# Radio Communication Handbook

Fourteenth Edition

Editor
Mike Browne, G3DIH

Radio Society of Great Britain

Published by the Radio Society of Great Britain, 3 Abbey Court, Fraser Road, Priory Business Park, Bedford. MK44 3WH. Tel 01234 832700. Website: www.rsgb.org

First published 2020



Design and Layout: Mike Browne, G3DIH

Cover design: Kevin Williams, M6CYB

Production: Mark Allgar, M1MPA

Printed in Great Britain by Short Run Press Ltd of Exeter



Hardback - ISBN 9781 9101 9393 8
Paperback - ISBN 9781 9101 9394 5

# Contents

# Acknowledgements

*The principal contributors to this book were:*

| | | |
|---|---|---|
| Chapter 1: | Principles | Alan Betts, G0HIQ |
| Chapter 2: | Passive components | Stuart Swain, G0FYX |
| Chapter 3: | Semiconductors | Alan Betts, G0HIQ |
| Chapter 4: | Building blocks 1: Oscillators | Peter Goodson, G4PCF |
| Chapter 5: | Building blocks 2: Amplifiers, mixers etc | Peter Goodson, G4PCF<br>Mike Stevens, G8CUL/F4VRB |
| Chapter 6: | HF receivers | Roger Wilkins, G8NHG |
| Chapter 7: | HF transmitters and transceivers | Peter Hart, G3SJX |
| Chapter 8: | Software Defined Radio (SDR) | Andrew Barron, ZL3DW |
| Chapter 9: | VHF/UHF receivers, transmitters and transceivers | Andy Barter, G8ATD<br>Chris Waters, 2E0UCW |
| Chapter 10: | Low frequencies: Below 1MHz | Dave Pick G3YXM |
| Chapter 11: | Practical microwave receivers and transmitters | Andy Barter, G8ATD |
| Chapter 12: | Propagation | Peter Duffett-Smith, G3XJE |
| Chapter 13: | Antenna basics and construction | Mike Parkin, G0JMI |
| Chapter 14: | Transmission lines | Mike Parkin, G0JMI |
| Chapter 15: | Practical HF antennas | Mike Parkin, G0JMI |
| Chapter 16: | Practical VHF/UHF antennas | Peter Swallow, G8EZE |
| Chapter 17: | Practical microwave antennas | Peter Swallow, G8EZE |
| Chapter 18 | The great outdoors | Jamie Davies, MM0JMI<br>Richard Marshall, G4ERP |
| Chapter 19: | Morse code | Roger Cooke, G3LDI |
| Chapter 20: | Data communications | Andy Talbot, G4JNT |
| Chapter 21: | Computers in the shack | Andy Talbot, G4JNT |
| Chapter 22: | Electromagnetic compatibility | Robin Page-Jones, G3JWI |
| Chapter 23: | Power supplies | Stuart Swain, G0FYX |
| Chapter 24: | Measurement and test equipment | Philip Lawson, G4FCL |
| Chapter 25: | Construction and workshop practice | Eamon Skelton, EI9GQ |
| Chapter 26: | QRP (Low Power Operating) | Steve Hartley, G0FUW |

Thanks also go to the contributors to previous editions of this book and the authors of the published *RadCom* articles which provided some of the source material.

# Preface

Welcome to the 14th edition of the Radio Communication Handbook. This book has 832 pages, with over 600,000 words plus 2,000 illustrations and tables, and is therefore the one book you need if you have any interest at all in the technical side of amateur radio. Whether you are an operator who wants to know more about what goes on 'under the hood', an avid constructor or someone keen to go on learning about electronics and radio communication, there is plenty for you here.

Chapters included in this edition are on Practical HF Antennas which has been rewritten by Mike Parkin, G0JMI, Morse and Digital Communications, EMC (Electromagnetic Compatibility) and Measurement & Test equipment., also an additional chapter on Software Defined Radio (SDR) by Andrew Barron, ZL3DW. and an introduction into QRP (Low Power Operating) by Steve Hartley, G0FUW Many other chapters have been revised and updated.

New readers and old hands alike will find hundreds of pages packed with the distilled knowledge and experience of acknowledged experts on each topic. In amateur radio there is always plenty to learn and this book is the ideal way to expand your knowledge on your favourite activity, or to discover and explore something new.

Whatever you use this book for, it will become a valuable tool in your amateur radio activities, helping you get the very best out of our wonderful hobby.

*Mike Browne, G3DIH*

# Principles

# 1

A good understanding of the basic principles and physics of matter, electronics and radio communication is essential if the self-training implicit in amateur radio is to be realised. These principles are not particularly difficult and a good grasp will allow the reader to understand the following material rather than simply accepting that it is true but not really knowing why. This will, in turn, make more aspects of the hobby both attractive and enjoyable.

## STRUCTURE OF MATTER

All matter is made up of atoms and molecules. A molecule is the smallest quantity of a substance that can exist and still display the physical and chemical properties of that substance. There is a very great number of different sorts of molecule. Each molecule is, in turn, made up of a number of atoms. There are about 102 different types of atom which are the basic elements of matter. Two atoms of hydrogen will bond with one atom of oxygen to form a molecule of water for example. The chemical symbol is $H_2O$. The H stands for hydrogen and the subscript 2 indicates that two atoms are required; the O denotes the oxygen atom.

A more complex substance is $H_2SO_4$. Two hydrogen atoms, one sulphur atom and four oxygen atoms form a molecule of sulphuric acid, a rather nasty and corrosive substance used in lead-acid batteries.

Atoms are so small that they cannot be seen even under the most powerful optical microscopes. They can, however, be visualised using electronic (not electron) microscopes such as the scanning tunnelling microscope (STM) and the atomic force microscope (AFM). **Fig 1.1** shows an AFM representation of the surface of a near-perfect crystal of graphite with the carbon atoms in a hexagonal lattice.

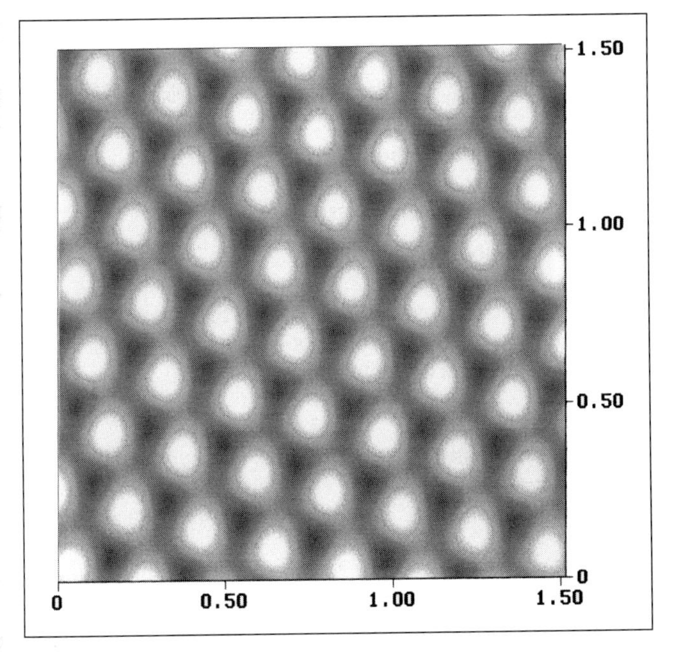


**Fig 1.1: Image of the atoms in a piece of high-purity graphite (distances in nanometres). The magnification is approximately 45 million times. Note that no optical microscope can produce more than about 1000 times magnification**

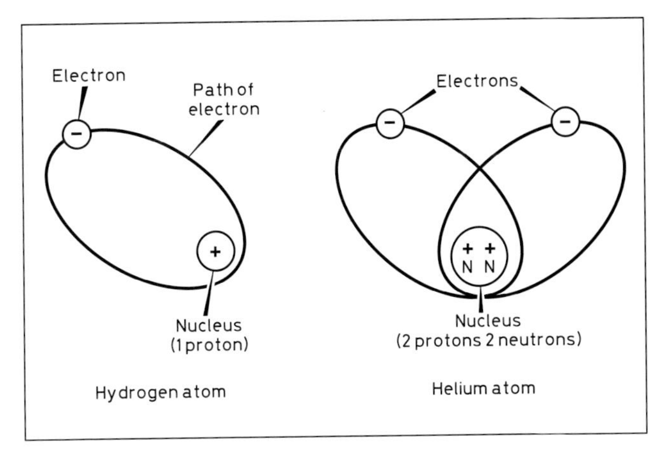


**Fig 1.2: Structure of hydrogen and helium atoms**

Atoms are themselves made up of yet smaller particles; the electron, the proton and the neutron, long believed to be the smallest things that could exist. Modern atomic physics has shown that this is not so and that not only are there smaller particles but that these particles, energy and waves are, in certain scenarios, indistinguishable from each other. Fortunately we need only concern ourselves with particles down to the electron level but there are effects, such as in the tunnel diode, where the electron seems to 'disappear' and 'reappear' on the far side of a barrier.

The core of an atom comprises one or more protons and may include a number of neutrons. The electrons orbit the core, or 'nucleus' as it is called, rather like the planets orbit the sun. Electrons have an electrical charge that we now know to be a negative charge. Protons have an equal positive charge. The neutrons are not charged.

A hydrogen atom has a single proton and a single orbiting electron. A helium atom has a nucleus of two protons and two neutrons with two orbiting electrons; this is shown in **Fig 1.2**. An atom is electrically neutral; the positive charge on the nucleus is balanced by the negative charge on the electrons. The magnitude of the charge is tiny; it would require 6,250,000,000,000,000,000 ($6{\cdot}25 \times 10^{18}$) electrons to produce a charge of 1 coulomb, that is 1A flowing for 1s.

### Conductors and Insulators

The ease with which the electrons in a substance can be detached from their parent atoms varies from substance to substance. In some substances there is a continual movement of electrons in a random manner from one atom to another and the application of an electrical force or potential difference (for example from a battery) to the two ends of a piece of wire made of such a substance will cause a drift of electrons along the wire called an electric current; electrical conduction is then said to take place. It should be noted that if an electron enters the wire from the battery at one end it will be a different electron which immediately leaves the other end of the wire.

To visualise this, consider a long tube such as a scaffold pole filled with snooker balls. As soon as another ball is pushed in one end, one falls out the other but the progress of any particular ball is much slower. The actual progress of an individual electron along a wire, the drift velocity, is such that it could take some minutes to move even a millimetre.

The flow of current is from a point of positive charge to negative. Historically, the decision of what represented 'positive' was arbitrary and it turns out that, by this convention, electrons have a negative charge and the movement of electrons is in the opposite direction to conventional current flow.

Materials that exhibit this property of electrical conduction are called conductors. All metals belong to this class. Materials that do not conduct electricity are called insulators, and **Table 1.1** shows a few examples of commonly used conductors and insulators.

## ELECTRICAL UNITS

### Charge (q)

Charge is the quantity of electricity measured in units of coulombs. **Table 1.2** gives the units and their symbol.

One coulomb is the quantity of electricity given by a current of one ampere flowing for one second.

Charge q = current (A) x time (s), normally written as: $q = I \times t$

### Current Flow, the Ampere (A)

The ampere, usually called amp, is a fundamental (or base) unit in the SI (System International) system of units. It is actually defined in terms of the magnetic force on two parallel conductors each carrying 1A.

### Energy (J)

Energy is the ability to do work and is measured in joules. One joule is the energy required to move a force of one Newton through a distance of one metre. As an example, in lifting a 1kg bag of sugar 1m from the floor to a table, the work done or energy transferred is 9·81 joules.

### Power (W)

Power is simply the rate at which work is done or energy is transferred and is measured in watts, W.

$$\text{Power} = \frac{\text{energy transferred}}{\text{time taken}}$$

For example, if the bag of sugar was lifted in two seconds, the power would be 9·81/2 or approximately 5W.

### Potential Difference, Voltage (V)

If a source of electrical energy has a Potential Difference of 1 volt, each coulomb of electricity, ie charge, that flows has an energy of 1 joule. If one coulomb of charge flows in, for example, a bulb, and 12 joules of energy are transferred into heat and light, the potential difference across the bulb is 12 volts.

The definition of the volt is the number of joules of energy per coulomb of electricity.

| Conductors | Insulators |
|---|---|
| Silver | Mica |
| Copper | Quartz |
| Gold | Glass |
| Aluminium | Ceramics |
| Brass | Ebonite |
| Steel | Plastics |
| Mercury | Air and other gasses |
| Carbon | Oil |
| Solutions of salts or acids in water | Pure water |

**Table 1.1: Examples of conducting and insulating materials**

| Quantity | Symbol | Unit | Abbreviation |
|---|---|---|---|
| Charge | q | coulomb | C |
| Conductance | G | Siemen | S |
| Current | I | Ampere (Amp) | A |
| Voltage* | E or V | volt | V |
| Time | t | second | s or sec |
| Resistance | R | ohm | Ω |
| Capacitance | C | farad | F |
| Inductance | L | henry | H |
| Mutual inductance | M | henry | H |
| Power | P | watt | W |
| Frequency | f | hertz | Hz |
| Wavelength | λ | metre | m |

** 'Voltage' includes 'electromotive force' and 'potential difference'.*

*Since the above units are sometimes much too large (eg the farad) and sometimes too small, a series of multiples and sub-multiples are used:*

| Unit | Symbol | Multiple |
|---|---|---|
| Microamp | μA | 1 millionth ($10^{-6}$) amp |
| Milliamp | mA | 1 thousandth ($10^{-3}$) amp |
| Microvolt | μV | $10^{-6}$V |
| Millivolt | mV | $10^{-3}$V |
| Kilovolt | kV | $10^{3}$V |
| Picofarad | pF | $10^{-12}$F |
| Nanofarad | nF | $10^{-9}$F |
| Microfarad | μF | $10^{-6}$F |
| Femtosecond | fs | $10^{-15}$s |
| Picosecond | ps | $10^{-12}$s |
| Nanosecond | ns | $10^{-9}$s |
| Microsecond | μs | $10^{-6}$s |
| Millisecond | ms | $10^{-3}$s |
| Microwatt | μW | $10^{-6}$W |
| Milliwatt | mW | $10^{-3}$W |
| Kilowatt | kW | $10^{3}$W |
| Gigahertz | GHz | $10^{9}$Hz |
| Megahertz | MHz | $10^{6}$Hz |
| Kilohertz | kHz | $10^{3}$Hz |
| Centimetre | cm | $10^{-2}$m |
| Kilometre | km | $10^{3}$m |

*Note: The sub-multiples abbreviate to lower case letters. All multiples or sub-multiples are in factors of a thousand except for the centimetre.*

**Table 1.2: Units and symbols**

$$1\,\text{volt} = \frac{1\,\text{joule}}{1\,\text{coulomb}}$$

Historically, voltage was viewed as a force but that is not strictly true although the term electromotive force (emf) is still in use. The better term is Electrical Potential.

### Resistance

Resistance restricts the flow of charge, the current. In forcing electrons through a conductor, some energy is lost as heat. A longer, thinner conductor will have a greater loss, that is a higher resistance.

Different materials have differing resistivities, that is, a wire of the same dimensions will have different resistances depending on the material. The conductors in the list in Table 1.1 are in conductivity (inverse of resistivity) order.

Materials such a Nichrome, Manganin and Eureka are alloys with a deliberately high resistivity and are used in power resistors and wire-wound variable resistors. Tungsten has a relatively high resistance but its key property is a high melting point and relative strength when white hot. It is used to make the filament of incandescent light bulbs.

Specific resistance: This is simply the resistance of a standard size piece of the subject material, normally a 1 metre cube and is quoted in units of ohm metre, Ωm and has the symbol ρ, the Greek letter rho. Its purpose is to compare the resistivity of different materials.

The unit of resistance is the ohm, symbol Ω, the Greek upper case letter omega. It is defined as the ratio of the applied EMF and the resulting current.

$$\text{Resistance R ohms} = \frac{\text{applied EMF}}{\text{current flowing}}$$

## Ohm's Law

Ohm's Law is simply a restatement of the definition of resistance. In words it reads:

In a circuit at constant temperature the current flowing is directly proportional to the applied voltage and inversely proportional to the resistance. The reference to temperature is important. In a practical test, the energy will be converted to heat and most materials change their resistance as the temperature changes.

In algebraic form, where V is the applied voltage or potential difference, I is the current flowing and R is the resistance:

$$V = I \times R \qquad I = \frac{V}{R} \quad \text{and} \quad R = \frac{V}{I}$$

*Example.*

Consider the circuit shown in **Fig 1.3** which consists of a 4V battery and a resistance R of 8Ω. What is the magnitude of the current in the circuit?

Here V = 4V and R = 8Ω. Let I be the current flowing in amperes. Then from Ohm's Law:

$$I = \frac{V}{R} = \frac{4}{8} = \frac{1}{2} = 0{\cdot}5\text{A}$$

It should be noted that in all calculations based on Ohm's Law care must be taken to ensure that V, I and R are in consistent units, ie in amperes, volts and ohms respectively, if errors in the result are to be avoided. In reality , typical currents may be in mA or μA and resistances in kΩ or MΩ.

## Conductance

Conductance is simply the inverse or reciprocal of resistance. Many years ago it was measured in a unit called the mho; today the correct unit is the Siemen. A resistance of 10 ohms is the same as a conductance of 0·1 Siemen.

## EMF, PD and Source Resistance

Sources of electrical energy, such as batteries, hold a limited amount of energy and there is also a limit to the rate at which this energy can be drawn, that is the power is limited. A battery or power supply can be considered as a perfect source (which can supply any desired current) together with a series resistance. The series resistance will limit the total current and will also result in a drop in the potential difference or voltage at the terminals. This is shown in **Fig 1.4**.

The source V has a voltage equal to the open circuit terminal voltage and is called the electromotive force (EMF) of the device. On load, that is when a current is being drawn, the potential difference at the terminals will drop according to the current drawn. The drop may be calculated as the voltage across the internal resistor 'r' using the formula:

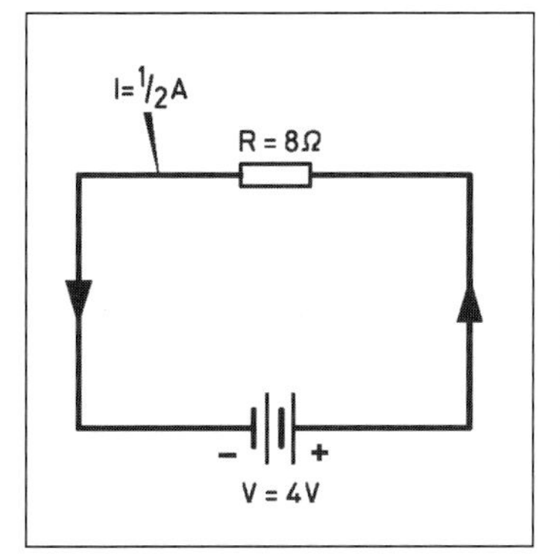


**Fig 1.3: Application of Ohm's Law**

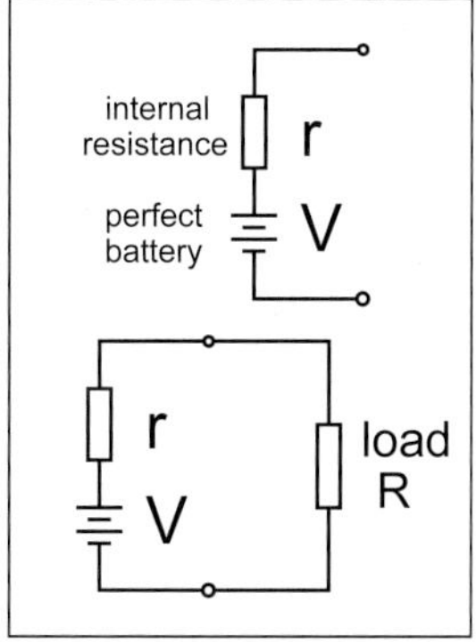


**Fig 1.4: A real battery**

Voltage drop = current drawn (I) x rΩ and the terminal voltage will be

$$V_{Terminal} = V_{Supply} - I \times r$$

## Maximum Power Transfer

It is interesting to consider what is the maximum power that can be drawn from a particular source. As the load resistance decreases, the current drawn increases but the potential difference across the load decreases. Since the power is the product of the load current and the terminal voltage, there will be a maximum point and attempts to draw more power will be thwarted by the drop in terminal voltage.

**Fig 1.5** shows the power in the load as the load resistance is varied from 0 to 10Ω, connected to a source of EMF, 10V and internal resistance 2Ω.

Maximum power transfer occurs when the load resistance is the same as the source resistance. For DC and power circuits this is never done since the efficiency drops to 50% but in RF and low level signal handling, maximum signal power transfer may be a key requirement.

## Sources of Electricity

When two dissimilar metals are immersed in certain chemical solutions, or electrolytes, an electromotive force (EMF or voltage) is created by chemical action within the cell so that if these pieces of metal are joined externally, there will be a continuous flow of electric current. This device is called a simple cell and such a cell, comprising copper and zinc rods immersed in diluted sulphuric acid, is shown in **Fig 1.6(a)**. The flow of current is from the copper to the zinc plate in the external circuit; ie the copper forms the positive (+) terminal of the cell and the zinc forms the negative (-) terminal.

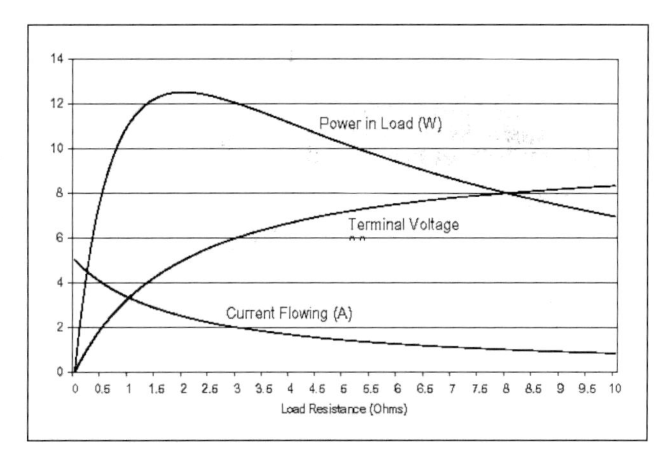


**Fig 1.5: Power dissipated in the load**

In a simple cell of this type hydrogen forms on the copper electrode, and this gas film has the effect of increasing the internal resistance of the cell and also setting up within it a counter or polarising EMF which rapidly reduces the effective EMF of the cell as a whole.

This polarisation effect is overcome in practical cells by the introduction of chemical agents (depolarisers) surrounding the anode for the purpose of removing the hydrogen by oxidation as soon as it is formed.

## Primary Cells

Practical cells in which electricity is produced in this way by direct chemical action are called primary cells; a common example is the Leclanché cell, the construction of which is shown diagrammatically in **Fig 1.6(b)**. The zinc case is the negative electrode and a carbon rod is the positive electrode. The black paste surrounding the carbon rod may contain powdered carbon, manganese dioxide, zinc chloride, ammonium chloride and water, the manganese dioxide acting as depolariser by combining with hydrogen formed at the anode to produce manganese oxide and water.

Alkaline cells are now more common and are interchangeable with zinc-carbon since both produce about 1.5V PD. They should not be mixed due to differing capacities. The steel can, the cathode (positive), is filled with a paste of manganese dioxide and carbon powder to improve conductivity. The centre has a porous separator filled with zinc powder in a potassium hydroxide paste electrolyte. This can leak a caustic gel at the end of its life forming a crystalline powdery coating which corrodes metal tracks and is a skin and respiratory irritant. Most cells are now leak proof, but attempts at recharging may accelerate any tendency to leak.

Silver oxide cells, typically button cells produce about 1.8V and have a silver oxide/ zinc chemistry, again with an alkaline electrolyte, typically sodium or potassium hydroxide.

Lithium batteries are now more common. These were relatively expensive and could suffer from catastrophic failure modes including fire, but technology improvements have largely overcome those difficulties. Their main advantage is much higher energy density, by volume or by weight. As with all batteries the manufacturer's instructions do need to be followed, particularly regarding removal when exhausted, safe disposal and the caution against attempted recharging.

Cells may be connected in series to form a battery with a higher voltage. The cells should be of the same type, capacity and age. With batteries of several cells it is quite possible for fresh cells to continue to cause a current which will reverse charge an old or dud cell. That may cause a chemical leak, swelling and physical damage and, possibly, overheating.

The symbol used to denote a cell in a circuit diagram is shown in **Fig 1.6(c)**. The long thin stroke represents the positive terminal and the short thick stroke the negative terminal. Several cells joined in series to form a battery are shown; for higher voltages it becomes impracticable to draw all the individual cells involved and it is sufficient to indicate merely the first and last cells with a dotted line between them with perhaps a note added to state the actual voltage. The amount of current which can be derived from a dry cell depends on its size and the life required, and may range from a few milliamperes to an amp or two.

## Secondary Cells

In primary cells some of the various chemicals are used up in producing the electrical energy - a relatively expensive and wasteful process. The maximum current available also is limited. Another type of cell called a secondary cell or accumulator offers the advantage of being able to provide a higher current and is capable of being charged by feeding electrical energy into the cell to be stored chemically, and be drawn out or discharged later as electrical energy again. This process of charging and discharging the cell is capable of repetition for a large number of cycles depending on the chemistry of cell.

A common type of secondary cell is the lead-acid cell such as that used in vehicle batteries. Vehicle batteries are of limited use for amateur radio for two reasons. Firstly they are liable to leak acid if tipped and give off hydrogen (explosive in confined spaces) and secondly, they are designed to float charge and start vehicle engines with very high current surges rather than undergo deep discharge.

Sealed or 'maintenance free' types are available but they must still be used the correct way up or leakage will occur as gas is generated and the pressure increases.

Deep cycle batteries (sometimes called leisure batteries for caravans) are available that are designed to be fully charge/discharge cycled but are considerably more expensive. These batteries are often available at amateur rallies. It is necessary to check which type they are and their origin. Those removed from alarm systems or uninterruptible power supplies (UPS) are likely to have been changed at the five-year maintenance review. They may well have a couple of years' service left and perhaps considerably more. It is a risk but can be a cheap way of obtaining otherwise expensive batteries.

Nickel-metal-hydride (NiMh) is now replacing Nickel-cadmium as the ubiquitous rechargeable battery, the European Union having banned NiCd imports due to the toxic nature of cadmium. Not that any battery technology should be considered entirely safe. They don't have quite the energy capacity per unit volume (energy density) but are much easier to handle. Lithium-ion batteries are still relatively expensive and offer a better energy density. All these types can be damaged by misuse, especially overcharging or the use of the wrong type of charger. As always, check with the manufacturer.

The chapter on Power Supplies discusses batteries in more detail.

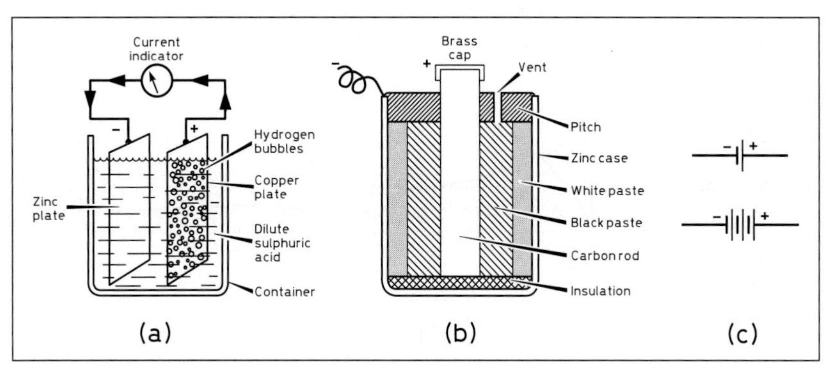


**Fig 1.6: The electric cell. A number of cells connected in series is called a battery. (a) A simple electric cell consisting of copper and zinc electrodes immersed in dilute sulphuric acid. (b) Sectional drawing showing construction of a dry cell. (c) Symbol used to represent single cells and batteries in circuit diagram**

## Mechanical Generators

Mechanical energy may be converted into electrical energy by moving a coil of wire in a magnetic field. Direct-current or alternating-current generators are available in all sizes but the commonest types likely to be met in amateur radio work are:

(a) Petrol-driven AC generators of up to 1 or 2kW output such as are used for supplying portable equipment; and

(b) Small motor generators, sometimes called dynamotors or rotary converters, which furnish up to about 100W of power and comprise a combined low-voltage DC electric motor and a high voltage AC or DC generator. These have two origins; ex military devices providing high voltage DC for use in valve transmitter/receiver equipments and those supplied for use in mobile caravans to provide mains voltage AC from the 12V DC battery. The latter function is now normally achieved by wholly electronic means (commonly known as inverters) but the waveform can be a compromise.

# ELECTRICAL POWER

When a current flows through a resistor, eg in an electric fire, the resistor gets hot and electrical energy is turned into heat. The actual rise in temperature depends on the amount of power dissipated in the resistor and its size and shape. In most circuits the power dissipated is insignificant but it is a factor the designer must consider, both in fitting a resistor of adequate dissipation and the effect of the heat generated on nearby devices.

The unit of electrical power is the watt (W). The amount of power dissipated in a resistor is equal to the product of the potential difference across the resistor and the current flowing in it. Thus:

Power (watts) = Voltage (volts) x Current (amperes)

$W = V \times I$

Ohm's Law states: V = I x R and I = V/R

Substituting V or I in the formula for power gives two further formulas:

$$W = I^2R$$

and

$$W = \frac{V^2}{R}$$

These formulas are useful for finding, for example, the power input to a transmitter or the power dissipated in various resistors in an amplifier so that suitably rated resistors can be selected.

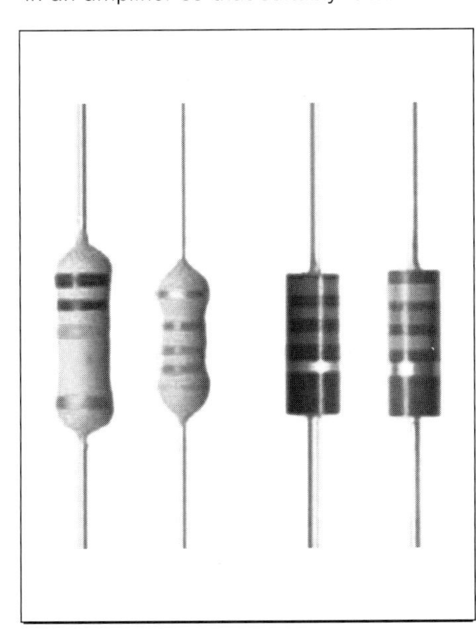

Fig 1.7: Metal film resistors revealed. The metal film is the dark coating on the outside of the white ceramic former, laser cut to make a broad spiral strip

| Colour | Value (numbers) | Value (multiplier) | Value (tolerance) |
|---|---|---|---|
| Black | 0 | 1 | - |
| Brown | 1 | 10 | 1% |
| Red | 2 | 100 | 2% |
| Orange | 3 | 1000 ($10^3$) | - |
| Yellow | 4 | $10^4$ | - |
| Green | 5 | $10^5$ | - |
| Blue | 6 | $10^6$ | - |
| Violet | 7 | - | - |
| Grey | 8 | - | - |
| White | 9 | - | - |
| Silver | - | 0.01 | 10% |
| Gold | - | 0.1 | 5% |
| No Colour | - | - | 20% |

*Note: A pink band may be used to denote a 'high-stability' resistor. Sometimes an extra band is added to give three figures before the multiplier.*

**Table 1.3: Resistor colour code**

To take a practical case, consider again the circuit of Fig 1.3. The power dissipated in the resistor may be calculated as follows:

Here V = 4V and R = 8Ω. so the correct formula to use is

$$W = \frac{V^2}{R}$$

Inserting the numbers gives

$$W = \frac{4^2}{8} = \frac{16}{8} = 2W$$

It must be stressed again that the beginner should always see that all values are expressed in terms of volts, amperes and ohms in this type of calculation. The careless use of megohms or milliamperes, for example, may lead to an answer several orders too large or too small.

# RESISTORS

## Resistors Used in Radio Equipment

As already mentioned, a resistor through which a substantial current is flowing will get hot. It follows therefore that in a piece of radio equipment the resistors of various types and sizes that are needed must be capable of dissipating the power as required without overheating.

Generally speaking, radio resistors can be divided roughly into two classes, (a) low power up to 3W and (b) above 3W. The low-power resistors are usually made of carbon film or metal film and may be obtained in a wide range of resistance values from about 10Ω to 10MΩ and in power ratings of 0·1W to 3W. The film usually has a spiral grove to narrow and lengthen the track to obtain the desired resistance, (see **Fig 1.7**) but the inductance introduced is low, typically a few nanohenries. Carbon composition resistors are no longer available. For higher powers, resistors are usually wire-wound on ceramic formers and the very fine wire is protected by a vitreous enamel coating. Typical resistors are shown in the Passive Components chapter.

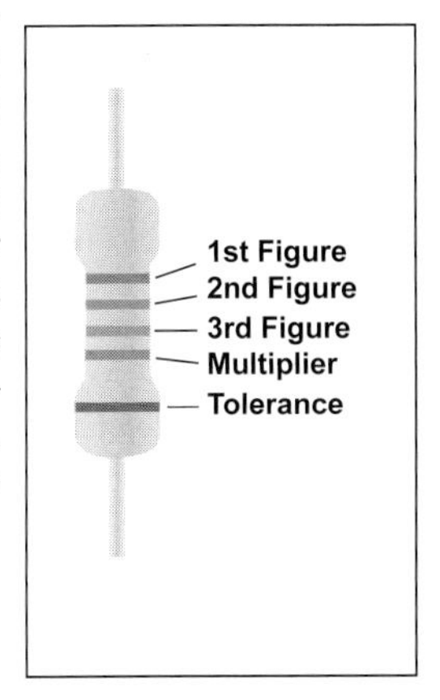


**Fig 1.8: Standard resistance value markings**

Resistors, particularly the small carbon types, are usually colour-coded to indicate the value of the resistance in ohms and sometimes also the tolerance or accuracy of the resistance. The standard colour code is shown in **Table 1.3**.

The colours are applied as bands at one end of the resistor as shown in **Fig 1.8** As an example, what would be the value of a resistor with the following colour bands: yellow, violet, orange, silver?

The yellow first band signifies that the first figure is 4, the violet second band signifies that the second figure is 7, while the orange third band signifies that there are three zeros to follow; the silver fourth band indicates a tolerance of ±10%. The value of the resistor is therefore 47,000Ω ±10% (47kΩ ±10%).

So far only fixed resistors have been mentioned. Variable resistors, sometimes called potentiometers or volume controls, are also used. The latter are usually panel-mounted by means of a threaded bush through which a quarter inch or 6mm diameter spindle protrudes and to which the control knob is fitted. Low-power high-value variable resistances use a carbon resistance element, and high-power lower-resistance types (up to 10,000Ω) use a wire-wound element. These are not colour coded but the value is printed on the body either directly, eg '4k7', or as '472' meaning '47' followed by two noughts.

## Resistors in Series and Parallel

Resistors may be joined in series or parallel to obtain a specific value of resistance. Some caution may be called for since the tolerance will affect the actual value and close tolerance resistors may be needed if the value is critical such as a divider used in a measuring circuit.

When in series, resistors are connected as shown in **Fig 1.9(a)** and the total resistance is equal to the sum of the separate resistances. The parallel connection is shown in **Fig 1.9(b)**, and with this arrangement the reciprocal of the total resistance is equal to the sum of the reciprocals of the separate resistances.

Series connection:

$$R_{total} = R1 + R2 + R3 + etc$$

Parallel connection

$$\frac{1}{R_{total}} = \frac{1}{R1} + \frac{1}{R2} + \frac{1}{R3} \quad etc$$

If only two resistors are in parallel, an easier calculation is possible:

$$\frac{1}{R_{total}} = \frac{1}{R1} + \frac{1}{R2} = \frac{R2}{R1 \times R2} + \frac{R1}{R1 \times R2} = \frac{R1 + R2}{R1 \times R2}$$

Inverting this and writing in 'standard form' (omitting multiplication signs) gives

$$R_{total} = \frac{R1\,R2}{R1 + R2}$$

This is a useful formula since the value of two resistors in parallel can be calculated easily.

It is also useful to remember that for equal resistors in parallel the formula simplifies to

$$R_{total} = \frac{R}{n}$$

where R is the value of one resistor and n is the number of resistors.

*Example 1.*

Calculate the resistance of a 30Ω and a 70Ω resistor connected first in series and then in parallel.

In series connection:
R = 30 + 70 = 100Ω

In parallel connection, using the simpler formula for two resistors in parallel:

$$R_{total} = \frac{R1\,R2}{R1 + R2} = \frac{30 \times 70}{30 + 70} = \frac{2100}{100} = 21\Omega$$

These two calculations are illustrated in **Fig 1.9(c)**.

*Example 2.*

Three resistors of 7Ω, 14Ω and 28Ω are connected in parallel. If another resistor of 6Ω is connected in series with this combination, what is the total resistance of the circuit?

The circuit, shown in **Fig 1.9(d)**, has both a series and parallel configuration. Taking the three resistors in parallel first, these are equivalent to a single resistance of R ohms given by:

$$\frac{1}{R_{total}} = \frac{1}{R1} + \frac{1}{R2} + \frac{1}{R3} = \frac{1}{7} + \frac{1}{14} + \frac{1}{28}$$

$$= \frac{4}{28} + \frac{2}{28} + \frac{1}{28} = \frac{7}{28}$$

Inverting (don't forget to do this!)

$$R_{total} = \frac{28}{7} = 4\Omega$$

This parallel combination is in series with the 6Ω resistor, giving a total resistance of 10Ω for the whole circuit.

*Note:* Often the maths is the most awkward issue and numerical mistakes are easy to make. A calculator helps but whether done with pencil and paper or by calculator, it is essential to estimate an answer first so mistakes can be recognised.

Let us consider the parallel calculation. The answer will be less than the lowest value. The 14Ω and 28Ω resistors will form a resistor less than 14Ω but greater than 7Ω since two 14Ω parallel resistors will give 7Ω and one of the actual resistors is well above 14Ω.

We now have this 'pair' in parallel with the 7Ω resistor. By the same logic, the answer will be less than 7Ω but greater than

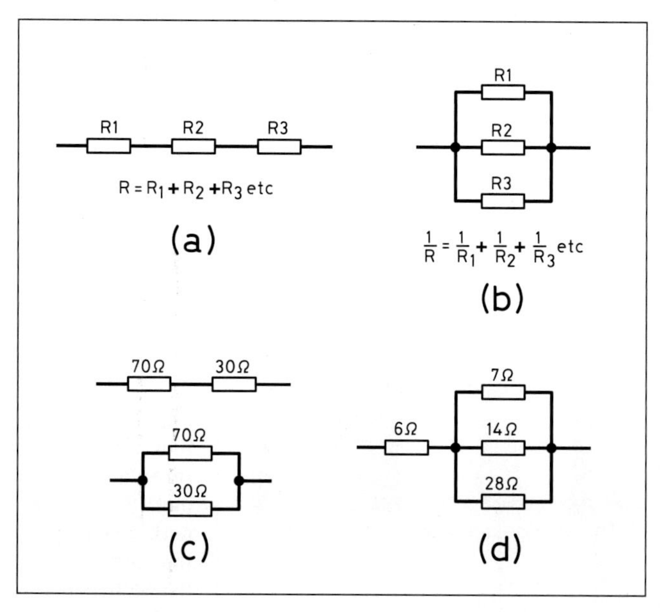


**Fig 1.9: Resistors in various combinations: (a) series, (b) parallel, (c) series and parallel, (d) series-parallel. The calculation of the resultant resistances in (c) and (d) is explained in the text**

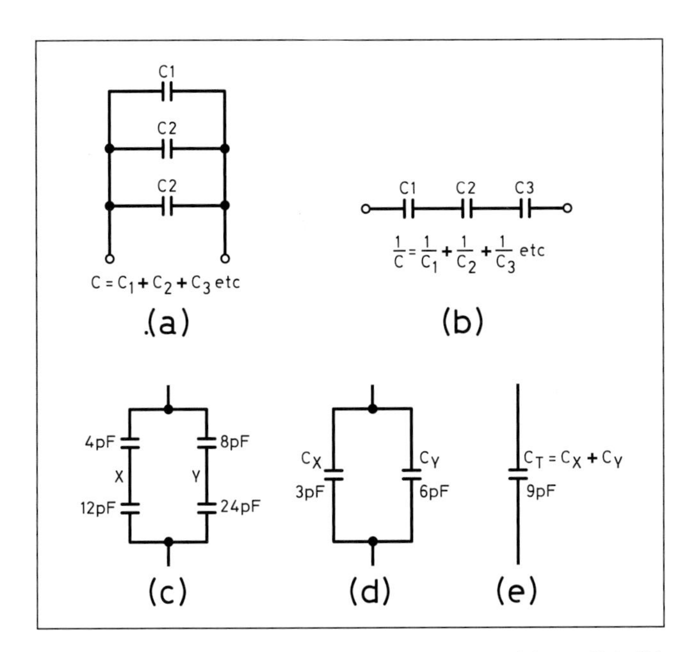


**Fig 1.10: Capacitors in various combinations: (a) parallel, (b) series, (c) series-parallel. The calculation of the resultant capacitance of the combination shown in (c) required first the evaluation of each series arm X and Y as shown in (d). The single equivalent capacitance of the combination is shown in (e)**

3·5Ω. The calculated answer of 4Ω meets this criterion so stands a good chance of being correct. Results outside the 3·5-7Ω range must be wrong.

This may seem a bit long winded. In words it is. In practice, with a little bit of experience it takes moments.

## Capacitors and Capacitance

Capacitors have the property of being able to store a charge of electricity. They consist of two parallel conducting plates or strips separated by an insulating medium called a dielectric. When a capacitor is charged there is a potential difference between its plates.

The capacitance of a capacitor is defined in terms of the amount of charge it can hold for a given potential difference. The capacitance 'C' is given by:

$$C = \frac{q}{V}$$

in units of Farads, F. (q in coulombs and V in volts)

A larger capacitor can hold a greater charge for a given voltage. Or to put it another way, a smaller capacitor will need a greater voltage if it is to store the same charge as a large capacitor.

The Farad is too large a unit for practical use, and typical values range from a few picofarads (pf) to some thousands of microfarads (μF). Unusually the convention is to write 20,000μF rather than 20mF which is, technically, correct notation.

The area of the two plates and the distance between them determines the capacitance. The material of the dielectric also has an effect; materials with a high dielectric constant can considerably enhance the capacitance.

The capacitance can be calculated from the formula

$$C = \frac{\varepsilon_0 \varepsilon_r A}{d}$$

where:

C is in Farads

$\varepsilon_0$ is a natural constant, the permitivity of free space

$\varepsilon_r$ is the relative permitivity (or dielectric constant K)

A is the area of the plates in $m^2$ and

d is their separation in m.

*Example:*

A tuning capacitor has six fixed plates and five movable plates meshed between them, with a gap of 1mm. Fully meshed at maximum capacitance the area of overlap is $8cm^2$. The dielectric is air, $\varepsilon_r$=1, and $\varepsilon_0$ is 8·85 x $10^{-12}$ F/m. What is the capacitance?

Firstly, it will be numerically easier to re-arrange the formula so the separation may be given in centimetres and the area in $cm^2$ and the answer in picofarads. The formula becomes

$$C = \frac{0 \cdot 0885\,A}{d} = \frac{0 \cdot 0885 \times 8 \times 10}{0 \cdot 1} = 71pF$$

The factor 10 above comes from the fact that there are 10 'surfaces' to consider in the meshed capacitor.

Such a capacitor will probably be acceptable for receiving and transmitting up to about 10-20W of power, depending in part on the matching and voltages involved. Higher powers and voltages will require considerably more separation between the plates, reducing the capacitance.

The factor by which the dielectric increases the capacitance compared with air is called the dielectric constant $\varepsilon_r$ (or relative permitivity K) of the material. Physicists tend to use the symbol $\varepsilon_r$ and engineers the symbol K. You may meet both, depending on which texts you are reading.

Typical values of K are: air 1, paper 2·5, glass 5, mica 7. Certain ceramics have much higher values of K of 10,000 or more. If the dielectric is a vacuum, as in the case of the inter-electrode capacitance of a valve, the same value of K as for air may be assumed. (Strictly, K = 1 for a vacuum and is very slightly higher for air.) The voltage at which a capacitor breaks down depends on the spacing between the plates and the type of dielectric used. Capacitors are often labelled with the maximum working voltage which they are designed to withstand and this figure should not be exceeded.

## Capacitors Used in Radio Equipment

The values of capacitors used in radio equipment extend from below 1pF to 100,000μF. They are described in detail in the Passive Components chapter.

## Capacitors in Series and Parallel

Capacitors can be connected in series or parallel, as shown in **Fig 1.10**, either to obtain some special capacitance value using a standard range of capacitors, or perhaps in the case of series connection to obtain a capacitor capable of withstanding a greater voltage without breakdown than is provided by a single capacitor. When capacitors are connected in parallel, as in **Fig 1.10(a)**, the total capacitance of the combination is equal to the sum of the separate capacitances. When capacitors are connected in series, as in **Fig 1.10(b)**, the reciprocal of the equivalent capacitance is equal to the sum of the reciprocals of the separate capacitances.

If C is the total capacitance these formulas can be written as follows:

Parallel connection

$$C = C1 + C2 + C3 \text{ etc}$$

Series connection

$$\frac{1}{C_{total}} = \frac{1}{C1} + \frac{1}{C2} + \frac{1}{C3} \quad \text{etc}$$

Similar to the formula for resistors in parallel, a useful equivalent formula for two capacitors in series is

$$C_{total} = \frac{C1\,C2}{C1 + C2}$$

The use of these formulas is illustrated by the following *example:*

Two capacitors of 4pF and 12pF are connected in series; two others of 8pF and 24pF are also connected in series. What is the equivalent capacitance if these series combinations are joined in parallel?

The circuit is shown in **Fig 1.10(c)**. Using the formula for two capacitors in series, the two series arms X and Y can be reduced to single equivalent capacitances $C_X$ and $C_Y$ as shown in **Fig 1.10(d)**.

$$C_X = \frac{4 \times 12}{4 + 12} = \frac{48}{16} = 3\text{pF}$$

$$C_Y = \frac{8 \times 24}{8 + 24} = \frac{96}{32} = 6\text{pF}$$

These two capacitances are in parallel and may be added to give the total effective capacitance represented by the single capacitor $C_T$ in **Fig 1.10(e)**.

$$C = C_X + C_Y = 3 + 6 = 9\text{pF}$$

The total equivalent capacitance of the four capacitors connected as described is therefore 9pF.

This is not just an academic exercise or something to do for an examination. Consider the circuit in **Fig 1.11**. C is a variable capacitor covering the range 5-100pF, a 20:1 range. It will be seen later that, in a tuned circuit, this will give a tuning range of about 4·5:1 which may well be rather greater than required and cramping the desired range over a relatively small part of the 180 degree rotation normally available. A 5pF capacitor C1 is connected in parallel with C, giving a capacitance range for the pair of 10-105pF by addition of the capacitor values. This is now in series with a 200pF capacitor, so the capacitance range becomes 9·5-69pF, a ratio of 7·3 and a tuning range of 2·7:1. Many variable capacitors have a small parallel trimmer capacitor included in their construction. Often that is used to set the highest frequency of the tuning range and a series capacitor or a variable inductor (see later in this chapter) used to set the lower end.

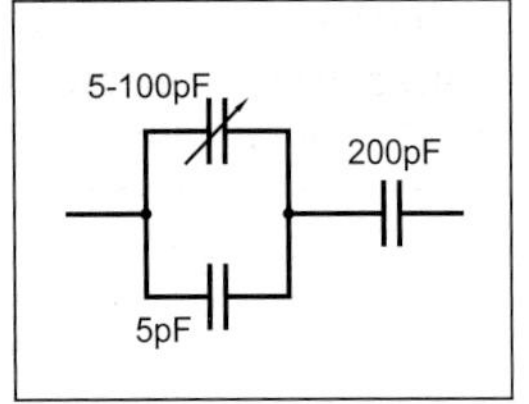


**Fig 1.11: Padding a variable capacitor to change its range**

# MAGNETISM

## Permanent Magnets

A magnet will attract pieces of iron towards it by exerting a magnetic force upon them. The field of this magnetic force can be demonstrated by sprinkling iron filings on a piece of thin cardboard under which is placed a bar magnet. The iron filings will map out the magnetic field as sketched in **Fig 1.12** and the photograph **Fig 1.13**. It will be seen that the field is most intense near the ends of the magnet, the centres of intensity being called the poles, and lines of force spread out on either side and continue through the material of the magnet from one end to the other.

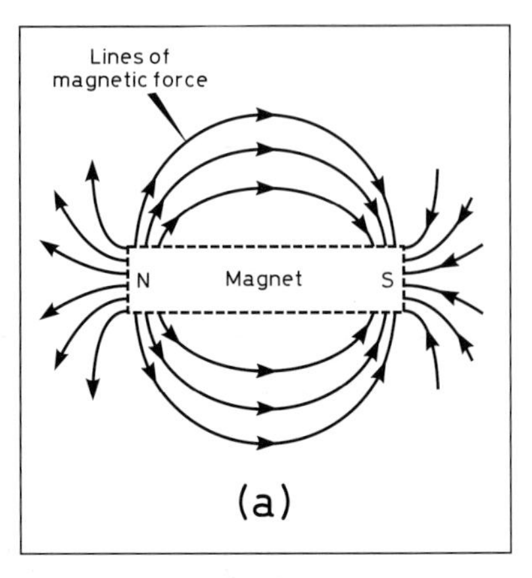


**Fig 1.12: magnetic field produced by a bar magnet**

If such a magnet is suspended so that it can swing freely in a horizontal plane it will always come to rest pointing in one particular direction, namely towards the Earth's magnetic poles, the Earth itself acting as a magnet. A compass needle is simply a bar of magnetised steel. One end of the magnet (N) is called the north pole, which is an abbreviation of 'north-seeking pole' and the other end (S) a south pole or south-seeking pole. It is an accepted convention that magnetic force acts in the direction from N to S as indicated by the arrows on the lines of force in Fig 1.12.

If two magnets are arranged so that the north pole of one is near the south pole of another, there will be a force of attraction between them, whereas if similar poles are opposite one another, the magnets will repel: see **Fig 1.14**.

Permanent magnets are made from certain kinds of iron, nickel and cobalt alloys and certain ceramics, the hard ferrites (see Passive Components chapter) and retain their magnetism more or less indefinitely. They have many uses in radio equipment, such as loudspeakers, headphones, some microphones, cathode-ray tube focusing arrangements and magnetron oscillators.

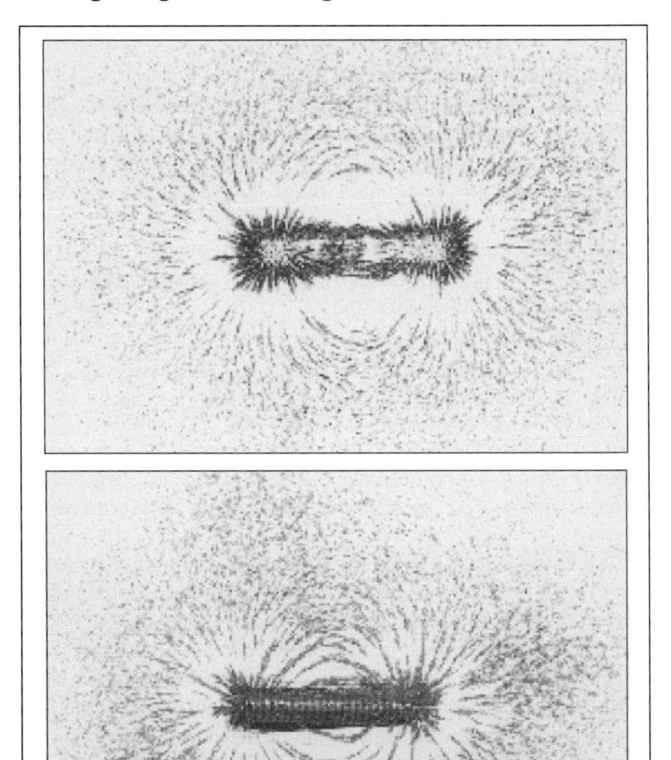

**Fig 1.13: Iron filings mapping out the magnetic field of (top) a bar magnet, and (bottom) a solenoid carrying current**

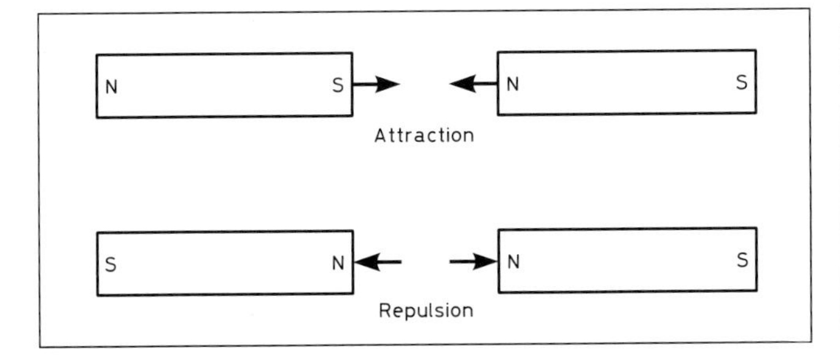


Fig 1.14: Attraction and repulsion between bar magnets

Other types of iron and nickel alloys and some ceramics (the soft ferrites), eg soft iron, are not capable of retaining magnetism, and cannot be used for making permanent magnets. They are effective in transmitting magnetic force and are used as cores in electromagnets and transformers. These materials concentrate the magnetic field by means of a property called permeability. The permeability is, essentially, the ratio of the magnetic field with a core to that without it.

## Electromagnets

A current of electricity flowing through a straight wire exhibits a magnetic field, the lines of force of which are in a plane perpendicular to the wire and concentric with the wire. If a piece of cardboard is sprinkled with iron filings, as shown in **Fig 1.15**, they will arrange themselves in rings round the wire, thus illustrating the magnetic field associated with the flow of current in the wire. Observation of a small compass needle placed near the wire would indicate that for a current flow in the direction illustrated the magnetic force acts clockwise round the wire. A reversal of current would reverse the direction of the magnetic field.

The corkscrew rule enables the direction of the magnetic field round a wire to be found. Imagine a right-handed corkscrew being driven into the wire so that it progresses in the direction of current flow; the direction of the magnetic field around the wire will then be in the direction of rotation of the corkscrew.

The magnetic field surrounding a single wire is relatively weak. Forming the wire into a coil will combine the field of each turn producing a stronger field. A much greater increase in the magnetic field can then be achieved by inserting a piece of soft iron, called a core, inside the coil.

**Fig 1.16** and the bottom photograph in Fig 1.13 show the magnetic field produced by a coil or solenoid as it is often called. It will be seen that it is very similar to that of a bar magnet also shown in Fig 1.13. A north pole is produced at one end of the coil and a south pole at the other. Reversal of the current will reverse the polarity of the electromagnet. The polarity can be deduced from the S rule, which states that the pole that faces an observer looking at the end of the coil is a south

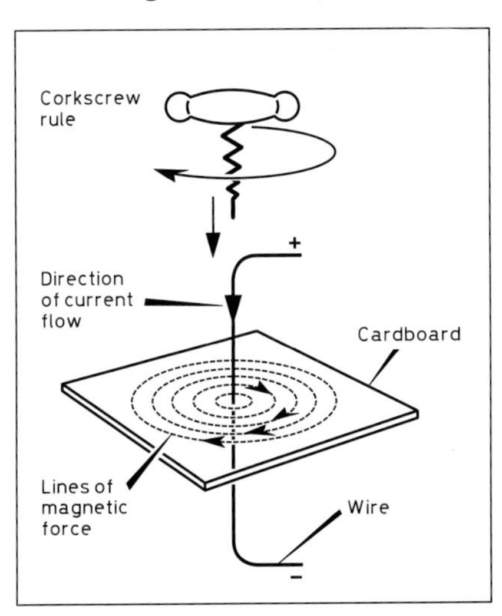


**Fig 1.15: Magnetic field produced by current flowing in a straight wire**

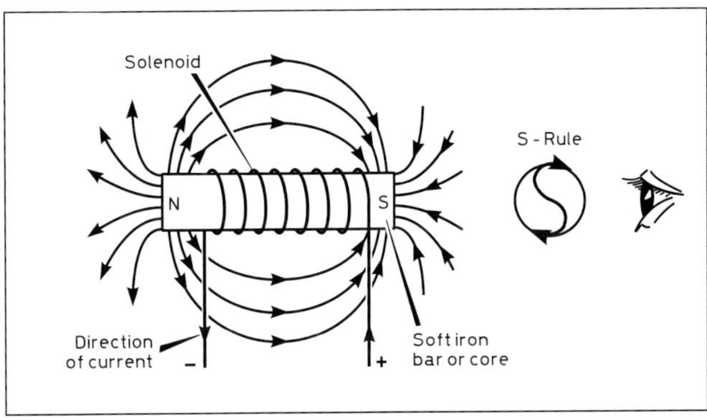


**Fig 1.16: The S rule for determining the polarity of an electromagnet**

pole if the current is flowing in a clockwise direction; see Fig 1.16. The current is the conventional current, not the direction of flow of electrons.

The strength of a magnetic field produced by a current is directly proportional to the current, a fact made use of in moving coil meters (see Test equipment chapter). It also depends on the number of turns of wire, the area of the coil, and the permeability of the core.

## Interaction of Magnetic Fields

Just as permanent magnets can attract or repel, so can electromagnets. If one of the devices, a coil for example, is free to move, then a current will cause the coil to move with a force or at a rate related to the magnitude of the current. The moving coil meter relies on this effect, balancing the force caused by the current against a return spring so that the movement or deflection indicates the magnitude of the current.

## Electromagnetic Induction

If a bar magnet is plunged into a coil as indicated in **Fig 1.17(a)**, the moving-coil microammeter connected across the coil will show a deflection. The explanation of this phenomenon, known as electromagnetic induction, is that the movement of the magnet's lines of force past the turns of the coil causes an electromotive force to be induced in the coil which in turn causes a current to flow through the meter. The magnitude of the effect depends on the strength and rate of movement of the magnet and the size of the coil. Withdrawal of the magnet causes a reversal of the current. No current flows unless the lines of force are moving relative to the coil. The same effect is obtained if a coil of wire is arranged to move relative to a fixed magnetic field. Dynamos and generators depend for their operation on the principle of electromagnetic induction.

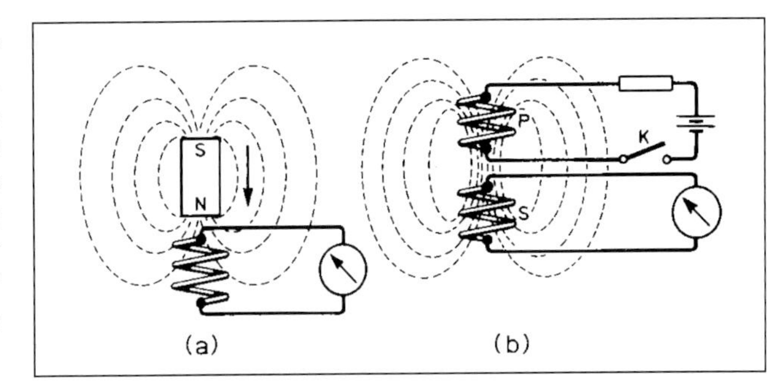


**Fig 1.17: Electromagnetic induction. (a) Relative movement of a magnet and a coil causes a voltage to be induced in the coil. (b) When the current in one of a pair of coupled coils changes in value, current is induced in the second coil**

Consider a pair of coils of wire are arranged as shown in **Fig 1.17(b)**. When the switch K is open there is no magnetic field from the coil P linking the turns of the coil S, and the current through S is zero. Closing K will cause a magnetic field to build up due to the current in the coil P. This field, while it is building up, will induce an EMF in coil S and cause a current to flow through the meter for a short time until the field due to P has reached a steady value, when the current through S falls to zero again. The effect is only momentary while the current P is changing.

The fact that a changing current in one circuit can induce a voltage in another circuit is the principle underlying the operation of transformers.

## Self-inductance

Above we considered the effect of a change of current in coil P inducing a voltage in coil S. In fact the changing field also induces an EMF in coil P even though it is the current in coil P that is causing the effect. The induced EMF is of a polarity such that it tends to oppose the original change in current.

This needs some care in understanding. On closing the switch K the EMF induced in P will tend to oppose the build up of current, that is it will oppose the voltage from the battery and the current will build up more slowly as a result. However if K is opened the current will now fall and the EMF induced in P will be of opposite polarity, trying to keep the current flowing. In reality the current falls more slowly.

This effect of the induced EMF due to the change in current is known as *inductance*, usually denoted by the letter L. If the current is changing at the rate of one amp per second (1A/s), and the induced EMF is one volt, the coil has an inductance of one Henry (1H). A 2H coil will have 2V induced for the same changing current. Since the induced voltage is in the coil containing the changing current this inductance is properly called *self-inductance*.

$$\text{Inductance } L = \frac{V}{dI/dt}$$

where dI is a small change in current and dt is the time for that change.

That is

$$dI/dt$$

is the rate of change of current in A/s

The inductance values used in radio equipment may be only a very small fraction of a henry, the units millihenry (mH) and microhenry (μH) meaning one thousandth and one millionth of a henry respectively are commonly used.

The inductance of a coil depends on the number of turns, on the area of the coil and the permeability of the core material on which the coil is wound. The inductance of a coil of a certain physical size and number of turns can be calculated to a fair degree of accuracy from formulas or they can be derived from coil charts.

## Mutual Inductance

A changing current in one circuit can induce a voltage in a second circuit: as in **Fig 1.16(b)**. The strength of the voltage induced in the second circuit depends on the closeness or tightness of the magnetic coupling between the circuits; for example, if both coils are wound together on an iron core practically all the magnetic flux from the first circuit will link with the turns of the second circuit. Such coils would be said to be tightly coupled whereas if the coils were both air-cored and spaced some distance apart they would be loosely coupled.

The mutual inductance between two coils is also measured in henrys, and two coils are said to have a mutual inductance of one henry if, when the current in the primary coil changes at a rate of one ampere per second, the voltage across the secondary is one volt. Mutual inductance is often denoted in formulas by the symbol M.

## Inductors in Series and Parallel

Provided that there is no mutual coupling between inductors when they are connected in series, the total inductance obtained is equal to the sum of the separate inductances. When they are in parallel the reciprocal of the total inductance is equal to the sum of the reciprocals of the separate inductances.

If L is the total inductance (no mutual coupling) the relationships are as follows:

Series connection

$$L_{total} = L1 + L2 + L3 \text{ etc}$$

Parallel connection

$$\frac{1}{L_{total}} = \frac{1}{L1} + \frac{1}{L2} + \frac{1}{L3} \quad \text{etc}$$

These two formulas are of the same format as the formula for resistors and the special case of only two resistors is also true for inductors. In reality paralleled inductors are uncommon but may be found in older radio receivers that are permeability tuned. That is where the inductance of a tuned circuit (see later) is varied rather than the more common case of variable capacitors.

## CR Circuits and Time Constants

**Fig 1.18(a)** shows a circuit in which a capacitor C can either be charged from a battery of EMF E, or discharged through a resistor R, according to whether the switch S is in position a or b.

If the switch is thrown from b to a at time ta, current will start to flow into the capacitor with an initial value E/R. As the capacitor charges the potential difference across the capacitor increases, leaving less PD across the resistor, and the current through the circuit therefore falls away, as shown in the charging portion of **Fig 1.18(b)**. When fully charged to the voltage E the current will have dropped to zero.

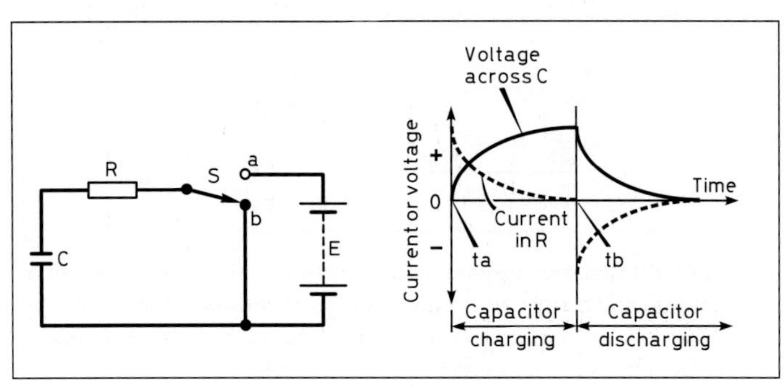


**Fig 1.18: In (a) a capacitor C can be charged or discharged through the resistor R by operating the switch S. The curves of (b) show how the voltage across the capacitor and the current into and out of the capacitor vary with time as the capacitor is charged and discharged. The curve for the rise and fall of current in an LR circuit is similar to the voltage curve for the CR circuit**

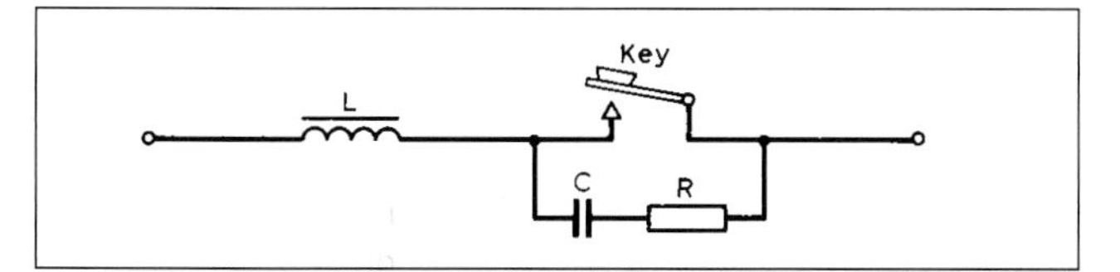


**Fig 1.19: Typical key-click filter. L serves to prevent a rise of current. C, charging through R, serves to continue flow of current briefly when key contacts open. Typical values: L = 0.01 to 0·1H, C = 0·01 to 0·1µF, R = 10 to 100 ohms**

At time tb, the switch is thrown back to b, the capacitor will discharge through the resistor R, the current being in the opposite direction to the charging current, starting at a value -V/R and dying away to zero. As the capacitor discharges, the PD across its plates falls to zero as shown in the discharge portion of Fig 1.18(b).

The voltage at any point during the charge cycle is given by the formula

$$V = V_b\left(1 - e^{-\frac{t}{CR}}\right)$$

Where $V_b$ is the battery voltage and t is the time, in seconds, after the switch is thrown. C and R are given in Farads and Ohms respectively and e is the base of natural logarithms.

This is known as an exponential formula and the curve is one of a family of exponential curves.

As the capacitor approaches fully charged, or discharged, the current is very low. In theory an infinite time is needed to fully complete charging. For practical purposes the circuit will have charged to 63% (or 1/e) in the time given by CR and this time is known as the *time constant* of the circuit. On discharge the voltage will have fallen to 37% of the initial voltage (1-1/e) after one time constant. Over the next time constant it will have fallen to 37% of its new starting voltage, or 14% of the original voltage. Five time constants will see the voltage down to 0·7%.

Time constant $\tau$ = CR seconds

As an example, the time constant of a capacitance of 0·01µF ($10^{-8}$F) and a resistance of 47kΩ ($4·7 \times 10^4$Ω) is:

$\tau = 10^{-8} \times 4·7 \times 10^4$

$= 4·7 \times 10^{-4}$s = 0·47ms

High voltage power supplies should have a bleeder resistor across the smoothing capacitor to ensure lethal voltages are removed before anyone can remove the lid after switching off, thinking it is safe! The time constant here may be a few tens of seconds. In audio detector circuits, the capacitor following the detector diode is chosen so that, along with the load resistance, the time constant is rather longer than the period of the intermediate frequency to give good smoothing or filtering. However it also needs to be rather shorter than the period of the highest audio frequency, or unwanted attenuation of the higher audio notes will occur.

A similar constraint applies in AGC (automatic gain control) circuits where the receiver must be responsive to variation caused by RF signal fading without affecting the signal modulation.

Digital and pulse circuits may rely on fast transient waveforms. Here very short time constants are required, unwanted or stray capacitance and inductance must be avoided.

## LR Circuits

Inductors oppose the change, rise or fall, of current. The greater the inductance, the greater the opposition to change. In a circuit containing resistance and inductance the current will not rise immediately to a value given by V/R if a PD V is applied but will rise at a rate depending on the L/R ratio. LR circuits also have a time constant, the formula is:

$\tau$ = L/R where L is in henries and R in ohms

Fig 1.18 showed the rise in capacitor voltage as it charged through a resistor; the curve of the rise in current in an LR circuit is identical. In one time constant, the current will have risen to 63%, (1 - 1/e) of its final value, or to decay to 37% of its initial value.

An example of the use of an inductor to slow the rise and fall of current is the Morse key filter shown in **Fig 1.19**. Slowing the rise of current as the key is depressed is simple enough but when the key is raised the circuit is cut and a high voltage can be developed which will cause sparking at the key contacts. The C and R across the key provide a path for the current to flow momentarily as it falls to zero. The reason for the use of this filter is discussed in the Morse chapter.

## ALTERNATING CURRENT

So far we have concerned ourselves with uni-directional current flow or direct current (DC). Audio and radio circuits rely heavily on currents (and voltages) that change their polarity continuously; alternating currents (AC).

**Fig 1.20(a)** shows a battery and reversing switch. Continually operating the switch would cause the current in the resistor to flow in alternate directions as shown by the waveform in **Fig 1.20(b)**. The waveform is known as a square wave.

Alternating current normally has a smoother waveform shown in **Fig 1.20(d)** and can be produced by an electrical generator shown in **Fig 1.20(c)**. Consider for a moment a single wire in the coil in Fig 1.19(c) and remember that if a wire moves through a magnetic field an EMF is induced in the wire proportional to the relative velocity of the wire.

At the bottom the wire is travelling horizontally and not cutting through the field; no EMF is induced. After 90 degrees of rotation the wire is travelling up through the field at maximum velocity and the induced EMF is at a maximum. At the top the EMF has fallen to zero and now reverses polarity as the wire descends, falling to zero again at the bottom where the cycle begins again. The vertical component of the velocity follows a sinusoidal pattern, as does the EMF; it is a sine wave. The precise shape can be plotted using mathematical 'sine' tables used in geometry.

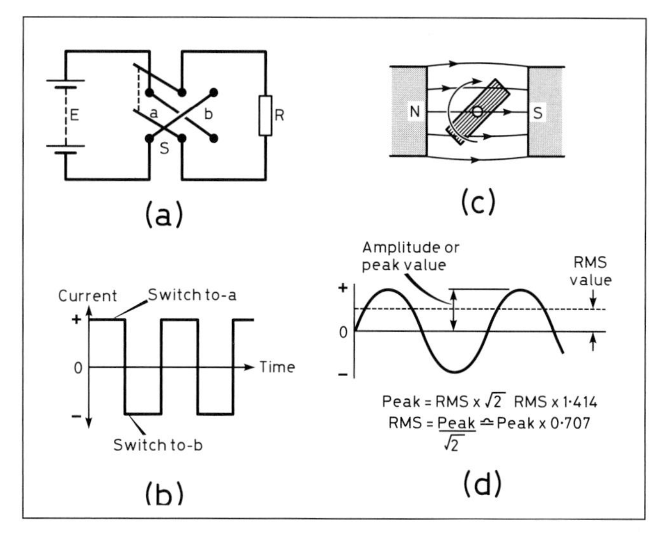


**Fig 1.20: Alternating current. A simple circuit with a current-reversing switch shown at (a) produces a square-wave current through the resistor R as shown in (b). When a coil is rotated in a magnetic field as in (c) the voltage induced in the coil has a sinusoidal waveform (d)**

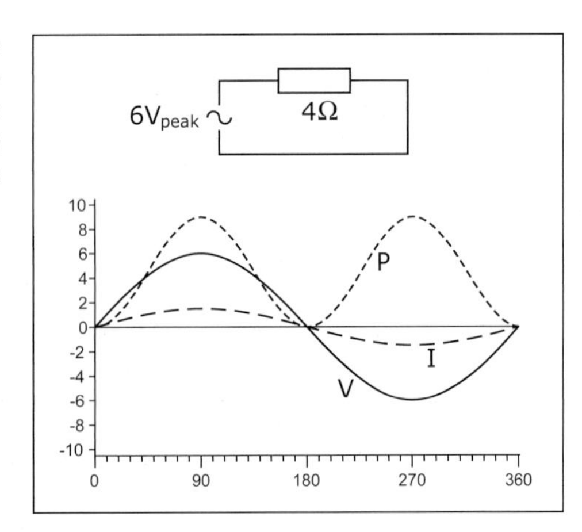


**Fig 1.21: The power dissipated in a resistor with a sine wave of current**

## Specifying an AC Waveform

For AC there are two parameters that must be quoted to define the current or voltage. Like DC we must give the magnitude or *amplitude* as it is called and we must also say how fast the cycles occur.

If t is the time for one cycle (in seconds) then 1/t will give the number of cycles occurring in a second; this is known as the frequency.

t = 1/f and f=1/t

The unit of frequency is cycles per second and is given the name Hertz, abbreviation Hz.

*Example:* the UK mains has a frequency of 50Hz, what is the periodic time?

f =50Hz so the time for 1 cycle t =1/f = 1/50 second or 0·02s.

## RMS Values

Specifying the amplitude is more interesting. It would seem sensible to quote the maximum amplitude and allow the reader to find the amplitude at other parts of the cycle by looking up the values in sine tables. In actual fact the RMS (root mean square) value is used but what does that mean?

Consider passing an AC current through a resistor. The power dissipated in the resistor will vary over the cycle as shown in **Fig 1.21**.

The peak current is 6V/4Ω = 1·5A and peak power is 6V x 1·5A = 9W. The power curve is symmetrical about 4·5W and the average power over the cycle is indeed 4·5W. The DC voltage that would produce 4·5W in a 4Ω resistor is just over 4V (4·24V). This is the RMS value of the 6V peak waveform. It is that value of DC voltage (or current) that would produce the same heating effect in a resistor.

The RMS value is related to the peak value by

$$\mathrm{RMS} = \frac{\mathrm{Peak}}{\sqrt{2}}$$

or

$$\mathrm{Peak} = \mathrm{RMS} \times \sqrt{2}$$

It may help to remember that

$$\sqrt{2} = 1 \cdot 4142$$

and

$$1/\sqrt{2} = 0 \cdot 707$$

*Example:* suppose your mains supply is 240V RMS, what is the peak value?

$$\mathrm{Peak} = \mathrm{RMS} \times \sqrt{2} \quad = 240 \times \sqrt{2} \quad = 340\mathrm{V}$$

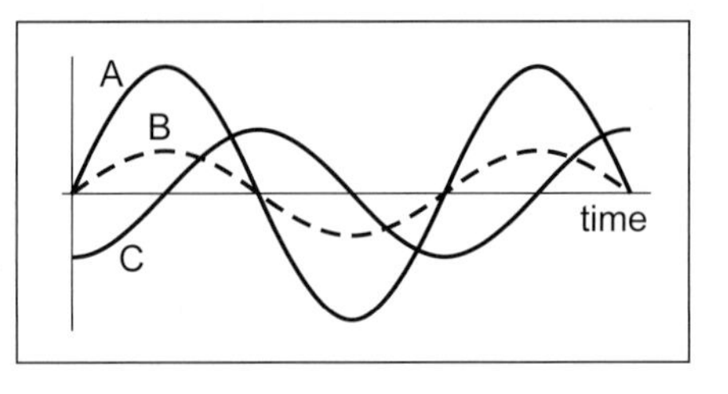


**Fig 1.22: Relative phase**

## Other Frequencies

Audible frequencies range from around 50-100Hz up to 20kHz for a young child, rather less for adults. Some animals can hear up to around 40-50kHz. Dogs will respond to an ultrasonic whistle (above human hearing) and bats use high pitch sounds around 50kHz for echo location.

Radio frequencies legally start at 9kHz, and 20kHz is used for worldwide maritime communication. Below 30kHz is known as the VLF band (very low frequency; LF is 30-300kHz; MF (medium frequencies) is 300-3000kHz. The HF, high frequency, band is 3 to 30MHz, although amateurs often refer to the 1·8MHz amateur band as being part of HF. VHF is 30-300MHz and UHF (ultra-high frequencies) is 300-3000MHz. Above that is the SHF band 3-30GHz, super-high frequencies. Also, by common usage, the 'microwave band' is regarded as being above 1GHz.

## Phase and Harmonics

Two waveforms that are of the same frequency are *in phase* if they both start at the same point in time. If one waveform is delayed with respect to the other then they are not in phase and the phase difference is usually expressed as a proportion of a complete cycle of 360°. This is shown in **Fig 1. 22**. Waveforms A and B are in phase but are of different amplitudes. Waveform C is not in phase with A, it lags A by 1/4 cycle or 90°, or A leads C by 90°. It lags because the 'start' of the cycle is to the right on the time axis of the graph, that is, it occurs later in time. The 'start' is conventionally regarded as the zero point, going positive.

If two waveforms are of different frequencies then their phase relationships are continuously changing. However, if the higher frequency is an exact multiple of the lower, the phase relationship is again constant and the pattern repeats for every cycle of the lower frequency waveform. These multiple frequencies are known as harmonics of the lower 'fundamental' frequency. The second harmonic is exactly twice the frequency and the third, three times the frequency.

**Fig 1.23** shows a fundamental and a third harmonic at 1/6 amplitude. The heavy line is the sum of the fundamental and the harmonic. It shows a phenomenon known as harmonic distortion.

If a sine wave is distorted, perhaps by over-driving a loudspeaker or by imperfections (often deliberate) in electronic circuits, then the distortion can be regarded as the original sine wave plus the right amplitudes and phases of various harmonics required to produce the actual distorted waveform. The procedure to determine the required harmonics is known as Fourier analysis.

It is important to appreciate that in distorting an otherwise clean sine wave, the harmonics really are created; new

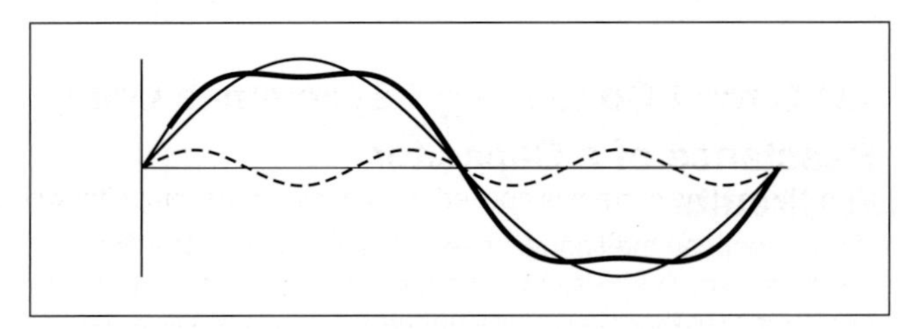

**Fig 1.23: Harmonics of a sine wave**

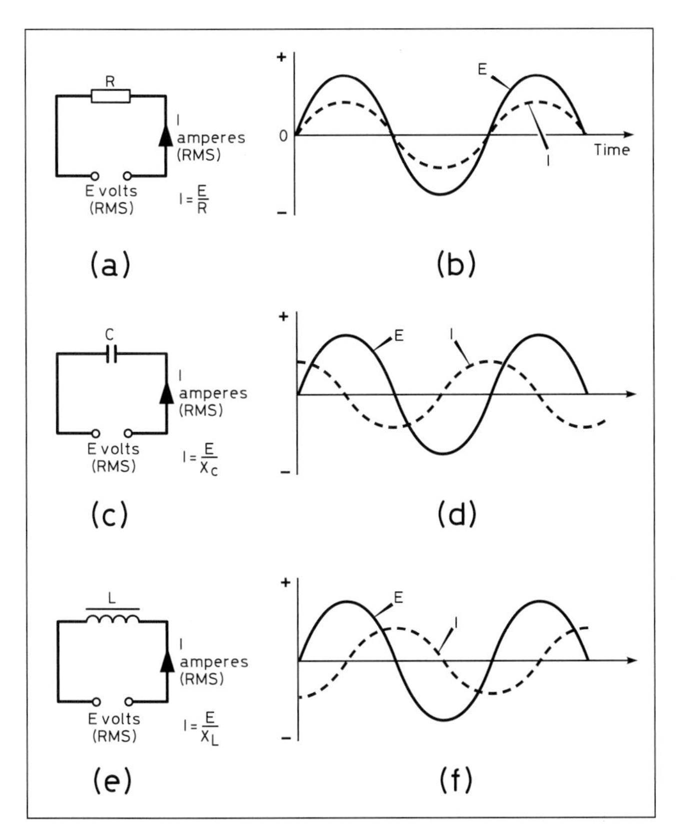


**Fig 1.24: Voltage and current relationships in AC circuits comprising (a) resistance only, (c) capacitance only and (e) inductance only**

frequencies are now present that were not there prior to the distortion. If a distorted sine wave is observed using a spectrum analyser, the harmonic frequencies can be seen at their relative amplitudes. Moreover, if they can be filtered out, the distortion will have been removed. This will prove a useful technique later in the discussions on radio receivers, intermediate frequencies and AGC (automatic gain control).

## AC Circuit Containing Resistance Only

Ohm's Law is true for a resistor at every point in a cycle of alternating current or voltage. The current will flow backwards and forwards through the resistor under the influence of the applied voltage and will be in phase with it as shown in **Fig 1.24(a)**.

The power dissipated in the resistor can be calculated directly from the power formulas, provided that RMS values for voltage and current are used.

$$P = VI \qquad P = \frac{V^2}{R} \qquad P = I^2R$$

If peak values of voltage and current are used, these formulas become:

$$P = \frac{V_P}{\sqrt{2}} \times \frac{I_P}{\sqrt{2}} = \frac{V_P I_P}{2} \qquad P = \frac{V_P{}^2}{2R}$$

## AC Circuit Containing Capacitance Only; Reactance of a Capacitor

If an alternating current is applied to a capacitor the capacitor will charge up. The PD at its terminals will depend on the amount of charge gained over the previous part cycle and the capacitance of the capacitor. With an alternating current, the capacitor will charge up first with one polarity, then the other. **Fig 1.24(c)** shows the circuit and **Fig 1.24(d)** shows the voltage (E) and current (I) waveforms. Inspection of **Fig 1.24(d)** shows that the voltage waveform is 90 degrees lagging on the current. This is explained by remembering that the voltage builds up as more charge flows into the capacitor. When the current is a maximum, the voltage is increasing at its maximum rate. When the current has fallen to zero the voltage is (momentarily) constant at its peak value.

There is no such thing as the resistance of a capacitor. Consider **Fig 1.24(d)** again. At time zero the voltage is zero yet a current is flowing, a quarter cycle (90°) later the voltage is at a maximum yet the current is zero. The resistance appears to vary between zero and infinity! Ohm's Law does not apply.

What we can do is to consider the value of $V_{rms}/I_{rms}$. It is not a resistance but it is similar to resistance because it does relate the voltage to the current. The quantity is called *reactance* (to denote to 90° phase shift between V and I) and is measured in ohms, Ω.

Remember that for the same charge, a larger value capacitor will have a lower potential difference between the plates. This suggests a larger value capacitor will have a lower reactance. Also, as the frequency rises, the periodic time falls and the charge, which is current x time, also falls. The reactance is lower as the frequency rises, it varies with frequency. The proof of this requires integral calculus which is beyond the scope of the book but the formula for the reactance of a capacitor is

$$X_C = \frac{1}{2\pi fC}$$

where f is the frequency in Hertz and C the capacitance in Farads. You may also meet this formula written using the symbol ω instead of 2πf. The lower case Greek letter omega, ω, gives the frequency in radians per second - remember there are 2π radians in 360° and ω is called the angular frequency.

*Example:*

A capacitor of 500pF is used in an antenna matching unit and is found to have 400V across it when the transmitter is set to 7MHz. Calculate the reactance and current flowing.

$$X_C = \frac{1}{2\pi fC} = \frac{1}{2\pi \times 7 \times 10^6 \times 500 \times 10^{-12}}$$

$$X_C = \frac{1}{7\pi \times 10^{-3}} = \frac{1000}{7\pi} = 45 \cdot 5\Omega$$

With 400V applied the current is

$$I = \frac{V}{X_C} = \frac{400}{45 \cdot 5} = 8 \cdot 8A$$

## AC Circuit Containing Inductance Only; Reactance of a Coil

The opposition of an inductance to alternating current flow is called the inductive reactance of the coil: **Fig 1.24(e)** shows the circuit and the current and voltage waveforms. If a potential difference is applied to a coil, the current builds up slowly, how slowly depending on the inductance. Similarly, if the PD is reduced, the current in the coil tries to keep flowing. With AC the current waveform lags the voltage waveform by 90°. As the frequency increases the current in the coil is being expected to change more rapidly and the opposition is greater, the magnitude of the current is less and the reactance higher.

This leads to the formula for the reactance of a coil as:

$$X_L = 2\pi fL$$

As before $X_L$ is in ohms, L in henries and f in Hertz.

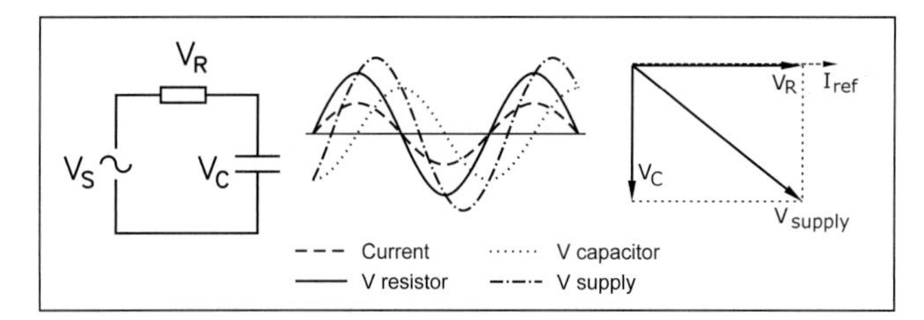


**Fig 1.25: Voltage and current in a circuit containing a resistor and capacitor**

*Example.* A coil has an inductance of 5μH, calculate its reactance at 7MHz.

$$X_L = 2\pi fL \quad = 2\pi \times 7 \times 10^6 \times 5 \times 10^{-6}$$
$$= 70\pi \quad = 220\Omega$$

## AC Circuits with R and C or R and L

Consider **Fig 1.25**, a resistor and capacitor in series. The same current flows through both R and C and is shown by the dashed waveform. The voltage across R will be in phase with the current, shown solid, and the voltage across C will lag by 90°, shown dotted. The supply voltage is the sum of the voltages across R and C. Since the voltages across R and C are not in phase, they cannot simply be added together; they must be added graphically, shown by the dash-dot line. For example when $V_R$ is at a maximum, $V_C$ is zero, so the sum should be coincident with $V_R$.

The waveform diagram is messy to say the least, it has been shown once to illustrate the situation. Fortunately there is a simpler way, the phasor diagram shown to the right in Fig 1.25. By convention the current, which is common to both components, is drawn horizontally to the right. The vector or phasor representing the voltage across the resistor is drawn parallel to it and of a length representing the magnitude of the voltage across R. $V_C$ lags by 90°, shown as a phasor downwards, again of the correct length to represent the magnitude of the voltage across C. The vector sum, the resultant, gives the supply voltage. The length of the arrow represents the actual voltage and the angle will give its phase.

Since $V_R$ and $V_C$ are at right angles we can use Pythagoras' Theorem to obtain a formula for the voltage.

$$V_{supply} = \sqrt{V_R^{\ 2} + V_C^{\ 2}}$$

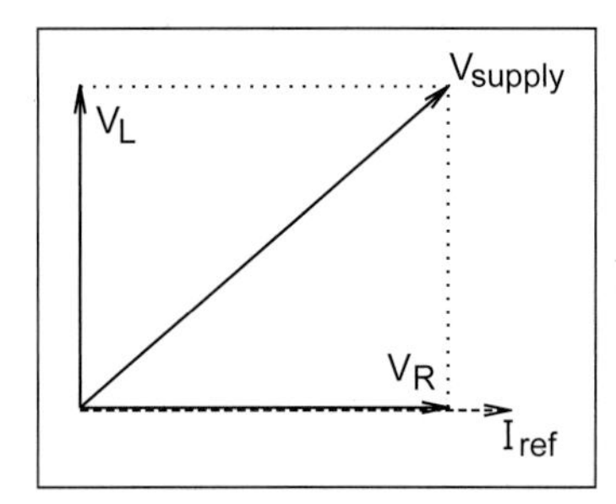


**Fig 1.26: Phasor diagram for R and L**

**(below) Fig 1.27: L and C in series**

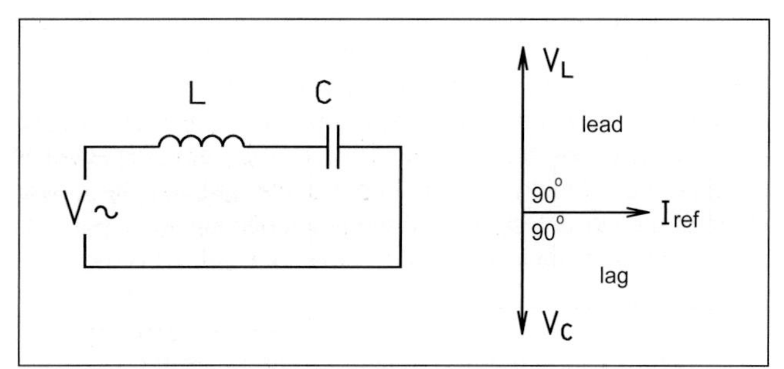


By a similar argument we can also obtain:

$$Z = \sqrt{R^2 + X_C^{\ 2}}$$

where Z is the impedance of the CR circuit, R is the resistance and $X_C$ the reactance.

Impedance (symbol Z) is the term used to denote the 'resistance' of a circuit containing both resistance and reactance and is also measured in ohms. The convention is that *resistance* means V and I are in phase; *reactance* means 90° (capacitive or inductive) and *impedance* means somewhere between, or indeterminate.

**Fig 1.26** shows the phasor diagram for R and L. Since the coil voltage now leads the current, it is drawn upwards but the geometry and the formula are the same.

$$V_{supply} = \sqrt{V_R^{\ 2} + V_L^{\ 2}}$$

$$Z = \sqrt{R^2 + X_L^{\ 2}}$$

## Capacitance and Inductance, Resonance

In the series circuit containing C and L the current is still the common factor and the voltage across C will lag the current by 90° whilst the voltage across the inductor will lead by 90°. Consequently these two voltages will be 180° out of phase; that is, in anti-phase. The two voltages will tend to cancel rather than add. **Fig 1.27** shows the circuit and the phasor diagram. As drawn, the voltage across the capacitor is less than the voltage across the inductor, indicating the inductor has the greater reactance. In general the total reactance is given by:

$X = X_L - X_C$.

However the reactance varies with frequency as shown in **Fig 1.28.** If the reactances of C and L are equal, the voltages $V_C$ and $V_L$ will also be equal and will cancel exactly. The voltage across the circuit will be zero. This occurs at a frequency where the curves intersect in Fig 1.28, it is called the resonant frequency.

At resonance the total reactance $X = X_L - X_C$ will be zero and if $X_L = X_C$ then:

$$2\pi fL = \frac{1}{2\pi fC}$$

Rearranging this equation gives:

$$f = \frac{1}{2\pi\sqrt{LC}} \quad \text{or} \quad C = \frac{1}{4\pi^2 f^2 L} \quad \text{or} \quad L = \frac{1}{4\pi^2 f^2 C}$$

**Fig 1.29** shows a circuit containing a capacitor, inductor and resistor. This is representative of real circuits, even if the resistance is merely that of the coil. This may be higher than expected due to 'skin effect'; a phenomenon explained later in the chapter. The overall impedance of the circuit is high when away from resonance but it falls, towards 'R' the value of resistance, as resonance is approached. The series resonant circuit is sometimes called an acceptor circuit because it accepts current at resonance. The current is also shown in Fig 1.29.

*Example 1:*

A 50μH inductor and a 500pF capacitor are connected in series; what is the resonant frequency?

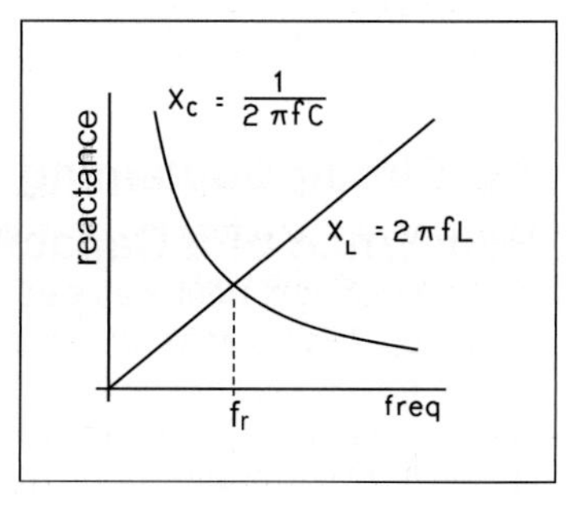


**Fig 1.28: Reactance of L and C**

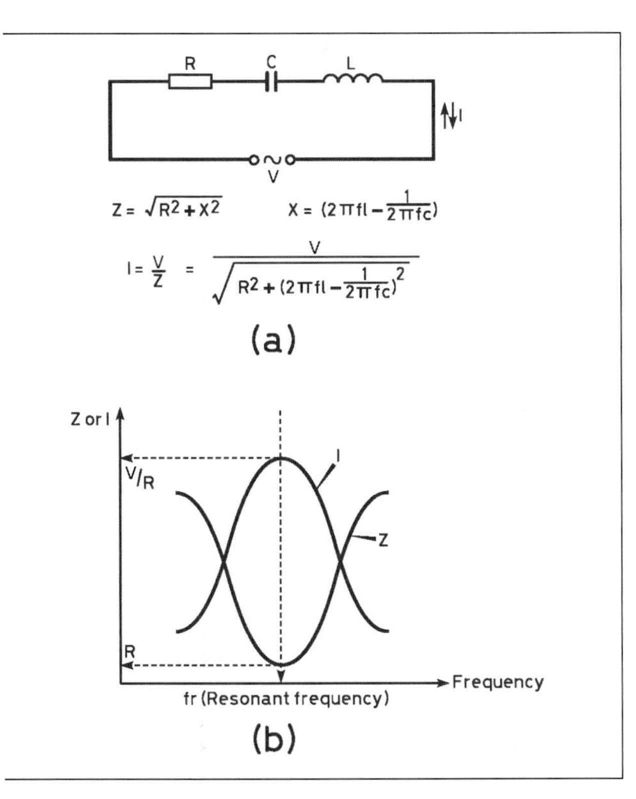


**Fig 1.29: The series-resonant circuit. The curves shown at (b) indicate how the impedance and the current vary with frequency in the type of circuit shown at (a)**

The formula is:

$$f = \frac{1}{2\pi\sqrt{LC}}$$

So inserting the values:

$$f = \frac{1}{2\pi\sqrt{50\times10^{-6}\times500\times10^{-12}}} = \frac{1}{2\pi\sqrt{5\times10^{-5}\times5\times10^{-10}}}$$

$$f = \frac{1}{2\pi\sqrt{25\times10^{-15}}} = \frac{1}{2\pi\sqrt{2\cdot5\times10^{-14}}}$$

$$f = \frac{1}{2\pi\times1.58\times10^{-7}} = \frac{10^7}{3\cdot16\pi} = 1\text{MHz}$$

*Note:* The calculation has been set out deliberately to illustrate issues in handling powers and square roots. In line 1 the numbers were put in standard form, that is a number between 1 and 10, with the appropriate power of 10. This then resulted in 25 x $10^{-15}$ inside the square root sign. The root of 25 is easy but the root of an odd power of 10 is not. Consequently the sum has been changed to 2·5 x $10^{-14}$. It so happens that this is also then in standard form but the real purpose was to obtain an even power of ten, the square root of which is found by halving the power' or exponent.

There is another method which may be easier to calculate; namely to use the formula for $f^2$ but remembering to take the square root at the end.

The formula is:

$$f^2 = \frac{1}{4\pi^2 LC}$$

An advantage is that $\pi^2$ is 9·87 which is close enough to 10, remembering the error is much less than the tolerance on the components.

*Example 2:*

A vertical antenna has a series inductance of 20μH and capacitance of 100pF. What value of loading coil is required to resonate at 1·8MHz?

The formula is

$$L = \frac{1}{4\eth^2 f^2 C} = \frac{1}{40\times\left(1\cdot8\times10^6\right)^2\times10^{-10}}$$

$$L = \frac{1}{40\times3\cdot24\times10^{12}\times10^{-10}} = \frac{1}{12960}$$

which is 77μH.

The antenna already has 20μH of inductance, so the loading coil needs to add another 57μH. A coil some 10cm diameter of 30 turns spread over 10cm could be suitable.

## Parallel Resonance

The L and C can also be connected in parallel as shown in **Fig 1.30**. Resistor 'r' is the internal resistance of the coil which should normally be as low as reasonably possible. The voltage is common to both components and the current in the capacitor will lead by 90° while that in the coil lags by 90°.

This time the two currents will tend to cancel out, leaving only a small supply current. The impedance of the circuit will increase dramatically at resonance as shown in the graph in Fig 1.30.

This circuit is sometimes called a rejector circuit because it rejects current at resonance; the opposite of the series resonant circuit.

## Magnification Factor, Q

Consider again the series resonant circuit in Fig 1.29. At resonance the overall impedance falls to the resistance R and the current is comparatively large. The voltage across the coil is still given by V = I x $X_L$ and similarly for the capacitor V = I x $X_C$. Thus the voltages across L and C are very much greater than the voltage across the circuit as a whole which is simply V = I x R. Remember $X_C$ and $X_L$ are very much larger than R.

The magnification factor is the ratio of the voltage across the coil (or capacitor) to that across the resistor.

The formula are:

$$Q = \frac{X_L}{R} \text{ or } \frac{X_C}{R}$$

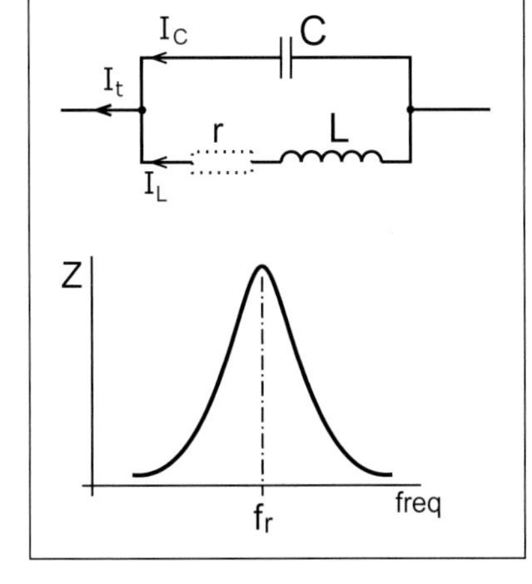


**Fig 1.30: Parallel resonance. The resistor 'r' may simply be the resistance of the coil L, but this is the resistance at the resonant frequency. See Skin Effect**

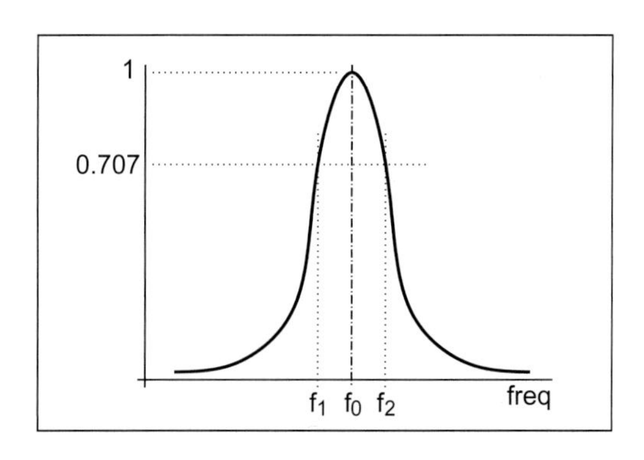


**Fig 1.31: Bandwidth of a tuned circuit**

that is:

$$Q = \frac{2\pi fL}{R} \quad \text{or} \quad \frac{1}{2\pi fCR}$$

This can also be written as:

$$Q = \frac{\omega L}{R} \quad \text{or} \quad \frac{1}{\omega CR} \qquad \text{where } \omega=2\pi f.$$

## Resonance Curves and Selectivity

The curves in Figs 1.29 and 1.30 showed how tuned circuits were more responsive at their resonant frequency than neighbouring frequencies. These resonance curves show that a tuned circuit can be used to select a particular frequency from a multitude of frequencies that might be present. An obvious example is in selecting the wanted radio signal from the thousands that are transmitted.

The sharpness with which a tuned circuit can select the wanted frequency and reject nearby ones is also determined by the Q factor. **Fig 1.31** shows a resonance curve with the centre frequency and the frequencies each side at which the voltage has fallen to $1/\sqrt{2}$ or 0·707 of its peak value, that is the *half power bandwidth*. The magnification or Q factor is also given by:

$$Q = \frac{\text{resonant frequency}}{\text{bandwidth}}$$

where the resonant frequency is $f_0$ and the bandwidth $f_2 - f_1$ in Fig 1.31.

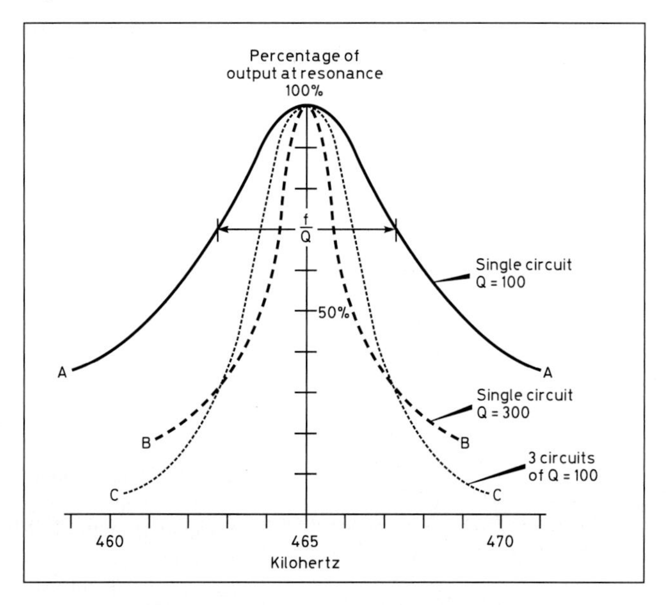


**Fig 1.32: Selectivity curves of single and cascaded tuned circuits**

| Relative bandwidth | Percentage output (voltage) | Loss (dB) |
|---|---|---|
| f/3Q | 95 | 0·45 |
| f/2Q | 90 | 0·92 |
| f/Q | 70 | 3 |
| 2f/Q | 45 | 6.9 |
| 4f/Q | 24 | 12.4 |
| 8f/Q | 12 | 18.4 |

**Table 1.4: Selectivity of tuned circuits**

*Example:*

A radio receiver is required to cover the long and medium wave broadcast bands, from 148·5kHz to 1606·5kHz. Each station occupies 6kHz of bandwidth. What Q factors are required at each end of the tuning range.

For long wave the Q factor is 148·5/6 ≈ 25. At the top of the medium wave broadcast band the required Q factor is 1606·6/6 ≈ 268. It is not realistic to achieve a Q factor of 268 with one tuned circuit.

**Fig 1.32** shows several resonance curves offering different selectivities. Curve A is for a single tuned circuit with a resonant frequency of 465kHz and a Q factor of 100. The width of the curve will be 4·65kHz at 70·7% of the full height. An unwanted signal 4kHz off-tune at 461kHz will still be present with half its voltage amplitude or one quarter of the power.

The dashed curve, B, has a Q of 300; the higher Q giving greater selectivity. In a radio receiver it is common to have several stages of tuning and amplification, and each stage can have its own tuned circuit, all contributing to the overall selectivity. Curve C shows the effect of three cascaded tuned circuits, each with a Q of 100. In practice this is better than attempting to get a Q of 300 because, as will be seen in later chapters, the usual requirement is for a response curve with a reasonably broad top but steeper sides. For a single tuned circuit the response, as a proportion of the bandwidth, is shown in **Table 1.4**. The values are valid for series circuits and parallel circuits where the Q-factor is greater than 10, as is normally the case.

## Dynamic Resistance

**Fig 1.33** shows a parallel tuned circuit with the resistance of the coil separately identified. This resistance means that the currents $I_L$ and $I_C$ are not exactly equal and do not fully cancel. The impedance of the circuit at resonance is high, but not infinite. This impedance is purely resistive, that is V and I are in phase, and is known as the dynamic impedance or resistance.

To find its value, the series combination of R and L must be transformed into its parallel equivalent.

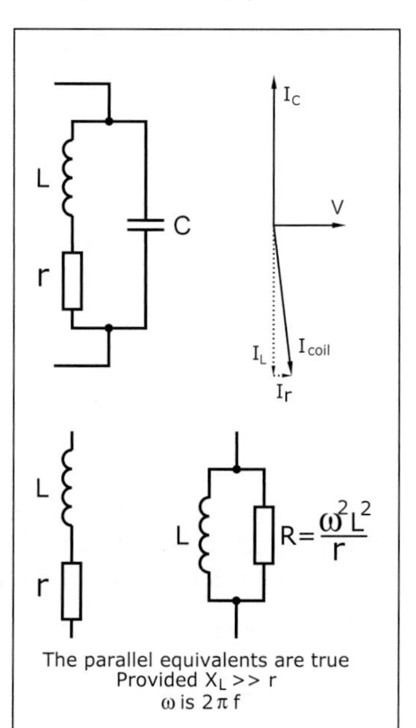


**Fig 1.33: A parallel LCR circuit. The phasor diagram shows that the current in the C and LR branches do not cancel exactly due to the phase change caused by r. At resonance the parallel tuned circuit behaves as a high value resistor called $R_D$, the dynamic resistance**

$$R_P = \frac{R_S^2 + X_S^2}{R_S} \quad \text{and} \quad X_P = \frac{R_S^2 + X_S^2}{X_S}$$

The proof of these transformations is outside the scope of this book.

If the resistance is considerably lower than the reactance of L; normal in radio circuits, then this simplifies to

$$R_P = \frac{X_S^2}{R_S} \quad \text{and} \quad X_P = X_S$$

Remembering that $X_S=2\pi fL$ and that, at resonance, $2\pi fL = 1/2\pi fC$, then the dynamic resistance, normally written $R_D$ can be written as

$$R_D = \frac{L}{Cr}$$

where L and C are the coil and capacitor values and r is the series resistance of the coil.

Remembering also that $Q=2\pi fL/r$, allows another substitution in the formula to get:

$Q=2\pi fCR_D$.

The significance of this is to note that a high $R_D$ implies a high Q.

## L/C Ratio

From the formula for $R_D$ above, it can be seen that as well as keeping r as low as practicable, the values of L and C will influence $R_D$. The value of L x C is fixed (from the formula for the resonant frequency) but we can vary the ratio L/C as required with the choice determined by practical considerations.

A high L/C ratio is normal in HF receivers, allowing a high dynamic resistance and a high gain in the amplifier circuits. Stray capacitances limit the minimum value for C. In variable frequency oscillators (VFOs) on the other hand a lower ratio is normal so the capacitor used swamps any changes in the parameters of the active devices, minimising frequency instability.

External components in the rest of the circuit may appear in parallel with the tuned circuit, affecting the actual $R_D$ and reducing overall Q factor. The loading of power output stages has a significant effect and the L/C ratio must then be a compromise between efficiency and harmonic suppression.

## Skin Effect

Skin effect is a phenomenon that affects the resistance of a conductor as the frequency rises. The magnetic field round the conductor also exists inside the conductor but the magnetic field at the surface is slightly weaker than at the centre. Consequently the inductance of the centre of the conductor is slightly higher than at the outer surface. The difference is small but at higher frequencies is sufficient that the current flows increasingly close to the surface. Since the centre is not now used, the cross-section of the conductor is effectively reduced and its resistance rises.

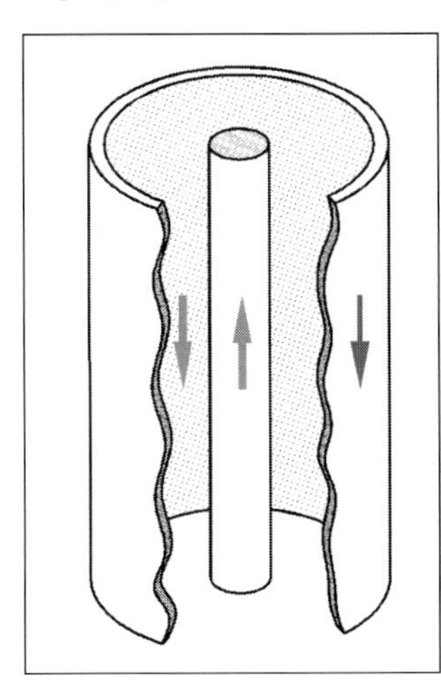

Fig 1.34: Currents in a coaxial cable

The skin depth is the depth at which the current has fallen to 1/e of its value at the surface. (e is the base of natural logarithms). For copper this works out as:

$$\text{skin depth } \delta = \frac{66 \cdot 2}{\sqrt{f}} \text{ mm}$$

At 1MHz this is 0·066 mm. 26SWG wire, which might be used on a moderately fine coil, has a diameter of 0·46mm, (0·23mm radius) so current is flowing in approximately half the copper cross-section and the resistance will be roughly doubled.

At UHF, 430MHz, however, a self-supporting coil of 16SWG coil (wire radius 0·8mm) the skin depth is 0·003mm and only the surface will carry a current, resulting in a considerable increase in resistance. If possible it will be better to use an even larger wire diameter which has a larger surface or use silver plated wire because the resistivity of silver is lower.

A coaxial cable has three conducting surfaces when carrying RF currents. This is shown in **Fig 1.34**. Current flows mainly along the surface of the inner conductor, depicted by the 'up' arrow. An exactly equal and opposite current flows along the inside surface of the outer conductor. The outside surface of the outer conductor is an entirely separate conducting surface and may have no current at all or even an extraneous interfering signal unrelated to the signal inside the coax.

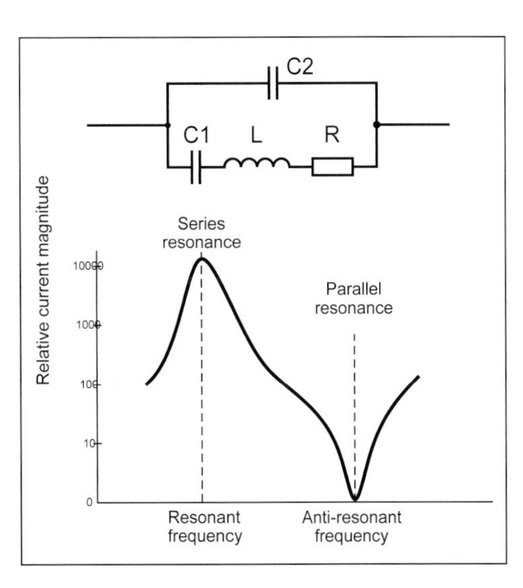


**Fig 1.35: Typical variation of current through a quartz crystal with frequency. The resonant and anti-resonant frequencies are normally separated by less than 0·1%; a few hundred hertz for a 7MHz crystal**

## Quartz Crystals

A quartz crystal is a very thin slice of quartz cut from a naturally occurring crystal. Quartz exhibits the piezo-electric effect, which is a mechanical-electrical effect. If the crystal is subjected to a mechanical stress, a voltage is developed between opposite faces. Similarly, if a voltage is applied then the crystal changes shape slightly. When an AC signal is applied to a crystal at the correct frequency, its mechanical resonance produces an electrical resonance. The resonant frequency depends on the size of the crystal slice. The electrical connections are made by depositing a thin film of gold or silver on the two faces and connecting two very thin leads.

Below 1MHz, the crystal is usually in the form of a bar rather than a thin slice. At 20kHz the bar is about 70mm long. Up to 22MHz the crystal can operate on its fundamental mode; above that a harmonic or overtone resonance is used. This is close to a multiple of the fundamental resonance but the term 'overtone' is used since the frequency is close to odd multiples of the fundamental but not exact.

**Fig 1.35** shows the equivalent circuit of a crystal and the two resonances possible with L and either C1 or C2. The two frequencies are within about 0·1% of each other. The crystal is supplied and calibrated for one particular resonance, series or parallel and should be used in the designed mode.

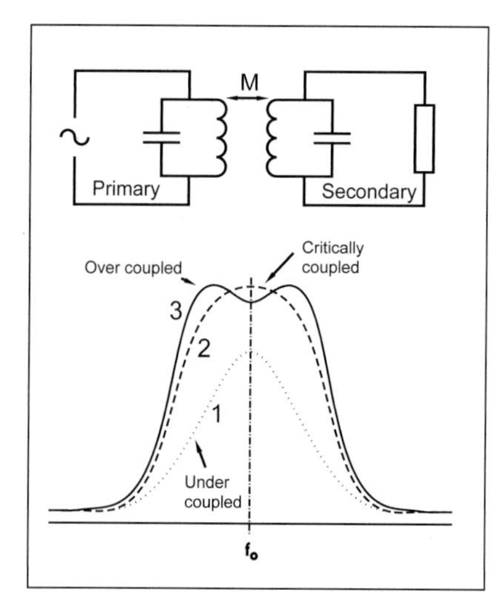


**Fig 1.36: Inductively coupled tuned circuits. The mutual inductance M and the coupling increase as the coils are moved together. The response varies according to the degree of coupling with close coupling giving a broad response**

The key advantage of using a crystal is that its Q factor is very much higher than can be achieved with a real LC circuit. Qs for crystals are typically around 50,000 but can reach 1,000,000. Care must be taken to ensure circuit resistances do not degrade this. There are more details in the oscillators chapter.

## Coupled Circuits

Pairs of coupled tuned circuits are often used in transmitters and receivers. The coupling is by transformer action (see later in this chapter), and the two tuned circuits interact as shown in **Fig 1.36**.

When the coupling is loose, that is the coils are separated, the overall response is as shown in curve 1. As the coils are moved closer, the coupling and the output increase until curve 2 is reached which shows *critical coupling*. Closer coupling results in curve 3 which is *over-coupling*. Critical coupling is the closest coupling before a dip appears in the middle of the response curve. Often some over-coupling is desirable because it causes a broader 'peak' with steep sides which rapidly attenuate signals outside the desired frequency range.

Two tuned circuits are often mounted in a screening can, the coils generally being wound the necessary distance apart on the same former to give the required coupling. The coupling is then said to be fixed.

Some alternative arrangements for coupling tuned circuits are shown in the Building Blocks chapter.

## Filters

Filters may be 'passive', that is an array of capacitors and inductors or 'active' where an amplifier and, usually, capacitors and resistors are used. Mostly active filters are used at lower frequencies, typically audio, and passive LC filters at RF. In both cases the aim is to pass some frequencies and block or attenuate others.

**Fig 1.37** shows the four basic configurations; low pass, high pass, band pass and band stop or notch; together with the circuit of a typical passive filter, and the circuits for passive filters. The Building Blocks chapter discusses the topic further.

## Transformers

Mutual inductance was introduced in Fig 1.17(b); a changing current in one coil inducing an EMF in another. This is the basis of the transformer. In radio frequency transformers the degree of coupling, the mutual inductance, was one of the design features. In power transformers, the windings, the primary and secondary, are tightly coupled by being wound on a bobbin sharing the same laminated iron core to share and maximise the magnetic flux (the name for magnetic 'current'). The size of core used in the transformer depends on the amount of power to be handled.

A transformer can provide DC isolation between the two coils and also vary the current and voltage by varying the relative number of turns. **Fig 1.38** shows a transformer with $n_P$ turns on the primary and $n_S$ turns on the secondary. Since the voltage induced in each coil will depend on the number of turns, we can conclude that:

$$V_S = V_P \times \frac{n_S}{n_P}$$

With a load on the secondary (the output winding) a secondary current. $I_S$ will flow. Recognising that, neglecting losses, the power into the primary must equal the power out of the secondary, we can also conclude that

$$I_P = I_S \times \frac{n_S}{n_P}$$

It is simple to say that if power is drawn from the secondary, the primary current must increase to provide that power, but that does not really provide an explanation. When current flows in the secondary the magnetic field due to the secondary current will weaken the overall field, resulting in a reduced back EMF in the primary. This allows more primary current to flow, restoring the field. The primary current does not fall to zero with no

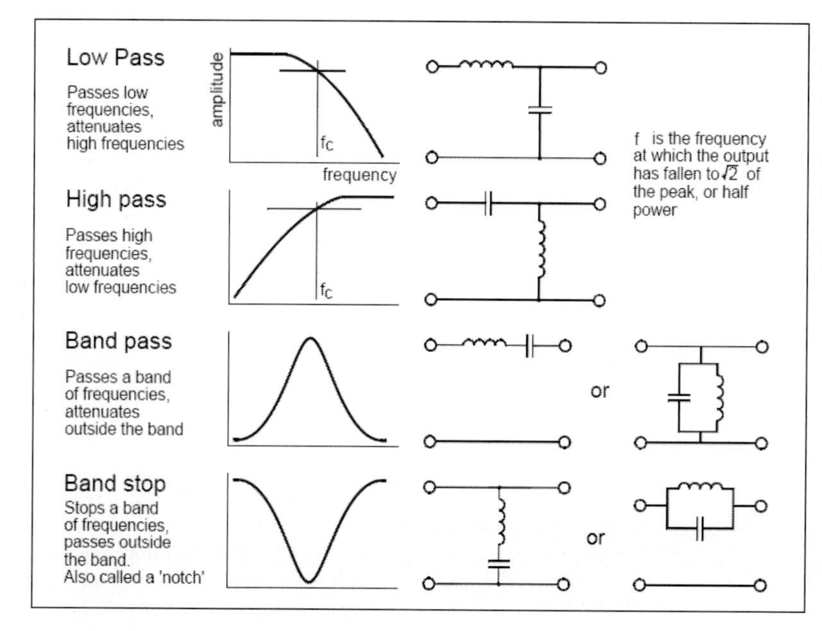


**Fig 1.37: The circuits and frequency response of low pass, high pass, band pass and band stop (or notch) filters**

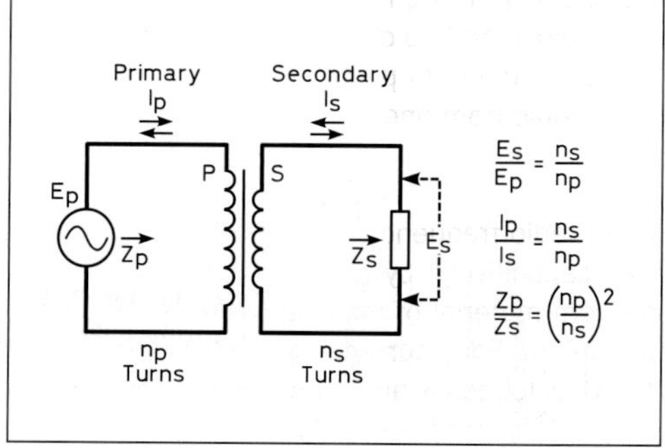


**Fig 1.38: The low-frequency transformer**

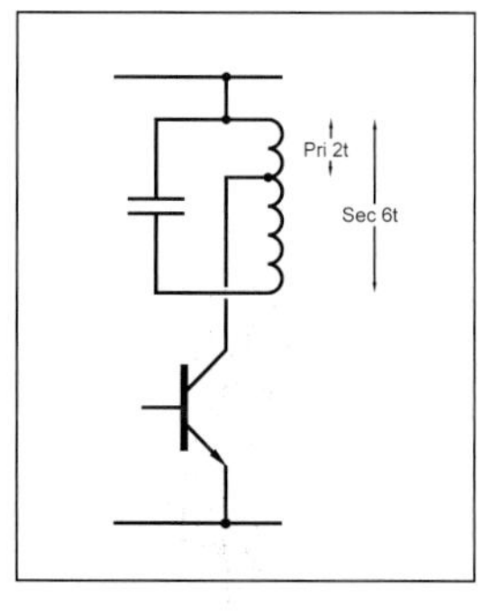


**Fig 1.39: An autotransformer in a parallel tuned circuit. Transformer tapping will reduce the loading of the transistor on the tuned circuit $R_D$**

secondary load, the residual current, known as the *magnetising current*, is that which would flow through the inductance of the primary. Normally this current and any small losses in the transformer can be neglected but it should be noted that the windings do rise in temperature during use and an overload could cause breakdown of the internal insulation or start a small fire before the fuse blows.

Since the current and voltages can be changed, it follows that the impedances, given by $Z_P = V_P/I_P$ and $Z_S = V_S/I_S$ will also change. If, for a step-down transformer, the secondary voltage is halved, the secondary current will be double the primary current. The impedance has reduced to a quarter of the primary value.

$$Z_P = Z_S\left(\frac{n_P}{n_S}\right)^2$$

The transformation of impedance is a valuable property and transformers are widely used for impedance matching. Several examples will be seen in the transmitters chapters.

## Auto-transformers

If DC isolation between the primary and secondary is not required, an auto-transformer can be used, which has a single winding. It is tapped at an appropriate point so that, for a step-down transformer, the whole winding forms the primary but only a few of those turns form the secondary. An example is a mains transformer with 1000 turns, tapped at 478 turns to give a 110V output from the 230V supply. The advantage of the auto-transformer is partly simplified construction but mainly space and weight saving.

**Fig 1.39** shows a step-up transformer in a tuned circuit in the collector of a transistor. The loading effect of the transistor will be considerably reduced, thereby preserving the Q of the unloaded tuned circuit.

## Screening

When two circuits are near one another, unwanted coupling may exist between them due to stray capacitance between them, or due to stray magnetic coupling.

Placing an earthed screen of good conductivity between the two circuits, as shown in **Fig 1.40(b)**, can eliminate stray capacitance coupling. There is then only stray capacitance from each circuit to earth and no direct capacitance between them. A useful practical rule is to position screens so that the two circuits are not visible from one another.

Stray magnetic coupling can occur between coils and wires due to the magnetic field of one coil or wire intersecting the other. At radio frequency, coils can be inductively screened (as well as capacitively) by placing them in closed boxes or cans made from material of high conductivity such as copper, brass or aluminium. Eddy currents are induced in the can, setting up a field which opposes and practically cancels the field due to the coil beyond the confines of the can.

If a screening can is too close to a coil the performance of the coil, ie its Q and also its inductance, will be considerably reduced. A useful working rule is to ensure the can in no closer to the coil than its diameter, see **Fig 1.40(c)**.

At low frequencies eddy current screening is not so effective and it may be necessary to enclose the coil or transformer in a box of high-permeability magnetic material such as Mumetal in order to obtain satisfactory magnetic screening. Such measures are not often required but a sensitive component such as a microphone transformer may be enclosed in such a screen in order to make it immune from hum pick-up.

It is sometimes desirable to have pure inductive coupling between two circuits with no stray capacitance coupling. In this case a Faraday screen can be employed between the two coils in question, as shown in **Fig 1.40(d)**. This arrangement is sometimes used between an antenna and a receiver input circuit or between a transmitter tank circuit and an antenna. The Faraday screen is made of stiff wires (**Fig 1.40e**) connected together at one end only, rather like a comb. The 'open' end may be held by non-conductive material. The screen is transparent to magnetic fields because there is no continuous conducting surface in which eddy currents can flow. However, because the screen is connected to earth it acts very effectively as an electrostatic screen, eliminating stray capacitance coupling between the circuits.

# SEMICONDUCTORS

Silicon forms the basis of most transistors and diodes. Specialist devices may be formed of the compound gallium arsenide. Early semiconductors were based on germanium and this is now making a come back in microwave transistors with silicon-germanium junctions (Si-Ge), offering similar performance to GaAsFets but at a lower cost. A simplified picture of the silicon atom is shown in **Fig 1.41**. Silicon has an atomic number of 14 indicating it has 14 electrons and 14 protons. The electrons are arranged in shells that must be filled before starting the next

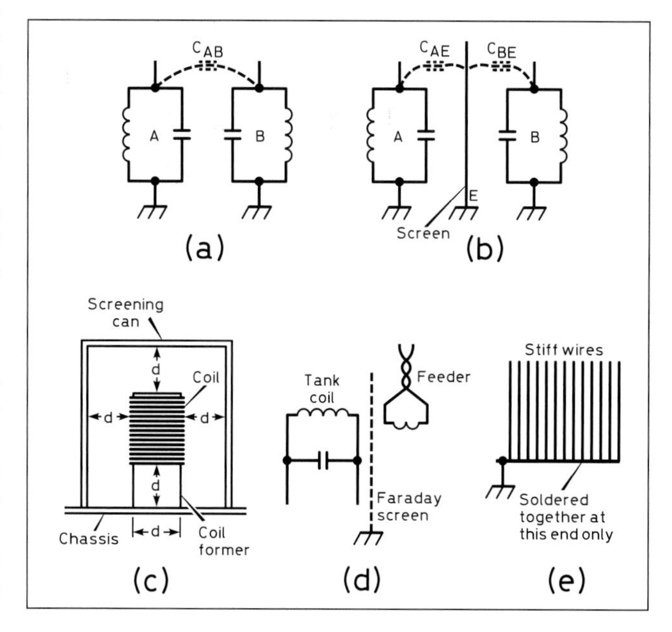


**Fig 1.40: (a) Stray capacitance coupling $C_{AB}$ between two circuits A and B. The introduction of an earthed screen E in (b) eliminates direct capacitance coupling, there being now only stray capacitance to earth from each circuit $C_{AE}$ and $C_{BE}$. A screening can (c) should be of such dimensions that it is nowhere nearer to the coil it contains than a distance equal to the diameter of the coil d. A Faraday screen between two circuits (d) allows magnetic coupling between them but eliminates stray capacitance coupling. The Faraday screen is made of wires as shown at (e)**

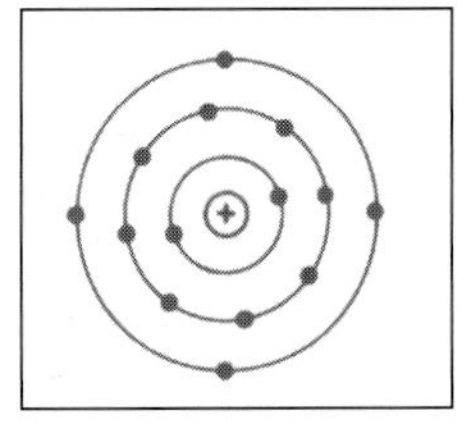

Fig 1.41: The silicon atom

outermost layer. It has four electrons in its outer shell, which are available for chemical bonding with neighbouring atoms, forming a crystal lattice. A single near perfect crystal can be grown from a molten pool of pure material. As a pure crystal, the four outer electrons are all committed to bonding and none are available to support the flow of current. Pure crystaline silicon is an insulator. To obtain the semiconductor, carefully controlled impurities are added. This is known as doping.

If a small quantity of an element with five outer electrons, for example phosphorus, antimony or arsenic, is added at around 1 part in $10^7$, then some of the atoms in the lattice will now appear to have an extra electron. Four will form bonds but the fifth is unattached. Although it belongs to its parent nucleus, it is relatively free to move about and support an electric current. Since this material has mobile electrons it is called an n-type semiconductor.

If an element with three outer electrons is added to the crystal, eg indium, boron or aluminium, these three electrons form bonds but a hole is left in the fourth place. An electron that is part of a nearby bond may move to complete the vacant bond. The new hole being filled, in turn, by another nearby electron. It is easier to consider the hole as being the mobile entity than to visualise several separate electrons each making single jumps. This material is a p-type semiconductor.

A *diode* is formed from a p-n junction and will only allow current to flow in one direction. This is used in rectification, changing AC into unidirectional half-cycles which can then be smoothed to conventional DC using a large capacitor.

A transistor uses a three layer semiconductor device formed either as n-p-n or, perhaps less common, as p-n-p sandwich. The transistor may act as an amplifier, increasing the amplitude or power of weak signals, or it can act as a switch in a control circuit. These uses, and more detailed discussion of the operation of transistors and diodes, are contained in this book.

## BALANCED AND UNBALANCED CIRCUITS

**Fig 1.42** shows a balanced and unbalanced wire or line. The balanced line (top) has equal and opposite signals on the two conductors, the voltages and currents are of equal magnitude but opposite in direction. For AC, this would mean a 180 degree phase difference between the two signals. The input and output from the line is taken between the two conductors. If each conductor had a 2V AC signal, the potential difference between the two conductors would be 4V.

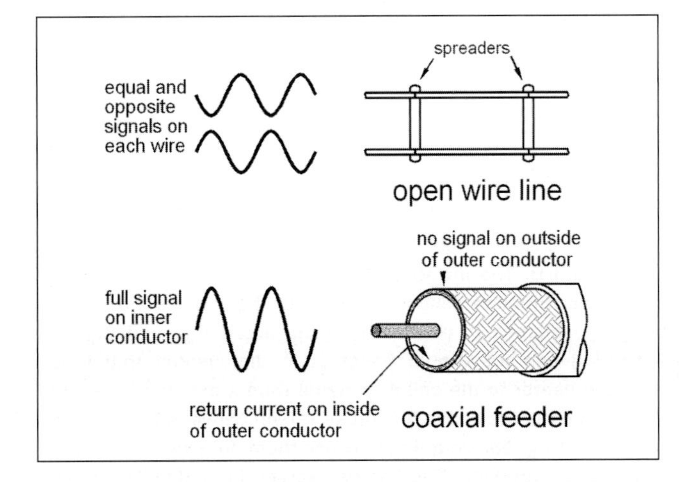


Fig 1.42: Signals on balanced and unbalanced lines

The unbalanced line may have a single conductor with an 'earth return', typically the chassis of an equipment or a real earth in the case of telegraph communication (see below) or may be a coaxial cable as shown in Fig 1.42. The centre conductor carries the signal and the return current path is the inside surface of the braid of the coaxial cable. With a true unbalanced termination at the load end, the braid will be at zero volts and the full magnitude of the signal is on the centre conductor. The outer surface of the braid will also be at zero volts, being earthed at one or both ends. In line telegraph use it was common for there to be a single wire, often alongside railway tracks and the earth literally was used as the return path to complete the circuit.

Any wire carrying a changing current has a tendency to radiate part of the signal as electromagnetic waves or energy. This is the principle of the antenna (aerial). The efficiency with which radiation occurs is a function of the length of the wire as a proportion of the wavelength of the signal. This is covered later in the chapters on HF and VHF/UHF antennas. A radiator can equally pick up any stray electromagnetic signals which will then get added to the wanted signal in the wire.

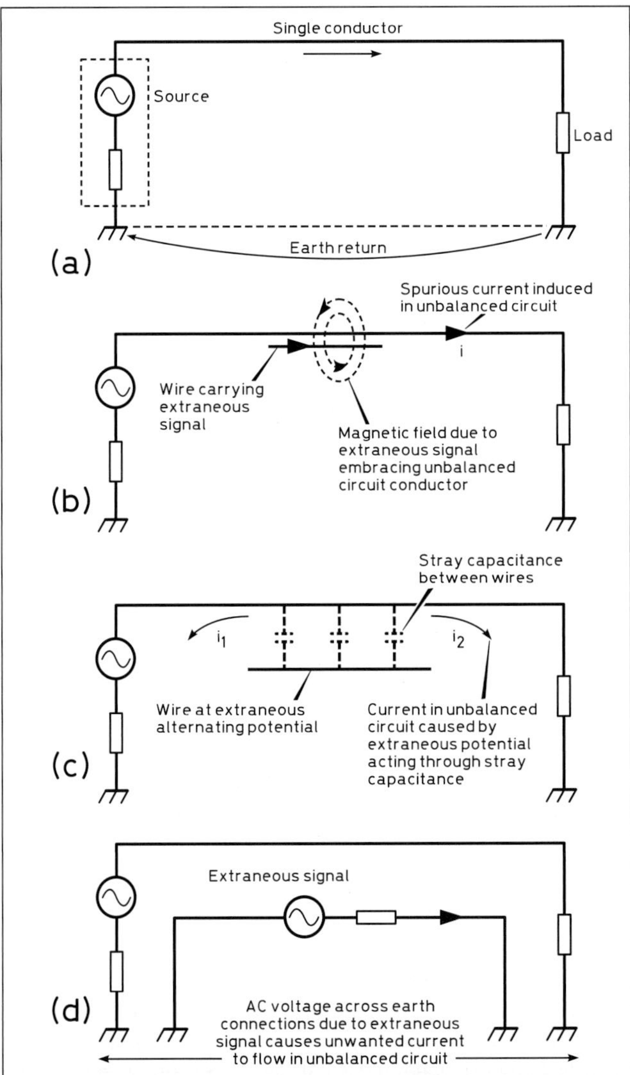


**Fig 1.43: The unbalanced circuit. (a) The basic unbalanced circuit showing earth return path. (b) How extraneous signals and noise can be induced in an unbalanced circuit by magnetic induction. (c) Showing how extraneous signals can be induced by stray capacitance coupling. (d) Showing how extraneous signals can be induced due to a common earth return path**

Unbalanced circuits are particularly prone to this effect. **Fig 1.43** shows an unbalanced wire with magnetic (mutual inductance) coupling and capacitive coupling to another nearby conductor. Circuits of this type are very commonly used in radio equipment and are perfectly satisfactory provided leads are kept short and are spaced well away from other leads. It is, however, prone to the pick-up of extraneous noise and signals from neighbouring circuits by three means: inductive pick-up, capacitive pick-up and through a common earth return path.

*Inductive pick-up*, **Fig 1.43(b)**, can take place due to transformer action between the unbalanced circuit wire and another nearby wire carrying an alternating current; a common example is hum pick-up in audio circuits due to the AC mains wiring.

*Capacitive pick-up*, **Fig 1.43(c)**, takes place through the stray capacitance between the unbalanced circuit lead and a neighbouring wire. Such pick-up can usually be eliminated by introducing an earthed metal screen around the connecting wire.

If the unbalanced circuit has an earth return path that is *common* to another circuit, **Fig 1.43(d)**, unwanted signals or noise may be injected by small voltages appearing between the two earth return points of the unbalanced circuit. Interference of this type can be minimised by using a low-resistance chassis and avoiding common earth paths as far as possible.

A balanced circuit is shown in **Fig 1.44**. As many signal sources, and often loads as well, are inherently unbalanced (ie one side is earthed) it is usual to use transformers to connect a source of signal to a remote load via a balanced circuit. In the balanced circuit, separate wires are used to conduct current to and back from the load; no current passes through a chassis or earth return path.

The circuit is said to be 'balanced' because the impedances from each of the pair of connecting wires to earth are equal. It is usual to use twisted wire between the two transformers as shown in Fig 1.44. For a high degree of balance, and therefore immunity to extraneous noise and signals, transformers with an earthed screen between primary and secondary windings are used. In some cases the centre taps of the balanced sides of the transformers are earthed as shown dotted in Fig 1.44.

The balanced circuit overcomes the three disadvantages of the unbalanced circuit. Inductive and capacitive pick-up are eliminated since equal and opposite currents are induced in each of the two wires of the balanced circuit and these cancel out. The same applies to interfering currents in the common earth connection in the case where the centre taps of the windings are earthed.

The argument also applies to radiation from a balanced circuit. The two conductors will tend to radiate equal and opposite signals which will cancel out. Some care is required however because if conducting objects are close to the feeder, comparable in distance to the separation of the two conductors, then the layout may not be symmetrical and some radiation due to the imbalance may occur.

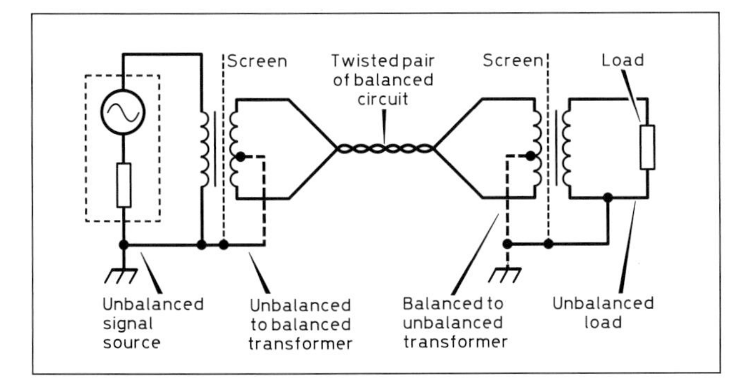


**Fig 1.44: The balanced circuit**

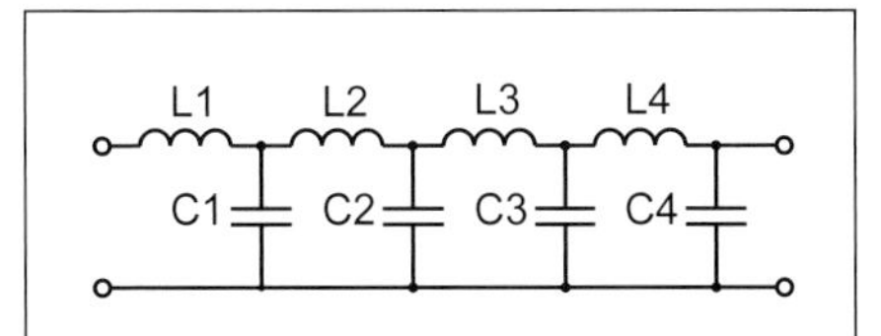


**Fig 1.45: A long feeder can be represented by series inductors and parallel capacitors**

Looking again at Fig 1.42, the balanced feeder must be symmetrically run, away from walls and other conductors, but the unbalanced feeder has an earthed conducting sleeve, typically an outer braid or metal tape to screen the centre conductor and minimise radiation outside the cable. This is known as coaxial cable. It may be run close to walls and conductors but the manufacturers advice on minimum bending radius should be heeded.

## FEEDERS

The feeder is the length of cable from the transmitter/receiver to the antenna. It must not radiate, using the properties of balance or screening outlined above and it must be as loss free as possible. Consider the circuit shown in **Fig 1.45.** Each short length of feeder is represented by its inductance, and the capacitance between the conductors is also shown. Assume now a battery is connected to the input. The first length of cable will charge up to the battery voltage but the rate of rise will be limited by the inductance of L1 and the need for C1 to charge. This will occur progressively down the cable, drawing some current from the battery all the while. Clearly with a short cable, this will take no time at all. However an infinitely long cable will be drawing current from the battery for quite some time. The ratio of the battery voltage to the current drawn will depend on the values of L and C and is the characteristic impedance of the cable.

It is given by:

$$Z_0 = \sqrt{\frac{L}{C}}$$

where L and C are the values per unit length of the feeder.

This argument and the formula assume the series resistance of the feeder is negligible, as is the leakage resistance of the insulation. This is acceptable from the point of view of $Z_0$ but the series resistance is responsible, in part, for feeder losses; some of the power is lost as heat. The other facet of the loss is the dielectric loss in the insulation, that is the heating effect of the RF on the plastic, easiest viewed as the heating of the plastic in a microwave oven. The loss is frequency dependant and, to a simple approximation, rises as the square root of the frequency. A very long length of feeder will look like a resistance of $Z_0$. So will a shorter length terminated in an actual resistor of value $Z_0$. Signals travelling down the feeder will be totally absorbed in a load of value $Z_0$ (as if it were yet more feeder) but if the feeder is terminated in a value other than $Z_0$, some of the energy will be absorbed and some will be reflected back to the source. This has a number of effects which are discussed in the chapters on antennas. Properly terminating the feeder in its characteristic impedance (actually a pure resistance) is termed correct matching. It is also, of course, necessary to use a balanced load on a balanced line (twin feeder) and an unbalanced load on an unbalanced line such as coaxial cable. Much more on feeders can be found in the chapter on transmission lines.

## THE ELECTROMAGNETIC SPECTRUM

Radio frequencies are regarded by the International Telecommunication Union to comprise frequencies from 9kHz to 400GHz but the upper limit rises occasionally due to advances in technology. This, however is only a small part of the electromagnetic spectrum shown in **Fig 1.46**.

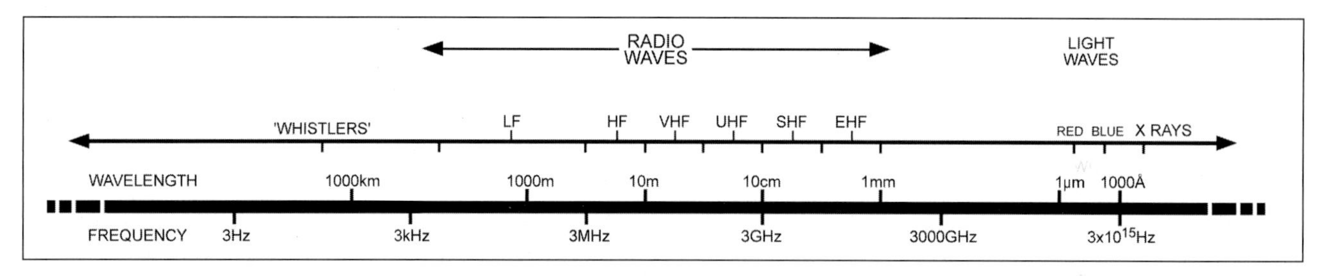


Fig 1.46: Graphical representation of the electromagnetic spectrum

The whole spectrum comprises radio waves, heat, light, ultraviolet, gamma rays, and X-rays. The various forms of electromagnetic radiation are all in the form of oscillatory waves, and differ from each other only in frequency and wavelength. They all travel through space with the same speed, approximately 3 x $10^8$ metres per second. This is equivalent to about 186,000 miles per second or once round the world in about one-seventh of a second.

As might be surmised from the name, an electromagnetic wave consists of an oscillating electric field and a magnetic field. It can exist in a vacuum and does not need a medium in which to travel. **Fig 1.47** shows one and a half cycles of an e-m wave. The electric (E) and magnetic (H) fields are at right angles and both are at right angles to the direction of propagation. E and H are in phase and their magnitudes are related by the formula:

$$\frac{E}{H} = \sqrt{\frac{\mu_0}{\varepsilon_0}} = 120\pi\,\Omega$$

where $\mu_0$ is the permeability of free space and $\varepsilon_0$ is the permitivity of free space, both natural physical constants. The quantity E/H is known as the impedance of free space and may be likened to the characteristic impedance of a feeder.

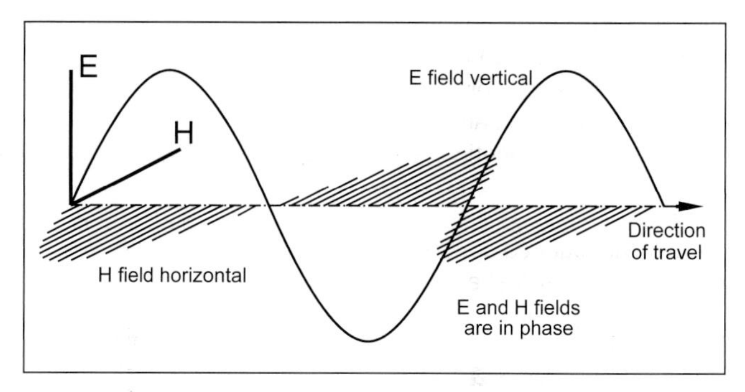


**Fig 1.47: An electromagnetic wave. By convention the polarisation is taken from the electric field, so this is a vertically polarised wave**

## Frequency and Wavelength

The distance travelled by a wave in the time taken to complete one cycle of oscillation is called the *wavelength*. It follows that wavelength, frequency and velocity of propagation are related by the formula

Velocity = Frequency x Wavelength, or c = f$\lambda$

where c is the velocity of propagation, f is the frequency (Hz), and $\lambda$ is the wavelength in metres.

*Example:*

What are the frequencies corresponding to wavelengths of (i) 150m, (ii) 2m and (iii) 75cm?

From the formula c = f$\lambda$, the frequencies are given by:

150m

$$f = \frac{c}{\lambda} = \frac{3\times10^8}{150} = \frac{300\times10^6}{150} = 2\cdot0\text{MHz}$$

2m:

$$f = \frac{c}{\lambda} = \frac{3\times10^8}{2} = \frac{300\times10^6}{2} = 150\text{MHz}$$

75cm:

$$f = \frac{c}{\lambda} = \frac{3\times10^8}{0\cdot75} = \frac{300\times10^6}{0\cdot75} = 400\text{MHz}$$

It is important to remember to work in metres and hertz in these formula. However, it will also be noticed that if the frequency is expressed in MHz throughout and $\lambda$ in metres, a simplified formula is:

$$f = \frac{300}{\lambda}\text{MHz}$$

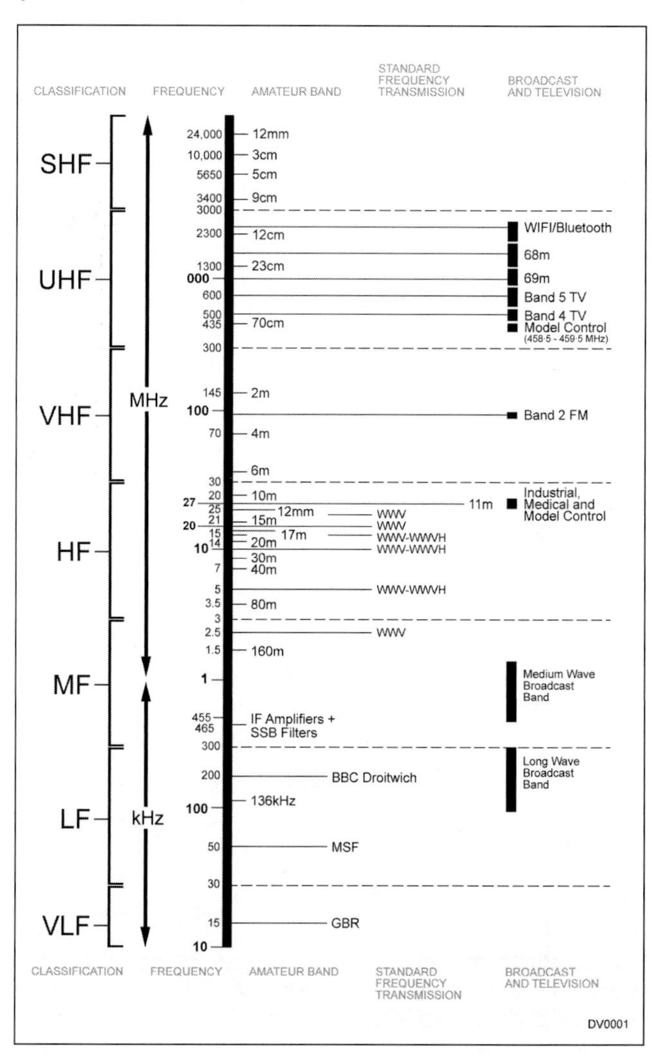


**Fig 1.48: The amateur frequency bands in relation to other user services**

or

$$\lambda = \frac{300}{f} \text{ metres}$$

**Fig 1.48** shows how the radio spectrum may be divided up into various bands of frequencies, the properties of each making them suitable for specific purposes. Amateur transmission is permitted on certain frequency bands in the LF, MF, HF, VHF and UHF, SHF/microwave ranges.

## ANTENNAS

An antenna (or aerial) is used to launch electromagnetic waves into space or conversely to pick up energy from such a wave travelling through space. Any wire carrying an alternating current will radiate electromagnetic waves and conversely an electromagnetic wave will induce a voltage in a length of wire. The issue in antenna design is to radiate as much transmitter power as possible in the required direction or, in the case of a receiver, to pick up as strong a signal as possible, very often in the presence of local interference.

### Isotropic Radiator

An isotropic radiator is one that radiates equally in all, three-dimensional, directions. In practical terms, it does not exist but it is easy to define as a concept and can be used as a reference.

The power flux density *p* from such an antenna, at a distance *r* is given by:

$$p = \frac{P}{4\pi r^2} \quad W/m^2$$

where P is the power fed to the antenna.

This leads to the electric field strength E, recalling the formula for the impedance of free space above:

$$p = \frac{E^2}{R} = \frac{E^2}{120\pi} = \frac{P}{4\pi r^2}$$

rearranging gives

$$E = \frac{\sqrt{30P}}{r} = \frac{5 \cdot 5\sqrt{P}}{r}$$

Note that this formula is for an isotropic radiator.

### Radiation Resistance

The radiation resistance of the antenna can be regarded as:

$$R_r = \frac{\text{total power radiated}}{(\text{RMS current at antenna input})^2}$$

This will vary from one antenna type to another. Electrically short antennas, much less than a wavelength, have very low radiation resistances leading to considerable inefficiency if this approaches the resistance of the various conductors. This resistance is not a physical resistor but the antenna absorbs power from the feeder as if it were a resistor of that value.

### Antenna Gain

Antennas achieve gain by focussing the radiated energy in a particular direction. This leads to the concept of Effective Radiated Power (ERP) which is the product of the power fed to the antenna multiplied by the antenna gain. The figure can be regarded as equivalent to the power required to be fed to an antenna without gain to produce the same field strength (in the wanted direction) as the actual antenna. It should be noted that this gain also applies on receive because antennas are passive, reciprocal devices that work identically on transmit or receive.

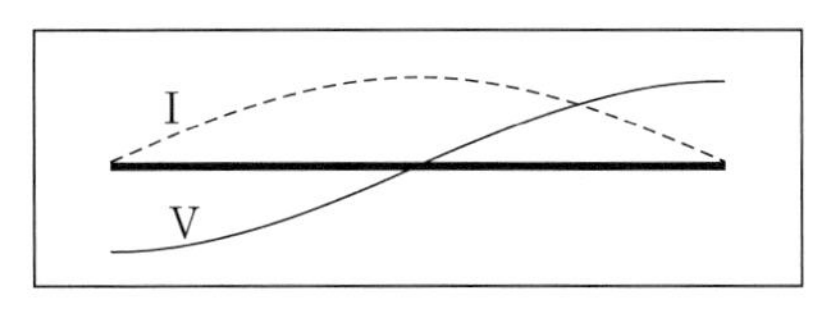


**Fig 1.49: Current and voltage distribution on a half-wave dipole**

### Effective Aperture

Consider a signal of field strength E arriving at an antenna and inducing an EMF. This EMF will have a source resistance of $R_r$, the radiation resistance and will transfer maximum power to a load of the same value. This power can be viewed as the area required to capture sufficient power from the incident radio wave.

An antenna of greater gain will have a larger effective aperture; the relationship is:

$$A_{eff} = \frac{\lambda^2}{4\pi} \times G$$

where λ is the wavelength of the signal and G is the antenna gain in linear units.

This aperture will be much larger that the physical appearance of a high gain antenna and is an indication of the space required to allow the antenna to operate correctly without loss of gain.

### A Dipole

The current and voltage distribution on a dipole are shown in **Fig 1.49**. The current is a half sine wave. By integrating the field set up by each element of current over the length of the dipole, it can be shown that the dipole has a gain of 1·64 or 2·16dB and a radiation resistance (or feed impedance) of 73 ohms (see below for a discussion of dB, decibels). The dipole is often used as the practical reference antenna for antenna gain measurements. It must then be remembered that the gain of an antenna quoted with reference to a dipole is 1·64 times or 2·16dB less than if the gain is quoted with reference to isotropic.

### Effective Radiated Power

The effective radiated power (ERP) of an antenna is simply the product of the actual power to the antenna and the gain of the antenna. Again it is necessary to know if this is referenced to isotropic (EIRP) or a dipole (EDRP). ERP is normally quoted with reference to a dipole. This figure indicates the power that would need to be fed to an actual dipole in order to produce the field strength, in the intended direction, that the gain antenna produces.

The total radiated power is still that supplied to the antenna, the enhancement in the intended direction is only as a result of focussing the power in that direction. A side benefit, but an important one, is that the power in other directions is much reduced.

The field from such an antenna is given by:

$$E = \frac{7 \cdot 01\sqrt{ERP}}{d} \quad V/m$$

where d is the distance from the antenna (shown as r above for the isotropic radiator).

The different coefficient of 7·01 rather than 5·5 accounts for the fact that the ERP is reference to a dipole, which already has some gain over isotropic.

### Feeding an Antenna

The concept of maximum power transfer applies to an antenna, but there is a additional complication if the antenna is not a resonant length. Fig 1.49 showed the current and voltage distribution on the dipole. The dipole is a half wavelength long and the

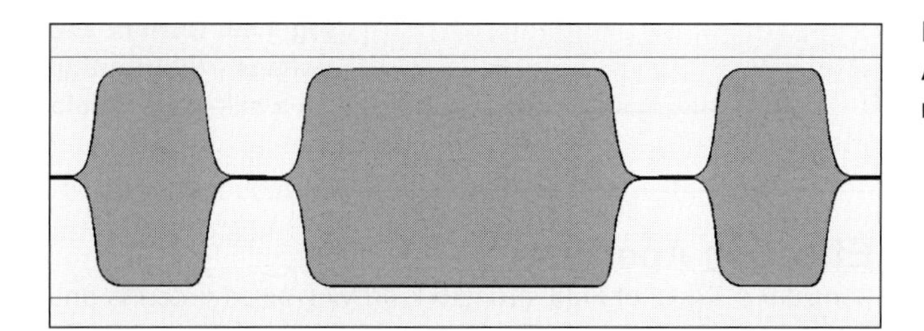

Fig 1.50: The Morse letter R

current can be viewed as millions of electrons 'sloshing' from one end to the other rather like water in a bath.

As a half wavelength (or a multiple) the antenna is resonant and can be regarded as an LC resonant circuit with a resistor, the radiation resistance $R_r$, to account for the power radiated. Below the resonant frequency the antenna will appear net capacitive (too short in length) and net inductive above the resonant frequency (too long in length).

This means that the antenna will no longer appear as a good match to the feeder and some power will be reflected back down the feeder. This, in turn, will result in the feeder being a poor match to the transmitter resulting in less than maximum power being transferred. To recover this situation, the opposite type of reactance must be inserted to offset that presented to the transmitter. The antenna/feeder/inserted reactance now represents a pure resistance to the transmitter. This function is usually performed by an antenna tuning (or matching) unit and is covered more fully in the chapter on HF antennas.

## MODULATION

A simple sine wave radio signal conveys no information except that it is there and perhaps the direction of its origin. In order to convey information we must change the signal in some agreed way, a process called modulation.

The simplest form of modulation is to turn the signal on and off and this was all that early pioneers of radio could achieve. This led to the adoption of the Morse code (originally invented for line telegraphy) which remains very effective at sending text messages and is still practiced by many amateurs although Morse has given way to other techniques in commercial applications. A Morse signal is shown in **Fig 1.50**. It will be noticed that the edges of the carrier waveform are rounded off; the reason will be explained shortly in the section below. It is called CW or carrier wave because the carrier (the sine wave signal) is all that is transmitted. However we now know that is not quite correct.

## Amplitude Modulation (AM)

In amplitude modulation, the amplitude of the carrier is varied according to the instantaneous amplitude of the audio modulating signal as shown in **Fig 1.51**. Only the carrier is transmitted in the absence of modulation, the audio (or other) signal increasing and decreasing the transmitted amplitude. The depth of modulation, m, is given by the quantity B/A or B/A x 100%.

It is desirable to achieve a reasonably high level of modulation but the depth must never exceed 100%. To do so results in distortion of the audio signal, the production of audio harmonics and a transmitted signal that is wider than intended and likely to splatter over adjacent frequencies causing interference.

### Sidebands

The process of modulation leads to the production of sidebands. These are additional transmitted frequencies each side of the carrier, separated from the carrier by the audio frequency. This is first shown mathematically and then the result used to demonstrate the effect in a more descriptive manner.

Fig 1.51: Amplitude modulation

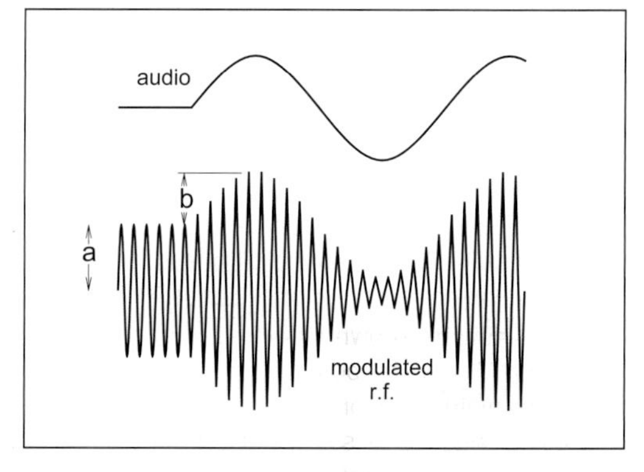


A single carrier has the form:

$$A \sin \omega_c t$$

where A is the amplitude and $\omega_c$ is $2\pi f_c$, the frequency of the signal. When amplitude modulated the formula becomes:

$$A(1 + m \sin \omega_a t) \sin \omega_c t$$

where m is the depth of modulation and $\omega_a$ the audio frequency used to modulate the carrier $\omega_c$.

Expanding this gives:

$$A \sin\omega_c t + Am \sin\omega_a t \times \sin\omega_c t$$

The second term is of the form sinA sinB which can be represented as ½(sin(A-B)+sin(A+B)), so our signal becomes:

$$A \sin \omega_c t + \frac{Am}{2}(\sin(\omega_c t - \omega_a t) + \sin(\omega_c t + \omega_a t))$$

Inspection will show that these three terms are the original carrier plus two new signals, one at a frequency $f_c$ - $f_a$ and the other at $f_c$+$f_a$ and both at an amplitude dependant on the depth of modulation but with a maximum amplitude of half the carrier. This is shown in **Fig 1.52**.

It still might not be all that clear just why the sidebands need to exist. Consider the effect of just the carrier and one of the sidebands but as if they were both high pitch audio notes close together in frequency. A beat note would be heard. Add the other sideband, giving the same beat note. The overall effect is the carrier varying in amplitude. The bandwidth needed for the transmission depends on the audio or modulating frequency. The two sidebands are equidistant from the carrier so the total bandwidth is twice the highest audio frequency.

Modulating the carrier also increases the transmitted power. The formula above gave the amplitude (voltage) of the signal. If the power in the carrier is $P_c$ then the total transmitted power is:

$$P_c + 2 \times P_C \left(\frac{m}{2}\right)^2 = P_c\left(1 + \frac{m^2}{2}\right)$$

For 100% modulation, the total transmitted power is 150% of the carrier power, that is the carrier plus 25% in each sideband.

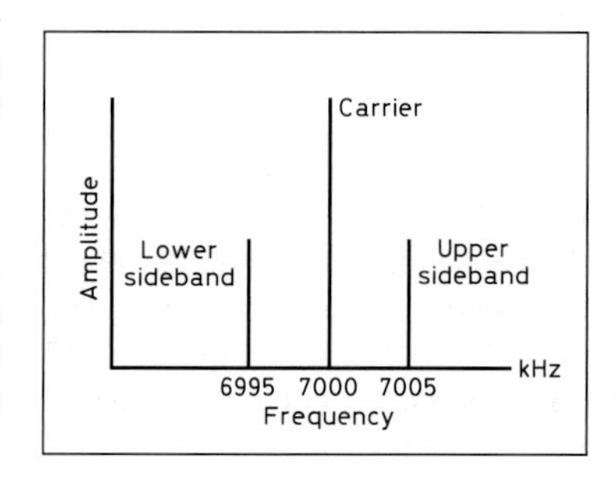


Fig 1.52: Sideband spectrum of a 7MHz carrier which is amplitude modulated by a 5kHz tone

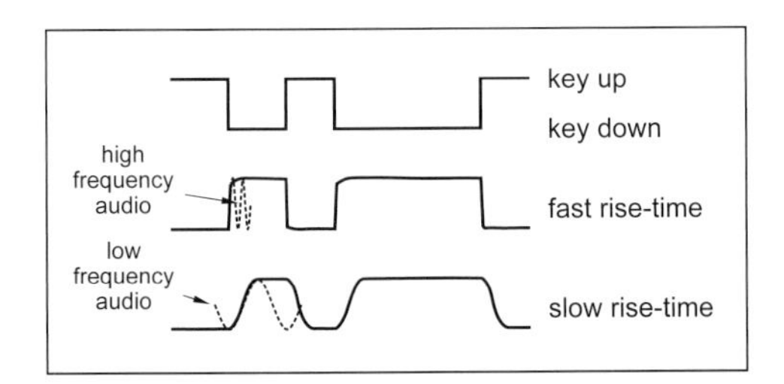


**Fig 1.53: Fast and slow rise times on a CW keying signal**

It can now be shown why the keying of the Morse signal was rounded. Morse can be regarded as amplitude modulation with only two values, 0 or 100% modulation. Simplistically, there would be sidebands separated from the carrier dependant on the keying speed. That is not the whole story however.

The rise and fall of the transmitted envelope greatly affect the overall bandwidth. **Fig 1.53** shows the Morse keying with fast and slow rises and falls of the keying and thus the transmitted RF envelope. The fast rise can be regarded as a small part of a high frequency audio signal; a slower rise time being part of a lower frequency audio signal. The sidebands at the moment of keying will be those applicable to the 'audio'. The bandwidth will be much wider than that assumed from the Morse speed alone. A well designed keying circuit is inaudible outside the tuning range over which the Morse tone can be heard. Bad keying causes 'key clicks' that can be heard several kHz each side of the Morse signal proper. It should be avoided.

## Single Sideband (SSB)

**Fig 1.54** shows an AM signal where the audio is speech containing a range of frequencies from just above zero to 3kHz. The upper sideband is a copy of the audio, albeit shifted up in frequency to just above the carrier, and the lower sideband is a mirror image just below the carrier. In fact all the audio information is contained in the sidebands; as stated at the very beginning, the carrier contains no information. We can, therefore, dispense with the carrier and either one of the sidebands. In the figure the lower sideband is shown as filtered out and the carrier has been suppressed.

The modulation and detection techniques needed to do this are shown in the chapters on receivers and transmitters. SSB transmission requires only half the bandwidth and represents a considerable saving in power, as evidenced by the formula for the power in the various components of the AM signal above.

It will be recalled that the frequency of the audio signal was retained in the transmitted signal by the frequency separation between the carrier and the sideband. If the carrier and the other sideband are removed the reference point for determining the audio frequency is lost. It is 're-inserted' at the receiver by the act of tuning in the signal. The receiver must be accurately tuned or all the received frequencies will be offset by the same amount, namely the error in the tuning. This is why the pitch of the audio of an SSB signal varies as the receiver is tuned.

The bandwidth of an SSB signal is simply the audio bandwidth, half that of the AM transmission. The power varies continuously from zero, when unmodulated, to its peak value depending on the audio volume and the designed power of the transmitter.

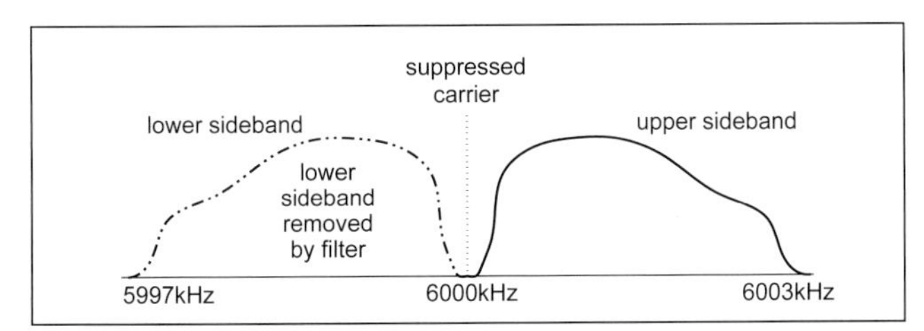


**Fig 1.54: SSB signal from a balanced modulator, with one sideband filtered out**

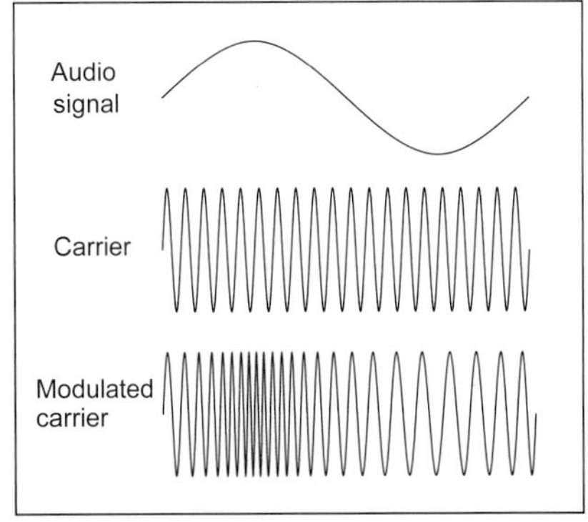


**Fig 1.55: Frequency modulation**

## Frequency Modulation (FM)

In FM, the instantaneous frequency of the carrier is varied according to the instantaneous amplitude of the carrier as shown in **Fig 1.55**. Unlike AM there is no natural limit to the amount by which the frequency may be deviated from the nominal centre frequency. The peak deviation is a system design parameter chosen as a compromise between audio quality and occupied bandwidth. The ability of an FM detector or discriminator to combat received noise increases as the cube of the deviation. However, the total received noise increases linearly with the bandwidth so the overall effect is a square law. For instance, doubling the deviation and bandwidth improves the recovered audio signal to noise ratio by a factor of four (6dB).

For FM the deviation ratio, corresponding to the depth of modulation of AM, is defined as:

$$\text{Deviation ratio} = \frac{\text{actual deviation}}{\text{peak deviation}}$$

The modulation index relates the peak deviation to the maximum audio frequency:

$$\text{Modulation index} = \frac{\text{peak deviation}}{\text{audio frequency}}$$

A narrow band FM system is one where the modulation index is about or less than unity. Amateur FM systems are narrow band (NBFM). At 2m, with 12·5kHz channels, the max audio frequency is around 2·8kHz and the peak deviation 2·5kHz; a mod index of 0·9. At 70cm, with 25kHz-spaced channels, the calculation is 5/3·5, a mod index of 1·4.

VHF broadcast stations have an audio range up to 15kHz and a peak deviation of 75kHz, giving a mod index of 5. This is a wide band FM system (WBFM). TV sound has a deviation of 50kHz but good quality TV sound systems now use the NICAM digital sound channel which also offers stereo reproduction. Simplistically, the mod index will give a guide to the audio quality to be expected.

The bandwidth of an FM transmission is considerably wider than twice the peak deviation, as might be expected. The modulation gives rise to sidebands but even for a single audio modulating tone, there may be several sidebands. The mathematical derivation is much more complex and the 'sin' function requires Bessel functions to calculate. The formula is:

$$A\sin(\omega_c t + m\sin\omega_a t)$$

where m is the modulation index.

This gives a transmitted spectrum shown in **Fig 1.56** which has, theoretically, an infinite number of sidebands of decreasing amplitude. The rule of thumb often applied, known as Carson's rule, is that the necessary bandwidth is given by:

Bandwidth = 2(peak deviation + max audio frequency)

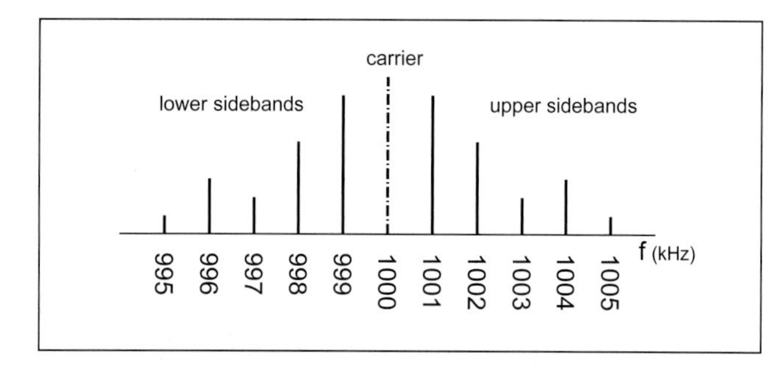


**Fig 1.56: The spectrum of an FM signal**

For a typical 2m system this works out as:

Bandwidth = 2(2·5 + 2·8) = 10·6kHz

Carson's rule tends to over-estimate the required bandwidth at the lower modulation indices used by 12·5kHz-channel systems. In practice, a receiver filter with a bandwidth around 7·5kHz will prove adequate, and also filter out more of the ever-present RF noise. The number of relevant sidebands depends on the modulation index, according to the graph in **Fig 1.57**. This shows the carrier and the first three sidebands. For those with access to *Microsoft Excel®*, the function BesselJ, available in the maths and statistics pack, will give the values. This pack is provided on the CD but frequently not loaded. The help menu gives guidance on how to load the pack if it is not loaded.

It is interesting to note that the amplitude of the carrier also varies with the modulation index, and goes to zero at an index of 2·4. This provides a very powerful method to set the deviation. Set the transmitter up with an audio signal generator at full amplitude at 1·042kHz (2·5/2·4). Listen to the carrier on a CW narrow bandwidth receiver and adjust the deviation on the transmitter for minimum carrier; this can also be done on a spectrum analyser if one is available. The deviation is now correctly set provided signals of greater amplitude are not fed into the microphone socket. Ideally the transmitter will have a microphone gain control, followed by a limiter and then followed by a deviation control. Thus excessive amplitude signals will not reach the deviation control and the modulator.

## Phase Modulation (PM)

This is a subtle variant of frequency modulation. In FM it is the instantaneous frequency of the transmitter that is determined by the instantaneous modulating input voltage. Clearly, as the frequency varies, the phase will differ from what it would have been had the carrier been unmodulated. There are phase changes but these are as a consequence of the frequency modulation. In phase modulation, it is the phase that is directly controlled by the input audio voltage and the frequency varies as a consequence.

Frequency modulation is usually achieved by a variable capacitance diode in the oscillator circuit. This has the side effect of introducing another source of frequency drift or instability. An RC circuit situated after the oscillator can achieve phase modulation by using a variable capacitance diode for C. This way the oscillator is untouched and more stable. Many transceivers do in fact use phase modulation and can pre-condition the audio so the transmitted signal appears frequency modulated. However, there is an advantage in retaining phase modulation because the higher audio frequencies are transmitted with a higher deviation than with FM, helping to combat any noise present in the receiver. Many data transmissions use phase modulation, the digital signal directly changing the carrier phase.

## Vector Representation of Modulation

Modulation can be shown on a vector diagram; first introduced in Fig 1.25. Remember the vector diagram is a 'freeze frame' of the signal showing the phase relationship between the various components. In 'reality' the diagram is rotating (by convention anti-clockwise) at the signal frequency.

Amplitude modulation is the easiest to visualise and is shown in **Fig 1.58**. The carrier ($f_c$) is the reference shown from the 'centre' to the 3 o'clock position. For a single audio tone the upper sideband is shown added to the carrier. Because it is a higher frequency it is rotating anticlockwise at the audio frequency ($f_a$) with reference to the carrier. The carrier and the whole drawing are already rotating anti-clockwise at the carrier frequency so the upper sideband is rotating at $f_c + f_a$ which is correct. Similarly the lower sideband is rotating clockwise with respect to the carrier, also a frequency $f_a$. That subtracts from the rotation of the carrier so represents a frequency of $f_c - f_a$.

The sum of the carrier plus these two sideband vectors is a signal at the carrier frequency and phase rising and falling in

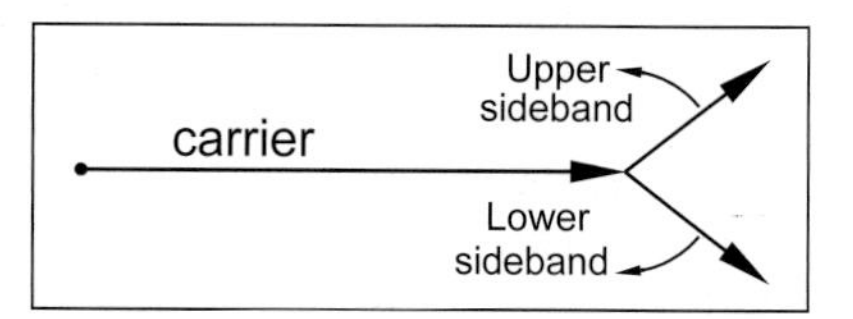


**(above) Fig 1.58: Phasor diagram showing a single audio tone amplitude modulating an RF carrier**

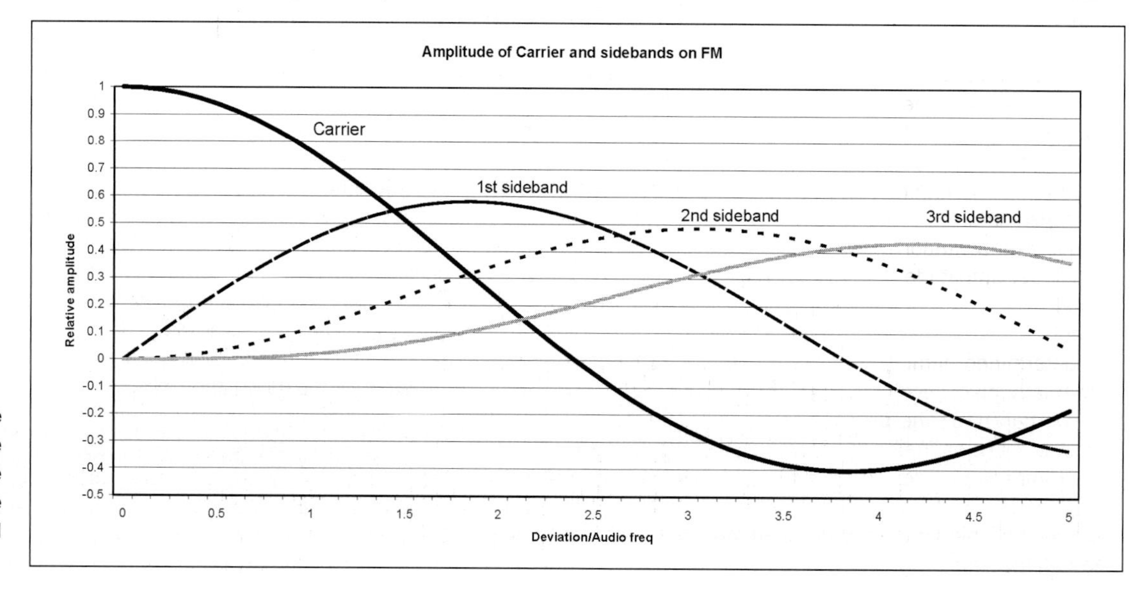


**Fig 1.57: The amplitude of the carrier and the sidebands are given by Bessel functions**

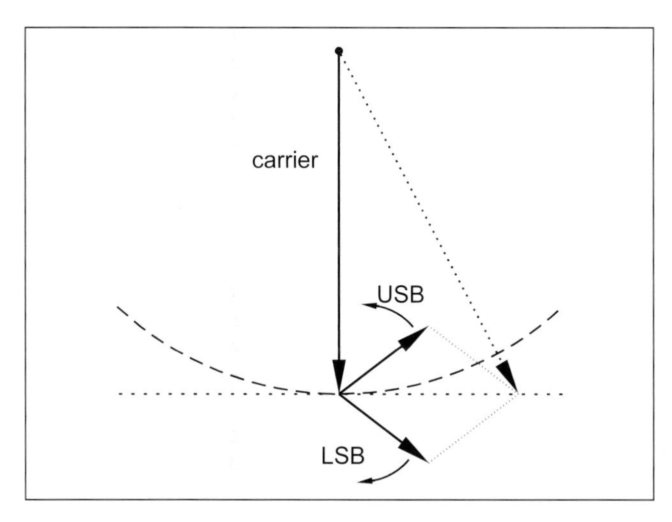


**Fig 1.59: Phasor diagram showing a single audio tone frequency modulating an RF carrier**

amplitude at the audio frequency. If the two sidebands have an amplitude of half the carrier amplitude then the resultant will be a signal just rising to double the unmodulated carrier and just falling to zero. That is 100% modulation as one would expect.

Now let's consider frequency modulation. First we will start with a narrow signal where only the first pair (or order) of sidebands, are relevant. Reference to Fig 1.57 shows for a low value of modulation index (deviation/audio frequency) the carrier is at full amplitude, the first order sidebands are of modest amplitude and the higher order ones are almost zero.

The situation is shown in **Fig 1.59**. Note that the two sidebands are shown as before for AM but we have rotated the carrier vector by 90°. Now as the two sidebands rotate, the resultant will move the head of the carrier arrow left and right but its length will remain substantially constant. As the head is moving to the right on the diagram (carrier vector rotating anti-clockwise) it is rotating slightly faster signifying a momentary frequency increase. As it moves to the left it is rotating slightly slower signifying a momentary frequency decrease. A complete 360° rotation of the two sideband vectors results in one cycle of increase and decrease of the carrier frequency as one would expect for FM.

As the modulation index increases the amplitude of the two sidebands increases. That suggests the length of the carrier vector will vary since it moves along the horizontal line shown dotted in Fig 1.59, not round the circle as it would need to do if its amplitude were to remain constant. We are rescued by looking again at Fig 1.57 which shows the second order sidebands are no longer zero. We must add them to our vector diagram. They will rotate at twice the audio frequency above and below the carrier. A few moments thought will show they add a vertical component to the locus of the head of the resultant vector. The resultant will indeed travel round the arc of the circle, not along the horizontal line.

This representation can be used to depict any modulation technique, amplitude, frequency or phase. This is also true of digital modulation methods discussed next. Simplistically digital modulation means that the resultant vector will only occupy a pre-determined number of set positions - assuming there is no unwanted noise.

If we can track the position and movement of the resultant vector in x-y co-ordinates then we can recover the original modulation. That can be done using two mixers each fed with a signal at the carrier frequency but with a 90° phase difference. That, of course, is the fundamental concept in SDR receivers where the two recovered signals are named I and Q, for the in phase and quadrature components. The technology for that is discussed elsewhere in this book.

## Digital Modulation

The techniques of digital modulation are little different, except the modulating input will normally comprise a data signal having one of two discrete values, normally denoted '0' and '1' but in practice two different voltages. In CMOS digital logic this may be 0V and 10V, or for RS232 signals from a computer ±6V.

Amplitude shift keying, that is two different amplitudes, either of the carrier or of an audio modulating signal is uncommon, partly because nearly all interference mechanisms are amplitude related and better immunity can be achieved by other methods.

Frequency shift keying is popular, the carrier having two frequencies separated by 170Hz for narrow shift or 450 and 750Hz wide shift typically used commercially. The occupied bandwidth is again greater than the frequency shift since each of the two 'carriers' will have sidebands depending on the keying speed and method. The mode can be generated either by direct keying of the modulator or feeding audio tones (1275 and 1455Hz for narrow shift amateur use) into an SSB transmitter.

Phase shift modulation is used in many amateur modes. These modes are discussed in more detail in the data communications chapter.

Two phases, 0° and 180° may be used. This is binary phase shift keying, BPSK. Another method, known as QPSK, quadrature phase shift keying has 4 states or phases; 0°, 90°, 180° and 270°. This allows two binary digits (each with two states 0 and 1, giving four in total) to be combined into a single character or symbol to be sent. By this means the symbol rate (the number of different symbols per second) is lower, half in this case, than the baud rate, defined as the number of binary bits per second.

Each symbol may be advanced in phase by 45° in addition to the phase change required by the data so that, irrespective of the data, there is always some phase change. This is needed to

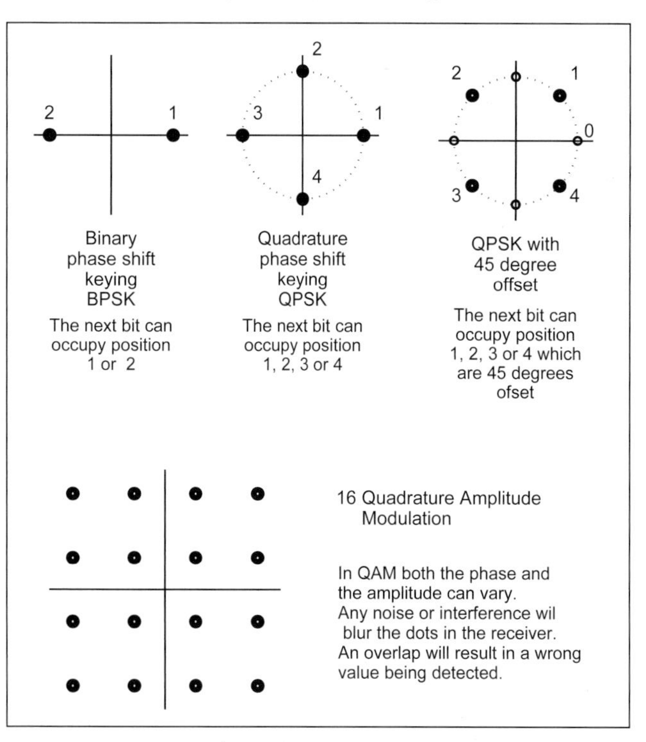


**Fig 1.60: Digital modulation schemes often vary the phase of the signal to represent the next transmitted symbol. 16 symbols can transmit four binary bits at once, but use amplitude as well as phase changes**

ensure the receiver keeps time with the transmitter. If this did not happen and the data happened to be a string of 20 zeros with no phase changes, there is a risk that the receiver would lose timing and output either 19, 20 or 21 zeros. It is then out of synchronisation with the transmitter and the output data signal may well be meaningless.

Higher level modulation schemes can combine even more binary digits into a symbol, allowing even higher data rates. Quadrature Amplitude Modulation (QAM) is used professionally but not often in amateur circles. This utilises both amplitude and phase changes to encode the data. Whilst higher data rates are achievable, the difference between each symbol gets progressively less and less as each symbol is used to contain more and more binary digits. **Fig 1.60** shows these systems diagrammatically. In isolation this is all very well but the effects of noise and interference get progressively worse as the complexity of the system increases.

Amateurs have tended to stick to relatively straightforward modulation schemes but have utilised then in novel and clever ways to combat interference or send signals with very low bandwidth, finding a slot in an otherwise crowded band. The data communications chapter has the details.

## Shannon limit

We saw above and in the QAM diagram in Fig 1.60 that a single link can use a more sophisticated multi-level modulation scheme to get more binary bits per symbol and thus a faster data (baud) rate for a given symbol rate. That is a higher bits/second/Hz. The downside is that the symbols are closer together, in amplitude, phase or both, so that the link is more susceptible to noise.

In professional planning the aim is to get many links in a given geographical area, maybe all converging on a single node, and the risk then is interference from other links, not natural noise. Higher capacity links need less noise and need to be more widely separated, leading to a trade off.

Amateur practice has been to optimise in a different direction, minimising the bandwidth required, eg PSK31, to fit in between other transmissions, or performing well at levels below the noise floor, eg WSJT. However that poses the question: how far can we go? Is there a theoretical maximum and how close to it are we? If we are within one decibel of the limit then trying to design for a 3dB improvement is doomed to failure.

We can try to combat bit errors by the use of error correcting codes. Very good codes exist but sending the code bits requires either a higher capacity channel or sacrifices user bit rates. Another trade off.

The limit is given by the Shannon formula:-

$$C = B\log_2\left(1+\frac{S}{N}\right)$$

Where C is the channel capacity in bits/sec (that is binary bits), B is the channel bandwidth in Hz, S is the received signal power in bandwidth B and N is the noise power in the same bandwidth. S/N is the signal to noise ratio in linear units. The signal and noise powers must be expressed as a power, not a voltage.

This formula assumes the noise has a Gaussian distribution, that is it is truly random noise as would be experienced from a hot resistor. If the ambient noise has a different statistical distribution, due typically to a semi-distant transmission or impulsive interference of man-made origin, then the S/N term will require slight modification.

Note that the log term is to the base 2, not the normal base 10. The $\log_{10}$ of 1000 is 3 because $10^3$ = 1000. Similarly $\log_2$ of 16 is 4 because $2^4$ = 16. Finding logs to the base 2 is simply achieved by remembering the definition of a logarithm.

The log of a number, x, is the power to which the base of the log must be raised to equal x.

$$\text{If} \quad 2^y = x \quad \text{then} \quad \log_2(x) = y$$

Take logs of both sides ($\log_{10}$)

$$y\log_{10}(2) = \log_{10}(x)$$

$$\text{so} \quad y = \frac{\log_{10}(x)}{\log_{10}(2)} = \frac{\log_{10}(x)}{0\cdot301}$$

Simply find the log of x (where x=1+S/N) and divide it by 0.3.

*Example:*

A radio channel has a 3 kHz bandwidth and a signal to noise ratio of 20dB or 100:1

$$C = B\log_2\left(1+\frac{S}{N}\right) = 3000\log_2\left(1+\frac{100}{1}\right) = 3000\log_2(101)$$

$$C = 3000\times 6\cdot 66 \approx 20\,kbit/\sec$$

The formula gives no indication how this may be achieved, that is for the designer; but the designer now knows the maximum bit rate in that bandwidth with that amount of noise (which includes receiver noise) which cannot be exceeded.

# Intermodulation

Mixing and modulation often use non-linear devices to perform the required function. Non-linear means that the output is not just a larger (or smaller) but otherwise identical copy of the input. Consider the equation y=4x. The output (y) is simply 4 times the input (x). Plotting the function on graph paper would give a straight line - linear.

Now consider $y=x^2$. Plotting this would give a curve, not a straight line; it is non-linear.

If the input was a sine wave, for example $\sin \omega_c t$, then the output would be of the form:

$$\sin^2 \omega_c t \quad \text{which is} \quad \tfrac{1}{2}(1-\cos 2\omega_c t)$$

The output will contain a DC component and a sine wave of twice the frequency. This is a perfect frequency doubler.

Similarly if two signals are applied, $\sin \omega_a$ and $\omega_b$, the output will be:

$$(\sin \omega_a + \sin \omega_b)^2$$

which is

$$\sin^2 \omega_a + 2\sin \omega_a \sin \omega_b + \sin^2 \omega_b$$

We already know the two $\sin^2$ terms are frequency doubling and we also know (from the amplitude modulation section) that sinA x sinB gives us sin (A+B) and sin (A-B), that is the sum and difference frequencies. We now have a mixer.

Unfortunately no device is quite that simple, and amplifiers are not perfectly linear. The errors are usually small enough to neglect but when badly designed, or overdriven, the non-linearities will produce both harmonics of the input frequencies and various sums and differences of those frequencies.

For two sine wave inputs A and B, the output contains frequencies m x A ± n x B where m and n are any integers from 0 upwards. These are called intermodulation products (IMPs). The frequencies 2A, 3A; 2B, 3B etc are far removed from the original frequencies and are usually easily filtered out. However, the frequencies m x A - n x B where m and n differ by 1, are close to the original frequencies.

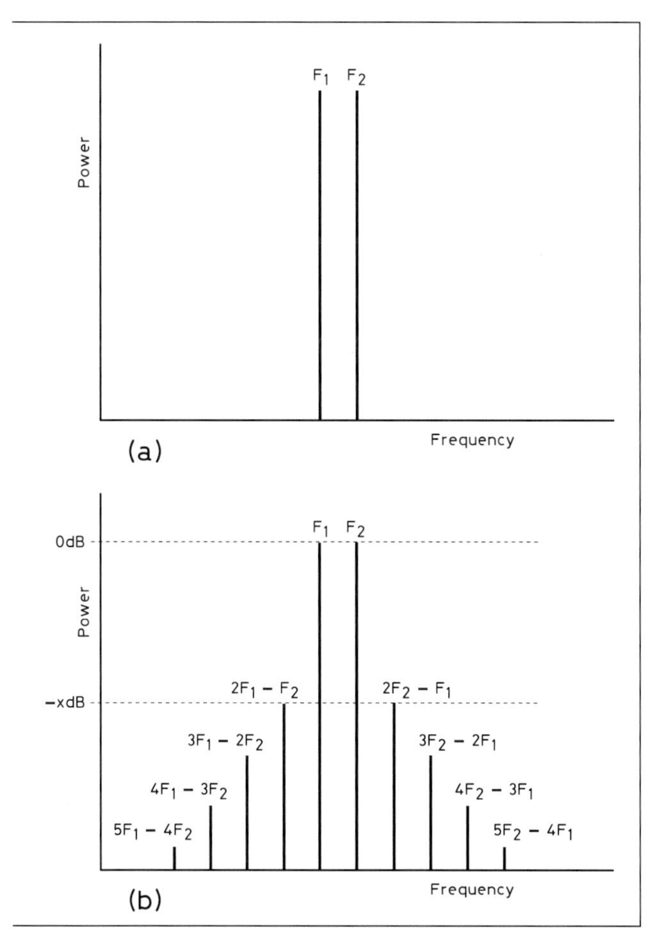


**Fig 1.61: Spectrum of two frequencies (a) before and (b) after passing through a non-linear device**

This can happen in an SSB power amplifier where two (or more) audio tones are mixed up to RF and then amplified. The unwanted intermodulation products may well be inside the pass-band of the amplifier or close to it such that they are not well attenuated by the filter. This, then represents distortion of the signal, or, worse, interference to adjacent channels. **Fig 1.61** illustrates this.

The value of m+n is known as the 'order' of the intermodulation product. It will be noticed that it is only the odd orders that result in frequencies close to the wanted signals. It is also the case that the lower order products (ie 3rd) will have a greater amplitude than higher order products. The power in these relative to the fundamentals, xdB in **Fig 1.61(b)**, should be at least 30dB down and preferably more.

Intermodulation can also occur in other systems. In receivers it can occur in the 'front-end' if this is operating at a non-linear level (due for example by being overloaded). For instance, in a crowded band such as 7MHz with high-powered broadcast signals close to (and even in!) the amateur band, intermodulation can occur between them. It should be noted that these signals may be outside the bandwidth of the later stages of the radio receiver but inside the band covered by the RF front end. The IMPs occur in the overloaded front end (normally in the mixer as this is a non-linear stage), producing extra signals which may well be close to, or on top of, the wanted signal.

This degrades the receiver performance if the products are within the pass-band of the receiver. The use of high-gain RF amplifiers in receivers can be a major source of this problem and may be countered by reducing the gain by using an input attenuator.

Alternatively, only just enough gain should be used to overcome the noise introduced by the mixer. For frequencies up to 14MHz, it is not necessary to have any pre-mixer gain provided the mixer is a good low-noise design (see the chapter on HF receivers).

Intermodulation can also occur in unlikely places such as badly constructed or corroded joints in metalwork. This is the so-called rusty-bolt effect and can result in intermodulation products which are widely separated from the original transmitter frequency and which cause interference with neighbours' equipment (see the chapter on electromagnetic compatibility).

## ANALOGUE TO DIGITAL CONVERSION

Frequently it is desirable to handle an analogue signal by digital means. Digital signals have discrete values so random noise, an ever present problem, must be sufficiently strong to exceed half the difference between the discrete values before there is risk of error. Furthermore, with digital systems, it is possible to have parity bits and other more complex error detection and correction systems. In many cases these will allow the RF signal to be successfully recovered even though it is well below the noise floor. Later chapters in this book discuss such systems.

Functions, such as filtering and demodulation can be performed as mathematical processes, often with better results than achievable by analogue circuitry. The wanted signal often had identifiable characteristics, whereas the noise (AWGN, Additive White Gaussian Noise) appears as random fluctuations. That allows the noise to be substantially suppressed, improving the perceived signal to noise ratio. Similarly, multipath reception can be either allowed for or its effects minimised. All this has the umbrella term DSP, Digital Signal Processing. However, before any of this can be carried out the analogue signal must be digitised by sampling it and representing the samples as digital values.

### Digital Sampling

Two decisions must be made when sampling an analogue signal. How accurately must we represent the amplitude of the signal, and how often must we do it?

#### Quantisation distortion

Analogue signals are continuously variable and may adopt any level from zero to a set maximum, and may be either positive or negative. A binary word of 8 bits, eg 10010011, can have $2^8$ different values, including zero. That is 256 different levels. The nearest of those 256 levels is chosen to represent the analogue signal.

There will be a small error, known as the quantisation error, as a result of digitising the signal. The reason for the error is that

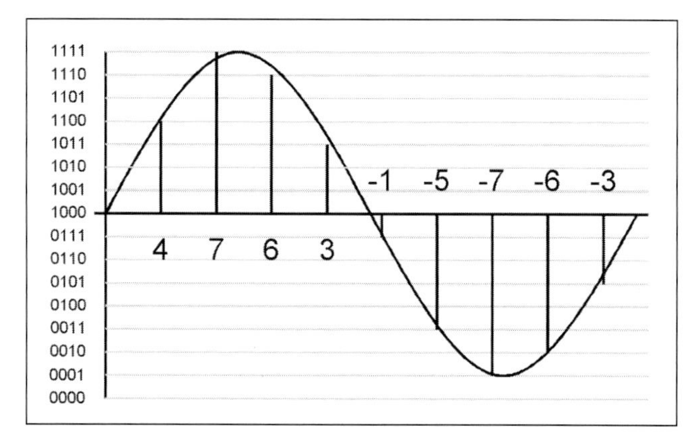


**Fig 1.62: Digitising the amplitude results in quantisation errors**

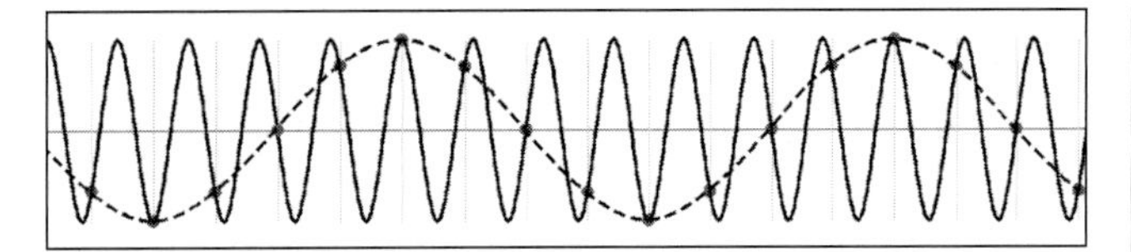

Fig 1.63: Under-sampling the signal

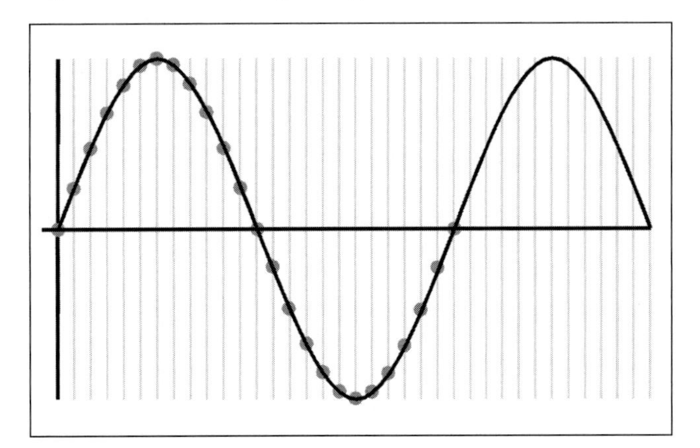

Fig 1.64: Over-sampling the signal results in a lot of data

there are only 256 different levels to choose from, rather than the infinite number needed to give an exact fit to the analogue signal. These errors can be regarded as a real signal plus quantisation 'noise', giving rise to the concept of a signal to quantisation noise ratio.

**Fig 1.62** shows part of an analogue signal and the errors in selecting the best fit digital representation. The vertical samples are restricted to one of eight levels each side of zero. A four bit binary word is being used, giving 16 values with 0000 representing the -8 level, 1000 representing the signal 0 point, and 1111 representing the +7 level.

To reduce the error, more binary bits are needed. 16 bits will give 65536 levels to choose from. That is much more accurate, but at the expense of needing more bits to represent the signal. That will translate into more DSP processing power required.

Assuming the quantisation noise has an amplitude uniformly distributed between plus or minus one half of the distance between the least significant bits, then the signal to noise ratio is given by

$$\mathrm{SNR} = 20\mathrm{Log}(2^{Q}) \quad \mathrm{dB}$$

where Q is the number of binary bits *(For explanation of dB see later in this chapter).*

For sine wave signals and assuming the full range of the converter is used, this formula can be simplified to

$$\mathrm{SNR} = 1 \cdot 76 + 6\mathrm{Q} \quad \mathrm{dB}$$

### Sampling rate

Clearly we need to sample sufficiently often to capture all the details in the signal; but how fast is that?

**Fig 1.63** shows a sine wave, the solid line, sampled 17 times in 15 cycles, that is just over once per cycle. The sample points are shown as dots. The sampled points suggest the signal was the dashed waveform. The sampling is too infrequent to capture all the detail in the signal.

**Fig 1.64** shows a signal being sampled very frequently. The exact shape of the signal is captured well. This results in a lot more data than needed and extra work for the digital processing system. Fewer samples would capture this signal perfectly well enough to reproduce it properly.

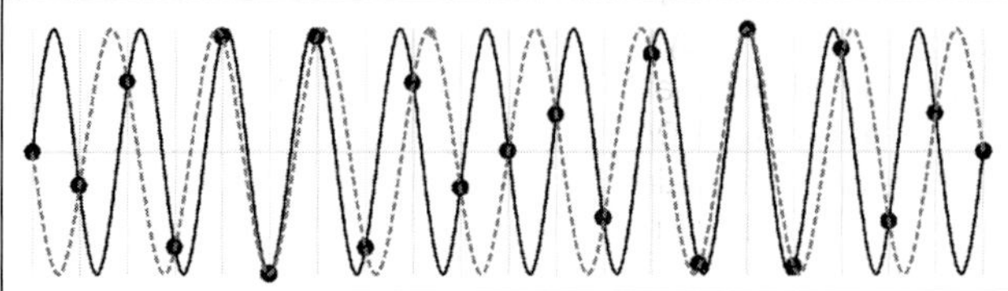

Fig 1.65: Sampling at the optimum rate with an unwanted higher frequency present

### Finding the best sampling rate

**Fig 1.65** shows 10ms of time divided up into a sample taken every half millisecond, 21 samples if we include time zero.

The solid waveform is a sine wave signal at 900Hz; there are nine cycles in 10ms. The 0.5ms sampling points are shown by the black dots. Also shown, dashed, is an 1100Hz signal. It will be noticed that the results of the sampling of that signal are identical to the 900Hz one. The sampling is failing to distinguish between a 900Hz signal and an 1100Hz signal. Which one is correct?

It can be shown that it is necessary to sample at least as fast as twice the highest frequency signal present. The proof is quite complex and the reader is referred to texts on the Nyquist Sampling Theorem for further information [6, 7, 8].

### Aliasing and anti-aliasing filter

In the example above, it would be necessary to have a good filter in front of the analogue to digital conversion process, to remove all frequencies above 1000Hz. If a small trace of an 1100Hz signal were allowed through the input filter, it would be processed as if it were a 900Hz signal. It is quite likely that it would interfere with the wanted signals.

The 1100Hz, unwanted, signal is called an *alias* and the effect of these higher frequency signals appearing as lower frequency ones inside the wanted frequency range, is called *aliasing*. Since the input filter cannot be perfect, in practice it is necessary to sample rather more often than twice the highest frequency present.

## Digital to Analogue Conversion

Having been processed, the digital signal must be converted back into analogue form. The output of a digital to analogue convertor (DAC or D2A) will be a series of steps both in amplitude and time. This is shown in **Fig 1.66**. The output of a DAC is shown as a solid black line. It is based on the samples of the dotted sine wave. At each conversion point the DAC changes its output to match the present value of that sine wave.

The DAC output is squared off into one of the 31 discrete levels with level 16 (binary 1000) being taken as the zero level allowing the signal to have both positive and negative values.

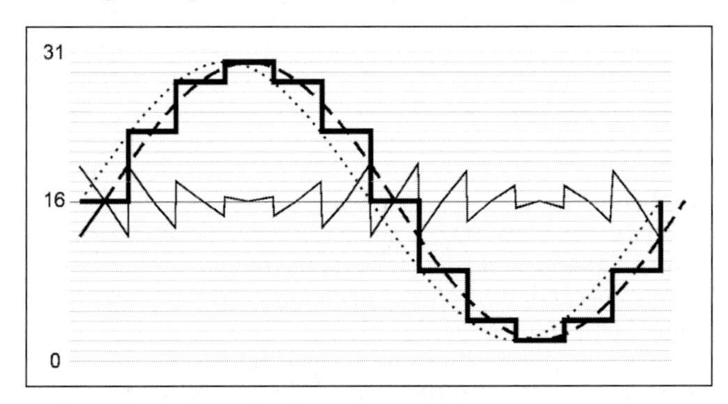


Fig 1.66: The DAC output of a sinewave

Negative values are signified by binary numbers below 1000, and positive values by binary numbers above 1000.

The dashed sine wave shows the sinewave representing the average value of the DAC output.

Two errors are noticed: Firstly, the output is delayed with respect to the digital representation. This is in addition to any delays occasioned by any mathematical processing in the DSP. Secondly, the square waveform is not a good representation of the smooth sinewave.

The timing error is of no consequence unless the original signal can be heard. On devices used to replay an earlier recording, this is irrelevant. Where the processing introduces extended delays, some interruption in smooth voice communication can be observed.

The fine black line shows the difference between the averaged sinewave and the actual output. The output can be regarded as the wanted sinewave and an additional signal represented by the fine black line. That is noise introduced by the digital to analogue conversion process.

Inspection of the noise waveform shows it is of much higher frequency. Consequently it can be removed by filtering. Passing the DAC output through a filter will remove all the higher frequency components and leave just the dashed waveform (further delayed by around half a sample period). The filter has smoothed out the rough edges of the DAC output.

In practice a 5-bit DAC is too coarse for practical use. A minimum of 8 bits is required and music systems are likely to utilise 14 or 16 bit DACs.

These devices are much more common than might be realised. CD and MP3 players all store the material, usually music, in digital form, which requires conversion to analogue. Mobile phones are now fully digital and require both A to D and D to A conversion.

## DECIBELS

Consider the numbers:

100 = $10^2$
1000 = $10^3$
10000 = $10^4$

The logarithm of a number is defined as the power to which the base of the logarithm (in this case 10) must be raised to equal the number. So the logarithm (log for short) of 1000 is 3, because $10^3$ is 1000. This is written as $\mathrm{Log}_{10}$ (1000) = 3. However it is usually understood that base 10 is used so the expression is often simplified to Log1000 = 3.

It should also be noted that $10^3$ (1,000) x $10^4$ (10,000) is 10,000,000 or $10^7$. In multiplying the two numbers the powers have been added.

It is possible to consider the Logs of numbers other than powers of 10. For example, the Log of 2 is 0·3010 and the Log of 50 is 1·6990. If these two Logs are added, we get 0·3010+1·6990 =2. We know the Log of 100 is 2, so the anti-Log of 2 is 100. So adding the Log of 2 and the Log of 50 gives us an answer equal to the Log of 100.

Adding the logs of two numbers and then finding the anti-log of the result gives us an answer equal to the product of the two original numbers.

Today this seems an incredibly awkward way of working out that 2 x 50=100. However, computers and calculators are relatively recent devices and Logs were a simple way of multiplying rather more cumbersome numbers.

The aim of the above is really to set out a principle. Multiplying and dividing can be achieved by adding and subtracting the Logs of the numbers concerned.

Consider the route from a microphone to the distant loudspeaker. A microphone amplifier provides some gain, the modulator may provide a gain or a loss, the RF power amplifier will provide gain, the feeder to the antenna will cause a loss and the antenna may well have some gain.

The radio path will have losses associated with distance, trees and building clutter, hills and weather conditions. The receive antenna will provide gain, the feeder a loss and then the many stages in the receiver will have their own gains and losses.

These gains and losses will not be simple round numbers and they all need multiplying or dividing.

It is normal in radio and line transmission engineering to express all these gains and losses in terms of units proportional to the logarithm of the gain (positive) or loss (negative). In this way gains and losses in amplifiers, networks and transmission paths can be added and subtracted instead of having to be multiplied together.

The unit is known as the decibel or dB. It is defined as:

$$\text{Gain (or loss)} = 10\mathrm{Log}_{10}\left(\frac{P_{out}}{P_{in}}\right)$$

It is always a power ratio, not a current or voltage ratio. If current or voltage ratios were used, a voltage step-up transformer would appear to have a gain, which is not the case. The power out of a transformer is never greater than the power in.

However, provided and only provided, the currents or voltages are being measured in the same resistance (so a voltage increase does signify a power increase) then it is possible to use the definition:

$$\text{Gain (or loss)} = 20\mathrm{Log}_{10}\left(\frac{V_{out}}{V_{in}}\right) \quad \text{or} \quad 20\mathrm{Log}_{10}\frac{I_{out}}{I_{in}}$$

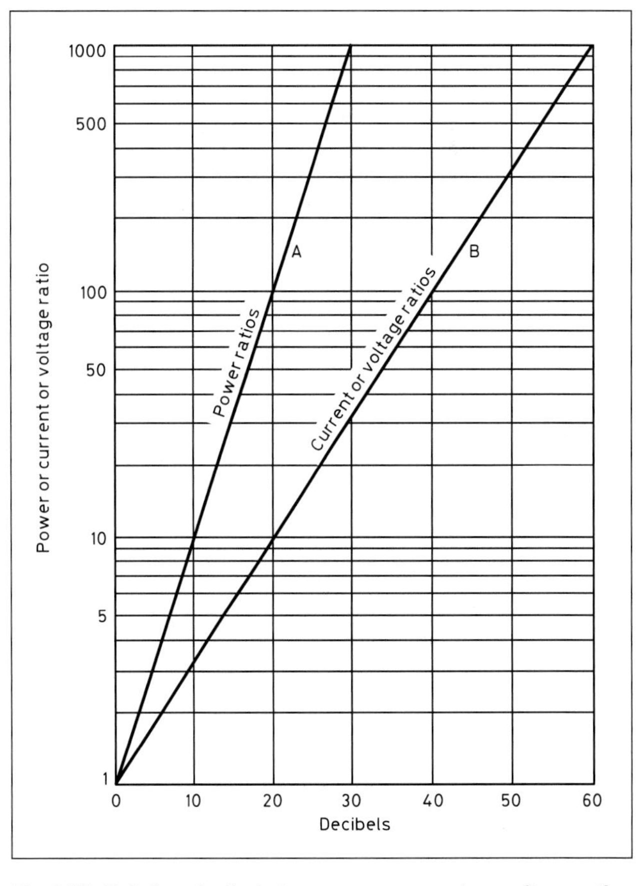


**Fig 1.67: Relating decibels to power or current or voltage ratio**

| multiplier | construction | dB construction | dB value |
|---|---|---|---|
| 2 | - | - | 3 |
| 4 | 2 x 2 | 3+3 | 6 |
| 8 | 2 x 2 x 2 | 3+3+3 | 9 |
| 16 | 2 x 2 x 2 x 2 | 3+3+3+3 | 12 |
| 10 | - | - | 10 |
| 100 | 10 x 10 | 10+10 | 20 |
| 20 | 2 x 10 | 3+10 | 13 |
| 40 | 2 x 2 x 10 | 3+3+10 | 16 |
| 5 | 10 ÷ 2 | 10-3 | 7 |
| 50 | 10 x 10 ÷ 2 | 10+10-3 | 17 |
| 25 | 10 x 10 ÷ 2 ÷ 2 | 10+10-3-3 | 14 |
| 1·26 | - | - | 1 |
| 1·58 | - | - | 2 |

**Table 1.5: Construction of dB values**

The change to 20x follows from the formula $P=V^2/R$ or $I^2R$. If a number is squared (ie multiplied by itself) then the Log will be doubled (added to itself). Line A in **Fig 1.67** can be used for determining the number of decibels corresponding to a given power ratio and vice versa. Similarly line B, shows the relationship between decibels and voltage or current ratios.

The use of Log tables (or the Log function on a calculator) can often be avoided by remembering two values and understanding how decibels apply in practice.

It was stated above that Log2 = 0·3010, so a doubling of power is 3·01dB, equally a halving is -3·01dB. The ·01 is usually omitted so doubling is simply assumed to be 3dB.

Quadrupling is double double and, like Logs, decibels are added, giving 6dB. The other one to remember is 10 times is 10dB. **Table 1.5** can be constructed from just those two basic values.

It can also be remembered that, with the same resistance (or impedance), the ratios in column 1 applied to voltage or current will double the dB value.

## REFERENCES AND FURTHER READING


[1] *Tables of Physical and Chemical Constants*, Kaye and Laby, 14th edition, Longman, London, 1972, pp102-105

[2] *Electronic and Radio Engineering*, Terman, 4th edn, McGraw Hill

[3] *Advance! The Full Licence Manual*, Alan Betts, 1st edn, RSGB

[4] *The RSGB Guide to EMC,* Robin Page-Jones, 2nd edn, RSGB

[5] *Four Figure mathematical tables*, Castle, Macmillan

[6] *Data Communications and Networks*, R L Brewster, pp 25 and 44

[7] *Telecommunications*, J Brown and E V D Glazier, pp 47

[8] http://en.wikipedia.org/wiki/Nyquist%E2%80%93Shannon_sampling_theorem

# Passive Components

2

*Stuart Swain, G0FYK*

While most passive components are very reliable, reliability is just as important for the amateur as for the professional. To us, a breakdown may mean a frustrating search with inadequate instruments for the defective component. This can be especially trying when you have just built a published circuit, and it does not work. Reliable components do tend to be expensive when bought new, but may sometimes be bought at rallies if you know what to look for. The aim of this chapter is to give some guidance.

If you test components before using, you save yourself from this troubleshooting.

Components should be de-rated as much as possible. In 1889 a Swedish chemist, Arrhenius found an exponential law connecting chemical reaction speed with temperature. As most failures are caused by high temperature, reducing the temperature will greatly reduce failure rate. Some components, particularly electrolytic capacitors, do not take kindly to being frozen, so take the above advice with proper caution. Similarly, operating at a voltage or current below the maker's maximum rating favours long life.

## WIRE

Copper is the conductor most often used for wire, being of low resistivity, easily worked and cheaper than silver. PVC or PTFE covered wire is most often used in equipment and for leads. Kynar covering is favoured for wire-wrap layouts. **Table 2.1** gives the characteristic of some coverings. Silicone rubber is cheaper than PTFE and can be used in high temperature locations but has not the strength of PTFE.

For inductors, the oleo enamel, cotton, regenerated cellulose and silk of yesteryear have given way to polyester-imide or similar plastics. These have better durability and come with a variety of trade names. They are capable of much higher temperature operation, and resist abrasion and chemicals to a greater degree. Either trade names or maximum temperature in degrees C are used to describe them. They are more difficult to strip, especially fine or 'Litz' stranded wires. Commercially, molten sodium hydroxide (caustic soda) is used, leaving bright easily soldered copper, but there is a hazard, not only is the material harmful to the skin but, if wet, is liable to spit when heated. Paint stripper is a good substitute, but is liable to 'wick' with Litz wire. Be sure to clean it off.

There are also polyurethane coatings, usually pink or, for identification purposes, green, purple or yellow. These can be used up to 130°C, and are self-fluxing, if a sufficiently hot iron is used. The fumes are irritating and not very good for you, especially if you are asthmatic.

Dual coatings of polyester-imide with a lower melting point plastic layer outside are made. When a coil has been wound with these, it is heated by passage of current and the outer layers fuse, forming a solid bond. Dual coatings are also found where the outer coating is made of a more expensive material than the inner.

Oleo enamel and early polyester-imide crack with age and kinking so old wire should be viewed with suspicion. The newer synthetic coatings can resist abrasion to a remarkable degree, but kinked wire, even of this type should not be used for inductors or transformers.

Wire size can be expressed as the diameter of the conductor in millimetres, Standard Wire Gauge (SWG) or American Wire Gauge (AWG). The sizes and recommended current carrying capacity are given in the general data chapter at the end of this book. Precise current carrying capacity is difficult to quote as it depends upon the allowable temperature rise and the nature of the cooling. Table 2.1 and **Table 2.2** give data and a guide to usage.

Radio frequency currents only penetrate to a limited depth; for copper this is $6.62/\sqrt{(F_{HZ})}$ cm. For this reason, coils for the higher frequencies are sometimes silver plated; silver has a marginally better conductivity than copper (see Table 2.2) and is less liable to corrosion, though it will tarnish if not protected. Generally the cost is not justified. For high currents, tubing may be used for reduced cost, as the current flows on the outside.

| Material | Temperature guide (maximum °C) | Remarks |
|---|---|---|
| PVC | 70-80 | Toxic fume hazard |
| Cross-linked PVC | 105 | Toxic fume hazard |
| Cross-linked polyolefine | 150 | |
| PTFE (Tefzel) | 200 | Difficult to strip |
| Kynar | 130 | Toxic fume hazard. $\varepsilon_r$=7* |
| Silicone rubber | 150 | Mechanically weak |
| Glassfibre composite | 180-250 | Replaces asbestos |

*If temperature range is critical, check with manufacturer's data.*
** $\varepsilon_r$ is the dielectric constant.*

**Table 2.1: Equipment wire and sleeving insulation**

| Metal | Resistivity at 20°C (Ω/m) | Temperature coefficient of resistivity per °C from 20°C (ppm/°C) | Notes |
|---|---|---|---|
| Annealed silver | $1.58 \times 10^{-8}$ | 4000 | |
| Annealed pure copper | $1.72 \times 10^{-8}$ | 3930 | |
| Aluminium | $2.8 \times 10^{-8}$ | 4000 | (1) |
| Brass | $9 \times 10^{-8}$ | 1600 | (1) |
| Soft iron | $11 \times 10^{-8}$ | 5500 | |
| Cast iron | $70 \times 10^{-8}$ | 2000 | |
| Stainless steel | $70 \times 10^{-8}$ | 1200 | (1) |
| Tungsten | $6.2 \times 10^{-8}$ | 5000 | |
| Eureka | $50 \times 10^{-8}$ | 40 | (1, 2) |
| Manganin | $43 \times 10^{-8}$ | ±10 | (3) |
| Nichrome | $110 \times 10^{-8}$ | 100 | (1) |

*Notes:*

*(1) Depends on composition and purity. Values quoted are only approximate.*

*(2) Copper 60%, nickel 40% approx. Also known as 'Constantan', 'Ferry' and 'Advance'. High thermal EMF against copper.*

*(3) Copper 84%, manganese 12%, nickel 4%. Low-temperature coefficient only attained after annealing at 135°C in an inert atmosphere and a low thermal EMF against copper. Cannot be soft-soldered.*

**Table 2.2: Resistivity and temperature coefficient for commonly used metals. Resistivity describes a material's ability to transmit electrical current that is independent of the geometrical factors**

From about 10kHz up to about 1MHz, a special stranded wire called 'Litz' gives lower radio frequency resistance due to the skin effect mentioned above. The strands are insulated from each other and are woven so that each strand takes its turn in the centre and outside of the wire. Modern Litz wire has plastic insulation, which has to be removed by one of the methods already described. Self-fluxing covering is also made. Older Litz wire has oleo enamel and silk insulation, needing careful heating in a meths flame for removal. There has been considerable dispute over the effect of failing to solder each strand; it is the writer's experience that it is essential to solder each strand, at least if the coil is to be used at the lower range of Litz effectiveness. Litz loses its usefulness at the high end of the range because of the inter-strand capacitance making it seem to be a solid wire. This may explain why some authorities maintained that it was unnecessary to ensure that each strand was soldered - at the top end of the range it may not make much difference.

Data on radio frequency cables will be found in the general data chapter.

For some purposes high resistivity is required, with a low temperature coefficient of resistivity. Table 2.2 gives values of resistivity and temperature coefficient of commonly used materials. For alloys, the values depend very much on the exact composition and treatment.

## FIXED RESISTORS

These are probably the components used most widely and are now among the most reliable if correctly selected for the purpose. As the name implies, they resist the passage of electric current - in fact the unit 'Ohm' is really the Volt per Amp.

Carbon composition resistors in the form of rods with radial wire ends were once very common. There is no excuse for using them now, as they are bulky, and suffer from a large negative temperature coefficient, irreversible and erratic change of value with age or heat due to soldering. Due to the granular nature of the carbon and binder, more than thermal noise is generated when current flows, another reason not to use them!

Tubes of carbon composition in various diameters with metallised ends are made for high wattage dissipation. These may bear the name 'Morganite' and are excellent for RF dummy loads (**Fig 2.1**). Recently developed film resistors are also suitable.

Carbon films deposited by 'cracking' a hydrocarbon on to ceramic formers are still used. To adjust the value, a helix is cut into the film, making the component slightly inductive. Protection is either by varnish, conformal epoxy or ceramic tube. Tolerance down to 5% is advisable, as the temperature coefficient (tempco) of between -100 and -900 parts per million per degree Celsius (ppm/°C) makes the value of closer tolerance doubtful.

In the 1930s, Dubilier made metal film resistors similar in appearance to carbon composition ones of that era, but they never became popular. In the 1950s, metal oxide film resistors were introduced, with tempco (temperature coefficient) of 300ppm/°C, soon to be replaced for lower wattage by the forgotten metal film types. These were more stable, less noisy and more reliable than carbon film, now only a little more expensive. They use the same helical groove technique as carbon film, and little difficulty should be experienced due to this up to about 50MHz. The tempco is of the order of ±15ppm/°C, allowing tolerance of 1% or better. Protection is either by conformal or moulded epoxy coating.

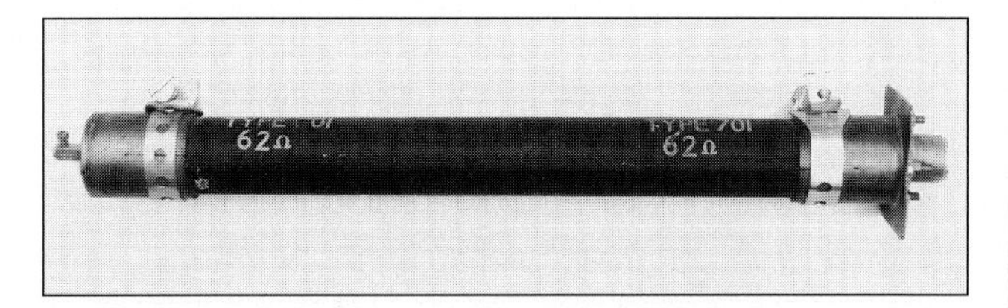


**Fig 2.1: An amateur-made 62-ohm dummy load with UHF coaxial connector. It is suitable for 200W for short periods**

**Fig 2.2: LTO-100 power resistor using thick film technology. It is rated at 100 watts at 25°C case temperature, heatsink mounted. Direct mounting ceramic on heatsink. [Info/picture from www.vishay.com]**

Metal film, 0.6W typically, is an ideal replacement for carbon film 1/4W, 1/3 W and 1/2W types. Body size is the same as 1/4W types. Superior replacement for most metal oxide and thick film resistors. Do not buy a closer tolerance resistor than is needed. Metering circuits, for instance, will demand ±1%, but bias resistors can use a wider tolerance such as ±5%.

Metal film resistors have excellent stability under load and severe environmental conditions. They exhibit very low noise current and voltage coefficients. They are available in a wide range of resistor values and are suitable for general purpose and precision applications.

Thick film power resistors (**Fig 2.2**) are non-inductive and are available from 0.015 to 1M ohms, with tolerances of 1, 2, 5 and 10%. A 50 ohm item would make a good dummy load at HF and below.

Surface mounting has now become universal in the commercial world, and as amateurs we need to embrace it. **Fig 2.3** shows a typical board layout. More about this later in this chapter, and in the construction chapter.

Resistor values are quoted from a series that allows for the tolerance. The series is named according to the number of members in a decade. For example, the E12 series has 12 values between 1 and 10 (including 1 but not 10). In this case, each will be the twelfth root of ten times the one below, rounded off to two significant figures (**Table 2.3**). The E192 series allows for

**Fig 2.3: Surface mounting components. The small dark components on the left labelled '473', '153', '333' and '243' are resistors, the larger grey one on the right is a capacitor and the light-coloured one labelled 'Z1' (bottom left) is a transistor. The scale at the bottom is in millimetres**

| E3 ±40% | E6 ±20% | E12 ±10% | E24 ±5% |
|---|---|---|---|
| 1.0 | 1.0 | 1.0 | 1.0 |
| - | - | - | 1.1 |
| - | - | 1.2 | 1.2 |
| - | - | - | 1.3 |
| - | 1.5 | 1.5 | 1.5 |
| - | - | - | 1.6 |
| - | - | 1.8 | 1.8 |
| - | - | - | 2.0 |
| 2.2 | 2.2 | 2.2 | 2.2 |
| - | - | - | 2.4 |
| - | - | 2.7 | 2.7 |
| - | - | - | 3.0 |
| - | 3.3 | 3.3 | 3.3 |
| - | - | - | 3.6 |
| - | - | 3.9 | 3.9 |
| - | - | - | 4.3 |
| 4.7 | 4.7 | 4.7 | 4.7 |
| - | - | - | 5.1 |
| - | - | 5.6 | 5.6 |
| - | - | - | 6.2 |
| - | 6.8 | 6.8 | 6.8 |
| - | - | - | 7.5 |
| - | - | 8.2 | 8.2 |
| - | - | - | 9.1 |

*The values shown above are multiplied by the appropriate power of 10 to cover the range.*

**Table 2.3: 'E' range of preferred values, with approximate values of tolerance from the next number in the range**

better than 0.5% tolerance, the figures now being rounded off to three significant figures. Unfortunately E12 is not a subset of E192, and some unfamiliar numbers will be met if E12 resistors are required from E192 stock. Values can either be colour coded (see the general data chapter) or for surface mounted ones, marked by two significant digits followed by the number of noughts. This is called the 'exponent system'. For example, 473 indicates 47k, but 4.7 Ohms would be 4R7. Metal film resistors can also be made into integrated packages either as dual in line, single in line or surface mount. By using these, space is saved on printed circuit boards.

For higher powers than metal oxide will permit, wire wound resistors are used. The cheaper ones are wound on a fibre substrate, protected by a rectangular ceramic tube or, rarely, moulded epoxy. More reliable (and expensive) ones are wound on a ceramic substrate covered by vitreous or silicone enamel. For even greater heat dissipation, it may be encased in an aluminium body designed to be bolted to the chassis or other heat sink. Vitreous enamelled wound resistors can be made with a portion of the winding left free from enamel so that one or more taps can be fitted. Wire wound resistors, even if so-called 'non-inductively' wound, do have more reactance than is desirable for RF use. However Ayrton Perry wound resistors in gas filled glass bulbs are suitable for use in the HF bands, but hard to come by.

Unless an aluminium-clad resistor is used for high power, care should be taken to ensure that the heat generated does not damage the surroundings, particularly if the resistor is mounted on a printed circuit board. The construction chapter gives practical advice about this. Marking is by printed value, perhaps including wattage and tolerance as well.

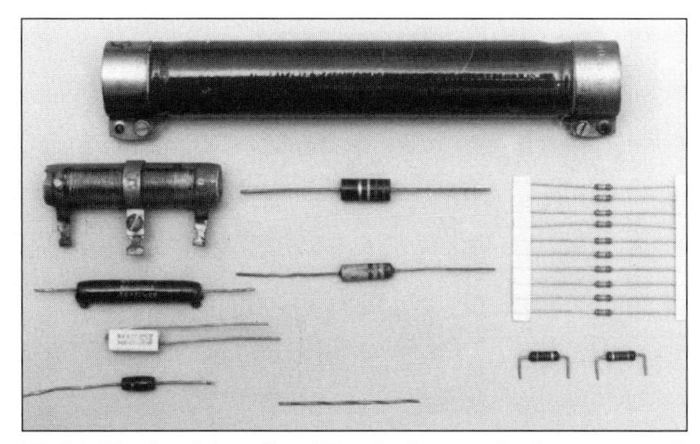

**Fig 2.4: Fixed resistors. Top: 50 watt wire wound, vitreous-enamel covered. Left, top to bottom: 10 watt wire wound with adjustable tapping, 12 watt vitreous enamelled, 4 watt cement coated and 2.5 watt vitreous enamelled. Centre: 2 watt carbon composition, 1 watt carbon film. Right, top: ten quarter-watt metal film resistors in a 'bandolier' of adhesive tape, two carbon film resistors with preformed leads. Bottom: striped marker is 50mm long**

**Fig 2.4** shows photographs of many of the fixed resistors that have just been described.

Single-in-line resistor networks provide the user with nominally equal resistors, each connected to a common pin (pin no. 1). A suitable application would be for a resistor-capacitor line. **Fig 2.5** shows a typical example.

## VARIABLE RESISTORS

Often a resistor has to be made variable, either for control or precise adjustment purposes; the latter is usually called a trimmer. Generally variable resistors are made with a moveable tapping point, and the arrangement is called a potentiometer or 'pot.' for short, the name coming from a laboratory instrument for use in measuring voltage. If only a variable component is required, it is advisable to connect one end of the track to the slider.

Tracks are made from carbon composition, cermet or conductive plastic, or they may be wire wound. Circular tracks with either multi-turn or single turn sliders are made, both for pots and trimmers, in varying degrees of accuracy and stability. The HiFi world likes linear tracks for control purposes, as do some rigs available to amateurs. Wire wound types have the disadvantage of limited resolution and higher price, but have the advantage of higher wattage and good stability. It is not always possible to tell the type from the external appearance as the photograph of **Fig 2.6** shows.

Single turn rotary pots for volume control and many other uses usually turn over a range of 250°-300°, with a log, anti-log or linear law connecting resistance with rotation. Ganged units and pots with switches are common. Precision components generally have a linear law (some for special purposes may have another law) and turn some 300°. For greater accuracy, multi-turn pots are used, some in conjunction with turn counting dials.

**Fig 2.5: Single-in-line resistor network**

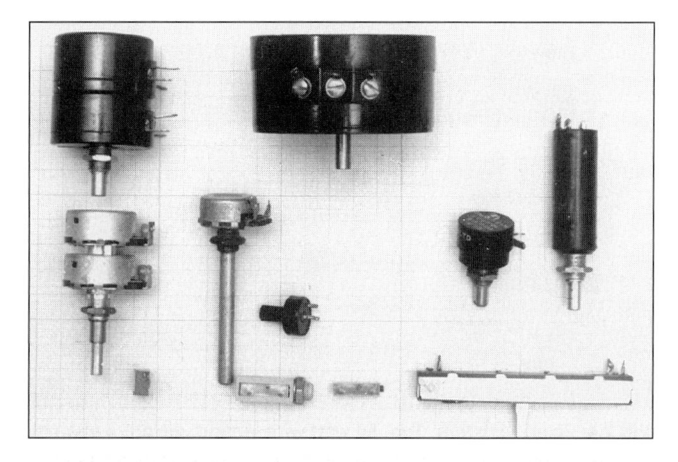

**Fig 2.6: Variable resistors. Top, left to right: double-gang 2 watt wire wound, 5 watt wire wound. Middle, left to right: two independent variable resistors with coaxial shafts, standard carbon track, single-turn preset, standard wire wound and 10-turn variable resistor. Bottom, left to right: 10-turn PCB vertical mounting preset, 10-turn panel and PCB horizontal mounting preset. Right: a slider variable resistor**

There are multi turn linear trimmers using a screw mechanism to move the slider over the linear track. Available values are frequently only in the E3 range, to 20% tolerance, the value being printed on the component, possibly in the exponent system.

## NON-LINEAR RESISTORS

All the resistors so far described obey Ohm's Law closely, that is to say that the current through them is proportional to the applied voltage (provided that the physical conditions remain constant, particularly temperature). Resistors, except metal film and wire wound, do change their value slightly with applied voltage, but the effect can generally be ignored. For some purposes, however, specialist types of resistor have been developed.

*Thermistors* are sintered oxides or sulphides of various metals that have a large temperature coefficient (tempco) of resistance, and are deliberately allowed to get hot. Both positive (PTC) and negative (NTC) tempco thermistors are made. Of all the forms of construction, uninsulated rods or discs with metallised ends are the most interesting to amateurs. Bead thermistors have special uses, such as stabilising RC oscillators or sensing temperature, one use being in re-chargeable battery packs.

PTC thermistors are used for current limitation. The thermistor may either operate in air or be incorporated into the device to be protected. Self-generated heat due to excess current or rise of ambient temperature will raise the resistance, and prevent further rise of current They should not be used in constant current circuits because this can result in ever increasing power dissipation and therefore temperature (thermal runaway), so destroying the device.

Unlike metals, PTC thermistors do not exhibit a positive tempco over the whole range of temperature likely to be encountered. **Fig 2.7(a)** shows this behaviour for a particular thermistor.

One type of PTC 'thermistor', though not usually classed as such, is the tungsten filament lamp. Tungsten has a large tempco of about 5000ppm/°C (Table 2.2) and lamps operate with a temperature rise of some 2500°C. An increase to about 12 times the cold resistance can be expected if the lamp is lit to full brilliance. These lamps are relatively non-inductive and can be used as RF loads if the change of resistance is acceptable.

NTC thermistors find a use for limiting in-rush currents with capacitor input rectifier circuits. They start with a high resist-

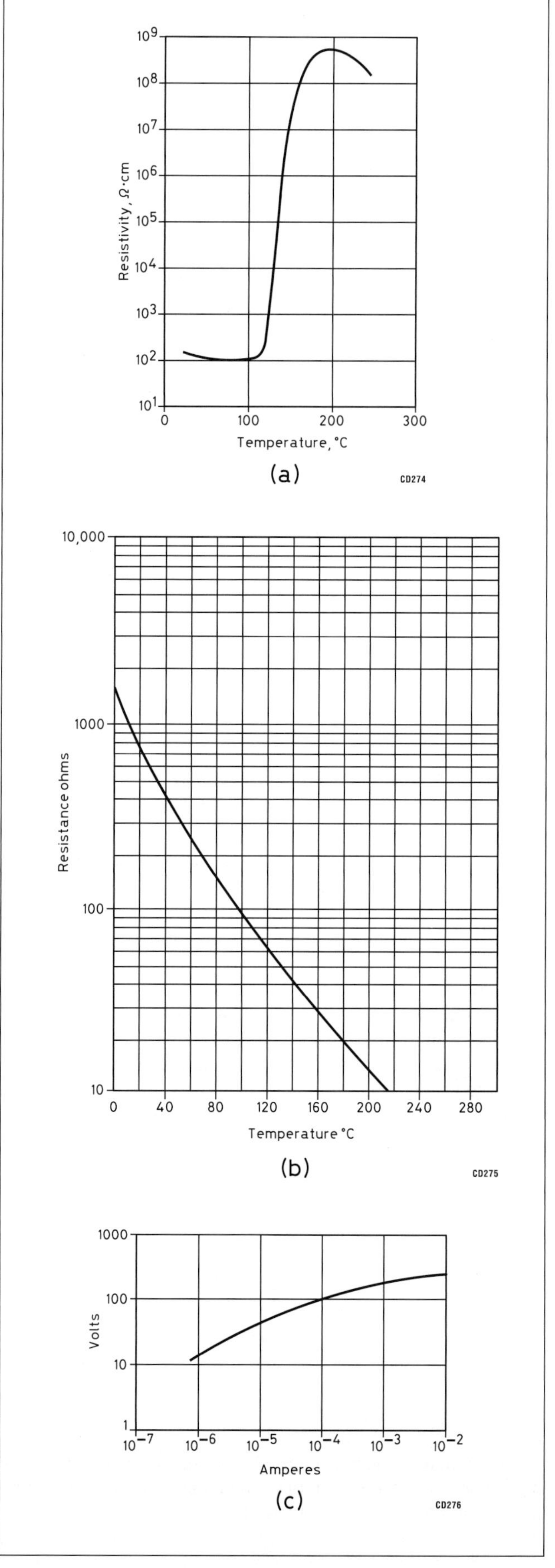


**Fig 2.7: Characteristics of thermistors**

ance, which decreases as they heat up. (**Fig 2.7(b)**). Remember that they retain the heat during a short break, and the in-rush current will not be limited after the break. When self-heated, the thermistor's temperature may rise considerably and the mounting arrangements must allow for this.

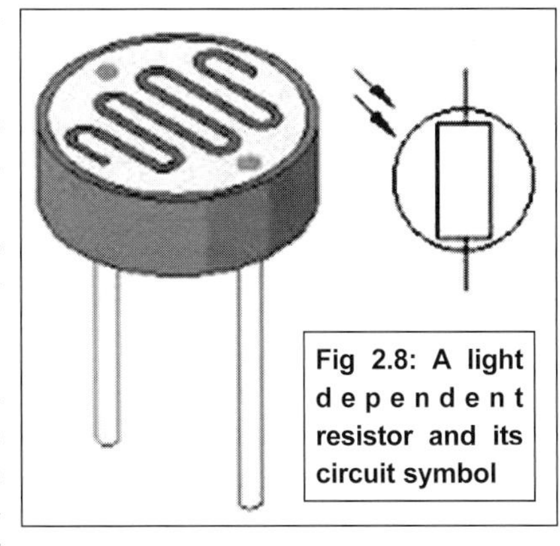

Fig 2.8: A light dependent resistor and its circuit symbol

PTC bead thermistors are the ones used to stabilise RC oscillators. Also, they are used in temperature compensated crystal oscillators where self-heating is arranged to be negligible and the response to temperature is needed to offset the crystal's tempco.

Non-ohmic resistors in which the voltage coefficient is large and positive are called voltage dependant resistors (VDRs) or varistors. VDRs having no rectification properties, ie symmetrical VDRs are known by various trade names (eg Metal Oxide Varistors). Doped zinc oxide is the basis, having better properties than the silicon carbide (Atmite, Metrosil or Thyrite) formerly used, as its change of resistance is more marked (**Fig 2.7(c)**).

VDRs are used as over-voltage protectors; above a certain voltage the current rises sharply. As the self-capacitance is of the order of nanofarads, they cannot be used for RF. Alternative over-voltage protectors are specially made zener diodes or gas discharge tubes.

Light dependent resistors (**Fig 2.8**) are an improvement on the selenium cell of yesteryear. Cadmium sulphide is now used, its resistance decreasing with illumination.

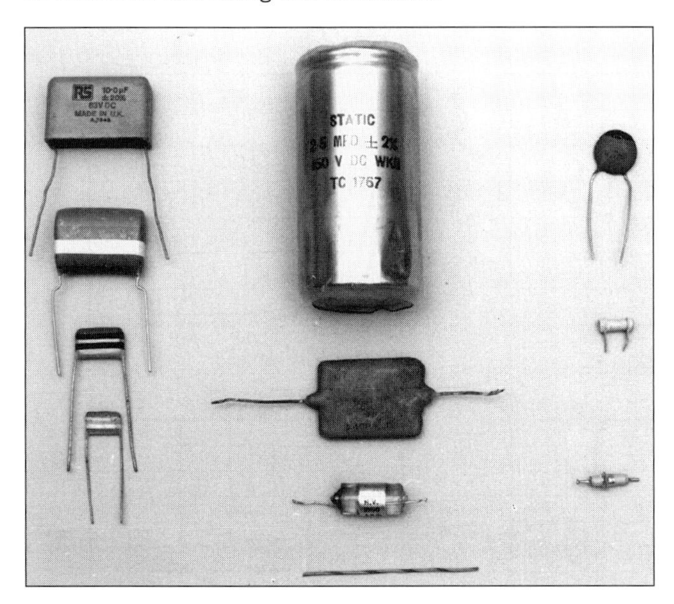

**Fig 2.9: Fixed capacitors. Left column (top to bottom): 10µF, 63V polycarbonate dielectric, 2.7µF 250V, 0.1µF, 400V, 0.033µF, 250V all polyester dielectric. Centre column from top: 2.5µF 150V ±2% polyester, 240pF ±1% silver mica, 1nF ±2% polystyrene (all close tolerance). Right column from top: 4.7nF disc ceramic, 220pF tubular ceramic, 3nF feedthrough. Striped marker is 50mm long**

Semiconductor photo-diodes, however, are preferred as they are cheaper to buy and are more sensitive.

## FIXED CAPACITORS

To get the values wanted for radio purposes into a reasonable space, fixed capacitors are made with various dielectrics, having a value of dielectric constant ($\varepsilon_r$) according to the intended use of the capacitor. **Fig 2.9** shows a selection of fixed capacitors.

### Ceramic Capacitors

The smallest capacitors have a ceramic dielectric, which may be one of three classes. Class 1 has a low $\varepsilon_r$ and therefore a larger size than the other two for a given capacitance. The loss factor for Class 1 is very low, comparable with silvered mica, and the value stays very nearly constant with applied voltage and life.

The designations 'COG' and 'NPO' refer to some members of this class of ceramic capacitor, which should be used where stability of value and low loss factor are of greater importance than small size. Other low loss ceramic capacitors are available with either N*** or P*** identification, where *** is the intentional negative or positive tempco in ppm/°C. These are for compensation to allow for the opposite temperature coefficient of other components in the circuit.

Class 2 comprises medium and high $\varepsilon_r$ capacitors. They both have a value depending on voltage, temperature and age, these effects increasing with $\varepsilon_r$. It is possible to restore the capacitance lost by ageing by heating above the Curie temperature, this being obtained from the manufacturer - not really a technique to be done at home!

'X7R' refers to a medium $\varepsilon_r$ material, Z5U, 2F4 and Y5V to higher $\varepsilon_r$ ceramic. **Fig 2.10** shows the performance of COG, X7R and Y5V types. This class should not be used where stability is paramount, but they are very suitable for RF bypassing, lead through capacitors being made where low inductance is required. Class 1 and 2 capacitors are made in surface mount, either marked in the exponent system, or unmarked.

Class 3 have a barrier layer dielectric, giving small size, poor loss factor and wide tolerance. Z5U dielectric has made this class obsolete, but some may be found as radial leaded discs in surplus equipment.

Large high voltage tubular or disc capacitors are made for

| Function | Type | Advantages |
|---|---|---|
| AF/IF coupling | Paper, polyester, polycarbonate | High voltage, cheap |
| RF coupling | X7R ceramic | Small, cheap but lossy |
| | Polystyrene | Very low loss, low leakage but bulky and not for high temperature |
| | COG ceramic or silver mica | Close tolerance |
| | Stacked mica | For use in power amplifiers |
| RF decoupling | X7R or Z5U ceramic disc or feedthrough | Very low inductance |
| Tuned circuits | Polystyrene | Close tolerance, low loss, negative temperature coefficient ( 150ppm) |
| | Silver mica | Close tolerance, low loss, positive temperature coefficient (+50ppm) |
| | COG ceramic | Close tolerance, low loss |
| | Class 1 ceramic | Various temperature coefficients available, more lossy than COG |

**Table 2.4: Use of fixed capacitors**

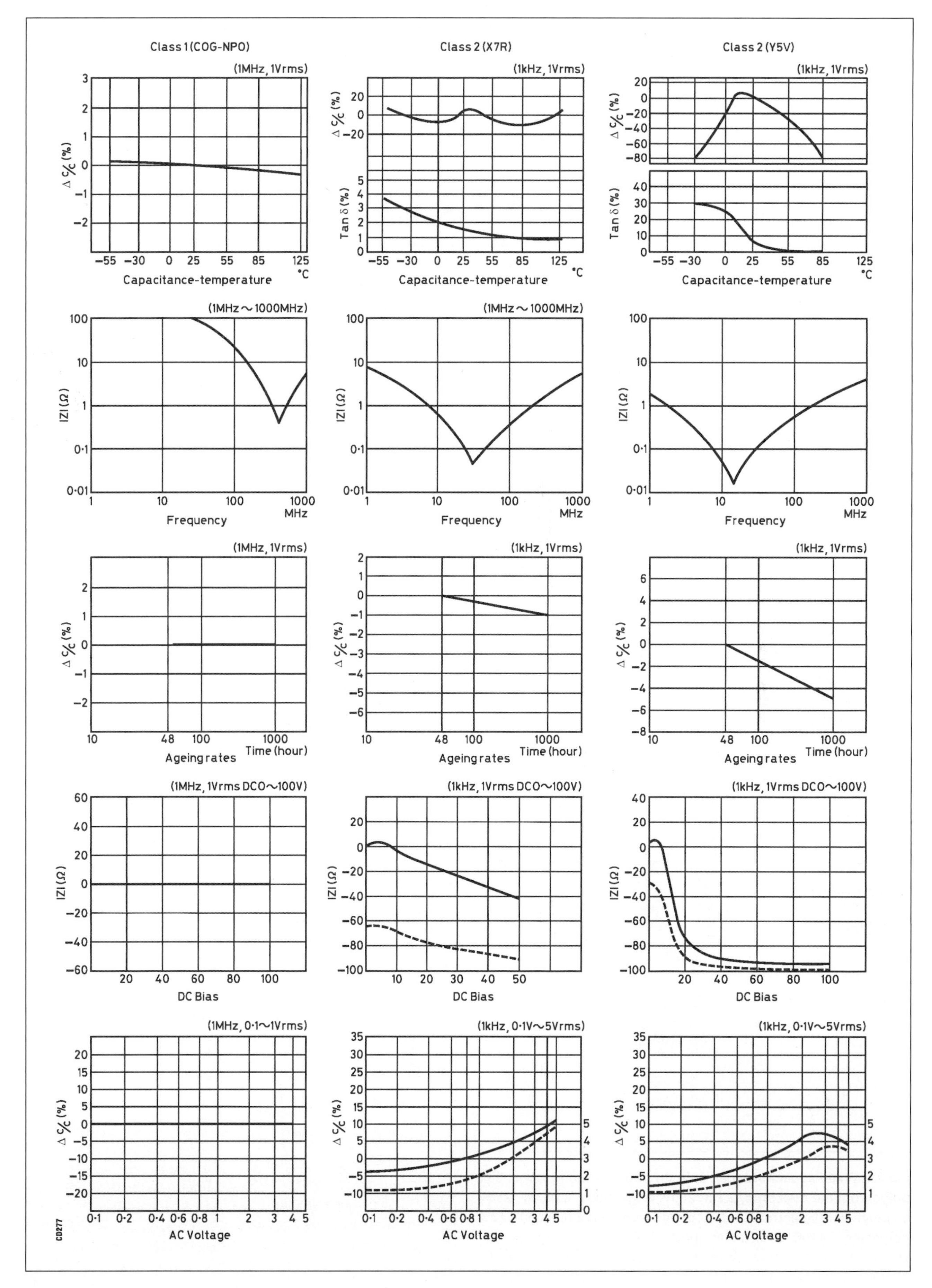


**Fig 2.10: Ceramic capacitor networks**

transmitter use which, if they can be found, are particularly useful for coupling the valves in a linear amplifier to the matching network.

The TDK company have recently introduced Multi-Layer Ceramic Capacitors (MLCC) with the following advantages particularly for switched-mode converters (see the chapter on power supplies):

- Lower ESR (equivalent series resistance)
- Lower impedance
- Better cap stability with frequency
- Higher voltage strength
- Higher reliability
- No polarity

**Table 2.4** summarises ceramic capacitor properties and recommended usage. The value colour coding is to be found in the general data chapter and **Table 2.5** gives the dielectric codes used.

## Mica Capacitors

These are larger than similar ceramic capacitors for a given value and have a tempco of between +35 and +75ppm/°C, the variation being caused by them being natural rather than synthetic. Today, silver electrodes are plated directly on to the mica sheets and the units encapsulated for protection. Because they generally remain stable over life and have low loss, they are popular for tuned circuits and filters. Occasionally however, mica capacitors will jump in value and this effect is unpredictable.

High voltage and current stacked foil types were used for coupling in transmitters, but are not easy to get now. The value and voltage rating is usually marked on the body of mica capacitors.

**Class 1 - r < 500**

COG, NPO temperature coefficient ±30ppm/°C
N*** Temperature coefficient: ***ppm/°C
P*** Temperature coefficient: +***ppm/°C

**Class 2 - r > 500**

***EIA coding***

| Working temperature range | | | | Capacitance change | |
|---|---|---|---|---|---|
| Lower | | Upper | | over range | |
| **Letter** | **Temp** | **Figure** | **Temp** | **Letter** | **Change** |
| Z | +10°C | 2 | +45°C | R | ±15% |
| Y | 30°C | 4 | +65°C | S | ±22% |
| X | 55°C | 5 | +85°C | T | +22 to 33% |
| | | 6 | +105°C | U | +22 to 56% |
| | | 7 | +125°C | V | +22 to 82% |
| | | 8 | +150°C | | |

*X7R and Z5U are commonly met. X7R was formerly denoted W5R.*

***CECC 32100 coding***

| Code | Capacitance change over range | | Temperature range (°C) | | | | |
|---|---|---|---|---|---|---|---|
| | | | -55 | -55 | -40- | -25 | -10 |
| | At 0V DC | At rated voltage | +125 | +85 | +85 | +85 | +85 |
| | | | Final code figure | | | | |
| | | | 1 | 2 | 3 | 4 | 6 |
| 2B* | ±10% | +10 to 15% | | * | * | * | |
| 2C* | ±20% | +20 to 30% | * | * | * | | |
| 2D* | +20 to 30% | +20 to 40% | | | | * | |
| 2E* | +22 to 56% | +22 to 70% | | * | * | * | * |
| 2F* | +30 to 80% | +30 to 90% | | * | * | * | * |
| 2R* | ±15% | | * | | | | |
| 2X* | ±15% | +15 to 25% | * | | | | |

*Reference temperature +20°C.*

*Example: X7R (EIA code) would be 2R1 in CECC 32100 code, tolerance ±15% over 55 to +125°C.*

**Class 3 barrier layer**

Not coded but tolerance approximately +50% to -25% over temperature range of -40 to +85°C. Refer to maker for exact details.

**Table 2.5: Ceramic capacitor dielectric codes**

## Glass Capacitors

These are used for rare applications where they have to work up to 200°C. Because of this, they do not appear in most distributors catalogues, but may be found in surplus equipment.

## Paper Dielectric Capacitors

At one time paper dielectric capacitors were much used, but now they may only be found in some high voltage smoothing capacitors and interference suppression capacitors. Paper capacitors are large and expensive for a given value, but are very reliable, especially when de-rated. There can be trouble if inferior paper is used, with voids in it. The capacitors tend to break down at the voids, where the electric stress is concentrated. Both foil and metallised electrodes are used and the units can be somewhat inductive if not non-inductively wound. A resistor can be included if the unit is to be used for spark suppression.

## Plastic Film Capacitors

These have replaced paper capacitors, and there are two distinct types of plastic used, polar and non-polar.

Polar plastics include polycarbonate, polyester and cellulose acetate. These are characterised by a moderate loss, which can increase at frequencies where the polar molecules resonate. They also suffer from dielectric absorption, meaning that after a complete discharge some charge reappears later. This is of some importance in DC applications. The insulation resistance is very high, making the capacitors suitable for coupling the lower frequencies of AC across a potential difference (within the voltage rating of course).

Commonly, values range from about a nanofarad to several microfarads at voltages from 30V up to kilovolts, thus replacing paper. Tolerances are not usually important as these capacitors are not recommended for tuned circuits, but as they may be used in RC oscillators or filters, 5% or better can be bought at increased cost. Polycarbonate is the most stable of this group and cellulose the worst and cheapest.

Polyethylene terephthalate (PETP) is used for polyester film and is also known as Mylar (in USA) and domestically as Terylene or Melinex. PETP exhibits piezo-electric properties, as can be demonstrated by connecting a PETP capacitor to a high impedance voltmeter and squeezing the capacitor. PETP capacitors therefore behave as rather poor microphones, which could introduce noise when used in mobile applications.

Non-polar plastics suitable for capacitors are polystyrene, polypropylene and polytetrafluoroethylene (PTFE). These exhibit very low loss, independent of frequency, but tend to be bulky. Polystyrene capacitors in particular have a tempco around -150 ppm/°C which enables them to offset a similar positive tempco of ferrites cores used for tuned circuits. Unfortunately polystyrene cannot be used above 70°C and care must be taken when soldering polystyrene capacitors particularly if you use lead free solder. The end connected to the outer foil may be marked with a colour and this should be made the more nearly grounded

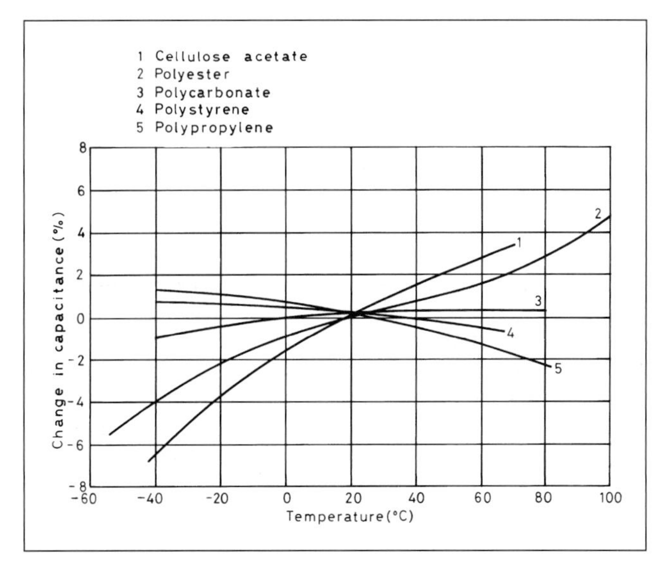


Fig 2.11: Temperature coefficients of plastic capacitors

electrode where possible.

Polypropylene capacitors can be used up to 85°C and are recommended for pulse applications. They are also used at 50Hz for power factor correction and motor starting. Again, take care when soldering.

Encapsulation of plastic foil capacitors is either by dipping or moulding in epoxy or it may be omitted altogether for cheap components where environmental protection is not considered necessary. The value is marked by colour code or printing along with tolerance and voltage rating. Small values are mostly tubular axial leaded, but single ended tubular and boxed are commonly available. You can also get surface mount ones.

**Fig 2.11** shows the effect of temperature on five different types.

PTFE fixed capacitors are not generally available, but variable ones are, see later in this chapter. Small lengths of PTFE or, if unobtainable polystyrene coaxial cable, make very good low loss high voltage capacitors for use as tuning elements in multiband HF antenna systems.

Cutting to an exact value is easy, if you know the type, as the capacitance per metre is quoted for cables in common use. The length must be much shorter than the wavelength for which the cut cable is to be used, to avoid trouble due to the inductance.

## Niobium Capacitors

Niobium Oxide capacitors are only available in surface mount package (**Fig 2.12**), but with a little ingenuity could be used as through hole on a PCB. Available values range from 4.7 to 330 microfarads in voltage range 4.7 to 10 Volts. They are particularly suitable as coupling capacitors for audio frequencies.

## Electrolytic Capacitors

This type is much used for values bigger than about 1µF. The dielectric is a thin film of either aluminium, tantalum, or niobium oxide on the positive plate

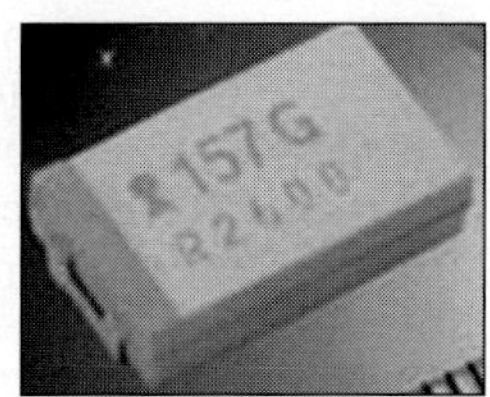


Fig 2.12: A niobium capacitor [info from AVX website]

Fig 2.13: Electrolytic capacitors. Left, from top: 220µF 385V, 150µF 63V, 2.5µF 15V, electrodes. Centre: 100,000µF, 10V, all with aluminium. Right, from top: 68µF 15V, 22µF 25V and 22µF 6.3V, all tantalum types. Striped marker is 50mm long

(**Fig 2.13**).

An electrolyte (liquid or solid) is used to connect to the negative plate. For this reason the electrolytic will be polarised, unless constructed with two oxide coated plates. This is not common in amateur usage, but can be used for motor starting on AC, and in the output or cross-over circuits of some HiFi amplifiers. Alternatively back-to-back capacitors can be used (see **Fig 2.14b & Fig 2.14c**).

Both tantalum and niobium oxide capacitors can stand a reverse voltage up to 0.3 volts. Aluminium capacitors cannot withstand any reverse voltage. The oxide film is formed during manufacture, and deteriorated if the capacitor is not used. Capacitors, unused for a long time, should be reformed by applying an increasing voltage with a high value resistor (approx 100k) in series until the rated voltage is reached, and leakage current falls.

Electrolytics are rated for voltage and ripple current at a specified maximum temperature, and it is here that de-rating is most advisable. At full rating some electrolytics have only 1000 hours advertised life. There is always an appreciable leakage current, worst with liquid aluminium ones.

Tantalum and niobium oxide capacitors should not be allowed to incur large in-rush currents. As the capacitor may fail, add a small resistance in series. Niobium oxide capacitors fail to an open circuit; all others go to a short circuit. For all types the maximum ripple current should not be exceeded to prevent rise of temperature. This is particularly important when the capacitor

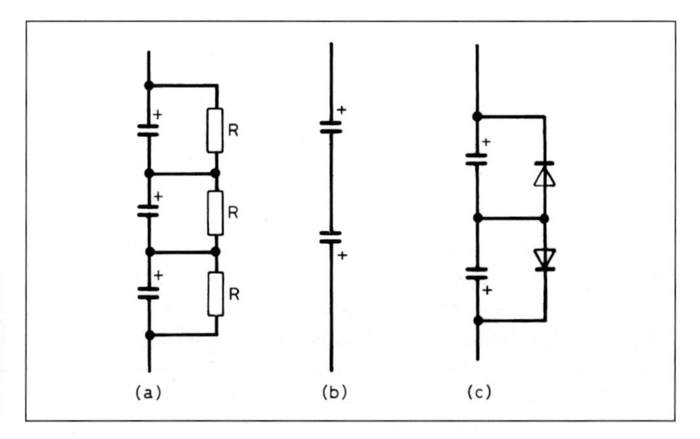


Fig 2.14: Circuits for electrolytic capacitors

Fig 2.15: Variable capacitors. Left: 500pF maximum, four-gang air-spaced receiver capacitor. Right: 125pF max transmitter capacitor with wide spacing. Bottom, left to right: 500pF max twin-gang receiver capacitor. Solid dielectric receiver capacitor. Single-gang 75pF max receiver capacitor. Twin-gang 75pF max receiver capacitor

Fig 2.16: Trimmer capacitors. Top, from left: 50pF max 'postage stamp', 70pF max air dielectric, 50pF max flat ceramic. Bottom, from left: 27pF max tubular ceramic, 30pF max plastic film, 12pF max miniature air dielectric, 30pF max 'beehive' air dielectric

is used after a rectifier (reservoir capacitor). There is more on this in the power supplies chapter.

Although capacitive reactance decreases with frequency, electrolytics are unsuitable for radio frequencies. The internal construction may give rise to considerable inductance and the equivalent series resistance increases with frequency. To avoid this, a small ceramic capacitor should be paralleled with the electrolytic, 0.1μF to 1nF being common. For switch-mode converters, special considerations apply. Again see the power supplies chapter for more information.

## VARIABLE CAPACITORS

These are most often used in conjunction with inductors to form tuned circuits, but there is another use, with resistors in RC oscillators - see the oscillators chapter. Small values (say less than 100pF) with a small capacitance swing are used as trimmers to adjust circuit capacitance to the required value (**Figs 2.15 and 2.16**).

Capacitors with vacuum as the dielectric are used to withstand large RF voltages and currents, for example in valve power amplifiers. The capacitance is adjusted through bellows, by linear motion. The range of adjustment usually allows one amateur band to be covered at a time. For the current UK power limit, the expense of new vacuum capacitors is normally not justified.

Air spaced variable capacitors are widely used both for tuning and trimming. In the Z-match ATU (**Fig 2.17**), a split-stator capacitor is required. More information is given in the chapter on Practical HF Antennas.

For precise tuning of a variable frequency oscillator (VFO) the construction must be such that the capacitance does not vary appreciably with temperature or vibration. This implies low loss, high grade insulating supports for the plates and good mechanical design. These are not factors that make for cheapness.

For valve power amplifiers, the plate spacing must be wide enough to withstand both the peak RF voltage, including the effects of any mismatch, and any superimposed DC. Rounded plate edges minimise the risk of flash over, and application of DC should be avoided by use of choke-capacity coupling. However, extreme stability is not required owing to the low Q of the circuit in which they are used.

As the moving plates are generally connected to the frame, it may be necessary to insulate this from the chassis in some applications. Ceramic shafts have been used to avoid this. It is possible to gang two or more sections, which need not have the same value, but it is usual for the moving vanes to be commoned. The plates need not be semi-circular, but can be shaped so as to give a wanted law to capacitance versus rotation angle. In early superhet receivers specially shaped vanes solved the oscillator-tracking problem.

For small equipment, plastic film dielectric variable capacitors are much used but do not have the stability for a VFO. Trimmers may have air or film dielectric, the latter being a low loss non-polar plastic or even mica. Both parallel plate and tubular types are used for precise applications, and mica compression ones now only occasionally where stability is of less importance than cost (mainly in solid state transmitter PA stages).

Junction diodes can be used as voltage variable capacitors - they are described in the semiconductors chapter.

## INDUCTORS

Inductors are used for resonant circuits and filters, energy storage and presenting an impedance to AC whilst allowing the passage of DC, or transforming voltage, current or impedance. Sometimes an inductor performs more than one of these functions simultaneously, but it is easier to describe the wide range of inductors by these categories.

### Tuned Circuits

Either air or magnetic cored coils are used - in the latter case different cores are selected for optimum performance at the operating frequency and power level. The criterion for 'goodness' for a

Fig 2.17: Split stator capacitor as used in some ATUs

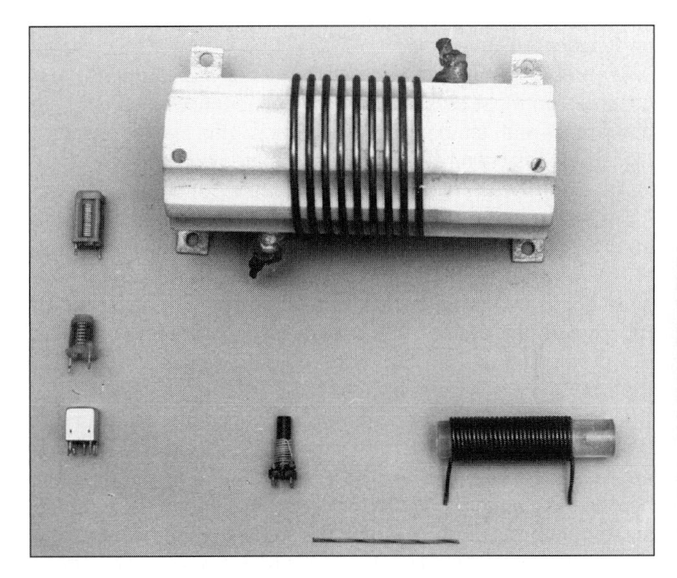

Fig 2.18: Air-cored inductors. Top: high-power transmitting on ceramic former. Left: VHF inductors moulded in polythene, the bottom one in a shielding can. Centre: small air-cored HF inductor (could have an iron dust core). Right: HF inductor wound on polystyrene rod. The striped marker is 50mm long

Fig 2.19: 'Roller-coaster' variable inductor

tuning coil is Q, the ratio of reactance to resistance at the operating frequency. With air-cored coils, the ratio of diameter to length should be greater than 1 for maximum Q, but Q falls off slowly as the length increases. It is often convenient to use a long thin coil rather than a short fat one, accepting the slight loss of Q.

Calculation of the inductance of air-cored coils is difficult and usually done by the use of charts or computer programs. However there is an approximate formula [1] which may be used:

$$L(\mu H) = \frac{r^2 N^2}{25.4\,(9r + 10l)}$$

where $r$ is the radius, $l$ is the length of the coil, both in millimetres and $N$ the number of turns. If you use inches, omit the 25.4 in the denominator. The formula is correct to 1% provided that $l > 0.8r$, ie the coil is not too short.

**Figs 2.18 and 2.19** show some different air-cored coils including a much sought after 'roller coaster' used for antenna and power amplifier tuning networks. **Fig 2.20** shows three versions of another type of variable inductor, the variometer.

Magnetic cores may be either ferrite or iron (carbonyl powder for RF, iron alloy for AF). Screw cores can be used for adjustment in nominally air-cored coils or in pot cores (**Fig 2.21**). For closed magnetic circuits, a factor called $A_L$ is quoted for extreme positions of the core. $A_L$ is the inductance in nanohenries that a one turn coil would have, or alternatively millihenries per thousand

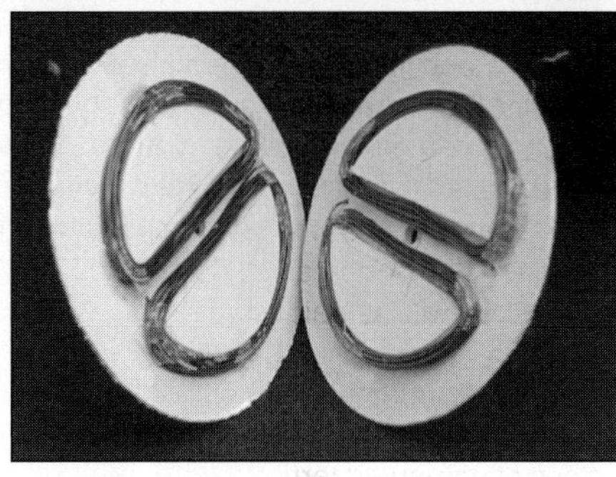

Fig 2.20: Variometers. On the left is an inductance in millihenries for use with a high power 136kHz transmitter. On the right are two ways of implementing low power variometers with inductances in the region of tens of microhenries. The flat coils, bottom right are connected in series and placed together with a non-conducting screw through the hole in the centre; one disc is rotated with respect to the other to give aiding and opposing mutual inductance

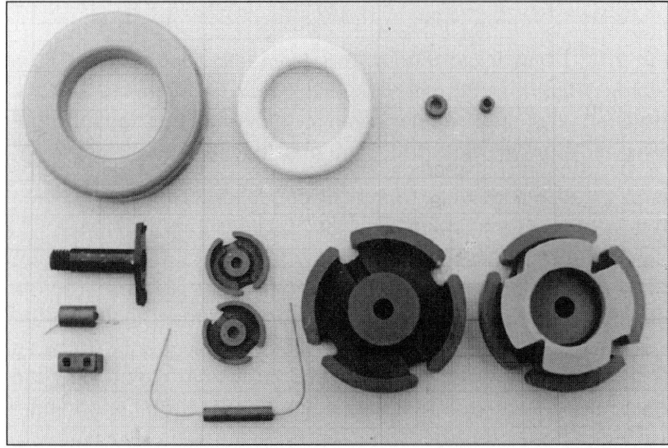

Fig 2.21: RF magnetic materials. Top, left to right: large, medium and small ferrite rings, and a ferrite bead with a single hole. Below, left top: tuneable RF coil former; middle: six-hole ferrite bead with winding as an RF choke; bottom: ferrite RF transformer or balun former; centre: small pot cores and an RF choke former with leads moulded into the ends; right: large pot core with former

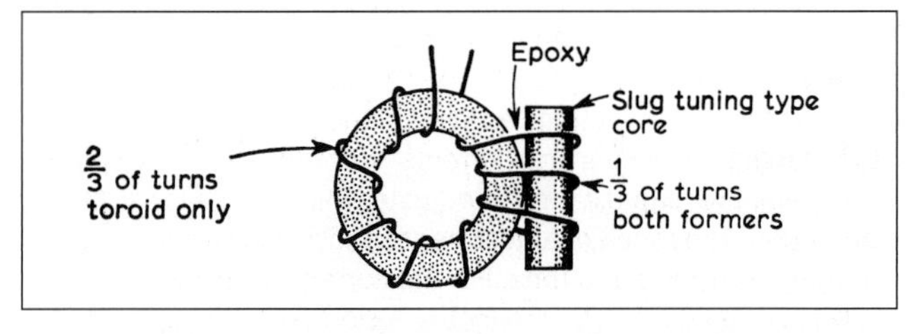


Fig 2.22: Tuneable toroid technique. About 10% variation in inductance can be achieved

turns. For a coil of N turns the inductance will be $N^2A_L$ nanohenries. Checking by grid dip oscillator is difficult because of the closed nature of the magnetic circuit. Also shown in Fig 2.21 are toroidal cores.

**Fig 2.22** shows a technique for adjusting the inductance on a toroidal core where it is not important to preserve a completely closed magnetic circuit. Trimming by removing turns is more easily done with the aid of a small crochet hook.

There are two common types of ferrite, manganese-zinc (Mn/Zn) and nickel-zinc (Ni/Zn). Mn/Zn has lower resistivity than Ni/Zn and lower losses due to hysteresis. The latter is caused by the magnetic flux lagging behind the magnetic field, so offering a resistive component to the alternating current in any coil wound round the material. As the applied frequency is increased, eddy currents and possibly dimensional resonances play a greater part and the higher resistivity Ni/Zn has to be used. Unfortunately, Ni/Zn has greater hysteresis loss. Both types of ferrite saturate in the region of 400mT, this being a disadvantage in high power uses (saturation is when increasing magnetic field fails to produce the same increase in magnetic flux as it did at lower levels). More will be said about this in non-tuned circuit applications.

There are many proprietary ferrites on the market and reference must be made to the catalogues for details. If you have a piece of unmarked ferrite, one practical test is to apply the prods of an ohmmeter to it. Mn/Zn ferrite will show some resistance on a 100k range, but Ni/Zn will not. Further tests with a winding could give you some idea of its usefulness.

Microwave ferrites such as yttrium iron garnet have uses for the amateur who operates on the Gigahertz bands.

Inevitably, coils will possess resistance and distributed capacitance as well as inductance. The distributed capacitance can confuse the measurement of inductance, and hence both confuses and limits the tuning range with a variable capacitor. When winding a tuning coil, it may be necessary to use a coating for environmental protection. All 'dopes' have an $\varepsilon_r$ greater than one and will increase the self-capacitance. Polystyrene with an $\varepsilon_r$ of only 2.5, dissolved in toluene (a hazardous vapour) or cellulose thinners is recommended. Polythene or PTFE would be attractive, but there is no known solvent for them.

## Energy Storage Inductors

The type of core will depend on the operating frequency. At mains supply frequency, silicon iron or amorphous iron are the only choices. To minimise eddy current loss caused by the low resistivity of iron, either thin laminations or grain orientated silicon strip (GOSS - the grains of which lie along the length of the strip in which the magnetic flux lies), is used. If magnetic saturation would be a problem, an air gap is left in the magnetic circuit. The calculation of core size, gap, turns and wire size is outside the scope of this book. Advice on winding is given later, under the heading 'Transformers'.

At higher frequencies such as for switch mode converter use, ferrites and amorphous alloys are widely used in the form of E shaped half cores, in pairs. As with silicon iron, an air gap is required to prevent saturation. The core manufacturers' data books give the information needed for design.

## Chokes

While energy storage inductors are most often called chokes, the term was originally applied to inductors used to permit the passage of DC, while opposing AC, and it still does. The cores used are as for energy storage for the lower frequencies, but at higher frequencies, where ferrites are too lossy, carbonyl powdered iron is used. At VHF and above, air-cored formers are used, surface mounted ones being available.

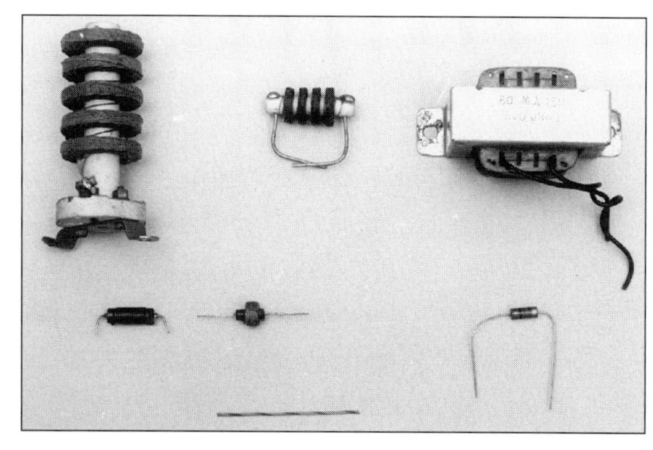

**Fig 2.23: RF and AF chokes. Top, left to right: 1mH large, 1mH small RF chokes and 15H AF type. Bottom, left to right: 10μH single layer, 300μH 'pile' wound, 22μH encapsulated. Striped marker is 50mm long**

There is a problem with self-resonance in RF chokes. It is impossible to avoid capacitance, and at some frequency it will resonate with the inductance to give a very high impedance - ideal - but at higher frequencies the so-called inductor will behave as a capacitor whose impedance decreases with frequency. The situation is complicated by the fact that the self-capacitance is distributed and there may be multiple parallel and series resonances. For feeding HT to valve anodes, the choke should not have series resonances in any of the bands for which it is designed. Where available, a wave winder should be used to wind different numbers of turns in separate coils which, being on the same former, are in series. A less efficient, but simpler alternative, is to sectionalise a single layer solenoid winding. **Fig 2.23** shows a variety of chokes.

The Coilcraft company make a range of inductors and transformers whose value and limitations are specified. Their use may save considerable calculation and trial. See **Fig 2.24** to see some examples.

## Transformers

The Principles chapter described mutual inductance, and this property enables transformers to be made. The magnetic flux from one coil is allowed to link one or more other coils, where it produces an EMF. Two types of transformer are met, tightly coupled and loosely coupled. With tight coupling the object is to make as much as possible of the flux from the first (primary) coil link the other (secondary). In this case the ratio of the applied EMF to induced EMF is that of the turns ratio secondary to pri-

**Fig 2.24: Ready-made inductors from Coilcraft**

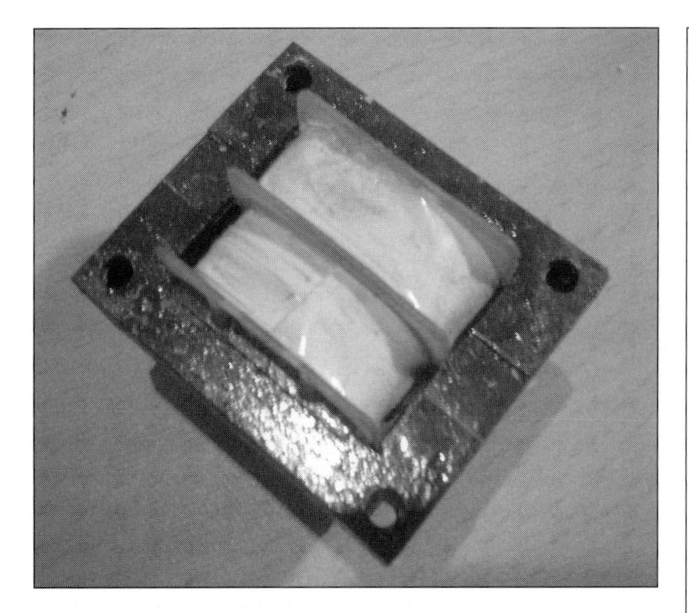

**Fig 2.25: A transformer with the primary and secondary wound separately, with an insulated divider**

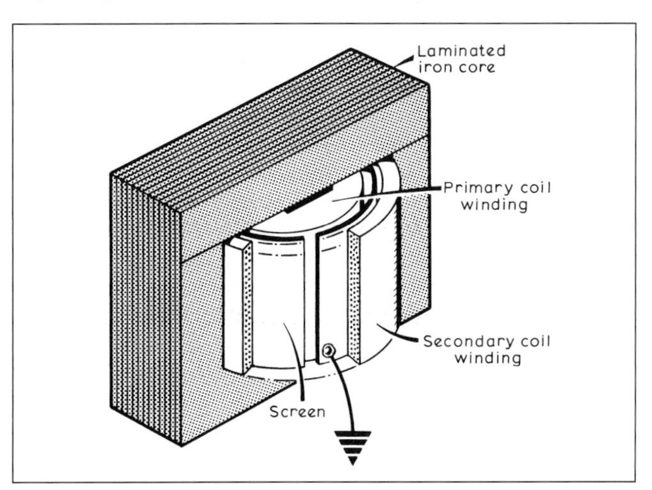


**Fig 2.26: A Faraday screen is an earthed sheet of copper foil between the primary and secondary windings. For clarity, a gap is shown here between the ends, but in reality they must overlap while being insulated from one another to avoid creating a shorted turn. As copper is non-magnetic, mutual induction is not affected. The screen prevents mains-borne interference from reaching the secondary winding by capacitive coupling. Some transformers may have another screen (shorted this time) outside the magnetic circuit to prevent stray fields**

mary. Formulas for calculating the required turns are given in the general data chapter.

Leakage inductance caused by incomplete flux linkage is minimised by interleaving the primary and secondary windings, however this does increase the self self-capacitance. This may not matter at 50Hz. Some modern transformers wind the primary and secondary separately, with an insulating divider instead of on top of each other (**Fig 2.25**). This provides greater safety at the cost of poorer regulation (the decrease of output voltage with current). An alternative is to use two concentric bobbins. Many small commercial transformers suffer from poor regulation due to this effect.

If capacitance between primary and secondary is to be minimised, a Faraday screen is inserted - see **Fig 2.26** for the way to do this. It is essential that the screen should not cause a

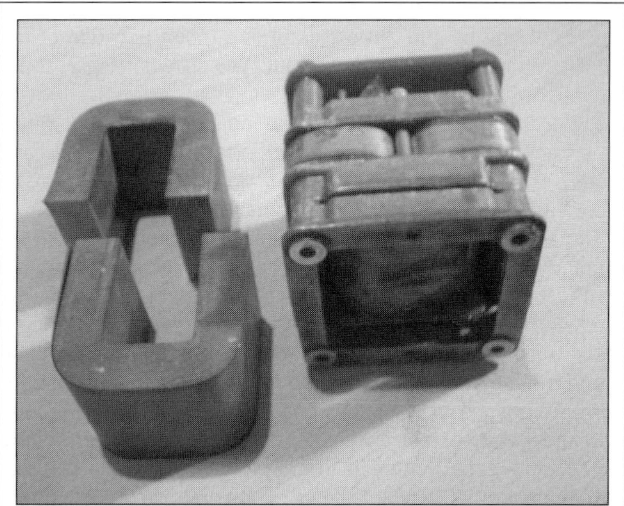

**Fig 2.27: A few examples of the use of 'C' cores and toroids**

shorted turn.

Insulation, if required, is best made with Kapton or Nomex tape, which will withstand as high a temperature as the wire.

Toroidal or 'C' cores (**Fig 2.27**) are used to reduce the external magnetic field and in Variacs$^{tm}$ (see later and **Fig 2.28**). Beware of making a shorted turn with the fixing arrangements. As toroidal transformers are worked near the maximum permissible flux density, a large in-rush current may be noticed when switching

Fig 2.28: A variac

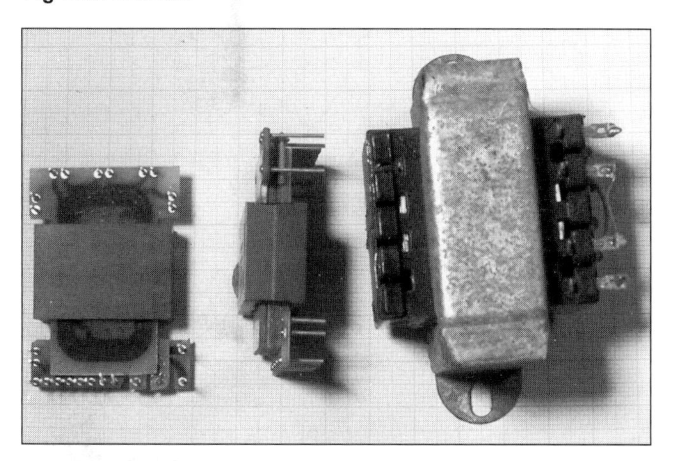

Fig 2.29: The two transformers on the left are 125W devices for 500kHz operation and weigh 40g each, while the one on the right is a 12W transformer for mains frequency (50Hz) and weighs in at 420g. The standard graph paper background has 1cm squares and 1mm subdivisions

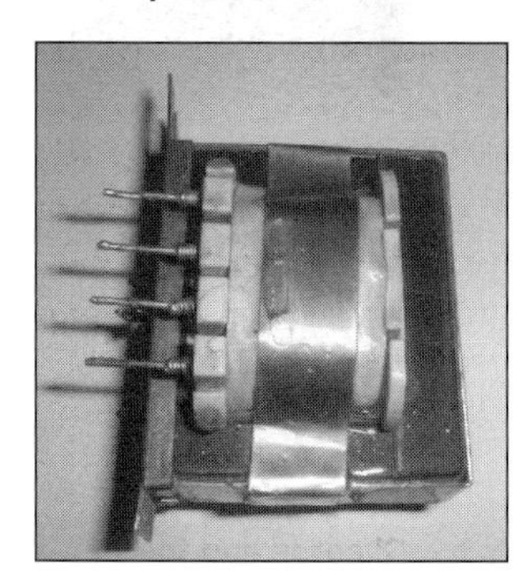

(left) Fig 2.30: Copper foil surrounds the winding and the outer limbs of the core to reduce radiation

on. If the supply is switched on as the voltage crosses zero, doubling of the current will occur as the voltage and current establish the correct phase relationship; it could even saturate the core momentarily. This is particularly noticeable with the Variac.

**Fig 2.29** shows a comparison between a switch mode transformer, which runs at 500kHz, and a standard 50Hz transformer. Where the stray magnetic field has to be reduced, a shorted turn of copper foil outside the whole magnetic circuit is used - see **Fig 2.30**.

Practical note: When winding on E and I or T and U laminations, remember that part of the bobbin's cheeks will later be covered by the core, so do not bring the leads out in this region! If the wires are very thin, skeining them before bringing them out makes a sounder job.

If isolation is not required between input and output, an autotransformer can be used. This will increase the power handling capacity of a particular core. The primary and secondary are continuous - the transformation ratio being still the ratio of the number of turns, as in **Fig 2.31.** Wire gauge is chosen to suit the current. Since the secondary is part of the primary (or vice versa), more use is made of the winding window. The tapping point can be made variable to allow adjustment of the output voltage, the device then being known as a Variac.

In loosely coupled transformers, if the windings are tuned by capacitors, the degree of coupling controls the bandwidth of the combination (**Fig 2.32**). Such transformers are widely used in intermediate frequency stages and wideband couplers. Again, ferrite, iron dust or brass slugs may be used for trimming in the appropriate frequency range. Loose coupling is used on mains in ballasts for sodium lamps and in microwave ovens, whose magnetron needs a constant current supply.

Beware of trying to use a microwave oven transformer to supply a valve linear amplifier, apart from loose coupling, the inner of the secondary is either earthed or not well insulated. Additionally, the output voltage falls rapidly with output current. This is not due to the winding's resistance but to what is called leakage reactance. This is why you should not try to use them in mains power supplies.

## MATERIALS

Earlier in this chapter conductors were described - here some of the insulators used in amateur radio will be considered first. These will be grouped into three categories: inorganic, polar and non-polar plastics.

Mica (mainly impure aluminium silicate), glass and ceramics are the most likely to be used (asbestos is potentially dangerous). Vitreous quartz is one of the best insulators, but is not generally available. It has nearly zero temperature coefficient of expansion, and would make a good former for VFO inductors. Mica is not generally used except in capacitors. Glass has a tendency to collect a film of moisture, but the leakage is hardly likely to be troublesome. If it is, the glass should be coated with silicone varnish. Porcelain antenna insulators do not suffer from this trouble. Ceramic coil formers leave little to be desired but are becoming increasingly more difficult to obtain.

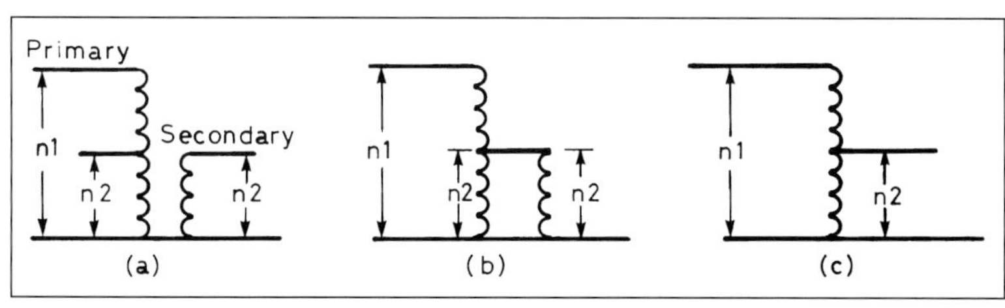


Fig 2.31: The auto-transformer

Non-polar plastics like polystyrene, polypropylene and PTFE are all very good. Polypropylene rope avoids the use of antenna insulators (except where very high voltages are encountered as when using a Marconi antenna on the 136kHz band) which is particularly useful in portable operation.

Polar plastics such as phenol-formaldehyde (Tufnol and Bakelite), Nylon, Perspex and PVC have higher losses that are frequency sensitive. The test using a microwave oven is not the best, as the loss

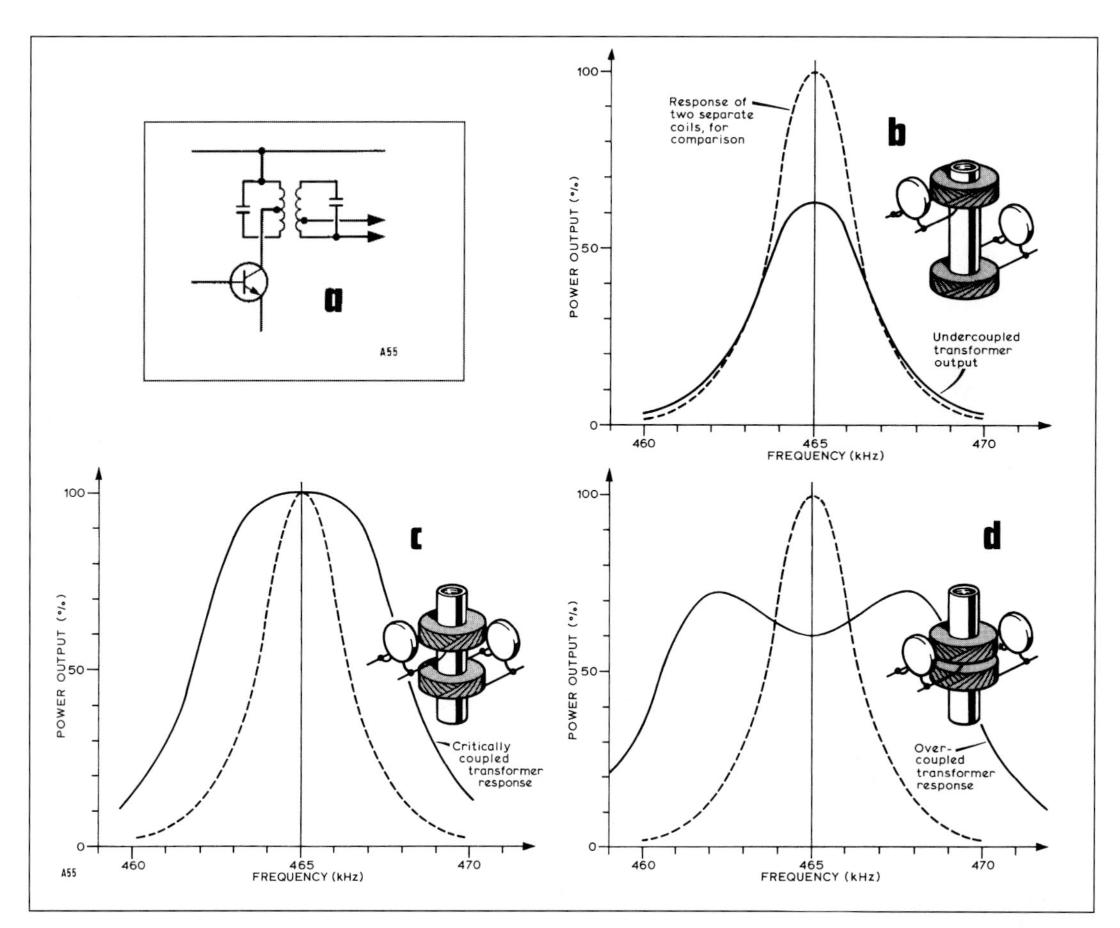


Fig 2.32: Bandwidth of coupled circuits

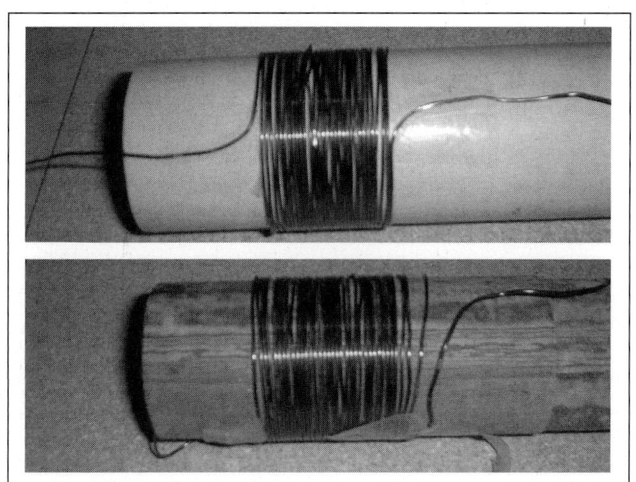

Fig 2.33: Similar coils wound on (a) polythene and (b) Tufnol have different Q values

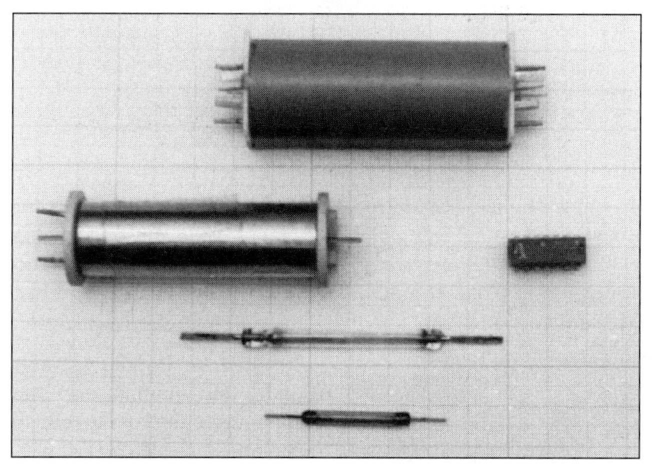

Fig 2.35: Reed relays. Top: four reeds in one shielded coil. Middle left: reed in open coil. Right: miniature reed relay in 14-pin DIL case. Below: large and small reed elements

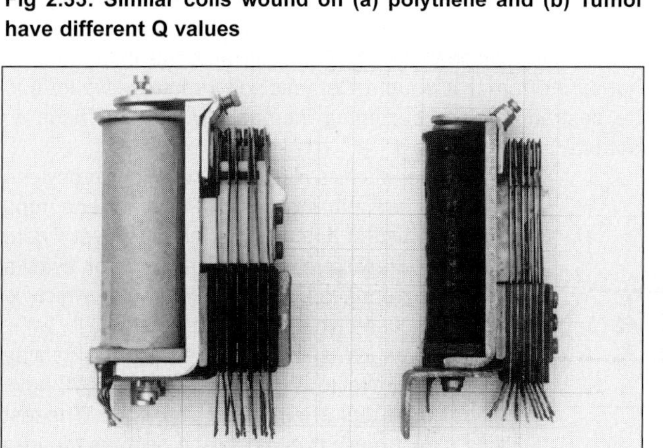

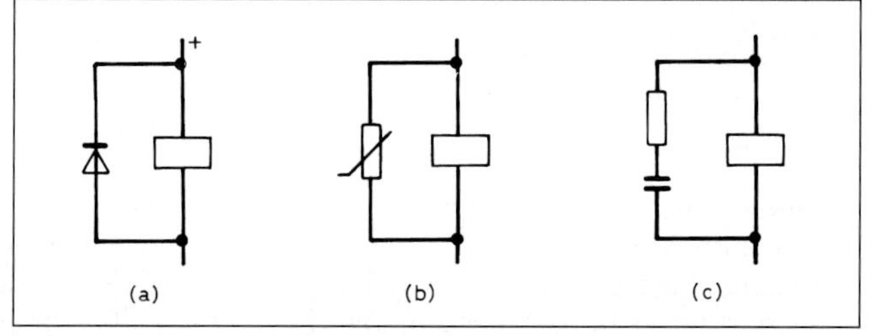


Fig 2.36: Relay coil suppression circuits

(left) Fig 2.34: Post Office relays. Left type 3000, right type 600

at 2450MHz may not be the same as at the required frequency. A better test is to build an oscillator for the frequency you intend to use, insert a slice of the unknown plastic between the capacitor plates and note the increase in supply current. Compare with similar slices of known plastics obtainable from your local sign manufacturer. Tests involving burning are not advised. Because of possible loss, they are not recommended to withstand high RF voltages. **Fig 2.33** shows two coils, one on polythene and the other on Tufnol. The one on the polythene former has a Q of 300, but the Tufnol one only 130 at 5MHz. Some protective varnishes are polar, being made for 50Hz. Only polystyrene dope previously described should be used where RF loss is important.

Permanent magnets may be a special ferrous alloy, ceramic rare earth alloy or neodymium-boron-iron. Ferrous alloys such as Alnico and Ticonal were much used in DC machines, headphones and loudspeakers. The others now replace them. Ceramic magnets are cheap and withstand being left without a keeper better than ferrous alloys. The other two store more magnetic energy, are very good on open magnetic circuits but are expensive.

## RELAYS

The 'Post Office' types 3000 and 600 are still available in amateur circles (**Fig 2.34**). However they are not sealed, and many sealed relays are now available. Newer relays need less operating current, some being able to operate directly from logic ICs. Contacts are made from different materials according to the intended use, such as very high or low currents.

The abbreviations 'NO', 'NC' and 'CO' apply to the contacts and stand for 'normally open', 'normally closed' and 'change over', the first two referring to the unenergised state. Some manufacturers use 'Form A' for NO, 'Form B' for NC and 'Form C' for CO. Multiple contacts are indicated by a number in front of the abbreviation.

Latching relays are stable in either position and only require a set or reset momentary energisation through separate coils or a pulse of the appropriate polarity on a single coil. These relays use permanent magnet assistance, meaning that correct polarity must be observed.

It is possible to get relays for operation on AC without the need for a rectifier, the range including 240V, 50Hz. Part of the core has a short circuited turn of copper around its face - this holds the relay closed while the current through the coil falls to zero twice per cycle.

If RF currents are to be switched, reed relays (**Fig 2.35**) or coaxial relays which maintain a stated impedance along the switched path should be used. Reed relays, either with dry or mercury wetted contacts, are among the fastest operating types, and should be considered if a relay is needed in a break-in circuit.

As a relay is an inductive device, when the coil current is turned off, the back EMF tries to maintain the current, possibly creating a spark. The back EMF may damage the operating transistor or IC, so it must be suppressed (**Fig 2.36**). A diode across the coil, connected so that it does not conduct during energisation will conduct when the back EMF is created, but in so doing it slows down the release of the relay. Varistors or a series RC combination are also effective. If rapid release is required, a high voltage transistor must be used with an RC combination which will limit the voltage created to less than the maximum of this transistor.

Another undesirable feature of relays is contact bounce, which may introduce false signals in a digital system. Mercury wetted reed relays are one way over this problem, but some will only work in a particular orientation.

## ELECTRO-ACOUSTIC DEVICES

Moving coil loudspeakers (and headphones) use the force generated when a magnetic flux acts on a current carrying coil (**Fig 2.37**). The force moves the coil and cone to which the coil is fastened. Sound is then radiated from the cone. Some means of preventing the entry of dust is provided in the better types.

## Sounders

The piezo-electric effect is used to move a diaphragm by applying a voltage. The frequency response has a profound resonance so the device is often used as a sounder at this frequency.

## Headphones

Small versions of the moving coil and piezo-electric loudspeakers are used as headphones. The in-ear headphone is a moving coil element, with a neodymium-boron-iron magnet to reduce the size. Deaf aid in-ear headphones use the piezo-electric effect, with a better response than the sounder has.

Another type has a magnetic diaphragm, which is attracted by a permanent magnet with the audio frequency field superimposed on it (**Fig 2.38**) The permanent field is necessary because magnetic attraction is proportional to the square of the flux (and current). Without the permanent flux, second har-

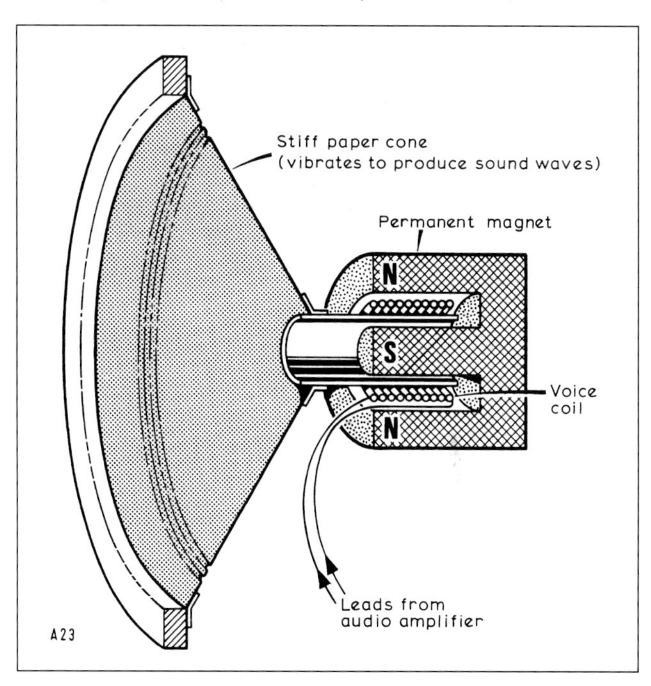


Fig 2.37: Moving-coil loudspeaker

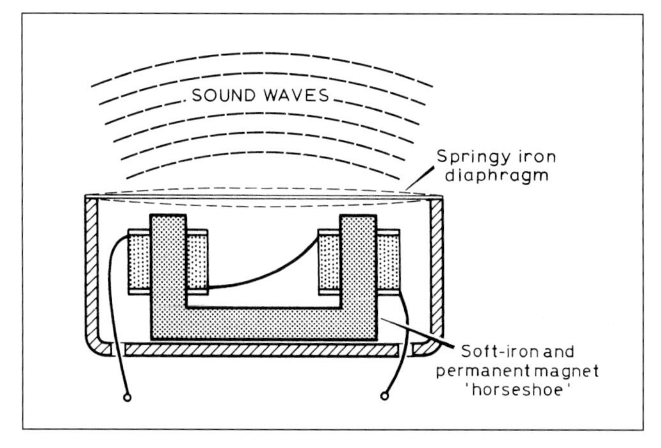


Fig 2.38: Moving iron headphone

Fig 2.39: Piezoelectric loudspeakers

Fig 2.40: Microphone inserts. Left: carbon mic. Right: dynamic mic

Fig 2.41: An electret microphone

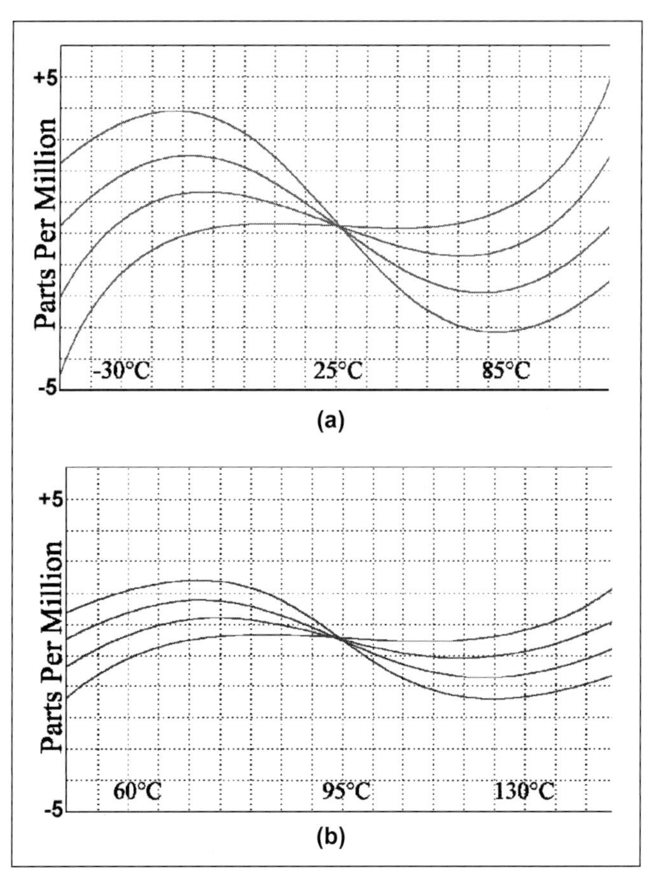


Fig 2.42: (a) Temperature curves for typical AT-cut crystals; (b) Temperature curves for typical SC-cut crystals

monic production would take place. The over-riding steady flux prevents this.

Noise-cancelling headphones reduce unwanted ambient sounds by means of active noise control (ANC). This process involves using one or more microphones placed near to the ear, and electronic circuitry which uses the signal from the microphone(s) to generate an 'antinoise' signal. When the anti-noise signal is produced by the speaker driver in the headphone, destructive interference cancels out the ambient noise as heard within the enclosed volume of the headphone. They require a power source (eg batteries) to operate.

## Microphones

The moving coil loudspeaker in miniature forms the dynamic microphone. It is possible to use the same unit both as a loudspeaker and microphone, often in hand-held equipment.

The piezo-electric elements shown in **Fig 2.39** are reversible, and form the basis of the crystal microphone. The equivalent circuit is a small capacitance in series with a small resistance. This causes a loss of low frequencies if the load has too low a resistance. FETs make a very suitable load. Moisture should be avoided, as some crystals are prone to damage.

It is possible to use the diaphragm headphone as a microphone since the effect is reversible, and some ex-service capsules are made this way.

Probably the carbon microphone is the oldest type still in telephone use; one type is shown in **Fig 2.40**. Sound waves alternatively compress and relieve the pressure on the carbon granules, altering the current when energised by DC. It amplifies, but is basically noisy and subject to even more unpleasant blasting noises if overloaded acoustically. It is rarely used nowadays in amateur radio. The impedance is some hundreds of ohms.

The above types can be made (ambient) noise cancelling by allowing noise to be incident on both sides of the diaphragm (or element) with little effect. Close speaking will influence the nearer side more, improving the voice to ambient noise ratio.

The so-called condenser microphone has received a new lease of life under the name of electret microphone. The condenser microphone had a conductive diaphragm stretched in front of another electrode and a large DC potential difference maintained, through a resistive load, across the gap. Variations in the position of the diaphragm by sound waves caused a variation of capacitance, and so an AF current flows through the load resistance. As the variation in capacitance is small, a large load resistance is required. A valve amplifier was included in the assembly to avoid loss of signal in leads and possible unwanted pick up. In the electret microphone, the need for a high polarising voltage is obviated by using an electret as the rear electrode and an FET in place of the valve.

An electret is the electrostatic counterpart of a permanent magnet, and maintains a constant potential across the electrodes without requiring any power. The FET provides impedance transformation to feed low impedance leads. A point worth remembering is that the field from the electret attracts dust! Two small ones are shown in **Fig 2.41**.

# QUARTZ CRYSTALS

Quartz is a mineral, silicon dioxide and the major constituent of sand. In its crystalline form it exhibits piezo-electric properties. The quartz crystal is hexagonal in section, and has hexagonal points. The useful piezo-electric effect takes place only at certain angles to the main axis from point to point. If a slice is cut at one of the correct angles [2], it can act like a bell and ring if struck. Ringing is an oscillation and with small slices this will

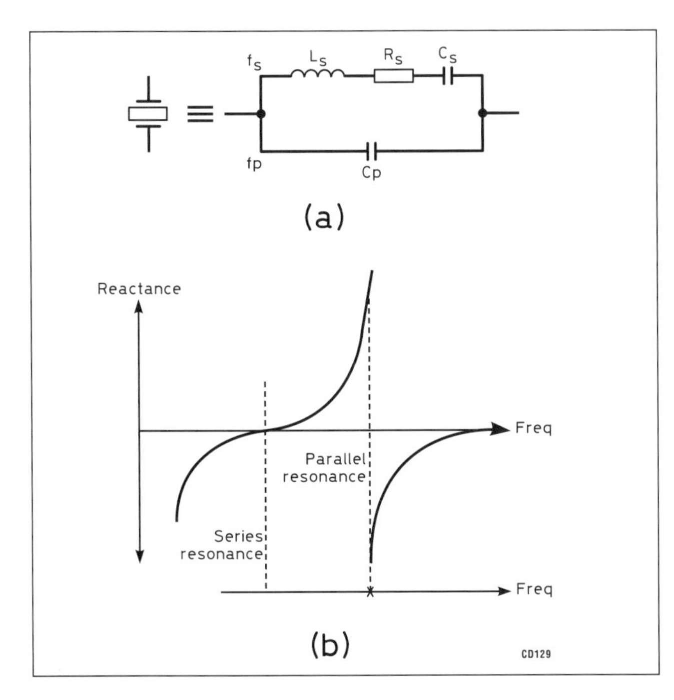


**Fig 2.43: (a) Equivalent circuit of a piezoelectric crystal as used in RF oscillators or in filter circuits. Typical values for a 7MHz crystal are given in Table 2.6. Cp represents the capacitance of the electrodes and the holder. (b) Typical variation of reactance with frequency. The series and parallel reactances are normally separated by about 0.01% of the frequencies themselves (eg about a kilohertz for a 7MHz crystal)**

| Parameter | 7MHz crystal | 7MHz LC |
|---|---|---|
| Ls | 42.5mH | 12.9μH |
| Cs | 0.0122pF | 40pF |
| Rs | 19 | 0.19 |
| Q | 98,000 | 300 |

**Table 2.6: Parameters of a crystal compared with an LC circuit**

take place in the MHz region. Mechanical ringing is accompanied by electrical ringing, and with proper amplifier connection this can be sustained (see the chapter on oscillators).

Charge is applied to the quartz element from metal plates sprung on to it, or better by silver electrodes plated on.

The type of deformation that the charge produces, stretching, shearing or bending, depends on the angle of cut and mode of mechanical oscillation to be excited. Various different cuts have different temperature coefficients. DT cut is for low frequencies such as those at 32.768 kHz, which are used in watches. It has an inverted parabolic curve of frequency against temperature. For amateur use at HF, AT and SC cuts are the most popular. Their frequency/temperature curves are shown in **Fig 2.42**. As the flat part of the SC curve occurs at about 105°C, the crystal needs to work in an oven and the ageing rate will increase. More detail of this can be found in the oscillators chapter.

Naturally occurring quartz always had defects in the crystal, which made for unpredictability. Nowadays crystals are made by dissolving raw impure quartz in a caustic soda solution under high pressure and temperature. The dissolved quartz grows onto a new seed slice of quartz by hydrothermal synthesis.

The Q of the crystal is much higher than can be obtained with LC circuits. As an example, a 7MHz crystal may have a Q of 100,000. The equivalent circuit is in **Fig 2.43** where $C_s$, $L_s$ and

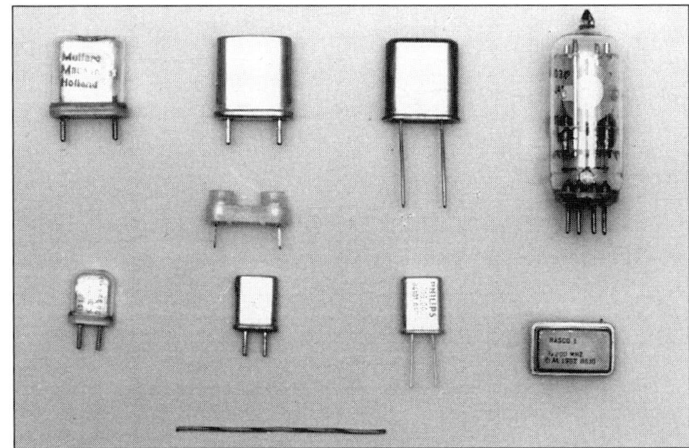

**Fig 2.44: Quartz crystals in different holders. Top, from left: HC27U, HC6U, HC47U (wire ended) and B7G. Holder for HC6U between the rows. Bottom, from left: HC29U, HC25U, HC18U (wire ended) and a packaged crystal oscillator. The bar is 50mm long**

$R_s$ represent the properties of the crystal and $C_p$ the shunt capacitance of the holder. The measured values of these parameters are strange when compared with the LC circuit, as **Table 2.6** reveals. Note the very large inductance of the crystal; also a Q of 300 for the LC circuit is good.

A quartz crystal is not a primary standard of frequency; it has to be compared with another. Not only this, all crystals age, even if not excited. Switching on and over excitation add to ageing, which is usually towards a higher frequency.

Crystals can normally be cut, ground or etched for the kilohertz region up to about 30MHz, above this they are too fragile to be of practical value. An overtone crystal can be used above this frequency, that is nearly equal to an odd number times the fundamental. It will still try to oscillate at the fundamental, but the external circuit must be made to prevent this.

Since crystals act as tuned circuits, they can also be used as filters; piezo-ceramics being much used. More information can be found in the chapters on receivers and transmitters. The use of cheap, readily available computer or colour TV crystals in filters has been well documented [3]. [4].

## Mechanical Design

Quartz crystals must be protected from the environment and are mounted in holders; these are usually evacuated. There are several types and sizes (**Fig 2.44**) but basically they are either in metal or glass. If in metal, the cheapest is solder sealed, next is resistance welded and the most expensive cold welded.

## Ceramic Resonators

Certain ceramics, mainly based on titanium and/or zirconium oxides, have similar properties to quartz and are used as oscillators or filters. They have a lower Q, a lower maximum frequency and a higher temperature coefficient, but these are offset by a lower price. They are also only available for a limited range of frequencies.

There is another type of ceramic resonator that works at microwave frequencies and is widely used in satellite TV receivers.

**Fig 2.45: MEMS resonator. This example measures just 0.6 x 0.8mm**

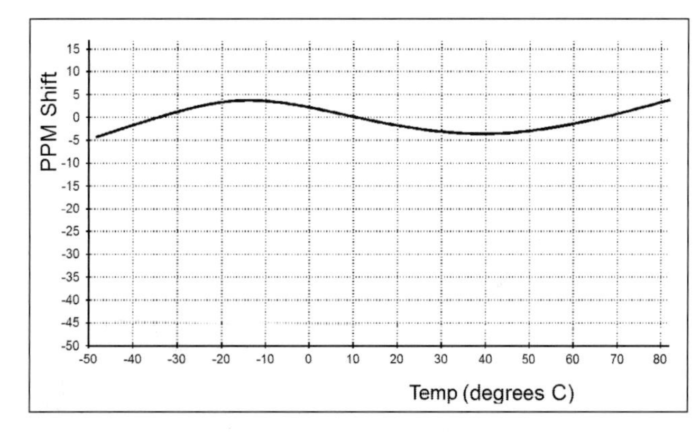


Fig 2.46: MEMS resonator frequency vs temperature

## MEMS Resonators

A recent invention is the silicon micro electronic mechanical system (MEMS) resonator, **Fig 2.45** This promises much smaller influence of temperature and ageing, as **Fig 2.46** shows. Factory programmable oscillators, including VCOs, containing a MEMS resonator are now available from the SiTime company; their accuracy is only relatively moderate.

# SURFACE MOUNT TECHNOLOGY

Although most people have seen surface mount technology (SMT), relatively few without a professional involvement in electronics have used it in construction. SMT dominates commercial technology where its high-density capability and automated PCB manufacture compatibility is of great value. The arrival of SMT combined with the increasing dominance of monolithic IC parts over discrete components means that amateurs will have to move with the technology or rapidly lose the ability to build and modify equipment. SMT has an image of being difficult but this is not the case and it is quite practical for amateurs to use the technology. An incentive is that traditional leaded components are rapidly disappearing with many recently common parts no longer being manufactured and new parts not being offered in through-hole format. The cost of SMT parts in small quantities has fallen dramatically and through-hole construction will become increasingly expensive in comparison when it is possible at all.

The variety of available SMT components is enormous and growing rapidly. While the fundamental component types are familiar, the packaging, benefits and limitations of SMT components are significantly different to their through-hole equivalents and a brief review is worthwhile (**Fig 2.47**).

Populated SMT boards should be handled very carefully. Resist the temptation to bend or flex the board when handling, as this will cause micro-fractures to result in the ceramic components.

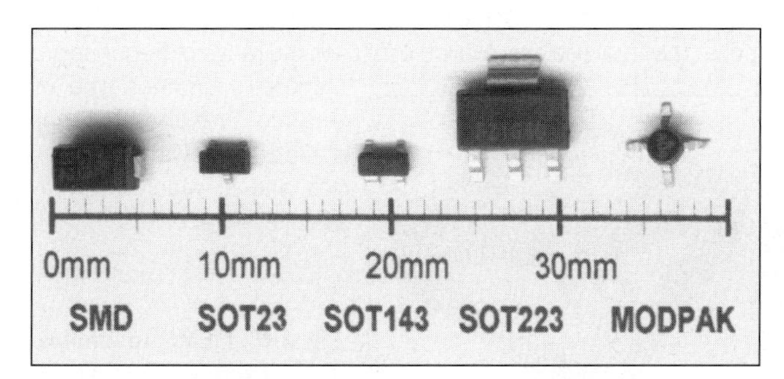


Fig 2.47: Discrete SMT packages

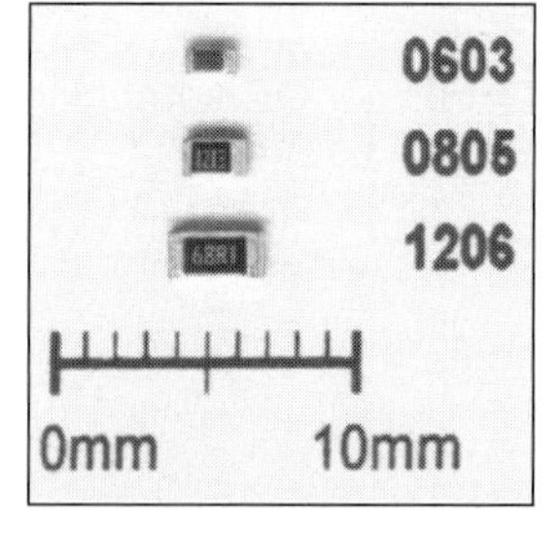


Fig 2.48: SMT resistors

| Style | Size (mm) L | Size (mm) W | Rating (mW) |
|---|---|---|---|
| 1206 | 3.2 | 1.60 | 250 |
| 0805 | 2.0 | 1.25 | 125 |
| 0603 | 1.6 | 0.80 | 100 |
| 0402 | 1.0 | 0.50 | 63 |
| 0201 | 0.6 | 0.30 | 50 |

Table 2.7: SMT resistors

## Surface Mount Resistors

Typical sizes of resistors continue to shrink due to market pressure for ever-higher densities and lower costs. The surface mount parts are often referred to by their nominal case sizes such as '1206' which is 12 hundredths of an inch long by 6 hundredths of an inch wide. **Fig 2.48**.

**Table 2.7** lists the commoner sizes but the largest volume is currently in the 0603 size. The 0805 is probably slightly better for the amateur as smaller parts tend not to bear any markings. One standard technique with surface mount is to use the space between the two pads under a component to run a surface track under the component and save a wire link. Below 0805 size this technique becomes demanding of amateur PCB processes. Tolerances of resistors have also improved with standard parts being 1% tolerance and 100ppm stability. Such has been the improvement in manufacturing processes looser tolerance parts are generally not available. The 0201 size is in restricted commercial use but is of limited relevance to the amateur. The power ratings are for comparison purposes only as in practice they are partly dependent on the PCB design and other factors.

## Surface Mount Capacitors

Capacitors are available in similar sizes to resistors but due to physical constraints it is not possible to settle on a single case size for all requirements (**Fig 2.49**). The dominant type is the monolithic multi-layer ceramic type with tantalum dielectric as the -next most common. Progress has been made with ceramic capacitors to decrease size and to increase maximum values to over 10µF, displacing electrolytic parts in some applications

| Dielectric type | Relative permittivity | Tolerance (%) |
|---|---|---|
| COG, NPO | low | ±5 |
| X7R | medium | ±10 |
| Z5U | high | +80/ -20 |
| Y5V | high | +80/ -20 |

Table 2.8: Ceramic capacitors. For typical temperature stability see Fig 2.7

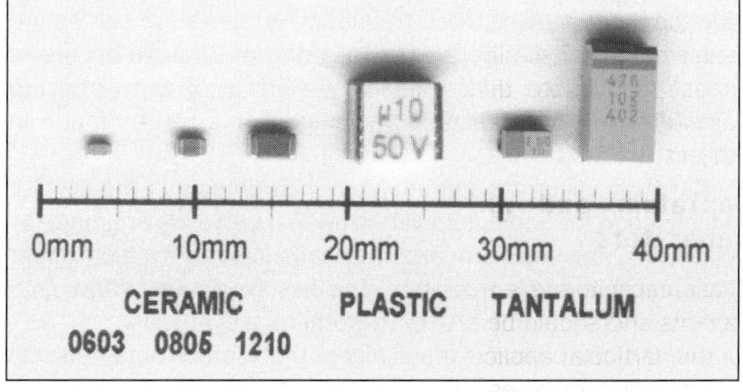


Fig 2.49: SMT capacitors

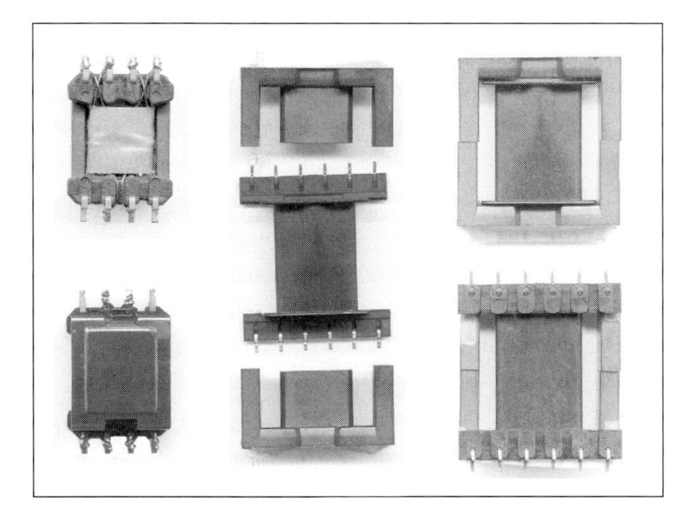

**Fig 2.50: SMT inductors. The left hand pair are 12 x 20mm and the other pictures show the same 30 x 32mm transformer**

although with a much lower ESR. Plastic film parts are also available but suffer from a problems (see below).

### Ceramic capacitors

When selecting these, choices must be made on tolerance, stability and voltage rating which will result in a number of size alternatives. The voltage rating is important since larger values may have quite low ratings. The choice of dielectrics is wide but some of the commoner choices are given in **Table 2.8** with some typical properties. The structure of these parts is a multilayer monolithically fabricated structure.

Although the parasitic inductances of these surface-mount devices are superior to leaded equivalents there are special parts available for microwave frequencies that have better-defined parasitic properties and high-frequency Q, albeit at a cost penalty. The monolithic multi-layer capacitors are more fragile than they may appear at first and unfortunately seldom give any external sign of failure.

Mechanical shock can damage the parts quite easily and at least one supplier warns that the parts should not be dropped onto a hard surface. When hand soldering, avoid any contact between the iron and the unplated body of the component. The damage will often take the form of shattering of some of the internal plates resulting in a significant capacitance error. Commercially an ultrasonic microscope or x-ray investigation can identify the problem.

The parts can suffer permanent value changes due to soldering and the values also change slightly with the DC voltage applied to them. The higher-permittivity dielectrics are particularly subject to these effects.

### Tantalum and niobium oxide capacitors

These capacitors are available in many variants and should be chosen carefully for the particular application. In addition to the standard types there are parts that provide lower effective series resistance (ESR), usually for PSU filter applications, and parts that provide either surge resistance or built-in fuses. It is important to consider safety, as an electrolytic failure on a power rail will often result in debris being blown off the PCB and even a small capacitor can cause eye damage. This means using appropriate voltage ratings, and using parts that will tolerate maximum current surges and have appropriate power dissipation and temperature ratings.

### Plastic dielectric capacitors

Although available, these have only made a limited impact due to the problem that the dielectric has to withstand direct solder temperatures and many manufacturers simply will not approve them for their solder processes.

Parts typically use polyester and care must be taken to avoid overheating them while soldering. Even with care the parts may change value significantly due to heating and the resultant mechanical distortion.

### Aluminium electrolytic capacitors

These are available in SMT, often as slight variations of a leaded package to allow surface soldering. Although they have been displaced to an extent by tantalum parts they still have a role to play.

## Surface Mount Inductors and Ferrites

A large range of inductive devices exists, based on either traditional wound structures or layered ferrite structures (**Fig 2.50**). The advantage of smaller size in many SMT parts may be offset by reduced performance, often with lower Q and poorer stability in comparison to large leaded parts. The wound parts come in a number of styles and an additional option is often whether the part is to be shielded, ie the magnetic field is mostly self-contained as in a toroid. The current rating needs to be carefully watched as the parts will saturate and lose performance long before any other effects are noticed. The ferrite parts are available in multi-layer structures, a little like a monolithic multi-layer capacitor and can be very useful for RF suppression. A vast

**Fig 2.51: Typical board populated with SMT components**

range of devices exist that integrate capacitors into the package to make various filters usually for EMC purposes, but monolithic low-pass filters are available for VHF and UHF transmitter harmonic suppression.

## Surface Mount Connectors

Surface mount connectors are now relatively common for both multi-way interconnects and for coaxial connections. There is little to say specifically about them since they are essentially similar to the equivalent through-hole parts. The small size and the reliance on solder connections for mechanical strength means that they should be treated with care as they can readily be torn from a PCB. Often additional mechanical methods will be needed to prevent damage, eg adhesive to glue the part to the board.

An additional problem is that due to their small size and low mating forces the connectors are very vulnerable to pollution on the contacts arising from flux or dirt and it is essential that they are kept very clean if problems are to be avoided. Although Pressfit connectors are through-hole components they are included here as they are common on surface-mount boards. These connectors typically have square pins that are inserted into slightly undersize round-plated through holes so that the pins cut into the copper hole plating, resulting in a cold weld requiring no soldering. Considerable force may be required for large multi-way connectors but the construction is attractive to PCB assemblers as it can eliminate an additional solder process, because parts such as connectors etc are generally fitted as a second stage operation after all the SMT devices have been soldered in place.

This technique is starting to appear on miniature coaxial PCB-mounting sockets. If unsoldered connectors are seen on a PCB it should not be automatically assumed that some mistake has been made! These parts can be used by the amateur - they should be treated as ordinary through-hole parts and hand soldered.

## SMT Component Markings

### Resistors

The small size of these parts (see **Fig 2.51**) means that often no markings are printed onto parts smaller than 0805. If they are marked, resistors are printed with their value using a three or four digit number, for example:

393 = $39 \times 10^3$ = 39kΩ
1212 = $121 \times 10^2$ = 12.1kΩ
180 = $18 \times 10^0$ = 18.0Ω
3R3 = 3.3 = 3.3Ω

Note: On some very small parts the value is often laser cut and is very difficult to see with the the naked eye.

### Capacitors

Unfortunately capacitors are seldom marked in this manner and if marked normally use a code system (EIA-198) of a letter followed by a number to indicate value. Access to a capacitance measurement device is extremely useful since the majority is unmarked. The EIA-198 capacitor marking system is shown in **Table 2.9**.

### Inductors

These are usually marked with their value in microhenries using a similar system to resistors, for example:

3u3 = 3.3µH
333 = 33mH

| Alpha | | | | Numeral | | | | | |
|---|---|---|---|---|---|---|---|---|---|
| **chr** | **9** | **0** | **1** | **2** | **3** | **4** | **5** | **6** | **7** |
| A | 0.10p | 1.0p | 10p | 100p | 1.0n | 10n | 100n | 1.0µ | 10µ |
| B | 0.11p | 1.1p | 11p | 110p | 1.1n | 11n | 110n | 1.1µ | 11µ |
| C | 0.12p | 1.2p | 12p | 120p | 1.2n | 12n | 120n | 1.2µ | 12µ |
| D | 0.13p | 1.3p | 13p | 130p | 1.3n | 13n | 130n | 1.3µ | 13µ |
| E | 0.15p | 1.5p | 15p | 150p | 1.5n | 15n | 150n | 1.5µ | 15µ |
| F | 0.16p | 1.6p | 16p | 160p | 1.6n | 16n | 160n | 1.6µ | 16µ |
| G | 0.18p | 1.8p | 18p | 180p | 1.8n | 18n | 180n | 1.8µ | 18µ |
| H | 0.20p | 2.0p | 20p | 200p | 2.0n | 20n | 200n | 2.0µ | 20µ |
| J | 0.22p | 2.2p | 22p | 220p | 2.2n | 22n | 220n | 2.2µ | 22µ |
| K | 0.24p | 2.4p | 24p | 240p | 2.4n | 24n | 240n | 2.4µ | 24µ |
| L | 0.27p | 2.7p | 27p | 270p | 2.7n | 27n | 270n | 2.7µ | 27µ |
| M | 0.30p | 3.0p | 30p | 300p | 3.0n | 30n | 300n | 3.0µ | 30µ |
| N | 0.33p | 3.3p | 33p | 330p | 3.3n | 33n | 330n | 3.3µ | 33µ |
| P | 0.36p | 3.6p | 36p | 360p | 3.6n | 36n | 360n | 3.6µ | 36µ |
| Q | 0.39p | 3.9p | 39p | 390p | 3.9n | 39n | 390n | 3.9µ | 39µ |
| R | 0.43p | 4.3p | 43p | 430p | 4.3n | 43n | 430n | 4.3µ | 43µ |
| S | 0.47p | 4.7p | 47p | 470p | 4.7n | 47n | 470n | 4.7µ | 47µ |
| T | 0.51p | 5.1p | 51p | 510p | 5.1n | 51n | 510n | 5.1µ | 51µ |
| U | 0.56p | 5.6p | 56p | 560p | 5.6n | 56n | 560n | 5.6µ | 56µ |
| V | 0.62p | 6.2p | 62p | 620p | 6.2n | 62n | 620n | 6.2µ | 62µ |
| W | 0.68p | 6.8p | 68p | 680p | 6.8n | 68n | 680n | 6.8µ | 68µ |
| X | 0.75p | 7.5p | 75p | 750p | 7.5n | 75n | 750n | 7.5µ | 75µ |
| Y | 0.82p | 8.2p | 82p | 820p | 8.2n | 82n | 820n | 8.2µ | 82µ |
| Z | 0.91p | 9.1p | 91p | 910p | 9.1n | 91n | 910n | 9.1µ | 91µ |
| a | 0.25p | 2.5p | 25p | 250p | 2.5n | 25n | 250n | 2.5µ | 25µ |
| b | 0.35p | 3.5p | 35p | 350p | 3.5n | 35n | 350n | 3.5µ | 35µ |
| d | 0.40p | 4.0p | 40p | 400p | 4.0n | 40n | 400n | 4.0µ | 40µ |
| e | 0.45p | 4.5p | 45p | 450p | 4.5n | 45n | 450n | 4.5µ | 45µ |
| f | 0.50p | 5.0p | 50p | 500p | 5.0n | 50n | 500n | 5.0µ | 50µ |
| m | 0.60p | 6.0p | 60p | 600p | 6.0n | 60n | 600n | 6.0µ | 60µ |
| n | 0.70p | 7.0p | 70p | 700p | 7.0n | 70n | 700n | 7.0µ | 70µ |
| t | 0.80p | 8.0p | 80p | 800p | 8.0n | 80n | 800n | 8.0µ | 80µ |
| y | 0.90p | 9.0p | 90p | 900p | 9.0n | 90n | 900n | 9.0µ | 90µ |

*The letter may be preceded by a manufacturer's mark such as a letter or symbol..*

**Table 2.9: EIA-198 capacitor marking system, showing the capacitance (pF, nF, µF) for various identifiers**

## REFERENCES

[1] 'Simple inductance formulae for radio coils. *Proc.IEE*, Vol. 16, Oct 1928, p 1398

[2] *Shortwave Wireless Communications*, Ladner and Stoner, 4th edition, Chapman & Hall, London 1946 pp 308-339

[3] G3JIR, *Radio Communication* 1976 p896, 1972, pp28, 122 and 687, RSGB

[4] G3OUR, quoted in 'Technical Topics' columns, *Radio Communication*, Dec 1980, and Oct 1982

*The chapter authors are grateful for the assistance given by the following: Advanced Crystal Technology; Automatic Windings; AVX Corporation of the USA; Charcroft Electronics; Coilcraft; 'Components in Electronics', Euroquartz; Lap-Tech Inc; Simon Tribe, G0IEY; SiTime; TDK; Vishay.com; the late Herr Professor J A W Zenneck.*

# Semiconductors

3

*Alan Betts, G0HIQ and Fred Ruddell, GI4MWA*

The development of semiconductor technology has had a profound impact on daily life by facilitating unprecedented progress in all areas of electronic engineering and telecommunications. Since the first transistors were made in the late 1940s, the number and variety of semiconductor devices have increased rapidly with the application of advanced technology and new materials. The devices now available range from the humble silicon rectifier diode used in power supplies to specialised GaAsFETs (gallium arsenide field effect transistors) used for low-noise microwave amplification. Digital ULSIICs (ultra large-scale integrated circuits), involving the fabrication of millions of tiny transistors on a single chip of silicon, are used for the low-power microprocessor and memory functions which have made powerful desktop personal computers a reality.

Within amateur radio, there are now virtually no items of electronic equipment that cannot be based entirely on semiconductor, or 'solid-state', engineering. However, thermionic valves may still be found in some high-power linear amplifiers used for transmission.

## SILICON

Although the first transistors were actually made using the semiconductor germanium (atomic symbol Ge), this has now been almost entirely replaced by silicon (Si). The silicon atom, shown pictorially in **Fig 3.1**, has a central nucleus consisting of 14 protons, which carry a positive electrical charge, and also 14 neutrons. The neutrons have no electrical charge, but they do possess the same mass (atomic weight) as the protons. Arranged around the nucleus are 14 electrons. The electrons, which are much lighter particles, carry a negative charge. This balances the positive charge of the protons, making the atom electrically neutral and therefore stable.

The 14 electrons are arranged within three groups, or shells. The innermost shell contains two electrons, the middle shell eight, and the outer shell four. The shells are separated by forbidden regions, known as energy band gaps, into which individual electrons cannot normally travel. However, at temperatures above absolute zero (0 degrees Kelvin or minus 273 degrees Celsius) the electrons will move around within their respective shells due to thermal excitation. The speed, and therefore the energy, of the excited electrons increases as temperature rises.

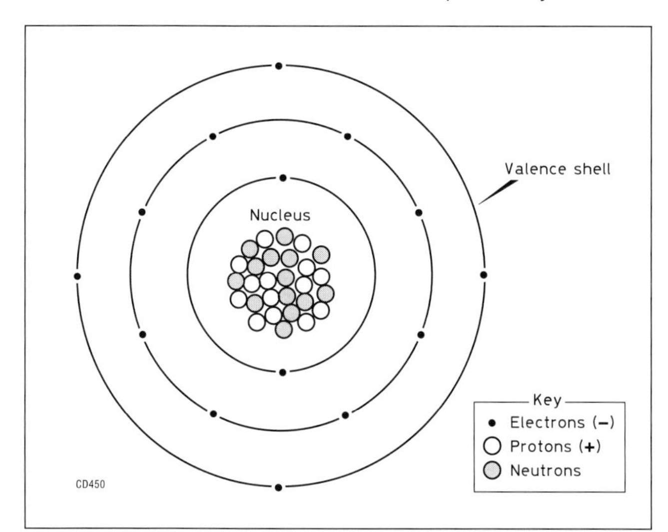


**Fig 3.1: The silicon atom, which has an atomic mass of 28.086. 26% of the Earth's crust is composed of silicon, which occurs naturally in the form of silicates (oxides of silicon), eg quartz. Bulk silicon is steel grey in colour, opaque and has a shiny surface**

It is the outermost shell, or valence band, which is of importance when considering whether silicon should be classified as an electrical insulator or a conductor. In conductors (see **Fig 3.2(a)**), such as the metals copper, silver and aluminium, the outermost (valence) electrons are free to move from one atom to another, thus making it possible for the material to sustain electron (current) flow. The region within which this exchange of electrons can occur is known, not surprisingly, as the conduction band, and in a conductor it effectively overlaps the valence band. In insulating materials (eg most plastics, glass and ceramics), however, there is a forbidden region (bandgap) which separates the valence band from the conduction band (**Fig 3.2(b)**). The width of the bandgap is measured in electron volts (eV), 1eV being the energy imparted to an electron as it passes through a potential of 1V. The magnitude of the bandgap serves as an indication of how good the insulator is and a typical insulator will have a bandgap of around 6eV. The only way to force an insulator to conduct an electrical current is to subject it to a very high potential difference, ie many thousands of volts. Under such extreme conditions, the potential may succeed in imparting sufficient energy to the valence electrons to cause them to jump into the conduction band. When this happens, and current flow is instigated, the insulator is said to have broken down. In practice, the breakdown of an insulator normally results in its destruction. It may seem odd that such a high potential is necessary to force the insulator into conduction when the bandgap amounts to only a few volts. However, even a very thin slice of insulating material will have a width of many thousand atoms, and a potential equal to the band gap must exist across each of these atoms before current flow can take place.

**Fig 3.2(c)** shows the relationship between the valence and conduction bands for intrinsic (very pure) silicon. As can be seen, there is a bandgap of around 1.1eV at room temperature. This means that intrinsic silicon will not under normal circumstances serve as an electrical conductor, and it is best described as a narrow bandgap insulator. The relevance of the term 'semiconductor', which might imply a state somewhere between conduction and insulation, or alternatively the intriguing possibility of being able to move from one to the other, is partly explained by the methods that have been developed to modify the electrical behaviour of materials like silicon.

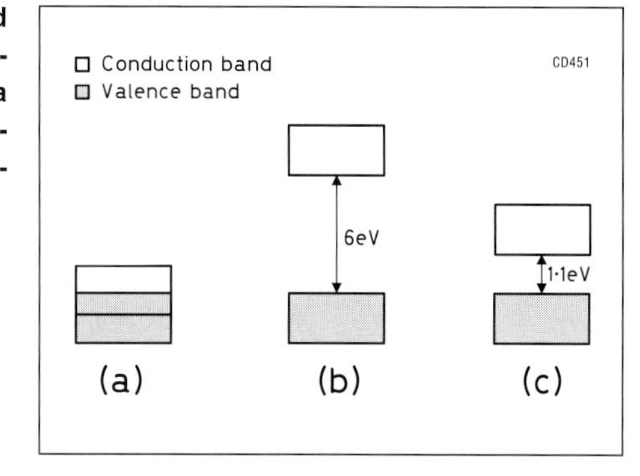


**Fig 3.2: Band gaps of conductors (a) a typical insulator (b) and silicon (c)**

## Doping

The atoms within a piece of silicon form themselves into a criss-cross structure known as a tetrahedral crystal lattice - this is illustrated three-dimensionally by **Fig 3.3**. The lattice is held together by a phenomenon called covalent bonding, where each atom shares its valence electrons with those of its four nearest neighbours by establishing electron pairs. For greater clarity, **Fig 3.4** provides a simplified, two-dimensional, representation of the crystal lattice.

It is possible to alter slightly the crystal lattice structure of silicon, and through doing so modify its conductivity, by adding small numbers of atoms of other substances. These substances, termed dopants (also rather misleadingly referred to as 'impurities'), fall into two distinct categories:

- Group III - these are substances that have just three valence electrons: one less than silicon. Boron (atomic symbol B) is the most commonly used. Material doped in this way is called P-type.
- Group V - substances with five valence electrons: one more than silicon. Examples are phosphorus (P) and arsenic (As). This produces N-type material.

The level of doping concentration is chosen to suit the requirements of particular semiconductor devices, but in many cases the quantities involved are amazingly small. There may be only around one dopant atom for every 10,000,000 atoms of silicon (1 in $10^7$). Silicon modified in this way retains its normal structure because the Group III and Group V atoms are able to fit themselves into the crystal lattice.

In the case of N-type silicon **(Fig 3.5(a)**, each dopant atom has one spare electron that cannot partake in covalent bonding. These 'untethered' electrons, are free to move into the conduction band and can therefore act as current carriers. The dopant atom is known as a donor, because it has 'donated' an electron to the conduction band. In consequence, N-type silicon has a far higher conductivity than intrinsic silicon. The conductivity of silicon is also enhanced by P-type doping, although the reason for this is less obvious. **Fig 3.5(b)** shows that when a Group III atom fits itself into the crystal lattice an additional electron is 'accepted' by the dopant atom (acceptor), creating a positively charged vacant electron position in the valence band known as a hole. Holes are able to facilitate electron (current) flow because they exert a considerable force of attraction for any free electrons. Semiconductors which utilise both free electrons and holes for current flow are known as bipolar devices. Electrons, however, are more mobile than holes and so semiconductor devices designed to operate at the highest frequencies will usually rely on electrons as their main current carriers.

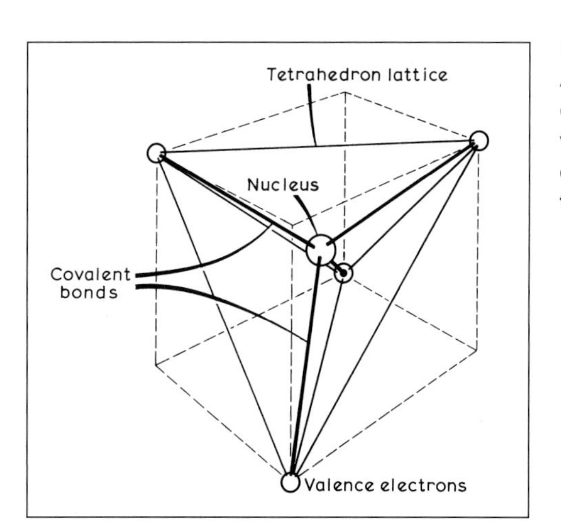


**Fig 3.3: A three-dimensional view of the crystal lattice**

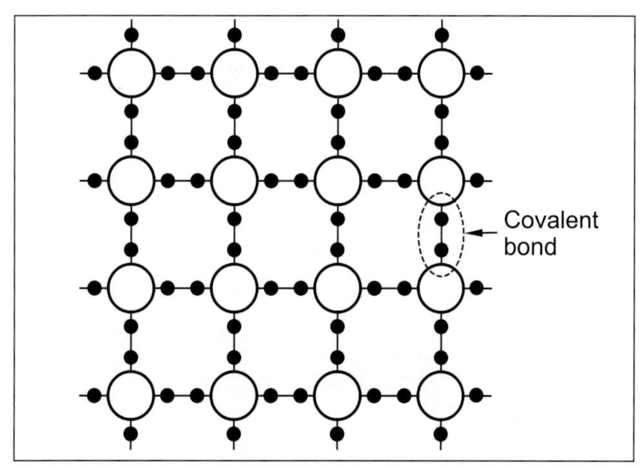


**Fig 3.4: Covalent bonding in the crystal lattice**

## THE PN JUNCTION DIODE

The PN junction diode is the simplest bipolar device. It serves a vital role both in its own right (in rectification and switching applications), and as an important building block in more complex devices, such as transistors. **Fig 3.6** shows that the junction consists of a sandwich of P- and N-type silicon. This representation suggests that a diode can be made by simply bonding together small blocks of P- and N-type material. Although this is now technically possible, such a technique is not in widespread use.

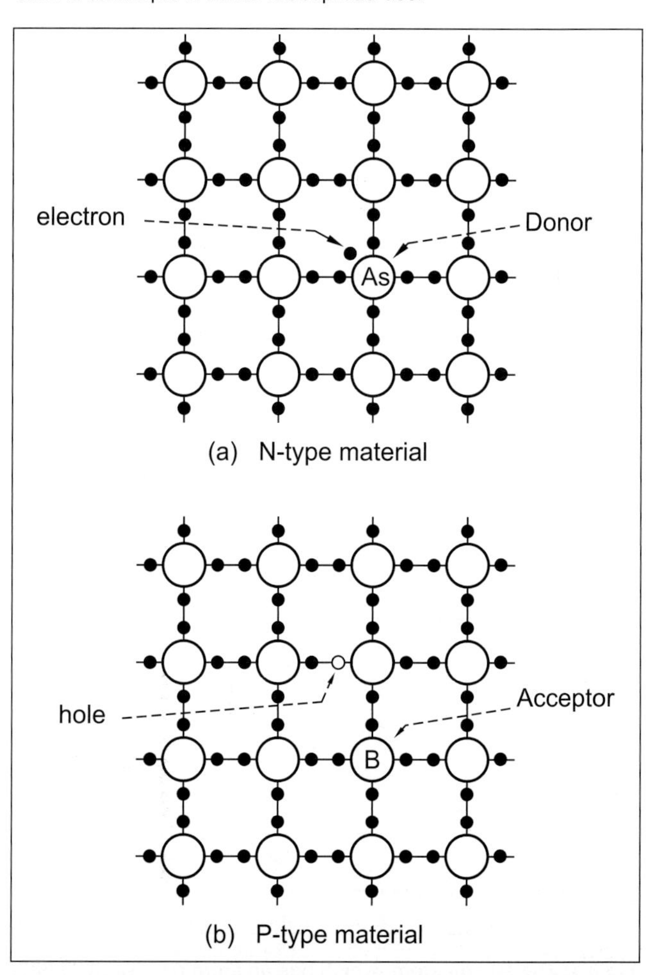


**Fig 3.5: (a) The introduction of a dopant atom having five valence electrons (arsenic) into the crystal lattice. (b) Doping with an atom having three valence electrons (boron) creates a hole**

**Fig 3.6: The PN junction and diode symbol**

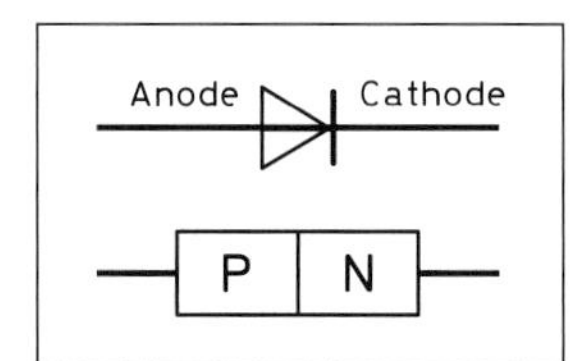


The process generally used to fabricate junction diodes is shown in **Fig 3.7**. This is known as the planar process, and variations of this technique are employed in the manufacture of many other semiconductor devices. The starting point is a thin slice of N-type silicon which is called the substrate (see **Fig 3.7(a)**). An insulating layer of silicon dioxide ($SiO_2$) is thermally grown on the top surface of the substrate in order to protect the silicon underneath. The next step (**Fig 3.7(b)**) is to selectively etch a small hole, referred to as a window, into the oxide layer - often using dilute hydrofluoric acid. A P-type region is then formed by introducing boron dopant atoms into the silicon substrate through the oxide window, using either a diffusion or ion implantation process (**Fig 3.7(c)**). This means, of course, that a region of the N-type substrate has been converted to P-type material. However, the process does not involve removing any of the N-type dopant, as it is simply necessary to introduce a higher concentration of P-type atoms than there are N-type already present. Ohmic contacts (ie contacts in which the current flows equally in either direction) are then formed to both the substrate and window (**Fig 3.7(d)**). In practice, large numbers of identical diodes will be fabricated on a single slice (wafer) of silicon which is then cut into 'chips', each containing a single diode. Finally, metal leads are bonded onto the ohmic contacts before the diode is encapsulated in epoxy resin, plastic or glass.

The diode's operation is largely determined by what happens at the junction between the P-type and N-type silicon. In the absence of any external electric field or potential difference, the large carrier concentration gradients at the junction cause carrier diffusion. Electrons from the N-type silicon diffuse into the P side, and holes from the P-type silicon diffuse into the N-side. As electrons diffuse from the N-side, uncompensated positive donor ions are left behind near the junction, and as holes leave the P-side, uncompensated negative acceptor ions remain. As **Fig 3.8(a)** shows, a negative space charge forms near the P-side of the junction (represented by '-' symbols), and a positive space charge forms near the N-side (represented by '+' symbols). Therefore this diffusion of carriers results in an electrostatic potential difference across the junction within an area known as the depletion region. The magnitude of this built-in potential depends on the doping concentration, but for a typical silicon diode it will be around 0.7V. It is important to realise, however, that this potential cannot be measured by connecting a voltmeter externally across the diode because there is no net flow of current through the device.

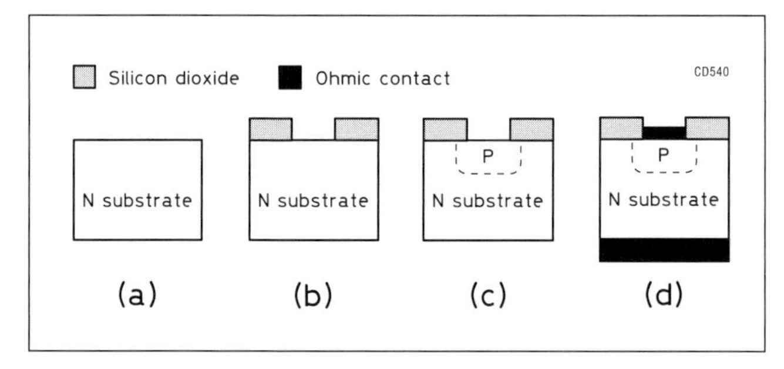


**Fig 3.7: Fabrication of a PN junction diode**

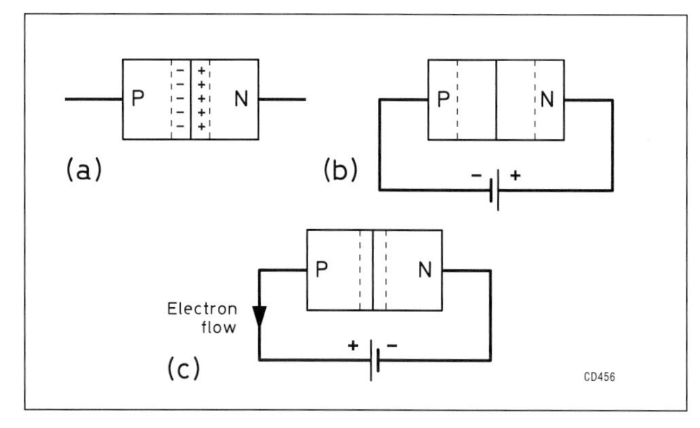


**Fig 3.8: Behaviour of the junction diode under zero bias (a) reverse bias (b) and forward bias (c)**

If a battery is connected to the diode as in **Fig 3.8(b)**, the external applied voltage will increase the electrostatic potential across the depletion region, causing it to widen (because the polarity of the external voltage is the same as that of the built-in potential). Under these conditions practically no current will flow through the diode, which is said to be reverse-biased. However, if the external bias is raised above a certain voltage the diode may 'break down', resulting in reverse current flow.

Reversing the polarity of the battery (**Fig 3.8(c)**) causes the external applied voltage to reduce the electrostatic potential across the depletion region (by opposing the built-in potential), therefore narrowing the depletion region. Under this condition, known as forward bias, current is able to flow through the diode by diffusion of current carriers (holes and electrons) across the depletion region.

The diode's electrical characteristics are summarised graphically in **Fig 3.9**, where a linear scale is used for both current and voltage. In Region A, where only a small forward bias voltage is applied (less than about 0·7V), the current flow is dominated by carrier diffusion, and in fact rises exponentially as the forward bias voltage is increased. Region B commences at a point often termed the 'knee' of the forward bias curve, and here series resistance (both internal and external) becomes the major factor in determining current flow. There remains, however, a small voltage drop which, in the case of an ordinary silicon diode, is around 0·7V .

Under reverse bias (region C) only a very small leakage current flows, far less than 1μA for a silicon diode at room temperature. This current is typically due to generation of electron-hole pairs in the reverse-biased depletion region, and does not vary with

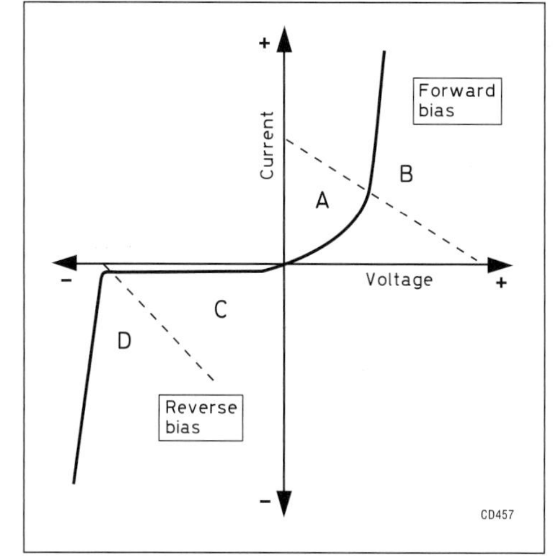


**Fig 3.9: Characteristic curve of a PN junction diode**

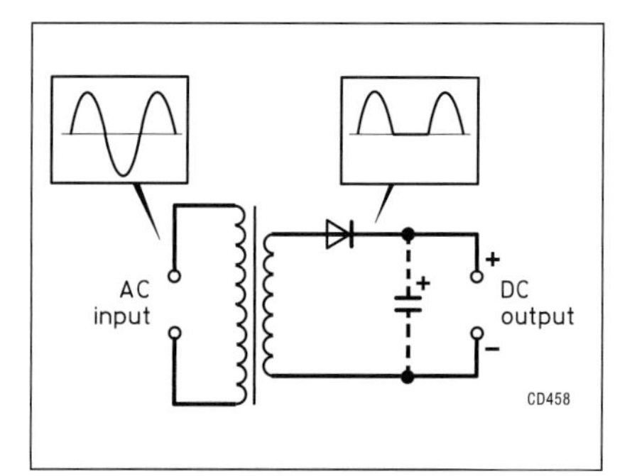

**Fig 3.10: Junction diode used as a half-wave rectifier**

reverse voltage. In Region D, where the reverse bias is increased above the diode's breakdown voltage, significant reverse current begins to flow. This is often due to a mechanism of avalanche multiplication of electron-hole pairs in the high electric field in the depletion region. Although this breakdown process is not inherently destructive, the maximum current must be limited by an external circuit to avoid excessive heating of the diode.

Diodes have a very wide range of applications in radio equipment, and there are many types available with characteristics tailored to suit their intended use. **Fig 3.10** shows a diode used as a rectifier of alternating current in a simple power supply. The triangular part of the diode symbol represents the connection to the P-type material and is referred to as the anode. The single line is the N-type connection, or cathode (Fig 3.6). Small diodes will have a ring painted at one end of their body to indicate the cathode connection (see the lower diode in **Fig 3.11**). The rectifying action allows current to flow during positive half-cycles of the input waveform, while preventing flow during negative half-cycles. This is known as half-wave rectification because only 50% of the input cycle contributes to the DC output. **Fig 3.12** shows how a more efficient power supply may be produced using a bridge rectifier, which utilises four diodes. Diodes A and B conduct during positive half-cycles, while diodes C and D conduct during negative cycles. Notice that C and D are arranged so as to effectively reverse the connections to the transformer's secondary winding during negative cycles, thus enabling the negative half-cycles to contribute to the 'positive' output. Because this arrangement is used extensively in equipment power supplies, bridge rectifier packages containing four interconnected diodes are readily available. When designing power supplies it is necessary to take account of the voltages and currents that the diodes will be subjected to. All rectifiers have a specified maximum forward current at a given case temperature. This is because the diode's forward voltage drop gives rise to power dissipation within the device - for instance, if the voltage across a rectifier diode is 0·8V at a current of 5A, the diode will dissipate 4W (0·8 x 5 = 4). This power will be converted to heat, thus raising the diode's temperature. Consequently, the rectifier diodes of high-current power supplies may need to be mounted on heatsinks. In power supplies especially, diodes must not be allowed to conduct in the reverse direction. For this reason, the maximum reverse voltage that can be tolerated before breakdown is likely to occur must be known. The term peak inverse voltage (PIV) is used to describe this characteristic (see the chapter on power supplies).

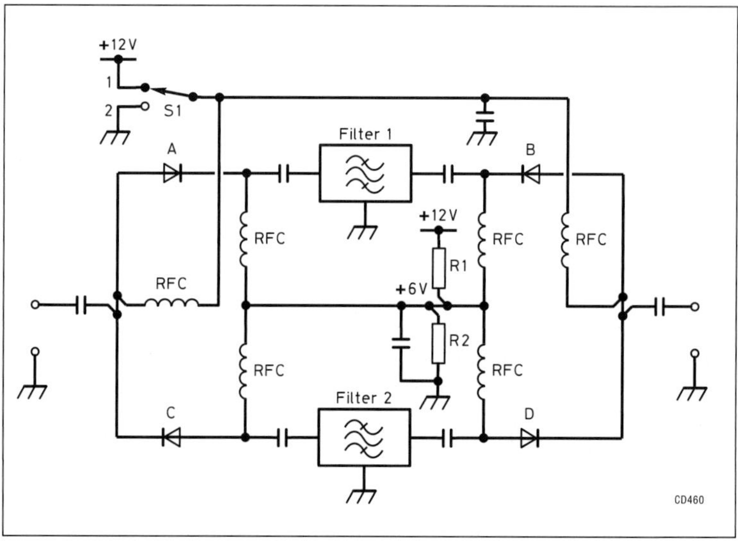

**Fig 3.13: Diodes are frequently used as switches in the signal circuitry of transceivers. The setting of S1 determines which of the two filters is selected. R1 and R2 will both have a value of around 2.2kΩ (this determines the forward current of the switching diodes). For HF applications the RFCs are 100µH**

**Fig 3.11: Rectifiers. From the left, clockwise: 35A, 100PIV bridge, 1.5A 50PIV bridge, 6A 100PIV bridge, 3A 100PIV diode. Note that the small squares are 1mm across**

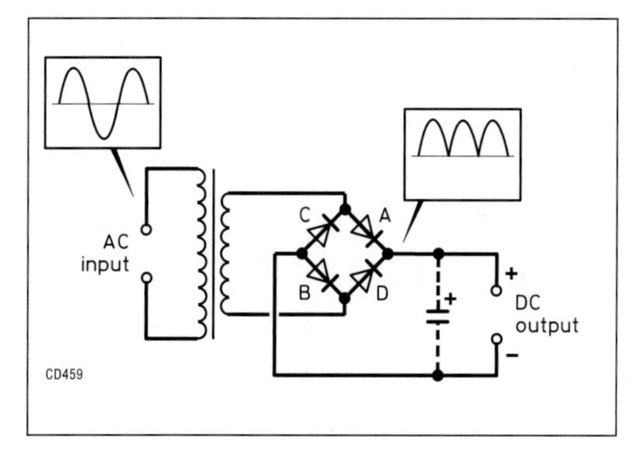

**Fig 3.12: A full-wave, or bridge, rectifier using four junction diodes**

Miniature low-current diodes, often referred to as small-signal types, are very useful as switching elements in transceiver circuitry. **Fig 3.13** shows an arrangement of four diodes used to select one of two band-pass filters, depending on the setting of switch S1. When S1 is set to position 1, diodes A and B are forward biased, thus bringing filter 1 into circuit. Diodes C and D, however, are reverse-biased, which takes filter 2 out of circuit. Setting S1 at position 2 reverses the situation. Notice how the potential divider R1/R2 is used to develop a voltage equal to half that of the supply rail. This makes it easy to arrange for a forward bias of 6V, or a reverse bias of the same magnitude, to appear across the diodes. Providing that the peak level of signals at the filter terminations does not reach 6V under any circumstances, the diodes will behave as almost perfect switches. One of the main advantages of diode switching is that signal paths can be kept short, as the leads (or PCB tracks) forming connections to the front-panel switches carry only a DC potential which is isolated

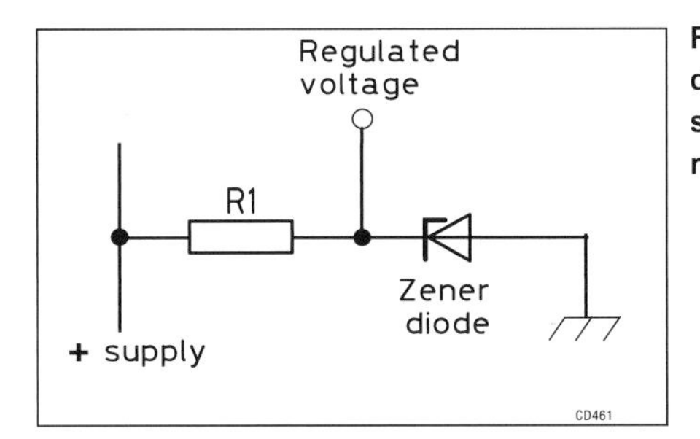


Fig 3.14: A zener diode used as a simple voltage regulator

from the signal circuitry using the RF chokes and decoupling capacitors shown.

## Zener Diodes

These diodes make use of the reverse breakdown characteristic discussed previously. The voltage at which a diode begins to conduct when reverse-biased depends on the doping concentration. As the doping level is increased, the breakdown voltage drops. This fact can be exploited during the manufacture of the diodes, enabling the manufacturer to specify the breakdown voltage for a given component. Zener diodes with breakdown voltages in the range 2·7V to over 150V are available and can be used to provide reference voltages for power supplies and bias generators.

**Fig 3.14** shows how a zener diode can be used in conjunction with a resistor to provide voltage regulation (note the use of a slightly different circuit symbol for the zener diode). When power is initially applied, the zener diode will start to conduct as the input voltage is higher than the diode's reverse breakdown value. However, as the diode begins to pass current, an increasingly high potential difference will appear across resistor R1. This potential will tend to rise until the voltage across R1 becomes equal to the difference between the input voltage and the zener's breakdown voltage. The net result is that the output voltage will be forced to settle at a level close to the diode's reverse breakdown potential. The value of R1 is chosen so as to limit the zener current to a safe value (the maximum allowable power dissipation for small zener diodes is around 400mW), while ensuring that the maximum current to be drawn from the regulated supply will not increase the voltage across R1 to a level greater than the difference between the zener voltage and the minimum expected input voltage (see the chapter on power supplies).

## Varactor Diodes

The varactor, or variable capacitance, diode makes use of the fact that a reverse-biased PN junction behaves like a parallel-plate capacitor, where the depletion region acts as the spacing between the two capacitor plates. As the reverse bias is increased, the depletion region becomes wider. This produces

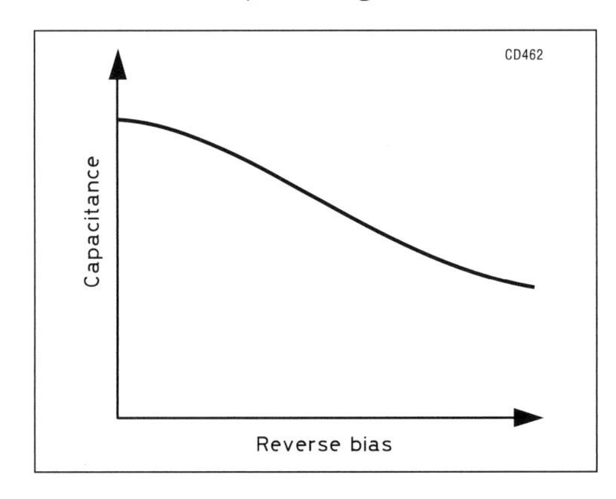


Fig 3.15: Relationship between capacitance and reverse bias of a varactor (variable capacitance diode)

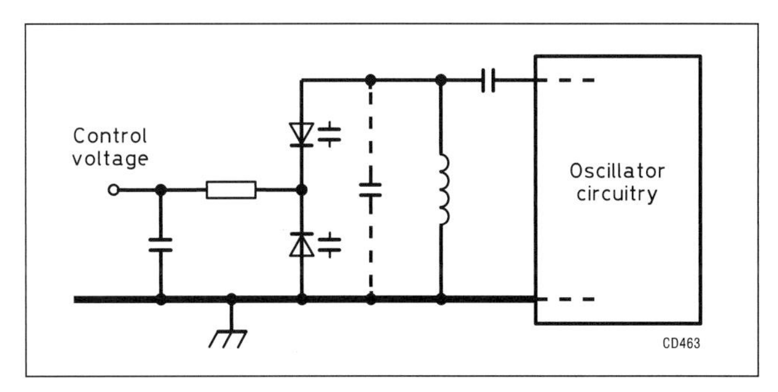


**Fig 3.16: Two varactor diodes used to tune a voltage-controlled oscillator. The capacitor drawn with dotted lines represents the additional component which may be added to the tank circuit in order to modify the LC ratio and tuning range**

the same effect as moving the plates of a capacitor further apart - the capacitance is reduced (see the fundamentals chapter). The varactor can therefore be used as a voltage-controlled variable capacitor, as demonstrated by the graph in **Fig 3.15**. The capacitance is governed by the diode's junction area (ie the area of the capacitor plates), and also by the width of the depletion region for a given value of reverse bias (which is a function of the doping concentration). Varactors are available covering a wide range of capacitance spreads, from around 0·5-10pF up to 20-400pF. The voltages at which the stated maximum and minimum capacitances are obtained will be quoted in the manufacturer's literature, but they normally fall in the range 2-20V. The maximum reverse-bias voltage should not be exceeded as this could result in breakdown.

Varactors are commonly used to achieve voltage control of oscillator frequency in frequency synthesisers. **Fig 3.16** shows a typical arrangement where two varactors are connected 'back to back' and form part of the tank circuit of a voltage-controlled oscillator (VCO). The use of two diodes prevents the alternating RF voltage appearing across the tuned circuit from driving the varactors into forward conduction, which is most likely to happen when the control voltage is low. Because the varactor capacitances appear in series, the maximum capacitance swing is half that obtainable when using a single diode. Three-lead packages containing dual diodes internally connected in this way are readily available. It is also possible to obtain multiple-diode packages containing two or three matched diodes but with separate connections. These are used to produce voltage-controlled versions of two-gang or three-gang variable capacitors.

When used as a capacitive circuit element, the varactor's Q may be significantly lower than that of a conventional capacitor. This factor must be taken into account when designing high-performance frequency synthesisers (see the chapter on oscillators).

## PIN Diodes

A PIN diode, **Fig 3.17(a)**, is a device that operates as a variable resistor at RF and even into microwave frequencies. Its resistance is determined only by its DC excitation. It can also be used to switch quite large RF signals using smaller levels of DC excitation.

The PIN diode chip consists of a chip of pure (intrinsic or I-type) silicon with a layer of P-type silicon on one side and a

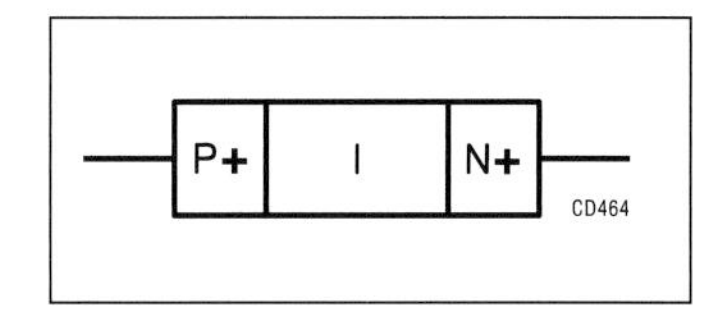


Fig 3.17(a): The PIN diode

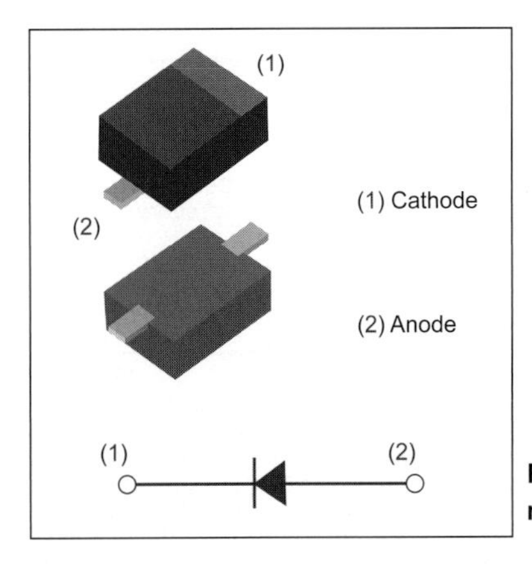


Fig 3.17(b): Surface mount Pin diode

layer of N-type on the other **Fig 3.17b)**. The thickness of the I-region (W) is only a little smaller than the thickness of the original wafer from which the chip was cut.

### Forward-biased PIN diodes

When forward-biased electrons and holes are injected into the I-region from the N- and P-regions respectively, these charges don't recombine immediately but a finite charge remains stored in the I-layer.

The quantity of stored charge, Q, depends on the recombination time (t, usually called the carrier lifetime) and the forward current I as:

$$Q = It$$

The resistance of the I-layer is proportional to the square of its thickness (W) and inversely proportional to Q, and is given by:

$$R = \frac{W^2}{(N+p)Q}$$

Combining these two equations, we get:

$$R = \frac{W^2}{(N+p)It}$$

Thus the resistance is inversely proportional to the DC excitation I. For a typical PIN diode, R varies from 0·1Ω at 1A to 10kΩ at 1μA. This is shown in **Fig 3.18**.

This resistance-current relationship is valid over a wide frequency range, the limits at low resistance being set by the parasitic inductance of the leads, and at high resistance by the junction capacitance. The latter is small owing to the thickness

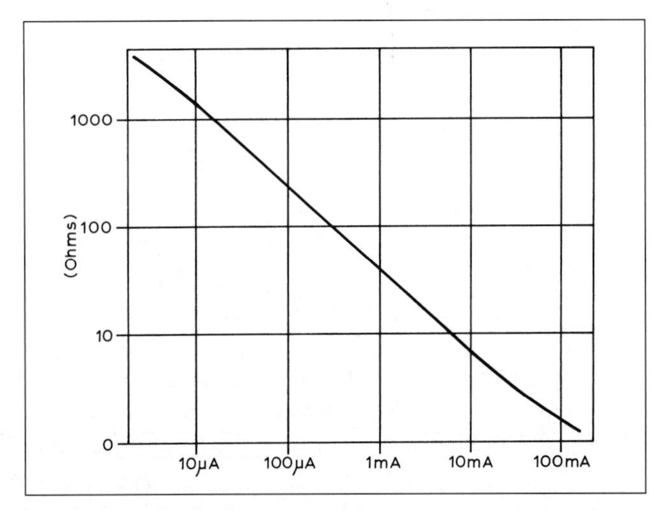


**Fig 3.18: Relationship between forward current and resistance for a PIN diode**

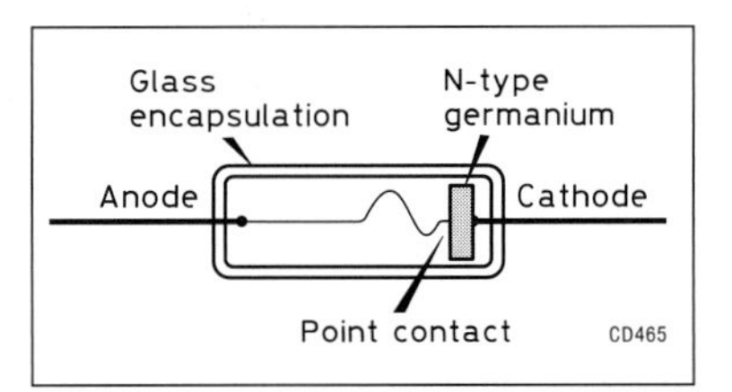


**Fig 3.19: The germanium point-contact diode**

of the I-layer which also ensures that there is always sufficient charge so that the layer presents a constant resistance over the RF cycle.

### Reverse-biased PIN diodes

At high RF, a reverse-biased PIN diode behaves as a capacitor independent of the reverse bias with a value of:

$$C = \varepsilon A/W \text{ farads}$$

where ε is the permittivity of silicon. Note that although the dielectric constant is quite high, the capacitance of the diode is small because of the small area of the diode and the thickness of the intrinsic layer. Therefore as a switch the isolation in the reverse bias state is good but gets slightly worse as the frequency increases. This is not as serious as it seems since nearly all PIN diode switches at VHF are operated in 50Ω circuits and the reactance is high compared to 50Ω.

## Germanium Point-contact Diodes

Ironically, one of the earliest semiconductor devices to find widespread use in telecommunications actually pre-dates the thermionic valve. The first broadcast receivers employed a form of envelope detector (RF rectifier) known colloquially as a cat's whisker. This consisted of a spring made from a metal such as bronze or brass (the 'whisker'), the pointed end of which was delicately bought into contact with the surface of a crystal having semiconducting properties, such as galena, zincite or carborundum.

The germanium point-contact diode is a modern equivalent of the cat's whisker and consists of a fine tungsten spring which is held in contact with the surface of an N-type germanium crystal (**Fig 3.19**). During manufacture, a minute region of P-type material is formed at the point where the spring touches the crystal. The point contact therefore functions as a PN diode. In most respects the performance of this device is markedly inferior to that of the silicon PN junction diode. The current flow under reverse bias is much higher - typically 5mA, the highest obtainable PIV is only about 70V and the maximum forward current is limited by the delicate nature of the point contact.

Nevertheless, this device has a number of saving graces. The forward voltage drop is considerably lower than that of a silicon junction diode - typically 0·2V - and the reverse capacitance is also very small. There is also an improved version known as the gold-bonded diode, where the tungsten spring is replaced by one made of gold. **Fig 3.20** shows a simple multimeter probe which is used to rectify low RF voltages. The peak value can then be read with the meter switched to a normal DC range. The low

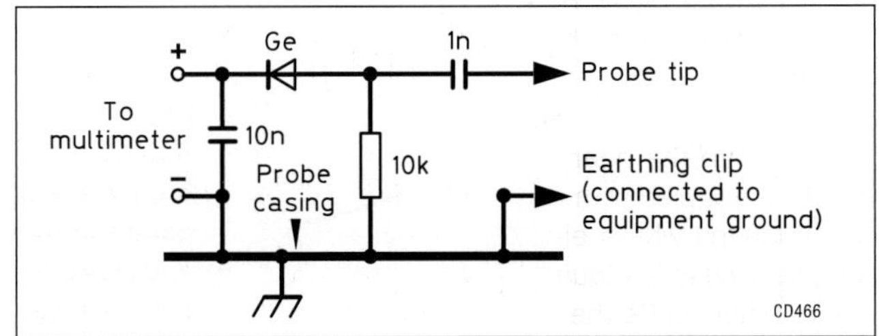


**Fig 3.20: A simple RF probe using a germanium (Ge) point-contact diode**

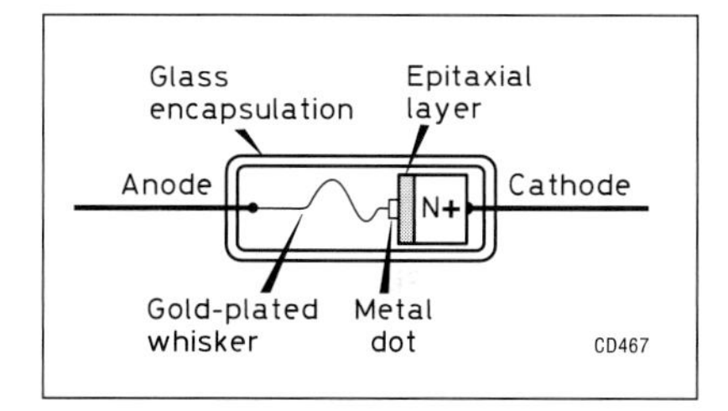


(above) **Fig 3.21: Hot-carrier (Schottky) diode**

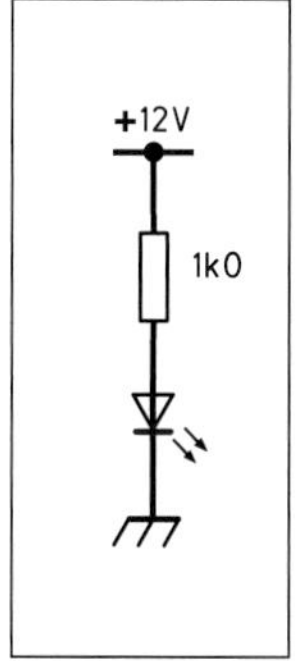


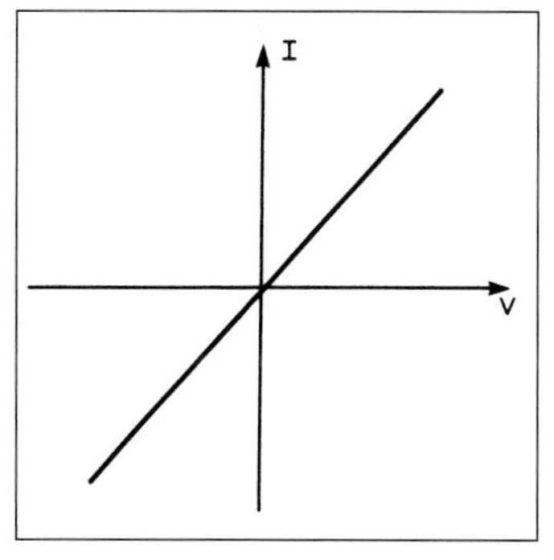


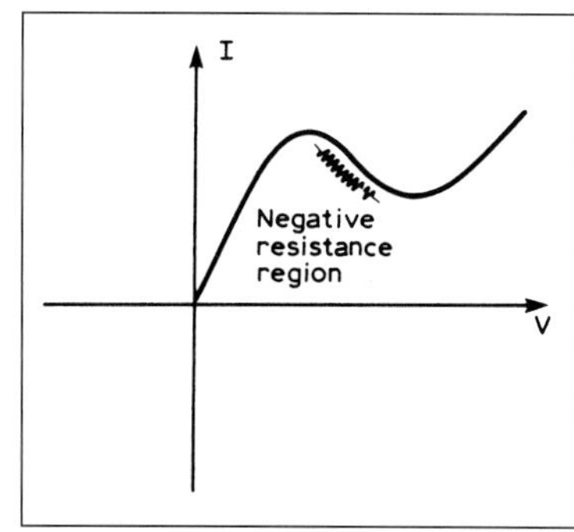


**Fig 3.22: An LED emits light when forward biased. The series resistor limits the forward current to a safe value**

**Fig 3.23: Voltage-current through a resistor**

**Fig 3.24: Voltage-current through a Gunn diode**

forward voltage drop of the point-contact diode leads to more accurate readings.

## The Schottky Barrier Diode

Ordinary PN junction diodes suffer from a deficiency known as charge storage, which has the effect of increasing the time taken for a diode to switch from forward conduction to reverse cut-off when the polarity of the applied voltage is reversed. This reduces the efficiency of the diode at high frequencies. Charge storage occurs because holes, which are less mobile than electrons, require a finite time to migrate back from the N-doped cathode material as the depletion region widens under the influence of reverse bias. The fact that in most diodes the P-type anode is more heavily doped than the N-type cathode tends to exacerbate matters.

The hot-carrier, or Schottky barrier, diode overcomes the problem of charge storage by utilising electrons as its main current carriers. It is constructed (**Fig 3.21**) in a similar fashion to the germanium point-contact type, but there are a few important differences. The semiconductor used is N-type silicon which is modified by growing a layer (the epitaxial region) of more lightly doped material onto the substrate during manufacture. The device is characterised by its high switching speed and low capacitance. It is also considerably more rugged than the germanium point-contact diode and generates less noise.

Hot-carrier diodes are used in high-performance mixers of the switching, or commutating, type capable of operating into the microwave region.

## Light-emitting Diodes (LEDs)

Light-emitting diodes (LEDs) utilise the recombination of the excess carriers injected in a forward-biased PN diode to produce light emission in the ultra-violet, visible, or infra-red spectrum. When electrons recombine with holes across the energy bandgap of a suitable semiconductor, photons of light energy are radiated. The energy, and therefore wavelength, of these photons is determined by the width of the bandgap. In indirect bandgap materials like Si and Ge, radiative carrier recombination is very unlikely. However, in direct bandgap materials like GaAs radiative recombination readily occurs, and therefore such semiconductors are used for light generation.

A GaAs LED produces photon emission outside the visible spectrum in the near infra-red (~900nm). This device may be optically coupled with a Si PIN detecting diode to replace relays in situations where electrical isolation between the input and output circuits is required.

For visible LEDs the compound semiconductor $GaAs_{1-y}P_y$ (gallium arsenide phosphide) is often used. This material has a bandgap which increases with y, the mole fraction, and so the LED colour can be fixed during manufacture by selecting the appropriate value of y. For example, red (y = 0.4), orange (y = 0.65), and yellow (y = 0.85) LEDs can be produced. Green LEDs may be made from GaP (gallium phosphide). High brightness blue LEDs may be made from wide bandgap materials based on GaN (gallium nitride). White LEDs which combine red, green and blue colours have also been produced.

Visible LEDs are highly efficient and reliable, and are commonly used for displays and indicators. A discrete LED indicator component contains an LED chip and a plastic lens, which is usually coloured to serve as an optical filter and to enhance contrast. More complex LED displays are in a seven segment numeric format, or in a 5x7 matrix alphanumeric array. These displays can be made by monolithic processes similar to those used to make silicon integrated circuits, or by using individual bar segment LED chips mounted on a reflector.

LEDs are operated in forward bias, and at a current of 10mA they will generate a useful amount of light without overheating. The forward voltage drop at this current is about 1·8V. **Fig 3.22** shows a LED operating from a 12V power rail (note the two arrows

**Fig 3.25: Current in Gunn diode**

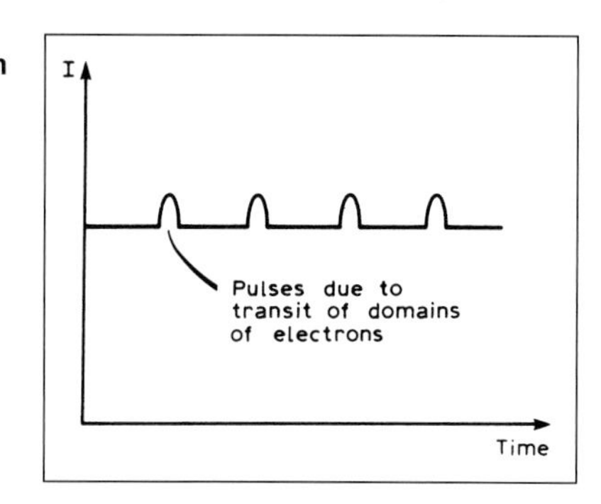


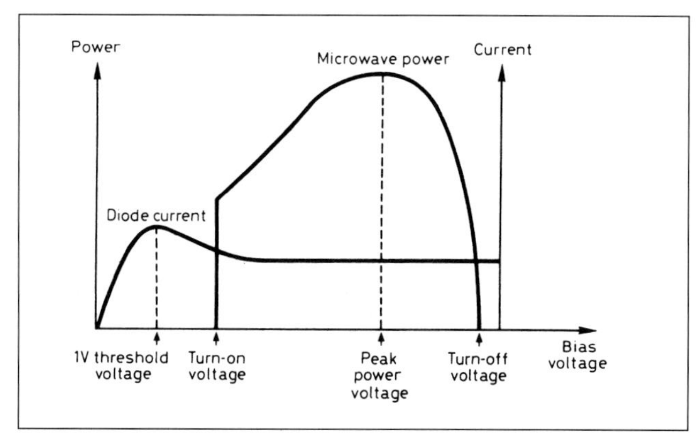


**Fig 3.26: Characteristic shape of a Gunn oscillator's bias power curve**

representing rays of light which differentiates the LED circuit symbol from that of a normal diode). The series resistor determines the forward current and so its inclusion is mandatory.

## The Gunn Diode

The Gunn diode, named after its inventor, B J Gunn, comprises little more than a block of N-type gallium arsenide. It is not, in fact, a diode in the normally accepted sense because there is no P-N junction and consequently no rectifying action. It is properly called the Gunn effect device. However, 'diode' has become accepted by common usage and merely explains that it has two connections.

It consists of a slice of low-resistivity N-type gallium arsenide on which is grown a thin epitaxial layer, the active part, of high-resistivity gallium arsenide with a further thicker layer of low-resistivity gallium arsenide on top of that. Since the active layer is very thin, a low voltage across it will produce a high electric field strength. The electrons in gallium arsenide can be in one of two conduction bands. In one they have a much higher mobility than in the other and they are initially in this band. As the electric field increases, more and more are scattered into the lower mobility band and the average velocity decreases. The field at which this happens is called the threshold field and is 320V/mm. Since current is proportional to electron velocity and voltage is proportional to electric field, the device has a region of negative resistance. This odd concept is explained by the definition of resistance as the slope of the voltage-current graph and a pure resistor has a linear relationship (**Fig 3.23**). On the other hand, the Gunn diode has a roughly reversed 'S'-shaped curve (**Fig 3.24**) and the negative resistance region is shown by the hatching. The current through the device takes the form of a steady DC with superimposed pulses (**Fig 3.25**) and their frequency is determined by the thickness of the active epitaxial layer. As each pulse reaches the anode, a further pulse is generated and a new domain starts from the cathode. Thus the rate of pulse formation depends on the transit time of these domains through the epitaxial layer. In a 10GHz Gunn diode, the layer is about 10μm thick, and a voltage of somewhat greater than 3·5V gives the high field state and microwave pulses are generated. The current through the device shows a peak before the threshold voltage followed by a plateau. See **Fig 3.26**. The power output reaches a peak at between 7·0 and 9·0V for 10GHz devices.

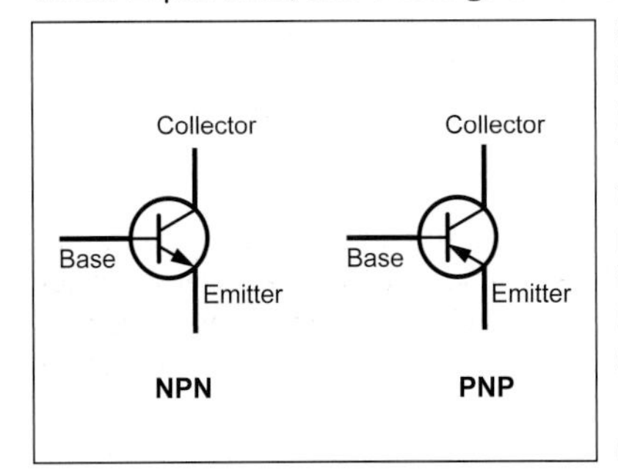


**Fig 3.27: Circuit symbols for the bi-polar transistor**

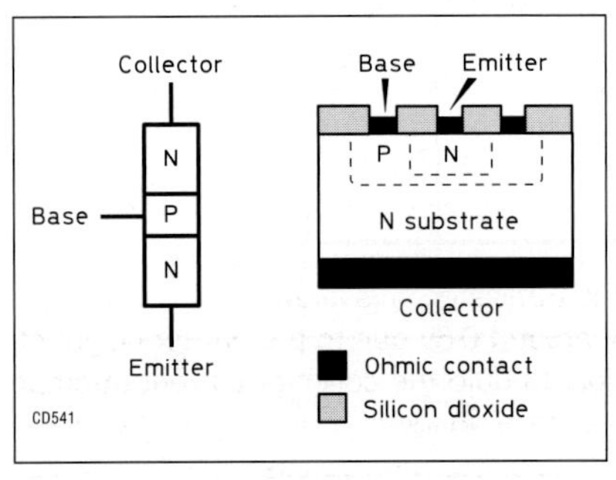


**Fig 3.28: Construction of an NPN transistor**

The Gunn diode is inherently a wideband device so it is operated in a high-Q cavity (tuned circuit) and this determines the exact frequency. It may be tuned over a narrow range by altering the cavity with a metallic screw, a dielectric (PTFE or Nylon) screw or by loading the cavity with a varactor. With a 10GHz device, the whole of the 10GHz amateur band can be covered with reasonable efficiency. Gunn diodes are not suitable for narrow-band operation since they are of low stability and have relatively wide noise sidebands. The noise generated has two components, thermal noise and a low frequency 'flicker' noise. Analysis of the former shows that it is inversely proportional to the loaded Q of the cavity and that FM noise close to the carrier is directly proportional to the oscillator's voltage pushing, ie to the variation of frequency caused by small variations of voltage. Clearly, the oscillator should be operated where this is a minimum and that is often near the maximum safe bias.

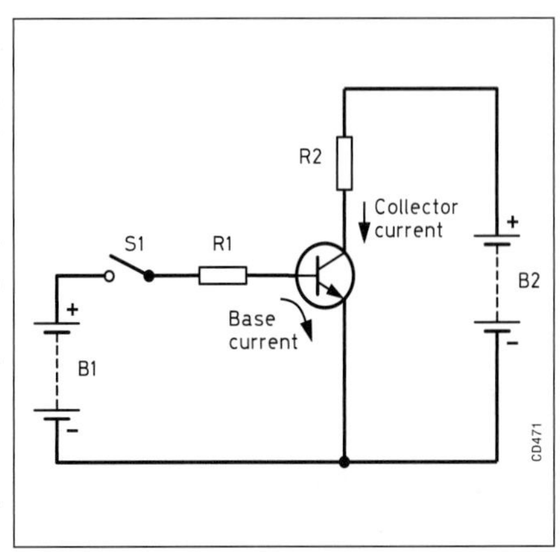


**Fig 3.29: Applying forward bias to the base-emitter junction of a bipolar transistor causes a much larger current to flow between the collector and emitter. The arrows indicate conventional current flow, which is opposite in direction to electron flow**

## THE BIPOLAR TRANSISTOR

The very first bipolar transistor, a point-contact type, was made by John Bardeen and Walter Brattain at Bell Laboratories in the USA during December 1947. A much-improved version, the bipolar junction transistor, arrived in 1950 following work done by another member of the same team, William Schockley. In recognition of their pioneering work in developing the first practical transistor, the three were awarded the Nobel Prize for physics in 1956.

The bipolar transistor is a three-layer device which exists in two forms, NPN and PNP. The circuit symbols for both types are shown in **Fig 3.27**. Note that the only difference between the symbols is the direction of the arrow drawn at the emitter connection.

**Fig 3.28** shows the three-layer sandwich of an NPN transistor and also how this structure may be realised in a practical device. The emitter region is the most heavily doped.

In **Fig 3.29** the NPN transistor has been connected into a simple circuit to allow its operation to be described. The PN junction between the base and emitter forms a diode which is forward biased by battery B1 when S1 is closed. Resistor R1 has been included to control the level of current that will inevitably flow, and R2 provides a collector load. A voltmeter connected between the base and emitter will indicate the normal forward voltage drop of approximately 0·6V typical for a silicon PN junction.

The collector-base junction also forms the equivalent of a PN diode, but one that is reverse-biased by battery B2. This suggests that no current will flow between the collector and the emitter. This is indeed true for the case where S1 is open, and no current is flowing through the base-emitter junction. However, when S1 is closed, current does flow between the collector and emitter. Due to the forward biasing of the base-emitter junction, a large number of electrons will be injected into the P-type base from the N-type emitter. Crucially, the width of the base is made less than the diffusion length of electrons, and so most of these injected carriers will reach the reverse-biased collector-base junction. The electric field at this junction is such that these electrons will be swept across the depletion region into the col-

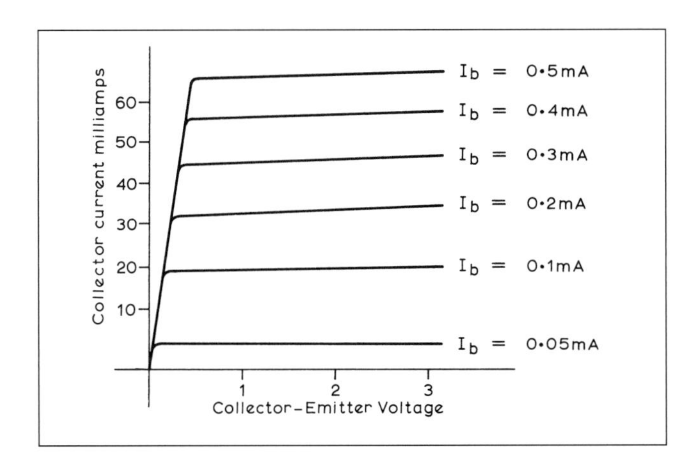


Fig 3.30: The relationship between base current and collector current for a bipolar transistor

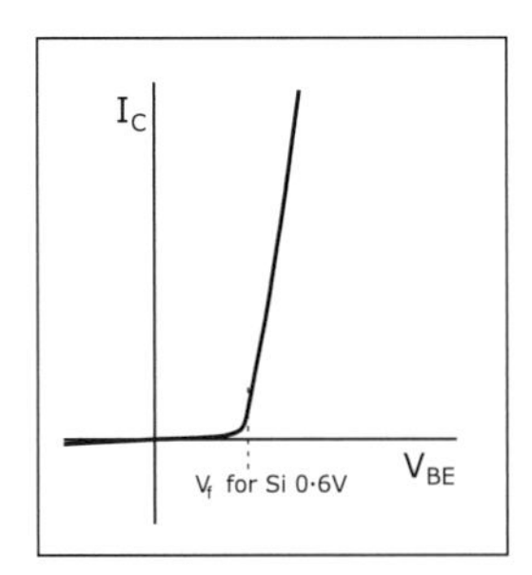


Fig 3.31: The relationship between base-emitter voltage and collector current for a bipolar transistor

lector, thus causing significant current to flow between the emitter and collector.

However, not all the electrons injected into the base will reach the collector, as some will recombine with holes in the base. Also, the base-emitter junction forward bias causes holes to be injected from the P-type base to the N-type emitter, and electrons and holes will recombine in the base-emitter depletion region. These are the three main processes which give rise to the small base current. The collector current will be significantly larger than the base current, and the ratio between the two (known as the transistor's ß) is related to the doping concentration of the emitter divided by the doping concentration of the base. The PNP transistor operates in a similar fashion except that the polarities of the applied voltages, and also the roles of electrons and holes, are reversed. The graph at **Fig 3.30** shows the relationship between base current ($I_B$) and collector current ($I_C$) for a typical bipolar transistor. The point to note is that the collector current is very much determined by the base current ($I_C = ßI_B$) and the collector-emitter voltage, $V_{CE}$, has comparatively little effect. The relationship between base-emitter voltage, $V_{BE}$, and collector current $I_C$ is shown in **Fig 3.31**. The graph should be compared with the diode characteristics previously shown in Fig 3.9. The graph is almost identical but the current axis is ß times greater due to the current gain of the transistor. It should be noted that the graph of transistor base current against applied voltage $I_B/V_{BE}$ is simply the graph of the base-emitter 'diode'.

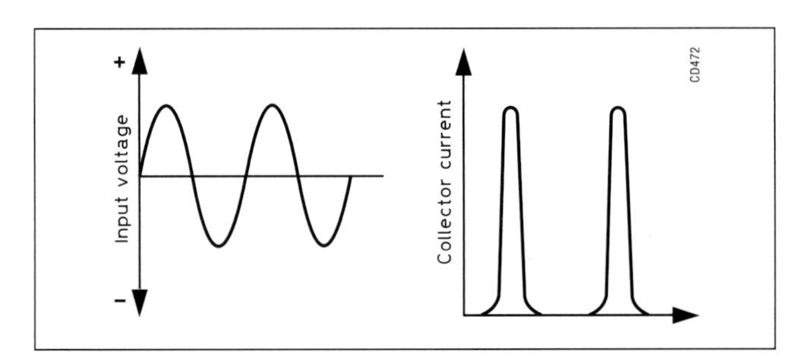


Fig 3.32: The circuit shown in Fig 3.29 would not make a very good amplifier

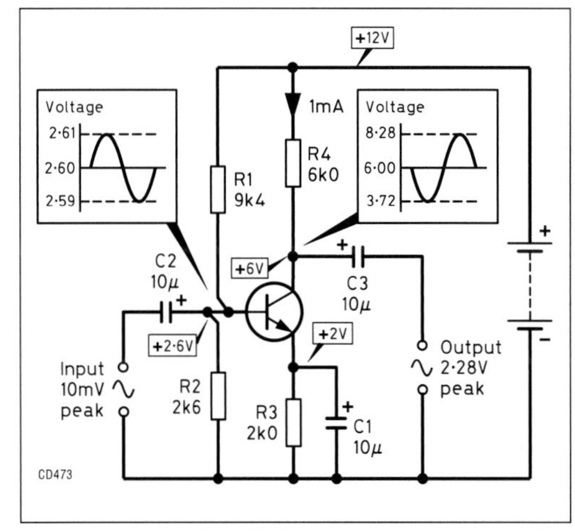


Fig 3.33: A practical common-emitter amplifier

The circuit shown in Fig 3.29 therefore provides the basis for an amplifier. Small variations in base current will result in much larger variations in collector current. The DC current gain (ß or $h_{FE}$), of a typical bipolar transistor at a collector current of 1mA will be between 50 and 500, ie the change in base current required to cause a 1mA change in collector current lies in the range 2 to 20µA. The value of ß is temperature dependent, and so is the base-emitter voltage drop ($V_{BE}$), which will fall by approximately 2mV for every 1°C increase in junction temperature.

## A Transistor Amplifier

If the battery (B1) used to supply base current to the transistor in Fig 3.29 is replaced with a signal generator set to give a sine-wave output, the collector current will vary in sympathy with the input waveform as shown in **Fig 3.32**. A similar, although inverted, curve could be obtained by plotting the transistor's collector voltage.

Two problems are immediately apparent. First, because the base-emitter junction only conducts during positive half-cycles of the input waveform, the negative half-cycles do not appear at the output. Second, the barrier height of the base-emitter junction results in the base current falling exponentially as the amplitude of the input waveform drops below +0·6V. This causes significant distortion of the positive half-cycles, and it is clear that if the amplitude of the input waveform were to be significantly reduced then the collector current would hardly rise at all.

The circuit can be made far more useful by adding bias. **Fig 3.33** shows the circuit of a practical common-emitter amplifier operating in Class A. The term 'common emitter' indicates that the emitter connection is common to both the input and output circuits. Resistors R1 and R2 form a potential divider which establishes a positive bias voltage at the base of the transistor. The values of R1 and R2 are chosen so that they will pass a current at least 10 times greater than that flowing into the base. This ensures that the bias voltage will not alter as a result of variations in the base current. R3 is added in order to stabilise the bias point, and its value has been calculated so that a potential of 2V will appear across it when, in the absence of an input signal, the desired standing collector current of 1mA (0·001A) flows (0·001A x 2000 = 2V). For this reason, the ratio between the values of R1 and R2 has been chosen to establish a voltage of 2·6V at the base of the transistor. This allows for the expected forward voltage drop of around 0·6V due to the barrier height of the base-emitter junction. Should the collector current attempt to rise, for instance because of an increase in ambient temperature causing $V_{BE}$ to fall, the voltage drop across R3 will increase, thus reducing the base-emitter voltage and preventing the collector current rising. Capacitor C1 provides a low-impedance

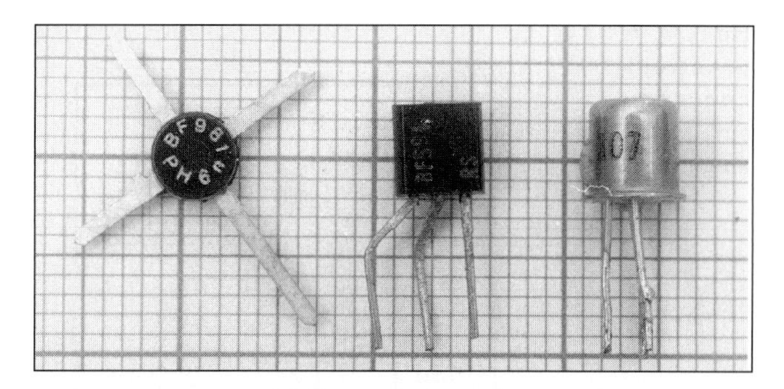

**Fig 3.34: Low-power devices. From the left: UHF dual-gate MOSFET, HF bipolar transistor (plastic-cased), AF bipolar transistor (metal-cased). The small squares are 1mm across**

**Fig 3.35: Power devices. From the left, clockwise: NPN bipolar power Darlington transistor, power audio amplifier IC, NPN bipolar power transistor. The small squares are 1mm across**

path for alternating currents so that the signal is unaffected by R3, and coupling capacitors C2 and C3 prevent DC potentials appearing at the input and output. The value of the collector load resistor (R4) is chosen so that in the absence of a signal the collector voltage will be roughly 6V - ie half that of the supply rail. Where a transistor is used as an RF or IF amplifier, the collector load resistor may be replaced by a parallel tuned circuit provided suitable precautions are taken to limit the collector current and achieve suitable biasing. This is covered in more detail in the Building Blocks chapters.

An alternating input signal will now modulate the base voltage, causing it to rise slightly during the positive half-cycles, and fall during negative half-cycles. The base current will be similarly modulated and this, in turn, will cause far larger variations in the collector current. Assuming the transistor has a ß of 100, the voltage amplification obtained can be gauged as follows:

In order to calculate the effect of a given input voltage, it is necessary to develop a value for the base resistance. This can be approximated for a low-frequency amplifier using the formula:

$$\text{Base resistance} = \frac{26 \times \beta}{\text{Emitter current in mA}}$$

The emitter current is the sum of the base and collector currents but, as the base current is so much smaller than the collector current, it is acceptable to use just the collector current in rough calculations:

$$\text{Base resistance} = \frac{26 \times 100}{1} = 2600\Omega$$

Using Ohm's Law, and assuming that the peak amplitude of the input signal is 10mV (0·01V), the change in base current will be:

$$\frac{0 \cdot 01}{2600} = 3 \cdot 8\mu A$$

This will produce a change in collector current 100 times larger, that is 380µA.

The change in output voltage (again using Ohm's Law) is:

$$380\mu A \times 6k\Omega\ (R4) = 2{\cdot}28V$$
$$(3{\cdot}8 \times 10^{-4} \times 6\ 10^{3} = 2{\cdot}28)$$

The voltage gain obtained is therefore:

$$\frac{2 \cdot 28}{0 \cdot 01} = 228 \quad \text{or } 47\text{dB}$$

Note that the output voltage developed at the junction between R4 and the collector is phase-reversed with respect to the input. As operating frequency is increased the amplifier's gain will start to fall. There are a number of factors which cause this, but one of the most significant is charge storage in the base region. For this reason, junction transistors designed to operate at high frequencies will be fabricated with the narrowest possible base width. A guide to the maximum frequency at which a transistor can be operated is given by the parameter $f_T$. This is the frequency at which the ß falls to unity (1). Most general-purpose junction transistors will have an $f_T$ of around 150MHz, but specialised devices intended for use at UHF and microwave frequencies will have an $f_T$ of 5GHz or even higher. An approximation of a transistor's current gain at frequencies below $f_T$ can be obtained by:

$$\beta = f_T \div \text{operating frequency}$$

For example, a device with an $f_T$ of 250MHz will probably exhibit a current gain of around 10 at a frequency of 25MHz.

## Maximum Ratings

Transistors may be damaged by the application of excessive voltages, or if made to pass currents that exceed the maximum values recommended in the manufacturers' data sheets. Some, or all, of the following parameters may need to be considered when selecting a transistor for a particular application:

**$V_{CEO}$** The maximum voltage that can be applied between the collector and emitter with the base open-circuit (hence the 'O'). In practice the maximum value is dictated by the reverse breakdown voltage of the collector-base junction. In the case of some transistors, for instance many of those intended for use as RF power amplifiers, this rating may seem impracticably low, being little or no higher than the intended supply voltage. However, in practical amplifier circuits, the base will be connected to the emitter via a low-value resistance or coupling coil winding. Under these conditions the collector-base breakdown voltage will be raised considerably (see below).

**$V_{CBO}$** The maximum voltage that can be applied between the collector and base with the emitter open-circuit. This provides a better indication of the collector-base reverse breakdown voltage. An RF power transistor with a $V_{CEO}$ of 18V may well have a $V_{CBO}$ rating of 48V. Special high-voltage transistors are manufactured for use in the EHT (extra high tension) generators of television and computer displays which can operate at collector voltages in excess of 1kV.

**$V_{EBO}$** The maximum voltage that can be applied between the emitter and base with the collector open-circuit. In the case of an NPN transistor, the emitter will be held at a positive

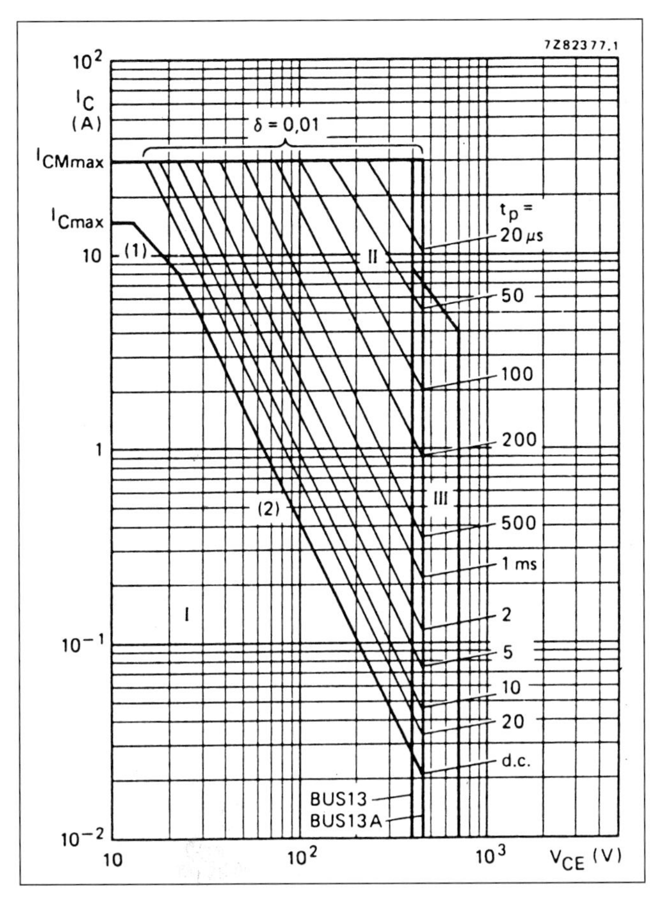


**Fig 3.36: The safe operating (SOAR) curves for a BUS13A power bipolar transistor. When any power transistor is used as a pass device in a regulated PSU, it is important to ensure that the applied voltage and current are inside the area on the graph marked 'I - region of permissible DC operation'. Failure to do so may result in the device having a very short life because of secondary breakdown. This is due to the formation of hot spots in the transistor's junction. Note that power FETs are not as prone to this limitation. Key to regions: I - Region of permissible DC operation. II - Permissible extension for repetitive pulse operation. III - Area of permissible operation during turn-on in single-transistor converters, provided $R_{BE} \leq 100\Omega$ and $t_p \leq 0.6\mu s$. (Reproduction courtesy Philips Semiconductors)**

potential with respect to the base. Therefore, it is the reverse breakdown voltage of the emitter-base junction that is being measured. A rating of around 5V can be expected.

**$I_C$** The maximum continuous collector current. For a small-signal transistor this is usually limited to around 150mA, but a rugged power transistor may have a rating as high as 30A.

**$P_D$** The maximum total power dissipation for the device. This figure is largely meaningless unless stated for a particular case temperature. The more power a transistor dissipates, the hotter it gets. Excessive heating will eventually lead to destruction, and so the power rating is only valid within the safe temperature limits quoted as part of the $P_D$ rating. A reasonable case temperature for manufacturers to use in specifying the power rating is 50°C. It is unfortunate that as a bipolar transistor gets hotter, its $V_{BE}$ drops and its ß increases. Unless the bias voltage is controlled to compensate for this, the collector current may start to rise, which in turn leads to further heating and eventual destruction of the device. This phenomenon is known as thermal runaway.

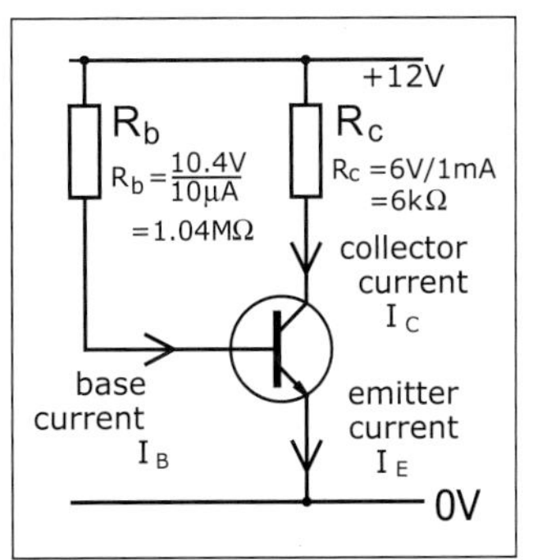


**Fig 3.37: Simple bias circuit offering no bias stability**

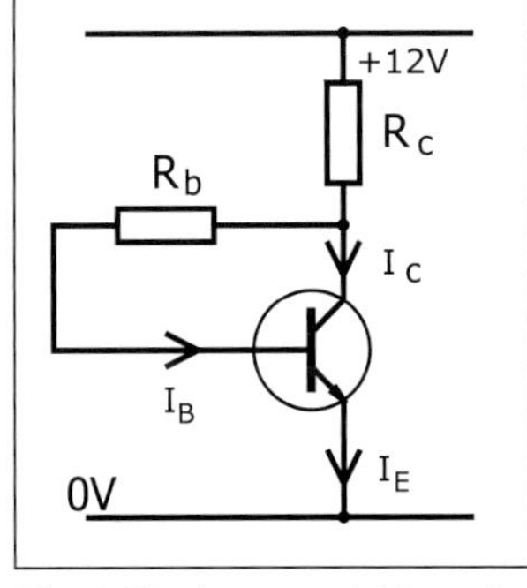


**Fig 3.38: Improved bias circuit with some bias stability**

The possibility of failure due to the destruction of a transistor junction by excessive voltage, current or heating is best avoided by operating the device well within its safe limits at all times (see below). In the case of a small transistor, junction failure, should it occur, is normally absolute, and therefore renders the device useless. Power transistors, however, have a more complex construction. Rather than attempting to increase the junction area of an individual transistor in order to make it more rugged, power transistors normally consist of large numbers of smaller transistors fabricated on a single chip of silicon. These are arranged to operate in parallel, with low-value resistances introduced in series with the emitters to ensure that current is shared equally between the individual transistors. A large RF power transistor may contain as many as 1000 separate transistors.

In such a device, the failure of a small number of the individual transistors may not unduly affect its performance. This possibility must sometimes be taken into account when testing circuitry which contains power transistors.

### Safe operating area - SOAR

All bipolar transistors can fail if they are over-run and power bipolars are particularly susceptible since, if over-run by excessive power dissipation, hot spots will develop in the transistor's junction, leading to total destruction. To prevent this, manufacturers issue SOAR data, usually in the form of a graph or series of graphs plotted on log-log graph paper with current along one axis and voltage along the other. A typical example is shown in **Fig 3.36.**

Most amateur use will be for analogue operation and should be confined to area I in the diagram (or, of course, to the corresponding area in the diagram of the transistor being considered). For pulse operation, it is possible to stray out into area II. How far depends on the height of the current pulses and the duty cycle. Generally speaking, staying well within area I will lead to a long life.

All bipolar transistors have this SOAR but, in the case of small devices, it is not often quoted by the manufacturers and, in any case, is most unlikely to be exceeded.

## Other Bias Circuits

Fig 3.33 showed a transistor amplifier biased to bring the operating point of the transistor, that is the standing or quiescent voltages and currents, to a mid-range value allowing a signal to vary those currents and voltages both up and down to provide a maximum available range or swing. It is important that these values are predictable and tolerant of the variation in characteristics, particularly gain, from one transistor to another, albeit of the same type and part number.

The circuit in **Fig 3.37** would provide biasing. The base current required is calculated as 1mA/ß (assuming $I_c$ = 1mA) and the resistor $R_b$ chosen accordingly. ß has been assumed to be 100. However the value of ß is not well controlled and may well

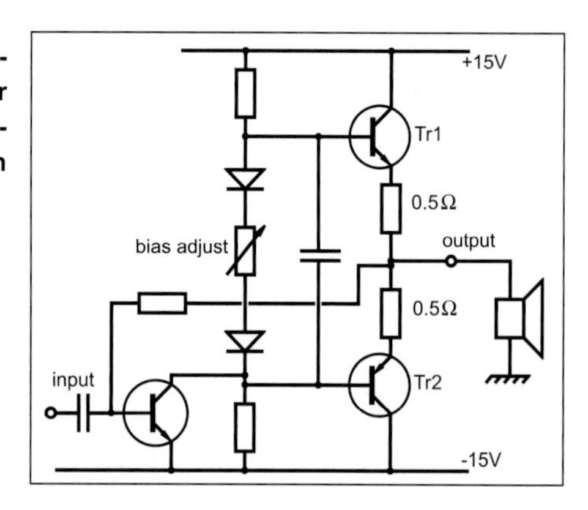


**Fig 3.39: A push-pull amplifier using two transistors biased in Class B**

vary over a 4:1 range. In this circuit, if ß actually was 200, the collector current would double, with 12V across $R_c$ and almost none across the transistor. Clearly unsatisfactory, the design is not protected against the vagaries of the transistor ß.

If the top of $R_b$ was moved from the supply rail to the collector of the transistor, as shown in **Fig 3.38**, then an increase in collector current (due to ß being higher than expected) would cause a fall in collector voltage and a consequent drop in the voltage across $R_b$. This would reduce the base current into the transistor, offsetting the rise in collector current. The circuit is reasonably stable against variations in gain and is sometimes used although $R_b$ does feed some of the output signal back to the base, which, being inverted or in anti-phase, does reduce the signal gain.

The method of bias stability in Fig 3.33 is better. If the collector current is higher than calculated, perhaps because ß is greater, the potential difference across R3 will rise. However the base voltage is held constant by R1 and R2, so the base-emitter voltage $V_{BE}$ will fall. Reference to Fig 3.31 will show that a reduction in $V_{BE}$ has a marked effect in reducing $I_C$ quickly offsetting the assumed rise.

## Bias Classes

The choice of base and collector voltages is normally such that an input signal is amplified without the distortion shown in Fig 3.32. However there are occasions where this is not intended. Take, for example, the circuit in **Fig 3.39**. The two output transistors may be biased just into conduction, drawing a modest quiescent current. If the signal voltage at the collector of the input transistor rises, then the $V_{BE}$ on transistor 1 will tend to increase and transistor 1 will conduct, amplifying the signal. The $V_{BE}$ on transistor 2 will reduce and it will not conduct.

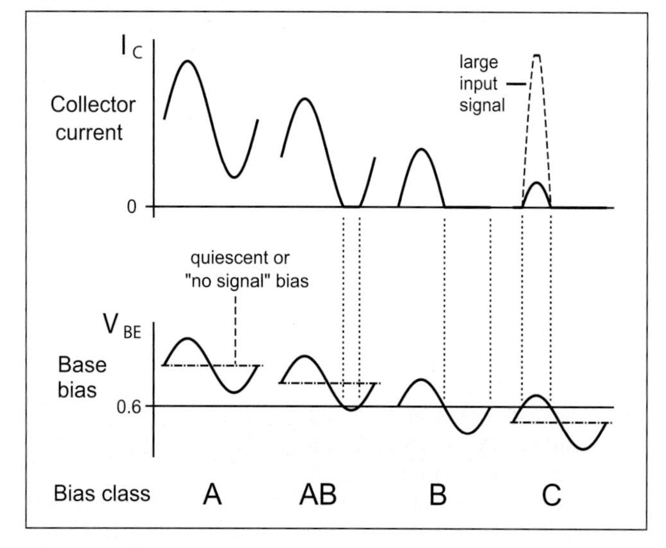


**Fig 3.40: Classes of bias relate to the proportion of the signal waveform for which collector current flows**

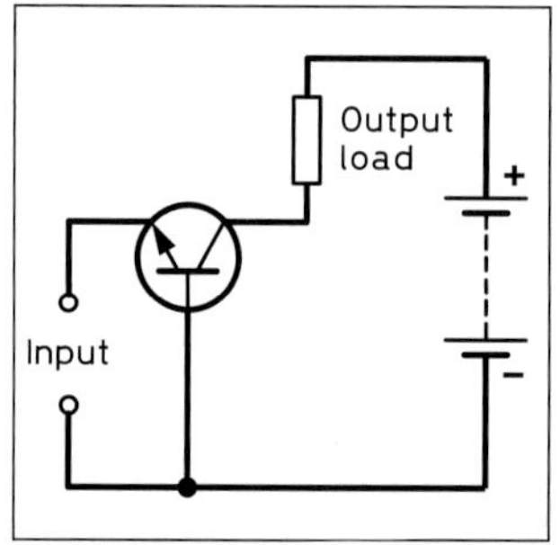


**Fig 3.41: The common-base configuration**

Conversely, if the voltage at the collector of the input transistor falls then transistor 1 will not conduct and transistor 2 will amplify the signal.

The purpose of this arrangement, know as a push-pull amplifier, is that it allows the quiescent current in the output transistors to be very low. Current only flows when the transistor is actually amplifying its portion of the signal. By this means the heat dissipation is minimised and a much higher output power can be realised than would be achieved normally. This is also desirable in battery-powered devices to prolong battery life.

The 'normal' bias of Fig 3.33 is known as Class A bias, where collector current flows over the entire cycle of the input signal waveform. The push-pull amplifier has Class B bias, where collector current flows over half the input signal waveform. Class AB bias denotes biasing such that collector current flows for more than half but less than the whole cycle of the input signal. Class C bias tells us that the collector current is flowing for less than half the input waveform. This is illustrated in **Fig 3.40**.

The distortion of class B bias is avoided simply by the two transistors sharing the task. In reality there is a small overlap between the two transistors to minimise any cross-over distortion where conduction swaps from one transistor to the other. It could be argued that, in reality, the transistors are biased in class AB, but close to class B. Some schools of thought use the notation AB1 and AB2 to denote closer to class A and closer to class B respectively. The design issue is distortion versus standing current in the transistor. The standing current having implications for heat dissipation/power handling and, especially in battery powered devices, current drain and battery life.

No such option is available for class C bias and distortion will occur. Class C can only be used at radio frequencies where tuned circuits and filters can be used. It is, however highly efficient, up to 66% or 2/3. That is to say 2/3 of the DC supply is transformed into RF signal. Modern RF semiconductors operating in Class-C can exhibit efficiency figures of up to 80% in narrow band circuits.

It will be recalled that any form of waveform or harmonic distortion introduces harmonic frequencies into the signal. Filtering and tuned circuits can remove the harmonics, thereby removing the distortion. This enables a transistor to produce even higher output powers, but, as will be seen in later chapters, the technique is not useable with any form of amplitude modulation of the input signal although a class C RF power amplifier stage can itself be amplitude modulated using collector or anode modulation. This is a technique where the RF input drive is unmodulated and the audio modulating signal is superimposed on the DC supply to the RF power amplifier transistor.

## Transistor Configurations

The transistor has been used so far with its emitter connected to the 0V rail and the input to the base. This is not the only way of connecting and using a transistor. The important things to remember are that the transistor is a current driven device, which responds to base current and the potential difference $V_{BE}$. The output is a variation in collector current, which is normally translated into a voltage by passing that current through a collector resistor. It must also be noted that $I_E=I_B+I_C$ and that $I_C>>I_B$.

The common-emitter configuration, already seen in Fig 3.33, is usually preferred because it provides both current and voltage

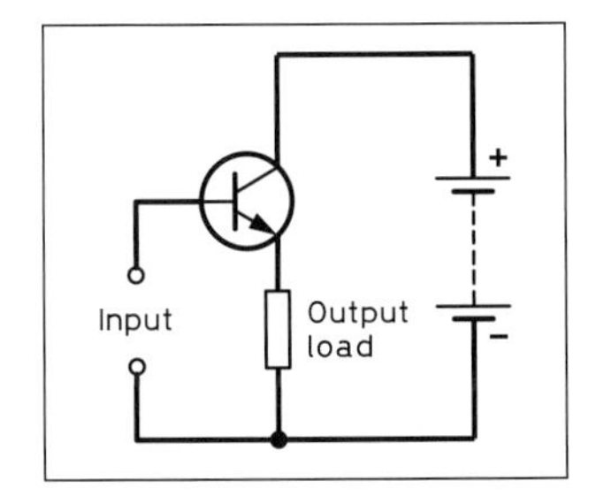


Fig 3.42: Common-collector, or emitter-follower, configuration

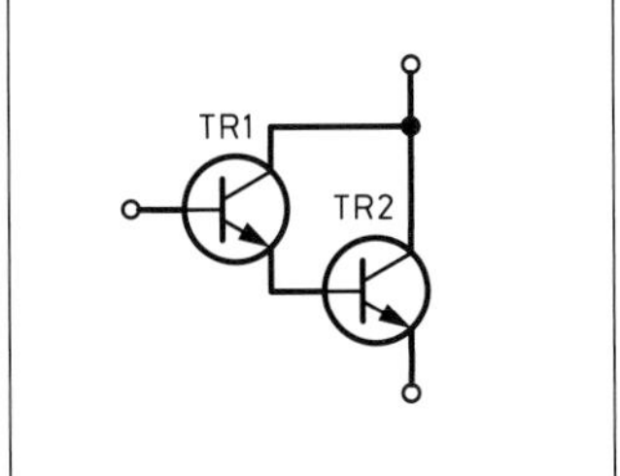


Fig 3.43: A Darlington pair is the equivalent of a single transistor with extremely high current gain

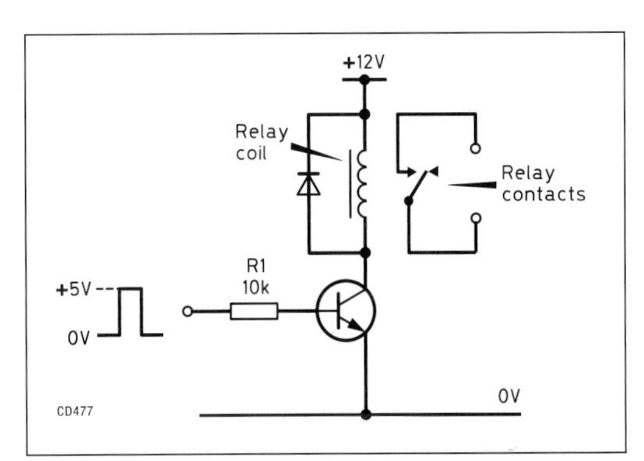


Fig 3.44: A transistor used to switch a relay

gain. The input impedance of a common-emitter amplifier using a small, general-purpose transistor will be around 2kΩ at audio frequencies but, assuming that the emitter resistor is bypassed with a capacitor, this will drop to approximately 100Ω at a frequency of a few megahertz. The output impedance will be in the region of 10kΩ. The chapter on Building Blocks provides more information on the design of common-emitter amplifiers.

## Common-base configuration

The common-base configuration is shown in **Fig 3.41**. The input and output coupling and bias circuitry has been omitted for the sake of clarity. As the emitter current is the sum of the collector and base currents, the current gain will be very slightly less than unity. There is, however, considerable voltage gain. The input impedance is very low, typically between 10 and 20Ω, assuming a collector current of 2mA. The output impedance is much higher, perhaps 1MΩ at audio frequencies. The output voltage will be in phase with the input. The common-base configuration will sometimes be used to provide amplification where the transistor must operate at a frequency close to its $f_T$.

## Common-collector configuration

The common-collector configuration is normally referred to as the emitter follower, shown in **Fig 3.42**. This circuit provides a current gain equal to the transistor's ß, but the voltage gain is very slightly less than unity. The input impedance is much higher than for the common-emitter configuration and may be approximated by multiplying the transistor's ß by the value of the emitter load impedance. The output impedance, which is much lower, is normally calculated with reference to the impedance of the circuitry which drives the follower, and is approximated by:

$$\text{Output impedance} = \frac{Z_{out}}{\beta}$$

where $Z_{out}$ is the output impedance of the circuit driving the emitter follower.

For example, if the transistor used has a ß of 100 at the frequency of operation and the emitter follower is driven by circuitry with an output impedance ($Z_0$) of 2kΩ, the output impedance of the follower is roughly 20Ω. As with the common-base circuit, the output voltage is in phase with the input. Emitter followers are used extensively as 'buffers' in order to obtain impedance transformation, and isolation, between stages (see the later chapters).

## The Darlington pair

**Fig 3.43** shows how two transistors may be connected to produce the equivalent of a single transistor with extremely high current gain (ß). A current flowing into the base of TR1 will cause a much larger current to flow into the base of TR2. TR2 then provides further current gain. Not surprisingly, the overall current gain of the Darlington pair is calculated by multiplying the ß of TR1 by the ß of TR2. Therefore, if each transistor has a ß of 100, the resultant current gain is 10,000. As an alternative to physically connecting two separate devices, it is possible to obtain 'Darlington transistors'. These contain a pair of transistors, plus the appropriate interconnections, fabricated onto a single chip.

## The transistor as a switch

There is often a requirement in electronic equipment to activate relays, solenoids, electric motors and indicator devices etc using control signals generated by circuitry that cannot directly power the device which must be turned on or off. The transistor in **Fig 3.44** solves such a problem by providing an interface between the source of a control voltage (+5V in this example) and a 12V relay coil. If the control voltage is absent, only a minute leakage current flows between the collector and emitter of the switching transistor and so the relay is not energised. When the control voltage is applied, the transistor draws base current through R1 and this results in a larger current flow between its collector and emitter. The relay coil is designed to be connected directly across a potential of 12V and so the resistance between the collector and emitter of the transistor must be reduced to the lowest possible value. The desired effect is therefore the same as that which might otherwise be obtained by closing a pair of switch contacts wired in series with the relay coil.

Assuming that R1 allows sufficient base current to flow, the transistor will be switched on to the fullest extent (a state referred to as saturation). Under these conditions, the voltage at the collector of the transistor will drop to only a few tenths of a volt, thus allowing a potential almost equal to that of the supply rail (12V) to appear across the relay coil, which is therefore properly energised. Under these conditions the transistor will not dissipate much power because, although it is passing considerable current, there is hardly any resistance between the emitter and collector. Assuming that the relay coil draws 30mA when energised from a 12V supply, and that the ß of the transistor is 150 at a collector current of this value, the base current will be:

$$\frac{30 \times 10^{-3}}{150} = 200\mu A$$

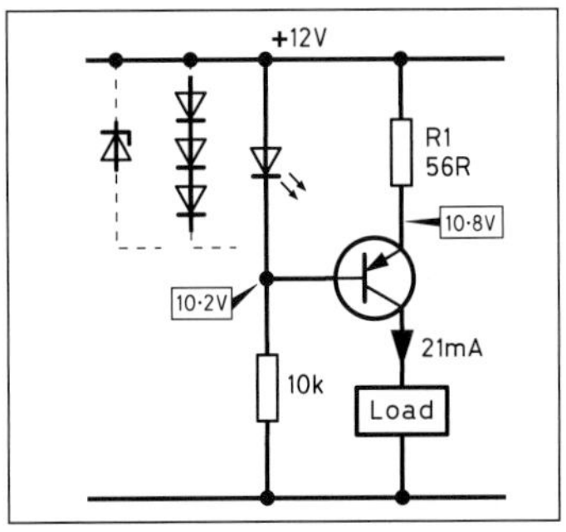

Fig 3.45: The constant-current generator

Using Ohm's Law, this suggests that the current limiting resistor R1 should have a value of around 25kΩ. In practice, however, a lower value of 10kΩ would probably be chosen in order to make absolutely sure that the transistor is driven into saturation. The diode connected across the relay coil, which is normally

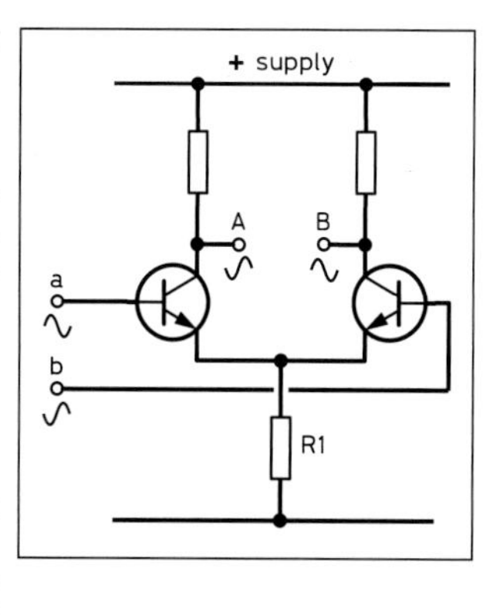


Fig 3.46: A differential amplifier or long-tailed pair

reverse-biased, protects the transistor from high voltages by absorbing the coil's back EMF on switching off.

### Constant-current generator

**Fig 3.45** shows a circuit that will sink a fixed, predetermined current into a load of varying resistance. A constant-current battery charger is an example of a practical application which might use such a circuit. Also, certain low-distortion amplifiers will employ constant-current generators, rather than resistors, to act as collector loads. The base of the PNP transistor is held at a potential of 10·2V by the forward voltage drop of the LED (12 - 1·8 = 10·2V). Note that although LEDs are normally employed as indicators, they are also sometimes used as reference voltage generators. Allowing for the base-emitter voltage drop of the transistor (approximately 0·6V), the emitter voltage is 10·8V. This means that the potential difference across R1 will be held at 1·2V. Using Ohm's Law, the current flowing through R1 is:

$$\frac{1\cdot 2}{56} = 0\cdot 021\text{A} \quad \text{or } 21\text{mA}$$

The emitter current, and also the collector current, will therefore be 21mA. Should the load attempt to draw a higher current, the voltage across R1 will try and rise. As the base is held at a constant voltage, it is the base-emitter voltage that must drop, which in turn prevents the transistor passing more current.

Also shown in Fig 3.45 are two other ways of generating a reasonably constant reference voltage. A series combination of three forward-biased silicon diodes will provide a voltage drop similar to that obtained from the LED (3 x 0·6 = 1.8V). Alternatively, a zener diode could be used, although as the lowest voltage zener commonly available is 2·7V, the value of R1 must be recalculated to take account of the higher potential difference.

### The long-tailed pair

The long-tailed pair, or differential amplifier, employs two identical transistors which share a common emitter resistor. Rather than amplifying the voltage applied to a single input, this circuit provides an output which is proportional to the difference between the voltages presented to its two inputs, labelled 'a' and 'b' in **Fig 3.46**.

Providing that the transistors are well matched, variations in $V_{BE}$ and $\beta$ will cause identical changes in the potential at the two outputs, A and B. Therefore, if both A and B are used, it is the voltage difference between them that constitutes the wanted output. The

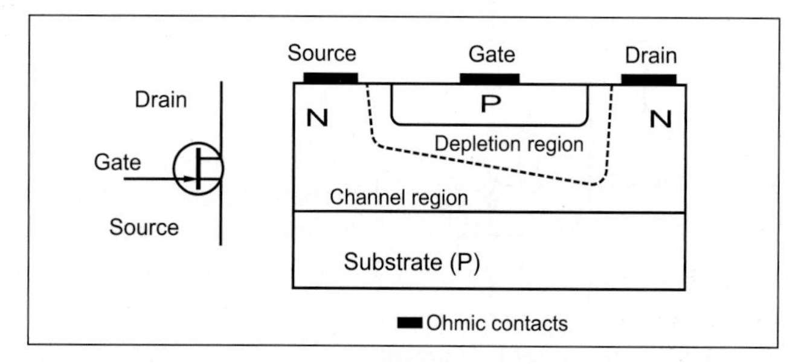


Fig 3.47: The circuit symbol and construction of the JFET

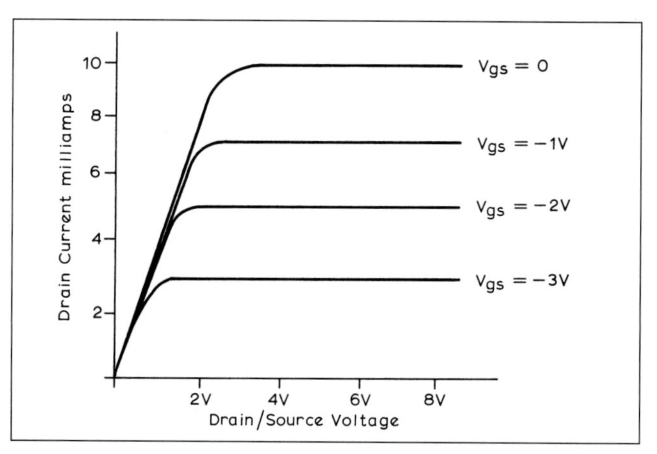


Fig 3.48: The relationship between gate voltage and drain current for a JFET

long-tailed pair is very useful as an amplifier of DC potentials, an application where input and output coupling capacitors cannot be used. The circuit can be improved by replacing the emitter resistor (R1) with a constant-current generator, often referred to in this context as a current source. The differential amplifier is used extensively as an input stage in operational amplifiers.

## JUNCTION FIELD EFFECT TRANSISTOR

Like its bipolar counterpart, the junction field effect transistor (JFET) is a three-terminal device which can be used to provide amplification over a wide range of frequencies. **Fig 3.47** shows the circuit symbol and construction of an N-channel JFET. This device differs from the bipolar transistor in that current flow between its drain and source connections is controlled by a voltage applied to the gate. The current flowing to the gate terminal is very small, as it is the reverse bias leakage current of a PN junction. Thus the input (gate) impedance of a JFET is very high, and the concept of current gain, as applied to bipolar devices, is meaningless. The JFET may be referred to as a unipolar device, since the current flow is by majority carriers only, ie electrons in an N-channel transistor. In such a device holes (which have lower mobility than electrons) have no part in the current transport process, thus offering the possibility of very good high-frequency performance.

If both the gate and source are at ground potential ($V_{GS}$=0), and a positive voltage is applied to the drain ($V_{DS}$), electrons will flow from source to drain through the N-type channel region (known as the drain-source current, $I_{DS}$). If $V_{DS}$ is small, the gate junction depletion region width will remain practically independent of $V_{DS}$, and the channel will act as a resistor. As $V_{DS}$ is increased, the reverse bias of the gate junction near the drain increases, and the average cross-sectional area for current flow is reduced, thus increasing the channel resistance. Eventually $V_{DS}$ is large enough to cause the gate depletion region to expand to fill the channel at the drain end, thus separating the source from the drain. This condition is known as pinch-off. It is important to note that current ($I_{DS}$) continues to flow across this depletion region, since carriers are injected into it from the channel and accelerated across it by the electric field present. If $V_{DS}$ is increased beyond pinch-off, the edge of the depletion region will move along the channel towards the source. However the voltage drop along the channel between the source and the edge of the depletion region will remain fixed, and thus the drain-source current will remain essentially unchanged. This current saturation is evident from the JFET current-voltage characteristics shown in **Fig 3.48** ($V_{GS}$=0).

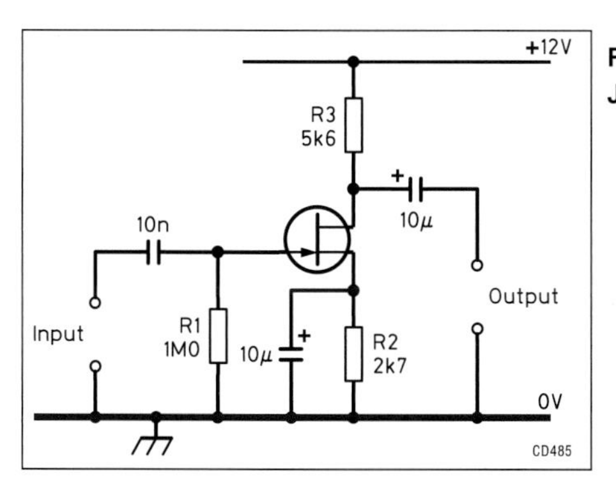


**Fig 3.49: A JFET amplifier**

A negative voltage applied to the P-type gate ($V_{GS}$) establishes a reverse-biased depletion region intruding into the N-type channel (see Fig 3.47). Thus for small values of of $V_{DS}$ the channel will again act as a resistor, however its resistance will be larger than when $V_{GS}$=0 because the cross-sectional area of the channel has decreased due to the wider depletion region. As $V_{DS}$ is increased the gate depletion region again expands to fill the channel at the drain end, causing pinch-off and current saturation. However, the applied gate voltage reduces the drain voltage required for the onset of pinch-off, thus also reducing the saturation current. This is evident from Fig 3.48 for $V_{GS}$ = -1V, -2V and -3V. Furthermore, Fig 3.48 also suggests that a varying signal voltage applied to the gate will cause proportional variations in drain current. The gain, or transconductance ($G_m$ or $Y_{fs}$), of a field effect device is expressed in siemens (see Chapter 1). A small general-purpose JFET will have a $G_m$ of around 5 milli-siemens.

The circuit of a small-signal amplifier using a JFET is shown in **Fig 3.49**. The potential difference across R2 provides bias by establishing a positive voltage at the source, which has the same effect as making the gate negative with respect to the source. R1 serves to tie the gate at ground potential (0V), and in practice its value will determine the amplifier's input impedance at audio frequencies. The inherently high input impedance of the JFET amplifier is essentially a result of the source-gate junction being reverse-biased. If the gate were to be made positive with respect to the source (a condition normally to be avoided), gate current would indeed flow, thus destroying the field effect. The value of R3, the drain load resistor ($R_L$), dictates the voltage gain obtained for a particular device transconductance (assumed to be 4 millisiemens in this case) as follows:

$$\begin{aligned}\text{Voltage gain} &= G_m \times R_L \\ &= 4\times10^{-3} \times 5\cdot6\times10^{3} \\ &= 4\times5\cdot6 \\ &= 22\cdot4 \quad \text{or } 27\text{dB}\end{aligned}$$

The voltage gain obtainable from a common-source JFET amplifier is therefore around 20dB, or a factor of 10, lower than that provided by the equivalent common-emitter bipolar amplifier. Also, the characteristics of general-purpose JFETs are subject to considerable variation, or 'spread', a fact that may cause problems in selecting the correct value of bias resistor for a particular device. However, the JFET does offer the advantage of high input impedance, and this is exploited in the design of stable, variable frequency oscillators. JFETs are also employed in certain types of RF amplifier and switching mixer.

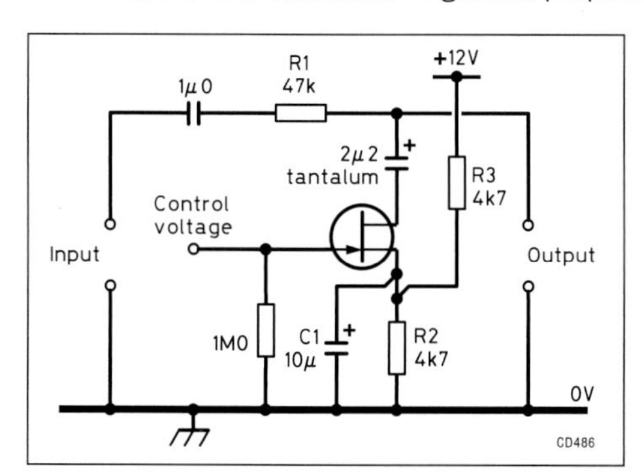


**Fig 3.50: The JFET may be used as a voltage-controlled attenuator**

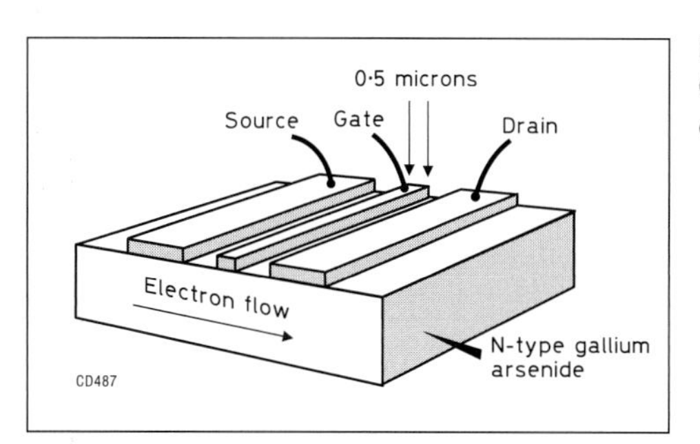


**Fig 3.51: Construction of a GaAsFET**

The JFET is often used as a voltage-controlled variable resistor in signal gates and attenuators. The channel of the JFET in **Fig 3.50** forms a potential divider working in conjunction with R1. Here R2 and R3 develop a bias voltage which is sufficient to ensure pinch-off. This means that the JFET exhibits a very high resistance between the source and drain and so, providing that the following stage also has a high input impedance, say at least five times greater than the value of R1, the signal will suffer practically no attenuation. Conversely, if a positive voltage is applied to the gate sufficient to overcome the effect of the bias, the channel resistance will drop to the lowest possible value - typically 400Ω for a small-signal JFET. The signal will now be attenuated by a factor nearly equal to the ratio between R1 and the channel resistance - 118, or 41dB (note that C1 serves to bypass R2 at signal frequencies). The circuit is not limited to operation at these two extremes, however, and it is possible to achieve the effect of a variable resistor by adjusting the gate voltage to achieve intermediate values of channel resistance.

## GaAsFETs

Although field effect devices are generally fabricated from silicon, it is also possible to use gallium arsenide. GaAsFETs (gallium arsenide field effect transistors) are N-channel field effect transistors designed to exploit the higher electron mobility provided by gallium arsenide (GaAs). The gate terminal differs from that of the standard silicon JFET in that it is made from gold, which is bonded to the top surface of the GaAs channel region (see **Fig 3.51**). The gate is therefore a Schottky barrier junction, as used in the hot-carrier diode. Good high-frequency performance is achieved by minimising the electron transit time between the source and drain. This is achieved by reducing the source drain spacing to around 5 microns and making the gate from a strip of gold only 0·5 microns wide (note that this critical measurement is normally referred to as the gate length because it is the dimension running parallel to the electron flow).

The very small gate is particularly delicate, and it is therefore essential to operate GaAsFETs with sufficient negative bias to ensure that the gate source junction never becomes forward biased. Protection against static discharge and supply line transients is also important.

GaAsFETs are found in very-low-noise receive preamplifiers operating at UHF and microwave frequencies up to around 20GHz. They can also be used in power amplifiers for microwave transmitters.

## MOSFETs

The MOSFET (metal oxide field effect transistor), also known as the IGFET (insulated gate field effect transistor), is a very important device, with applications ranging from low-noise preamplifi-

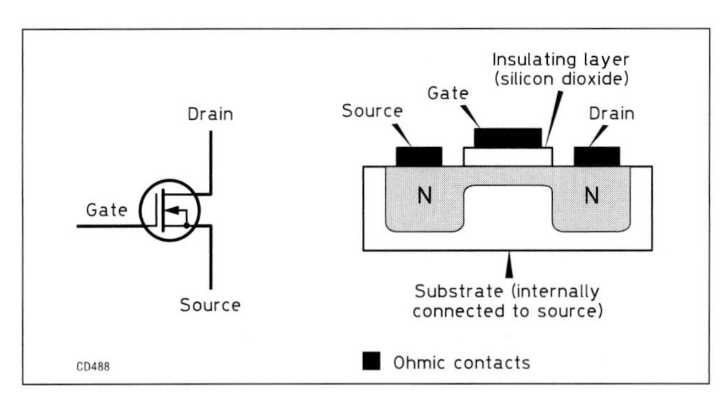


**Fig 3.52: Circuit symbol and method of construction of a depletion-mode N-channel MOSFET (metal oxide semiconductor field effect transistor). The circuit symbol for the P-channel type is the same, except that the direction of the arrow is reversed**

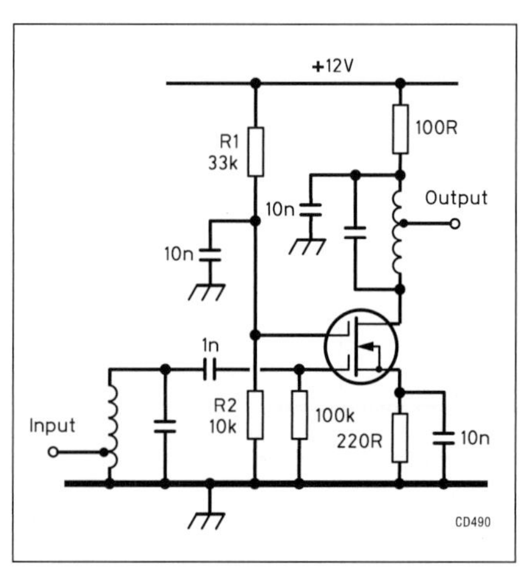


**Fig 3.54: An RF amplifier using a dual-gate MOSFET**

cation at microwave frequencies to high-power amplifiers in HF and VHF transmitters. Ultra large scale integrated circuits (ULSICs), including microprocessors and memories, also make extensive use of MOS transistors.

The MOSFET differs from the JFET in having an insulating layer, normally composed of silicon dioxide ($SiO_2$), interposed between the gate electrode and the silicon channel. The ability to readily grow a high quality insulating material on silicon is a key reason for the dominance of silicon device technology. The source and drain are formed by diffusion into the silicon substrate. This insulation prevents current flowing into, or out of, the gate, which makes the MOSFET easier to bias and guarantees an extremely high input resistance. The insulating layer acts as a dielectric, with the gate electrode forming one plate of a capacitor. Gate capacitance depends on the area of the gate, and its general effect is to lower the impedance seen at the gate as frequency rises. The main disadvantage of this structure is that the very thin insulating layer can be punctured by high voltages appearing on the gate. Therefore, in order to protect these devices against destruction by static discharges, internal zener diodes are normally incorporated. Unfortunately, the protection provided by the zener diodes is not absolute, and all MOS devices should therefore be handled with care.

MOSFETs have either N-type or P-type channel regions, conduction being provided by electrons in N-channel devices, and holes in P-channel devices. However, as electrons have greater mobility than holes, the N-channel device is often favoured because it promises better high-frequency performance.

The current-voltage characteristics of MOSFETs are analagous in form to JFETs. For low values of drain-source voltage ($V_{DS}$) the channel acts as a resistor, with the source-drain current ($I_{DS}$) proportional to $V_{DS}$. As $V_{DS}$ increases, pinch-off eventually occurs, and beyond this point $I_{DS}$ remains essentially constant.

The gate-source voltage ($V_{GS}$) controls current flow in the channel by causing carrier accumulation, carrier depletion or carrier inversion at the silicon channel surface. In accumulation, $V_{GS}$ causes an enhanced concentration of majority carriers (eg electrons in N-type silicon), and in depletion the majority carrier concentration is reduced. In inversion the applied gate voltage is sufficient to cause the number of minority carriers at the silicon surface (ie holes in N-type silicon) to exceed the number of majority carriers. This means that, for example, an N-type silicon surface becomes effectively P-type. The value of $V_{GS}$ required to invert the silicon surface is known as the threshold voltage.

There are basically two types of N-channel MOSFETs. If at $V_{GS}$=0 the channel resistance is very high, and a positive gate threshold voltage is required to form the N-channel (thus turning the transistor on), then the device is an enhancement-mode (normally off) MOSFET. If an N-channel exists at $V_{GS}$=0, and a negative gate threshold voltage is required to invert the channel surface in order to turn the transistor off, then the device is a depletion-mode (normally on) MOSFET. Similarly there are both enhancement and depletion-mode P-channel devices. Adding further to the variety of MOSFETs available, there are also dual-gate types.

**Fig 3.52** shows the circuit symbol and construction for a single-gate MOSFET, and **Fig 3.53** features the dual-gate equivalent. Dual-gate MOSFETs perform well as RF and IF amplifiers in receivers. They contribute little noise and provide good dynamic range. Transconductance is also higher than that offered by the JFET, typically between 7 and 15 millisiemens for a general-purpose device.

**Fig 3.54** shows the circuit of an RF amplifier using a dual-gate MOSFET. The signal is presented to gate 1, and bias is applied separately to gate 2 by the potential divider comprising R1 and R2. Selectivity is provided by the tuned circuits at the input and output. Care must be taken in the layout of such circuits to prevent instability and oscillation. A useful feature of the dual-gate amplifier is the

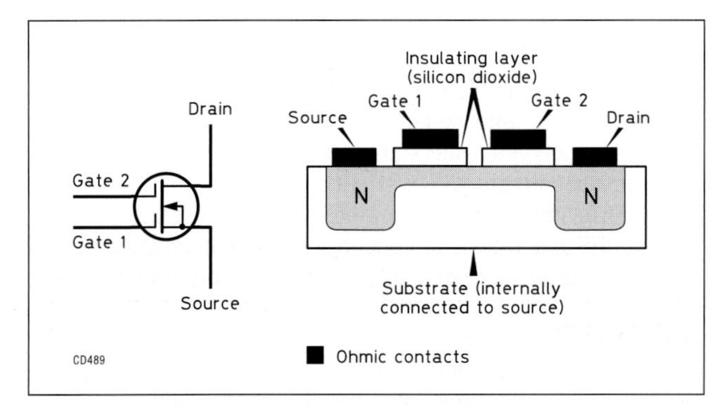


**Fig 3.53: A depletion-mode dual-gate MOSFET**

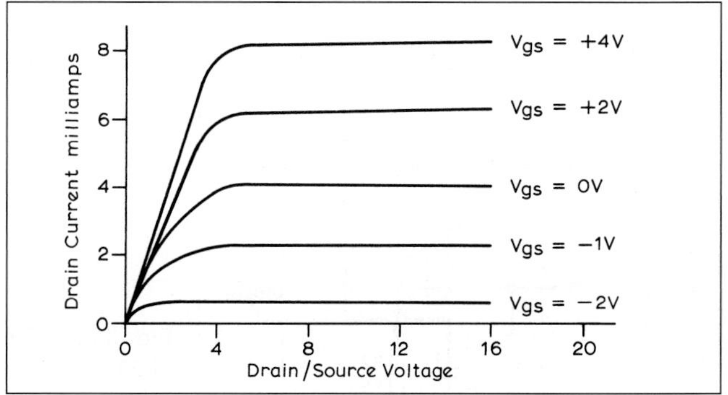


**Fig 3.55: The relationship between gate voltage and drain current for an N-channel depletion mode MOSFET**

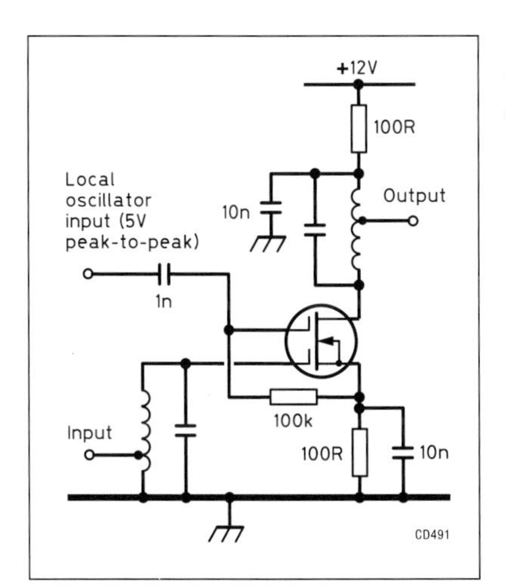


Fig 3.56: A mixer circuit based on a dual-gate MOSFET

ability to control its gain by varying the level of the gate 2 bias voltage. This is particularly useful in IF amplifiers, where the AGC voltage is often applied to gate 2. The characteristic curve at **Fig 3.55** shows the effect on drain current of making the bias voltage either negative or positive with respect to the source.

Dual-gate MOSFETs may also be used as mixers. In **Fig 3.56**, the signal is applied to gate 1 and the local oscillator (LO) drive to gate 2. There is a useful degree of isolation between the two gates, and this helps reduce the level of oscillator voltage fed back to the mixer input. For best performance, the LO voltage must be sufficient to turn the MOSFET completely off and on, so that the mixer operates in switching mode. This requires an LO drive of around 5V peak to peak. However, as the gate impedance is high, very little power is required.

## Power MOSFETs

As already discussed, MOSFETs exhibit very low gate leakage current, which means that they do not require complex input drive circuitry compared with bipolar devices. In addition, unipolar MOSFETs have a faster switching speed than bipolar transistors. These features make the MOSFET an attractive candidate for power device applications. The VMOS™ (vertical metal oxide semiconductor), type of power MOSFET, is constructed in such a way that current flows vertically between the drain, which forms the bottom of the device, to a source terminal at the top (see **Fig 3.57**). The gate occupies either a V- or U-shaped groove etched into the upper surface. VMOS devices feature a four-layer sandwich comprising N+, P, N- and N+ material and operate in the enhancement mode. The vertical construction produces a rugged device capable of passing considerable drain current and offering a very high switching speed. These qualities are exploited in power control circuits and transmitter output stages. **Fig**

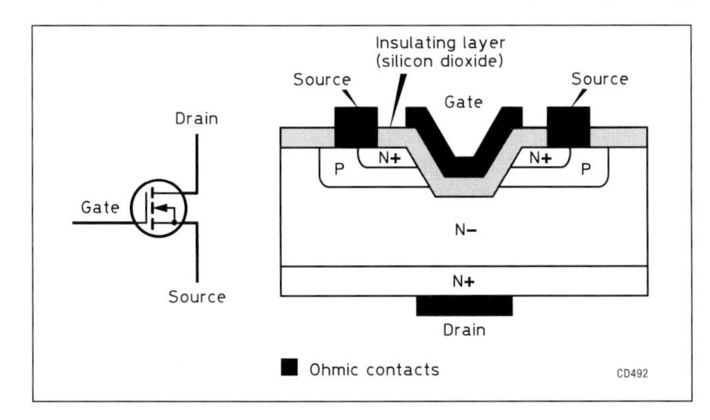


Fig 3.57: Circuit symbol and method of construction of a VMOS transistor

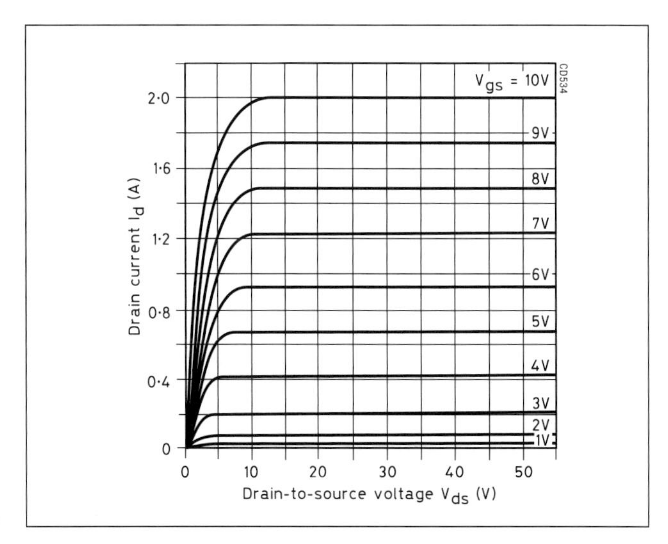


Fig 3.58: Relationship between gate voltage and drain current for a VMOS power transistor

**3.58** shows the characteristic curves of a typical VMOS transistor. Note that the drain current is controlled almost entirely by the gate voltage, irrespective of drain voltage. Also, above a certain value of gate voltage, the relationship between gate voltage and drain current is highly linear.

Although the resistance of the insulated gate is for all intents and purposes infinite, the large gate area leads to high capacitance. A VMOS transistor intended for RF and high-speed switching use will have a gate capacitance of around 50pF, whereas devices made primarily for audio applications have gate capacitances as high as 1nF.

A useful feature of these devices is that the relationship between gate voltage and drain current has a negative temperature coefficient of approximately 0·7% per degree Celsius. This means that as the transistor gets hotter, its drain current will tend to fall, thus preventing the thermal runaway which can destroy bipolar power transistors.

**Fig 3.59** shows the circuit of a simple HF linear amplifier using a single VMOS transistor. Forward bias is provided by the potential divider R1, R2 so that the amplifier operates in Class AB. In this circuit R3 and C3 provide a small amount of negative feedback to help prevent instability.

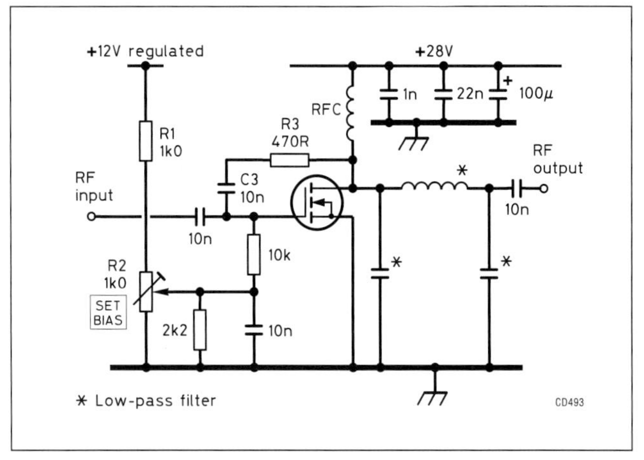


Fig 3.59: A linear amplifier utilising a VMOS power transistor. Assuming a 50Ω output load, the RFC value is chosen so that it has an inductive reactance ($X_L$) of approximately 400Ω at the operating frequency

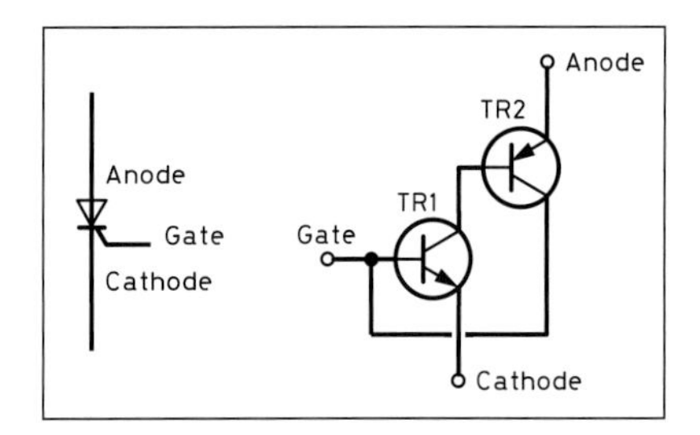


**Fig 3.60: The symbol for a thyristor (silicon controlled rectifier) and its equivalent circuit**

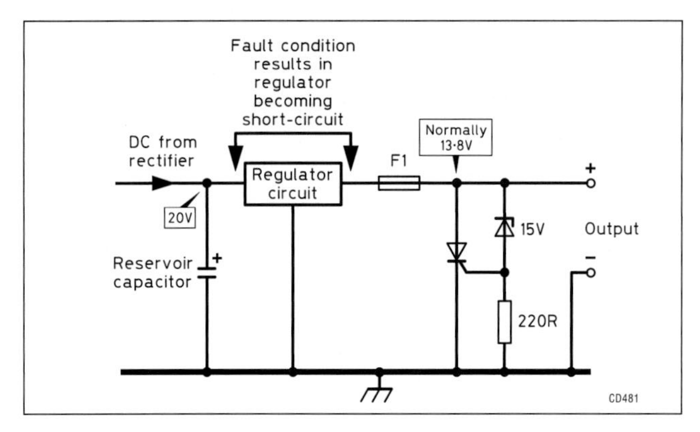


**Fig 3.61: Crowbar over-voltage protection implemented with a thyristor**

More complex high power push-pull amplifiers operating from higher voltage supplies are now available. The most recent high power MOSFETs are designed for DC supply voltages of up to 450V, with a pair of such devices capable of delivering 1.5kW at HF with good IMD performance. A pioneer of this technology is Richard Frey, K4XU, senior design engineer at Advanced Power Technology (Microsemi Corporation acquired APT in May 2006). An advantage of high voltage transistors is that the drain impedance is quite high, typically around 12.5Ω , which means that a straightforward 4:1 transmission line transformer can be used to obtain a wide bandwidth match to 50Ω. However, voltage transients which may cause device failure by breakdown of the drain-source junction must be eliminated. At such high power levels careful attention to heat sinking is also required. An amplifier delivering 1.5kW output with a typical efficiency level of 50% will dissipate 1.5kW in the MOSFETs. Another consideration is that these power MOSFETs are not self-protecting as junction temperature rises. The negative temperature coefficient that is characteristic of VMOS devices is not apparent in higher voltage designs and some protective circuitry is essential to prevent thermal runaway under high power dissipation. This may be achieved by using a thermistor bonded to the heat sink to reduce the gate voltage as the temperature increases.

## THYRISTORS

The thyristor, or silicon controlled rectifier (SCR), is a four-layer PNPN device which has applications in power control and power supply protection systems. The thyristor symbol and its equivalent circuit is shown in **Fig 3.60**. The equivalent circuit consists of two interconnected high-voltage transistors, one NPN and the other PNP. Current flow between the anode and cathode is initiated by applying a positive pulse to the gate terminal, which causes TR1 to conduct. TR2 will now also be switched on because its base is forward biased via TR1's collector. TR1 continues to conduct after the end of the trigger pulse because collector current from TR2 is available to keep its base-emitter junction forward biased. Both transistors have therefore been latched into saturation and will remain so until the voltage between the anode and cathode terminals is reduced to a low value.

**Fig 3.61** shows an over-voltage protection circuit for a power supply unit (PSU) based on a thyristor. In the event of regulator failure, the PSU output voltage rises above the nominal 13·8V, a situation that could result in considerable damage to any equipment that is connected to the PSU. As this happens, the 15V zener diode starts to conduct, and in doing so applies a positive potential to the thyristor gate. Within a few microseconds the thyristor is latched on, and the PSU output is effectively short-circuited. This shorting, or crowbar action, will blow fuse F1 and, hopefully, prevent any further harm.

The thyristor will only conduct in one direction, but there is a related device, called the triac (see **Fig 3.62** for symbol), which effectively consists of two parallel thyristors connected anode to cathode. The triac will therefore switch currents in either direction and is used extensively in AC power control circuits, such as the ubiquitous lamp dimmer. In these applications a trigger circuit varies the proportion of each mains cycle for which the triac conducts, thus controlling the average power supplied to a load. Having been latched on at a predetermined point during the AC cycle, the triac will switch off at the next zero crossing point of the waveform, the process being repeated for each following half cycle.

## INTEGRATED CIRCUITS

Having developed the techniques used in the fabrication of individual semiconductor devices, the next obvious step for the electronics industry was to work towards the manufacture of complete integrated circuits (ICs) on single chips of silicon. Integrated circuits contain both active devices (eg transistors) and passive devices (eg resistors) formed on and within a single semiconductor substrate, and interconnected by a metallisation pattern.

ICs offer significant advantages over discrete device circuits, principally the ability to densely pack enormous numbers of devices on a single silicon chip, thus achieving previously unattainable functionality at low processing cost per chip. Another advantage of integrated, or 'monolithic', construction is that because all the components are fabricated under exactly the same conditions, the operational characteristics of the transistors and diodes, and also the values of resistors, are inherently well matched. The first rudimentary hybrid IC was made in 1958 by Jack Kilby of Texas Instruments, just eight years after the birth of the bipolar junction transistor. Since then, advances in technology have resulted in a dramatic reduction in the minimum device dimension which may be achieved, and today MOSFET gate lengths of only 50nm ($5 \times 10^{-6}$ cm) are possible. This phenomenal rate of progress is set to continue, with 10nm gate lengths expected within the next decade.

The earliest ICs could only contain less than 50 components, but today it is possible to mass-produce ICs containing billions of components on a single chip. For example, a 32-bit microprocessor chip may contain over 42 million components, and a 1Gbit dynamic random access memory (DRAM) chip may contain over 2 billion components.

ICs fall into two broad categories - analogue and digital. Analogue ICs contain circuitry which responds to finite changes in the magnitude of voltages and currents. The most obvious example of an analogue function is amplification. Indeed, virtually all analogue ICs, no matter what their specific purpose may be, contain an amplifier. Conversely, digital ICs respond to only two voltage levels, or states. Transistors

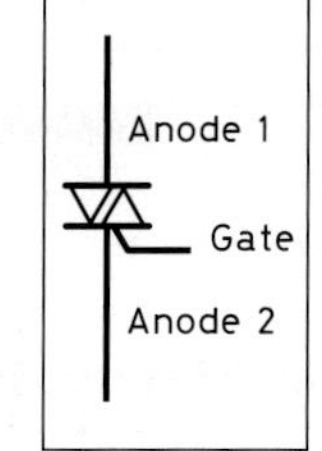


**Fig 3.62: The triac or AC thyristor**

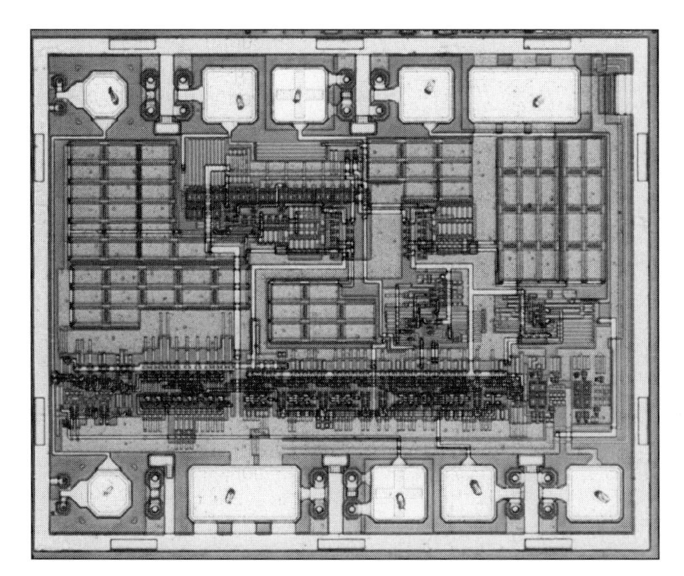

**Fig 3.63: Photograph of the SP8714 integrated circuit chip showing pads for bonding wires (GEC-Plessey Semiconductors)**

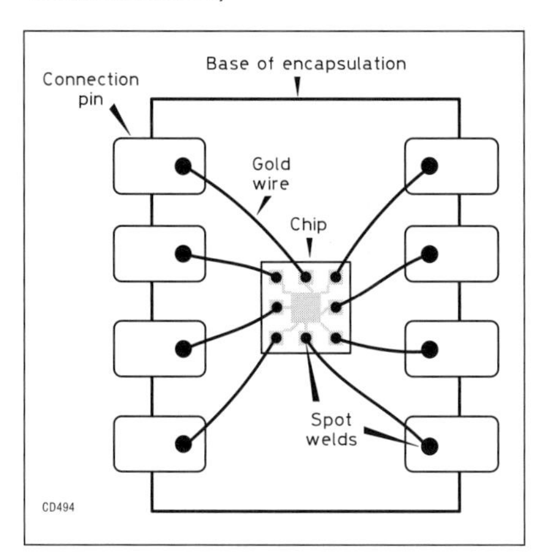


**Fig 3.64: Internal construction of an integrated circuit**

within the IC are normally switched either fully on, or fully off. The two states will typically represent the ones and zeros of binary numbers, and the circuitry performs logical and counting functions.

The main silicon technologies used to build these ICs are bipolar (NPN), N-channel MOSFET, and complementary MOS (CMOS) transistors (incorporating P-channel and N-channel MOSFET pairs). While silicon is by far the dominant material for IC production, ICs based around gallium arsenide MESFETs have also been developed for very high frequency applications.

ICs are produced using a variety of layer growth/deposition, photolithography, etching and doping processes, all carried out under dust-free cleanroom conditions. All chemicals and gases used during processing are purified to remove particulates and ionic contamination. The starting point is a semiconductor wafer, up to 300mm in diameter, which will contain many individual chips. The patterns of conducting tracks and semiconductor junctions are defined by high resolution photolithography and etching. Doping is carried out by diffusion or ion implantation. These processes are repeated a number of times in order to fabricate different patterned layers. Strict process control at each stage minimises yield losses, and ensures that the completed wafer contains a large number of correctly functioning circuits. The wafer is then cut into individual chips and automatically tested to ensure compliance with the design specification. Each selected chip is then fixed to the base of its encapsulation, and very fine gold wires are spot welded between pads located around the chip's periphery and the metal pins which serve as external connections - see **Figs 3.63 and 3.64**. Finally, the top of the encapsulation is bonded to the base, forming a protective seal. Most general-purpose ICs are encapsulated in plastic, but some expensive devices that must operate reliably at high temperatures are housed in ceramic packages.

## LINEAR INTEGRATED CIRCUITS

### Operational Amplifiers

The operational amplifier (op-amp) is a basic building-block IC that can be used in a very wide range of applications requiring low-distortion amplification and buffering. Modern op-amps feature high input impedance, an open-loop gain of at least 90dB (which means that in the absence of gain-reducing negative feedback, the change in output voltage will be at least 30,000 times greater than the change in input voltage), extremely low distortion, and low output impedance. Often two, or even four, separate op-amps will be provided in a single encapsulation. The first operational amplifiers were developed for use in analogue computers and were so named because, with suitable feedback, they can perform mathematical 'operations' such as adding, subtracting, logging, antilogging, differentiating and integrating voltages.

A typical op-amp contains around 20 transistors, a few diodes and perhaps a dozen resistors. The first stage is normally a long-tailed pair and provides two input connections, designated inverting and non-inverting (see the circuit symbol at **Fig 3.65**). The input transistors may be bipolar types, but JFETs or even MOSFETs are also used in some designs in order to obtain very high input impedance. Most op-amps feature a push-pull output stage operating in Class AB which is invariably provided with protection circuitry to guard against short-circuits. The minimum value of output load when operating at maximum supply voltage is normally around 2kΩ, but op-amps capable of driving 500Ω loads are available. Between the input and output circuits there will be one or two stages of voltage amplification. Constant-current generators are used extensively in place of collector load resistors, and also to stabilise the emitter (or source) current of the input long-tailed pair.

In order to obtain an output voltage which is in phase with the input, the non-inverting amplifier circuit shown at **Fig 3.66** is used. Resistors R1 and R2 form a potential divider which feeds a proportion of the output voltage back to the inverting input. In most cases the open-loop gain of an op-amp can be considered

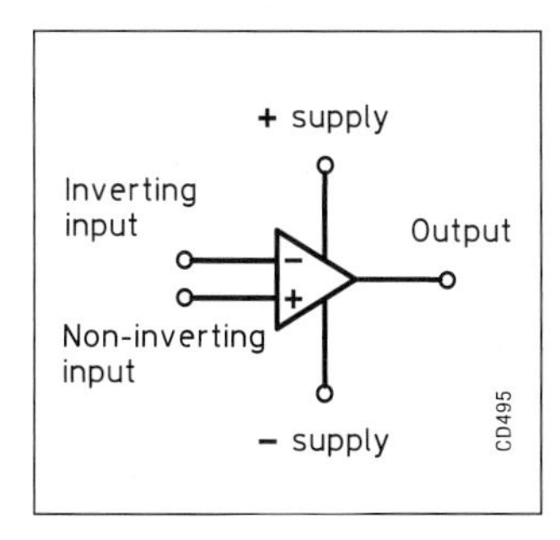

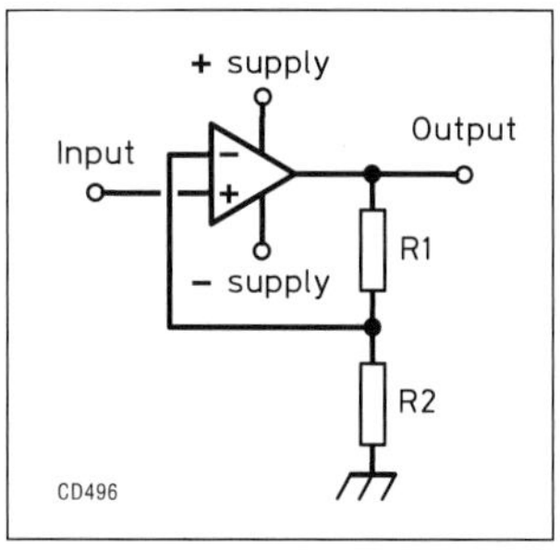

**Fig 3.65: Circuit symbol for an operational amplifier (op-amp)**

**Fig 3.66: A non-inverting amplifier**

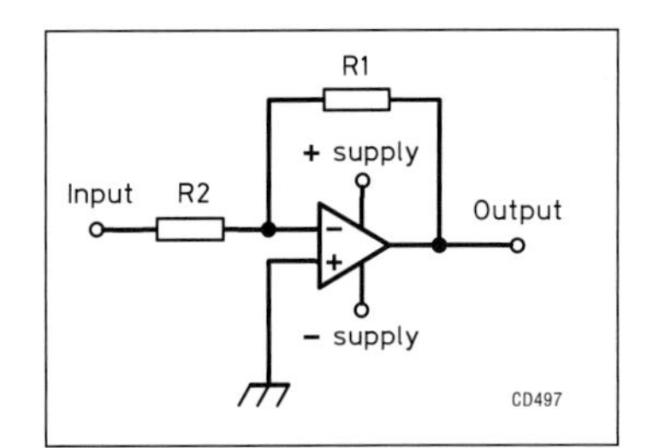


Fig 3.67: An inverting amplifier

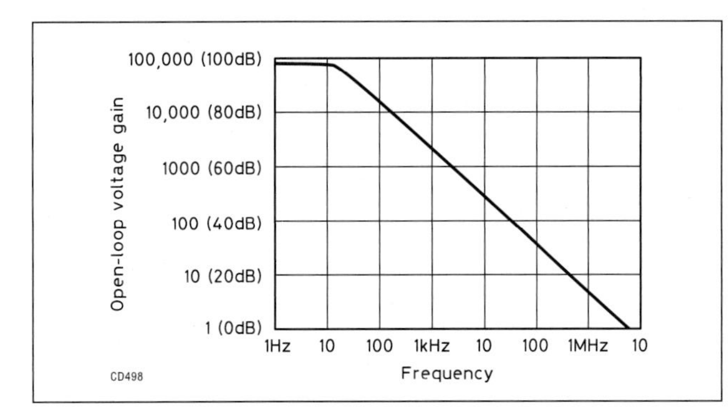


Fig 3.68: The relationship between open loop gain and frequency for a typical internally compensated op-amp

infinite. Making this assumption simplifies the calculation of the closed-loop gain obtained in the presence of the negative feedback provided by R1 and R2. For example, if R1 has a value of 9k and R2 is 1k , the voltage gain will be:

$$\frac{R_1 + R_2}{R_2} = \frac{9000 + 1000}{1000} = 10 \text{ or } 20\text{dB}$$

If R2 is omitted, and R1 replaced by a direct connection between the output and the inverting input, the op-amp will function as a unity gain buffer. **Fig 3.67** shows an inverting amplifier. In this case the phase of the output voltage will be opposite to that of the input. The closed-loop gain is calculated by simply dividing the value of R1 by R2. Therefore, assuming the values are the same as used for the non-inverting amplifier:

$$\frac{R_1}{R_2} = \frac{9000}{1000} = 9 \text{ or } 19\text{dB}$$

At low frequencies, the operation of the negative feedback networks shown in Figs 3.66 and 3.67 is predictable in that the proportion of the output voltage they feed back to the input will be out of phase with the input voltage at the point where the two voltages are combined. However, as the frequency rises, the time taken for signals to travel through the op-amp becomes a significant factor. This delay will introduce a phase shift that increases with rising frequency. Therefore, above a certain frequency - to be precise, where the delay contributes a phase shift of more than 135 degrees - the negative feedback network actually becomes a positive feedback network, and the op-amp will be turned into an oscillator. However, if steps are taken to reduce the open-loop gain of the op-amp to below unity (ie less than 1) at this frequency, oscillation cannot occur. For this reason, most modern op-amps are provided with an internal capacitor which is connected so that it functions as a simple low-pass filter. This measure serves to reduce the open-loop gain of the amplifier by a factor of 6dB per octave (ie for each doubling of frequency the voltage gain drops by a factor of two), and ensures that it falls to unity at a frequency below that at which oscillation might otherwise occur. Op-amps containing such a capacitor are designated as being 'internally compensated'.

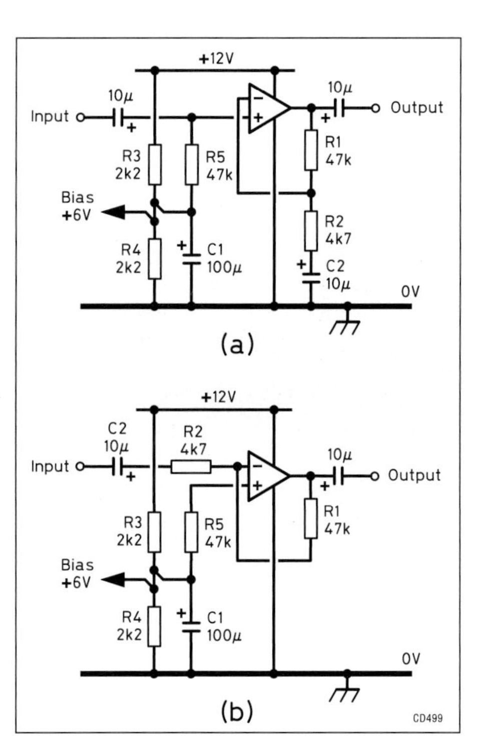


Fig 3.69: Single supply rail versions of the non-inverting (a) and inverting (b) amplifier circuits

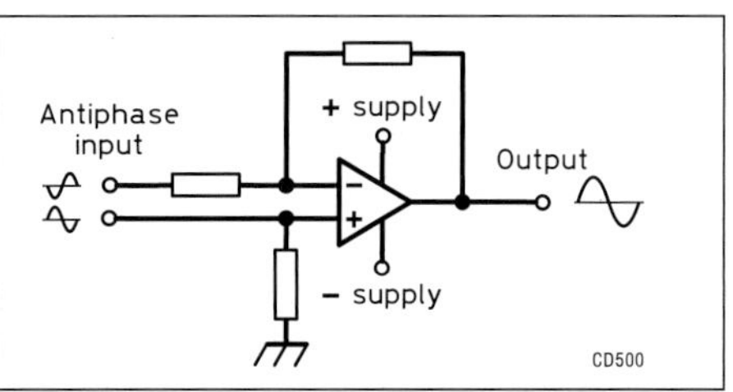


Fig 3.70: A differential amplifier

Compensated op-amps have the advantage of being absolutely stable in normal use. The disadvantage is that the open-loop gain is considerably reduced at high frequencies, as shown by the graph at **Fig 3.68**. This means that general-purpose, internally compensated op-amps are limited to use at frequencies below about 1MHz and they will typically be employed as audio frequency preamplifiers, and in audio filters. There are, however, special high-frequency types available, usually featuring external compensation. An externally compensated op-amp has no internal capacitor, but connections are provided so that the user may add an 'outboard' capacitor of the optimum value for a particular application, thus maximising the gain at high frequencies.

In Figs 3.66 and 3.67 the op-amps are powered from dual-rail supplies. However, in amateur equipment only a single supply rail of around +12V is normally available. Op-amps are quite capable of being operated from such a supply, and **Fig 3.69** shows single-rail versions of the non-inverting and inverting amplifiers with resistor values calculated for slightly different gains.

A mid-rail bias supply is generated using a potential divider (R3 and R4 in each case). The decoupling capacitor C1 enables the bias supply to be used for a number of op-amps (in simpler circuits it is often acceptable to bias the op-amp using a potential divider connected directly to the non-inverting input of the op-amp, in which case C1 is omitted). The value of R5 is normally made the same as R1 in order to minimise the op-amp input current, although this is less important in the case of JFET op-amps due to their exceptionally high input resistance. C2 is incorporated to reduce the gain to unity at DC so that the output voltage will settle to the half-rail potential provided by the bias generator in the absence of signals.

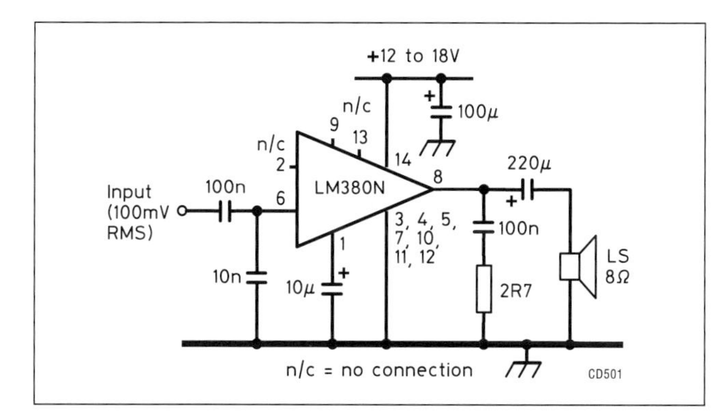


**Fig 3.71: An audio output stage using the LM380N IC. The series combination of the 2.7Ω resistor and the 100nF capacitor at the output form a Zobel network. This improves stability by compensating for the inductive reactance of the loudspeaker voice coil at high frequencies**

**Fig 3.70** shows a differential amplifier where the signal drives both inputs in antiphase. An advantage of this balanced arrangement is that interference, including mains hum or ripple, tends to impress voltages of the same phase at each input and so will be eliminated by cancellation. This ability to reject in-phase, or common-mode, signals when operating differentially is known as the common-mode rejection ratio (CMRR). Even an inexpensive op-amp will provide a CMRR of around 90dB, which means that an in-phase signal would have to be 30,000 times greater in magnitude than a differential signal in order to generate the same output voltage.

Further information on op-amps and their circuits is given in the Building Blocks chapter.

## Audio Power Amplifiers

The audio power IC is basically just an op-amp with larger output transistors. Devices giving power outputs in the range 250mW to 40W are available, the bigger types being housed in encapsulations featuring metal mounting tabs (TO220 for example) that enable the IC to be bolted directly to a heatsink. **Fig 3.71** shows a 1W audio output stage based on an LM380N device.

The LM380N has internal negative feedback resistors which provide a fixed closed loop voltage gain of approximately 30dB. A bias network for single-rail operation is also provided, and so very few external components are required. Not all audio power ICs incorporate the negative feedback and single-rail bias networks 'on-chip', however, and so some, or all, of these components may have to be added externally.

## Voltage Regulators

These devices incorporate a voltage reference generator, error amplifier and series pass transistor. Output short-circuit protection and thermal shut-down circuitry are also normally provided. There are two main types of regulator IC - those that generate a fixed output voltage, and also variable types which enable a potentiometer, or a combination of fixed resistors, to be connected externally in order to set the output voltage as required. Devices capable of delivering maximum currents of between 100mA and 5A are readily available, and fixed types offering a wide variety of both negative and positive output voltages may be obtained. **Fig 3.72** shows a simple mains power supply unit (PSU), based on a type 7812 regulator, providing +12V at 1A maximum.

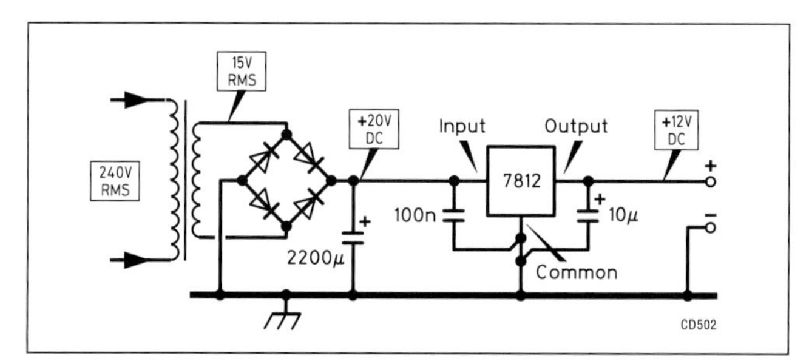


**Fig 3.72: A simple 12V power supply unit using a fixed voltage IC regulator type 7812**

Switched-mode power regulator ICs, which dissipate far less power, are also available. Further information on regulator ICs is given in the chapter on power supplies.

## RF Building Blocks

Although it is now possible to fabricate an entire broadcast receiver on a single chip, this level of integration is rarely possible in amateur and professional communications equipment. To achieve the level of performance demanded, and also provide a high degree of operational flexibility, it is invariably necessary to consider each section of a receiver or transceiver separately, and then apply the appropriate technology to achieve the design goals. In order to facilitate this approach, ICs have been developed which perform specific circuit functions such as mixing, RF amplification and IF amplification.

An example of an IC mixer based on bipolar transistors is the Philips/Signetics NE602AN (featuring very low current consumption as demanded in battery-operated equipment). A useful wide-band RF amplifier is the NE5209D, and there is a similar dual version (ie containing two RF amplifiers within the same encapsulation), the NE5200D. There are also devices offering a higher level of integration - often termed sub-system ICs - which provide more than one block function. For more information on how to use such ICs, see chapters on receivers and transmitters. Amplification at UHF and microwave frequencies calls for special techniques and components. A wide range of devices, known generically as MMICs (microwave monolithic integrated circuits) are available, see the microwaves chapter.

# DIGITAL INTEGRATED CIRCUITS

## Logic Families

In digital engineering, the term 'logic' generally refers to a class of circuits that perform relatively straightforward gating, latching and counting functions. Historically, logic ICs were developed as a replacement for computer circuitry based on large numbers of individual transistors, and, before the development of the bipolar transistor, valves were employed. Today, very complex ICs are available, such as the microprocessor, which contain most of the circuitry required for a complete computer fabricated on to a single chip. It would be wrong to assume, however, that logic ICs are now obsolescent because there are still a great many low-level functions, many of them associated with microprocessor systems, where they are useful. In amateur radio, there are also applications which do not require the processing capability of a digital computer, but nevertheless depend upon logic - the electronic Morse keyer and certain transmit/receive changeover arrangements, for example.

By far the most successful logic family is TTL (transistor, transistor logic) although CMOS devices have been replacing them for some while. Originally developed in the 'sixties, these circuits have been continuously developed in order to provide more complex functions, increase speed of operation and reduce power consumption. Standard (type 7400) TTL requires a 5V power rail stabilised to within ±250mV. Logic level 0 is defined as a voltage between zero and 800mV, whereas logic 1 is defined as 2·4V or higher. **Fig 3.73** shows the circuit of a TTL NAND gate ('NAND' is an abbreviation for 'NOT AND' and refers

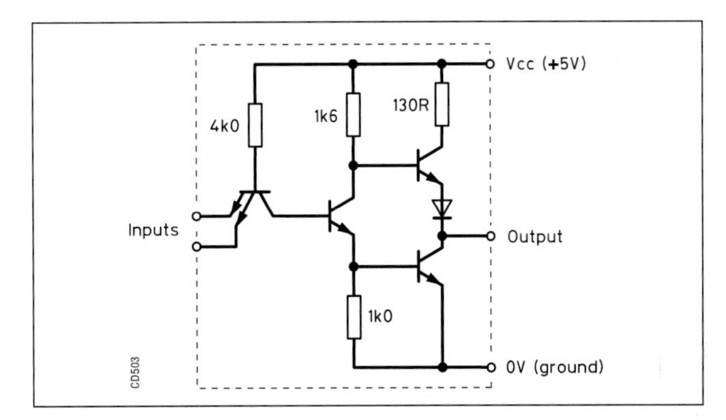


**Fig 3.73: The internal circuit of a standard TTL two-input NAND gate. 74LS (low-power Schottky) TTL logic uses higher-value resistors in order to reduce power consumption, and the circuitry is augmented with clamping diodes to increase the switching speed of the transistors**

to the gate's function, which is to produce an output of logic 0 when both inputs are at logic 1. Really this is an AND function followed by an inverter or NOT function. Note the use of a dual-emitter transistor at the input which functions in the same way as two separate transistors connected in parallel.

One of the first improvements made to TTL was the incorporation of Schottky clamping diodes in order to reduce the turn-off time of the transistors - this enhancement produced the 74S series. Latterly, a low-power consumption version of Schottky TTL, known as 74LS, has become very popular. The latest versions of TTL are designed around CMOS (complementary metal oxide silicon) transistors. The 74HC (high-speed CMOS) series is preferred for general use but the 74HCT type must be employed where it is necessary to use a mixture of CMOS and 74LS circuits to implement a design. A 74HC/HCT counter will operate at frequencies up to 25MHz. The 'complementary' in 'CMOS' refers to the use of gates employing a mixture of N-channel and P-channel MOS transistors.

**Fig 3.74** shows the simplified circuit of a single CMOS inverter. If the input of the gate is held at a potential close to zero volts (logic 0), the P-channel MOSFET will be turned on, reducing its channel resistance to approximately 400Ω, and the N-channel MOSFET is turned off. This establishes a potential very close to the positive supply rail ($V_{dd}$ - logic 1) at the gate's output. An input of logic 1 will have the opposite effect, the N-channel MOSFET being turned on and the P-channel MOSFET turned off.

As one of the transistors is always turned off, and therefore has a channel resistance of about 10,000MΩ, virtually no current flows through the gate under static conditions. However, during transitions from one logic state to another, both transistors will momentarily be turned on at the same time, thus causing a measurable current to flow. The average current consumption will tend to increase as the switching frequency is raised because the gates spend a larger proportion of time in transition between logic levels. The popular CD4000 logic family is based entirely on CMOS technology. These devices can operate with supply voltages from 5 to 15V, and at maximum switching speeds of between 3 and 10MHz.

One of the most useful logic devices is the counter. The simplest form is the binary, or 'divide-by-two' stage shown at **Fig 3.75**. For every two input transitions the counter produces one output transition. Binary counters can be chained together (cascaded) with the output of the first counter connected to the input of the second, and so on. If four such counters are cascaded, there will be one output pulse, or count, for every 16 input pulses. It is also possible to obtain logic circuits which divide by 10 - these are sometimes referred to as BCD (binary coded decimal) counters. Although originally intended to perform the arithmetic function of binary division in computers, the counter can also be used as a frequency divider. For instance, if a single binary counter is driven by a series of pulses which repeat at a frequency of 100kHz, the output frequency will be 50kHz. Counters can therefore be used to generate a range of frequencies that are sub-multiples of a reference input. The crystal calibrator uses this technique, and frequency synthesisers employ a more complex form of counter known as the programmable divider.

Counters that are required to operate at frequencies above 100MHz use a special type of logic known as ECL (emitter coupled logic). ECL counters achieve their speed by restricting the voltage swing between the levels defined as logic 0 and logic 1 to around 1V. The bipolar transistors used in ECL are therefore never driven into saturation (turned fully on), as this would reduce their switching speed due to charge storage effects. ECL logic is used in frequency synthesisers operating in the VHF to microwave range, and also in frequency counters to perform initial division, or prescaling, of the frequency to be measured.

## Memories

There are many applications in radio where it is necessary to store binary data relating to the function of a piece of equipment, or as part of a computer program. These include the spot frequency memory of

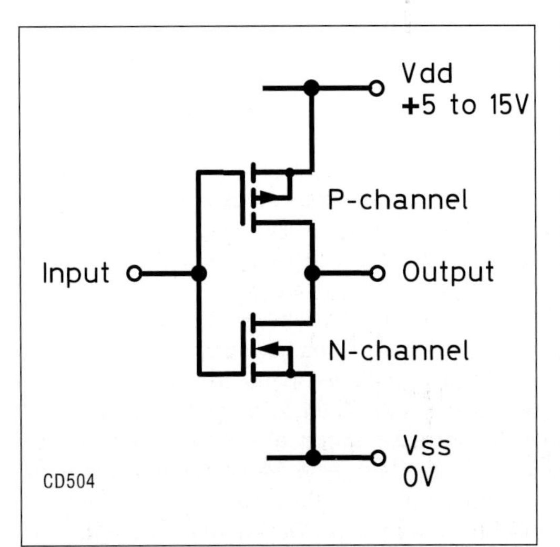


**Fig 3.74: Simplified circuit of a CMOS inverter as used in the CD4000 logic family**

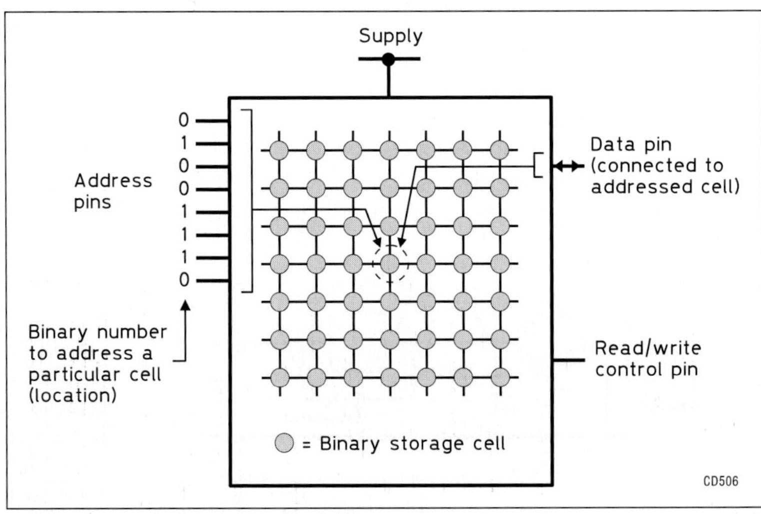


**Fig 3.76: Representation of an IC memory**

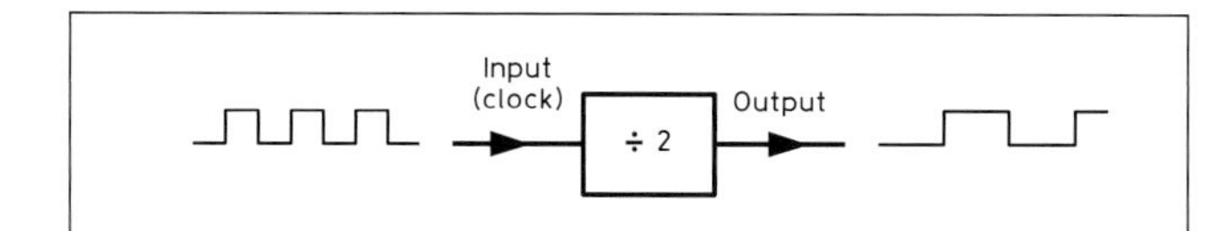


**Fig 3.75: A binary counter may be used as a frequency divider**

a synthesised transceiver, for instance, or the memory within a Morse keyer which is used to repeat previously stored messages.

**Fig 3.76** provides a diagrammatic overview of the memory IC. Internally, the memory consists of a matrix formed by a number of rows and columns. At each intersection of the matrix is a storage cell which can hold a single binary number (ie either a zero or a one). Access to a particular cell is provided by the memory's address pins. A suitable combination of logic levels, constituting a binary number, presented to the address pins will instruct the IC to connect the addressed cell to the data pin. If the read/write control is set to read, the logic level stored in the cell will appear at the data pin. Conversely, if the memory is set to write, whatever logic level exists on the data pin will be stored in the cell, thus overwriting the previous value. Some memory ICs contain eight separate matrixes, each having their own data pin. This enables a complete binary word (byte) to be stored at each address.

The memory described above is known as a RAM (random access memory) and it is characterised by the fact that data can be retrieved (read) and also stored (written) to individual locations. There are two main types of RAM - SRAM (static RAM) in which data is latched within each memory cell for as long as the power supply remains connected, and DRAM (dynamic RAM) which uses the charge, or the absence of a charge, on a capacitor to store logic levels. The capacitors within a DRAM cannot hold their charge for more than a few milliseconds, and so a process known as refreshing must be carried out by controlling circuitry in order to maintain the stored levels. The method of creating addresses for the memory locations within a DRAM is somewhat complex, in that the rows and columns of the matrix are dealt with separately. DRAM memory chips are used extensively in desktop computers and related equipment because the simple nature of the capacitor memory cell means that a large number of cells can easily be fabricated on a single chip (there are now DRAMs capable of storing 16 million binary digits, or bits). SRAMs fabricated with CMOS transistors are useful for storing data that must be retained while equipment is turned off. The low quiescent current consumption of these devices makes it practicable to power the memory from a small battery located within the equipment, thus providing an uninterrupted source of power.

The ROM (read-only memory) has data permanently written into it, and so there is no read/write pin. ROMs are used to store computer programs and other data that does not need to be changed. The ROM will retain its data indefinitely, irrespective of whether it is connected to a power supply. A special form of ROM known as the EPROM (eraseable programmable ROM) may be written to using a special programmer. Data may later be erased by exposing the chip to ultra-violet light for a prescribed length of time. For this reason, EPROMs have a small quartz window located above the chip which is normally concealed beneath a UV-opaque protective sticker. The EEPROM (electrically eraseable programmable ROM) is similar to the EPROM, but may be erased without using UV light. The PROM (programmable ROM) has memory cells consisting of fusible links. Assuming that logic 0 is represented by the presence of a link, logic 1 may be programmed into a location by feeding a current into a special programming pin which is sufficient to fuse the link at the addressed location. However, once a PROM has been written to in this way, the cells programmed to logic 1 can never be altered.

## Analogue-to-digital Converters

Analogue-to-digital conversion involves measuring the magnitude of a voltage or current and then generating a digital numeric value to represent the result. The digital multimeter works in this way, providing an output in decimal format which is presented directly to the human operator via an optical display. The analogue to digital (A/D) converters used in signal processing differ in two important respects. Firstly, the numeric value is generated in binary form so that the result may be manipulated, or 'processed' using digital circuitry. Secondly, in order to 'measure' a signal, as opposed to, say, the voltage of a battery, it is necessary to make many successive conversions so that amplitude changes occurring over time may be captured. In order to digitise speech, for instance, the instantaneous amplitude of the waveform must be ascertained at least 6000 times per second. Each measurement, known as a sample, must then be converted into a separate binary number. The accuracy of the digital representation depends on the number of bits (binary digits) in the numbers - eight digits will give 256 discrete values, whereas 16 bits provides 65536.

Maximum sampling frequency (ie speed of conversion) and the number of bits used to represent the output are therefore the major parameters to consider when choosing an A/D converter IC. The fastest 8-bit converters available, known as flash types, can operate at sampling frequencies of up to 100s of MHz, and can be used to digitise television signals. 16-bit converters are unfortunately much slower, with maximum sampling rates of only 100s of kHz. It is also possible to obtain converters offering intermediate levels of precision, such as 10 and 12 bits.

Having processed a signal digitally, it is often desirable to convert it back into an analogue form - speech and Morse are obvious examples. There are a variety of techniques which can be used to perform digital-to-analogue conversion, and ICs are available which implement these.

## Microprocessors

The microprocessor is different from other digital ICs in that it has no preordained global function. It is, however, capable of performing a variety of relatively straightforward tasks, such as adding two binary numbers together. These tasks are known as instructions, and collectively they constitute the microprocessor's instruction set. In order to make the microprocessor do something useful, it is necessary to list a series of instructions (write a program) and store these as binary codes in a memory IC connected directly to the microprocessor. The microprocessor has both data and address pins (see **Fig 3.77**), and when power is first applied it will generate a pre-determined start address and look for an instruction in the memory location accessed by this. The first instruction in the program will be located at the starting address. Having completed this initial instruction, the microprocessor will fetch the next one, and so on. In order to keep track of the program sequence, and also provide temporary storage for intermediate results of calculations, the microprocessor has a number of internal counters and registers (a register is simply a small amount of memory). The manipulation of binary numbers in order to perform arithmetic calculations is carried out in the arithmetic logic unit (ALU). A clock oscillator, normally crystal controlled, controls the timing of the program-driven events. The microprocessor has a number of control pins, including an interrupt input. This allows normal program execution to be suspended while the microprocessor responds to an external event, such as a keyboard entry.

Microprocessors exist in a bewildering variety of forms. Some deal with data eight bits at a time, which means that if a number is greater than 255 it must be processed using a number of separate instructions. 16-bit and 32-bit and more recently 64-bit micropro-

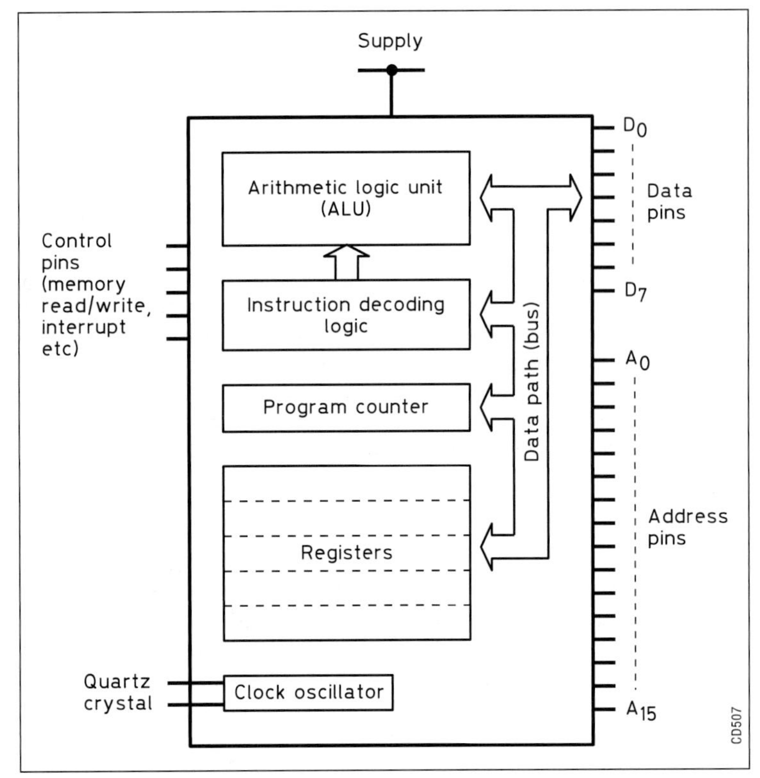


Fig 3.77: An 8-bit microprocessor

more quickly, and by a processor using a smaller number of transistors.

Computers developed using the 86 series followed briefly by the 186, 286 and 386 devices. The 586 series was renamed the Pentium™ and following a ruling at that time that a number could not be trademarked the P1, P2, P3, P4 series had a variety of names for a number of foundries.

Amateurs today make frequent use of the 'PIC' series of microprocessors and these are discussed in more detail elsewhere in the handbook.

DSP (digital signal processing) ICs are special microprocessors which have their instruction sets and internal circuitry optimised for fast execution of the mathematical functions associated with signal processing - in particular the implementation of digital filtering.

Fig 3.78: Surface mounting components

cessors are therefore generally faster. A special class of microprocessor known as the microcontroller is designed specifically to be built into equipment other than computers.

As a result, microcontrollers tend to be more self-sufficient than general microprocessors, and will normally be provided with internal ROM, RAM and possibly an A/D converter. Microcontrollers are used extensively in transceivers in order to provide an interface between the frequency synthesiser's programmable divider, the tuning controls - including the memory keypad - and the frequency display.

Following the development of the first microprocessors in the 1970s, manufacturers began to compete with each other by offering devices with increasingly large and ever-more-complex instruction sets. The late 1990s saw something of a backlash against this trend, with the emergence of the RISC (reduced instruction set microprocessor). The rationale behind the RISC architecture is that simpler instructions can be carried out

## SURFACE MOUNT DEVICES

As the semiconductor industry has developed, all packaging has been reduced in size and manufacturing techniques have changed so that devices are mounted on the copper-side surface of the printed circuit board. These SMDs (surface mounting devices) are now the norm and we as amateurs have to live with them. An example is shown in **Fig 3.78** further details on SMDs and how to handle them are given in the Passive Components, and Construction and Workshop Practice chapters.

*Past contributors to this chapter were **Alan Betts** and **Fred Ruddell***

# Building Blocks 1: Oscillators

4

*Peter Goodson, G4PCF*

Oscillators are fundamental to radio and are used in almost all types of radio equipment. The purpose of an oscillator is to generate an output at a specific frequency. For most applications, an oscillator would ideally generate a pure sine wave. If a spectrum of voltage against frequency were to be plotted, it would consist of a single line at the required frequency. **Fig 4.1(a)** shows the spectrum of the 'ideal' oscillator.

No oscillator produces this ideal output, and there are always harmonics, noise and often sidebands in addition to the wanted output frequency. A more realistic plot is shown in **Fig 4.1(b)**. Tuning a receiver across this would show where the problems are. The noise floor (the background level of noise which is equal at all frequencies) is apparent, as is the '1/*f* ' noise, which is an increase in noise close to the centre (wanted, or carrier) frequency. The harmonics also show these effects, as do the sidebands and spurious oscillations which often occur in synthesisers because of practical limitations in loop design.

Noise and sidebands are always undesirable, and can be very difficult to remove once generated, so one purpose of this chapter is to indicate ways of minimising them. Harmonics are less of a problem. They are a long way in frequency terms from the wanted signal, and can often be simply filtered out. In some cases, they may even be wanted, since it is often convenient to take an output frequency from an oscillator at a harmonic instead of the fundamental. Examples of this are shown below for crystal oscillators.

Oscillators may be classed as variable frequency oscillators (VFOs), crystal oscillators (COs) (including the class of variable crystal and ceramic resonator oscillators (VXOs)), phase-locked loop synthesisers (PLLs), which include VFOs as part of their system and, more recently, direct digital synthesisers (DDSs). Any or all of these may be used in a particular piece of equipment, although recent trends commercially are to omit any form of 'free-running' (ie not synthesised) VFO. The name is retained, however, and applied to the sum of the oscillators in the equipment. Since these are digitally controlled, it is possible to have two (or more) virtual 'VFOs' in a synthesised rig, where probably only a single variable oscillator exists physically. It can be retuned in milliseconds to any alternative frequency by the PLL control circuit.

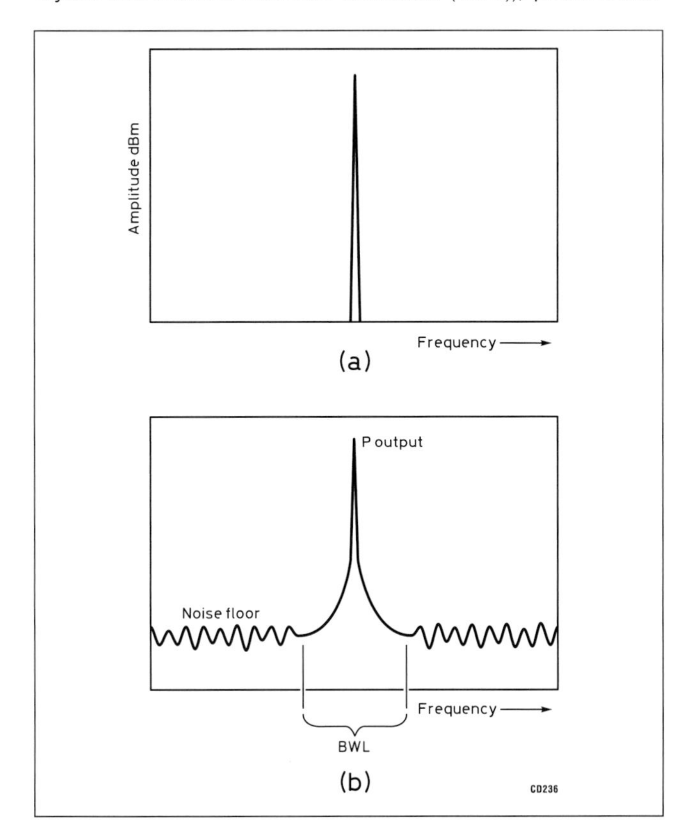


**Fig 4.1: (a) Spectrum of 'ideal' oscillator. (b) Spectrum of local oscillator showing noise**

All oscillators must obey certain basic design rules. The first essential for an oscillator is an amplifying element. This will be an active device such as a bipolar transistor (sometimes called a 'BJT' for bipolar junction transistor), a junction field-effect transistor (JFET), a metal-oxide semiconductor FET (MOSFET) or a gallium arsenide FET (GaAsFET). There are lesser-used devices such as gallium arsenide bipolars, and gallium arsenide HEMTs (high electron mobility transistors), a variant of the GaAsFET with lower noise.

Operational amplifiers are rarely used at RF, since very few are capable of high-frequency operation. Some integrated circuits do contain oscillator circuits, but they are very often variants of standard discrete circuits. Valves, now long obsolete in the context of oscillator design, will be neglected in this text, but it is worth observing that the fundamental circuit configurations owe their origins to valve designs from the early part of the 20th century. The Colpitts and Hartley oscillators were both first published in 1915. The Colpitts oscillator is named after its developer, Edwin Henry Colpitts (1872-1949) and the Hartley oscillator was developed by Ralph Hartley (1889-1970). For those interested in valve design, reference should be made to earlier editions of this handbook, and to references [1] and [2]. A more recent source is Section 4 of reference [3].

The basic requirements for an oscillator are:

1. There must be gain over the whole frequency range required of the oscillator.
2. There must be a feedback path such that the product of the forward gain and the feedback attenuation still leaves a net loop gain greater than 1.
3. The feedback must occur in such a way that it is in phase with the input to the gain stage.
4. There must be some form of resonant circuit to control the oscillator frequency. This is most often an inductor and capacitor (L-C). More precise frequency control can be achieved using electromechanical structures such as crystals, ceramic resonators or surface acoustic wave (SAW) resonators. This latter category is only really feasible in specialist applications where the high cost can be justified.

At low frequencies, a resistor-capacitor (RC) or inductor-resistor (LR) network may be used for frequency control. These cannot be said to be truly resonant themselves but, with active gain stages, resonant peaks which are sometimes of very high Q can be achieved.

A potential successor to the crystal and ceramic resonator is the electrochemically etched mechanical resonator, usually in the same silicon chip as the oscillator circuit. These devices are still in the development stage, but may well come into common use soon.

An oscillator therefore needs gain, feedback and a resonant circuit. However, there are more criteria that can be applied:

5. For stable oscillations, the resonant circuit should have high Q. This tends to rule out RC and LR oscillators for high frequencies, although they are often used for audio applications. More information on low-frequency oscillators can be found in audio texts [4].
6. For stable oscillations, the active components should also show a minimum of loading on the resonant circuit, ie the loaded Q should be as high as possible. Although an oscillator could be said to be the ultimate in Q-multipliers, high inherent Q gives better stability to the final product.
7. The choice of active device is critical in achieving low-noise operation. This will be covered extensively below, since in almost all cases low-noise oscillations are required, and the difference between 'noisy' and 'quiet' oscillators may be only in the choice of a transistor and its matching circuit.
8. The output stage of an oscillator can be very important. It must drive its load adequately, without changing the loading of the oscillator and thereby changing the oscillator frequency (an effect known as load pulling).
9. The power supply to an oscillator should be very carefully considered. It can affect stability, noise performance and output amplitude. Separate supply decoupling and regulation arrangements are usually essential in good oscillator design.

## EFFECT OF VFO PERFORMANCE ON RECEIVER PERFORMANCE

When an oscillator is used as the first local oscillator in a receiver, its performance has a bearing on the receiver performance. The most obvious of these is frequency drift, but other oscillator parameters include:

- The oscillator noise sideband performance will affect the ability of the receiver to resolve a weak signal when there are nearby strong signals.
- The oscillator noise floor may affect the ability of the receiver to resolve a weak signal when there are strong signals far removed (in frequency) from the wanted signal.
- In an FM receiver, the local oscillator phase noise gets added to the received signal in the first mixer. The FM detector will then demodulate the local oscillator phase noise. If the phase noise is excessive, it may limit the maximum signal to noise ratio available from the receiver.
- In an SSB receiver, the local oscillator phase noise gets added to the received signal in the first mixer. This noise gets demodulated in the SSB detector and, if excessive, will give a 'fuzzy' sound to the demodulated audio.
- A high level of harmonics from an oscillator may increase the spurious responses in the receiver.
- The value of VFO buffer output impedance may be important to enable some mixers to give their best performance.
- If the oscillator signal is picked up by the antenna it may cause a degradation of receiver performance. For this reason it may be necessary to screen the oscillator.

For a high performance receiver (a receiver which has excellent intermodulation performance) an oscillator with good noise performance will be required or else the work and cost put into achieving the intermodulation performance will be wasted.

However, if the receiver is one of fairly modest performance, it is not necessary to spend a lot of time or money on a high performance oscillator. This shows that the performance of the receiver signal path and oscillator need to be matched. This subject is covered in more detail in the chapter on HF Receivers.

## VARIABLE-FREQUENCY OSCILLATORS

A VFO is a type of oscillator in which the oscillation frequency is adjustable by the operator. Normally this is done by the tuning control on the front panel of the equipment. This control operates through a variable capacitor or variable inductor to control the oscillation frequency. Most VFOs consist of a single active device with a tuned circuit and a feedback network to sustain the oscillation.

Two such examples are shown in **Figs 4.2 and 4.3**. Some comments on these diagrams will show the design compromises in action. Fig 4.2 uses a tapped inductor and is known as the Hartley oscillator. The diagram shows the basic circuit and does not include DC bias components. In the Hartley oscillator of Fig 4.2, the tuned circuit is designed for high Q. The active device is an FET, chosen for good gain at the intended frequency of oscillation. A dual-gate MOSFET could be used or, especially at UHF and microwave frequencies, a GaAsFET. The high input impedance of the FET puts little loading onto the tuned circuit, while the FET source provides feedback at low impedance. The fed-back voltage is transformed up to a higher impedance and appropriate phase, so the oscillation loop is complete. This is one example where the active device provides only impedance matching to the oscillator loop; no voltage gain or phase shift in the FET is involved.

Fig 4.3 shows the capacitively tapped version of basically the same circuit. This is known as the Colpitts oscillator. Capacitive transformers are less easy to understand intuitively but, if one thinks of the whole as a high-impedance resonant circuit, then a capacitive tap is a reasonable alternative. While in principle the two circuits are very similar, in practice the additional capacitors of the Colpitts make the circuit less easy to tune over a wide range. In the Hartley circuit, with careful

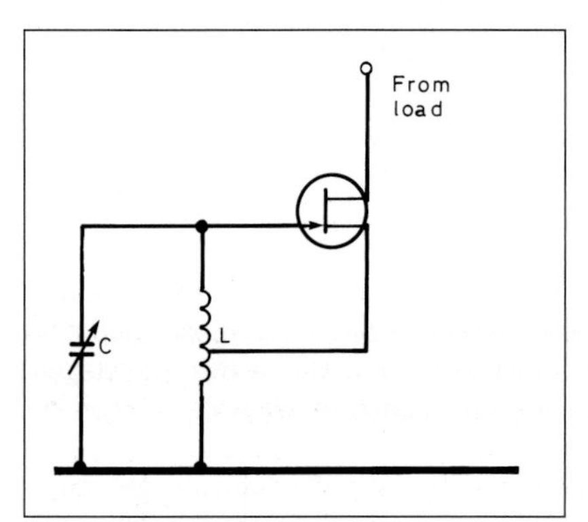


**Fig 4.2: Hartley oscillator**

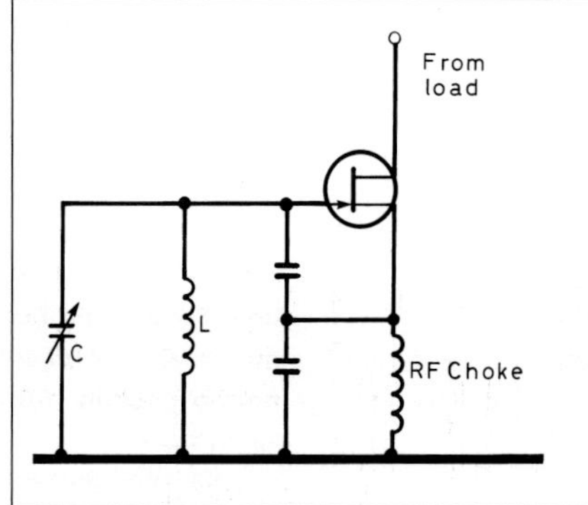


**Fig 4.3: Colpitts oscillator**

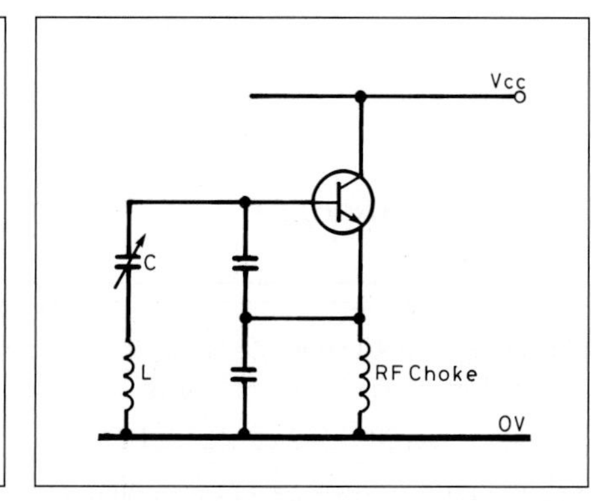


**Fig 4.4: Clapp oscillator**

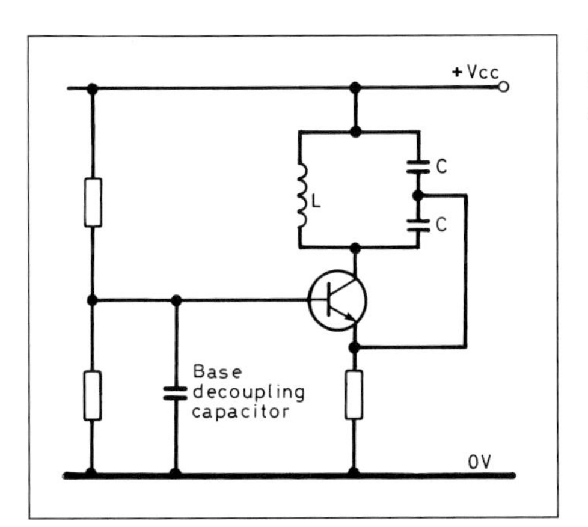


**Fig 4.5: 180 degree phase-shift oscillator**

design and a low-input-capacitance FET, a very wide range of total circuit capacitance variation (and hence wide frequency range) can be achieved. The Colpitts circuit is preferred at higher frequencies, since the transformer action of the inductor is much reduced where the 'coil' is in the form of a straight wire, say at 500MHz or above.

Sometimes described as a variant of the Colpitts circuit is the Clapp oscillator: **Fig 4.4**. This shows a series-resonant circuit, which is more easily matched into the bipolar transistor shown. This circuit is especially suitable to UHF and low microwave work, where it is capable of operating very close to the cut-off frequency (Ft) of the transistor.

Variants of both the Hartley and Colpitts circuits have been derived for feedback around the active device including a 180° phase shift. An example is shown in **Fig 4.5** and this is again more suited to a bipolar transistor.

In the technical press, there has been much discussion on whether the best choice for the active device in an oscillator is a bipolar transistor or an FET. There are many factors involved in this choice, but some simple rules can be derived. First, since the loaded Q must be maintained as high as possible, an FET is attractive for its high input impedance. A bipolar transistor could be used as an emitter follower with an undecoupled emitter resistor but a noise analysis of the circuit would put all of that resistance in the noise path. As an example, **Figs 4.6 and 4.7** compare noise sources in oscillator (or amplifier) input circuits. The FET, provided it is operated well within its frequency range, has a high input impedance. Any input capacitance is absorbed into the tuned circuit fairly directly, since it looks just like a capacitor with at most an ohm or two of input series resistance and perhaps 1 or 2nH of series inductance from the bondwire on the chip. The output impedance at the source, ie in a follower, is $1/g_m$, (plus a small ohmic resistance term) where $g_m$ is the mutual conductance at the operating frequency. As a gain stage, the output impedance of most FETs is very high, so the gain is determined by gm and the load impedance. An exception to this is the GaAs MESFET, where the output impedance is typically a few hundred ohms, so there is a serious limit to the gain available per stage. Some silicon junction FETs (JFETs) have low $g_m$, so it is worth choosing the device carefully for the frequency used. GaAsFETs tend to have much higher $g_m$, especially at high frequencies, and so are recommended for microwave work, with the proviso that they are rather prone to $1/f$ noise. This will be described below.

Turning now to the bipolar device, it can be seen that the input impedance is inherently low. At low frequencies, the input impedance is approximately $h_{fe}$ (the AC current gain) multiplied by the emitter resistance ($R_e + r_e$). This term is comprised of $R_e$, the $1/g_m$ term as in the FET, and $r_e$, the ohmic series resistance.

This latter can be several ohms, while the former is determined by the current through the transistor; at room temperature, $R_e = 26/I_e$, where $I_e$ is the emitter current in milliamps. Thus at, say, 10mA, the total emitter resistance of a transistor may be an ohm or two, and the input impedance say 50 to 100Ω. This will severely degrade the Q of most parallel-tuned circuits. There is no advantage in using a smaller emitter current, since this leads to lower $F_t$ and hence lower gain; as the frequency approaches $F_t$, the AC current gain is degraded to unity by definition at $F_t$. For completeness, the bipolar noise sources are included. In this respect, the best silicon bipolar transistors compare roughly equally with the best silicon FETs. GaAs devices tend to be better again, especially above 1GHz, but the effect of $1/f$ noise has to be considered, so that in an oscillator (but not an amplifier) a silicon device has many advantages right into the microwave region. GaAs devices only predominate because most of them are designed for amplifier service; if sub-0.5 micron geometry, silicon discrete FETs became commercially available, they could become the mainstay of oscillators to 10GHz and beyond.

So what is this $1/f$ noise, and how does noise affect an oscillator? Noise is more familiarly the province of the low-noise amplifier builders, but the principles are the same for oscillators. Where noise comes into the receiver context is that if the signal to be generated is the local oscillator (LO) in a receiver, then most or all

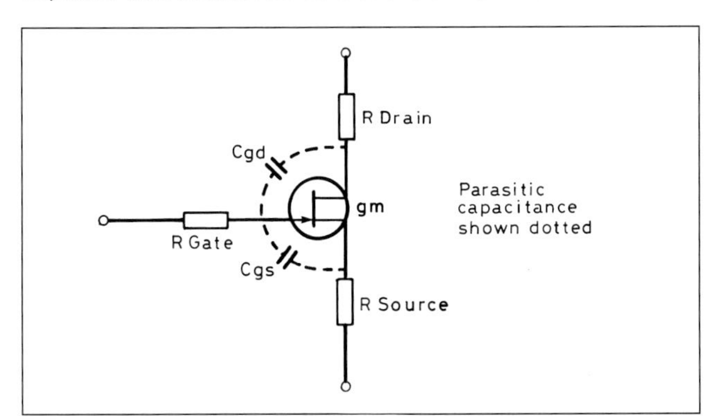


**Fig 4.6: FET input characteristics. Input resistance is very high. Input capacitance is approximately equal to $C_{gs}$ + (voltage gain) $C_{gd}$. Input equivalent noise resistance is approximately equal to $R_{GATE} + R_{SOURCE} + 1/g_m$. Typically $R_{GATE}$ is less than 10 and $R_{SOURCE}$ is approximately equal to $R_{DRAIN}$ and less than 100. Power devices may be less than 1**

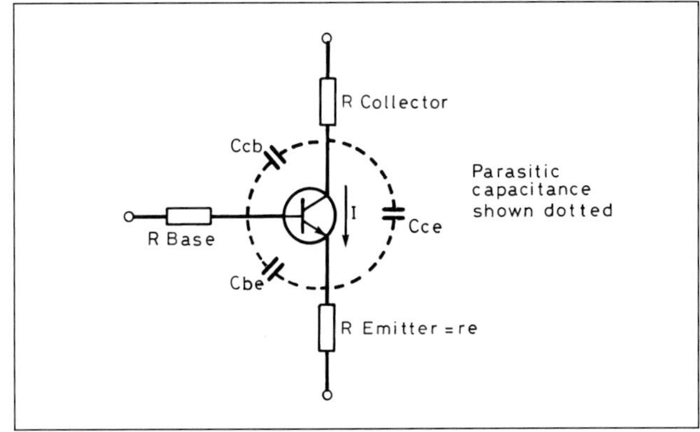


**Fig 4.7: Bipolar input characteristics. Input resistance is approximately $R_{BASE} + h_{fe}$ x $(r_e + R_e)$ where $R_e = 1/g_m$ and $g_m = (q/kT)$ x $I$. Input capacitance is very approximately equal to $C_{be} + V_{gain}$ x $C_{cb}$ but transistor effects will usually increase this dynamically. Input equivalent noise resistance ($R_{IN}$) is equal to $R_{BASE} + (R_e + r_e)/2$. Typically, for a small device, $R_{BASE}$ is about 100Ω, $R_e$ is 26Ω at 1mA, and $r_e$ is 3Ω. Therefore $R_{IN}$ is approximately 129Ω**

of the local oscillator noise is modulated onto the wanted signal at the intermediate frequency (IF). There are several mechanisms for this, including the straightforward modulation of the wanted signal, and the intrusion in the IF of reciprocal mixing (see the chapter on HF receivers), where a strong but unwanted signal, which is close to the wanted one within the front-end passband, mixes with LO noise. A low-noise oscillator is therefore a major contributor to a low-noise receiver. On transmit, the effects may not be so obvious to the operator, but the transmission of noisy sidebands at potentially high powers is inconvenient to other band users, and may in extreme cases cause transmissions outside the band. In practice, an oscillator with low enough noise for reception is unlikely to be a problem on transmit.

All devices, active and passive, generate noise when not at absolute zero temperature. Inductors and capacitors only do so through their non-ideal resistive terms, so can be neglected in all practical cases. Resistors and transistors (FET and bipolar) are the real sources of noise in the circuit. The noise power generated by a resistor is:

$$\text{Noise power} = kTB$$

where $k$ is Boltzmann's constant, a fundamental constant in the laws of physics; $T$ is the absolute temperature (room temperature is usually approximated to 300K, ie 27°C); and B is the measurement bandwidth.

This equation is sufficiently accurate for most modern resistors like metal film and modern carbon film resistors. Note that some older types, notably carbon-composition types, do generate additional noise. Active devices are less so, and an equivalent noise resistance ($R_{IN}$ in Fig 4.6 and 4.7) can be derived or measured for any active device which approximates the device to a resistor.

This may have a value close to the metallic resistance around the circuit, or it may be greater. For many applications, this resistance can be used to estimate the noise contribution of the whole circuit. However, there are other noise sources in the device, most noticeably in the oscillator context those due to $1/f$. Fig 4.1(b) shows $1/f$ noise diagrammatically. Noise sources of this type have been studied for many years in many different types of device. Classically, this noise source increases as frequency is reduced. It was thought that there had to be a turnover somewhere in frequency, but this need not be the case.

If we take a DC power supply, it has a voltage at the output, but it has not always been so. There was a time before the supply was switched on, so the 'noise' is infinite, ie $1/f$ holds good at least in qualitative terms. This gives rise to the concept of noise within a finite bandwidth, usually 1Hz for specification purposes. Practical measurements are made in a sensible bandwidth, say a few kilohertz, and then scaled appropriately to 1Hz. The goodness of an oscillator is therefore measured in noise power per hertz of bandwidth at a specified offset frequency from the carrier. A good crystal oscillator at 10MHz would show a noise level of -130dBc (decibels below carrier) at 10kHz offset. A top-class professional source might be -150dBc. There are several possible causes of $1/f$ noise and, when examining the output of an oscillator closely, there is invariably a region close to the carrier where the noise increases as $1/f$. This is clearly a modulation effect, but where does it come from?

The answer is that most (but not all) oscillators are very non-linear circuits. This is inherent in the design. Since it is necessary to have a net gain around the oscillation loop (which is the gain of gain stage(s) divided by the loss of feedback path), then the signal in the oscillator must grow. Suppose the net gain is 2. Then the signal after one pass round the loop is twice what it was, after a further pass four times and so on. After many passes, the amplitude should be 'infinite'. There clearly must be a practical limit. This is usually provided by voltage limitations due to power supply rails, or gain reduction due to overdrive and saturation of the gain stage(s). Occasionally, in a badly designed circuit, component breakdown can occur where breakdown would not be present under DC conditions. This can be disastrous as in a burn-out or, more sinister, as in base-emitter breakdown of bipolar transistors, which will lead to permanent reduction of transistor gain and eventual device failure in service.

So, almost all practical oscillators are highly non-linear and have sources of noise, be they wide-band (white noise) or $1/f$ (pink noise). The non-linearity thus modulates the noise onto the output frequency. In contrast, a properly designed amplifier is not non-linear, has very little modulation process, and is therefore only subject to additive noise.

The exceptions to this are the very few oscillators which have a soft limitation in the output amplitudes, ie they are inherently sine wave producers. This requires very fast-acting AGC circuitry, and is very rarely used because it adds much cost and complication. Most practical oscillators are inherently square-wave generators but a tuned circuit in the output will normally remove most harmonic energy so that the output looks like a sine wave on an oscilloscope. A spectrum analyser normally gives the game away, with harmonics clearly visible. The DDS devices are an exception to all this for reasons explained below.

Generally, a silicon device will have very much lower levels of $1/f$ noise than a gallium arsenide device. This is because silicon is a very homogeneous material, of very high purity, and in modern processes with very little surface contamination or surface states. Gallium arsenide is a heterogeneous material which is subject to surface states. The $1/f$ knee is the frequency below which the noise levels associated with the surface states start to increase, typically at 6dB/octave of frequency. In, say, 1980 this frequency would have been 100MHz or more for most gallium arsenide processes. In 1990 the figure was as low as 5MHz in some processes. Silicon typically has a $1/f$ knee of less than 100Hz, sometimes less than 10Hz, so $1/f$ noise is unimportant for most practical cases. It means that there is a possibility of increased close-to-carrier noise in GaAs-based oscillators. Whether this makes the oscillator better or worse than a silicon design depends on the exact choice of devices and circuits. More information on these topics can be found in [5].

## PRACTICAL VFOs

A VFO design should start with a set of clear objectives, such as:

- The required frequency tuning range.
- The exact means of tuning.
- The requirement for frequency stability.
- The RF power output required from the oscillator. The RF level is important because most mixers have a minimum oscillator power requirement. This is the oscillator power that will give the specified mixer performance.
- For a high performance receiver it may be necessary to define the oscillator noise performance.

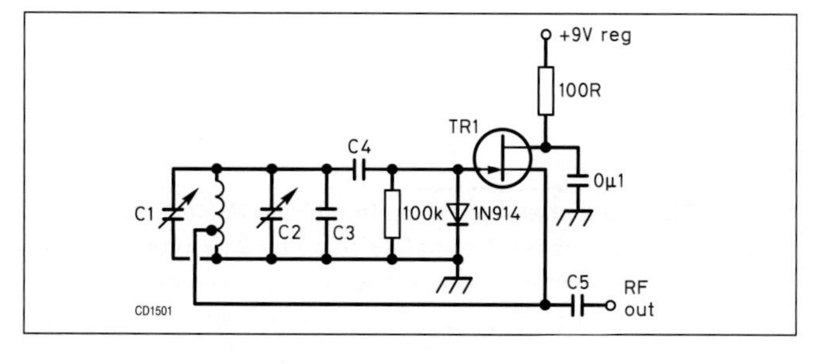


**Fig 4.8: Hartley VFO (*W1FB's QRP Notebook*)**

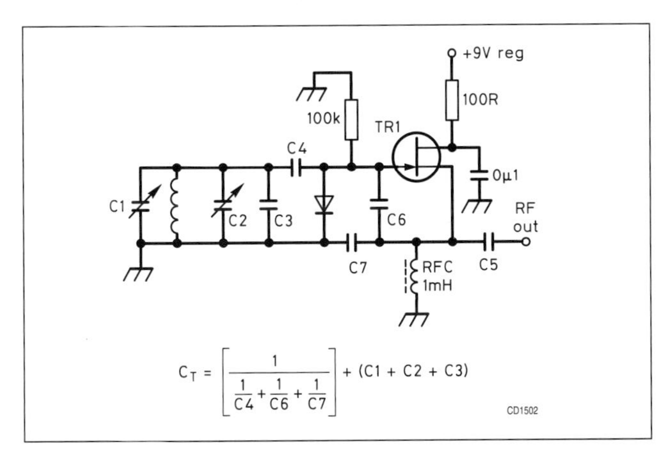


**Fig 4.9: Colpitts VFO (*W1FB's QRP Notebook*) Suggested values for reactances are: $X_{C1}$ 1200; $X_{C2}$ 3500; $X_{C3}$, $X_{C4}$, $X_{C5}$ 880; $X_{C6}$, $X_{C7}$ 90; $X_{L1}$ 50 (*W1FB's QRP Notebook*)**

A typical VFO design objective is as follows:

1. Frequency range 5.0 to 5.5 MHz. The VFO therefore tunes over 500kHz
2. The VFO is required to be tuned by a variable capacitor.
3. The VFO is required to tune an SSB/CW transceiver. Therefore this requires that the frequency should not drift by more than 100Hz in an hour, and should always be repeatable to within 200Hz.
4. RF Power output needs to be 10mW.

The frequency range required is determined by the band that is required to be covered, and the intended IF. In this example the use of a VFO covering 5 - 5.5MHz allows the 80m and 20m bands to be tuned, using an IF of 9MHz.

The third requirement is defined by the fact that an SSB receiver needs to be tuned to within about 50Hz of the correct frequency. This puts a difficult requirement on VFO stability because small changes in temperature caused by, for example switching on a heater in the room, can cause considerably more change than 50Hz.

**Figs 4.8 to 4.10** show a range of practical VFO circuits, which include the DC bias components. In each circuit, C1 is the main tuning capacitor and C2 provides bandspread. The circuit in Fig 4.8 is a Hartley oscillator. The feedback tap should attach at about 25% of the coil from the earthy end; actually, the lower down the better, consistent with easy starting over the whole tuning range. A bipolar transistor could be used in this circuit with bias modification, but it will tend to reduce circuit Q and hence be less stable than a FET. The FET should be chosen for low noise, and JFETs are usually preferred. However this circuit will also work well with modern dual-gate MOSFETs. Its second gate should be connected to a bias point at about 4V. Fig 4.9 shows the Colpitts version of the circuit. The additional input capacitance restricts the available tuning range but this is rarely significant. Again, a dual-gate MOSFET version is possible. A voltage-tuned version of the Hartley is shown in Fig 4.10. This is actually a VCO (see later) whose tuning voltage is derived from a potentiometer, so to the operator it operates like a VFO.

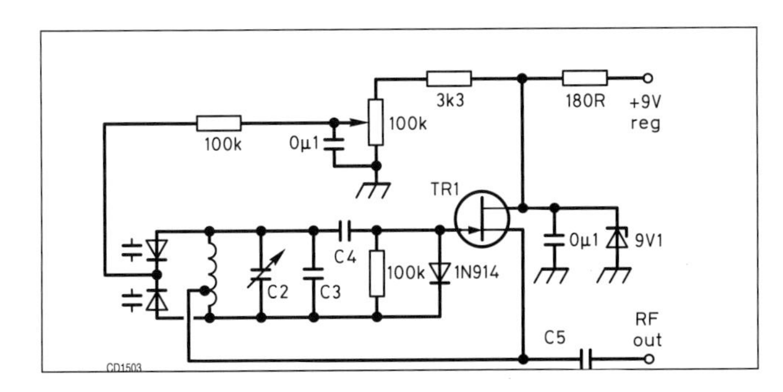


**Fig 4.10: Varactor-tuned Hartley VFO.**

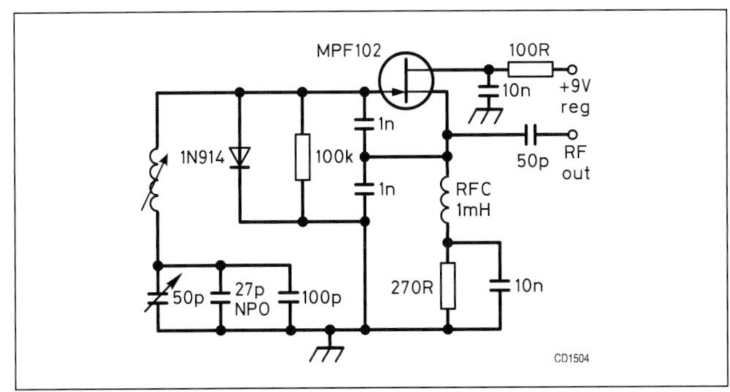


**Fig 4.11: A JFET series-tuned Clapp VFO for 1.8MHz (*ARRL Electronic Data Book*)**

Many other variants on these basic circuits have been published over the years; some examples are shown in **Figs 4.11**, a Clapp oscillator and **4.12**, a Colpitts oscillator using a dual-gate MOSFET as the active device. In Fig 4.12, the biasing of the dual-gate MOSFET in oscillator service is typical; G1 is at source DC voltage level and G2 at 25% of the drain voltage.

As well as the Colpitts. Clapp and Hartley oscillator, other designs described in this chapter are the Franklin and Vackar oscillators.

**Fig 4.13** illustrates a Franklin oscillator. This uses two active devices which in this example are GaAs MESFETs. As mentioned above, the MESFET tends to have relatively low gain per stage, so there is an advantage in using two gain stages in series to provide positive gain over a very wide frequency range. The devices are

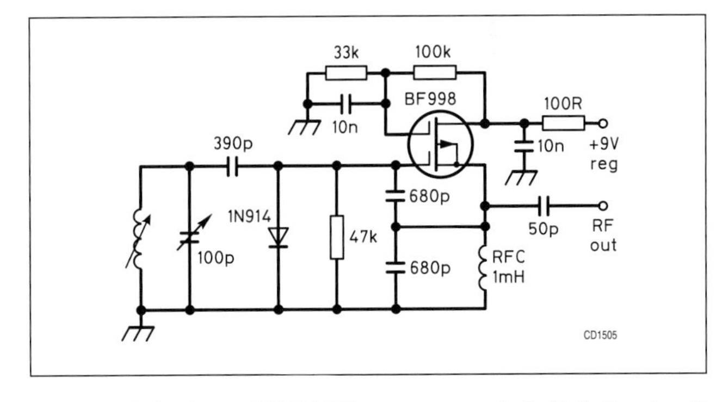


**Fig 4.12: A dual-gate MOSFET in a common-drain Colpitts circuit (*ARRL Electronic Data Book*)**

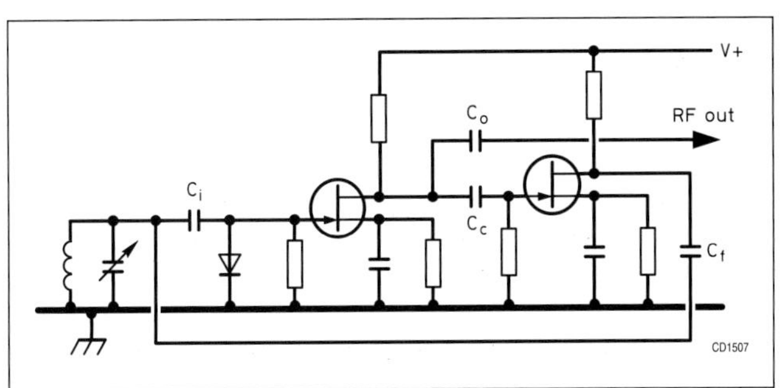


**Fig 4.13: The Franklin oscillator**

shown running with self-bias on the gates. Ideally, they should be run close to $I_{dss}$, where the gain is usually greatest. The feedback path is arranged to give 360° phase shift, ie two device inversions. The feedback is then directed into the tuned circuit.

The output impedance of the MESFET is relatively low, but it is prevented from damping the resonant circuit by very light coupling; $C_i$ and $C_f$ typically are 1-2pF in the HF region, smaller still at higher frequencies. The high gain ensures oscillation even in this lightly coupled state, so that very high circuit Q is maintained, and the oscillation is stable and relatively noise-free. A particular feature of this configuration is the insensitivity of the circuit to supply variation; but the supply should still be well decoupled. A disadvantage of the circuit is the total delay through two stages, which limits the upper frequency, but with gallium arsenide devices this can still yield 5GHz oscillators.

## DESIGNING / BUILDING A VFO

The following are recommendations for building variable frequency oscillators:

- For best stability, operate the VFO within the range 5 - 10MHz.
- Build the VFO in a sturdy box, eg a die-cast box. This will provide mechanical stability.
- Screen the oscillator to prevent the transfer of energy to/from other electronic circuits nearby. The screening box mentioned above will achieve this, but note that a box will not provide screening against magnetic fields.
- The VFO should be positioned away from circuits which generate high magnetic fields eg transformers.
- Protect the circuit from radiant and convected heat sources. Placing the VFO in a box will achieve this, but minimise the number of heat sources inside this box. For example, voltage regulators should be placed outside the box.
- Avoid draughts across the tuned circuit.
- Provide adequate decoupling.
- Use a buffer / amplifier stage to isolate the oscillator from the load.
- Provide regulated voltage supplies to the oscillator and buffer amplifier.
- Use a FET as the active device. The FET should have reasonably high gm (> 5mS). A J310 FET is a good example.
- Resistors should be metal film or carbon film types.
- Resistors should be rated at 0.5W or 1W dissipation. This reduces the temperature rise caused by currents flowing in the resistors.
- The active device will experience a temperature rise due to the operating current flow. This will cause a warm-up drift in the oscillator frequency. For this reason a conventional leaded package should be used, not a surface mounted package, and a small heat sink should be fitted.
- If the VFO is built on a PCB the board should be single-sided. This is because double-sided boards may show a capacitance change with temperature due to the dielectric material characteristics changing with temperature.
- Use single-point earthing of frequency-determining components in the tuned circuit.
- All components should be as clean as possible.
- The oscillator should be fitted with a good quality slow-motion drive to give the required tuning rate on the main tuning control. A typical figure chosen here might be 5kHz/turn of the tuning control for SSB mode.
- The receiver/transceiver operating frequency is most easily indicated using a digital counter/display. This should have a resolution of 0.1kHz or better, to allow easy re-setting to a certain frequency. The frequency display may use a software-based frequency counter (using a PIC) or it may be a all-hardware design using digital counters.
- The oscillator supply should be de-coupled to ground using a good quality electrolytic capacitor (eg 100µF). Low-Impedance types are recommended. For example, types that are intended for high quality audio equipment are Ideal.

### Oscillator Power Level

In order to avoid excessive temperature rise In the VFO after switch-on, It Is desirable to operate the VFO at low power level. However, If a good noise floor performance It Is required, It Is necessary that the oscillator Is operated at a high power level. This is shown by Leeson's equation (see chapter 6).
This shows that there is sometimes a conflict between requirements for oscillator design.

### Oscillator Tuning Components

The oscillation frequency of an oscillator is primarily determined by the value of the inductor and capacitor(s) in the tuned circuit. Undesired changes in the reactance of these components (due to temperature change or some other factor) will have a direct effect on the oscillator frequency. It can be easily shown mathematically that if the inductor reactance changes by +100ppm then the oscillator frequency will move by -50ppm. A similar result is obtained if the capacitor reactance changes by +100ppm. This shows that there is a direct relationship between component value and oscillator frequency. For example, consider a 5MHz oscillator which employs a silvered mica capacitor with a temperature coefficient of +100ppm/°C. The capacitor would cause the oscillator to drift by -250Hz/°C. This is significant drift, and if the VFO was used in an SSB receiver, would require frequent retuning as the room temperature changed.

All components in the tuned circuit chosen should therefore be as stable as possible, and this is normally achieved by making the right choice of the type of component for each capacitor and inductor. A single poor quality component in an otherwise good oscillator design may ruin oscillator stability.

#### Capacitors

The following is a description of various capacitor types and their suitability for oscillator use.

- X7R dielectric ceramic capacitors must be avoided. These have low Q and are microphonic. They also exhibit capacitance change with applied voltage.
- COG dielectric ceramic capacitors are significantly better than X7R. They have higher Q and better temperature stability. These are available in surface-mounted (leadless) form and these are recommended. This is because they can be fitted directly onto a PCB and are then mechanically very stable.
- Silvered Mica. These are excellent for oscillators. They have high Q, and are temperature stable. However, the quantities used in the electronics industry are declining and they are becoming difficult to obtain. Typical temperature coefficient is -20 to +100 ppm/°C
- Polystyrene. These have a high Q at lower radio frequencies and are temperature stable. These are also becoming difficult to obtain. Typical temperature coefficient is typically -150 ppm/°C
- Porcelain capacitors. These are specialised capacitors, which are designed for use in high power microwave amplifiers. These offer very high Q and have typical temperature coefficient of +90 ppm/°C.

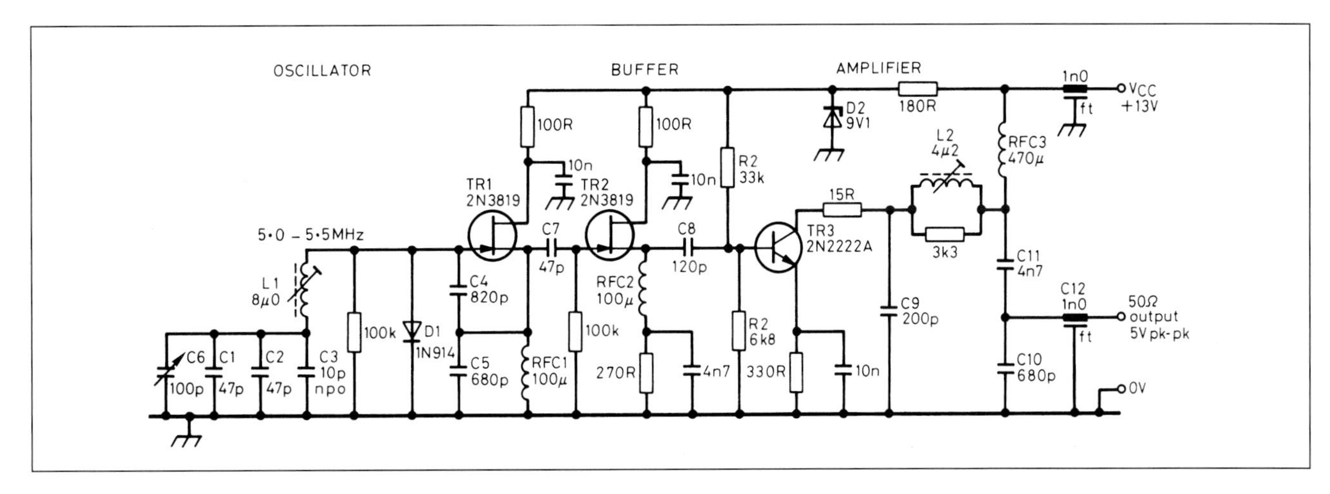


**Fig 4.14: Clapp VFO with amplifiers for 5.0-5.5MHz. Reactance values are L1 = 265Ω, L2 = 140Ω, C1 = 690Ω, C2 = 690Ω, C3 = 2275Ω, C4 = 33Ω, C5 = 48Ω, C6 = 303Ω min, C7 = 690Ω, C8 = 227Ω, C9 = 152Ω, C10 = 48Ω, C11 = 4.5Ω, C12 = 23Ω, RFC1 = 4400Ω, RFC2 = 4400Ω**

- Some trimmer capacitors show very poor temperature stability. Air dielectric trimmers on a ceramic base are preferred.
- The main VFO tuning capacitor should be a double bearing type to aid mechanical stability. Those with silver-plated brass vanes show better temperature stability than those with aluminium plates. The capacitor should be securely mounted to avoid vibration.
- If the Q of a capacitor is rather low, a useful technique is to make up the value required by putting several small value capacitors in parallel. This usually produces a capacitor with a greater Q. It also reduces the RF current flowing in each capacitor, which reduces the heating effect of that current.
- Ceramic plate capacitors are available with positive and negative temperature coefficients as well as COG / NPO. Commonly available ceramic plate capacitors have designations such as N150, meaning it decreases its value by 150 ppm per degree Celsius. Conversely, a P100 capacitor increases the capacitance by 100 ppm for every degree Celsius. As the inductor in the VFO normally has a strong positive temperature coefficient the use of negative temperature coefficient capacitors (eg N150) in parallel with the tuning capacitor can reduce the drift with temperature to practically zero.

### Inductors

If the guidelines (above) about capacitors are followed, the Q of an HF oscillator coil will usually be less than that of the capacitors. Therefore, the coil is the limiting factor on oscillator Q.

- The oscillator Q can usually be increased by raising the Q of the coil, and this can be done by winding the coil from Litz wire. Litz wire can be made by twisting a number of thin wire strands together. Each individual strand should have enamel insulation. Better Q may be obtained if the strands are plaited, rather than twisted.
- If the oscillator is operated on VHF, coil Q is usually high (typically 400) and the tuned circuit Q may be limited by the capacitors.
- Ceramic or fused silica formers are preferred for coil formers.
- Plastic formers should be avoided unless there is no other alternative.
- The coil former should not be too large. This is because a large coil will have a large magnetic field and it will be very susceptible to movement of objects which are within its field. 2 - 3cm diameter is ideal.
- If the coil is fitted with an adjustable screw slug the slug should be of a low permeability type. The number of turns on the coil should be adjusted so that the final adjustment position of the slug is just into one end of the coil.
- Toko coils are sometimes used in projects that must be easily reproducible. These are slug-tuned coils, and do not provide the best stability, but are commercially available as a ready-made part. For use in a VCO (corrected by a PLL) they offer good performance into the VHF range.
- If the coil is wound on a toroidal core the type of core material should be chosen carefully because some core materials have a high temperature coefficient of permeability. The resulting oscillator would show very high temperature drift. Powdered iron cores are more stable than ferrite.

Many of the specialised parts mentioned above may be obtained at radio rallies.

## TWO PRACTICAL VFOS

### Gouriet-Clapp Oscillator

The series-tuned Colpitts oscillator, often referred to as the Clapp or Gouriet-Clapp (devised by G C Gouriet of the BBC), has been a favourite with designers for many years. The series-tuned oscillator enables a higher value of inductance to be used than would normally be required for a parallel-tuned design, resulting in claims of improved stability. The circuit of a practical VFO, **Fig 4.14**, has component values selected for a nominal operating frequency of 5MHz, but reactance values are tabulated for frequency-determining components to allow calculation of optimum values for operation on other frequencies in the range 1.8 to around 10MHz. Simply substitute the desired VFO frequency into the reactance formula and the required L and C values can be deduced. Round off the calculated value to the nearest preferred value.

L1 should be wound on a ceramic low-loss former and the inductance is made variable by fitting a low-permeability, powdered-iron core. The series capacitors C1, C2 and C3 are the most critical components in the circuit; they carry high RF currents and the use of three capacitors effectively decreases the current through each one. A single capacitor in this location may cause frequency-jumping brought about by dielectric stress as a result of

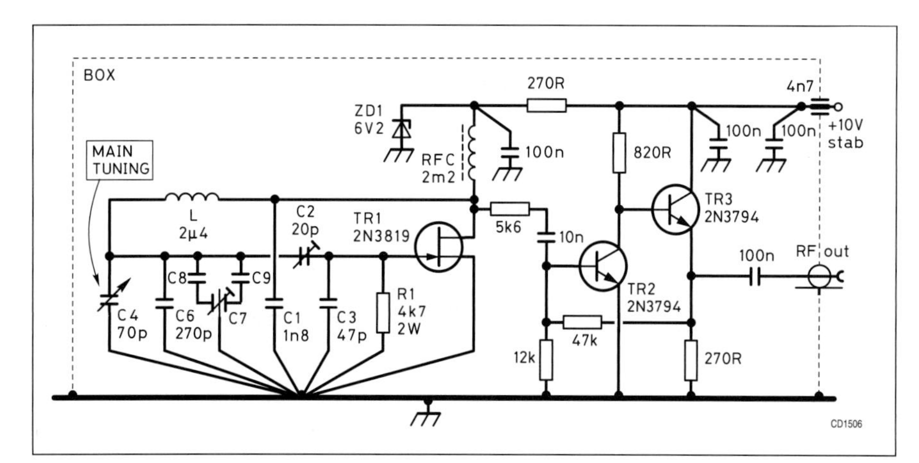


**Fig 4.15: This Vackar oscillator covers 5.88 - 6.93MHz with two-stage buffering. It is based on a G3PDM design. A temperature compensation scheme consisting of C7, C8 and C9 is described in the text. A lower-noise option would be to derive the supply from a LM78L09 regulator with a liberal amount of electrolytic capacitors to kill any low frequency noise on the supply rail**

heat generated by the RF current. C3 is selected to counteract the drift of the circuit. In most circuits the tuning capacitor C6 will be located in series with the inductor and in parallel with C1-C3; a trimmer capacitor may also be placed in parallel with the main tuning capacitor. By placing the tuning capacitor in parallel with C5, a smaller tuning range is possible. This may be desirable for a 40m VFO where the frequency range required is only 200kHz.

The active device is a JFET and another JFET, TR2, acts as a buffer amplifier. This is lightly coupled to the source of TR1. Both transistors operate from a zener-stabilised 9V power supply. A better, lower-noise option would be to derive the supply from a LM78L09 regulator with a liberal amount of electrolytic capacitors to kill any low frequency noise on the supply rail.

TR3 is a voltage amplifier operating in Class A and uses a bipolar device such as the 2N2222A. Bias for Class A operation is established by the combination of the emitter resistor and the two base bias resistors which are also connected to the stabilised 9V supply.

The collector impedance of TR3 is transformed to 50Ω using a pi-network which also acts as a low-pass filter to attenuate harmonics. The loaded Q of the output network is reduced to approximately 4 by adding a parallel resistor across the inductor, and this broadens the bandwidth of the amplifier. For optimum output the network must be tuned either by adjusting L2 or by altering the value of C9. The 15Ω series resistor is included to aid stability.

The VFO described in Fig 4.14 is capable of driving a solid-state CW transmitter, and it is ideally suited to provide the local oscillator drive for a diode ring mixer. It also has sufficient output to drive a valve amplifier chain in a hybrid design.

## Vackar Oscillator

This is an oscillator configuration which was first published in 1949 by Jiri Vackar. The circuit is a modification of the Colpitts oscillator, and is capable of extremely high stability. A notable example of the Vackar oscillator was a design by P G Martin, G3PDM (**Fig 4.15**) which is described more fully in [6]. This design has adjustable temperature-compensation and employed most of the measures described earlier to produce a stable oscillator. The design used a silver-plated copper wire for the tank coil (wound on a ceramic former), and included a two-transistor buffer stage.

The temperature compensation components are the 100 + 100pF differential air-spaced preset capacitor C7, the 100pF ceramic C8 with negative temperature coefficient and the 100pF ceramic C9 with positive temperature coefficient. Turning C7 effects a continuous variation of temperature coefficient of the combination between negative and positive with little variation of the total capacitance of about 67pF. Differential capacitors are hard to find but can be constructed by ganging two singles with semicircular plates, offset by 180 degrees. Adjustment for, and confirmation of, zero temperature coefficient of the whole oscillator is a time-consuming business. An updated version of this circuit would probably use a dual-gate MOSFET, which should offer marginally better performance. The buffer stages could use more easily obtainable devices such as the 2N3904.

VFOs are generally not used commercially because every VFO needs setting-up. This is costly and it is cheaper to use a synthesiser system, even if the synthesiser is fairly complex. However, for the amateur, building and adjusting a VFO for an SSB receiver/transceiver is a viable option to obtain variable frequency operation. It requires good mechanical skills to build a good quality VFO. It requires a significant amount of time and effort to build the VFO and adjust the temperature compensation. However, time spent will be rewarded with an oscillator which is essentially drift-free.

In this design, it is not recommended that the tuning capacitor C4 Is replaced with tuning diodes. There are two reasons for this:

1. The RF voltage across C4 be may fairly high (depending upon the Q of the Inductor L). It may be too high for tuning diodes to work effectively.
2. The tuning voltage connection produces a port by which unwanted noise can be introduced into the oscillator.

An exception to this Is that a small value tuning diode may be used to provide a very limited tuning range RIT function.

# EVALUATION OF A NEWLY-BUILT VFO

It is essential that a VFO is properly tested for frequency stability before being used, particularly if it is to be used for transmitting. The process of testing the VFO is described below.

The oscillator should be located in the chassis in which it will finally be used and the output connected to a dummy load. The normal supply voltages supplied to the oscillator and buffer should be applied and the total current consumption checked. This should be no more than 30mA. Check that the circuit is oscillating by looking at the output on an oscilloscope, checking the frequency on a frequency counter, or listening on a receiver tuned to the expected frequency of oscillation. You may have to tune around on the receiver slightly because the oscillator probably won't be on the exact frequency you expected.

Adjust the oscillator to the required frequency range. This is normally done by adjusting the value of the inductor with the tuning capacitor set to mid-travel. The inductor is adjusted to set the oscillation frequency to the middle of the required tuning range. The padding capacitors are then adjusted to set the upper and lower frequency range required.

Frequency measurements should then be done to check the temperature stability. Temperature drift can be adjusted by exchanging fixed capacitors for ones having the same capacitance value, but different temperature coefficients (see the chapter on passive components).

## VOLTAGE-CONTROLLED OSCILLATORS

These are a class of oscillator which are similar to a VFO. However a tuning capacitor or tuning inductor is not used and the oscillation frequency is controlled by an applied voltage. This is achieved by applying the control voltage to a component (or components) in the oscillator tuned circuit whose reactance varies with the applied voltage. Having the oscillator frequency controlled by a voltage allows the oscillator to be used in applications where electronic control of the frequency is required. The usual element used for this is a tuning diode. This is a junction diode which is operated in reverse bias, and exhibits a capacitance which varies with the applied bias voltage. Diodes are available which are designed specifically for operation as a tuning diode. These are optimised for wide capacitance variation with applied voltage, and good Q. Other means of electronic tuning are possible in some circuits, but the use of tuning diodes is most common. The important parameter of the VCO is its tuning rate in MHz/volt and a typical design figure might be 1MHz/volt. VCO design is not without problems, particularly if the design needs a high tuning rate.

A VCO is normally used in a closed-loop control system. The system is designed to control the frequency of oscillation by means of the DC input control voltage. The requirements for short term stability (ie phase noise) are similar to VFOs. Therefore, most of the design rules presented above are equally applicable for VCOs as for VFOs. There is one exception to this: Because of the closed-loop nature of the VCO control system, the design rules for long-term stability of the oscillator no longer apply or can be significantly relaxed. VCOs can be built for virtually any frequency range from LF to microwave.

## CRYSTAL OSCILLATORS

A crystal oscillator is a type of oscillator which is designed to operate on one frequency only. The frequency of oscillation is determined by a quartz crystal and the crystal must be manufactured specifically for the frequency required. Quartz shows the property of the Piezo-electric effect and can be made to oscillate at radio frequencies. The main virtue of quartz is that it has a very low temperature coefficient so the resulting oscillator will be very temperature stable. It also has a very high Q. The types of crystal are similar to the ones used in quartz crystal filters.
Crystal oscillators exhibit the following characteristics.

- Good temperature stability.
- The high Q gives good oscillator phase noise.

They are used where an accurate frequency source is needed. Examples of this are in electronic instruments, frequency synthesisers and RF signal generators.

For a description of the equivalent electrical circuit of a crystal see the next building blocks chapter.

Crystal oscillators employ one of two modes of oscillation:

- Fundamental mode, which is normally used up to about 24MHz.
- Overtone mode, which is normally used above about 24MHz.

There really are no user serviceable parts inside a crystal, but it is interesting to open up an old one carefully to examine the construction. The crystal plate is held in fairly delicate metal springs to avoid shocks to the crystal. The higher-frequency crystals are very delicate indeed.

In the last few years, crystals have become available which are intended to be operated in fundamental mode up to about 250MHz. They are manufactured using ion beam milling techniques to produce local areas of thinning in otherwise relatively robust crystal blanks. This maintains some strength while achieving previously unattainable frequencies. These are intended for professional applications and are generally too expensive for the amateur, and not easily available.

Crystals intended for fundamental-mode operation are generally specified at their anti-resonant or 'parallel' frequency with a specific 'load' capacity, most often between 12 and 30pF. When a crystal is operated in fundamental mode, it is possible to tune, or trim, the oscillation frequency slightly by adjustment of a reactance in the oscillator circuit. This is usually done with a variable capacitor and is done to adjust for the initial tolerance of the crystal frequency and can be done subsequently to adjust the frequency as the crystal ages.

When a crystal is operated in the overtone mode, it is in a series resonant, mode. Overtones available are always odd integers and the normal overtones used are 3rd and 5th. Overtone operation allows an oscillator to operate up to about 130MHz (5th overtone). A trimmer capacitor can be used to effect a slight adjustment of the frequency but the ability to adjust the frequency is much less than with the fundamental mode of operation. The 'pullability' of an overtone crystal is approximately $R = 1/N^2$, where R is the ratio in number of Hz it can be pulled and N is the overtone number eg 3 or 5 etc compared to a fundamental (parallel resonant) crystal of the same frequency. Hence, a 10MHz fundamental might be able to be pulled, say, 200Hz and a 30MHz third overtone only about 70Hz at 10MHz and hence a total of 210Hz at 30MHz. The 10MHz fundamental when multiplied up to 30MHz would allow about 600Hz variation. For FM transmitters a fundamental crystal oscillator provides a more linear deviation method than overtone oscillators.

### Crystal Oscillator Circuits

The operation of a crystal oscillator is basically the same as the principles of any oscillator (VFO or VCO) described so far in this chapter. The best option for the amateur is to follow one of the circuits described here. Some crystal oscillator circuits do not need to use inductors, so the Hartley circuit is less common than the Colpitts type of circuit. Most circuits use a single active device.

For fundamental-mode crystal oscillators, any low power HF transistor is adequate, even a BC109. For overtone oscillators any RF type is recommended, eg NPN transistor types 2N2222A (metal can), BF494 or 2N3904 (plastic); N-channel plastic JFETs include J310, MPF102 and 2N3819. Virtually any dual-gate MOSFETs can also be used. A good rule of thumb is to choose a device with an Ft that is at least three to five-times the oscillator frequency. The BC109 and 2N2222 both have an Ft of about

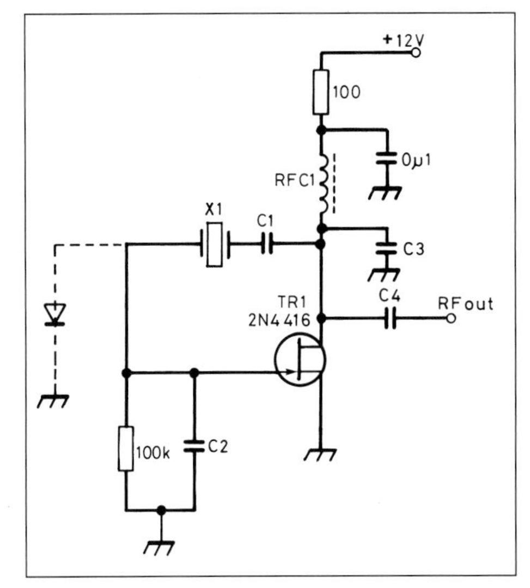


**Fig 4.16: Pierce oscillator. Reactance values are: $X_{C1}$, $X_{C3}$ 230; $X_{C2}$, $X_{C4}$ 450. (*W1FB's QRP Notebook*)**

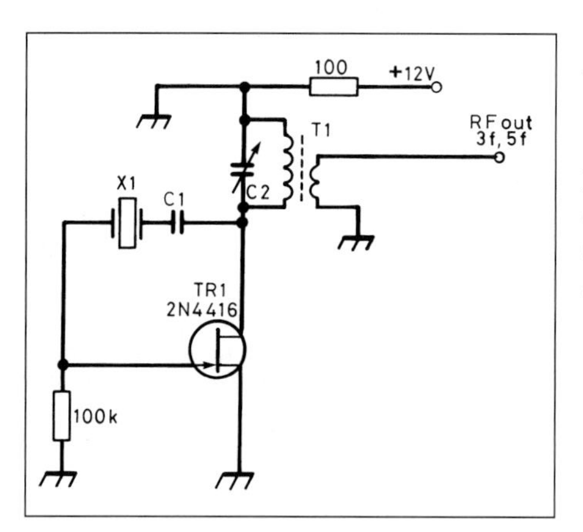


**Fig 4.17: Overtone oscillator. Reactance values are: $X_{C1}$ 22., $X_{C2}$ 150. at resonance. (*W1FB's QRP Notebook*)**

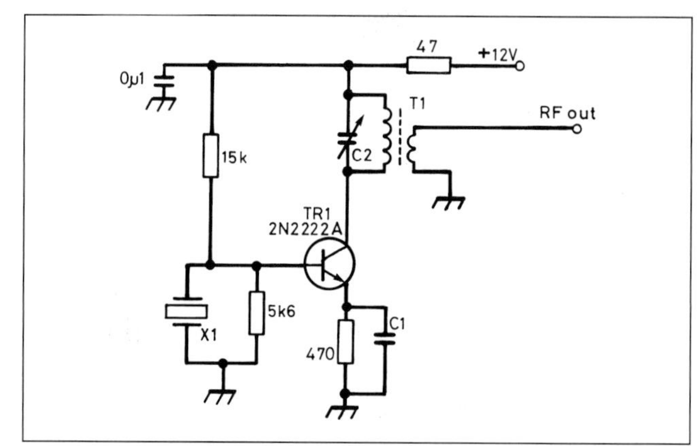


**Fig 4.19: The tuning of the LC collector circuit determines the mode of oscillation: fundamental or overtone**

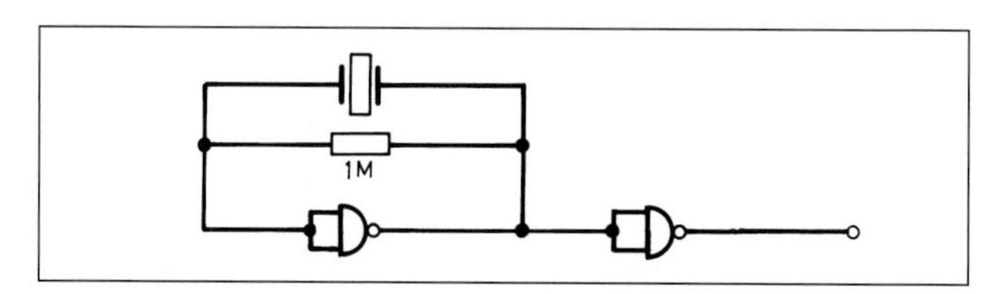


**Fig 4.20: CMOS oscillator circuit. With modern 'HC' CMOS, this will oscillate to over 20MHz with a good square-wave output**

100MHz so would be suitable for oscillators up to about 30MHz.

Excessive feedback should be avoided, primarily because it tends to reduce stability. Excessive oscillation levels at the crystal can lead to damage, but this is very unlikely in circuits operating on a supply of 12V or less. Sufficient feedback is needed to guarantee starting of the oscillator. Some experimentation may be needed, especially if the crystal is of poor quality. Pressure-mounted crystals and those in epoxy-sealed aluminium cans, eg colour-burst crystals from early colour television sets, are often poor.

This section describes a range of fixed-frequency crystal oscillators with practical circuit details. Crystal oscillator circuits are not hard to build, and most reasonable designs work fairly well since the crystal itself provides extremely high Q, and in such a way that it is difficult to degrade the Q without deliberately setting out to do so.

Note that an overtone crystal is more active on its fundamental frequency than the overtone. If no method is used to suppress the fundamental mode then spurii can be generated. A common method with overtone crystal oscillators is to use either an inductor in shunt with the crystal resonating with the holder capacitance or a low value resistor, the value should be about ten-times the ESR of the series resonant crystal. A third overtone crystal at 50MHz has an ESR of about 40 ohms worst case, so a 390 ohm resistor or slightly higher would be suitable.

**Figs 4.16 to 4.19** show four basic crystal oscillator circuits. Fig 4.16 is a Pierce oscillator using a FET as the active device. The circuit operates as a fundamental oscillator and the crystal is connected in the feedback path from drain to gate. The only frequency setting element is the crystal. For this to be the case, the RFC must tune with the circuit parasitic capacitances to a frequency lower than the crystal. Most small JFETs will work satisfactorily in this circuit. Note that the FET source is grounded, which gives maximum gain.

Fig 4.17 shows an overtone version of Fig 4.16. In this, a tuned circuit is inserted in the drain of the FET. The tuned circuit is tuned to the overtone frequency required and hence the FET only has gain at that frequency. This causes the circuit to operate only on the wanted overtone.

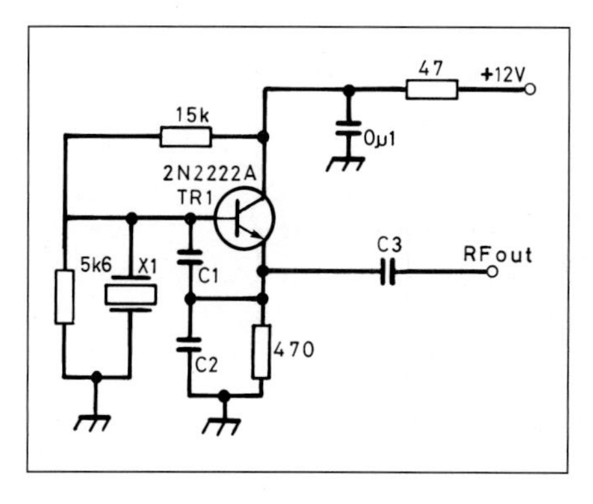


**Fig 4.18: Colpitts oscillator. Reactance values are: $X_{C1}$, $X_{C2}$, $X_{C3}$ 450Ω.. (*W1FB's QRP Notebook*)**

Fig 4.18 shows a Colpitts fundamental oscillator circuit, which uses a bipolar transistor. The chief advantage of this configuration is that the crystal is grounded at one end. This facilitates switching between crystals, which is much less easily achieved in the Pierce circuits.

Fig 4.19 shows an overtone oscillator using a bipolar transistor. One side of the crystal is earthed, which makes switching crystals easier. However, switching crystals in an overtone oscillators is not recommended, since the stray reactance associated with the switching can lead to loss of stability or moding of the crystal. This causes some crystals to oscillate at an unintended overtone, or on another frequency altogether. If switching of overtone crystals is required, it is better to build a separate oscillator for each crystal and then switch oscillators.

Oscillators can also be built using an integrated circuit as the active device. **Fig 4.20** shows an oscillator using a standard CMOS gate package. Almost any of the CMOS logic families will work in this circuit. The primary frequency limitation is the crystal. This oscillator will generate a very square output waveform, and with the faster CMOS devices can be a rich source of harmonics to be used, for example, as marker frequencies. Note that a square wave contains only odd harmonics.

Self-contained crystal oscillators can be purchased which are intended for use as clocks for digital circuits. They frequently contain the crystal plate, the gates and any resistors on a substrate in a hermetic metal case. These are unsuitable for most amateur receiver or transmitter applications because there is no way of tuning out manufacturing frequency tolerance or subsequent ageing. Also, the phase noise performance is usually poor.

Where an oscillator with very high temperature stability is required, a temperature-compensated crystal oscillator (TCXO) can be used. These are normally supplied in a sealed package and they were developed to maintain a precise frequency over a wide temperature range for long periods without consuming the stand-by power required for a crystal oven. These are suitable for use as a frequency reference in frequency synthesisers and are used by the mobile communication industry. They are now included in top-of-the-line amateur transceivers and offered as an

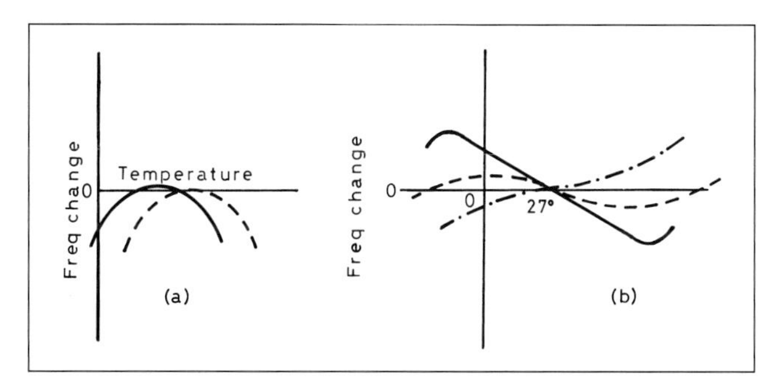


Fig 4.21: Crystal temperature coefficients. Frequency / temperature curves of 'zero-temperature coefficient' crystals. (a) Typical BT and DT-cut crystals. (b) Typical AT-cut crystals

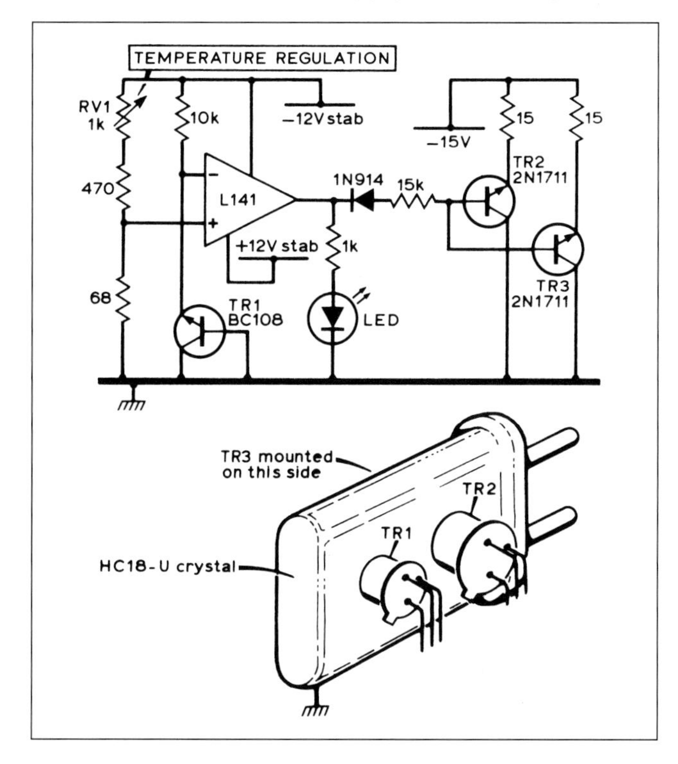


Fig 4.22: Proportional temperature control system for HC/16U crystals as proposed by I6MCF and using a pair of transistors as the heating elements and a BC108 transistor as the temperature sensor

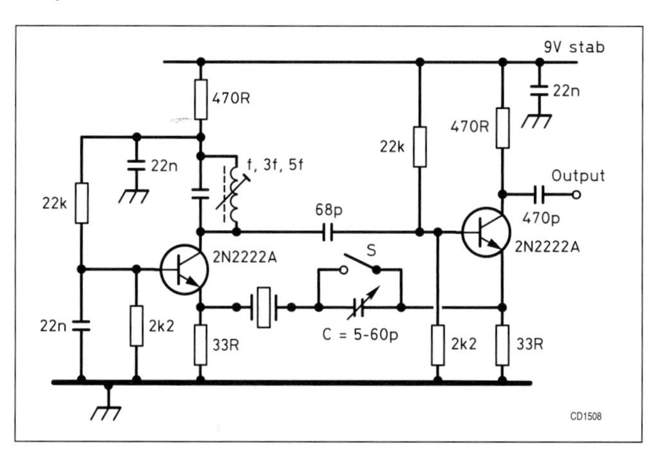


Fig 4.23: A crystal test oscillator using the Butler circuit. With S closed, the resonant frequency is generated. With S open and C set to the specified load capacitance, the anti-resonant frequency can be measured

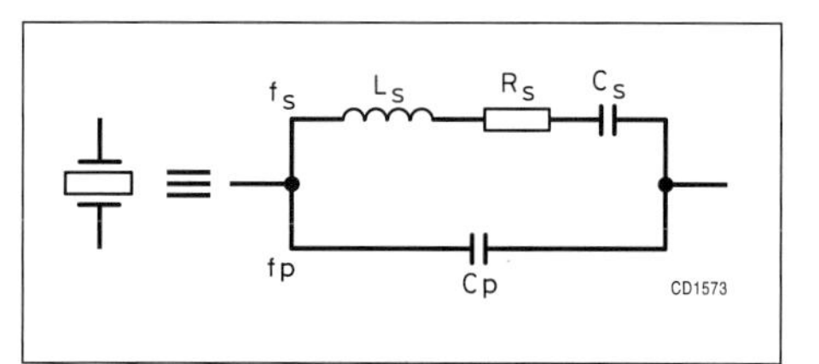


Fig 4.24: Crystal equivalent circuit

accessory in the mid-price field.

A TCXO is pre-aged by temperature cycling well above and below its normal operating range. During the last cycle, the frequency-vs-temperature characteristic is computer logged; the computer then calculates a set of temperature compensation values, which are stored in an on-board ROM. A good TCXO might have a tolerance of 1ppm over 0 - 60°C and age no more than 1ppm in the first year after manufacture, less thereafter. There is normally a provision to make small adjustments to the output frequency so that it can be adjusted from time to time.

## Buying Crystals

Crystals must be bought for the frequency and mode that you require. Many common frequencies can be bought ready-made (eg exact frequencies like 5MHz or 10MHz). If you need to get a crystal made you will need to specify (as a minimum):

- Frequency
- Mode
- Loading
- Operating temperature

Crystals are normally available in HC6/U, HC18/U, HC25/U or similar holders. If crystals are bought at a rally, much older, larger packages may be found.

Crystals made for commercial use are similar in specification to amateur devices; they may be additionally aged or, more commonly, ground slightly more precisely to the nominal frequency. For most purposes this is not significant unless multiplication into the microwave region is required.

Occasionally, crystals are specified at a temperature other than room temperature (see **Fig 4.21**). Ideally, the crystal should have a zero temperature coefficient around the operating temperature, so that temperature variations in the equipment or surroundings have little effect. In some cases, the crystal is placed in a crystal oven at 70°C. It is important to note, however, that the best result will only be obtained if the crystal is cut for the oven temperature. Although any temperature stabilisation is useful, a room-temperature-cut crystal may have little advantage in an oven. A scheme for thermal stabilisation of crystals is shown in **Fig 4.22**.

**Fig 4.23** shows a Butler oscillator which is a favourite for testing oscillator crystals. With the switch S closed, the circuit will oscillate at the crystal's resonant frequency; the output amplitude is then measured. Next, a 100Ω variable composition resistor is substituted for the crystal and adjusted to produce the same output. Its final resistance value is equal to the crystal resistance, $R_s$ in **Fig 4.24**.

# VARIABLE CRYSTAL (VXOS) AND CERAMIC-RESONATOR OSCILLATORS

A VXO is a type of crystal oscillator which is designed to have a much greater frequency adjustment range than a standard crystal oscillator. A typical use for a VXO is to obtain crystal control on an HF amateur band. The frequency can be adjusted by several kilohertz to avoid causing interference to other band users.

VXOs are used frequently in portable equipment where the virtues of simplicity and stability are particularly useful.

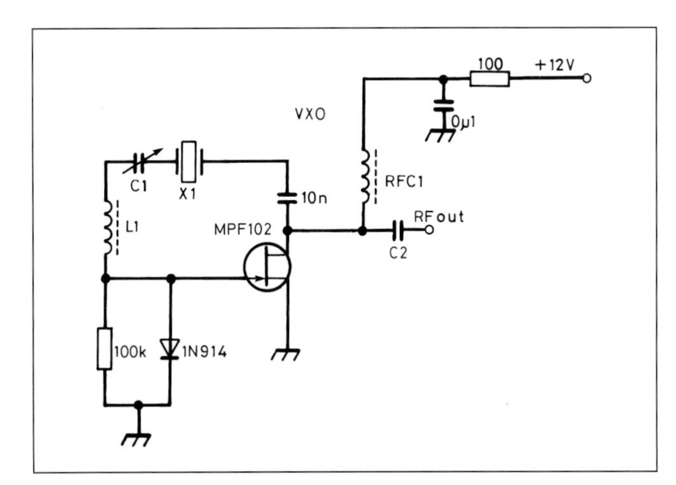


**Fig 4.25: A Pierce VXO for fundamental-mode crystals. The diode limits the output amplitude variation over the tuning range. Reactances at the crystal frequency are: C1 = 110Ω, C2 = 450Ω, L = 1.3Ω, RFC = 12.5kΩ**

Fundamental mode oscillators should be used because these can be pulled in frequency much more than overtone oscillators. The principle can be understood if one replaces the piezoelectric crystal by the electrical equivalent circuit shown in Fig 4.24. If a variable inductor is placed in series with that equivalent circuit, the resonant frequency of the combination can be 'pulled' somewhat below that of the crystal alone; for convenience, tuning is usually carried out with a variable capacitor in series with this coil which is over-dimensioned for that purpose.

When building VXOs the precautions described for VFO design should be used, but VXOs are less sensitive to external influences since the oscillation frequency is controlled primarily by the crystal. The two problems of VXOs are the very limited tuning range, typically of 0.1%, and the tendency to instability if this range is exceeded. A good VXO should be very close to crystal stability, with the added ability to change frequency just a little. Of course, multiplication to higher frequencies is possible, and several of the older types of commercial equipment successfully used VXOs on the 2m band. Typical VXO circuits are shown in **Figs 4.25 and 4.26**. Fig 4.25 is a Pierce oscillator using an FET as the active device. Stray capacitance in the circuit should be minimised. Ideally, the VXO should be able to shift from below to just slightly

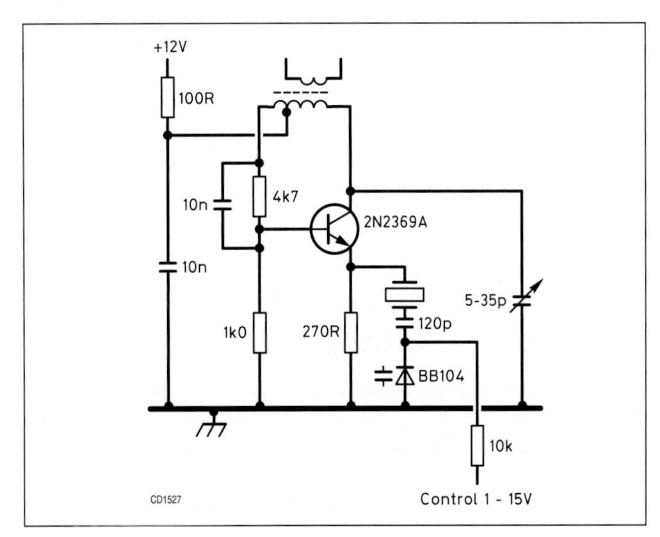


**Fig 4.27: Voltage-controlled 21.4 and 20MHz oscillators used by G3MEV in his 'heterodyne' 1.4MHz VXO system**

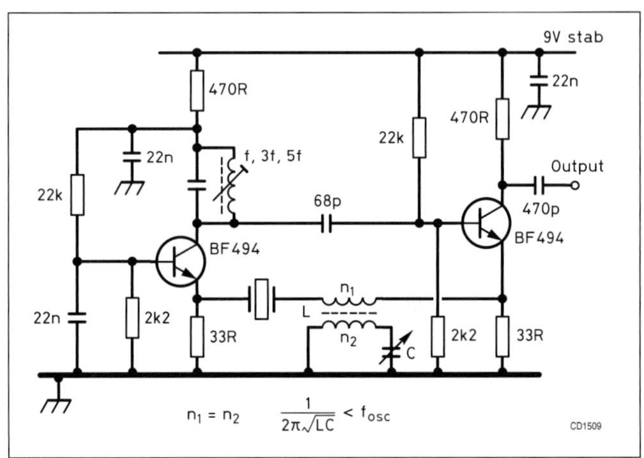


**Fig 4.26: A Butler VXO for fundamental or overtone crystals**

above the nominal crystal frequency. Fig 4.26 shows a two-transistor Butler oscillator.

Crystal switching in VXOs should be avoided and it is better to switch oscillators. Non-linear operation of capacitance against frequency should be expected, so the tuning scale will be non-linear. Note that some linearisation is possible using combinations of series and parallel trimmer capacitors with the crystal. Instead of a tuning scale, a better approach is to fit an LCD frequency readout or to use the station digital frequency meter, thus avoiding altogether the need for a tuning scale.

One way to achieve a wider tuning range is to use two VXOs ganged to shift in opposite directions at two high frequencies differing by the desired much lower frequency, as proposed by G3MEV in reference [7]. **Fig 4.27** shows G3MEV's VXO circuit. Another idea is the use of one VXO to interpolate between closely-

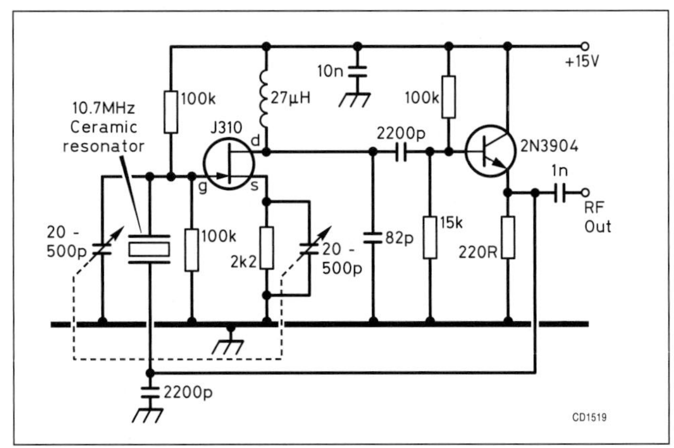


**Fig 4.28: K2BLA's variable ceramic-resonator oscillator covers a 2% frequency range**

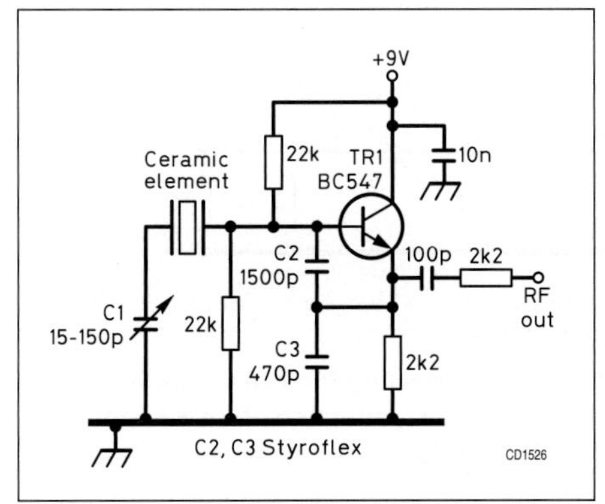


**Fig 4.29: LA8AK's oscillator can tune 3.5-3.6MHz with a 3.58MHz ceramic resonator**

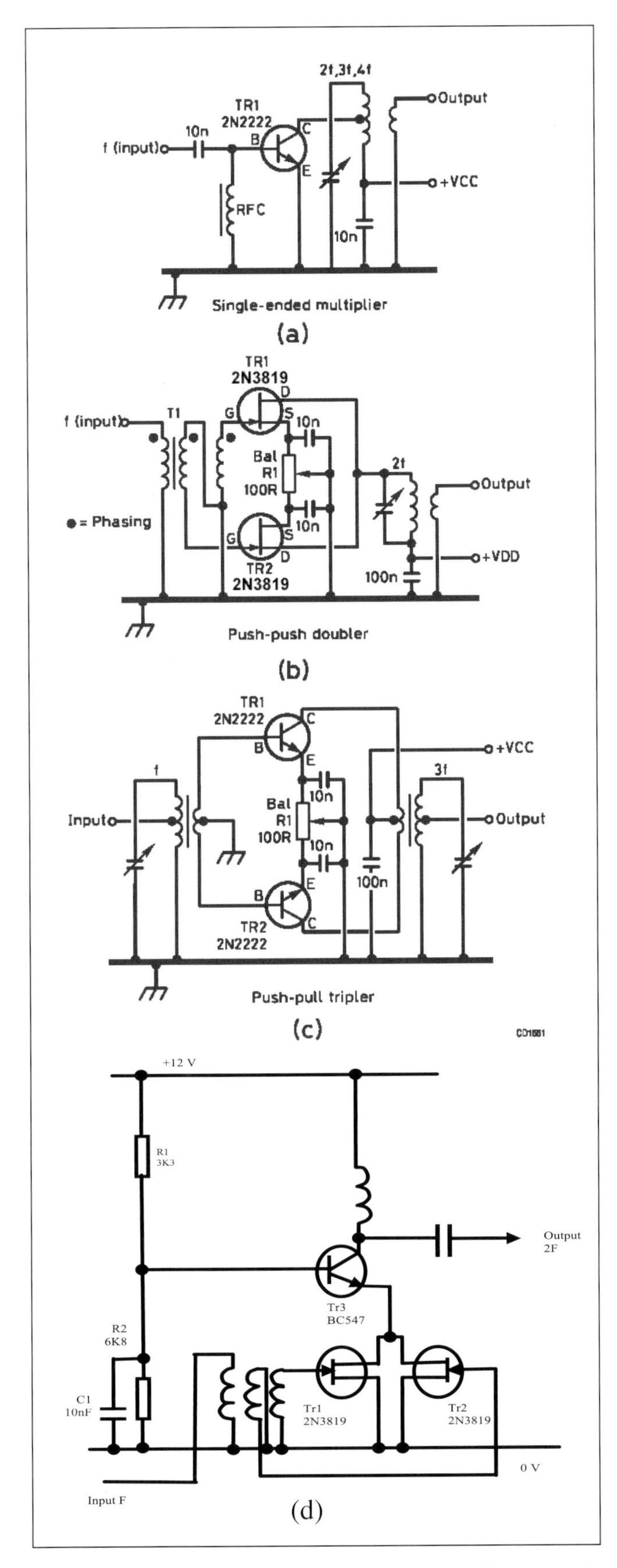


**Fig 4.30: Frequency multipliers. (a) Single-ended. (b) Push-push doubler. (c) Push-pull tripler**

spaced multi-channel synthesiser frequencies; The difference frequencies between a VXO tuning 24.10125 to 24.11125MHz (only 10kHz in 24MHz) and the 40 synthesised channels of a CB transmitter will cover 3.5 - 3.9MHz.

Ceramic resonators are now inexpensively available in many standard frequencies. They permit the construction of stable variable-frequency oscillators with a wider pulling range than VXOs. See reference [8].

Like quartz crystals, they are piezo-electric devices; they have a worse temperature coefficient than crystals and a lower Q, but for most HF applications it is adequate and the lower Q permits pulling over a wider frequency range. The oscillator circuits are similar to those for VXOs. **Figs 4.28 and 4.29** are examples.

## USING CRYSTAL CONTROL ON VHF

To obtain a crystal controlled source on VHF an overtone oscillator could be used. For example, if a crystal-controlled AM or CW transmitter is to be built for the 50MHz or 70MHz amateur bands, then there is significant benefit in using an overtone oscillator to directly generate the frequency required. Compared to using a fundamental oscillator with frequency multiplier stage(s), the overtone circuit is much simpler, and reduces the likelihood of spurious outputs occurring from the transmitter.

Crystals for fundamental use are, however, much cheaper than those intended for overtone use. Also if the source is intended for use in an FM transmitter, it will be easier to achieve the frequency deviation required if a fundamental oscillator is used. It is necessary to multiply the frequency and this can be achieved using frequency multiplier stages.

Normally an RF amplifier is biased and driven so that it does not generate a high level of harmonics at its output. However, when a frequency multiplier is designed, its bias and drive level are set so that it will deliberately generate harmonics of the the input frequency.

A frequency multiplier has its output tuned to a multiple of its input frequency, eg x2, x3 or x4. These multipliers are known as doublers, triplers and quadruplers, respectively. VHF/UHF bipolar transistors work well as frequency multipliers. FETs do not generally make good frequency multipliers, except where a frequency doubler is required. The square-law characteristic of FETs means that they make very effective frequency doublers. **Fig 4.30(a)** illustrates a typical multiplier.

If a higher multiplication factor is required, it is better to use two or more multipliers, eg for x9 multiplication use two tripler stages. The multiplier may be operated with a small amount of forward bias dependent on the transistor forward transfer characteristic but the stage must be driven hard enough (with RF) for it to function correctly. The bias is adjusted for maximum efficiency (output) after the multiplier has been tuned to the desired output frequency.

The multiplier can be single-ended, as in **Fig 4.30(a)**, or with two FETs configured as a push-push doubler as in **(b)**, or two transistors as a push-pull tripler as in **(c)**. The efficiency of **(b)** and (c) is typically higher than **(a)**. The push-push doubler discriminates against odd-order multiples and the push-pull tripler dis-

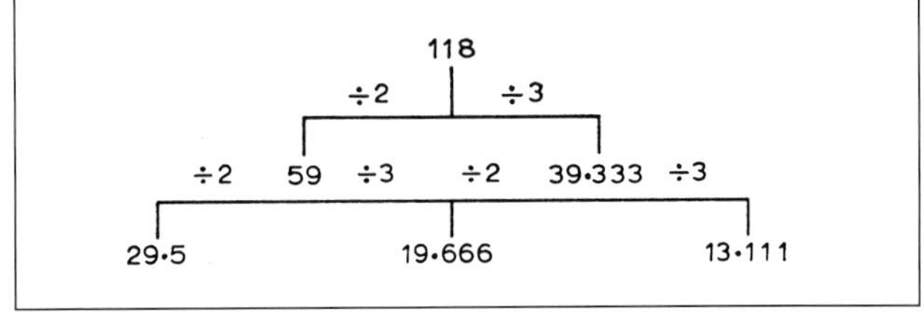


**Fig 4.31: Type of chart for determining the fundamental crystal frequency required in an oscillator multiplier chain**

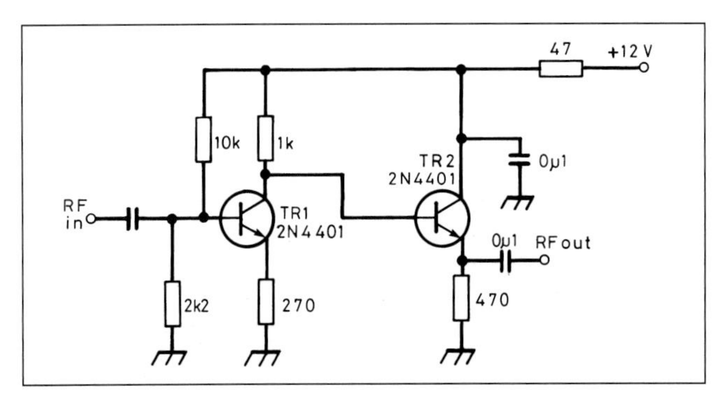


Fig 4.32: Direct-coupled buffer circuit (*W1FB's QRP Notebook*)

criminates against even-order multiples, so additional attenuation of unwanted outputs is much improved over the single-ended multiplier. **Fig 4.30(d)** shows an Improved version of **Fig 4.30(b)** that uses a cascode configuration to give a significantly increased output level. R1 in **(b)**, and **(c)** and **(d)** is adjusted for optimum dynamic balance between TR1 and TR2, ie maximum minimum unwanted multiplier output.

It is always good practice to start with a reasonably high oscillator frequency in a receiver LO multiplier chain, thus requiring fewer multiplier stages. This is because there is the distinct possibility that one of the unwanted multiples will reach the mixer and cause a spurious response (or responses) within the receiver tuning range. The magnitude of any unwanted injection frequencies depends on the operating conditions of the final multiplier, including the working Q of its output circuit. To minimise this problem it is advisable to include a buffer amplifier stage with input and output circuits tuned to the LO frequency between the final multiplier and the mixer.

These circuits may consist of a single high-Q tuned circuit called a high-Q break or two loosely coupled circuits. Helical filters provide an ideal solution to this problem. Two filters may be cascaded to give greater attenuation.

Adequate filtering is essential in transmitter frequency multipliers, particularly between the final multiplier and the first RF amplifier, to attenuate all unwanted multiplier output frequencies from the oscillator to the final multiplier. In order to establish the crystal frequency it is useful to draw a chart such as **Fig 4.31**. This shows the combination of frequencies and multiplication factors which could be used to obtain an output on 118MHz.

It is important to understand that any frequency change in the final multiplier output is the frequency change (or drift) of the oscillator multiplied by the multiplication factor. As an example, if the frequency change in a 6MHz crystal oscillator was 50Hz

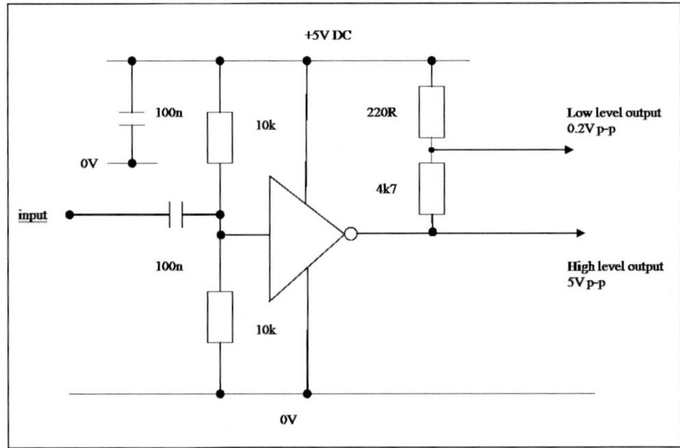


Fig 4.34: Buffer using a Schmitt trigger

and the multiplication factor is 24, then at 144MHz the change will be 1.2kHz. This change is not too serious for NBFM but would not be acceptable for SSB equipment.

## BUFFER STAGE

The output of an oscillator is not normally connected directly to the load. This is because changes in load impedance will affect the oscillator frequency. Changes in load impedance may be caused by:

- Changes in temperature
- Switching from receive to transmit
- Changes in supply voltage

A buffer amplifier must always be used to isolate the oscillator from the load. A buffer should be used with VFOs, VCOs and crystal oscillators. It is an RF amplifier which is designed for high reverse isolation. This greatly reduces the effect of load changes on the oscillator frequency.

The following needs to be known when designing a buffer amplifier:

- The expected load impedance.
- The gain required.
- The output level required.

Some of the oscillator circuits in this chapter show a buffer amplifier. Examples are:

- Fig 4.15. In this circuit the buffer is formed by TR2 and TR3.
- **Fig 4.32**. In this circuit the buffer is formed by TR2 which operates as an emitter follower.

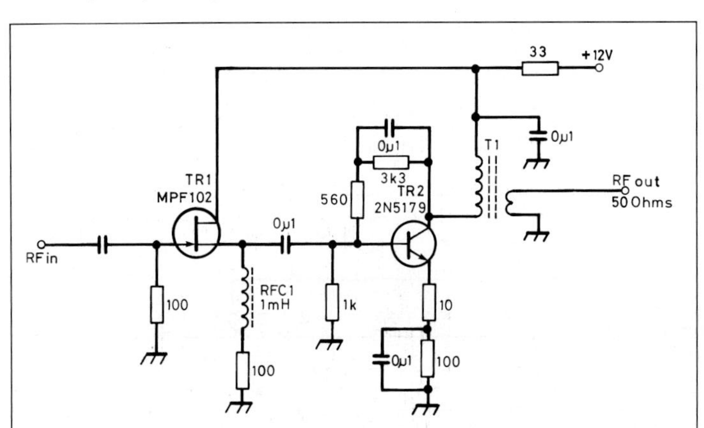


Fig 4.33: Transformer output buffer circuit (*W1FB's QRP Notebook*)

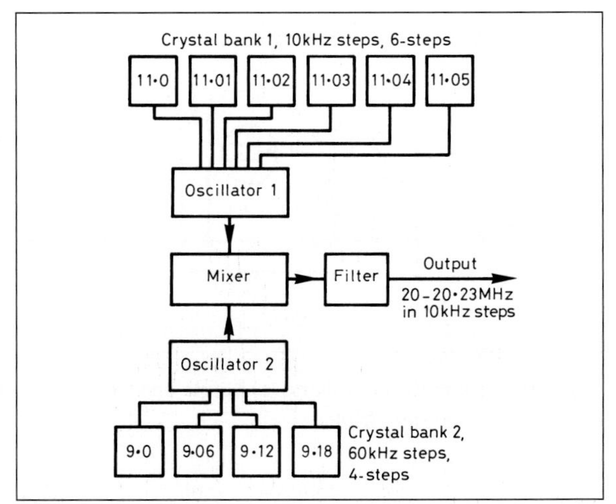


Fig 4.35: Simplified crystal bank sythesiser

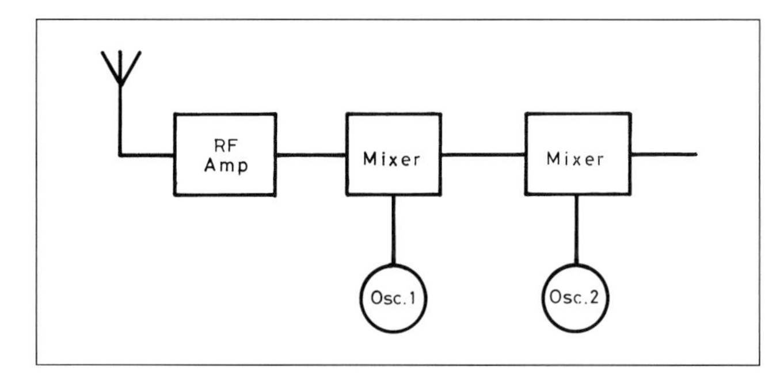


**Fig 4.36: Multiple oscillators/mixers, an alternative to frequency synthesis**

- **Fig 4.33** shows a two-stage buffer amplifier which is designed to drive a 50 ohm load.
- **Fig 4.34** shows a buffer using a CMOS Schmitt trigger invertor. This is suitable for mixers that require a large voltage swing (5V p-p). There is also an alternative low-level output that can be used to drive Gilbert Cell type mixers.

Amplifiers employing feedback should not be used on their own as a buffer because these have poor reverse isolation. Buffers should be built using high quality components and should be solidly built like a VFO.

# FREQUENCY SYNTHESIS

The description 'frequency synthesiser' is generally applied to a circuit in which the output frequency is non-harmonically related to the synthesiser reference frequency. Therefore, this excludes circuits which use a crystal oscillator and frequency multiplier stage to generate the output frequency. This section contains a description of the operation of analogue synthesis, PLL frequency synthesis and Direct Digital synthesis (DDS). Early forms of synthesis are no longer used commercially, but are described here because they are still of relevance to the radio amateur.

## Analogue Synthesis Systems

Analogue (or Direct) synthesis was the first form of synthesis. It involves the production of an output frequency, by the addition of several (sometimes many) oscillators, usually by a combination of mixing, multiplying, re-mixing etc with many stages of filtering. This was difficult and expensive to set up really well. Examples were primarily limited to specialist, often military, synthesisers in the 'fifties. The output spectrum could be made very clean, especially in respect of close-to-carrier phase noise, but only by significant amounts of filtering. Direct synthesisers are still used in specialist areas, because they can give very fast frequency hopping and low spurious levels well into the microwave region.

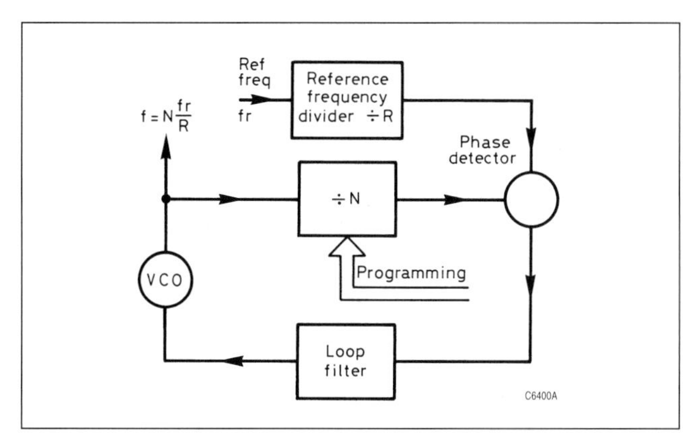


**Fig 4.37: Basic PLL synthesiser**

A simple version did go into quantity production, in the early US-market CB radios. This was the crystal bank synthesiser shown in **Fig 4.35**. This arrangement used two arrays of crystals, selected in appropriate combinations to give the required coverage and channel spacing. It worked very well over a restricted frequency range, could offer very low phase noise (because crystal oscillators are inherently 'quiet'), and could be made acceptably (for the time) compact and low power. An example on the UK market in the 1970s, was the Belcom Liner 2, which was an SSB transceiver operating in the 2m (144-146MHz) band. This used crystal bank synthesis with a front-panel-tuned VXO on a further crystal oscillator for the conversion up and down from the effective 'tuneable IF' at 30MHz to the working frequency of 145MHz.

The need for the VXO illustrates the weakness of this type of synthesiser; close channel spacing is essentially prohibited by the very large number of crystals which would be needed. Triple bank versions were described in the professional literature, but were not produced for amateur purposes. The other disadvantage of this class of synthesiser is the high cost of the crystals, but it remains an interesting technique capable of very high performance with reasonable design.

A popular alternative to synthesis for many years used a carefully engineered VFO and multiple mixer format. Examples are in the KW range, the Yaesu/Sommerkamp FT-DX series and, in homebrew form, the G2DAF designs (see earlier editions of this *Handbook*). This technique survived in solid-state form, with among others, the FT101 [9]. The technique is illustrated in **Fig 4.36** [10].

## Phase-Locked Loop (PLL) Frequency Synthesisers

In parallel with the analogue synthesis systems described above, came early attempts at combining the best VFO characteristics with PLL synthesis. A PLL synthesiser employs a closed-loop control system, which uses analogue and/or digital circuits to generate the required output frequency. This allows a single-conversion receiver, and is able to achieve greater receiver dynamic range. An example of this was G3PDM's classic phase-locked oscillator, as used in his receiver and described in earlier editions of this *Handbook* [6]. This was an analogue synthesiser which used mainly valves, but with a FET VFO. This design is still regarded as the standard to beat. The FET VFO was discussed earlier in this chapter (Fig 4.15). A demonstration of this receiver was given where a signal below 1μV suffered no apparent degradation from a 10V signal 50kHz away. This amazing >140dB dynamic range

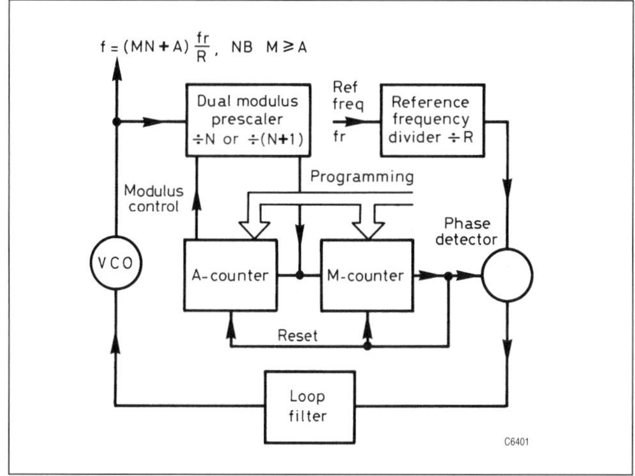


**Fig 4.38: Dual modulus prescaler**

was made possible by the extremely low phase noise of this synthesiser, and the superb linearity of a beam-deflection mixer valve.

Most modern PLLs contain digital circuits and the PLL output frequency is normally driven by digital control inputs. The first digital synthesisers appeared in about 1969, chiefly in military equipment. An outstanding example was the UK-sourced Clansman series of military radios.

A basic PLL digital synthesiser block diagram is shown in **Fig 4.37**. The VCO output is divided in a programmable divider down to a frequency equal to that of the crystal-controlled reference source; this may be at the crystal frequency or more usually at some fraction of it. The phase comparator compares the two frequencies, and then produces a DC voltage which is proportional to the difference in phase. This is applied to the VCO in such a way as to drive it towards the wanted frequency (negative feedback). When the two frequencies at the phase comparator input are identical, the synthesiser is said to be in-lock. Ideally, the phase comparator is actually a phase and frequency comparator, so that it can bring the signals into lock from well away. The 'DC' term referred to is a varying voltage determined by the frequency and phase errors. The bandwidth available to this voltage is said to be the loop bandwidth and it determines the speed at which lock can be achieved.

The programmable divider is a frequency divider using digital counters. Fully programmable dividers of appropriate frequency range were not immediately available to the early PLL designers. Two solutions were proposed for this:

- The dual-modulus pre-scaler.
- A variant of the crystal mix scheme.

The dual-modulus counter (**Fig 4.38** uses a dual-modulus pre-scaler and is a particularly clever use of circuit tricks to achieve the objective of high-speed variable ratio division. Dividers were designed which could be switched very rapidly between two different division ratios, for example 10 and 11. If the divider is allowed to divide by 11 until the first programmable counter (the A-counter) reaches a preset count (say, $A$) and then divides by 10 until the second counter reaches $M$, the division ratio is:

$$(11 \times A) + 10 \times (M - A) \text{ or } (10M + A)$$

The advantage in the system is that the fully programmable counters M and A need only respond, in this example, to one-tenth of the input frequency, and can therefore use lower cost, slower logic. The disadvantage is that there is a minimum available count of A x (11), where $A$ is the largest count possible in the A-counter. This is rarely a problem in practice.

Fig 4.38 shows the complete loop consisting of the VCO, the dual-modulus counter, a main counter, a reference oscillator and divider chain, a phase detector and a loop amplifier/ filter.

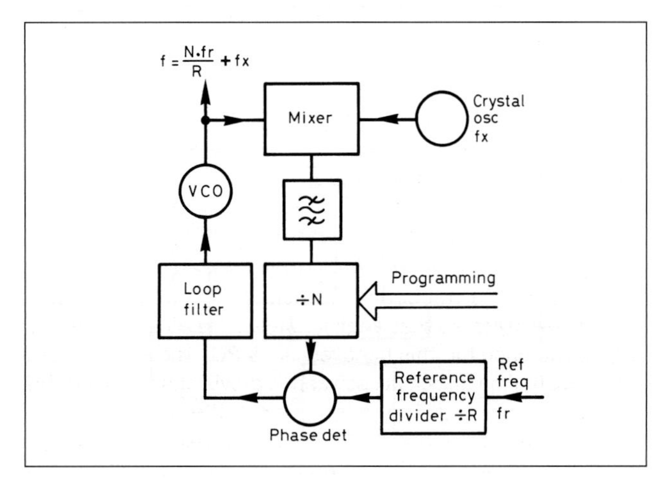


**Fig 4.39: Mixer-loop PLL**

This type of synthesiser is relatively easy to design when a synthesiser is needed with large step sizes. It has been described many times in the professional and amateur radio fields [11]. An example of this is a synthesiser for an FM receiver, which is required to step in 25kHz or 12.5kHz steps. This is also used in TV receivers, satellite receiver and Band II FM tuners. Most commercially available PLL synthesiser chips contain the digital parts of the PLL. That is the main divider, the reference divider and the phase detector, and the counters are normally programmed by serial input. Most modern PLL chips will operate up to VHF without the need for a dual-modulus pre-scaler. They also provide two (or more) phase detectors to cover the far-from-lock and locked-in cases.

The second method, that of crystal mixing with a digital synthesiser, is probably the most widely used technique in amateur equipment, where wide operating frequency range is not needed. It offers advantages of simplicity and low component count, especially when the digital functions of the system can be contained in a single chip; see **Fig 4.39**. Unlike the dual modulus scheme, no pre-scaler is needed, but a high-frequency mixer is necessary, usually with two crystals, one for mixing down and one for the reference.

The PLL is a closed-loop system, which employs negative feedback. Like any feedback system, the loop has a finite speed at which it can operate and this is defined by the closed-loop bandwidth. There is a (usually) wider bandwidth called the lock-in bandwidth, which is the range of starting frequencies from which lock can be achieved.

The closed loop bandwidth is under control of the designer and is set by the open-loop parameters. These are:

- VCO gain
- Divider ratios
- Phase detector gain
- Loop filter amplitude/frequency and phase/frequency responses

The control loop is able to control and hence reduce the effects of VCO noise within the closed-loop bandwidth, this includes long term drift in the VCO. Outside the loop bandwidth, VCO noise is unaffected by the loop and the closed-loop bandwidth is normally set to about one tenth of the reference frequency. The closed loop is a second-order system and so it is possible for the loop to exhibit loop instability. The designer controls the loop stability primarily by controlling the response of the loop filter. The lock-up time is inherently related to the closed-loop bandwidth. For an FM-only rig, with say 12.5kHz minimum frequency step, lock-up time can be very quick.

For SSB receiver and transceivers, there are strong commercial pressures to replace the VFO with a full digital synthesis system because this avoids many of the problems and costs of building and setting up a good VFO. However, trying to reproduce the action of a good VFO in digital synthesiser form is difficult. A PLL synthesiser can only tune in steps and so to reproduce the VFO action for SSB requires very small steps, typically 20Hz. This cannot be done in a single loop because the loop bandwidth would have to be very narrow, and the loop would be very slow. Professional systems overcome this by complex, multiple-loop synthesis, which can be expensive to design. They can also be expensive to manufacture and are sometimes bulky.

The operating frequency of a digital synthesiser is normally controlled by pulses from a shaft encoder system. This allows control of the frequency by a tuning knob in the same way as a VFO. Frequently, the tuning rate of the knob is programmable so

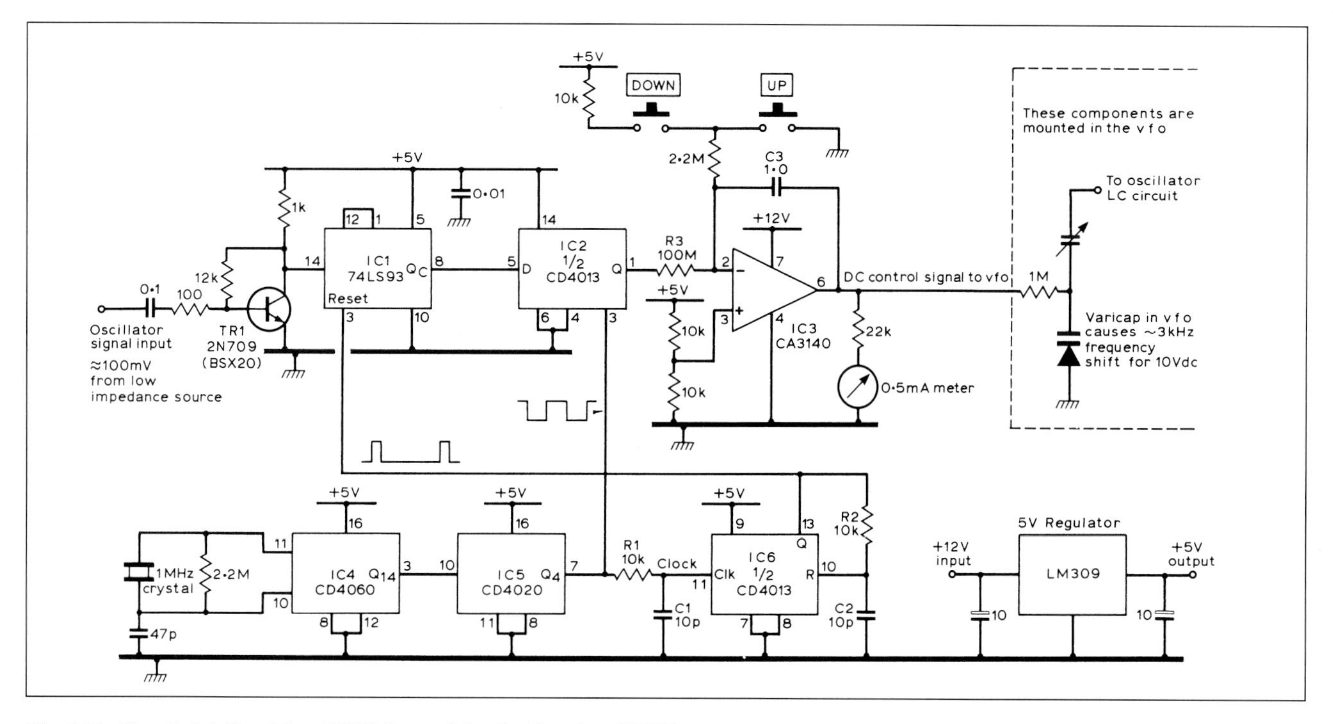


**Fig 4.40: Circuit details of the CMOS form of the 'huff-and-puff' VFO stabilising system - a version of PA0KSB's system as described in *Ham Radio***

that the tuning rate (in kHz per turn) is adjustable for different modes of signal. AM or FM modes can use faster tuning rates than SSB. The equipment uses a digital frequency display and this normally has resolution of 100Hz or 10Hz. It should be noted that this does not mean that the frequency is accurate to 100Hz or 10Hz. (Resolution doesn't equal accuracy). The frequency accuracy of the synthesiser is determined entirely by the accuracy of the reference oscillator.

PLL synthesiser design is a very demanding and essentially analogue task. The VCO is particularly difficult to design because very small unwanted voltages induced onto the tuning voltage will generate noise or spurious outputs from the synthesiser. Nevertheless, truly excellent performance can be achieved by the professional engineer in factory-built equipment. Properly designed, PLL synthesis works very well indeed, and most commercial amateur equipment uses some form of PLL synthesiser. However, the problems of providing narrow frequency increments, with the demand for complete HF band coverage (at least on receive) have lead to rather compromised synthesiser designs. This has been seen in particular in reviews of otherwise excellent radios limited in performance by synthesiser noise and spurious outputs causing reciprocal mixing.

A PLL digital frequency synthesiser can be designed to cover a much wider frequency range than a typical good VFO, but still achieve good frequency stability. It offers an alternative to the amateur who doesn't have the skills or tools to build a VFO. Some synthesisers (eg those that use serial control) require software to enable them to operate, but synthesisers can be built using hardware only, and hence require no software. The frequency display is normally digital, and it is generally easy to obtain the frequency display information from the synthesiser control circuits.

It is possible for the very dedicated and well-equipped amateur to design a PLL system for SSB use, particularly as he does not cost his time into the project. It should be said that a spectrum analyser, or at least a continuously tuneable receiver covering the frequency of interest and a harmonic or two, is essential for any synthesiser work. Loop stability and noise should be checked at several frequencies in the synthesiser tuning range.

Amateur equipment practice has tended to take a different route. A synthesiser is employed which has a minimum step size of typically 1kHz. Increments between the 1kHz steps are achieved either with a separate control knob, on older equipment, or by a digital-to-analogue converter (DAC) operated by the last digit of the frequency setting control. The analogue voltage tunes a VXO in the rig, which may either be the conversion crystal in the synthesiser or the reference crystal itself. Care must be taken to ensure that the pulling range is accurate or the steps will show a jump in one direction or the other at the 1kHz increments. Actually, many rigs do show this if observed carefully; the reason is that until recently the DAC was a fairly simple affair consisting of a resistor array and switching transistors.

More recently, commercial DACs have been used, but this is not a complete solution, since the VXO is unlikely to be linear to the required degree, so some step non-linearity is inevitable. At least if the end points are not seriously wrong, this should not be a problem. Some rigs do tune in 10Hz steps, especially on HF; this is an extension of the technique to 100 steps instead of just 10. Again, there is the issue of the 1kHz crossover points; but with 10Hz steps this can be made less noticeable on a well-designed and adjusted rig.

One variant on PLL synthesis which appeared some years ago, and which is particularly suitable to amateur construction, is the so-called huff-and-puff VFO [12]. This could be described as a frequency-locked loop and consists (**Fig 4.40**) of a VFO, and a digital locking circuit. This has the same frequency reference chain and phase-sensitive detector, but the divider chain is largely omitted. The VFO is tuned using a variable capacitor, and the loop locks the VFO frequency to the nearest multiple of the reference frequency possible. The reference is typically below 10Hz, so that analogue feel is retained in the tuning. To the operator, this behaves like a PLL. The loop can take a significant time to lock up, so various provisions to speed this up can be employed. The loop must not jitter between different multiples of the reference, hence the need for very high initial stability in the VFO. As a commercial proposition, this scheme is unattractive compared to a true PLL, but as an

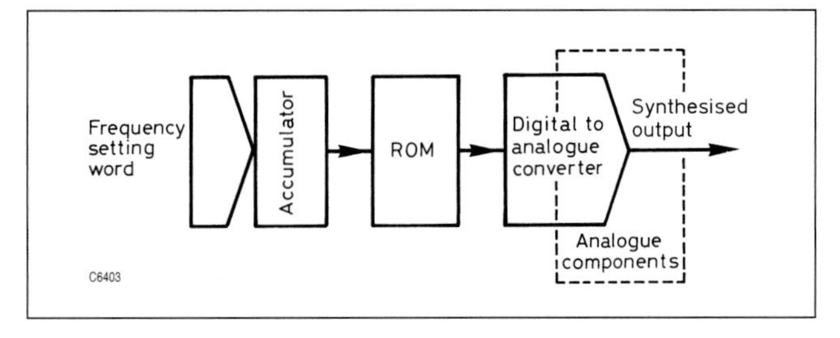


**Fig 4.41: The DDS concept**

amateur approach it can overcome the VFO tuning problem in an elegant way.

## Direct Digital Synthesis (DDS)

Direct Digital Synthesis is a method of frequency synthesis which differs from the methods described so far. A DDS is virtually all-digital and the only analogue part is a digital-to-analogue convertor (DAC) at the DDS output. In fact many modern DDS devices now contain two DACs, supplying two separate outputs. A DDS does not contain a negative feedback closed-loop system like a PLL. The DDS offers the possibility of generating an output which has very small frequency steps but, unlike a PLL, very rapid tuning to new frequency. Compared to a digital synthesis PLL type system, a DDS is relatively easy to design because the complex, critical electronics are contained within the DDS chip. Construction is relatively easy because it is built up as a conventional logic PCB and there are none of the problems of setting-up which occur with a conventional VFO. The DDS approach to synthesis is most suited to the amateur who is familiar with building digital systems. It may require some software to be written to drive the DDS chip. The output frequency stability (both short-term and long-term) is determined by the characteristics of the reference oscillator. For the amateur, the best option for the reference oscillator is to use a quartz crystal oscillator and it is worthwhile using a high quality oscillator circuit to obtain low phase noise and good frequency stability. The alternative is to purchase a ready-made high- performance crystal oscillator module.

Direct digital synthesis has been frequently mentioned in *RadCom*, and is a feature of current amateur equipment [13, 14]. The basic direct digital synthesiser consists of an arrangement to generate the output frequency directly from the clock and the input data. The simplest conception is shown in **Fig 4.41**. This consists of a digital accumulator, a ROM containing, in digital form, the pattern of a sine wave, and a digital-to-analogue converter. Dealing with the accumulator first, this is simply an adder with a store at each bit. It adds the input data word to that in the store. The input data word only changes when the required frequency is to be changed. In the simplest

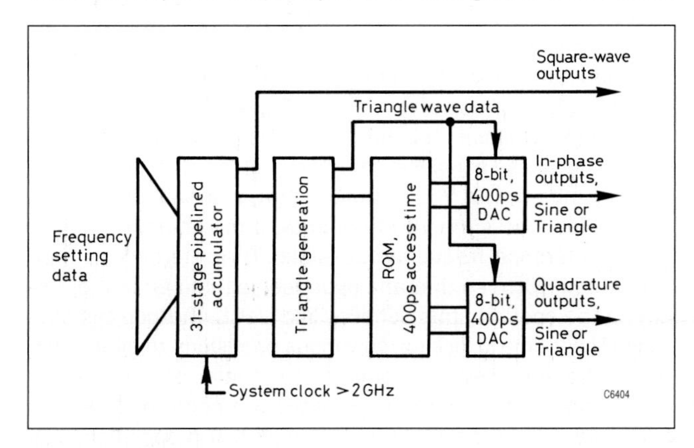


**Fig 4.42: More complex DDS**

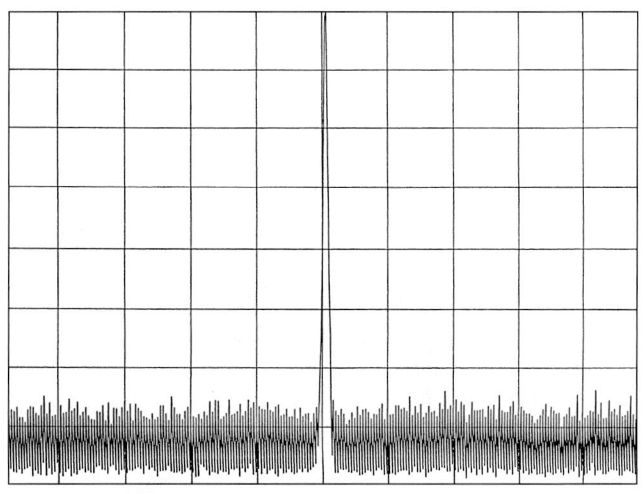

**Fig 4.43: DDS spectrum at 250MHz with a 1GHz clock. 10MHz/div, X axis; 10dB/div, Y axis**

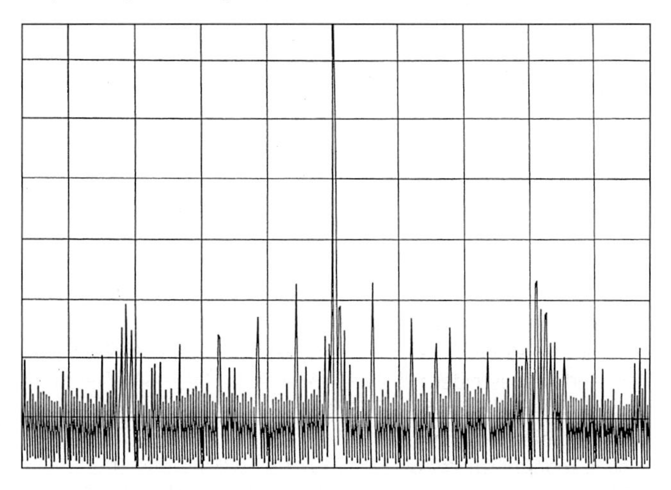

**Fig 4.44: DDS spectrum at 225MHz with a 1GHz clock. 10MHz/div, X axis; 10dB/div, Y axis**

case the length of accumulator is the clock frequency divided by the channel spacing, although it is usually calculated the other way round, ie if a 5kHz channel spacing up to 150MHz is needed, and since at least two clock pulses are required per output cycle, a clock frequency of at least 300MHz will be required. More conveniently, a 16-bit accumulator gives 65,536 steps. If the step size is to be 5kHz, then a clock frequency to the accumulator of 65,536 x 5kHz is needed, ie 327.68MHz. Such a DDS would produce any frequency in the range covered, ie 5kHz to over 100MHz in a single range without any tuned circuits.

Other frequency increments are available; any multiple of 5kHz by selection of input data, and others by choice of clock frequency, eg 6.25kHz requires a clock at 204.8MHz with a two-channel (2 x 3.125kHz) program word. Of course, frequency multiplication or mixing can be used to take the output to higher frequencies. Programming is very easy, DDS devices have either a parallel-input data format or are designed to be driven from a serial input. The control circuits which drive the DDS can also be used to drive a digital frequency display.

A more advanced device is shown in **Fig 4.42** [15]. This has on-chip DACs and facilities for square, triangle and sine outputs. Two DACs are used, because both phase and quadrature output signals are available in true and complement form. The most significant bit (MSB) from the accumulator feeds the square-wave output buffer direct. In parallel, the next seven bits from the accumulator, which digitally represent a sawtooth waveform, feed a set of XOR gates, under control of the MSB, so that a triangle output is generated, in digital form, at this point in the circuit. Actually, two triangles in quadrature are generated. These can be digitally steered to the output DACs or can be used to address a ROM containing data for sine and cosine waves; only 90° is needed, since all four quadrants can be generated from one. Finally, if selected, the digital sine/cosine is fed to the DACs for conversion to analogue form.

This device was designed to operate at up to 500MHz output frequency with 1Hz steps, so the full range is 1Hz - 500MHz, again in a single 'range' with no means of or requirement for tuning. The clock frequency is necessarily very high; the quadrature requirement adds a further factor of two, so nominal clock frequency is $2^{31}$ Hz, ie 2.147483648GHz. This must be supplied with crystal-controlled stability, although since phase noise is effectively divided down in the synthesiser and the frequency is fixed, it can be generated by a simple multiplier from a crystal source.

Output spectra are shown in **Figs 4.43 and 4.44**. Fig 4.43 shows a clean output at exactly one quarter of the clock frequency. Fig 4.44 shows a frequency not integrally related to the clock frequency. Here the spurious sidebands have come up to a level about 48dB below the carrier. This is the fundamental limitation of the DDS technique, and where most development work is going on. The limit comes from the finite word size and accuracy of the DAC, and incidentally the ROM, although this could have been made bigger fairly easily.

Fast DACs are difficult to make accurately, for two reasons. The first is technological. IC processes in general have a limit to a component matching accuracy of about 0.1%, ie 9 or 10 bits accuracy in a careful design. Other techniques, such as laser trimming of the resistors, can be used on some processes. However, this tends to use older, slower processes, and especially requires large resistors for trimming which slow the DAC settling time. The second problem is simply the requirements on fast settling: the DAC is required to get to the final value quickly and this will in general happen with the smallest number of bits. An 8-bit system as in this example was chosen to fit the process capabilities and the device requirements, but this leads to a limitation in the high level of spurious signals present.

Basically, although some frequencies are very clean, in the worst case the spurious level is 6*N* dB below the carrier, where *N* is the number of effective DAC bits. For an 8-bit system, this gives -48dB. Oversampling, ie running the clock at more than twice the output frequency, gives some improvement (at 6dB per octave) by improvement of the DAC resolution, but only up to a limit of the DACs accuracy. Typically this is about 9 bits or -54dB. In some applications, this may not be serious, since filtering can remove all but the close-in spurs. Phase-locked translation loops into the microwave region also act as filters of relatively narrow bandwidth, while retaining the 1Hz step capability.

In the amateur rigs using DDS, the synthesisers operate at relatively low frequencies and are raised to the working frequency by PLL techniques. This is complicated, but makes fine frequency increments available without the compromise of PLL design. The devices used are CMOS types, with DAC accuracies of 8 and 10 bits. This gives spurious signals of theoretically -60dB referred to the carrier, which is adequate with filtering from the PLL.

## Direct Digital Synthesiser for Radio Projects

This project was written for *RadCom* by Andy Talbot, G4JNT [16]. In this article, a DDS chip is included in a small stand-alone module controlled by straightforward text-based messages from a PC. The module has been designed to be used as a drop-in component in larger projects such as receivers or transceivers.

An onboard PIC microcontroller allows control of the DDS chip from a standard PC via the serial port, and a straightforward command syntax has been developed so that standard software commands, written in any language, can be used to set the

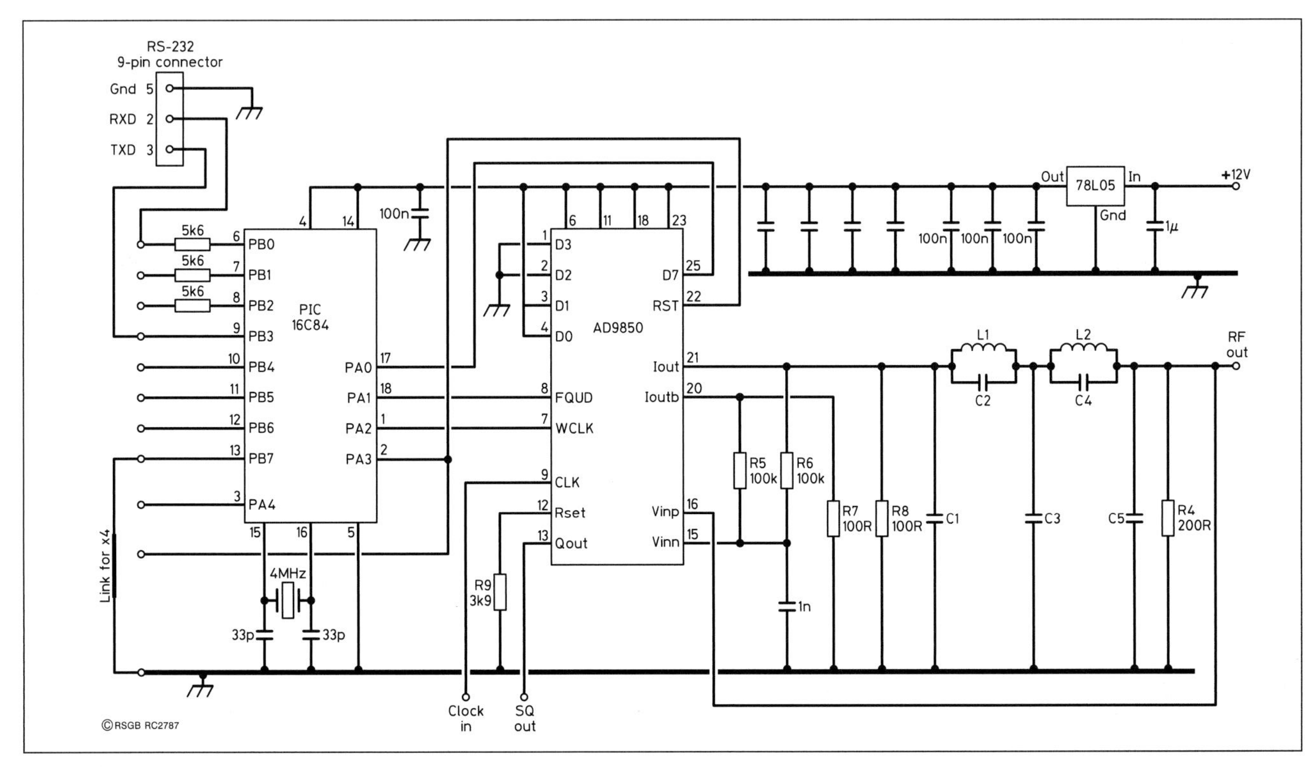


**Fig 4.45: Circuit diagram of the AD9850 DDS module. Filter components and some decoupling components are dependent upon clock frequency**

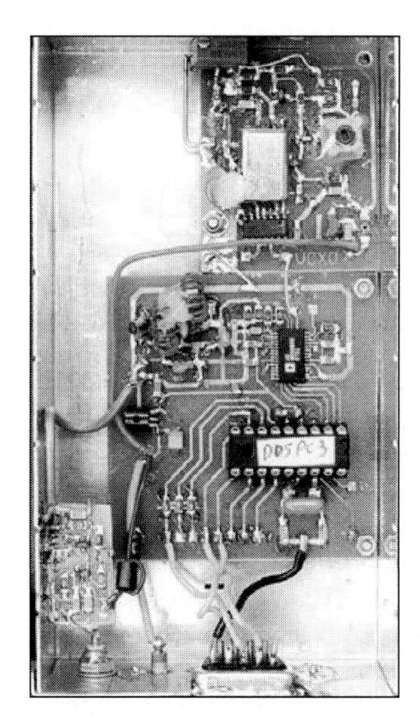

**Fig 4.46: A prototype version of the G4JNT DDS board**

operating parameters. This PIC can easily be reprogrammed to suit any user's requirements in a stand-alone project and enough spare Input / Output lines are provided to allow for this.

This article is intended to provide an overview of the DDS module as a component for larger projects and, therefore, only limited details are included.

## DDS module design

The module is based around an Analog Devices AD9850 DDS chip, full details of which are available from Analog Devices [17]. The device will accept a clock signal up to 120MHz, although a suitable source is not provided within the module. This is best left to individual constructors.

The DDS can generate an output up to approximately one third of the clock frequency with a resolution of over 4 billion so, for a 120MHz clock, frequencies from DC to 40MHz can be produced in steps of approximately 28 millihertz (mHz), and so the actual frequency of the clock is unimportant, provided of course it is known with reasonable accuracy. The RF output is at a level of 1V p-p (0.35V RMS) and the output impedance is 100Ω.

The supplied PIC controller translates text based messages received from the serial port into command codes for the AD9850, and (optionally) stores these in non-volatile RAM for immediate setting at switch-on. Spare memory in the PIC allows user information, such as the exact clock frequency, to be readout on request, allowing common software to be written that can drive individual modules, each having different clock frequencies.

Another option available in the firmware is a 'times-four' output, where the output from the module is designed to drive a quadrature frequency generator which performs the final frequency division.

A single hex-digit module address is included as part of the command syntax, to allow multiple modules to be driven in a multi-drop arrangement from a common controller with one COM port.

## Output spurii

Spurious output signals from a DDS are complex in their nature and are not harmonically related - the device data sheet [17] gives more details.

All spurii from the design here are at levels of -60dB or better. This figure is obtained from the manufacturer's specification and is affected by the design of the output filter. One useful facet about DDS circuitry is that the spurious levels below the output filter cut-off frequency are inherently dependent on the circuits internal to the DDS chip, and are not affected by any poor circuit layout; this aspect is often trouble-some in other synthesiser designs, as filtering cannot remove close-in products.

60dBc spurious levels may be considered a bit excessive if this module were to be used alone as the local oscillator of a high-performance wide-band receiver, but there are additional techniques such as Phase Locked Loops that can clean up this signal.

## Construction

The circuit diagram is shown in **Fig 4..45**, from which it can be seen that the module actually has very few components, most of them being for decoupling and output filtering. Apart from the integrated circuits, all other components are either of the surface-mount 1206 or 0805 style; for the output filter, wire-ended components are used. The PIC controller is mounted in a socket for easy re-programming.

The photograph **Fig 4.46** shows a prototype version of the DDS board, together with a high-stability 94MHz source on the PCB top right and an output buffer amplifier bottom left.

Values of components for the output filter depend on the clock frequency chosen, as do the values of decoupling capacitors. With a clock input that can range from a few kHz up to 120MHz, the optimum values of these can vary over a wide range.

## Software control

Commands are sent using ASCII / Hex characters over a bi-directional RS-232 link with no handshaking. Parameters are 19,200 baud, 8 data bits, no parity, 1 stop bit. A simple terminal programme such as HYPERTERM (Windows®) or PROCOMM (DOS) can be used to command the frequency source. Set this to 19200 N 8 1, full-duplex, no flow control and all start up, and modem commands set to null.

Alternatively, custom software can be written to drive the COM port with the commands.

The first character sent is a board address which precedes all commands. This is a single Hex character sent as ASCII 0 - F and potentially allows up to 16 modules to be driven from the same COM port.

The next character is a command which may have hex data following it.

- Q followed by eight hex digits for the frequency command word terminated by a carriage return [CR]
- P followed by two hex digits for phase word and [CR]
- U writes the data sent above to the AD9850 DDS chip
- W as for U, and also stores all data in the PIC's non-volatile EEPROM memory for switch-on next time
- Y followed by one Hex digit, changes the board address and stores in EEPROM. No [CR] needed
- K followed by 10 hex digits and [CR]. User data, not used for driving the DDS. (In practice, read as decimal number for user data, typically clock frequency)
- R read back current data values - not necessarily those in EEPROM

The 32 bit or 8 hexadecimal character, value N (required for frequency-setting), can be derived from:

$$N = F_{out} / F_{clock} \times 2^{32}$$

Phase can be set to any one of 32 values in increments of 11.25 degrees. These form the five highest significant bits of the phase word Pxx. The lowest three bits are ignored.

Data is sent back from the DDS in text strings which can be read directly by application software.

An example of a command to set the output frequency is:

| | |
|---|---|
| 5Q03D70A3D [CR] | Board address 5, set frequency word N = hex 03D70A3D |
| 5U | Programme the DDS to this value (with a 120MHz clock, this gives an output at 1.8MHz). |

### Applications

Any experiments or testing that needs an agile frequency source is a candidate. Just about any programming language that includes commands to drive the serial port can be employed, and does not even have to be PC-based. The only requirement is that you can actually write suitable software!

One application written to demonstrate the functionality of the module is for generating a narrow-band Multi-Tone Hellschreiber signal for LF use. SMT Hell transmits visible text as an image, and can be received on a frequency / time plot, commonly known as a spectrogram or waterfall display, using public domain commonly-available audio analysis software. The software generates SMT-Hell signals by directly commanding the DDS module (in real time) to set the frequencies that make up the vertical elements of each character sent. The horizontal components are made up by appropriately setting software delays. Transmissions as narrow as 2.5Hz bandwidth have been sent on 137kHz and successfully decoded as visible letters even where the signal is completely inaudible below the noise.

The ability to set the output signal phase to one of 32 values means that slow Phase Shift Keying is also possible by direct command. An additional command code (T) allows frequency and phase updates to be synchronised to an external trigger input on port B1 such as that from a GPS receiver. The DDS chip is updated within 3μs of the trigger signal rising edge and will allow, for example, precisely-timed low data rate signalling experiments.

By using the DDS output to drive the reference input to a conventional Phase Locked Loop synthesiser, the best of both worlds becomes possible.

The high frequency capability of PLL synthesisers, up to many GHz, can be coupled with the tiny step size of the DDS. For example, a PLL operating with an output at 2.4GHz could be made with a step size of 0.55Hz. DDS frequency resolution is considerably better than the stability of most crystals will allow. In fact, the actual crystal frequency can be measured, stored in the user data area of memory and then used in subsequent high-accuracy output frequency calculations.

DDS devices are evolving rapidly, to operate at higher frequencies and with more bits, but at reasonable cost. This trend will continue, and will make DDS more and more attractive for amateur applications. Web sites are the best source of up-to-date information on DDS chips.

At the time of writing, the fastest devices available from Analog Devices are the AD9858, AD9910 and AD9912. These will operate with a clock frequency up to 1000MHz, and can operate with either series or parallel control (except the AD9912 which uses serial control only). The AD9858 is a 10-bit device and the AD9910 and AD9912 are 14-bit devices.

For other examples of the use of DDS techniques, see [18] and the chapter in this handbook on software defined radios.

## REFERENCES AND BIBLIOGRAPHY


[1] *Electronic and Radio Engineering*, F E Terman, McGraw-Hill, any older edition, eg the fourth, 1955

[2] *Radio Receiver Design*, K R Sturley, Chapman and Hall, 1953

[3] *Amateur Radio Techniques*, 7th edn, Pat Hawker, RSGB

[4] *Special Circuits Ready-Reference*, J Markus, McGraw-Hill

[5] *GaAs Devices and their Impact on Circuits and Systems*, ed Jeremy Everard, Peter Peregrinus

[6] 'A receiver with noise immunity and frequency synthesis', Peter Martin, G3PDM, *Radio Communication Handbook*, 5th edn, RSGB, 1976, pp10.104-10.108

[7] 'Technical Topics', *Radio Communication* May 1991, p31

[8] See 'Using ceramic resonators in oscillators', Ian Braithwaite, G4COL, *RadCom*, February 1994, pp38-39

[9] *Maintenance Service Manual*, FT-101 series, Yaesu Musen Co Ltd

[10] 'An easy-to-set-up amateur band synthesiser', Ian Keyser, G3ROO, *Radio Communication* December 1993, pp33-36

[11] The Plessey Company, part of GEC, have, when they were still making telecommunications ICs, published several applications books describing PLL synthesis; any of these are well worth acquiring, although they are only obtainable second-hand. Examples are *Radio Communications Handbook*, 1977; *Professional Radio Communications*, 1979; *Radio Telecoms IC Handbook*, 1987; *Professional Data Book*, 1991; and *Frequency Dividers And Synthesisers IC Handbook*

[12] Klaas Spaargaren, PA0KSB, in the ARRL publication *QEX* (Feb 1996, pp19-23), Chas F Fletcher, G3DXZ in *RadCom* December 1997, as well as corrections, improvements, additions and comments in 'Technical Topics' in *RadCom*, Jul, Aug, Sep 1996, Feb, Dec 1997, and Feb, Apr 1998

[13] 'Technical Topics', Pat Hawker, G3VA, *Radio Communication* Dec 1988, pp957-958

[14] 'Direct digital synthesis, what is it and how can I use it?', P H Saul, *Radio Communication* Dec 1990, pp44-46

[15] Plessey Semiconductors Data Sheet *SP2002*

[16] 'Direct Digital Synthesis for Radio Projects'. By Andy Talbot G4JNT. *RadCom* Nov 2000

[17] Analog Devices web site. http://www.analog.com

[18] 'CDG 2000 HF Transceiver'. *RadCom* Jun 2002 to Dec 2002


***Peter Goodson*** *is qualified with a Hons Degree in Communication Engineering. He began his career working on the component-level design of HF (SSB) and VHF (FM) transceivers. He now works on the system level design of RF systems, which includes the issues of communications security, EMC, co-siting of radio equipment, RF hazard assessments and Immunity to lightning strikes.*

# Building Blocks 2: Amplifiers, Mixers etc

# 5

Peter Goodson, G4PCF and Mike Stevens, G8CUL/F4VRB

For an amateur designing a complex circuit, it is normally a good idea first to draw the proposed circuit in block diagram form. For example, **Fig 5.1** shows the block diagram of a proposed 80m SSB receiver.

- The block diagram shows the intended signal flow. Normally, inputs are shown on the left of the block diagram and outputs are on the right. Therefore, the signal flows from left to right.
- The block diagram does not show the 0V (ground) connections between the stages. It is assumed that all stages have a common 0V connection.
- Each block in the diagram has an intended function, eg mixer, amplifier, oscillator.
- At the block diagram stage of designing, the designer may have no idea about what circuits will be used to 'fill' the blocks.
- The block diagram can be used to set the gain distribution.
- For complex circuits, the block diagram is easier to understand than the complete circuit diagram.
- The block diagram can be used to check for potential spurious responses.
- The block diagram can be used to set the noise figure/gain requirements of the front-end.

See other chapters in this handbook for many examples of block diagrams. Once the designer is happy that the diagram represents a viable design, each of the blocks can be 'filled in'. The purpose of these Building Blocks chapters is to present examples of circuits which can be used to fill the blocks.

Some building blocks are standard circuits using discrete semiconductors, and some are made up from application-specific integrated circuits. The performance of building blocks can be described by the following parameters.

- **Gain** describes the available AC power gain of the circuit from input to output. The source and load impedance may also need to be specified for the gain specification to be useful. Gain is normally expressed in dB.
- **Isolation** (normally between the ports of a mixer) refers to the input applied to one port affecting whatever is connected to another port. In the simplest of receivers that uses a mixer without much isolation, peaking the antenna tuning circuit connected to the signal input may 'pull' the frequency of an unbuffered free-running oscillator connected to the local oscillator (LO) port. Conversely, the LO signal may unintentionally be transmitted from the antenna. Isolation is normally expressed in dB.
- **Noise** is generated in all circuits. It is quantified as a 'noise figure', expressed in decibels over the noise power generated by a resistor at the prevailing temperature, eg 50Ω at 27°C. For example, if front-end mixer noise is significant as compared to the smallest signal to be processed, the signal-to-noise ratio will suffer. The same noise level or 'noise floor', in terms of dBm, is shown in **Fig 5.2**. Below 10MHz, however, atmospheric noise will override receiver noise and most mixers will be adequate in this respect.
- **Overload** occurs if an input signal exceeds the level at which the output is proportional to it. The overloading input may be at a frequency other than the desired one. Overload is normally expressed in dBm.
- **Compression** is the gain reduction which occurs when the signal input magnitude exceeds the maximum the circuit can handle in a linear manner. The input level for 1dB of gain reduction is often specified. Note the bending of the 'fundamental component' line in Fig 5.2. The signal input level should be kept below the one causing this bending.
- **Intermodulation** products are the result of non-linearity in the circuit, and are generated when the input contains two or more signals. Intermodulation performance is often specified as the power level in dBm of the 'third-order intercept point'. This is the fictitious intersection on a signal input power versus output power plot, Fig 5.2, of the extended (dashed) fundamental (wanted) line and the third-order intermodulation line. Note that the third-order line rises three times steeper than the fundamental line.
- **Harmonic distortion** is a result of non-linearity in the circuit. The result of harmonic distortion is to produce, at the output, harmonics of the input fundamental frequency.

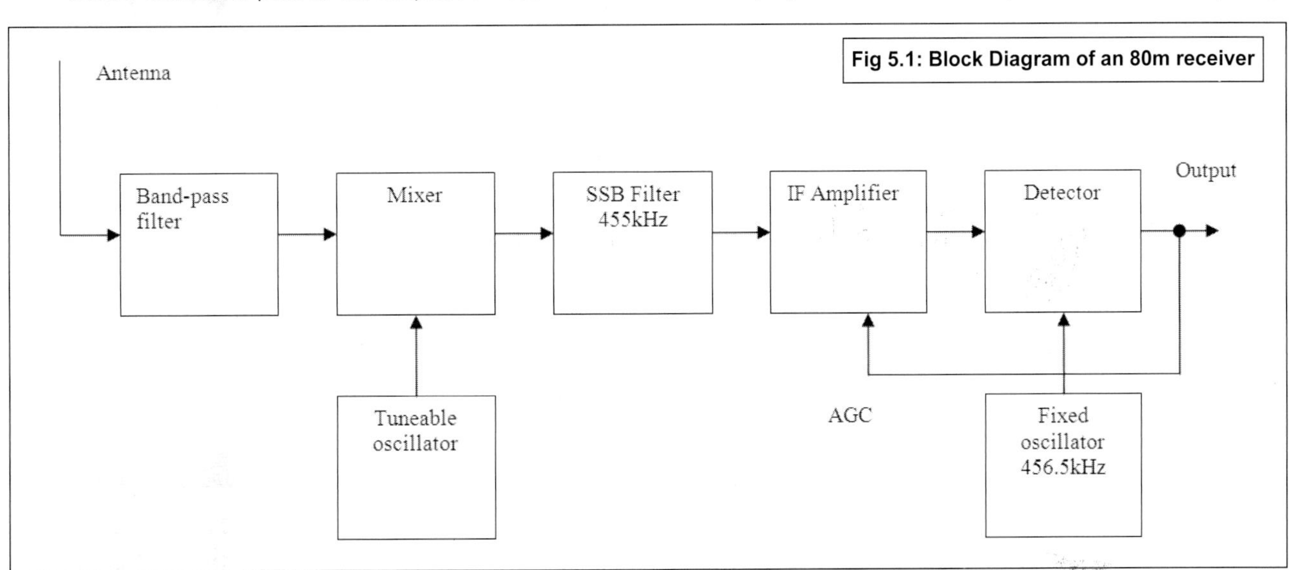


Fig 5.1: Block Diagram of an 80m receiver

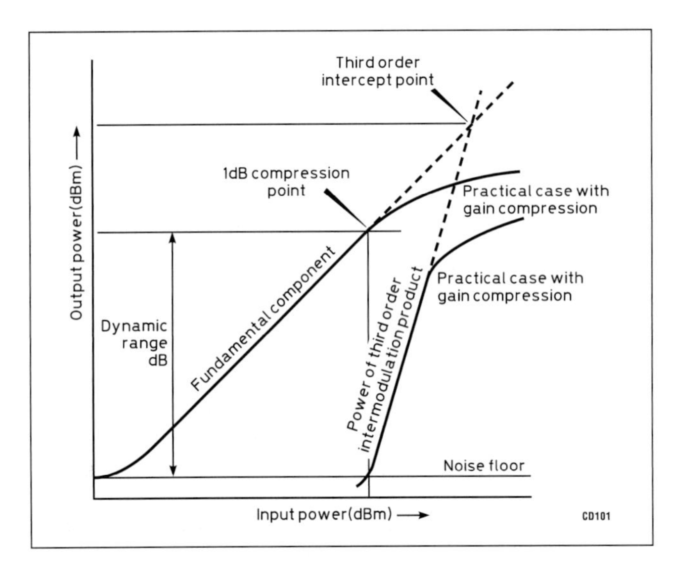


**Fig 5.2: Noise floor, 1dB compression point, dynamic range and 3rd-order intercept indicated on a mixer output vs input plot (*GEC-Plessey Professional Products IC Handbook*)**

Even-order harmonics (ie 2nd, 4th, 6th etc) can be virtually eliminated by the use of push-pull circuits.

- **Operating impedance**. This defines the input and output impedances of the circuit, or the source and load impedances for which the circuit is designed. Circuits which operate at low impedance (typically 50 ohms) can be designed to have wide bandwidth. They normally operate at low voltage, high current. Circuits which operate at high impedance (typically 1000 ohms) tend to have reduced bandwidth. They tend to operate at a higher voltage, but with lower current consumption than low-impedance circuits.
- **Bandwidth** defines the range of frequencies over which the circuit is designed to function. A wideband circuit design is one which typically covers several octaves and would be used, for example, for a circuit required to operate over 1.8 to 52MHz. Wideband design is not always necessary for amateur use, and is of no value for those who only operate on one amateur band.
- **Power consumption** is the power which the circuit draws from the DC power supply. The power drawn is given by the product of the DC supply voltage multiplied by the operating current. Power consumption is most significant when the equipment is operated from batteries, because high power consumption will reduce battery life.

Building blocks may be classed as passive or active. Passive circuits require no power supply and therefore provide no gain. In fact they may have a significant insertion loss. Examples of passive circuits are diode mixers, diode detectors, LC (inductor + capacitor) filters, quartz filters and attenuators. Examples of active circuits are amplifiers, active mixers, oscillators and active filters.

## MIXERS

### Terms and Specifications

A mixer is a three-port device, of which two are inputs and the third is the output. The output voltage is the mathematical product of the two input voltages. In the frequency domain, this can be shown to generate the sum and difference of the frequencies on the two input ports. The following circuits are better known by the specific function that they perform, but they are all examples of circuits which operate as a mixer, or multiplier.

- Front-end mixer
- Detector
- Product detector
- Synchronous detector
- Demodulator
- Modulator
- Phase detector
- Quadrature FM detector
- FM Stereo decoder
- TV Colour demodulator
- Analogue multipliers
- Multiplexer
- Sampling gate

These all use the principle of multiplying the two input signals. However, there are differences in the performance of the circuits. For example, a mixer required to operate as the front-end mixer in a high-performance receiver would be designed for high dynamic range, whereas a mixer operating as a detector in the back-end of a receiver only needs very limited dynamic range. This is because the level of signal into the demodulator is constant (due to the receiver AGC system).

It should be noted that there is a distinction between mixers (or multipliers) in this context and mixers which are used in audio systems. The mixers used in audio systems operate by adding the input voltages, and the function is therefore mathematically different.

Of the two inputs to a mixer, one, $f_s$, contains the intelligence, the second, $f_o$, is specially generated to shift that intelligence to (any positive value of) $\pm f_s$, $\pm f_o$, of which only one is the desired output. In addition, the mixer output also contains both input frequencies, their harmonics, and the sum and difference frequencies of any two of all those. If any one of these many unwanted mixer products almost coincides with the wanted signal at some spot on a receiver dial, there will be an audible beat note or birdie at that frequency. Similarly, in a transmitter, a spurious output may result.

In a general-coverage receiver and multi-band transmitter

| **Unbalanced mixer** | | | | | |
|---|---|---|---|---|---|
| $f_o$ | $2f_o$ | $3f_o$ | $4f_o$ | $5f_o$ | |
| fs | $f_o \pm f_s$ | $2f_o \pm f_s$ | $3f_o \pm f_s$ | $4f_o \pm f_s$ | $5f_o \pm f_s$ |
| 2fs | $2f_s \pm f_o$ | $2f_o \pm 2f_s$ | $3f_o \pm 2f_s$ | $4f_o \pm 2f_s$ | $5f_o \pm 2f_s$ |
| 3fs | $3f_s \pm f_o$ | $3f_s \pm 2f_o$ | $3f_o \pm 3f_s$ | $4f_o \pm 3f_s$ | $5f_o \pm 3f_s$ |
| 4fs | $4f_s \pm f_o$ | $4f_s \pm 2f_o$ | $4f_s \pm 3f_o$ | $4f_o \pm 4f_s$ | $5f_o \pm 4f_s$ |
| 5fs | $5f_s \pm f_o$ | $5f_s \pm 2f_o$ | $5f_s \pm 3f_o$ | $5f_s \pm 4f_o$ | $5f_o \pm 5f_s$ |
| **Balanced mixer–half the number of mixer products** | | | | | |
| $f_o$ | $2f_o$ | $3f_o$ | $4f_o$ | $5f_o$ | |
| fs | $f_o \pm f_s$ | $2f_o \pm f_s$ | $3f_o \pm f_s$ | $4f_o \pm f_s$ | $5f_o \pm f_s$ |
| 3fs | $3f_s \pm f_o$ | $3f_s \pm 2f_o$ | $3f_o \pm 3f_s$ | $4f_o \pm 3f_s$ | $5f_o \pm 3f_s$ |
| 5fs | $5f_s \pm f_o$ | $5f_s \pm 2f_o$ | $5f_s \pm 3f_o$ | $5f_s \pm 4f_o$ | $5f_o \pm 5f_s$ |
| **Double-balanced mixer–one quarter the number of mixer products** | | | | | |
| $f_o$ | $2f_o$ | $3f_o$ | $4f_o$ | $5f_o$ | |
| fs | $f_o \pm f_s$ | - | $3f_o \pm f_s$ | - | $5f_o \pm f_s$ |
| 3fs | $3f_s \pm f_o$ | - | $3f_o \pm 3f_s$ | - | $5f_o \pm 3f_s$ |
| 5fs | $5f_s \pm f_o$ | - | $5f_s \pm 3f_o$ | - | $5f_o \pm 5f_s$ |

*$f_O$ is the local oscillator. Note that a product such as $2f_O$ +-$f_s$ is known as a third-order product, $3f_s$+-$3f_O$ as a sixth-order product and so on*

**Table 5.1: Mixing products in single, balanced and double-balanced mixers**

design, the likelihood of spurious responses occurring somewhere in the tuning range is very high, but this problem can be reduced by the use of balanced mixers. When a signal is applied to a balanced input port of a mixer, the signal, its even harmonics and their mixing products will not appear at the output. If both input ports are balanced, this applies to both $f_0$ and $f_s$, and the device is called a double-balanced mixer. **Table 5.1** shows how balancing reduces the number of mixing products. Note that products such as $2f_0 \pm f_s$ are known as third-order products, $3f_0 \pm 3f_s$ as sixth-order and so on. Lower-order products are generally stronger and therefore more bothersome than higher-order products. It should be noted that the term 'balanced' (and 'double balanced') in this context does not necessarily mean that the inputs are in a balanced form (ie with differential inputs) but rather refers to the internal configuration of the mixer circuit.

## Practical Mixer Circuits

Virtually any device which has a non-linear characteristic can be used as a mixer, which probably explains why so many different mixer circuits and configurations have been published. However, this chapter concentrates on using common semiconductor devices like diodes, transistors and FETs. Even with a limited range of devices, there is still a wide range of circuit configurations which can be built. None of them is expensive to build, which enables the amateur to experiment with circuit configurations.

### Passive mixers

**Fig 5.3** is a single-balanced version of the diode mixer. It is popular for direct-conversion receivers where the balance reduces the amount of local oscillator radiation from the receiving antenna. The RC output filter reduces local oscillator RF reaching the following AF amplifier.

**Fig 5.4** shows a diode ring double-balanced mixer. This design has been used for many decades and its performance can be second to none. Diode mixers are low-impedance devices, typically 50 ohms. This means that the local oscillator must deliver power, and typically 5mW (+7dBm) is required for best dynamic range. The low impedances mean that the circuit is essentially a wideband design. The balance of this mixer may be further improved by adding external transformers as shown in the chapter on HF receivers.

Diode mixers can be built by the amateur, and for very little cost. The four diodes should be matched for both forward and reverse resistance. Inexpensive silicon diodes like the 1N914 can be used, but Schottky barrier types such as the BA481 (for UHF) and BAT85 (for lower frequencies and large-signal applications) are capable of higher performance. The bandwidth of the mixer is determined largely by the bandwidth of the transformers. For wide-band operation, these are normally wound on small ferrite toroids. For HF and below, the Amidon FT50-43 is suitable, and 15 trifilar turns of 0.2mm diameter enamelled copper wire are typical. The dots indicate the same end of each wire.

Complete diode mixers can be bought ready-made and are now available quite cheaply. These are usually manufactured as double-balanced mixers and one of the cheapest is the Mini-Circuits SBL-1. They contain four matched diodes and two transformers and are available in a sealed metal package. The advantage of these is that their mixing performance is specified by the manufacturer. When studying mixer data sheets, it can be seen that the main difference between mixer models is the intermodulation performance. The models which have the best intermodulation performance also require much greater local oscillator input power. Mixers can be purchased which have local oscillator requirements from about 0dBm up to about +27dBm. Commercial diodes mixers are typically specified to operate over 5–500MHz, but should still offer reasonable performance beyond these frequencies.

Passive mixers have a conversion loss of about 7dB. If such a mixer is used as a receiver front-end mixer, this loss will contribute significantly to the noise figure of the receiver. It is recommended that the mixer is followed immediately by an IF amplifier. If the receiver is being used above about 25MHz, a low noise-figure receiver is important, and this IF amplifier must, therefore, be a low-noise type.

The performance of passive diode mixers is normally specified with the three ports all terminated with 50 ohms. Failure to do this on any port may increase third-order intermodulation products. On the LO port, as it is difficult to predict the output impedance of an oscillator circuit over a wide frequency range, it is best to generate more LO power than that required by the mixer and insert a resistive attenuator, 3-6dB being common. Using this method tends to force the other two ports to more closely adopt an impedance of 50Ω.

Reactive termination of the output port can increase conversion loss, spurious responses and third-order IMD. Most mixers work into a filter to select the desired frequency. At that frequency, the filter may have a purely resistive impedance of the proper value but at the frequencies it is designed to reject it certainly does not. If there is more desired output than necessary, an attenuator is indicated; if not, the filtering may have to be done after the following amplifier, which may be the one making up for the conversion loss. This problem can be overcome by use of the 'constant-R' matching network [2].

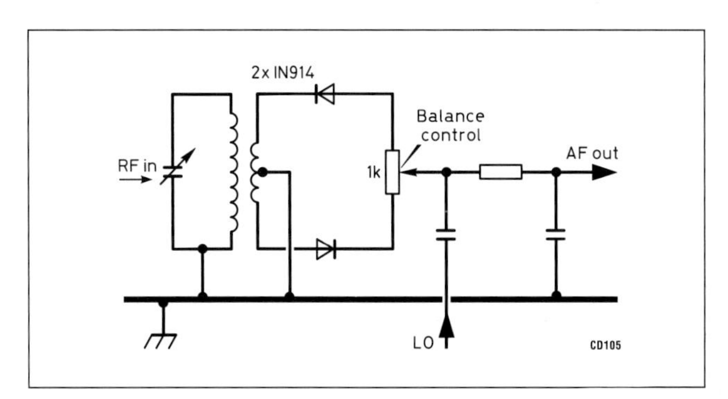


**Fig 5.3 In this two-diode balanced mixer for direct conversion receivers, balancing helps to keep LO drive from reaching the antenna**

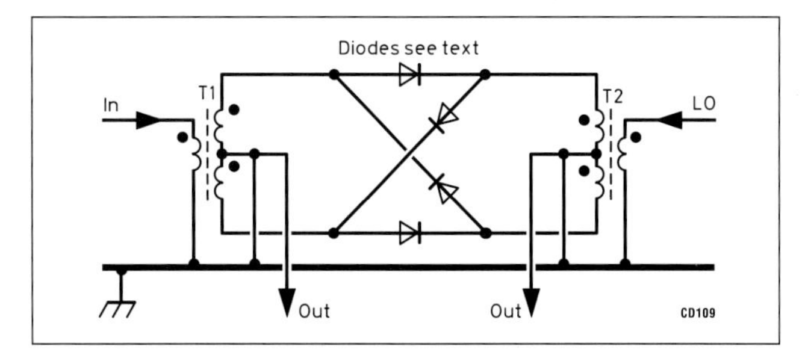


**Fig 5.4: Diode ring mixers are capable of very high performance at all the usual signal levels but several sometimes costly precautions have to be taken to realise that potential; see text. Further improvement of the balance can be obtained by the use of push-pull feed to the primaries of T1 and T2 and feeding the output from between the T1 and T2 centre-taps into a balanced load**

## Active mixers

An active mixer is able to provide gain at the same time as providing the mixing action. It is normal to use active devices like bipolar transistors, junction FETs, single and dual-gate MOSFETs. FETs are particularly good as mixers because they have a strong second-order response in their characteristic curve, and it is the second-order response which gives the mixing action in active mixers (see **Table 5.2**). Active devices like transistors have the advantage that they have three terminals (E-B-C) so they are very easy to connect up as a three-port circuit. Some devices, for example the dual-gate MOSFET, are often shown in their data sheet being operated as a mixer. However, for many devices the data sheet doesn't describe their use as a mixer, but this does not necessarily preclude them from being used.

**Fig 5.5** shows what is probably the simplest active mixer. This employs a single transistor, and is a non-balanced mixer, but is able to provide significant gain.

**Fig 5.6** shows what is probably the simplest FET mixer circuit. It is a non-balanced mixer but does provide some isolation between ports. It also provides low noise, high conversion gain and a reasonable dynamic range. The latter can be improved by the use of an FET, such as the J310, biased for high source current, say 20mA, depending on the device. JFETs require careful adjustment of bias (source resistor) and local oscillator input level for best performance.

The signal is applied to the high-impedance gate; hence, loading on the resonant input circuit is minimised and gain is high. The local oscillator, however, feeds into the low-impedance source; this implies that power is required from the local oscillator, which may need buffering. One could reverse the inputs, thereby reducing the local oscillator power requirements, but that would reduce the gain. The output is taken from the drain with a tuned circuit selecting the sum or difference frequency.

**Fig 5.7** shows a dual-gate MOSFET in a mixer circuit in which both inputs feed into high-impedance ports, but the dynamic range of these devices is somewhat limited. In this circuit, the G2 voltage equals that of the source. The local oscillator injection must be large, 1-3V p-p. Alternatively, G2 could be biased at approximately 25% of the supply voltage. Figs 6.53 and 6.55 (in Chapter 6) show examples of single-balanced mixers using FETs.

| Device | Advantages | Disadvantages |
|---|---|---|
| Bipolar | Low noise figure | High intermodulation |
| transistor | High gain | Easy overload |
| Low DC powerSubject to burn-out | | |
| Diode | Low noise figure | High LO drive |
| High power handling | | Interface to IF |
| High burn-out level | | Conversion loss |
| JFET | Low noise figure | Optimum conversion gain |
| Conversion gain | | not at optimum square-law |
| Excellent square-law | | response level |
| characteristic High LO power | | |
| Excellent overload | | |
| High burn-out level | | |
| Dual-gate | Low IM distortion | High noise figure |
| MOSFET | AGC | Poor burn-out level |
| Square-law | Unstable | |
| characteristic | | |

**Table 5.2: Comparison of semiconductor performance in mixers**

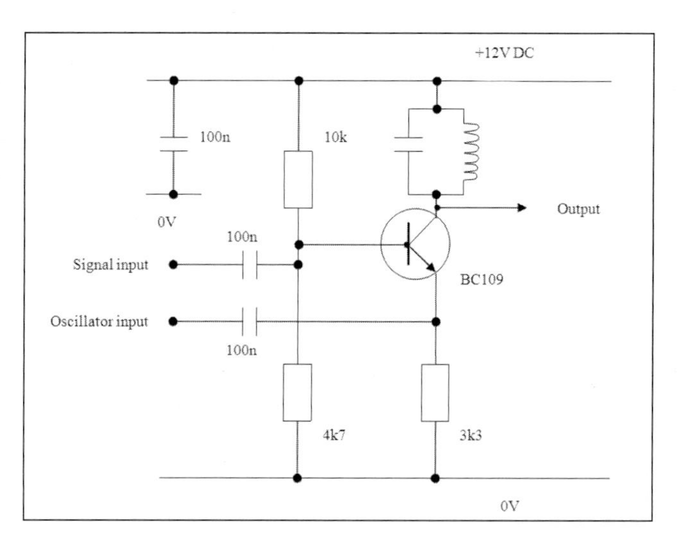


**Fig 5.5: A transistor mixer**

The circuit of **Fig 5.8** shows a single-balanced mixer, which is balanced with respect to the RF input. The circuit consists of a Dual-gate MOSFET (TR1), which amplifies the RF input and drives an RF current into the bipolar transistors TR1 and TR2. These two transistors achieve the mixing action and need to be reasonably well matched (a matched pair is ideal). The two bipolar transistors can be virtually any low to medium $f_T$ devices. They should not have a very high $f_T$ because they may tend to be unstable at VHF or UHF. For the MOSFET, virtually any Dual-gate FET can be used. The output inductor L1 was bifilar wound on a ferrite core and is resonant with C5 at the intermediate frequency. The local oscillator input transformer T1 was trifilar wound on a ferrite two-hole bead.

Bipolar ICs can also be configured as single-balanced or double-balanced mixers for use in the VHF range and below. ICs developed for battery-powered instruments such as portable and cellular telephones have very low power consumption and are used to advantage in QRP amateur applications. Often, a single IC contains not only a mixer but other functions such as a local oscillator and RF or IF amplifiers.

Monolithic technology relieves the home constructor of the task of matching components and adjusting bias, while greatly simplifying layout. Also, the results are more predictable, if not always up to the best obtainable with discrete components.

The NE602N mixer/local oscillator IC serves as an example.

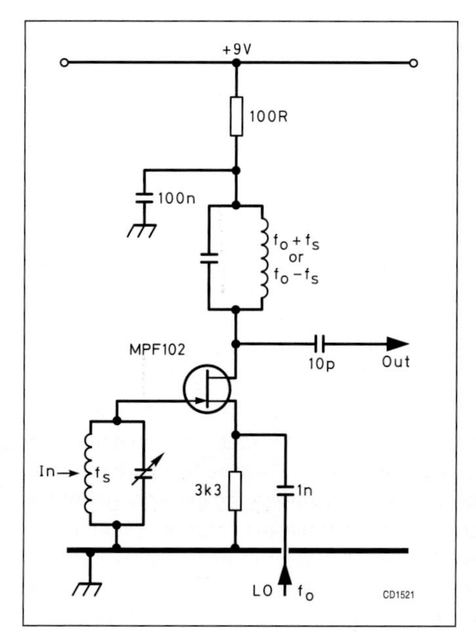


**Fig 5.6: A single JFET makes a simple inexpensive mixer. The LO drives a low-impedance port, requiring it to supply some power**

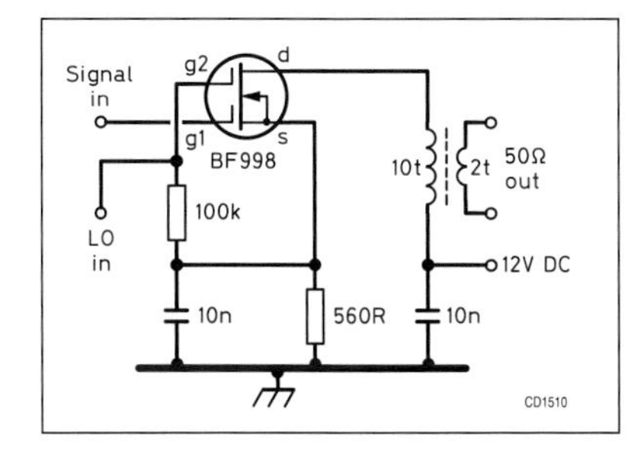


Fig 5.7: The two gates of a dual-gate MOSFET present high impedance to both the signal input and the local oscillator. A self-biasing configuration is shown here

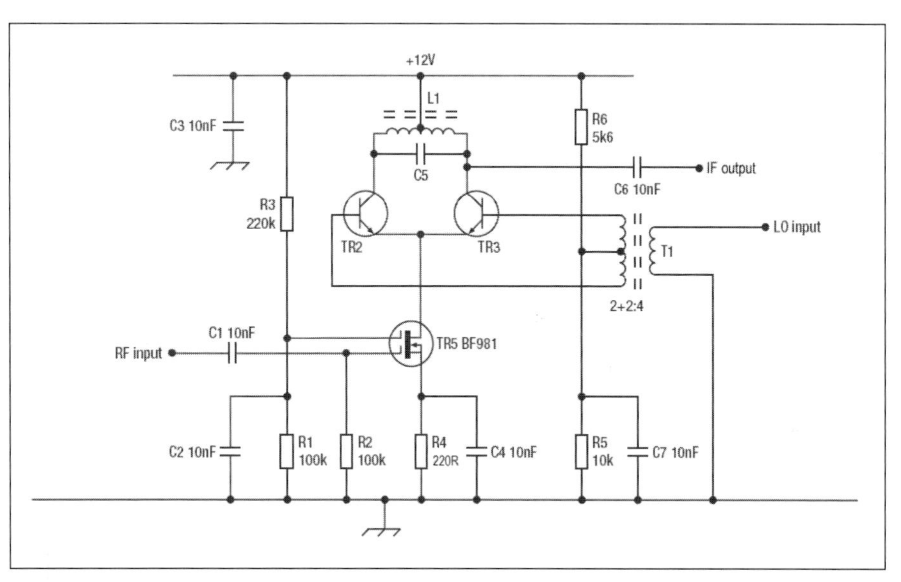


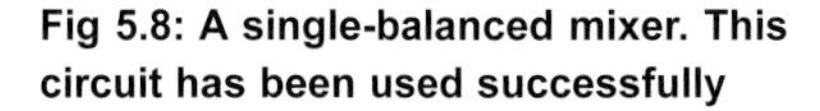

Fig 5.8: A single-balanced mixer. This circuit has been used successfully

**Fig 5.9** shows a circuit of a mixer using an NE602A. A similar device is available under the number SA602A or SA612A.

**Fig 5.10** the equivalent circuit of the IC and the following description of how it works were taken from reference [3]. The IC contains a Gilbert cell (also called a transistor tree), an oscillator/buffer and a temperature-compensated bias network. The Gilbert cell is a differential amplifier (pins 1 and 2) which drives a balanced switching cell. The differential input stage provides gain and determines the noise figure and the signal-handling performance.

Performance of this IC is as follows. Its conversion gain is typically 18dB at 45MHz. The noise figure of 5dB is good enough to dispense with an RF amplifier in receivers for HF and below. The large-signal handling capacity is not outstanding, as evidenced by a third-order intercept point of only -15dBm (typically +5dBm referred to output because of the conversion gain). This restricts the attainable dynamic range to 80dB, well below the 100dB and above attainable with the preceding discrete-component mixers, but this must be seen in the light of this IC's low power consumption (2.4mA at 6V) and its reasonable price. The on-chip oscillator is good up to about 200MHz, the actual upper limit depending on the Q of the tuned circuit. The NE602 provides great flexibility on the input and output connections. Figures 6.88 and 6.89 (in Chapter 6) show various methods of connecting up the inputs and outputs of the NE602A.

The following double-balanced mixers can be used where low noise and good dynamic range must be combined with maximum suppression of unwanted outputs, eg in continuous-coverage receivers and multiband transmitters. Two different approaches are presented.

**Fig 5.11** shows a double-balanced mixer using four JFETs. This is capable of high performance and is used in the Yaesu FT-1000 HF transceiver. The circuit offers some conversion gain and the local oscillator needs to supply little power as it feeds into the high-impedance gates.

**Fig 5.12** shows a double-balanced mixer, which was derived from the single-balanced mixer in Fig 5.9. The circuit is designed on the principle of the Gilbert Cell. In this design however, the lower pair of transistors in the Gilbert Cell have been replaced with two dual-gate MOSFETs. The two MOSFETs do not operate as mixers but as amplifiers. In the prototype, matching was done by selection, finding two devices (from a batch of BF981s) that would operate at the same DC conditions when fitted into the circuit. However, pairs of MOSFETs in a single SMD package are available, eg Philips BF1102, and the use of these devices is recommended. The top four bipolar

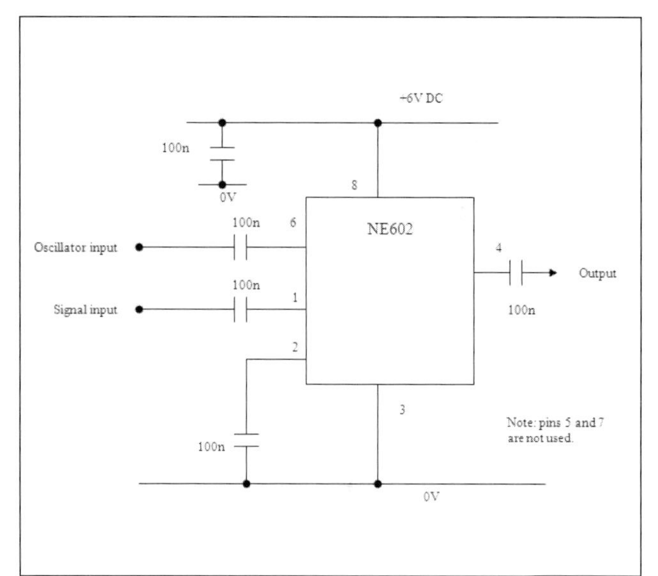


Fig 5.9: Circuit of Mixer using NE602

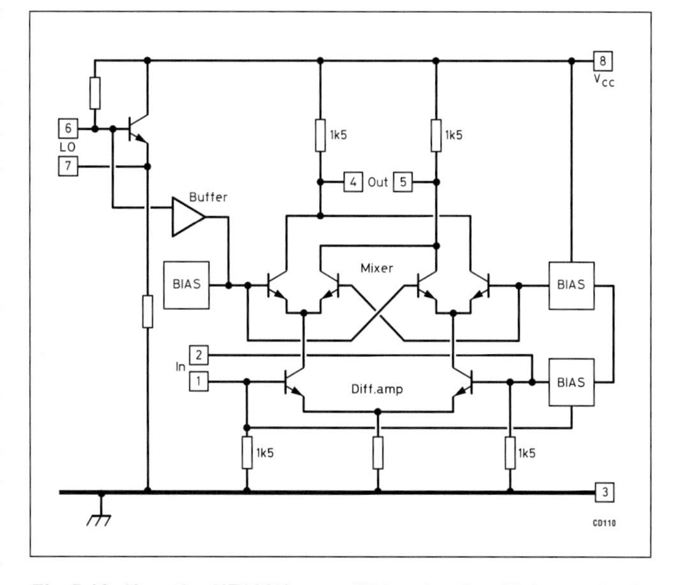


Fig 5.10: How the NE602N monolithic mixer/oscillator works is described in the text with reference to this equivalent circuit (*Signetics RF Communications Handbook*)

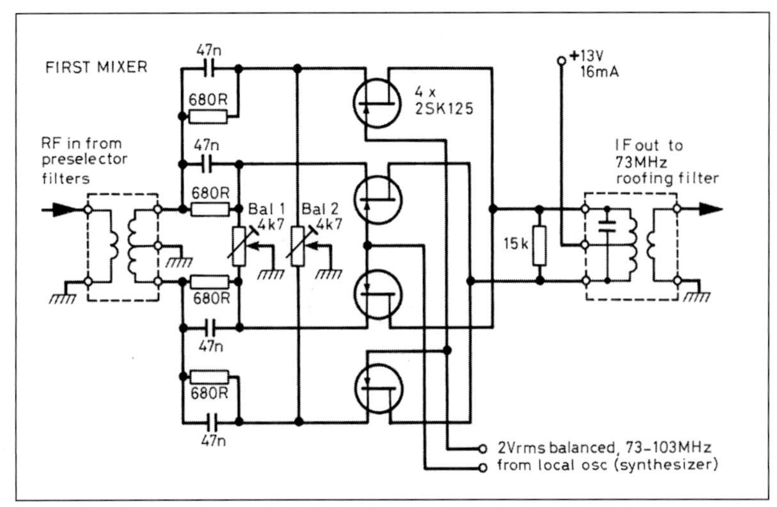


**Fig 5.11: This four-JFET double-balanced mixer is as good as any seen in amateur receivers but does not have the problems of conversion loss, termination and high LO drive associated with diode ring mixers**

transistors achieve the mixing action in the same way as in the Gilbert Cell. The four transistors (TR1 to TR4) were contained within a CA3046 integrated circuit so they are very well matched. C3 and L1 were selected to be resonant at the intermediate frequency. T1 is similar to transformer T1 in Fig 5.9. The circuit gives extremely good balance, good noise figure, reasonable intermodulation performance, and a useful amount of gain.

The 3rd order intercept is about 25dB higher than the NE602, and the circuit is suitable as a front-end mixer in a medium performance HF receiver and would not require an RF amplifier. The circuit operates with a total supply current of about 20mA.

One of the advantages of the Gilbert Cell type of mixer is that the circuit will operate extremely well with a low level of local oscillator input. The circuit will give full performance with just 250mV p-p of local oscillator input. A square wave is best.

Further information and examples of mixer circuits can be found in chapters 6 and 7 of this handbook.

# AMPLIFIERS

Amplifiers are an essential part of radio frequency electronics, and are used anywhere it is necessary to raise the power of a signal. Examples of this are described below.

- An RF amplifier is used on the antenna input of a VHF or UHF receiver. The input signals here may be significantly less than 1uV.
- In a receiver, an IF amplifier is used to provide most of the overall gain.
- A transmitter contains a power amplifier to raise the power of the signal up to the level required for transmission. The output power may be many hundreds of watts.

The following section considers amplification of signals from just above audio frequency up to UHF. This includes examples of amplifiers for receiver input stages through to amplifiers which will transmit the normal power limit for UK amateurs (400W PEP), especially those which might be considered for home construction projects. Integrated circuits are available which require few extra components but they are often application-specific and so the performance is fixed. The advantage of using discrete components is that the performance of the circuit is under control of the designer. Therefore, included in this section are small-signal amplifiers with discrete semiconductors and ICs and power amplifiers with discrete semiconductors and hybrids (semi-integrated modules).

## Intermediate Frequency (IF) Amplifiers

The primary requirement of an IF amplifier is to provide most of the gain needed in a receiver. In addition, it will be required to provide all (or most) of the gain control range. When choosing a device for use as an IF amplifier the main requirement is to choose one that will give a high gain. Other factors such as noise figure are usually secondary.

If discrete devices are to be used, the best choice is to use a bipolar transistor in common emitter configuration. **Fig 5.13** shows the circuit of an IF amplifier operating at 455kHz. For this frequency the transistor does not need to be a high

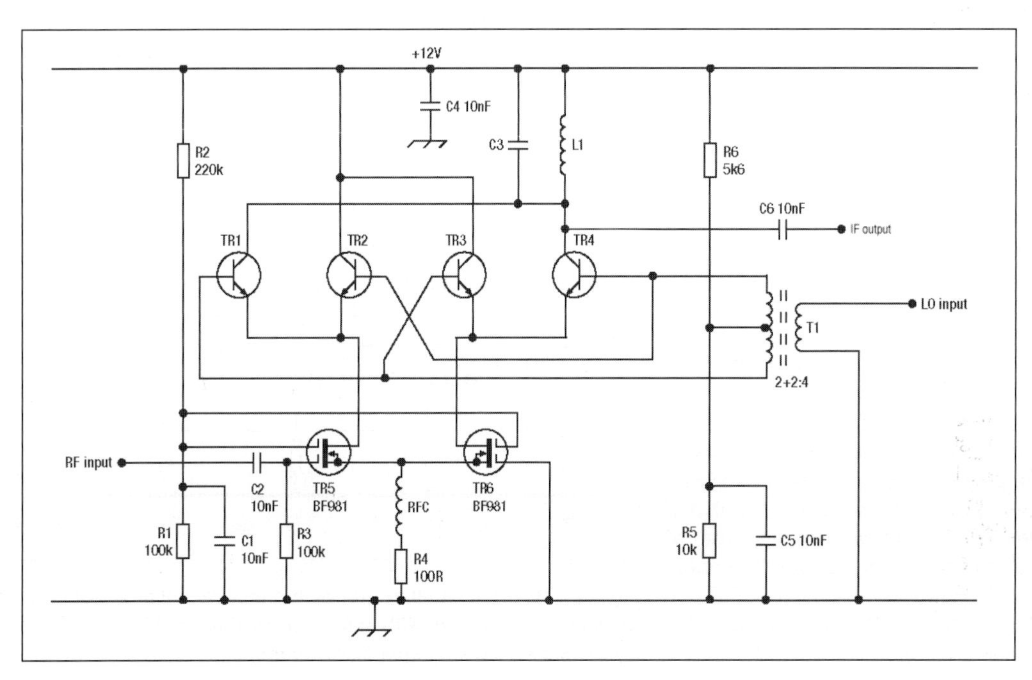


**Fig 5.12: A double-balanced version of the mixer shown in Fig 5.9**

frequency one. An audio type of transistor is adequate, and is also very cheap. The circuit shown operates at a current of about 1mA.

Even in common-emitter configuration, the maximum gain available from this circuit is limited, and if a large amount of gain is required in the IF strip then two or three such stages may be needed. The maximum gain in Fig 5.13 is limited by the following transistor parameters:

- The output conductance (expressed in transistor h-parameters as $h_{OE}$).
- The reverse voltage transfer ratio (expressed in transistor h-parameters as $h_{RE}$).
- The miller capacitance (base-collector capacitance).

These can be overcome, and hence much more gain available, by using the cascode configuration as shown in **Fig 5.14**. This has only three extra components compared to Fig 5.13 but is able to provide much higher gain. Measurements on a prototype of this circuit showed a voltage gain of 2000 from the input of TR1 to the collector of TR2. This is an extremely useful IF amplifier that, in many cases, is able to provide all of the receiver IF gain.

A dual-gate MOSFET may be used as the amplifying element in an IF amplifier (**Fig 5.15**). Although the Dual-Gate MOSFET is configured internally rather like a cascode amplifier, it is not able to provide as much gain as Fig 5.14. The main virtue of the Dual-Gate MOSFET is that it has a good gain adjustment range. The circuit in Fig 5.15 is based upon a circuit from the BF981 data sheet, and will give about 50dB range of gain adjustment.

## Integrated circuits

In the past there were many application-specific integrated circuits which provided this function, but most of these are no longer manufactured, although they may still be available from surplus suppliers. Although now obsolete, one IC which is still generally available is the MC1350 (also available in a metal package as the MC1590).

The MC1350 provides the function of a high-gain amplifier, and has an input that can be used to control the gain. The gain control range is typically 68dB. **Fig 5.16** shows the circuit of an IF amplifier employing an MC1350. The current consumption of the circuit is typically 14mA. If the gain control pin is left unconnected the IC will operate close to full gain.

It should be noted that any IF amplifier which has a high gain is prone to instability. This is because even a small amount of coupling from the output back to the input will provide a path for the amplifier to become unstable. For this reason, the layout of such a circuit is critical and it may require screening, particularly at higher IFs like 9MHz, 10.7MHz or 45MHz.

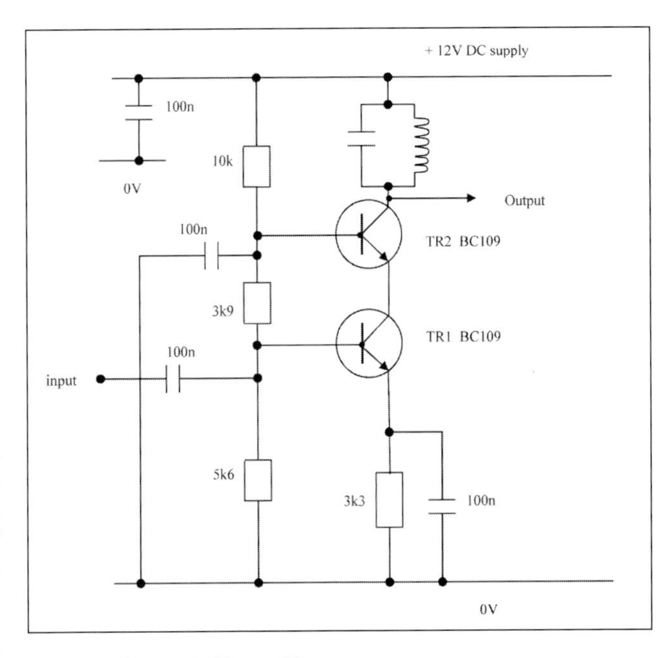


**Fig 5.14: Cascode IF amplifier**

## Methods of gain control

For reception of CW, SSB or AM signals, it is normal for the IF strip to have adjustable gain. The gain control may be derived from an AGC circuit, or it may be part of a manual gain control system or a combination of the two. The gain control is normally achieved by varying an input voltage (or current). There are many circuits that provide adjustment of gain. Ideally a circuit should provide a very wide range of gain adjustment, and the gain adjustment should be linear (ie have a constant dB per volt) over the full gain range.

Gain control using the MC1350 and a dual-gate MOSFET has already been described in this section. It should be noted that the gain control characteristic for the dual-gate FET (Fig 5.16) is not particularly linear, which may cause gain control problems if the dual-gate FET is included in an automatic gain control circuit.

A FET, when operated with zero drain-source bias voltage, acts as a resistor, the resistance value of which can be adjusted by

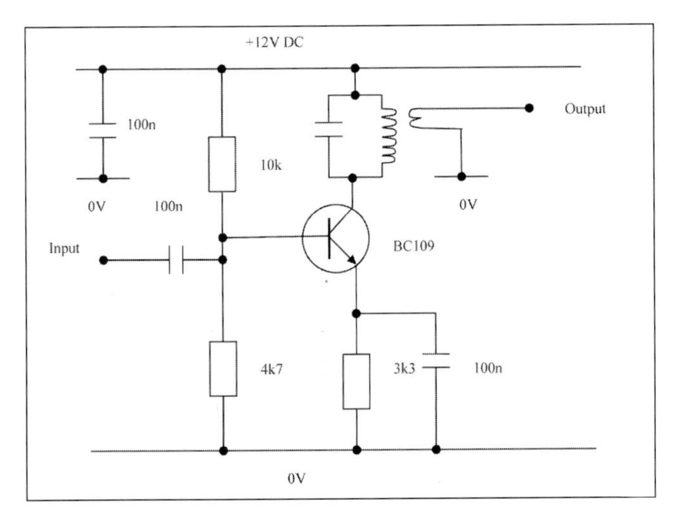


**Fig 5.13: Transistor IF amplifier**

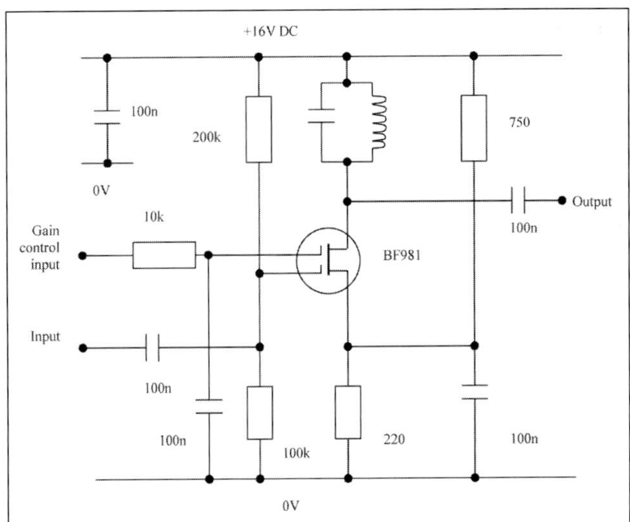


**Fig 5.15: Dual gate MOSFET IF amplifier**

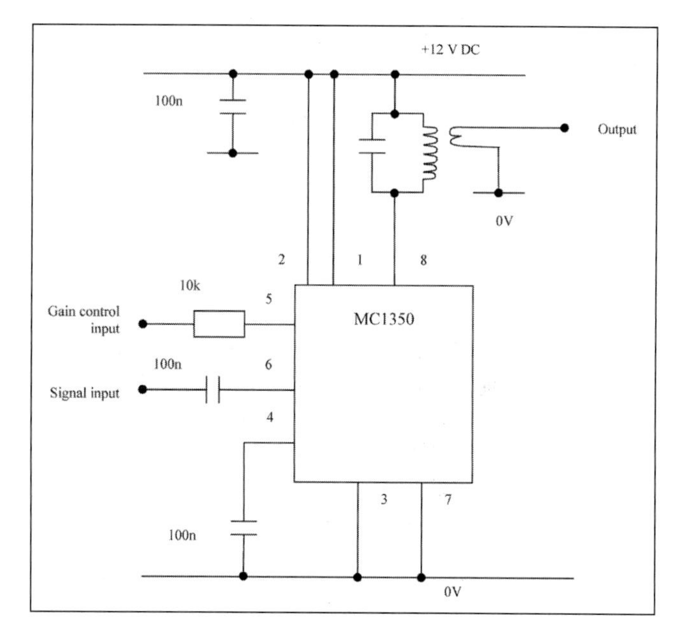


Fig 5.16: Circuit of IF amplifier using MC1350

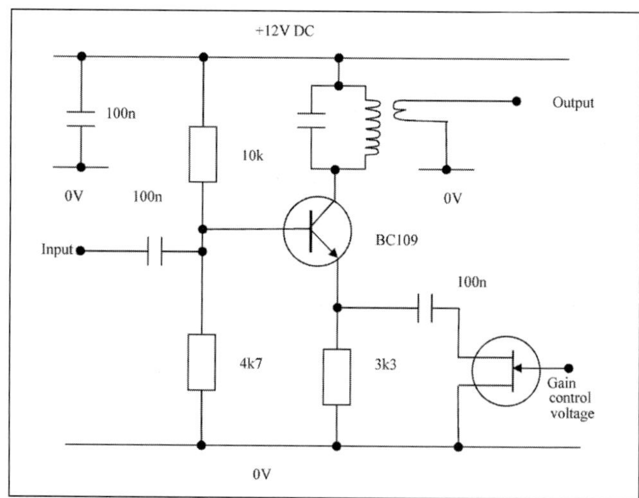


Fig 5.17: Transistor IF Amplifier with gain control applied to the transistor emitter

the gate-source voltage. Examples of this are shown in **Fig 5.17 and 5.18**. In Fig 5.17 the FET is fitted in the emitter circuit of a transistor amplifier to adjust the transconductance of the amplifier. In Fig 5.18 the FET is connected across the collector load of the transistor to act as a variable load.

In **Fig 5.19** a light-dependant-resistor (LDR) is connected across the tuned circuit load of the transistor. The resistance of the LDR can adjusted by varying the light falling on it from the LED. The LDR and the LED should be very close together and fitted inside a light-proof container. This circuit has been used by the author at 455kHz, and gave a very smooth control of gain. The dynamic characteristics of the LDR give a fast attack / slow decay, which makes it particularly useful for use in an SSB receiver.

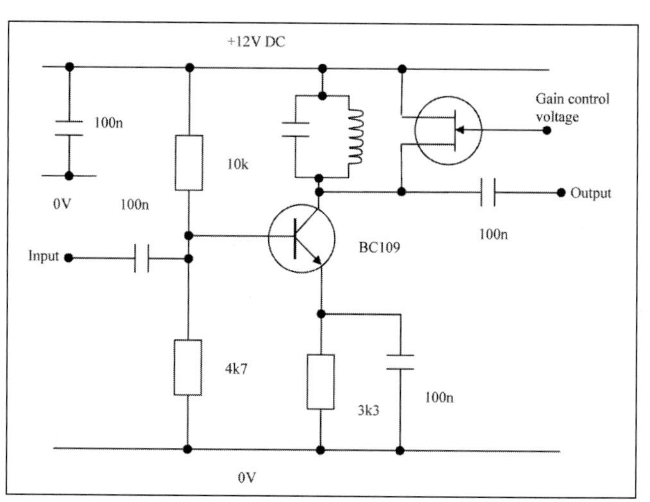


Fig 5.18: Transistor IF Amplifier with gain control applied to the collector load

## Radio Frequency (RF) Amplifiers

### Low-level discrete-semiconductor RF amps

Discrete semiconductors are indicated when the function requires no more gain than that which can be realised in one stage, generally between 6 and 20dB, or when the required noise level or dynamic range cannot be met by an IC. In receiver input stages, ie ahead of the first mixer, both conditions may apply.

Modern discrete devices for use in RF amplifiers are extremely cheap, and the cost is therefore not a deciding factor in the decision on which device to use. Modern design of the first stage in a receiver demands filters rejecting strong out-of-band signals, a wide dynamic range, low-noise IF at VHF and above, and only enough gain to overcome the greater noise of an active first mixer or the conversion loss of a passive mixer, typically a gain of 10dB. Bipolar, field-effect and dual-gate MOSFET transistors can provide that. AGC can easily be applied to FET amplifiers, but the trend is to switch the RF amplifier off when not required for weak-signal reception.

In most of the following low-level RF circuits inexpensive general-purpose transistors such as the bipolar 2N2222A, FETs BF245A, J310, MPF102 and 2N3819, and the SMD dual-gate MOSFET BF998 can be used. However, substitution of one device by another or even by its equivalent from another manufacturer frequently requires a bias adjustment. For exceptional dynamic range, power FETs are sometimes used at currents up to 100mA. Similarly, for the best noise figure above 100MHz,

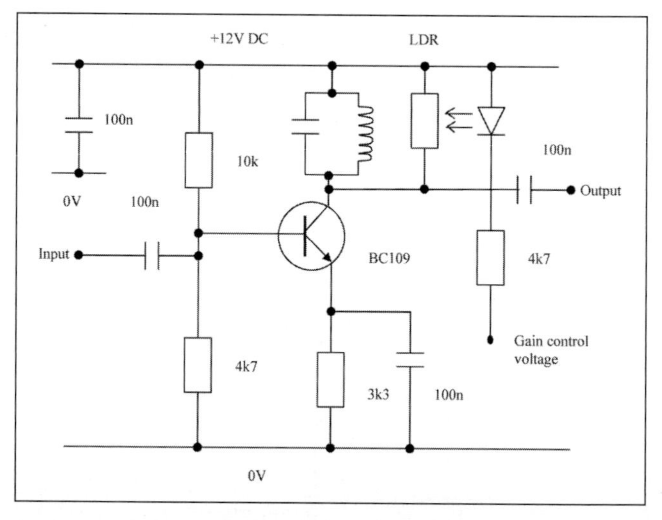


Fig 5.19: Transistor IF Amplifier with gain control using an LDR

gallium-arsenide FETs (GaAsFETs) are used instead of silicon types.

**Fig 5.20** shows the circuit of a 50MHz amplifier using an FET in grounded-gate configuration. The circuit has networks on the input and output which provide impedance matching and tun-

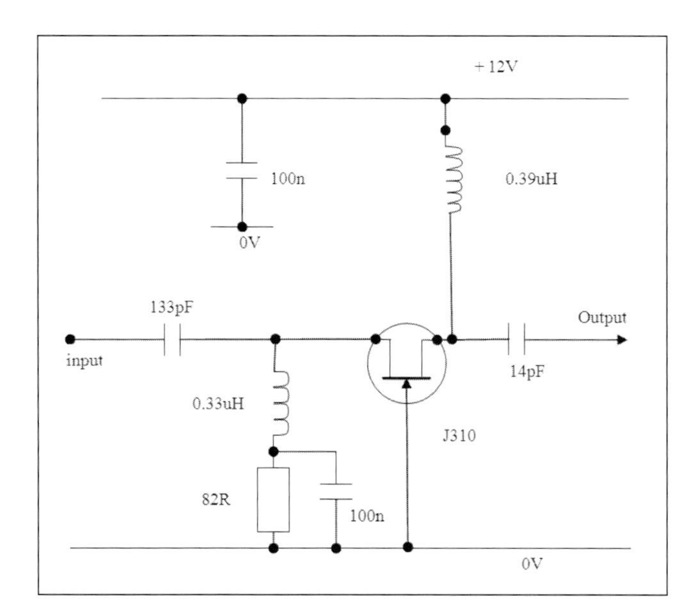


**Fig 5.20: 50MHz FET amplifier**

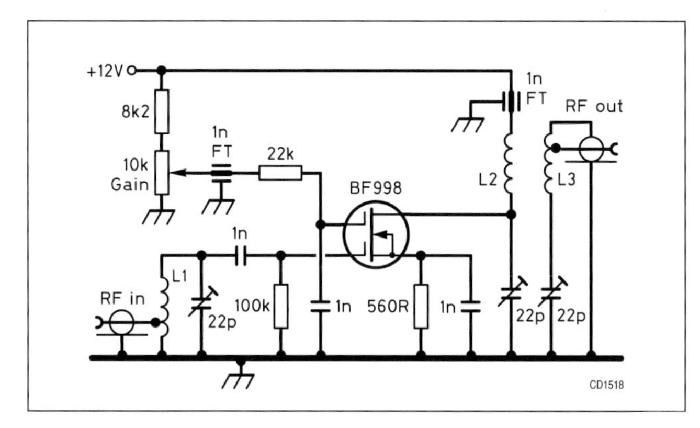


**Fig 5.21: A 144MHz preamp using a BF998 SMD dual-gate MOSFET. L1: 5t 0.3mm tinned copper tapped 1t from earthy end, 10mm long, 6mm dia. L2, L3: 6t 1mm tinned copper 18mm long, 8mm dia. L3 tapped 1t from earthy end. L2, L3 mounted parallel, 18mm between centres**

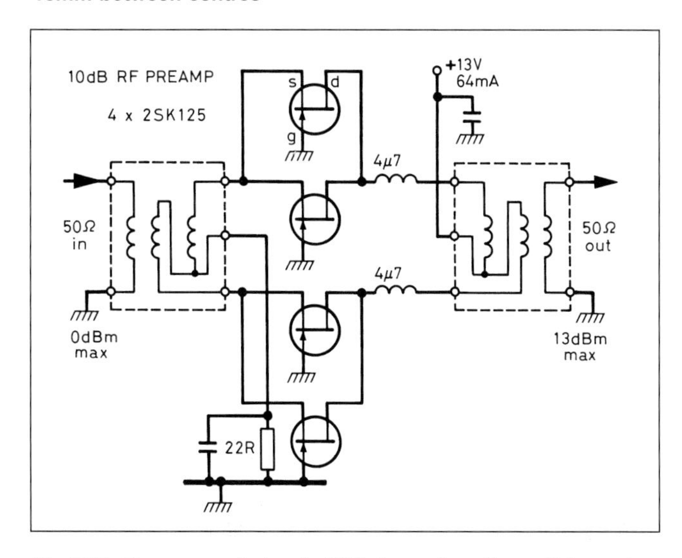


**Fig 5.22: Four grounded-gate FETs in push-pull parallel running high drain currents give this RF stage of the Yaesu FT-1000 HF transceiver its excellent dynamic range, suppression of second-order intermodulation, and proper termination of preceding and following bandpass filters**

ing. The circuit operates between source and load impedances of 50 ohms.

The VHF amplifier in **Fig 5.21** uses a dual-gate MOSFET.

Four FETs in a push-pull parallel grounded-gate circuit, **Fig 5.22**, are used in the RF amplifiers of some top-grade HF receivers. Push-pull operation reduces second-order intermodulation. Several smaller FETs in parallel approach the wide dynamic range of a power FET, and the source input provides, through a simple transformer, proper termination for the preselector filters.

## Low-level IC Amplifiers

Monolithic integrated-circuit RF amplifiers come in a great variety, sometimes combined on one chip with other functions such as a mixer, detector, oscillator or AGC amplifier. A linear IC, though more expensive than its components in discrete form, has the advantage of well-specified performance in the proven PCB layouts given in the manufacturer's data sheet.

Some ICs are labelled 'general purpose', while others are optimised for a specific application. If intended for a VHF hand-portable FM transceiver, low power consumption might be a key feature, but linearity would be unimportant. However, an IC for fixed-station SSB use would be selected for good linearity, but the current consumption would be unimportant. Any potential user would do well to consult the data sheets or books on several alternative ICs before settling on any one. New devices are being introduced frequently and the price of older ones, often perfectly adequate for the intended function, is then reduced.

The following elaborates on a few popular devices, but only a fraction of the data sheet information can be accommodated here.

The Philips-Signetics NE604AN contains all the active components for a NBFM voice or data receiver IF system [3]-see **Fig 5.23**. It contains two IF amplifiers, which can either be cascaded or used at different frequencies in a double-conversion receiver, a quadrature detector, a 'received signal-strength indicator' (RSSI) circuit with a logarithmic range greater than 90dB, a mute switch (to cut the audio output when transmitting in simplex operation) and a voltage regulator to assure constant operation on battery voltages between 4.5V (at 2.5mA) and 8V (at 4mA). The IC is packaged in a 16-pin DIP. The operating temperature range of the NE604AN is 0 to 70°C.

The Toko TK10930V IC is an AM-FM 2nd IF-detector system in a 24-pin DIP. In it, the 10.7MHz 1st IF input is converted to a 455kHz 2nd IF and processed in separate AM and FM amplifier and detector channels to simultaneously provide audio output in each. AVC, RSSI and squelch circuitry is included. DC requirements are only 3V @ 4mA (FM only) or 7mA (both channels on). Applications are in portable scanners, airband and marine receivers but, unfortunately, there is no provision for SSB reception. Internal and external components are shown in **Fig 5.24**. The Philips NE592N is an inexpensive video amplifier IC with sufficient bandwidth to provide high gain throughout the HF region. Its differential input (and output) enabled G4COL to use it, without a transformer, in a very sensitive antenna bridge in which both the RF source and the unknown impedance are earthed [4] The circuit is shown in **Fig 5.25**. For video response down to DC this IC requires both positive and negative supply voltages; for AC-only, ie RF, a single supply suffices, as explained for operational amplifiers later in this chapter.

## RF Solid-state Power Amplifiers

Representative professional designs of high-power amplifiers are given and explained in later chapters.

The reader will notice the many measures required to protect the often very costly transistors under fault conditions, some of which may occur during everyday operation.

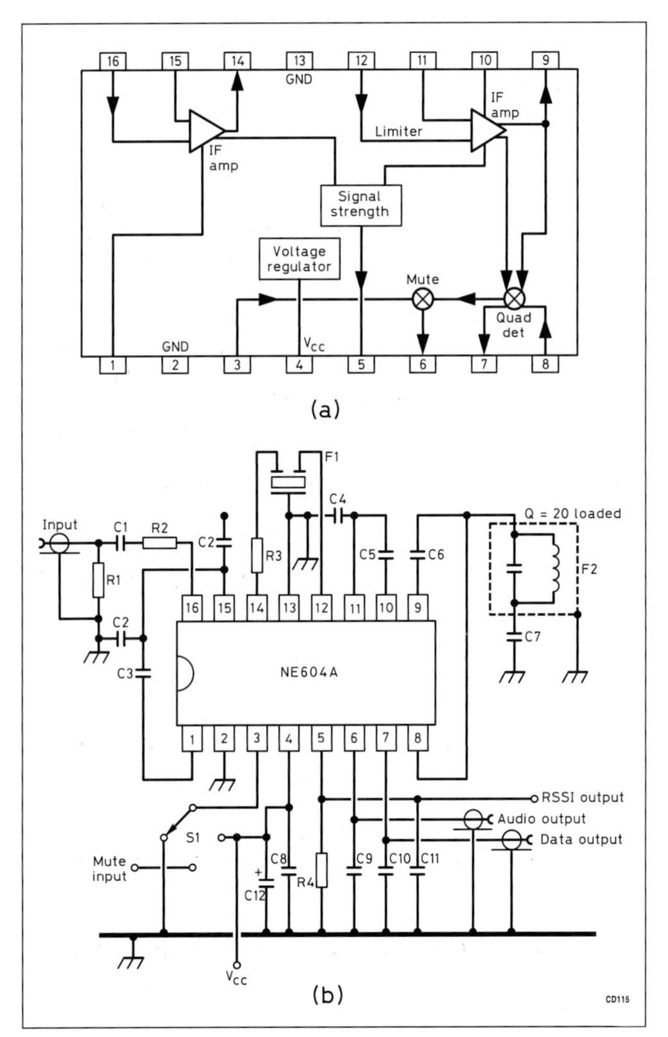


**Fig 5.23: The NE604A FM receiver IFIC. (a) Internal block diagram. (b) External connections**

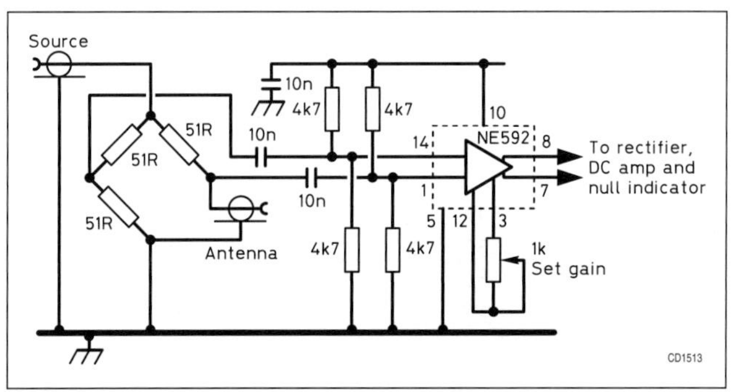


**Fig 5.25: An inexpensive video amplifier IC, NE592N, can serve as a wide-band HF amplifier, as it does in G4COL's sensitive HF antenna bridge**

That does not encourage experimentation with those designs but amateurs have discovered that devices either not designed for RF service or widely available on the surplus market cost much less and can be made to perform in homebrew amplifiers.

One might be tempted to use VHF transistors salvaged from retired AM PMR (VHF or UHF) transmitters, in HF amplifiers. This is attractive because of the very high power gain of VHF transistors used a long way below their design frequency. However, stability problems must be expected.

## Use of Feedback in Amplifiers

Frequently in both receiver and transmitter design, it is found that one particular amplifier stage is limiting the overall linearity performance. In this case, there is a requirement to improve the linearity of the particular stage. This can be achieved by the application of negative feedback. In RF circuits feedback is normally applied around individual stages. This is in contrast to AF amplifiers, where feedback is normally applied around the whole amplifier.

There are various methods of applying negative feedback, but the easiest for the amateur to implement is to apply feedback to a common-emitter amplifier as shown in **Fig 5.26**. Resistors Rsh and Rse are responsible for applying the feedback. The feedback takes two forms:

- Rsh applies shunt feedback.
- Rse applies series feedback.

The application of feedback using resistors means that the amplifier characteristics are set primarily by the values of the feedback resistors, rather than other components in the circuit. As well as improving the linearity, feedback has

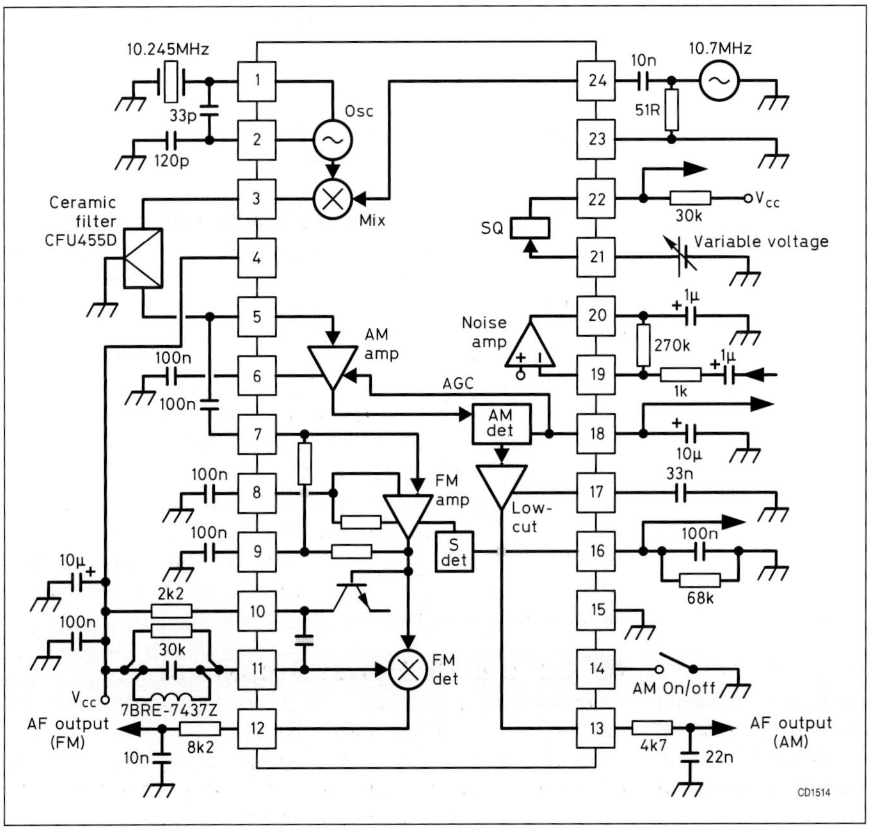

**Fig 5.24: The Toko TK10930V IC comprises a mixer/LO from a 10.7MHz 1st IF to 455kHz and separate AM and FM IF-detector channels, plus auxiliary circuitry. Low drain, from a 3V battery, invites portable applications**

the following effects on the amplifier performance (compared to the amplifier without feedback).

- Amplifier gain is reduced.
- Bandwidth is increased.
- Gain is stabilised against variations in temperature.
- Gain is stabilised against changes in operating volts/current.
- Gain is stabilised against variations of transistor parameters.
- Feedback allows the designer to control the input and output impedance of the amplifier.
- The noise figure is usually degraded.

Examples of the use of feedback can be seen in the chapter on HF Transmitters. See particularly the G3TXQ transceiver..

An example of amplifiers involving feedback are modern commercial MMIC (Microwave Monolithic Integrated Circuit) devices. These have built-in feedback resistors to control the amplifier characteristics. Many such devices are available and an example is shown in **Fig 5.27**. Avago Semiconductors make several types of MMIC which provide gain from DC up to the 2.3GHz band with very few external components [5, 6]. They have 50 ohms input impedance, can be cascaded without interstage tuning, have noise figures as low as 3dB at 1GHz and are capable of 10-20mW of output into 50 ohms . Packaging is for direct soldering to PCB tracks. The only external components required are a choke and series resistor in the DC supply lead and DC blocking capacitors at their input and output as shown in Fig 5.27. No listing is given here, because new models appear frequently. The suppliers' data sheet should be consulted for a full set of data on these devices.

## Switching MOSFETs at HF

As 'real' HF power transistors remain expensive, amateurs have found ways to use selected inexpensive MOSFETs intended for audio, digital switching, switch-mode power supplies or ultrasonic power applications at ever-increasing frequencies. Because these are not designed for RF use, they may not perform well in linear amplifiers, and internal capacitance and inductance may be an issue.

MOSFETs, as compared with bipolar power transistors, facilitate experimentation for several reasons. Their inherent reduction of drain current with increasing temperature contrasts with bipolar power transistors where 'thermal run-away' is an ever-present

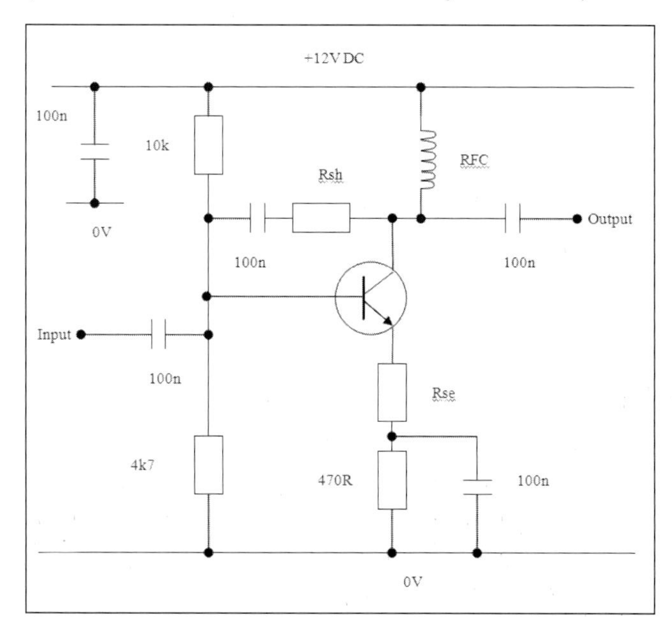


**Fig 5.26: Applying negative feedback to a common-emitter RF amplifier**

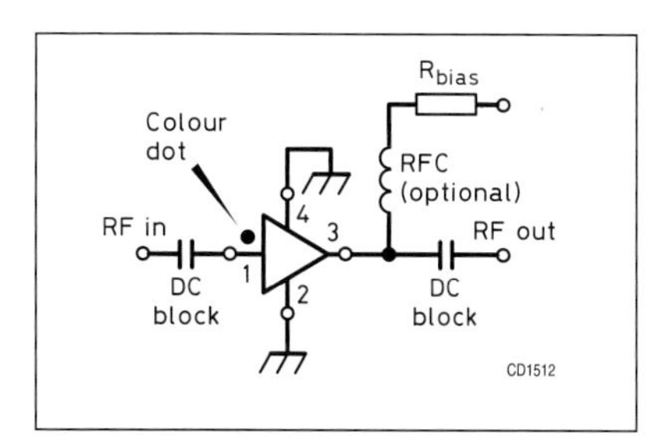


**Fig 5.27: MMICs can supply high wideband gain up to 2.3GHz with a reasonable noise figure. Input and output are 50 ohms and no tuned circuit is required**

danger. MOSFET gates require only driving voltage, not power; though developing that voltage across the high gate-to-source capacitance requires some ingenuity, it greatly simplifies biasing.

Where driving power is available, a swamping resistor across the gate-to-source capacitance helps, especially if that capacitance is made part of a pi-filter dimensioned to match the swamping resistor to the desired input impedance, eg 50Ω. A swamping resistor also reduces the danger of oscillation caused by the considerable drain-gate capacitance of a MOSFET. The most important limitation is that MOSFETs require a high supply voltage for efficient operation, 24-50V being most common; on 13.8V, output is limited.

Here follow some such designs. Note that most use push-pull circuits. This suppresses even harmonics and thereby reduces the amount of output filtering required. With any wide-band amplifier, a proper harmonic filter for the band concerned must be used; without it even the third or fifth harmonic is capable of harmful interference. See **Table 5.3**.

Parasitic oscillations have also dogged experimenters. They can destroy transistors before they are noticed. A spectrum analyser is the professional way to find them but with patience a continuous coverage HF/VHF receiver can also be used to search for them. It is useful to key or modulate the amplifier to create a 'worst case' situation during the search. If found, a ferrite bead in each gate lead and/or a resistor and capacitor in series between each source and drain are the usual remedies. W1FB found negative feedback from an extra one-turn winding on the output transformer to the gate(s) helpful. VHF-style board layout and bypassing, where the back of the PCB serves as a ground plane, are of prime importance [7].

The first example is aimed at constructors without great experience; those following represent more difficult projects.

## 1.8-10.1MHz 5W MOSFET Amplifier

A push-pull broad-band amplifier using DMOS FETs and providing 5W CW or 6W PEP SSB with 0.1W of drive and a 13.8VDC supply has been described by Drew Diamond, VK3XU [8]. He used two inexpensive N-channel DMOS switching FETs which make useful RF amplifiers up to about 10MHz. This amplifier had a two-tone IMD of better than -30dBc (-35dBc typical) and, with the output filter shown in **Fig 5.28**, all harmonics were better than -50dBc. The amplifier should survive open or shorted output with full drive and remain stable at any load SWR.

The drain-to-drain impedance of the push-pull FETs is 2 x 24 = 48Ω so that no elaborate impedance transformation is needed to match into 50Ω. T3 serves as a balun transformer. T2 is a balanced choke to supply DC to the FETs. Negative RF feedback is provided by R3 and R4, stabilising the amplifier and helping to keep the frequency response constant throughout the range. The bias zener ZD1 is positioned against the heatsinks of TR1 and TR2 with a small blob of heatsink compound so that it tracks the temperature of the FETs, causing the bias voltage to go down when the temperature goes up. The amplifier enclosure must have adequate ventilation.

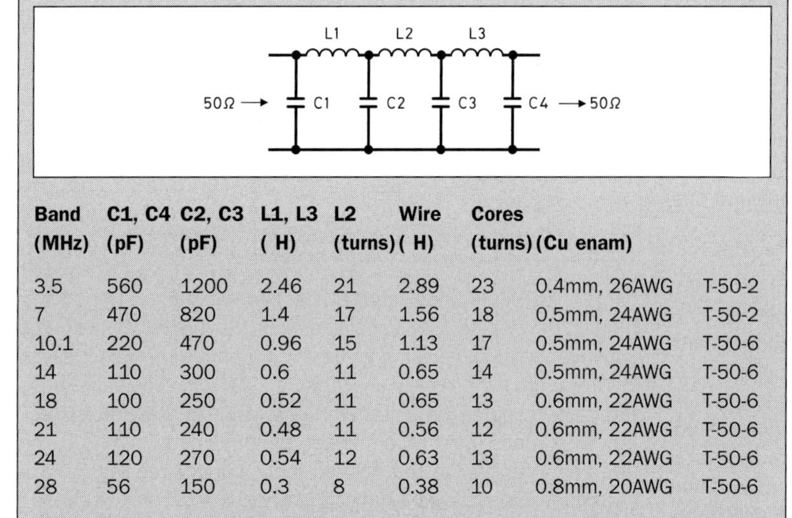


| Band | C1, C4 | C2, C3 | L1, L3 | L2 | Wire | Cores | | |
|---|---|---|---|---|---|---|---|---|
| (MHz) | (pF) | (pF) | ( H) | (turns) | ( H) | (turns) | (Cu enam) | |
| 3.5 | 560 | 1200 | 2.46 | 21 | 2.89 | 23 | 0.4mm, 26AWG | T-50-2 |
| 7 | 470 | 820 | 1.4 | 17 | 1.56 | 18 | 0.5mm, 24AWG | T-50-2 |
| 10.1 | 220 | 470 | 0.96 | 15 | 1.13 | 17 | 0.5mm, 24AWG | T-50-6 |
| 14 | 110 | 300 | 0.6 | 11 | 0.65 | 14 | 0.5mm, 24AWG | T-50-6 |
| 18 | 100 | 250 | 0.52 | 11 | 0.65 | 13 | 0.6mm, 22AWG | T-50-6 |
| 21 | 110 | 240 | 0.48 | 11 | 0.56 | 12 | 0.6mm, 22AWG | T-50-6 |
| 24 | 120 | 270 | 0.54 | 12 | 0.63 | 13 | 0.6mm, 22AWG | T-50-6 |
| 28 | 56 | 150 | 0.3 | 8 | 0.38 | 10 | 0.8mm, 20AWG | T-50-6 |

*Notes. Various cut-off frequencies and ripple factors used to achieve preferred-value capacitors. Coil turns may be spread or compressed with an insulated tool to peak output. Cores are Micrometals toroids. Capacitors are silver mica or polystyrene, 100V or more.*

**Table 5.3: Low pass output filters for HF amplifiers up to 25W**

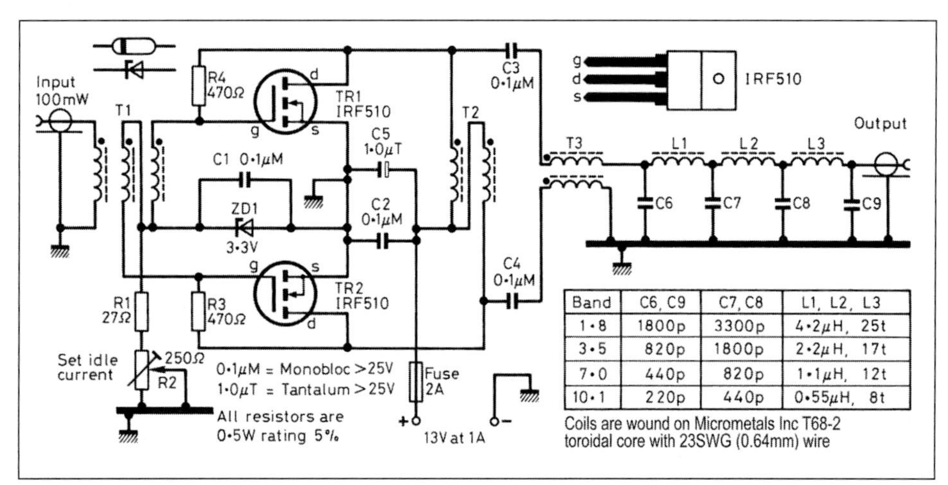


| Band | C6, C9 | C7, C8 | L1, L2, L3 |
|---|---|---|---|
| 1·8 | 1800p | 3300p | 4·2μH, 25t |
| 3·5 | 820p | 1800p | 2·2μH, 17t |
| 7·0 | 440p | 820p | 1·1μH, 12t |
| 10·1 | 220p | 440p | 0·55μH, 8t |

Coils are wound on Micrometals Inc T68-2 toroidal core with 23SWG (0.64mm) wire

**Fig 5.28: Circuit diagram of the 1.8-10.1MHz amplifier providing 5W CW or 6W PEP SSB using switching FETs. T1 comprises three 11t loops of 0.5mm enameled wire on FT50-43 core. T2, T3 are three 11t loops of 0.64mm enam wire on Amidon FT50-43. * Indicates start of winding**

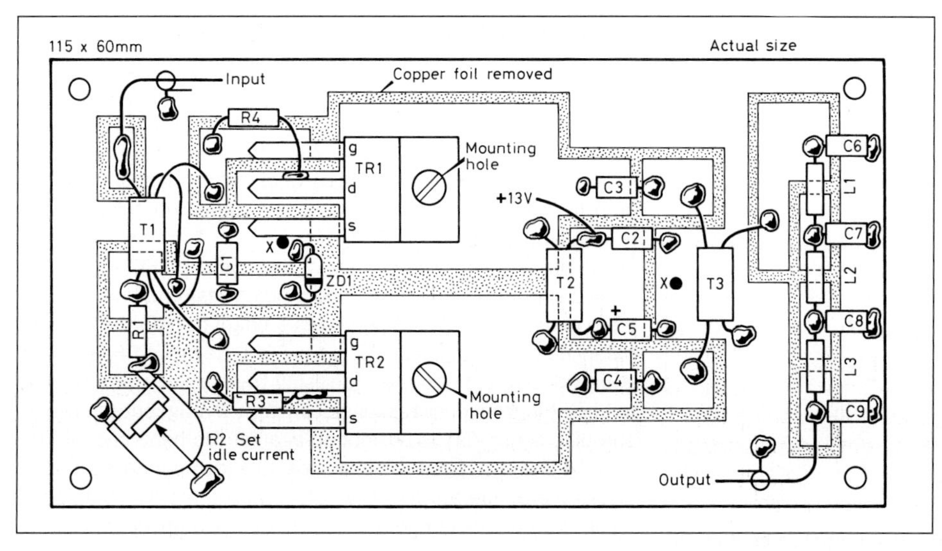


**Fig 5.29: Layout of the 5W FET amplifier on double-sided PCB**

The complete amplifier, with an output filter for one band, can be built on a double-sided 115 x 60mm PCB.

For stability, the unetched 'ground plane' should be connected to the etched-side common/earth in at least two places marked 'X' in **Fig 5.29**. Drill 1mm holes, push wires through and solder top and bottom.

If multiband operation is required, the filter for the highest band should be accommodated on the amplifier board and kept in the circuit on all bands. Lower-frequency filters can then be mounted on an additional board. Polystyrene or silvered-mica capacitors should be used in the filters. Hard-to-get values can be made up of several smaller ones in parallel.

When setting up, with R2 at minimum resistance the desired no-signal current is 200-300mA. With 100mW drive and a 50Ω (dummy) load, the supply current should be about 1A. After several minutes of operation at this level and with suitable heat-sinks, the latter should not be uncomfortably hot when touched lightly. While 100mW drive should suffice on the lower bands, up to 300mW may be needed at 10.1MHz, which is about the limit. Overdriving will cause flat-topping. With larger heatsinks and higher supply voltage, more output would be possible.

## 100W Multiband HF Linear

David Bowman, G0MRF, built and described [9] in great detail the design, construction and testing of his amplifier, complete with power supply, T/R switching and output filtering. With a pair of inexpensive 2SK413 MOSFETs, it permits an output of 100W from topband through 14MHz and somewhat less up to 21MHz. The RF part is shown in **Fig 5.30**.

## A Linear 50MHz Amplifier

When selecting cheap switching FETs it is usually found that the drain is connected to the metal tab. Clamping this to ground with an insulating washer adds unwanted capacitance. The usual solution is to use two insulating washers, thus halving the capacitance. There is, of course, a slight reduction of the heat transfer efficiency.

PA0KLS found that useful 50MHz output could be obtained from a single IRF610 MOSFET in the circuit of **Fig 5.31** [10]. With a 50V supply, the forward bias for Class AB operation was adjusted for an idling current of 50mA. A

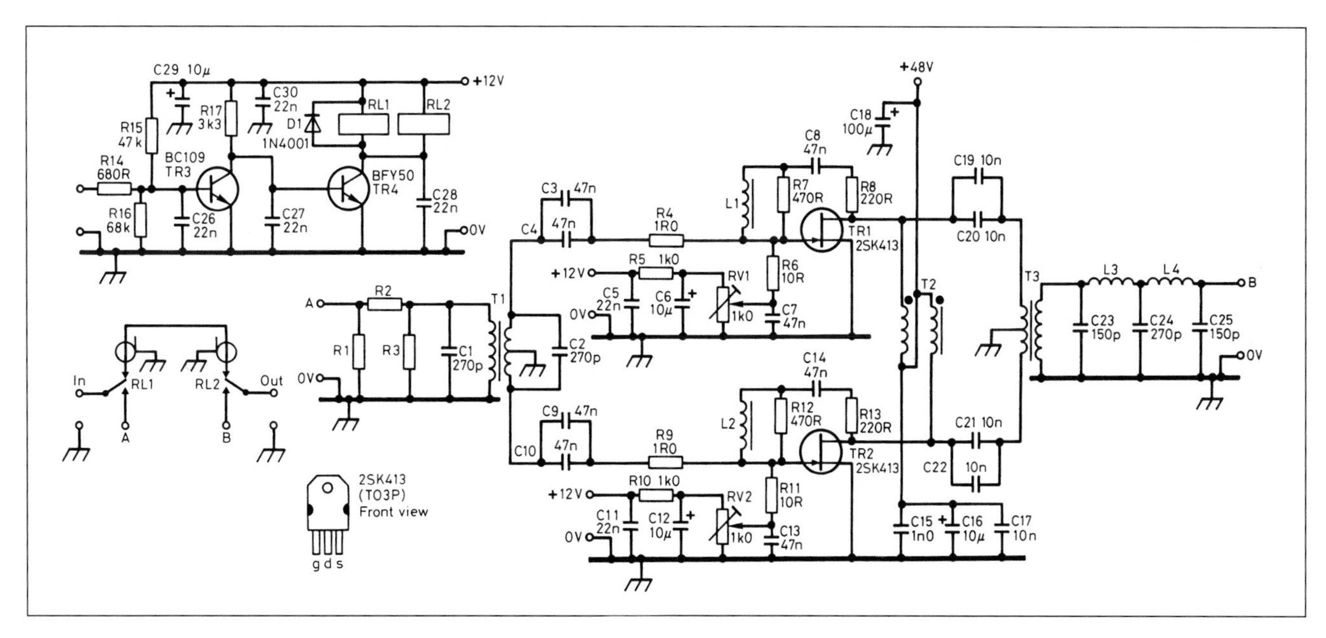


Fig 5.30: G0MRF's 100W HF linear amplifier uses inexpensive MOSFETs

two-tone input of 0.2-0.3W then produced an output of 16W at a drain current of 0.5A. Third-order distortion was 24dB below either tone, not brilliant but adequate. Voice quality was reported to be good. On CW, the output exceeded 20W.

Best output loading of $(V_d)^2/2P_o$ = 78Ω was obtained with another pi-network; while this provides only single-band matching, it reduces harmonics. The drain choke was made roughly resonant with the IRF610's drain-to-source capacitance of 53pF, so that the pi-coupler only has to match near-pure resistances.

As in a single-ended amplifier there is no cancellation of even harmonics and the second harmonic of 50MHz is in the 88-108MHz FM broadcast band, additional low-pass filtering is mandatory. Set up for 28MHz, this amplifier produced 30-40W CW; at this power level, with an insufficiently large heatsink, the MOSFET got very hot but it did not fail due to the inherent reduction of drain current mentioned above.

The difference between 28 and 50MHz output shows that 50MHz is about the upper limit for the IRF610.

Peter Frenning, OZ1PIF, has published a design [11] using eight IRF510 MOSFETs to produce 250W at 50MHz.

## VHF Linear Amplifiers

For medium-power VHF linear amplifiers there is another cost-saving option. Because transistors intended for Class C (AM, FM or data) cost a fraction of their linear counterparts, it is tempting and feasible to use them for linear applications. If simple diode biasing alone is used, however, the transistors, such as the MRF227, go into thermal runaway. Two solutions to this problem were suggested in the ARRL lab for further experimentation [12]. Both methods preserve the advantages of a case earthed for DC.

The current-limiting technique of **Fig 5.32(a)** should work with any device, but is not recommended where current drain or power dissipation are important considerations. The approach is simply to use a power supply which is current limited, eg by a LM317 regulator, and to forward bias the transistor to operate in Class B. Forward bias is set by the values of RB1 and RB2.

The active biasing circuit of **Fig 5.32(b)** is not new but has not been much used by amateurs. The collector current of the transistor is sampled as it flows through a small-value sensing resistor $R_s$; the voltage drop across it is amplified by a factor $R_f/R_i$, the gain of the op-amp differential amplifier circuit. If the collector current

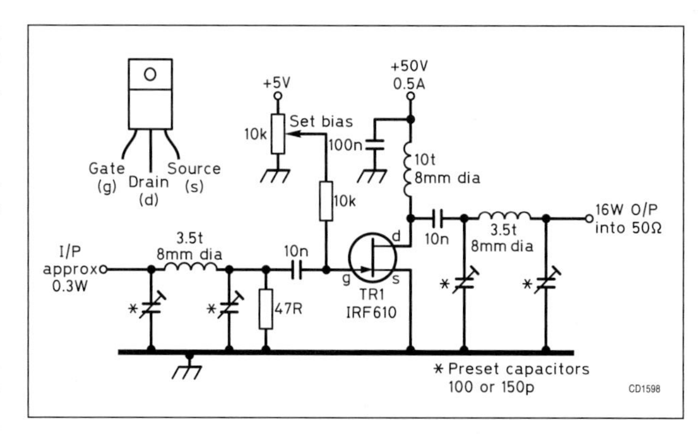


Fig 5.31: This 50MHz linear amplifier using an inexpensive switching MOSFET will deliver 16W output with less than 0.3W of drive

goes up, the base voltage is driven down to restore the balance. The gain required from the op-amp circuit must be determined empirically. While experimenting, the use of a current-limiting power supply is recommended to avoid destroying transistors.

### 50-1296MHz power amplifier modules

Designing a multi-stage UHF power amplifier with discrete components is no trivial task. More often than not, individually tuned interstage matching networks are required. Duplicating even a proven design in an amateur workshop has its pitfalls. One solution, though not the least expensive, is the use of a sealed modular sub-assembly. These are offered by several semiconductor manufacturers, including Mitsubishi (see **Table 5.4**), Motorola and Toshiba.

For each band, there is a choice of output power levels, frequency ranges, power gain and class of operation (linear in Class AB for all modes of transmission or non-linear in Class C for FM or data modes only). All require a 13.8V DC supply, sometimes with separate external decoupling for each built-in stage, and external heatsinking.

In general, their 50Ω input provides a satisfactory match for the preceding circuitry, and a pi-tank (with linear inductor on the higher bands) is used for antenna matching and harmonic suppression. They are designed to survive and be stable under any

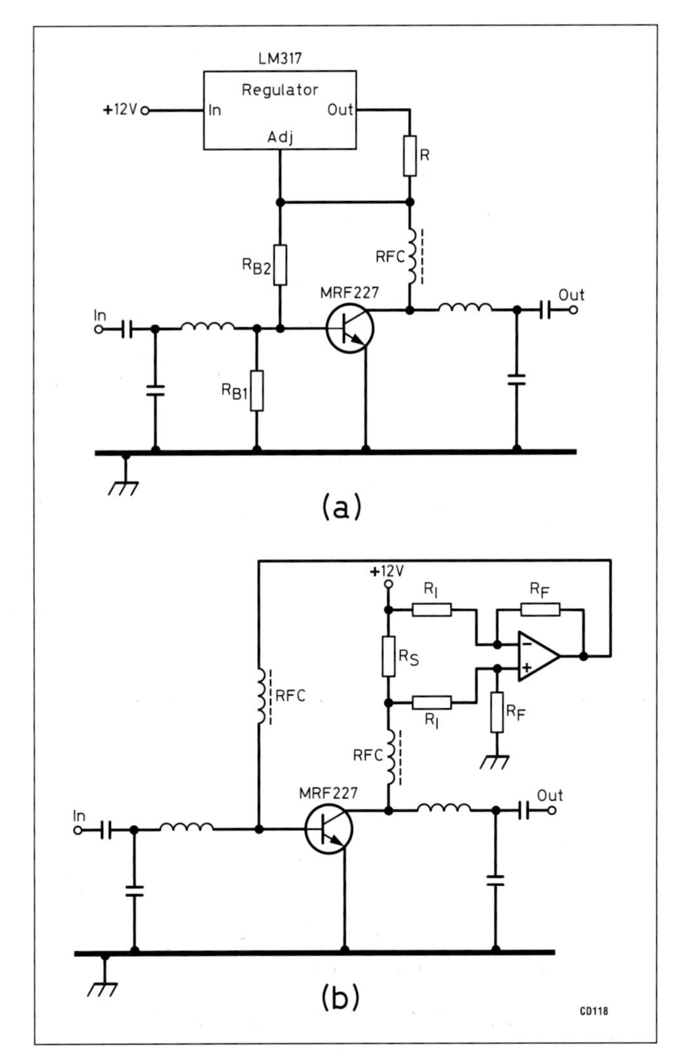


**Fig 5.32: Two methods to prevent thermal runaway of medium-power VHF transistors designed for Class C only in linear amplifiers. (a) A supply current regulator. (b) An active biasing circuit (*QST*)**

| Part No. (W) | Po (mW) | Pin (MHz) | Band | Mode |
|---|---|---|---|---|
| M57735 | 19 | 200 | 50 | lin |
| M57706L | 8 | 200 | 145 | - |
| M57719L | 14 | 200 | 145 | - |
| M57732L | 7 | 20 | 145 | lin |
| M57741UL | 28 | 200 | 145 | - |
| M57796L | 7 | 300 | 145 | lin |
| M67748L | 7 | 20 | 145 | lin |
| M67781L | 40 | 300 | 145 | - |
| M57704M | 13 | 200 | 435 | - |
| M57714M | 7 | 100 | 435 | - |
| M57716 | 17 | 200 | 435 | lin |
| M57729 | 30 | 300 | 435 | - |
| M57745 | 33 | 300 | 435 | lin |
| M57788M | 40 | 300 | 435 | - |
| M57797MA | 7 | 100 | 435 | lin |
| M67709 | 13 | 10 | 435 | lin |
| M67728 | 60 | 10 | 435 | lin |
| M67749M | 7 | 20 | 435 | lin |
| M57762 | 18 | 1000 | 1296 | lin |

**Table 5.4: Some Mitsubishi RF power modules. All operate on 12V mobile systems, but those rated over 25W cannot be supplied from a 5A (cigar lighter) circuit and generally require forced air cooling of the heat sink. Suitable for SSB only if marked 'lin' (Extracted from a 1998 Mainline Electronics pamphlet)**

load, including open- and short-circuit. As each model has its peculiarities, it is essential to follow data sheet instructions to the letter.

### A 435MHz 15W mobile booster

PA0GMS built this 70cm power amplifier with RF VOX to boost the FM output from his hand-held transceiver to a respectable 15W for mobile use [13]. The circuit is shown in **Fig 5.33.** Its top half represents the amplifier proper, consisting of (right to left) the input T/R relay, input attenuator to limit input power to what the module requires for rated output, the power module with its power lead decoupling chokes (ferrite 6 x 3mm beads with three turns of 0.5mm diameter enamelled copper wire) and bypass capacitors (those 0.1μF and larger are tantalum beads with their negative lead earthed; lower values are disc ceramics, except for the 200pF which is a feedthrough type). The output tank (two air trimmers and a 25mm long straight piece of 2mm silvered wire) feeds into the output T/R relay. Both relays are National model RH12 shielded miniature relays not designed for RF switching but adequate for UHF at this power level. The choke between the relay coils is to stop RF feedback through the relays.

The relays are activated by what is sometimes misnamed a VOX (voice-operated control) but is in fact an RF-operated control. Refer to the diagram at the bottom of Fig 5.33. A small fraction of any RF drive applied to the amplifier input is rectified and applied to the high-gain op-amp circuit here used as a comparator. Without RF input, the op-amp output terminal is at near-earth potential, the NPN transistor is cut off, and the relays do not make, leaving the amplifier out of the circuit as is required for receiving. When the operator has pressed his PTT (push-to-talk) switch, RF appears at the booster input, the op-amp output goes positive, the transistor conducts, and the relays make, routing the RF circuit through the booster.

This kind of circuit is widely used for bypassing receiving amplifiers and/or inserting power boosters in the antenna cable when transmitting; at VHF or UHF, this is often done to make up for losses in a long cable by placing these amplifiers at masthead or in the loft just below. Though not critical, the 1pF capacitor may have to be increased for lower frequencies and lower power levels, and reduced for higher frequencies and higher input power. The NPN transistor must be capable of passing the coil currents of the relays. While not an issue with 0.5W in and 15W out, at much higher power levels proper relay sequencing will be required.

No PCB is used for the amplifier proper. It is built into a 110 x 70 x 30mm tin-plate box, **Fig 5.34**. The module is bolted through the bottom of the box to a 120 x 70mm heatsink which must dissipate up to 30W. Heat transfer compound is used between the module and the box and also between the box and the heatsink; both must be as flat as possible. Care is required when tightening the bolts as the ceramic substrate of the module is brittle. The other components are soldered in directly. The RF-operated relay control is a sub-assembly mounted on a 40 x 30mm PCB.

Adjustments are simple. Set the input attenuator to maximum resistance and the two air trimmers to half-mesh. Apply 0.6 or 0.7W input from the hand-held and verify that both relays make. A few watts of output should be generated at this stage. Adjust the two air trimmers to peak the output, then reduce the input

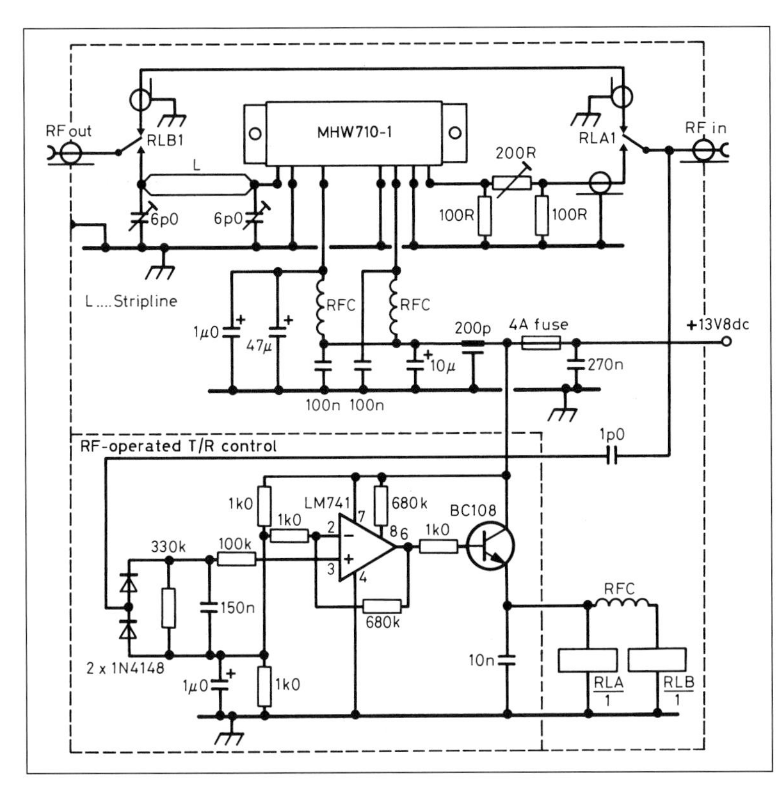


Fig 5.33: The PA0GMS 15W 435MHz mobile booster. The original MHW710-1 power module is obsolete, but model M57704M (different pin-out) is suitable

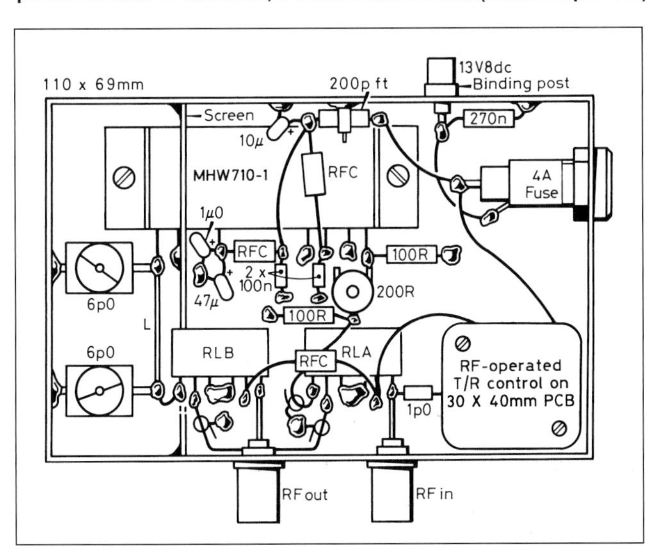


Fig 5.34: Construction of the PA0GMS 15W 435MHz mobile booster. The RF-operated T/R control is a sub-assembly on a PCB

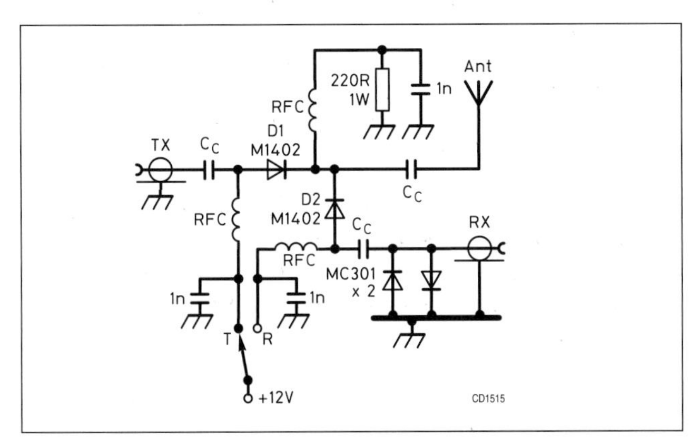


attenuation until the output reaches 15W with a supply current of about 3A. This completes the adjustment. Harmonics were found to be -60dBc and intermodulation -35dBc.

## PIN Diode T/R Switching

If the same antenna is to be used both for transmitting and receiving, change-over switching is required. This normally is accomplished by means of an electromechanical relay. If, however, the receiver must function as quickly as possible after the end of each transmission, as in high-speed packet operation, the release plus settling time of a relay, several milliseconds, restricts the data throughput and solid-state switching is indicated.

This is done by using PIN diodes as switches. When such a diode is 'biased on' by (typically) 50mA DC in its pass direction, its RF resistance is almost as low as that of closed relay contacts; when biased by a voltage in the opposite direction, it acts like a pair of open relay contacts.

To separate bias and RF circuits, bias is applied through RF chokes; capacitors pass only the RF. **Fig 5.35** shows the basic circuit. The resistor limits the bias current and provides the 'off bias' voltage. The back-to-back diodes limit the transmitter power reaching the receiver under fault condition. The insertion loss and isolation (the fraction of transmitter power getting through to the receiver) achievable with discrete components is tabulated.

At UHF and above, less-effective chokes and stray capacitances limit performance. For the 432 and 1296MHz bands, there are hybrid modules available containing the whole circuit optimised for one band.

## DC AND AF AMPLIFIERS

This section contains information on the analogue processing of signals from DC (ie zero frequency) up to 5kHz. This includes audio amplifiers for receivers and transmitters for frequencies generally between 300 and 3000Hz as well as auxiliary circuitry, which may go down to zero frequency.

## Operational Amplifiers (Op-amps)

An op-amp is an amplifier which serves to drive an external network of passive components, some of which function as a feedback loop. They are normally used in IC form, and many different types are available. The name operational amplifier comes from their original use in analogue computers where they were used to perform such mathematical operations as adding, subtracting, differentiating and integrating.

**(left) Fig 5.35: Solid-state antenna switch using pin diodes for transmitter power up to 25W. The diode numbers are Mitsubishi models. Typical data are:**

| **Frequency (MHz)** | **Isolation Tx to Rx (dB)** | **Insertion loss (dB)** |
|---|---|---|
| **29** | **40** | **0.3** |
| **50** | **39** | **0.3** |
| **144** | **39** | **0.3** |
| **220** | **38** | **0.4** |
| **440** | **36** | **0.5** |

**The off-state isolation can be improved by providing a DC return when the diodes are reverse-biased. Connect a resistor (1–10k) across both 1n capacitors**

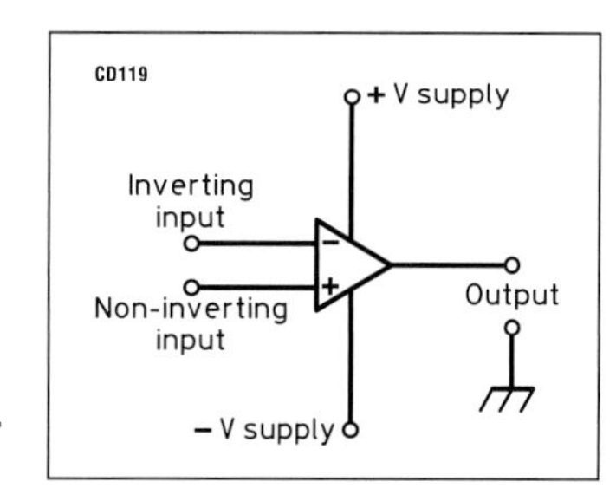


**Fig 5.36: Symbol of an operational amplifier (op-amp)**

The response of the circuit (for example the gain versus frequency response) is determined entirely by the components used in the feedback network around the op-amp. This holds true provided the op-amp has sufficient open-loop gain and bandwidth.

Why use IC op-amps rather than discrete components? Op-amps greatly facilitate circuit design. Having established that a certain op-amp is adequate for the intended function, the designer can be confident that the circuit will reproducibly respond as calculated without having to worry about differences between individual transistors or changes of load impedance or supply voltages. Furthermore, most op-amps will survive accidental short-circuits of output or either input to earth or supply voltage(s). General-purpose op-amps are often cheaper than the discrete components they replace and most are available from more than one maker. Manufacturers have done a commendable job of standardising the pin-outs of various IC packages. Many devices are available as one, two or four in one DIL package; this is useful to reduce PCB size and cost but does not facilitate experimenting.

Many books have been written about the use of op-amps; some of these can be recommended to amateur circuit designers as most applications require only basic mathematics. In the following, some basics are explained and a few typical circuits are included for familiarisation.

The symbol of an op-amp is shown in **Fig 5.36**. There are two input terminals, an inverting or (-) input and a non-inverting or (+) input, both with respect to the single output, on which the voltage is measured against earth or 'common'. The response of an op-amp goes down to DC and an output swing both positive and negative with respect to earth may be required; therefore a dual power supply is used, which for most op-amps is nominally ±15V.

Where only AC (including audio) signals are being processed, a single supply suffices and blocking capacitors are used to permit the meaningless DC output to be referred to a potential halfway between the single supply and earth established by a resistive voltage divider. In application diagrams, the power supply connections are often omitted as they are taken for granted.

To understand the use of an op-amp and to judge the adequacy of a given type for a specific application, it is useful to define an ideal op-amp and then compare it with the specifications of real ones. The ideal op-amp by itself, open loop, ie without external components, has the following properties:

- Infinite gain, ie the voltage between (+) and (-) inputs is zero.
- Output is zero when input is zero, ie zero offset.
- Infinite input resistance, ie no current flows in the input terminals.
- Zero output resistance, ie unlimited output current can be drawn.
- Infinite bandwidth, ie from zero up to any frequency.

No real op-amp is ideal but there are types in which one or two of the specifications are optimised in comparison with general-purpose types, sometimes at the expense of others and at a much higher price. The most important of these specifications will now be defined; the figures given will be those of the most popular and least expensive of general-purpose op-amps, the µA741C.

## Maximum ratings

Values which the IC is guaranteed to withstand without failure include:

- Supply voltage: ±18V.
- Internal power dissipation: 0.5W.
- Voltage on either input: not exceeding applied positive and negative supply voltages
- Output short-circuit: indefinite.

## Static electrical characteristics

These are measured at DC ($V_s$ = ±15V, T = 25°C).

- Input offset voltage ($V_{oi}$): the DC voltage which must be applied to one input terminal to give a zero output. Ideally zero. 6mV max.
- Input bias current ($I_b$): the average of the bias currents flowing into the two input terminals. Ideally zero. 500nA max.
- Input offset current ($I_{os}$): the difference between the two input currents when the output is zero. Ideally zero. 200nA max.
- Input voltage range ($V_{cm}$): the common-mode input, ie the voltage of both input terminals to power supply common. Ideally unlimited. ±12V min.
- Common-mode rejection ratio (CMRR): the ratio of common-mode voltage to differential voltage to have the same effect on output. Ideally infinite. 70dB min.
- Input resistance ($Z_i$): the resistance 'looking into' either input while the other input is connected to power supply common. Ideally infinite. 300kΩ min.
- Output resistance ($Z_o$): the resistance looking into the output terminal. Ideally zero. 75Ω typ.
- Short-circuit current ($I_{sc}$): the maximum output current the amplifier can deliver. 25mA typ.
- Output voltage swing (±Vo): the peak output voltage the amplifier can deliver without clipping or saturation into a nominal load. ±10V min ($R_L$ = 2kΩ).
- Open-loop voltage gain ($A_{OL}$): the change in voltage between input terminals divided into the change of output voltage caused, without external feedback. 200,000 typ, 25,000 min. ($V_o$ = ±10V, $R_L$ = 2kΩ).
- Supply current (quiescent, ie excluding $I_o$) drawn from the power supply: 2.8mA max.

## Dynamic electrical characteristics

- Slew rate is the fastest voltage change of which the output is capable. Ideally infinite. 0.5V/µs typ. ($R_L$ = 2kΩ).
- Gain-bandwidth product: the product of small-signal open-loop gain and the frequency (in Hz) at which that gain is measured. Ideally infinite. 1MHz typical.

Understanding the gain-bandwidth product concept is basic to the use of op-amps at other than zero frequency. If an amplifier circuit is to be unconditionally stable, ie not given to self-oscillation, the phase shift between inverting input and output must be kept below 180° at all frequencies where the amplifier has gain. As a capacitive load can add up to 90°, the phase shift within most op-amps is kept, by internal frequency compensation, to 90°, which coincides with a gain roll-off of 6dB/octave or 20dB/decade. In **Fig 5.37**, note that the open-loop voltage gain from zero up to 6Hz is 200,000 or 106dB. From 6Hz, the long slope is at -6dB/octave until it crosses the unity gain (0dB) line at 1MHz. At any point along that line the product of gain and frequency is the same: $10^6$. If one now applies external feedback to achieve a signal voltage or closed-loop gain of, say, 100 times, the -3dB

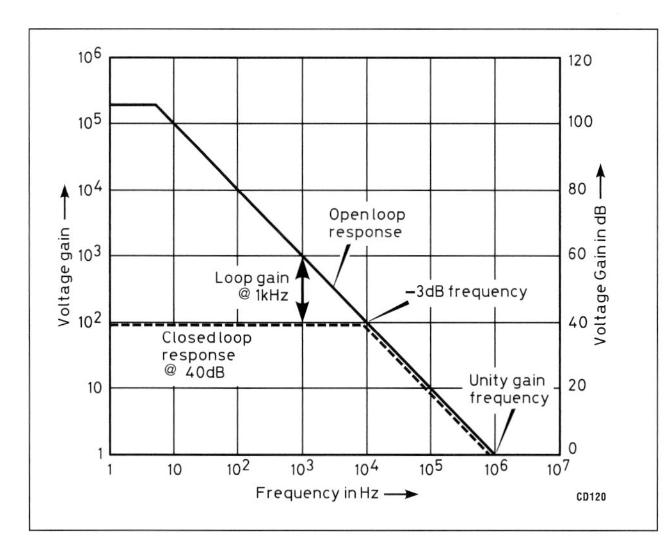


**Fig 5.37: Open-loop, closed-loop and loop gain of an op-amp, internally compensated for 6dB/octave roll-off**

point of the resulting audio amplifier would be at $10^6/100$ = 10kHz and from one decade further down, ie 1kHz to DC, the closed-loop gain would be flat within 1%. At 3kHz, ie the top of the communication-quality audio range, this amplifier would be only 1dB down and this would just make a satisfactory 40dB voice amplifier. If one wished to make a 60dB amplifier for the same frequency range, an op-amp with a gain-bandwidth product of at least $10^7$ = 10MHz should be used or one without internal frequency compensation.

## Op-amp types

The specifications given above apply to the general-purpose bipolar op-amp µA741C. There are special-purpose op-amps for a variety of applications, only a few are mentioned here.

**Fast op-amps:** The OP-37GP has high unity gain bandwidth (63MHz), high slew rate (13.5V/µs) and low noise, (3nV/√Hz) which is important for low-level audio.The AD797 has a gain bandwidth product of 110MHz. It has a slew rate of 20V/µs and has an extremely low noise. The input noise voltage is 0.9nV/√Hz. This device is fairly expensive but has sufficient bandwidth that it can be used as an IF amplifier at 455kHz. The low input noise voltage means that it will give a good signal/noise when driven from a low impedance source (eg 50 ohms).

The AD797 also has very good DC characteristics. The input offset voltage is typically 25µV. if this device is used, the manufacturers data sheet should be consulted [19]. This shows how the op-amp should be connected and decoupled to ensure stability.

**FET-input op-amps:** The TL071-C has the very low input current characteristic of FET gates ($I_{os} \leq$ 5pA) necessary if very-high feedback resistances (up to 10MΩ) must be used, but it will work on a single battery of 4.5V, eg in a low-level microphone amplifier, for which it has the necessary low noise. The input impedance is in the teraohm ($10^{12}$Ω) range.

**Power op-amps:** The LM383 is intended for audio operation with a closed-loop gain of 40dB and can deliver up to 7W to a 4Ω speaker on a single 20V supply, 4W on a car battery, and can dissipate up to 15W if mounted on an adequate heatsink.

## Basic op-amp circuits

**Fig 5.38** shows three basic configurations: inverting amplifier (a), non-inverting amplifier (b) and differential amplifier (c).

The operation of these amplifiers is best explained in terms of the ideal op-amp defined above. In (a), if $V_o$ is finite and the op-

**Fig 5.38: Three basic op-amp circuits: (a) inverting; (b) non-inverting; and (c) differential amplifiers**

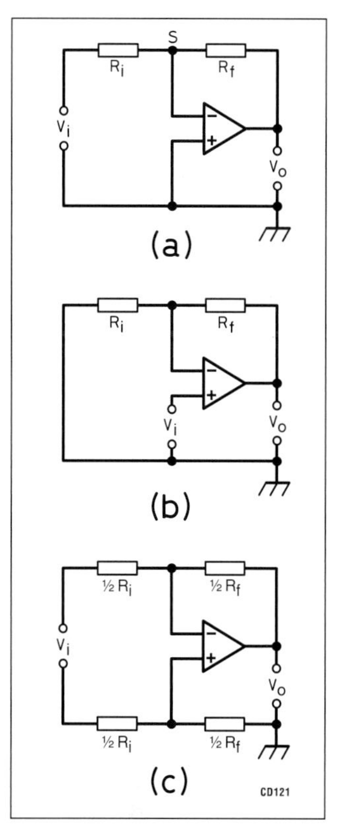


amp gain is infinite, the voltage between the (-) and (+) input terminals must approach zero, regardless of what $V_i$ is; with the (+) input earthed, the (-) input is also at earth potential, which is called virtual earth. The signal current Is driven by the input voltage $V_i$ through the input resistor $R_i$ will, according to Ohm's Law, be $V_i/R_i$. The current into the ideal op-amp's input being zero, the signal current has nowhere to go but through the feedback resistor $R_f$ to the output, where the output voltage $V_o$ must be $-I_sR_f$; hence:

$V_o = -V_i \frac{R_f}{R_i}$ in which $-\frac{R_f}{R_i} = A_{cl}$, the closed-loop gain

The signal source 'sees' a load of $R_i$; it should be chosen to suitably terminate that signal source, consistent with a feedback resistor which should, for general purpose bipolar op-amps, be between 10kΩ and 100kΩ; with FET-input op-amps, feedback resistors up to several megohms can be used.

The junction marked 'S' is called the summing point; several input resistors from different signal sources can be connected to the summing point and each would independently send its signal current into the feedback resistor, across which the algebraic sum of all signal currents would produce an output voltage. A summing amplifier is known in the hi-fi world as a mixing amplifier; in amateur radio, it is used to add, not mix, the output of two audio oscillators together for two-tone testing of SSB transmitters.

For the non-inverting amplifiers, **Fig 5.38(b)**, the output voltage and closed-loop gain derivations are similar, with the (-) input assuming Vi:

$V_o = V_i \frac{R_i + R_f}{R_i}$ in which $\frac{R_i + R_f}{R_i} = A_{cl}$, the closed loop gain

The signal source 'sees' the very high input impedance of the bare op-amp input. In the extreme case where $R_i = \infty$, ie left out, and $R_f = 0$, Vo = Vi and the circuit is a unity-gain voltage follower, which is frequently used as an impedance transformer having an extremely high input impedance and a near-zero output impedance. Non-inverting amplifiers cannot be used as summing amplifiers.

A differential amplifier is shown in **Fig 5.38(c)**. The differential input, ie $V_i$, is amplified according to the formulas given above for the inverting amplifier. The common-mode input, ie the average of the input voltages measured against earth, is

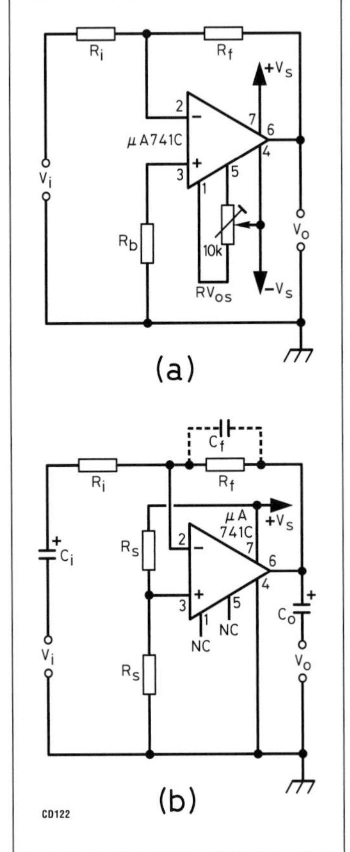


**Fig 5.39: DC to audio amplifier (a) requires dual power supplies; it has input bias current and voltage offset compensation. The audio-only amplifier (b) needs only a single power supply**

rejected to the extent that the ratio of the resistors connected to the (+) input equals the ratio of the resistors connected to the (-) input. A differential amplifier is useful in 'bringing to earth' and amplifying the voltage across a current sampling resistor of which neither end is at earth potential; this is a common requirement in current-regulated power supplies; see also Fig 5.32. **Power supplies for op-amps**

Where the DC output voltage of an op-amp is significant and required to assume earth potential for some inputs, dual power supplies are required. Where a negative voltage cannot be easily obtained from an existing mains supply, eg in mobile equipment, a voltage mirror can be used. It is an IC which converts a positive supply voltage into an almost equal negative voltage, eg type Si7661CJ. Though characterised when supplied with ±15V, most op-amps will work off a wide range of supply voltages, typically ±5V to ±18V dual supplies or a single supply of 10V to 36V, including 13.8V. One must be aware, however, that the output of most types cannot swing closer to either supply bus than 1-3V.

In audio applications, circuits can be adapted to work off a single supply bus. In **Fig 5.39**, the AC-and-DC form of a typical inverting amplifier with dual power supplies (a) is compared with the AC-only form (b) shown operating off a single power supply.

The DC input voltage and current errors are only of the order of millivolts but if an amplifier is programmed for high gain, these errors are amplified as much as the signal and spoil the accuracy of the output.

In **Fig 5.39(a)**, two measures reduce DC errors. The input error created by the bias current into the (-) input is compensated for by the bias current into the (+) input flowing through $R_b$, which is made equal to the parallel combination of $R_i$ and $R_f$. The remaining input error, $I_{os}$ ' $R_b$ + $V_{os}$, $i_s$ trimmed to zero with $RV_{os}$, which is connected to a pair of amplifier pins intended for that purpose. Adjustment of $RV_{os}$ for $V_o$ = 0 must be done with the $V_i$ terminals shorted.

Note that some other amplifier designs use different offset arrangements such as connecting the slider of the offset potentiometer to $+V_S$. Consult the data sheet or catalogue.

In audio-only applications, DC op-amp offsets are meaningless as input and output are blocked by capacitors $C_i$ and $C_o$ in **Fig 5.39(b)**. Here, however, provisions must be made to keep the DC level of the inputs and output about half-way between the single supply voltage and earth. This is accomplished by connecting the (+) input to a voltage divider consisting of the two resistors $R_S$. Note that the DC closed loop gain is unity as $C_i$ blocks $R_i$.

In the audio amplifier of **Fig 5.39(b)**, the high and low frequency responses can be rolled off very easily. To cut low-frequency response below, say 300Hz, $C_i$ is dimensioned so that at 300Hz, $XC_i = R_i$. To cut high-frequency response above, say 3000Hz, a capacitor $C_f$ is placed across Rf; its size is such that at 3000Hz, $XC_f = R_f$.

For more sophisticated frequency shaping, see the section on active filters.

## A PEP-reading module for RF power meters

The inertia of moving-coil meters is such that they cannot follow speech at a syllabic rate and even if they could, the human eye would be too slow to follow. The usual SWR/power meter found in most amateur stations, calibrated on CW, is a poor PEP indicator.

GW4NAH designed an inexpensive circuit [20], **Fig 5.40**, on a PCB small enough to fit into the power meter, which will rise to a peak and hold it there long enough for the meter movement, and the operator, to follow.

The resistance of RV1 + RV2 takes the place of the meter movement in an existing power meter; the voltage across it is fed via R1 and C1 to the (+) input of op-amp IC1a. Its output charges C3 via D2 and R6 with a rise time of 0.1s, but C3 can discharge only through R7 with a decay time constant of 10s. The voltage on C3 is buffered by the unity-gain voltage follower IC1b and is fed via D3 to the output terminals to which the original meter movement is now connected.

The input-to-output gain of the circuit is exactly unity by virtue of R5/R1 = 1. C2 creates a small phase advance in the feedback loop to prevent overshoot on rapid transients. The LM358 dual op-amp was chosen because, unlike most, it will work down to zero DC output on a single supply.

The small voltage across D1 is used to balance out voltage and current offsets in the op-amps, for which this IC has no built-in provisions, via R3, R4 and RV3. D4 protects against supply reversals and C4 is the power supply bypass capacitor. D5 and C5 protect the meter movement from overload and RF respectively.

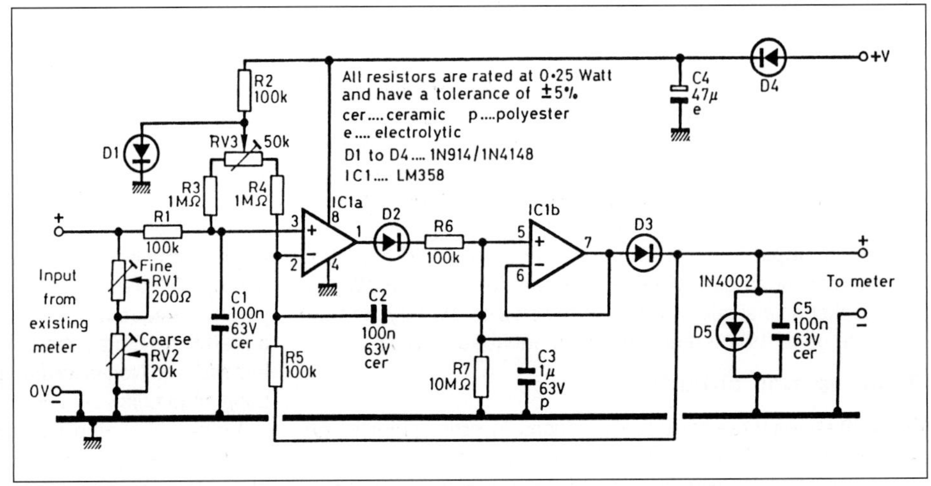


**Fig 5.40: A PEP-reading module for RF power meters using a dual op-amp (GW4NAH)**

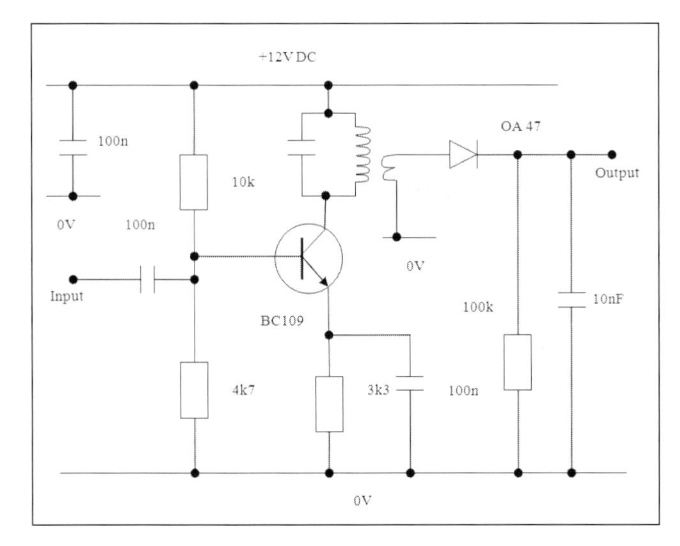


**Fig 5.41: Transistor IF amplifier with diode detector**

## AM Detectors

The advantage of AM (double sideband / full carrier) is that transmitters and receivers are generally much less complicated than for SSB. For the beginner it is relatively easy to build AM equipment. The detector stage of an AM receiver may consist of a single diode operating as a rectifier. **Fig 5.41** shows an example.

This detector has a voltage below which it will not detect. This is because of the threshold voltage of the diode. It is desirable to have a low threshold voltage and this can be achieved by using a germanium diode. It should be noted, however, that not all germanium diodes have the same threshold voltage and it is beneficial to select a diode which has a low forward voltage. A good choice is the OA47, and if more than one is available, it may be useful to select the one with the lowest threshold voltage because there is always a small variation between diodes of the same type. This is easily done by using a modern multimeter, most of which will show the forward voltage of a diode. Typical voltage for an OA47 type diode is approximately 240mV.

If a lower threshold than this is required, then an active rectifier can be used. An example of this is shown in **Fig 5.42**. This circuit uses a diode in a feedback circuit. This uses an op amp to reduce the threshold of the detector, and gives a value of typically 25mV.

## Receiver Audio

The audio signal obtained from the demodulator of a radio receiver generally requires filtering and voltage amplification; if a loudspeaker is to be the 'output transducer' (as distinct from headphones or an analogue-to-digital converter for computer processing) some power amplification is also required. For audio filtering, see the section on filters. For voltage amplification, the op-amp is the active component of choice for reasons explained in the section on them.

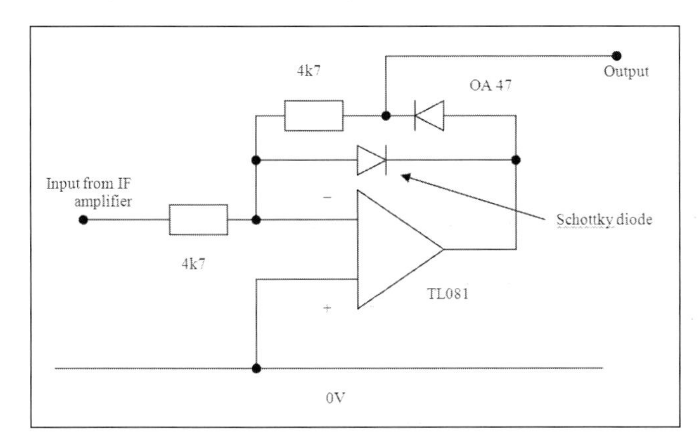


**Fig 5.42: Active rectifier detector**

There is also a great variety of ICs which contain not only the op-amp but also its gain-setting resistors and other receiver functions such as demodulator, AGC generator and a power stage dimensioned to drive a loudspeaker. A fraction of a watt is sufficient for a speaker in a quiet shack but for mobile operation in a noisy vehicle several watts are useful. Design, then, comes down to the selection from a catalogue or the junk box of the right IC, ie one that offers the desired output from the available input signal at an affordable price and on available power supply voltages.

The data sheet of the IC selected will provide the necessary details of external components and layout. **Fig 5.43** is an example of voltage and power amplification in an inexpensive 14-pin DIL IC; the popular LM380 provides more than 1W into an 8Ω speaker on a single 13.8V supply. Even the gain-setting resistors are built-in.

Note that many ICs which were popular as audio amplifiers are now being withdrawn by the manufacturers. The newer devices which replace them are often switched-mode amplifiers. These offer advantages for the user because they are more efficient than linear amplifiers and hence require less heatsinking. They are not recommended however, for use in a receiver because of the significant amount of radio-frequency energy generated by the amplifier.

If a receiver is used for long periods, it can become fatiguing for the operator to listen to receiver audio. This effect can be reduced by the use of good quality components in the receiver audio stages (particularly resistors and capacitors), thus improving the receiver audio and reducing fatigue. One solution is to use a good quality loudspeaker, and many hi-fi loudspeakers designed as mid-range speakers provide a bandwidth which is optimum for communications (300–3000Hz). The optimum enclosure for a mid-range speaker is typically one litre, which is quite small.

## Transmitter Audio

The audio processing in amateur as in commercial and military transmitters consists of low-level voltage amplification, compression, limiting and filtering, all tasks for op-amp circuitry as described in the section on them. The increase in the consumer usage of radio transmitters in cellular telephones and private mobile radios has given incentive to IC manufacturers to integrate ever more of the required circuitry onto one chip.

First, an explanation of a few terms used in audio processing: clipper, compressor, VOGAD and expander.

The readability of speech largely resides in the faithful reproduction of consonant sounds; the vowels add little to the readability but much to the volume. Turning up

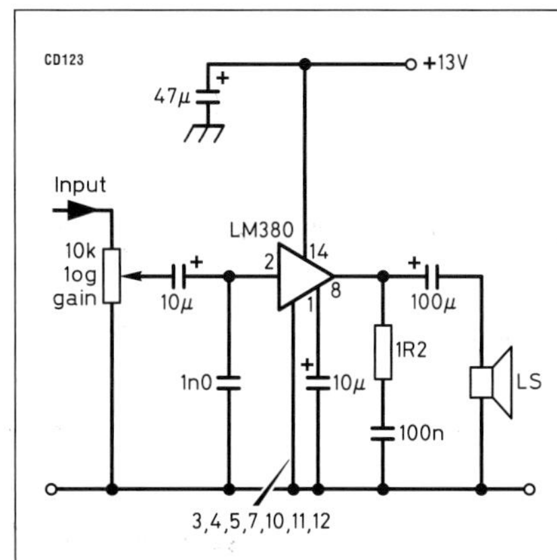


**Fig 5.43: This simple receiver audio IC provides up to 40dB voltage gain and over a watt of low-distortion output power to drive an 8-ohm loudspeaker**

the microphone gain does enhance the consonant sounds and thereby the readability, but the vowel sounds would then overload the transmitter, distort the audio and cause RF splatter. One remedy is to linearly amplify speech up to the amplitude limit which the transmitter can process without undue distortion and remove amplitude peaks exceeding that limit. This is called clipping.

If a waveform is distorted, however, harmonics are created and as harmonics of the lower voice frequencies fall within the 300-2700Hz speech range where they cannot subsequently be filtered out, too much clipping causes audible distortion and reduces rather than enhances readability. One way to avoid this, at least in a single-sideband transmitter, is to do the clipping at a higher frequency, eg the IF where the SSB signal is generated; the harmonics then fall far outside the passband required for speech and can be readily filtered out.

VOGAD stands for 'voice-operated gain adjustment device'. It automatically adjusts the gain of the microphone amplifier so that the speech level into the transmitter, averaged over several syllables, is almost independent of the voice level into the microphone. It is widely used, eg in hand-held FM transceivers.

A compressor is an amplifier of which the output is proportional to the logarithm of the input. Its purpose is to reduce the dynamic range of the modulation, in speech terms the difference between shouting and whispering into the microphone, and thereby improve the signal to noise ratio at the receiving end. If best fidelity is desired, the original contrast can be restored in the receiver by means of an expander, ie an exponential amplifier; this is desirable for hi-fi music but for speech it is seldom necessary.

A compander is a compressor and an expander in one unit, nowadays one IC. It could be used in a transceiver, with the compressor in the transmit and the expander in the receive chain.

## An RF speech clipper

As explained above, clipping is better done at a radio frequency than at audio. Analogue speech processors in the better pre-DSP SSB transceivers clip at the intermediate frequency at which the SSB signal is generated.

Rigs without a speech processor will benefit from the stand-alone unit designed by DF4ZS [21]. In **Fig 5.44** the left NE612 IC (a cheaper version of the NE602 described above) mixes the microphone audio with a built-in 453kHz BFO to yield 453kHz + audio, a range of 453.3 to 455.7kHz. The following filter removes all other mixing products. The signal is then amplified, clipped by the back-to-back diodes, amplified again, passed through another filter to remove the harmonics generated by the clipping process, and reconverted to audio in the right-hand NE612. A simple LC output filter removes non-audio mixing products.

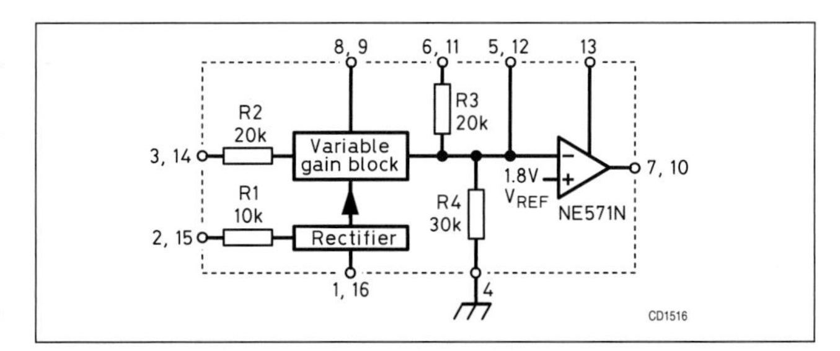


**Fig 5.45: Block diagram of the Philips NE571N two-channel compander. Basic input-to-output characteristics are as follows:**

| Compressor input level or expander output level (dBm) | Compressor output level or expander input level (dBm) |
|---|---|
| +20 | +10 |
| 0 | 0 |
| -20 | -10 |
| -40 | -20 |
| -60 | -30 |
| -80 | -40 |

## A compander IC

The Philips NE571N Compander IC contains two identical channels, each of which can be externally connected as a compressor or as an expander. Referring to **Fig 5.45**, this can be explained (in a very simplified way) as follows:

To expand, the input signal to the device is also fed into the rectifier which controls a 'variable gain block' (VGB); if the input is high, the current gain of the VGB is also high. For example, if the input goes up 6dB, the VGB gain increases by 6dB as well and the current into the summing point of the op-amp, and hence the output, go up 12dB, R3 being used as the fixed feedback resistor.

To compress, the output is fed into the rectifier and the VGB is connected in the feedback path of the op-amp with R3 being connected as the fixed input resistor. Now, if both the output and the VGB gain are to go up 6dB, the input must rise 12dB.

Having two channels enables application as a stereo compressor or expander, or, in a transceiver, one channel can compress the transmitted audio while the other expands the receiver output.

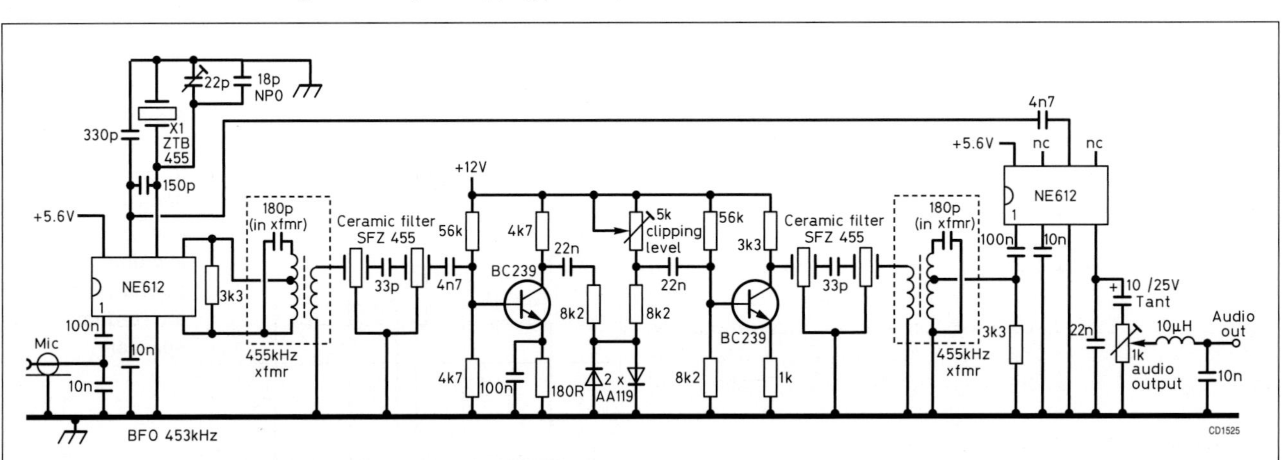


**Fig 5.44: The DF4ZS RF speech clipper combines high compression with low distortion**

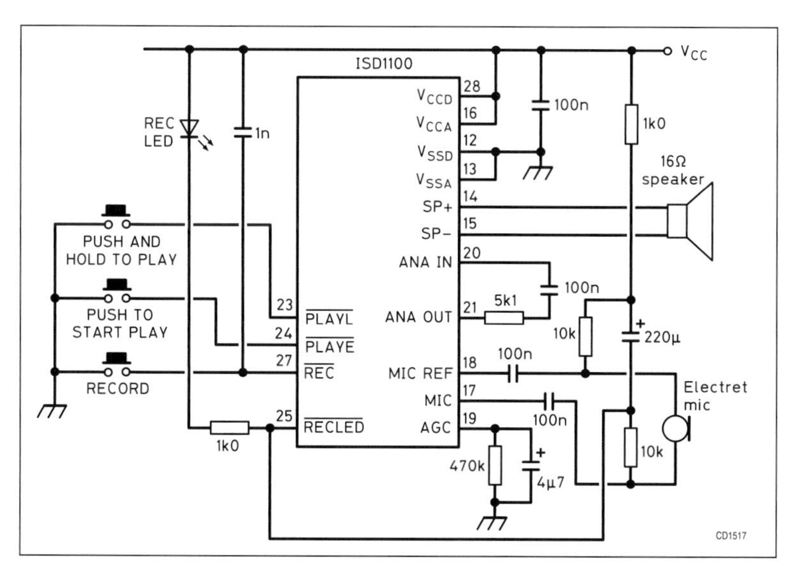


**Fig 5.46: A record-playback IC, with its associated circuitry, will record 10-20 sec of speech, store it in non-volatile memory, and play it back at the push of a button**

### A voice record-playback device

Contesters used to get sore throats from endlessly repeated 'CQs'. Repeated voice messages can now be sent with an IC which can record into non-volatile erasable analogue memory a message of 10-20 seconds in length from a microphone, and play it back with excellent fidelity through a loudspeaker or into the microphone socket of a transmitter. Playback can be repeated as often as desired and then instantly cleared, ready for a new message. The US company Information Storage Devices' ISD1100-series ICs sample incoming audio at a 6.4kHz rate, which permits an audio bandwidth up to 2.7kHz. **Fig 5.46** shows a simple application diagram.

Velleman, a Belgian manufacturer, makes a kit, including a PCB, on which to assemble the IC and the required passive components and switches. ON5DI showed how to interface this assembly with a transceiver and its microphone [22].

## FILTERS AND LC COUPLERS

Filters are circuits designed to pass signals of some frequencies and to reject or stop signals of others. Amateurs use filters ranging in operation from audio to microwaves. Applications include:

- Preselector filters which keep strong out-of-band signals from overloading a receiver.
- IF (intermediate frequency) filters which provide adjacent-channel selectivity in superheterodyne receivers.
- Audio filters which remove bass and treble, which are not essential for speech communication, from a microphone's output. This minimises the bandwidth taken up by a transmitted signal.
- A transmitter output is filtered, using a low-pass filter, to prevent harmonics being radiated.
- Mains filters, which are low-pass filters, used to prevent mains-borne noise from entering equipment.

Filters are classified by their main frequency characteristics. High-pass filters pass frequencies above their cut-off frequency and stop signals below that frequency. In low-pass filters the reverse happens. Band-pass filters pass the frequencies between two cut-off frequencies and stop those below the lower and above the upper cut-off frequency. Band-reject (or band-stop) filters stop between two cut-off frequencies and pass all others. Peak filters and notch filters are extremely sharp band-pass and band-reject filters respectively which provide the frequency characteristics that their names imply.

A coupler is a unit that matches a signal source to a load having an impedance which is not optimum for that source. An example is the matching of a transmitter's transistor power amplifier requiring a 2-ohm load to a 50-ohm antenna. Frequently, impedance matching and filtering is required at the same spot, as it is in this example, where harmonics must be removed from the output before they reach the antenna. There is a choice then, either to do the matching in one unit, eg a wide-band transformer with a $1:\sqrt{(50/2)} = 1:5$ turns ratio and the filtering in another, ie a 'standard' filter with 50Ω input and output, or to design a special filter-type LC circuit with a 2Ω input and 50Ω output impedance. In multiband HF transceivers, transformers good for all bands and separate 50Ω/50Ω filters for each band would be most practical. For UHF, however, there are no satisfactory wide-band transformers; the use of LC circuits is required.

## Ideal Filters and the Properties of Real Ones

Ideal filters would let all signals in their intended pass-band through unimpeded, ie have zero insertion loss, suppress completely all frequencies in their stop-band, ie provide infinite attenuation, and have sharp transitions from one to the other at their cut-off frequencies (**Fig 5.47**). Unfortunately such filters do not exist. In practice, the cut-off frequency, that is the transition point between pass-band and stop-band, is generally defined as the frequency where the response is -3dB (down to 70.7% in voltage) with respect to the response in the pass-band; in very sharp filters, such as crystal filters, the -6dB (half-voltage) points are frequently considered the cut-off frequencies. There are several practical approximations of ideal filters but each of these optimises one characteristic at the expense of others.

## LC Filters

If two resonant circuits are coupled together, a band-pass filter can be made. The degree of coupling between the two resonant circuits, both of which are tuned to the centre frequency, determines the shape of the filter curve (**Fig 5.48**). Undercoupled, critically coupled and overcoupled two-resonator filters all have their applications.

Four methods of achieving the coupling are shown in **Fig 5.49**. The result is always the same and the choice is mainly one of convenience. If the signal source and load are not close

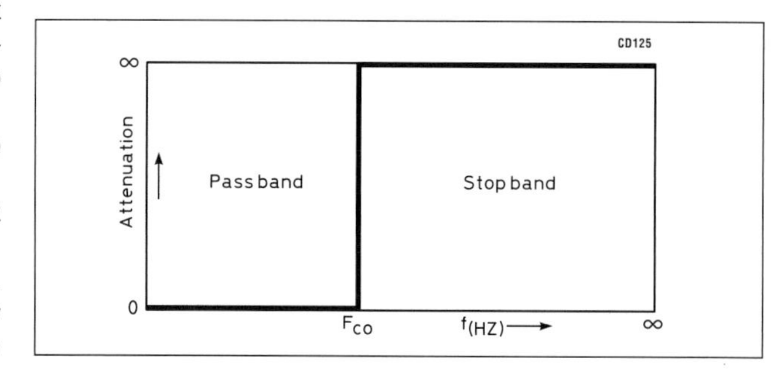


**Fig 5.47: Attenuation versus frequency plot of an ideal low-pass filter. No attenuation (insertion loss) in the pass-band, infinite attenuation in the stop-band, and a sharp transition at the cut-off frequency**

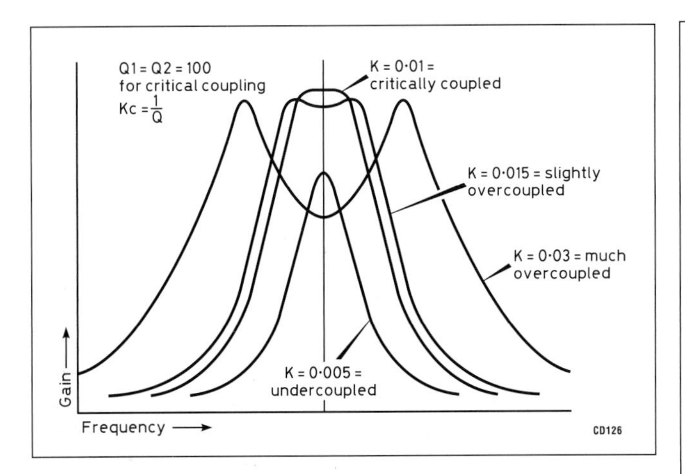


**Fig 5.48: Frequency response of a filter consisting of two resonant circuits tuned to the same frequency as a function of the coupling between them (after Terman). Under-coupling provides one sharp peak, critical coupling gives a flat top, mild overcoupling widens the top with an acceptably small dip and gross overcoupling results in peaks on two widely spaced frequencies. All these degrees of coupling have their applications**

together, eg on different PCBs, placing one resonant circuit with each and using link coupling is recommended to avoid earth loops. If the resonant circuits are close together, capacitive coupling between the 'hot ends' of the coils is easy to use and the coupling can be adjusted with a trimmer capacitor. Stray inductive coupling between adjacent capacitively coupled resonant circuits is avoided by placing them on opposite sides of a shield and/or placing the axes of the coils (if not on toroids or pot cores) at right-angles to one-another. The formulas for the required coupling are developed in the General Data chapter.

All filters must be properly terminated to give predictable bandwidth and attenuation. **Table 5.5** gives coil and capacitor values for five filters, each of which passes one HF band. The 7MHz and 14MHz filters are wider than the others to permit their use in frequency multiplier chains to the FM part of the 29MHz band. Included in each filter input and output is a 15kΩ termination resistor, as is appropriate for filters between a very high-impedance source and a similar load, or the drain of one dual-gate MOSFET amplifier and the gate of the next.

Frequently, a source or load has itself an impedance much lower than 15kΩ, eg a 50Ω antenna. These 50Ω then become the termination and must be transformed to 'look like' 15kΩ by tapping down on the coil. As the circulating current in high-Q resonant circuits is many times larger than the source or load current and if the magnetic flux in a coil is the same in all its turns, true in coils wound on powdered iron or ferrite toroids or pot cores, the auto-transformer formula may be used to determine where the antenna tap should be: at √(50/15,000) ≈6% up from the earthy end of the coil. In coils without such cores, the flux in the end turns is less than in the centre ones, so the tap must be experimentally located higher up the coil.

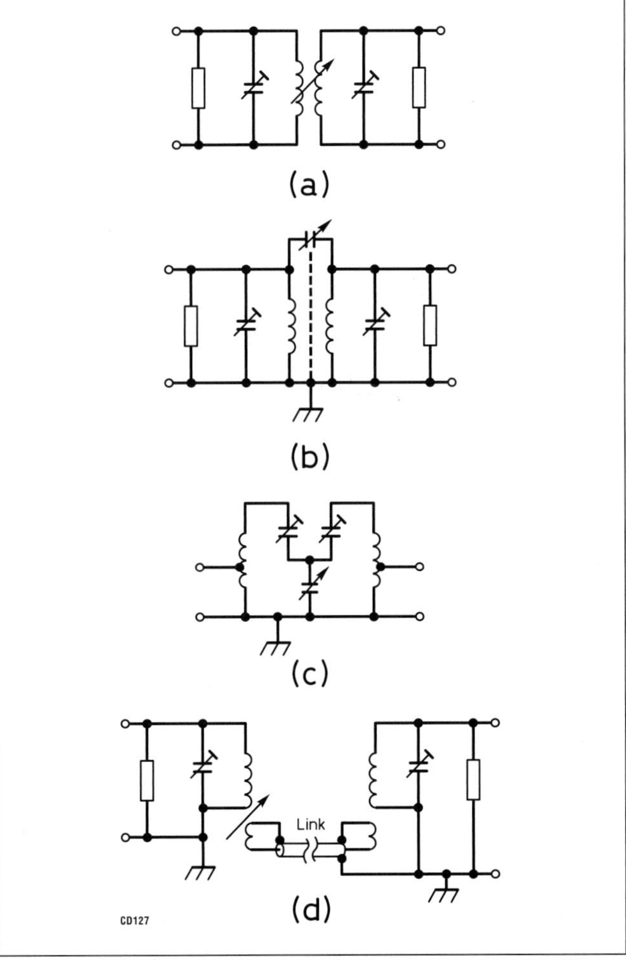


**Fig 5.49: Four ways of arranging the coupling between two resonant circuits. Arrows mark the coupling adjustment. (a) Direct inductive coupling; the coils are side-by-side or end-to-end and coupling is adjusted by varying the distance between them. (b) 'Top' capacitor coupling; if the coils are not wound on toroids or pot cores they should be shielded from each other or installed at right-angles to avoid uncontrolled inductive coupling. (c) Common capacitor coupling; if the source and/or load have an impedance lower than the proper termination, they can be 'tapped down' on the coil (regardless of coupling method). (d) Link coupling is employed when the two resonant circuits are physically separated**

| Lowest frequency (MHz)(MHz) | Centre frequency (MHz) | Highest frequency (pF) | Coupling (pF) | Parallel capacitance L (μH)(formers ¾in long, 3/8in dia) | Winding details |
|---|---|---|---|---|---|
| 3.5 3.65 | 3.8 | 6 | 78 | 2460t 32SWG close-wound | |
| 7 7.25 | 7.5 | 3 | 47 | 1040t 28SWG close-wound | |
| 14 14.5 | 15 | 1.5 | 24 | 527t 24SWG close-wound | |
| 21 21.225 | 21.45 | 1 | 52 | 112t 20SWG spaced to ¾in | |
| 28 29 | 30 | 0.6 | 10 primary | 321t 24SWG spaced to ¾in | |
| 30 secondary | 112t 20SWG spaced to ¾in | | | | |

*The use of tuning slugs in the coils is not recommended. Capacitors can be air, ceramic or mica compression trimmers. Adjust each coupler to cover frequency range shown.*

**Table 5.5: Band-pass filters for five HF bands and 15kΩ input and output terminations as shown in Fig 5.49(b)**

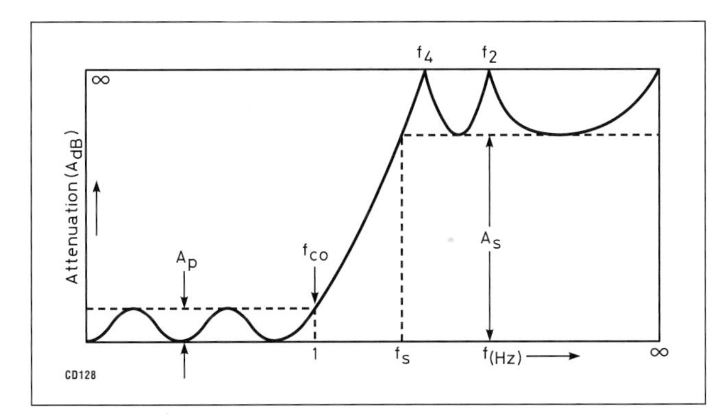


**Fig 5.50: Attenuation vs frequency plot of a two-section elliptic low-pass filter. A is the attenuation (dB); $A_p$ is the maximum attenuation in the pass-band or ripple; $f_4$ is the first attenuation peak; $f_2$ is the second attenuation peak with two-section filter; $f_{co}$ is the frequency where the attenuation first exceeds that in the pass-band; $A_s$ is the minimum attenuation in the stop-band; and $f_s$ is the frequency where minimum stop-band attenuation is first reached**

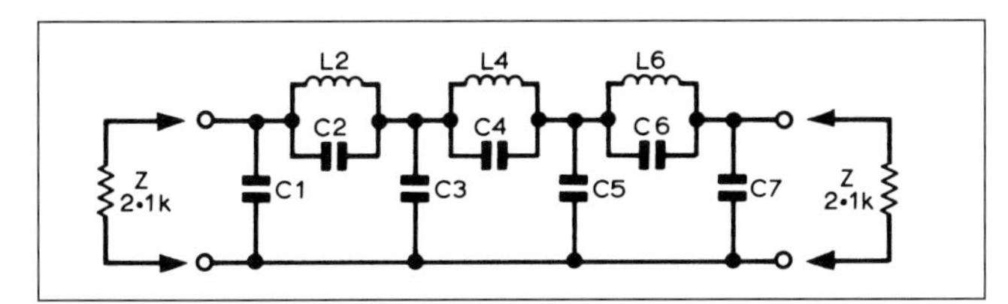


**Fig 5.51: A three-section elliptic low-pass filter with 3kHz cut-off. C1 = 37.3nF, C2 = 3.87nF, C3 = 51.9nF, C4 = 19.1nF, C5 = 46.4nF, C6 = 13.5nF, C7 = 29.9nF, mica, polyester or polystyrene. L2 = 168mH, L4 = 125mH, L6 = 130mH are wound on ferrite pot cores**

Impedances lower than 50Ω can be accommodated by placing that source or load in series with the resonant circuit rather than across all or part of it.

DC operating voltages to source and load devices are often fed though the filter coils. If properly bypassed, this does not affect filter operation. To avoid confusion, DC connections and bypass capacitors are not shown in the filter circuitry in this chapter.

Several more sophisticated LC filter designs are frequently used. All can be configured in high-pass, low-pass, band-pass and band-reject form. Butterworth filters have the flattest response in the pass-band. Chebyshev filters have a steeper roll-off to the stop-band but exhibit ripples in the pass-band, their number depending on the number of filter sections. Elliptic filters have an even steeper roll-off, but have ripples in the stop-band (zeroes) as well as in the pass-band (poles): see **Fig 5.50**. Chebyshev and elliptical filters have too much overshoot, particular near their cut-off frequencies, for use where pulse distortion must be kept down, eg in RTTY filters.

The calculation of component values for these three types would be a tedious task but for filter tables normalised for a cut-off frequency of 1Hz and termination resistance of 1Ω (or 1MHz and 50Ω where indicated). These can be easily scaled to the desired frequency and termination resistance. See the General Data chapter for more information.

M-derived and constant-k filters are older designs with less-well-defined characteristics but amateurs use them because component values are more easily calculated 'long-hand'. The diagrams and formulas to calculate component values are contained in the General Data chapter.

From audio frequency up to, say, 100MHz, filter inductors are mostly wound on powdered-iron pot cores or toroids of a material and size suitable for the frequency and power, and capacitors ranging from polystyrene types at audio, to mica and ceramic types at RF, with voltage and current ratings commensurate with the highest to be expected, even under fault conditions.

**Fig 5.51** shows an audio filter to provide selectivity in a direct-conversion receiver. It is a three-section elliptic low-pass filter with a cut-off frequency of 3kHz, suitable for voice reception.

To make any odd capacitance values of, say, 1% accuracy, start with the next lower standard value (no great accuracy required), measure it precisely (ie to better than 1%), and add what is missing from the desired value in the form of one or more smaller capacitors which are then connected in parallel with the

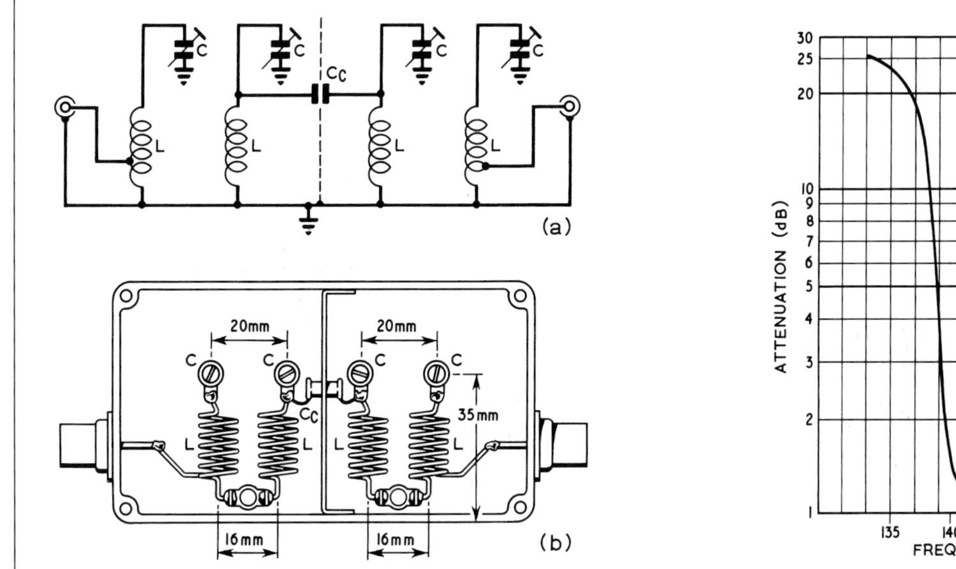


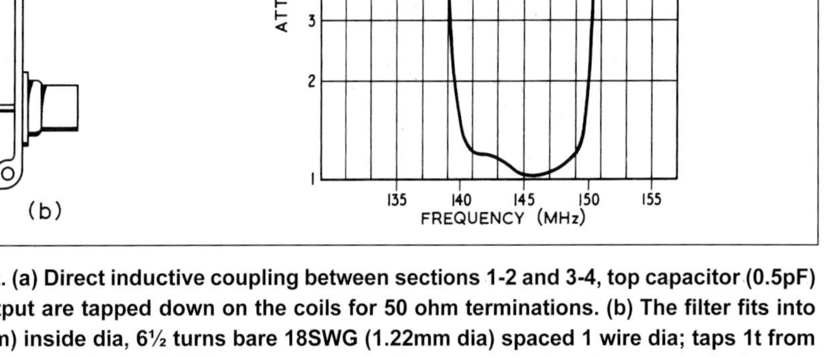


**Fig 5.52: A four-section band-pass filter for 145MHz. (a) Direct inductive coupling between sections 1-2 and 3-4, top capacitor (0.5pF) coupling between sections 2-3; both input and output are tapped down on the coils for 50 ohm terminations. (b) The filter fits into an 11 x 6 x 3cm die-cast box; coils are 3/8in (9.5mm) inside dia, 6½ turns bare 18SWG (1.22mm dia) spaced 1 wire dia; taps 1t from earthy end; C are 1–6pF piston ceramic trimmers for receiving and QRP transmitting; for higher power there is room in the box for air trimmers. (c) Performance curve**

first one. The smaller capacitors, having only a small fraction of the total value, need not be more accurate than, say, ±5%.

## Filtering at VHF and Above

At HF and below, it may be assumed that filters will perform as designed if assembled from components which are known to have the required accuracy, either because they were bought to tight specifications or were selected or adjusted with precise test equipment. It was further assumed that capacitor leads had negligible self-inductance and that coils had negligible capacitance.

At VHF and above these assumptions do not hold true. Though filter theory remains the same, the mechanics are quite different. Even then, the results are less predictable and adjustment will be required after assembly to tune out the stray capacitances and inductances. A sweep-generator and oscilloscope provide the most practical adjustment method. A variable oscillator with frequency counter and a voltmeter with RF probe, plus a good deal of patience, can also do the job.

At VHF, self-supporting coils and mica, ceramic or air-dielectric trimmer capacitors give adequate results for most applications. For in-band duplex operation on one antenna, as is common in repeaters, bulky and expensive very-high-Q cavity resonators are required, however.

The band-pass filter of **Fig 5.52** includes four parallel resonant circuits [15]. Direct inductive coupling is used between the first two and the last two; capacitive coupling is used between the centre two, where a shield prevents stray coupling. The input and output connections are tapped down on their respective coils to transform the 50Ω source and load into the proper terminations. This filter can reduce harmonics and other out-of-band spurious emissions when transmitting and suppress strong out-of-band incoming signals which could overload the receiver.

At UHF and SHF, filters are constructed as stripline or coaxial transmission line sections with air-dielectric trimmer capacitors. One type of stripline, sometimes called microstrip, consists of carefully dimensioned copper tracks on one side of high-grade (glass-filled or PTFE, preferably the latter) printed circuit board 'above' a ground plane formed by the foil on the other side of that PCB. The principle is illustrated in **Fig 5.53** and calculations are given in the General Data chapter.

Many amateurs use PCB strip lines wherever very high Q is not mandatory, as they can be fabricated with the PCB-making skills and tools used for many other home-construction projects.

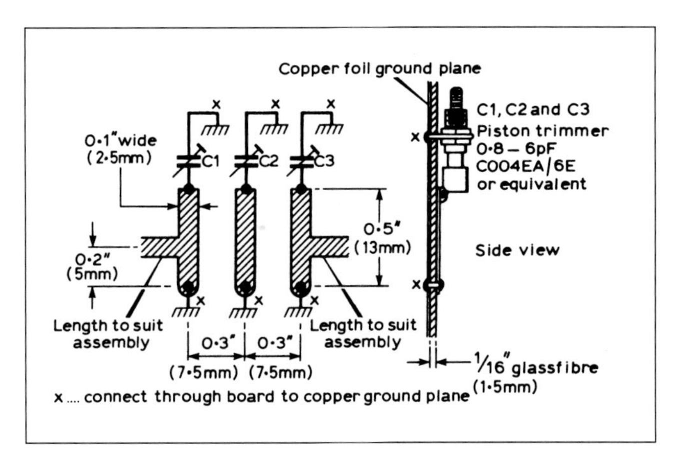


**Fig 5.54: A 1.3GHz microstrip filter consisting of three quarter-wave resonators. Coupling is by the stray capacitance between trimmer stators. The input and output lines of 50 ohm microstrip are tapped down on the input and output resonators**

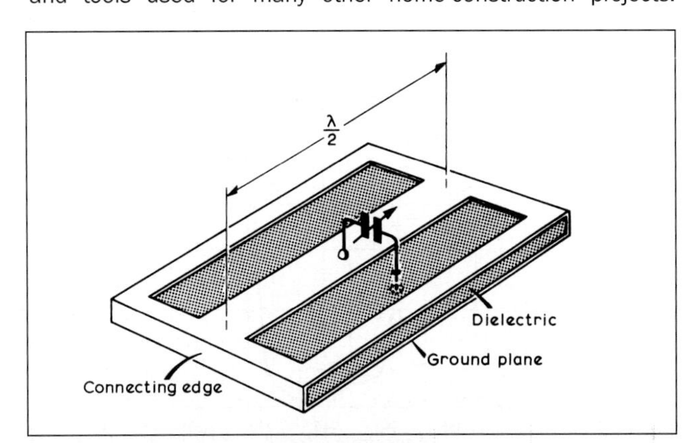


**Fig 5.53; Basic microstrip half-wave resonator. The electrical length of the strips is made shorter than their nominal length so that trimmers at the voltage maxima can be used to tune to resonance; the mechanical length of the strips is shorter than the electrical length by the velocity factor arising from the dielectric constant of the PCB material**

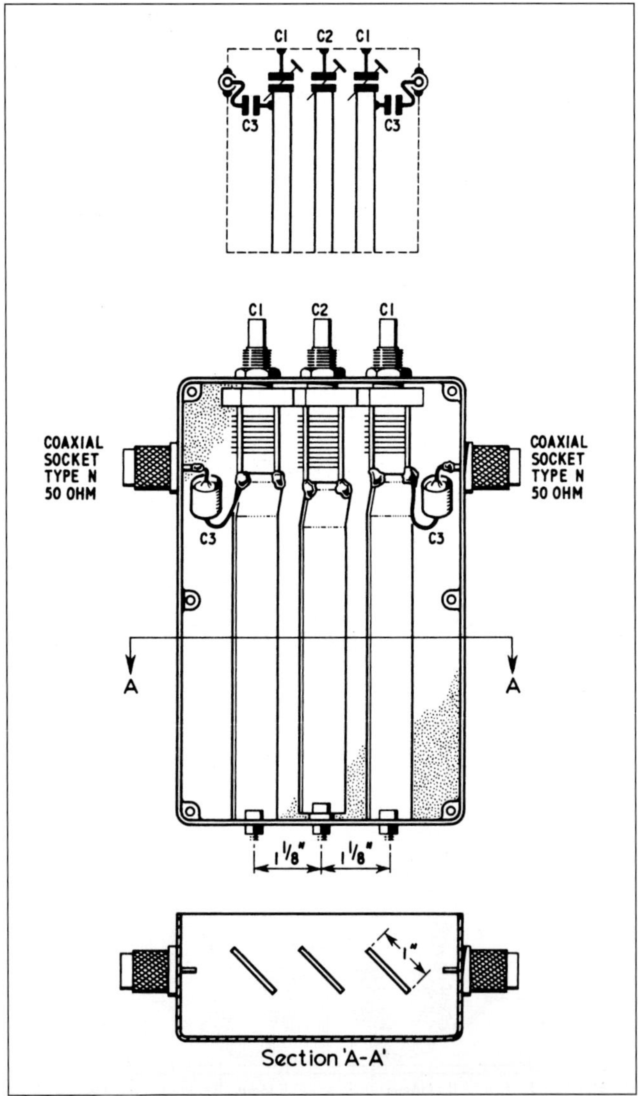


**Fig 5.55: The circuit and layout for a 100W 145MHz slab-line filter. The strips are 1 x 1/16in (25 x 1.5mm) sheet copper, offset at 45° to avoid overcoupling. Input and output lines are 6½in (165mm) long; the centre resonator is slightly shorter to allow for the greater length of C2 and a rib in the cast box. C1 = 50pF, C2 = 60pF, C3 = 4.4pF**

Frequently, in fact, such filters are an integral part of the PCB on which the other components are assembled.

**Fig 5.54** exemplifies a PCB band-pass filter for the 1.3GHz band. It consists of three resonators, each of which is tuned by a piston-type trimmer. It is essential that these trimmers have a low-impedance connection to earth (the foil on the reverse side of the PCB). With 1-5pF trimmers, the tuning range is 1.1-1.5GHz. The insertion loss is claimed to be less than 1dB. The input and output lines, having a characteristic impedance of 50Ω, may be of any convenient length. This filter is not intended for high-power transmitters.

For higher power, the resonators can be sheet copper strip-lines, sometimes called slab lines, with air as the dielectric. **Fig 5.55** gives dimensions for band-pass filters for the 144MHz band [15]. The connections between the copper fingers and the die-cast box are at current maxima and must have the lowest possible RF resistance.

In the prototype, the ends of the strips were brazed into the widened screwdriver slots in cheesehead brass bolts. The 'hot' ends of the strips are soldered directly to the stator posts of the tuning capacitors. After the input and output resonators are tuned for maximum power throughput at the desired frequency with the centre capacitor fully meshed, the latter is then adjusted to get the desired coupling and thereby pass-band shape (**Fig 5.56**).

For the highest Q at UHF and SHF, and with it the minimum insertion loss and greatest out-of-band attenuation, coaxial cavity resonators are used. Their construction requires specialised equipment and skills, as brass parts must be machined, brazed together and silver plated.

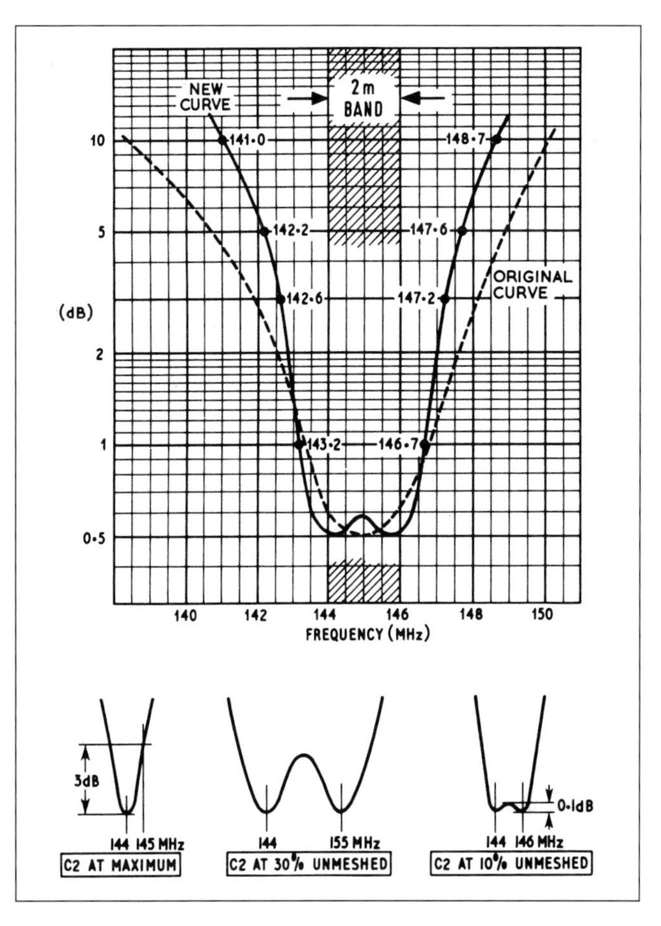


**Fig 5.56: Performance curves of the 145MHz slab-line filter as affected by the setting of C2. Note that the insertion loss is only 0.6dB. Also, compare the 10dB bandwidth of this filter with that of the four-section LC filter: 7.7MHz for this filter when tuned to 'new curve', 13MHz for the four-section filter with small, lower-Q coils**

## LC Matching Circuits

LC circuits are used to match very low or very high impedances to the 50 or 75-ohm coaxial cables which have become the standard for transporting RF energy between 'black boxes' and antennas. The desired match is valid only at or near the design frequency. The calculations for L, pi and L-pi circuits are given in the chapter on HF transmitters.

## High-Q Filter Types

The shape factor of a band-pass filter is often defined as the ratio between its -60dB and -6dB bandwidth. In a professional receiver, a single-sideband filter with a shape factor of 1.8 could be expected, while 2.0 might be more typical in good amateur equipment. It is possible to make LC filters with such performance, but it would have to be at a very low intermediate frequency (10-20kHz), have many sections, and be prohibitively bulky, costly and complicated. The limited Q of practically realisable inductors, say 300 for the best, is the main reason. Hence the search for resonators of higher Q. Several types are used in amateur equipment, including mechanical, crystal, ceramic, and surface acoustic wave (SAW) filters. Each is effective in a limited frequency range and fractional bandwidth (the ratio of bandwidth to centre frequency in percent). See **Fig 5.57**.

### Mechanical filters

There are very effective SSB, CW and RTTY filters for intermediate frequencies between 60 and 600kHz based on the mechanical resonances of small metal discs (**Fig 5.58**). The filter comprises three types of component: two magnetostrictive or piezo-electric transducers which convert the IF signals into mechanical vibrations and vice versa, a number of resonator discs, and coupling rods between those disks. Each disc represents the mechanical equivalent of a high-Q series resonant circuit, and the rods set the coupling between the resonators and thereby the bandwidth. Shape factors as low as 2 can be achieved. Mechanical filters were first used in amateur equipment by Collins Radio (USA). They still are offered as options in expensive amateur transceivers but with the advent of digital signal processing (DSP) in IF systems, typically between 10 and 20kHz, similar performance can be obtained at lower cost.

### Crystal filters

Piezo-electric crystals, cut from man-made quartz bars, are resonators with extremely high Q, tens of thousands being common. Their electrical equivalent, **Fig 5.59(a)**, shows a very large inductance $L_s$, an extremely small capacitance $C_s$ and a small

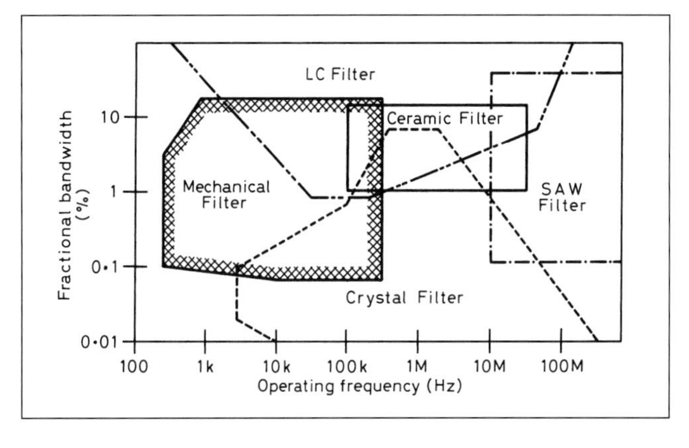


**Fig 5.57: Typical frequency ranges for various filter techniques**

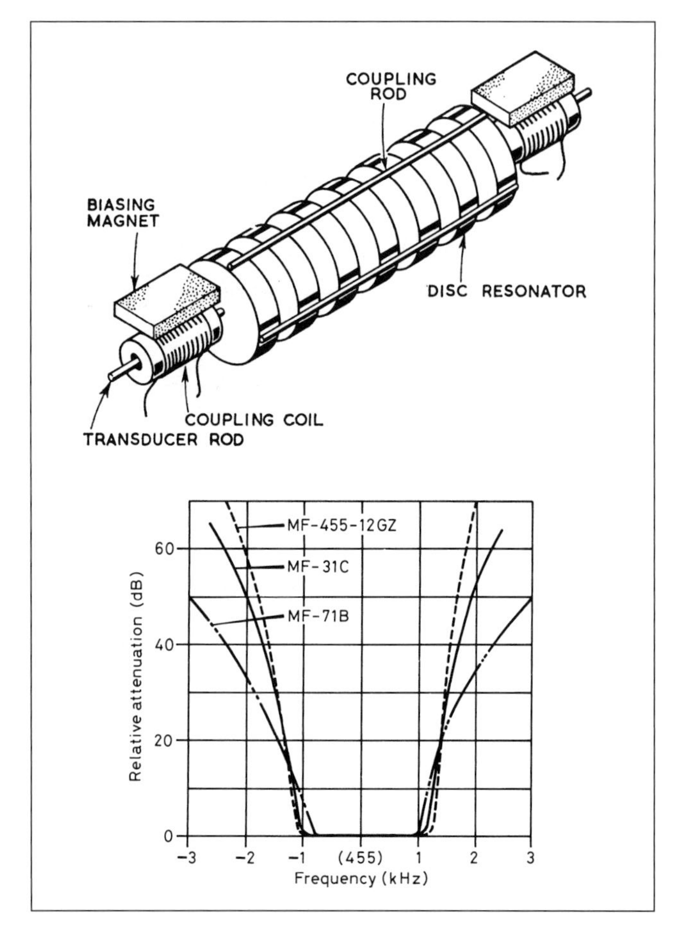


**Fig 5.58: The Collins mechanical filter. An IF signal is converted into mechanical vibrations in a magnetostrictive transducer, is then passed, by coupling rods, along a series of mechanical resonators to the output transducer which reconverts into an electrical signal. Below are response curves of three grades of Goyo miniature mechanical filters**

loss resistance $R_s$, in a series-resonant circuit. It is shunted by the parallel capacitance $C_p$, which is the real capacitance of the crystal electrodes, holder, socket, wiring and any external load capacitor one may wish to connect across the crystal.

At frequencies above the resonance of the series branch, the net impedance of that branch is inductive. This inductance is in parallel resonance with $C_p$ at a frequency slightly above the series resonance: **Fig 5.59(b)**. The essentially mechanical series resonance, the zero, may be considered user-immovable but the parallel resonance or pole can be pulled down closer to the series resonant frequency by increasing the load capacitance, eg with a trimmer.

Though their equivalent circuit would suggest that crystals are linear devices, this is not strictly true. Crystal filters can therefore cause intermodulation, especially if driven hard, eg by a strong signal just outside the filter's pass-band. Sometimes, interchanging input and output solves the problem. The two most common configurations are the half-lattice filter, **Fig 5.60**, and the ladder filter, **Fig 5.61**. Half-lattice filter curves are symmetrical about the centre frequency, an advantage in receivers, but they require

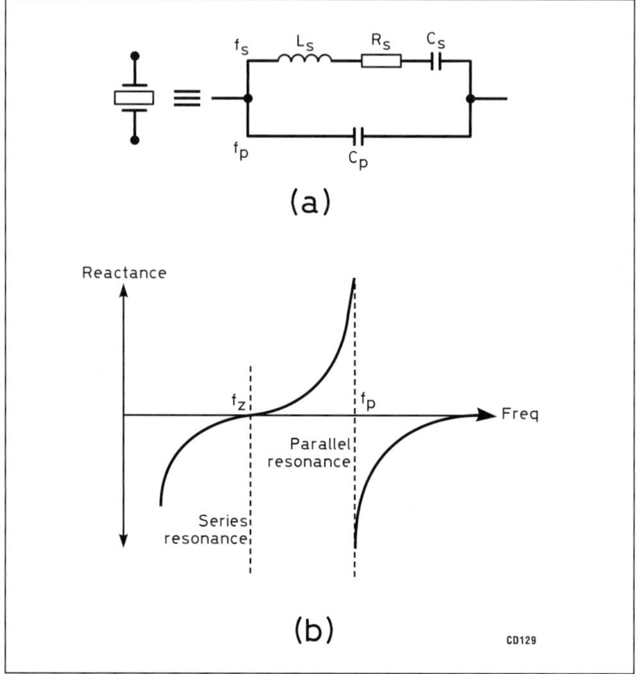


**Fig 5.59: The crystal equivalent circuit (a), and its reactance vs frequency plot (b)**

crystals differing in series resonant frequency by somewhat more than half the desired pass-band width and an RF transformer for each two filter sections. Two to four sections, four to eight crystals, are required in a good HF SSB receiver IF filter. Model XF-9B in **Table 5.6** is a high-performance filter which seems to be of this type. Half-lattice filters make good home construction projects but only for the well-equipped and experienced.

Ladder filters require no transformers and use crystals of only one frequency but they have asymmetrical pass-band curves; this creates no problem in SSB generators but is less desirable in a receiver with upper and lower sideband selection. With only five crystals a good HF SSB generator filter can be made. Model XF-9A in Table 5.6 seems to be of this type.

Ladder crystal filters using inexpensive consumer (3.58MHz telephone and 3.579, 4.433 or 8.867MHz TV colour-burst) crystals have been successfully made by many amateurs. A word of

| **Filter type** | **XF-9A** | **XF-9B** | **XF-9C** | **XF-9D** | **XF-9E** | **XF-9M** |
|---|---|---|---|---|---|---|
| **Application** | SSB TX | SSB TX/RX | AM | AM | NBFM | CW |
| **Number of poles** | 5 | 8 | 8 | 8 | 8 | 4 |
| **6dB bandwidth (kHz)** | 2.5 | 2.4 | 3.75 | 5.0 | 12.0 | 0.5 |
| **Passband ripple (dB)** | <1 | <2 | <2 | <2 | <2 | <1 |
| **Insertion loss (dB)** | <3 | <3.5 | <3.5 | <3.5 | <3 | <5 |
| **Termination** | 500 /30pF | 500 /30pF | 500 /30pF | 500 /30pF | 1200 /30pF | 500 /30pF |
| **Shape factor** | 1.7 (6-50dB) | 1.8 (6-60dB) | 1.8 (6-60dB) | 1.8 (6-60dB) | 1.8 (6-60dB) | 2.5 (6-40dB) |
| 2.2 (6-80dB) | 2.2 (6-80dB) | 2.2 (6-80dB) | 2.2 (6-80dB) | 4.4 (6-60dB) | | |
| **Ultimate attenuation (dB)** | >45 | >100 | >100 | >100 | >90 | >90 |

**Table 5.6: KVG 9MHz crystal filters for SSB, AM, FM and CW bandwidths**

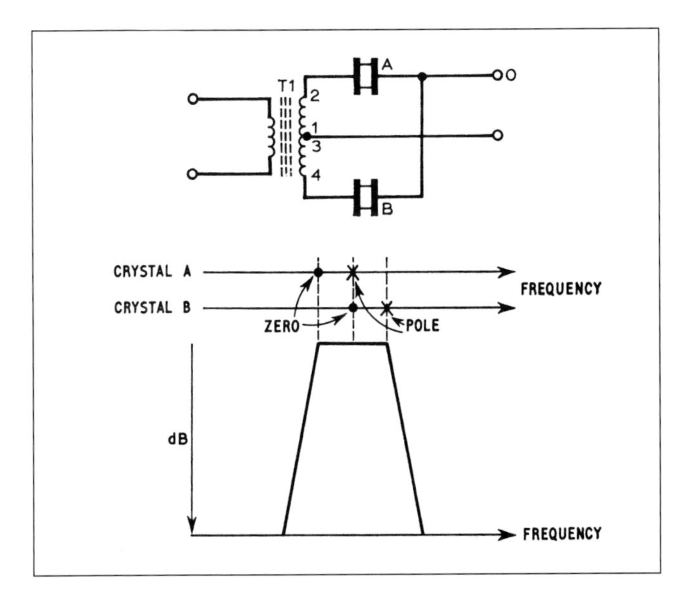


Fig 5.60: The basic single-section half-lattice filter diagram (top) and its idealised frequency plot (bottom). Note the poles and zeros of the two crystals in relation to the cut-off frequencies; if placed correctly, the frequency response is symmetrical. The bifilar transformer is required to provide balanced inputs

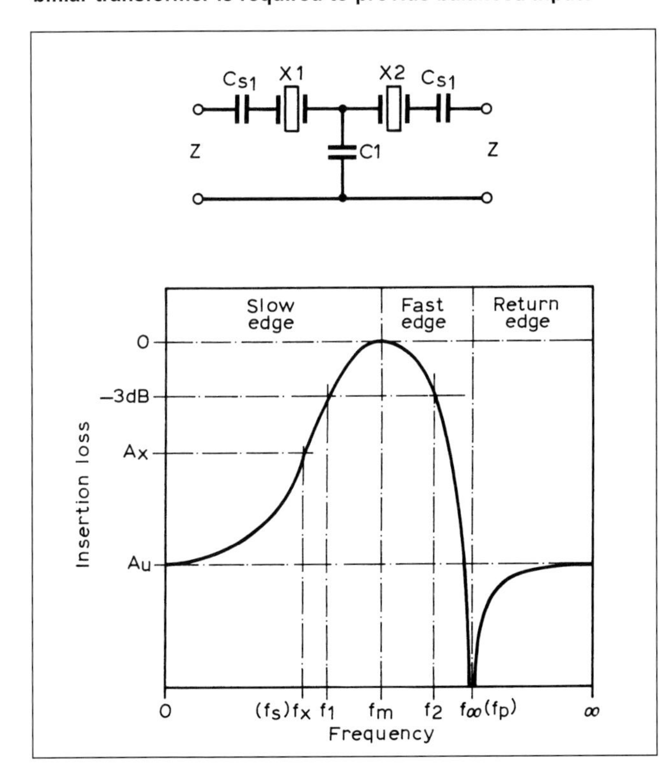


Fig 5.61: The basic two-pole ladder filter diagram (top) and its generalised frequency plot (bottom). Note that the crystals are identical but that the resulting frequency response is asymmetric

warning: these crystals were made for parallel-resonant oscillator service; their parallel-resonant frequencies with a given load capacitance, typically 20pF, were within 50ppm or so when new. Their series-resonant frequencies, unimportant for their original purpose but paramount for filter application, can differ by much more. It therefore is very useful to have many more crystals than needed so that a matched set can be selected.

To design a filter properly around available crystals one must know their characteristics. They can be measured with the test circuit of **Fig 5.62**, where the three positions of S1 yield three equations for the three unknowns $L_s$, $C_s$ and $C_p$. $R_s$ can be measured after establishing series-resonance with S1 in position 3 by substituting non-inductive resistors for the crystal without touching the frequency or output level of the signal generator. A resistor which gives the same reading on the output meter as the crystal is equal to $R_s$. Note that the signal generator must be capable of setting and holding a frequency within 10Hz or so. All relevant information and PC programs to simplify the calculations are given in [23].

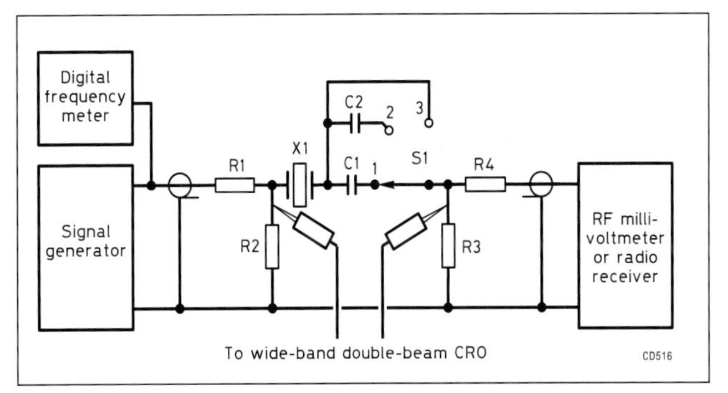


Fig 5.62: G3JIR's crystal test circuit. At resonance, both series (switch position 3) and parallel (switch positions 1 and 2), the two 'scope traces are in-phase. For 9MHz filter crystals, R1 = R4 = 1k , R2 = R3 = 220 ohms. The signal generator and frequency meter should have a frequency resolution of 10Hz

If a sweep generator and 'scope are used to adjust or verify a completed filter, the sweep speed must be kept very low: several seconds per sweep. Traditionally this has been done by using 'scopes with long-persistence CRTs. Sampling 'scopes, of course, can take the place of long-persistence screens.

The home-made filter of **Fig 5.63**, described by Y27YO, uses six 4.433MHz PAL colour TV crystals [24]. The filter bandwidth can be changed by switching different load capacitors across each crystal. See **Fig 5.64**. Note that at the narrower bandwidths the ultimate attenuation is reduced because of the greater load capacitors shunting each crystal. Each crystal with its load capacitors and switch wafer should be in a separate shielding compartment.

The input impedance of any crystal filter just above and below its pass-band is far from constant or non-reactive; therefore it is unsuitable as a termination for a preceding diode-ring mixer, which must have a fixed purely resistive load to achieve the desired rejection of unwanted mixing products. The inclusion in the Y27YO filter of a common-gate FET buffer solves this problem.

Virtually all available crystals above about 24MHz are overtone crystals, which means that they have been processed to have high Q at the third or fifth mechanical harmonic of their fundamental frequency. Such crystals can be used in filters. The marking on overtone crystals is their series resonant frequency, which is the one that is important in filters.

| Centre frequency (f0) | 48.0012MHz |
|---|---|
| 3dB bandwidth | 2.6kHz |
| 6dB bandwidth | 3kHz |
| 40dB bandwidth | 9kHz |
| 60dB bandwidth | 15.1kHz |
| Spurious responses | < 70dB |
| Pass-band ripple | <0.2dB |
| Insertion loss | 2.1dB |

Table 5.7: Performance of the 48MHz crystal filter

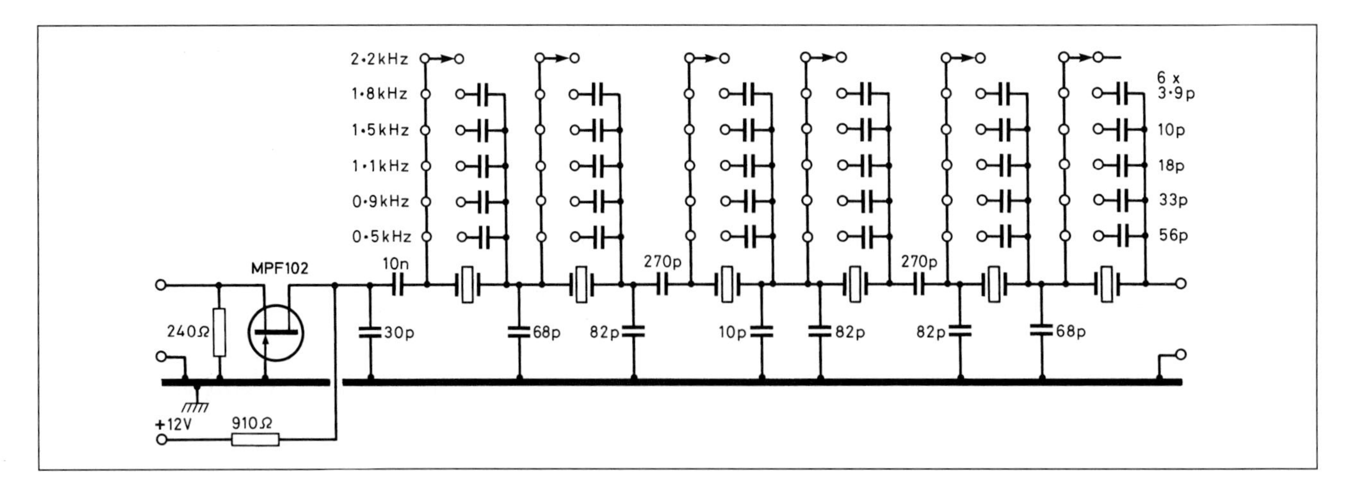


**Fig 5.63: Y27YO described this six-section 4.433MHz ladder filter with switch-selectable bandwidth in *Funkamateur* 1/85. The FET buffer is included to present a stable, non-reactive load to the preceding diode mixer and the proper source impedance to the filter. While it uses inexpensive components, including PAL TV colour-burst crystals, careful shielding between sections is required and construction and test demand skill and proper instrumentation**

48MHz is a common microprocessor clock frequency and third-overtone crystals for it are widely available and relatively inexpensive. 48MHz also is a suitable first IF for dual- or triple-conversion HF receivers, which then require a roofing filter at that frequency. PA0SE reported on the 48MHz SSB filter design shown in **Fig 5.65**. Prototype data are shown in **Table 5.7** -these were measured with a 50Ω source and 150Ω//7pF as shown in Fig 5.65.

Adequate shielding of the whole filter and between its sections is essential at this high frequency.

## Monolithic crystal filters

Several pairs of electrodes can be plated onto a single quartz blank as in **Fig 5.66**. This results in a multi-section filter with the coupling between elements being mechanical through the quartz. Monolithic crystal filters in the 10-100MHz range are used as IF filters where the pass-band must be relatively wide, ie in AM and FM receivers and as roofing filters in multimode receivers.

## Ceramic filters

Synthetic (ceramic) piezo-electric resonators are being made into band-pass filters in the range of 400kHz to 10.7MHz. Monolithic, ladder and half-lattice crystal filters all have their ceramic equivalents. Cheaper, ceramic resonators have lower Q than quartz crystals and their resonant frequencies have a wider tolerance and are more temperature dependent. While ceramic band-pass filters with good shape factors are made for bandwidths commensurate with AM, FM and SSB, they require more sections to achieve them, hence their insertion loss is greater. Care must

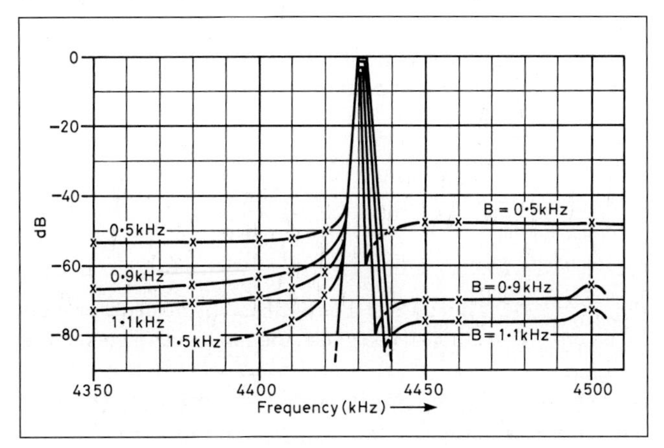


**Fig 5.64: Response curves of the the Y27YO filter**

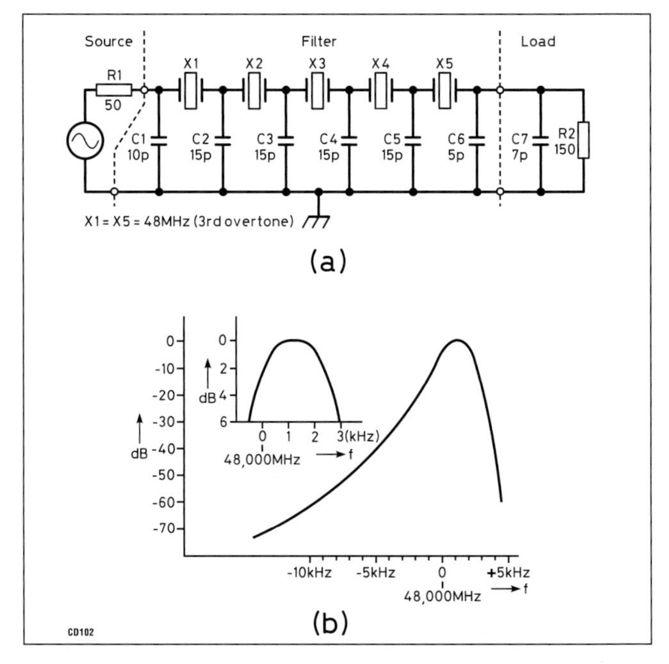


**Fig 5.65: Overtone crystals, always marked with their series-resonant frequency, are used in this 48MHz ladder filter. This design by J Wieberdink from the Dutch magazine *Radio Bulletin* 10/83 makes a good roofing filter for an HF SSB/CW receiver, but its top is too narrow to pass AM or NBFM and its shallow skirt slope on the low side requires that it be backed up by another filter at a second or third IF**

be taken that BFOs used with them have sufficient frequency adjustment range to accommodate the centre frequency tolerance of a ceramic filter, eg 455±2kHz at 25°C. Input and output matching transformers are included in some ceramic filter modules. **Fig 5.67** shows a Murata ladder filter and a Toko monolithic filter.

## Active filters

When designing LC audio filters, it is soon discovered that the inductors of values one would wish to use are bulkier, more expensive and of lower Q than the capacitors. Moreover, this lower Q requires the use of more sections, hence more insertion loss and even greater bulk and cost. One way out is the active

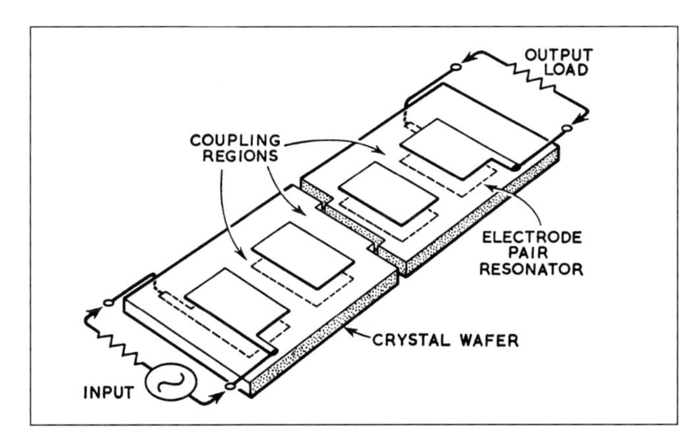


**Fig 5.66: A monolithic crystal band-pass filter consists of several resonators, ie pairs of electrodes, on a single quartz plate. The coupling between the resonators is essentially mechanical through the quartz. Monolithic filters are mounted in three or four-pin hermetically sealed crystal holders. At HF, they are less expensive though less effective than the best discrete-component crystal filters, but they do make excellent VHF roofing filters**

filter, a technique using an amplifier to activate resistors and capacitors in a circuit which emulates an LC filter. Such amplifiers can be either single transistors or IC operational amplifiers, both being inexpensive, small, miserly with their DC supply, and capable of turning insertion loss into gain. Most active audio filters in amateur applications use two, three or four two-pole sections, each section having an insertion voltage gain between one and two. Filter component (R and C) accuracies of better than 5% are generally adequate, polystyrene capacitors being preferred.The advantage of op-amps over single transistors is that the parameters of the former do not appear in the transfer (ie output vs input) function of the filter, thereby simplifying the calculations.

Note that most IC op-amps are designed to have a frequency

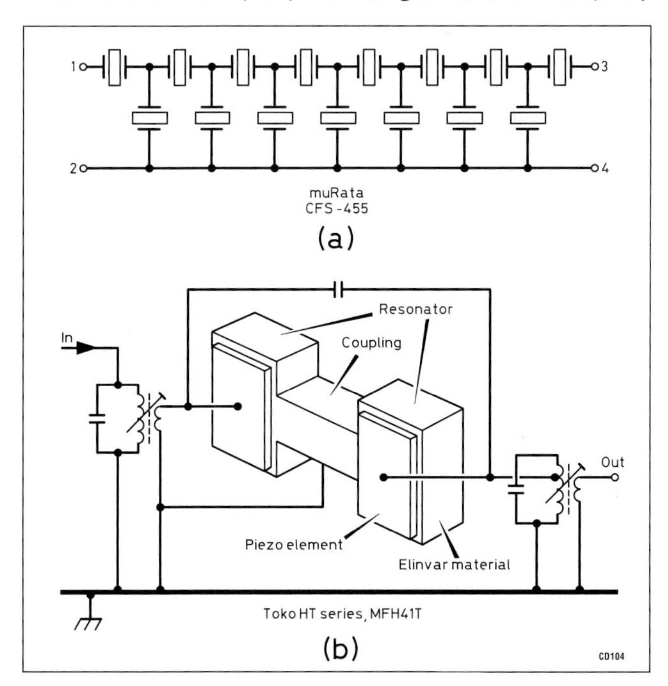


**Fig 5.67: Ceramic band-pass filters are made in several configurations resembling those of crystal filters. While cheaper than the latter, they require more sections for a given filter performance, hence have greater insertion loss. Centre-frequency tolerances are 0.5% typical. Shown here are a ladder filter (a) and a monolithic filter (b)**

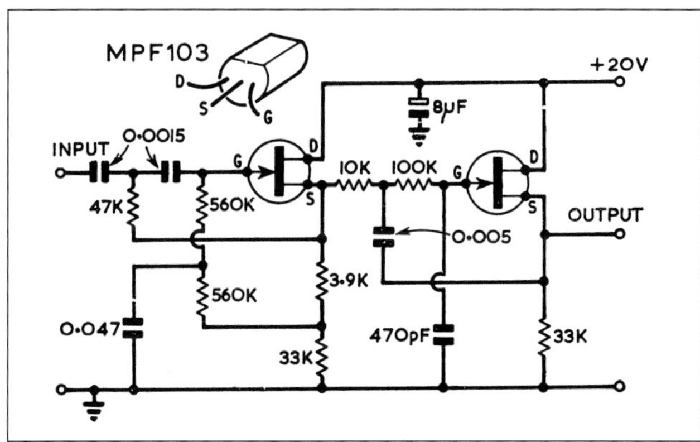


**Fig 5.68: The ZL2APC active band-pass audio filter using FETs and a single DC supply**

response down to DC. To allow both positive and negative outputs, they require both positive and negative supply voltages. In AC-only applications, however, this can be circumvented. In a single 13.8V DC supply situation, one way is to bridge two series-connected 6.8kΩ 1W resistors across that supply, each bypassed with a 100µF/16V electrolytic capacitor; this will create a three-rail supply for the op-amps with 'common' at +6.9V, permitting an output swing up to about 8V p-p. The input and output of the filter must be blocked for DC by capacitors.

A complete active filter calculation guide is beyond the scope of this chapter but some common techniques are presented in the General Data chapter. Here, however, follow several applications, one with single transistors and others using op-amps.

A discrete-component active filter, which passes speech but rejects the bass and treble frequencies which do not contribute to intelligibility, is shown in **Fig 5.68**. This filter, designed by ZL2APC, might be used to provide selectivity for phone reception with a direct-conversion receiver (though it must be pointed out that active filters do not have sufficient dynamic range to do justice to those very best DC receivers, which can detect microvolt signals in the presence of tens of millivolts of QRM on frequencies which the audio filter is required to

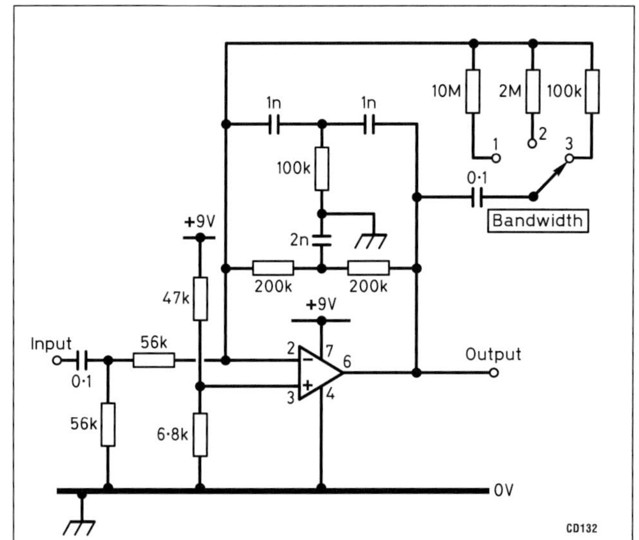


**Fig 5.69: G3SZW put an 800Hz twin-T filter in the feedback loop of an op-amp to obtain a peak filter, then widened the response to usable proportions with switch-selectable resistors: to 60 and 180Hz (positions 1 and 2) for CW or 300-3500Hz (position 3) for voice. This filter does not require dual DC supplies**

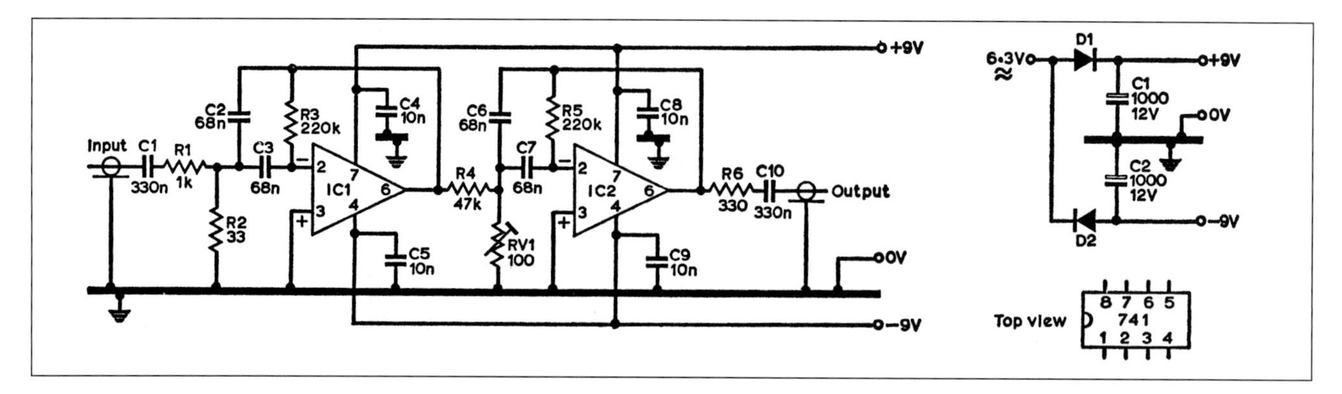


**Fig 5.70: This CW band-pass filter using two multiple-feedback stages comes from LA2IJ and LA4HK. The centre frequency of the second stage can be equal to or offset from the first. If the two frequencies coincide, the overall bandwidth is 50Hz at 6dB (640Hz at 50dB), if staggered 200Hz (1550Hz). The dual power supply is derived from 6.3VAC, available in most valved receivers or transmitters**

reject).

A twin-T filter used in the feedback loop of an op-amp is shown in **Fig 5.69**. A twin-T filter basically is a notch filter which rejects one single frequency and passes all others. With the R and C values shown, that frequency is about 800Hz. Used in the feedback loop of an op-amp, as in this design by G3SZW, the assembly becomes a peak filter which passes only that one frequency, too sharp even for CW. G3SZW broadened the response by shunting switch-selectable resistors across the twin-T; to 60Hz with 10MΩ, 180Hz with 2MΩ, and from 300Hz to 3500Hz, for phone, with 100kΩ.

Twin-T filters require close matching of resistors and capacitors. That would, in this example, be best accomplished by using four identical 1nF capacitors and four identical 200kΩ resistors, using two of each in parallel to make the 2nF capacitors and 200k resistors.

This circuit also demonstrates another technique for the use of op-amps on a single supply, here 9V. The DC level of both inputs is set by the voltage divider to which the (+) input is connected:

$$[6.8/(6.8 + 47)] \times 9 = 1.14\text{V}$$

The op-amp's DC output level is set by its DC input voltage and inverting gain:

$$1.14 + [(200 + 200)/(56 + 56)] \times 1.14 = 5.2\text{V}$$

roughly half-way between +9V and earth. The capacitor in series with the bandwidth switch is to prevent the lowest bandwidth resistor from upsetting the DC levels. The input blocking capacitor has the same purpose with respect to any DC paths through the signal source or load.

A CW filter with two 741 or 301A op-amps in a multiple-feedback, band-pass configuration was described by LA2IJ and LA4HK [25], and is shown in **Fig 5.70**. It would be a worthwhile accessory with a modern transceiver lacking a narrow crystal filter.

The first stage is fixed-tuned to about 880Hz, depending partly on the value of R2. The corresponding resistor in the second stage is variable and with it the resonant frequency can be adjusted to match that of the first stage, or to a slight offset for a double-humped band-pass characteristic. In the first state the filter has a pass-band width of only 50Hz at -6dB (about 640Hz at -50dB). When off-tuned the effective pass-band can be widened to about 200Hz (1550Hz).

Note that a bandwidth as narrow as 50 Hertz is of value only if the frequency stability of both the transmitter and receiver is such that the beat note does not drift out of the pass-band during a transmission, a stability seldom achieved with free-running home-built VFOs. Few analogue filters can compete with the ears of a skilled operator when it comes to digging out a weak wanted signal from among much stronger QRM.

A scheme to provide second-order CW band-pass filtering or a tuneable notch for voice reception was described by DJ6HP [26]. It is shown in **Fig 5.71**, and is based on the three-op-amp so-called state variable or universal active filter. The addition of a fourth op-amp, connected as a summing amplifier, provides a notch facility.

The resonant Q can be set between 1 and 5 with a single variable resistor and the centre frequency can be tuned between 450 and 2700Hz using two ganged variable resistors.

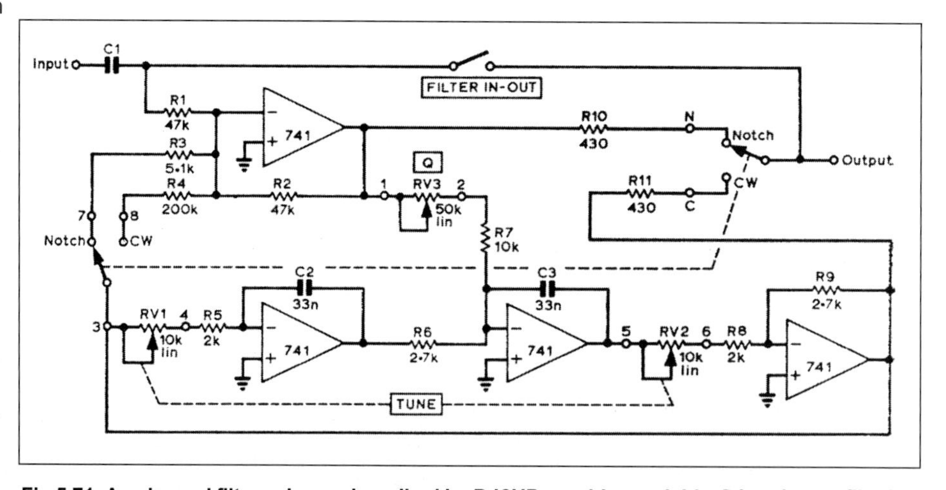


**Fig 5.71: A universal filter scheme described by DJ6HP provides variable-Q band-pass filtering for CW or a tunable notch for 'phone. Dual DC supplies between ±9V and ±15V are required**

## Switched-capacitor filters

The switched-capacitor filter is based on the digital processing of analogue signals, ie a hybrid between analogue and digital signal processing. It depends heavily on integrated circuits for its implementation. It pays to study the manufacturers' data sheets of devices under consideration before making a choice.

WB4TLM/KB4KVE describe the operation of switched-capac-

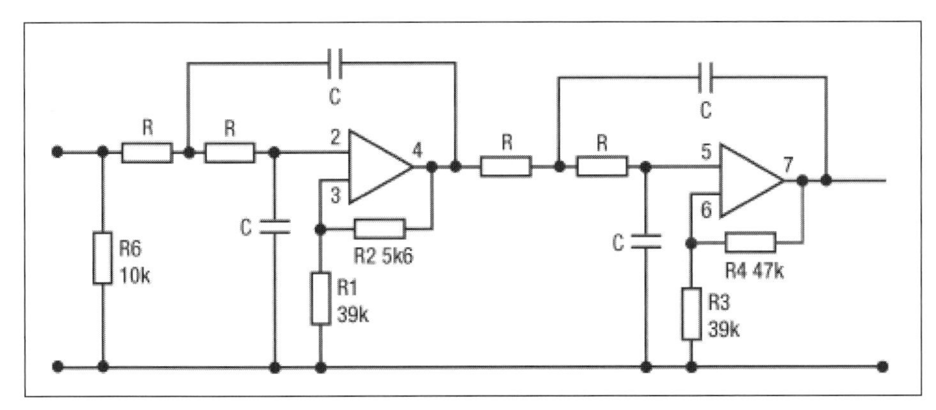


Fig 5.72: Audio low-pass filter, fourth-order Butterworth design

itor filters as follows: "The SCF works by storing discrete samples of an analogue signal as a charge on a capacitor. This charge is transferred from one capacitor to another down a chain of capacitors forming the filter. The sampling and transfer operations take place at regular intervals under control of a precise frequency source or clock. Filtering is achieved by combining the charges on the different capacitors in specific ratios and by feeding charges back to the prior stages in the capacitor chain. In this way, filters of much higher performance and complexity may be synthesised than is practical with analogue filters".

**Fig 5.72** shows the circuit of an active low-pass filter which has a Butterworth (maximally flat) type of response and is a fourth-order design. Outside the passband, the attenuation of a fourth-order filter increases at 24dB/octave. The filter was designed using an active filter design reference book [27].

The values of R and C were set to R = 10kΩ and C = 5n6. This gives a cut-off frequency (where the amplitude vs frequency response curve is 3dB down) of 2.8kHz. The 5n6 capacitors should be polystyrene types because this gives the best audio quality. The filter was designed to improve the overall selectivity of a receiver which has poor IF selectivity, but could also be used in a transmitter to limit the bandwidth of the radiated signal. The op-amps were contained within a NE5532 type dual op-amp package, and pin numbers are given for this device. The resistor R6 is a bias resistor, and is required only if the source does not have a DC path to supply bias current for input pin 2 of the NE5532. The power supply used was ±12V.

## VOLTAGE REGULATORS

The voltage regulator is an important building block. Ideally, it provide a constant output DC voltage, which is independent of the input voltage, load current and temperature. If the voltage regulator is used to regulate the supply to a VFO (for example) then it will not be affected by variations in supply voltage. For sensitive circuits like the VFO, this is a significant benefit. This allows circuits to be designed to operate at one voltage, which eases the design. As well as the DC parameters of regulators,

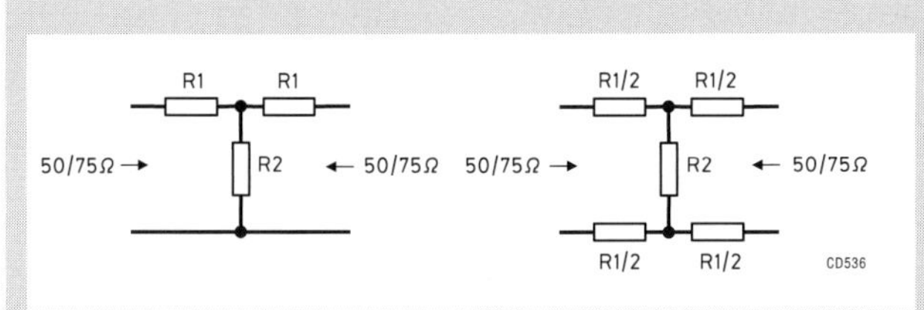


| Attenuation (dB) | R1 | 50Ω T-pad R2 | R1 | 75Ω T-pad R2 |
|---|---|---|---|---|
| 1 | 2.9 | 433 | 4.3 | 647 |
| 2 | 5.7 | 215 | 8.6 | 323 |
| 3 | 8.5 | 142 | 12.8 | 213 |
| 4 | 11.3 | 105 | 17.0 | 157 |
| 5 | 14.0 | 82 | 21.0 | 123.4 |
| 6 | 16.6 | 67 | 25.0 | 100 |
| 7 | 19.0 | 56 | 28.7 | 83.8 |
| 8 | 21.5 | 47 | 32.3 | 71 |
| 9 | 23.8 | 41 | 35.7 | 61 |
| 10 | 26.0 | 35 | 39.0 | 52.7 |
| 11 | 28.0 | 30.6 | 42.0 | 45.9 |
| 12 | 30.0 | 26.8 | 45.0 | 40.2 |
| 13 | 31.7 | 23.5 | 47.6 | 35.3 |
| 14 | 33.3 | 20.8 | 50.0 | 31.2 |
| 15 | 35.0 | 18.4 | 52.4 | 25.0 |
| 20 | 41.0 | 10.0 | 61.4 | 15.2 |
| 25 | 44.7 | 5.6 | 67.0 | 8.5 |
| 30 | 47.0 | 3.2 | 70.4 | 4.8 |
| 35 | 48.2 | 1.8 | 72.4 | 2.7 |
| 40 | 49.0 | 1.0 | 73.6 | 1.5 |

| Attenuation (dB) | R3 | 50Ω π-pad R4 | R3 | 75Ω π-pad R4 |
|---|---|---|---|---|
| 1 | 5.8 | 870 | 8.6 | 1305 |
| 2 | 11.6 | 436 | 17.4 | 654 |
| 3 | 17.6 | 292 | 26.4 | 439 |
| 4 | 23.8 | 221 | 35.8 | 331 |
| 5 | 30.4 | 179 | 45.6 | 268 |
| 6 | 37.3 | 151 | 56.0 | 226 |
| 7 | 44.8 | 131 | 67.2 | 196 |
| 8 | 52.3 | 116 | 79.3 | 174 |
| 9 | 61.6 | 105 | 92.4 | 158 |
| 10 | 70.7 | 96 | 107 | 144 |
| 11 | 81.6 | 89 | 123 | 134 |
| 12 | 93.2 | 84 | 140 | 125 |
| 13 | 106 | 78.3 | 159 | 118 |
| 14 | 120 | 74.9 | 181 | 112 |
| 15 | 136 | 71.6 | 204 | 107 |
| 20 | 248 | 61 | 371 | 91.5 |
| 25 | 443 | 56 | 666 | 83.9 |
| 30 | 790 | 53.2 | 1186 | 79.7 |
| 35 | 1406 | 51.8 | 2108 | 77.7 |
| 40 | 2500 | 51 | 3750 | 76.5 |

Table from *Datacom*.

Table 5.8. Resistor values for 50-ohm and 75-ohm T- and pi-attenuators

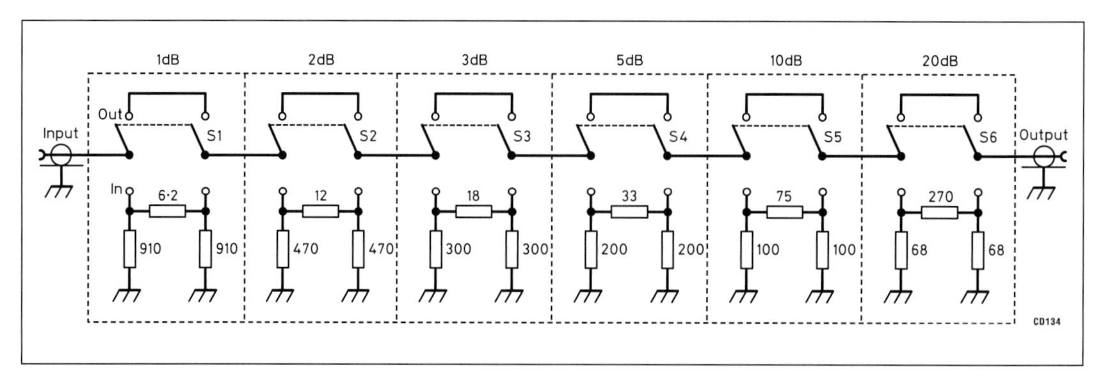


Fig 5.73: 50Ω attenuator with a 0-41dB range in 1dB steps. At RF, it is unwise to try to get more than 20dB per step because of stray capacitances (*ARRL Handbook*)

most circuits also provide a significant isolation of high frequencies. This isolation is provided partly by the decoupling capacitors which accompany the regulator, and partly by the electronic isolation of the regulator itself. IC type regulators are now so cheap (typically 50p for an LM317) that it is economical to use one for each stage of a complex circuit. For example, in a frequency synthesiser separate regulators could be used for:

- The VCO
- The VCO buffer/amplifier stage
- The reference oscillator
- The digital divider circuits

More on regulators can be found in the power supplies chapter.

## ATTENUATORS

Attenuators are resistor networks which reduce the signal level in a line while maintaining its characteristic impedance. **Table 5.8** gives the resistance values to make up 75Ω and 50Ω unbalanced RF T- and pi-attenuators; the choice between the T- and pi- configurations comes down to the availability of resistors close to the intended values, which can also be made up of two or more higher values in parallel; the end result is the same. Attenuators are also discussed in the chapter on test equipment. Here are some of the applications:

### Receiver Overload

Sensitive receivers often suffer overload (blocking, cross-modulation) from strong out-of-band unwanted signals which are too close to the wanted signal to be adequately rejected by preselector filters. This condition can often be relieved by an attenuator in the antenna input. In modern receivers, one reduction of sensitivity is usually provided for by a switch which removes the RF amplifier stage from the circuit; and another reduction by switching in an attenuator, usually 20dB.

There is an additional advantage of an attenuator. However carefully an antenna may be matched to a receiver at the wanted frequency, it is likely to be grossly mismatched at the interfering frequency, leaving the receiver's preselector filter poorly terminated and less able to do its job. A 10dB attenuator keeps the nominally 50Ω termination of the preselector filter between 41 and 61Ω under all conditions of antenna mismatch.

### An S-meter as a Field Strength Meter

When plotting antenna patterns, the station receiver is often used as a field strength indicator. As the calibration of S-meters is notoriously inconsistent, it is better to always use the same S-meter reading, say S9, and to adjust the signal source or the receiver sensitivity to get that reading at a pattern null. All higher field strengths are then reduced to S9 by means of a calibrated switchable attenuator.

A suitable attenuator, with a range of 0–41dB in 1dB steps, is

| Nominal frequency (Hz) | FX365 frequency (Hz) | Δfo (%) | D0 | D1 | D2 | D3 | D4 | D5 |
|---|---|---|---|---|---|---|---|---|
| 67.0 | *A* 67.05 | +0.07 | 1 | 1 | 1 | 1 | 1 | 1 |
| 71.9 | *B* 71.90 | 0.0 | 1 | 1 | 1 | 1 | 1 | 0 |
| 74.4 | 74.35 | 0.07 | 0 | 1 | 1 | 1 | 1 | 1 |
| 77.0 | *C* 76.96 | 0.05 | 1 | 1 | 1 | 1 | 0 | 0 |
| 79.7 | 79.77 | +0.09 | 1 | 0 | 1 | 1 | 1 | 1 |
| 82.5 | *D* 82.59 | +0.10 | 0 | 1 | 1 | 1 | 1 | 0 |
| 85.4 | 85.38 | 0.02 | 0 | 0 | 1 | 1 | 1 | 1 |
| 88.5 | *E* 88.61 | +0.13 | 0 | 1 | 1 | 1 | 0 | 0 |
| 91.5 | 91.58 | +0.09 | 1 | 1 | 0 | 1 | 1 | 1 |
| 94.8 | *F* 94.76 | 0.04 | 1 | 0 | 1 | 1 | 1 | 0 |
| 97.4 | 97.29 | 0.11 | 0 | 1 | 0 | 1 | 1 | 1 |
| 100.0 | 99.96 | 0.04 | 1 | 0 | 1 | 1 | 0 | 0 |
| 103.5 | *G* 103.43 | 0.07 | 0 | 0 | 1 | 1 | 1 | 0 |
| 107.2 | 107.15 | 0.05 | 0 | 0 | 1 | 1 | 0 | 0 |
| 110.9 | *H* 110.77 | 0.12 | 1 | 1 | 0 | 1 | 1 | 0 |
| 114.8 | 114.64 | 0.14 | 1 | 1 | 0 | 1 | 0 | 0 |
| 118.8 | *J* 118.80 | 0.0 | 0 | 1 | 0 | 1 | 1 | 0 |
| 123.0 | 122.80 | 0.17 | 0 | 1 | 0 | 1 | 0 | 0 |
| 127.3 | 127.08 | 0.17 | 1 | 0 | 0 | 1 | 1 | 0 |
| 131.8 | 131.67 | 0.10 | 1 | 0 | 0 | 1 | 0 | 0 |
| 136.5 | 136.61 | +0.08 | 0 | 0 | 0 | 1 | 1 | 0 |
| 141.3 | 141.32 | +0.02 | 0 | 0 | 0 | 1 | 0 | 0 |
| 146.2 | 146.37 | +0.12 | 1 | 1 | 1 | 0 | 1 | 0 |
| 151.4 | 151.09 | 0.20 | 1 | 1 | 1 | 0 | 0 | 0 |
| 156.7 | 156.88 | +0.11 | 0 | 1 | 1 | 0 | 1 | 0 |
| 162.2 | 162.31 | +0.07 | 0 | 1 | 1 | 0 | 0 | 0 |
| 167.9 | 168.14 | +0.14 | 1 | 0 | 1 | 0 | 1 | 0 |
| 173.9 | 173.48 | 0.19 | 1 | 0 | 1 | 0 | 0 | 0 |
| 179.9 | 180.15 | +0.14 | 0 | 0 | 1 | 0 | 1 | 0 |
| 186.2 | 186.29 | +0.05 | 0 | 0 | 1 | 0 | 0 | 0 |
| 192.8 | 192.86 | +0.03 | 1 | 1 | 0 | 0 | 1 | 0 |
| 203.5 | 203.65 | +0.07 | 1 | 1 | 0 | 0 | 0 | 0 |
| 210.7 | 210.17 | 0.25 | 0 | 1 | 0 | 0 | 1 | 0 |
| 218.1 | 218.58 | +0.22 | 0 | 1 | 0 | 0 | 0 | 0 |
| 225.7 | 226.12 | +0.18 | 1 | 0 | 0 | 0 | 1 | 0 |
| 233.6 | 234.19 | +0.25 | 1 | 0 | 0 | 0 | 0 | 0 |
| 241.8 | 241.08 | 0.30 | 0 | 0 | 0 | 0 | 1 | 0 |
| 250.3 | 250.28 | 0.01 | 0 | 0 | 0 | 0 | 0 | 0 |
| No tone | No tone | - | 0 | 0 | 0 | 0 | 1 | 1 |

From *Consumer Microelectronics Ltd IC Data Book*, 1st edn

**Table 5.9: CTCSS (continuous tone coded squelch system) EIA-standard frequencies. The letters behind some frequencies refer to the tones used by UK voice repeaters. The eight right-hand columns refer to the FX365 LSI chip (see text)**

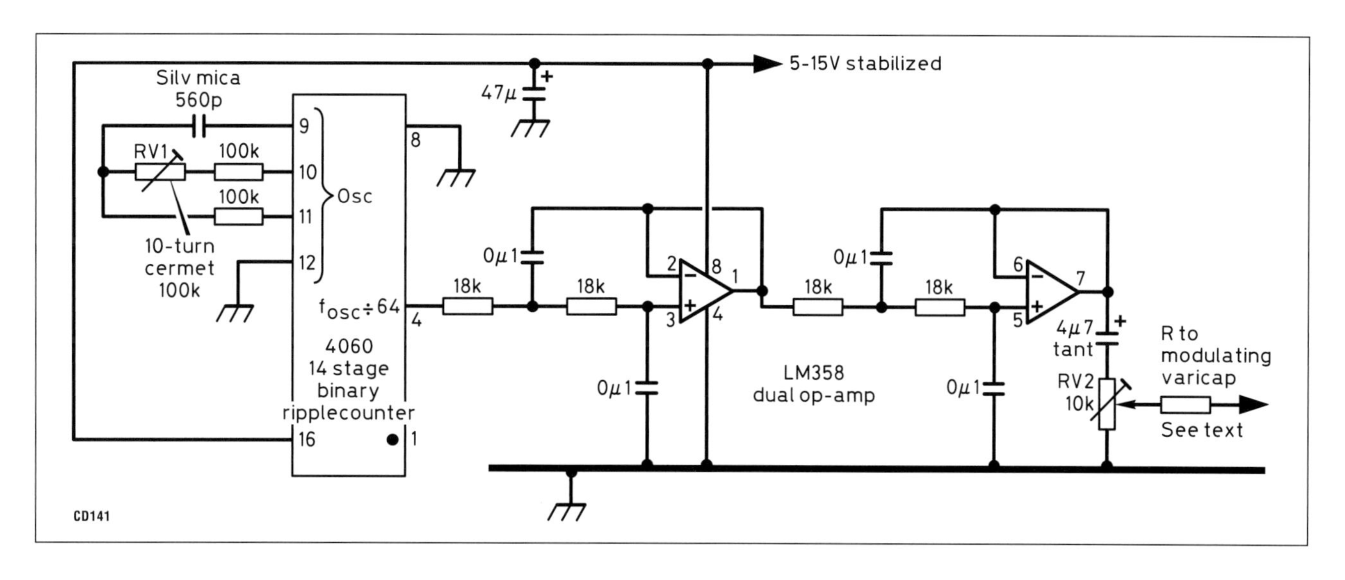


**Fig 5.74: The Tuppenny simple CTCSS encoder (G0CBM) consists of an RC oscillator and ÷64 divider in one CMOS IC, together with a two-stage active LP filter built around a dual op-amp (G8HLE) (*Kent Repeater Newsletter*)**

shown in **Fig 5.73.** Such an attenuator bank can be constructed from 5% carbon composition or film resistors of standard values and DPDT slide switches, eg RS 337-986. The unit is assembled in a tin-plate box; shields between individual attenuators reduce capacitive coupling and can be made of any material that is easy to cut, shape and solder, ie tin plate, PCB material or copper gauze. At VHF, the unit is still useful but stray capacitances and the self-inductance of resistors are bound to reduce accuracy. Verification of the accuracy of each of the six sections can be done at DC, using a volt or two from a battery or PSU and a DVM; do not forget to terminate the output with a 50Ω resistor.

## Driving VHF Transverters

RF attenuators may sometimes be used at higher powers, eg where an HF transmitter without a low-level RF output is to drive, but not to overdrive, a VHF transverter. Care must be taken to ensure that the resistors can handle the power applied to them.

# TONE SIGNALLING: CTCSS

If an FM receiver requires, for its squelch to open and remain open, not only a carrier of sufficient strength but also a tone of a specific frequency, much co-channel interference can be avoided. If a repeater transmits such a tone with its voice transmissions but not with its identification, a continuous-tone coded squelch system (CTCSS) equipped receiver set for the same tone would hear all that repeater's voice transmissions but not its idents. Its squelch would not be opened either, eg during lift conditions, by a repeater on the same channel but located in another service area and sending another CTCSS tone. Conversely, a mobile, positioned in an overlap area between two repeaters on the same channel but which have their receivers set for different CTCSS tones, could use one repeater without opening the other by sending the appropriate tone.

Similar advantages can be had where several groups of stations, each with a different CTCSS tone, share a common frequency. With all transceivers within one group set for that group's tone, conversations within the group would not open the squelch of stations of other groups monitoring the same frequency.

The Electronic Industries Association (EIA) has defined 38 CTCSS standard sub-audible tone frequencies. They are all between 67 and 250.3Hz, ie within the range of human hearing but outside the 300-3000Hz audio pass-band of most communications equipment and their level is set at only 10% of maximum deviation for that channel; hence sub-audible.

The left column of **Table 5.9** gives the frequency list. CTCSS encoders (tone generators) and also decoders (tone detectors)

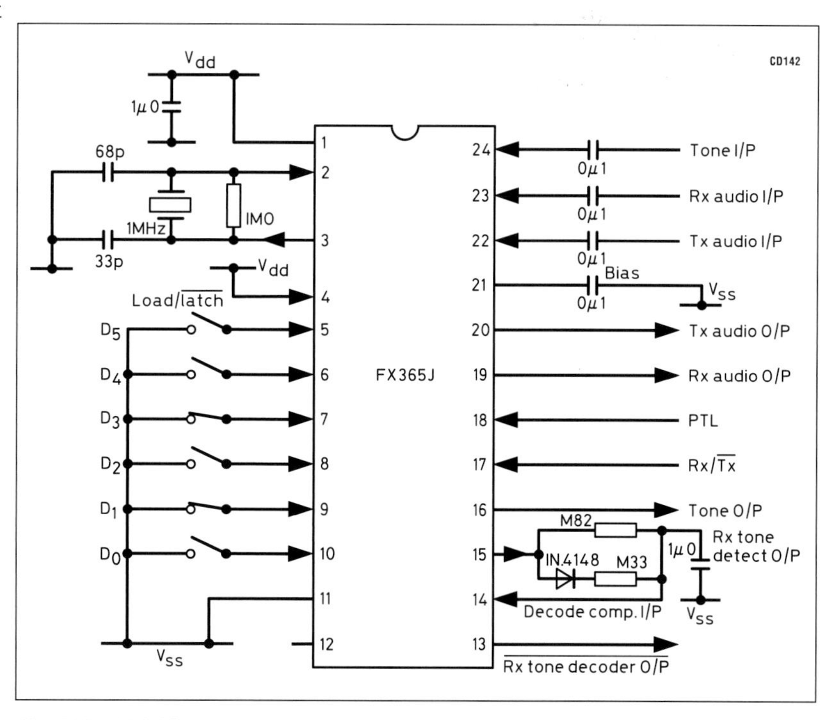


**Fig 5.75: CTCSS encoder/decoder using a CML CMOS LSI device. The chip also contains 300Hz-cut-off HP and LP filters to separate tones and speech and a crystal reference oscillator. Tone selection can be by microprocessor or hard-wired switches (*Consumer Microcircuits IC Data Book*)**

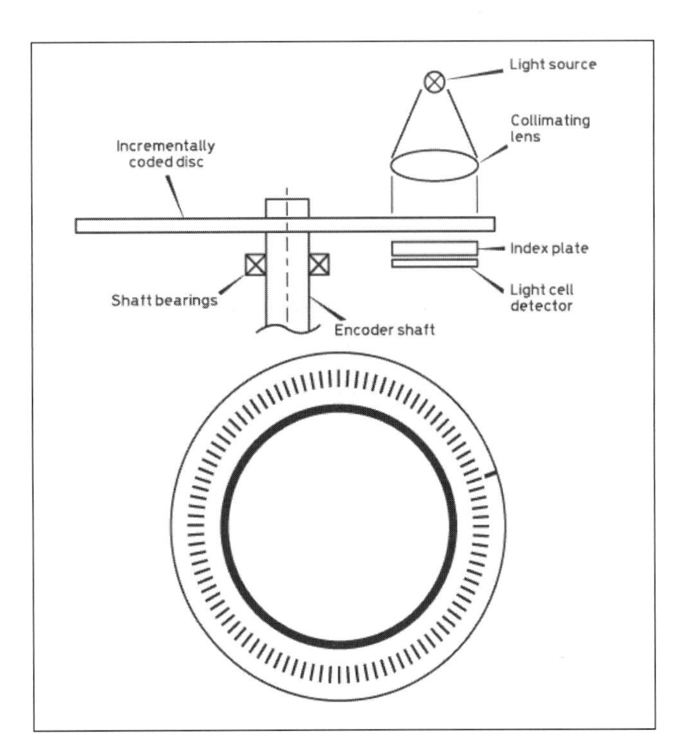


**Fig 5.76: The incremental shaft encoder. In some radios with digital frequency synthesis, the tuning knob turns the shaft which turns the disc which alternatingly places transparent and opaque sectors in the light path; each pulse from the light detector increases or decreases the frequency by one step. (Taken from *Analog-Digital Conversion Handbook*, 3rd edn, Prentice-Hall, 1986, p444, by permission. Copyright Analog Devices Inc, Norwood, MA, USA)**

are offered as options for many earlier mobile and hand-held VHF and UHF FM transceivers. Most current transceivers have CTCSS encode built-in as standard.

Commercial standards require encoder frequencies to be within 0.1% of the nominal tone frequency under all operating conditions, attainable only with crystal control; most amateur repeaters are more tolerant, however, and RC-oscillators have been used successfully. The encoder tone output must be a clean sine wave lest its harmonic content above 300Hz becomes audible; this requires good filtering.

G0CBM's very simple Tuppenny CTCSS tone generator was designed for retrofitting in a surplus PMR transmitter to access local repeaters: **Fig 5.74** [28]. It is built around the inexpensive CMOS oscillator-frequency divider IC 4060. The parts connected to pins 9, 10 and 11 are the frequency-determining components

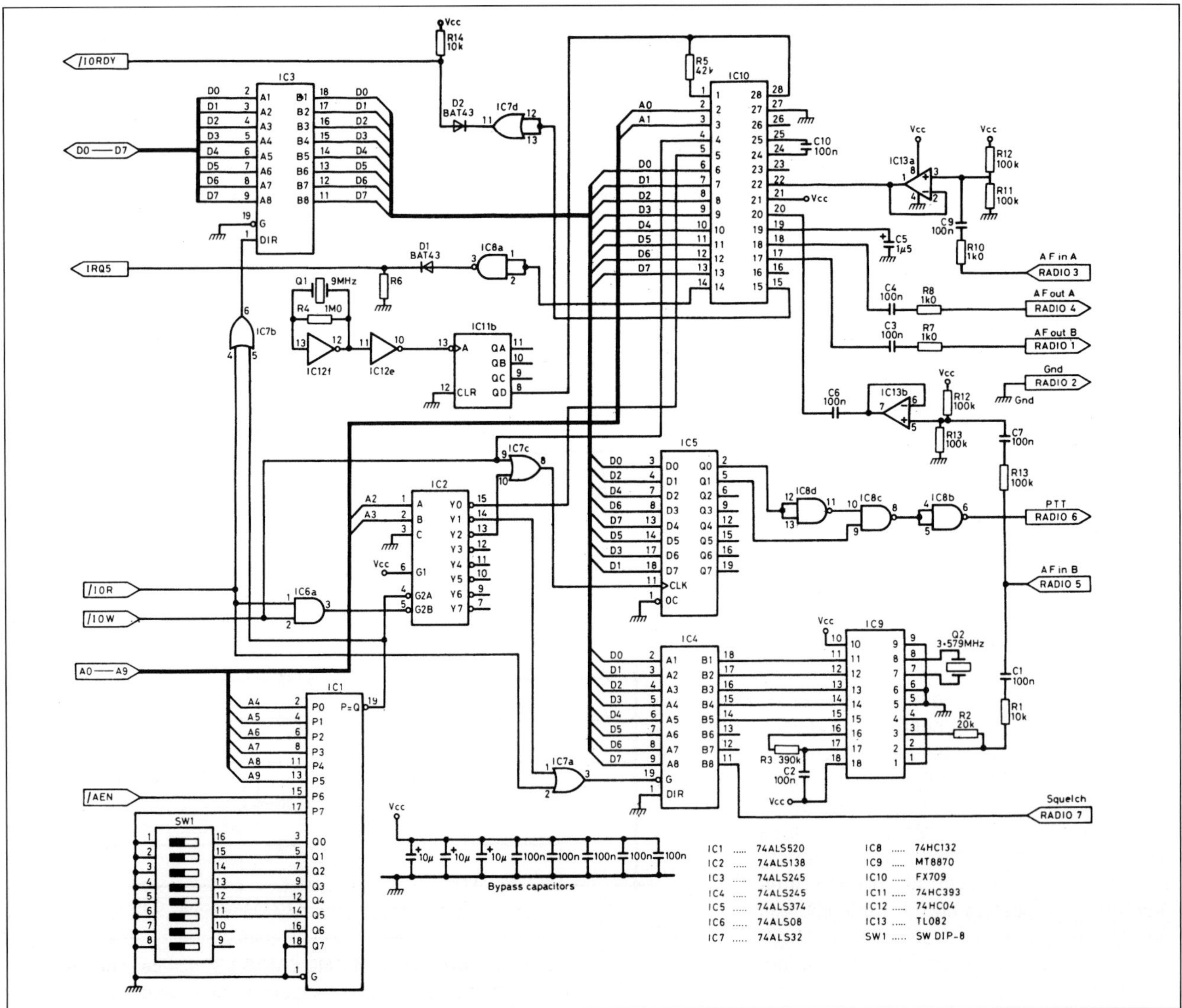


**Fig 5.77: Speech encoder/decoder designed by DG3RBU and DL8MBT for voice mailboxes on UHF FM repeaters**

of the RC oscillator; they have a tolerance of 5% but parts with the lowest possible temperature coefficient should be chosen to obtain adequate frequency stability, especially under mobile operation. RV1 allows frequency adjustment over the range 4288-7603Hz, which, after dividing by 64, yields at pin 4 any CTCSS frequency between 67 and 119Hz; this includes tones A-J assigned to UK repeaters.

A two-stage active low-pass filter was designed by G8HLE to get sufficient suppression of any harmonics above 300Hz of all tones. The LM358 dual op-amp was chosen because it is small, cheap and works on a low, single supply voltage. The -6dB frequency was chosen at 88Hz, ie the higher tones to be passed fall outside the pass-band.

As the tone amplitude is far greater than required, this attenuation is no disadvantage, but each tone tuned in with RV1 requires a different setting of RV2 to get the same deviation. The output resistor R depends on the modulator circuitry in the transmitter. It should be dimensioned to get the proper CTCSS deviation at the highest tone frequency to be used with RV2 set near maximum.

Encoders/decoders are more complicated and the LSI CMOS device used, eg CML FX365, is expensive: **Fig 5.75** [29]. It contains not only the encoder and decoder proper, but also a high-pass filter which prevents any received CTCSS tone becoming audible, a low-pass filter to suppress harmonics of CTCSS tones and a crystal-controlled reference oscillator from which the tones are are derived; see Table 5.9. Tone selection is according to the six right-hand columns of that table, either by microprocessor or by hard-wired switches. Another feature is transmit phase reversal on release of the PTT switch; this shortens the squelch tail at the receiver.

## ANALOGUE-DIGITAL INTERFACES

Most real-world phenomena are analogue in nature, meaning they are continuously variable rather than only in discrete steps; examples are a person's height above ground going up a smooth wheel-chair ramp, the shaft angle of a tuning capacitor, the temperature of a heatsink and the wave shape of someone's voice. However, some are digital in nature, meaning that they come in whole multiples of a smallest quantity called least-significant bit (LSB); examples are a person's height going up a staircase (LSB is one step), telephone numbers (one cannot dial in between two numbers) and money (LSB is the penny/cent and any sum is a whole multiple thereof).

Frequently, it is useful to convert from analogue to digital and vice versa in an analogue-to-digital converter (ADC) or a digital-to-analogue converter (DAC). If anti-slip grooves with a pitch of, say, 1cm are cut across the wheelchair ramp, it has, in fact, become a staircase with minuscule steps (LSBs) which one can count to determine how far up one is, with any position between two successive steps being considered trivial. Without going into details of design, a number of conversions commonly employed in amateur radio equipment will now be explained.

### Shaft Encoders & Stepper Motors

Amateurs expect to twist a knob, an analogue motion, when they want to change frequency. Traditionally, that motion turned the shaft of a variable capacitor or screwed a core into or out of a coil.

Now that frequencies in many radios are generated by direct digital synthesis, a process more compatible with keyboard entry and up/down switches, amateurs not only still like to twist knobs, they also like them to feel like the variable capacitors of yesteryear; equipment manufacturers comply but what the knob actually does is drive an optical device called an incremental shaft encoder: **Fig 5.76** [30].

The encoder is often a disc divided into sectors which are alternately transparent and opaque. A light source is positioned at one side of the disc and a light detector at the other. As the tuning knob is spun and the disc rotates, the output from the detector goes on or off when a transparent or an opaque disc sector is in the light path. Thus, the spinning encoder produces a stream of pulses which, when counted, indicate the change of angular position of the shaft. A second light source and detector pair, at an angle to the main pair, indicates the direction of rotation. A third pair sometimes is used to sense the one-per-revolution marker seen at the right on the disc shown.

Available encoder resolutions (the number of opaque and transparent sectors per disc) range from 100 to 65,000. The SSB/CW tuning rate of one typical radio was found to be 2kHz per knob rotation in 10Hz steps; this means an encoder resolution of 200. The ear cannot detect a 10Hz change of pitch at 300Hz and above, so the tuning feels completely smooth, though its output is a digital signal in which each pulse is translated into a 10Hz frequency increment or decrement.

Stepper motors do the opposite of shaft encoders; they turn a shaft in response to digital pulses, a step at a time. Typical motors have steps of 7.5 or 1.8 degrees; these values can be halved by modifying the pulse sequence. Pulses vary from 12V at 0.1A to 36V at 3.5A per phase (most motors have four phases) and are applied through driver ICs for small motors or driver boards with ICs plus power stages for big ones. The drivers are connected to a control board which in turn may be software-programmed by a computer, eg via an RS232 link (a standard serial data link). In amateur radio they are used on variable capacitors and inductors in microprocessor-controlled automatic antenna tuning units and on satellite-tracking antenna azimuth and elevation rotators.

### Digitising Speech

While speech can be stored on magnetic tape, and forwarded at speeds a few times faster than natural, each operation would add a bit of noise and distortion and devices with moving parts, ie tape handling mechanisms, are just not suitable for many environments. Digital speech has changed all that. Once digitised, speech can be stored, filtered, compressed and expanded by computer, and forwarded at the maximum speed of which the transmission medium is capable, all without distortion; it can then be returned to the analogue world where and when it is to be listened to.

If speech were to be digitised in the traditional way, ie sampled more than twice each cycle on the highest speech frequency, eg at 7kHz for speech up to 3kHz, and 10-bit resolution of the amplitude of each sample were required for reasonable fidelity, a bit rate of 70kbit/s would result, about 20 times the 3.5kHz bandwidth considered necessary for the SSB transmission of analogue speech. Also, almost a megabyte of memory would be required to store each minute of digitised speech. Differential and adaptive techniques are used to reduce these requirements.

DG3RBU and DL8MBT developed hardware and software for the digital storage and analogue retransmission of spoken messages on normal UHF FM voice repeaters and the digital forwarding of such messages between repeaters via the packet network. They described their differential analogue-digital conversion as follows.

"The conversion is by continuously variable slope decoding (CVSD), a form of delta modulation; in this process it is not the instantaneous value of an analogue voltage that is being sampled and digitised but its instantaneous slope at the moment of sampling. Binary '1' represents an increasing voltage, '0' a decreasing voltage.

"Encoding the slope of the increase or decrease depends on the prior sample. If both the prior and present samples are '1', a steeper slope is assumed than when a '1' follows a '0'. Decoding does the same in reverse. This system of conversion is particularly suitable for speech; even at a relatively modest data rate of 16kbit/s it yields good voice quality. An FX709 chip is used.

"The analogue-to-digital converter (ADC) for speech input and digital-to-analogue converter (DAC) for speech output are assembled on a specially designed plug-in board for an IBM PC computer, **Fig 5.77** [31]. It provides all the required functions, starting with the address coding for the PC (IC1 and 2). In the FX709 (IC10), the signal passes through a bandpass filter to the one-bit serial encoder; after conversion to an eight-bit parallel format the data pass to the PC bus via IC3; in the other direction, voice signals pass through a software-programmable audio filter; registers for pause and level recognition complete the module.

"The FX709 has a loopback mode which permits a received and encoded speech signal to be decoded and retransmitted, an easy way to check the fidelity of the loop consisting of the radio receiver-encoder-decoder-radio transmitter. The built-in quartz clock oscillator (IC12f) and divider (IC11) permit experiments with different externally programmable clock rates.

"The maximum length of a text depends on the data rate. We used 32kbit/s, at which speed it is hard to tell the difference between the sound on the input and an output which has gone through digitising, storage, and reconversion to analogue.

"The maximum file length is 150s. The reason for this time limit is that the FX709 has no internal buffer; this requires that the whole file must be read from RAM in real time, ie without interrupts for access to the hard disk."

## Digital Panel (volt) Meters (DPMs)

Analogue methods for DC voltage measurements, say with a resolution of a millivolt on a 13.8V power supply, are possible but cumbersome. With a digital voltmeter they are simple and comparatively inexpensive. Digital panel meters are covered in the Test Equipment chapter.

## SOFTWARE BUILDING BLOCKS

The PIC [32] is a microcontroller which is being used more and more in amateur radio projects (for instance the one shown in Fig 5.82).

A PIC is an economical way of providing software control of functions in home-built amateur radio equipment. Examples of the use of PIC devices are:

- ATU control [33]
- DDS controller [34] (see also the oscillators chapter of this Handbook)
- Frequency-dependant switch [35]
- Bug key [36]
- Transceiver control [37]
- Keyer [38]
- Morse code speed calibrator [39]
- Morse code reader [40]

Note that the references above also have links to internet sites.

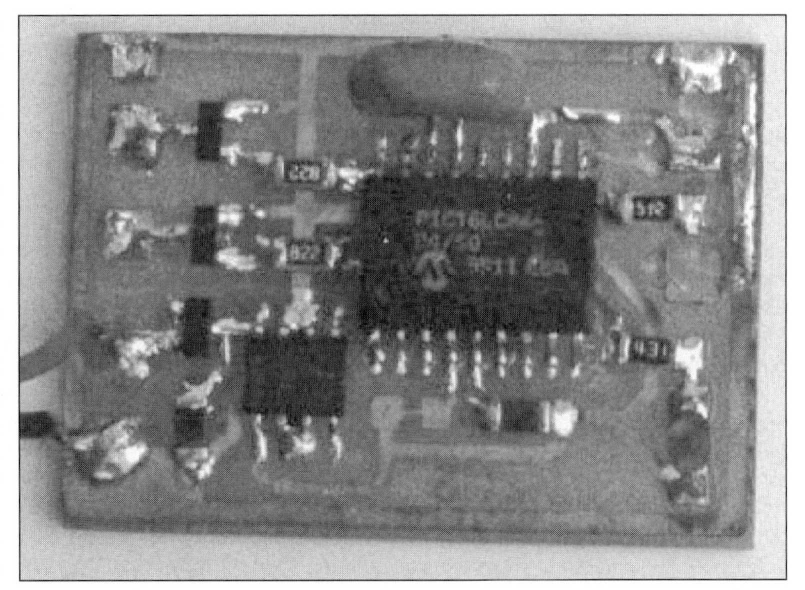

**Fig 5.78: Postage stamp sized beacon keyer using a PIC chip**

## PIC Code

In some of the above projects, the PIC can be purchased ready programmed and this is an easy solution to reproduce a published design.

However, there are strong reasons for the constructor to write his or her own code:

- The constructor may wish to modify existing code to either modify an existing function or add extra requirements to an existing design.
- The constructor may wish to write all of the code from scratch.
- The existing design may be suitable, but further functions will be added in the future. For example, this may be caused by a change in amateur bands used, or a change of mode.

The ideal environment in which to develop PIC software is the MPLAB integrated package, which is available from the manufacturers of the PIC [32]. This runs on a PC, but the PC doesn't need to be sophisticated. An old, unused PC may be adequate.

For the purpose of developing software, it is recommended that PIC devices with on-board EEPROM are used. An example of a small PIC project is shown in **Fig 5.78**.

Further information on PIC programming can be found in the chapter on computers in the shack and from [41].

## REFERENCES


[2] Constant-R network. *RF Design Basics*, John Fielding, ZS5JF. RSGB, 2006
[3] *Signetics RF Communications Handbook*
[4] Ian Braithwaite, G4COL, *RadCom*, Jul 1997, pp38-39
[5] http://www.avagotech.com
[6] http://www.mini-circuits.com
[7] Doug DeMaw, W1FB, *QST*, Apr 1989, pp30-33
[8] Drew Diamond, VK3XU, *Amateur Radio*, 10/88
[9] David Bowman, G0MRF, in *RadCom*, Feb 1993, pp28-30, and Mar 1993, pp28-29
[10] Klaas Spaargaren, PA0KLS, in 'Eurotek', *RadCom*, Nov 1998 (corrections in *RadCom*, Jan 1999, p20
[11] http://www.frenning.dk/OZ1PIF_HOMEPAGE/50MHz_IRF 510.htm
[12] Zack Lau, KH6CP, *QST*, Oct 1987
[13] PA0GMS in 'Eurotek', *RadCom*, Dec 1992, p49
[19] Analog Devices web site, http://www.analog.com
[20] GW4NAH, *Radio Communication*, Jan 1989, p48
[21] DF4ZS in 'Eurotek', *RadCom*, Oct 1998
[22] ON5DI in 'Eurotek', RadCom, Dec 1998
[23] 'Computer-aided ladder crystal filter design', J A Hardcastle, G3JIR, *Radio Communication*, May 1983
[24] H R Langer, Y27YO, *Funkamateur*, Jan 1985; *Radio Communication*, Jun 1985, p452
[25] LA2IJ and LA4HK, *Amator Radio*, Nov 1974
[26] DJ6HP, *cq-DL*, Feb 1974
[27] *Active Filter Cookbook,* Don Lancaster. Howard W. Sams and Co. ISBN 0-672-21168-8
[28] *Kent Repeater Newsletter,* Jan 1993
[29] *IC Data Book*, 1st edition, Consumer Microcircuits Ltd, p2.40
[30] *Analogue-Digital Conversion Handbook*, 3rd edition, Prentice-Hall, 1986, p215
[31] *RadCom*, May 1992, p62
[32] www.microchip.com
[33] 'picATUne–the intelligent ATU', Peter Rhodes, G3XJP, *RadCom*, Sep 2000 to Jan 2001
[34] 'Use of PICs in DDS Design' in 'Technical Topics' column, Pat Hawker G3VA. *RadCom*, Jan 2001
[35] 'Pic-A-Switch A Frequency Dependent Switch', Peter Rhodes, G3XJP, *RadCom*, Sep–Dec 2001
[36] 'Bugambic: Son of Superbug', Chas Fletcher, G3DXZ, *RadCom*, Apr 2002
[37] 'CDG2000 HF Transceiver', Colin Horrabin, G3SBI, Dave Roberts, G3KBB, and George Fare, G3OGQ, *RadCom*, Jun 2002 to Dec 2002
[38] 'The programmer and the keyer', Ed Chicken, G3BIK, *RadCom*, Nov 2004
[39] 'A Morse Code Speed Calibrator'. Jonathan Gudgeon, G4MDU, *RadCom*, Aug 2004
[40] 'A Talking Morse Code Reader', Jonathan Gudgeon, G4MDU, *RadCom*, Jun 2001
[41] *Command*, Andy Talbot, G4JNT, RSGB, 2003



***Peter Goodson*** *is qualified with a Hons Degree in Communication Engineering. He began his career working on the component-level design of HF (SSB) and VHF (FM) transceivers. He now works on the system level design of RF systems, which includes the issues of communications security, EMC, co-siting of radio equipment, RF hazard assessments and Immunity to lightning strikes.*

***Mike Stevens****, (G8CUL/F4VRB) has ONC, HNC qualifications and an honours degree in Engineering Science, the latter from Exeter University. He is also an MIET. He is now retired from over 40 years designing embedded systems for the Nuclear Industry and the MoD but still exercises his brain with the design of embedded systems but now for (mainly) radio-based equipment. Although equipped for activity on all bands from 160m to 9cm, his main interests lie in the VHF/UHF and microwave bands.*

# HF Receivers

# 6

*Roger Wilkins, G8NHG*

Amateur HF operation, whether for two-way contacts or for listening, imposes stringent requirements on the receiver. The need is for a receiver that enables an experienced operator to find and hold extremely weak signals on frequency bands often crowded with much stronger signals from local stations or from the high-power broadcast stations using adjacent bands. The wanted signals may be fading repeatedly to below the external noise level, which limits the maximum usable sensitivity of HF receivers.

Although the receivers now used by most amateurs form part of complex, factory-built HF transceivers, the operator should understand the design parameters that determine how well or how badly they will perform in practice, and appreciate which design features contribute to basic performance as HF communications receivers, as opposed to those which may make them more user-friendly but which do not directly affect the reception of weak signals. This also applies to dedicated receivers that are factory-built.

Ideally, an HF receiver should be able to provide good intelligibility from signals which may easily differ by up to 10,000 times and occasionally by up to one million times (120dB) - from less than 1μV from a weak signal to nearly 1V from a near-neighbour. **Table 6.1** shows the relationship between the various ways of measuring the input signals; pd (potential difference) and dBm (input power) are most commonly used, together with the ITU standard S-point scale.

To tune and listen to SSB or to a stable CW transmission while using a narrow-band filter, the receiver needs to have a frequency stability of within a few hertz over periods of 15 minutes or so, representing a stability of better than one part in a million. It should be capable of being tuned with great precision, either continuously or in increments of at most a few hertz.

| for a 50 ohms power matched system | emf | pd | dBm | S-points |
|---|---|---|---|---|
| | 1 V | 0.5 V | 7 | |
| | 100 mV | 50 mV | -13 | |
| | 10 mV | 5 mV | -33 | |
| | 1 mV | 0.5 mV | -53 | S9 + 20 |
| | | | -73 | S9 |
| | | | -79 | S8 |
| possible range of signal levels | 100 uV | 50 uV | -80 | |
| | | | -85 | S7 |
| | | | -91 | S6 |
| | 10 uV | 5 uV | -93 | |
| | | | -97 | S5 |
| | | | -103 | S4 |
| | | | -109 | S3 |
| | 1 uV | 0.5 uV | -113 | |
| typical receiver sensitivity | | | -115 | S2 |
| | | | -121 | S1 |
| thermal noise floor (3KHz bandwidth) | 0.1 uV | 50 nV | -133 | |

**Table 6.1: The relationship between emf, pd and dBm**

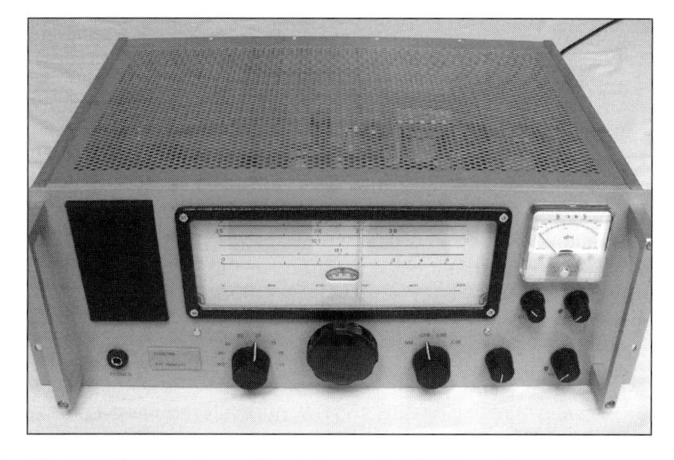

**Fig 6.1: The chapter author's home-built HF receiver**

A top-quality receiver may be capable of receiving transmissions on all frequencies from 1.8MHz to 30MHz (or even 50MHz) to provide 'general coverage' or only on the bands allotted to amateurs. Such a receiver may be suitable for a number of different modes of transmission - SSB, CW, AM, NBFM, data (RTTY/ packet) etc - with each mode imposing different requirements in selectivity, stability and demodulation (decoding). Such a receiver would inevitably be complex and costly to buy or build. On the other hand, a more specialised receiver covering only a limited number of bands and modes such as CW-only or CW/SSB-only, and depending for performance rather more on the skill of the operator, can be relatively simple to build at low cost.

As with other branches of electronics, the practical implementation of high-performance communications receivers has undergone a number of radical changes since their initial development in the mid-1930s, some resulting from the improved stability needed for SSB reception and others aimed at reducing costs by substituting electronic techniques in place of mechanical precision.

It is still possible to build reasonably effective HF receivers, particularly those for limited frequency coverage, on the kitchen table with the minimum of test equipment. With more effort and test gear, a more complex receiver can be home-constructed (**Fig 6.1**). Furthermore, since many newcomers will eventually acquire a factory-built transceiver but require a low-cost, stand-alone HF receiver in the interim period, this need can be met either by building a relatively simple receiver, or by acquiring, and if necessary modifying, one of the older type receivers that were marketed for amateur operation in the years before the virtually universal adoption of the transceiver. Even where an amateur has no intention of building or servicing his or her own receiver, it is important that he or she should have a good understanding of the basic principles and limitations that govern the performance of all HF communications receivers.

## BASIC REQUIREMENTS

The main requirements for a good HF receiver are:-

1. Sufficiently high sensitivity, coupled with a wide dynamic range and good linearity to allow it to cope with both the very weak and very strong signals that will appear together at the input; it should be able to do this with the minimum

impairment of the signal-to-noise ratio by receiver noise, cross-modulation, blocking, intermodulation, reciprocal mixing, hum etc.

2. Good selectivity to allow the selection of the required signal from among other (possibly much stronger) signals on adjacent or near-adjacent frequencies. The selectivity characteristics should 'match' the mode of transmission, so that interference susceptibility and noise bandwidth should be as close as possible to the intelligence bandwidth of the signal.
3. Maximum freedom from spurious responses - that is to say signals which appear to the user to be transmitting on specific frequencies when in fact this is not the case. Such spurious responses include those arising from image responses, breakthrough of signals and harmonics of the receiver's internal oscillators.
4. A high order of stability, in particular the absence of short-term frequency drift or jumping.
5. Good read-out and calibration of the frequency to which the set is tuned, coupled with the ability to reset the receiver accurately and quickly to a given frequency or station.
6. Means of receiving SSB and CW, normally requiring a stable beat frequency oscillator preferably in conjunction with product detection.
7. Sufficient amplification to allow the reception of signals of under 1μV input; this implies a minimum voltage gain of about one million times (120dB), preferably with effective automatic gain control (AGC) to hold the audio output steady over a very wide range of input signals.
8. Sturdy construction with good-quality components and with consideration given to problems of access for servicing when the inevitable occasional fault occurs.

A number of other refinements are also desirable: for example it is normal practice to provide a headphone socket on all communications receivers; it is useful to have ready provision for receiver 'muting' by an externally applied voltage to allow voice-operated, push-to-talk or CW break-in operation; an S-meter to provide immediate indication of relative signal strengths; a power take-off socket to facilitate the use of accessories; an IF signal take-off socket to allow use of external special demodulators for NBFM, FSK, DSBSC, data etc.

In recent years, significant progress has continued to be made in meeting these requirements - although we are still some way short of being able to provide them over the entire signal range of 120dB at the ideal few hertz stability. The introduction of more and more semiconductor devices into receivers has brought a number of very useful advantages, but has also paradoxically made it more difficult to achieve the highly desirable wide dynamic range. Professional users now require frequency read-out and long-term stability of an extremely high order (better than 1Hz stability is needed for some applications) and this has led to the use of frequency synthesised local oscillators and digital read-out systems; although these are effective for the purposes which led to their adoption, they are not necessarily the correct approach for amateur receivers since, unless very great care is taken, a complex frequency synthesiser not only adds significantly to the cost but may actually result in a degradation of other even more desirable characteristics.

So long as continuous tuning systems with calibrated dials were used, the mechanical aspects of a receiver remained very important; it is perhaps no accident that one of the outstanding early receivers (HRO) was largely designed by someone whose early training was that of a mechanical engineer. Some circuit blocks continue to benefit from good mechanical practice eg oscillators that require mechanical stability, screening and thermal management to reduce frequency drift.

It should be recognised that receivers which fall far short of ideal performance by modern standards may nevertheless still provide entirely usable results, and can often be modified to take advantage of recent techniques. Despite all the progress made in recent decades, receiver designs dating from the 'thirties and early 'forties are still capable of being put to good use, provided that the original electrical and mechanical design was sound. Similarly, the constructor may find that a simple, straightforward and low-cost receiver can give good results even when some aspects of its specification eg large signal handling, LO stability and selectivity are well below that now possible. It is ironical that almost all the design trends of the past 30 years have, until quite recently, impaired rather than improved the performance of receivers in the presence of strong signals!

## THE HF RADIO ENVIRONMENT

The radio environment is a combination of background noise and potentially interfering signals that exists at a particular location. Radio noise is made up of natural noise eg lightning, together with man-made noise coming from electrical machinery, switched mode power supplies and modems. Background radio noise varies:-

(a) From place to place; rural areas tend to have less radio background noise than towns and cities because of the reduced component of man-made noise.
(b) With frequency; the lower frequencies are noisier than the higher frequency bands.
(c) With the time of day; as propagation changes following the sun, so does the amount of more distant signals.

**Fig 6.2** shows day and night time background noise levels across the HF band measured in a particular rural location. Below 10MHz, the background noise level rises rapidly.

As well as external background noise coming in from the antenna, the receiver front end generates its own noise that gets added to the external background noise and it is this internally generated noise that limits the small signal performance of the receiver. All electronic components generate noise power as a function of temperature and the measurement bandwidth, which is why radio telescopes immerse their front end amplifiers in liquid helium. The self generated noise of a receiver is

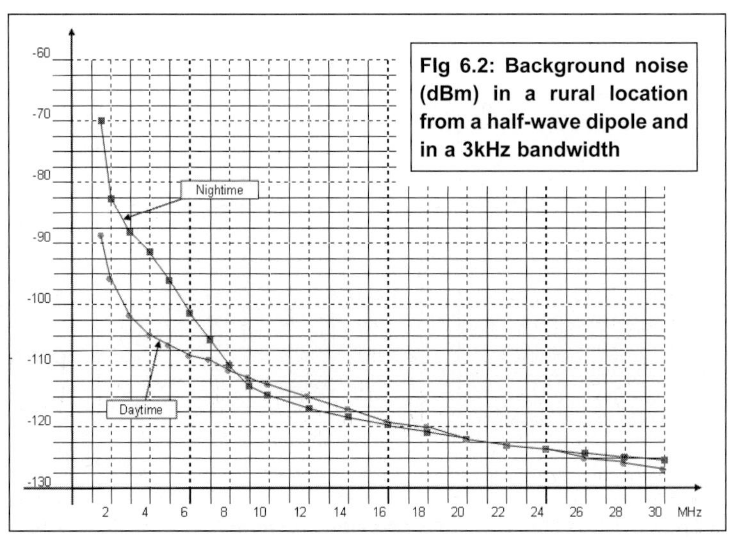


**Fig 6.2: Background noise (dBm) in a rural location from a half-wave dipole and in a 3kHz bandwidth**

described by its Noise Factor ($NF_a$) and its related dB version, Noise Figure ($NF_i$).

## Noise Factor and Noise Figure

Noise Factor defines the maximum sensitivity of a receiver without regard to its bandwidth or its input impedance, and is largely determined by the front-end of the receiver. It is defined as the ratio:

*Noise factor,* $NF_a = (S/N\ in) / (S/N\ out)$
where the S/N is noise power ratio.

When expressed logarithmically:

*Noise Figure,* $NF_i = 10 \log10 (NF_a)$ dB

The minimum equivalent input noise power for a receiver at room temperature (290°K) is -174dBm/Hz. This is given by:

$P = kTB$
where P is the noise power in watts, k is Boltzmann's constant ($1.38 \times 10^{-23}$ J/K) and B is the bandwidth in hertz.

The noise floor of the receiver is then given by:

*noise floor* $= -174 + NF_i + 10 \log_{10} B$ dBm
where $NF_i$ is the Noise Figure in dB and B is the receiver noise bandwidth in hertz.

For CW, the bandwidth is approximately 300Hz and for SSB it is approximately 3kHz. So a receiver with a 3kHz bandwidth and a 10dB noise figure will have a noise floor of -130dBm. That is to say, if the receiver was perfect, it would behave as if the input noise level was -130dBm. The relationship between sensitivity and the various ways of describing noise is shown in **Fig 6.3**.

Because of galactic, atmospheric and man-made noise always present on HF there is little need for a receiver noise figure of less than 15-17dB up to about 18-20MHz, or less than about 10dB up to 30MHz, even in quiet sites. It may, however, be an advantage if the first stages (preamplifier or mixer or post-mixer) have a lower noise figure since this will permit good reception with an electrically short antenna. It should be noted that excessive sensitivity is likely to impair the strong-signal handling capabilities of the receiver.

Signal-to-noise ratio (or, as measured in practice, more accurately signal-plus-noise to noise ratio) gives the minimum antenna input voltage to the receiver needed to give a stated output S/N ratio with a specified noise bandwidth. The input voltage for a given input power depends on the input impedance of the receiver; for modern sets this is invariably 50 or 75 ohms but for older (though still useful) receivers this may be 400 ohms and such sets may thus appear wrongly to be less sensitive unless the impedance is taken into account. It should be noted that where sensitivity is defined in terms of S/N ratio, this depends not only on input impedance but also on the output S/N ratio which, while usually 10dB, may occasionally be 6dB; it will also depend as noted above on the noise bandwidth of the receiver.

It is useful to note also that the Minimum Discernible Signal (MDS) measured in dBm is defined as the signal level equal to the noise floor. In practice MDS is measured by determining the signal that is perceptably different from the noise floor, such as by 1dB. This introduces very little error in the measurement of MDS. The MDS is thus the noise floor of the receiver and represents the power required from a signal generator to produce a 3dB (S+N)/N output, which means that the input signal equals the receiver noise.

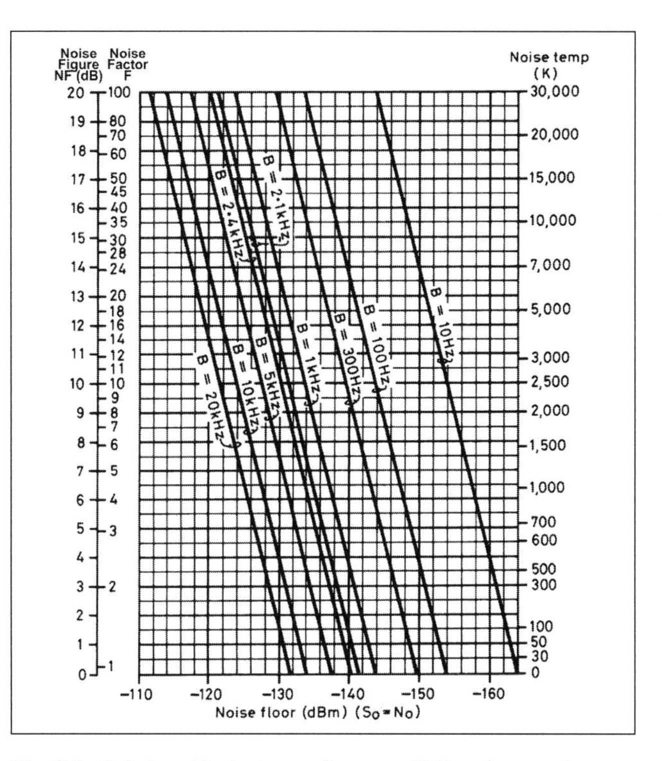


**Fig 6.3: Relationship between the sensitivity of a receiver as defined in terms of noise factor, noise figure (dB) or noise temperature and the noise floor (noise = signal) in dBm for various receiver bandwidths. Note how the minimum detectable signal reduces with narrower bandwidths for the same noise factor**

*MDS (dBm)* $= -174\text{dBm} + 10 \log B + NF_i$
(at room temperature)

The noise floor of HF receivers is conventionally given in dBm in a 3kHz noise bandwidth and in practice may range from about -120dBm to about -145dBm. Alternatively, receiver noise in a 3kHz bandwidth at 50 ohm impedance may be represented by an equivalent signal expressed in dBmV EMF, which will numerically equal the Noise Figure less 26dB.

With modern forms of mixers, it is possible to achieve a Noise Figure better than 10dB so that no preamplification is required to achieve maximum usable sensitivity on HF up to about 21MHz; however, low-gain amplification may still be advisable in order to minimise radiation of the local oscillator output from the antenna and to permit the use of narrow-band or resonant input filtering. Oscillator radiation is limited for amateur equipment to a maximum level of -57dBm conducted to the antenna for emissions below 1GHz, and to -47dBm above 1GHz, by the European Radiocommunications Committee Recommendation on Spurious Emissions.

To calculate the overall Noise Factor of cascaded stages, the equation is:

$NF_t = NF_1 + [(NF_2 - 1)/g_1] + [(NF3 - 1)/g_1 g_2] + \ldots$
where (all in power ratios and *not* decibels) $NF_t$ is the overall Noise Factor, $NF_2$, $NF_3$ are the individual stage noise factors and $g_1$, $g_2$ are the individual stage gains.

To convert back to Noise Figure in dB we take the $\log_{10}$ and multiply by 10, ie $10 \log_{10} NF_t$. Obviously, the series can be extended to cover as many stages as necessary. For design work, this can usefully be incorporated into a spreadsheet approach for rapid evaluation of changes; however, the values must be converted from decibels to absolute power ratios.

The relationships between noise factor, noise figure, noise temperature, noise resistance, minimum discernable signal and sensitivity are as follows:

From the above, noise factor, $NF_a$ = (S/N in) / (S/N out) and noise figure, $NF_i = 10 \log_{10}$ ((S/N in) /(S/N out)), expressed in dB

From the previous text we have seen that thermal noise power is derived from;

$P = kTB$

where P is the power in watts, k is Boltzmann's constant (1.38 x 10-23 J/K), T is the temperature in K, and B is the bandwidth in hertz.

If this is converted into its voltage form it becomes:

$v^2 = 4kTBR$

where v is the resulting RMS noise voltage created across a resistance of R ohms.

This formula can be rearranged as:

$T = v^2/(4kBR)$

which is known as noise temperature, normally used at microwave frequencies. Alternatively it can be rearranged as:

$R = v^2/(4kTB)$

that is known as noise resistance, previously used in valve specifications.

Sensitivity, is an input signal level that produces a given S/N ratio output, usually 10dB for SSB. Sensitivity is a practical meaningful measurement from which all the other ways of defining noise can be derived. So for a measured sensitivity of -113dBm (0.5µV PD) giving a 10dB S/N ratio, the MDS and the actual receiver noise floor is -123dBm.

Assuming a 3kHz bandwidth and room temperature, you get a theoretical thermal noise level of -133dBm. The difference between this theoretical figure and the actual receiver noise floor is the Noise Figure, $NF_i$ which in this example is 10dB.

## Large Signal Handling

Table 6.1 indicates the range of larger signals to be expected with S9+20dB equivalent to -53dBm or 0.5mV. Signals from broadcast stations found adjacent to the top end of the 40m amateur band can be up to 50mV, and these represent a threat to successful reception of an S3 amateur signal.

**Fig 6.4** is a graph showing the relationship between the antenna input signal and the output signal from a receiver front end. At low and modest input levels, the output follows the input. At some point, the receiver front end cannot produce any more output, it overloads and any further increase in input is no longer matched by an equivalent change in output. The straight line curves into a flat line; the slope becomes non-linear. As we shall see later in this chapter, this non-linearity creates an unexpected mixer that potentially produces three detrimental effects; blocking, crossmodulation and intermodulation. Of the three, intermodulation, often referred to as IP3, is the most prevalent.

So, self generated noise limits a receiver's small signal performance and intermodulation limits its large signal performance. The difference between the two is the receiver's dynamic range.

**Fig 6.5** brings together and shows the relationship between noise, S/N ratio, MDS, IP3 and spurious free dynamic range.

Large signal performance has been driving HF receiver design for the last decade. Ironically in the valve era, large signal performance was adequate and attracted little design attention. However when semiconductors came into use, intermodulation became an issue. Not surprisingly since valves have a supply voltage of 200V whilst semiconductors run off 12V, so a 1V signal is trivial for a valve but challenging for a transistor.

Crossmodulation and intermodulation (**Table 6.2**) are both

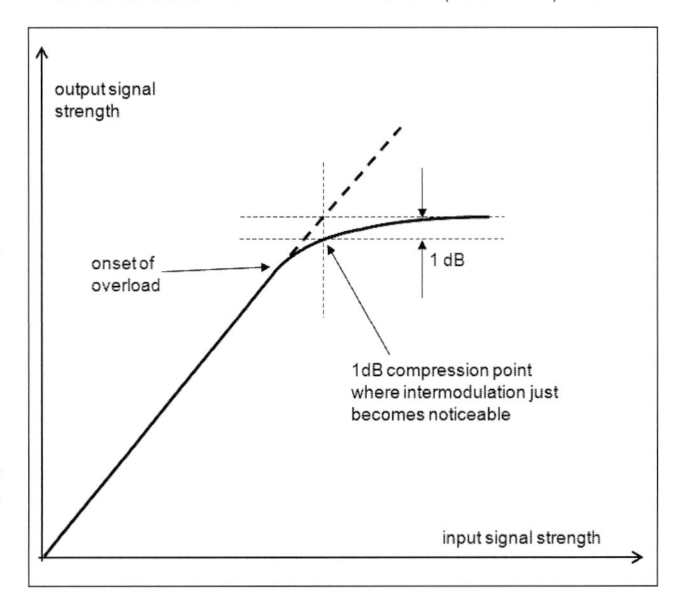


**Fig 6.4: Antenna input signal versus output signal from a receiver front end, showing the point at which overload occurs**

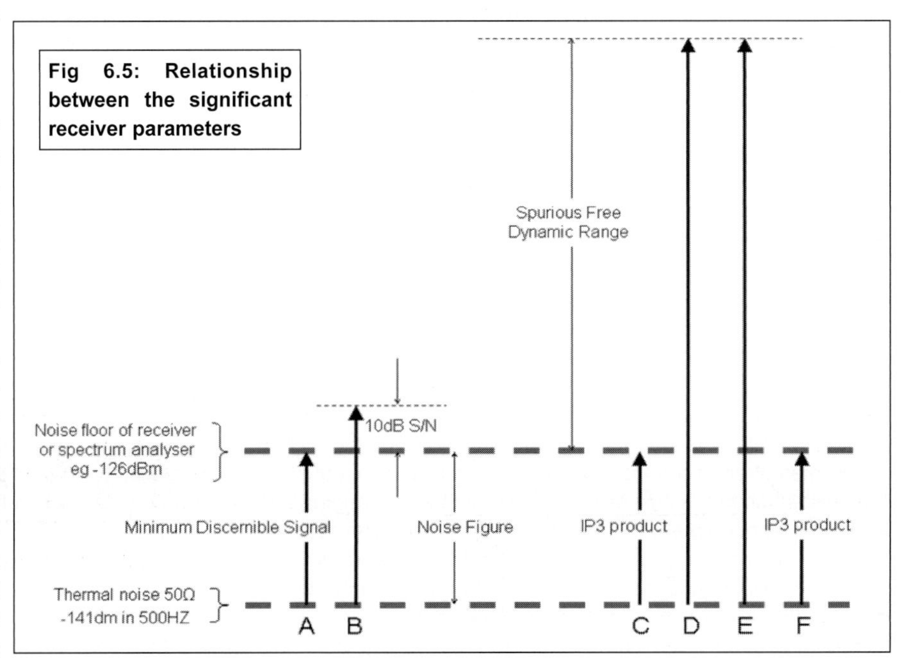


**Fig 6.5: Relationship between the significant receiver parameters**

**Blocking** is when a large unwanted out of band signal causes the receiver to de-sensitise.

**Crossmodulation** is when the modulation of a large unwanted out of band signal is transferred onto a weak wanted signal.

**Intermodulation** happens when two large unwanted out of band signals mix to produce a range of additional smaller unwanted signals, one of which occurs at the wanted frequency. If this particular signal rises above the noise floor of the receiver, it will be heard as phantom signal, ie it is not actually there.

**Table 6.2: Non-linear effects in receivers allow large, unwanted, out of band signals to impact on the performance of the receiver**

effects of non-linearity in the presence of large unwanted signals. The mixer is the most non-linear stage and consequently produces most of the unwanted effects. IP3 and spurious free dynamic range (SFDR) are both measures of the amount of intermodulation.

The compression point is the overload point of a circuit when a further increase in input signal produces no further increase in output signal.

IP3, more correctly called the third order intercept point, is the theoretical point on a graph where the increase in wanted signal meets the corresponding increase in the intermodulation product, see **Fig 6.6**. As the input signal rises, the wanted output rises at the same rate whilst the unwanted intermodulation product rises at three times the rate (in dB). Low level measurements can be extrapolated to a point where the two lines cross and this point is known as the third order intercept point, or IP3 for short. It is a key measurement of non-linearity and it crops up in all sorts of analysis and calculations. IP3 is measured in dBm.

### Relationships between IP3 and gain

There is input IP3 (IIP3) and output IP3 (OIP3) and gain (or loss) in between.

*OIP3 = gain (in dB) + IIP3*

The figure for a receiver is its IIP3. Alternatively if you are measuring a circuit block, for instance a mixer, the more useful form of the relationship is:

*IIP3 = OIP3 - gain (in dB)*

Spurious Free Dynamic Range (SFDR) is 2/3 of the difference

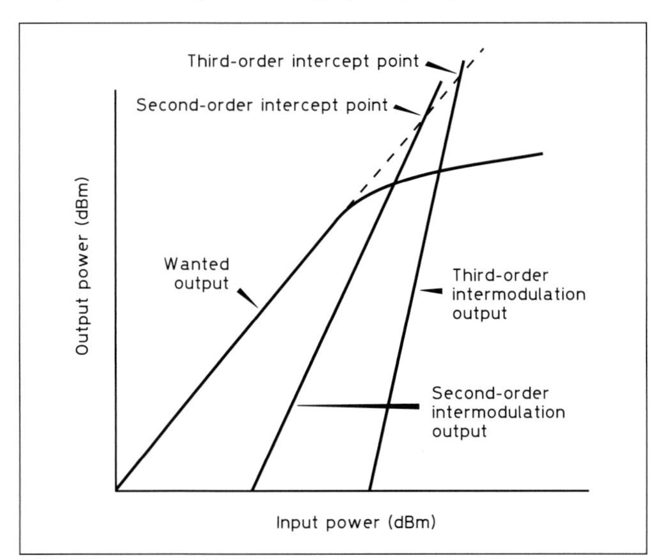


**Fig 6.6: Graphical representation of receiver dynamic range performance**

between the IIP3 and the Minimum Discernible Signal (MDS). So a receiver with an IIP3 of 30dBm and an MDS of -115dBm (two S points) will have an SFDR of 97dB.

The compression point of a switching mixer is approximately 10 to 15dB below its IP3, so knowing the compression point gives a good idea of the expected IP3.

The level of crossmodulation transferred onto a wanted signal can be calculated from the IIP3, the power of the unwanted signal and its percentage modulation using:

$M_W = 4M_uP / (IIP3 + 2P)$ where:

$M_W$ is the percentage of modulation transferred to the wanted signal, $M_u$ is the percentage of modulation on the unwanted signal, P is the power of the unwanted signal in watts and IIP3 is the receiver third-order input intercept point power in watts

## RECEIVER ARCHITECTURES

The simplest receiver is the crystal set, see **Fig 6.7**. It has very few components and does not require power, but it does need a long aerial. Energy is transferred from passing radio waves into the aerial and delivered to the tuned circuit. This provides some selectivity and voltage magnification due partly to the tuned circuit Q and partly from the transformer step-up. The diode rectifies the signal and the capacitor removes the remaining RF component leaving the recovered audio as heard in the headphones.

The weaknesses of this very simple receiver are inadequate selectivity because there is only one tuned circuit and this has relatively poor Q, and the recovered audio is very weak because there is no power gain in the receiver. An obvious next step is to add an audio amplifier eg a transistor; now there is more volume but still poor selectivity.

The next step in receiver evolution was the regenerative design (shown in a modern version in **Fig 6.8**) where the incoming RF signal enters through the tuned circuit, is amplified and then fed back though the same tuned circuit and so on until the signal is sufficiently large to be rectified. Finally, an audio amplifier is

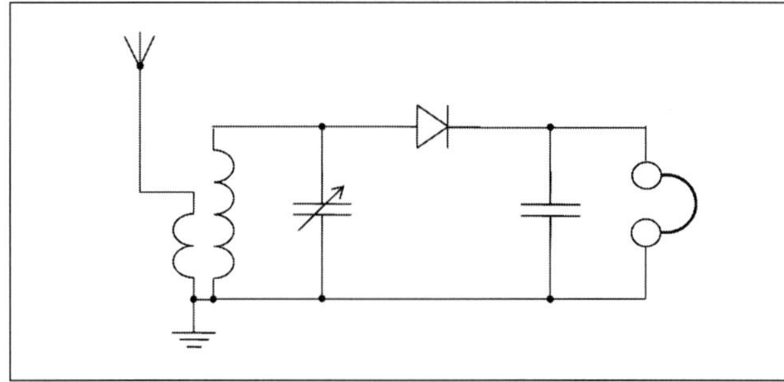

**Fig 6.7: Circuit of a crystal set, the simplest receiver**

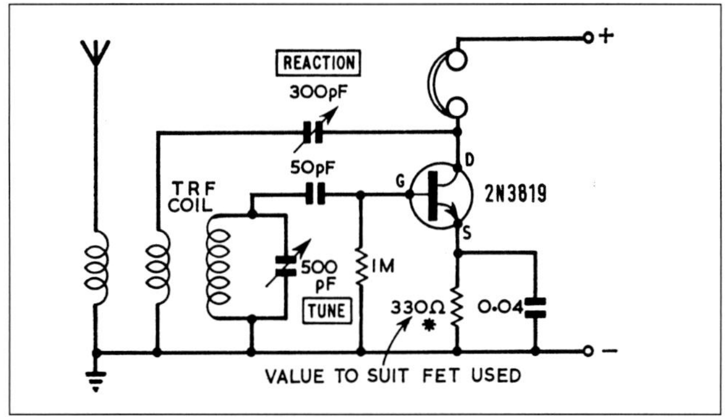


**Fig 6.8: A single-FET receiver**

added to raise the recovered audio to a reasonable level. Now we have greater selectivity and gain because the signal passes through the tuned circuit numerous times.

The principal weakness of this type of receiver is it requires careful adjustment of the feedback, different for each incoming signal. Too much feedback and the circuit oscillates, and being connected to an aerial, it becomes a transmitter! Nevertheless, such circuits can be very rewarding to build, producing quite a lot from very little.

## TRF

In 1916 a new type of receiver design appeared; the TRF (Tuned Radio Frequency) where gain and selectivity were achieved by cascading several stages of tuned circuit and amplification before the detector see **Fig 6.9**. If these stages have too much gain the overall RF amplifier chain will oscillate. Also ensuring that the cascade of tuned circuits all stay tuned to the same frequency is challenging so careful adjustment is required.

## Superhet

The trade off between gain, the number of tuned circuits and their adjustment was solved two years later when the superhetrodyne, or superhet was introduced, see **Fig 6.10**. This design recognised that a large amount of RF gain was required, but that it was better to achieve this at a relatively low fixed intermediate frequency, the IF.

The incoming RF signal was shifted down to the fixed IF using a frequency changer or mixer stage where a local oscillator (LO) was mixed with the incoming RF signal to produce the IF according to the relationship:-

> The IF equals the sum or difference between the RF and the LO frequency.

Tuning was now achieved by setting the frequency of the local oscillator and selectivity was achieved in the IF amplifier using multiple tuned circuits. The maths is simple; mixing the LO and RF signals together produces amongst other frequencies the sum and difference between the two, the IF being the LO minus RF, or alternatively the RF minus the LO depending if the LO is above or below the RF frequency.

If we assume a wanted RF signal of 21MHz and an IF of 1MHz, then the LO could be 20MHz. However an unwanted signal at 19MHz will also produce a 1MHz IF, and this 19MHz signal is known as the image frequency. The image frequency must be removed before the mixer and so the TRF survived and became the RF amplifier albeit with one stage of amplification and two tuned circuits. The combined selectivity of the RF amplifier tuned circuits reduces the level of the unwanted image frequency whilst the RF gain overcomes the noise introduced by the subsequent mixer.

Selectivity in the IF amplifier could produce an almost ideal response being made up of several double tuned coupled pairs. The gain could be high and gain control could be added thus maintaining the output level irrespective of the input level. When this is done automatically, you get automatic gain control, or AGC (see later).

## Choice of IF

Choice of the intermediate frequency (IF) or frequencies is a most important consideration in the design of any superhet receiver. The lower the frequency, the easier it is to obtain high gain and good selectivity and also to avoid unwanted leakage of signals round the selective filter. On the other hand, the higher the IF, the greater will be the frequency difference between the wanted signal and the 'image' response, so making it simpler to obtain good protection against image reception of unwanted signals and also reducing the 'pulling' of the local oscillator frequency. These considerations are basically opposed, and the IF of a single-conversion receiver is thus a matter of compromise. However, in recent years it has become easier to obtain good selectivity with higher-frequency band-pass crystal filters and it is no longer any problem to obtain high gain at high frequencies.

The very early superhet receivers used an IF of about 100kHz; then for many years 455-470kHz was the usual choice - many modern designs use between about 3 and 9MHz, and SSB IF filters are now available to 40MHz. A good rule of thumb is that the IF should be no less than 5%, and preferably 10% of the signal frequency. Where the IF is higher than the signal frequency the action of the mixer is to raise the frequency of the incoming signal, and this process is now often termed up-conversion (a term formerly reserved for a special form of parametric mixer). Up-conversion, in conjunction with a low-pass filter at the input, is an effective means of reducing IF breakthrough as well as image response.

A superhet receiver, whether single- or multi-conversion, must have its first IF outside its tuning range. For general-coverage HF receivers tuning between, say, 1.5 and 30MHz, this limits the choice to below 1.5MHz or above 30MHz.

To reduce image response without having to increase pre-mixer RF

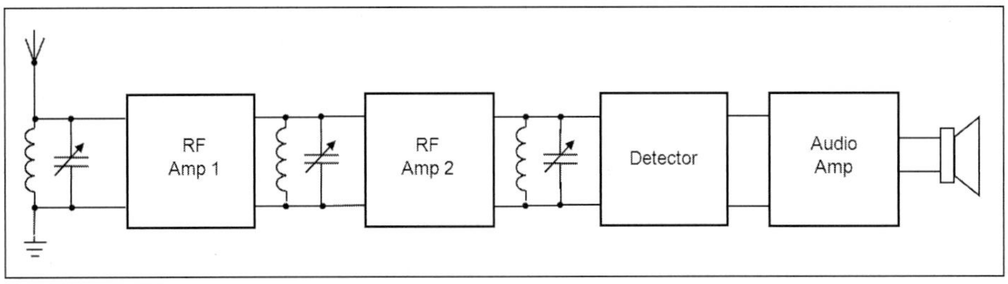

**Fig 6.9: Block diagram of a tuned radio frequency (TRF) receiver**

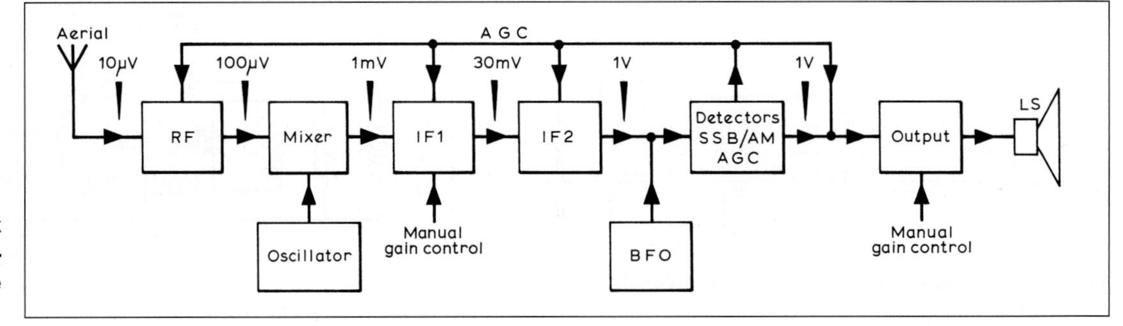


**Fig 6.10: Block outline of representative single conversion**

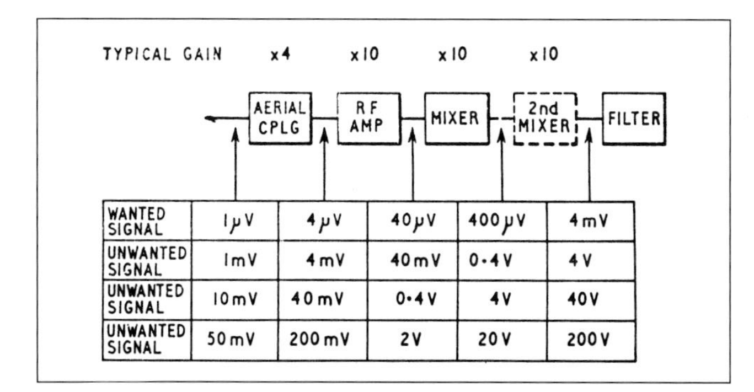


| | | | | | |
|---|---|---|---|---|---|
| WANTED SIGNAL | 1μV | 4μV | 40μV | 400μV | 4mV |
| UNWANTED SIGNAL | 1mV | 4mV | 40mV | 0·4V | 4V |
| UNWANTED SIGNAL | 10mV | 40mV | 0·4V | 4V | 40V |
| UNWANTED SIGNAL | 50mV | 200mV | 2V | 20V | 200V |

**Fig 6.11: This diagram shows how unwanted signals are built up in high-gain front-ends to levels at which cross-modulation, blocking and intermodulation are almost bound to occur**

selectivity (which can involve costly gang-tuned circuits) professional designers are increasingly using a first IF well above 30MHz. This trend is being encouraged by the availability of VHF crystal filters suitable for use either directly as an SSB filter or more often with relaxed specification as a roofing filter.

The use of a very high first IF, however, tends to make the design of the local oscillator more critical (unless the Wadley triple-conversion drift-free technique is used - when the intermodulation problems are more difficult because of the number of stages prior to the selectivity determining filters). For amateur-bands-only receivers the range of choice for the first IF is much wider, and 3.395MHz and 9MHz are typical.

A number of receivers have adopted 9MHz IF with 5.0-5.5MHz local oscillator: this enables 4MHz-3.5MHz and 14.0-14.5MHz to be received without any band switching in the VFO; other bands are received using crystal-controlled converters with outputs at 3.5 or 14MHz. With frequency synthesisers, up-conversion is commonly found, with the first IF between 40-70MHz.

## Gain distribution

In many receivers of conventional design, it has been the practice to distribute the gain throughout the receiver in such a manner as to optimise signal-to-noise ratio and to minimise spurious responses.

So long as relatively noisy mixer stages were used, it was essential to amplify the signal considerably before it reached the mixer. This means that any strong unwanted signals, even when many kilohertz from the wanted signal, pass through the early unselective amplifiers and are built up to levels where they cause cross-modulation within the mixer: see **Fig 6.11**.

Today it is recognised that it is more satisfactory if pre-mixer gain can be kept low to prevent this happening. Older multigrid valve mixers had an equivalent noise resistance as high as 200,000 ohms (representing some 4-5mV of noise referred to the grid). Later types such as the ECH81 and 6BA7 reduced this to an ENR of about 60,000 ohms (about 2.25mV of noise) while the ENR of pentode and triode mixers was lower still (although these may not be as satisfactory for mixers in other respects).

The noise contribution of semiconductor mixers is also low - for example an FET mixer may have a noise factor as low as 3dB - so that generally the designer need no longer worry unduly about the requirement for pre-mixer amplification to overcome noise problems. Nevertheless a signal frequency stage may still be useful in helping to overcome image reception by providing a convenient and efficient method of coupling together signal-frequency tuned circuits, and when correctly controlled by AGC it becomes an automatic large-signal attenuator. Pre-mixer selectivity limits the number of strong signals reaching the mixer. The use of double-balanced mixers is generally advisable to reduce the spurious response possibilities.

**Fig 6.12** shows a typical gain distribution as found in a modern design in which the signal applied to the first mixer is much lower than was the case with older (valved) receivers.

The significant conversion losses of modern diode and FET ring mixers (6-10dB), the losses of input band-pass filtering and the need for correct impedance termination means that in the highest-performance receivers it is desirable to include low-gain, low-noise, high dynamic range pre- and post-mixer amplifiers and a diplexer ahead of the (main or roofing) crystal filter to achieve constant input impedance over a broad band of frequencies.

If the diplexer is a simple resistive network this will introduce a further loss of some 6dB. Stage-by-stage gain, noise figure, third-order intercept, 1dB compression point and 1dB desensitisation point performance of the N6NWP high-dynamic-range MF/HF front-end of a single-conversion (9MHz IF) receiver ([1] and see later) is shown in **Fig 6.13**. More complex diplexers may be used to divert the local oscillator feedthrough signal from entering the IF strip.

For double- and multiple-conversion receivers, the gain distribution and performance characteristics of all the stages preceding the selective filter need to be considered. In general the power loss in any band-pass filter will decrease as the bandwidth increases: for example a 75MHz, 25kHz-bandwidth filter might have a power loss of 1dB whereas a 75MHz, 7kHz-bandwidth filter might have a power loss of 3.5dB.

### Automatic gain control (AGC)

The fading characteristics of HF signals and the absence of a carrier wave with SSB make the provision of effective AGC an important characteristic of a communications receiver, although it should be stressed that no AGC may be preferable to a poor AGC system that can degrade overall performance of the receiv-

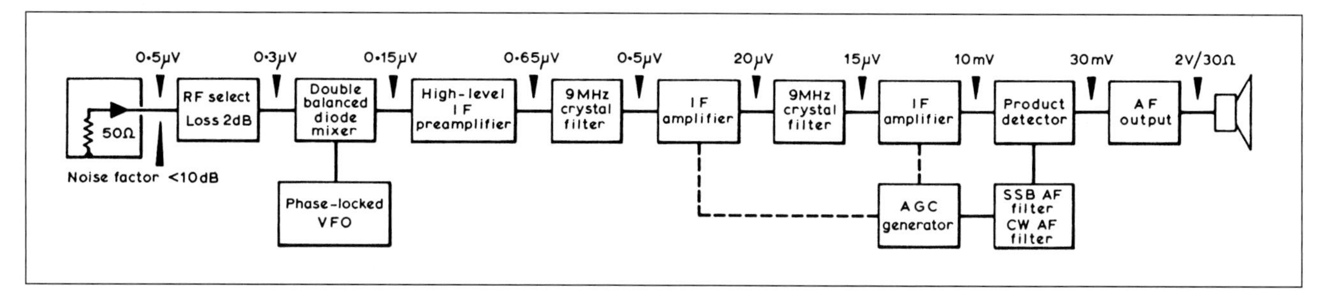


**Fig 6.12: Gain distribution in a high-performance semiconductor single-conversion receiver built by G3URX using seven integrated circuits, 27 transistors and 14 diodes. 1μV signals can be received 5kHz off-tune from a 60mV signal and the limiting factor for weak signals is the noise sidebands of the local oscillator, although the phase-locked VFO gives lower noise and spurious responses than the more usual pre-mixer VFO system**

Fig 6.13: Stage by stage gain and characteristics of the N6NWP front-end of single-conversion 14MHz HF receiver with 9MHz IF

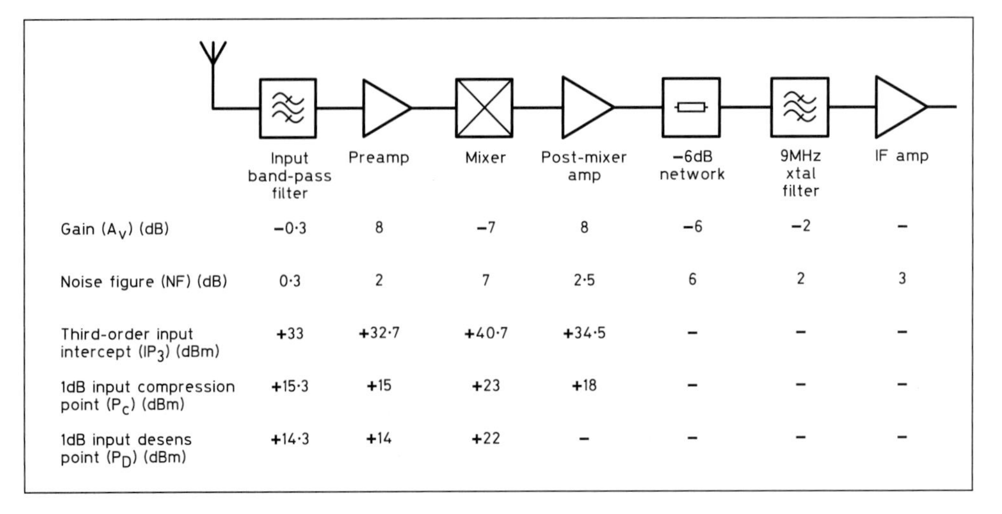

| | Input band-pass filter | Preamp | Mixer | Post-mixer amp | −6dB network | 9MHz xtal filter | IF amp |
|---|---|---|---|---|---|---|---|
| Gain ($A_V$) (dB) | −0·3 | 8 | −7 | 8 | −6 | −2 | – |
| Noise figure (NF) (dB) | 0·3 | 2 | 7 | 2·5 | 6 | 2 | 3 |
| Third-order input intercept ($IP_3$) (dBm) | +33 | +32·7 | +40·7 | +34·5 | – | – | – |
| 1dB input compression point ($P_C$) (dBm) | +15·3 | +15 | +23 | +18 | – | – | – |
| 1dB input desens point ($P_D$) (dBm) | +14·3 | +14 | +22 | – | – | – | – |

er. Unfortunately, unwanted dynamic effects of an AGC system cannot be deduced from the usual form of receiver specification which usually provides only limited information on the operation of the AGC circuits, indicating only the change of audio output for a specified change of RF input. For example, the specification may state that there will be a 3dB rise in audio output for an RF signal input change from 1mV to 50mV. This represents a high standard of control, provided that the sensitivity of the receiver is not similarly reduced by strong off-tune signals, and that the control acts smoothly throughout its range without introducing intermodulation distortion etc.

Basically, AGC is applied to a receiver to maintain the level of the wanted signal output at a more or less constant value, while ensuring that none of the stages is overloaded, with consequent production of IMPs etc. The control voltages, derived usually at the end of the IF amplifying stages, are applied to a number of stages in the signal path while usually ensuring that the IF gain is reduced first, and the RF/first mixer later, in order to preserve the SNR.

When AGC is applied to an amplifier, this shifts the operating point and may affect both its dynamic range and the production of intermodulation distortion. One partial solution is the use of an AGC-controlled RF attenuator(s) ahead of the first stage(s).

All AGC systems are designed with an inherent delay based on resistance-capacitance time-constants in their response to changes in the incoming signal; too-rapid response would result in the receiver following the audio envelope or impulsive noise peaks. The delays are specified as the attack time, ie the time taken for the AGC to act, and the decay time for which it continues to act in the absence or fade of the wanted signal (**Fig 6.14**).

For AM signals, now only rarely encountered in amateur radio, the envelope detector may be used directly to generate the AGC voltage. In this case the attack and delay times need to be fast enough to allow the AGC system to respond to fading, but slow enough not to respond to noise pulses or the modulation of the carrier. Typically time-constants are about 0.1 to 0.2s.

For amateur SSB signals there is no carrier level and the AGC voltage must be derived from the peak signal level. This is sometimes done by using the AF signal from a product detector but it is more satisfactory to incorporate a dedicated envelope detector to which a portion of the final IF signal is fed. This avoids the effects of the AGC signal disappearing at low audio beat notes. For SSB, a fast attack time is needed but the release (decay) time needs to be slow in order not to respond to the brief pauses of a speech transmission. Such a system is termed hang AGC and can be used also for CW reception. Typically the attack time needs to be less than 20ms and the release time some 200 to

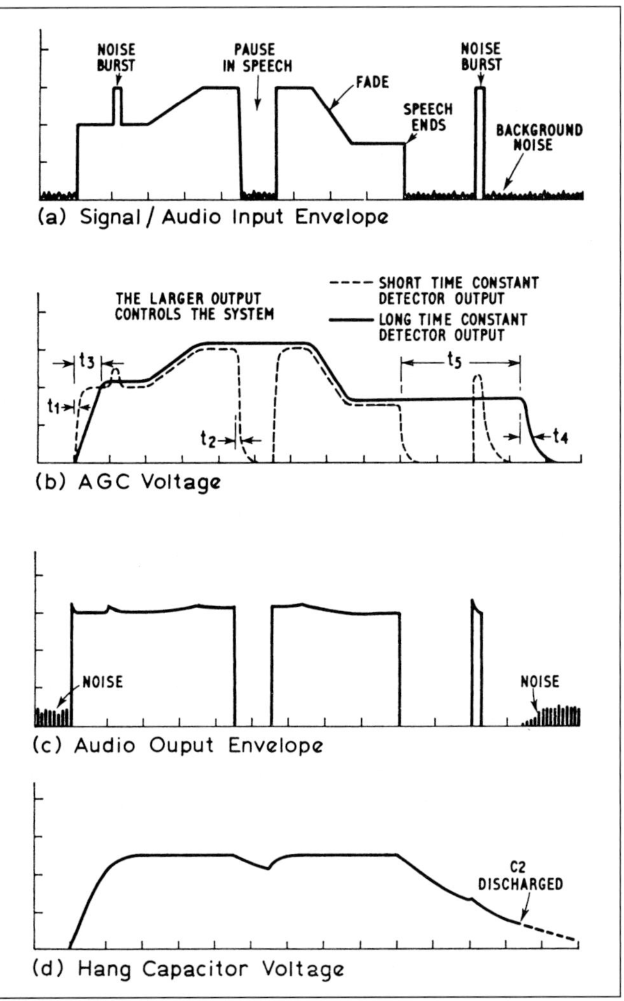


Fig 6.14: The behaviour of dual time-constant AGC circuits under various operating conditions, where $t_1$ is the fast detector rise time, $t_2$ the fast detector decay time, $t_3$ the slow detector rise time, $t_4$ the slow detector fall time, and $t_5$ the hang time

1000ms; with a receiver intended also for AM reception it is useful to have a shorter release time (say 25ms) available for fast AGC. Hang AGC systems are often based on two time-constants to allow the system to be relatively unaffected by noise pulses while retaining fast attack and hang characteristics.

For both SSB and CW reception, it is important that the BFO or carrier insertion oscillator (CIO) should not affect the AGC system and for many years this problem was sidestepped by turning the AGC system off during CW reception. Even with modern AGC systems, it may be useful to be able to turn the system off, particularly where final narrow-band CW selectivity depends on post-demodulation filtering, with the result that the AGC system will react to signals within the bandwidth of the final IF system which may be about 3kHz unless narrow-band IF crystal filters are fitted.

The distribution of gain control is an important, but often neglected function. Obviously, gain could be controlled by an antenna attenuator, but the result would be that the signal-to-noise ratio was that obtained for the weakest input signal. Ideally, an increase of input signal of 20dB would provide an increase in SINAD of 20dB, but there is usually an ultimate receiver SINAD limit of 40 to 50dB. A test of SINAD improvement ratio is usefully but rarely done: a signal is fed into the receiver to provide a 20dB SINAD ratio, and is increased in 10dB steps. A reasonable receiver will show about 28 to 29dB SINAD for the first step, and no less than 37dB SINAD for the second.

## Selectivity

The ability of a receiver to separate stations on closely adjacent frequencies is determined by its selectivity. The limit to usable selectivity is governed by the bandwidth of the type of signal which is being received.

For high-fidelity reception of a double-sideband AM signal the response of a receiver would need ideally to extend some 15kHz either side of the carrier frequency, equivalent to 30kHz bandwidth; any reduction of bandwidth would cause some loss of the information being transmitted. In practice, for average MF broadcast reception the figure is reduced to about 9kHz or even less; for communications-quality speech in a double-sideband system we require a bandwidth of about 6kHz; for single-sideband speech about half this figure or 3kHz is adequate, and filters with a nose bandwidth of 2.7 or 2.1kHz are used, providing, in the case of a 2.7kHz filter, audio frequencies from 300 to 3000Hz with little loss of intelligibility. For CW, at manual keying speeds, the minimum theoretically possible bandwidth will reduce with speed from about 100Hz to about 10Hz for very slow Morse.

**Fig 6.15** shows how 'single signal' reception is achieved with a narrow filter. Excessively narrow filters with good shape factors make searching difficult - and many operators like to have some idea of signals within a few hundred hertz of the wanted signal.

Ideally, again, we would like to receive just the right bandwidth, with the response of the receiver then dropping right off as shown in **Fig 6.16**, to keep the noise bandwidth to a minimum. Although modern filters can approach this response quite

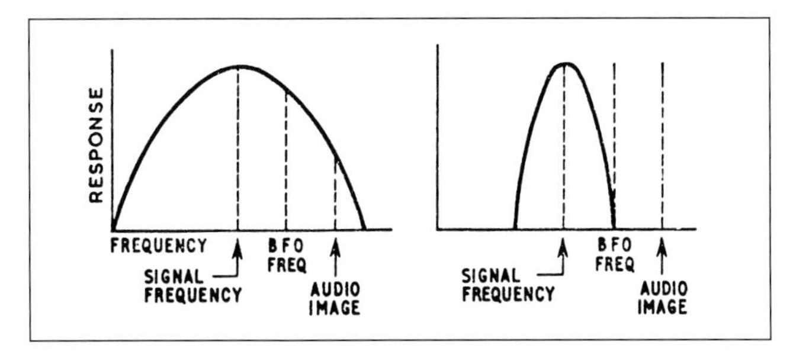


**Fig 6.15: How a really selective receiver provides 'single-signal' reception of CW. The broad selectivity of the response curve on the left is unable to provide substantial rejection of the audio 'image' frequency whereas with the more selective curve the audio image is inaudible and CW signals are received only on one side of zero-beat**

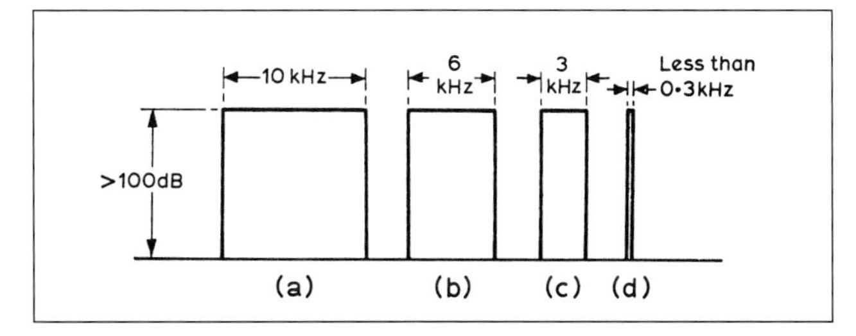


**Fig 6.16: The ideal characteristics of the overall band-pass of a receiver are affected by the type of signals to be received. (a) This would be suitable for normal broadcast reception (DSB signals) permitting AF response to 5kHz. (b) Suitable for AM phone (AF to 3kHz). (c) For SSB the band-pass can be halved without affecting the AF response (in the example shown this would be about 300 to 3300Hz). (d) Extremely narrow channels (under 100Hz) are occupied by manually keyed CW signals but some allowance must usually be made for receiver or transmitter drift and a 300Hz bandwidth is typical - by selection of the carrier insertion oscillator frequency any desired AF beat note can be produced**

closely, in practice the response will not drop away as sharply over as many decibels as the ideal.

To compare the selectivity of different receivers, or the same receiver for different modes, a series of curves of the type shown in **Fig 6.17** may be used.

There are two ways in which these curves should be considered: first the bandwidth at the nose, representing the bandwidth over which a signal will be received with little loss of strength; the other figure - in practice every bit as important - is the bandwidth over which a powerful signal is still audible, termed the skirt bandwidth.

The nose bandwidth is usually measured for a reduction of not more than 6dB, the skirt bandwidth for a reduction of one thou-

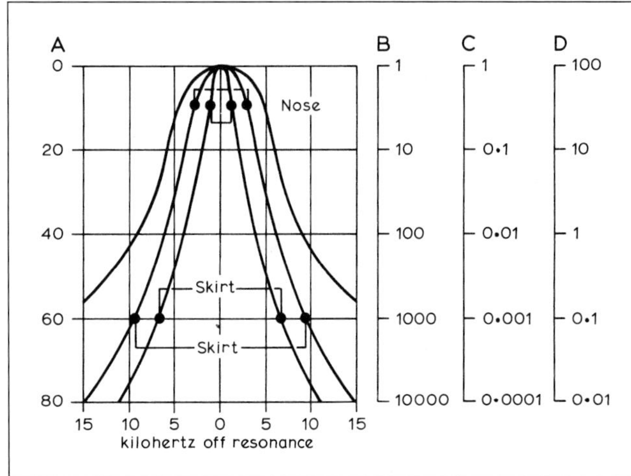


**Fig 6.17: The ideal vertical sides of Fig 6.16 cannot be achieved in practice. The curves shown here are typical. These three curves represent the overall selectivity of receivers varying from the 'just adequate' broadcast curve of a superhet receiver having about four tuned IF circuits on 470kHz to those of a moderately good communications receiver. A, B, C and D indicate four different scales often used to indicate similar results. A is a scale based on the attenuation in decibels from maximum response; B represents the relative signal outputs for a constant output; C is the output voltage compared with that at maximum response; D is the response expressed as a percentage**

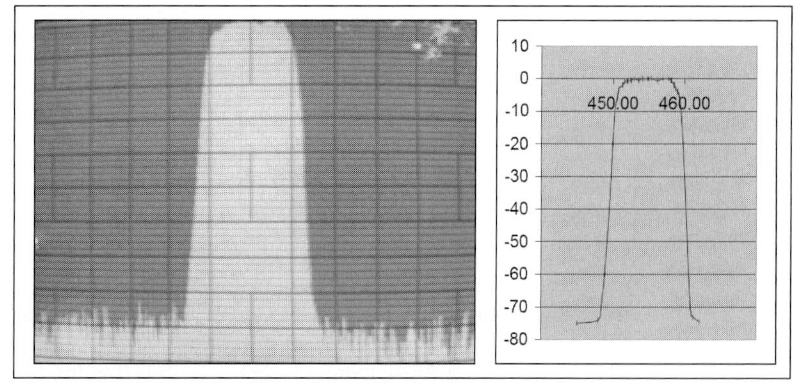


Fig 6.18: (left) Measured and (right) theoretical selectivity of a LF-D6 ceramic filter

sand times on its strength when correctly tuned in, that is 60dB down. These two figures can then be related by what is termed the shape factor, representing the bandwidth at the skirt divided by the bandwidth at the nose.

The idealised curves of **Fig 6.16**, which have the same bandwidth regardless of signal strength, would represent a shape factor of 1; such a receiver cannot be designed at the present state of the art, although it can be approached by digital filters and some SSB filters; the narrower CW filters, although much sharper at the nose, tend to broaden out to about the same bandwidth as the SSB filter and thus have a rather worse shape factor, although this may not be a handicap.

Typically a high-grade modern receiver might have an SSB shape factor of 1.2 to 2 with a skirt bandwidth of less than 5kHz. Note that, although 6 and 60dB are conventionally used for nose and skirt, some radios are specified at 3 and 30dB, or even 6 and 30dB - it is important to be aware of what figures are used when comparing one filter or receiver with another.

**Fig 6.18** shows both the specification and the measured response of an LF-D6 ceramic filter indicating a shape factor of 1.14.

It should be noted that such specifications are determined when applying only one signal to the input of the receiver; unfortunately in practice this does not mean that a receiver will be unaffected by very strong signals operating many kilohertz away from the required signal and outside the IF pass-band; this important point is considered elsewhere in this chapter.

It therefore needs to be stressed that the effective selectivity cannot be considered solely in terms of static characteristics determined when just one test signal is applied to the input, but rather in the real-life situation of hundreds of signals present at the input: in other words it is the dynamic selectivity which largely determines the operational value of an amateur HF receiver.

With the advent of block IF amplifiers, it has become common practice to put one high quality filter in front of the IF gain block, but this does nothing to limit the amount of noise generated within the block from reaching the detectors. Before gain blocks, selectivity was provided progressively along the amplifier as part of the interstage coupling, distributed selectivity in today's parlance. Ideally, some additional selectivity is required prior to the detectors to eliminate this additional source of IF noise.

## Demodulation

Demodulation, or detection as it used to be called, is the process of recovering the intelligence being conveyed by a radio signal.

Most modes used on the amateur HF bands use a form of amplitude modulation (AM). CW is on-off keying, ie switching the signal's amplitude from zero to 100%. SSB is a derivative of amplitude modulation and digital modes are usually received as amplitude signals before being separately demodulated (see the chapter on digital communication).

AM can be considered in two ways; (a) where the amplitude of the carrier wave changes in sympathy with the modulation and (b) where the modulation mixes with the carrier wave to produce components at the carrier frequency plus and minus the modulation frequency (the two sidebands) together with the original carrier (the carrier) and the original modulation (which is discarded).

The simplest AM demodulator is the diode half wave rectifier, or envelope detector. Since the modulated carrier is symmetrical about the centre line, throwing half of it away results in a series of half sine waves at the carrier frequency varying in amplitude with the modulation. Simple RF filtering discards the RF component leaving only the modulation.

Because half the power in an AM signal is vested in the carrier, it can be eliminated in the transmitter thereby improving the efficiency of the transmission; this is double sideband, suppressed carrier (DSBSC). Since both sidebands contain the same intelligence and each share the remaining 50% of the transmitted power, one of them can be filtered out creating single sideband suppressed carrier (SSBSC), more commonly called SSB. This reduces the transmitted occupied bandwidth and further increases the transmission efficiency.

At the receive end, the demodulation process requires that the carrier is effectively re-inserted (using the Beat Frequency Oscillator, or BFO, alternatively known as the Carrier Insertion Oscillator, or CIO) thus re-creating an AM signal. Whilst this can be demodulated using a simple envelope detector, it can also be demodulated using mixing techniques where the BFO mixes with the incoming sideband producing the other sideband, the BFO and the original modulation which is easily filtered out. This form of demodulator is known as a product detector.

## Superhet Refinements

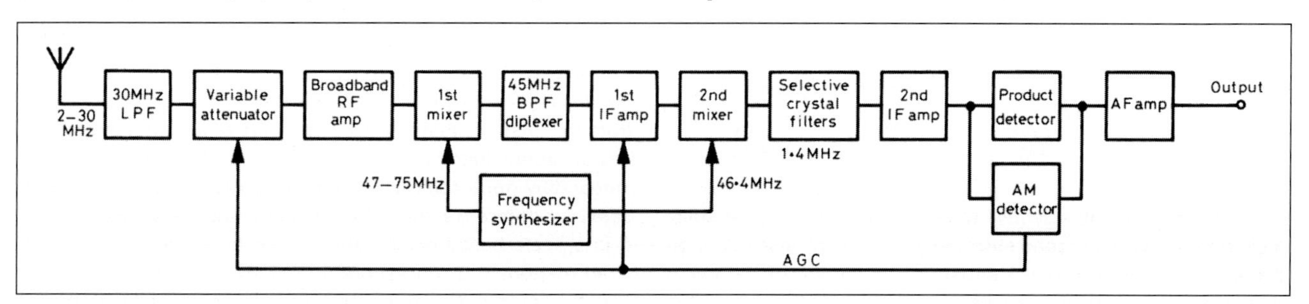


Fig 6.19: Representative architecture of a modern communications receiver, a general-coverage double-conversion superhet with up-conversion to 45MHz first IF and 1.4MHz second IF

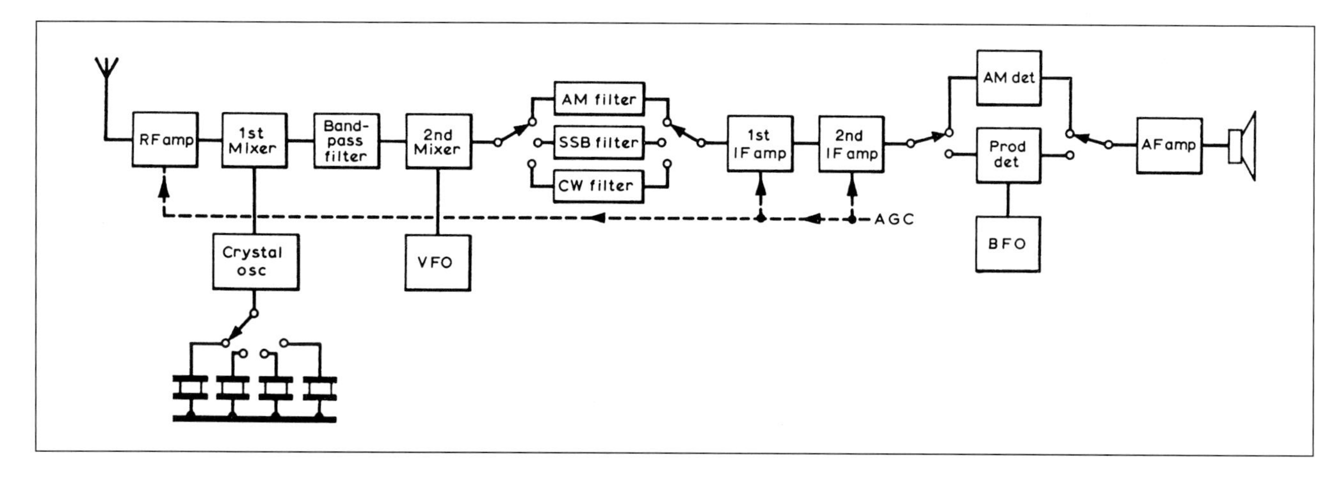


Fig 6.20: Block diagram of a double-conversion receiver with crystal-controlled first oscillator - typical of many late 20th-century designs

### Triple conversion

Such was the success of the superhet that it has prospered and remains the principle receiver architecture today.

Over the years several refinements to the architecture have been developed, most notably the multiple conversion superhet, see **Fig 6.19**. The early IF amplifiers achieved selectivity through the inter-stage coupling at frequencies of typically less that 500kHz. Whilst this was adequate for medium wave reception, a wanted RF signal at say 21MHz required an LO frequency of 20.5MHz producing an image response at 20MHz. This 1MHz difference (twice the IF) between the wanted signal and the unwanted image is very hard to suppress with just two RF tuned circuits.

One answer is to change the IF to say 5MHz, now the image is 10MHz away from the wanted and adequate suppression is achievable with two RF tuned circuits. However, obtaining good selectivity at 5MHz was difficult and so the double conversion superhet was born.

The function of the first IF at higher frequency is to enable the suppression of the image response. The first IF amplifier does not need much gain; it is primarily a filter whose output is then mixed down to the second IF where the bulk of the selectivity and gain take place as before.

The ultimate solution to the image response problem is to move the first IF above the highest RF frequency, up conversion typically to 45MHz so the image will be 90MHz away.

This is not the perfect architecture as there is now an additional image problem from the second mixer because going from 45MHz down to 0.455MHz produces a second image only 0.910MHz away and the 45MHz filter will need to work hard to eliminate this second image. The answer was a third conversion from 45MHz down to say 5MHz and then down again to 0.455MHz. Triple conversion receivers have been built, but the choice of first, second and third LO frequencies becomes critical because of the multiplicity of spurious products generated in the less than perfect mixers.

Another variation of the double conversion superhet was the tunable IF where the first mixer is fed from a fixed frequency LO, see **Fig 6.20**. This produces a band of frequencies determined by the front end filtering. The wanted frequency is selected by making the second LO variable so that it tunes across the band of frequencies coming out of the first mixer. So the architecture looks like a receiver with a converter in front.

## Frequency stability

In the early development of receivers, sensitivity and selectivity issues had been solved. The next problem to be tackled was tuning drift where the frequency of the LO shifted slowly because it was temperature dependant.

The first solution, appearing in 1954, was the Wadley drift cancelling loop, see **Fig 6.21** that made use of the tunable IF architecture and some clever LO mixing to eliminate the effects of LO drift. The first commercial receiver to use this approach was the Racal RA17 followed by several others including the Yaesu FRG-7.

The next evolution regarding tuning drift was the frequency synthesiser that used digital frequency divider integrated circuits, phase detectors and voltage controlled oscillators (VCOs) to create highly stable oscillators with predictable frequency output. These were not continuously variable oscillators (VFOs), instead they jumped from one frequency to the next, the step size being governed by the digital division ratio. The circuit technique is known as phase locked loop, or PLL (see **Fig 6.22**) with the division ratios being controlled by a microprocessor.

This was the first appearance of a microprocessor in a receiver

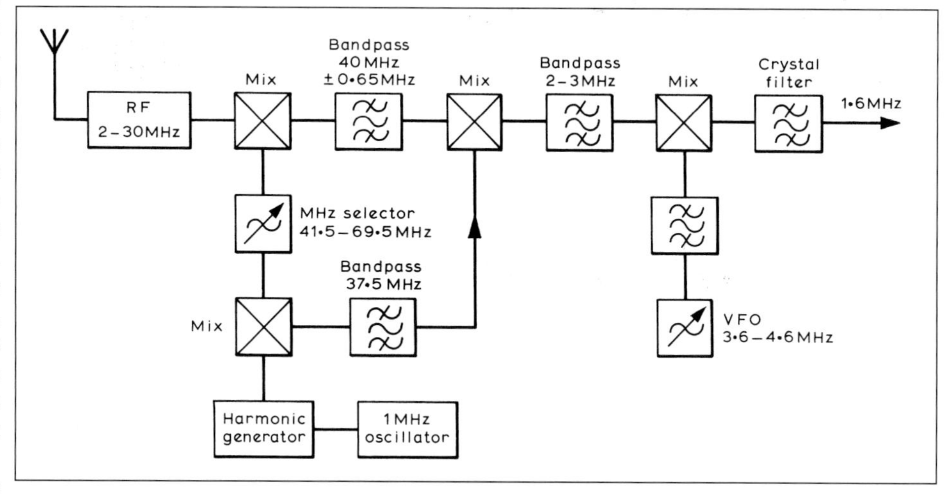


Fig 6.21: The Wadley drift-cancelling loop system as used on some early Racal HF receivers but requiring a considerable number of mixing processes and effective VHF band-pass filters

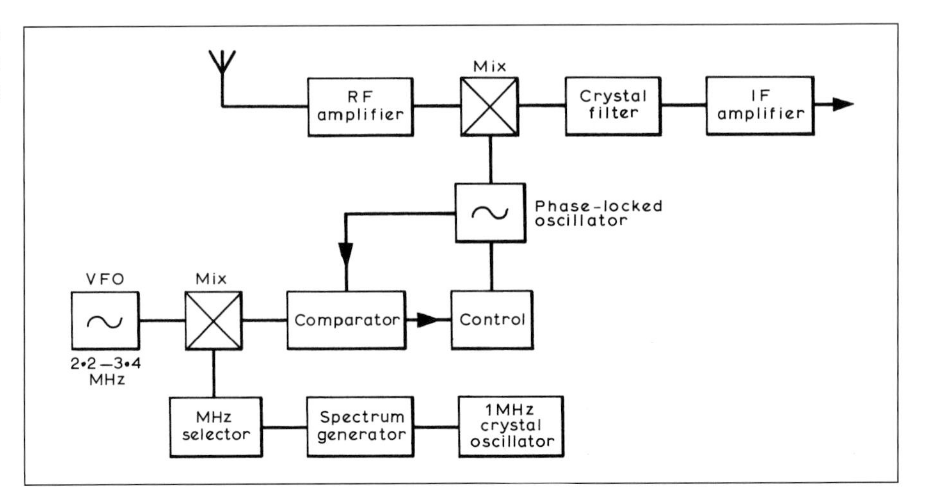


**Fig 6.22: Practical frequency synthesis using fixed-range VFO with megahertz signals derived from a single 1MHz crystal**

and its use was expanded to provide other control and user interface features. It was the beginning of Software Controlled Radio that evolved into today's Software Defined Radio.

## Shortcomings

By the 1970s, the most important performance shortcomings were blocking, crossmodulation and intermodulation.

It is interesting to note that these problems became much more apparent when valves were replaced by transistors. At the time the excitement and emphasis in receiver design was for smaller, low power (battery powered), portable receivers and little attention was paid to the shortcomings.

Eventually, however, the industry woke up to the deficiencies which were attributed to non-linear circuits in the presence of large signals, particularly the mixer. Early mixers were non-linear, the non-linearity produces the wanted products but the attendant unwanted products produce blocking, crossmodulation and intermodulation by default. The search was on for linear mixers and designers started to focus on large signal performance. Of the mixing techniques available, the switching mixer was identified as the way forward and in the last ten years they have become the norm.

One other deficiency has been identified and is now attracting a lot of development focus. **Fig 6.23** shows the close-in spectrum of an LO where its noisy sidebands can be seen. This noise can mix with a high level off frequency signal to produce an unwanted signal in the receiver passband, a process known as reciprocal mixing, see **Fig 6.24**. This is not a shortcoming of the mixer, the problem is the noisy oscillator, so today's design focus is for very low noise oscillators.

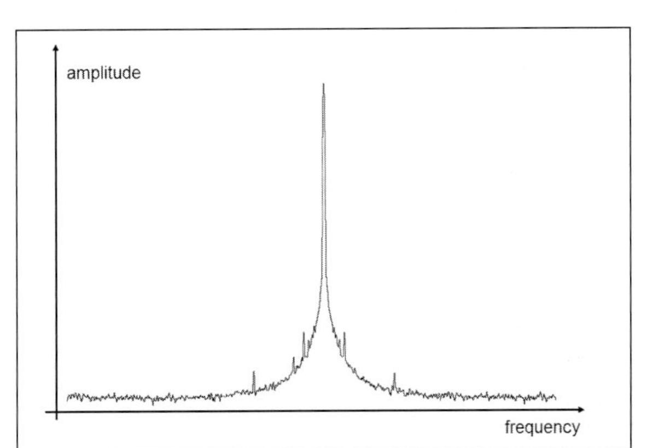


**Fig 6.23: Close-in oscillator noise can lead to unwanted signals in the receiver**

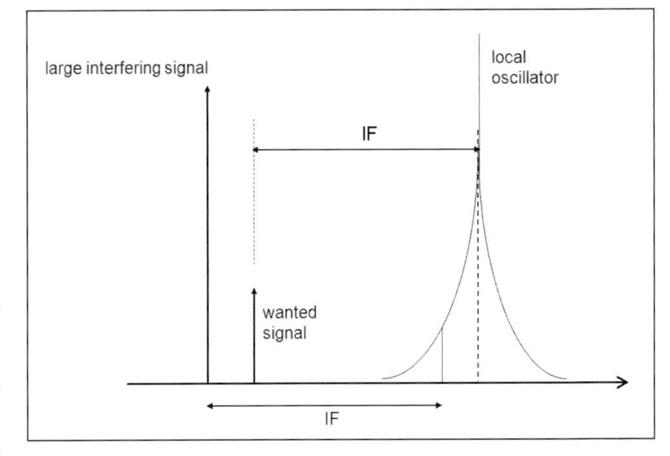


**Fig 6.24: Reciprocal mixing**

## Direct Conversion

One interesting derivative of the superhet is the synchrodyne, homodyne, zero IF or direct conversion receiver devised in the 1930s. The principle is that the incoming RF signal is mixed down to zero Hz using an LO on the same frequency as the RF signal. The amplification and selectivity is then achieved in the very high gain audio amplifier that follows the mixer, see **Fig 6.25**.

There are three weaknesses in the concept (a) the image, which is the other sideband and is not suppressed (b) the very high gain audio amplifier is difficult to keep stable and (c) its susceptibility to mains hum and other pickup. As is often the case that concepts developed years ago eventually becoming possible with modern circuit techniques and devices.

What made the direct conversion receiver popular a few years ago was the development of image cancelling mixers. The image cancelling mixer is derived from the phasing technique used to generate SSB. The same circuit generates two outputs known as I and Q which if added together produce one sideband and if subtracted produce the other sideband. Such

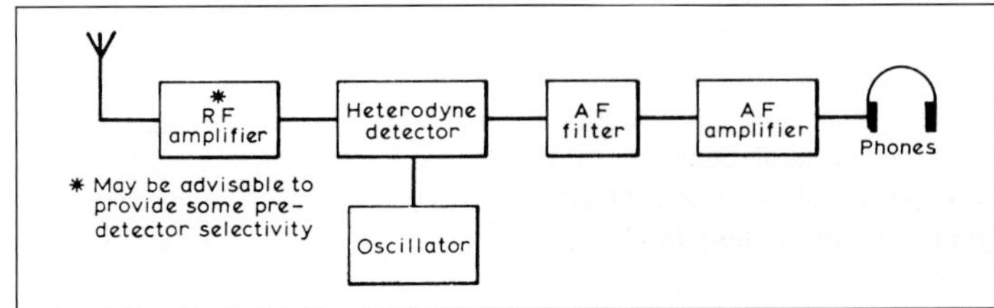


**Fig 6.25: Outline of a simple direct-conversion receiver in which high selectivity can be achieved by means of audio filters**

circuits require very careful adjustment, but can produce 50dB sideband suppression.

Direct conversion receivers are capable of reasonable results and are very suitable for home construction largely because not much test equipment is needed.

# DIGITAL TECHNIQUES

The availability of general-purpose, low-cost digital integrated circuit devices made a significant impact on the design of communication receivers although, until the later introduction of digital signal processing, their application was primarily for operator convenience and their use for stable, low-cost frequency synthesisers rather than their use in the signal path.

## Display

By incorporating a digital frequency counter or by operation directly from a frequency synthesiser, it is now normal practice to display the frequency to which the receiver (or transceiver) is tuned directly on matrices of light-emitting diodes or liquid crystal displays. This requires that the display is offset by the IF from the actual output of the frequency synthesiser or free-running local oscillator. Such displays have virtually replaced the use of calibrated tuning dials.

## Tuning

Frequency synthesisers are commonly 'tuned' by a rotary shaft-encoded switch which can have the 'feel' of mechanical capacitor tuning of a VFO, but this may be supplemented by push-buttons which enable the wanted frequency to be punched in.

Common practice is for the frequency change per knob revolution to be governed by the rate at which the knob is spun, to speed up large changes of frequency. With a synthesiser the frequency may change in steps as small as 100Hz, 10Hz or even 1Hz, and it is desirable that this should be free of clicks and should appear to be almost instantaneous.

Many of the factory-built equipments incorporate digital memory chips which can be programmed with frequencies to which it is desired to return to frequently. It should be noted that a phase-locked oscillator has an inherent jitter that appears as phase and amplitude noise that can give rise to reciprocal mixing and an apparent raising of the noise floor of the receiver. Digital direct frequency synthesis (DDS) can reduce phase noise and is being increasingly used. More on DDS can be found in the oscillators chapter in this book.

Digital techniques may be used to stabilise an existing free-running VFO by continuously 'sampling' the frequency over pre-determined timing periods and then applying a DC correction to a varactor forming part of the VFO tuned circuit. The timing periods can be derived from a stable crystal oscillator and the technique is capable of holding a reasonably good VFO to within a few hertz. Here again some care is needed to prevent the digital pulses, with harmonics extending into the VHF range, from affecting reception.

## User Interface

Microprocessor control of user interfaces can include driving the tuning display, memory management (including BFO and pass-band tuning), frequency and channel scanning; data bus to RS-232 conversion.

As stressed by Dr Ulrich Rohde, DJ2LR, several key points need to be observed:

1. Keypad and tuning-knob scanning must not generate any switching noises that can reach the signal path.
2. All possible combination of functions such as frequency steps, operating modes, BFO frequency offset and pass-band tuning should be freely and independently programmable and storable as one data string in memory.
3. It is desirable that multilevel menus should be provided for easy use and display of all functions. This includes not only modes (USB, LSB, CW etc) but also AGC attack and decay times. Such parameters should be freely accessible and independently selectable.

It should be understood that most of the above microprocessor functions are for user convenience and do not add to the basic performance (apart from frequency stability) of a receiver; and unless care is exercised may in practice degrade performance. **Fig 6.26** shows an example of the computer control architecture of a modern receiver.

## Software Defined Radio

In recent years, Software Defined Radio (SDR) has become established. Whilst it began by digitally processing the recovered audio using sound cards, it quickly evolved to replace the IF and demodulators as well.

**Fig 6.26** shows the structure of a SDR where the RF is converted down to the I and Q signals suitable for the analogue to digital converter (ADC), prior to it being dealt with in the digital domain.

A/D conversion is where the analogue signal is digitised thus enabling most of the receiver functions to be performed through software; the ADC is therefore a key component of the SDR.

There are two principle components of the ADC specification; the resolution of the conversion i.e. the number of bits the analogue signal is converted into, and the speed at which the conversion can take place, known as the sampling rate. These two components are mutually exclusive i.e. an ADC can have high resolution but sample at a slow rate, or vice-versa – not both.

Resolution is described by the device's number of bits where each bit is equivalent to 6dB of dynamic range, so a 24 bit ADC has an in-principle 144dB's of dynamic range. Assuming a 0 to 5V input range, this corresponds to a step size of 30uV which is within the noise floor of the device, so the actual dynamic range is reduced to 107dB.

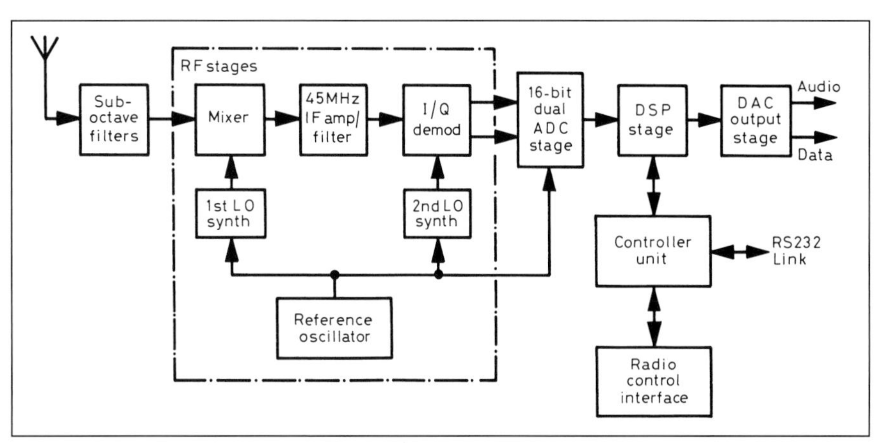


**Fig 6.26: Outline of prototype high-performance analogue / digital communications receiver with baseband digitisation following two-phase I/Q demodulation (Roke Manor Research Ltd)**

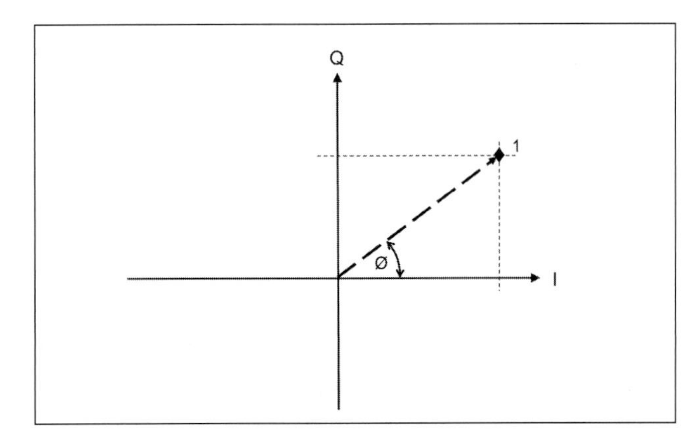


Fig 6.27: A single digital bit on the I and Q plane.

Sampling rate is the number of times per second the input waveform is captured and converted into digits, so a 24 bit ADC may have a sampling rate of 125K samples/second, delivered serially.

Nyquist sampling theory says that to A/D a waveform and then D/A it back again satisfactorily, the sampling rate must be at least twice the highest frequency being sampled. Anything less gives rise to what is called aliasing, the digital equivalent to the mixer image. To guard against aliasing, the bandwidth of the input signal must be restricted (by filtering) to less than half the sampling rate. So for a sampling rate of 125K samples/sec, the highest frequency that can be sampled is 62.5KHz. It follows then that the IF frequency (the I and Q channels) must be less that 62.5KHz. The HF7070 receiver described at the end of this chapter used a 24 bit ADC at an I and Q frequency of 44KHz and produces a dynamic range of 115dB.

But what is I and Q all about? Digital modulation techniques convey information through the phase of the carrier (Phase Shift Keying or PSK) and often the amplitude as well (Quadrature Amplitude Modulation or QAM). **Fig 6.27** shows a single digital bit (1) located on the I and Q plane. Its position can be defined by its vector length and phase angle or by it's I and Q values.

**Fig 6.28** shows an 8 bit PSK on the I and Q plane. [This type of diagram is known as a constellation.] All the vectors have the same length it's only the phase angle that changes with a 45° step size, hence 8 bits. An FM receiver is best suited to this type of encoding.

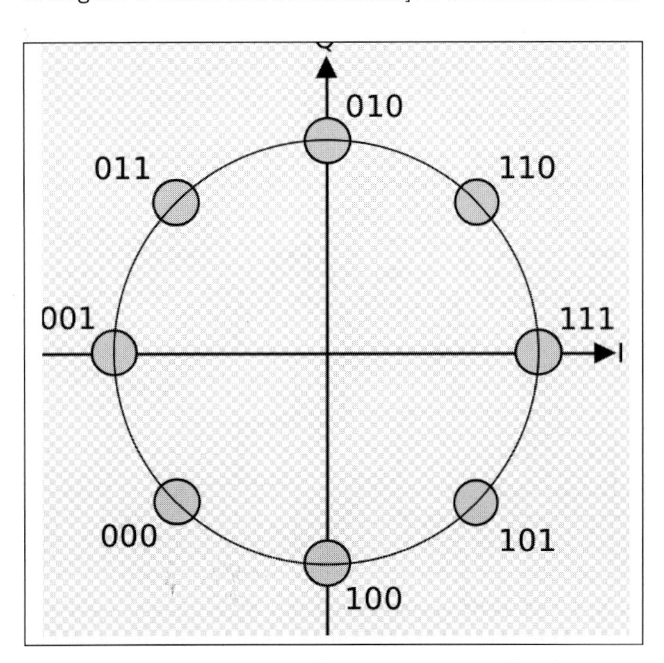


Fig 6.28: 8 bit PSK shown on the I & Q plane.

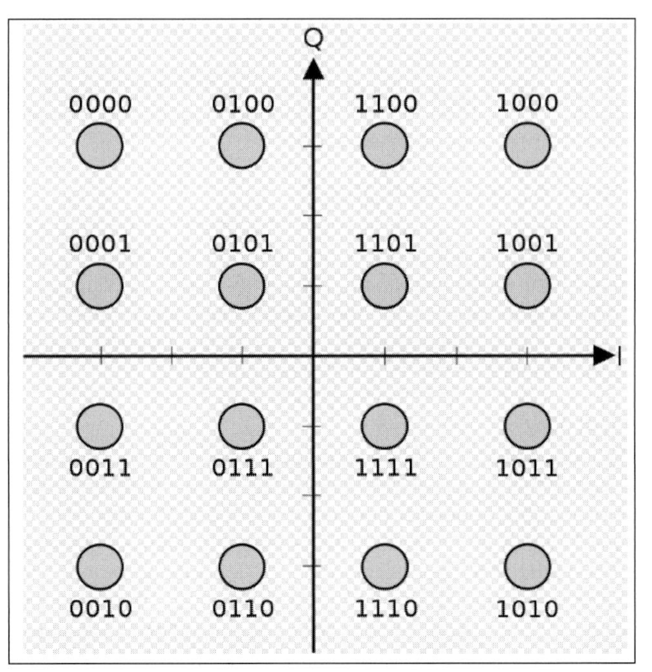


Fig 6.29: 16 bit QAM shown on the I & Q plane.

**Fig 6.29** shows a 16 bit QAM constellation where each bit is defined by its phase angle plus its vector length, so an AM/SSB receiver is required.

Incidentally, the diagrams show the ideal case where each bit has an exact location. As reception becomes noisy, the location of each bit starts to become fuzzy, but as long as there is clear space between each fuzzy bit location, the software can successfully decode the data. This is the digital equivalent to signal to noise ratio.

The I and Q approach enables the demodulation of digitally encoded signals and as a bonus, also enables the demodulation of analogue signals. I and Q followed by A to D plus appropriate software results in a universal demodulator.

## Direct digital synthesis

As part of the digital evolution PLL synthesisers have/are being replaced by direct digital synthesis (DDS) where the integrated circuit creates a sine wave using clocks and lookup tables under the control of a microprocessor, see **Fig 6.30**. The software can then command the production of any frequency with a very small step size and no settling time.

### Software controlled radio

The microprocessor used to control the LO, its spin-wheel or keyboard input and LCD display went on to control circuit blocks such as filters, demodulators and AGC systems. This management of the receiver is known as Software Controlled Radio or SCR.

Virtually all of a receiver's circuit functions can be described mathematically and assuming the input signal has been digitized, these algorithms can be carried out inside a microprocessor. A powerful microprocessor is required to carry out all the multiple near-simultaneous operations necessary to become the receiver, so to satisfy these kinds of applications an additional specialist device appeared, known as a Digital Signal Processor; it and digital signal processing has become known as DSP.

Sound studios and audio processing were amongst the early adopters of DSP, hence the availability of sound cards for PCs. For

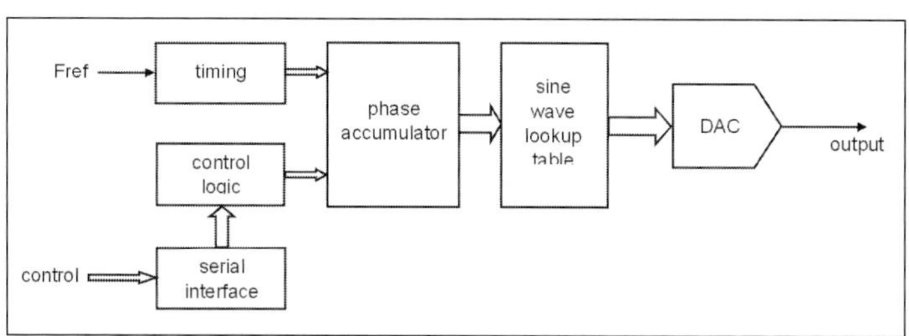


**Fig 6.30: Block diagram of a DDS oscillator**

SDR receivers, the DSP has been replaced by the much more powerful Field Programmable Gate Array (FPGA) that can process several hundred simultaneous arithmetic activities.

## Practical SDRs

The difference between Software Controlled Radio and what is now called Software Defined Radio, or SDR, was the adoption of digital signal processing. Previous analogue functions like filtering can now be achieved as a mathematical process producing spectacular results, because the process is lossfree. The bandscope is another notable success of SDR. A number of commercial designs are now available for the amateur market and have become extremely popular, such as the Flex Radio SDR1000.

In many cases, these designs use a Direct Digital Synthesis (DDS) local oscillator and a mixer direct to baseband (zero IF), creating separate 90° phased I and Q signals. The computing power is provided by the operator's desktop PC, which accepts the I and Q signals from the analogue front end and demodulates them into single sideband audio. Advanced on-screen user interfaces provide the operator with numerous features such as panoramic display of a large segment of band, and incredible control over the filtering and characteristics of the receiver.

The ultimate goal of SDR is the direct digitisation of a wide band of frequencies, such as the HF Spectrum from 0 - 30MHz, direct from the antenna, with the entire receiver implemented digitally and defined/controlled by software. The only analogue stage in such a system would be a low-pass filter to prevent deterioration of ADC performance due to interfering signals higher than 30MHz. Until recently, the dynamic range available from ADC devices and the computing power requirements were not able to meet the demands of a true all-digital SDR.

Almost all SDRs have an RF front end that that filters and then mixes the RF down to a very low IF, typically 40kHz, or zero IF as in the direct conversion receiver. This low IF or baseband signal is then digitized using a high bit A/D converter that drives the FPGA. The overall system is controlled by either a microcontroller or a PC. See **Fig 6.31a.**

However, several all-digital receivers are now available which boast high performance over the entire HF range. Whilst it may be argued that the highest performance in terms of dynamic range, sensitivity and immunity from cross-modulation is still the domain of analogue receivers, or at least analogue front ends, there is no doubt that continued advances will soon allow all-digital receivers to overtake their analogue counterparts.

The range of SDRs available today all have the same building blocks, what makes them different is the way the blocks are partitioned. Put everything in one box and you have the SDR receiver such as the HF7070, see **Fig 6.31b**. Put the front end, the A/D and the FPGA into one box and plug it into a PC and you get the Flexradio, Winradio, Perseus or Funcube, for instance, see **Fig 6.31c**. Alternatively, Softrock connects its front end into the PC through the soundcard, see **Fig 6.31d**.

Whilst SDR has provided a wealth of new functions, in most cases the basic performance of the receiver remains dependent on the analogue front end. Even state of the art direct digitization of the incoming signal is reliant on the same silicon technology that today's H-mode mixers are built from. Reference to the Sherwood Engineering receiver measurement comparison site [2] and the transceiver reviews carried out by Peter Hart in [3] show that SDR radio offers high performance and has now become the established way of building receivers and transceivers.

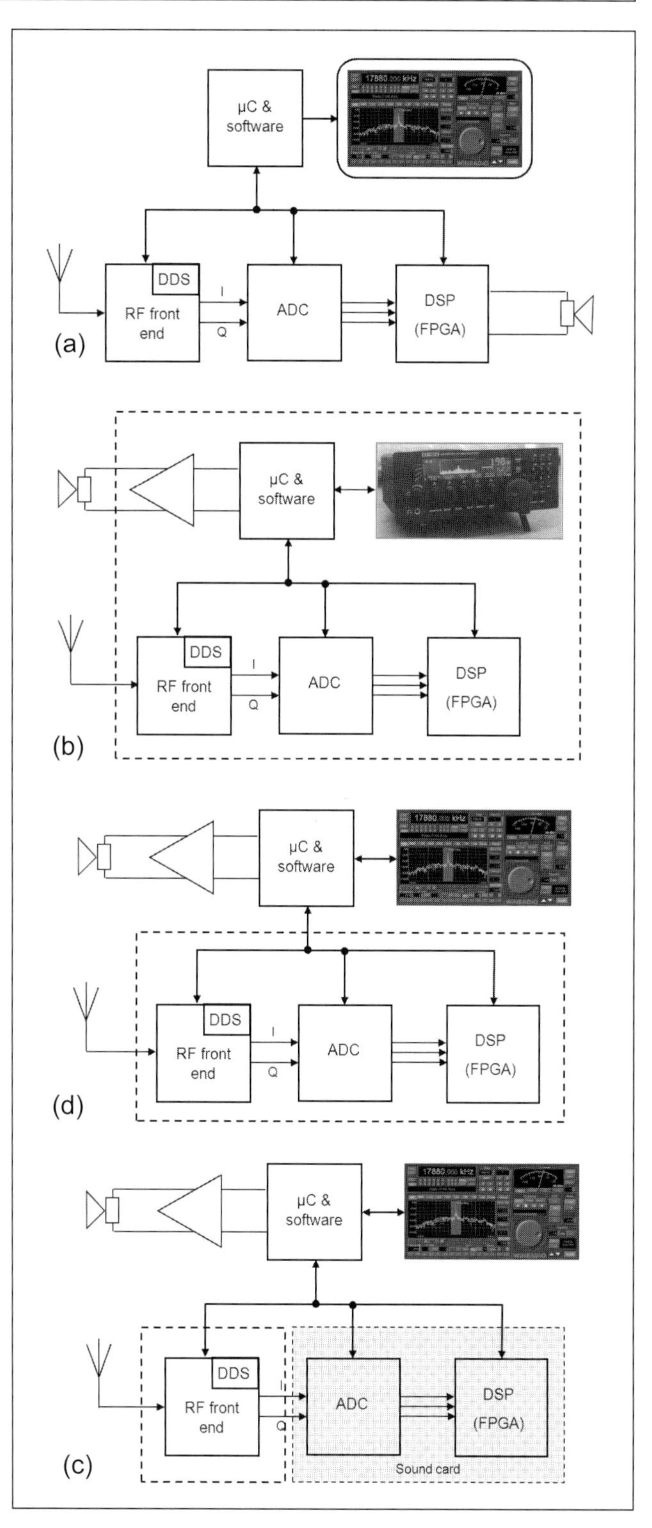


**Fig 6.31: Four SDR variants: (a) Basic setup, (b) Self-contained SDR based transceiver, (c) an SDR receiver such as the WinRadio or Perseus and (d) the simplest SDR, eg a Softrock**

# PERFORMANCE

## Specifying Performance

Few international standards specify the performance of amateur radio receivers. Standards makers are only concerned with the interference that such equipment might cause to other co-located equipment and there is no interest in how well the receiver performs its primary function. Therefore manufacturers are largely free to set their own specifications, which include the cost. We also have the customers of such equipment who at some point have to make a choice, probably based on cost, recommendations, reviews and advertising. Finally there is the designer/constructor who strives for high performance at any cost or the best performance for a minimum cost or technical novelty or any number of personal goals.

Often the performance is a function of the architecture or devices chosen at the beginning and the final specification emerging at the end of the project, and this approach is entirely in the spirit of amateur radio. However, when considering major projects there is the other way where the specification comes first and the design evolves to meet it. The challenge for the designer/constructor is to set a specification that is meaningful, affordable and achievable.

The basic job of a radio is to receive and demodulate a signal at a prescribed frequency, so the top level performance parameters are sensitivity and frequency. All the other receiver performance parameters are related to how well it doesn't do things, usually referred to as rejection ratios eg adjacent channel rejection and image rejection. Consequently, with the exception of sensitivity and frequency, a list of such parameter measurements will each be several tens of decibels. **Table 6.3** shows a list of receiver parameters and their definitions.

If you measured the performance of a cheap short wave radio, a lot of its parameters would be at the 40dB level, whilst measuring a professional receiver would probably produce results in the 100dB range. The cheap short wave radio is unlikely to have any feature that achieves 100dB similarly the professional receiver would not be spoilt with a 40dB feature. So to an approximation you can define the performance of a receiver by the number of dBs delivered by most of its parameters: the cheap short wave receiver is a '40dB radio' whilst the professional receiver is a '100dB radio'. For the designer/constructor, deciding to build a 70dB radio or a 90dB radio is all that is necessary because the rest ie performance, specifications, complexity and cost are inherent in that decision. Similarly, commercial receivers could be classified in the same way so as to give the prospective buyer a better way of choosing.

| PARAMETER | PROPOSED DEFINITION |
|---|---|
| **Reference sensitivity** | The input level (dBm) required to produce a 10dB S/N output |
| **Reference output** | 100mW into 8 ohms |
| **AGC range** | The change in input (dB) above the reference sensitivity that produces a change of 3dB in the reference output |
| **Spurious free dynamic range** | Derived from the IIP3 minus the minimum discernible input signal range |
| **Blocking** | The number of decibels above the reference sensitivity an unwanted off channel (by 20kHz) CW signal must be to cause a 3dB drop in a wanted reference output |
| **Cross-modulation** | The number of decibels above the reference sensitivity that an unwanted off channel signal* must be to cause a 3dB change in the wanted reference output. [* the unwanted signal is 20kHz off-channel and 50% amplitude modulated] |
| **Image rejection** | The input level (dB) above the reference sensitivity level that an input at the image frequency(s) must be to give the reference output level |
| **IF rejection** | The input level (dB) above the reference sensitivity level that an input at the IF must be to give the reference output level |
| **Local oscillator leakage** | Conducted power measurement at the LO frequency(s) coming out of the antenna socket |
| **Adjacent channel selectivity** | The input level (dB) above the reference sensitivity that an adjacent channel signal must be to produce the reference output |
| **Internally generated spurious signals** | The input level (dBm) for the equivalent to the highest internally generated spurious signal |

**Table 6.3: Receiver parameter definitions**

## Measuring Performance

Standards for measuring the performance of radio equipment are usually simple in concept although normally written in standards language. So, for example, measuring the sensitivity of a receiver is in essence connect up a signal generator, tune the receiver into the signal and slowly reduce the signal level until you can only just hear it. See **Fig. 6.32**. At this point you have measured the receiver sensitivity. Standards will define 'until you can only just hear it' in terms of Signal to Noise (S/N) ratio in decibels, whereas the amateur can decide that point for himself by listening.

The complications come with the modulation type and the receiver bandwidth. If the operator prefers CW, the receiver bandwidth is reduced to a few hundred Hz, whereas an SSB operator would use a receiver bandwidth of say 3kHz. As the saying goes, "the wider you open the window, the more dirt flies in", which is why the narrower CW bandwidth gives greater receiver sensitivity. Again the measurement regime can be decided by the amateur.

Selectivity and shape factor can be measured using the same test setup used for sensitivity. In this case, the receiver is tuned into the CW signal generator whose level is adjusted to give an S2 (-115dBm) reading on the receiver S-meter. The CW signal is then moved progressively to either side of the IF centre frequency and its level re-adjusted upwards to maintain the S2 reading. When the CW signal has gone up to -55dBm, this marks the -60dB bandwidth.

The measurements can be continued to say ±100kHz either side of the IF centre frequency to give a complete picture of the

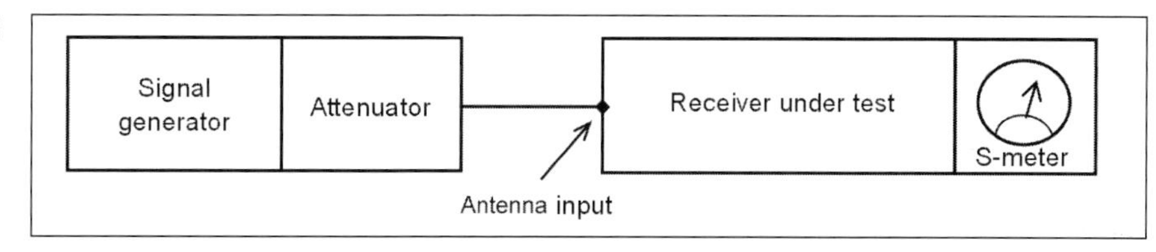


**Fig 6.32: Basic method of measuring the sensitivity of a receiver**

selectivity that includes the effects of LO noise (reciprocal mixing). It will also capture any throw-ups that some filters exhibit.

The S2 signal level can be considered as the reference receiver sensitivity against which all its other performance parameters can be judged.

So for example, if the CW signal generator is moved to the image frequency and its level re-adjusted to give an S2 reading, then the difference between the CW signal generator level at the wanted frequency that gives the reference level (S2 or -115dBm) and its level that gives S2 at the image frequency is the image rejection. In a receiver with more than one IF frequency there will be more than one image rejection figure, but they will all be measured using the same method.

By moving the CW signal generator to say ±20kHz away from the wanted frequency, a measure of adjacent channel selectivity can be achieved using the same technique.

IF rejection is the ability of the receiver to reject a signal transmitted at the receiver's IF frequency. If there is such a signal, which is more likely in multi-conversion receivers with several IFs, then that signal will be heard irrespective of band-switching or tuning.

The measurement is made by setting the CW signal generator to the IF frequency and adjusting its output level until the reference input level (S2 or -115dBm) is measured on the S-meter. The ratio of the CW signal generator output level at the IF frequency to its output level at the wanted signal frequency, both giving the same S2 signal level gives the IF rejection.

AGC range is measured by plotting the AM modulated signal generator output level against the receiver audio output signal. The AGC should hold the audio output level constant over a wide range of RF input levels. So starting at an input level corresponding to S2 and noting the audio level, slowly increase the RF input level until the audio level goes up by 50%, ie 3dB. The difference between the two RF input levels is the AGC range.

Blocking, crossmodulation and intermodulation all require two signal generators.

For blocking, the wanted signal is set to give the reference sensitivity (S2, -115dBm) and the unwanted signal is set 20kHz away and its level adjusted so that the signal level of the wanted signal drops half an S point (3dB). The ratio of the two signal generator output levels is the blocking.

Crossmodulation is the same measurement setup except that the unwanted signal is now modulated. The objective is to determine the level of the unwanted modulated signal necessary to cause the apparent modulation of the wanted signal as heard in the receiver output. This will be an AM measurement as few, if any, signal generators can generate SSB.

Intermodulation is measured by setting the two CW signal generators, one offset by 20kHz and the other by a further 20kHz from the wanted frequency. In this case there is no wanted signal as it is the two unwanted signals that will generate a false wanted signal in the front end of the receiver. So the two unwanted signals are kept at the same level which is then increased until a wanted signal is just detected, say S2.

## MODIFYING AN EXISTING RECEIVER

There are several reasons for modifying an existing receiver:-

- A repair is required and one or some of the original components are no longer available, and the damaged part of the circuit has to be replaced with an equivalent or improved circuit
- One or more shortcomings have been identified whilst using the receiver and these can be overcome by modifying the circuit
- Improvements have been posted on the internet
- Adding functionality, such as providing a connection to a computer

Before carrying out any changes to the receiver there are a few notes of caution. Firstly, assuming mains input, the receiver's power supply is dangerous. Secondly, if the receiver is valve based, it contains high voltages that can hurt. Thirdly, if the receiver is relatively new there may be guarantee implications to be considered. Upgrades, modifications and repairs are not for the newcomer.

The very first thing is to obtain and study the circuit details and diagram of the receiver. This information can usually be found on the internet, as can any previous attempt at similar modifications. If this is the case then the modification can proceed following the advice or instructions discovered.

Overcoming shortcomings to give a greater user experience depends of course on the shortcoming identified. The following suggestions are a good starting point and several of the helpful circuits are shown later in this chapter.

### Improving Frequency Setting

This might be the requirement where the frequency is read from a mechanical dial. It can be achieved by tapping into the local oscillator (LO) and connecting up a frequency counter. The tap should be a high impedance connection so as not to disturb the amplitude or frequency of the LO. Since the measurement is of the LO, the actual signal frequency is calculated by adding or subtracting the known IF frequency.

### Improving Tuning Stability

Drift can be cured by adding a huff and puff stabilising circuit. Again, a high impedance tap is required to feed a sample of the LO into the huff and puff stabiliser [4].

The stabiliser locks the LO to the nearest of many closely spaced steps typically 30Hz apart. The circuit creates a slowly varying DC voltage in proportion to the frequency error which then controls the capacitance of a tuning diode placed across the main LO tuning capacitor (which may also be a tuning diode). This change in overall tuning capacitance causes the LO frequency to drift back to where it was thus achieving a much improved long term stability.

Since we need only a very small change in capacitance then one of the VHF tuning diodes is ideal because it will not cause much change in LO frequency just by being there and only a tiny change in overall tuning capacitance is required.

### Improving Sensitivity

This rarely needs improvement. If a receiver lacks sensitivity, it should first be investigated for a fault or the antenna needs attention.

A receiver has sensitivity at the small signal end and an overload point at the high signal end. The difference between the two is the spurious free dynamic range, the SFDR, and the bigger the better. So if the SFDR is, say, 80dB and the corresponding sensitivity is -110dBm then the maximum signal level for the two tone test signal not to cause interference is -30dBm. If the sensitivity is improved to -113dBm, a change of 3dB, the corresponding maximum two tone test signal not to cause interference drops to -39dBm and the resulting SFDR falls to 74dB. So in this case an improvement to the sensitivity of 3dB reduces the large signal capability by 6dB. It is obvious that caution is required before improving a receiver's sensitivity, especially when bearing in mind the limiting factor of background noise (see **Fig 6.2**).

## Improving Selectivity

A receiver's selectivity is normally achieved in the IF amplifier. Older receivers had their final IF in the range 455kHz to 470kHz. Even older receivers achieved their selectivity through a series of double tuned IF transformers, whilst later receivers normally used a single ceramic block filter.

Ultimately, the selectivity is set by the noise performance of the LO through a process known as reciprocal mixing (see above), so by the time the IF filter skirts are about 80dB down, it is the LO noise that starts to dominate.

Ceramic block filters made by Murata can still be found through the internet; they come centred on 455kHz in various bandwidths and ultimate attenuations. A good approach is to fit the additional ceramic block filter on the output of the IF amplifier; several filters with different bandwidths could be switched in.

Care must be taken with impedance matching. Failure to correctly match a ceramic filter will result in disappointing results as the expected shape factor and/or attenuation will not be achieved. Matching will require the laborious use of a signal generator and oscilloscope, or better test equipment.

If using ceramic filters in a valve based receiver, note that they will not tolerate large voltages across them.

If the receiver uses double tuned IF transformers, another technique worth investigating is the Q-multiplier. This adds a limited amount of positive feedback to one of the IF amplifier stages thus increasing the apparent Q and hence peaking the bandwidth of the IF transformer. A video on fitting a Q-multiplier to a WW2 receiver can be found at [5].

## Improving Image Response

Suppose the receiver has an IF of 455kHz, the LO is on the low side and it is tuned to a station transmitting on 7.10MHz. The LO will be tuned to 6.945MHz causing an image susceptibility at 6.19MHz, which is in the 49m broadcast band. With inadequate image rejection, the broadcast station on 6.19MHz will be heard on top of the amateur transmission being listened to on 7.1MHz.

Improvement to image performance has to be carried out prior to the mixer. In a single conversion superhet, there is usually an RF amplifier with tuned circuits on its input and output. It is this selectivity that controls the image response. Consequently, it is difficult to improve this selectivity; however an external pre-selector is an excellent alternative, particularly on the lower bands.

## Reducing Overload

Symptoms of receiver overload come in four flavours:-

- In the absence of other signals, the wanted signal is just too large for the receiver and the recovered audio ends up distorted. Since most of the HF band is crowded with other signals, this effect is not that common.
- Blocking - this effect is difficult to identify; you just think the wanted signal is weak, not weaker than it should be.
- Crossmodulation - when the modulation of a strong unwanted signal is transferred to a weak wanted signal.
- Intermodulation - when two large unwanted signals mix to produce a range of additional smaller unwanted signals, one of which may be at the receiver frequency.

All these effects come from the non-linear parts of the receiver, primarily the first mixer.

The simple answer to overload effects is to attenuate all the incoming signals together, so that although the wanted signal drops a bit, the unwanted effects drop a lot more and the intelligibility of the wanted signal improves. Many receivers come with front end signal attenuators that can be switched in. Equally, many transceivers come with additional built-in preamps that can be switched out under overload conditions.

## Improving Audio Quality

The audio quality of a receiver is often overlooked by the designers and can benefit from some improvements.

If the loudspeaker is mounted on the metal cabinet or front panel then using an external loudspeaker in a suitable enclosure will produce immediate results. Replacing the audio output stage with a modern integrated audio power amplifier will reduce the distortion and probably the background noise as well.

Finally, additional audio filtering can be inserted prior to the audio output stage to reduce the out of band noise, with one bandwidth for voice and a narrower bandwidth for CW.

## Other Improvements

Repairing an existing receiver starts with the circuit diagram and any other information available, the internet being an excellent source.

As a general rule, invest in a can of switch cleaner and expect to replace dried out electrolytic capacitors and scratchy potentiometers, particularly those with switches.

Beware of positive earth receivers based on PNP transistors and take obvious care with power supply polarity! Also cross-connecting test equipment, power supplies and positive earth radios can easily create short circuits across the external power supply.

# BUILDING YOUR OWN RECEIVER

There are many published designs for partial or complete receivers and it becomes an issue of personal choice.

The governing factors behind the choice are the resources available in terms of knowledge, cost, time, mechanical capability and test equipment as these will set the complexity and hence performance of the finished project. Cost and time are obvious; mechanical capability requires the ability to cut, drill, form and fasten materials and possibly the production of printed circuit boards. Some test equipment is required for almost anything to do with radio/electronics; a multimeter, an oscilloscope, a signal generator and a power supply are the bare minimum. It is these limitations, particularly knowledge and test equipment that govern the performance of the finished receiver.

Knowledge, experience and confidence are all related and in terms of home construction are somewhat interchangeable; the important thing is to have a go. Try to avoid getting too complex too quickly as finding out why something doesn't work escalates with complexity. Break the project down into small circuit blocks, build them and test them progressively before adding more.

**Fig 6.33: The chapter author's shack showing his range of test equipment**

Start with the audio output stage and work towards the antenna so that you can hear progress as you add more and more. This approach will give a much greater insight into how a circuit works and what it is capable of. In the end you will have a stock of circuit blocks that you have confidence in and can incorporate into other projects.

Test equipment is expensive but there is a great deal of second-hand equipment available through the clubs, magazines and internet auction sites. Alternatively there are still dealers that have a good range of equipment [6]. As with all second-hand goods, guarantees are rare and usually associated with the more expensive end of the market. **Fig 6.33** shows the author's home facility capable of HF receiver designs up to 95dB. There is much more on this subject in the chapter on Test Equipment.

Variable Ls & Cs underpin almost all radios. Many years ago when radio was king you could readily buy coil formers with slugs and cans to wind your own coils and variable capacitors; sadly no longer.

A limited range of ready wound 10mm variable inductors made by TOKO can be found at [7] and on *eBay*, as can some coil formers. However without the means to measure inductance, the coil formers are of limited use. Today's inductor technology is based on the toroid (also available from [7]) which provides a predictable, but fixed result. Whilst it is possible to make minor adjustments to a toroid based inductor by bunching or stretching the turns, the adjustment range is very limited.

Toroids come in three flavours; the T range for tuned circuits, the FT range for wideband transformers and the BN range which is a binocular version of the FT range. Within each range there are different ferrite mixes for different frequency ranges and different sizes for different powers. In the case of receivers, sizes relate to IP3 and to winding length; the bigger the better.

Design guides are available on the internet (a good one being [8]) which translate inductance into turns, thereby reducing the dependency on measuring equipment.

Variable capacitors are hard to find, other than the Jackson range available from several sources eg [9]. Many years ago, the tuning capacitor found in receivers was replaced by the varicap diode, a useful component inside a feedback loop, but with an inherently poor temperature coefficient when not.

Air-spaced single and multi-ganged tuning capacitors can be found at rallies, on *eBay* and at surplus stores such as [10]. With the supply situation being as it is, the user must expect to change the overall value of the tuning capacitor using series and occasionally padding capacitors.

The ability to measure capacitance is very helpful and most DVMs have this facility. For more on measuring inductors and capacitors, see the chapter on Test Equipment.

## Crystal Set

This is the simplest of all receivers, it can receive AM only and is powered by energy gathered by the aerial, see Fig 6.7. It is of very limited use in amateur radio because of its poor selectivity, low gain and inability to demodulate CW or SSB. However there is an extreme crystal set enthusiast community. The website at [11] is very enlightening and informative.

## Regenerative Receiver

These excellent value-for-money receivers can demodulate CW and SSB, they have good gain and selectivity, the downside is they are tricky to adjust. The principle is that the incoming signal is amplified at RF by the transistor and a portion of the output is then fed back into the input. This positive feedback creates high gain and high Q, ie good selectivity. As the output amplitude of the signal rises, the transistor begins to act as a detector and so demodulated audio is obtained.

The tricky adjustment is in maintaining the level of feedback so that the circuit is just short of oscillation, thereby giving maximum gain and selectivity. Since the transistor is effectively connected directly to the aerial, bad adjustment that allows oscillation can create a transmitter. It is therefore essential to add an RF amplifier between the aerial and the regenerative device to isolate the potential oscillator from the aerial, and at the same time, beneficial to add an audio amplifier.

A good source of material on the subject of building regenerative receivers can be found at [12].

## Direct Conversion Receiver

Direct conversion receivers require little test equipment, but they do need a lot of constructional skill for optimum performance. Because the receiver is based on a very high gain audio amplifier circuit, layout, screening, decoupling and freedom from mains hum pickup are all crucial elements of the construction. One other effect heard is the twang sound associated with mechanically disturbing the LO, correctly called microphony; any shock or shaking of the LO (usually the inductor) causes a momentary shift in frequency which beats with the incoming signal producing a short lived non-zero IF, hence the twang noise.

The fundamental issue with the direct conversion receiver is its inability to discriminate between the upper and lower sidebands because both sidebands are mixed down to zero IF. The unwanted sideband is at the image frequency in direct conversion receivers, hence the interest in image rejection mixers. There are three methods of suppressing the unwanted sideband, two derived from SSB generating techniques; see the chapter on HF Transmitters.

The first is the phasing technique that uses two balanced modulators and two wideband quadrature generators (**Fig 6.34**). Sideband suppression is directly related to the amplitude and phase balance of the two wideband quadrature generators, and the wider the band, the greater the difficulty in maintaining the 90° phase shift and balanced amplitude. With the right circuits and through careful adjustment 40 to 50dB sideband suppression can be achieved.

The second technique is known as the Weaver or 3rd method and uses four balanced modulators, one narrow band and one

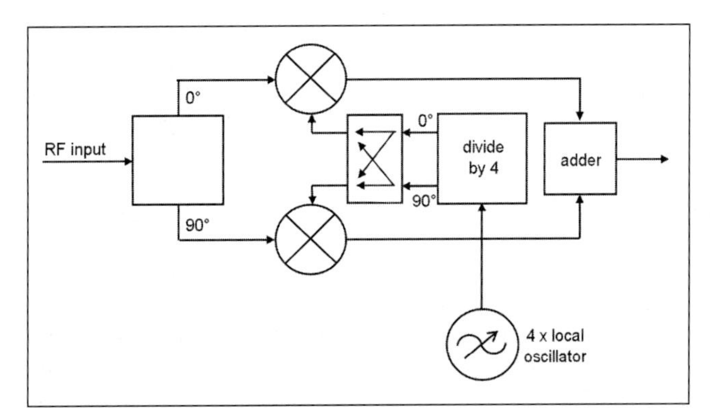


**Fig 6.34: Phasing mixer**

wideband quadrature generator and two low pass filters (**Fig 6.35**). The circuit is not particularly intuitive but it has two advantages over the phasing method. Firstly, only one wideband quadrature generator is required - the LO, and secondly, the sideband suppression is not dependant on the phase or amplitude balance of the quadrature generators.

The third technique, known as the Tayloe product detector, translates the incoming RF signal into four phases of baseband signal using an LO at four times the RF frequency. The four output phases can then be combined in a polyphase network to select the wanted sideband. **Fig 6.36** shows a practical Tayloe detector that feeds the polyphase circuit shown in **Fig 6.37**, whilst **Fig 6.38** shows the theoretical sideband suppression obtained. The Tayloe detector is a remarkable circuit giving low noise, low conversion loss and high IP3. The downsides are an LO at four times the RF frequency is required as is the supporting polyphase network.

## Superhet Receiver

Fewer people build superhets these days. One of the reasons is that as the complexity of the radio increases, so does the requirement for test equipment. A lot can be achieved with an RF signal generator, an oscilloscope and a frequency counter. But for superior results a spectrum analyser to see filter responses and measure gains, and a Q meter to measure inductors add a great deal. All this equipment is available second hand.

The excellent series of pragmatic articles written by Eamon Skelton for *RadCom* include all sorts of receiver circuit blocks. Of particular interest is the receiver part of his HF transceiver project published throughout 2011 which is reproduced in this book.

## Software Defined Receiver

It could be assumed that those interested in SDR homebrewing are interested in the software rather than the hardware. The hardware is just a platform for the software and as such becomes something that is bought in ready made. Also considerable software knowledge required to program and run applications on FPGAs.

This high entry level knowledge limits the scope for DiY SDR development. However there are open source code suppliers such as GNU Radio [13] that allow the user to incorporate their own programs into various pieces of hardware, including Perseus and Funcube.

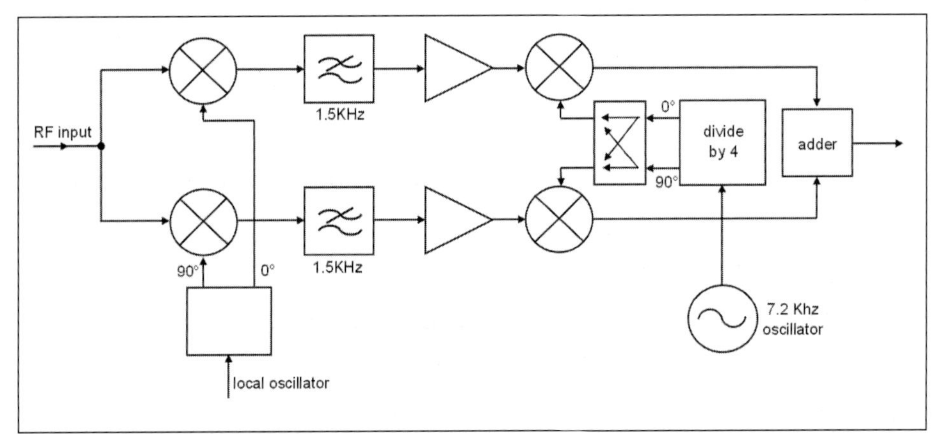


**Fig 6.35: The Weaver or third method**

# SOME USEFUL CIRCUITS

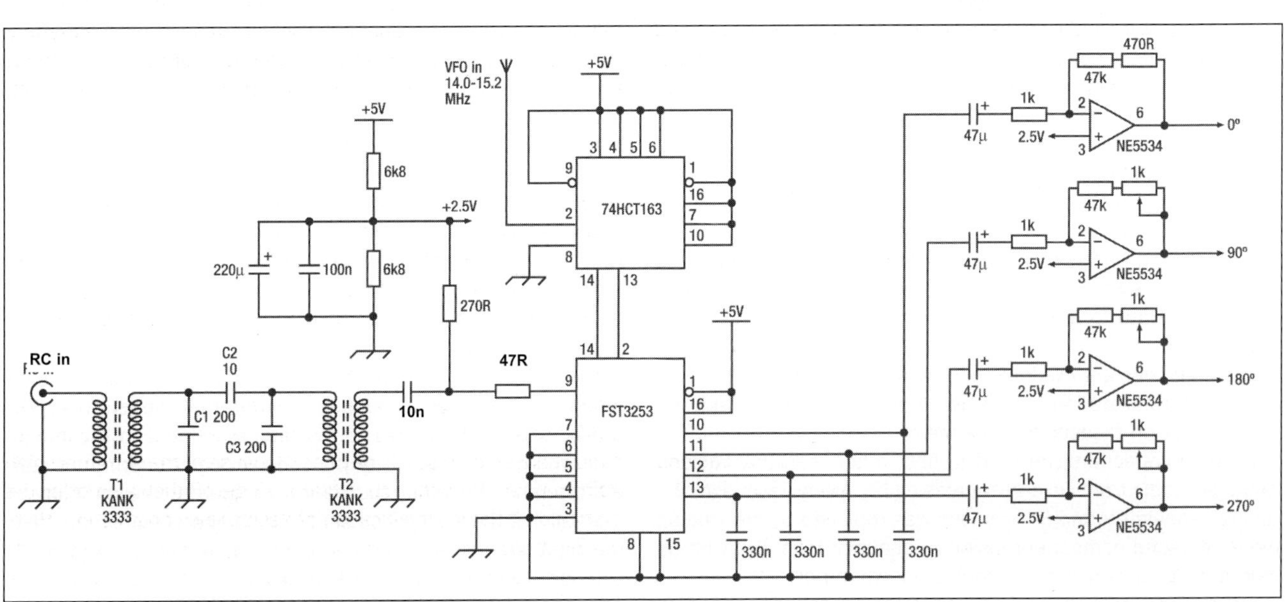


**Fig 6.36: Front-end: RF filter and Tayloe Detector. Op-amps pin 4 is grounded and pin 7 is +12V**

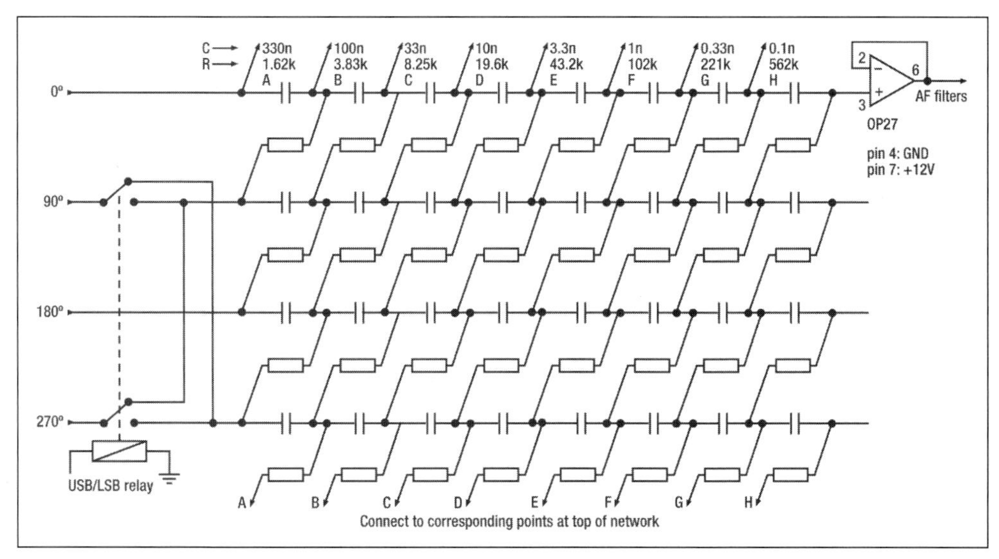


Fig 6.37: Polyphase network. Resistors and capacitors should be high tolerance types

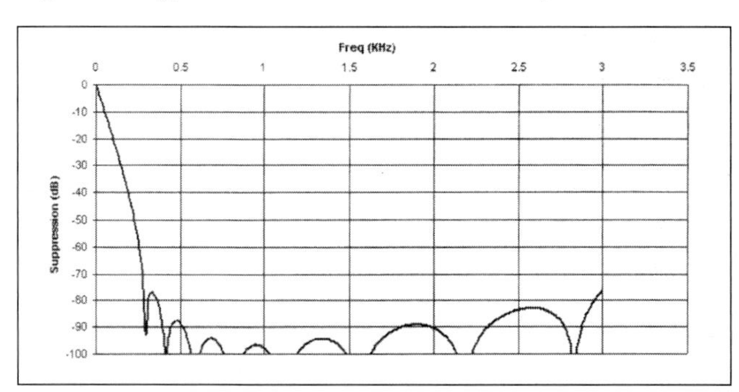


Fig 6.38: Theoretical opposite sideband suppression (dB) versus frequency (kHz)

Radio amateurs today are faced with a wide and sometimes puzzling choice of active devices around which to design a receiver: valves, bipolar transistors, field-effect transistors including single and dual-gate mosfets and junction FETs, special diodes such as Schottky (hot-carrier) diodes, and an increasing number of integrated circuits, many designed specifically for receiver applications.

Each possesses advantages and disadvantages when applied to high-performance receivers, and most recent designs tend to draw freely from among these different devices. Characteristics of these are compared below.

## Active Devices

### Valves

In general, valves are bulky, require additional wiring and power supplies for heaters, generate heat, and are subject to ageing in the form of a gradual change of characteristics throughout their useful life. However, they are not easily damaged by high-voltage transients; were manufactured in a wide variety of types for specific purposes to fairly close tolerances; and are capable of handling large signals with good linearity. Mixers, however, are noisy. The variable-mu devices, eg the EF183, have excellent AGC characteristics.

### Bipolar transistors

These devices can provide very good noise performance with high gain and are simple to wire. They need only low-voltage supplies, consuming very little power and so generating very little heat (except power types needed to form the audio output stage). On the other hand, they are low-impedance, current-operated devices, making the interstage matching more critical and tending to impose increased loading on the tuned circuits; they have feedback capacitances that may require neutralising; they are sensitive to heat, changing characteristics with changing temperature; and they can be damaged by large input voltages or transients. Their main drawback in the signal path of a receiver is the difficulty of achieving wide dynamic range and satisfactory AGC characteristics. However, bipolar transistors are suitable for most AF applications.

The bipolar transistors developed for CATV (wire distribution of TV) such as the BFW17, BFW17A, 2N5109 etc used with a heatsink, can form excellent RF stages or mixers, especially where feedback is used to enhance linearity. Not all transistors are equal in linearity, and those developed for CATV show advantages.

### Field-effect transistors

FETs can offer significant advantages over bipolar devices for the low-level signal path. Their high input impedance makes accurate matching less important; their near square-law characteristics make them comparable with variable-mu valves in reducing susceptibility to cross-modulation; at the same time, second-order intermodulation and responses to signals 'half the IF' away necessitate careful consideration of front-end filtering.

They can readily be controlled by AGC systems, although attempts to obtain too much gain reduction can lead to severe distortion. The dual-gate form of device is particularly useful for small-signal applications, and forms an important device for modern receivers. They tend, however, to be limited in signal-handling capabilities. Special types of high-current field-effect transistors have been developed capable of providing extremely wide dynamic range (up to 140dB) in the front-ends of receivers.

A problem with FETs is the wide spread of characteristics between different devices bearing the same type number, and this may make individual adjustment of the bias levels of FET stages desirable. The good signal-handling capabilities of mosfet mixers can be lost by incorrect signal or local oscillator levels. mosfets must be treated as fragile devices because their very high input resistance means they are subject to damage from static electricity that can be developed just through handling.

### Integrated Circuits

Special-purpose ICs use large numbers of bipolar transistors in configurations designed to overcome many of the problems of circuits based on discrete devices. Because of the extremely high gain that can be achieved within a single IC they also offer the home-constructor simplification of design and construction. High-performance receivers can be designed around a few special-purpose linear ICs, or one or two consumer-type ICs may alternatively form the 'heart' of a useful communications receiver.

It should be recognised, however, that their RF signal-handling capabilities are less than can be achieved with special-purpose discrete devices and their temperature sensitivity and heat generation (due to the large number of active devices in close proximity) usually make them unsuitable for oscillator applications. They also have a spread of characteristics which may make it desirable to select devices from a batch for critical applications.

The electronics industry has tended towards digital techniques where possible, and generally to application specific integrated circuits (ASICs) for both analogue and digital applications. In general, devices such as operational amplifiers and some audio amplifiers will continue to be available, but many of the small-scale integration (SSI) devices that were the mainstay of the amateur throughout the 1970s and 1980s have now disappeared. Of the circuits manufactured for radio applications, many are aimed at markets such as cellular radio, and have become so dedicated that their use (and availability in small quantities) for amateur radio is doubtful.

#### Integrated-circuit precautions

As with all semiconductor devices it is necessary to take precautions with integrated circuits, although if handled correctly high reliability may be expected. Recommended precautions include:

- Do not use excessive soldering heat. Ensure that the tip is at earth potential.
- Take precautions against static discharge - eg don't walk across a synthetic fibre carpet on a cold day and then touch a device. It is advisable if handling CMOS devices to use a static bleed wriststrap, connected to earth.
- Check and re-check all connections several times before applying any voltages.
- Keep integrated circuits away from strong RF fields.
- Keep supply voltages within ±10% of those specified for the device from well-smoothed supplies.

Integrated circuit amplifiers can provide very high gains (eg up to 80dB or so) within a single device having input and output leads separated by only a small distance: this means that careful layout is needed to avoid instability, and some devices may require the use of a shield between input and output circuits. Some devices have earth leads arranged so that a shield can be connected across their underside.

Earth returns are important in high-gain devices: some have input and output earth returns brought out separately in order to minimise unwanted coupling due to common earth return impedances, but this is not true of all devices.

Normally IC amplifiers are not intended to require neutralisation to achieve stability; unwanted oscillation can usually be traced to unsatisfactory layout or circuit arrangements. VHF parasitics may generally be eliminated by fitting a 10 ohm resistor in series with either the input or output lead, close to the IC.

With high-gain amplifiers, particular importance attaches to the decoupling of the voltage feeds. At the low voltages involved, values of series decoupling resistors must generally be kept low so that the inclusion of low-impedance bypass capacitors is usually essential. Since high-Q RF chokes may be a cause of RF oscillation it may be advisable to thread ferrite beads over one lead of any RF choke to reduce the Q.

As with bipolar transistors, IC devices (if based on bipolar transistors) have relatively low input and output impedances so that correct matching is necessary between stages. The use of an FET source follower stage may be a useful alternative to step-down transformers for matching.

Maximum and minimum operating temperatures should be observed. Many linear devices are available at significantly lower cost in limited temperature ranges which are usually more than adequate for operation under normal domestic conditions. Because of the relatively high temperature sensitivity and noise (especially LF noise) of bipolar-type integrated circuits, they are not generally suitable as free-running oscillators in high-performance receivers.

For the very highest grade receivers, discrete components and devices are still required in the front-end since currently available integrated circuits do not have comparable dynamic range.

The IC makes possible extremely compact receivers; in practice miniaturisation is now limited - at least for general-purpose receivers - by the need to provide easy-to-use controls for the non-miniaturised operator.

## Selective Filters

The selectivity characteristics of any receiver are determined by its filters: these filters may be at signal frequency (as in a straight receiver); intermediate frequency; or audio frequency (as in a direct-conversion receiver). Filters at signal frequency or IF are usually of band-pass characteristics; those at AF may be either band-pass or low-pass. With a very high first IF (112-150MHz) low-pass filters may be used at RF. A number of different types of filters are in common use: LC (inductor-capacitor) filters as in a conventional IF transformer or tuned circuit; crystal filters; mechanical filters; and ceramic filters; RC (resistor-capacitor) active filters (usually only at AF but feasible also at IF).

### Roofing filters

If the main selective filter is placed early in the signal path, for instance immediately following a diode mixer or low-gain, low-noise, post-mixer amplifier, where the signal voltage is very low, subsequent amplification (of the order of 100dB or more) will introduce considerable broad-band noise unless further narrow-band filtering is provided (which may be AF filtering). Another answer is to use an initial roofing filter with the final, more-selective filter(s) further down the signal path.

### Crystal filters

The selectivity of a tuned circuit is governed by its frequency and by its Q (ratio of reactance to resistance). There are practical limits to the Q obtainable in coils and IF transformers. In 1929, Dr J Robinson, a British scientist, introduced the quartz crystal resonator into radio receivers. The advantages of such a device for communications receivers were appreciated by James Lamb of the American Radio Relay League and he made popular the IF crystal filter for amateur operators.

For this application a quartz crystal may be considered as a resonant circuit with a Q of from 10,000 to 100,000 compared with about 300 for a very-high-grade coil and capacitor tuned circuit.

From earlier chapters, it will be noted that the electrical equivalent of a crystal is not a simple series or parallel-tuned circuit, but a combination of the two. It has (a) a fixed series resonant frequency ($f_s$) and (b) a parallel resonant frequency ($f_p$). The parallel frequency $f_p$ is determined partly by the capacitance of the crystal holder and by any added parallel capacitance and can be varied over a small range.

The crystal offers low impedance to signals at its series resonant frequency, a very high impedance to signals at its parallel resonant frequency, and a moderately high impedance to signals on other frequencies, tending to decrease as the frequency increases due to the parallel capacitance.

#### Ladder crystal filters

Single or multiple crystal half-lattice filters were used in the valve era. The modern filter of choice is the ladder filter, which can provide excellent SSB filtering at frequencies around 4-11MHz. This form of filter uses a number of crystals of the same (or nearly the same) frequency and so avoids the need for accurate crystal etching or selection. Further, provided it is correctly terminated, it does not require the use of transformers or inductors. Plated crystals such as the HC6U or 10XJ types are more likely to form good SSB filters, although virtually any type of crystal may be used for CW filters.

A number of practical design approaches have been described by J Pochet, F6BQP, [14] and in a series of articles by J A Hardcastle, G3JIR, [15]. **Fig 6.39** outlines the F6BQP approach. By designing for lower termination impedances and/or lower frequency crystals excellent CW filters can be formed.

A feature of the ladder design is the very high ultimate out-of-band rejection that can be achieved (75-95dB) in three- or four-section filters.

For SSB filters at about 8MHz a suitable design impedance would be about 800 ohms with a typical 'nose' band-pass of 2.0-2.1kHz. Intercept points of SSB and roofing filters can be over +50dBm with low insertion losses, and over +45dBm for narrow-band CW filters.

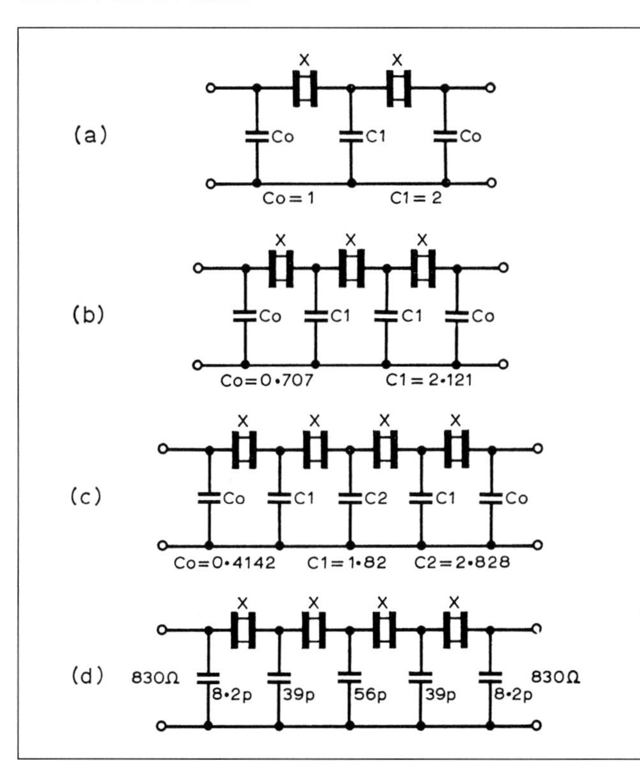


**Fig 6.39: Crystal ladder filters, investigated by F6BQP, can provide effective SSB and CW band-pass filters. All crystals (X) are of the same resonant frequency and preferably between 8 and 10MHz for SSB units. To calculate values for the capacitors multiply the coefficients given above by 1/(2πfR) where f is frequency of crystal in hertz (MHz by $10^6$), R is input and output termination impedance and 2π is roughly 6.8. (a) Two-crystal unit with relatively poor shape factor. (b) Three-crystal filter can give good results. (c) Four-crystal unit capable of excellent results. (d) Practical realisation of four-crystal unit using 8314kHz crystals, 10% preferred-value capacitors and termination impedance of 820Ω. Note that for crystals between 8 and 10MHz the termination impedance should be between about 800 and 1000Ω for SSB. At lower crystal frequencies use higher design impedances to obtain sufficient bandwidth. For CW filters use lower impedance and/or lower frequency crystals**

The ladder configuration is particularly attractive for home-built receivers since they can be based on readily available, low-cost 4.43MHz PAL colour-subcarrier crystals produced for use in domestic television receivers to P129 or P128 specifications. In NTSC countries, including North America and Japan, 3.58MHz TV crystals can be used, although this places the IF within an amateur band and would be unsuitable for single-conversion superhet receivers. Low-cost crystals at twice the PAL sub-carrier frequency (ie 8.86MHz) are also suitable.

Virtually any combination of crystals and capacitors produces a filter of some sort, but published equations or guidance should be followed to achieve optimum filter shapes and desired bandwidth. If this is done, ladder filters can readily be assembled from a handful of nominally identical crystals (ideally selected with some small offsets of up to about 50 or 100Hz) plus a few capacitors, yet providing SSB or CW filters with good ultimate rejection, plus reasonably low insertion loss and pass-band ripple.

Ladder filters have intercept points significantly above those of most economy-grade, lattice-type filters.

A valuable feature of the ladder configuration is that it lends itself to variable selectivity by changing the value of some, or preferably all, of the capacitors. The bandwidth can be varied over a restricted but useful range simply by making the middle capacitor variable, using a mechanically variable capacitor or electronic tuning diode (which may take the form of a 1W zener diode). **Fig 6.40** shows an 8MHz filter which has a 'nose' bandwidth that can be varied from about 2.8kHz down to 1.1kHz.

**Fig 6.41** illustrates a 4.43MHz nine-crystal filter that was built by R Howgego, G4DTC, for AM/SSB/CW/RTTY reception; the 3dB points can be varied from 4.35kHz down to 600Hz. In development, he noted that the bandwidth is determined entirely by the 'vertical' capacitors. If, however, these are reduced below about 10pF, the bandwidth begins to narrow rather than widen. The maximum bandwidth that could be achieved was about 4.5kHz. This could be widened by placing resistors (1kΩ to 10kΩ) across the capacitors but this increases insertion loss. Terminating impedances affect the pass-band ripple, not the bandwidth.

The filter gives continuously variable selectivity yet is relatively easy to construct. It is basically a six-pole roofing filter followed by a variable three-pole filter. It was based on low-cost Philips HC18-U type crystals and these were found to be all within a range of 80Hz. Crystals in the large case style (eg HC6-U) tend to be about 200Hz lower. C1, C2 is a 60 + 142pF miniature tuning capacitor as found in many portable analogue broadcast receivers.

The integral trimmers are set for maximum bandwidth when capacitor plates are fully unmeshed. Set R1 for best compromise between minimum bandwidth and insertion loss, 1.2kΩ nominal.

The following specification should be achievable.

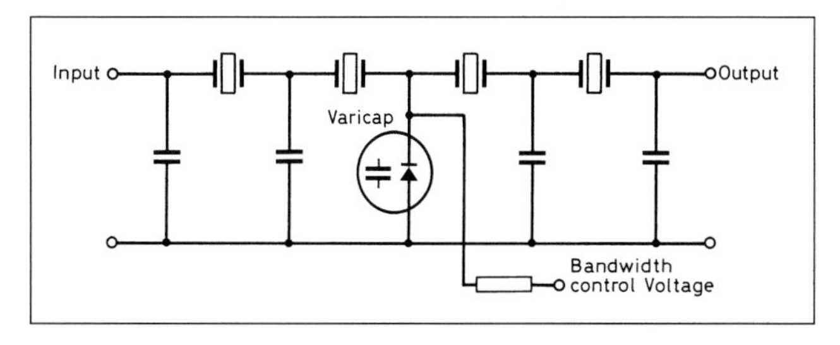


**Fig 6.40: A simple technique for varying the bandwidth of an 8MHz crystal ladder filter from about 12.8kHz down to 1.1kHz by using a varicap diode**

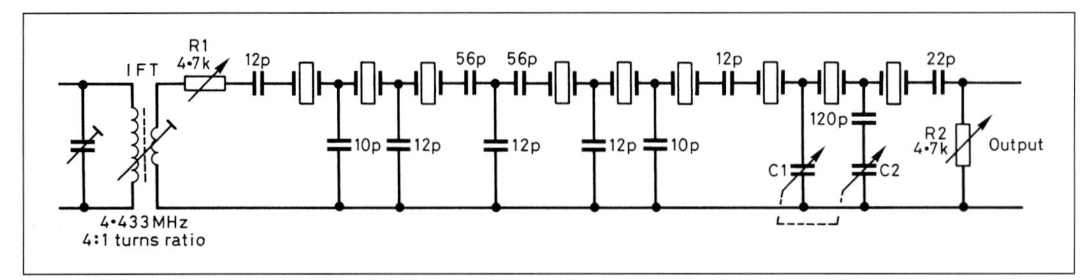


**Fig 6.41: Variable-selectivity ladder filter using low-cost PAL colour-TV crystals for AM/SSB/CW/RTTY reception**

- C1, C2 plates unmeshed: 3dB points at 4437.25kHz and 4432.90kHz, bandwidth 4.35kHz.
- C1, C2 plates half-meshed: 3dB points at 4434.0kHz and 4432.90kHz, bandwidth 1.10kHz.
- C1, C2 plates meshed: 3dB points at 4433.5kHz and 4432.90kHz, bandwidth 600Hz.
- Insertion loss in pass-band: maximum (R1 2.5kΩ, R2 1.2kΩ) 10dB, minimum (R1 0kΩ , R2 1.2kΩ) 6dB.
- Pass-band ripple 1-3dB (dependent on R1). Ripple reduces with bandwidth.
- Stop-band attenuation better than 60dB. -20dB bandwidth typically 1kHz wider than the -3dB bandwidth.

The filter shown in **Fig 6.42**, designed by D Gordon-Smith, G3UUR, provides six different bandwidths suitable for both SSB and CW operation, switching the value of all capacitors, and ideally preceded by a roofing filter. In order to reduce the number of switched components, the terminating resistors remain constant and the ripple merely decreases with bandwidth. This also reduces the variation in insertion loss.

The 2.4kHz position has a 1dB ripple Chebyshev response, and the 500Hz position represents a Butterworth response. A 5:1 bandwidth change is possible if the maximum tolerable ripple is 1dB. This range is fixed by design constraints: the ratio of 1dB to 0dB ripple response terminating resistance is approximately 5:1 for the same bandwidth, and therefore the same terminating resistance satisfies bandwidths that have a ratio of about 5:1.

The main disadvantage is that the pass-band moves low in frequency as it is narrowed. This could be compensated for by moving the carrier crystals down in sympathy with the filter centre frequency; a total shift of 1kHz or less would probably be adequate, and could be corrected at the VFO by the RIT shift control.

A different approach to ladder filters is to put the crystals in shunt with the signal rather than in series. The filter then has its steeper slope on the low-frequency side instead of the high-frequency side, and this may be preferable for narrow-bandwidth filters (eg CW filters), particularly when using relatively low-Q plated crystals in HC-18 holders. **Fig 6.43** shows a shunt-type crystal filter designed by John Pivnichy, N2DCH. The filter response is shown in **Fig 6.44**.

## Ceramic filters

Piezoelectric effects are not confined to quartz crystals; in recent years increasing use has been made of certain ceramics, such as lead zirconate titanate (PZT). Small discs of PZT, which resonate in the radial dimension, can form economical selective filters in much the same way as quartz, though with considerably lower Q.

Ceramic IF transfilters are a convenient means of providing the low impedances needed for bipolar transistor circuits. The simplest ceramic filters use just one resonator but numbers of resonators can be coupled together to form filters of required bandwidth and shape factor. While quite good nose selectivity is achieved with simple ceramic filters, multiple resonators are required if good shape factors are to be achieved. Some filters

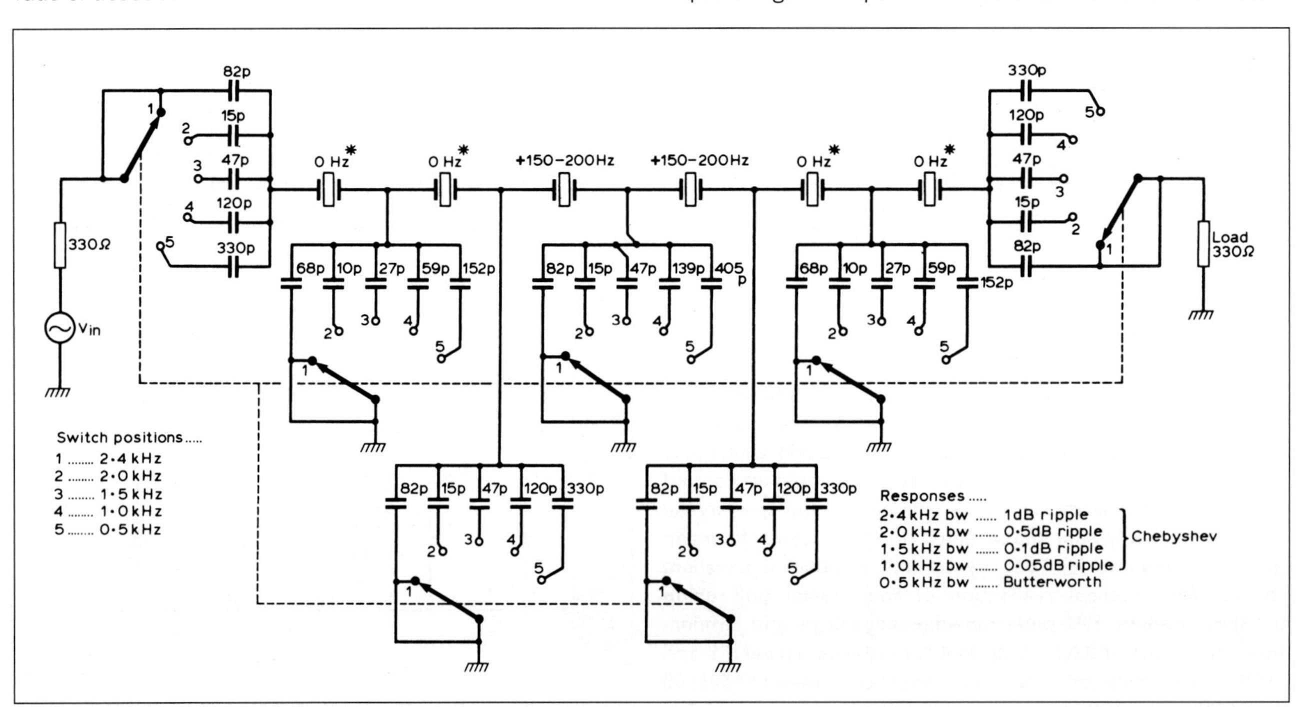


**Fig 6.42: G3UUR's design for a switched variable-bandwidth ladder filter using colour-TV crystals. Note that crystals shown as 0Hz offset can be in practice ±0.5Hz without having too detrimental an effect on the pass-band ripple**

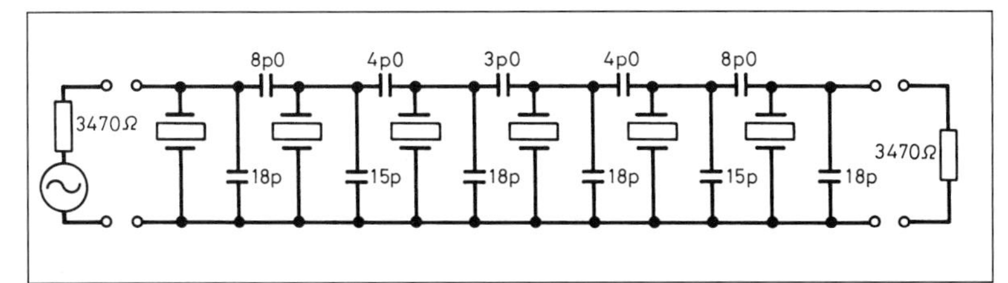


**Fig 6.43: Shunt-type crystal ladder CW filter using six 3.58MHz NTSC crystals designed for 3470Ω terminations. Crystals should usually be matched to within 100Hz. (TV crystals are often only specified as within 300Hz but are usually within 200Hz)**

are of 'hybrid' form using combinations of inductors and ceramic resonators.

Examples of ceramic filters include the Philips LP1175 in which a hybrid unit provides the degree of selectivity associated with much larger conventional IF transformers; a somewhat similar arrangement is used in the smaller Toko filters such as the CFT455C which has a bandwidth (to -6dB) of 6kHz. A more complex 15-element filter is the Murata CFS-455A with a bandwidth of 3kHz at -6dB, 7.5kHz at -70dB and insertion loss 9dB, with input and output impedances of 2kΩ and centre frequency of 455kHz. In general ceramic filters are available from 50kHz to about 10.7MHz centre frequencies. **Fig 6.45** provides more details.

Ceramic filters tend to be more economical than crystal or mechanical filters but have lower temperature stability and may have greater pass-band attenuation.

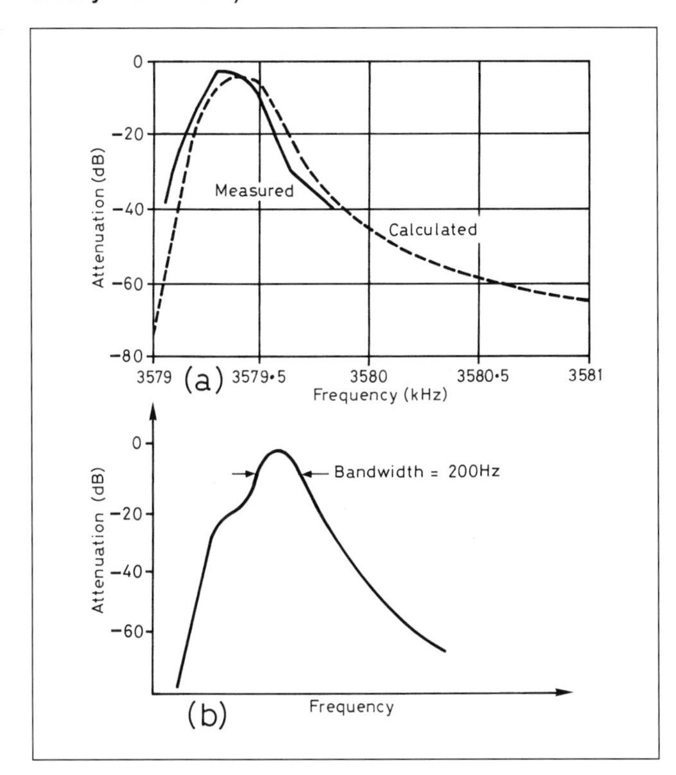


**Fig 6.44: (a) Calculated and measured response curves for the CW filter. (b) Filter response resulting from use of poorly matched crystals**

## Mechanical and miscellaneous filters

Very effective SSB and CW filters at intermediate frequencies from about 60 to 600kHz depend on the mechanical resonances of a series of small elements usually in the form of discs. The mechanical filter consists of three basic elements: two magneto-striction transducers which convert the IF signals into mechanical vibrations and vice versa; a series of metal discs mechanically resonated to the required frequency; and disc coupling rods.

Each disc represents a high-Q series resonant circuit and the bandwidth of the filter is determined by the coupling rods. 6-60dB shape factors can be as low as about 1.2, with low pass-band attenuation. The limitation of mechanical filters to frequencies of about 500kHz or below has led to their virtual disappearance from the amateur markets and they are now found only in older models despite their excellent performance below 500kHz.

Other forms of mechanical filters have been developed which include ceramic piezoelectric transducers with mechanical coupling: they thus represent a combination of ceramic and mechanical techniques. These filters may consist of an H-shaped form of construction; such filters include a range manufactured by the Toko company of Japan. Generally the performance of such filters is below that of the disc resonator type, but can still

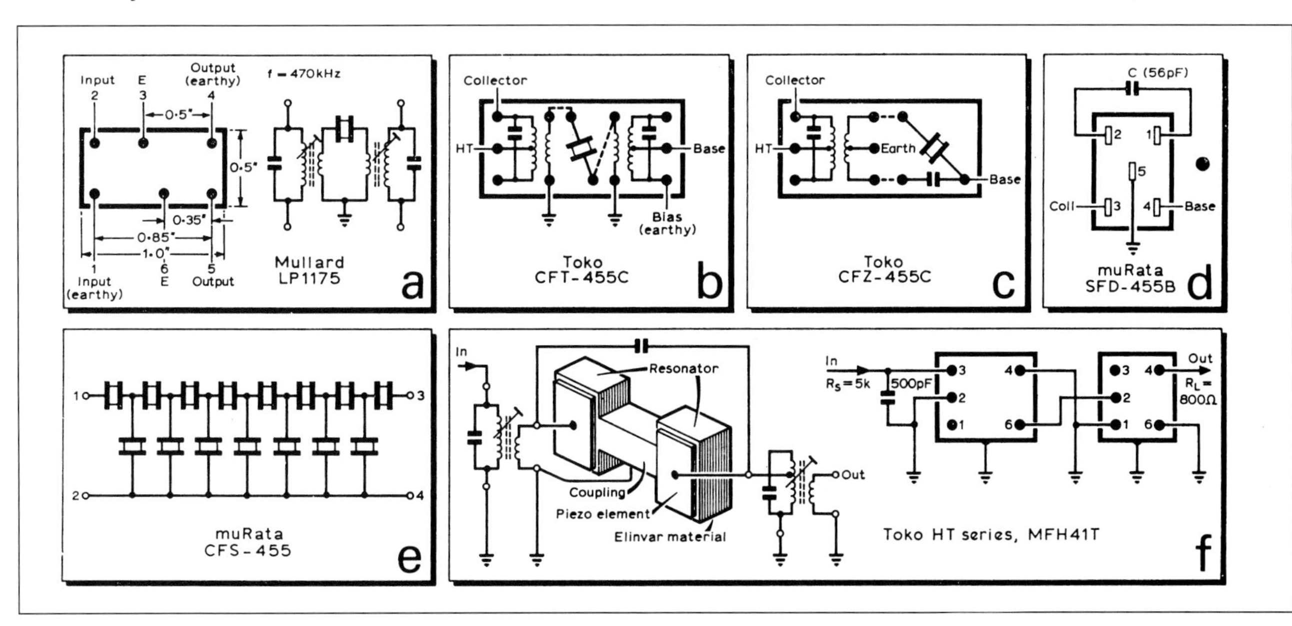


**Fig 6.45: Representative types of ceramic filters**

be useful.

Surface acoustic wave (SAW) filters are available for band-pass filter applications where discrete-element filters have previously been used, including IF filters. Filters in the 80 to 150MHz region are manufactured for cellular telephone applications, and those designed for the American AMPS and IS136 standards may be narrow enough (<30kHz) to be useful as roofing filters. It should be noted that the mechanism of signal propagation in the SAW filter is such that they are extremely resistant to intermodulation difficulties.

## Receiver Protection

Receivers, particularly where they are to be used alongside a medium- or high-powered transmitter, need to be protected from high transient or other voltages induced by the local transmitter or by build-up of static voltages on the antenna. Semiconductors used in the first stage of a receiver are particularly vulnerable and invariably require protection. The simplest form of protection is the use of two diodes in back-to-back parallel configuration. Such a combination passes signals less than the potential hill of the diodes (about 0.3V for germanium diodes, about 0.6V for silicon diodes) but provides virtually a short-circuit for higher-voltage signals. This system is usually effective but has the disadvantage that it introduces non-linear devices into the signal path and may occasionally be the cause of cross- and inter-modulation.

The mosfet devices are particularly vulnerable to static puncture and some types include built-in zener diodes to protect the 'gates' of the main structure. Since these have limited rating it may still be advisable to support them with external diodes or small gas-filled transient suppressors.

## Input Circuits

It has already been noted that with low-noise mixers it is now possible to dispense with high-gain RF amplification. Amplifiers at the signal frequency may, however, still be advisable: to provide pre-mixer selectivity; to provide an AGC-controlled stage which is in effect a controlled attenuator on strong signals; and to counter the effects of conversion loss in diode and FET-array mixer stages.

In practice, semiconductor RF stages are often based on junction FETs as shown in **Fig 6.46** or dual-gate mosfets, or alternatively integrated circuits in which large numbers of bipolar transistors are used in configurations designed to increase their signal-handling capabilities.

Tuned circuits between the antenna and the first stage (mixer or RF amplifier) have two main functions: to provide high attenuation at the image frequency and to reduce as far as possible the amplitude of all signals outside the IF pass-band.

### Coupled tuned circuits

Most amateur receivers still require good pre-mixer selectivity; this can be achieved by using a number of tuned circuits coupled through low-gain amplifiers, or alternatively by tuneable or fixed band-pass filters that attenuate all signals outside the amateur bands (**Fig 6.47**). The most commonly used input arrangement consists of two tuned circuits with screening between them and either top-coupled through a small-value fixed capacitor, or bottom-coupled through a small common inductance.

In order that the coupling is maintained constant over the tuning range, the coupling element should be the same sort as the fixed element, as in **Fig 6.47b**.

Critical coupling is achieved when the coupling coefficient, k, multiplied by the working Q is equal to 1. For **Fig 6.47a**, k is given by:

$$k = Cc/\surd C_1C_2$$

where $C_1$, $C_2$ are the tuning capacitors, and Cc is the top coupling capacitor. In Fig 6.47b:

$$k = Lc/\surd L_1L_2$$

where $L_1$, $L_2$ are the tuning inductors, and Lc is the coupling inductor.

Critical coupling can be hard to achieve, and for amateur use, it is frequently the case that slight undercoupling is preferable to overcoupling. Slightly more complex, but capable of rather better results, is the minimum-loss Cohn filter; this is capable of reducing signals 10% off-tune by as much as 60dB provided that an insertion loss of about 4dB is acceptable. The Cohn filter is a bottom C coupled pair. This compares with about 50dB (and rather less insertion loss) for an undercoupled pair of tuned circuits. The Cohn filter is perhaps more suited for use as a fixed band-pass filter which can be used in front of receivers having inadequate RF selectivity.

Before integrated circuits, receiver IF amplifiers were built up with a chain of valves or transistors coupled together with transformers. Each transformer was a usually a double tuned circuit

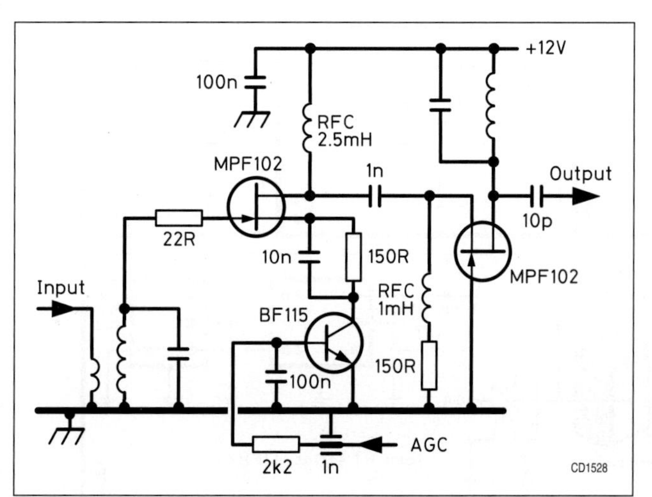


**Fig 6.46: Cascode RF amp using two JFETs with transistor as AGC**

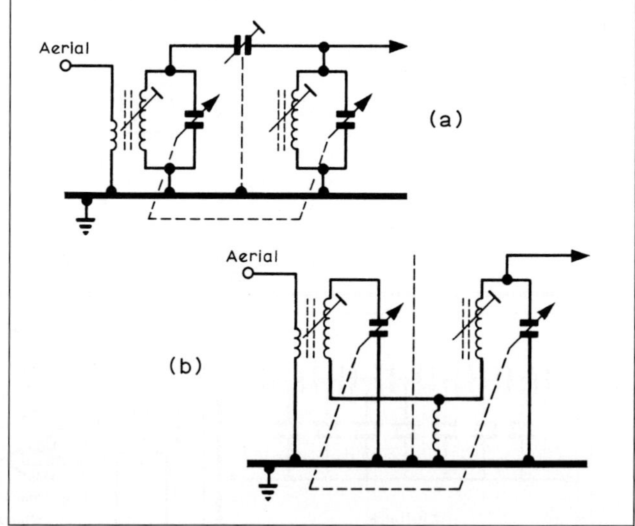


**Fig 6.47: Typical RF input circuits used to enhance RF selectivity and capable of providing more than 40dB attenuation of unwanted signals 10% off tune**

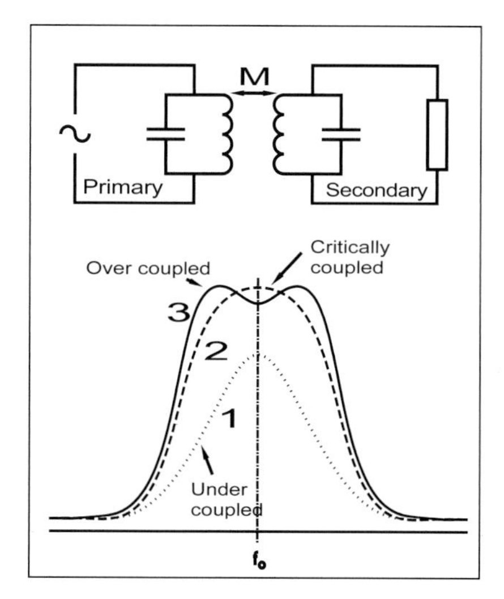


**Fig 6.48: Inductively coupled tuned circuits. The mutual inductance M and the coupling increase as the coils are moved together. The response varies according to the degree of coupling with close coupling giving a broad response**

providing a proportion of the overall selectivity. With the advent of integrated circuits, all of the IF gain became a single block with the selectivity now in front between the mixer and the input to the IF amplifier. This was good because of the protection given to the IF amplifier against strong off channel signals, however, the detector now experienced all the wideband noise generated within the IF amplifier. Ideally the overall IF selectivity should be provided by both input filtering where most of the selectivity takes place, plus output filtering to eliminate the wideband noise generated within the IF amplifier.

**Fig 6.48** illustrates the principle of the coupled tuned circuit IF transformer and **Fig 6.49** shows the traditional embodiment of the device (these two figures are reproduced from chapters 1 and 2 which provide more information on this subject). In essence, there are two tuned circuits whose physical spacing determines the coupling and hence the bandwidth of the transformer. There are several other ways of providing the all important coupling, two of which are illustrated in **Fig 6.50**.

Top C coupling is where a small value capacitor connects the two hot ends of the identical tuned circuits, as illustrated in **Fig 6.50a**. A feature of top C coupling is that the attenuation falls away more rapidly on the low frequency side of resonance. Bottom C coupling is achieved by connecting the cold ends of the two capacitors together and then shunting away most of the energy by use of a large value capacitor as shown in **Fig 6.50b**. Bottom C coupling is characterised by the attenuation falling off more rapidly on the high side of resonance. One advantage of bottom C coupling is if the two halves of the pair are physically separated, they can be joined via a piece of coax that forms part of the bottom C.

Whilst the traditional type of IF transformer can still be found at rallies, they are too big for modern construction practice. Most home built receivers will use the TOKO 10mm 10K and 10EZ range of coils that are out of production but still available from a few stockists, eg [7]. However this range does not include double tuned transformers and so some way of fabricating the double tuned device is required. **Fig 6.51** illustrates a way of assembling an IF transformer centered on 455KHz from two TOKO coils using top C coupling. Also shown is the measured frequency response illustrating the characteristic dip in the middle of the passband due to slight overcoupling and the reduced attenuation on the high side of resonance. The particular TOKO coil inductance was 680µH tuned with an external 180pF capacitor and a coupling capacitor of 5.6pF.

## RF amplifiers and attenuators

Broad-band and untuned RF stages are convenient in construction but can be recommended only when the devices used in the

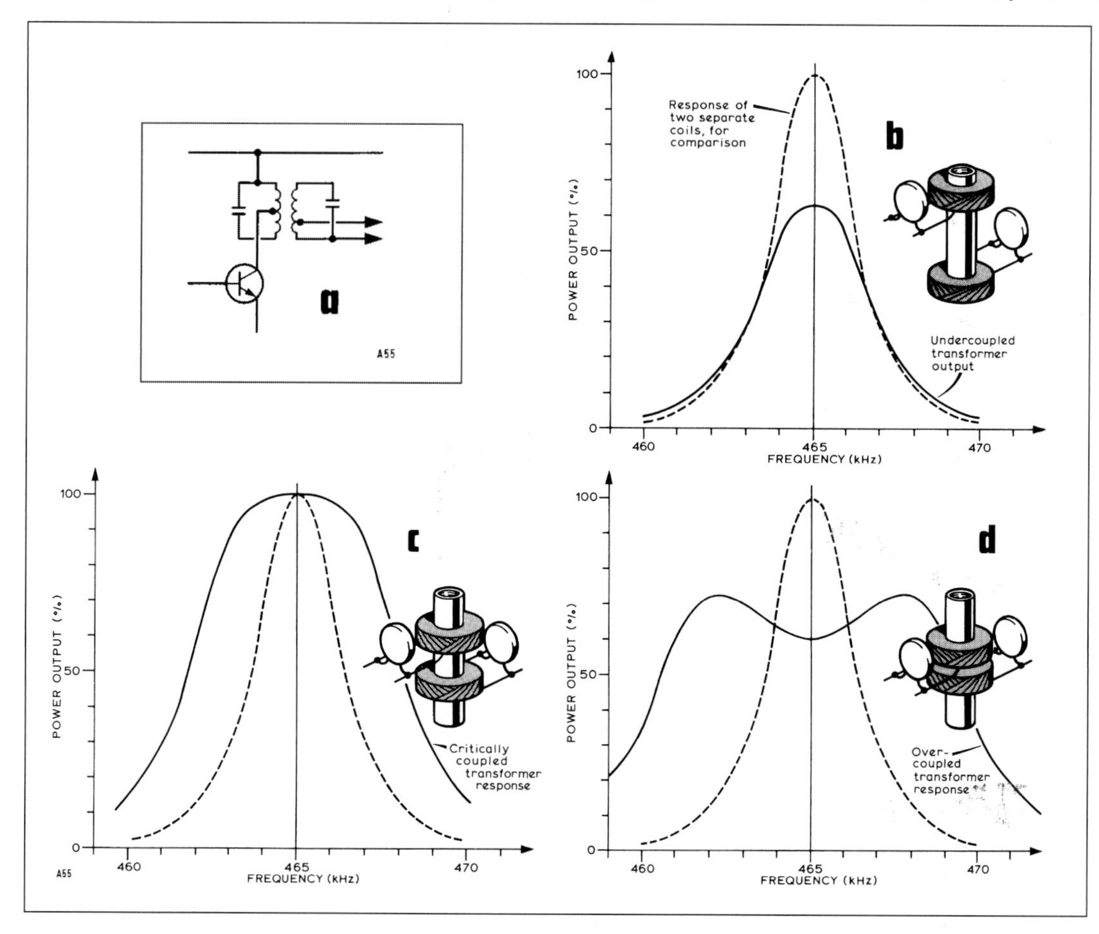


**Fig 6.49: Bandwidth of coupled circuits**

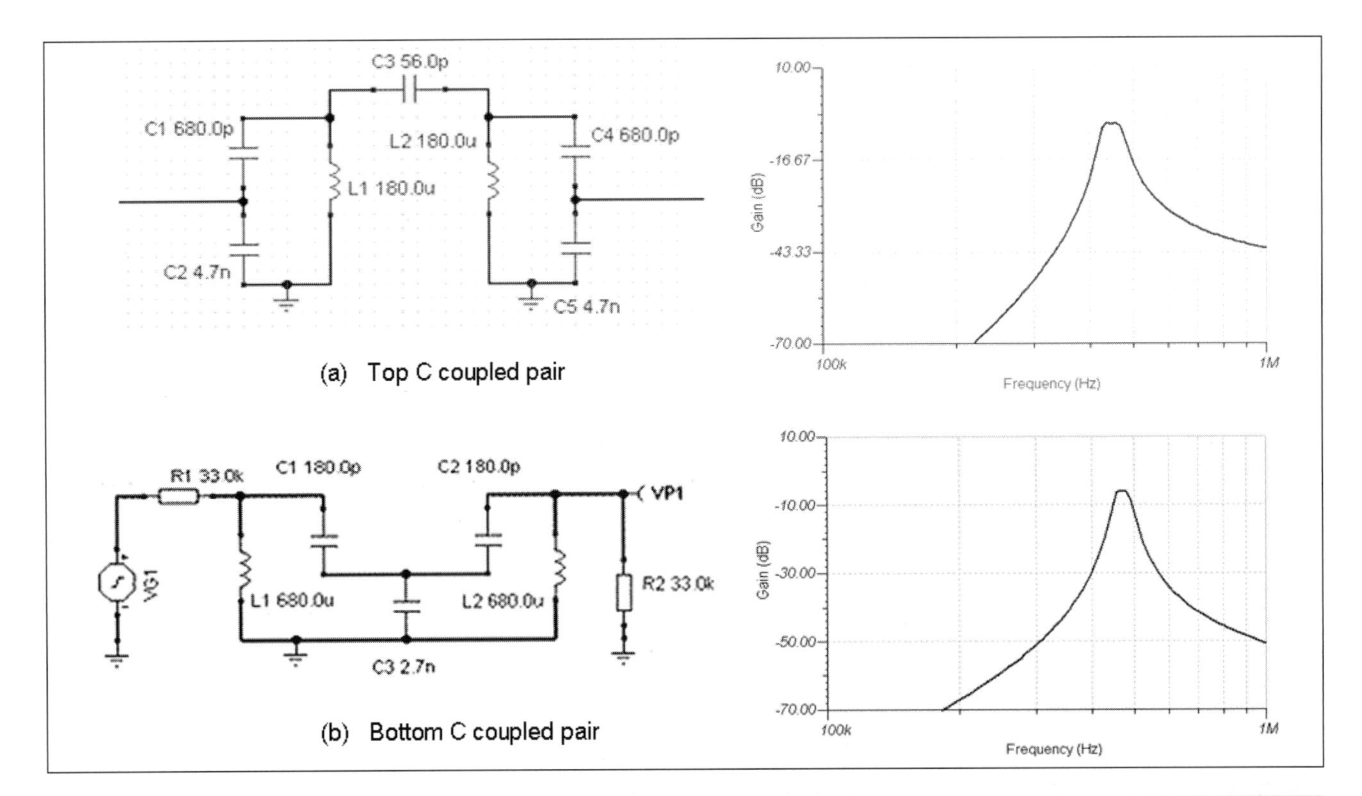


Fig 6.50: Two methods of connecting tuned circuits

front-end of the receiver have wide dynamic range.

Unless the front-end of the receiver is capable of coping with the full range of signals likely to be received, it may be useful to fit an attenuator working directly on the input signal. **Fig 6.52** shows simple techniques for providing manual attenuation control; **Fig 6.53** is a switched attenuator providing constant impedance characteristics.

A wide-band amplifier placed in front of a mixer of wide dynamic range must itself have good spurious free dynamic range. Amplifiers based on power FETs can approach 140dB dynamic range when operated at low gain (about 10dB) and with about 40mA drain current. This compares with about 90 to 100dB for good valves (eg E810F), 80 to 85dB for small-signal FET devices; and 70 to 90dB for small-signal bipolar transistors. A high-dynamic-range bipolar amplifier can exhibit a third-order input intercept point of +33dBm with a noise figure of 3dB, using suitable transistors and noiseless feedback techniques, representing about 160dB of dynamic range.

**Fig 6.54** shows two wideband RF amplifiers, both capable of better than +40dBm OIP3. The dynamic range of an amplifier can be increased by the operation of two devices in a balanced

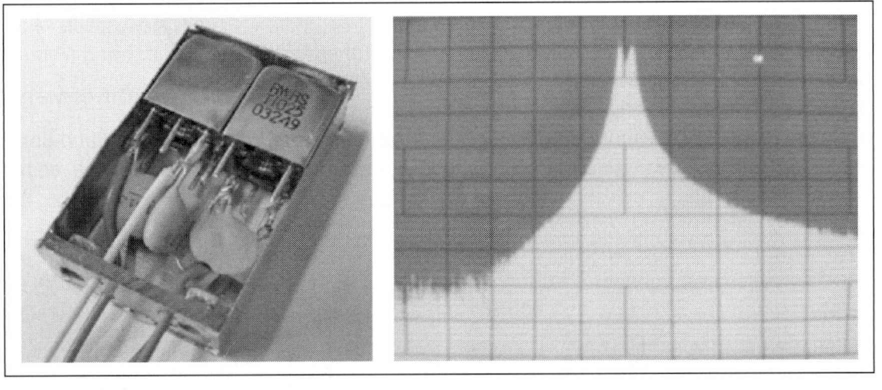

Fig 6.51: (left) Assembling two TOKO coils into a 455kHz transformer; (right) The resulting passband [50kHz per div and 10dB per div]

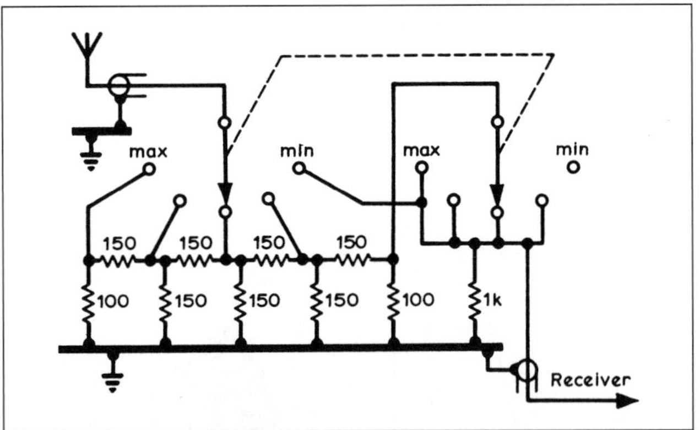


(above) Fig 6.53: Switched antenna attenuator for incorporating in a receiver

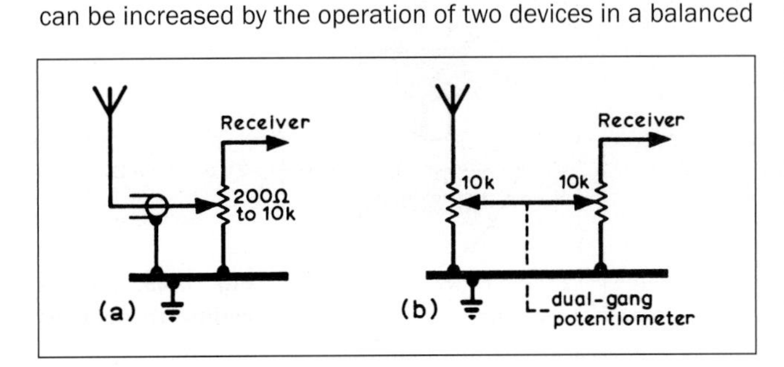


(left) Fig 6.52: Simple attenuators for use in front of a receiver of restricted dynamic range. (a) No attempt is made to maintain constant impedance. (b) Represents less change in impedance

**Fig 6.54: Two wideband RF amplifiers**

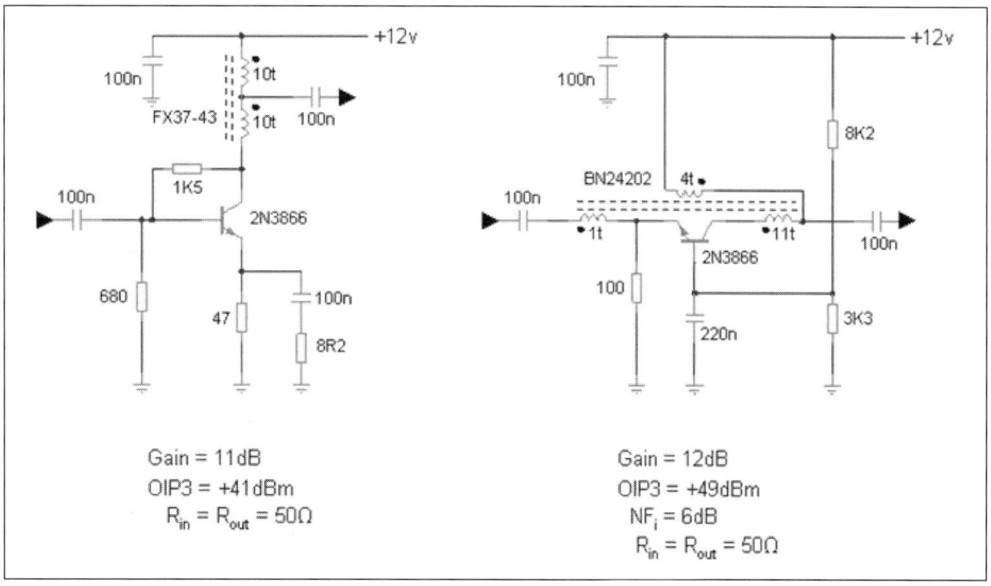

Gain = 11dB
OIP3 = +41dBm
$R_{in} = R_{out} = 50\Omega$

Gain = 12dB
OIP3 = +49dBm
$NF_i$ = 6dB
$R_{in} = R_{out} = 50\Omega$

(push-pull) mode, but at a penalty of a 3dB increase in noise figure.

Various forms of attenuators controlled from the AGC line are possible. **Fig 6.55** shows a system based on PIN diodes; **Fig 6.56** is based on toroid ferrite cores and can provide up to about 45dB attenuation when controlled by a potentiometer. **Fig 6.57** shows an adaptation with a mosfet control element for use on AGC lines, although the range is limited to about 20dB.

The tuned circuits used in front-ends may be based on toroid cores since these can be used without screening with little risk of oscillation due to mutual coupling. However, the available Q cannot always compete with that of air-cored coils of thick wire, or of bunched conductor wire and pot cores at the lower frequencies.

It is important to check filters and tuned circuits for non-linearity in iron or ferrite materials. Intermodulation can be caused by the cores where the flux level rises above the point at which saturation effects begin to occur, although this is not usually a problem with dust iron cores.

Since the optimum dynamic range of an amplifier is usually achieved when the device is operated at a specific working point (ie bias potential) it may be an advantage to design the stage for fixed (low) gain with front-end gain controlled by

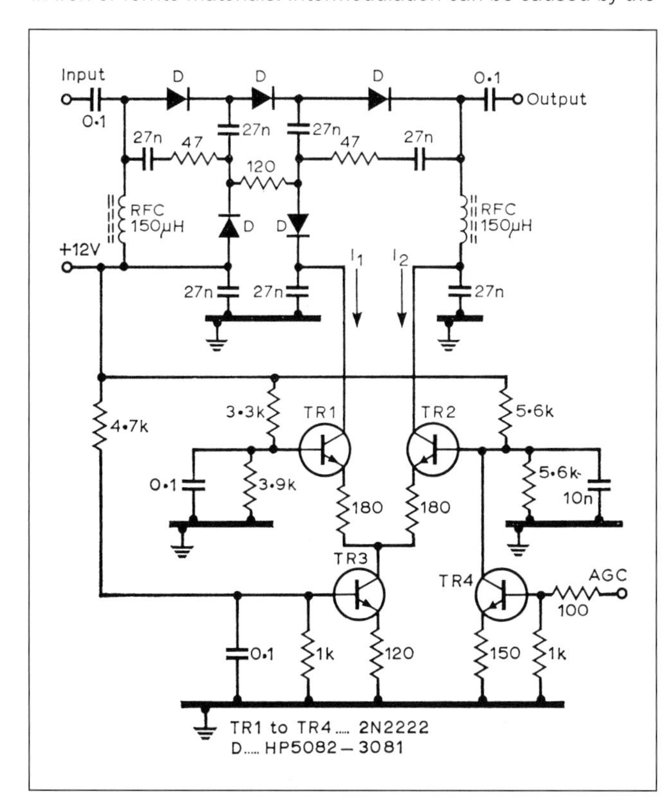


**Fig 6.55: Five PIN diodes in a double-T arrangement form an AGC-controlled attenuator. The sum of the transistor collector currents is maintained constant to keep input and output impedances constant**

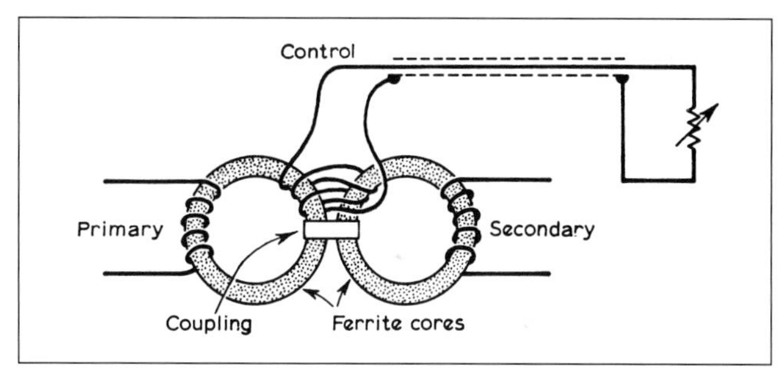


**Fig 6.56: Basic form of RF level control using two toroidal ferrite cores**

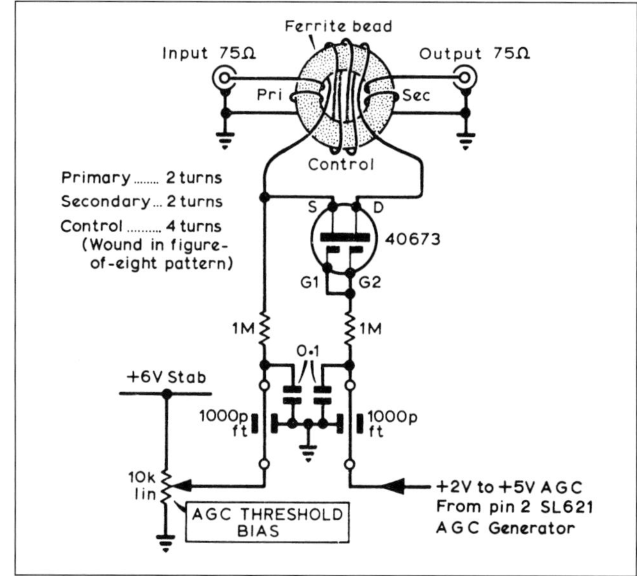


**Fig 6.57: Automatic antenna attenuator based on the technique shown in Fig 6.53**

means of an antenna attenuator (manual or AGC-controlled). Attenuation of signals ahead of a stage subject to intermodulation is often beneficial since 1dB of attenuation reduces the potential for intermodulation by 3dB.

For semiconductor stages using FETs and bipolar transistors the grounded-gate / grounded-base configuration is to be preferred. Special types of bipolar transistors (such as the BF314, BF324 developed for FM radio tuners) provide a dynamic range comparable with many junction FETs, a noise figure of about 4dB and a gain of about 15dB with a collector current of about 5mA. Generally, the higher the input power of the device, the greater is likely to be its signal-handling capabilities: overlay and multi-emitter RF power transistors or those developed for CATV applications can have very good intermodulation characteristics.

For general-purpose, small-signal, un-neutralised amplifiers the zener-protected dual-gate mosfet is probably the best and most versatile of the low-cost discrete semiconductor devices, with its inherent cascode configuration. If gate 2 is based initially at about 30 to 40% of the drain voltage, gain can be reduced (manually or by AGC action) by lowering this gate 2 voltage, with the advantage that this then increases the signal-handling capacity at this stage. This type of device can be used effectively for RF, mixer, IF, product detector, AF and oscillator applications in HF receivers.

## Mixers

Mixers operate either in the form of switching mixers (the normal arrangement with diode ring mixers) or in what are termed continuous non-linear (CNL) modes. Switching mixers work by chopping up the signal and the resulting components include the wanted output. CNL mixers use their non-linear transfer characteristic to mathematically multiply the input and LO together, one of the results being the wanted output. Generally, switching mixers can provide better performance than CNL modes but require more oscillator injection, preferably as a square wave.

Because of their near square-law transfer characteristics field-effect devices make successful mixers provided that care is taken on the LO drive level and the operating point (ie bias resistor); preferably both these should be adjusted to suit the individual device used.

Junction FETs used as mixers can be operated in three ways:

**RF signal applied to gate, oscillator signal to source.** This provides high conversion gain but requires high LO power and may result in oscillator pulling.

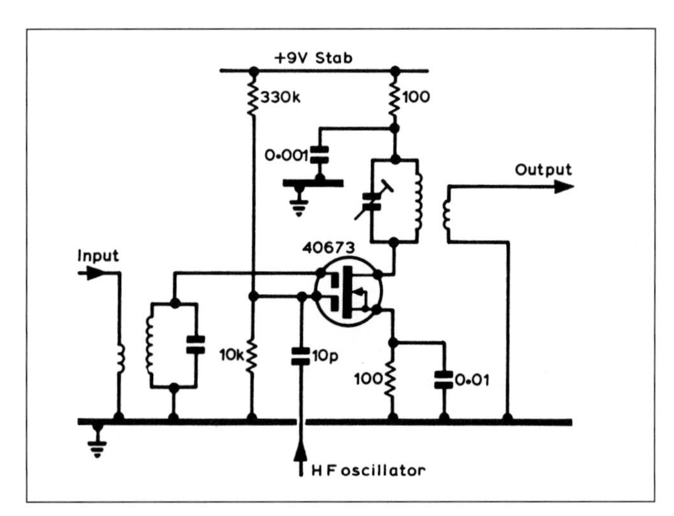


**Fig 6.58: Typical dual-gate MOSFET mixer - one of the best 'simple' semiconductor mixers providing gain and requiring only low oscillator injection, but of rather limited dynamic range**

**RF signal to source, oscillator signal to gate.** This gives good freedom from oscillator pulling and requires low LO power, but provides significantly lower gain.

**RF and oscillator signals applied to gate.** This gives fairly high gain with low LO power and may often be the optimum choice.

For most applications the dual-gate mosfet mixer (**Fig 6.58**) probably represents the best of the 'simple' single-device arrangements.

During the 1980s and '90s, much attention was given to improving the input intercept point of receivers in order to cope better with the range of strong signals that reach low-noise broad-band mixers with a minimum of pre-mixer selectivity. A number of high-performance, double-balanced switching mixers were developed which were available in packaged form for home-construction.

It has been noted that a double-balanced mixer offers advantages over single-device and single-balanced mixers in reducing the number of IMPs, and also the oscillator radiation from the antenna and the oscillator noise entering the IF channel.

Double-balanced mixers using four hot-carrier diodes (eg the one shown in **Fig 6.59**) when driven at a suitable level and with correct output impedance can provide third-order intercept points (IP3) of up to about +40dBm. Conversion loss is likely to

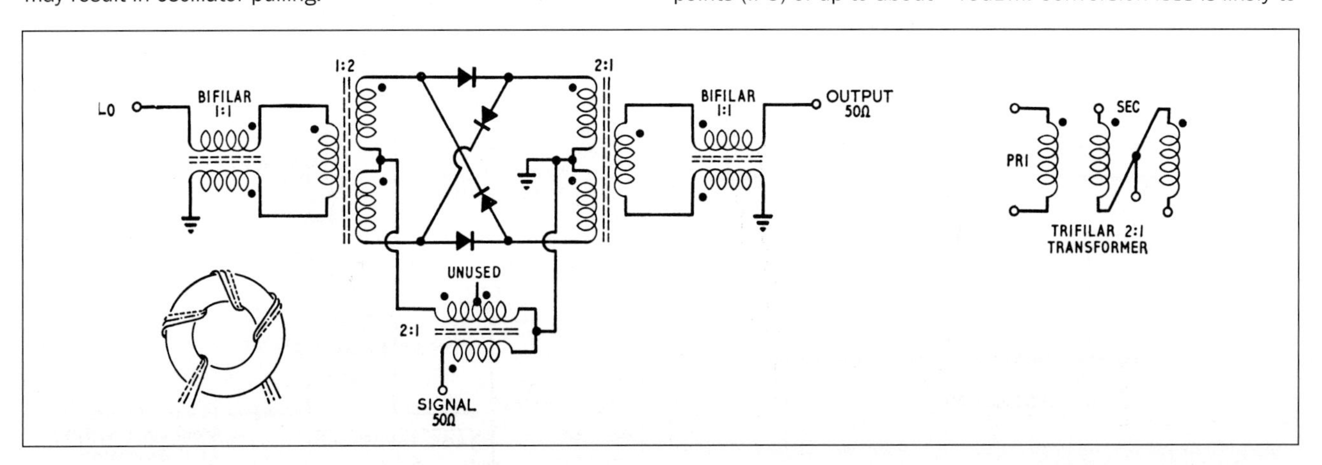


**Fig 6.59: Double-balanced diode ring mixer showing how additional bifilar-wound transformers can be added to improve balance, with details of the transformers. The three strands of wire should be twisted together before winding; each winding consists of 12 to 20 turns (depending on frequency range) of No 32 enamelled wire. Injection signal should be 0.8 to 3V across 50 ohms (4-12mW). The toroid material is dependent on the frequency range**

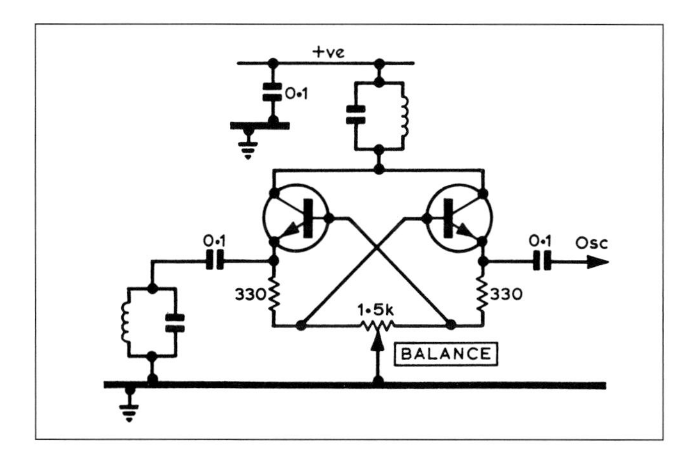


**Fig 6.60: Use of cross-coupled transistors to form a double-balanced mixer without special balanced input transformers**

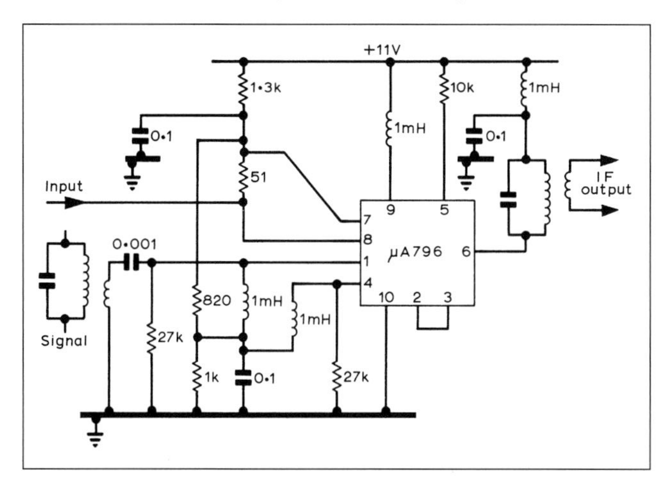


**Fig 6.61: Double-balanced IC mixer circuit for use with µA796 or MC1596G etc devices**

be at least 6dB, and optimum performance requires a high-level of LO injection (+7dBm for the popular SBL-1) with a near-square waveform. For optimum IP3 performance attention must be paid to the broadband output termination which should be 50 ohms.

High-level ring-type mixers can also be based on the use of medium-power bipolar transistors in cross-coupled 'tree' arrangements (**Fig 6.60**). With such an arrangement there need be no fundamental requirement for push-pull drive or balanced input/output transformers, although their use minimises local oscillator conducted emissions. This approach is used at low-level in the Motorola MC1596 and now-obsolete Plessey SL640 IC packages, and at high-level in the obsolete Plessey SL6440. The SL6440 was specifically developed as a high-level, low-noise mixer and is capable of +30dBm intercept point, +15dBm 1dB compression point with a conversion 'gain' of -1dB. A circuit for the µA796 and similar integrated circuits is shown in **Fig 6.61**.

Switching mixers, if they are to achieve a high IP3 require a drive that must approach an ideal square wave and have sufficient amplitude to switch the devices fully on and off. Further, to maintain good overall performance in terms of minimum conversion loss, maximum dynamic range (ie taking into account the noise figure) and maximum strong signal performance, it is desirable to incorporate image-frequency termination.

Details of the performance of various mixer devices are shown in **Fig 6.62**.

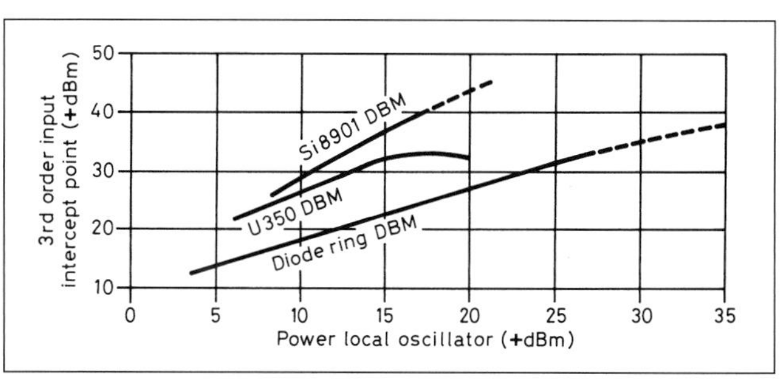


**Fig 6.62: Performance comparison between Si8901 DBM, U350 active FET DBM and diode ring DBM**

## The H-Mode mixer

This is an extreme perfomance version of the double balanced switching mixer, developed by G3SBI and it takes its name from the shape of its circuit diagram. What makes the H-Mode mixer different is that its switching elements operate in shunt rather than series.

In essence, if the switching elements go from open circuit to short circuit in an infinitely short time then you get the perfect mixer and modern very fast bus switches and logic allow the H-Mode mixer to be the closest we have come. For more, see later in this chapter under Super-linear front-ends.

The handicap for the home constructor is that the physical device packages are surface mount and small, consequently, they do not lend themselves to dead-bug construction.

## Image-rejection mixer

In recent years, there has been a marked trend towards multi-conversion general-coverage receivers with the first IF in the VHF range since this facilitates the use of frequency synthesisers covering a single span of frequencies with a broad-band up-conversion mixer. It has long been considered that subsequent down-conversion mixer stages should not change the frequency by a factor of more than about 10:1 in order to minimise the 'image' response.

Thus a receiver with a first IF of the order of 70MHz or 45MHz and a final IF of, say, 455kHz or 50kHz (to take advantage of digital signal processing) normally requires an intermediate IF of, say, 9 or 10.7MHz and possibly a further IF of about 1MHz. With triple or even quadruple conversions, it becomes increasingly difficult to achieve a design free of spurious responses and of wide dynamic range.

It is possible to eliminate mid-IF stages by the use of an image-rejecting two-phase mixer akin to the form of demodulator used in two-phase direct conversion receivers; this is best implemented using two double-balanced diode-ring mixers or the equivalent.

**Fig 6.63** outlines the basic arrangement of an image-rejection mixer. Image rejection mixers offer slightly better intermodulation performance, all other considerations being equal, because the input signal level is reduced by 3dB. This reduces the absolute power of the 3rd order intermodulation products by 9dB, and summing the products increases this level by a maximum of 3dB. However, there is a 3dB greater noise figure, and so the absolute improvement in dynamic range disappears. Image rejection mixers are generally limited to about 30dB rejection: with careful trimming, possibly 40dB may be attained over time and temperature variations. Such a mixer can convert a VHF signal directly to, say, 50kHz while maintaining image response at a low level.

More about mixers can be found in the one of the Building

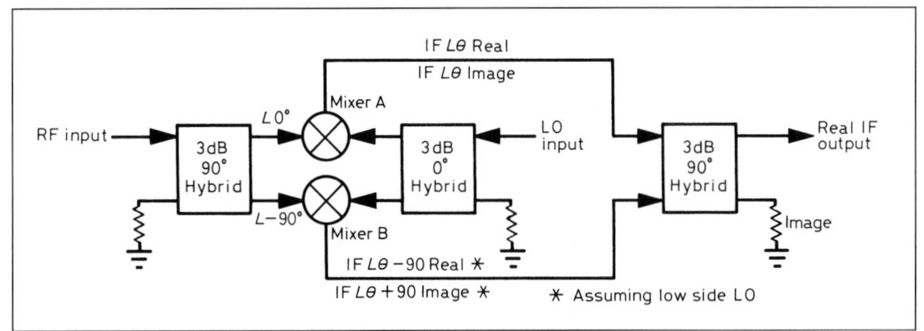


**Fig 6.63: Image rejecting mixer from *Microwave Solid-state Circuit Design* by Bahl and Bhartia (1988)**

Blocks chapters earlier in this book.

## HF Oscillator(s)

The frequency to which a superhet or direct-conversion receiver responds is governed not by the input signal frequency circuits but by the output of the local oscillator. Any frequency variations or drift of the oscillator are reflected in variation of the received signal; for SSB reception, variations of more than a few tens of Hertz will render the signal unintelligible unless the set is retuned. Much more on this topic can be found in the chapter Building Blocks: Oscillators.

Mentioned earlier in this chapter is the unwanted effect known as reciprocal mixing which is directly related to oscillator noise. This aspect represents the next evolutionary step in HF receiver design.

## IF Amplifiers

The IF remains the heart of a superhet receiver, for it is in this section that virtually all of the voltage gain of the signal and the selectivity response are achieved. Whereas with older superhets (which had significant front-end gain) the IF gain was of the order of 70-80dB, today it is often over 100dB.

Where the output from the mixer is low (possibly less than 1μV) it is essential that the first stage of the IF section should have low-noise characteristics and yet not be easily overloaded. Although it is desirable that the crystal filter (or roofing filter) should be placed immediately after the mixer, the very low output of diode and passive FET mixers may require that a stage of amplification takes place before the signal suffers the insertion loss of the filter.

Similarly it is important that where the signal passes through the sideband filter at very low levels the subsequent IF amplifier must have good noise characteristics. Further, for optimum CW reception, it will often be necessary to ensure that the noise bandwidth of the IF amplifier after the filter is kept narrow. The noise bandwidth of the entire amplifier should be little more than that of the filter. This can be achieved by including a further narrow-band filter (for example a single-crystal filter with phasing control) later in the receiver, or alternatively by further frequency conversion to a low IF.

To achieve a flat AGC characteristic it may be desirable for all IF amplifiers to be controlled by the AGC loop, and it is important that amplifier distortion should be low throughout the dynamic range of the control loop.

The dual-gate mosfet (**Fig 6.64**) with reverse AGC on gate 1 and partial forward AGC on gate 2 has excellent cross-modulation properties but the control range is limited to about 35dB per stage. Integrated circuits with high-performance gain-controlled stages are available.

Where a high-grade SSB or CW filter is incorporated it is vital to ensure that signals cannot 'leak' around the filter due to stray coupling; good screening and careful layout are needed.

In multiple-conversion receivers, it is possible to provide continuously variable selectivity by arranging to vary slightly the frequency of a later conversion oscillator so that the band-pass of the two IF channels overlap to differing degrees. For optimum results this requires that the shape factor of both sections of the IF channel should be good, so that the edges are sharply defined.

With double-tuned IF transformers, gain will be maximum when the product kQ is unity (where k is the coupling between the windings). IF transformers designed for this condition are said to be critically coupled; when the coupling is increased beyond this point (over-coupled) maximum gain occurs at two points equally spaced about the resonant frequency with a slight reduction of gain at exact resonance: this condition may be used in broadcast receivers to increase bandwidth for good-quality reception.

If the coupling factor is lowered (under-coupled) the stage gain falls but the response curve is sharpened, and this may be useful in communications receivers.

## FETs as Small-signal Amplifiers

Field effect transistors make good small-signal RF or IF amplifiers and offer advantages when used properly. They can provide low-noise, rugged amplifiers once wired in circuits having an easy, low-resistance path between their gate(s) and earth, though they are vulnerable to electrostatic discharge when out of circuit, unless protected by an internal diode(s). However, despite being, like thermionic valves, voltage-controlled devices, they should not be used as replacements for valves in similar circuits.

FETs have very high slope of up to 30mA/V (much higher than most valves) but also much greater drain-to-gate capacitance than the inter-electrode capacitances of triode valves. This is a sure prescription for self-oscillation if connected in a typical pentode-type amplifier circuit.

However, the FET can be used with a very low output load impedance and thus can be used as a stable low-gain device, yet providing excellent stage gain because of the voltage step-up that can be readily achieved with a resonant input transformer.

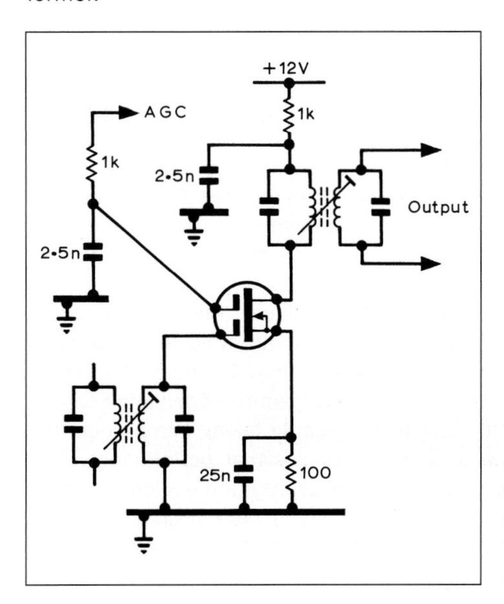


**Fig 6.64: Typical automatic gain-controlled IF amplifier, a dual-gate mosfet**

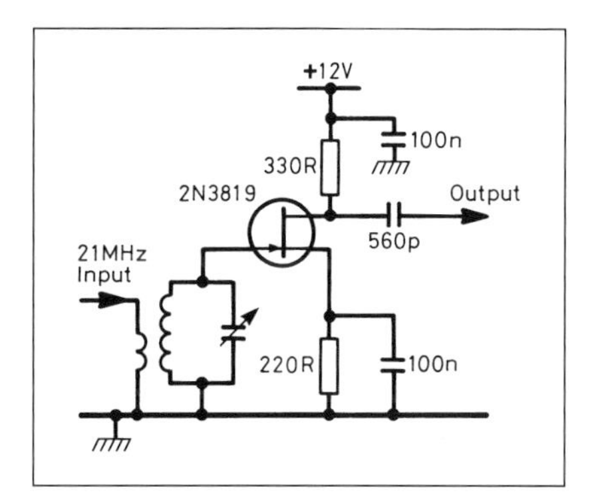


Fig 6.65: Stable FET preamp with low-impedance output load and with the main part of the gain coming from the step-up input transformer

For example, the 21MHz preamplifier shown in **Fig 6.65** using the 2N3819 FET (10mA/V) with a 330 ohm resistor as load has a device gain of only three, but the input tuned circuit can provide a voltage gain of about seven, resulting in a stable voltage gain of about 21, with the FET's very high input impedance presenting only light loading of the input transformer. This provides an unconditionally stable stage gain of over 20dB. **Figs 6.66, 6.67 and 6.68** show typical FET and dual-gate FET small-signal amplifiers.

The most critical amplifier in a modern, high-performance receiver is usually the post-mixer amplifier, ie the IF amplifier that follows the mixer, either directly or after a crystal filter. It needs to be of low noise in order to cope with the conversion loss of a ring mixer, and if preceding the crystal filter will have to cope with large off-frequency signals, requiring a high intercept characteristic.

For the highest-performance receivers, push-pull bipolar transistors with a noise figure of about 2dB and a third-order intercept point equal or better than that of the mixer are required.

Possibly the simplest post-mixer amplifier for high-performance receivers is a power-FET common-gate stage, such as that shown earlier. With a 2N5435 FET this can provide a 2dB noise figure with a 50 ohm system gain of 9dB and an output third-order intercept point of +30dBm when biased at $V_{DD}$ of 12-15V at 50mA.

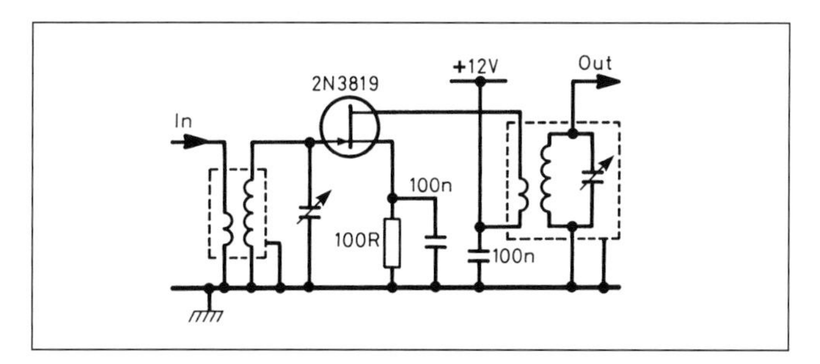


Fig 6.66: Stable FET IF amplifier using two bipolar transistor-type IF transformers in reverse configuration

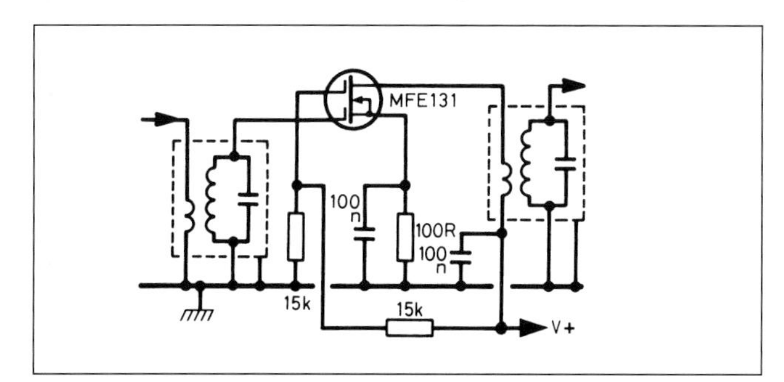


Fig 6.67: Dual-gate FET IF amplifier

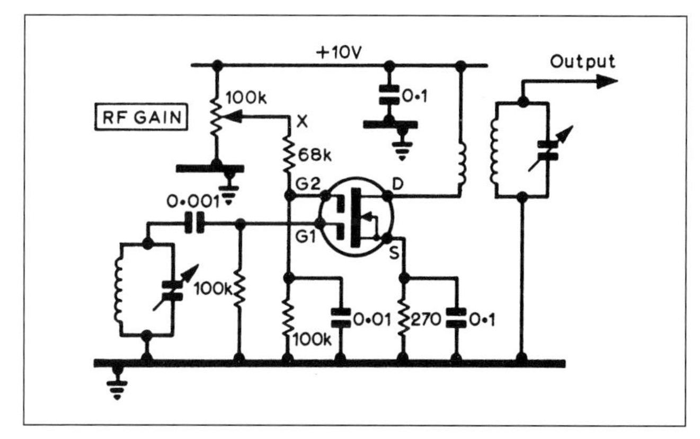


Fig 6.68: Typical dual-gate MOSFET RF or IF amplifier. G2 is normally biased to about one-third of the positive voltage of drain. In place of manual gain, control point X can be connected to a positive AGC line

## Demodulation

For many years, the standard form of demodulation for communications receivers, as for broadcast receivers, was the envelope detector using diodes. Envelope detection is a non-linear process (part mixing, part rectification) and is inefficient at very low signal levels.

On weak signals this form of detector distorts or may even lose the intelligence signals altogether. On the other hand, synchronous or product detection preserves the signal-to-noise ratio, enabling post-detector signal processing and audio filters to be used effectively (**Fig 6.69**).

Synchronous detection is essentially a frequency conversion process and the circuits used are similar to those used in mixer stages.

The IF or RF signal is heterodyned by a carrier at the same frequency as the original carrier frequency and so reverts back to the original audio modulation frequencies (or is shifted from these frequencies by any difference between the inserted carrier and the original carrier as in CW where such a shift is used to provide an audio output between about 500 and 1000Hz). Typical product detector circuits are shown in **Fig 6.70**.

It should be noted that a carrier is needed for both envelope and product detection: the carrier may be radiated along with the sidebands, as in AM, or locally generated and inserted in the receiver (either at RF or IF - usually at IF in superhets, at RF in direct-conversion receivers).

Synchronous or product detection has been widely adopted for SSB and CW reception in amateur receivers; the injected carrier frequency is derived from the beat frequency oscillator, which is either LC or crystal controlled. By using two crystals it is possible to provide selectable upper or lower sideband reception.

The use of synchronous detection can be extended further to cover AM, DSBSC, NBFM and RTTY but for these modes the injected carrier really needs to be identical to the original carrier,

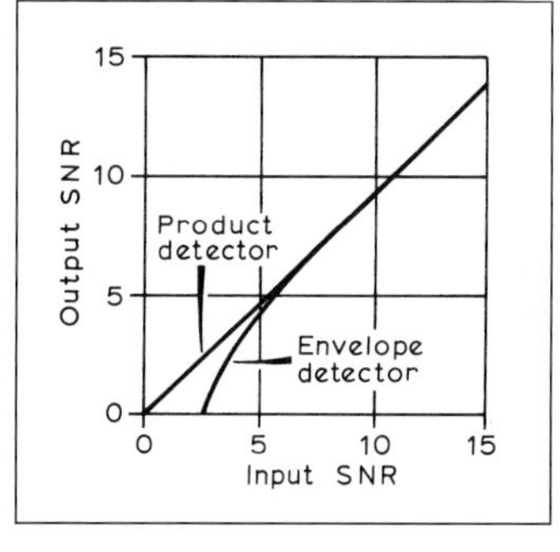


Fig 6.69: Synchronous (product) detection maintaining the SNR of signals down to the lowest levels whereas the efficiency of envelope detection falls off rapidly at low SNR, although as efficient on strong signals

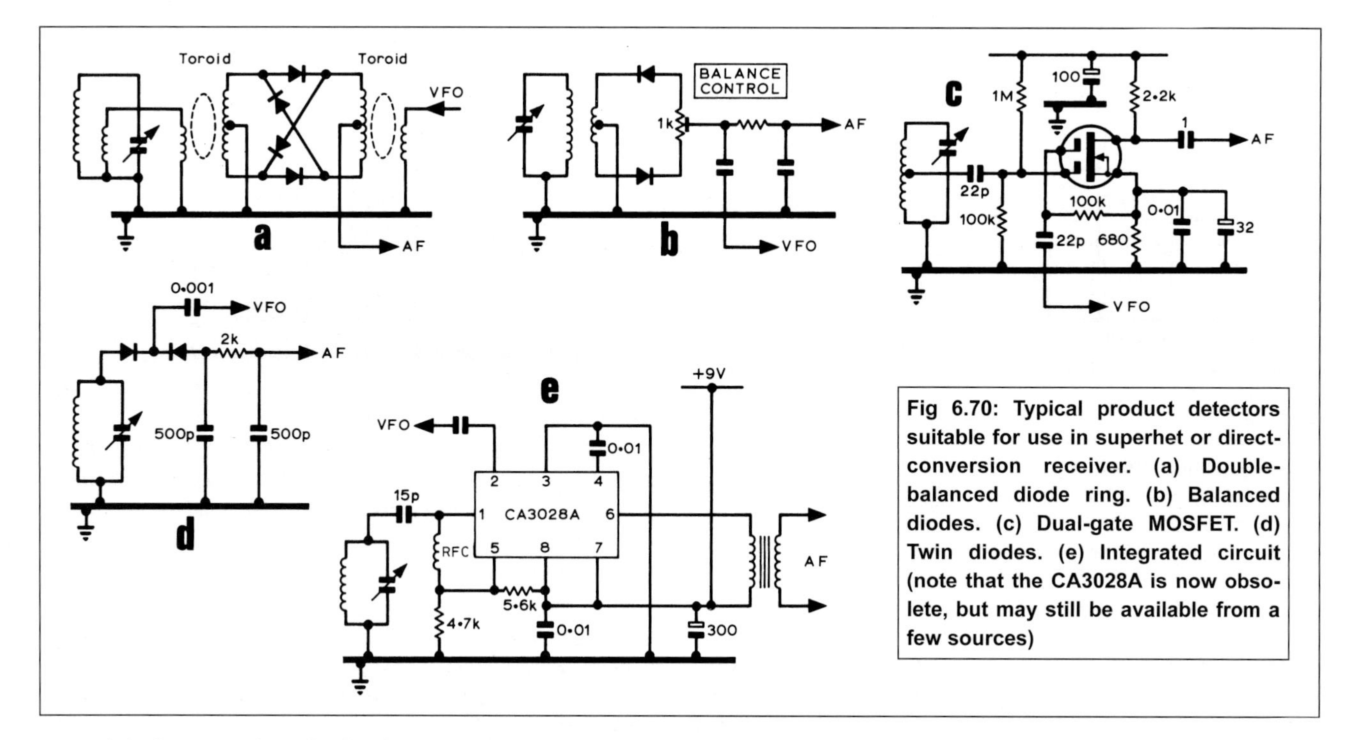


**Fig 6.70: Typical product detectors suitable for use in superhet or direct-conversion receiver. (a) Double-balanced diode ring. (b) Balanced diodes. (c) Dual-gate MOSFET. (d) Twin diodes. (e) Integrated circuit (note that the CA3028A is now obsolete, but may still be available from a few sources)**

not only in frequency but also in phase: that is to say the local oscillator needs to be in phase coherence with the original carrier (an alternative technique is to provide a strong local carrier that virtually eliminates the original AM carrier - this is termed exalted carrier detection).

Phase coherence cannot be achieved between two oscillators unless some effective form of synchronisation is used. The simplest form of synchronisation is to feed a little of the original carrier into a local oscillator, so forcing a phase lock on a free-running oscillator; such a technique was used in the synchrodyne receiver. The more usual technique is to have a phase-lock loop. At one time such a system involved a large number of components and would have been regarded as too complex for most purposes; today, however, complete phase-lock loop detectors are available in the form of a single integrated circuit, both for AM and NBFM applications. Apart from the phase-lock loop approach, a number of alternative forms of synchronous multimode detectors have been developed. One interesting technique which synthesises a local phase coherent carrier from the incoming signal is the reciprocating detector.

## Noise Limiters, Null-steerers and Blankers

The HF spectrum, particularly above 15MHz or so, is susceptible to man-made electrical impulse interference stemming from electric motors and appliances, car ignition systems, thyristor light controls, high-voltage power lines and many other causes. Static and locally generated interference from appliances, TV receivers etc can be a serious problem below about 5MHz, and may be maximum at LF/MF.

These interference signals are usually in the form of high-amplitude, short-duration pulses covering a wide spectrum of frequencies. In many urban and residential areas this man-made interference sets a limit to the usable sensitivity of receivers and may spoil the reception of even strong amateur signals.

Because the interference pulses, though of high amplitude, are often of extremely short duration, a considerable improvement can be obtained by 'slicing' off all parts of the audio signal which are significantly greater than the desired signal. This can be done by simple AF limiters such as back-to-back parallel diodes. For AM reception, more elegant noise limiters develop fast-acting biasing pulses to reduce momentarily the receiver gain during noise peaks. The ear is much less disturbed by 'holes of silence' than by peaks of noise. Many limiters of this type have been fitted in the past to AM-type receivers.

Unfortunately, since the noise pulses contain high-frequency transients, highly selective IF filters will distort and broaden out the pulses. To overcome this problem, noise blankers have been developed which derive the blanking bias potentials from noise pulses which have not passed through the receiver's selective filters. In some cases a parallel broadly tuned receiver is used, but more often the noise signals are taken from a point early in the receiver.

For example, the output from the mixer goes to two channels: the signal channel which includes a blanking control element which can rapidly reduce gain when activated; and a wide-band noise channel to detect the noise pulse and initiate the gain reduction of the signal channel.

To be most effective it is necessary for the gain reduction to take place virtually at the instant that the interference pulse begins. In practice, because of the time constants involved, it is difficult to do this unless the signal channel incorporates a time delay to ensure that the gain reduction can take place simultaneously with or even just before the noise pulse. One form of time delay which has been described in the literature utilises a PAL-type glass ultrasonic television delay line to delay signals by 64 microseconds. It is, however, difficult to eliminate completely transients imposed on the incoming signal.

One possible approach, which has been investigated at the University College, Swansea, is to think in terms of receivers using synchronous demodulation at low level so that a substantial part of the selectivity, but not all of it, is obtained after demodulation. This allows noise blankers to operate at a fairly low level on AF signals.

A control element which has been used successfully consists of a FET gate pulsed by signals derived from a wide-band noise amplifier. The noise gate is interposed between the mixer and the first crystal filter, with the input signal to the noise amplifier taken off directly from the mixer.

Noise limiters and noise blankers are suitable for use only on

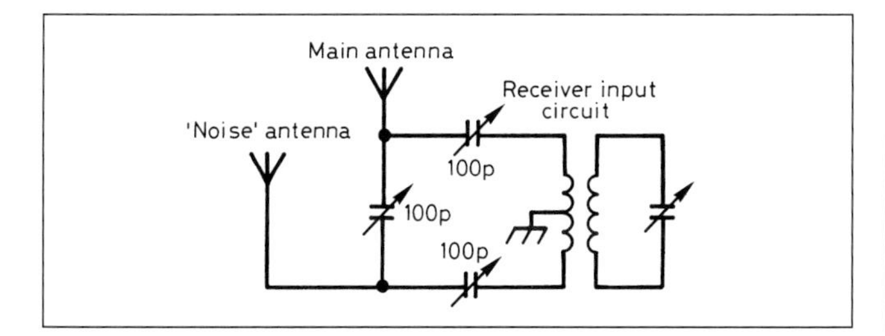


**Fig 6.71: The original Jones noise-balancing arrangement as shown in early editions of *The Radio Handbook*. Local electrical interference could be phased out by means of pick-up on the auxiliary 'noise' antenna. Although it could be effective it required very careful setting up**

pulse-type interference in which the duty cycle of the pulse is relatively low. An alternative technique, suitable for both continuous signals and noise pulses, is to null out the unwanted signal by balancing it with anti-phase signals picked up on a short 'noise antenna'.

This has led to the revival of the 1930s Jones noise-balancing technique in which local interference could be phased out by means of pick-up on an auxiliary noise antenna. This system (**Fig 6.71**) was capable of reducing a specific unwanted source of interference by tens of decibels while reducing the wanted signal by only about 3-6dB, but required critical adjustment of the controls.

In the early 1980s John K Webb, W1ETC, developed a more sophisticated method of phasing out interference using coiled lengths of coaxial cable as delay lines to provide the necessary phase shifts: **Fig 6.72**. This included a compact null-steerer located alongside the receiver/transceiver, capable of generating deep nulls against a single source of interference, resembling the nulls of an efficient MW ferrite-rod antenna. The two controls adjusted phase and amplitude of the signals from the auxiliary noise antenna. W1ETC summarised results [16] with such a unit as follows:

1. The available null depth in signals propagated over short paths of up to 20 miles is large and stable, limited only by how finely the controls are adjusted.
2. Nulls on signals arriving over short skywave paths of up to a few hundred miles are in the order of 30dB, provided there is a single mode of propagation and one direction of arrival. Such nulls are usually stable.
3. Signals propagated over paths of 10 to 100 miles may arrive as a mixture of ground-wave and skywave. A single null is thus ineffective.
4. Signals propagated by skywave over long distances frequently involve several paths, each having a different path length so that a single null has little effect on what is usually the 'wanted' signal.
5. Broad-band radiated noise can be nulled as deeply as any radio signal. This seems to be a more effective counter to noise than blanking or limiting techniques with local electrical noise deeply nulled, and with little effect on wanted long-distance signals.

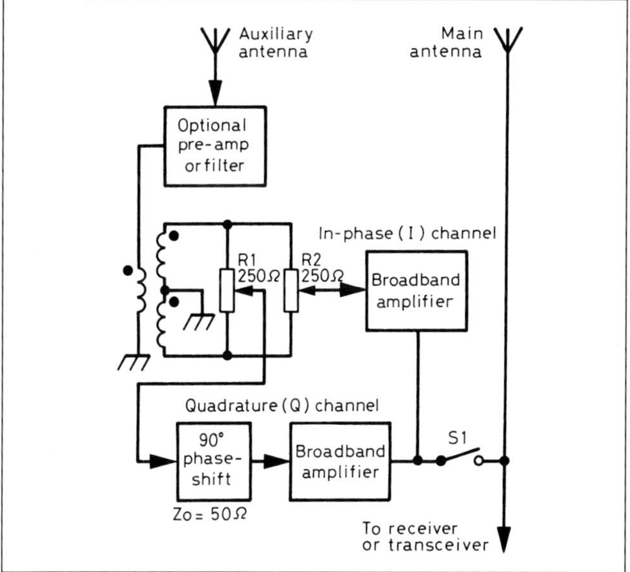


**Fig 6.72: Functional diagram of the electronic null steering unit for use in conjunction with an HF receiver or transceiver (S1 is a relay contact to disconnect the system during transmission) as described by John Webb, W1ETC, of the Mitre Corporation in 1982**

To meet result (5) the interference has to be directly radiated to the receiver antennas and not enter the receiver in a less-directional manner (for example re-radiation from mains cabling etc). Various methods of implementing null-steering have been described: **Figs 6.73-6.75** show a design by Lloyd Butler, VK5BR, utilising the phase shifts of off-tune resonant circuits.

## AF Stages

The AF output from an envelope or product detector of a superhet receiver is usually of the order of 0.5 to 1V, and many receivers incorporate relatively simple one- or two-stage audio amplifiers, typically using an IC device and providing about 2W output. On the other hand the direct-conversion receiver may require a high-gain audio section capable of dealing with signals of less than 1μV.

Provided that all stages of the receiver up to and including the product detector are substantially linear, many forms of post-demodulation signal processing are possible: for example band-pass or narrow-band filtering to optimise signal-to-noise ratio of the desired signal, audio compression or expansion; the removal of audio peaks, AF noise blanking, or (for CW) the removal by

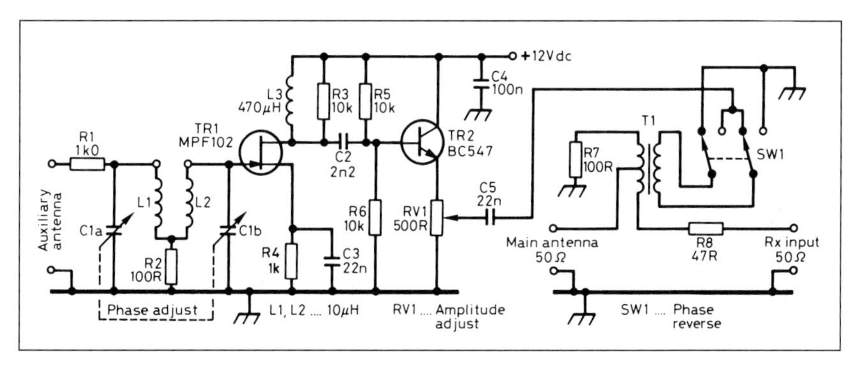


**Fig 6.73: VK5BR's Mk 2 interference-cancelling circuit as described in the January 1993 issue of *Amateur Radio*. As shown this covers roughly 3.5 to 7MHz. C1 ganged 15-250pF variable capacitor or similar. L1, L2 miniature 10μH RF chokes. T1 11 turns quadfilar wound on Amidon FT-50-75m toroidal core**

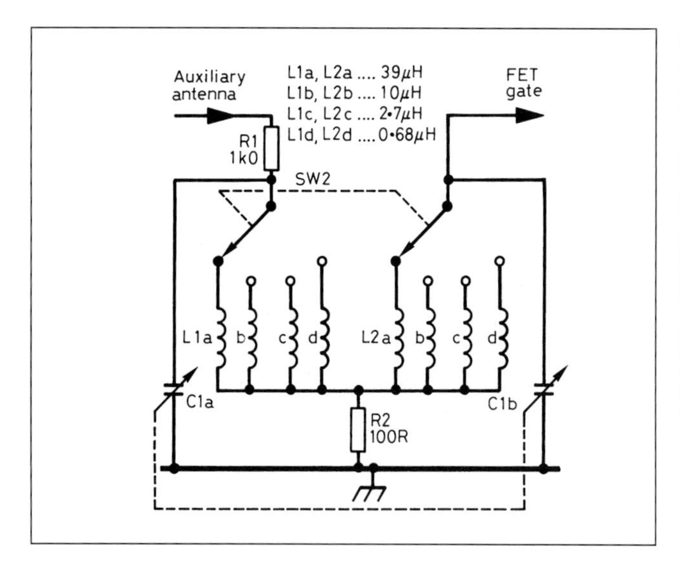


**Fig 6.74: Bandswitching modification to provide 1.8 to 30MHz in four ranges**

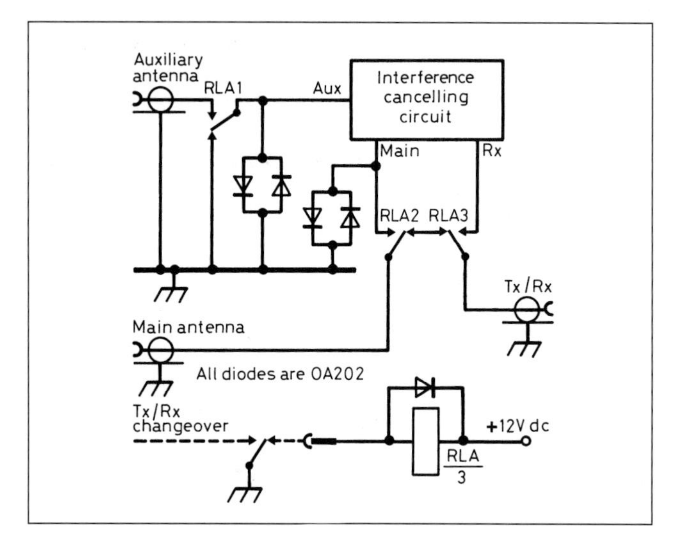


**Fig 6.75: Transmit-receive switching with protection diodes for use with the VK5BR interference-cancelling circuit with a transceiver**

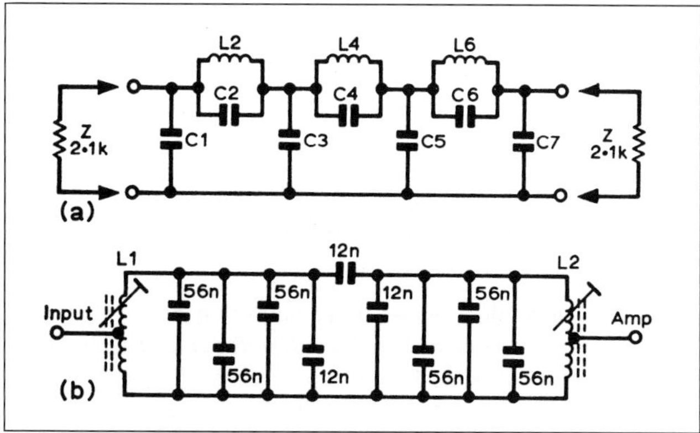


**Fig 6.76: (a) Phone and (b) CW AF filters suitable for use in direct-conversion or other receivers requiring very sharply defined AF responses. The CW filter is tuned to about 875Hz. Values for (a) can be made from preferred values as follows: C1 37.26nF (33,000 + 2200 + 1800 + 220pF); C2 3.871nF (3300 + 560pF); C3 51.87nF (47000 + 4700 + 150pF); C4 19.06nF (18,000 + 1000pF); C5 46.41nF (39,000 + 6800 + 560pF); C6 13.53nF (12,000 + 1500pF); C7 29.85nF (27,000 + 2700 + 150pF). All capacitors mica or polyester or styroflex types. L2 168.2mH (540 turns), L4 124.5mH (460 turns); L6 129.5mH (470 turns) using P30/19 3H1 pot cores and 0.25mm enam wire. Design values based on 2000Ω impedance**

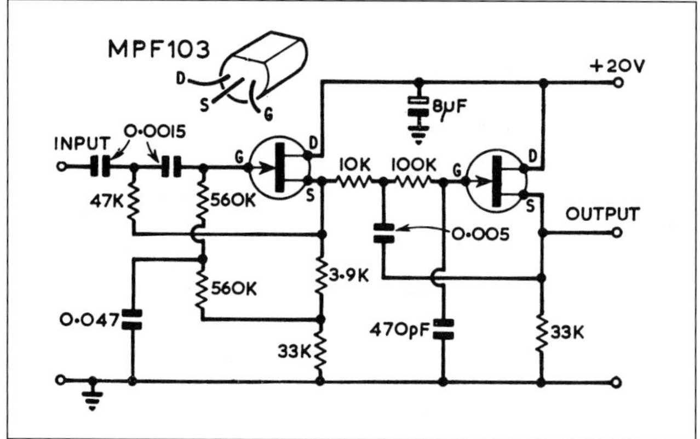


**Fig 6.77: Active band-pass AF filter for amateur telephony. -6dB points are about 380 and 3200Hz, -18dB about 160 and 6000Hz**

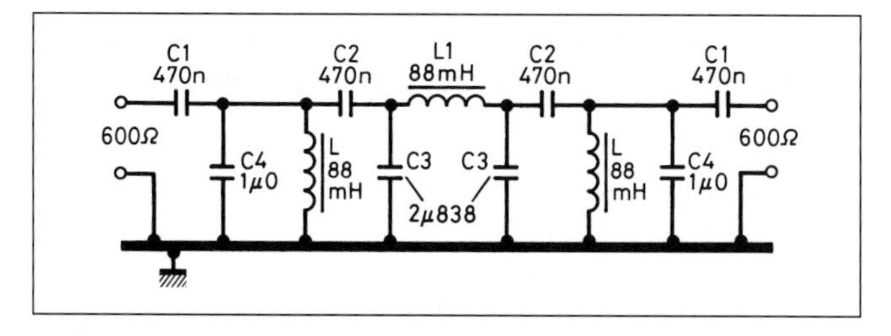


**Fig 6.78: Passive AF filter by DJ1ZB using standard 88mH toroids and with a centre frequency of about 420Hz and bandwidth about 80Hz. Note that this design is for 600-ohm input/output impedance [*Sprat* No 58]**

gating of background noise. Audio phasing techniques may be used to convert a DSB receiver into an SSB receiver (as in two-phase or third-method SSB demodulation) or to insert nulls into the audio pass-band for the removal of heterodynes. Then again, in modern designs the AGC and S-meter circuits are usually operated from a low-level AF stage rather than the IF-derived techniques used in AM-type receivers.

It should be appreciated that linear low-distortion demodulation and AF stages are necessary if full advantage is to be taken of such signal processing, since strong intermodulation products can easily be produced in these stages. Thus, despite the restricted AF bandwidth of speech and CW communications, the intermodulation distortion characteristics of the entire audio section should preferably be designed to high-fidelity audio standards. Very sophisticated forms of audio filtering, notching and noise reduction are now marketed in the form of add-on digital-signal-processing (DSP) units, or may be incorporated into a receiver or transceiver.

Audio filters may be passive, using inductors and capacitors, or active, usually with resistors and capacitors in conjunction with op-amps or FETs (see **Figs 6.76-6.79**). Many different circuits have been published covering AF filters of variable bandwidth, tuneable centre frequencies and for the insertion of notches.

The full theoretical advantage of a narrow-band AF filter for CW reception may not always be achieved in operational use: this is because the human ear can itself provide a 'filter' bandwidth of about 50Hz with a remarkably large dynamic range and the ability to tune from 200 to 1000Hz without introducing

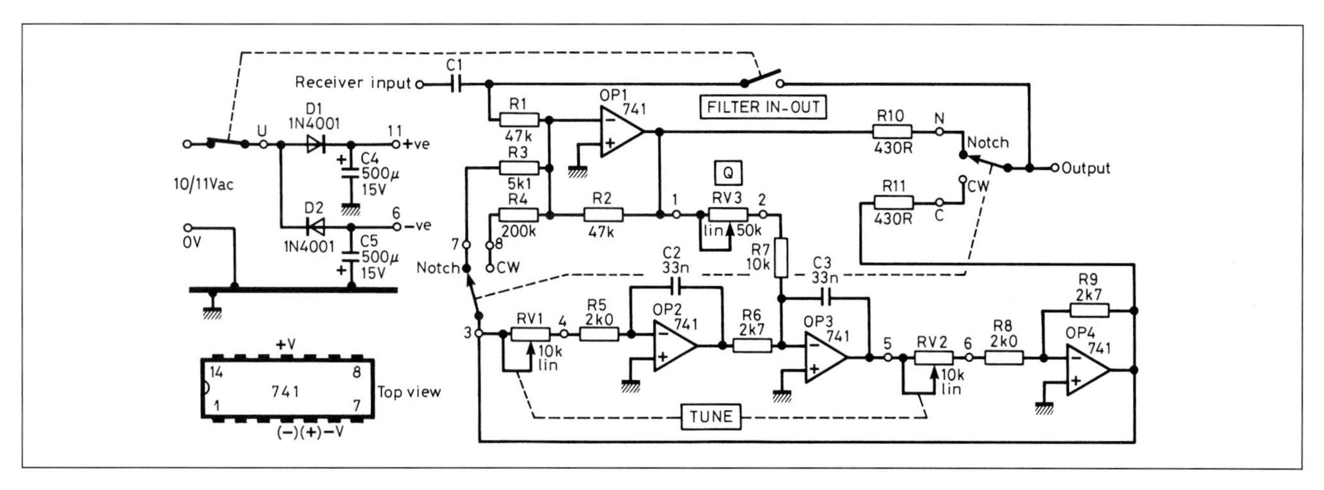


**Fig 6.79: Versatile active analogue AF filter for speech or CW reception as described originally by DJ6HP in 1974 and which continues to represent an effective design. It provides a CW filter tuneable over about 450 to 2700Hz with the Q (bandwidth) variable over a range of about 5:1. For speech the filter can be switched to a notch mode. Although modern digital audio filters could provide more precisely shaped tuneable filtering, this analogue filter has received many endorsements over the years**

'ringing'.

Finally, there is the loudspeaker, often the cheapest device on a dreadful baffle resulting in a tinny rattley sound that throws away a lot of the earlier achievements. Choose the loudspeaker carefully and fit it in its own rigid, ie wooden, enclosure even if it then goes into a larger enclosure.

## DESIGNS FOR HOME CONSTRUCTION

The simplest types of HF receiver suitable for home construction 'on the kitchen table' without the need for high-grade measuring equipment etc are undoubtedly the various forms of 'straight' (direct-conversion) receivers using either a regenerative detector, Q-multiplier plus source-follower detector, or the now more popular homodyne form of direct conversion, preferably using a balanced or double-balanced product detector. In practice, to avoid the complications of band-switching, most of the simple models tend to be either single-band receivers or may still use the once-popular 'plug-in' coils.

**Figs 6.80 - 6.85** show a representative selection of circuit diagrams for simple receivers, suitable for use directly with high-impedance headphones, or with output transformers permitting the use of modern low-impedance headphones.

As mentioned in the previous section, simple regenerative receivers of the type shown in Figs 6.80 and 6.81 can re-radiate the local oscillator signal which can cause interference over a sometimes considerable area. Care may need to be taken to ensure this does not occur.

For simple receivers where a very high dynamic range is not sought, construction can be simplified by the use of SA602 type IC devices which contain a double-balanced mixer, oscillator and isolator stages.

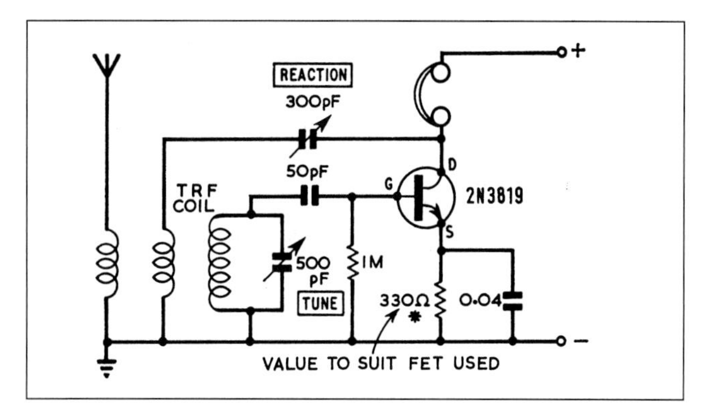


**Fig 6.80: Single-FET receiver**

This device, originally developed for VHF portable radiophones, has been widely adopted by amateurs for use as frequency converters, complete front-ends for direct-conversion receivers, and as product detectors etc.

**Fig 6.86** outlines the SA602 and some ways it can be used, while **Fig 6.87** shows its use as the front-end of a 28MHz direct-conversion receiver which could be adapted for lower HF bands.

The SA602 is equally suitable for use as a crystal-controlled frequency converter to provide extra bands in front of an exist-

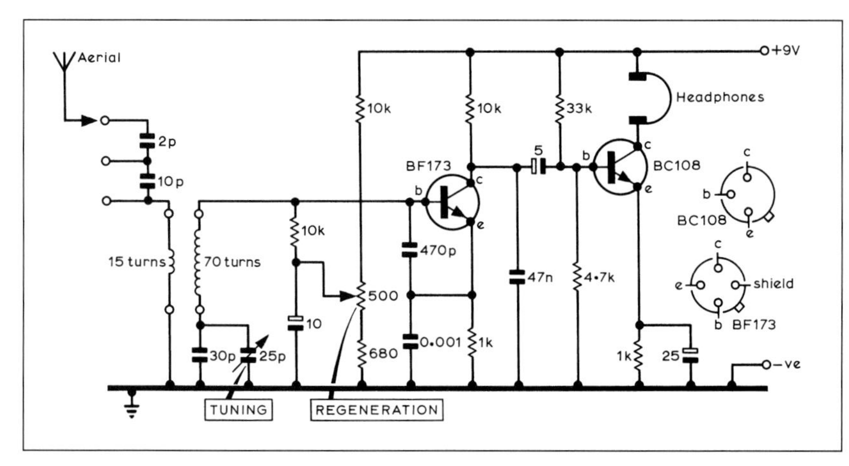


**Fig 6.81: A simple 'straight' receiver intended for 3-5MHz SSB/CW reception and using Clapp-type oscillator to improve stability**

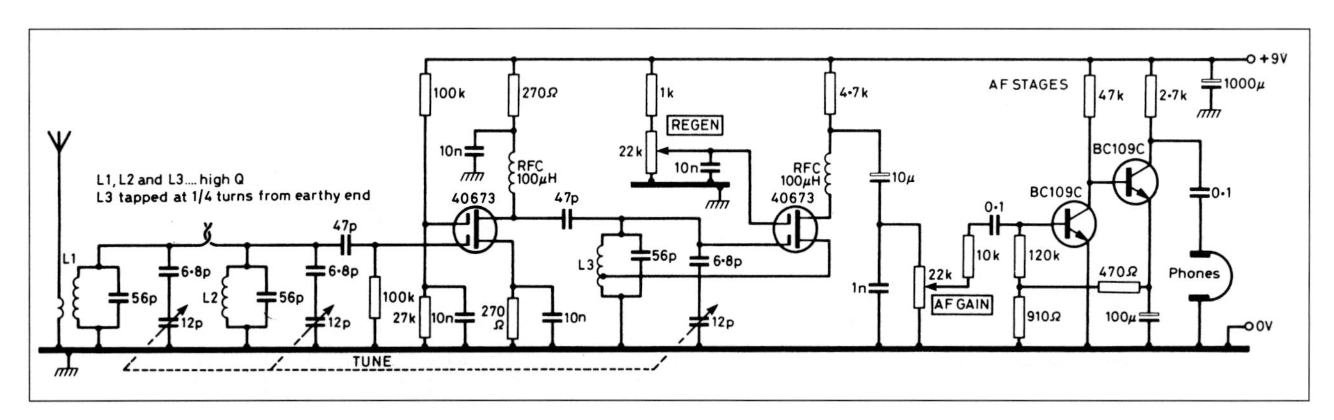


**Fig 6.82: Solid-state 14MHz TRF 'straight' receiver originally described by F9GY in *Radio-REF* in the 1970s and intended primarily as a monoband CW receiver**

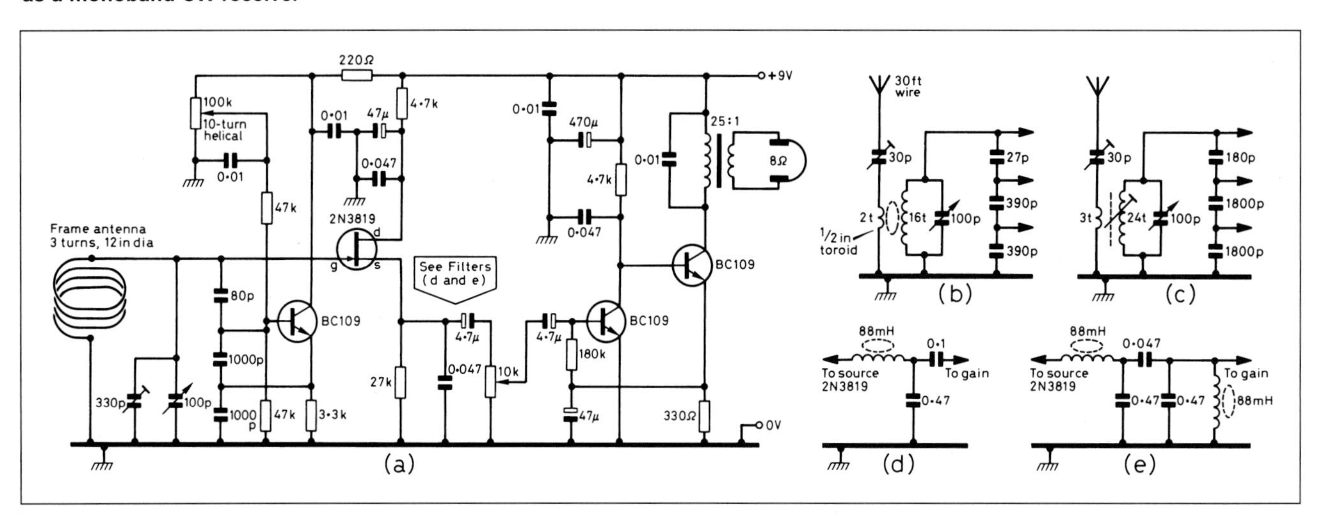


**Fig 6.83: GI3XZM's solid-state regenerative 'blooper' receivers. (a) 3.5MHz version with frame antenna mounted about 12in above chassis with miniature coaxial-cable 'downlead'. (b) Input circuit for 9-16MHz version. (c) Conventional input circuit for 3.5MHz receiver using wire antenna (coil 26SWG close-wound on 0.5in slug-tuned former). (d) Audio filter that replaces the 4.7µF capacitor shown in (a). (e) CW filter 14MHz Mk2 version**

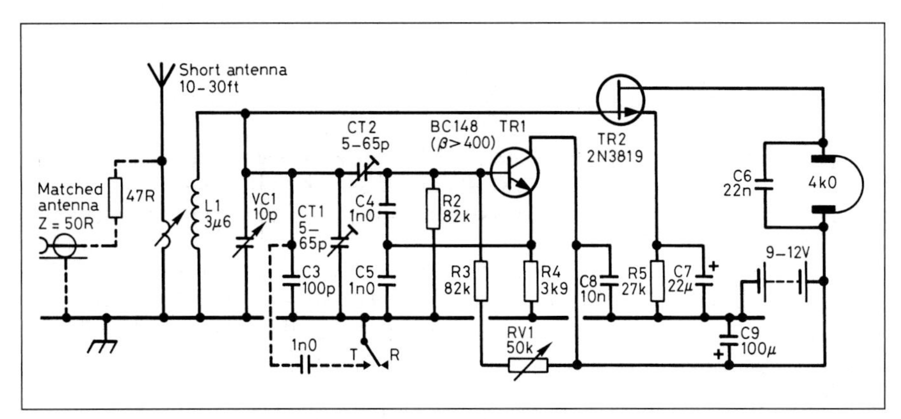


**Fig 6.84: G3RJT's 'two-transistor communications receiver' or 'active crystal-set receiver', based on the design approach of GI3XZM but using a higher-gain drain-bend detector that permits the omission of the two-transistor AF amplifier provided that high-impedance headphones of good sensitivity are used**

ing receiver or for the 'super-gainer' form of simple superhet or as the mixer/oscillator stages of a conventional superhet, possibly using a second SA602 as a product-detector.

Receivers based on standard IC devices have particular application for 'listening' or as the receiver section of compact low-power transceivers.

**Fig 6.88** shows how a Motorola MC3362 IC can form the complete front-end of a single-band superhet including IF and detector stages as part of a compact transceiver. The chip pinout is in **Fig 6.89**.

## Super-linear Front-ends

The front-end of a superhet or direct-conversion receiver comprises all circuitry preceding the main selectivity filter. For a superhet this includes the passive preselector, the RF amplifier(s), the mixer(s) and heterodyne oscillator(s), the diplexer between mixer and post-mixer amplifier, the roofing filter, and any IF stages up to an including the main (crystal) filter. For a direct-conversion receiver, the front-end comprises all stages up to the selective audio-filter(s), including the product detector which for a high-performance receiver may be of the two-phase, audio-image-rejecting type.

For any receiver, superhet or direct-conversion, in which digital signal processing at IF or audio baseband is used to determine

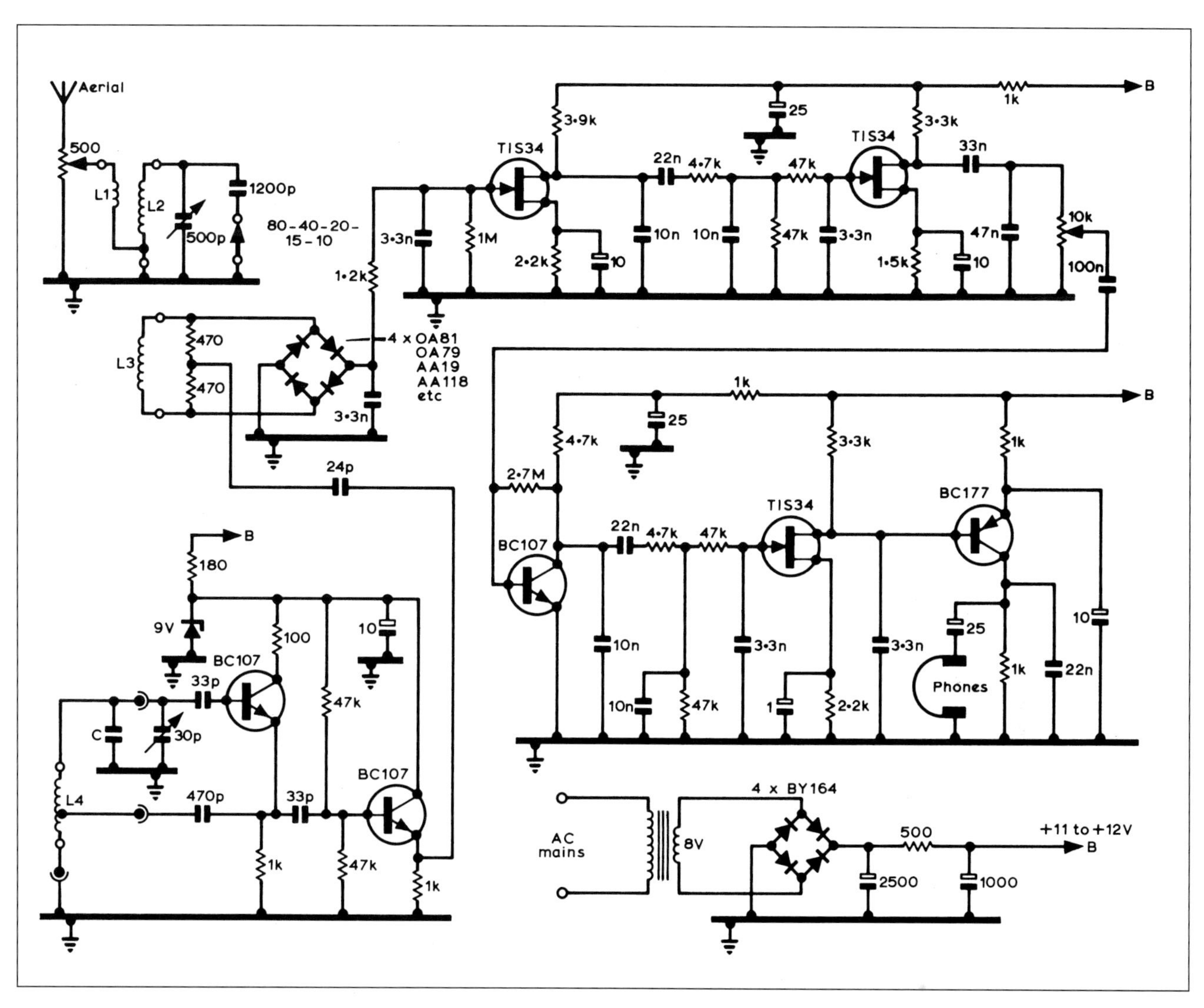


**Fig 6.85: Complete circuit diagram of a multiband direct-conversion receiver using diode ring demodulator and plug-in coils for oscillator section**

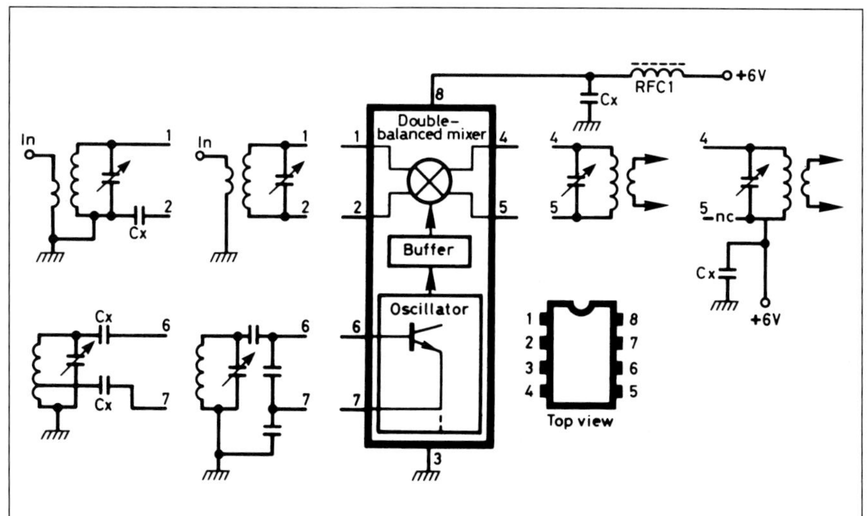


**Fig 6.86: Typical configurations of the SA602. Balanced circuits are to be preferred but may be more difficult to implement. Cx blocking capacitor 0.001 to 0.1µF depending on frequency. RFC1 (ferrite beads or RF choke) recommended at higher frequencies. Supply voltage should not exceed 6V (2.5mA). Noise figure about 5dB. Mixer gain 20dB. Third-order intercept +15dBm (do not use an RF preamplifier stage). Input and output impedances are both 2 x 1.5kΩ.**

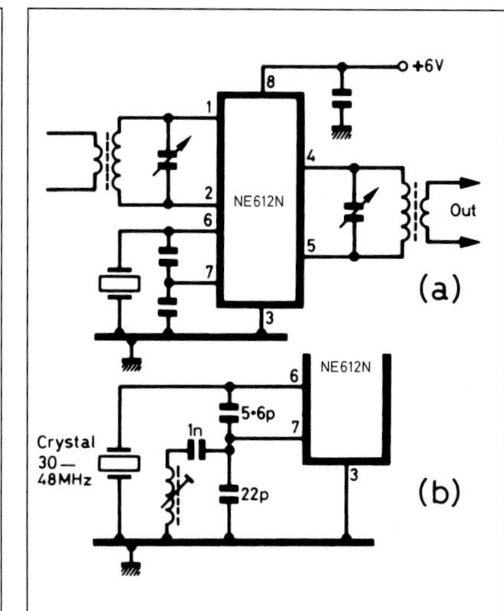


**Fig 6.87: The use of an SA602 IC as a crystal-controlled converter. The 5dB noise figure is low enough to achieve optimum sensitivity right up to about 50MHz without having to build an RF amplifier**

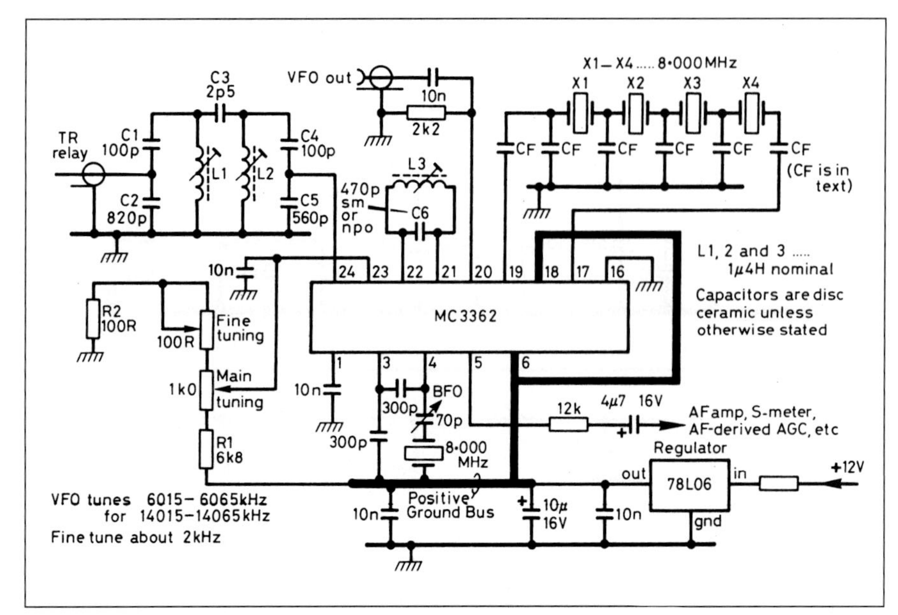


Fig 6.88: K9AY uses the MC3362 as the complete front-end of the 14MHz superhet receiver

Fig 6.89: The MC3362 chip showing pin-out and basic functions

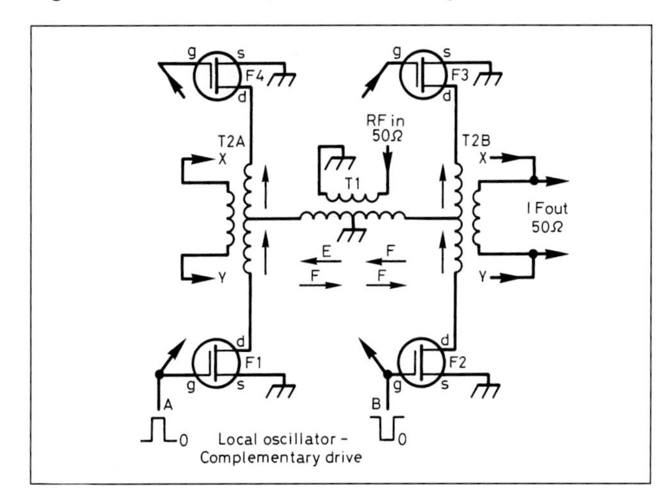


Fig 6.90: G3SBI 'H-mode' mixer

the selectivity, the A/D converter and digital filtering must be considered in determining the front-end performance in terms of linearity and dynamic range.

In designing a receiver for the highest possible front-end performance, attention must be paid to all of the circuitry involved, to the gain distribution, to the noise characteristics, to both the strong-signal handling characteristics and to the intermodulation intercept points.

The ability to hold and copy an extremely weak signal, barely above the atmospheric noise level, adjacent to a strong local signal or close to signals from super-high-power broadcast signals places heavy demands on the active and passive components available within amateur budgets, including any ferrite-cored transformers and the crystal filters.

The mixer developed by G3SBI (intellectual title held by SERC) is shown in **Fig 6.90** which illustrates why it has been given the name 'H-mode'. Operation is as follows: Inputs A and B are complementary square-wave inputs derived from the sine-wave local oscillator at twice the required frequency. If A is 'on' then FETs F1 and F3 are 'on' and the direction of the RF signal across T1 is given by the 'E' arrows. When B is 'on', FETs F2 and F4 are 'on' and the direction of the RF signal across T1 reverses (arrows 'F'). This is still the action of a commutation mixer, but

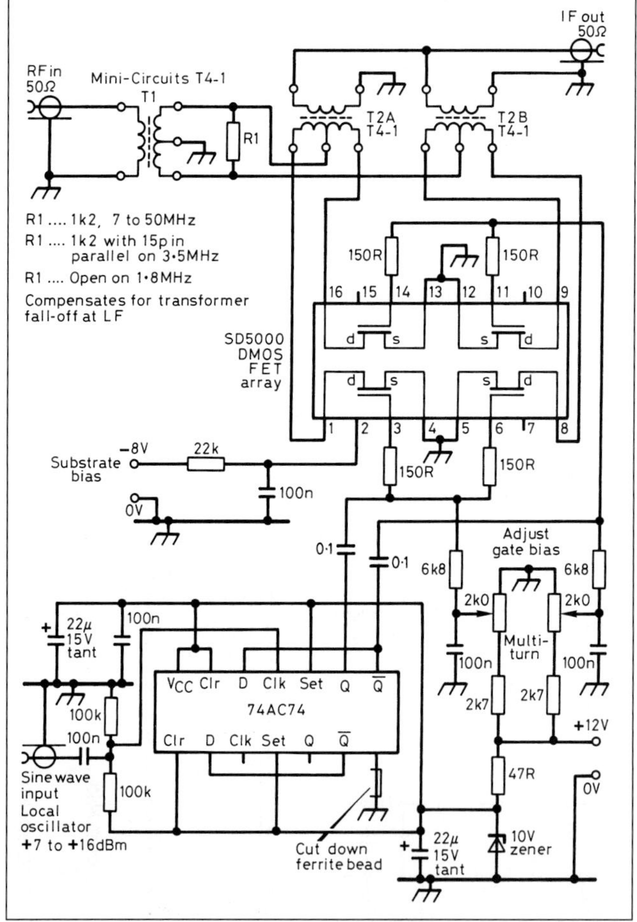


Fig 6.91: Test assembly for 'H-mode' mixer

now the source of each FET switch is grounded, so that the RF signal switched by the FET cannot modulate the gate source voltage.

In this configuration the transformers are important: T1 is a Mini-Circuits type T4-1; T2 is two Mini-Circuits T4-1 transformers with their primaries connected in parallel. The parallel-connected transformers give good balance and perform well.

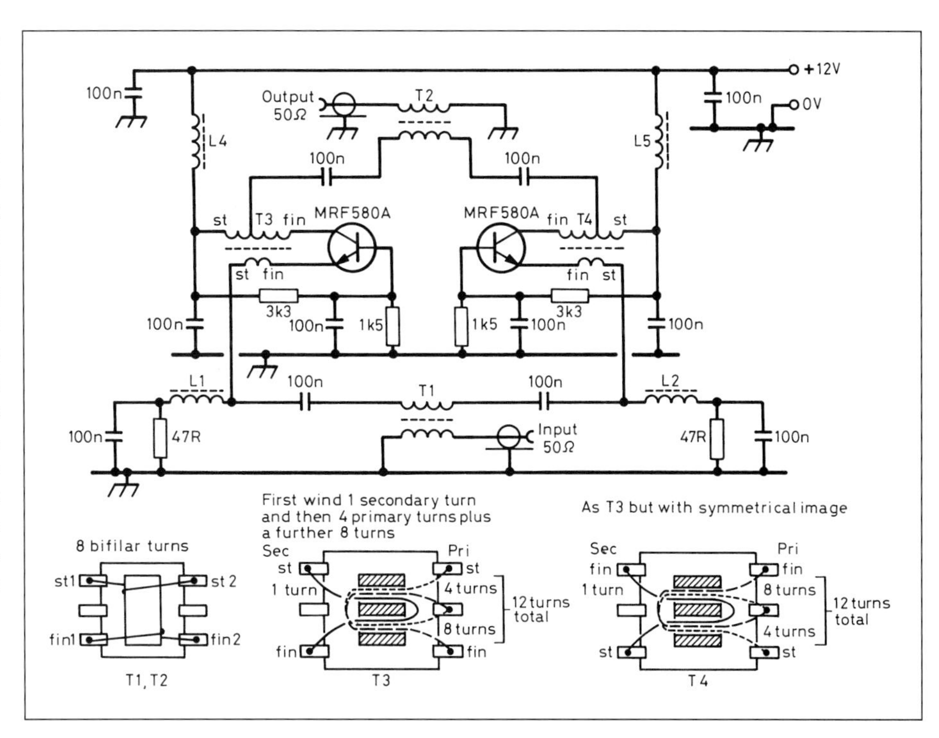


Fig 6.92: G3SBI's modified post-mixer test amplifier adapted from the N6NWP design but using the MRF580A devices. All resistors 0.25W metal-film RS Components. All 0.1µF capacitors monolithic ceramic RS Components. L4, L5 4t of 0.315mm dia bicelflux wire on RS Components ferrite bead. L1, L2 5t 0.315mm dia bicelflux wire (RS Components). T1-T4 use 40SWG bicelflux wire. Take two glassfibre Cambion 14-pin DIP component headers, cut each into two parts and bend the tags 90° outwards. Stick a piece of double-sided tape onto the header and mount the balun cores on this. Wind the transformers as shown above. The amplifier is constructed with earth-plane layout

A practical test circuit of the H-mode mixer is shown in **Fig 6.91**. It was constructed on an earth plane board with all transformers and ICs mounted in turned-pin DIL sockets. The printed circuit tracks connecting T1 to T2 and from T2 to the SD5000 are kept short and of 0.015in width to minimise capacitance to ground. The local oscillator is divided by two in frequency and squared by a 74AC74 advanced CMOS bistable similar to that used in the N6NWP-type mixer. However, the bistable is run from +10V instead of +9V and a cut-down RS Components ferrite bead is inserted over the ground pin of the 74AC74 to clean up the square wave.

The preferred method of setting the gate-bias potentiometers with the aid of professional-standard test equipment is as follows. One potentiometer is set to the desired bias voltage for a specific test run, the other is then set by looking at the RF-to-IF path feedthrough on the spectrum analyser at 14MHz and adjusting the potentiometer for minimum IF feedthrough. The setting is quite sharp and ensures good mixer balance. An RF test signal of 11dBm (0.8V RMS) was used for each test signal for the two-tone IMD tests. The gate bias level chosen enabled an input level of +12dBm to be reached before the IMD increased sharply.

The performance of the H-mode test mixer was as follows. With an input RF test level of +11dBm (spaced at 2kHz or 20kHz) the conversion loss was 8dB; RF to IF isolation -68dB; LO to IF isolation -66dB. Input intercept points: 1.8 to 18MHz +53dBm; 21 to 28MHz +47dB, or better; 50MHz +41dB. These results were achieved with a gate-to-source DC bias of +1.95V and -8V substrate bias, a square-wave local oscillator amplitude of 9V and IF at 9MHz.

**Fig 6.92** shows a 9MHz post-mixer amplifier developed by G3SBI, again based on the N6NWP approach but with changes that provide improved performance in terms of gain and output intercept point, with a noise figure of 0.5dB. Whereas N6NWP used the MRF586 device, G3SBI used the MRF580A device, giving a lower noise figure at a collector current of 60mA. Measured performance showed a gain of 8.8dB, output intercept +56dBm, noise figure 0.5dB. However, a crystal filter driven by the amplifier would present a complex impedance, particularly on the slope and near the stop-band, and would seriously degrade performance of the amplifier.

G3SBI also investigated the performance of quadrature hybrid 9MHz crystal-filter combinations, and he found that the performance of budget-priced crystal lattice-type filters is a serious limitation (home-made ladder filters appear to have higher intercept points and lower insertion loss although the shape factor may not be quite so good). The problem with budget-priced lattice filters can be reduced by eliminating the post-mixer amplifier with the mixer going immediately to a quadrature hybrid network 2.4kHz-bandwidth filter, followed by a low-noise amplifier. The 2.4kHz filter is then used as a roofing filter. Although this is not an ideal arrangement, it can result in an overall noise figure of about 13.5dB (5dB noise figure due to the filter and amplifier, another 7dB from mixer loss, and a further 2.5dB loss due to the antenna input band-pass filter). This would be adequate sensitivity on 7MHz without a pre-mixer amplifier.

## Experimental DC Polyphase Receiver

This design by Hans Summers, G0UPL, [17] combines new and old techniques to produce a simple but high performance direct conversion receiver. As discussed above, the most obvious disadvantage of direct conversion is that both sidebands are detected. This can be solved via the use of audio phasing networks such that the unwanted sideband is mathematically cancelled by summation of correctly phased signals. Conventionally this required the use of two mixers, fed by phase shifted RF inputs.

A recent commutative mixer design by Dan Tayloe, N7VE, produces four audio outputs at 90-degree phase angles using a very simple circuit [18, 19] Despite its simplicity the detector boasts impressive performance as shown in **Table 6.4**.

The circuit is not really a mixer producing both sum and difference frequencies: it might more accurately be described as a switching integrator. It possesses a very useful bandpass filter characteristic, tracking the local oscillator frequency with a Q of typically 3000.

The quadruple phased audio output of the Tayloe Detector is ideally suited to drive a circuit of much older heritage: the passive polyphase network.

This consists of a resistor-capacitor network with resonant

| | |
|---|---|
| Conversion Loss: | 1dB |
| Minimum Discernable Signal (MDS): | -136dBm |
| Two Tone Dynamic Range (2TDR): | 111dB |
| Third Order Intercept (IP3): | +30dBm |
| Bandpass RF filter characteristic: | Q ~3000 |

**Table 6.4: N7VE's performance figures for his commutative mixer**

frequencies (poles) tuned such that they occur at evenly spaced intervals across the audio band of interest. The effect is a quite precise 90-degree phase shift throughout the audio band. In the current experimental receiver, the polyphase network connection is such that the unwanted sideband is mathematically completely cancelled inside the network, resulting in a single sideband audio output that is then amplified and filtered in a conventional way. **Fig 6.93** shows the block diagram of the experimental receiver design.

## Input Filter and detector

The high performance characteristics of the Tayloe Detector make a preceeding RF amplifier unnecessary. In this receiver, the RF signal is filtered by a simple bandpass filter consisting of two TOKO KANK3333 canned transformers (**Fig 6.94**). Circuit values are shown for 80m, but the circuit can readily be adapted to any HF amateur band, see **Table 6.5**.

The input to the Tayloe Detector is biased to mid-rail (2.5V) in

| BAND MHz | T1 - T2 TYPE | T1 - T2 Inductance | C1 - C3 [pF] | C2 [pf] |
|---|---|---|---|---|
| 1.8 - 2.0 | 3333 | 45µH | 150 | 12 |
| 3.5 - 3.8 | 3333 | 45µH | 39 | 3.3 |
| 7.0 - 7.1 | 3334 | 5.5µH | 100 | 8.2 |
| 10.1 - 10.15 | 3334 | 5.5µH | 47 | 6.8 |
| 14.0 - 14.35 | 3334 | 5.5µH | 22 | 3.3 |
| 18.07 - 18.17 | 3335 | 1.2µH | 68 | 6.8 |
| 21.0 - 21.17 | 3335 | 1.2µH | 47 | 4.7 |
| 24.89 - 24.99 | 3335 | 1.2µH | 33 | 3.3 |
| 28.0 - 29.7 | 3335 | 1.2µH | 22 | 3.3 |

**Table 6.5: Details of using the receiver on all HF amateur bands**

order to obtain maximum dynamic range. The FST3253 IC is a dual 1-4 way analogue multiplexer designed for memory bus switching applications and possessing high bandwidth and low ON-resistance.

A local oscillator signal at four times the reception frequency is required to accurately generate the necessary switching signals: a synchronous binary counter type 74HC163 accomplishes this easily. The four audio outputs are buffered by low noise NE5534 op-amps configured for a gain of 33dB. The gain of three of the op-amps is made adjustable by mult-turn 1kΩ preset potentiometers to allow the amplitude of each of the four paths to be matched precisely.

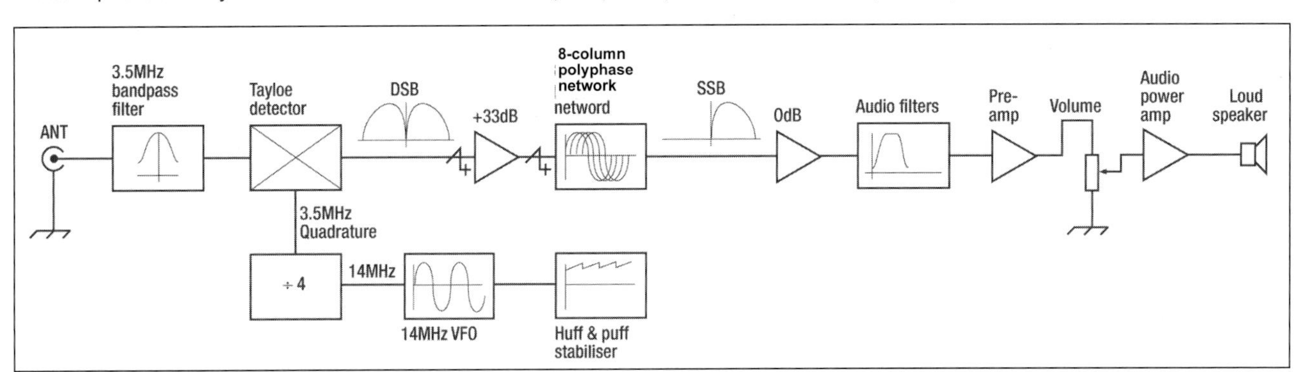


**Fig 6.93: Block diagram of experimental Direct Conversion Polyphase receiver**

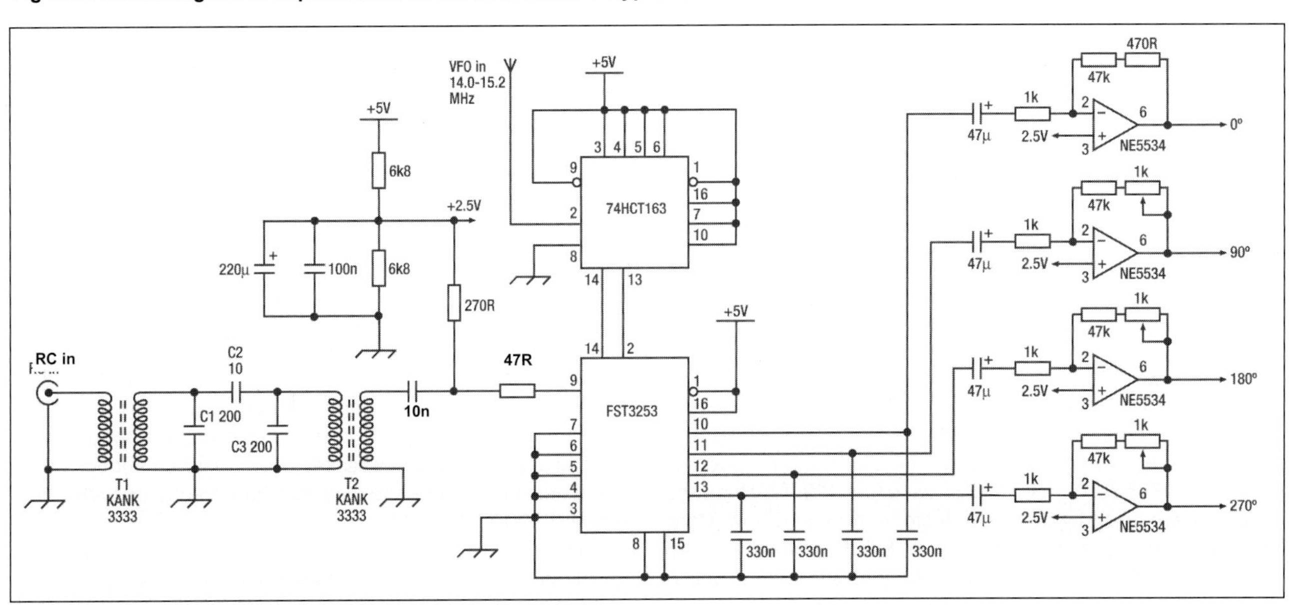


**Fig 6.94: Front-end: RF filter and Tayloe Detector. Op-amps pin 4 is grounded and pin 7 is +12V**

**Fig 6.95: Prototype of the experimental direct conversion polyphase receiver by Hans Summers, G0UPL**

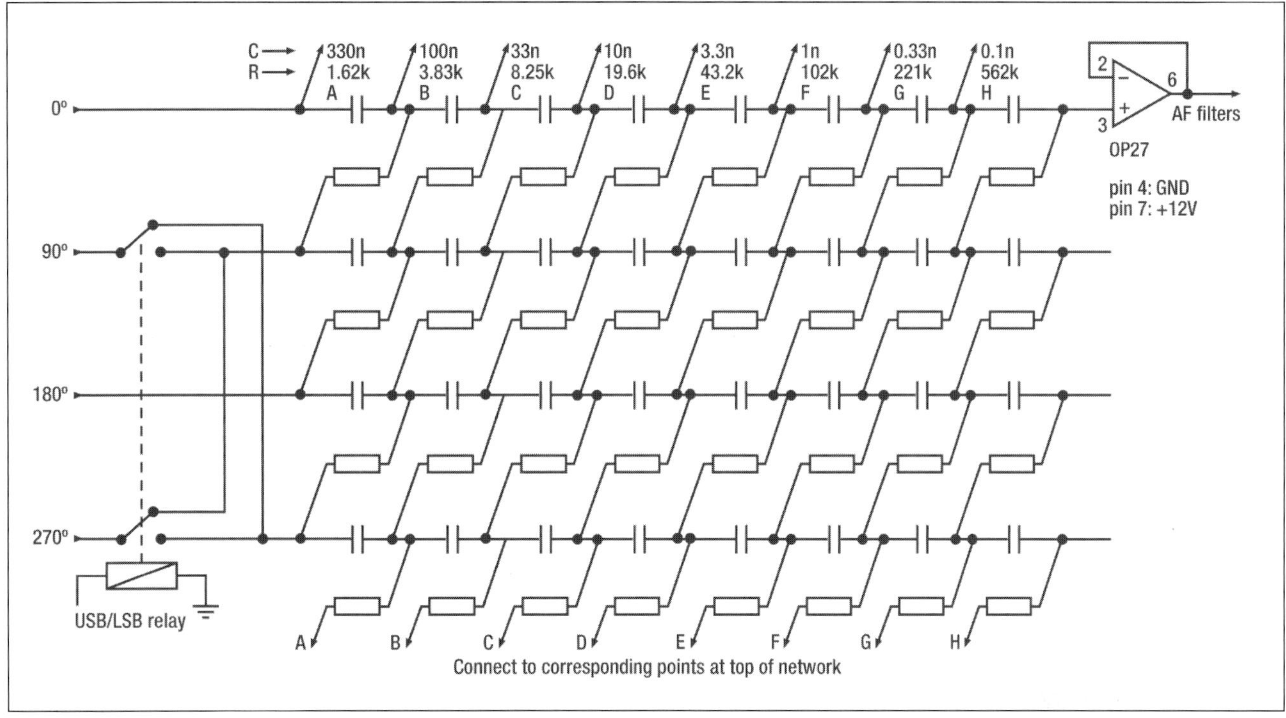


**Fig 6.96: The polyphase network. Resistors and capacitors should be high tolerance types**

## Polyphase network

Polyphase networks are usually designed using either a constant capacitance value throughout the network, or constant resistance. Since capacitors are available in less-closely specified tolerances compared to resistors, a constant value capacitance made it possible to choose sets of four closely matching capacitors for each column in the network. However such a network can result in considerable losses, which are not constant across the audio band of interest. These losses must be compensated by higher amplification elsewhere in the signal chain, which generally degrades noise performance and dynamic range of the receiver.

This component values for this design were calculated by reference to an excellent article by Tetsuo Yoshida JA1KO [20]. Tetsuo's unique design process increases the resistance value from column to column in such a way as to produce a lossless passive polyphase network. This counterintuitive result has been verified by measurement.

An eight-column polyphase network was designed (**Fig 6.96**). The theoretical opposite sideband suppression of this network (**Fig 6.97**) assumes precise component values. In practice, this is impossible to achieve and the real world performance will be degraded somewhat compared to the ideal curve.

This degradation can be mitigated as far as possible by the use of high accuracy (0.1 percent) resistors, together with 'padding' capacitors by connecting smaller capacitors in parallel until meas-

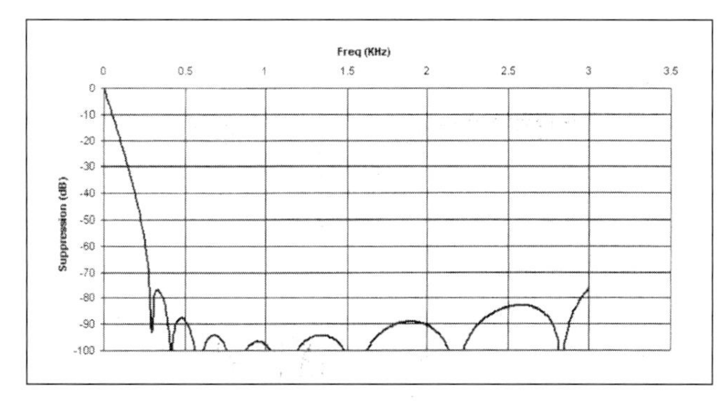


**Fig 6.97: Theoretical opposite sideband suppression (dB) versus frequency (kHz)**

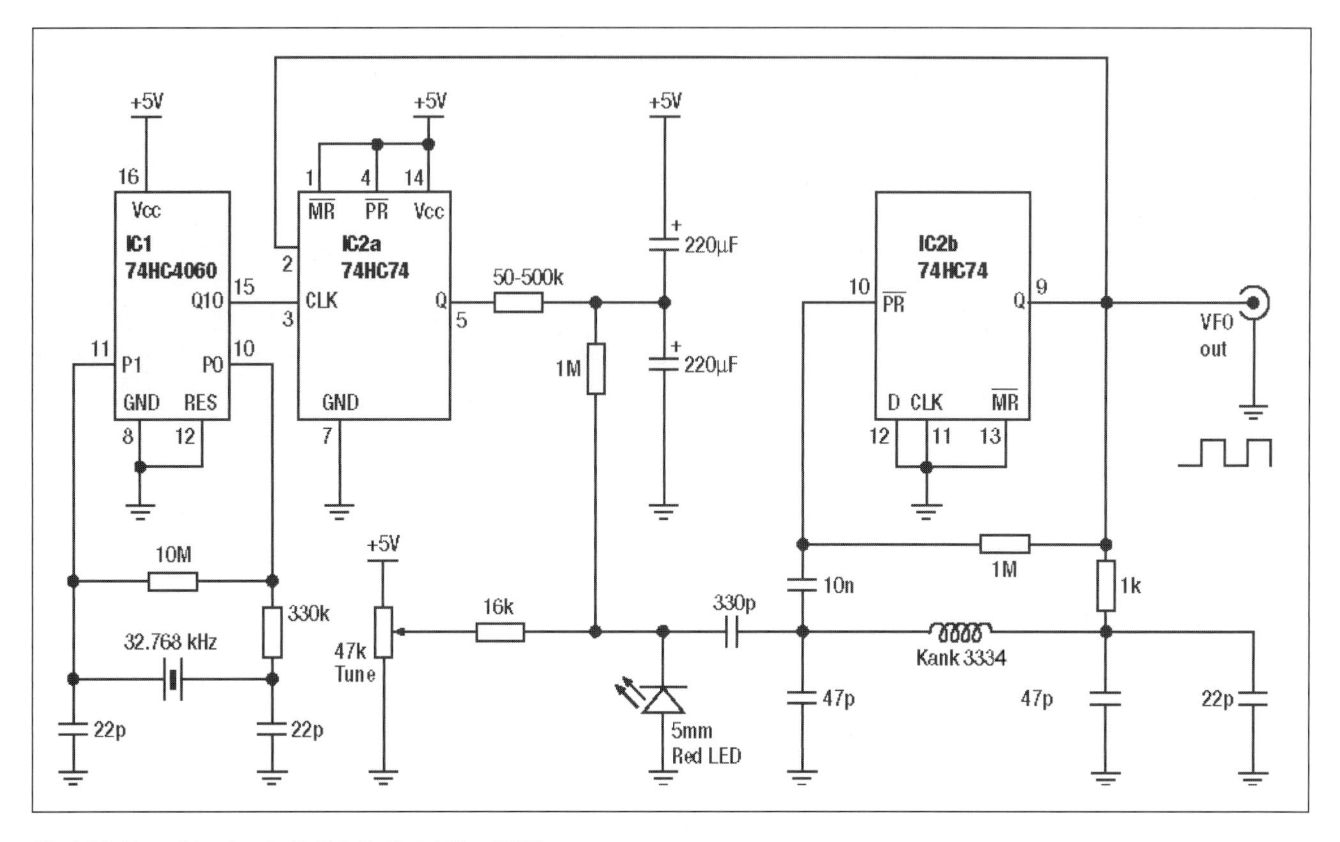


Fig 6.98: Two-chip simple Huff & Puff stabilised VFO

urement indicates four matched capacitors for each column.

Note that selection of upper or lower sideband is simply a matter of swapping the 90- and 270-degree inputs to the network. This could be accomplished by a DPDT relay or switch; or alternatively for single-band use the circuit could be hard-wired. A single unity-gain high impedance low-noise op-amp (OP27) follows the polyphase network.

### Oscillator - the Huff & Puff stabiliser

The Tayloe switching detector requires a VFO at four times the receive frequency. One way to obtain a stable VFO is by use of a Huff & Puff stabiliser. This simple circuit was developed initially by Klaas Spargaren PA0KSB in the early 1970s and many subsequent modifications have appeared in *RadCom*'s 'Technical Topics' column and elsewhere (see also the chapter on Building Blocks: Oscillators and ref [4]).

The stabiliser compares the VFO to a stable crystal reference frequency and locks the VFO to this reference in small frequency steps. It might be described as a 'frequency locked loop' rather than a 'phase locked loop'. The locking process is a slow loop,

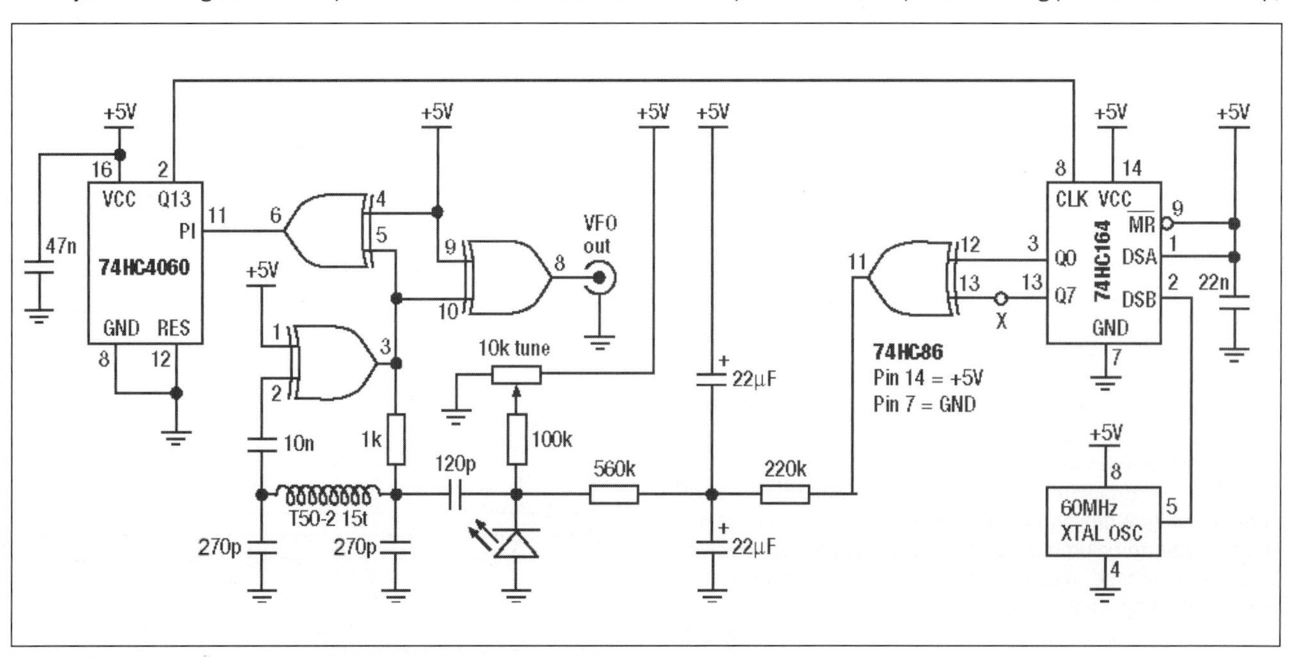


Fig 6.99: Three-chip 'fast' Huff & Puff stabilised VFO, offering improved performance

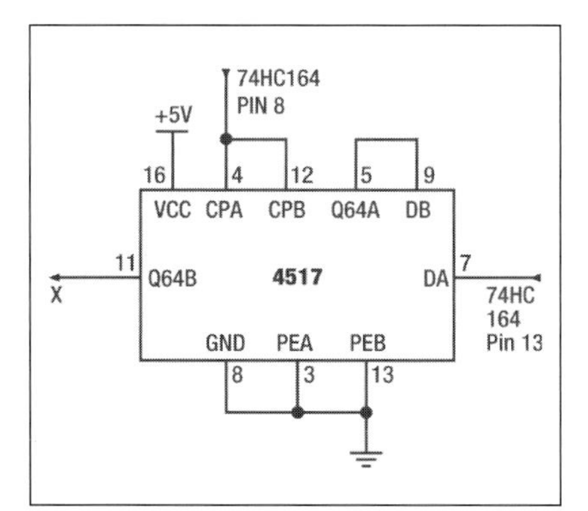


**Fig 6.100: For higher frequencies, a 4517 CMOS IC may be added to improve the Huff & Puff stabiliser even further**

and lacks the complication of phase locked loops and frequency synthesisers, whilst easily achieving a low phase noise output.

Two simple, minimalist designs (**Figs 6.98 and 6.99**) developed subsequently by G0UPL make it even simpler to build a Huff & Puff stabilised VFO. These designs combine the VFO and stabiliser, resulting in a stable output frequency at TTL-logic levels, perfectly suited for driving the Tayloe Detector.

**Fig 6.98** shows an implementation of PA0KSB's original stabiliser. This design uses one half of a 74HC74 D-type flip flop uniquely forced to behave as a simple inverter gate, and pressed into service as an oscillator. The crystal reference frequency is generated from a cheap 32.768kHz watch crystal. The frequency step size of the resulting VFO is determined by the division ratio of the 32.768kHz reference frequency, eg 32Hz. Remember that in the Tayloe Detector the VFO is divided by four, which will also divide the tuning step by the same factor.

A later development was the 'fast' stabiliser by Peter Lawton, G7IXH, which was described in an article in *QEX* magazine [21]. He used a shift register as a n-stage delay line and compared the input and output of the delay line using an exclusive OR-gate (XOR).

The effect is a statistical averaging of the output control signal. The 'fast' method makes it possible to stabilise a worse VFO compared to the standard method, or stabilise a comparable VFO with much less frequency ripple. The frequency step-size is given by:

$$\text{Step} = 10^6 \times \text{VFO}^2 / ( z \times M \times \text{xtal})$$

where VFO is the VFO frequency in MHz, z is the number of stages of delay, xtal is the crystal reference frequency in MHz, and $M = 2^n$ where n is the number of divide-by-2 stages in the VFO divider.

The minimalist design in **Fig 6.99** uses only three ICs to implement G7IXH's 'fast' Huff & Puff method. One XOR gate is used as the VFO. The shift 74HC164 register effectively provides a seven-stage delay line.

To increase this further, a 4517 CMOS IC (128-stage shift register) could be cascaded in series with the 74HC164 to provide a 135-stage delay line (**Fig 6.100**). Note that the 4517 (part numbers HEF4517, CD4517 etc) is a member of the original CMOS 4000-series and was not produced in later, higher speed families such as the 74HC-series.

Therefore it must be connected *after* the 74HC164 so that the '164 is responsible for detecting the fast edges of the 60MHz reference oscillator.

This 'fast' design is recommended for higher frequency VFOs such as might be used to build this experimental receiver for higher HF bands.

### Audio stages

The remainder of the experimental receiver is relatively conventional and non-critical. Low-noise NE5534 op-amps are used to construct high-pass, low-pass filters to restrict the SSB bandwidth to 300Hz - 2.8kHz.

A switchable narrow filter at 800Hz could be added for CW reception. A standard TDA2002 audio power amplifier, produces sufficient output power to drive a 4 ohm loudspeaker for comfortable 'arm-chair' copy.

### Conclusions and further development

The receiver described has been found very satisfactory in use. Other designers of similar receivers have implemented parts of the circuit slightly differently. Many of these modifications would make the circuit more complex, but the following points might suggest avenues for further experiment:

1. The switching order of the divide-by-four circuit in the Tayloe detector can be altered to a 'gray code' sequence such that only one of the two-bit outputs changes state at each clock pulse. This is said to produce lower switching noise, though on 80m the atmospheric noise probably swamps any such effects anyway.
2. A clock-squarer circuit can be employed to generate a precise 50% duty cycle from a VFO at two times the reception frequency; this obviously imposes less stringent requirements on the VFO, which is harder to construct for higher frequencies.
3. The VFO could be replaced by one generated by Direct Digital Synthesis or other precise oscillator methods (see the Building Blocks: Oscillators chapter).
4. It is possible to use the other half of the FST3253 switch in parallel to halve the ON-resistance; alternatively a double-balanced Tayloe detector may be constructed using the second switch and a phase-splitting transformer at the input. This would make the detector more immune to noisy VFO signals such as might be produced by digital methods, eg DDS.
5. Many designers combine 0, 180 and 90, 270 degree outputs of the Tayloe detector prior to feeding the polyphase network. This can reduce certain common-mode noise sources.
6. Instead of using just one output from the polyphase network, all four can be combined thereby averaging errors and improving the signal-to-noise ratio.
7. For transmit operation, it is possible to connect the Tayloe Detector in 'reverse' as a high-performance SSB modulator to make a simple, high performance direct conversion SSB transceiver.

# STATE OF THE ART

## The Front End

Aside from features and functionality, the trend in HF receiver design effort in early 2014 remains focused on high signal level performance, and today that means the receiver front end which is everything from the aerial input to the A to D converter (ADC). A 24 bit ADC has a dynamic range of 112dB, so the preceding front end should exceed this, and the gain, noise and IP3 distribution along the front end circuit blocks is key to achieving this. The use of RF system design tools such as SysCalc [22] allows the front end to me modelled from the individual circuit block measurements, and vice versa, enabling the design trade-offs to be understood.

Because of the background noise present in the HF spectrum, an excessively low noise factor is unnecessary and given a sufficiently low noise first mixer means that the traditional RF amplifier is no longer required. Because there is no gain between the aerial and the 1st mixer, the demands on the mixer's linearity are eased to the point where an H-mode mixer

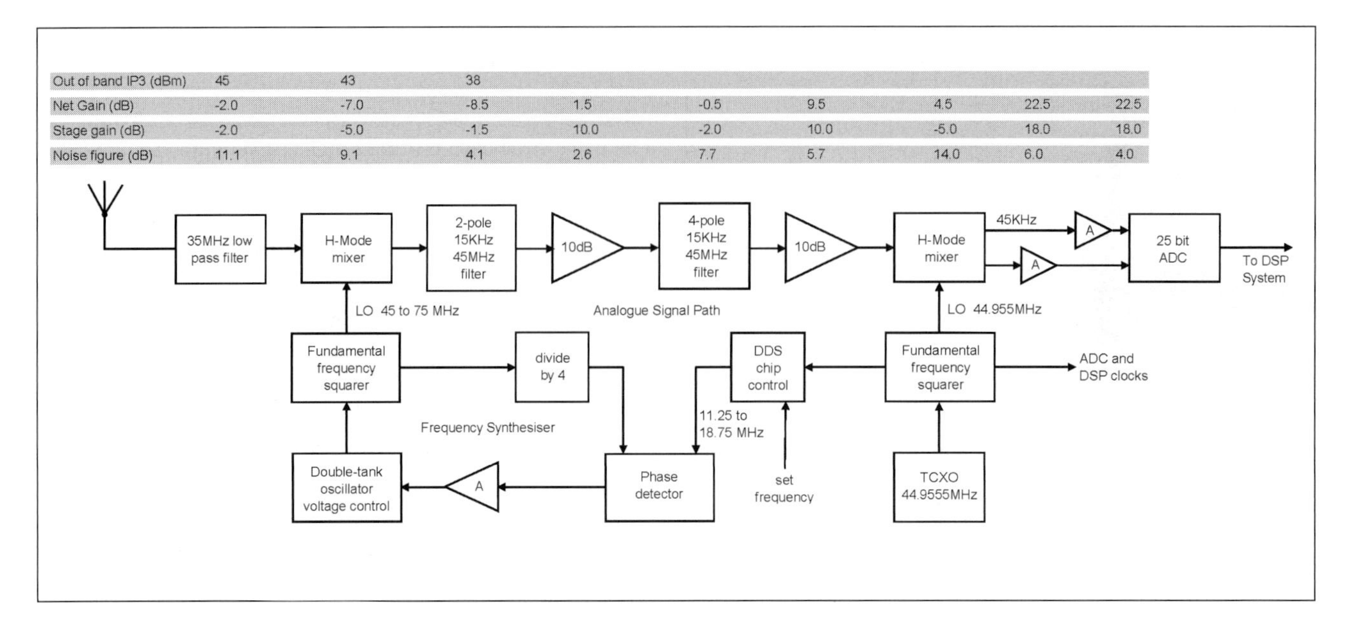


**Fig 6.101: Block diagram of the HF7070 high performance HF/LF receiver**

preceded by no filtering does the job.

The HF7070 [23] is a good example of a high performance HF/LF receiver. It has no sub-octave filtering at the mixer input so in order to achieve good intermodulation performance both 2nd order and 3rd order effects have to be well suppressed. Filtering was omitted to improve noise figure and reduce complexity. Also the inductor requirements for high dynamic range make the filters large and expensive - most standard sized inductors would add more intermodulation than they would filter out! This is a staggering vindication of the H-Mode mixer performance where the ferrites in the input filter produce more IP3 than the mixer.

Because there is no RF amplifier and because of the design of the subsequent IF amplifiers, there is no AGC in the front end. Therefore the entire front end has to handle signal levels between -115dBm (0.4uV) and -5dBm (130mV) ie 110dB of dynamic range.

Intermodulation is now understood, as is mixer performance. Essentially if the switching mixer could go from complete open circuit to complete short circuit in zero time, you would have the perfect mixer. The H-mode mixer devised by Colin Horrabin, G3SBI, and developed by Martein Bakker, PA3AKE, using the new range of fast analogue switches eg the FSA3157 is now state of the art. To maximise the 2nd order performance of the H-mode mixer an active drive system with an integrator is used to maintain an accurate 50:50 square wave drive to the signal switches and a capacitive trim is applied to balance the two combining transformers in the mixer.

It is also understood that the mixer must be correctly terminated across and far beyond the band of interest whilst at the same time provide sufficient roofing filtering to protect the subsequent first amplifier. This has lead to the search for high IP3 crystal filters and flat diplexers, and designs based on quadrature hybrids with two identical crystal filters have emerged with IP3s reaching +50dBm and corresponding SFDRs of 120dB.

Assuming up-conversion, the roofing filter has two functions; to protect the 1st IF amplifier against out of band high level signals and to provided 2nd image rejection. This has now been achieved by the use of six poles of crystal filtering. When it comes to IP3, crystal filters vary considerably and very few manufacturers offer the exceptional IP3 figures required in state of the art designs.

Following the diplexer and the first two poles of roofing filter is the 1st IF amplifier. This is the first gain block in the system and its noise factor and IP3 are of great significance, particularly its IP3. Good linearity comes with high current as you push further up the devices' exponential transfer characteristic. This can be achieved by using a single transmitting power device or a number of smaller devices in parallel. This latter approach has been developed by Bill Carver, W7AAZ, with four FETs in parallel with feedback that deliver 10dB of gain, a noise figure of 1.3dB with an IP3 of 40dBm.

The following 24 bit ADC has an input range from approximately 0 to 5V, and if the peak to mean modulation is 3:1, that is 1.7v mean input for a 130mV aerial input signal, implying a front end gain of 22dB. This requires two IF amplifiers which conveniently allows for the other four poles of crystal filtering.

A second H-mode mixer translates the 45MHz 1st IF down to 45kHz to feed the ADC via a further 6dB of gain. See **Fig 6.101** for a block diagram.

One of the predecessors to the HF7070, the CDG2000, achieved a lot in terms of IP3, sufficient for selectivity to have attracted further design attention. Whilst selectivity down to about 80dB is set by the actual IF filter(s), below that level it becomes governed by the noise of the LO. There are two components to the noise level; the wideband noise floor determined by the active devices and the power supply, and the close-in phase noise controlled by a combination of synthesizer loop filter and the Q of the actual oscillator. Direct digital synthesis or DSS has had an impact, but the accompanying unwanted spurs produced by current devices have to be removed by a tracking filter ie a phase locked loop (PLL) frequency synthesizer, and that brings us back to square one where the DDS acts as a reference frequency for the PLL.

Close attention to power supply regulation and filtering together with a careful choice of active device produce noise floors in the region of -160dBc/Hz.

The development of the double-tank LO by Colin Horrabin, and refined by John Thorpe, has produced an improvement to the close in LO phase noise fall off rate from 20dB/decade to 30dB/decade.

All these developments have been adopted in the HF7070. The low-frequency IF at 44kHz is all balanced with high-speed op-amps as the final gain / low-pass filter stage. Phase compensation feedback is used to keep the amplifiers

**Fig 6.102: The high performance HF7070 receiver [pic ARRL]**

stable at the low gain setting used. Impedances in this section are not 50 ohms, but 200 ohms balanced to connect to the 2nd mixer without transformers, thereby eliminating the requirement for high performance from the transformer cores at low frequencies.

During the development of the HF7070 it became clear that the excellent IP3 figures of +50dBm was only part of the IP3 picture, and there is the associated in-band IP3 that contributes to the quality of the final audio. Equipment reviewers such as Sherwood Engineering Labs [2] and Peter Hart have been measuring the in-band IP3 performance of receivers for some time. Contributary developments by Martein Bakker, PA3AKE have showed that with up-conversion front ends, the use of wider bandwidth roofing filters can improve in-band IP3 so that the front end is no longer the critical point in the signal path. Listening tests have confirmed the exceptionally clear recovered audio from the HF7070 using the wider bandwidth roofing filter.

## A-D Converter

The ADC uses two channels of 24-bits which are summed to give an extra bit of resolution. The sampling rate is high at 176kHz which is later decimated to give an extra two bits, thus potentially the digital noise floor is 27 bits (160dB) below peak level. The whole digital/DSP exercise in the HF7070 was to make the receiver seem as analogue as possible, but just to work better. To this end John Thorpe, the managing designer, felt the noise signature of the radio should be analogue and for this reason the gain distribution was set to give five or six bits of analogue noise from the IF stages in the digital signal. This seems to effectively mask the digital noise and spurii from the converter giving a very smooth, uncoloured noise floor, with 19-20 bits (115dB) of available dynamic range above the noise floor. The advantage of this approach has paid off since you can easily hear signals below the noise floor level coming clearly out of 'clean' noise rather than being distorted by digital noise artefacts.

## Digital IF

The digital final IF system is made in a very similar way to an analogue radio but constructed with algorithms and arithmetic rather than electronic components. The advantage of the maths is that some devices can work perfectly whereas their electronic counterparts are compromised.

For example it is possible to build two filters that have exactly the same gain and phase response, irrespective of temperature. Also you can make a perfect mixer that has no RF-IF feedthrough. With these building blocks a high-performance, variable bandwidth IF system can be created. There are, however, some pitfalls that can disrupt the perfection. The one issue that does not occur in the electronic version or in the mathematical models but appears in a big way to disrupt signal fidelity is rounding error. Small errors at each stage of the calculations compound to add grittiness and distortion to the final sound.

The digital IF system can be implemented in the time domain - signals processed step by step as they would be in analogue electronics, or in the frequency domain - signals converted into their frequency components by Fourier transform, processed to allow the required frequencies to pass, and then converted back to a signal to listen to.

The two techniques are mathematically equivalent and can produce the same results. Although the latter process sounds more complex it actually needs less computational power to achieve, thanks to a very efficient algorithm called the Fast Fourier Transform (FFT). This method seems to be adopted by most of the software radios on the market and it runs well on PC-type processors.

The HF7070, however, has a dedicated digital signal processor and its architecture is tuned to give high performance for an operation called convolution. This is the mainstay of filtering signals in the time domain using Finite Impulse Response (FIR) filters. The HF7070 processor has a separate, dedicated filter engine that spends 100% of its time computing the IF bandpass filters.

So the 7070 bucks the trend in digital IF by processing the whole signal path in the time domain - why? The processing is less efficient but the DSP is fast enough to do the job if the program code is written well. The advantages are:-

1) The intermediate calculation results can be kept to a very high precision (56 bits used for the main filter calculations).
2) The latency of the signal passing through the IF is minimised. In the HF7070 the signal takes less than 25ms from antenna to speaker. Although this is a small delay, it does matter because it impacts search tuning and for break-in CW.
3) All of the IF controls are implemented directly in the signal path and have an effect as soon as they are changed. This gives the real 'hands-on' feel of an analogue receiver.

The last part of the HF7070 that is important for its overall sound and feel is the AGC system. Although not unique, this takes advantage of the precision available with a digital IF and uses feed-forward gain correction. A simple explanation is that the signal level is measured as it leaves the bandwidth filters and then amplified to give a constant level output. All the normal dual rate, hold, attack and decay rates are applied to the control, but the important difference to a conventional AGC system is that the gain control is done after the level measurement and therefore there can be no AGC instability.

The HF7070 was designed as a successor to the AR7030 produced by AOR UK Ltd. Although this receiver was in production for several years, it eventually came to an end because several key components were withdrawn due to the RoHS Directive. Sadly AOR UK Ltd. was dissolved in June 2010 and with it went any possibility of the HF7070 being manufactured other than a few pre-production models, see **Fig 6.102**. The ill-fated HF7070 embodies all the above techniques in its front end developed by Colon Horrabin and together with its SDR section created by John Thorpe can be considered as state of the art (in 2014).

This product design is a particularly notable achievement from the British radio amateur community.

## REFERENCES

[1] 'High-dynamic-range MF/HF front-end of a single-conversion receiver', N6NWP, *QST*, Feb 1993, ARRL
[2] Sherwood Engineering Inc: www.sherweng.com/table.html.
[3] *RadCom*, the monthly journal of the Radio Society of Great

Britain, see www.rsgb.org
[4] www.hanssummers.com/huffpuff.html
[5] www.youtube.com/watch?v=BFGkkk72iNO
[6] Stewart of Reading: www.stewart-of-reading.co.uk
[7] Jabdog Components web site: http://stores.ebay.co.uk/JabDogElectronicComponents
[8] http://toroids.info
[9] Mainline Electronics: http://mainline-group.com
[10] Surplus Sales of Nebraska: www.surplussales.com
[11] www.crystal-radio.eu/index.html
[12] www.arrl.org/files/file/Technology/tis/info/pdf/9811qex026.pdf
[13] http://gnuradio.org/redmine/projects/gnuradio/wiki
[14] 'Technical Topics' column, *RadCom*, September 1976 and *Wireless World*, July 1977
[15] 'Some experiments with HF ladder crystal filters', J A Hardcastle, G3JIR, *RadCom*, Dec 1976, Jan/Feb/Sep 1977
[16] *QST*, Oct 1982, ARRL
[17] 'Experimental Direct Conversion Polyphase Receiver' *Radio Communication Handbook*, 11th edition. ch 6
[18] United States patent 6230000
[19] 'A Software Defined Radio for the Masses, Part 1', Gerald Youngbloood, AC5OG, *QEX*, Jul/Aug 2002, ARRL
[20] *QEX* magazine, Nov 1995, ARRL
[21] *QEX* magazine, Nov 1998, ARRL
[22] SysCalc: www.Ardentech.com
[23] www.arrl.org/files/file/QEX_Next_Issue/Jul-Aug_2013/QEX_7_13_Horrabin.pdf

**Roger Wilkins, G8NHG,** *has been a radio engineer for many years, firstly designing TV sets, then mobile phones and then developing mobile phone radio networks. He now works as a forensic expert. However his major lifelong passion has been the design and construction of HF receivers. He is a great believer in hands-on experience supported nowadays by the wealth of knowledge, tools, components and material available on the internet. He aspires to join the club of 100dB radio designers.*

# HF Transmitters and Transceivers

# 7

*Peter Hart, G3SJX*

The purpose of a transmitter is to generate RF energy which may be keyed or modulated and thus employed to convey intelligence to one or more receiving stations. This chapter deals with the design of that part of the transmitter which produces the RF signal, while keying, data modulation etc are described separately in other chapters.

Transmitters operating on frequencies between 1.8 and 30MHz only are discussed here; methods of generating frequencies higher than 30MHz are contained in other chapters. Where the frequency-determining oscillators of a combined transmitter and receiver are common to both functions, the equipment is referred to as a transceiver; the design of such equipment operating in the HF spectrum is also included in this chapter.

One of the most important requirements of any transmitter is that the desired frequency of transmission shall be maintained within fine limits to prevent interference with other stations and to ensure that the operator remains within the allowed frequency allocation.

Spurious frequency radiation capable of causing interference with other services, including television and radio broadcasting, must also be avoided. These problems are considered in the chapter on electromagnetic compatibility.

The simplest form of transmitter is a single-stage, self-excited oscillator coupled directly to an antenna system: **Fig 7.1(a)**. Such an elementary arrangement has, however, three serious limitations:

- The limited power which is available with adequate frequency stability;
- The possibility of spurious (unwanted) radiation;
- The difficulty of securing satisfactory modulation or keying characteristics.

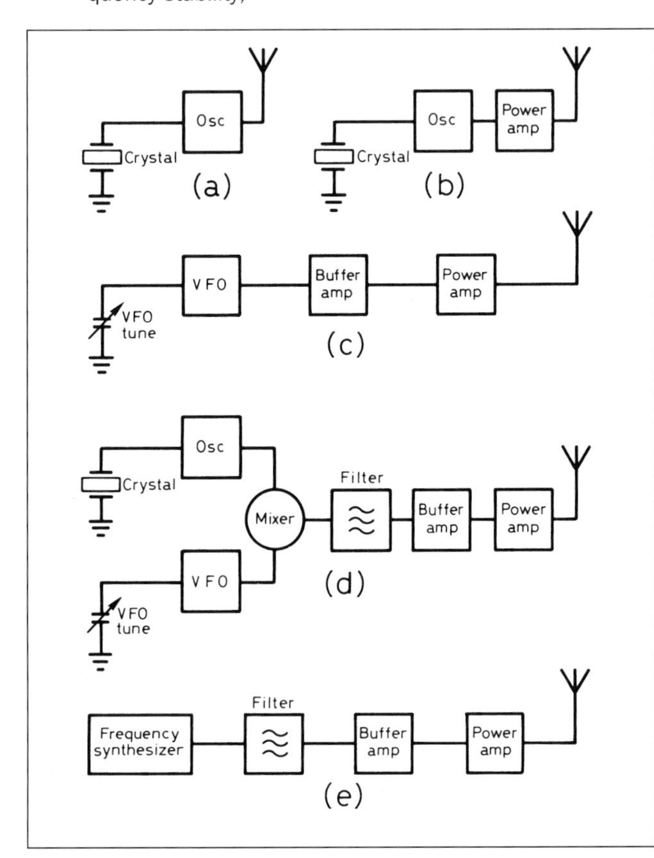


**Fig 7.1: Block diagrams of basic transmitter types**

In order to overcome these difficulties, the oscillator must be called upon to supply only a minimum of power to the following stages. Normally an amplifier is used to provide a constant load on the oscillator: **Fig 7.1(b)**. This will prevent phase shifts caused by variations in the load from adversely affecting the oscillator frequency.

A FET source follower with its characteristic high input impedance makes an ideal buffer after a VFO. In practice, a two-stage buffer amplifier is often used, **Fig 7.1(c)**, incorporating a source follower followed by a Class A amplifier.

Transmitters designed for use on more than one frequency band often employ two or more oscillators; these signals are mixed together to produce a frequency equal to the sum or difference of the original two frequencies, the unwanted frequencies being removed using a suitable filter. This heterodyne transmitter, **Fig 7.1(d)**, is very similar in operation to the superhet receiver. More recently, frequency synthesisers have become popular for the generation of RF energy, **Fig 7.1(e)**, and have almost entirely replaced conventional oscillators in commercial amateur radio equipment. Synthesisers are ideally suited for microprocessor control and multi-frequency coverage. Oscillators are covered in detail in the earlier chapter, Building Blocks: Oscillators.

## INTERSTAGE COUPLING

Correct impedance matching between a stage and its load is important if the design power output and efficiency for the stage are to be achieved. The load may be a succeeding stage or an antenna, and correct matching is essential for achieving efficient operation. In order to keep dissipation in the stage within acceptable limits, the load impedance is often higher than that needed for maximum power transfer. Thus a transmitter designed to drive a 50Ω load usually has a much lower source impedance than this. The output power available from a stage can be calculated approximately from the formula:

$$P_{out} = \frac{V_{cc}^2}{2_{ZL}}$$

where ZL is the load impedance in ohms.

Determination of the base input impedance of a stage is difficult without sophisticated test equipment, but if the output of a stage is greater than 2W its input impedance is usually less than 10Ω and possibly even as low as 1 to 2Ω. For this reason some types of LC matching do not lend themselves to this application. With the precise input impedance of a stage unknown, adjustable LC networks are often employed as these lend themselves best to matching a wide range of impedances. Additionally, a deliberate mismatch may be introduced by the designer to control power distribution and aid stability.

In the interest of stability it is common practice to use low-Q networks between stages in a solid-state transmitter, but the penalty is poor selectivity and little attenuation of harmonic or spurious energy. Most solid-state amplifiers use loaded Qs of 5 or less compared to Qs of 10 to 15 found in valve circuitry. All

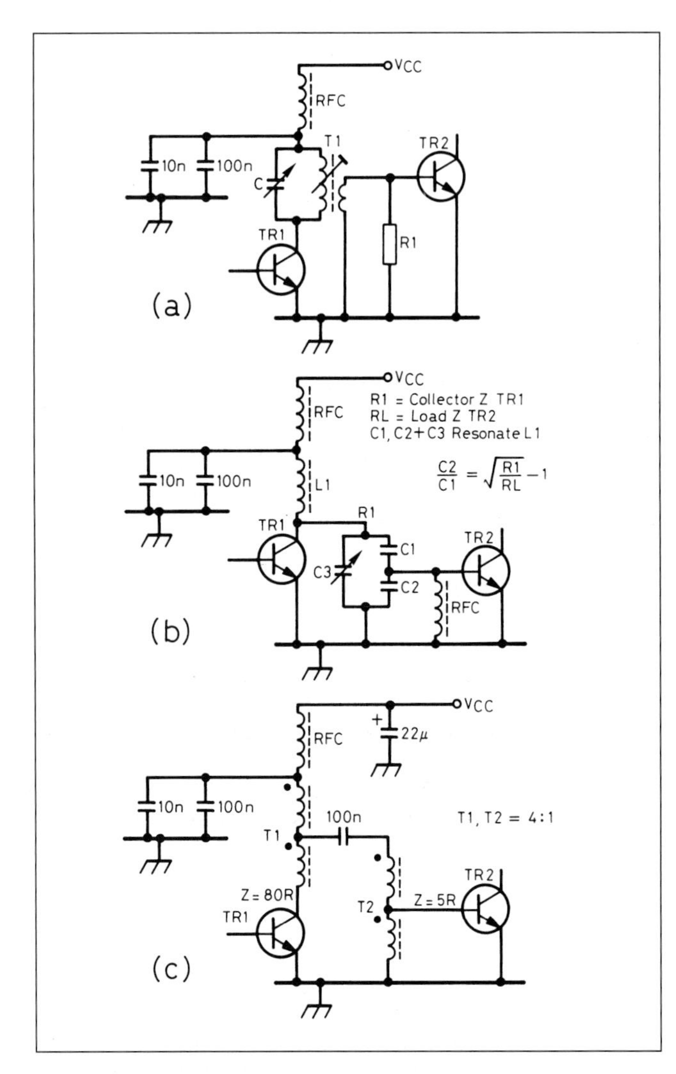


**Fig 7.2: Interstage coupling. (a) Transformer coupling; (b) capacitive divider coupling; (c) broad-band transformer matching**

calculations must take into account the input and output capacitance of the solid-state devices, which must be included in the network calculations. The best source of information on input and output capacitance of power transistors is the manufacturers' data sheets. Impedances vary considerably with frequency and power level, producing a complex set of curves, but input and output capacitance values are independent of these parameters.

Transformers have always been a popular choice for interstage coupling with solid-state devices: **Fig 7.2(a)**. Toroidal transformers are the most efficient and satisfactory up to 30MHz, the turns ratio of T1 being arranged to match the collector impedance of TR1 to the base impedance of TR2. In some circumstances only one turn may be required on the secondary winding. A low value of resistor is used to slug this secondary winding to aid stability.

An alternative to inductive coupling is to use a capacitive divider network which resonates L1 as well as providing a suitable impedance tap: **Fig 7.2(b)**. Suitable RF chokes wound on high-permeability (μ = 800) ferrite beads, adequately decoupled, are added to aid stability. This circuit is less prone to VHF parasitics than the one in Fig 7.2(a), especially if C2 has a relatively high value of capacitance.

When the impedance values to be matched are such that specific-ratio broad-band transformers can be used, typically 4:1 and 16:1, one or more fixed-ratio transformers may be cascaded as shown in **Fig 7.2(c)**. This arrangement also exhibits a lack of selectivity due its low Q value, but has the advantage of offering broad-band characteristics. Dots are normally drawn on to transformers to indicate the phasing direction of the windings (all dots start at the same end).

Simple LC networks provide practical solutions to matching impedances between stages in transmitters. It is assumed that normally high output impedances will be matched to lower-value input impedances. However, if the reverse is required, networks can be simply used in reverse to effect the required transformation.

Networks 1 and 2, **Figs 7.3(a) and 7.3b)**, are variations on the L-match and may used fixed or with adjustable inductors and capacitors. Network 3, **Fig 7.3(c)**, is used by many designers because it is capable of matching a wide range of impedances. It is a low-pass T-network and offers harmonic attenuation to a degree determined by the transformation ratio and the total network Q. For stages feeding an antenna, however, additional harmonic suppression will normally be required. The value of Q may vary from 4 to 20 and represents a compromise between bandwidth and attenuation.

Network calculations are shown in **Table 7.1**.

Conventional broad-band transformers are very useful in the construction of solid-state transmitters. They are essentially devices for the transformation of impedances relative to the ratio of the transformer and are not specific impedance devices, eg a solid-state PA with a nominal 50Ω output impedance may employ a 3:1 ratio broad-band output transformer. The impedance ratio is given by the square of the turns ratio and will be 9:1. The impedance seen by the collector is thus:

$$Z_c = \frac{50}{9} = 5.55\Omega$$

The power delivered to the load is then determined by:

$$P_{out} = \frac{V_{cc}^2}{2Z_c}$$

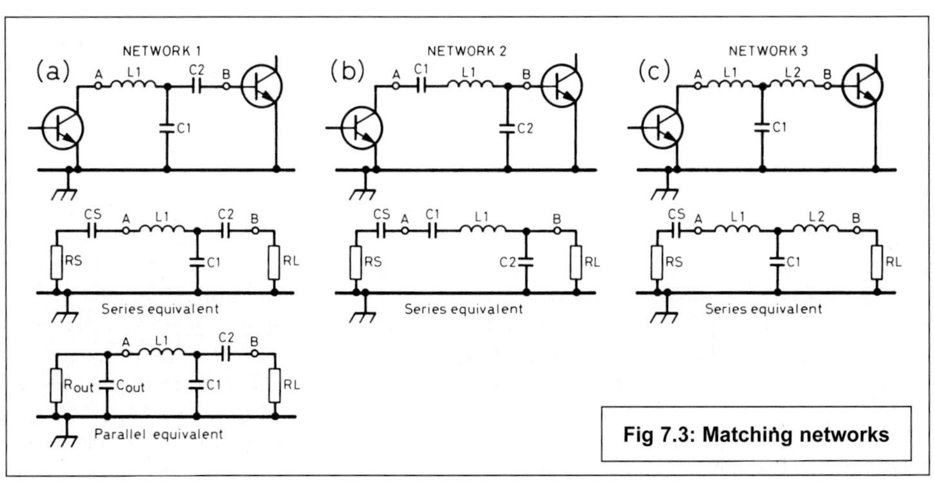


**Fig 7.3: Matching networks**

Another type of broad-band transformer found in transmitting equipment is the transmission-line transformer. This acts as a conventional transformer at the lower frequencies but, as the frequency increases, the core becomes less 'visible' and the transmission-line properties take over. The calculations are complex, but it is well known that a quarter-wavelength of line exhibits impedance-transformation properties. When a quarter-wave line of impedance $Z_0$ is terminated with a resistance $R_1$, a resistance $R_2$ is seen at the other end of the line.

$$Z_0 = \sqrt{R_1 R_2}$$

Transmission-line transformers are often constructed from twisted pairs of wires wound onto a ferrite toroidal core having a initial permeability (μ) of at least 800.

This ensures relatively high values of inductance with a small number of turns. It should be borne in mind that with high-power solid-state transmitters some of the impedances to be transformed are very low, often amounting to only a few ohms.

## ANODE TANK CIRCUITS

Whilst solid state devices have replaced valves in most amateur equipment, high power valve linear amplifiers are likely to be in use for some time into the future.

### Pi Network

The pi-tank (**Fig 7.4**) has been the most popular matching network since the introduction of multiband transmitters in the 1950s. It is easily bandswitched and owes its name to its resemblance to the Greek letter $\pi$.

The anode tank circuit must meet the following conditions:

1 The anode circuit of the valve must be presented with the proper resistance in relation to its operating conditions to ensure efficient generation of power.

2. This power must be transferred to the output without appreciable loss.

3 The circuit Q must be sufficient to ensure good flywheel action in order to achieve a close approximation of a sinusoidal RF output voltage. This is especially important in Class C amplifiers where the drive from the valve is in the form of a series of pulses of RF energy.

#### Tank circuit Q

In order to quantify the ability of a tank circuit to store RF energy (essential for flywheel action), a quality factor Q is defined. Q is the ratio of energy stored to energy lost in the circuit.

$$Q = 2\pi \frac{W_S}{W_L} = \frac{X}{R}$$

where $W_S$ is the energy stored in the tank circuit; $W_L$ is the energy lost to heat and the load; X is the reactance of either the inductor or capacitor in the tank circuit; R is the series resistance.

Since both circulating current and Q are inversely proportional to R, then the circulating current is proportional to Q. By Ohm's Law, the voltage across the tank circuit components must also be proportional to Q.

When the circuit has no load the only resistance contributing to R are the losses in the tank circuit. The unloaded $Q_U$ is given by:

$$Q_U = \frac{X}{R_{loss}}$$

where X is the reactance in circuit and $R_{loss}$ is the sum of the resistance losses in the circuit.

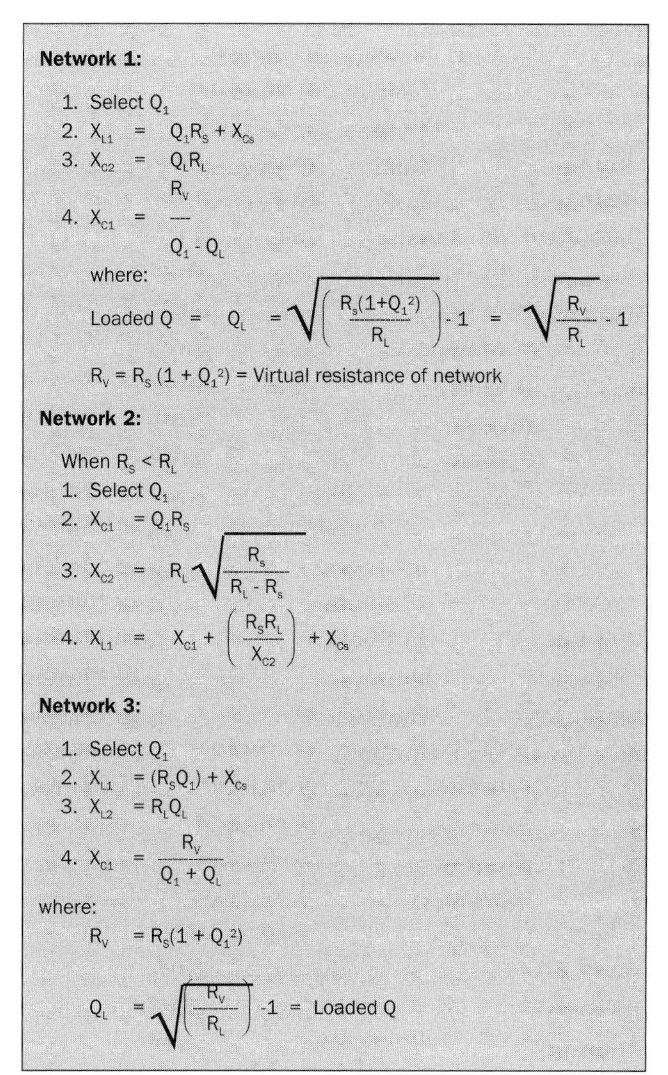

**Network 1:**

1. Select $Q_1$
2. $X_{L1} = Q_1R_S + X_{CS}$
3. $X_{C2} = Q_LR_L$
4. $X_{C1} = \dfrac{R_V}{Q_1 - Q_L}$

where:

$$\text{Loaded Q} = Q_L = \sqrt{\left(\frac{R_S(1+Q_1^2)}{R_L}\right) - 1} = \sqrt{\frac{R_V}{R_L} - 1}$$

$R_V = R_S(1 + Q_1^2)$ = Virtual resistance of network

**Network 2:**

When $R_S < R_L$

1. Select $Q_1$
2. $X_{C1} = Q_1R_S$
3. $X_{C2} = R_L\sqrt{\dfrac{R_S}{R_L - R_S}}$
4. $X_{L1} = X_{C1} + \left(\dfrac{R_SR_L}{X_{C2}}\right) + X_{CS}$

**Network 3:**

1. Select $Q_1$
2. $X_{L1} = (R_SQ_1) + X_{CS}$
3. $X_{L2} = R_LQ_L$
4. $X_{C1} = \dfrac{R_V}{Q_1 + Q_L}$

where:

$R_V = R_S(1 + Q_1^2)$

$$Q_L = \sqrt{\left(\frac{R_V}{R_L}\right)} - 1 = \text{Loaded Q}$$

**Table 7.1: Calculations for the three networks shown in Fig 7.3**

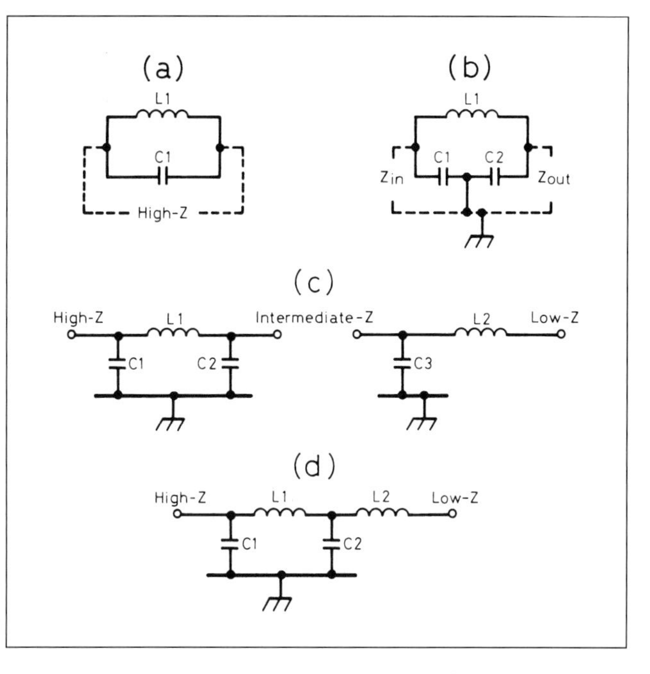


**Fig 7.4: Pi-network and derivation. (a) Parallel-tuned tank circuit; (b) parallel-tuned tank with capacitive tap; (c) pi-network and L-network; (d) pi-L network**

**Table 7.2: Pi-network values for selected anode loads ($Q_L$ = 12) and load (antenna impedance) of 50 ohms**

| | MHz | 1500 | 2000 | 2500 | 3000 | 3500 | 4000 | 5000 | 6000 | 8000 |
|---|---|---|---|---|---|---|---|---|---|---|
| | 1.8 | 708 | 531 | 424 | 354 | 303 | 264 | 229 | 206 | 177 |
| | 3.5 | 364 | 273 | 218 | 182 | 156 | 136 | 118 | 106 | 91 |
| **C1** | 7.0 | 182 | 136 | 109 | 91 | 78 | 68 | 59 | 53 | 46 |
| **(pF)** | 14.0 | 91 | 68 | 55 | 46 | 39 | 34 | 30 | 27 | 23 |
| | 21.0 | 61 | 46 | 36 | 30 | 26 | 23 | 20 | 18 | 15 |
| | 28.0 | 46 | 34 | 27 | 23 | 20 | 17 | 15 | 13 | 11 |
| | 1.8 | 3413 | 2829 | 2415 | 2092 | 1828 | 1600 | 1489 | 1431 | 1392 |
| | 3.5 | 1755 | 1455 | 1242 | 1076 | 940 | 823 | 766 | 736 | 716 |
| **C2** | 7.0 | 877 | 728 | 621 | 538 | 470 | 411 | 383 | 368 | 358 |
| **(pF)** | 14.0 | 439 | 364 | 310 | 269 | 235 | 206 | 192 | 184 | 179 |
| | 21.0 | 293 | 243 | 207 | 179 | 157 | 137 | 128 | 123 | 119 |
| | 28.0 | 279 | 182 | 155 | 135 | 117 | 103 | 96 | 92 | 90 |
| | 1.8 | 12.81 | 16.60 | 20.46 | 24.21 | 27.90 | 31.50 | 36.09 | 39.96 | 46.30 |
| | 3.5 | 6.59 | 8.57 | 10.52 | 12.45 | 14.35 | 16.23 | 18.56 | 20.55 | 23.81 |
| **L1** | 7.0 | 3.29 | 4.29 | 5.26 | 6.22 | 7.18 | 8.12 | 9.28 | 10.26 | 11.90 |
| **(µH)** | 14.0 | 1.64 | 2.14 | 2.63 | 3.11 | 3.59 | 4.06 | 4.64 | 5.14 | 5.95 |
| | 21.0 | 1.10 | 1.43 | 1.75 | 2.07 | 2.39 | 2.71 | 3.09 | 3.43 | 3.97 |
| | 28.0 | 0.82 | 1.07 | 1.32 | 1.56 | 1.79 | 2.03 | 2.32 | 2.57 | 2.98 |

**Table 7.3: Pi-L network values for selected anode loads ($Q_L$ = 12) and load (antenna impedance) of 50 ohms**

| | MHz | 1500 | 2000 | 2500 | 3000 | 3500 | 4000 | 5000 | 6000 | 8000 |
|---|---|---|---|---|---|---|---|---|---|---|
| | 1.8 | 784 | 591 | 474 | 397 | 338 | 297 | 238 | 200 | 152 |
| | 3.5 | 403 | 304 | 244 | 204 | 174 | 153 | 123 | 103 | 78 |
| **C1** | 7.0 | 188 | 142 | 114 | 94 | 81 | 71 | 57 | 48 | 36 |
| **(pF)** | 14.0 | 93 | 70 | 56 | 47 | 40 | 35 | 29 | 24 | 18 |
| | 21.0 | 62 | 47 | 38 | 32 | 27 | 23 | 19 | 16 | 12 |
| | 28.0 | 48 | 36 | 29 | 24 | 21 | 18 | 15 | 13 | 9 |
| | 1.8 | 2621 | 2355 | 2168 | 2026 | 1939 | 1841 | 1696 | 1612 | 1453 |
| | 3.5 | 1348 | 1211 | 1115 | 1042 | 997 | 947 | 872 | 829 | 747 |
| **C2** | 7.0 | 596 | 534 | 493 | 468 | 444 | 418 | 387 | 368 | 337 |
| **(pF)** | 14.0 | 292 | 264 | 240 | 222 | 215 | 204 | 186 | 172 | 165 |
| | 21.0 | 191 | 173 | 158 | 146 | 136 | 137 | 125 | 117 | 104 |
| | 28.0 | 152 | 135 | 127 | 115 | 106 | 107 | 95 | 87 | 86 |
| | 1.8 | 14.047 | 17.933 | 21.730 | 25.466 | 29.155 | 32.805 | 40.011 | 47.118 | 61.119 |
| | 3.5 | 7.117 | 9.086 | 11.010 | 12.903 | 14.772 | 16.621 | 20.272 | 23.873 | 30.967 |
| **L1** | 7.0 | 3.900 | 4.978 | 6.030 | 7.070 | 8.094 | 9.107 | 11.108 | 13.081 | 16.968 |
| **(µH)** | 14.0 | 1.984 | 2.533 | 3.069 | 3.597 | 4.118 | 4.633 | 5.651 | 6.655 | 8.632 |
| | 21.0 | 1.327 | 1.694 | 2.053 | 2.406 | 2.755 | 3.099 | 3.780 | 4.452 | 5.775 |
| | 28.0 | 0.959 | 1.224 | 1.483 | 1.738 | 1.989 | 2.238 | 2.730 | 3.215 | 4.171 |
| | 1.8 | 8.917 | *The value of L2 remains constant for all values of anode impedance.* | | | | | | | |
| | 3.5 | 4.518 | | | | | | | | |
| **L2** | 7.0 | 2.476 | | | | | | | | |
| **(µH)** | 14.0 | 1.259 | | | | | | | | |
| | 21.0 | 0.843 | | | | | | | | |
| | 28.0 | 0.609 | | | | | | | | |

To the tank circuit, a load acts in the same way as circuit losses. Both consume energy but only the circuit losses produce heat energy. When energy is coupled from the tank circuit to the load, the loaded Q ($Q_L$) is given by:

$$Q_L = \frac{X}{R_{load} + R_{loss}}$$

It follows that if the circuit losses are kept to a minimum the loaded Q value will rise.

Tank efficiency can be calculated from:

$$\text{Tank efficiency (\%)} = \left(1 - \frac{Q_L}{Q_U}\right) \times 100$$

where $Q_U$ is the unloaded Q and $Q_L$ is the loaded Q.

Typically the unloaded Q for a pi-tank circuit will be between 100 and 300 while a value of 12 is accepted as a good compromise for the loaded Q. In order to assist in the design of anode tank circuits for different frequencies, inductance and capacitance values for a pi-network with a loaded Q of 12 are provided in **Table 7.2** for different values of anode load impedance.

## Pi-L Network

The pi-L network, **Fig 7.4(d)**, is a combination of the pi-network and the L-network. The pi-network transforms the load resistance to an intermediate impedance, typically several hundred ohms, and the L-network then transforms this intermediate impedance to the output impedance of 50Ω. The output capacitor of the pi-network is in parallel with the input capacitor of the L-network and is combined into one capacitor equal to the sum of the two individual values.

The major advantage of the pi-L network over a pi-network is considerably greater harmonic suppression, making it particularly suitable for high-power linear amplifier applications. A table of values for a pi-L network having a loaded Q of 12 for different values of anode load impedance is given in **Table 7.3**. Both Table 7.2 and Table 7.3 assume that source and load impedances are purely resistive; the values will have to be modified slightly to compensate for any reactance present in the circuits to be matched. Under certain circumstances matching may be compromised by high values of external capacitance, in which case a less-than-ideal value of Q may have to be accepted.

## VOICE MODULATION TECHNIQUES

A description of the theory of Amplitude Modulation (AM), Single Sideband (SSB), and Frequency and Phase Modulation (FM and PM) can be found in the Principles chapter.

Although currently still common for broadcasting, AM is now rarely used on the amateur bands. The use of FM and PM is confined to the bands above 30MHz (with the exception of a little activity in the upper part of the 28MHz band), so they are covered in the chapter on VHF/UHF transmitters and receivers. This chapter will deal with the practical aspects of SSB modulation.

An SSB transmitter has a similar architecture to an SSB receiver as shown in **Fig 7.5**. This makes it very suitable for combining into a transceiver, as can be seen later.

## BALANCED MODULATORS

Balanced modulators are essentially the same as balanced mixers, balanced demodulators and product detectors; they are tailored to suit different circuit applications especially with respect to the frequencies in use. The balanced modulator is a circuit which mixes or combines a low-frequency (audio) signal with a higher frequency (RF) signal in order to obtain the sum-and-difference frequencies (sidebands); the original RF frequency is considerably attenuated by the anti-phase or balancing action of the circuit.

A *singly* balanced modulator is designed to balance out only one of the input frequencies, either $f_1$ or $f_2$, normally the higher frequency. In a *doubly* balanced modulator, both $f_1$ and $f_2$ are balanced out, leaving the sum and difference frequencies $f_1 + f_2$ and $f_1 - f_2$. In addition, intermodulation products (IMD) will appear in the form of spurious signals caused by the interaction and mixing of the various signals and their harmonics.

Balanced modulators come in many different forms, employing a wide variety of devices from a simple pair of diodes to complex ICs. In their simplest form they are an adaptation of the bridge circuit, but it should be noted that diodes connected in a modulator circuit are connected differently to those in a

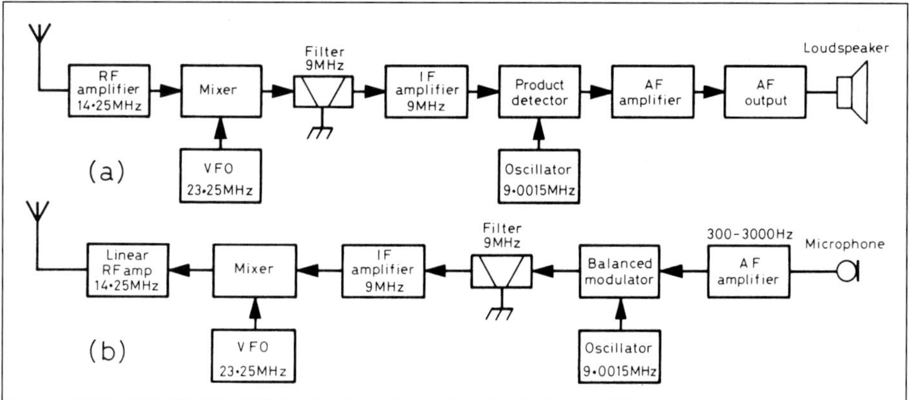


**Fig 7.5: (a) Typical single-conversion superhet receiver. (b) Single conversion SSB transmitter**

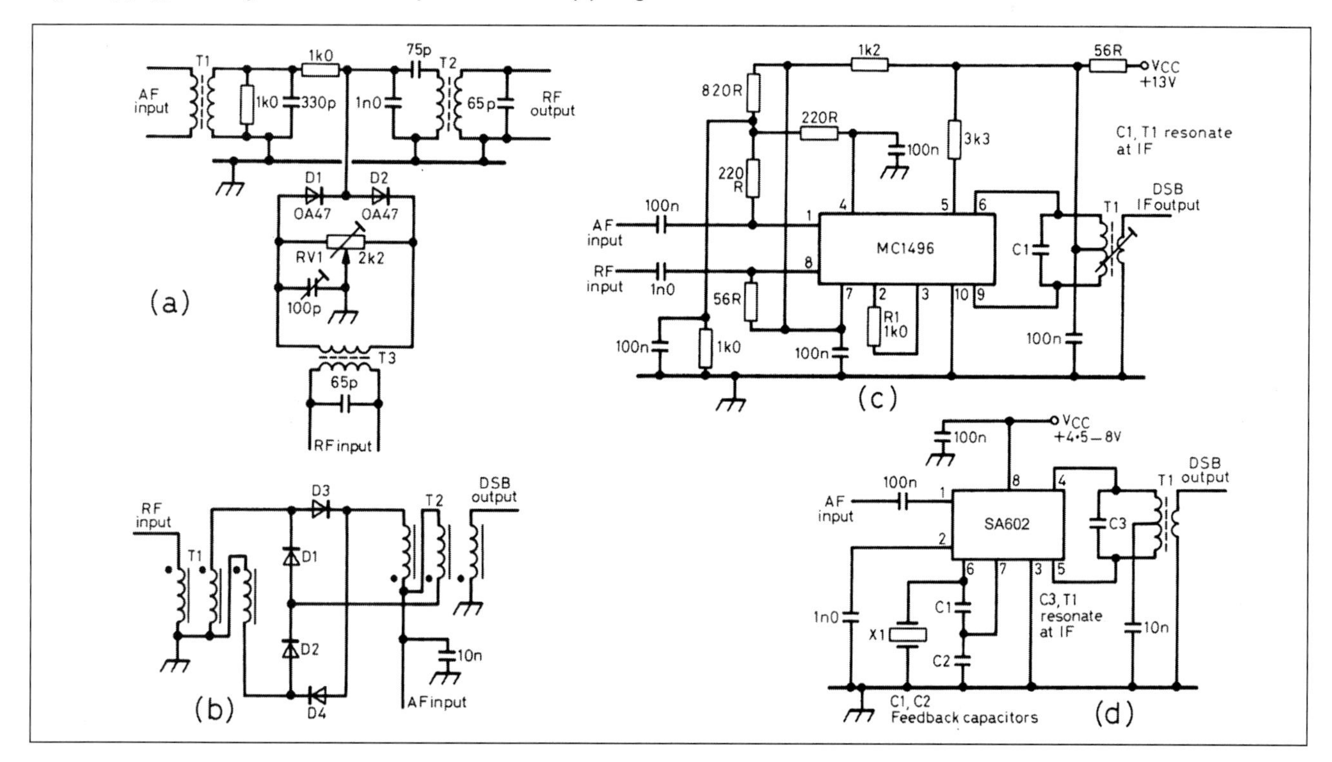


**Fig 7.6: Balanced modulators: (a) Singly balanced diode modulator; (b) Doubly balanced diode ring modulator; (c) MC1496 doubly balanced modulator; (d) SA602 doubly balanced modulator with internal oscillator**

bridge rectifier. Simple diode balanced modulators can provide high performance at low cost. Early designs used point-contact germanium diodes while more recent designs use hot-carrier diodes (HCD). The HCD offers superior performance with lower noise, higher conversion efficiency, higher square law capability, higher breakdown voltage and lower reverse current combined with a lower capacitance. In practice, almost any diode can be used in a balanced modulator circuit, including the ubiquitous 1N914.

In the early days of SSB, simple diode balanced modulators were very popular, easy to adjust and capable of good results. Doubly balanced diode ring modulators have subsequently proved very popular because of their higher performance. However, they incur at least a 6dB signal loss while requiring a high level of oscillator drive. Doubly balanced modulators offer greater isolation between inputs as well as between input and output ports when compared to singly balanced types.

The introduction of integrated circuits resulted in a multitude of ICs suitable for use in balanced modulator and mixer applications. These include the popular Philips SA602 which combines an input amplifier, local oscillator and double balanced modulator in a single package, and the MC1496 double balanced mixer IC from Motorola [1]. The majority of IC mixers are based upon a doubly balanced transistor tree circuit, using six or more transistors on one IC. The major difference between different types of IC lies in the location of resistors which may be either internal or external to the IC.

IC mixers offer conversion gain, lower oscillator drive requirements and high levels of balance, but IMD performance can be inferior to that of diode ring modulators. Some devices, such as the Analog Devices AD831, permit control of the bias current. This allows the designer freedom to improve IMD performance at the expense of power consumption.

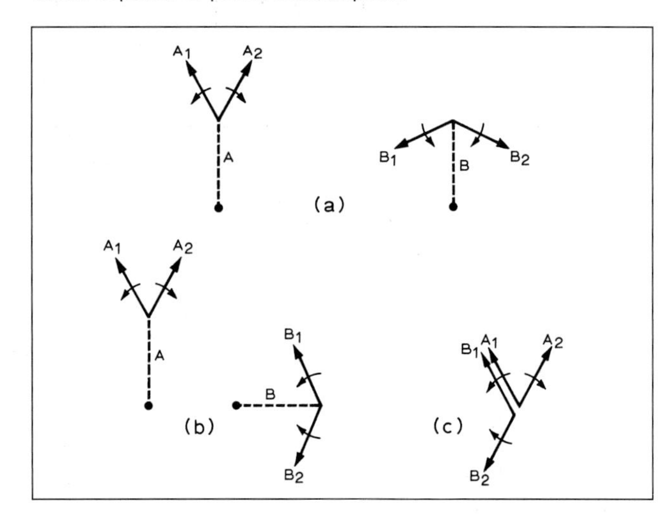


**Fig 7.7: Phasing system vectors**

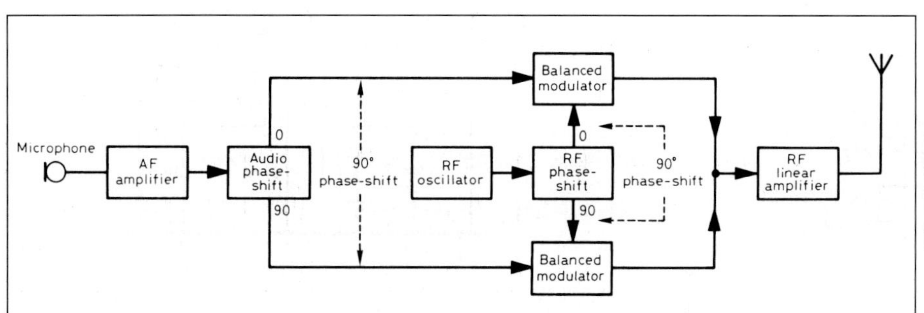


**Fig 7.8: Phasing-type exciter**

**Fig 7.6** illustrates a range of practical balanced modulator circuits. The shunt-type diode modulator, **Fig 7.6(a)**, was common in early valve SSB transmitters, offering a superior balance to that achievable with conventional valve circuitry. **Fig 7.6(b)** shows a simple diode ring balanced modulator capable of very high performance - devices such as the MD108 and SBL1 are derivations of this design. The designs in **Figs 7.6(c)** and **7.6(d)** illustrate the use of IC doubly balanced mixers. The SA602, primarily designed for very-low-power VHF receiver mixer applications, offers simplicity of design and a low external component count. Although most of the devices mentioned are no longer in production they remain popular for homebrew construction and can still be occasionally be found for sale at rallies and on the internet.

## GENERATING SSB

The double sideband (DSB) signal generated by the balanced modulator has to be turned into SSB by attenuating one of the sidebands to an acceptable level. A figure of 30-35dB has come to be regarded as the minimum acceptable standard. With care, suppression of 50dB or more is attainable but such high levels of attenuation are of questionable benefit.

The unwanted sideband may be attenuated either by phasing or filtering. The two methods are totally different in conception, and will be discussed in detail.

### The Phasing Method

The phasing method of SSB generation can be simply explained with the aid of vector diagrams. **Fig 7.7(a)** shows two carriers, A and B, of the same frequency and phase, one of which is modulated in a balanced modulator by an audio tone to produce contra-rotating sidebands A1 and A2, and the other modulated by a 90° phase-shifted version of the same audio tone. This produces sidebands B1 and B2 which have a 90° phase relationship with their A counterparts. The carrier vector is shown dotted since the carrier is absent from the output of the balanced modulators. **Fig 7.7(b)** shows the vector relationship if the carrier B is shifted in phase by 90° and **Fig 7.7(c)** shows the addition of these two signals. It is evident that sidebands A2 and B2 are in antiphase and therefore cancel whereas A1 and B1 are in phase and are additive. The result is that a single sideband is produced by this process.

A block diagram of a phasing-type transmitter is shown in **Fig 7.8** from which will be seen that the output of an RF oscillator is fed into a network in which it is split into two separate components, equal in amplitude but differing in phase by 90°.

Similarly, the output of an audio amplifier is split into two components of equal amplitude and 90° phase difference. One RF and one AF component are combined in each of two balanced modulators. The double-sideband, suppressed-carrier signal from the two balanced modulators iare added together. The relative phases of the sidebands produced by the two balanced modulators are such that one sideband is balanced out while the other is reinforced. The resultant is an SSB signal.

The main advantages of a phasing exciter are that sideband suppression may be accomplished at the operating frequency and that selection of the upper or lower sideband may be made by reversing the phase of the audio input to one of the balanced modulators. These facilities are denied to the user of the filter system.

If it were possible to arrange for

absolute precision of phase shift in the RF and AF networks, and absolute equality in the amplitude of the outputs, the attenuation of the unwanted sideband would be infinite. In practice, perfection is impossible to achieve, and some degradation of performance is inevitable. Assuming that there is no error in the amplitude adjustment, a phase error of 1° in either the AF or the RF network will reduce the suppression to 40dB, while an error of 2° will produce 35dB, and 3.5° will result in 30dB suppression. If, on the other hand, phase adjustment is exact, a difference of amplitude between the two audio channels will similarly reduce the suppression. A difference between the two voltages of 1% would give 45dB, and 4% 35dB approximately. These figures are not given to discourage the intending constructor, but to stress the need for high precision workmanship and adjustment if a satisfactory phasing-type SSB transmitter is to be produced.

The early amateur phasing transmitters were designed for fundamental-frequency operation, driven directly from an existing VFO tuning the 80m band, and used a low-Q phase-shift network. This low-Q circuit has the ability to maintain the required 90° phase shift over a small frequency range, and this made the network suitable for use at the operating frequency in single-band exciters designed to cover only a portion of the chosen band. The RF phase-shift network is incapable of maintaining the required accuracy of phase shift for operation over ranges of 200kHz or more, and the available sideband suppression deteriorates to a point at which the exciter is virtually radiating a double sideband signal.

For amateur band operation a sideband suppression of 30-35dB and a carrier suppression of 50dB should be considered the minimum acceptable standard. Any operating method that is fundamentally incapable of maintaining this standard should not be used on the amateur bands. For this reason, the fundamental type of phasing unit is not recommended. For acceptable results, the RF phase shift must be operated at a fixed frequency outside the amateur bands. The SSB output from the balanced modulator is then mixed to the required bands by using an external VFO.

### Audio phase-shift network

Achieving the audio phase shift necessary for SSB generation in a phasing exciter traditionally required the use of high-tolerance components, often necessitating the use of a commercially made phase-shift network. Such devices can prove more costly than the crystal filter required for a filter-type exciter.

An alternative method of deriving the required 90° phase shift using off-the-shelf values was devised by M J Gingell and is referred to as the polyphase network (**Fig 7.9**). Standard 10% tolerance resistors and capacitors are used in the construction of a six-pole network capable of providing four outputs of equal amplitude, all lagging one another by 90°. The network is designed to phase shift audio signals between 300Hz and 3000Hz, and it is therefore necessary to limit the bandwidth of the audio input using a filter or clipper circuit which can be either active or passive. Audio input to the polyphase network is derived using a simple phase splitter to provide the two phase inputs required by the network. Resistors used in the network are of one common value and Mylar audio-grade capacitors are suitable for the capacitive elements.

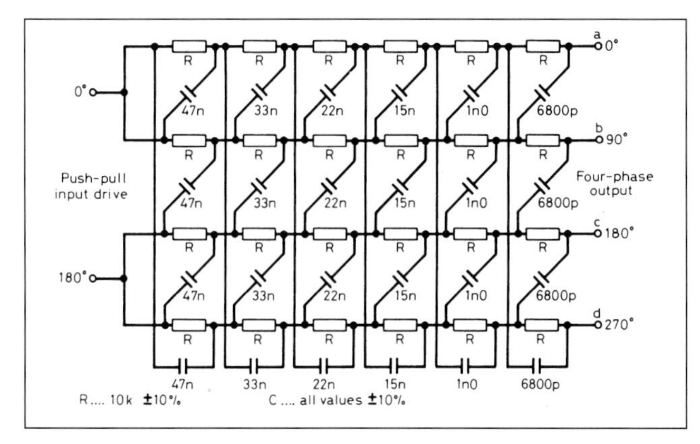


**Fig 7.9: Gingell polyphase network**

### RF phase-shift network

Traditionally the most satisfactory way to produce a 90° RF phase shift was to employ a low-Q network comprising of two loosely coupled tuned circuits which exhibits a combination of inductance, resistance and capacitance.

**Fig 7.10(a)** illustrates such a network in the anode circuit of a valve amplifier. The primary coils are inductively coupled while the link couplings are connected in series. When both circuits are tuned to resonance there will be exactly 90° phase shift between them. Difficulties occur when the frequency is changed, and the network has to be retuned, restricting the bandwidth to no more than 200kHz.

With the advent of digital ICs, it became relatively easy to obtain the required phase shift by dividing the output of an oscillator using a flip-flop IC. **Fig 7.10(b)** shows the RF circuitry for a 160m phasing exciter, in which the signal from a VFO tuning 7.2-8MHz is divided by four using a 74HC73 (J-K flip-flop), providing a square-wave output between 1.8 and 2MHz.

The VFO is conveniently implemented by one inverter of a 74HC00 (quad NAND-gate), which directly drives the flip-flop producing both 0 and 90 degree outputs. The fundamental square-wave signal will be phased out in the balanced modulator.

### Four-way phasing method

The four-way phasing method is an adaptation of the conventional phasing method and can be simply described as a double two-way method. **Fig 7.11** illustrates a four-way phasing exciter. The major requirement for acceptable carrier and sideband suppression is a good audio phase shifter. The polyphase network (Fig 7.9) is ideal and provides the required four output signals at 90° phase intervals. The RF output from the carrier generator must also provide four RF outputs phase-shifted by 90° from one another, and this is achieved by using a dual J-K flip-flop which also divides the input frequency by a factor of four. The phase-shifted AF and RF signals are fed to four modulators, the outputs of which are summed in a tuned adder, resulting in an SSB output signal.

A practical 9MHz four-way SSB generator is illustrated in **Fig 7.12**. The TTL oscillator is operated at 12MHz, providing a 3MHz signal at the output of the flip-flop. As this square wave signal is rich in odd-order harmonics, the tuned adder can be adjusted to tune to the third harmonic in preference to the fundamental signal, and the result is a 9MHz output SSB signal.

One major disadvantage of the digital phase shifter is the necessity to operate the oscillator on four times the output frequency. The technique used in Fig 7.12 represents one solution to the problem, but an alternative would be to heterodyne the output to the desired frequency using a VFO.

## The Filter Method

Since the objective is to transmit only a single sideband, it is necessary to select the desired sideband and suppress the unwanted sideband. The relationship between the carrier and sidebands is shown in the Principles chapter. Removing the unwanted sideband by the use of a selective filter has the

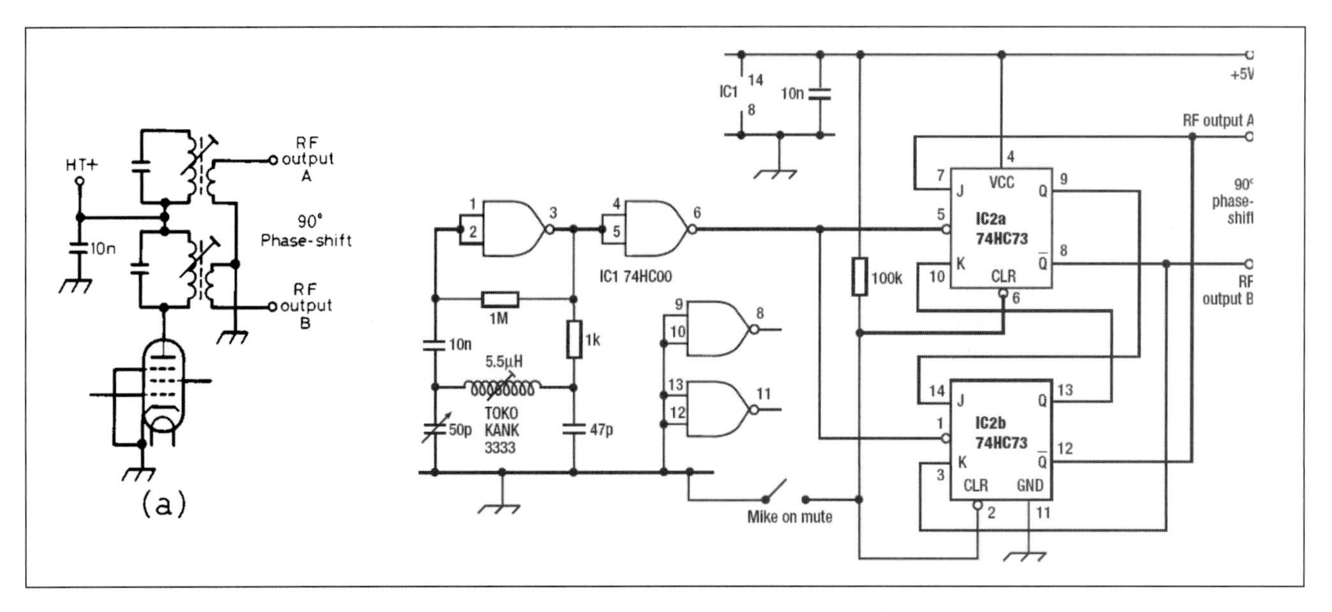


Fig 7.10: Methods of obtaining RF phase shift. (a) Traditional method of obtaining 90° phase shift using loosely coupled tuned circuits. (b) Active RF phase shifter 7.2MHz VFO providing 1.8MHz output with 90° shifts

Fig 7.11: Four-phase SSB generator

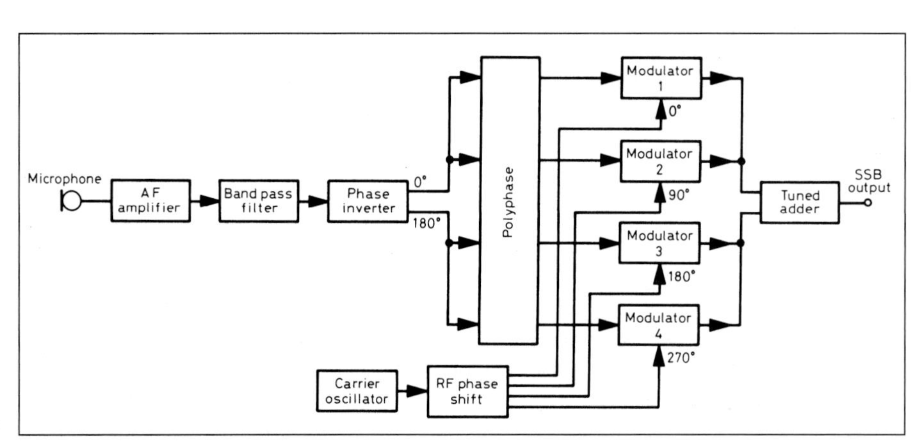


(below) Fig 7.12: 9MHz phasing exciter

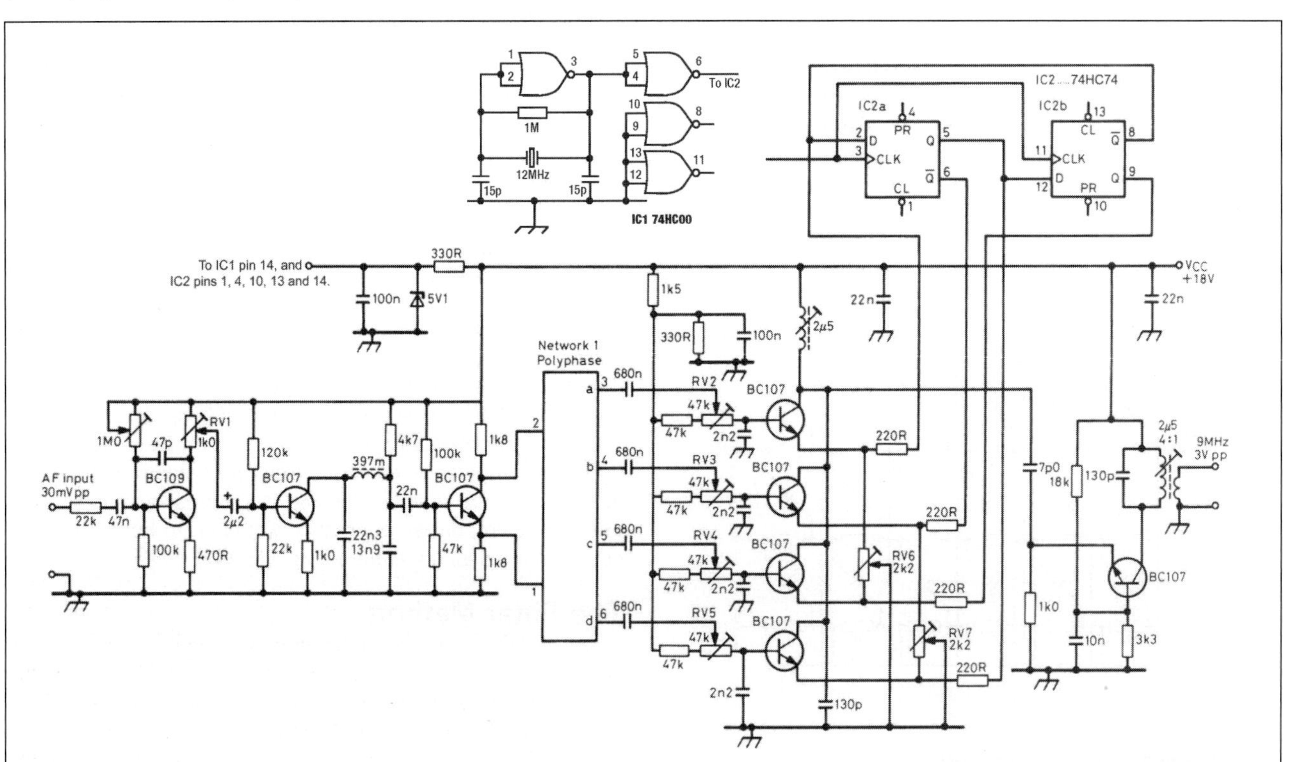

advantage of simplicity and good stability, and this is therefore the most widely adopted method of generating SSB. The unwanted sideband suppression is determined by the attenuation of the sideband filter, while the stability of this suppression is determined by the stability of the elements used in constructing the filter. High stability can be achieved by using materials that have a very low temperature coefficient. Commonly used materials are quartz, ceramic and metallic plates.

The filter method, because of its proven long-term stability, has become the most popular method used by amateurs. At present three types of selective sideband filters are in common use:

- High-frequency crystal filter
- Low-frequency mechanical filter
- Low-frequency ceramic filter

## Crystal filters

The crystal filter is the most widely used type of filter found in SSB transmitters. In a transceiver, one common filter can be used for both transmit and receive functions. Generation of a SSB signal in the 9MHz range permits single-frequency conversion techniques to be employed to cover the entire HF spectrum, and for this reason 9MHz and 10.7MHz filters have virtually dominated the market. A number of other frequencies have also been employed for filters, including 5.2MHz, 3.18MHz and 1.6MHz, the latter mainly for commercial applications.

The principle of operation of a crystal or quartz filter is based upon the piezo-electric effect. When the crystal is excited by an alternating electric current, it mechanically resonates at a frequency dependent upon its physical shape, size and thickness. A crystal will easily pass current at its natural resonant frequency but attenuates signals either side of this frequency (see the chapter on Passive Components).

By cascading a number of crystals having the same, or very closely related, resonant frequencies it is possible to construct a filter having a high degree of attenuation either side of a band of wanted frequencies, typically 40-60dB with a six-pole filter, 60-80dB using an eight-pole filter, and 80-100dB with a 10-pole filter. The characteristic bandwidth of a SSB filter is selected to pass a communications-quality audio spectrum of typically 300-3000Hz. For SSB transmission the best-sounding results can be achieved using a 3kHz wide filter, but for receiver applications a slightly narrower filter is preferable and 2.4kHz has become the accepted compromise. SSB transmissions made using narrower filters have a very restricted audio sound when received. Filter bandwidths are normally quoted at the 6dB and 60dB attenuation levels, and the ratio of the two quoted bandwidths is referred to as the shape factor, 1:1 being the ideal but not realistically achievable. Anything better than 2.5:1 may be regarded as acceptable.

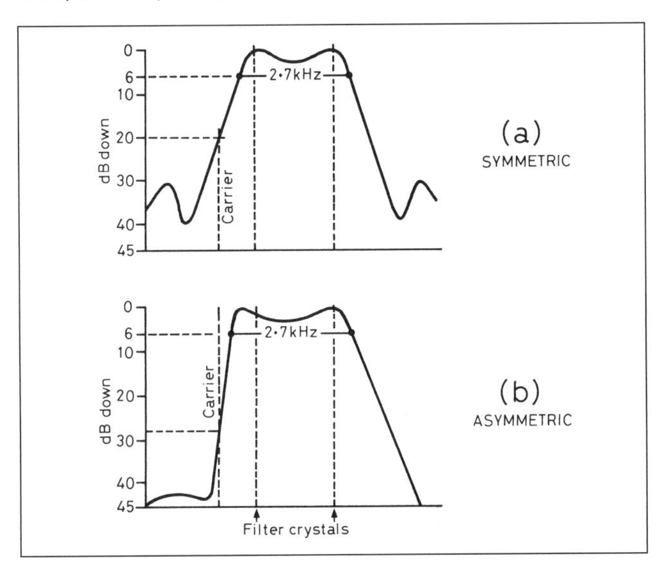


Fig 7.13: Response curves of two- and four-pole symmetric and asymmetric crystal filters

The purpose of the crystal filter is to attenuate the unwanted sideband but it conveniently has a secondary function: to attenuate further the already-suppressed carrier. A balanced modulator seldom attenuates the carrier by more than 40dB, which equates to 1mW of carrier from a 100W SSB transmitter. A further 20dB of carrier suppression is available from the sideband filter, making a carrier suppression of 60dB possible.

The passband of a crystal filter may be symmetrical in shape, **Fig 7.13(a)**, or asymmetric as in **Fig 7.13(b)**. Home-constructed filters are invariably asymmetric to some degree whereas commercially made filters will be designed to fall into either of the two categories. Assuming that we wish to generate a LSB signal with a carrier frequency of 9MHz, a SSB filter will be required to pass the frequency range 8.9975MHz to 9.0MHz, whereas if we wished to change to USB, the filter would be required to pass the frequency range 9.0MHz to 9.0025MHz. At first sight it would appear that two filters are required. In commercial equipment, the use of two filters has been common practice, most probably asymmetric and annotated with the sideband that they are designed to generate and the intended carrier frequency. In amateur radio equipment a much cheaper technique is adopted: it is easier to use a symmetrical filter with a centre frequency of 9.000MHz and move the carrier frequency from one side of the filter to the other in order to change sidebands.

Typically 8.9985MHz for USB and 9.0015MHz for LSB are used; note the filter frequency is above the carrier frequency to give USB and below the carrier frequency to give LSB. Asymmetric filters are invariably marked with the carrier frequency and have the advantage of higher attenuation of the unwanted carrier and sideband, due to the steeper characteristic of the filter on the carrier frequency side. This is the primary reason why they are used in commercial applications where they are required to meet a higher specification. One minor disadvantage of the symmetrical filter and switched carrier frequency method is that changing sideband causes a shift in frequency approximately equal to the bandwidth of the filter. This can be compensated for by an equal and opposite movement of the frequency-conversion oscillator.

Radio amateurs have adopted the practice of operating LSB below 10MHz and USB above 10MHz. It is therefore necessary to be able to switch sidebands if operation on all bands is contemplated.

Home construction of crystal filters is only to be recommended if a supply of cheap crystals is available. Fortunately there are now several sources. Clock crystals for microprocessor applications are manufactured in enormous quantities and cost pence rather than pounds. Another source of crystals is those intended for TV colour-burst; the UK and continental frequency of 4.43MHz is the more suitable as the USA colour-burst frequency is in the 80m amateur band. **Fig 7.14(a)** shows an eight-pole ladder filter designed by G3UUR and constructed using colour-burst crystals. The frequency of individual crystals is found to vary by as much as 200Hz, but by careful selection of crystals it is possible to find those that are on frequency and those that are slightly above or below the nominal frequency.

For optimum results it is recommended that the on-frequency crystals are located at either end of the filter while the centre four crystals should be slightly higher in frequency. **Fig 7.14(b)** shows the frequency response of the G3UUR filter constructed from TV colour-burst crystals.

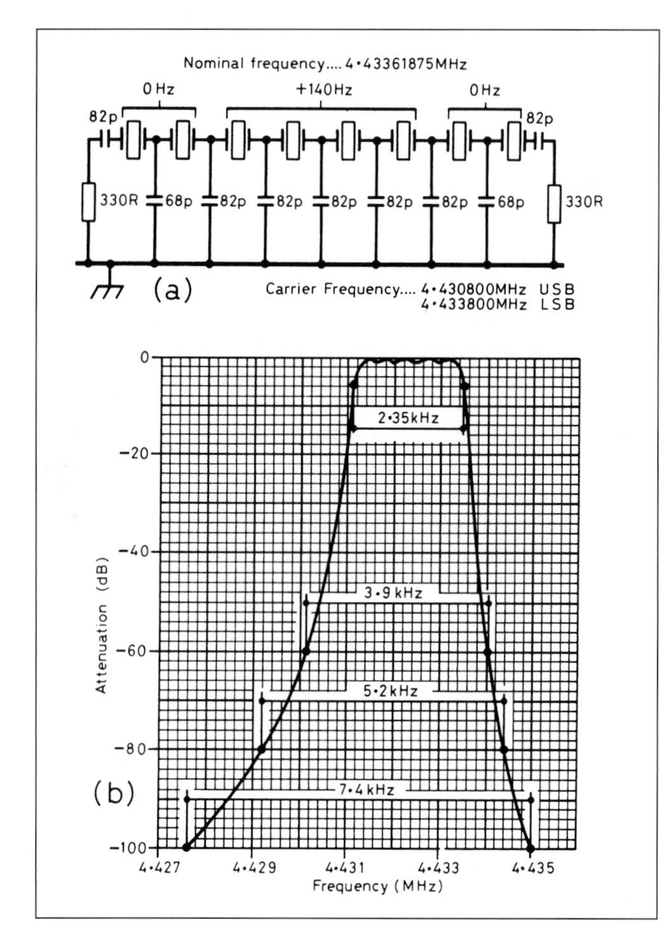


**Fig 7.14: (a) Ladder crystal filter using 4.43MHz TV crystals. (b) Ladder crystal filter using P129 specification colour-TV 4.43MHz crystals. It provides a performance comparable with the ladder filter in the Atlas 180 and 215 transceivers. Insertion loss 4-5dB, shape factor (6/60dB) 1.66. Note that the rate of attenuation on the low-frequency side of the response is as good as an eight-crystal lattice design; on the HF side it is better**

Commercially available crystal filters are primarily confined to 9MHz and 10.7MHz types, though the sources seem to come and go. Numerous filters are available new, and surplus filters can often be found at bargain prices. **Table 7.4** lists the KVG range of filters - alternative filters are listed in **Table 7.5**.

## Mechanical filters

The mechanical filter was developed by the Collins Radio Company for low-frequency applications in the range 60-500kHz. The F455 FA-21 was designed specifically for the amateur radio market with a nominal centre frequency of 455kHz, a 6dB bandwidth of 2.1kHz and a 60dB bandwidth of 5.3kHz, providing a shape factor of just over 2:1.

The mechanical filter is made up from a number of metal discs joined by coupling rods. The discs are excited by magnetostrictive transducers employing polarised biasing magnets and must not be used in circuits where DC is present. The input and output transducers are identical and are balanced to ground so the filter can be used in both directions. Mechanical filters have always been expensive but provide exceptional performance, and examples of the Collins filters can often be found in surplus equipment. Japanese Kokusai filters were available for a number of years and offered a more cost-effective alternative. Mechanical filters are now seldom used in new equipment.

## Ceramic filters

Ceramic filters have been developed for broadcast radio applications; they are cheap, small and available in a wide range of frequencies and bandwidths. Narrow-bandwidth ceramic filters with a nominal centre frequency of 455kHz have been manufactured for use in SSB receiver IF applications and provide a level of performance making them ideally suited for use in SSB transmitters. Bandwidths of 2.4 and 3kHz are available with shape factors of 2.5:1 and having attenuation in excess of 90dB (Table 7.5).

## SSB filter exciter

The SSB filter exciter is inherently simpler than the phasing-type exciter. **Fig 7.15(a)** illustrates a 455kHz SSB generator requiring just two ICs and using a ceramic filter. The SA602 doubly balanced modulator is provided with a 455kHz RF input signal

| Filter type | XF-9A | XF-9B | XF-9C | XF-9D | XF-9E | XF-9M |
|---|---|---|---|---|---|---|
| Application | SSB TX | SSB TX/RX | AM | AM | FM | CW |
| Number of poles | 5 | 8 | 8 | 8 | 8 | 4 |
| 6dB bandwidth (kHz) | 2.5 | 2.4 | 3.75 | 5.0 | 12.0 | 0.5 |
| Passband ripple (dB) | <1 | <2 | <2 | <2 | <2 | <1 |
| Insertion loss (dB) | <3 | <3.5 | <3.5 | <3.5 | <3 | <5 |
| Termination | 500 /30pF | 500 /30pF | 500 /30pF | 500 /30pF | 1200 /30pF | 500 /30pF |
| Shape factor | 1.7 (6-50dB) | 1.8 (6-60dB) | 1.8 (6-60dB) | 1.8 (6-60dB) | 1.8 (6-60dB) | 2.5 (6-40dB) |
| | | 2.2 (6-80dB) | 2.2 (6-80dB) | 2.2 (6-80dB) | 2.2 (6-80dB) | 4.4 (6-60dB) |
| Ultimate attenuation (dB) | >45 | >100 | >100 | >100 | >90 | >90 |

**Table 7.4: KVG 9MHz crystal filters for SSB, AM, FM and CW applications**

| Filter type | 90H2.4B | 10M02DS | CFS455J | CFJ455K5 | CFJ455K14 | QC1246AX |
|---|---|---|---|---|---|---|
| Centre frequency (MHz) | 9.0000 | 10.7000 | 0.455 | 0.455 | 0.455 | 9.0000 |
| Number of poles | 8 | 8 | Ceramic | Ceramic | Ceramic | 8 |
| 6dB bandwidth (kHz) | 2.4 | 2.2 | 3 | 2.4 | 2.2 | 2.5 |
| 60dB bandwidth (kHz) | 4.3 | 5 | 9 (80dB) | 4.5 | 4.5 | 4.3 |
| Insertion loss (dB) | <3.5 | <4 | 8 | 6 | 6 | <3 |
| Termination | 500 /30pF | 600 /20pF | 2k | 2k | 2k | 500 30pF |
| Ultimate attenuation (dB) | >100 | - | >60 | >60 | 60 | >90 |
| Manufacturer/UK supplier | IQD | Cirkit | Cirkit | Bonex | Bonex | SEI |

**Table 7.5: Popular SSB filters**

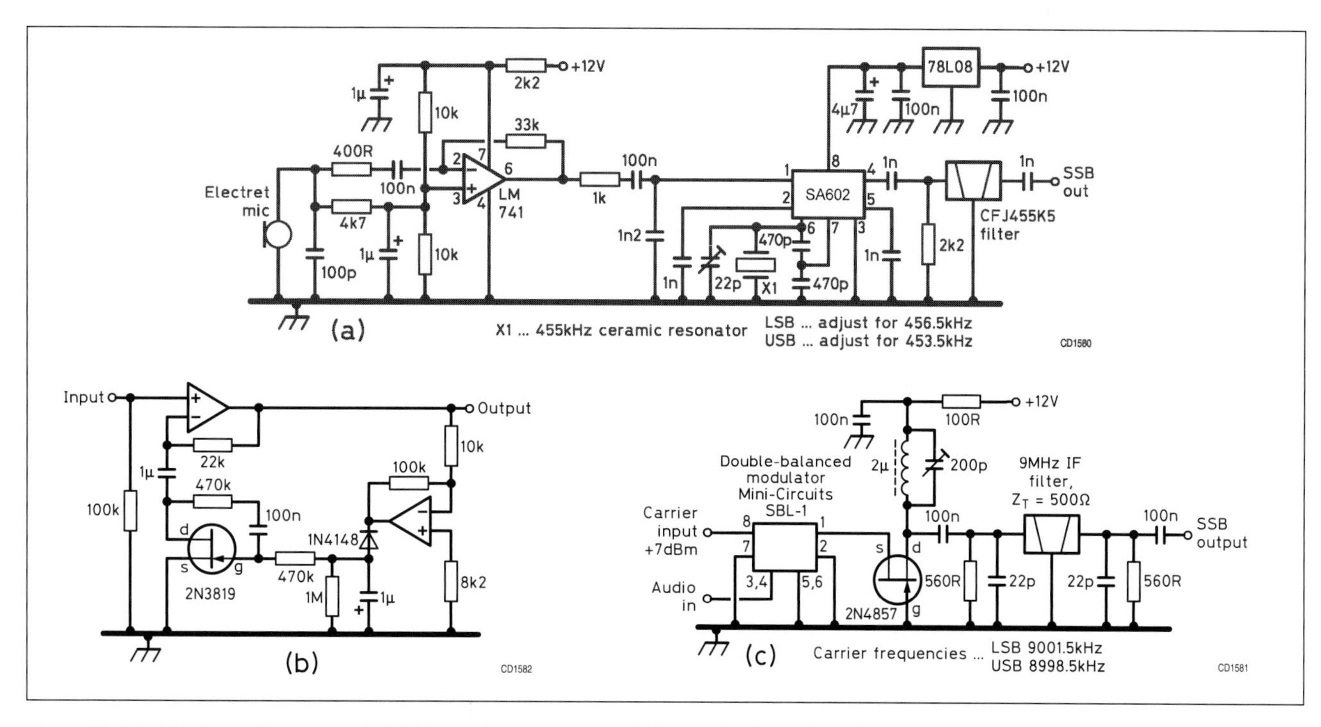


Fig 7.15: (a) 455kHz SSB generator. (b) Example of an audio AGC circuit. (c) 9MHz SSB generator

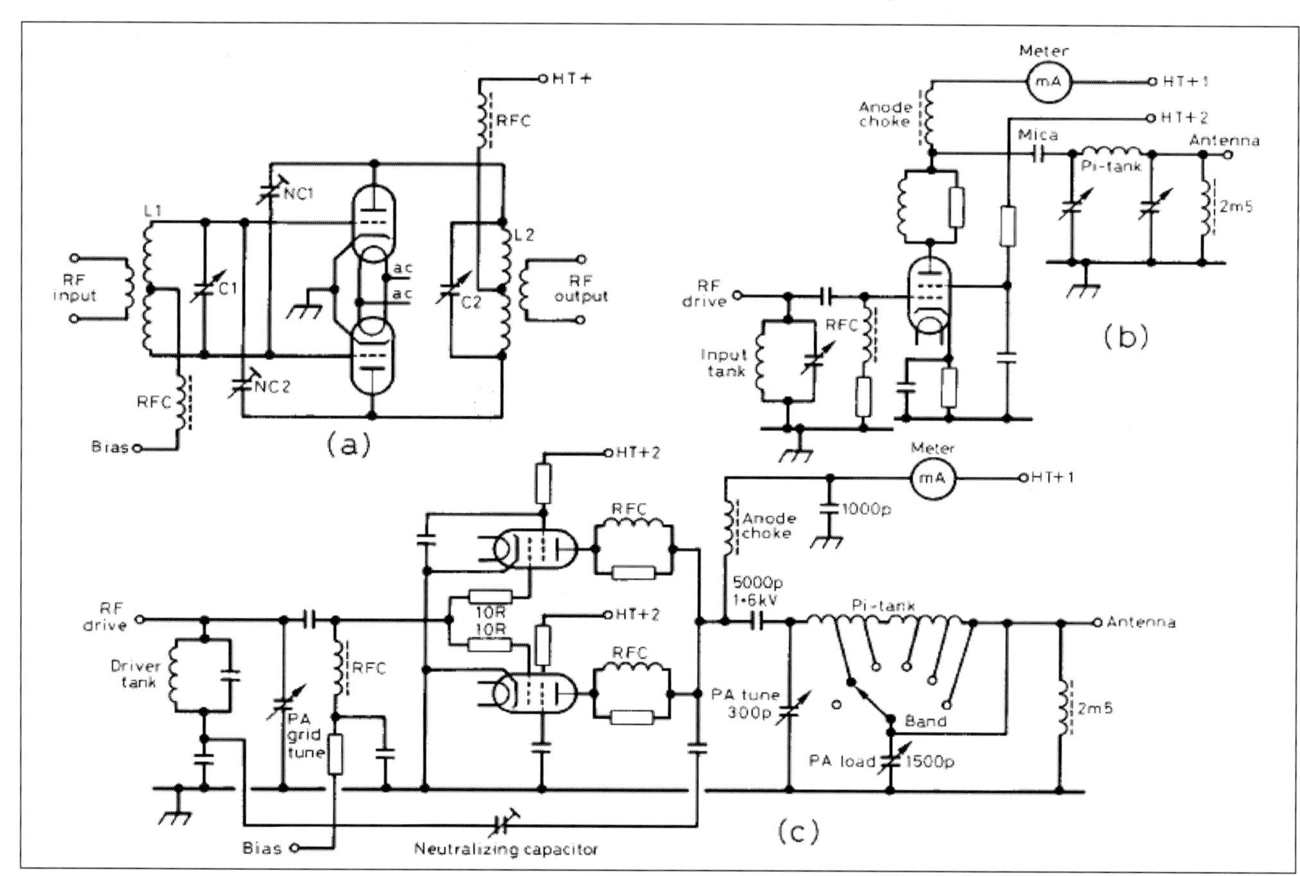


Fig 7.16: Valve power amplifiers. (a) Push-pull amplifier; (b) Single-ended Class-C amplifier; (c) Parallel-pair RF amplifier

from its own internal oscillator. This can use a 455kHz crystal or the cheaper ceramic resonator, and the latter can be pulled in frequency to generate either a LSB or USB carrier. A DSB signal at the output of the balanced modulator is fed directly to the ceramic filter for removal of the unwanted sideband, resulting in a 455kHz SSB output signal. This can be heterodyned directly to the lower-frequency amateur bands or via a second higher IF to the HF bands. The circuit performance can be improved by the addition of an audio AGC circuit, such as the one shown in **Fig 7.15(b)**. **Fig 7.15(c)** illustrates a design for a 9MHz SSB generator. Microphone audio would need to pass through a suitable audio amplifier before being applied to the balanced modulator at pins 3 and 4. Output from the 9MHz SSB exciter can be heterodyned directly to any of the LF and HF amateur bands. The only adjustment required in either circuit Fig 7.15(a) or (c) is to adjust the oscillator to the correct frequency in relation to the filter; this frequency is normally located 20dB down the filter response curve. Quite often the oscillator can simply be adjusted for the best audio response at the receiver.

## POWER AMPLIFIERS

The RF power amplifier is normally considered to be that part of a transmitter which provides RF energy to the antenna. It may be a single valve or transistor, or a composite design embodying numerous devices to take low-level signals to the final output power level. RF amplifiers are also discussed in detail in the Building Blocks chapter. Push-pull valve amplifiers, **Fig 7.16(a)**, were popular until the 1950s and offered a number of advantages over single-ended output stages as in **Fig 7.16(b)**. The inherent balance obtained when two similar valves having almost identical characteristics are operated in push-pull results in improved stability, while even-order harmonics are phased out in the common tank circuit. One major shortcoming of the push-pull valve amplifier is that switching of the output tank circuit for operation on more than one frequency band is exceedingly difficult due to the high RF voltages present.

During the 1960s, push-pull amplifiers were superseded by the single-ended output stage, often comprising two valves in parallel, **Fig 7.16(c)**, coupled to the antenna via a pi-output network and low-impedance coaxial cable. The changes in design were brought about by two factors. First, the rapid expansion in TV broadcasting introduced problems of harmonically related TVI. This demanded greater attenuation of odd-order harmonics than was possible with the conventional tank circuit used in push-pull amplifiers. Second, there was a trend towards the development of smaller, self-contained transmitters capable of multiband operation, ultimately culminating in the transceiver concept which has almost totally replaced the separate transmitter and receiver in amateur radio stations.

The pi-output tank circuit differs from the conventional parallel-tuned tank circuit in that it uses two variable capacitors in series: Figs 7.4(a) and 7.4(b). The junction of the two capacitors is grounded and the network is isolated from DC using a high-voltage blocking capacitor. The input capacitor has a low value, while the output capacitor has a considerably higher value. This provides high-to-low impedance transformation across the network, enabling the relatively high anode impedance of the power amplifier to be matched to an output impedance of typically 50Ω. The pi-tank circuit performs the function of a low-pass filter and provides better attenuation of harmonics than a link-coupled tank circuit. The low-impedance output is easily band-switched over the entire HF spectrum by shorting out turns, and facilitates direct connection to a dipole-type antenna, an external low-pass filter for greater reduction of harmonics or an antenna matching unit for connection to a variety of antennas.

The Class C non-linear power amplifiers commonly used in CW and AM transmitters produced high levels of harmonics, making compatibility with VHF TV transmissions exceedingly difficult. Fortunately the introduction of amateur SSB occurred at much the same time as the rapid expansion in television operating hours, and the introduction of 'linear' Class AB amplifiers, essential for the amplification of SSB signals, greatly reduced the problems of TVI with TV frequencies that were often exact multiples of the 14, 21 and 28MHz amateur bands. The parallel-pair valve power amplifier became the standard output stage for amateur band transmitters and is still in common use in many external linear amplifiers and a few amateur radio transceivers .

Solid-state power amplifiers first appeared in the late 1970s and were capable of up to 100W RF output. Initially the reliability was poor but improved rapidly, and within 10 years virtually all commercially made amateur radio transmitters were equipped with a 100W solid-state PA unit. Transistors for high-power RF amplification are specially designed for the purpose and differ considerably internally from low-frequency switching transistors. Initial attempts by amateurs to use devices not specifically designed for RF amplification resulted in a mixture of success and failure, probably giving rise to initial claims that solid-state power amplifiers were less reliable than their valve counterparts.

Due to the very low output impedances of solid-state bipolar devices, it is almost impossible to match the RF output from a solid-state power amplifier to a resonant tank circuit with its characteristic high impedance. As a result, low-Q matching circuits (Fig 7.2 and 7.3) are used to transform the very low (typically 1 or 2Ω) output impedance encountered at the collector of an RF power amplifier to the now-standard output impedance of 50Ω.

As the matching circuits are not resonant, they are broad-band in nature, making it possible to operate the amplifier over a wide range of frequencies with no need for any form of tuning. Unfortunately, the broad-band amplifier also amplifies harmonics and other unwanted products, making it essential to use a low-pass filter immediately after the amplifier in order to achieve an adequate level of spectral purity. It is also essential that the amplifier itself should not contribute to the production of unwanted products. For this reason solid-state amplifier designs have reverted to the balanced, push-pull mode of operation, with its inherent suppression of even-order harmonics. As band switching and tuning of the amplifier is neither possible or necessary, construction of the amplifier is relatively simple. Typically, HF broad-band amplifiers will provide an output over the entire HF spectrum from 1.8 to 30MHz.

Power amplifiers can be categorised into two basic types, valve and solid-state, and then further into sub-groups based upon the output power. Low-power amplifiers can be regarded as 10W or less, including QRP (usually regarded as less than 5W output power), medium power up to 100W, and high power in excess of 100W.

The class of operation of a power amplifier is largely determined by its function. Class C amplifiers are commonly used for CW, AM and FM transmissions because of their high efficiency. Class C is also a pre-requisite for successful high-level modulation in an AM transmitter. Due to the non-linear operation of Class C amplifiers, the harmonic content is high and must be adequately filtered to minimise interference.

Single sideband transmission demands linear amplification if distortion is to be avoided, and amplifiers may be operated in either Class A or B. Class A operation is inefficient and normally confined to driver stages, the high standing current necessary to achieve this class of operation usually being unacceptable in amplifiers of any appreciable power rating. Most linear amplifiers designed for SSB are operated in between Class A and B, in what are known as Classes AB1 and AB2, in order to achieve a compromise between efficiency and linearity.

### Solid-state Versus Valve Amplifiers

The standard RF power output for the majority of commercially produced amateur radio transmitters/transceivers is 100W, and at this power level solid-state amplifiers offer the following:

- Compact design.
- Simpler power supplies requiring only one voltage (normally 13.8V) for the entire equipment.
- Broad-band, no tune-up operation, permitting ease of operation.
- Long life with no gradual deterioration due to loss of emission.
- Ease of manufacture and reduced cost.

There are of course some disadvantages with solid-state amplifiers and it is for this reason that valves have not disappeared entirely. However, valve amplifiers have considerably more complex power supply requirements and great care must be taken with the high voltages involved if home construction is contemplated. Although solid state amplifiers only require relatively low voltages, they require very high currents for the generation of any appreciable power, placing demands upon the devices and their associated power supplies. Heatsinking and voltage stabilisation become very important. While the construction of solid-state amplifiers up to 1kW is feasible using modern devices operated from a high-current 50V power supply, valve designs offer a simpler and more cost-effective alternative. However, the development of VMOS devices with much higher input and output impedances is beginning to bridge the gap. Quite possibly within a few years VMOS devices will offer a cheaper alternative to the valve power amplifier at the kilowatt level.

## Impedance Matching

All types of power amplifier have an internal impedance made up from a combination of the internal resistance, which dissipates power in the form of heat, and reactance. As both source and load impedances are fixed values and not liable to change, it is necessary to employ some form of impedance transformation or matching in order to obtain the maximum efficiency from an amplifier. The power may be expressed as:

$$P_{input} = P_{output} + P_{dissipated}$$

where $P_{input}$ is the DC input power to the stage; $P_{output}$ is the RF power delivered to the load and $P_{dissipated}$ is the power absorbed in the source resistance and dissipated as heat.

$$\text{Efficiency} = \frac{P_{output}}{P_{input}} \times 100\%$$

When the source resistance is equal to the load resistance, the current through either will be equal as they are in series, with the result that 50% of the power will be dissipated by the source and 50% will be supplied to the load. The object of a power amplifier is to provide maximum power to the load. Design of a power amplifier must also take into account the maximum dissipation of the output device as specified by the manufacturer. An optimum load resistance is selected to ensure maximum output from a power amplifier while not exceeding the amplifying device's power dissipation. Efficiency increases as the load resistance to source resistance ratio increases and vice versa. The optimum load resistance is determined by the device's current transfer characteristics and for a solid-state device is given by:

$$R_L = \frac{V_{cc}^2}{2P_{out}}$$

Valves have more complex current transfer characteristics which differ for different classes of operation; the optimum load resistance is proportional to the ratio of the DC anode voltage to the DC anode current divided by a constant which varies from 1.3 in Class A to approximately 2 in Class C.

$$R_L = \frac{V_a}{KI_a}$$

The output from a RF power amplifier is usually connected to an antenna system of different impedance, so a matching network must be employed. Two methods are commonly used: pi-tank circuit matching for valve circuits and transformer matching for solid-state amplifiers. The variable nature of a pi-tank circuit permits matching over a wide range of impedances, whereas the fixed nature of a matching transformer is dependent upon a nominal load impedance of typically 50Ω, and it is therefore almost essential to employ some form of antenna matching unit between the output and the final load impedance. Matching networks serve to equalise load and source resistances while providing inductance and capacitance to cancel any reactance.

## Valve Power Amplifiers

Valve power amplifiers are commonly found in the output stages of older amateur transmitters. Usually two valves will be operated in parallel, providing twice the output power possible with a single valve. The output stage may be preceded by a valve or solid-state driver stage.

The valve amplifier is capable of high gain when operated in the tuned input, tuned output configuration often referred to as TPTG (tuned plate tuned grid). Two 6146 valves are capable of producing in excess of 100W RF output with as little as 500mW of RF drive signal at the input. Operation of two valves in parallel increases the inter-electrode capacitances by a factor of two and ultimately affects the upper frequency operating limit. Fig 7.16(c) illustrates a typical output stage found in amateur transceivers.

### Neutralisation

The anode-to-grid capacitance of a valve provides a path for RF signals to feed back energy from the anode to the grid. If this is sufficiently high, oscillation will occur. This can be overcome by feeding back a similar level of signal, but of opposite phase, thus cancelling the internal feedback. This process is referred to as neutralisation.

Once set, the neutralisation should not require further adjustment unless the internal capacitance of the valve changes or the valve has to be replaced.

While there are numerous ways of achieving neutralisation, the most common method still in use is series-capacitance neutralisation. A low-value ceramic variable capacitor is connected from the anode circuit to the earthy end of the grid input circuit as in Fig 7.16(c).

The simplest way to adjust neutralisation is firstly to tune up the amplifier into a dummy load and then to remove all the high voltage supplies from the valves, leaving the filaments powered. Connect a sensitive RF voltmeter across the dummy load and drive the input of the amplifier with rated power. When the neutralising capacitor is adjusted, the RF voltage at the output will dip to a minimum at the optimum tuning point. This method is safe and accurate. An alternative is to reverse the amplifier connections, that is apply the drive power to the antenna port and measure the RF voltage at the grid network and similarly tune for a dip. Neutralising capacitors should be adjusted with non-metallic trimming tools as metal tools will interfere with the adjustment.

### Parasitic oscillation

It is not uncommon for a power amplifier to oscillate at some frequency other than one in the operating range of the amplifier. This can usually be detected by erratic tuning characteristics and a reduction in efficiency. The parasitics are often caused by the resonance of the connecting leads in the amplifier circuit with the circuit capacitance. To overcome problems at the design stage it is common to place low-value resistors in series with the grid, and low-value RF chokes, often wound on a resistor body, in series with the anode circuit. Ferrite beads may also be strategically placed in the circuit to damp out any tendency to oscillate at VHF.

## HIGH-POWER AMPLIFICATION

Output powers in excess of 100W are invariably achieved using an add-on linear amplifier. In view of the high drive power available, the amplifier can be operated at considerably lower gain, with the advantages of improved stability and no requirement for neutralisation. Input circuits are usually passive, with valves operated in either passive-grid or grounded-grid modes. The grounded-grid amplifier has a cathode impedance ideally suited to matching the pi-output circuit of a valve exciter. Pi-input networks are usually employed to provide the optimum 50-ohm match for use with solid-state exciters. Passive grid amplifiers often employ a grid resistor of 200 to 300Ω, which is suitable for connecting to a valve exciter but will require a matching network such as a 4:1 auto-transformer for connection to a solid-state exciter.

### Output matching

There are only two output circuits in common use in valve power amplifiers - the pi-output network is by far the most common and suitable values for a range of anode impedances are provided in Table 7.2. More recently, and especially for applications in high-power linear amplifiers, an adaptation of the circuit has appeared called the pi-L output network, and here the conventional pi-network has been combined with an L-network to provide a matching network with a considerable improvement in attenuation of unwanted products. The simple addition of one extra inductor to the circuit provides a considerable improvement in performance. Suitable values for a pi-L network are given in Table 7.3 for a range of anode impedances.

Valve amplifiers employing pi-output networks require a suitable RF choke to isolate the anode of the power amplifier from the high voltage power supply. This choke must be capable of carrying the anode current as well as the high anode voltages likely to be encountered in such an amplifier. The anode choke must not have any resonances within the operating range of the amplifier or it will overheat with quite spectacular results. Chokes are often wound in sections to reduce the capacitance between turns and may employ sections in varying diameters. Ready-made chokes are available for high-power operation from a number of sources on the internet. A typical example is the very reasonably priced amateur bands transmitter choke rated at 4kV (part number RFC-3) offered by RF Parts Company in the USA [2].

## Solid-state Power Amplifiers

Solid-state power amplifiers are invariably designed for 50Ω input and output impedances, and they employ broad-band transformers to effect the correct matching to the devices. While single-ended amplifiers are often shown in test circuits, Unless they are tuned (**Fig 7.17(a)**), their use is not recommended for the following reasons:

(a) higher levels of second harmonic may be present than when using push-pull. These are difficult to attenuate using standard low-pass filters;

(b) multiband operation is more difficult to achieve due to the more complex filter requirements.

The broad-band nature of a solid-state amplifier with no requirement for bandswitching enables push-pull designs to be used,

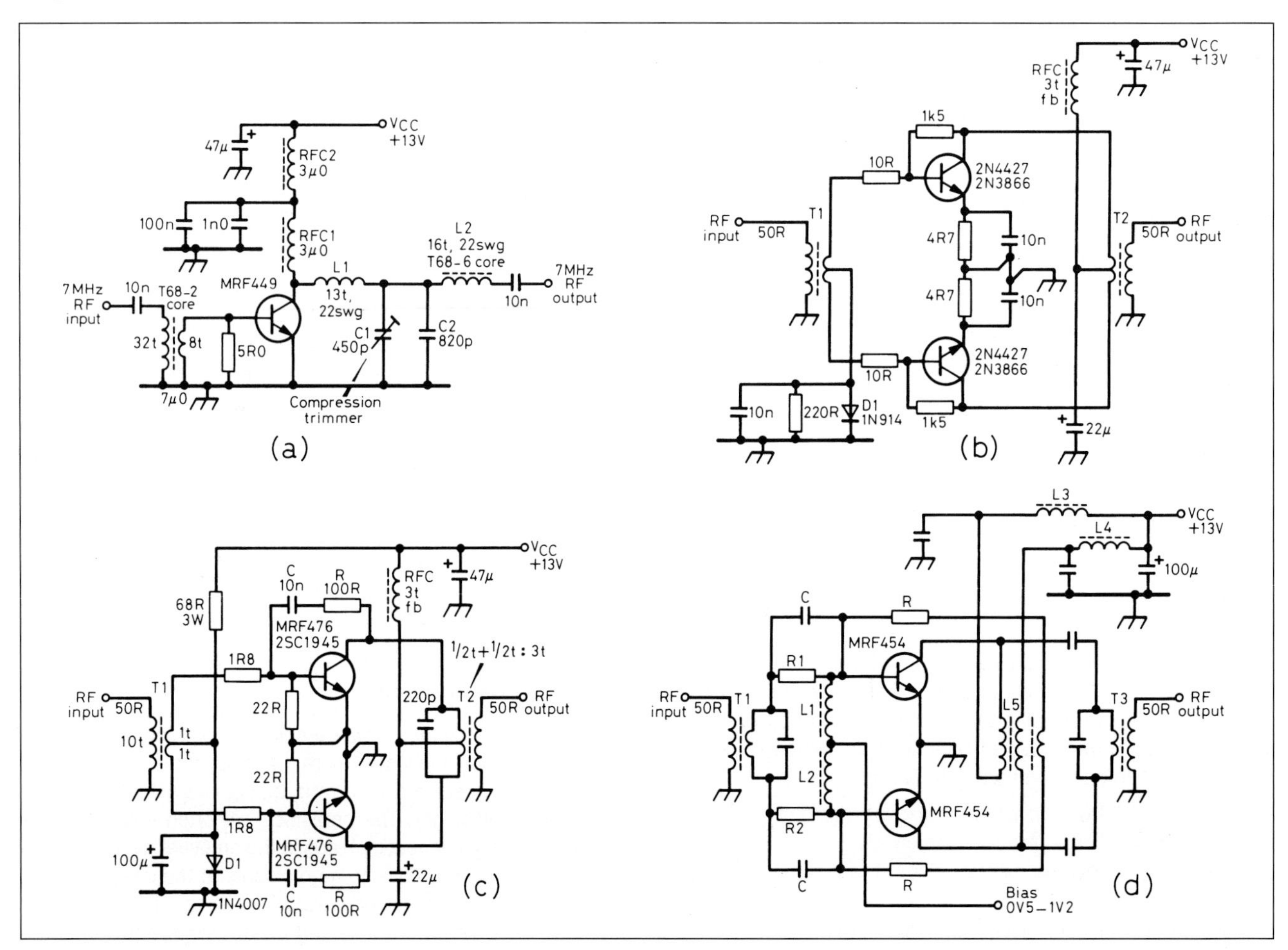


Fig 7.17: Solid-state power amplifiers. (a) Single-ended Class C amplifier (15W); (b) low-power amplifier driver; (c) low-power push-pull amplifier (10-25W); (d) medium-power push-pull amplifier (100W)

taking advantage of their improved balance and natural suppression of even-order harmonics (though care should be taken to reduce odd-order harmonics). The gain of a solid-state broadband amplifier is considerably lower than that of a valve power amplifier (typically 10dB) and may necessitate cascading a number of stages in order to achieve the desired power level. This is easily achieved using common input and output impedances.

The reduced gain has the advantage of aiding stability, but the gain rises rapidly with a reduction in frequency and demands some form of frequency-compensated gain reduction. The latter is achieved using negative feedback with a series combination of R and C: Figs 7.17(c) and (d). Good decoupling of the supply down to audio frequencies is essential. The use of VHF power transistors is not recommended in the HF spectrum as instability can result even when high levels of negative feedback are employed.

Solid-state amplifiers can be operated in Class C for use in CW and FM transmitters. However, their use for AM transmission should be treated very carefully because device ratings must be capable of sustaining double the collector voltage and current on modulation peaks, ie four times the power of the carrier. Additional safety margins must be included to allow for high RF voltages generated by a mismatched load.

For AM transmission it is recommended that the signal be generated at a low level and then amplified using a linear amplifier. Again, allowance should be made for the continuous carrier and the power on modulation peaks.

### Output filters

Solid-state amplifiers must not be operated into an antenna without some form of harmonic filtering. The most common design is the pi-section filter, comprising typically of a double pi-section (five-element) and in some cases a triple pi-section (seven-element) filter. Common designs are based upon the Butterworth and Chebyshev filters and derivations of them. The purpose of the low-pass filter is to pass all frequencies below the cut off frequency (f0), normally located just above the upper band edge, while providing a high level of attenuation to all frequencies above the cut-off frequency. Different filter designs provide differing attenuation characteristics versus frequency, and it is desirable to achieve a high level of attenuation by at least three times the cut-off frequency in order to attenuate the third harmonic.

The second harmonic, which should be considerably lower in amplitude due to the balancing action of the PA, will be further attenuated by the filter which should have achieved approximately 50% of its ultimate attenuation.

Elliptic filters are designed to have tuned notches which can provide higher levels of attenuation at selected frequencies such as 2f and 3f. Filters are discussed in detail in the chapters on Building Blocks and General Data.

The desire to achieve high levels of signal purity may tempt constructors to place additional low-pass filters between cascaded broad-band amplifiers, but this practice will almost certainly result in spurious VHF oscillations. These occur when the input circuit resonates at the same frequency as the output circuit, ie when the filter acts as a short-circuit between the amplifier input and output circuits. Low-pass filters should only be employed at the output end of an amplifier chain. If it is essential to add an external amplifier to an exciter which already incorporates a low-pass filter, it is important either to use a resistive matching pad between the exciter and the amplifier, or modify the output low-pass filter to ensure that it has a different characteristic to the input filter. If a capacitive input filter is used in the exciter, an inductive input filter should be employed at the output of the linear amplifier. The resulting parasitics caused by the misuse of filters may not be apparent without the use of a spectrum analyser, and the only noticeable affect may be a rough-sounding signal and warm low-pass filters.

### Amplifier matching

Broad-band transformers used in solid-state amplifiers consist of a small number of turns wound on a stacked high-permeability ferrite core. The secondary winding may be wound through the primary winding which may be constructed from either brass tube or copper braid. The grade of ferrite is very important and will normally have an initial permeability of at least 800 (Fairite 43 grade). Too low a permeability will result in poor low-frequency performance and low efficiency. Some designs use conventional centre-tapped transformers for input and output matching (Fig 7.17(b) and (c)); these transformers carry the full DC bias and PA currents.

Other designs, Fig 7.17(d), include phasing transformers to supply the collector current while the output transformer is blocked to DC by series capacitors. The latter arrangement provides an improvement in IMD performance of several decibels but is often omitted in commercial amateur radio equipment on grounds of cost.

### Amplifier protection

The unreliability of early solid-state PAs was largely due to a lack of suitable protection circuitry. ALC (automatic level control) has been used for controlling the output of valve amplifiers for many years - by sampling the grid current in a valve PA it is possible to provide a bias that can be fed back to the exciter to reduce the drive level. A similar system is used for solid-state amplifiers but it is usually derived by sampling some of the RF output present in a SWR bridge circuit.

ALC is very similar to the AGC system found in a receiver. One disadvantage of sampling the RF output signal is that excessive ALC levels will cause severe clipping of the RF signal with an associated degradation of the IMD performance of the amplifier. For optimum performance the ALC system should only just be operating.

One of the major differences between solid-state and valve amplifying devices is that the maximum voltage ratings of solid-state devices are low and cannot be exceeded without disastrous consequences.

Reverse ALC is provided to overcome this problem and works in parallel with the conventional or forward ALC system. High RF voltages appearing at the PA collectors are attributable to operating into mismatched loads which can conveniently be detected using a SWR bridge. The reverse or reflected voltage can be

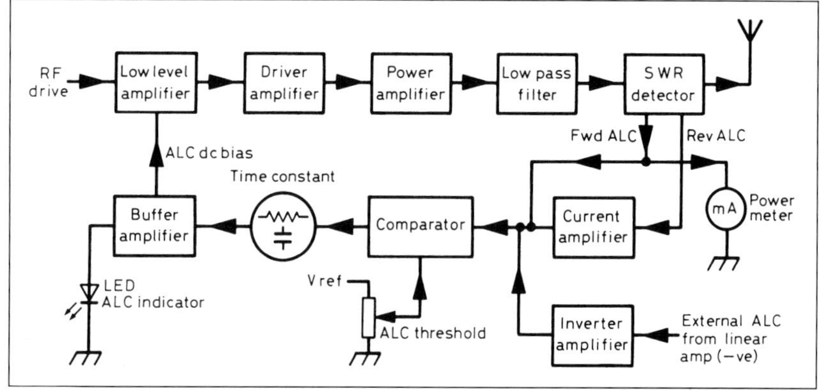


Fig 7.18: ALC system incorporating forward and reverse protection

sampled using the SWR bridge and amplified and used to reduce the exciter drive by a much greater level than that used with the forward ALC. The output power is cut back to a level which then prevents the high RF voltages being generated and so protects the solid-state devices. Some RF output devices are fitted internally with zener diodes to prevent the maximum collector voltage being exceeded. **Fig 7.18** illustrates a typical ALC protection system.

## Heatsinking for solid-state amplifiers

The importance of heatsinking for solid-state amplifiers cannot be overstressed. On no account should a power amplifier be allowed to operate without a heatsink, even for a short period of time.

Low-power amplifiers can often be mounted directly to an aluminium chassis if there is adequate metalwork to dissipate the heat, but a purpose-made heatsink should ideally be included for power levels in excess of 10W. For SSB operation the heatsinking requirements are less stringent than for CW or data operation due to the much lower average power dissipation. Nevertheless, an adequate safety margin should be provided to allow for long periods of key-down operation.

Solid-state power amplifiers are of the order of 50% efficient, and therefore a 100W output amplifier will also have to dissipate 100W of heat. It can be seen that for very-high-power operation heatsinking becomes a major problem due to the compact nature of solid-state amplifiers. The use of copper spreaders is advisable for powers above 200W. The amplifier is bolted directly to a sheet of copper at least 6mm (0.25in) thick, which in turn is bolted directly to the aluminium heatsink. The use of air blowers to circulate air across the heatsink should also be considered.

The actual mounting of power devices requires considerable care; the surface should ideally be milled to a flatness of ±0.012mm (0.5 thousandths of an inch). For this reason, diecast boxes must not be used for high-power amplifiers as the conductivity is poor and the flatness is nowhere near to being acceptable. Heatsinking compounds are also essential for aiding the rapid conduction of heat away from the device. Motorola Application Note AN1041/D [1] provides guidance in the mounting of power devices in the 200-600W range.

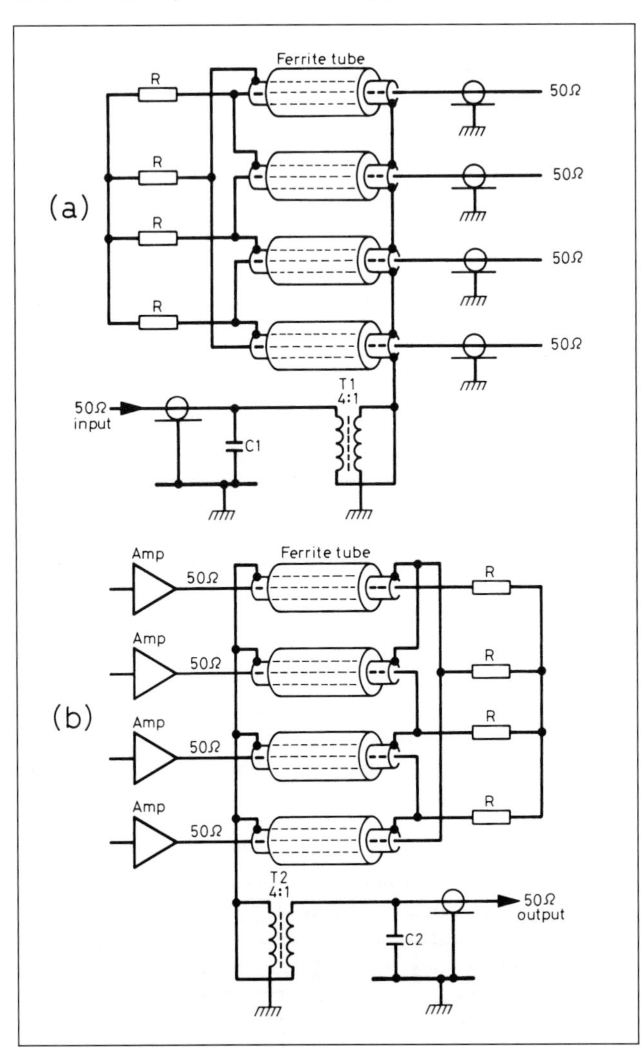


**Fig 7.19: Combining multiple power amplifiers. (a) Four port power divider. (b) Four-port output combiner**

## Power dividers and combiners

It is possible to increase the power output of a solid-state amplifier by combining the outputs of a number of smaller amplifiers. For instance it is practical to combine the outputs of four 100W amplifiers to produce 400W, or two 300W amplifiers to produce 600W. Initially the drive signal is split using a power divider and then fed to a number of amplifiers that are effectively operated in parallel. The outputs of the amplifiers are summed together in a power combiner that is virtually the reverse of the divider circuit. **Fig 7.19** illustrates a four-way divider and combiner.

The purpose of the power divider is to divide the input power into four equal sources, providing an amount of isolation between each. The outputs are designed for 50Ω impedance, which sets the common input impedance to 12.5. A 4:1 step-down transformer provides a match to the 50Ω output of the driver amplifier. The phase shift between the input and output ports must be zero and this is achieved by using 1:1 balun transformers.

These are loaded with ferrite tubes to provide the desired low-frequency response without resort to increasing the physical length. In this type of transformer, the currents cancel, making it possible to employ high-permeability ferrite and relatively short lengths of transmission line. In an ideally balanced situation, no power will be dissipated in the magnetic cores and the line loss will be low. The minimum inductance of the input transformer should be 16µH at 2MHz - a lower value will degrade the isolation characteristics between the output ports and this is important in the event of a change in input VSWR to one of the amplifiers. It is unlikely that the splitter will be subjected to an open- or short-circuit load at the amplifier input, due to the base-frequency compensation networks in the amplifier modules. The purpose of the balancing resistors R is to dissipate any excess power if the VSWR rises.

The value of R is determined by the number of 50Ω sources assumed unbalanced at any one time. Except for a two-port divider, the resistor values can be calculated for an odd or even number of ports as:

$$R = \left( \frac{R_L - R_{IP}}{n + 1} \right) n$$

where $R_L$ is the impedance of output ports (50Ω); $R_{IP}$ is the impedance of input port (12.5Ω); n is the number of correctly terminated output ports.

Although the resistor values are not critical for the input divider, the same formula applies to the output combiner where mismatches have a larger effect on the total power output and linearity.

The power divider employs ferrite sleeves having a µ of 2500 and uses 1.2in lengths of RG-196 coaxial cable; the inductance is approximately 10µH. The input transformer is wound on a

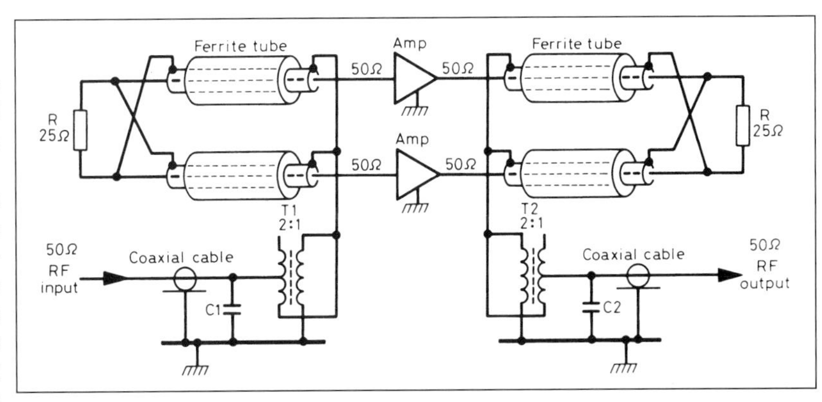


Fig 7.20: Two-port divider and combiner

63-grade ferrite toroid with RG-188 miniature coaxial cable. Seven turns are wound bifilar and the ends are connected inner to outer, outer to inner at both ends to form a 4:1 transformer.

The output combiner is a reverse of the input divider and performs in a similar manner. The ballast resistors R must be capable of dissipating large amounts of heat in the event that one of the sources becomes disabled, and for this reason they must be mounted on a heatsink and be non-inductive. For one source disabled in a four-port system, the heat dissipated will be approximately 15% of the total output power, and if phase differences occur between the sources, it may rise substantially. The resistors are not essential to the operation of the divider/combiner; their function is to provide a reduced output level in the event of an individual amplifier failure. If they are not included, failure of any one amplifier will result in zero output from the amplifier combination. The output transformer must be capable of carrying the combined output power and must have sufficient cross-sectional area. The output transformer is likely to run very warm during operation. High-frequency compensating capacitors C1 and C2 may be fitted to equalise the gain distribution of the amplifier but they are not always necessary.

A two-port divider combiner is illustrated in **Fig 7.20**; operation is principally the same as in the four-port case but the input and output transformers are tapped to provide a 2:1 ratio. Detailed constructional notes of two- and four-port dividers and combiners are given in Motorola Application Notes AN749 and AN758 respectively [1]

## G4JNT DIRECT AUDIO UP-CONVERTER

I/Q or quadrature upconverters are ideally suited for translating the audio from a soundcard directly up to RF. Image cancelling or I/Q upconversion is possible by making use of the two stereo channels from a soundcard carrying the I and Q audio signals - without the need for a separate audio phasing network. Alternatively, for narrow band datamodes, a simple phasing network working over no more than a few hundred Hz bandwidth is quite straightforward, and useable where stereo I/Q output have not been provided by the software author.

The design shown in **Fig 7.21** is a direct-from-audio upconverter suitable for the LF to low HF bands. The only part not shown is the actual RF source itself which could be a crystal oscillator, a VFO or a frequency synthesizer such as a DDS.

The upconverter design is based around fast CMOS bus switches as used in the Softrock and several similar designs of Software Defined Radios for direct conversion to audio. By turning the mixer round a high performance image cancelling direct upconverter can result.

At the left hand side of the circuit, the RF input is buffered by a pair of logic gates biased as a high gain limiting amplifier then applied to a ring counter made from a pair of flip flops. One oddity of this divider making it different from a conventional binary counter is that the outputs count 0, 2, 3, 1 rather than 0, 1, 2, 3. This changes the order of the connections needed to the two poles of the CMOS switch. A ring counter is used as it is easier to ensure both output drives are exactly synchronous than it is with a conventional counter.

The output filter is essential. As the RF drive takes the form of a square wave, all odd order harmonics are present, and the relative power of these rolls off following a $1/F^2$ law. So the third harmonic is only 10.LOG(1/9) or approximately 10dB down, the fifth at -14dB and so on. The values shown give a cut off of 505kHz. This is followed by a simple broadband amplifier raising the output level to around +6dBm maximum. The filter component values can be scaled for other frequency ranges.

The I/Q drive is generated from a single audio input with an all-pass network made from a pair of op-amps. Each channel of the all-pass network maintains a constant gain of unity whatever the input frequency, with a phase shift over the audio band depending on the CR product connected to each op-amp positive input. This goes from 0 degrees at DC, 90 degrees from input to output at F = 1/2.π.C.R and 180 degrees at high frequencies. By choosing a pair of differing CR time constants for each channel, the difference in phase between the two outputs can be kept sufficiently close to 90 degrees over a limited frequency band as shown in **Fig 7.22**.

The simple arrangement can only generate I/Q signals with a sufficiently accurate 90 degree phase difference over a limited range of frequencies. As the phase shift degrades from the ideal quadrature values the sideband isolation falls with its value, in dB, given by 20 * LOG (TAN(ØI - ØQ)) and assumes the amplitudes of I and Q are identical; if this is not so, sideband rejection is further degraded. At 90 degrees it would yield an infinite value (perfect rejection) but at 89 degrees phase shift (1 degree error) an isolation of 35dB results. At 5 degree of error sideband isolation becomes a rather poor 21dB. **Fig 7.23** shows the resulting sideband rejection that can be achieved. Only single I and Q channel signals are supplied and we need differential drive, so a second pair of op-amps are used as unity gain buffers to generate minus-I and minus-Q. If I and Q signals can be generated by software instead of having to be synthesized by analogue networks, the connections shown by the dotted lines (in Fig 7.21) can be used to configure the first two pair of op-amps as buffers rather than all-pass networks. Several components need to be shorted out or removed to do this. They are marked on the schematic diagram.

Setting up involves little more than adjusting the I/Q networks for optimum sideband rejection and checking / trimming carrier rejection. As an absolute minimum a receiver that can be tuned either side of the output band of interest to look at sideband rejection is adequate, with a suitable clean sinusoidal drive signal. Ideally the receiver should have a spectral display several kHz wide - any soundcard based SDR will do admirably.

For the source, it is helpful if the drive can be mixed with white noise to show sideband rejection over the whole output band simultaneously. A suitable test signal can come from another SSB receiver tuned to a weak carrier. The audio out from this contains reasonably flat noise typically from 300Hz to 2700Hz and a tone. First display its audio output by feeding directly into the soundcard input while running spectral analysis software such as Spectran [3]. Tune to any single carrier and adjust input

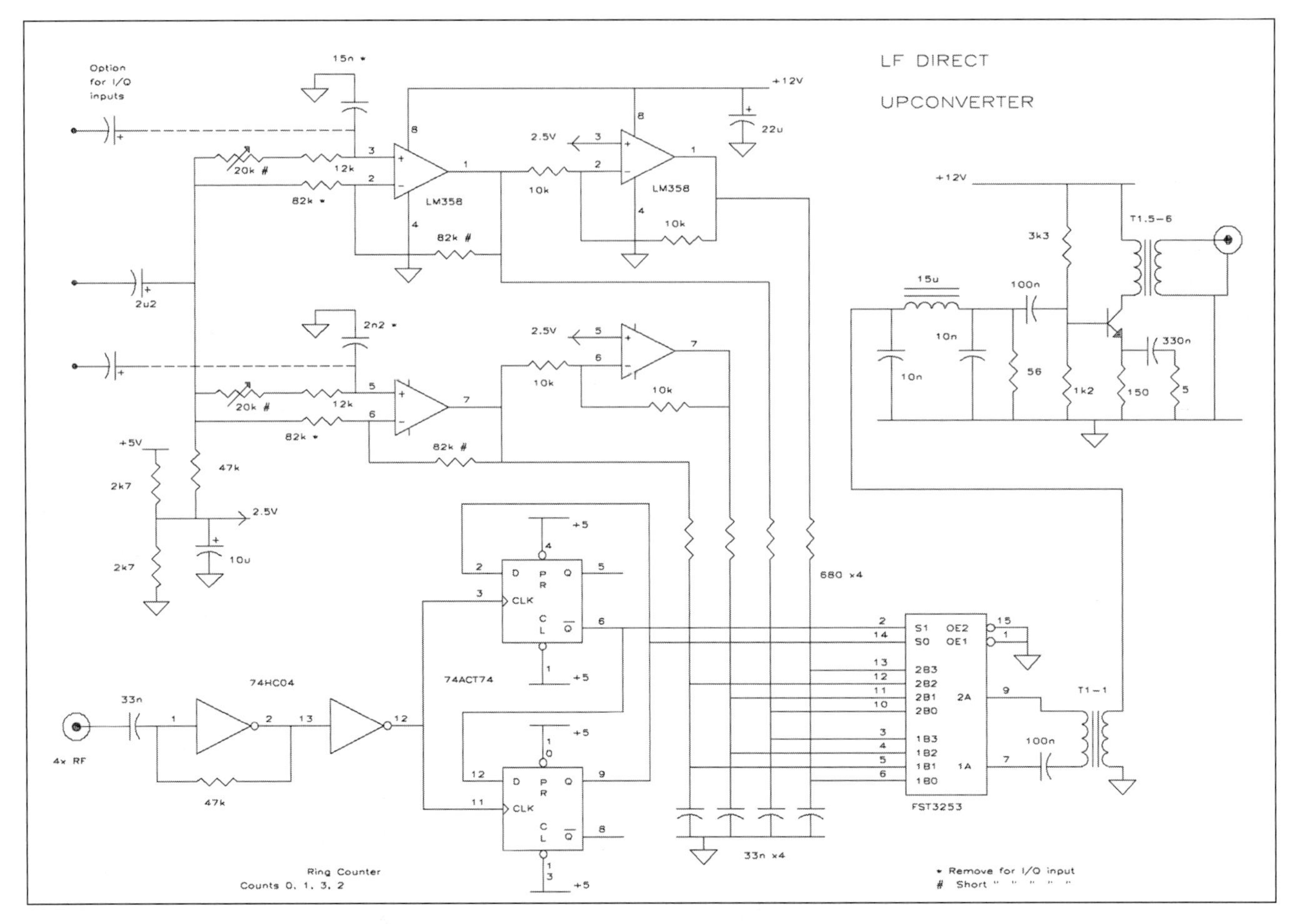


Fig 7.21: Circuit diagram of an upconverter for generating RF direct from soundcard outputs

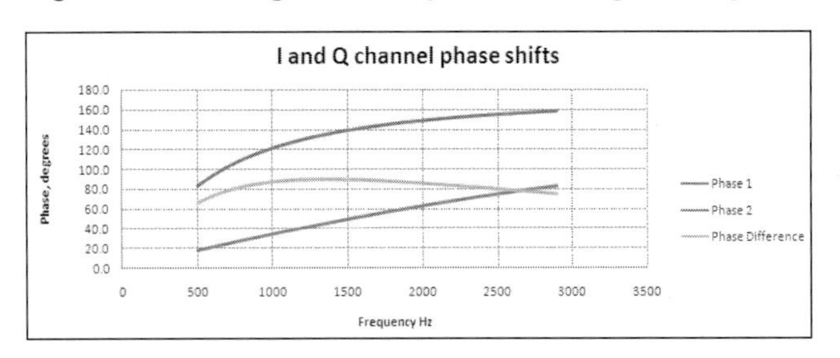


Fig 7.22: Phase response from the all-pass network in Fig 7.21

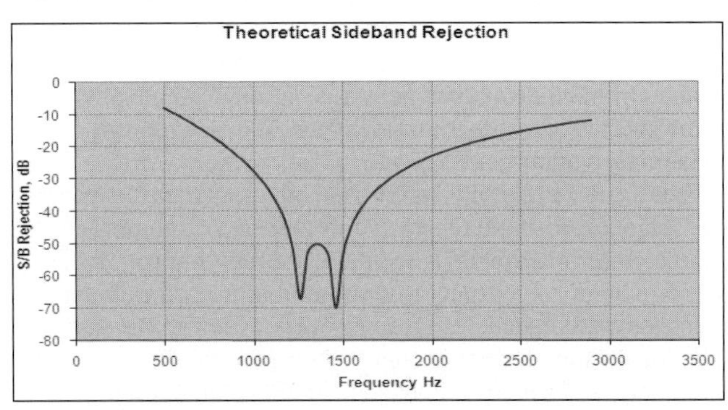


Fig 7.23: Theoretical sideband rejection from the all-pass network in Fig 7.21

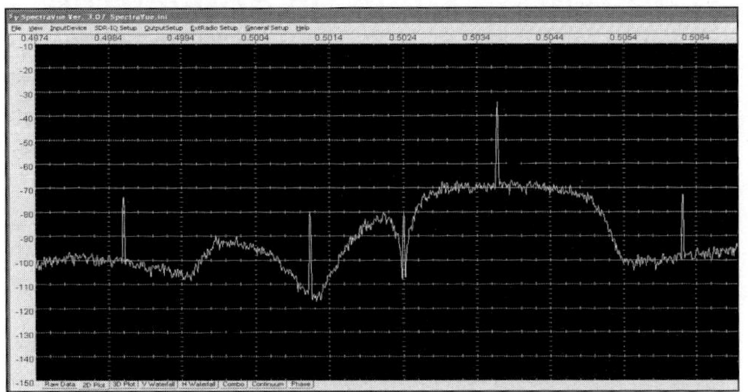

Fig 7.24: Measured plot of the up-converter output when driven by a mixture of white noise and a tone. For this plot the RF input was 2.0096MHz (4 * 502.4kHz) with the output from the up-converter fed via a 20dB attenuator directly to my SDR-IQ antenna input. The SDR-IQ was tuned to 502.4kHz centre frequency and 10kHz span. Leakage from the carrier appears exactly in the middle of the display with the upconverted sidebands on either side. Audio drive was generated by tuning my IC-746 to a weak carrier while monitoring the audio output using *Spectran*. The RF was attenuated until the carrier sat about 35dB above noise on the spectral display then the audio drive was transferred to the up-converter input

attenuation / signal pickup until this sits a few tens of decibels above the background noise. Ensure things stay reasonably constant and that you can change the tone frequency with the tuning knob, and the noise stays flat. Transfer the audio so it becomes the drive to the upconverter, with a way of controlling the level from zero to maximum - such as by adjusting the receiver's volume control.

Apply a four-times RF source to the up-converter RF input buffer, and connect the RF output via suitable attenuation to a test receiver tuned to the wanted frequency. With the audio drive set to minimum some carrier leakage should be observed.

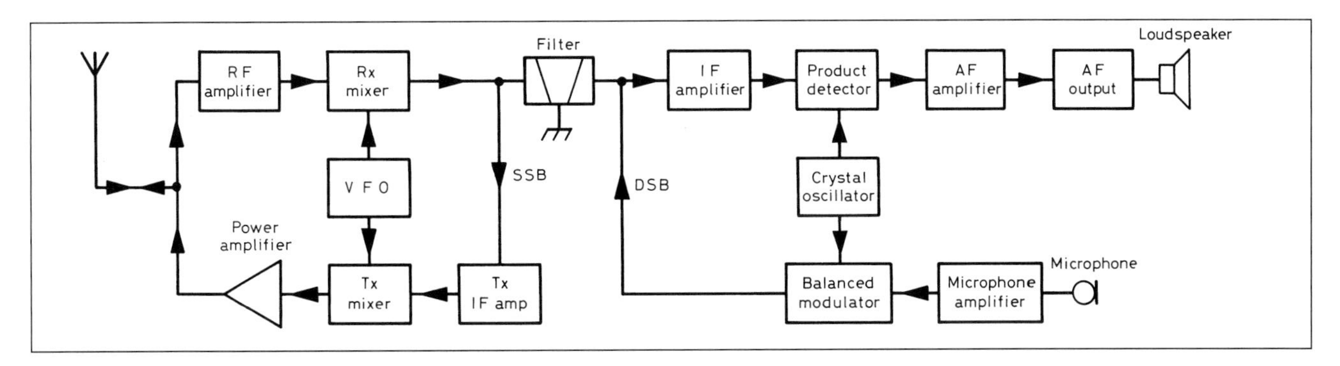


**Fig 7.25: Block diagram of a single sideband (SSB) transceiver**

Slowly increase audio drive and observe the upconverted RF increase in level by several tens of dB above the carrier leakage which should remain constant. Audio sidebands either side of the carrier will be seen, but a section of the audio band, one sideband, should be noticeably different from the other. It may be the upper or lower sideband - it doesn't matter which at this stage. **Fig 7.24** shows the sort of plot that can be expected with a tone plus noise drive.

Adjust the two input preset resistors to optimise the rejection. 40dB - 50dB can be achieved in a narrow band, with 35 - 40dB over a range of a few hundred hertz. If it is impossible to reach a particularly high rejection at even a single point, then it is likely that the amplitudes of the I and Q channel are not matched. With 1% resistors used throughout, little trimming should be needed. RF carrier rejection of 50dB ought to be possible, and there is a bit of scope for optimising this by adjusting the inverting buffer gain. But note that this will then affect I/Q balance and it could get very time consuming trying to get the ultimate all round performance. The pragmatic solution is to match resistors using a DVM to around 0.3 - 0.5% then leave well alone, just accepting the result. To change between upper and lower sideband, swap over I and Q drive signals.

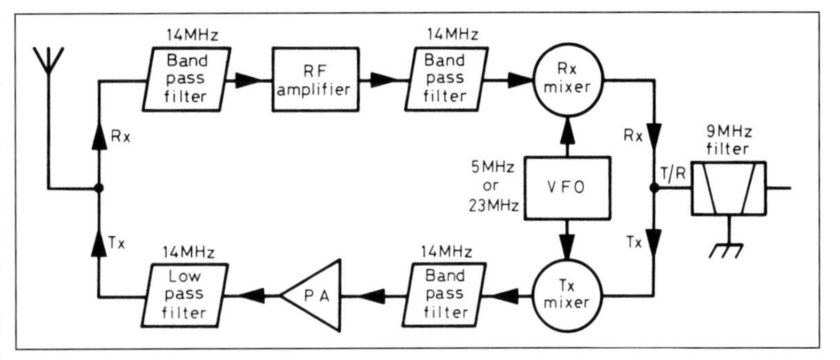


**Fig 7.26: HF transceiver radio frequency filtering**

## TRANSCEIVERS

Separate transmitter and receiver combinations housed in one cabinet may be referred to as a 'transceiver'. This is not strictly correct as a transceiver is a combined transmitter and receiver where specific parts of the circuit are common to both functions. Specifically, the oscillators and frequency-determining components are common and effectively synchronise both the transmitter and receiver to exactly the same frequency. This synchronisation is a pre-requisite for SSB operation and transceivers owe their existence to the development of SSB transmission. While many early attempts at SSB generation used the phasing method, the similarity of the filter-type SSB generator circuit to a superhet SSB receiver circuit (Fig 7.3(b)) makes interconnection of the two circuits an obvious development (**Fig 7.25**). True transceive operation is possible by simply using common oscillators, but it is also advantageous to use a common SSB filter in the IF amplifier stages, which provides similar audio characteristics on both transmit and receive as well as providing a considerable saving in cost.

Initially low-frequency SSB generation necessitated double-conversion designs, often employing a tuneable second IF and a crystal-controlled oscillator for frequency conversion to the desired amateur bands. This technique was superseded by single-conversion designs using a high-frequency IF in the order of 9MHz, with a heterodyne-type local oscillator consisting of a medium-frequency VFO and a range of high-frequency crystal oscillators. With the advent of the phase-locked loop (PLL) synthesiser and the trend towards wide-band equipment, modern transceivers typically up-convert the HF spectrum to a first IF in the region of 40-70MHz. This is then mixed with a synthesised local oscillator and converted down to a working IF in the order of 9MHz. Often a third IF in the order of 455kHz will be employed, giving a total of three frequency conversions.

In order to simplify the construction of equipment, designers have attempted to combine as many parts of the transceiver circuit as possible. Front-end filtering can be cumbersome and requires a number of filters for successful operation. **Fig 7.26** illustrates the typical filtering requirements in a 14MHz transceiver. Traditionally, the receiver band-pass filter and even the transmit band-pass filter would have employed a variable tuning capacitor, often referred to as a preselector, to provide optimum selectivity and sensitivity when correctly peaked. To arrange for a number of filters to tune and track with one another requires careful design and considerable care in alignment, especially in multiband equipment. The introduction of the solid-state PA with its wide-band characteristics has lead to the development of wideband filters possessing a flat response across an entire amateur band. By necessity these filters are of low Q and consequently must have more sections or elements if they are to exhibit any degree of out-of-band selectivity. By employing low-Q, multi-section, band-pass filters it is possible to eliminate one filter between the receiver RF amplifier and the receive mixer. The gain of the RF amplifier should be kept as low as possible and in most cases can be eliminated entirely for use below 21MHz. By providing the band-pass filters with low input and output impedances, typically 50Ω, switching of the filters can simplified to the extent that one filter can be used in both the transmit and receive paths, thus reducing the band-pass filter requirement to one per band.

**Fig 7.27(a)** illustrates a typical band-pass filter configuration for amateur band use, and suitable component values are listed in **Table 7.6**. A transmit low-pass filter is essential for attenuation of all unwanted products and harmonics amplified by the

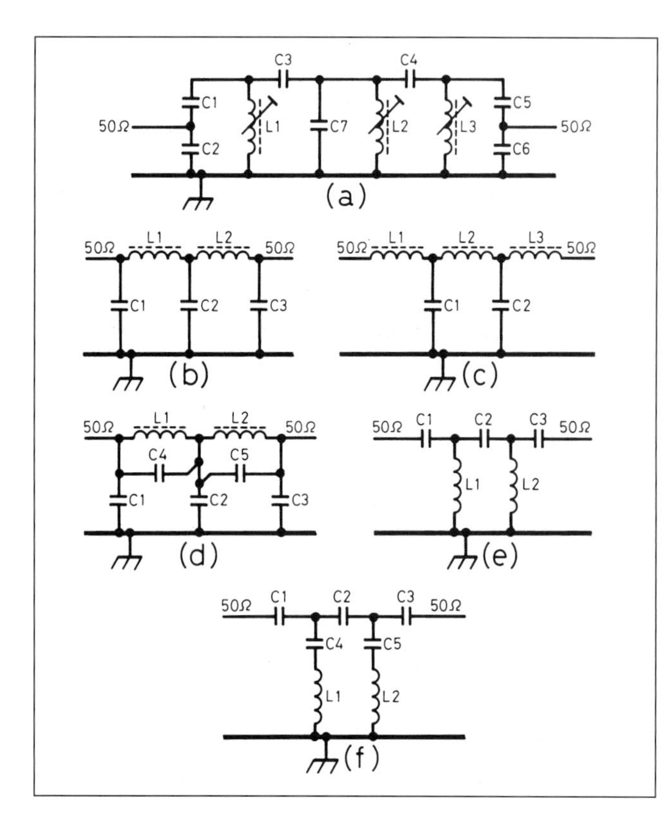


**Fig 7.27: Band-pass, low-pass and high-pass filters. (a) Band-pass filter; (b) Chebyshev low-pass filter capacitive input/output; (c) Chebyshev low-pass filter inductive input/output; (d) elliptic function low-pass filter; (e) Chebyshev high-pass filter; (f) elliptic high-pass filter**

broad-band amplifier chain. Suitable values for a typical Chebyshev filter, **Fig 7.27(b)**, suitable for use at the output of a power amplifier chain providing up to 100W output, are given in **Table 7.7**. One disadvantage of the standard low pass filter designs is the limited attenuation in the stop band and how quickly the attenuation increases for increasing frequency. The Butterworth filter is the worst, the Chebyshev far superior and the so-called elliptic filter (Cauer-Chebyshev) **Fig 7.27(d)** the best for the same number of poles.

The addition of two capacitors placed in parallel with the two filter inductors results in the circuits L1, C4 and L2, C5 being resonant at approximately two and three times the input frequency respectively to provide peaks of attenuation at the second and third harmonic frequencies. Where possible, the use of elliptic filters is recommended - typical values for amateur band use are given in **Table 7.8**.

For an external solid-state amplifier being driven by an exciter that already contains a capacitive-input low-pass filter such as that in Fig 7.23(b), the inductive input design, **Fig 7.27(c)**, may be necessary at the final amplifier output.

The combination of the two different types of filter should eliminate parasitic oscillations which will almost certainly occur if two filters with similar characteristics are employed. Values for the inductive input filter are given in **Table 7.9** for use at powers up to 300W.

The development of HF synthesisers has led to the development of HF transceivers providing general-coverage facilities and requiring yet further changes in the design of band-pass filters for transceiver front-ends. There is a finite limit to the bandwidth that can be achieved using conventional parallel-tuned circuit filters. For general-coverage operation from 1 to 30MHz, approximately 30-40 filters would be required, and this is obviously not a practical proposition. While returning to the mechanically tuned filter might reduce the total number of filters required, the complexity of electronic band-changing would be formidable. By combining the characteristics of both low-pass and high-pass filters, **Figs 7.27(d) and (f)**, the simple action of cascading two such filters will result in a band-pass filter, **Fig 7.28(a)**, having a bandwidth equal to the difference between the two filter cut-off frequencies. The limiting bandwidth will be one octave, ie the highest frequency is double the lowest frequency. Unless electronically tuned using varicap diodes, filter bandwidth is restricted to slightly less than one octave. Typical filters in a general-coverage HF transceiver might cover the following bands:

(a) 1.5-2.5MHz
(b) 2.3-4.0MHz
(c) 3.9-7.5MHz
(d) 7.4-14.5MHz
(e) 14.0-26.0MHz
(f) 20.0-32.0MHz

| Band (m) | L1, L3 (µH) | L2 (µH) | L-type | C1, C5 (pF) | C2, C6 (pF) | C3, C4 (pF) | C7 (pF) |
|---|---|---|---|---|---|---|---|
| 160 | 8.0 | 8.0 | 27t KANK3335R | 1800 | 2700 | 180 | 750 |
| 80 | 5.8 | 5.8 | KANK3334R | 390 | 1800 | 47 | 270 |
| 40 | 2.8 | 2.8 | KXNK4173AO | 220 | 1000 | 10 | 150 |
| 30 | 1.3 | 1.3 | KANK3335R | 220 | 1000 | 10 | 180 |
| 20 | 1.2 | 1.2 | KANK3335R | 120 | 560 | 4.7 | 100 |
| 17 | 0.29 | - | Toko S18 Blue | 220 | 750 | 8.2 | - |
| 15 | 0.29 | - | Toko S18 Blue | 180 | 560 | 6.8 | - |
| 12 | 0.29 | - | Toko S18 Blue | 100 | 560 | 2.7 | - |
| 10 | 0.29 | - | Toko S18 Blue | 82 | 390 | 2.7 | - |

*All capacitors are polystyrene except those less than 10pF which are ceramic. L2 and C7 are not used on the 10-17m bands.*

**Table 7.6: Band-pass filter of Fig 7.23(a). 50 ohms nominal input/output impedance**

| Band (m) | L1, L2 | Core | C1, C3 (pF) | C2 (pF) |
|---|---|---|---|---|
| 160 | 31t/24SWG | T50-2 | 1200 | 2500 |
| 80 | 22t/20SWG | T50-2 | 820 | 1500 |
| 40 | 18t/20SWG | T50-6 | 360 | 680 |
| 30/20 | 12t/20SWG | T50-6 | 220 | 360 |
| 17/15 | 10t/20SWG | T50-6 | 100 | 220 |
| 12/10 | 9t/20SWG | T50-6 | 75 | 160 |

*All capacitors are silver mica or polystyrene - for 100W use 300VDC wkg; for <50W use 63VDC wkg. Cores are Micrometals Inc T50-2 Red or T50-6 Yellow from Amidon.*

**Table 7.7. Chebyshev low-pass filter of Fig 7.23(b)**

| Band (m) | L1 | L2 | Core | C1 (pF) | C2 (pF) | C3 (pF) | C4 (pF) | C5 (pF) |
|---|---|---|---|---|---|---|---|---|
| 160 | 28t/22SWG | 25t/22SWG | T68-2 | 1200 | 2200 | 1000 | 180 | 470 |
| 80 | 22t/22SWG | 20t/22SWG | T50-2 | 680 | 1200 | 560 | 90 | 250 |
| 40 | 18t/20SWG | 16t/20SWG | T50-6 | 390 | 680 | 330 | 33 | 100 |
| 30/20 | 12t/20SWG | 11t/20SWG | T50-6 | 180 | 330 | 150 | 27 | 75 |
| 17/15 | 10t/20SWG | 9t/20SWG | T50-6 | 120 | 220 | 100 | 12 | 33 |
| 12/10 | 8t/20SWG | 7t/20SWG | T50-6 | 82 | 150 | 68 | 12 | 39 |

*Capacitors 300VDC wkg silver mica up to 200W. All cores Micrometals Inc from Amidon.*

**Table 7.8: Elliptic low-pass filter of Fig 7.23(d)**

| Band (m) | L1, L3 | L2 | Core | C1,C2 (pF) |
|---|---|---|---|---|
| 160 | 8.1µH/24t | 11.4µH | T106-2 | 1700 |
| 80 | 4.1µH/17t | 5.8µH/21t | T106-2 | 860 |
| 40 | 2.3µH/13t | 3.2µH/15t | T106-2 | 470 |
| 30/20 | 1.18µH/10t | 1.65µH/12t | T106-6 | 240 |
| 17/15 | 0.79µH/8t | 1.11µH/10t | T106-6 | 160 |
| 12/10 | 0.57µH/7t | 0.8µH/8t | T106-6 | 120 |

*This filter is for use with a high-power external amplifier, when capacitive input is fitted to exciter. Capacitors silver mica: 350VDC wkg up to 200W; 750VDC wkg above 300W. Use heaviest possible wire gauge for inductors. All cores Micrometals Inc from Amidon.*

**Table 7.9: Chebyshev low-pass filter inductive input of Fig 7.27(c)**

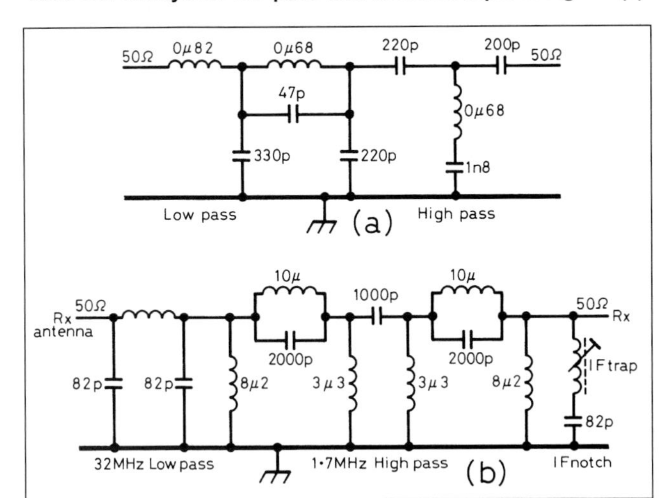


**Fig 7.28: Multifunction filters. (a) Band-pass filter using high/low pass filters; (b) composite receiver input filter**

It can be seen that six filters will permit operation on all the HF amateur bands as well as providing general coverage. Hybrid low/high-pass filters invariably use fixed-value components and require no alignment, thus simplifying construction. The transmit low-pass filter may also be left in circuit on receive in order to enhance the high-frequency rejection; it has no effect on low-frequency signals. The use of separate high-pass filters in the receiver input circuit prior to the band-pass filter serves to eliminate low-frequency broadcast signals. Typically, a high-pass filter of the multi-pole elliptic type, **Fig 7.28(b)**, having a cut-off of 1.7MHz, is fitted to most commercial amateur band equipment.

## TRANSVERTERS

Transverters are transmit/receive converters that permit equipment to be operated on frequencies not covered by that equipment. Traditionally HF equipment was transverted to the VHF/UHF bands but, with the increase in availability of 144MHz SSB equipment, down-conversion to the HF bands has become popular. **Fig 7.29** shows the schematic of a typical 144/14MHz transverter providing HF operation with a VHF transceiver.

A transverter takes the output from a transmitter, attenuated to an acceptable level, heterodynes it with a crystal-controlled oscillator to the desired frequency and then amplifies it to the required level. The receive signal is converted by the same process in reverse to provide transceive capabilities on the new frequency band. The techniques employed in transverters are the same as those used in comparable frequency transmitting and receiving equipment. Where possible it is desirable to provide low-level RF output signals for transverting rather than having to attenuate the high-level output from a transmitter with its associated heatsinking requirements.

Occasionally transverters may be employed from HF to HF in order to include one of the 'WARC bands' on an older transceiver, or to provide 160m band facilities where they have been omitted. In some cases it may prove simpler to add an additional frequency band to existing equipment in preference to using a transverter. Transverters have also been built to convert an HF transceiver to operate on the 136kHz band.

## PRACTICAL TRANSMITTER DESIGNS

### QRP + QSK - A Novel Transceiver with Full Break-in

This design by Peter Asquith, G4ENA, originally appeared in *Radio Communication* [4].

#### Introduction

The late 20th century saw significant advances in semiconductor development. One such area was that of digital devices. Their speed steadily improved to the point which permitted them to be used in low band transceiver designs. One attractive feature of these components is their relatively low cost.

The QSK QRP Transceiver (**Fig 7.30**) employs several digital components which, together with simple analogue circuits, provide a small, high-performance and low-cost rig. Many features have been incorporated in the design to make construction and operation simple.

One novel feature of this transceiver is the switching PA stage. The output transistor is a tiny IRFD110 power MOSFET. This device has a very low 'ON' state resistance which means that very little power is dissipated in the package, hence no additional heatsinking is required. However, using this concept does mean that good harmonic filtering must be used.

The HC-type logic devices used in the rig are suitable for operation on the 160m and 80m bands. The transmitter efficiency on 40m is poor and could cause overheating problems. Future advances in component design should raise the top operating frequency limit.

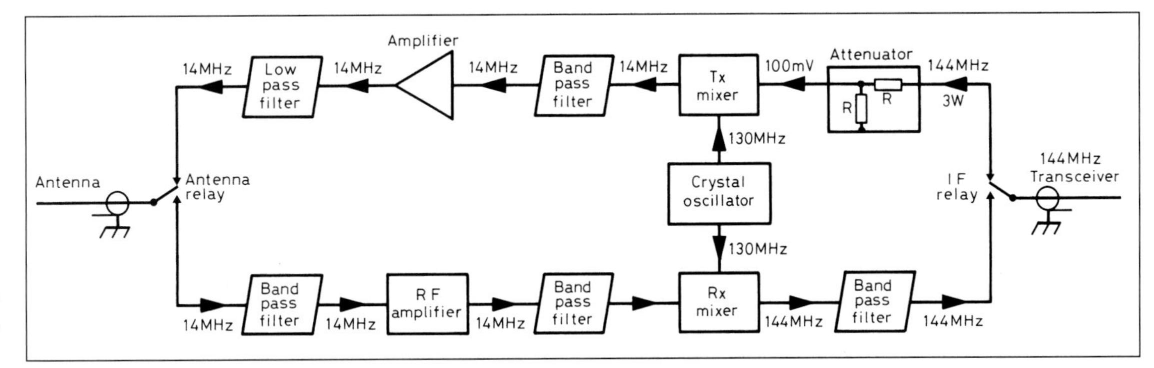


**Fig 7.29: 144MHz to 14MHz transverter**

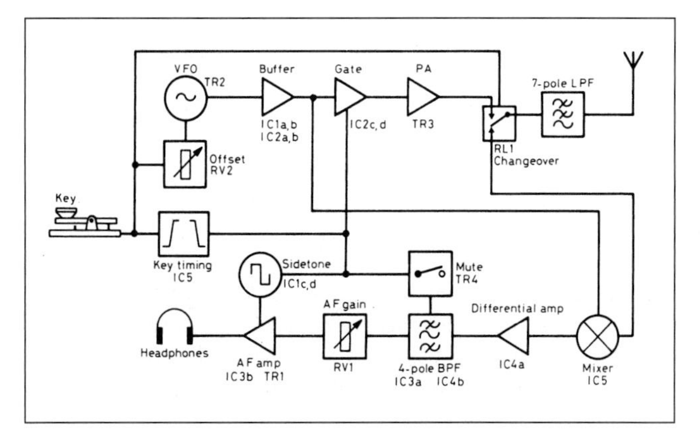


Fig 7.30: Block diagram of the G4ENA transceiver

## Circuit description

**VFO:** The circuit diagram of the transceiver is shown in **Fig 7.31**. TR2 is used in a Colpitts configuration to provide the oscillator for both receive and transmit. The varicap diode D1 is switched via RV2/R1 by the key to offset the receive signal by up to 2kHz, such that when transmitting, the output will appear in the passband of modern transceivers operating in the USB mode. C1 controls the RIT range and C29/30 the band coverage. IC1a, IC1b and IC2a buffer the VFO. It is important that the mark/space ratio of the square wave at IC1b is about 50:50. Small variations will affect output power.

**Transmitter:** When the key is operated, RL1 will switch and, after a short delay provided by R19/C15, IC2c and IC2d will gate the buffered VFO to the output FET, TR3. TR3 operates in switch mode and is therefore very efficient. The seven-pole low-pass filter after the changeover relay removes unwanted harmonics, which are better than -40dB relative to the output.

**Receiver:** The VFO signal is taken from IC2a to control two changeover analogue switches in IC5, so forming a commutating mixer and providing direct conversion to audio of the incoming stations. IC4a is a low-noise, high-gain, differential amplifier whose output feeds the four-pole CW filter, IC4b and IC3a, before driving the volume attenuator, RV1. IC3b amplifies and TR1 buffers the audio to drive headphones or a small speaker. When the key is down TR4 mutes the receiver, and audio oscillator IC1c and d injects a sidetone into the audio output stage, IC3b. The value of R5 sets the sidetone level.

## Constructional notes

The component layout is shown in **Figs 7.32 and 7.33**, and the Components List is in **Table 7.10**.

1. Check that all top-side solder connections are made.
2. Wind turns onto toroids tightly and fix to PCB with a spot of glue.
3. Do not use IC sockets. Observe anti-static handling precautions for all ICs and FETs.
4. All VFO components should be earthed close to VFO.
5. Fit a 1A fuse in the supply line.
6. Component suppliers: Bonex, Cirkit, Farnell Electronic Components, JAB Electronic Components etc.

## Test and calibration

Before TR3 is fitted, the transceiver must be fully operational and calibrated. Prior to switch-on, undertake a full visual inspection for unsoldered joints and solder splashes. Proceed as follows.

1. Connect external components C29/30, RV1/2, headphones and power supply.
2. Switch on power supply. Current is approx 50mA.
3. Check +6V supply, terminal pin 7. Voltage is approx 6.3V.
4. Select values for C29 to bring oscillator frequency to CW portion of band (1.81-1.86MHz/3.50-3.58MHz). Coverage should be set to fall inside the band limits of 1.81/3.50MHz.

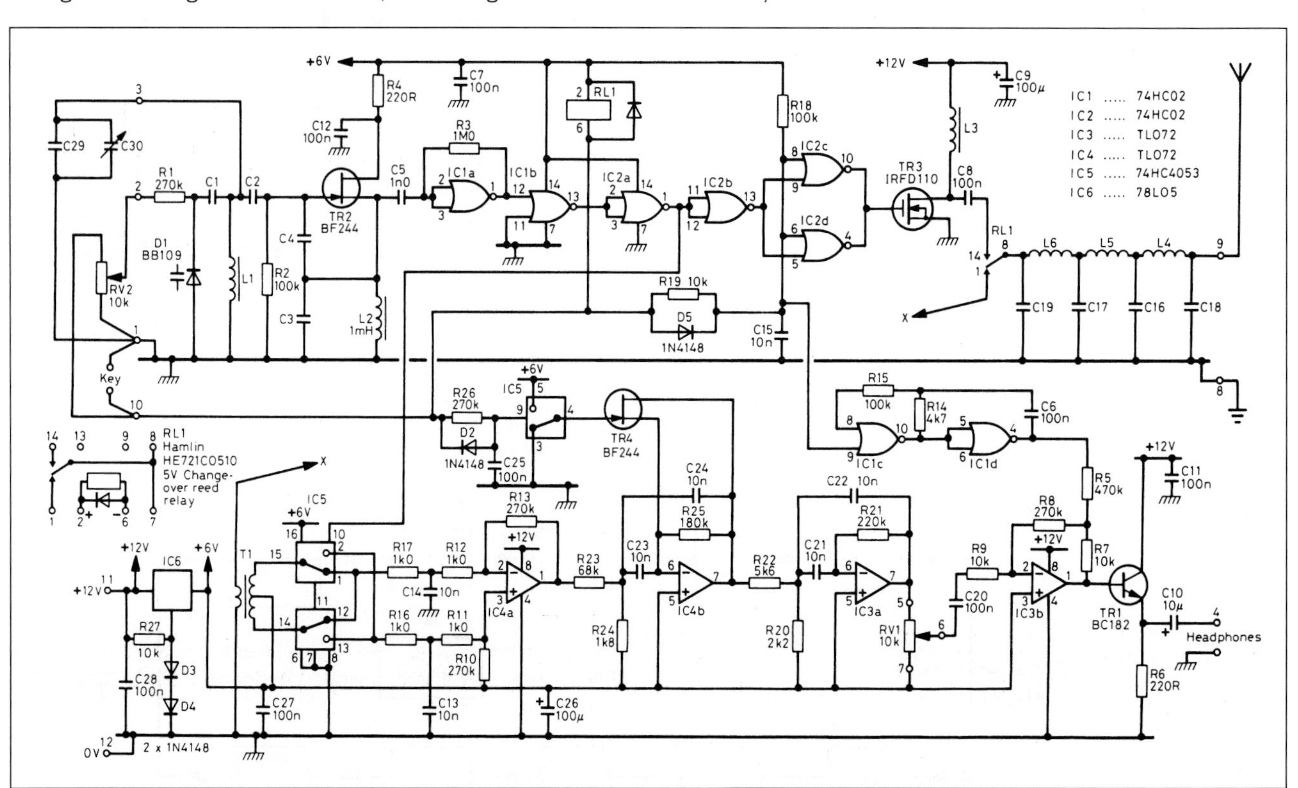


Fig 7.31: The switching PA stage requires a seven-pole filter, as shown in the circuit diagram above

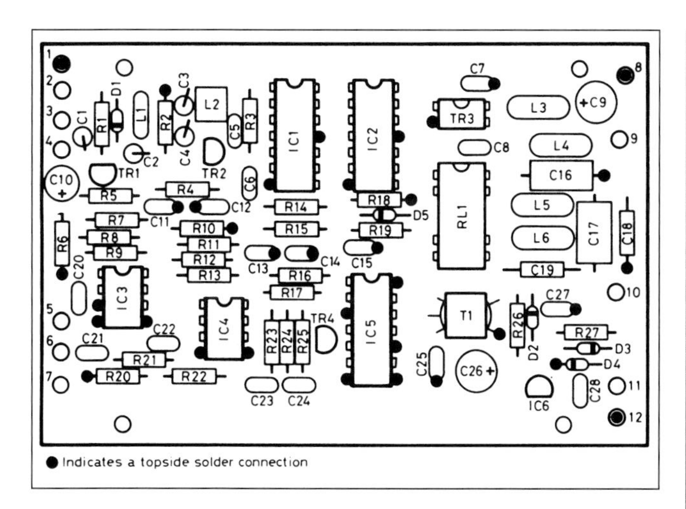


**Fig 7.32: A neat component layout results in a compact unit suitable for portable operation (see also Fig 7.34)**

**Fig 7.33: Inside view of the completed G4ENA novel QRP + QSK transceiver**

5. Connect an antenna or signal generator to terminal pin 9 and monitor the received signal on headphones. Tuning through the signal will test the response of the CW filter which will peak at about 500Hz.
6. Connect key and check operation of sidetone and antenna changeover relay. Sidetone level can be changed by selecting value of R5.
7. Monitor output of IC2c and IC2d (TR3 gate drive) and check correct operation. A logic low should be present with key-up, and on key-down the VFO frequency will appear. This point can be monitored with an oscilloscope or by listening on a receiver with a short antenna connected to IC2c or IC2d.
8. When all checks are complete fit TR3 (important! - static-sensitive device) and connect the transceiver through a power meter to a dummy load. On key-down the output power should be at least 5W for +12V supply, rising to 8W for 13.8V supply. Note: should the oscillator stop when the key is pressed it will instantly destroy TR3. Switch off power when selecting VFO components.
9. Connect antenna and call CQ. When a station replies note the position of the RIT control. The average receive offset should be used when replying to a CQ call.

## On air

The QSK (full break-in) concept of the rig is very exciting in use. The side tone is not a pure sine wave and is easily heard if there is an interfering beat note of the same frequency. One important note is to remember to tune the receiver into a station from the high-frequency side so that when replying your signal falls within his passband.

Both the 160 and 80m versions have proved very successful on-air. During the 1990 Low Power Contest the 80m model was operated into a half-wave dipole and powered from a small nicad battery pack. This simple arrangement produced the highest

**RESISTORS** *(All fixed resistors 0.25W 2%)*

|  |  |  |  |
|---|---|---|---|
| R1, 8, 10, 13, 26 | 270k | R14 | 4k7 |
| R2, 15, 18 | 100k | R20 | 2k2 |
| R3 | 1M | R21 | 220k |
| R4, 6 | 220R | R22 | 5k6 |
| R5 | 470k | R23 | 68k |
| R7, 9, 19, 27 | 10k | R24 | 1k8 |
| R11, 12, 16, 17 | 1k | R25 | 180k |
| RV1, 2 | 10k lin | | |

**CAPACITORS** ( **Select on test component)*

| Ref | Type | Pitch | Value (80m) | Value (160m) |
|---|---|---|---|---|
| C1 | Ceramic plate 9 | 2.54 | 4p7 | 15p |
| C2 | Polystyrene | - | 47p | 100p |
| C3, 4 | Polystyrene | - | 220p | 470p |
| C5 | Ceramic monolithic | 2.54 | 1n | 1n |
| C6, 7, 8, 11, 12, 20, 25, 27, 28 | Ceramic monolithic | 2.54 | 100n | 100n |
| C9, 26 | Aluminium radial 16V | 2.5 | 100 | 100µ |
| C10 | Aluminium radial 16V | 2.0 | 10µ | 10µ |
| C13, 14, 15, 21, 22, 23, 24 | Ceramic monolithic | 2.54 | 10n 10% | 10n 10% |
| C16, 17 | Polystyrene | - | 1n5 | 2n7 |
| C18, 19 | Polystyrene | - | 470p | 1n |
| C29* | Polystyrene | - | 470p | 820p |
| C30* | Air-spaced VFO | - | 25p | 75p |

**INDUCTORS**

| Ref | Type | 80m | 160m |
|---|---|---|---|
| L1 | T37-2 (Amidon) | 31t 27SWG (0.4mm) | 41t 30SWG (0.315mm) |
| L2 | 7BS (Toko) | 1mH | 1mH |
| L3 | T37-2 | 2.2µH 23t 27SWG | 4.5µH 33t 30SWG |
| L4, 6 | T37-2 | 2.9µH 26t 27SWG | 5.45µH 36t 30SWG |
| L5 | T37-2 | 4.0µH 31t 27SWG | 6.9µH 41t 30SWG |
| T1 | Balun | 2t primary, 5+5t secondary, 0.2mm 36SWG (28-43002402) | |

**SEMICONDUCTORS**

|  |  |  |  |
|---|---|---|---|
| D1 | BB109 | IC1, 2 | 74HC02* |
| D2, 3, 4, 5 | 1N4148 | IC3, 4 | TL072* |
| TR1 | BC182 (not 'L') | IC5 | 74HC4053* |
| TR2, 4 | BF244 | IC6 | 78L05 |
| TR3 | IRFD110* | | |

**Static-sensitive devices*

**MISCELLANEOUS**

|  |  |
|---|---|
| RL1 | 5V change-over reed relay, Hamlin, HE721C0510; PED/Electrol, 17708131551-RA30441051 |
| PCB | Boards and parts are available from JAB Electronic Components [5] |

**Table 7.10: G4ENA transceiver components list**

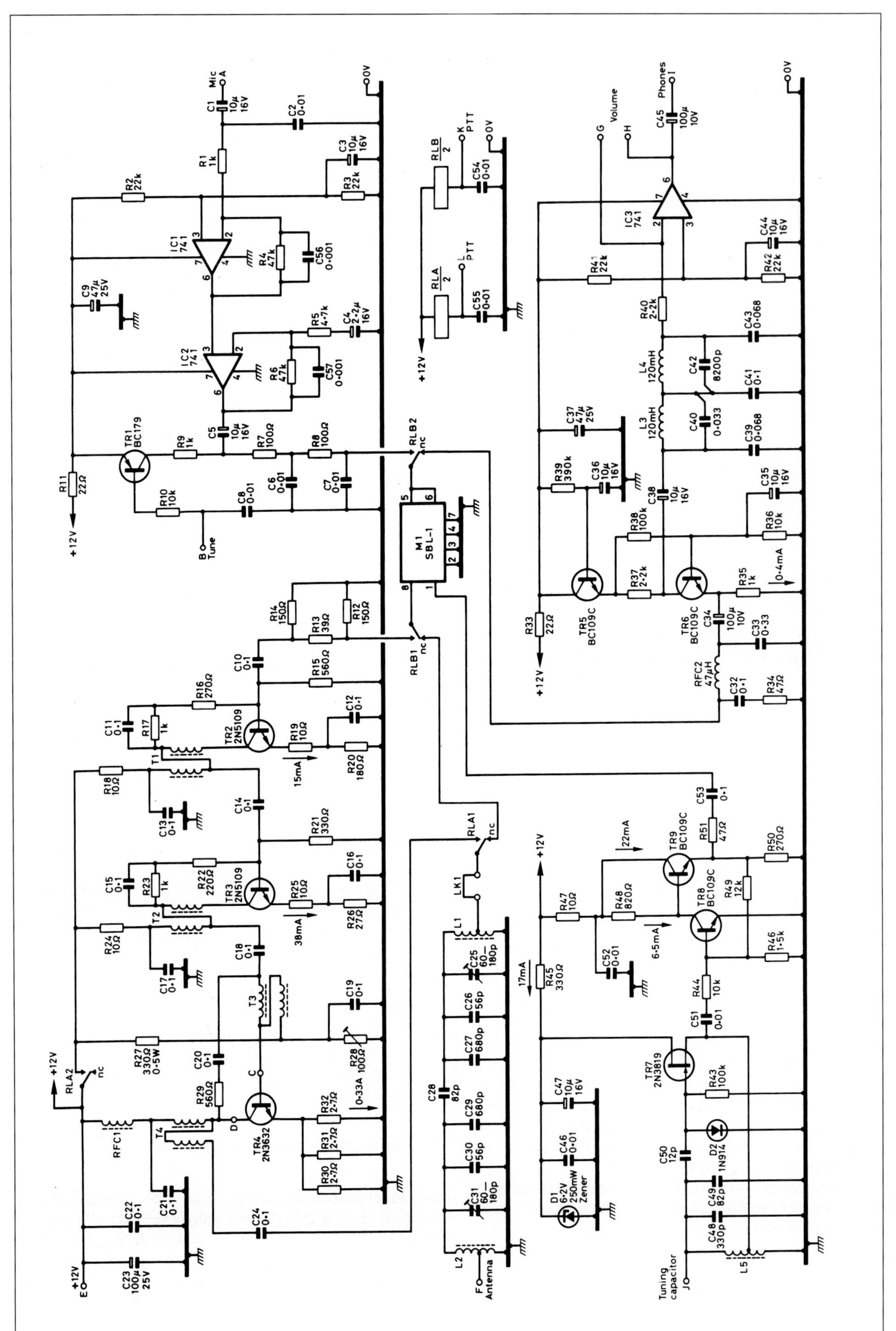


Fig 7.34: Circuit of the G3TXQ transceiver

80m single band score! Its small size and high efficiency makes this rig ideal for portable operation. A 600mAh battery will give several hours of QRP pleasure - no problem hiding away a complete station in the holiday suitcase!

## A QRP Transceiver for 1.8MHz

This design by S E Hunt, G3TXQ, originally appeared in *Radio Communication* [6].

### Introduction

This transceiver was developed as part of a 1.8MHz portable station, the other components being a QRP ATU, a battery-pack and a 200ft kite-supported antenna. It would be a good constructional project for the new licensee or for anyone whose station lacks 1.8MHz coverage. The 2W output level may seem a little low, but it results in low battery drain and is adequate to give many 1.8MHz contacts.

The designer makes no claim for circuit originality. Much of the design was adapted from other published circuitry; however, he does claim that the design is repeatable - six transceivers have been built to this circuit and have worked first time. Repeatability is achieved by extensive use of negative feedback; this leads to lower gain-per-stage (and therefore the need for more stages) but makes performance largely independent of transistor parameter variations.

### Circuit description

The transceiver circuit (**Fig 7.34**) comprises a direct-conversion receiver together with a double-sideband (DSB) transmitter. This approach results in much simpler equipment than a superhet design, and is capable of surprisingly good performance, particularly if care is taken over the mixer circuitry.

During reception, signals are routed through the band-pass filter (L1, L2 and C25-C31) to a double-balanced mixer, M1, where they are translated down to baseband. It is vital for the mixer to be terminated properly over a wide range of frequencies, and this is achieved by a diplexer comprising, RFC2 and C32-R34. Unwanted RF products from the mixer, rejected by RFC2, pass through C32 to the 47-ohm terminating resistor R34. The wanted audio products pass through RFC2 and C34 to a common-base amplifier stage which is biased such that it presents a 50-ohm load impedance. The supply rail for this stage comes via an emitter-follower, TR5, which has a long time-constant (4s) RC circuit across its base. This helps to prevent any hum on the 12V rail reaching TR6 and being amplified by IC3.

The voltage gain of the common-base stage (about x20) is controlled by R37 which also determines the source resistance for the following low-pass filter (L3, L4 and C39-C43). This filter is a Chebyshev design and it determines the overall selectivity of the receiver. The filter is followed by a single 741 op-amp stage which give adequate gain for headphone listening; however, an LM380 audio output stage can easily be added if you require loudspeaker operation.

On transmit, audio signals from the microphone are amplified in IC1 and IC2, and routed to the double balanced mixer where they are heterodyned up into the 1.8MHz band as a double-sideband, suppressed-carrier signal. Capacitors C56 and C57 cause some high-frequency roll-off of the audio signal and thereby restrict the transmitted bandwidth.

A 6dB attenuator (R12-R14) provides a good 50Ω termination for the mixer. The DSB signal is amplified by two broad-band feedback amplifiers, TR2 and TR3, each having a gain of 15dB. TR3 is biased to a higher standing current to keep distortion products low.

The PA stage is a single-ended design by VE5FP [7]. The inclusion of unbypassed emitter resistors R30-R32 establishes the gain of the PA and also helps to prevent thermal runaway by stabilising the bias point. Additional RF negative feedback is provided by the shunt feedback resistor R29. The designer chose to run the PA at a moderately high standing current (330mA) in order to reduce distortion products, thinking that at some stage he might use the transceiver as a 'driver' for a 10-15W linear amplifier.

The PA output (about 2W PEP) is routed through the band-pass filter to the antenna. The designer used a 2N3632 transistor in the PA because he happened to have one in the junk-box; the slightly less expensive 2N3375 would probably perform just all well. VE5FP used a 2N5590 transistor but this would need different mounting arrangements.

At the heart of the transceiver is a Hartley VFO comprising TR7 and associated components. The supply of this stage is stabilised at 6.2V by zener diode D1 and decoupled by C46 and C47. It is important for best stability that the Type 6 core material is used for L5 as this has the lowest temperature coefficient. Output from the VFO is taken from the low impedance tap on L5.

The VFO buffer is a feedback amplifier comprising TR8 and TR9. The input impedance of this buffer is well-defined by R44 and presents little loading of the VFO. Its gain is set by the ratio R49/R44 and R51 has been included to define the source resistance of this stage at approximately 50Ω. Changeover between transmit and receive is accomplished by two DPDT relays which are energised when the PTT lines are grounded.

A CW signal for tuning purposes can be generated by grounding the TUNE pin - this switches on TR1, which in turn unbalances the mixer, allowing carrier to leak through to the driver and PA stages.

### Construction

The transceiver, pictured in **Fig 7.35**, is constructed on a single 6 by 5in PCB. The artwork, component layout and wiring diagram are shown in **Figs 7.36** (see Appendix B)**, 7.37 and 7.38** respectively. The PCB is double-sided - the top (component) surface being a continuous ground plane of unetched copper. The Components List is shown in **Table 7.11**.

**Fig 7.35: The top view of the G3TXQ low power 1.8MHz transceiver**

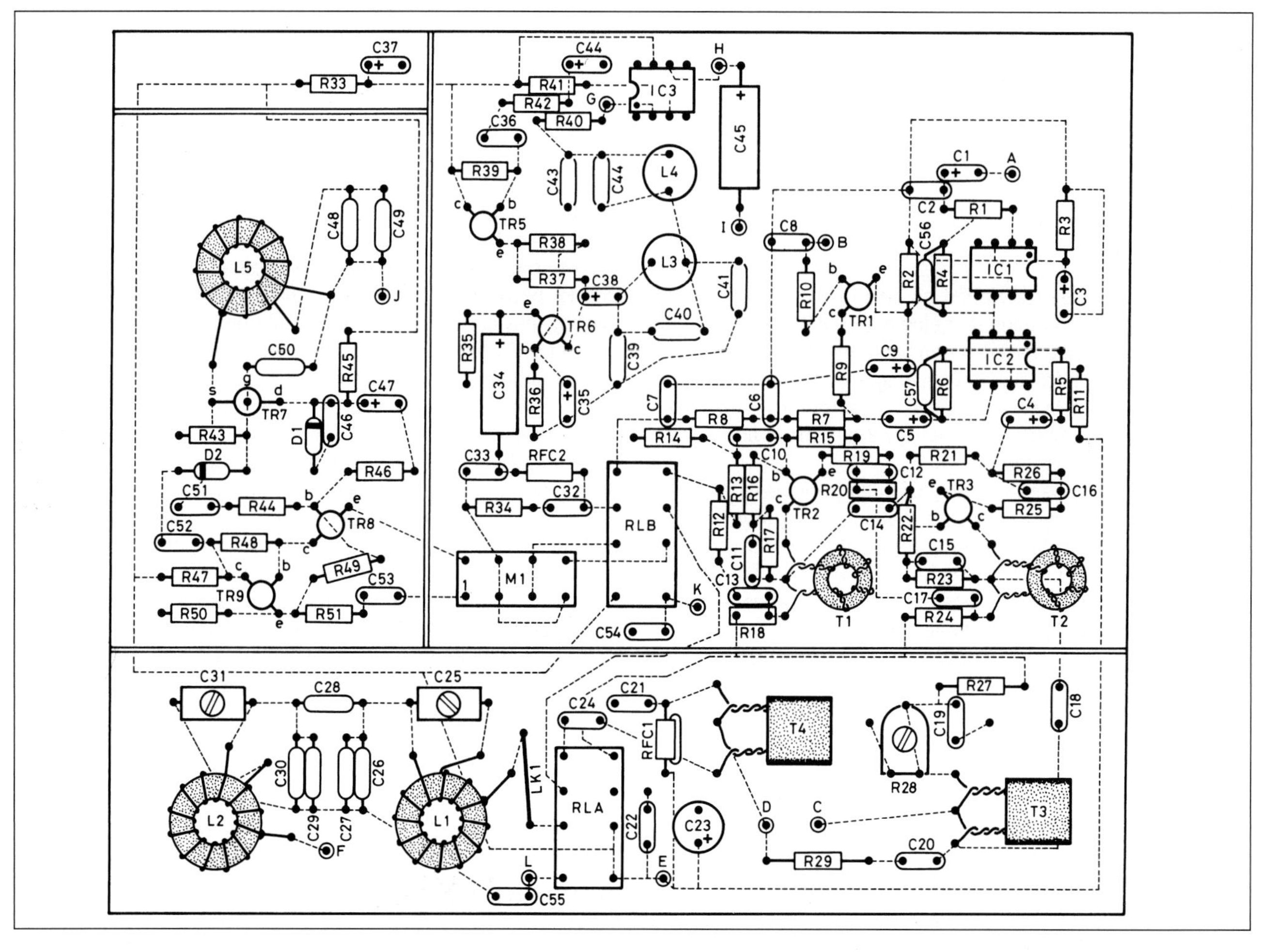


Fig 7.37: Component layout for the G3TXQ transceiver. The PCB details can be found in Appendix B

Without the facility to plate-through holes, some care needs to be taken that components are grounded correctly. Where a component lead is not grounded, a small area of copper must be removed from the ground plane, using a spot-cutter or a small twist drill. Where a component lead needs to be grounded, the copper should not be removed and the lead should be soldered to the ground plane as well as to the pad on the underside. This is easy to achieve with axial-lead components (resistors, diodes etc) but can be difficult with radial-lead components. In most cases the PCB layout overcomes this by tracking radial leads to ground via nearby resistor leads. A careful look at the circuit diagram as each component is loaded soon shows what is needed.

Remember to put in a wire link between pins L and K, and in position LK1. Screened cable was used for connecting pins G and H to the volume control - connect the outer to pin H.

There are no PCB pads for C56 and C57, so these capacitors should be soldered directly across R4 and R6 respectively. TR4 must be adequately heatsinked as it dissipates almost 4W even under no-drive conditions. TR4 was bolted through the rear panel to a 1.5 by 2.5in finned heatsink. Resistors R30-R32 are soldered directly between the emitter of TR4 and

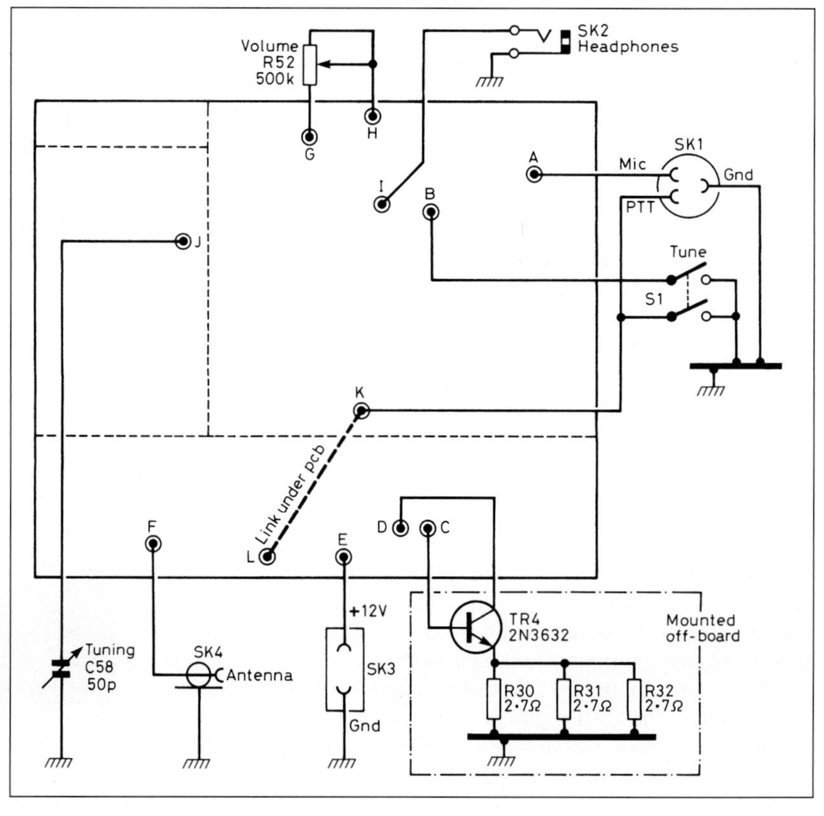


Fig 7.38: Wiring diagram for the G3TXQ transceiver

|  |  |
|---|---|
| R1, 9, 17, 23, 35 | 1k |
| R2, 3, 41, 42 | 22k |
| R4, 6 | 47k |
| R5 | 4k7 |
| R7, 8 | 100R |
| R10, 36, 44 | 10k |
| R11, 33 | 22R |
| R12, 14 | 150R |
| R13 | 39R |
| R15, 29 | 560R |
| R16, 50 | 270R |
| R18, 19, 24, 25, 47 | 10R |
| R20 | 180R |
| R21, 45 | 330R |
| R22 | 220R |
| R26 | 27R |
| R27 | 330R 0.5W |
| R28 | 100R preset |
| R30, 31, 32 | 2R7 |
| R34, 51 | 47R |
| R37, 40 | 2k2 |
| R38, 43 | 100k |
| R39 | 390k |
| R46 | 1k5 |
| R48 | 820R |
| R49 | 12k |

|  |  |
|---|---|
| C1, 3, 5, 35, 36, 38, 44, 47 | 10μ 16V tant bead |
| C2, 6, 7, 8, 46, 51, 52, 54, 55 | 0.01μ ceramic |
| C4 | 2μ2 16V tant bead |
| C9, 37 | 47μ 25V tant bead |
| C10-22, 24, 32, 53 | 0.1μ ceramic |
| C23 | 100μ 25V elect |
| C25, 31 | 60-180p trimmer (Cirkit 06-18006) |
| C26, 30 | 56p silver mica |
| C27, 29 | 680p silver mica |
| C28 | 82p silver mica |
| C33 | 0.33μ |
| C34, 45 | 100μ 10V elect |
| C39, 43 | 0.068μ |
| C40 | 0.033μ |
| C41 | 0.1μ polystyrene |
| C42 | 8200p silver mica |
| C48 | 330p silver mica |
| C49 | 82p silver mica |
| C50 | 12p silver mica |
| C56, 57 | 0.001μ ceramic |
| C58 | 50p air-spaced variable, SLC law (Maplin FF45Y) |

|  |  |
|---|---|
| R52 | 500k log pot |
| L1, 2 | 37t on T68-2 core tapped at 7t from ground |
| L3, 4 | 120mH (eg Cirkit 34-12402) |
| L5 | 57t on T68-6 core tapped at 14t from ground |
| RFC1 | 2t on small ferrite bead |
| RFC2 | 47μH choke |
| T1, 2 | 10t twisted wire on 10mm OD ferrite toroidal core Al = 1μH/t (eg SEI type MM622). See fig. |
| T3, 4 | 4t twisted wire on two 2-hole ferrite cores. Al = 4 H/t (eg Mullard FX2754). See fig. |
| TR1 | BC179 |
| TR2, 3 | 2N5109 or 2N3866 |
| TR4 | 2N3632 (see text) |
| TR5, 6, 8, 9 | BC109C |
| TR7 | 2N3819 |
| D1 | 6.2V 250mW zener |
| D2 | 1N914 |
| IC1, 2, 3 | 741 op-amp |
| M1 | Mini-circuits SBL-1 double-balanced mixer |
| RLA, B | DPDT 12V relay (eg RS Electromail 346-845) |
| SK1 | Microphone socket |
| SK2 | Headphone socket |
| SK3 | DC power socket (eg Maplin YX34M) |
| SK4 | Antenna socket |
| S1 | DPDT toggle switch |
| Slow-motion drive for C58 (eg Maplin RX40T) | |
| Heatsink approx 1.5 by 2in | |
| Knob for R52 | |

**Table 7.11: 1.8MHz QRP transceiver components list**

the ground plane.

It is important that the VFO coil L5 is mechanically stable. Ensure that it is wound tightly and fixed rigidly to the PCB; the coil was 'sandwiched' between two Perspex discs and bolted through the discs to the PCB (**Fig 7.39**). Also, be sure to use rigid heavy-gauge wire for connecting to C58.

The designer used a 6:1 vernier slow-motion drive which, with the limited tuning range of 100kHz, provides acceptable bandspread; the 0-100 vernier scale (0 = 1.900MHz, 100 = 2.000MHz) gives a surprisingly accurate read-out of frequency, the worst-case error being 1kHz across the tuning range.

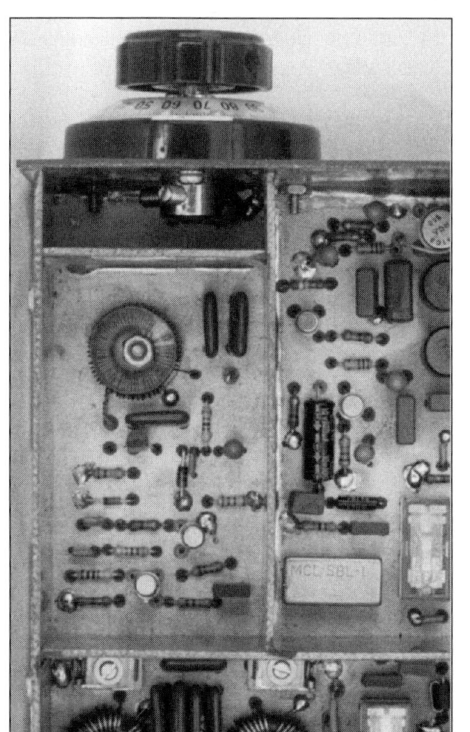

**Fig 7.39: Detail of top view with C58 removed to show mounting arrangements**

The broad-band transformers, T1-T4, are wound by twisting together two lengths of 22SWG enamelled copper wire. The twisted pair is then either wound on a ferrite toroidal core (T1 and T2), or wound through ferrite double-holed cores (T3 and T4). Identify the start and finish of each winding using an ohmmeter - connect the start of one wire to the finish of the other to form the centre tap (see **Figs 7.40 and 7.41** for more details). All transformers and the band-pass filter coils were secured to the PCB with adhesive. The designer made all of the transceiver, other than the top and bottom panels, by soldering together double-sided PCB materials.

It is vital to have a good screen between the PA and the VFO, otherwise the transmitter will frequency modulate badly. Two-inch high screens were used around the PA and VFO area, and

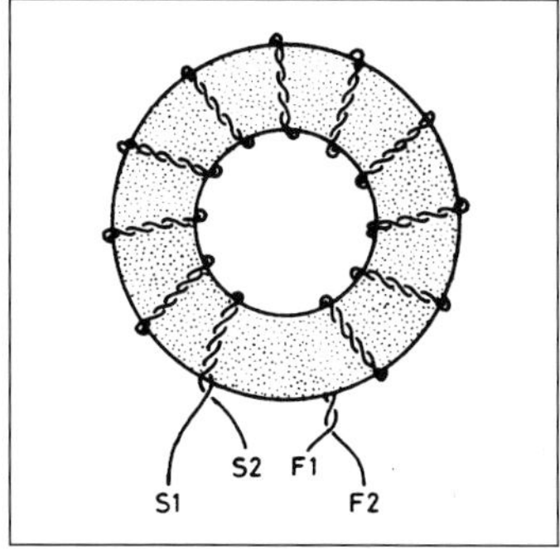


**(above left) Fig 7.40: Winding details of T1 and T2. Connect S2 and F1 to form the centre tap. Note that the two wires are twisted together before winding. S1, F1: start and finish respectively of winding 1. S2, F2: start and finish respectively of winding 2. Core: 10mm OD ferrite toroid**

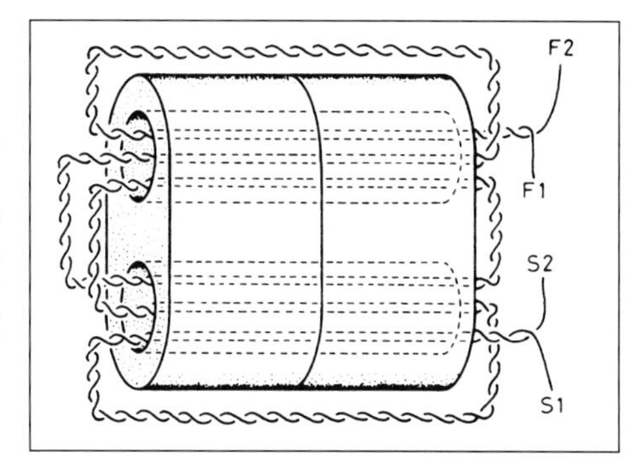


**(above right) Fig 7.41: Winding details of T3 and T4. Connect S2 and F1 to form the centre tap. Note that the two wires are twisted together before winding. S1, F1: start and finish respectively of winding 1. S2, F2: start and finish respectively of winding 2. Core: two 2-hole cores stacked end-to-end**

| | Emitter | Base | Collector | Note |
|---|---|---|---|---|
| TR1 | 12.2 | 11.6 | 11.8 | Tune switch operated |
| TR2 | 2.85 | 3.6 | 12 | Transmit |
| TR3 | 1.4 | 2.15 | 11.6 | Transmit |
| TR4 | 0.3 | 1 | 12.2 | Transmit |
| TR5 | 11.2 | 11.8 | 12.2 | |
| TR6 | 0.4 | 1 | 10.3 | |
| TR8 | 0 | 0.65 | 6.75 | |
| TR9 | 6 | 6.75 | 12 | |

**Table 7.12: Bipolar transistor DC voltages (with 12.2V supply)**

| | Source | Gate | Drain |
|---|---|---|---|
| TR7 | 0 | 0 | 6.2 |

**Table 7.13: FET DC voltages (with 12.2V supply)**

| Circuit node | AC voltage | Notes |
|---|---|---|
| TR7 source | 2.6V p/p | 1.8MHz RF |
| TR9 emitter | 2.6V p/p | 1.8MHz RF |
| Mic input | 4mV p/p | Transmit audio |
| IC1 pin 6 | 200mV p/p | Transmit audio |
| IC2 pin 6 | 2.2V p/p | Transmit DSB RF |
| TR2 base | 200mV p/p | Transmit DSB RF |
| TR4 collector | 15V p/p | Transmit DSB RF |
| Ant (50 ) | 30V p/p | Transmit DSB RF |

**Table 7:14: AC voltages**

a screen was included at the front of the VFO compartment on which to mount C58. If you use lower screens you may need to put a lid over the VFO; cut a tightly fitting piece of PCB material and bolt it in position to four nuts soldered into the corners of the VFO compartment.

### Alignment

Check the PCB thoroughly for correct placement of components and absence of solder bridges.

Turn the volume control fully counter-clockwise, the TUNE switch to the off position and R28 fully counter-clockwise. Connect the transceiver to a 12V supply and switch on. Check that the current drawn from the supply is about 50mA.

Check the frequency of the VFO either by using a frequency counter connected to the source of TR7, or by monitoring the VFO on another receiver. With C58 set to mid-position, the frequency should be about 1.95MHz; if it is very different, you can adjust L5 slightly by spreading or squeezing together the turns. Alternatively, major adjustments can be made by substituting alternative values for C49. Check that the range of the VFO is about 1.9 to 2.0MHz. Plug in a pair of headphones and slowly advance the volume control; you should hear receiver noise (a hissing sound). If you have a signal generator, set it to 1.95MHz and connect it to the antenna socket; if not, you will have to connect the transceiver to an antenna and make the next adjustment using an off-air received signal. Tune to a signal at 1.95MHz and alternately adjust C25 and C31 for a peak in its level.

Connect the transceiver to a 50Ω power meter or through an SWR bridge to a 50Ω load. Plug in a low-impedance microphone and operate the PTT switch. Note the current drawn from the supply - it should be about 200mA. Slowly turn R28 clockwise and note that the supply current increases; adjust R28 until the supply current has increased by 330mA. Release the PTT switch and operate the TUNE switch; the power meter should indicate between 1 and 2W. At this stage, final adjustments can be made to C25 and C31. Swing the VFO from end to end of its range and note the variation in output power. The desired response is a slight peak in power at either end of the VFO range with a slip dip at mid-range. It should be possible to achieve this by successive adjustments to C25 and C31. For those of you lucky enough to have access to a spectrum analyser and tracking generator, LK1 was included to allow isolation of the band-pass filter.

If you have any problems, refer to **Tables 7.12 to 7.14** which show typical DC and AC voltages around the circuit. If necessary, you can tailor the gain of IC2 to suit the sensitivity of your microphone by changing the value of R5.

### Final thoughts

In retrospect it would have been useful to have included the low-pass filter (L3, L4, C39-C43) in the transmit audio path to restrict the bandwidth further. Normally the roll-off achieved by C57 / C56 combined with the low output power means that you are unlikely to cause problems for adjacent contacts. But when using a 200ft vertical antenna during portable operation, the transceiver puts out a potent signal and a bandwidth reduction would then be more 'neighbourly'.

A CW facility could be added fairly easily using the TUNE pin as a keying point. You would need to add RIT (receive independent tune) facilities - probably by placing a varactor diode between TR7 source and ground. You might also consider changing to a band-pass audio filter rather than a low-pass audio filter in the receiver.

The transceiver can be adapted for other bands by changing the VFO components and the band-pass filter components - all other circuitry is broad-band. You will need to worry more about VFO stability as you increase frequency, and you may find the gain of the buffer falls - you can overcome this by decreasing the value of R44. The noise figure of the receiver is adequate for operation on the lower frequency bands but on 14MHz and above a preamplifier will probably be needed. Those who enjoy experimentation might try changing the VFO to a VXO, adding a preamplifier to the receiver, and seeing if operation on 50MHz is possible!

Finally, it has been interesting to note that, despite theory, with careful tuning it is quite possible to resolve DSB signals on the direct-conversion receiver.

## The FOXX2 Transceiver

The FOXX2 is based on a design by George Burt, GM3OXX, and modified by Rev George Dobbs, G3RJV. This article originally appeared in *SPRAT* [8] and shows just how simple a CW transceiver can be.

### Introduction

The FOXX transceiver is an elegant little circuit which uses the same transistor for the transmit power amplifier and the receive mixer. It is capable of transceiver operation on several bands and generates around 1W of RF power out. The original FOXX circuit has been revised with a few design changes.

### Circuit design

The circuit diagram is shown in **Fig 7.42**. TR1 is a VXO (variable crystal oscillator) stage. The feedback loop formed by the crystal and the trimmer capacitor (C1) tunes the circuit to the desired frequency. C1 provides a small amount of frequency shift. The output is coupled to a power amplifier stage. This stage is unusual in that a PNP transistor is used with the emitter connected to the positive supply and the output taken from the collector load, which goes to ground. The output of the transmitter may be adjusted by a resistor (Rx - a few hundred ohms) to around 1W. TR2 should be

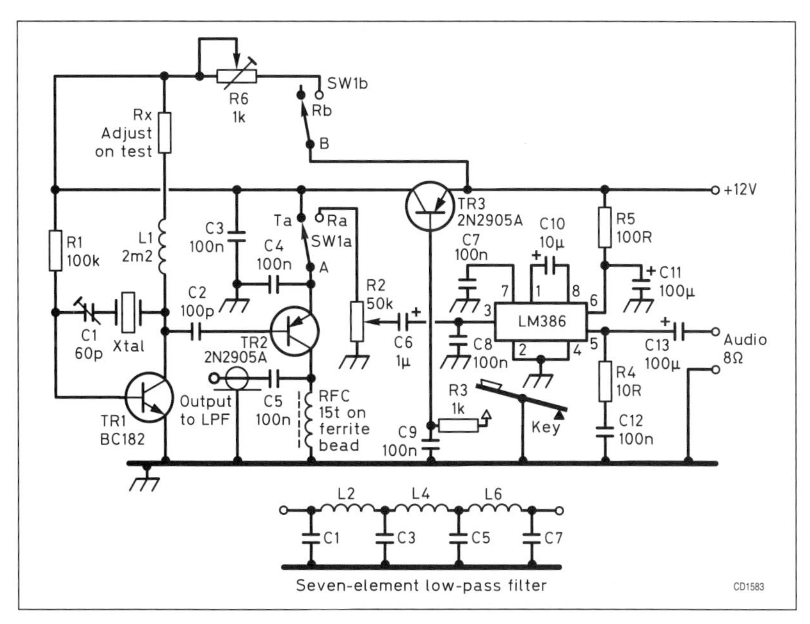


**Fig 7.42: FOXX2 circuit diagram**

fitted with a clip-on heatsink. TR3, another PNP transistor, allows the transmitter to be keyed with respect to ground. TR3 and TR2 are both 2N2905A PNP switching transistors.

The low-pass filter is a seven-element circuit based on the circuit and constants described by W3NQN. The transmit-receive function is performed by a double-pole, double-throw switch, SW1A and SW1B. The receive position has two functions. It bypasses the keying transistor, TR3, to ensure that the oscillator TR1 remains on during the receive position to provide the local oscillator. It also switches the supply line away from the power amplifier TR3 and connects the latter to the audio amplifier. In this position TR3 functions as a diode mixer, mixing the signals from the antenna which appear at the emitter and the signal from TR1.

The audio amplifier is an LM386 working in maximum gain mode. The supply for the LM386 is taken directly from the 12V supply line which means it is on during both transmit and receive functions. This has the advantage of providing a rudimentary sidetone to monitor the keying. 'Sidetone' is an over-statement because all it does is produce clicks in time with the keying.

A preset potentiometer is added in series with TR1 supply on receive. This is a very simple form of RIT (receiver incremental tuning). If the supply voltage to TR1 is reduced enough, it shifts the frequency of the oscillations. Assuming the value of Rx to be in the order of a few hundred ohms (just to reduce the drive from TR1 a little on transmit), a 1kΩ preset at R6 can be set to shift the frequency by around 700-800Hz, giving a comfortable offset for CW reception. Values for the seven-element low pass filter are shown in **Table 7.15**. The wire gauge is not critical but wind the coil so as to comfortably fill the core over about three-quarters of its full circumference. The PCB layout is shown in **Fig 7.43** (in Appendix B), the component layout is in **Fig 7.44**, and the Components List is in **Table 7.16**.

| Band (MHz) | C1, 7 (pF) | C3, 5 (pF) | L2, 6 (turns) | L4 (turns) | Core | Wire (SWG) |
|---|---|---|---|---|---|---|
| 3.5 | 470 | 1200 | 25 | 27 | T37-2 | 28 |
| 7.0 | 270 | 680 | 19 | 21 | T37-6 | 26 |
| 10.1 | 270 | 560 | 19 | 20 | T37-6 | 26 |
| 14.0 | 180 | 390 | 16 | 17 | T37-6 | 24 |

**Table 7.15: FOXX2 low-pass filter values**

## The Epiphyte-2

This section is based on articles by Derry Spittle, VE7QK [9] and Rev George Dobbs, G3RJV [10]. The Epiphyte is a remarkable little transceiver which has introduced many QRP operators to the pleasure of building their own SSB equipment. It was designed by Derry Spittle, VE7QK, who needed reliable radio communication when journeying into wilderness areas in British Columbia. Beyond the range of VHF repeaters, simple battery-operated HF equipment offers the only practical means of communication. The objective was to build a small transceiver capable of providing effective voice communication with the British Columbia Public Service Net on 3729kHz from anywhere in the province. The Epiphyte began as a project of the QRP Club of British Columbia and the design was first published in 1994 by the G-QRP Club and the QRP Club of Northern California. The Epiphyte-2 now includes many modifications and suggestions since made by their members.

### Circuit description

The block diagram of the EP-2 is shown in **Fig 7.45**. The circuit is based around a pair of SA602 double-balanced mixers (IC2, IC3), and a MuRata miniature 455kHz ceramic SSB filter (F1). IC1 switches the LF and HF oscillators, permitting the same mixers to be used for both transmitting and receiving. On transmit, the modulated signal passes though a band-pass filter to remove the image, before being applied to the driver (IC5). Transmit/receive switching is accomplished with a DPDT relay (K1) and IC1. When receiving, B+ is disconnected from the microphone preamplifier (IC6) and the driver (IC5), and forward bias is removed from the PA (TR4). When transmitting, B+ is removed from the audio amplifier (IC6) and the RF input to the first mixer is disconnected from the low-pass filter and grounded. At the same time, the relay provides switching voltages to

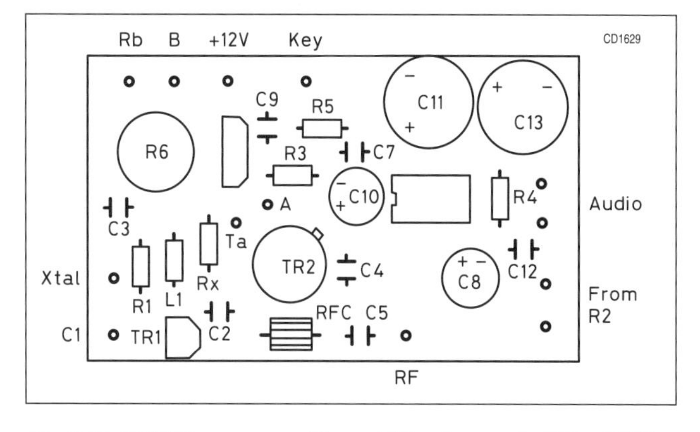


**Fig 7.44: FOXX2 component layout (See Appendix B for PCB)**

| | | | |
|---|---|---|---|
| R1 | 100k | C3-5, 7-9, 12 | 100n |
| R2 | 50k | C6 | 1µ |
| R3 | 1k | C10 | 10µ |
| R4 | 10R | C11, 13 | 100µ |
| R5 | 100R | L1 | 2.2mH |
| R6 | 1k | RFC | 15t on ferrite bead |
| Rx | See text | TR1 | BC182 |
| C1 | 60p trimmer | TR2, 3 | 2N2905A |
| C2 | 100p | IC1 | LM386 |

**Table 7.16. FOXX2 component list**

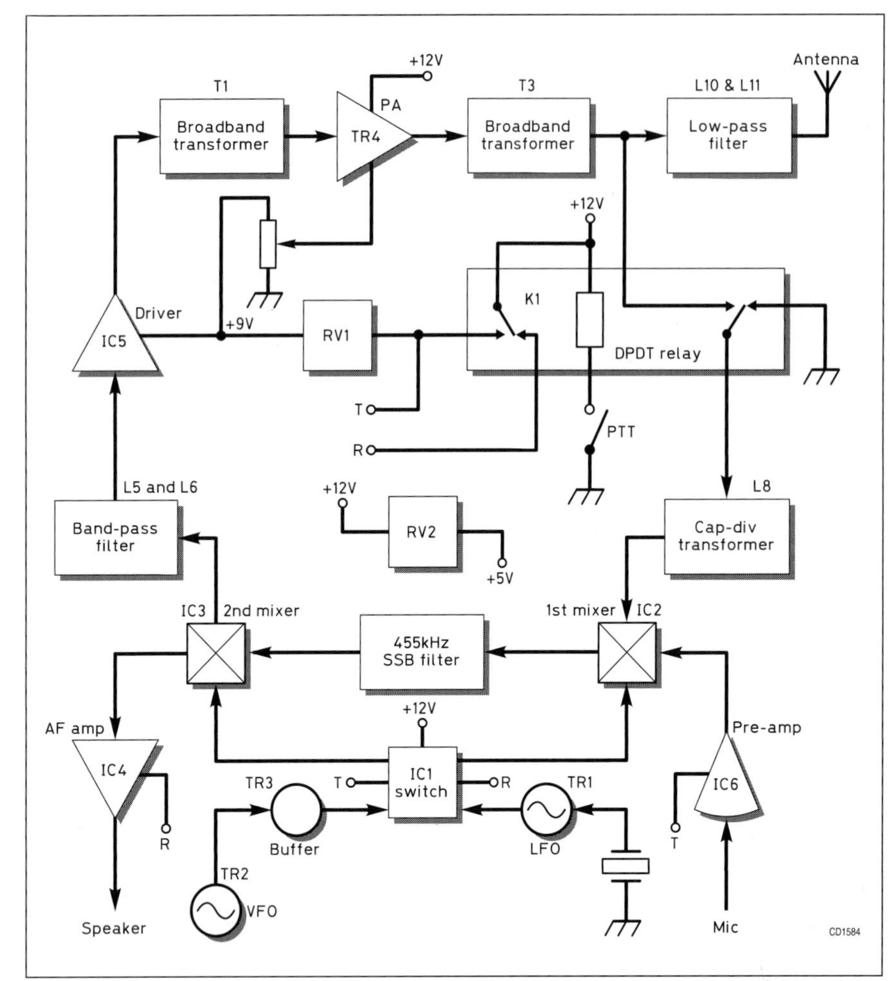


Fig 7.45: Block diagram of Epiphyte-2

IC1. The antenna remains connected to the low-pass filter and PA at all times. The circuit is shown in **Figs 7.46 and 7.47. Table 7.17** shows the Components List.

The VFO (TR2) is a varactor-tuned Vackar circuit using a Toko 3.3µH variable coil. This is buffered by an emitter follower (TR3). The LF oscillator (TR1) uses a 455kHz ceramic resonator adjusted to 452.8kHz. An MPF102 FET is used for both oscillators.

The RF bandpass filter was modelled in a series-tuned configuration to give a reasonably flat response over 200kHz and a sharp roll-off on the high-frequency side for rejection of the image frequency. It is designed for an input impedance of 1500Ω (to match the SA602 mixer) and to terminate in a 100Ω resistive load to ensure stability in the driver. It uses a pair of Toko 4.7µH coils and the fixed capacitors are standard values.

The driver stage (IC5) is a CA3020A differential amplifier. Operating from a 9V supply, this stage has a power gain of 60dB and is capable of 500mW output. The broad-band transformer (T1) has a bifilar-wound primary link-coupled to a 47Ω resistive load (R23) at the gate of TR4. The power amplifier (TR4) is an IRF510 MOSFET with an RF output of 5W PEP. A broad-band transformer (T3) matches the output to a conventional 50Ω low-pass filter.

The receiver RF input from the low-pass filter is a capacitance-divider matching circuit to the first mixer and is tuned to the centre of the phone band. The AF amplifier (IC4) is a LM386 with a balanced input. The receiver is quite sensitive and with a dipole or inverted-V antenna at 25ft it provides adequate speaker volume from all but the weakest signals. Band noise is usually the limiting factor.

R19 provides the polarising voltage for an electret microphone (two-terminal type) and should be omitted if a dynamic microphone is used. The speech amplifier (IC6) is an LM741. The value of R20 should match the impedance of the microphone and the stage gain may be set by adjusting the value of R17.

The component layout is shown in **Fig 7.48** and the PCB layout is in **Fig 7.49** (in Appendix B). A ground loop present in earlier versions (eg **Fig 7.50)** has now been eliminated.

## Assembly

Assembly is fairly straightforward. Ensure that the Toko coils (L3, 5, 6, 8), SSB filter (F1), ceramic resonator (X1) and trimmer capacitor (C1) fit the PCB and enlarge the holes if necessary. Install the CA3020A (IC5) first as it is easier to align the 12 pins without the other components in place. The tab is over pin 12. Be sure to solder in the two bare jumper wires on top of the PCB before installing the socket for IC1. Some fairly large value polystyrene capacitors are specified and their physical size should be ascertained before ordering if they are to fit comfortably on the board.

Alternatively, NPO/COG ceramic capacitors may be substituted. Cut off the centre pin (drain) before mounting the IRF510 (TR4) and heatsink with a 4-40 machine screw, nut and star washers. Output from TR4 is taken from the tab. Remove or cut-off unused terminals from the relay socket. Finally, don't bother soldering the three unconnected pins on the Toko coils to the ground plane - you may need to remove them one day!

## Alignment and testing

This must be carried out with the single-sided PCB fastened to a ground plane with four metal stand-offs.

1. Remove both metering jumpers. Install the relay (K1) and PTT switch if not built into the microphone. Remove all socketed ICs and connect to a 12-14V FUSED supply. Verify that RV2 is delivering 5V and that RV1 is delivering 9V on transmit. With an RF probe, check that both oscillators are functioning. Install IC1 and verify that the VFO is switching between transmit and receive.
2. Set the LFO to 453kHz with the trimmer (C10). Monitor the ninth harmonic at 4.075MHz on a communications receiver. It may be necessary to change the value of the padder (C11).
3. Set R25 to mid position and adjust L3 until the frequency at the 'test point' (Con9) measures approximately 4.2MHz. To avoid having the ferrite core protrude above the case of L3, screw it completely into the coil and tune it upwards (anti-clockwise). The core will remain firmly in position without further 'fixing'. R24 limits the overall frequency coverage and, at the same time, ensures that the varactor diode is always biased positive. The value of R24 may be changed to alter the bandspread.
4. Install IC4 and connect the antenna, the speaker and the

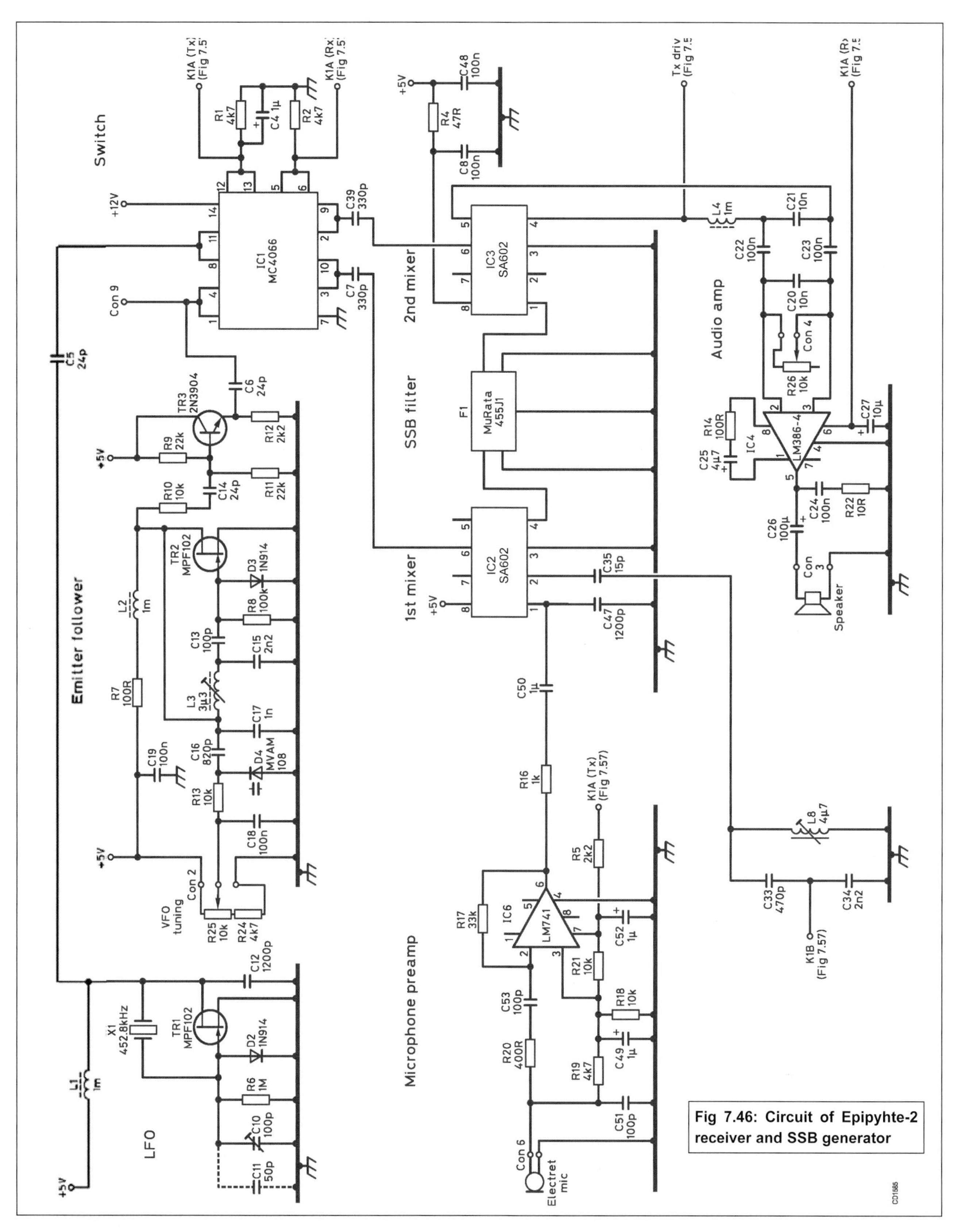


Fig 7.46: Circuit of Epipyhte-2 receiver and SSB generator

volume control (R26). The leads to R26 should be kept as short as possible. Test the receiver and adjust L8.

5. The RF voltage at pin 6 on IC2 and IC3 should read 100 to 150mV RMS. If necessary, change the value of C5 and/or C6 to adjust.
6. Set the RF drive control (R15) to minimum. The transmit idling current in the driver (IC5) should measure 25mA +10% at Con 7. If not then TR1 or IC5 has probably been installed incorrectly.
7. Adjust R3 to set the transmit idling current in the power amplifier (TR4) to read 50mA at Con8. Install IC6, microphone, both metering jumpers and a 50-ohm dummy load at the antenna terminal. Advance the RF driver (R15) until RF voltage appears across the dummy load while modulating

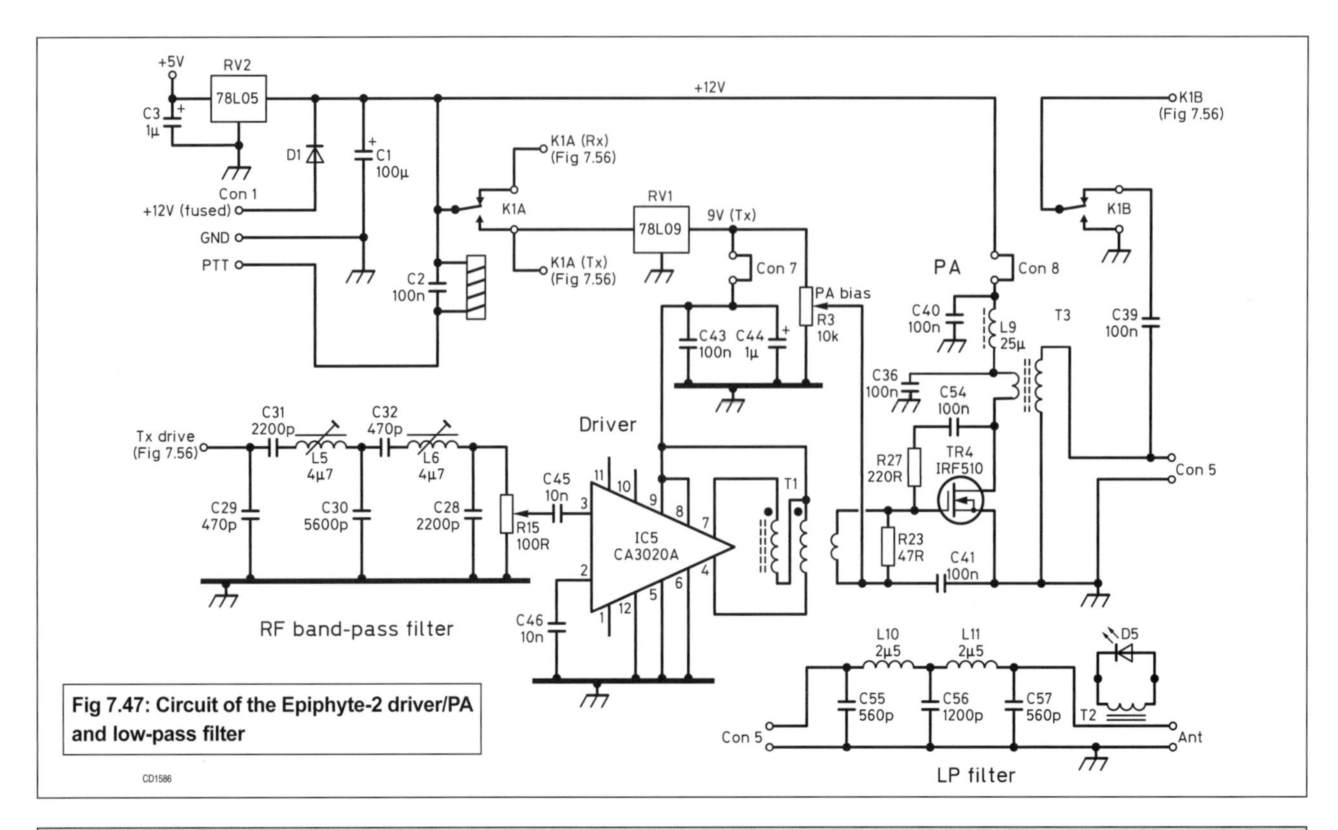


**Fig 7.47: Circuit of the Epiphyte-2 driver/PA and low-pass filter**

| | | | |
|---|---|---|---|
| R1, 2, 19, 24 | 4k7 | C27 | 10µ tantalum |
| R3 | 10k lin (multi-turn trim pot) | C29, 32, 33 | 470p axial polystyrene |
| R4, 23 | 47R | C30 | 5600p axial polystyrene |
| R5, 12 | 2k2 | C35 | 15p NPO |
| R6 | 1M | C55, 57 | 560p axial polystyrene |
| R7, 14 | 100R | C56 | 1200p axial polystyrene |
| R8 | 100k | L1, 2, 4 | 1mH choke |
| R9, 11 | 22k | L3 | 3.3µH (Toko BTKANS9445) |
| R10, 13, 18, 21 | 10k | L5, 6, 8 | 4.7µH (Toko 154AN-T1005) |
| R15 | 100R lin (multi-turn trim pot) | L9 | RFC (7t on Amidon FB-43-801) |
| R16 | 1k | L10, 11 | 2.5µH (25t on Amidon T-32-2) |
| R17 | 33k | T1 | 4:1 broad-band transformer (5t bifilar on FB-43-801 with 3t link) |
| R20 | 400R | | |
| R22 | 10R | T2 | 2t on Amidon FB-43-2401 |
| R25 | 10k (precision 10 turn pot) | T3 | binocular broadband transformer (2t pri, 5t sec on Amidon BM-43-202) |
| R26 | 10k log | | |
| R27 | 270R | IC1 | MC14066 |
| RV1 | 78L09 | IC2, 3 | SA602 |
| RV2 | 78L05 | IC4 | LM386-4 |
| C1, 26 | 100µ electrolytic | IC5 | CA3020A |
| C3, 4, 44, 49, 52 | 1µ tantalum | IC6 | LM741 |
| C50 | 1µ non-polar ceramic | D1 | polarity protection diode, eg 1N4001 |
| C5, 6, 14 | 24p NPO | D2, 3 | 1N914 |
| C7, 9 | 330p ceramic | D5 | LED |
| C2, 8, 18, 19, 22, 24, 36, 39-43, 48, 53, 56 | 100n monolithic ceramic | D4 | MVAM108 |
| | | TR1, 2 | MPF102 |
| | | TR3 | 2N3904 |
| | | TR4 | IRF510 |
| | | F1 | 455kHz SSB filter (MuRata 455J1) |
| C15, 28, 31, 34 | 2200p axial polystyrene | K1 | Miniature DPDT relay |
| C10 | 100p trimmer | X1 | 455kHz ceramic resonator |
| C11 | 50p NPO | Two 3-pin, four 2-pin polarised Molex terminals | |
| C12, 47 | 1200p ceramic | Two metering jumpers and terminals | |
| C13, 51 | 100p NPO | One 1-pin test point | |
| C32 | 470p NPO | Heatsink for TR4 | |
| C16 | 820p axial polystyrene | Four 8-pin, one 14-pin, one 16-pin IC sockets | |
| C17 | 1n axial polystyrene | Four ¼in metal stand-offs | |
| C20, 21, 45, 46 | 10n | Printed circuit boards and a kit of parts for the project are available from JAB Electronic Components [3] | |
| C25 | 4µ7 | | |

**Table 7.17: Epiphyte-2 component list**

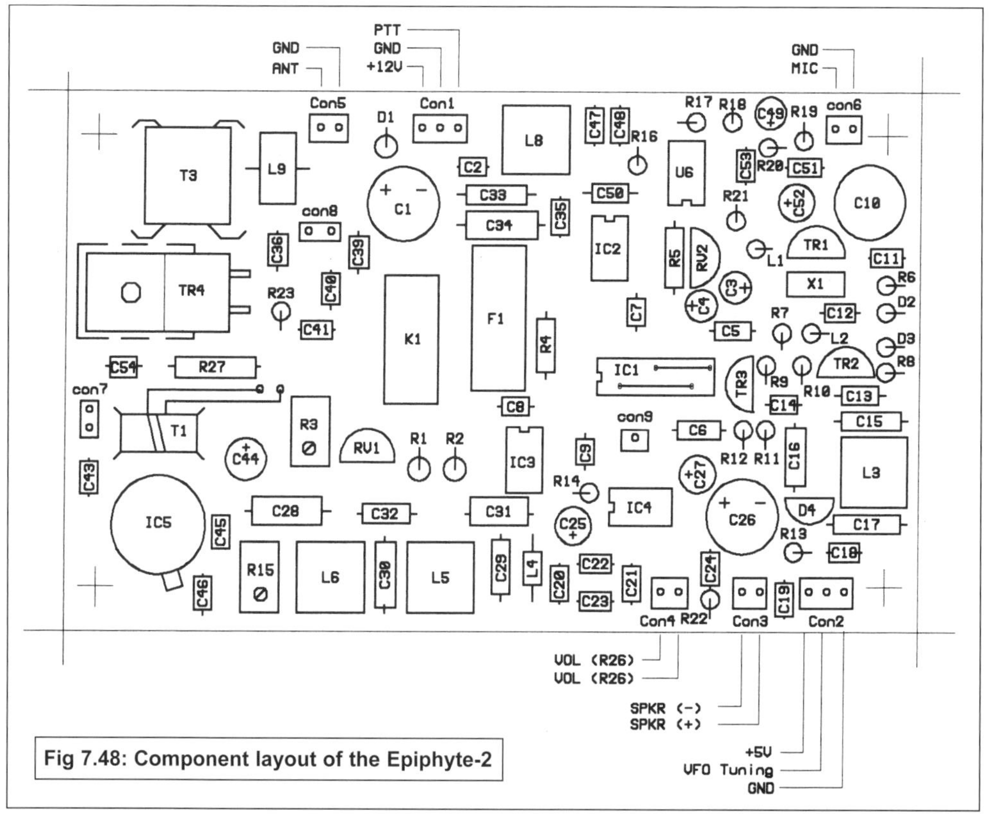


Fig 7.48: Component layout of the Epiphyte-2

with a constant tone (whistle!). Adjust the band-pass filter (L5, L6) to maximise, but do not 'stagger tune' them. Continue increasing the drive until it peaks at around 16V RMS with modulation. The driver current should rise to 60 or 70mA and the PA current to around 600mA with a continuous heavy tone. The 'average' current with normal speech modulation will be considerably less. Monitoring your own signal with phones on a receiver will disclose any serious problem before testing on the air with a local amateur.

Fig 7.50: An assembled Epiphyte-2 which uses a earlier version of the PCB

## The GQ-40 CW Transceiver

This section describes the GQ-40 (**Fig 7.51**) a high-performance, single-band CW transceiver by GW8ELR which first appeared in *SPRAT* [11].

The main Components List is shown in **Table 7.18**, and component values are provided for 7, 10, and 14MHz versions (**Tables 7.19, 7.20 and 21**). **Table 7.22** shows transformer winding details.

### General

The receiver is a conventional superhet design at an IF of 4.4MHz, with a high-dynamic-range, double-balanced, mixer, a six-pole 500Hz crystal IF ladder filter and high-power audio stage. The transmitter is of the mixer type, utilising the receive mixer in a bi-directional mode. The driver and PA are MOSFETs of the inexpensive commercial switching type; adjustable gate bias allows user selection of operating class. The PA is run in push-pull to give a low harmonic content and when in Class AB1 an output of 7W is typical. Power output in normal operation is fully variable via a power control (drive) potentiometer.

Full QSK is achieved by electronic timing of the antenna pin diodes, positive supply lines and IF gain control. Conventional rectifier diodes 1N4007 are used for the antenna changeover system as these are inexpensive and have a similar doping profile to more expensive PIN diodes. Insertion loss is less than 0.1dB at 10MHz with an IP3 of +50dBm when biased at 5mA minimum.

Frequency control is by a Colpitts VFO with a high-quality variable capacitor. Due to the high frequency required for the local oscillator on 14MHz, the VFO is pre-mixed with a crystal oscillator.

### Receiver front-end

Signals from the antenna are routed through the LPF L5-7/C74-77 to the diode changeover system (**Fig 7.52**). D5/6 are biased from the permanent 12V line when TR6 is switched on by the receive 12V line. Bias is regulated by R26/25 with RFC2-4 and C55, 56 keeping RF from entering the DC supply. D8 prevents

Fig 7.51: GQ-40 interior view

| R1, 41 | 47k | 40, 42, 45, | | T1-3, 6, 7 | 37KX830 (matt black) |
|---|---|---|---|---|---|
| R2, 4 | 180R | 55-60, 62, | | T4, 5, 10 | BLN43002402 (two hole |
| R3, 5 | 10R | 66, 68,79, | | | balun core) |
| R6 | 560R | 80, 83, 84, | | T8, 9 | 56-61001101 (matt black) |
| R7, 9, 24, 51, | | 86-91, 100, | | X1-8 | 4.4336MHz |
| 52, 65 | 4k7 | 101, 110 | 100n | RFC1 | 100µH (Toko 7BS) |
| R8, 19, 27, | | C9-11, 43, 61, | | RFC2-5 | 1mH (Toko 7BS) |
| 45, 46, 50, | | 65, 71, 72 | 1n | RFC6,7 | 15µH (Toko 8RBSH) |
| 53-56 | 10k | C13, 15, 16, | | RFC8 | 10µH (Toko 7BS) |
| R15,16 | 68k | 18, 19, 21, | | D1, 2, 4, 9, | |
| R17, 20, 23, | | 31 | 180p | 10, 11 | 1N4148 |
| 25, 26, 28 | 1k | C17 | 220p | D3, 8 | BA243 |
| R18 | 220R | C13a,15a, | | D5-7 | 1N4007 |
| R21 | 5R6 | 9a, 21a | 22p | ZD1 | BZY88C4V7 |
| R22 | 220k | C14a, 20a | 4p7 | ZD3 [varicap | |
| R29, 42, 44, | | C16a, 18a | 8p2 | D12] | BB105 |
| 47-49, 57, | | C17a | 18p | TR1, 2, 5, 16 | J310 |
| 58, 67 | 100k | C22, 32, 38, | | TR3, 12 | 2N3904 |
| R31 | 3k | 39, 49, 53, | | TR4 | 2N2222 |
| R32 | 27R | 63, 64, 69, | | TR6 | PN2222/MPS3392 |
| R33, 35, 36 | 22k | 70, 73, 78, | | TR7 | MFE201 |
| R34 | 2k2 | 97, 111 | 10n | TR8 | VN66AFD |
| R10, 38-40 | not used | C30, 48 | 47p | TR9, 10 | IRF510 |
| R59 | 390R | C33 | 1µ | TR11 | 2N3906 |
| R63 | 680R | C37, 67 | 10µ | TR13, 14 | BD140 |
| R64 | 12R | C41 | 4n7 | TR15 | BS170 |
| RFC2-5 | 1mH | C44 | 47µ | TR17 | 2N3866 |
| VR1 | 1k preset | C46, 47, 81 | 100µ | IC1 | HPF-505X/SBL-1 |
| VR2, 3 | 10k preset | C50 | 150p | IC2 | MC1350P |
| RV1, 2 | 10k log | C51 | 560p | IC3, 9 | SA612/602 |
| RV3 | 22k log switched | C52 | 68p | IC4 | LF351 |
| RV4 | 100k lin | C82 | 4µ7 | IC5 | LM380 |
| RV5, 6 | 22k lin | C85 | 22µ tantalum | IC6 | 4093 |
| C14, 20, 28 | 100p | C92 | 6p8 | IC7, 8 | 78L08 |
| C8, 12, 23-27, | | TC1, 2 | 60p (brown) | IC10 | 78L05 |
| 29, 34, 35, | | TC3 | 6-10p (blue) | IC11 | 78L06 |

**(above) Table 7.18: GQ transceiver parts list - all bands**

| R61 | not fitted |
|---|---|
| R37 | 820R |
| C1, 4, 6 | 100p |
| C2, 7 | 1n |
| C3, 5 | 3p9 |
| C74, 77 | 47p + 220p polystyrene 63V |
| C75, 76 | 680p polystyrene |
| C93 | 47p + 47p + 33p |
| C94 | 470p polystyrene |
| C95, 96 | 1800p polystyrene |
| C99 | 82p |
| C102-109 | not fitted - install bypass link |
| L1-3 | Toko KANK 3334 (yellow) |
| L4 | 40t 28SWG on T68-6 or T50-6 (yellow) |
| L5, 7 | 21t 26SWG on T37-6 (yellow) |
| L6 | 24t 26SWG on T37-6 (yellow) |
| RFC9 | 1mH |
| TC4 | not fitted |

*VFO frequency 2566-2666kHz for 7000-7100kHz coverage*

**Table 7.19: GQ transceiver - additional parts for 40m version**

signal loss during receive through T6. Signals now enter the three-pole input filter formed by C1-7/L1-3 (**Fig 7.53**). The input filter is a low-loss Butterworth design with a bandwidth of 300kHz. Capacitive dividers C1/2 and C6/7 match the characteristic filter impedance to 50Ω. The filter output feeds IC1, a double-balanced hot-carrier mixer, type SBL-1/HPF-505 etc. LO drive for the mixer is provided from a J310 buffer amplifier, and to ensure a correct termination a 50-ohm T pad is used. The mixer is terminated by a simple active termination which uses a step-up transformer to a

| R61 | not fitted |
|---|---|
| R37 | 820R |
| C1, 6 | 47p |
| C2, 7 | 470p |
| C5 | fit wire link |
| C74, 77 | 270p polystyrene |
| C75, 76 | 560p polystyrene |
| C93 | 33p + 39p NPO |
| C94 | 220p + 220p polystyrene |
| C95 | 180p polystyrene |
| C96 | 220p + 220p polystyrene |
| C99 | 82p |
| C102-109 | not fitted - install bypass link |
| C104 | 47p |
| TC4 | not fitted |
| L1 | not fitted |
| L2, 3 | Toko KANK 3334 (yellow) |
| L4 | 32t 28SWG on T68-6 or T50-6 (yellow) |
| L5, 7 | 19t 26SWG on T37-6 (yellow) |
| L6 | 20t 26SWG on T37-6 (yellow) |
| RFC9 | 1mH |

*VFO frequency 5666-5766kHz for 10,100-10,200kHz coverage. Adjust VC1S series cap to bandspread*

**Table 7.20: GQ transceiver parts list - additional parts for 30m version**

J310 FET amplifier with a gate resistor. Whilst this is not as good as the more common diplexer arrangement, it requires no setting up.

## IF and crystal filter

The filter (Fig 7.53) is a six-pole ladder design with a centre frequency of 4.433MHz. The unit uses high-quality, high-volume crystals designed for TV colour burst timing. The filter has a 3dB

| | |
|---|---|
| R61 | 100R |
| R37 | 4k7 |
| C1, 6 | 120p |
| C2, 7 | 1n |
| C5 | 4p7 |
| C3 | fit wire link |
| C4 | not used |
| C74, 77 | 180p polystyrene |
| C75, 76 | 390p polystyrene |
| C93 | 33p + 39p NPO |
| C94 | 220p + 220p polystyrene |
| C95 | 180p polystyrene |
| C96 | 220p + 220p polystyrene |
| C98, 102 | 10n |
| C99 | 82p |
| C103 | not used |
| C104 | 47p |
| TC4 | 18p fixed |
| C105 | 56p |
| C106 | 100n |
| C107 | 1p8 |
| C108 | 56p |
| C109 | 10n |
| X9 | 24.00MHz |
| L1 | not fitted |
| L2, 3 | Toko KANK 3335 (pink) |
| L4 | 32T 28SWG on T68-6 or T50-6 (yellow) |
| L5, 7 | 16T 26SWG on T37-6 (yellow) |
| L6 | 17T 26SWG on T37-6 (yellow) |
| L8, 9 | 1.2µH Toko KANK 3335 (pink) |
| RFC9 | 1mH |
| IC9 | SA612/602 |

*VFO frequency 5566-5466kHz (LO = 24 VFO) for 14000-14100kHz coverage*

**Table 7.21: GQ transceiver - additional parts for 20m version**

| | | | | | | |
|---|---|---|---|---|---|---|
| T1 | K37X830 | 3t IC1 | | 13t TR2 | | 32SWG |
| T2 | - | 15t TR2 | 10t C13 | | | 32SWG |
| T3 | - | 6t C21 | 24t IC2 | | | 32SWG |
| T4 | BLN 43002402 | 2t TR5 | Ct D3 | 6t 12VT | | 32SWG |
| T5 | - | 2t TR1 | Ct C11 | 6t 12VP | | 32SWG |
| T6 | K37X830 | 4t D8 | | 15t TR7 | | 32SWG |
| T7 | - | 14t TR7 | 3t R31 | | | 32SWG |
| T8 | 59-61001101 | 12t PA | | 12t TR8 | | bifilar 28SWG |
| T9 | - | 12T | | | | trifilar 28SWG |
| T10 | BLN 43002402 | 6t R66 | | 2t R3 | | 32SWG |

**Table 7.22: GQ transceiver transformer winding details**

bandwidth of 500Hz and a Gaussian shape by use of Butterworth design constants. T2, 3 match the input and output impedance of the IF amplifier and filter. IC2, a MC1350P, provides up to 60dB gain at the IF frequency. During transmit the amplifier is muted by applying bias to the AGC control, pin 5. This bias voltage is switched by TR3 controlled from the T3 line.

A front-panel IF gain control is fed to the AGC control input via D1 and external AGC may also be applied via D2. The exclusion of an AGC system was a deliberate design policy. Many of the audio-derived AGC systems found in current QRP transceiver designs are far from satisfactory and the inclusion of AGC is often counter-productive to the reception of weak CW signals. If you must add an AGC system, add a good one.

The amplified IF frequency is converted to audio frequency in IC3, an SA602 balanced mixer. LO (BFO) injection for the conversion is provided by an on-chip crystal oscillator. The BFO frequency is adjusted by TC2. The SA602 requires a 6 volt power supply and this is provided from IC11, a three-terminal regulator.

## AF preamp and amplifier

A low-noise LM351 preamplifier, IC4, with some frequency tailoring drives an LM380 audio output amplifier (Fig 7.53). The output is capable of about 1W of audio, and has a Zobell network to ensure stability.

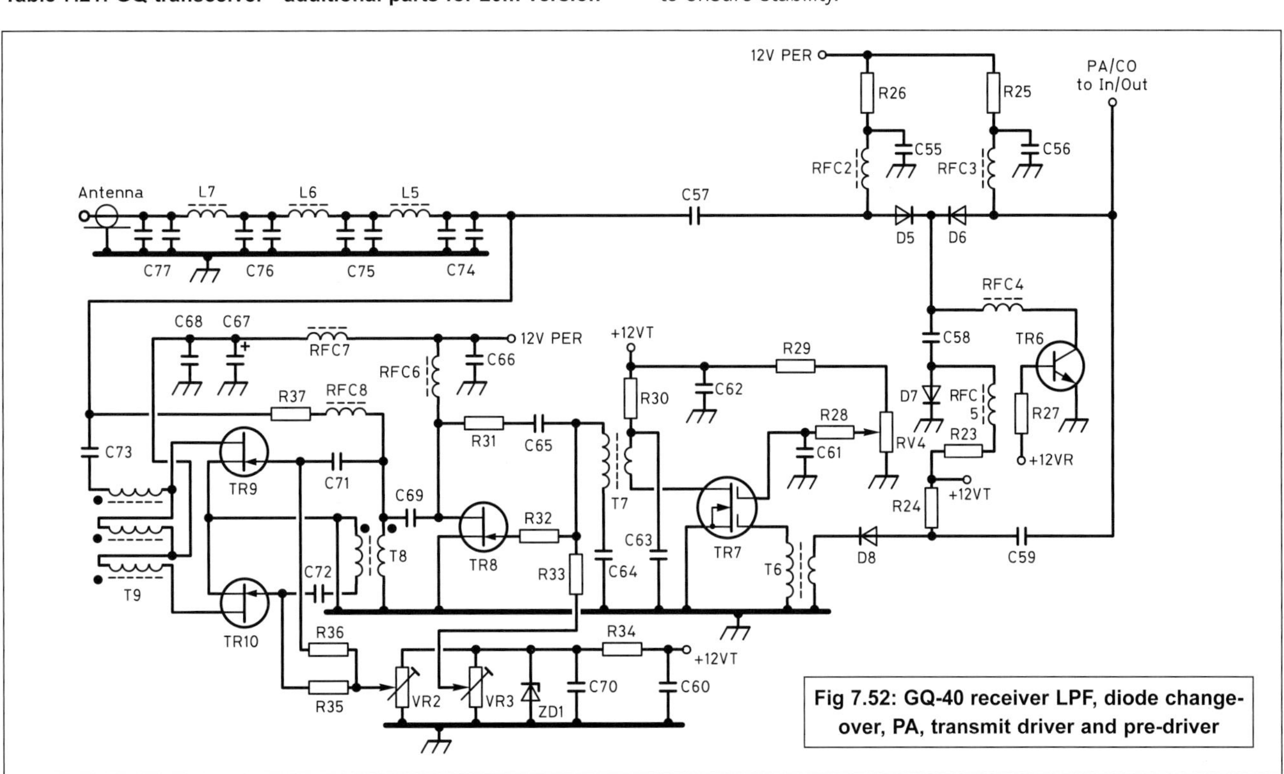


Fig 7.52: GQ-40 receiver LPF, diode change-over, PA, transmit driver and pre-driver

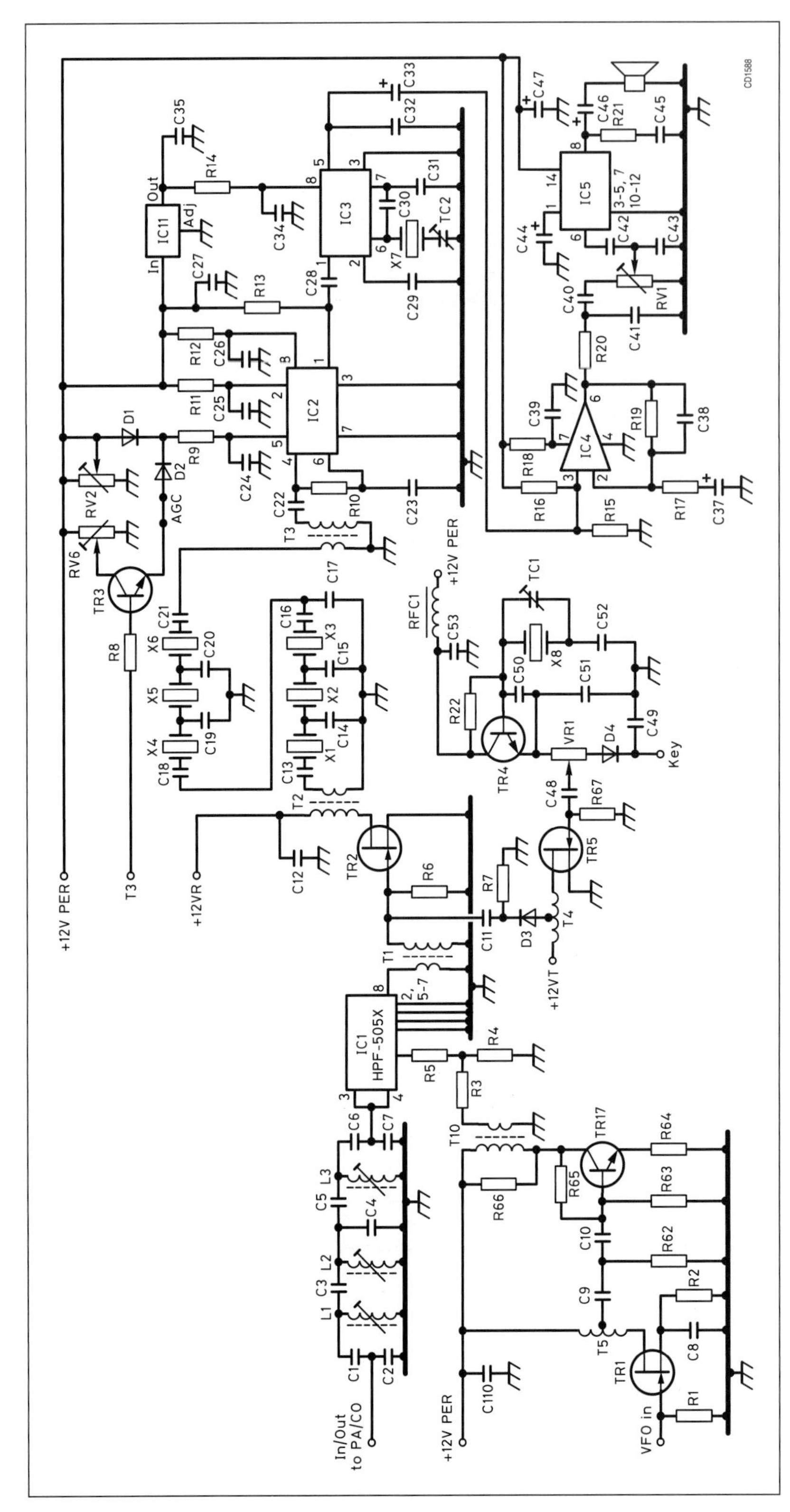


Fig 7.53: GQ-40 crystal filter, CO, buffer and mixer

## Transmit carrier oscillator, buffer and mixer

TR4 is the transmit oscillator (Fig 7.53) and is crystal controlled at 4.433MHz. TC1 allows the frequency to be offset to match the receive BFO audio note. The oscillator is keyed directly at the keying rate via the key line.

TR5 is a J310 buffer amplifier. This feeds the mixer via D3 which is biased on during transmit, via the 12VT line. The output is coupled to the mixer transformer T1 via C11, a DC blocking capacitor. The transmit signal is mixed in IC1 to the output frequency and routed back through the band-pass filter to the pre-driver, TR7.

## Transmit pre-driver and driver

When in transmit, D8 is biased on by the 12VT line, coupling the transmit signal to TR7 by T6 (see Fig 7.52). D7 is also biased on to provide an RF ground between D5/6 to improve the input/output receive bypass isolation. C58 provides a DC block to prevent D5/6 being switched on.

The pre-driver is a dual-gate MOSFET; gate 2 is used as a control element by varying its voltage by RV4, a panel-mounted potentiometer. This stage is transformer coupled to the next by T7, a broad-band matching transformer. C64 on the secondary provides an RF ground but provides DC isolation.

The driver, TR8, a VN66 FET, is biased as a Class A amplifier. Feedback is applied to the driver to ensure stability; this is controlled by R31. Bias adjustment is by VR3. The bias voltage is stabilised by ZD1 and supplied from the 12VT line. As the amplifier has no bias during the receive period, TR8 is connected to the 12V permanent line.

## Power amplifier

TR9, 10 are run as a push-pull amplifier (Fig 7.52). To maintain stability. R37/RFC8 provide feedback. Push-pull drive for the amplifier is provided from a phase-splitter transformer T8. The amplifier is provided with bias to allow the operating class to be varied. The best balance between output power and efficiency will be in Class B. VR2 is the bias control potentiometer and is adjusted for the required quiescent current.

T9, a trifilar transformer matches the transistors' output impedance to the load presented by the output filter. LPF L5-7 and C74-77 form a seven-element Chebyshev low-pass filter. This has a low SWR at the operating frequency, with the cut-off just above the maximum bandwidth upper frequency. The filter is used bi-directionally to cascade with the bandpass filter when in receive mode.

## VFO and pre-mix

The VFO is a standard Colpitts oscillator (**Fig 7.54**). The voltage to the oscillator is stabilised by a three-terminal regulator at 5V. C93 is the main fixed capacitance. This is made up from three values with NPO dielectric, and a small 10pF trimmer for calibration.

The inductance is wound on type 6 material which is a good compromise between thermal stability and the winding size. Receiver incremental tuning is provided by a variable capacitance diode. To maintain a reasonable voltage swing on the diode, its effective capacitance is reduced by series capacitor C92.

Output from the oscillator is coupled by C99, either to the pre-mixer, IC9, or direct to the main mixer buffer amplifier TR1. When the transceiver model requires a LO frequency greater than 6MHz, IC9 pre-mixer is employed. This is a SA602A used as

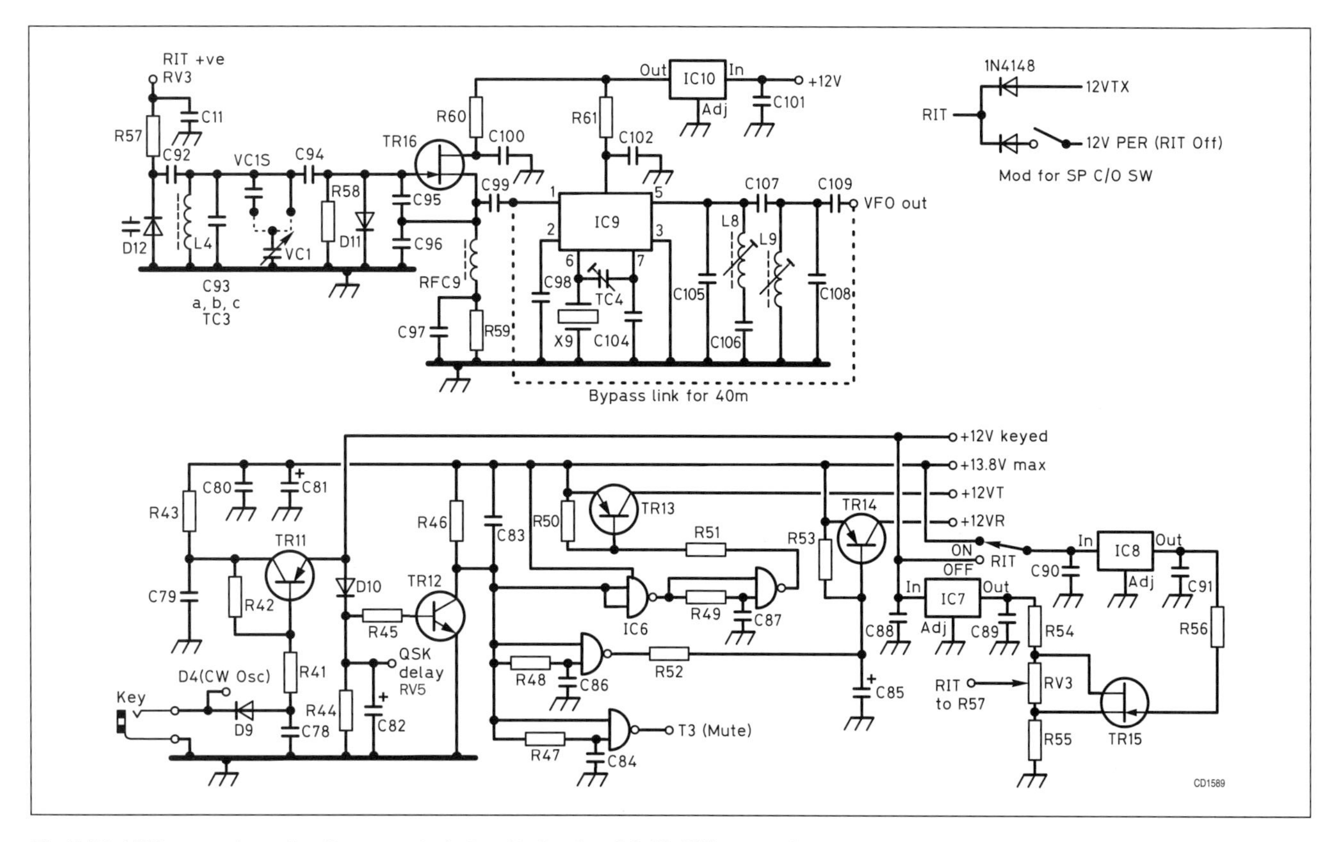


Fig 7.54: VFO, pre-mix and voltage control circuits for the GQ-40 CW transceiver

a crystal oscillator/mixer. The LO pre-mixer output is filtered by a two-pole bandpass filter, C105-109/L8-L9.

### LO buffer and drive amplifier

TR1, a J310 FET, buffers either the pre-mix or VFO frequency dependent on the model (Fig 7.53). Its output is transformer coupled via T5 to a Class A driver, TR17. This is a 2N3866, and the amplifier is designed as a 50-ohm line driver feeding the mixer IC1 T-pad, via T10

### Voltage and control switching

The 12VT and 12VR power supply lines are switched by PNP power transistors controlled from a time-and-delay circuit (Fig 7.58). The timing switch IC6 is a quad NAND Schmitt trigger package, gate timing being adjusted with small ceramic capacitors C84-87. Timing is initiated by TR11 keyer which directly keys the CW oscillator and the delay generator TR12. TR12 will hold on, dependent on the charge across C82. The charge is adjustable via RV5 which is mounted on the rear panel.

RIT voltage is supplied through a divider network controlled by RV3, a front-panel potentiometer. Voltage to the divider is stabilised at 8V by a three-terminal regulator IC7. During transmit, or when the RIT is switched off, gate voltage is applied to TR15 from IC8. This turns on the FET to short RV3 and centre the voltage equivalent to the mid-position of RV3.

### Test and alignment

The minimum requirements for test and alignment are a high-impedance multimeter, power meter, 50Ω dummy load and a general-coverage receiver. The power meter can be a simple peak diode detector across the dummy load used in conjunction with the multimeter. If available, a signal generator or in-band test oscillator are very useful, as is a frequency counter and oscilloscope.

Before starting, carry out a resistance check across the 13.8V input. This should be about 200Ω. If all is OK, connect the multimeter in series with the DC+ lead on its highest amperage range, and switch on. With no signal present and minimum volume, expect the current to be 100-200mA. If the current is appreciably more, switch off and check for faults. Once this test is complete, remove the meter and connect normally to the supply. Turn up the volume control and check for some background noise.

Next test and set up the VFO. Check the VFO output frequency using a counter or a general-coverage receiver with a short antenna near the VFO. Adjust TC3 for the LF band edge. If you are unable to correct the frequency with TC3, then spread or compress the turns on L4. If still outside the range, remove or add turns to L4, or add capacitance using 5mm NPO discs at C93a, C93b, C93c. Next unmesh VC1 and note the HF frequency. If the coverage is more than required, reconnect VC1 through a fixed capacitor to the VC1 pin. VC1 is now in series with the capacitor and its effective capacitance swing will be reduced dependent on the value. Check the HF and LF limits again and readjust TC3 as necessary. If necessary alter the fixed value and repeat the checks. Once you are happy with the coverage, cement the turns on L4, then cement it to the board using Balsa or plastic adhesive.

Once the VFO is up and running correctly, any temperature drift may be corrected. The capacitance of C93 can be made up with three capacitors using the A and B pads. If required, it should be possible to correct long-term drift by mixing dielectric types.

If a VFO crystal mixer is installed, check for the correct output frequency. If an oscilloscope is available, check the output at R5 or use a general-coverage receiver. Peak the output by adjusting L8/L9.

Attach an antenna to the input pin and adjust the cores of L1-3 for maximum received signal strength - do not use a screwdriver for adjustment as the core is very brittle. If you do not have the correct tool, an old plastic knitting needle with the end filed is ideal. Tune to a test signal and adjust the BFO frequency using TC2. Adjust for the best filter response and opposite sideband suppression.

Test the transmitter next. Connect a 50-ohm dummy load and a power meter or diode detector and multimeter to the antenna output. Fit a temporary heatsink to TR8-10; a clip-on TO220 type is ideal (the tab is part of the drain so be careful that during the checks it is not grounded). Set VR2/3 to the earthy end of their travel. (If you have connected RV4, this must also be at the earth end of its range.) Set VR1 wiper to maximum (the TR4 end of its travel). Connect a multimeter on its 'amps' range in the 13.8V supply line, and key a dash length by grounding the KEY pin. If all is well, the transceiver should change to transmit and hold for a short time. With a multimeter on its DC voltage range, check at R27 that the 12VR line switches correctly. Re-check that 4.7V appears across ZD1 on transmit.

Current consumption will be around 200mA providing there is no drive from TR7. If the consumption is excessive, remove D8 to isolate the driver and PA, and re-check. Possible faults include reversed diodes, ICs or transistors, incorrect values on decoupling or bias resistors, or shorted or incorrectly installed transformers. Once any errors have been corrected, reconnect D8.

Adjust TC1 to obtain a 700Hz beat note from the sidetone. If the audio level is excessive, check that the mute voltage to the IF amp of 8V plus appears at the junction of R9/D1.

Attach a multimeter on its current range in series with the DC supply line. Key the transmitter and set TR8 quiescent current by adjusting VR3 so that the meter current rises by 200mA (ie TR8 standing current is 200mA). Now adjust VR2 so that the current increases by a further 100mA (this sets TR9/10 quiescent to 100mA). The adjustments should be smooth with no power output on the meter. Jumps on the current or the power meter would indicate instability at some level of drive. In this case, remove D8 to isolate the drive and re-check. If there is no improvement, check that T9 is correctly wired. Also check the antenna isolation: confirm that TR6 is switching and that D7 is biased on by the 12VT line. Finally, attach RV4 and adjust it to give 2-3V on the gate of TR7. Key the transmitter and check for power output. This completes the tests.

### GQ transceiver PCB

The PCB, a complete kit of parts and a manual containing the latest revisions can be obtained from Hands Electronics, Tegryn, Llanfyrnach, Dyfed SA35 0BL. Tel 01239 698427.

## MEDIUM AND HIGH-POWER SOLID-STATE HF LINEAR AMPLIFIERS

The construction of a solid-state linear amplifier now represents a realistic alternative to valve designs even up to the normal maximum UK power level of 400W. The construction of solid-state amplifiers has often been discouraged in the amateur press with claims that such projects can only be made with the use of expensive test equipment, and in particular, a spectrum analyser. The 'simple spectrum analyser' described by Roger Blackwell, G4PMK, in [12] is ideal for the purpose, and fully justifies the small amount of money necessary for its construction.

*Motorola Applications Notes AN762, AN758 and EB104*, available from Communications Concepts Inc [13] and on the internet [1, 14], provide the designers of sold-state linear amplifiers with outputs in the range 140W to over 1kW. The application notes (AN) and engineering bulletins (EB) describe the construction and operation of suitable amplifiers using Motorola devices, and include printed circuit foil information, making construction relatively easy. It is interesting to note that the PA units fitted to virtually all of the currently available commercial amateur radio equipment are based upon these designs by Motorola, with only a few individual differences to the original design.

With the availability of suitable application notes it is perhaps surprising that few amateurs seem to have embarked upon such a project. High prices for solid-state power devices, combined with a lack of faith in their reliability, may well account for the reluctance to construct solid-state linear amplifiers. Suitable components have been readily available in the USA for some time: Communications Concepts Inc of Xenia, Ohio [13] have offered a complete range of kits and components for the Motorola designs.

The kits include the PCB, solid-state devices, and all the components and the various ferrite transformers already wound, so that all the constructor has to do is solder them together. A large heatsink is necessary and is not supplied.

Home construction of PCBs for an amplifier is possible from the foil patterns available in the application notes, but it should be borne in mind that the thickness of copper on much of the laminate available to amateurs is unknown and inadequate for the high current requirements of a linear amplifier. A ready-made PCB is therefore a very sensible purchase and the kit of parts as supplied by CCI represents a very-cost-effective way of building any of the Motorola designs.

Whilst these HF linear designs by Motorola have now been in existence for more than two decades, they remain a useful standard and a very viable design for home construction. The Motorola Semiconductors business was split up in around the year 2000 and the specified devices may be harder to find today, but the general principles apply and in many cases substitutions are available. A matched pair of MRF454 is available in the US from CCI [13]. A suitable substitution for the MRF454 is the SD1487 and a matched pair is available from Birketts in the UK [15].

### Choosing an Amplifier

Perhaps the most useful amplifier for the radio amateur is described in AN762. It operates from a 13V supply and is capable of providing up to 160W output with only 5W of drive. This design is the basis for the majority of commercial 100W PA units and lends itself to both mobile and fixed station operation using readily available power supplies. **Table 7.23** illustrates a number of alternative designs.

AN758 describes the construction of a single 300W output amplifier operating from a 50V supply and further describes a method of using power combiners to sum the outputs of a number of similar units to provide power outputs of 600 and 1200W. Two such units would be ideal for a full-power linear amplifier for UK use, giving an output comparable to the FL2100 type of commercial valve amplifier. It is recommended that any prospective constructor reads the relevant applications note before embarking upon the purchase and construction of such an amplifier.

After constructing a number of low-power transceivers it was decided to commence the construction of the AN762 amplifier. The EB63 design is similar, but uses a slightly simpler bias circuit.

| Number | Power out (W) | Supply voltage (V) |
|---|---|---|
| AN762 | 140 | 12-14 |
| EB63 | 140 | 12-14 |
| EB27A | 300 | 28 |
| AN758 | 300-1200 | 50 |
| EB104 | 600 | 50 |

**Table 7.23: Motorola RF amplifiers**

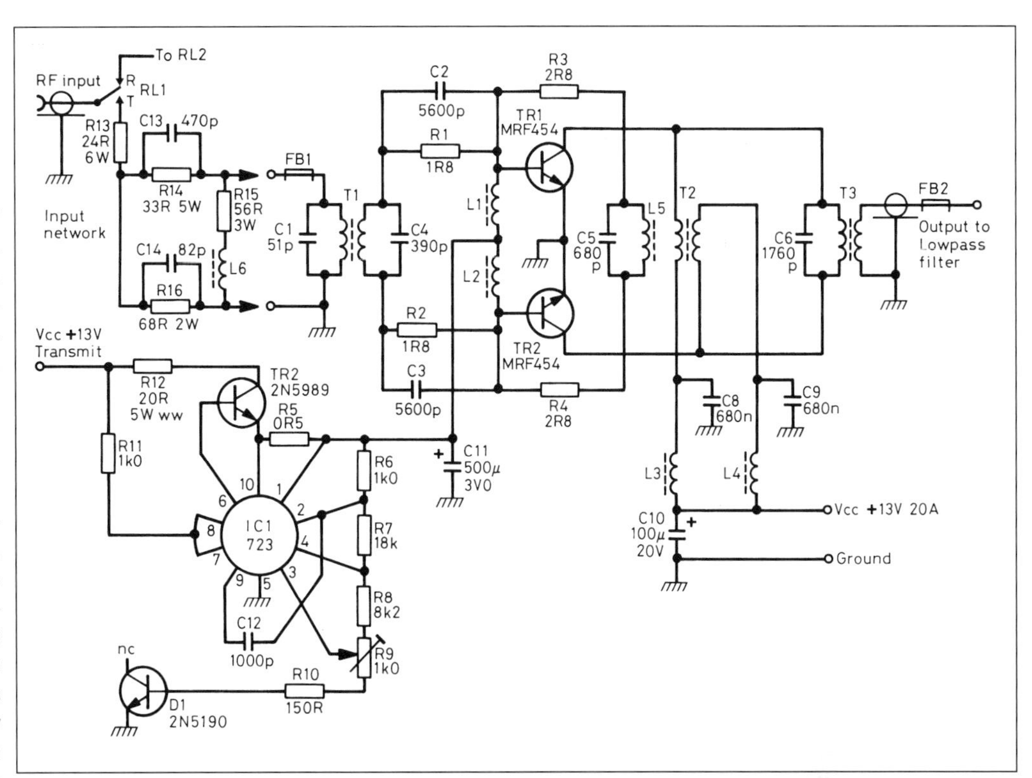


Fig 7.55: The HF SSB 140-300W linear amplifier circuit

AN762 describes three amplifier variations: 100W, 140W and 180W using MRF453, MRF454 and MRF421 devices respectively, and the middle-of-the road MRF454 140W variant was chosen (**Fig 7.55**).

## Constructing the Amplifier

Construction of the amplifier is very straightforward, especially to anyone who has already built up a solid-state amplifier. The PCB drawings (**Fig 7.56**, see Appendix B) and layouts are very good and a copy of the relevant application note was included with the kit. **Figs 7.57, 7.58 and 7.59** show how the amplifier was constructed. A Components List is in **Table 7.24**.

Fixing the transformers to the PCB posed a slight problem: on previous amplifiers that the author has built they were soldered directly to the PCB, but not this one. Approximately 20 turret tags (not supplied) are required: they are riveted to the board and soldered on both sides.

A modification is required if you wish to switch the PA bias supply to control the T/R switching: the bias supply must be brought out to a separate terminal rather than being connected to the main supply line. This is achieved by fitting a stand-off insulator to the PCB at a convenient point; the bias supply components are then soldered directly to this stand-off rather than directly to the PCB.

Fig 7.57: The HF linear amplifier

Fig 7.58: Low-pass filter and ALC circuits

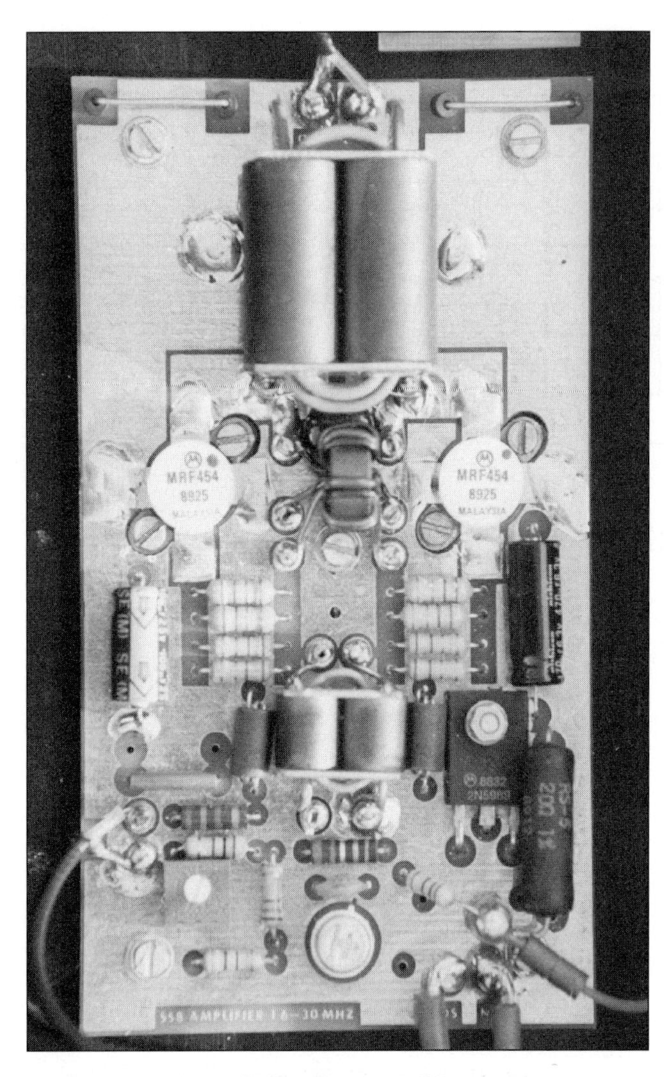


**Fig 7.59: Circuit board of the linear amplifier**

A number of small ceramic chip capacitors have to be soldered directly to the track on the underside of the PCB; it should be borne in mind when doing this that the clearance between the PCB and heatsink is slightly less than 1/8in and the capacitors must not touch the heatsink.

It is advisable to mount the PCB to the heatsink before attempting to make any solder connections to it, as it is necessary to mark the mounting holes accurately, ideally drilling and tapping them either 6BA or 3mm. The power transistor mounting is very critical in order to avoid stress on the ceramic casing of the devices. The devices mount directly onto the heatsink and should ideally be fitted by drilling and tapping it. The PCB is raised above the heatsink on stand-offs made from either 6BA or 3mm nuts, so that the tabs on the transistors are flush with the PCB - they must not be bent up or down. When the PCB and transistors have been mounted to the heatsink correctly, they may be removed for the board to be assembled. The transistors should not be soldered in at this stage. The nuts to be used as the stand-offs can be soldered to the PCB, if required, to make refitting the latter to the heatsink a little simpler, but ensure the alignment of the spacers is concentric with the holes.

Assembly should commence with the addition of the turret tags and stand-offs. Then the ceramic chip capacitors should be added under the PCB. D1, which is really a transistor, is also mounted under the PCB; only the emitter and base are connected - the collector lead is cut off and left floating. This transistor is mounted on a mica washer and forms a central stand-off when

**HF LINEAR AMPLIFIER**

| | |
|---|---|
| C1 | 51p chip |
| C2, 3 | 5600p chip |
| C4 | 390p chip |
| C5 | 680p chip |
| C6 (C7) | 1760p (2 x 470p chips plus 820p silver mica in parallel) |
| C8, 9 | 0.68μ chip |
| C10 | 100μ 20V |
| C11 | 500μ 3V |
| C12 | 1000p disc |
| C13 | 470p silver mica |
| C14 | 82p silver mica |
| R1, 2 | 2 x 3.6R in parallel |
| R3, 4 | 2 x 5.6R in parallel |
| R5 | 0.5R |
| R6 | 1k |
| R7 | 18k |
| R8 | 8k2 |
| R9 | 1k trimpot |
| R10 | 150R |
| R11 | 1k |
| R12 | 20R 5W WW |
| R13 | 24R carbon 6W* |
| R14 | 33R carbon 5W* |
| R15 | 56R carbon 3W* |
| R16 | 68R carbon 2W* |

* Make up from several higher values in parallel.

| | |
|---|---|
| L1, 2 | VK200 19/4B choke (6-hole ferrite beads) |
| L3, 4 | Fairite beads x 2 (2673021801) on 16SWG wire |
| L5 | 1t through T2 |
| L6 | 0.82μH (T50-6) |
| T1 | 2 x Fairite beads 0.375in x 0.2in 0.4in, 3:1 turns |
| T2 | 6t 18SWG ferrite 57-9322 toroid |
| T3 | 2 x 57-3238 ferrite cores (7d grade) 4:1 turns |
| FB1, 2 | Fairite 26-43006301 cores |
| RL1 | OUD type |
| TR1, 2 | MRF454 |
| TR3 | 2N5989 |
| D1 | 2N5190 |
| IC1 | 723 regulator |

**LOW-PASS FILTER**

| | |
|---|---|
| C1 | 1200p |
| C2, 16 | 180p |
| C3 | 2200p |
| C4 | 470p |
| C5 | 1000p |
| C6, 13 | 680p |
| C7 | 90p |
| C9 | 250p |
| C10 | 560p |
| C11 | 390p |
| C12, 24 | 33p |
| C14 | 100p |
| C15, 18 | 330p |
| C17 | 27p |
| C19 | 75p |
| C20, 28 | 150p |
| C21 | 120p |
| C22, 27 | 12p |
| C23 | 220p |
| C25 | 100p |
| C26 | 82p |
| C29 | 39p |
| C30 | 68p |
| C31, 32 | 10n ceramic |
| C33 | 10p trimmer |
| C34 | 220p silver mica |

Note: C1 C30 silver mica 350VDC

| | |
|---|---|
| L1 | 28t 22SWG T68-2 |
| L2 | 25t 22SWG T68-2 |
| L3 | 22t 22SWG T50-2 |
| L4 | 20t 22SWG T50-2 |
| L5 | 18t 20SWG T50-6 |
| L6 | 16t 20SWG T50-6 |
| L7 | 12t 20SWG T50-6 |
| L8 | 11t 20SWG T50-6 |
| L9 | 10t 20SWG T50-6 |
| L10 | 9t 20SWG T50-6 |
| L11 | 8t 20SWG T50-6 |
| L12 | 7t 20SWG T50-6 |
| T1 | 18t bifilar T50-43 pri: 1t |
| R1 | 68R |
| R2, 3 | 22k trimpot |
| R4 | 1k8 |
| L13 | 1mH RFC |
| RL2 | OM1 type |
| RL3-14 | 2A SPCO PCB mtg, 6V coil |
| D1, 2 | OA91 or OA47 |

**SWITCHING AND ALC**

| | |
|---|---|
| R1, 17, 20 | 10k |
| R2 | 4k7 |
| R3-7 | 1k |
| R8, 9 | 33k |
| R10, 13, 21, 22 | 1M |
| R11 | 3k3 |
| R12, 14, 18 | 47k |
| R15 | 47k trimpot |
| R16 | 390R |
| R19 | 220k |
| R23 | 8k2 |
| TR1 | BC212 |
| TR2 | BC640 |
| IC1 | LM3900 |
| C1, 14-26 | 100n |
| C2-5, 7, 9-11, 13 | 10n |
| C6 | 0.22μ |
| C8 | 10 x 16V |
| C12 | 1μ 16V |
| D1-3 | LEDs |
| D4, 5, 7-18 | 1N914 |
| D6 | 10V zener |
| S1 | 1-pole 6-way |
| S2, 3 | SPCO |
| Meter | 500μA or similar |

**Table 7.24: HF linear amplifier components list**

the PCB is finally screwed down to the heatsink. The mounting screw passes through the device which must be carefully aligned with the hole in the PCB. A number of holes on the PCB are plated through and connect the upper and lower ground planes together. The upper-side components can be mounted starting with the resistors and capacitors, and finally the transformers can be soldered directly to the turret tags. Soldering should be to a high standard as some of the junctions will be carrying up to 10A or more.

When the board is complete it should be checked at least twice for errors and any long leads removed from the underside to ensure clearance from the heatsink. Mount the PCB to the heatsink and tighten it down. Now apply heatsink compound and mount the power transistors which should fit flush with the upper surface of the PCB. Tighten them down by hand, ensuring that there is no stress on the ceramic cases. If any of the connections need to be slightly trimmed to fit, cut them with metal cutters. Ensure the collector tab is in the correct place. Once the transistors fit correctly, they can be removed again and very lightly tinned. The PCB should also be lightly tinned. The transistors can now be refitted and tightened down. Now they can be soldered in but, once in, they are very difficult to remove, so take great care at this stage. The amplifier board is now complete.

Construction of the amplifier takes very little time but requires considerable care to avoid damage to the output devices; the metal work may take a little longer. A large heatsink is required for 140W and an even larger one for 300W. It is recommended that the higher-power amplifiers are mounted onto a sheet of 6mm (0.25in) copper which is in turn bolted to the main heatsink. (Note: on no account should the transistors be mounted onto a diecast box due to surface imperfections and poor thermal conductivity.) Blowing may be necessary at the higher powers or if a less-than-adequate heatsink is used.

## Setting Up and Testing the Amplifier

There is only one adjustment on the amplifier, making setting up relatively simple. Before connecting any power supplies, check and re-check the board for any possible errors. The first job is to test the bias supply. This must always be done before connecting the collector supply to the amplifier as a fault here could destroy the devices instantly. With +13V connected to the bias supply only, it should be possible to vary the base bias from approximately 0.5V to 0.9V. Set it to the lowest setting, ie 0.5V. Disconnect the bias supply from the 13V line. When conducting any tests on the amplifier always ensure that it is correctly terminated in a 50-ohm resistive dummy load. Apply +13V to the amplifier and observe the collector current on a suitable meter. It should not exceed a few milliamps - if it does something is wrong, so stop and check everything. Assuming that your amplifier only draws 3 to 4mA, connect the bias supply to the +13V supply and observe an increase in current, partly caused by the bias supply itself and also by the increased standing current in the output devices. The current can be checked individually in each of the output devices by unsoldering the wire links L3 and L4 on the PCB. Set the bias to 100mA per device by adjusting R9. The current should be approximately the same in each device: if it is not it could indicate a fault in either device or the bias circuitry to it. Increase the standing current to ensure that it rises smoothly before returning it to 100mA per device. Once the total standing current is set to 100 + 100 = 200mA, the amplifier is ready for operation.

With a power meter in series with the dummy load, apply a drive signal to the input, steadily increasing the level. The output should increase smoothly to a maximum of about 160W. It will go to 200W but will exceed the device specification. Observing the output on a spectrum analyser should reveal the primary signal, together with its second, third and higher harmonics. Check that there are no other outputs. Removal of the input signal should cause the disappearance of the other signals displayed. It is helpful during initial setting up to monitor the current drawn by the amplifier. At full output, efficiency should be in the order of 50%, perhaps lowering slightly at the upper and lower frequency limits and increasing a little somewhere in the 20MHz range. The maximum current likely to be drawn by the 140W amplifier is in the order of 24A.

## Putting the Amplifier to Use

Building and setting up the amplifier is undoubtedly a simple operation, and may lead one into a false sense of security. Before the amplifier can be used it must have a low-pass filter added to the output to remove the harmonics generated. For single-band operation only one filter would be required, but for operation on the HF amateur bands a range of filters is required with typically six switched filters covering the range 2 to 30MHz. For most applications a five-pole Chebyshev filter will provide all the rejection required, but the majority of commercial designs now use the elliptic type of filter providing peaks of rejection centred around the second and third harmonics. Such filters can be tuned to maximise the rejection at specific frequencies. An elliptic function filter was decided upon as it only requires two extra components over and above the standard Chebyshev design and setting up is not critical.

The construction of a suitable low-pass filter (**Fig 7.60**) may take the form of the inductors and capacitors mounted around

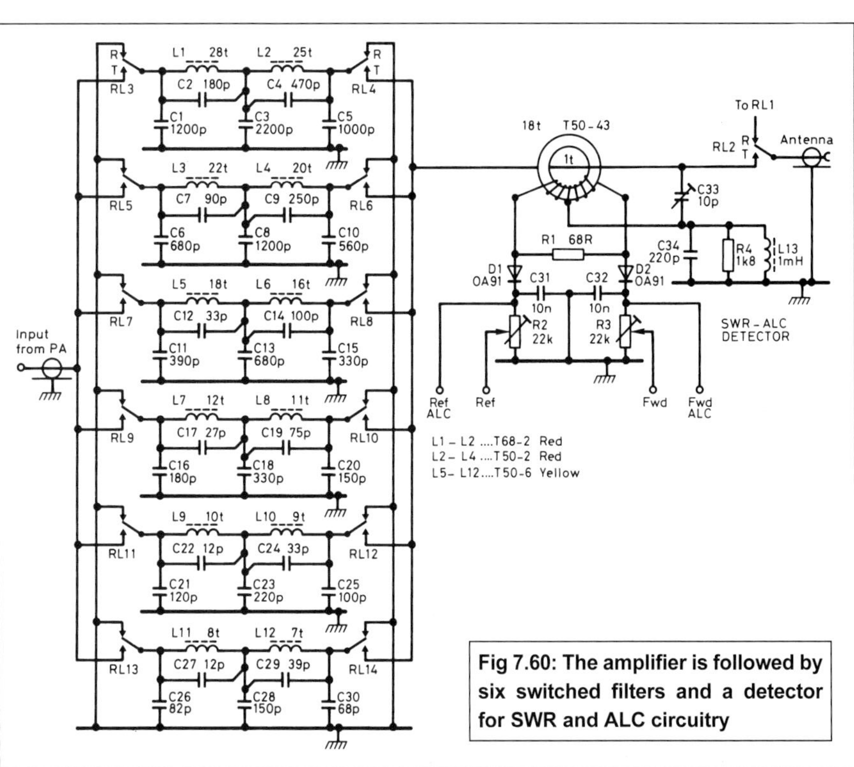


**Fig 7.60: The amplifier is followed by six switched filters and a detector for SWR and ALC circuitry**

a suitable wafer switch, or they may be mounted on a PCB and switched in and out of circuit using small low-profile relays. This makes lead lengths shorter and minimises stray paths across the filter. Unused filters may be grounded easily using relays permitting only one filter path to be open at a time. The relays need only to be able to carry the output current; they are not required to switch it. 2A contacts are suitable in the 100-140W range.

Micrometals Inc cores ensure the duplication of suitable inductors while silver mica capacitors should be used to tune the filters. The voltage working of the capacitors should be scaled to suit the power level being used; ideally, 350V working should be used in the 100-150W range and, for powers in the region of 400-600W, 750V working capacitors should be used, the latter being available from CCI.

The antenna change-over relay may be situated at either end of the low-pass filter. If it is intended to use the filter on receive then the relay will be placed between amplifier and filter, but if the amplifier is an add-on unit then it may not be necessary to use the filter on receive and the relay may be located at the filter output. Filter performance is enhanced if it is mounted in a screened box with all DC leads suitably decoupled.

## SWR Protection and ALC

One of the major shortcomings of early solid-state amplifiers was PA failure resulting from such abuse as overdriving, short-circuited output, open-circuited output and other situations causing a high SWR. A high SWR destroys transistors either by exceeding the collector-base breakdown voltage for the device or through overcurrent and dissipation.

ALC (automatic level control) serves two functions in a modern-day transmitter: it controls the output power to prevent overdriving and distortion and can be combined with a SWR detector to reduce the power if a high SWR is detected, which reduces the voltages that can appear across the output device and so protects it.

A conventional SWR detector provides indication of power (forward) and SWR (reflected power) which can be amplified and compared with a reference. If the forward power exceeds the preset reference an ALC voltage is fed back to the exciter to reduce the drive and hence hold the power at the preset level. A high SWR will produce a signal that is amplified more than the forward signal and will reach the reference level more quickly, again causing a reduction in the drive level.

The circuit (**Fig 7.61**) has been designed to work in conjunction with the ALC system installed in the G3TSO modular transceiver [16] and produces a positive-going output voltage. The LM3900 IC used to generate the ALC voltage contains two unused current-sensing op-amps which have been used as a meter buffer with a sample-and-hold circuit providing a power meter with almost a peak reading capability. In practice it reads about 85% of the peak power compared to the 25% measured on a typical SWR meter.

## T/R Control

There are many ways of controlling the T/R function of an amplifier. It was decided to make this one operate from the PTT line but unfortunately direct connection resulted in a hang-up when relays in the main transceiver remained activated after the PTT line was released. The buffer circuit comprising TR1 and TR2 simply switch the input and output relays from receive to transmit and provide a PA bias supply on transmit.

## Interfacing Amplifier to Exciter

The G3TSO modular transceiver was used as the drive source for the AN762 solid-state amplifier, 5W of drive producing 140W output from the linear. A little more drive and 200W came out. This was rapidly reduced by setting the ALC threshold. Initially the spectrum analyser showed the primary signal, harmonics reduced by the action of the elliptic low-pass filter, but alas a response at 26MHz and not many decibels down on the fundamental. A check with the general-coverage transceiver revealed that there really was something there, while a finger on the 80m low-pass filter showed quite a lot of heat being generated.

Investigations revealed quite clearly that this type of broad-band amplifier cannot be operated with a capacitive-input, low-pass filter at either end without it going into oscillation at some frequency, usually well above the cut-off frequency of the filter. The filter input impedance decreases with frequency, and with two such filters located at either end of the amplifier there comes a point where the input circuit and output circuit resonate at the same frequency and a spurious oscillation occurs. Removal of either filter solves the problem.

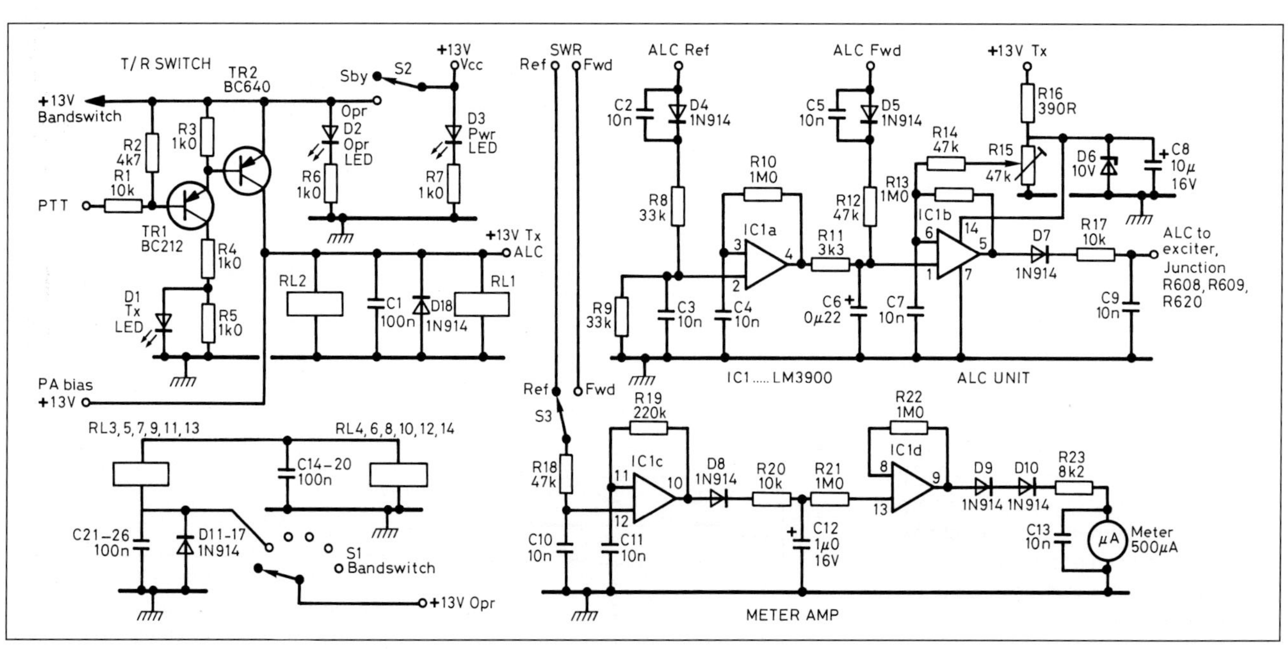


**Fig 7.61: Transmit/receive switching, ALC and SWR circuits**

A direct connection between the exciter and linear amplifier is the preferred solution, but is not always practicable in the case of add-on amplifiers where there is already a low-pass filter installed in the exciter. Another solution is to provide a resistive termination at the input of the amplifier; this is far simpler to effect and is used in several commercial designs.

The network used comprises a 56Ω carbon resistor across the input of the amplifier which reduces the input impedance to 25Ω, so a 24Ω resistor is placed in series with the drive source to present a near-50Ω impedance to the exciter. Power from the exciter will be absorbed in these resistors which must be built from a number of lower-wattage resistors in parallel, ie five 150Ω resistors make a 30Ω resistor with five times the power rating.

In addition it was found necessary to add a ferrite bead to the input and output leads to the amplifier to effect a complete cure to the parasitic problem which was at its worst on 21MHz, the parasitics occurring above 40MHz. An alternative solution would be to use an inductive-input low-pass filter (Fig 7.27(c) & Table 7.9).

## Conclusion

The construction of a solid-state high-power amplifier is very simple, especially as the parts are obtainable in kit form via international mail order from the USA. The use of a spectrum analyser, no matter how simple, greatly eases the setting up of such an amplifier and ensures peace of mind when operating it.

Amplifiers are available for a number of different power levels, and can be combined to provide higher power levels. The use of 13V supplies practically limits powers to about 180W and below, while 28 or 50V facilitates higher-power operation without the need for stringent PSU regulation. The full-power solid-state linear is now possible at a price showing a considerable saving on the cost of a commercially made unit.

There is a great similarity between all the Motorola designs: the 300W amplifier described in AN758 is virtually the same as

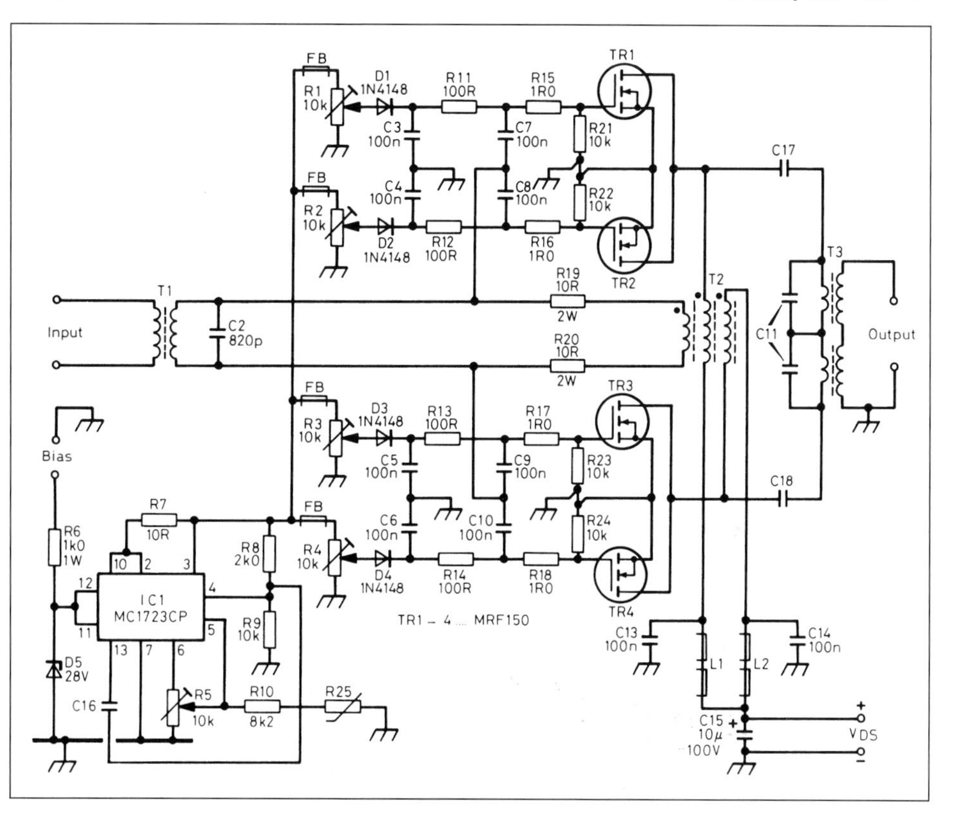

Fig 7.62: 600W output RF amplifier

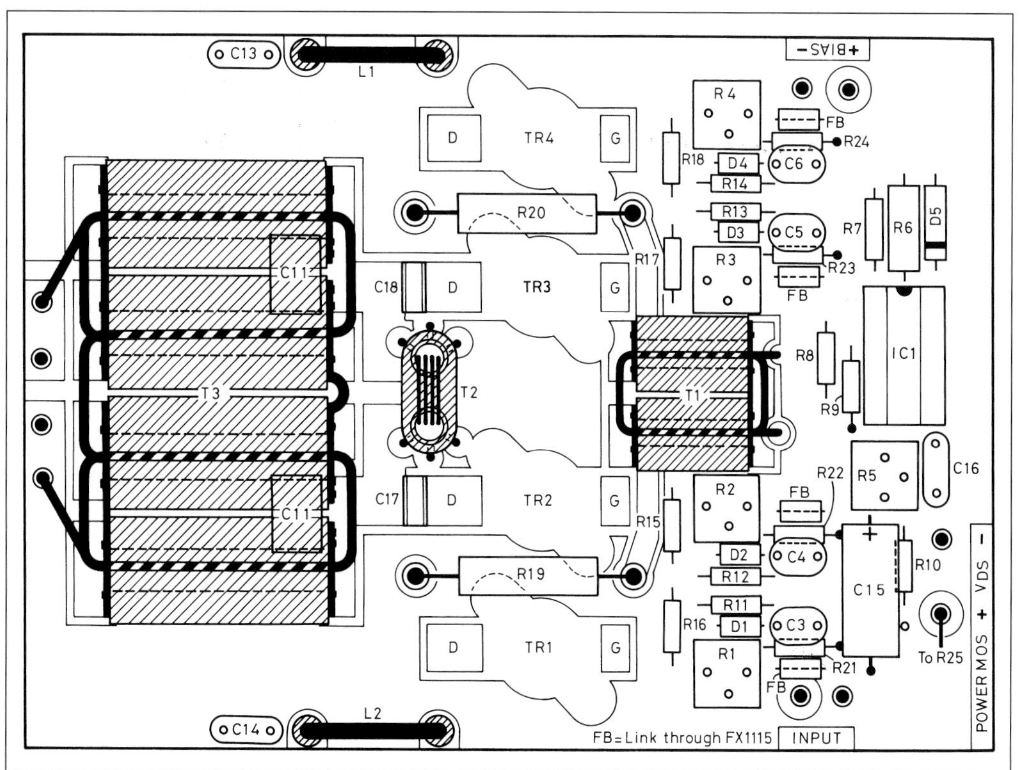

Fig 7.64: 600W amplifier component layout

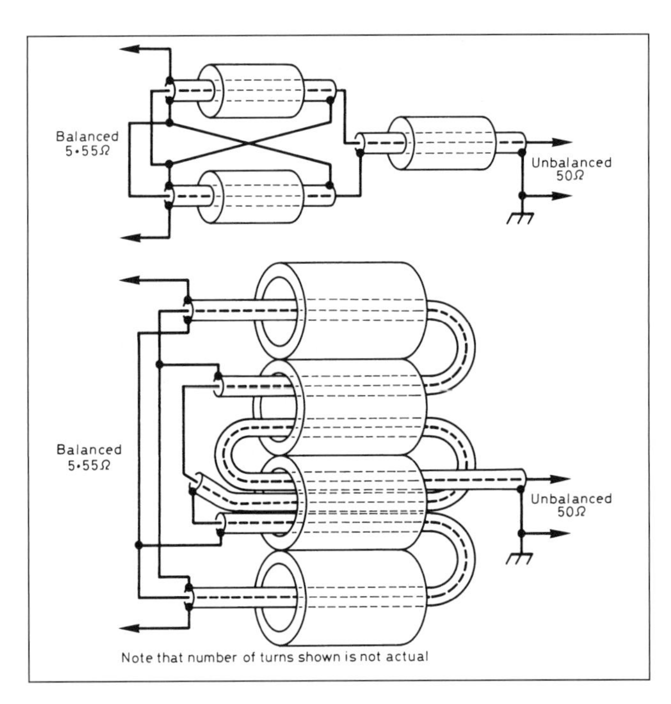


**Fig 7.65: Winding the transformers**

| | |
|---|---|
| R1-R5 | 10k trimpot |
| R6 | 1k, 1W |
| R7 | 10R |
| R8 | 2k |
| R9, 21-24 | 10k |
| R10 | 8k2 |
| R11-14 | 100R |
| R15-18 | 1R |
| R19, 20 | 10R, 2W carbon |
| R25 | Thermistor, 10k (25°C), 2k5 (75°C) |
| *All resistors ½W carbon or metal film unless otherwise noted.* | |
| C1 | Not used |
| C2 | 820p ceramic chip |
| C3-6, 13, 14 | 100n ceramic |
| C7-10 | 100n ceramic chip |
| C11 | 1200p each, 680p mica in parallel with an Arco 469 variable or three or more small mica capacitors in parallel |
| C12 | Not used |
| C15 | 10µ, 100V elec |
| C16 | 1000p ceramic |
| C17, 18 | Two 100n, 100V ceramic each (ATC 200/823 or equivalent) |
| D1-4 | 1N4148 |
| D5 | 28V zener, 1N5362 or equivalent |
| L1, 2 | Two Fair-Rite 2673021801 ferrite beads each or equivalent, 4 H |
| T1 | 9:1 ratio (3t:1t) |
| T2 | 2µH on balun core (1t line) |
| T3 | See Fig 7.64 |
| TR1-4 | MRF150 |
| IC1 | MC1723CP |

**Table 7.25: Components list for 600W output RF amplifier**

the AN762 amplifier, the fundamental difference being the 28V DC supply voltage.

The 600W MOSFET amplifier (**Figs 7.62, 7.63** in Appendix B, **7.62 and 7.64**, plus **Table 7.25**) described in EB104 represents a real alternative to the valve linear amplifier and operates from a 50V supply, but considerably more attention must be paid to the dissipation of heat.

## CHOOSING A COMMERCIAL TRANSCEIVER

*Apart from low power radios, the majority of transceivers used in the shacks of UK amateurs are commercial, usually manufactured in the far east. Many of these are reviewed in the RSGB members' magazine* RadCom, *by* ***Peter Hart, G3SJX****. The following is his guide to finding your way round the bewildering range of choices available. Further information and detailed reviews are included in [17].*

Buying an HF transceiver can be one of the bigger purchases which the amateur is likely to make. The large number of radios on the market and the various features and functions which they provide can be rather daunting, especially to the newcomer and the following guidelines may help in the decision making process.

### Evolution of the Modern Transceiver

The SSB/CW transceiver, as distinct from a separate receiver and transmitter first started to make an appearance during the 1960s. At the beginning of the 1980s, the first models to include a microcontroller were introduced. This had a profound effect on the architecture of the radio, possibly unsurpassed since Edwin Armstrong invented the superhet a century ago. The microcontroller made possible the single tuning knob digital frequency synthesiser and with this the ability to make stable oscillators at VHF. This in turn made viable the up-conversion broadband superhet architecture which forms the basis of virtually all successive designs up until the present day.

This was a very timely introduction as in 1979 we had gained three more HF bands, the so-called WARC bands, adding three more band positions and complexity to the traditional bandswitched design.

These new generation radios featured memories, twin VFOs and variable IF bandwidth. Valved power amplifiers were replaced by broadband semiconductor designs as power transistor technology was becoming more mature and cheaper.

As we progressed into the 1990s, the Direct Digital Synthesiser made its appearance, significantly reducing the complexity of the multi-loop PLL and improving performance. Digital Signal Processing started to be used in the audio stages giving degrees of filtering and noise reduction facilities previously unattainable by analogue technologies. With a greater realisation of the importance of dynamic range we started to see radios with substantially improved performance.

As the decade progressed the level of features provided snowballed. DSP started to be used in the IF stages giving a far greater range of bandwidth options and removing the need to fit expensive extra IF filters. LCD panels of increasing size and complexity began to dominate over LED and fluorescent display technology. HF radios also started to appear with extensions into the VHF and UHF bands and some even catered for full duplex satellite operation and coverage up as far as the 23cm band.

Entering the 21st century, radios were just as well-featured for the data modes as for the traditional modes with a growing number of models including RTTY and packet decoders. A fresh look at performance and features has resulted in a new range of really high-end designs and currently the state of the art in dynamic range.

At the other end of the scale, a range of lightweight models appeared aimed at the mobile user or traveller and currently the only true HF (plus VHF and UHF) hand portable. Computer interfacing had for some time featured strongly and now black boxes started to appear using the PC as the sole user interface.

The close affinity with PCs continues and we are now seeing radios sporting VGA connectors for external displays, LAN connectors and USB ports for control and interface lines, and upgradeable firmware downloadable from the Internet. Touch

**Table 7.26: Comparison of the categories of commercially manufactured amateur radio transceivers**

| Type of radio | Characterised by | Principle use | Typical models |
|---|---|---|---|
| Base stations mains and 12V | Largest size, most features, highest performance, versatile, highest cost, may need external mains PSU | Home use, DXing, contesting, general use, all modes | Yaesu FTDX5000<br>Icom IC-7851<br>Kenwood TS-990<br>Ten-Tec Orion |
| Mid size 12V | Mid size, needs external mains PSU, many features, mid/high performance, mid cost | General all round use principally at home. | Yaesu FTDX3000<br>Icom IC-7600<br>Kenwood TS-590SG<br>Elecraft K3S |
| Small size 12V | Small size, multifunction knobs, mid performance, mid/low cost, external PSU for mains use | General all round use at home, portable, transportable, lightweight DXpeditions | Yaesu FT-450D<br>Icom IC-7200<br>Kenwood TS-480 |
| Mobile | Dash mounting or detachable panel, 12V operated, multifunction knobs, mid cost, mid performance | Mobile, transportable | Yaesu FT-857<br>Icom IC-7100<br>Kenwood TS-50S |
| Battery operated portables | Small size, lightweight, low power, fewer features, internal or external batteries | Hand portable, take anywhere | Yaesu FT-817<br>Icom IC-703<br>Elecraft KX3 |
| Software Defined Radios (SDR) | Unlimited features, downloadable software, needs PC, keyboard and mouse control, spectrum display with point and click tuning | Home use, an experimenters radio | FlexRadio 5000A<br>Apache ANAN-100D |

screen displays are now starting to be used and add a new dimension to the way in which we use our radios. Remote operation is becoming fairly common and radios are now being supported with appropriate software to allow full remote operation via a LAN or via the internet.

The current generation of radios use DSP for all signal processing functions with comprehensive channel filtering, tailoring of the passband shape and noise reduction modes. To achieve the very highest close-in dynamic ranges for the toughest of crowded conditions there is a move away from up-conversion architectures back to lower frequency IFs with narrow roofing filters as low as 300Hz in some radios.

Software Defined Radio (SDR) is the latest major milestone in the evolution of the radio transceiver with all processing functions performed in software and with minimal hardware. First implementations used down-conversion and the PC soundcard as the radio 'engine'.

More recently direct digital sampling using A to D conversion at signal frequency is used removing the need for a high quality sound card. The latest designs use very high speed digital processing to provide multiple slice receivers and spectrum displays with outstanding levels of resolution and dynamic range. Feature implementation is limited only by imagination, and with open source code and enthusiastic developers this continues to be a fast moving area.

| ALL RADIOS | HIGHER END RADIOS | TOP CLASS RADIOS |
|---|---|---|
| Multimode | Selectable IF filters | Excellent RF performance |
| Twin VFOs | Variable bandwidth | Excellent audio performance |
| Memories | Notch filters | Top class channel filtering and filtering armoury |
| Clarifier | AF Filters | Ability to operate in multi station environment |
| Switchable front end | Dual watch capability | Uncompromised dual receiver capability |
| Variable Power | CW keyer | Interfacing to QSK linears |
| Interfacing | Auto ATU | Extensive interfacing to accessories |
| Switchable AGC | 50MHz and higher coverage | Ergonomic use in contest and expedition environments |
| Many software features | More comprehensive displays | Upgradeable software |
| Noise blankers | More user customisation | Use with receive only antennas |
| Computer interface | DSP filters and noise reduction<br>Improved data capability<br>More features<br>Spectrum display | |

**Table 7.27: Features provided on HF transceivers**

Turning our attention now to choosing a radio, **Table 7.26** summarises the different types of radio available for HF use. There are several main factors to consider when choosing a transceiver, and these are described below.

## Intended Use

Will the radio be used at home, portable, in the car or on a DXpedition, or a multiple of these uses? This determines the overall size and weight. Is the main use for general operating or for competitive activities such as contesting and DXing? A higher performance radio is desirable for contesting and DXing, whereas a lower performance and cheaper radio is entirely satisfactory for more casual operating.

Is the main use on a specific mode such as SSB, CW or Data modes? Some radios have more features suited to different modes, such as full break-in and built-in keyers for CW and extensive PC/audio interfacing and dedicated data mode selection and filters for RTTY, PSK and other specialist modes.

## Features and Functions

Most modern radios are very well equipped with all the features you are likely to need for most general purpose operation. See **Table 7.27**. Filtering functions and noise reduction capabilities are better implemented on the higher and top end models. DXing frequently uses split frequency operation and the ability to tune and receive on both channels is important. Dual receiver models are ideal for this purpose. The larger base station radios provide more extensive interfacing capabilities to multiple and receive-

| RECEIVE PERFORMANCE | |
|---|---|
| ON-CHANNEL | Sensitivity<br>Distortion / AF quality<br>Bandwidth<br>AGC |
| OFF-CHANNEL | Stopband selectivity<br>Spurious responses<br>Non-linearity / intermod<br>Phase noise |

**Table 7.28: Key receive performance parameters**

| TRANSMIT PERFORMANCE | |
|---|---|
| ON-CHANNEL | Power output<br>Distortion / AF quality<br>Bandwidth<br>Keying characteristics |
| OFF-CHANNEL | Spurious outputs<br>Sideband splatter/clicks<br>Non-linearity<br>Phase noise<br>Wideband noise |

**Table 7.29: Key transmit performance parameters**

only antennas, linear amplifiers, PC and audio lines etc. Radios at home are more likely to be used in combination with a linear amplifier, PC control and sound card for data, CW and voice keyers, transverters etc. If using a small radio at home make sure that it has the interfaces that are needed.

The new breed of software defined radios (SDR) offers limitless possibilities. SDR inherently provides high resolution spectrum displays and panadaptors with 'point and click tuning' and in conjunction with further processing software, such as CW Skimmer, intelligent analysis and multiple decoding of CW and data signals across a wide band is possible. However, using the PC as the user interface with keyboard control and mouse tuning is not to everyone's liking. We are now starting to see standalone radios with conventional controls and displays but employing the latest SDR techniques for all internal functions. Expect rapid progress to continue in this area.

## Ease of Use

This is most important if you are to obtain maximum enjoyment from the use of your radio. Large well-spaced control knobs and buttons, dedicated controls rather than menu or context switched controls, and clear displays all make for easy operation.

Compromises are, however, inevitable on the smaller sized radios. Ease of use is particularly important in minimising fatigue in extended operating periods such as in contests. Try to check out the radio on air before buying.

## Performance

The most important performance parameters for both the receiver and the transmitter are shown in **Tables 7.28 and 7.29**. A high performance receiver is most beneficial in contesting and DXing where wanted signals can be weak, bands crowded and unwanted signals strong. A 100dB dynamic range in SSB bandwidths represents a target for the very best receivers, 95dB is very good, 85-95dB is typical for mid price radios and 80-85dB for budget priced radios.

Third order intermodulation and reciprocal mixing are the principal performance measuring parameters of significance for dynamic range. Achieving a respectable dynamic range at frequency offsets greater than 20kHz from the receive frequency is achieved by most commercial amateur receivers these days, but closer in, particularly inside the roofing filter bandwidth, the performance of most receivers degrades sharply. There are a relatively few number of radios which achieve 90dB dynamic range at 5kHz offset, and cost is not necessarily an indicator of best performance at these close spacings.

**Table 7.30** summarises the performance of receivers measured for *RadCom* equipment reviews over the last 20 years in respect of their dynamic range range in 2.4kHz bandwidth due to third order intermodulation at 5kHz and 50kHz offset (ILDR - intermodulation limited dynamic range). Reciprocal mixing figures are quoted in terms of phase noise in 1Hz bandwidth (dBC/Hz) at 10kHz offset. Subtract 34dB to give the phase noise limited dynamic range in 2.4kHz bandwidth or 27dB to give this figure in 500Hz bandwidth. Table 7.30 is in order of close-in dynamic range with the highest performance radios appearing at the top.

## Cost

Cost is generally related to performance and features. Second hand purchases can be a good buy as radios do not normally wear out and significant savings can be made. Some useful guidance on buying second hand is contained in the RSGB Publication *The Rig Guide*, edited by Steve White. G3ZVW.

## Frequency Coverage

All HF transceivers cover the bands 1.8 to 30MHz but many also provide coverage of the VHF and UHF bands. Virtually all recent introductions now include 50MHz and a lesser number cover 144, 432 and even 1296MHz. If your interests cover VHF and UHF as well as the HF bands then a multiband radio may be of particular interest. Some of these radios allow simultaneous reception and transmission (full duplex) between main band groupings eg between HF, 144 and 432MHz and those with frequency tracking are particularly suitable for satellite working.

## RF Power Output

Most HF transceivers provide a transmit output power of 100W unless they are intended for use on internal batteries in which case the power is significantly less.

A few provide a higher power of 200W and even 400W and this extra power can be an advantage. However where a linear amplifier is also used the extra power is unnecessary, and indeed unwanted where there is a danger of overdrive of the linear amplifier. Some of the higher power radios allow the output amplifier to run in low distortion class-A mode and help minimise the bandwidth occupied on the bands, always a desirable feature.

## Available Radios

The "big three" Yaesu, Icom and Kenwood have over the years produced the greatest range of available models of HF transceivers. However the US suppliers Ten-Tec and more recently Elecraft and FlexRadio have also produced popular models, in many cases specialising in good RF performance. **Table 7.31** lists the key features of many of the most popular models produced over the last 25 years.

## REFERENCES


[1] Datasheets available from www.datasheetarchive.com/

[2} www.rfparts.com/

[3] Spectran software. www.weaksignals.com/

[4] 'QRP + QSK - a novel transceiver with full break-in', Peter Asquith, G4ENA, *Radio Communication*, May 1992

[5] Kit and component supplier JAB Electronic Components,

PO Box 5774, Great Barr, Birmingham B44 8PJ. Tel: 0121 682 7045. Web www.jabdog.com/. E-mail jabdog@blueyonder.co.uk
[6] 'A QRP transceiver for 1.8MHz', S E Hunt, G3TXQ, *Radio Communication*, Sep 1987
[7] 'Wideband linear amplifier', J A Koehler, VE5FP, *Ham Radio*, Jan 1976
[8] 'The FOXX 2 - an old favourite revisited', George Dobbs, G3RJV and George Burt, GM3OXX, *SPRAT,* Summer 1997
[9] 'The EP-2 portable 75m SSB transceiver', Derry Spittle, VE7QK, *SPRAT,* Winter 1995/96
[10] 'Epiphytes for the Third World', George Dobbs, G3RJV, *Radio Communication*, Jul 1997
[11] 'The GQ-40 (GQ-20) CW transceiver', Sheldon Hands, GW8ELR, *SPRAT*, Summer 1985
[12] *Radio Communication,* November 1989 (also described at www.g3pho.free-online.co.uk/microwaves/ssa.html)
[13] Communications Concepts Inc, 508 Millstone Drive, Xenia, Ohio 45385, USA. Tel (513) 426 8600. Website www.communication-concepts.com
[14] EB104 at http://oh8jep.kotinet.com/2eb104.html
[15] J Birkett, Radio Components, 25 The Strait, Lincoln, LN2 1JD. Telephone (UK) 01522 520767. Website: www.zyra.org.uk/birkett.htm
[16] 'A modular multiband transceiver', Mike Grierson, *Radio Communication*, Oct/Nov 1988
[17] *Twenty Five years of Hart Reviews,* Peter Hart, G3SJX, 2007, RSGB
[18] *Hart Reviews,* Peter Hart, G3SJX, 2015, RSGB

| TRANSCEIVER | | Phase Noise @10kHz | ILDR @5kHz | ILDR @50kHz |
|---|---|---|---|---|
| Kenwood | TS-990 | -144dBC/Hz | 109dB | 109dB |
| Yaesu | FTDX3000 | -131dBC/Hz | 107dB | 107dB |
| Yaesu | FTDX5000 | -142dBC/Hz | 105dB | 105dB |
| Icom | IC-7851 | -151dBC/Hz | 104dB | 114dB |
| Apache | ANAN-100D | -148dBC/Hz | 104dB | 104dB |
| FlexRadio | FLEX-6000 | -146dBC/Hz | 104dB | 104dB |
| Hilberling | PT-8000A | -153dBC/Hz | 101dB | 106dB |
| Elecraft | K3S | -147dBC/Hz | 101dB | 101dB |
| Elecraft | K3 | -134dBC/Hz | 101dB | 101dB |
| Elecraft | KX3 | -159dBC/Hz | 98dB | 98dB |
| Ten-Tec | Eagle | -131dBC/Hz | 98dB | 98dB |
| Yaesu | FTDX1200 | -130dBC/Hz | 97dB | 97dB |
| FlexRadio | FLEX-5000A | -124dBC/Hz | 97dB | 97dB |
| Kenwood | TS-590SG | -144dBC/Hz | 96dB | 104dB |
| FlexRadio | SDR-1000 | -135dBC/Hz | 94dB | 94dB |
| Yaesu | FTDX9000D | -137dBC/Hz | 93dB | 98dB |
| Ten-Tec | ORION 1&2 | -129dBC/Hz | 93dB | 93dB |
| Elecraft | K2/100 | -128dBC/Hz | 91dB | 95dB |
| Ten-Tec | CORSAIR | -132dBC/Hz | 90dB | 90dB |
| Yaesu | FT-2000D | -125dBC/Hz | 89dB | 98dB |
| Yaesu | FT-991 | -114dBC/Hz | 89dB | 100dB |
| Icom | IC-7800 | -134dBC/Hz | 88dB | 111dB |
| Ten-Tec | OMNI-VI | -130dBC/Hz | 88dB | 88dB |
| Icom | IC-7600 | -124dBC/Hz | 87dB | 105dB |
| Ten-Tec | OMNI-VII | -127dBC/Hz | 84dB | 89dB |
| Kenwood | TS-950 | -136dBC/Hz | 83dB | 102dB |
| Yaesu | FT-1000MP | -128dBC/Hz | 82dB | 97dB |
| Yaesu | FT-950 | -128dBC/Hz | 82dB | 97dB |
| Icom | IC-737 | -138dBC/Hz | 80dB | 102dB |
| Yaesu | FT-747 | -119dBC/Hz | 80dB | 97dB |
| JRC | JST-245 | -115dBC/Hz | 80dB | 92dB |
| Icom | IC-725 | -117dBC/Hz | 79dB | 95dB |
| Yaesu | FT-990 | -127dBC/Hz | 78dB | 97dB |
| Icom | IC-707 | -127dBC/Hz | 77dB | 91dB |
| Icom | IC-746 | -124dBC/Hz | 77dB | 98dB |
| Kenwood | TS-930 | -123dBC/Hz | 77dB | 95dB |
| Kenwood | TS-940 | -132dBC/Hz | 76dB | 94dB |
| Icom | IC-736/8 | -128dBC/Hz | 76dB | 100dB |
| Yaesu | FT-450D | -123dBC/Hz | 76dB | 94dB |
| Drake | R8E | -123dBC/Hz | 76dB | 92dB |
| Icom | IC-751A | -139dBC/Hz | 75dB | 104dB |
| Kenwood | TS-850 | -134dBC/Hz | 75dB | 98dB |
| Kenwood | TS-480 | -123dBC/Hz | 75dB | 98dB |
| Ten-Tec | JUPITER | -112dBC/Hz | 75dB | 89dB |
| Alinco | DX-70TH | -112dBC/Hz | 75dB | 92dB |
| Icom | IC-756PROIII | -127dBC/Hz | 74dB | 105dB |
| Icom | IC-756 | -125dBC/Hz | 74dB | 89dB |
| Icom | IC-7000 | -125dBC/Hz | 74dB | 88dB |
| Yaesu | FT-1000MP mk5 | -129dBC/Hz | 73dB | 95dB |
| Icom | IC-7400 | -127dBC/Hz | 73dB | 97dB |
| Icom | IC-756PROII | -126dBC/Hz | 73dB | 98dB |
| Icom | IC-756PRO | -126dBC/Hz | 73dB | 92dB |
| Icom | IC-703 | -118dBC/Hz | 73dB | 91dB |
| Icom | IC-775DSP | -134dBC/Hz | 72dB | 93dB |
| Yaesu | FT-920 | -131dBC/Hz | 72dB | 96dB |
| Kenwood | TS-50 | -129dBC/Hz | 69dB | 96dB |
| Yaesu | FT-900 | -125dBC/Hz | 69dB | 97dB |
| Yaesu | FT-890 | -125dBC/Hz | 69dB | 95dB |
| Kenwood | TS-2000 | -124dBC/Hz | 68dB | 95dB |
| Icom | IC-7100 | -123dBC/Hz | 68dB | 95dB |
| Yaesu | FT-100 | -121dBC/Hz | 68dB | 90dB |
| Yaesu | FT-817 | -119dBC/Hz | 68dB | 90dB |
| Yaesu | FT-847 | -124dBC/Hz | 67dB | 94dB |
| Yaesu | FT-1000 | -125dBC/Hz | 65dB | 96dB |
| Yaesu | FT-857 | -118dBC/Hz | 65dB | 93dB |
| Yaesu | FT-897 | -116dBC/Hz | 65dB | 92dB |
| Kenwood | TS-870 | -131dBC/Hz | 61dB | 94dB |

**Table 7.30: Dynamic range receive performance of HF transceivers**

| Maker | Model | Type | Bands | Introduced | Supply | HF Power | *RadCom* Review | Other key features |
|---|---|---|---|---|---|---|---|---|
| Alinco | DX-70 | mobile | HF 50 | 1995 | 13V | 100W | August 1995 | 10W on 6m |
| Alinco | DX-70TH | mobile | HF 50 | 1999 | 13V | 100W | August 1999 | 100W on 6m |
| Apache | ANAN-100D | SDR | HF 50 | 2013 | 13V | 100W | October 2014 | direct digital, multi RX, open HPSDR |
| Elecraft | K2 | base small | HF | 1999 | 13V | 15/100W | | Kit construction. Optional PA |
| Elecraft | K3 | base small | HF 50 | 2008 | 13V | 100W | July 2008 | many options inc twin receiver |
| Elecraft | K3S | base small | HF 50 | 2015 | 13V | 100W | April 2016 | upgraded performance and functions |
| Elecraft | KX3 | portable | HF 50 | 2012 | 8-15V | 10W | April 2013 | RF performance, QRP, many features |
| FlexRadio | FLEX-1500 | SDR | HF 50 | 2010 | 13V | 5W | April 2011 | USB interface to external PC |
| FlexRadio | FLEX-3000 | SDR | HF 50 | 2009 | 13V | 100W | August 2009 | External PC needs fireware interface |
| FlexRadio | FLEX-5000A | SDR | HF 50 | 2007 | 13V | 100W | January 2008 | External PC needs fireware interface |
| FlexRadio | FLEX-6000 | SDR | HF 50 | 2013 | 13V | 100W | October 2014 | 3 models, direct digital, multi RX |
| FlexRadio | SDR-1000 | SDR | HF 50 | 2005 | 13V | 1/100W | June 2006 | External PC needs soundcard |
| Hilberling | PT-8000A | base large | HF 50 70 144 | 2013 | mains | 200W | November 2013 | Professional quality |
| Icom | IC-7000 | mobile | HF 50 144 430 | 2005 | 13V | 100W | April 2006 | spectrum scan. 2m 50W, 70cm 35W |
| Icom | IC-703 | mobile | HF 50 | 2003 | 9-15V | 5/10W | October 2003 | transportable, detachable front panel |
| Icom | IC-706 | mobile | HF 50 144 430 | 1995 | 13V | 100W | June 1997 | 70cm on G version 20W, 2m 50W |
| Icom | IC-707 | base small | HF | 1993 | 13V | 100W | April 1994 | front speaker |
| Icom | IC-7100 | multi-use | HF 50 70 144 432 | 2013 | 13V | 100W | February 2014 | touch screen, DSTAR, ext controller |
| Icom | IC-7200 | base small | HF 50 | 2008 | 13V | 100W | January 2009 | water resistant. USB interface |
| Icom | IC-736 | base mid | HF 50 | 1994 | mains | 100W | May 1995 | 100W on 6m |
| Icom | IC-737 | base mid | HF | 1993 | 13V | 100W | September 1993 | auto ATU, keyer |
| Icom | IC-738 | base mid | HF | 1994 | 13V | 100W | May 1995 | As IC-736 without mains PSU or 6m |
| Icom | IC-7300 | base mid | HF 50 70 | 2016 | 13V | 100W | August 2016 | As HF, 50 & 70MHz full SDR technology |
| Icom | IC-7400 | base mid | HF 50 144 | 2002 | 13V | 100W | October 2002 | 100W 6m,2m. spectrum scan, IF DSP |
| Icom | IC-7410 | base mid | HF 50 | 2011 | 13V | 100W | January 2012 | feature packed. USB interface |
| Icom | IC-746 | base mid | HF 50 144 | 1997 | 13V | 100W | March 1998 | 100W 6m,2m. spectrum scan, AF DSP |
| Icom | IC-756 | base large | HF 50 | 1997 | 13V | 100W | May 1997 | AF DSP, dual watch, Spectrum display |
| Icom | IC-756PRO | base large | HF 50 | 1999 | 13V | 100W | March 2000 | IF DSP, dual watch, feature packed |
| Icom | IC-756PROII | base large | HF 50 | 2002 | 13V | 100W | June 2002 | IF DSP, dual watch, feature packed |
| Icom | IC-756PROIII | base large | HF 50 | 2004 | 13V | 100W | February 2005 | IF DSP, dual watch, feature packed |
| Icom | IC-7600 | base large | HF 50 | 2009 | 13V | 100W | June 2009 | successor to IC-756PRO series |
| Icom | IC-7700 | base large | HF 50 | 2008 | mains | 200W | June 2008 | IF DSP, single RX, feature packed |
| Icom | IC-775DSP | base large | HF | 1995 | mains | 200W | January 1996 | AF DSP, daul watch, auto-ATU, keyer |
| Icom | IC-7800 | base large | HF 50 | 2004 | mains | 200W | August 2004 | IF DSP, dual RX, feature packed |
| Icom | IC-7851 | base large | HF 50 | 2015 | mains | 200W | November 2015 | IC-7800 upgrade |
| Icom | IC-9100 | base mid | HF 50 144 432 1.3G | 2011 | 13V | 100W | May 2012 | Duplex satellite, DSTAR, 23cm options |
| JRC | JST-245 | base large | HF 50 | 1994 | mains | 150W | October 1997 | 150W on 6m |
| Kenwood | TS-2000 | base mid | HF 50 144 432 1.3G | 2000 | 13V | 100W | April 2001 | Data TNC, duplex satellite, 23cm option |
| Kenwood | TS-480 | mobile | HF 50 | 2003 | 13V | 100/200W | March 2004 | Remote panel. 100W / 200W versions |
| Kenwood | TS-50S | mobile | HF | 1993 | 13V | 100W | May 1993 | small size |
| Kenwood | TS-570D | base mid | HF | 1996 | 13V | 100W | December 1996 | auto ATU, keyer, AF DSP |
| Kenwood | TS-590S | base mid | HF 50 | 2010 | 13V | 100W | January 2011 | twin architecture |
| Kenwood | TS-590SG | base mid | HF 50 | 2014 | 13V | 100W | March 2015 | TS-590S upgrade |
| Kenwood | TS-690S | base mid | HF 50 | 1992 | 13V | 100W | November 1992 | 50W on 6m |
| Kenwood | TS-850S | base large | HF | 1991 | 13V | 100W | October 1991 | options for ATU, message stores, DSP |
| Kenwood | TS-870S | base large | HF | 1995 | 13V | 100W | April 1996 | IF DSP, ATU |
| Kenwood | TS-990 | base large | HF 50 | 2013 | mains | 200W | June 2013 | Top performance and features |
| Ten-Tec | EAGLE | base small | HF 50 | 2010 | 13V | 100W | July 2011 | RF performance, simple controls |
| Ten-Tec | JUPITER | base mid | HF | 2000 | 13V | 100W | January 2004 | IF DSP. Spectrum scope. |
| Ten-Tec | OMNI-VI | base large | HF | 1992 | 13V | 100W | January 1994 | Ham bands only |
| Ten-Tec | OMNI-VII | base mid | HF 50 | 2007 | 13V | 100W | September 2007 | Remote LAN operation. Spectrum scan |
| Ten-Tec | ORION 1 | base large | HF | 2003 | 13V | 100W | June 2004 | Dual RX, IF DSP, narrow roof, top end |
| Ten-Tec | ORION 2 | base large | HF | 2006 | 13V | 100W | August 2006 | Dual RX, IF DSP, narrow roof, top end |
| Yaesu | FT-100 | mobile | HF 50 144 430 | 1999 | 13V | 100W | June 1999 | spectrum scan. 2m 50W, 70cm 20W |
| Yaesu | FT-1000MP | base large | HF | 1995 | mains/13V | 100W | January 1996 | Dual RX, AF DSP, auto-ATU, keyer |
| Yaesu | FT-1000MP mk5 | base large | HF | 2000 | mains | 200W | October 2000 | Separate PSU, Class AB/A PA, Dual RX |
| Yaesu | FT-2000 | base large | HF 50 | 2006 | mains/13V | 100W | | Dual RX, IF DSP, DMU display features |
| Yaesu | FT-2000D | base large | HF 50 | 2007 | mains | 200W | March 2008 | Separate PSU, Class AB/A PA, Dual RX |
| Yaesu | FT-450 | base small | HF 50 | 2007 | 13V | 100W | October 2007 | IF DSP, optional ATU |
| Yaesu | FT-450D | base small | HF 50 | 2011 | 13V | 100W | November 2011 | style upgrade. ATU included |
| Yaesu | FT-817 | portable | HF 50 144 430 | 2001 | 9-13V | 5W | June 2001 | int batteries/ext supply |
| Yaesu | FT-840 | base small | HF | 1993 | 13V | 100W | February 1994 | Optional FM |
| Yaesu | FT-847 | base mid | HF 50 144 430 | 1998 | 13V | 100W | August 1998 | 50W on 2m/70cm, duplex satellite, 4m |
| Yaesu | FT-857 | mobile | HF 50 144 430 | 2003 | 13V | 100W | June 2003 | detachable panel. 50W/2m 20W/70cm |
| Yaesu | FT-890 | base small | HF | 1992 | 13V | 100W | September 1992 | optional ATU |
| Yaesu | FT-897 | base small | HF 50 144 430 | 2003 | mains/13V | 100W | April 2003 | 20W with int batteries. |
| Yaesu | FT-900 | base small | HF | 1994 | 13V | 100W | November 1994 | optional ATU |
| Yaesu | FT-920 | base large | HF 50 | 1997 | 13V | 100W | August 1997 | ATU, keyer, voice store, AF DSP |
| Yaesu | FT-950 | base mid | HF 50 | 2007 | 13V | 100W | December 2007 | IF DSP, ATU, DMU display features |
| Yaesu | FT-991 | base small | HF 50 144 430 | 2015 | 13V | 100W | February 2016 | touch screen, ATU, spectrum scan |
| Yaesu | FTDX1200 | base mid | HF 50 | 2013 | 13V | 100W | March 2014 | crisp display, easy menu access |
| Yaesu | FTDX3000 | base mid | HF 50 | 2013 | 13V | 100W | January 2014 | down conversion RX, crisp display |
| Yaesu | FTDX5000 | base large | HF 50 | 2010 | mains | 200W | June 2010 | 3 variants, top end, feature packed |
| Yaesu | FTDX9000D | base large | HF 50 | 2005 | mains | 100W | December 2006 | Dual RX, microtune filters, top end |

**Table 7.31: Some of the most popular models of HF transceiver over the last 25 years**

# Software Defined Radio (SDR)

8

*Andrew Barron, ZL3DW*

Software Defined Radio (SDR) has become one of the most talked about and perhaps most misunderstood technologies to be adopted by amateur radio enthusiasts since the introduction of single sideband back in the 1940s. In many respects, SDR is just a natural progression from the I.F. DSP (intermediate frequency digital signal processing) stages that have been incorporated into amateur radio HF transceivers since the late 1970s. But SDR also encompasses a wide range of small receivers and transmitters that can cover a huge section of the radio spectrum, often capable of receiving signals above 1.5GHz and sometimes as high as 6GHz. This has opened access to the VHF, UHF and microwave bands at a relatively low cost. Where previously you were committed to buying or building narrowband receiver converters or transverters, you can now receive any frequency in the range. However, you should still use an LNA (low noise preamplifier), or a LNB (low noise block converter), mounted at the antenna end of the feeder cable, for microwave frequencies, because it is important to keep the overall receive noise figure as low as possible.

Software defined radio is a merging of radio technology and computer technology. Some of the radio is still made of hardware circuits but the operating controls, display, and digital signal processing are managed by a software application, running on a personal computer or computer chips within the radio. The panadapter spectrum and waterfall display is the most readily identifiable feature of software defined radio receivers, but not all SDRs have one, and not all radios equipped with a band scope display are SDRs.

Software defined radios have a fundamentally different architecture to the traditional superheterodyne radios that we have been using. They offer performance gains, extended frequency coverage, and sometimes cost savings, compared to older technologies.

The aim of software defined radio developers is to create radios that convert a section of the received RF spectrum into a digital signal as early as possible in the receiver chain and in SDR transmitters to keep the signal in digital format for as long as possible. Digital signals are manipulated by computer software to perform the functions that are performed by hardware circuitry in a conventional radio without adding any noise or distortion. Software algorithms change the numbers in the digital data streams, to apply filters, shift frequencies, and modulate or demodulate the signals. This ability to use computer power rather than complex and expensive crystal filters, mixers, oscillators, I.F. amplifiers, modulators and demodulators makes software defined radios cheaper to manufacture and easier to tune up than older technologies. It also makes them easy to upgrade and update. For example, a manufacturer may decide to incorporate a new digital mode or perhaps a PSK decoder. It is just a matter of downloading some new software. In the case of SDR hardware that uses a PC program for DSP functions, control of the radio, and the display of the controls and panadapter, you may have a choice of two or more software applications. With different software, your SDR may look and operate completely differently.

The following are some of the features and benefits that you can expect from a software defined radio. Not all functions are available with all software defined radios, or SDR software. But these eight items will give you a taste of what you can gain by buying an SDR.

1. Generally, SDRs have very good receiver performance. This is mostly due to the fundamental difference in technology between an SDR and a superheterodyne receiver. Digital down conversion HF receivers have no mixers (in hardware) so they don't suffer from the intermodulation distortion effects and 'birdies' that mixers and oscillators contribute to standard superheterodyne receivers. Of course, the manufacturers of conventional radios do go to great lengths to try to eliminate these problems by using excellent designs, quality components and by picking I.F. frequencies that place image signals outside the ham bands. But great performance comes at the cost of increased complexity and a higher price.
2. The panadapter screen of a software defined radio displays the signals across an amateur radio, commercial, or shortwave band. A simple mouse click tunes the radio to any signal that you can see. The panadapter will also show other interesting things such as interfering signals or the effect of poor-quality transmissions. CW key clicks, overmodulated digital mode signals, and linear amp splatter are easy to spot.
3. Tuning around a band with a conventional receiver you might easily miss a station making occasional CQ calls. With an SDR, you will see the signal pop up on the waterfall or spectrum display. You can click on the waterfall or spectrum display to move directly to the next station without having to tune up and down the band. This is fantastic during contests because you can select each station in turn as you work up or down the band.
4. With some software defined radios, you can listen to two receivers at the same time, to hear a DX station and simultaneously the station they are working in the pileup. The panadapter will show both the DX station and the pileup of calling stations. This lets you pick a relatively quiet spot to transmit or work out which way the DX station is progressing through the calling stations. You can easily see when the DX station is transmitting and time your call appropriately.
5. Being able to monitor several bands at the same time is quite useful. You can quickly see which bands are open. If you are changing bands, to work a station on another band you can see their signal come up on the new frequency, without moving off the old frequency. No more changing bands only to find the frequency is in use or you can't hear the other station. With a software defined radio, you can watch and hear your old frequency and your new frequency at the same time. You can watch a frequency for a Net to begin, or for a DX station while working or listening to another station on a different band.
6. If you are using SDR software running on your PC, you won't need cables to connect the PC sound card to your transceiver. Audio cross connections are done with special PC utilities. You will not need a CAT cable for computer control either since the radio software is already running on the PC. You may have to load some shareware or freeware utility programs to perform these tasks.
7. You won't need external audio equalizers either since the SDR software usually includes microphone and receive audio, equalizers.

8. The CW Skimmer program is very popular with CW DX stations and contesters. It can decode as many as 700 Morse Code signals at the same time. A software defined radio can send the entire CW band segment, or even multiple bands, to CW Skimmer, so that all of the CW signals can be identified. With conventional receivers, only signals within the relatively narrow 3kHz receiver bandwidth can be seen by the CW Skimmer application.

Many articles in amateur radio magazines now feature screenshots of SDR receiver displays. The people who are interested in the development of new digital modes, EME, propagation, shortwave signals, weather satellites, ham satellites, contesting, interference detection, and monitoring intruders on our bands are finding that software defined radios are very useful for their particular interest. An SDR receiver can be used as a spectrum analyser to observe different types of modulation or interference signals. You can even see the frequency response of the audio modulation on a single sideband signal and you can use the panadapter to monitor your own signal while you are transmitting to ensure that your transmission is clean and legal.

One of the joys of software defined radio is that it is evolving rapidly. New features are being developed continuously. This is great for those who like to tinker with new technology and be at the bleeding edge in learning what is possible with these new radios. Others may be more interested in the competitive edge that using an SDR can provide.

## TYPES OF SDR

There are many different types of SDR. USB dongles, small box direct sampling receivers, QSD/QSE (quadrature sampling detector or exciter) based receivers and transceivers, SDR kits, VHF UHF and SHF boards that use a traditional front-end oscillator and mixer to get an I.F. frequency low enough for direct sampling, HF transceivers using direct digital synthesis, SDRs with knobs and onboard DSP, superheterodyne radios with audio frequency IQ outputs from their DSP stages, and commercial or military radios that use digital sampling and DSP chips.

Some software defined radios have the usual knobs and buttons on the front panel. They don't need to be connected to a computer. Software defined radios that do not have front panel controls do require connection to a computer that is running suitable SDR software. The underlying hardware technology is the same in both types and it is radically different to conventional radios. The SDR just has a few integrated circuits. There are none of those little adjustable tuning coils and variable capacitors that have to be carefully aligned at the factory.

Some software defined radios have dedicated software specifically written to match the hardware. Others can be used with a variety of different software applications. This is one of the advantages of SDR technology. You can completely change how your radio looks and operates simply by using a different software package.

Different software may offer different modes. For example, one program may include DRM digital broadcast mode, and another might include FM and SSB, or a PSK decoder. Using an SDR that has front panel controls is much the same as using a receiver or transceiver based on the conventional superheterodyne architecture. Using an SDR connected to a PC is no more difficult than using any other computer program.

### Dongle SDR Receivers

Dongle SDR receivers (**Fig 8.1**) have a similar physical format to USB memory 'flash drives.' They are designed to be plugged directly into the USB 'Type A' port on a computer or USB extender hub. You will need to load a free SDR program such as SDR#

**Fig 8.1: A typical dongle type SDR receiver**

(SDR sharp), HDSDR, or SDR Console, and you may have to load a USB device driver called Zadig and/or a data format interface utility called EXTIO.dll. Most SDR dongle receivers use the R820T/T2 as the tuner chip, and the Realtek RTL2832U chip as the analogue to digital converter, digital down converter, and USB port interface. They typically cover frequencies from 24 to 1766MHz or up to 2200MHz in some models. Some are bundled with cheap up-converters to provide coverage of the HF bands. The maximum panadapter bandwidth is typically about 2.5MHz.

RTL dongles have a filter in front of the preamplifier, but they do not have separate bandpass filters for individual bands, so they are prone to overload and ADC gain compression. The 8 bit ADC means that they don't have a particularly large dynamic range, to begin with. However, they are very sensitive, and they cover a huge frequency range. Most importantly they are very cheap. They provide a great introduction to software defined radio and they are extremely popular.

The FUNcube Pro+ dongle receiver uses a different chipset. It also has better filters, including SAW (surface acoustic wave) filters for the 2m and 70cm amateur radio bands, a more stable TXCO reference clock, and a 32 bit ADC. The radio has a frequency range of 150kHz to 260MHz and 410MHz to 2.05GHz. Covering every ham band from the 1800m band to the 23cm band. That is L.F to L Band. Seventeen amateur radio bands, from a receiver you can carry in your pocket! However, it only supports a maximum panadapter bandwidth of 192kHz, compared to about 2.5MHz using an RTL dongle.

### Small Box SDR Receivers

There's a big range of small box SDR receivers and a few

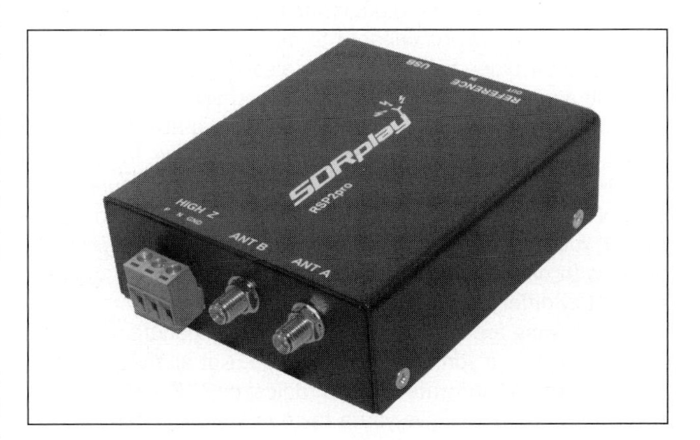


**Fig 8.2: A typical small "black box" SDR receiver**

small box transceivers on the market (**Fig 8.2**). Some use QSD technology but most use direct digital sampling. Some cover the LF, MF, and HF bands while others such as the SDRplay RSP2 shown in the picture include coverage of the VHF and UHF bands as well. The RSP2 covers 10kHz - 2GHz.

The HF models usually use direct sampling with DSP performed in an FPGA. The VHF/UHF models typically use a tuner chip such as the Mirics MSi001 RF tuner with the MSi2500 'digital interface' chip. The interface to the computer is usually a USB 2.0 or USB 3.0 connection. But some radios use an Ethernet connection or PCIe.

These radios usually have a 12 bit, 14 bit, or 16 bit ADC and they can support panadapter bandwidths of 8MHz or more. They are often bundled with their own free SDR software and many can also be used with SDR# (SDR sharp), HDSDR, SDR Console or other programs.

## SDR Transceivers

**Fig 8.3: Expert Electronics HF and VHF transceiver**

**Fig 8.4: The Apache Labs ANAN-100D 100W transceiver**

The SunSDR2 pro (**Fig 8.3**) from Expert Electronics is a 20 Watt HF, and 8 Watt VHF, SDR transceiver, covering a frequency range from 90kHz to 66MHz, or 95MHz to 148MHz. It is a direct sampling radio with a 16 bit ADC sampling at 160MHz. Interestingly, it uses the ADCs second Nyquist zone when it is operating in the VHF mode. Using the free bundled SDR software, the radio supports two 312kHz wide panadapters and a lower resolution wideband display to 80MHz. CW generation is performed in 'hardware' for low latency. The radio uses an Ethernet connection to the controlling PC.

The Apache Labs ANAN-8000DLE (**Fig 8.4**) is a 200 Watt HF +6m transceiver. It has two synchronous 16-bit ADCs. With the PowerSDR MRx or Thetis software, it can support up to seven 1.5MHz wide panadapters. The dual ADCs allow true diversity reception. When using the 'PureSignal' function the transmitter IMD performance is outstanding at better than -72 dB @ 200W on the 20m band. The radio uses an Ethernet connection to the controlling PC.

The Icom IC-7610 (**Fig 8.5**) is a 100 Watt HF +6m transceiver with two identical receivers and a large touchscreen. Each receiver can have its own panadapter and waterfall display. It is a direct sampling SDR which does not require connection to a PC, although you can connect it to a computer for CI-V control and digital mode operation.

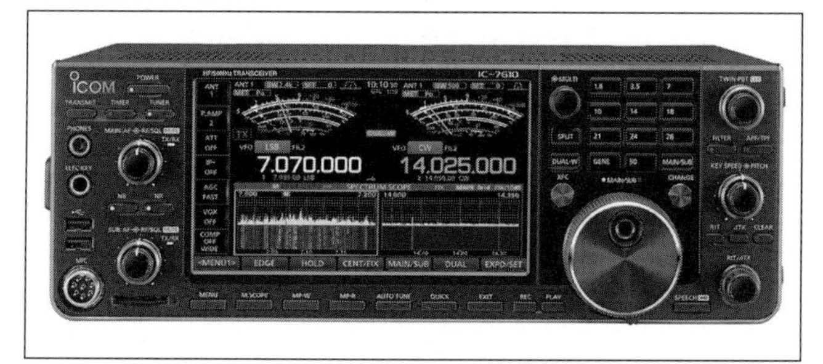


**Fig 8.5: The Icom IC-7610 transceiver**

## Technical Variety

Radios for the LF (low frequency), HF (high frequency), and in some cases VHF (very high frequency) bands are typically 'direct sampling,' meaning that the RF (radio frequencies) arriving at the receiver input are sampled and converted into a digital signal using an ADC (analogue to digital converter).

The signals are then processed using DSP (digital signal processing) software, running on a programmable device such as an FPGA (field programmable gate array) or possibly dedicated DSP chips. Receivers that cover higher UHF (ultra-high frequency) and microwave bands use a conventional oscillator and mixer arrangement to shift a band of radio frequencies down to a frequency that the SDR part of the receiver can sample. The SDR section is often a dedicated receiver or tuner chip, but it can be an ADC-based radio as described above.

The big advantages of SDR are the panadapter display, fully adjustable receiver bandwidth, great noise and interference suppression, and excellent receiver performance. The panadapter display lets you see signals across the band of interest. This means that you can jump directly to stations rather than carefully tuning across the band straining to hear an elusive weak signal.

A panadapter is different to the band scope found on some conventional superheterodyne receivers. A band scope displays signals above and below the frequency that the receiver is tuned to. A panadapter typically displays a much wider range of frequencies and you can place one or more receivers at any point on the display.

Most panadapters also include a waterfall display. A waterfall display indicates signals over a period of time. It can show you signals that have a short duration such as a CQ call, air traffic voice channel traffic, land-mobile radio, or burst transmissions. You can see the traces of signals that could easily be missed if you were tuning across a band using a conventional receiver.

SDR receivers feature a completely adjustable receiver bandwidth. This means that users can adjust the receiver to exactly match the bandwidth of the signal they are listening to. On AM and SSB, this means operators are not listening to noise that is outside of the wanted signal range. You can set the receiver to a wide bandwidth for broadcast AM or FM signals and to a narrow bandwidth for narrowband signals. Adjustable bandwidth is a real advantage if you want to receive weather images from the NOAA weather satellites on 137MHz. A conventional FM receiver is too narrow. You can decode the received images much better using an SDR receiver.

## Signal Processing

Most SDR software or firmware has very good noise blankers and filters because these are created using DSP software code. The filter parameters are often user-adjustable, and some extremely sharp filter responses can be achieved without the 'ringing' artefacts typical of sharp hardware filters. It is true that very aggressive DSP filters can make the audio sound weird and fluttery, but this may be preferable than listening to a very noisy signal. In some SDR software, the filter taps can be adjusted for the best compromise between a distorted fluttery signal and a noisy signal. Some software supports tracking notch filters which stay on a selected frequency. When you tune the receiver, the notch stays over the interfering signal and if you return to the frequency later, the notch filter will still be engaged.

SDR receivers tend to have excellent receiver performance. Because of the different way that these radios work, they do very well in the receiver performance tests published in magazines and on websites. Generally, there is no receiver blocking on strong signals until the incoming signal reaches the clipping point of the ADC. At that stage, the receiver performance will crash. But this only occurs with extremely large signals and it can easily be cured by switching in some front-end attenuation.

SDR receivers do not suffer from 'AGC pumping,' which is the effect where a strong signal within the receiver's passband causes the receiver AGC (automatic gain control) to operate, desensitizing the receiver and causing the weak signal that you want to receive to disappear. SDR receivers are unaffected and their intermodulation performance can, in fact, be improved, by the presence of nearby signals.

Many people comment that SDR receivers 'sound better' or 'quieter' that their old conventional radios. This is because SDR receivers don't have the multiple oscillators, mixers, and I.F. amplifiers that are inside a standard double or triple conversion receiver. Each amplifier and oscillator in a conventional receiver contributes noise to the output audio signal and the mixing process adds intermodulation distortion products and image signals. Those 'birdies' you hear as you tune across the band.

Direct conversion SDR receivers sample the signal close to the antenna, usually following a preamplifier, step attenuator, and bandpass filter. They don't have hardware mixers, oscillators, and I.F. amplifiers, so you don't get the noise and intermodulation that they produce. The result is much cleaner, less hissy audio. This fundamental difference in architecture creates one of the major benefits of software defined radios over conventional radios and it is the reason that a low cost SDR can sound better than an expensive double or triple conversion receiver.

## SDR Downsides

While SDR offers a lot of performance and cost-saving benefits there are some things that they are not as well suited to. Firstly, SDR receivers do not make good scanning receivers. It is certainly possible to program SDR software so that the receiver will operate as a scanning receiver, but I am not aware of any software that has that functionality. The reason for this is that the SDR panadapter display lets you see channels as they are activated. You can click on the panadapter to hear the activity on that channel. To do that successfully, on the high VHF and the UHF bands you need a radio that supports a wide panadapter display. At least 10MHz.

The problem here is that you end up looking at a lot of bandwidth that does not contain the channels you want to hear. You can, of course, program memories and scan through them. But it is not the same as using a scanning receiver. Fundamentally the objectives are different. A scanning receiver sweeps a band, stopping on active channels. An SDR panadapter lets you see which channels are active, but you have to click the panadapter or waterfall display in order to hear the signal.

Secondly, SDR receivers and transceivers are not well suited to mobile operation. They rely heavily on the ability to display signals on a panadapter or waterfall display. You click on the signals that you want to hear. Of course, this is not practical while you are driving a vehicle. You need to keep your eyes on the road, not on the screen of your radio.

## SDR Receivers

HF software defined radio receivers are based on one of two commonly used architectures. Some SDR receivers are QSD designs which use a Tayloe detector as a direct conversion mixer to create I and Q audio streams that are passed on to two analogue to digital converters. The others are direct sampling SDR receivers that use digital down conversion (DDC). In these, the RF signal is sampled with a single very high-speed analogue to digital converter and then the digital signal is split into I and Q signals. This is often done in an FPGA, but it can be done in a dedicated receiver chip, or by an embedded microprocessor, or a DSP chip. Some designs take advantage of the computing power available from SBCs (single board computers) such as the Raspberry Pi or the Beagle Bone.

Receivers working at VHF and higher frequencies use a slightly different design. Typically, a hardware oscillator and mixer are used to mix the RF signal down to an I.F. channel and then a digital down conversion SDR performs the analogue to digital conversion and DSP functions. Most of these radios use a dedicated 'tuner' or receiver chip rather than stand-alone oscillator and mixer.

The tuner chips have internal hardware oscillators and mixers followed by an analogue to digital conversion. A variety of methods can be used to produce the IQ data output streams for the DSP stage. The RTL dongles use a Realtek 2832U tuner chip and the Afedri receivers use a Texas Instruments front end chip. The HackRF board uses a Xilinx CPLD and an ARM Cortex processor. The BladeRF board uses an ARM9 general purpose processor and an FPGA.

A CPLD is a Complex Programmable Logic Device. It is another field programmable logic device like an FPGA. CPLD chips tend to have fewer programmable gates than FPGA chips and their internal structure is different. The main practical difference apart from size is that an FPGA does not usually have any non-volatile memory. The gate configuration program is loaded from a separate EPROM or a connected computer each time the FPGA is powered up. A CPLD has internal non-volatile memory so it is ready to go as soon as power is applied. A big CPLD will have around the same number of programmable logic gates as a very small FPGA.

## SDR Transmitters

There are basically three classes of SDR transmitter. Almost all commercially available models are QRP (low power) radios. Radios like the FLEX-1500, FLEX-3000, SoftRock, SDR Cube and the Genesis G59 use a Tayloe detector in reverse as a quadrature sampling exciter (QSE). The digital (IQ) signals carry the modulated audio frequency signal and then the QSE mixes the audio up to the final RF frequency, which is amplified and passed through low pass harmonic filters to the antenna. Some transceivers use a QSE plus onboard digital to analogue conversion and DSP. Often these are the software defined radios with knobs like the Elecraft KX3, SDR Cube, mcHF, and RS-918 transceivers. Next, there are digital up-conversion (DUC) transmitters. The DSP stage creates a modulated digital IQ signal, which is up-converted to the final frequency as a digital signal which is then converted to an analogue signal using a fast digital to analogue converter. This is followed by an RF power amplifier and a low pass filter.

VHF transmitters might be straight digital up-conversion designs, or they may be hybrids. For UHF and SHF frequencies the transmitters are typically similar to the equivalent receivers. A direct up-conversion SDR creates an I.F. frequency, which is mixed up to the final RF frequency by a hardware mixer and oscillator. Almost all of the software defined radios that work at UHF and above are very low power 'sub-QRP' transceivers and boards.

Nearly all hand-held radios and most mobile radios are fully digital inside. But digital does not necessarily mean software defined. The term 'software defined' implies an ability to reconfigure the functions of the radio, not just change the frequencies stored in the memories, by changing the software within the radio.

There are some 100 Watt and 200 Watt, DDS (direct digital synthesis) radios using digital down conversion receivers and digital up-conversion transmitters. Some are 'no knob' designs that require a PC or other computer device to run the SDR display, control and DSP software. Radios in this class include the Apache Labs ANAN and some of the FlexRadio Signature radios. The Expert Electronics MB1 is similar but it has a display and controls plus a computer built into the box so it can be used as a stand-alone radio. The ANAN-7000DLE MKII has no front panel controls, but it also has a computer built into the radio. Finally, there is the new breed of SDR and SDR hybrid radios from FlexRadio, TenTec, Alinco, Icom, and Yaesu. These radios look just like other high-end amateur radio transceivers, but they use varying degrees of SDR technology. The FlexRadio 'M series' and the Icom IC-7300 and IC-76-10 transceivers are true direct sampling radios. The others use some SDR technology making them 'Hybrid SDRs.'

## Commercial (Non-Amateur Radio) SDRs

Software defined radio is well established in the commercial radio world. SDR technology is very common in cellular base stations and military radios. It is rapidly becoming common for mobile and hand-held radios as well. These types of radios are probably better described as 'digital' rather than software defined, because the internal firmware is not usually user upgradeable.

Many of the manufacturers of software defined radios for the amateur radio and hobby market also make radios for the commercial and military markets.

## Hybrid SDRs

There are quite a few radios that advertise 'SDR' as a feature. Some manufacturers are now using terms like 'superheterodyne + SDR,' or 'SDR technology.' This partly due to the fact that SDR features are fashionable and partly because it is quite difficult to define exactly what software defined radios are. These radios are often called hybrids because you do get some SDR functionality, even if it is only a band scope. It is not the knobs or the lack of them that makes the difference.

Some examples of + SDR radios are the Elecraft K3S, the Alinco DX SR9T/E, and the AOR AR-2300 / AR-5001D. These radios all output IQ streams at audio frequencies that can be used for a band scope display with a bandwidth of up to around 200kHz. But they all have standard superheterodyne receivers with I.F. DSP. One recently released transceiver uses an SDR receiver to drive the panadapter display, but a hybrid architecture for the main receivers and the transmitter.

Less easy to quantify are radios like the Comm Radio CR1a, the Elecraft KX3, and the SDR Cube, which use QSD technology. They are not traditional superheterodyne receivers. These radios have onboard DSP, their modulation, demodulation and filtering are performed inside the radio and they have limited or no ability to display a band-scope or panadapter. Most people do refer to these radios as software defined radios, but in some respects, they are hybrids. It seems that using QSD rather than superheterodyne architecture is enough to ensure their status as being software defined.

Not many people would be brave enough to suggest that the new FlexRadio Systems 6000 series radios are hybrids. They use direct digital synthesis and can support several very wide panadapters. In fact, they are at the pinnacle of SDR design. But they too have all of their DSP performed in hardware by DSP chips and embedded processors. Certainly, this is field upgradable firmware but then so is the firmware used in conventional receivers.

I prefer the term 'superheterodyne + SDR' to 'hybrid SDR' or 'SDR technology.' At least you know what you are dealing with. The fact is. Direct sampling SDR and superheterodyne 'conventional' technologies are going to continue to merge and fairly soon, "a radio will be a radio." It might have knobs, or it might not. It will probably use some direct sampling technology and if it is a base station radio it will probably offer a panadapter spectrum scope and waterfall display.

## Conventional Superheterodyne Radios with IF Outputs

Some non-SDR receivers feature an output connector specifically to allow connection to an external SDR receiver. These usually provide a tap off the receiver's I.F. (intermediate frequency) or directly from the antenna via a coupler. This is actually quite useful, as you can connect a 'black box' or dongle type SDR and get the advantages of a second receiver and a panadapter band display. Some software, such as HDSDR for example, can be set to an 'I.F. mode' so that the SDR receives the I.F. frequency from your radio but displays the incoming RF frequencies on the screen. CAT or CI-V control via 'Omni-Rig' software tunes your main receiver as you click on the SDR receiver's panadapter display, integrating the operation of the SDR and the conventional receiver. If your SDR is connected to your receiver's I.F. output, the bandwidth that you can display on the SDR panadapter will be restricted to the I.F. bandwidth of the receiver. If the SDR is connected to a port that shares the main receiver's antenna, you will probably be able to tune across the full bandwidth of the antenna. But this depends on where the output port is in the main receiver chain. It may follow preselecting filters, in which will restrict the possible bandwidth but also offer protection from strong 'out of band' signals.

## SDRs as a Test Instrument

Some SDRs are marketed specifically for use as spectrum analysers rather than as general-purpose receivers. A good example is the Signal Hound range of spectrum analysers and SDR receivers. Micropower SDR transceivers can be used as vector network analysers and as spectrum analysers. Examples include the Pocket VNA *http://www.pocketvna.com and the miniVNA from Mini Radio Solutions http://miniradiosolutions.com. These devices come bundled with excellent PC software and they perform very well. The OSA103 mini SDR and its bundled free software can be used as an oscilloscope, function generator, frequency meter, spectrum analyser, vector network analyser, antenna analyser, LC meter, or with come constraints as an RF signal generator, www.osa103.ru/en/main-page.*

## Web SDR

You can get a taste of what it is like to operate an SDR by logging on to one of the online web SDR receivers at www.websdr.org/ or https://sdr.hu/?q=kiwisdr. These allow you to control and listen to an SDR receiver that is located in another area or country. They allow you to listen to HF signals when the bands are closed at your place and you can see band activity on the panadapter. Several listeners can use the same receiver at the same time.

WebSDR is a worldwide network of more than 300 SDR receivers that are available for you to use for free as remote receivers. Each radio is made available to anyone at no cost to the user, even though there is a significant amount of internet bandwidth used at each receiver site. There is no 'sign up' or 'login' required, so I encourage you to try out some of these remote receivers. You can tune to any frequency in the HF spectrum and in some cases on VHF or UHF frequencies as well. One advantage is that by picking a receiver located on the far side of the world you can listen to bands when there is no propagation on that frequency to your location. If you are an amateur radio operator, you can transmit from home and monitor your signal as it is heard in another country. Note that this does not work particularly well for making contacts over amateur radio as there is a significant latency (delay) on the received signals. It is also not 'legal' for amateur radio contests or awards. Just looking at the signals on the waterfall at various locations can give you an indication of what bands are 'open' to various countries.

## The QSD/QSE Method

Designing and producing high-performance amateur radio receivers and transceivers is an exceptionally difficult task. Every amplifier stage in a receiver adds noise to the signal and this degrades the noise figure. Non-linear devices like mixers cause intermodulation distortion and local oscillators can add noise and harmonic related signals. The radio manufacturers have built on many years of experience and work very hard to minimise these effects. There is fierce competition to produce transceivers with excellent receiver performance. Receiver I.F. frequencies are chosen to reduce the image frequencies and place them outside the ham bands. New receivers have very stable local oscillators with low phase noise. Some new receivers have roofing filters, variable pre-selectors, or external microprocessor controlled Hi-Q front end filters to make the receiver selectivity track the wanted receive signal rather than just having a wide roofing filter.

A large part of the reason that SDR receivers have very good performance is that the SDR design eliminates the need for multiple mixers, local oscillators and I.F. amplifiers. Eliminating these receiver components removes the noise and distortion they cause. All of the 1st generation 'sound card' and 2nd generation (A/D conversion in the radio) SDR receivers use a QSD (quadrature sampling detector) which acts as a direct conversion mixer. In most cases, they use a circuit known as a Tayloe Detector which was patented by Dan Tayloe N7VE in 2001. The same design can be used in reverse as a quadrature sampling exciter (QSE) to make a signal which can be amplified to create a transmit signal.

In any discussion about software defined radio, it is not long before I and Q streams are mentioned. It is at this point people's eyes glaze over and they decide that SDR is too complicated for them. Don't panic, it is not really too difficult. But before we get into how the QSD design works I should explain about direct conversion receivers and 'the phasing method'. QSD based SDRs are direct conversion receivers that use the 'phasing' method of image rejection. This is not a new idea. Direct conversion 'phasing' receivers such as the Central Electronics CE 100v were very popular in the mid-1950s until improvements in filter design made superheterodyne receivers the preferred option.

Direct conversion receivers are similar to superheterodyne receivers except the I.F. output is directly at audio frequencies extending from 0 Hz up to the bandwidth of the receiver. The local oscillator is at the same frequency as the wanted receive frequency, so in the mixer, an RF signal extending to 3kHz above the LO (local oscillator) frequency becomes an audio signal between 0 to 3kHz. This is great except the mixer works equally well with input signals below the LO frequency. An RF signal 1kHz below the LO frequency cannot become -1kHz, so it is reflected into the audio range as a 1kHz signal with a phase change of 180 degrees. This reverses the audio signal. For example, if the signal that is received below the LO frequency is an upper sideband signal, it becomes a lower sideband signal at the output of the mixer. It will appear as a lower sideband signal on the panadapter display and it will sound distorted, exactly like receiving an upper sideband signal with your radio set to LSB by mistake. Worse than that, this "image" signal will interfere with the audio from the wanted RF signal above the LO frequency. You will see both signals superimposed on the panadapter and hear both signals at once!

Traditionally this image frequency problem is managed in one of two ways; either the signals below (or above) the LO frequency are filtered out before the mixer, in the same way, that image signals are filtered out before a mixer in a superheterodyne receiver, or the 'phasing method' is used to eliminate the image frequencies. SDR receivers use the phasing method for image frequency cancellation.

In the phasing method, the RF signal is split into two streams before the mixer and one stream is delayed by 90 degrees. They are called "I" incident and "Q" quadrature signals. After the mixer, a 90-degree phase change is applied to the other stream so that when the two streams are combined the wanted signals are back in phase and add together. But the unwanted image signals have been given an additional 180-degree phase change due to being received from below the LO frequency. When the two streams are combined after the mixer, they are out of phase and so they cancel each other out, eliminating the image signals from the audio output.

## The Tayloe Detector

The Tayloe QSD detector is very simple and cheap to make. It uses three very basic integrated circuits. A dual Flip Flop D type latch configured to divide the local oscillator signal by four, a multiplex switch, and a dual low noise Op-Amp.

Take a look at the schematic diagram of the 'double balanced' Tayloe detector **Fig 8.6**. The two flip-flops in a 74LC74 chip are configured to divide the 14MHz clock signal by four. The two outputs generate a 00, 01, 10, 11 binary pattern at 3.5MHz which is used to switch the two multiplex switches in the FST3253 chip, to each of the four outputs in turn. If the VFO signal is from a variable device like a Si570 chip, the receiver can be made to cover a wide range of frequencies. The Tayloe design can be used up to around 1GHz, but in this example, the local oscillator signals to the switch are at 3.5MHz and the received signals are also centred around 3.5MHz. This means that the signal from the input transformer is switched to all four outputs during every cycle of the input frequency. A pulse of the input signal voltage is applied to each of the four capacitors in turn.

The capacitors store the voltage on the inputs to the Op-amps and end up averaging the input levels over time. This is called a sample and hold circuit. It works as an envelope detector in the same way that the capacitor following the diode in a crystal set

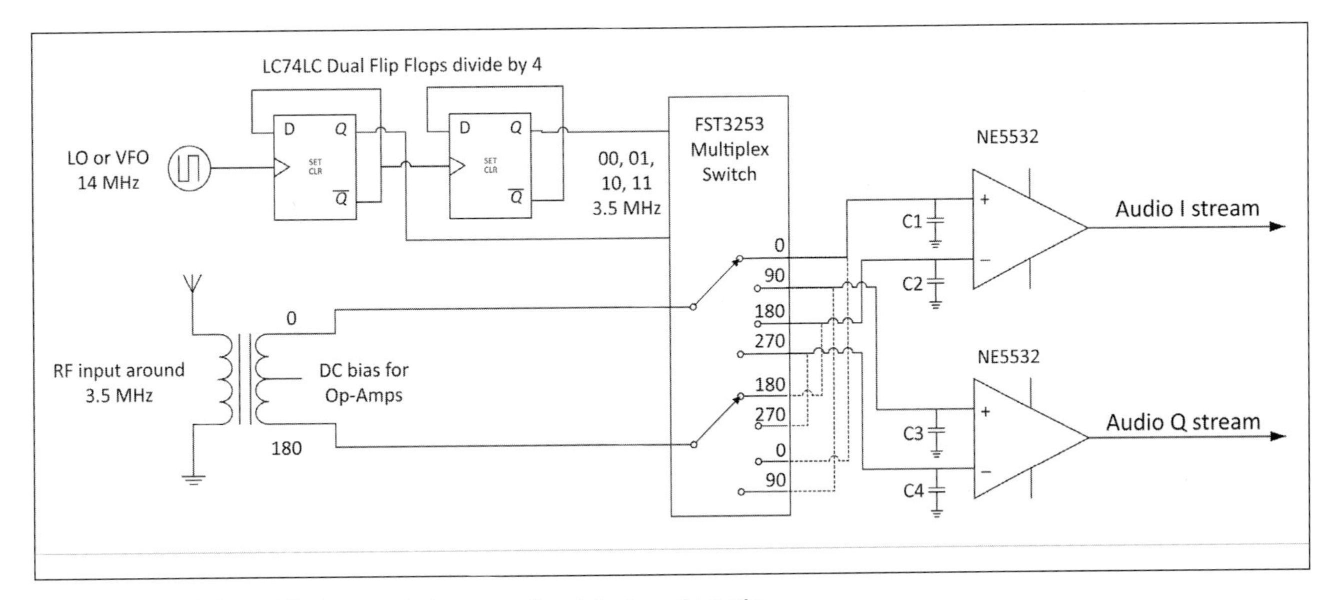


**Fig 8.6: Double balanced Tayloe quadrature sampling detector schematic**

does. You get an audio frequency output. In a double balanced Tayloe detector, the signal from the other side of the input transformer is not wasted. Because it is anti-phase it is used to top up the capacitor that is 180 degrees out of phase. So, for each of the four switch positions two of the capacitors are charged up with the input signal. As Dan Tayloe states in his article titled 'Ultra Low Noise, High Performance, Zero I.F. Quadrature Product Detector and Preamplifier,' "... two separate detectors are driven with the inputs 180 degrees apart using an input transformer. The two detectors use a common set of four detector caps. Since the outputs of the two detectors are 180 degrees out of phase, the detector capacitors are now driven two at a time."

A DC bias voltage for the Op-Amps is applied to the transformer centre tap. It passes through the switches to the Op-Amp inputs.

Because the $0^{0}$ and $180^{0}$ capacitors are charged up at the same time, (during switch positions 1 and 3), the signal on the $180^{0}$ capacitor is the same as the 00 capacitor but with reverse polarity. This means that the $0^{0}$ and $180^{0}$ audio signals can be combined using the differential inputs of the Op-amp, providing a handy 6 dB increase in the I stream signal level without adding any noise to the signal. Likewise, the $90^{0}$ and $270^{0}$ audio signals are combined in the other Op-Amp to create the Q signal.

Just like the 1950s phasing receivers, both audio outputs contain essentially the same signals, but the Q signal is delayed by 90 degrees. The audio output extends from 0 Hz up to the bandwidth of the receiver. The bandwidth of a Tayloe detector is limited by the RC time constant (Ohms x Farads) of the capacitors and the overall resistance of the network. Because in the double balanced version of the Tayloe detector the capacitors get topped up twice as often. They can have a smaller value as they only have to hold their charge for half as long. Less capacitance means that the bandwidth of the double balanced Tayloe detector is wider than the original single balanced detector. The other factor limiting the bandwidth is the sample rate of the ADC (analogue to digital converters) following the detector.

In a 1st generation SDR the ADC is the PC sound card. The I and Q signals are connected to the Left and Right line input on the PC. Don't use the microphone input because it may not be stereo.

## The Quadrature Sampling Exciter

The QSE or quadrature sampling exciter **Fig 8.7** is the transmitter version of a Tayloe detector. It is virtually the same circuit in reverse, and it works in much the same way. Audio signals

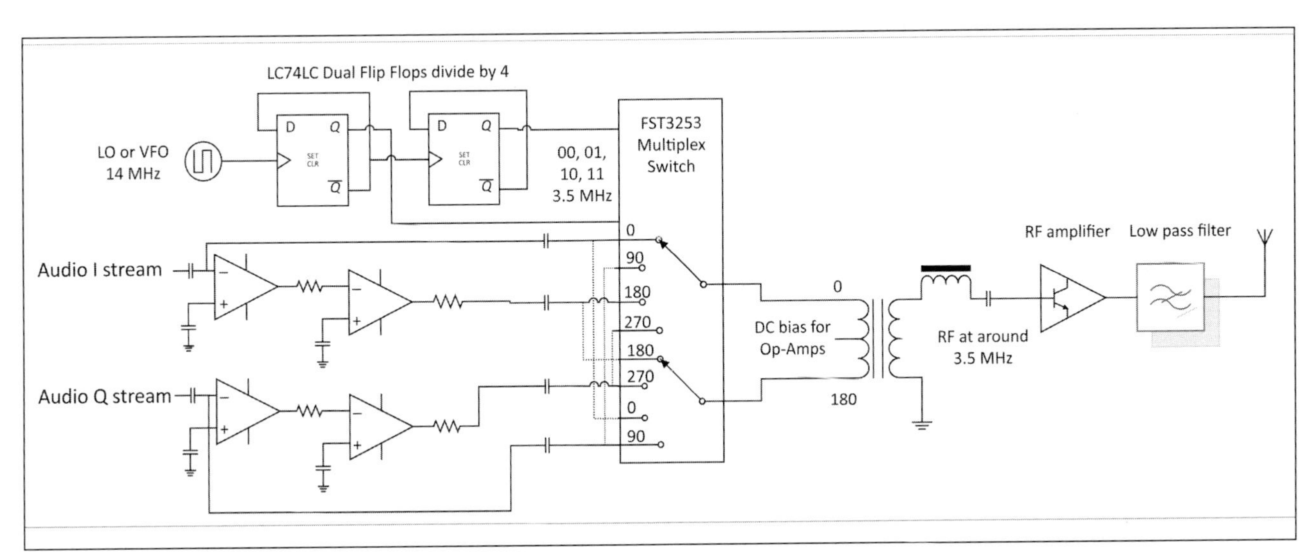


**Fig 8.7: QSE quadrature sampling exciter**

that have been created from the data are sent to the QSE via the sound card DAC (digital to analogue converter). The digital signals feeding the DAC carry the modulation added by the DSP software. In the QSE, the I stream is split into $0^{o}$ and $180^{o}$ signals. The Q stream, which already has a 900 degree phase lag, is split into $90^{o}$ and $270^{o}$ signals. The FST3253 is switched at the 3.5MHz rate causing a 3.5MHz RF signal to be created. This is amplified and passed through a low pass filter, then connected to the antenna.

The QSE circuit is simple but effective. Similar designs are used in all generation 1, 2, and 3 SDR transmitters. The Elecraft KX3 and the SoftRock Ensemble use the circuit above. The Genesis design uses more Op-amps and eight individual 74HC4066N switches rather than the FST3253 multiplex switch, but at a functional block level, the design is the same.

## Solving the Image Problem

The Tayloe detector is a 'direct conversion' mixer meaning the RF signal at the input is converted directly to an audio output signal. The problem with all direct conversion receivers is that 'image' signals from below the LO (local oscillator) frequency end up in the audio signal as well as the signals from above the LO frequency. We want to be able to hear signals from above and below the LO frequency and see them on the panadapter display without them appearing on top of each other. Luckily the I and Q data allows us to separate the image signals from the wanted signals.

The local oscillator (LO) is at the same frequency as the receiver centre frequency (Fo). In the mixer, an upper sideband signal that is 10kHz above the local oscillator frequency becomes an upper sideband audio signal at 10kHz. See the lighter coloured signal on **Fig 8.8** below.

An RF signal that is 5kHz below the local oscillator frequency cannot become -5kHz. It becomes reflected into the audio range with a phase change of 180 degrees. See the darker coloured signal on the diagrams. So now, we have a wanted signal at 10kHz and an unwanted 'image' signal at 5kHz. The image signal becomes a lower sideband audio signal due to the $180^{o}$ reflection. Problems arise when an image signal from below the local oscillator frequency appears right on top of a wanted signal.

The left side of the drawing **Fig 8.9** represents the hardware part of the receiver, usually a Tayloe detector. The right side shows what happens inside the PC. Note that one output has the signals from above the LO frequency with the cancellation of image signals from below the LO frequency. The other output has the signals from below the LO frequency with the cancellation of image signals from above the LO frequency. Combining the output onto the same panadapter shows all of the signals above and below the receiver centre frequency with no image signals.

The Tayloe QSD detector creates I and Q audio frequency signals in quadrature because the RF signal charging the capacitors on the Q stream occurs 90 degrees later. The two audio signals are converted to digital data using the PC sound card or a dedicated analogue to digital converter (ADC) chip. Software in the PC applies a $90^{o}$ degree phase shift to the I stream. This realigns the I and Q streams, so they become 'in-phase.' The two data streams are then added together to make a new audio stream with twice the original amplitude. Subtracting the two data streams creates a second data stream containing the signals from below the local oscillator frequency and cancelling the signals from above the local oscillator frequency. The image cancellation is achieved using mathematics applied to the digital data streams rather than circuitry. By combining these two streams into a single stream via a data buffer, the radio can display signals below and above the LO frequency while suppressing the image signals.

The process could be done in hardware like the old 1950s phasing receiver, but in a software defined radio, everything after the analogue to digital converters is done using computer programming.

The drawing **Fig 8.10** shows how the panadapter display is created from the incoming RF signal. First, there is the single conversion receiver front end, which creates I and Q audio signals. Both streams are virtually the same and both contain unwanted image signals. However, the phase relationships are different.

We want to create a panadapter that displays the signals above and below the nominal centre frequency Fo, with no pesky image signals. We use the I+Q signal to supply the high end of the panadapter and cancel the image signals from below Fo. The I-Q signal provides the lower half of the panadapter and cancels the images from above Fo.

By selecting the data from a different part of the FFT output buffer, we get a reversal of the I-Q part of the signal. This corrects the sideband reversal and places the signal at the proper frequency offset from the centre of the panadapter. Then the data can be combined to display a panadapter.

The image cancellation process can also be depicted using vector diagrams to show the vector addition and subtraction in action. (see **Fig 8.11** )

To get good image cancellation the audio level of the I and Q streams must be the same and the phase difference must be exactly 90 degrees. Achievement of 40 dB of image cancellation requires the difference between the I and Q levels to be within 0.1 dB and the phase error to be less than 1 degree. For 60 dB of image cancellation, the levels must be within 0.01 dB and the phase error less than 0.1 degrees. The PC software is able to compensate for phase and amplitude errors to minimize the display and reception of image signals, but the cor-

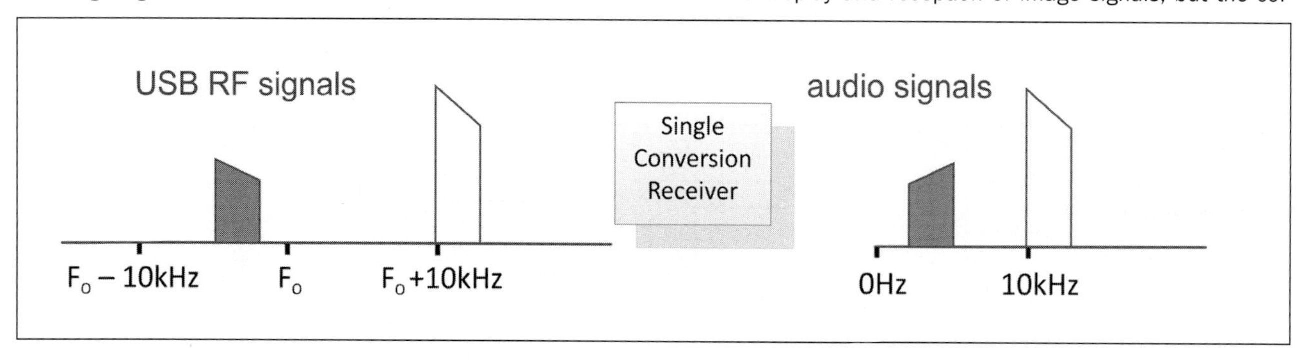


**Fig 8.8: The single conversion process creates image signals in the audio output**

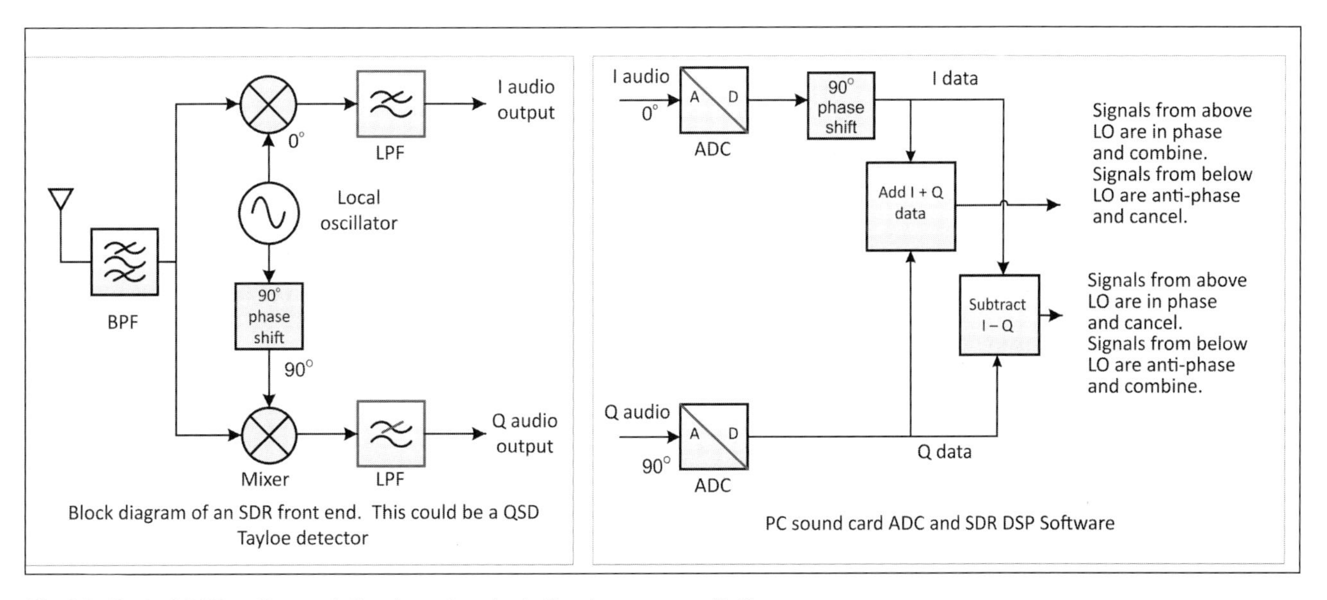


**Fig 8.9: Typical QSD software defined receiver including image cancellation**

rection is not perfect across the whole spectrum.

If you have a QSD type receiver, you can see the effect of the image cancellation by disconnecting or turning down the level of either the I or the Q stream, thus disrupting the IQ balance. With the streams unbalanced, each wanted signal will have a mirror image on the other side of the spectrum display, equidistant from the centre frequency. The mirror images will have the sideband(s) reversed, which is not a problem and not visible for CW, AM or PSK signals but obvious with SSB signals. Using a mono rather than a stereo sound card input will have the same effect i.e. no image cancellation.

To make the image cancellation as good as possible the PC software looks at signals you are receiving, works out where the image frequency would be and adjusts the phase between streams and the audio levels to minimise the image signal. It does this all the time the receiver is operating and over time it learns the characteristic of your particular receiver. In most amateur radio SDR applications the software code used for this adaptive correction is based on a very clever and remarkably simple algorithm invented by Alex Shovkoplyas, VE3NEA, the author of the Rocky SDR software. He writes; "The algorithm

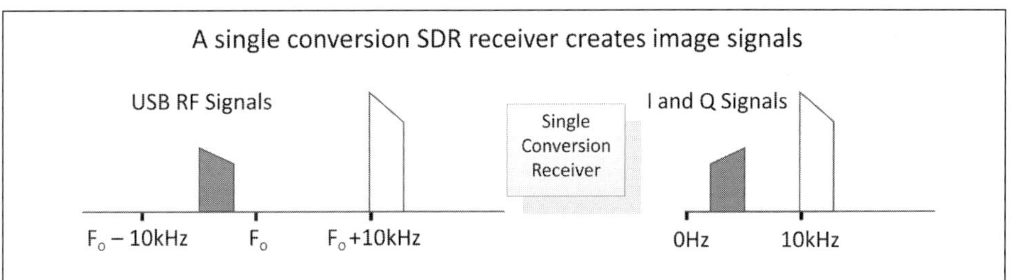

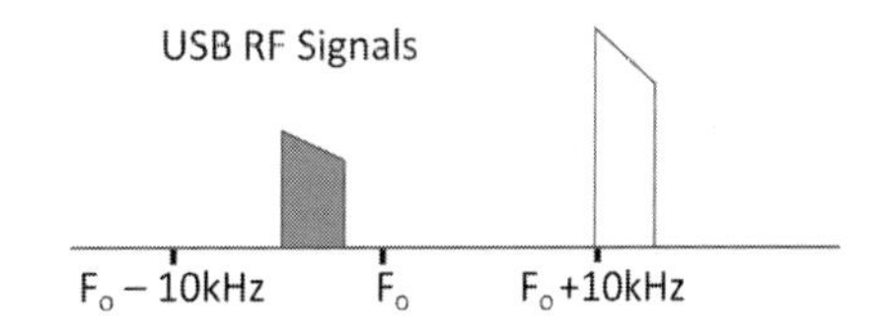


**Fig 8.10: Using the I and Q streams for image cancellation**

works as follows - the receiver pass spectrum is scanned for signals that are at least 30 dB above the noise. For each signal, synchronous detection of the image is performed, using the main signal as a reference oscillator. The synchronous detector has very high sensitivity and can detect the image signal, even if is below the noise."

This technique allows QSD SDR receivers to achieve around 90 dB of image suppression which is comparable to a very good superheterodyne receiver. SDR guru Phil Harman VK6PH reverently referred to these lines as, "The four lines of code that changed the SDR world".

Direct digital conversion receivers use a mixer and local oscillator created in software running on the FPGA chip to create the I and Q audio streams. This creates a perfect phase and level relationship, so the image cancellation is excellent, at least as good as the 100 dB or so achievable in the best conventional architecture receivers.

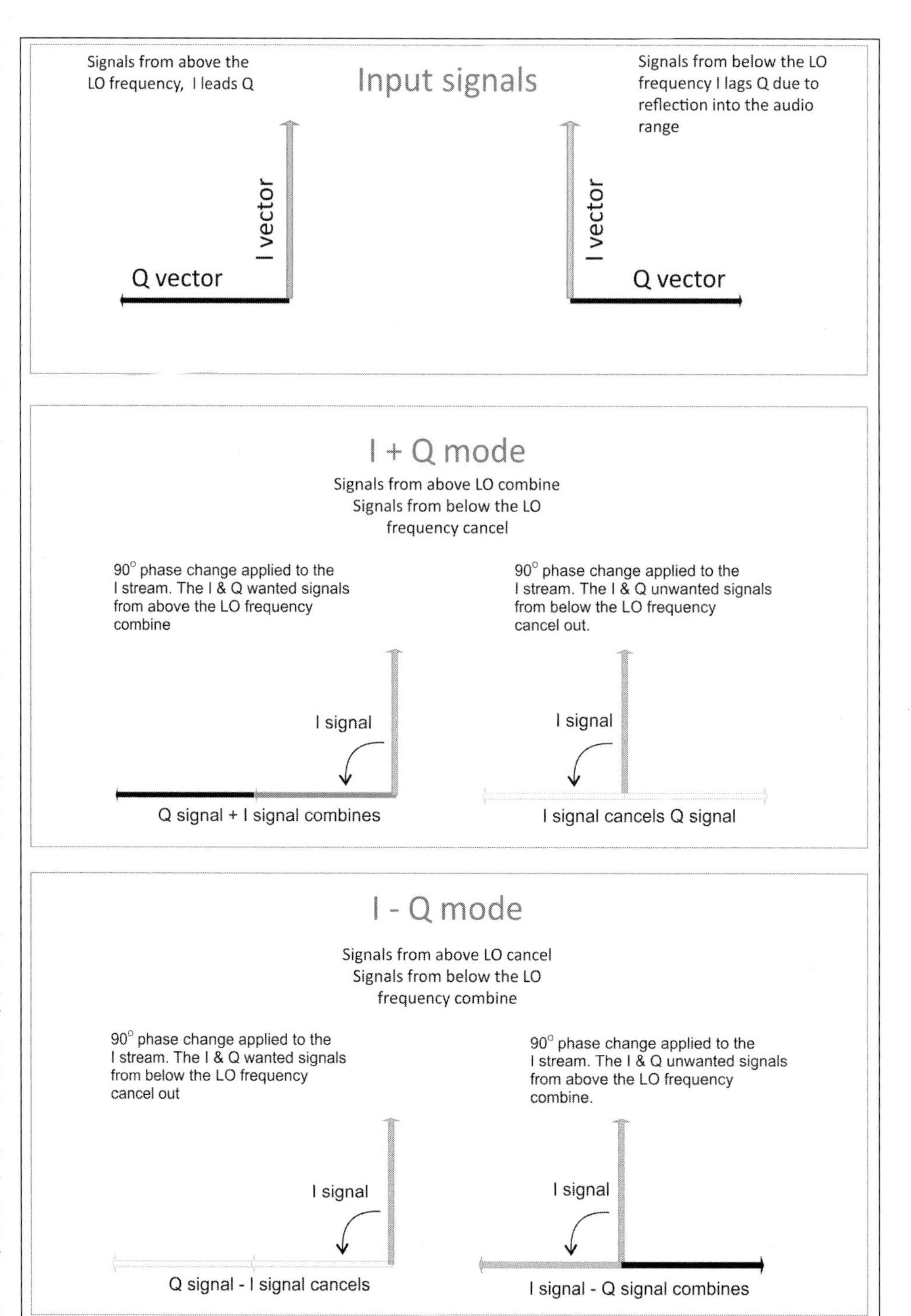


**Fig 8.11: Vector diagrams illustrate the image cancellation process**

## The QSD Receiver

As well as creating the data for the panadapter display, the IQ digital signals are used to determine the amplitude and phase of the received signals at the times they were sampled. The phase differences are used to decode FM or other phase modulated signals and the amplitude information is used to decode amplitude modulated signals like CW, AM and SSB. If you have I and Q signals you can demodulate any type of modulation.

A 48kHz or 96kHz bandwidth is enough to display quite a few SSB signals or the entire CW or digital mode section of the band. You can click your computer mouse on any displayed signal and hear the QSO. Another function unique to SDR receivers is that you can record the full panadapter bandwidth and play it back later. You can listen to QSOs anywhere on the recorded band segment even signals you didn't listen to earlier.

Some other statistics for the Tayloe detector include a conversion loss of 0.9 dB (a typical conventional receiver mixer has 6 to 8 dB), a low Noise Figure around 3.9 dB, (the NF of a typical HF receiver can be anything up to 20 dB), and a 3rd order intercept point of +30 dBm. It also has 6 dB of gain without adding noise and it is very cheap.

The Tayloe QSD SDR provides very good receiver performance at a much lower cost than conventional receivers and it achieves the goal of making the modulator and demodulator a part of the digital signal processing, bridging the gap between I.F. DSP and A.F. DSP. But it still has one mixing process, which is a potential source of intermodulation distortion. The next logical step is to eliminate the QSD mixer and sample the RF spectrum directly at the antenna. The process is known as direct digital synthesis (DDS). It uses digital down conversion (DDC) for the receiver and digital up-conversion (DUC) in the transmitter.

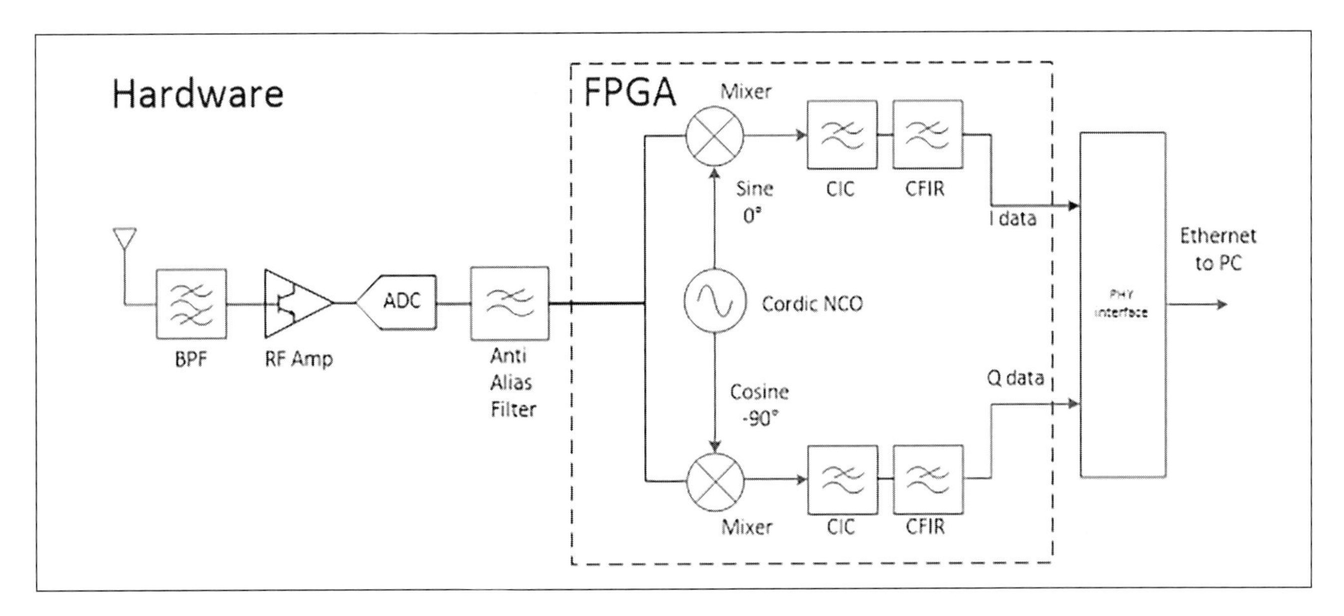


Fig 8.12: Typical digital down conversion (DDC) HF receiver

## DIRECT DIGITAL SYNTHESIS

### The DDC Receiver

In a digital down conversion receiver (**Fig 8.12**), the RF signal is directly sampled by a single very fast analogue to digital converter. The digital data output is then split into the I and Q streams by the SDR software or by 'firmware' running on an FPGA (field programmable gate array), or on a microprocessor, a single board computer (SBC), a tuner chip, or a dedicated DSP chip.

The block diagram of the mixer in the FPGA looks similar to the mixer in a QSD Tayloe detector, but it is constructed purely in software code. The difference is that in a QSD receiver the mixer is before the analogue to digital conversion and works at audio frequencies. In the direct sampling receiver, the mixer is after the analogue to digital conversion and it works on the high-speed digital signals.

The NCO is a numerically controlled oscillator. It creates data streams of numbers that are equivalent to a Sine wave and a Cosine wave. A Cosine wave is a Sine wave with a $90^{0}$ phase lag.

A CIC filter is a type of FIR filter that either interpolates or decimates. What? I hear you gasp! OK, CIC stands for cascaded integrating comb filter. They are multi-stage filters with the advantage that the mathematics governing the filter's operation is easy and efficient to code in software. The drawbacks are a certain amount of latency (delay) and they are not completely flat. They have a slight roll off at the top end. The more CIC stages employed, the more roll off occurs and that is why the last filter in each of the two receiver chains are CFIR (compensating finite impulse response) filters. They are trickier to implement in the FPGA because they use more FPGA elements and resources. The CFIR acts as a filter that has a rise at the high-frequency end which compensates for the droop in the earlier filters. The result is a flat panadapter across the band. Without the CFIR filter, you would see a drop off at the far left and right ends of the panadapter. You sometimes see this drop-off in level at the edges of 'sound card' SDR panadapters.

Decimation in the DDC receiver involves throwing away a whole lot of the ADC output bits. In some radios, the data rate from the ADC is too fast to send to the PC. The combination of the CORDIC mixer and decimation selects a subset of the input spectrum and sends it to the PC at a much-reduced data rate. The CIC filters shown in the receiver block diagram, slow the data rate down from the ADC rate to the sampling rate that is sent to the PC. For example, in an ANAN radio, the CIC filters slow the data from the ADC sample rate of 122.88 Msps (122.88 million samples per second) to 384 ksps (384 thousand samples per second) or lower if a narrower panadapter is selected. As the sampling rate is reduced, you must apply a low pass filter at the Fs/2 frequency to stop alias signals appearing in the wanted pass-band. The CIC filters do both the decimation and the filtering. Note that not all direct sampling receivers use decimating filters. Onboard DSP may process the entire output of the ADC.

### The DUC Transmitter

The direct up conversion transmitter (**Fig 8.13**) is quite similar to the receiver design. A modulated IQ signal is sent from the PC via Ethernet or a USB cable. The IQ data from the PC is at a much lower speed than the rate required for the DAC. It only has to be at least twice the highest modulating frequency. The CIC Interpolation filters speed the data up to the DAC sampling rate. Then the CORDIC NCO mixes the resulting data up to the RF frequency. The DAC converts the high-speed digital data stream into an analogue signal at the RF frequency, which is amplified and filtered before being sent to the antenna. There will usually be several stages of amplification and filtering because the output from the DAC is only about 1 mW of RF power.

Interpolation in the DUC transmitter is a mathematical way of filling in the gaps. The data rate of the IQ signals carrying the modulation is much lower than the data rate required by the DAC. This means that every 16-bit sample applied to the CIC interpolation filter results in a whole lot of 16-bit output samples. The output could just repeat each input sample two thousand five hundred and sixty times to interpolate between the 48 ksps transmit IQ stream and the 122.88 Msps DAC stream of an ANAN transmitter, but that would cause sudden signal level changes. So, the interpolation filter uses mathematics to smooth the rate of change.

In the diagram of the Icom IC-7610 receiver (**Fig 8.14**), the software running inside the FPGA is not illustrated. However, we

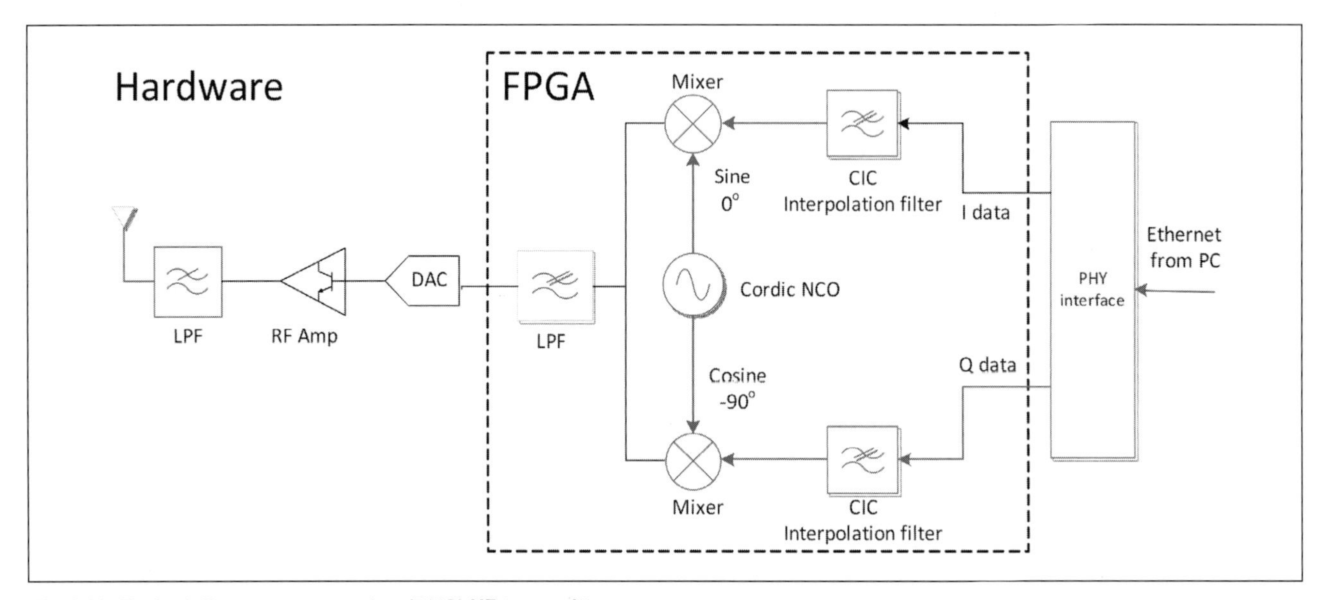


**Fig 8.13: Typical direct up-conversion (DUC) HF transmitter**

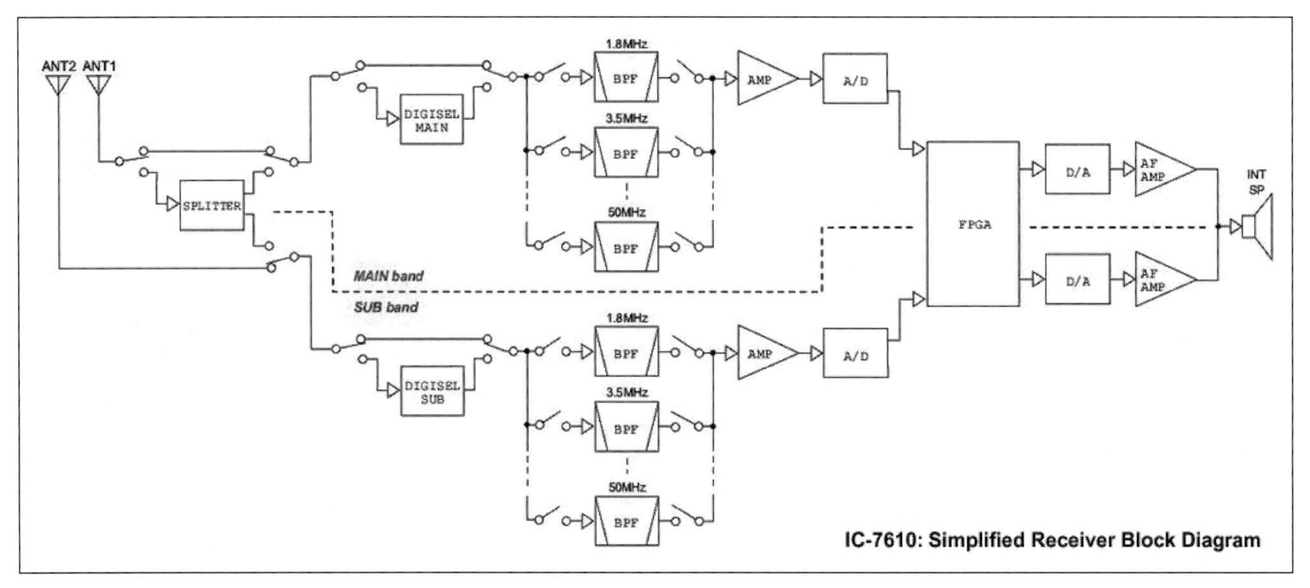


**Fig 8.14: Block schematic drawing of the Icom IC-7610 SDR transceiver**

can see the SDR architecture, including the two identical receivers. The optional 'DigiSel' preselector is actually a digitally switched hardware filter made using inductors and capacitors. It is configured to follow the receiver frequency by switching in or out the inductors and capacitors using relays. Rather like a modern antenna tuner. The DigiSel' preselector is followed by a selection of twelve bandpass filters followed by a selection of twelve bandpass filters (not all shown), which are also automatically selected according to the receiver frequency. Next, there is an adjustable preamplifier and a software selectable attenuator (not shown). Each receiver has an analogue to digital converter (A/D) which outputs a 16-bit high-speed data stream to the FPGA. The FPGA performs all the DSP functions and contains a software emulated computer CPU which takes care of the front panel controls and sends data to the touchscreen display.

## The Panadapter

The most immediately noticeable difference between conventional radio receivers and software defined radios is the panadapter spectrum and waterfall display. At first glance, the panadapter display on one of the new SDR based transceivers looks very similar to the band scope on some of the more expensive conventional superheterodyne architecture radios. But there is a world of difference. A band scope displays signals above and below the frequency that the receiver is tuned to. A panadapter can be made to emulate a band scope, but generally, they display a much wider range of frequencies. You can place one or more receivers at any point across the display. A simple click of the mouse or a touch on the touchscreen allows you to jump from signal to signal. Most panadapters and band scopes also include a waterfall display. A waterfall display indicates signals over a period of time. It can show you signals that have a short duration such as a station intermittently calling CQ. You can see the traces of signals that could easily be missed if you were tuning across a band using a conventional receiver.

Radios that use a computer monitor to display the panadapters often have more flexibility, having the space to be able to display wider bandwidths, multiple panadapters, and often two or more receivers. For example, the FlexRadio transceivers can show up to seven receivers on seven bands with a maximum panadapter bandwidth of 14MHz. You simply can't display that much information on the small screen on a transceiver. The band scopes in the top end conventional radios and new SDR transceivers can only display one or two spectrum and waterfall panadapter displays with a maximum span of 1MHz.

In a direct sampling HF receiver the ADC samples the entire RF spectrum up to around 60MHz so potentially the entire HF +6m spectrum can be displayed on the panadapter. Some PC SDR software can display the whole HF spectrum as a low-resolution image, but generally, the maximum panadapter bandwidth is reduced due to the requirement to handle very high-speed data streams. The problem is not so much managing the data within the computer as carrying the data over WiFi, Ethernet, or USB from the radio to the computer. Radios that have onboard digital signal processing are able to display a wider panadapter than radios that require an Ethernet or USB connection to a computer or SBC (single board computer).

Most of the small box and dongle type SDRs only support one panadapter and one receiver but you can still see activity across a wide band of frequencies. The maximum panadapter size of these radios varies from 48kHz up to more than 10MHz depending on the sampling rate of the ADC and any decimation occurring in the radio before the data is sent to the computer. Decimation is a way of reducing the data rate of the I and Q digital data streams so that they can be carried over the circuit between the SDR and the computer. Reducing the data rate has the effect of reducing the bandwidth that the digital stream can carry. For example, if you have a 100 Msps 16-bit data stream, you generate a data rate of 1,600 Mbits (1,600 million bits per second) and get a Nyquist bandwidth of 50MHz. If you throw away every second 16-bit sample, you will halve the data rate, resulting in a 50 Msps 16-bit data stream. The data rate is reduced to 800 Mbits (800 million bits per second) and the Nyquist bandwidth is reduced to 25MHz. Once that data is available on the computer, you can display a 25MHz wide panadapter and place several receivers placed within that range of frequencies.

You might wonder what advantages a band scope or panadapter gives you. The answer depends on how you operate. If your radio is always used for chatting to locals or on your favourite net, the panadapter offers little value. But for those chasing DX or participating in contests, a band scope, or panadapter is rapidly becoming indispensable. During a contest can see each contest station and click on the screen to work each one without having to tune slowly across a noisy band. It is easy to avoid stations that you have already worked without having to waste time listening for their call sign each time you tune up or down the band. You can see weak stations that are very close to strong stations and use the SDR's variable filters to pull them out of the noise. When you are trying to beat the pile-up and work a rare DX station, you can see and hear both the DX station and the stations in the pileup. This allows you to work out how the DX station is working. Is the operator sitting on one receive frequency? Or are they moving higher or lower after each call? It also allows you to choose a transmit frequency that is in a quiet spot in the pile-up, or a frequency that the DX station is moving towards. These techniques can give you a significant edge. Increasing your contest score or working that rare DX station that you might otherwise miss. The ability of some SDRs to show two or more bands at the same time can be useful if you want to monitor a frequency while working stations on another band. For example, you might check out the 20m band while waiting for a friend to pop up on the 40m band. Or you could watch beacons frequencies on 6m to see when the band opens while operating FT8 on another band.

The panadapter has other uses as well. It is essentially a spectrum analyser display and it can be used in that role. The Y axis, showing the level of the incoming signals is usually calibrated in dBm rather than S points and it is normally quite accurate. The X axis displays the range of frequencies that you have selected. The noise level seen between stations as 'grass' on the bottom of the screen indicates the level of the background RF noise level in the area that you live in. Rural locations typically have a noise floor better than -110 dBm while in an urban location it will be closer to -100 dBm. On a panadapter, you may be able to see interference from ADSL routers, plasma TVs, power lines, electric fences, LED lighting, or other sources. You can compare the signal level received from different antennas very accurately, or compare the signals received from different stations. The panadapter spectrum display can also reveal the quality of the signals that you are receiving. Splatter and over-modulation are immediately obvious because you can see the shape and bandwidth of the other station's transmission. You can even make a guess as to what digital mode is being transmitted. Each digital mode has a distinctive look on the spectrum and waterfall display. The waterfall display can reveal the individual dits and dahs of a Morse Code transmission. You can see any key clicks if you study the waterfall trace carefully. The slope across an SSB signal spectrum shows the flatness of the modulating audio. A higher level near the carrier frequency indicates more low-frequency bass in the signal. You tend to see this when the station has been set up for rag chewing and local contacts. A higher level at the edge furthest from the carrier frequency indicates more high audio frequencies in the signal. You tend to see this where the station has been set up for DX operation. The width of the signal on the spectrum display and especially the waterfall indicates the bandwidth of the transmission. You can tell if the other station is transmitting a 2.4kHz wide DX or contest grade SSB signal, or a wider bandwidth of 2.7 to 3kHz. If an SSB station is transmitting a bit off frequency, you can see that on the panadapter as well. You can also use the panadapter to monitor your own transmitted signal to ensure that your transmissions are clean and legal.

Another feature of panadapters, as opposed to band scopes, is that most panadapters support computer mouse operation. It might feel strange at first, but you quickly get used to clicking on the signal you want rather than turning a VFO knob. The SDR software makes operating your radio as easy as using any other computer program.

The panadapter is not a new idea. The first use of a spectrum display attached to a receiver is attributed to Marcel Wallace a French engineer and 'radio ham,' F3HM, who invented the idea around 1932. He filed his patent application on 17th March 1938. Similar cathode ray tube displays were fitted to some receivers used during WWII. Interestingly Wallace used a motor driving a generator to produce the sawtooth sweep signal for the display. The term 'Panadapter' is apparently a contraction of 'Panoramic Adapter.' The first device to be marketed as a 'Panadapter' was the Model PCA-2, Type T-200 Panoramic Adaptor, made in the mid-1940s by Panoramic Radio Products of Mount Vernon, New York. It was a separate box that sat on or beside the receiver. From 1946, they produced the same unit under contract to Hallicrafters where it was re-badged as the SP-44.

(sourced from *http://portabletubes.co.uk/boats/pca2.htm).*

All software defined radio receivers use an analogue to digital converter to sample the incoming RF signals, or audio in QSD receivers, and represent them as a series of digital numbers. The analogue to digital converter samples the level of the incoming wideband received signal at instants in time and each sample is represented by a digital number appearing on the ADC output pins. On the next clock cycle, another sample is read and its level becomes the next number on the output pins. Since the data coming from the analogue to digital converter represents the signal at specific 'sample' times, it is in the 'time domain.' This serial data is ok for decoding the modulation and for DSP functions like filtering and noise reduction, but it is no good for the spectrum display. We want to see a display showing signals represented at their operating frequencies across the band, not signals at different times. Fast Fourier Transformation (FFT) converts the data into a form that can be displayed on the panadapter.

Instead of a series of numbers representing the signal level at different times, FFT creates a series of numbers representing the signal level at different frequencies.

After the FFT process, each digital number in the output buffer represents the signal level of a narrow section of the frequency spectrum called a bin. As well as being able to display the spectrum on your panadapter, having the signal represented in terms of frequency is useful for some DSP functions. For example, to create something like a low pass filter, you could progressively reduce the level of the bins above the cut off frequency. By altering the level in individual bins, you can make fine adjustments to the shape, or the high and low cut off frequencies of a filter. Frequency translation can be done simply by moving the position of bins in the software buffer. Amplification or equalization can be achieved by multiplying the level in each bin by a calculated amount. Or, you could create a notch filter by simply reducing the level in a few bins.

The spectrum display is made by plotting the level of each bin across the bandwidth of the panadapter. Because each bin is only a few Hertz wide, they each have a large dynamic range, due to process gain. Stacking thousands of bins side by side gives you a wideband display with a low noise floor. The same technique is used in modern spectrum analyzer test instruments. Without FFT, a wide panadapter display would have less process gain and therefore a higher noise floor than the noise level within the receiver bandwidth. Using FFT, the panadapter can display a lower noise floor than the noise being heard within the receiver bandwidth.

A 'sample' is a number represented as a digital word that describes the level of the signal at a particular instant in time. In the case of the output of an ADC, it is the time that the input level was measured and recorded as a number. A bin is a number represented as a digital word that describes the level of the signal inside a very narrow bandwidth of frequencies. A bit like an 'S meter' measuring the level inside a 100 Hz receiver bandwidth.

I should point out that the image you see on the panadapter spectrum and waterfall display is not the same as the signal that you are listening to on the receiver. Although both the receiver and the panadapter use the same I and Q data streams, the digital signal processing (DSP) software process is completely different. It is easy to be fooled by this. You may be able to see a signal on the waterfall that is too weak to be heard in the receiver. Again, the reason for this is the relative bandwidth. The receiver is wider than the bins used to create the panadapter, so the noise level will be higher within the receiver passband, reducing the dynamic range and the signal to noise ratio.

The waterfall display is created from the same data samples as the spectrum display and usually updated at the same rate. If we assume a panadapter screen that is 1024 pixels wide, the value of each of the 1024 FFT bins is recorded as a dot across the top line of the waterfall display. Large value numbers representing big signals are displayed in bright colours and small values representing weak signals, are displayed in dark colours. Every time a line of spectrum information is displayed on the spectrum display, a new line is added to the top of the waterfall display. In PowerSDR, the waterfall is stored as an image file that is 1024 pixels wide and 256 lines long. Unlike the spectrum image, which is drawn across the panadapter dot by dot, like the trace on an oscilloscope, each waterfall line is added as a complete line.

## The ADC

The heart of any software defined radio, or indeed the DSP stage in a conventional receiver or transceiver, is the conversion of the analogue signals that we receive on our antenna into digital signals that we can process with computer technology. We do it because manipulating the digital signal adds none of the noise and distortion that happens in the amplifiers and mixers of a conventional radio. The ADC (analogue to digital converter) reads the voltage at its input, decides what the closest voltage step is, and then outputs a number representing that voltage step. This happens very fast. For example, the ADC used in the new FlexRadio FLEX-6500 samples the input voltage 245,760,000 times per second.

When we need to convert a signal from digital back to analogue, we use a DAC (digital to analogue converter). This integrates voltages between samples to create a smoothed analogue output signal.

## ADC Bits

The output of an 8 bit ADC can represent 256 different input voltage steps because that is the maximum number of different numbers that can be fitted into 8 bits. If the input range is an analogue signal with a minimum level of 0 and a maximum level of 1 Volt, then the voltage steps are 3.9 mV apart. That means that every sample has an error, which could be as much as 3.899 mV. An input level just over the 1000 0001 level will be represented as 1000 0010, which is the next step up. The error is known as quantization error because it is a mistake in quantifying the input level. Quantization errors cannot be recovered. When you convert the signal back to an analogue level, the output for that particular sample will be wrong by the amount of the quantization error. This manifests itself as noise on the recovered signal. The only way to reduce quantization errors is to make the sampling voltage steps smaller. To do that you need to use more bits. However, that adds to the complexity of the device. Using 14 or 16 output pins instead of 8 means a larger chip size with more internal buffers.

Actual quantization errors are minimized by pipeline sampling in the ADC. Pipeline sampling is a successive approximation methodology which reduces the quantisation error by comparing the sample against previous samples.

A 16 bit ADC can represent 65,536 individual voltage levels, so if the input is in the same 0 to 1 Volt range, the voltage steps are now 0.015 mV apart. This means much higher sampling accuracy i.e. much less quantization noise. Adding more bits increases the data rate that we need to transfer over the serial USB or Ethernet interface between the radio and the SDR software running on the PC. Doubling the number of bits from an 8-bit data packet to a 16-bit data packet almost doubles the data rate required. The actual serial data rate is not quite, double because the number of overhead bits in each data frame remains constant.

## Is The Number of ADC Bits Important?

The answer is yes, but not as important as some people think. The number of bits that the ADC uses to output each sample affects the receiver's dynamic range and signal to noise ratio. The analogue to digital converter in the SDR receiver creates a stream of data words, which represent the signal at the ADC input at very specific instants in time. In other words, the ADC looks at the input signal and outputs an 8, 12, 14, or 16-bit number, depending on its type, which represents the voltage step nearest to the input voltage at that time. This is called sampling. An 8 bit ADC can describe 255 voltage levels and a 16-bit digital number can describe 65,536 voltage levels, although in practice internal noise performance reduces the number of usable bits. The noise causes the ADC to confuse the reading of very small input signals and this reduces the effective number of bits. For example, a 16 bit ADC may only be able to output 12 bits of usable data. This is known as the 'effective number of bits' or ENOB. It is the ENOB of the ADC, and the process gain from bandwidth reduction (decimation) that determines the maximum dynamic range of an SDR receiver, not the number of bits in the output stream. This is why radios with 12 bit or 14-bit ADCs often have a dynamic range nearly as good as radios with 16-bit ADCs. With radios that have a low noise amplifier or a tuner before the ADC, the noise performance is dominated by the noise factor of the front-end device rather than the ADC. A dongle type SDR with an 8 bit ADC can provide very good performance provided the preamplifier in the tuner preceding the ADC stage has a good noise figure and is not overloaded by strong input signals. In QSD based SDRs, the ADC is done on audio signals after the Tayloe detector. A soundcard or audio chip that uses 24-bit sampling might give a very marginal improvement over a standard 16-bit device, but it is unlikely you could tell the difference unless you are listening to a high fidelity music station.

## The ADC Sampling Rate

Harry Nyquist (1889-1976) determined the fundamental rules for analogue to digital conversion. He found that as long as the sample rate was at least twice the rate of the highest frequency in the analogue signal, the original analogue signal could be recreated accurately from the digital data. This means that to accurately, sample and then recreate a band of radio signals from a few kHz up to 50MHz, the ADC should sample the radio spectrum at a minimum of 100 Msps, (one hundred million samples per second). Higher ADC sampling rates mean that more bandwidth can be sampled, and a wider spectrum of frequencies can be displayed on the panadapter. Most direct sampling HF receivers use an ADC sample rate between 60 Msps and 150 Msps. A sample rate of 60 Msps allows the receiver to cover 0-30MHz. A sample rate of 150 Msps means that the receiver can cover 0-75MHz.

By the way, an ADC reads one sample for every clock cycle, so the ADC sample rate in Msps is the same as the ADC clock frequency in MHz. Therefore, a direct sampling HF radio has a frequency range equal to half the ADC clock rate and also half of the ADC sample rate.

Dongle and other SDRs that work above the HF bands, use a receiver chip or some other frequency translation stage, possibly an ordinary mixer and oscillator, to shift a band of frequencies down to a range that the ADC can sample. Often the ADC sample rate is around 10 Msps, so this type of SDR can display a band of frequencies that is 10MHz wide. The frequencies displayed could be in the 2GHz range or even higher, but the panadapter display is limited to the bandwidth set by the ADC sample rate. Eagle-eyed readers will have spotted an anomaly here. The HF receiver could only display a maximum bandwidth of one half of the ADC sample rate, as set by the Nyquist theorem. But the receivers operating at higher frequencies can display a panadapter that is equal to the sample rate and this seems to break the Nyquist rule.

The reason for this is that the data from the ADC is split into two 16 bit data streams known as the I and Q signals. The I and Q data is used to create the panadapter spectrum and the waterfall display. It is also used to feed the DSP stage so that you can demodulate and hear the receiver or receivers that are displayed on the panadapter. If you only had a single data stream you would only be able to display 5MHz of spectrum from a 10 Msps ADC, but by using both the I and the Q data streams it is possible to accurately display 5MHz above the nominal centre frequency, and 5MHz below it. This does not work for the HF receiver because being direct conversion the nominal centre frequency is zero. You can only display the frequencies above zero and up to the Nyquist frequency which is half the ADC sample rate. Having a higher ADC sample rate means that the receiver can display a wider bandwidth of the received spectrum.

## Nyquist Zones

A direct sampling HF SDR can receive signals from a few kHz above zero to one half of the ADC sample rate. This range of frequencies is called the 'first Nyquist zone.' The range of frequencies from one half of the ADC sample rate up to the full sampling rate is called the 'second Nyquist zone' and there are more zones going up in frequency. For example, if the ADC is sampling at 100 Msps, then the first Nyquist zone is 0-50MHz. The second Nyquist zone is 50-100MHz and the third is 100-150MHz.

The ADC can sample frequencies well above the top of the first Nyquist zone. For example, the ADC in an ANAN radio can sample frequencies up to around 700MHz. So why don't we build receivers that do that? The answer is that you can, but there are constraints. The main one being that you can only use one Nyquist zone at a time.

## Alias Frequencies

Take a piece of blank paper or plastic sheet, something semi-transparent is ideal. Place it on a table with the long side facing up (landscape format). Using a ruler and a pen, draw a line right across the page. Label the left edge of the line '0' and label the right edge '200MHz.' Fold the paper in half and at the fold, label the line with '100MHz.' Fold both sides back so that you have a fan fold 'W' pattern. Label the line on the new folds with '50MHz' and '150MHz.' On the first quarter draw a computer screen that almost fits the range 0-50MHz. Fold the paper flat and draw an RF spectrum with interesting radio signals like you have seen on images of panadapter displays, all the way across the paper. You now have a model of an SDR system. Each quarter page is a Nyquist zone. The signals in the 0-50MHz zone are displayed on the panadapter. Fold the paper so that the second Nyquist zone overlays the first and hold the paper up to a bright light.

The RF signals in the second zone also appear on the panadapter, but they are reversed. A signal at 80MHz appears at 20MHz on the panadapter. A signal at 60MHz appears at 40MHz on the panadapter. The actual radio signals are reversed as well, an upper sideband signal at 80MHz appears on the panadapter as a lower sideband signal at the 20MHz mark. This is not a problem for modes like AM and FM, but it is a concern for SSB stations.

If you overlay the third Nyquist zone you find that the signals are normal (not inverted) and the fourth zone signals are inverted again. With all four zones overlaid you will see a horrible mess of interference signals from 50 - 200MHz overlaying the wanted frequencies. The unwanted frequencies are called alias frequencies. Obviously, it is very important that you stop the ADC from sampling the alias frequencies above the first Nyquist band. Once a signal on an alias frequency has been sampled it is impossible to remove it, so all good HF SDR receivers will have a low pass 'anti-alias' filter before the ADC.

Here is the trick! If you replace the 50MHz low pass filter with a 50-100MHz bandpass filter, you can use the receiver to pick up signals in the 50-100MHz range. The panadapter will be backwards and the sidebands reversed but this is very easy to fix in the SDR software. If you change to a 150-200MHz bandpass filter the receiver can be used to receive 150-200MHz. This effect is used on the Expert SUNSDR2 pro transceiver so that it can receive the 80-160MHz range, but it is not commonly used. Probably because it is more effective to use a front-end frequency shifter or a receiver chip like the Dongle and other small SDR receivers do. That technique offers a smaller panadapter bandwidth over a much wider frequency range, which is generally more useful.

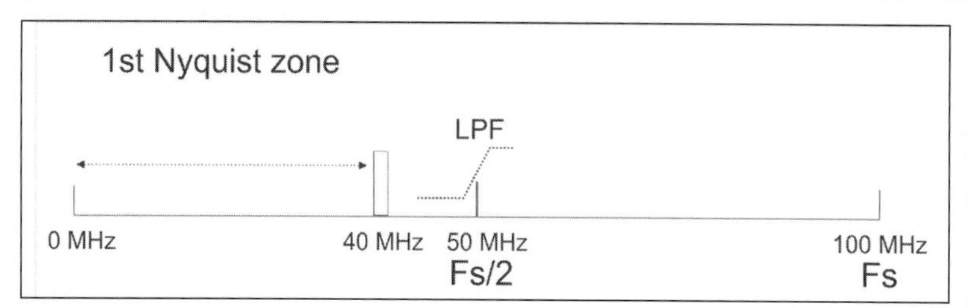


Fig 8.15: Receiving signals in the first Nyquist zone

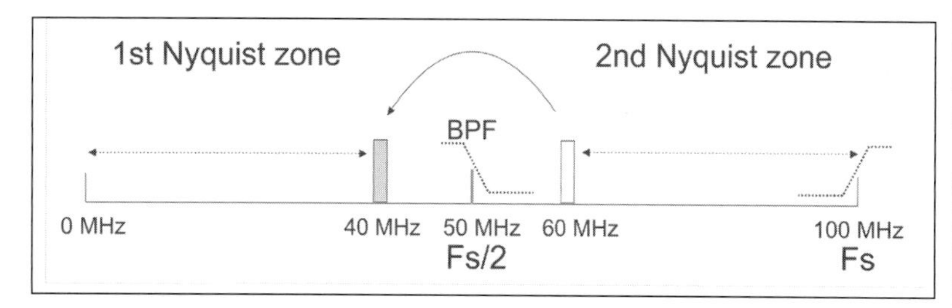


Fig 8.16: Receiving signals in the second Nyquist zone

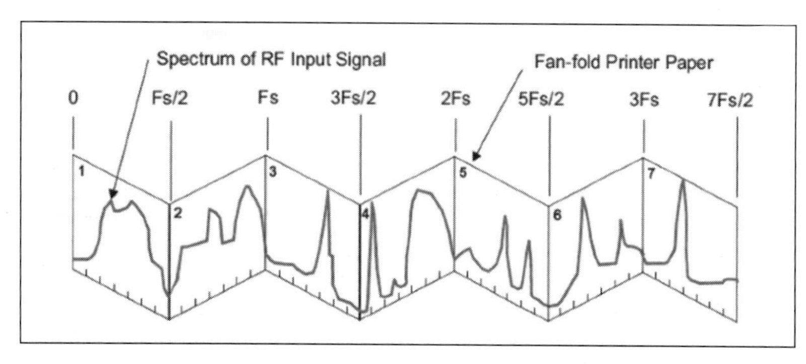


Fig 8.17: Nyquist zones fold back. Every second zone has the spectrum and signals inverted.

## Front-End Filters

Conventional superheterodyne HF receivers have a bandpass filter before the first mixer stage. The filter is there to stop image signals appearing within the I.F. (intermediate frequency) passband where they would interfere with the wanted band of signals.

Direct sampling SDR receivers for the HF bands don't have an 'image problem.' The direct conversion performed by the analogue to digital converter means that direct sampling SDRs don't have a 'hardware' mixer and oscillator so there are no image frequencies. But all SDR receivers should have a low pass or a bandpass filter before the ADC to prevent it from sampling signals from higher Nyquist zones.

The standard configuration for a direct sampling HF receiver is to have a low pass filter at around the Nyquist frequency placed just before the ADC. Signals in the 1st Nyquist zone are sampled by the ADC and end up displayed on the panadapter. Signals from above the 1st Nyquist zone are not sampled because they are attenuated by the anti-alias low pass filter.

If you want to receive signals in the 2nd Nyquist zone you can replace the low pass filter with a bandpass filter that covers the second Nyquist zone. In the example drawing, the filter would pass frequencies from 50MHz to 100MHz. The 60MHz signal is reflected back and it appears on the panadapter at the 40MHz position. In a real SDR receiver, the panadapter would be adjusted (flipped) to show 50 to 100MHz and the inverted sideband would be corrected as well.

The Nyquist zones are rather like a semi-transparent, fan folded, paper map. If you look through page one, the second page folds over and is reversed but the third page folds over with the spectrum the right way around again.

The drawing showing the Nyquist zones (**Fig 8.17**) illustrates the way that 'even' numbered Nyquist zones are reflected or folded back in an inverted way where the panadapter would appear to be reversed. The 'odd' numbered zones are folded the same way as zone one so the panadapter and sidebands are not inverted.

Note that except for modes that only use a single sideband, or which carry different information on each of the sidebands, the sideband reversal does not matter. AM and FM are unaffected by the sideband reversal since the sidebands are identical. SSB, digital modes carried as audio on a single sideband, ISB and some phase shift modes would need to be corrected because USB becomes LSB. PSK31 will decode OK because the information is carried by the phase changes, but the operating frequency would be offset below rather than above the nominal 'carrier' frequency. RTTY would be affected as it uses high and low tones, which would become reversed.

## Anti-Alias Filter Frequency And Response

The design of the front-end filter and the choice of its 'roll-off' frequency is quite important. Especially when you are talking about an HF receiver. The radio must be protected from the very high-power FM broadcast stations operating in the 88-104MHz range. Those signals are usually in the second Nyquist zone and without a good anti-alias filter before the ADC they can break through and be displayed on the panadapter display as wideband interference.

There is a temptation for SDR manufacturers to use as high a sample rate as they can, in order to get a receiver that can cover a wide frequency range. But if, for example, you use an ADC with a 160MHz clock, the Nyquist zone ends at 80MHz and that is perilously close to those large FM broadcast transmissions. Unless the anti-alias filter has a sharp cut-off, there is a strong chance of getting interference from the FM band on the panadapter. This break-through of unwanted signals can be avoided if the anti-alias filter is tuned for a frequency that is lower than the Nyquist frequency. Yes, you lose some of the receiver's possible frequency coverage, but any large signals above the Nyquist frequency are attenuated to a much greater degree and they probably won't appear on the panadapter.

## ADC Dither

Some SDR software and SDR radios allow you to select an option called 'Dither.' Dither improves the intermodulation performance of the radio by ensuring the input signals are well spread across the analogue to digital transfer characteristic of the ADC. Noise is added to the analogue signal and then removed after the analogue to digital conversion. In the LTC2208 ADC, the noise is generated as a pseudorandom digital code inside the ADC. The code is converted into an analogue signal and mixed with the incoming signal from the antenna. After the combined signal has been sampled and converted to digital data the original digital code is mathematically subtracted back out again. The result is better IMD performance with only a small degradation in noise performance.

The RF that the ADC samples, is the vector sum of everything at its input, so unless you have a particularly poor antenna there will usually be enough medium-size signals to ensure that the ADC input signal will transverse most of the ADCs range. Dither could be helpful if you are using external narrow band filters for contest operation and it does improve the IMD performance when a 'two-tone' test signal is used as part of lab testing. But in the real world, with the antenna connected, there are many signals being processed by the ADC so using the dither function is usually unnecessary.

## ADC Random Function

Some SDR software and SDR radios allow you to select an option called 'Random.' "The ADC is a small device with a big problem." On the antenna side, there are very small sub microvolt signals which we want to hear in our receiver and on the other side, there are very high speed 3.3 V square waveform data and clock streams. Inside and around the chip both signals are in very close proximity. The ADC samples at a high rate such as 122.88 Msps and the same clock drives the FPGA. At 16 bits per sample, the output stream is a square wave at around 1.97GHz. Capacitive and radiated RF coupling of the large 3.3 V signal into the sensitive analogue input stage is a problem even with very good circuit board design. The effect, particularly when there are a limited number of input signals such as during testing, is that the coupling causes fixed signal spikes called spurs which can be demodulated and are visible on the spectrum display. The generation of spurs is a particular problem when the radio is being used in a situation where the receiver signal(s) are at fixed levels or frequencies such as a fixed radio link or cellular base station. The random nature of the signals on the ham bands means the input signal is fairly random and so the output signal is quite variable. In most cases, this means the internal random function does not need to be used while listening on the ham bands.

The problem with fixed levels and signals on the ADC input is the repetitive nature of the data coming out of the ADC. This causes harmonics and related intermodulation products which could be coupled into the ADC input. Signals up to 700MHz can be aliased back into the wanted bandwidth and they are being generated after our front end filters. The answer is to randomise the data coming out of the ADC so any coupling becomes noise rather than visible and audible spurs. The LTC2208 ADC does this by applying an Exclusive OR function to bits 1 to 15 of the output signal based on the value of bit 0. This produces a randomised output bit stream resulting in a clean spectrum display with no spurs. As soon as the scrambled signal gets to the FPGA it can be converted back into usable data by simply applying another Exclusive OR function to bits 1 to 15 based on the value of bit 0.

The output of bit 0 is random due to the noise performance of the ADC so the scrambling function really does produce a randomised output. ADCs used in other SDR receivers may not include the internal application of dither and random. The functions may be added with other hardware or not used at all.

## The FPGA

After the signals have been converted into a digital data stream by the ADC they are passed on to the DSP part of the radio inside the FPGA (or some other computer device). The functions of the FPGA can be described in terms of analogue signals and depicted as hardware gates and devices, but all the signal processing is actually achieved by the manipulation of digital bits. The internal gates, registers, and memory are built out of simple logic elements and are programmed connections rather than existing as physical devices inside the chip.

The Altera Cyclone IV FPGA in an ANAN radio performs the tasks of down conversion and filtering. It does this job very well and very fast. Some FPGAs can be configured to act like a 'software' or microprocessor CPU, allowing them to run programs as well as performing the DSP functions. The 'softcore' CPU can be used to control functions in the radio. FPGA chips have between 39,600 and 150,000 logic elements which can be configured to do many tasks simultaneously. An FPGA may be a little slower than a microprocessor, but it has a much higher capacity for parallel processing, and it can be reconfigured internally at any time.

Like a microprocessor in a computer, the FPGA does not retain any internal connections or information when it is not powered. The code which governs how it works is stored in external non-volatile memory and is loaded into the FPGA when you turn the radio on. Firmware updates overwrite the information in the non-volatile memory and then the new file is loaded into the FPGA when the chip is reset.

If the random function has been turned on in the ADC, the first thing the FPGA software does is remove the randomisation of the incoming data stream. The input signal to the FPGA is a 16-bit digital data stream containing the sampled spectrum from 0 Hz up to the Nyquist frequency (half of the ADC sample rate). If we assume an ADC sample rate of 122.88 Msps, the data is coming out of the ADC at a rate of 1.966 Gbits per second. It is not practical to send such a lot of data over a USB or Ethernet connection to the PC, so many current SDR models use a mixer plus decimation and filtering in the FPGA to select only a part of the HF spectrum. The new FlexRadios get around this problem by performing the DSP functions inside the radio, dramatically reducing the amount of data that needs to be sent to the PC. Only the data required for the demodulated receiver audio and the panadapter display has to be sent to the radio. Of course, radios 'with knobs' like the Icom IC-7300 perform all of the DSP and the display and control functions inside the radio.

## Factors Affecting The Dynamic Range of SDR Receivers.

Some people wonder how the panadapter can display a much wider dynamic range than the quoted dynamic range of the analogue to digital converter (ADC) chip. They complain wrongly that this can't be true and that the SDR manufactures are faking the dynamic range quoted in specifications and shown on the panadapter display. The dynamic range of a software defined radio receiver is the difference between the noise level displayed on the panadapter display when the antenna is not connected and the level at which the ADC becomes overloaded and clipping occurs. The achievable dynamic range is affected by the number of bits that the ADC uses (8, 14, or 16 bits), but in most SDRs, it is at least 100 dB. This is much more than the dynamic range of superheterodyne receivers which must use AGC to compensate. AGC reduces the RF gain when there are large signals in the receiver passband making the superheterodyne receiver seem to have a wider dynamic range than it actually has.

Narrow bandwidth receivers have a better signal to noise ratio than wide bandwidth receivers. This is the main reason that CW signals received on a receiver with a narrow 500 Hz filter are easier to copy than SSB signals received on a receiver with a wider 2500 Hz filter. Think of leaving a soup bowl and a beer bottle out in a heavy rainstorm. After an hour pour the water out of the soup bowl into a second beer bottle and compare the water in each. The wide bandwidth soup bowl captures much more noise (rainwater) than the narrow bandwidth beer bottle. The amount of air between the water level and the top of each bottle is the dynamic range of each device. When you apply this theory to the panadapter there is an immediate problem. The panadapter has a very wide bandwidth. Not as wide as the spectrum being sampled by the ADC, but much wider than the 2.5kHz bandwidth of an SSB receiver. That implies that the panadapter will have a poor dynamic range. The receiver will be more sensitive than the panadapter and you would be able to hear signals in the receiver that are not visible on the panadapter. That is not what you want.

The trick to improving the dynamic range of the panadapter display is a manipulation of bandwidth. The panadapter is created using the same method as used in digital storage spectrum analysers. Instead of displaying the spectrum as a whole. The panadapter displays the level of a thousand or more narrow frequency 'bins' arranged side by side. Each frequency bin has a very narrow bandwidth, only a few hertz wide, so each bin has a high dynamic range. Put them side by side and you can create a wide spectrum display with excellent dynamic range. This ensures that the panadapter can show any signal that you are able to hear in the receiver.

The dynamic range of a direct down-conversion SDR receiver is predominately affected by;

1. The effective number of bits (ENOB) calculated from the published SINAD, not the number of bits the ADC uses to output the data
2. Process gain arising from the ratio of the sampling frequency to the final receive bandwidth, and
3. The number of large signals being received. Actually, the sum power of all signals arriving at the input to the ADC

The noise floor displayed on the spectrum display with no antenna connected is affected by

1. All of the above, plus
2. The bandwidth of the FFT (fast Fourier transformation) bins displayed, and
3. A constant (k) determined by the type of 'Windowing' calculation in use

As long as the noise floor displayed on the spectrum scope is lower than the noise level within the receiver bandwidth, as indicated by the S meter, you will be able to see any signal you can hear.

## Process Gain

Process gain is a trade-off between the dynamic range of the radio and the bandwidth of the signal. The ADC might have an SFDR (spurious free dynamic range) of around 95 dB, but that is while sampling the entire 700MHz bandwidth of the ADC. The noise level is spread out. When you filter out the frequencies you don't want and just look at the signal in a narrow bandwidth such as the 2.4kHz receiver bandwidth required for an SSB signal, or a 500 Hz wide bandwidth for CW there is much less noise and so the dynamic range and SNR is much improved. The improvement in signal to noise ratio is directly proportional to the reduction in bandwidth. It is normally expressed in decibels, as a gain in signal level, because an increase in the signal to noise ratio is equivalent to an increase of the dynamic range. Process gain = 10 * log10(BW1/BW2) dB.

Process gain is the reason that an SDR can deliver more than 120 dB of dynamic range from an ADC that might only have around 77 dB SINAD.

- The theoretical maximum dynamic range of an ADC is (N bits x 6.02) +1.76.
- For a 16 bit ADC, it is (16 x 6.02) +1.76 = 98 dB.

However, the ADC's internal noise performance reduces the number of usable bits. The noise causes the ADC to confuse the reading of very small input signals and this reduces the effective number of bits. For example, a 16 bit ADC may only be able to output 12 bits of usable data. This is known as the 'effective number of bits' or ENOB.

- For a 16 bit LTC2208 ADC, the published ENOB is 12.6.
- The 'real world' dynamic range of an ADC is (ENOB bits x 6.02) +1.76.
- The dynamic range is (12.6 x 6.02) +1.76 = 77.6 dB.

But this dynamic range extends over the entire 700MHz range of the ADC including over the bandwidth of the first Nyquist zone. So we can add some 'process gain' to the receivers dynamic range.

- Process gain for an SDR is 10 x log (Fs/2 / B), where Fs is the ADC sample rate, Fs/2 is the Nyquist bandwidth and B is the receiver bandwidth.

As an example, for an ANAN radio with a sample rate of 122.88 Msps the process gain is

- 10 x log (122,880,000 / 2 / 2400) = 44 dB for a receiver bandwidth of 2.4kHz.
- 10 x log (122,880,000 / 2 / 500) = 51 dB for a receiver bandwidth of 500 Hz.

The overall receiver dynamic range is 77.6 + 44 = 121.6 dB for a receiver bandwidth of 2.4kHz and 77.6 + 51 = 128.5 dB for a receiver bandwidth of 500 Hz. This compares well with the measured performance of the ANAN-100 receiver which is better than 120 dB in a 2.4kHz bandwidth and better than 128 dB in a 500 Hz bandwidth.

## ADC Gain Compression:

When you connect the antenna to your receiver the noise level shown on the spectrum display increases. This happens for two reasons. Firstly, the antenna picks up all kinds of; atmospheric noise, electrical noise, splatter and harmonics from transmitters, and in my case interference from my ADSL modem. Secondly, the dynamic range of the ADC is affected by the total power of all the signals presented to it. Receiving a lot of large signals reduces the dynamic range and the displayed noise floor rises. We don't want any signals from above the Nyquist frequency and we also don't want unwanted signals, such as AM broadcast stations, reducing our receiver's dynamic range and signal to noise ratio. Having a few large signals, from within the Nyquist bandwidth will not affect the noise floor by too much and will act to reduce the effects of intermodulation distortion.

In many SDRs, hardware bandpass filters are installed before the ADC to protect the receiver from gain compression and to combat possible out of band interference. They must be designed and made with care so that they do not add any new noise or cause intermodulation problems.

SDR dongles that don't have bandpass filters are quite susceptible to ADC gain compression. You can see the noise floor rise indicating a reduction in the dynamic range when very strong signals such as FM broadcast stations are within the sampled bandwidth.

## SDR Transmitters

Almost all of the benefits and advancements achieved from using software defined radios relate to receiver performance. In most cases, the transmitter performance of software defined radio transmitters is exactly the same as older architecture transmitters because it is mostly governed by the linearity of the transmitter power amplifier. There are some promising advancements under development including very efficient transmitters using 'class E' modulation. CESSB (controlled envelope SSB modulation) is an enhancement that is available on a few of the latest HF SDR transceivers. CESSB is a way of maximizing the talk power of your SSB signal. It works like a compressor, but it is impossible to overdrive the modulator. There is no clipping and overshoot is controlled. Your average transmitted power is almost doubled. That is like adding another Yagi on top of your existing antenna. Both of these advancements are possible due to the fact that the signals to be transmitted are digital signals, which can be manipulated by software before being converted to analogue and transmitted. CESSB could be implemented in conventional radios with DSP, provided there is sufficient number crunching capability in the radio.

## Adaptive Transmitter Pre-Distortion

Amplifier nonlinearity in the transmitter power amplifier or a linear amplifier results in poor IMD (intermodulation distortion) performance, which is bad. It may distort your transmitted signal making it harder to copy and it causes the radiated signal to be wider than the wanted bandwidth creating splatter that interferes with other band users. Adaptive transmitter pre-distortion modifies the audio modulation signal to correct for the nonlinearity of the amplifier so that the net result is a flat linearity response. This improves the transmit IMD performance a lot. It sounds easy, but it is not! Both phase and amplitude linearity errors need to be corrected and the nonlinearity characteristics change dynamically with changes in frequency, antenna load, temperature, power supply voltage and all sorts of other factors.

The PureSignal adaptive transmitter pre-distortion software was written by Warren Pratt, NR0V. It has been incorporated into the PowerSDR mRX PS SDR software package used for OpenHPSDR radios, Hermes Lite, and the Apache Labs ANAN transceivers. It is able to dramatically improve the transmitter's IMD performance, achieving results that were previously unheard of.

What is transmitter nonlinearity? All RF power amplifiers introduce some nonlinearity into the transmitted signal. 'Nonlinearity' just means that the relationship between the input power and the output power is not entirely flat. In other words, the output signal is never a perfect but louder copy of the input signal. The power amplifier in your transceiver may have a little less gain when the input signal is small, or it might not put out as much power as it should when the input signal is large. This is usually due to the power supply voltage drooping under a high current load. Components heat up under high current conditions. This can cause them to have more resistance or to change their capacitance or inductance, which might de-tune the transmitter slightly. All of these factors combine to introduce two kinds of nonlinearity. Amplitude nonlinearity where the gain of the amplifier does not remain the same for all input levels and phase nonlinearity is when the phase of the output signal is being affected by the input level or frequency.

Using a 'linear' amplifier after your transceiver adds an additional source of nonlinearity. If you place a feedback coupler at the output of the linear amplifier instead of the output of the transceiver, the adaptive pre-distortion software can actually compensate for the overall nonlinearity of the transceiver and linear amp combination.

Why do we want good transmitter linearity? When we transmit a digital mode or SSB HF transmission, the last thing we want is complaints that our transmission sounds bad, or that it is causing splatter interference across the band. We are legally required to ensure that we do not cause interference to other band users and this includes making sure that our signal is mostly contained within the wanted bandwidth, usually less than 3kHz for an SSB signal.

SDR transceivers are in general no better and no worse than any other kind of ham transceiver and using the standard 'two-

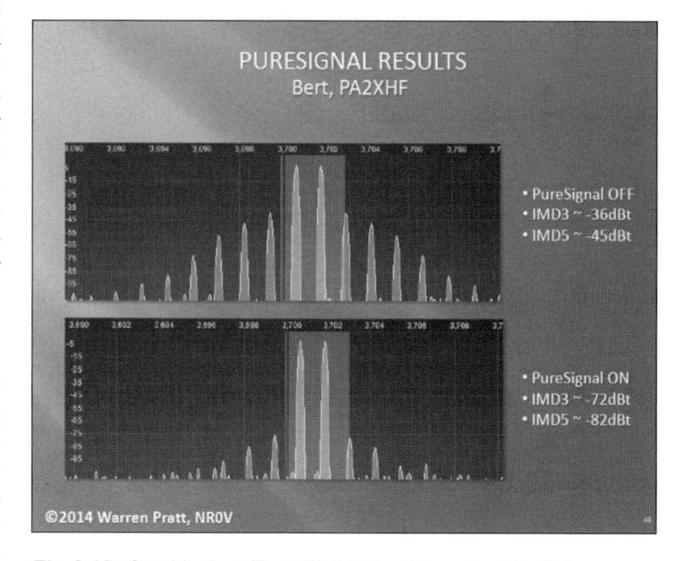


**Fig 8.18: On this test Bert PA2XHF achieved a 36 dB improvement in transmitter IMD performance.**

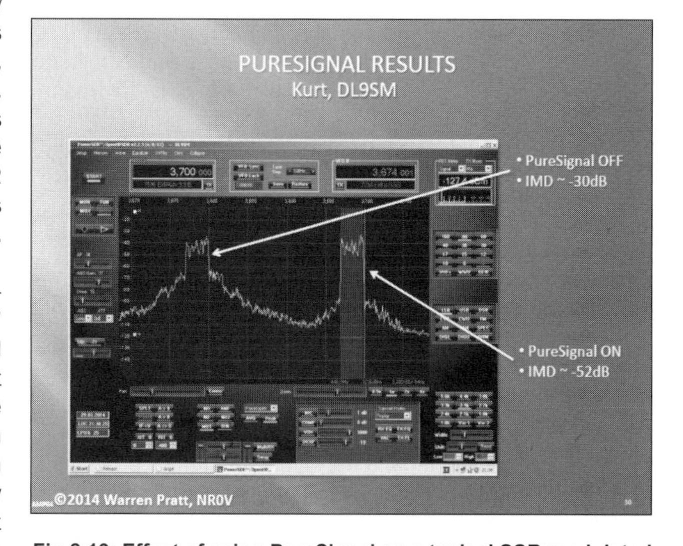


**Fig 8.19: Effect of using PureSignal on a typical SSB modulated signal**

tone' test, the 3rd order IMD product is usually between 30 dB and 38 dB below the test tone level. In some cases, this level of interference could be heard at the receiver's location and would interfere with a station operating very close to the transmit frequency. For example, if you were transmitting USB on 14.200MHz and a station was receiving your signal at 30 dB over 9, then a station in the same locality listening on 14.203MHz would hear interference at around S9. Improving the transmit IMD performance of your transmitter is a way of being "nice to the neighbours." Transceivers with higher voltage 50V output stages tend to have slightly better 3rd order transmit IMD performance than 12V transceivers. The improvement is typically 5 dB - 8 dB. Transmitter pre-distortion is able to improve the 3rd order transmit IMD performance by a massive 30 dB to 35 dB (**Fig 8.18**).

PureSignal is an innovation for PowerSDR mRX PS that makes use of the full duplex capability of the software defined radios. It also requires the wide bandwidth capability of the transmitter and the calculating power of the PC application.

Full duplex is the ability to receive and transmit simultaneously. Most conventional architecture HF radios are not able to operate in a duplex mode. They tend to use the same local oscillator and I.F. amplifiers for transmitting and receiving and their receivers may not be able to cope with the high levels from the transmitter. Many VHF and UHF radios are able to operate in full duplex mode, but that is more often used for cross band operation. So far, there are very few ham radio transceivers with the ability to use 'adaptive transmit pre-distortion.'

The program was developed by Warren Pratt NR0V. It dramatically improves the transmitter's intermodulation performance by adjusting the modulated RF signal going into the RF power amplifier so that the signal at the output of the amplifier is a more accurate but amplified representation of the input signal. It is so effective that when PureSignal is employed, it currently gives the ANAN radios better transmit intermodulation performance figures than any other HF ham radio transceiver tested by ARRL labs, (based on reviews published in QST). The ANAN-8000DLE claims a 'third order IMD' typically -72dB below PEP @ 200W output on the 20m band (**Fig 8.19**).

How does it work? A low-level sample of the transmitter's output power is taken, usually from a coupler at the antenna port. It is received and demodulated by the receiver, which in an ANAN or Hermes radio is always running because the radio is full duplex. It is extremely important that this feedback signal is attenuated by at least 50 dB to avoid damage to the sensitive receiver front end.

The demodulated sample signal is compared with the audio input signal being sent to the modulator. Phase and amplitude changes are applied to the modulated signal in order to dynamically reduce the IMD products. This is not straightforward because the amount of adjustment required changes continuously with the changing input signal. The level and phase adjustment needs to adapt. Also, the amount of nonlinearity caused by the amplifier will be different at different operating frequencies and when the transceiver is connected to different antenna impedances. If the radio was just sending an un-modulated carrier signal the phase and level adjustment would be easy. But then life would be boring because you can't convey any information with an un-modulated carrier signal. No, not even CW, which is an amplitude modulated digital mode, "carrier on, carrier off".

Inside the SDR software both the audio signal being used to modulate the transmitter and the audio from the received feedback signal are digital signals. The PureSignal adjustments can be calculated mathematically and applied by changing bits in the 'transmit' audio data stream. Over a short period, the software adjusts the modulation signal to compensate for the IMD created by the amplifier. It continuously compares the transmitted signal with the modulation and adapts to changing conditions.

The idea of using pre-distortion to improve the transmitter's linearity has been around for quite a while, but it is tricky to implement with conventional electronics.

In a software defined radio, the problem can be resolved using a mathematical algorithm and that allows fast, dynamic, adjustment of the transmitted signal. The receiver needs to be working at the same time as the transmitter because you need to continuously receive and compare samples of the transmitted signal. The process needs to be able to apply adjustments over a wider frequency range than the normal SSB transmission bandwidth so that harmonics of the modulation frequencies are included. This is another problem if you are trying to implement adaptive transmitter pre-distortion on conventional architecture radios.

Adaptive transmitter pre-distortion could be implemented in the DSP chips of conventional transceivers. But, they do not usually have enough transmitter bandwidth and most are not able to transmit and receive on the same band at the same time. The reuse of circuit blocks such as the local oscillator limits the ability for many conventional radios to operate in a duplex mode. Future designs may use an embedded SDR receiver or a sub receiver to receive the required feedback signal.

The technical background: Amateur radio transceivers use linear power amplifiers because the voice signals that we want to transmit vary in amplitude, causing the transmitted RF signal to range from zero to the full output power of the radio. You see this as the peaks on your transmit power meter during SSB operation. It is important that the output power of the amplifier accurately follows the changing level of the input signal. This is called the amplifier's amplitude linearity.

During AM or SSB transmission, the phase of the output signal should not be affected by changes in the level or frequency of the audio input signal, this is called the amplifier's phase linearity. Phase nonlinearity can be caused by components changing their inductance or capacitance when they are stressed due to high voltage or current causing heating. The change of value causes de-tuning of the RF power amplifier resulting in a phase shift. Phase nonlinearity effectively causes a small amount of phase modulation when we want amplitude modulation.

Severe nonlinearity causes the different frequency components of the RF signal to mix together creating unwanted sum and difference frequencies. The result is known as 'intermodulation distortion.' The amount of degradation is dependent on the design and tuning of the output stage of the amplifier, the output filters, load impedance, drive level, the frequency of operation, modulation signal, the stiffness of the power supply and components like toroids, and the output transistors.

Linear amplifiers tend to be less linear at low input levels and the output power tapers off near full drive level. If you chart the output power vs input power you would see a slight S shape rather than the desired linear line.

Intermodulation products mixed into the wanted transmitter signal sound bad because they interfere with the wanted speech. Intermodulation products that are created outside of the wanted transmit bandwidth are heard as 'splatter' and make the transmitted signal wider than intended. If your transmitter is only used for CW and digital modes like PSK and RTTY the audio modulation level does not vary. The modulation is either on, or off. With these modes, the RF amplifier does not have to be linear. They generate much less IMD but PureSignal can still be used.

What it can't do: PureSignal can improve amplitude and phase linearity problems. But is less successful with 'memory effects' because they are related to past events, not the input signal that is currently being compared. 'Memory effects' are phase or amplitude distortions caused by earlier signal events affecting components in the radio. For example, a large input signal could cause the RF transistors and heatsink to get hot resulting in an amplitude and /or phase change in the amplifier output signal. When the input signal level drops, the devices take some time to cool so the distortion persists even though the input signal has reduced. Another example is a high input level causing the amplifier to run at near full power, drawing maximum DC current. This may cause the power supply voltage to droop and the lower supply voltage will affect the amplifier linearity. When the input signal level reduces, the DC supply voltage takes some time to return to full voltage, so again the distortion persists even though the input signal has reduced.

Adaptive pre-distortion can only improve IMD caused by non-linearity that is attributable to the input modulation signal. It may not be able to adjust for fixed sources of IMD such as saturated Toroids in the output filter, badly tuned linear amp stages, or IMD occurring in the antenna system after the feedback coupler.

## SDR For Radio Listening

A lot of software defined radios are used for listening to the shortwave bands and for listening to interesting signals on the higher bands, such as the aircraft band or the marine band. You can also listen to commercial traffic and unencrypted military communications. There is a wealth of other transmissions, which are of interest to some people. Aircraft ADS-B positioning information can be presented on maps showing commercial flights. ACARS text data has information sent from aircrew to the ground stations. Maritime VHF AIS beacons indicate the ship's name, cargo, and position information.

Software defined radios are great for listening. You get great performance and a much wider frequency range than was available on older receivers. Most SDR software has good noise suppression filters, which are important when you want to listen to broadcast quality audio like music or news. Another big advantage is that you can set the bandwidth to match the signal that you are listening to. The AM filters in older receivers that were primarily designed for SSB may be too narrow and not adjustable. Being able to adjust the receiver's bandwidth precisely means that you don't have to listen to noise from outside of the transmitted signal. Being able to increase the bandwidth past 3kHz means that you can listen to the full fidelity of the AM signal. The audio sounds much nicer. Also, software defined radios have a better ability to drag out weak signals that are near large signals they are less prone to intermodulation distortion.

There are a dwindling number of commercial AM broadcast stations on the HF short wave bands. Many of the remaining ones broadcast religious programmes. Some countries still broadcast general music, news, weather, and current affairs. The BBC, Radio Australia, and Radio New Zealand International are still broadcasting 24 hours a day on several frequencies. They change frequencies at intervals to maximize the reach into various regions as daily propagation changes.

## FAX and Weather FAX

You can decode unencrypted commercial radio fax and weather fax pictures. Some of these are on the HF bands and some use satellite downlinks. Software defined radios are excellent for receiving the 137MHz NOAA weather satellites because you need a wideband FM receiver. Unlike software defined radios, most VHF communications receivers and amateur band transceivers capable of receiving 137MHz only have narrow band FM.

There are weather fax transmissions on many HF frequencies. Radiofax is also known as HF FAX, or WEFAX, although WEFAX is also used to describe satellite downlink based systems. Reception of WEFAX requires an SSB receiver. Note that the published frequencies for weather fax refer to the frequencies marked on dedicated weather receivers. An SSB radio should be tuned for upper sideband at 1.9kHz below the nominated frequency.

## APRS

APRS (automatic packet reporting system) beacons include position and other data relating to; ham radio operators, experimental balloon and rocket flights, vessels using inland waterways and seaports, and even satellites. APRS is also commonly used by amateur weather station operators, so you can use it to check the real-time weather conditions at various locations around your city or elsewhere.

## SDR for CW

CW operators can expect excellent performance from SDR transceivers. There are significant advantages in using an SDR for CW operation. Using the waterfall and panadapter spectrum displays you can see the strength and quality of each CW signal on the band. Allowing you to easily, skip from station to station across the band. Tuning each signal is a breeze. Just click on each station on the waterfall. The very narrow filters allow you to listen to only a single CW signal without hearing any other QSOs. At the same time, you can run CW Skimmer or other software like MRP40 and see all the activity on the band, or on several bands at once.

## CW Skimmer

CW Skimmer by Afreet Software (Alex Shovkoplyas, VE3NEA), has been developed with wideband receivers like software defined radios in mind. When used with a conventional radio, Skimmer is restricted to decoding CW signals within the radio's 3kHz passband. But when using the IQ signal from a software defined radio, Skimmer can decode up to 700 signals in parallel. You can download CW Skimmer for free for a 30-day trial but after that, the program must be registered.

CW Skimmer is widely used by CW enthusiasts, DXpeditions and contest stations. It decodes many CW signals simultaneously allowing the operator to quickly identify stations without having to listen to them. You can easily move up the band working each station in turn, without duplicating contacts or wasting contest time listening for call signs.

## Built in CW

Some software defined radios and some + SDR radios include a CW decoder as well as the more normal CW keyer. The Elecraft KX3 (SDR) and K3S (+ SDR) radios can decode and display CW, RTTY, and PSK. If you are proficient with a paddle, the KX3 can also convert the CW that you send via the 'key' into RTTY or PSK transmissions.

## SDR for Digital Modes

The main advantages of using an SDR for digital modes are the ability to adjust the filter bandwidth so that you can see and decode the whole of each digital mode band segment, the ability to operate on several bands at the same time, and the ability to operate several digital modes at the same time. You also get the same advantages as you do when using an SDR for SSB operation such as a range of noise reduction filters and blankers.

Using software defined radios for digital modes is not much different from using a conventional radio. You do get the advantage that you don't need to connect an audio cable between the radio and the PC for digital mode operation. The audio signal between the SDR software and the digital mode software is carried by a 'virtual audio cable,' which is a software utility program. If you have a FlexRadio Signature range transceiver, you will use the built-in DAX (digital audio exchange). Rig control is performed using CAT or CI-V commands sent to the SDR software over a virtual com port. You can use com0com or Virtual Serial Port 'VSP' to connect the digital mode software Com port to the SDR software Com port.

In the future, we will see more digital mode software that supports IQ input signals, or which include a virtual audio cable and com port manager into the package.

One difference when you are using an SDR is that the bandwidth of the signals sent to the digital mode software can be changed. You can view much more than the normal 2.5kHz of PSK signals. When you have decided to concentrate on one signal, you can narrow the receive filter down so that only one signal can be heard, or 'seen,' and decoded.

## Dolly or Twin Pass Band Filters

Some SDR transceivers and PC software feature double lobe filters especially for receiving RTTY signals. They are very good. They will allow you to copy with 100% accuracy signals that are just random strings of letters without the filter.

## SDR for Contesting

I love using my software defined radio in contests. It is one of the areas of the hobby where using a software defined radio really helps a lot. In an SSB contest, most stations are spaced at 1kHz or possibly 500 Hz apart. If you set the mouse wheel tuning for a 500 Hz step, you can click directly onto any station in the panadapter and jump to the exact frequency. This is great for 'search and pounce' contest operation. You simply click on each signal in turn and work up and down the band. You will quickly know where the 'big gun' stations are operating because you can see them on the panadapter. You get to recognize the "look" of each station, so you won't waste time carefully tuning to them over and over again, through the night.

If you are in 'run mode,' you will be able to see if your frequency is being squashed by big stations nearby. You can check the status of other bands by opening a second panadapter. No more wasting time and possibly losing your run frequency while you tune around a dead band.

In a busy contest, it is inevitable that there will be stations overlapping. You will be able to see this on the waterfall and panadapter displays. With careful filtering, you can usually eliminate enough of the interference to work both stations. Receivers without I.F. DSP don't have that advantage. Being able to see exactly where the two signals are, really helps you to set the filters in a way that accentuates the wanted signal and masks the unwanted one. You can watch the relative levels of each station as you turn the beam as well.

DSP receiver filters are sharper than non-digital filters and they can be adjusted to exactly the right place. Other radios may have good filters and they may be adjustable, but without a panadapter or a band scope, you lack the ability to place them exactly where you need them.

There is no point in having a receiver passband that is any wider than the transmitted signal. All that you get when listening outside that bandwidth is unwanted noise. The waterfall display will show you exactly how wide each transmitter is operating, and you can adjust the receive filter width to achieve the best signal to noise.

The AGC action in conventional receivers will reduce the receiver gain if there are strong signals within the I.F. passband. This desensitizing makes it hard to work weak stations that are operating near strong signals. This desensitizing does not occur in software defined radios. So, it is much easier to work weak stations. The blocking dynamic range of a software defined radio is equal to the clipping level of the radio. SDRs do not normally exhibit any loss of sensitivity when there are big signals within the receiver passband or displayed on the panadapter.

During a contest, the band is packed with many signals operating very close together. Receiving a lot of signals in the receiver pass-band at the same time is a problem for conventional radios. The signals mix together and mix with the local oscillators inside the receiver causing intermodulation products, which can be heard. This distortion adds to the cacophony of noise that is so typical in a contest situation. A lot of old-timers who complain that they "can't hear anything" during contests are experiencing this kind of intermodulation distortion. They believe that it is poor transmitters causing splatter, when in fact it is intermodulation distortion occurring within their own receiver. Dealing with this intermodulation problem is why we prize receivers with very good 'close in,' 2kHz spacing, IMD performance. SDR receivers usually have the same IMD performance at narrow spacing as they do with signals spaced further apart. So, they perform very well when there are many stations operating very close together.

Most SDR software has a selection of different noise reduction filters and noise blankers. You should be able to find one to combat any kind of noise interference.

Another advantage of using a software defined radio for contesting is that you will probably be using a contest logging program and you might be using digital mode software or an application like CW skimmer as well. It is convenient to be operating the radio on the same computer. The best way is to have the SDR displayed on a second computer monitor. PowerSDR mRX has just introduced a function that will return active control to the N1MM logging program or other nominated software as soon as you have finished making any adjustment on the radio. This is specifically intended for use during contest operation. It gives the logging program priority over the radio to stop you inadvertently entering keystrokes into the SDR program when you want them in the logger or CW keyer.

With the latency and delayed side tone problems now fixed, SDRs are great for CW contesting. The same advantages mentioned above also apply to CW contesting. You can click directly on each CW signal and net automatically to the transmitter offset tone. With an SDR, you can use super narrow 'brick wall' filters, which can isolate one CW signal even when the band is frantic. With careful filtering, you can make a signal sound as clean as a code practice oscillator in the shack.

## SDR for Interference Monitoring

Some people like to monitor the bands for interesting signals and intruders on the amateur radio bands. Software defined radios offer many advantages for this type of activity. For a start, a software defined radio is effectively a spectrum analyzer, so fairly accurate measurements of the signal level and frequency can be made. You can monitor a wide range of frequencies on the panadapter and waterfall display so you will see bursts of illegal transmission that might be missed if you were just tuning around with a conventional receiver.

## Ionosondes And Signals That Change Frequency

'Whistlers,' Ionosondes, and other signals that are swept across a frequency range may only show up as a blip on a conventional radio, but they are displayed as a diagonal line on a waterfall display. Since you know the rate that the waterfall is updating in frames per second and its width, you can calculate how fast the signal is changing frequency. Fast sweeps cause a flat diagonal line on the waterfall and slow sweeps cause a steeper diagonal line.

An Ionosonde is a Radar that examines the Ionosphere by sweeping the HF band and receiving the echoes. Ionosondes sweep from 1-40Mhz. They send a line of very short pulses up the band. It looks like a slanted line when observed on a waterfall display. *http://www.sigidwiki.com/wiki/Ionosonde.*

## Recording

Many SDR programs include the ability to record signals or even the whole panadapter. If you have recorded the whole panadapter IQ stream, you can listen to signals that you were not tuned to when the recording was made. This means that you can leave the receiver recording a band overnight, then observe, and listen to any of the signals at a later time. This is a really neat feature if you are watching a band for intruders or waiting for a signal. You could use this feature to find out when a band is open to a DXpedition, to watch the downlink from a satellite that is passing over at an inconvenient time, or to catch a pirate who is operating intermittently.

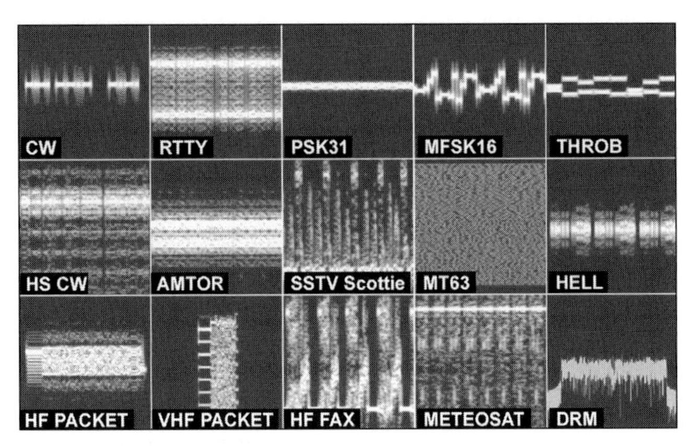


**Fig 8.20: Digital modes have distinctive patterns**

## Identification

The waterfall and spectrum displays can help you to identify what mode is being transmitted. Digital modes and signals like fax and packet have distinctive patterns, which can be observed on the screen (**Fig 8.20**). You can see and hear what some modes sound like at:

*http://www.astrosurf.com/luxorion/qsl-ham-history15.htm*

## Interference

You can use the panadapter to see what kind of interference you are experiencing. Wideband noise, wobbly frequency variations, and multiple interfering signals show up very well. Combs of signals and swept signals are also easy to spot. Sometimes you can see interference that is generated inside your building. Likely culprits are plasma TVs, switch mode power supplies especially cheap ones, powering computers and other portable devices, and low voltage lighting systems. ADSL routers used for Internet connections are another source of local interference. They typically radiate a collection of closely spaced carriers at intervals right across the HF spectrum.

## Diversity Reception

Software defined radios which have two phase coherent receivers can be used in some very interesting ways. 'Phase coherent' just means that both receivers are clocked from the same source. This means that if both receivers are connected to the same antenna and are tuned to the same frequency the receiver outputs will be at the same level and exactly in phase. If you use two antennas, the phase difference that results can be used to enhance a wanted signal or reject an unwanted one. The SDR software can delay the signal from one receiver in relation to the other and alter the gain of each receiver.

You can use the diversity mode to make the signals from two antennas add together. For example, on the low bands, you might have a Beverage receive antenna and a long wire or V beam transmit antenna. If you are receiving two stations and you only want one of them, you can change the phase relationship to null out the unwanted signal. Say for example you want to hear a signal from New Zealand, but there is a European station on or near the same frequency. The phase relationship of the ZL station will be different from the EU station. So, you can manipulate the diversity mode phase and level to pull the ZL out. Then if you want to work the other person, just change the phasing to emphasize the European station. This technique is called 'beam steering' because you are emphasizing signals from a particular direction. Beam steering can also be used to find the direction that a signal is from without having steerable antennas.

If you are suffering from received noise or interference, you can use diversity reception as a noise filter. It works best if the second antenna is picking up signals at around the same level as your wanted signal. But a wire dipole or a long wire 'noise antenna' can be surprisingly effective. The idea is to use the phase cancellation mode to null out the high noise level.

To reduce the noise, you tune the diversity phase and level control on the SDR software for a null in the noise level rather than for a peak in the wanted signal. Diversity noise suppression can be very effective and make the difference between barely hearing a signal and an easily heard station. Up to 40dB of noise suppression should be achievable. For this application, your secondary noise antenna should be placed not too close to the transceiver's main antenna. You don't want to couple too much energy back into your sensitive receiver. Avoid purposely placing the noise antenna near a noise source as that would just make the situation worse by coupling more noise into the system. The idea is for the noise antenna to be receiving the same noise level but with a different phase relationship to the wanted signal.

## SDR for Eme, Microwave and Satellites

Software defined radios are becoming popular for VHF, UHF and microwave operation. Many software defined radios can be locked to a GPS disciplined or another very stable clock source, which makes them good platforms for use with receive converters or transverters. Many of the new SDR boards work well up into the microwave region so you can work microwave frequencies directly, without a transverter. Previously an LNB (low noise block) would be connected at the feed point of a microwave dish to convert the signal down to a lower frequency. Now you can use an LNA (low noise amplifier) and an SDR directly at the wanted receive frequency.

Receivers like the RTL and FUNcube dongles and transceivers like the HackRF and the Ettus USRP are much cheaper than conventional radios covering the same range of frequency bands.

Very stable oscillators are important when you are using a converter or transverter. Any frequency error becomes multiplied. For example, a good HF radio may have an oscillator with a 0.5 ppm (parts per million) frequency stability. At a frequency of 30MHz, this is a respectable frequency error of plus or minus 15 Hz. At 6GHz the same oscillator has a frequency error of plus or minus 3000 Hz. Frequency stability is normally measured after the radio has been operating for at least an hour because oscillators are often very sensitive to temperature variations.

The panadapter is great for weak signal work such as EME (Earth - Moon - Earth) operation or VHF DX. You can see signals that are outside of the passband of a conventional receiver. For example, if you were working 2m EME with a conventional receiver you might easily miss a station transmitting a few kilohertz up or down the band. As well as displaying a wider bandwidth, the waterfall can show signals that are very weak or on a different frequency to what you expect.

The same is true when working amateur radio satellites with linear transponders. You can display the entire transponder width and see any downlink signals. You can see the relative strength of the downlink signals and compare them to the level from the beacon if there is one. Sometimes working out if a signal is actually from the satellite is confusing. You can get spoofed by local signals. The waterfall display from a software defined radio will show a slanted signal from a satellite because the frequency of the signal will be moving due to Doppler shift. A signal from a terrestrial source will not show any slant. You will even be able to tell if the satellite is coming towards your location or heading away. If you are using software control to correct the receiver frequency for Doppler shift, it will be the terrestrial signals that show as slanted.

Cross polarization loss is a big problem, particularly for EME operations. Big EME stations often have both horizontal and vertical polarized antenna arrays, but they don't usually use both at the same time. Software defined radios with two-phase coherent receiver front ends can combine signals from the vertical and horizontal polarized arrays and adjust the phase relationship to enhance the received signal. This will make use of both antenna arrays at the same time providing gain and eliminating the deep nulls that often result from cross polarization.

The height that your antenna is above the ground can also cause nulls in the received signal strength as the Moon rises. This is called ground gain (or loss). It happens when the signal coming directly from the Moon is enhanced or partially cancelled by a signal reflected from the foreground surface. If you have two antennas at different heights, a software defined radio with two receiver front ends can be used to combine them in order to maximize the received signal. The advantage over having a fixed antenna 'stack' connected together with coax cable matching sections, is that the phase relationship between the signals can be dynamically adjusted.

## SDR for Radio Astronomy

Some people are using software defined radio receivers for other weak signal work including radio astronomy and propagation studies. One project is Sudden Ionospheric Disturbances (SID monitoring).

This topic is way out of my area of expertise, but there is plenty of information available online for those that are interested in using a software defined radio for this kind of activity. If you are interested in using an SDR for radio astronomy there is an interesting paper by Dr David Morgan www.dmrdas.co.uk.

There is special software for astronomy at www.radio-astronomy.org/rasdr. More on the topic, at; *http://www.britastro.org/radio/projects/An_SDR_Radio_Telescope.pdf.*

'A 21cm Radio Telescope for the Cost-Conscious' by Marcus Leech, of Science Radio Laboratories, Inc. is another interesting article on using software defined radio for radio astronomy. The pdf is available online at; *http://www.sbrac.org/files/budget_radio_telescope.pdf.*

There is an article by Pieter-Tjerk de Boer, PA3FWM on using the WebSDR at Twente University to create 'Ionograms' from data obtained by tracking the ionospheric reflections of signals generated by 'Chirp' transmitters at *http://websdr.ewi.utwente.nl:8901/chirps/article/* .

## Summary

People should not become too concerned about the definition of what radios are and are not SDRs. New receivers and transceivers will increasingly employ SDR technology or a mix of SDR and conventional architecture. The important thing is how these new radios enhance your enjoyment of the amateur radio hobby. Direct sampling HF transceivers offer technical performance enhancements over older technology radios and panadapters promote new ways to operate. There are definite advantages for contest stations and anyone trying to work a DX station when there is a pileup and people that call CQ will be rewarded with more contacts because stations using an SDR can see the whole band at once. Rather than calling endlessly and hoping that someone happens to tune across the band. You can call CQ in the knowledge that anyone using an SDR has the chance of seeing your signal appear on the spectrum and waterfall display. Some radios can monitor two (or more) bands, or even allow you to operate on two bands at the same time, switching the transmitter between bands as required. SDRs integrate well with digital mode software. It is likely that future SDR transceivers will include digital voice, and digital modes such as FT8 and PSK decoders, 'in the radio.' Conversely, digital mode programs are starting to take advantage of the wideband IQ output from SDR receivers.

The dongle, small 'black box' receivers, and sub-QRP transceiver boards have made it easy to access the UHF and microwave amateur bands, and also the LF bands, at a much lower cost than before. You can use them to receive all sorts of interesting signals; from satellite transmissions, beacons from aircraft and ships, APRS, mobile radio, aircraft band, marine band, and even noise from distant galaxies. Every amateur radio magazine features screenshots of panadapters showing signals such as; intruders on the amateur bands, radar, interference, contest activity, band activity, poor quality signals, and good quality signals. Being able to see these signals and show them to others by taking a screenshot has been revolutionary.

## Latest SDR technology

The IC-7300 [1] and the IC-9700 are the latest HF+ and VHF/UHF radios from Icom and is quite a revolutionary design. They are the first stand-alone transceiver's from the main Japanese suppliers to adopt full direct-digital sampling SDR technology. All processing is on-board, no controlling computer is needed and the radio sports the usual buttons, controls and displays seen on traditional designs. Here we look an the IC-7300 covering the HF bands plus 50 and 70MHz, the radio is packed with features and functions including a multicolour touch-screen display, a high-resolution real-time spectrum scope and much more.

## Basic functions

The IC-7300 is a compact midi-sized radio weighs about 4.2kg. The receiver tunes from 30kHz to 74.8MHz and the transmitter is enabled in the amateur bands at a maximum of 100W output power (1.8 to 50MHz). 4m is enabled (70.0 to 70.5MHz) the UK and also in Europe where 70MHz is an allocation, and the transmit output on this band is 50W maximum. 60m transmit coverage extends continuously from 5.255MHz to 5.405MHz and includes all operating modes.

The usual modes, SSB, CW, RTTY, AM and FM are provided with reverse sidebands selectable on SSB, CW and RTTY and with AFSK data on SSB, FM and AM.

The radio requires the usual nominal 13.8V supply and draws a maximum current of 21A. CE marked models marketed in the UK and Europe have a separate EMC filter box incorporated into the power supply lead. The full manual is supplied on CD ROM as a PDF file together with a full set of circuit diagrams. Running to around 170 pages it is very detailed, book-marked and full of cross-links, with many pictorial representations. A subset of the full manual is also provided on paper, 72 pages covering the initial set-up and basic operating instructions. The radio is provided with a standard hand electret microphone, the HM-219, but other microphones are, of course, also suitable.

## Radio design and architecture

Direct digital sampling is used for both the receive and transmit signal paths. Incoming receive signals pass through a diode-switched bandpass filter unit where one of 15 narrowband filters is selected to cover the tuning range of the receiver. A pre-amplifier with a switchable high gain or a low gain setting and / or an input attenuator allows the receiver to accommodate differing signal levels. An AGC controlled PIN attenuator is also included in the signal path. The RF signal is then sampled by the A/D converter and passed in parallel digital format to the FPGA, a fast field-programmable gate array. This extracts a slice of input signal at an IF of 36kHz by a process of down-conversion decimation and passes the result to a DSP device for all further processing of the received signal. This is the same DSP chip as is used in the IC-7100 receiver, except in that receiver the down-conversion process to 36kHz is done as an analogue superhet radio. Hence with the same, or at least similar, DSP code the various filtering and processing functions are the same for both radios. A separate down-conversion process in the FPGA simultaneously produces the spectrum scope signal path. Direct sampling tends to produce low-level distortion products, which can be noticeable under certain situations, particularly on quiet bands. Dithering in the A/D converter can reduce or eliminate this and is enabled by the IP+ function.

On transmit the process operates in reverse. The DSP generates the transmit signal, which is up-converted in the FPGA to the desired output frequency and converted to analogue format by the D/A converter. The usual amplifier chain follows to the PA and the relay switched low-pass output filters. A high stability TCXO reference oscillator is built in, which achieves 0.5ppm stability.

## Front panel

Operation of the radio centres around the touch-screen display and the associated buttons and controls. The multicolour display has a high resolution and is particularly clear and bright and retains readability well under bright lighting. Band, mode, filter selection, meter selection and VFO/memory functions are all selected by touching the appropriate areas on the display which brings up a grid of selectable options. Five hardware buttons along the bottom of the display select top level functions for the display such as spectrum scope, RTTY decoder, keyers (via the MENU key) or the receive parameters such as AGC, preamps, noise reduction (via the FUNCTION key). Dedicated buttons on the front panel also provide a fast alternative way to set the various receive, VFO and memory functions. Clicking the MULTI rotary control sets adjustable functions such as transmit power level and microphone gain.

The display shows a large number and complete set of status indicators and function values. Both frequencies are shown in split frequency operation with a single bargraph style meter for signal strength and various selectable transmit functions. Simultaneous display of multiple transmit functions can also be selected. A very comprehensive set mode allows tailoring of an enormous number of functions. These are all accessed via the touch screen display with nested menu items and many are set using MULTI or the rotary tuning knob. A keyboard is displayed when alphanumeric data needs to be entered. This can be in either a full QWERTY or a 10-key format and makes data entry very straightforward.

Tuning is very smooth and easy using the 50mm diameter rotary control. Tuning is in 10Hz or 1Hz steps at 6kHz or 600Hz per knob revolution with auto speed-up on fast tuning. A quarter rate is also selectable on CW and data modes. A higher rate for faster navigation is also selectable, with a variety of mode-dependant step sizes. AF gain combined with RF squelch and Twin PBT (passband tuning) are given separate rotary controls on the front panel and there is the usual 8-pin DIN microphone connector and a 3.5mm headphone jack.

An SD memory card slot is also provided for storing various items such as received and transmitted audio files, RTTY, CW and voice memory stores, RTTY decode logs, memory contents and set-up data. Screen images can also be captured. The SD card is also the route for transferring firmware updates from a PC to the radio. SD or SDHC cards up to 32GB can be used.

## Rear panel

The rear panel contains a single SO239 antenna socket; a pity a second wasn't included for flexibility as the radio covers two VHF bands as well as HF. A 13-pin socket, as used on several other Icom radios, is used for connecting various accessories, including control of linear amplifiers, audio input and output for data modes, and band data for external control such as for ATUs. A separate socket interfaces to the Icom AH-4 ATU and AH-740 antenna. Phono connectors provide alternative switching and ALC feedback for linear amplifiers. The usual external speaker connection is provided, and a single key jack accommodates CW paddles, straight keys or computer keying. Both the CI-V interface and a USB port are provided for computer interfacing. The USB port can also be used in conjunction with data mode software on the PC to transfer receive and transmit audio, PTT, CW and RTTY keying. A 12kHz IF output may also be enabled for DRM decoding.

## Receiver features

The usual receiver functions are provided. Twin VFOs (A/B) allow split operation with the ability to easily check and tune the trans-

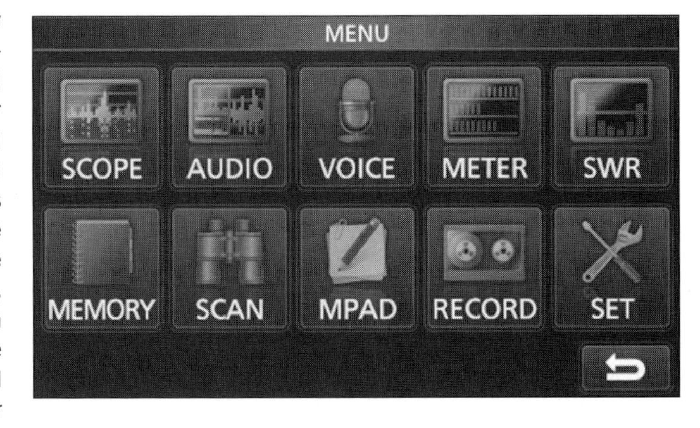


**Fig 8.21: Menu display**

**Fig 8.22: IC-7300 front view**

mit frequency via XFC. Incremental tuning (RIT, XIT) functioning on both receive and transmit, pitch control and auto-tuning on CW are all provided. Band selection via the touch screen buttons allows three separate frequencies to be stored per band (ie a band stacking register). 5MHz and 70MHz bands do not have dedicated buttons but are included on the GENE general coverage button. It is a good idea to store several 5MHz and 70MHz frequencies into memory as the GENE band stacked frequencies are easily overwritten when tuning outside the amateur bands, eg to broadcast frequencies. Frequencies can also be keyed in directly.

There are a total of 99 regular memory channels and 2 scan edge channels. Memory channels can be selected, saved and copied in several different ways, making access very straightforward. Memory channels can be assigned names up to 10 characters in length and this is quick and easy with the on-screen keyboard. A separate quick access memopad stack for 5 or 10 stores is also included. A host of scanning functions are also provided.

Filtering functions are very comprehensive, as with all Icom radios. A touch button on the display scrolls around three preset IF filter bandwidths with separate settings for each mode from a menu of over 40 different bandwidths. Both sharp and soft passband shapes are available. Twin PBT (passband tuning) allows either side of the filter passband to be shifted independently, shifting or narrowing the overall shape to assist in combating adjacent channel interference. A manual notch filter operates at IF inside the AGC loop and hence prevents desensitisation with strong carriers. It has excellent depth with wide, medium or narrow width settings. A separate auto-tuning notch filter operating at audio removes multiple tones effectively but does not prevent strong carriers from desensitising the receiver.

A noise reduction system reduces background noise and improves readability in certain situations and a separate noise blanker eliminates pulse-type noise from car ignition systems. Both systems are adjustable. Three separate AGC time constants are selectable from a menu of 13 different values (0.1 to 6s) and are set separately for all modes except FM. The AGC can also be switched off.

The receiver audio response can be tailored independently for each mode. The high-pass and low-pass roll-offs can be adjusted separately and the bass and treble responses cut or enhanced, so there is much to play with here. There is no CW audio peak filter but on RTTY a sharp twin peak filter is provided.

## Transmit features

Transmit functions for SSB include the usual speech compressor, VOX and a transmission monitor. The audio transmit filter bandwidth may be set to wide, mid or narrow, where the upper and lower bandwidth points are adjustable. By default the wide setting is 100Hz to 2900Hz, mid setting 300Hz to 2700Hz and narrow is 500Hz to 2500Hz. In addition, the bass and treble responses can be cut or enhanced separately for each voice mode in a similar fashion to the receive audio.

On CW there is the usual provision for full and semi break-in with adjustable drop back delay. The keying envelope rise and fall times are adjustable between 2 and 8ms and an additional delay is selectable to accommodate slow-switching linear amplifiers or other accessories. Different delays may be set for HF, 50 and 70MHz. An automatic antenna tuner is built in, covering all bands including 50 and 70MHz. The tuner matches antennas up to a VSWR of 3:1 at full power or to a higher VSWR in an emergency mode when the power output is then limited to 50W. The tuner includes the usual memories to enable rapid retuning when the frequency changes.

A full CW message keyer is included operating over the speed range 6 – 48 WPM with adjustable weighting and a variety of keying paddle arrangements. Eight memories will each store up to 70 characters, with a provision to send automatically incrementing serial numbers and auto-repeat after a time delay. The message stores are programmed in text via the touch screen keyboard display and may be sent either from display buttons or via an external homebrew keypad connected to the microphone socket. However, the external keypad only allows for sending stores 1-4. Stores 5-8 need to be sent from the display buttons.

FM mode operation includes CTCSS access and tone squelch, and repeater split frequency operation. Repeater offsets are stored separately for HF and 50MHz and are programmable over wide limits.

## RTTY

A built-in Baudot decoder for standard 45-baud RTTY signals displays 4 lines of 35 characters in standard mode or 9 lines of

35 characters in wide mode. An audio spectrum and waterfall display is provided for tuning purposes and there are a host of user setups and options, easy to access. There are 8 message stores, each holding up to 70 characters for pre-programmed transmitted messages and these are accessed and stored in a similar fashion to the CW message stores. Receive and transmit messages can be time stamped and saved to SD card.

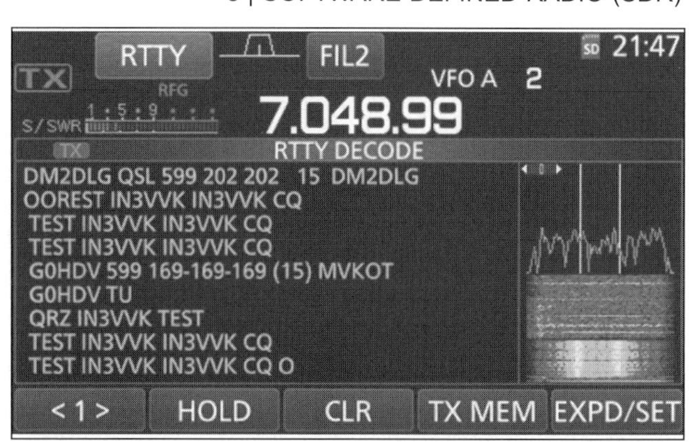


**Fig 8.23: RTTY decoder display in expanded mode.**

## Spectrum and scope displays

One of the key features that SDR brings to a radio at relatively low cost is the high performance real-time spectrum scope display. This operates simultaneously with normal receiver operation. Two displays are provided: a spectrum of the signals for the band currently selected and an audio scope showing the waveform and spectrum of the receiver or transmitter audio. Both spectrums can show an associated waterfall display. Various combinations of spectrum and audio display are possible in expanded or in mini-scope mode used in conjunction with other displays such as the RTTY decoder. The displays have excellent resolution.

The spectrum display has two modes of operation. The centre mode will display the spectrum on either side of the on-tune frequency with spans selectable from ±2.5kHz to ±500kHz. The Fixed mode will display the spectrum between two fixed points, and three fixed edge bands can be set for each amateur band. In both modes there are a number of settings that select sweep speed, colours, VFO markers, peak hold, averaging, reference level etc. The displayed vertical range is 80dB.

One of the features of the spectrum display is touch-screen tuning. When the spectrum area or the waterfall area is touched either with a finger or a stylus, the frequency span immediately adjacent to the touched area is zoomed. Then a second touch will precisely tune the radio to the wanted frequency. Zooming greatly improves the accuracy of tuning by this method.

## Auxiliary features

The IC-7300 includes a digital voice recorder for transmitting messages such as CQ calls. These are only stored on the SD card, which must be in place. Eight channels are available, each with 90 seconds recording time, and these can be tagged with labels up to 16 characters long for easy identification on the display screen. Messages can also be set to repeat after a time delay. Messages are sent from the display buttons or from an external keypad in a similar fashion to the CW and RTTY stores.

As well as providing message stores on transmit, the voice recorder can also store the receive and transmit audio. Files are automatically named and placed in folders together with time and date, frequency, mode etc. The recording time is limited only by the amount of available memory on the SD card. Individual files are limited to 2GB in length but new files are created automatically if necessary. Stored as .WAV files at a data rate of 128kbps, 2GB corresponds to around 35 hours of recording time. The audio can be transferred to a PC or played back on the radio where the usual CD audio-style navigation buttons (fast forward, pause etc) are provided.

The antenna SWR can be plotted graphically against frequency, which can be useful to check antenna performance over the band. The measuring step is selectable from 10 to 500kHz and the number of steps from 3 to 13. **Fig 8.25** shows the match on my 160m dipole across the band.

Other features include a built-in calendar and 24-hour clock, transmit timeout timer, screensaver, and a voice synthesiser for audible readout of frequency, mode and S-meter. Remote control over the internet or home network is possible using the optional RS-BA1 IP remote control software. With an appropriate monitor, this provides the possibility of a much larger spectrum and waterfall display together with mouse driven point and click tuning.

## On-the-air performance

I was very impressed with how well the radio performed and how easy it was to use. Much thought has obviously gone into achieving an excellent ergonomic design; the controls, the display and overall handling is close to ideal. The touchscreen display is very responsive and undemanding even with large fingers and is crisp and clear. Tuning is easy to navigate and memories quick and easy to access. Touch tuning from the spectrum display worked well and the zoom feature generally ensured that the frequency was fairly accurately set.

The performance was excellent. Sensitivity was good but dropped markedly on the lower frequency broadcast and time-code bands although the response was clean. The strong signal performance was generally very good. Signal overload and intermodulation effects were observed with excessive use of preamplifier gain but in all cases could be avoided by correct use of the preamplifier / attenuator / RF gain controls. No low-level intermodulation effects were observed on the quiet bands but the radio was not used when large signals were around such as in the more popular contests on 50 or 70MHz. The channel filters, notches, noise blanker and noise reduction system all per-

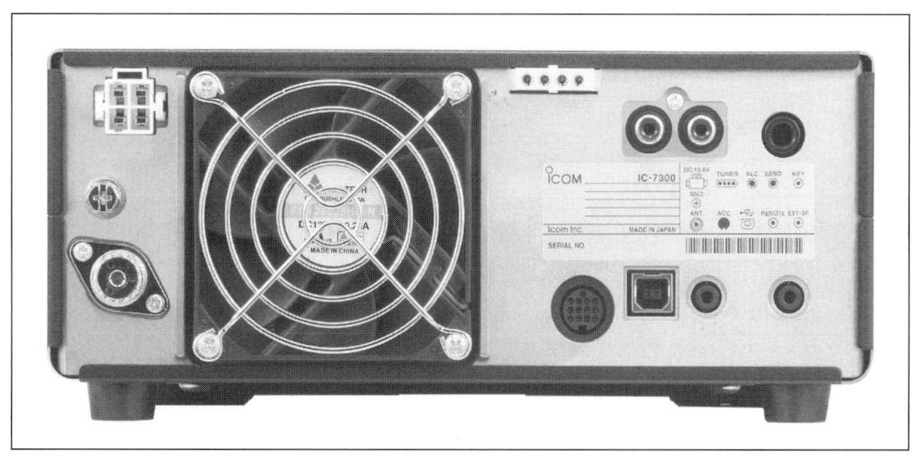


**Fig 8.24: Rear panel of the IC-7300**

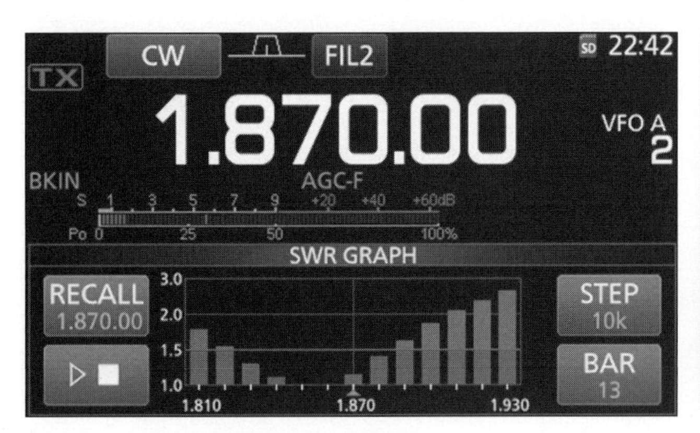


**Fig 8.25: Antenna SWR graph display.**

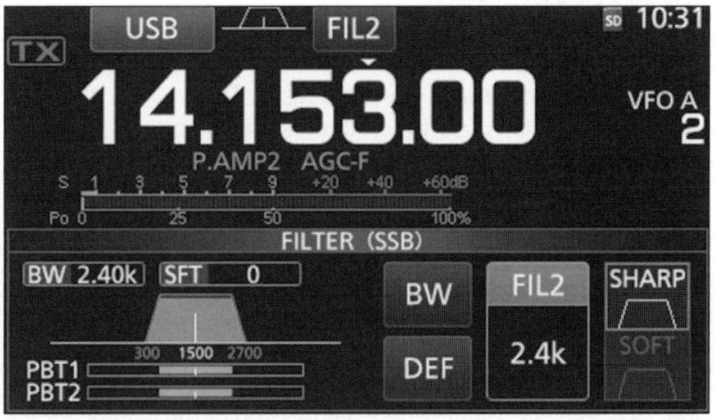

**Fig 8.26: Filter bandwidth display.**

formed extremely well. The audio quality and volume from the internal speaker was remarkably good with no rattles, although on CW a distinct resonance was heard. The quality on headphones was excellent.

Transmit operation was very well behaved. The fan operates continuously on transmit even when cold and is fairly quiet except on initial start. It does not operate on receive and keeps the temperature well in check even during long periods of continuous transmission. Audio quality reports were excellent with the supplied HM-219 microphone. CW break-in was clean, with just a hint of clicks on the sidetone. On full break-in the changeover relay was slightly noisy but allowed listening between characters up to about 25 WPM.

The spectrum scope, waterfall and audio screens were very effective with high resolution and no appreciable delay, a great improvement on previous radios. The RTTY decoder was very effective and easy to use for rubber-stamp contacts. RTTY message stores are best to access from an external keypad as two key presses are necessary when using the front panel buttons, which doesn't lend itself to slick operation. It is about time that Icom produced a keypad accessory along the lines of the Yaesu FH-2. During the course of the review a firmware upgrade was released, which I installed without problems

The IC-7300 is a superb radio with some great features and a good performance. Currently priced at around £1050 it is outstanding value for money and the SDR approach adopted will surely pave the way for the future for this style of radio.

*Full Measured Performance for the ICOM IC-7300 can be found in Peters review in the August 2016 edition of the RadCom on the RSGB website.*

## Creative Credits

Fig 1. https://www.nooelec.com
Fig 2: https://www.sdrplay.com/rsp1a/
Fig 3: https://sunsdr.eu/product/sunsdr2-pro
Fig 4: https://apache-labs.com
Fig 5: http://www.icom.co.jp/world/products/amateur/hf/ic-7610/
Fig 6: image by author
Fig 7: image by author
Fig 8: image by author
Fig 9: image by author
Fig 10: image by author
Fig 11: image by author
Fig 12: image by author
Fig 13: image by author
Fig 14: http://www.icom.co.jp/world/products/amateur/hf/ic-7610/
Fig 15: image by author
Fig 16: image by author
Fig 17: drawing from Pentek document, 'Putting under-sampling to work' www.pentek.com
Fig 18: from Digital pre-distortion. Linearizing our Amplifiers. Presentation for Ham Radio Friedrichshafen 2014 by Dr Warren C Pratt, NR0V. Test image by 'Bert' Meijer PA2XHF.
Fig 19: from Digital pre-distortion. Linearizing our Amplifiers. Presentation for Ham Radio Friedrichshafen 2014 by Dr Warren C Pratt, NR0V. Test image by Kurt Tausch DL9SM.
Fig 20: http://www.astrosurf.com/luxorion/qsl-ham-history15.htm
[1] Latest SDR Technology on page 8.24 is abridged from a review by Peter Hart, G3SJK published in the August 2016 edition of RadCom.

***Andrew Barron*** *is the author of 'SDR Software Defined Radio,' The Radio Today guide to the Icom IC-7610, and 'Amsats and Hamsats.' All available from the RSGB bookshop.*

# VHF/UHF Receivers, Transmitters & Transceivers 9

*Andy Barter, G8ATD and Chris Waters, 2E0UCW*

The purpose of this chapter is to give the reader an insight into what is available to the radio amateur on the VHF and UHF bands in the UK.

There is some theory about the choice of the equipment to use but the emphasis is on the practical aspects of choosing equipment and enhancing it with preamplifiers, and power amplifiers to give you a station that can be used to its maximum effect.

## GETTING THE BEST OUT OF YOUR VHF/ UHF STATION

One of the great attractions of operating on the VHF/UHF bands is that there are so many different aspects of the hobby that can be utilised at these frequencies. Interested in voice communications? You can use the VHF bands for both local and international contacts. Perhaps your interest lies in digital communications, packet radio (AX25) to access mailboxes or the DX Cluster.

A further aspect of this technology is the automatic packet reporting system (APRS) that allows real-time tracking of mobile (or fixed) stations. Image communication such as slow scan television (SSTV) is also popular, especially now that most of the processing is achieved using a computer and sound card. And don't forget Morse! This 'digital' mode is still very much used on the VHF bands by the DX community. Once you get hooked on working DX you'll then discover exotic propagation modes such as trans-equatorial propagation (TEP), Sporadic-E (Sp-E or Es), Aurora and meteor scatter (MS). And it doesn't have to be two-way terrestrial contacts.

You can also make use of amateur satellites or even bounce your VHF/UHF signals off the moon (earth moon earth EME) to make world-wide contacts. You can operate from home, in the car or go out back-packing from the hill tops. Other activities include low power or high power, rag chews or contesting. The VHF/UHF bands really do have something for everyone.

### Prime Mover

The one piece of equipment that determines exactly what facilities you can ultimately use on the VHF bands is the station transceiver. This will either be a single-mode or multi-mode base station, mobile unit or portable hand-held radio. Most single-mode transceivers available today are mobile units (often pressed into service for home use) and portable hand-helds. These are designed to operate exclusively on FM and are very popular, as they can be used for short-range telephony (either direct or via a repeater) and for data communications such as packet radio. FM transceivers can be obtained from amateur radio retailers, but that's not the only source of this type of equipment.

**Fig 9.1: The VHF/UHF station of the Handbook editor**

Commercial operators regularly upgrade their private mobile radio (PMR) equipment, and this can be obtained from traders who specialise in electronic surplus. It will get you operational very quickly and at a price that will suit most pockets. Indeed, for many fixed station applications, I would recommend that you use dedicated PMR equipment as it does possess many advantages. It is designed for use by a wide range of operators in varying environments. Because of this the equipment is normally of rugged construction. Drop it and it will probably keep working. The majority of PMR equipment has to be built to a high technical performance and reliability. Spectral purity of the transmitted signal is very good. Some amateur band allocations are very close to the commercial PMR bands. By looking around you should find equipment suitable for the 50MHz, 70MHz, 144MHz and 430MHz bands. Most equipment is relatively easy to modify and in some instances may not need any modification at all. However, before you hand over your money there are a few points to note. Is the equipment working on a frequency range close to an amateur band? What transmission mode does it use? Is it AM or FM? What is the channel spacing of the equipment? Is it 50kHz, 25kHz or 12.5kHz? The latter two are preferable, whereas the 50kHz channel spacing would indicate that the equipment is many years old and uses wide bandwidth filters which may not be suitable for use on the VHF bands today.

### The VHF/UHF Bands

The three UK VHF_amateur bands are 50MHz, 70MHz and 144MHz, and the only UHF band is 430MHz. Profiles of these bands, their different propagation characteristics and the bandplans can be found in the *Amateur Radio Operating Manual*, by Mike Dennision G3XDV & Steve Telenius-Lowe, PJ4DX, and the *RSGB Yearbook* both available from the RSGB.

### Long Distance

As mentioned, the use of FM equipment is for short-range communication links. If you want to broaden your horizons and contact stations much further away then you'll need to procure a multi-mode rig which in addition to FM includes CW and SSB transmission modes. Unfortunately you won't be able to find surplus PMR equipment that can be pressed into service as a multi-mode rig, so it really is a case of digging deep into your pockets and buying a suitable transceiver.

If you already possess a multi-mode HF transceiver. you may wish to consider the use of a transverter. A transverter is a transmitting converter, a receiving converter and a local oscillator source all combined into one unit. It connects to the antenna socket of an existing transceiver that provides the driving signal, typically at 28MHz. The transverter then mixes the IF drive from the transceiver with its own local oscillator to produce an output on the VHF/UHF band of your choice. On receive a similar process takes place, the VHF signals being down-converted to provide an output signal in the 28MHz band.

In practice, transverters are available for all VHF/UHF bands and for a variety of IF drive frequencies. Although the majority will be at 28MHz you'll also find models that will accept drive at 144MHz. So if you already have a transceiver on this VHF/UHF band you should have no problem finding a transverter that will allow you to operate on the 50MHz band. The advantage of using a transverter is that it allows all the functions and performance of the driving transceiver to be used on the VHF/UHF band of your choice; more on this topic later.

## Optimisation

Now it's time to take a look at what a VHF/UHF station and how you can make simple improvements. **Fig 9.1** shows the editors station with the Kenwood TS-2000x covering 50, 70, 144, 430MHz and 1200MHz as well as HF, also shown is the Icom IC-7300 which in addition to HF covers 50 and 70MHz. New on the scene for VHF/UHF is the IC-9700 SDR radio covering VHF/UHF/1200MHz all mode including D-Star DV/DD. No matter what VHF/UHF band or transmission mode you wish to use, the basic system will always be the same. It's a transceiver feeding an antenna via a length of coaxial cable.

So why do some stations consistently perform better than others? One of the most important factors is the site on which the VHF/UHF station is located. Ideally, a hill-top location is best, but good results can be obtained in low lying areas that are clear of local obstructions. Results depend very much on the band used, obstructions having considerably less effect at 50MHz than at 144MHz or 430MHz.

We cannot all live at 250m above sea level with a clear take-off, so you need to pay special attention to the most significant item in your station. That of course is the antenna; **Fig 9.2** shows a typical VHF/UHF antenna array covering 6m, 2m & 70cm.

Convention dictates that FM operation, for both telephony and digital communications, an antenna with vertical polarisation is required. If you want to make local contacts then you'll probably need omni-directional coverage. For packet radio you will require a similar vertically polarised antenna, although you might consider using a small 4 or 5-element beam for a fixed link.

For serious VHF DX work, using CW or SSB, a horizontally polarised directional Yagi is recommended. There are many types of beam antennas available, some very good and some, well, not so good. But the difference between the poorest designs to that of the very best may only amount to 4dB or so.

The point here is that if you are only interested in working occasional DX when the band is open, what 'real' difference do a few decibels make when propagation conditions can vary by many tens of dB? So, unless you really want to eke out the very last vestige of antenna gain, the most important criterion is not ultimate gain but build quality. After all, a long boom antenna is no good if it folds in half during the winter gales. Similarly, the longer the antenna boom the sharper the directivity of the array becomes.

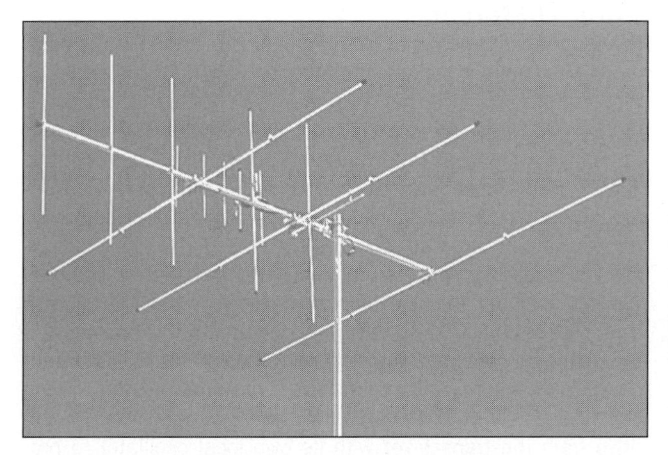

**Fig 9.2: 6m, 2m & 70cm VHF/UHF antenna system**

The possibility of missing stations away from the main antenna lobe becomes increasingly likely. So you might consider trading off some gain for an increase in beamwidth. Taking all these factors into account you might find that a pair of stacked 9-element Yagis (on the 144MHz band) will provide a more practical solution than using a single 18-element Yagi.

Much more information can be found in the chapter on VHF/UHF_antennas.

## Siting and Cabling

The siting of an antenna is just as important as the type of antenna used. A ground-plane antenna located on a chimney top, clear of any obstructions, may give better results than a beam antenna located in a loft space. Unless you have restrictions imposed at your home, the best place for a VHF/UHF antenna is always outside in an uncluttered location. If possible, mount it on a suitable pole, elevating it above the roof and away from nearby television aerials.

The coaxial feeder connecting the antenna to the transceiver should have a low loss at the frequency in use and this is especially important on the VHF/UHF bands. A poor quality cable will lose valuable transmit and receive signal power, so be prepared to spend more money on the main feeder than on the antenna. It really will be an investment.

Finally, make sure that the connectors you use are of the highest quality. Although the use of N-type plugs and sockets is recommended, they are not essential, especially on the lower VHF bands.

## Background Noise

Having paid attention to the antenna and feeder, it's now time to look at the receiver. The background sky noise arriving at the antenna effectively limits the maximum receiver sensitivity required for normal communications. On the lower VHF bands of 50MHz and 70MHz man-made noise often exceeds the background noise by 10dB or so. Consequently, receiver noise figures as high as 12dB and 10dB respectively are quite adequate for these bands. At 144MHz and 430MHz however, the sky noise is much less and a receiver noise figure of around 2.5dB will be quite adequate for most types of terrestrial communication. Unfortunately you probably won't find out the overall noise figure of your commercially made transceiver because it is rarely given. Normally the specification is given in terms of so many μV for a signal to noise ratio of so many dB. For example, one 144MHz transceiver quotes "better than 0.5μV for 11dB s/n", making the most favourable assumptions, this translates to a noise figure of 11dB.

Now you can see how little some manufacturers are really offering the VHF/UHF enthusiast. Much effort seems to be exerted in producing rigs with 100 memories, air-band receive facilities, computer control and displays that say "Hello", when what is really required is a VHF/UHF transceiver with a low noise figure, a dynamic range in excess of 100dB, switchable filters, IF shift, notch filtering, adjustable noise blankers and full CW break-in. All these features can be found on a modern HF radio, which brings us nicely back to the original suggestion of using a VHF/UHF transverter with an HF transceiver. You really do get the best system performance by adopting this technique.

## Pre-amps

Another way of overcoming the basic lack of sensitivity is to use an external pre-amplifier and, if this is mounted at the antenna, it will also eliminate the effect of feeder loss in the receive direction. Unfortunately, the receive sensitivity is only improved if the pre-amplifier has sufficient gain, but this extra gain also decreases the strong-signal handling capability of the receiver. Therefore the use of a pre-amplifier may show overload effects on some signals that originally didn't cause any problems. Try to use a pre-amplifier that has adjustable gain, so that you can adjust it to suit your receiver. Typically, a gain of between 6 and 15dB will be sufficient for most needs.

## Summary

The biggest improvements to your VHF/UHF station always come first. Changes to the antenna system, coaxial feeder, making the receiver more sensitive and increasing your transmit power will easily improve your system performance. After that it becomes a little bit more difficult. The rewards are still available but each improvement will be less significant.

## D-STAR

We are at the start of the digital revolution in amateur radio so it is important to understand the new modes that are becoming available. One that has the support of a major amateur radio equipment supplier, Icom, is likely to gain a significant part of the market. The following section by Gavin Nesbitt, M1BXF, describes how D-Star works and how to use the mode.

Imagine a radio system that finds your friends for you; A system where you are able to call CQ on a handheld and making the contact exactly the same whether your friend is next door or on another continent; an amateur radio system that lets you send position reports in a similar way to analogue APRS and other messages without nasty noises on the audio channel.

Until recently, this sort of connectivity has been the stuff of dreams. But now, D-Star makes it all a reality. It transparently links mobiles, handhelds, home stations and repeaters world-wide using the Internet. In normal use it simply doesn't matter if your friend is on the local repeater or on holiday miles away, D-Star sorts it all out and lets you communicate effortlessly.

The great thing about D-Star is that you don't have to learn about the technical stuff to operate using it - unless you want to.

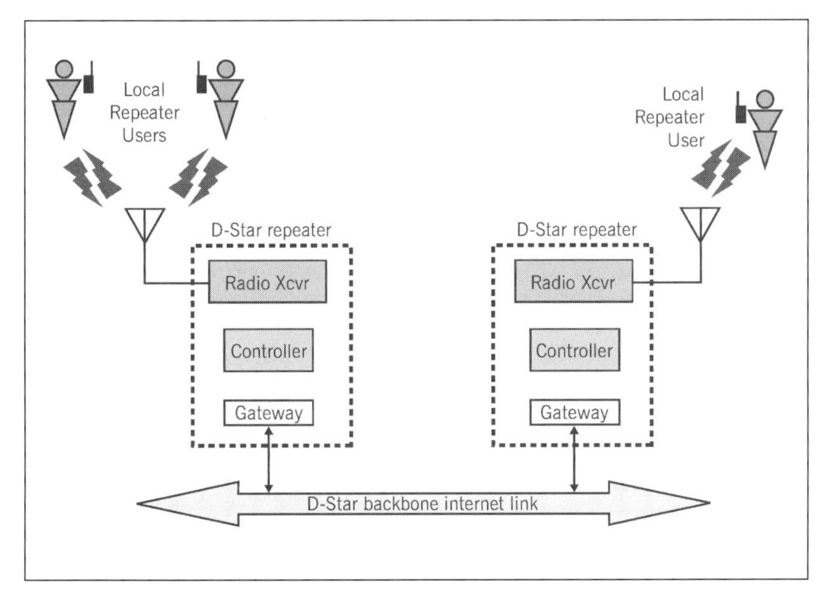


**Fig 9.3: D-Star can connect individuals directly, via a local repeater or over large distances via the Internet**

## Overview

The D-Star system consists of individual radios, repeaters, servers known as gateways and Internet connections. **Fig 9.3** shows a diagram of the D-Star network. D-Star repeaters work like standard analogue voice repeaters, with some added bonuses. For a start, they can operate on several bands at once, with 2m and 70cm being most common.

So, what are the fundamentals of getting active on D-Star? First, you'll need a D-Star compatible radio (or DV-Dongle, more of which later). The only people currently making D-Star radios are Icom, although there is nothing to stop others climbing aboard the bandwagon because D-Star is an open standard so that anyone can make equipment.

You need to tell the D-Star radio who you are, which identifies you on the D-Star network. Next, you need to work out who you want to talk to. If you haven't got any D-Star equipped friends yet, just set it to *CQCQCQ* and someone will come back, just like any other mode.

Finally, you need to make the radio talk to your local repeater by entering its frequency and callsign.

So, the radio knows who you are, who you want to talk to, and the name of the repeater you want to work through. What are you waiting for? Call CQ!

## The Basics

D-Star stands for Digital Smart Technology for Amateur Radio and was developed by the Japanese Amateur Radio League (JARL) back in 1999, in conjunction with Japanese universities and amateur radio suppliers. Their goal was to advance the hobby into the digital age by developing an open standard, using off-the-shelf parts, such as the AMBE 2020 voice encoder chip, which would allow anyone to design and produce digital radios or interfaces for radio amateurs. So D-Star was born.

The D-Star standard has two modes. The one causing the most excitement, and the main one used in the UK, is Digital Voice. DV mode allows simultaneous voice and serial data and easily fits in a 12.5kHz channel. The other mode is a high-speed digital data mode, DD, which gives a 128kHz half-duplex data connection. DD requires wide channels and is only used on 23cm.

Digital Voice uses a 4800 baud data stream. This is split into 3600 baud for voice, including 1200 baud of error correction, and 1200 baud for non-error-corrected data. When DV mode signals are marginal, the first noticeable effects are errors on the 1200 baud serial data stream. The voice holds up reasonably well as signals degrade further, eventually failing quickly as the signal quality falls below the error correction capability.

Think of D-Star as the transport mechanism, just like the meaning of FM or SSB. It's what you send over it, in DV or DD mode, which makes it interesting. A little like packet radio, packet by itself is useful but using packet for APRS or the DX cluster is what makes packet what it is today. D-Star uses GMSK (Gaussian Minimum Shift Keying) modulation to send the 4800 baud data stream over the air. As it is digital, the data stream includes headers such as the route the data should take and where it came from. There are four values that are used for routing and are easily programmed in all the radios available. These are MYCALL, URCALL (sometimes seen as YOURCALL), RPT1 and RPT2. In real life, these are used as follows:

**MYCALL:** Station/Terminal identifier, this is usually your own callsign but can be any name you wish such as CAR 1, USER A, etc. There is no licence requirement for it to be your callsign and this entry is what is used to authenticate and track you on the D-Star network (more later) so using your callsign is favoured.
**URCALL:** The destination address. Can be another users callsign, a repeater or, if there is no intended recipient, *CQCQCQ* is used.
**RPT1:** The callsign of the repeater being accessed, this is used in the same way as CTCSS on analogue repeaters - meaning if you are in range of two D-Star repeaters only the one corresponding to the RPT1 value will trigger.
**RPT2:** This value is optional; it is only used if you wish to route traffic to a different place such as another local port on a different band, through the gateway if you are connecting to a remote repeater or finding the location of a user and the echo test server - more about this later.

All the D-Star radios have these values that can be updated and, in the case of Icom radios, each memory channel supports having unique URCALL, RPT1 and RPT2 saved alongside other memory data. This means you can save a channel with the correct values to route to a remote repeater or user, and name the channel appropriately.

## Advantages

Being digital is what makes D-Star an interesting mode. It brings all manner of new and enhanced features to radio amateurs such as data and call routing, simultaneous voice and data traffic, web reporting of stats and a real-time last heard list of users on the network. But, it's when in range of a D-Star repeater on the gateway, ie connected to the Internet, that the system provides most of the advantages over analogue repeaters, being able to form a wide area network.

A key feature is that the D-Star network can track a user's last known location in the form of the last repeater and port the user was heard on.

This can be used like a cellular system and automatically route calls so, if someone wants to contact another station without knowing where that station is located, the gateway can do a lookup of where he was last heard and can route the call there automatically.

Other features of gateways are that they are able to connect to reflectors, GPS (DPRS) position reporting to the web and echo test, which is just a simple way to test your transmission. It's worth saying that the gateway has user access control and users must be registered to use its features, there is no access control on using the repeater for local contacts.

The main reason for registration is to add your callsign, a MYCALL value, to the gateway that assigns it an ID (an IP address) that is used by all the gateways for tracking and routing calls. Contact your local D-Star repeater keeper or group to be added to the gateway once you get a D-Star radio.

## Repeaters

Currently only Icom have a D-Star repeater system available but there are open source repeaters in the pipeline. Icom's repeater controller can support four repeaters, simultaneously associated with three ports, simply known as Port A (same for DV or DD), B or C. The ports are harmonised across the network so the user knows which repeater/band they are connecting to. Normal aliases are Port A for 23cm DV and/or DD, B is 70cm DV and C is 2m DV. There are other ports such as G for the gateway (internet connection) or E for echo test. When routing, you always need to specify the port you are connecting to on a remote repeater and, in Icom's case, the port number always has to be in the 8th position in the URCALL, RPT1 and RPT2 fields. If you start the URCALL with a forward slash '/' this indicates the callsign is that of a repeater and not a user.

Let's look at a few examples of how routeing works. If I wanted to connect to another repeater from the 70cm port on GB7PI to, say, the 70cm port on GB7NM (north Manchester) I would set my values as:

*MYCALL = M1BXF,*
*RPT1 = GB7PI B,*
*RPT2 = GB7PI G,*
*URCALL =/GB7NM B.*

When I transmit with these settings I would be heard locally on the 70cm port of GB7PI and would also be heard, almost instantly, on the 70cm port on GB7NM. If I used this connection a lot, I could save these values in a memory channel and call it something like GB7NM.

If I wanted to call a user directly, I would simply change the URCALL from the above example to the callsign of the station I wish to contact like URCALL = G7LWT. What happens now is the gateway PC does a lookup of G7LWT's last known repeater/port and then routes my transmission to that repeater/port. It's that easy.

If I were just calling locally, I would just need to set RPT1 = GB7PI B and URCALL = CQCQCQ. However, with the advent of reflectors and DV Dongle users it is suggested that everyone put the local gateway into RPT2. This is because DV Dongles and reflectors connect to repeaters as a gateway and not a user. So, for your data to be routed back to those connected via reflectors or DV Dongles, you must use the gateway, hence you need to set RPT2 = GB7PI G.

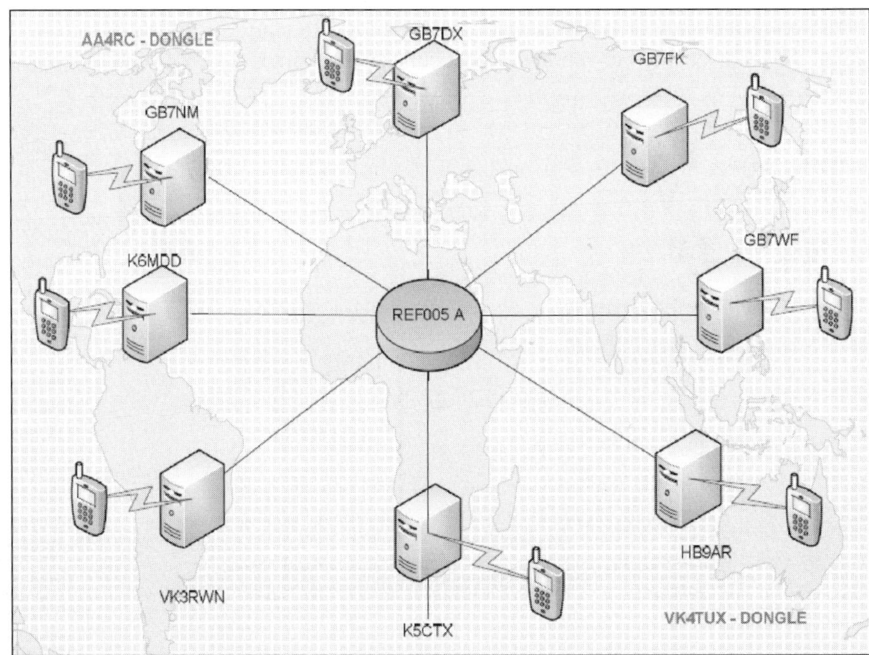


**Fig 9.4: Reflectors link D-Star repeaters, RF stations and DV dongle users from around the world into a single meeting place**

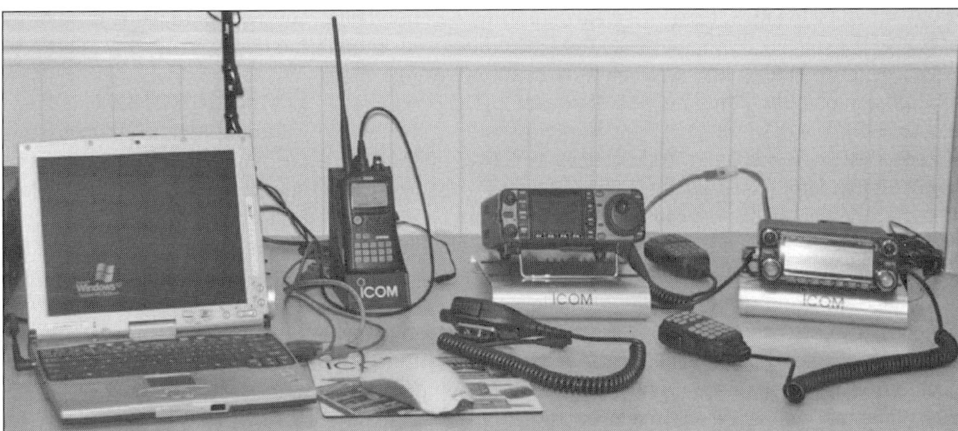

Fig 9.5: A collection of D-Star radios plus D-Star, DMR HotSpot based on the Rasberry Pi

## Reflectors

I keep mentioning reflectors. Users of *Echo Link*, *IRLP* and *eQSO* will be familiar with conference rooms, and reflectors are the D-Star equivalent. D-Star reflectors were developed by Robin, AA4RC, who has written many enhancement applications for the D-Star gateways. His *dPlus* is the code that reflectors use. Reflectors link D-Star repeaters, RF stations and DV Dongle users from around the world into a single meeting place as shown in **Fig 9.4**. There are reflectors located in many parts of the world. Two of these are in London: REF005 and REF006. Each has three ports, A, B and C, giving a total of six conference rooms that are used by different European countries or groups.

Connecting to reflectors is simple and only requires the name and port of the required reflector plus an L for 'link' or U for 'unlink' in the URCALL field, eg REF005AL to connect and REF005AU to disconnect. Once connected, an audible announcement is given and the users should then change their URCALL back to CQCQCQ so *dPlus* doesn't keep trying to connect to the reflector every time you transmit. Again these settings can be saved as a memory channel that can be recalled whenever needed.

## DV Dongle

The DV Dongle is an AMBE2020 chip connected to a computer via a USB interface. *DV Tool*, the DV Dongle control software, is supported in Windows, Mac and Linux and connects to repeater gateways or reflectors to allow DV Dongle users to speak with D-Star radio users from their computer. As mentioned earlier, it does require remote RF users to have their RPT2 value configured correctly to the gateway port. The DV Dongle is not the same as an *Echo Link* user sitting at their PC as it connects to the gateway and not a user. In a comparative sense, it's like a link connecting to a repeater. At present, there is no radio used at the DV Dongle end but I believe there are plans to make DV Tool interface into a 9k6 port of an existing radio that will give the user the joy of being RF again.

## Radios and Interfaces

D-Star is not Icom's own standard or technology, but they are the first and so far only company currently to make D-Star equipment commercially available. There are various different Icom radios available, including mobiles and handhelds, some examples are shown in **Fig 9.5**. These comprise ID-1 for 23cm DD and DV mode, ID-800 for 2m and 70cm DV, IC-2200 for 2m DV, IC-2820 2m and 70cm DV mobiles and IC-E91, IC-E92 and ID-51E Plus 2m and 70cm DV handhelds. Also shown is one of many Hotspots available to link into your WiFi network. All radios are effectively normal FM radios with D-Star being an optional extra on some and fitted as standard in the other. I've had the chance to play with many of these radios and they are all feature rich even if just used on FM.

If buying a radio is not for you then there are many projects appearing in the form of adapters that connect to a 9k6 port in an existing FM radio and allow users to join in on D-Star. A word of caution: not all 9k6 radios are built the same and some have problems with transmitting GMSK. Have a look at [1] for project details. A final word if you are thinking of D-Star, there are many forums available where you can ask questions or just monitor if you prefer. The UK forum can be found at [2].

## DIGITAL MOBILE RADIO [DMR]

DMR has been around for a long time in the commercial world and is used by numerous companies. The start of DMR in the amateur radio community has its origins in the DMR-MARC Network [MARC representing the Motorola Amateur Radio Club] made up of members that were, in the main, associated with Motorola. This was the beginning of a network that would quickly develop on a world-wide basis comprising of Motorola repeaters, commercial C-bridge systems and most users utilised Motorola radios as they were quite popular with commercial users and had the advantage of being available for the amateur market too. Over time, the use of DMR in the amateur radio world has progressed – we now have various brands of repeaters [including home brew], a wide variety of radios, several networks and the use of hotspots.

To get started on DMR, you need to apply for an ID number, this can be done on this site - https://register.ham-digital.org/ - note that you only need one ID regardless of the number of radios you use and the registration system may limit the number of ID's it will issue to a call sign. Hotspot software such as pi-star allows you to add a prefix number to your ID so that the networks can determine the flow of traffic to each unit should you have multiple hotspots on the same network.

## Terms & Technical Explainations

DMR comes with a different technical side as well as different terminology to analogue. Many will hear terms such as code plug, talk group, time slot, promiscuous mode and many others. Whilst it's not possible to explain all of them in a short article, I shall explain a few of the more regular terms whilst covering some of the basics on the technical side of DMR to give some understanding of the mode.

Both analogue and DMR use 12.5kHz of bandwidth however, DMR has the advantage that it can effectively split this bandwidth in two by using fast digital switching between 2 channels of 6.25kHz (typically 30Ms intervals) thus generating two 'Time Slots'. In more simple terms, it could be seen as having "two repeaters in one box", thus we have Time Slot 1 [TS1] and Time Slot 2 [TS2].

Each DMR network has their own layout of talk groups [TG] which are not inter linked except for a few minor exceptions. This is like a switchboard system routing a call, this could be to users on the same repeater to users across the entire global network as well as covering numerous topics. There are also a few guidelines regarding talk group usage in order to minimise congestion and unnecessary data usage as many repeaters are connected to the internet via wireless systems. TG9 is referred to as local - the QSO does not route via the network but may still be logged on monitors. Some talk groups are referred to as "always on" which allows you to hear them if there is traffic and others are referred to as "user activated" [UA] where the user needs to press the PTT to "open" the talk group. This is another route used to reduce congestion across the network and reduce data usage.

Every radio requires a code plug which contains your DMR ID, various settings, a talk group list, digital contacts [DMR ID's with user's names, call signs and possibly location], zones, scan lists and simplex/repeater details. All this is done via a CPS [customer programming software] which used to read/write/edit the code plug. The code plug can be seen as a spreadsheet and the CPS as MS Excel/Libre Office. The CPS is available free to download except for the Motorola CPS - this is not readily available via legal channels to users outside of the USA.

When registering for a DMR ID, your details are entered into a digital contacts database. Your ID number is transmitted when you press the PTT, this can be seen on last heard logs, live monitors and on other radios. The ID database is used by networks, software providers and within radios to display your details instead of just your ID number. This is useful to be able to see the call sign, name and possibly the town/country a person is from as well as showing on live monitors so you can see the activity and who is in a QSO.

Many may have come across or used Reflectors [REF] on D-Star. Within DMR, they are fairly much the same as D-Star and Fusion, allowing you to manually connect to rooms or talk groups by going onto TG9 then using the manual dial system to type in the reflector number then pressing the PTT for around 3 seconds to create the link. The numbering system is from 4001 to 4999, with 4000 being used to disconnect and 5000 to get the connection status on a repeater. As with talk groups, reflectors are not interlinked across networks however if the reflector links to a talk group, the talk group may be interlinked. Reflectors can be easier to use on many of the older radios as they do not need to be programmed into the radio.

Promiscuous mode is a feature not available on many of the older and commercial radios. When setting up a repeater in a code plug, a channel is required for each talk group which is also linked to the talk group [and the applicable time slot that the talk group is used on] within a list. Promiscuous mode allows you to hear traffic that is not specific to the talk group of the channel that you are on, if the frequency and colour code [CC] is the same.

Firmware updates are important as they sort out bugs in the radio as well as add features or create more memory space for items such as the DMR database (which is constantly increasing and currently contains around 161000 ID entries. Some manufacturers such as Anytone release both a new Firmware and CPS simultaneously usually as one package. It's important to do both the Firmware update and install the new CPS. Note that any other files included in the package should be read as they may contain instructions on the updates as well as how to perform the updates and any other requirements as part of the update.

## DMR Simplex

Many people also utilise DMR simplex just as would be done on analogue, on both 2m and 70cm. In fact, most of the frequencies used for DMR simplex are the same as the analogue frequencies except for the calling channels. DMR has its own calling channels as per the RSGB Band Plan [https://rsgb.services/public/bandplans/html/rsgb_band_plan_2020.htm] therefore leaving the standard calling channels specifically for analogue users. The standard settings found in most radios for simplex are to use TG9, Slot 1 and CC1, which makes it easier to contact other users.

## DMR Networks

There are several DMR networks available with the two main networks providing international coverage.

Phoenix UK [http://www.dmr-uk.net/] is predominantly a repeater network which was originally the European arm of DMR-MARC [https://dmr-marc.net/]. This is run by a group called OpenDMR which still manages the European connection to the DMR-MARC network in the USA. There are around 51 repeaters across the UK which have a set talk group structure on each time slot. There are currently 39 talk groups available.

Phoenix UK [and DMR-MARC] makes use of a German reflector based network called DMR Plus to allow hotspot connections. Phoenix UK users can connect via reflectors to numerous UK talk groups offered by Phoenix UK or utilise most of the talk groups that are available on repeaters.

Connected to Phoenix UK is the Northern DMR Cluster [https://www.northerndmrcluster.com/] which is an independently managed group in the north west of the UK, which is links into Phoenix UK to offer the same national, international and hotspot access as Phoenix UK users. There are currently around 9 repeaters connected to the cluster.

Connected to Phoenix UK and DMR Plus, is DV Scotland [http://dvscotland.net/] which is an independently managed group of around 12 repeaters. They use Phoenix UK talk groups on time slot 1 and reflectors on time slot 2 (which requires TG9).

Brandmeister [http://www.bm-dmr.uk/dash/] is an international repeater and hotspot network which has around 67 repeaters across the UK and a large number of hotspot users. Globally, around 30% of the connections are hotspots. Brandmeister uses both talk groups and reflectors [some of which are linked to talk groups] however, they plan to phase out the use of reflectors at the end of 2020. Brandmeister has around 1400 talk groups [https://brandmeister.network/?page=talkgroups]. Repeater keepers decide which talk groups are always on for each time slot of their repeater. This can be found by locating the repeater on the Brandmeister website then clicking on the website to show various details for the repeater.

Connected to Brandmeister in the UK is the SALOP Cluster [https://salop-repeater.weebly.com/] which is an independently managed group in the Shropshire, Welsh Borders, Hereford, Worcester, Gloucestershire, Staffordshire, West Midlands and the Cotswolds. There are currently around 9 repeaters connected to the cluster. They utilise TG75 on TS1 for the cluster [which can also be accessed via Brandmeister using TG23575] and TG9 on TS2 to link to Brandmeister reflectors.

The South West Cluster [https://gb7bs.com/gb7bs/the%20southwest%20cluster.html] is an independently managed group of around 9 repeaters in the south west. The cluster only utilises two talk groups - TG9 on TS1 for local QSO's and TG950 on TS2 for the cluster. This cluster has no link outside of their group

Fig 9.6: DMR Networks

## DMR RADIOS

Commercial DMR users mainly utilised UHF and there was no commercial requirement for dual band therefore earlier radios are mono band covering analogue and DMR. As time has passed, radios have evolved to become more amateur friendly, offering more memory capacity, additional features to make using the networks easier as well as a range of dual band radios being launched.

There are many manufacturers of DMR radios - Motorola, Hytera, Connect Systems, Retevis/TYT, Radioditty, Anytone, Baofeng, Wouxan to name the majority. Choice of radio is important but can be difficult for new users and the wrong choice can lead to frustration and leaving the mode. Memory, features and the ability to get a code plug [especially for new users to the mode and if not supplied by the seller] are the important points. Digital APRS is also offered as an option for many manufacturers and is added as standard for the Anytone range.

Memory requirements are important as it affects the usability of the radio. Channels are important - each simplex and analogue repeater added requires a channel. DMR repeaters can require a lot more as a channel is required for each talk group you wish to use [a Phoenix UK repeater can use 39 channels]. Channels per zone affects the layout. Earlier radios only allow for 16 channels in a zone so a Phoenix repeater could be spread over a few zones whereas newer radios can hold 50-100+ channels allowing you to add an entire repeater into one zone. Scan lists can also be important depending on the user's preferences. Promiscuous mode helps reduce the number of channels for a repeater that you need to place in the scan list - all that is required is one channel from each time slot.

Manufacturers such as Motorola, Hytera and Connect Systems are no longer major players in the UK amateur market. Motorola and Hytera have kept their focus on the commercial market whilst Connect System found it difficult to maintain there "foot in the door" following the launch of their newer radios.

**Retevis / TYT** launch the same radios under different brand/model numbers. Their earlier models [RT3/MD380 & RT8/MD390] are still extremely popular mono band radios. The launch of a few options of "experimental" firmware increased their popularity as the firmware added more amateur friendly options and allowed the radios to hold more of the DMR ID database. A 3rd party CPS was also launched via a UK amateur which was easier to used and allowed users to export/import data to a spreadsheet therefore making it quicker to create or modify a code plug. The manufacturer then went on to launch the first dual band radios, starting with the RT82/MD2017 [the RT92 was also sold under the Moonraker badge as the HT-500D] handhelds. These radios were met with mixed feelings mainly due to a trackball [which could be disabled]. Ty Weaver [KG5RKI] who created Ty Toolz for their mono band radios, also released an "alternative" firmware and User DB Manager for the RT82 which is available from the Retevis website [https://www.

Fig 9.7: A collection of DMR radios

retevis.com/resources-center/]. The RT90/MD9600 was the first dual band mobile to be launched. There were some teething issues with the earlier hardware versions which were fixed. G4EML also came up with fixes and modifications such as changing the display light colour. Retevis also launched the Ailunce HD1 which has the feel of a Motorola. Whilst a little difficult to get around in the earlier stages, I believe that the radio has improved with firmware updates.

**Radioditty** launched the GD77 which like the Ailunce HD1, was not an easy radio to get started with, which has also improved with firmware updates. There has also been a project called OpenGD77 by Roger Clarke which connects the radio to a raspberry pi running pi-star to turn it into a hotspot without any additional hardware.

**Anytone** are one of the newer played in the DMR market having been better known for analogue radios and their DMR range have become the most popular due to memory capacity, functionality, the ability to export/import most parts of the code plug and being able to do many functions via the radio. They started with a handheld range launching the AT-D868UV which became extremely popular. Then the AT-D878UV was launched adding roaming and analogue APRS followed by the AT-D878UV Plus which added a bluetooth module which can connect with most bluetooth earpieces and car audio systems. They then launched the AT-D578UV Pro mobile based on the AT-D878UV Plus with a few additional features such as cross band repeat.

**Baofeng** has also jumped into the DMR market. Their first model, the DM-5R was not very successful as it transmitted on both time slots simultaneously regardless of the channel settings. Baofeng resolved this issue with further product launches such as the DM-9HX. Unfortunately, memory capacity has not been expanded compared to the analogue range therefore these radios are not suitable for wide area use.

### Hotspots

Hotspots for digital use is not a new concept as they have been utilised for some time to connect to D-Star. Now, they can connect to a variety of digital modes such as DMR, D-Star, YSF, NXDN and P25. Just as with radios, hotspots have evolved and there are even devices that do not require a radio. Hotspots are essentially the equivalent of having a low power repeater however, many of them are simplex devices.

The DV Mega is one of the earlier hotspots that were originally used for D-Star and now can be used with a variety of digital modes. There are now a variety of hotspots available from DV Mega using various hardware and base boards including the DV Stick which does not require a radio and can be used via PC or mobile phone. DV Mega products can use various software depending on the hardware - BlueDV [http://www.pa7lim.nl/bluedv/] and pi-star [https://www.pistar.uk/] are the most popular brands of software.

Another earlier hotspot is known as the DV4 Mini which were extremely popular in the early days, but they seem to have faded off the market with very few global vendors.

There are a variety of hotspots available using different models of raspberry pi and pi-star software. This includes the JumboSpot, duplex hotspots and some of the DV Mega units. Some of the units also make use of OLED or Nextion displays. These can be purchased from a variety of vendors as well as the usual online sites. Depending on where you shop, you can get assembled units, components that require final assembly or the boards with electrical components that require some further assembly work.

SharkRF has several models - the first model released was the openSpot 1, requires a LAN connection, the openSpot 2 was their first portable unit and now the upgraded portable unit, the openSpot 3 which includes cross modes. These are only available direct from SharkRF [USA] via their website and are quite pricy compared to other hotspots.

## Further Information

There is a wealth of information on the mode and various hardware available on the internet from websites, YouTube videos and Facebook groups. Even if you aren't into social media or Facebook, it may be worth joining just to find information, ask questions and keep up to date on the mode, hardware and updates. These are run by numerous amateurs who take an interest in helping others with the mode and the numerous types of hardware that are available.

## Fusion System

Using C4FM 4-level FSK Technology to transmit digital voice and data over the Amateur radio bands, Fusion is Yaesu's implementation of Digital Amateur Radio.

System Fusion is an Amateur Friendly Digital Operating mode, providing a simpler interface with features that meet the demands and needs of the Amateur enthusiast.

There are Four Operating Modes to System Fusion

**V/D** (Digital Narrow) - Voice+Digital or "**V/D**" mode

Voice FR (VW) Mode - Utilizes all available bandwidth for high-fidelity voice operation, providing the most crystal clear of voice communications.

Highspeed Data

Transfer data such as images or text messages at full rate with speeds up to 9600 Bits-per-second

## Analog FM Mode

Maintains backwards compatibility with existing Analog FM Equipment, allowing a wide range of users to experiment with System Fusion Digital.

Capabilities such as Automatic Mode Select (AMS) On the DR-1X Repeater allow an even wider range of users to communicate, by running the repeater in "Fixed FM" mode on Transmit, and "Automatic Mode Select" on receive the repeater will automatically detect the incoming signal and convert it to an Analog FM Transmission. This mode allows digital users to communicate with existing Analog FM users without the need to switch their radios into FM Mode, allowing crystal clear Digital reception into the repeater that is converted into a conventional FM Signal.

Automatic Mode Select - Complete Digital Co-existence

This function instantly recognizes whether the received signal is C4FM digital or conventional FM. The communication mode automatically switches to match the received mode. Even if a digital signal is being used, you can switch to FM communication if radio signals are received from a FM station. This function enables stress-free operation by removing the need to manually switch the communication method each time.

AMS function breaks down into the following operating modes that are fully selectable by the radio user:

**AUTO**: The both RX/ TX mode is automatically selected from one of the four operating modes (DN, VW, DW and FM) to match the characteristics of the received signal. This is the current AMS operation.

**TX MANUAL**: The both RX/ TX mode is automatically selected from four operating modes to match the characteristics of the received signal. And if desired to change the TX mode, by pressing the Microphone PTT switch momentarily, the TX mode switch between DN and FM modes.

**TX FM FIXED**: The RX mode is automatically selected from one of the four operating modes, but the TX mode is fixed to FM.

**TX DN FIXED**: The RX mode is automatically selected from one of the four operating modes, but the TX mode is fixed to the 'DN' (Digital Voice Narrow) mode.

**TX VW FIXED**: The RX mode is automatically selected from one of the four operating modes, but the TX mode is fixed to the 'VW' (Digital Voice Wide) mode.

## Enhanced Communication Functions of System Fusion

Yaesu incorporated a suite of features within the System Fusion product line that are designed specifically for Amateur Radio Use. These features allow the operator to transmit High quality digital voice simultaneously along with Digital Data, and a High Rate (Data FR Mode) dedicated Digital Data mode that provides a method of transmitting Images, Text message and Telemetry Data at a high rate of speed.

## Group Monitor (GM)

Digital Group Monitor automatically checks whether users within a communication group are in or out of range, and displays information such as distance and orientation on the screen of the client radio for up to 24 Stations.

Each individual group can share Text and Picture messages between themselves, allowing intelligent control of how content is distributed amongst a large operating group.

Group Monitor is almost an invaluable feature when an operator or group of operators needs to track resources, such as in an emergency communication operation. Resources can easily be tracked and controlled, letting operators know when they are going to fall out of range and may need to return to the coverage area, or providing invaluable telemetry data for locating and tracking individual operators.

## Snapshot Function (Picture Messaging)

By simply connecting the optional MH-85A11U Speaker Microphone with Camera, an operator can quickly take advantage of the high speed data functions of any System Fusion C4FM radio and can easily transmit images to other C4FM users.

Image data which sent from a group member is displayed on the full-color screen of the FTM-400DR or Monochromatic display of the FT2DR. This image data also retains a time record and the GPS location data of the snapshot. The pictures and data files may be easily viewed and edited by using a personal computer by simply inserting the SD Card into any SD Card reader.

## Smart Navigation Functions/Backtrack Function

The Smart/Real-time navigation function enables location checking at any time. In digital V/D mode, information such as position data is transmitted together with voice signals so the distance and direction to the other stations can be displayed in real-time while communicating with them.

The Backtrack Function enables navigation to a registered location at the touch of a button. When hiking or camping, simply register your starting point or campsite before departure, and the distance and orientation from the current location are displayed on the screen.

## Text Messaging

Text messaging could not be simpler, with direct entry via T9 Text input or the On Screen Keyboard (FTM-400), messages can be sent quickly to an individual operator or group of operators (GM Mode).

## Easy Infrastructure Migration

Migrating to System Fusion could never be easier! Since System Fusion enables the operator to quickly select operating modes using the revolutionary AMS (Automatic Mode Select) System, backwards-compatibility is fully maintained. Using the Yaesu DR-1X Repeater enables seamless integration with existing Analog systems.

## WiRES-X – Wide-Coverage Internet Repeater Enhancement System

WIRES-X (Wide-Coverage Internet Repeater Enhancement System) is a comprehensive and easy-to-use system for linking repeaters and/or home stations together, using Internet voice technology. Now you can talk to old friends, or make new ones, around the world.

# DESIGN THEORY

This section will explore some of the theory and practice of designing receiving and transmitting equipment for the 50, 70, 144 and 432MHz amateur bands.

## Receivers

Standards for VHF/UHF receivers are strongly based on the performance expected from HF receivers, in particular the ability of the receiver to detect, without any deterioration in performance, a weak signal in the presence of one or more unwanted strong signals present at the same time.

Above 50MHz background noise is much lower, so with a good receiver it is possible to realise a performance superior in terms of sensitivity and signal-to-noise ratio. An HF signal of a few microvolts is often down in the noise but at VHF and UHF, communication between stations can be achieved with signal levels as low as a few nanovolts (1nV = 1V x $10^{-9}$). On the HF bands a limit is imposed by both man-made and natural interference, beyond which any attempt to recover signals is fruitless. In VHF/UHF signal reception, there is no appreciable atmospheric noise with the exception of that caused by lightning discharges or from electrically charged rain drops. The limiting factor, when the receiver (and antenna) is in a good location, is extraterrestrial noise but the receiver can be designed to respond to signals only slightly above this level.

### Definition of noise

Broadly, noise is unwanted signal of a more or less random nature within the pass-band of the receiver. It may be natural or man-made. Examples of natural noise are the radiation from the Sun or, as described earlier, that from electrical storms and charged rain drops. These can only be avoided by excluding the Sun or the electrical storms from the 'field of view' of the antenna. Also, there is the inescapable noise generated in a resistor at any temperature above absolute zero, and shot noise produced in semiconductors, caused by the random generation and recombination of electron-hole pairs in their operations.

Examples of man-made noise are the radiation from switches and thermostats when they break current, and the radiation from computers caused by their processing pulses with very fast rise and fall times.

In the design of VHF/UHF receivers only the inescapable natural noise needs to be considered. Resistors introduce thermal noise, due to the random motion of charge carriers that produce random voltages and currents in the resistive element. There is unfortunately no resistor that will not produce these random products unless the receiver is operated at a temperature at absolute zero (0K). However, resistor noise generation can be minimised, particularly in the receiver front-end, by the correct choice of resistor. Metal film types are recommended. Thermal noise is also known as Johnson or white noise.

Shot noise in semiconductors is due to charge carriers of a particle-like nature having fluctuations at any one instance of time when direct current is flowing through the device. The random fluctuations cause random instantaneous current changes. Shot noise is also known as Schottky noise.

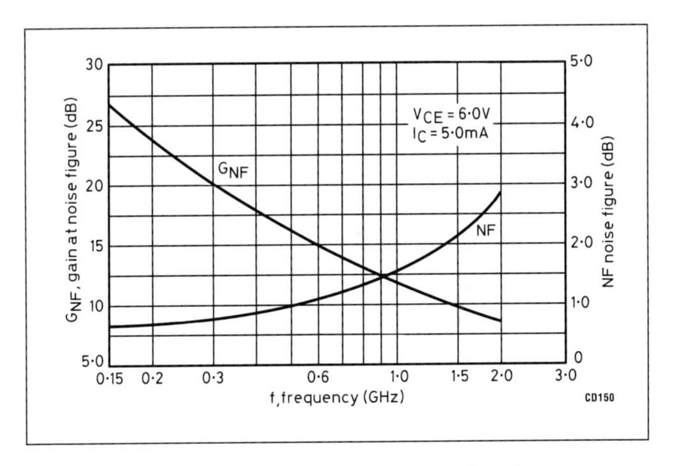


**Fig 9.8: Showing gain at noise figure and noise figure versus frequency**

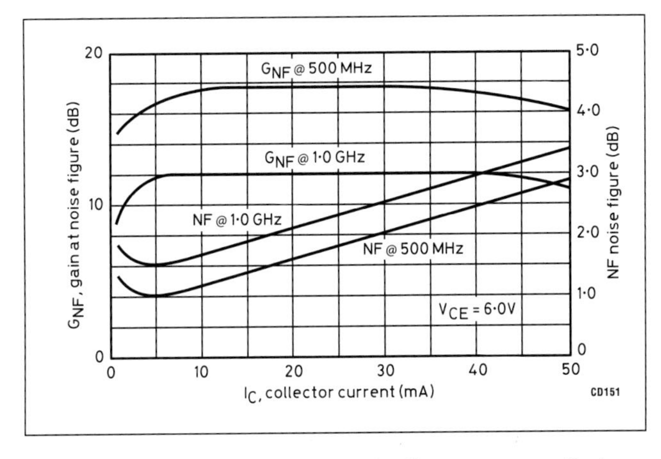


**Fig 9.9: Gain at noise figure and noise figure versus collector current**

## Noise factor and noise figure

The noise factor is the ratio of the input signal-to-noise ratio to the output signal-to-noise ratio. The noise figure is the noise factor expressed in decibels and is used as a figure of merit for VHF and UHF circuits:

$$f = \frac{\text{Input S/N}}{\text{Output S/N}}$$

$$NF = 10\log_{10} f$$

It is measured as the noise power present at the receiver output assuming a conventional S/N ratio of 1 at the input. An ideal noiseless receiver does not produce any noise in any stage. Thus the equation becomes 1/1 or a noise factor of 1 or 0dB. The noise factor of a practical receiver that will generate noise in any stage, particularly the front-end, is the factor by which the receiver falls short of perfection.

Amateur communication receiver manufacturers usually rate the noise characteristics with respect to the signal input at the antenna socket. It is commonly expressed as: (signal + noise) / noise, or "signal-to-noise ratio".

The sensitivity is usually expressed as the voltage in microvolts at the antenna terminal required for a (signal + noise) / noise ratio of 10dB. Sensitivity can also be specified as the minimum discernible signal or noise floor of the receiver.

An important point to remember in VHF/UHF receivers is that the optimum noise figure of an RF amplifier does not necessarily coincide with the highest maximum usable gain from that stage.

**Figs 9.8 and 9.9** illustrate this feature. The transistor is capable of operation up to 2.0GHz. As Fig 9.8 shows, however, the gain falls with increasing frequency but the actual noise also increases with increasing frequency. This characteristic is also shown in Fig 9.9 but here the gain at noise figure and the actual noise figure are plotted for 500MHz and 1GHz against variations in collector current. Note that the maximum gain occurs with good input matching but minimum noise does not. For example, a GaAsFET preamplifier may have an input VSWR as high as 10:1 when tuned for the lowest noise figure.

The definition of noise figure and degradation of receiver performance due to noise implies that the front-end stages, namely the RF amplifier and mixer, must use active devices, either bipolar or field effect types, with a low inherent noise figure. The noise figure quoted for the transistor illustrated in Figs 9.8 and 9.9 applies only when the device is connected to a 50 ohm source. As can be seen the noise figure increases by almost two times between 1 and 2GHz. This figure can be reduced by mismatching to the source at 1GHz and above.

Modern design theory and practice now employ S-parameters (scatter parameters) to obtain maximum performance from an RF amplifier and mixer while maintaining the noise figure at low levels. However, S-parameters require an advanced knowledge of design which includes the use of the Smith chart and the availability of some sophisticated equipment such as a network analyser. Manufacturers' data on RF devices includes tables of S-parameters. This method of design of amplifiers and mixers is outside the scope of this handbook.

## Intermodulation

Intermodulation occurs when two or more signals combine to produce additional (spurious) signals that were not originally present at the receiver input and possibly causing interference to a weak wanted signal. The receiver front-end is handling many incoming signals of different strengths but only those signals passing through the selective (IF) filters will eventually be detected. The RF circuits can have a bandwidth of several megahertz but the selective IF filters reduce the bandwidth to that required for adequate resolution of signals, dependent on the method of modulation of the wanted carrier. At low signal levels the front-end will have optimum linearity, ie there is no unwanted mixing between signals. However, as stated previously, very strong signals will cause the front-end to go into its non-linear region of operation and then these signals will mix together and produce new signals which can appear in the IF pass-band. Second-order intermodulation products (IPs) are caused by two signals mixing, viz $f_1$ and $f_2$, and generating new frequencies which appear as ($f_1 + f_2$) and ($f_1 - f_2$) and the second harmonics of each signal ($2f_1$ and $2f_2$), generating the second-order IPs. However, if $f_1$ and $f_2$ are close spaced, their second-order IPs will be well spaced and can be easily filtered out by the selective IF filters.

However, if the $f_1$ and $f_2$ signals are increased in strength then another set of IPs is generated. These are third-order intermodulation products, due to the fact that mixing occurs between three signals. The three signals can be independent but the same products can be generated by $f_1$ and $f_2$ by themselves. These frequencies $f_1$ and $f_2$ can add or subtract to produce the following third-order IPs:

| | |
|---|---|
| Third harmonics: | ($f_1+f_1+f_1$) and ($f_2+f_2+f_2$) |
| Sum products: | ($f_1+f_1+f_2$) and ($f_2+f_2+f_1$) |
| Difference products: | ($f_1+f_1-f_2$) and ($f_2+f_2-f_1$) |

It is clear that if $f_1$ and $f_2$ are equally spaced above and below the wanted frequency, interference will be severe. When $f_1 + f_2$ are close to the wanted frequency the third-order sum products will appear in the third harmonic area of this frequency and will be attenuated by the selective filters in the receiver. However, when the difference products containing a minus sign are close to $f_1$ and $f_2$ and are generated by the receiver, the filters, however selective, will not remove these spurious signals. These products could cause unwanted interference to a wanted weak signal.

When third-order intermodulation products are generated in the receiver, they will increase in level by 3dB for every 1dB increase in the levels of $f_1$ and $f_2$. Thus the appearance of intermodulation products above the receiver noise floor is quite usual. When further levels of $f_1$ and $f_2$ occur higher odd-order intermodulation products are generated, eg fifth, seventh etc, which can interfere with a weak wanted signal. These higher-order products will appear even more quickly than third-order products but require stronger $f_1$ and $f_2$ signals, eg fifth-order products will be generated five times as fast as $f_1$ and $f_2$ as the level of $f_1$ and $f_2$ is increased. Significant intermodulation products can only result from $f_1$ and $f_2$ when their strength is high. If either the $f_1$ or $f_2$ signal disappears, leaving only one signal, the intermodulation product will disappear. Optimising linearity in the receiver front-end will minimise generation of these unwanted intermodulation products and hence interference to wanted signals.

## Gain compression

This occurs when a strong incoming signal appearing at the antenna socket causes one (or more) stages in the front-end to be driven into the non-linear region of its output characteristic. As an example, when an amplifier stage is operating in its linear region, an increase by 3dB in signal level at its input will cause

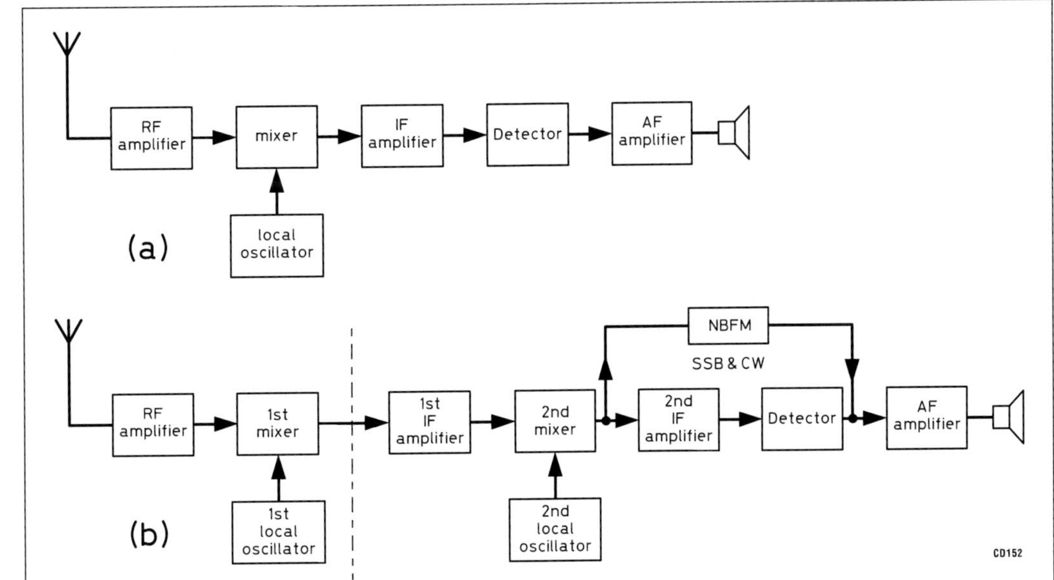


**Fig 9.10: Two conventional superheterodyne configurations: (a) Self-contained single superhet with tuneable oscillator. (b) Self-contained double superhet or converter in front of a single superhet. Either or both oscillators may be tuneable**

by linear transfer, a 3dB increase in signal level at the output. However, a further increase in input signal level could cause non-linear transfer and limit the output level increase to 1dB. A very strong signal could drive the stage into extreme non-linearity, making the stage degenerative (gain less than 1) and desensitising the receiver. Background noise will decrease in level together with all other signals, including wanted weak signals. As with intermodulation, optimising front-end linearity will minimise receiver desensitisation by strong signals.

### Reciprocal mixing

This phenomenon occurs when the receiver local oscillator produces excessive sideband noise on its carrier and a strong off-channel RF carrier mixes with this noise to produce the IF. Reciprocal mixing causes an increase in receiver noise level when a strong carrier appears, the opposite effect of gain compression. The receiver selectivity is not necessarily defined by front-end RF filters. The 'cleanest' local oscillators are LC (VFOs) and crystal-controlled oscillators. Some of the `noisiest' oscillators are found in receivers employing a synthesised system, for example phase-locked-loop synthesised oscillators. Some earlier receivers suffered from reciprocal mixing effects, ie generation of an unwanted spurious signal in the IF pass-band, for example, noise on the voltage-controlled oscillator control line leading to FM noise sidebands. However, modern receivers now employ 'quiet' synthesised local oscillators that minimise reciprocal mixing.

### Dynamic range

The main problems of front-end overload are gain compression, intermodulation and reciprocal mixing. Each phenomenon has its own characteristic and level at which strong unwanted signal(s) cause degradation in receiving wanted signals. Just one strong signal causes gain compression or reciprocal mixing, whereas two are required to cause intermodulation products. The gain of the front-end, ie the RF stage and the mixer, should be kept as low as possible, consistent with good sensitivity and signal/noise performance. Gain compression and intermodulation are caused when either or both stages are driven beyond their linear transfer range. The front-end should be designed so it cannot be overloaded by even the strongest amateur band signals. Intermodulation products will not be a problem if they are restricted to the level of the background noise and gain compression and reciprocal mixing effects are not a problem if they do not significantly change the system noise level.

The lowest end of the dynamic range will be designed for the lowest power audible signal and, conversely, the highest end will be designed for the unwanted signal of the highest power level, ie signals without any overload effects degrading the front-end performance. This principle is called spurious-free dynamic range but the range will change according to the differing power levels of unwanted signals.

## Receiver Front-End Stages

Thus the requirements for front-end stage design in a VHF/UHF receiver are:

- Low noise figure
- Large dynamic range
- Power gain consistent with good sensitivity and signal-to-noise ratio

The noise figure and dynamic range requirements have already been described in detail. However, 'power gain' must be brought into the equation to complete the design philosophy. 'Power gain' needs some explanation because sheer power gain is not sufficient in itself or even desirable. In a multistage receiver with, say, eight stages of gain, input noise originating in the first stage, normally an RF amplifier, will be amplified by the eight gain stages, that in the second by seven and so on. If the effective noise voltages are denoted by $V_1$, $V_2$, $V_3$, ... $V_8$ and the stage gains by $G_1$, $G_2$, $G_3$,... $G_8$, the total noise present at the receiver detector will be:

$$V_1(G_1 G_2 G_3 .. G_8) + V_2(G_2 G_3 .. G_8) + V_3(G_3 G_4 .. G_8) \text{ and so on.}$$

If the voltage gain of the RF amplifier ($G_1$) is high, for example 20dB (10 times) or more, the important noise contribution is due to $G_1$. Provided the remaining gain stages are correctly designed and provide evenly distributed gain, the overall noise contribution from them will be very small.

Additional noise generated by any stage that is regenerative or is actually oscillating will degrade the overall receiver noise performance, and might actually cause receiver desensitisation and consequent poor weak-signal performance.

However, at the output of $G_3$ (eg the first IF stage) an amplified signal will be large compared to the noise contributed and $G_4$ ... $G_8$ should not degrade noise performance. The function of the RF stage is to provide just sufficient gain to overcome the noise

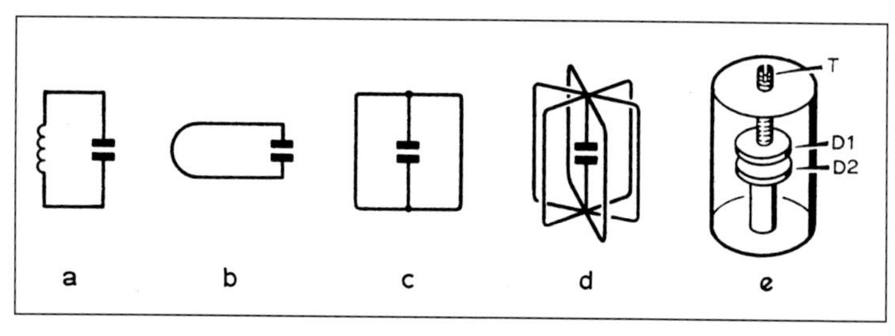


**Fig 9.11: Progressive development of tuned circuits from a coil to a cavity as the frequency is increased**

contribution of the mixer stage. The mixer is by definition a non-linear device and normally contributes more noise than any other stage no matter how well designed. The RF stage therefore considerably improves the receiver signal-to-noise ratio, improving weak-signal performance. If the RF stage gain is too high then the problems of strong signals, ie intermodulation, gain compression and reciprocal mixing products, will, as described, degrade receiver performance.

A system of gain control on the RF amplifier may appear to be the answer. However, the use of AGC must be carefully considered, otherwise the weak-signal performance will be degraded if the onset of AGC is not delayed. Reducing the amplifier gain, using an AGC current for bipolar transistors or a voltage for field effect transistors, can degrade the RF stage linearity. The noise factor of a transistor amplifier is also dependent on emitter (source) current. A large variation from the manufacturer's data given for noise factor V $I_E$ ($I_S$) will also degrade weak-signal performance. Circuit layout and correct shielding for VHF/UHF RF amplifiers is of paramount importance for stable operation. Instability and spurious oscillation can be produced via RF feedback through the amplifier transistor. Even the latest designs of transistor, both bipolar and field effect types, can give rise to these effects. Regeneration or actual oscillation can be prevented by neutralising the internal feedback, using an external circuit from output to input of the amplifier. This will feed back an equal amount of out-of-phase signal and, providing there is no other feedback path, the amplifier will be stable.

## Circuit noise

Noise due to devices other than transistors is produced solely by the resistive component; inductive or capacitive reactances do not produce noise. Inductors of any form have negligible resistance at VHF and UHF but the leakage resistance of capacitors and insulators is important. It is imperative to choose high-quality capacitors such as silver mica, ceramic, polycarbonate, or polystyrene types with negligible leakage current and high-Q properties. Tantalum capacitors should be used where it is necessary to decouple at LF, as well as VHF and UHF. Attention should be given to the use of low-noise resistors. Carbon or metal film types (of adequate power dissipation) must be used. The old-style carbon composition resistors are very good noise generators. 25dB noise difference between film and composition resistors has been observed when a direct current was passed through the two types, both being the same value.

Circuit noise due to regeneration has already been discussed. Common causes of regeneration are:

(a) Insufficient decoupling of voltage supplies (at each stage) and particularly emitter (source) and collector (drain) circuits.

Suitable capacitor values and types are as follows:

| | |
|---|---|
| 10MHz | 47nF (0.047µF) ceramic disc or plate |
| 10-20MHz | 22nF (0.022µF) ceramic disc or plate |
| 20-30MHz | 10nF (0.01µF} ceramic disc or plate |
| 30-100MHz | 4.7nF (0.0047µF) ceramic disc or plate |
| 100-200MHz | 2.2nF (0.0022µF) ceramic disc or plate |
| 200-500MHz | 1nF (0.001µF) ceramic disc or plate |
| 500-1000MHz | 220pF (0.00022µF) ceramic disc or plate |

Leadless capacitors should be used above about 400MHz.

(b) Closely sited input and output circuits. Every attempt should be made to build the amplifier(s) in a straight line and where possible, to use shielded coils, particularly for 6m, 4m and 2m equipment.

(c) Insufficient or wrongly placed screening between input and output circuits. In transistor amplifiers where a screen is required, it should be mounted close fitting across the transistor with input and output circuits (normally base/gate and collector/drain) on opposite edges of the screen. The screen can conveniently be made of double-sided copper laminate and soldered to the main PCB.

(d) Circulating IF currents in the PCB due to multipoint grounding. Decoupling capacitors for each stage should be grounded at a single point very close to the emitter (source). It is preferable to use a double-sided PCB using one side as the ground plane.

## Effect of bandwidth on noise

If the noise factor of a receiver is measured with a noise generator it is independent of receiver bandwidth. Generator noise has the same characteristics as circuit noise so, for instance, if the bandwidth doubled, the overall noise is doubled. For the reception of a signal of finite bandwidth however, the optimum signal-to-noise ratio is obtained when the bandwidth of the receiver is only just sufficient to accommodate the signal. Any further increase in bandwidth results in increased noise. The signal-to-noise ratio at the receiver detector therefore depends on the power per unit bandwidth of the transmitted signal.

As an example, a receiver may generate 0.25µV of noise for each 2.5kHz of bandwidth. Assuming an SSB transmitter radiates a sideband signal of 2.5kHz bandwidth and produces 2.5µV of signal at the receiver detector, the signal-to-noise ratio is therefore 10, provided the receiver overall bandwidth is 2.5kHz.

If the transmission bandwidth is reduced to 1.25kHz for a CW

| Parameter | HRW (231MT-10001A) | HRQ (232MT-1001A) |
|---|---|---|
| Centre frequency (MHz) | 435 | 435 |
| Bandwidth at 3dB (MHz) | 12 min | 11 |
| Attenuation (dB) | 20 min at ±30MHz | 25 min at ±15MHz |
| Max ripple (dB) | 1.5 | 2 |
| Max insertion loss (dB) | 2.5 | 4 |
| Impedance (Ω) | 50 | 50 |

**Table 9.1: Electrical characteristics of Toko HRW and HRQ helical resonators**

**Fig 9.12: (left) Toko resonators. (left) HRW and (right) HRQ**

signal and the radiated power is unchanged, the receiver input will remain at 0.25µV but if the bandwidth is also reduced to 1.25kHz, the receiver detects only 0.125µV of noise and the signal-to-noise ratio will increase to 20.

Using receiver bandwidths that exceed transmission bandwidths is therefore undesirable when optimum signal-to-noise ratio is the prime factor. Transmitters with poor frequency stability will either require the receiver to be retuned or the use of wider bandwidth, resulting in a degraded signal-to-noise ratio. Fortunately, well designed transmitters with PLL synthesised oscillators or crystal-controlled oscillators are now employed in the majority of the VHF/UHF bands.

## Choice of Receiver Configuration

Receivers using other than superheterodyne techniques are rare on VHF or UHF. Modern superheterodyne receivers may have one, two or three frequency changes before the final IF, each with its own oscillator which may be tuneable (by the receiver tuning control) or of fixed frequency. Receivers may have a variety of configurations; two are illustrated here in **Fig 9.10**. **Fig 9.10(a)** shows a conventional single superheterodyne for use on the HF bands. The local oscillator will be partially synthesised, ie use a pre-mixer driven by an HF crystal-controlled oscillator and a LF VFO to produce the local oscillator for the main mixer. **Fig 9.10(b)** shows a double superheterodyne. This can be a purpose-built receiver (or as illustrated in Fig 9.10a)) to which is added (to the left of the dotted line) a VHF/UHF converter. The first local oscillator is crystal controlled and tuning is accomplished using the HF receiver (second) oscillator.

The main disadvantage of this method of using an HF receiver as a 'tuneable IF' amplifier preceded by a converter is again the problem of overloading the first amplifier and second mixer with strong signals. This will result in intermodulation and reciprocal mixing products, if not gain compression, particularly if the converter gain is high, say, 20 to 30dB.

A superior arrangement is to build a tuneable IF amplifier containing all the refinements of a normal HF receiver, including an NBFM IF amplifier and detector, and restrict the tuning range to a few megahertz to cover the VHF/UHF ranges of the converters. The HF receiver gain in front of the second mixer must be low. The first IF amplifier can be omitted but the pre-mixer selectivity should be retained. The converter gain (from VHF/UHF to first IF) should also be low: 10 to 14dB. This will result in a VHF/UHF receiver with a very good noise factor and dynamic range, and demodulation of NBFM in addition to CW and SSB signals.

### Choice of the first IF

In any superheterodyne receiver it is possible for two incoming frequencies to mix with the local oscillator to give the IF; these are the desired signal and the image frequency. A few figures should make the position clear. It will be assumed that the receiver is to cover the 144 to 146MHz band, and that the first IF is to be 4 to 6MHz. The crystal oscillator frequency must differ from that of the signal by this range of frequencies as the band is tuned and could therefore be 144 - 4MHz = 140MHz or, alternatively, 144 + 4MHz. However, the choice of 144 + 4MHz would invert the sideband being received. If the signal transmitted were USB, the tuneable IF would need to be set to LSB to resolve the signal correctly. As this is inconvenient the best choice is 144 - 4MHz. Assuming that the lower of the two crystal frequencies is used, a signal on 136MHz would also produce a difference of 4MHz and unless the RF and mixer stages are selective enough to discriminate against such a signal, it will be heard along with the desired signal on 144MHz. From the foregoing, it will be appreciated that the image frequency is always removed from the signal frequency by twice the IF and is on the same side as the local oscillator.

It should be noted that even if no actual signal is present at the image frequency, there will be some contributed noise which will be added to that already present on the desired signal. It is usual to set the RF and mixer tuned circuits to the centre of the band in use so that on the 144MHz band they should be at least 2MHz wide in order to respond to signals anywhere in the band. This bandwidth only represents approximately 1.4% of the mid-band frequency and it is not surprising that appreciable response will be obtained over the image frequency range of 134 to 136MHz unless additional RF filtering is employed. Naturally the higher the first IF, the greater the separation between desired and image frequencies. However, an IF as low as 4 to 6MHz is feasible, provided some attempt is made to restrict the bandwidth of the converter by, for example, employing two inductively-coupled tuned circuits between the RF and mixer stages, thus providing a band-pass effect.

The choice of the first IF is also conditioned by other factors. Firstly, it is desirable that no harmonic of the oscillator in the main receiver should fall in the VHF band in use and secondly, there should be no breakthrough from stations operating on the frequency or band of frequencies selected for the first IF.

Many HF receiver oscillators produce quite strong harmonics in the VHF bands and, although these are high-order harmonics and are therefore tuned through quickly, they can be distracting when searching for signals in the band in question. The problem only exists when the converter oscillator is crystal controlled, as freedom from harmonic interference is then required over a band equal in width to the VHF band to be covered. This also applies of course to IF breakthrough.

As it is practically impossible to find a band some hundreds of kilohertz wide which is unoccupied by at least some strong signals, it is necessary to take steps to ensure that the main receiver does not respond to them when an antenna is not connected. Frequencies in the range 20 to 30MHz are often chosen, since fewer strong signals are normally found there than on the lower frequencies but this state of affairs may well be reversed during periods of high sunspot activity.

With the greatly increased use of general-coverage receivers covering 100kHz to 30MHz, the best part of the spectrum for 6m, 4m and 2m is from 28 to 30MHz. Full coverage of the 70cm band will require the receiver to be tuned from 10 to 30MHz. IF breakthrough is minimised and frequency calibration is simple.

## Tuned Circuits

Tuning is readily achieved at HF by lumped circuits, ie those in which the inductor and capacitor are substantially discrete components. At VHF the two components are never wholly separate, the capacitance between the turns of the inductor often being a significant part of the total circuit capacitance. The self inductance of the plates of the capacitor is similarly important. Often the capacitance required is equal to, or less than, the necessary minimum capacitance associated with the wiring and active devices, in which case no physical component identifiable as 'the capacitor' is present and the circuit is said to be tuned by the 'stray' circuit capacitance.

As the required frequency of a tuned circuit increases, obviously the physical sizes of the inductor and capacitor become smaller until they can no longer be manipulated with conventional tools. For amateur purposes the limits of physical coils and capacitors occur in the lower UHF hands: lumped circuits are often used in the 432MHz band but are rare in the 1.3GHz band.

## Distributed circuits

**Fig 9.11** illustrates how progressively lower inductances are used to tune a fixed capacitor to higher frequencies. In **Fig 9.11(b)** the 'coil' is reduced to a single hairpin loop, this configuration being commonly used at 432MHz. Two loops can be connected to the same capacitor as in **Fig 9.11(c)**. This halves the inductance and can be very convenient for filters.

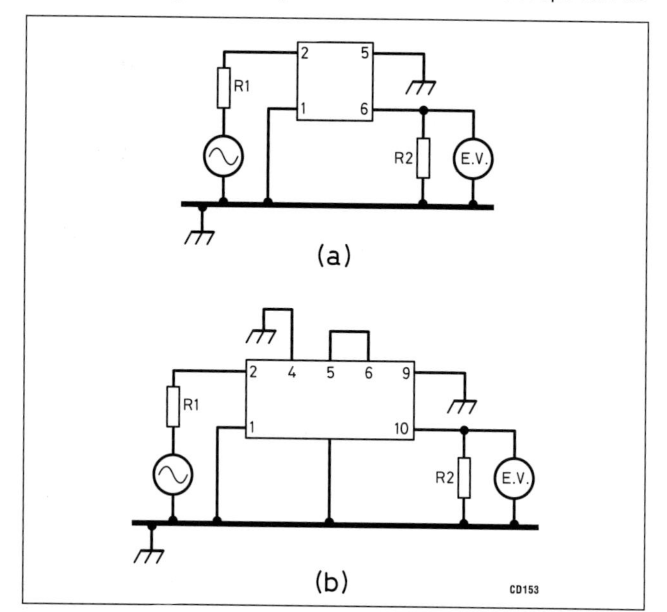


**Fig 9.13: Test circuit for (a) Toko HRW and (b) Toko HRQ filters. The case lugs must be grounded. R1 = R2 = 50Ω**

**Fig 9.11(d)** represents a multiplication of this structure and in **Fig 9.11(e)** there are in effect an infinite number of loops in parallel, ie a cylinder closed at both ends with a central rod in series with the capacitor. If the diameter of the structure is greater than its height it is termed a rhumbatron, otherwise it is a coaxial cavity. The simple hairpin, shown at Fig 9.11(b), is a very convenient form of construction: it can be made of wide strip rather than wire and is especially suitable for push-pull circuits. It may be tuned by parallel capacitance at the open end, or by a series capacitance at the closed end.

In a modification of the hairpin loop, the loop can be produced from good-quality double-sided printed circuit board and such an arrangement is known as microstripline. The loop is formed on one side; the ground-plane side of the PCB, through the dielectric, makes the stripline. When the PCB is very thin, the result is called microstrip and that is used in many commercial receivers.

## Bandpass circuits

Tuning of antenna and RF circuits to maintain selectivity in the front-end of a VHF/UHF receiver cannot be undertaken with normal ganged tuning capacitors, not only due to the effects of strong capacitance coupling between circuits, which become prominent in these frequencies, but also due to the difficulty of procuring small-swing (say 20pF) multi-ganged capacitors. The varicap diode can be used to replace mechanical capacitors, but can degrade the receiver performance when strong signals are present by rectifying these signals and introducing intermodulation products into the mixer, thus seriously reducing the receiver's dynamic range. Fortunately modern construction techniques have enabled coil manufacturers to introduce a band-pass circuit in a very small screened unit, namely the helical filter.

The helical filter in simple terms is a coil within a shield. However, a more accurate description is a shielded, resonant section of helically wound transmission line, having relatively high characteristic impedance. The electrical length is approximately 94% of an axial quarter-wavelength. One lead of the winding is connected to the shield; the other end is open-circuit. The Q of the resonator is dictated by the size of the shield, which can be round or square. Q is made higher by silver plating the shield. Resonance can be adjusted over a small range by opening or closing the turns of the helix. The adjustment is limited over the small frequency range to prevent degradation of Q. Modern miniature helical resonators can be obtained in a shield only 5mm square, but the minimum resonant frequency is normally 350MHz. Maximum $F_r$ can be 1.5GHz. Large-size resonators (10mm square) will resonate down to 130MHz.

The band-pass filter is obtained by combining two to four resonators in one unit with slots cut in each resonator screen of defined shape to couple the resonators. This forms a high-selectivity tuned circuit with minimum in-band insertion loss and maximum out-of-band attenuation.

Helical filters can be cascaded to increase out-of-band attenuation. As an example, a quadruple filter with a centre frequency of 435MHz might have a 3dB bandwidth of 11MHz and 25dB attenuation at plus or minus 15MHz; with a ripple factor of 2dB and insertion loss of 4dB (see **Table 9.1**). **Fig 9.12** shows pictures of the helical filters tested. The test circuit is shown in **Fig 9.13**, and **Figs 9.14 and 9.15** show the test results.

Tuning is accomplished by brass screws in the top of the screen, one for each helix. The nominal input/output impedance is 50 ohms formed by placing a tap on the helix. This impedance is ideal for matching to antennas and to RF amplifiers and mixers designed using S-parameters. Thus the helical filter replaces conventional tuned circuits in the receiver front-end, resulting in a considerable improvement in selectivity. One note of caution;

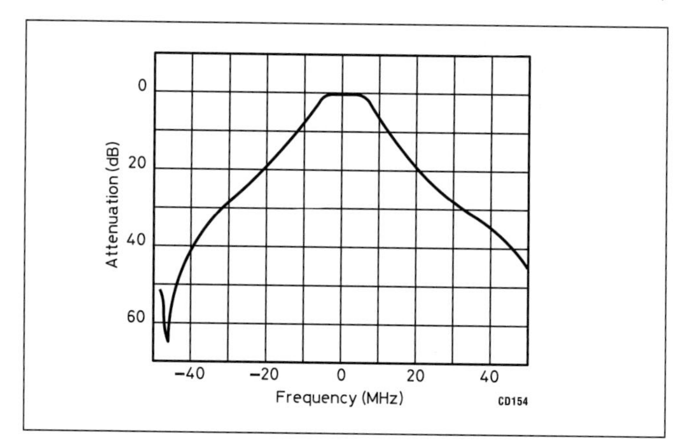


**Fig 9.14: The Toko HRW frequency response**

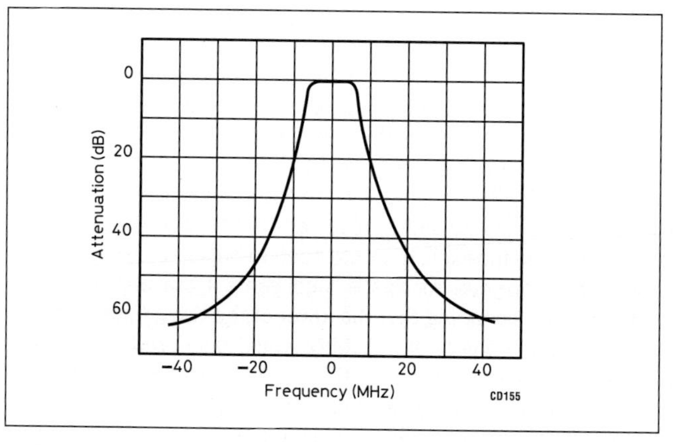


**Fig 9.15: The Toko HRQ frequency response**

the screening can lugs must be soldered perfectly to the PCB, otherwise the out-of-band attenuation characteristics of the filters will be degraded. These filters can be used for 2m, 70cm and 23cm receiver front-ends.

## Preamplifiers

As available devices improve and new circuit designs are published, it will become apparent that a receiver that may have been considered a first-rate design when built is no longer as good as may be desired. Specifically, a receiver using early types of transistor may not be as sensitive as required, although the local oscillator may perform satisfactorily. The sensitivity of such a receiver can be improved without radical redesign by means of an additional separate RF amplifier, usually referred to as a preamplifier. Such an amplifier should have the lowest possible noise figure and just sufficient gain to ensure that the overall performance is satisfactory.

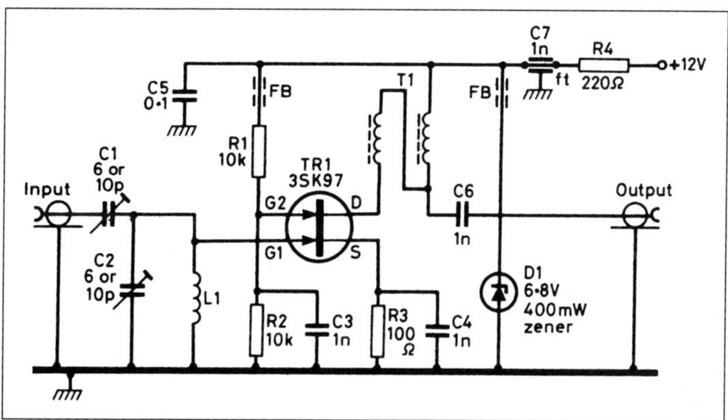


Fig 9.18: GaAsFET preamplifier for 144 or 432MHz

**Fig 9.16** shows the improvement to be expected from a preamplifier given its gain and the noise factor of the preamplifier alone and the main receiver.

An example will suffice to show the application. An existing 145MHz receiver has a measured noise figure of 6dB and is connected to its antenna via a feeder with 3dB loss. It is desired to fit a preamplifier at the masthead; what is the performance required of the preamplifier?

Suppose a BF180 transistor is available: this has a specified maximum noise figure of 2.5dB at 200MHz and will be slightly better than this at 145MHz. The main receiver and feeder can be treated as having an overall noise figure of 3 + 6 = 9dB. From Fig 9.16, if the preamplifier has a gain of 10dB, the overall noise figure will be better than 4.1dB. Increasing the gain of the preamplifier to 15dB will only reduce the overall noise figure to 3.6dB and may lead to difficulty due to the effect of varying temperatures on critical adjustments. The addition of so much gain in front of an existing receiver is also very likely to give rise to intermodulation products from strong local signals. If it is desired to operate under such conditions, it is essential that provision is made for disconnecting the preamplifier when a local station is transmitting.

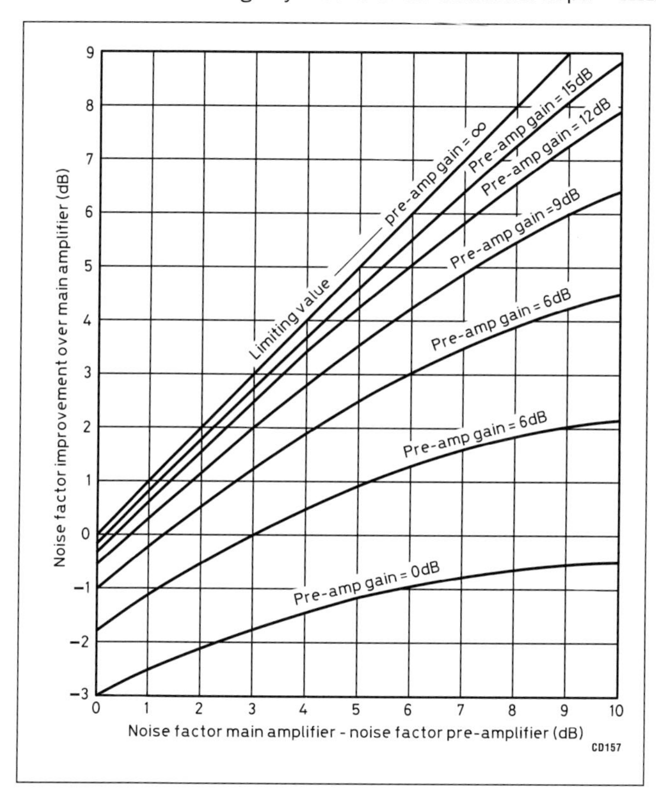


Fig 9.16: Receiver noise figures

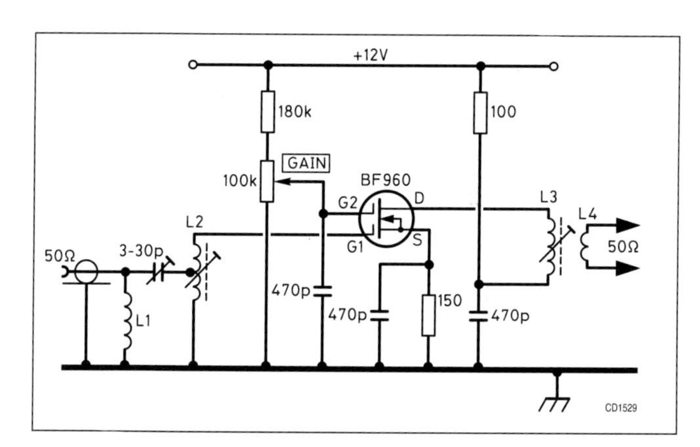


**Fig 9.17: Dual-gate MOSFET RF amplifier for 2m with gain control. L1: 3 turns 1.0mm enamelled wire, 6.0mm ID, 8.0mm long. L2: 2 turns 1.0mm tinned copper wire on 8.0mm former 10mm long, tapped at 1½ turns and tuned with dust core. L3: 6 turns 1.0mm enamelled wire on 8mm former tuned with dust core and coupled to L4. L4: 2 turns 1.0mm enamelled wire on same former, close spaced to capacitor end of L3**

## RF Amplifier Design

The advent of the field effect transistor (FET) eased the design of VHF RF amplifiers in two ways. The relatively high input resistance at the gate permits reasonably high-Q tuned circuits providing protection against strong out-of-band signals such as from broadcast or vehicle mobile stations. Also the drain current is exactly proportional to the square of the gate voltage; this form of non-linearity gives rise to harmonics (and the FET is a very efficient frequency doubler) but a very low level of intermodulation.

The use of a square-law RF stage is not, however, as straightforward as it first appears. Second-order products are still present, such as the sum of two strong signals from transmitters outside the band, typically Band II broadcasts, mixing to generate the unwanted signals on the 144MHz band. It follows, of course, that if two FET amplifiers are operated in cascade a band-pass filter is required between them to ensure that the distortion products generated in the first stage are not passed to the next stage where they will be re-mixed with the wanted signal.

A development of the FET was the metal oxide semiconductor FET (MOSFET), the gate is insulated by a very thin layer of silica.

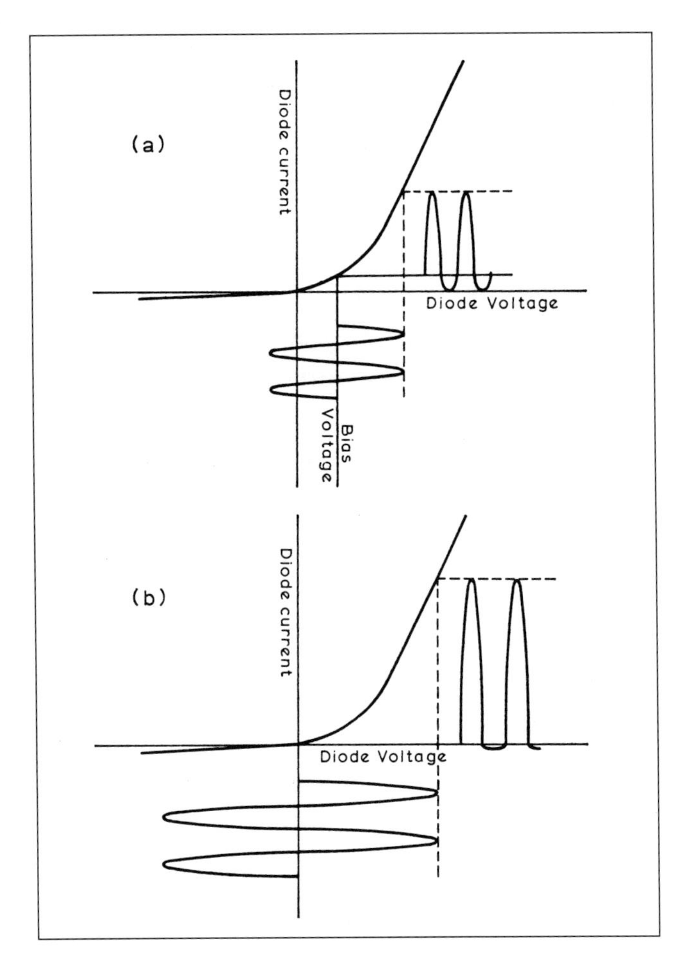


Fig 9.19: Working conditions of diodes mixers. In (a) forward bias is required to provide an optimum point. In (b) the local oscillator is higher and no bias is required

The gate therefore draws no current and a high input resistance is possible, limited only by the losses in the gate capacitance. These devices may be damaged by static charges and must be protected against antenna pick-up during electrical storms and also by RF from the transmitter feeding through an antenna changeover relay with an excessively high contact capacitance. MOSFETs now have protective diodes incorporated in the device which limit the input voltage to a safe level and these devices are thus much more rugged.

In the dual-gate MOSFET the drain current is controlled by two gates, resulting in various useful circuit improvements. If it is desired to apply gain control to a bipolar transistor stage the control voltage is applied to the same electrode as the signal but the result can be a reduction in signal handling capacity, showing up as intermodulation and/or blocking. The dual-gate FET avoids this problem and automatic or manual gain control can be applied to gate 2 without reducing the signal-handling capability at gate 1. **Fig 9.17** shows how such an RF stage for a 2m converter is arranged. When a strong local station causes intermodulation at the mixer or an early stage of the main receiver, the RF gain can be reduced until interference-free reception is again possible. Note that reducing the gain by this method increases the noise figure. For a gain reduction of more than a few dB the noise figure will begin to degrade.

The GaAsFET is similar to the MOSFET but is based on gallium arsenide rather than silicon. Gallium arsenide has larger electron mobility than silicon and therefore has a better performance at UHF. GaAsFETs were originally designed for use in tel-

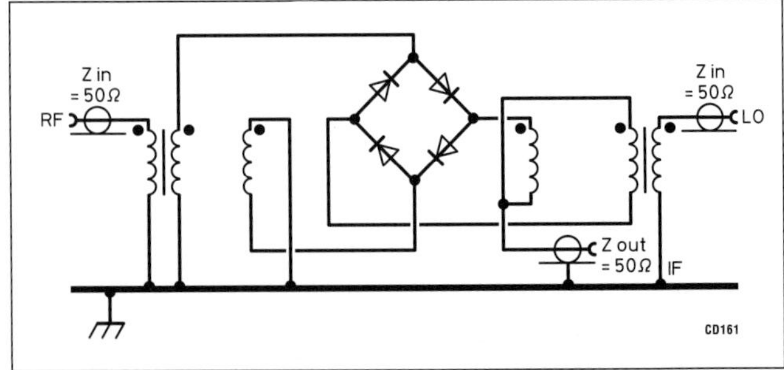


Fig 9.20: Typical diode ring mixer circuit (*ARRL Handbook*)

evision receiver tuners and are entirely suitable for 144MHz and 432MHz preamplifiers or as a replacement for an existing RF amplifier. These devices are available from major semiconductor manufacturers and include the 3SK97, 3SK112, C3000 and the CF739 (Siemens).

In a correctly designed amplifier the GaAsFET will provide excellent performance at these frequencies. The quoted noise figure for the 3SK97 is 1dB at 900MHz. Silicon diodes are mounted in the GaAsFET chip to protect the gates against ESD (electrostatic discharge) breakdown. The noise figure obtainable is related to the manufacturer's data sheet. Biasing is the same as for silicon MOSFETs, ensuring, however, that the gates are never positively biased with respect to the channel.

The circuit for a GaAsFET preamplifier [3] that incorporates a self-biasing arrangement is shown in **Fig 9.18**. Construction of L1 and T1 for 144MHz and 432MHz is given below.

L1 144MHz: 6 turns 2.0mm tinned copper wire 6mm ID 13mm long
432MHz: copper line 15mm wide 57mm long spaced 4mm above ground plane

T1 144MHz: 12 turns bifilar wound 0.5mm enam copper wire.
432MHz: 5 turns 0.5mm enam copper wire centre tapped as a 4:1 transformer on an Amidon T-20-12 toroid Core

Double-sided copper laminate is used for mounting most of the components with a vertical screen of tinplate forming a screen between input and output circuits. Care must be taken against static when mounting the GaAsFET in the circuit. After careful checks for constructional errors the current should be checked before alignment. This should be between 25-30mA.

Power gain alignment will be in the order of 26dB at 144MHz and 23dB at 432MHz. Any tendency towards instability can be cured by fitting a ferrite head over the drain lead close to the FET.

An attenuator must be used between the preamplifier and input to an existing receiver or converter to prevent degradation of the strong-signal performance.

## Mixers

The most common types of mixers in use today are designed

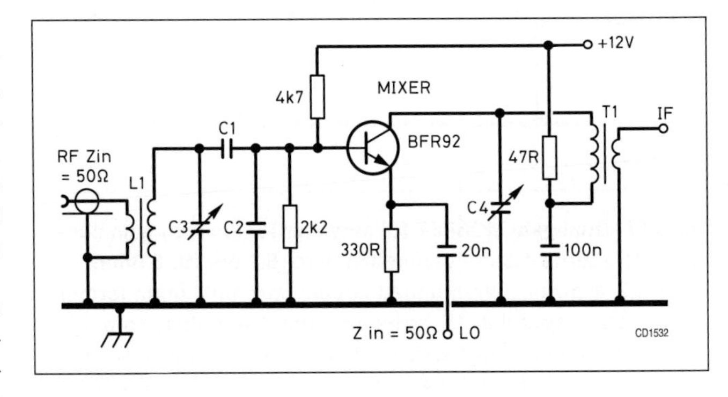


Fig 9.21: Bipolar mixer circuit (*ARRL Handbook*)

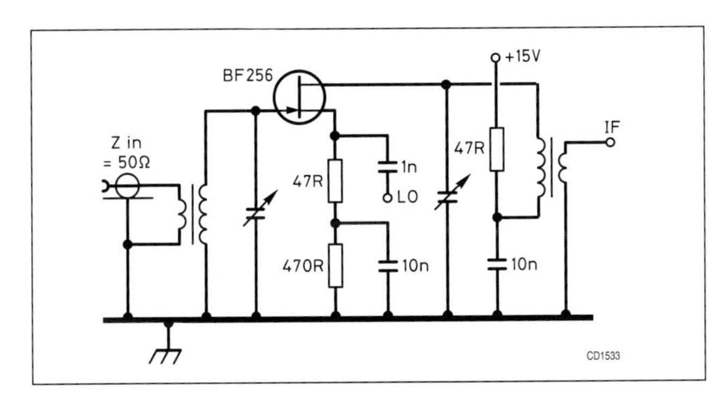


Fig 9.22: A JFET mixer (*ARRL Handbook*)

around diodes, bipolar and field effect transistors; integrated circuits are now available for use in mixer applications.

## Diode mixers

A diode mixer operates non-linearly, either around the bottom bend of its characteristic curve where the current through the diode is proportional to the square of the applied voltage - see **Fig 9.19(a)**, or by the switching action between forward and reverse conduction as shown in **Fig 9.19(b)**. For optimum working conditions in (a), it is usually necessary to apply DC forward bias to the diode of typically 100 to 200mV; this will vary with the type of diode.

The switching diode mixer (Fig 9.18(b)) is used where a high overload level (strong signals) is required. Signal levels approaching one tenth of the local oscillator power can be handled successfully, and the oscillator level is limited only by the power handling capacity of the diode. The noise generated in the mixer rises with increasing diode current, however, and this sets the limit on the usable overload level if maximum sensitivity is required. The LO power can be adjusted to select a compromise between sensitivity and overload capacity.

Diode mixers display high intercept points and almost all of them are balanced. Conversion loss is an inherent characteristic of diode mixers. In practice this is usually between 3 and 6dB. It is therefore essential that the stage following the mixer, eg an IF amplifier, has the lowest possible noise figure.

## Balanced-diode mixers

Noise component transfer from the LO and the mixer to the post-mixer stage can be reduced to a low level by using a balanced two-diode design employing modern low-noise diodes.

However with modern diodes it is not usually necessary to provide adjustment of balance at the LO frequency for best noise performance.

Conversion loss is minimised with as little injection as 1mW

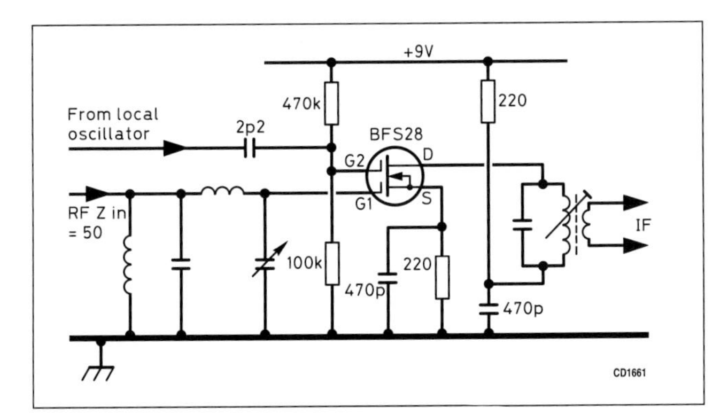


Fig 9.23: A dual-gate MOSFET mixer

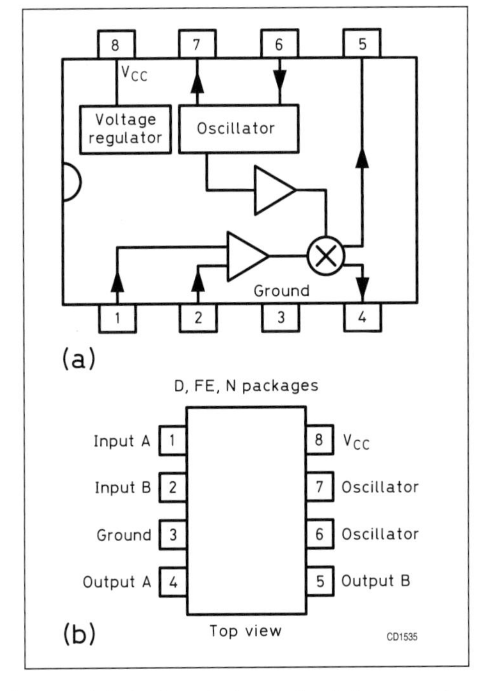


**Fig 9.24: NE612 double-balanced mixer and oscillator. (a) Block diagram. (b) Pin configuration (Signetics)**

(OdBm), but IMD (intermodulation) product rejection is improved by increasing the LO drive level to +7 to +10dBm (5mW to 10mW ).

Diodes suitable for VHF and UHF receiver mixers are the hot-carrier, high-speed silicon switching types and Schottky barrier diodes. The latter are the only types used in mixers designed specifically to have a high overload capability. A Schottky barrier diode such as the HP 2800 is probably the best choice for mixers operating at these frequencies. The diode ring mixer as shown in **Fig 9.20** is capable of very high performance at VHF and UHF at all the usual signal levels. The transformer cores are normally a low-permeability ferrite with a μ value of typically 125. Toroids are not the best type to use but multi-hole ferrite beads are ideal and make transformer construction simple. For optimum balance the diodes themselves must be dynamically balanced, and for this reason it is easier to purchase a ready-built diode ring mixer, which is usually built into a screened and potted assembly.

## Bipolar transistor mixers

A bipolar transistor can be used as a mixer at VHF and will provide a reasonable performance. From the circuit shown in **Fig 9.21** it will be seen that RF input is made to the base and LO injection is made to the emitter. The injection is made at low impedance - the LO coupling capacitor has a low reactance. Bipolar transistor mixers require a fairly low value of LO input power. This reduces the output power requirement of the LO. A typical level is -10dBm. Conversion gain is moderate and correct choice of the transistor will minimise the noise figure.

However, the intrinsic exponential forward-transfer characteristic of the bipolar transistor severely limits the large-signal handling capabilities. Blocking and IMD products become evident even with moderate signal input levels to the mixer. The gain of the preceding RF stage (if used) must be kept low, eg 6dB, to minimise these unwanted products.

## Junction field effect transistor mixers

Junction FETs can provide very good performance as mixers at VHF. A typical JFET mixer circuit is shown in **Fig 9.22**. The input impedance is high but the conversion gain is only 25% of the gain of the same device used as a VHF amplifier. Bias is critical, and is normally chosen so the gate-source voltage is 50% of the pinch-off voltage of the device. The LO voltage is injected into the source, the level being chosen to avoid the pinch-off region, but not sufficient to cause the source-gate diode to be driven into conduction. Normally the LO peak-to-peak amplitude should be kept a little below the pinch-off voltage.

The JFET has moderate output impedance, typically 10kΩ. This eases impedance matching to the following IF filter, which is usually performed via a step-down transformer as shown in Fig 9.22. The JFET does not have an ideal square-law input characteristic due to the effect of bulk resistance associated with the source. However, the generation of unwanted IMD products under strong signal conditions is much lower than the bipolar mixer, although the noise figure is similar to that of the latter.

### Dual-gate MOSFET mixers

MOSFET mixers can provide a superior performance compared with both bipolar and junction field transistors. They have excellent characteristics including a low noise figure, almost perfect square-law forward transfer, together with high input and output impedances. Conversion gain can be high and at the same time there is very low generation of IMD products at the IF output under large signal input levels. Overall performance is extremely good at both VHF and UHF.

**Fig 9.23** shows the dual-gate MOSFET in a mixer circuit. The signal input is applied to gate 1 as for an RF amplifier, while the LO is applied to gate 2, the IF output at the drain is controlled by the input levels. Optimum conversion gain is obtained with about 5V peak-to-peak of LO. This mixer has the advantage of having inherent isolation between the signal and LO gate inputs. Next to the ring diode mixer the dual-gate MOSFET mixer has the highest 'overload' level, at the same time giving conversion gain instead of loss.

Balanced mixers, whether they use bipolar transistors, junction FETs or MOSFETs, will give a much improved performance where low noise and the largest dynamic range must be achieved at the same time as maximum suppression of unwanted mixer outputs (typically IMD products and LO feedthrough) either to the antenna or to the IF amplifier.

### Integrated circuit mixers

Using IC types can obviate the problems of matching transistors and components in balanced mixers. These are available in the form of a monolithic bipolar double-balanced mixer intended for use at VHF and UHF. External circuit layout is simplified, bias adjustment is eliminated and results are more predictable.

The NE612 is an example of this type of mixer. The block diagram is shown in **Fig 9.24** and the equivalent circuit in **Fig 9.25**. Input signal frequencies can be as high as 500MHz. The mixer is a Gilbert cell multiplier configuration which can provide 14dB or more conversion gain. The Gilbert cell is a differential amplifier that drives a balanced switching cell. The differential input stage provides gain and determines the noise figure and signal handling performance. The mixer noise figure at 50MHz is typically <6dB.

The NE612 contains a local oscillator. This can be configured as a CCO or a VFO. The oscillator can also be reconfigured as a buffer amplifier for an external oscillator (the latter is to be preferred). The low power consumption, typically 2.4mA at Vcc = 6V, makes this type of device well suited for use in battery operated receivers.

## IF Filters

The performance of a purpose-built double-conversion VHF/ UHF receiver with a 'tuneable' first local oscillator can be enhanced by using a crystal filter between the first mixer and IF amplifier. Commercial crystal filters are now readily available, not only for the well-known IFs of 10.7 and 21.4MHz, but also for 45 and 75MHz. Until recently, crystal filters above about 25MHz were only available with third-overtone-mode crystal elements. Now 45 and 75MHz filters are available with fundamental crystal elements. Recommended filters for VHF/UHF bands are:

| | |
|---|---|
| 6m and 4m | 10.7MHz |
| 2m | 21.4MHz |
| 70cm | 45.0MHz |

Modern crystal filters are available between 2-poles and 10-poles, the higher the number of poles the greater the attenuation to out of band signals.

Crystal filters need to be correctly matched to the circuitry to attain the specified response. Some types of filters have built-in matching transformers and others do not. Generally types with built-in matching networks have low impedance, typically about 470Ω, whereas types without matching networks have higher impedance, typically 3k3Ω. Often the filter will require a reactive impedance and this is shown by, for example, "3k3Ω//2pF".

The value of the terminating impedance quoted by the manufacturer in the data sheet needs to be achieved for the correct pass-band response, skirt attenuation and minimum insertion loss. Note that the "terminating impedance" is the value the circuitry needs to present to the filter terminals and is the conjugate of the filter impedance seen when you look into the filter with a Vector Network Analyser or other test equipment.

The preferred matching mechanism to the first mixer is the 'constant-R' network which ensures constant filter terminating impedance.

### Bandwidth

Crystal filters must be correctly chosen for the type of modula-

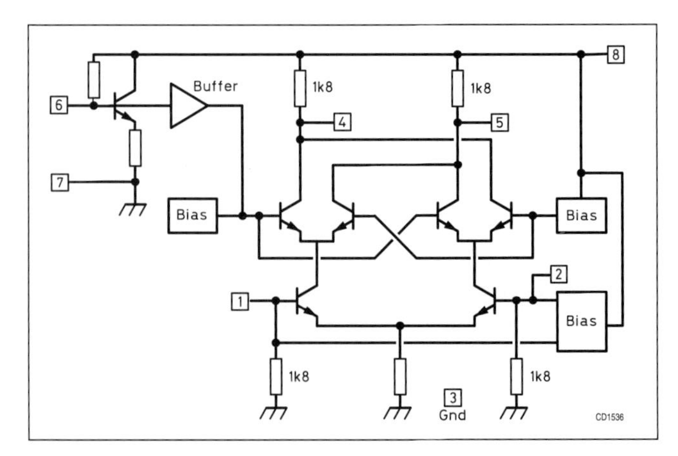


**Fig 9.25: Equivalent of NE612 double-balanced mixer and oscillator (Signetics)**

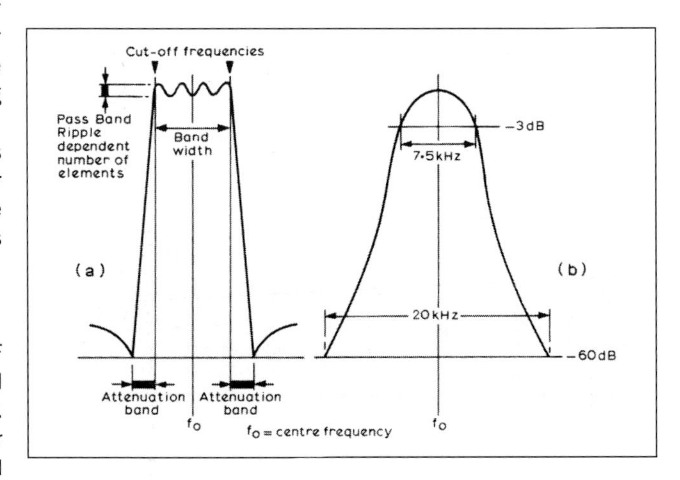


**Fig 9.26: Block filter characteristics (a) SSB, (b) NBFM**

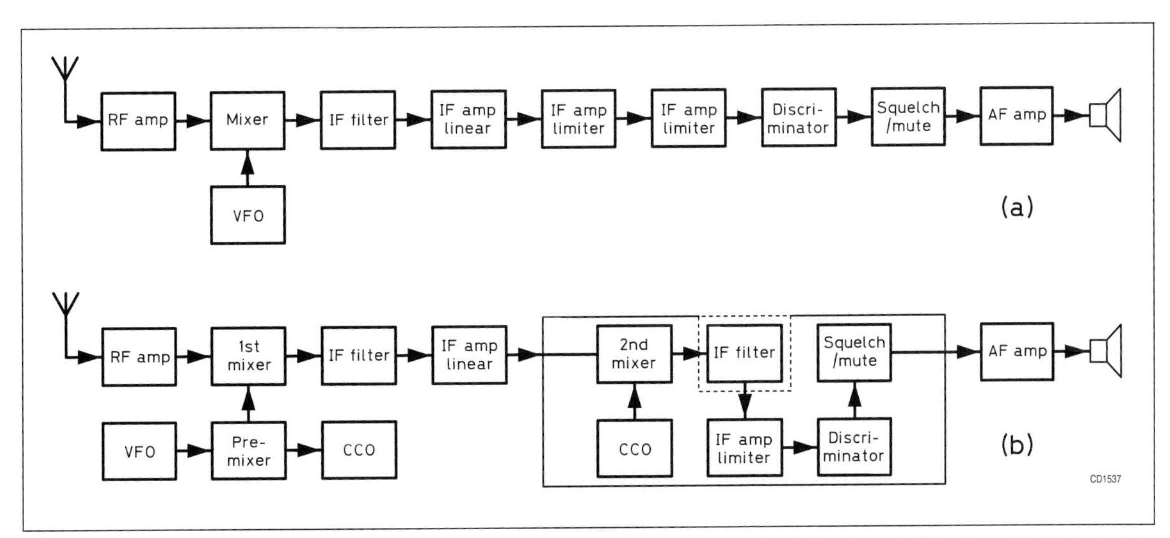


Fig 9.27: Comparison of the essential stages of receivers for (a) NBFM with discrete circuits and (b) NBFM with integrated circuits

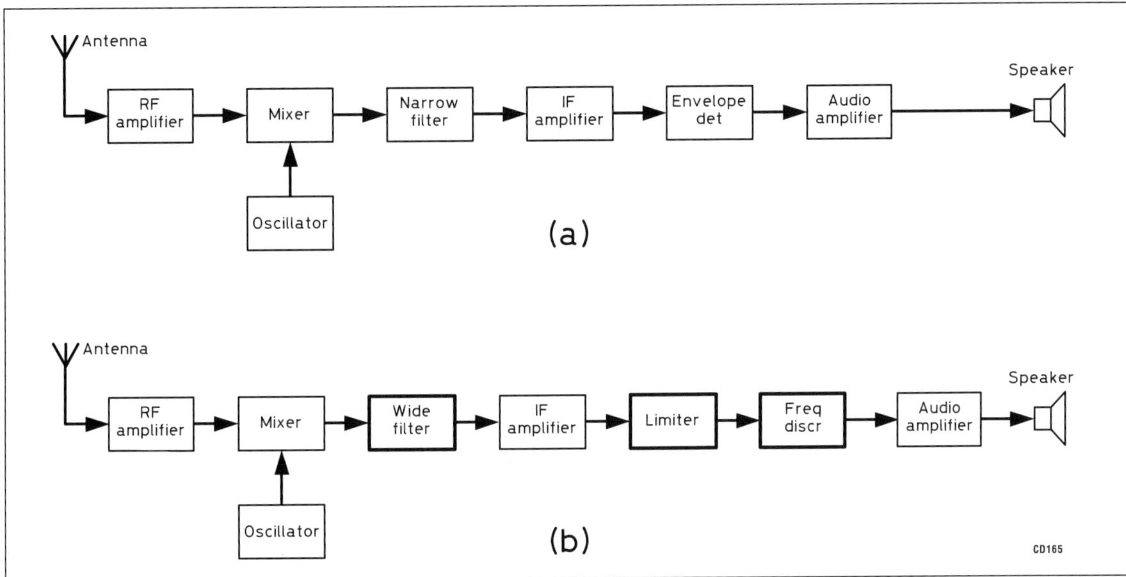


Fig 9.28: Block diagrams of (a) an AM and (b) an FM receiver. Dark borders outline the sections that are different from the AM set (ARRL Handbook)

tion to be detected by the receiver. The -3dB bandwidth for SSB filters can be as little as 2.1kHz (500Hz for CW receivers). The -60dB bandwidth can be 4.2kHz, giving a shape factor of 2:1.

However, filters for NBFM receivers must have a -3dB bandwidth of ±3.75kHz for 12.5kHz channel spacing and ±7.5kHz for 25kHz channel spacing for linear 'detection' of NBFM transmitters using maximum frequency deviation of the modulator.

Pass-band ripple must be minimal (not more than 1dB) to avoid phase modulation distortion. The -60dB bandwidth for a typical six-pole NBFM crystal filter can be 16-20kHz for 12.5kHz channel spacing and 32-40kHz for 25kHz channel spacing, giving a shape factor of around 2.5:1 All crystal filters have some insertion loss and it is usual to build a post-filter IF amplifier in the receiver to negate this loss. **Fig 9.26** illustrates the difference in SSB and NBFM IF filter characteristics.

## RECEPTION OF FM SIGNALS

There are two principal features in receivers designed to receive FM signals, namely limiting rather than linear amplifiers precede the detector and the latter is designed to convert IF variations into AF signals of varying amplitude, dependent on the degree of frequency variation in the transmitter carrier.

### The FM Receiver

The block diagrams of an FM receiver and AM/SSB receiver are shown in **Figs 9.27 and 9.28**. The principal difference between the receivers are the IF filter bandwidths (see above) and the IF amplifier gains required before the detector.

It is necessary to provide sufficient gain between the antenna and detector of an FM receiver to ensure receiver quieting; ie optimum signal-to-noise ratio with the weakest signal. Usually this is less than 0.35µV PD or -116dBm (into 50 ohms).

Thus it is necessary to use the double superheterodyne principle to achieve the required voltage gain, usually greater than 1 million or 120dB, whilst ensuring optimum stability independent of the input frequency. Other receiver stages, particularly the RF amplifier, mixer, oscillator and audio stages, can be identical to those employed in AM/SSB/CW receivers.

In a multimode receiver designed for reception and detection

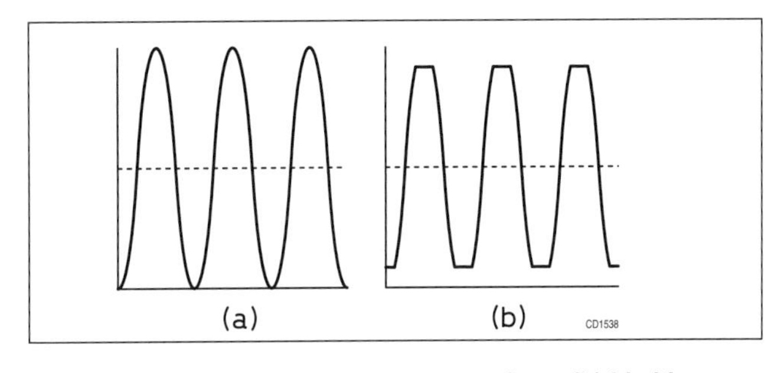


**Fig 9.29: (a) Linear amplifier output waveform. (b) Limiting IF amplifier 'clipped' output waveform**

of all principal methods of modulation, the difference in signal-to-noise ratio and effect of interference is very noticeable between FM and AM/SSB signals. The limiter and detector (discriminator) for FM signals reduce interference effects, usually impulse noise, to a very low level, thus achieving a high signal-to-noise ratio. However, it is necessary to align the detector correctly and in use tune the receiver accurately to achieve noise suppression.

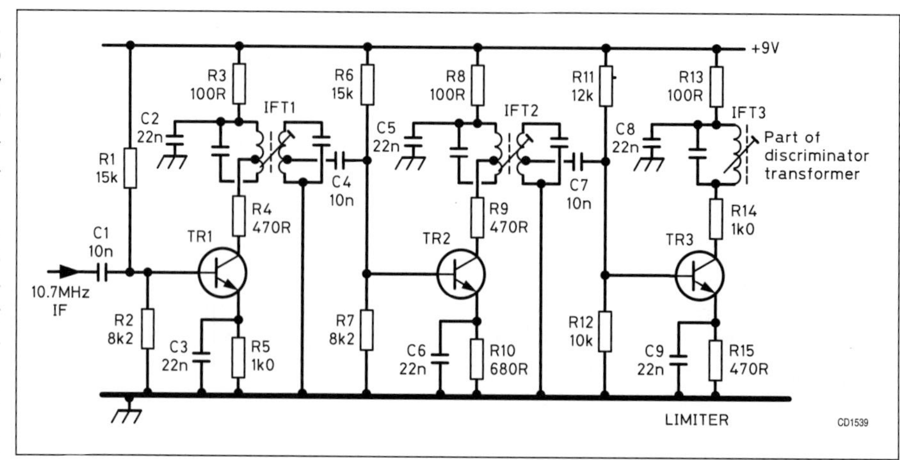


**Fig 9.30: A three-stage amplifier and limiter. Transistors are BF194 or similar**

An unusual effect peculiar only to FM receivers, and known as capture effect, occurs when a strong signal appears exactly on the frequency to which the receiver is tuned. If this strong signal has a carrier amplitude more than two to three times that of the wanted signal, the strong signal will be detected. This effect can be a problem in mobile operation, particularly in a geographical area between two repeater outputs on the same frequency.

Weak-signal reception in AM and FM receivers can be degraded by a much stronger carrier on or near the frequency of the weak carrier.

## Selectivity

As already stated in the previous section on filters, it is essential to choose the IF filters designed for NBFM reception. A narrow-bandwidth filter will introduce unwanted harmonic distortion. Too wide a bandwidth will degrade adjacent channel selectivity. However, transmitters exceeding the recommended limit on frequency deviation, will aggravate distortion effects and may cause 'squelch-blocking' where a squelch momentarily closes on peaks of the deviation. With modern NBFM transmitters this is less of a problem. Poor adjacent channel selectivity can cause receiver desensitisation, particularly when strong signals are present on either or both adjacent channels.

A transmitter with poor adjacent channel power rejection can also degrade weak-signal reception. These effects again cause receiver desensitisation but the transmitter modulation will not appear on the wanted signal (cross-modulation cannot in theory occur in FM systems).

## Limiters

Limiting IF amplifiers are specifically designed to introduce gain compression into their forward-transfer characteristics. If an amplifier is driven into limiting, its output signal level remains unchanged as the input signal level is varied. This effectively removes any sudden amplitude change, which is important as it is necessary to remove any impulse noise and AM on the carrier prior to an FM detector.

**Fig 9.29** shows the difference between a linear and a limiting IF amplifier output waveform. The clipping action removes the AM component. The overall amplifier gain must be high enough to ensure the limiting stages are limiting even with weak signals or with large changes of signal level IF to the receiver input. With an IF input of typically 5.0μV (equivalent to 0.25μV RF input to a receiver front-end with a conversion gain of 26dB) to the IF amplifier input, a minimum of three stages are required to raise the level of the signal for limiting action to commence. As the IF carrier level increases above 5.0μV the limiting action starts.

Now the signal-to-noise ratio improves until at a certain level the noise disappears. This is known as the receiver quieting characteristic referred to earlier in this section, usually the input for 20dB signal-to-noise ratio.

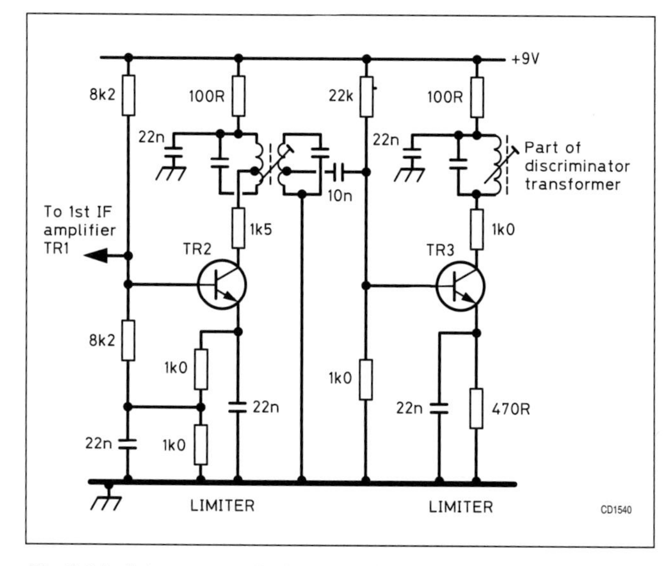


**Fig 9.31: A two-stage limiter developed from the circuit of Fig 9.30**

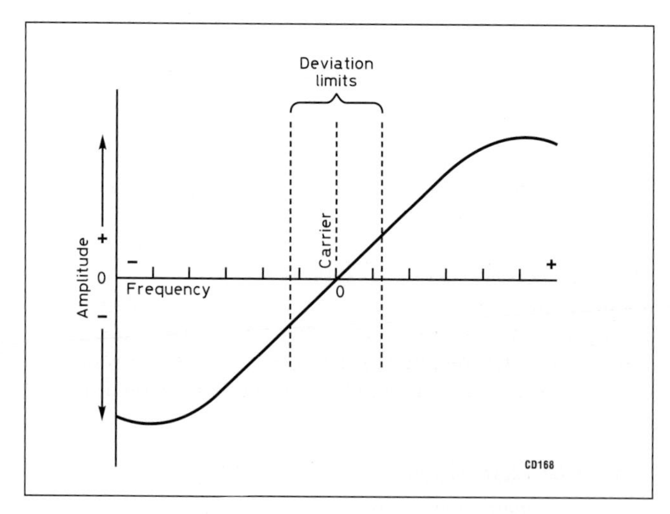


**Fig 9.32: The characteristic of an FM discriminator**

Discrete limiting amplifier(s) preceded by linear amplifier(s) with interstage IF transformers are still to be found in some early designs but are not now employed in modern NBFM receivers. Examples are shown in **Figs 9.30 and 9.31**. Linear amplifiers precede the limiting amplifiers. The base bias on the final IF amplifier in Fig 9.30 is varied to set the required limiting input level. In Fig 9.31 the base bias on the two final amplifiers is varied. This sets the limiting knee characteristic of the transistors to a point at which, for an increasing input, there will be no further increase in collector current. The amplifiers saturate, giving the required limiting and consequent noise level reduction for good receiver quieting. The circuit in Fig 9.31 gives an improved limiting performance compared to the circuit in Fig 9.30.

Some FM receiver manufacturers incorporated an IC limiting amplifier containing two or more stages, as they gave superior limiting action compared to discrete designs such as the MC1590 and the CA3028A. These ICs became obsolete with the introduction of devices containing six or more differential DC coupled IF amplifiers with improved and consistent limiting characteristics, operating at a relatively low IF, typically 455kHz. This system removed the requirement for IF transformers between each stage, but caused layout problems. However, it considerably simplified alignment. These ICs are called NBFM IF subsystems, and will be described after the section on FM demodulators.

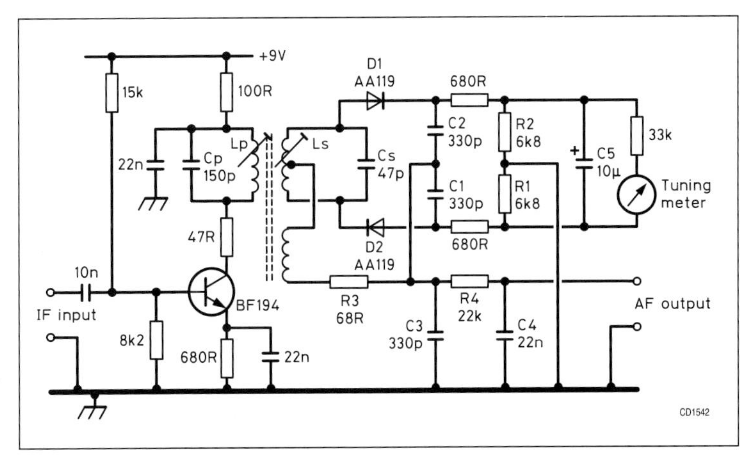


**Fig 3.34: A practical ratio detector circuit**

## FM demodulators

The FM detector, or more correctly the FM discriminator, was evolved to be able to respond only to changes in frequency as received from an FM transmitter and not to amplitude variations as received from an AM transmitter when the carriers are modulated. The degree of frequency change (frequency deviation) corresponds to the amplitude of the audio signal applied to the transmitter modulator. For example, if the transmitted carrier is deviated by ±2.5kHz by the modulator, the received carrier will be changing in frequency by ±2.5kHz.

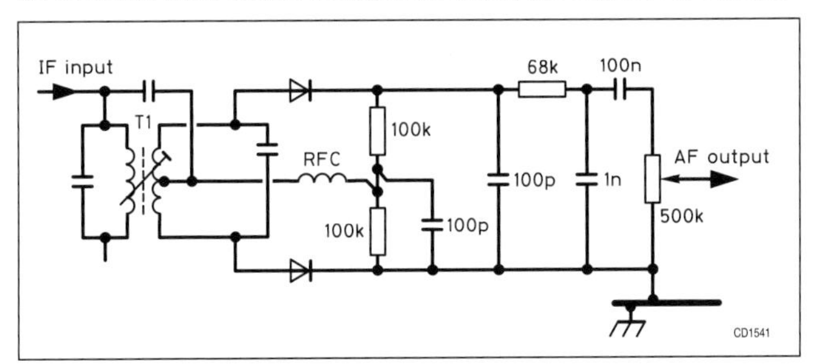


**Fig 9.33: The Foster-Seeley discriminator**

This frequency 'swing' is detected by an FM discriminator. **Fig 9.32** shows clearly the 'S' curve characteristic of the discriminator. Provided the swing is in the linear (straight) portion of the curve, ie about ±6kHz, very little distortion will be present in the recovered audio signal.

Maximum frequency deviation in amateur FM equipment is limited to ±5kHz compared to ±75kHz for Band 2 broadcast transmitters and receivers. This explains why FM voice and data communications systems are normally known as narrow-band FM systems.

A practical discriminator, known as the Foster-Seeley discriminator, after its inventors, is shown in **Fig 9.33**. T1 is the discriminator transformer. Voltage, due to the IF carrier, is developed across the primary of T1. Primary-to-secondary inductive coupling induces a current in the secondary 90° out of phase with the primary current. The IF carrier is also coupled to the secondary centre tap via a coupling capacitor.

The voltages on the secondary are combined in such a way that these lead and lag the primary voltage by equal amounts (degrees) when an unmodulated carrier is present. The resultant rectified voltages are of equal and opposite polarity.

When the received carrier is deviated the phase is changed between primary and secondary, resulting in an increased output level on one side and a decreased output on the other side. These voltage-level differences, after rectification by D1 and D2, represent recovered audio. This discriminator responds to carrier amplitude variations (AM) unless driven hard by the limiting IF amplifier stages and, for this reason, it fell from favour.

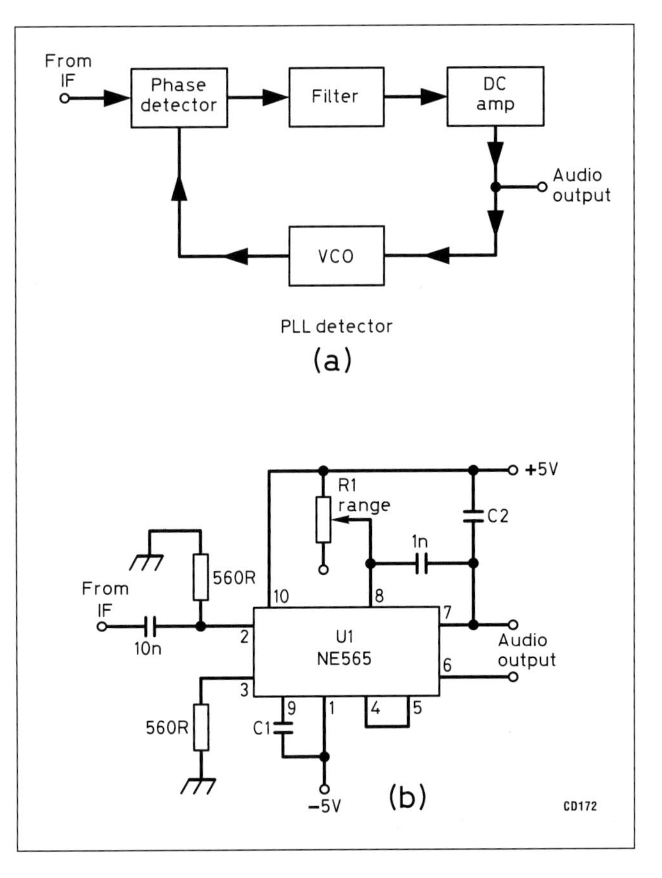


**Fig 9.35: (a) Block diagram of a PLL demodulator; (b) Complete PLL circuit**

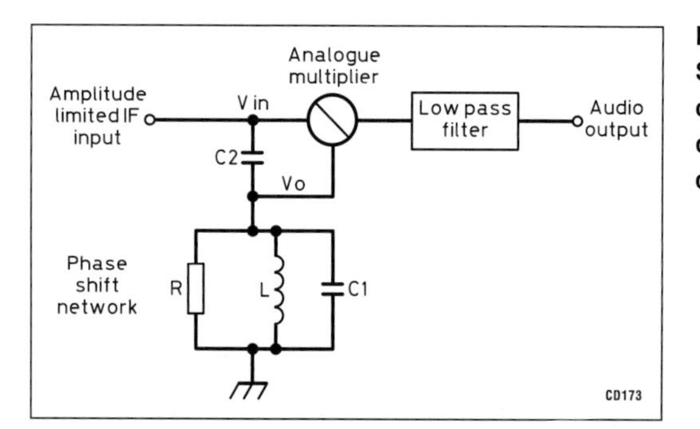


**Fig 9.36: Symbolic circuit of quadrature detector**

## Ratio detectors

This type of FM detector was developed from the frequency discriminator and became very popular in broadcast FM receivers - it has also been widely used in amateur receivers. The ratio detector is far less susceptible to carrier amplitude variations, hence less limiting is required before it. The detector is shown in **Fig 9.34**. T1 is a ratio detector transformer comprising of a primary, centre-tapped secondary and a tertiary winding, tightly coupled to the primary.

As its name implies, it functions by dividing the rectified DC voltages from D1 and D2 appearing across R1 and R2 into a ratio equal to the amplitude ratio present on each side of transformer secondary ($L_S$). The DC voltage sum appears across the electrolytic capacitor C5. This has sufficient capacitance to maintain this voltage at a constant level during fluctuations in carrier levels, caused for example by AM or noise signals being present on the carrier. Hence the ratio detector has its own inherent limiting characteristic. With a detector that responds only to ratios, the strength of the IF carrier can vary considerably without causing the output level to change. Therefore only FM can be detected and not AM. The carrier level should not fall below a level to cause D1 and D2 to become partially or fully non-conductive.

When the carrier is deviated by FM the audio signal is recovered from the tertiary winding L1, The IF carrier is filtered out by R3 and C3. The diode load resistors are lower in value than for the discriminator. It will be noted the diodes are connected in series rather than series-opposing as in the discriminator. This makes the ratio detector 6dB less sensitive. The diodes such as the gold-bonded AA 119 should preferably be matched dynamically.

## Phase-lock-loop detectors (PLL)

With the advent of the single-chip PLL it has been possible to design a reliable NBFM detector without tuned circuits and, therefore, without the necessity for alignment.

An example of an IC PLL detector is shown in **Fig 9.35**. The block diagram is shown in **Fig 9.35(a)** and the circuit including external components in **Fig 9.35(b)**.

Referring to Fig 9.35(a), the VCO oscillates close to the carrier frequency, in most cases 455kHz. The phase detector produces an error voltage when the VCO frequency and the carrier frequency are not identical. This error voltage is a DC voltage that is amplified after filtering, and 'corrects' the VCO frequency. When the carrier is deviated by audio modulation the frequency change is sensed by the phase detector and the resultant error voltage corrects the VCO frequency, causing it to remain locked to the carrier frequency, The system bandwidth is controlled by the loop filter. As the error voltage corresponds exactly to the frequency deviation, the PLL circuit functions as a precise FM detector.

It has a high sensitivity, requiring typically a 1mV carrier level for the PLL circuit to function correctly. Referring to Fig 9.35(b), R1 and C1 set the VCO frequency close to the carrier frequency. C2 controls the loop filter bandwidth that in turn controls the PLL capture range. The capture range is the maximum deviation from the carrier frequency to which the loop will gain and maintain lock on the carrier frequency.

Because the VCO runs at the same frequency as the IF and is pulled into frequency and phase lock with the signal to be demodulated, any signal leakage from the VCO into the earlier IF stages will cause the system to self limit or 'jam' the VCO signal. This is often detected as a low amplitude 'whistle' or beat note which varies in frequency when the input signal is absent. For this reason the PLL Detector has largely fallen out of favour and today the preferred type is the Quadrature Detector.

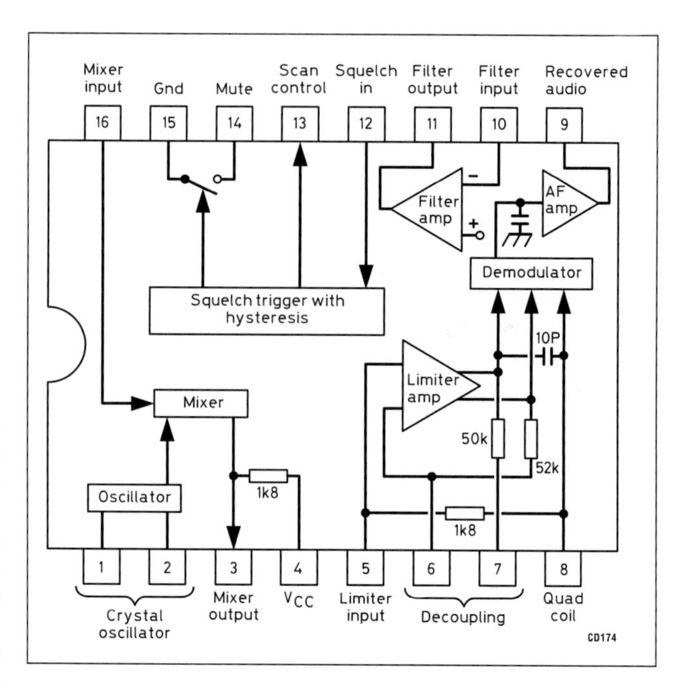


**Fig 9.37: Internal block diagram of the MC3361 NBFM IC**

## Quadrature detectors

This is also known as the quadrature discriminator (sometimes as the coincidence detector). The symbolic circuit is shown in **Fig 9.36**. It is found in virtually all NBFM subsystem ICs. Alignment is simple, there being only one coil to align for maximum audio output.

The IF input is fed to the detector, both directly and via a 90° (quadrature) phase shift. The quadrature input is fed to the detector using an appropriate value capacitor (C2) and a phase-shift network C1 and L which resonates at the IF centre frequency. The detector itself is an analogue multiplier. The phase-shifted signal is multiplied by the IF signal deviation from centre frequency (when modulated with an audio signal). The direct signal is multiplied with the phase-shifted signal to recover the audio signal, via the low-pass filter from the multiplex output spectrum.

For small frequency deviations (as in NBFM systems) the phase shift, controlled by the quadrature network, is sufficiently linear to give acceptable audio quality, ie with very little distortion.

The working Q of L in the network can be controlled by shunting it with a resistor (R). The lower the resistor value, the better the linearity, as this will increase the peak deviation capability of the detector. However, the audio recovery level is reduced.

## De-emphasis

It is normal practice to insert a de-emphasis network, usually a resistor and capacitor combination, in the post-detector section of the receiver, irrespective of the type of detector employed, to attenuate noise and audio frequencies above 3kHz by 6dB/octave.

## NBFM IF subsystems

These are used in modern dual-conversion receivers. The IC integrates the limiting amplifiers with a second mixer and oscillator for converting to a low IF, typically 455kHz, a quadrature detector, an active filter for driving a squelch circuit and a post-detector AF preamplifier in a single monolithic block. This results in considerable space saving, particularly for hand-held and mobile equipment. Power consumption is low and setting up and alignment is simple. Early examples of this type of IC are the Motorola MC3357 and the MC3359; these ICs can be found in many professional and amateur NBFM receivers.

More recently improved performance versions have appeared, eg the MC3361 and MC3362. The MC3361 is shown in block diagram form in **Fig 9.37** and in a practical circuit in **Fig 9.38**.

The mixer is balanced to reduce spurious radiation. It converts the first IF input signal to the second IF of 455kHz. After passing through an external band-pass filter, normally a multi-pole ceramic type, the IF signal is fed to the five-stage limiting amplifier and then to the quadrature detector where the audio signal is recovered. The 10pF on-chip capacitor produces the 90° phase shift between one output port of the final limiting amplifier and one input port of the quadrature detector. The quadrature coil is shunted by parallel resistor R2. This controls the linearity of the detector swing from centre frequency, hence the harmonic distortion in the recovered AF output which is buffered by the internal amplifier. R3 and C7 form the de-emphasis network. R4, R5 and C9 apply attenuated output to the filter amplifier. C10 and R6 peak the filter response to approximately lOkHz, ie above the normal audio pass-band of 300Hz to 3kHz.

In the absence of a carrier, only the noise signal is amplified this is detected by D1, which conducts, causing the DC voltage at the junction of' R9/R10 to fall. This level change is applied to the squelch switch input. The internal switch that stops noise output appearing across RV1 grounds the mute output pin. R10 and C12 filter the DC output from D1 anode. R7 and R8, with RV2, the squelch control, set the initial biasing of D1 and hence the squelch threshold level. When a carrier is detected and audio signals are recovered, D1 will not conduct and the squelch mute switch will remain open, removing the muting of the audio signal across RVl. The scan control at pin 13 can be used in conjunction with a digital PLL tuning system for 'locking' onto channels where the receiver has scanning facilities.

Fig 9.38 shows the external components required to complete the circuit of a practical NBFM receiver IF system. An IF block filter, as described earlier, must be inserted between the 1st mixer output and the IC input (2nd mixer) to provide adequate adjacent selectivity. The oscillator is an internally biased Colpitts type. The circuit shows a 10.245MHz fundamental-mode crystal oscillator for converting the 1st IF of 10.7MHz into the 2nd IF of 455kHz. However, a 20.945MHz crystal can be used for a 21.4MHz IF input and a 44.454MHz third overtone crystal for a 45MHz IF input. C1 and C2 form the crystal load capacitance.

For 10.7MHz to 455kHz mixers the alternative crystal is 11.155MHz and this will be used if the harmonics of the 10.245MHz crystal fall closer than 100kHz to the input signal. For example, 102.45MHz as an input signal would be blocked by the harmonics from the 10.245MHz crystal. Changing the second conversion crystal obviates this problem. The other IFs have similar alternative crystals, 21.855MHz and 45.455MHz are commonly available from crystal manufacturers as standard items.

The doubly balanced mixer has a high input impedance of about 3kΩ. This characteristic enables crystal filters to be matched easily to the mixer input. Similarly the output impedance is fixed by the internal 1.8kΩ resistor for correct matching to the 455kHz ceramic bandpass filter (FL1). The filter -3dB bandwidth should be 7.5kHz for 12.5kHz and 15kHz for 25kHz channel spacing respectively. Ultimate adjacent-channel selectivity is a function of the filter stop-band attenuation. A six-pole filter will provide sufficient attenuation. Measured at ±100kHz the attenuation should be 40dB minimum: this is adequate for NBFM receivers. The filter is matched at the IF input by an internal 1.8kΩ resistor.

The Motorola MC3371 and MC3372 represent the two low-power NBFM sub-system ICs. They are basically very similar to their predecessors, the MC3361 and MC3362. The principal difference between the MC3371 an MC3372 is in the limiter and quadrature circuits. The MC3371 has internal components connecting the final limiter to the quadrature detector for use with parallel LC discriminator. In the MC3372 these components are omitted and must be added externally The MC3372 can be used with a ceramic discriminator. Application circuits are shown in **Figs 9.39 and 9.40**.

Both ICs have a meter output at pin 13 which indicates the

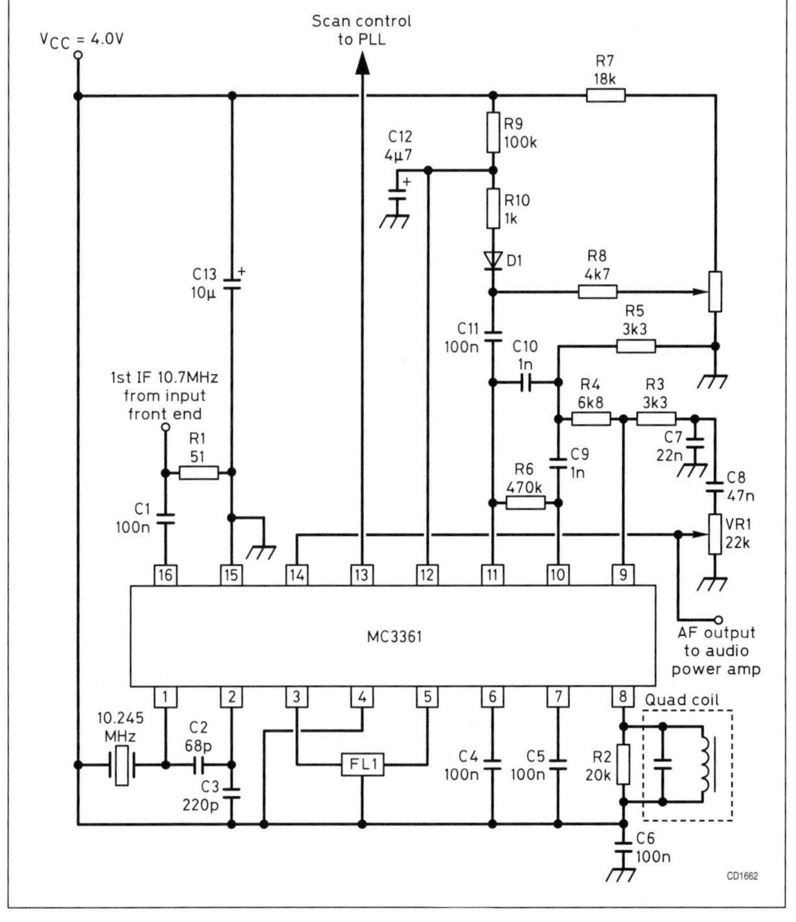


Fig 9.38: Practical circuit using the MC3361 (Motorola)

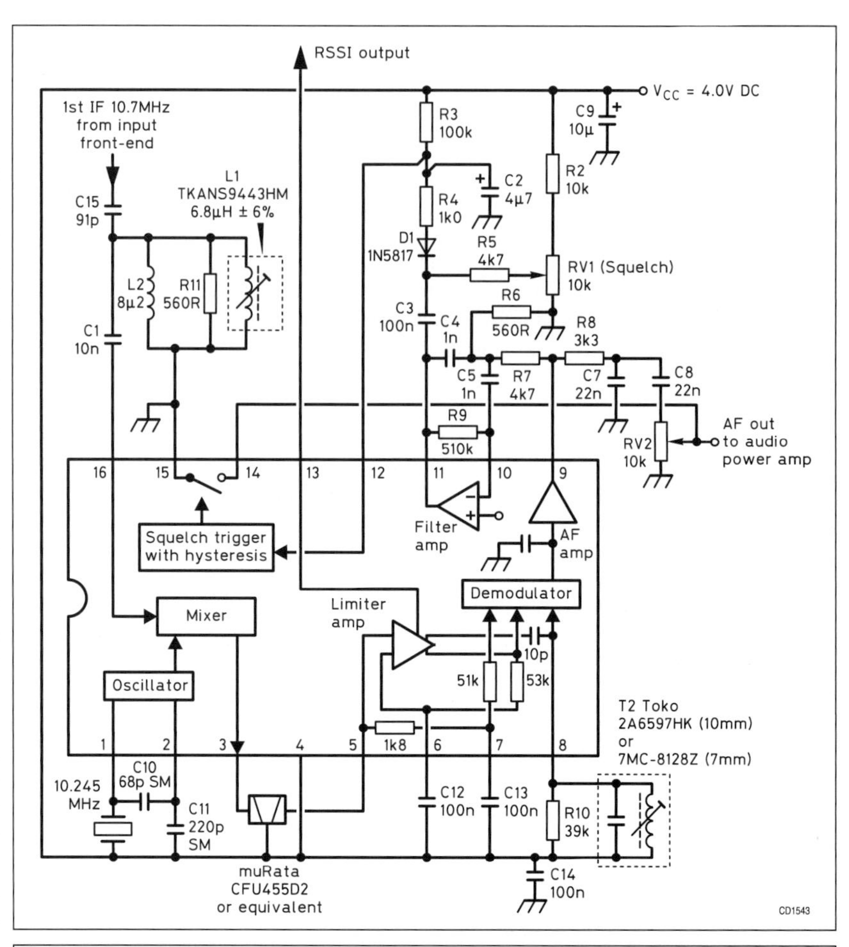


**Fig 9.39: Typical application for the MC3371 at 10.7MHz with a parallel LC discriminator (Motorola)**

strength of the IF level and the output current is proportional to the logarithm of the IF input signal. A maximum of 60µA is available to drive an S-meter and to detect the presence of an IF carrier. This feature is known as a received signal strength indicator (RSSI) or S-meter. Pin 13 is resistively terminated (to ground) to provide a DC voltage proportional to the IF signal level. The resistor value is estimated by $V_{CC}$ - 1.0V/60µA, so for $V_{CC}$ = 4.0V a 50kΩ resistor will provide a maximum swing of 3.0V.

# PREAMPLIFIERS

Modern semiconductors have made it possible for amateurs to build high quality preamplifiers to improve the performance of their existing transceivers or transverters for the VHF and UHF bands. The examples included in this chapter have been designed by amateurs to meet the specific needs of operation on the VHF and UHF bands. The three designs for the lower bands are from Dragoslav Dodricic, YU1AW [4] and show that bipolar transistors rather than FET transistors still have their place. State of the art commercial preamplifiers are reviewed by Sam Jewell, G4DDK to show what can be obtained off the shelf. The design for 23cm comes from Sam Jewell, G4DDK and has a novel two stage design where the second stage can be bypassed to improve performance under local strong signal conditions.

## Preamplifiers for 6m, 4m and 2m

A new type of low noise preamplifier is described here, which is recommended for its exceptional noise and inter-modulation characteristics not only for normal DX operation, but also for operation under difficult conditions when there are a significant number of powerful local stations, for example during competitions. The amplifiers are designed to have low noise, unconditional stability and exceptional linearity, thanks to the use of special ultra-linear, bipolar, low noise transistors designed for TV signals amplifiers. Since they are widely used, they are readily available at low cost. Construction is extremely simple with a small number of components, very simple adjustments and a high repeatability. This has been achieved using extensive computer non-linear and statistical optimisation. Designs are available for all amateur bands from 6m to 23cm [4]; watch that web

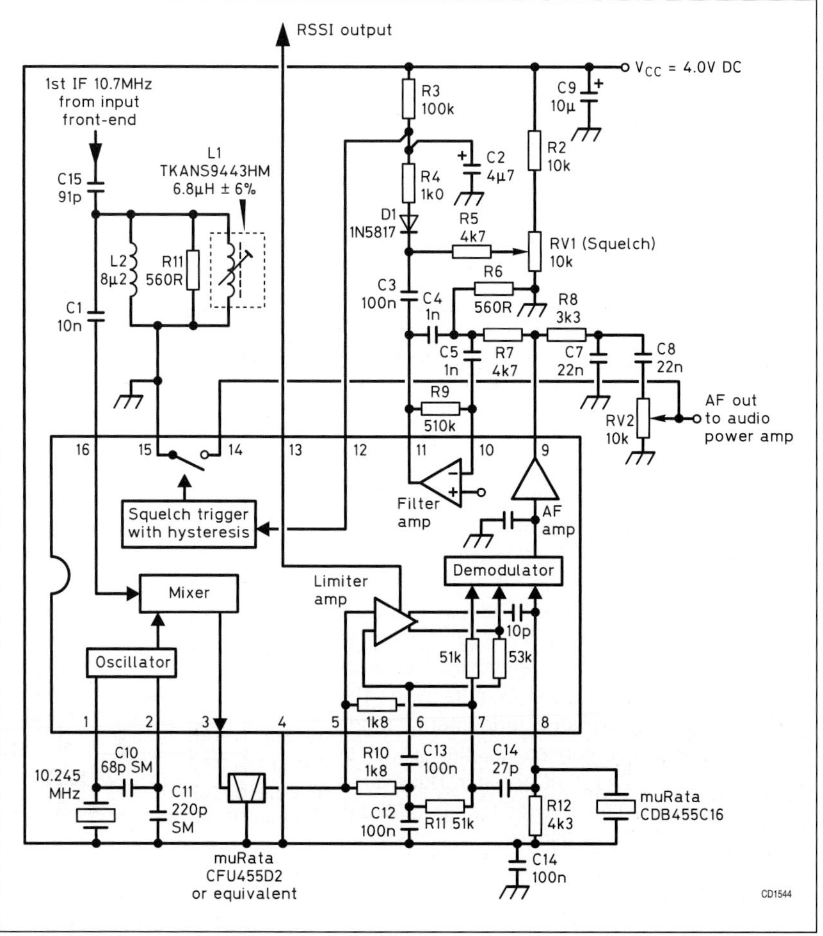


**Fig 9.40: Typical application for the MC3372 at 10.7MHz with a ceramic discriminator (Motorola)**

site for latest updates to these designs.

For three decades, MOSFET or GaAsFET transistors have been used almost exclusively in preamplifier designs. The reason for this is their superior noise performance and amplification. What we inevitably encounter when using GaAsFETs is a stability problem due to their conditional stability on VHF and UHF frequencies [5 - 8]. However, with an increasing number of stations using greater output powers, especially during competitions, the majority of these low noise preamplifiers that are successfully used for DX, MS or EME activity, become overloaded. This is manifested by a large number of inter-modulation products that contaminate the band, this is attributed to other stations and especially those that use powerful amplifiers.

The problem, of course, could be in non-linear power amplifiers due to excess input power causing saturation. This generates a high level of inter-modulation products. However, in practice it is more frequently due to the receiver's excessively high amplification and insufficient linearity, ie its input stage is overloaded causing it to generate products that look as though they really exist on the band.

In order to understand how to cure this problem, it is necessary to know how, where and under what conditions it occurs. It turns out that the source of this problem is very high amplification, a feature that the majority of amateurs praise the most and should be praised the least or even avoided. Technically it is much more difficult to achieve the two other important properties of an amplifier: low noise factor and strong signal performance, ie linearity. These are the most important properties for an amplifier to those for whom decibels are not just numbers that cover ignorance.

How do we determine which amplifier is of good quality? In order to resolve this dilemma a measure of an amplifier's quality has been introduced which encompasses all of an amplifier's three characteristics: noise factor, amplification and the output level of a signal for a determined level of non-linear distortions. This measure of quality is called the dynamic range of an amplifier and represents a range in which the level of a signal on an amplifier's input can be changed, while the output signal degradation stays within defined limits. The lower limit of this range is determined by the minimum allowable signal/noise ratio of the output signal and it is directly determined by the amplifier's noise factor, and the upper limit is the allowable level of non-linear distortion.

The lower limit of a dynamic range is the level of the input signal that gives a previously determined minimal signal/noise ratio (S/N) at the output. If the lower limit value is a S/N = 0 (incoming signal and following noise are equal) and if the upper limit of this range is limited by the maximum output signal voltage at which the amplifier, due to non-linear distortion, generates products equal to the level of noise on the output of the amplifier, then this is the so-called SFDR (spurious free dynamic range) or a dynamic range free from distortion, ie products of intermodulation distortion or IMD.

Since the third order inter-modulation distortion (IMD3) is dependant on the cube of the input signal, that is with each increase or decrease of the input signal by 1dB, the third order inter-modulation products increase or decrease by 3dB. It is therefore possible to calculate the maximum output level for different values of relations between products and the signal that is being used, or the value of IMD3 products, at different output signal levels. Using an attenuator enables us to also check whether an amplifier is overloaded, ie to recognise whether an audible signal on our receiver really exists on the band or whether it is simply the 'imagination' of our overloaded receiver. This enables us to dispose of overload and IMD3.

Since the level of products rise faster than the basic signal, by increasing the input signal we reach the point at which third order inter-modulation products, IMD3, reach the level of a useful signal at the output and that point is known as IP3 (Intercept Point). When the IP3 value is quoted it is necessary to state if it is referenced to the input or output of the amplifier. These values naturally differ by the value of the amplifier's amplification. Occasionally, it is stated as the TOI (third order intercept). This point is often taken as a measure of an amplifier's linearity and is highly convenient when comparing different amplifiers. Knowing the value of an amplifier's IP3 enables us to precisely calculate the value of IMD3 products at some arbitrarily chosen

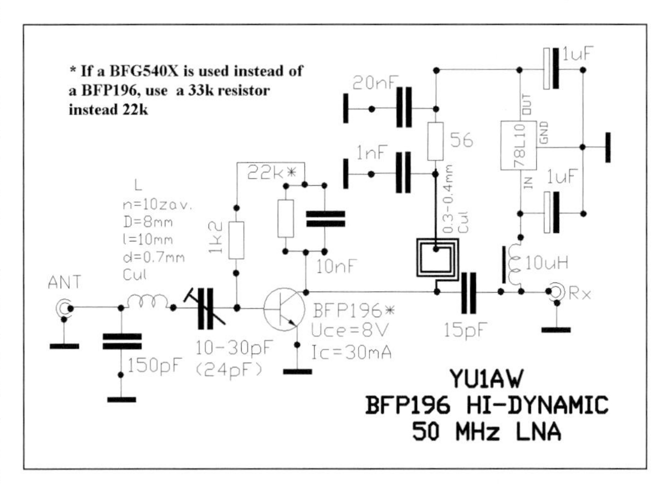


**Fig 9.44: Circuit diagram for the 6m low noise amplifier**

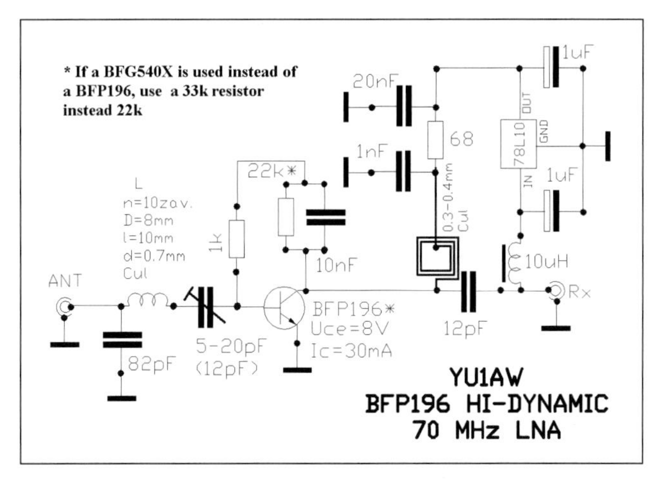


**Fig 9.45: Circuit diagram for the 4m low noise preamplifier**

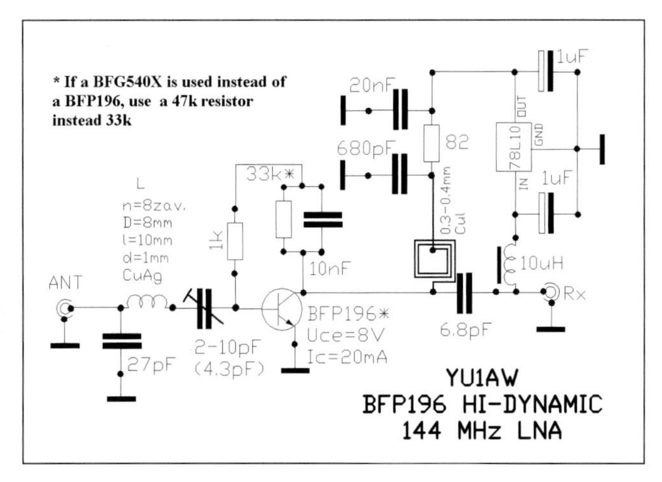


**Fig 9.46: Circuit diagram for the 2m low noise amplifier**

output or input signal level.

If excessive amplification is used, for example in a multi stage amplifier, a danger exists where antenna noise and the noise of the first amplifier are amplified to such an extent that they exceed the limit of linear operation of the last transistor, at which point the amplifier is saturated with the noise itself without any signal.

The conclusion is clear: An amplifier is worth as much as its dynamic range value, rather than how great its amplification is!

Therefore, if we want to construct an amplifier with the maximum amount of SFDR we have to fulfil the following conditions:

- the noise factor is as low as possible
- the IP3 is as high as possible
- it has acceptable amplification

On the one hand, amplification should be as large as possible, to prevent second degree influence on noise factor, and on the other hand it should be as small as possible so that the IP3 input is as high as possible, ie so that the amplifier should withstand the highest possible input signals without distortion. Compromise is essential and it usually ranges between 13-20dB amplification, depending on which parameter is more important for us.

If we want a low noise amplifier with a high dynamic range, then the choice of a corresponding transistor is extremely important. It is necessary to choose the type of transistor that besides low noise and sufficient amplification on the given frequency fulfils the condition of good linearity, that is high IP3 along with unconditional stability. Hitherto, MOSFET and GaAsFET transistors did not fulfil this condition in a satisfactory manner. Specially built transistors for ultra linear working, primarily for CATV do fulfil these criteria. For that reason, Siemens BFP196 bipolar transistors in SMD packaging were chosen. The Philips transistor BFG540/X corresponds closely to the Siemens device, it requires only slightly different base bias resistors. This Philips transistor should be used on 1296MHz because it gives several decibels greater amplification. It should be stressed that BFG540 without /X could be used, but the layout of pins is different, - it is not pin-to-pin compatible with the BFP196 - therefore the printed circuit board has to be changed, which is not recommended.

Since we are talking about a broadband transistor whose Znf and S11 values are relatively close to 50 ohms, the input circuit has been chosen to optimally match the transistor with regards to noise, while at the same time it provides some selectivity at the input. By varying the circuit values a compromise is found which provides the highest selectivity with minimal degradation of the noise factor. On lower bands where the noise factor is not as important, the compromise was in favour of selectivity which is more important than noise on these bands. The operating point of the transistor was also chosen as a compromise between minimal noise and maximum IP3. The output circuit is

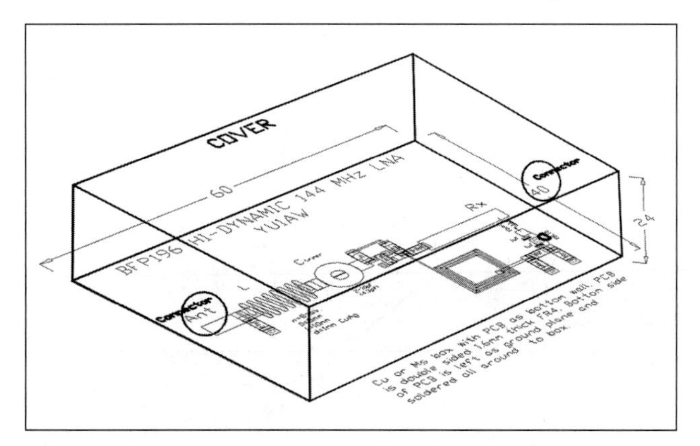


**Fig 9.47: Mechanical layout of the low-noise amplifier**

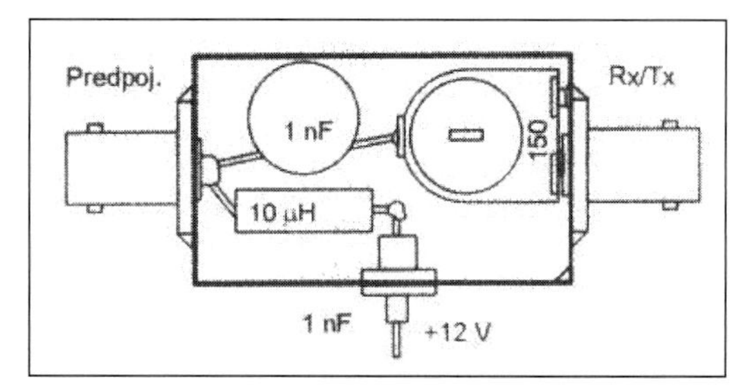


**Fig 9.48: Power feed for the pre-amplifiers**

relatively broadband and it is implemented using a printed inductor to reduce coupling with the input and to provide high repeatability. In order to maintain optimal output matching that gives minimal IMD, any matching by trimmer capacitors or by variable inductances is forbidden on the output. In order to achieve unconditional stability, minimal IMD, optimal amplification and minimal noise, negative feedback is applied which cannot be changed arbitrarily.

The printed circuit board is made with the dimensions shown in the relevant figure (**Fig 9.41** for 6m, **Fig 9.42** for 4m and **Fig 9.43** for 2m - all located in the Appendix B). Double sided board, type G10 or FR4 is suitable. The bottom copper surface is an un-etched ground plane.

SMD components are the 1206 type and the ground connections are made using through plated holes or with wire links through the holes soldered on both sides. The parallel resistor

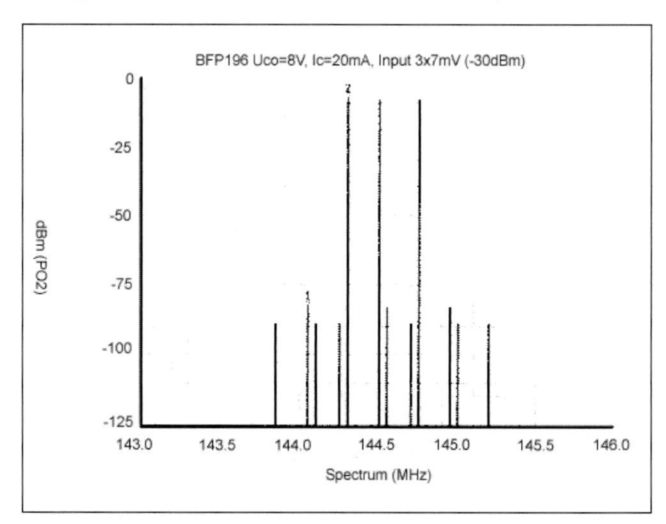


**Fig 9.49: IMD products from a BFP196 preamplifier**

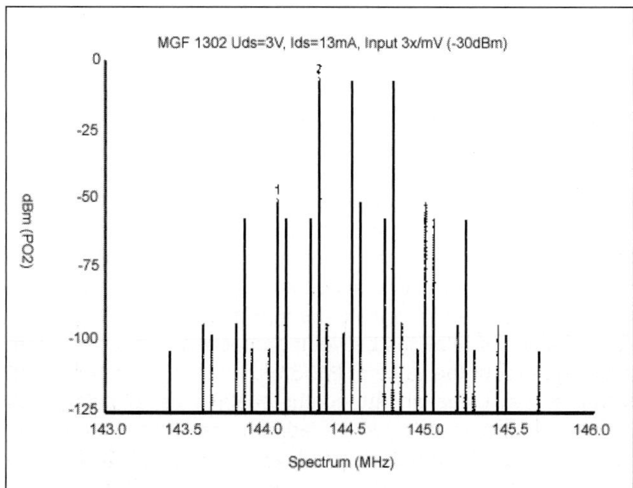


**Fig 9.50: IMD products from a MGF1302 preamplifier**

and capacitor in the base bias circuit are soldered on top of each other and not next to each other. The transistor collector is connected to the wider track.

The trimmer used is either of the air or PTFE foil type, although a ceramic one can also be used if it has a suitable capacity range. It is especially important for the higher band amplifiers that the trimmer capacitor has a low enough minimum capacity.

The coil is wound, as shown in the relevant circuit diagram (**Fig**

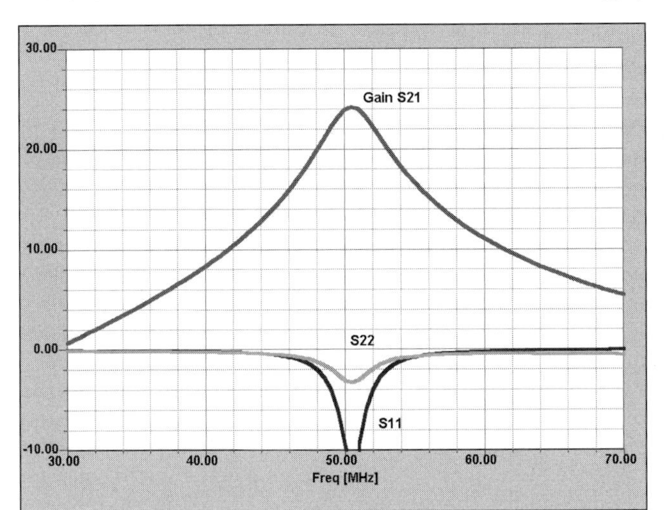


**Fig 9.54: Amplification, input and output adjustment for the 6m low noise preamplifier**

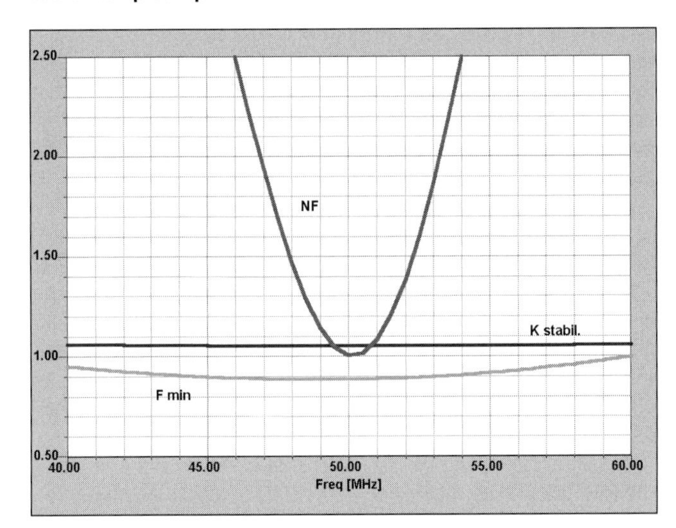


**Fig 9.55: Noise figure, minimum noise and stability of the 6m low noise preamplifier**

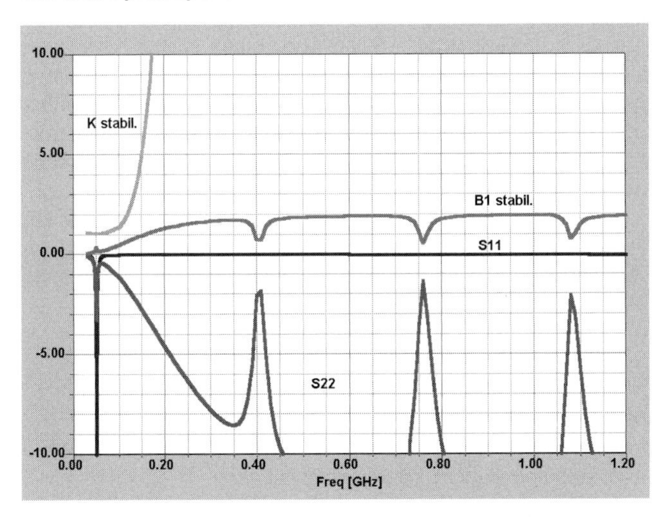


**Fig 9.56: Stability factor and adjustment for the 6m low noise preamplifier**

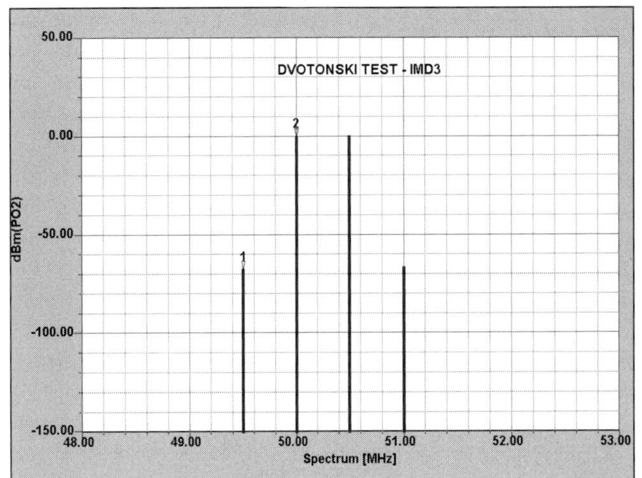


**Fig 9.57: Two tone test and IMD products for the 6m low noise preamplifier**

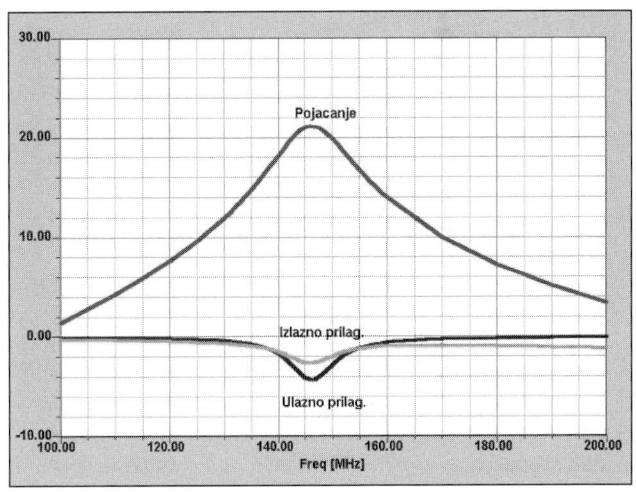


**Fig 9.58: Amplification, input and output adjustment for the 2m low noise amplifier**

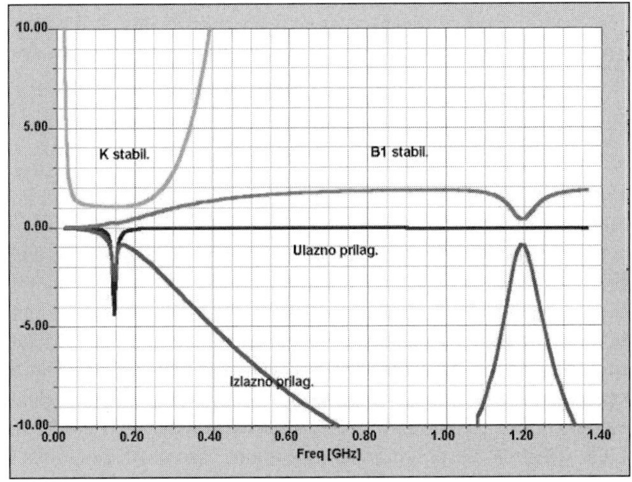


**Fig 9.59: Noise figure, minimum noise and stability of the 2m low noise preamplifier**

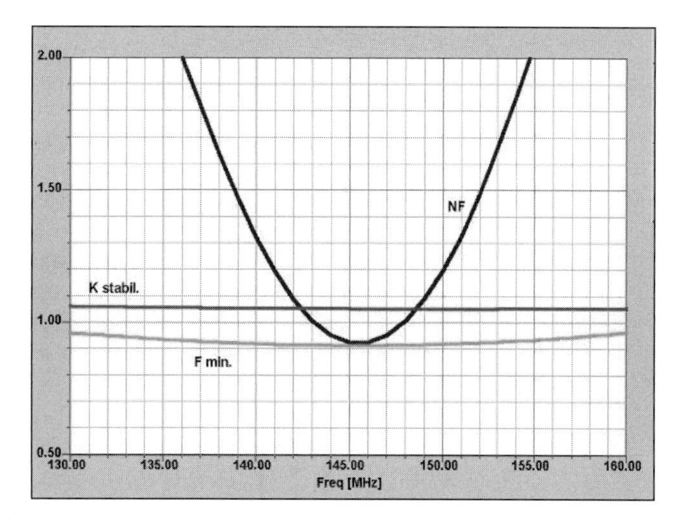


**Fig 9.60: Stability factor and adjustment for the 2m low noise preamplifier**

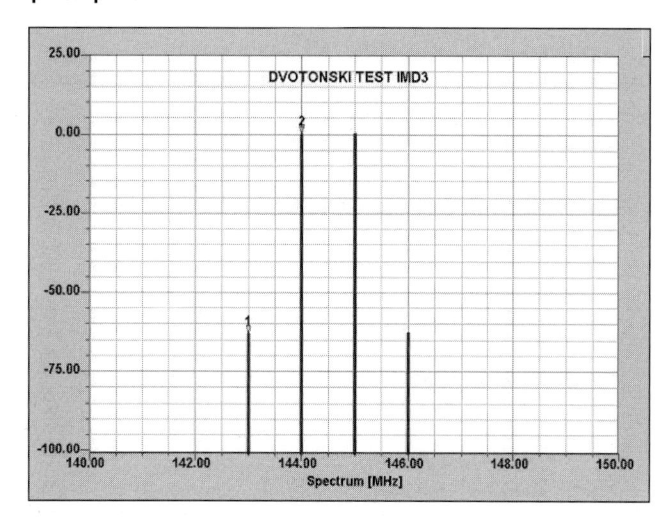


**Fig 9.61: Two tone test and IMD products for the 2m low noise preamplifier**

**9.44** for 6m, **Fig 9.45** for 4m and **Fig 9.46** for 2m), with silver plated copper wire, thickness 'd' and 'n' turns with a body diameter 'D'. The coil is to be expanded to length 'l'. When the coil is fitted it has to be positioned so that the bottom is approximately 3mm above the printed circuit board.

The box for the amplifier is made so that the printed board is the bottom side of the box, as can be seen in **Fig 9.47**. The easiest way to do this is to solder 25-30mm wide copper or brass strip, 0.3mm thick, around the edges of the printed board. Connections are mounted onto the box created, and a lid is made out of the same kind of sheet metal.

Once everything is carefully soldered, check for any possible mistakes such as short circuits, then connect the DC supply voltage and measure the collector current and voltage. If everything is correct and properly connected and the transistor is functioning properly, the values should be close to the ones given in the circuit diagram. If the differences are within 10%, everything is OK. If the differences are greater, check the supply voltage and then reduce the value of the base bias resistor, which is in parallel with the capacitor. Make it lower to raise the collector current and vice-versa. Do not change the value of the other resistor in the base bias circuit. If the collector voltage is not correct at the correct value of collector current, adjust the value of the resistor in the supply line. Such corrections are extremely rare and are necessary only if the particular transistor used has different characteristics from the common characteristics for that transistor type.

When both collector voltage and current are within the expected range, connect an antenna to the amplifier input and the receiver to the output and adjust the trimmer capacitor for maximum received signal using a weak station. This completes the final adjustment; the performance will be very close to the predicted values. With higher band amplifiers, especially 23cm, there is a slight difference in matching for maximum amplification and for minimum noise. The amplifier should be set to maximum amplification and then adjusted to a slighter lower frequency, ie slightly raise the trimmer capacity, until the amplification falls by 1-2dB.

Any further changes or modifications except the ones stated above are absolutely not recommended, because the amplifier is optimised so that it immediately reaches the required characteristics. Any modification would prevent that and would produce much worse results than the expected.

The amplifier should always be mounted as close to the antenna as possible and connected to the antenna with the shortest possible cable, using coaxial relays to switch the anrtenna from receive to transmit. Its power supply should be fed through the coaxial cable that connects it to the receiver, using the adapter shown in **Fig 9.48**.

As expected, very good noise characteristics have been proved in practical use, which mainly satisfy every requirement for serious DX work. Only for EME work at 432 and 1296MHz you might try using lower noise value amplifiers, ie the GaAsFET amplifiers [5 - 8], but in all other cases the amplifier satisfies even the most rigorous noise requirements. These amplifiers have shown exceptional linearity with IP3 values far exceeding +30dBm on all bands, except at 1296MHz where it is 3-4dB lower.

As a comparison, **Figs 9.49 and 9.50** show the performance of this amplifier and a common amplifier which uses an MGF1302 GaAsFET. Both amplifier inputs have three signals of 7.1mV (-30dBm) to simulate three strong stations on the band. The graphs show what would be heard with an ideal receiver without its own IMD. With a real receiver, because of its possible IMD, things would look even worse! Before you accuse someone of band wasting check with an input attenuator whether your receiver might actually be creating IMD due to a strong input signal!

The amplifier using the BFP196 is superior to the one using the MGF1302. The difference in the IMD products appearing on the output of an ideal receiver was over 30dB! Of course, in both cases the amplifiers had approximately the same amplification.

The component layouts for the amplifiers are shown in **Figs 9.51 - 9.53** (all in Appendix B) and the predicted performance is shown in **Figs 9.54 - 9.61**. The values shown have been simulated on a computer. Also, in real life the values have been proven on a sufficient number of built and measured amplifiers that they do not differ more than usual for this type of construction. Strict adherence to the guidelines given here will produce amplifiers with performance very close to those shown.

The final results achieved with these amplifiers in real life conditions largely depend on the IMD characteristics of the receiver used. If it has weaker characteristics, then the results may even be worse in respect to IMD because when signals, amplified in the preamplifier, reach the input of a bad receiver they cause overload and IMD and the results are poor. That is why the minimum necessary amplification is recommended between this amplifier and the first mixer in the receiver or the transverter in order to preserve as much of the dynamic range of the whole receiving system as possible.

If IMD is apparent in the receiver, put a variable attenuator between the amplifier and the receiver, define the lowest attenuation at which it disappears, replace it with a fix attenuation of the same value and work with that in circuit. This method is highly efficient because the IMD products are attenuated three times faster than the wanted signal, so that it is possible to weaken the products to the level where they are not heard whilst the useful signal has very little attenuation!

Don't be afraid that you will not hear the desired signal, there is too much amplification as soon as IMD appears - feel free to lower it!

A miniature 100-500 ohm trimmer potentiometer connected to the receiver input can be used in place of a variable attenuator. It can be built into the amplifier supply adaptor box as shown in Fig 9.48. This represents a very practical and rather elegant solution at least on of the lower bands. You can also use a variable 20dB attenuator used in CATV.

## SSB 2m, 70cm and 23cm Preamplifiers

### Introduction

Serious VHF/UHF weak signal operators use low noise amplifiers (LNA) to improve the sensitivity of their receiver systems. External noise, whether from earth or galactic in origin, tends to be at a very low level above a few hundred megahertz. This usually means that the limit to receiving weak signals is set by the internal noise of the receiver and by the loss of the signal in any antenna to receiver feeder.

Using an LNA immediately after the antenna, but before the feeder and the main receiver can often make a marked improvement in sensitivity. This is known as a preamplifier. In general, the lower the noise figure of the preamplifier, and the higher its gain, the better results will be. However, too much gain can result in the receiver system suffering from strong, unwanted, signal problems.

Low noise figure, adjustable gain, preamplifiers are a good solution. This way the optimum gain can be set for any particular situation, so as to minimise strong signal effects.

During transmit the preamplifier needs to be switched away from the antenna in order to protect its sensitive input stage from high transmit power. When a single feeder system is used the preamplifier needs to be bypassed so that a low loss path exists between the transmitter output and the antenna. In the SSB Electronics SP series preamplifiers this is done with the use of two coaxial changeover relays.

It is not advisable to allow transmit power to arrive at the relay contacts before they have closed. This greatly increases their power handling capacity. The relay contacts that do the path switching should ideally be allowed time to settle before the next stage of switching commences. The device that controls the timing sequence of the relays and stage switching is known as a sequencer.

SSB Electronics manufacture a range of high quality, weatherproof, preamplifiers for the VHF and UHF bands from 50MHz to 2.3GHz. Not only do these preamplifiers have low noise figures, but they also have user adjustable (preset) gain. They are also configured to use a single coaxial feeder for both transmit and receive and to be both powered and switched over the same cable using a sequencer, in what is known as the failsafe mode. Power can be supplied to the masthead preamplifier via an S0239 socket if preferred. **Fig 9.62** shows the SP2000 preamplifier. Externally the SP7000 and SP23 look identical.

In this review the results of RF measurements on the SSB Electronics SP2000 (144MHz band), SP7000 (432MHz band) and SP23 (1.3GHz band) preamplifiers, together with their complementary DCW2004B and DCW2004 SHF sequencers are presented.

### SP2000 and SP7000 preamplifiers

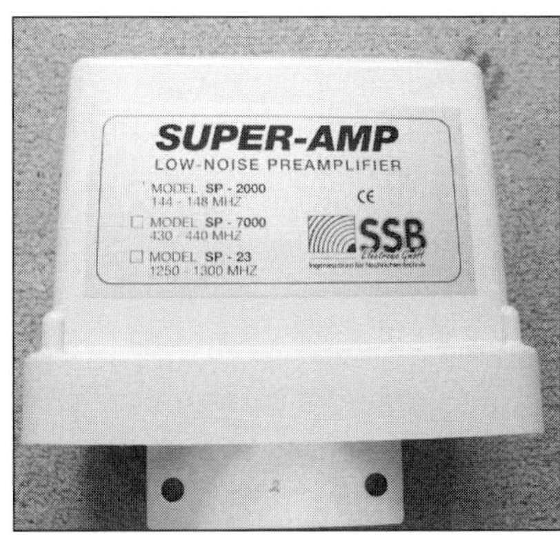


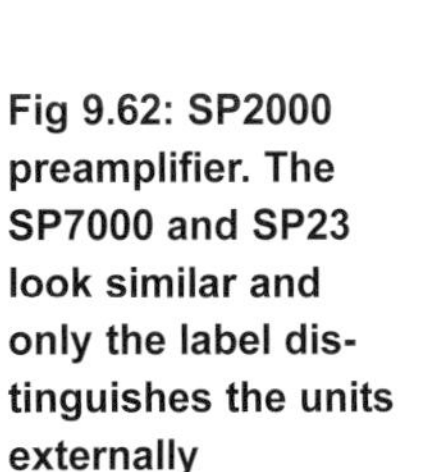

**Fig 9.62: SP2000 preamplifier. The SP7000 and SP23 look similar and only the label distinguishes the units externally**

These two preamplifiers use very similar circuits, consisting of a Mitsubishi MGF1302 Gallium Arsenide Field Effect Transistor (GaAsFET) low noise first stage followed by an Agilent (Avago) MSA1104 Silicon Microwave Integrated Circuit (MMIC) second stage. Band pass filtering is incorporated in both amplifiers. In the SP2000 preamplifier the band pass filter consists of a simple top-coupled tuned filter pair whilst in the SP7000 preamplifier the filter is a commercial two-stage helical block filter. Low noise input matching to the GaAsFET is by means of a capacitor and inductor/capacitor tuned "L" match.

Microwave GaAsFETs, such as the MGF1302, can have very low inherent noise figures at VHF, but the noise matching input circuit must itself have low loss if it is not to dominate the overall noise figure, since it adds directly to the device noise figure. This requires very high Q parts in the matching circuit. SSB Electronics appear to have used components of an acceptable quality in these two amplifiers.

The amplifiers are gain adjustable with a variable resistive attenuator, accessible through the lid of the tin plate box that is housed within the plastic weatherproof outside housing.

Both the SP2000 and the SP7000 use type N connectors for the antenna and transceiver connections. A connector provides for external DC power, if this is preferred over powering via the coaxial feeder cable.

Two methods of switching the preamplifier out of circuit, for transmit, are incorporated. It can be either RF detection switching (RF VOX) when RF is detected on the cable from the transceiver/transverter, or removal of the DC powering (via transceiver PTT) on the coaxial cable. The sequencer allows use of higher transmit powers due to the carefully controlled relay timing. The manufacturer claims that the RF VOX method results in a significant reduction in power handling capability.

### SP23 preamplifier

The SP23 1.3GHz band preamplifier also uses an MGF1302

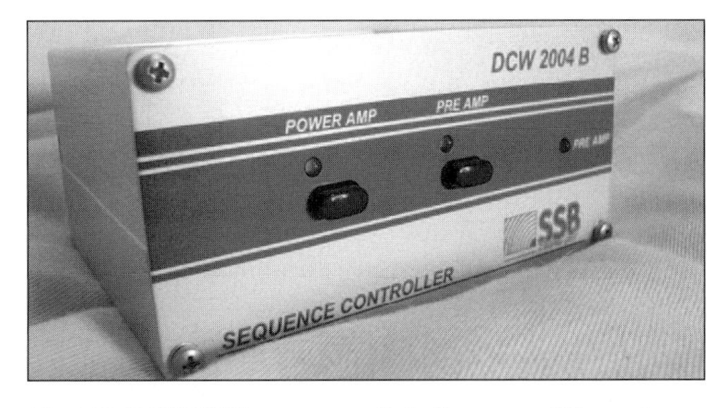


**Fig 9.63: DCW2004B sequencer, Note the preamplifier and power amplifier enable switches and indicator LEDs**

| Claimed Noise figure and gain (dB) | Noise figure and gain (dN) (High gain) | Noise figure and gain (dB) (Low gain) | 3rd order input intercept (dBm) | Input return loss (dB) | 3/20dB bandwidth (MHz) | Preamplifier off insertion loss (dB) |
|---|---|---|---|---|---|---|
| ***SP2000 (144MHz)*** | | | | | | |
| 144MHz 0.8/20 | 0.95/20.5 | 1.23/11.5 | -8 | <1 | 6/19 | 0.11 |
| 145MHz | 0.9/21.6 | 1.14/12.5 | | | | |
| 146MHz | 0.88/21.8 | 1.11/12.8 | | | | |
| ***SP7000 (430MHz)*** | | | | | | |
| 430 | 1.5/19.4 | 1.92/9.6 | -3 | 3.1 | 14/37 | 0.23 |
| 432 0.9/20 | 1.33/21.4 | 1.66/11.4 | | | | |
| 434 | 1.31/21.4 | 1.6/12.0 | | | | |
| 436 | 1.29/21.4 | 1.58/11.8 | | | | |
| 438 | 1.3/21.2 | 1.58/11.8 | | | | |
| 440 | 1.33/21.3 | 1.6/11.0 | | | | |
| ***SP23 (1200MHz)*** | | | | | | |
| 1240 | 2.34/14.53 | | | | | |
| 1260 | 1.28/19.65 | | | | | |
| 1280 | 1.05/21.0 | | | | | |
| 1300 0.9/20 | 1.2/18.9 | 0.5 | 7.2 | 65/195 | 0.41 | |
| 1320 | 2.09/13.5 | | | | | |

**Table 9.2: Results of measurements on the SP2000 (144MHz), SP7000 (430MHz) and the SP23 (1.3GHz) preamplifiers**

GaAs FET front end, but with a revised low noise, air spaced matching circuit. A commercial helical filter is used to shape the pass band. The second stage uses an ERA55 GaAsFET MMIC, providing enough additional amplification to raise the overall gain to around 20dB. No user adjustable gain control is provided.

The SP23 uses the same connector arrangement as the VHF preamplifiers.

## DCW2004B and DCW2004B SHF sequencer

Some multiband transceivers, such as the FT847 and TS2000, can provide the necessary DC powering and switching for the SP2000 and SP7000. The alternative is to use an SSB Electronics DC2004 sequencer. The DCW2004B is specified for 50MHz, 144MHz and 432MHz and not only provides the required DC supply to the preamplifier, via a suitable power injector (bias tee), it also provides suitable time delays to allow sequenced switching of the preamplifier, power amplifier (if used), and transceiver/transverter. Independent selection of preamplifier on or off and power amplifier on or off is provided for by front panel push (momentary) switches with LED indication of mode. **Fig 9.63** shows the front panel of the DCW2004 and **Fig 9.64** shows the DCW2004B SHF rear panel connectors.

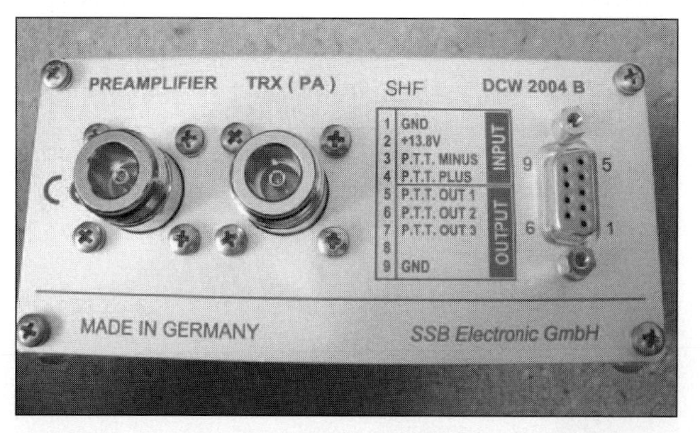


**Fig 9.64: Rear panel view of the DCW2004B SHF sequencer showing the N connectors and 9 pin D connector used for powering and interconnections to the transceiver and power amplifier (if used). The non-SHF version looks similar**

The DCW2004B SHF is specified for use with the SP23 (1.3GHz) and SP13 (2.3GHz) preamplifiers. The difference in the two types of sequencer appears to be in the frequency range of the bias tee used for DC power injection.

## Measurements

Accurate noise figure measurement is notoriously difficult for a number of reasons including external interference, noise source impedance changes, LNA input match and poor equipment calibration. These preamplifiers have been measured using a Hewlett Packard (HP) HP8970A and HP346A (5dB ENR) noise head.

This equipment has been used extensively to measure literally hundreds of low noise preamplifiers and transverters at a number of UK VHF and microwave events. Cross-checking between measurements performed on a number of the same LNAs, using different measurement equipment over a period of several years provides confidence that the noise figure results presented here are as accurate as is likely to be achieved using similar equipment.

The third order input intercept (IIP3) measurements were made using an Agilent E4432 vector signal generator providing multi-tone output (50kHz, 2-tone separation) signal drive and an Agilent E4405 spectrum analyser for the measurements, whilst the input return loss was measured using a HP8754 vector network analyser and HP8502A reflection test set.

The estimated uncertainty for noise figure is 0.1dB based on

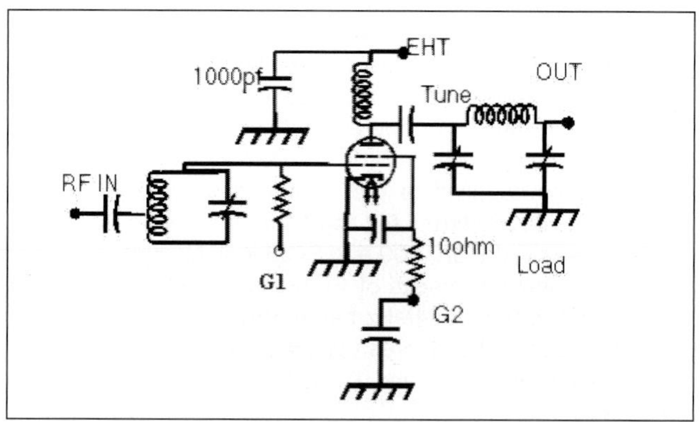


**Fig 9.65: Circuit diagram of the 6 metre power amplifier**

| | |
|---|---|
| **Anode Section:** | |
| Anode Tuning Capacitor | Jackson C804 25pf (wide-spaced) |
| Loading Capacitor | Jackson C804 150pf |
| Anode Isolating Capacitor | 1000pf 20kV 'door knob' |
| Anode Coil | Connects between Tune and Loading capacitors, 5t 12SWG, 1.375in dia |
| EHT RFC | 36t 22SWG enamelled wire on 5/8in dia PTFE rod |
| **Grid Section:** | |
| Grid Tuning | Jackson C804 50pf |
| Grid Tuning Coil | 6t 14SWG 1/2 in diameter |
| C1 input capacitor | Connects between input connector and Grid Tuning Coil, 1000pf mica |

**Table 9.3: Parts list for 6m power amplifier**

measurement of 'golden' (ie known noise figure) LNAs. Measurement uncertainty is -0.5dB for gain, return loss and IIP3. All cable and sequencer losses have been taken into account in the overall preamplifier gain results.

### Results

The measured results for the three preamplifiers are shown in **Table 9.2**, together with the manufacturers claimed figures.

Measurements on the sequencers gave the following results:-

DCW2004B Insertion loss (144MHz and 432MHz) = <0.1dB

DCW2004B SHF Insertion Loss= <0.1dB

### Conclusions

The SSB Electronics preamplifiers provide a convenient and cost effective solution to the problem of what to use to improve your VHF/UHF receive system.

Within the accuracy limits of the measurement system, the SP2000 noise figure and gain measured very close to the manufacturers typical figures, whilst the SP23 noise figure was a little higher than claimed by the manufacturer. To put this in context, an increase of 0.3dB will probably not be noticed in terrestrial operation.

The SP7000 noise figure was a little disappointing but again, putting the measured noise figure in context, would give a big improvement over most transceivers in common use. The 3rd order input intercept measured a very respectable -3dBm.

The SP23 measured IIP3 of +0.5dBm is remarkably good and should allow this preamplifier to be used in some strong signal situations. A good preamplifier cannot compensate for an indifferent transverter or transceiver, so do not expect miracles.

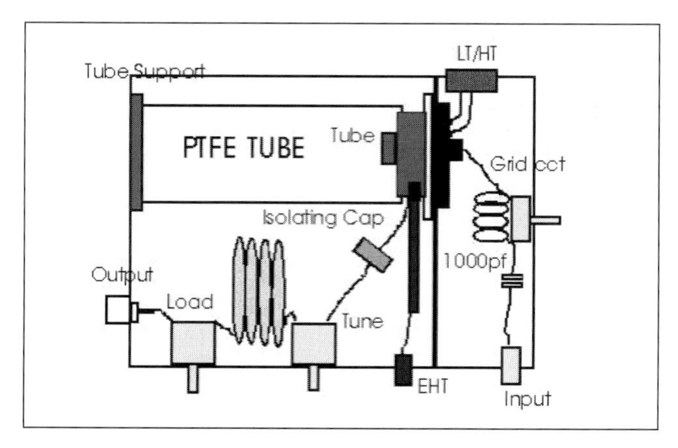


**Fig 9.66: Component layout for the 6m power amplifier**

**Fig 9.67: The 6m power amplifier**

English language data sheets are available on the Diode Communications [9] web site.

## POWER AMPLIFIERS

There is a simple decision to be made when you are thinking of making a power amplifier: should it use valves or should it use semiconductors?

There is no doubt that you can get more power for less money with valves, the down side is that high power valve amplifiers require very high voltages, 1- 2kV, that can be very dangerous if you don't take the correct precautions. There are many designs for valve power amplifiers for all of the VHF and UHF bands, in this chapter two more novel designs are included. The single 4CX250B 6m power amplifier was designed by Geoffrey Brown, G(J)4ICD [10] some time ago,

The 70cm valve power amplifier is from *VHF Communications Magazine* 4/1998 [11]. Semiconductor amplifiers fall into two types, those that use hybrid modules and those that use discrete semiconductor devices. Hybrid modules are easier to use because they require very few external components to get them working. Unfortunately the most common modules used by radio amateurs, from Mitsubishi, have been discontinued. They are still available

**Fig 9.68: The A200 amplifier. (Picture supplied by the Pye Museum [23])**

**Fig 9.69: The identification plate on the A200 amplifier (Picture supplied by the Pye Museum [38])**

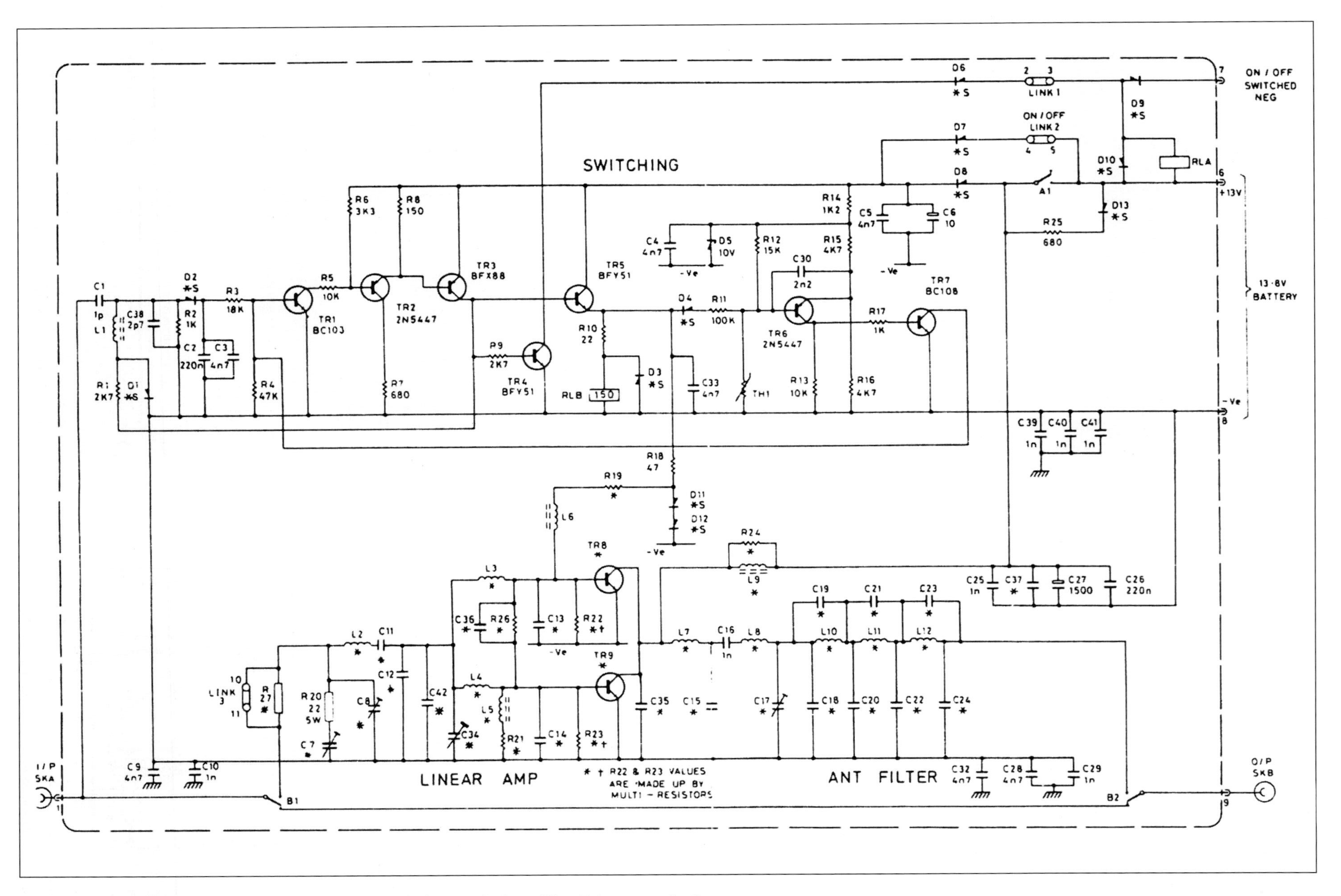


Fig 9.70: Circuit of the A200 amplifier. Reproduced with the permission of Pye Telecommunications

form some suppliers and on the second hand market and there are other modules available that can be used. The designs for transverters shown later in this chapter have optional amplifiers using hybrid modules. The 400W amplifier for 2m shows how to design and construct a reliable semiconductor power amplifier.

If you don't want to embark on a big constructional project you can get more power on the VHF and UHF bands by modifying a commercial amplifier. Modification of the ex PMR A200 amplifier is shown as an example of this possibility.

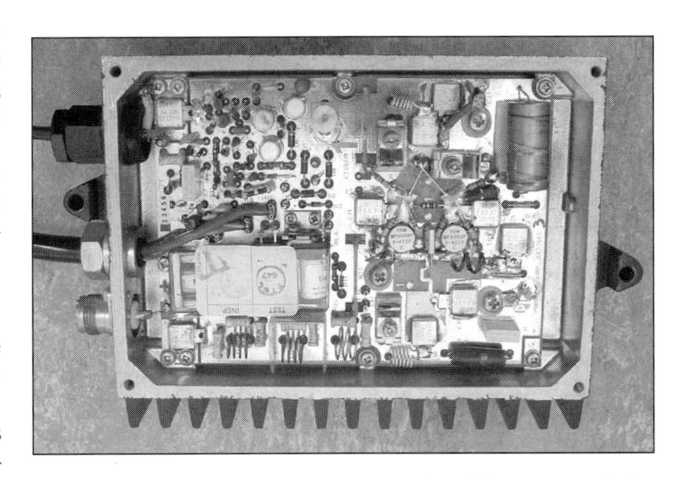

Fig 9.71: Internal view of the A200 amplifier. (Picture supplied by the Pye Museum [38])

## 6m Amplifier

This 6m amplifier can be built in under a day and will provide 250+ watts out for about a couple of watts in. The circuit diagram is shown in **Fig 9.65** and the parts list is shown in **Table 9.3.**

You will need a diecast box, fan, SK600 or SK610 surplus socket for the valve, a 4CX250B valve, a couple of tuning capacitors, some PTFE sheet to make the anode cooling chimney, and a multiple connector for the various voltages, plus a high voltage connector such as a PET 100 or TNC for the EHT of 1kV to 1.5kV.

Remember that any valve type amplifier has to operate with high voltages in order to work. The typical voltages applied to the 4CX250B series of valves is as follows:

- EHT (anode) is between 1 and 2kV. For this design 1.8kV at 250mA is ideal
- Screen grid + which MUST be regulated by zeners or stabilising valves, this should be 300 volts with a current capacity of 50mA
- A bias supply for the G1, this should be variable to -75 volts
- A heater voltage is also required which is 6.3 volts at a couple of amps
- A relay voltage is also required for the bias circuit and the antenna changeover relays

This amplifier was built in a die-cast box. The box measures 9 x 5 x 5 inches and a plate is fitted across the right hand end about 2.5 inches in. This plate (made of aluminium) has the SK600/SK610 socket fitted on it but offset towards the back (see **Fig 9.66**). The tuning capacitor and loading capacitor are fitted towards the front of the box along with the EHT choke, isolating capacitor and tuning coil. Cooling is via the anode compartment, so no manufacturer's chimney is used. A chimney made of PTFE sheet bonded together with silicon rubber is fabricated to fit onto the 4CX250B and run to the left hand end of the box. A small printed circuit board is used to support the chimney, it has a hole cut in it for the exhaust and a brass shim soldered to support the chimney. The blower is fitted to the lid of the box.

Anode and loading capacitors are fitted in the front panel of the box, as can be seen in Fig 9.65 and **Fig 9.67**. A band of brass, or an electrolytic capacitor mounting clamp is utilised to connect the isolating capacitor and the EHT choke onto the 4CX250B. RF out is fitted on the right hand end of the box. The grid circuit is straightforward with the tuned circuit being fitted behind the SK600/610 socket. A power connector is fitted to the rear

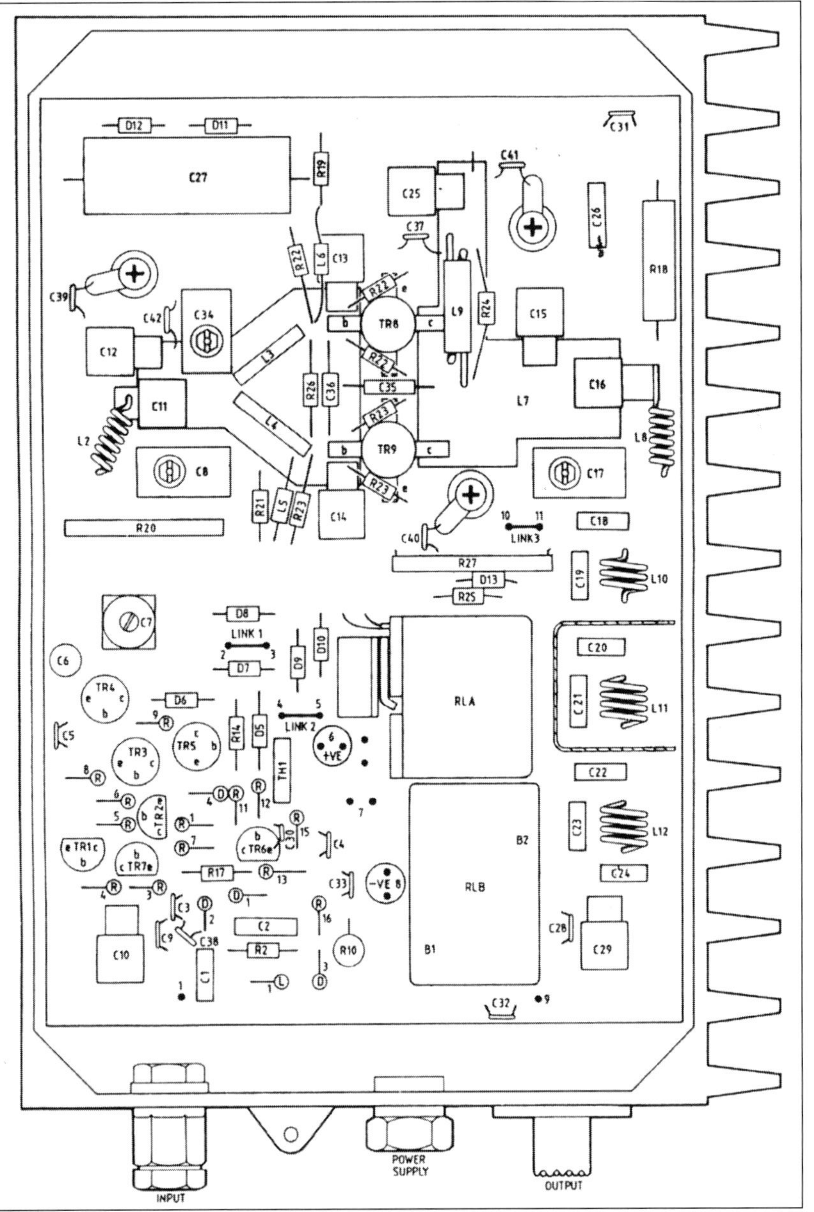


Fig 9.72: Component layout of the A200 amplifier. Reproduced with the permission of Pye Telecomm

wall of the grid compartment.

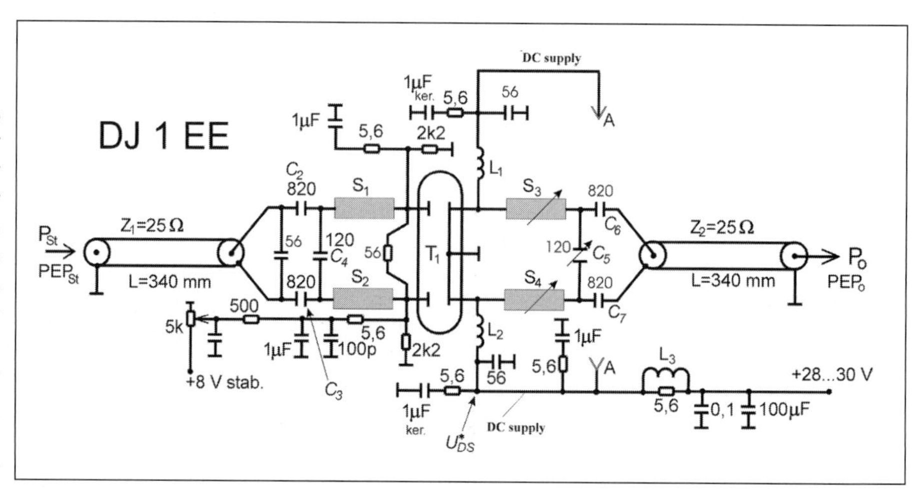


Fig 9.73: Circuit diagram of the 400 watt 2m power amplifier

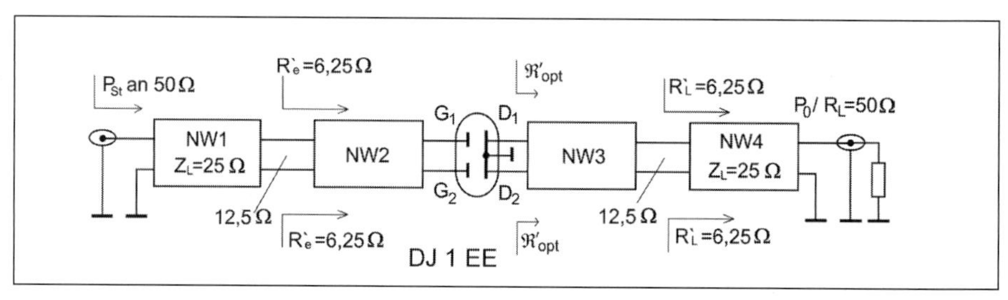


Fig 9.74: Block diagram of the 400 watt 2m power amplifier

## 4m Amplifier

The Pye (later Philips) A200 was designed as a boot mounting linear amplifier to give more output power for their range of mobile radios. They are still available on the second hand market but many radio amateurs do not realise the potential of these units to add a useful amount of extra power to a 4m station, they can also be used on 2m and 6m. The A200 is built to last in a heavy weatherproof case with automatic RF sensing for transmit/receive switching, so it is unlikely that you will buy one that does not work.

It is easy to spot the A200 by the chunky black case shown in **Fig 9.68**. There are three connections at one end, these are RF input, RF output and a thick DC power lead. The DC power lead is actually heavy duty mains cable with brown being the positive 13.8V supply, blue is negative and green/yellow is for switching. Do not connect this lead to a mains supply, this is a sure fire way to destroy your new acquisition. Also be careful not to confuse a VR200 24V to 12V converter for an A200, it has a similar case but two DC cables coming out of the side.

Two types of A200 were manufactured; early models had a TNC connector for the RF input and an N-Type RF output socket. Later models had a flying lead for the RF input and an SO239 RF output socket. Both models are very similar inside.

To decide if the unit is suitable for 4m, look at the identification plate on the side, see **Fig 9.69**. They are marked "Cat No. A200" but the aligned frequency is often blank. Fortunately the 'Code' should be marked, something like "01 E0", this will tell you the frequency range:

| | |
|---|---|
| E0: | 68 - 88MHz |
| M1: | 105 - 108MHz |
| B0: | 132 - 156MHz |
| A0: | 148 - 174MHz |

The E0 model is suitable for 4m and the A0 or B0 models will work on 2m. The E0 model can be modified to work on 6m, this is described in [12].

**Fig 9.70** shows the circuit diagram of the A200. A pair of MPX085P or BLW60 transistors are used in the output stage with bias derived using a wire wound resistor and two forward biased diodes. Printed circuit inductors are used for input and output circuits, tuned with compression trimmers. There is a three-stage low pass filter in the output. The RF sensing circuit switches the amplifier into circuit if DC power is applied to the A200.

The amplifier is well protected including a thermal cut out to shut down the unit if the output transistors are overheating. **Fig 9.71** shows an internal picture of the amplifier and **Fig 9.72** shows the component layout.

As an initial check, ensure that links between 2 and 3 plus 4 and 5 are fitted. This will ensure that the RF sensing is enabled. This switches power to the amplifier via relay A and the RF path through the amplifiers via relay B when RF is sensed on the input. If you want to use direct switching, remove these two links and switch the green/yellow wire to 0V to enable the amplifier.

The amplifier requires about 10 watts of drive to produce 60 - 70W output and will draw 10 - 15A from a 13.8V supply. To align for 4m the following steps should be used:

- Set C7 to minimum, this reduces the input drive to the amplifier
- With 2 - 15W input power, check that the relays operate
- Tune C8 and C17 to achieve maximum output power. It may be necessary to repeat adjustment of these two capacitors to achieve optimum tuning.
- If an SWR bridge is available insert it between your transmitter and the A200 and tune C8 for minimum SWR. This should coincide with maximum power output.

Because the amplifier is linear, it will operate on AM, FM or SSB.

For AM operation C7 should be set for a maximum output of 25W with no modulation to prevent over driving the amplifier.

For SSB operation either direct switching should be used or the 'hang time' of the RF sensing circuit should be increased to prevent chatter. Fitting a 0.68µF across C2 and C3 will give a 'hang time' of approximately 0.75 seconds. It is also necessary to increase the sensitivity of the circuit by fitting a 4.7pF capacitor in parallel with C1. C7 should be adjusted to reduce the maximum power by about 10% from the maximum, to prevent overdriving the amplifier. This will still mean that you get 45-50W PEP output with the third order IMD products at least 28db down.

## 400W Power Amplifier for 2m

Because the 2m band is enjoying an increase in popularity a power amplifier with the maximum output of 400 - 450W with a supply voltage of 28 – 32V is described below. The active device was chosen as a proven 'VHF workhorse' from the manufacturer SEMELAB. It is very robust; the load VSWR may vary up to 20:1 thus giving more scope for output network optimisation without destroying the semiconductor. Full data for the transistor can be found in the datasheet [13]. The circuit is not complicated yet has good characteristics.

## The circuit

The critical part of a transistor power amplifier circuit (**Fig 9.73**) is the output network. A mismatch at the output represents a danger to the semiconductor. Therefore the matching circuits shown in **Fig 9.74** will be described in more detail.

A narrow band solution would be sufficient for the 144MHz to 146MHz frequency range; it is nevertheless advisable to use a wider band solution in order to allow for adjustment tolerances and alignment sensitivity.

The input and output matching circuits NW1 and NW4 consists of quarter-wave matching transformers (25 ohm coax cable), they also provide a 50 ohm asymmetric to 12.5 ohm symmetric transformation. The networks NW2 and NW3 are very simple L/C circuits with inductances made from individual striplines.

## Design of the circuit for network NW3

The quarter-wave matching transformer NW4 is made from 25 ohm coaxial cable that transforms the asymmetric 50W load RL to 2 x R´L = 2 x 6.25W = 12.5W. Flexible or semirigid cable should be used with an outside diameter not less than 3mm because of the 400W output power that it must handle)

To make the computation a bit clearer, **Fig 9.75** shows that the real load is divided from 12.5 ohms to 2 x 6.25 ohms. The network NW3 must be designed to match the optimal load resistance $R'_{opt}$ at half of the total output of 200W with the internal values of the transistor shown in **Fig 9.76**. The value R´opt is shown in the SEMELAB data sheet. The internal load resistance $R'_{DS}$ can be determined from the effective RF Drain voltage $V'_{RF}$ at 200W. The effective RF voltage is close to the supply less 'the bottoming voltage' $V_K$ (assumed to be 3V):

$$R'_{DS} = \frac{(V - V_K)^2}{2 \cdot P_o} = \frac{(28-3)^2}{400} = 1.56\Omega$$

In parallel with this is an output capacitance $C_{ob}$ = 190pF (see data sheet). The effective capacity is:

$$C'_{ob} = 1.3 \cdot C_{ob} = 1.3 \cdot 190 = 250 pF$$

The inductance of the drain connection is taken from the SEMELAB data and is 0.63nH.

## Calculating the value of L1 in Fig 9.76

L1 is made from a small low impedance stripline. The transformation procedure can be seen by using a Smith chart. The Smith chart program by Fritz Dellsperger [14] was used for this process and found to be extraordinarily helpful for this task. The Smith chart in **Fig 9.77** is standardised for the terminal resistance R'2 = 6.25 ohms. The internal load resistance $R'_{DS}$ = 1.56 ohms and the goal value $R'_L$ = 6.25 ohms can be seen. The chart is simplified to make it easier to understand. Point 1 (Z1) on the real axle represents $R'_{DS}$ = 1.56 ohms. $C'_{ob}$ = 250pF is point 2 on the conductance circle. The bond

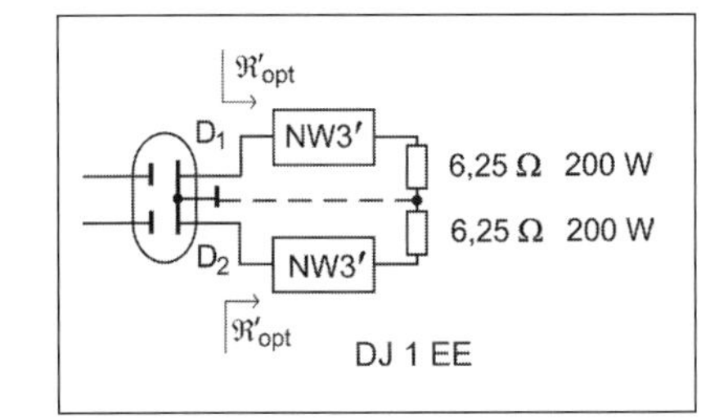


Fig 9.75: Network NW3

inductance is a series inductor L = 0.63nH and is represented as Z2 = 1.4 + j0.5 ohms giving point 3. This is the impedance at the drain connector lug; Z3 = 1.4 + j0.1 ohms. This 'transistor connection resistance' is now transformed with an inductance, formed by striplines, S3 and S4, on the circle around the centre of the diagram to point 4 giving Z4 = 1.4 + j2.6 ohms. This is the intersection with the conductance circle corresponding to the target resistance Z5 =

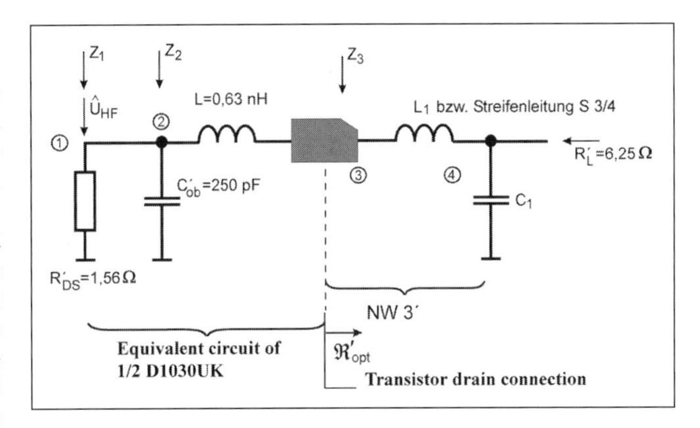


Fig 9.76: The equivalent circuit of half D1030UK

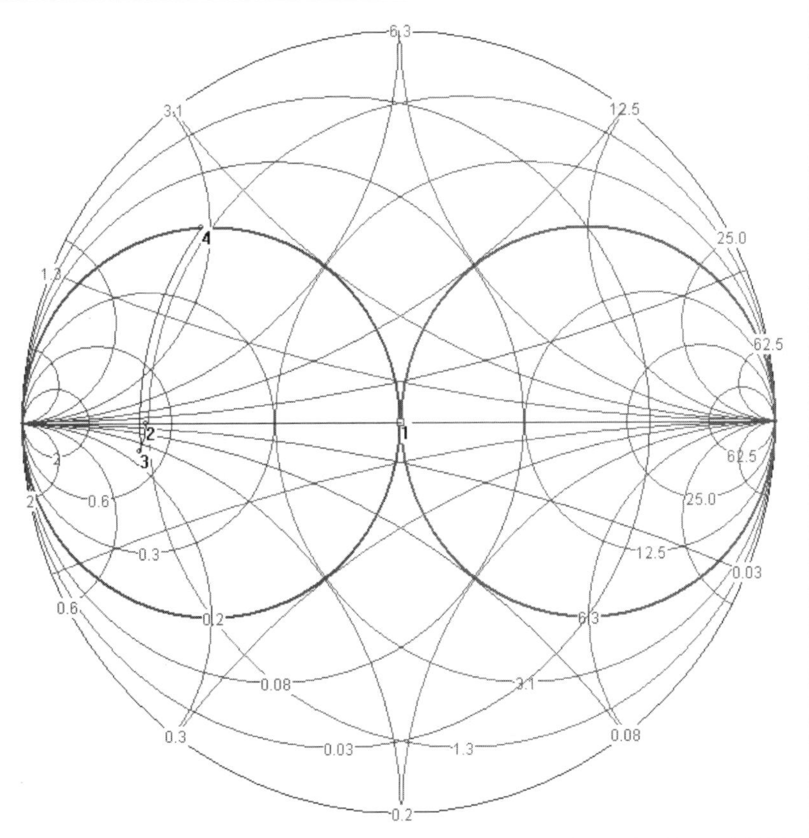

Fig 9.77: Smith chart showing the transformation to 6.25 ohms

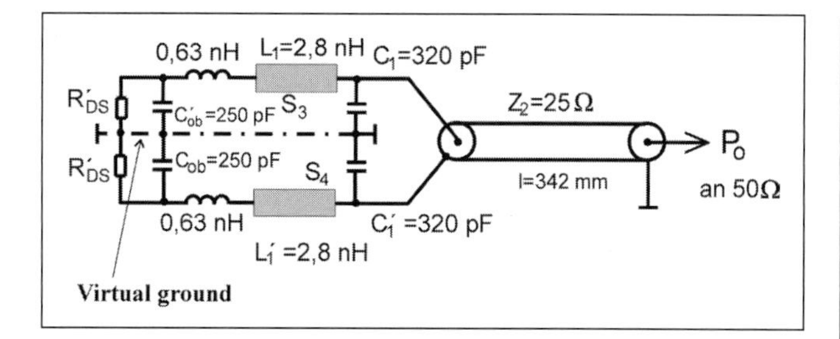


**Fig 9.78: Circuit showing the two halves of the push-pull amplifier**

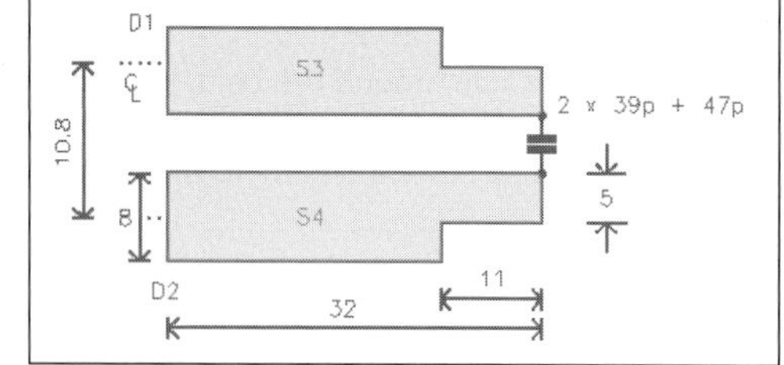


**Fig 9.79: Dimensions of striplines S3 and S4 after optimisation**

6.25 + j0 ohms.

The stripline width W was selected as 8mm and the substrate thickness, H, used was 0.83mm. The characteristic impedance $Z_0$ is given by:

$$\frac{W}{H} = \frac{8mm}{0.83mm}$$

This is for a dielectric constant $\varepsilon_R = 1$. In reality the characteristic impedance of the striplines S3 and S4 are 16.5 ohms because the substrate has a dielectric constant $\varepsilon_R = 3.3$. The inductance of this stripline is 2.8nH and is 'quasi-stable' the length of the line is l = 30 mm. The inductance of this very short line is independent of $\varepsilon_R$!

If a 320pF capacitor is connected from point 4 to ground then a real resistance of 6.25 ohms is achieved (point 5 on the Smith chart). This matches one half of the power transistor to 6.25 ohms. For the push-pull circuit twice this value is used, see **Fig 9.78**. Because the 'virtual ground' does not exist, the two single capacitors in Fig 9.78 are combined into a single capacitor (2 x 320pF in series = 160pF).

If the output networks are used with the calculated values the output power will be approximately 250 to 300W. The transistor equivalent circuit used do not exactly correspond to the actual values partly because of differences in the mounting of the transistor. In order to achieve the maximum output power of 400W with good efficiency (~70%), changes in the striplines S3 and S4 as well as the capacitor C5 are required. This is shown in Fig 9.73 by the variable arrow on these components. This tuning was carried out while watching the efficiency and the values shown in **Fig 9.79** were the result.

The characteristics of the output stage with a supply voltage $V_{DS}$ of 28V and 32V are shown in **Fig 9.80**.

## Measured performance

The performance of the amplifier was measured with the equipment shown in **Fig 9.81**. The results for a supply voltage of 28V are shown in **Table 9.4** and for a 32V supply voltage in **Table 9.5**.

## Input matching

The input impedance of a half D1030UK has the typical values shown in **Fig 9.82**.

$R'_{GS}$ is found, like $R'_{DS}$, in the data sheet [13]. $R'_{GS}$ can be transformed to 6.25 +j0 ohms using a simple L/C circuit. The inductor is again made from a stripline. The transformation to 50 ohms is achieved with quarter-wave matching transformers made from 25 ohm coax cable; thin cable is sufficient here because the power is low. From the values calculated using a Smith chart, the input matching from 5:1 to 2:1 is achieved. The striplines and capacitors must be optimised as shown in **Fig 9.83**. Unlike the output the input can be trimmed to give a return of zero.

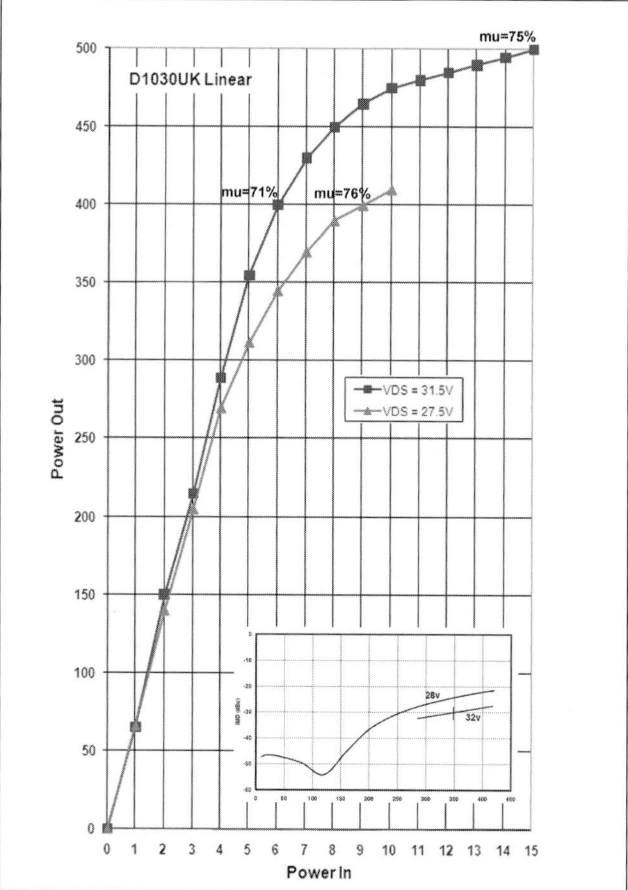


**Fig 9.80: Output characteristics of the D1030UK**

## Mechanical construction

**Fig 9.84** shows the construction of the amplifier. The 50 x 150mm baseboard is fitted to the large 150 x 120 x 80mm heatsink as shown in **Fig 9.85**. The baseboard is 1.6mm thick copper clad FR4 material; the upper surface is the RF and DC ground. In order that fitting the power transistor does not interrupt this ground, a very thin copper foil (~ 0.1mm) is fitted into the cutout for the transistor. The striplines are individually cut from 0.83mm thickness material and soldered to the baseboard ground surface. The DC wiring can take place as desired.

Good heat transfer between the foil, transistor and heatsink is extremely important. Spread thermal compound on the individual surfaces very thinly because only the pores in the metal are to be filled! The heatsink should be well cooled with the aim of a maximum flange temperature of 60°C with a PEP output of 300W.

# 70cm Amplifier

The 70cm linear power amplifier described here can be built in little more than a weekend. It delivers the power required for satellite working, small antennas and short cables or larger antennas and longer cables. Three readily available 2C39 disc-seal triodes are used in parallel delivering 300W output for an input drive of 15W.

As can be seen in **Fig 9.86**, the three triodes operate in a grounded grid circuit with the cathodes being driven in parallel. The amplifier requires only two supply voltages for reliable operation: the anode and the filament voltages. The anode voltage may be between 1.3 and 1.5kV and the filament between 5.8 and 6.0VAC (at 3A).

With 1.3kV on the anode, the anode current can be driven up to 400mA giving an RF output power of some 300W for 15W drive power. It is quite possible that, if good tubes are used, the output power will be even more but they should not be overdriven. A good axial air-blower should be used for the anode cooling.

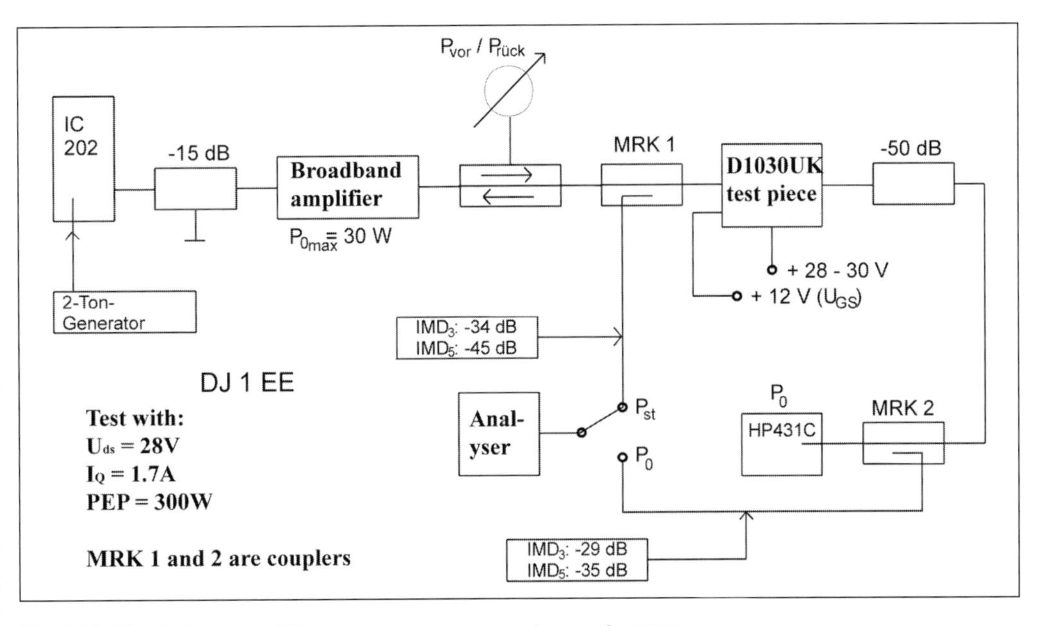


Fig 9.81: Block diagram of the performance measuring equipment

The RF circuits were computed with the aid of a computer program. Special attention was given to optimise the half-wave anode line so that with the given impedance the lowest possible loaded Q was obtained. Through a careful selection of the parameters, a unloaded Q of 39 was achieved which, for this application, is the lowest possible value. This ensures that the anode tank circuit and the amplifier work with the maximum efficiency.

The variable capacitors, shown in Fig 9.86, should have the following calculated values for optimum operation:

| | |
|---|---|
| C1 | 4.3pF |
| C2 | 5.3pF |
| C3 | 1.4pF |
| C4 | 5.3pF |

The dimensions for the cathode ($L_K$) and the anode ($L_A$) line resonators as well as the coupling and tuning plates (C1 to C4) are shown in **Fig 9.87**.

The construction is very simple as can be seen from **Figs 9.88 to 9.90**. A few special parts of the circuit should be explained.

In order that the anode resonator ($L_A$) can be properly connected to the valves, the latter should be modified in the following manner:

- The cooling-fins are taken off and tapped, 4mm. The strip line can then be held tightly between the cooling-fin body and the tube's anode. A 25mm ceramic pillar supports the other end of the strip line

| | | |
|---|---|---|
| **CW:** | $P_o$ | 400W |
| | $P_{ST}$ | 8.9W |
| | Efficiency | 75% |
| | $V'_{DS}$ | 58V |
| **Harmonics:** | 2 x $f_0$ | -49dB |
| | 3 x $f_0$ | -44dB |
| **SSB power:** | $PEP_0$ | 300W |
| | $PEP_{ST}$ | 4.7W |
| | $FMD_3$ | -29 to 31dB |
| $FMD_5$ | -35dB | |
| Efficiency | 49% | |

**Table 9.4: Measured parameters at 28V ($V_{DS}$ = 28V; $I_{DQ}$ = 1.7A; f = 145MHz)**

| $P_o$/W | $P_{ST}$/W | $I_D$/A | G/dB | Effic. % |
|---|---|---|---|---|
| 50 | 0.7 | 6.8 | 18.7 | 24 |
| 100 | 1.35 | 9.2 | 18.9 | |
| 200 | 2.7 | 12.7 | 18.7 | |
| 300 | 4.3 | 15.9 | 18.5 | |
| 350 | 5.0 | 17.0 | 18.5 | |
| 400 | 5.9 | 18.1 | 18.3 | 71 |
| 450 | 8.2 | 19.6 | 17.4 | 74 |
| **SSB Power** | $PEP_0$ | 350W | | |
| | $IM_3$ | -30dB | | |
| Maximum drain voltage: (DC + RF) ≅ 65V for $P_o$ = 480W | | | | |

**Table 9.5: Measured parameters at 32V ($V_{DS}$ = 32V; $I_D$ ≅ 1.8A; f=145MHz)**

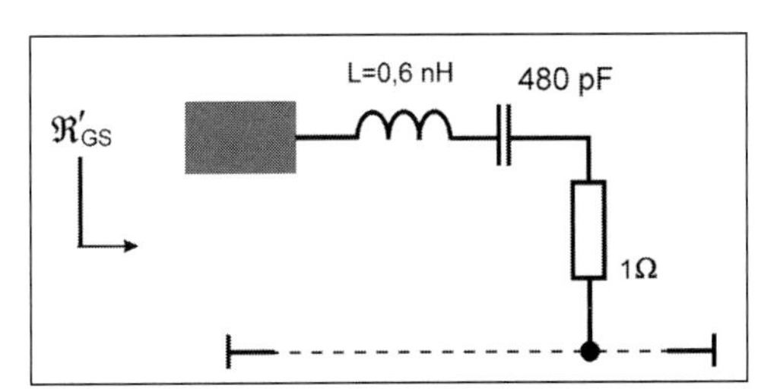


**Fig 9.82: The input equivalent circuit**

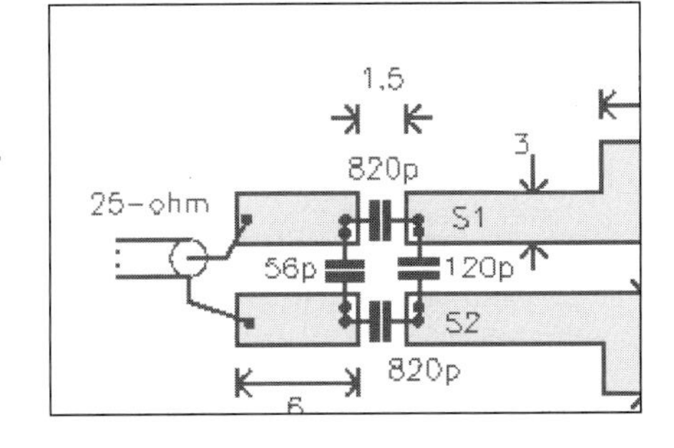


**Fig 9.83: Dimensions of striplines S1 and S2 after optimisation**

- Strip contact fingers are used to make the grid-ring contact. The cathode contact, on the other hand, can be fashioned from 10mm outer diameter copper tube of 0.5mm wall thickness. This tube is 12mm long and slit longitudinally down to the middle. The slotted half is then press-fitted over the cathode contact and the other end soldered to the cathode strip line $L_K$. The remote end of the strip line is secured to a PTFE or ceramic pillar.

The strip lines for the anode and cathode resonators are cut from 1 to 1.5mm stock and silvered, if at all possible.

The amplifier is built into a housing made from 1mm thick brass plate, see Figs 9.88, 9.89 and 9.90. The sides are soldered together.

Tuning capacitors C2 and C4 are made from 0.5mm thick brass plate (Fig 9.87) and hinged and rotated using nylon fishing line. A piece of insulating material - PTFE or polystyrene - is positioned as a stop to prevent direct contact with the opposite electrode. A couple of thick knots tied in the fishing line serve the same purpose.

These tubes require a lot of cooling air if they are to work reliably over a long period. The air blast must also be powerful in order to achieve sufficient cooling over all the surface of the cooling fins. The forced air comes in from above via C4 and cools both the anode resonator and the anode itself and is then vented out of the anode area. It is recommended that a couple of not too small holes be provided in the screening wall between anode and cathode enclosures (Fig 9.88) in order to allow a weak flow of air from the mainstream to flow over the cathode resonator and cathode.

The HT supply as with the drive power is connected to the amplifier by BNC panel sockets. An N socket is used for the RF output.

The valve heaters are connected in parallel. Between the inner heater contact and the cathode lead of every tube, a 1nF disc ceramic (C8) is fitted using the shortest possible connections. The RF chokes (RFCs) are wound using a 6 to 8mm shaft with 0.8 to 1mm diameter copper wire. They are 6 to 7 turn coils, supported from their soldered ends.

Tuning the amplifier is very straightforward, simply tune for maximum output power. This may be accomplished with the aid of a UHF SWR meter or by using the detector circuit shown in Fig 9.86. The coupling (C9) to the detector is adjusted by varying the distance of the silicon diode to the N socket centre pin. The first tuning attempt should take place with very low input drive power and then gradually increase it to maximum.

When the amplifier is in tune the following conditions should exist:

| | |
|---|---|
| Anode voltage | 1300V |
| Grid voltage | -10 to -12V |
| Filament voltage | 5.8 to 5.9V |
| Filament current | 3A |
| Quiescent anode current | 120mA (40mA per valve) |
| Maximum anode current | 400mA 130mA per valve) |
| Maximum grid current | 100mA (32mA per valve) |
| Output power | 280 to 300W |
| Power dissipation | 210W (70W per valve) |
| DC Input power | 520W |
| Efficiency | 60% |
| Gain | 13dB |

It has been found that the anode voltage can remain on during transmit breaks and receive periods. If noise interference can be heard in the receiver, a 10kΩ resistor can be included in the circuit at the point marked X. This resistor must, of course, be short-circuited during transmit. Any type of available relay will do this job.

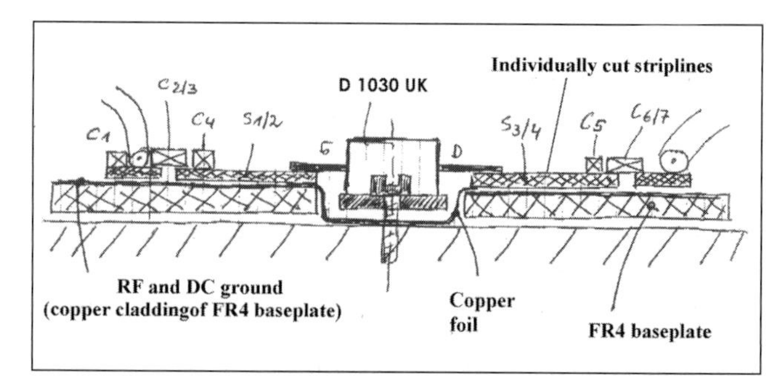


Fig 9.84: Rough sketch of the mechanical construction of the 400 watt 2m power amplifier

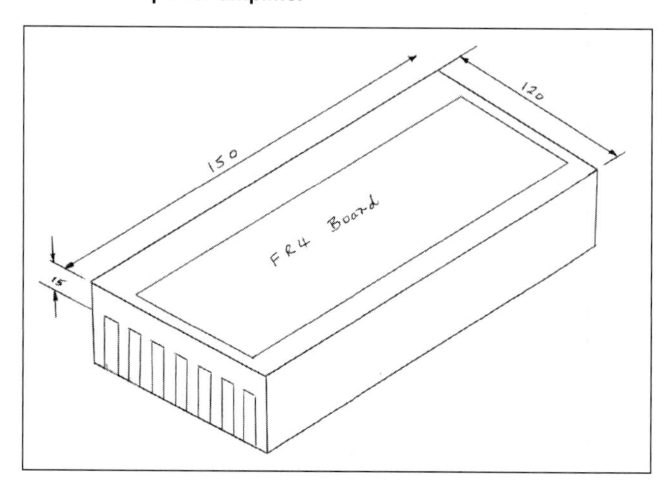


Fig 9.85: The heatsink baseboard

## RECEIVERS

With the large number of commercial receivers available for the VHF and UHF bands, not many amateurs build their own. This design from *RadCom* by Andy Talbot, G4JNT, shows a new type

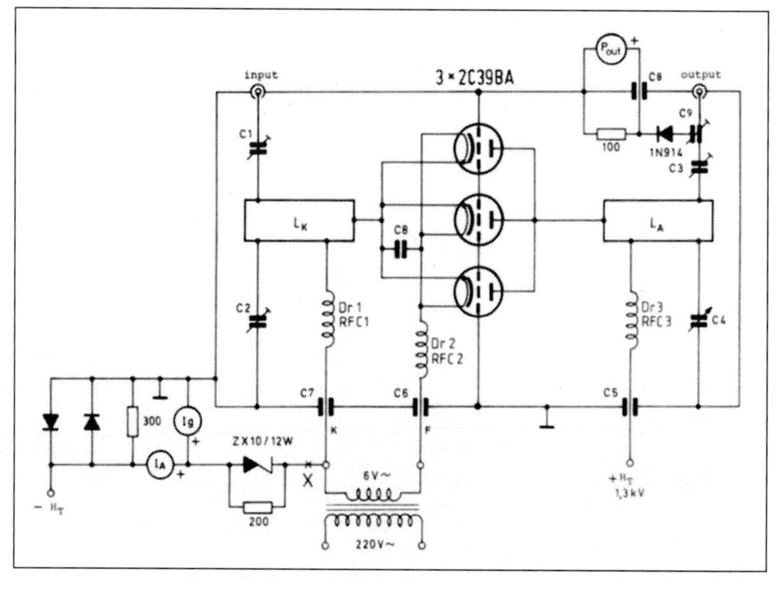


Fig 9.86: Circuit diagram of the 300 watt 12dB gain, 2C39A power amplifier for the 70cm band

of receiver [15].

The converter was designed with the primary aim of using it for the IF stage on microwave transverters. A linear receiver was needed with no AGC, but with a calibrated gain control to make accurate relative measurements of microwave beacons using a PC soundcard-based system for the actual level and signal-to-noise ratio measurements. A straightforward gain calibration could then be used to convert these into absolute readings, making this a useful piece of test equipment for propagation studies.

There is nothing inherently narrowband in the design - filtering limits the RF bandwidth to around 8MHz to eliminate strong signals from broadcast and PMR and the audio bandwidth is kept to about 20kHz, wide enough for the normal maximum soundcard sampling rate of 44100Hz. Any subsequent audio filtering for listening purposes is performed by the software or in separate audio processing circuitry.

The circuit diagram is shown in **Fig 9.91**. In the RF path two MMICs, a MAR-6 and a MAR-3, amplify the RF; there is a two stage bandpass filter between them with 10MHz bandwidth. The output feeds into two SRA-1 type DBMs via a resistive splitter, with the quadrature local oscillator (LO) signal generated using a MiniCircuits PSCQ-2-160 90° power splitter. This device guarantees less than 3° phase error over 100 to 160MHz; as 144MHz is near the middle of the range, we can expect better performance here.

The local oscillator is an AD9851 DDS, currently clocked at 100MHz, generating 16 to 16.67MHz followed by a x9 RF multiplier. The DDS source is not described here, but the module in a basic form is described in reference [16]. The active stages in the multiplier consist of MAR-6 MMICs configured as a pair of cascaded tuned x3 stages with a final MAR-6 as amplifier/limiter, this combination forming probably the simplest tuned RF multiplier possible! There are a couple of CW spurii generated by the DDS, but once you know where they are they can be ignored. All filtering is designed to allow the LO to tune over 144 to 150MHz to cover more than the normal 2MHz narrowband segments on the microwave bands, and allow for odd LO frequencies. The multiplier output level is +l0dBm drive to the quadrature hybrid.

By using the internal x6 option in the AD9851 DDS chip the LO could be driven from a 10MHz frequency reference, producing a clock of 60MHz, but this has not been tried.

The mixer outputs drive a pair of identical NE5532 op-amps with a voltage gain approaching 300 (the exact value is a bit uncertain due to the internal impedance of the mixer IF port). No clever matching is used, just the mixer feeding the inverting input, giving 800 ohm input resistance at audio, and low-pass filtering to get rid of RF leakage. The I/Q outputs feed another pair of op-amps with precisely switchable gain from 0 to 40dB in 10dB steps. Audio bandwidth is not especially tailored, but rolls off gently from about 20kHz to allow for 44100Hz sampling rate in a soundcard.

The total system gain and dynamic range is based on 16-bit digitisation, and is sufficient at maximum (+40dB) to place its own thermal noise at least 10dB above the quantisation noise pedestal. Strong signals and extra RF gain in transverters are catered for by backing off the audio gain. For signals too strong even for this (80db S/N in 20kHz) an external (calibrated) RF attenuator can be added.

No attempt was made to put this on a proper PCB. The con-

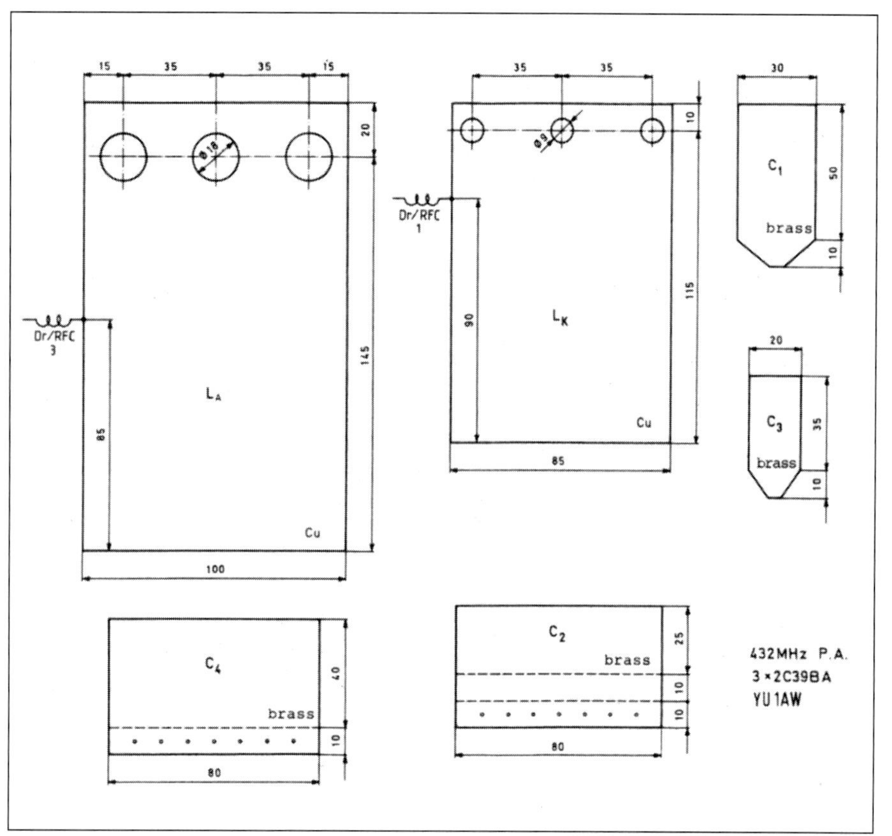


Fig 9.87: Dimensions of the housing parts for the 2C39A power amplifier for the 70cm band

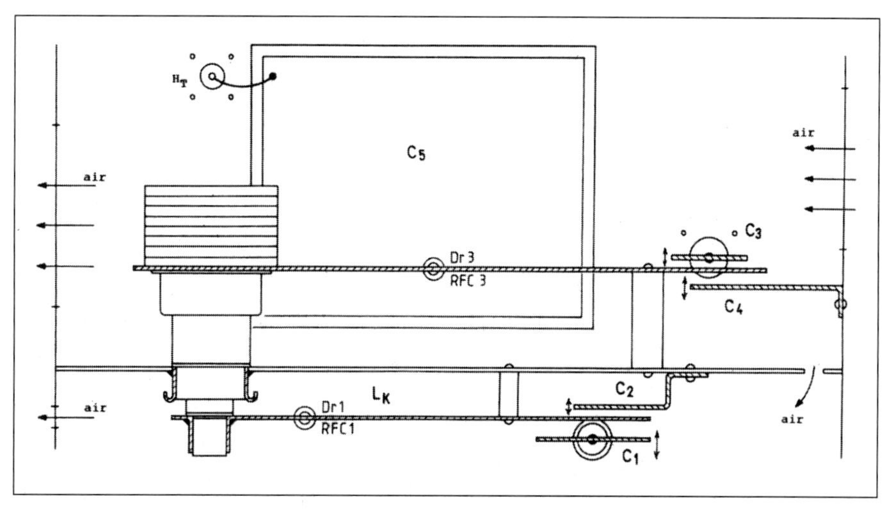


Fig 9.88: Side view of the 2C39A power amplifier for the 70cm band

verter and audio stages were built birds-nest style on a piece of un-etched copper clad PCB as can be seen in the photograph. Plenty of decoupling and short direct wires ensure stable performance. As there is a lot of gain - particularly at audio - the whole unit was built into a tinplate box for screening

Using parallel and series 1% resistors for the switchable gain stage, no special trimming or adjustment was necessary, the traces looked well enough matched on an oscilloscope and, as the aim was only 20 - 25dB sideband rejection to make opposite sideband noise insignificant, tweaking was not necessary. 3° phase error will give 25dB rejection, assuming the amplitude is correct, which is about equivalent to 5% amplitude imbalance. All power rails are regulated and well-filtered for operation from a portable 12V supply.

The LO multiplier was made by cutting a 50 ohm microstrip line into a double-sided PCB. To make a 50 ohm line quickly without etching, score two lines 2.8mm apart through the copper on the top face of the PCB for the full width; use a Stanley knife or similar, making sure you penetrate the copper fully. A 2.8mm width on normal 1.6mm-thick fibreglass PBC gives about 50 ohm characteristic impedance. Then, score two more lines about 1mm from each of these.

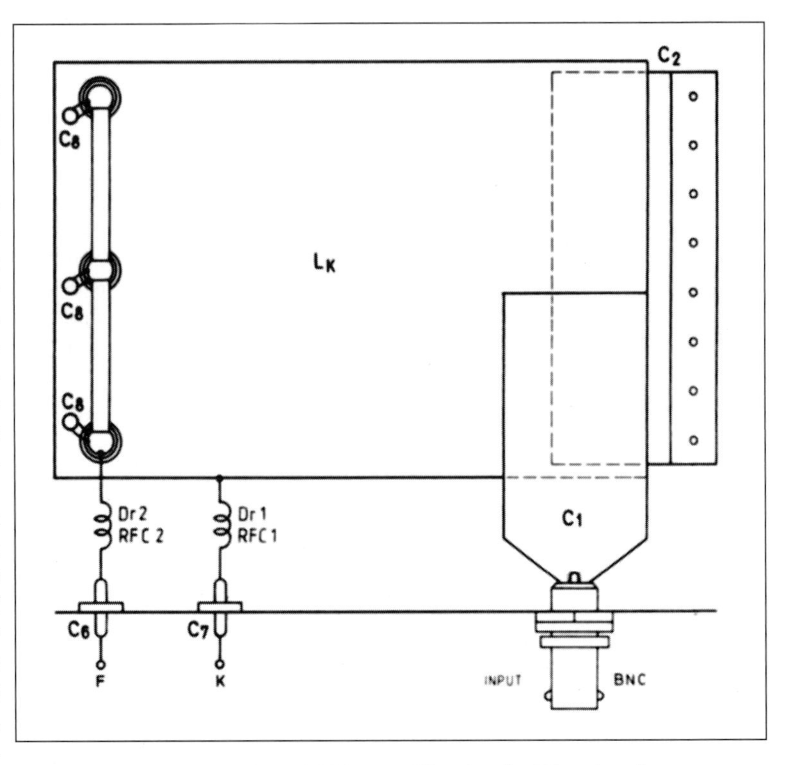

Fig 9.89: Bottom view of the 2C39A amplifier for the 70cm band

Using a hot soldering iron, use this to soften the adhesive and with a pair of tweezers, lift up and remove the two 1mm wide strips, which will give a single 50 ohm line surrounded by a copper ground-plane. Drill a number of 0.8 to 1mm holes through the top ground plane to the underside and fit wire links to give a solid RF ground structure. Wire links are best fitted close to where grounding and decoupling components are connected.

Cut the 50 ohm line into segments with gaps for the MMICs, DC blocking capacitors and filters. Other connections around the filters are made up bird's-nest style. When completed and aligned, coils can be held in place with glue (a hot glue gun is a useful accessory to have around).

For the stand alone unit for use as a receiver in the field, a simple quadrature network and loudspeaker amplifier can be added to make a complete receiver. A high/low pass pair of all-pass networks will give 15dB sideband rejection over 400Hz to 2kHz, which is good enough for listening to beacon signals on hill tops. Alternatively, look at [17] for phasing-type SSB networks to give an improved SSB performance.

A meter driven from the audio level via a precision rectifier circuit can be added to allow quite precise signal strength measurements to be made in conjunction with the calibrated attenuator. Alternatively, take at look at the Software-Defined Radio software [18] from I2PHD, for another solution

The DDS module, described in [16], has new PIC software, along with a rotary encoder and LCD display to give a user friendly interface. For anyone who has the original DDS board, G4JNT can supply PIC software for this modification. However, the AD9850 and AD9851 chips are in short supply now - they have been replaced in most cases by larger, faster, new devices in a different package. G4JNT has also developed a rotary encoder / display for the AD9852 DDS which gives a better route for a local oscillator as it can generate up to 100MHz. He can be reached at [19].

Alternatively, emulate the venerable IC-202 transceiver and build a VCXO to supply the signal to the multiplier. Or use a VFO/mixer, or a synthesiser - the choice is yours!

## TRANSCEIVERS

Building a transceiver may seem to be a thankless task but seasoned contester André Jamet, F9HX has some different views as he explains below:

### A 144MHz Transceiver – for SHF

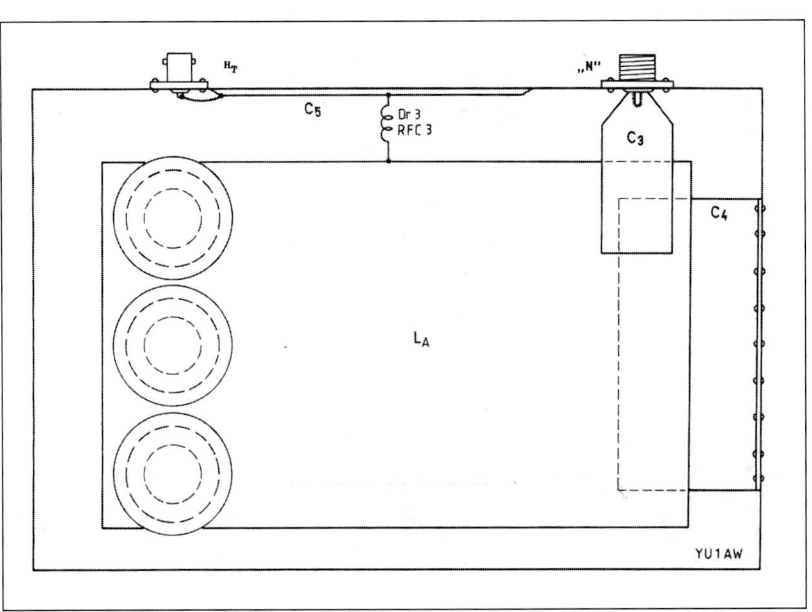

Fig 9.90: View of the anode enclosure of the 2C39A power amplifier for the 70cm band

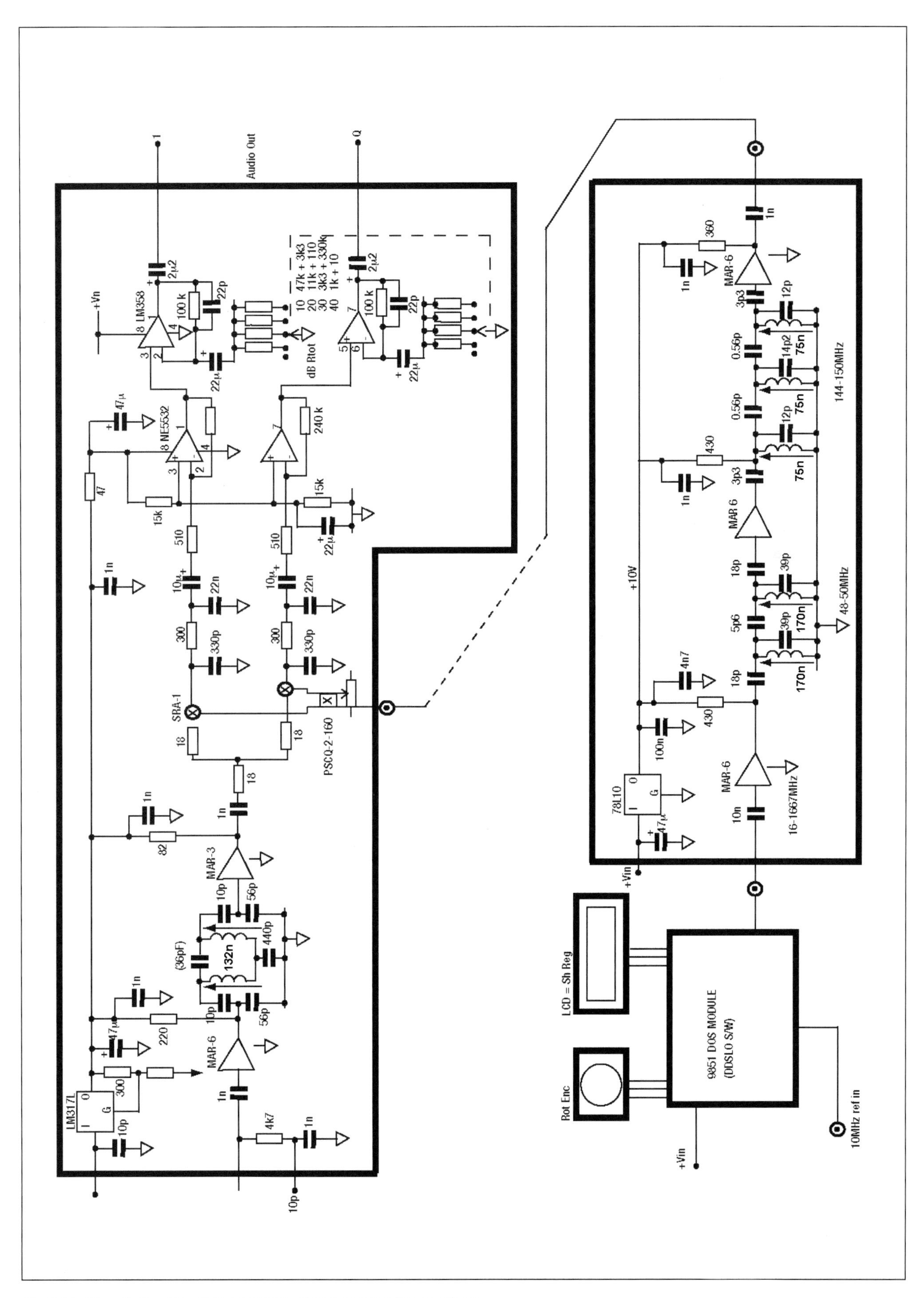

**Fig 9.91: Circuit diagram of the 144MHz direct conversion receiver**

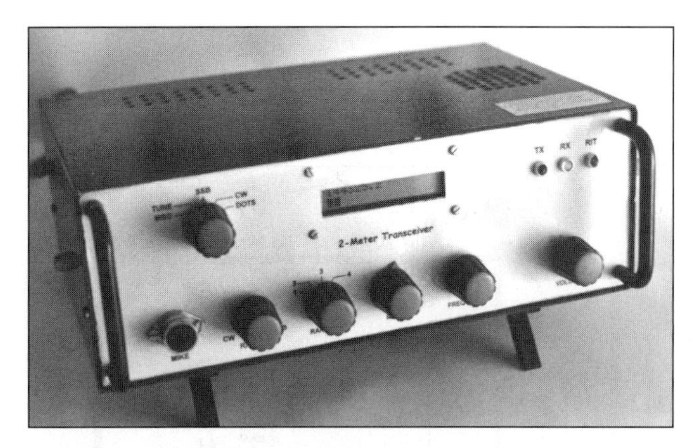


Fig 9.92: The completed 144MHz transceiver

## Typical equipment for SHF operation

For operating on the 5.7,10, 24 and 47GHz bands and beyond a transverter is usually used to reduce the signal to be received or transmitted to lower frequencies. The 144MHz band is in frequent use as an intermediate frequency up to 10GHz, but the 432MHz and 1,296MHz bands are also used for higher frequencies.

We therefore need a VHF or UHF transceiver with the characteristics required to work in combination with the transverter – ie one that can generate SSB and CW, and also has certain accessories which are very useful for SHF use.

One 144MHz transceiver which is very widely used for this application is the famous IC-202. In spite of its faults:

- Imprecise frequency display
- An S-meter which is just as imprecise
- No receive selectivity adjustment (to adjust the pass band in order to improve the signal-to-noise ratio)
- No transmit power control
- No pip generator (to make it easier to get into contact)
- Very frequently poor health

Also bear in mind the great age of those in service and the amount of travelling they have had to endure.

So among SHF enthusiasts a wish has often been expressed to replace this old companion with a more modern transceiver that performs better. Unfortunately, tests carried out using modern transceivers fitted with a very large number of accessories have not always given the expected results. If the various faults mentioned above have disappeared, a new one has seen the light of day. The spectrum purity of their local oscillator is not up to that of the older equipment! This is a hindrance to the reception of weak signals [20, 21], when high amplitude signals are received, and to the generation of a 'narrow' transmission. Modern transceivers use PL's and, above all, DDSs, and their spectrum purity close to the carrier frequency (and also at a distance, in spite of numerous filters) does not attain that of a simple crystal oscillator, even when pulled in frequency in a VXO, as used in the IC-202.

All this is perhaps slightly exaggerated, but the 10GHz specialists (and not only in France) have a lot of trouble in replacing their IC-202s, and several have reconditioned them to give them a new lease of life, adding on the new equipment required.

## What if we replaced our IC-202s?

F9HX has written several articles for the French SHF magazine [22]. The original idea was to create a 144MHz transceiver that would have precisely the characteristics required, without any unnecessary accessories.

## The transceiver principle implemented

It would have been simple to retain the IC-202 structure, ie

- A simple intermediate-frequency conversion receiver operating around 10MHz, comprising a quartz filter to obtain

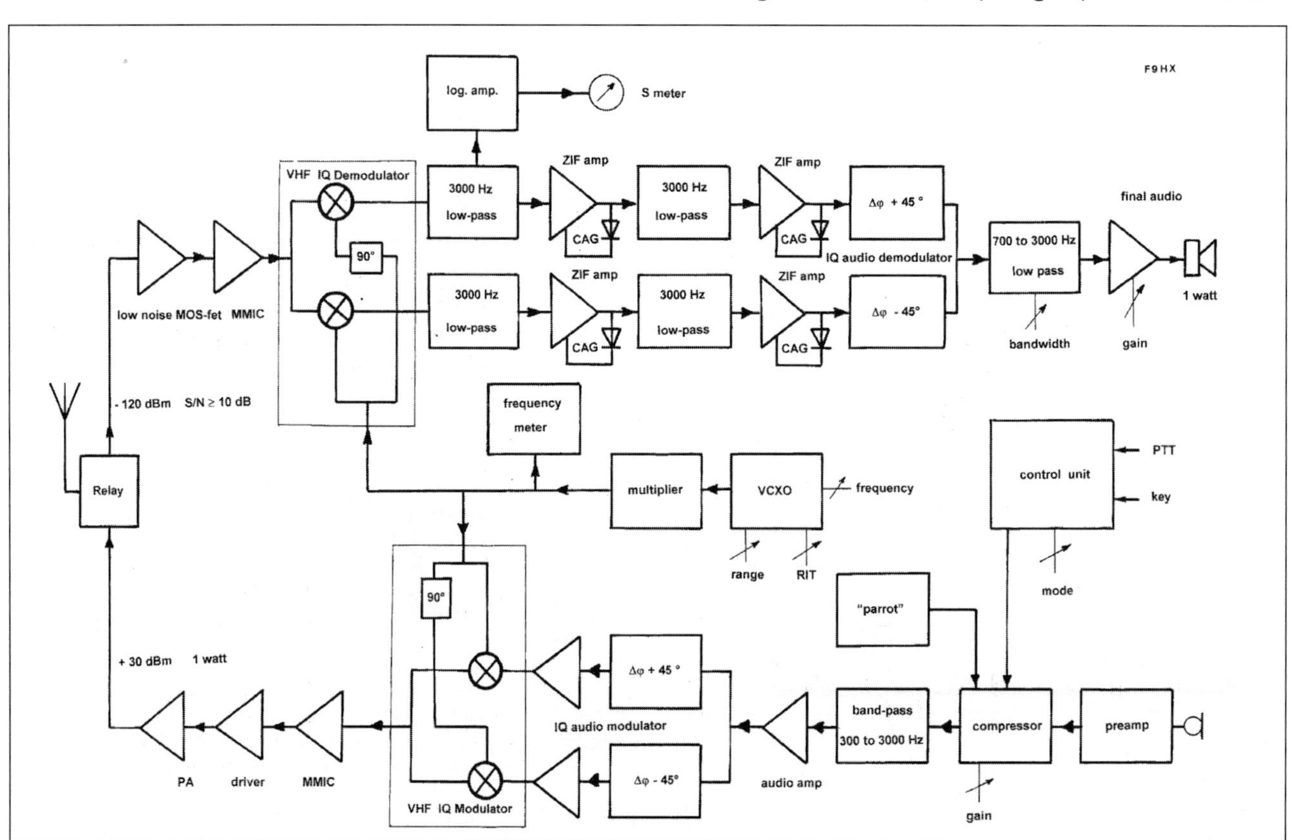


Fig 9.93: Block diagram of the 144MHz transceiver

the desired selectivity followed by a product detector for the demodulation of the CW and SSB.

- A transmitter using the same quartz filter to reject the unwanted sideband and generate SSB.

This solution was adopted by F1BUU, and has been described in several articles [23]. But we can also generate and demodulate SSB using the phasing method [24], ie using phase converters to cancel the unwanted sideband. (For more on methods of SSB generation, see the chapter on HF Transmitters in this Handbook)

Reception is based on a single frequency change. But, since the local oscillator is on the same frequency as the signal received, the intermediate frequency is directly in the audio range. This is referred to as being at 'zero intermediate frequency', since if the modulation signal is at zero frequency, the intermediate frequency is as well (and not at 10MHz).

In English publications, the expression 'direct conversion' refers simultaneously to the single frequency change and to the zero intermediate frequency [25, 26], whereas in France some assume that direct conversion corresponds to the single frequency change, without the intermediate frequency being at zero.

The resulting transceiver is shown in **Fig 9.92** and the block diagram is shown in **Fig 9.93**.

In the receiver, the antenna is matched to a low noise FET by a simple LC circuit. The output is fed to an MMIC through a bandpass filter. This feeds a Mini Circuits double balanced mixer to demodulate the signal into I and Q audio signals. These very low level audio signals (in the order of a microvolt for VHF reception in nanovolts) are amplified by two identical channels of amplifiers fitted with automatic gain controls. They also have active low-pass and high-pass filters in order to limit the pass band received. When the level is sufficient, the two square-wave signals are phase-converted in what are known as Hilbert filters, in such a way that, when they are subsequently added together, the signals from the wanted sideband are added and those from the other are cancelled out. An elliptic 8th order filter using a capacitor switching IC gives an adjustable bandwidth from one to three kilohertz. A one watt audio amplifier ensures a loud signal from the speaker. The demodulator is fed by the LO which comprises four VCXOs switched from the front panel and multiplied to give VHF reception. A logarithmic high dynamic range IC is used for the S-meter.

In the transmitter, modulation is obtained from a microphone or a one kilohertz signal for CW, tune or dots. This signal is initially amplified, then rigorously filtered to allow only the band needed for SSB to pass.

Two Hilbert filters produce two square wave signals to feed the double balanced modulator that produces a VHF SSB signal that only needs to be amplified up to the desired power. A voice record and playback IC stores a twenty second message for calling CQ.

For those interested in the theory, articles have been published in the amateur press [27, 28] explaining mathematically the functioning of this method and also the Weaver method, which is a refinement of it. Some intermediate-frequency direct conversion UHF and even SHF Weaver transceivers, up to 10GHz, are described in [29]. These are models of application of modern techniques. A direct conversion zero IF decametric receiver is described in [30].

## Review of various functions of transceiver

To study the behaviour of the transceiver, modules were created to handle one or more related functions, each on a printed circuit. This also proved to be useful for the final design, and the idea of a single printed circuit was set aside for the final assembly.

Starting from the antenna in **Fig 9.94**, we first find a 50Ω relay, which handles the transmit receive switching for the VHF section.

### Receive section

**VHF module (Fig 9.94):** The VHF signal is amplified by a low noise selective stage fitted with a robust BF 998 dual gate FET transistor, with a performance level at least equal to that of the CF 300, which is well known for its voltage fragility. A filter limits the pass band to the limits of the 2m band and feeds an untuned amplifier fitted with an MMIC. The amplified VHF signal feeds a Mini Circuits quadrature demodulator, which also receives the signal from the local oscillator described below. The output signals from the demodulator are two square wave audio signals referred to as I and Q.

**Intermediate frequency amplifier (Figs 9.95 and 9.96):** The printed circuit comprises two identical amplification channels with a low noise transistor at the input of each of them, followed by a low pass filter and a variable gain amplifier acting as an automatic gain control. Next is another low pass stage and another variable gain stage. The outputs from this module are thus always two square wave audio signals, but amplified, calibrated for the pass band and amplitude compressed.

**Audio demodulator (Figs 9.97 and 9.98)**: On another printed circuit, there are two channels with different phase conversion. These are the Hilbert circuits that bring the signals from the desired sideband into phase and those from the other sideband into opposition. A passive circuit combines the two channels to obtain only the desired sideband. A first order active high pass filter and an eighth order elliptical low pass filter actively limit the pass band and play the major role in defining the transceiver band. A knob on the front panel can control the low pass filter. This adjusts the cut off frequency from 700 to 3,000Hz to cover the SSB and CW requirements.

Finally, a power amplifier stage feeds the internal loudspeaker and/or a headset.

### Transmit section:

As shown in **Fig 9.99**, the signal from the microphone, which can be ceramic, electret or magnetic, is amplified by a stage followed by an adjustable compressor. Then high pass and low pass filters, as efficient as those used in the receiver, limit the pass band to 300 to 3,000Hz.

An input is provided for the signal from the 'parrot' and 800Hz generator for CQ calls, CW and the generation of pips to assist when aligning parabolic antennas.

The signal is then fed to two channels, each including Hilbert phase converters to generate square wave audio signals (**Fig 9.100**).

On the same printed circuit as the receiver section, we find the transmit section (Fig 9.94). It receives square wave audio signals and feeds a Mini Circuits modulator, which is also fed by the signal from the local oscillator. The local oscillator is on a separate module, and is divided into two outputs by a 3dB resistive divider to feed the receive demodulators and transmit modulator. The output from the modulator is amplified by an MMIC, followed by two temperature stabilised class AB stages, each having a diode thermally linked to its casing. The output power can be adjusted using a knob located on the rear face of the transceiver by controlling the level of I and Q signals feeding the modulator.

### Local oscillator

This consists of a 24MHz VXO, the frequency is adjusted by means of a varicap diode, and a ten turn potentiometer (**Fig**

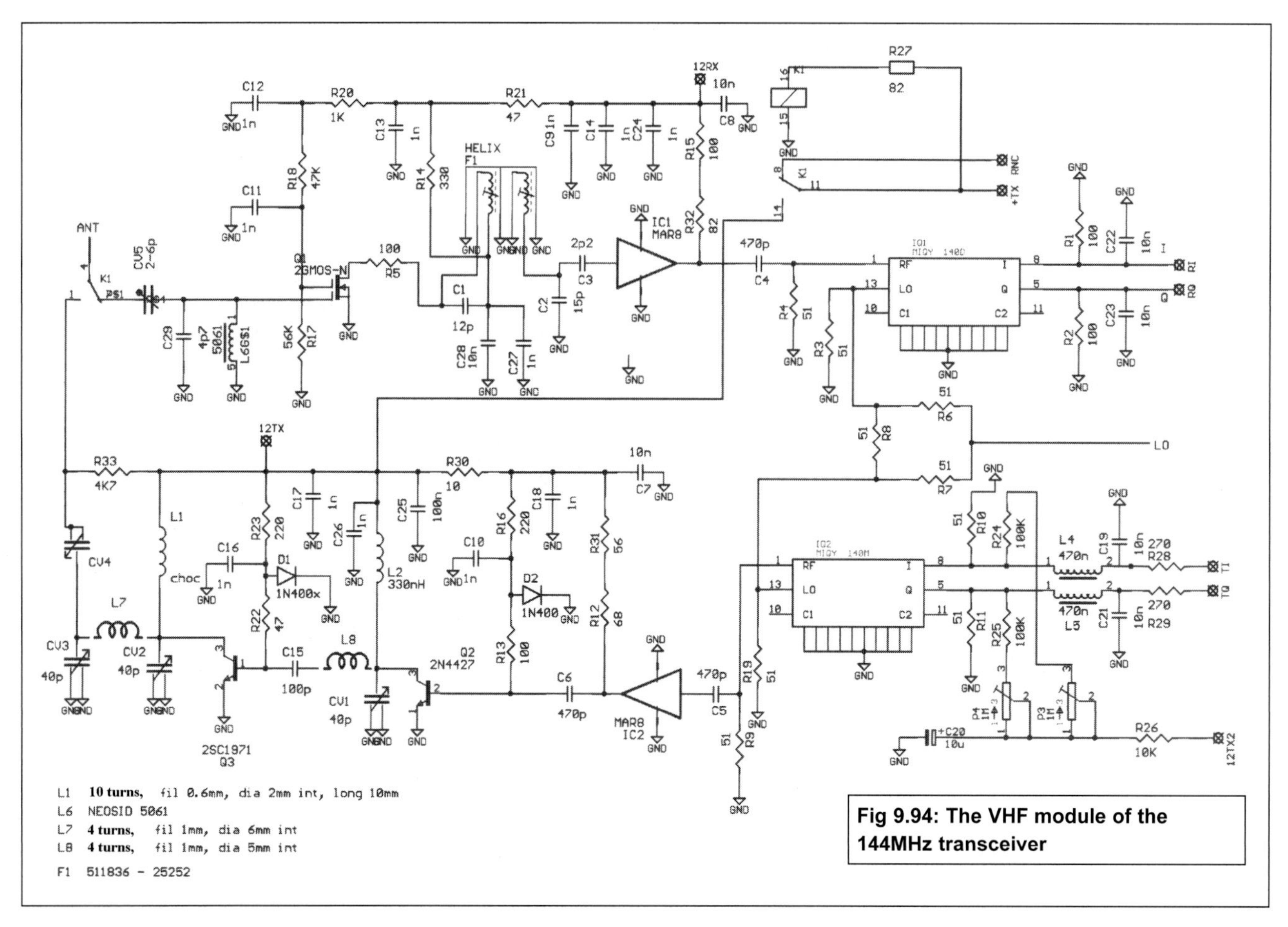

Fig 9.94: The VHF module of the 144MHz transceiver

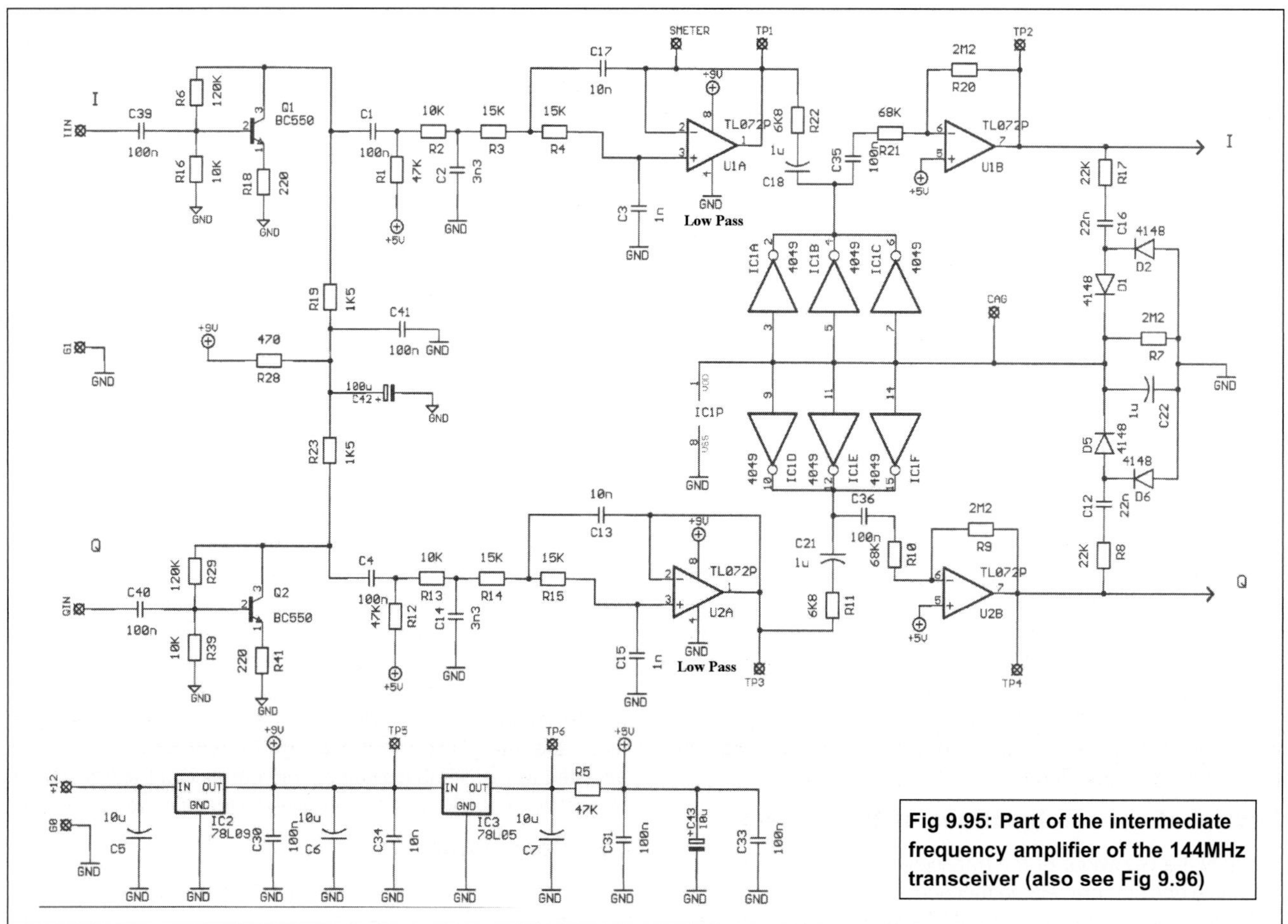

Fig 9.95: Part of the intermediate frequency amplifier of the 144MHz transceiver (also see Fig 9.96)

**9.101**). An RIT can be used for reception using a potentiometer with a notch at the central position, thus a click can be felt when the knob is rotated. A switch makes it possible to select one of four oscillators to cover four ranges of at least 200kHz within the 144-146MHz band. In contrast to the VXO of the IC-202, the switching is not effected via VHF, but through the DC feed of the selected oscillator. This avoids interference from other capacities, which would reduce the range covered by the varicap diode. The crystals used on the equipment and the ranges covered are:

- 24.038MHz crystal:- 144 to 144,200MHz
- 24.071MHz crystal:- 144.271 to 144.400MHz
- 24.133 MHz crystal:- 144.600 to 144.800MHz
- 24.172 MHz crystal:- 144.800 to 145.000MHz

The oscillator is followed by the multiplier stages and an amplifier stage, to provide the level required in the 144MHz band (**Fig 9.102**). In the same module there is a divider (x 10) supplying the signal for the frequency meter.

## Auxiliary circuits

The following auxiliary functions are on a single printed circuit (**Figs 9.104 and 9.105**):

A DC voltage regulator, with reverse polarity protection, limits the voltage applied to various modules to 12V. The other modules include second regulation if necessary, for example for the VXO.

The PIC-based circuit controls the selection of the type of transmission and its generation: CW, SSB, pips, message, tune. It also controls the switching from transmit to receive, with a 'K' at the end of the message. Signals in CW, tune and pips are at approximately 800Hz.

An S-meter, using a logarithmic amplifier, receives one of the audio signals, taken from the output of the first IF amplifier stage, before the automatic gain control. This allows a linear deviation, in decibels, from the signal received (scale 100dB).

## Frequency meter

The 24MHz signal generated by the local oscillator is divided by ten using an ECL divider to feed the frequency meter module. This is made up of a gate, two counters (16 bit counting), a PIC and a two line by 16 character back lit display (Fig 9.105 and **Fig 9.106**).

The PIC and its 20MHz crystal control the frequency meter, generating the gate opening time (0.1 seconds) and all the signals required for LCD display. The PIC code is optimised in order to measure frequencies in the 2m band, with a refresh time in the order of 120ms. The PIC clock frequency can be adjusted using a capacitor for an accurate display.

The analogue/digital converter function of the PIC is used to produce an S-meter display as a bar graph with a length proportional to the logarithm of the signal received.

## Parrot

A recorder repeater makes it possible to modulate the transmitter, using a single or repeated message (**Fig 9.107**). The recording uses an electret microphone, mounted at the back of the transceiver, with knobs for the various operations necessary.

## Switching operations

Two switches make it possible to select the range of frequencies received and the functioning mode. **Figs 9.108 to 9.110** show how they are connected up to the various modules.

## Assembly

The transceiver is housed in a metal box with only the controls strictly necessary for SHF operating on its front panel:

- Selection of range covered at 144MHz
- Frequency control
- RIT
- Receive audio volume
- Receive pass band
- Transmission type selection: message, tune, SSB, CW, pips
- Frequency and S-meter display

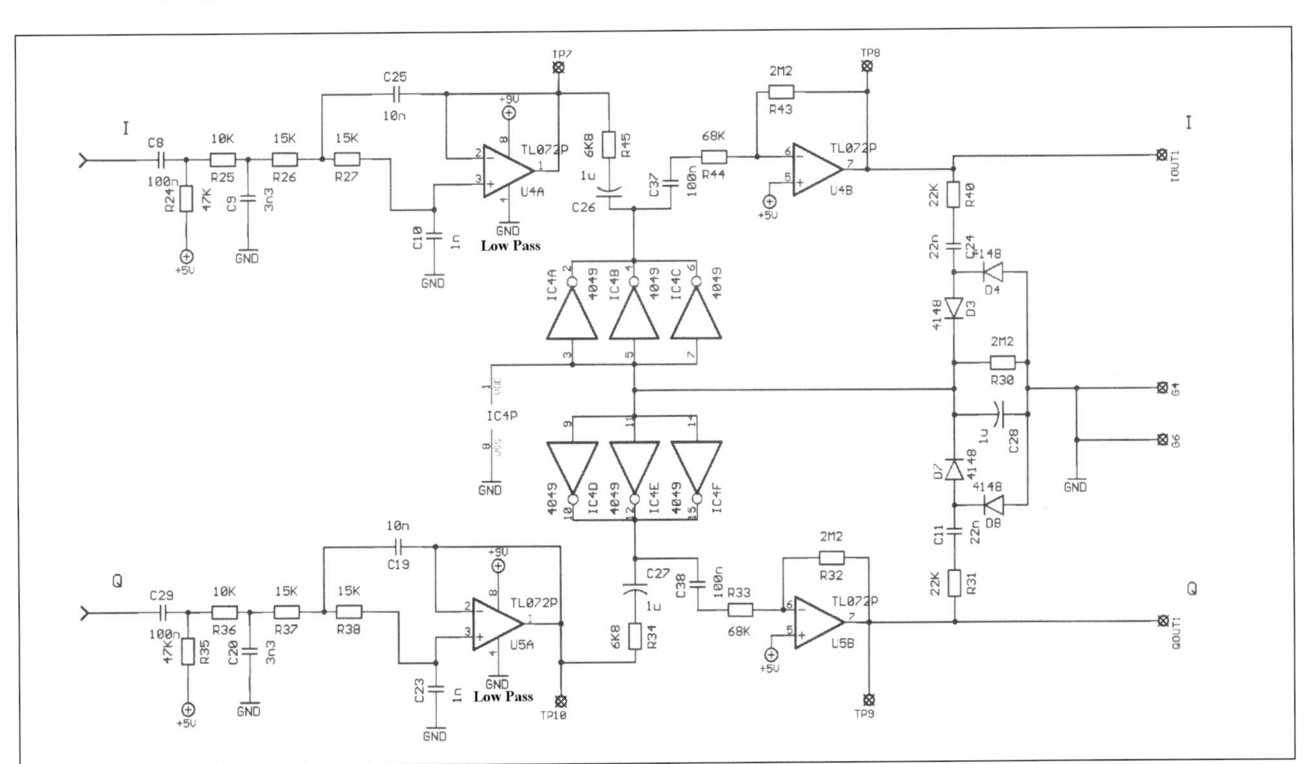


Fig 9.96: Part of the intermediate frequency amplifier of the 144MHz transceiver (see also Fig 9.95)

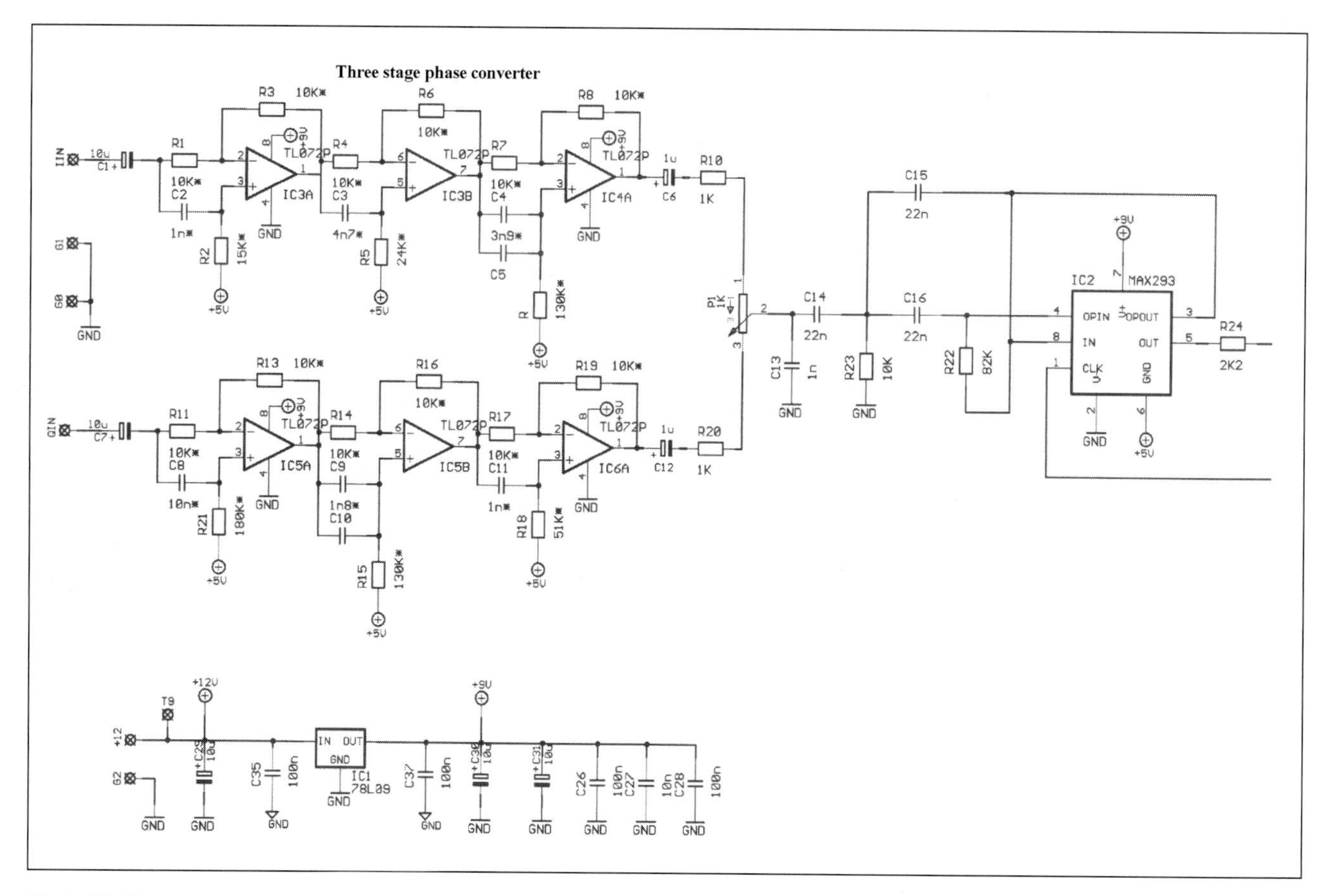


**Fig 9.97: The audio demodulator of the 144MHz transceiver**

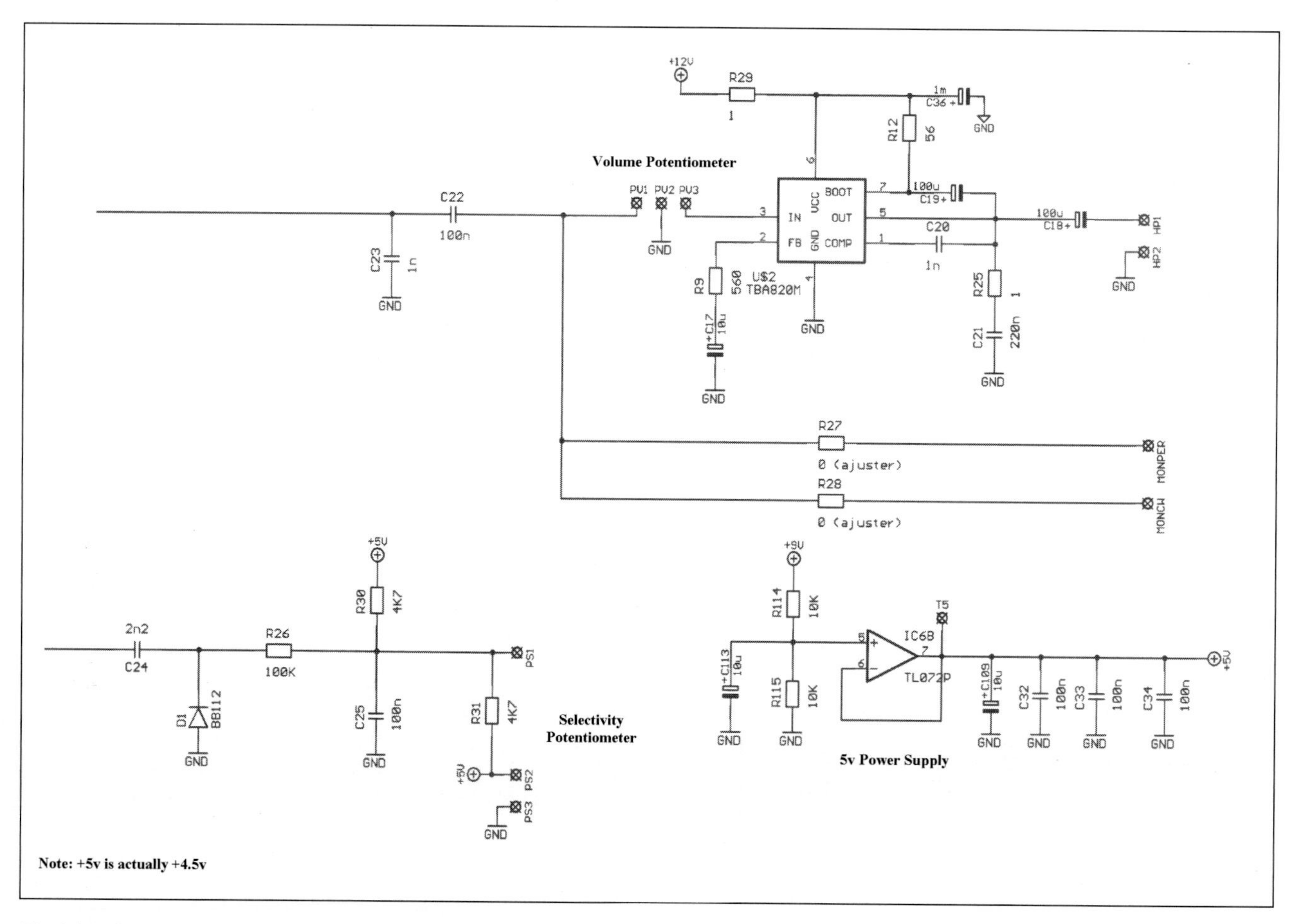


**Fig 9.98: The audio amplifier of the 144MHz transceiver**

- Green indicator light: reception
- Flashing red indicator light: transmission
- Microphone socket

The other knobs and jacks are at the rear, since they do not have to be operated during a contact:

- 12 Volt input
- Fuse
- Connector for 144MHz input and output
- Parrot microphone
- Indicator light and operating knobs for same
- Jack for loudspeaker or headset, without internal loudspeaker cut-off
- Jack as above, but with cut-off (can also be wired up for insertion of a DSP)
- Jack for key
- Transmit power adjust knob

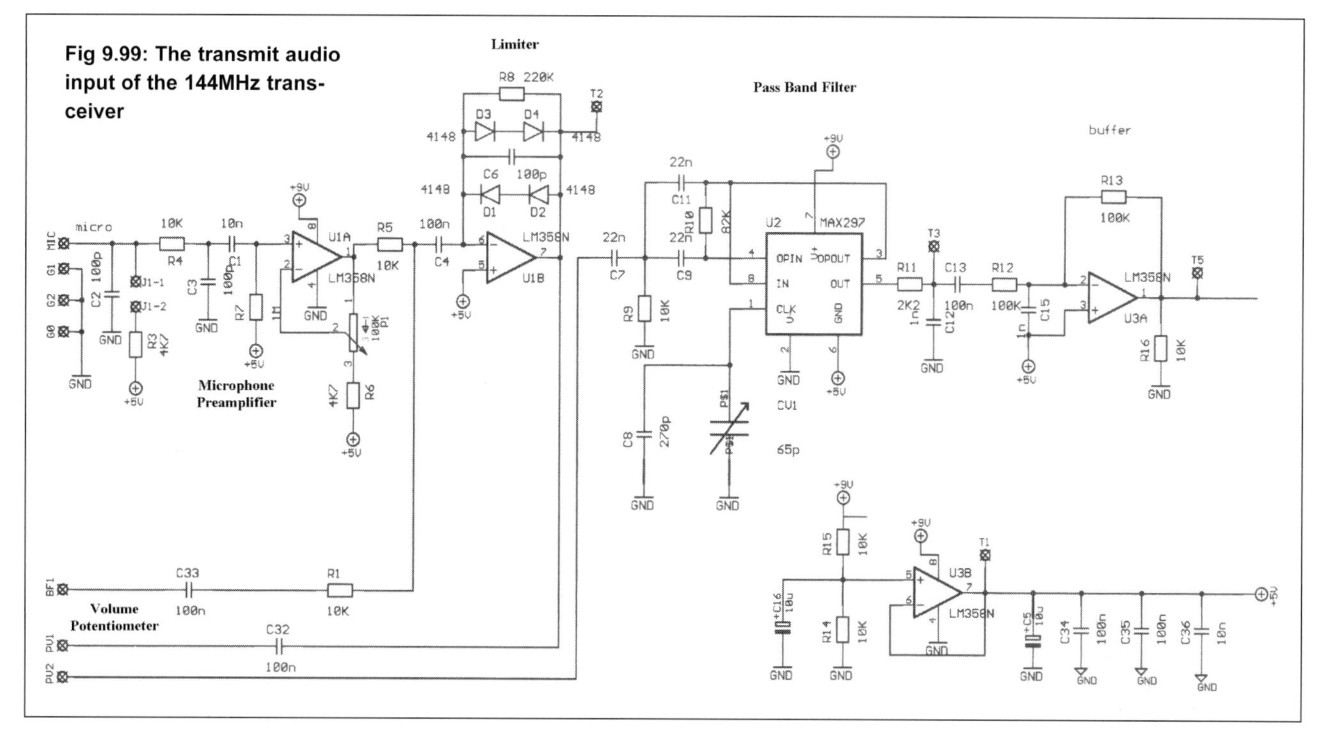


**Fig 9.99: The transmit audio input of the 144MHz transceiver**

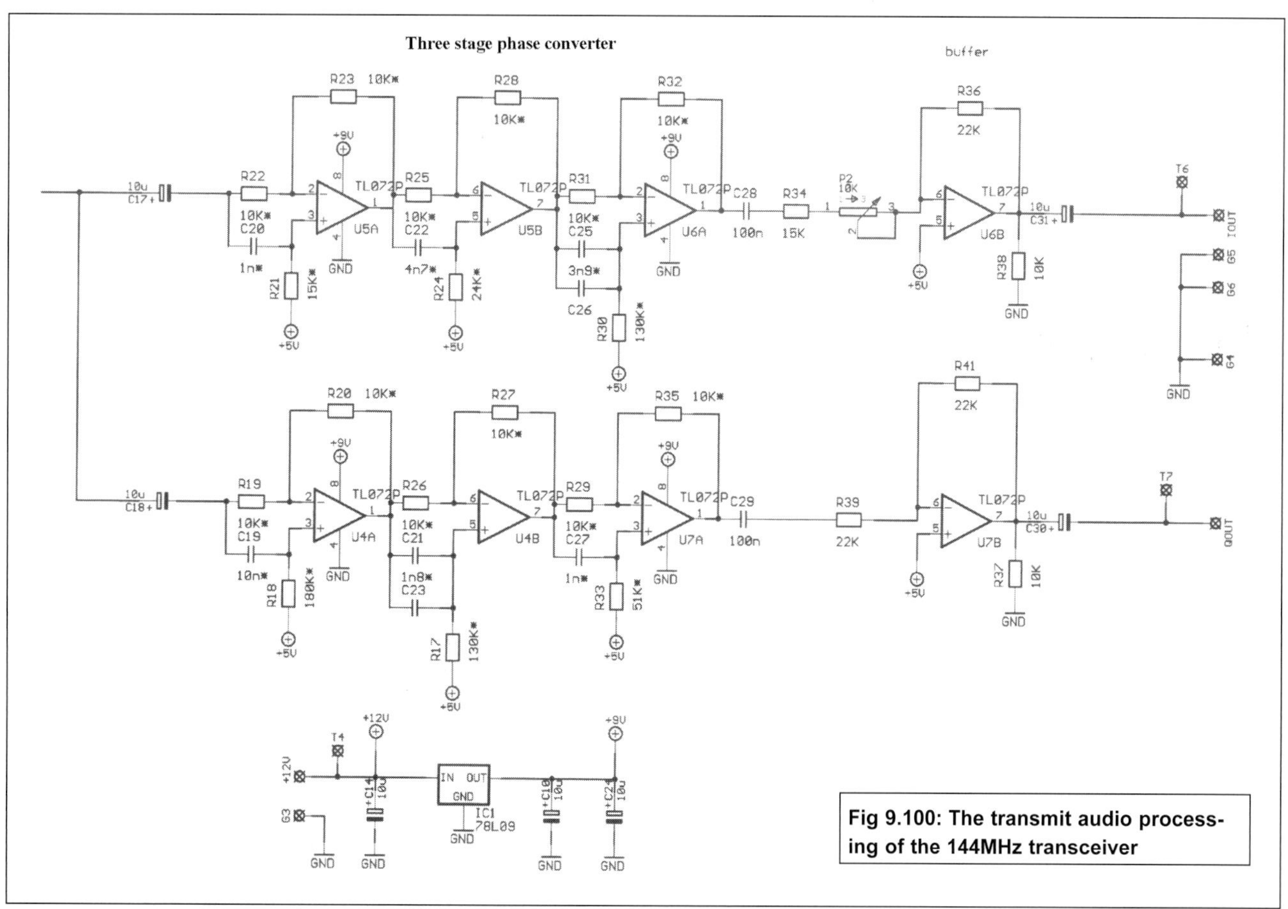


**Fig 9.100: The transmit audio processing of the 144MHz transceiver**

All the modules are mounted on printed circuit boards (single sided or double sided, depending on requirements). The finished prototype is shown in **Figs 9.112 and 9.113**.

The RF and local oscillator modules are housed in tinned sheet metal enclosures, which measure 74 x 111 x 50mm. (reduced to 40mm.) and 74 x 111 x 30mm. respectively. 1nF feed through capacitors are used for all the connections unless they are high frequency links. The antenna is connected via a TNC connector, providing for a very reliable contact, as with an N type, but taking up less space. Professional standard BNCs can also be used, but the mass market models should not be used, since they generate crackling due to earth contact resistance variations. The lower frequency connections and the frequency meter use SMA connectors. They are assembled using CMS components, except the RF power circuits. The interconnection is in the form of a star, using normal wire for DC circuits, shielded wires for the audio circuits, and small diameter coaxial for RF.

## Using the transceiver without an SHF transverter

As with all receivers based on the principle of direct conversion to zero intermediate frequency there is a phenomenon that can be a real nuisance. This is the reception of interference from very strong signals, AM or SSB even if they are located outside the band. This is due to their demodulation by the demodulator's diodes, as soon as their detection threshold is reached. In a receiver with an intermediate frequency of about 10MHz, this detection produces an audio signal that is removed by the intermediate frequency amplifier. By contrast in this design the signal is amplified. We hear powerful stations from the 144MHz band, and even outside it, if they are strong enough to get through the filters located on the VHF module. This does not cause problems when the transceiver is used with an SHF transverter, but using it on 144MHz from a good location can quickly lead to problems due to the reception of interference. Trials now in progress appear to show that a solution requiring only some modest additions to the intermediate frequency/AGC circuit would lead to an appreciable reduction in this fault.

Another problem has been noted with radiation from the local oscillator. In a traditional receiver, this is not in the band of the frequencies received, but offset by the the intermediate frequency, if it emits radiation it is not noticed. By contrast, our transmitter emits radiation at the reception frequency, and can be heard by another receiver if it is nearby. This can be a possible source of conflict with neighbours!

Some direct conversion receiver makers have been worried by microphony. By avoiding any ceramic capacitors in the low level audio circuits, we can exclude the microphonic behaviour of these components.

## Tests and adjustments

Each printed circuit board should be carefully inspected, preferably under an illuminating magnifying glass, to detect any poor soldering or any bridges between connections.

Each module should be individually tested before being fitted into the housing – initially, without putting the integrated circuits in place on their sockets. The modules run at 12 volts, apart from auxiliary equipment circuits, which operate at 13 volts. We

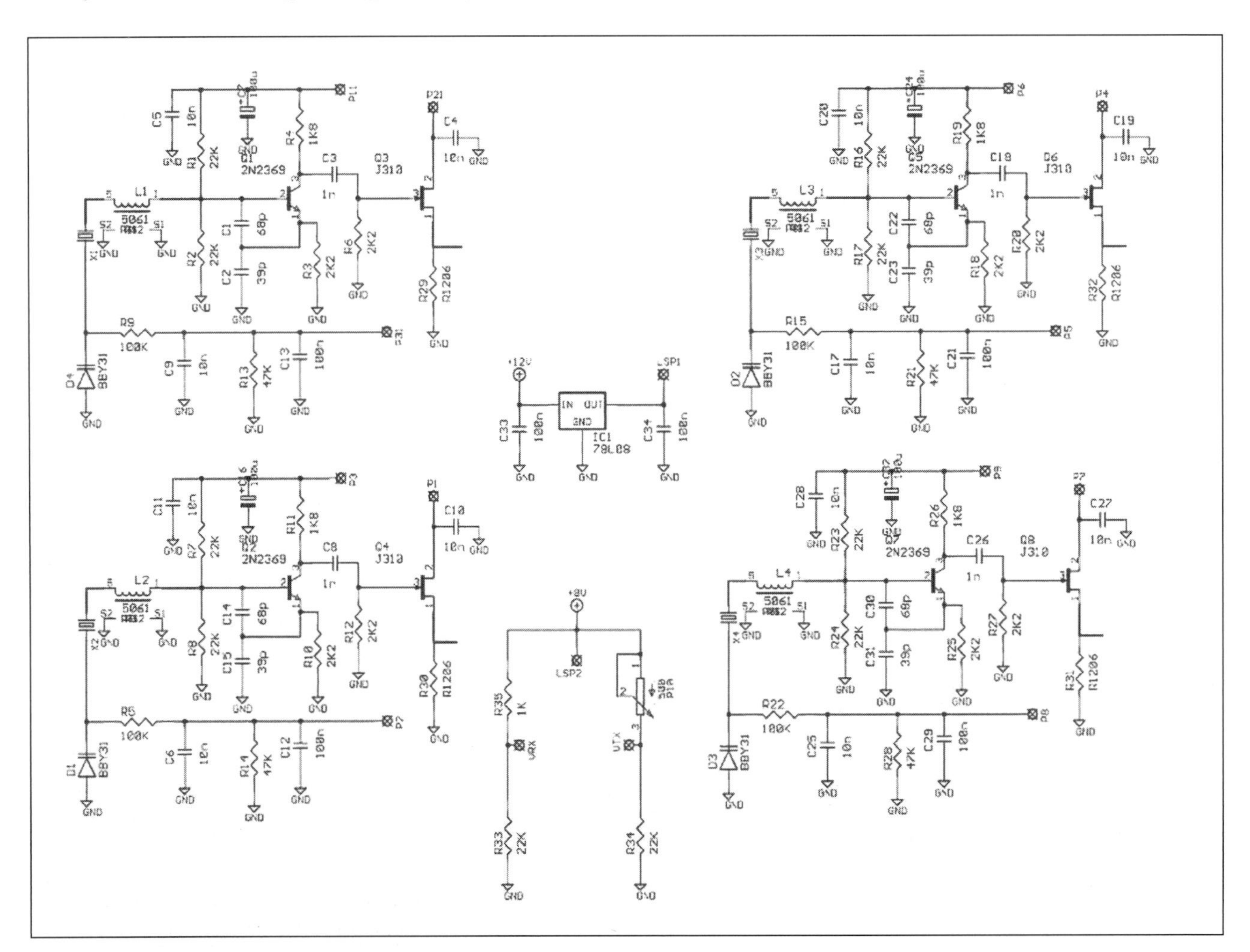


Fig 9.101: The local oscillator of the 144MHz transceiver

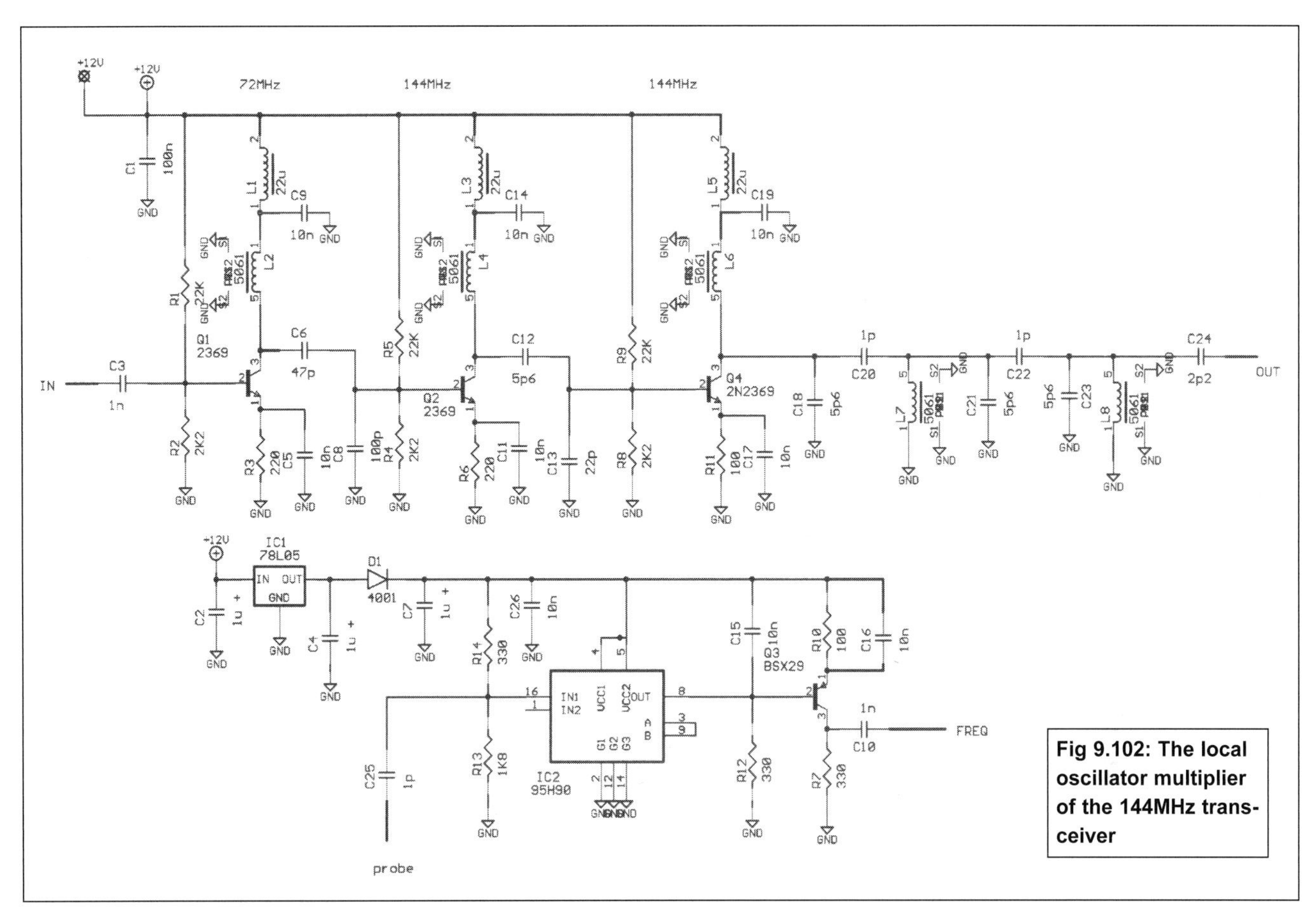

Fig 9.102: The local oscillator multiplier of the 144MHz transceiver

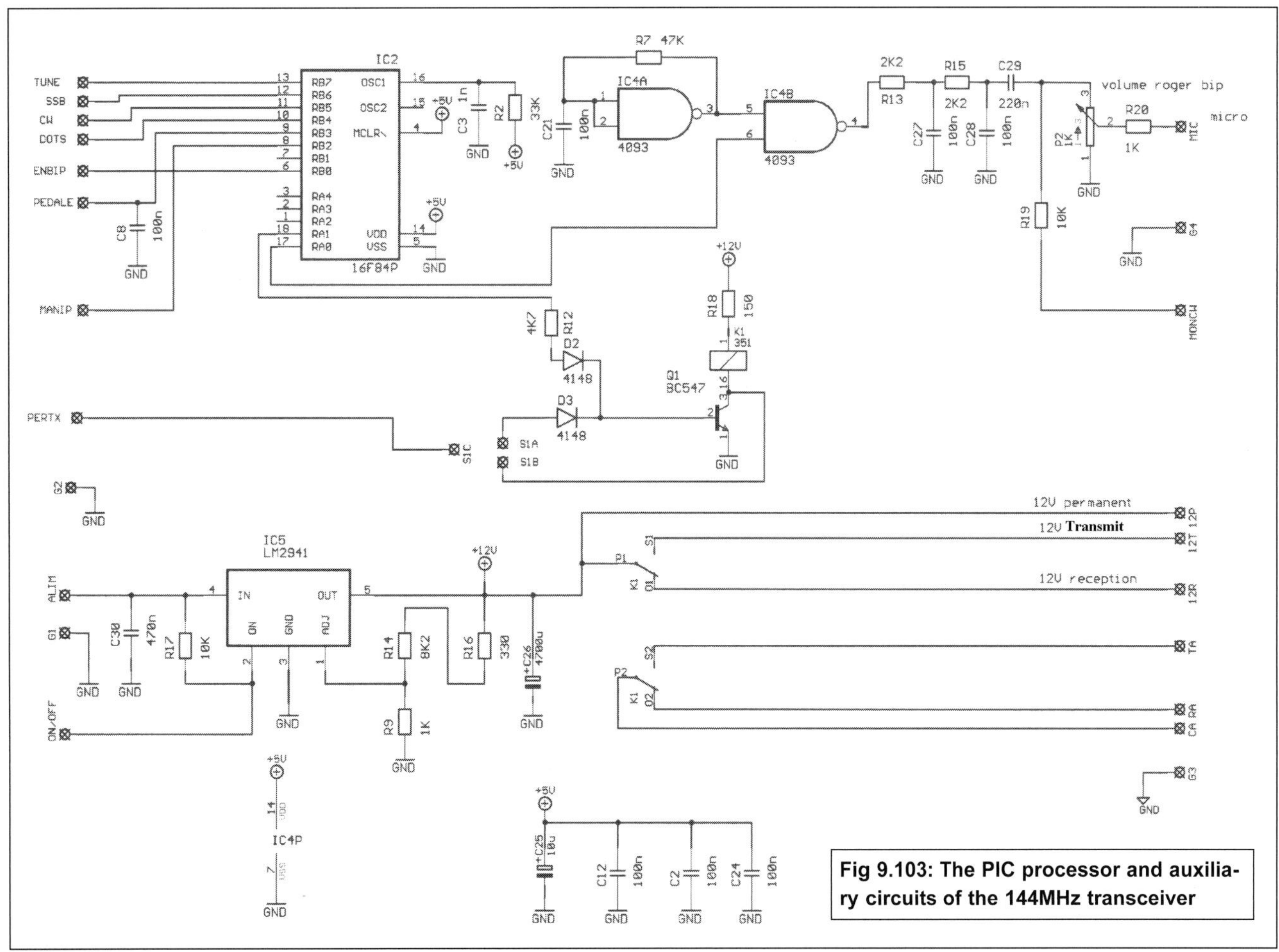

Fig 9.103: The PIC processor and auxiliary circuits of the 144MHz transceiver

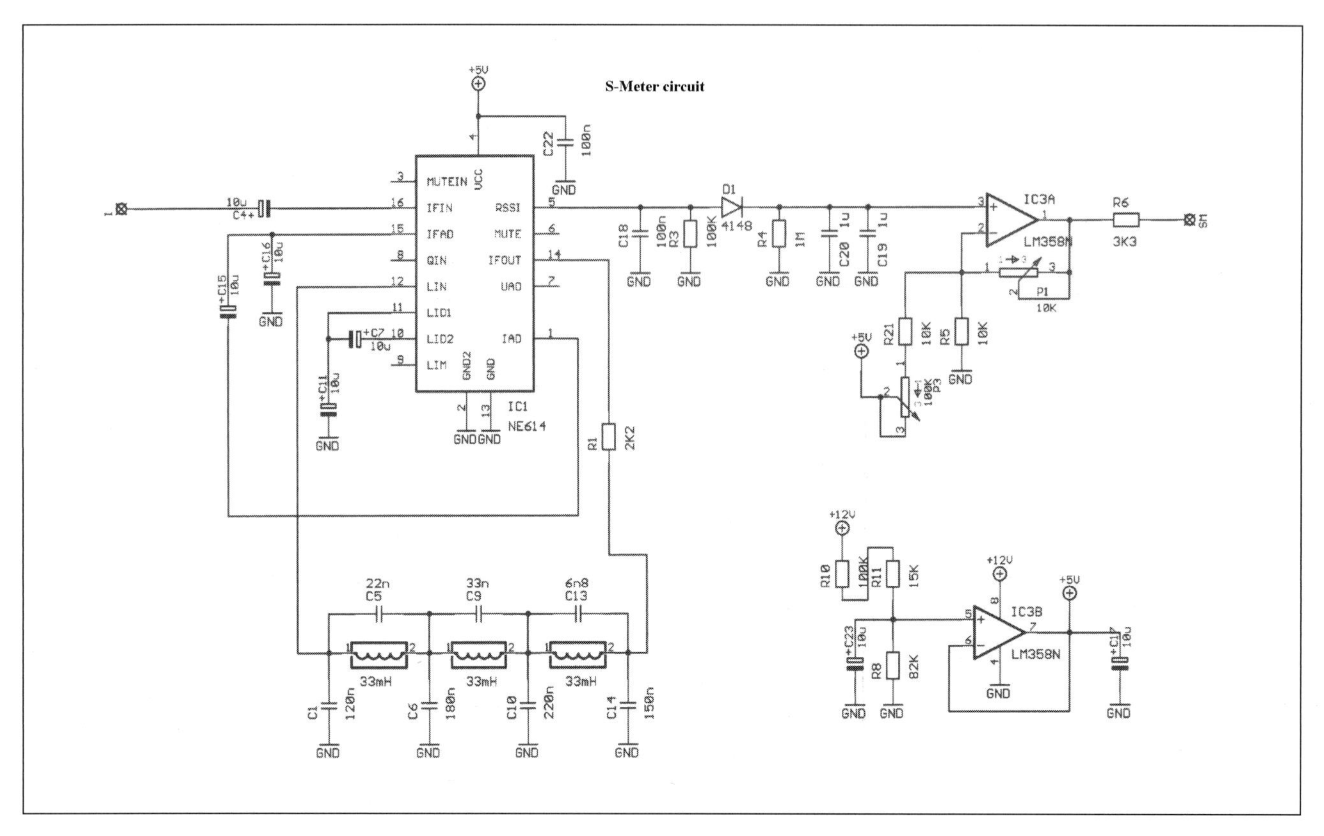


Fig 9.104: The S meter of the 144MHz transceiver

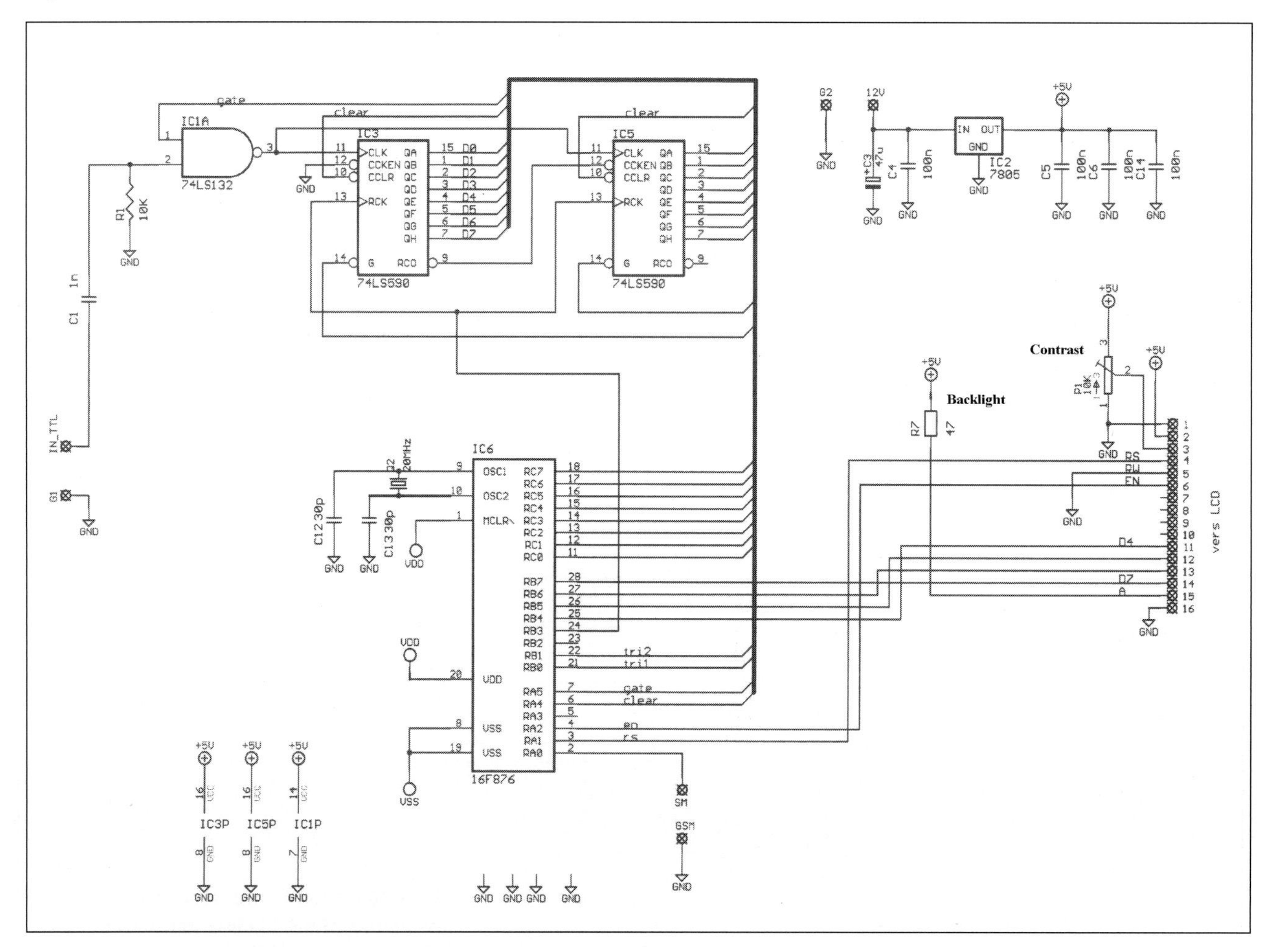


Fig 9.105: The frequency counter of the 144MHz transceiver

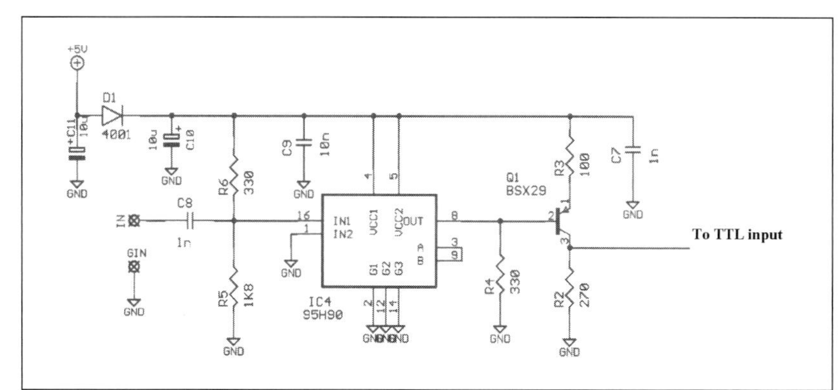


**Fig 9.106: The divider for feeding the frequency counter**

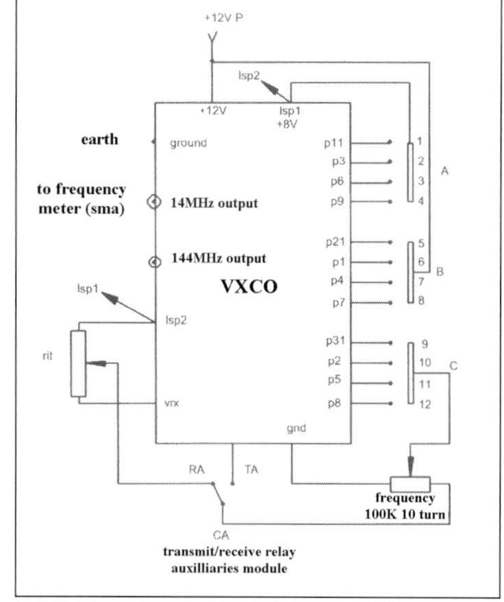


**Fig 9.108: Local oscillator interconnections**

can check whether the voltages are correct at various significant points on the circuit. Then, with the integrated circuits in place, we can proceed to the functioning tests.

The VXO should be tested and adjusted for each of the crystals to obtain the band coverage required. The multiplier section should be set for maximum output at 145MHz. The intermediate frequency module is tested using an audio generator by putting the two inputs in parallel.

The two outputs should supply the same amplified and limited signal, thanks to the AGC. The second receiver module can be tested using an audio generator, preferably followed by a buffer stage supplying two square wave signals. The transmit modulator should be driven by an audio generator, to check that the outputs are square wave. The compressor should be adjusted as desired.

The VHF module should be tested in receive mode, with the local oscillator connected. Applying a 145MHz signal to the TNC socket will make it possible to adjust the input stage by measuring the output voltages I and Q, the input VHF signal being offset by approximately 1.5kHz in relation to the local oscillator frequency. It is possible to adjust the three adjustments, antenna and coupling, to obtain a flat pass band of between 144 and 146MHz, or to favour the most frequently used range, for example, from 144 to 144.200MHz for use with a 10GHz transverter. The phase difference of outputs I and Q can be checked by creating a lissajous figure with a two channel oscilloscope, which should show a circle.

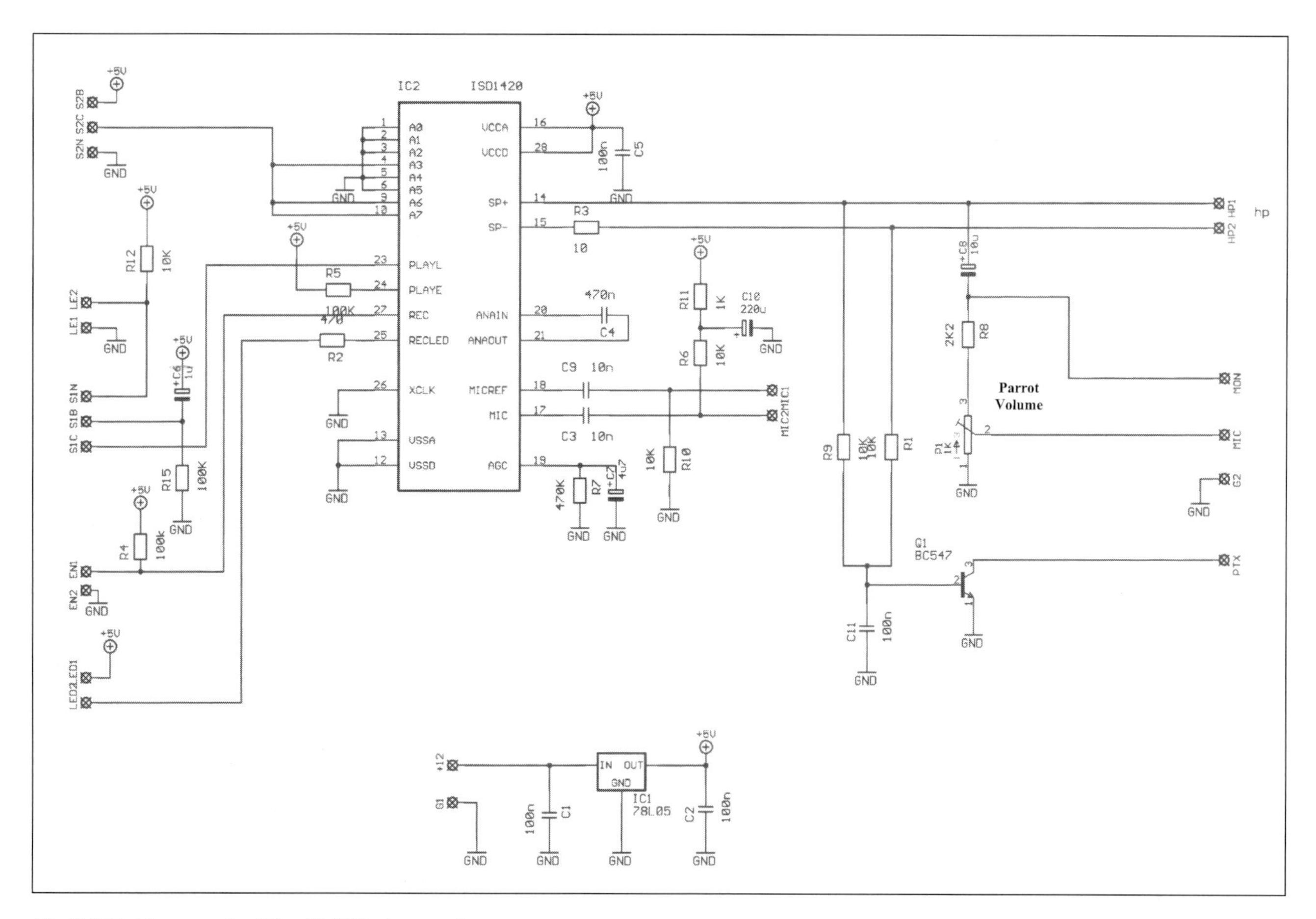


**Fig 9.107: The parrot of the 144MHz transceiver**

### 144MHz Transceiver Specification

| | |
|---|---|
| **Dimensions:** | 292 x 230 x 103mm, weight 1.8kg. |
| **Power supply:** | 12 to 15 volts for full transmit power, reduced power operation from 10 to 12 volts |
| **Transmit power at output:** | 1 Watt (+ 30dBm) in tune, CW, peak SSB positions |
| **Sensitivity:** | 0.16µV (-120dBm) with S/N = 10dB for 3kHz pass band for a signal modulated at 1,500Hz |
| **Receive AGC:** | For 100dB variation in the received signal, the output varies by only 30dB |
| **Pass band:** | |
| **Transmit:** | 300-3,000Hz, with -50dB drop at 4,000Hz (**Fig 9.111**) |
| **Receive:** | Adjustable on front panel from 700 to 3,000Hz, with -50dB drop at 4,000Hz for setting at 3,000Hz |
| **Audio power:** | 1 Watt into 8 ohms |

During transmit, with the local oscillator connected, two square wave I and Qs, produced as referred to for the receive section will make it possible to generate a 145MHz signal. This can be tuned for a maximum using the adjustable capacitors.

When all the modules are functioning correctly, it is time to install them into the housing, to check that the assembly is operating properly, and to fine tune the settings.

### Conclusion

This long project took two years to complete After several months of use on activity days, some very satisfactory results were achieved at 10GHz with numerous contacts at distances of close to 500 kilometres, whilst using a very modest parabola with a diameter of 48 centimetres.

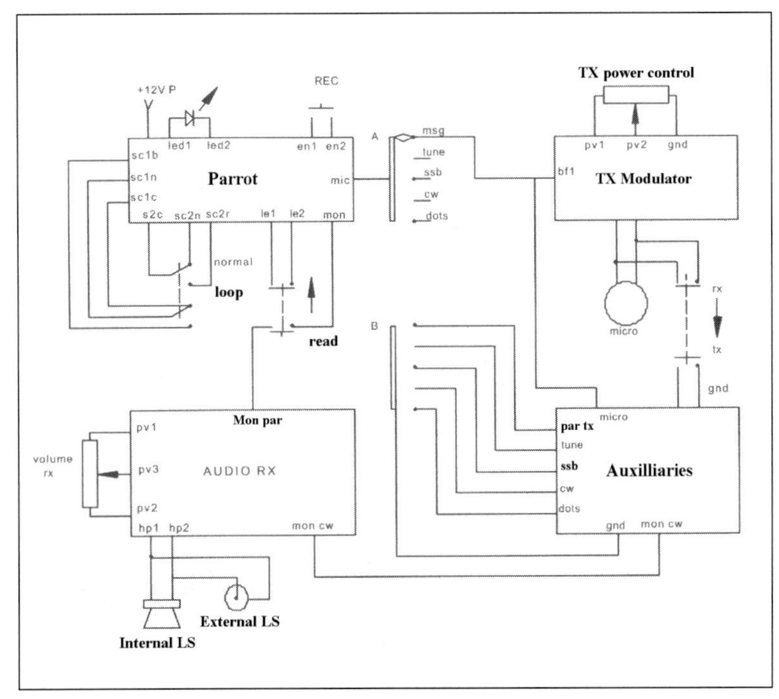


Fig 9.109: Main interconnections for the 144MHz transceiver

### F5CAU's contribution

F5CAU carried out the work of designing printed circuits to professional standards, and also designed the frequency meter, the 800Hz generator and the PIC for the auxiliary modules. In addition, as the assembly instructions are too extensive to be published in full here, he has posted it on his Internet site [31], where the printed circuit drawings and some other useful documents can be download free of charge.

## TRANSVERTERS

If you already have an HF band transceiver, one of the easiest ways of getting onto the VHF and UHF bands is to use a transverter. This takes the output of your transceiver, usually the 28 - 30MHz band, converts it to the chosen VHF or UHF band and converts received signals on the VHF or UHF band so that they are received on your transceiver. The transceiver output and input is commonly called a tuneable IF. The advantage of this approach is that all of the facilities of your HF transceiver are available on the VHF or UHF band.

This section contains transverter designs for all of the VHF and UHF bands. The first set of designs cover 6m, 2m and 70cm using similar circuits. The 4m transverter design was used as a club project by the Andover Radio Amateurs Club [32].

## 2m Transverter

In 1990, Wilhem Schüerings, DK4TJ, and Wolfgang Schneider, DJ8ES, presented a paper at the 35th VHF Congress in Weinheim on a universal transverter concept [33] and [34]. The following design is the resulting 28/144MHz transverter [35]. It should be possible for the transverter to directly feed a standard power amplifier, the design of a suitable amplifier using a hybrid amplifier module is shown.

Transverters for the 2m band are always of interest, in an attempt to match the current state of art in amateur radio technology this transverter was developed using modern components to convert the 144 - 146MHz range into the 10m band.

Concepts such as high-level signal strength and oscillator signal spectral purity have taken on increasing significance. It is also important that the equipment can be reproduced easily. The transverter described below represents a circuit that corresponds to today's requirements.

**Fig 9.114** shows the circuit of the 28/144MHz transverter. The local oscillator uses a tried and tested U310 crystal oscillator at 116MHz (note: the U310 has been superseded by the J310).

This signal is amplified by the next stage, an MSA1104 MMIC, giving an output level of 50mW. The SRA1H high-level ring mixer requires a local oscillator level of +17dBm (50mW) and can be used at up to 500MHz.

A pi attenuator, consisting of R1 to R3 is used to the control the output from the driving transmitter (IF). For a 'clean' signal (intermodulation products <50 dB), the ring mixer must be driven by a maximum of 1mW (0dBm) at the IF port. **Table 9.6** shows the resistance values needed for the attenuator for various IF input power levels. The attenuator uses standard value resistors. The attenuator also acts as a 50 ohm termination for the ring mixer.

The converted receiver signal is fed from the mixer by L2 and C1 to a high impedance amplifier using a BF992 low-noise transistor to give the required intermediate frequency amplification.

The 2m received signal is fed to the gate of the

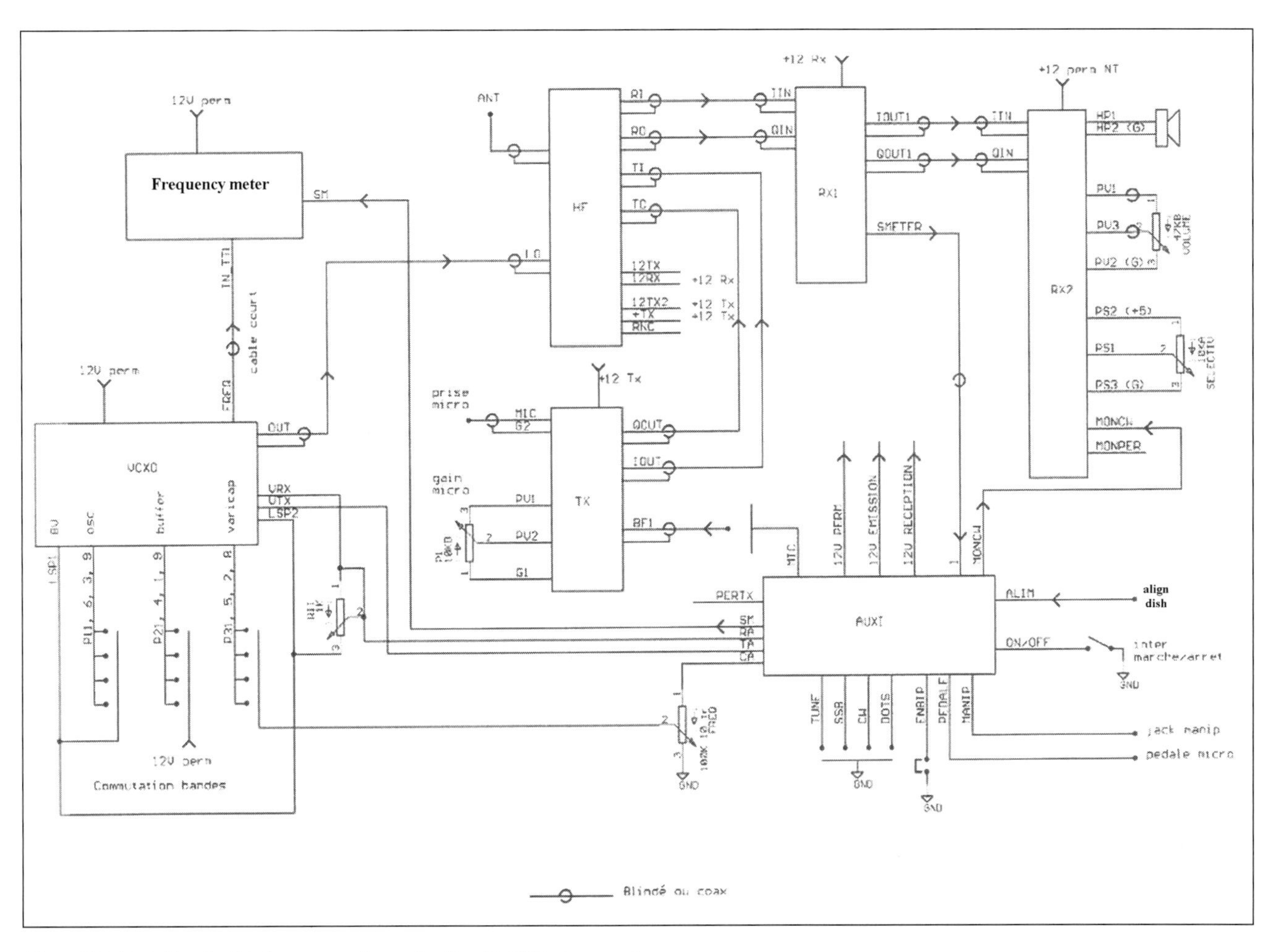


Fig 9.110: Details of all interconnections for the 144MHz transceiver

BF992 RF amplifier through a pi filter from the 50 ohm aerial input. The RF amplifier is followed by a two-pole filter. The received signal is switched to the ring mixer by the +12V receiver supply voltage through the PIN diode, D1 (BA886).

In transmit mode, diode D2 is activated. The 2m signal from the ring mixer first passes through a three-pole filter. The signal is then amplified by three MMIC amplifiers (IC3, IC4, IC5). The combination of MSA0104, MSA0304 and MSA1104 guarantees an output level of 50mW (+17dBm). The transverter can be used with any power amplifier but additional harmonic filtering is recommended.

The 28/144MHz transverter is assembled on a double sided PCB measuring 54mm x 108mm. The board can be mounted in standard tinplate housing of 55.5mm x 111mm x 30mm, **Fig 9.115** shows pictures of the completed transverter. The board should either be made as a through-plated PCB, or copper rivets used to make the earth connections for the coils and ring mixer. The parts list is shown in **Table 9.7** and component layouts for both sides of the PCB are shown in **Figs 9.116 and 9.117** (both in Appendix B). Suitable holes for the crystal, trimmers and Neosid coils, etc are drilled on the earth side of the boards (fully coated side) using a 2.5mm drill. Holes that are not used for earth through connections should be countersunk using an 8mm drill. Suitable slots are to be sawn

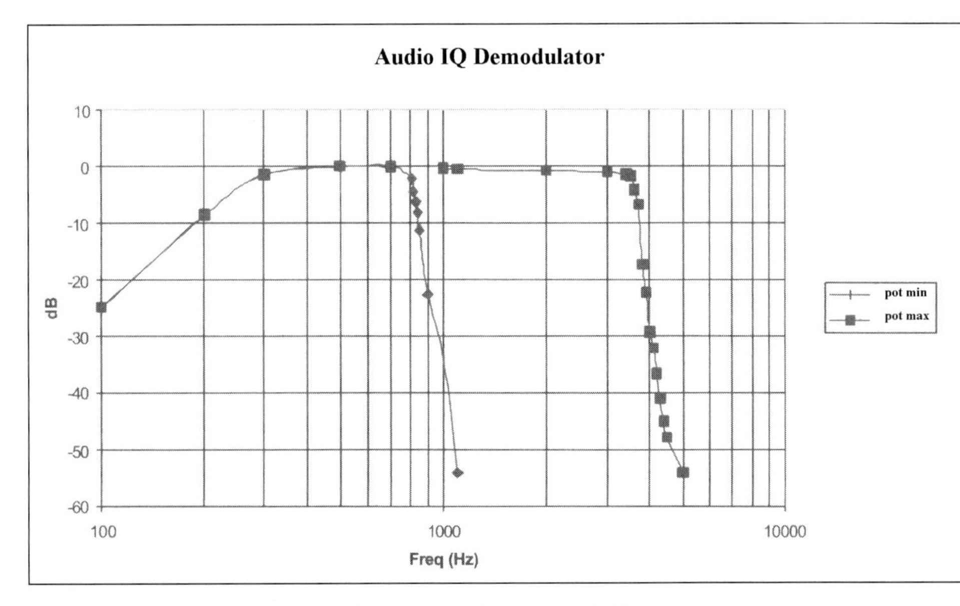


Fig 9.111: Frequency response of the audio filter of the 144MHz transceiver

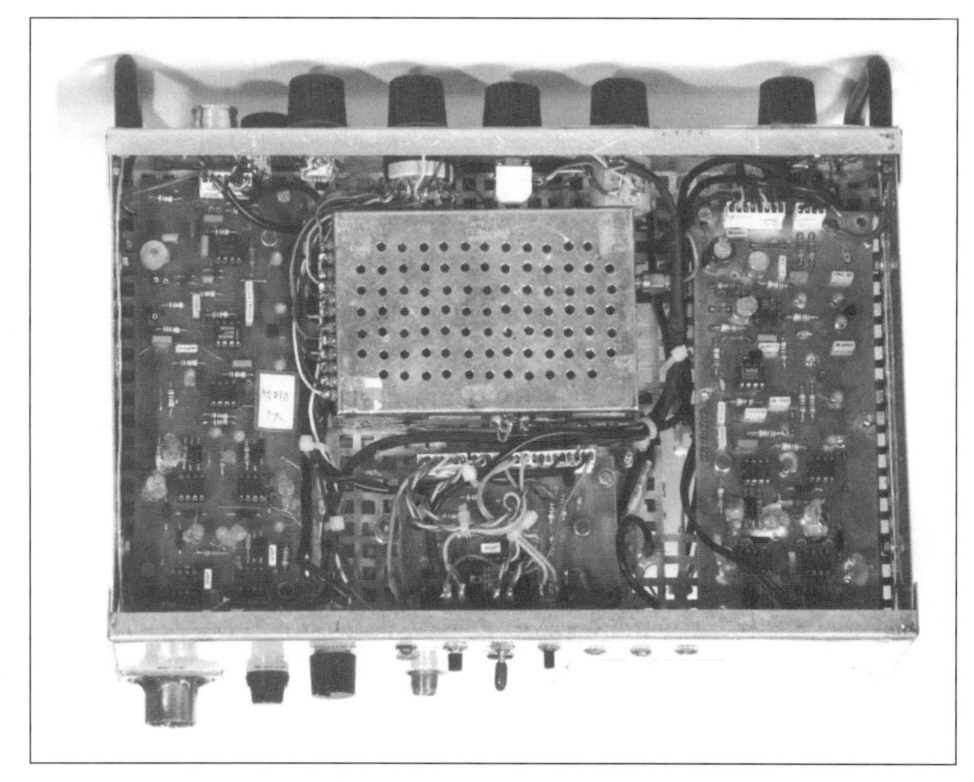

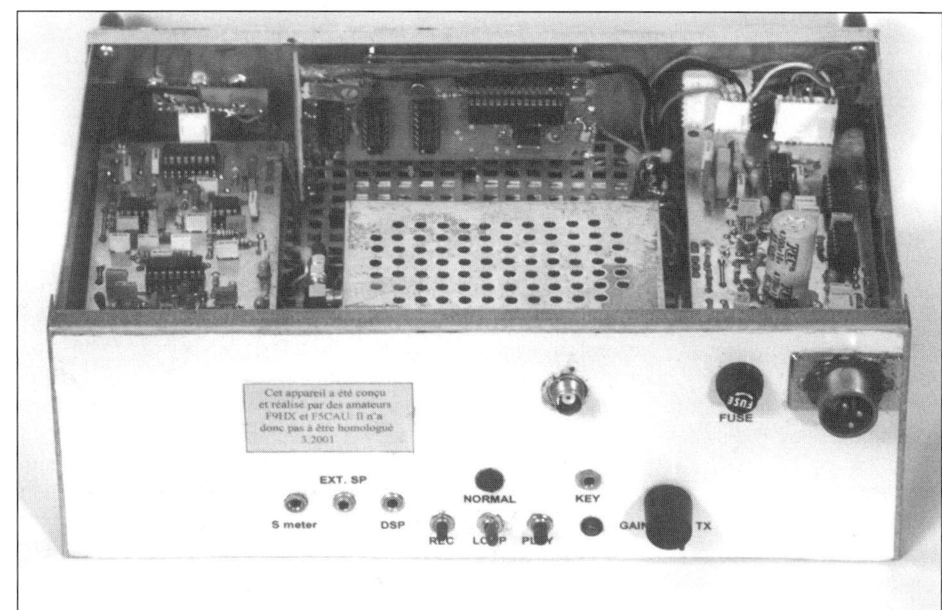


**(Top) Fig 9.112: Underside of the completed 144MHz transceiver**

**(Bottom) Fig 9.113: The top side of the completed 144MHz transceiver**

out in the printed circuit board for the BNC connectors. When the board has been soldered to the sides of the housing, the actual assembly can be undertaken. The boards are fitted into the housing so that the connector pins of the RF connectors are level with the surface of the PCB (cut off projecting Teflon collars with a knife first). When the mechanically large components (filter coils, trimmers, crystal and ring mixer) have been fitted it must be possible to fit the housing cover without any obstruction.

When the equipment is used for the first time, the following test equipment should be available: multimeter, frequency counter, diode probe, wattmeter and received signal (eg beacon). First the crystal oscillator is tuned by adjusting L1, the power consumption should be approximately 65mA. In transmit modes the only adjustment required is to tune the three-pole filter, the power consumption should be approximately 130mA. In receive mode the input filter should be tuned using a weak signal eg a beacon, then the 28MHz filter, C1 / L2, should be tuned.

A power amplifier can be added to the 2m transverter to increase the output power to 20W. The amplifier uses a Mitsubishi hybrid module, these are still available from some suppliers and can be found in surplus equipment.

**Fig 9.118** shows the relatively simple circuit for the 144MHz power amplifier. The core of the circuit is a Mitsubishi M57727 hybrid module (IC1). This module operates at a working voltage of 12 volts. With exactly 27dB amplification, the transverter signal is raised to an output voltage of 20W. The output power to input power ratio is shown in **Fig 9.119**. The current consumption of the module is directly proportional to this ratio.

Such amplifier modules are constructed using thick film technology. This module is designed for the 144 - 148MHz frequency range, and the amplification is achieved in two stages. **Fig 9.120** shows what a typical module looks like from the inside. The 50 ohm input and output matching circuits are clearly visible.

A filter (**Fig 9.121**) on the output provides the harmonic suppression required. Amazing suppression is obtained using only two pi filters wired together. **Fig 9.122** shows the output of the PUFF CAD design software used to design the filter.

The amplifier is assembled on a double-sided printed circuit board measuring 54mm x 108mm. The board can fit into a standard tinplate housing (55.5mm x 111 mm x 30mm), the parts list is shown in **Table 9.8**. A suitable sized hole is sawn out for the hybrid module. Fastening holes are drilled along the edge as shown in **Fig 9.123** (in Appendix B). Good earth connections are essential for the circuit to operate correctly. The through contacts required are made by the M3 screws that secure the assembly to the heat sink. The BNC connectors are placed at suitable points on the side wall of the housing. Also positioned in the side wall is the feed-through capacitor for the power supply. The components are not inserted until the board has been soldered to the sides of the housing. The board should be fitted at the edges of the sides, so that the amplifier module will lie flat on the heat sink.

The two coils (L1, L2) and the coupling capacitor, CK, are hand

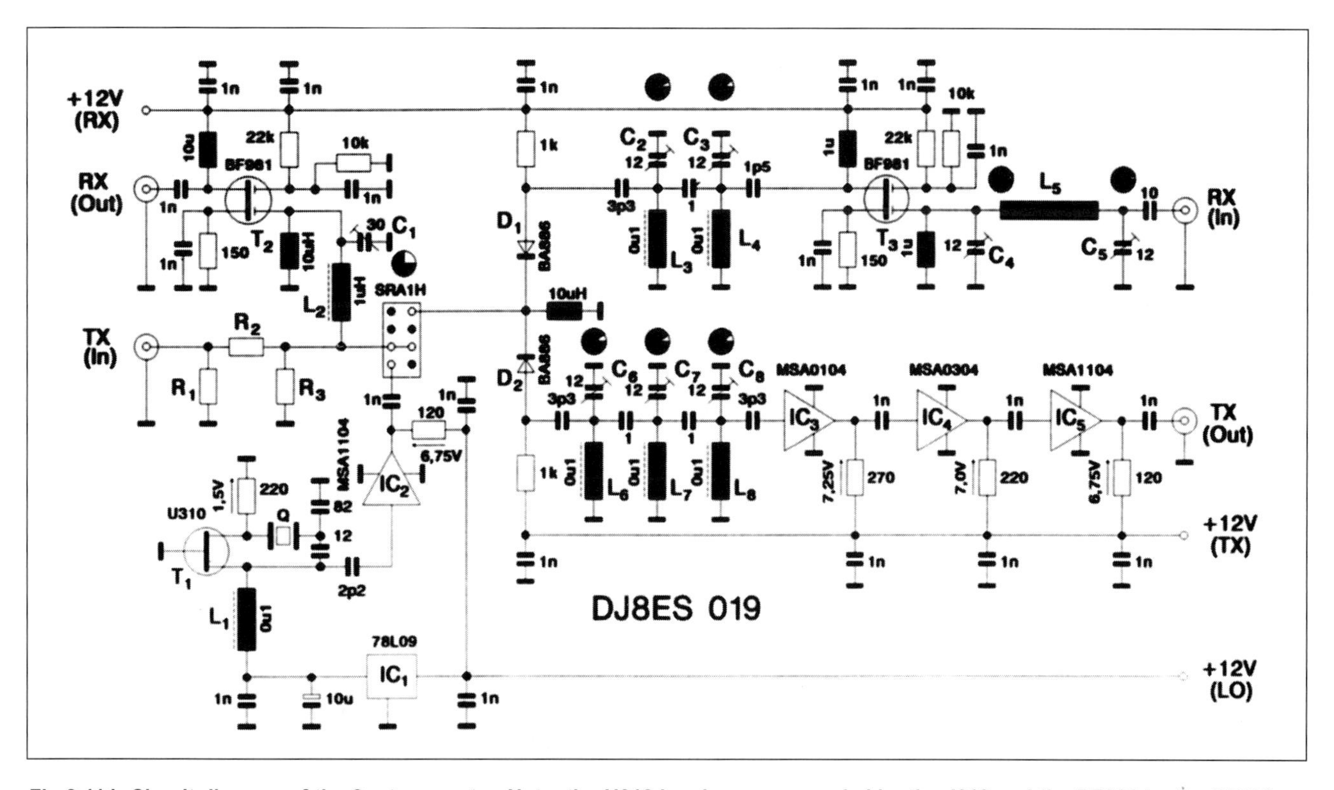


**Fig 9.114: Circuit diagram of the 2m transverter. Note: the U310 has been superseded by the J310 and the BF981 by the BF992**

made. The coils are 8.5 turns of silvered copper wire with a diameter of 1mm. The wire is wound around a 6mm mandrel (eg a 6mm drill shank) and soldered on with a 1mm clearance from the PCB. The coupling capacitor, $C_K$, is made from a 1cm long piece of coaxial cable (RG174), the length is chosen to give the required 1pF capacitance. A standard chip capacitor cannot be used here, due to the relatively high power level. A thin copper plate is soldered between the two pi filters for screening, see Fig 9.123 (in Appendix B). Finally, the remaining components are added. The module is screwed directly onto the heatsink using two M4 screws after applying heat conducting paste.

A power meter and a multimeter are required for putting the equipment into operation. The quiescent current should be approximately 400mA, which rises to around 2.5A under full drive with an input power of 60mW. This gives an output power of the order of 18W.

Only the low-pass filter (C1, C2) requires tuning in the hybrid amplifier. The trimmers are normally screwed about half way in when the unit is correctly tuned. In order protect the hybrid module, carry out this tuning procedure with only a low drive power level (max 10mW).

## 6m Transverter

| Pin | dB | R1 in | R2 in | R3 in |
|---|---|---|---|---|
| 1mW | 0 | – | 0 | 51 |
| 2mW | 3 | 300 | 18 | 300 |
| 5mW | 7 | 120 | 47 | 120 |
| 10mW | 10 | 100 | 68 | 100 |
| 20mW | 13 | 82 | 100 | 82 |
| 50mW | 17 | 68 | 180 | 68 |
| 100mW | 20 | 62 | 240 | 62 |

**Table 9.6: Resistance values for attenuator used in 6m, 2m, and 70cm transverters**

A transverter for the 6m band can be produced based on the 28/144MHz transverter described above [36]. All that is required is modification of the oscillator and the filter.

**Fig 9.124** shows the complete circuit for the 28/50MHz transverter. The circuit can be assembled using the printed circuit board used for the 2m transverter. The pi filter at the input of the receiver needs to be altered; **Fig 9.125** (in Appendix B) shows details of the modification. All the coils and some of the

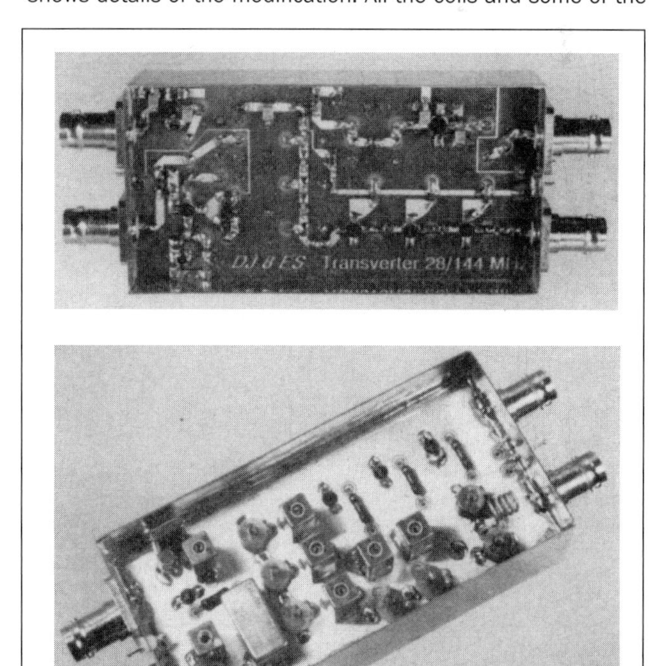

**Fig 9.115: The completed 2m transverter**

capacitors have different values for the lower frequency range, **Table 9.9** show the parts list. To make it easier to produce the 6m version of the transverter, the layout of the printed circuit board with the appropriate components for the 50MHz version is illustrated in **Fig 9.126 and Fig 9.127** (both in Appendix B).

To increase the power output an M57735 hybrid module is used in a separate amplifier stage for the 6m band, the circuit diagram is shown in **Fig 9.128**. The M57735 module was developed for use around 50MHz and is still available from some suppliers or in surplus radio equipment. About 10W can be expected at the output of the PA from the 50mW output from the transverter.

The low-pass filter provides the harmonic filtration required. Only components of appropriate quality (eg air-core coils and air-spaced trimmers) should be used here. The 50MHz amplifier can be assembled on the printed circuit board used for the 2m version using the same construction techniques. The parts list is shown in **Table 9.10**.

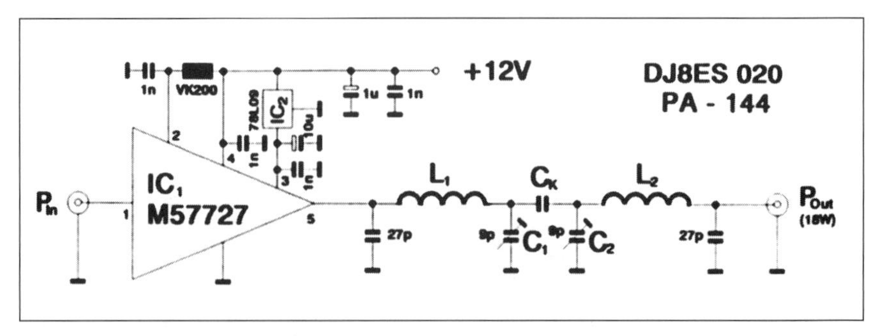


Fig 9.118: Circuit of the 2m power amplifier to be used with the 2m transverter

## 70cm Transverter

The following design for a 28/432 MHz transverter [37] is similar to the 28/144MHz transverter described above. It uses two boards; the oscillator and the transverter **Fig 9.129** shows a picture of the completed units. It should be possible for the transverter to directly feed a standard power amplifier.

Using wide-band amplifier ICs and a ring mixer makes the circuit very flexible, by just changing the filters and the crystal oscillator, the tuning range can be changed to suit the requirements.

**Fig 9.130** shows the circuit of the local oscillator, it uses a U310 as the crystal oscillator at 101MHz. The 404MHz required for the local oscillator is produced using a quadrupler. A printed circuit 2-pole filter provides the necessary filtering. Two wide-band integrated amplifiers, MSA0404 (IC1) and MSA1104 (IC2) supply the desired output of 50mW. The correct level of amplification is important, only the amplification, which is actually necessary,

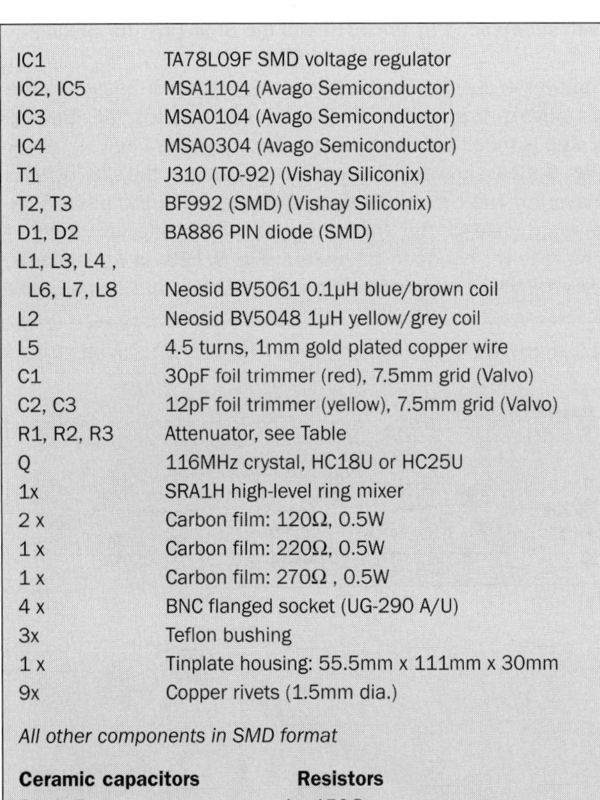

| | |
|---|---|
| IC1 | TA78L09F SMD voltage regulator |
| IC2, IC5 | MSA1104 (Avago Semiconductor) |
| IC3 | MSA0104 (Avago Semiconductor) |
| IC4 | MSA0304 (Avago Semiconductor) |
| T1 | J310 (TO-92) (Vishay Siliconix) |
| T2, T3 | BF992 (SMD) (Vishay Siliconix) |
| D1, D2 | BA886 PIN diode (SMD) |
| L1, L3, L4, L6, L7, L8 | Neosid BV5061 0.1µH blue/brown coil |
| L2 | Neosid BV5048 1µH yellow/grey coil |
| L5 | 4.5 turns, 1mm gold plated copper wire |
| C1 | 30pF foil trimmer (red), 7.5mm grid (Valvo) |
| C2, C3 | 12pF foil trimmer (yellow), 7.5mm grid (Valvo) |
| R1, R2, R3 | Attenuator, see Table |
| Q | 116MHz crystal, HC18U or HC25U |
| 1x | SRA1H high-level ring mixer |
| 2 x | Carbon film: 120Ω, 0.5W |
| 1 x | Carbon film: 220Ω, 0.5W |
| 1 x | Carbon film: 270Ω , 0.5W |
| 4 x | BNC flanged socket (UG-290 A/U) |
| 3x | Teflon bushing |
| 1 x | Tinplate housing: 55.5mm x 111mm x 30mm |
| 9x | Copper rivets (1.5mm dia.) |

*All other components in SMD format*

| Ceramic capacitors | Resistors |
|---|---|
| 3 x 1pF | 1 x 150Ω |
| 1 x 1.5pF | 2 x 220Ω |
| 1 x 2.2pF | 2 x 1kΩ |
| 4 x 3.3pF | 2x 10kΩ |
| 1 x 10pF | 2 x 22kΩ |
| 1 x 12pF | **Inductors** |
| 1 x 82pF | 2 x 1µH choke |
| 17 x 1nF | 3 x 10µH choke |
| 1 x 10µF / 20V Tantalum | |

Table 9.7: Parts list for 2m transverter

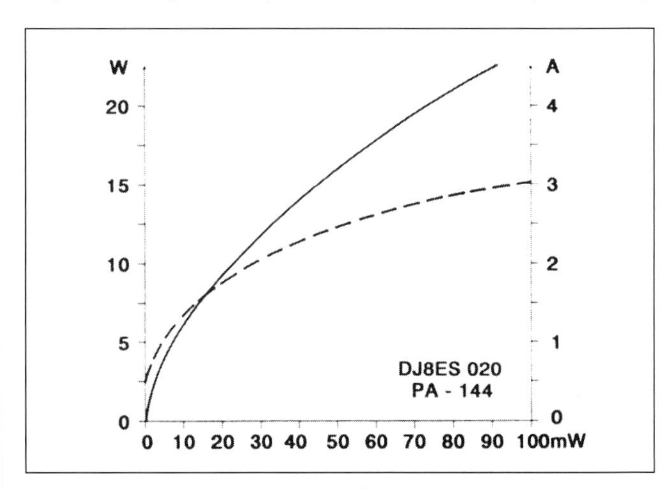


Fig 9.119: Power transfer characteristics of the Mitsubishi M57727 hybrid module

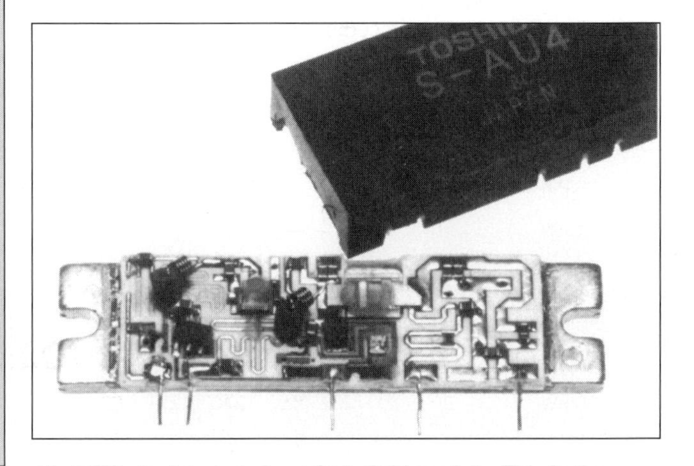

Fig 9.120: An internal view of a hybrid module. This is the Toshiba S-AU4 which is a 70cm amplifier

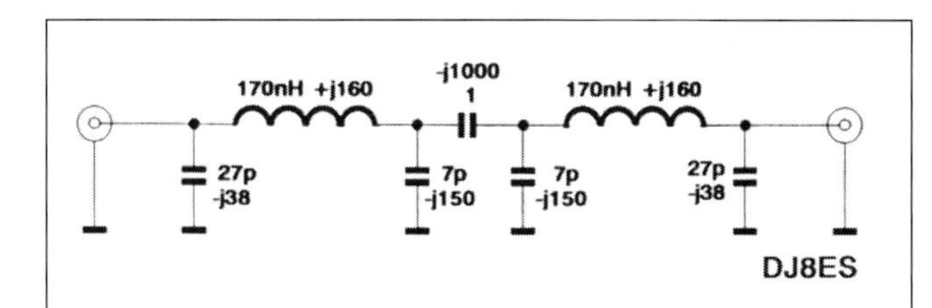


Fig 9.121: Circuit of the filter used in the two metre power amplifier

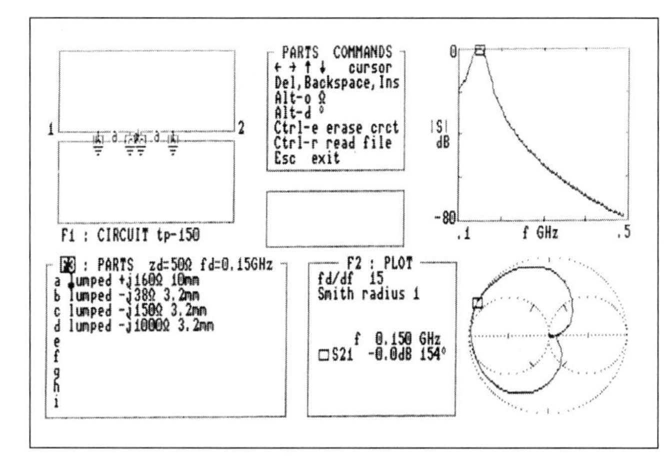


Fig 9.122: PUFF CAD software output used to design the two metre filter

should be used. Any excess increases the spurious outputs.

**Fig 9.131** shows the circuit diagram of the transverter. The SRA1H ring mixer used in the transmit/receive converter is suitable for use up to 500MHz, and requires a local oscillator level of 50mW. The mixer is controlled using an attenuator, which should provide an intermediate frequency (IF) level of no more than 1mW at the ring mixer. The attenuator must be designed on the basis of the IF output available.

Table 9.6 shows the resistor values required for the attenuator

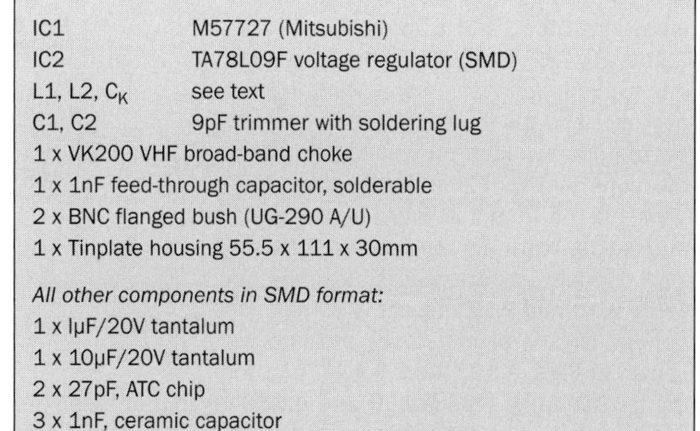

| | |
|---|---|
| IC1 | M57727 (Mitsubishi) |
| IC2 | TA78L09F voltage regulator (SMD) |
| L1, L2, $C_K$ | see text |
| C1, C2 | 9pF trimmer with soldering lug |
| 1 x VK200 VHF broad-band choke | |
| 1 x 1nF feed-through capacitor, solderable | |
| 2 x BNC flanged bush (UG-290 A/U) | |
| 1 x Tinplate housing 55.5 x 111 x 30mm | |
| *All other components in SMD format:* | |
| 1 x lµF/20V tantalum | |
| 1 x 10µF/20V tantalum | |
| 2 x 27pF, ATC chip | |
| 3 x 1nF, ceramic capacitor | |

Table 9.8: Parts list for 2m hybrid amplifier

in relation to the IF power level. All the values are based on the standard values from the E12 to E24 ranges. The attenuator also serves as a wide-band 50-ohm termination for the ring mixer (SRA1H). The received signal is matched at high impedance to the CF300 (T3) using L4 and C3. This low-noise transistor stage provides the necessary intermediate frequency amplification.

The 70cm received signal is passed to the gate of the BF994 (T4) through a pi filter (aerial impedance 50 ohms) that is followed by an MSA0304 amplifier. When the receive +12V power supply is connected, the PIN diode D1 (BA479) is biased on and the signal passed through.

The printed circuit 3-pole 70cm filter is used for both receive and transmit. The transmit signal initially passes through the filter and diode D2 which is biased on. The subsequent amplifier uses integrated wide-band amplifiers (IC6, IC7, IC8). The combination of MSA0104, MSA0304 and MSA1104 provides an output of 50mW (+17dBm) with 40dB of amplification.

In practical operation, such transverters are used with the same driving unit; an additional filter for harmonics and spurious transmissions is recommended. **Fig 9.132** shows a possible two-pole bandpass filter. It can be assembled as an air core con-

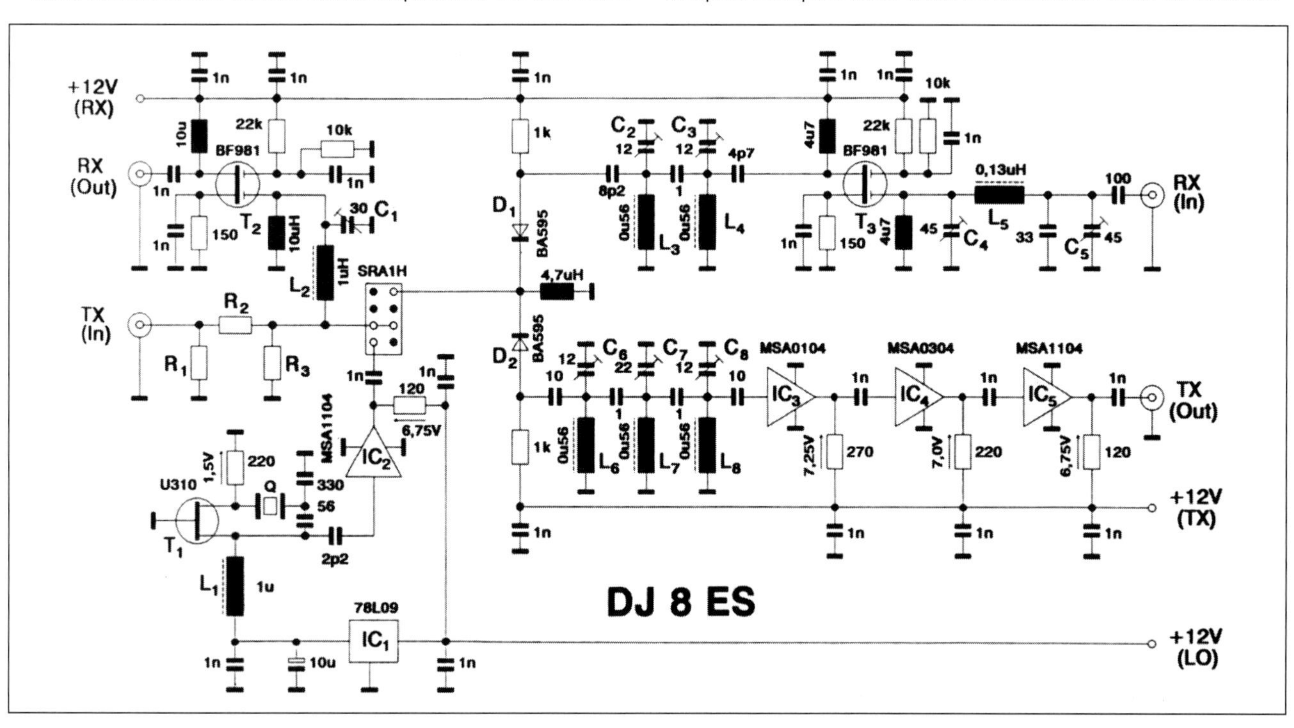


Fig 9.124: Circuit diagram of the 6m transverter. Note: the U310 has been superseded by the J310 and the BF981 by the BF992

struction using a standard tinplate housing measuring 55.5 x 111 x 30mm.

The 28/432 MHz transverter is divided into two independent assemblies: the local oscillator and the transmit/receive converter. The double sided printed circuit boards measure 54mm x 72mm for the local oscillator and a 54mm x 108mm for the transmit/receive converter. The parts list for the local oscillator and transverter are shown in **Table 9.11 and 9.12** respectively. The component layout for the local oscillator is shown in **Figs 9.133 and 9.134**, and for the transverter in **Figs 9.135 and 9.136** (all four of these are in Appendix B). The PCB are mounted in standard tinplate housings, suitable holes are drilled for the stripline transistors and the wide-band amplifiers; these components are mounted level with the surface of the board. The holes for the crystal, trimmers and Neosid coils, etc are drilled on the earth side of the boards (fully coated side) using a 2.5mm drill. Suitable slots are to be sawn out in the printed circuit board for the SMC or BNC connectors. The same applies to the 1nF capacitors at the source connection of the amplifier transistors, T3 and T4. When the individual boards have been soldered to the sides of the housing, the actual assembly can be undertaken. The boards are fitted into the housing so that the connector pins of the RF connectors are level with the surface of the PCB (cut off projecting Teflon collars with a knife first). When the mechanically large components (filter coils, trimmers, crystal and ring mixer) have been fitted it must be possible to fit the housing cover without any obstruction.

When the equipment is used for the first time, the following test equipment should be available: Multimeter, Frequency counter, Wattmeter and Received signal (eg beacon). The assemblies switched on one after another.

Firstly, the crystal oscillator is set to its operating frequency of 101MHz by adjusting coil, L1. The onset of oscillation results in a slight increase in the collector current of T2 (monitoring voltage drop across 100 ohm resistor). A frequency counter is loosely coupled and the oscillator frequency measured.

The two-pole filter after the quadrupler T2 (BFR90a) filters out the 404MHz frequency required. To adjust this, the two trimmers, C1 and C2 are adjusted one at a time for maximum output. The local oscillator should supply an output of at least 50mW. The current consumption for an operating voltage of +12V is about 120mA.

The transmit branch of the transmit/receive converter is put into operation first. Only the three-pole filter (C4, C5, C6) has to be adjusted. A current of approximately 130mA should be measured for an operating voltage of +12V. This is an indication that the amplifier stages are operating satisfactorily. If the input attenuator is selected as described in Table 9.6, an output greater than 50mW can be expected. Possible spurious outputs (oscillator, image frequency, etc.) are suppressed by better than 50dB.

The receiver can be calibrated using a strong received signal (eg a beacon). Because the same filter is used as in the transmit branch, the beacon signal should be audible immediately. Another filter is used at the intermediate frequency (28MHz) after the mixer. The trimmer C3 should be adjusted to give maximum signal output. Optimising the signal-to-noise ratio using the pi filter, C7, C8 and L8 completes the calibration. The current consumption of the receive converter is very low, only 50mA. The noise figure is approximately 2dB with a conversion gain of approximately 30dB.

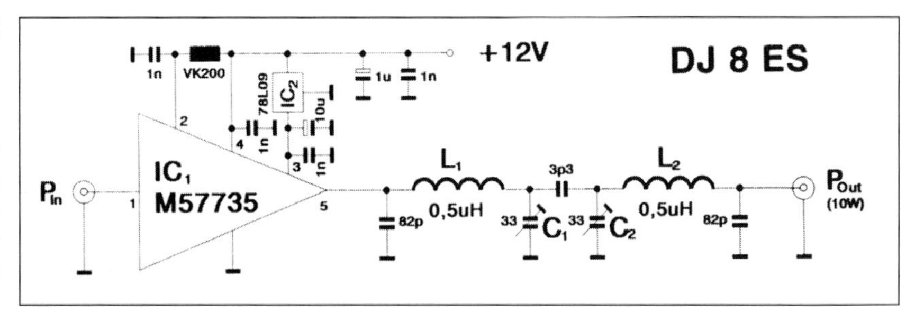


Fig 9.128: Circuit of the 6m power amplifier to be used with the 6m converter

The designer uses the transverter described in association with an external preamplifier and power amplifier. Modern hybrid modules are just the thing for power amplifier stages. The output signal can be increased from 50mW to 10 - 20W in one go using such components. **Fig 9.137** shows the circuit for such a module using a Mitsubishi M55716. A 2C39 valve PA can be fully driven using this 10W output.

## 4m Transverter

This transverter was designed as a club project for the Andover Radio Amateur Club [32]. Various schemes such as transverting from two meters were considered but after some discussion, it was decided to transvert from 27MHz or optionally 28MHz. Using 27MHz as the drive source was particularly attractive for a number of reasons. Many members own or could cheaply obtain a 27MHz CB rig which would mean that the completed transverter could be permanently connected to a dedicated drive source without 'tying-up' and restricting the use of equipment used regularly on other bands. This would make it more practical to establish a club frequency that could be monitored whenever one is in the shack. It was felt this would help to build up band activity. Although CB radios have 10kHz channel spacing whilst the 4m band uses 25kHz, the two spacings coincide on a number of frequencies including all the calling frequencies (70.45 FM, 70.26 All Mode and 70.2MHz SSB) and two of the most commonly used simplex FM working frequencies 70.35 and 70.40MHz.

The design of this transverter was not intended to push the frontiers of technology, but to provide a simple, repeatable design based on readily available components many of which could be found in the 'junk box' thus keeping down costs. The circuit diagram for the transverter is shown in **Figs 9.138 and 9.139.**

The receive converter uses Dual Gate MOSFET RF and mixer stages which provide good gain, noise performance and stability. L1 is resonated by the series combination of C1 and C2 to provide the first stage of input filtering, while the ratio C2:C1 provides matching from the 50 ohm input to the higher impedance of Q1 Gate 1. The RF and mixer stages are band-pass coupled (L2 and L3) to improve rejection of unwanted, out of band signals.

The local oscillator uses a third overtone crystal and is shared between the receive and transmit mixers. The choice of crystal frequency depends on the drive source to be used. For 28MHz the crystal frequency is 42MHz, for a CB 27/81 driver a frequency of 42.49875MHz is required while for the newer PR 27/94 CB Rigs a frequency of 43.0850MHz is required. The zener diode ZD1 is an option which may be required to improve stability if the transverter is to be used for SSB or CW operation (Initial tests indicate that it is not necessary so long as a regulated supply is used).

The transmit mixer uses a proprietary doubly balanced mixer to provide additional suppression of the oscillator and driver frequencies and their products. There are a number of different

| | |
|---|---|
| IC1 | TA78L diameter 9F voltage regulator (SMD) |
| IC2, IC5 | MSA1104 (Avago Semiconductor) |
| IC3 | MSA0104 (Avago Semiconductor) |
| IC4 | MSA0304 (Avago Semiconductor) |
| T1 | J310 (TO-92) (Vishay Siliconix) |
| T2, T3 | BF992 (Siemens) |
| DI, D2 | BA595 PIN diode (SMD) |
| L1, L2 | BV5048 Neosid coil,1μH, yellow/grey |
| L3, L4 | BV5036 Neosid coil, 0.58μH, orange/blue |
| L5 | BV5063 Neosid coil, 0.58μH, blue-orange |
| L6, L7, L8 | BV5063 Neosid coil, L8 0.58μH, orange/blue |
| C1 | 30pF foil trimmer (red) 7.5mm grid (Valvo) |
| C2, C3 | l2pF foil trimmer (yellow) 7.5mm grid (Valvo) |
| C4, C5 | 45pF foil trimmer (violet) 7.5mm grid (Valvo) |
| C6, C8 | 12pF foil trimmer (yellow) 7.5mm grid (Valvo) |
| C7 | 22pF foil trimmer (green) 7.5mm grid (Valvo) |
| Q | 22MHz crystal, HC18U or HC25U |
| 1 x SRA1H ring mixer | |
| 2 x 120Ω / 0.5W Carbon film | |

1 x 220Ω / 0.5W Carbon film
1 x 270Ω / 0.5 W Carbon layer
4 x BNC flanged connector (UG-290 A/U)
3 x Teflon bushing
1 x Tinplate housing 55.5 x 111 x 30mm
9 x 1.5mm dia. Copper rivets

*All other components are SMD format:*

| **Ceramic capacitors** | **Resistors etc** |
|---|---|
| 3 x 1pF | 1 x 150Ω |
| 1 x 2.2pF | 2 x 220Ω |
| 1 x 4.7pF | 2 x 1kΩ |
| 1 x 8.2pF | 2 x 10kΩ |
| 2 x 10nF | 2 x 22kΩ |
| 1x 33pF | 1x 10μF / 20V Tantalum |
| 1 x 56pF | 3x Choke, 4.7μH |
| 1 x 330pF | 2x Choke, 10μH |
| 17 x 1nF | |

**Table 9.9: Parts list for the 6m transverter**

mixer units that will operate satisfactorily, but if substituting a different device beware of its pin connections as two different pin layouts are in common use and only the right one will work! Also, some types are not as shown in **Fig 9.140** but have pins 2, 5, 6 and 7 bonded directly to the case. This is not a problem. Note that pin 1 is identified by a different coloured bead that may be lighter or darker than its neighbours.

These doubly balanced mixers require about +7dbm of local oscillator injection and less than 0dBm (1mW) of Drive (IF) for optimum IMD performance. The output of the oscillator is loosely coupled to L6 providing a band-pass arrangement to reduce unwanted harmonics. C19 and C20 in series resonate L6 while their ratios provide an impedance transformation to match the oscillator output to the mixer (U1) at 50 ohms.

The input from the driving 'rig' is first coupled through C42 to a voltage doubling detector D3/D4 which acts as an RF sensor driving Q7 to operate the change over relays RL1 and RL2. Note that the choice of device for Q7 was based largely on the need for a high current gain ($H_{fe}$), the specified BC548C having an $H_{fe}$ min. of 420. Once a sufficient level of RF is present at the input, Q7 turns ON and relays RL1 (and RL2) close routing the RF through an attenuator, R29, R30 and R31 to reduce its level to about 0dbm (1mW). Please note that RF switching is generally acceptable for FM but can be problematic for SSB operation so direct switching is recommended for this mode. There are two options for external switching: a positive voltage on Transmit applied to the EXT PTT + input or a contact closure to ground applied to EXT PTT LO. Note that the latter should be from voltage free contacts or possibly through a series diode.

Assuming that a standard CB rig is being used with an output of 4 watts, the attenuation required is 36dB. The values shown for these resistors produce about 36dB of attenuation. If you wish to use a different input level, you will need to change these values, Table 9.13 shows some common values. Bear in mind that R29 will dissipate the bulk of the power output from the driving rig and should be rated accordingly. This design uses a TO220 style non-inductive power resistor for R29, and this is bolted to the front panel and hence chassis of the transverter to dissipate the heat. Although rated at 20W, a maximum drive level of 10 watts should not be exceeded and if you intend to have long overs using FM, you should keep to a maximum input level of 4 watts or the case will get mightily hot!

The mixer is band pass coupled (L7/L8) to a pre driver stage Q4 which is in turn band-pass coupled (L9/L10) to the driver Q5. Note that space has been provided for a trap (L17/C43) from the base of Q4 to ground. This trap which is expected to be necessary only if a 144 or 145MHz IF is used should be resonated at the LO frequency (74/75MHz). With a 28MHz IF and no trap, the local

| | |
|---|---|
| IC1 | M57735 (Mitsubishi) |
| IC2 | TA78L09F voltage regulator (SMD) |
| L1, L2 | 0.5μH air-core coil |
| C1, C2 | 33pF trimmer with soldering lugs |
| 1 x VK200 VHF wide-band choke | |
| 1 x 1nF feed through capacitor, solderable | |
| 2 x BNC flanged bush (UG-290 A/U) | |
| 1x Tinplate housing 55.5 x 111 x 30mm | |
| 1x 1μF / 20V Tantalum | |
| 1x 10μF / 20V Tantalum | |
| 1x 3.3pF, ATC chip | |
| 2x 82pF, ATC chip | |
| 3x 1nF, ceramic capacitor | |

**Table 9.10: Parts list for 6m hybrid amplifier**

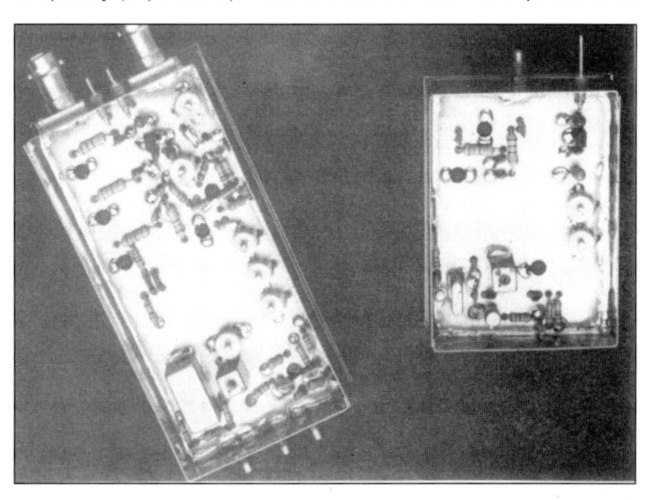

**Fig 9.129: The completed local oscillator and transverter units of the 70cm transverter**

**Fig 9.130: Circuit of the local oscillator used in the 70cm transverter**

**(below) Fig 9.131: Circuit diagram of the 70cm transverter. Note that the CF300 has been replaced by a BF994 and the U310 by a J310 (see parts list)**

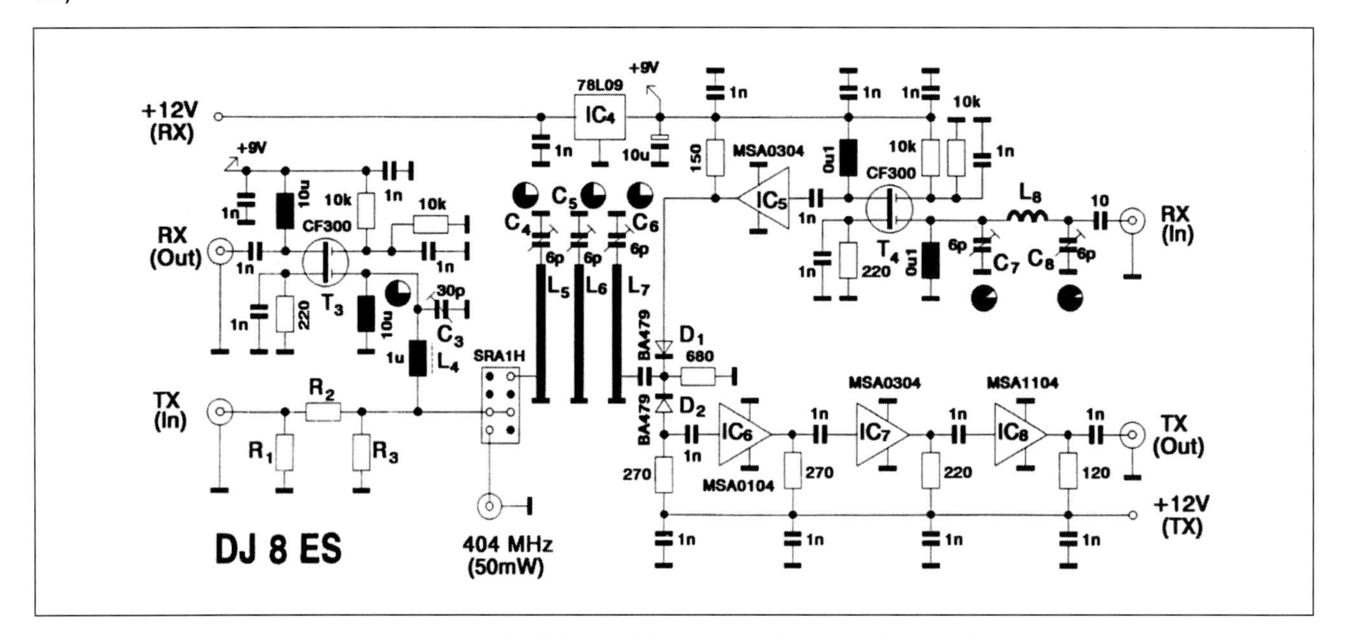


oscillator output level was measured at about 47dB below full output power which was felt to be acceptable. As in the receive part of the transverter, split capacitance is used (C21/C22, C24/C25 and C30/C31) to effect impedance matching.

The Driver and PA devices were chosen on the basis of price and availability. The PA is characterised for FM (Class C) operation (5W at 150MHz) but it was felt that there was a good chance that it could be made to operate satisfactorily in Class B and that would allow the transverter to be used for SSB (and AM) if required.

The first prototype appeared to provide about 4W and this power level was considered entirely adequate for the local activity that this project was designed to stimulate.

Subsequent improvements to the layout yielded the full 5W. For those wanting higher power, the layout provides for connection of an external power amplifier within the RF switching provided. Please note, however, that relays with a higher contact current rating may be needed to perform the DC switching of such a PA as their absolute maximum limit is 2A DC and in practice this should not be closely approached.

Space on the PCB has been left free for additional output filtering which it was thought may be necessary to reduce the harmonic output from the -35dB level observed on the prototype. Although not absolutely necessary at the 5 watt level, it would nonetheless represent good practice and tests suggest that this filter reduces harmonics to about 60dB down on the wanted signal at full power. Tests on a sample filter suggest a loss of about 0.85dB reducing the output from 5W to 4W. In real terms this is insignificant. If a PA is to be fitted a filter is essential and should be fitted to its output.

PCB assembly commences with the smallest components first as placement of the larger components makes it difficult to reach and inspect the smaller ones. Start with the pins, whose

**Fig 9.132: 70cm bandpass filter that can be used with the 70cm transverter**

| | |
|---|---|
| IC1 | MSA0404 (Avago Semiconductor) |
| IC2 | MSA1104 (Avago Semiconductor) |
| IC3 | 78L09 voltage regulator |
| T1 | J310 (TO-92) (Vishay Siliconix) |
| T2 | BFR90a (Valvo) |
| L1 | Neosid BV5061 0.1µH blue/brown coil |
| L2, L3 | λ/4 stripline, etched |
| C1, C2 | 6pF foil trimmer (grey), 7.5mm grid (Valvo) |
| Q | 101MHz crystal, HC18U or HC25U |
| 1 x | Carbon film: 180Ω, 0.5W |
| 1 x | Carbon film: 120Ω, 0.5W |
| 1 x | SMC or BNC flanged socket (UG-290 A/U) |
| 1 x | Teflon bushing |
| 1 x | Tinplate housing: 55.5mm x 74mm x 30mm |
| 2 x | 1nF trapezoid capacitor |
| 2 x | IOµF 20V tantalum capacitor |
| *Ceramic Capacitors (2.5mm grid)* | *Resistors (1/8W, IOmm)* |
| 1 x 2.7pF | 1 x 100 |
| 1 x 18nF | 1 x 220 |
| 1 x 82pF | 1 x 2.2k |
| 6 x 1nF | 1 x22k |
| 1 x 100nF | |
| *SMD Capacitor (model 1206 or 0805)* | |
| 2 x 1nF | |

**Table 9.11: Parts list for 70cm transverter local oscillator**

| | |
|---|---|
| IC4 | 78L09 voltage regulator |
| IC6 | MSA0104 (Avago Semiconductor) |
| IC5, IC7 | MSA0304 (Avago Semiconductor) |
| IC8 | MSA1104 (Avago Semiconductor) |
| T3, T4 | BF994 (Vishay Siliconix) |
| DI, D2 | PIN diode BA479 |
| L4 | BV5048 Neosid coil, 1µH, yellow/grey |
| L5, L6, L7 | λ/4 stripline, etched |
| L8 | 1.5 turns, 1mm CuAg wire |
| 1x | SRA1H high-level ring mixer |
| C3 | 30pF foil trimmer (red) 7.5mm grid (Valvo) |
| C4, C5, C6 | 6pF foil trimmer (grey) 7.5mm grid (Valvo) |
| C7, C8 | 6pF foil trimmer (grey) 7.5mm grid (Valvo) |
| R1, R2, R3 | Attenuator, see Table |

1 x Carbon film: 120Ω , 0.5W
1 x Carbon film: 150Ω, 0.5W
1 x Carbon film: 220Ω, 0.5W
I x Carbon film: 270Ω, 5W
5 x SMC sockets (some of which may be BNC flanged: UG-290 A/U) (see photo of specimen assembly)
2 x Teflon bushing
1 x Tinplate housing 55.5 x 111 x 30mm
4 x 1nF trapezoid capacitor
2 x 0.1 µH choke, IOmm grid, axial
2 x 10µH choke, IOmm grid, axial
1 x 10µF 20V tantalum

| *Resistors (1/8W, 10mm)* | *Ceramic Capacitors (2.5mm grid)* |
|---|---|
| 2 x 220Ω | 1 x 10pF |
| 1 x 270Ω | 12 x 1nF |
| 1 x 680Ω | SMD Capacitor (model 1206 or 0805) |
| 4 x 10kΩ | 6 x1n |

**Table 9.12: Parts list for 70cm transverter**

| Input Level | Required Attenuation | R30 |
|---|---|---|
| 1W | 30dB | 820 |
| 4W | 36dB | 1.5K |
| 10W | 40dB | 2.7K |

**Table 9.13: Attenuator values for 4m transverter**

| | | | |
|---|---|---|---|
| C3, 4, 6, 8, 13, | | R14 | 33k |
| 14 ,26, 27, | | R1, 5 | 100k |
| 32, 33, 38, | 0.1µF Multi | R2, 6 | 330k |
| 36, 37, 39, | layer | Q3, 4 | 2N2369 |
| 40, 41, 44 | ceramic | Q1, 2 | BF981 |
| C7 | 1pF/1.5pF | Q5 | 2N3866 |
| C23, 29 | 2.7pF | Q6 | 2SC1971 |
| C9 | 3.3/3.9pF | Q7 | BC548C |
| C18 | 5.6pF | U1 | DB Mixer |
| C42 | 10pF | SBL-1 | |
| C16 | 22pF | D1, 2, 3, 4, 5 | 1N4148 |
| C15 | 33pF | ZD1 | Zener 9V1 |
| C17, 19 | 47pF | RL1, 2 | DPCO Relays |
| C5, 10, 12, 28, 30 | 56pF | L1, 2, 3, 7, 8, 9, 10 | Toko style |
| C2, 22, 24 | 82pF | | MC119 3.5T |
| C31 | 120pF | L4 | Toko style |
| C11, 20, 21, 25, 35 | 150pF | | MC119 15.5T |
| C1 | 220pF | | without can |
| C34 | 22 F 25V | L5, 6 | Toko style |
| VC1, 2,3,5 | Variable 22pF | | MC119 9.5T |
| VC4,6,7 | Var 5-60pF | | without can |
| R24,25 | 1R0 2W | 10mm screening cans | |
| R21 | 10R | L11, 12, 13, | |
| R20 | 22R | 14, 15, 16 | Airspaced coils |
| R29 | 47R 20W | XT1 | Crystal, see text |
| R31 | 56R | Enclosure AB10 | Maplin LF11M |
| R4, 8, 12, 16 | 100R | Heatsink | Famell 170-071 |
| R11 | 150R | Sockets BNC | |
| R3, 7, 17 | 220R | Misc. Nuts bolts wire | |
| R22 | 470R | M3 x 6mm PH Screw | |
| R30 | 1k5 | M3 x 10mm Spacer | |
| R18, 23, 27 | 2k2 | M3 x 10mm Screw | |
| R10, 15, 19, 26 | 10k | Nuts + washer + spw | |
| R9, 28 | 22k | Feet | |

**Table 9.14: Parts list for 4m transverter**

positions are marked on the PCB component layout overlay as circles, the PCB is shown in **Figs 9.141 and 9.142** in Appendix B. Push them firmly into place and this is most easily done before any other components have been fitted. The ridges on the pins hold them in place until soldering which need not begin until almost all the components are fitted. Leave the fitting of the following components until much later: MOSFETS Q1 and Q2. The power resistor R29 is not fitted until the unit is assembled in its box, the parts list for the transverter is shown in **Table 9.14**.

When fitting components, pre-form all leads so that the component will sit as close to the PCB as possible except for the self supporting coils which should sit 2mm above the board. Some capacitors such as C3, 7, 8, 9 must be pre-formed to a slightly wider pitch (than 0.1in) as conductors pass between their pins and more clearance is required. Don't cut the leads until you have fitted the component and bent its leads over to about 45 degrees from the board. Leave 1.5-2mm protruding. This will hold them in place until soldering which should not start until all the components except the MOSFETS, Inductors and R29 have been fitted. Note that the PCB layout shows VC5 as a large capacitor, in fact a small (green) one should be fitted in this position. The circuit diagram and parts list are correct. Note that Q5, the driver transistor should be mounted on a T05 transipad.

After fitting the small components, carry out a careful inspection to confirm that everything is correctly placed before soldering. Carefully inspect your soldering to ensure that all the joints are good. Remove all surplus flux using a PCB cleaning solvent and a stiff brush (such as a half-inch paint brush with its bristles cut short) before inspection if possible as surplus flux has been known to mask a bad joint. Modern fluxes are hygroscopic and will absorb moisture from the atmosphere and cause corrosion if not removed.

When all the small components have been fitted it is time to fit the pre-wound inductors. These should be fitted one at a time and one pin 'tack soldered' merely to hold them in place. Note that L17 and C43 are only required if a 2m IF is used.

Next, the self supporting coils should be wound using a mandrel such as a drill bit with the correct diameter (6mm). All coils are specified for close (no) spacing between turns, details for winding the coils are shown in **Table 9.15**. While each coil is still on its mandrel, carefully scrape off the insulation enamel at the ends, then remove it from the mandrel and pre-tin the ends. Next, pre-form the leads to fit the PCB in the appointed space observing the orientation as shown on the component layout. Fit the coils so that they lie 2mm above the PCB (a matchstick makes a useful spacing tool) and bend the leads over at the rear

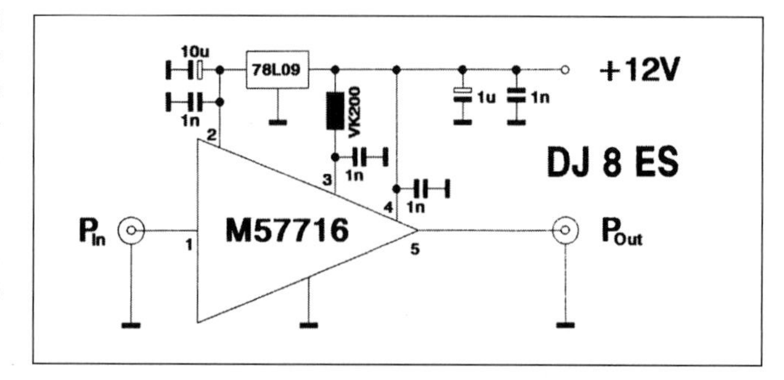


**Fig 9.137: Circuit diagram of the 70cm power amplifier to be used with the 70cm transverter**

of the PCB by about 45 degrees to hold them in place as you did with the other components. Be sure to observe the correct orientation of the axis of the self supporting coils. The sense (clockwise/anticlockwise) in which they are wound doesn't matter. When all inductors are fitted, inspect the PCB again for correct placement then solder them. With the pre-wound coils in particular, solder the unsoldered pin first to avoid them dropping out of the PCB!

Next fit the wire link LK1 and, if you have opted to fit the low pass filter, place a link from the transmitter output to the filter input which is nearby. If you are not fitting the filter, run a miniature co-axial link from the PA stage to the change-over relay RL2 using additional pins where CF4 would have been fitted.

After that, fit and solder the MOSFETS Q1 and Q2 keeping the

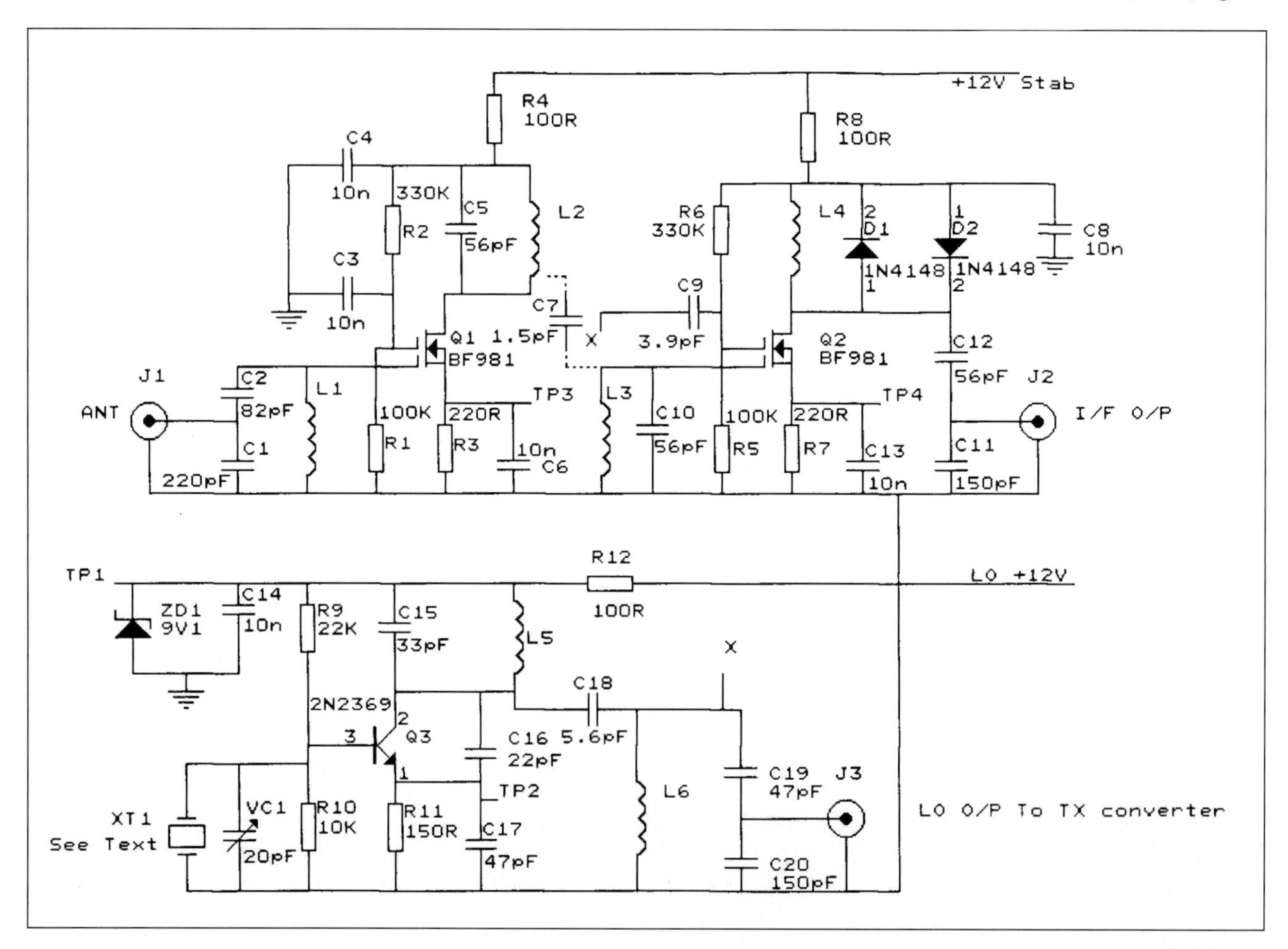


**Fig 9.138: Receive converter circuit diagram for the 4m transverter.** ***[Note: BF981 and 2N2369 are now obsolete devices. A suitable DG mosfet would be the BF992 in SMD and the 2N2369 can be replaced with the MPSH-10 in TO-92 with a slightly different pin out.]***

| | | | |
|---|---|---|---|
| L1,2,3,7,8,9,10 | Cirkit | 35-11934 | |
| L4 | Cirkit | 35-13415 | |
| L5,6 | Cirkit | 35-13492 | |
| L11 | 5t | 22SWG | 6mm dia close wound |
| L12 | 6t | 22SWG | 6mm dia - ditto - |
| L13 | 1t | 22SWG | 6mm dia - ditto - |
| L14 | 7t | 22SWG | 6mm dia - ditto - |
| L15 | 7t | 22SWG | 6mm dia - ditto - |
| L16 | 6t | 22SWG | 6mm dia - ditto - |
| L17 | 2m IF version only. | | |
| LF1,2,3 | 6t | 22SWG | 6mm dia close wound |

**Table 9.15: Coil winding data for 4m transverter**

leads as short as possible and taking sensible anti-static precautions (see below). First pre-form the leads downwards 2mm from the body so that they will pass through the holes on the PCB.

Fit 'tails' of 22SWG Tinned copper wire about 1in (25mm) long to the pins adjacent to the IF socket and 50mm long to the ANT and GND Pins adjacent to RL2 by making one turn tightly around the pin then soldering. Bend these tails up at right angles to the PCB. They will be used to connect to the IF and antenna sockets when the PCB is fitted in its box.

Finally, make a trial assembly of the PCB into its case, using the spacers provided and fit the IF BNC socket. Leave the other BNC socket off at this stage as it will prevent removal of the PCB from the box once fitted. Next, fit R29 in place, bolt it firmly in

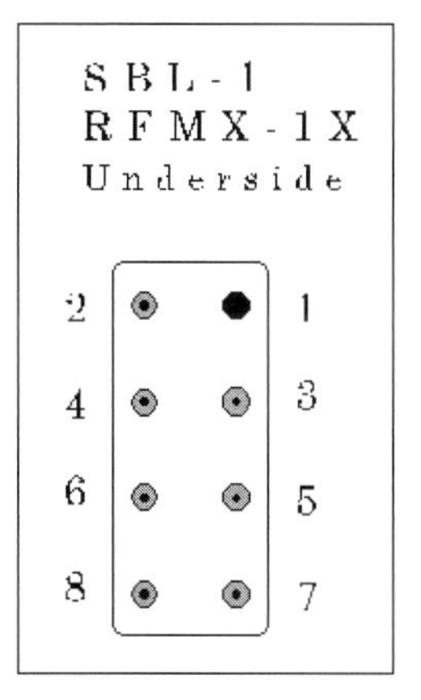


**Fig 9.140: Pin connections for the mixer U1 used in the 4m transverter**

position and 'tack solder' its leads on the top of the PCB. Remember that it will need to be un-bolted from the box when the PCB is removed. Now remove the PCB from the box and trim and solder R29's leads on the underside of the PCB. It is recommended that initial testing is carried out before final assembly into the box.

To test the transverter you will need a multi-meter with an input impedance of 20kΩ/V or better and either an oscilloscope with an input sensitivity of from 10mV per division (bandwidth immaterial) or a millivolt meter. A frequency counter, a general coverage receiver and a 70MHz signal source are also helpful, but not essential. In addition to the basic equipment mentioned, you will need two very simple and useful pieces of test equipment, an RF 'sniffer' and a diode probe to turn your oscilloscope or millivolt meter into a wideband RF level indicator.

Before final assembly into the box, connect a red 14/0.2mm lead to one of the pins at either end of LK1 and a black lead to any convenient point on the ground plane. Carefully apply 13.8 volts DC from a current limited supply. If possible limit to 100mA to avoid damage if there are any serious errors in the construc-

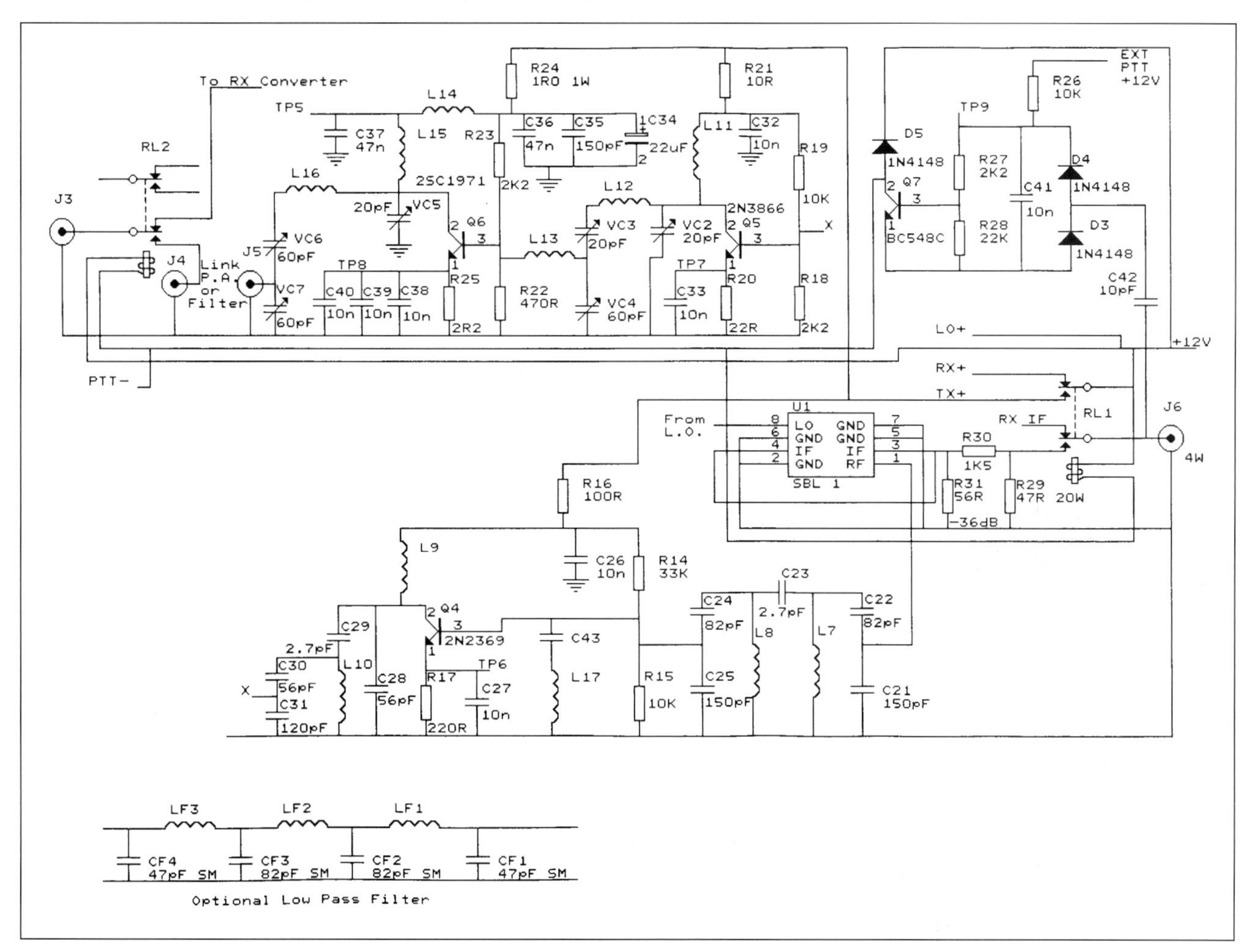


**Fig 9.139: Transmitter converter circuit diagram for the 4m transverter** ***[Note: 2N2369 is now obsolete but it can be replaced with the MPSH-10 in TO-92 with a slightly different pin out. The 2N3866 is now only made in the SMD version (SO-8 package)]***

tion. If you do not have a current limit on your supply or if it cannot be set to 1 amp or less, connect the supply through a resistor of 47 to 100 ohms.

With a voltmeter having an input impedance of 20kΩ/V or more, set to a range which extends to at least 15 volts, check the following points:

| | | |
|---|---|---|
| Local Oscillator supply. | TP1 | 12V |
| Local Oscillator Emitter. | TP2 | 1.8V |
| Q1 Source | TP3 | 0.55V |
| Q2 Source. | TP4 | 0.8V |

With a diode sniffer, tune L5 for maximum deflection, then tune L6. There will be slightly less deflection for L6 but the meter should move significantly. These two circuits interact so repeat the process. Set VC1 so that it is about half engaged and re-tune L5 for the correct frequency. If you have a frequency counter, couple it loosely to L6 using a single or two turn loop and re-tune L5 for the correct crystal frequency. If you do not have a counter don't worry as the oscillator is unlikely to be far out and can be trimmed to frequency using an off air signal later. For fine trimming adjust VC1.

Now increase the current limit on your power supply to 500mA, or reduce the series resistor to about 22 ohms and connect the EXT PTT input (R26) to the +ve supply. You should hear the relays click. Now check the following voltages:

| | | |
|---|---|---|
| TX Supply | TP5 | 13V |
| Q4 Emitter | TP6 | 1.8V |
| Q5 Emitter | TP7 | 0.9V |
| Q6 Emitter | TP8 | 0.3V |

Satisfactory results to these tests give us some confidence that resistors and semiconductors are correctly placed and there are no disastrous short circuits between tracks!

As taking the PCB out of the box to correct errors is very tedious, it was found it best to run through the tune-up procedure with the PCB out of the box first and carry out a final re-tune later after fitting the fully tested PCB into the box.

First, however, fit the 10mm spacers in the box and screw the PCB in place temporarily, Next, fit R29 in place and bolt it firmly to the box before soldering it in place. Now un-bolt R29 and remove the PCB for initial set-up and test. Fit 'tails' of 22SWG tinned copper wire to the IF and Antenna pins and adjacent ground pins then connect the loose BNC sockets to these tails with the ground 'tails' soldered to the large solder tags supplied with the sockets.

Having already tuned the oscillator, the front end, mixer and IF stages (L1, L2 and L3 and L4) can be tuned by connecting a receiver (or transceiver) to the IF socket, and using a strong local signal.

Before tuning the transmit section, you should adjust the output of the driving transceiver to a level compatible with the attenuator R29, R30 and R31. If you are using a CB transceiver, set it to the low power (0.4W) power setting.

The input level to the transmit mixer is just 0dBm (1mw). If you are using an HF transceiver which might be capable of 100 watts or more, test the power level into a dummy load first. Once you have got the level about right, a simple way to check it is by measuring the voltage at TP9. For 4W input it should be about 15-20V and for 0.4W about 6-8V. Remember however that if you apply much too much power you could damage the transverter and you definitely will produce a poor quality signal.

With enough power applied to operate the relays and switch the transverter to transmit, work through the tuned circuits in the transmit path using the diode probe and an oscilloscope, millivolt meter or a 50µA meter movement. Start at the junction of L7, C22 and C23 and tune L7 for maximum. Next move on to the base of Q4 and tune L8 for Maximum. Repeat the process for the base of Q5, tuning L9 and L10 for maximum.

After this, sufficient power should be present in the self supporting coils to give a reading on the RF sniffer. Hold its loop near the circuit to be tuned and tune for 'maximum smoke'. At this stage, a dummy load should be fitted to the output. If you have a power meter, put this in the output circuit. To tune the driver and PA stages a number of variable capacitors need to be tuned and some will interact with others so you should expect to adjust each several times to achieve maximum output. Next, check that the output is at the correct frequency within the 4m band ie between 70.000 and 70.500MHz. This can be done using a frequency counter and a wavemeter, taking a wide sweep on the later to ensure there are no measurable spurious products. Also, check that the output falls to zero when the drive is removed while the transverter is held in transmit mode using the PTT. If it does not, there is self oscillation. Under no circumstances transmit until this problem is solved!

Assuming all is well, you are ready to go on the air for a test.

## REFERENCES


[1] http://www.d-star.asia/index.html.en

[2] http://www.d-staruk.co.uk

[3] 'Gallium arsenide FETs for 144 and 432MHz' John Regnault, G4SWX, *Radio Communication*, Apr 1984

[4] These designs are shown on the web site: www.qsl.net/yu1aw/Misc/engl.htm. The author, Dragoslav Dobricic, can be contacted on: dobricic@eunet.yu

[5] 'Low noise aerial amplifier for 144 MHz', *Radioamater*, Dragoslav Dobricic, YU1AW, 10/1998, pp 12-14 (part I) and 11/1998, pp 12-15 (part II). Also: 'Low noise aerial amplifier for 144MHz', KKE Lecture text, Dec 1998

[6] 'Low Noise aerial amplifier for 144MHz', Dragoslav Dobricic, YU1AW, *CQ ZRS*, Dec 1999, pp 26-31

[7] 'Low noise aerial amplifier for 432MHz', Dragoslav Dobricic, YU1AW, *Radioamater*, 1/2001 and 2/2001

[8] 'Low noise aerial amplifier for 432MHz', Dragoslav Dobricic, YU1AW, *CQ ZRS*, 6/2000, pp 27-31

[9] Diode Communications: www.diodecomms.co.uk

[10] Geoffrey Brown, G4ICD, http://www.qrz.com/db/G4ICD

[11] *VHF Communications Magazine* – www.vhfcomm.co.uk

[12] *Surplus 2-Way Radio Conversion Handbook*, Chris Lorek, pp 85 - 93

[13] SEMELAB, web: www.semelab.co.uk. The data sheet for the D1030UK can be downloaded from www.semelab.co.uk/pdf/rf/D1030UK.pdf

[14] Fritz Dellsperger, Smith chart program and Smith chart tutorials, E-mail: fritz.dellsperger@isb.ch, web: www.fritz.dellsperger.net/

[15] '144MHz direct-conversion receiver with I/Q outputs for use with software defined radio', Andy Talbot, G4JNT, *RadCom*, Nov 2004, pp 102 - 103

[16] 'AD9850 DDS Module', *RadCom*, Nov 2000

[17] *RadCom* series on SSB phasing net works, Feb-June 2004

[18] http://www.weaksignals.com/

[19] Update firmware available from the author, Andy Talbot,

G4JNT: ac.talbot@btinternet.com
[20] *RF Design Guide*, Peter Vizmuller, Artech House
[21] 'Radio amateurs' equipment', F6AWN, *Mégahertz* 2 / 1997
[22] 'What if we replaced our IC-202's?' F9HX, *Hyper*, no. 37, pp 46, 48, 54
[23] 'Simple direct-conversion VHF transmitter-receiver', F1BBU, *Mégahertz* 7 / 2000, 2 / 2001, *Radio-REF* 4 / 2001, 6 / 2001 (sold as kit: see REF-Union shop)
[24] 'What's phasing-system SSB?', Uncle Oscar's notebooks, *Mégahertz* 7 / 2000
[25] 'Direct Conversion Prepares for Cellular Prime Time', Patrick Mannion, *Electronic Design*, 11 / 1999
[26] 'On the Direct Conversion Receiver – A Tutorial', A.Mashour, W.Domenico, N.Beamish, *Microwave Journal*, June 2001
[27] 'The Generation and Demodulation of SSB Signals using the Phasing Method', DB2NP, *VHF Communications* 2 and 3 / 1987
[28] 'Weaver Method of SSB-Generation', DJ9BV, *Dubus*, 3 / 1997
[29] 'No-Tune SSB Transceivers for 1.3, 2.3, 5.7 and 10GHz', S53MV, *Dubus*, 3, 4 / 1997 and 1, 2 / 1998 and *International Microwave Handbook*, ISBN 1-872309-83-6
[30] 'Study of a very low-priced decametric transceiver', F61WF, *Proceedings of CJ*, 1995 and 1996
[31] http://perso.wanadoo.fr/f5cau. Note: the documents on this web site are in French but the PIC code can also be found on the *VHF Communications* web site: http://www.vhfcomm.co.uk
[32] The transverter is described on the 70MHz organisation web site: www.70MHz.org/transvert.htm. The 4 metre transverter design is on the web site: http://myweb.tiscali.co.uk/g4nns/ARACTVT.html
[33] 'Transverters for the 70, 23 or 13cm tuning areas', *1992 Weinheim Congress proceedings*, Wolfgang Schneider, DJ8ES
[34] '28/432MHz Transverter instructions, tips and improvements', *1993 Weinheim Congress proceedings*, Wolfgang Schneider, DJ8ES
[35] '28/144MHz Transverter', Wolfgang Schneider, DJ8ES, *VHF Communications Magazine*, 4/1993, pp 221 - 226
[36] '28/50MHz Transverter', Wolfgang Schneider, DJ8ES, *VHF Communications Magazine*, 2/1994, pp 107 - 111
[37] '28/432MHz Transverter', Wolfgang Schneider, DJ8ES, *VHF Communications Magazine*, 2/1994, pp 98 - 106
[38] Pye museum, run by G8EPR, www.qsl.net/g8mgk/pye/Pye.htm
[39] https://icomuk.co.uk/IC-9700/Amateur_Radio_Ham_Base_Stations

***Andy Barter*** *was an apprentice at ICL in the 1960s where he trained as an electronics engineer. He spent 15 years working in the electronics industry before becoming a computer consultant for a further 20 years. He is now self-employed, publishing the popular VHF Communications magazine and carrying out consultancy work. Andy is also the editor of The International Microwave Handbook, VHF/UHF Handbook, Microwave Projects and Microwave Projects 2.*

*He was licensed as G8ATD in 1966 and has for many years spent his time constructing UHF and microwave equipment for use in contests. Andy lives in Luton, Bedfordshire, UK and is an active member of the Shefford & District Amateur Radio Society, where he is the contest organiser, and The Luton VHF Group, G3SVJ who take part in VHF, UHF and Microwave contests.*

***Chris Waters*** *is an aviation enthusiast who as part of the hobby, got into radio scanning and amateur radio, obtaining his licence in mid-2014. Shortly after, he took an interest in DMR learning about code plug programming as well as taking an interest in the DMR network, which at that time was still in its infancy. He has been writing code plugs as well as assisting amateurs with questions on the mode for many years. He is also in the progress of setting up a website – www.dmr-guide.uk – to provide a central source of information for DMR users in the UK*

# Low Frequencies: Below 1MHz

# 10

*Dave Pick, G3YXM*

The 136kHz band, 135.7kHz - 137.8kHz was introduced in January 1998 and is unique in being in the LF frequency range (Low Frequency, defined as 30kHz - 300kHz). It is available to holders of all classes of UK licence.

In February 2007, the UK licensing authority Ofcom agreed to issue some amateurs with licence Notices of Variation to experiment on frequencies between 501 and 504kHz. Following internationally coordinated negotiations, this allocation was replaced by an internationally-agreed band from 472 to 479kHz which in the UK is available only to Full licensees.

The 136kHz and 472kHz bands each have unique characteristics and are different to all the higher frequency amateur bands. Propagation on 136kHz and 472kHz is very different to the HF bands (for more on this, see the chapter on propagation).

Due to the narrow bandwidth available at 136kHz (a total of only 2.1kHz), the low radiated power level permitted (1W ERP) and the high noise levels present on this band, digital modes like JT9 and Opera are prevalent, and CW has fallen into disuse except for "QRSS" and "DFCW". Some LF DX modes are described in detail in the Digital Communications chapter.

The 472kHz band has a slightly higher radiated power limit of 5W EIRP, but it offers its own challenges, particularly the very deep fading that occurs at intermediate and long distances. There is still some CW operation on this band but digital modes are now the most popular.

## MODES IN USE

Although conventional CW is still used at the lower end of the 472kHz band, most activity takes place using a digital mode of some kind. New modes come and go but the ones in use at the time of writing are:

**QRSS** Not really a digital mode but it does use a computer to send and receive it. It is merely Morse code sent at a very slow rate. For DX each dot might be 60 seconds long and the dashes 180 seconds long! The received signals are displayed on a spectrum display program such as Argo by I2PHD [8] and read from the screen.

**DFCW** A version of QRSS using frequency shift keying to differentiate dots from dashes. Conventionally dashes are sent at a slightly higher frequency than dots. Using this technique the dashes can be the same length as the dots and, within a character, there are no spaces between them. This makes DFCW considerably faster than QRSS.

**WSPR** This is the same mode as used on HF but there is also a slower version suitable for long distance LF work. In the experimental WSPR-X software package you will find WSPR-15 which uses 15 minute send and receive periods instead of the normal WSPR's 2 minutes. It is claimed to be 9 dB more sensitive than WSPR-2, decoding signals as weak as -37 dB in the standard 2500 Hz reference bandwidth. The deep fading on 472kHz can reduce this advantage somewhat. WSPR is a beacon mode, not a QSO mode.

**Opera** This mode is easy to send because it merely keys the transmitter on and off on a single frequency and so doesn't use the sound card or require a linear PA. It was written by EA5VHK and can be found on his web-site [1]. At the receiving end the software can work from an expected list of callsigns and attempt to correlate against that. It indicates whether a decode has been made unambiguously or with use of this information. Opera is intended as a beacon mode.

**JT-9** Part of the WSJT-X suite of programs from Joe Taylor K1JT, JT-9 works in a similar way to FT-8 but with a one minute transmit period. This makes more sensitive than FT-8 and thus better suited for LF QSOs.

**Slow JT9** In the original WSJT-X package it was possible to select various different sub-modes of JT-9. By increasing the time period to as much as 10 minutes it is possible to realise a considerable signal-to-noise gain. Unfortunately this facility has been dropped in the new versions but ON7YD has revived it in his "Slow JT9" program. [2] Rik's version can be set to 1, 2, 5 and 10 minute time slots.

**EbNaut** Written by Paul Nicholson, EbNaut is a UT synchronous coherent BPSK mode for use at VLF and low LF, enabling communications close to the channel capacity. It has enabled trans-Atlantic messages to be received on 8.9kHz using amateur antennas. At faster rates it is also effective on 136kHz. [3]

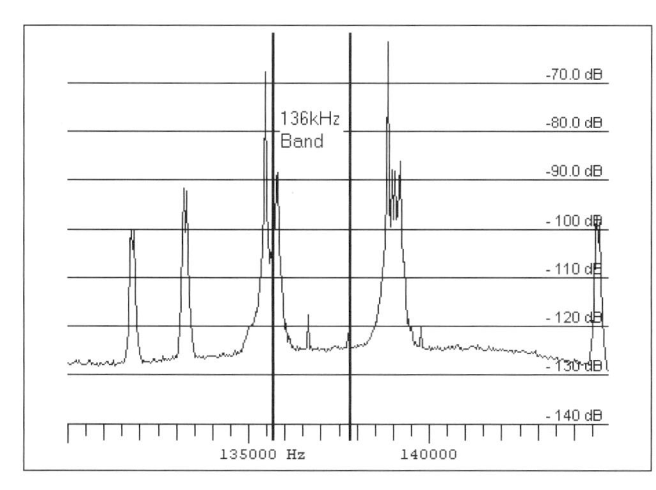


**Fig 10.1: Spectrum in the vicinity of the 136kHz band. The very strong signals just above and below the band are utility stations in Germany and Hungary respectively. Sidebands of their data bursts are audible inside the amateur allocation**

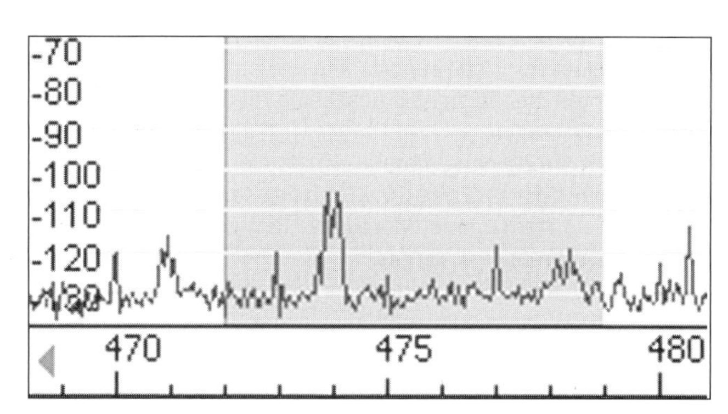


**Fig 10.2: Spectrum in the vicinity of the 472kHz band. From this dusk view it is obvious that there are no strong adjacent signals. Inband, two Polish airport beacons can be seen with carriers at 474.0 and 474.5kHz and CW idents appearing approximately 1kHz either side. Notice the amateur WSPR stations clustered around 475.7kHz**

## RECEIVERS FOR 136KHZ AND 472KHZ

The majority of stations use commercially available receivers. Many amateur HF receivers and transceivers include general coverage that extends to 472kHz and 136kHz, and in many cases these can be successfully pressed into service; however, unlike HF reception, where reasonable results are often achieved simply by connecting a 'random' wire antenna to the receiver input, successful reception on these bands is a bit more difficult, for a number of reasons. First, there is usually a very large mismatch between the impedance of a wire antenna at these frequencies and the typically 50-ohm receiver input impedance, which leads to a large reduction in signal level at the receiver input. This is often exacerbated by degraded receiver sensitivity at 472kHz and below, particularly in amateur-type equipment. Secondly, amateurs with their relatively tiny radiated signals share the spectrum with vastly more powerful broadcast and utility signals; unless effective filtering is provided, this can result in severe problems with overloading at the receiver front end. Fortunately, very satisfactory results can usually be achieved by using quite simple antenna matching, preamplifier, and/or preselector arrangements, as will be seen later in this section.

## LF AND MF RECEIVER REQUIREMENTS

Requirements for LF receivers depend on the type of operation that is envisaged. Adequate sensitivity is obviously required; the internal receiver noise level should be well below the natural band noise at all times. As a guide, similar figures to those for HF receivers (a few tenths of a microvolt in a CW bandwidth) will suffice. If a large, transmitting-type antenna is used, the signal level will be high enough to allow a receiver with considerably poorer sensitivity to be used. Small loop and whip receiving antennas usually require a dedicated low noise preamplifier (see section on receiving antennas and noise reduction).

Because of the narrowness of the bands - 136kHz is only 2.1kHz wide, and 472kHz is 7kHz wide - almost all modes in use occupy less than SSB bandwidth so good IF filtering is important. A 500Hz bandwidth is adequate for CW, but narrower bandwidths can be used to advantage. On 136kHz, there are several strong utility signals just outside (and sometimes inside) the amateur band (**Fig 10.1**), and so a good filter shape factor is important since the utility signals can be 60dB or more above the level of readable amateur signals, with a frequency separation of less than 1kHz.

For specialised extremely narrow-bandwidth modes such as JT9, QRSS and WSPR (see the chapters on Morse and Digital Communications), selectivity is provided at audio frequencies using DSP techniques in a personal computer, but good basic receiver selectivity is still important to prevent strong out-of-band and unwanted inband signals desensing the receiver and entering the audio stages.

In the UK strong non-amateur signals close to, or inside, the 472kHz - band are rare (**Fig 10.2**), making selectivity less critical than on the 136kHz band. More important is rejection of medium wave AM broadcast signals.

Frequency stability requirements depend on the operating mode. For CW operation, maintaining frequency within 100Hz during a contact is not usually a problem. Narrow band digital modes require much better stability, and frequency setting accuracy. The best stability and accuracy is obtained by GPS disciplining of the receiver's master oscillator [6]. This is required for the narrowest bandwidth modes such as EbNaut but a receiver with a decent TCXO type master oscillator will be fine for JT9 and OPERA.

Most modern fully synthesised receivers are equipped with a TCXO, or offer one as an option, and generally exhibit a setting accuracy within 1 or 2Hz and drift a fraction of a Hertz over an extended period. This degree of frequency stability is adequate for the vast majority of applications, including the reception of intercontinental 136kHz beacon signals using QRSS60 and QRSS120 speeds, with bandwidths of as little as 0.01Hz, along with narrowband, weak-signal digital modes such as WSPR and JT9.

## RECEIVERS IN AMATEUR TRANSCEIVERS

Many amateur HF receivers and transceivers can tune to 136kHz and below, and since they are already available in many shacks, probably a majority of 136kHz and 472kHz amateur stations use receivers of this type. All modern equipment is fully synthesised, so frequency accuracy and stability are good. Older receivers using multiple crystal oscillators or an interpolating VFO have relatively poor stability and are not suitable for anything other than CW use. Modern crystal or mechanical CW filters have excellent shape factors, giving good rejection of strong adjacent signals. In some receivers, multiple filters can be cascaded, giving further enhanced selectivity.

Since reception at frequencies below 1.8MHz is generally included as an afterthought, manufacturers rarely specify sensitivity of amateur receivers at 136kHz and 472kHz. Unfortunately it can often be poor. There is little relation between the HF performance, cost, or sophistication of a particular model, and the sensitivity at lower frequencies. Therefore it may well be that older, cheaper models perform better at LF than their newer successors.

Few laboratory-quality sensitivity measurements are available for the LF and MF sensitivity of amateur transceivers and receivers, but the following lists some models which have been used successfully as LF receivers.

Classified as "good" are most recent Kenwood transceivers, the Icom IC-7100 and IC-756 ProIII. Classified as "adequate" are the Icom IC-706, IC-718, IC-7610, IC-761, IC-765, and IC-781, TS-870, Yaesu FTDX-3000D, FTDX-5000MP, FT-817 and FT-1000MP. These may require either a large antenna and/or an external preamplifier to achieve adequate sensitivity. The popular IC-7300 is useable on LF, having good stability and accuracy but poor sensitivity and a high noise floor at MF means that a preselector-preamplifier stage will be required between the antenna and the receiver input.

Some transceivers have a transverter port, or even a low-level transmit output port which will work at LF, making it easy to get a fully-featured LF/MF radio in conjunction with an external home-made transverter or amplifier.

The reason for poor sensitivity lies within the receiver front-end design. The inter-stage coupling components, in particular the first mixer input transformer, are optimised for operation at HF, and often have high losses at LF, reducing the signal level. Internally generated synthesiser noise may also be higher at LF. The front end filter used when the receiver is tuned to 472kHz or below is normally a simple low-pass filter with a cut-off frequency of 1 - 2MHz, often including an attenuator pad to reduce overloading due to medium wave broadcast signals; this further reduces sensitivity, without eliminating the broadcast signals. Some LF operators have improved receiver performance substantially by replacing the mixer input transformer with one having extended low frequency response [7]; this component must be carefully designed if receiver HF performance is to be maintained. A simpler and more common approach is to use an external preamplifier, and provide additional signal frequency selectivity, as described later in this section.

## PROFESSIONAL AND VINTAGE EQUIPMENT FOR LF

Many professional communications receivers made by such firms as Racal, Plessey, Harris, Collins, Eddystone, Rohde & Schwarz and others include coverage of the LF and MF spectrum, and surplus prices are often competitive with the amateur-type equipment discussed above. Ex-professional equipment is usually fully specified at LF and MF frequencies, so sensitivity and dynamic range are usually good at 136kHz and 472kHz. Fully synthesised professional receivers often have precision reference oscillators with excellent stability; they also usually have inputs for an external frequency reference. These features are not often found on amateur-type equipment, making them attractive if the more specialised LF communications modes are to be explored. A drawback is that affordable examples are usually fairly old, so servicing and repairs may be required from time to time. Also, they have a rather Spartan feel, with few of the 'bells and whistles' operator facilities found on modern amateur rigs. The Racal RA1792 has been popular with UK amateurs on 136kHz and 472kHz. The older RA1772 also performs well.

A few amateurs have used vintage receivers, including the HRO, Marconi CR100 and AR88LF for 136kHz or 472kHz. Valve-era equipment designed for marine service also often includes LF and MF coverage. The antenna input circuit of this type of equipment is generally designed to be operated in the lower frequency ranges using an un-tuned wire antenna, and usually gives good sensitivity on these bands without requiring additional antenna tuners or preamplifiers. The major disadvantage of most vintage receivers is their relatively poor stability making them unsuitable for anything other than CW use. Their single-pole crystal filters have poor skirt selectivity compared to modern IF filters and this can result in strong utility signals several kilohertz from the receive frequency reaching the IF and detector stages of the receiver, causing blocking and heterodyne whistles which swamp the weak amateur signals. Unmodified vintage receivers are therefore usually poor performers at 136kHz. As noted above, selectivity is less critical for 472kHz operation, and vintage receivers can perform quite well on CW.

Selective level meters (SLMs), also called selective measuring sets or selective voltmeters, are instruments designed for measuring signal levels in the now-obsolete frequency division multiplex telephone systems; consequently, they are sometimes available surplus at low cost. SLMs are designed for precision measurement of signals down to sub-microvolt levels; their frequency range extends from a few kilohertz into the MF or HF range, so can make effective LF and MF receivers. Well-known manufacturers are Hewlett-Packard (HP3625) and the German companies Wandel and Golterman (the SPM- selektiver pegelmesser series) and Siemens.

SLMs are not purposely designed as receivers and therefore do not have many normal receiver features, such as AGC and selectable operating modes, or sometimes even an audio output. Filter bandwidths are designed for telephony systems and are not always suited for amateur radio operating modes. Normally the 'CW pitch' is fixed at around 2kHz, so they are not well suited to CW operating, although this presents few obstacles for 'sound card' operating modes. The area where SLMs excel is in signal measurement; they have been used by a number of amateurs for 136kHz and 472kHz field strength measurements (see LF Measurements and Instrumentation section). They are often available with a tracking level generator, which is very useful for measurements on filters, or bridge-type impedance measurements.

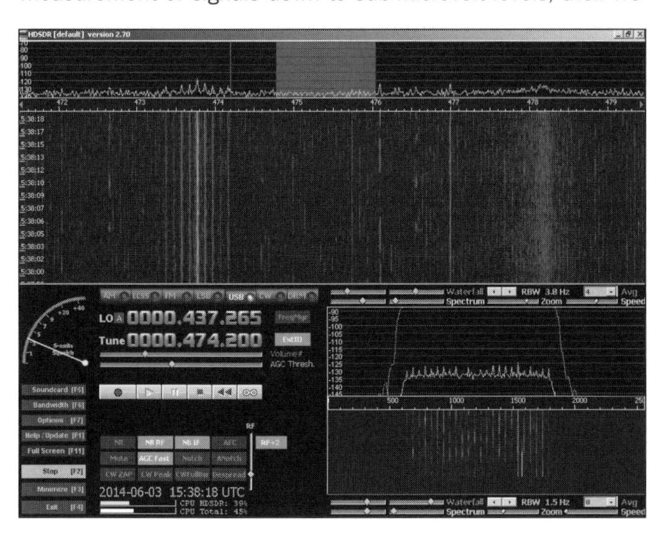


**Fig 10.3: A typical SDR screen shows the entire 472kHz band (top) and the audio spectrum of the WSPR slot (bottom right)**

## MODERN DEDICATED RECEIVERS

### Superhet receivers

Stand-alone receivers are likely to perform well at low frequencies although some don't go down to 136kHz. The Yaesu FRG100 is reported to be sensitive and stable. Tuning well below the 136kHz band, the AOR7030 is useful for LF work with good sensitivity, 10Hz tuning steps and optional 300Hz or 500Hz IF filters. JRC's NRD-345 is sensitive and stable, and has several optional narrow filters; the NRD-525, 535 and 545 are described as very good and the NRD-91 tunes down to 10kHz. The Icom IC-R75 can be fitted with an oven controlled oscillator and is very sensitive; the AGC can be turned off. The Kenwood R75 is getting rather old now but is still a good performer. The AOR AR5000 is a good receiver and can be locked to an external frequency reference. Most will need a preselector or band-pass filter in line to prevent cross-modulation from strong out-of-band signals. Many professional grade receivers cover LF and are available on the surplus market, such as the Racal RA1792 or the RFT EKD-300. They are very good performers right down to the bottom end of their frequency coverage and have high quality reference oscillators.

### Software defined radio receivers

Software defined radios (SDRs) are now part of the mainstream of amateur radio, see the chapter on Receivers. PC-based spectrogram software has been used for several years in conjunction with conventional receivers for the 'visual' LF/MF operating modes such as QRSS and narrow-bandwidth data modes; SDR is the natural extension of this trend, marrying hardware and PC software [8, 9]. Because the 136kHz and 472kHz bands are so narrow, they can easily be monitored in their entirety (**Fig 10.3**) with even the simplest SDR.

General coverage, direct-digitising SDR receivers are now available at reasonable prices and many cover the LF/MF range. For instance the Perseus SDR receiver [11] and the Airspy HF+ Discovery perform extremely well at LF and MF. The relatively low-priced Afedri SDR-NET has produced excellent results on both 136kHz and 472kHz with a resonant antenna and a low pass filter to reduce medium wave broadcast signals. The Elad FDM DUOr is unusual in that it can be used in "stand-alone" mode or with SDR software via USB. It can be locked to an external frequency standard. For less than £100 the SDR-Play RSP-1A receiver includes a MW broadcast band filter which helps with their LF use and even the simplest 'dongle' SDRs are capable of adequate performance, though dynamic range may be an issue.

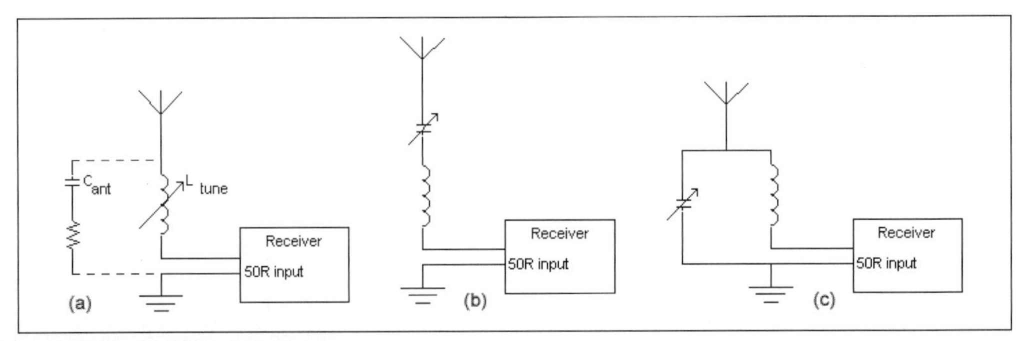


**Fig 10.4: Receive antenna tuning circuits. Note that the arrangements in (b) and (c) are not suitable for transmitting, except for very low power levels**

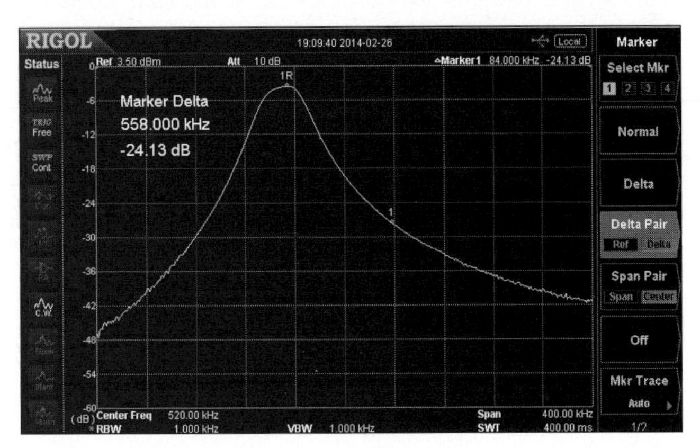


**Fig 10.5: Typical performance of the G0MRF preamp at 474kHz (plot 370kHz to 770kHz)**

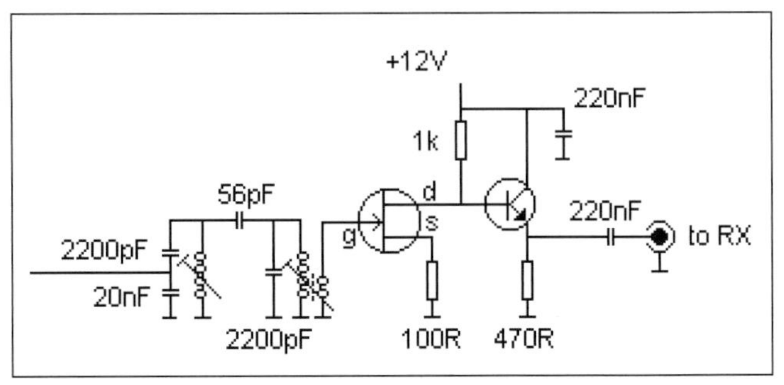


**Fig 10.6: The 136kHz version of the G0MRF preamp**

## RECEIVE ANTENNA TUNING

The impedance of a typical long-wire antenna at LF or MF can be modelled as a series resistor and capacitor. Taking the example in the Transmitting Antennas section of this chapter, a typical long-wire antenna, the capacitance might be 287pF in series with 40 ohms at 136kHz. At 472kHz, the capacitance will be almost the same, but the resistance could be lower, perhaps 20 ohms. Assuming a receiver input impedance of 50 ohms, the SWR at the feed point of the antenna is about 8200:1 at 137kHz! This mis-match results in an unacceptable signal loss of about 32dB. The loss due to mismatch at 472kHz is less severe, but still more than 20dB.

Most of the loss is caused by the capacitive reactance; signal levels can be greatly improved by resonating the antenna at the operating frequency with a series inductance. In the example above, the antenna with resonating inductance form a tuned circuit with Q around 40 and bandwidth of only a few kilohertz, which is very effective in filtering out powerful broadcast band signals.

The practical effect of resonating the antenna is dramatic. Normally with a long-wire antenna connected directly to the receiver, the only signals heard in the 136kHz range are numerous intermods. With the antenna tuned, these disappear and the band noise is audible above the noise floor of reasonably sensitive receivers. Attempts to receive signals at 472kHz with un-tuned wire antennas are more successful at locations where broadcast signal levels are fairly low, but an antenna tuner still yields substantial improvements.

Typical circuits used to tune wire antennas for reception are shown in **Fig 10.4**. **Fig 10.4**(a) is a simple series inductor; the value required is:

$$L_{tune} = \left( \frac{1}{2\pi f \sqrt{C_{ant}}} \right)$$

with $L_{tune}$ in henries, Cant in farads and f in hertz. A useful rule of thumb is that the antenna capacitance Cant will be roughly 6pF for each metre of wire, typically $L_{tune}$ of a few millihenries will be required for 136kHz and a few hundred microhenries at 472kHz. Because of the high Q the inductance must be adjustable; this can be done using the same techniques as for transmitting antennas, or a slug-tuned coil can be used. It is often more convenient to use a fixed

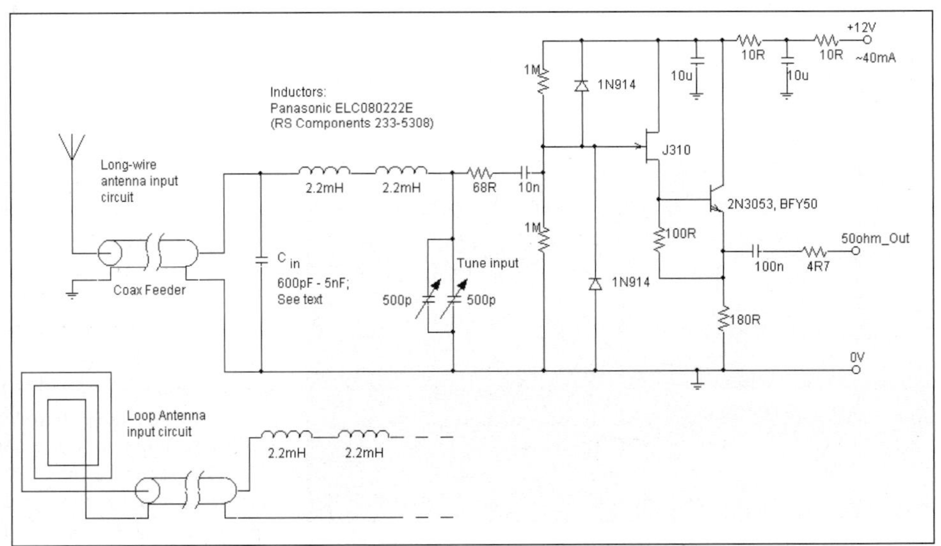


**Fig 10.7: LF Antenna tuner / preamp**

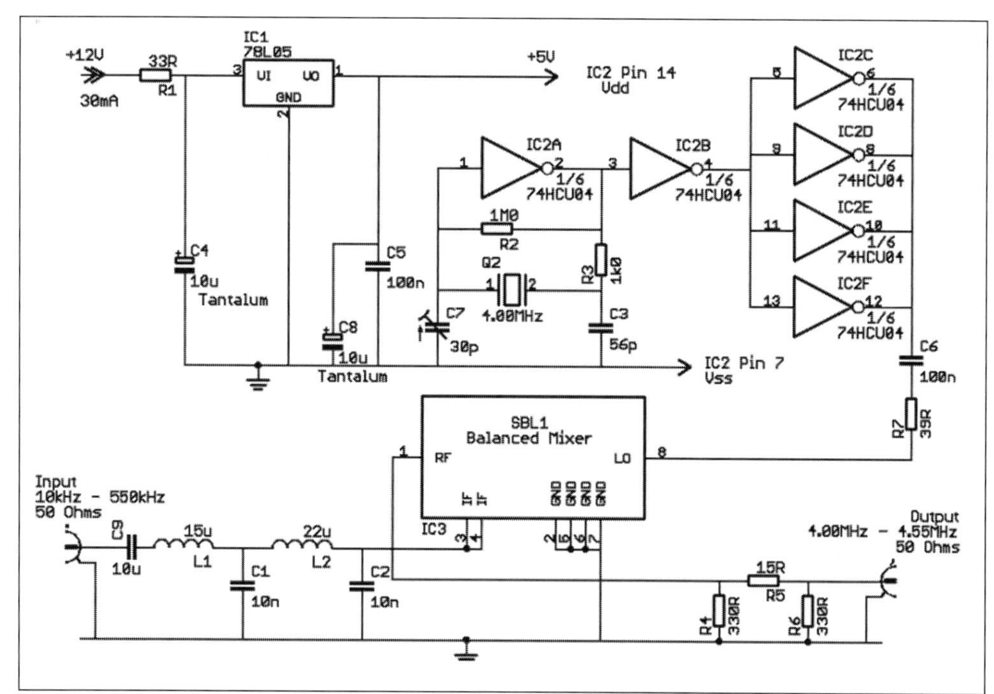


Fig 10.8: M0BMU's LF/MF to HF converter

inductor, and adjust to resonance using a variable capacitor, as shown in **Fig 10.4**(b). This can be a broadcast-type variable, with both sections paralleled to give about 1000pF maximum.

A higher tuning inductance is required to make up for the overall reduction in capacitance. The shunt-tuned circuit of **Fig 10.4**(c) has the convenience of one side of the tuning capacitor being grounded. The impedance match will not be quite as good, although normally perfectly adequate.

## RECEIVING PREAMPS

To overcome reduced sensitivity at lower frequencies, many amateur-type receivers require a preamplifier. Because of the strong broadcast signals in the LF and MF frequency ranges, it is important that adequate selectivity is provided at the signal frequency. To obtain a good S/N ratio, it is also necessary to pay attention to impedance matching between antenna and preamplifier.

Both amplification and selectivity are provided by a design, originally by G3YXM, available ready-built from G0MRF [12] in versions for 136kHz (**Fig 10.5**) and 472kHz. These incorporate double-tuned input filters providing a bandwidth of a few kilohertz centred on the amateur band (**Fig 10.6**). Gain is around 12dB. The preamp is designed for 50-ohm input impedance, so antenna matching as described in the previous section will normally be required.

The LF antenna tuner/preamp circuit of **Fig 10.7** combines the antenna matching, filtering and preamplifier functions. It is quite flexible and can be used with a wide range of long-wire and loop antenna elements. It can easily be modified for other frequencies, including 472kHz (see below). It has been used successfully with an IC-718 transceiver, which has fairly poor sensitivity at 136kHz. The preamp is a compound follower, with a high-impedance JFET input, and a bipolar output to drive a low impedance load with low distortion. The gain of the follower is about unity, but the high Q, peaked low-pass filter input circuit provides voltage gain, and also gives substantial attenuation of unwanted broadcast signals at higher frequencies.

The gain of the circuit depends on the type of antenna element used, of the order of 10dB with a long wire element and 30 - 40dB with a loop element. The 2.2mH inductors are the type wound on small ferrite bobbins with radial leads, and have a Q around 80 at 136kHz; other types of inductor with similar or greater Q could also be used. For wire antennas, Cin should be in the range 600pF - 5000pF, with large values giving a reduced signal level with longer wires, and smaller values suiting short wire antennas.

The antenna can be fed with coaxial cable, in which case the distributed capacitance of the coax (about 100pF/m for 50-ohm cable) makes up part or all of Cin. This allows the receiving antenna to be located remote from the shack, which is often useful in reducing interference pick-up. This circuit has given good results with wire antennas ranging from a 5m vertical whip to a 55m long wire. For loop antennas, Cin is omitted. Satisfactory sensitivity was obtained using a 1m2 loop with 10 turns of 1mm2 insulated wire, and also with larger single-turn loops with area around 10m2 (see the Receiving Antennas section of this chapter). Loop antennas can also be fed via quite long lengths of coax cable. The frequency range of the input circuit can easily be extended by using different values of inductance in the input tuning circuit; for example, the circuit was used to receive the 17.2kHz VLF broadcasts from SAQ by using 200mH inductance, and has also been used at 472kHz with 330µH inductance.

## CONVERTERS

For HF receivers without coverage of 136kHz, or those that have very poor performance at LF and MF, a converter can offer

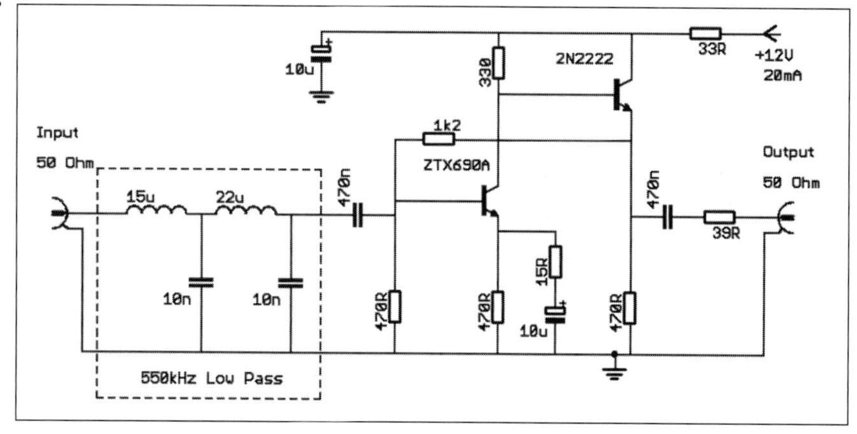


Fig 10.9: 20dB preamplifier

excellent reception. A number of LF/VLF - HF converters have been manufactured in the past (Datong VLF converter, Palomar VLF-S, VLF-A) and may still be available second-hand. However., these have mainly been aimed at broadcast reception, and may not give good results in the more demanding circumstances of the LF and MF amateur bands. The Juma LF/MF transmitters include receive converters with IFs in the HF amateur bands. Unfortunately Juma don't make these any more but you may find a second-hand one.

**Fig 10.8** shows a simple homebrew LF/MF to HF converter used successfully by M0BMU, mostly for portable reception in conjunction with a Yaesu FT-817. This uses a 4MHz crystal oscillator and broadband diode mixer module to up-convert input signals from a few kilohertz up to the 550kHz cut-off frequency of the input low-pass filter, to an output range of 4.00 - 4.55MHz. The LF/MF input signal is fed into the DC-coupled IF port of the SBL-1 mixer, and the HF output is taken from the RF port - this allows input frequencies below the 500kHz minimum of the RF port.

The crystal oscillator uses one gate of a 74HCU04 hex CMOS inverter IC, with the remaining five gates used as a buffer amplifier to drive the diode mixer. A wide range of other crystal frequencies could also be used to obtain different output frequency ranges if preferred; due to the broad-band nature of the mixer, no further modification is needed other than to change the crystal. An oscillator frequency below 2MHz makes the circuit more susceptible to IF breakthrough and image responses, while frequencies much over 10MHz will lead to reduced frequency stability, which may be a problem when narrow-band modes are being received.

The converter output has a simple -3dB attenuator pad to reduce the effect of output impedance variations on the mixer; the circuit therefore has an overall loss of about 10dB. No post-mixer amplifier stage is included, since most HF receivers include a low-noise preamplifier that can be switched in to perform this function.

The crystal frequency can be adjusted by setting the HF receiver to exactly 4MHz (or other crystal frequency), and adjusting the trimmer capacitor so that the oscillator signal is exactly centred in the CW passband of the receiver. The received input frequency is then the value displayed by the receiver, minus the crystal frequency. This converter gives good performance from 550kHz down to very low frequencies, and for receivers with very poor sensitivity at LF or MF will often give better results than the addition of high-gain preamplifiers. It is also an effective way of extending the low frequency capability of HF-only receivers.

The prototype converters were either built on ground-plane prototyping board, or 'dead bug style' on un-etched PCB board. The only setting up required is adjusting the oscillator frequency. The frequency drift of the prototypes was of the order of 1 or 2Hz over a period of hours, which is adequate for all except the most extreme narrow band modes. This converter has been used with wire antennas and the tuner/preamp of **Fig 10.7**. For small loop antennas, the preamp circuit of **Fig 10.9** has been used to increase the signal level by about 20dB.

## TRANSMITTERS FOR 136KHZ AND 472KHZ

Although many amateur HF transceivers can receive 136kHz and 472kHz signals, they cannot generate useful transmitter power at these frequencies due to PA limitations. Surprisingly there are no commercially available stand-alone transmitters for these bands. A couple of transverters are available, from Monitor Sensors and Mini-Kits, but most LF/MF amateur stations use home-made transmitters or transverters.

Due to the very low efficiencies of typical amateur antennas for the LF and MF, transmitters are usually required to produce outputs between 50W and a few kilowatts.

One approach to LF transmitter design is to use HF circuit techniques, with appropriate scaling of components for a lower operating frequency [15]. However, most LF operators are currently using transmitters with switching-mode output stages, operating in class D or class E modes. These can achieve high output powers using quite simple circuits, along with very good efficiency, which considerably simplifies cooling problems associated with high-power linear amplifiers. These circuits are also well suited to inexpensive power MOSFETs and other components intended for switch-mode power supplies operating in a similar frequency range.

Switching mode circuits are, however, more difficult to key or modulate satisfactorily for so-called 'linear' operating modes. Fortunately, most LF operation uses simple on-off keying of the transmitter, or frequency or phase-shift keying. Class D and E PA stages have seen relatively little use for the higher frequency amateur bands; therefore, a description of their design and operation is given below, together with practical designs for complete transmitters. Further LF/MF transmitter designs can be found at [12, 16].

## CLASS D TRANSMITTERS

Class D amplifiers (sometimes referred to as 'tuned class D' to differentiate from class D audio amplifiers) fall into two distinct types, voltage-switching, **Fig 10.10**(a), or current-switching, **Fig 10.10**(b). In each case, the load is connected to the output stage via a resonant tank circuit. The voltage-switching type has a series-tuned tank circuit; the switching MOSFETs develop a square-wave voltage at the input side of the series tank circuit, however the tank circuit ensures the current flowing in the load is almost a pure sine wave.

The current-switching type has a parallel-tuned tank circuit; the supply to the output devices is a constant current which is applied to the tank circuit in alternate directions depending on which MOSFET is switched on. The resulting square wave current applied to the tank circuit again results in an almost sinusoidal voltage across the load. Since a constant-current DC

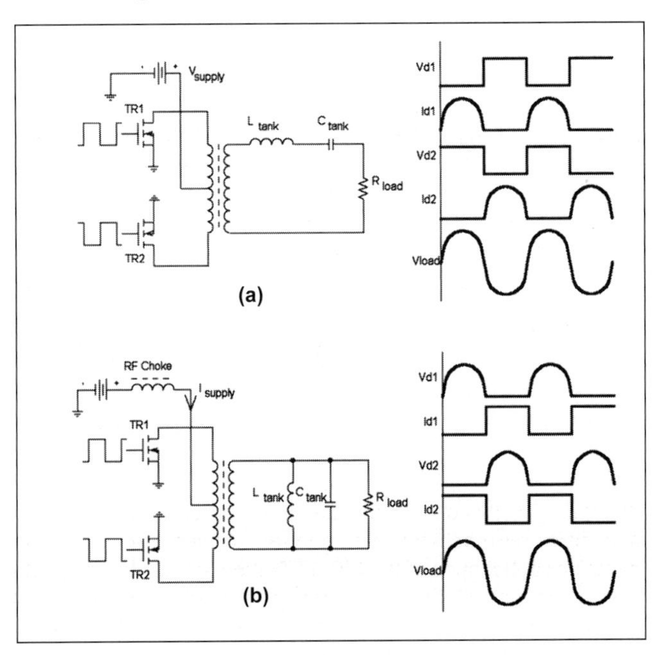


**Fig 10.10: (a) Voltage-switching class D amplifier, and (b) Current-switching class D amplifier with MOSFET drain waveforms**

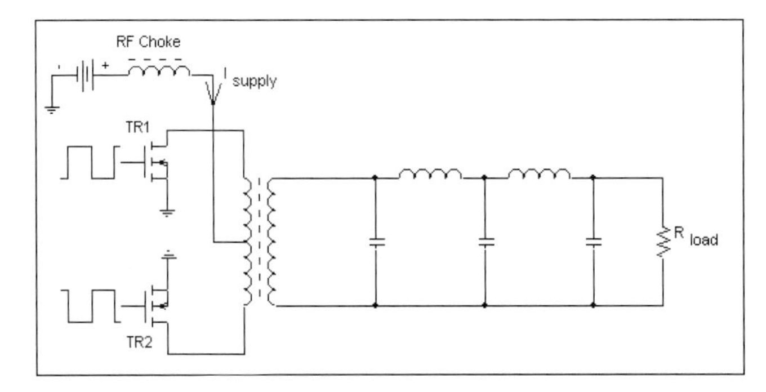


Fig 10.11: Practical form of class D output stage

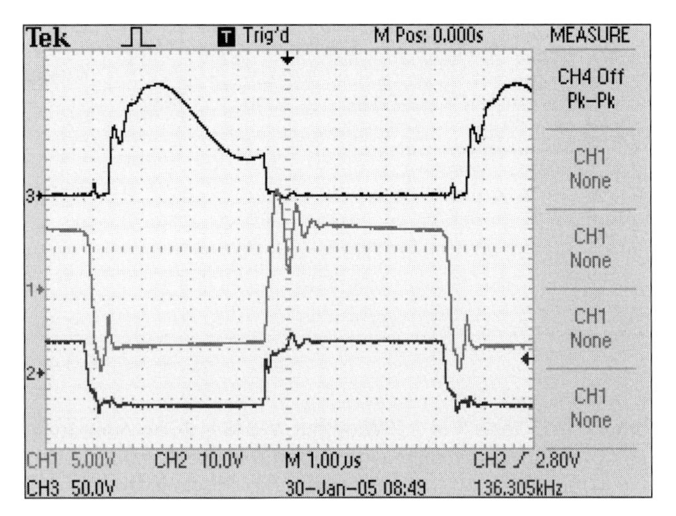


**Fig 10.12: Class D waveforms. Upper trace, drain voltage; middle trace, drain current; lower trace, gate drive voltage**

supply is not very practical, a constant voltage supply is used with a series RF choke. Provided the impedance of the choke is much greater than the load resistance, the supply current is almost constant. The major advantage of class D is that the MOSFETs are either fully 'on', in which case the only power loss is due to the MOSFET 'on' resistance, rDS(on), or fully off, with essentially zero power dissipation. In practice, there are additional losses, but these are small compared to linear amplifiers, and efficiency can exceed 90%.

In many amateur circuits, the tank circuit is replaced by a low-Q low-pass filter, **Fig 10.11**. This circuit is 'quasi-parallel resonant'; it provides a resistive load at the output frequency, but a low shunt impedance at the harmonics. The low Q leads to non-ideal class D operation in that the voltage waveform is not a perfect sine wave, but has the advantages that smaller inductors and capacitors are required, tolerances are less critical, and better rejection of higher harmonics is provided by the multiple filter sections. The voltage and current waveforms of a real-world class D output stage using this circuit are shown in **Fig 10.12**. Compared to the idealised waveforms, some high frequency 'ringing' is visible. This is due to stray capacitance and inductance which inevitably exists in the circuit, and is undesirable since it causes increased losses, as well as the potential for generating high-order harmonics. It is therefore important to minimise stray reactance, two important causes of which are the parasitic capacitance of the MOSFETs themselves, and the leakage inductance of the output transformer. Adding damping RC 'snubber' networks can also usefully reduce the level of ringing.

As with other types of amplifier, the output power of class D amplifiers is defined by the supply voltage Vcc, output transformer turns ratio n and load impedance RL. For the voltage switching amplifier:

$$P_L = \frac{8n^2V_{cc}^2}{\pi^2 R_L}$$

While for the current-switching class D:

$$P_L = \frac{\pi^2 n^2 V_{cc}^2}{8R_L}$$

These formulas assume that losses in the circuit are negligible; in practice, some losses do occur but since they are small, the results given by the formulas are reasonably accurate.

## CLASS D PA DESIGN EXAMPLE

The design process for a class D transmitter output stage is best illustrated by an example. The following design is for a LF transmitter with about 200W output, using a current-switching class D circuit. This is a modest power level for 136kHz, but the principles discussed have been applied equally well to designs with 1kW or more output using this circuit configuration, which is probably the most popular in use at present. The complete circuit is shown in **Fig 10.13**.

The first design decision is what DC supply voltage to use, since the power supply is normally the most expensive and bulky part. In this case 13.8V was selected; it can use the standard DC supply found in many amateur shacks. The DC input power required will be about 10% greater than the RF output due to losses, so the expected supply current will be 220W 13.8V = 16A, a level that most 13.8V supplies can readily deliver. For higher power designs, 40 - 60V is often a good compromise, since the problem of large DC and RF currents is then reduced. It is perfectly possible to use an 'off line' directly rectified AC mains supply with no bulky mains transformer, as has been done by G4JNT [17]; note that design for electrical safety is absolutely critical in this case. Inexpensive switching MOSFETs are available suitable for any of these supply voltages. In the ideal push-pull current switching circuit, the peak MOSFET drain-source voltage will theoretically be π times the DC supply voltage, in practice about four times the 13.8VDC supply is likely. MOSFETs should be selected so that only a few percent of the DC input power will be dissipated in their 'on' resistance, rDS(on). This condition also ensures the MOSFET will have adequate drain current rating.

STW60NE10 devices were used for TR2, TR3, with BVDS of 100V, and a typical rDS(on) of 0.016 ohms, leading to about 4W dissipation due to the 'on' resistance and 16A supply current (I2 x rDS(on)). Additional dissipation occurs during the transient period where the device is switching 'on' or 'off'. This can be determined from measurement of circuit waveforms, but can be assumed to be similar to that due to rDS(on). In normal operation therefore, each MOSFET will only dissipate a few watts; however with a severe mismatch, power dissipation can be much higher, especially without DC supply current limiting. For a robust design, the MOSFETs and their heatsink should be able to dissipate of the order of 50% of the total DC input power, at least during a short overload period. The STW60NE10 devices have a TO-247 package and can dissipate 90W each at a case temperature of 100 degrees C, which is adequate.

The output transformer is the most important part of the design. It is normally wound on a core using the same ferrite grades that are used for switch-mode power supplies. These may be large toroidal cores, pot cores, or 'E' cores with plastic bobbins. Several manufacturers produce suitable materials; these include Ferroxcube (Philips) 3C8, 3C85, 3C90, Siemens

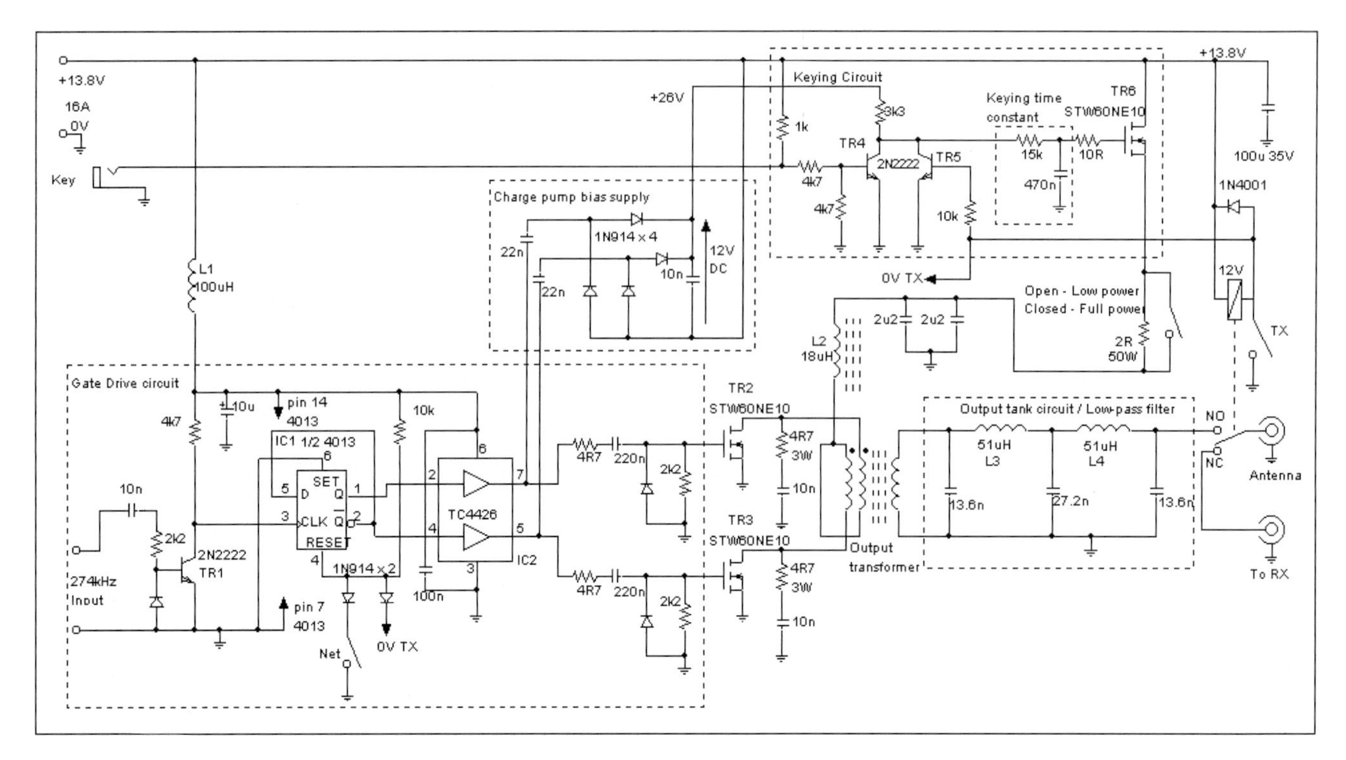


**Fig 10.13: 200W class D transmitter circuit**

N27, N87, Neosid F44, and Fair-Rite #77 grades. All these ferrites have permeability around 2000, and have reasonably low loss at 136kHz. They are available in a variety of forms, such as EE, EC, and ETD styles, and sizes; a designation such as ETD49 means an ETD style core that is 49mm wide. A good selection of different core types is available from component distributors [18, 19] at reasonably low cost.

Transformer design is a complex topic in its own right, but a simplified procedure usually gives satisfactory results for amateur purposes, as follows. Given the supply voltage and load impedance (usually 50 ohms), the turns ratio of the output transformer determines the output power. Rearranging the formula for current-switching class D given in the previous section gives:

$$n = \sqrt{\frac{8P_L R_L}{\pi^2 V_{cc}^2}}$$

For Vcc = 13.8V, RL = 50Ω, and PL = 220W, this gives n = 1:6.8. Next, a suitable sized core is chosen. As a guide, using the types of ferrite listed above, an ETD34 core is suitable for powers up to 250W, an ETD44 core for 500W, and an ETD49 for up to 1kW. Similar sizes in different styles have similar power handling. If in doubt, use a bigger core!

The number of turns N in the secondary winding can then be determined. N must be large enough to keep the peak magnetic flux Bpeak to a value well below the saturation level at the expected output voltage level, VRMS:

$$B_{peak} = \frac{V_{RMS}}{4.44 fNA_e}, \quad V_{RMS} = \sqrt{P_L R_L}$$

Where Ae is the effective area of the core in m2. The number of turns must also be large enough so that the inductance of the winding has a large reactance compared to the load impedance. A value of XL about 5 - 10 times the load impedance is desirable. An ETD34 core and bobbin of 3C85 ferrite material was available, which according to the manufacturer's data has Ae of 97.1mm2 (97.1 x 10-6 m2), and AL of 2500nH/T2. A suitable maximum value of Bpeak for power-grade ferrite materials is around 0.15 tesla. For 220W output, VRMS is 105V. A few trials using the formulas resulted in n = 14 turns, Bpeak = 0.127T and L = 490uH, XL = 422Ω, which meets the criteria given.

Two turn primary windings result in a turns ratio of 1:7, close enough in practice to the 1:6.8 design value. The primary windings were 4 x 2 turns, quadrifilar wound of 1mm2 enamelled copper wire, using two windings in parallel for each half of the primary winding. The secondary of 14 turns of 0.8mm2 enamelled copper was wound on top of the primaries, and insulated from them with polyester tape.

The DC feed choke L2 must be capable of handling the full DC supply current without saturation and also have a high reactance at 137kHz compared to the load impedance at the transformer primary, which is (50Ω / 72), about 1Ω. A reactance of 10Ω or greater is adequate, requiring at least 12uH. A high Q is not required, since only a small RF current flows in the choke. The 18uH choke used a Micrometals T-106-26 iron dust core. Iron dust cores of similar types to this can often be salvaged from defunct PC switch-mode PSUs. The winding used 2 x 17 turns in parallel of 1mm2 enamelled copper wire. An air-cored inductor would also be feasible, if more bulky.

The output filter consists of two identical cascaded pi-sections. The filter should provide a resistive load at the 136kHz output frequency, but a low capacitive reactance at harmonics. This can be achieved by designing the pi-sections as low-Q matching networks, with equal source and load resistances. This yields a circuit with two equal capacitors. The standard pi-section design formulas can be used, modified for Rin = Rout = R:

$$X_C = \frac{R}{Q}, \quad X_L = \frac{2QR}{Q^2+1}$$

$$C = \frac{1}{2\pi f X_C}, \quad L = \frac{X_L}{2\pi f}$$

Most designers select Q between 0.5 and 1. Metallised polypropylene capacitors are a good choice, since they have low losses at 136kHz, and are available with large values and high voltage ratings. The DC voltage rating should be several times larger than the RMS RF voltage present; the main limitation is the heating effect of the RF current causing internal heating of the capacitor. Several 6.8nF, 1kV polypropylene capacitors were available, so C = 2 x 6.8nF = 13.6nF was used. This has reactance of 85.4Ω at 137kHz, forcing Q = 0.585, and giving XL = 43.6Ω, L = 50.6uH.

The inductors must have low loss at 136kHz to avoid excessive heating. Micrometals T130-2 iron dust cores were used, wound with 68 turns of 0.7mm2 enamelled wire. It is a good idea to check the capacitance and inductance of the filter components using an LCR meter or bridge. However, the main effect of small errors is only to slightly alter the output power from the circuit, without greatly affecting the efficiency.

The drive signal applied to the class D output stage is a 50% duty cycle square wave. The 137kHz gate drive signal is obtained from a 274kHz input using a D-type flip-flop in a divide-by-2 configuration, guaranteeing an accurate 50% duty cycle. When the circuit is switched to receive, the flip-flop is disabled by pulling the reset input high, preventing a 137kHz signal leaking to the receiver input and causing interference. For netting the transmitter, the 'net' switch enables the flip-flop. The MOSFETs require zero or negative gate voltage to switch the transistor fully off, and +10 to +15 volts to bias them fully on. The MOSFET gates behave essentially as capacitors, requiring transient charging and discharging currents as the drive voltage switches on and off, but drawing no current while the gate voltage remains stable. In order to achieve fast MOSFET switching, a TC4426 gate driver IC is used.

These driver ICs accept a TTL-compatible logic level input signal, and are designed to produce peak output currents of 1A or more which charge and discharge the MOSFET gate in a fraction of a microsecond. A disadvantage of using a flip-flop to generate the drive signal is that, if the input signal is lost, one MOSFET will remain switched on and act as a virtual short across the supply. To avoid this, the gate driver is capacitively coupled to the MOSFETs; the shunt diodes perform a DC restoration function, making the full positive peak voltage available to drive the gate. If drive is lost, the gates discharge through the 2.2k resistors, switching both MOSFETs off. The 4.7 ohm resistors in series with the gate drivers help to reduce ringing.

Each MOSFET has a series RC damping network from drain to source, reducing high frequency 'ringing' superimposed on the drain waveform. The component values are best determined experimentally, since they depend on the individual circuit. A good starting point is to make the capacitor about five times larger than the MOSFET output capacitance. A resistance between 2 and 20 ohms is usually effective. Effectiveness is best checked by examining the MOSFET drain waveforms with an oscilloscope, and compromising between minimising high-frequency ringing and excessive power dissipation in the resistors. Larger capacitors and smaller resistors normally result in reduced ringing, but increased dissipation. These components should be appropriate for high frequency use; in this circuit, 4.7 ohm, 3W metal film resistors, and 10nF, 250V polypropylene capacitors were satisfactory.

The transmitter is keyed using series MOSFET TR6. The MOSFET should have low rDS(on) to minimise loss when switched on. A third STW60NE10 was used, although since the maximum voltage applied to this device is the 13.8V DC supply, a lower voltage device could be used instead. During the rise and fall of the keying waveform, dissipation in the keying MOSFET peaks at about 25% of the maximum DC input, 55W in this case. However, when the MOSFET is fully on, it dissipates only a few watts due to rDS(on), and when fully off dissipation is practically zero, so the average power dissipated is small, under 10W in this circuit, provided it is not keyed very rapidly. In order to bias this MOSFET fully on, a voltage around 10V higher than the 13.8V DC rail must be applied to the gate.

Only a few milliamps bias are required; a small auxiliary DC supply could be used in a mains-powered transmitter, but in this case the bias voltage was obtained by rectifying the capacitively-coupled gate drive waveform using a charge-pump circuit. The bias is controlled by the key input via transistor TR4, and the keying waveform is shaped by the RC time constant to give around 10ms rise and fall times. TR5 maintains the keying 'off' when the circuit is switched to receive. The 15V zener across the MOSFET gate and source limits the gate voltage to prevent damage. The output power of a class D transmitter can be varied either by changing the supply voltage, or by having multiple taps on the output winding; both these techniques are used in the G0MRF and G3YXM designs described later.

In this design, a resistor can be switched in series with the DC supply, reducing the supply voltage to the output stage to around 4V, and RF output to 18W, for tuning-up purposes. The 2 ohm wirewound resistor dissipates nearly 40W in low power mode, so greatly reduces efficiency, but does make it very difficult to damage the PA due to its inherent current limiting, useful when using a battery supply or when initially testing the circuit.

Construction of this type of transmitter is reasonably non-critical. The low-power parts of the transmitter can be assembled using Veroboard or similar, but the gate driver IC must have a 0.1µF ceramic decoupling capacitor directly across the supply pins due to the large transient currents present. Also for this reason, the gate leads, and the ground return from the MOSFET sources should be kept very short (<30mm). The MOSFETs, output transformer, keying circuit and DC feed carry heavy cur-

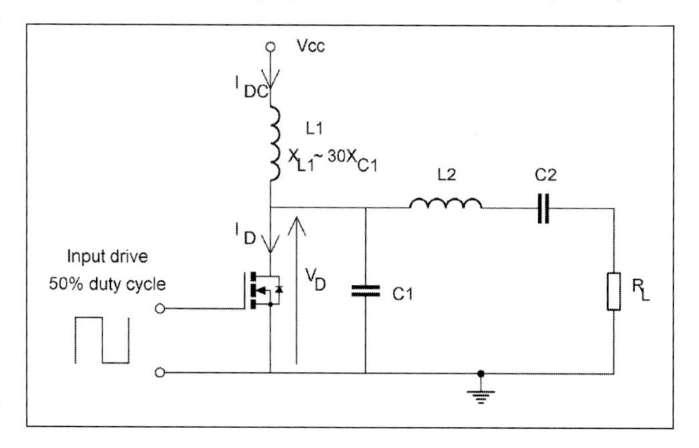


Fig 10.14: Basic class E PA circuit

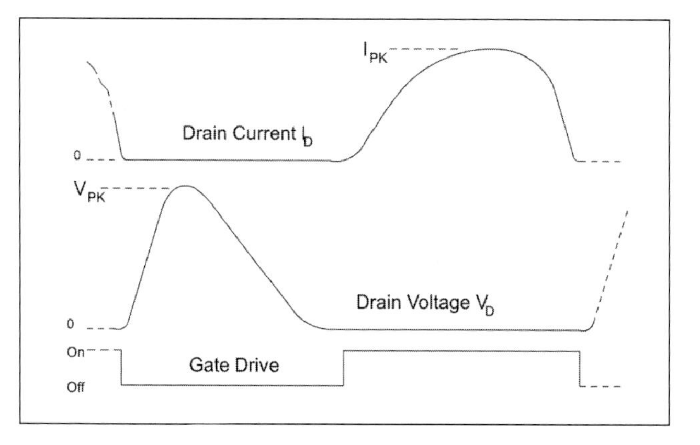


Fig 10.15: Class E waveforms

rents, so connections should be as short as possible and use thick wire (at least 2.5mm2). The RC damping components should be mounted directly across the MOSFET drain and source pins, and the connections to the output transformer kept short. The circuit described above was assembled on an aluminium plate about 160 x 200 x 3mm, which provided ample heatsinking for the three MOSFETs when air was allowed to circulate freely. In an enclosed box, a fan would probably be desirable.

Testing a class D circuit should start by checking operation of the gate-drive circuit, ensuring that complementary 12Vp-p square waves are present at the MOSFET gates at the correct frequency. A dummy load is almost obligatory for testing LF transmitters (see section on LF measurements). If possible, apply a reduced DC supply voltage to the output stage (but not to the gate drive circuit!), or use a series resistor to reduce the supply voltage, as included in this circuit. An oscilloscope is the ideal tool to check the correct waveforms are present. A useful check is the efficiency; the ratio of RF output power to DC input power should be well over 80% if the circuit is working correctly.

## CLASS E TRANSMITTERS

The class E power amplifier is another form of switch-mode output stage that has been successfully applied to 136kHz and 472kHz amateur transmitter construction. The circuit was invented by Nathan Sokal, WA1HQC [20]. The basic class E circuit is single-ended, as shown in **Fig 10.14**. The single switching MOSFET drives a tank circuit C1, L2, C2 with series and shunt capacitors, and is fed with DC supply current via a high impedance choke L1.

As in class D PAs, the active device (typically a power MOSFET) is used as a switch that is either fully 'off' or 'on' and thus, in an idealised circuit, no power is dissipated. In switch mode circuits in general, some small losses will occur, partly due to the unavoidable 'on' resistance of the MOSFET, but also partly due to the finite time taken for switching to occur. During switch on, the current is rising while the voltage falls and the reverse occurs during switch off. The instantaneous power dissipation can be high during these short transition periods, since both voltage and current in the MOSFET is simultaneously quite large. The class E tank circuit design aims to minimise these losses by shaping the voltage and current waveforms in the MOSFET so that:

- The MOSFET current decreases to zero before switch-off occurs, so the MOSFET current is zero while the voltage rises
- The voltage across the MOSFET reaches zero before switch-on occurs, so current in the MOSFET does not rise until the voltage is zero

The approximate waveforms of the voltage and current are shown in **Fig 10.15**. The output waveform delivered to RL is effectively filtered by the series-tuned circuit formed by L2, C2 and is an almost pure sine wave. Analysis of this circuit is complex due to its non-linear nature and the non-sinusoidal waveforms. A good description, together with detailed design formulas has been provided in [20]. An approximate design procedure is as follows. Initially, the operating frequency f, supply voltage VCC, output power PL and tank circuit loaded Q, QL are selected. Typically QL is between 5 and 10. Referring to **Fig 10.14**, the component values are approximately:

$$R_L \approx 0.577\frac{V_{CC}^2}{P_L};\ C_1 \approx 0.184\frac{1}{2\pi f R_L}$$

$$C_2 \approx \frac{1}{2\pi f Q_L R_L};\ L_2 \approx \frac{Q_L R_L}{2\pi f}$$

The DC current supplied to the PA, and the peak values of the

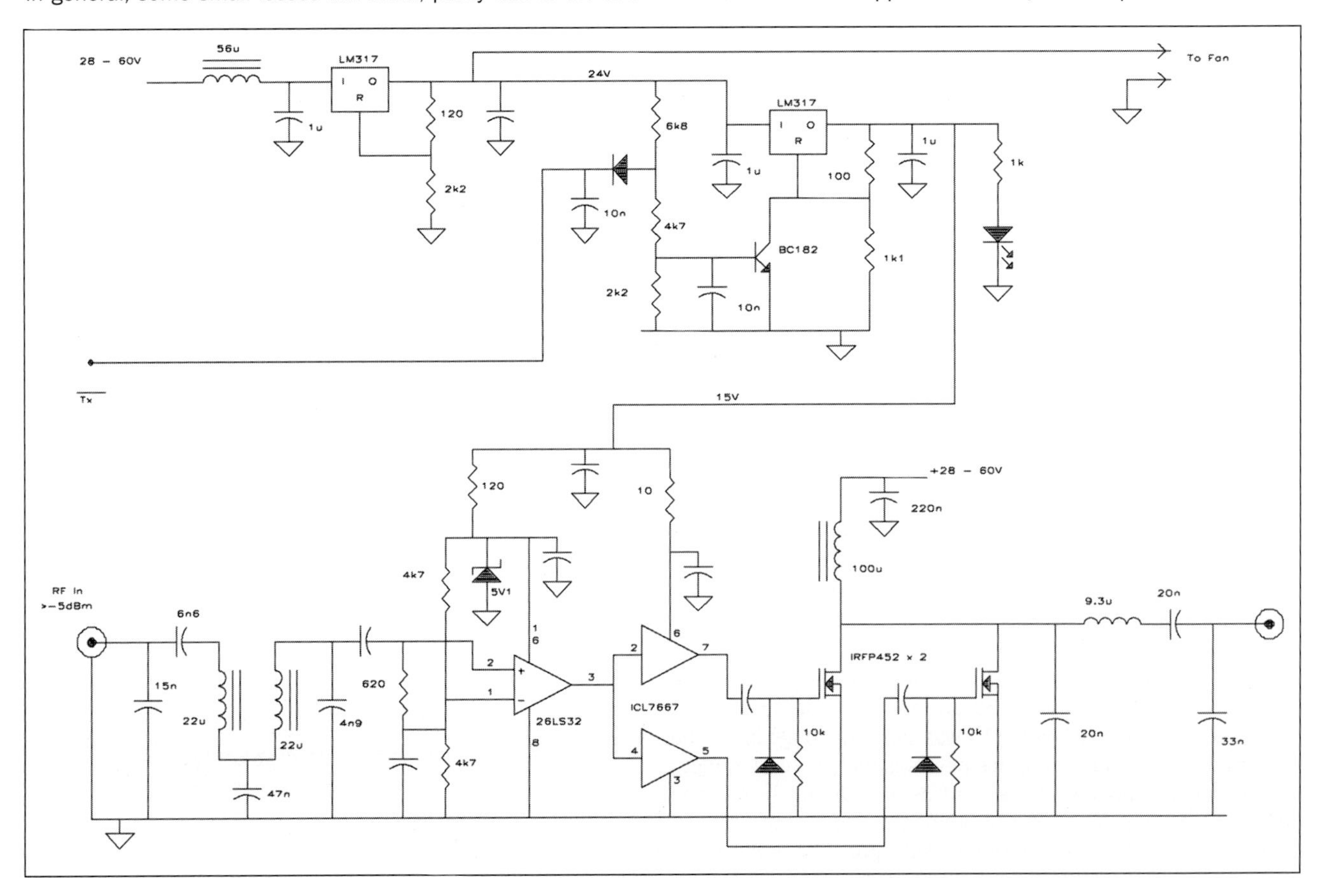


**Fig 10.16: G4JNT QRO class E MF PA. All unmarked capacitors are 100nF. Diodes are 1N4145**

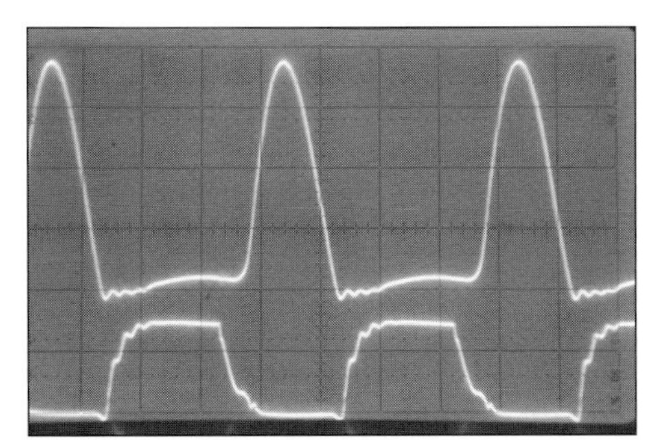

**Fig 10.17: Class E PA waveforms: Upper trace - Drain voltage (50V/div); Lower trace - Gate drive voltage (10V/div)**

voltage and current waveforms of **Fig 10.15** are approximately given by:

$$I_{DC} \sim \frac{P_L}{V_{CC}};\quad V_{PK} \sim 3.56\,V_{CC};\quad I_{PK} \sim 2.86\,I_{DC}$$

These approximations neglect the effect of losses, finite tank circuit QL, and the impedance of the feed choke L1, but yield values within typically 20%. To obtain more accurate values, the more complex formulas in [20] should be used; design spreadsheets are available from [21] and [22] to simplify this process. In any case, due to component tolerances and parasitic capacitance and inductance in the circuit, it is generally necessary to adjust the tank circuit components experimentally to their final values while observing the voltage waveform with an oscilloscope. A detailed description of this process is also given in [20]. The value of RL will generally not be 50 ohms; a ferrite-cored matching transformer may be used to obtain the desired load impedance as for the class D circuit, or the tank circuit may be extended to include an additional L-network as described in the G4JNT design in [23] and below. Additional low-pass filtering for harmonic suppression may also be required.

Compared to class D PAs, perhaps the main attractions of class E designs for amateur LF/MF use include a simpler circuit with single-ended input drive and output, and 'cleaner' waveforms with less possibility of high frequency ringing and switching transients, especially compared to the push-pull current-switching class D circuit. This is probably mostly due to the elimination of the push-pull transformer with its inevitable parasitic inductance and capacitance. At higher frequencies, switching losses in class E PAs are significantly reduced compared to class D, but at 136kHz and 472kHz, efficiencies of well over 80% are possible with both types. The main disadvantages of class E are the need for relatively bulky high Q tank circuit inductors and capacitors, and the attendant requirement for a more precise tuning up procedure.

## G4JNT High Power MF Class E Amplifier

Originally designed for 500kHz, this project is a low cost, high power class E switch mode amplifier. A power output of 500W running from a 50V supply was decided upon as an aiming point. References [20] and [21] provided all the design details. A spreadsheet was developed from one by G3NYK that added a series L / shunt C output matching network to raise $R_{load}$ to 50 ohms. The series L is absorbed into either the tank circuit L or C values.

Several IRFP462 FETs were available, rated at 500V, 170W dissipation and RDS(on) of 0.4 ohms. The RDS(on) value is a bit high for the projected power level, so two devices were used in parallel.

The air-cored tank coil was wound with 2.5mm2 Litz wire. Ceramic capacitors proved to have high losses when used in the tank circuit, and the final breadboard uses 3.3nF, 1700V metallised polypropylene capacitors built up in parallel combinations to give the required values and share the RF current.

The completed circuit in **Fig 10.16** has input circuitry consisting of a bandpass filter followed by a line receiver to get a close-to-50% duty cycle square wave from a low level input. For driving the FETs, ICL7667 gate drivers were already available. Since the

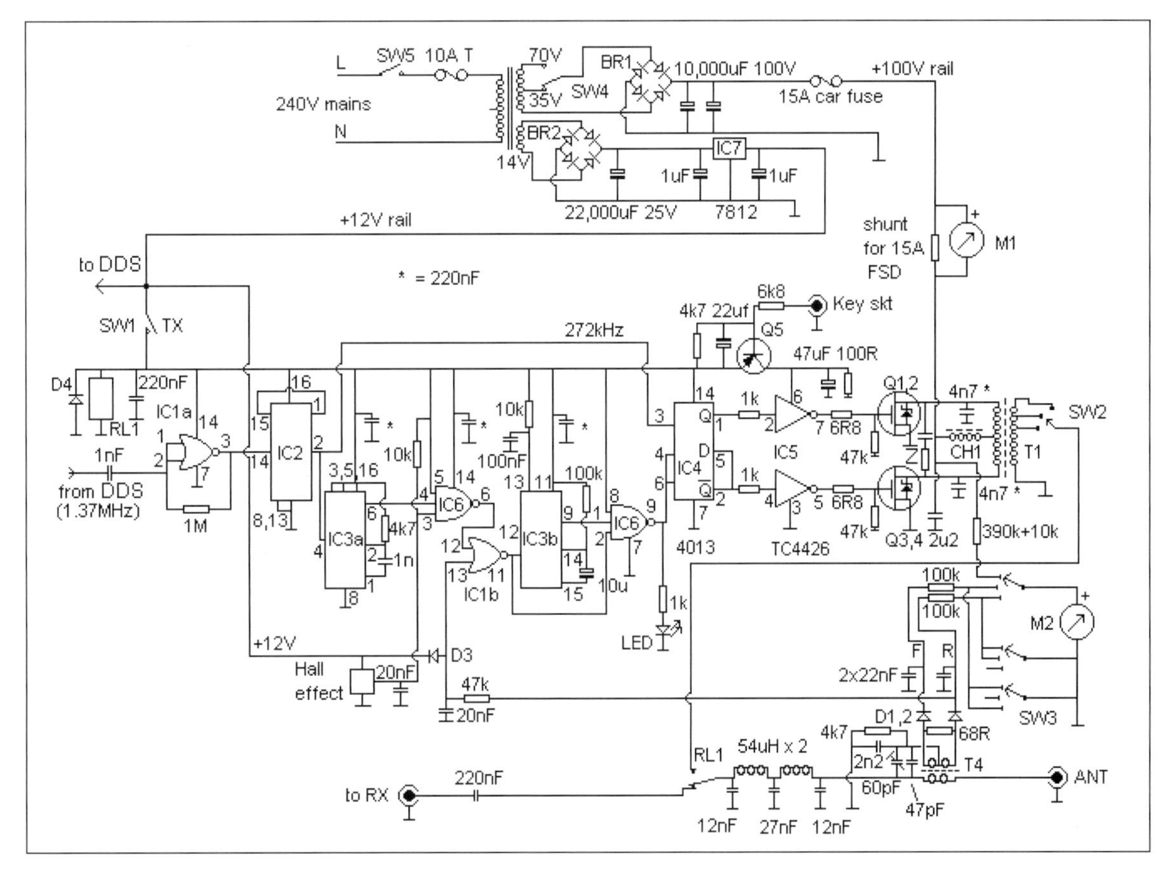


**Fig 10.18: G3YXM one kilowatt transmitter**

| | |
|---|---|
| IC1 | HEF4001 |
| IC2 | HEF4017 |
| IC3 | HEF4538 |
| IC4 | HEF4013 |
| IC5 | TC4426 |
| IC6 | HEF4023 |
| IC7 | 7812 |
| Q1,2,3,4 | IRFP450 |
| Q5 | IRFP260 |
| D1, D2 | 1N4936 |
| D3, D4, D4, D6 | 1N4006 |
| Hall effect device | OHN3040U (Farnell 405-656) |
| BR1, BR2 | 35A, 600V |
| R (for Zobel network "Z") | 22Ω, 25W (Farnell 345-090) |
| Mains transformer | 2 x 35V, 530W |
| T1 Primary | 2 x 8 turns, secondary 20 turns, tapped at 12 and 16 turns |
| CH1 | 20 turns on 50mm length of antenna-type ferrite rod |
| T4 | Primary 1 turn, secondary 2 x 18 turns bifilar |
| Output filter inductors | 54µH. 65 turns 1mm enamelled wire on Micrometals T200-2 toroid core |

**Table 10.1: Component details for the G3YXM 1kW transmitter**

ICL7667 contains two identical drivers, one was used for each parallel FET to reduce the loading due to the FET's 2000pF of input capacitance amplified by Miller feedback. The 100 micro-henry DC feed choke is a large toroid, and must be capable of passing the 10A DC feed current without excessive heating.

The unit was powered initially with a 12V supply to the PA to allow tuning safely without over-stressing PA components. The PA was tuned by adding or removing 3.3nF capacitors from the parallel combinations making up the tank circuit capacitors, while viewing the switching waveform and power output (reference [20] contains a description of the tuning procedure and expected waveforms).

The quantisation of the capacitance values in 3.3nF steps is a bit coarse, but combinations could be found giving close to the right waveform shape (**Fig 10.17**) and output power. With 50V, 10.1A DC input, RF output of 400W was achieved, with the heat-sink running cool at this level. The peak drain voltage of 190V is also well within the FET's rating so it may be possible to safely increase the DC supply voltage to achieve the 500W target.

Efficiency at 79% is not the highest that can be obtained from class E PA stages - values in excess of 90% have been claimed. The major power loss in the circuit is due to the relatively high 'on' resistance of the IRFP462 FETs, as can be seen from the approximately 8V Vds voltage drop visible in the waveform of **Fig 10.17**. More modern devices are available with much lower Rds(on). Also the ICL7667 gate drivers may be marginal in this circuit; again more recent devices are available with higher drive capability. A more detailed article describing this project and the development process can be found at [23], and the tank circuit design spreadsheet incorporating the output matching network can also be downloaded from [22].

## G3YXM 136kHz 1kW Transmitter

G3YXM set out to design a transmitter that is reasonably small, produces around 1kW RF output, and will withstand antenna mis-match and other mishaps. The description here is an abridged version of the original article available via G3YXM's web pages [24]. The circuit is shown in **Fig 10.18** and the major components are listed in **Table 10.1**.

The original design had a VFO at 1.37MHz but these days it would be more appropriate to use a DDS module such as GW4GTE's Multi-Rock [55]. The 1.37MHz signal is squared and divided by 10, the output at IC4 being a symmetrical square wave, driving the output MOSFETs via gate driver IC6. Each MOSFET shown is actually two devices in parallel. The output transformer ratio is set by switch S2. Higher output is obtained with more secondary turns selected. Across the primary of the transformer the Zobel network marked 'Z' (22 ohms and 4n7 in series) reduces ringing. The output is fed to the antenna via a low-pass filter. The cut-off frequency is quite high at about 220kHz as virtually no second harmonic is produced.

The SWR bridge consists of T4 and associated components. It is a bifilar winding of 2x18 turns which forms the centre-tapped secondary and the coax inner passing through the toroid core forms the single-turn primary. The protection circuit which cuts the drive for about a second is triggered by high SWR via IC1B, or over-current signal from the Hall-effect device, which is triggered by the magnetic field of CH1, which is made from a 50mm piece of ferrite antenna rod wound with 20 turns of 1.5mm enamelled copper wire. The Hall-effect detector is placed near one end of CH1 and the spacing adjusted to trip at about 20 amps. The receive pre-amp uses coupled tuned circuits giving a band-pass response over the 135 to 138kHz range. A single JFET (Q7) makes up for the filter loss.

The mains transformer used in the power supply has two series 35V windings. The DC voltage is either 50V from the centre-tap or 100V from both windings. An auxiliary 12V winding was added by winding 30 turns of 16SWG wire through the toroid. At full output the HT will drop to about 80V. The keying circuit uses a series MOSFET with shaping of rise and fall times to prevent key-clicks. To turn this MOSFET fully on, its gate must be at least 5V positive of its source which is close to the main supply voltage. Diodes D5 and 6 are supplied via a high voltage capacitor from the 12V winding to produce an extra 20V bias for this purpose.

The low-level circuitry was built on strip-board, taking care to keep the tracks short and earth unused inputs. The TC4426 chip IC5 is capable of driving 1.5A into the gate capacitances of the MOSFETs and the decoupling capacitors must be fitted close to the chip with short leads. The 6R8 series resistors are mounted on the gate pins of the MOSFETs, the resistor leads forming the connections to the strip board. It is probably best to use one resistor for each gate. The strip board should be grounded to the earth plane as near as possible to the MOSFETs, which should have the source leads soldered to the ground plane. The two 4n7 capacitors should be connected directly across the MOSFETs. Output transformer T1 should be

**Fig 10.19: Prototype of GW3UEP's VFO**

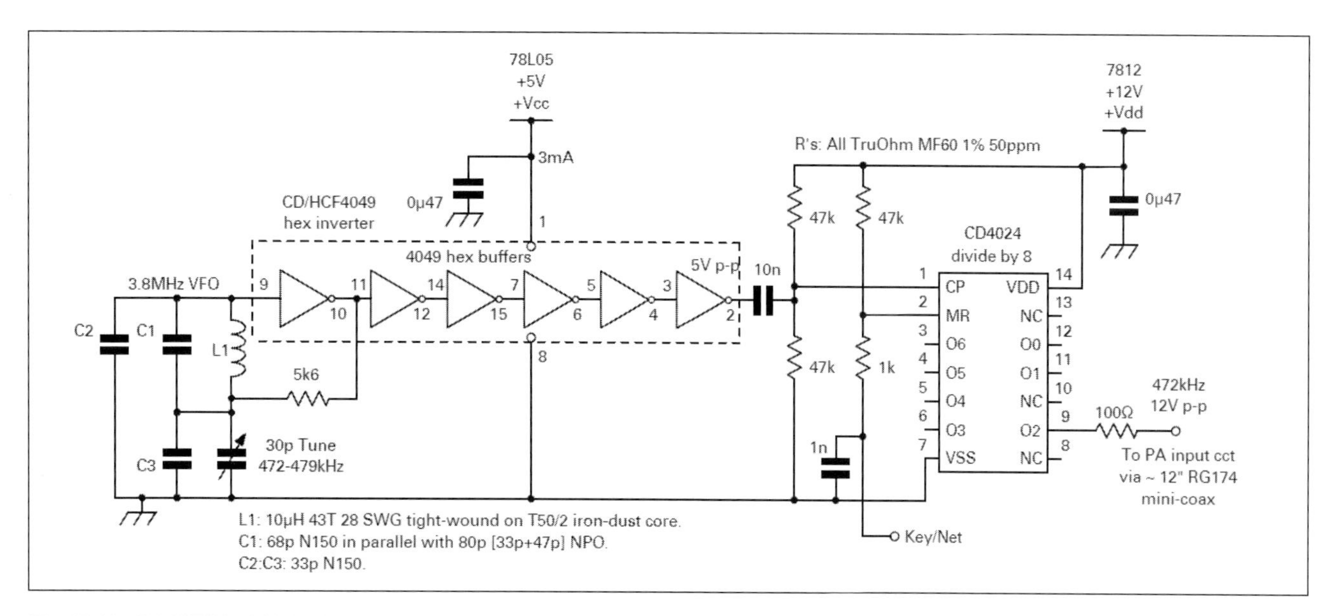


**Fig 10.20: GW3UEP's 3.8MHz VFO divides down to the 472kHz band to provide a stable signal**

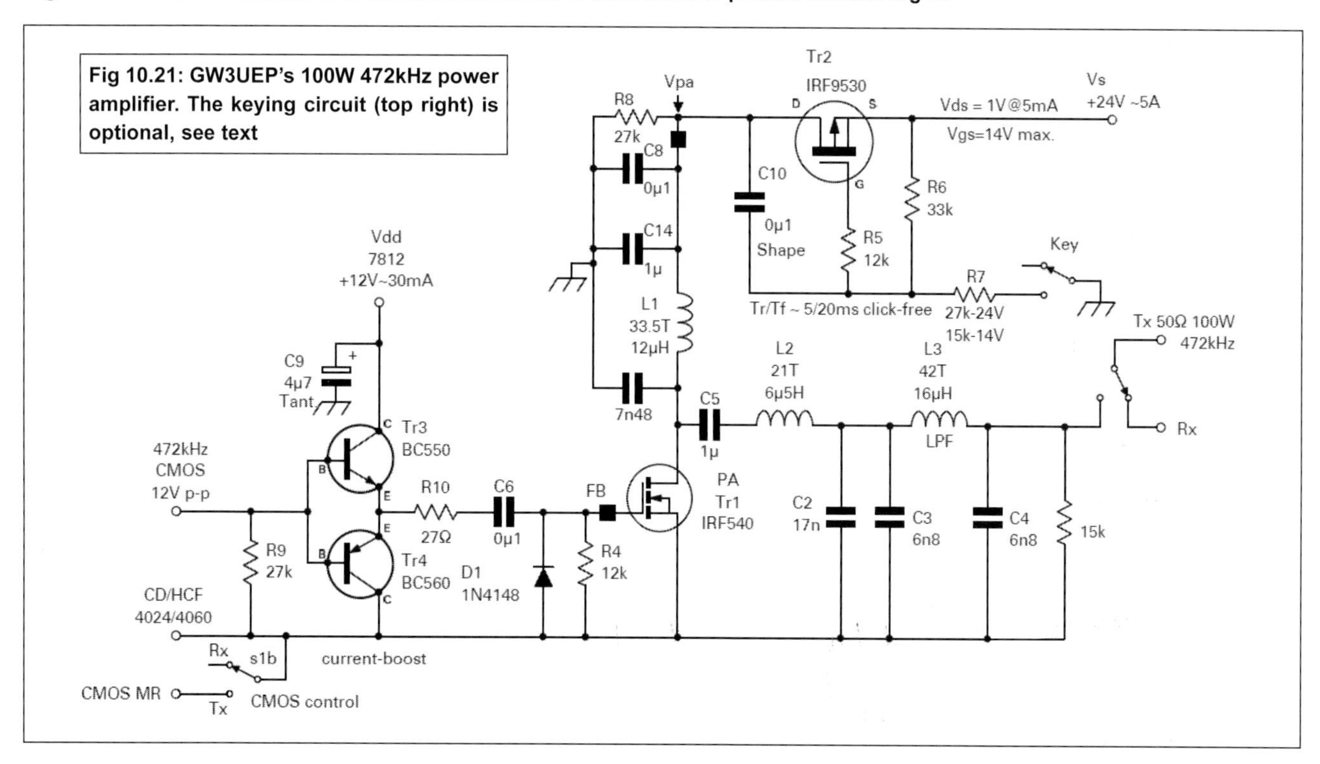


**Fig 10.21: GW3UEP's 100W 472kHz power amplifier. The keying circuit (top right) is optional, see text**

constructed from two-core 'figure of eight' speaker cable wound eight times through the ferrite toroid, connected as a centre-tapped primary by connecting one end of one winding to the opposite end of the other. The secondary is wound over it with 20 turns of thin wire tapped at 12 and 16 turns. The Zobel network should be wired from drain to drain with short wires.

Get the PSU, oscillator and CMOS stages working first. Check with a scope that you have complementary 12V square waves on the gates, the waveform will be slightly rounded off due to the gate capacitance. Connect the transmitter to a 50 ohm load and, having selected the first tap on SW2, apply 50V (SW4 in low position) with a resistor in place of the fuse to limit the current. The MOSFETs should draw no current without drive. Press the key and the output stage should draw a few amps and produce a few watts into the dummy load. If the shut-down LED comes on, either the load is mis-matched, the SWR bridge is connected backwards or the 60pF capacitor needs adjustment. If all seems well, remove the current limiting resistor and increase

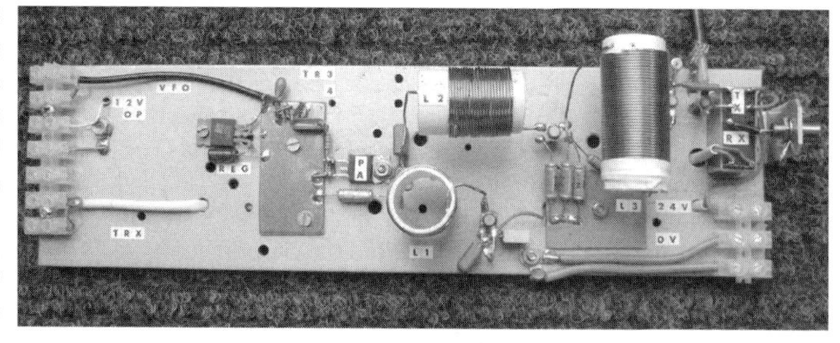

**Fig 10.22: The 100W 472kHz amplifier ready for testing. This is the version without PA keying**

the power by selecting taps, key the rig in short bursts and check for overheating of MOSFETs and cores. When you are happy that the transmitter is working OK, load it up to 15A PA current and slowly move the Hall device nearer to the end of the ferrite rod (CH1) until the protection circuit trips. Move it just a tiny bit further away and fix in position with silicone rubber. The receive preamplifier filter inductors can be aligned using a signal near 137kHz; the tuning is very sharp.

## GW3UEP 100W 472kHz CW Transmitter

This simple 100W MF CW transmitter and the 25W QTX ('Quick-TX', see [25]) were developed by Roger Plimmer, GW3UEP. Many of his designs are in use on the 472kHz band.

### Stable VFO

The 3.8MHz VFO is housed separately (**Fig 10.19**) to reduce heat coupling from the PA affecting stability. It includes division by eight to the 472kHz band (**Fig 10.20**) and has a low component count yet will produce a stable enough signal for CW, Opera and QRSS modes. The oscillator runs continuously.

It is important to take care to ensure maximum stability of the 3.8MHz oscillator. Inductor L1 is laquered then hot glued in place 4mm above the ground plane and clear of metalwork. Note the temperature coefficients specified for the capacitors in **Fig 10.20**. Non-ferrous (eg brass) screws/fixings should be used in the vicinity of the toroid. The VFO box is separate from the PA unit in order to avoid thermal coupling and temperature change. More details can be found at [25].

### Keying

Simple CMOS keying of the VFO is achieved using the 4024 divider reset-line. When used with the companion 100W transmitter, the keyed RF envelope is free from spikes and glitches, reducing key-clicks. For modes where there is a requirement for a coherent signal, or for minimal key clicks, keying of the PA is recommended (see below).

## 100W POWER AMPLIFIER

This amplifier produces 100 watts at high efficiency from a 24V regulated supply (operation over the range 14-24V is recommended, though power output will be reduced at lower voltages). It is nevertheless simple to build. **Fig 10.21** shows the circuit of the amplifier and a prototype can be seen in **Fig 10.22**. The IRF540 MOSFET was chosen for operation as the power switch, TR1. Note that if other devices are used, for instance an IRF640, resistor R10 must be fitted directly on the gate pin. The PA operates in switch mode (Class E) with drain efficiency in the 80% range.

TR3 and TR4 form a zero-biased complementary voltage follower, buffering the IC2 output stage and providing adequate source/sink current for the IRF540 gate charge (alternative devices for TR3 and TR4 are BC549/559, BC337/327, BC109/BCY71 or 2N3904/3906). The gate is AC-coupled to the buffer and DC-restored to ground, to prevent high DC current flow in TR1 should a fault occur in the 472kHz drive.

The output circuit provides matching and low-pass filter functions, and presents a clean sine wave into the 50 ohm load. C1/L1 forms a resonant MF tank circuit. L-match C2/L2 transforms the 50-ohm output load to a lower impedance at the drain. The output inductors are air-cored and wound on 22mm diameter plastic 'waste pipe' available from plumbing suppliers.

PA keying (if required – see above) is achieved with P-channel MOSFET TR2, which also shapes the keyed RF envelope and eliminates key-clicks. The key input switches TR2 gate via R6 and R7, which along with C10 also set the rise and fall times. R5 ensures stability by rolling off the frequency response of TR2, forming a LPF with its input capacitance. If keying is to be done in the oscillator stage instead, this circuitry can be omitted and TR2 replaced by additional transmit-receive relay contacts, which switch-off the PA drain supply on receive for key-down netting.

The maximum supply voltage for the transmitter is 25V; this allows for a voltage drop across TR2, which has VDS of 1V at 5A supply current. R6/R7 reduce the gate-source voltage of TR2 to 14V with a 25V supply. Heat sinks are required - at maximum output TR1 dissipates 20-25W, and TR2 5W. Ideally, a stabilised 24V PSU with current limiting set to approximately 5A should be used. Additionally, a 5A quick-blow fuse should be incorporated. An un-regulated PSU should deliver 24-25V maximum on load.

A typical setting-up procedure includes the following steps and approximate values (assumes a 24V DC supply. Measurements made using a DVM, plus oscilloscope with a 10:1 probe for the RF tests):

- Terminate the transmitter output with a 100W dummy load/power meter and observe the DC supply current
- With no VFO input, check PA current 20mA when switched to transmit and receive [excludes relay current]
- Apply VFO input
- With Vpa = 0, check >10Vpk-pk at 472kHz across R4
- Apply 14V or 24V supply Vpa
- With transmitter key-down: check 100W RF output, 5A DC supply current (Vpa = 24V) or 30W RF output and 3A (Vpa=14V)
- Check that PA drain waveform is a clean pulse and that efficiency is >80%. The drain voltage waveform should be approximately 100Vpk-pk. The drain voltage waveform should look similar to that in **Fig 10.17**

It is important that the transmitter is connected to a properly matched load with an SWR of no more than 1.25:1. A simple QRP LF/MF reflectometer and power meter can be found at [48].

| Mode types | Efficiency |
|---|---|
| A1A (CW), FSK (no envelope modulation): | 100% |
| BPSK with envelope shaping (eg PSK31): | 89% |
| Two tone 'IFK' modulation (eg Throb): | 78% |

**Table 10.2: Theoretical efficiencies achievable using EER. Practical efficiencies may be some 20% lower**

## M0BMU's 200W Multi-Mode EER Transverter for 136kHz and 472kHz

The two main objectives of this design are to provide a convenient way of transmitting and receiving digital mode signals on the 136kHz and 472kHz amateur bands, and to act as a test bed for transmission of different modes using the EER (envelope elimination and restoration) technique for power amplification.

The transverter signal source is an HF SSB transceiver with audio input and output from a PC sound card. This enables the use of a wide range of 'sound card mode' software, together with the convenient VFO and filter facilities of the HF rig. The basic transverter mixes this signal with a 4MHz local oscillator to obtain a low-level output in the range 20kHz - 550kHz. For reception, the signal path is reversed to convert the LF/MF input signal up to the 4.0 - 4.5MHz range. Transmit output at 200W PEP is achieved using a class D PA and modulator using the EER technique, although the transverter also has a low-level output for use with a conventional linear PA.

Envelope elimination and restoration (Kahn technique) is a method of generating amplitude- and phase-modulated signals

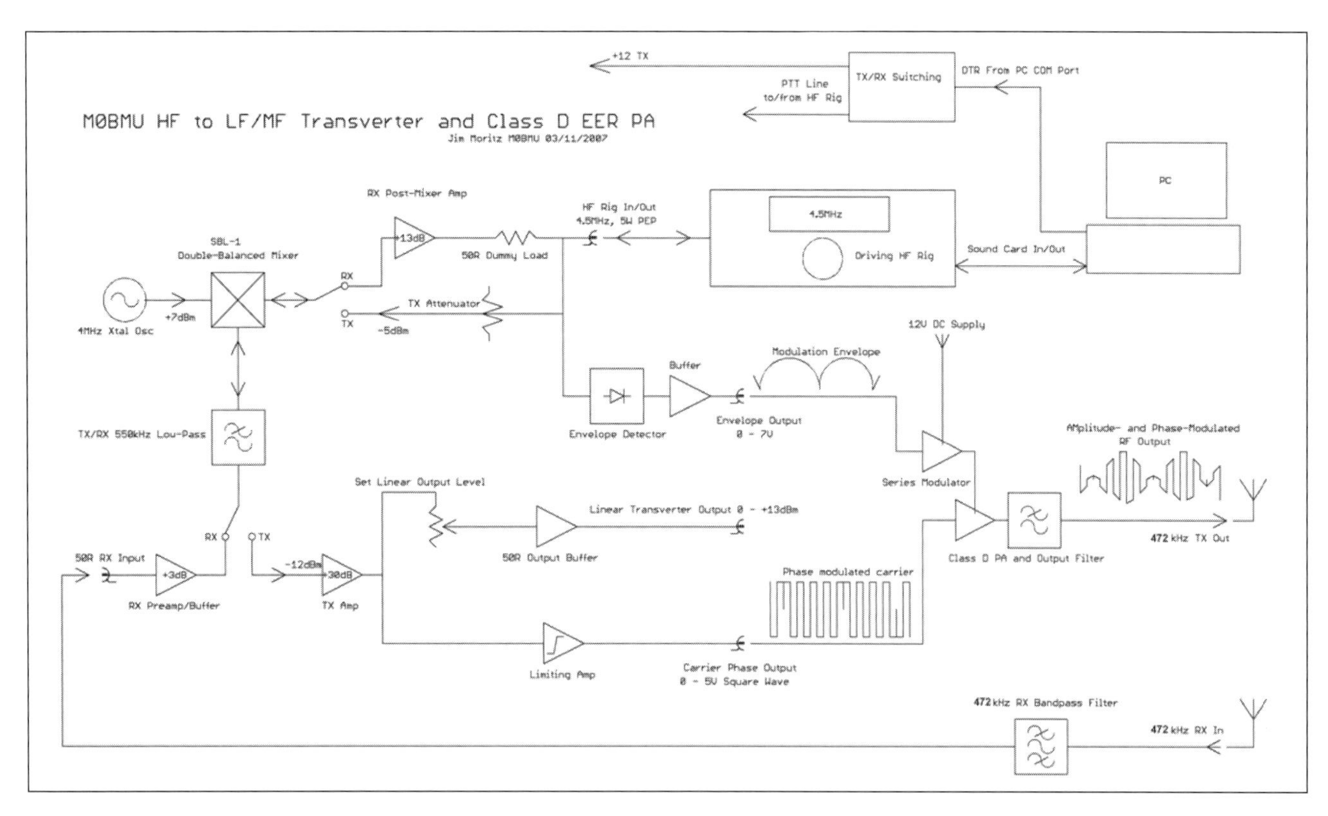


**Fig 10.23: EER Transverter block diagram**

using a high efficiency, non-linear PA, without introducing excessive distortion. The signal to be amplified is divided into a carrier phase channel and an amplitude envelope channel.

The constant-amplitude, phase-modulated carrier frequency is applied to the PA input, which can be a class D or E switching mode type to achieve high efficiency. The amplitude modulation signal is used to modulate the DC supply voltage to the PA. The output from the PA retains the phase modulation in the carrier signal, but now also varies in amplitude proportional to the envelope modulation signal. Since any type of signal is effectively a carrier frequency with a combination of amplitude and phase modulation, the EER technique can in principle be used with any form of modulation. If the high efficiency amplifier is used in combination with a switch-mode modulator, very high efficiencies are possible, making the technique popular for high-power AM/SSB/digital broadcast transmitters.

For these relatively high power, wide-band transmitters, complicated techniques are required to ensure alignment between amplitude and phase channels. But for LF/MF amateur applications at moderate power levels and using narrow-band modulation, satisfactory signal quality can be achieved quite easily.

This design uses a simple linear modulator which dissipates a significant amount of the DC input power. For types of modulation with a reasonably high crest factor (ie ratio between average power and PEP) the losses in the modulator are quite low. For example the efficiency calculated for some common types of modulation, assuming an idealised system with a 100% efficient PA and a series modulator that delivers the full supply voltage to the PA at modulation peaks gives the results shown in **Table 10.2**.

In practice, there are additional losses in both PA and modulator, so in this design actual efficiency is perhaps 20% less than these figures, but this is still better than a class AB linear PA transmitting the same modes. The linear design is simpler than a switch-mode modulator, and power dissipation in the modulator is quite manageable at the 200W output level.

A block diagram of the transverter system is shown in **Fig 10.23**. The driving IC-718 HF transceiver is operated at 4.1 - 4.5MHz IF frequency. This frequency was chosen to utilise some 4MHz crystals that were available; the IC718 can easily be modified to operate on frequencies outside HF amateur allocations.

The transverter IF input/output circuit is wide-band, so other input frequency ranges could be used just by changing the crystal frequency. The low-level transverter output is also broad band, and is about 3dB down at 20kHz and 550kHz, allowing it to be used anywhere in this range, although the PAs described are narrow band and limited to frequencies around 136kHz and 472kHz. The receive channel covers a similar frequency range, and achieves about 13dB SNR with a 0.1μV input signal in 250Hz bandwidth. This level of sensitivity allows the use of separate small receiving antennas to combat interference problems.

In the transverter, the carrier phase and amplitude signals are separated from the modulated signal from the HF transceiver. The carrier phase signal is obtained by feeding the down-converted signal into a limiting amplifier. The limiter output is a logic-level square wave at constant amplitude. The envelope modulation signal is obtained by rectifying the HF signal with a diode envelope detector, buffered by an op-amp follower.

The output from the HF transceiver is set to about 5W PEP; this was found to give a clean signal, and the level is high enough to be adjusted easily using the HF rig drive level control. Also, good linearity is obtained from the envelope detector with this high level signal. A buffered linear output at up to +13dBm is also provided.

The modulator and PA are designed for use with 13.8V DC. The objective was to produce a rig for possible future portable battery operation, although it is also convenient for use with the high current 13.8V PSUs present in many shacks. The PA circuits draw about 19A at full output.

Separate PA circuits are used for 136kHz and 500kHz; since

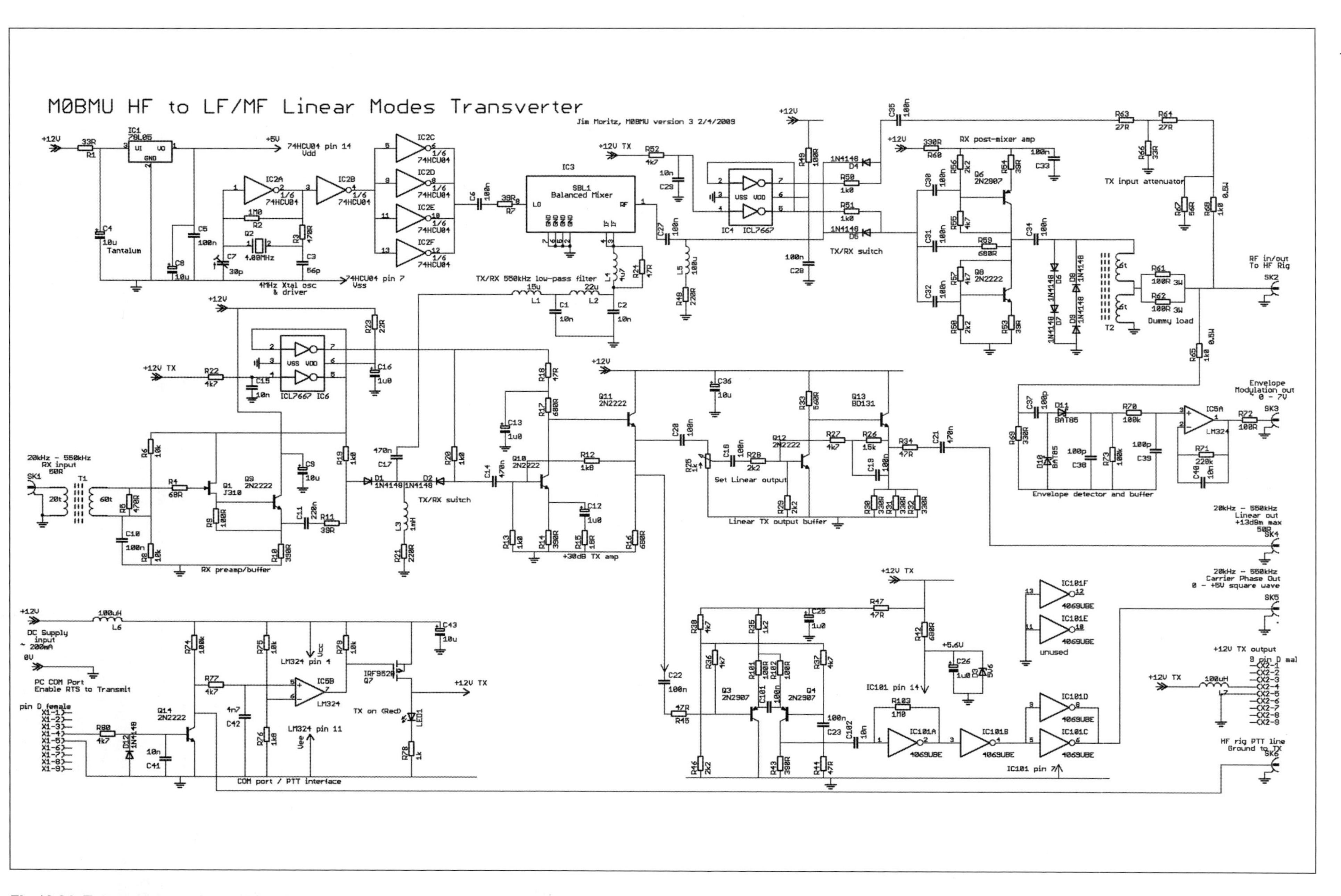


**Fig 10.24: Transverter receive and low-level transmit sections**

the MOSFET and driver components are cheap, and band switching would be quite awkward due to the high currents and low impedances involved, it was felt better to have separate PAs for each band rather than trying to produce a dual band design.

The schematic of the low-level sections of the transverter is shown in **Fig 10.24** The output from the driving HF transceiver is connected permanently to a 50-ohm dummy load, R61, R62.

On transmit, diodes D6 - D9 conduct, effectively grounding the other end of the load resistor via T2. An attenuated portion of the transmitter output is fed to the diode mixer via the diode transmit/receive switch.

On receive, D6 - D9 do not conduct, and the load resistor is in series with the output of the post-mixer amplifier Q6, Q8. This circuit was designed to make the transverter robust against being damaged by accidentally transmitting into the receive circuits, and can tolerate a 5W level indefinitely, and 100W for short periods.

The transmit/receive diode switch uses an ICL7667 driver IC to drive 10mA forward bias, or 2.5V reverse bias into switching diodes D4, D5. IC3, a SBL-1 double-balanced diode mixer is used for down/up conversion. The LF/MF signal enters or leaves via the DC-coupled IF port, to allow operation at low frequencies. The LO signal for the mixer is from a 4MHz crystal oscillator using one gate of CMOS inverter IC2. The remaining gates form a driver for the 50-ohm input impedance of the mixer. The LF/MF mixer input/output is filtered by a 550kHz low-pass filter to remove LO and image components.

The LF signal is switched between transmit and receive paths by another diode switch (D1, D2) - in this case the switch driver also switches the DC bias to the receive preamp and the transmit amplifier. This effectively increases the switch isolation, reducing possible problems with feedback between transmit and receive paths. If an external receive preamplifier was used with this transverter, it would be advisable to arrange that this was also switched off on transmit.

The preamp/buffer Q1, Q9 provides a 50 ohm termination at the receive input, and for the mixer and low-pass filter. The gain of this amplifier, and the overall receive gain, is only around 3dB. This minimises problems due to overloading by strong signals, whilst maintaining a reasonably low noise figure.

On transmit, the signal from the mixer is amplified by Q10, Q11. An output is taken from this point via a level-setting pot and buffer Q12, Q13 for optional use with a conventional linear PA. The signal is also applied to a limiting amplifier to generate the carrier phase output. The long-tailed pair Q3, Q4 provide a well-defined symmetrical limiting action. This output is applied to a biased CMOS inverter, IC101a, which gives additional gain at low signal levels, and additional inverters bring levels up to a logic-compatible 0V, +5V.

The output square wave is symmetrical within a few percent over the likely input signal range, which minimizes unwanted AM to PM conversion. The envelope modulation is extracted from the HF input signal by envelope detector D10, D11 and buffered by op-amp IC5A. The high signal frequency allows small detector time constants, providing adequate filtering without introducing large phase shifts. Schottky diodes are used for their lower forward voltage drop. The modulation envelope output is about +7V at the modulation peaks.

Transmit/receive switching is controlled via the PC COM port DTR handshake line by most digital modes software. The IC-718 and most other rigs are switched to transmit by pulling their PTT line to ground, which is done by Q14, which also switches MOSFET Q7 on via op-amp IC5B, providing a switched +12V Tx line, which is used for T/R switching in the transverter, PA and elsewhere if needed. This arrangement also operates the +12V Tx line if the PTT line is switched to transmit internally, for instance when operating CW. The only adjustment required to the transverter circuit is to trim the conversion oscillator frequency. The frequency can be set using an accurate frequency counter; preferably the HF rig frequency should also be trimmed. If a suitable counter is not available, the best approach is to set the HF rig to 4.000MHz CW receive, and trim the transverter oscillator so that the receive audio note is equal to the nominal CW pitch. This can be set within a few Hertz using the spectrogram display in many digital mode programs. This procedure makes any frequency offset in the transverter equal to the offset of the HF rig, minimising the error in the output frequency

The 200W series modulator schematic is shown in **Fig 10.25**. The modulator has to be capable of dissipating significant power - about 55W peak during normal operation, but reaching nearly 100W during worst-case conditions with the PA output shorted.

Four MOSFETs (Q3 - Q6) in parallel are used. To achieve accurate current sharing, each MOSFET has a current sensing resistor (R14, 18, 22, 26). The op-amps IC2a-d control the gate bias

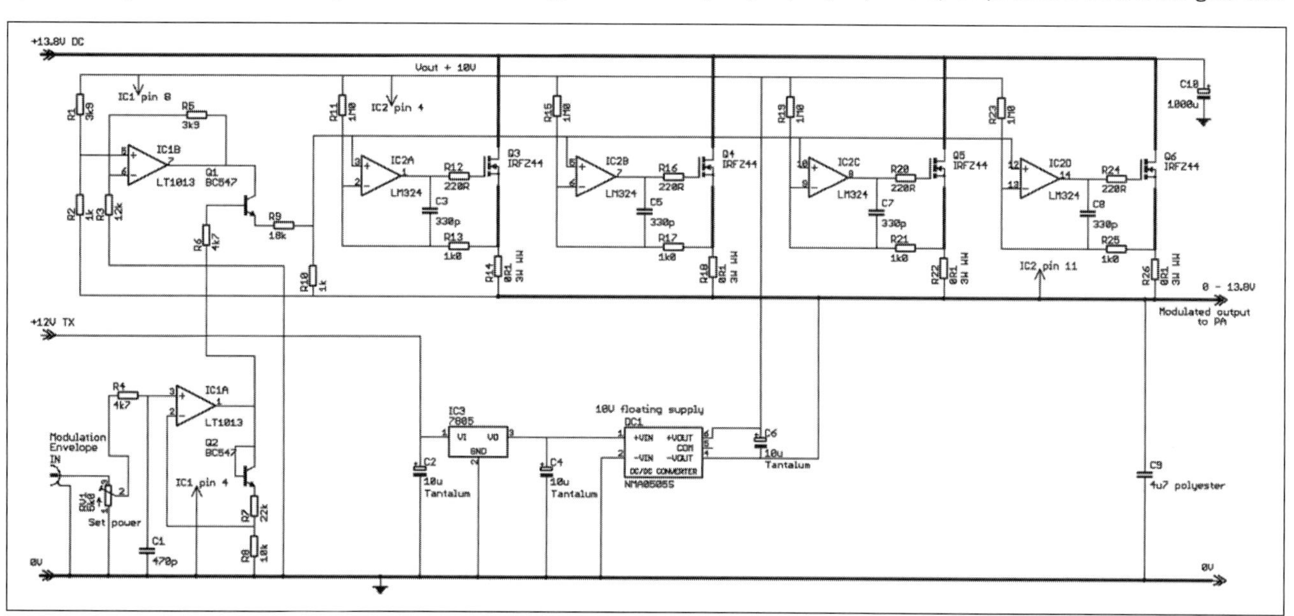


**Fig 10.25: 200W series modulator**

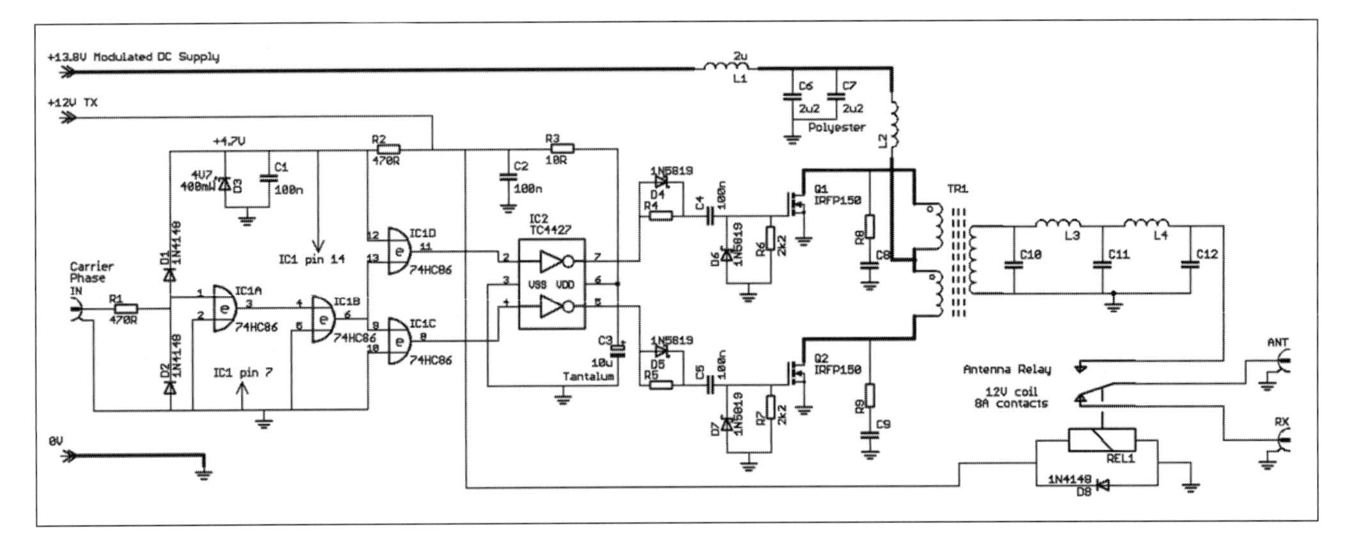


Fig 10.26: 200W class D PA

on each MOSFET so that the voltage across the current sensing resistor is equal to the control voltage applied to the non-inverting inputs of the op-amps; thus the current in each MOSFET is the same, and determined by the control voltage between the junction of R9, R10 and the output voltage rail. Since this control voltage is sensed between the terminal of R9 and the output voltage terminal, the overall circuit behaves as a voltage follower, with the output terminal voltage following the voltage applied to R9 with somewhat less than unity gain.

The modulation envelope input is amplified by IC1a and drives R9 via follower Q1. To prevent excessive power dissipation in both modulator and PA under fault conditions, a 'fold-back' current limiting characteristic is desirable, where the maximum output current is reduced at low output voltages, reducing the maximum dissipation that can occur in the modulator. Current limiting occurs because the maximum control voltage is equal to the collector voltage of Q1. This voltage therefore determines the maximum voltage across the current sensing resistors and the maximum output current. The collector voltage of Q1 is generated by op-amp IC1B summing a fraction of the output voltage with a DC offset produced by R1, R2. The voltage varies from about 2.6V with zero output to about 6.8V at maximum output voltage. This sets the current limit; with the given values of R9, R10 and the 0.1 current-sensing resistors to about 8A at zero output voltage, up to about 23A at maximum output. Thick lines in **Fig 10.25** represent high-current paths - up to 20A at 200W out.

A small 5V to +5/-5V DC - DC converter provides a 10V floating supply for the MOSFET driving op-amps. The DC-DC converter is supplied from the +12V TX line via IC3, so that it is switched off on receive to prevent possible QRM - the internal switching frequency is about 100kHz.

IC3 has a small clip-on heatsink. The current-limiting op-amps

| | |
|---|---|
| **Q1, Q2** | Mounted on 2.2°C/W Heatsink (RS components 490-7191) with silicone insulating washers |
| **Thick lines** | Represent high-current paths - up to 20A at 200W out |
| **C3** | Should be mounted very close to IC2 supply pins; IC2 should be positioned close to MOSFETs with short connections to gates and ground. |
| **TR1** | Primary connections and MOSFET source connections should be kept as short as possible |

**136kHz version**

| | |
|---|---|
| R4, R5 | 39R |
| R8, R9 | 10R, 3W metal film |
| C8, C9 | 2n2, 630V polypropylene |
| TR1 | ETD34 core and bobbin assembly, no air gap. Core material 3C90, F44, N67 or similar with mu around 2000, available from RS, Farnell etc. Primary 4 x 2 turns 1mm enamelled wire, each primary has 2 wires in parallel. Secondary 16 turns of 1mm enamelled wire. |
| C10, C12 | 23n; e.g. 22n + 1n in parallel, 1kV metallised polypropylene |
| C11 | 46n; e.g. 2 x 22n + 2n2 in parallel, 1kV metallised polypropylene |
| L1 | Approx. 2uH e.g. 6 turns 2 x 1mm enamelled wire on Micrometals T90-26 toroid core |
| L2 | Approx. 18uH e.g. 17 turns 2 x 1mm enamelled wire on Micrometals T106-26 toroid core |
| L3, L4 | 58uH nominal - 63 turns 0.8mm enamelled copper wire on Micrometals T106-26 toroid core |

**472kHz version**

| | |
|---|---|
| R4, R5 | Direct connection (D4, D5 can be omitted) |
| R8, R9 | 4R7 3W metal film |
| C8, C9 | 4n7, 630V polypropylene |
| TR1 | ETD29 core and bobbin assembly, no air gap, Core material 3C90, F44, N67 or similar with mu around 2000, available from RS components, Farnell, etc. Primary 4 x 2 turns 1mm enamelled wire, each primary has 2 wires in parallel. Secondary 16 turns 1mm enamelled wire. |
| C10, C12 | 6.2n; e.g. 4n7 + 1n5 in parallel, 1kV metallised polypropylene |
| C11 | 12.4n; e.g. 2 x 4n7 + 2 x 1n5 in parallel, 1kV metallised polypropylene |
| L1 | Approx. 2μH - e.g. 6 turns 2 x 1mm enamelled wire on Micrometals T90-26 toroid core |
| L2 | Approx. 7μH - e.g. 10 turns 2 x 1mm enamelled wire on Micrometals T90-26 toroid core |
| L3, L4 | 15.6μH nominal - 32 turns 0.8mm enamelled wire on Micrometals T106-2 toroid core |

Table 10.3 200W class D PA component notes

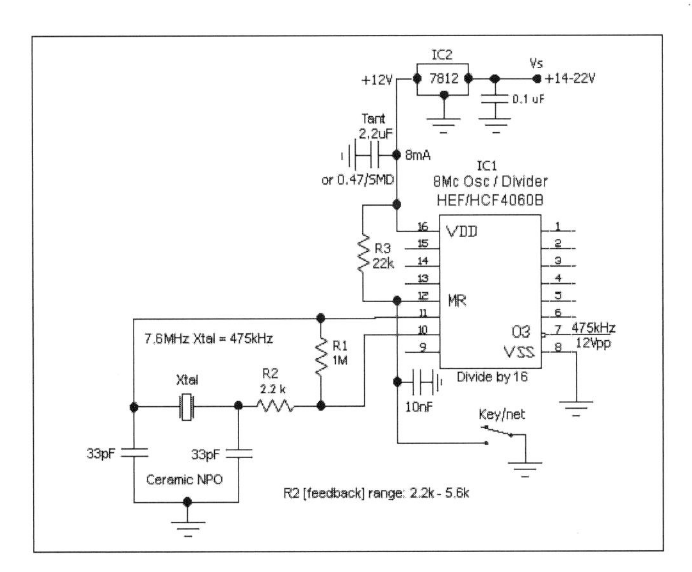


**Fig 10.27: GW3UEP's 472kHz crystal divider oscillator. Any crystal between 7.552 and 7.664MHz will produce a signal within the band by using this division ratio**

IC2 are only required to operate with inputs and outputs positive with respect to the modulator output, and the LM324 is a 'single supply' type, so the negative supply of this IC is connected to the output rail. IC1 inputs must operate down to zero, so its negative supply pin is connected to 0V, and it must also be a single-supply type. The total supply voltage of this IC is thus about 23.8V maximum, depending on the modulator output. In order to allow possible future use of the circuit with a higher supply voltage such as 24V DC, the LT1013 dual op-amp was chosen, as it allows up to 44V total supply voltage.

The heatsink used for the modulator depends on the type of operation required. For normal transmit/receive use, average power dissipation is low and a 1°C/W heatsink provides adequate cooling, but if long-duration beacon transmissions are envisaged, with possible long periods of overload resulting, a bigger heatsink will be needed. A fairly cheap and compact solution to this requirement is a large 'CPU cooler' heatsink with fan as available from PC parts suppliers. The one used for the prototype (Akasa AK862) was more than adequate, with only about 20°C temperature rise after a prolonged short circuit. Note that the fan has to be running for this type of heatsink to be effective. If a smaller heatsink is used, it would be advisable to attach a normally-closed bimetallic thermal switch with a trip temperature of around 70°C, to disconnect the +12V Tx line if the heatsink temperature is excessive. It was found that the Eltec SMPS 60/20 power supply used with the prototype modulator showed large fluctuations in output voltage when the modulator was keyed on and off at maximum output, apparently due to poor transient regulation - a large (68000µF) electrolytic capacitor across the PSU output improved regulation. Short, thick power cables are essential to minimise voltage drops.

The 200W PA circuit diagram is shown in **Fig 10.26**, with component details in **Table 10.3**. The push-pull current-switching type of class D output stage described in the design example above is used. It is well suited to low supply voltage, high current designs. The transverter carrier phase output signal is applied to a 'phase splitter' using the 74HC86 exclusive-or gates to generate antiphase square waves. These drive the gates of the output MOSFETs via the TC4427 gate driver IC. A 'half wave' output filter/tank circuit is used. Using the IRFP150 MOSFETs, about 85% efficiency is achieved. The MOSFETs are mounted on a 2.2°C/W heatsink attached vertically to the prototype board. For normal operation this runs quite cool, but for continuous beacon operation at full power, or if ventilation is restricted, a small fan is desirable to cool the heatsink and other output circuit components on the board. In the prototype, the air current produced by the modulator heatsink fan nearby gave plenty of cooling.

This transverter has so far been used successfully with a wide range of digital modes, including RTTY, PSK31, 'Olivia' and weak-signal modes such as WSPR and JT65. Also, it has been used for conventional CW and 'visual' modes such as QRSS and DF6NM's Chirped Hellschreiber using DL4YHF's Spectrum Lab software. All that is necessary to change modes is to load the appropriate software into the PC and 'follow the instructions'. For more on these modes, see the Digital Communications chapter.

The sound card output level is set to a point just below where the HF rig ALC starts to operate. The PA modulator level can be set by setting the sound-card software to produce a CW tuning tone and adjusting the output power pot to a point just below where saturation is reached. Or better, monitor the RF output on an oscilloscope, and adjust the pot to a point just below where the modulation peaks are clipped. To operate in CW, or to generate a CW output for tuning up, etc, the HF rig is just switched to CW mode. A longer article describing this transverter system can be found at [26].

## TRANSMITTER DRIVE SOURCES

Several different types of frequency source are in use for 136kHz and 472kHz transmitters. As in the case of receivers, frequency stability requirements vary depending on operating mode. For straightforward CW operation, a simple VFO could be used, the low output frequency tends to mean low drift also. Higher stability is needed for the extreme narrow band modes; frequency synthesisers of various forms are usually used. Crystal oscillators are sometimes used, usually in conjunction with a frequency divider, but the crystal frequency can usually only be pulled a few tens of hertz at the LF or MF output frequency, and being able to change frequency is very desirable even in a narrow band.

## OSCILLATORS

It is possible to make a fairly stable VFOs for 136kHz or 472kHz, usually operated at a high frequency divided down, which makes the components less cumbersome and can give quite good performance on CW. It is easier to find suitable components for higher frequency VFOs; high stability inductors suitable for LF are difficult to produce. An example of this type of VFO is used in the GW3UEP transmitter described above, where the VFO operates around 3.8MHz and is divided by 8. This is convenient for use with a frequency counter.

A simple fixed frequency oscillator for 472kHz is shown in **Fig 10.27**. It was designed for the GW3UEP 100W transmitter described earlier, and is capable of very high stability. Other division ratios could be used with available surplus crystals. Note that although this will provide a simple signal source, crystal control can restrict the number of contacts available.

Several amateurs have used a so-called 'crystal mixer' scheme. Two crystal oscillators are operated in the HF range, the difference in frequency being the desired output frequency. The output of the two oscillators are mixed together, and a low-pass filter at the mixer output selects the difference frequency component. Because the crystals operate at relatively high frequency, their pulling range is relatively large, and by making one or both oscillators a VXO, the whole band may be covered.

In this circuit, **Fig 10.28**, the crystal oscillators are both tuned by varicap diodes, and the oscillators and mixer are implement-

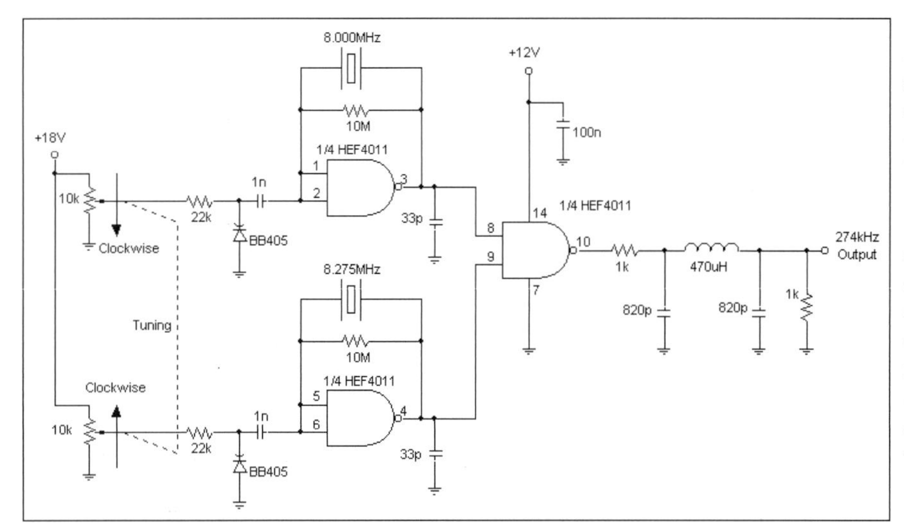


Fig 10.28: 'Crystal mixer' VFO from G0MRF transmitter

ed using a CMOS quad NAND gate IC. This type of frequency source is less stable than a simple crystal oscillator, since drift is due to differential changes in frequency between the two oscillator frequencies, but typically the output frequency is maintained within a few hertz over a considerable period, which is adequate for many applications.

None of the techniques outlined above are suitable for the very narrow-band modes increasingly used today and some kind of frequency synthesiser based upon a high-stability master oscillator is recommended unless your interest is solely in CW.

For generating modulated digital mode signals directly at LF or MF, an I/Q up-converter can be used. A modulated signal is generated at audio frequencies using PC software, with in-phase and quadrature components of the signal either being provided at the stereo sound card outputs by the software, or using hardware audio phase shift networks.

The modulated audio I/Q signal is up-converted to the desired RF frequency by a fixed oscillator, with quadrature channels providing cancellation of the unwanted sideband. Tuning over a narrow range is provided by varying the audio carrier frequency in software. A simple up-converter designed by G4JNT suitable for 136kHz or 472kHz use is described in [27] (see also the HF Transmitters chapter in this book).

Several types of frequency synthesiser have been used as LF or MF transmitter drive sources. A number of DDS (Direct digital synthesis - see the chapter on oscillators) synthesisers have been produced suitable for LF/MF operation [28, 29, 30]. These are well suited to extreme narrow-band application, since they are capable of high tuning resolution, often tuning in steps of small fractions of a hertz.

Many kits and modules are available such as the "Multi-Rock II" DDS from GW4GTE [55] **Fig 10.29** and the ARDU-5351 VFO from EA3GCY [56]. These designs are usually based on a DDS chip such as the AD9850 or the SI5351, controlled by a micro-controller of some kind. The Multi-Rock II uses a PIC processor which comes pre-programmed in the kit and the ARDU-5351 uses an Arduino. The oscillators on the DDS modules are usually cheap 50 or 20ppm computer types but these can be swapped for a high stability "TCXO" (temperature controlled crystal oscillator) unit if required.

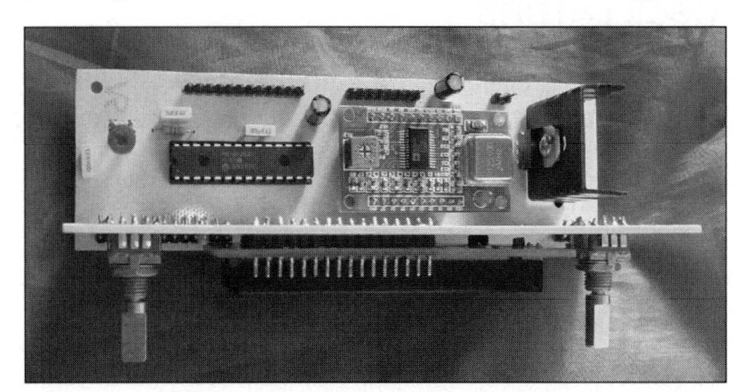

Fig 10.29: The Multi-Rock. A DDS VFO based on the AD9850

Some older synthesised signal generators are available as surplus quite cheaply, especially ones that do not cover the UHF range. These often have precision reference oscillators, capable of very high frequency stability, and are obviously useful for other purposes in the shack.

Another synthesised signal source that has been widely used is an HF transceiver, whose output is digitally divided down to the LF or MF range. A typical example of this scheme due to G3KAU is shown in **Fig 10.30**. Output from an HF rig at 13.6MHz is applied to two cascaded decade counters, giving output at 136kHz. This design also includes filtering to produce a sinusoidal output waveform for a linear transmitter. It was originally designed to include the now defunct 73kHz band in addition to 136kHz, but could be modified to include the 472kHz band instead. Note that most modern HF rigs inhibit transmission outside the amateur bands at frequencies such as 13.6MHz. However, in most cases only simple modification is required to enable transmission at out-of-band frequencies; contact the manufacturers for information.

## ULTIMATE3S QRSS/WSPR KIT

Hans Summers, G0UPL, markets a very reasonably priced kit for a DDS-controlled driver/transmitter running 100mW or so on any frequency from a few kilohertz to a couple of hundred megahertz (**Fig 10.31**). It is intended as a low power beacon transmitter with pre-programmed messages being transmitted

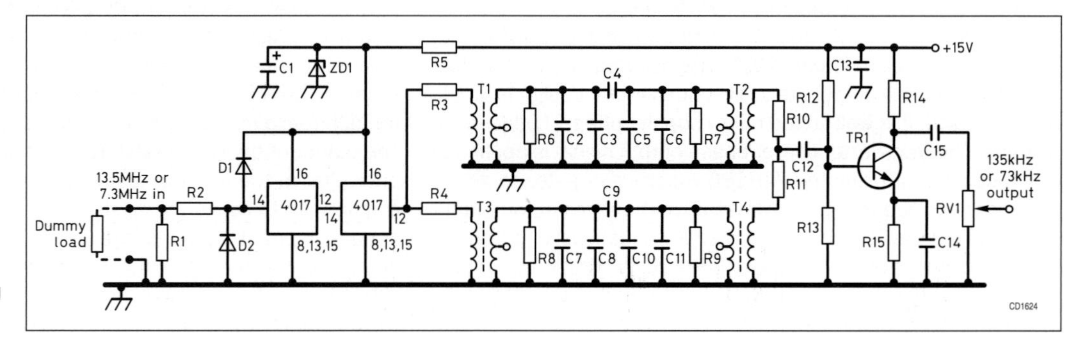


Fig 10.30: G3KAU divide-by-100 circuit

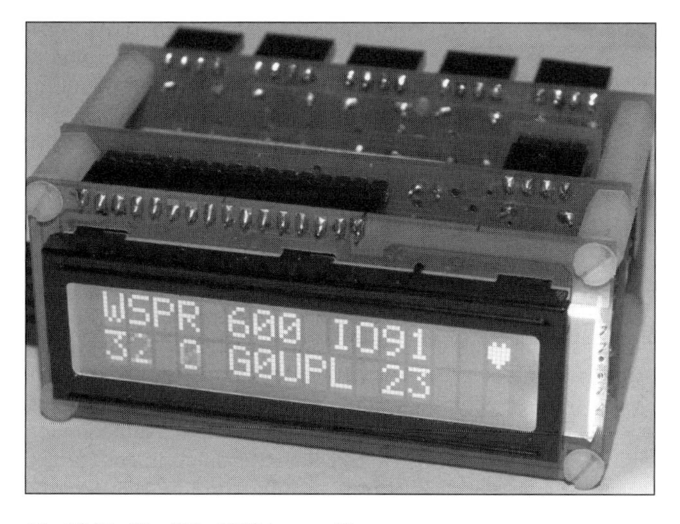


**Fig 10.31: The U3s DDS transmitter**

in various built-in modes including QRSS, DFCW, CW, FSK-CW, WSPR, JT9 and OPERA. Facilities are included for GPS locking if required. A range of inexpensive low pass filters are available from the same source [31].

Several 472kHz operators have used this kit as a stand-alone low power WSPR transmitter. It also makes a good driver for an LF or MF transmitter. By adjusting the settings appropriately, it can transmit an unmodulated constant carrier, which can turn it into a useful signal generator, or even as the driver for a keyed CW transmitter such as those described above.

## COMMERCIAL TRANSCEIVERS

For many digital modes, eg WSPR, it is convenient to use an SSB transmitter to generate the tones required and an existing HF transceiver can be used in conjunction with a transverter such as the one described above to produce a few watts on LF/MF. There are a few commercial transverters available such as the Monitor Sensors TVTR1 (for 472kHz) [1] and TVTR2 (for 136kHz) or the Minikits EME223 which covers 472kHz [2]. Other transverters are described in [3].

A few modern amateur transceivers, including some SDRs, have transmit outputs directly on the 136 and/or 472kHz bands but these are typically at the milliwatt level and will require considerable amplification. It is possible to modify some transceivers, such as the IC-706 and IC-7300, to produce some low frequency output but again this is at low power. Modifying a commercial radio should of course be carried out with great care and will invalidate any remaining guarantee.

Note that some digital modes (but not WSPR) will require the use of linear amplifying stages.

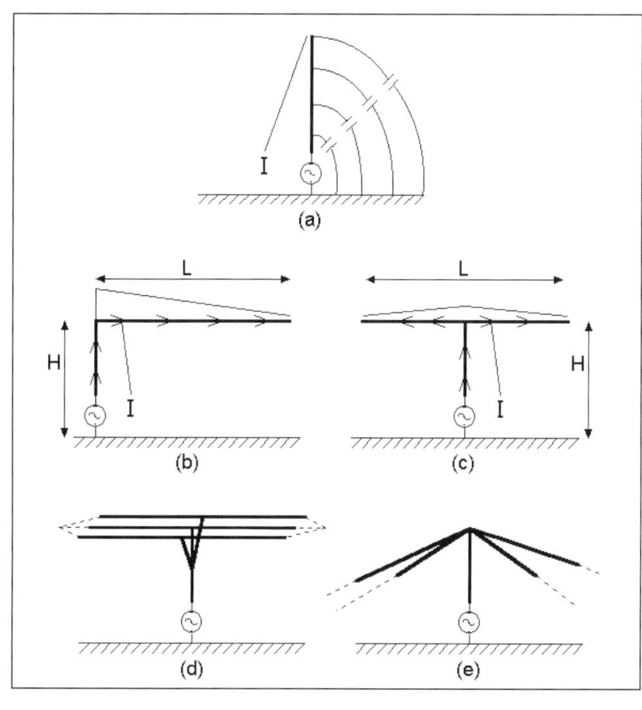


**Fig 10.32: Vertical antenna configurations**

## TRANSMITTING ANTENNAS & MATCHING

The underlying principles of antennas are of course the same at 136kHz and 472kHz as at other frequencies, but with operating wavelengths around 2200m and 635m respectively, most amateur antennas can only be a small fraction of a wavelength in length. This means they are always operated far below their self-resonant frequency, and require a large amount of inductive and/or capacitive loading in order to present a resistive load to the transmitter.

'Electrically small' antennas of this type are very inefficient, typically 0.1% efficiency would be a 'good' figure for a back-garden amateur antenna on 136kHz, rising to a few percent at 472kHz. However, in spite of this, successful amateur operation at LF and MF is done using quite ordinary wire antennas of similar dimensions to those used for HF operation.

Electrically small antennas fall into two types, vertical or loop. In the UK and Europe, the vast majority of amateurs have used vertical antennas at 136kHz and 472kHz, however LF loops have been popular with operators in North America, for reasons that will be discussed in the section on loop antennas. A well-known reference for information on amateur LF/MF antennas is ON7YD's 'Antennas for 136kHz' web page [32], and his dedicated 472kHz web-site [34] which contains extensive additional information.

## VERTICAL ANTENNAS

The vertical or 'Marconi' antenna is the most widely used LF transmitting antenna. It consists of a vertical monopole element a small fraction of a wavelength long, driven against a ground plane, **Fig 10.32**(a). The voltage on the element is nearly equal at all points, while the current is a maximum at the feed point, tapering to zero at the end, as the current flows to the ground plane through the distributed capacitance of the antenna element.

A figure of merit for a vertical antenna is the effective height, Heff. This is the height of a notional 'ideal' vertical antenna element with a uniform current all the way along its length that would generate the same radiated field as the real antenna when fed with the same current. Because of the non-uniform current distribution, the effective height of a real antenna is always less than the physical height.

The effective height can be increased by adding top loading to the basic vertical element, as is done with the T and inverted-L shown in **Figs 10.32**(b) and (c).

These can often be existing HF dipole or long wire antennas. A large proportion of the antenna current flows to ground through the distributed capacitance of the top loading wire, increasing the current flowing in the upper part of the vertical section. In either the T or inverted-L, the distributed capacitance of the vertical section, CV, and the horizontal section CH is approximately:

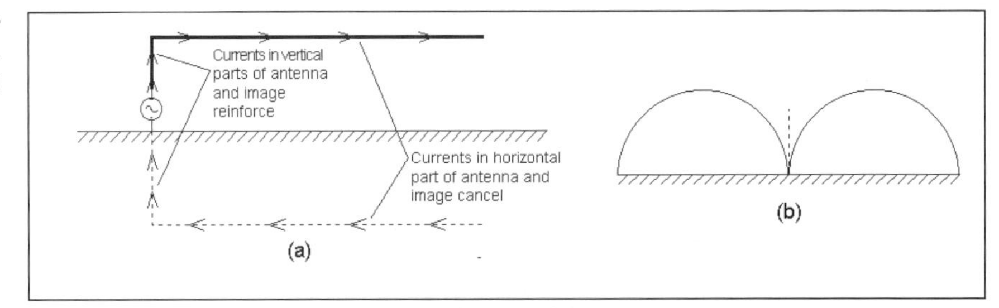


**Fig 10.33: (a) Cancellation of horizontally-polarised radiation from vertical. (b) Radiation pattern of short vertical antenna**

$$C_V = \frac{24H}{\text{Log}_{10}\left(\frac{1.15H}{d}\right)}, \quad C_H = \frac{24L}{\text{Log}_{10}\left(\frac{4H}{d}\right)}$$

Where CV, CH are in picofarads. H is the height of the vertical section, L the length of the horizontal section and d the wire diameter, all in metres. For plastic covered wire, d can be taken as the overall diameter of the insulation. A handy approximation for CV is 6pF per metre of height, and CH 5pF per metre of length.

To maximise the amount of top loading in a limited amount of space, multiple top-loading wires are sometimes used, as in the 'flat-top' T antenna of **Fig 10.32**(d). Due to proximity effects, multiple parallel wires have less capacitance than a single wire of the same total length. As a guide, two 1mm wires spaced 100mm apart will have about 39% greater capacitance than a single wire over the same span, spacing 1m apart will increase the capacitance by 68% compared to the single wire.

The effective height of the top-loaded vertical depends on the physical height H, and the relative values of CV and CH:

$$H_{eff} = H\frac{\frac{C_H}{C_V} + \frac{1}{2}}{\frac{C_H}{C_V} + 1}$$

With no top loading, ie with CH = 0, Heff = 1/2H. With a very large top loading capacitance, CH >> CV and Heff is nearly equal to H. Therefore, adding a large amount of top loading can nearly double the effective height of the basic vertical element. The actual shape of the top loading is not very important; the main objective is to maximise its capacitance.

Another form of vertical antenna often used at LF and MF is the umbrella, **Fig 10.32**(e). In this case, the top loading consists of sloping wires, with the advantage that only a single tall support is required. With only two top loading wires, the umbrella becomes an inverted-V, or with one wire just a sloping long wire. The drawback is that the current flowing in the sloping wires will have a 'downwards' component, partly cancelling the 'upwards' current of the vertical section. Bringing the ends of the loading wires close to ground level is therefore likely to reduce the effective height of the antenna considerably. Many combinations of length and angle of slope are possible, but provided the lower ends of the loading wires are at least half the height of the central vertical element, the overall effect of the umbrella top loading will be beneficial. The formulas for the T and inverted-L antennas can still be used to calculate the approximate effective height of the umbrella, with the modification that H is now the average height of the sloping wires, rather than the highest vertical point of the antenna, and L is the horizontal length of the sloping wires.

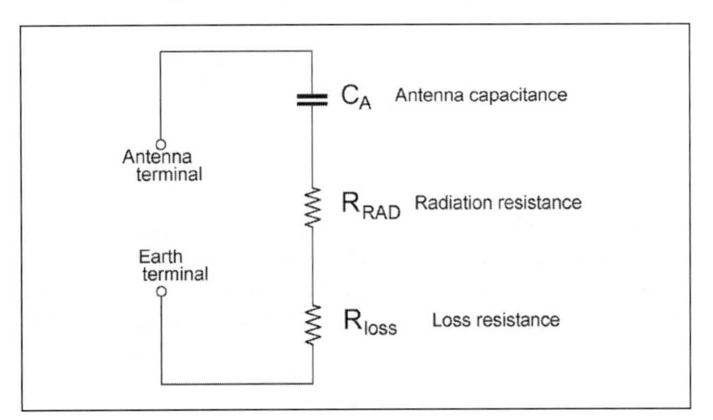


**Fig 10.34: Equivalent circuit of vertical antenna**

Although the horizontal parts of the antenna wire are often much longer than the vertical section, little horizontally polarised radiation is generated. This is because, when the height of the horizontal section is a tiny fraction of the wavelength, the effect of the horizontally-flowing current components are almost completely cancelled out by the 'image' currents reflected in the ground plane, **Fig 10.33**(a). Thus, these antennas are still classified as verticals, and the radiation produced is almost entirely vertically polarised.

The radiation pattern of any electrically short vertical antenna is virtually the same. It is omnidirectional in the azimuth plane, and has field strength proportional to the cosine of elevation, giving rise to maximum radiation towards the horizon, and a null vertically upwards, **Fig 10.33**(b). The directional gain of all electrically short vertical antennas is close to 2.62dB with respect to a dipole, or 1.83 as a power ratio, irrespective of their shape or size.

As an example, consider a typical antenna that might be used for 136kHz or 472kHz operation, a 40m long horizontal wire 10m above ground, made from 2mm diameter wire. Using the formulas given above, CV = 64pF and CH = 223pF. The antenna capacitance CA is the sum of CV and CH , 287pF. Heff becomes 8.9m, as expected somewhat less than the physical height.

The impedance of this and other electrically short vertical antennas can be represented by a series combination of two resistances, $R_{RAD}$ and $R_{loss}$, and CA (**Fig 10.34**). $R_{RAD}$ is the radiation resistance, which represents the conversion of transmitter power into radiated electromagnetic waves. $R_{RAD}$ is related to the effective height Heff, and the wavelength of the radiated signal λ in metres, by the formula:

$$R_{RAD} = 160\pi^2 \frac{H_{eff}^2}{\lambda^2}$$

For 137kHz, this becomes:

$$R_{RAD} = 0.0004 \times H_{eff}^2$$

For 475kHz:

$$R_{RAD} = 0.00449 \times H_{eff}^2$$

So $R_{rad}$ is typically very small; for the example antenna with 8.9m effective height, $R_{rad}$ is only 0.026 ohms at 137kHz, and 0.32 ohms at 475kHz. The power radiated from the antenna as electromagnetic waves is $I2R_{rad}$, so quite large antenna currents are required for appreciable power to be radiated.

The other resistive component of the impedance is the loss resistance, $R_{loss}$, which represents the power losses in the antenna and its surroundings. These include the resistance of the antenna wires and the ground system, and power dissipated due to dielectric losses in the ground under the antenna, and in objects near the antenna, such as trees and buildings. Power dissipated in $R_{loss}$ is converted to heat, and is therefore wasted. The same antenna current flowing in $R_{rad}$ also has to flow through the loss resistance, resulting in a power $I2R_{loss}$ being wasted, so the ratio $R_{rad}/R_{loss}$ is a measure of the antenna efficiency. $R_{loss}$ depends greatly on the environment around the antenna.

There is no known reliable way of determining $R_{loss}$ by theoretical calculation; measured values for typical amateur antennas in the LF and MF range vary from perhaps 10 to 200 ohms, with larger antennas having lower loss resistance, and decreasing loss resistance at higher frequency. For our example antenna an optimistic figure would be 40 ohms at 136kHz, decreasing to 20 ohms at 472kHz. The efficiency of this antenna is therefore 0.026/40 = 0.00065, or only 0.065% at 136kHz. Efficiency at 472kHz is 1.6%, nearly 25 times greater than the lower frequency, but still very low compared to antennas on the higher frequency amateur bands.

In many cases, actual efficiency is even lower, due to environmental effects; see the Practical Antenna Considerations section below.

## ESTIMATING RADIATED POWER

The Schedule contained in the Amateur license conditions specifies a maximum power level for the 136kHz band of 1W (0dBW) Effective Radiated Power (ERP), rather than the transmitter output power limit specified on most bands.

The power limit in the Notice of Variation issued for the 472kHz band is also differently specified. It is 5W (7dBW) Effective Isotropic Radiated Power (EIRP). EIRP is a rather like ERP, but the calculation also includes the difference in gain between the antenna in use and an isotropic radiator. The box below gives more detail.

On most amateur bands, we are permitted a maximum RF power, which we can then use with any antenna no matter how much gain it has. However, on the 136kHz band, the power limit is 1W ERP, the ERP indicating the power radiated from the antenna relative to a dipole in free space. Thus if the antenna has one hundredth of the gain of a dipole (-20dBd), the maximum power permitted from the transmitter is

1 x 100 = 100W.

This same idea applies on the 472kHz band, but instead of the radiated power being compared to a dipole, it relates to an entirely theoretical 'isotropic antenna' (the 'I' in EIRP) which radiates equally in every direction. A dipole has a directional gain of 2.1dB compared to an isotropic antenna. This is the first time that EIRP has been used in the UK licence. The result of this is that we need to factor in the gain of our antenna compared to an isotropic radiator in any calculation of the actual radiated power.

Using radiated power (ERP or EIRP) as the limit provides a certain equality, whereby someone with a small antenna can (theoretically at least) increase his transmitter power until he radiates the same power as a person with a much larger antenna.

It is therefore important to have an understanding of the concept of radiated power, and methods of determining ERP (or EIRP), when assembling a 136kHz (or 472kHz) station.

The gain of the antenna can be considered to be the product of the antenna's efficiency, and its directivity (directional gain) compared to a theoretical reference antenna - a half-wave dipole in free space for the 136kHz band and an isotropic radiator on 472kHz.

### Radiated Power Limits: ERP and EIRP

On most amateur bands, we are permitted a maximum RF power, which we can then use with any antenna no matter how much gain it has. However, on the 136kHz band, the power limit is 1W ERP, the ERP indicating the power radiated from the antenna relative to a dipole in free space. Thus if the antenna has one hundredth of the gain of a dipole (-20dBd), the maximum power permitted from the transmitter is 1 x 100 = 100W.

This same idea applies on the 472kHz band, but instead of the radiated power being compared to a dipole, it relates to an entirely theoretical 'isotropic antenna' (the 'I' in EIRP) which radiates equally in every direction. A dipole has a directional gain of 2.1dB compared to an isotropic antenna. This is the first time that EIRP has been used in the UK licence. The result of this is that we need to factor in the gain of our antenna compared to an isotropic radiator in any calculation of the actual radiated power.

Using radiated power (ERP or EIRP) as the limit provides a certain equality, whereby someone with a small antenna can (theoretically at least) increase his transmitter power until he radiates the same power as a person with a much larger antenna.

## ESTIMATING ERP BY CALCULATION

The ERP can in principle be determined from a knowledge of the transmitter power, antenna efficiency and directivity. Antenna efficiency can be estimated by calculating the radiation resistance $R_{rad}$ as described in the previous section, and measuring the loss resistance $R_{loss}$ using an RF bridge or similar. Multiplying the transmitter output power Pout by the efficiency gives the total radiated power, $P_{rad}$:

$$P_{rad} = P_{out} \frac{R_{RAD}}{R_{loss}}$$

An easier way to find $P_{rad}$ is to measure the RF current flowing at the antenna feed point, Iant. The radiated power is then just:

$$P_{rad} = I_{ant}^2 R_{RAD}$$

The ERP is the radiated power multiplied by the gain with respect to a dipole (but see below for converting this figure to EIRP), 2.62dB or a power ratio of 1.83:

$$P_{ERP} = 1.83\, I_{ant}^2 R_{RAD}$$

In summary, estimating ERP involves the following steps:

- Using the formulas given previously, calculate the radiation resistance $R_{RAD}$ using the measured dimensions of the antenna.
- Measure the antenna current (see LF/MF Measurements section for details of suitable RF ammeter designs).
- Calculate the ERP:

$$P_{ERP} = 1.83\, I_{ant}^2 R_{RAD}$$

Using the previous example of a 10m high, 40m long wire antenna with radiation resistance of 0.026 ohms at 137kHz, it can be seen that an antenna current of 4.6A is required to obtain 1W ERP. The power wasted in the 40 ohm loss resistance of the antenna in producing this current is 846W, emphasising the need for large transmitter powers on this band!

## ESTIMATING ERP BY FIELD STRENGTH

An equivalent definition of ERP is that it is the amount of power fed to a reference antenna that would produce the same received field strength as the actual antenna does, with the conditions that both reference and actual antennas are the same distance from the receiver, and that the direction from the antennas to the receiver is the direction of maximum gain. The reference dipole in free space is obviously a theoretical abstrac-

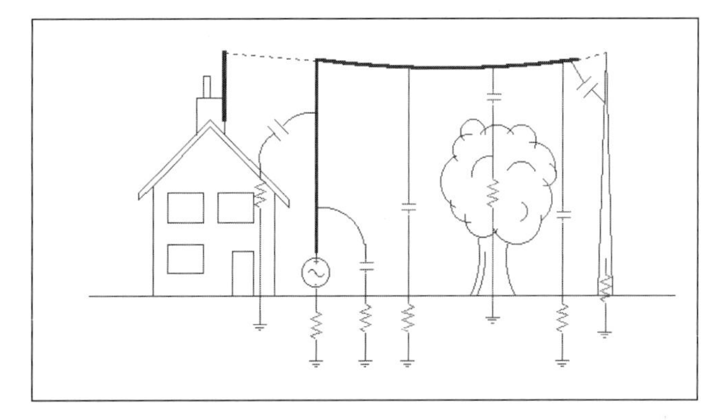

**Fig 10.35: Loss in the environment around an antenna**

tion (especially at these low frequencies), however it is easy to calculate what field strength E (in volts per metre) it would produce if it really existed:

$$E = 7\frac{\sqrt{P}}{d}$$

where P is the power fed to the ideal dipole and d is the distance in metres.

For example, at a distance of 10km from the dipole, with 1W feeding it, the field strength would be 700μV/m. Therefore, any transmitter and antenna combination producing a received field strength of 700μV/m at 10km distance is radiating 1W ERP. Thus, a definitive way of determining ERP is to measure the field strength at a known distance from the station. This has the advantage that it does not rely on the assumptions made regarding calculation of antenna radiation resistance and gain, but is by no means a simple measurement to make (see Field Strength Measurement section).

## ENVIRONMENTAL FACTORS

For antennas in 'near-ideal locations', ie located on open ground, free from obstructions such as buildings, trees and metal structures, it is generally found that there is close agreement between the value of ERP determined from field-strength measurements and that estimated using the antenna dimensions and antenna current. But many amateur antennas are in far from ideal locations, being in domestic gardens surrounded by many obstructions. In these conditions it is invariably found that the measured ERP is lower than the estimated value, typically by 3 - 6dB, but several decibels more for stations in urban or heavily wooded environments. This is due to reduction of the antenna's radiation resistance by the screening effects of the surroundings (see next section). It is not possible to determine this environmental loss from measurements on the antenna itself. However, since the effect of these losses is always to reduce ERP, the simple estimate above is a reliable method of setting an upper limit on the possible ERP of the station, and thereby ensuring compliance with licence conditions.

## CONVERTING ERP TO EIRP FOR 472KHZ

Having estimated the ERP using one or more of the methods described above, it must be multiplied by a factor of 1.4 (2.15dB) to get the power in EIRP as specified in the UK licence. This is the gain of the free space dipole used in the calculations above over an isotropic radiator (which theoretically radiates equally in all directions). The formula above now becomes:

$$EIRP = 2.56 \times 1^{1}_{ant} \times R_{rad}$$

For example, we have already determined that our typical antenna has a radiation resistance of 0.32 ohms at 475kHz, so in that case a current of just under 2.47A is required to reach the 5W EIRP limit. The transmitter power needed to achieve this with a typical 20 ohm loss resistance is only 122W.

Note that the same transmitter and antenna will produce 1.4 more times EIRP as ERP. Of course the actual radiated power is the same, but the power is simply defined differently.

**Fig 10.36: Similar antennas in different environments: (a) in an open field location; (b) in a domestic garden**

Fig 10.37. Running high power on 136kHz requires care. This burn was caused by VO1NA's antenna wire coming into contact with his ATU box on a wet and windy night

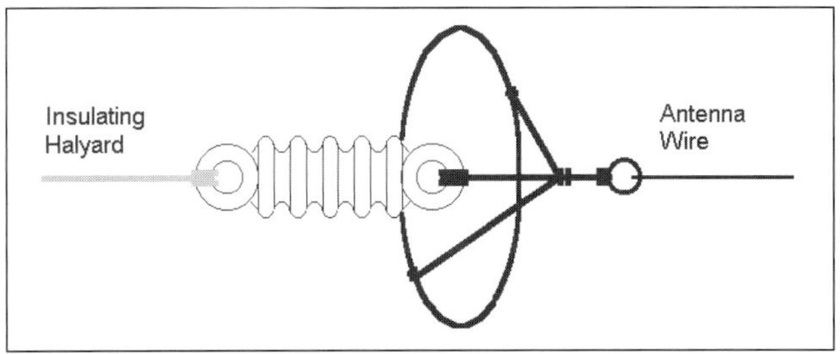


Fig 10.38: Insulator fitted with 'corona ring'

## PRACTICAL ANTENNA CONSIDERATIONS

"As much wire as possible as high as possible" is a good guiding principle for LF and MF transmitting antennas. As seen in the previous section, the radiated power is proportional to the radiation resistance of the antenna, and the radiation resistance is proportional to the square of the effective height, Heff. Therefore, for a given value of antenna current, and other things being equal, the amount of power radiated is proportional to the square of the height of the antenna. The most important dimension of an LF or MF antenna is therefore always its height.

The effective height can also be increased by maximising the capacitance of the top loading section, so the second most important dimension is the length of wire making up the top loading section of the antenna; ensuring this covers as much ground as possible will make the effective height of the antenna as close as possible to the physical height. However, as noted in the discussion of umbrella antennas and sloping wires, it is the average height of the top loading that is important, so having long loading wires that have a large sag, or droop close to the ground is counterproductive.

After achieving the maximum possible effective height, the next most important goal is to minimise antenna losses. The vertical antenna can be thought of as a capacitor, one plate of the capacitor being the antenna wire, and the other plate being the ground system. Losses occur due to the resistance of the 'plates', and also due to the 'dielectric', made up of the air in between, the ground underneath, and the other objects surrounding the antenna, such as buildings and trees (**Fig 10.35**). The major source of losses depends on the shape and size of the antenna. For large commercial antennas, most of the loss occurs in the resistance of the ground system, but for much smaller amateur antennas, experiments have shown that most of the loss occurs in the 'dielectric'. Part of the electric field of the antenna penetrates the ground, which is far from a perfect conductor. The electric field also induces RF currents to flow in the structure of buildings, and the wood and leaves of trees and plants, leading to further loss.

Measurements taken a few years ago showed that antenna loss resistance can be greatly increased due to these effects; the loss resistance of a particular inverted L antenna erected at M0BMU, in a domestic back garden surrounded by trees, was measured as 61 ohms at 136kHz, and 26 ohms at 502kHz, while a near-identical antenna erected in an open field for comparison had loss resistances of only 8.5 ohms at both 136kHz and 502kHz (see **Fig 10.36**).

A further environmental effect is the reduction in effective height, Heff, caused by the screening effect of the objects near the antenna. The fields close to the antenna induce RF currents flowing to ground in these surrounding objects, which therefore behave as parasitic antenna elements. Since the direction of parasitic element current flow is opposite to that in the antenna itself, and the element spacing is a small fraction of a wavelength, the radiation from the antenna is partly cancelled. Measurements show a substantial fraction of the antenna current can return to ground via these paths, significantly reducing ERP. The effective height of the antennas in **Fig 10.36** calculated from the geometry of the antenna wires was 8.4m, and measurements of the radiated field strength of the open-field antenna of of **Fig 10.36**(a) yielded similar values. However, similar measurements on the home antenna of **Fig 10.36**(b) gave Heff as only 5.8m at 502kHz, and 4.5m at 136kHz. Therefore the environment can be the dominant effect on overall antenna efficiency, both due to increased resistive losses and reduced radiation resistance due to reduced effective height.

Maximising the height of the antenna obviously keeps the antenna wires as far as possible from the ground and other lossy materials. Also, maximising the size of the top loading will reduce losses; a higher antenna capacitance will lead to a reduced voltage (see next section). This results in a reduced electric field intensity, reducing dielectric losses, which are proportional to the square of the field strength.

The capacitance between the antenna wires and poorly-conducting objects such as trees and buildings should be kept to a minimum by keeping the antenna, including the downlead, as far from them as possible. If the antenna is supported by a metal mast, it is desirable to maximise the clearance between mast and antenna, supporting the antenna wires some metres clear using insulating halyards, or perhaps replacing the top part of the mast with fibreglass.

The effect of the mast will be reduced if it is well insulated from ground, but if it is not possible to adequately insulate the mast, ensure it has a good earth connection in order to minimise power loss due to the circulating RF current.

## ANTENNA VOLTAGE AND SAFETY

An important practical consideration, particularly at 136kHz, is the antenna voltage. The RF voltage Vant is almost the same at all points on the antenna, and is approximately equal to the antenna current multiplied by the reactance of the antenna capacitance:

$$V_{ant} = I_{ant} \times \frac{1}{2\pi f C_A}$$

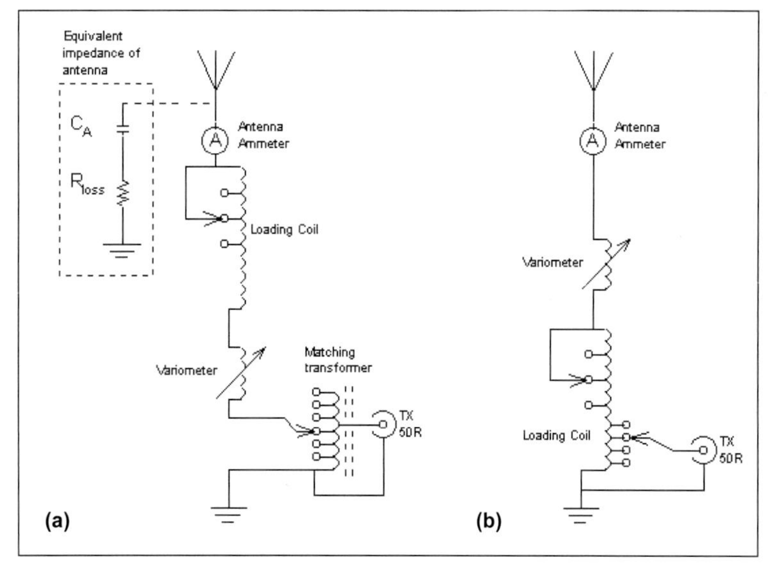


**Fig 10.39: (a) LF antenna tuner; (b) Alternative antenna tuner circuit**

At 137kHz, and with C in picofarads, this formula reduces to:

$$V_{ant} = 1.16 \times 10^6 \times \frac{I_{ant}}{C_A}$$

For our previous 10 metre high, 40 metre long wire example, Cant = 287pF, Iant = 4.6A, Vant becomes 18,600V!

Very large voltages are typical, particularly when using high powers with small antennas at 136kHz. Special attention must therefore be paid to the insulation of LF antennas. At 472kHz, with reduced antenna impedance, voltages are much smaller (typically a few kV) and high voltage breakdown is much less of an issue.

A particular problem experienced by many LF operators is corona discharge. Corona occurs where the electric field around the antenna is most intense; typically at the ends of wires where they attach to insulators, at sharp bends, where loose ends of wire project, or where the antenna passes near another object that is at ground potential. Corona manifests as a continuous, diffuse electrical discharge that produces a hissing sound; it is often very hard to see, even in the dark it may appear only as a faint glow. However, it absorbs substantial RF power, and so generates a lot of heat, which can ignite plastic insulators, support ropes and antenna tuning components, leading to the collapse of the antenna. Several instances have occurred where plastic insulators, ignited by corona discharge, have dripped burning plastic on to the ground beneath.

If plastic insulators are used, ensure that they are in a position where burning debris cannot fall on a building or people. If insulators must be positioned over a building, glass or ceramic ones are safer. Even glass or ceramic insulators are eroded and cracked by corona over a period of time.

The downlead of the antenna is equally prone to corona, and this represents a particular hazard if the antenna wire is brought directly into the shack; a number of minor fires have resulted from corona discharges igniting nearby woodwork (**Fig 10.37**). Measures to prevent corona include the following:

- Make the capacitance of the antenna top loading as large as possible to minimise the antenna voltage
- Use 'corona rings' (**Fig 10.38**) at ends of antenna wires, or at sharp corners, to reduce the field gradient. These can be made from loops of stiff wire, 100mm or more in diameter.
- Dress the ends of the wire so there are no sharp projecting points.
- Use insulating rope halyards, rather than conducting wires, to support insulators; this will reduce the voltage gradient across the insulator
- Keep antenna wires well clear of buildings and trees, and other antennas. Locate the antenna downlead and antenna tuner away from the house or shack

It will be seen that these are mostly the same guidelines as those given for reducing antenna losses, so taking precautions against corona also help to improve efficiency.

High antenna voltages are obviously also hazardous to people. Coming too close to an LF antenna wire whilst transmitting at high power can result in a severe RF burn, even without direct contact. Be sure that it is not possible for a person to come within a few metres of any part of the antenna while operating. It is also possible for large metal objects near the antenna, such as ladders, garden furniture or other antennas, to have sufficient RF voltage induced on them to cause burns to a person touching them. This can be prevented by earthing all such objects.

## GROUND SYSTEMS

A ground system is required for any Marconi antenna. In professional MF, LF and VLF antenna systems, the ground system is normally the dominant source of losses, and earth mats containing many kilometres of buried wire are used. But for amateur LF and MF antennas, it is usually easy to produce a ground system that has a negligible contribution to overall antenna loss, since other losses in the antenna system are much higher.

The most widely used ground system consists of a number of ground rods connected to a common point close to the antenna tuner. This type of ground works well over much of the UK, where soil conductivity is quite high. As a very rough guide, a single 1m long ground rod will have a resistance of the order of 20 ohms; where several rods are used, spaced a few metres from each other, this figure is roughly divided by the number of rods. The losses in other parts of the antenna system are normally tens of ohms or more, so a point of diminishing returns is quickly reached where the ground system resistance is only a few ohms, and further improvements to the ground system yield little reduction in overall loss resistance. If a large number of ground rods are used, it is found that relatively little RF current flows in the rods that are further away from the feed point. This appears to be because the distributed inductance of the longer connecting wire has a large impedance compared to the rest of the ground system. This may not be true where the ground conductivity is very low. In this situation, a ground system distributed over a wider area could be expected to give a useful improvement, although little practical data is available.

Ground rods intended for domestic mains earthing are ideal; these are designed to be rigid enough to be hammered into the ground. It is also possible to use copper water pipe in very soft ground, or inserted into pre-made holes, but this material quickly buckles when hammered. The rods should be as long as possible, and in contact with permanently damp soil.

Systems of buried radial wires as used for HF verticals have also proved effective. Except where ground conductivity is very low, radials tend to be more useful on 472kHz than on 136kHz.

## MATCHING VERTICAL ANTENNAS

In principle, you could use the same matching networks used for HF antenna matching, such as the pi- or T networks, to match vertical antennas at 136kHz or 472kHz. But in practice it

Fig 10.40: Typical loading coils for 136kHz

Fig 10.41: Some of the loading coils of Table 10.4

is found that the component values, particularly for capacitors, are impracticably large, and for 136kHz require very high ratings due to the high antenna voltage. The two most popular LF/MF antenna matching circuits are shown in **Fig 10.39**.

In **Fig 10.39**(a), a series loading coil has an inductive reactance that cancels out the capacitance, CA of the antenna. The resistive component of the impedance (practically equal to the loss resistance) is then matched to 50 ohms, or other value of transmitter output impedance, using a ferrite-cored transformer. The capacitance of back garden amateur antennas, typically hundreds of picofarads, corresponds to a loading inductance of a few millihenries at 136kHz, and a few hundred microhenries at 472kHz. Because the antenna reactance is much larger than the resistance, the loading inductor must be capable of fine adjustment to obtain resonance accurately. Coarse adjustment is usually achieved by a series of taps on the loading coil. For fine tuning, the inductance is made variable over a narrow range using a variometer, described below.

For most antennas, $R_{loss}$ is between perhaps 10 and 200 ohms, requiring a transformer with turns ratios between about 1:2 and 2:1 to match to 50 ohms. One design of 136kHz matching transformer, satisfactory at power levels up to 1kW, uses an ETD49 transformer core in 3C90 ferrite material, wound with 32 turns of 1.5mm enamelled copper wire, tapped every two turns. A much smaller transformer can be used at 472kHz, for example an ETD29 core assembly. The 50-ohm transmitter output is connected at the 16 turn tapping point, and the 'cold' end of the loading coil connected to the tap that gives optimum matching. This matching arrangement is very straightforward to use, since the adjustment of antenna resonance and resistance loading are almost completely independent.

Another popular matching network uses a tapped loading coil as shown in **Fig 10.39**(b). The low potential end of the coil is equipped with closely-spaced taps, so the loading coil also performs the function of the matching transformer. Although this is physically simpler than **Fig 10.39**(a), the electrical behaviour of this circuit is more complicated.

The primary and secondary of the transformer are not tightly coupled, so the transformer impedance ratio will not closely correspond to the turns ratio and the adjustment of the antenna to resonance and the selection of the impedance-matching tap will be somewhat interdependent. However, it is not difficult to find a suitable tapping point by trial and error, and this will not often then need to be changed.

The range of antenna loss resistance that can be matched using the tapped loading coil depends on the coil geometry. In general, if the coil has a relatively small diameter and coarse winding pitch (ie it is wound with thick wire), the maximum value of $R_{loss}$ that can be matched is quite low. If the coil has large diameter and fine winding pitch, much higher $R_{loss}$ can be matched. However, the coil is then less suitable for low resistance antennas, because the required tap is only a few turns from the grounded end of the coil, giving very coarse steps in matching adjustment. A Microsoft Excel spreadsheet is availa-

| | Former | Winding length | Wire | Turns | L, mH | Series R @137kHz | Q @ 137kHz |
|---|---|---|---|---|---|---|---|
| 1 | 110mm dia PVC tube | 280mm | 0.9mm Enamelled | 281 | 2.75 | 9.4 | 250 |
| 2 | 156mm dia PVC tube | 190mm | 0.9mm enamelled | 190 | 3.50 | 15 | 220 |
| 3 | 200mm dia PVC tube. | 435mm | 2mm dia (1.25mm dia conductor) Teflon insulated | 225 | 3.75 | 10.3 | 330 |
| 4 | 395mm dia Polythene bucket | 300mm | 2.7 dia 729 strand Litz | 109 | 3.89 | 3.0 | 1100 |
| 5 | 485mm dia ribbed "sectional manhole", polypropylene | 200mm (multi-layer) | 4.1mm dia Polythene insulated 729 strand Litz | 79 | 3.83 | 5.1 | 650 |
| 6 | ditto | 440mm | 3.8mm dia PVC insulated | 95 | 3.37 | 9.6 | 300 |

Table 10.4: Practical loading coil data

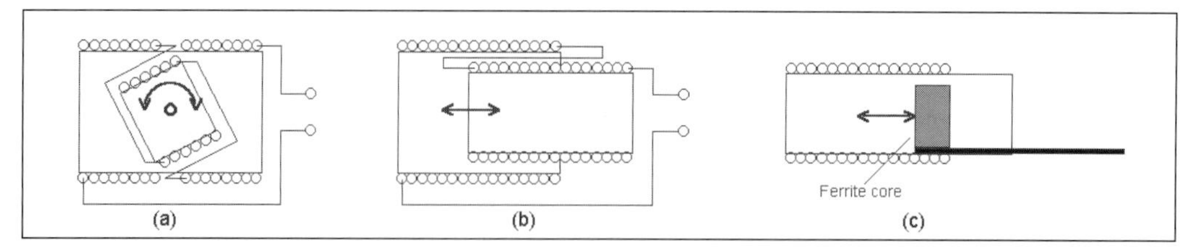


Fig 10.42: Three forms of variometer

ble at [33] for calculating the range of antenna loss resistance that can be matched in this way with any coil diameter and winding pitch.

Another disadvantage of the tapped coil matching arrangement is that the variometer must be at the 'hot' (very high voltage) end of the coil so any adjustments must be done with low power and attention must be given to the insulation on the adjuster.

## LOADING COIL CONSTRUCTION

The loading coil is the most critical component in the vertical antenna tuning system. It must have low losses (ie a high Q factor) if it is not to substantially reduce antenna efficiency. The coil may have to handle several amps of RF current. At 136kHz, the coil must also withstand large RF voltages. With high power 136kHz transmitters, the loading coil may need to dissipate a few hundred watts.

To meet these requirements, the most practical form of loading coil is a single layer solenoid of large dimensions. The former for the coil may be large diameter plastic tubing; many other roughly cylindrical plastic objects such as buckets and bins have been used to good effect. A visit to a garden or DIY centre may yield useful materials (**Fig 10.40**). Wooden coil forms have been tried, but these result in disappointingly high losses, probably due to the poor dielectric properties of most wood.

Enamelled or plastic insulated wire is usually used for winding. If available, Litz wire has considerably lower loss. Litz wire is composed of many fine strands of enamelled copper wire twisted together in the form of a rope. The purpose of this construction is to reduce the 'skin effect', where at high frequencies current flow is confined to the surface of a conductor, increasing its resistance. The strands of Litz wire weave in and out of the bundle, forcing current to flow throughout the thickness of the wire. This reduces the RF resistance typically by a factor of 2 or 3 at 137kHz but is less useful on 472kHz. For best efficiency it is important to ensure that all of the individual wires are properly soldered at each end.

Fig 10.43: Combined loading coil and variometer for 472kHz built by F6HCC

The best form of loading coil usually depends on the materials available. Data on some practical 136kHz loading coils (see **Fig 10.41**) is given in **Table 10.4**. Coils 1 and 2 are close-wound with enamelled copper wire on PVC drain pipe formers. This gives a compact coil for a given inductance, but a relatively low Q. Coil 3 is wound with teflon-insulated, stranded core equipment wire. Teflon is an excellent dielectric, and withstands high temperatures and voltages, so produces a robust, higher rated coil with lower losses. This type of wire is quite expensive, but surplus bargains sometimes appear. Coils 4 and 5 are wound with Litz wire, the 'Rolls-Royce' of loading coil materials, but hard to obtain and relatively very expensive. Coil 6 is wound with wire obtained by stripping the outer sheathing from inexpensive 2.5mm2 'twin and earth' cable available for domestic mains wiring.

The resistance of the loading coil results in the loss of a proportion of the transmitter power in the coil. The percentage loss of power is given by:

$$\%\,\text{Loss} = \frac{R_{coil}}{R_{loss} + R_{coil}} \times 100\%$$

$$\text{or, in decibels, Loss(dB)} = 10Log_{10}\left(\frac{R_{coil}}{R_{loss} + R_{coil}}\right)$$

where Rcoil is the coil resistance, and $R_{loss}$ is the loss resistance of the antenna.

Taking the inverted L antenna discussed above as an example, with $R_{loss}$ of 40Ω at 136kHz, using coil 2 in the table with a Q of 220 will result in 27% of the transmitter power being dissipated in the coil, while if coil 4 with a much higher Q of 1100 is used, only 6% of the power is dissipated in the coil. The loss in radiated power in decibels is 1.4dB for coil 2 and 0.25dB for coil 4. In either case, this loss amounts to only a fraction of an S-point at the receiving station, so the effect on overall system performance is minimal.

What is more significant is the power-handling capability of the coil. Physically small coils such as 1 and 2 are suitable for transmitter power levels of up to a few hundred watts. Larger coils with higher Q dissipate a smaller proportion of the trans-

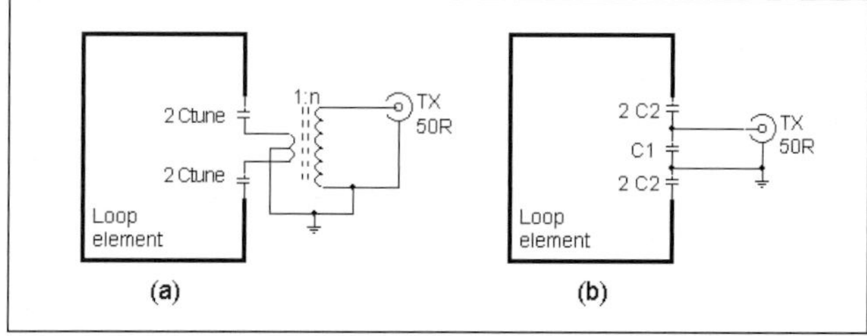


Fig 10.44: Transmitting loop matching circuits

mitter power, and also have greater surface area for cooling. Coils 3,4 and 5 are suitable for kilowatt power levels, as is coil 6, which in spite of being relatively low Q has a very large surface area.

It is wise to make the loading coil inductance somewhat higher than that calculated to resonate the antenna, and provide several taps (preferably at the end of the coil closest to the matching unit) to accommodate changes in antenna capacitance. The winding should be attached to the former in several places, otherwise there is a tendency for wire to spill off when the former expands and contracts with changes in temperature. Loading coils are often located out-of-doors, and require protection from the elements by some sort of housing.

The housing may itself cause considerable loss, especially if made of materials with high RF loss such as wood. A good protective housing is a large plastic dustbin. It is advisable to lift the housing at least a metre clear of the ground, to reduce dielectric losses in the ground underneath.

## VARIOMETERS

A variometer is essentially a variable inductor, which is invaluable for fine adjustment of the loading inductance. The most common type of variometer is shown in **Fig 10.42**(a), and consists of two coils, one rotating inside the other. When the coils are aligned so their magnetic fields add together, the mutual inductance between the coils adds to the self inductance of each coil, resulting in maximum overall inductance. Rotating the inner coil by 180 degrees results in the fields opposing, and the mutual inductance subtracting, giving minimum inductance. An adjustment range of about 2:1 in inductance is possible. Another form of variometer is shown in **Fig 10.42**(b), in this case the two coils 'telescope' together to vary the mutual inductance.

A very simple arrangement is shown in **Fig 10.42**(c); this uses a ferrite core for permeability tuning of an air-cored coil. This may simply be a coil of a few hundred microhenries inductance wound on a horizontal tube former, with the ferrite core, glued to a plastic handle, free to slide inside the tube. It is important to use a short, thick piece of ferrite rather than a long rod, since in the latter case the flux density can easily become very high, leading to saturation and severe heating of the core. A large switch-mode power supply transformer core works well.

A variometer constructed by F6HCC [35] is in **Fig 10.43**, and other examples can be seen at [36, 37]. Reference [38] includes an on-line calculator for designing the **Fig 10.42**(a) form of variometer.

## TRANSMITTING LOOP ANTENNAS

A large, single-turn loop is the 'alternative' electrically small LF/MF transmitting antenna. While the vertical is a high voltage, relatively low current device, the loop features relatively low voltage, and high current. Like smaller receiving loop antennas, the transmitting loop has a figure-of-eight radiation pattern in the azimuth plane, with deep nulls at right angles to the plane of the loop.

The impedance of the loop can be modelled as the loop inductance in series with a radiation resistance $R_{rad}$ and a loss resistance $R_{loss}$. The value of $R_{rad}$ depends on the area of the loop A in square metres:

$$R_{rad} = 2 \times 320\pi^4 \frac{N^2A^2}{\lambda^4}$$

where N is the number of turns, normally one. The factor of 2 is included in the formula due to the effect of the ground plane

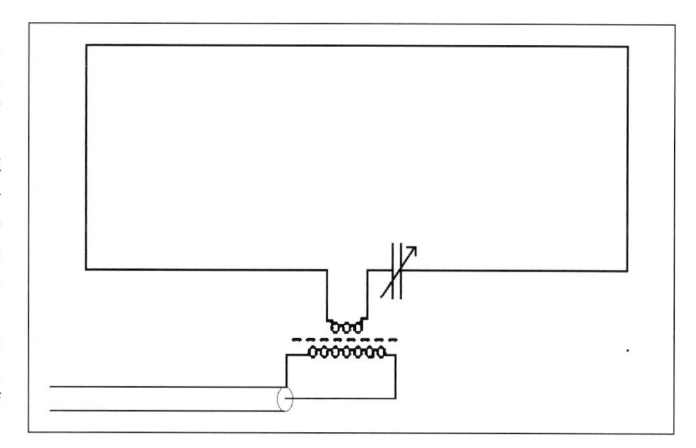

Fig 10.45: Practical loop feed and tuning circuit

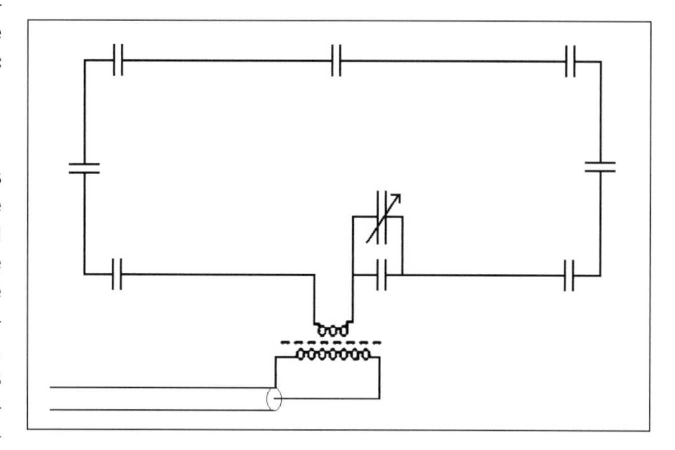

Fig 10.46: Distributed capacitance loop

underneath the loop, which in principle doubles $R_{rad}$ due to the 'image' antenna reflected in the ground plane.

$R_{rad}$ for back-garden sized loop antennas, which typically have areas of hundreds of square metres, is normally below a milliohm at 136kHz. This requires loss resistance $R_{loss}$ to be below a couple of ohms to achieve efficiency comparable with vertical antennas. The dominant source of losses for loop antennas is the AC resistance of the loop conductor. To achieve reasonable efficiency, very thick wire, or multiple parallel wires are required; several builders have resorted to using the braid of UR67 coax, or even copper water pipe!

At 472kHz, a similar loop can be expected to have a much higher radiation resistance, of the order of tens of milliohms, making efficiency comparable with vertical antennas.

Matching a loop antenna normally uses one of the circuits shown in **Fig 10.44**. **Fig 10.44**(a) uses a step-down transformer to match the loop $R_{loss}$ to the 50-ohm transmitter output, and a series capacitance to resonate the loop inductance $L_{ant}$.

$L_{ant}$ in henries is given approximately by the formula:

$$L_{ant} = 2 \times 10^{-7}\, P.Log_e\{3440A/dP\}$$

where P is the overall length of the loop perimeter (m), A is the loop area (m2), and d is the conductor diameter (mm). Ctune is therefore:

$$C_{tune} = \left(\frac{1}{2\pi f\sqrt{L_{ant}}}\right)^2$$

Ctune is often divided into two series capacitors as shown, to make the loop voltages approximately balanced with respect to ground. The required transformer turns ratio is $\sqrt{R_{load}/R_{loss}}$.

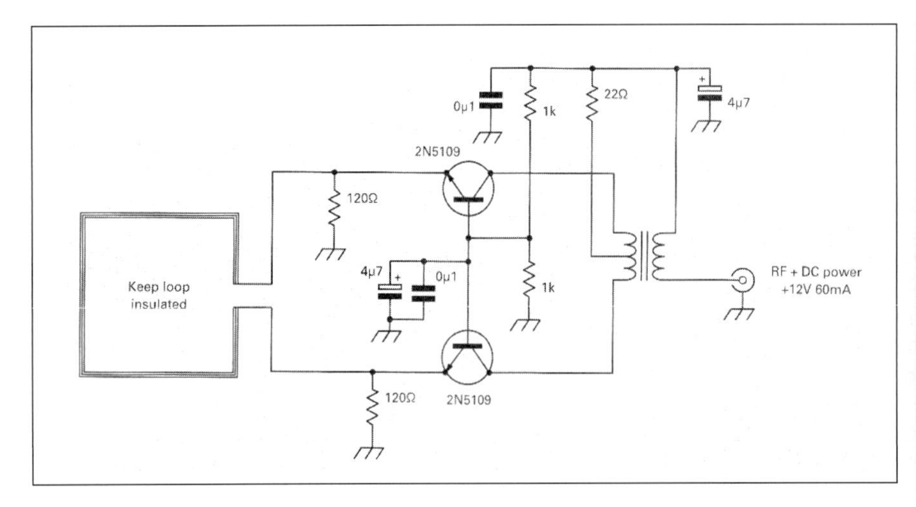


**Fig 10.47: Active loop preamplifier suitable for a single-turn loop**

**Fig 10.48: F6HCC's receiving loop wound on plastic plumbing pipe**

An alternative matching scheme uses a capacitive matching network, **Fig 10.44**(b). The values of C1 and C2 are:

$$C_1 = \frac{\sqrt{\frac{R_{load} - R_{loss}}{R_{loss}}}}{2\pi f R_{load}}, \quad C_2 = \frac{1}{2\pi f\left(2\pi f L_{ant} - \sqrt{R_{loss}(R_{load} - R_{loss})}\right)}$$

A simple practical way to build a transmitting loop is a variation of **Fig 10.44**(a) using just one capacitor which could be a vacuum variable or other high voltage component, perhaps with some fixed capacitors in parallel if required (**Fig 10.45**). The transformer is formed by passing the loop conductor through a large ferrite core and winding a few turns around this for the 50-ohm primary. Start with about 8 turns and experiment.

Because the bulky tuning and matching components are heavy it is usual to place them at the bottom of the loop. This gives rise to some losses as the high Voltages present on the bottom run are close to the ground. One idea to combat this problem is shown in **Fig 10.46**. The tuning capacitor, instead of being one component is split into, say eight, fixed components around the loop. As these are in series they must be of a correspondingly higher value and must be rated to carry the loop current of perhaps 10A or so, but they will be subject to less Voltage stress and the leakage to ground should be less noticeable. A variable capacitor across one C will still be needed to trim the resonance.

As with vertical antennas, achieving good performance from loop antennas depends mostly on size. The radiation resistance is proportional to the square of the loop area, so every attempt should be made to make the loop as long and as high as possible. In 'open field' sites, loops are usually less efficient than similarly sized verticals for 136kHz operation due to their very low radiation resistance.

At 472kHz, a loop antenna's efficiency is over ten times better than at 136kHz and, because it doesn't rely on an earth system, tends to be better than a vertical of the same height. Loops are relatively easy to put up if you have a couple of supports - such as trees - available. G3YXM tested a square loop made from 50 metres of "Twin and Earth" mains cable which gave 472kHz WSPR reports from the USA almost every night in January 2020.

The main advantage of transmitting loops is that the loop voltages are much lower than for the vertical, resulting in lower dielectric losses in objects around the antenna. This makes a loop a good choice for wooded surroundings, where many trees close to the antenna would lead to very poor efficiency with a vertical. This seems to be a common situation in North America, where several LF loop antennas have been constructed using branches of tall trees to support the loop element. Loops also do not rely on a low resistance ground connection, so may be an improvement where there is very dry or rocky soil. A disadvantage is that stronger antenna supports are required to hold the thick loop conductor. A further drawback is the directional pattern of the antenna; the radiated signal will be reduced in some directions due to the nulls in the radiation pattern and options for changing the orientation of a large loop are usually limited.

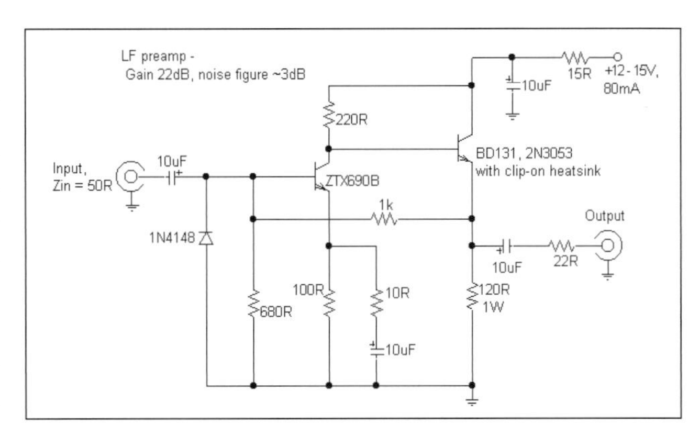


**Fig 10.49: 50-ohm preamp suitable for loop antennas**

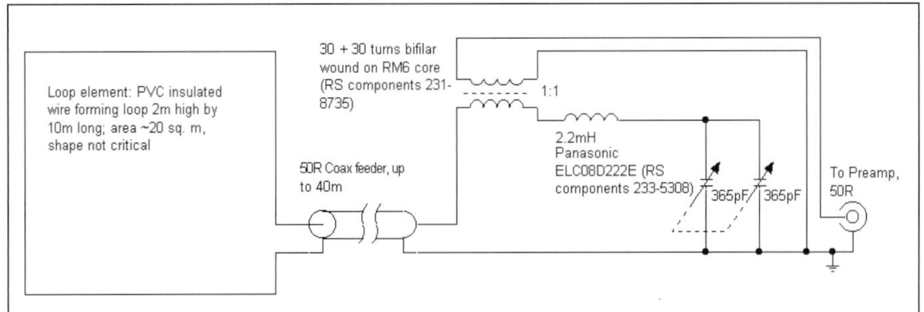


**Fig 10.50 'Lazy loop' and tuning arrangement**

A detailed article on LF transmitting loop antennas can be found at [39]; descriptions of practical 136kHz transmitting loops are given at [40, 41].

## RECEIVING ANTENNAS AND INTERFERENCE REDUCTION

Both the 136kHz and 472kHz bands are subject to high levels of naturally occurring and man-made noise. The major source of naturally occurring noise is thunderstorms, which give rise to the characteristic crackling lightning static (QRN) heard at most times during the night on either band, but reduced or absent during the day.

**(above / right) Fig 10.51: Band-switched loop construction. The loop connections are the two bolts at the back of the case. The loop tuning capacitors C1 and C2 are mounted on the toggle band switch at the bottom**

When QRN is low, the audible 136kHz band noise floor in the UK is usually dominated by low-level sidebands from some of the high-power utility stations operating in mainland Europe and transmitting FSK data bursts in the vicinity of (see **Fig 10.1**).

In contrast, the 472kHz band is relatively little affected by unwanted signals, the main ones being aeronautical beacons (NDBs) which use MCW and appear on an amateur receiver as a carrier and two sidebands, each 400Hz either side of the carrier and comprising a CW callsign.

A more serious problem for many amateur stations is locally generated, man-made noise. This has many sources, most associated with mains electrical wiring. Many devices use switch-mode power supplies, operating at switching frequencies in the LF/MF range. Equipment with rectifiers or triac-based phase-control circuits can generate significant levels of harmonics of the mains frequency throughout the range. Digital systems can also generate wide-band noise in the LF/MF spectrum, a particular problem if the source is a computer within the shack. Broadband internet delivered via telephone lines gives rise to wide band noise that can affect the 472kHz band. Plasma televisions can also be a problem at MF.

A good first stage in isolating a local noise source is to switch off all mains-operated equipment. This usually requires unplugging the equipment completely, since often the same or sometimes greater noise level can be generated when switched to 'standby'. The most certain method is to switch off the house mains supply at the main switch, whilst using a battery-operated receiver to monitor the noise level. If the offending device is on the premises, the noise can be eliminated by simply switching it off while operating. Unfortunately, mains-related noise can propagate considerable distances along the mains wiring, so often the amateur has no control over the noise source. Useful advice on tracing and dealing with local man-made noise can be found in [42], though this deals primarily with interference to HF reception.

Many low-frequency amateur stations use their transmit antenna for reception. However, there are also many situations, especially where QRM is a problem, where it is advantageous to use a separate receiving antenna. Due to the high band noise level at low frequencies, it is possible to design very small antennas that yield perfectly satisfactory signal-to-noise ratios for reception in spite of their inevitably low efficiency.

Another option is to use a noise cancelling system. In this

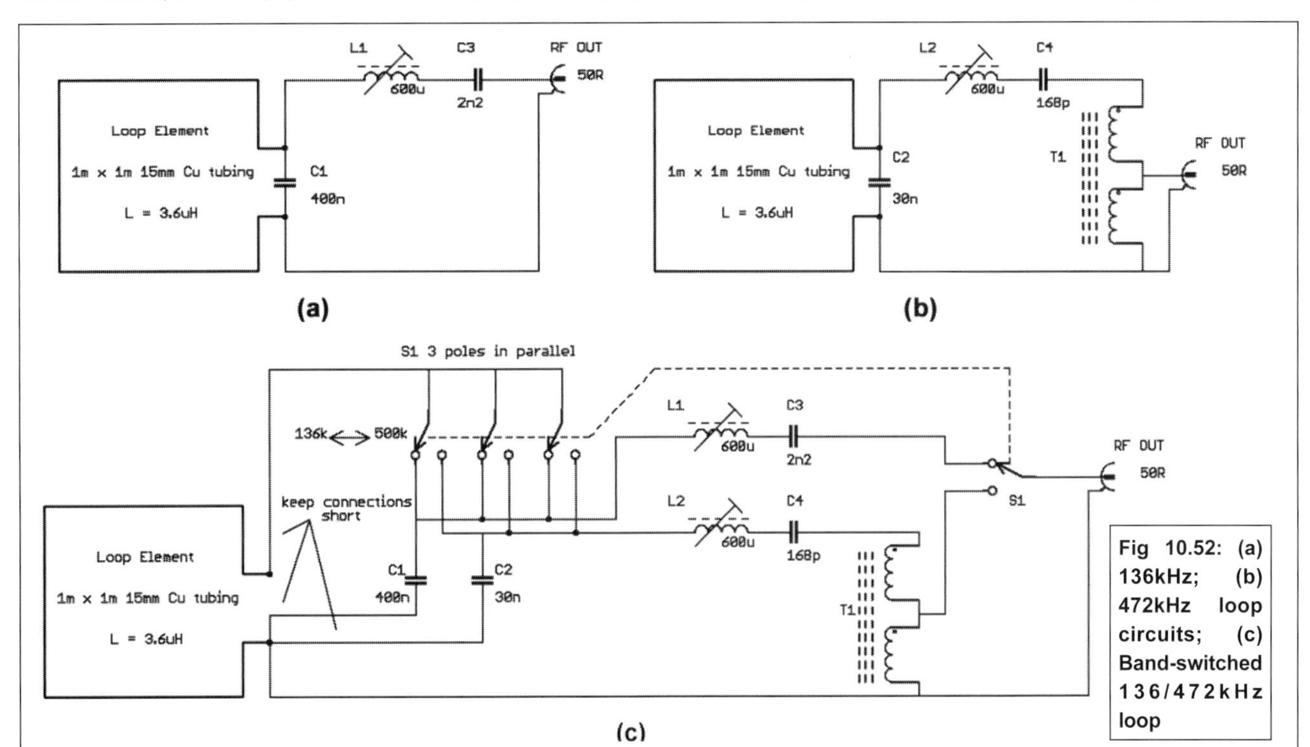


**Fig 10.52: (a) 136kHz; (b) 472kHz loop circuits; (c) Band-switched 136/472kHz loop**

situation the small antenna is positioned to pick up as much of the noise as possible, this is then subtracted from the main antenna's receive signal (see Noise Reduction below).

In the case of local man-made noise, the noise level may vary greatly over short distances, and often moving an antenna only a few metres can result in substantial noise reduction. This is easily done with a compact receiving antenna where trying to move the much larger transmitting antenna to a different location is usually impractical. Many 136kHz and 472kHz operators therefore use separate transmit and receive antennas, and in one case a solar powered remote receiving station. Where the noise originates from a localised point, noise-cancelling schemes are sometimes very effective.

Even if attempting to reduce the noise level is not effective, it is often found that the noise level varies a lot during the day, and is sometimes low enough for satisfactory operation. Some modes, especially those with long integration times such as QRSS, Opera or WSPR, can be operated successfully under conditions that may be too noisy for CW. Very occasionally, the noise may abruptly disappear for good, when one noisy appliance is replaced by a quieter one, but unfortunately, the reverse is also very possible!

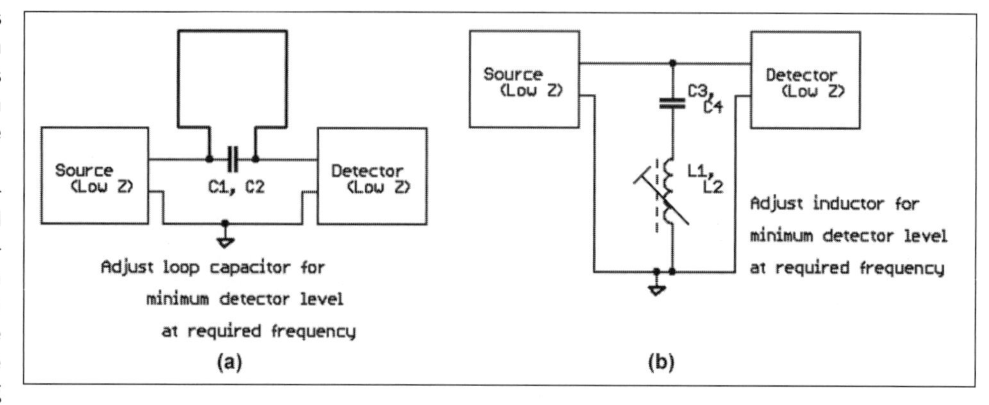


Fig 10.53: How to adjust the resonant frequency of (a) loop, (b) auxiliary tuned circuit

## RECEIVING LOOP ANTENNAS

Much has been said about the 'noise reducing' properties of receiving loops. Electrically small loop antennas respond essentially to the magnetic field component of an electromagnetic wave, and so reject noise that exists as a local electric field. Unfortunately, most local noise sources in the LF/MF range involve common-mode noise currents flowing through mains wiring, giving rise to magnetic fields which the loop will pick up. Using a loop antenna inside or near a building therefore usually yields poor results. However, these fields rapidly decrease in strength as the antenna is moved away from the offending wiring; often moving the receive antenna by only a few metres results in a substantial reduction in noise. It is therefore most important to experiment with different positions for receiving loops, often a quiet spot can be found even where high noise levels exist all around.

Loop receiving antennas have a figure-of-eight directional pat-

| Capacitors | Stable, low-loss types: |
|---|---|
| C1 | 4 x 100nF, 100V metallised polypropylene in parallel (capacitors selected during alignment) |
| C2 | 2 x 15nF polystyrene in parallel (capacitors selected during alignment) |
| C3 | 2.2nF polypropylene |
| C4 | 150pF polystyrene + 18pF in parallel |
| L1, L2 | Primary winding of 455kHz 10mm 'Toko' IF transformer with 180pF tuning capacitor or similar; capacitor removed |
| S1 | 4-pole double-throw miniature toggle switch |
| T1 | 18 bifilar turns on FT-50-43 (5943000301) 12.7mm dia, = 850 toroid, or similar |

Table 10.5: Component notes for bandpass loops

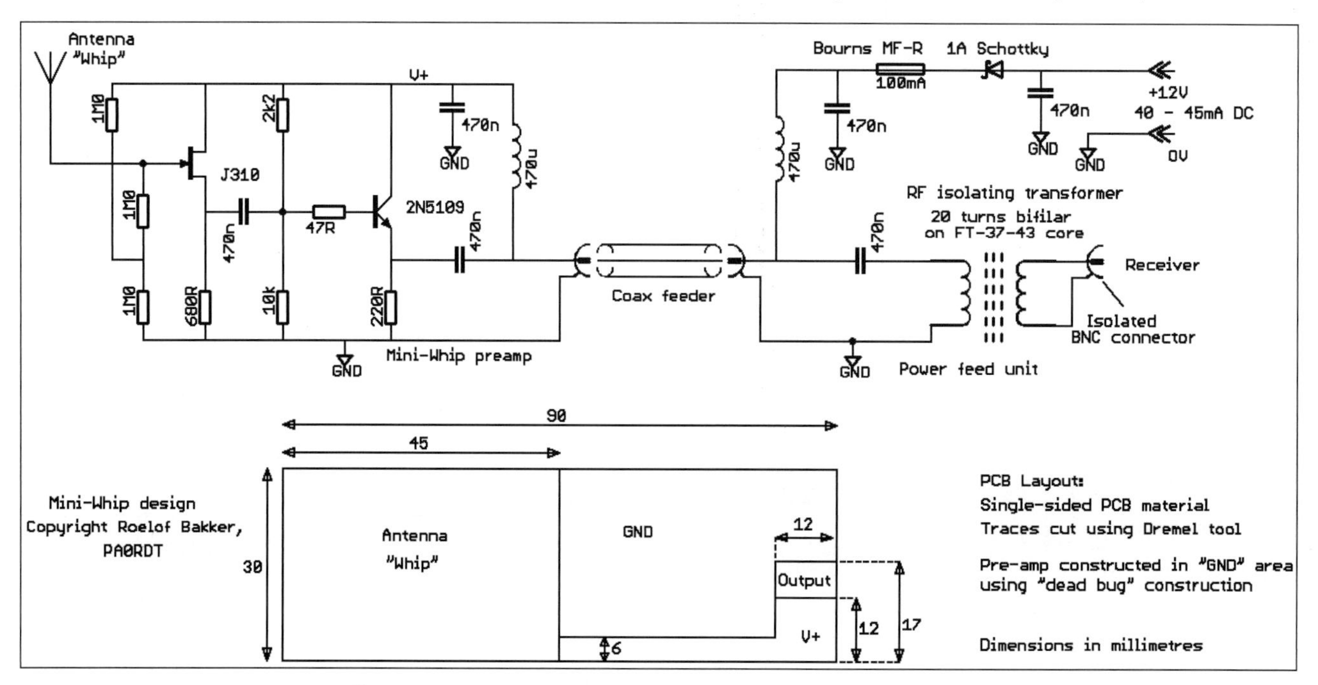


Fig 10.54: The PA0RDT-Mini-Whip©

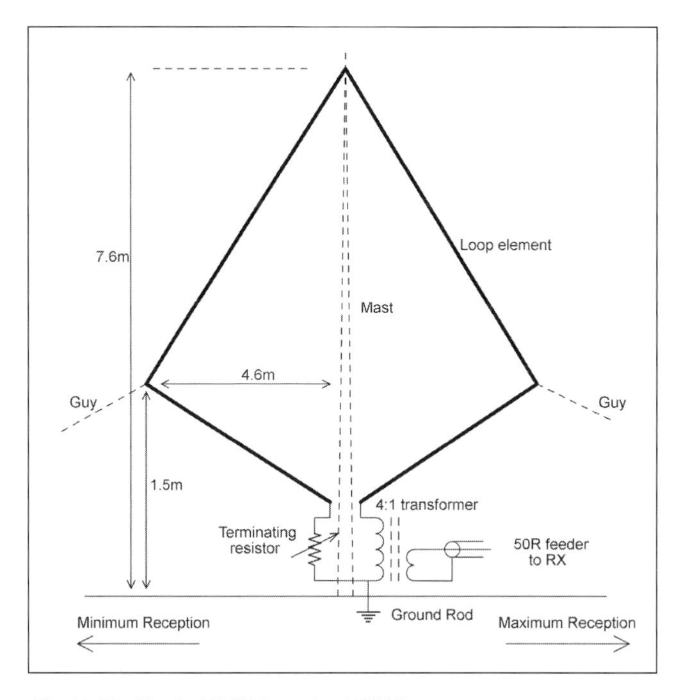


Fig 10.55: Single K9AY loop for LF/MF use

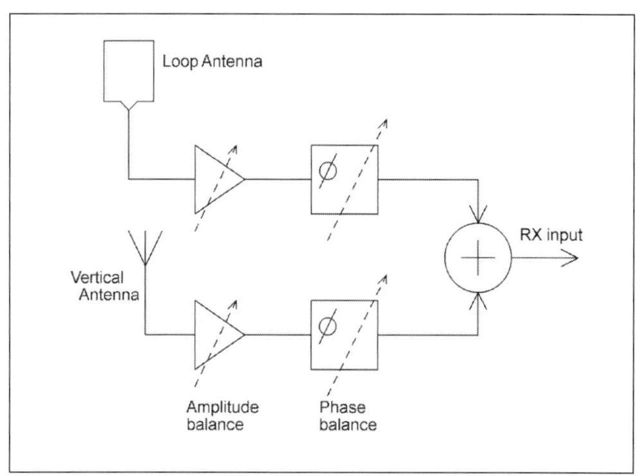


Fig 10.56: Loop/vertical directional array for low frequency reception

tern, with nulls at right angles to the plane of the loop, and maximum sensitivity along the plane of the loop. The directional null of a loop is often very effective in eliminating distant noise sources such as utility stations. Also, the loop null can sometimes be used to suppress local noise.

Loops can be of the single-turn broadband active type or the multi-turn tuned type. Active loops are usually about 1m diameter and made from aluminium tube. Rather than a capacitor tuning the loop there is a low-noise differential amplifier which is often powered by DC sent back down the coax cable from the shack end. These loops can be bought commercially, the most famous being the Wellbrook ALA1530 which works well at LF, or home constructed. They will generally have a response from below 100kHz up to 30MHz or more so offer no protection from strong broadcast transmissions. A simple design from the RadCom Design Notes column is shown in **Fig 10.47**.

A typical tuned loop antenna is shown in **Fig 10.48**, and for 136kHz can consist of about 30 turns of wire wound onto a wooden or plastic frame, with an area of about 1m2, tuned by a 1000pF variable capacitor. For 472kHz, about 10 turns is sufficient, with a 500pF tuning capacitor. The receiver input is fed via a low impedance single-turn link winding. The output of the loop is small, and a low-noise preamplifier will normally be required, such as the one shown in **Fig 10.49**. The Q of the loop is typically 100 or more, so re-tuning will be required within the narrow amateur bands. This selectivity is very useful in reducing intermodulation due to strong out-of-band signals. A number of amateurs have used much larger tuned loops for reception, which achieve higher output signal levels and so can dispense with the preamplifier, at the expense of being more bulky.

An alternative is the 'lazy loop' of **Fig 10.50**. This uses a large single-turn loop, the area of which is around 10 - 20m2. The shape is not at all important, and it can be normal insulated wire slung from bushes or fence posts, etc, hence the name! The loop can be fed through a coax feeder, allowing it to be positioned remote from the shack to reduce noise, whilst having the tuning components easily accessible at the receiver end of the feeder. This tuning arrangement is not optimal from the point of view of minimising losses, but due to the large loop area the signal-to-noise ratio is more than adequate. It is also possible to use somewhat smaller, multi-turn loops. Again, a low-noise preamp, such as **Fig 10.49** will be required. This type of loop element also gives good results with the LF Antenna tuner/preamp circuit of **Fig 10.7**.

## BANDPASS LOOPS FOR 136KHZ AND 472KHZ

The frequent need to re-tune a high Q loop is something of a drawback. The Q can be reduced by adding resistive loading, but unfortunately this also reduces the signal level available to the receiver. An alternative approach is to combine the tuned circuit formed by the loop with other capacitors and inductors to form a bandpass filter. This results in a wider bandwidth, while at the same time improving rejection of out-of-band signals. Another drawback of conventional loops is the multi-turn winding, which is hard to weatherproof. A single-turn loop made of tubing is

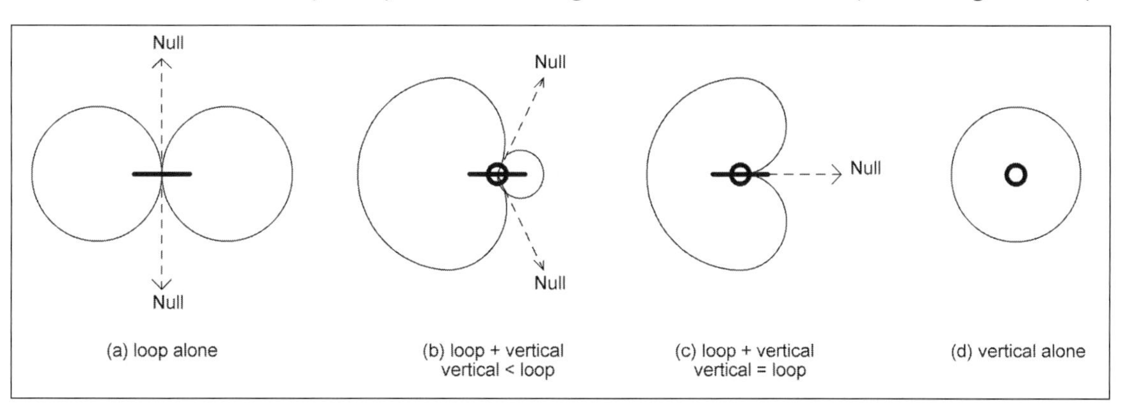


Fig 10.57: Directional patterns obtainable with loop / vertical array

more convenient and robust, and has quite high Q, even though very large tuning capacitance is needed. The following designs are based on 1m x 1m loop elements made of 15mm copper water pipe (as shown in **Fig 10.51**) and using the bandpass filter principle to achieve coverage of the full band without retuning.

**Figs 10.52** (a) and (b) show similar circuits used for both 136kHz and 472kHz loops, with the same inductance values for both. The transformer in the 472kHz design is not critical; any transformer giving a low loss and 200:50 ohm transformation at 472kHz could be used. The -3dB bandwidths of the prototypes were approximately 10kHz for the 136kHz version, and 37kHz for the 472kHz version, more than enough to cover each band.

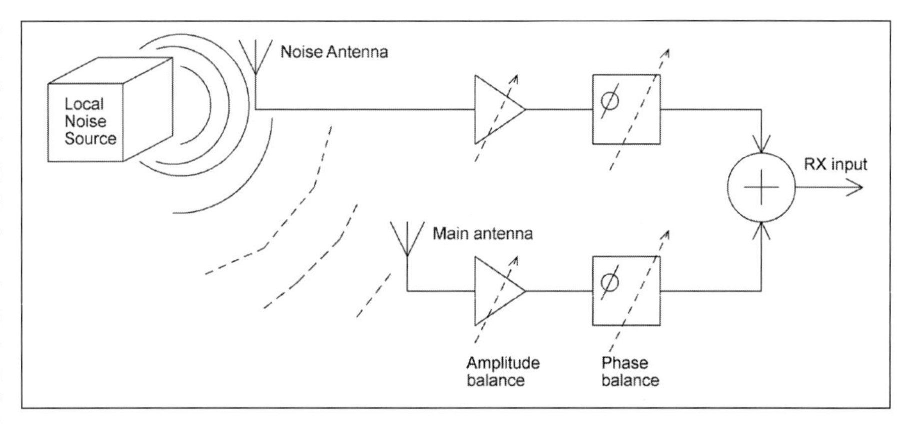


**Fig 10.58: Noise-cancelling array**

A band-switched version has also been built, see **Fig 10.52**(c). This simply switches one loop element between the two circuits. The band-change switch selecting the loop resonating capacitor must have low contact resistance in order not to increase the losses in the loop, so a 4-pole, double throw toggle switch with three poles connected in parallel was used to select the loop capacitor.

**Table 10.5** shows details of the components. The 600μH adjustable inductors are the primary windings of 'Toko' or similar 455kHz IF transformers; the types that have 180pF capacitors have a suitable inductance value that is adjustable over a fairly wide range. The internal ceramic capacitor must be disconnected or removed; in the Toko types, this is most easily done by carefully breaking up the ceramic capacitor in the moulded plastic base with a pointed implement. The primary winding is usually connected to the two end pins of the row of three pins on the IFT base - check with an ohmmeter to find the largest winding resistance. Other coils with similar inductance and a Q >50 could be used.

The loop element is made from 4 x 1m lengths of 15mm copper water pipe, joined in a square with 90 degree solder elbows. One side of the loop is cut in the middle, the cut ends flattened and drilled, and brass bolts passed through and soldered into place. The bolts pass through the wall of a plastic box containing the tuning components, and connections are made using solder tags. This gives a good low-resistance connection. The connections to the loop tuning capacitors should be as short as possible and direct to the loop terminals, especially for the band-switched version - see **Fig 10.51**. The loop element is attached to a wooden support using plastic pipe clips.

Alignment consists of adjusting the resonant frequencies of the loop and auxiliary tuned circuits. This is best done during assembly by temporarily configuring the tuned circuits as parallel or series traps as in **Fig 10.53**; when the resonant frequency is equal to the source frequency, a sharp dip in detector level will be seen. First, the parallel resonant frequency of the loop itself is adjusted to be close to the centre of the band of interest by selecting a suitable parallel combination of capacitors (**Fig 10.53**(a)). The auxiliary tuned circuit is then adjusted to an identical series resonant frequency by itself (**Fig 10.53**(b)). Then the connections between the tuned circuits and the output are made to complete the circuit; no further adjustment should be required. Nominal resonant frequencies for the two bands are 137kHz and 475kHz, although deviations of a couple of percent from these values are not serious due to the fairly wide

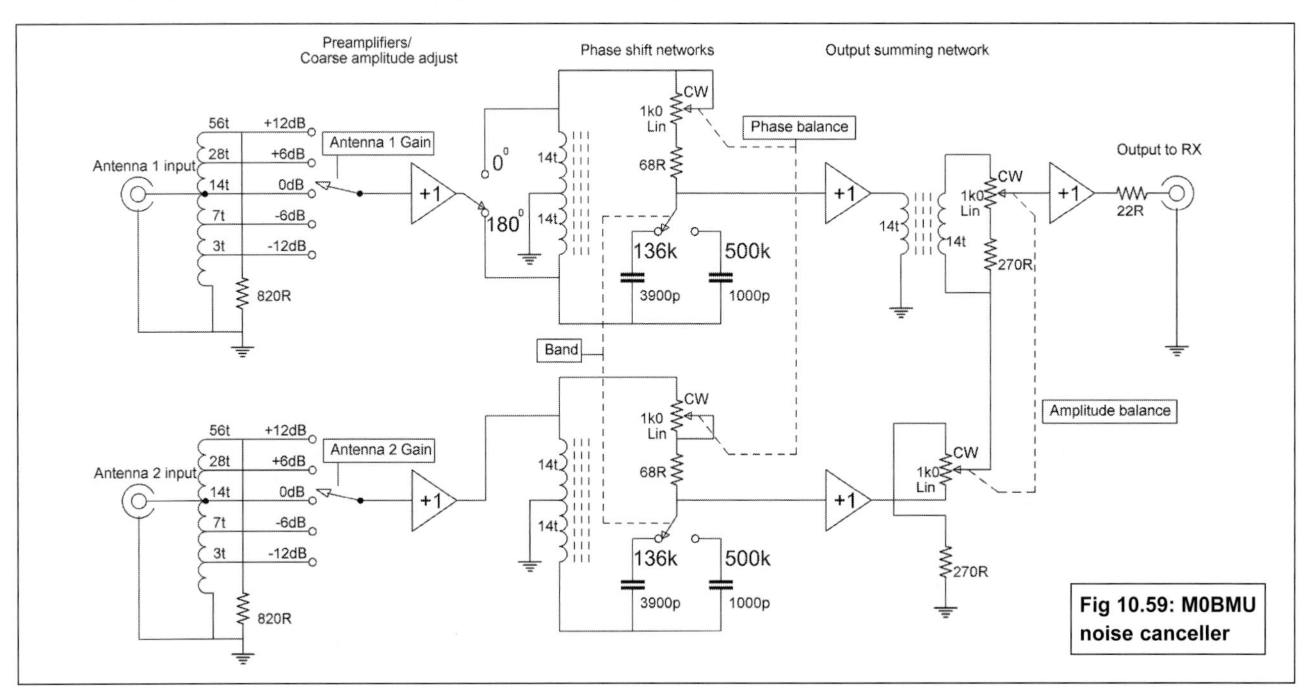


**Fig 10.59: M0BMU noise canceller**

passband - the main thing is that both tuned circuits are set to the same frequency. Note that, when tuning the Toko inductors, these tiny ferrite cores can be saturated by quite low signal levels; about -20dBm from the source is safe.

The output signal level is quite small, and a low-noise preamplifier such as the circuit of **Fig 10.49** is used. Sensitivity is then more than adequate to hear the band noise levels. The loops are not very sensitive to de-tuning; even a large metal object like a step-ladder has little effect unless it is nearly touching the loop.

These designs were originally published as a longer article which can be found at [43].

## ACTIVE WHIP ANTENNAS

Quite short wire or whip antennas provide adequate signal to noise ratio at LF when matched to the receiver input by a high impedance buffer amplifier. The response of the resulting Active Whip antenna, or E-field antenna is broad-band, and can extend from the VLF range to the VHF range, depending on the amplifier used. The preamplifier is located at the base of the antenna, and its DC supply is usually fed up the coax.

The whip element is often around 1 - 2m long, and the small size makes the antenna easy to site in an electrically quiet location. As with the loop antennas, it is important to experiment with the antenna location to find a site that has a low noise level. Since all signals over a wide frequency range are presented to the buffer amplifier input, good dynamic range is important, and overload problems may occur if there are high-power broadcast stations nearby.

Unlike loop receiving antennas, the signal output level from an active whip depends strongly on its position, partly due to the screening effect of surrounding buildings and other objects, and partly on the height of the whip element above the ground plane. Greater output will be obtained if the whip is mounted on a mast, or on the roof of a building, rather than at ground level.

### The PA0RDT Mini-Whip antenna

The PA0RDT-Mini-Whip© was designed by Roelof Bakker, PA0RDT, and has been built and used successfully for LF and MF reception by numerous amateurs. Good receive performance extends from 10kHz to over 20MHz. The compact size of this antenna also makes it very suitable for portable operation.

The Mini-Whip 'whip' element is in fact a small piece of copper-clad board. The shape is not important provided capacitance to ground is similar, and other whip element construction can also be used, eg a small metal box with the preamp inside. Tests were performed to optimise the size of the whip element, and the design achieves good sensitivity while maintaining maximum overall output at about -20dBm to prevent receiver overload. The buffer amplifier is optimised for good strong-signal performance. Second order output intercept has been measured by AA7U as being greater than +70dBm, and third order intercept greater than +30dBm. Power is fed from a 12 - 15V DC supply to the Mini-Whip via the power feed unit and the coaxial feed line, which can be up to 100m long. The power feed unit includes an RF isolating transformer, which reduces noise due to ground loops, but this is not always essential.

The buffer amplifier is constructed 'dead bug' style on the ground plane half of the board next to the whip element, which is formed by the other half of the board (see **Fig 10.54**). The complete circuit board is mounted inside a 100mm long section of 40mm plastic drain pipe, with two end-caps. One end-cap carries an insulated BNC connector to which the circuit board is soldered.

PA0RDT notes that the electric field from most interference sources is largely confined within the building. For best results therefore, the Mini-Whip should be mounted on a non-conducting pole in a position well clear of buildings. Grounding the outer braid of the coaxial feeder to a ground rod installed close to the point where the feeder enters the shack also helps to reduce interference generated by noise currents flowing on the feeder.

A more detailed article can be found at [44]. A ready-made Mini-Whip can also be purchased from PA0RDT [45].

**Fig 10.60: High impedance buffer used in M0BMU noise canceller**

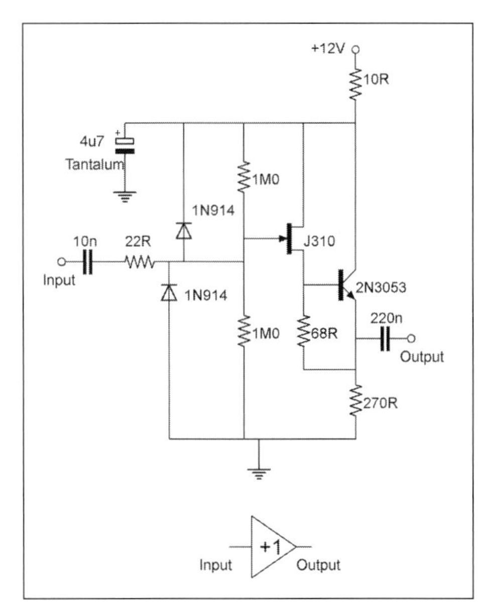


## TERMINATED LOOPS

Terminated loop antennas such as the K9AY and EWE [47, 48] designs have been quite popular directional receiving antennas for the lower HF bands, and also are effective in the VLF, LF and MF ranges. These antennas use large wire loop elements simultaneously in a loop receiving mode and as a vertical antenna. The proper summation of loop and vertical signals at the feed point gives rise to a unidirectional pattern with a single directional null. The relative levels of loop and vertical signals can be adjusted to obtain the deepest null by varying the terminating resistor. A single loop of the K9AY type was used successfully at M0BMU (**Fig 10.55**). About 20dB reduction in noise level radiated from nearby broadcast transmitters was readily obtained. It was found that a terminating resistor of around 200 ohms was required, rather lower than values quoted for the full two-loop K9AY HF array. This may be due to the use of only a single loop, or environmental factors, such as ground conductivity. Using a 1k ohm pot as the terminating resistor will allow a wide variation to be accommodated.

The received signal level from these loops is quite low, and a preamplifier will be required with most receivers. The size and shape of the loop is not particularly critical; larger loops will generally give more output.

The Beverage antenna is another well-known form of unidirectional receiving antenna. At low frequencies, the 0.9λ length of the classical Beverage is usually impractical (about 2km at 136kHz!), but much shorter lengths have been used successfully. These probably function in a similar way to the terminated loops described above.

## NOISE REDUCTION

By combining the signals from two or more separate antennas with suitable adjustment of amplitude and phase, it is sometimes possible to cancel distant or local noise sources and so improve signal to noise ratio of wanted signals. For distant signals, this amounts to creating a directional receiving array with a null in the direction of the unwanted signal source.

## DIRECTIONAL ANTENNA

At low frequencies, multi-element directional arrays as used at HF and above are generally not practical due to the large spacings that would be needed between the elements. For the amateur, the most practical form of low frequency directional array consists of a rotatable loop and a vertical antenna (**Fig 10.56**). This type of array was widely used in the past for radio direction finding by adjusting for a null of a beacon signal, but can equally well be used to null unwanted noise.

By summing the omnidirectional vertical pattern with the figure-of-eight pattern of the loop, a skewed figure-of-eight pattern results (**Fig 10.57**(b)). Varying the amplitude of the vertical signal relative to the loop signal allows the skewing of the pattern to be controlled.

With zero vertical signal one obtains the basic loop pattern with two nulls at right angles to the plane of the loop (**Fig 10.57**(a)); as the signal from the vertical is increased, the pattern becomes asymmetrical with a smaller angle between nulls, until the limiting case of **Fig 10.57**(c) is reached, where the two nulls coincide to produce a cardioid pattern with a single null. Thus, the two nulls of the loop/vertical array can be 'electrically steered' to any angle between 0 and 180 degrees, whilst wanted signals in substantially different directions are received with little attenuation.

## NOISE CANCELLING

For local noise sources, noise cancelling usually relies on being able to position two antennas so that one is relatively much closer to the noise source as in **Fig 10.58**. Distant signals will received by both antennas at a similar level, but the 'noise' antenna will have a relatively higher level of the local noise present. Suitably attenuating and adjusting the phase of the noise antenna signal before summing the outputs of both antennas results in cancellation of the local noise, with relatively little change to distant signals. To be successful, this scheme requires that the noise originates predominantly from a single source; it is unlikely to be practically possible to arrange that multiple noise sources will have the correct amplitude and phase to all be cancelled simultaneously.

It can be seen that the overall system required for obtaining directional reception or local noise cancellation is essentially the same; two receiving antennas, and a combiner that allows adjustment of relative gain and phase of the antenna signals to achieve cancellation. Very simple passive combining networks are possible, although these place restrictions on the type of antennas that can be used.

A few commercial units are available such as the Timewave ANC-4 [46] which claims to function down to 100kHz. Some SDRs have twin antenna inputs and can use their software to perform a noise-cancelling function.

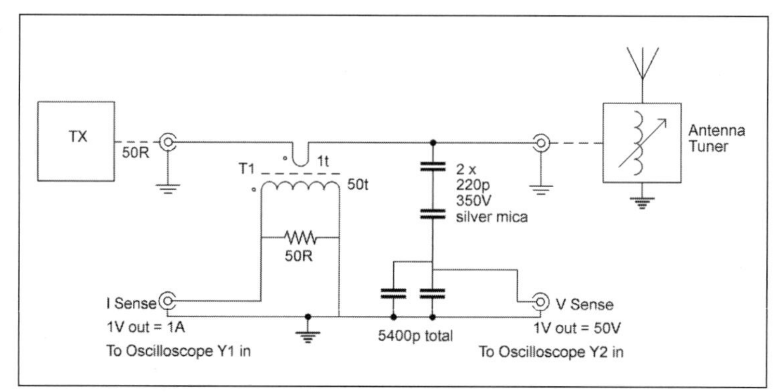


Fig 10.62: 'Scopematch' tuning aid

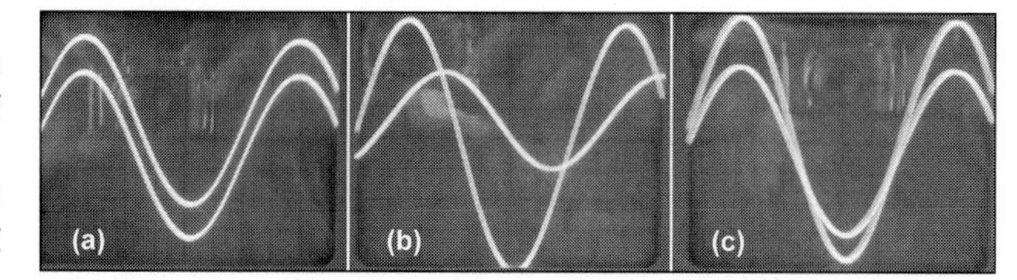


**Fig 10.63: Scopematch displays - upper trace voltage, lower trace current: (a) V and I in phase and equal magnitude; load is 50Ω resistive; (b) V lags I in phase, V less than I, so capacitive load of magnitude < 50Ω (about 20Ω here); (c) V and I in phase, but V greater than I, so resistive load >50Ω (about 80Ω here)**

A versatile noise canceller with adjustable gain, and phase shift that can be varied over a full 360 degree range, is shown in **Fig 10.59**. Buffer amplifiers are used to isolate the gain- and phase-adjusting networks, making these adjustments independent of one another. A noise canceller of this type was in use at M0BMU with a variety of receiving antennas at LF and MF.

The circuit is based on five identical high input impedance unity gain buffer circuits (**Fig 10.60**). Coarse adjustment of gain between -12dB and +12dB is provided at the two antenna input channels via tapping points on step-up/step down auto-transformers before being applied to the input buffers. Resistive loading of the transformers ensures a fixed input impedance close to 50 ohms independent of gain.

The input buffers each drive RC variable phase shift networks. The dual-gang 'phase balance' pot is wired differentially so that as the phase shift in one channel increases, the other channel decreases. This gives an overall phase adjustment of about ±120 degrees. A further switched 0/180 degree phase shift is provided by inverting one channel, so that a full 360 degree range is covered with overlap.

Output from the phase shift networks is buffered and applied to a gain adjustment network. The 'amplitude balance' pot provides approximately -10dB to 0dB gain variation in each channel, and again is wired differentially so that as the pot is rotated, the signal amplitude in one channel increases while the other decreases.

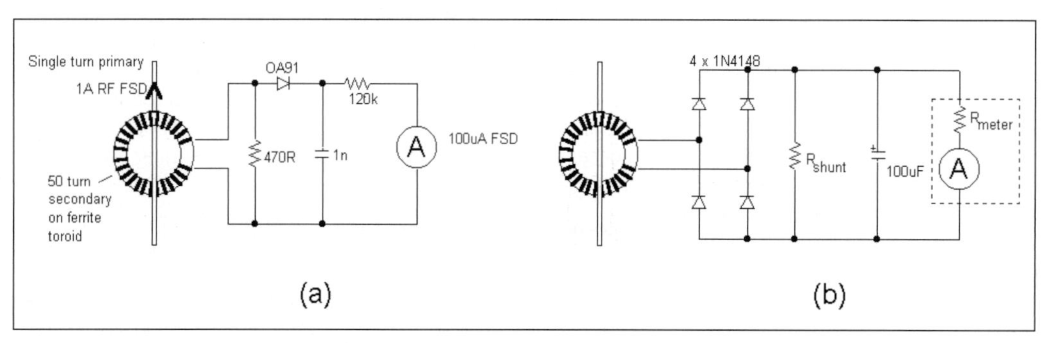


**Fig 10.61: Two types of RF ammeter**

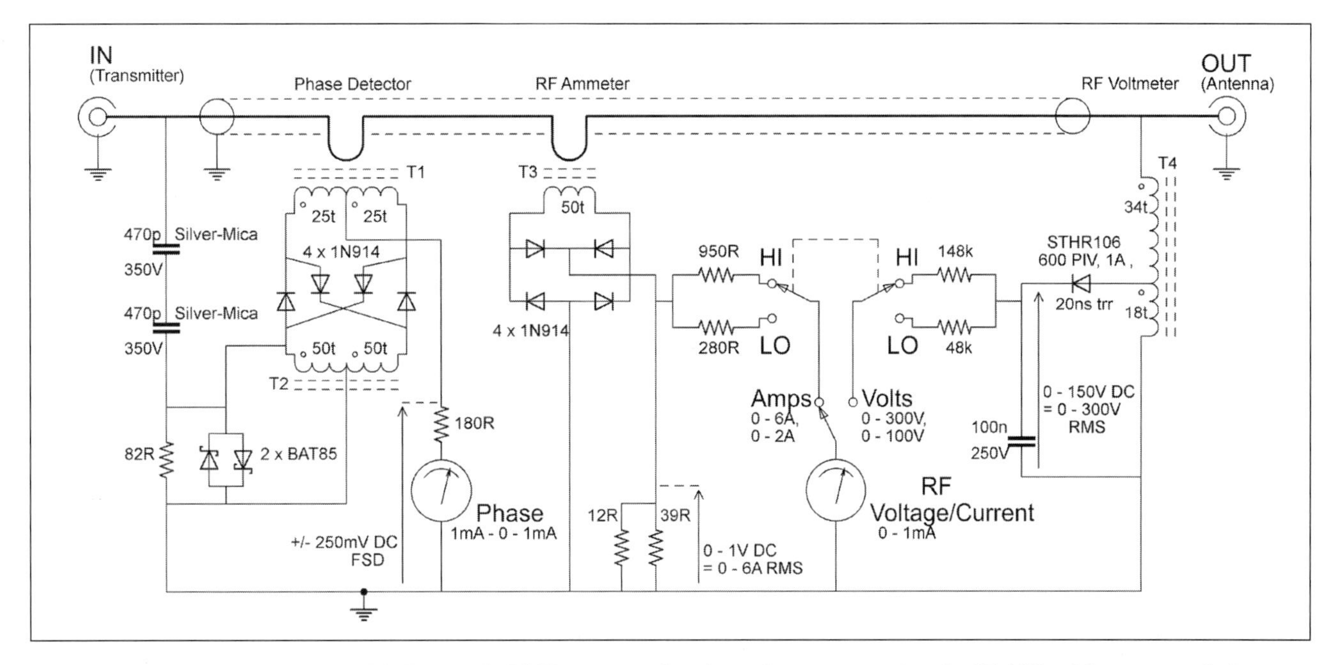


**Fig 10.64: LF Tuning meter circuit. T1: Primary is RG58 coax passing through core, secondary 2 x 25t bifilar 0.3mm enamelled copper. T2: 2 x 50t bifilar 0.25mm enamelled copper. T3: Primary is RG58 coax passing through core, secondary 50t 0.5mm enamelled copper (see text for details of transformer construction)**

Together with the coarse input gain adjustment, unwanted signals differing by over 30dB at the antenna inputs can be adjusted to a null, permitting a wide range of receiving antennas to be used. Signals in the two channels are summed by connecting the gain-adjusted outputs in series, one being made floating with respect to RF ground by an isolating transformer. The combined output passes through a final buffer to a low-impedance receiver input. The transformers shown in **Fig 10.59** are all wound on high permeability ferrite toroids with AL of approximately 4000nH/turn, similar to RS components part number 232-9561.

The noise canceller has been used for suppressing local noise sources, and to produce directional nulls on distant signals, mostly using the bandpass loop antennas described above, together with an un-tuned vertical antenna. Active whip and terminated loop type receiving antennas also work well. High Q resonant antennas are not well suited to noise cancelling schemes, since the signal phase and amplitude, and hence the depth of the null, is very sensitive to any alteration of the resonant frequency. Also, the bandwidth of the null is extremely small, and the controls must be re-adjusted for even a slight change in receiver frequency.

The system can be set up as follows. For nulling distant interference using the loop/vertical combination, the loop is first oriented for maximum received interference level (plane of the loop directed towards the source), or, if two sources are to be nulled, on a bearing mid-way between the interference sources. For cancelling local noise sources, one antenna is positioned to maximise noise pick-up, and the other for minimum noise. The amplitude controls of the noise canceller are then adjusted with each antenna individually connected in turn to obtain as close as possible to equal unwanted noise levels at the receiver from each antenna. Then both antennas are connected to the canceller, and the phase balance control adjusted for a null in the noise level. A few iterations of adjusting the amplitude and phase balance control should then result in a deep null.

## LF MEASUREMENTS & INSTRUMENTATION

### RF Ammeters

As was seen in the section on antennas, the most straightforward way of estimating the effective radiated power, or determining the efficiency of an LF station requires measurement of the antenna current. Most LF stations have some form of RF ammeter so that antenna current can be measured. The traditional RF ammeter uses a thermojunction (sometimes called thermocouple); these are difficult to obtain, and are easily damaged by overload. Fortunately, it is easy to make a rectifier-based RF ammeter that is much more robust.

The RF ammeter of **Fig 10.61**(a) has a full-scale deflection of 1 amp, and uses a ferrite toroid as an RF current transformer. The single-turn primary is the current-carrying wire threaded through the toroid, which is wound with a 50 turn secondary. The secondary current, which is thus 1/50 times the primary current, or 20mA maximum, develops a voltage of 9.4V RMS across the 470 ohm load resistor. This voltage is measured by a simple diode voltmeter; the peak voltage across the smoothing capacitor is 1.414 x VRMS, or 13.3V, less around 0.5V diode forward voltage, giving approximately 12.8V DC which the DC series resistor sets as full scale deflection of the 100μA meter. The current range can be increased by proportionally reducing the value of the load resistor; it is desirable to maintain the voltage across the load at around 10V in order to maintain the linearity of the diode voltmeter. Note that with higher maximum currents, the power dissipation in the load resistor can reach a few watts, so the resistor should be appropriately rated.

A slightly different ammeter circuit is shown in **Fig 10.61**(b). In this circuit, the output current from the secondary of the transformer is fed directly to a bridge rectifier, whose mean output is measured by a DC milliameter. If IRF is the RMS RF current, and the secondary has N turns, the mean DC current at the rectifier output is 0.90 x IRF/N. The meter movement requires a shunt resistor to read the desired full-scale value. For example, if the current transformer has a 50 turn secondary, and 6A RMS full scale is required:

$$I_{mean} = 0.90 \times \frac{I_{RF}}{N} = 0.90 \times \frac{6}{50} = 0.108A$$

A 1mA meter with 75 ohm resistance was used, so the required shunt resistor Rshunt was:

$$R_{shunt} = R_{meter} \times \frac{I_{meter}}{(I_{mean} - I_{meter})} = 75 \times \frac{1mA}{(108mA - 1mA)} = 0.70\Omega$$

The shunt resistor was made up of larger value resistors in parallel. This circuit has good linearity down to low currents because the diodes are fed from the high impedance of the transformer secondary winding.

The voltages in this circuit are essentially just the forward voltage drop of the rectifier diodes, resulting in somewhat less power dissipation than the previous circuit, and also leading to less error due to the shunting effect of the transformer inductance at lower frequencies.

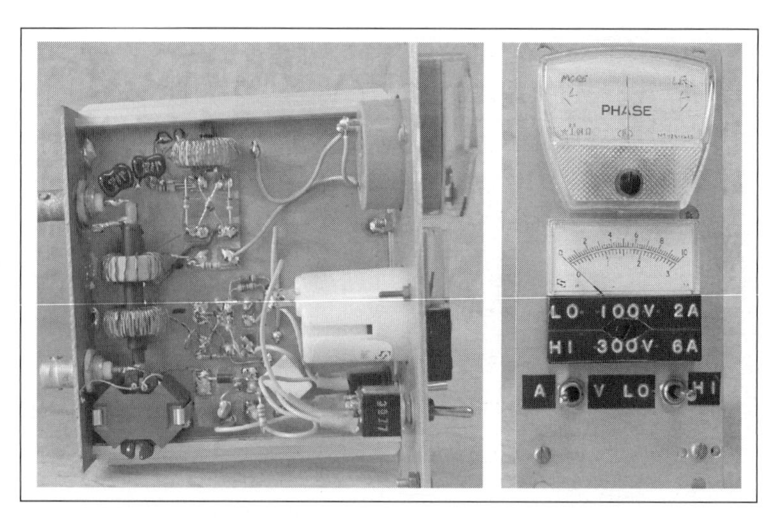


Fig 10.65: LF tuning meter prototype

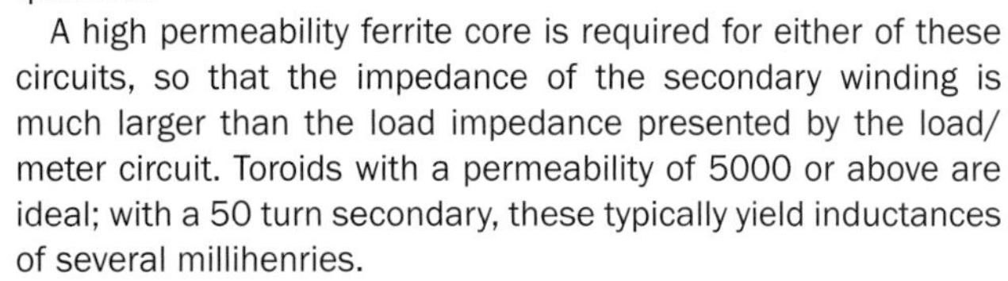

A high permeability ferrite core is required for either of these circuits, so that the impedance of the secondary winding is much larger than the load impedance presented by the load/meter circuit. Toroids with a permeability of 5000 or above are ideal; with a 50 turn secondary, these typically yield inductances of several millihenries.

A split ferrite core can be used to allow the ammeter to be put into the circuit without disconnecting the wire. If the RF ammeter is to be used at a high voltage point, such as at the feed point of the antenna wire, a screening metal case enclosing the transformer and meter circuit is advisable, to prevent the possibility of stray currents flowing in the meter circuit due to capacitive coupling, and causing errors.

## The Scopematch Tuning Aid

136kHz or 472kHz antenna tuning is critical; the loaded Q is usually of the order of 100, giving a very sharp resonance and rapidly varying impedance as the antenna is tuned. The 'scopematch' displays the voltage and current waveforms on a dual-trace oscilloscope, which provides a convenient and intuitive visual aid to adjusting antenna tuning. If voltage and current are out of phase, the antenna impedance is partly inductive or capacitive, if in-phase the impedance is resistive. The antenna resistance can then easily be determined from the ratio of voltage to current.

The circuit diagram of the scopematch is shown in **Fig 10.62**. Any high permeability ferrite core of about 18mm diameter will be suitable for the current-sensing transformer, which has a single turn primary winding and 50 turn secondary. Adequate insulation is required between primary and secondary to withstand the applied RF voltage. The 1:50 transformer and 50 ohm load give a scale factor of 1V = 1A at the 'I' output, the 50:1 capacitive potential divider gives a scale factor of 1V = 50V at the 'V' output. Thus when the load resistance is 50 ohms, the amplitudes at the V and I outputs will be equal.

In use, transmitter power is applied to the antenna tuner, and the loading coil adjusted until V and I waveforms are in phase. If the amplitude of V is greater than I, loading is adjusted to reduce the resistance, or to increase the resistance if V is less than I; see **Fig 10.63** for some example displays. A more detailed description of the Scopematch can be found in [49].

## LF Tuning Meter

The LF tuning meter was designed as an aid to tuning transmit antennas at LF and MF which provides more useful information than the traditional SWR bridge. It has two meters, one indicating phase, and thus antenna resonance, while the other can be switched between measuring RF voltage and current, allowing the magnitude of the load impedance and the transmitter output power to be determined.

The circuit diagram is shown in **Fig 10.64**, and a picture of the prototype in **Fig 10.65**. The phase meter indicates the DC output of a passive double balanced diode mixer, whose inputs are the antenna current, sampled by current transformer T1, and a sample of the antenna voltage phase shifted by 90 degrees obtained using the potential divider formed by the series 470pF capacitors and 82 ohm resistor.

With a resistive load, antenna current and voltage are in phase, and the DC output of the mixer is zero. With an inductive or capacitive load, the phase difference between voltage and current results in a positive or negative DC output from the mixer. In this simple phase meter, output voltage depends somewhat upon power level as well as load reactance, however a resistive load gives a zero indication over a wide power range.

The RF voltage is measured by a diode rectifier voltmeter driven from step-down auto-transformer T4. A toroidal current

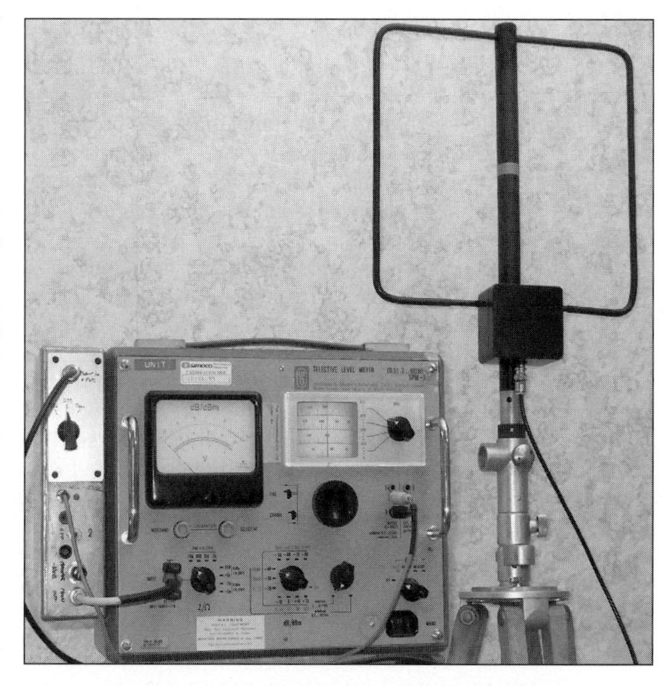

Fig 10.66: Field strength measuring system at M0BMU

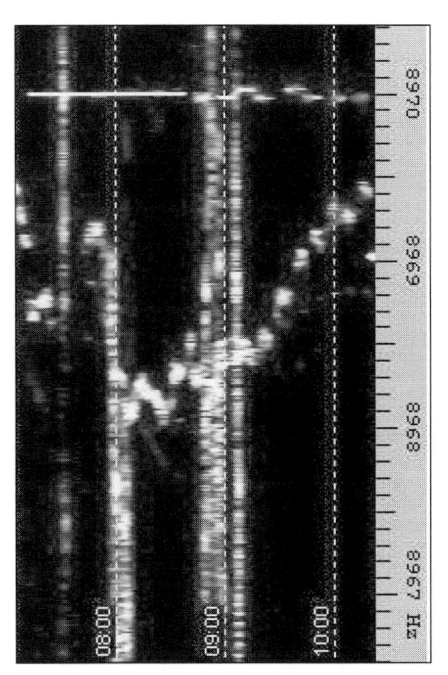


**Fig 10.67: 8.97kHz signal from DK7FC received over a three hour period at M0BMU, together with local noise. The signal consists of a fixed carrier, followed by "FB" encoded in DFCW**

transformer T3 is used with a bridge rectifier to measure current (see RF Ammeters section above).

'Hi' (300V, 6A) and 'Lo' (100V, 2A) voltage and current ranges are provided, which makes the meter usable over a range of about 20W to 200W (Lo range) or 200W to 1.8kW (Hi range) transmitter power. The voltage and current scales are chosen so that the voltage scale is 50 times the current scale. Therefore, with a 50Ω load connected, the same deflection is seen on the meter scale when the meter is switched between voltage and current ranges; the operator can immediately see if the load is matched to 50Ω (V/A readings equal deflection), greater than 50Ω (V>A), or less than 50Ω (V<A).

The back-to-back BAT85 schottky diodes limit the phase detector input signal to a reasonably constant 500mV pk-pk with changing power levels. Other switching or small rectifier schottky diodes rated at 200mA or more will also be satisfactory. The phase detector diodes should ideally be matched to minimise the error at zero phase - this can be done using the 'diode check' range on most digital multimeters. 1N914, 1N4148 or similar diodes work well. Note the phase detector diodes are connected as a 'rat-race', while the ammeter section diodes are connected as a bridge rectifier.

The voltmeter section uses a high voltage 'ultra-fast' rectifier diode rated at 600V, 1A; similar diodes intended for SMPS applications will be suitable. The transformers T1, T2, T3 are wound on high-permeability ferrite cores about 22mm diameter with AL of about 4000nH/t, and permeability of about 5000. TThe RS components 212-0910 and numerous others should be suitable.

The coax links the input and output sockets, and T1 and T3 are threaded onto the coax. Note that only one end of the coax braid is grounded! T4 is wound on an RM10 pot core in 3C85 or similar 'power' ferrite grade. The main winding has to withstand the 300V maximum RF voltage - 'Kynar' insulated wire-wrapping wire was used instead of standard enamelled wire.

The meters and range-setting resistors will depend on what you have in the junk box, and what transmitter power is to be used. Meters of different sensitivity, and different full-scale ranges can be used by changing the range-setting resistors. The values shown were assembled from series combinations during testing; presets could be used instead. Accuracy is not critical, since we are mainly just interested in the sign of the phase, and the relative level of voltage and current.

In use, select the 'Hi' or 'Lo' range to suit the power level in use. Apply transmitter power to the antenna, and note the phase meter reading. If the phase is negative (capacitive), increase the loading coil inductance, and reduce the inductance if the phase is positive (inductive). When zero phase is obtained, switch between 'V' and 'A' ranges. The ratio V/I is the resistance - if it is greater or less than 50 ohms, as indicated by the meter deflections not being equal, adjust the matching transformer, loading coil tap, etc. as appropriate. If the 'V' deflection is greater than the 'A' deflection, adjust to give a reduced load resistance; if 'V' is less than 'A', adjust for a higher load resistance. The transmitter output power when the load is matched is V2/50 or 50 x I2. A more detailed description of the tuning meter can be found in [49]

## Field Strength Measurement

As pointed out in the section on antennas, the definitive way of determining the effective radiated power of a station is to measure the field strength at a known distance from the antenna. The relation between field strength E (volts/metre) at a distance d metres, and ERP is given by the formula:

$$P_{ERP} = \frac{E^2 d^2}{49}$$

For this relationship to be valid, the distance d must be in the far field region of the antenna, where the field strength falls away in inverse proportion to the distance from the antenna. The near field region is closer to the antenna, where the field strength decreases more rapidly with distance, and the formula above does not apply. For small antennas at 136kHz and 472kHz, a safe minimum distance is about 1km. At distances much greater than a few tens of kilometres, the formula also becomes invalid, due to the effects of ground loss on the propagating wave close to the ground, and ionospheric reflections. For amateur stations, the signal is also likely to be too weak to accurately measure at larger distances.

Two pieces of equipment are required to measure field strength, a calibrated receiver and a calibrated 'measuring' antenna (**Fig 10.66**). The calibrated receiver must be capable of accurately measuring signal levels down to a few microvolts, and have sufficient selectivity to reject unwanted adjacent signals; the ideal amateur equipment for this purpose is the selective level meter (see section on receivers).

Calibrated antennas have a specified antenna factor (AF), which is the number of decibels which must be added to the signal voltage measured at their terminals to obtain the field strength. Quite good accuracy in the LF and MF ranges can be obtained using a simple single turn loop antenna. Such loops have a low feed point impedance, so the received signal level is little affected by the load impedance. The output voltage of an N-turn loop with area A square metres at a frequency f hertz is given by:

$$V = 2.1 \times 10^{-8} \times fNAE$$

From this, the antenna factor of a single turn loop at 137kHz is:

$$AF(dB) = 20Log_{10}\left(\frac{1}{2.1 \times 10^{-8} \times 137 \times 10^3 \times A}\right) = 20Log_{10}\left(\frac{350}{A}\right)$$

At 475kHz, AF = 20Log10(100/A). A square or circular loop made of tubing is usually used, with an area between 0.5m2 and 1m2. As an example, suppose a signal level of 7.5dBμV (ie 7.5 decibels above 1μV, or 2.4μV; selective level meters usually

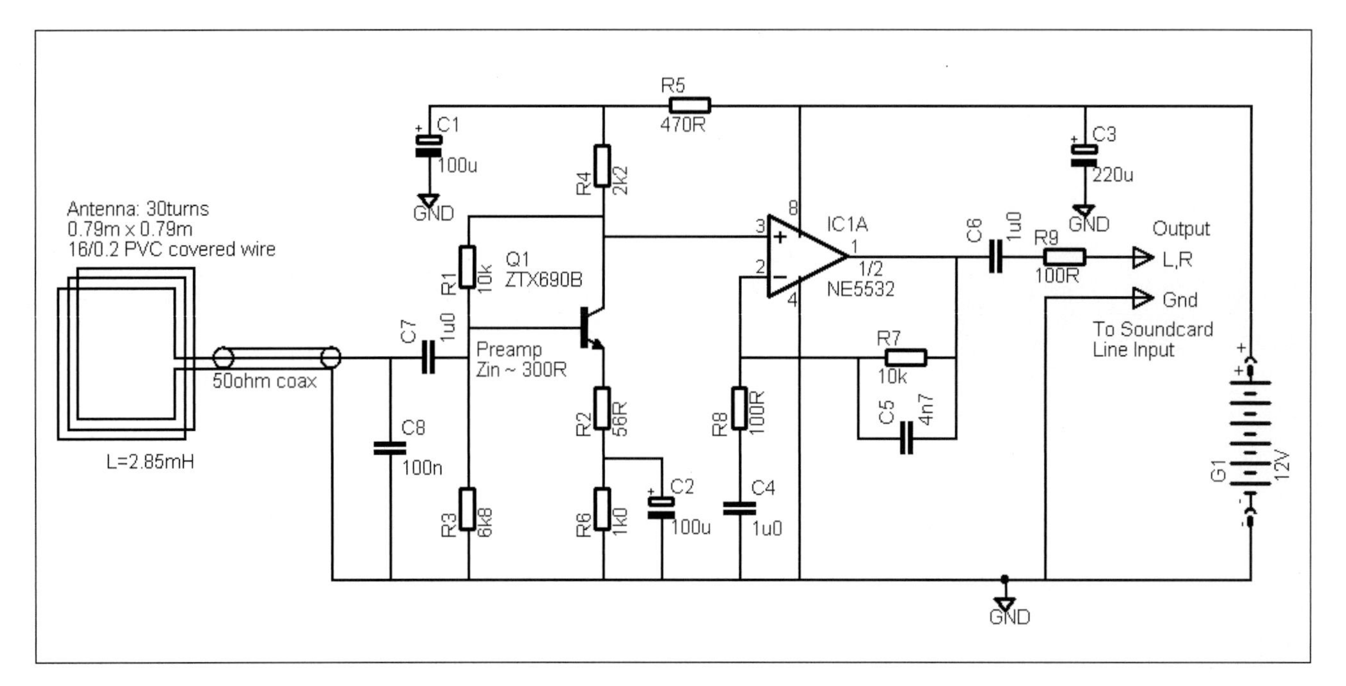


**Fig 10.68: (above right) A VLF loop antenna together with preamplifier for use with an ordinary PC sound card. The input transistor, Q1 can be replaced by a BC337 with little change in performance**

give a decibel-scaled reading) is measured at a distance of 5km from the transmitting antenna, using a 1m2 loop at 137kHz. From the formula above, AF is 51dB, so the field strength is 58.5dBμV/m, or 840μV/m. Using the ERP formula gives PERP = 350mW. When making measurements on the 472kHz band, it may be convenient to multiply the ERP figure by 1.4 to convert to EIRP (see the section on Estimating Radiated Power).

A more compact alternative to the loop is a tuned ferrite rod antenna, however this requires calibration with a known field strength to determine the antenna factor. A field strength measuring system, including ferrite rod antenna, measuring receiver, and calibration set-up has been described by PAOSE [50].

Field strength measurements are prone to errors caused by environmental factors. The measured field strength is particularly affected by conducting objects giving rise to parasitic antenna effects. Such parasitic antennas can be large steel-framed structures such as buildings and road bridges, overhead power and telephone wires, even such things as fence wires and shallow buried cables. Such factors are difficult to avoid entirely, so several field strength measurements should be made at different locations over as wide an area as possible. Locations giving widely different values of ERP can then be rejected; it will be found that a few decibels of variation still exists between different measurement sites, so the ERP should be taken as an average of several measuring sites.

## VERY LOW FREQUENCIES

The very low frequency (VLF) radio spectrum is defined as 3kHz - 30kHz. For many years it has been the nearly exclusive preserve of military communications and navigation utilities, but in recent years there have been several successful transmission and reception tests by amateurs in this range.

In some countries, frequencies below 9kHz are not regulated, so are freely available to amateur experimenters; in the UK, a Notice of Variation to the amateur licence must be obtained from Ofcom for transmission experiments in this frequency range; a few UK stations have obtained such NoVs in the past.

The use of large antennas and fairly high-power transmitter have resulted in European signals between 9kHz and 5kHz being widely received across the continent and as far afield as Iceland and Israel. Several US stations have now made successful transmissions below 30kHz, including crossing the Atlantic on 9kHz.

Transmitting and receiving techniques used at VLF are broadly similar to those used at LF, although even lower efficiencies from antennas of feasible size (the wavelength at 8.97kHz is 33.4km!) and much higher atmospheric and man-made noise levels make generating and detecting any signal very challenging.

Transmission has generally used the largest possible wire vertical antennas, with RF power generated by class D PAs, or modified audio amplifiers. Very large loading coils of the order of hundreds of millihenries are required. Efficiency is extremely low, with ERP in the micro-watt range for back-garden sized antennas, reaching milliwatts for larger kite or balloon supported wires.

Signals are received using digital signal processing to generate extremely narrow-band spectrograms with milli- or even micro-hertz resolution and integration periods of minutes to hours to give a detectable signal to noise ratio. Therefore, transmissions are usually fixed-frequency carriers, or keyed with frequency shifts of a few millihertz in simple patterns to facilitate identification (**Fig 10.67**). Transmitting single Morse characters in this way can take hours, so communications capability is negligible, but it is possible to make technically interesting observations of propagation changes, noise, etc.

A recent innovation has been the development of the EbNaut mode by Paul Nicholson [3] allowing more than just a detection of a carrier on a particular frequency. EbNaut uses coherent BPSK with both transmitter and receivers being GPS disciplined so that the timings are exact at each end of the link. The long duration phase-stable paths which occur at VLF can be exploited to send messages within 1 dB of the theoretical limit. EbNaut has been used to achieve the first transatlantic amateur communication below 9kHz.

Setting up an amateur VLF transmitting station represents a major investment of time and effort, so reception reports are always very welcome. Successful reception of amateur VLF signals can be achieved with quite modest equipment, and does not require large antennas, or the obtaining of a NoV. Reception has mostly been achieved by feeding the signal from a suitable antenna / preamplifier directly into the sound card input of a PC. The standard 48kHz sound card sampling rate allows direct processing of VLF signals up to 24kHz. Most receiving stations have used DL4YHF's Spectrum Lab free software suite [8], which performs down-conversion, filtering, noise blanking and frequency stabilisation functions, as well as generation of the spectrogram display, entirely using software. Therefore, the ordinary shack PC can be used as a remarkably sophisticated and economical software-defined radio for VLF reception.

Several types of receiving antenna have been used [52]. Large wire antennas can be effective if in a reasonably noise-free location; normally these are resonated in the VLF range (see Receive Antenna Tuning section above), and some input protection for the sound card is advisable. Small active whip and loop antennas are also successfully used, with the usual benefits of portability and the possibility of positioning to minimise local noise levels. For example, the PA0RDT active whip design described above can be modified for extended low-frequency response by increasing the values of coupling / decoupling capacitors and inductors, and increasing the input FET gate bias resistor. A simple active loop design that has been successfully used by M0BMU to receive several VLF transmissions is shown in **Fig 10.68**. Although response is broadly peaked at 9kHz, the bandwidth of this antenna is wide enough to receive other signals up to 24kHz, for example the occasional CW broadcasts to amateurs by Grimeton Radio SAQ on 17.2kHz.

VLF amateur radio is at an early stage in development and, although never likely to be a 'mainstream' activity, what has been called the 'Dreamers' Band' has already shown some surprising possibilities. For current up-to-date information on VLF activities, the reader is referred to the various on-line resources dedicated to this subject [53, 54].

## REFERENCES


[1] https://rosmodem.wordpress.com/
[2] http://www.472khz.org/SlowJT9/
[3] http://abelian.org/ebnaut/
[4] LF Today, 3rd edition, Mike Dennison, G3XDV, RSGB, 2013
[5] The Amateur Radio Operating Manual, 8th Edition, Mike Dennison, G3XDV, and Steve Telenius-Lowe, PJ4DX, RSGB, 2015
[6] www.g4jnt.com/freqlock.htm
[7] www.qsl.net/i2phd/TS950/change.html - TS950 mods for improved LF sensitivity
[8] Spectrum Lab and other radio-related software. DL4YHF's web pages: www.qsl.net/dl4yhf/.
[9] Winrad SDR software, Argo and other radio-related software. I2PHD's Weaksignals site: www.weaksignals.com
[10] www.rfspace.com - RFSpace Inc. website, including CloudIQ receiver
[11] http://microtelecom.it/index.html - Microtelecom S.r.l. Website, including Perseus SDR receiver
[12] G0MRF preamps and amplifiers: http://g0mrf.com
[13] http://www.monitorsensors.com/ham-radio/630m-transverter
[14] https://www.minikits.com.au/eme223
[15] http://w5jgv.com/400w-ssb-amp/index.htm - 400W LF linear amplifier
[16] A 1kW class D transmitter for LF and MF is described at www.w1vd.com/137-500-KWTX.html
[17] www.wireless.org.uk/jnt.htm - G4JNT off-line 600W PA design
[18] RS Components web pages and on-line catalogue: https://uk.rs-online.com/web/
[19] Farnell web pages and on-line catalogue: www.farnell.com
[20] www.cs.berkeley.edu/~culler/AIIT/papers/radio/Sokal%20AACD5-poweramps.pdf - description of class E PA operation, including design formulas and tuning procedure
[21] G3NYK's design paper on Class E amplifiers is at: http://g3nyk.ham-radio-op.net/classepa.htm.
[22] www.g4jnt.com/DownLoad/classe_match.xls - design spreadsheet for class E PA, including impedance matching section
[23] www.g4jnt.com/QRO_500kHz_PA_Breadboard.pdf - G4JNT class E PA for MF
[24] www.wireless.org.uk/136rig.htm - G3YXM 136kHz transmitter designs
[25] Variants of GW3UEP's transmitter design and associated test gear: http://www.gw3uep.ukfsn.org/472kHz.htm
[26] https://sites.google.com/site/uk500khz/members-files/files - download file 'EER_Transverter_v3.pdf'.
[27] www.g4jnt.com/LFUpconv.pdf - I/Q transmitting upconverter for 136kHz or 472kHz
[28] 'Build a PIC controlled DDS VFO, 0 to 6MHz' Johan Bodin, SM6LKM, http://home.swipnet.se/~w-41522/minidds/minidds.html
[29] 'A software based DDS for 137kHz', http://wireless.org.uk/swdds.htm
[30] DDS exciter by ZL1BPU, www.qsl.net/zl1bpu/MICRO/EXCITER/
[31] https://qrp-labs.com
[32] www.wireless.org.uk/on7yd/index.htm - ON7YD's extensive information on LF transmitting antennas
[33] www.wireless.org.uk/tap_coil.xls - Spreadsheet for tapped loading coil design
[34] http://www.472khz.org/
[35] F6HCC's 472kHz project. http://f6hcc.free.fr/tx472khz.htm
[36] http://g0mrf.com/variometer.htm - simple variometer for 136kHz
[37] www.oz8nj.dk/spoler.htm - variometers constructed for 136kHz
[38] www.qsl.net/in3otd/variodes.html - on-line variometer design calculator
[39] http://njdtechnologies.net/wp-content/uploads/2017/04/Bill-Ashlock-loops-CQ-July-2017.pdf
[40] www.wireless.org.uk/loopy.htm - transmitting loop used by G3YXM
[41] www.w1tag.com/XESANT.htm - transmitting loop used by WD2XES
[42] Elimination of Electrical Noise by Don Pinnock, G3HVA. A Radio Today publication available from the RSGB
[43] www.wireless.org.uk/BPloops2.pdf - Bandpass receiving loop antennas for LF and MF
[44] www.radiopassioni.it/pdf/pa0rdt-Mini-Whip.PDF - PA0RDT 'Mini Whip' design
[45] E-mail PA0RDT: roelof@ndb.demon.nl

[46] https://timewave.com/product/timewave-anc-4-antenna-noise-canceler/#
[47] www.hard-core-dx.com/nordicdx/antenna/loop/k9ay/k9ay_orig.pdf - K9AY receiving antenna array article
[48] http://www.iw5edi.com/ham-radio/files/qst_1_95_ewe.pdf - 'Is this Ewe for you?' article on EWE antenna
[49] https://sites.google.com/site/uk500khz/members-files/files - download file 'LF_tuning_aids.pdf', including details of LF tuning meter, 'scopematch' tuning aid and simple resistive SWR bridge.
[50] www.wireless.org.uk/pa0se.htm - PA0SE field strength measurement system
[51] 'Experimental investigation of very small low frequency transmitting antennas', J R Moritz, IEE 9th International Conference on HF Radio Systems and Techniques, June 2003, IEE Conference Publication n. 493 pp 51 - 56
[52] https://sites.google.com/site/sub9khz/antennas - VLF receiving antenna designs
[53] https://groups.io/g/vlf - VLF amateur radio discussion group
[54] https://sites.google.com/site/sub9khz/home - Information and news on VLF amateur radio activity.
[55] https://s9plus.com/
[56] https://www.qrphamradiokits.com



***Dave Pick G3YXM****, has been involved with amateur LF experiments since 1998 when UK amateurs were first allowed access to the 136kHz band. He is still active on 472 and 136khz. He has written the RadCom LF column since 1999.*
*Dave has taken over producing this chapter from Mike Dennison and we thank Mike for all his hard work in the past.*

# Microwave Receivers & Transmitters 11

*Andy Barter, G8ATD*

Many of the techniques for generating and receiving microwave frequencies were investigated and developed more than 70 years ago, in the 1930s. Microwave usage was given added impetus by the development of radar and the advent of the Second World War. Before 1940, the definition of the higher parts of the radio frequency spectrum [1, 2] read like this:

| | |
|---|---|
| *30 to 300Mc/s* | *Very high frequencies (VH/F)* |
| *300 to 3000Mc/s* | *Decimetre waves (dc/W)* |
| *3000 to 30.000Mc/s* | *Centimetre, waves (cm/W)* |

Radio frequencies above 30,000Mc/s (now 30GHz) apparently did not exist! Various definitions have appeared in the intervening years. These have included terms such as super-high frequencies (SHF) and extra-high frequencies (EHF).

In the course of time, the unit of frequency cycles per second (c/s), its decimal multiples, kilocycles per second (kc/s) and megacycles per second (Mc/s), have been replaced by the unit hertz (Hz), its decimal multiples kilohertz = $10^3$Hz (kHz), megahertz = $10^6$Hz (MHz), gigahertz = $10^9$Hz (GHz) and terahertz = $10^{12}$Hz (THz).

Today, the term microwave has come to mean all radio frequencies above 1000MHz (1GHz). The division between radio frequencies and other electromagnetic frequencies, such as infra-red, visible (light) frequencies, ultra-violet and X-rays, is still not well defined since many of the techniques overlap, just as they do in the transition between HF and VHF or UHF and microwaves.

There has been keen interest in amateur communications using infra-red and visible light, over the last few years, so a section on it has been included in this chapter. To a large extent the divisions are artificial insofar as the electromagnetic spectrum is a frequency continuum, although there are several good reasons for these divisions.

Around 1GHz (30cm wavelength) the lumped circuit techniques used at lower frequencies are replaced by distributed circuit techniques such as resonators and microstrip. Conventional components, such as resistors and capacitors, become a significant fraction of a wavelength in size so surface-mount devices (SMDs) are used which are very small and leadless. These require special techniques for constructors; these are described later.

Conventional valves (vacuum tubes) and silicon bipolar solid state devices are usable beyond 1GHz - perhaps to about 3.5GHz - and, as frequencies increase, these devices are replaced by special valves such as klystrons, magnetrons and travelling-wave tubes. The first semiconductor devices to be used at the higher microwave frequencies were varactor diodes, PIN diodes and Gunn diodes. Because of the massive development in semiconductors for use by commercial telecommunications systems there is now a wide range of transistors available to radio amateurs including gallium arsenide field effect transistors (GaAsFETs), metal epitaxial semiconductor field effect transistors (MESFETs), pseudomorphic high electron mobility transistors (pHEMTs) and many more.

The other device that has revolutionised microwave designs is the microwave monolithic integrated circuit (MMIC); these are usually designed to work in matched 50-ohm networks, making them easy to use and (usually) free from instability problems. They are used as building blocks, in a similar way that the operational amplifier (OP-amp) is used in DC and audio designs, but are extremely wide band and often require bandpass filtering to obtain the desired results.

Many of the more exotic semiconductor devices are now appearing on the surplus market making them more acceptable to the amateur's pocket. These can be used to build new amateur equipment, or whole surplus units can be modified to work on amateur bands. Quite a few articles have appeared, describing the modification of commercial equipment, so there are example articles later in this chapter.

Other advantages of operation in the microwave spectrum are compact, high-gain antennas and available bandwidth. None of these advantages is attainable in an amateur station operating on the lower frequency amateur bands. High gain antennas are impossibly large below VHF, and the levels of spectral pollution from man-made and natural noise are such that low noise receivers, needed to handle weak signals, cannot now be effectively used, even at VHF. Communication on the many available microwave bands over distances of hundreds of kilometres is now quite common (sometimes over thousands of kilometres, given favourable tropospheric propagation conditions or the use of amateur satellites or moonbounce). This destroys the perception that microwaves are useful only over limited line-of-sight paths! A good place to find details of the latest records and operating conditions is *Dubus Magazine*; all of the main amateur bands are reported with details of activity using various types of propagation.

There is an increasing amount of commercially produced equipment available from amateur radio retailers for all of the microwave bands and plenty of designs for the constructor. It is true that attaining really high transmitter power output above about 3 or 4GHz is still difficult and expensive for most amateurs. Many successful amateur operators settle for comparatively low power output, ranging from perhaps 50 to 100W in the lower-frequency bands, to milliwatts in the 'centimetre' bands, or even microwatts in the 'millimetre' bands. This is compensated for by using very high antenna gain and, as already mentioned, receivers with very low noise figures.

Since the first essential requirement of microwave construction is easy availability of designs and components, many leading microwave amateurs have launched small-quantity component sources or have designed and can supply either kits of parts for home construction or ready-made equipment to these designs. Most microwave equipment now uses printed circuit boards (PCBs) and surface-mount components. This avoids the use of the precision engineering usually associated with older, waveguide based designs and means that construction of microwave equipment is not restricted to the amateur who has his own mechanical workshop. Conventional tools can be used together with some fairly simple test equipment to construct some very sophisticated equipment that produces excellent results.

There are many examples of designs that can be purchased in kit form or ready built; suppliers are listed in the Bibliography at the end of this chapter. The most widely used designs in the UK come from Michael Kuhne (DB6NT) who has equipment for all bands up to 241GHz; most have been described in the pages of *Dubus Magazine* or the Dubus Technik publications.

Free and easy access to practical information is important to the microwave amateur enthusiast. The question most often asked by amateurs new to microwaves is: "Where do I get reliable information and (possibly) help?" Many microwave designs have appeared in the amateur press, in published books or magazines and in the various national amateur radio societies' journals. Some of the more prolific or rewarding titles are given in the bibliography. In addition to these sources, obtaining up-to-date designs, component information and design tools is extremely easy using the internet. The number of suppliers of microwave components has mushroomed with the expansion of the mobile phone networks, so has the sophistication of the design tools available.

Many suppliers of design tools have student or 'Lite' versions of their software free to download from their websites. These generally have reduced functionality compared with the full versions of the software, which may cost several thousand pounds, but are more than adequate for most amateur use. A quick search with one of the popular search engines, using the relevant key words, should find information about the software.

The range of current amateur microwave allocations offers scope to try out all of the modes and techniques available to amateurs. All amateurs are encouraged to try out some of these which will help retain our allocations. The lowest microwave frequency amateur allocation, the so called 23cm band, (1240MHz to 1325MHz in the UK), can be regarded as the transition point from 'conventional' radio techniques and components to the 'special' microwave techniques and components to be reviewed here.

In the space of a single chapter it will only be possible to give a flavour of some of the practical techniques involved, by outlining a few representative designs for most of the bands currently used by amateurs. If you need more detail, there are plenty of pointers to other sources of information shown in the bibliography.

The microwave bands support a wide range of activities such as:

- All narrow band modes
- Amateur TV, including wide band colour transmission
- Moon bounce (EME)
- Amateur satellite operation
- Meteor scatter

Since a significant amount of amateur microwave interest centres on the use of narrow band modes to achieve long distance, weak signal communication, the majority of the designs outlined here will concentrate on such equipment. More details of components and techniques (including wide band modes) are available in other publications [3, 4, 5].

In some instances construction and alignment procedures are described in some detail, again to illustrate the techniques used by amateurs in the absence of elaborate or costly test equipment, such as microwave noise sources, power meters, frequency counters or spectrum analysers. Most of the designs described are capable of being home constructed without elaborate workshop facilities (most can be constructed using hand tools, a generous helping of patience and some basic knowledge and skills!) and aligned with quite ordinary test equipment such as matched loads, directional couplers, attenuators, detectors, multimeters and calibrated absorption wavemeters.

## AMATEUR MICROWAVE ALLOCATIONS

Most countries in the world have amateur microwave allocations extending far into the millimetre wave region, ie above 30GHz. Many of these allocations are both 'common' and 'shared Secondary', ie they are similar in frequency in many countries but are shared with professional (in this case 'Primary') users who take precedence. Amateur usage must, therefore, be such that interference to Primary users is avoided and amateurs must be prepared to accept interference from the Primary services, especially in those parts of the spectrum designated as Industrial, Scientific and Medical (ISM) bands.

The UK Amateur Service allocations are summarised in **Table 11.1** and the UK Amateur Satellite Service allocations are shown in **Table 11.2**.

All the familiar transmission modes are allowed under the terms of the amateur licence: in contrast to the lower frequency bands, most of the microwave bands are sufficiently wide to

| Allocation (MHz) | Amateur Status | Narrow band segment centre of activity (MHz) |
|---|---|---|
| 1,240 - 1,325 | Secondary | 1,296.200 |
| 2,310 - 2,450 | Secondary | 2,320.200 |
| 3,400 - 3,475 | Secondary | 3,410 (EME) |
| 5,650 - 5,680 | Secondary | 5,668.200 |
| 5,755 - 5,765 | Secondary | 5,760.100 |
| 5,820 - 5,850 | Secondary | |
| 10,000 - 10,125 | Secondary | |
| 10,225 - 10,475 | Secondary | 10,368.100 |
| 10,475 - 10,500 | Secondary | Amateur Satellite Service only |
| 24,000 - 24,050 | Primary | 24,048.200 |
| 24,050 - 24,250 | Secondary | |
| 47,000 - 47,200 | Primary | 47,088.200 |
| 75,500 - 75,875 | Secondary | |
| 75,875 - 76,000 | Primary | 75,976.200 |
| 76,000 - 77,500 | Secondary | |
| 77,500 - 78,000 | Primary | 77,500.200 |
| 78,000 - 81,00 | Secondary | |
| 122,250 - 123,000 | Secondary | |
| 134,000 - 136,000 | Primary | 134,928.0 - 134,930.0 |
| 136,000 - 141,000 | Secondary | |
| 241,000 - 248,000 | Secondary | |
| 248,000 - 250,000 | Primary | |

**Table 11.1: UK Amateur Service allocations 2014**

| Allocation (MHz) | Amateur Status | Comments |
|---|---|---|
| 1,260 - 1,270 | Secondary | ETS |
| 2,400 - 2.450 | Secondary | ETS/STE |
| 5,650 - 5,668 | Secondary | ETS |
| 5,830 - 5,850 | Secondary | STE |
| 10,475 - 10,500 | Secondary | ETS/STE |
| 24,000 - 24,050 | Primary | ETS/STE |
| 47,000 - 47,200 | Primary | ETS/STE |
| 75,500 - 75,875 | Secondary | ETS/STE |
| 75,875 - 76,000 | Primary | ETS/STE |
| 76,000 - 77,500 | Secondary | ETS/STE |
| 77,500 - 78,000 | Primary | ETS/STE |
| 78,000 - 81,000 | Secondary | ETS/STE |
| 134,000 - 136,000 | Primary | ETS/STE |
| 136,000 - 141,000 | Secondary | ETS/STE |
| 241,000 - 248,000 | Secondary | ETS/STE |
| 248,000 - 250,000 | Primary | ETS/STE |

*ETS = Earth to Space, STE = Space to Earth*

**Table 11.2: UK Amateur Satellite Service allocations 2014**

**Table 11.3: Some harmonic relationships for the microwave bands**

| Starting frequency | Multiplication | Output frequency |
|---|---|---|
| 144MHz | x3 | 432MHz |
| | x9 | 1296MHz |
| | x16 | 2304MHz |
| | x24 | 3456MHz |
| | x46 | 5760MHz |
| | x72 | 10,368MHz |
| | x108 | 24,192MHz |
| 432MHz | x3 | 1296MHz |
| | x8 | 3456MHz |
| | x24 | 10,368MHz |
| | x56 | 24,192MHz |
| 1152MHz | +144* | 1296MHz |
| | x2 | 2304MHz |
| | x3 | 3456MHz |
| | x5 | 5760MHz |
| | x9 | 10,368MHz |
| | x21 | 24,192MHz |

* Note: additive mixing, not multiplication

support such modes as full-definition fast-scan TV (FSTV) or very high speed data transmissions as well as the more conventional amateur narrow-band modes, such as CW, NBFM and SSB.

Many of the bands are so wide (even though they may be Secondary allocations) that it may be impracticable for amateurs to produce equipment, particularly receivers that cover a whole allocation without deterioration of performance over some part of the band. Most amateur operators do possess a high-performance multimode receiver (or transceiver) as part of their station equipment and this will frequently form the 'tuneable IF' for a microwave receiver or transverter. Commonly used intermediate frequencies are 144-146MHz or 432-434MHz. either of which are spaced far enough away from the signal frequency to simplify the design of good image and local oscillator carrier sideband noise rejection filters. An intermediate frequency of 1296-1298MHz is often used for the millimetre bands, ie 24GHz and higher.

There are 'preferred' sub-bands in virtually all of the amateur allocations where the majority of narrow band (especially weak signal DX) operation takes place. Typically 2MHz wide sub bands, often harmonically related to 144MHz as shown in **Table 11.3**, were originally adopted for this purpose.

Some of these harmonic relationships are no longer universally available or usable because the lower microwave bands are rapidly filling up with Primary user applications. Indeed, the position is changing particularly rapidly at the time of publication and the reader should refer to the latest ITU/IARU band plans (see RSGB web site) to get up-to-date information on current amateur usage, even though the current narrow band segment centres of activity are indicated in Table 11.1.

## MODERN MICROWAVE COMPONENTS AND CONSTRUCTION TECHNIQUES

### Static Precautions

Some types of microwave components, for example Schottky diodes (mixers and detectors), microwave bipolar transistors and GaAsFETs can be damaged or destroyed by static charges induced by handling, and thus certain precautions should be taken to minimise the risk of damage.

Such sensitive devices are delivered in foil lined, sealed envelopes in conductive (carbon filled) foam plastic or wrapped in metal foil. The first precaution to be taken is to leave the device in its wrapping until actually used. The second precaution is to ensure that the device is always the last component to be soldered in place in the circuit. Once in circuit the risk is minimised since other components associated with the device will usually provide a 'leakage' path of low impedance to earth that will give protection against static build up.

Before handling such devices the constructor should be aware of the usual sources of static. Walking across nylon or polyester carpets and the wearing of clothes made from the same materials are potent sources of static, especially under cold dry conditions. The body may carry static to a potential of several thousand volts although much lower leakage potentials existing on improperly earthed mains voltage soldering irons are still sufficient to cause damage. Some precautions are listed below:

- Avoid walking across synthetic fibre carpets immediately before handling sensitive devices.
- Avoid wearing clothes of similar materials.
- Ensure that the soldering iron is properly earthed whilst it is connected to its power supply. This is a common sense precaution in any case. Preferably use a low voltage soldering iron.
- Use a pair of crocodile clips and a flexible jumper wire to connect the body of the soldering iron to the earth plane of the equipment into which the device is being soldered.
- If the component lead configuration allows (and the usual flat pack will), place a small metal washer over the device before removing it from its packing in such a position that all leads are shorted together before and during handling. Alternatively, it might be possible to use a small piece of aluminium foil to perform the same function, removing the foil once the device has been soldered in place.
- A useful precaution that will minimise the risk of heat damage, rather than static damage, is to ensure that the surfaces to be soldered are very clean and preferably pre tinned.
- Immediately before handling the device, touch the earth plane of the equipment and the protective foil to ensure that both are essentially at the same potential.
- Place the device in position, handling as little as possible.
- Disconnect the soldering iron from its power supply and quickly solder the device in place. It may be necessary to repeat some of the operations if the soldering iron has little heat capacity.

Finally, when assembling items of equipment to form a complete operating system, for instance when installing a masthead pre-amplifier and associated transmit/receive switching, it is important to keep leads carrying supply voltages to the sensitive devices well away from other leads carrying appreciable RF levels or those leads which might carry voltage transients arising from inductive (relay) switching. Such supply lines should be well screened and decoupled in any case, but physical separation can minimise pick up, thus making the task of decoupling easier.

### PCB Materials

A printed circuit board (PCB) in a microwave design is not like the one you find in HF equipment. The printed tracks are an integral part of the circuit not just there to interconnect the components. The tracks are microstrip transmission lines, used to form matching circuits, tuned circuits and filters. The design process takes into account the base material of the board used, the thickness of the copper deposit and the dimensions of the track etched. It is therefore important to use the material specified in the design that you are using otherwise the circuit may not perform as expected. Conventional Epoxy/glass PCB board is usable, with care, up to about 3GHz. Most designs use special PCB materials such as Rogers RO 4003 or RT/duroid 5870.

If possible try to use PCBs that have been professionally produced using the correct PCB material and good artwork. If that is not possible you can produce your own PCBs using conventional etching techniques. Microwave PCBs always have an earth plane on one side and etched tracks on the other side; the earth connections from one side to the other are important and are formed by plating through the appropriate holes on professionally manufactured boards. For home made PCBs the best technique is to use small rivets to make these connections, they are fitted and soldered in place forming a good, low inductance, interconnection.

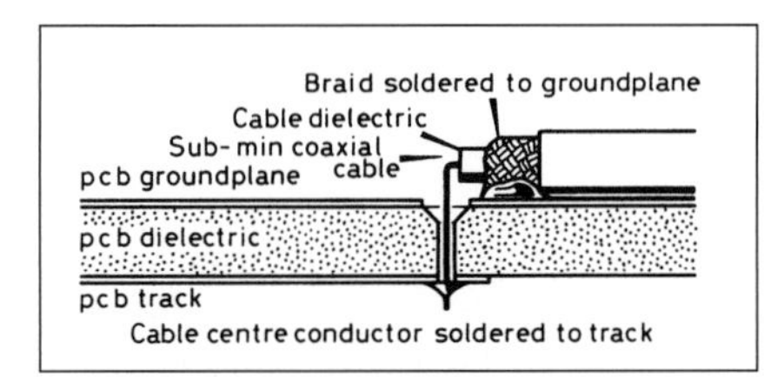


Fig 11.1: Correct coaxial cable connection technique

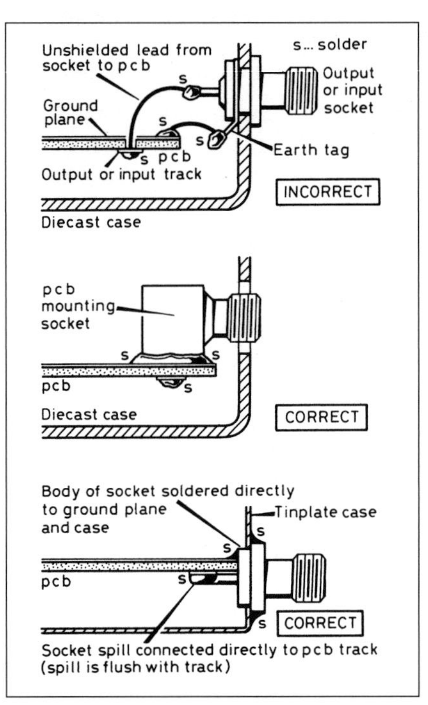


Fig 11.2: Correct and incorrect methods of connecting coaxial sockets to microwave PCBs

## Earthing and Interconnections

Earthing is a very important topic for the microwave constructor. As mentioned above, the earth connections from one side of a PCB to the other must be as good as possible to reduce the effects of stray inductance. The same is true for all other earth connections; they must be as short and solid as possible. It is important to house finished PCBs properly in order to screen the circuitry from stray pick-up and provide a good earth. Small die-cast boxes can be used but they are quite expensive and difficult to use.

Piper Communications stock tinplate boxes of various sizes that are an acceptable substitute for the die-cast boxes. These are widely used in Europe for housing such PCBs and are much less expensive. They consist of two L-shaped side pieces, and top and bottom lids. It is intended that the PCB be put into the box, joining the edges of the ground plane to the sides of the box.

To interconnect circuits it is necessary to use RF connectors. N-type connectors are too large and BNC connectors can be unreliable. UHF sockets must not be used (they are absolutely useless at UHF, despite their name, and significantly mismatched even at 144MHz) and the only really reliable types are SMA, SMB or SMC, all of which are expensive. You might like to consider taking the outputs away by directly connecting miniature 50-ohm coaxial cable as shown in **Fig 11.1**. Do not take the output away as shown in **Fig 11.2**, this is disastrous as it will almost certainly cause mismatch, stray inductive losses and may detune the output lines so that they will not resonate properly.

## Surface Mount Components

Surface mount components are ideally suited for microwave construction because there are no leads to introduce extra inductance in series with the actual component being used. This means that circuit performance can be reproduced more easily. It does introduce a new construction challenge for the newcomer to microwaves. Dealing with tiny components and solder-

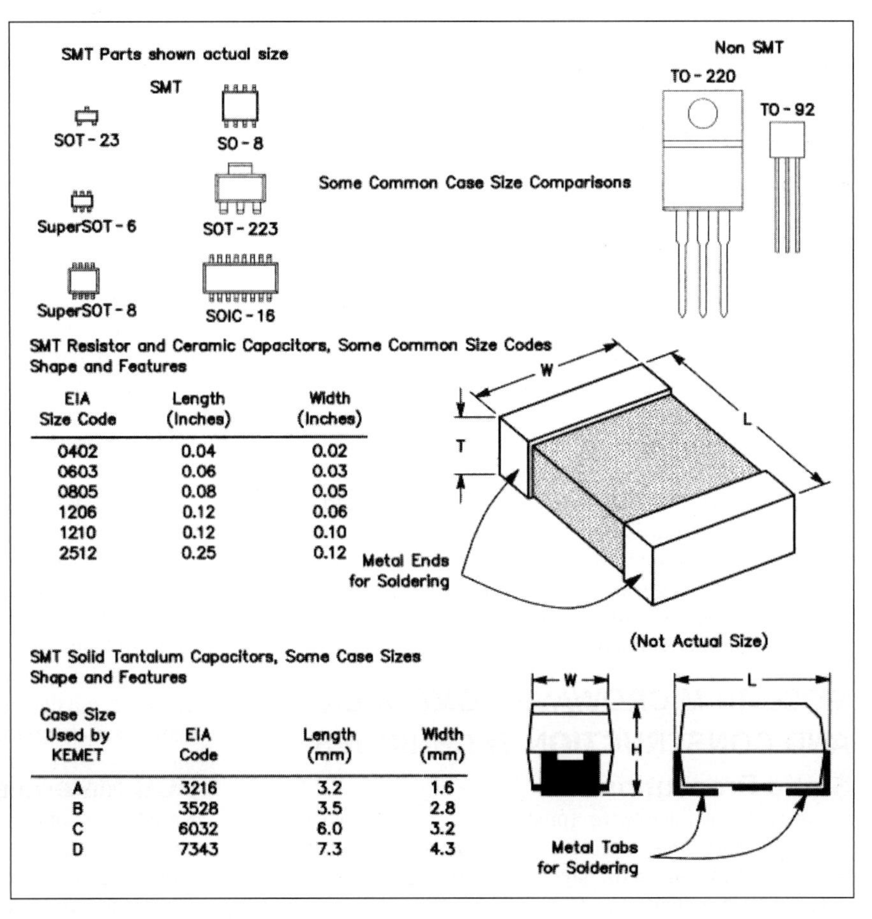


| EIA Size Code | Length (Inches) | Width (Inches) |
|---|---|---|
| 0402 | 0.04 | 0.02 |
| 0603 | 0.06 | 0.03 |
| 0805 | 0.08 | 0.05 |
| 1206 | 0.12 | 0.06 |
| 1210 | 0.12 | 0.10 |
| 2512 | 0.25 | 0.12 |

| Case Size Used by KEMET | EIA Code | Length (mm) | Width (mm) |
|---|---|---|---|
| A | 3216 | 3.2 | 1.6 |
| B | 3528 | 3.5 | 2.8 |
| C | 6032 | 6.0 | 3.2 |
| D | 7343 | 7.3 | 4.3 |

Fig 11.3: Size comparison of some surface mount devices and their dimensions

| Part number | Description | Bias voltage (V) | Bias current (mA) | NF (dB) |
|---|---|---|---|---|
| MGA-13516 | High gain, high linearity, active bias, low noise amplifier | 5 | 54 | 0.66 |
| MGA-14516 | High gain, high linearity, active bias, low noise amplifier | 5 | 45 | 0.66 |
| MGA-30116 | 750MHz - 1GHz 1/2W high linearity amplifier | 5 | 202.8 | 2.0 |
| MGA-30216 | 1.7 - 2.7GHz 1/2W high linearity amplifier | 5 | 206 | 2.8 |
| MGA-30316 | 3.3 - 3.9GHz 1/2W high linearity amplifier | 5 | 198 | 2.7 |
| MGA-52543 | 5V LNA, 32dBm OIP3, 0.4 - 6GHz, SOT343(SC-70) | 5 | 53 | 1.9 |
| MGA-53543 | 5V high linearity LNA, 39dBm OIP3, 0.44 - 6GHz, SOT343(SC-70) | 5 | 54 | 1.5 |
| MGA-53589 | 50MHz to 6GHz high linearity amplifier | 5 | 54 | 1.66 |
| MGA-545P8 | Low current 22dBm medium power amplifier in LPCC2x2 for 5 - 6GHz | 3.3 | 92 | 4.4 |
| MGA-565P8 | 20dBm Psat high isolation buffer amplifier | 5 | 67 | |
| MGA-61563 | Current adjustable low noise amplifier | 3 | 41 | 1.2 |
| MGA-62563 | Current adjustable low noise amplifier | 3 | 60 | 0.9 |
| MGA-631P8 | Low noise, high linearity amplifier with active bias | 4 | 60 | 0.5 |
| MGA-632P8 | Low noise, high linearity amplifier with active bias | 4 | 60 | 0.6 |
| MGA-665P8 | 0.5 - 6GHz low noise amplifier | 3 | 21 | 1.5 |
| MGA68563 | Current adjustable low noise amplifier | 3 | 11 | 1.0 |
| MGA-685T6 | Current adjustable low noise amplifier | 3 | 10 | 0.9 |
| MGA-71543 | 3V LNA with bypass switch, 0 to 9dBm adjustable IIP3, SOT343(SC-70) | 2.7 | 10 | 0.8 |
| MGA-72543 | 3V LNA with bypass switch, 2 to 14dBm adjustable IIP3, SOT343(SC-70) | 2.7 | 20 | 1.4 |
| MGA-725M4 | 3V LNA with bypass switch, 2 to 14dBm adjustable IIP3, SOT343(SC-70) | 2.7 | 20 | 1.3 |
| MGA-785T6 | Low noise amplifier with bypass switch | 3 | 7 | 1.1 |
| MGA-81563 | 3V driver amplifier. 14dBm P1dB, low noise, 0.1 - 6GHz, SOT363(SC-70) | 3 | 42 | 2.8 |
| MGA-82563 | 3V driver amplifier. 14dBm P1dB, low noise, 0.1 - 6GHz, SOT363(SC-70) | 3 | 84 | 2.2 |
| MGA-83563 | 3V PA/driver amplifier. 22dBm PSAT, 0.5 - 6GHz, SOT363(SC-70) | 3 | 152 | |
| MGA-85563 | 3V LNA, 12 to 17dBm adjustable OIP3, 0.8 - 6GHz, SOT363(SC-70) | 3 | 15 - 30 | 1.9 |
| MGA-86563 | 5V LNA. 20dB high gain, 0.5 - 6GHz, SOT363(SC-70) | 5 | 14 | 1.5 |
| MGA-86576 | 5V LNA. 23dB high gain, 1.5 - 8GHz, SOT363(SC-70) | 5 | 16 | 1.6 |
| MGA-87563 | 3V LNA, 4.5mA low current, 0.5 - 4GHz, SOT363(SC-70) | 3 | 4.5 | 1.6 |

**Table 11.4: Data for a selection of Avago GaAs MMICs (Copyright Avago, reproduced with their permission)**

ing them in place can seem a daunting task but after some practice it is easy.

**Fig 11.3** shows some of the more common SMD component sizes, obviously some special tools are needed to cope with these small components. The essential tools are:

- An illuminated magnifying glass. For instance, a bench-mounted magnifier with a five inch glass.
- A low power temperature-controlled soldering iron. This should have its tip earthed to reduce problems with static. A fine conical tip is best.
- Thin flux cored solder, preferably 26SWG (0.5mm). Larger diameter solder is difficult to use because it floods the solder pads and tends to create short circuits between solder pads.
- De-soldering braid, for use when too much solder has been applied.
- A flux pen to apply a small amount of flux before components are mounted.
- A good pair of non-magnetic tweezers.
- A PCB frame or some other method to hold your printed circuit board whilst soldering. If you don't hold the PCB down you will run out of hands to hold everything else!

Before you start to mount components, the PCB should be lightly tinned, just enough solder for the solder to flow onto the component but not too much, otherwise short-circuits will be a problem. As with wired components the assembly sequence should start with the low value parts such as resistors and capacitors. Position the component in position and apply heat just long enough for the solder to flow; you will need to hold the component in place otherwise it will move as the solder flows and may well land up on the end of your soldering iron rather than on the PCB!

Multi-leaded devices, like ICs, should be tacked in place by soldering leads at opposite corners and then flowing solder to all the other legs. Some of the latest ICs have an earthing pad underneath which makes life very difficult. The only successful technique that I have heard of is to mount such components first by heating the complete PCB over an electric cooker to flow the solder under the IC.

## Monolithic Microwave Integrated Circuit (MMIC) Amplifiers

MMICs are now widely used in amateur radio designs and are available from several manufacturers including Mini-Circuits and Avago (formerly Agilent or Hewlett Packard) at a price that makes them very attractive for many applications.

Keeping track of the devices that are available can be a problem; two useful sources of information are the Avago website [6] and the Minicircuits website [7]. Some useful information from these suppliers is reproduced in **Tables 11.4 and 11.5**.

The Avago website has a wide range of application notes to show how to use their MMICs. They also supply evaluation boards so that the complete circuit can be built and tested. One such application note is for a 2,400MHz LNA, designed to provide an optimum noise match from 2,400 - 2,500MHz, making it useful for applications that operate in the 2,400 to 2,483MHz ISM band. The component labels appearing in the following paragraphs refer to positions shown in **Fig 11.4**. The input match consists of a shunt inductor at L1 and a series inductor at L2. Both of these inductors use the tracks as originally etched

| Model Number | Frequency range (MHz) Low | High | Gain (dB). Typ | Max. Power output @ 1dB comp. (dBm) Typ. | N.F. (dB) Typ. | IP3 (dBm) Typ. | VSWR (:1) Typ. In | Out | Device DC operating power Voltage (V) | Current (mA) |
|---|---|---|---|---|---|---|---|---|---|---|
| ERA-1+ | DC | 8000 | 10.9 | 12.0 | 4.3 | 26.0 | 1.5 | 1.5 | 3.4 | 40 |
| ERA-2+ | DC | 6000 | 14.4 | 13.0 | 4.0 | 26.0 | 1.3 | 1.2 | 3.4 | 40 |
| ERA-3+ | DC | 3000 | 18.7 | 12.5 | 3.5 | 25.0 | 1.5 | 1.4 | 3.2 | 35 |
| ERA-4+ | DC | 4000 | 13.4 | 17.3 | 4.2 | 34.0 | 1.2 | 1.3 | 4.5 | 65 |
| ERA-5+ | DC | 4000 | 18.5 | 18.4 | 4.3 | 32.5 | 1.3 | 1.2 | 4.9 | 65 |
| ERA-6+ | DC | 4000 | 12.2 | 17.9 | 4.5 | 36.0 | 1.3 | 1.6 | 5.0 | 70 |
| ERA-1SM+ | DC | 8000 | 10.9 | 12.0 | 4.3 | 26.0 | 1.5 | 1.5 | 3.4 | 40 |
| ERA-2SM+ | DC | 6000 | 14.4 | 13.0 | 4.0 | 26.0 | 1.3 | 1.2 | 3.4 | 40 |
| ERA-21SM+ | DC | 8000 | 12.2 | 12.6 | 4.7 | 26.0 | 1.1 | 1.3 | 3.5 | 40 |
| ERA-3SM+ | DC | 3000 | 18.7 | 12.5 | 3.5 | 25.0 | 1.5 | 1.4 | 3.2 | 35 |
| ERA-33SM+ | DC | 3000 | 17.4 | 13.5 | 3.9 | 28.5 | 1.6 | 1.25 | 4.3 | 40 |
| ERA-4SM+ | DC | 4000 | 13.4 | 17.3 | 4.2 | 34.0 | 1.2 | 1.3 | 4.5 | 65 |
| ERA-4XSM+ | DC | 4000 | 13.5 | 17.0 | 4.2 | 35.0 | 1.3 | 1.3 | 4.5 | 65 |
| ERA-5SM+ | DC | 4000 | 17.6 | 18.4 | 4.3 | 32.5 | 1.3 | 1.2 | 4.9 | 65 |
| ERA-5XSM+ | DC | 4000 | 17.6 | 17.8 | 3.5 | 33.0 | 1.3 | 1.3 | 4.9 | 65 |
| ERA-50SM+ | DC | 2000 | 19.4 | 17.2 | 3.5 | 32.5 | 1.3 | 1.2 | 4.4 | 60 |
| ERA-51SM+ | 0C | 4000 | 16.1 | 18.1 | 4.1 | 33.0 | 1.1 | 1.2 | 4.5 | 65 |
| ERA-6SM+ | DC | 4000 | 12.2 | 17.9 | 4.5 | 36.0 | 1.3 | 1.6 | 5.0 | 70 |
| ERA-8SM+ | DC | 2000 | 19.0 | 12.5 | 3.1 | 25.0 | 1.4 | 1.8 | 3.7 | 36 |
| GALI-1+ | DC | 8000 | 11.8 | 10.5 | 4.5 | 27.0 | 1.3 | 1.4 | 3.4 | 40 |
| GALI-19+ | DC | 7000 | 11.6 | 9.0 | 6.5 | 23.7 | 1.6 | 1.5 | 3.6 | 40 |
| GALI-2+ | DC | 8000 | 14.8 | 11.0 | 4.6 | 27.0 | 1.6 | 1.6 | 3.5 | 40 |
| GALI-21+ | DC | 8000 | 13.1 | 10.5 | 4.0 | 27.0 | 1.1 | 1.3 | 3.5 | 40 |
| GALI-24+ | DC | 6000 | 16.6 | 19.3 | 4.3 | 35 3 | 1.4 | 2.0 | 5.8 | 80 |
| GALI-29+ | DC | 7000 | 19.7 | 10.0 | 6.0 | 24.7 | 1.5 | 1.5 | 3.6 | 40 |
| GALI-3+ | DC | 3000 | 19.1 | 10.5 | 3.5 | 25.0 | 1.5 | 1.2 | 3.3 | 35 |
| GALI-33+ | DC | 4000 | 17.5 | 11.4 | 3.9 | 28.0 | 1.6 | 1.2 | 4.3 | 40 |
| GALI-39+ | DC | 7000 | 19.7 | 9.0 | 2.4 | 22.9 | 1.6 | 1.5 | 3.5 | 35 |
| GALI-4+ | DC | 4000 | 13.5 | 16.0 | 4.0 | 34.0 | 1.2 | 1.4 | 4.6 | 65 |
| GALI-4F+ | DC | 4000 | 13.4 | 13.8 | 4.0 | 32.0 | 1.2 | 1.5 | 4.4 | 50 |
| GALI-49+ | DC | 5000 | 13.6 | 15.0 | 5.5 | 33.3 | 1.7 | 1.5 | 5.0 | 65 |
| GALI-5+ | DC | 4000 | 17.5 | 16.0 | 3.5 | 35.0 | 1.2 | 1.4 | 4.4 | 65 |
| GALI-5F+ | DC | 4000 | 17.4 | 14.2 | 3.5 | 31.5 | 1.2 | 1.4 | 4.3 | 50 |
| GALI-51+ | DC | 4000 | 16.1 | 16.5 | 3.5 | 35.0 | 1.3 | 1.5 | 4.5 | 65 |
| GALI-51F+ | DC | 4000 | 15.9 | 14.4 | 3.5 | 32.0 | 1.2 | 1.5 | 4.4 | 50 |
| GALI-52+ | DC | 2000 | 17.8 | 15.5 | 2.7 | 32.0 | 1.35 | 1.4 | 4.4 | 50 |
| GALI-55+ | DC | 4000 | 18.5 | 15.5 | 3.3 | 28.5 | 1 25 | 1.3 | 4.3 | 50 |
| GALI-59+ | DC | 5000 | 18.3 | 17.6 | 4.3 | 33.3 | 1.6 | 1.5 | 4.8 | 65 |
| GALI-6+ | DC | 4000 | 11.3 | 18.2 | 4.5 | 35.5 | 1.5 | 1.8 | 5.0 | 70 |
| GALI-6F+ | DC | 4000 | 11.6 | 15.8 | 4.5 | 35.5 | 1.5 | 1.9 | 4.8 | 50 |
| GALI-74+ | DC | 1000 | 21.8 | 18.3 | 2.7 | 38.0 | 1.2 | 1.6 | 4.8 | 80 |
| GALI-84+ | DC | 6000 | 16.7 | 21.0 | 4.4 | 37.4 | 1.4 | 2.1 | 5.8 | 100 |
| GALI-S66+ | DC | 3000 | 17.3 | 2.8 | 2.7 | 18.0 | 1.25 | 1.7 | 3.5 | 16 |
| MAR-1+ | DC | 1000 | 16.5 | 2.5 | 3.5 | 14.0 | 1.3 | 1.2 | 5.0 | 17 |
| MAR-1SM+ | DC | 1000 | 16.5 | 2.5 | 3.3 | 14.0 | 1.3 | 1.2 | 5.0 | 17 |
| MAR-2SM+ | DC | 2000 | 12.0 | 7.0 | 3.7 | 22.0 | 1.3 | 1.3 | 5.0 | 25 |
| MAR-3+ | DC | 2000 | 12.0 | 10.0 | 6.0 | 23.0 | 1.5 | 1.7 | 5.0 | 35 |
| MAR-3SM+ | DC | 2000 | 12.0 | 10.0 | 3.7 | 28.0 | 1.3 | 1.3 | 5.0 | 35 |
| MAR-4+ | DC | 1000 | 8.0 | 12.5 | 7.0 | 25.5 | 1.5 | 1.9 | 5.25 | 50 |
| MAR-4SM+ | DC | 1000 | 8.0 | 12.5 | 6.0 | 25.5 | 1.6 | 2.0 | 5.3 | 50 |
| MAR-6+ | DC | 2000 | 20.0 | 3.0 | 3.0 | 14.5 | 1.7 | 1.7 | 3.5 | 16 |
| MAR-6SM+ | DC | 2000 | 20.0 | 3.0 | 3.0 | 14.5 | 1.3 | 1.3 | 3.5 | 16 |
| MAR-7SM+ | DC | 2000 | 12.5 | 3.5 | 5.0 | 19.0 | 1.3 | 1.3 | 4.0 | 22 |
| MAR-8A+ | DC | 1000 | 25.0 | 12.5 | 3.1 | 25.0 | 1.4 | 1.8 | 3.7 | 36 |
| MAR-8ASM+ | DC | 1000 | 25.0 | 12.5 | 3.1 | 25.0 | 1.4 | 1.8 | 3.7 | 36 |
| MAR-8SM+ | DC | 1000 | 22.5 | 12.5 | 3.3 | 27.0 | 3.1 | 3.1 | 7.8 | 36 |

**Table 11.5: Extract from Mini Circuits [7] web page showing MMIC characteristics**

on the circuit board without modification. The output is matched with a simple shunt open circuited stub (S1) on the output 50-ohm microstripline. 22pF capacitors were used for both the input (C1) and output (C2) blocking capacitors.

A 16-ohm chip resistor placed at R1 and decoupled by a 100pF capacitor at C3 provides a proper termination for the device power terminal. An additional bypass capacitor (100 to 1000pF) placed further down the power supply line at location C4 may be required to further decouple the supply terminal, especially if this stage is to be cascaded with an additional one. Proper decoupling of device VCC terminals of cascaded amplifier stages is required if stable operation is to be obtained. If desired, a 50-ohm resistor placed at R2 will provide low frequency loading of the device. This termination reduces low frequency gain and enhances low frequency stability. The MGA-87563 has three ground leads, all of which need to be well grounded for proper RF performance. This can be especially critical at 2.4GHz where common lead inductance can significantly decrease gain.

The performance of the LNA as measured on the HP 8970 Noise Figure Meter is shown in **Table 11.6**. At 2.4GHz, the loss of the FR-4/G-10 epoxy glass material can add several tenths of a dB to noise figure and lower gain by double the amount. The swept plots, **Figs 11.5 - 11.7** were taken on a scalar analyser and show the performance of the amplifier.

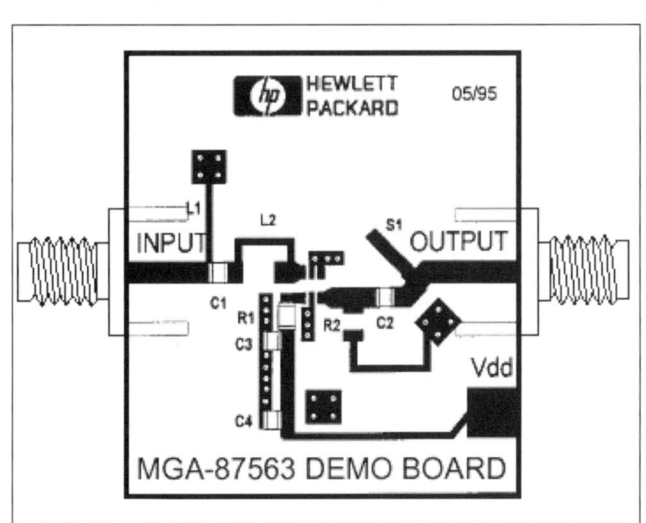


**Fig 11.4: Agilent** PCB for 2,400MHz LNA. For more details see [6]**

| Frequency (MHz) | Gain (dB) | Noise Figure (dB) |
|---|---|---|
| 1,700 | 10.4 | 2.60 |
| 1,800 | 13.8 | 2.57 |
| 1,900 | 11.3 | 2.45 |
| 2,000 | 11.9 | 2.38 |
| 2,100 | 13.3 | 2.06 |
| 2,200 | 12.8 | 2.02 |
| 2,300 | 12.9 | 2.12 |
| 2,400 | 11.5 | 2.05 |
| 2,500 | 11.5 | 2.14 |
| 2,600 | 10.5 | 2.25 |
| 2,700 | 10.9 | 2.29 |
| 2,800 | 10.3 | 2.33 |
| 2,900 | 9.8 | 2.35 |
| 3,000 | 9.6 | 3.42 |

**Table 11.6: 2,400MHz LNA noise figure and gain with Vd = 3V**

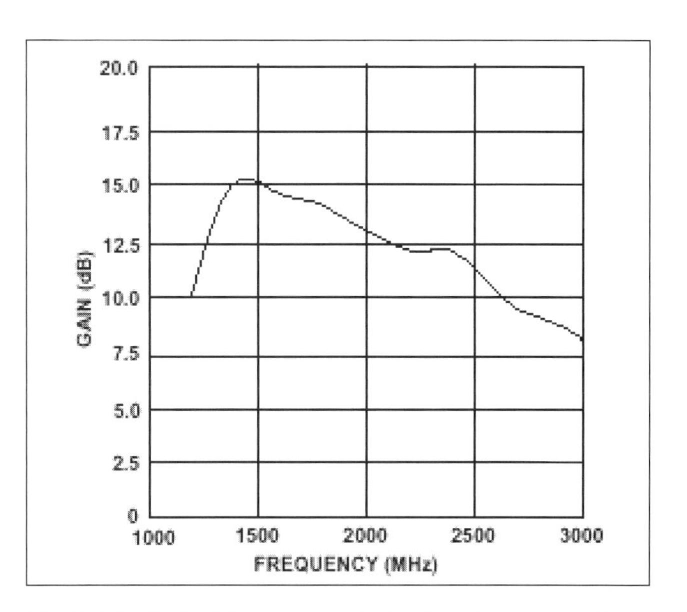


**Fig 11.5: 2,400MHz LNA associated gain at maximum noise figure, with Vd = 5V**

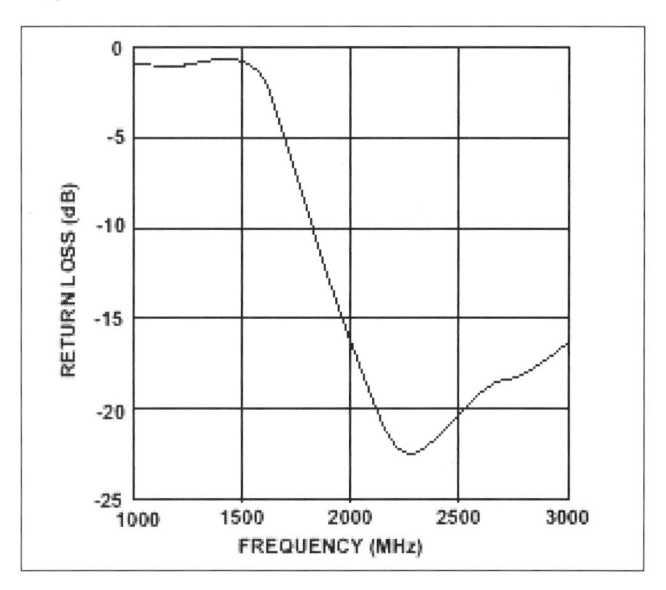


**Fig 11.6: 2,400MHz LNA output return loss, with Vd = 5V**

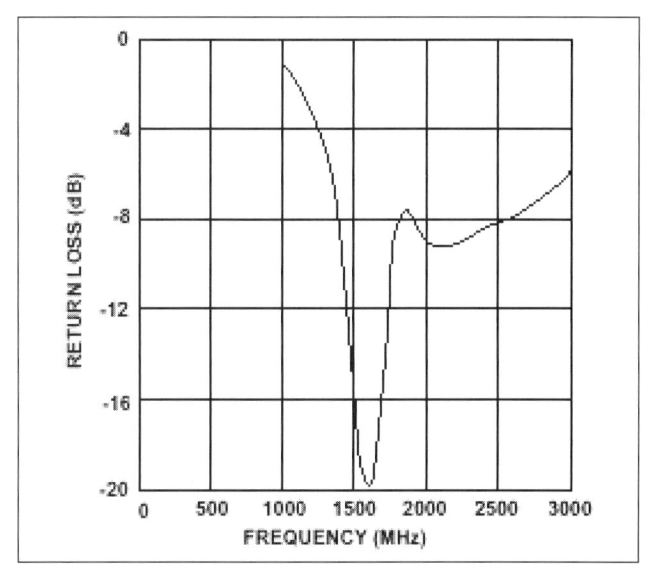


**Fig 11.7: 2,400MHz LNA input return loss, with Vd = 5V**

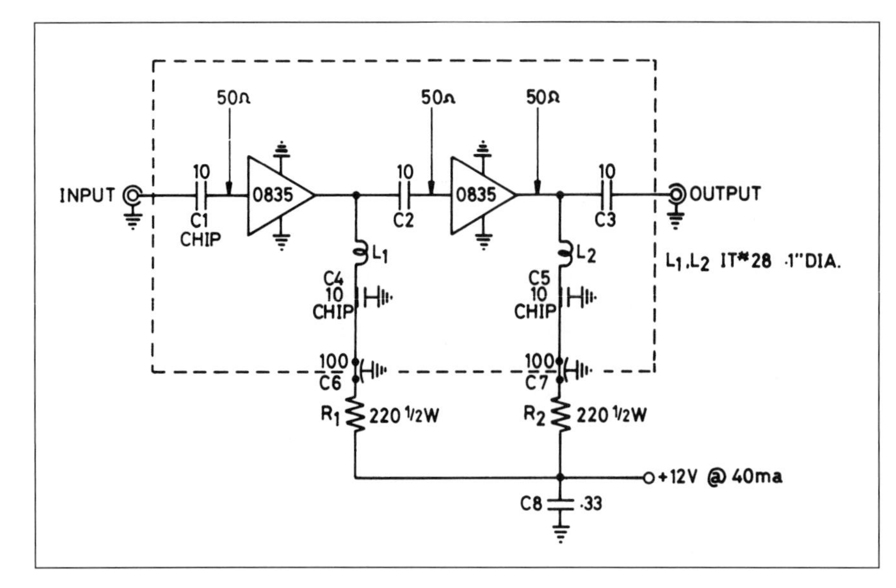


Fig 11.8: Circuit of the two stage MSA0835 wide band amplifier

Another simple design that is usable up to 5.7GHz, using MMIC amplifiers as gain blocks, is shown here. Two designs are presented, a two and a three stage amplifier using similar circuitry [8]. The first amplifier is shown in **Fig 11.8**. It uses two MSA0835 devices, in cascade, with a gain that varies from 27dB at 2GHz down to 7dB at 6GHz. Output power is only a few milliwatts at the upper end of the range but the amplifier is not intended to form the output stage of a transmitter. This function is better provided by a GaAsFET power amplifier above 3GHz. and by a bipolar amplifier below 3GHz.

Construction of the two stage amplifier is straightforward using a simple microstrip line track of 50-ohm impedance, on either glass fibre or PTFE double sided board, as shown in **Fig 11.9**. PTFE board material is preferred if operation above 3GHz is required.

Small 10pF ceramic chip capacitors prevent the bias supply to the MSA0835 devices being shorted out by the source and load. Another capacitor prevents the collector supply to the first stage shorting out the second stage base bias. Simple resistor current limiting from the amplifier 12V supply provides bias. Single-turn chokes in the collector leads of the MMICs prevent the bias resistors shunting the output signal to ground. The frequency response of the two stage amplifier is shown in **Fig 11.10**.

The three stage amplifier shown in **Fig 11.11** offers more gain than the two stage amplifier right across the frequency range. Construction is similar to the latter but uses a slightly longer board as shown in **Fig 11.12**. Its frequency response is shown in **Fig 11.13**.

More output can be achieved in this design if a Siemens CGY40 MMIC is used in the output stage in preference to the MSA0835. Gain will be slightly less and the maximum frequency of operation will fall to 3.4GHz. With this modification the output power at 3.4GHz can be as high as 50mW. This power level should be satisfactory for short links or longer line-of-sight paths.

These amplifiers have very high gain and are only conditionally stable when used between good 50-ohm load and source impedances. In practice, the insertion of a 3 or 6dB attenuator at the input and output of the amplifier should ensure stable operation, although in many cases the attenuators will not be required.

Replacing the MSA0835 MMIC with the lower gain 0735 device results in unconditional stability and much less gain variation across the frequency range. The penalty for this is lower gain and output power. About 32dB gain at 2GHz, falling to 22dB at 3.4GHz can be expected with this design.

It cannot be stressed enough how important the grounding is around the emitter leads of the MMIC devices on this board. In the absence of a plated through-hole board, through-board wire links must be provided where shown, and especially under the emitter lead connections to the top ground plane of the board. Soldering copper foil along the edges of the board is just not sufficient to ensure good ground integrity and the amplifier will self oscillate if links are not used.

## DETERMINING THE PERFORMANCE OF A TERRESTRIAL MICROWAVE RECEIVER SYSTEM

What constitutes acceptable performance for an amateur microwave terrestrial receiver system? This can be a very difficult question to answer, as it will depend on location and operating aspirations.

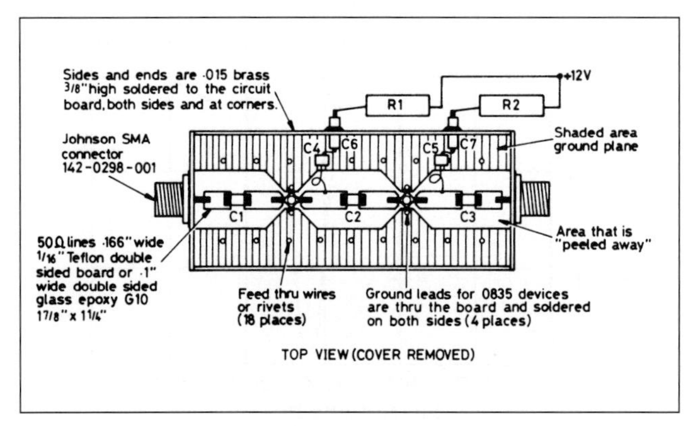


Fig 11.9: Layout of components in the two stage MSA0835 wide band amplifier

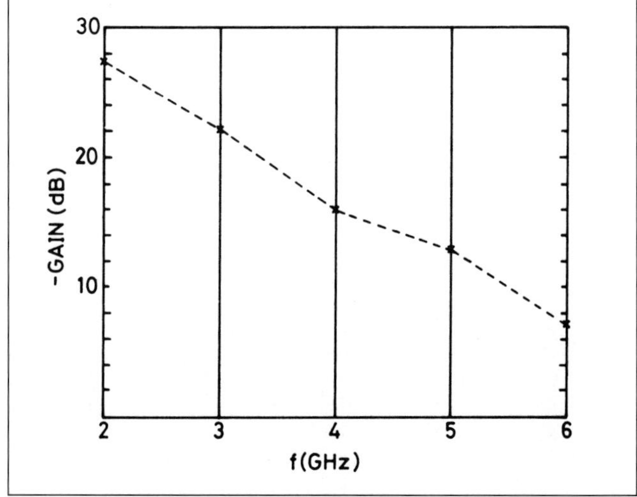


Fig 11.10: Frequency response of the two stage MSA0835 wide band amplifier

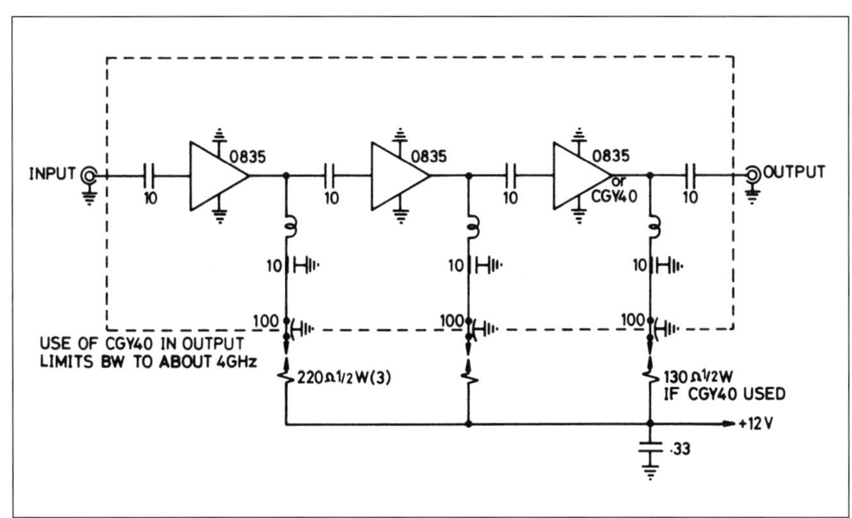


**Fig 11.11: Circuit of the three stage MSA0835 wide band amplifier**

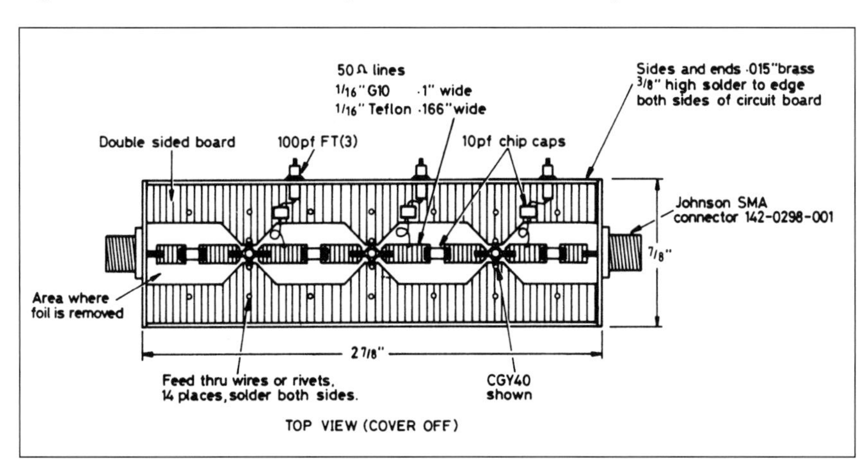


**Fig 11.12: Layout of the components in the three stage MSA0835 wide band amplifier**

The usual amateur radio approach is to use a high-gain, low-noise preamplifier, preferably at masthead, into the station transverter or transceiver. While this will undoubtedly provide a very sensitive receiver system, an ever-growing number of commercial digital radio-based systems means that our receivers are now being bombarded with many high level out-of-band signals as well as strong in-band signals from other amateurs and primary users. These signals may cause the appearance of intermodulation products as well as odd noises due to reciprocal mixing right on top of a weak station you are trying to work. No longer is it advisable to "pile on the preamplifiers" as we have done in the past, to achieve an arbitrarily low system noise figure. A good receive system design must now take account of the strong signal performance of the entire receiver as well as its noise figure. But how do you decide what is a good strong signal performance?

All amplifiers are non-linear. Low-noise preamplifiers are especially poor in this respect. The onset of non-linear operation is normally associated with the output signal level no longer increasing at the same rate as the input signal level increases. When the amount of increase is 1dB less than it should be, this is known as the 1dB compression point and is a common way of expressing the linearity of an amplifier, especially a power amplifier. While the compression point can be very useful, a better measure is something called the intercept point (IP). All amplifiers have an infinite number of intercept points, depending on the number of input signals and the order of the product generated from these. In general, only the second- (IP2) and third-order (IP3) are required to characterise strong signal performance. In this section, we will be looking mainly at IP3, as this is the parameter used in the *AppCAD* software programme. It is also important to realise that intercept points can refer to both input (IIP2, IIP3) and output (OIP2, OIP3).

Intercept points are theoretical values, which cannot be directly measured, but which can be extrapolated from measurements of the input or output signal level and the level of the products generated within the amplifier from these signals. There are many variations on this depending on the number of input signals, and the order of the harmonic mix. The one of interest here is the mixing of the second harmonic of one signal plus or minus the fundamental of a second signal. This is the classic third order product. See **Fig 11.14** for a graphical explanation of third-order product.

$f_1$ = frequency of signal 1

$f_2$ = frequency of signal 2

Due to the non-linear nature of the amplifier, a pair of unwanted signals at frequency $f_3$ and $f_4$ are generated at the amplifier output. Their frequencies are given by:

$$f_3 = 2 \times f_1 - f_2 \text{ and } f_4 = 2 \times f_2 - f_1$$

For example if

$f_1$ = 2320.200MHz, $f_2$ = 2320.250MHz,

then

$f_3$ = (2 x 2320.200) - 2320.250 = 2320.150MHz

and

$f_4$ = (2 x 2320.250) - 2320.2 = 2320.300MHz

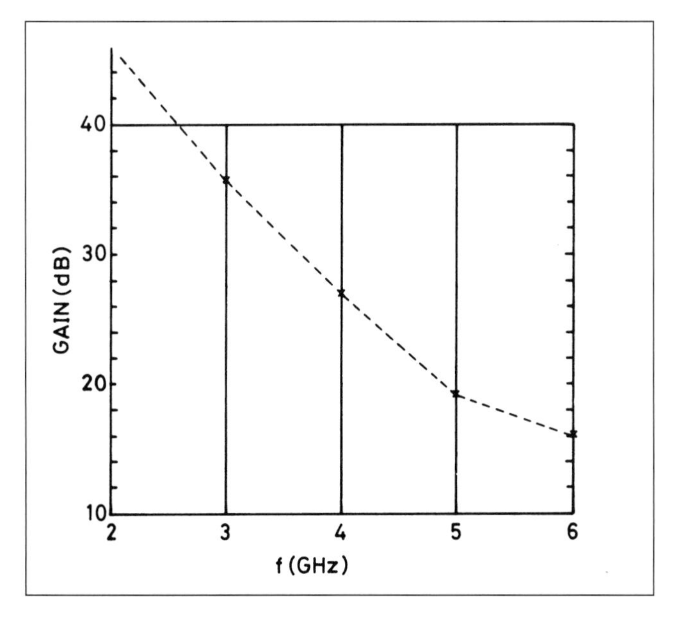


**Fig 11.13: The frequency response of the three stage MSA0835 wide band amplifier**

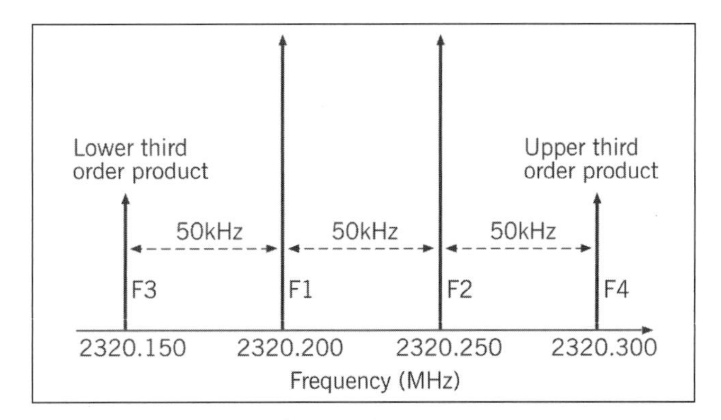


**Fig 11.14: Third order products generated from $f_1$ and $f_2$**

These two products are symmetrically spaced from $f_1$ and $f_2$ by the difference frequency (50kHz) between $f_1$ and $f_2$. The significance of these unwanted signals is that they are within the frequency band of interest, and cannot practically be removed by filtering. Because they are third-order products, their levels will increase at three times the rate of the level of the wanted signal. That is, for every 1dB increase in the level of the wanted signal $f_1$ and $f_2$, the unwanted third-order products $f_3$ and $f_4$ will increase by 3dB. Clearly, if the unwanted products increase at three times the rate at which the wanted signals increase there will be a point at which the levels of $f_1 = f_2 = f_3 = f_4$. In practice, the amplifier will saturate long before this point, so it cannot practically be measured. However, with the knowledge that the unwanted signal levels increase at three times the input rate, it is possible to extrapolate the point at which the unwanted signals, $f_3$ and $f_4$, equal $f_1$ and $f_2$. This theoretical point is called the intercept point and, when extrapolated at the output, is called the output third-order intercept (OIP3) and if extrapolated at the amplifier input is called the input third-order intercept (IIP3).

It is usually easier to measure the level of the wanted signals $f_1$ and $f_2$ and the unwanted products $f_3$ and $f_4$ at the output of an amplifier rather than the input. The output intercept point can be extrapolated from these values. The input intercept point is then the output intercept minus the amplifier gain.

The third order intercept can be calculated from:

$$IIP3 = P_{in} + 0.5\,(P_{in} - IM_{3in})$$

A graphical extrapolation of the intercept point usually requires that the output level of $f_1$ or $f_2$ is plotted on linear graph paper and then the unwanted product at $f_3$ or $f_4$ is plotted on the same graph. The slope of $f_3$ or $f_4$ should be three. Extrapolating the $f_3$ or $f_4$ lines to the point where either crosses the $f_1$ or $f_2$ lines gives the intercept point. This is shown in **Fig 11.15**.

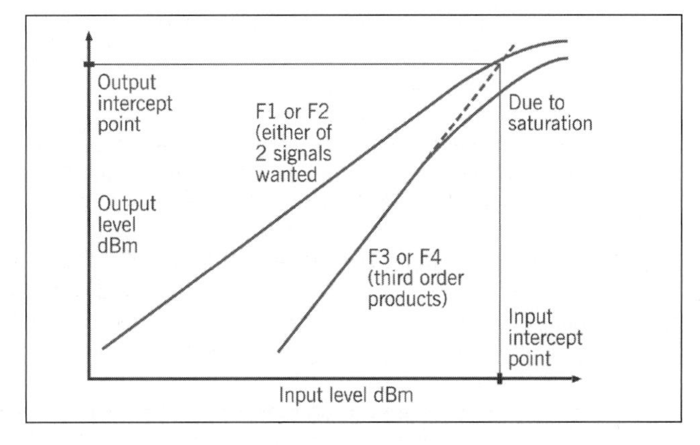


**Fig 11.15: Extrapolation of the third order intercept point**

Practically, it can be difficult to determine the actual line to extrapolate, and there may be a few dB of uncertainty in the resulting intercept point determination using this technique. This can be avoided by using a useful piece of software called *Intercept Point* within the AppCAD programme suite, which is available to download from Avago Technologies [6].

The great value of knowing the intercept point is that the level of unwanted products can be determined accurately from the level of the wanted signals and the intercept point. When the two equal-level input signals just produce unwanted products at the amplifier noise floor, the difference in the level between the noise floor and either of the two input tones (in dB) gives the spurious-free dynamic range (SFDR) of the amplifier. This is a good measure of the amplifier (and receiver system) strong-signal performance.

Before moving on to use this information to determine system performance, it is necessary to understand a bit about noise figure (NF) and noise temperature.

Noise figure is the most common way to express the sensitivity of amplifiers and receivers. It is a measure of the excess noise (always) added by a circuit or component, such that at the output of a circuit there is always a worse signal to noise ratio compared with the signal going into the circuit. Noise figure is normally measured and quoted with a source temperature of 290 Kelvin (290K) or 17 degrees C. In determining the performance of a microwave radio system it is often more convenient to work with system noise temperature rather than NF To convert from noise figure to noise temperature the formula is:

$$T = 290 \times (10^{NF/10} - 1)$$

where T= noise temperature in Kelvin, and NF = noise figure in dB. For example, 1dB noise figure = 70K

Conversely, to convert noise temperature to noise figure:

$$NF = 10 \log (1 + (T/290))$$

A terrestrial microwave band antenna would normally be orientated towards the horizon. In this position, the antenna would "see" a complex noise temperature consisting of approximately half cold sky at perhaps 10 - 30K and half warm earth at near 290K. See **Fig 11.16** for a graphical explanation. In practice, the resulting mean noise temperature will be somewhere between 150 and 200K due to a less-than-perfect horizon line, often consisting of trees and buildings protruding into the cold sky area, and unwanted antenna lobe noise pick-up.

The noise temperature of the complete system, consisting of antenna and receiver noise temperatures limits receiver system sensitivity. The lower $T_{sys}$, the more sensitive the receiver system.

$$T_{sys}\,(K) = T_{antenna} + T_{receiver}$$

Clearly, if the contribution from the receiver (including feeders etc) can be minimised, the overall sensitivity can be increased. Playing with *AppCAD* will show that attempting to reduce the receiver contribution too far may well result in a very sensitive system, but with poor dynamic range. Compromise is usually necessary.

**Fig 11.16: Possible antenna noise temperatures as seen by a microwave antenna**

## Using *AppCAD*

- Download the executable file AppCAD-302.exe from the website and install it on your PC.
- Open AppCAD. Select Signals-systems from the menu on the left-hand side of the window.
- Select NoiseCalc from the menu on the left hand side of the window.
- The open page displays an example worksheet for a CDMA handset receiver.
- Select and press Clear (next to Main Menu [F8]). The graphical boxes will clear of information. Don't worry; the graphics will re-appear next time you open AppCAD. You are now ready to begin entering data.

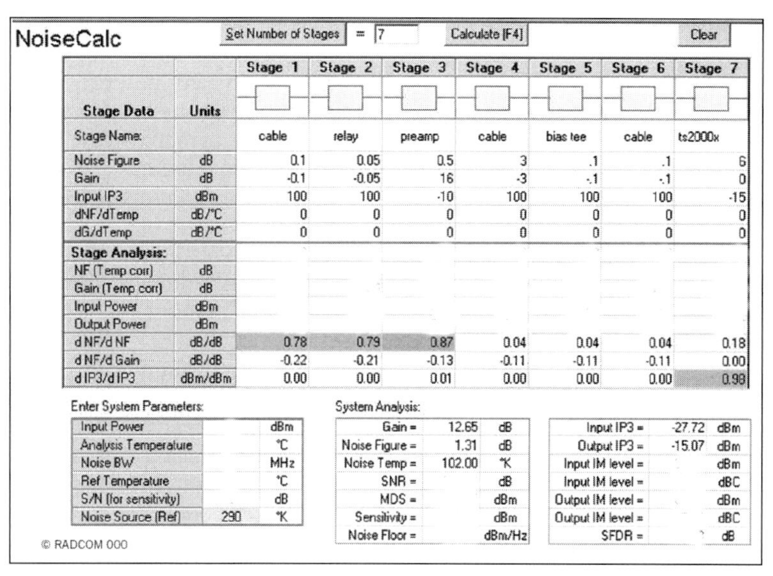
NoiseCalc — Set Number of Stages = 7 — Calculate [F4] — Clear

| Stage Data | Units | Stage 1 | Stage 2 | Stage 3 | Stage 4 | Stage 5 | Stage 6 | Stage 7 |
|---|---|---|---|---|---|---|---|---|
| Stage Name: | | cable | relay | preamp | cable | bias tee | cable | ts2000x |
| Noise Figure | dB | 0.1 | 0.05 | 0.5 | 3 | .1 | .1 | 6 |
| Gain | dB | -0.1 | -0.05 | 16 | -3 | -.1 | -.1 | 0 |
| Input IP3 | dBm | 100 | 100 | -10 | 100 | 100 | 100 | -15 |
| dNF/dTemp | dB/°C | 0 | 0 | 0 | 0 | 0 | 0 | 0 |
| dG/dTemp | dB/°C | 0 | 0 | 0 | 0 | 0 | 0 | 0 |
| **Stage Analysis:** | | | | | | | | |
| NF (Temp corr) | dB | | | | | | | |
| Gain (Temp corr) | dB | | | | | | | |
| Input Power | dBm | | | | | | | |
| Output Power | dBm | | | | | | | |
| d NF/d NF | dB/dB | 0.78 | 0.79 | 0.87 | 0.04 | 0.04 | 0.04 | 0.18 |
| d NF/d Gain | dB/dB | -0.22 | -0.21 | -0.13 | -0.11 | -0.11 | -0.11 | 0.00 |
| d IP3/d IP3 | dBm/dBm | 0.00 | 0.00 | 0.01 | 0.00 | 0.00 | 0.00 | 0.98 |

Enter System Parameters:

| Parameter | Value | Units |
|---|---|---|
| Input Power | | dBm |
| Analysis Temperature | | °C |
| Noise BW | | MHz |
| Ref Temperature | | °C |
| S/N (for sensitivity) | | dB |
| Noise Source (Ref) | 290 | °K |

System Analysis:

| | | |
|---|---|---|
| Gain = | 12.65 | dB |
| Noise Figure = | 1.31 | dB |
| Noise Temp = | 102.00 | °K |
| SNR = | | dB |
| MDS = | | dBm |
| Sensitivity = | | dBm |
| Noise Floor = | | dBm/Hz |

| | | |
|---|---|---|
| Input IP3 = | -27.72 | dBm |
| Output IP3 = | -15.07 | dBm |
| Input IM level = | | dBm |
| Input IM level = | | dBC |
| Output IM level = | | dBm |
| Output IM level = | | dBC |
| SFDR = | | dB |



**Fig 11.17: AppCAD worksheet for the 1.3GHz example receiver system. Unused numbers have been removed for clarity**

The following paragraphs will work through an example of a 1296MHz receive system using a masthead preamplifier and a Kenwood TS-2000X, but it could equally be any equipment and band of choice.

Under Options, at the top of the window, select Automatically Calculate after Edit and Intercept Point specified at Input.

It is necessary to set the number of stages you need. This should include one each for antenna cable, transmit/receive relay, preamplifier and receive feeder, bias-T (if used), second connection cable (if used) and radio. That is seven stages. It is better to have too many than too few. Type "7" to the Set Number of Stages box and press the similarly marked button. The worksheet will show seven stages. You cannot enter any graphics into these boxes now, so use the Stage Name worksheet cells to enter "ant cable", "relay" etc. If you are using, say, four antennas, enter the antenna cable to combiner cable once including the excess loss for the combiner. Don't forget the combiner to preamplifier cable loss, in this configuration. This would now be eight boxes.

Starting with the antenna to preamplifier cable, enter its loss in negative dB in the Gain cell. Enter its noise figure as the loss, eg if the loss of this cable is 0.1dB, enter -0.1 as the gain and +0.1 as the noise figure. Do the same for the relay and following stages. You only need enter the noise figure for the final (TS-2000X) stage. Leave the gain as "0". You will note that we have not entered the Input IP3 yet. These should be left as "100" for now.

The program has been calculating the system noise figure as you have been entering the data. Look at the System Analysis box at the bottom of the window. You will see that the cumulative gain is shown as 12.65dB up to the TS-2000X input, whilst the system NF is 1.31dB. Under NF, the Noise Temp cell shows the equivalent noise temperature of 102K.

Now go back and add the Input IP3 data as shown in **Fig 11.17** The passive components are entered as "100" as it is assumed they contribute little intermodulation. The preamplifier is entered as -10dBm, but obviously you would enter the data for your particular preamplifier. The average is about -10dBm for a small-signal HEMT device. The TS-2000X is entered as -15dBm from the ARRL measurements on this radio.

In the System Analysis boxes, look at the box on the lower right. The Input IP3 is shown as -27.72dBm. This is the Third Order Input Intercept (IIP3) for the complete receiver system. It is not particularly good. Below the Input IP3 is the Output IP3. This is the figure at the input to the TS-2000X and is shown as 15.07dBm. This is about the same as the TS-2000X on its own. Try increasing the preamplifier Input IP3 to -3dBm. The system Input IP3 doesn't improve very much so there is little to be gained by fitting this higher IIP3 preamplifier to the system as shown. A much higher preamplifier IIP3 is needed to have much impact. The TS-2000X is the limitation in this system.

You will notice that the Stage Analysis box has the stage 7 (TS-2000X) cell highlighted in red and showing 0.98. This is a sensitivity analysis where the closer the number is to "1" the weaker the stage is in terms of the parameter shown. In this case, d(IP3)/d(IP3) which is the sensitivity of the system IIP3 to the stage IIP3, is shown as 0.98 for the rig. The lower the better. The d(NF)/d(NF) line shows some numbers in blue. Here, the sensitivity of the system NF to the NF of the stage is shown. The loss of the relay and cable are shown to be sensitive but less so than the preamplifier (0.87).

You will learn a lot by playing with the numbers and hopefully gain a better understanding of how your system works and to what you should pay most attention and what to leave until later. The worksheet shows lots of other numbers to aid system analysis. As a guide, for the middle microwave bands including 1.3GHz and probably down to 430MHz, a good target for a terrestrial system noise figure is about 1dB, as the antenna noise temperature is about 170K (2dB NF) for a good terrestrial antenna in the clear. Try entering 19dB as the gain for the preamplifier in the worksheet, holding the NF at 0.5dB. Notice the system IIP3 reduces to around -30dBm. A 1dB system NF means that antenna noise will be the main contributor to system sensitivity. If you strive for much better than about 1dB, it may result in poor system dynamic range, but this is very dependent on other system parameters.

Ideally, the IIP3 for the system should be 0dBm or better. An acceptable figure is -10dBm, while -27dBm probably means some strong-signal problems will be experienced.

Receiver system design usually involves a trade-off between strong-signal performance and noise figure. AppCAD (and other system analysis programs such as TCALC) are essential tools in analysing and designing better systems. AppCAD does not deal with parameters such as Blocking Dynamic Range (BDR) or the effects of local oscillator phase noise on reciprocal mixing.

# GETTING STARTED ON 3CM

Brian Coleman, G4NNS, and Ian Lamb, G8KQW, describe how to become operational on the 10GHz band with minimum effort.

10GHz is the most popular of the higher amateur radio microwave bands and provides abundant opportunity for challenging operation. The current UK (terrestrial) distance record of 1347km was made between two home stations. This might be a surprise for those who don't know the band or who perhaps thought DX could only be worked between mountaintops. There are, of course, many opportunities for those who want to work from portable locations and the compact size and lightweight of the system described here makes this not only possible but also very enjoyable and rewarding. This section describes how to become operational on the 10GHz band with minimum effort, but with a system that provides a sound basis for future development. Even in its basic form hilltop contacts to 300 or 400km should be possible, and even from the average home station location such distances are possible during enhanced propagation conditions that include rain scatter and thermal inversions.

The system consists of a DB6NT transverter and a suitable relay with associated drive circuit, all housed in a weatherproof plastic enclosure. The transverter is small and light enough to be mounted, with the appropriate feed horn, on the support arm of a small, surplus, TVRO (Receive Only) offset dish. The feed horn is mounted in place of the original LNB (**Fig 11.18**). Thus the need for waveguide (plumbing) is minimised.

While 200mW output might sound like low power (QRP), a small TVRO dish will provide about 30dBi of gain and thus achieve an Effective isotropic Radiated Power (EiRP) of about 200W.

## The Transverter

Kuhne Electronic (DB6NT) offers the modern MKU 10 G2, high performance, 10GHz transverter, either as a ready-built module or as a kit. Whilst the kit is easy to build, this route should only be taken by experienced home constructors who are well versed in working with surface mount components. While the kit comes with adequate instructions, John Hazell, G8ACE, has written some additional notes, a link to which can be found in the hardware section of the UK Microwave Group web site [9]. At time of publishing this edition, a new 'G3' version is available and details can be found at [10].

Any 144MHz band transceiver that can provide between 200mW and 2W output, can be used to drive the transverter. A multimode transceiver is preferred as most communications on 10GHz takes place using either SSB or CW. A transceiver used for this purpose is known as an Intermediate Frequency (IF) rig. The actual transceiver output power is not too critical as an adjustable attenuator inside the transverter allows the user to set full 10GHz output for any input in this range. The small, battery powered, Yaesu FT290 or FT817 are commonly used as if transceivers by microwave operators. The transverter has separate connectors (SMA) for RF in, RF out and IF.

## The Changeover Relay

Whilst there is no requirement for an rf relay between the transceiver and the transverter, an external coaxial change over relay is required between the transverter and the antenna.

Coaxial relays that have reasonably low loss (<1dB) and sufficient port to port isolation at 10GHz may sound expensive. They are, in fact, quite common and appear frequently at mobile radio rallies and on eBay. Prices are normally in the range £10 to £25 depending on type and condition. Make sure that the relay you buy is useable at 10GHz before you part with your money. These relays invariably have SMA connectors.

An example is shown in **Fig 11.19** but they come in a variety of shapes, sizes and colours and most commonly have a 28V coil. These will operate satisfactorily down to about 20V, but the simple circuit shown in **Fig 11.20** can be used to operate them from a 12V supply. If, at a later date, you add a power amplifier you will need to think about sequencing the relay to avoid damage to the PA or transverter front end. You should check the relay for correct operation at DC before connecting it to the transverter. C1 may need some adjustment. Too small a value and the relay won't close, too large a value and it has an excessive hang time when switching from transmit to receive the feed horn.

Coaxial cable is lossy at 10GHz so its use is kept to an absolute minimum. Waveguide is much less lossy but can put some

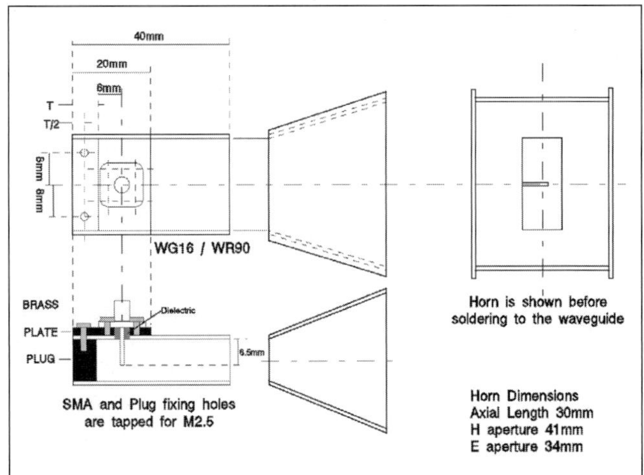


Fig 11.21: SMA to waveguide transition and feed horn

Fig 11.18: Quickstarter for 3cm mounted on the feed arm of a surplus TVRO offset dish

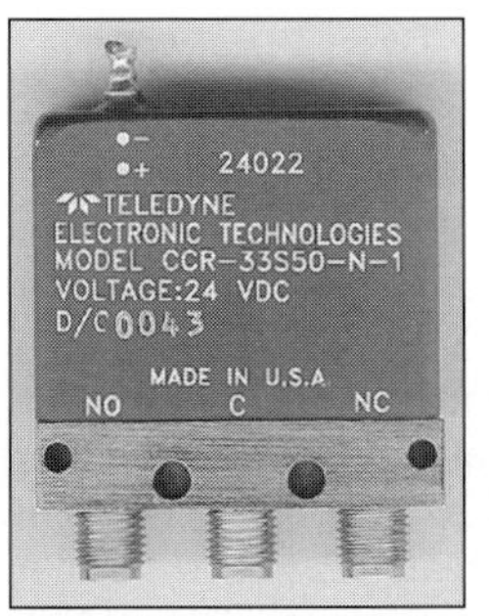


Fig 11.19: A typical SMA relay suitable for use at 10GHz

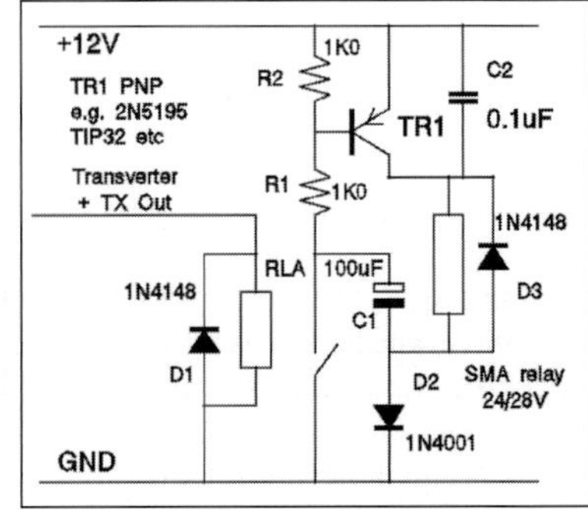


Fig 11.20: A circuit to drive a 24V or 28V relay from a 12V supply

constructors off. In the Quickstarter, the transverter is purposely located very close to the dish feed point to minimise the feeder length. The feed system consists of two components, combined into one assembly. The components are a coax to waveguide transition and a feed horn designed to provide efficient illumination of the dish. An SMA to waveguide transition is shown in **Fig 11.21**. These are easier to make than you might think, but for those unwilling to try, the complete transition can be supplied by, for example, the UK Microwave Group or a number of commercial suppliers. The critical dimensions are the centring of the SMA connector, the length of probe within the waveguide and the distance from the probe to the waveguide plug (short circuit) to the rear. If you have access to the necessary test equipment the probe can be made over sized and trimmed down for minimum return loss (best SWR). If you follow the dimensions shown you will have a satisfactory transition. Construction details and a template for making the 10GHz feed horn from copper laminate can be found on the G4NNS webpage [11].

## Metalwork

Many of us find metalwork a challenge so the metalwork for this project has been kept very simple. It should be possible to fashion the various parts from surplus aluminium sheet of 1.0 to 1.5mm thickness, using basic tools such as a hacksaw, drill and vice.

Essentially there are six basic components: A pair of right angle brackets to clamp the horn in place (**Fig 11.22**), a pair of right angle brackets to clamp the unit onto the feed support arm of the dish (**Fig 11.23**), and two plates to clamp the transverter in place (**Fig 11.24**). The pairs of clamps are folded in opposite directions to form mirror images of each other. The plastic enclosure is available from Maplin [12], type MB6, Code YN39. For portable operation the enclosure needs only a few drilled holes. For a more permanent installation, it should be painted with a reflective paint such as silver, and thoroughly sealed against rain ingress. Some people prefer to add a small drain / breather hole at the lowest point of the enclosure.

I have not shown the hole positions as they will vary according to the dish used and its feed support arm in particular. What is important is that the centre line of the feed support arm and the feed horn are aligned. The brackets allow adjustment of the whole assembly on the feed support arm and alignment of the feed horn to place it at the focus and pointing at the centre of the dish.

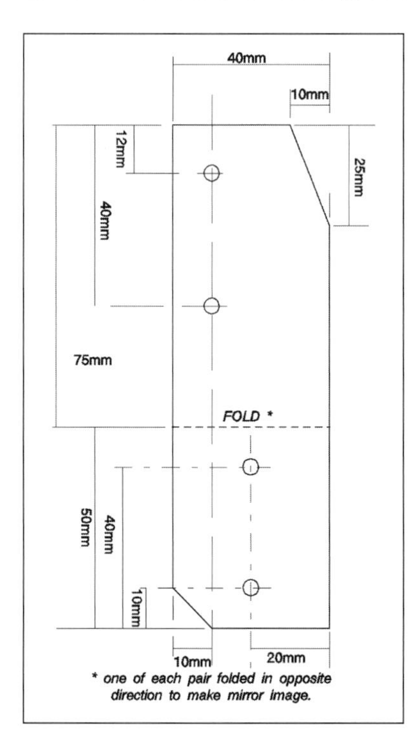


Fig 11.22: Feed horn bracket

## The Antenna

Surplus TVRO dishes often appear at mobile rallies and can often be found as scrap. It is important that the profile is in good condition and that the feed support arm is included. If it still has the Low Noise Block (LNB) converter, or its fixing clamp attached, so much the better as this will give a clear idea of where the focus is located. If this is unknown it can be calculated using the W1GHZ program *HDL_Ant* [13]. Use the "Offset dish calculation" option, enter the frequency (10368MHz), the large "diameter" of the dish, the small "diameter" of the dish, the depth at the deepest point and the distance of the deepest point from what is described as the "bottom" edge of the dish. As we will be using the dish mounted on its side, this point is perhaps best described as the edge nearest the feed arm and I will refer to it as the "feed point edge". The software will then provide the location of the focus in terms of distance from the feed point edge and the opposite edge of the dish. It also provides a calculated gain.

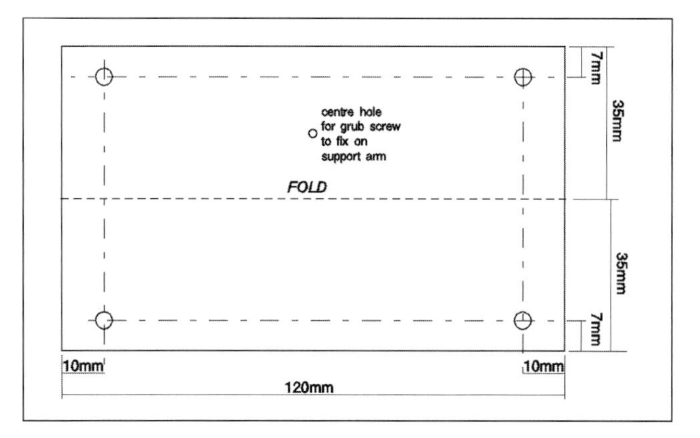


Fig 11.23: Support arm bracket

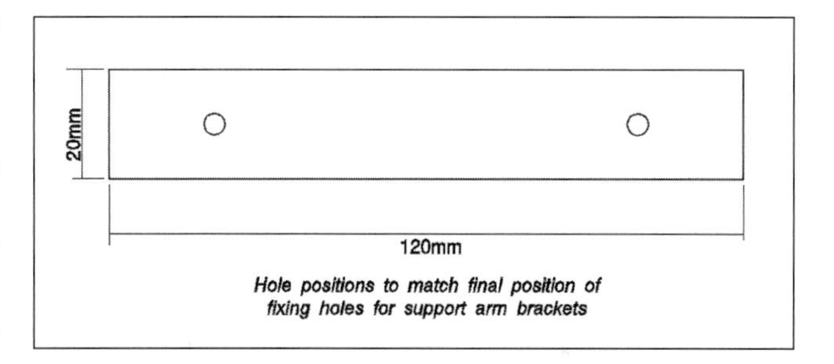


Fig 11.24: Clamping plates for the transverter

For terrestrial operation I prefer to use offset dishes mounted on their side as this makes mounting them on a vertical mast much simpler and avoids the critical adjustment of elevation that is otherwise needed. The exact method of mounting depends upon the construction of the dish. Usually there is a bracket at the rear of the dish that fixes the feed support arm and the mounting brackets. The screws fixing this to the dish can be replaced by studding (usually M6) and a plate or angle sections fixed to the rear see **Figs 11.25 and 11.26**. The plate or angle sections then have holes drilled for standard mast clamps. These holes should be positioned to give the best balance point for the dish, transverter and cables.

The direction in which the dish points in terms of azimuth, when mounted on its side, may appear a bit of a mystery but in fact is usually described by the line from the feed point edge of the dish to the centre of the horn, see **Fig 11.27**. This is because offset dishes are usually sections of a normal parabola that includes the centre at the feed point edge. Calibration can be confirmed by listening to a signal on a known bearing and unlike elevation this can be easily compensated using a rotator or by hand. With the offset dish mounted on its side like this it is also easy to add an additional feed horn for another band (such as 5.7GHz), but both feed horns cannot occupy the same place so there will be some offset in azimuth between the two (or more) bands.

Fig 11.25: Mounting arrangement for a surplus TVRO dish using angle sections

Fig 11.26: Mounting arrangement for a surplus TVRO dish using a plate

## Antenna Support

The arrangement described is ideal for fixing the dish to a mast either for home station or portable operation. For portable operation a tripod, if a sufficiently sturdy one is available, is useful. Many potential sites have hedges or small trees that cannot be cleared with a tripod so a short mast is a good choice. With a 144MHz talk back antenna at the top and the dish mounted below the guys even a small mast can provide good support for a dish at a greater height than with a tripod. Some means of locking the azimuth will be necessary.

## Coaxial Connections

As mentioned previously any coaxial cable will be lossy at these frequencies, as are most connectors. The preferred cable is RG402 (also available as UT141) semi rigid or the more flexible and easier to use equivalents such as Sucoflex and Quickform (available from Farnell [14] stock code 157 995). All these cables have an outside diameter (OD) of about 3.5mm and are used with direct solder SMA connectors such as Farnell 105-6352. These cables frequently appear at mobile rallies with the connectors already fitted. They are unlikely to be the right length but can often be cut so that only one additional connector is needed. Avoid right angle connectors as these can present a mismatch and can be very lossy. Although few of us have the correct tools to work with these cables, they can be worked quite satisfactorily with standard hand tools. Cut the cable to length using a junior hacksaw. Remove the outer conductor, where necessary, by heavily scoring it all the way around with a sharp knife. Flex the cable holding the piece of outer you wish to remove, using pliers so that it breaks away quite easily. Then cut the inner conductor and insulation in accordance with the instructions for the connector you are using. The total amount of coaxial cable you need for this project is 150mm or less.

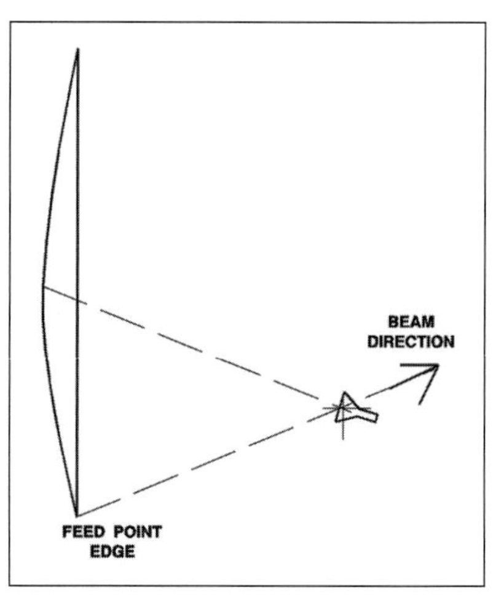


Fig 11.27: Dish beam pattern geometry

## IF Transceiver

The most popular choices of if transceiver are the FT290 Mk1 and the FT817 but other multi mode transceivers that can provide a suitably low level output can be used.

The FT290 Mk1 can be found second hand for around £100. It provides a DC output on the antenna line when in transmit mode and this is used to set the transverter to transmit.

It is also possible to do this by grounding the PTT line but if the DC on the IF line is used an additional connection is not needed. Without internal modification this is slightly harder to do for the FT817. However the circuit shown in **Fig 11.28** can be used. It has the slight disadvantage that if the FT817 is switched off while power is supplied to the transverter it will be switched to the transmit state. This is not likely to result in damage and as most people will turn off power to the transverter when the IF transceiver is off this is not likely to be a problem. The miniature DIN DATA plug is the same as on some old computer mice so you probably don't need to buy one! Whichever transceiver you choose, set up your transverter to be driven with the minimum power.

## Conclusion

The system described above will provide a quick and easy means of getting going on 10GHz and although not the cheapest route will provide good results and a sound basis for further development. With the growing popularity of microwave operation it is likely to retain its value. If you need help with the project contact a member of the UK Microwave Group Technical Support Team [9]. See you on 10GHz soon I hope.

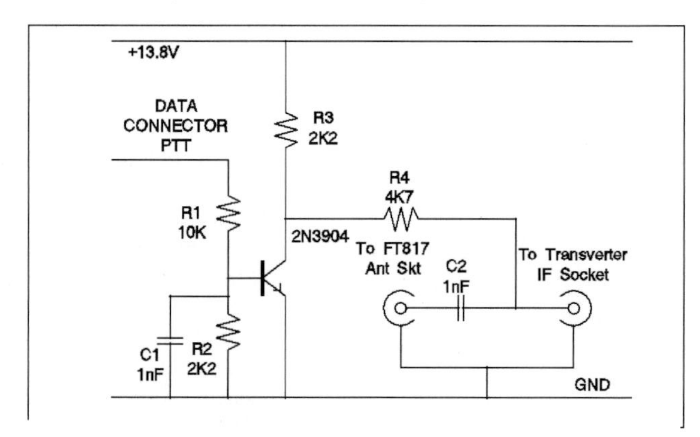


Fig 11.28: A Bias T for the FT817

Fig 11.29: A typical Solfan 10GHz transceiver

## 10GHZ WIDEBAND ACTIVITY REVIVAL

Sam Jewell, G4DDK noticed in 2006 that the UK Microwave Group [9] was receiving a lot of correspondence indicating renewed interest in wideband frequency modulation (WBFM) equipment and operation within the 10GHz amateur band allocation. During the 1960s and 1970s, WBFM was virtually the only way to become active on the band, with klystrons (723A/B and similar) providing an easy way to generate a few tens of milliwatts just inside the lower edge of the band.

Perhaps more popular, from the mid 1970s, were Gunn diode oscillators. These used surplus, low-power, low-voltage, Gunn effect semiconductor devices to generate 10mW or so at 10GHz. Early Gunn oscillators were often home-constructed using waveguide-based cavities and, as I recollect, were not particularly stable and were often highly temperamental.

The introduction of the commercial Gunnplexer enabled more stable transceivers to be assembled although, for many of us, the Gunnplexer was just too expensive and we sought an alternative. This turned out to be the commercial Doppler intruder alarm module, the best known was the Solfan unit. These were (and still are) also available with other markings, such as C&K, but are otherwise identical. The Solfan unit could easily be retuned from its normal operating frequency of around 10.6GHz down to typically 10.2 to 10.3GHz. The Solfan unit contained not only the Gunn oscillator but also a mixer diode as part of the Doppler receiver.

Solfan head Gunn transceivers use a single oscillator for the transmitter and the receiver. This means that a pair of these transceivers needs to operate with an agreed oscillator frequency offset such as 10.7, 30 or 100MHz, this being the intermediate frequency or IF. Transceivers with different IFs are still able to communicate, but one operator has to change frequency between transmit and receive in order to keep signals tuned-in. This is a rather hit-or-miss affair at the best of times and tends to lead to more complex solutions using separate transmitter and receiver Gunn oscillators. The almost unique feature of the common IF, single -source transceiver is that full-duplex (simultaneous transmit and receive) operation is possible. **Fig 11.29** shows a typical Solfan 10GHz transceiver, using an old Ambit 10.7MHz IF strip.

**Fig 11.30** shows a pair of full-duplex Gunn transceivers together with an example of some typical full-duplex frequencies. When used with a wideband FM receiver (such as a scanner or broadcast FM receiver) as the IF, a simple but effective wideband 10GHz transceiver can still be put together for a few tens of pounds. Solfan heads are still regularly seen on the surplus market and a few are still in amateur use in 10GHz transceivers.

The growth in popularity in the late 1980s of stable, narrowband, systems based on GaAsFET amplifiers, mixers and multipliers, used in 144MHz to 10GHz transverters, meant that system performance improved dramatically. It became possible to work from home station to home station at distances of over 200km under normal conditions and over 1000km under enhanced propagation conditions.

Faced with the reality of what could be achieved on the higher bands with more advanced equipment, wideband equipment operation has fallen away in recent years, except for the dedicated few who enjoyed the challenge of working long distances with modest equipment. Whilst this is not necessarily a bad thing from the aspect of equipment development and having sufficient performance to investigate new propagation phenomena, it does mean that many prospective newcomers to the microwave bands no longer go through the worthwhile learning process that simple and low-cost equipment can provide. I wonder how many potential newcomers have been put off by the relatively high cost of the modern 10GHz transverter and therefore never give microwaves a go?

Sams own interest in simple wideband FM was again sparked when he read the microwave group messages and later was given a pair of Solfan Doppler intruder alarm heads.

Andrew, MOBXT, had measured the sensitivity of the Solfan-based transceiver shown in Fig 11.29. It was found possible to detect a usable signal (tone modulated FM) down to about -87dBm. A typical WBFM scanner (Fair Mate HP100) was used as a WBFM IF, has a measured sensitivity of about -100dBm at both 10.7MHz and 30MHz. The conversion loss of a typical Solfan head, when optimally adjusted for mixer diode current, is about 13dB, giving a usable overall sensitivity of about -87dBm.

Putting this into perspective; a 10mW (+10dBm) output Gunn oscillator transmitter, Solfan based receiver, directly connected to dishes with +30dBi gain (typical of measured 60cm diameter dishes with penny feeds) at each end the path, will have a path loss capability (plc) given by:

plc(dB) = eirp (dBm) - ers (dBm)

where: Transmit effective isotropic radiated power (eirp) = +10dB +30dB = +40dB, and Effective receive sensitivity (ers) = -87dB - (+30dB) = -117dB, so Path loss capability (plc) = +40 - (-117) = 157dB

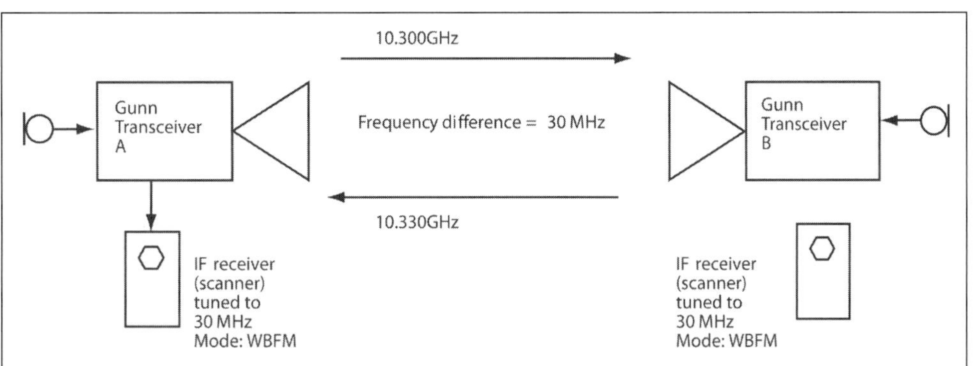


Fig 11.30: A simple Gunn device based duplex transceiver pair with 30MHz IF

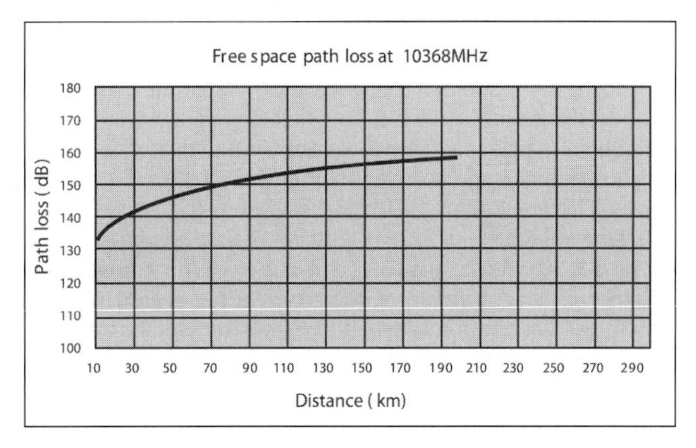


**Fig 11.31: A plot of free space path loss at 10.368GHz**

Free-space loss is given by:

Path loss (dB) = 32.45 + 201og(f) + 201og(d)

where f is in MHz and d is in km.

Transposing for distance, a free-space path loss of 157dB gives a working distance of approximately 150km as can be seen in **Fig 11.31** This is also the distance that must be worked on the 10GHz band in order to achieve the coveted RSGB 10GHz minimum distance award. This simple equipment must be working well to achieve this award.

10GHz WBFM system performance can be improved quite dramatically by adding GaAs FET preamplifiers or moving from the use of Gunn oscillators to more stable Dielectric Resonant Oscillator (DRO) sources as used in satellite TV Low-Noise Block (LNB) converters.

An LNB based receive converter can be stable enough to receive quasi-narrowband FM modulation (20 - 30kHz) instead of the typical 50 - 100kHz deviation used with Gunn-only systems. The reduction in bandwidth in the receiver will give a small but worthwhile receive system improvement while the low noise figure (<1dB) of a modern LNB front-end, even at 10GHz, would further improve the sensitivity by 12dB or so. Of course we are starting to get more complex than is possible with the simple Gunn transceiver, and it then becomes a question of where to put the work in order to get the best return for that effort.

## The Boomerang

A simple but effective way to test a Solfan or similar transceiver is called the boomerang and requires that, in addition to the transceiver to be tested, you need a signal generator or simple crystal oscillator at the wanted IF frequency and a suitable waveguide mixer. It is possible to use the mixer part of a second Solfan waveguide head.

The signal generator is set to the receiver IF frequency and connected to the boomerang waveguide mixer. A signal generator output of between 1 and 10mW is required at IF. If the Gunn transceiver has a built-in modulation source, the signal generator need not be modulated, otherwise set the modulation to FM and about 50kHz deviation at, say, 1kHz modulating frequency. With the Gunn transceiver aimed at the mixer, the modulated IF signal applied to the mixer will be heard in the transceiver receiver.

If the transceiver is self modulated with a tone, this should be switched on and will be heard when the mixer and transceiver antenna are lined up. A good transceiver will detect a good mixer at up to several hundred metres, although the exact distance will depend on several factors including the gain of the antenna on the transceiver and if any antenna is used on the mixer waveguide.

The boomerang works by the radiated RF from the transmitter mixing in the boomerang mixer diode with the locally generated IF signal and then being reradiated by the mixer diode. If the transceiver uses a 30MHz IF and the signal generator is set to 30MHz, then the reradiated 10,300MHz signal will contain 30MHz sidebands and these will appear at 10,270 and 10,330MHz as a received signal at the transceiver receive frequency. See **Fig 11.32** for a diagrammatic explanation.

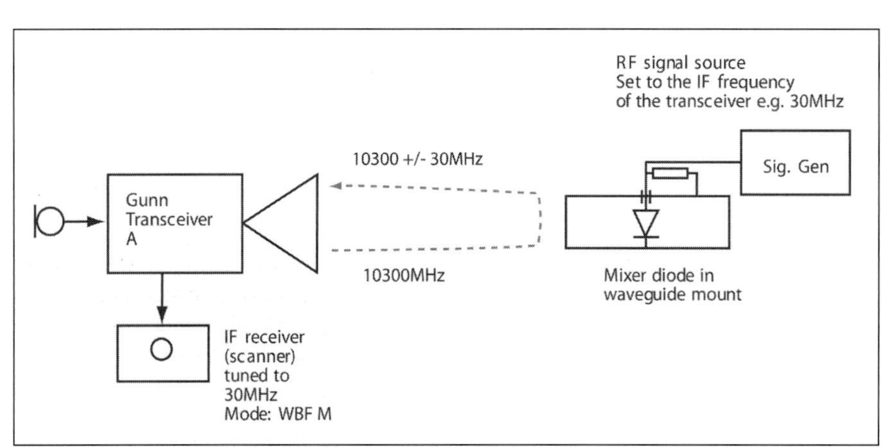


**Fig 11.32: A simple 'Boomerang' system for testing a full duplex transceiver**

Care must be taken to ensure that the IF receiver does not receive the signal generator by direct pick-up. This is most easily arranged by aiming either the mixer or transceiver antenna away from the other unit and checking that the received signal disappears.

A timely reminder of the temperamental nature of the Solfan (and other) Gunn transceivers was experienced when measuring the sensitivity of the 10GHz transceiver. As the Gunn oscillator voltage was varied, in order to fine-tune the Gunn frequency, numerous spurious signals were heard. Investigation with a spectrum analyser showed these to be due to unwanted moding or spurious oscillations, in the Gunn at certain critical tuning voltages. Simple Gunn transceivers are not always problem-free, but can offer a simple route to 10GHz operation.

## MICROWAVE LOCAL OSCILLATOR SOURCES

Early experimenters on the amateur microwave bands used wide band modes, so frequency accuracy was not that important. Many designs in the 1960s used free running oscillators at the operating frequency using Gunn diodes. Then improvements were made and the free running oscillators were locked to a known stable source. The quest for communications using narrow band modes, such as SSB, necessitated a different approach.

A stable crystal controlled oscillator, followed by a multiplier chain to produce the required local oscillator frequency became commonplace. The local oscillator is mixed with the output of a commercial transceiver to produce the required signal on the amateur band to be used (see **Fig 11.33**). The local oscillator and mixer are usually combined into a single unit - a transverter. This still remains the technique of choice used by serious microwave operators. Any

144MHz transceiver can be used but the IC202 is still often used because of its clean and stable output. The main design criteria for a good local oscillator source for microwave use are:

- Good short term frequency stability. Short term frequency variations may be caused by such things as the type of crystal used, the type of oscillator circuit, stability of supply voltages and temperature changes. Small changes in the frequency of the crystal oscillator are multiplied, eg if a local oscillator is used to generate an output on the 10GHz band, a 106.5MHz crystal frequency will be multiplied by 96 to give a local oscillator frequency of 10,244MHz. Thus a change of 25Hz in crystal frequency will change the frequency at 10GHz by about 2.5kHz.
- Good long term frequency stability. Long term frequency variations may be caused by ageing of the crystal used. Also crystals suffer from a hysteresis effect that causes them to operate on a slightly different frequency each time they are started.
- Good signal purity. This is governed by the design of the oscillator. It is important to keep phase noise and spurious outputs of the crystal oscillator to a minimum because the multiplier chain magnifies these.

Other techniques, such as Phase Locked Loops (PLL) and Direct Digital Synthesisers (DDS) are used to generate the local oscillator signal, but these can suffer from poor signal purity.

## 1.0 - 1.3GHz High Quality Microwave Source

Sam Jewell's design for the DDK001 L band source was first published in the *RadCom* 'Microwaves' column in 1987 and in volume two of the *RSGB Microwave Handbook*. Supplies of the PCBs dried up many years ago after a key component became unavailable, effectively making the design obsolete.

Ongoing demand for sources for use as local oscillators, test sources and small 1.3GHz transmitters encouraged Sam to update the original design. The new, compact, version 2 DDK001 local oscillator source (2001 LO) covers the frequency range from about 1150MHz to 1305MHz, determined by the available Toko helical filters used in the final multiplier stage. Output level is typically +7dBm at 1152MHz and the output spectrum is cleaner than the original 001 source. The following description assumes an output frequency of 1152MHz (LO frequency for 2m IF at 1296MHz, among other applications).

Although the new 2001 LO follows the same architecture as the older 001 design, there are several significant changes. The side coupled stripline filters have been replaced by discrete component lumped element band pass filters in the first two positions and by a Toko 2-pole helical filter in the output stage. This final filter determines the frequency range of the source. An external high stability source can be connected in place of the internal crystal oscillator if required.

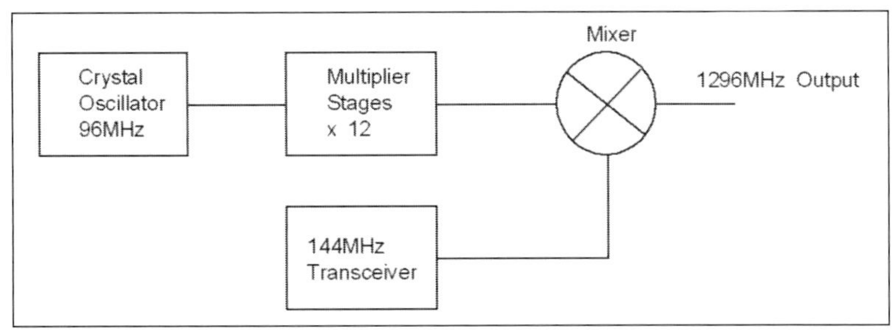


**Fig 11.33: Local oscillator mixed with transceiver to give microwave output**

**Fig 11.34: One of the prototype microwave signal sources. This version uses different trimmer capacitors to those specified. The trimmers shown are red and now replaced by black bodied 4 to 25pf trimmers**

The two-stage Butler bipolar transistor overtone crystal oscillator has been used again. This design has been endlessly analysed with respect to its phase noise performance and stability. Whilst there can be little doubt that the Driscoll and some other Butler variants can produce better phase noise performance, it is still a versatile and low noise oscillator design that is easy to align, forgiving of component variation and reliable in operation. The oscillator has proven itself over the years and even when multiplied up to 10GHz the phase noise performance is still good. Butler oscillators have been operating in the GB3MHL and GB3MHX (J002PB) beacons for over 20 years without failure. A finished 2001 LO source is shown in **Fig 11.34**.

### Circuit description

An overtone crystal oscillator drives a series of frequency multiplier stages with inter-stage filtering to define the output frequency. The overall multiplication is 12, implemented as x3 in the Butler oscillator stage and then two frequency doubler stages. The circuit is shown in **Fig 11.35**.

Common base amplifier, TR1, is the oscillator maintaining stage. Its collector tank circuit is tuned to the crystal overtone frequency (96MHz). The tuned circuit is heavily damped by R6. Trimmer C5 resonates the circuit. This stage is also used as the input buffer amplifier when an external source is to be connected.

TR2 is an emitter follower with its output feeding the overtone crystal. Its collector is tuned to 288MHz, the third harmonic of the crystal overtone frequency. Soft limiting is used to reduce phase noise degradation, but this results in low harmonic output levels. A compromise has had to be made here.

Since the crystal is connected between the emitters of TR1 and TR2, the feedback is in phase, which allows oscillation. Capacitors C9 and C10 allow the crystal to be pulled over a few hundred Hz at 96MHz. L2 may not be necessary but can be used if required in order to allow the crystal to be pulled onto frequency. In practice, C10 should be used to set the oscillator frequency and C5 used to peak the output, although there will be strong interaction between the settings of these capacitors.

The quality of the crystal oscillator is largely determined by the quality of the crystal used. Cheap crystals may not prove to be so economical in terms of performance.

L3, C13, C15, L4 and C14 form a 288MHz band pass filter. This is sharply tuned and only the recommended trimmer capacitors should be used to ensure resonance is achieved at 288MHz. Top capacitive coupling is provided by C15.

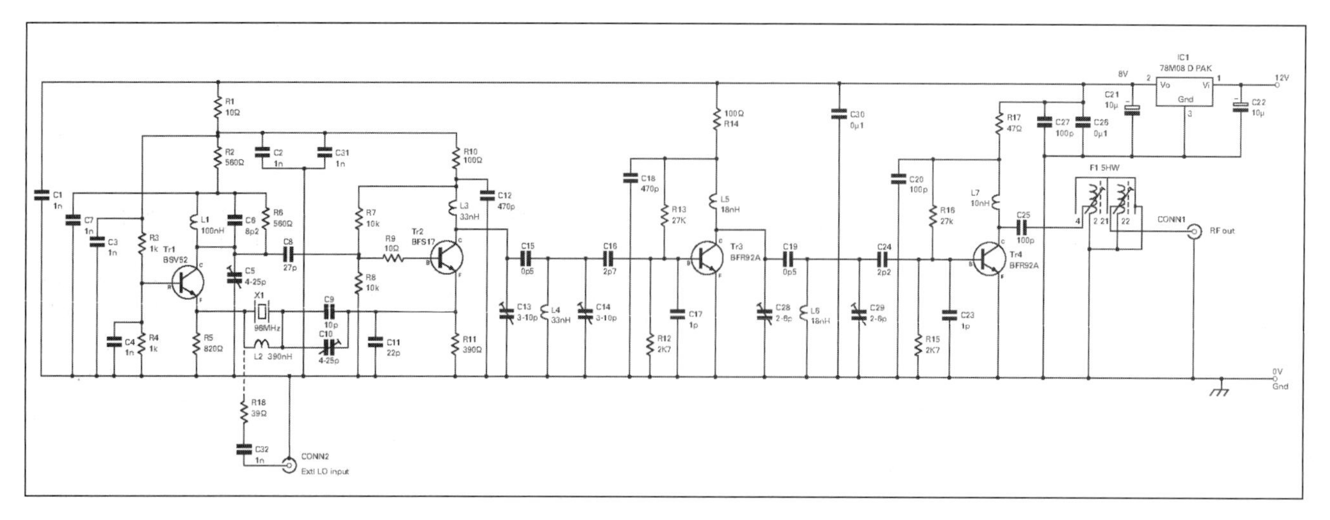


**Fig 11.35: Circuit diagram of the 20011.0 microwave source**

TR3 is the frequency doubler from 288MHz to 576MHz with L5, C28, C19, L8 and C29 forming the band pass filter at this frequency. Bias stabilisation is provided by returning R13 to the collector side of R14.

TR4 is the frequency doubler from 576 to 1152MHz and uses a similar bias arrangement to that of TR3. A Toko 5HW series helical filter selects the second harmonic of the 576MHz drive. The PCB connections are for the Toko F pin-out configuration filters (5HW 115045A-1195, 5HW 120050F-1225 and 5HW 125055F-1305). A 78M08 surface mount voltage regulator provides a stabilised 8V to drive the source.

No attempt has been made to temperature stabilise the crystal oscillator as it is felt that where this is required, an external, low noise, stabilised crystal oscillator (OCXO), direct frequency synthesiser (DFS) or PLL could be used. However, the small housing used for the 2001 LO source means that it is practical to enclose the entire unit within a temperature stabilised oven to control the whole of the unit and not just the crystal.

Where an external source is to be used, remove crystal X1 and add R18 and C32 as shown in Fig 11.35. Since the input impedance of the common base stage is low at about 7 ohms it is necessary to increase the impedance, to achieve a decent match, by adding series 39 ohm resistor, R18. The required drive is in the range -10 to +6dBm, with optimum drive at about -3dBm.

With the component values shown in Fig 11.35 the oscillator will operate satisfactorily between approximately 95 and 109MHz, giving an output of between 1150 and 1305MHz respectively. One of the three versions of the helical filter will be required to cover the whole frequency range.

## Construction

The source is constructed on a 37 x 74 x 1.6mm, FR4, double sided PCB, shown in **Fig 11.36**. All the components used are surface mount devices (SMD) with the exception of the crystal and the helical filter. 0805 size passive components are used, with SOT23 packaged transistors, D-Pak voltage regulator and B case size tantalum polarised decoupling capacitors. The trimmer capacitors are Murata TZB4 series.

The source will fit into a Schubert 37 x 74 x 30mm size tin plate box, or it can be fitted into convenient die-cast aluminium or other housing as required. The Schubert tin plate box is available from G3NYK [15]. If the tin plate box is used then it should be marked with a line around the inside 16.5mm down from the rim. This marks the position of the ground plane (copper side) of the PCB, with the component side of the board now 11.3mm below the rim of the box. The area above the PCB ground plane side is sufficient to clear the helical filter and crystal, which are both mounted on the ground plane side of the PCB.

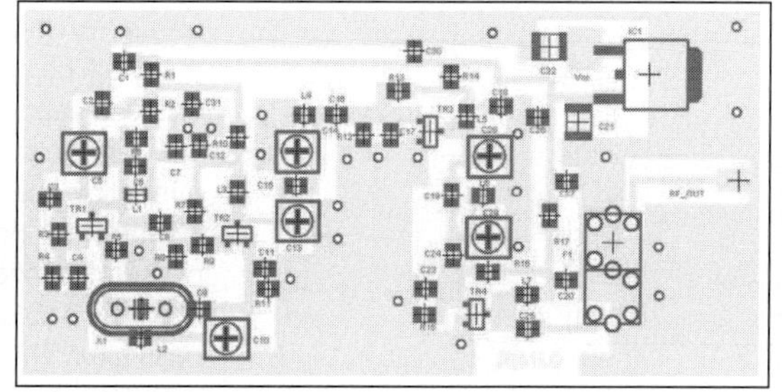

**Fig 11.36: Component overlay, slightly smaller than full size (71.6 x 34.6mm)**

Next, mark the position where the two or 4 hole gold plated SMA RF output connector will be soldered to the outside of the box. Drill a 4mm diameter hole through the box in a position where the connector spill will lie flat to the PCB RF output track. If an external source is to be used, an SMA connector can be fitted at the other end of the box, near the crystal location.

Drill a hole in the end of the box for the supply feedthrough capacitor. This should be above the voltage regulator and on the short edge of the box near the RF output SMA connector.

It is preferable to solder the PCB into the box before soldering the SMD parts into place. This avoids accidental damage to these small and sometimes fragile parts during the seam soldering process.

If you are making your own PCB for the source, it will be necessary to use thin wire through-board links to connect the ground plane to areas of the component side of the PCB. Normally these are plated through holes (PTH). Use 0.3 - 0.4mm diameter tinned wire through 0.5mm diameter holes. Use only 28SWG solder, both for the through board links and for soldering the SMD parts. Standard size 22SWG solder is unacceptable and will make a mess of the PCB. The 2001 LO PCB component overlay is shown in Fig 11.36.

## Alignment

Apply +12V and check that the current draw is no more than about 50mA. If it is significantly higher, check for faults. Check for +8 volts at the output of IC1.

The source can be aligned with little more than a multimeter. However, this is not recommended, as it can be difficult to ascertain if intermediate stages have been tuned to the correct frequency. A far better technique is to use a spectrum analyser with the probe described in [16], see **Fig 11.37**. Tune the analyser to 96MHz with

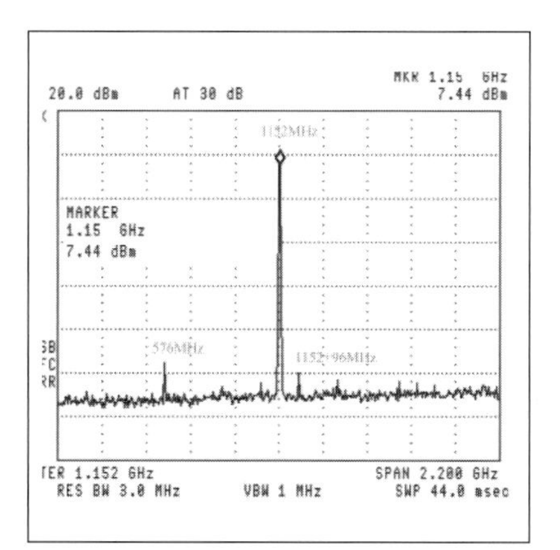


**Fig 11.37: Output spectrum of the 20011.0 microwave source from 50 to 2250MHz. The measured output is +9dBm at 1152MHz on narrow span. Half frequency (576MHz) is >45dB down on the 1152MHz output. All other non-harmonic outputs are over 50dB down on the 1152MHz output**

1MHz span and set to the amplitude reference to 0dBm. Place the probe on the base of TR2 and adjust C5 for a strong 96MHz output indication (approximately -15dBm with the -20dB probe). C5 should be set around mid capacitance. Re-tune the analyser to 288MHz and place the probe on the base of TR3. Tune C13 and C14 for maximum 288MHz output (approximately -15dBm). C13 and C14 should be set to near maximum capacitance.

Re-tune the analyser to 576MHz. Place the probe on the base connection of TR4 and adjust C28 and C29 for maximum 576MHz output (approximately -15dBm). C28 and C29 should be set around mid capacitance.

Finally, re-tune the analyser to 1152MHz and set the amplitude reference to +20dBm. Connect the analyser to the output connector and adjust the cores of Fl for maximum output. This should be in the range +7 to +11dBm.

With a suitable means to accurately measure the output frequency, adjust C10 for 96.000000MHz. It may be necessary to remove C9 with some crystals. L2 is not normally required, but provision is made to add this inductor for use with some crystals that will not otherwise adjust onto frequency.

There will be some frequency interaction with the adjustment of C13 and C14 as the TR2 frequency tripler will tend to pull the frequency. To a lesser extent, so will adjustment of C28 and C29.

### Work in progress

By the time this is in print, small changes may have been made to the circuit. It is advisable to check Sam Jewell's web page [17] for the latest information before commencing construction.

RSGB members can download the PCB foil pattern and comprehensive components list from the RSGB *RadCom Plus* website [18]

## HIGH PRECISION FREQUENCY 10MHZ STANDARD

A high stability frequency standard for 10MHz can be created using only three system components, a voltage controlled crystal oscillator (VCXO), a counter controlled by the signal from the GPS satellite system and a D/A converter for fine control of the VXCO. The short term and long term frequency stability that can be obtained by this simple means far exceed the requirements for practical amateur radio operations. The 10MHz standard can thus be used as the basis of a highly accurate method of generating local oscillator signals for microwave use.

The most commonly used method for precision time comparisons nowadays makes use of the satellites of the Global Positioning System (GPS). The GPS satellites carry atomic clocks of the highest accuracy, the operation of which is carefully monitored by the ground stations. A stable quartz oscillator regulated with the aid of the GPS ensures that its maximum frequency deviation always remains better than $1 \times 10^{-11}$. This is a precision of 0.0001Hz in 10MHz! Or, for the microwave amateur, 1Hz in 100GHz.

The frequency control of a 10MHz oscillator using GPS, shown in **Fig 11.38**, was designed by Wolfgang Schneider, DJ8ES, and Frank-Peter Richter, DL5HAT, and published in *VHF Communications* [19]. It uses an HP10544A VXCO (**Fig 11.39**), these are now available on the surplus market and often found for sale on eBay. In practice an accuracy of approximately $4 \times 10^{-10}$ can be achieved or, in other words, 4Hz in 10GHz. This value results from the inaccuracies of the counting process built into the system. In all frequency counters, the last bit should be taken with a pinch of salt. Depending on the phase position of the gate time to the counting signal, an error occurs of ±1 bit (phase error ±100ns). For a gate time of 1s, that would be 1Hz for the measuring frequency 10MHz ($\pm 1 \times 10^{-7}$).

The first practical measurements were based on a gate time of 8s, which corresponds to a resolution of 0.125Hz. Together with the phase jitter of the GPS signal, there should have been uniform distribution and thus a levelling off of the reading over a relatively long period of time (max. 64 measurements). This turned out to be wishful thinking. On investigation, it was established that the oscillator frequency varied very slowly around the required value of 10MHz. The absolute frequency here was 10.0MHz ±0.0305Hz.

If the gate time is increased to 128s, in theory the reading improves to at least ±0.0078125Hz. However, the influence of the GPS phase jitter is now reduced. This results in an effective usable precision for the 10MHz signal of approximately $4 \times 10^{-10}$ or 4Hz at 10GHz.

The control stage (**Fig 11.40**) operates like a frequency counter with an additional numerical comparator. The 10MHz output of the HP10544A oscillator is counted. The gate time of the counter is generated from the 1pps signal of the GPS receiver with a 74LS393. For control operation a gate time of 128 seconds is used, and 8s is used in the comparison mode for the OCXO.

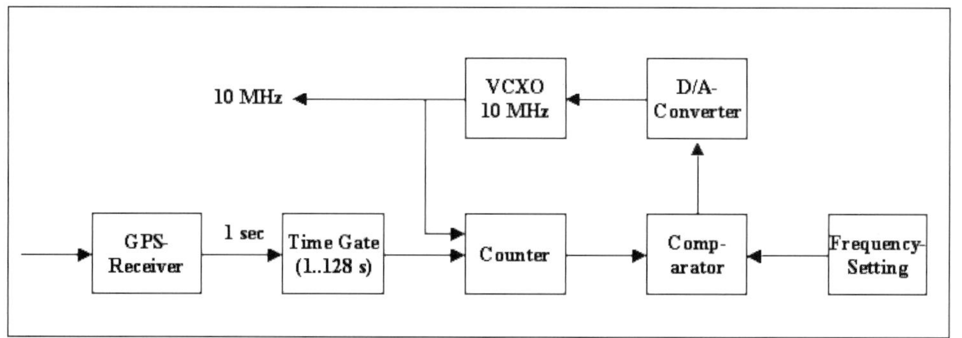


**Fig 11.38: Block diagram of frequency control via GPS**

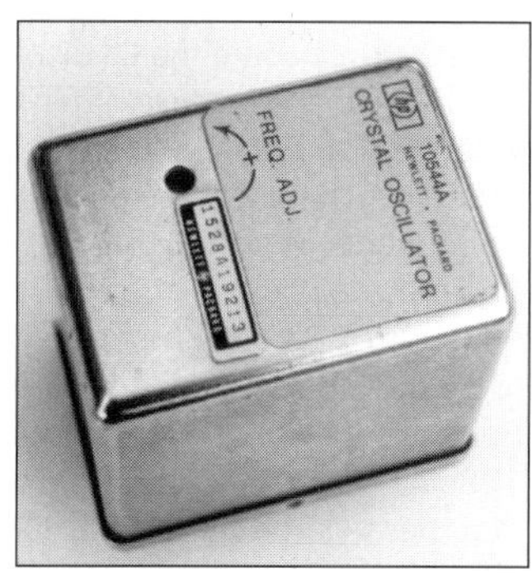


**Fig 11.39: HP10544A VXCO**

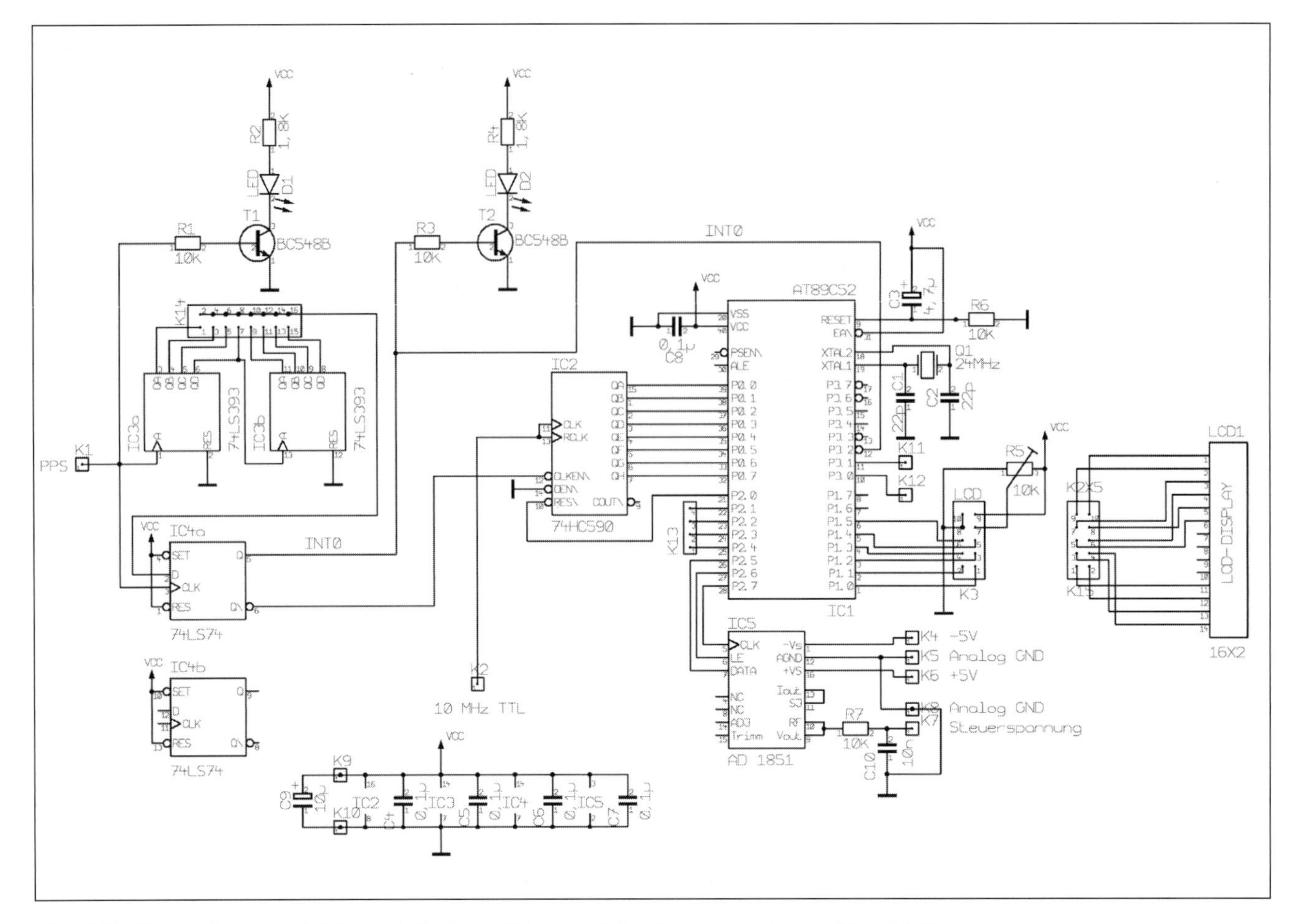

**Fig 11.40: Circuit diagram of GPS control stage of high precision frequency standard for 10MHz**

The 74HC590 8-bit counter can be used, with a gate time of 8s, to measure the input frequency 10MHz ±16Hz. The minimum resolution here is 0.125Hz. In control operation (gate time 128s), this is improved by a factor of 16, resulting in the system determined precision of 0.78 x $10^{-9}$, based on the 10MHz frequency oscillator.

The frequency of the HP oscillator can be finely adjusted using a tuning voltage of ±5V. This is done by the digital to analogue converter (AD 1851). It has a resolution of 16 bits for a control voltage range of ±3V. This gives a setting range for the OCXO of approximately ±0.5Hz.

The AT89C52 micro-controller controls all the functions including the D/A converter and the status in the LC display. The software in the micro-controller performs two tasks. Firstly, it enables a rough comparison operation to be carried out, and secondly it will continuously carry out the final fine adjustment using the GPS signal.

If pin 4 of K13 is open, then when the voltage is applied to the control circuit board the LC display shows "Warming Up". In order to eliminate any artificial jitter in the GPS signal, a mean value is formed and displayed from 64 readings from the 74HC590 counter. A change in the oscillator frequency using its frequency adjustment control will thus not display any effect for some time. So after a change of the control we must just wait for approximately 64 x 8s until the next adjustment takes place. If a value of ≥ ±0.250Hz is attained, we can switch over to the basic control by earthing Pin 4 of K13 and selecting the long gate time of 128s by means of a bridge between pins 15 and 16 at K14. The first message on the display is "Warming Up", the first value is displayed after approximately 15 minutes. It is not the deviation in Hertz but the value that is written in the AD 1851 D/A converter. This value can reach a maximum of ±32767, which means approximately ±3V.

The software assesses the output of the counter and calculates the value for the AD 1851 D/A converter. From the present value of the counter and the mean of the last 64 counter results, the figure is determined which is to be added to or subtracted from the current D/A converter value. This can be seen on the display.

Theoretically, with the gate time of 128s, and with a mean value formed over 64 readings, the precision of 4x$10^{-10}$ Hz (ie 4Hz in 10GHz) is achieved after 4.5 hours. However, it has been demonstrated in practice that this value has already been reached after approximately two hours.

The frequency controller circuit is assembled on 100mm x 100mm double sided PCB (**Fig 11.41** in Appendix B), the component layout can be see in **Fig 11.42** and the component list is shown in **Table 11.7**.

In DL5HAT's prototype (**Fig 11.43**) the HP10544A oscillator was used to drive a buffer stage designed by DJ8ES [20]. This gives TTL outputs at 1, 5 and 10MHz and 3 separate 10MHz sine wave outputs. The 1pps signal was generated from a GPS receiver manufactured by Garmin (GPS 25-LVS receiver board). The control assembly output supplies the control voltage for the HP oscillator. The tuning voltage, ±5V, must be separately generated in the frequency controller power supply. The following power supplies are required:

| | |
|---|---|
| +24V | HP oscillator |
| +5V | GPS receiver |
| +5V | control assembly |
| ±5V | control voltage |

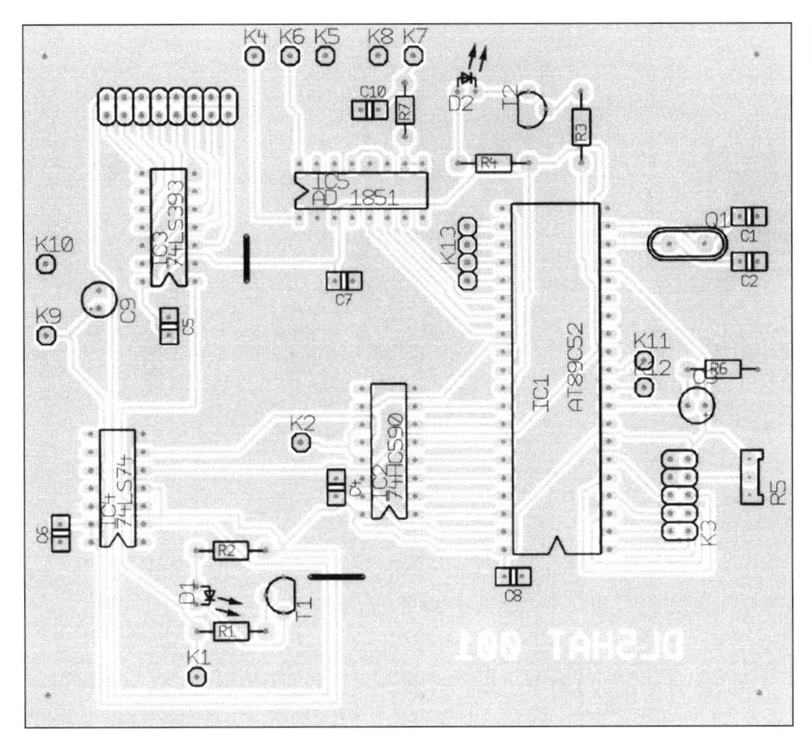


Fig 11.42. Component layout for GPS control stage of high precision frequency standard for 10MHz

Table 11.7: (right) Component list for GPS controller for 10MHz frequency standard

| Miscellaneous components | |
|---|---|
| 1 x micro-controller | AT89C52 |
| 1 x A/D converter | AD1851 |
| 1 x TTL-IC | 74LS74 |
| 1 x TTL-IC | 74LS393 |
| 1 x TTL-IC | 74HC590 |
| 2 x Transistor | BC848B |
| 2 x LED | green, low current |
| 1 x crystal | 24MHz |
| 1 x potentiometer | 10k |
| 1 x socket terminal strip, 10-pin | |
| 1 x plug strip, 10-pin | |
| 1 x stud strip, 14-pin | |
| 1 x jumper | |
| 1 x circuit board | DL5HAT 001 |
| **Resistors** | |
| 2 x | 1.8k |
| 4 x | 10k |
| **Ceramic capacitors** | |
| 5 x | 0.1µF |
| 2 x | 22pF |
| 1 x | 10nF |
| **Tantalum capacitors** | |
| 1 x | 4.7µF/25V |
| 1 x | 10µF/25V |

Long-term observations of the 10MHz frequency standard over approximately four weeks confirmed the design criteria.

## RECEIVE PREAMPLIFIERS

Receive preamplifiers are an important part of the microwave station. They are often used as masthead preamplifiers to overcome feeder loss that can be quite considerable for the higher frequency bands. There are many suitable semiconductors available to produce very low noise preamplifiers for all the amateur bands up to 10GHz. Above that, the devices are available but they are fairly expensive. Three designs are shown here to illustrate the technology available.

### 23cm Preamplifier

Developments in the early 2000s of cellular radio component technology have provided the VHF, UHF and microwave enthusiast with access to low-cost, low-noise, high-dynamic-range parts such as the Agilent ATF54143 PHEMT FET. (Pseudomorphic High Electron Mobility Transistor) As a result, if you are looking for an effective low-noise 23cm preamplifier there have been several designs published in the amateur press. While there have been many such designs optimised for the low-noise amplifier stage, there has been a dearth of designs for the equally-important receive converter second-stage amplifier. Using some of this new technology, a switchable, low-noise, second-stage amplifier has been developed for use with a masthead preamplifier.

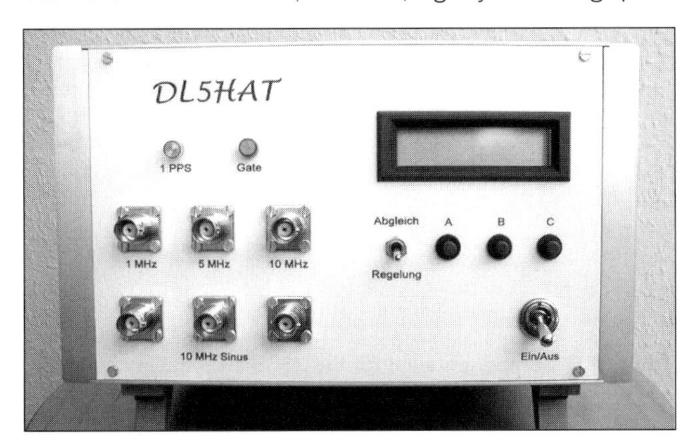


Fig 11.43: Picture of the prototype high precision frequency standard for 10MHz controlled by GPS

Sam Jewell, G4DDK had been developing a new 23cm transverter and one of his design objectives was to enable the receive side to maintain high dynamic range when using a masthead preamplifier, yet still have adequate sensitivity if used as a stand-alone transverter. The key to achieving this was discovered when the data sheet for the Agilent MGA71543 was studied. This GaAs Microwave Monolithic Integrated Circuit (MMIC) amplifier device is designed for use in Code Division Multiple Access (CDMA) and W-CDMA cellular receivers, where the level of received signal is critical to maintaining traffic capacity. This device incorporates a simple, internal, switching system where the amplifier can be bypassed whilst maintaining a good match at the input and output. This type of device is called a low noise amplifier mitigated bypass switch. When used with a good masthead preamplifier, the system receive noise figure can be maintained below 1dB, with relatively good dynamic range, but when a very strong local signal appears, the MGA71543 bypass switch can be operated to reduce system gain, so that operation can continue, although with a higher system noise figure, and with reduced blocking or intermodulation problems.

The transverter design required 30dB of pre-mixer gain without the masthead preamplifier. An MGA71543, on its own, can provide 16dB gain with a noise figure below 1dB at 23cm. A second amplifier stage is obviously required to achieve 30dB

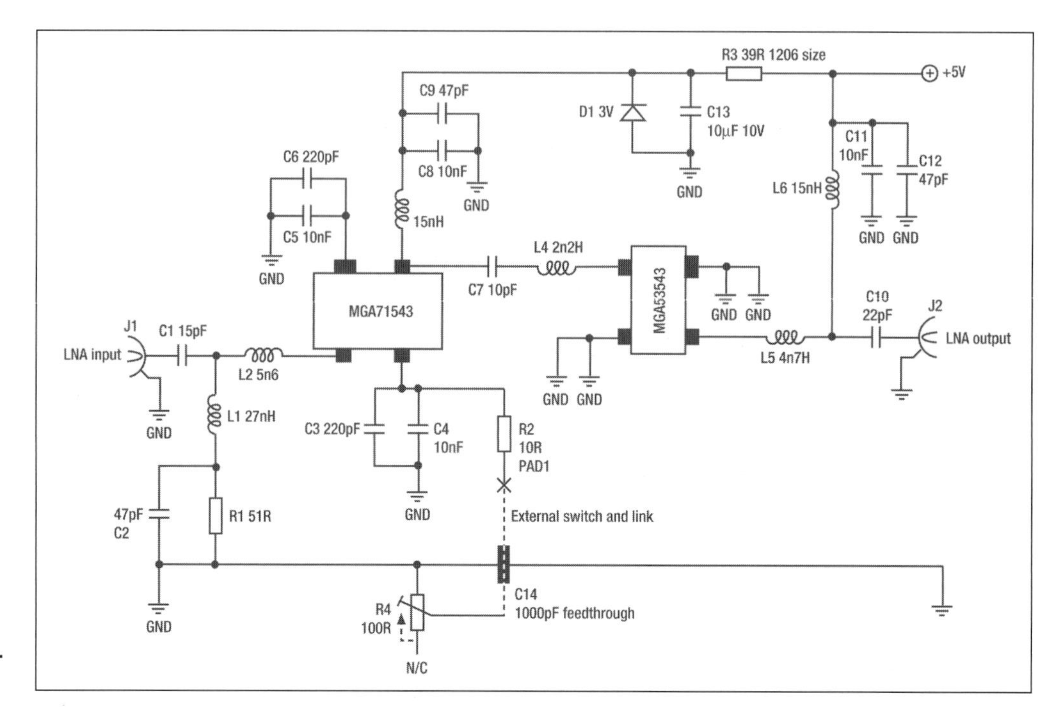


**11.44: Circuit diagram of the 23cm switchable amplifier**

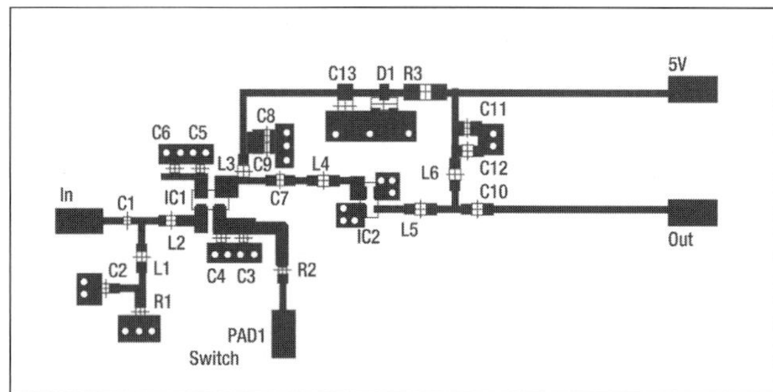


**Fig 11.45: PCB layout of the 23cm switchable amplifier (not exact size - see text)**

gain. A good second-stage candidate is the MGA53543. Using the MGA71543 as the first stage and the MGA53543 as the second stage, the two-stage amplifier measured 29.6dB gain at a noise figure of 1.1dB and an input intercept (IIP3) of -3dBm (calculated from two-tone measurements). This is a slightly lower IIP3 than expected, although it is adequate for current requirements. Switching out the MGA71543 reduces the overall amplifier gain to 8.7dB, with 8.2dB noise figure, but provides a measured 11P3 of +11.5dBm. Depending on the gain and noise figure of the masthead preamplifier and receive feeder loss, the overall system noise figure can still be maintained below 2dB with the MGA71543 switched out and below 1dB with it in circuit.

**Fig 11.44** shows the circuit schematic of the two-stage amplifier and the components list is in **Table 11.8**. The layout (**Fig 11.45**) has been derived from the application notes for each device, with matching for 1.3GHz rather than the published 900MHz or 1800MHz cellular allocations. The MGA71543 first stage device is noise-matched with a "T" circuit using a series capacitor and inductor and a shunt inductor. It is important that the indicated values are used or noise figure will suffer, as will gain.

The input device has two source connections. One of these is double decoupled with two capacitors. The second source connection is also decoupled with a similar arrangement, but here the source connection is connected through a 10 ohm resistor and then a feedthrough capacitor to an external 100 ohm potentiometer and toggle switch, used to set the bias and bypass the stage respectively. The value of the potentiometer should be adjusted for best noise figure or best intercept, depending on

| Component | Value | Type | Supplier |
|---|---|---|---|
| R1 | 51R | SMD 0603 | Rapid Electronics |
| R2 | 10R | SMD 0603 | Rapid Electronics |
| R3 | 39R | SMD 1206 | Rapid Electronics |
| R4 | 100R | 10 Turn trimmer | Rapid Electronics |
| C1 | 15pF | SMD 0603 NPO | Farnell InOne |
| C2 | 47pF | SMD 0603 NPO | Farnell InOne |
| C3 | 220pF | SMD 0805 NPO | Farnell InOne |
| C4 | 10nF | SMD 0805 NPO | Farnell InOne |
| C5 | 10nF | SMD 0805 NPO | Farnell InOne |
| C6 | 220pF | SMD 0805 NPO | Farnell InOne |
| C7 | 10pF | SMD 0603 NPO | Farnell InOne |
| C8 | 10nF | SMD 0805 NPO | Farnell InOne |
| C9 | 47pF | SMD 0805 NPO | Farnell InOne |
| C10 | 22pF | SMD 0603 NPO | Farnell InOne |
| C11 | 10nF | SMD 0805 NPO | Farnell InOne |
| C12 | 47pF | SMD 0805 NPO | Farnell InOne |
| C13 | 10ìF | SMD 10V wkg Tantalum | Farnell InOne |
| C14 | 1000pF | Screw- or solder-in feedthrough | Mainline Electronics |
| L1 | 27nH | SMD 0603 | Farnell InOne |
| L2 | 5n6H | SMD 0603 | Farnell InOne |
| L3 | 15nH | SMD 0603 | Farnell InOne |
| L4 | 2n2H | SMD 0603 | Farnell InOne |
| L5 | 4n7H | SMD 0603 | Farnell InOne |
| L6 | 15nH | SMD 0603 | Farnell InOne |
| D1 | 3V | SOT23 Zener | Rapid Electronics |
| IC1 | MGA71543 | | Agilent BFI Optilas |
| IC2 | MGA83543 | | Agilent BFI Optilas, Farnell InOne |
| Tinplate box | | E-Kafka 1000102 | Eisch-Kafka Electronics |
| J1 | | SMA 4 hole mounting socket | Farnell InOne |
| J2 | | SMA 4 hole mounting socket | Farnell InOne |

**Table 11.8: Components list for the 23cm switchable amplifier**

requirements. Opening the switch removes the bias and causes the MGA71543 to bypass itself internally. A further improvement in noise figure can be achieved with the use of better (lower loss) decoupling capacitors on the source connections of the

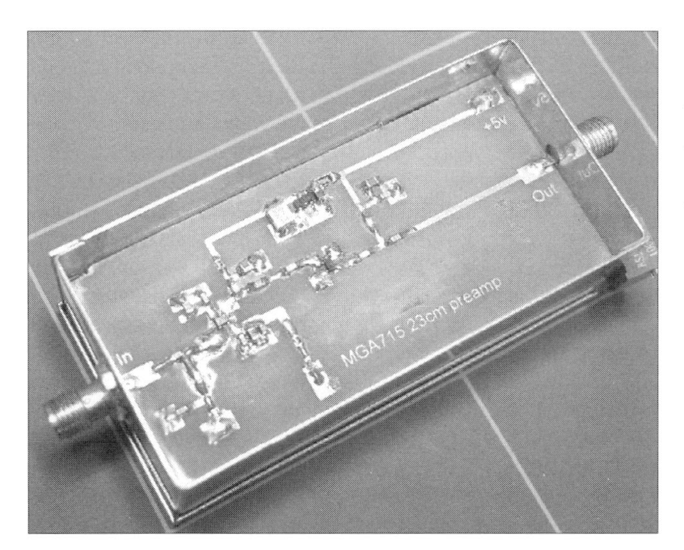


Fig 11.46: The prototype 23cm switchable amplifier in its tinplate box. This view shows the component side of the PCB

MGA71543.

The first stage 3V drain voltage is derived from the 5V supply by using a well-bypassed 3V Zener diode and dropping resistor. The resistor value is chosen to pass 10mA Zener current when the MGA71543 is set to draw 40mA. If the first stage is operated at much less than 30mA, to improve noise figure, the series resistor R3 must be increased accordingly to ensure that the Zener diode current does not exceed about 10mA.

An MGA53543 is capable of a +38dBm output intercept at 15.4dB gain, if the output match is optimised. In this amplifier, the output match is not fully optimised, resulting in a lower than expected IIP3.

The second stage is matched at its input using surface mount inductors and capacitors. The MGA53543 requires 5V drain-to-source bias and therefore operates directly from the +5V supply.

**Fig 11.46** shows the amplifier built into in a standard Schuberth tinplate box, with dimensions 37 x 74 x 30mm, obtainable from Eisch-Kafka Electronics in Germany [21]. The amplifier is built on FR4 double-sided 1.6mm thick fibreglass PCB. A 1:1 PCB foil for the amplifier can be obtained from G4DDK on request [17].

Home produced double-sided PCBs with good quality through board connections can be difficult to engineer. G4DDK's solution to this problem is to drill 0.5 - 0.6mm diameter holes, where through board ground connections are required, using a sharp high-speed drill bit. A single strand of thin tinned or silver plated wire is then threaded though the holes and soldered both sides using very thin (28SWG) solder to make a small size soldered joint over the wire. The small diameter hole seems to produce a capillary action, sucking the solder down into the hole and effectively making an excellent through board connection. This technique avoids the unsightly and often-ineffective wire `worms' formed by off cuts of old 0.25W resistors.

The two MGA devices are mounted as shown, taking care to ensure that the larger (wider) source lead is orientated as shown in the circuit schematic. SMA connectors are used at the input and output.

The 5V supply should be connected into the box through a second feedthrough capacitor.

G4DDK has become a devotee of surface mount inductors and capacitors in place of the more popular, but ungainly, microstrip matching arrangements. He uses 0603 size surface mount components where possible although, at 23cm, the larger 0805 size is quite usable. 0603 size inductors and capacitors are often cheaper than the 0805 size alternatives.

## 13cm Preamplifier

This design is by Rainer Bertelsmeier, DJ9BV, and originally appeared in *Dubus Magazine*. A preamp equipped with a PHEMT provides a top-notch performance in noise figure and gain as well as unconditional stability for the 13cm amateur band. The noise figure is 0.35dB at a gain of 15dB. It utilises the C band PHEMT, NEC NE42484A and provides a facility for an optional second stage on board. The second stage with the HP GaAs MMIC MGA86576 can boost the gain to about 40dB in one enclosure. The preamplifier is rather broadband and usable from 2300 to 2450MHz.

The construction of this LNA follows the proven design of the 23cm HEMT LNA [22] by using a wire loop with an open stub as an input circuit (**Fig 11.47**). The FET's grounded source requires a bias circuit to provide the negative voltage for the gate. A special active bias circuit (**Fig 11.48**) is integrated into the RF board that provides regulation of voltage and current for the FET. The component list is in **Table 11.9**.

Stub ST and inductance L1 provide a match for optimum source impedance for minimum noise figure. L1 is as a dielectric transmission line above a PTFE board and has somewhat lower loss than a microstripline. L3 and L4 provide inductive feedback to increase the stability factor and input return loss. R1, R2, L9 and R3 increase the stability factor.

The system of C2/L5/C6/L7 and L8 is specially designed to match the output of the single stage version to 50 ohms and to allow easy insertion of the GaAs MMIC for the two stage version. In the two stage version (**Fig 11.49**) it provides the appropriate input and output match to the MMIC. This solution was found by doing some hours of design work with the software design package *Microwave Harmonica*. It allows the two versions to have the same PCB. C4 provides a short on 2.3GHz, because it is in series resonance at this frequency. On all frequencies outside the operating band the gate structure is terminated by R1. Dr1 is a printed λ/4 choke to decouple the gate bias supply.

The two stage version utilises a HP GaAs MMIC, MGA86576 in the second stage. It provides about 2dB noise figure and 24dB gain. Input is matched by a wire loop for optimum

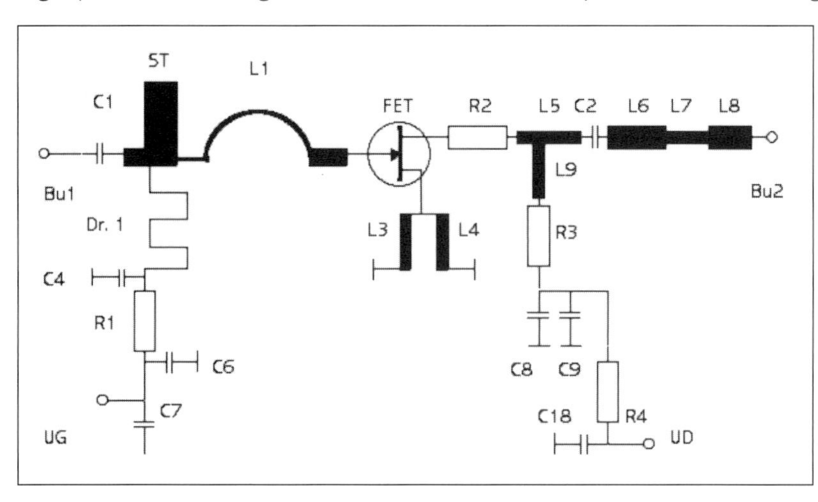


Fig 11.47: Circuit diagram of single stage 13cm PHEMT preamplifier

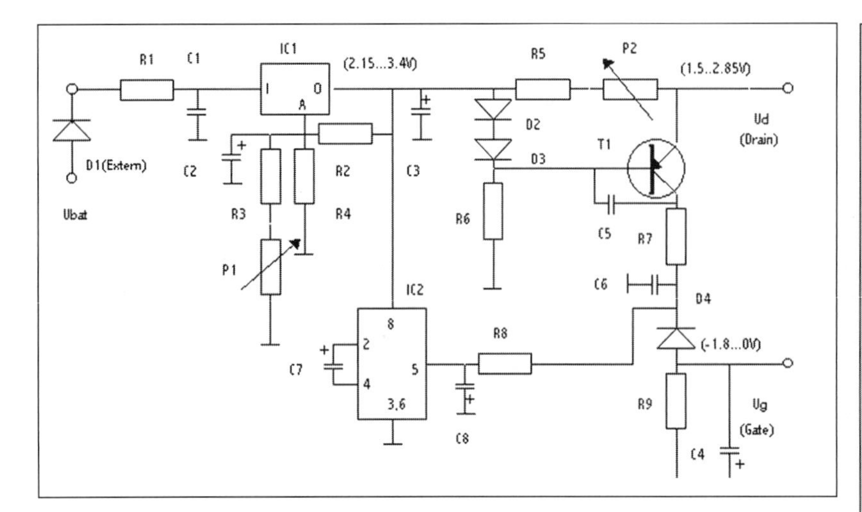


Fig 11.48: Circuit diagram of bias circuit for 13cm PHEMT preamplifier

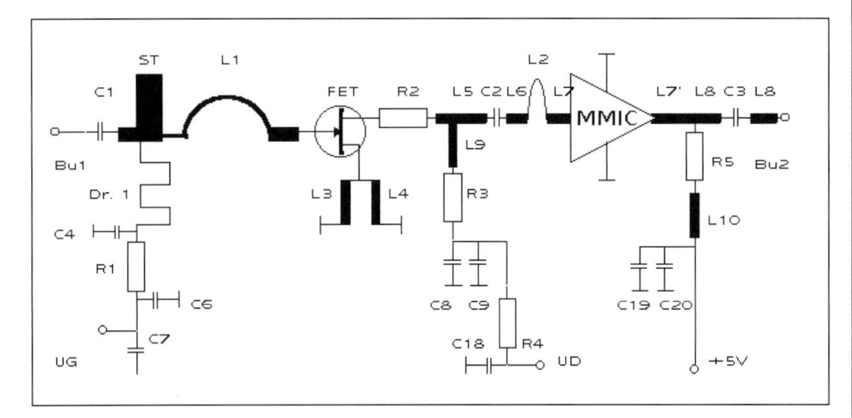


Fig 11.49: Circuit of the two stage 13cm PHEMT preamplifier

noise figure. Output is terminated by a resistor R5 and a short transmission line L10. Together with L7/L8 and C3 a good output return loss is measured. The source pads have to provide a very low inductance path to the ground plane, to preserve the MMIC's inherent unconditional stability.

To achieve unconditional stability four ground connections are needed on each source. Appropriate source pads are provided on the PCB. Simulation indicates a minimum K factor of 1.2 in this arrangement on a 0.79mm thick substrate. A thicker substrate is prohibitive. The MMIC typically adds 0.07dB to the noise figure of the first stage. This is somewhat difficult to measure, because most converters will exhibit gain compression, when the noise power of the source, amplified by more than 40dB, will enter the converter.

The construction uses microstripline techniques on glass PTFE substrate Taconix TLX with 0.79mm thickness with the PCB having dimensions of 34 x 72mm. An active bias circuit, which provides constant voltage and current, is integrated into the PCB (**Fig 11.50** in Appendix B). **Fig 11.51** shows a top view of FETs and **Fig 11.52** shows the component layout.

The construction process is:

- Prepare tinned box (solder side walls).
- Prepare PCB to fit into box.
- Prepare holes for N connectors. Note, Input and output connector are asymmetrical. Use PCB to do the markings.
- Drill holes for through connections (0.9mm diameter) and use 0.8mm gold plated copper wire at the positions indicated.
- Solder all resistors onto PCB.
- Solder all capacitors onto PCB.

| **Capacitors** | |
|---|---|
| C1 | 4.7pF Chip-C 50mil (500 DHA 4R7 JG) |
| C2, 3, 6 | 100pF SMD-C, size 0805 |
| C4 | 5.6pF SMD-C, size 0805 |
| C7, 8, 19 | 1000pF SMD, size 0805 |
| C9, 18, 20 | 10nF SMD-C, size 0805 |
| C10, 12, 17 | 0.1µF SMD-C, size 1206 |
| C11, 14, 15 | 10µF SMD-Electro, size 1210 |
| C13 | 1µF SMD-Electro, size 1206 |
| C16 | 1000pF Feedthrough |
| **Resistors** | |
| R1, 3 | 470 SMD-R size 1206 |
| R2, 14 | 390 SMD-R size 1206 |
| R4, 5 | 100 SMD-R size 1206 |
| R12 | 6.8kO SMD-R size 1206 |
| P1 | 1000 SMD-Pot, Murata 4310 |
| **Miscellaneous** | |
| Dr.1 | Printed λ/4 |
| L1 | Wire loop, 0.5mm gold plated copper wire 18mm long, 1mm above board |
| L2 | Wire loop, 0.5mm gold plated copper wire 8mm long, on PCB |
| D1 | IN4007 |
| FET | NE42484A, NEC |
| MMIC | MGA-86576, HP |
| T1 | PNP eg BC807, BC856, BC857, BC858, BC859, SOT-23 |
| IC1 | LTC1044SN8 |
| Bu1, 2 | N small flange or SMA |
| PCB | Taconix TLX, 35 x 72mm, 0.79mm er = 2.55 |
| Box | Tinplate 35 x 74 x 30mm |

Table 11.9: Components for the 13cm PHEMT preamp

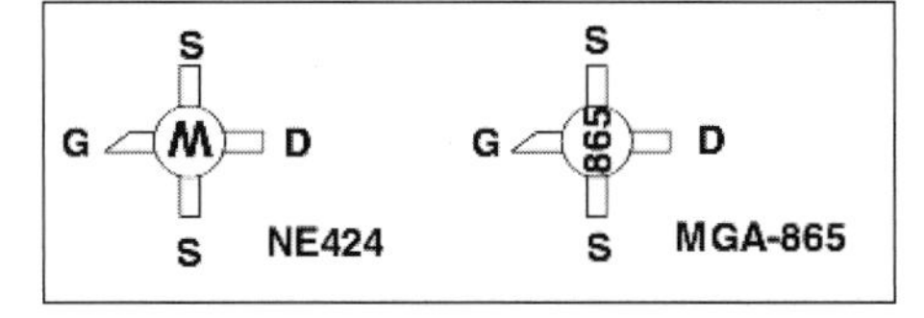


Fig 11.51: Top view of FET and MMIC used in the 13cm preamplifier

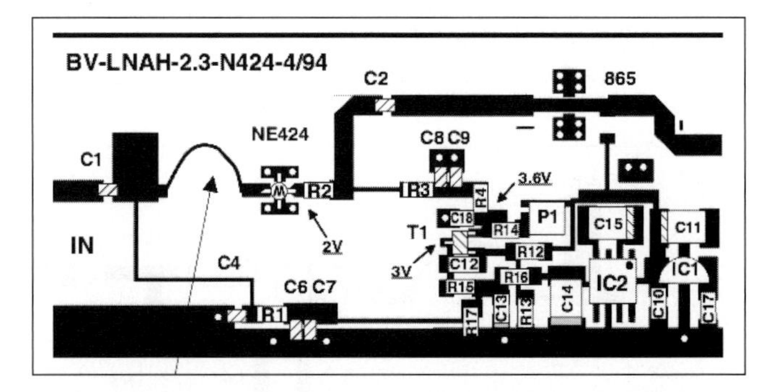


Fig 11.52: Component layout for single stage 13cm PHEMT preamplifier

- For L1, cut a 17mm length of gold plated copper 0.5mm diameter wire. Bend down the ends at 1mm length to 45 degrees. Form wire into a half circle loop and solder into the circuit with 1mm clearance from the PCB. The wire loop has to be flush with the end of the gate stripline and should be soldered at right angles to it. The wire loop has to be oriented flat and parallel to the PCB.

- Verify the open circuit function of bias circuit. Adjust P1 to 45 ohms. Solder a 100 ohm test resistor from the drain terminal on the PCB to ground. Apply +12V to IC1 and measure +5V at output of IC1, -5V at IC2 Pin5,_-2.5V> at collector of T1, +3.6V at emitter of T1, +3V at base of T1, -2.5V at R17 and +2.0V across the 100 ohms. If OK, remove 100-ohm test resistor.
- Solder PHEMT onto PCB. Ground the PCB, your body and the power supply of the soldering iron. Never touch the PHEMT on the gate, only on the source or the drain, when applying it to the PCB and solder fast (much less than 5 seconds).
- Solder N Connectors into sides of the box.
- Solder the finished PCB into the box, solder both sides of the PCB at the sides of the box and solder centre pins of the connectors to the microstriplines.
- Solder feed-through capacitors into box.
- Connect D1 between feed-through capacitor and PCB.
- Connect 12V and adjust P1 for 16mA drain current (measure 160mV across R4 on RF Board). Voltages should be around +2.0V at the drain terminal, -0.4V at the gate, +3.6V at emitter of T1.
- Connect LNA to a noise figure meter, if you have one, and adjust input wire loop, adjust the clearance to PCB as well as drain current by adjusting P1 for minimum noise figure. Even without tuning, the noise figure should be within 0.1dB of minimum because of the limited tuning range of the wire loop.
- Glue conducting foam inside the top cover and slip into the top of the box.

To add the MMIC Amplifier, refer to **Fig 11.53** for Construction.

- Prepare PCB by cutting slits into the microstriplines around the MGA865. These are a 2mm slit for L2, a 1.8mm slit for the MMIC and a 0.8mm slit for C3.
- For L2 cut an 8mm length of gold plated copper 0.5mm diameter wire. Form wire into a half circle loop and solder wire loop into the circuit. The wire loop has to lie flat on the PCB, flush with the end of the gate stripline and should be soldered in a right angle to it.
- Follow other instructions given above

Measurements were taken using an HP8510 network analyser and HP8970B/HP346A noise figure analyser, transferred to a PC and plotted. **Figs 11.54 and 11.55** show the results for gain and noise figure for the one stage and two stage version respectively. Using a special PHEMT, NEC NE42484A optimised for C Band, a typical noise figure of 0.35dB at a gain of 15dB can be measured on 2.32GHz. An optional second stage on the same PCB using the GaAs MMIC MGA86576 from HP will boost the gain from 15db to 41dB. The two stage version measures with a noise figure of 0.45dB. This version can be used for satellite operation. For EME, where lowest noise figure is at premium, a cascade of two identical one stage LNAs may be more appropriate. Both versions are broadband. They can cover the various portions of the 13cm amateur allocation from 2300 to 2450MHz without re-tuning.

The real surprise is the performance of the C band PHEMT NE424. It performs better than several other HEMTs (FHX35, FHX06, NE324, and NE326) tried in this circuit, and it measures 0.15dB better than its published noise figure. In fact the *Microwave Harmonica* simulation predicts a 0.5dB noise figure based on the data sheet value. The lower noise figure measured seems to be due to a special bias current and the lower value of

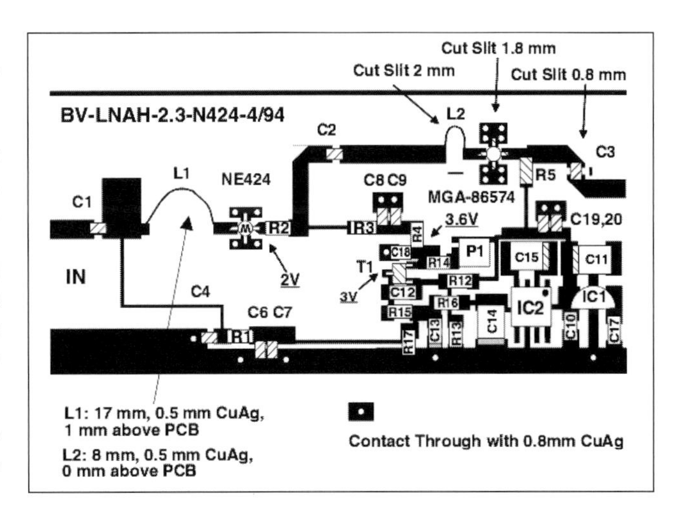


**Fig 11.53: Component layout for the two stage 13cm PHEMT preamplifier**

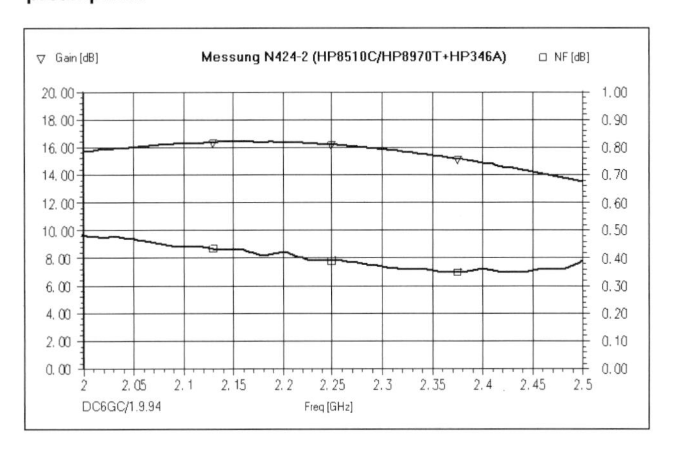


**Fig 11.54: Noise figure and gain measurements for single stage 13cm preamplifier**

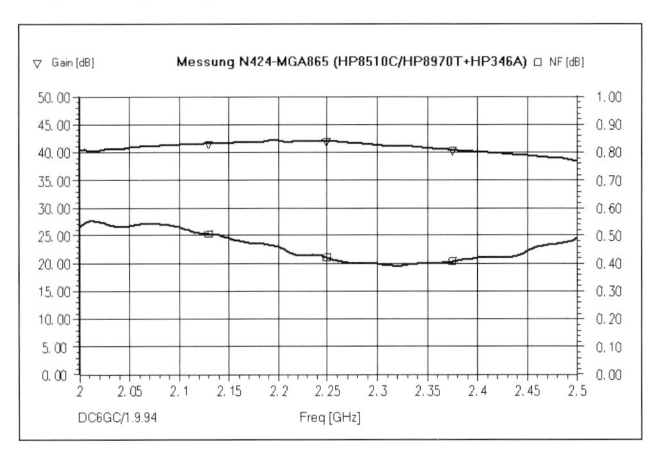


**Fig 11.55: Noise figure and gain measurements for two stage 13cm preamplifier**

gamma at approximately 0.75 which is due to the gate length of 0.35 micrometres. This provides optimum properties for application in 2 - 4GHz LNAs. Stability is excellent. This has been achieved by a carefully controlled combination of inductive source feedback, resistive loading in the drain and non resonant DC feed structures for drain and gate. A broadband sweep from 0.2 to 20GHz showed a stability factor K of not less than 1.2 and the B1 measure was always greater than zero. These two properties indicate unconditional stability. At the operating frequency of 2.3GHz, stability factor is about 1.6. The two-stage version with the MGA865 measures with K>>4 at all frequencies.

The preamplifier provides quantum leap towards the perfect noiseless preamplifier. It uses a low cost and rugged C band PHEMT instead of relying on expensive X band HEMTs. An improvement of about 0.2dB in noise figure has been achieved in comparison to the no tune HEMT preamplifier described in [23]. This improvement provides roughly 1.5dB more S/N in EME or satellite operation but is not noticeable in terrestrial links. However, the new preamplifier has to be tuned. This requires a noise figure meter for alignment. For those who like a no tune device, the HEMT preamplifier in [23] provides adequate performance with a typical NF of 0.55dB.

## Low Priced 10GHz Preamplifiers

### Described by Gerard Galve, F6CXO

Franco Rota runs an RF component supply company in Italy called R F Elettronica [24]. His main objective is to sell bulk components such as SMD parts to the electronics industry. He attends some radio rallies in Europe and often has interesting items for sale that can be used or adapted by radio amateurs for use on the amateur bands. Franco sells the PCB described in this article (**Fig 11.56**) for 3 Euros. The board was initially purchased to salvage the 4 NE 32584s. After examination, I realised it was fairly rare to find satellite boards with preamplifiers that are so well aligned and so suitable for modifications.

**Fig 11.57** shows a close up of the two preamplifier circuits. Preamplifier No. 1 can be adapted to a wider range of enclosures. The input track can be cut at any point throughout the black area.

**Fig 11.58** shows that if you are careful, you can even salvage two preamplifier circuits on this board; use the output capacitor of one as the input for the other. You could just salvage transistors (that was the original idea).

**Fig 11.59** shows how the output capacitor of the right hand preamplifier is connected to the input of the left hand one. The negative supply potentiometer can be fitted on the existing printed circuit board.

It is that easy, the hardest part is to find a housing that suits the length of the printed circuit board. Just use your imagination and see what might be available at the back of some old drawer.

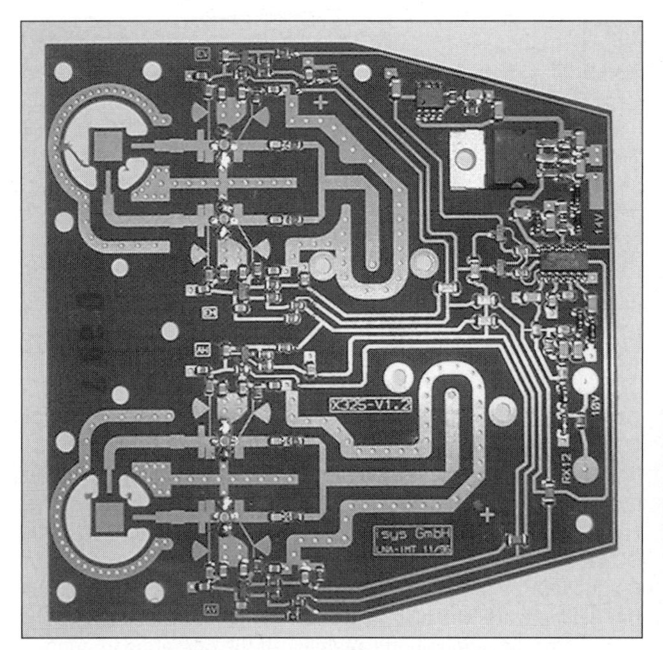


**Fig 11.56: The amplifier PCB available from Franco Rota**

**Figs 11.60 and 11.61** show the author's example.

### Performance

This box of tricks produced a noise figure of 0.7dB without the lid, 0.8dB with the lid, and a gain of 14dB from 10,368 to 10,450MHz. That's pretty fantastic for 1.5 Euros.

### Alternative use

Circuit No. 2 is a little too short to position a capacitor at the input. It can be used very easily with the input line directly in the waveguide as used in the original design (**Fig 11.62**).

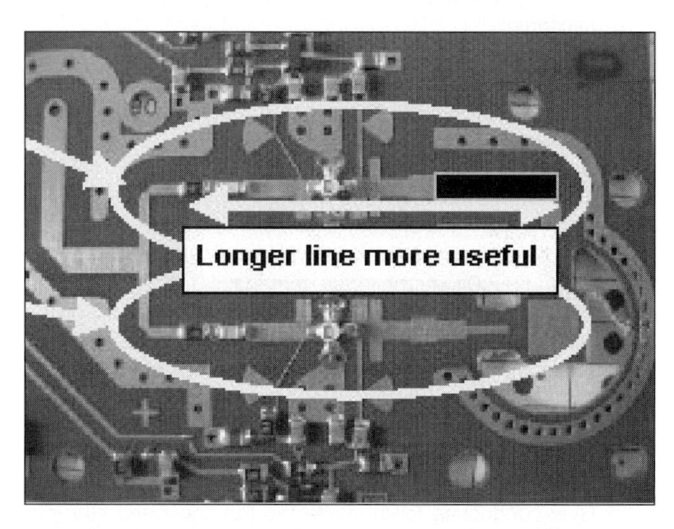


**Fig 11.57: One pair of amplifiers. The top circuit is referred to as Preamplifier No 1; the bottom one is Amplifier No 2**

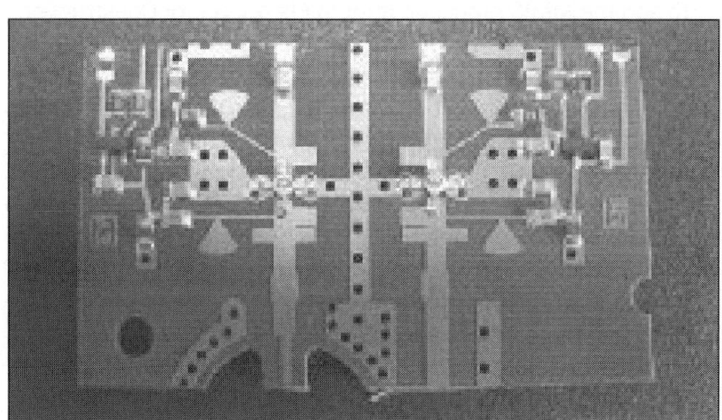

**Fig 11.58: How to use two amplifier circuits**

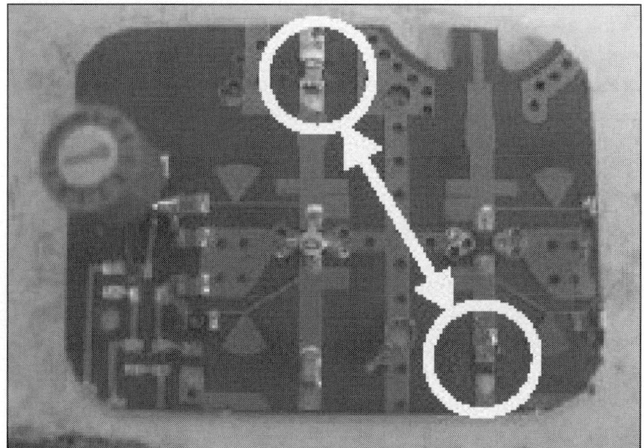

**Fig 11.59: Connecting two amplifiers together**

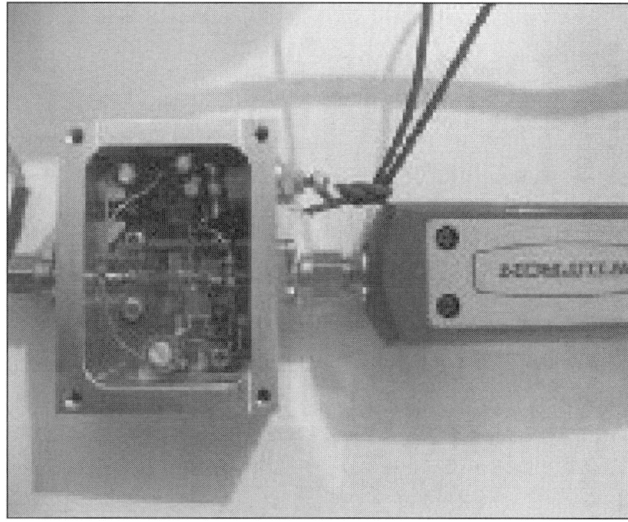

**(Above right and left) Figs 11.60 and 11.61: Example housing for the 10GHz amplifier**

**(Left) Fig 11.62: Using Preamplifier No 2 with direct waveguide input**

### Circuit diagram

The original diagram is in the box shown in **Fig 11.63**. The power supply was made very simple with wiring "in the air". The ICL 7660 is upside down and the components are soldered directly onto the pins. The drain current should be set to 10mA.

## TRANSMIT POWER AMPLIFIERS

Generating any reasonable amount of power on the microwave bands was the speciality of the valve. On the lower bands the trusty 2C39A was well used in anything up to eight valve designs.

On the higher bands the travelling wave tube was used but this needed special powers supplies. These are being replaced with semiconductor devices, with their compact design and more manageable power requirements. These can also be mast-head mounted.

On 23cm, the Mitsubishi M57762 hybrid amplifier has been used for many years but this is now going out of production and being replaced by the MOSFET hybrid amplifier RA18F1213G. Discrete semiconductors are also replacing hybrid modules, an example is shown here.

As the frequency increases it becomes more difficult to find, or afford, semiconductors for power amplifiers. Fortunately the gain from the antennas used comes to the rescue reducing the input power needed to achieve the required radiated power. On the higher bands special techniques are still required, such as direct bonding to semiconductor substrates, and an example of these techniques is shown below.

## L Band Power Amplifier for AO-40 Uplink

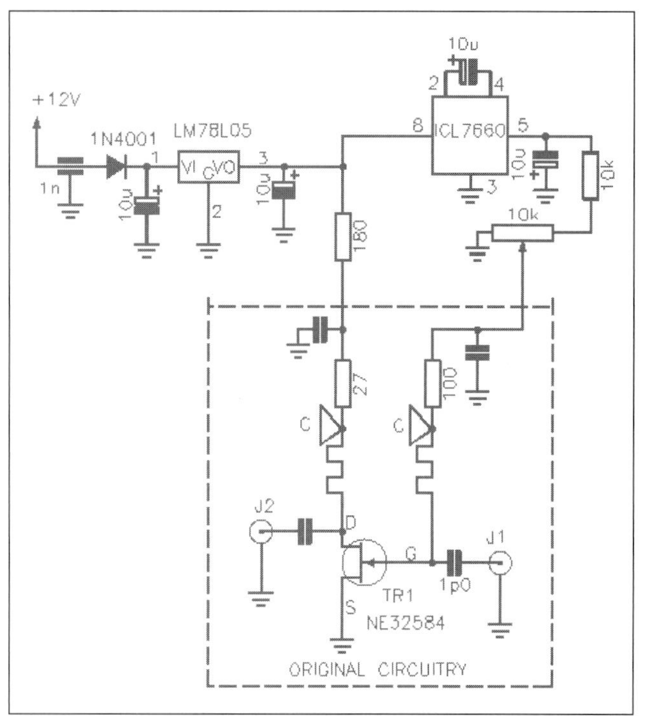


**Fig 11.63: Circuit diagram of the preamplifier and power supply used**

Amateur radio satellites offer many options for experimentation as can be seen at the end of this chapter on *page 11.67*.

The AMSAT-OSCAR 40 (AO-40) satellite is currently in orbit after surviving an explosion on board when the orbit was being established. In addition to operating on other bands, it has an uplink in the 1296MHz L Band. However, the power required for satisfactory radio contact is above the output of a normal transceiver, which makes it necessary to use a power amplifier.

A suitable RF power amplifier has been designed by Konrad Hupfer, DJ1EE. The experiences of many AO-40 users have demonstrated that a PEP of approximately 50W at the input of a circular radiating antenna with approximately 20dB gain is sufficient for the L Band uplink, even using squint angles. Looked at from the point of view of cable losses it makes most sense to generate the power at the point where it is used: directly at the antenna power feed. The amplifier described here has the following characteristics:

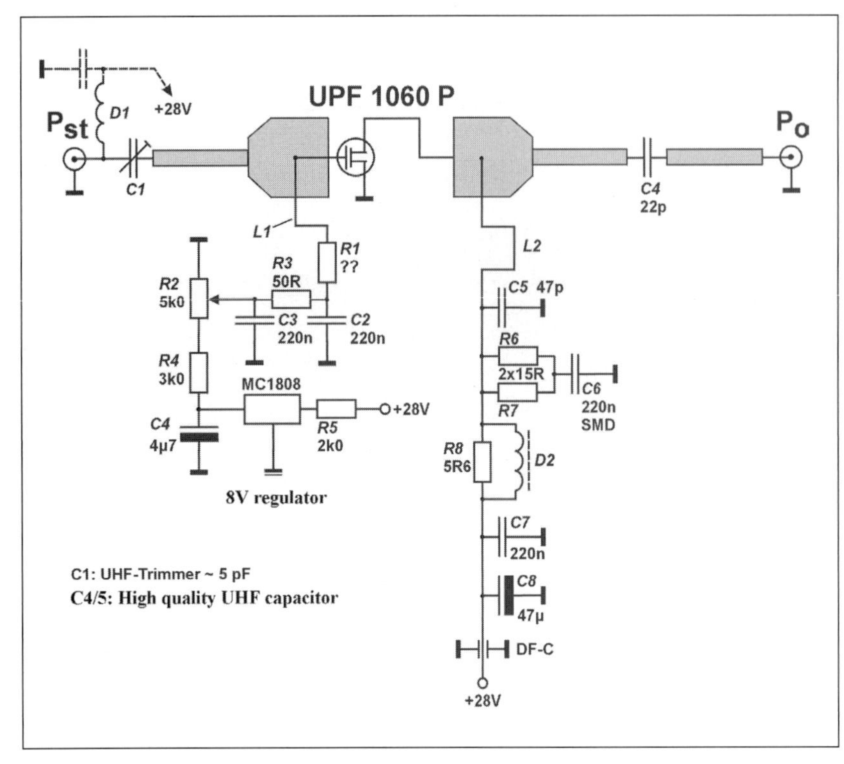


Fig 11.64: Circuit of L band amplifier

- $P_{opep}$ = 50W
- $U_{ds}$ = 28V
- G = 12dB
- $I_q$ = 300mA

Directly mounting an L Band helix or a patch antenna on the rear of the reflector offers the ideal solution. The reflector plate of an antenna (eg made from 3mm aluminium) having the normal area of approximately 400cm², is not quite adequate for a heatsink at ambient temperatures of >25°C. One remedy can be an additional 'chimney', consisting of approximately 1mm thick aluminium plate, with an associated cover. This additional cooling makes it possible to obtain thermally stable functioning, even in summer, for normal SSB mode with a PEP of approximately 55 watts.

## The amplifier

The L Band amplifier uses a fairly standard circuit (**Fig 11.64**). Printed line transformers are used to transform the relatively low complex input and output impedances of the L-DMOS transistor used (UPF 1060P from ULTRA RF). These transformers are made on the familiar material RO 4003, substrate thickness 0.79mm, r = 3.35, using stripline technology.

The general calculations for transformation networks of this type can be found, among other places, in [25]. The printed matching networks are shown in **Figs 11.65 and 11.66**. Since the most important dimensions of the lines are shown, you can easily construct them yourselves.

The first specimens of the amplifier circuit boards were designed with Indian ink, using the old technology, and then etched in the usual way. The undersides of the circuit boards naturally have a copper coating.

In the present circuit, for the sake of simplicity, no stabilisation is provided for the 300mA quiescent current. It would also be expedient to incorporate temperature compensation, to be prepared for 'extreme cases'. A cut out when excessive temperatures arise is also recommended.

The DC wiring can be laid out as you wish in accordance with the mechanical size of the components used. It can be seen from **Figs 11.67 and 11.68** that, in addition to the transformer lines, some small areas are provided. These are the so-called trim elements, used to fine-tune the amplifier for the best input matching or for maximum output power and linearity. This is needed to compensate for, for example, when similar transistors are used, differences in the input and output impedances occur or there are tolerances in the transistor mounting.

## Mechanical and electrical assembly

The semiconductor type selected, UPF 1060 P, is a flangeless model, because of price. Fig 11.67 and **Figs 11.69 and 11.70** show one possible assembly, using a 3mm thick carrier plate made from copper as a heat spreader. This mounting plate should be as level as possible on both sides; it should preferably be finished by surface milling on both sides. The individual circuit boards are soldered onto the copper plate using a hotplate. The distance of 6.5mm between the circuit boards should be adhered to; the copper surface must not be tinned in this area.

When the circuit boards have been assembled, the transistor is put in place, but initially without its connections soldered to the striplines.

The module (copper plate with soldered-on circuit boards) is screwed into the housing, the base of which has been finished by surface milling. A uniform application of heat conducting paste should be used between the transistor, the copper plate and the housing base. The transistor is now pressed onto the copper plate using a flexible implement (un-coated printed circuit material, fibreglass etc). You must treat the equipment carefully. Finally the connecting lugs can be soldered (Figs 11.69 and 11.70).

The amplifier must now be mounted flat onto the chimney/

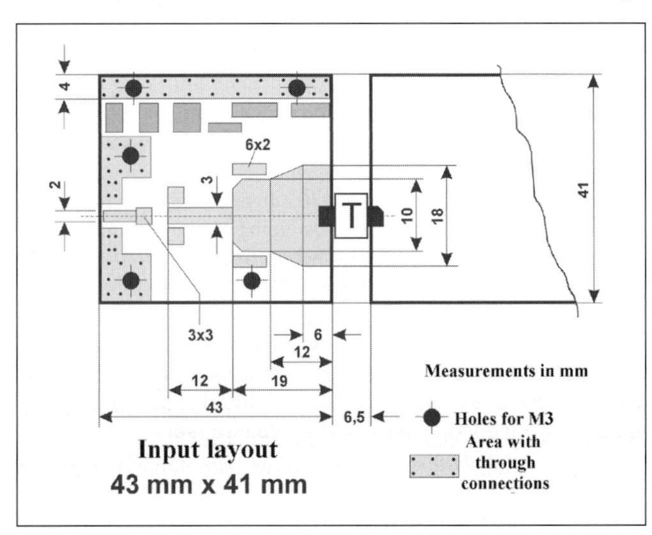


Fig 11.65: The input circuit layout showing measurements of matching networks

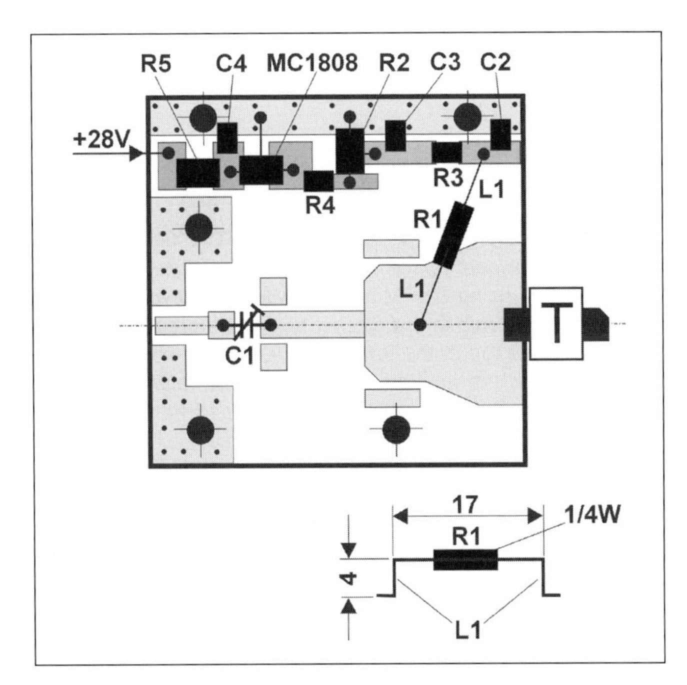


Fig 11.66: Component layout for input circuit showing details of L1 and R1

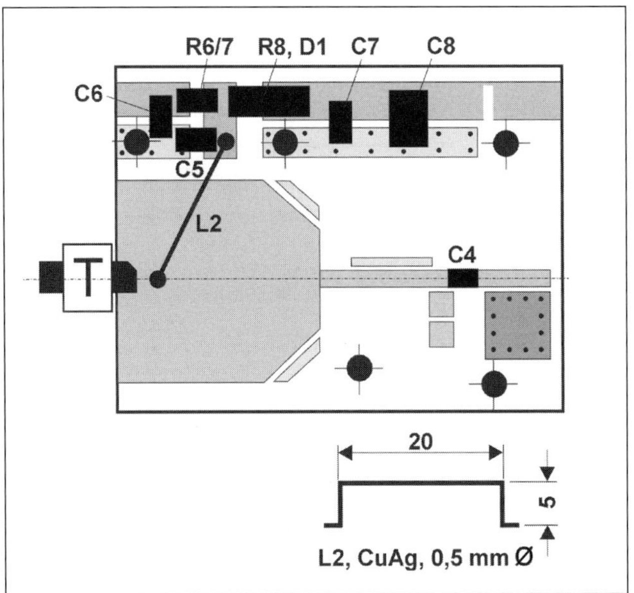


Fig 11.68: Component layout for output circuit showing details of L2

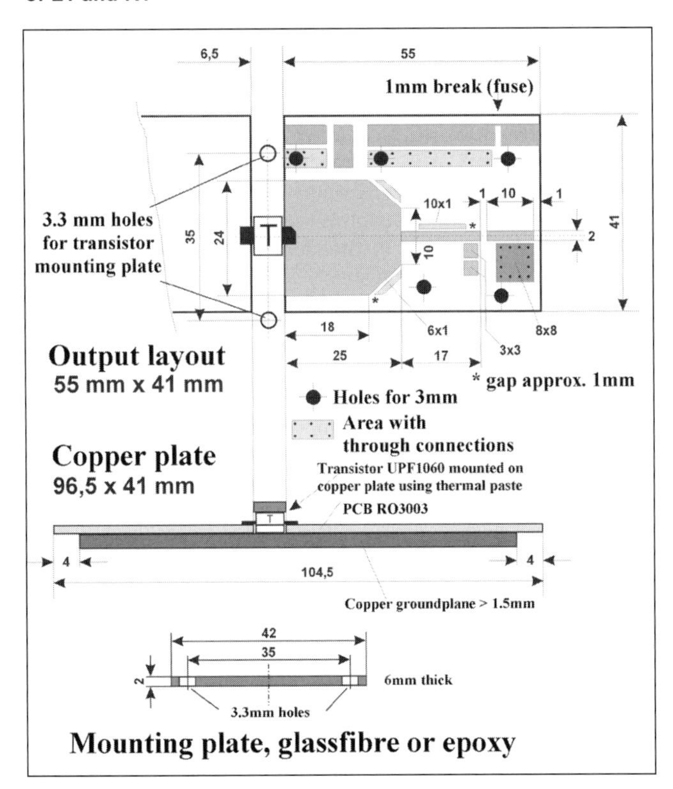


Fig 11.67: Output circuit layout showing measurements of matching networks and details of transistor mounting

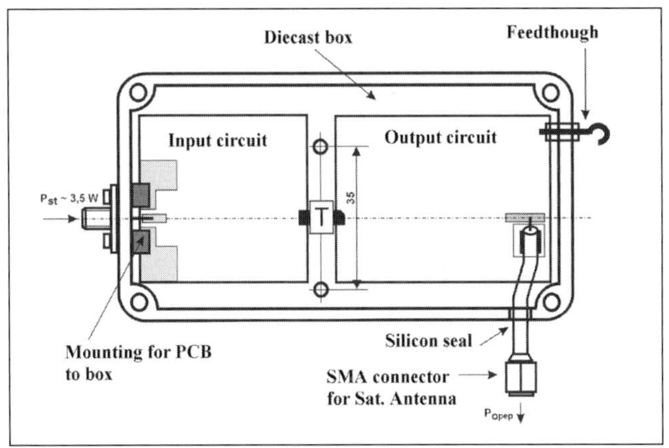


Fig 11.69: Installing the amplifier in an Eddystone diecast box measuring 111mm x 61mm x 30mm

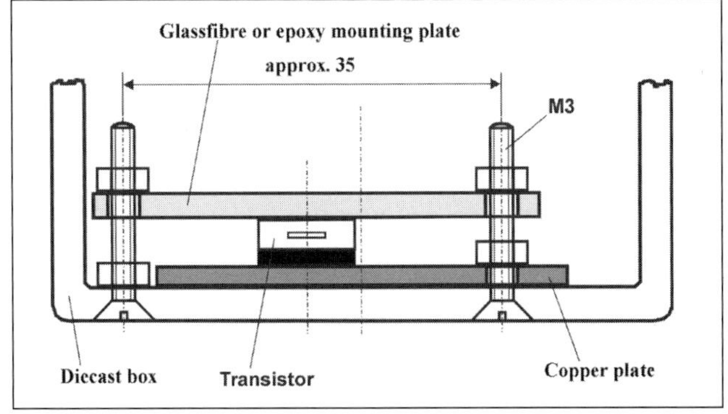


Fig 11.70: Details of transistor mounting using mounting plate

radiation reflector installation in its housing for best heat dissipation. The holes required for M3 countersunk screws can be made in the reflector. Another solution is to tap M3 threads into the reflector and then screw the entire unit onto the reflector from the amplifier side (**Figs 11.71 and 11.72**).

Various mechanical solutions can be used depending on the application. For simplicity's sake, the power supply in this layout is provided using a feed-through filter mounted in the housing wall. The supply voltage of 28V can be fed through the coaxial feeder cable. The entire structure should be painted white to keep the temperature as low as possible, even when the sun is shining!

The circuit has been assembled and tested many times. The characteristics, interpolated from the worst values of three amplifiers are shown in **Fig 11.73**. The typical measurements of the RF power amplifier for single-tone test are as follows:

- $P_0$ = 50W (with compression

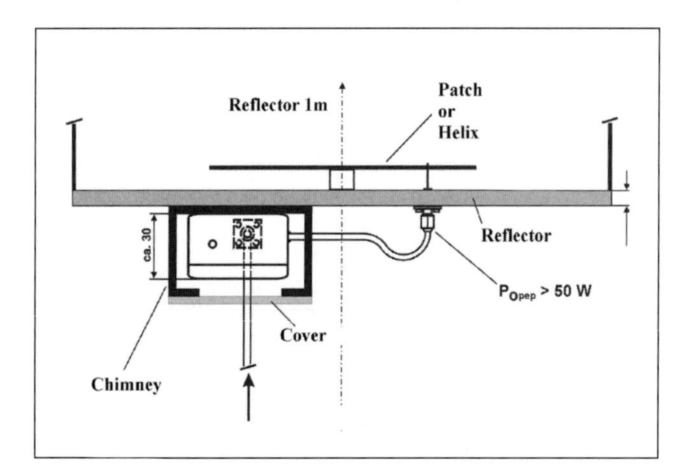


Fig 11.71: Mounting the amplifier on an antenna with cooling chimney approximately 150 x 200mm

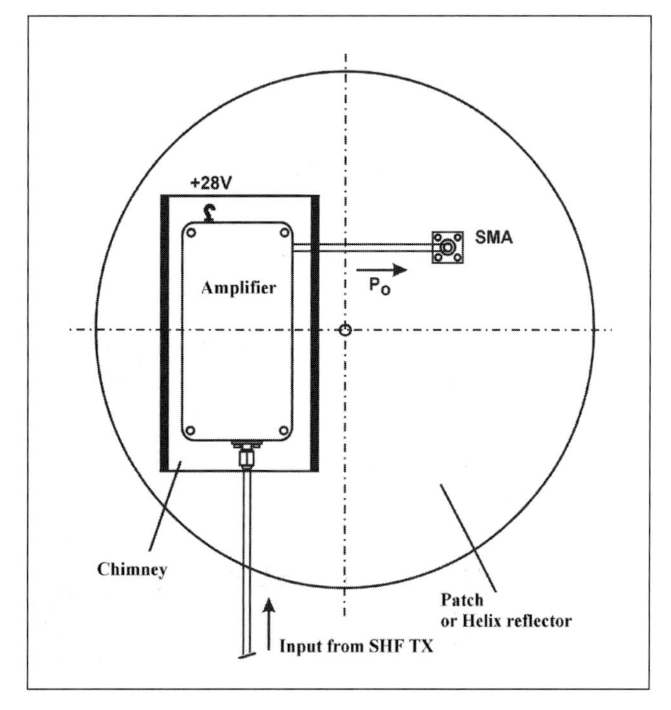


Fig 11.72: Rear view of mounting the amplifier on an antenna with cooling chimney

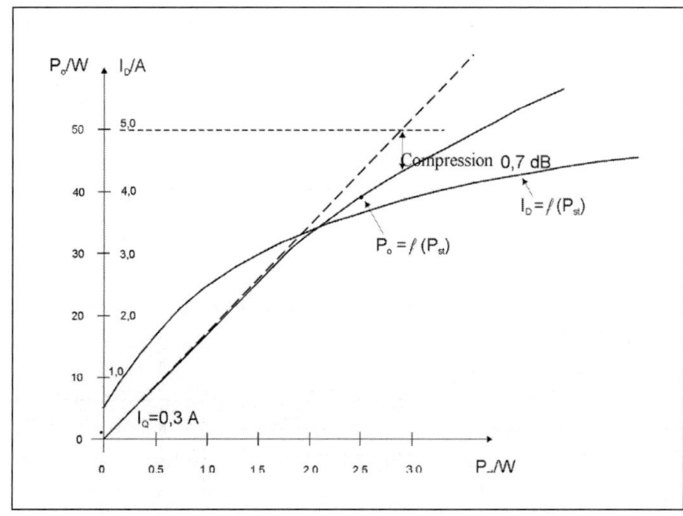


Fig 11.73: Transfer characteristics of the L band amplifier

approximately 0.6dB)

- $J_D$ = 4.5A
- $J_{DQ}$ = 300mA
- $P_{st}$ = 3.2W
- G = 12dB
- Efficiency ≥ 49 %

## G H Engineering PA1.3-100 23cm Power Amp

For the last 10 years I have been using a 23cm power amplifier for contesting using two Mitsubishi M57762 power modules. The amplifier is built into a diecast box with a preamplifier and mounted at the top of the tower to reduce the feeder loss. The power is fed to the amplifier from a 12V gel battery mounted near the amplifier, the battery is float charged from the shack 60 feet below. This arrangement has worked very well but I decided that it was time to try and increase the power output and chose the new PA1.3-100 amplifier from G H Engineering [26]. This is available as a mini kit or a ready built unit; I decided to go for the mini kit and placed an order as soon as the kits became available early in 2008. The following article describes my construction and testing efforts.

### Circuit description (from the GH Engineering documentation)

The PA1.3-100 is a linear amplifier for the 1.3GHz band. It is capable of producing at least 100W in the frequency range 1240 - 1300MHz with a drive level of 3W. The amplifier operates in linear mode, which makes it suitable for SSB operation as well as FM. For best linearity, it is recommended that the input power does not exceed 2W PEP, which will give an output power of approximately 80W at 1296MHz. The amplifier uses four Mitsubishi RA18H1213G PA modules that require no tuning being internally matched to 50 ohms at the input and output. An external 13.8V DC power supply is required that is able to supply at least 45A. The amplifier is rated for continuous operation with an input power of 3W and a supply voltage of 13.8V

The PA1.3-100 is intended for use with a transverter of 3W output or similar. It can be used with a 10W transceiver with the use of an external 6dB attenuator. The amplifier incorporates a 7.5dB input attenuator on the PCB; this can be changed or removed if required, thus allowing the amplifier to be used with an input signal down to approximately 750mW.

**Fig 11.74** shows a photograph of the circuit diagram. The original is printed on an A3 sheet so it is not really suitable for reproduction on an A4 page in this handbook but the main elements can be seen to help understand the following description of the RF distribution.

The RF input signal is split with a Wilkinson splitter and each of the two signals is then split again with two more splitters. This gives four signals of equal amplitude and phase that are attenuated by approximately 7.5dB by pi-networks.

Each RA18F1213G module provides a small signal gain of approximately 23 - 25dB depending on frequency. The input and output microstrip lines have been designed to ensure that the overall amplifier gain is as flat as possible over the 1240 - 1300MHz band.

The output signal is fed to a 4-way Gysel combiner, which is a derivative of the N-way Wilkinson combiner. It consists of four quarter-wave transmission lines, each 100 ohm impedance. Therefore the output signal is transformed to 200 ohms, which gives an output of 50 ohms when all four are connected together in parallel, The isolating network is somewhat more complex than for the input splitters. three-quarter wavelength transmission lines are connected to each of the junctions of the 50 ohm and 100 ohm lines. The ends of these lines are terminated with 50 ohm high power terminat-

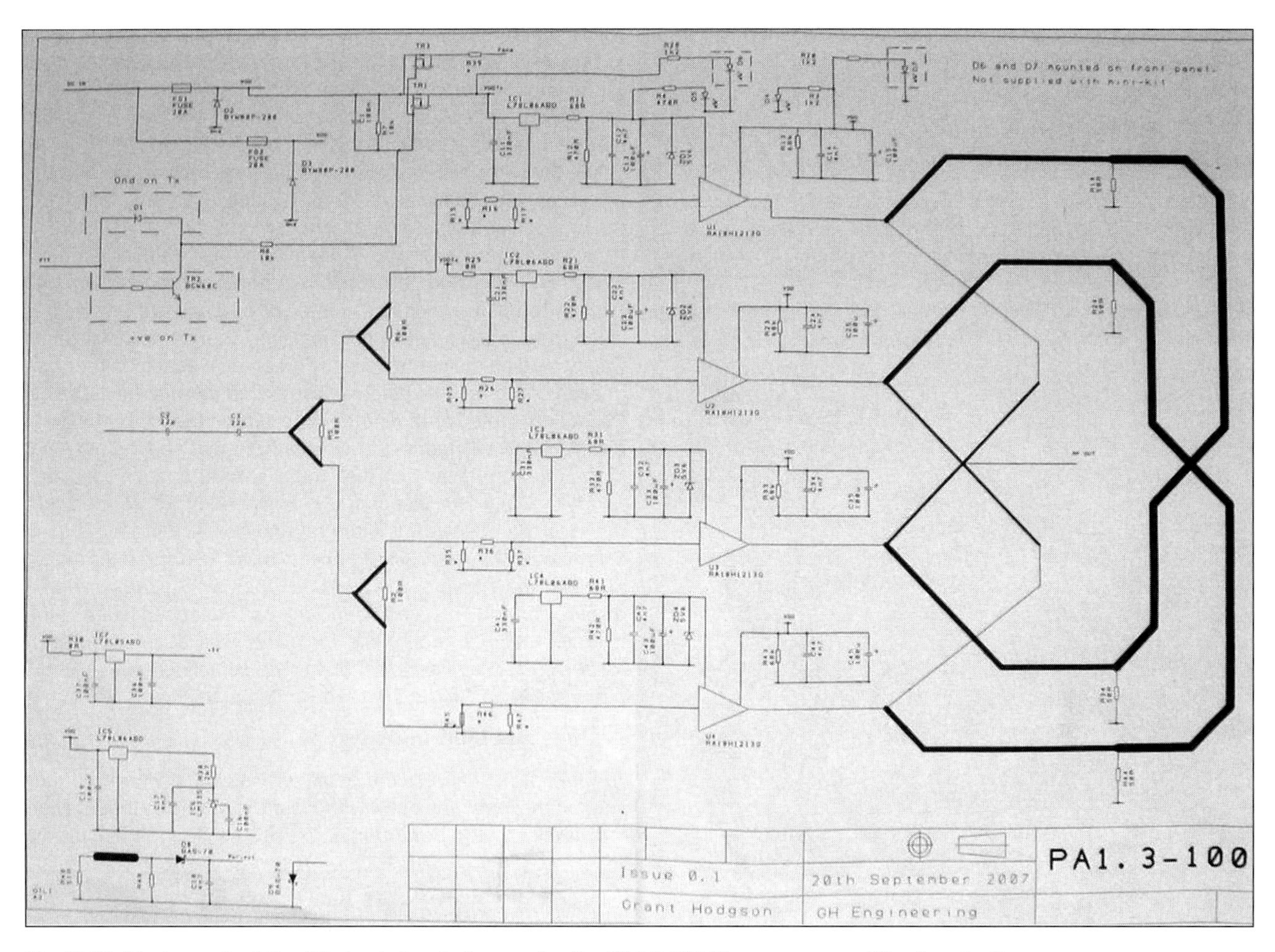


**Fig 11.74: Photograph of the A3-sized circuit diagram for the PA1.3-100 23cm power amplifier (see text)**

ing resistors. This is also connected to the end of another quarter wavelength line, of 25 ohm impedance. These four 25 ohm lines are connected together. At this point there is a virtual ground, in that no current flows across the junction, but the impedance is actually very high, and approaches an open circuit. Therefore the impedance at the termination resistor is very low (approaching a short circuit) and no power flows into the resistors. Therefore, the impedance looking into the end of the three-quarter wavelength lines is very high. Consequently, the three-quarter wavelength line places no load on the output, and no current flows into the isolating network.

The PA1.3-100 is fitted with a PCB temperature sensor giving an output voltage that is proportional to temperature. The voltage gradient is 10mV/ C and the voltage is directly proportional to absolute temperature in Kelvin. Therefore at 20 degrees C the output voltage would be 2.93V and at 50 degrees C would be 3.23V.

There is an output power detector using a sampling line and a BAS-70 diode. The output voltage developed is dependant on temperature, to compensate for this dependency a second diode is fitted. A simple power meter can be made using a operational amplifier that compares the output voltage of the two diodes.

Both the temperature sensor and power detector are provided as unsupported features for the constructor to use as they see fit.

## The kit

The kit comprises all the electronic components, a pre drilled heatsink and a set of documentation. **Fig 11.75** shows the kit of parts and **Fig 11.76** shows the documentation and parts list. The first task was to check that everything was present. Thoughtfully the SMD were split into several self-seal bags with the values on a list stuck to the bag. Because some components are difficult, or impossible, to identify the bags are split so that the components can be identified by the quantity of each in the bag. I made a note to keep the components in their bags until I used them so they did not get mixed up.

The next thing to do was to read the assembly instruction to make sure that I knew of any special requirements during assembly. There were a few modifications to do to the PCB and a decision to be made about the sense of the PTT switching. Fortunately I wanted PTT switched to ground for transmit since one of the PCB modifications made it impossible to fit the additional components for PTT going high for transmit. The remaining instructions looked straightforward SMD assembly followed by some care required when mounting the four power modules.

## Construction

The heatsink is a hefty piece of aluminium and made a suitable place to mount the PCB and hold it still for assembly of the SMD components (**Fig 11.77**). Because there were a few errors on the circuit diagram and the component layout care had to be taken to double-check each part as it was assembled but assembly of all the SMD parts was straightforward. The only point to note is that any component connected to the ground surface needed an extra bit of heat to make a good solder joint, so I used two different bits on my old Weller soldering iron. If you have a more modern soldering station this may not even be noticeable.

Once all the SMD components are fitted the instructions require some voltage checks before the power modules are fitted. In my case the voltage measurement of the top power module bias supply was not correct. This was quickly traced to the monitor LED that placed an extra load on the circuit. A quick wire link to a different supply point and all was well.

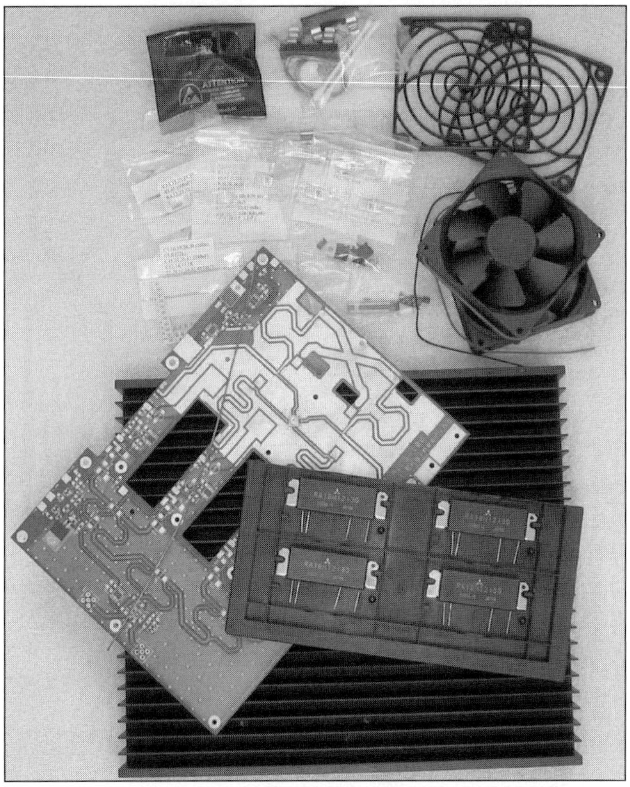

Fig 11.75: The kit for the PA1.3-100 23cm power amplifier

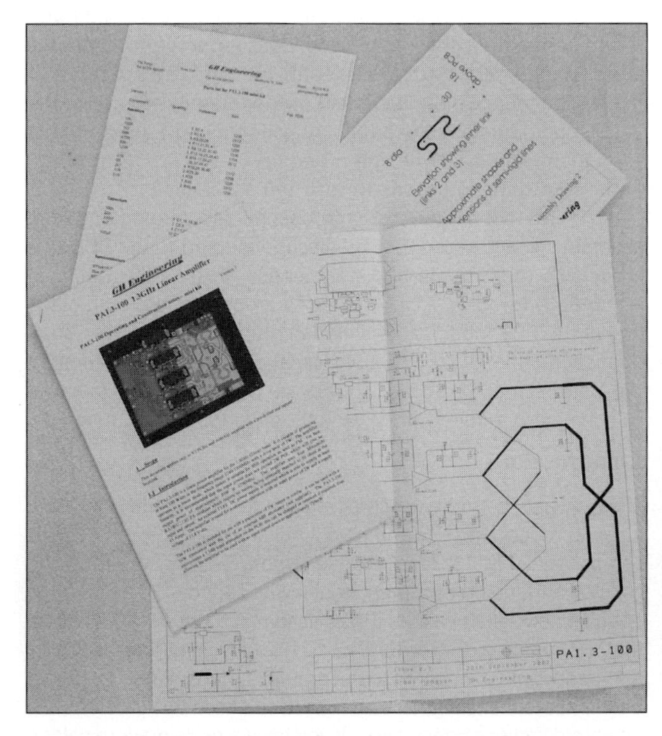


Fig 11.76: The documentation for the PA1.3-100 23cm power amplifier

The next step was to fit the two three-quarter wavelength lines that are made from semi rigid cable. Some care was required to make these and shape them as shown on the diagram in the instructions. My attempt at this is shown in **Fig 11.78**.

The final step was to bolt the PCB to the heatsink and fit the power modules. Static precautions must be taken when fitting the power modules because they are sensitive to static electricity. The instructions mention that the mounting holes for the power modules must be as clean as possible to provide a good DC and RF path to the power modules, some cotton buds are provided to clean the tapped holes in the heatsink. Thermal transfer paste was spread on the centre (slightly recessed) section on the base of each power module before it was bolted down to the heatsink Because I wanted to mount the amplifier on the top of the mast I wanted the finished unit to be weather proof and did not want to mount the two fans on the heatsink. After a quick conversation with Grant Hodgeson at G H Engineering we decided that the fans are needed if the power amplifier is going to be used continuously for FM or ATV, but because my use was intermittent SSB and the amplifier will be in the open air the fans could be dispensed with. This meant blocking up the tapped holes in the heatsink used to mount the fans; this was done using Allen screws coated with silicon sealant. The completed power amplifier is shown in **Fig 11.79**, ready for initial testing.

## Testing and fault finding

To make an initial test the amplifier was fitted with a dummy load on the input and output. The instructions state that the current drawn by the power amplifier with no RF input should be

Fig 11.77: The PCB mounted on the heatsink as a workbench to hold the PCB still while fitting the SMD components

Fig 11.78: The three-quarter wavelength lines fitted to the output circuit

Fig 11.79: The completed PCB ready for initial testing

approximately 15A so I intended to use my 25A bench power supply for the initial test. When I switched on, the power supply went into current overload, so I knew there was a problem.

The final method of powering the amplifier was to be a 38AH sealed lead acid battery so that was connected up and the amplifier was actually drawing 26A with no RF input. A quick check showed that the amplifier was producing about 10W output on random frequencies (**Fig 11.80**); it was self-oscillating. An email to G H Engineering indicated that the mounting holes for the power amplifiers were probably not clean enough or some thermal paste had oozed under the mounting flanges.

The modules were removed, holes cleaned and a small spot of thermal paste removed from one of the modules. The method that I used to clean the holes was to shave down a cotton bud and spray it with degreasing cleaner then 'screw' it into the tapped hole. When the amplifier was switched on again it still drew 26A but the output was now on a single frequency. By grounding the control, pin (Pin 2) on each power module in turn the oscillation was isolated to power modules 3 and 4. Again the problem was thought to be poor grounding of the power module so some brass strips were fitted to improve the grounding of modules 3 and 4 (**Fig 11.81**) but this did not solve the problem.

After carrying out more checks it was noticed that pressing the PCB around the general area of power module 3 and 4 stopped the oscillation; this was the clue. The mounting screws for the PCB to the heatsink were loosened on one side and re-tightened with the PCB pressed firmly against the heatsink. The amplifier then drew 15A and no oscillation.

Next the amplifier was fed with 23cm RF and the output measured, with 12.5V at the battery terminals 87W output was measured. To get full output the full 13.8V supply would be required. To do this, the power supply wiring had to be tidied up; it is surprising how much volt drop there can be across a thick piece of

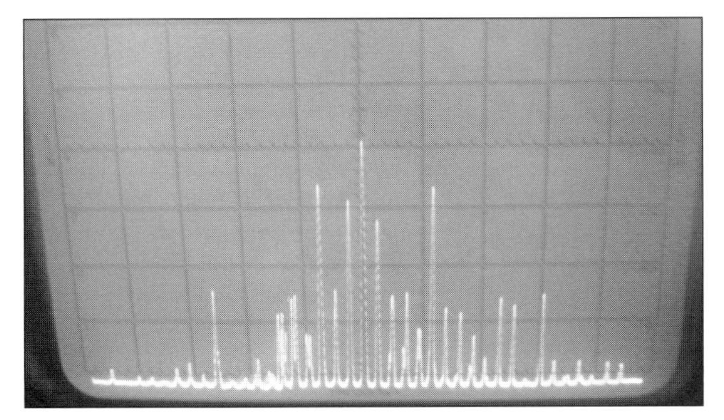

Fig 11.80: The output of the amplifier when it was self-oscillating

Fig 11.81: The additional grounding on modules 3 and 4

cable when it is carrying the 40 or so amps drawn at full power.

The amplifier was then fitted into its box to tidy up the wiring and fit the final input and output connections to the changeover relays. I spent some time researching the possibilities for an enclosure; my first thought was diecast aluminium with a fully sealed lid. However, having seen the price and weight of suitably sized enclosures, I realised there had to be a better option. While searching the Farnell web site I found that there is a range of steel terminal enclosures [27] that are fully sealed at a reasonable price and weighing less than the equivalent aluminium enclosure (the aluminium enclosures are built to support a tank). The only problem with this enclosure is cutting out a 200mm x 250mm hole to fit the power amplifier. The result is shown in **Fig 11.82**. Final testing with everything tidied up gave nearly 100W output.

In conclusion, the PA1.3-100 kit is well designed and easy to build by anyone with a reasonable level of experience. The amplifier performs well and does exactly what it says on the tin.

The main thing to note when using the amplifier is that a 45A 13.8V supply is difficult to provide at the top of a mast but I think the advantages of mounting the amplifier near the antenna for contest working outweigh the problems involved. If I were using the amplifier for a permanent installation this would need careful thought.

## Surplus High Power Amplifiers for 2.3GHz

Activity on the 2.3GHz band was at a low level until mobile

Fig 11.82: The amplifier mounted in its metal enclosure with a sequencer and preamplifier

quency section and the earth in the input and output areas, silver conducting lacquer can be smeared there (very sparingly, of course).

The partly assembled board is now fitted into the suitably prepared aluminium housing and screwed down by five M2 screws. When the connections to the feed-through capacitor have been completed, it is possible to check the DC function. For this purpose, the two trimmers are pre set to a gate voltage of about -1.5V.

The trickiest stage in the procedure is the soldering of the GaAsFET into the milled grooves. To this end, the aluminium housing is first heated, with the board inside, to precisely 150°C. Each milled groove is then pre tinned, using low temperature solder with a melting temperature of 140°C.

Excess tin is then removed using a de-soldering pump. The transistors are next placed in the grooves; all the relevant safety measures known must be taken. Normally, the tin binds very well with the gold plated flanged base, something that can easily be tested by a visual check of the flanged holes. Naturally, this soldering process should be carried out as rapidly as possible. The housing is then immediately placed on a cold copper block or a large cooling body, so that the temperature quickly falls.

Drain and gate connections are soldered onto the striplines; all the relevant safety measures must be taken. The two 10µF capacitors on the drain side are fitted and the SMA flanged bushes are screwed on. The power amplifier is ready for tuning (**Fig 11.90**).

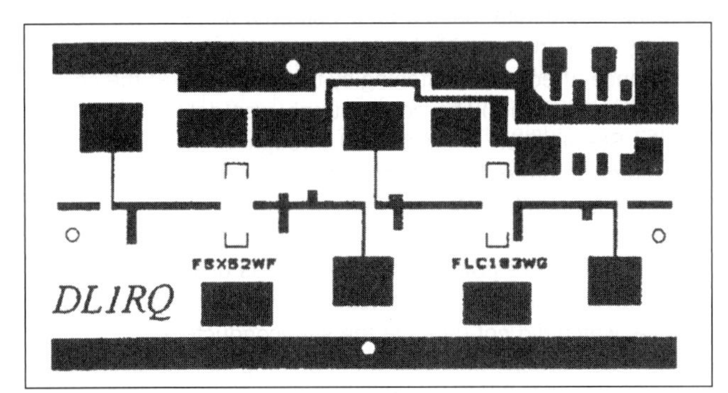

**Fig 11.87: PCB layout for the 1W GaAsFET amplifier for 3cm**

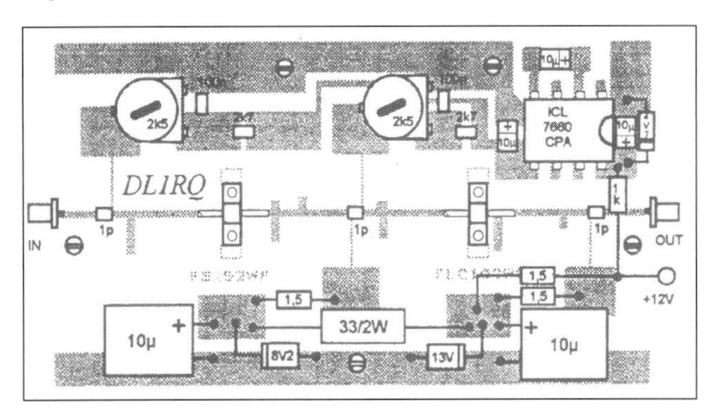

**Fig 11.88: Component layout for the 1W GaAsFET amplifier for 3cm**

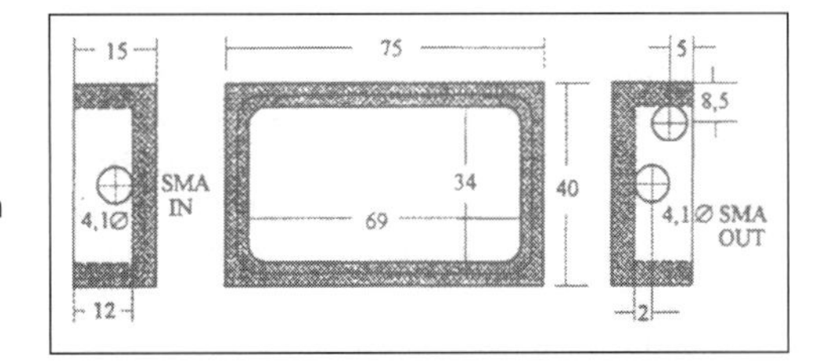

**Fig 11.89: Dimensions of a milled housing for the 1W, 3cm GaAsFET amplifier**

First, the no-signal currents are set as follows:

- For the FSX52WF at approximately 70mA, this corresponds to a voltage drop of 105mV, across a 1.5-ohm protective resistor. For the FLC103WG, at approximately 240mA, it corresponds to a voltage drop of 360mV, across a 1.5-ohm protective resistor.
- With 30mW drive at the desired frequency, an output of approximately 400mW (in the worst case) and of 1W (in the ideal case) should be measurable.

The "small disc method" is normally of assistance when tuning the amplifier. You will need small discs, measuring about 2 - 4mm$^2$, a few toothpicks to press down and push and a lot of patience. Above all, the greatest care in watching out for short circuits will (hopefully) soon lead you to achieve full output power. After tuning, an aluminium cover plate 1mm thick can be fitted.

In the overwhelming majority of the power amplifiers measured, almost no influence from the cover could be detected.

Of course, there were just a few cases in which minimal self excitation were detected when the cover was fitted. This is caused by astonishingly stable housing resonance, slightly above the calibration frequency, with a few milliwatts of power at the output.

Even this undesirable oscillation disappeared with a low powered drive. A strip of absorbent material about 5mm wide and about 10mm long, glued to the inside of the cover in the area above the FSX52WF, provided a reliable remedy here.

A comparison of the output data from the semiconductors (**Figs 11.91 and 11.92**) with the readings from a typical power amplifier (**Figs 11.93 - 95**) makes clear how successful the project is in practice.

**Fig 11.90: The 1W GaAsFET amplifier for 3cm**

## 76GHz Amplifier

This design [36] is part of a series of articles by Sigurd Werner, DL9MFV, published in *VHF Communications* magazine. It describes an amplifier that uses two MMICs (IAF-MPA7710) connected in series, and originates from development work by the Fraunhofer Institute for Applied Solid-State Physics in Freiburg. The gain of the amplifier is 24dB at 76,088MHz.

The first difficulty lies in the procurement (selection would be something of an exaggeration) of suitable MMICs for this frequency band.

Siemens manufacture a two-stage GaAs amplifier chip (T602B-MPA-2) for use in its car collision radar, which amplifies small signals by approximately 9.5dB [37].

One genuine alternative to this is an MMIC that has been developed at the Fraunhofer Institute. This is another two-stage amplifier chip (1.5mm x 1mm x 0.635mm), with the designation IAF-MPA7710. It has the following characteristics:

- Frequency range: 73 - 80GHz
- Gain: >11dB
- Output >+14dBm at 1dB compression, and with a power consumption of approximately 800mW [38] and [39].

Since gate 1 is earthed, only two positive voltages of approximately 1.5V (for Gate 2) and approximately 4V each for all drain connections are required. In order to attain a usable level of amplification, two MMIC's are wired in series, without any additional matching circuit between the two chips.

The housing with dimensions of 27.5mm x 39.8mm x 13.5mm was milled from brass and subsequently gold-plated (**Fig 11.96**). The two chips were mounted in a 0.5mm hollow in the 4.1mm machined cavity. This balances out the difference in height between the MMIC's (0.635mm!) and the connection substrates (0.254mm). Strict attention was paid to ensure that the distance between the two MMIC's and their distance from the substrates were as small as possible (approximately 60 or 75μm). The connection substrates that link the chips with the two WR-12 waveguides (amplifier input and output) are made from aluminium nitride (approximately 1.1mm wide). They were sawn out of an existing ceramic, a substrate designed for other purposes by R&S. Since the chips have co-planar RF connections, the substrates likewise have a short co-planar section (0.5mm), which then goes into a 50-ohm stripline (8mm). This line projects approximately λ/8 into the

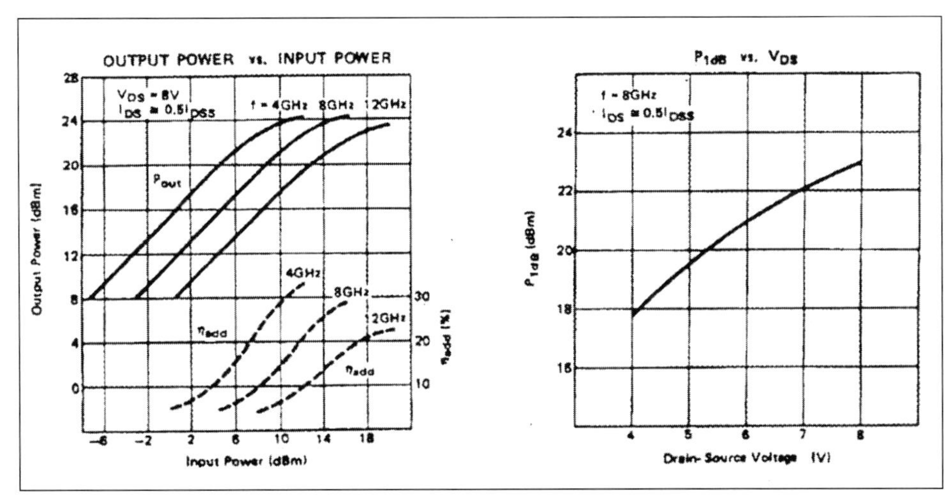


**Fig 11.91: Data for the Fujitsu FSX53WF used in the 1W GaAsFET amplifier for 3cm**

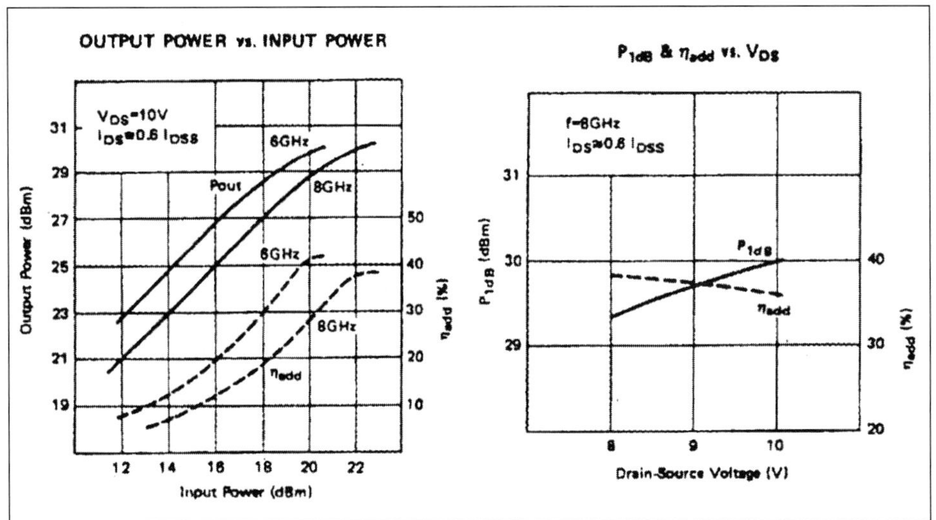


**Fig 11.92: Data for the Fujitsu FSX53WF used in the 1W GaAsFET amplifier for 3cm**

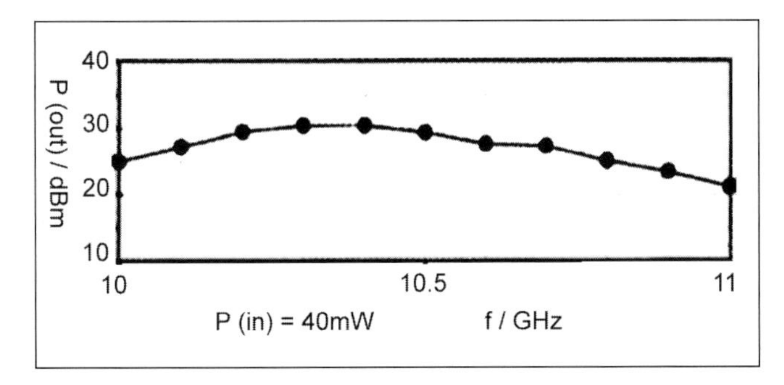


**Fig 11.93: Power bandwidth of the 1W GaAsFET amplifier for 3cm**

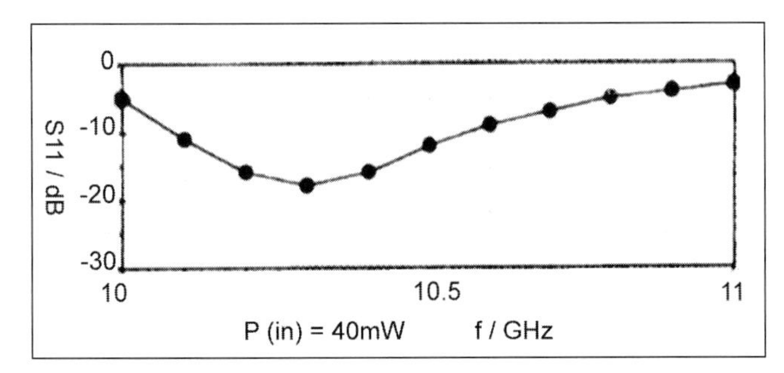


**Fig 11.94: Input matching of the 1W GaAsFET amplifier for 3cm**

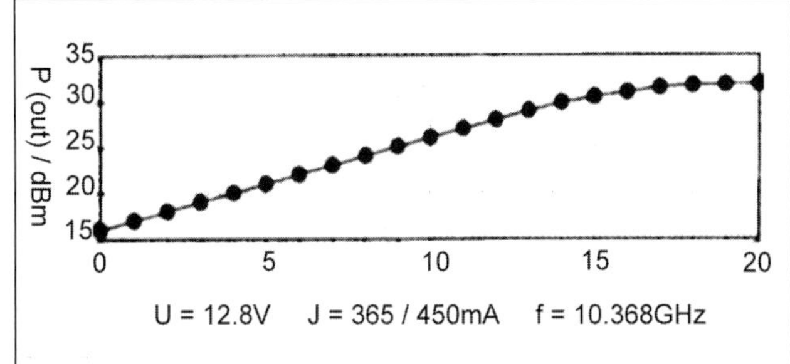


**Fig 11.95: Linearity of the 1W GaAsFET amplifier for 3cm**

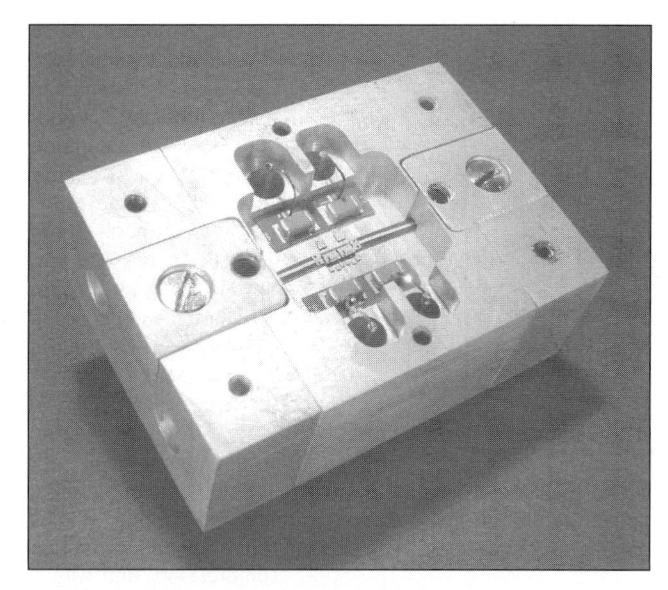

Fig 11.96: Picture of the completed 76GHz amplifier

Fig 11.97: Picture of the completed 76GHz amplifier with WR-12 waveguide connections fitted

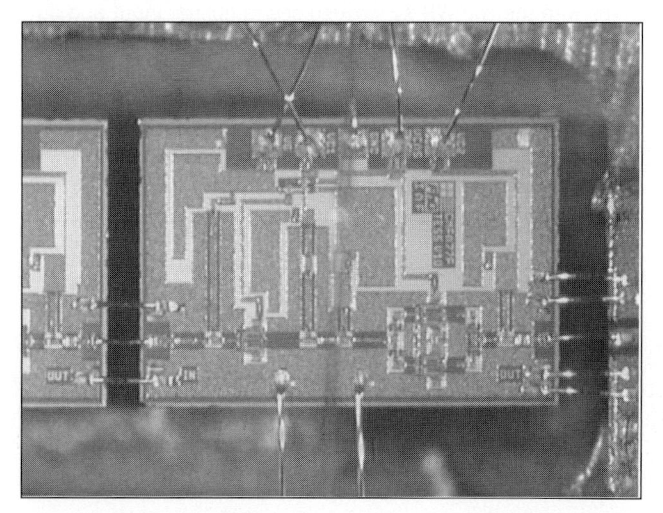

Fig 11.98: The 1AF-MPA7710 chip, used in the 76GHz amplifier, magnified about 120 times

waveguide. This construction technique is described in more detail in [40]. The ground plane of the ceramic projecting into the waveguide was milled off. The power supplies for the chips are initially blocked with 100pF single layer capacitors, and subsequently with 100nF ceramic capacitors, and are then fed out via feedthrough filters through the housing base (see **Fig 11.97**). The connections between chips and substrate (or capacitors) were created using wedge-wedge bond technology [40].

The following problem arose here: The RF connection pads of the MMICs, which were actually designed for flip-chip installation, are extremely small (see **Fig 11.98**). Directly behind these pads there are air bridges running to the chip circuit. These fragile structures are very easily caught and destroyed during the bonding by the tool that feeds the 17µm gold thread. The chip is then naturally unusable. This difficulty was avoided through the use of a still thinner needle and a correspondingly finer gold thread of 12µm. However, all other connections were created using a conventional 17µm thread.

The amplifier was initially operated at low power levels (approximately 50µW) at 76,032MHz. The gain observed was initially very disappointing (in the region of 8dB). Even after the fine adjustment of the waveguide short circuit screws the gain reading was scarcely 10dB. On the basis of experience of mismatching of the waveguide couplings obtained during the transverter project, another series of gold threads was attached and fastened to the striplines using a UV activated adhesive (see [41]). The input power during matching amounted to -13dBm.

After a laborious sequence of nine pennants, the work was rewarded by an amplification of 24dB. That means 12dB per amplifier stage, a value which tallies well with the specifications in the data sheet [38]. A value of > 9dB (SWR < 2.1) was measured at the amplifier input, with > 25dB (SWR < 1.1) at the output. The gain and the output power for various input levels (f = 76,032GHz) is shown in **Fig 11.99**. It can be seen that for an input power of > -15dBm the amplification is already decreasing, a behaviour to be expected. At an input power of 100µW, there is still 10mW measured at the output anyway (20dB gain). The saturation power of +12.6dBm remains unsatisfactory (approximately 18mW). Approximately 15dBm would have been expected! There could be several reasons for this behaviour. The large number of threads certainly increased the matching and thus the amplification, but at the same time a lot of energy is lost with each stub attached. Secondly, the length of the striplines at these frequencies leads to additional losses. The power

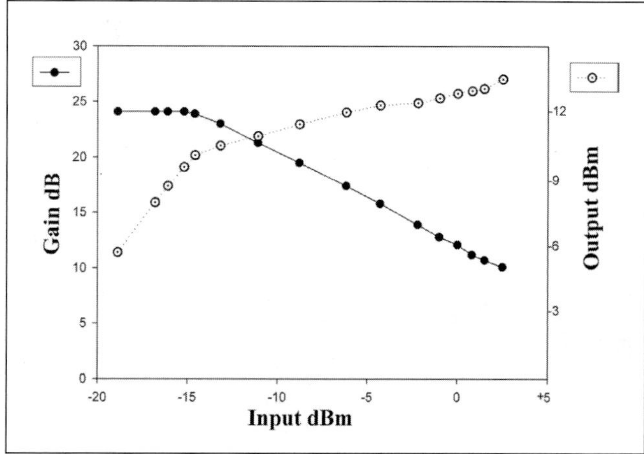


Fig 11.99: Transfer characteristics of the 76GHz amplifier at 76,032MHz

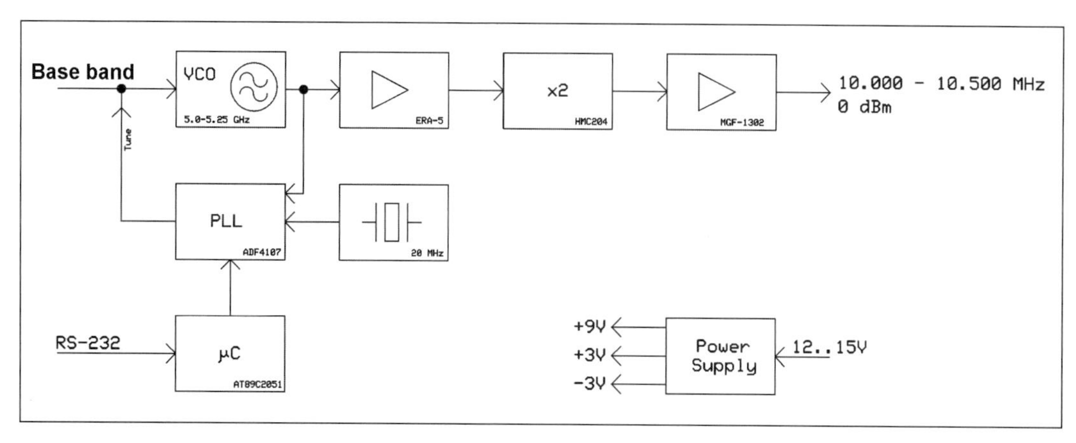


**Fig 11.100: Block diagram of the 10GHz ATV transmitter**

supply voltages were varied in a further attempt to attain a higher saturation power on the individual MMIC's. The optimal setting yielded the following values:

- The driver chip, gate 2 +1.4V, drains 3.0V (190mA);
- The output chip, gate 2 + 1.2V, drains 3.7V (220mA).

The result was an increase in the power of only approximately 20%. The MMIC (IAF-MPA7710) can provide good service in the manufacture of a power amplifier. Better performance could be achieved by using at least two chips operated in parallel. Possible solutions for the addition of the outputs are "magic tee" or a 3/4λ Wilkinson coupler on a quartz substrate (0.127mm!). A project showing these techniques was published in [42].

## ATV TRANSMITTER WITH PLL FOR 10GHZ

The ATV mode is very popular because many ATV repeaters have become active in recent years. This has meant that the 3cm band is becoming more and more important, in addition to links; it is increasingly used for operator input and output. For more on ATV, see the chapter on imaging techniques.

With this in mind, a modern, frequency-stabilised ATV transmitter has been developed by Alexander Meier, DG6RBP, the transmit frequency can be simply and accurately selected at any point on the entire 10GHz amateur radio band. This makes it suitable for use as a transmitter in a repeater or for direct operation.

A transmitter in an ATV repeater, eg as operator output or for a link, is subjected to temperature variations and is in operation for long periods of time. So the transmit frequency should change as little as possible. For direct operation, on the other hand, simple frequency adjustment is important, particularly if it is also used to work with repeaters. The PLL stabilised transmitter has a few advantages here: the transmit frequency can be selected simply and easily anywhere in the 3cm. band, by means of a microcontroller, and frequency deviations are prevented by the Phase Locked Loop (PLL).

The transmitter with a PLL is based on Voltage Controlled Oscillators (VCOs), that have recently become available at reasonable prices up to and beyond 10GHz [43]. You won't find a suitable VCO that will cover the entire 3cm band with an acceptable tuning rate. But there is an option to use a VCO of 5.00 to 5.25GHz then doubling and amplifying the output signal. There is, however, a disadvantage of this approach; compared with free running oscillator or DRO transmitters, the circuit complexity is considerably greater.

**Fig 11.100** shows the block diagram of the project. A VCO of 5.00 to 5.25GHz is stabilised using a PLL and after a passive doubler (x2) covers the entire 3cm band from 10.00 to 10.50GHz. The losses that this generates are balanced out again using two amplifier stages with an output of approximately 1mW at 10GHz. This is sufficient to drive an additional power amplifier, for example a small TWT or a 200mW FET amplifier.

The PLL is controlled by means of an 8051 microcontroller from ATMEL. The transmit frequency can thus be conveniently selected anywhere on the 3cm band via the RS232 interface from a PC or through a separate control module. It is not obligatory to use the interface for fixed frequency operation in a repeater because the transmitter starts at a fixed frequency when it is switched on (pre-programmed in the microcontroller). An LED signals when the PLL is locked. If this is not the case, the microcontroller switches the amplifier stages off and sends an acknowledgment via the RS232 interface.

As an option, the ATV transmitter can also monitor an externally connected power amplifier (PA); if the voltage at the monitor output of the external PA falls below a selected threshold value, the microcontroller reports this via the RS-232 interface. A picture of the completed ATV transmitter is shown in **Fig 11.101**.

### Circuit description

The circuit diagram of the transmitter is shown in **Fig 11.102** and **Table 11.10** shows the parts list.. The VCO, U2, can be tuned at pin 22, using a tuning voltage of between 0 and 10V giving between 5.0 and 5.5GHz at Pin 22. Since only a frequency range of 5.0 to 5.25GHz is required before doubling, a

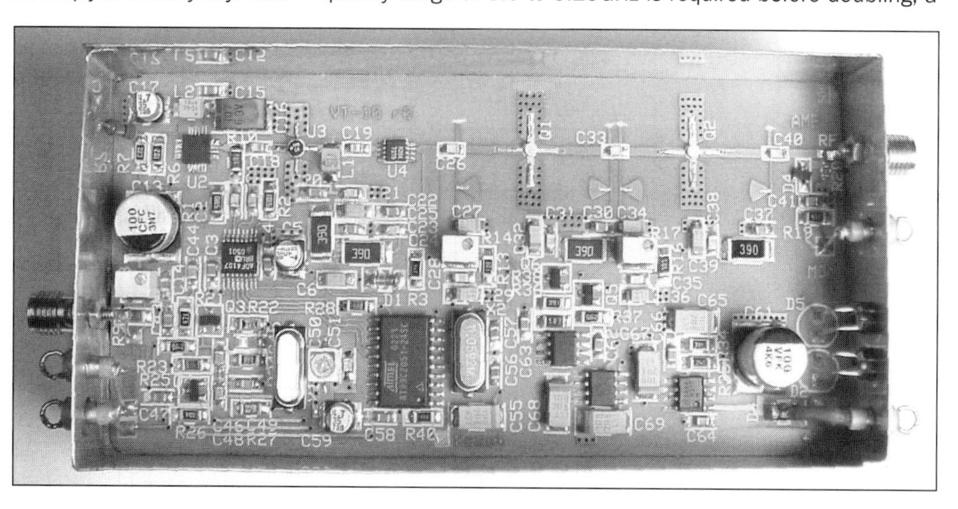

**Fig 11.101: The completed 10GHz ATV transmitter**

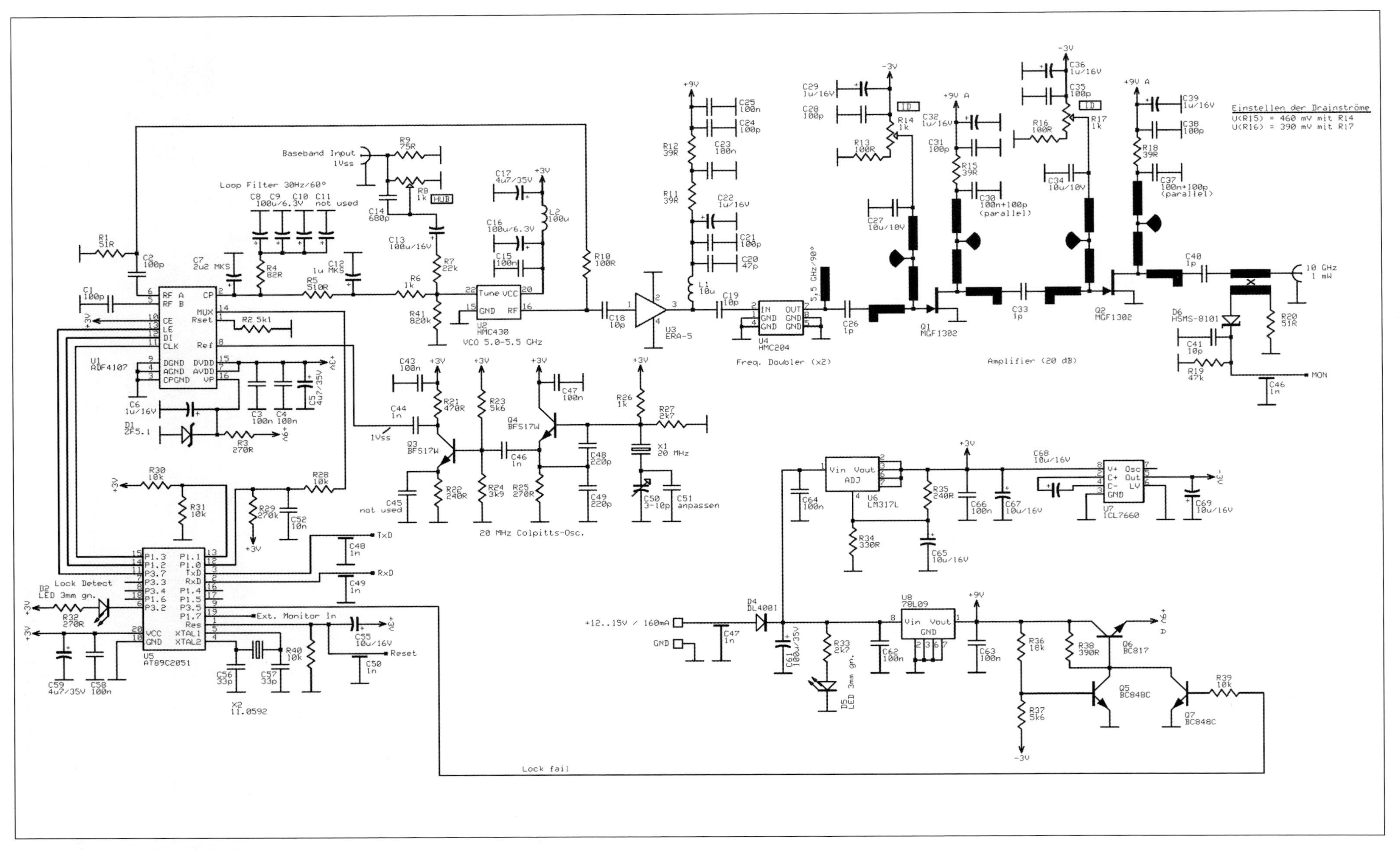


Fig 11.102: Circuit diagram of the 10GHz ATV transmitter

| **Resistors** | | |
|---|---|---|
| R1, R20 | 51R | 1206 |
| R9 | 75R | 1206 |
| R4 | 82R | 1206 |
| R10, R13, R16 | 100R | 1206 |
| R22, R35 | 240R | 1206 |
| R3, R25, R32 | 270R | 1206 |
| R34 | 330R | 1206 |
| R38 | 390R | 1206 |
| R21 | 470R | 1206 |
| R5 5 | 10R | 1206 |
| R6, R26 | Ik | 1206 |
| R27, R33 | 2k7 | 1206 |
| R24 | 3k9 | 1206 |
| R2 | 5k1 | 1206 |
| R23, R37 | 5k6 | 1206 |
| R28, R30, R31, R39, | | |
| R40 | 10k | 1206 |
| R36 | 18k | 1206 |
| R7 | 22k | 1206 |
| R19 | 47k | 1206 |
| R29 | 270k | 1206 |
| R41 | 820k | 1/4W |
| R11, R12, R15, | | |
| R18 | 39R | 2510 |
| R8, R14, R17 | 1k | SMD Trimmer |
| | | |
| **Capacitors** | | |
| C26, C33, C40 | IpF | 0805 |
| C18, C19, C41 | 10pF | 0805 |
| C51 | 18pF | 0805 |
| C56, C57 | 33pF | 0805 |
| C20 | 47pF | 0805 |
| C1, C2, C21, C24, | | |
| C28, C30, C31, | | |
| C35, C37, C38 | 100pF | 0805 |
| C48, C49 | 220pF | 0805 |
| C14 | 680pF | 0805 |
| C44, C46 | 1 nF | 0805 |
| C52 | 10nF | 0805 |
| C3, C4, C15, C23, | | |
| C25, C30, C37, C43, | | |
| C47, C58, C62, C63, | | |
| C64, C66 | 100nF | 0805 |

| | | |
|---|---|---|
| C6, C22, C29, C32, | | |
| C36, C39 | IμF/16V | SMD Tantalum |
| C12 | 1 μF | MKS-2 |
| C7 | 2.2μF | MKS-2 |
| C5, C17, C59 | 4.7μF/35V | SMD Electrolytic |
| C27, C34 | 10μF/10V | SMD Tantalum |
| C55, C65, C67-C69 | 10μF/16V | SMD Tantalum |
| C8-C10, C16 | 100μF/6.3V | SMD Tantalum |
| C13 | 100μF/16V | SMD Electrolytic |
| C61 | 100μF35V | SMD Electrolytic |
| C50 | C-Trimmer 3 - 10pF | SMD |
| C46-C50 | 1nF | Feedthrough |
| | | |
| **Inductors** | | |
| L1 | 10μH SIMID | 1210 |
| L2 | 100μH, SIMID | 1210 |
| | | |
| **Crystals** | | |
| XI | 20MHz, | SMD CL = 16pF |
| X2 | 11.0592MHz | SMD |
| | | |
| **Semiconductors** | | |
| D1 | ZF5.1 | |
| D4 | DL4001 | |
| D2, D5 | 3mm green LED | low current |
| D6 | HSMS-8101 | |
| QI, Q2 | MGF 1302 | |
| Q3, Q4 | BFS 17W | |
| Q5, Q7 | BC 848C | |
| Q6 | BC 817 | |
| | | |
| U1 | ADF4107 BRU | |
| U2 | HMC 430 LP4 | |
| U3 | ERA-5 | |
| U4 | HMC 204 MS8G | |
| U5 | AT 89C2051 | SMD, programmed |
| U6 | LM 317L | SMD |
| U7 | ICL 7660 | SMD |
| U8 | 78L05 | SMD |
| | | |
| **Hardware** | | |
| 1 x | tinplate housing | 111 x 55 x 30mm |
| 1 x | PCB DG RBP-VT 10 | |
| 2 x | SMA socket | |

**Table 11.10: Components list for the 10GHz ATV transmitter**

tuning voltage of between 0 and 5V is satisfactory. The operating voltage is fed to the VCO via a low pass filter (L2 and C15-C17) in order to suppress noise and other residual frequencies in the supply voltage.

The VCO output is approximately 1mW, one part of this is fed to the PLL via the resistor, R10. The majority goes to the amplifier, U3. This gain block increases the power to a few mW to feed the passive doubler, U4. A stub cable at the output of the doubler suppresses the first harmonic at 5GHz. The second harmonic at 10GHz (approximately -20dBm) corresponds to the desired frequency and is amplified, using the Mitsubishi FETs Q1 and Q2, to an output of approximately 0dBm (1mW). The quiescent currents of the amplifier are set by means of the trimmer potentiometers, R14 and R17. If required, the printed circuit board provides the option of a monitor circuit with a directional coupler and a Schottky diode (R19, R20, C41, C46, D4).

The PLL module (U1) from Analog Devices has the output frequency of the VCO (5.00 - 5.25GHz) fed into it via pin 6, and the reference frequency (20MHz) via pin 8. The phase comparison is carried out at 100kHz, so that there is a more than satisfactorily small step size of 0.2MHz in the 10GHz band. The loop filter of the PLL (C7, R4, C8 to C10, R5, C12) has been designed for the transmission of video signals. The loop bandwidth is 30Hz. The low frequency of the phase comparison makes it possible to use practical component values. High-quality foil capacitors must be used for C7 and C12, and good tantalum capacitors for C8 to C10.

The tuning voltage is combined with the modulation signal (base band) via R6 and R7 before the VCO. The trimmer, R8, serves to set the frequency deviation. The capacitor, C14, is used to provide for a slight increase in the high modulation frequencies.

A simple quartz oscillator in a Colpitts circuit (Q4) is used as a reference oscillator. The frequency can be varied somewhat using the trimmer capacitor, C50. A subsequent amplifier stage (Q3) serves mainly as a buffer. If required, the capacitor, C45, could be used to increase the AC voltage amplification of the stage.

The PLL's programming is carried out by an 8051 microcontroller (U5) from ATMEL via pins 11, 14 and 15. The controller operates with a quartz crystal frequency of 11.0592MHz to generate a baud rate of 9600 Baud. The lock-detect signal of the PLL is compared with a pre-set voltage, using an integrated comparator (Pin 12, 13). If the current at pin 12 exceeds the reference voltage of 2.5V, the controller recognises that the PLL has locked and switches the amplifier stages on via pin 9. In addition, an LED signals to pin 6 that the PLL has locked.

For the optional supervision of an external output amplifier stage, the corresponding monitor signal is compared using a comparator (external circuit as and when required). The output signal of the comparator is then fed to the microcontroller via pin 19. A voltage level of 0V means the PA has failed, while a level of 5V means it is operating. The status of pin 19 is continuously monitored in the controller, and should any change occur a message is transmitted via the RS232 interface.

The supply voltage for the module (12 to 15 Volts) is fed in through a reverse battery protection diode (D4). The fixed voltage regulator (U8) uses it to generate the operating voltage for the amplifier stages (9V). This is applied, through the protective circuit of Q5-Q7, only if the PLL is locked and the negative gate voltage is correct.

An easily adjustable voltage regulator, type LM 317 (U6) is used for the power supply for both the PLL and the VCO (3V). The charge pump (U7) uses this to generate a negative voltage for the gates of the amplifier stages.

## Assembly and tuning

The entire circuit is built on a 109 mm x 54 mm x 0.51 mm printed circuit board. The inexpensive RO-40003C from Rogers is used as the printed circuit board material. To make assembly easier, particularly for the VCO (U2) and the doubler (U4), the printed circuit board is through-hole plated and coated with a solder resist.

After a coating of solder paste has been applied, the VCO is melted on, using a reflow soldering unit or a hot air station. This component cannot be mounted using a soldering iron because of the VCO package connections are on the underside. This is the same for the doubler (U4), it has an earth side on its underside, which must be soldered correctly.

Finally, all other components apart from Q6 are mounted, with no particular instructions. A soldering iron can be used here. Good eyes and steady hands are required for the PLL (U1) when the components are mounted by hand. The wired resistor, R41 is soldered 'quick as a flash' by hand, depending on the earth (eg housing wall). The PCB layout is shown in **Figs 11.103** and **11.104** and the component layout is in **Figs 11.105 and 11.106** (all four of these drawings are in Appendix B).

Prepare the tinplate housing by drilling the appropriate holes. Clean the flux residue off of the printed circuit board and fit it into the housing, soldering it all around the underside. Then the printed circuit board is cleaned again.

Once the printed circuit board has been connected to the feedthrough capacitors, the first function test can be carried out. When the supply voltage has been applied, the voltage on the gates of, Q1 and Q2, is initially set to -3V, using the trimmers, R14 and R17. Switch off and fit transistor Q6. Switch on again and check the important parts of the circuit. The most important elements are the operating voltages (3V, -3V, 9V, 5.1V at pin 16 of U1) and the 20MHz quartz oscillator. Next the tuning begins.

When the supply voltage is applied, the module starts at the default, pre-programmed, frequency. Now the quiescent currents of the amplifier stages are set. First, the voltage is measured across resistor R15 and adjusted to 460mV using the trimmer R14. The voltage across R18 is set to 390mV using R17. A spectrum analyser is used to check whether the PLL has locked correctly. The output will still be somewhat below 1mW. Finally, the module is connected to a PC via the RS232 interface and the frequency is altered from 10.000 to 10.500GHz in steps of 50MHz and check that the PLL has locked correctly. The stubs at the input and output of the transistors, Q1 and Q2, are shortened slightly with a scalpel to find the maximum output in the middle of the frequency range or at any preferred frequency. The frequency shift of the transmitter is set to ±3.5MHz (as per DL2CH ATV standard [44]) using a spectrum analyser and the carrier minimum method. A 15MHz signal is applied at the base band input, with an amplitude of 1Vpp. Increase the shift until the first time that the carrier frequency reaches a minimum. The frequency deviation can also be set to a higher value if required.

After a final function test with a test image and the transmitter is ready for operation.

## Programming

The programming of the default frequency of the module, ie the frequency at which the module starts when the supply voltage is applied, is pre-set once and for all in the software of the microcontroller (U5). If the transmitter were used in an ATV repeater, this would be the output frequency. The transmitter could then be used without utilising the RS232 interface.

Otherwise, the output frequency can be changed at any time via the RS232 interface in steps of 1MHz. The PLL would allow for smaller steps, but these are not expedient for ATV applications. To set the frequency, we can use a PC with a USB programming adapter [45], another microcontroller with an RS232 interface, or a frequency input module developed individually for the transmitter [46]. Direct connection to the RS232 interface of a PC is possible, using a level converter (0/5V, e.g. MAX232).

**Table 11.11** lists the interface commands. Thus, when the supply voltage is applied, the transmitter transmits a 'P' for 'Power On'. As soon as the PLL is locked, the transmitter reports this with an 'L'. If we now wish, for example, to set the frequency to 10.450MHz, we transmit the character string 'F10450' to the transmitter. The new frequency is confirmed and selected using 'OK'. As the PLL is not locked during the frequency change, the transmitter reports this with 'U'. As soon as the PLL is locked, we receive an 'L' as a response.

| | | |
|---|---|---|
| Controller to ATV transmitter | F10xxx | New transmit frequency 10xxx = frequency in MHz |
| Controller to ATV transmitter | S | Request transmitter status |
| ATV transmitter to Controller | P | Power on |
| ATV transmitter to Controller | L | PLL locked |
| ATV transmitter to Controller | U | PLL un-locked |
| ATV transmitter to Controller | A0 | External monitor signal fail |
| ATV transmitter to Controller | A1 | External monitor signal OK |
| ATV transmitter to Controller | E | Error |
| ATV transmitter to Controller | ERR | Error |
| ATV transmitter to Controller | OK | New frequency confirmed |
| ATV transmitter to Controller | F10xxxyAz | Transmitter status 10xxx = frequency in MHz; y = L,U (locked, un-locked); z = 0,1 (external monitor 0,1) |

**Table 11.11: Programming control by the RS-232 port on the ATV transmitter**

**Table 11.12: Technical data for ATV transmitter**

| | |
|---|---|
| Frequency range | 10.000 to 10.500GHz |
| Steps | 1MHz |
| Frequency stabilisation | PLL |
| Long term stability | ±5ppm / year |
| Temperature stability | ±50ppm (+10 - +40°C) |
| Modulation input | Baseband I$V_{pp}$ (positive input gives positive modulation) |
| Modulation type | FM |
| Modulation depth | ±3.5MHz for I $V_{pp}$ (internally adjusted with reserve) |
| Modulation freq. range | 30Hz to >7MHz |
| Interface | RS-232, 0/3V signalling, 9600 baud, 8 data bits, 1 stop bit, no parity |
| Interface cable length | 20m |
| RF output impedance | 50 ohms |
| Typical output | 1mW / 0dBm (±3dB) |
| Typical harmonic output | <-40dBc (5 - 5.25GHz), <-40dBc (20 - 21GHz) |
| Other harmonics | <-50dBc |
| Supply voltage | 12 to 15V DC |
| Supply current | Approximately 160mA |
| Temperature range | +10 to +40°C |
| Connectors | SMA socket for RF output and baseband input. Feedthrough capacitors for RS-232 |
| Indicators | Power - green LED, Locked - green LED |
| Case | Tinplate housing |
| Size | 111 x 55 x 30mm approximately |

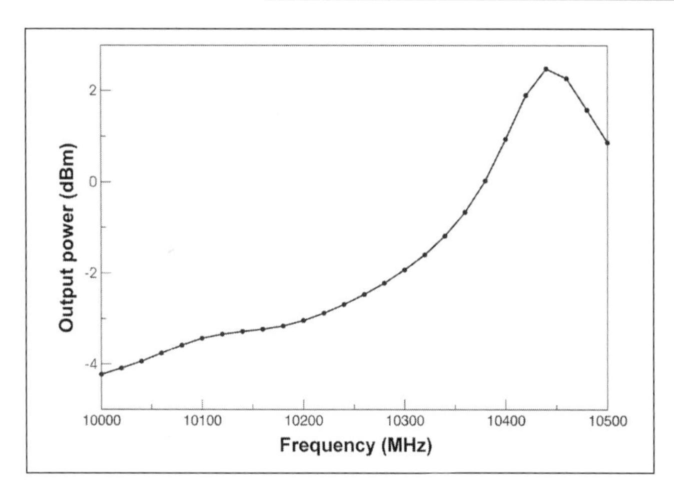


**Fig 11.107: Graph of output power against frequency for the ATV transmitter**

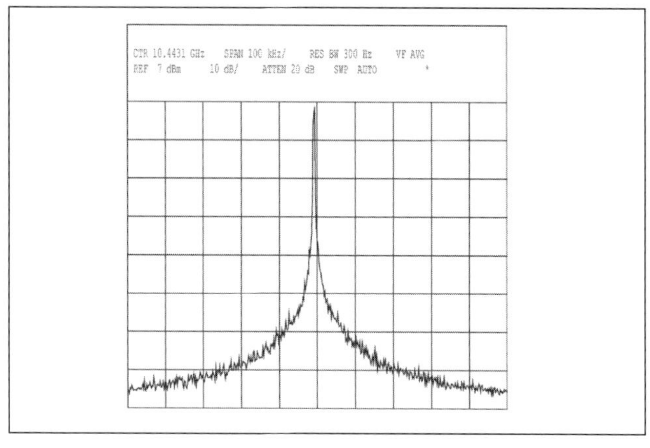

**Fig 11.108: Output spectrum of the ATV Tx with no modulation**

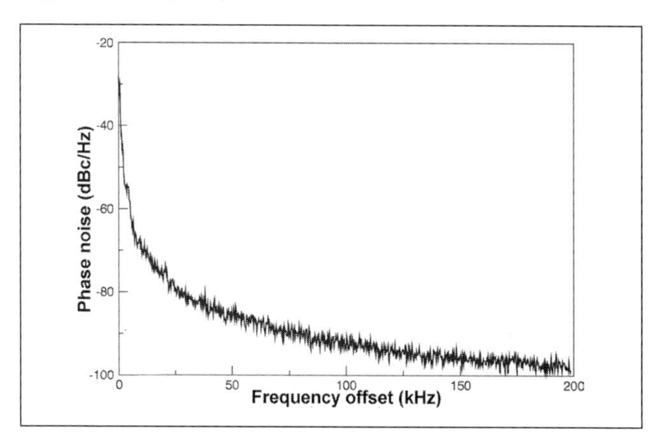


**Fig 11.109: Phase noise performance of the ATV transmitter**

We can check the transmitter status by sending the character 'S' to the transmitter. For example, if the transmitter answers 'F10200LA1', this means: frequency 10200MHz, PLL locked (L) and monitor signal for external amplifier stage OK (A1).

The technical data of the ATV transmitter are summarised in **Table 11.12**. As the transmitter modulates the output frequency towards higher frequencies for positive voltages on the base band input, the video signal must be inverted before being applied to the base band input to obtain the inverted image usually required for ATV repeater outputs.

The output power at different frequencies can be seen in **Fig 11.107**. The frequency of the maximum power is determined during tuning. Tuning was carried out here at 10.440GHz. **Fig 11.108** shows the output spectrum without modulation and the phase noise of the transmitter is shown in **Fig 11.109**.

## TRANSVERTERS

The most popular method of becoming active on the microwave bands today is to use a commercial transceiver and a transverter. The transverter performs two functions, it down-converts the incoming signal, on the microwave band being used, to the chosen input frequency of the commercial transceiver (tuneable IF) and up-converts the output of the transceiver to the microwave band. The most common tuneable IF is 144MHz; using lower frequencies makes it difficult to filter out the unwanted signals. The transverter will use a common local oscillator chain so that transmit and receive frequencies on the microwave band will be the same, making it possible to use narrow band modes such as SSB.

Designs of transistorised transverters for 23cm started to appear in the mid 1970s. As semiconductor technology improved, designs for the higher amateur bands became available. Most of these were quite tricky to build and persuade to work properly. In the mid 1980s 'No Tune' transverter designs started to appear. These overcame many of the construction and set-up problems. Some of the amateurs who designed these transverters now produce them as kits and ready built units, these are all listed in the bibliography. Sam Jewell has reviewed the DB6NT 23cm and 3cm transverters.

Amateurs have always enjoyed modifying commercial equipment to work on the amateur bands. This is now possible for most of the microwave bands because there is a lot of surplus microwave link equipment on the market. The problem with this route is to recognise the potential of the equipment that we all see at rallies and equipment sales. The Internet is a wonderful source of knowledge about this, including pictures of the equipment and modification details, one such modification is shown later in this chapter.

## Kuhne Electronics MKU13 G3 1.3GHz Transverter Review by Sam Jewell, G4DDK

DB6NT, in conjunction with DF9LN, published a 1.3GHz transverter design in the European magazine *Dubus* [47] 4/1992. This design was very popular and when DB6NT formed Kuhne Electronics the design was made available commercially as the MKU 13 G. It was later upgraded to the G2. The obsolescence of several key parts, together with the usual commercial imperative to update designs, has resulted in the release of the new G3 transverter. An external photograph of the new G3 is shown in **Fig 11.110**.

Although the G2 was sold as both a kit and a ready-built module, the G3 is only being made available in ready-built form. Many of the small surface mount components used in the G3 now make it impractical for most home constructors to build.

The most significant features of the new transverter are more output power and the addition of an external 10MHz local oscillator locking input. This latter feature will be much appreciated by those who fought the thermally-induced frequency drift that the older G2 was noted for. It is not necessary to use the locking facility if it is not required. **Fig 11.111** shows the top side of the PCB with the frequency locking board, helical filters and gain setting controls.

### Circuit description

In common with the previous design from DB6NT, the G3 uses

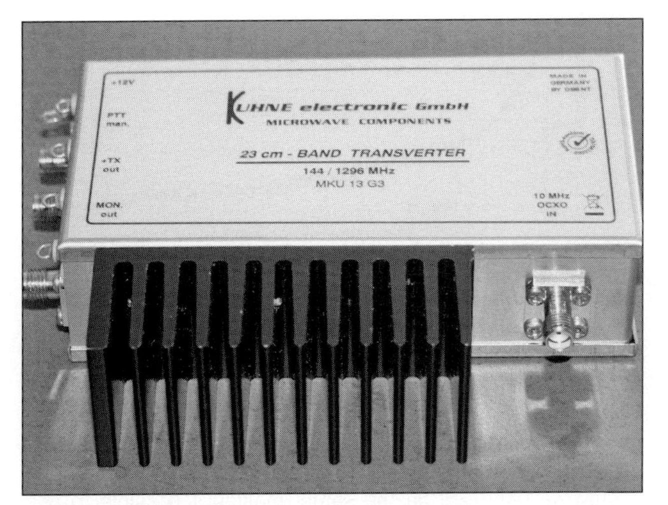


Fig 11.110: Outside of the G3 1.3GHz transverter

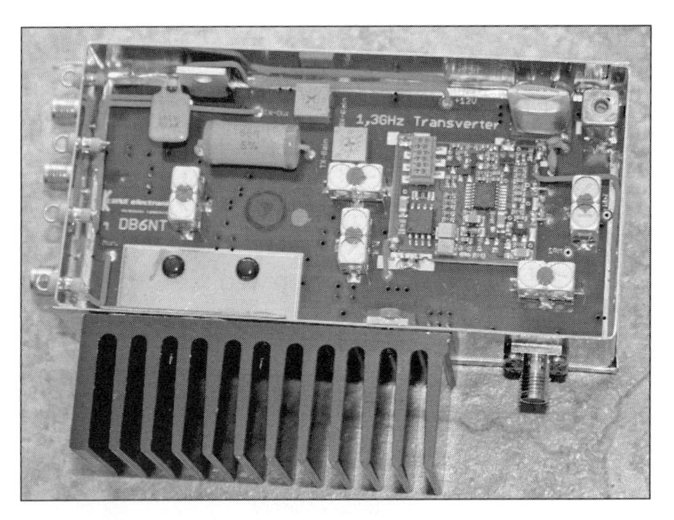


**Fig 11.111: Top side of the G3 unit. The frequency locking board, helical filter, crystal (with heater) and the 50-ohm IF load can all be seen**

GaAsFETs in many of the critical RF stages. An NE32584 GaAsFET low noise amplifier (LNA) provides a claimed noise figure of under 1dB. This is followed by a two-pole helical filter into an integrated circuit gain block ERA3. A second helical filter feeds into a high level (+17dBm) ADE 17 diode ring mixer. Following the 144MHz diplexer a miniature relay switches the receiver IF output via an adjustable attenuator to the common IF connector. The gain is adjustable over a large range to allow optimisation of the receive system according to cable lengths etc.

DB6NT told me, at Weinheim, that they had decided to revert to using miniature relays rather than PIN diodes because of greater isolation and reliability at the higher 144MHz IF transmit level that can now be used. This has become necessary because many newer VHF multimode rigs cannot be adjusted to below 5W RF output at 144MHz.

On transmit the 144MHz IF signal is connected, via the relay contacts, to the mixer. A transmit variable attenuator allows the transmit drive to the mixer to be set to the correct level so as not to over drive the mixer (a common problem).

The wanted 1296MHz product, at the mixer output, is selected by a helical filter and then PIN diode switched to the first amplifier stage, again filtered and then amplified to 3.4W output. This is a linear process, up to the point where the transmit chain starts to compress, and therefore is suitable for use with all modes including SSB, CW, digital (eg JT65C) and FM.

The G3 uses a Watkins Johnson FP31 HFET [48], rated at 2W for 1dB compression, in the transmitter output. This device is in a plastic 28 pin QFN package and this must be one of the reasons why Kuhne Electronics believe it is no longer practical for most radio amateurs to solder assemble units like the G3. **Fig 11.112** shows the component side of the G3 PCB. The FP31 can be seen clearly.

The transmitter and receiver chains are separate and unless separate transmit and receive antennas are used, an external antenna changeover relay must be used with a single antenna or array system. Most operators will want to use an external masthead preamplifier on receive and an external high power amplifier on transmit. If this is the case then the antenna changeover relay must be situated between these components and the antenna. It is also desirable to use an external low pass filter to reduce harmonics from the transmitter and this should be situated after the power amplifier.

Changeover, from receive to transmit, can be initiated by either applying a ground connection to the manual PTT input or,

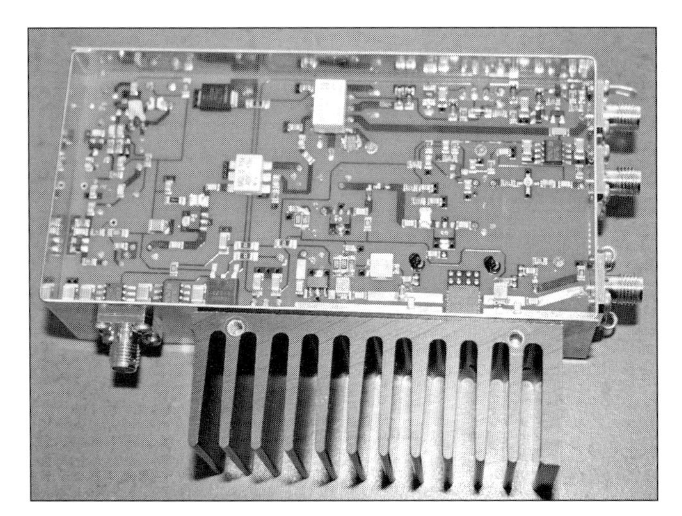

**Fig 11.112: Surface mount component side of the PCB. The mixer, IF relay and FP31 output are readily identified**

with a suitable IF rig, applying a small positive voltage on the IF coaxial cable connection. Details of where to find how to modify a number of popular rigs are given in the accompanying instruction sheet.

### Frequency locking

Transverter local oscillators usually consist of an overtone crystal oscillator followed by a chain of frequency multipliers to the required injection frequency. Whilst many overtone oscillators are capable of better than 1kHz accuracy (at the final frequency) over a temperature range of around 0°C to +40°C, this is no longer regarded as adequate with the growing use of digital modes like JT65C and the desire to monitor the transmission from amateur beacons to an accuracy adequate to determine the amount of Doppler shift being introduced by the propagation path.

There are several ways to achieve the desired accuracy, including locking to GPS, an Oven Controlled Crystal Oscillator (OCXO) or a high stability atomic source (such as Rubidium) as the reference. Possibly the most popular way is to use what is known as a Reflock [49] board to lock the existing on-board overtone oscillator. This method is capable of giving very low phase noise because it uses the inherently high Q overtone crystal oscillator rather than a low Q voltage controlled oscillator (VCO).

An external, high stability, low phase noise, 10MHz signal connected to the Reference input port will cause the in-built 96MHz crystal oscillator to phase lock to the 10MHz reference. This is subsequently multiplied to the required 1152MHz local oscillator injection frequency.

The technique used is limited by the very small frequency range over which an overtone oscillator can be pulled. If the oscillator drifts more than a few tens of Hz from the nominal frequency, the Reflock cannot maintain control. This drift may be caused by local heating of the crystal, for example when the transverter is in transmit mode, and the output amplifier is therefore generating considerable local heat inside the transverter box. To reduce this effect Kuhne Electronic has also used one of their small, proportional control, crystal heater boards attached to the crystal by a short length of heat shrink tubing. This is most effective.

Kuhne Electronics have used a National Semiconductor LMX2306 synthesiser chip in a conventional Phase Locked Loop (PLL) circuit to control the 96MHz crystal oscillator. The external 10MHz input acts as the reference frequency for the PLL. An Atmel eight-bit microcontroller ensures that the synthesiser is loaded with the correct frequency data each time the G3 is powered up.

### Construction and interfaces

The transverter is built on a double sided PCB contained in folded metal box plated with what Kuhne Electronics describes as German Silver (typically 12.5% nickel, 50-65% copper and the balance zinc). This should not rust like the more common tin plate boxes, but it is more expensive to produce.

A small heat sink is attached to the side of the case and heat from the power amplifier device is conducted to the heat sink through an L shaped bracket. I do wonder if the copper sheet bracket will prove adequate in conducting away the heat. To be fair, the instructions do specify that the small heat sink must be attached to a larger heat sink. The heat sink is suitably drilled and tapped and two screws are provided to facilitate this.

SMA connectors are provided for the 144MHz IF, 1.3GHz transmitter output, 1.3GHz receiver input and the 10MHz external oscillator input. The power supply connection is via a feed-through capacitor. Feed-through capacitors are also provided to bring out an active connection on transmit, to drive the antenna changeover relay, a DC output representing the RF level and a manual press to talk (PTT) input.

### On-air impressions

I was able to test the G3 in the heat of battle during the October 2007 IARU multiband contest. Used with my FT-817 as the IF and my two 44-element Wimo Yagis at 10m, this was an ideal opportunity to see how the transverter would perform with a mixture of strong local signals and a number of much weaker DX signals. I used an FXLabs 1500MHz low pass filter and CX540 antenna relay. With this arrangement I was not going to be a very big signal, so I allowed myself the luxury of using the ON4KST chat system to advertise my presence.

During the 1.5 hours I operated with the G3 I worked seven stations on the 1.3GHz band. The best DX was G3CKR/P at 252km and P14Z at 186km. Reports from local stations were complimentary when I told them what I was using.

Conditions were pretty average, but a good number of German and Dutch stations were heard, although with just 3W it was difficult to attract attention. The G3 acquitted itself well in the contest and the few receive problems I experienced were probably due to the performance of the FT-817. With about 20dB gain from the receive converter in front of the FT-817 this is not too surprising with strong local contest stations on the band. To get the best from the G3 a high performance 144MHz IF receiver must be used.

### Comments

The G3 is a well constructed unit and should provide many years of reliable service. It meets or exceeds its claimed performance, allowing for the accuracy of the measurements.

One side effect of the use of the HFET output device, with its very high gain at high frequencies, is the relatively high level of harmonic output. This transverter should not be used directly into an antenna without a transmitter low pass filter to reduce the level of the 2nd and higher harmonics. Mention is made of this in the accompanying instruction sheet.

Now that transverters have a microcontroller inside I wonder how long it will be before we can expect a USB socket on the side of the unit, and the ability not only to interrogate the transverter for operating conditions and failures but to be able to programme operating parameters such as LO offsets and system gains?

My thanks to Kuhne Electronics for the loan of the test unit.

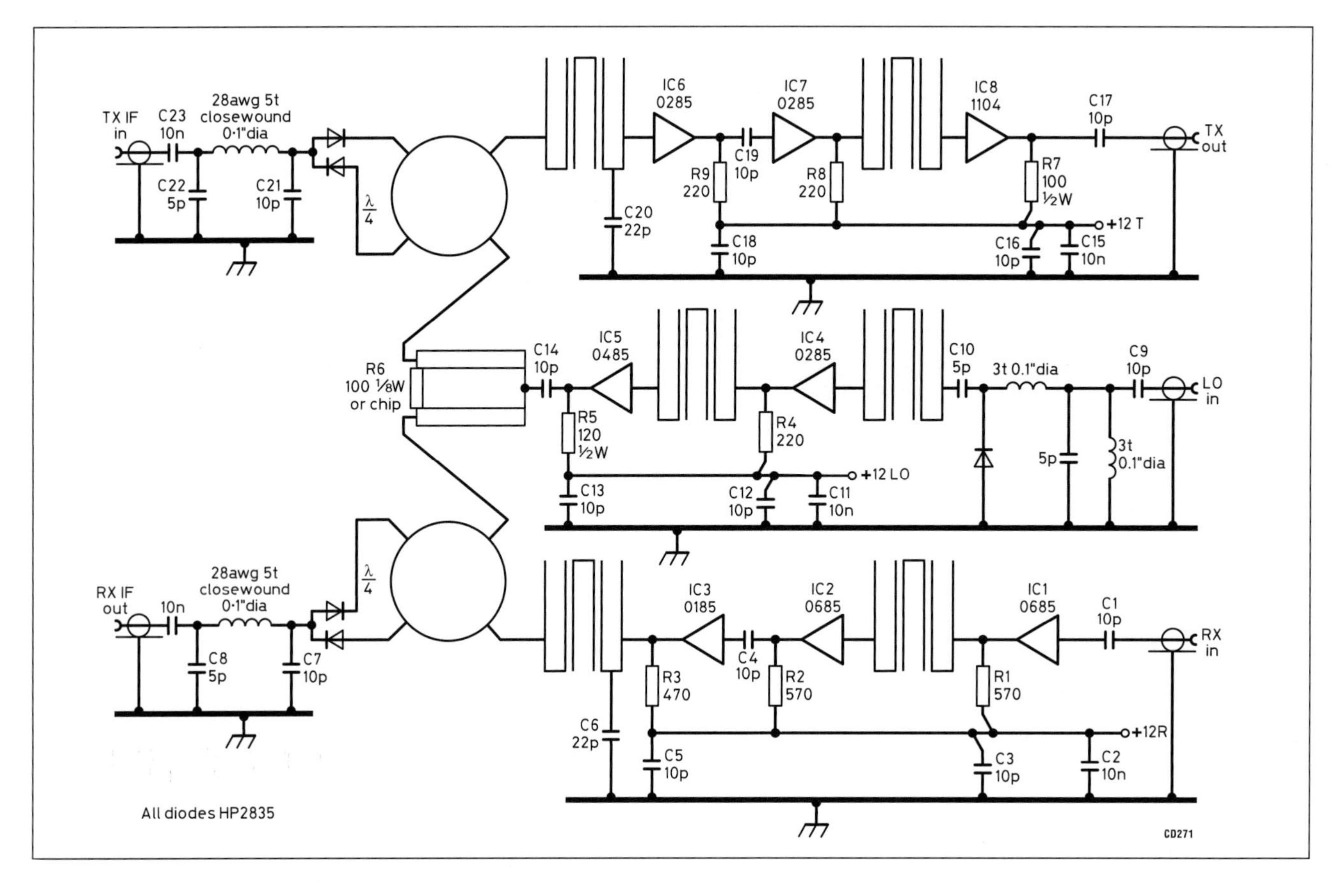


Fig 11.113: Circuit diagram of the KK7B transverter. BA481 Schottky diodes can be used instead of the HP5082-2835 diodes. Other, more modern MMICs can be substituted for the specified types, provided that the supply resistors are calculated and adjusted in value to suit the devices chosen. See reference [52]

## A Single-Board, No-tune 144MHz/1296MHz Transverter

The comparatively recent development of economically priced and readily available microwave monolithic integrated circuits (MMICs) has allowed the development of a number of broadband (no-tune) low-power transverters from the 144MHz amateur band to the lower microwave bands, typically 1.3, 2.3 and 3.4GHz. Such designs use microstrip technology, including no-tune inter-stage band pass filtering in the LO, receive and transmit chains.

A 144MHz to 1296MHz transverter circuit was described by KK7B in [50]. This circuit and layout, although it does not give 'ultimate' performance in terms of either receive noise figure or transmit output power (nor is it particularly compact in terms of board size), is probably one of the simplest and most cost-effective designs available. Its simplicity also makes it suitable for novice constructors. It is also flexible enough to allow the constructor to substitute new, improved MMICs as these become available, without major re-engineering.

The receive performance can be enhanced by means of an external (possibly mast-head) low-noise amplifier (LNA), such as a high-performance GaAsFET or PHEMT design. The transmit output level, at +13dBm (20mW), is ideal for driving a linear PA module such as the G4DDK-002 design [51]. This, in turn, could drive either a solid-state power block amplifier or a valve linear amplifier.

Precision printed circuit boards for this design and a similar design for the 2.3GHz band [52] have been available for some time, produced and marketed by Down East Microwave [53] in the USA. They are also available from a number of sources in the UK. A similar design concept was adopted for a transverter for the US 900MHz amateur band [54]. The Down East Microwave website [53] is worth visiting, it has a number of useful designs and application notes that can be freely downloaded

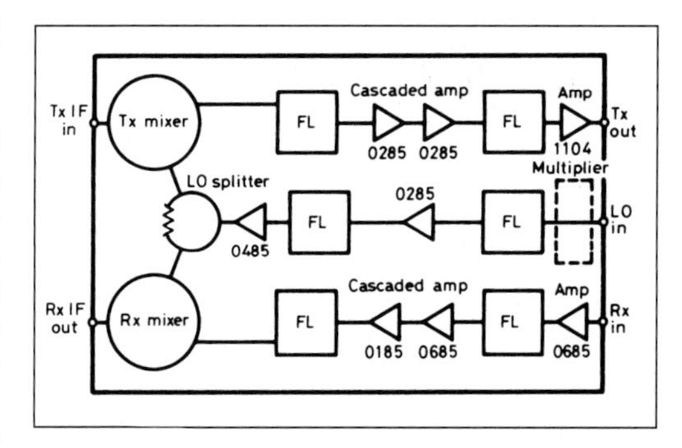


Fig 11.114: Layout for the KK7B transverter

The original circuit diagram of the 1296MHz version is given in **Fig 11.113** and the physical layout of the circuit is shown, not to scale, in **Fig 11.114**. Wide use is made of hairpin-shaped, self resonant, printed microstripline filters in the LO, receive (RX) and transmit (TX) chains, together with printed microstripline transmit and receive balanced mixers and 3dB power splitter for the LO chain. An external LO source at any sub-harmonic frequency of the required injection frequency (1152MHz for the 1296-1298MHz narrow-band communications segment of the 23cm band when using an IF of 144-146MHz) was used and a simple, on-board diode multiplier produced the required injection frequency from the LO input. Although direct injection of 1152MHz was mentioned in the original description, little guidance was given as to how to achieve this. With a simple modification to the LO chain and a few changes to circuit values and devices, without need for PCB changes, it is easily possible to use the G4DDK-001 1152MHz

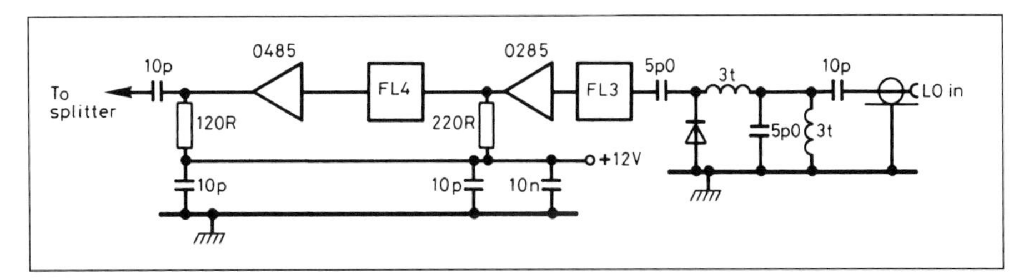


Fig 11.115: Original KK7B LO circuit used with his no-tune transverter

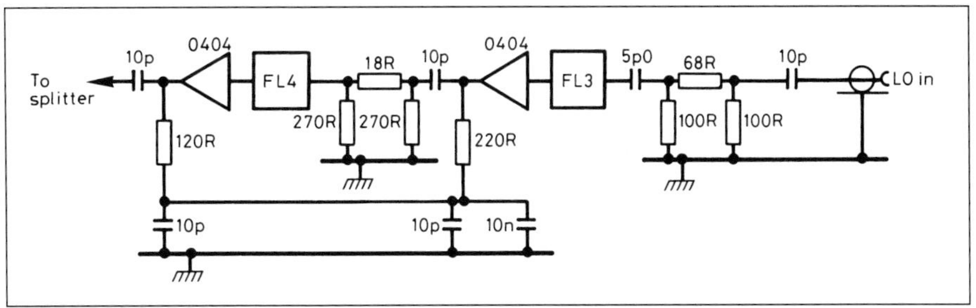


Fig 11.116: Modification to the KK7B LO circuit to allow use of the G4DDK-001 1152MHz source and the higher output level available from this circuit

source, already described, as the LO for this design.

**Fig 11.115** gives the circuit and component values for the original LO chain, while **Fig 11.116** shows the modifications to allow the correct mixer injection levels to be attained when using the single +13dBm output option of the G4DDK-001 1152MHz source. **Fig 11.117** shows the layout of the modified circuit using the existing PCB pads and tracks.

Construction is straightforward, using surface-mount techniques, ie all components, whether SMD or conventional, are mounted on the track side of the board unlike conventional construction. All non-semiconductor components - connectors, resistors, capacitors and inductors - should be mounted first, the MMICs and mixer diodes last, taking adequate precautions to avoid both heat and static damage. Note that the values of the bias resistors, which set the working points of the MMICs, were chosen for a supply rail of +12V DC. Higher supply voltages will require recalculation of these values and the constructor should refer to either the maker's data sheets for the particular devices used or to the more general information given in reference [55]. When construction is complete and the circuit checked out for correct values and placing of components, there is no alignment as such, assuming that the LO source has already been aligned! It should simply be a matter of connecting the transverter to a suitable 144MHz (multimode) transceiver via a suitable attenuator and switching interface such as that by G3SEK [56].

## Kuhne Electronics MKU10 G3 10GHz Transverter Review by Sam Jewell, G4DDK

Kuhne Electronic GmbH (DB6NT) produces a range of transverters, frequency converters, preamplifiers, power amplifiers and accessories for the amateur radio market. This is a review of the new MKU 10 G3 10GHz transverter. The 23cm MKU13 G3 1296MHz transverter, reviewed above has many circuit block similarities, so some of the following description may sound familiar if you read the MKU 13 G3 review above.

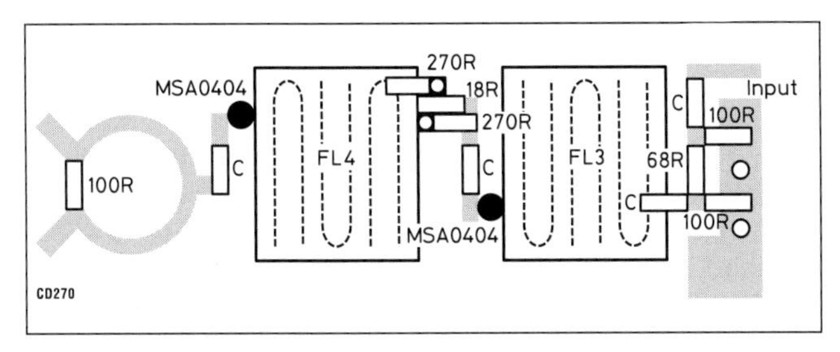


Fig 11.117: Layout of components for the modified LO chain in the KK7B transverter

### New features

Possibly the most significant feature of the new transverter is the addition of an external 10MHz local oscillator reference input. With the transverter local oscillator locked to a high quality 10MHz reference source, frequency stability is exceptional and an accuracy of better than 10Hz becomes practical on the 10GHz amateur band.

**Fig 11.118** shows the top side of the PCB with the frequency locking board, helical filters and gain setting controls. The other side, with all the surface mount components, is shown in Fig **11.119**.

### Construction

The transverter is constructed on a double sided printed circuit board soldered into a folded metal box. The main PCB carries surface mount components (SMD) on one side of the board and larger, leaded, parts such as the regulators and helical filters on the top side. A small daughter board, with SMD parts, carries the reference frequency locking circuit. A second daughter board carries the crystal heater controller. GaAsFETs are used in many of the critical RF stages. The remainder of the stages use bipolar transistors.

The RF inputs and outputs all use SMA connectors. Power supplies, Press to Talk (PTT), TX supply and Monitor outputs connect via feed through capacitors. Press fit top and bottom box lids allow access to both sides of the main PCB. The lower lid carries a piece of electric field absorber foam to aid stability. Two short lengths of similar absorber foam are glued to the SMD

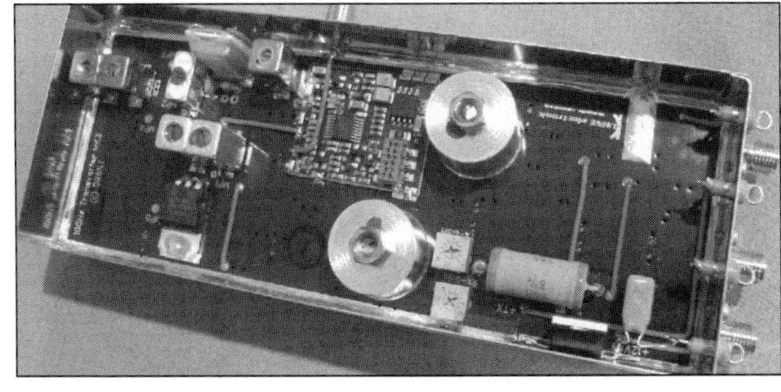

Fig 11.118: The top side of the G3 10GHz transverter, showing the frequency locking board, helical filters, crystal (with heater) and 50 ohm IF load

side of the main PCB in the vicinity of the transmitter and receiver amplifier chains.

## Receive path

An NEC NE32584 GaAsFET front end provides a noise figure of under 1.5dB. This is followed by two further GaAsFET amplifier stages using quarter wave length coupled microstrip capacitors. A microstrip Wilkinson splitter (combiner) separates the transmitter and receiver signal path. A substantial silver plated cavity filter is located between the splitter and the dual diode balanced mixer. This filter provides the sole RF selectivity on receive. A single mixer is used for both transmit and receive and is followed by a miniature relay that switches the mixer IF output to an adjustable attenuator (RX Gain) which is used to set the receive converter gain.

## Transmit path

On transmit, the relay switches the 144MHz IF to a power attenuator consisting of a 5W (+37dBm) rated resistive load in a Pi configuration, with a series resistor and a variable resistor (TX Gain) to allow a range of adjustment. It is compatible with many of the modern 144MHz transceivers that cannot be adjusted to below 5W RF output.

With the TX Gain control set to give maximum 10GHz output (saturation), with 5W of 144MHz transmit input, the 10GHz power output was slightly in excess of the claimed 180mW output.

Following the mixer, the wanted 10GHz transmit product is filtered in the cavity filter, separated from the receive path in the Wilkinson divider and then amplified in three GaAsFET stages to around 280mW (+25dBm), at saturation output. There is a second cavity filter located between the first and second stage transmit amplifier to improve spectral purity. As the spectrum plot of **Fig 11.120** shows, this is just adequate to reduce the local oscillator feed-through and the image frequency to acceptable levels, although an extra stage of filtering would be desirable.

Both the receiver and transmitter chains are linear (up to the point where they start to compress) and therefore the transverter is suitable for use with all modulation modes including SSB, CW and FM. Most of the GaAsFET amplifier stages use active bias to maintain the bias operating point over temperature and device spread.

The transmitter and receiver paths are separate and an external antenna changeover relay must be used to connect to the antenna.

Changeover from receive to transmit can be initiated by either applying a ground connection to the manual PTT input or, with a suitable IF rig, applying a small positive voltage on the IF coaxial cable connection. The transverter incorporates a simple but effective sequencer circuit for the internal transmit and receiver timing.

## Local oscillator

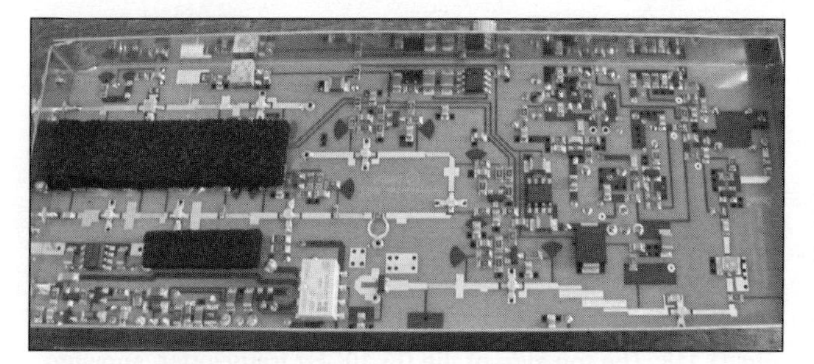

**Fig 11.119: Component side of the G3 10GHz transverter PCB. The black absorber foam can clearly be seen**

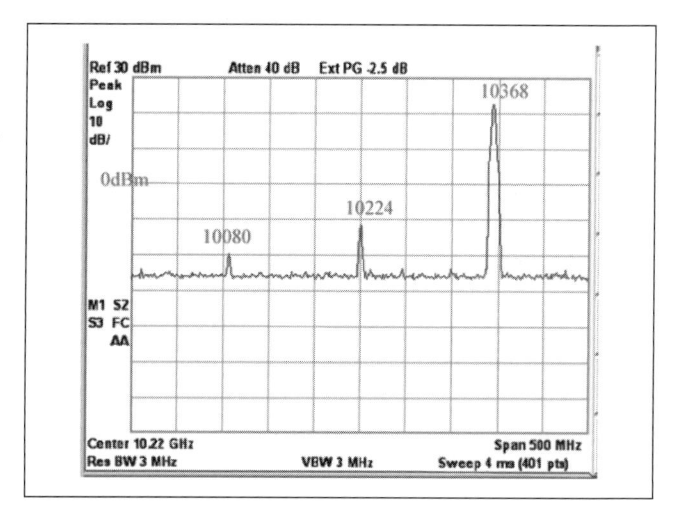


**Fig 11.120: Plot of the output spectrum showing the level of the local oscillator and the image frequencies**

In common with most other 10GHz transverters the MKU10 uses a crystal controlled overtone oscillator at 106.500MHz followed by a chain of frequency multipliers to the final injection frequency of 10224MHz. Bipolar transistor multipliers are used up to the final 2556MHz multiplier stage. From 2556MHz to 10224MHz a GaAsFET multiplier, followed by an LO amplifier is used. Intermediate stages use commercial helical filters whilst the final 10224MHz selection filter is a multistage side coupled microstrip circuit. The 2256MHz stage uses a miniature ceramic block filter.

## Frequency locking

In order to provide exceptional frequency stability the MKU 10 G3 uses an optional phase locking circuit. Whilst most overtone oscillators are capable of better than 1kHz accuracy (at the final 10GHz frequency) over a temperature range of around 0°C to +40°C, when suitably conditioned, this is no longer regarded as adequate, even for 10GHz.

There are several ways to achieve better accuracy using a reference source derived from GPS satellite, an Oven Controlled Crystal Oscillator (OCXO) or a high stability atomic source such as Rubidium. Possibly the most popular way to provide the desired locking is to use what is known as a Reflock [49] board overtone oscillator to lock one of these reference sources. This method also gives exceptionally low phase noise because it uses the inherently high Q overtone crystal oscillator rather than a low Q voltage controlled oscillator (VCO).

An external, high stability, low phase noise, 10MHz signal, connected to the Reference input connector, will cause the inbuilt 106.5MHz crystal oscillator to phase lock to the 10MHz reference.

The Reflock technique used is limited by the very small frequency range over which an overtone oscillator can be pulled. If the oscillator drifts more than a few tens of Hz from the nominal frequency, the Reflock cannot maintain control. To improve basic stability Kuhne Electronic has used one of their small, proportional control, crystal heater boards attached to the crystal by a short length of heat shrink tubing. The heater operates at 40°C and is very effective at holding the crystal at that temperature over a large range of ambient temperature up to 40°C.

Kuhne Electronic has used a National Semiconductor LMX2306 synthesiser chip in a conventional Phase Locked Loop (PLL) circuit to control the 106.5MHz crystal oscillator. The external 10MHz input acts as the reference frequency for the PLL. An Atmel 8 bit microcontroller ensures that the synthesiser is loaded with the correct frequency data each time the G3 is powered up. The PLL circuit is assembled on a small daughter PCB that is soldered onto the top of the main PCB.

### On-air impressions

I was only able to try the transverter on air in receive mode due to time constraints. I used a 1 metre length of low loss coaxial cable to connect a 17dBi gain horn antenna to the receive connector of the transverter. With the horn pointed out of the shack window, towards the dock cranes at Felixstowe, I was able to hear the 10GHz Martlesham GB3MHX beacon, by reflection, at strengths of up to 539. This is a remarkably good result and by comparison a barefoot G3JVL transverter (a popular design 20 years ago) gives a barely detectable signal when similarly connected.

I used my HP Z3801 10MHz GPS disciplined oscillator (+9dBm output) as the reference input and this worked extremely well, providing a frequency accuracy on 10GHz that was ultimately limited by my TS2000 IF transceiver and not by the transverter local oscillator chain.

### Comments

The MKU10 G3 is a well constructed unit and should give many years of service. It may appear to be an expensive way to get onto the amateur 10GHz band. However, many amateurs think nothing of spending over £1000 on an HF transceiver and often hundreds of pounds on a commercial antenna. By comparison, the 10GHz band is also capable of providing a great deal of fun and access to propagation modes such as rain scatter, tropospheric ducting and aircraft reflection that are not experienced on HF. My thanks to Kuhne Electronics for the loan of the test unit.

## A 10GHz Transverter from Surplus Qualcomm OmniTracks Units

These modifications were produced by Kerry Banke, N6IZW, of the San Diego Microwave Group and presented at The Microwave Update in 1999. The project offers an economical route to 10GHz, the unmodified transceiver, 10MHz TXCO and unmodified 1W PA can, at the time of writing, be ordered from Chuck Houghton for about £100 [57].

An earlier Qualcomm X-Band conversion project required considerable mechanical changes as well as electrical modifications and was based on replacing the original stripline filters with pipecap filters. These filters were required to provide sufficient LO and image rejection at 10GHz that the original stripline filters could not provide for a two meter IF. This version uses a somewhat smaller, more recent OmniTracks unit that contains the power supply and synthesiser on the same assembly as the RF board, and utilises dual conversion high side LO to allow use of the stripline filters. The filter modification has been proven to work well by extending the filter elements to specified lengths. Some additional tuning of the transmit output stages appears to be required for maximum output.

The synthesiser VCO operates at 2.272GHz, and when multiplied by five it becomes 11.360GHz for the first LO. The first IF frequency is 992MHz which is near the original internal IF frequency of 1GHz. The second LO is derived from the synthesiser pre-scaler, this divides the VCO frequency by two to produce 1,136MHz. Other second IF frequencies may be calculated using the relationship (RF-IF2)/0.9 = LO1 where RF is the 10GHz operating frequency (10,368MHz), IF2 is the second IF frequency, and LO1 is the first LO frequency. The synthesiser output frequency is then LO1 divided by five. **Table 11.13** shows the Excel spread sheet used to calculate the synthesiser programming.

The second conversion stage consists of a second LO amplifier (1,136MHz) and SRA-11 mixer converting the 992MHz 1st IF to the 144MHz 2nd IF. A 992MHz filter is required between the two conversion stages. Both Evanescent Mode and Coaxial Ceramic filters have been used. The conversion yields a reasonably high performance transverter with a noise figure of about 1.5dB and a power output of +8dBm, frequency locked to a stable 10MHz reference. Power required is +12VDC with a current consumption of about 0.5A on receive and 0.6A on transmit (about 1.5A total on transmit when including the 1W PA). **Fig 11.121** is a block diagram of the modified unit.

The unmodified circuit has a synthesiser output of 2,620MHz providing an LO of 13.1GHz. The original transmit frequency was around 14.5GHz with one watt output, and the receiver was near

| **3216 PLL Calculations for X Band Transverter with 144MHz 2nd IF; 1st LO=11,360MHz; 1st IF=992MHz** | | | | | | | | |
|---|---|---|---|---|---|---|---|---|
| Ref MHz | 2 | Ref MHz can be 10MHz divided by any integer from 1 - 16 | | | | | | |
| VCO MHz | 2,272 | | | | | | | |
| PLL MHz | 1,136 | PLL in MHz is VCO/2 and must be an integer multiple of Ref MHz | | | | | | |
| N | 568 | | | | | | | |
| | | M6(Pin15) | M5(Pin14) | M4(Pin13) | M3(Pin10) | M2(Pin9) | M1(Pin8) | M0(Pin7) |
| M | 55 | 0 | 1 | 1 | 0 | 1 | 1 | 1 |
| Board as is | | 0 | 0 | 0 | 0 | 0 | 0 | 0 |
| | | A3(Pin21) | A2(Pin20) | A1(Pin19) | A0(Pin18) | | | |
| A | 8 | 1 | 0 | 0 | 0 | | | |
| Board as is | | 0 | 0 | 0 | 0 | | | |
| | | R2(Pin5) | R2(Pin4) | R1(Pin3) | R0(Pin2) | | | |
| R | 4 | 0 | 1 | 0 | 0 | | | |
| Board as is | | 0 | 0 | 0 | 0 | | | |
| Lift pin22 | | | | | | | | |
| **Reference suppression filter modifications, parallel these capacitors with the following values** | | | | | | | | |
| Ref MHz | C1 | C2,C3 | | Add 1pF to VCO | | | | |
| 5 | None | None | | | | | | |
| 2 | 1000pF | 3000pF | | | | | | |
| 1 | 4700pF | 6800pF | | | | | | |

**Table 11.13: Synthesiser calculations for Qualcomm OmniTracks unit. 3216 PLL calculations for X Band Transverter with 144MHz 2nd IF; 1st LO=11,360MHz; 1st IF=992MHz**

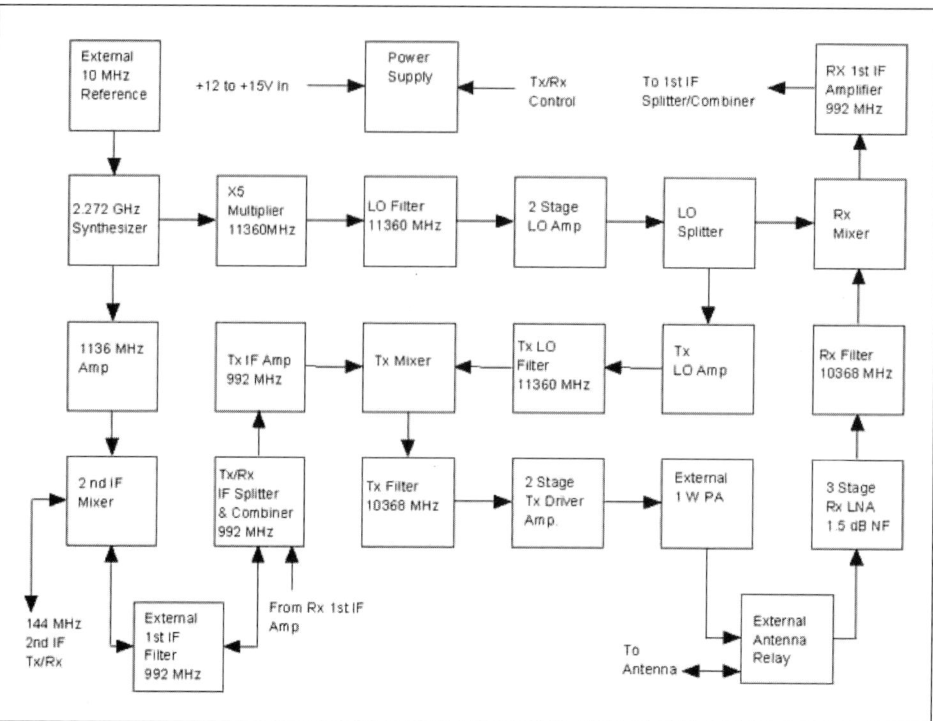


**Fig 11.121: Block diagram of Qualcomm X band transverter conversion**

12GHz. Unfortunately, the integrated PA in the original configuration provides no useful output below 12GHz and is not modifiable, so it has been removed for the 10GHz conversion. The transmit and receive IF preamplifiers make the transmit input requirement low (-10dBm) and provides high overall transverter receive gain.

**Fig 11.122** shows a picture of the modified transverter, 1W amplifier and 10MHz TCXO. **Fig 11.123** shows a picture indicating the locations of the various functions. The following is an outline of the conversion procedure [58]:

1. Marking location of RF connectors and removal of circuit boards.
2. Base plate modification for mounting two SMA connectors (10GHz receive and transmit) plus four SMA connectors installed (2 RF + 1 IF and 10MHz Reference input).
3. Clearing of SMA connector pin areas in PCB ground plane.
4. Remounting of PCBs.
5. Cuts made to PCB and coupling capacitors installed.
6. Stripline filter elements extended and tuning stubs added.
7. Synthesiser reprogrammed and 4 capacitors added.
8. Add tuning stubs to the x5 Multiplier stage
9. 2nd LO amplifier, mixer and 1st IF filter added.
10. Power and transmit/receive control wires added.
11. Test of all biasing.
12. Synthesiser and receiver test.
13. Transmitter test and output stage tuning.

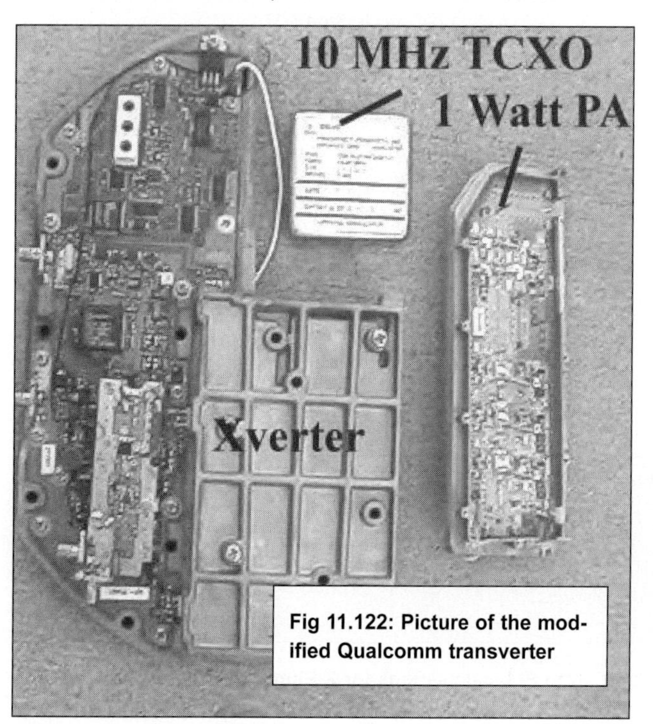


**Fig 11.122: Picture of the modified Qualcomm transverter**

Step 1. Mark the location of RF connectors and board cuts for coupling capacitors: Before removing the boards from the base plate, carefully drill through the board in the two places shown using a 0.050 inch diameter drill just deep enough to mark the base plate. These are the locations for receive and transmit RF SMA connectors. The upper connector hole (transmit) is located 0.5 inches to the left of the transistor case edge. The lower hole (receive) is located 0.4 inches to the left of the transistor case edge. Make the cuts as shown in **Fig 11.124** using a sharp knife.

Step 2. Base plate removal, modification, and connector installation: After making the holes and cuts, remove all screws and lift the boards off of the base plate. (Note: the original antenna connector pin must be de-soldered to remove the board. Once the boards are removed, drill through the plate in the 2 locations marked using a 0.161 inch drill to clear the teflon insulator of the SMA connectors. Use a milling tool to remove enough material on the back side of the base plate (see **Fig 11.125**) to clear the two SMA connector locations, taking the thickness down to about 0.125 inches (may vary depending on available SMA connector pin length). Locate, drill and tap the base plate for two 2-56 mounting screws at each connector. Mount the SMA connectors on the base plate and cut the Teflon insulator flush with the top side of the base plate (circuit board side).

Carefully clear the ground plane around the two connector holes on the bottom side of the circuit board to prevent the SMA probe from being shorted (using about a 0.125 inch drill rotated between your fingers). Reinstall the circuit boards onto the base plate.

Step 3. Add coupling capacitors: Add the three capacitors along with the additional microstrip pieces to modify as in **Fig 11.126**.

Step 4. Extend the transmit LO filter elements to the total length shown in **Fig 11.127**: Filter extensions are made by cutting 0.003 - 0.005 inch copper shim stock into strips about 0.07 inches wide and tinning both sides of the strip, shaking off excess solder. No additional solder is normally needed when attaching the extensions as the tinning re-flows when touched by the soldering iron. The length of the top element (0.21 inches) is measured between the marks as shown.

Step 5. Extend the LO filter elements as shown in **Fig 11.128**: Again, total element lengths are shown except for the right-most element that has additional dimensions.

Step 6. Extend the receive filter elements as shown in **Fig 11.129**: Dimensions shown are total element length.

Step 7. Extend the transmit filter elements as shown in **Fig 11.130**: Dimensions shown are total element length.

Step 8. Add the tuning stubs to the x5 Multiplier stage: This stage is located directly to the left of the LO filter which is shown in **Fig 11.128**. The gate of the x5 Multiplier stage requires addition of two stripline stubs, as shown in Fig 11.131.

Step 9. Modify the 2nd LO amplifier board, mount onto transverter and connect 1,136MHz LO input through 1pF coupling capacitor as shown in **Figs 11.132 - 11.134**. Fig 11.132 shows the overall second IF converter which is mounted using two grounding lugs soldered to the top edge of the LO amplifier board and secured by two of the screws which mount the main transverter board. Fig 11.133 shows the coax connected to the 1,136MHz point on the synthesiser through a series 1pF capacitor. Fig 11.134 shows the mounting and wiring of the SRA-11 mixer onto the LO amplifier board. Note the cut on the original amplifier output track after the connecting point to the mixer. The mixer case is carefully soldered directly to the LO amplifier board ground plane. The IF SMA connectors are mounted by carefully soldering them directly to the top of the mixer case.

Step 10. Program the synthesiser as shown in **Fig 11.135** by carefully lifting the pins shown with a knife. Ground pin 10, connecting it to pin 6 that is ground. Add the two 3000pF and 1000pF in parallel

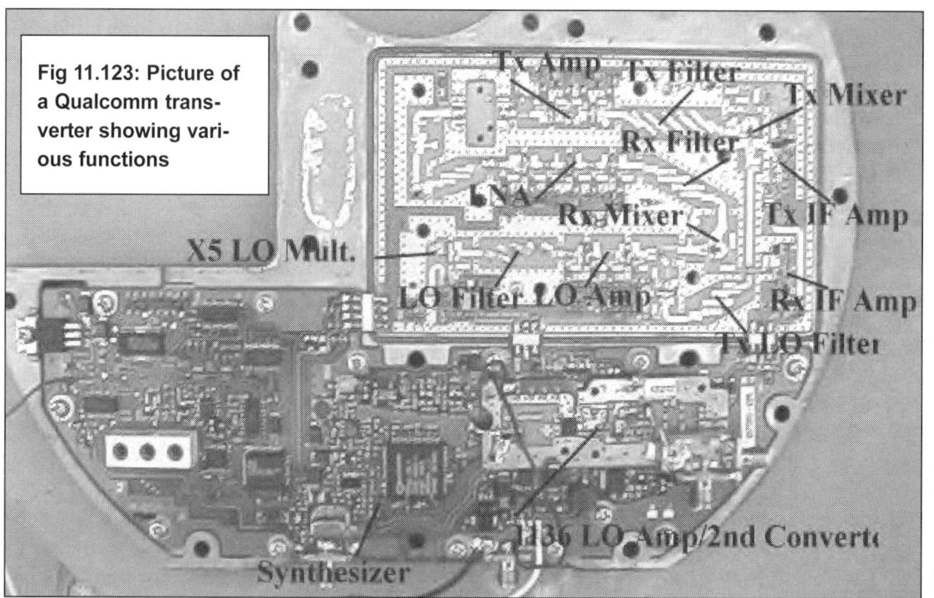


**Fig 11.123: Picture of a Qualcomm transverter showing various functions**

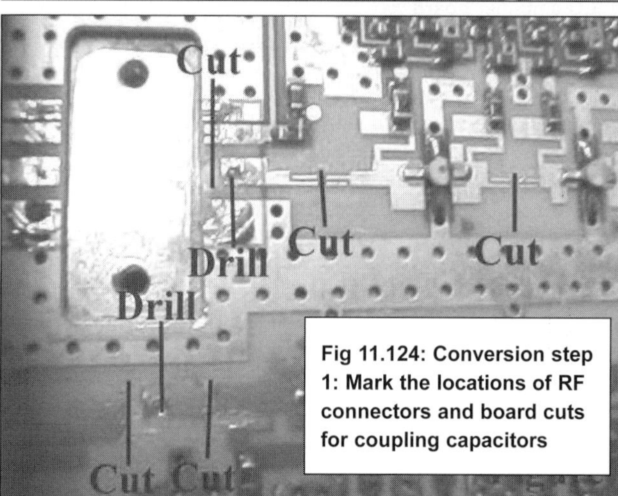


**Fig 11.124: Conversion step 1: Mark the locations of RF connectors and board cuts for coupling capacitors**

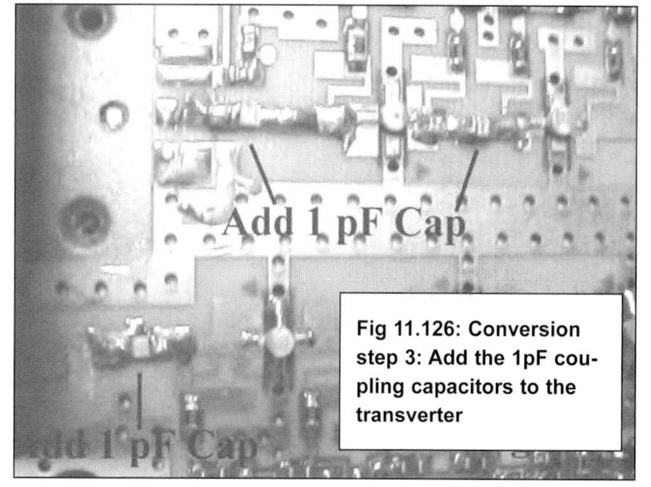


**Fig 11.126: Conversion step 3: Add the 1pF coupling capacitors to the transverter**

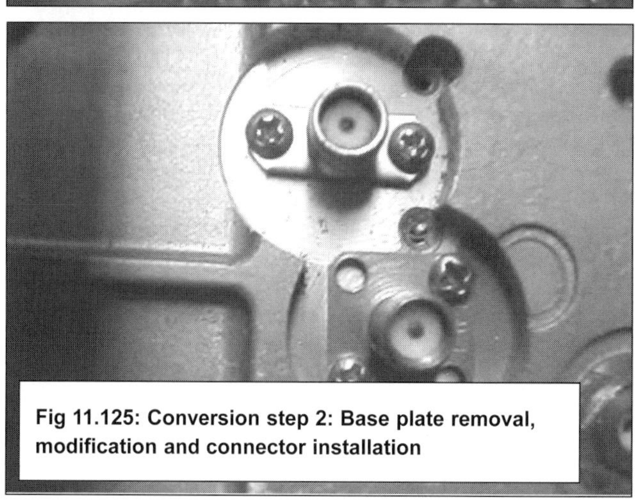

**Fig 11.125: Conversion step 2: Base plate removal, modification and connector installation**

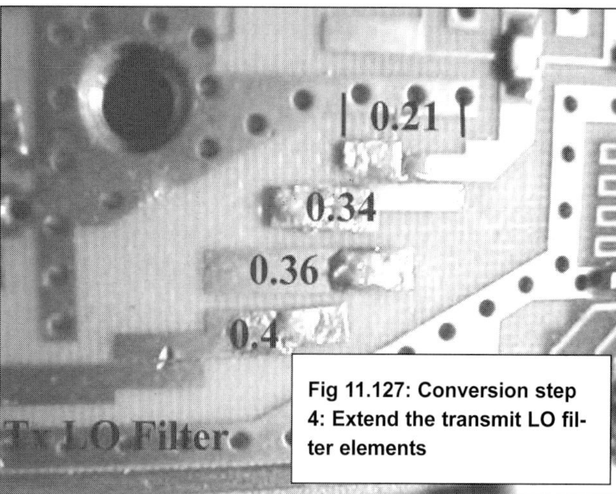


**Fig 11.127: Conversion step 4: Extend the transmit LO filter elements**

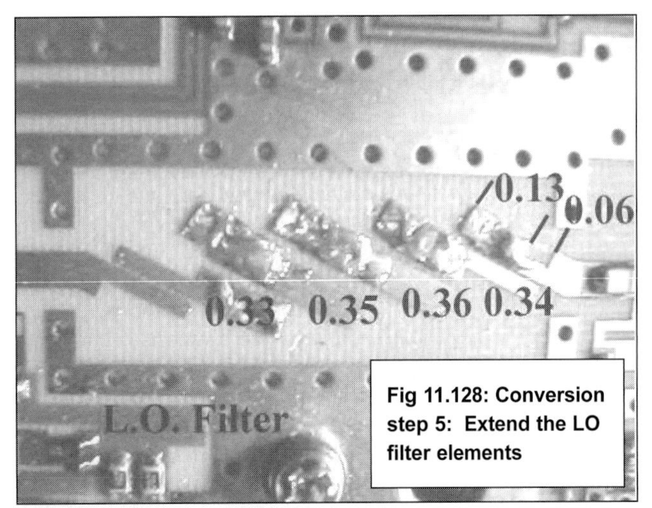


Fig 11.128: Conversion step 5: Extend the LO filter elements

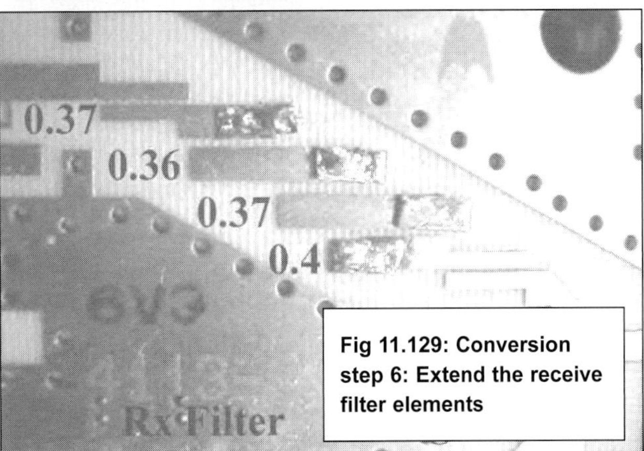


Fig 11.129: Conversion step 6: Extend the receive filter elements

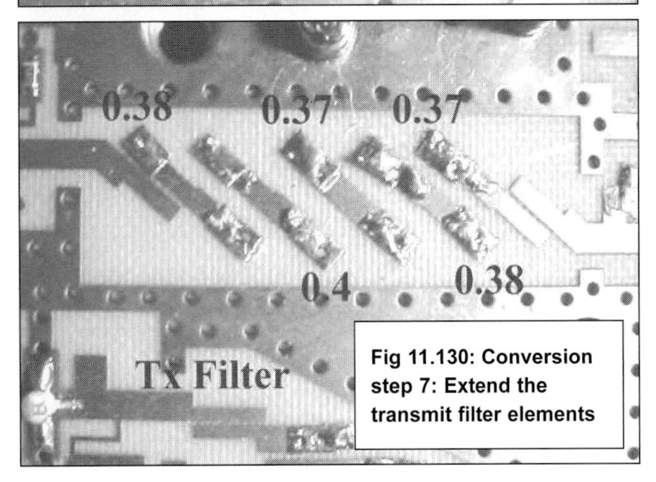


Fig 11.130: Conversion step 7: Extend the transmit filter elements

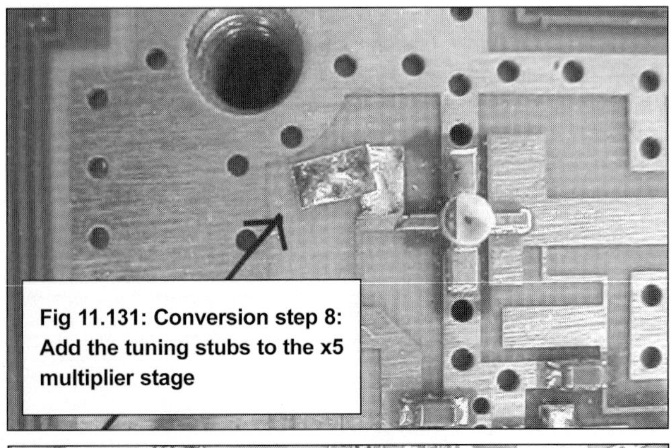

Fig 11.131: Conversion step 8: Add the tuning stubs to the x5 multiplier stage

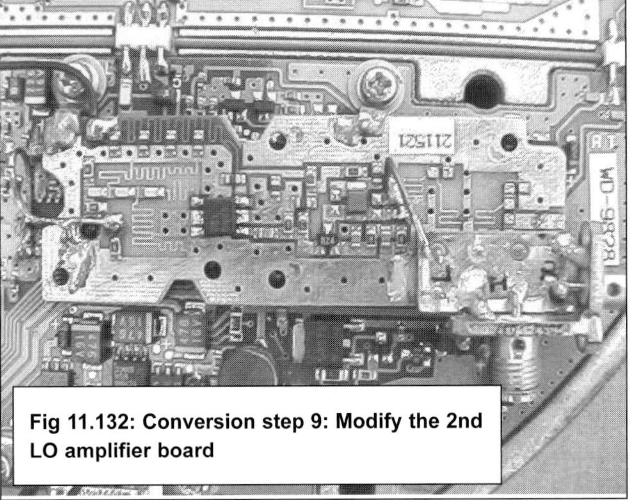

Fig 11.132: Conversion step 9: Modify the 2nd LO amplifier board

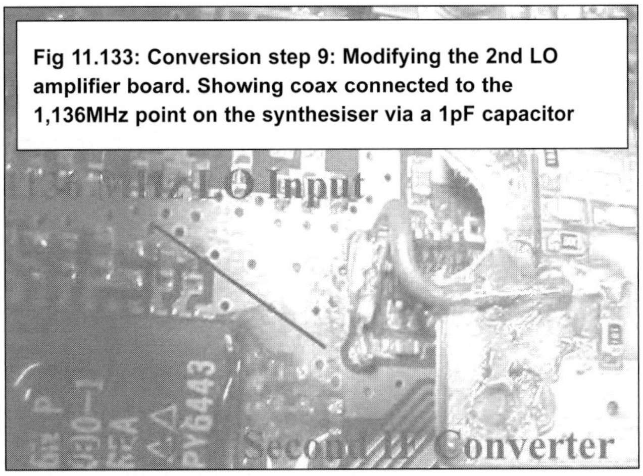

Fig 11.133: Conversion step 9: Modifying the 2nd LO amplifier board. Showing coax connected to the 1,136MHz point on the synthesiser via a 1pF capacitor

with the existing reference filter capacitors as shown in **Fig 11.136**.

Step 11. Add a 1pF capacitor as shown in **Fig 11.137** to lower the VCO frequency

Step 12. Add three transmit mixer tuning stubs as shown in **Fig 11.138**.

Step 13. The transmit/receive control is connected as shown in **Fig 11.139**. Grounding the control line places the transverter in transmit mode. The control can be open or taken to +5V to place the transverter in receive mode.

Step 14. The +12VDC power input is connected to the point shown in **Fig 11.140**. The original air core coil, with one end connected to that point, has been removed from the board. (This choke was originally used to supply +12V to the transverter through the 1st IF port).

Step 15. Powering up the Transverter: Apply +12V to the power connector and verify that the current drawn in receive mode is about 0.5A. Connect the 10MHz reference to the transverter board. Pin 43 of the synthesiser IC should be high when locked. If available, use a spectrum analyser to check (sniff using a short probe connected by coax) the synthesiser output frequency and spectrum. The synthesiser should be operating on 2,272MHz and no 2MHz or other spurs should be visible. Carefully probe the drain of each FET in the LO multiplier, LO amplifier, and LNA to verify biases are approximately +2 to +3VDC. A drain voltage of near 0V or 5V probably indicates a problem with that stage.

Place the transverter in receive mode and verify the biasing on the transmit LO amplifier and transmit output amp stages. Tune the 992MHz 1st IF filter (not part of the transverter board)

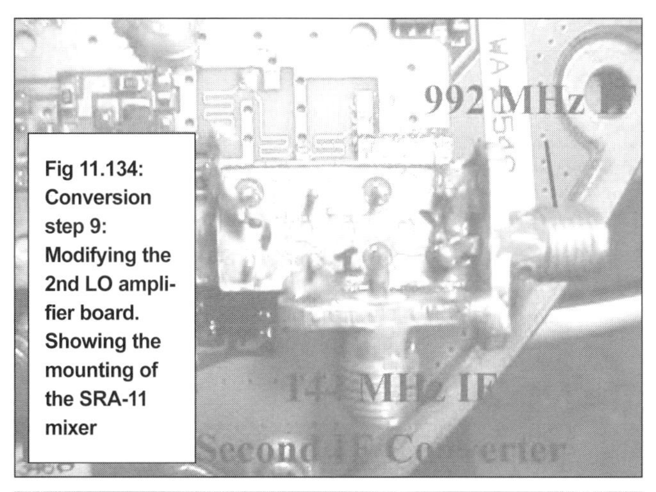

Fig 11.134: Conversion step 9: Modifying the 2nd LO amplifier board. Showing the mounting of the SRA-11 mixer

Fig 11.135: Conversion step 10: Program the synthesiser

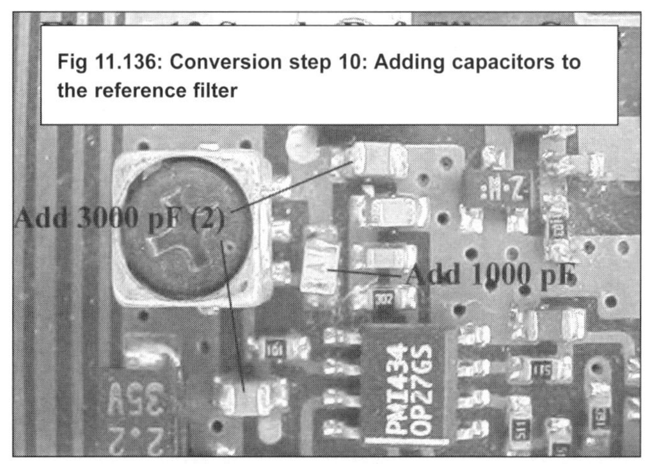

Fig 11.136: Conversion step 10: Adding capacitors to the reference filter

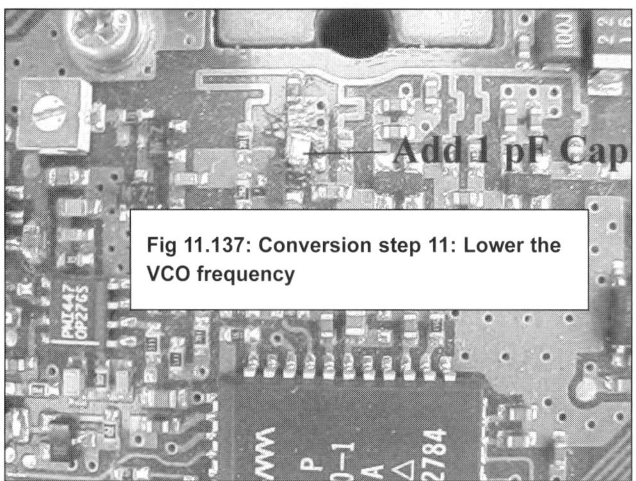

Fig 11.137: Conversion step 11: Lower the VCO frequency

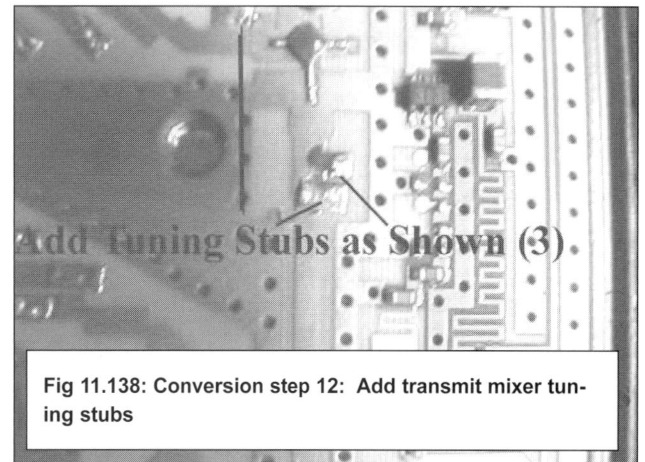

Fig 11.138: Conversion step 12: Add transmit mixer tuning stubs

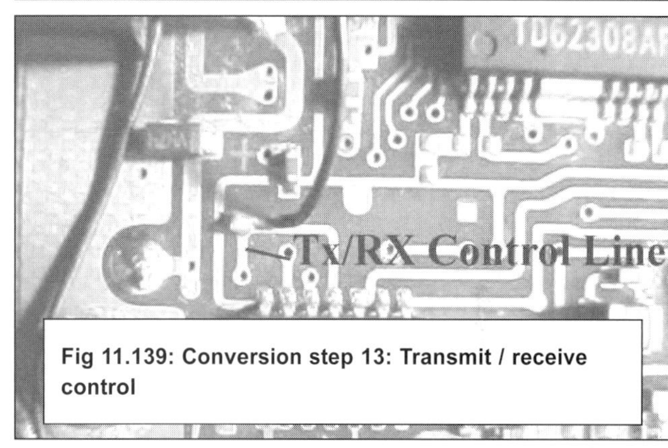

Fig 11.139: Conversion step 13: Transmit / receive control

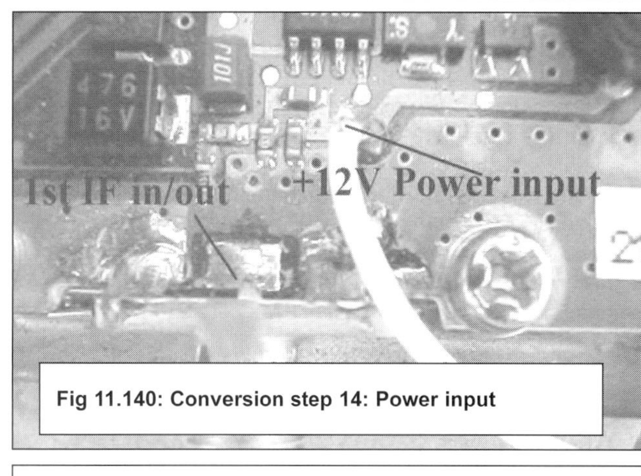

Fig 11.140: Conversion step 14: Power input

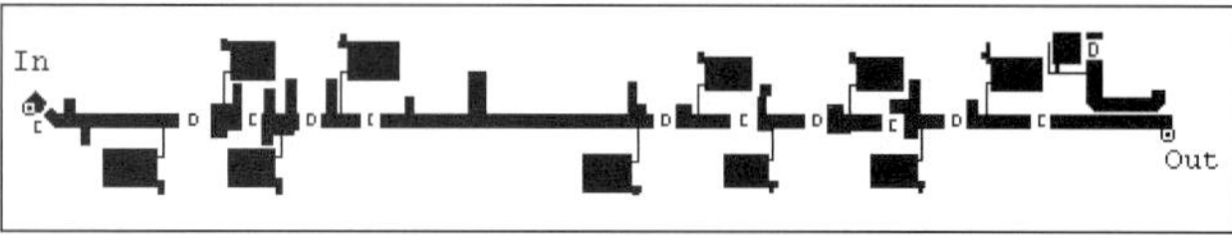

Fig 11.141: 1W PA board prior to tuning. -15dBm input gives +5 to +10dBm output with 10V at approximately 1A

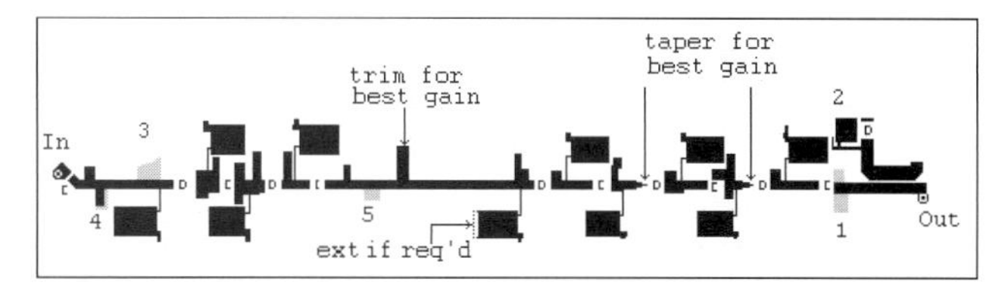

Fig 11.142: 1W PA board after tuning. The shaded tabs were added and tuned in the sequence shown. Results vary slightly from board to board. Key: [ = coupling capacitor, D = devices, Input coupled with 2pF

**** Devices from Agilent, Avantek or HP**

Originally Avantek Semiconductors manufactured the devices. Hewlett-Packard bought finished units (for example LNAs and VCOs) from Avantek for its range of test equipment as well as discrete transistors. Many of these were custom items although still bearing the Avantek logo. As the volumes supplied grew Hewlett-Packard purchased Avantek and the devices then became branded as HP ones. A little later during the HP reshuffle of its markets and operating divisions, the decision was made that the HP name would only be used for the computing systems, and a new name Agilent was coined for the semiconductor division and other allied divisions. In 2006 the name of Agilent was being used both for the old HP test equipment range and the semiconductor devices. HP decided that Agilent would only be used for the test equipment divisions and so Avago Semiconductors was invented for the semiconductor divisions.

For information on devices shown in this chapter as originating from Agilent, Avantek or HP see http://www.avagotech.com.

and connect it between the 1st IF ports on the transverter board and second IF converter. The receiver noise level at the 2nd IF port on the 2nd converter should be very noticeable on a 2m SSB receiver. A weak 10,368MHz signal can then be connected to the receiver RF input connector and monitored on the 2m receiver. The overall gain from receiver RF input to 2nd IF output should be roughly 35 to 45 dB.

Place the transverter into transmit mode and connect about -10dBm at 144MHz to the 2nd IF port. Monitor the power level at the transmit RF output port and add/move the transmit amplifier tuning stubs shown in **Fig 11.141** as required for maximum output. Typical transmit output will be about + 8dBm. This is considerably more than required to drive the one watt amp to full power.

### To convert the 1W PA

These conversion notes were produced by Ken Schofield, W1RIL [59]: Many PA boards have been successfully re-tuned for 10GHz operation. No two boards are exactly alike and each will tune a little differently from its apparent twin. The numbered steps in **Fig 11.142** will in many cases get your PA up into the gain range stated. You will find that numbered step 3 to be the most sensitive to gain increase. Unfortunately it is also one of the 'busiest' areas on the board.

Be careful! A few dos and don'ts are shown to help you bypass some of the many pitfalls that can be encountered.

- **Do** use a low voltage grounded soldering iron, and work in a static-free area.
- **Do** check for negative bias on all stages prior to connecting Vcc voltage.
- **Do** use good quality 50 mil chip caps - in and out approximately 1 to 2pF.
- **Do** remove all voltages prior to soldering on board.
- **Don't** work on board tracks when tired, shaky or after just losing an argument with your partner.
- **Don't** touch device inputs with anything that hasn't been just previously grounded.
- **Don't** apply Vcc to any stage lacking bias voltage.
- **Don't** shoot for 45dB gain - you won't get it! Be happy with 25 to 30dB

## OPTICAL COMMUNICATION

*The following contribution is by **Stuart Wisher, G8CYW.***

Communication by light has been used for many centuries, from beacons being used to warn of advancing invaders through the use of the Aldis Lamp to transmit Morse code messages, to advanced laser communications used in today's high speed telecommunications links. For amateurs, the use of light is an extension of the frequency spectrum into the Terahertz (THz) region. Visible light is in the range 380 - 750THz and infra red from 100THz - 380THz. Use of these frequencies introduces some new challenges, not least being much more dependency on weather conditions. Amateurs in Germany and America have been actively operating in the light spectrum for many years.

### Latest Developments

The use of small lasers as the active element in the transmitter has been superceded by the use of power LEDs. The author has been greatly aided by a small group of radio amateurs without whom much of this development would not have been possible.

### The AM Laser System

An economic solution is to use a separate photodiode and low noise op-amp. A SFH2030 photodiode and a NE5534 op-amp are used in the first receiver developed by G8CYW. The use of a precise 488Hz Morse tone for a modulated carrier wave (MCW) output and a very narrow filter in the receiver are continued, but with the inclusion of a 20kHz pulse width modulator (PWM) and a switch in the receiver filter circuit to broaden the bandwidth to permit voice communications.

The photodiode and amplifier are mounted in a 4 x 2 x 1 inch diecast box (**Figs 11.143 and 11.144**), acting as a receive head. The final stage of the receiver, a standard LM386 audio amplifier IC, volume control and loudspeaker, are contained in a separate box. A 3.5mm stereo jack plug, socket and lead are used to connect the boxes, carrying power, signal and ground. The transmitter circuit also has its own separate box, although there is no reason why it could not be accommodated in the receive audio amplifier and speaker box. **Fig 11.145** shows the circuit diagram of the receive head.

The Morse circuit is altered with the addition of an inverter to give maximum light output on key up (no modulation), which helps when aligning over a distance. The driver transistor pro-

**Fig 11.143: Original AM receive head. The switch adjusts bandwidth between CW and voice**

**Fig 11.144: The completed AM receive head, showing the photo diode and mounting tube**

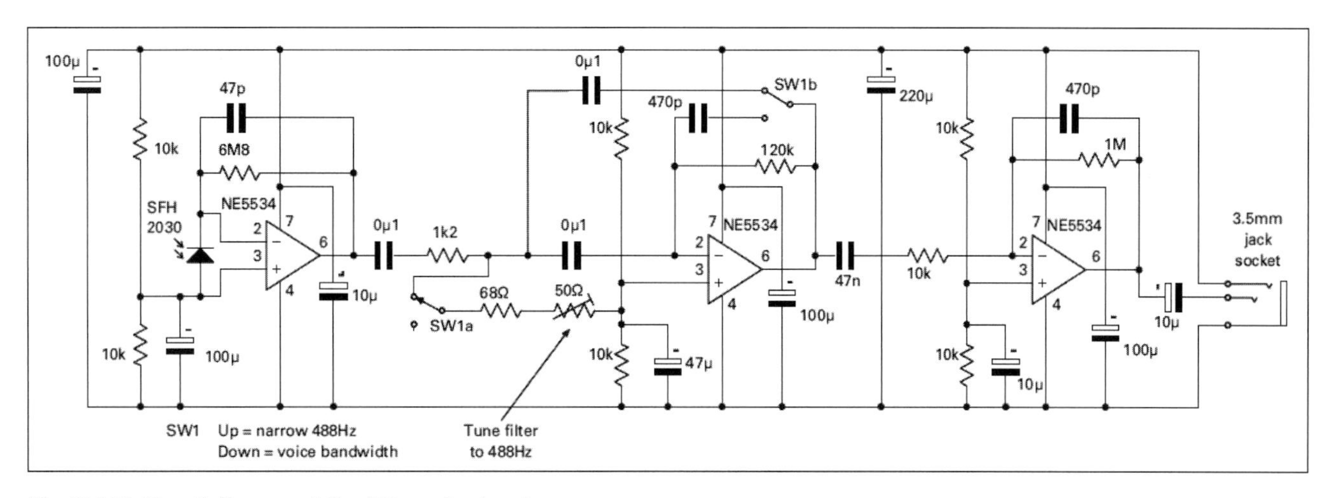


Fig 11.145: Circuit diagram of the AM receive head

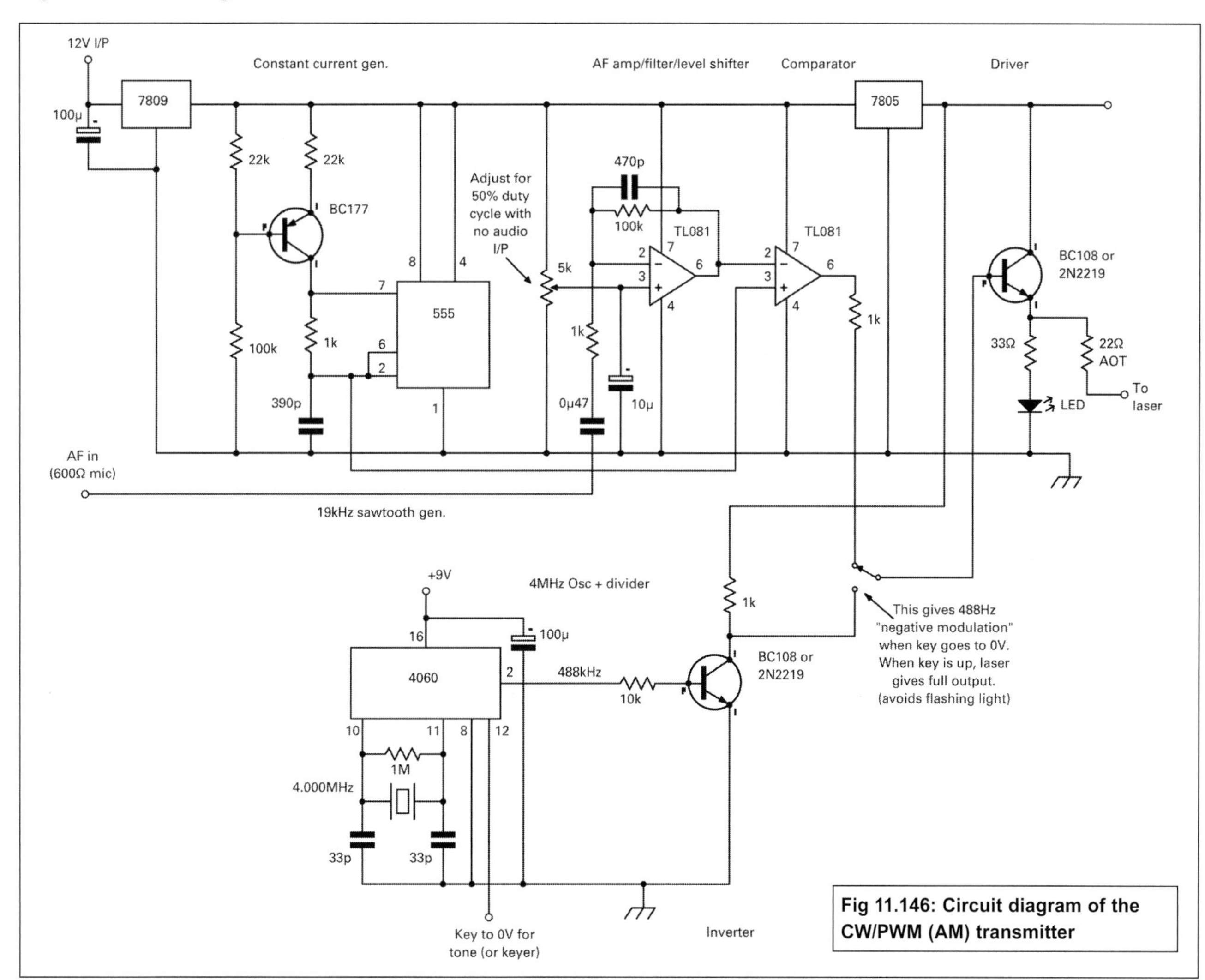


Fig 11.146: Circuit diagram of the CW/PWM (AM) transmitter

duces an output of 3V at around 30mA, ideal for the average laser pen; laser pointer or laser spirit level. Also included is an output for a front panel LED so that it would not only indicate when switched on, but it would act as a low power transmitter beacon for short range testing (which is very useful). **Fig 11.146** shows the CW/PWM transmitter circuit diagram and **Fig 11.147** shows the completed unit.

The performance of this system is such that the MCW tone can be received at a distance of 6km with no lens on the receiver; at 15km, a one inch lens is required.

## Optimum optics

All that is required of the optical part of the system is to collect as much of the available beam as possible and concentrate it onto the photodiode in much the same way as a microwave dish focuses radio waves on to the feed.

The optical system that evolved used a length of 110mm waste pipe, an end cap and a joining piece (all available from DiY warehouses). A Blue Spot 100mm magnifying glass was located at a pound shop, the handle was cut off and the protru-

Fig 11.147: Completed CW/PWM transmitter

Fig 11.148: Parts used in constructing the receive optics

sion where the handle met the rim was then filed smooth. This was found to be a tight push fit into the end of the tube. (Be careful at this point: the band around the lens is not parallel. Place it narrow end upwards on a sturdy table and push the tube over it. When the tube rim touches the table top, the lens is properly installed.) **Fig 11.148** shows the constituent parts.

The cylinder on the end cap is cut down to 20mm and the centre of the blanking disc drilled out to allow light to pass through. A 38mm diameter hole cut with a hole punch was found to be convenient. A 110mm pipe joining piece was cut exactly in half and one half is used to push the end cap, the cut end then is slid over the tube on the far end from the lens, making a loose fit that can be improved by using a turn or two of PVC insulating tape on the tube. The length of the tube depends on the lens. You need to be able to produce a focused image of a distant street lamp on the photo diode. The lenses were found to vary in focal length slightly; most required around 285mm from lens centre to the diode surface.

## Lasers

Originally, lasers were obtained from a cheap German supermarket, where a laser level kit was often available for £10. The kit includes a tripod, adjustable head and a spirit level with the laser in it. The spirit level can be modified to take two connections out to a 3.5mm socket to use it as the transmit head. A second kit provides everything needed to mount the receive tube. Simply bolt a section of the aluminum spirit level to two 110mm pipe wall clips (also useful for marking an accurate line around the tube before cutting) and slot the tube in place through the hoops. This means that a laser level is sacrificed to science, but this is the cheapest and easiest way. A further wall clip or two can be used to support a finder telescope or half a pair of cheap binoculars - an essential aid to lining up. This assembly then clamps in the adjustable head. A completed unit is shown in **Fig 11.149**, which also shows some 40mm waste pipe fittings used to hold the receive head. This arrangement also enables different heads to be slid in and out and for fine tuning the focus of the system, although once the focus is set it doesn't normally need adjusting.

The other half of the joining piece is used as a lens hood. This also seemed to finish the system off well. It has also proved useful as a holder for irises - cardboard discs with holes in - that act as signal attenuators. These can be used to assess how little signal is needed at a given distance, as an aid to calculating the potential range of the system. A 50mm diameter hole gives 6dB attenuation and a 25mm hole gives 12dB attenuation. Ignoring atmospheric effects, these equate to signal levels at twice and four times the range respectively.

### WARNING: Lasers are Dangerous

- The dangers from lasers are essentially from the amount of energy contained in a very small area. If a narrow beam is used, then all of the power can be directed into the typical 5mm diameter of the human eye causing significant damage.
- Whenever possible, ensure the beam created in home constructed or converted equipment has its beam expanded. If the beam is expanded to larger diameter then its energy is contained over a larger area and the danger of accidental eye damage is much reduced. Including a beam expander to your laser system also has the advantage of reducing the divergence of the beam, increasing the distance potential of the transmitter.
- Never leave a laser transmitter on and unattended as you are not in control of where the laser is pointing. If you move away from the transmitter always turn the laser off.
- Consider any beam, even that generated by a cheap laser pointer, as potentially dangerous until its beam has diverged to a minimum of 50mm diameter.
- Night-time eye response: At night and in dark conditions, the pupil in the human eye will dilate to increase sensitivity to light. This significantly increases the danger from lasers.
- Visible wavelength lasers can be seen by the operators and others, and their danger anticipated. However, special care should be taken when using invisible wavelengths. Infra red lasers are generally more powerful but the risks from accidentally leaving a laser source on are significantly increased. By the time you realise the source is on, it may already be too late. Consider a beam expander as mandatory with non visible wavelengths.

*Further information on laser safety can be obtained from the Health and Safety Executive we site http://www.hse.gov.uk*

## Operation

AM works well at short range, but there are, however, some issues with the use of this system. There is a lot of QRM from street lights. The signal flutters at long distance due to atmospheric scintillation (twinkling), and aiming the laser accurately can be quite difficult. To address these issues, a high power LED (and lens arrangement similar to the receive lens) is now used which eventually got to the point where the signal could be detected 34km away.

## Power LED

The advantage of a LED transmitter is that it is much easier to

(above) Fig 11.149: The complete receiver

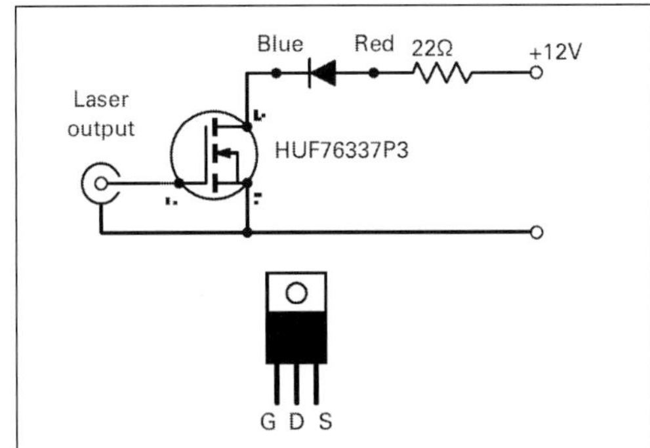


Fig 11.150: High power LED driver

aim because it has a broader beam than the laser. Also, it avoids the issues involved in aiming a laser over the countryside and, since the power density is much lower, it is safer. That said, a 1W LED using these optics still looks very bright, even when lined up over a distance of several kilometers.

Changing from a laser to a power LED is an easy move. Another tube and lens system is required - so there is a use for the second tripod, head and spirit level: to hold the second tube. At the rear of the end cap, a diecast box makes a good mount and heatsink for the power LED. The completed assembly looks rather like the detector in Fig 11.149.

The LED drive circuit (**Fig 11.150**) is an N-channel power MOSFET, which requires a small heatsink. The gate goes to the transmit electronics, the source to 0V and the drain to the LED cathode. The anode goes to +12V.

LED operation reduces the operating noise somewhat, because the wider (optical) beamwidth doesn't suffer from the 'speckle' and scintillation of signals that is common with lasers. But this remains an AM-based system, which has pronounced issues with fluttering signals plus QRM from streetlights and road traffic.

These issues caused much thought to see if a solution could be found to get round these problems. A web search, looking for 'laser DX' and 'optical communication' revealed a wealth of material out there. The progress made by various groups of optical communication enthusiasts can be found mainly in Australia, Czech Republic, Finland, Germany and USA, to name just some of the major contributions. Of special note are VK7MJ and group who have communicated by voice over 160km and KA7OEI and group who have exceeded even this. To date, most have now progressed into weak signal modes and the Australia/Tasmania groups have spanned the Bass Strait between Australia and Tasmania by cloudbounce, a distance of some 288km.

Since in the UK we do not have any huge mountains or dry flat deserts to provide long optical paths and our atmosphere is cloudy and misty most of the time, we cannot really compete on distance, so we re-defined our aims to involve immediate real-time microphone to loudspeaker communications.

## FM System

Web searches produced a receive head design by VK7MJ that had a frequency response from audio to 50kHz and beyond. This could be used in an FM subcarrier system, centred on approximately 25kHz. At this frequency the QRM from street lights was significantly reduced (if not absent) and the effect of the limiter in an FM receiver helped to overcome the fluttering of signals.

The FM transmitter shown in **Fig 11.151** uses an op-amp based microphone amplifier/filter connected to a 4046 PLL IC VCO generator running at 25kHz.

The audio signal produces frequency modulation around the main carrier frequency. A MOSFET driver connected directly to the oscillator then provides adequate drive to the power LED head.

The receive head (**Fig 11.152**) starts with VK7MJ's design, followed by an amplifier tuned to the sub-carrier frequency feeding a NE566 PLL demodulator. All of this fits into a 4 by 2 by 1in diecast box and uses the same 3.5mm stereo jack system for power and signal connections that connect to the audio amplifier/speaker box from the AM system. All the optics remain as they were for the AM system. A later modification is to make this head switchable between AM and FM by tapping directly into the output of the original VK7MJ circuit. The first dual-mode optical receiver was thus created. Just for good measure the transmitter box included a linear (rather than PWM), AM transmitter circuit to complete the dual-mode setup.

### FM results

Short range tests at 6km showed FM to hold much promise. Very strong signals with no QRM or flutter were achieved. Going on to the 15km path gave similarly good results even in near proximity to powerful lights. Since then FM has been used over all paths tried up to 34km and strong signals and clear communi-

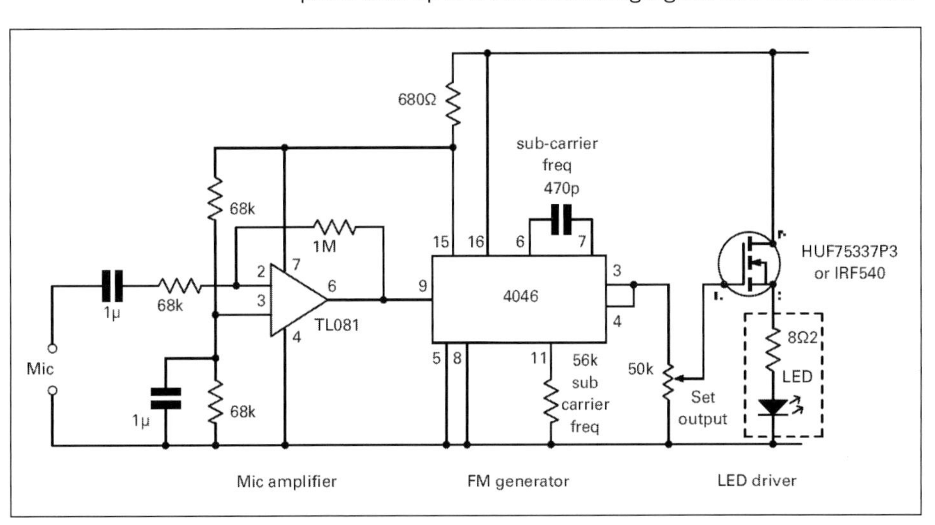


Fig 11.151: Basic FM transmitter operating at ~25kHz

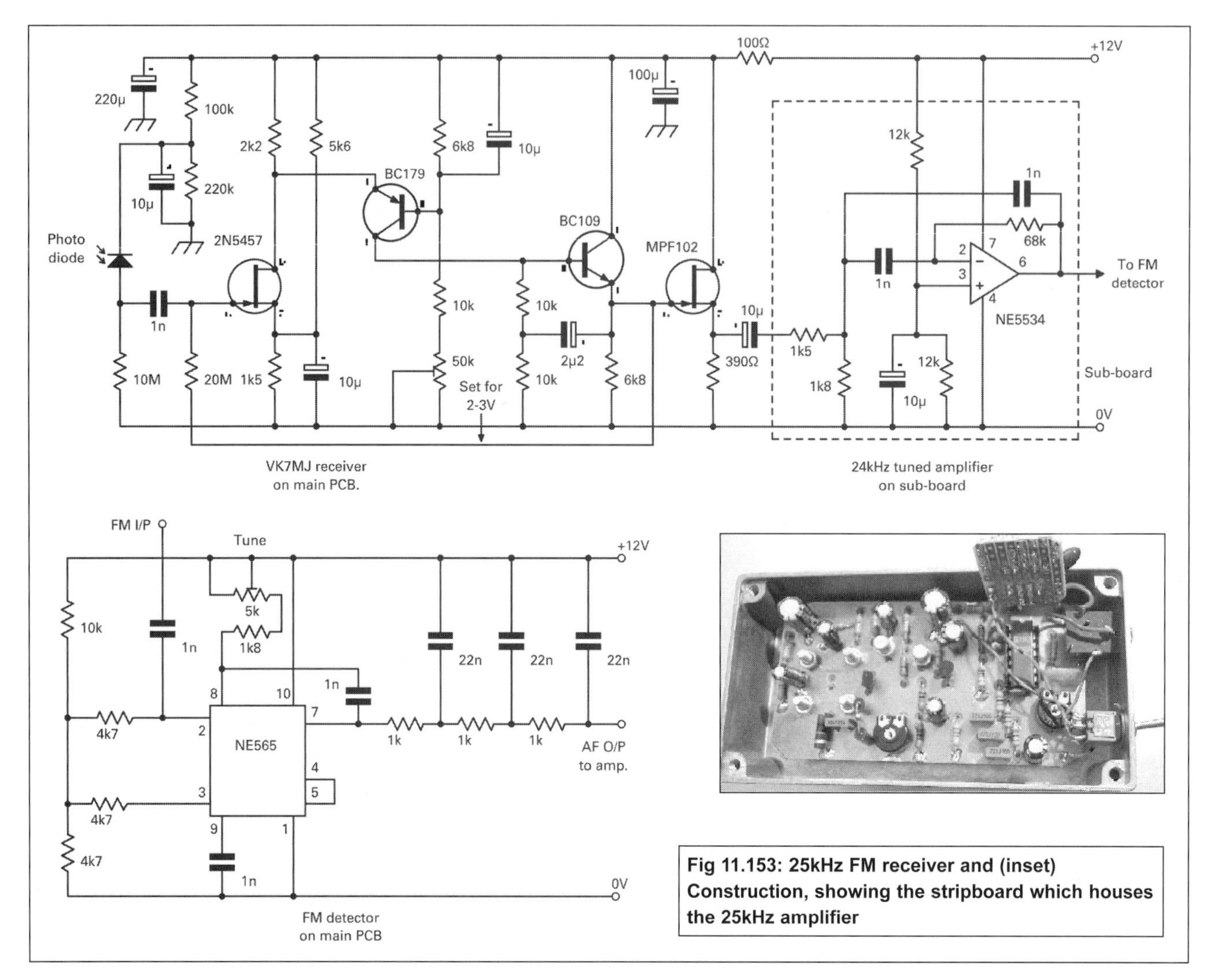


**Fig 11.153: 25kHz FM receiver and (inset) Construction, showing the stripboard which houses the 25kHz amplifier**

cation has always been possible.

## Optical Transverter

After building and testing the FM system, G8CYW wondered what advantage, if any, might be gained by going to SSB.

This mode is paramount for long distance voice communication over most, if not all, amateur bands where it is allowed, so why not on light? The thought of building stand-alone systems for receive and transmit for SSB was eclipsed by one of those *eureka!* moments: why not build a transverter? All of the gain and signal processing power of a small HF amateur radio transceiver (eg FT-817) could be utilised, converting the HF signals to light and vice versa.

Since the group were already operating on 20kHz PWM AM and 25kHz FM, thoughts turned to the possibility of sending and receiving single sideband (or any other mode for that matter) on around the same frequency. Crystals for 3.58MHz (actually 3.579545MHz) are readily available, enabling the 80m band to be used for the intermediate frequency. The chosen operating frequency is 3.6MHz RF, thus producing or receiving an optical signal around 20kHz. This is known as 'radio over light'. It even looks and feels like you are operating a real radio when making optical contacts.

### The circuit

The transverter circuit is shown in **Fig 11.153**. A relay switches the HF transceiver between the receive and transmit paths in the transverter (via resistive attenuators). The transmit attenuator reduces the 0.5W output from the transmitter to a few millivolts. This signal is mixed with the 3.58MHz local oscillator signal in the MC1496 balanced mixer to provide the 20kHz signal. This is amplified by a NE3354 opamp which also filters out the unwanted 7.2MHz mixer product, and the signal is then fed to the gate of a power MOSFET. The MOSFET has adjustable bias via R28 to set the quiescent gate voltage (and hence LED idling current). The LED and resistor are contained in a separate box (the transmit head, described later), mounted at the focus of a lens in a similar manner to the previous system.

On receive, the amplified 20kHz signal from a receive head (of which more later) is connected to the input of the mixer, which up-converts the signal to the 80m band again thanks to the 3.58MHz local oscillator. The output of the mixer is switched to an attenuator to protect the HF transceiver's front end. The attenuator also protects the mixer IC against inadvertent transmission into the mixer output if the PTT fails. The local oscillator can be heard at a low level on the HF receiver if you tune down to the region of 3.58MHz. It is not strong enough to de-sense the receiver if you keep at least 10kHz away from it.

This linear transverter is useable on any mode (although FM and SSB are best) and on a range of frequencies. 3.600 MHz RF gives 20 kHz optical; 3.650 MHz gives 70kHz and so on. Lower frequencies should be better because they are less demanding of the electro-optical system. It is best to keep to frequencies around 3.600 MHz (20 kHz optical).

The prototype transverter PCB and overlay are shown in **Figs 11.154 and 11.155** (in Appendix B). Note that the board should be double sided - the top is a ground plane that doesn't require

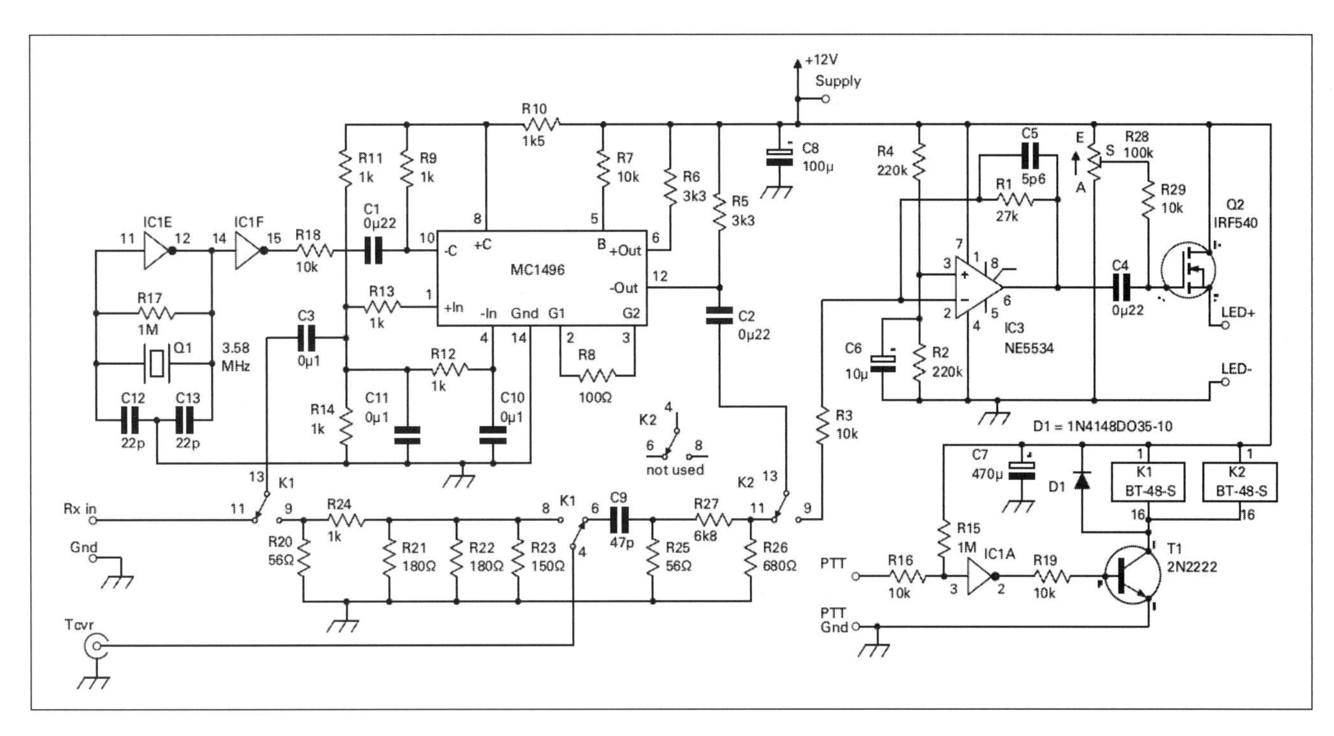


**Fig 11.153: Circuit diagram of the transverter to convert 3.5MHz SSB to 20kHz for 'radio over light'**

etching. After you drill the board, identify all the ground connections and mark them (a marker pen pressed against the holes will usually be visible form the other side). Then turn the board over and, using a hand-held drill bit of about 3mm, clear the copper around the non-earth holes (eg the diode in the foreground of **Fig 11.156**). When you populate the board, solder all earth connections on both sides of the board (eg the four resistors near the middle). This will result in a good quality of screening.

## Setting up

Before connecting power for the first time, turn R28 so the wiper is down to 0V (fully anticlockwise if you're using this PCB). Check the board current consumption - it should be around 20mA at 12V. Connect the LED head and, while monitoring current drawn, turn up R28 until the supply current is about 100mA more than you started, eg 120mA. This is akin to the bias setting on a conventional solid-state linear amplifier. It also gives adequate light for a distant receiver to line up on. Surprisingly, it is

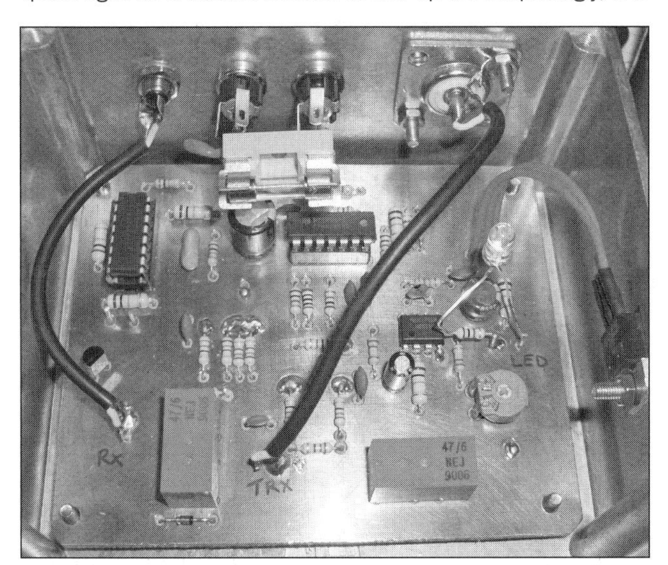


**Fig 11.156: Linear transverter (built & photographed by M0DTS)**

also left on when switched to receive, giving the distant station something to aim at.

## Mods for optimum performance

If you intend using this on FM only, or over short distances, it performs perfectly well, but there have been some more recent modifications to the design above that squeeze a little more performance out of the system.

Several recent tests over long distances revealed an odd situation with receiver response. FM signals were end-stopping on the FT-817 S-meter, but when switched to SSB, the S-meter would not indicate more than S9 with equally strong signals. This was traced to the receive head (and transceiver, the same circuit on receive) putting out so much signal that it was causing the transverter to limit. Of course this is fine on FM, but it would be nice to have an equal response to SSB. The cure is to reduce the gain in the head by removing the second op-amp and connecting the output of the first op-amp via a 0.1µF capacitor to the output socket, this can easily be done on the existing two op-amp circuit board design. In the transverter the receive output attenuator was modified as follows: change C9 from 47pF to 1nF, remove R25 altogether, and swap R26 with R27. If the receiver S-meter is deflected by noise after this modification, simply fit a 220 ohm potentiometer as an input level control at the point where the Rx signal from the head connects to the transverter board. The signal on the pot wiper is fed to the Rx in port on the board. Adjust the potentiometer to the point where the S-meter shows no indication when the head is in total darkness. This modification slightly lowers the noise figure of the receiver and enables the S-meter to have full range on SSB signals.

## Operating principle

It is easy to imagine what happens to an FM signal on transmit through the LED: the MOSFET is effectively in Class D and being driven hard, switching frequency modulated pulses to the LED at around 20 kHz.

It was problematic at first envisaging exactly what was happening with a single sideband signal, as only half of it will be conducted by the LED. This must be rather like putting a rectifier

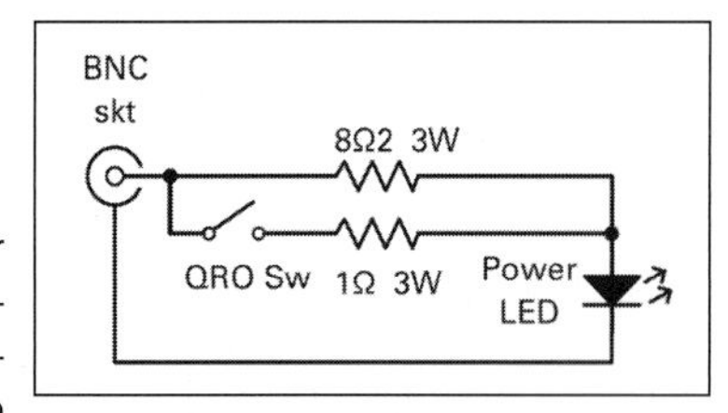


**Fig 11.157: Circuit diagram of the transmit head**

diode in series with your HF antenna! Only the positive half of the SSB signal gets converted into light (the LED is a diode after all). These half-signal pulses travel to the distant receiver. If you could pick this up directly, without a tuned circuit at signal frequency, it would sound like the most awful, overdriven distorted signal you have ever heard. But it does not sound distorted at all on an HF receiver because the tuned circuits in the receiver restore the waveform due to the flywheel action of a high Q tuned circuit.

Unlike many other transverters, there is nothing to tune up or adjust other than the LED bias potentiometer. Most who have built this have not even padded the crystal down to its design frequency (although there are spaces for capacitors on the board for this), just leaving it about 1 - 2kHz high is fine.

Do not exceed the LED current ratings. By monitoring the total transverter current, you can keep an eye on the average current through the LED. A 1A fuse is installed in the power supply line and it has blown several times on speech peaks.

## Transmit head

This consists of a power LED and switch mounted on a diecast box and current limiting resistors mounted inside the box. **Fig 11.157** shows the circuit diagram. In normal use the QRO switch is left open. When using SSB it is possible to close the switch, which lets a lot more current flow through the LED on speech peaks. Do it at your own risk - it gives about one extra S-point and several LEDs have been blown this way. Trying it on FM or CW is almost certain to blow the LED.

The LED is mounted on a 25mm square piece of 0.4mm fibreglass PCB on the base of the diecast box (see **Fig 11.158** and **Fig 11.159**). Power LEDs dissipate a few watts of heat so it is important to use heatsink paste between the LED and the board and again between the board and the box, which acts as a heatsink. Do not use normal-thickness PCB! The two power resistors are mounted in the box and a BNC socket is used for the drive connection. The box is mounted on a pipe-stop end with a central hole cut in it, then placed at the focus of a lens as described earlier.

An Osram Golden Dragon LR W5SM HYJY-1 LED, RS part number 665-6189 has been used here. It is quite important that you use this one because, although many others will work on transmit, It will be explained later how to make this particular LED also work as a photodiode on receive. It is quite likely that the original LED will have been discontinued as the development of LEDs is proceeding at a rapid pace, so it will be a case of finding an equivalent

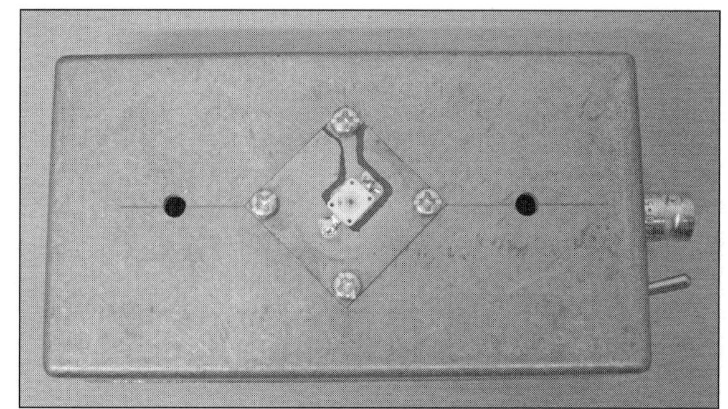

**Fig 11.158: Power LED on transmit head.**

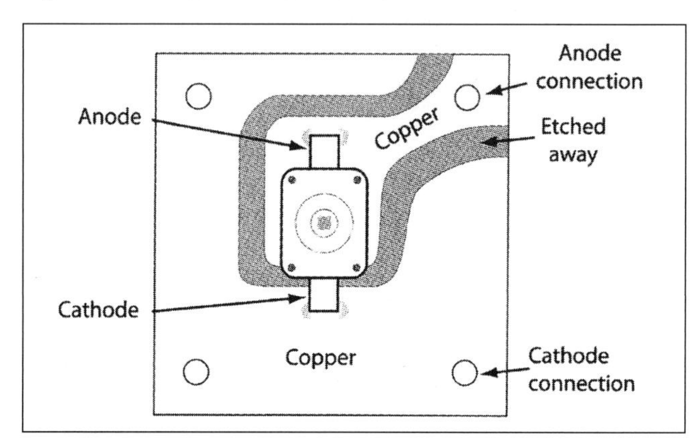


**Fig 11.159: Power LED mounting**

and testing it.

## Latest and most sensitive front end

**Fig 11.160** shows the best front end used to date. It is closely based on a design by Clint Turner, KA7OEI, that is found on his website [60]. The first stage has been altered to be a reverse biased SFH2030 photodiode RC coupled to the gate of the FET. The relatively elaborate bias circuit is to make best use of the low-noise performance of the FET. Some of KA7OEI's low pass filtering has been removed to increase the bandwidth of the circuit which now responds up to about 100kHz. T3 forms a

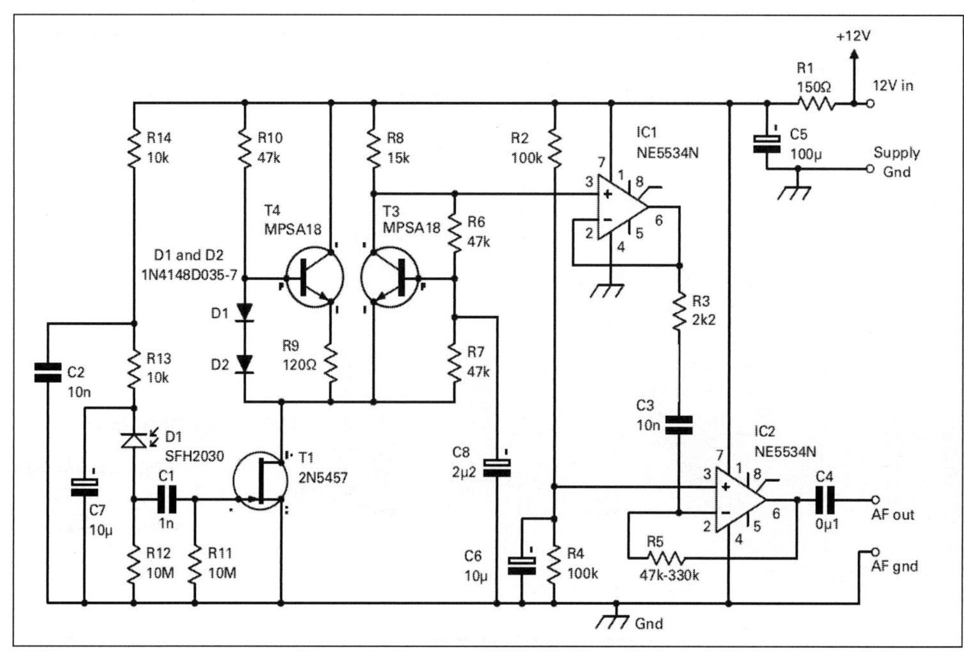


**Fig 11. 160: Receive head circuit diagram (this is the two op-amp version)**

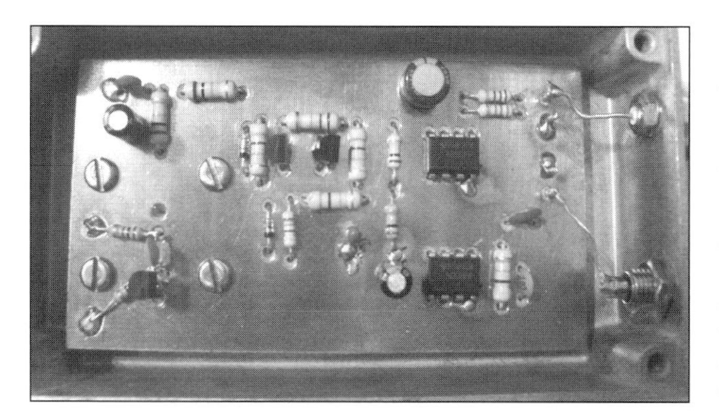

**Fig 11.161: Receive head (two op-amp version)**

cascode amplifier with the FET while T4 controls the FET drain current. The low noise op-amp(s) provide further gain.

The following information on the 2N5457 FET front end bias applies to both the receive head and transceiver head to be described later. Variations in the characteristics of the 2N5457, bias generator MPSA18 and its diodes can cause an incorrect FET drain voltage. It is vital for maximum sensitivity that the drain voltage sits at around half the power supply voltage. If the drain voltage is higher than 6V, a quick fix is to connect a resistor from drain to 0V, which will greatly improve matters. Some constructors have used a 4.7kΩ fixed resistor whilst others have used a 10kΩ variable resistor and adjusted it until the drain voltage is at 6V. A more careful approach is to adjust the value of the resistor (currently 120Ω) connected to the drain of the FET. This has affected around half the circuits built so far.

This circuit when biased correctly is so exquisitely sensitive that there is no alternative to a diecast box for shielding: even the power is fed down a shielded cable from the transverter. Use a 3.5mm stereo jack plug, socket and screened cable to bring power in and take the signal back to the transverter on separately shielded cables. The SFH2030 photodiode just peeps out of a 6mm diameter hole in the box. This is normally placed at the focus of a lens on a pipe end cap, drilled as before. Any daylight reaching the photodiode will vastly increase its noise output so, while testing, keep the head, mounted in the pipe end cap, face down on the bench to exclude the light. Sliding it slowly to the edge of the bench instantly reveals when the tiniest amount of light enters because the noise level rises so much. You end up skulking around in the dark with this one! Without a lens, this head gives a noticeable increase in noise from moonlight and, with a lens, Jupiter and the brighter stars are easily detected.

**Fig 11.161** shows the completed receive head, usable with one or two op-amps, with the PCB foil pattern and overlay in **Figs 11.162 and 11.163** (in Appendix B). As with the transverter, the PCB is double sided for screening; use the same drilling technique.

## Test Beacons

It is useful to have a small beacon transmitter that can output an optical signal for bench testing. Nothing more complicated than a pair of 555 timer ICs is required (**Fig 11.164**). The right hand 555 oscillates at about 25kHz and drives a red LED via a series resistor. 1kΩ gives a bright light for long distance testing up to 500m or so; 10kΩ is quite dim and suitable for indoor use. This beacon tunes in as a carrier a little above 3.605MHz on the FT-817. The left hand 555 is optional; when switched in, it keys the 25kHz oscillator, making the signal more easily identified.

### Extreme beacon

One of the longer distance contacts, at 65km, was nearly a failure because the stations could not locate each other for some considerable time. White light from powerful torches looks just like car headlamps; red lights look like car tail lamps. Then a Xenon strobe was used, identification is much easier with a regularly flashing lamp but, understandably, it was looking a little dim at 65km.

Enter the extreme beacon, **Fig 11.165**. In principle, it is similar to the test bench beacon described above, but uses a power FET on the output to drive a 20W LED. This has 25 individual LED chips arranged in a 5 by 5 matrix on a substrate. It runs on about 12V at around 2A. The LED does require a substantial heatsink - reckon on dissipating about 15W as heat.

With this LED and its heatsink at the focus of a 100mm lens in the optic tube (**Fig 11.166**), it produces a 2° wide pattern of light flashing at either 2.5Hz (for visual identification) or 25 kHz for receiver alignment. Do not look directly into the beam: the intensity is enough to cause eye damage, even at some considerable distance.

## The LED Transceiver

Tim Toast, of the Optical Links website, regularly scans the web for optical communication-related material. In October 2010 he provided a link to a paper, LED used as APD, written by a team at the University of Salerno, Italy. They had discovered that some GaP and GaAsP/GaP power LEDs, when reverse biased to large voltages, acted as photo sensitive diodes and as avalanche photo diodes, the latter being very expensive.

After reading this paper, G8CYW set about to repeat their experiments. He used the test bench beacon, several power supply units, a few resistors and capacitors, a selection of red LEDs and an oscilloscope. It was found that a signal could be recovered from a particular high brightness 5mm LED without

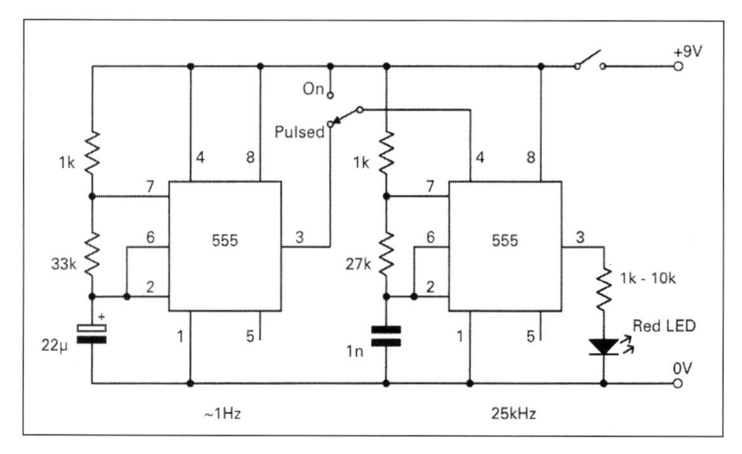


**Figure 11.164: Circuit diagram of the test bench beacon**

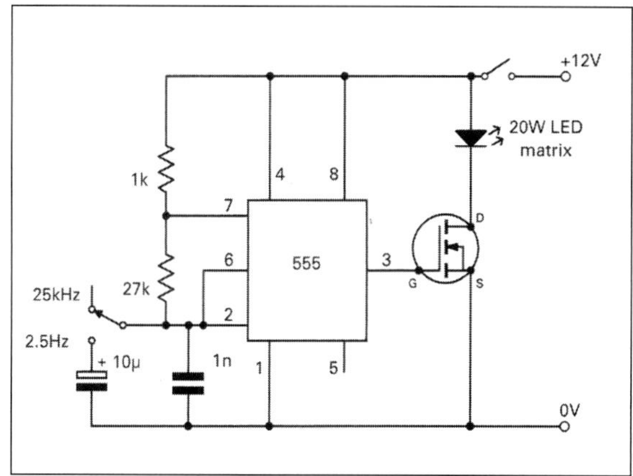


**Fig 11.165 Circuit diagram of the extreme beacon**

**Fig 11.166: The extreme beacon LED mounted in one of the now-standard housings**

any reverse bias; then as the reverse bias was increased to over 30V, the recovered signal was enhanced. Another receive head was rapidly put together, with a relay to switch the LED from forward bias (so that It could be used as a transmit head), then to reverse bias it in an attempt to use it as a photodiode.

There is also one very obvious advantage using an LED in this way as a transceiver, you only need to aim only one set of optics: if you can hear the other station; you are ready to work the other station.

This low power LED transceiver was tested over a 6.5km path and found to receive and transmit as well as the separate units used before. This may well have been the first occasion where a single LED was used to receive a free space optical signal over a distance of several kilometres.

## LED transceiver Mk2

The concept of the LED transceiver was proved, but the light level on transmit was way down on that from the power LEDs. So, the Golden Dragon LED used in the separate transmitter was looked at again. It is an InGaAlP device, a type not covered in the original research paper. Also, it could not operate in reverse bias due to the presence of a protection diode in reverse-parallel with the LED chip. There is even a warning in its data sheet that the LED is not intended for reverse bias operation. It was noticed that both diodes were set in a silicone gel and the minute gold leads to both diodes were separately visible within the gel. Enter the concept of 'microsurgery' on the LED.

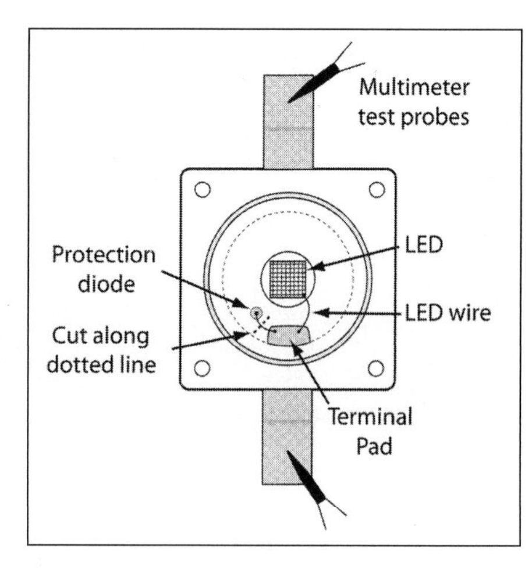


**Fig 11.167: Where to cut the protection diode wire**

A sharp knife was used to cut through the gold wire to the protection diode by simply pressing the blade down vertically on the wire. Checking with a multimeter on the diode range indicated the operation had been successful: with the meter leads one way round the LED lit up and the meter indicated about 1.6V; the other way round was now open circuit, having previously indicated the 1.065V forward voltage drop of the protection diode. In short, the LED now has a dual use; one way round it is a LED, the other way round it is now a functioning photodiode. Several operations have since been performed and an LED has not been lost yet. **Fig 11.167** shows the LED, diode and cut point.

This LED encouragingly developed a photovoltaic output of over 1.1V when strongly illuminated. Using the test bench beacon as a signal source, it gave an enhanced output voltage when reverse biased to 43V. This was then the centre of a redesigned front end which incorporated the same modified KA7OEI circuit as before, plus relay switching to use the LED on transmit as well, as shown in **Fig 11.168**. The LED series resistors were also included on the board design to make the complete transceiver head.

The transceiver LED is mounted on a thin fibreglass PCB **(Fig 11.169)**, this time in a design to minimise capacitance to ground on receive while retaining adequate heat conductivity on transmit. Use heatsink paste as before.

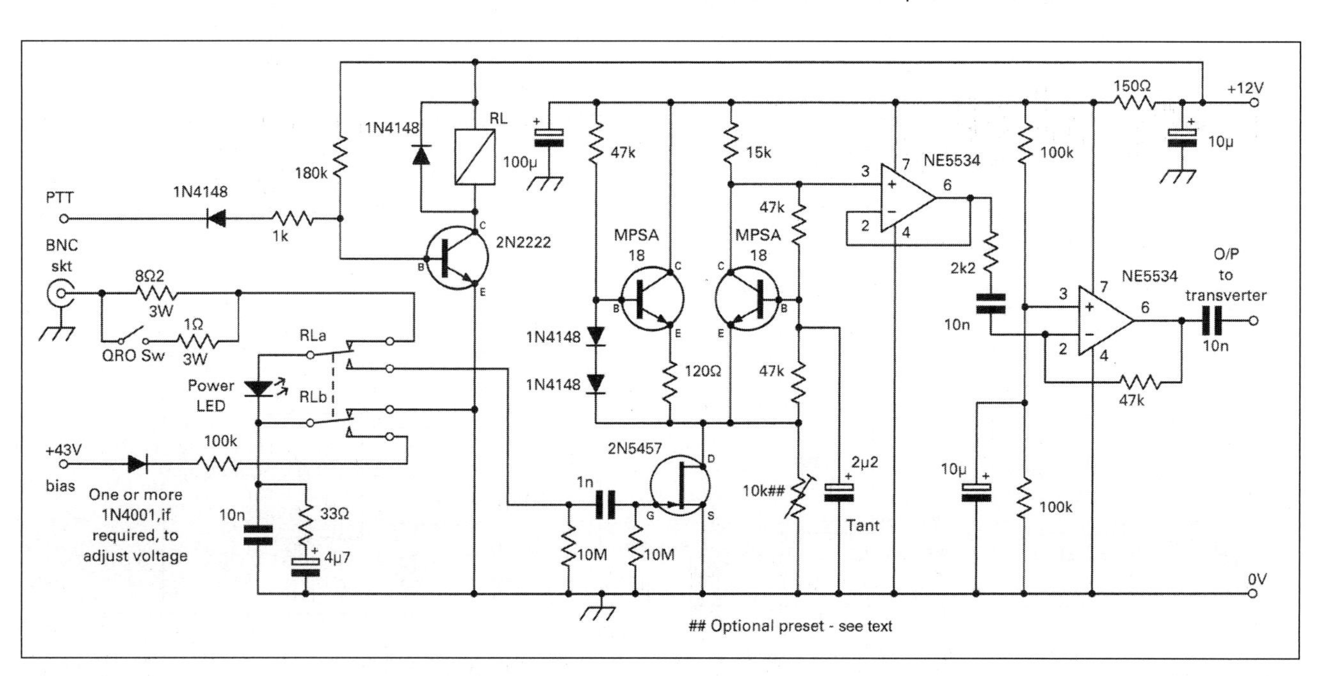


**Fig 11.168: Circuit diagram of the LED transceiver head (two op-amp version)**

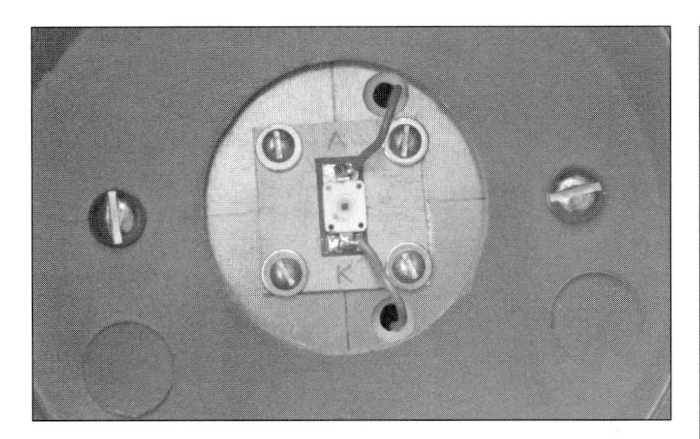


**Fig 11.169: Transceiver LED mounted on the head assembly**

It is thought that the exact value of reverse voltage will be an 'adjust on test' item, as one version of the transceiver required 48V to bring it to optimum performance. The bias current is extremely low (about 100nA) and is switched off by the relay when not in use (the reason for the PTT arrangement, it defaults to transmit when not powered). A combination of 12V keyfob batteries and/or button cells is used to achieve the best voltage for the particular transceiver. You can adjust the actual voltage by placing 1N4001 diodes in series to drop the voltage if required: it seems that you need to hit the optimum voltage to an accuracy of about half a volt. Do not use zener diodes, as they would generate noise.

The batteries are contained within a piece of 15mm copper water pipe attached inside the box using Terry clips, as can be seen in **Fig 11.170**. Others have used N cell holders, which are a good match for the keyfob batteries. Don't try to use an inverter or voltage multiplier to supply the bias voltage; noise would drown out the wanted opto signal. The PTT line can be connected to the FT817 (which grounds a pin on transmit) as well as the transverter PTT input and it should operate as required.

## The Fresnel Lens Rig

This uses flat A4 page magnifiers made from acrylic sheet. They are available from stationery shops or, rather cheaper, from many pound shops. They are actually Fresnel lenses which are of quite good quality. They are not quite the same size as an A4 sheet of paper, being about 21cm by 28cm. Compared to the Blue Spot lenses that fit the plumbing pipes, they have over 8dB further gain when used on receive or transmit. This figure was arrived at by simple calculation of the increased area and confirmed using the FT-817 S-meter on receive and a light meter on transmit. The transmit beamwidth of the Fresnel lens system works out at about a fifth of a degree - which is sharp, but not impossible to aim. It is however too sharp to aim if you do not know exactly where the other station is, hence the extreme beacon (with its 2° spread) described earlier. Later versions of this also employed a secondary meniscus lens of 28mm diameter and 28mm focal length mounted close up to the LED (about 10mm from the LED). This extra lens captures more of the light from the LED and directs it toward the Fresnel lens. This extra lens broadens the beam slightly, to about 0.3 degrees, but the beam is still brighter as a result.

This Fresnel lens system was introduced before the LED transceiver was developed, so G8CYW's version has two lenses side by side (**Fig 11.171**). This makes it simpler to operate because the separate receiver and transmitter, aligned side by side, automatically look at the same point in the distance. It is simply necessary to line up the receiver on at a distant (beacon) signal and you're ready to transmit back. With the advent of the transceiver, it is only necessary to use a single lens, but the second lens has been successfully used to house the extreme beacon. This simplifies the alignment procedure when operating from a hilltop.

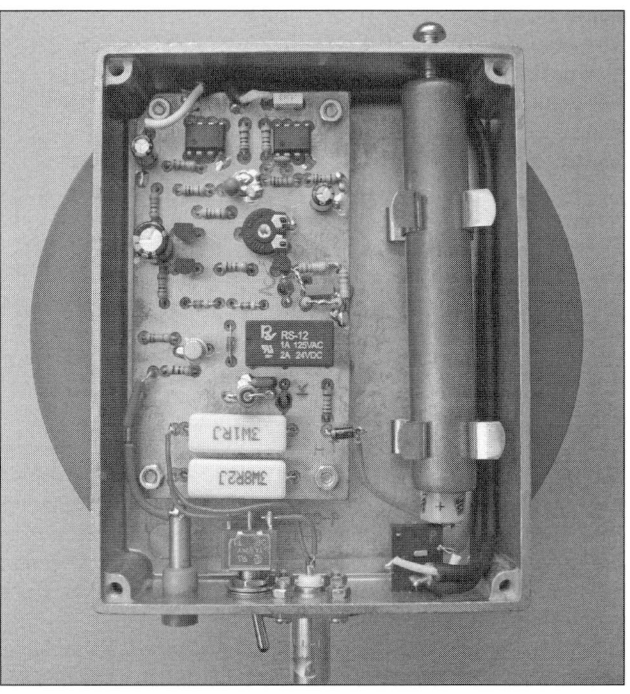

**Fig 11.170: General arrangement of the transceiver (two op-amp version). The copper tube contains the 43V bias batteries**

The Fresnel lenses tested have a focal length of 350mm, but it would be wise to test each lens individually by measuring the lens to image distance when producing a focused image of a distant streetlamp. The LED (or photodiode) is simply supported at this point. The ridged surface of the Fresnel lens must face towards the distant station. All these parts are mounted in a box, shown in Fig 11.171.

Two fixed cabinet feet are installed under the front of the black-painted lens box and one adjustable foot, on a screw thread, centrally at the rear (**Fig 11.172**). This gives adjustment in the vertical plane; for horizontal (azimuth) adjustment, it is simply necessary to move the whole box around on a square of MDF clamped on to a folding workbench or large tripod. This has proved adequate for contacts up to nearly 120km distance.

SSB has been proven ultimately win out over FM on weak signals as demonstrated when the UK distance record was increased to 117.6km, (M0DTS at Danby Beacon near Whitby and G8CYW near Alnwick, Northumberland), by operation over a difficult path over the North Sea. This path is technically non-line-of-sight, requiring help from atmospheric refraction. This gear seems to be a good addition to those who climb up hills for contests etc, to add 623nm (the wavelength of the red light produced by the LED), to the

**Fig 11.171: The Fresnel lens rig, suitable for a transmitter and receiver, or a transceiver and beacon**

**Fig 11.172: The Fresnel lens rig, rear view, showing sliding mounts for Rx, Tx, LED transceiver or extreme beacon**

**Fig 11.173: the IR LED and secondary lens, together with the longpass filter (left)**

selection of wavelengths available for communication.

## Infrared Communication

Analysis of several contacts has revealed that the maximum light level in which the system will operate giving the lowest noise level on receive is during nautical twilight when the sun is at least 12 degrees below the horizon. This places a constraint on the operating time in the summer months.

A recent development has been to use the near infrared band (IR), instead of visible red light. Several IR LEDS are readily available which produce radiation at wavelengths of 850nm or 940nm. Atmospheric data suggest that the shorter wavelength may be attenuated less over long paths, but both will be investigated. The LED transceiver is easily adapted for use with either type of IR LED, and the response as a photodiode (once the protection diode has been removed) to visible light seems to be reduced. The high bias voltage is not necessary and can be connected to the main LED transceiver supply of around 12V.

In a recent test, the types of IR LEDs used were drawn from the Osram Golden Dragon range, notably the SFH4230 and SFH4233 types which emit strong radiation at 850nm and 940nm respectively. Both these types will also act as sensors and have been used in single optic transceivers as noted earlier. More recently, the Osram range has been augmented by the more powerful Platinum Dragon range. The SFH4235 type has been used, although due to its internal structure it has not been found useful on receive and a separate photo diode has to be used.

An initial test on 940nm over a 6.3km path was successful in sunlight conditions, with strong signals reported between G8CYW and G8KPD/P. They were both using LED transceiver circuits connected to the transverter and FT 817, as has been used for all contacts, on sub-carrier SSB. After this contact, efforts were shifted to 850nm as information about atmospheric transparency indicated that this might be better at the slightly shorter wavelength.

It is worth noting here that cheap photographic filters, of the longpass type. which will pass the IR signal but block visible light are readily available. The author uses the 52mm diameter screw-on type as these conveniently fit in front of the secondary meniscus lens and they appear to have a very low transmission loss. The effect on receive is to drop the noise level by some two S-points even above the fact that using an IR LED on receive seems to have an already lowered response to visible light anyway. **Fig 11.173** shows the filter arrangement on the transceiver head, the filter has been removed from its mount making the IR LED and secondary lens visible.

Moving on to 850nm, a number of tests at increasing range were made until a range of 46km was attempted on a day with the sun visible between rain showers. This was carried out by G8CYW/P and G8KPD/P, between locations at around the 300m ASL mark. Both stations were using LED transceivers and single optics.

Deciding not to use aids such as beacons or strobes, both stations aligned their optics to a high accuracy hoping simply to aim accurately at each other. A 20kHz subcarrier on 850nm from one station was almost immediately picked up and confirmed by keying it on and off. Both rigs were adjusted for optimum and signal reports exchanged at 59 both ways on SSB, FM was also tried but fluttering made copy difficult. The wind was at least 30mph and this caused the gear to vibrate. In addition there was much turbulence in the atmosphere. This was another case where SSB gave a better result. Contact was maintained on various tests from about noon for around 90 minutes, during which it rained and there was even sunshine at times.

The next trial, which was to use an iris plate attenuator to reduce the signal by a known amount. Put simply, a 6dB attenuator (which reduced the area of the Fresnel lens by a factor of four to roughly a rectangle 10cm by 14cm) reduced signals on SSB to 57 one way and 58 the other. This roughly simulates the signal level at twice the distance without taking account of the larger extinction coefficient of the atmosphere at that distance.

A 12dB reduction in signal strength was tried next, and despite some technical problems it appeared to indicate that a further signal margin was available. The received noise level due to daylight was around S3-4 using 850nm long-pass filters in the rigs. Removing the filter increased the noise to S6-7 so the use of a long-pass filter is a benefit, particularly as it had no effect on the wanted signal.

The test result was that subcarrier communication 'in the middle of the day' is possible using 850nm NIR at 46km distance, with some margin in reserve.

## Ultra Violet Communication

The only part of the visible light spectrum area not yet investigated was ultra violet (UV). The borderline between visible light and ultra violet is at 400nm and there are available some 5mm through-panel LEDs that operate around 395nm - 400nm. These only radiate some 10mW of UV and were thought much safer to use than some industrial high power LEDs (see box 'WARNING: Lasers are Dangerous' earlier).

Compared to these industrial LEDs, some 20dB or so less power was available. The UV LEDs did not appear to be usable as sensors necessitating a return to the photo diode receivers, which had very poor response to UV, having a sensitivity of only 10% of the peak level at 850nm. It looked like the link budget on any path would down by some 30dB compared to IR.

A further limiting factor was the lack of atmospheric transparency at 400nm, so much UV is scattered by the atmosphere that our research seemed to indicate we may be wasting our time.

The usual series of tests were set up, with initial contact being made over 6km, no mean feat due to the power reduction. Extensions to 10km and 20km were routine, until the final (to date) test over 32.5km where a signal was received successfully at 57 on sub-carrier SSB. UV is certainly a challenge, but eventually has been demonstrated at distances much further than believed possible.

## The Phlatlight

A number of Nanowavers have recently obtained a very powerful LED, heatsink, Peltier effect cooler and even also a suitable fan from sellers on eBay [61]. The LED, a Phlatlight (photonic lattice) CBT-54 can run at up to 9.1A continuous and 13.5A at 50% duty cycle according to its data sheet. Startling for an LED! Even more interesting was the news from KA7OEI that the LED would survive at over 20A for a few seconds!

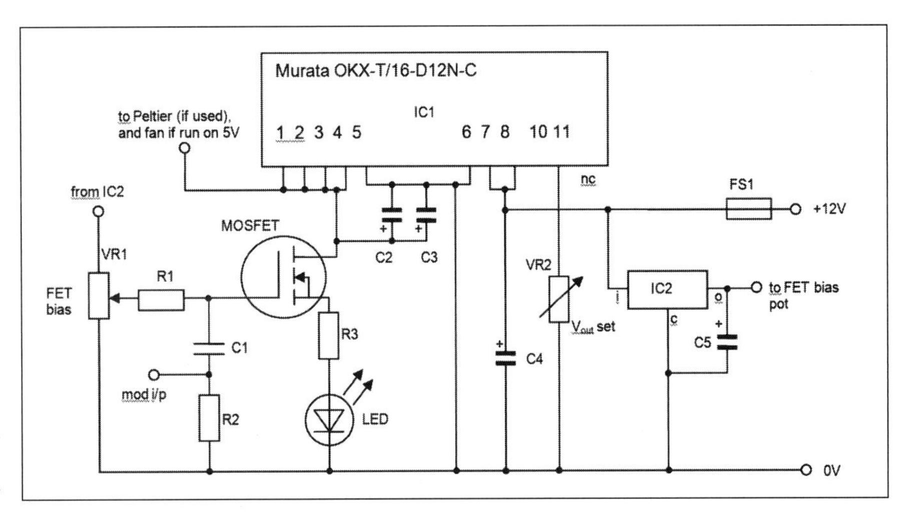


Fig 11.174: Phlatlight driver voltage converter, MOSFET driver, and LED circuit

### A simple Phlatlight LED driver for optical communications

The latest development has been a driver circuit for this LED capable of 20A peak on SSB. An extraordinary amount of red light is produced which has been received at 59 at some 30km distance with no lens!

Some of the following design is rather idiosyncratic, but it simply connects straight to my transverter with no modification to the transverter itself, and will run the LED at up to 20A peak current. Whilst designed for sub-carrier operation, it is equally suited to baseband AM with an easy adjustment of FET bias as this circuit will operate in class A for AM as well as class B for SSB.

The design uses a Murata DC-DC converter to obtain a low voltage for the LED from the more usual 12V vehicle supply or gel battery. The advantage here is that 5A at 12V from the battery is converted into 12A at 5V for the LED, and the Murata chip is rated at 16A maximum output with no heatsink required due to its greater than 90% efficiency. A low ESR capacitor is used to comfortably obtain peak output currents for SSB use in excess of this current level.

The next element of the design uses a suitably rated MOSFET to drive the LED in a source-follower circuit (**Fig 11.174**). It feels a little strange using a series resistor of just 0.05 ohms to limit the LED current, but it is all a matter of proportion and working at the level of current involved. This is a simple circuit that uses the MOSFET to bias the LED to provide whatever standing current is decided upon by the user. A modulating signal, capacitatively coupled to the gate, then appears superimposed on the source. The distortion products, even in class B, second and third harmonics were measured at -30dB or (much) better.

The MOSFET is bolted and thermally well connected to the diecast box, the low value resistor is Araldited to a corner of the box giving it plenty of thermal contact. Lead lengths are intentionally short in this area so as not to lose too much voltage at high currents. The bias arrangement makes use of a 78L08 regulator to make sure the MOSFET bias is stabilised against variations in battery voltage, the 5V output of the Murata device is insufficient here. The net result is that this system will keep going at a stable level until the nominal 12V supply is well below 11V.

It could well be argued that the overall thermal aspect of my design is rather 'belt and braces' with the Peltier effect cooler, heatsink and fan, but there is a point to this. The LED seems to be characterised for light output at 40°C and the light output increases slightly as the temperature is reduced. The implication is that up to another 20% light output is obtained by running at 20° rather than 40°. The Peltier device is run from the 5V supply and consumes 1.5A at this level. This translates to about an extra half amp from the 12V battery. There are no markings on the Peltier device and it has proved hard to obtain any information on it, but it is likely that it was run at this level in the equipment the LEDs were intended for. This could be missed out if you wish. The result of my efforts in cooling is that when tested in a room at just over 20°, the LED temperature as indicated by its integrated thermistor, is -6° when the LED runs at the 1A quiescent level I run for class B SSB. When 'talked up', the temperature hardly goes over 10°C.

This design runs somewhat 'on the edge', but to date, my spare Phlatlight remains unused. A 5A resettable circuit breaker limits the total input power to the whole driver to 60 watts averaged by another low ESR capacitor on the 12V side of the circuit.

### Thermal indication and protection

The third element in my design is a 'thermo chromic indicator', the integrated thermistor on the Phlatlight is coupled to a quad op-amp circuit that illuminates one of five small LEDs that convey information about the large LED chip temperature by effectively changing the colour of a little dot visible in the dark when out in the countryside. A line of 3mm LEDs is used with blue, green, yellow, orange and red used to indicate below 10°, between 10° and 20°, between 20° and 30°, between 30° and 40°, and above 40° respectively (**Fig 11.175**). The final state of the indicator can easily be used to switch the Phlatlight off if required. This circuit on its own may be of interest to anyone wanting an easy way to interpret temperature for any high power device. The interlink would also work by simply grounding any gate/base bias at the final temperature threshold and thereby prolonging the working lifetime of the device.

In starting the design of the indicator, I first of all noticed a simple mathematical relationship between the Phlatlight thermistor resistance and temperature (available on [62]), at the previously mentioned threshold levels, and designed a sort of cascade window comparator. No setting up is necessary providing the resistor values specified are adhered to and you don't go searching in the junk box for 'something that will do'. It is pleasing on switch-on to hear the fan start up, see the yellow led light initially, rapidly switching to green, and finally after a few more

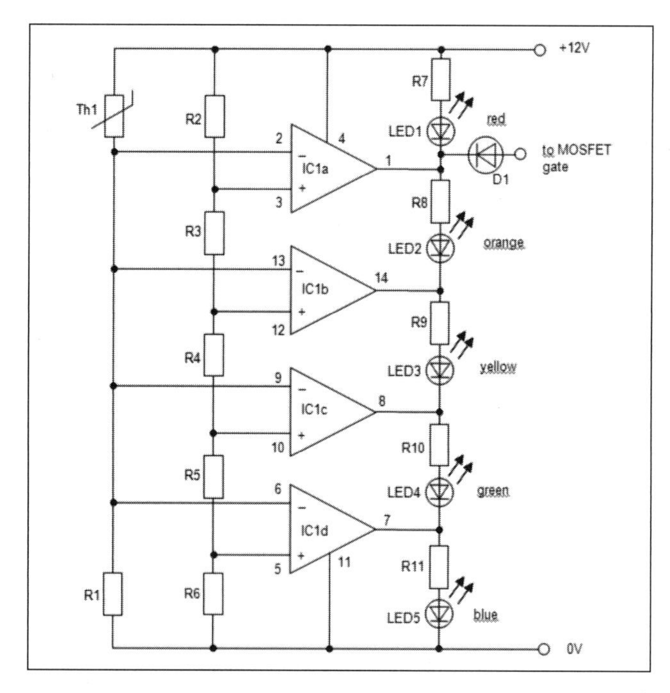


Fig 11.175: The temperature indicator designed for the Phlatlight could be used for other projects

| | |
|---|---|
| R1 | 10k |
| R2 | 1k, only required if connecting to my transverter design (it completes the output circuit in the transverter) |
| R3 | 0.05R 5W (in addition, the original LED leads add to this) |
| VR1 | 100k pot (MOSFET bias, set to 0V initially) |
| VR2 | 10k pot. set output volts) |
| C1 | 0.1µF, modulation coupling capacitor |
| C2 | 2.2µF, tantalum (decoupling for IC1, its switching frequency is around 300kHz) |
| C3 | 100µF low ESR (increase this to 1000µF for baseband use) |
| C4 | 100µF low ESR (increase this to 1000µF for baseband use) |
| C5 | 2.2µF tantalum, decoupling for IC2 |
| MOSFET: | I used a HUF 75337P3, but an IRF540 or similar would be OK, just bear in mind ID max and $RDS_{on}$. |
| IC1 | Murata OKX-T/16-D12N-C |
| IC2 | 78L08 |
| FS1 | 5A resettable fuse |
| LED | CBT54 |

Table 11.13: Components list for the Phlatlight driver

seconds, the blue led comes on, cool!

I have chosen to mount the LED in its original black plastic holder that forms a sort of clamp to hold the LED and cooler tightly together on the outside of a die cast box. It will be seen (**Fig 11.176**) that the heatsink is located inside the box with the fan blowing through the fins. Very small amounts of heatsink paste were used just to fill in the valleys in the metal-to-metal contact. The spare air from the fan circulates in the box cooling all the other high dissipation components before being vented to outside.

## Optics

I have removed the two lenses the LED came with as they rather under-illuminate the main reflector. Luckily, the dimensions of the black plastic mount supplied with the LED, after trimming off some projections and filing flat, is close enough to the best distance for mounting the meniscus lens used for optimally illuminating the A4 page magnifier Fresnel lens (46 dB gain antenna!).

The 28mm focal length 29mm diameter PMN lens from Surplus Shed [63] is glued flat on to the black plastic. Do not use super glue as I did at first - it fell off; get the Araldite out again. The resultant circular patch of red light produced by the meniscus lens just touches the short sides of the Fresnel lens leaving the corners dark. It should be added that using no lens at all here over-illuminates the Fresnel and results in some 10dB less focussed output. Another important issue is the size of the resultant LED image at distance, in this case the correct secondary lens keeps this (and the beamwidth) to a minimum of just less than a degree.

Fig 11.176: The completed Phlatlight driver

## Construction notes:

The components list is shown in **Table 11.13**. Make very solid connections to the voltage converter pins, bearing in mind the current levels increase considerably at its output. The sense input pin 3 is wired directly to the output pins as I used only 1cm of thick wire to the MOSFET drain.

Make sure the MOSFET bias pot is set to 0V before you connect up to the power supply. The original LED leads are used as part of the LED series resistance, do not shorten them. There are three wires in parallel for each LED connection, cut off the wires directly at the plug, strip and twist together before connecting to the resistor, and use a common earth point for the LED cathode and all other 0V connections. Wire the MOSFET source directly to the resistor using short direct wiring (I managed a 1cm distance here), cut off the thinner part of the MOSFET leads and connect close to the MOSFET body.

## Setting up

Make sure that the MOSFET pot is at the 0V end of the track, you have been warned! Connect a nominal 12V-13.8V power supply unit via a 10A meter and switch on. Adjust VR2 until you measure 5V on IC1 output (you could be cautious and use 4.5V to start with). Check the 78L08 output voltage; it should give its 8V output even when the supply drops to 11V. Slowly advance the MOSFET bias pot (use one of the nice new enclosed ones with a smooth action) to the point where the LED just lights.

Take a breather here, and just touch the MOSFET gate capacitor with a screwdriver in contact with your hand (the 'hum test' for a newly constructed audio amplifier), you should see the LED spring into life and brighten up just on the induced mains hum.

If you have the same Peltier cooler and fan on the 5V side as I have, increase the bias pot for sub-carrier SSB until you get an input current of 1A total, this will result in about 1A or so standing current through the LED. Or for other arrangements, monitor the voltage across the LED series resistor to achieve the same

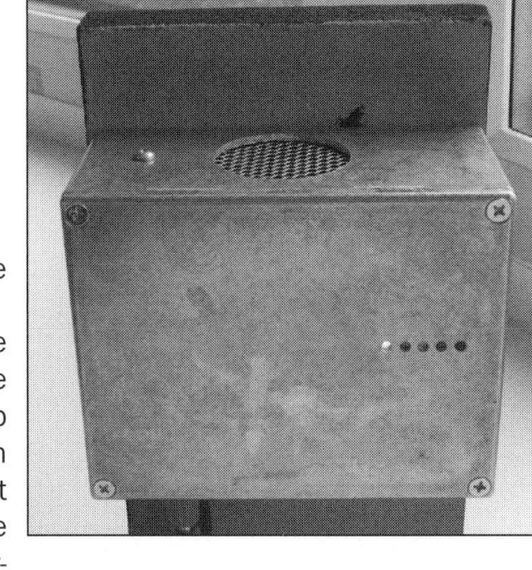

**Fig 11.177: The back of the box showing the temperature indicator LEDs**

current through the LED.

Operating from the transverter will see the current peak up on speech peaks on SSB. Be aware that you get more drive the lower the sub-carrier frequency. Test on 20kHz or so, go lower in frequency to increase power, after all you have the whole band to yourself! At 13kHz LSB I see 5A into the whole unit (this is averaged by the low ESR capacitors, now go and work out the current that must be coming out of the converter chip and then guess at the peak current through the LED!).

For baseband operations, having fitted the larger capacitors, decide for yourself where you want the 'half current' level for class A operation to be, try 5A if you dare! A volt or so of audio from an op-amp will drive the LED to 10A peak, so fit a pot on the drive to control it, and probably use a compressor/limiter as well.

If you have a variable supply, you should note that the input current to the whole system goes up as the input voltage goes down, showing the Murata chip is doing its job. The Murata converter is so efficient that it does not need to be connected to a heatsink, there is no provision for doing so, and it is simply cooled by the circulating air in the box as mentioned earlier. The MOSFET and series resistor are thermally well coupled to the box.

### LED temperature indicator

This is driven from the thermistor on the Phlatlight LED substrate. There are eight wires from the substrate, three each for the LED anode and cathode and two for the thermistor that is isolated from the LED.

The thermistor has a resistance close to 6kΩ at 40°C, 8kΩ at 30°C, 12kΩ at 20°C, and 18kΩ at 10°C, it is placed in a voltage divider circuit with a 12kΩ resistor connected between the thermistor and earth which gives a voltage across the fixed resistor of 0.67 (two-thirds), 0.6, 0.5 and 0.4 of the supply voltage at the temperatures given above.

Investigation revealed that a series combination of 33kΩ, 6.8kΩ, 10kΩ, 10kΩ, and 39kΩ would give the required threshold voltages to within a small margin of error for the comparators. By running both divider chains from the same supply neatly gives independence from the actual power supply voltage as the whole thing works in proportion.

Finally, the comparators (4 of them in a single LM324N IC) are connected so their outputs go low at the set temperatures which mean that if the +40°C limit is reached, the action of the relevant comparator going low can be used to steal the bias on the MOSFET and switch off the LED until things cool down.

I spaced the five small 3mm LEDs at 0.2 inch centres to match the holes drilled in the back of the box, using a piece of stripboard as a template to get the holes spaced correctly and in a straight line (**Fig 11.177**). Stripboard was also used also for the construction of the circuit, unconventionally using both sides of the board to obtain a very compact layout, about the same size as a large postage stamp. The symmetrical pin-out of the LM324N was a help here.

| | | | |
|---|---|---|---|
| Th1 | on-LED thermistor | IC1 | LM324N, pin numbers on circuit diagram |
| R1 | 12k | | |
| R2 | 33k | LED1 | 3mm red |
| R3 | 6.8k | LED2 | 3mm orange |
| R4 | 10k | LED3 | 3mm yellow |
| R5 | 10k | LED4 | 3mm green |
| R6 | 39k | LED5 | 3mm blue |
| R7 - 11 | all 1k, LED series resistors | D1 | 1N4148 or 1N4001 type |

**Table 11.14: Component list for the LED temperature indicator**

## AMATEUR SATELLITES

Since soon after the launch of the first artificial satellite, "Sputnik-1", radio amateurs have been constructing and operating amateur satellites. There have been more than one hundred amateur satellites launched since then, of which more than twenty are currently operating and available to amateurs. In the early days, only a few well-equipped stations with operators well versed in orbital mechanics were able to make QSOs consistently.

Nowadays, modern technology takes the strain off the mathematics and physics, making listening to and communicating via amateur satellites far easier than it used to be. Indeed it's often the case that amateurs find that they already have all the equipment and knowledge required to start operating amateur satellites.

*All radio amateurs licensed in the UK can communicate via satellites, Foundation licensees are actively encouraged to join in.*

### How do I find a satellite?

To figure out when a satellite is passing over, and where it will be in the sky at a given moment in time, prediction software is used. If the station has directional antennas, most prediction software can also steer the antennas in real time. It's important to ensure that prediction software has up-to-date orbital parameters, called Keplerian Elements, or often simply 'Keps'. These parameters are downloadable from the Internet, and most software can be configured to obtain these updates automatically.

Because satellites are moving relatively rapidly compared to the ground station, there will be some degree of Doppler shift in the receiving and transmitting frequencies. Doppler shift is experienced in daily life when, for example, the tone of an emergency vehicle's siren drops as it passes. When operating amateur satellites, Doppler shift may be adjusted by manual tuning, or alternatively most prediction software can update the radio's frequency in real time.

### Amateur satellite orbits

Although, these days, amateur satellite operators certainly don't need an understanding of orbital mechanics, knowing some basics of a satellite's orbit will help when operating.

Some satellites orbit higher in space than others. The Low Earth Orbit (LEO) satellites have coverage areas (or 'footprints') of between 3,000

**Fig 11.178: Lance Ginner, K6GSJ, poses with the flight model of Amateur Radio's first satellite, OSCAR I. The blue, stick-on label on the top of the spacecraft reads: "OSCAR I – AMATEUR RADIO BEACON SATELLITE"**

and 4500 miles range. Because of their low orbit, they complete an entire orbit in 90+ minutes, so they are only "visible" for a maximum of about fifteen minutes each orbit. Because of the short 'pass', QSOs and overs tend to be of short duration.

The 'Phase 3' High Earth Orbit (HEO) satellites such as AO-40 **Fig 11.179** had the benefit of a much larger footprint because they were designed to operate from a much higher altitude. They also appeared to the observer on the ground to be hovering around for several hours at a time, perfect for a ragchew. These spacecraft were in a highly elliptical orbit but, at the time of writing, there are no operational Phase 3 orbiting satellites.

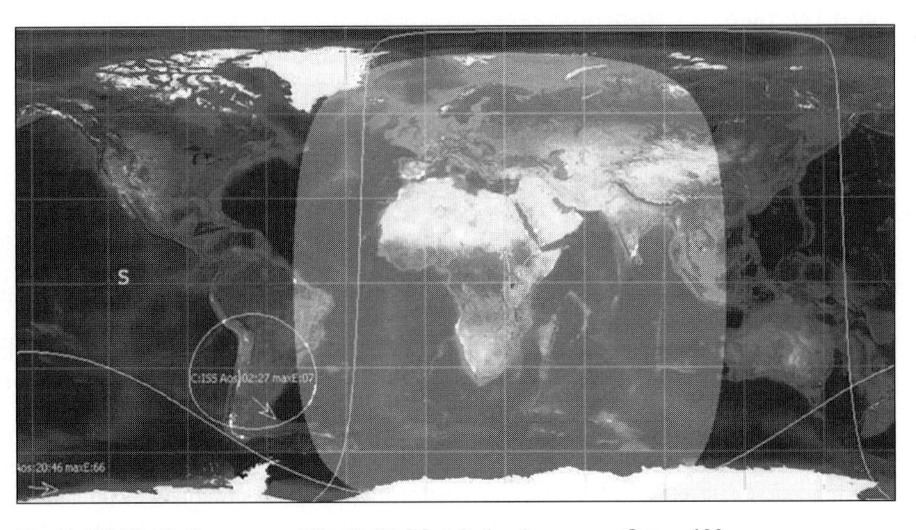

Fig 11.180: Earth Coverage of the Es'HailSat-2 also known as Oscar 100

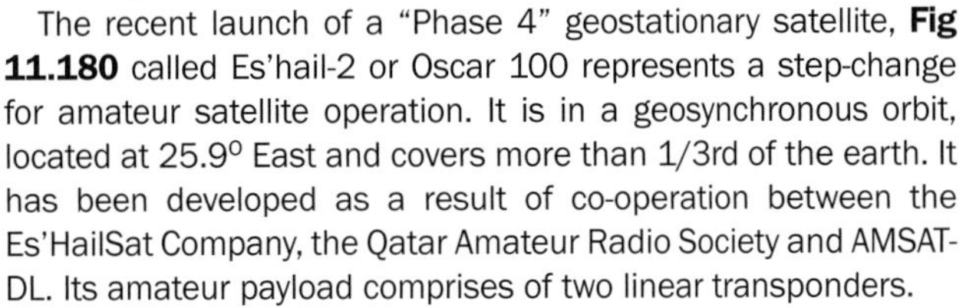

The recent launch of a "Phase 4" geostationary satellite, **Fig 11.180** called Es'hail-2 or Oscar 100 represents a step-change for amateur satellite operation. It is in a geosynchronous orbit, located at 25.9° East and covers more than 1/3rd of the earth. It has been developed as a result of co-operation between the Es'HailSat Company, the Qatar Amateur Radio Society and AMSAT-DL. Its amateur payload comprises of two linear transponders.

These transponders operate with 2400MHz uplinks and 10450MHz downlinks. They comprise of a 500kHz bandwidth linear transponder intended for conventional analogue operations and an 8MHz bandwidth transponder for experimental digital modulation schemes and amateur television using DVB-S and DVB-S2 modulation at a variety of symbol rates.

This spacecraft appears stationary in the sky and is therefore available for use 24/7. It also does not require tracking antennas so the ground station implementation can be quite simple. More information about this unique amateur spacecraft, and how to use it, can be found at *https://eshail.batc.org.uk*

## Ground Stations for operating amateur satellites

To figure out when a LEO satellite is passing over, and where it will be in the sky at a given moment in time, prediction software is used. If the station has directional antennas, most prediction software can also steer the antennas in real-time. It's important to ensure that prediction software has up-to-date orbital parameters, called Keplerian Elements, or often simply as 'Keps' or TLEs (Two Line Elements). These parameters are downloadable from the Internet, and most software can be configured to obtain these updates automatically.

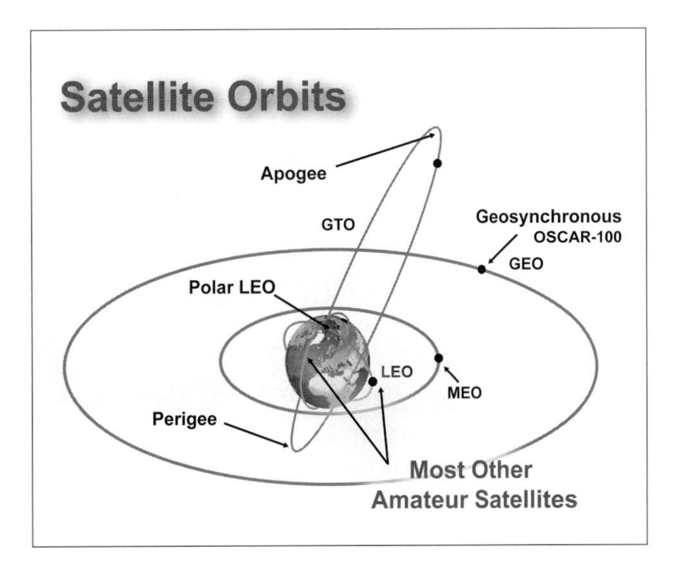


Fig 11.179: Typical orbits used by the Amateur Satellite Service

Because satellites are moving relatively rapidly compared to the ground station, there will be some degree of Doppler shift in the receiving and transmitting frequencies. Doppler shift is experienced in daily life when, for example, the tone of an emergency vehicle's siren drops as it passes. When operating amateur satellites, Doppler shift may be adjusted by manual tuning, or alternatively most prediction software can update the radio's frequency in real-time.

A well-equipped satellite station (**Fig 11.181**) has antennas of similar size to a standard domestic TV and FM Band II antenna configuration, with perhaps eight elements on 70cm and four on 2m. This makes it surprisingly easy to make a neighbour-friendly and capable amateur satellite station. These antennas are usually crossed dipoles and can either be steerable in azimuth only or, for more consistent QSOs, in elevation too.

Rather than purchasing expensive Az/El rotators, antenna pointing is very often done manually with the antennas at ground level, especially in temporary or portable configurations (**Fig 11.182**). Because the satellite is above the horizon in the sky, locations often considered ineffective for terrestrial radio communication can be effective for amateur satellites.

As technology progresses, more satellites will be operating on the higher bands. For microwave use, a small dish is employed and because of their narrower beamwidths, it's essential to be able to point the antennas accurately both in azimuth and in elevation.

Perhaps the most important rule of thumb in satellite operation is to concentrate on the station's receiving equipment before investing time, money and effort in the transmitting side. The nature of satellite communications means that the old adage 'if you can't hear them, you can't work them' is especially true. It is often tempting to improve your signal by increasing your station's ERP. It is likely that the transponder may already be limiting (eg, in linear transponders, the transponder's AGC has started to attenuate the passband so that its output can be maintained in the linear region), so more ERP is not going to be beneficial. Masthead preamps are always beneficial.

## Who makes amateur satellites?

Amateur satellites continue to be made by many organisations throughout the world, by individual national AMSAT societies, or

**Fig 11.181: Picture from N1FD showing a typical, well equipped amateur satellite ground station.**

a collection of national AMSAT societies. Many educational establishments have also launched amateur satellites.

AMSAT-UK is at the forefront of satellite building and AMSAT-UK has already created the FUNcube-1 CubeSat in collaboration with AMSAT-NL which was successfully launched in late 2013. This also carries the name Oscar AO73 and acts as a linear 70cms to 2m transponder at night and during weekends and holidays. At other times it provides telemetry data for educational outreach for schools and colleges. The sharing of this resource is intended to encourage the uptake of STEM (Science Technology Engineering & Mathematics) subjects as well as increasing knowledge about and interest in amateur radio.

This spacecraft continues to operate nominally and has provided more than 1.6GB of data which is stored on a central Data Warehouse and which is available for research purposes.

**Fig 11.182 Picture from Makezine.com showing Diana Eng with her portable ground station and homemade yagi antenna.**

AMSAT-UK also provided a similar FUNcube-2 payload for the UKube-1 spacecraft which was developed for the UK Space Agency. This spacecraft is also in orbit and although it has completed its science mission the transponder continued to function until mid-2018. Additionally, early 2017 saw the successful launch of Nayif-1 which provides a further addition to the FUNcube constellation.

Their latest launch, in late 2018, was JY1Sat. This is based on the FUNcube mission of having a transponder for radio amateurs and educational out-reach, with images in SSDV format, for schools and colleges. It was developed under the auspices of the Crown Prince Foundation in Jordan.

All artificial satellites have a limited lifetime. Solar panels gradually become less efficient as they are bombarded by the radiation in space, from which we are protected here on Earth. On other occasions, it might be that the rechargeable batteries fail first. Because of their inherent limited lifetime, there is a continual need for replacement satellites.

The longest surviving amateur satellite is AO-7, which was launched in 1974. AO-7 went silent for over two decades until Pat Gowen, G3IOR, heard it again in 2002. It is believed that AO-7 originally stopped functioning due to the battery going short circuit. After many years of the cumulative electrical and chemical stresses on the battery from the solar cells attempting to charge them, one of the cells in the battery went open circuit and now AO-7 is available again - although only when it is in sunlight. It is understood to be THE oldest earth orbiting spacecraft still in active use!

## REFERENCES


[1] CCIR, 1937

[2] *Admiralty Handbook of Wireless Telegraphy*, HMSO, London, 1938

[3] *International Microwave Handbook - 2nd edition*, editor - Andy Barter, G8ATD, RSGB 2008

[4] *Microwave Projects, and Microwave Projects 2*, editor - Andy Barter, G8ATD, RSGB 2003

[5] *UHF/Microwave Experimenter's Manual*, 'Antennas, components and design', ARRL

[6] Avago (see box on page 11.54) Web: http://www.avagotech.com/

[7] Minicircuits. Web: www.minicircuits.com

[8] *ARRL Proceedings of Microwave Update* 1987

[9] www.microwavers.org/

[10] www.kuhne-electronic.de/en/products/transverter/mku-10-g3.html

[11] http://myweb.tiscali.co.uk/g4nns/FeedHorn.html

[12] Maplin Electronics, www.maplin.co.uk

[13] Paul Wade, W1GHz HDL_Ant, - see www.w1ghz.org/antbook/contents.htm

[14] Farnell Electronics, http://uk.farnell.com/

[15] Boxes from G3NYK: http://www.xeropage.co.uk/g3nyk/componen.htm [note: "componen" is correct!]

[16] 'GHz Bands' Column, *RadCom*, June 2008

[17] Sam Jewell, G4DDK web page, www.g4ddk.com

[18] RSGB RadCom Plus website at www.rsgb.org/membersonly/publications/radcomplus/index.php

[19] 'High precision frequency standard for 10MHz. Part II: Frequency control via GPS', Wolfgang Schneider, DJ8ES and Frank-Peter Richter, DL5HAT, *VHF Communications* 1/2001 pp2-8

[20] 'High precision frequency standard for 10MHz', Wolfgang Schneider, DJ8ES, *VHF Communications* 4/2000 pp177-90

[21] Eisch-Kafka Electronics web: www.eisch-electronic.com

[22] 'HMET LNAs for 23cm', R Bertelsmeier, DJ9BV, *Dubus Technik IV*, pp191-197

[23] 'No tune HEMT preamp for 13cm', R Bertelsmeier, DJ9BV, *Dubus Technik IV*, pp191-197

[24] RF Elettronica di Rota Franco, www.rf-microwave.com/en/home.html

[25] 'Matching an LDMOS power transistor at 1.3GHz with an output of 120W'; *Proceedings of Congress on VHF-UHF 2002; Munich Technical College, 9th. - 10th March, 2002*

[26] G H Engineering, The Forge, West End, Sherborne St. John, Hants, RG24 9LE, Tel: 01256 889295, Fax 01256 889294, email: sales@ghengineering.co.uk, web: www.ghengineering.co.uk/products/

[27] Terminal boxes from Farnell: www.farnell.com/datasheets/22345.pdf, I used the KL1508 400 x 300 x 120mm box

[28] http://www.pamicrowaves.nl/website/technica_articles/PA1G2mod.pdf

[29] www.qsl.net/dl4mea/13ss/13ss.htm

[30] ON4IY, www.qslnet.de/member/on4iy/spectrian.html

[31] Garry C Hess, K3SIW: http://users.elite.net/k7xq/Spectrian%202304%20SSPA%2018%20September%202007.pdf

[32] 'GaAsFET power amplifier stage up to 5W for 10GHz', Paeter Volg, DL1RQ, *VHF Communications* magazine, 1/1995, pp.52-63

[33] '6cm transverters in modern stripline technology', Peter Vogl, DL1RQ, *VHF - UHF Munich* 1990, pp49-66

[34] '6cm transverters in stripline technology, part 2', Peter Vogl, DL1RQ, *VHF Communication Magazine* 2/1991, pp69-73

[35] 'Determination of parasitic inductances on capacitors', Roland Richer, *Special project at Dominicus vo Linprun Grammar School Viechtach*, 1994

[36] '76GHz amplifier', Sigurd Werner, DL9MFV, *VHF Communications Magazine*, 3/2003, pp163-169

[37] 'Medium Power Amplifier @ 77GHz', Infineon Technologies Data Sheet, T 602B_MPA_2; 05.05.99

[38] Data Sheet IAF-MAP7710, Fraunhofer Institut for Applied Solid-State Physics, Freiburg

[39] 'Frequency multiplier for 76GHz with an integrated amplifier', Sigurd Werner, DL9MFV, *VHF Communications Magazine*. 1/2002, pp35-41

[40] 'Amplifier for 47GHz using chip technology', Sigurd Werner, DL9MFV, *VHF Communications Magazine*, 3/2002, pp160-164

[41] 'A simple concept for a 76GHz transverter', Sigurd Werner, DL9MFV, *VHF Communications Magazine*, 2/2003, pp77-83

[42] 'Combining power at 76GHz: Three possible solutions discussed', Sigurd Werner, DL9MFV, *VHF Communications Magazine*, 1/2004, pp13-19

[43] Hittite Microwave Corporation, 20 Alpha Rd. Chelmsford, MA01824, www.hittite.com

[44] FM ATV standard for IARU region 1, Lillehammer conference report 1999

[45] Convenient procedure for filling VX-2 memories, *CQ-DL* 10/04, DARC-Verlag, Baunatal, 2004

[46] Frequency input module for 10GH ATV transmitter module, Alexander Meier, DG6RBP, *VHF Communications Magazine*, 4/2005, pp217 - 221

[47] Dubus www.dubus.org/

[48] www.triquint.com/products/p/FP31QF-F

[49] www.qsl.net/ct1dmk/reflock.html

[50] 'A single-board, no-tuning 23cm transverter', Richard L Campbell, KK7B, *23rd Conference of the Central States VHF Society, Rolling Meadows, Illinois, 1989*, pp44-52 and 'Engineering Notes', ibid, pp53-55. Subsequently re-published in *The ARRL Handbook for the Radio Amateur*, 69th edition, 1992

[51] 'A 1W linear amplifier for 1152MHz', Sam Jewell, G4DDK, *RSGB Microwave Handbook Volume 2*, pp8.21-8.23

[52] Richard L Campbell, KK7B, *Proceedings of the Microwave Update*, ARRL, 1988

[53] Down East Microwave, 954 Rt. 519 Frenchtown, NJ 08825, USA. Web: http://www.downeastmicrowave.com.

[54] Richard L Campbell, KK7B, *Proceedings of the Microwave Update*, ARRL, 1989

[55] 'VHF and microwave applications of monolithic microwave integrated circuits', Al Ward, WB5LUA, *ARRL UHF/Microwave Experimenter's Manual*, ARRL, 1990. pp7.32-7.47

[56] *Microwave Handbook, Vol 3,* edited by M W Dixon, G3PFR, RSGB, 1992, pp 18.116 - 18.118

[57] For further conversion and materials availability information contact Chuck Houghton, WB6IGP, E-mail: clhough@pacbell.net, or Kerry Banke N6IZW of the San Diego Microwave Group, E-mail: kbanke@qualcomm.com

[58] Additional conversion information articles and sources: 'Microwave GaAs FET Amps for Modification to 10GHz', C Houghton, WB6IGP & Kerry Banke, N6IZW, *Nts Feedpoint Newsletter*, December, 1993. 'UP, UP & Away to 10GHz Semi-Commercial Style', Bruce Wood, N2LIV, *Proceedings of the 20th Eastern VHF/UHF Conference*, August, 1994, p133. '10GHz Qualcomm Modification Notes' by Dale Clement, AF1T

[59] 'Suggestions for Modifications of Qualcomm LNA Board for 10GHz', Ken Shofield, W1RIL, *Proceedings of the 21st Eastern VHF/UHF Conference, August, 1995,* p63 and 'Modification Update of Omnitrack PA Board for 10GHz'. p65

[60] KA7OEI Optical Comms: www.modulatedlight.org/optical_comms/optical_index.html

***Andy Barter*** *was an apprentice at ICL in the 1960s where he trained as an electronics engineer. He spent 15 years working in the electronics industry before becoming a computer consultant for a further 20 years.*

*He was licensed as G8ATD in 1966 and has for many years spent his time constructing UHF and microwave equipment.*

*Thanks also go to AMSAT-UK*

# Propagation

# 12

*Dr Peter Duffett-Smith, G3XJE*

The work of Faraday, Maxwell, Hertz, Lodge and others in the 19th century led to demonstrations of the propagation of radio waves, and then to practical applications for communication. Indeed, the way in which radio waves can propagate over large distances without the need for wires, and the fact that the waves may be modulated to carry information, is the foundation of the radio communication industry.

In the early days (more than a century ago) the use of radio was confined to maritime and government communications with just a few amateur stations operating at frequencies above 1.5MHz. But the discovery in the 1920s, that communication was possible to great ranges with modest powers at frequencies up to 30MHz, led to major growth in the use of radio, together with a push for new technologies to explore the potential of still higher frequencies with greater available bandwidths. The stimulus for that growth was the existence of the ionosphere. This part of the upper atmosphere was first conjectured by Balfour Stewart in 1878, based on studies of the variations in the Earth's magnetic field, and was shown to exist by Appleton and others in the mid 1920s. In 1932 Robert Alexander Watt (later called Watson-Watt) gave the region a general name - he called it the ionosphere. Even today the ionosphere and its complex variations are not completely described, and there is still a place for careful experimentation and detailed record-keeping by radio amateurs of observed propagation behaviours. In this chapter, we may only scratch the surface of a very wide-ranging topic. Don't be put off by the appearance of the few equations. There is nothing very difficult here and a scientific calculator or spreadsheet can handle the tasks with ease.

As you read, compare the text with your own experiences on the bands. In all probability you know already much about the practicalities of radio propagation. Here, we hope to provide you with some underlying explanations. Be warned, though. This subject is strongly addictive!

In free space, power radiated from a radio antenna spreads out into space in a simple way. Our interest is in the effects on propagation caused by the Earth - obstructions of both the topography of the Earth and man-made structures - and particularly the atmosphere and ionosphere.

This chapter provides an introduction to propagation characteristics as they vary with frequency. It touches on the various effects caused by the ever-changing ionosphere at frequencies up to 30MHz. At higher frequencies, we consider how waves may be diffracted by obstacles, and how refraction in the troposphere and rain in the atmosphere affect things.

## ELECTROMAGNETIC WAVE SPECTRUM

Radio waves are a part of the whole spectrum of electromagnetic waves and are arbitrarily defined by the International Telecommunication Union (ITU) Radio Regulations as electromagnetic waves with frequencies lower than 3,000GHz (3 THz). In fact, frequencies are internationally allocated to radio services in the range 8.3kHz to 275GHz. The atmosphere becomes increasingly opaque for frequencies above about 500GHz, but it becomes nearly transparent again at infra-red and optical frequencies (as we know because we can see things).

The position of radio waves in the electromagnetic wave spectrum is shown in **Fig 12.1**. A logarithmic scale is used so as to fit the whole spectrum into one diagram, exaggerating the bandwidth available at the lower frequencies and under- representing it at the higher frequencies. Electromagnetic waves of all frequencies travel in free space with a speed of 2.99790 x $10^8$ m $s^{-1}$ (generally taken as about 300,000km $s^{-1}$). This is popularly known as the 'speed of light' although visible light forms but a fraction of the whole range.

The radio wave portion of the spectrum has been conveniently divided by the ITU into a series of bands based on successive orders of magnitude in wavelength. **Table 12.1** gives the recognised acronyms used to describe each decade of the frequency range of radio waves. These are useful for general discussions, although, in most cases, propagation characteristics change with frequency within each decade so the decades do not fit the purpose very well.

## RADIATION

Electromagnetic radiation is caused by the acceleration of charges. A static electric charge in free space has an isotropic radial electric field, or E-field, associated with it. The E-field at a point is a mathematical concept which describes the magnitude

| Acronym | Frequency range | Wavelength range |
|---|---|---|
| VLF | 3 - 30kHz | 100km - 10km |
| LF | 30 - 300kHz | 10km - 1km |
| MF | 300kHz - 3MHz | 1km - 100m |
| HF | 3 - 30MHz | 100m - 10m |
| VHF | 30 - 300MHz | 10m - 1m |
| UHF | 300MHz - 3GHz | 1m - 10cm |
| SHF | 3 - 30GHz | 10cm - 1cm |
| EHF | 30 - 300GHz | 1cm - 1mm |

**Table 12.1: Acronyms used to refer to the sections of the radio spectrum**

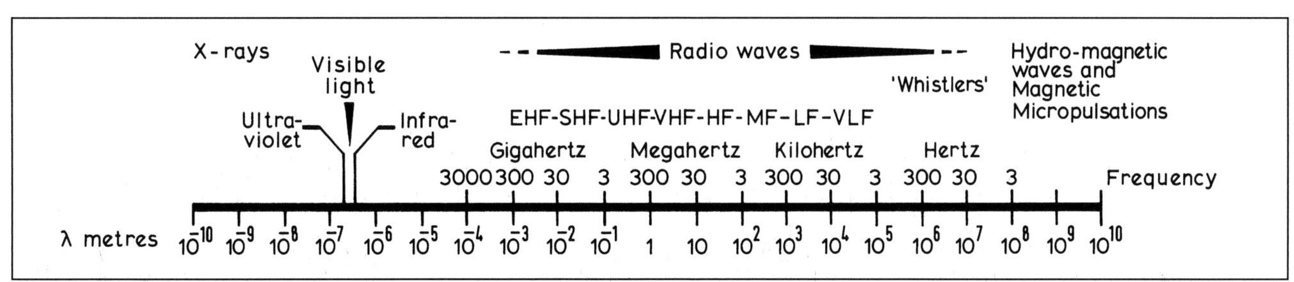


**Figure 12.1: The spectrum of electromagnetic waves. This diagram shows on a logarithmic scale the relationship between X-rays, ultra violet and visible light, heat (infrared), radio waves and the very long waves associated with geomagnetic pulsations, all of them similar in basic character.**

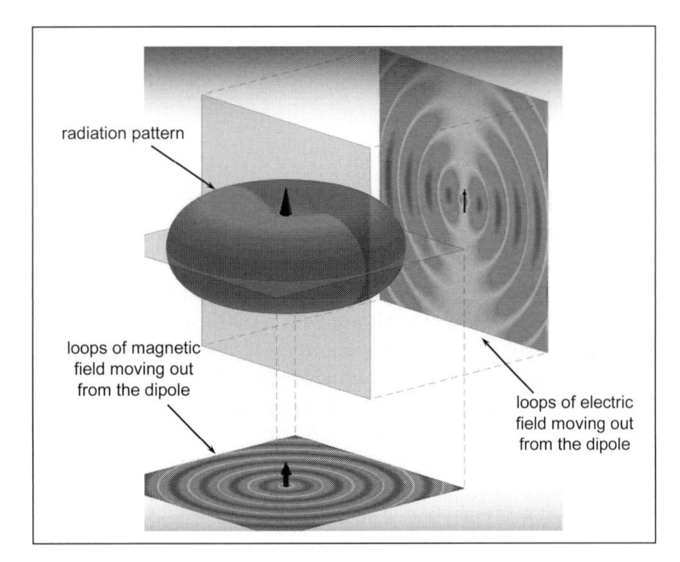


**Figure 12.2: Radiation from a dipole, illustrating the loops of electric and magnetic field propagating outwards. See the text for details.**

and direction of the force exerted on a hypothetical unit positive test charge placed at that point. The static electric charge has no magnetic field associated with it, but if you now move the electric charge at a constant speed in a straight line, a magnetic field appears in addition to the electric field, which takes the form of circles centred on the line of movement of the charge. Neither the electric nor the magnetic fields carry any energy away with them, and do not form part of an electromagnetic wave. However, if you now *accelerate* the charge, for example by oscillating it backwards and forwards, additional electric and magnetic fields are created which break free and propagate as loops of ever-increasing size moving at the speed of light away from the oscillating charge. These are electromagnetic waves. Their frequency is the same as the frequency of oscillation of the charge.

The simplest radiator of electromagnetic waves is the dipole. A monopole, for example a static electric charge, does not radiate as discussed above. You need acceleration in a given direction to cause radiation, and the simplest case is that of a charge oscillating sinusoidally backwards and forwards in a line. Figure 12.2 illustrates the loops of the electric and magnetic field components of the electromagnetic waves radiated by an oscillating dipole. The dipole is at the centre of the figure and is pointing upwards as shown by the tip of the arrow. The doughnut-shape around the dipole illustrates the radiation pattern of the dipole, showing that there is maximum radiation at right-angles to the line of the dipole in all horizontal directions, and zero radiation in the line of the dipole. The loops of electric field, breaking away from the dipole and propagating outwards, are shown projected from the grey plane square intersecting the doughnut in line with the dipole onto the plane 'screen' at the back of the diagram. The circular loops of magnetic field, breaking away from the dipole and propagating outwards, are shown projected from the grey plane square intersecting the doughnut at right-angles to the dipole onto the plane 'screen' at the bottom of the diagram. (This diagram has been adapted from an excellent animation which you are strongly encouraged to view at *https://youtu.be/UOVwjKi4B6Y*). The electric and magnetic components are tightly coupled together, varying in phase with each other and with their amplitudes in a constant ratio. When expressed in consistent units, the ratio of the electric field strength to that of the magnetic field is known as the 'impedance of free space', equal to approximately 377 Ohms.

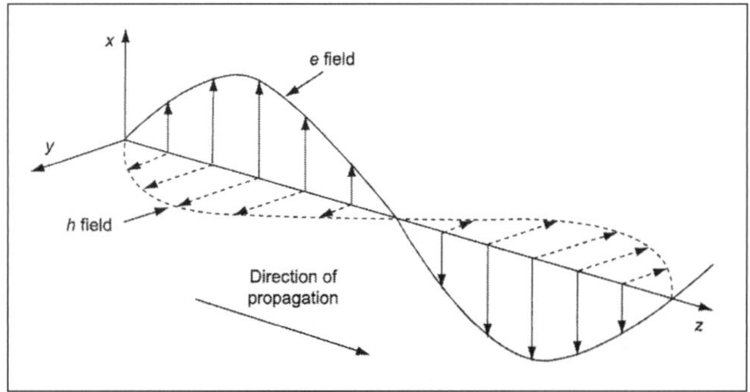


**Figure 12.3: Vector fields in an advancing plane wave [©IET]**

Note that the radiation is not uniform in all directions, as it would be if it were isotropic. In particular, you see that there is no radiation at all in line with the dipole, and maximum radiation perpendicular to it. This means that, even in this simplest of cases, there is directionality associated with the radiation, and you would need to specify the orientation of the dipole with respect to a given direction of propagation in order to describe the relative strength of the wave. This directionality is often referred to as the 'directivity' of the antenna, or the 'gain' when losses are also included and is usually expressed as a power ratio in dB. Even though an isotropic radiator does not exist, it is nevertheless convenient to imagine that it does and to use the power that would be in the wave from this hypothetical construct as the reference. Thus a dipole has a gain of +2.15 dB in a direction perpendicular to its line and minus infinity dB in its line, and its gain (i.e. in the direction of maximum propagation) is said to be 2.15 dB*i* (the *i* stands for the case of using an *i*sotropic radiator as the reference). Sometimes, antenna gains are referred to the dipole. In this case, a dipole itself would have a gain of 0 dB*d* (the *d* indicating a *d*ipole).

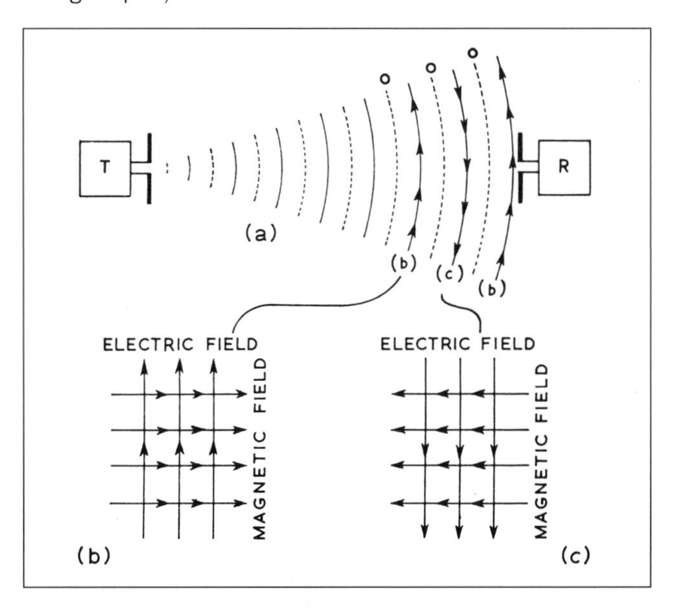


**Figure 12.4: (a) A vertical section through the fields radiated from a vertically-polarised transmitting antenna. The expanding spherical wave front consists of alternate reversals of electric field, shown by the arrowed arcs; nulls are shown dotted. At right angles to the plane of the paper, but not seen here, would be simultaneous alternate reversals of magnetic field. (b) and (c) These two 'snapshots' must be rotated through 90° in imagination, so that the magnetic field lines run into and out of the plane of the paper. The view is from T towards R.**

The electric and magnetic field strengths each vary as the inverse of the distance from the antenna. The power density (W $m^{-2}$) is proportional to the product of the fields, and hence varies as the inverse square of the distance. This is to be expected for a radially-expanding wave where the same amount of power must be spread over the area of a sphere of expanding radius. At large-enough distances, the small part of the wave front as seen by an observer appears to be a plane surface.

The two fields of an electromagnetic wave constantly change in magnitude and reverse their directions with every half-cycle of the transmitted carrier. **Figs 12.3 and 12.4** are two views of an advancing wave. **Fig 12.4** shows successive portions of wave fronts passing a receiving antenna - another dipole in this case - in which is induced a received signal which follows all the changes carried by the field and therefore reproduces the current in the transmitting dipole, but at a much lower amplitude. All of the characteristics of the transmitted signal are therefore reproduced in the receiving antenna.

The lines of the electric field are parallel to the line of the transmitting dipole. By convention, the direction of the electric field defines the direction of polarisation of the radio waves relative to the local plane of the Earth's surface. Thus horizontal dipoles propagate horizontally-polarised waves and vertical dipoles propagate vertically-polarised waves. In free space, remote from ground effects and the influence of the Earth's atmosphere, these polarisation senses do not change with distance and a receiving antenna aligned with the electric field would respond to the whole of the incident field. Another turned through 90 degrees would receive nothing as its direction would then be orthogonal to the incoming wave's polarisation. Besides these plane-polarised waves (horizontal or vertical), it is also possible to generate elliptically-polarised waves in which the instantaneous electric field rotates as it passes by, making one revolution every wave period, with the electric field vector tracing out the form of an ellipse (circularly-polarised waves are a special case).

Propagation between two points on the Earth is further complicated by variations in the conductivity of the Earth itself, the sense of polarisation, obstacles in the way such as hills, the characteristics of the atmosphere and troposphere, and the presence of the ionosphere. All of these things can cause changes in the sense of the received polarisation. Plane polarisation may be rotated or a degree of cross-polarisation may be introduced which results in signals arriving at the receiving antenna with elliptical polarisation.

## Propagation in Free Space

Imagine that we have a mythical isotropic radiator in free space. This emits power uniformly in all directions. The radiation spreads outwards, and at a distance $d$ metres the power is spread over the surface of a sphere of that radius, so that the power flux density $S$ (W $m^{-2}$) is given by

$$S = \frac{P_t}{4\pi d^2}$$

where $P_t$ is the total power radiated (W).

In a similar way to Ohm's law, the field strength $E_i$ (V $m^{-1}$) at that location is given by

$$S = \frac{E_i^2}{Z_0}$$

where $Z_0$ is the impedance of free space, 377 Ohms. Combining these two equations together and eliminating $S$ we get

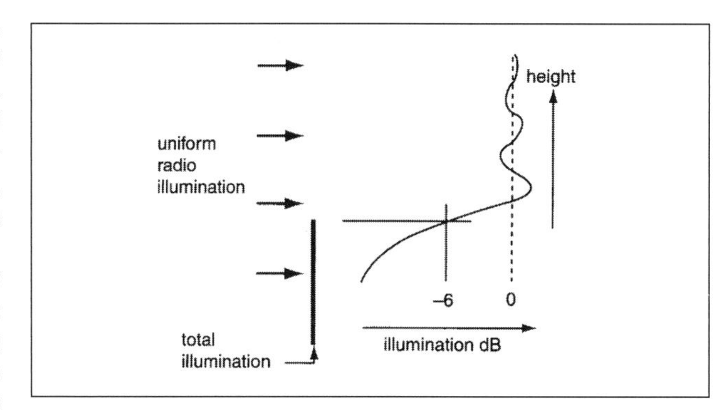


**Figure 12.5: Variation of signal strength at the edge of a radio shadow. [©IET]**

$$E_i = \sqrt{\frac{30P_t}{d^2}}$$

where the units of $E_i$ are V $m^{-1}$ if $P_t$ is expressed in W.

A half-wave dipole has a maximum gain, broadside on to the dipole, of 1.64 times (2.15 dB) compared to an isotropic radiator, so that with such a transmitting antenna the maximum field strength is approximately

$$E_d = 7\sqrt{\frac{P_t}{d^2}}$$

volts $m^{-1}$, which may be a convenient relationship to keep in mind as a rule of thumb. Note that this equation applies to the so-called *far field*, so $d$ should be greater than about 10 wavelengths or so.

The relationship between the gain, $G$ (expressed in natural units, not in dB), the wavelength, $\lambda$ (m), and the 'capture area', $A$ (i.e. the 'effective aperture', $m^2$), is

$$G = \frac{4\pi A}{\lambda^2}$$

.

A mythical isotropic receiving antenna has unity gain and therefore an effective aperture of $\lambda^2/4\pi$ , so that the power available in a matched receiver input is:

$$P_r = S\frac{\lambda^2}{4\pi} = \frac{P_t\lambda^2}{16\pi^2 d^2}$$

The ratio between the power radiated and the power received is the free space transmission loss, $L_{bf}$ , is therefore

$$L_{bf} = \frac{P_t}{P_r} = \frac{16\pi^2 d^2}{\lambda^2} = \frac{16\pi^2 d^2 f^2}{c^2}$$

where $f$ is the frequency in Hz and $c$ is the speed of light in $ms^{-1}$, since $f\lambda = c$ .

Expressing this in dB by taking $10\log_{10}(L_{bf})$, and changing the units of $f$ to MHz and $d$ to km, we get

$L_{bf} = 32.4 + 20\log_{10}(f) + 20\log_{10}(d)$dB.

This expression gives the basic transmission loss in free

| Type of ground | Permittivity | Conductivity | Penetration depth (m) | | |
|---|---|---|---|---|---|
| | ε | σ (S/m) | 100kHz | 1MHz | 10MHz |
| Sea water | 70 | 5 | 0.8 | 0.25 | 0.08 |
| Fresh water | 80 | 0.003 | 33 | 15 | 10 |
| Very moist ground | 30 | 0.01 | 17 | 5 | 3 |
| Average ground | 22 | 0.003 | 28 | 12 | 8 |
| Dry ground | 7 | 0.0003 | 80 | 50 | 40 |

**Table 12.2: Ground characteristics**

| Distance km | Inverse distance field* | Sea σ = 4 S/m 136 Hz | 475 Hz | 1.8 MHz | 3.5 MHz | Land σ = 0.003 S/m 136 Hz | 475 Hz | 1.8 MHz | 3.5 MHz |
|---|---|---|---|---|---|---|---|---|---|
| 3 | 100 | 0 | 0 | 0 | 1 | 0 | 1 | 8 | 18 |
| 10 | 90 | 0 | 0 | 1 | 1 | 1 | 3 | 18 | 29 |
| 30 | 80 | 0 | 0 | 1 | 2 | 2 | 6 | 28 | 41 |
| 100 | 70 | 1 | 1 | 2 | 7 | 3 | 16 | 44 | 56 |
| 300 | 60 | 1 | 3 | 8 | 23 | 4 | 32 | 68 | 85 |

**dB relative to 1μV/m for 1 kW radiated power*

**Table 12.3: The number of decibels to be subtracted from the calculated free-space field in order to take into account various combinations of ground conductivity and distance. The values are shown in each case for four LF/MF bands. Vertical polarisation is assumed**

space, taking account of the frequency and the distance. On the Earth we need to add the losses caused by propagation effects in the troposphere, and ionosphere, and reflections or scatter from the Earth's surface or obstructions to the radio path. It is also necessary to take account of the gains of the transmission and receiving antennas, and losses in the feeders. Hence we have

$$P_r = P_t + G_r + G_t - L_{ft} - L_{fr} - L_{bf} - L_m,$$

where $P_r$ is the received power available in the receiver and $P_t$ is the output power from the transmitter terminals, both expressed either in dBm or in dBW, $G_r$ is the gain of the transmitting antenna in the direction of the path relative to an isotropic radiator (dBi), $G_r$ is the gain of the receiving antenna in the direction of the path relative to an isotropic radiator (dBi), $L_{ft}$ is the loss in the feeder system connecting the transmitter terminals to the transmitting antenna (dB), $L_{fr}$ is the loss in the feeder system connecting the receiver terminals to the receiving antenna (dB), $L_{bf}$ is the free-space transmission loss derived above (dB), and $L_m$ is the additional propagation loss (dB).

## Reference Antennas and Radiated Power

The discussion above has used the concept of an isotropic antenna that radiates or receives equally in all directions in space. We saw previously that such an antenna is an idealised concept that cannot be made in practice. It is however a convenient and simple reference against which real antennas may be compared. Most antenna manufacturers quote antenna gains using an isotropic comparison, using dBi (the gain in dB in the direction quoted as compared with an isotropic antenna at the same location).

Another reference antenna commonly used, particularly at VHF, is a tuned $\lambda/2$ dipole in the direction of maximum radiation, at right angles to the dipole itself; this antenna has a gain of 2.15 dBi.

It is often useful to describe the power radiated from an antenna in a particular direction. This may be done by estimating or measuring the power received at a location with the real system, and then to quote the power that would be needed to be radiated from an isotropic antenna at the transmitter location to give the same received power. This is known as the Effective (or Equivalent) Isotropically-Radiated Power, EIRP.

In the same way if the real antenna is replaced by a $\lambda/2$ dipole, oriented to give maximum signal in the direction of propagation, the necessary power would be the Effective (or Equivalent) Radiated Power, ERP.

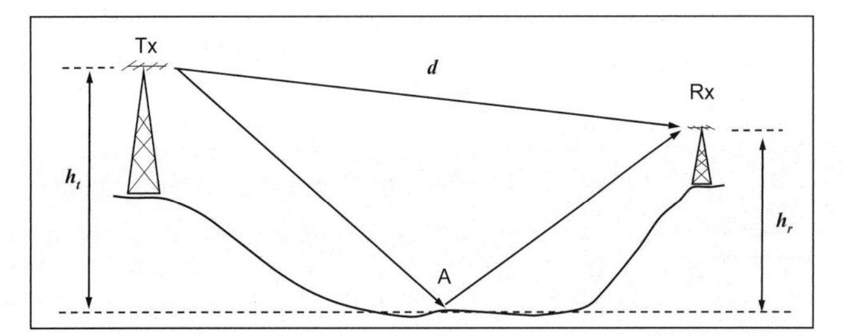


**Figure 12.6: The direct and reflected rays. Note that the antenna heights (shown greatly exaggerated in this diagram) are above the reflecting surface at point A.**

# MODES OF PROPAGATION

There are several different mechanisms which affect radio wave propagation. They are:

- travel in free-space, affected only by spreading with distance;
- travel in the Ionosphere, where the actions of free electrons in the upper levels of the Earth's atmosphere may yield terrestrial propagation by reflection etc. in the ionosphere, and which may modify the trans-ionospheric propagation to or from satellites or the Moon;
- travel in the troposphere, where variations in the refractive index structure of the atmosphere and meteorological conditions (such as rain) may alter the directions of radio waves;
- travel past obstacles and over surfaces, which may diffract, reflect, scatter or absorb radio waves; and
- travel as surface waves, often called ground waves, where the waves follow the surface of the Earth and are modified by the nature of the terrain over which they travel.

# DIFFRACTION

Diffraction is an alteration in direction of the propagation of a wave around an obstacle, so that some of the energy appears in the geometric shadow of the obstacle. We have already discussed propagation in free space where the waves spread out radially from the transmitting antenna. It is well known amongst radio amateurs working at frequencies above 100MHz that radio waves may be received when there is no clear path between the transmitter and receiver, over hills and roof tops, around obstructions, in road cuttings, etc. These waves may have been diffracted around the obstacles. Radio waves also diffract around the bulge of the Earth.

## Knife Edge Diffraction

The variation in the signal received beyond a knife edge is illustrated in **Fig 12.5**. Waves passing just above the edge show some variation with height. At the geometric edge there is a loss of 6dB - the field strength is halved - while in the geometric shadow the signal decreases. The effect varies with frequency. At VHF the signal spreads far into the shadow region but as the frequency increases towards SHF the shadow becomes much sharper.

A true knife edge may not often occur in practice, but results are similar over a ridge line, around the corner of a building, etc. On long routes, the diffracted signal over a prominent hill may be better than that over a smooth path of similar length. Where hill tops are rounded or irregular, as often occurs, diffraction still occurs but with greater losses. Techniques have been developed to predict the signal on long paths passing over a succession of terrain obstacles.

# SURFACE WAVE PROPAGATION

In contrast to the diffraction at higher frequencies described above, surface wave propagation, another mode, enables propaga-

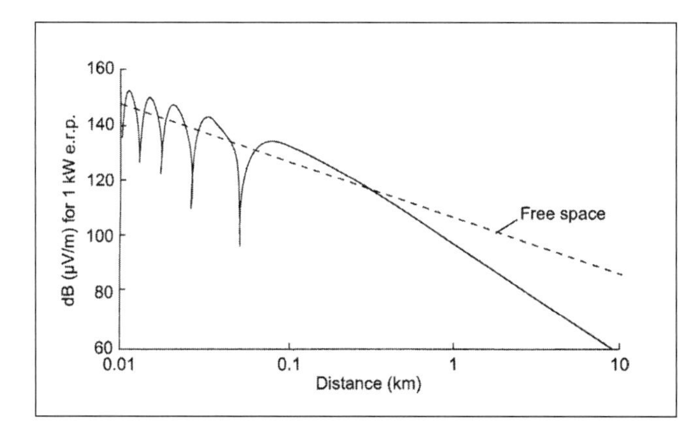


**Figure 12.7: An example of the variation of field strength with distance for a path with a single ground reflection. In this example, the power is one kW ERP, the frequency is 900MHz, and the antenna heights are both 3 m above the reflecting surface. The reflection coefficient is taken to be -1 [©IET]**

tion at lower frequencies around the curve of the Earth to long ranges. This mode is only usable with vertical polarisation and is attenuated because of the absorption of energy in the ground. The electrical characteristic of the ground are the conductivity, $\sigma$ (sigma, measured in units of Siemens per metre, S m$^{-1}$) and the relative permittivity, $\varepsilon$ (epsilon).

The values of $\sigma$ and $\varepsilon$ for various types of ground are given in **Table 12.2**, valid at frequencies of 100kHz, 1MHz, and 10MHz. This Table also gives the depth of penetration into the ground, d (delta), defined as the depth in metres at which the signal is attenuated by a factor of $1/e$ (37 per cent) of its value at the surface. At very low frequencies the field strength decreases inversely with the distance and **Table 12.3** gives the additional attenuation for various frequencies and ground types. However, it should be borne in mind that at the longer ranges there will also be an additional ionospheric component to the signal.

## REFLECTION AND SCATTER

The surface wave, described above, is important at low frequencies with low antenna heights, but at higher frequencies, particularly at VHF and above, antennas will often be several wavelengths above ground and there may be a direct line-of-sight path from transmitter to receiver, more-or-less with the characteristics of free space propagation, or a direct path from the transmitter to an obstacle beyond which the signal is diffracted.

In the line-of-sight case, there may be a second path with a reflection from the ground. **Fig 12.6** shows the geometry. Where the path lengths are much greater than the antenna heights, the angle of the reflected ray to the surface is very small and then, over real Earth, for both vertical and horizontal polarisation, the reflection coefficient at the reflecting surface is approximately -1. The reflected ray has a longer path length and so will have a phase delay with respect to the direct ray.

The resultant received signal power, $P_r$, combining the direct and reflected rays, is given approximately by

$$P_r = \frac{\lambda^2}{(4\pi d)^2}\left(2\sin\left(\frac{2\pi h_t h_r}{\lambda d}\right)\right)^2 G_t G_r P_t$$

where l is the wavelength, $d$ is the horizontal distance between the two antennas, $h_t$ and $h_r$ are the heights of the transmitting and receiving antennas respectively above the *reflecting surface* (not necessarily the local ground level - see **Fig 12.6**), $G_t$ and $G_r$ are the gains of the transmitting and receiving antennas respectively in their line-of-sight directions, and $P_t$ is the transmitter power. The heights, the distance, and the wavelength must all be expressed in the same units (usually metres). This expression is a good approximation when the distance between the antennas is much greater than the heights of the antennas above the reflecting surface. A flat Earth is also assumed.

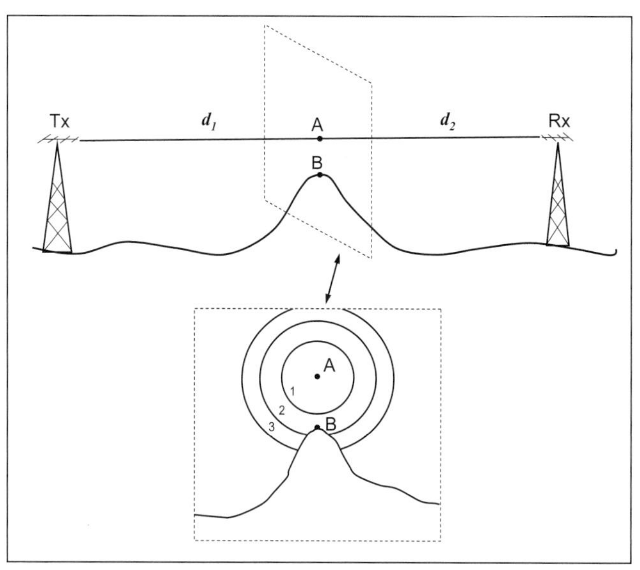


**Figure 12.8: Fresnel half-period zones around an obstacle. You can generally ignore Fresnel effects provided that the distance AB is at least 0.6 of the radius of the first Fresnel zone**

**Fig 12.7** shows an example of this combination. The nulls occur when the two components are out of phase, and the peaks, 6 dB above the free space level, are when they are in phase. At longer ranges the phase difference between the two rays decreases and does not reach the out-of-phase condition. At lower frequencies, or for greater antenna heights, the nulls would be farther apart.

A similar result would also be obtained for a fixed distance if one of the antennas is progressively raised in height. There would be a succession of peaks and nulls in the received signal. With a good reflecting surface and an open area, the maxima would have a transmission loss 6 dB smaller than that of free space; from this a method of measuring the combined gains of the antennas at VHF and UHF may be developed.

However, in a real environment such as a built-up area the situation would be much more complicated, since there would not usually be a line of sight path. Prediction methods may take, as a basis, the field strength at roof-top height, assumed to be 10 m in suburban areas.

Lower antenna heights, amongst the 'clutter' of the environment, would generally receive a weaker signal. This would be variable, dependent on the nearby environment and would vary with antenna height. As a rule of thumb, the expected reduction in median field strength at three example frequencies, when the antenna height is reduced from 10 to 3 metres, may be taken as

- 9 - 10 dB at 50 - 70MHz;
- 8 dB in flat terrain and 5 dB in urban or hilly areas at 145MHz; and
- 6 - 7 dB in suburban areas at 430MHz, although the median values are dependent on terrain irregularity.

### Multipath

The preceding section deals with the combination of a direct ray and one reflected from the ground. This does occur in practice but is a simple depiction that may not paint a realistic pic-

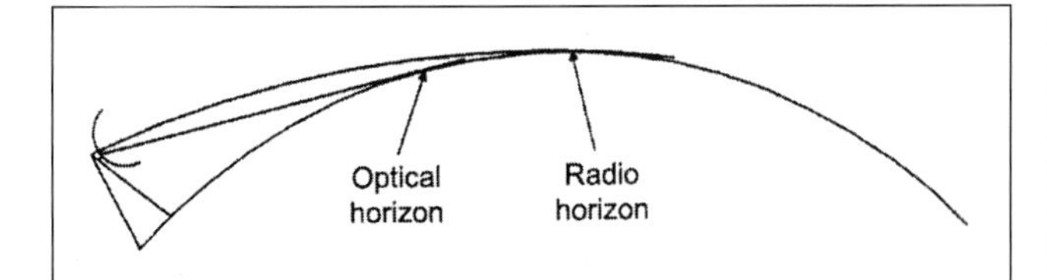


**Figure 12.9: Illustrating the extension of radio horizon caused by tropospheric bending (greatly exaggerated). [©IET]**

ture of the way in which the signals received at a distant location depend on the radio frequency and the distance from the transmitter, excluding other effects caused by the troposphere and the ionosphere.

The reason for this is that the wave incident on the receiving antenna is rarely only the one that has arrived by the most direct path but is more often the resultant of several waves which have travelled by different routes and have covered different distances in doing so. If these waves should eventually arrive in phase they would act to reinforce one another, but should they reach the receiving antenna in antiphase they would interfere with one another and, if they happened to be equal in amplitude, would cancel one another completely.

These alternative paths may arise as a result of reflections from the sides of buildings or other structures, or from the ground as discussed above. If the reflecting surfaces and the radio locations are stationary, the phase differences (whatever their values) would be constant and a steady signal would result.

It may happen that a surface of reflection is in motion as it would be if it was part of an aeroplane flying near the transmission path. In that case the distance travelled by the reflected wave would change continually and the relative phases would progressively advance or retard through successive cycles, leading to alternate enhancements and diminutions as the waves aid or oppose one another. (One can also regard this in terms of the Doppler effect. The reflected wave has a slightly higher or lower frequency and 'beats' occur between the reflected and stationary components.) If you remember the days of analogue TV signals, you might recall the fluttering of the picture as an aircraft passed by. UHF and low microwave band operators have used reflections from aircraft to make radio contacts.

The waves along the reflected path repeat themselves after intervals of exactly one wavelength, so it is only the portion of a wavelength 'left over' which determines the phase relationship in comparison with the direct-path wave. This suggests that relatively small changes in the position of a receiving antenna could have profound effects on the magnitude of the received signal when multiple paths are present, and this is indeed found to be the case, particularly where the reflection location is near at hand.

## Fresnel Zones

So far, the radio path has been considered as a narrow ray between the transmitter and receiver. This simple idea is often sufficient but it is more realistic to envisage the received signal as being the result of combining a bundle of rays.

Imagine that you are at the receiver and 'looking' towards the transmitter, as in Figure 12.8. There is an obstacle in the way, but not tall enough to block the direct path. The circles centred on the direct path are called 'Fresnel half-period zones' and are of radius given approximately by

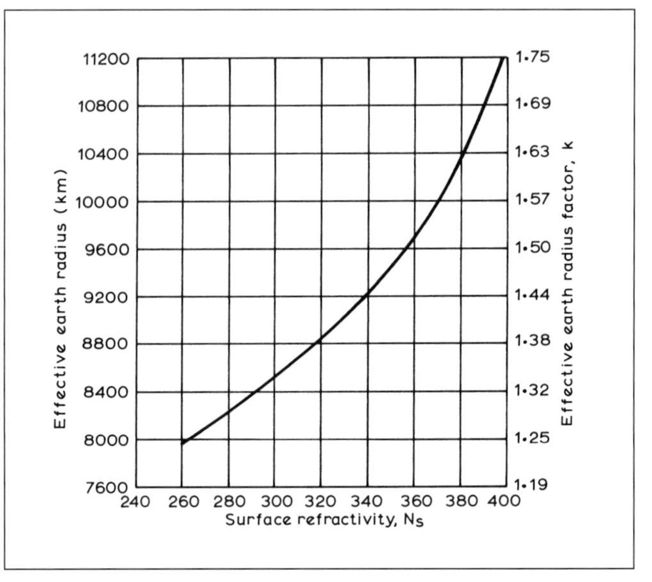


**Figure 12.10: The effective Earth radius corresponding to various values of the surface refractivity.**

$$R_n = \sqrt{\frac{n\lambda d_1 d_2}{(d_1+d_2)}}$$

where $d_1$ is the length of the path from the transmitter to the obstacle, $d_2$ is the length of the path from the obstacle to the receiver, $\lambda$ is the wavelength, and $n$ is one for the first half-period zone, two for the second and so on. (This approximation is valid for $d_1$ and $d_2$ both being very much larger than $n\lambda$.) The bundle of rays within the first zone arrive at the receiver more-or-less in phase with the direct ray, whilst those arriving from the second zone are more-or-less out of phase with the direct ray and tend to cancel those in the first bundle, thus reducing the resultant amplitude. The rays from the third zone again arrive (more or less) in phase with those in the first zone, and tend to add, and so on. The signal amplitude at the receiver is the vector sum of all the rays from all the unobscured zones and parts of zones. In practice, it is rarely necessary to consider more than the first three zones. This is the explanation for the periodic fluctuations in signal above an obstruction as shown in **Fig 12.4**.

The change in signal amplitude caused by Fresnel effects tends to be rather small as long as there is sufficient clearance above an obstruction; as a rule of thumb, you can ignore Fresnel effects provided that the clearance is at least 0.6 of the radius of the first Fresnel zone.

## Scatter

Real environments tend to be complicated: there may be many reflecting surfaces which may also be 'rough', that is they have irregularities which are of the same order of size as the wavelength. There may also be small component signals from surface features of buildings and other structures, and reflections from moving vehicles, etc. All these may result in a spread of weak signals each with its own time delay and shift. Combining all these together gives a resultant signal which is often called 'scatter'.

# PROPAGATION IN THE TROPOSPHERE

## The Standard Atmosphere

The propagation mechanisms already described - reflection, diffraction and surface waves - do not require the presence of the atmosphere. But the presence of the lower atmosphere enables propagation to longer ranges and is of special interest. The *troposphere* is that lower portion of the atmosphere in which the general tendency is for air temperature to decrease with height. It is separated from the stratosphere, the region immediately above, where the air temperature does not change much with height, by a boundary called the tropopause, usually at around 10 km. The troposphere contains all the well-known cloud forms and is responsible for nearly everything loosely grouped under the general heading of 'weather'.

In the general case, the atmosphere becomes less dense with increasing height, so that the refractive index of the air becomes closer and closer to unity. The effect upon radio waves is to bend them, generally in the same direction as that taken by the Earth's curvature, as a result of successive changes in the refractive index of the air through which the waves pass. In extreme cases this includes the optical phenomenon of the appearance of mirages, where objects beyond the horizon are brought into view by ray-bending, in that case resulting from temperature changes along the line-of-sight path. In the case of radio signals the distribution of water vapour also plays a part, often a major one where anomalous propagation events are concerned.

The refractive index, $n$ , of a sample of air is close to unity and the departure from unity is called the *refractivity*, $N$ , where $n = 1 + 10^{-6} N$ . An approximate value of $N$ is given by

$$N = \frac{77.6}{T}\left(P + 4810\frac{e}{T}\right)$$

where $P$ is the atmospheric pressure and $e$ is the water-vapour pressure, both expressed in millibars or hecto-Pascals (hPa), and $T$ is the temperature in degrees Kelvin (273.16 plus degrees Celsius).

In a standard atmosphere, the refractivity decreases by about 40 units in the first kilometre of altitude resulting in the situation where horizontal rays are bent down towards the Earth. Hence, the radio horizon is extended beyond the geometrical optical horizon (see Fig.12.9).

Drawing curved ray paths is not easy and for simplification it is possible to make a geometric transformation by increasing the real Earth radius until horizontal ray paths in the troposphere appear to be straight. This can be done for a standard atmosphere by using an *effective* Earth radius of $kr$ , where $r$ is the true radius and $k = 4/3$ . The horizon distance in the absence of hills or obstructions is given by $\sqrt{2krh}$ where $h$ is the antenna height.

The atmosphere varies so that for some world-wide locations and for different percentages of the time the four-thirds Earth concept will be inappropriate. In these cases, an estimate of the value of $k$ can be obtained from the surface refractivity $N_s$ (obtained from **Fig 12.10**), using for $N$ the value obtained from ground-level readings of pressure, temperature and vapour pressure. However, when marked anomalies are present in the vertical refractive index structure and the steady decrease of refractivity with height does not apply, then the effective Earth radius approach may not be helpful.

## Tropospheric Pressure Systems and Fronts

Weather systems in the troposphere are shown on weather maps as contours of equal pressure, isobars. There are two closed systems of isobars involved, known as *anticyclones* and *depressions* within or around which may appear ridges of high pressure, troughs of low pressure, and cols, which are slack regions of even pressure, bounded by two opposing anticyclones and two opposing depressions. The most important consideration about these pressure systems, in so far as it affects radio propagation at VHF and above, is the direction of the vertical motion associated with them.

Depressions are closed systems with low pressure at the centre. They vary considerably in size, and so also in mobility, and frequently follow one another in quick succession across the North Atlantic. They are accompanied by circulating winds that tend to blow towards the centre of the system in an anti-clockwise direction viewed from above in the northern hemisphere.

The air so brought in has to find an outlet, so it rises, whereupon its pressure falls, the air cools, and the relative humidity increases. When saturation is reached, clouds form and further rising may cause water droplets to condense out and fall as rain. *Point one: depressions are associated with rising air.*

Anticyclones are generally large closed systems, which have high pressure in the centre. Once established they tend to persist for a relatively long time, moving but slowly and effectively blocking the path of approaching depressions which are forced to go around them. Winds circulate clockwise (viewed from above in the northern hemisphere), spreading outwards from the centre as they do, and to replace air lost from the system in this way there is a slow downflow, called subsidence, which brings air down from aloft over a very wide area. As the subsiding air descends its pressure increases, and this produces dynamical warming by the same process that makes a bicycle pump warm when the air inside it is compressed.

The amount of water vapour that can be contained in a sample of air without saturating it is a function of temperature, so that if the air was originally near saturation to begin with, by the time the subsiding air has descended from, say, 5 km to 2 km, it arrives considerably warmer than its surroundings and by then contains much less than a saturating charge of moisture at the new, higher, temperature. In other words, it has become warmer and dryer compared to the air normally found at that level. *Point two: anticyclones are associated with descending air.*

In addition to pressure systems, weather maps are complicated by the inclusion of fronts, which are the boundaries between two air masses having different characteristics. They generally arrive accompanied by some form of precipitation, and they come in three varieties: warm, cold and occluded.

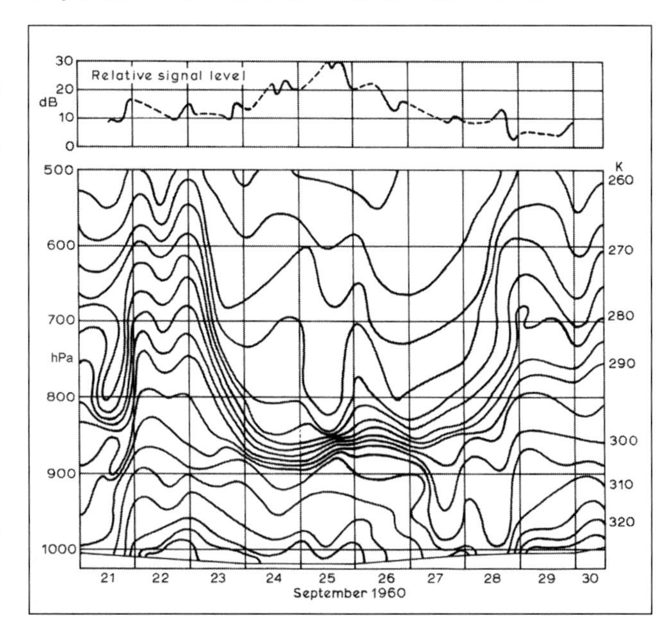


**Figure 12.11: The relationship between variations in potential refractive index in the atmosphere and signal strengths over a long-distance VHF tropospheric path (see text for details).**

Warm fronts (indicated on a chart by a line edged with rounded 'bumps' on the forward side) are regions where warm air meets cold air and the warm air is being forced to rise above it, causing precipitation on the way.

Cold fronts (indicated by triangular 'spikes' on the forward side of a line) are regions where cold air is undercutting warm air. The front itself is often accompanied by towering clouds and heavy rain (sometimes thundery), followed by the sort of weather described as 'showers and bright intervals'.

An occluded front (shown by alternate 'bumps' and 'spikes') [s really the boundary between three air masses and is, in effect, a cold front that has overtaken a warm front and one or the other has been lifted up above the ground.

It is a simple matter of observation that there is some correlation between VHF signal levels and surface pressure readings, but the pressure is generally found to be only a coarse indicator of radio propagation, sometimes showing little more than the fact that high signal levels tend to accompany high pressure and low signal levels tend to accompany low pressure. The reason that it correlates at all is because of the fact that high pressure generally indicates the presence of an anticyclone that, in turn, heralds the likelihood of descending air.

The reason that subsidence is so important stems from the fact that it causes dry air to be brought down to lower levels where it is likely to meet cool moist air which has been stirred up from the surface by turbulence. The result then is the appearance of a narrow boundary region in which refractive index falls off very rapidly with increasing height - the conditions needed to bring about the sharp bending of high-angle radiation - causing the radio waves to return to the ground many miles beyond the normal radio horizon. Those of us who remember the days of analogue TV may recall that so-called 'continental interference' often accompanied high-pressure systems over the UK and near continent.

The essential part of the process is that the descending air must meet turbulent moist air before it can become effective as a boundary. If the degree of turbulence declines, the boundary descends along with the subsiding air above it, and when it reaches the ground all the abnormal conditions rapidly become subnormal, and a sudden drop-out occurs. Occasionally this means that operators on a hill suffer the disappointment of hearing others below them still working DX that they can no longer hear themselves. Note, however, that anticyclones are not uniformly associated with descending air, nor is the necessary moist air always available lower down, but a situation such as a damp foggy night in the middle of an anticyclonic period is almost certain to be accompanied by a strong boundary layer. Ascending air on its own never leads to spectacular conditions. Depressions therefore result in situations in which the amount of ray bending is controlled by a fairly regular fall-off of the refractive index. The passage of warm fronts is usually accompanied by declining signal strength, but occasionally cold fronts and some occlusions are preceded by a short period of enhancement.

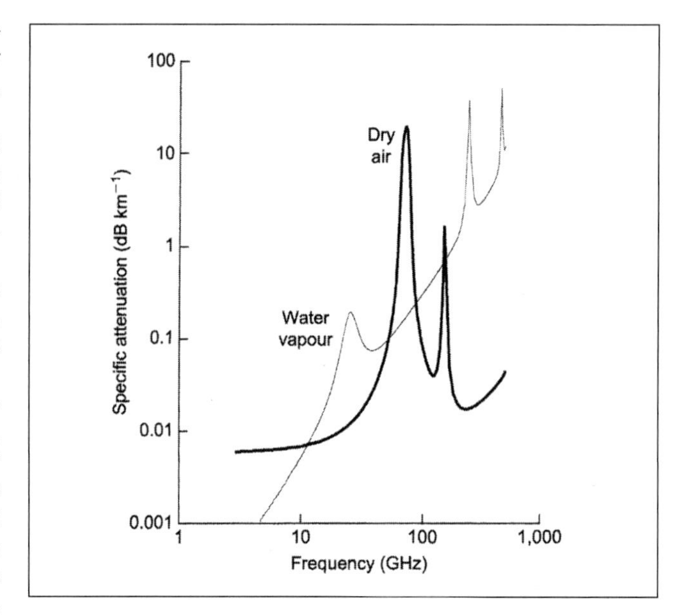


**Figure 12.12: The specific attenuation at sea level caused by atmospheric gases.**

To sum up, there is very little that can be deduced about propagation conditions from surface observations of atmospheric pressure. The only reliable indicator is a knowledge of the vertical refractive index structure in the neighbourhood of the transmission path.

It remains now to consider how the vertical distribution of refractive index can be displayed in a way that gives emphasis to those features which are important in tropospheric propagation studies. Obviously the first choice would be the construction of atmospheric cross-sections along paths of interest, at times when anomalous conditions were present, using values calculated by the normal refractive index formula. The results

| Frequency band (GHz) | Precipitation (mm/hour) | | | | | Fog or cloud water content (g/m³) (at 0°C) | | |
|---|---|---|---|---|---|---|---|---|
| | 100 | 50 | 25 | 10 | 1 | 2.35 | 0.42 | 0.043 |
| 3.4 | 0.1 | 0.02 | 0.01 | - | - | - | - | - |
| 5.6 | 0.6 | 0.25 | 0.1 | 0.02 | - | 0.09 | - | - |
| 10 | 3.0 | 1.5 | 0.6 | 0.2 | 0.01 | 0.23 | 0.04 | - |
| 21 | 13.0 | 6.0 | 2.5 | 1.0 | 0.1 | 0.94 | 0.17 | 0.02 |
| 24 | 17.0 | 8.0 | 3.8 | 1.5 | 0.1 | 1.41 | 0.25 | 0.03 |
| 48 | 30.0 | 17.0 | 9.0 | 4.0 | 0.6 | 4.70 | 0.84 | 0.09 |

*100mm/h = tropical downpour; 50mm/h = very heavy rain; 25mm/h = heavy rain: 10mm/h = moderate rain, 1mm/h = light rain.*

*2.35g/m³ visibility of 30m; 0.42g/m³ 100m; 0.043g/m³ 500m.*

**Table 12.4: The attenuation in decibels per km to be expected from various rates of rainfall and for various degrees of cloud intensity**

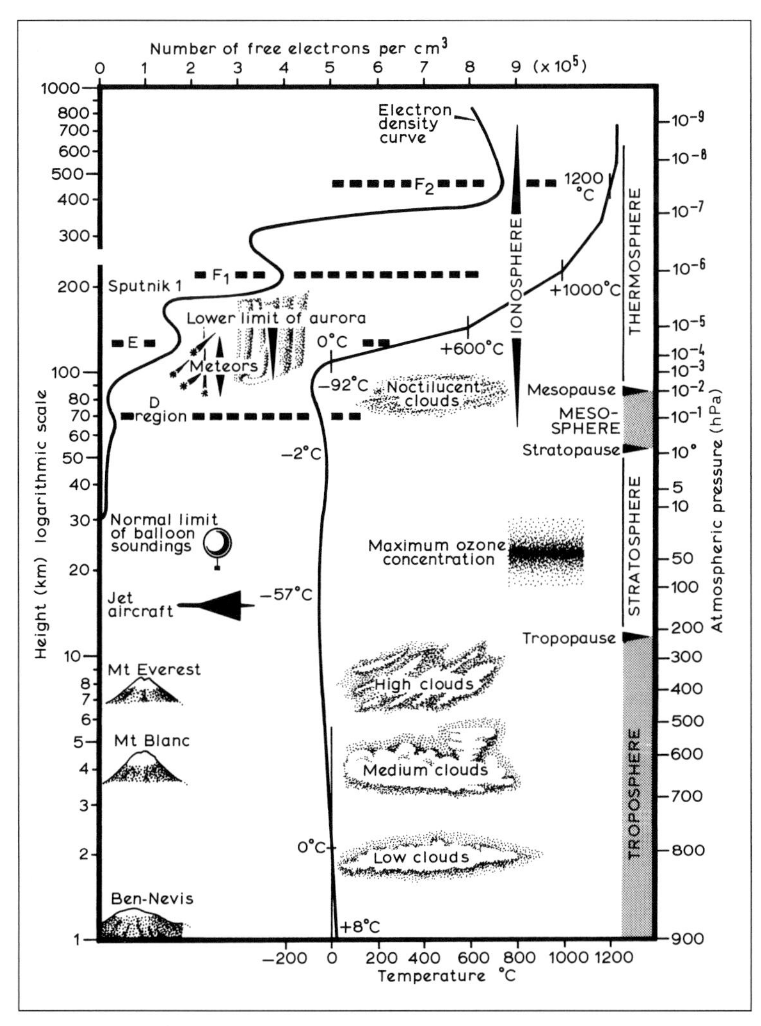


**Figure 12.13: Some features of the Earth's atmosphere. The height scale is logarithmic, beginning at 1 km above sea level. The equivalent pressure scale on the right is not regularly spaced because the relationship between pressure and height depends on temperature, which does not change uniformly with height. The figure sketches the variation of the atmosphere with height. Note that the logarithmic height scale unduly emphasises the lower heights. The atmospheric pressure has been scaled in hectopascals (hPa), the replacement term for millibars.**

are often disappointing, however, because the general decrease of refractive index with height is so great compared to the magnitude of the anomalies looked for that, although they are undoubtedly there, they do not strike the eye without a search.

A closely-related function of refractive index overcomes this difficulty, with the added attraction that it can be computed graphically and easily from published data obtained from upper-air meteorological soundings. It is called the *potential refractive index, $K$*, and may be defined as being the refractive index that a sample of air at any level would have if brought adiabatically (ie without gain or loss of heat or moisture) to a standard pressure of 1000 hPa.

This adiabatic process is the one that governs (among other things) the increase of temperature in air which is descending in an anticyclone, so that, besides the benefits of the normalising process (which acts in a way similar to that whereby it is easy to compare different-sized samples of statistics when they have all been converted to percentages) there is the added attraction that the subsiding air tends to retain its original value of potential refractive index all the time it is progressing on its downward journey.

This means that low values of $K$ are carried down with the subsiding air, in sharp contrast to the values normally found there. A cross-section of the atmosphere during an anticyclonic period, drawn up using potential refractive index, gives an easily recognisable impression of this.

The lower part of **Fig 12.11** shows the way in which the vertical potential refractive index distribution over Crawley, Sussex, varied during a 10-day period in September 1960. For a standard atmosphere, the potential refractive index would appear as horizontal lines but there is no mistaking the down-coming air from the anti-cyclone and the establishment of the boundary layer around 850 hPa (at height of about 1.5km).

Note how the signal strength of the Lille television transmission on 174MHz varied on a pen-chart recording (the top part of **Fig 12.11**) made near Reading, Berkshire, during the period, with peak amplitudes occurring around the time when the layering was low and well-defined. Observe also the marked decline which coincided with the end of the anticyclonic period. Time-sections such as these also show very clearly the ascending air in depressions (although the value of $K$ begins to alter when saturation is reached) and the passage of any fronts which happen to be in the vicinity of the radiosonde station at the time of ascent.

Radio waves of widely different wave lengths are liable to be disturbed by the troposphere in some way or other, but it is generally only those shorter than about 10 m (over 30MHz) which need be considered. There are two reasons for this; one is that ionospheric effects are usually so pronounced, and ray paths tend to be at a significant angles above the horizon, that tropospheric effects are negligible; and the other that anomalies in the refractive index structure of the troposphere, when they occur, do not extend over a sufficient range of heights to accommodate radio waves as long as 10 m or more.

For example, it might be that the decrease of refractive index with height becomes so sharp in a layer of the troposphere, that the radius of bending of the waves is smaller than the Earth's radius and waves become trapped in an atmospheric duct, within which they remain confined for abnormally long distances. The maximum wavelength which can be trapped completely in a duct of, say, 100 m thickness is about 1m (corresponding to a frequency of 300MHz), for example, so that the most favourable conditions are generally found in the VHF and UHF bands or above.

The relationship between maximum wavelength $\lambda$ and duct thickness $t$ is given by the expression

$$t = 500\lambda^{2/3},$$

where both $t$ and $\lambda$ are expressed in centimetres.

Advection ducts occur when a mixed warm air mass moves over a cooler sea. This situation might arise, for example, when there is an anti-cyclone over continental Europe with warm dry air carried over the North Sea.

Subsidence inversion ducts are elevated and extended radio ranges may occur as the subsidence moves towards the ground particularly at night.

Evaporation ducts occur frequently over the sea because of the very rapid decrease in water vapour pressure above the sea

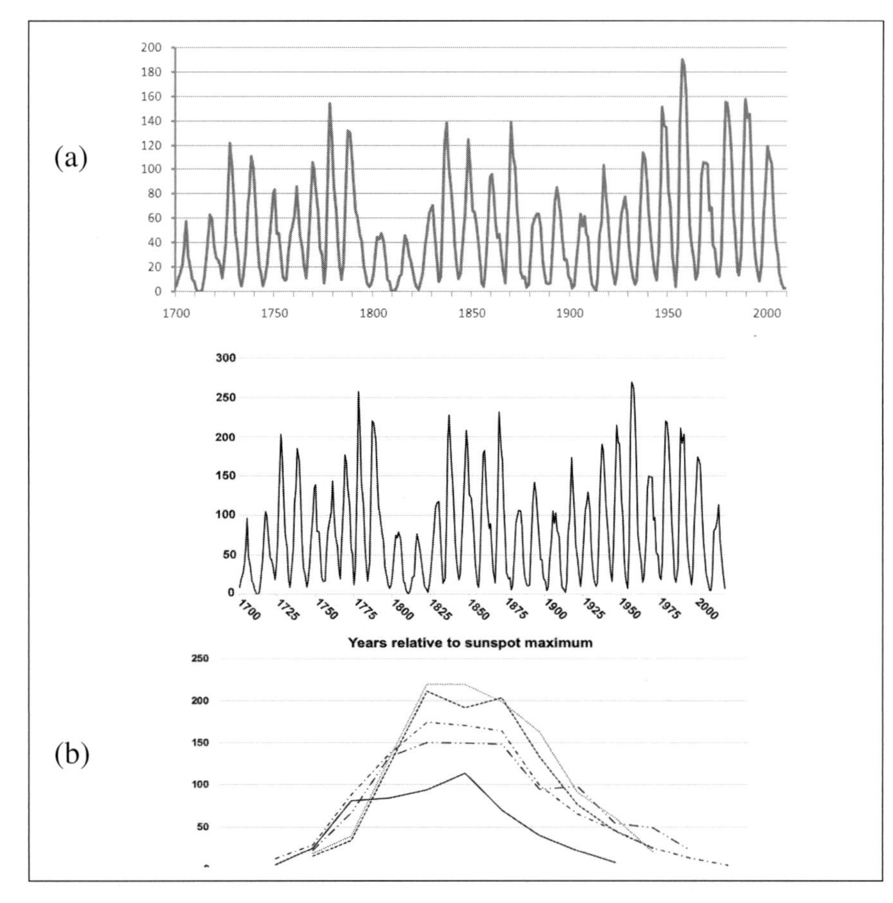


**Figure 12.14: The upper panel shows the annual relative sunspot numbers, the averages of monthly means of daily values. The lower panel shows a comparison of the six solar cycles recently completed. The steep rise to maximum and the relatively slow decline thereafter are characteristic of all sunspot cycles.**

surface. In the North Sea, the mean duct thickness is 5 to 6 m and can permit extended ranges for antennas on the shore within a duct.

Nocturnal radiation ducts may occur on clear, still nights when the air is dry, particularly in hot, dry climates.

## Scintillation

At centimetre wavelengths, signals propagating through the troposphere suffer rapid fluctuations in amplitude and phase due to irregular small-scale variations in refractive index which give rise to continuous changes known as scintillations (akin to the twinkling of stars), and they are also attenuated by water in the form of precipitation (rain, snow, hail, etc) or as fog or cloud. As **Table 12.4** shows, this effect increases both with radio frequency and with either the rate of rainfall or the concentration of water droplets. Precipitation causes losses by absorption and by random scattering from the liquid (or solid in the case of ice) surfaces and this scattering becomes so pronounced as to act as a 'target' for weather radars, which use these precipitation echoes to detect rain areas.

## Atmospheric Attenuation

At centimetre and shorter wavelengths resonances occur within the molecules of some of the gases, that make up the atmosphere. **Fig 12.12** shows the attenuation due to dry air and an example of the additional attenuation due to water vapour. The water vapour peak at about 22GHz is close to the 24GHz amateur band. The increasing gas attenuation at still higher frequencies is likely to prove to a limiting factor for paths passing through the atmosphere, as opposed to working in outer space.

## Rain scatter operation in the GHz bands

Rain showers, referred to by meteorologists as rain "cells", can be very effective in scattering signals, both forwards and backwards, at frequencies above 1GHz. This so-called "rain scatter" affects all theGHz bands to a greater or lesser extent but it is generally accepted that the most-affected bands are 5.7 and 10GHz. Why is that? Water forms in the atmosphere over a wide range of particle sizes; raindrops vary in size from about 3 mm (a tropical downpour) to 0.5 mm (drizzle), cloud particles are between 100 and 1 µm and fog or mist particles range from 10 µm down to 1 nm. The scattering from particles (in this case water particles) increases as the fourth power of the frequency, true for particles up to about a tenth of a wavelength across, after which the scattering remains more or less constant with frequency. For the lowest microwave bands at 1.29, 2.3, 3.4, 5.7, 10 and 24GHz, a tenth of a wavelength corresponds to about 23, 13, 9, 6, 3, and 1.2 cm respectively. One tenth of a 10GHz wavelength falls just below typical rain droplet sizes and one tenth of a 5.7GHz wavelength is slightly longer than a tropical downpour droplet size. RF power is more easily generated than at higher frequencies and water absorption is much less. With the fourth power law, the expected signal strength at 5.7GHz would be $(0.57)^4$ times the signal strength at 10GHz, i.e about 10 dB lower. Repeating this calculation for the lower GHz bands illustrates why there are fewer rain scatter QSOs there.

In practice is it quite easy to operate using rain scatter. Both stations need to be in range of the same rain cell, (referred to as a scatter-point or "scpt") and point their antennas towards it. Signals will be forward or backscattered between them. All modes work, with CW the easiest. SSB and CW have a characteristic "rasping" sound, not unlike Auroral signals, caused by spreading from droplets moving with different velocities, but if signals are strong enough, NBFM gives almost perfect audio quality.

Rain scatter QSO range is limited by the height of the scatter-point and your local take-off angle; some scatter points will be below your horizon. Typically rain cells are 2-4 km above ground and QSOs can be from a few km to several hundred km. You can use one of a number of rain-radar websites to find where a scatter points might be. A local beacon can also be useful: tune to it and rotate your antenna and see where signals come from, other than on the direct path. When you have heavy rain, signals can come from all directions, making multi-way QSOs possible. (More information may be found at *http://www.mike-willis.com/Tutorial/rainscatter.htm and http://www.wa1mba.org/10grain.htm.)*

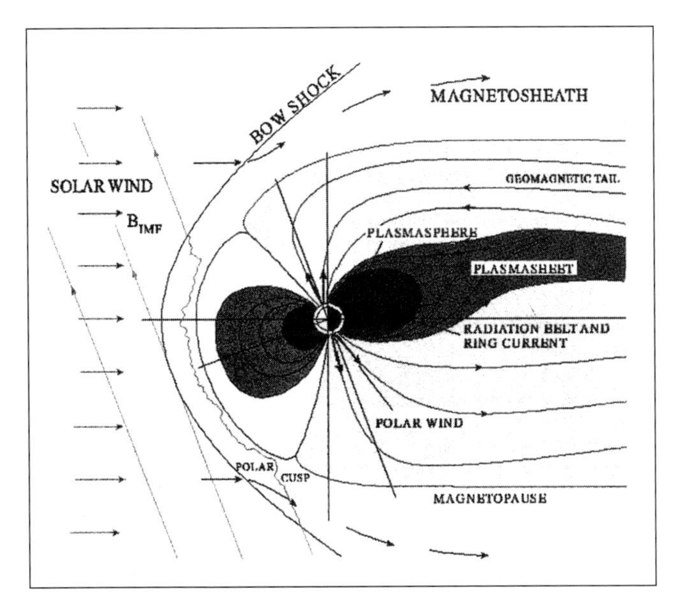


**Figure 12.15: The Earth in the solar wind.**

## THE EARTH'S ATMOSPHERE

In the troposphere, nearest the ground, temperature tends to decrease with height. At the tropopause, around 10 km (although all these heights vary from day to day and from place to place), it becomes fairly uniform at first and then begins to increase again in the stratosphere up to the stratopause at around 50 km. Above this is the mesosphere, with another temperature minimum at the mesopause (at about 80 km). Above this is the thermosphere, with temperatures increasing up to about +1200 °C around 700km. Note that atmospheric pressure decreases with height and that the atmosphere is extremely tenuous in the thermosphere. See **Fig 12.13**.

Incoming solar radiation and, to a much lesser extent, cosmic radiation interacts with the upper atmosphere and ionises a small percentage of the gas creating free electrons and ions. This region lies between about 60 and 700km and is called the ionosphere. Because of different ionising mechanisms, varying atmospheric conditions, and changes in the rates at which recombination between ions and electrons occur, the electron density shows a number of 'ledges' or layers, identified by the letters D, E, F1 and F2.

## SOLAR AND MAGNETIC INFLUENCES

### The Sun

Our Sun, at the centre of our solar system of planets, is a huge sphere of incandescence of a size that is equivalent to about double our moon's orbit around the Earth. However, despite appearances to the contrary, it has no true 'edge' because nearly the whole of the Sun is gaseous and the part we see with apparently sharp boundaries is merely a layer of the solar atmosphere called the photosphere which has the appearance of a bright surface, preventing us from seeing anything that lies beneath.

Above the photosphere is a relatively cooler, transparent layer called the chromosphere, so named because it has a bright rose tint when visible as a bright narrow ring during total eclipses of the Sun. From the chromosphere great fiery jets of gas, known as prominences, extend. Some are slowly changing and remain suspended for weeks, while others, called eruptive prominences, are like narrow jets of fire moving at high speeds and for great distances.

Outside the chromosphere is the corona, extending a distance of several solar diameters before it becomes lost in the general near-vacuum of interplanetary space. At the moment of totality in a solar eclipse it has the appearance of a bright halo surrounding the Sun, and at certain times photographs of it clearly show it being influenced by the lines of force of the solar magnetic field.

The visible Sun is not entirely featureless - often relatively dark sunspots appear and are seen to move from east to west, changing in size, number and dimensions as they go. They are of interest for two reasons: one is that they provide reference marks by which the angular rotation of the Sun can be gauged, the other that by their variations in number they reveal that the solar activity waxes and wanes in fairly regular cycles. The Sun's rotation period has been found to vary with latitude, with its maximum angular speed at the equator. The mean synodic rotation period, the time required for the Sun to rotate until the same part faces the Earth, is 27.3 days.

Each rotation of the Sun is given a unique number called the Carrington Rotation Number, starting from 9 November 1853. (Carrington Rotation number 2224 began on 12$^{th}$ November 2019.) When looking at the mid-day Sun from the UK, north is at the top, and astronomers conventionally label the east limb as that on the left side and the west limb on the right. A long-persistent feature which first appears on the east limb is visible for about 13.5 days before it disappears from sight around the west limb.

For many purposes, it is more convenient to refer the positions of noteworthy features to a related, but Sun-centred and fixed, set of co-ordinates, the heliocentric latitude and longitude, in which locations are described with respect to the centre of the visible disc. An important statistic relating to sunspots is the time of their central meridian passage.

Radio telescopes detect features which are usually situated in the vicinity of the solar corona, and some of them reveal disturbances beyond the limbs of the visible disc. They often travel across the face of the Sun at a faster rate than any spots beneath. The apparent diameter of the Sun varies with the choice of radio frequency. Because the lower frequencies come from the outer parts of the corona, the width of their Sun appears to be larger. At high frequencies the sources are situated below the level which provides the visible disc so that the width of the Sun then appears to be smaller. The optical and radio frequency diameters are similar at a frequency of 2800MHz (10.7cm wave- length), so that was chosen to provide the daily solar flux measurements. The daily flux figure, which can vary between about 65 at solar minimum and 300 at maximum, may be used, instead of sunspot number, as a measure of the Sun's ionising capability. Higher flux figures correspond to greater ionisation and thus to higher frequencies being available in the HF band.

### Sunspots

Sunspots are the visible manifestation of very powerful magnetic fields; adjacent spots often having opposing polarities. These intense magnetic fields also produce solar flares, which are eruptions of hydrogen gas. They are responsible also for the ejection of streams of charged particles and X-rays.

Sunspot numbers have been recorded for over 200 years and it has been found that their totals vary over a fairly regular cycle occupying around 22 years taking account of the solar magnetic polarity, or 11 years if only the magnitude of activity is considered. The 11-year peaks are known as sunspot maxima; the intervening troughs are sunspot minima.

The rise and fall times are not equal, though. Four years and seven years respectively are typical, although each cycle differs from the others in both timing and maximum value, as may be seen in **Fig 12.14**. At sunspot minimum the Sun may be completely spotless for weeks or months - or even, during the Maunder minimum in the 17th century, for years.

Tables of daily relative sunspot numbers are prepared monthly at the Sunspot Index Data Centre (SIDC) in Brussels, from information supplied by a network of participating observatories. In fact this is not the number of visible spots, which would vary with the observing system concerned and with the daily visibility, but is standardised in accordance with a formula devised by Dr Wolf in Zurich (hence the description Wolf number, still sometimes used professionally). The relative sunspot number, $R$, is found from the expression

$$R = k(10g + t),$$

where $k$ is a regulating factor that keeps the series to a uniform standard, $g$ is the number of spot groups, and $t$ is the total number of spots.

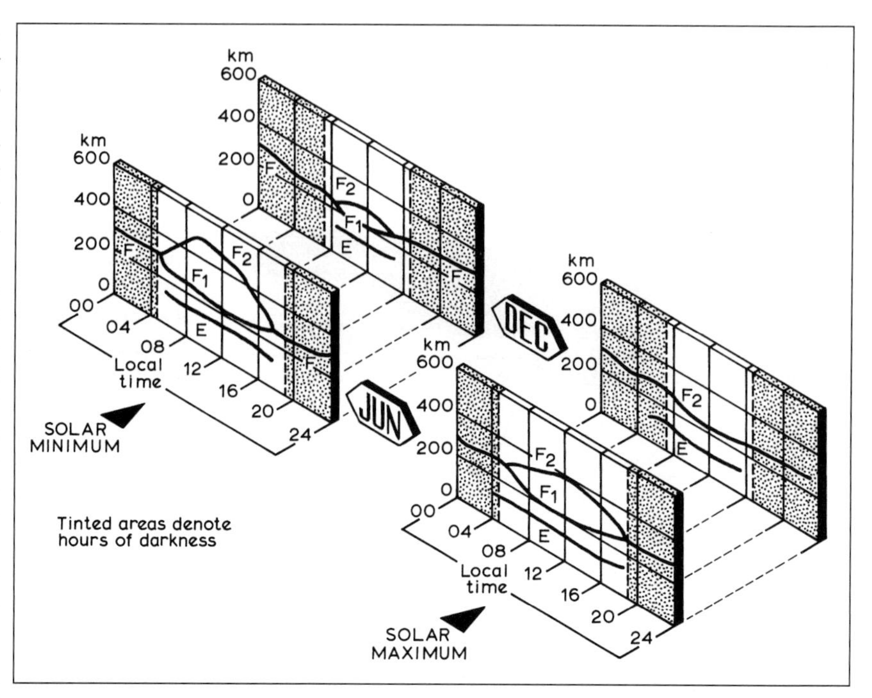


**Figure 12:16: Typical diurnal variations of layer heights in summer and winter at minimum and maximum points of the solar cycle.**

Daily figures obtained at an individual observatory use unity as the value of $k$. These daily figures are provisional because they will have been prepared in haste to meet a deadline. When the figures from the participating observatories have been combined at the SIDC, a value of $k$ which is less than unity will have been applied to correct for the "seeing" at each station and to maintain continuity with the past series of numbers. That figure does not appear in the tables, but it is currently around 0.6.

The SIDC figures are issued twice, first provisionally as soon as possible, then definitively after more careful scrutiny. A smoothed index, $R12$, is obtained from the monthly means of the definitive values. This is the arithmetic mean of 12 successive monthly means, the result being ascribed to the period at the centre of the sample. In order to make that fall in the middle of the month, rather than between months, 13 months are taken but the first and the last are given only half weight in the calculations. From the nature of this 12-month running average (as called for in ionospheric prediction programs) it must be evident that it never reaches the peaks and troughs of the individual monthly means and it falls far short of the maxima of the daily values.

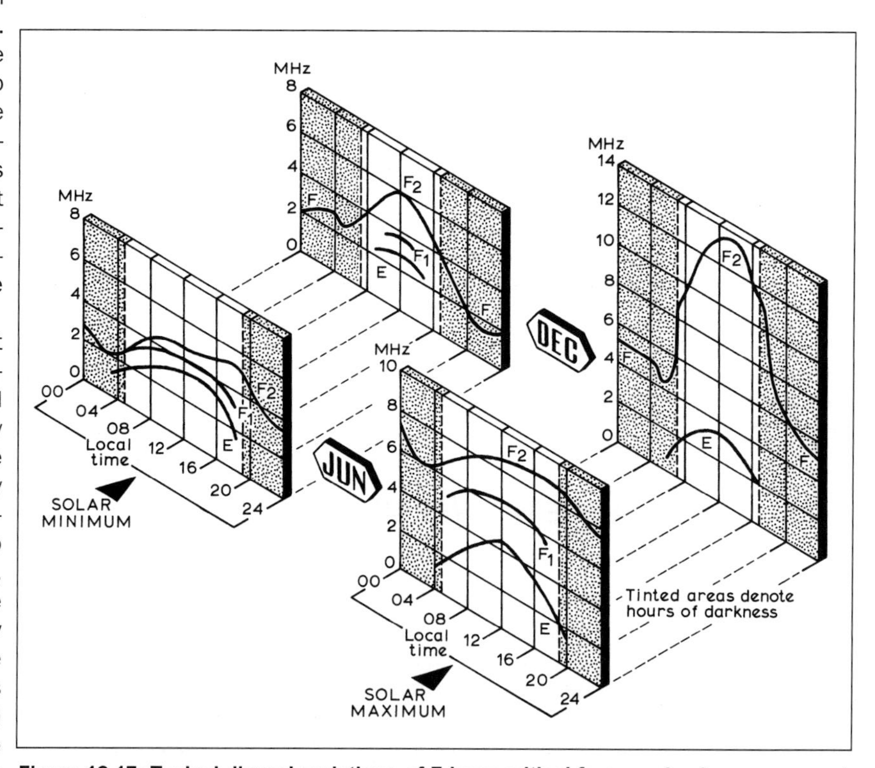


**Figure 12.17: Typical diurnal variations of F-layer critical frequencies in summer and winter at the extremes of the solar cycle.**

To put all these different versions of the 'sunspot number' into perspective we have only to look at a specific example, say the month which contained the peak of solar cycle 22, June 1989. The daily figures reached 401, and their monthly mean was 297. The SIDC definitive figure had a daily maximum of 265, with a monthly mean of 196. However, the smoothed figure $R12$ for the month was only 158 and, remember, this is the one that you need for prediction programs, not the 401 obtained as a maximum value of one day recorded at one station. You should be aware also that the latest smoothed figure available is always six months behind the current date, so a figure for the current month has been a forecast made six months ahead.

It is also of interest to observe that the peak Boulder figure was made up of 18 groups which, between them, contained a total of 221 spots. Put those figures in the formula and you come up with 401, the figure reported. The three largest groups accounted for 86, 53 and 26 spots respectively, and none of the others contained more than 7.

## The Solar Wind

The solar corona was described in the last section as extending outwards until it becomes lost in interplanetary space. In fact it turns into a tenuous flow of Ionised gas which expands outwards through the solar system known as the solar wind.

Near the Sun the corona behaves as a static atmosphere, but farther away it gradually accelerates with increasing distance to speeds of hundreds of kilometres per second. The gas particles normally take about nine days to travel the $1.5 \times 10^8$ km to the Earth (less when there is a high-speed coronal stream and less still after a major flare), carrying with them a magnetic field (because the gas is ionised) which assumes a spiral form because of the Sun's rotation. It is the solar wind, rather than visible light pressure alone, which is responsible for comets' tails flowing away from the Sun, causing them to take on the appearance of celestial wind-socks.

The existence of the solar wind was first detected and measured by space vehicles. They showed that its speed and turbulence are related to solar activity. Regular measurements of solar wind velocity are now routine: see Fig.12.15 for a representation of Earth in the solar wind.

There is thus a direct connection between the atmosphere of the Sun and the atmosphere of the Earth. In the circumstances it is hardly surprising that solar events, remote though they may at first seem, soon make their effects felt here on Earth.

## The Earth's Magnetosphere

It is well known that the Earth possesses a magnetic field and that the field appears to be concentrated at points somewhere near the north and south poles. Popular science articles have familiarised us with a picture of field lines surrounding the Earth like a section of a ring doughnut made up of onion-like layers.

Because the particles carried by the solar wind are charged, their movement produces a magnetic field, which interacts with the geomagnetic field. A bow shock is formed around the Earth; the shockwave forms as the supersonic particles comprising the solar wind abruptly slow to subsonic speeds. The region of the Earth's magnetosphere where these solar wind particles slow down and are redirected around the Earth's magnetic field is known as the magnetosheath. The solar wind flows around the magnetosheath rejoining behind where the field on the far side is stretched in the form of a long tail, the overall effect being reminiscent of the shape of a pear with its stalk pointing away from the Sun. The region within the magnetosheath, into which the wind does not pass, is called the magnetosphere.

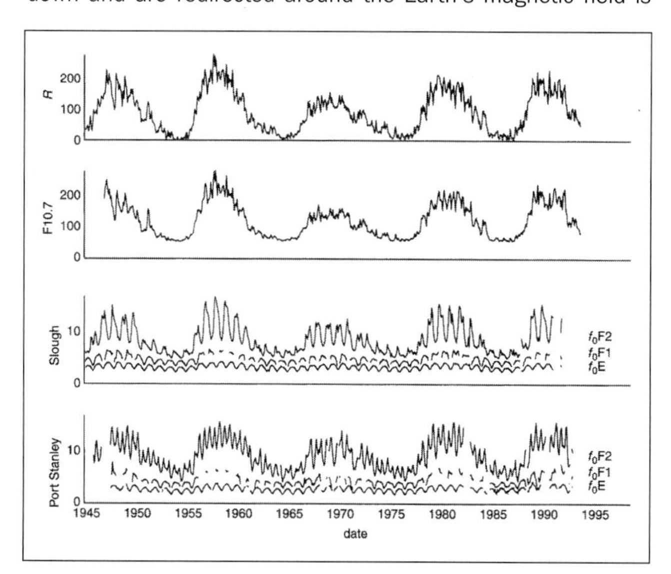


**Figure 12.18: The variation of monthly median noon values of the critical frequencies of the F2, F1 and E layers and of the sunspot number. [©IET**

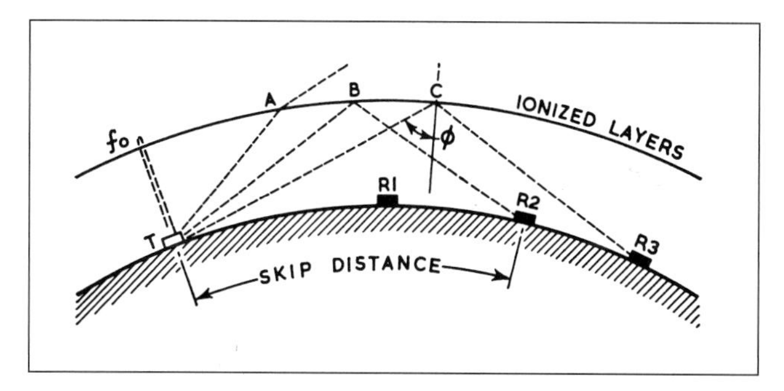


**Figure 12.19: Wave propagation via the ionosphere. T is the site of a transmitter, R1, R2 and R3 are the sites of three receivers. The significances of the ray at vertical incidence and the three oblique rays are explained in the text.**

On the Earthward side the magnetosphere merges into the ionosphere. Inside the magnetosphere there are regions where charged particles can become trapped by geomagnetic lines of force in a way that causes them to oscillate back and forth over great distances. Particles from the solar wind can enter these regions (often called Van Allen belts after their discoverer) in some way, as yet not perfectly understood.

The concentration of electrons in the magnetosphere can be gauged from the ground by observations of whistlers, naturally-occurring audio-frequency, electromagnetic oscillations of descending pitch which are caused by waves radiated from the electric discharge in a lightning flash. These travel north and south through the ionosphere and magnetosphere from one hemisphere to another along the magnetic lines of force. The various component frequencies propagate at different speeds so that the original flash (which appears at higher frequencies as an sharp crack) arrives at the observer considerably spread out in time. The interval between the reception of the highest and lowest frequencies is a function of the concentration of electrons encountered along the way.

These trapped particles move backwards and forwards along the geomagnetic field lines within the Van Allen belts and some collide with atoms in the ionosphere near the poles where the belts approach the Earth most closely. Here they yield up energy either as ionisation or illumination and are said to have been dumped.

These dumping regions surround the two poles, forming what are called the auroral zones. The radius of the circular motion in the spirals (they are of a similar form to that of a helical spring) is a function of the strength of the magnetic field, being small when the field is strong. Electrons and protons perform their circular motions in opposite senses, and the two kinds of spiralling columns drift sideways in opposite directions, the electrons eastward, the protons westward, around the world. Because of the different signs on the two charges these two drifts combine to give the equivalent of a current flowing in a ring around the Earth from east to west.

This ring current creates a magnetic field at the ground that combines with the more-or-less steady field produced from within the Earth. We shall see later the sort of effects that solar disturbances have on the magnetosphere, the ionosphere, and the total geomagnetic field.

| Layer | Distance (km) 1000 | 2000 | 3000 | 4000 |
|---|---|---|---|---|
| Sporadic E* | 4.0 | 5.2 | - | - |
| E | 3.2 | 4.8 | - | - |
| F1 | 2.0 | 3.2 | 3.9 | - |
| F2 winter | 1.8 | 3.2 | 3.7 | 4.0 |
| F2 summer | 1.5 | 2.4 | 3.0 | 3.3 |

** See section below on Sporadic E*

**Table 12.5: MUF (maximum usable frequency) factors for various distances assuming representative heights for the principal layers**

## The Quiet Ionosphere

The normal day-to-day working of the ionosphere is dependent for its chemistry on another form of incoming solar radiation. The gas molecules in the Earth's upper atmosphere are normally electrically neutral, that is to say the overall negative charges carried by their orbiting electrons exactly balance the overall positive charges of their nuclei. Under the influence of ultraviolet radiation from the Sun, however, some of the outer electrons can become detached from their parent atoms, leaving behind overall positive charges due to the resulting imbalance of the molecular structure. These ionised molecules are called ions, from which of course stems the word 'ionosphere'.

This process, called disassociation, tends to produce layers of free electrons brought about in the following manner. At the top of the atmosphere where the solar radiation is strong there are very few gas molecules and hence very few free electrons. At lower levels, as the numbers of molecules increase, more and more free electrons can be produced, but the action progressively weakens the strength of the radiation until it is unable to take full advantage of the increased availability of molecules and the electron density begins to decline. Because of this there is a tendency for a maximum (or peak) to occur in the production of electrons at the level where the increase in air density is matched by the decrease in the strength of radiation. A peak formed in this way is known as a Chapman layer, after the scientist who first outlined the process.

The height of the peak above ground level is determined not by the strength of the radiation, but by the density/height distribution of the atmosphere and by its capability to absorb the solar radiation (which is a function of the UV wavelength), so that the layer is lower when the radiation is less readily absorbed. The strength of the radiation affects the rate of production of electrons at the peak, which is also dependent on the direction of arrival. The electron density is greatest when the radiation arrives vertically and it falls off as a function of zenith distance, or zenith angle, the angle between the vertical and the direction of the incoming radiation. This is usually given the Greek letter $\chi$, and the intensity is proportional to $\cos(\chi)$.

Experimental results show that the E-layer (at about 110km) and the F1-layer (at about 200km) are formed according to Chapman's theory as a result of two different kinds of radiation with perhaps two different atmospheric constituents involved. At E layer heights the atmospheric density is such that electrons typically recombine with an ion after about 30 minutes so that the resultant electron density peaks just after midday, when the sun is highest, and falls to low levels at night.

At F2 layer heights (around 400 km), where the atmosphere is much more tenuous, the typical time for recombination may be about twelve hours and so the layer will persist through the night, with ionised layers being subject to movement by winds and the magnetic field. This gives rise to complicated geographic distributions.

When first observed, these effects were described as a series of anomalies, although they may now all be accounted for with current modelling techniques. The anomalies include: the diurnal anomaly, when the peak occurs at an unexpected time during the day; the night anomaly, when the intensity of the layer increases during the hours of darkness when no radiation falls upon it; the polar anomaly, when peaks occur during the winter months at high latitudes, when no illumination reaches the layer at all; the seasonal anomaly, when magnetically quiet days in summer (with a high Sun) sometimes show lower penetration frequencies than quiet days in winter (with a low Sun); and a geomagnetic anomaly where, at the equinoxes, when the Sun is over the equator at midday, the F2-layer is most intense at places to the north and south separated by a minimum along the magnetic dip equator.

The D region, at about 80 km, is generally not a distinct layer but is important since the relatively high atmospheric density at that height means that energy in free electrons may be absorbed causing attenuation of radio waves.

Measurements taken during a solar eclipse showed that recombination can take place over a period of just a few seconds.

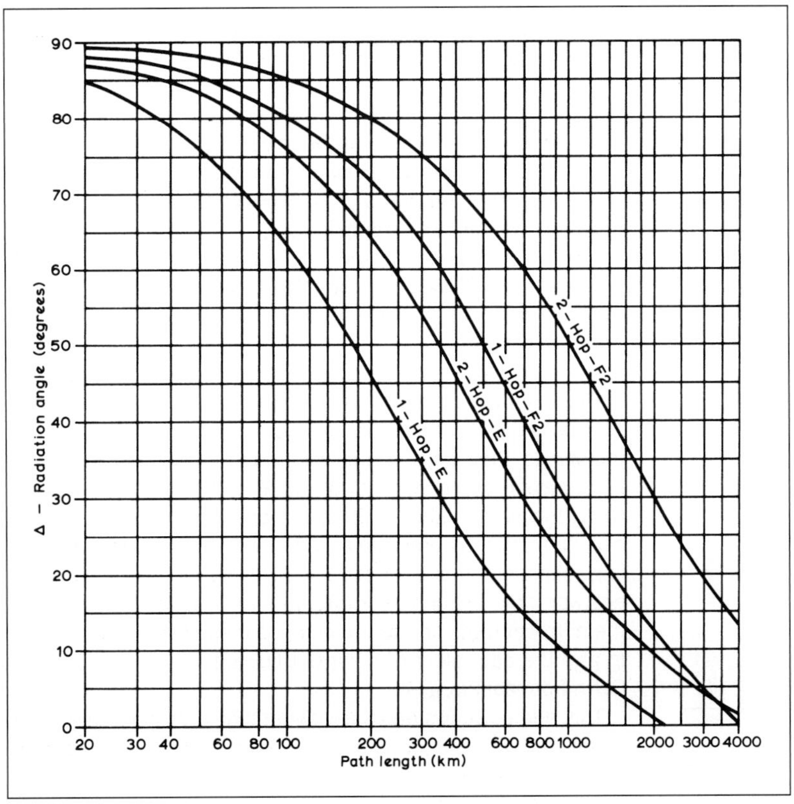

**Figure 12.20: Radiation angles involved in one-hop and two-hop paths via the E and F2-layers.**

## Regular Ionospheric Layers

From comments already made, it will be appreciated that the regular ionospheric layers exhibit changes which are basically a function of day and night, season and solar cycle.

Most of our knowledge of the ionosphere

| Tropospheric propagation | Sporadic-E propagation |
|---|---|
| May occur at any season | Mainly May, June, July and August |
| Associated with high pressure, or with paths parallel to fronts | No obvious connection with weather patterns |
| Gradual improvement and decline of signals | Quite sudden appearance and disappearance |
| Onset and decay times similar over a wide range of frequencies | Begins later and ends earlier as radio frequency increases |
| Observed at VHF, UHF, SHF | Rarely above 150MHz |
| Area of enhancement relatively stable for several hours at a time | Area of enhancement moves appreciably in a few hours |
| May last a week or more | Duration minutes or hours, never days |
| Wide range of distances with enhanced signals at shorter ranges | Effects mainly at 1000-2000km. No associated enhancement at short ranges |

**Table 12.6: Comparative characteristics of tropospheric and sporadic-E propagation**

comes from frequent soundings made at vertical incidence, using a specialised form of radar called an ionosonde that is continuously varied in frequency from MF up to about 20MHz, and beyond if conditions warrant.

Reflections from the various layers are recorded digitally and may be displayed in the form of a graphical ionogram, which displays virtual height (the apparent height from which the sounding signal is returned) as a function of signal frequency.

*Critical frequency*

The highest frequency reflected from a given layer is called the critical frequency. There are a number of ionosondes in the world; the one serving the United Kingdom is located at Chilton, Oxfordshire. It is under the control of the Rutherford Appleton Laboratory, which houses one of the World Data Centres for solar terrestrial physics (*www.ukssdc.ac.uk*) to which routine measurements of the ionosphere are sent from most parts of the world.

The two sets of diagrams (**Fig 12.16** and **Fig 12.17**) summarise the forms taken by the diurnal variations in height and critical frequencies for two seasons of the year at both extremes of the sunspot cycle. The actual figures vary very considerably from one day to the next, but an estimate of the expected monthly median values of maximum usable frequency and optimum working frequency between two locations at any particular year, month and time of day can be obtained from predictions. The critical frequencies of the E and F1-layers, $f_0E$ and $f_0F1$ respectively, are functions of $R12$, the smoothed SIDC relative sunspot number, and the cosine of the Sun's zenith angle, $\chi$, and are given inMHz approximately by the empirical expressions

$$f_0E = 0.9\left[(180 + 1.44R12)\cos(\chi)\right]^{0.25},$$

usually correct with about 0.2MHz, and

$$f_0F1 = (4.3 + 0.01R12)(\cos(\chi))^{0.2},$$

which is less accurate because of the uncertainty in the value of the exponent which varies with location and season.

The F2-layer is the most important for HF communication at a distance, but, as has been said, it is also the most variable. The F2 critical frequency, $f_0F2$, varies with the solar cycle, as shown in **Fig 12.18**, which indicates the sunspot number, $R$, and the monthly median noon values of $f_0F2$, $f_0F1$ and $f_0E$ at Slough in the Northern hemisphere and Port Stanley, Falklands Islands, in the South over five solar cycles.

## Wave Propagation in the Ionosphere

At ionospheric heights, the increase in electron concentration with height has the effect that the refractive index decreases with height so that radio waves are refracted away from the vertical. The magnitude of effect depends on the frequency as well as the electron density. At low-enough frequencies, the waves are bent downwards and, as a simplification, may be thought of as being reflected. At higher frequencies, the waves will be deflected to some extent but then continue into space.

Consider the circumstances outlined in **Fig 12.19**, where T indicates the site of a transmitting station and R1, R2 and R3 three receiving sites. For a given electron concentration, there is a critical frequency, $f_0$, which is the highest to return to the transmitter location from radiation directed vertically upward. Frequencies higher than this penetrate the layer completely and are lost in space. At oblique incidence, waves travel a greater distance but do not have to be refracted through as big an angle, and at R2 the ray bending becomes just sufficient to return signals to the ground, making R2 the nearest location to the transmitter at which the sky-wave is received. At point A the refraction is not sufficient so that no signals are received at R1.

The range over which no signals are received via the ionosphere is known as the skip distance, and the roughly circular area described by it is called the skip zone. At frequencies less than the critical frequency there is no skip zone.

Lower-angle radiation results in longer ranges, for example to point R3 from a reflection at C, and a second 'hop' may result from a further reflection from the ground. The longest ranges at HF are achieved this way - and it is possible for an HF signal to travel right round the world using a succession of hops. However, over long distances, signal paths can be complex and variable - not least because they are often asymmetrical. They may be affected by irregular ionospheric tilts and gradients, particularly around sunrise and sunset, which can alter the propagation geometry.

For oblique incidence on a particular path (eg the ray from T to R3 via point C in the ionosphere), there is a maximum usable frequency ($MUF$), higher than the critical frequency, given by

$$MUF = m_d f_0,$$

where $m_d$ is called the *$MUF$ factor* and is greater than one. It is a function of the path length and of the equivalent reflection height of the layer. If the Earth were flat, it would be the secant of the angle of incidence in the layer, but for real curved Earth typical values are given in **Table 12.5**.

The ionosphere is varying continually, so that operating at

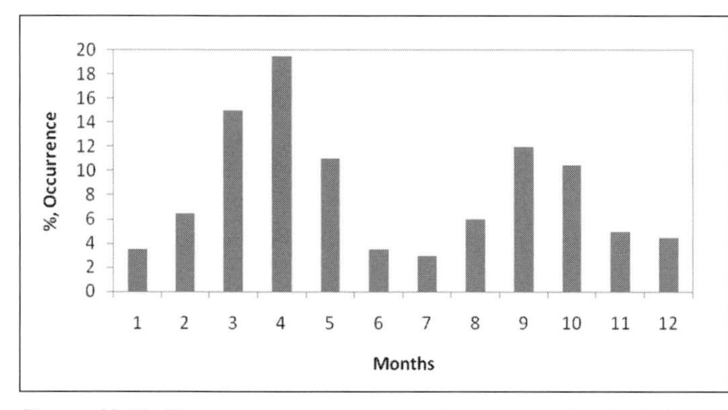


**Figure 12.21: The average occurrence of geomagnetically disturbed days.**

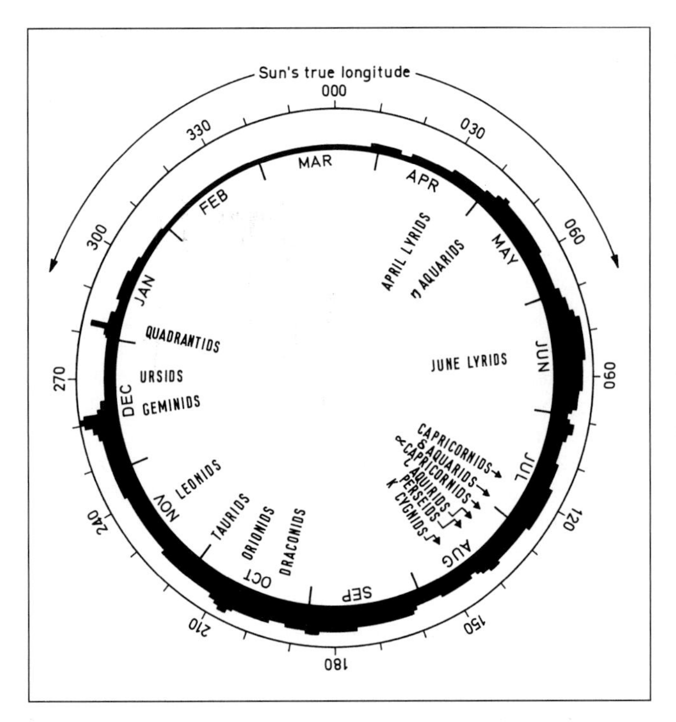


**Figure 12.22: The seasonal variation of meteor activity, based on a daily relative index. Prepared from tables of 24 h counts made by Dr Peter Millman, National Research Council, Ottawa. The maximum rate corresponds to an average of about 300 echoes per hour corresponding to an equivalent visual magnitude of six or brighter.**

the *MUF* is hard to maintain. The most reliable frequency for use on a given path is somewhat less than the *MUF*.

The optimum working frequency (*OWF*, also called the Frequency of Optimum Transmission, *FOT*) is the highest (of those available) which does not exceed the *MUF*. In the context of ionospheric predictions, the *OWF* is the highest frequency which may be expected to be available taking account of day to day variations in *MUF*, so that, on most days of the month, a somewhat higher frequency would be better still.

We have not yet mentioned the effect of the Earth's magnetic field in this brief summary. The field causes the ray paths in the ionosphere to be split into two parts, the *ordinary ray* and the *extraordinary ray*. At short ranges, the extraordinary ray will have an *MUF* about 0.7MHz higher than that predicted for the ordinary ray. This difference reduces as the range increases, becoming negligible beyond about 800 km.

There is also a lower limit to the band of frequencies which can be selected for a particular application. This is set by the lowest usable frequency (*LUF*), below which the circuit becomes either unworkable or uneconomical because of the effects of absorption in the D region and the increasing level of radio noise at the lower frequencies. Its calculation is quite a complicated process beyond the scope of this book.

It is sometimes useful to be able to estimate the radiation angle involved in one or two-hop paths via the E and F2-layers. Curves based on average heights are shown in **Fig 12.20**. A rule-of-thumb is that the maximum one-hop E range is 2000 km, the useful two-hop E range is 4000 km, and this is also approximately the one-hop F2 range. Of course, all extreme ranges require very low angles of take-off, not usually achieved by radio amateurs.

The treatment above is appropriate for frequencies greater than about 100kHz, for which the electrons appear to be concentrated in a succession of layers of increasing density, having the effect of progressively bending the rays as the region is penetrated.

At medium frequencies, including 475kHz, daytime absorption in the D region becomes so great that daytime propagation is only by surface wave. MF broadcast signals are thus of limited range during the day, but ranges are extended by ionospheric reflection during the night hours.

At lower frequencies, such as 136kHz, the daytime absorption decreases to some extent, since the waves do not penetrate so far into the D region.

Below about 50kHz, the wavelength is a significant fraction of the height of the bottom of the ionosphere and it becomes inappropriate to consider the wave propagation as 'rays'. Instead the space between two concentric spheres, one being the lower edge of the layer and the other the surface of the Earth, may be considered as a waveguide and propagation extends to great distances, although it becomes increasingly difficult to make an efficient antenna.

## IRREGULAR IONISATION

Besides the usual E, F1 and F2 layers, other more-localised concentrations of ionisation can occur, and these can contribute to radio-wave propagation. They generally occur at around the heights associated with the E-layer and their effects often extend well into the VHF range - unlike the usual E-layer, which cannot sustain propagation at frequencies above 30MHz.

### Sporadic-E

*Sporadic-E* (Es) was first observed at HF using ionospheric sounding apparatus in the early 1930s. Its characteristics differ according to whether it occurs in the polar regions, at mid latitudes or near the magnetic equator. At mid-latitudes, it takes the form of relatively thin but highly-ionised 'clouds', typically only a kilometre or so deep and around 100 km across, at a height of about 100-120 km and drifting at speeds of up to about 140 m $s^{-1}$. The distribution of ionisation within them, and thus the frequencies they can reflect, can vary considerably, at times causing sudden, deep fades, also affected by the motion of the clouds relative to transmit and receiving sites.

The mechanism behind Es is quite different from normal E layer ionisation. Es is caused by the ionisation of metallic ions, principally Iron, which enter the atmosphere from meteor dust. These ions have a much longer lifetime than atmospheric ions (Oxygen and Nitric Oxide) in the E region. This allows Es clouds to persist for hours. Only normal levels of solar radiation are

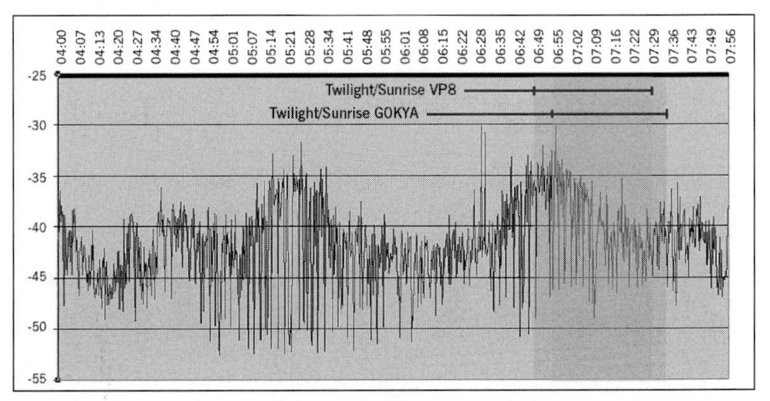


**Figure 12.23: Signals received at G0KYA from the Falkland Islands on the 40 metre band on 27 November 2002, showing a peak about 30-40 minutes before sunrise.**

required for the ionisation process, and there is little variation in Es propagation with the solar cycle.

On 28MHz, where Es occurs almost every day during the summer, operators may be able to maintain contacts for several hours and stations running low power may be received at great strength. At both 28 and 50MHz, multi-hop openings over long distances have been recorded, and exceptionally even exceeding 15000km..

The higher the frequency, the fewer and briefer the openings tend to be until, at 144MHz, they may last only a few minutes. One report from the US suggested that Es was responsible for a QSO at 220MHz - but even that is not necessarily the absolute limit. The $MUF$ factors for Es given in **Table 12.5** are for cases where the ionised layer may be treated as a thick layer, like the usual layers. For thinner layers, partial reflections may occur at much higher frequencies. Every year, radio amateurs make 2-metre contacts using sporadic-E. There is still much to learn about this mode and radio amateurs have long played a useful role in studying it. In individual cases, it may be hard to distinguish between VHF Es and extended range tropospheric modes. These are quite separate and a list of their distinguishing features will be found in **Table 12.6**.

At mid-latitudes, Sporadic-E is mainly a May to August phenomenon, though there is a weaker sub-peak in December in the northern hemisphere. Distances worked are not often less than 500 km and most contacts are at ranges of 1,000 to 2,000 km. The maximum range for a single hop is about 2,300km, although sometimes this can be extended by tropospheric enhancement at one or both ends. At low latitudes Es is generally weaker and displays no marked seasonal pattern.

*Auroral propagation*, by contrast, is found mainly at high latitudes and is closely associated with geomagnetic disturbances. At such times, the regions around the north and south poles where visual and radio aurora are commonplace, the so-called auroral ovals, expand towards the equator, taking with them a capacity to return VHF signals directed at them. The reflecting regions are usually around 110 km in height and are composed of highly- charged irregularities arrayed near vertically along the earth's magnetic field lines.

The greater the geomagnetic disturbance, the nearer the equator a QSO becomes possible. Thus, an operator in Orkney will have many more opportunities than one in Cornwall. **Fig 12.21** shows the seasonal variation in geomagnetic disturbance.

*Auroral backscatter* can occur at any season and any time of day. However, it is rather more frequent around the equinoxes. Typically (but not invariably) an event begins in the afternoon and may appear to be over by the end of the afternoon. But many events feature a second phase in the evening, which can be better in terms of DX worked - though this may partly be due to a greater number of stations being active at that time. These events can finish suddenly, just as some of the longest paths are being worked. Some of the best results occur well after local midnight.

It used to be thought that antennas had to be directed within a few degrees of due north or magnetic north. However, it is now known that optimum bearings can and do change in the course of an event. Finding the optimum reflecting area is crucial. While points somewhere between north-west and north-east are usually most productive for UK operators, particularly during major storms, bearings of over 90° or less than 270° have on occasion brought the best results. The wise operator will vary the beam heading from time to time.

To make a successful auroral contact, the stations involved have to direct their signals at a common reflection point. This will normally be at higher latitude than one or both of them. An aurorally-scattered signal is unmistakable, with a raw, rasping tone arising from the fact that it is being reflected by a highly disturbed medium and may have been subjected to a doppler shift or spreading.

At times this can make speech unintelligible - so CW is usually to be preferred. Auroral working is reported on 28MHz from time to time, but most contacts are on 50, 70 and 144MHz, with 430MHz available to well-equipped stations during major events.

In the course of geomagnetic disturbances, during a transition between phases in the circulating current system, auroral-E (AuE) may appear at 28MHz and VHF. This is a high-latitude phenomenon. Contacts are made on a direct beam heading and signals have the 'clean' tone characteristic of 'normal' mid-latitude Es rather than the rusty note of backscattered signals.

## Trans-equatorial Propagation

The study and use of trans-equatorial propagation (TEP) is one of the success stories of amateur radio research. Much of the pioneering work was carried out by amateurs, initially at 50MHz and subsequently at 144 and 430MHz - though most TEP working is at 50MHz. It is more frequent in years of high solar activity, though it occurs even at solar minimum, and it favours the months around the equinoxes. Example paths are Japan to northern Australia, the southern USA and the Caribbean to Argentina, and southern Europe to southern Africa. Britain lies to the north of the optimum area for TEP working though UK amateurs have occasionally experienced it, especially when signal paths are extended by, for example, sporadic-E. The stations in contact are usually (not invariably) symmetrically located with respect to the magnetic dip equator, with the path between them ideally being perpendicular to it.

Two types of TEP have been recognised. Afternoon-type TEP peaks at 1500 to 1700 local time with strong, clear signals and low fading rates at frequencies up to about 70MHz, typically at ranges of 5000 - 6500 km. Signals travel between the ionisation peaks each side of the equator without an intermediate ground reflection (and without the absorption in the D region) and thence to ground. This mode is also called chordal-hop propagation. The evening type is found mostly at frequencies above 30MHz but has been shown to occur as high as 430MHz. It tends to peak around 2000 to 2300 local time and is favoured by low geomagnetic activity. Ranges are between 3000 and 8000 km. Signals can be strong, but may be severely disturbed, with deep and rapid flutter fading and marked distortion caused by multipath and spreading, sounding a bit like auroral backscatter.

After sunset, the ionosphere over the magnetic equator breaks up into intense small-scale irregularities, seen on ionograms as spread-F, with characteristic rapid fading and distortion.

## Meteor Scatter

Meteoric ionisation is caused by the heating to incandescence by friction of small solid particles entering the Earth's atmosphere. This results in the production of a long pencil of ionisation extending over a length of 15 km or more, chiefly in the height range 80 to 120 km. This trail expands by diffusion and rapidly distorts due to vertical wind shears. Most trails detected by radio are effective for less than one second, but some last for longer periods, occasionally up to a minute and very occasionally for longer.

There is a diurnal variation in activity, most trails occurring between midnight and dawn when the Earth sweeps up the particles travelling towards it. There is a minimum around 1800 local time, when only meteors overtaking the Earth are observed.

Sporadic meteors are present throughout the year, but shower meteors have definite orbits and predictable dates. **Fig 12.22** shows the daily and seasonal variation in meteor activity. Intermittent communication is possible using meteoric ionisation between stations with antennas aligned to the optimum headings of 5° to 10° to one side of the great-circle path between them.

Small bursts of signal, referred to as pings, can be received by meteor scatter from distant broadcast (or other) stations situated 1000-1200 km away.

### Grey-line Propagation

The grey line is the ground-based boundary around the world that separates day from night, sunlight from shadow.

Many operators believe, and with some justification (**Fig 12.23**), that signals beamed along the grey line near sunrise or sunset will reach distant locations that are also experiencing sunrise or sunset for relatively short periods of time when conventional predictions may appear to be pessimistic. Sunrise and sunset are the periods when $MUF$ may be expected to rise or fall through their greatest range of the day.

This is a mode for the lower HF bands, 1.8, 3.5 and, perhaps, 7MHz because the operating frequency has to be below the $MUF$ for the path. On the ground, the grey line may be considered to be a great circle, but a rather badly defined one. It is not the ground-based shadow that determines the state of the ionosphere. The F2 region is illuminated for longer periods than the ground and other factors also govern the ionisation. Nevertheless, ground-based data is easy to obtain and antenna beam widths will be large at these frequencies, so that with a generous allowance for timing the circumstances may be favourable.

You may also get grey line enhancements for paths crossing the terminator at or near right angles near to the path ends. This is thought to be due to ionospheric tilting, although more research is needed for a definitive answer.

The Sun's azimuth, $A_S$ , when viewed from the ground at sunrise is given approximately by

$$\sin(A_S) = \sin(\delta_S)/\cos(l_G),$$

where $d_S$ is the Sun's declination, and $l_G$ is the station's geographical longitude, Similarly, at sunset we have

$$\sin(A_S) = -\sin(\delta_S)/\cos(l_G).$$

The sunrise and sunset azimuths are also shown in **Fig 12.26**.

### Ionospheric Scatter

At frequencies above the normal $MUF$ , propagation is still possible with much reduced signal levels by scattering from small irregularities in the ionosphere. This mode has been used commercially with high powers and very low angles of radiation using frequencies around 35MHz. Paths of some 2000 km are possible and this mode has the advantage of being workable in auroral regions where conventional HF methods are often unreliable.

## IONOSPHERIC MULTI-PATH AND FADING

When multi-path effects occur (as they frequently do) they may be between the ground wave and an ionospheric wave, between several ionospheric waves which may have travelled through different layers or, in the case of long range transmissions, by signals which have followed different paths entirely in different directions around the Earth's curvature, perhaps in several 'hops' between the ionosphere and the ground. Whatever the cause the result is inevitably a fading signal.

The ionosphere is not a perfect reflector; it has no definite boundaries and it is subject to frequent changes in form and intensity. These changes result in continual small alterations in phase or frequency of the received signal and, when only a single ionospheric wave is present, can pass almost unnoticed by the average HF listener, who is remarkably tolerant of imperfections on distant transmissions. However, when a second signal from the same source is present, which may be either the relatively steady ground wave or another ionospheric component, these phase or frequency changes become further emphasised by appearing as quasi-periodic changes in amplitude as the waves alternately reinforce and interfere, and by distortion of modulated signals if the various sideband frequencies do not resolve back into their original forms.

Occasionally, very long-distance transmissions may be heard with a marked echo on their modulation. The two signal components responsible have travelled routes of markedly differing length and they probably arrive from different azimuths at the receiving antenna, or are launched into different azimuths at the transmitter end of the path. This effect is most noticeable on omni-directional broadcast transmissions and may be minimised by the use of narrow-beam antennas both for transmission and reception.

### Fading

Fading is generally a consequence of interference between signals arriving at the receiving antenna along different paths which may be changing with respect to each other, or of changes in the received polarisation. It can also be caused by changes in attenuation. Generally the fading rate increases with frequency because a particular motion in the ionosphere causes a greater phase shift at the shorter wavelengths.

Interference fading, as its name implies, is caused by wave-interference - the vector addition of two or more component waves when some of the path lengths are changing, perhaps because of fluctuations in the ionosphere, or because of reflections from a moving surface. The fading period is relatively short, usually up to a few seconds. Fast interference fading is often called flutter. Auroral flutter comes in this category, being caused by motion of the reflecting or scattering surfaces.

Polarisation fading is brought about by continuous changes in polarisation due to the effect of the Earth's magnetic field on the ionosphere. Signals are at a maximum when they arrive with the same polarisation as the receiving antenna. The period is again up to a few seconds.

Absorption fading, generally of fairly long period, is caused by inhomogeneities in the ionosphere, with periods up to an hour, or longer.

Skip fading occurs when a receiver is on the edge of a skip zone and changes in the $MUF$ cause the skip distance to shorten and lengthen. The effect is to produce fading of an irregular nature without any well-defined period.

Selective fading occurs when the fading is a function of frequency. Transmission loss on one frequency may be low while on a nearby frequency it may be much greater. The effect will depend on the modulation system in use, i.e. the bandwidth of the signal.

## IONOSPHERIC DISTURBANCES

Like so much that affects radio propagation, ionospheric disturbances have their genesis in the Sun. From time to time flares occur, powerful explosions that hurl vast amounts of highly-charged particles out of the Sun along with electromagnetic radiation over a wide band of frequencies.

Initially, the radiation is in the form of X-rays, ultraviolet, visible light and radio waves which will reach the sun-lit side of the Earth within about eight minutes. The X-rays and UV light cause immediate increases in the D-layer ionisation, leading to absorption, described as a *Sudden Ionospheric Disturbance (SID) or a Short-Wave (or Dellinger) fade-out,* which may persist for anything up to two hours. However, for frequencies below 500kHz, the increased D-region ionisation *enhances* signals. The signal strength profile closely follows the flare X-ray flux plot shown on the NOAA web-site. As it is not possible to predict the timing of flares accurately, they are of limited use in achieving contacts.

The effects of an SID, experienced only on the daylight side of the Earth, appear first on the lower bands, then move up in frequency, sometimes wiping out even high-powered transmissions. Recovery works in the opposite direction, with the higher frequencies regaining propagation first. Sometimes prolonged bursts of radio noise also occur. Other effects observed are a sudden enhancement of atmospherics (SEA), a sudden absorption of cosmic noise (SCNA), and sudden phase anomalies (SPA) on VLF transmissions.

The ionised particles from a solar eruption may also escape the Sun, travelling somewhat more slowly. If the disposition of the solar corona allows, and if the solar wind and the earth's magnetic field are in the right orientations, the effects may reach Earth.

A Polar Cap Absorption event (PCA) is the intense absorption of radio waves in the polar regions caused by the arrival of high-energy solar protons concentrated in this region by the lines of force of the Earth's magnetic field, causing a radio blackout over trans-polar circuits. The interval between a major flare being observed on the Sun and the onset of these blackouts varies between a few minutes and several hours. They can last for anything up to ten days, though three days is more normal.

The main stream of particles arrives after an interval of 20-40 hours and consists of plasma clouds of protons and electrons borne by the solar wind. High-speed coronal streams, not necessarily originating in flare activity) can travel at speeds which sometimes exceed 1,000 km $s^{-1}$. What happens next greatly depends on the strength and orientation of the interplanetary magnetic field (IMF) - the intensity of which is expressed in nano-Tesla (nT) units. A southerly orientation of the IMF (that is, a minus nT value) favours a coupling with the Earth's magnetic field and thus the development of the disturbance. (A northerly, or positive, nT value does not couple so easily.) Where this applies, particles reaching the Earth's magnetosphere manifest themselves in visible displays of aurora and in auroral backscatter propagation at VHF, occasionally extending into the UHF range. Also, a strong polar electrojet current may flow in the lower ionosphere. Changes in the make-up of the trapping regions leads to variations in the circulating ring-current causing violent alterations in the strength of the geomagnetic field, which is the first indication of a magnetic storm.

Associated with magnetic storms are ionospheric storms, and both may persist for several days. The most prominent features are the reduction in F2 critical frequencies and an increase in D-region absorption. During the storm period, signal strengths remain very low and are subject to flutter fading. The effects of an ionospheric storm are most pronounced on paths which approach the geomagnetic poles. Conversely, while the *MUF* is falling at high and mid-latitudes, enhancements of twenty per cent or more may be experienced in low latitudes and over trans-equatorial circuits. LF signal levels are usually enhanced by up to about 10 dB at the peak of a solar flare.

The effects of magnetic storms can linger for some time, depending in large measure on their severity. Thus, enhanced D-layer absorption does not immediately disappear when the magnetic *K* index falls back to normal values. Electrons and ions trapped in the ring current exchange with the ionosphere at the daylight edge of the magnetosphere which becomes

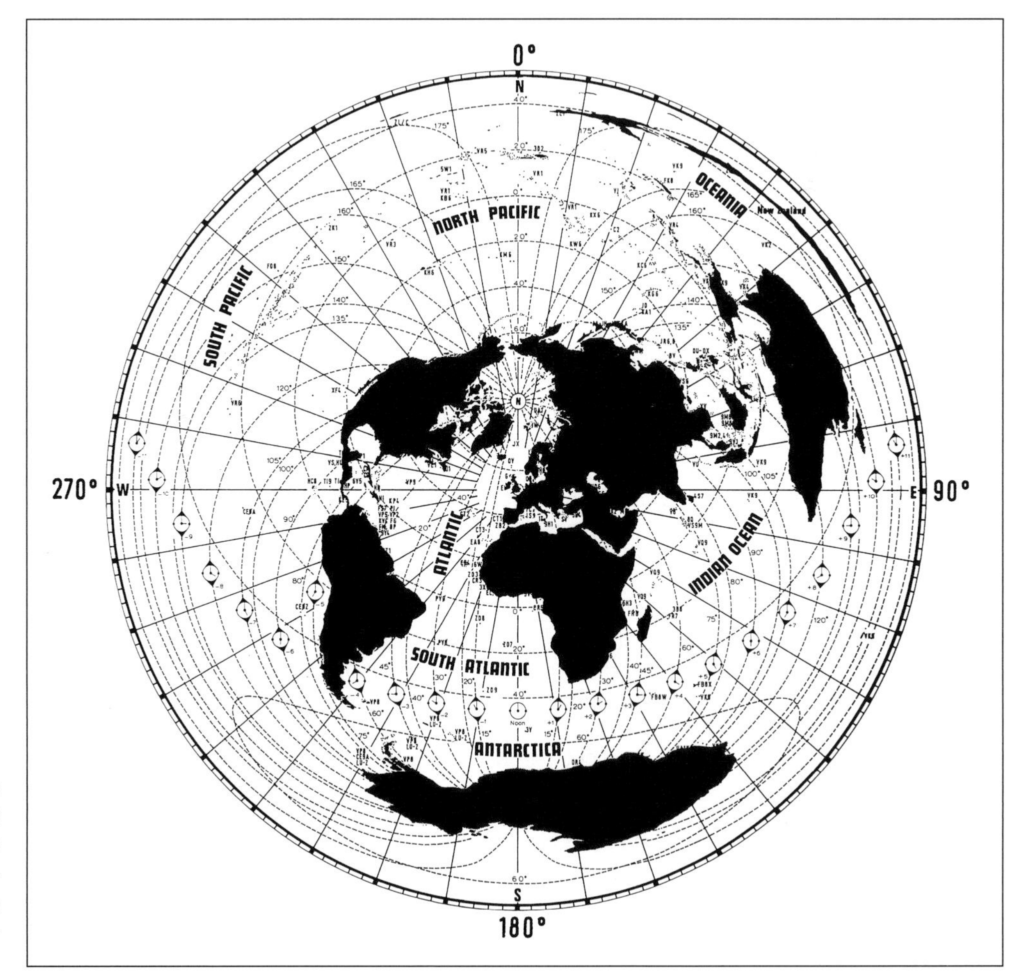


**Figure 12.24: An example of an azimuthal equidistant (or great-circle) map. This map shows the true bearing and distance from London of any place elsewhere in the world**

distorted by the pressure of the solar wind. This lets hot electrons out of their 'magnetic bottles' to be precipitated into the D-region, where they can persist even through the darkness hours. LF night-time signals do not return to normal levels until one or two days after the equatorial ring current has returned to near zero (<-20 nT). After a very severe storm, absorption can result in LF signals remaining below normal values for as long as 28 days.

## Geomagnetism

The Earth's magnetic field is the resultant of two components, a main field originating within the Earth, roughly equivalent to the field of a centred magnetic dipole inclined at about one degree to the Earth's axis, and an external field produced by changes in the electric currents in the ionosphere.

The main field is strongest near the poles and exhibits slow changes of up to about 0.1% a year. It is believed to be due to self-exciting dynamo action in the molten metallic core of the Earth. The field originating outside the Earth is weaker and very variable, but it may amount to more than 5% of the main field in the auroral zones, where it is strongest. It fluctuates regularly in intensity according to annual, lunar and diurnal cycles, and irregularly with a complex pattern of components down to micropulsations of very short duration.

Certain observatories around the world are equipped with sensitive magnetometers which record changes in the field on three different axes, the total field being a vector quantity having both magnitude and direction. Daily records of the field, (magnetograms), are obtained. From these the highest positive and negative departures from the 'normal' daily curve are determined for successive three-hourly periods, using a quasi-logarithmic scale ranging from 0 (quiet) to 9 (very disturbed). The various observatories do not all use the same scale factors in determining $K$-indices, but are chosen so as to make the frequency distributions similar at all stations.

Most large magnetic disturbances are global in nature and appear almost simultaneously all over the world. The more frequently used planetary $K$ index, $K_p$, is formed by combining the $K$ figures for a dozen selected observatories. It is more finely graded: $0, 1-$, $1o, 1+, 2-, 2o, 2+$, and so on up to $9-, 9o, 9+$. Indices of 5 or more may be regarded as being indicative of magnetic storm conditions.

The quasi-logarithmic scales of the $K$- and $K_p$-indices place more emphasis on small changes in low activity than in high. For some purposes, it is more convenient to work with a linear scale, particularly if the values are to be combined to derive averages, as of the day's activity, for example. The $A$-index, is one such linear scale, recording the daily equivalent amplitude on a scale running from 0 to 400 - the maximum figure for the most severe storms. As with $K_p$, the daily $A_p$ figure combines results from a selected group of observatories. Both indices are widely used as a shorthand description of geomagnetic activity.

Periods of high geomagnetic activity tend to occur somewhat more frequently around the equinoxes, while periods around the summer and winter solstice are more likely to be quieter. During some parts of the solar cycle, particularly in the years when the sunspot number is decreasing, there is a tendency for disturbances to repeat over several solar rotations and such events may not be due to solar flares but to coronal irregularities.

# PRACTICAL CONSIDERATIONS

## Map Projections

Maps are very much a part of the life of a radio amateur, yet how few of us ever pause to wonder if we are using the right map for our particular purpose or take the trouble to find out the reason why there are so many different forms of map projection.

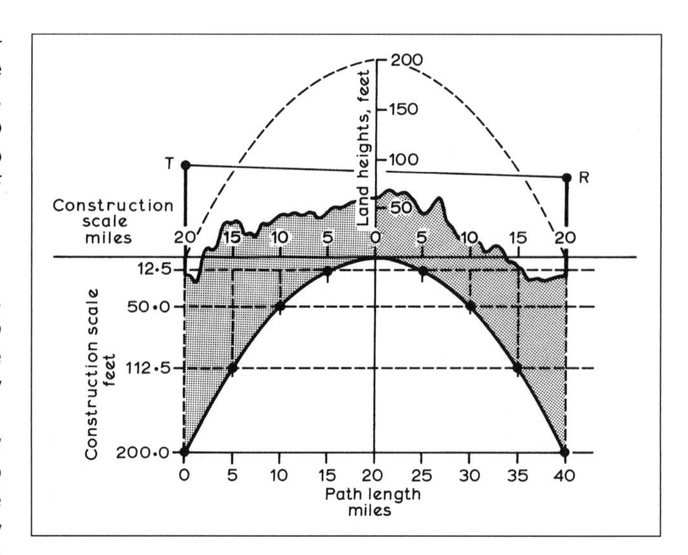


**Figure 12.25: An example of a 'four-thirds-earth' profile. Land heights are measured upward from the lower sea- level curve. On this diagram, rays subjected to 'normal' variations of refractive index with height may be represented by straight lines.**

| D (miles) | Horizontal scale (inches) | D² | D²/2 = h | Vertical scale (inches) |
|---|---|---|---|---|
| 5 | 0.5 | 25 | 12.5 | 0.125 |
| 10 | 1.0 | 100 | 50.0 | 0.500 |
| 15 | 1.5 | 225 | 112.5 | 1.125 |
| 20 | 2.0 | 400 | 200.0 | 2.000 |

**Table 12.7: Points on the curve drawn in Fig 12.25**

The cartographer is faced with a basic problem, namely that a piece of paper is flat and the Earth is not. For that reason his map, whatever the form (or *projection*) it may take, can never succeed in being faithful in all respects - only a globe achieves that. The amount by which it departs from the truth depends not only on how big a portion of the globe has been displayed at one viewing, but on what aspect the mapmaker has wanted to keep correct at the expense of all others.

Projections can be divided into three groups: those which show areas correctly, described, logically enough, as equal-area projections; those which show the shapes of small areas correctly, known as orthomorphic or conformal projections; and those which represent neither shape nor area correctly, but which have some other property which meets a particular need. The conformal group, useful for atlas maps generally, weather charts, satellite tracks etc includes the following members:

*Stereographic projections*: these are projections in which lines of latitude and longitude are all either straight lines or arcs of circles. The projections are made by projecting the surface of the Earth on to a plane surface, tangent either to one of the poles (polar), the equator (equatorial), or somewhere intermediate (oblique). Small circles on the globe remain circles on the projection, but the scale of the map increases with increasing radius from the centre of projection.

*Lambert's conformal conic projection*: this is formed by projection of the surface of the Earth on to a right circular cone whose axis passes through the Earth's poles. All meridians (lines of longitude) are straight lines and all parallels (lines of latitude) are circles.

*Mercator's projection*: this is based on a projection of the Earth's surface on to a cylinder which touches the globe at the equator. Meridians and parallels are straight lines in this projection and they intersecting at right-angles. The meridians are equidistant, but the parallels are spaced at intervals which rapidly increase with latitude. Any straight line on a Mercator's projection is a line of constant bearing, a so-called *rhumb line*. This is not the same thing as a *great circle*, which is the shortest path between two points on the Earth's surface. A great circle has its centre at the centre of the Earth. There is a scale distortion which gets progressively more severe with distance from the equator, to such an extent that it becomes impossible to show the poles, but most people accept these distortions as being "normal", because this is the best-known of all the projections.

*Transverse Mercator projection:* this is a modification of the 'classic' Mercator system, and is formed by projection of the Earth's surface on to a cylinder which touches the globe along selected opposing meridians. It therefore corresponds to an ordinary Mercator projection turned through ninety degrees, and is of value for displaying an area which is extensive in latitude but limited in longitude. A variant is the *Universal Transverse Mercator projection*, which forms the basis of a number of reference grids, including the one used by the British Ordnance Survey.

The equal-area group is used when it is necessary to display the relative distribution of something, generally on a worldwide scale. It includes:

*Azimuthal equal-area projection*: this has radial symmetry about the centre of the map, which may be at either pole (polar), at the equator (equatorial), or intermediate (oblique). With this system, the entire globe can be shown in a circular map, but there is severe distortion towards its periphery.

The final group, true direction projections, includes two which are of particular interest in propagation studies:

*Azimuthal equidistant projection*: this is centred on a particular place, whence all straight lines are great circles at their true azimuths. The scale is constant and linear along any radius. This is more commonly known as a great-circle map. See **Fig 12.24**.

*Gnomonic projection*: this is constructed by projection from a point at the Earth's centre on to a tangent plane touching the globe. Any straight line on the map is a great circle. The scale of the map expands very rapidly with increasing distance from the centre, so they do not normally cover large areas. They are sometimes produced as skeleton maps on which great circles can be drawn and are used to provide a series of latitudes and longitudes by which the path can be re-plotted on a more detailed map based on a different projection.

## Beam Heading and Locators

The shortest distance between two points on the surface of the Earth lies along the great circle that passes through them. This is easily done if you have a globe. All you need is to join the two locations with a tightly stretched thread. The shortest length of thread that can do this will be the great circle route and indicates the beam heading along which you need to direct your signal for best results. Sometimes, however, the best path actually lies in the opposite direction as ionospheric conditions allow for propagation along the *Long Path*. Irregularities in the distribution of ionisation can sometimes result in signals being diverted off the great circle - but this is an exception to the general rule. Alternatively, a great circle map along the lines of **Fig 12.24** indicates optimum beam headings. For most working purposes the map works well for most UK operators. Operators far removed from London, or who simply wish to be more exact, can readily find freeware programs on the Internet that will supply beam headings and distances for any location.

## Great circle calculations

It is sometimes useful to be able to calculate the great-circle bearing and the distance of one point from another, and the expressions which follow enable this to be done.

First label the two points A and B.

Then set:

$L_A$ to be the latitude of point A;

$L_B$ to be the latitude of point B;

$L_0$ to be the difference in longitudes of points A and B;

$C$ to be the direction of point B from point A, in degrees East or West from North in the northern hemisphere, or from South in the southern hemisphere; and

$D$ to be the angle of arc between points A and B (the angle subtended at the centre of the Earth).

The equations are:

$$\cos(D) = \sin(L_A)\sin(L_B) + \cos(L_A)\cos(L_B)\cos(L_0)$$ , and

$$\cos(C) = \frac{\sin(L_B) - \sin(L_A)\cos(D)}{\cos(L_A)\sin(D)}$$

$D$ can be converted to distance using 1 degree of arc = 111.2 km and 1 minute of arc = 1.853 km.

**Note that**:

latitudes are positive for stations in the northern hemisphere;

latitudes are negative for stations in the southern hemisphere;

$\cos(L_A)$ and $\cos(L_B)$ are always positive;

$\cos(L_0)$ is positive between 0 and 90°, negative between 90° and 180°;

$\sin(L_A)$ and $\sin(L_B)$ are negative in the southern hemisphere; and

the bearing for the reverse path can be found by transposing the two locations.

It is advisable to make estimates of the bearings on a globe, wherever possible, to ensure that they have been placed in the correct quadrant.

## THE IARU LOCATOR

It is always nice to know where your contact lives and operators have been asking that question since the earliest days of operating. But all too often the answer has been something like '16 km south' of some place your atlas does not deign to mention. That is particularly unsatisfactory if points in a contest or a personal distance record are at stake.

Happily, the International Amateur Radio Union [5] developed a system, originally called the *Maidenhead Locator* but now termed just the *Locator,* that can be used anywhere in the world, in which information about the location is expressed in a group of six characters. The result defines the location to within 0.04 degrees of latitude and 0.08 degrees of longitude. This is sufficiently precise for most purposes; on the relatively rare occasions where greater exactitude is required an eight-character or even a ten-character version is available.

The *RSGB Yearbook* [6] contains a map of grid squares in Europe. Several programs for converting latitude and longitude data into 'grid locators' (and conversely) anywhere in the world, as well as the distance between any two points, can be found on the Internet.

## Plotting VHF/UHF Path Profiles

For a tropospheric propagation study, it is standard practice to construct a path profile showing the curvature of the Earth as though its radius was four-thirds of its true value This is so that, under standard conditions o refraction, the ray path may be shown as a straight line relative to the ups and downs of the intervening terrain. To prepare a path profile, firs draw the sea level curve as shown by the lowe parabola in **Fig 12.25**. The sea level drops below the centre high point by a value of $h$ given by

$$h = s^2 / 2kR ,$$

where $s$ is the horizontal distance from the mid point, $R$ is the actual radius of the Earth, and $k$ is the effective Earth radius factor.

The antenna locations and heights above sea level may then be added at the appropriate dis tances as well as terrain heights along the path as found from height contours on a map.

It will be found that, as in **Fig 12.25**, if heights are plotted in feet and distances in miles the expression $h = s^2 / 2$ gives a very good approxima tion for a four-thirds Earth. It is simple to con struct the baseline of the chart using feet fo height and miles for distance, for the reason that by a happy coincidence, the various factors then cancel, leaving a very simple relationship tha demands little more than mental arithmetic to handle.

Suppose that a profile is required for a given path 40 miles in length. First, decide on suitable scales for your type of graph paper (**Fig 12.25** was drawn originally with 1 inch on paper repre senting 100 ft in height, and 10 miles in distance Then construct the sea-level datum curve, taking the centre of the path as zero and working down wards and outwards for half the overall distance in each direction, using the expression for $h$ to calculate points on the curve (**Table 12.7**). Remember that this works only if $h$ is expressed in feet and $D$ is expressed in miles.

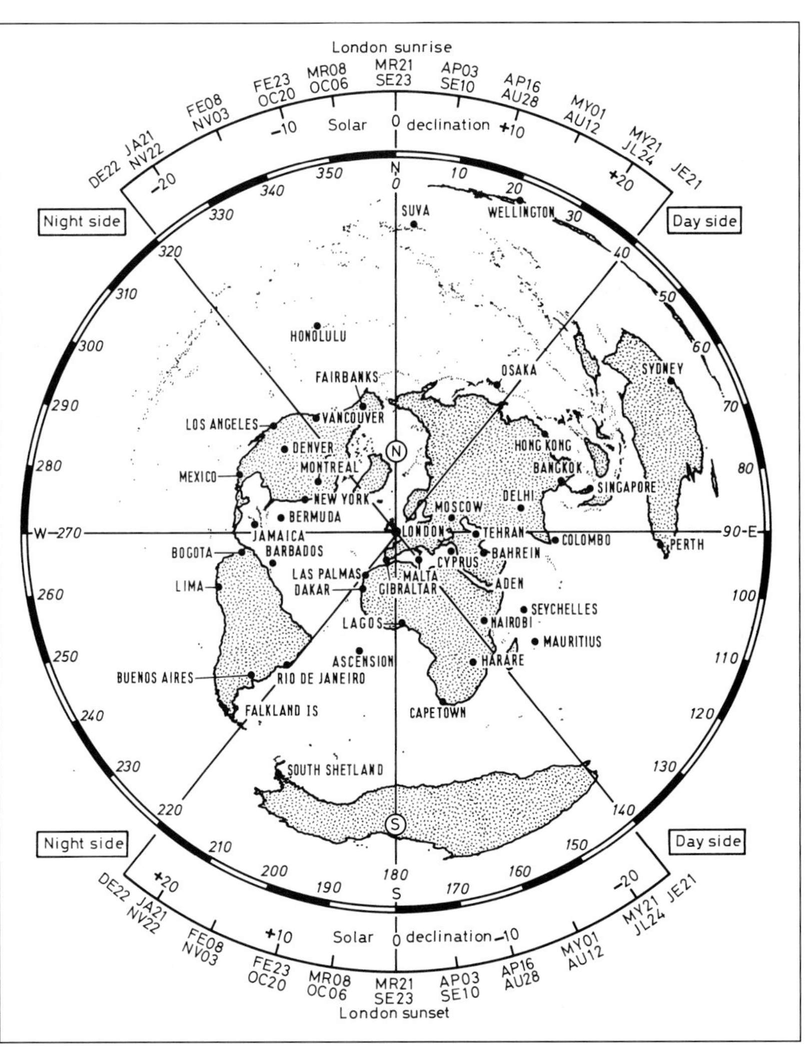


**Figure 12.26: The locations of the places used for the RadCom HF predictions.**

Next mark in the significant terrain height points. Usually it will be sufficient to limit this to just the contour values encountered along the path. Add the antenna heights at the two terminal points. Draw a line between the antenna locations. If the straight line from transmitting antenna to receiving antenna has sufficient Fresnel zone clearance (see page 12.6) over any obstacles along the way, then the path should be clear under normal conditions of refraction.

## Ionospheric Predictions

The quality of a radio circuit is usually highest when it is operated at a frequency just below the maximum usable frequency ( $MUF$ ) for the path. Three regions, E, F1 and F2, are considered in the determination of $MUF$ .

For the F2-layer, which is responsible for most HF long-distance contacts, the $MUF$ for paths less than 4000 km is taken as being the value that applies at the mid-point. For paths longer than 4000 km, the $MUF$ for the path is the lower of the values at the two "ends" of the path. For this purpose, the end point locations are not those of the transmit/receive antennas, but of their associated *control points* where low-angle radiation from a transmitter would reach the F2-layer. Those points are taken to be 2000 km from each antenna, along the great circle joining them.

To put this into perspective, a control point for a station located in the midlands of England would lie somewhere above a circle passing close by Narvik, Leningrad, Minsk, Budapest, Sicily, Algiers, Tangier, mid-Atlantic and NW Iceland, the place depending on the direction of take-off. It means that (for example) the steep rise in the $MUF$ associated with UK-end sunrise will occur something like four hours earlier on a path to the east than on a path to the west.

Taken over a month, the day-by-day path $MUF$ value at a given time can vary over a considerable range. A ratio of 2:1, as between maximum and minimum, could be considered as typical. Monthly predictions are based on median values, that is to say, on those values which have as many cases above them as below. Therefore the median values in predictions represent 50% probabilities because, by definition, for half of the days of the month the operating frequency would be too high to be returned by the ionosphere. Note that, however certain you might be that a given circuit at a given frequency at a given time might be open 15 days in a particular month (the meaning of 50% probability in this context), there is no way of knowing *which* 15 days they might be. Nor can you be sure that 'good days' on one circuit will be equally rewarding on another.

If you want to ensure the most reliable communication over a particular path, say to make a daily contact at a given time, you will need to operate below the median $MUF$ . Conventionally this optimum working frequency ( $OWF$ ), at which contact should be possible on 90 per cent of days in the month, is found by multiplying the $MUF$ figure by 0.85. Thus, for a monthly median $MUF$ of 20MHz, the $OWF$ will be around 17MHz.

| Name of band and limits of frequency | UK amateur bands by frequency (and usual wavelength description) | Principal propagation modes | Principal limitations |
|---|---|---|---|
| Very low frequency (VLF) 3-30kHz | | Extensive surface wave. Ground to ionosphere space acts as a waveguide | Very high power and very large antennas required. Few channels. |
| Low frequency (LF) 30-300kHz | 135.7-137.8kHz | Surface wave and reflections from lower ionosphere | High power and large antennas required. Limited number of channels available. Subject to fading where surface wave and sky wave mix |
| Medium frequency (MF) 300-3000kHz | 472-479kHz (NOV required)<br>1810-2000kHz (160m band, also known as topband) | Surface wave only during daylight. At night reflection from decaying E-layer | Strong D-region absorption during day. Long ranges possible at night but signals subject to fading and considerable co-channel interference |
| High frequency (HF) 3-30MHz | 3.50-3.80MHz (80m)<br>7.00-7.20MHz (40m)<br>5MHz (60m)<br><br>10.10-10.15MHz (30m)<br>14.00-14.35MHz (20m)<br>18.068-18.168MHz (17m)<br>21.00-21.45MHz (14m)<br>24.89-24.99MHz (12m)<br>28.00-29.70MHz (10m) | Nearly all long distance-working via F2-layer. Short-distance working via E-layer NVIS | Daytime attenuation by D-region, E and F1-layer absorption. Signal strength subject to diurnal, seasonal, solar-cycle and irregular changes |
| Very high frequency (VHF) 30-300MHz | 50.00-52.00MHz (6m)<br>70.00-70.50MHz (4m)<br>144.00-146.00MHz (2m) | F2 occasionally at LF end of band around sunspot maximum.- Irregularly by sporadic-E and auroral-E. Otherwise maximum range determined by temperature and humidity structure of lower troposphere | Ranges generally only just beyond the horizon but enhancements due to anomalous propagation can exceed 2000km |
| Ultra high frequency (UHF) 300-3000MHz | 430-440MHz (70cm)<br>1240-1325MHz (23cm)<br>2300-2450MHz (13cm) | Line-of-sight enhanced by tropospheric effects | Atmospheric absorption effects noticeable at top of band |
| Super high frequency (SHF) 3-30GHz | 3.400-3.475GHz (9cm)<br>5.650-5.850GHz (6cm)<br>10.00-10.50GHz (3cm)<br>24.00-24.25GHz (12mm) | Line-of-sight, enhanced by tropospheric and scattering effects | Attenuation due to oxygen, water vapour and precipitation becomes increasingly important |
| Extra high frequency (EHF) 30-300GHz | 47.00-47.20GHz (6mm)<br>75.50-76.00GHz (4mm)<br>142.00-144.00GHz (2mm)<br>248.00-250.00GHz (1.2mm) | Line-of-sight, enhanced by tropospheric and scattering effects | Atmospheric propagation losses create pass and stop bands. Background noise sets a threshold |
| 300-3000GHz | - | Line-of-sight, enhanced by tropospheric and scattering effects | Present limit of technology |

**Table 12.8: A survey of the radio-frequency spectrum**

At the other end of the scale, there will be days when the monthly $MUF$ figure is exceeded. This highest path frequency ( $HPF$ ) is found by multiplying the $MUF$ by 1.175 - in the example given earlier this would give a figure around 23.5MHz. Always remember that signals will propagate best if you are operating on the band closest to the operational $MUF$ on any particular day.

All this may seem to suggest that the lower the operating frequency, the greater would be the chance of success, but that is not necessarily true. The reason is that the further down from the median $MUF$ that one operates, the greater become the losses caused by absorption, and eventually these become the dominant factor. Noise levels tend to increase with decreasing frequency as well. For any given path there is a lowest useful frequency ( $LUF$ ) below which the quality of the circuit becomes too low.

The tables printed at the back of each issue of RadCom provide information in the form of percentage probabilities for each of the HF amateur bands, taking both the $HPF$ and the $LUF$ into account. A figure of 1 represents 10% (or three days a month), a figure of 5 represents 50% (or 15 days a month), and so on. Multiply by three the figure given in the appropriate column and row and you will have the expected number of days in the month on which communication should be possible at a given time on a given band.

Many operators believe that if it is known that solar activity is higher (or lower) than was expected when the predictions were prepared, then they can raise (or lower) all the probabilities by a fixed amount to compensate for the changed circumstances. That is not correct. The highest probability will always appear against the band closest to the $OWF$ (as defined in the predictions sense). If the $OWF$ is altered by fresh information, then the probabilities on one side of it will rise but on the other side will fall. The only sure way of establishing amended figures would be to enter the revised particulars into the computer program and to run off a complete set of new predictions. It is beyond the scope of this chapter to provide detailed descriptions of how to prepare predictions and relate them to the amateur bands.

For those wishing to prepare their own predictions, several useful freeware or shareware programs are available on the Internet. These prediction programs customarily assume a quiet geomagnetic field.

Monthly prediction programs usually require an index of solar activity comparable to the 12-month smoothed relative sunspot number $R12$ , which is derived from ionospheric data. There is nothing to be gained by substituting the latest unsmoothed monthly sunspot figure.

The ionosphere does not respond directly to fluctuations of daily sunspot numbers. On a daily time scale, signal performance is much more responsive to changes in the level of geomagnetic activity. Information about current levels of geomagnetic activity and short-term forecasts of the 'radio weather'' is available from several Internet sites.

The locations of places for which predictions are given every month in RadCom may be seen against their relative beam headings in the great-circle outline map of **Fig 12.26**. Note that this map is centred on London, but the predictions in Radcom are centred on the middle of the UK (54.5N, 2.00W)

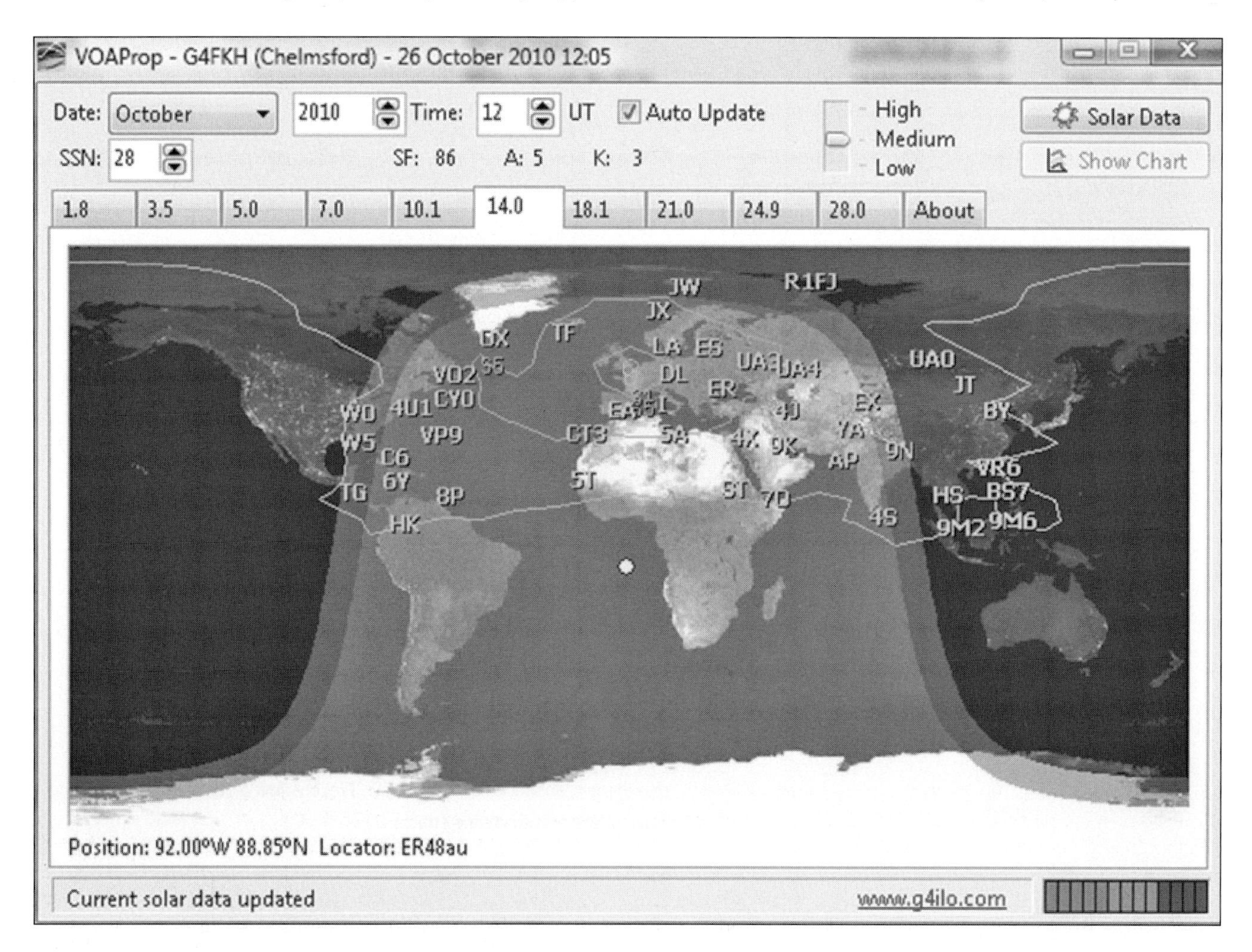


**Figure 12.27: A screen-grab from the VOAProp software.**

## HF Propagation Prediction Aids

There are several useful computer programs available free from the Web, for example the one used by Gwyn Williams, G4KFH, to produce the HF Propagation Predictions for *RadCom*. You can find this at the following Web address: *http://rsgb.org/main/operating/propagation-predictions/.*

Another useful tool is a little program called VOAProp by Julian Moss, G4ILO. It is important to read the instructions prior to installation. The program uses the VOACAP propagation engine. **Fig 12.27** shows a VOAProp screen indicating the call areas that it may be possible to work into, with the prevailing ionospheric conditions. Note the magnitude slider is set for medium. You can adjust this for your own station. Here, "medium" refers to a radio station delivering 100 W into a dipole. The inner contour shows a probable S-meter reading of '5', whilst the outer one suggests a considerably lower reading. The day/night terminator is also indicated, which can be a very useful guide. Right clicking on the map brings up a menu with various options including one to set the DX location to produce point-to-point predictions. The real value of this program is as a visualisation tool, which illustrates the general way that propagation changes from hour to hour, season to season, etc.

The RSGB has two online propagation prediction tools that can be used to calculate the probability of a contact being possible. Both use the same target locations used in RadCom and allow you to tweak the input to better match your station in terms of its power, antenna gain and mode.

The first at www.rsgb.org/voacap is based on the VOACAP engine and can produce charts for a given month. The second at www.rsgb.org/proppy uses the newer ITURHFPROP engine and does the same.

General prediction programs such as these cannot show exactly what should be heard at any particular time, but do provide good indications of what can be expected on average. The sunspot numbers used are those recommended by the VOACAP/ITURHFPROP authors and use 13-month running average sunspot numbers.

## The Beacon Network

One of the most useful aids for understanding actual propagation conditions, and exploiting whatever possibilities may be open, is offered by worldwide networks of beacon transmitters. These also offer a useful basis for personal propagation research projects. You can monitor these transmissions directly yourself, or you can use an automatic monitoring program available on the internet.

Formal permanent beacons are found mainly on the higher HF bands and upwards. For the most part they operate continuously with a simple message in Morse. This always includes their call sign and a long dash to facilitate signal strength measurements. Many also include their grid locator, town and power. Because, unlike other amateur stations, they are always on the air, they offer a useful indication of the state of a particular path.

This is particularly true at HF of the network of beacons created by the North California DX Foundation. This consists of 18 beacons, strategically dispersed around the world, transmitting in sequence on 14.100, 18.110, 21.150, 24.930 and 28,200MHz in turn in the course of a three-minute cycle. The power is stepped down from 100 W to 100 mW in four stages, giving a further indication of the state of the circuit.

There are hundreds more beacons worldwide at HF and on the VHF and UHF bands. Details of those likely to be heard in the UK are listed in the *RSGB Yearbook* [6] and changes between editions can be found on the Internet.. The RSGB web-site also has a link to G3USF's list at https://rsgb.org/main/technical/propagation/propagation-beacons/. For VHF and microwave beacons, see also the beaconspot website at https://www.beaconspot.uk/index2.php ,

## *GB2RS* News

Finally, the weekly GB2RS news bulletin, posted every Friday on the RSGB website [7] and broadcast every Sunday, contains items of propagation interest, including a summary of solar-geophysical events over the preceding week. Times and frequencies of these broadcasts vary according to location; details are printed in the *RSGB Yearbook* [6].

# BAND BY BAND

**Table 12.8** is an attempt to outline the principal propagation characteristics of each band, but it must be emphasised that there are no clear-cut boundaries to the various effects described. The practical uses for the information given so far have been discussed in a series of *RadCom* articles by Steve Nichols, G0KYA, and Alan Melia, G3NYK, detailing propagation band by band. The enquiring reader is referred to [8] and [9]. This is also available as a free downloadable e-book at [10].

# REFERENCES


[1] International Telecommunication Union (ITU): www.itu.int/

[2] SIDC, Brussels: *http://sidc.oma.be/*

[3] Boulder: www.ngdc.noaa.gov/stp/iono/ionohome.html

[4] Rutherford Appleton Laboratory: www.stfc.ac.uk/

[5] International Amateur Radio Union (IARU): www.iaru.org

[6] *RSGB Yearbook*, published annually by the RSGB

[7] *GB2RS* weekly news bulletin: *http://rsgb.org/main/news/gb2rs/*

[8] 'Understanding LF Propagation', by Alan Melia, G3NYK series in *RadCom*, September to November 2009, RSGB

[9] 'An introduction to HF Propagation', G Williams, G4FKH, *RadCom*, May 2007, pp 56-62

[10] *http://g0kya.blogspot.com*


# BIBLIOGRAPHY

## Print


*Radio Auroras*, Charlie Newton, G2FKZ and Neil Carr, G0JHC, RSGB, 2012

*Radio Propagation - Principles & Practice*, Ian Poole, G3YWX, RSGB, 2004

*LF Today*, Mike Dennison, G3XDV, RSGB, 2013

*The VHF/UHF Handbook*, edited by Andy Barter, G8ATD, RSGB, 2007

*Ionospheric Radio*, K Davies, P Peregrinus, IEE, 1990

*Propagation of Radio Waves*, ed L Barclay, IET, 2013

'Understanding Propagation', by Steve Nichols, G0KYA. series in *RadCom*, 2008 and 2009, RSGB

'5MHz Beacon Sounder Experiment', P Martinez, *RadCom*, January 2006, pp 64-68

'Aircraft Reflections', W F Blanchard, RadCom, March 2006, pp 78-79

'Patterns in Propagation', R G Flavell, *Journal of the IERE*, Vol 56 No 6 (Supplement), pp175-184

'The Twilight Zone Revisited - Recent Grey-line Research', S Nichols, RadCom, May 2006

'Re: The Twilight Zone Revisited - Recent Grey-line Research', Steve Telenius-Lowe, G4JVG, *RadCom*, August 2006

## Web

*http://rsgb.org/main/operating/propagation-predictions/*

G4ILO's Shack: www.g4ilo.com/voaprop.html

Current Solar Data: www.n3kl.org/sun/noaa.html

The Radio Propagation Page is at *http://rsgb.org/psc/. Run by the RSGB Propagation Studies Committee this carries links to a wide range of sources, ranging from the explanation of basic terms in radio propagation through the various forms of propagation to sites carrying data affecting propagation, as well as HF band prediction programs*

Amateur Radio Propagation Studies (VHF): www.df5ai.net

'DX Radius in Aurora and FAI Radio Propagation', Volker Grassman, DF5AI, August 2002: www.df5ai.net/ArticlesDL/FAIRadius/FAIRadius.html

Glossary on Solar-Terrestrial Terms: www.ips.gov.au/Educational/1/2/1

Great Circle bearings and Distances: www.gb3pi.org.uk/great.html

Introduction to the Ionosphere: www.ngdc.noaa.gov/stp/IONO/ionohome.html

Meteor Showers: www.amsmeteors.org/showers.html

Meteor Showers Online: *http://meteorshowersonline.com*

Real-time Space Weather: www.spacew.com

SWPC Radio Users' Page: www.swpc.noaa.gov/radio/

Solar and Geophysical Forecast: www.swpc.noaa.gov/forecast.html

Space Environment Center: www.swpc.noaa.gov

Spaceweather.com: www.spaceweather.com

Transequatorial Radio Propagation: www.ips.gov.au/category/Educational/Other%20Topics/Radio%20Communication/Transequatorial.pdf

Worldwide HF Beacon List (G3USF): www.keele.ac.uk/depts/por/28.htm

University of Colorado, Dst Index: *http://lasp.colorado.edu/space_weather/dsttemerin/dsttemerin.html*

## ACKNOWLEDGEMENTS

**Fig 12.3** is from the 3rd edition of *Propagation of Radiowaves*, edited by Les Barclay, IET, 2013

Figs 12.5 and 12.6 are from the 2nd edition of *Propagation of Radiowaves*, edited by Les Barclay, IET, 2012

Figs 12.8, 12.9, 12.12 and 12.18 are from the 1st edition of *Propagation of Radiowaves*, ed Hall, Barclay & Hewitt,IEE 1996

**Fig 12.8** is from *Radiowave Propagation* by L Boithias, North Oxford Academic Publishers Ltd, 1984

**Fig 12.11** is from J Atmos, *Terrest Phys*. Pergamon Press

**Fig 12.20** is from NBS Circular 462 *Radiowave Propagation*, 1948

*This chapter was revised for the 14th edition by **Dr Peter Duffett-Smith, G3XJE,** with contributions by members of the PSC. **Les Barclay, G3HTF, SK** who contributed to this chapter early in 2019, sadly passed away in July 2019. His contrubution to the Handbook and to amateur radio will be missed.*

# Antenna Construction Techniques 13

*Mike Parkin, G0JMI*

## INTRODUCTION

Operating a station on several bands involves installing a number of antennas, usually a mast and potentially several feeder cables between the antennas and the radio equipment. Guidance on how to construct, install and source material for amateur band antennas and their supporting structures is summarised within this chapter.

## ANTENNA MATERIALS

The two main types of conductive material used when building antennas are wire and metal tubing. Wire antennas are generally more straightforward to construct, although some arrays of wire elements can become complex. When tubing is required, most often aluminium tubing is used because it is relatively strong and lightweight.

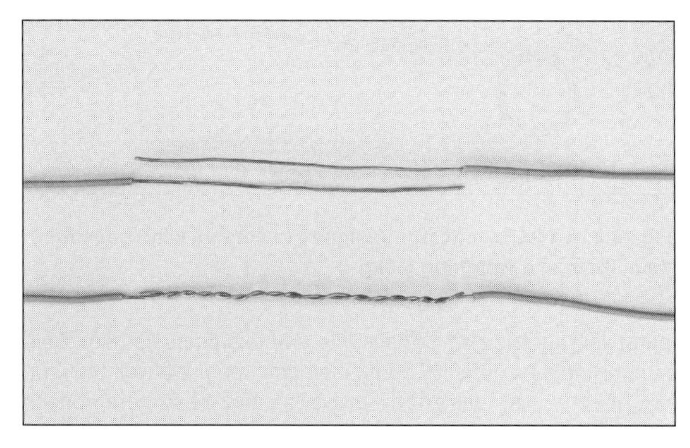

**Fig 13.2: Example of the wire spliced jointing technique before the joint was soldered and weatherproofed**

### Antenna Wire

Wire antennas can be made from any copper wire. The RF resistance of copper wire increases as the diameter of the wire decreases. However, in most types of antennas that are constructed from wire, the radiation resistance (R) will be much higher than the conductor's resistance (Rc) allowing the antenna to be an efficient radiator of RF energy. Even when quite thin wire is used, where Rc can have more of an effect, this gives a reasonable efficiency for the antenna. Wire diameter sizes as small as 0.3mm have been used quite successfully in the construction of antennas to help make them inconspicuous in areas where more conventional antennas cannot be installed. In most cases, the selection of wire for an antenna will be based primarily on the physical properties of the wire.

For long wire antennas the preferred material is 14 Standard Wire Gauge (SWG) hard-drawn copper wire and this is ideal for applications where significant stretch cannot be tolerated. Care is required when handling this wire because it has a tendency to spiral when it is unrolled. Make sure that kinks do not develop when handling this type of wire because this wire can have a tendency to break at a kink.

Most antenna material suppliers have various types of antenna wire available as shown in **Fig 13.1**. Antenna wire can often be obtained from scrap metal yards. However, scrap electrical wire is usually heavily insulated and can be too weighty for antenna elements although its use is fine for radials. Other potential sources of suitable wire that could be used for antenna applications are from some electrical retailers and online suppliers.

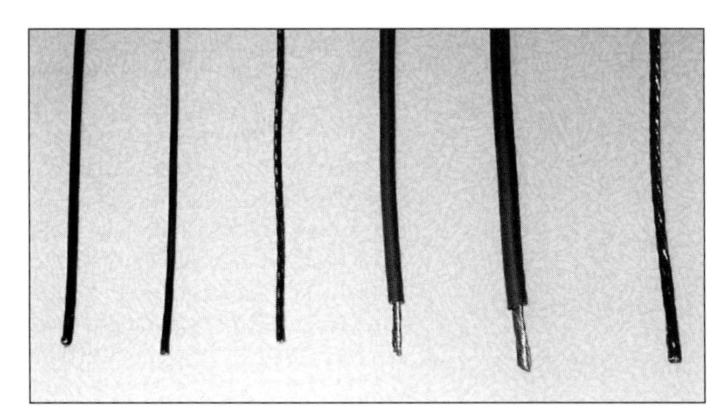

**Fig 13.1: Various types of antenna wire. From the left to right (1) 16SWG hard drawn single strand, (2) 14SWG hard drawn single strand, (3) 14SWG multi-strand, (4) Plastic covered 1.5mm multi-strand (5) Plastic covered 2mm multi-strand (6) Multi-strand FLEXWEAVE antenna wire (wire samples supplied by WH Westlake Electronics)**

The most practical material for wire beams is insulated 14 to 16SWG stranded flexible tinned copper wire. Single-core wire having an enamel coating is preferable compared to bare wire because the enamel coating resists oxidation and corrosion of the conductor that can tend to increase Rc. Wire antennas should preferably be made with unbroken lengths of wire. In instances where this is not possible, wire sections should be spliced. The insulation should be removed for a distance of about 100mm (4inches) from the end of each wire section. If the insulation is enamel coating rather than plastic, then this can be removed by scraping with a knife or rubbing with sandpaper until the copper conductor underneath is revealed. To form the spliced joint, the turns of wire should be brought up tight around the standing part of the wire and laterally twisted using broad-nose pliers. The twisted joint that is made should lock both the wires together, so stopping them from moving. The joint should be soldered with the crevices formed by the wire completely filled with resin-core solder. A large wattage soldering iron will be required to melt solder when working outside, or a propane torch could be used. The joint should be heated sufficiently so the solder flows freely into the joint when the source of heat is removed momentarily. After the joint has cooled completely, it should be wiped clean with a cloth, and then sprayed generously with acrylic, or a similar material, to prevent the joint from corroding. An example of the wire spliced jointing technique described is shown in **Fig 13.2** before the joint was soldered and weatherproofed.

### Aluminium Tubing

Many of the beam antennas described in the antenna chapters are constructed from aluminium tubing. In reality, most tube is an aluminium alloy whose other constituents can include, for example, copper, zinc, iron, magnesium and silicon (amongst others) to improve the mechanical and structural properties of the material. This is because pure aluminium, although com-

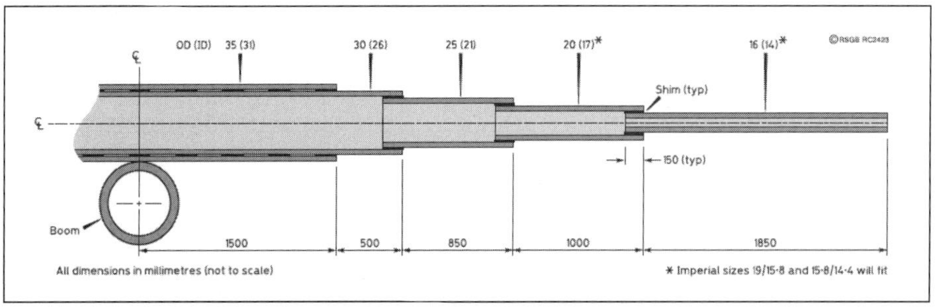

Fig 13.3: A 14MHz reflector designed to survive wind speeds of up to 159km/h, sagging less than 16cm and weighing 5.6kg

paratively lightweight, is malleable and this limits its use. Self-supporting horizontal HF beam elements require a well thought-out mechanical design to arrive at the best compromise between surviving storms, weight and sagging of the elements. This can be done by 'tapering' the elements through using tubes that telescope into each other at the ends. This technique allows thicker tubes to be utilized towards the centre of the antenna with gradually thinner tubes used in several steps towards the tips as shown in **Fig 13.3**.

G4LQI [1] has investigated the availability of aluminium tubing and the following is a summary of his work:

The most common range of American tubing comes in outside diameter (OD) steps of 3.18mm (1/8inch), with a wall thickness of 1.47mm (0.058inch). This means that each tube size conveniently slides into the next larger size. Amateurs in Britain who have tried to copy proven American designs have found that this could be difficult to do using the alloy tubing available in the UK. This is because the tube sizes easily obtainable in the UK tend to be Imperial, regardless whether designated in inches or millimetres. These tubes come in OD steps of 3.18mm (1/8inch) but with a wall thickness of 1.63mm (0.064inch), so the next smaller size does not fit into the larger one. This means the smallest taper step is 6.35mm (1/4inch) leaving a gap. To use these tube sizes requires this gap to be filled using aluminium shims and this method of construction is described later.

Metric size tubing, where all inside and outside tube diameters are in whole millimetres, are standard in mainland Europe and provide good compromise. This tubing can be arranged to fit, however some combinations of tube sizes may be need shims to provide a good fit. Table 1 summarises metric size tubing sizes for reference. Within the UK, there are now available several online aluminium suppliers who stock a range of tubing, angled sections and plate.

Aluminium tubing is also available from scrap yards. However, with the exception of scaffolding poles, aluminium tubing can be one of the scarcest materials to find. This is because there are not many applications that tend to use this material in quantity. Where aluminium tube is available it tends to come as one-off items such as tent poles for example.

There is another type of material available called duralumin that is commonly used for aircraft construction and boat masts. It has the advantage of being lighter and stronger than aluminium but is more brittle. Aluminium or duralumin tubing is useful when making up a lightweight mast and several sections can be joined together to give the desired height. Lightweight sections of thin-wall tubing can be joined together using a short joining section. A joining section can be made by longitudinally slicing a 30 to 45cm (12 to 18inches) length of the same tube with a hacksaw and then springing it open using a screwdriver. The two sections to be joined are forced into the joining section and clamped tight using hose clips as shown in **Fig 13.4**. Aluminium scaffold poles are useful for masts and booms for larger HF beams. This material is thick walled, is strong and relatively lightweight. Aluminium scaffold poles have the advantage of having clips, clamps and extension sleeves available that are used when building scaffolding platforms, an example of an extension sleeve is shown in **Fig 13.5**. An example of a fold-over mast that is constructed from scaffold poles is described later in the chapter.

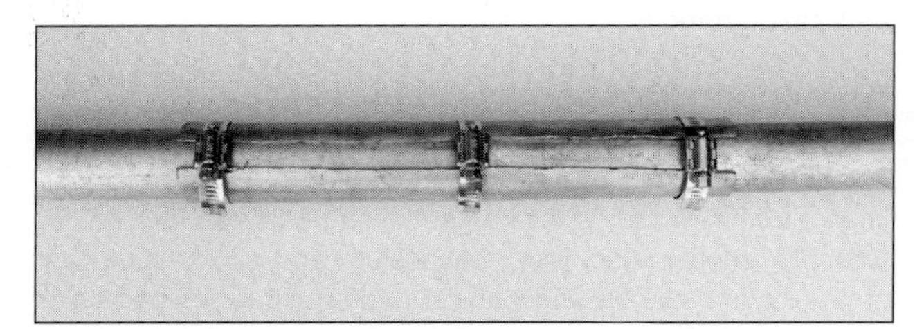

Fig 13.4: Method of joining lightweight tube of equal diameters

## Steel tubing

Steel tubing is an excellent material for constructing antenna masts and is often available in scrap metal yards. Steel tubing used for antenna masts should be free from damage and excessive corrosion. The lower sections of a self supporting steel mast of around 12m (40ft) in height should be at least 10cm (4inches) in diameter, with a wall thickness of at least 5mm (1/5inch).

Steel tubing is often available with the ends threaded to enable sections to be joined using screw couplers. This type of tube and their couplers are intended for use for the liquid or gas duct applications that are buried in the ground where bending stresses tend not to be encountered. The tube and screw couplers only have a short length of screw thread and this makes them a potential source of weakness and point of failure when used to join sections to form an antenna mast. This is because when the mast is put under load, the mast's coupled sections may have to withstand significant bending stresses that can tend to lead to the mast failing at these coupled points. Therefore, these couplers should not be used to extend masts.

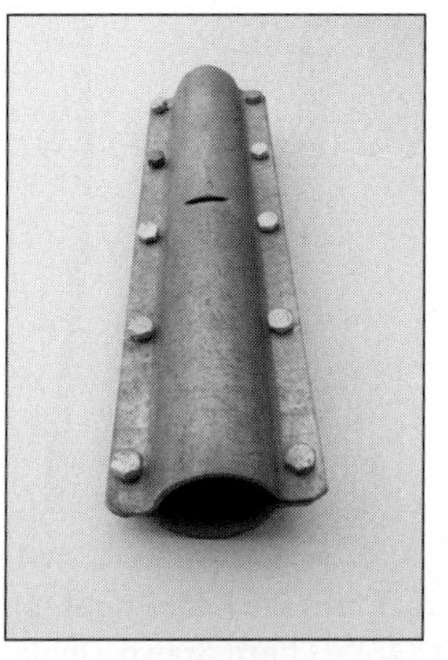

Fig 13.5: Two-piece steel sleeve of the type used to join scaffolding tubes together

Steel tubes should be joined by using lengths that telescope into each other, with at least 30cm (12inches) of overlap. These should be secured with a suitable nut and bolt passed through an appropriately sized hole drilled through both tubes (two securing nuts and bolts are preferable). It is not advisable to weld the sections together to form a mast because of the weight that this can involve. For example, even a 12m (40ft) steel mast made of welded sections can be very heavy making it difficult for an individual to raise or lower it on their own. Details on how to construct a steel mast are described later within this chapter. When a small diameter

**Fig 13.6: Method of fixing mast-to-boom and boom-to-element using an aluminium plate and U-clamps**

pole is joined to larger diameter pipe, eg scaffolding pole into 80mm (3inches) pipe, metal strip or angle iron shims can be used to pack any space between the differing diameters before securing with a nut and bolt passed through a suitable hole drilled through the tubes.

## Copper tubing

Copper has a very good conductivity but is comparatively heavy making it not really suitable for large HF antennas. Copper is suitable for small compact HF antennas, mobile and VHF antennas. A further advantage of copper for constructing antennas is that there is a good selection of couplings available. Builders' merchants, plumbing suppliers and some do-it-yourself (DIY) stores are good sources of supply with a range of copper tubing and couplings usually stocked. Copper tubing can be relatively plentiful at scrap metal yards because of central heating system applications. In the UK the most common copper tubing diameters available are 16 and 22mm. However, some older scrap copper tubing may still have imperial dimensions, so these should be checked if the tubing is to be integrated with an existing structure.

## Metal plates

Aluminium plates are particularly useful for making mast-to-boom and boom-to-element fixings. To make a sturdy fixing, aluminium plate of at least 5mm (1/5inch) thickness should be used, with holes drilled to take U-clamps and fixing bolts. An example of an HF boom-to-element clamp is shown in **Fig 13.6** and an arrangement used to clamp a UHF beam to a mast is shown in **Fig 13.7**.

## Scrapyards as a Source of Material

Some useful materials for constructing antennas can be obtained from a scrap metal yard. The best yards seem to be those located near an industrial estate because these often contain a selection of suitable materials for antenna construction including:

- Steel tubing (for antenna masts)
- Steel casings (for mast foundations)
- Steel angle material (for gin poles, clamps and guy rope anchors)
- Copper and aluminium tubing (for elements and booms)
- Paxolin, Bakelite or plastic sheet (Insulators)
- Electrical wire (antenna elements)
- Electric motors and gear-boxes (for rotators)
- Aluminium angle or L-section (for quad and Double-D antenna spreaders)
- Aluminium plate (to make couplers for joining elements to booms and booms to masts)

## Other Materials

The following materials and fixings are very useful for antenna construction.

## Hose Clips

Hose clips, or jubilee clamps, can be used for joining a variety of items including:

- Different diameter sections of elements
- Sections of mast
- Wire to metal elements
- Quad spreaders to angle or L-section.

This type of clamp is readily available at all hardware, DIY and car (auto) part stores. When a clamp is used as part of an outdoor antenna structure it is advisable to coat it with a film of grease to prevent corrosion. However, these clamps should not be painted or varnished because this can make it very difficult to dismantle the structure.

## Insulators

Antenna insulators should be made of material that will not absorb moisture. The best insulators for antenna use are those made of glass or glazed porcelain. Depending on the type of wire antenna, the insulator may need to be capable of taking the same strain as the antenna wire.

**Fig 13.7: Example of an UHF antenna to mast clamp made using an aluminium plate, U-clamps and fixing bolts**

'Egg' insulators designed for electric fence installations can be useful. They are moulded from fibreglass-reinforced plastic and are designed to withstand at least 10kV. Often egg insulators were used at each end of a wire antenna to support it under a tensile load. However, commercially available 'dog-bone' type insulators now tend to be used for such higher impedance antenna situations. A disadvantage of using egg insulators is that they can introduce excessive end capacitance or even leakage that can affect the performance of a dipole for example. It is possible to reduce the mechanical strain on an egg insulator by placing the insulator under a compressive load through crossing the wires as could be encountered with lower impedance antenna situations.

Many radio equipment suppliers stock antenna insulators that are usually of the smaller or larger 'dog-bone' type. Other sources of supply of this type of insulator include radio rallies and from online suppliers. An example of the use of a 'dog-bone' insulator is described later in this chapter and is shown in **Fig 13.23**.

## Pulleys

Several types of pulleys are readily available at most hardware stores or from online suppliers. Among these are small galvanised pulleys designed for tent or caravan awnings. There are also several styles and sizes of pulleys available for clotheslines. Heavier and stronger pulleys are those used in maritime applications. The amount of stress that a pulley is capable of handling depends upon the diameter of the shaft, how securely the shaft is fitted into the sheath and the size and material that the frame

**Fig 13.8: (a) Mast top guy ring bearing suitable for mast diameters up to 50mm (2in) This is an upside-down view to show guy connection holes. (b) Bearing suitable for base and mast top, for mast diameters up to 9cm (3.5in). Supplied with fittings (not shown) to connect guys**

is made from. Another important factor to be considered in the selection of a pulley is its ability to resist corrosion. Most good quality clothesline pulleys are made of alloys that do not readily corrode. These pulleys are designed to carry at least a 15m (50ft) length of line that could be loaded with wet clothing in strong winds. This means these pulleys should be adequate for normal spans of 30 to 40m (100 to 130ft) between stable supports. Choose a pulley to suit the line used because, if the line is too thin, it is likely to get trapped between the pulley wheel and the sheath causing the pulley to jam.

### Exhaust Pipe Clamps

In the USA these clamps are called muffler clamps. These clamps can be used to construct boom-to-mast and element-to-boom fittings. They usually comprise a U-bolt and brace that is secured using two nuts. Most auto-stores stock exhaust pipe clamps and they can also be sourced from online suppliers.

### Spreaders for Wire Beams

Lightweight bamboo cane or fibreglass rod can be used to make spreader-insulators for wire antennas. Fibreglass rods are preferred because they are lightweight, weather well and have excellent insulating properties.

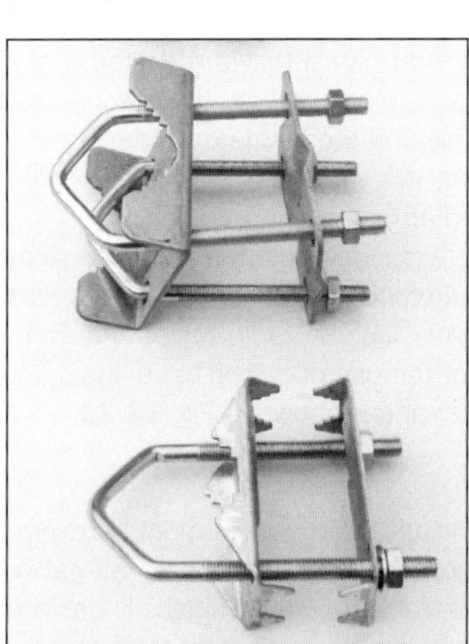

**Fig 13.9: Examples of commercially available antenna fittings typically used by TV antenna installers**

### Polypropylene Rope

Strong polypropylene rope is suitable as a halyard for pulling up masts and antenna structures. This type of rope is also good to use as guys to support such structures once upright to maintain them in place. Sources of this rope include marine suppliers, some DIY stores and online suppliers.

### Specialised Antenna Fittings

The importance of guying masts correctly is discussed later. The problem of how to guy a rotating mast can be overcome by using the appropriate fittings and there are various specialised commercial fittings available. Examples of two mast-top guy ring bearings suitable for mast diameters up to 50mm and 90mm are shown in **Fig 13.8**. These bearings are supplied with fittings to connect the guys.

### TV Type Fittings

There are several clamps that are used by the TV antenna industry that can be useful for amateur radio antenna applications. Some typical examples of these clamps are shown in **Fig 13.9** for reference. They can be useful when fitting smaller beam antennas to masts or the elements to the boom of an HF antenna for example. These fittings can be sourced from some DIY stores, TV antenna installation companies or from online suppliers.

## BEAM ANTENNA CONSTRUCTION

Beam antennas can be built using all-metal construction or with wire elements supported on spreaders or between insulators.

### All-metal Construction

The antenna's boom can be fixed to a tubular mast with a metal plate and car exhaust U-clamps as previously shown in Figures 6 and 7. These techniques can also be used to attach Elements to booms in a similar manner. Tapered elements can be constructed from lengths of aluminium alloy tubing with different diameters so that the lengths can be telescoped into each other. Often a section is not a snug-fit into the end of the adjoining section and needs to be modified using shims as shown in **Fig 13.10**. If there is a relatively large difference between the two joining sections, the shim may need to be made from a short section of tubing that is slit longitudinally. Any corrosion on any of the metal surfaces that make up the joint should be removed with fine sandpaper to ensure a good electrical connection between the sections. Four longitudinal slots of about 40mm (1.5inch) in length should be cut using a hacksaw down the outer section tube's end that will act to clamp the inner section in place by tightening a hose clip. Any burrs along the saw cuts should be removed using a suitable file. The surfaces are then wiped clean with a cloth, the sections intersected and the joint clamped tight using a hose clip. The exterior of the joint should then be coated with a thin film of grease to prevent corrosion.

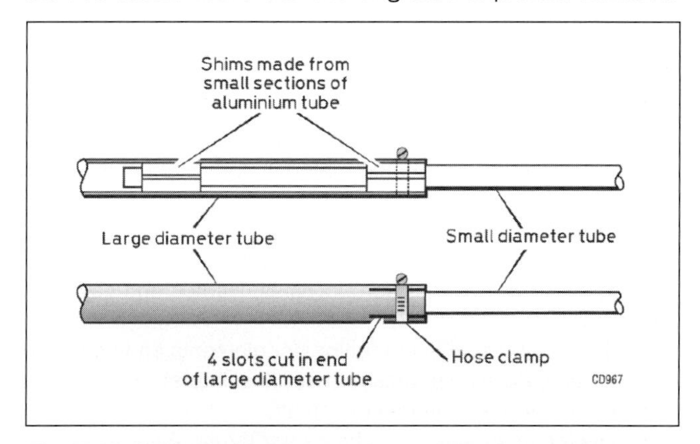


**Fig 13.10: Method of joining sections of aluminium tube where the tube diameters present a poor fit (top drawing depicts a cross section of the joint)**

Fig 13.11: Example of an element clamp made using four L-shaped aluminium sections used for a 6m beam antenna

This method has advantages over using a nut and bolt where a new set of holes has to be drilled every time an adjustment to length is made. The hose clip method also gives the joint a lower contact resistance.

Often the construction of a quad antenna involves joining tubing and insulated copper wire to make up its elements. To attach a wire to the end of a metal element, first remove a short length of the plastic insulation off the end of wire revealing the conductor. The conductor is then fixed to the end of the metal element using a hose clip forming a joint. It is particularly important that a copper wire/aluminium tube joint is protected with grease (or similar) to prevent the joint from corrosion. This is because, if not protected, exposure of the joint to the weather could make it become highly resistive and so have an effect on the antenna's performance.

There has been some discussion as to whether the contact between aluminium and copper should be avoided with a small stainless steel washer used to provide isolation between the two [2]. Tinning the copper wire may also help to lower the possibility of corrosion through electrolytic action. The HF Gamma match used by G3LDO and described in the Antenna Principles chapter is an example of where a copper conductor was connected to an aluminium element. This connection used a coating of grease to protect the joint against corrosion.

## All-metal element clamps

For smaller beam antennas, element clamps can be made from four aluminium L-sections that are held together using nuts and bolts as shown in **Fig 13.11**. This clamp was built for use on a 6m beam antenna using a 12mm element diameter and 25mm square boom. Two 60mm long L-sections were spaced 25mm apart to provide a snug fit onto the boom. Then two 90mm long L-sections were situated across these and spaced 12mm apart allowing a snug fit for the element. The assembly was held together using M3 nuts and bolts passed through suitably drilled holes as shown. A 6mm hole was drilled through the element's centre and the boom with a M6 bolt and nut securing everything in place.

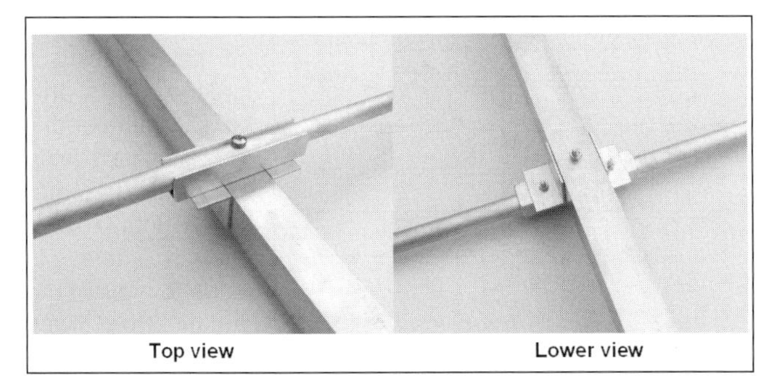


Fig 13.12: Example of an element clamp made using a trough-shaped and two L-shaped aluminium sections used for a 2m beam antenna

A similar clamp used for a 2m beam antenna is shown in **Fig 13.12**. This clamp was built to accommodate a 12mm element diameter and 25mm square boom. However, the technique has also been used with a 25mm round boom. Two 30mm long L-sections were spaced 25mm apart to provide a snug fit onto the boom. An 80mm length of U-shaped aluminium trough was situated across these and secured using M3 nuts and bolts passed through suitably drilled holes. The U-shaped trough had a 12mm internal dimension giving a snug fit with the element. A 5mm hole was drilled through the reflector's centre and the boom with a M5 bolt and nut securing everything in place. Several of these clamps were made up to support the elements for a 2m Yagi beam antenna. The aluminium L-section and trough used to make clamps was sourced from an online aluminium supplier.

## Wire Beam Construction

Insulating spreaders for wire beam antennas or helically wound elements, can be constructed using lightweight bamboo cane or fibreglass rod. The main disadvantage of these materials is that they can be easily damaged by crushing at the clamping point. To avoid damaging the ends of the spreaders, a sepa-

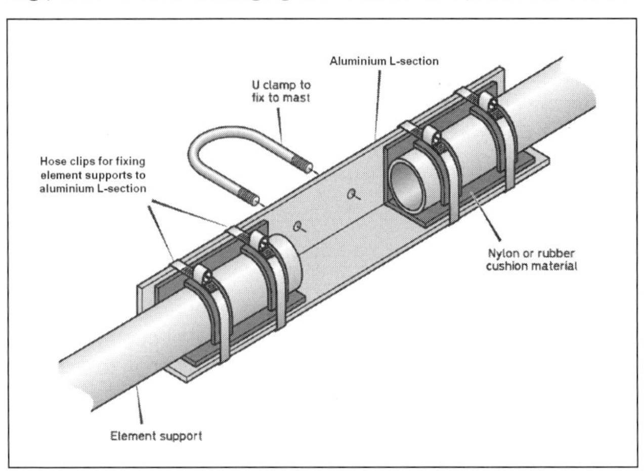


Fig 13.13(a): Method of fixing cane or fibreglass supports for wire beam elements to a boom or mast. The length of the aluminium L-section material and the spacing between the hose clip supports depends on the size of the antenna structure (not to scale)

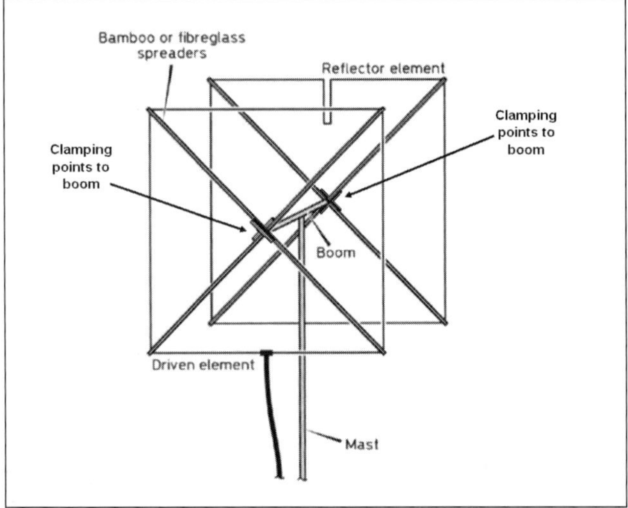


Fig 13.13(b): Concept of an HF 2 element quad constructing using the clamps illustrated in Fig 13.13(a)

rate clamp can be made up using aluminium L-section. The length of the aluminium L-section depends on the physical size and the frequency range of the antennas to be supported. For a conventional 2 element HF quad a 1m (3ft) long length of aluminium L-section is usually suitable. Four clamps are required for a 2 element HF quad, with two clamps used for each element to hold the supporting spreaders in place. Two holes are drilled at the centre of each clamp, their distance apart will be determined by the size of the U-bolts to be used to attach each clamp to the mast or boom. The cane or fibreglass spreaders are fixed to the ends of the aluminium L-section using hose clips as shown in **Fig 13.13**(a). Rubber or plastic tubing cushions can be used to prevent the clamps damaging the cane or fibreglass spreader ends. **Fig 13.13**(b) illustrates the concept of a 2 element HF quad using this technique for reference.

## ANTENNA SUPPORTS

### Using the House as an Antenna Support

Placing a large antenna 10 or 20m (30 to 60ft) above the ground, with access for adjustment and tuning can be major task in itself. Backyard locations often do not have sufficient space for a free-standing mast or tower. In this case a solution is to fix the antenna to the house. The usual method of doing this is to fix the antenna to the chimney (if the house has one) using a chimney bracket or to fix it to the side of the building with a wall bracket.

### Wall brackets

Useful advice for fixing wall brackets comes from G3SEK [3] and is summarised as follows:

Even a small antenna installation can generate considerable wind forces on the support structure depending upon the direction of the wind. If the wind is pushing the bracket on to the wall, the force is spread over several bricks, and the fixing is as strong as the wall itself. If the wind is blowing parallel to the wall and the bracket is strong enough, most kinds of wall fixings will be extremely secure against the sideways forces. The difficulty arises when the wind is blowing away from the wall and trying to pull the bolts straight out of the bricks (or the bricks straight out of the wall). This latter possibility is a serious one, unless the wall is well built. Older houses with mortar that has weakened over the years, and bricks made before factory quality control was introduced, are simply not a good prospect for a mast bolted to the wall.

**Fig 13.14: Example of a typical well-braced wall bracket**

Assuming the house has reasonably sound brickwork, the aim should be to mount the top bracket as high as possible to reduce the wind forces. At least three courses of bricks should be left between the bricks that are drilled and the top of the wall. Also leave plenty of sideways clearance from upstairs window openings because the brickwork can be considerably weakened. The best place to mount the top bracket is quite high on a gable end wall. This allows the unsupported upper section of mast to be shorter so reducing the wind forces. The bracket itself is important because it will be expected to provide a reliable fixing for potentially a number of years. A cheap and poorly made wall bracket that is really intended for UHF TV antennas is not very suitable. It is better to obtain a substantial, well made and preferably galvanise bracket from an amateur radio dealer. A suitable bracket will look something like the example shown in **Fig 13.14**, with a T-shaped piece that bolts to the wall and a well-braced arm for fixing the mast to. All the component parts should be solidly double-seam welded. Typically there should be two or more bolt holes in the horizontal member of the T, and one or two more in the vertical member. The top row of fixings will bear almost all the load and **Fig 13.15** shows a typical drilling pattern for this type of bracket.

To fix the bracket, it is best to use an expanding type wall anchor and there are several types available. However, they all work by expanding outwards and gripping the sides of the holes. The traditional 'Rawlbolts' are best because they give a very secure fixing. There are other anchor methods available, however the 'DIY' fixings using large plastic wall plugs and 'coach bolts' are not recommended.

This is a safety-critical application, so investing in some properly engineered fixings that are designed to work together as a system and following the manufacturer's instructions exactly is recommended. The holes for the 'Rawlbolt' anchors should be drilled in the centre of the brick as shown in **Fig 13.15**. The optimum size for ordinary brickwork is M10, which requires a 16mm diameter hole to be drilled. Three or four 'Rawlbolt' anchors should be more than adequate to withstand the wind forces envisaged, provided they are installed correctly. Choose a sound set of bricks for drilling, free from any hairline cracks. If necessary, be prepared to move the mast a little from its planned location. Drill into the exact centre of each brick, not near the edges, and never into mortar. If necessary, make new mounting holes in the bracket to suit the brickwork. The holes in the bracket should be 10mm diameter to take M10 bolts, with some extra clearance to help line the bolts up. Hold the bracket to the wall, level it with a spirit level and mark the centres of the holes. Place the point of a centre-punch, or similar, exactly on each drilling mark made and tap gently with a hammer to chip out a small dimple as a guide mark. This will prevent the drill-bit point from wandering when drilling is started using an electric drill. Initially, start drilling with a small masonry drill-bit before using a larger masonry drill-bit. If a hammer-drill is used, start without the hammer action until a deep enough hole is made to prevent the drill-bit from wandering. Do the same at each change of drill-bit as the hole is gradually opened

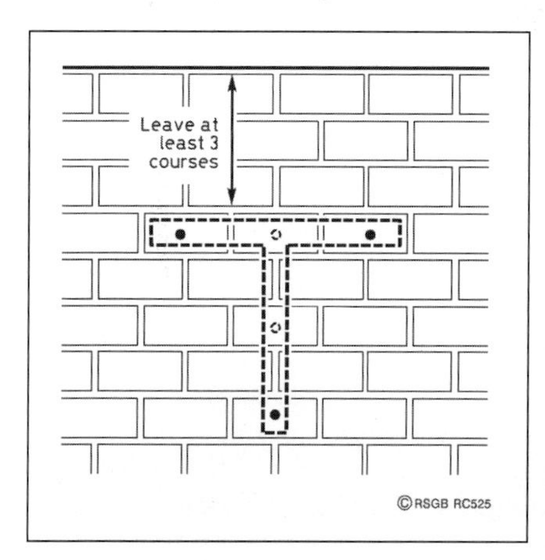


**Fig 13.15: Typical drilling pattern for fixing a wall bracket. Patterns may vary, but always drill into the centre of the bricks**

out, using progressively larger masonry drill-bit sizes. Taking time to do this, it is possible to drill good cylindrical holes that are square to the wall and exactly where they are required for the bracket.

The final holes must be exactly the right diameter as specified by the manufacturer. For example, the hole for an M10 Rawlbolt should be 16mm diameter. The correct hole diameter is very important because the strength of the wall fixing comes from the physical contact between anchor sleeve against the inside of the hole. The sleeve should be a gentle tap fit, so that when the bolt is tightened the anchor will immediately start to grip the brick. Tap the anchor sleeve into place just below the surface of the wall. This is to ensure that, when the bolts are tightened up, that the bracket contacts with the wall and not with the end of the sleeve so leaving a gap. Do this without the bolt inserted and then fit the bracket. Leave the bolts slightly loose to allow the bracket to be levelled, then tighten up the Rawbolts using a suitable spanner. The difficult part is to tighten the bolts to the correct torque to sufficiently expand the anchor sleeve enough so that the fixing develops strength, but not so much that it splits the brick and ruins the fixing.

Although G3SEK uses Rawlbolts, he notes that they can be very prone to split the bricks if over-tightened. It is worth considering alternative types, such as the Fischer bolts that use a softer plastic sleeve to grip the inside of the hole.

Nevertheless, it is best to use the type of anchor that has a free bolt that screws in, rather than the type with a stud that takes a nut.

If the house is built using modern bricks that have holes through the middle, conventional expanding anchors are not really feasible and other fixing systems need to be investigated. Wall anchors using a chemical adhesive fixing system are available and have the advantage of not stressing the bricks. These adhesive anchoring systems are claimed to have higher strengths compared to conventional expanding anchor systems. They can also be used for fixing into hollow bricks, but the strength of the bricks themselves may then become an issue. As with any adhesive bonding system, success depends on how well everything is prepared and through following the instructions. These adhesives usually make a bond that penetrates into the masonry and this can make it difficult to remove a bracket or fixture at a later date. When removing a bracket that has been attached to the masonry using an adhesive, care is needed not to damage the mortar and brickwork as the fixture is removed. Another technique when using conventional Rawlbolts in ordinary brickwork is to use epoxy resin to provide addition support to the fixing.

When fixing to a gable end wall, another option could be to drill through the wall into the loft space. Then use long bolts, or studs, to secure the bracket to a steel plate that spreads the load over the inside wall.

The lower bracket tends to bear much less of a load compared to the upper bracket. The lower bracket's main purpose is to steady the mast and prevent it from bowing below the upper bracket. After fixing the upper bracket in place, mark out and drill for the lower bracket, lining the two up with a plumb line. The bottom of the mast should also be fixed to prevent it from moving sideways. In the longer term, wall anchors can work loose due to a combination of frost action, thermal expansion/contraction cycles and the effect of the wind on them over time. Regularly check the fixings a couple of times a year, for example every spring and autumn. If it is intended to mount a commercial mast or antenna against the wall, obtain and follow the manufacturer's specific advice on how this should be installed and maintained.

### House chimney

The house chimney can be used for an antenna support as shown in **Fig 13.16**. The main advantage of this method is that chimney-mounting brackets are relatively easy to obtain. Some of these mounting brackets can be seen supporting some tall TV antenna structures in marginal TV signal areas. When mounting an antenna on a chimney of an older house, where the mortar may have weakened over the years, the mortar should be examined and if necessary re-pointed before installing the antenna. The single wire lashing kits used for TV antenna installation are not really suitable for use with amateur radio antennas. However, a heavier-duty double TV antenna chimney lashing kit can support a large VHF array or a small sized HF beam.

**Fig 13.16: Example of a chimney lashing used to support a compact 2-element HF beam antenna**

## ROUTING CABLES INTO THE HOUSE

Installing cables between the antenna and the station can be a lengthy process that should not be underestimated. Taking time to plan everything is a worthwhile exercise with the object of identifying where potential problems could arise before encountering them in practice. To plan an installation may seem to be a time consuming process, however once the cables have been installed, they could be in use for potentially many years. The installation at a station often involves:

- A multi-band HF beam with its coaxial cable
- VHF and UHF antennas and their coaxial cables
- Rotator control cables for the antennas
- A long wire antenna for the lower frequency bands and its feeder cable.

One way of overcoming this problem was to drill a lot of holes in the window frame. However, nowadays many houses use double-glazing with plastic and metal window frames. This type of window frame is not really a feasible option for routing cables through and another method needs to be used. G3SEK used a method to route the cables through a wall using a length of plastic drainpipe that was inserted through the wall [4]. To follow is a summary of how this was done:

Many British brick houses have cavity outside walls and the technique to be described uses a length of 40mm diameter plastic pipe passed through the cavity wall. Usually, a 40mm pipe is sufficient to route most cables with enough provision to route other cables at a later date if necessary.

**The tools required to take on this task include:**

- An electric hammer drill capable of taking larger diameter masonry drill-bits
- A masonry drill-bit of 10mm diameter that is long enough to go right through the double wall and cavity
- A shorter masonry drill-bit of about the same diameter

- A fairly large hammer
- A long but narrow cold chisel.

When starting the work, plan carefully to find the best place to drill through the cavity wall. It is best to leave at least one whole brick (or more) between the hole and doors or window frames. Remember that the frame has a solid lintel across the top, extending outwards on both sides that could cause a problem if drilled into. Check both the inside and the outside of the wall with a live cable metal detector to ensure that mains electrical cables are avoided before drilling. If the house is timber-framed with a brick outer skin, take care to avoid structural timbers.

First a pilot hole should be drilled right through the wall at a slight upwards angle from the outside of the wall (see later). This hole is going to be enlarged from each side of the wall. Start from the outside and drill through the mortar, halfway along a brick, as shown in **Fig 13.17**. Use one of the shorter masonry drill-bits to start this hole, taking care to drill it accurately at right angles to the wall (the mortar joint tends to act as a guide). When the drill-bit breaks through into the wall cavity, change to the longer drill-bit and carry on drilling. To avoid the drill-bit pushing a large patch of plaster off the inside wall, stop the drilling within a few centimetres of breaking through the wall. Switch off the drill's hammer action and continue with a very gentle pressure until the drill-bit has gone through both walls. Now mark out a circle on each side of the wall that is slightly larger than the diameter of the plastic pipe as shown in **Fig 13.17**. The pilot hole is at the top of each circle.

Working separately from each side of the cavity wall and using the shorter masonry drill-bit, drill a ring of holes as close together as possible, stopping when the drill-bit is through to the cavity. Each hole should be started beginning with a smaller drill-bit, using the drill at slow speed with the hammer action off. When the hole is well established into solid brickwork (or blockwork on the inside wall), change to the larger drill-bit and use the faster hammer action.

Next the hole is opened out from each side using the hammer and cold chisel, until the plastic pipe will slide right through. Chip away a little-at-a-time so as not to crack the outside brick or do any unnecessary damage to the interior plasterwork. Try to pull the rubble out rather than letting it fall into the cavity.

Make sure that the pipe will slide through with it sloping a few degrees upward from the outside so that rainwater will not run in (this is why previously it was suggested that the initial pilot hole was drilled at a slight upward angle). Set the end of the pipe just proud of the inside wall, and leave any overhang outside. Fill the gaps around the pipe with mortar or exterior filler on the outside and plaster or interior filler on the inside, letting it all set solid. After a few hours, saw off the outside end of the pipe, a few centimetres away from the wall. Now the cables can be threaded through the pipe into the building. It is a good idea to use a 45° or right-angled pipe elbow, facing downwards, fixed to the pipe where it emerges on the outside wall to help keep the rainwater out.

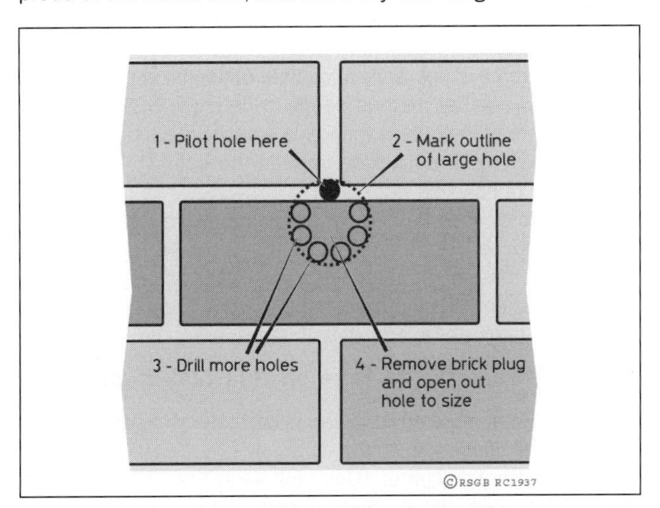


**Fig 13.17: A method to drill a 40mm diameter hole through bricks. This technique can be used for larger holes**

When all the cables are in place, push in plastic foam for draught proofing, or use aerosol-expanding foam. This method of installing a pipe is quite easily reversible before moving house. The pipe will pull out from the outside with a bit of effort. The interior hole can be plastered and the outside hole replaced with a brick if the drilling pattern in **Fig 13.17** has been followed.

## CONSTRUCTION OF FOLD-OVER MASTS

Many radio amateurs use commercial lattice construction masts that have a telescoping and fold-over capability. These masts have the advantage of having well-defined manufacturers' data regarding heights and wind loading capabilities. However, they can have a fairly high visual impact and this may be a problem in some locations. A drawback to trying to adjust an antenna using these types of support is that they take some time and effort to raise and lower. Additionally, the winch cables are not designed for continually raising and lowering the mast that can be required when a lot of antenna work is needed over a period of time. For this reason, a homemade structure can have the advantage of being designed to be easily to raised and lowered. This also has the additional advantage that the mast can be quickly lowered if severe gale force winds are forecasted. Two straightforward mast designs for single-handed construction are summarised as follows.

### The G2XK Lightweight Fold-over Mast

Eric Knowles, G2XK, used the method described as follows to support a 6 element 10 metre beam on an 11m boom at a height of 12m above the ground, using 80mm diameter thin-wall duralumin tubing. This large structure has weathered many gales that have swept across the Vale of York. The details of this mast and its layout are shown in **Fig 13.18** for reference.

There is nothing new in this method of supporting, raising and lowering a mast using guy ropes. The description that follows is of a mast that is suitable for supporting a medium sized beam antenna. No special tools or welding equipment are required to construct this structure and it is a suitable support for experimental antennas provided the space for the guy ropes is available. Do not use steel tubing for this mast design because it is too heavy. Aluminium scaffolding pole is not really suitable for this mast design either because of the weight-to-length ratio, although it could be used for shorter masts of up to 8m high for example.

This structure can be used with a fixed mast and rotator, or the mast could be rotated. In this case provision has to be made to allow the mast to rotate and be folded over. A minimum of four guy ropes is used to support this structure.

Referring to **Fig 13.18**, the anchor point for guy rope 4 must be above ground level and, for example, the chimney to the house could be suitable for this as shown. Guy ropes 1 and 2 are anchored along the baseline so that they retain the same tension when the mast is being raised or folded over. The length of guy rope 3 is adjusted so that it is under tension when the mast is in the vertical position. The original G2XK version used two sets of guy ropes but only one set is illustrated in **Fig 13.18** for clarity.

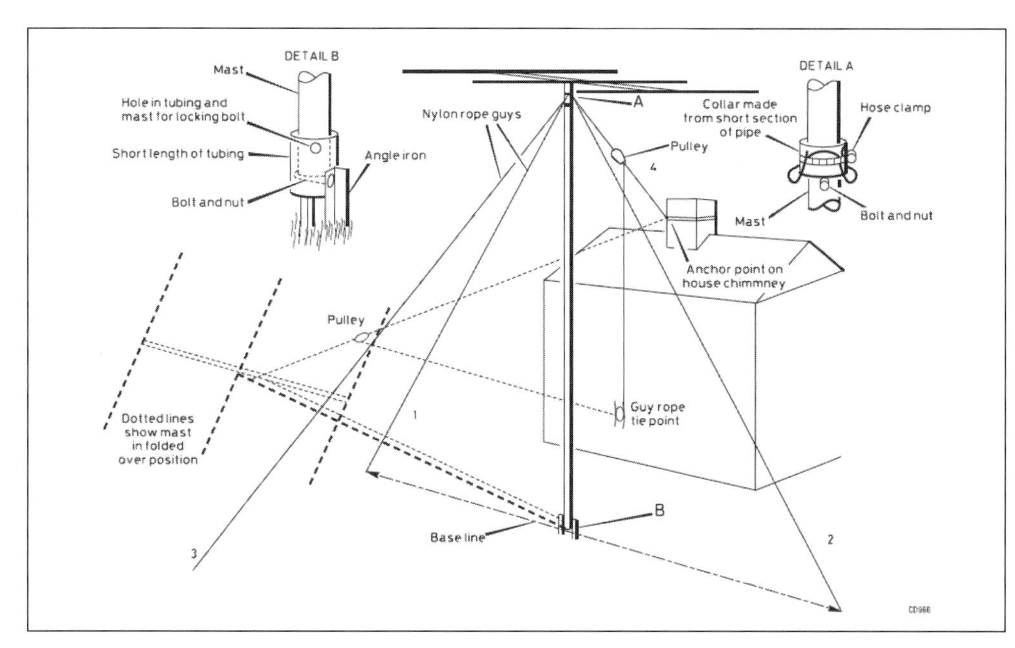


**Fig 13.18: Constructional details of the G2XK type of lightweight fold-over mast**

This structure gains all its strength from its guy ropes, so it is important that the guy ropes are strong and are connected securely, both at their anchorage points and at the top of the mast. Polypropylene rope (6mm diameter) is a suitable material for the guy ropes, which should ideally be at 45o to the mast. This angle can be reduced if space is limited but this increases the downward pressure on the mast in high winds, increasing the chances of the mast buckling and the structure failing. Commercial guy rope to mast bearings are available for the top of the mast and are recommended for this type of application. An example of two types of mast bearings were previously illustrated as **Fig 13.8**. The guy ropes should be connected to the bearings with D-clamps.

If a commercial rotatable guy rope support bearing is unavailable, one can be constructed using a short length of steel tube that is slightly larger in diameter than the mast. Very thick wire loops can be fixed to this tube using a long hose clip (or using two shorter hose clips joined together). The top of the mast is inserted into this tubing. A bolt, secured with a nut, is passed through the appropriate point on the mast to hold the guy support collar in position. **Fig 13.18** Detail A illustrates this assembly for reference.

The guy anchorage can be constructed from a length of angle iron of about 1m long. This is cut to a point at one end and a hole drilled in the other to make an anchorage stake. This stake can be driven into the ground at 90$^{o}$ to the angle of pull. The guy anchorage may need to be more substantial for very large masts or if the soil is light and sandy. The guy rope should be connected to the guy anchorage with a D-clamp. A pulley is required for the halyard to enable the mast to be hauled up. A good quality clothesline pulley is suitable for this.

The base pivot point comprises two lengths of angle iron, cut and drilled in a

similar way to the guy anchorage stakes. The two angle iron pieces are driven into the ground, with the holes aligned so that the pivot bolt can be fitted. If the design calls for a rotatable mast then a small section of tubing, whose internal diameter is slightly larger than the outside diameter of the mast, is pivoted to the angle iron. The mast fits inside this section of tubing and is free to rotate. Holes can be drilled through the base tubing and the mast to enable the structure to be locked on any particular heading. **Fig 13.18** Detail B of shows how the mast rotation assembly was made. To make the mast up to the desired height, lightweight sections of thin-wall tubing can be joined together using a suitable joining section as previously shown in **Fig 13.4**.

## Counter Weighted Fold-over Masts

This type of support is of a heavier construction and requires more effort to build it. This mast's main advantage is that guy ropes are not absolutely necessary. This design is based on an 18m fold-over support originally designed by Alfred W.Hubbard KOOHM [5]. The original mast was designed to support a 3-element tri-band beam and a rotator. In this design of fold-over mast, all sections of steel tubing of the mast were welded together and the partially counterweighted by filling the lower half of the tilt-over section with concrete. A pulley was used to manage the remaining 160kg of pull.

The design of the base comprises a section of casing fixed in the ground with a concrete foundation. The gap between the

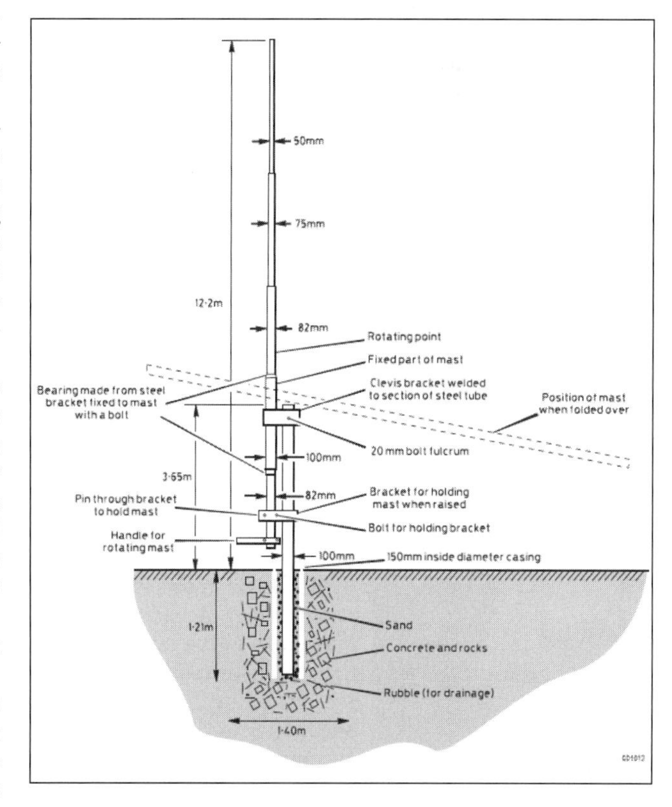


**Fiigure 19: Constructional details of the counterweighted fold-over rotatable 12m mast**

mast and the casing was filled with sand. This reduces the high-stress point at ground level that normally exists if the mast is set directly into concrete. The sand acts as a buffer and allows the mast to flex within the base during high winds. The internal casing diameter should be around 50mm (2inches) greater in diameter than the lowest section of the mast.

Peter Dodd, G3LDO, has built several of these masts. The largest of these was 18m (59ft) and supported an all metal quad antenna. The mast and payload should not be fully counterweighted. A top weight imbalance of around 45kg, controlled with a winch, will enable the momentum of the structure to be manageable for most users.

## Medium Size 12m Fold-over Steel Mast

The mast, described above, is too large for most suburban sites and a smaller version of the mast is described in the following. However, this mast will require a garden at least 12m (40ft) long to accommodate it. If necessary, the design could be scaled down still further if required. This mast is counterweighted with approximately 15kg of top weight so a winch is not required. It takes about 15 seconds to raise the mast and antenna into the vertical position. The mast is relatively lightweight with the top third of its length using 50mm diameter scaffolding pole.

**Fig 13.20: 12m version of the fold-over mast in upright position, supporting a 14MHz metal quad antenna**

**Fig 13.21: 12m mast and 14MHz quad in the folded over position**

The whole mast is manually rotated using a handle fixed to the bottom of the mast. The sections of steel tubing making up the mast are telescoped into each other for about 30cm and secured using a nut and bolt passed through a suitably drilled hole. This allows the mast to be assembled, modified or repositioned much more readily than if the sections were welded. The details of this mast can be seen in **Fig 13.19** and more general views can be seen in **Fig 13.20** and **13.21**. Two weights initially were used as the counterweight. However, these weights were eventually dispensed with by making the lower section of the mast out of solid 82mm steel rod.

Although these structures can be built single-handed the following are areas where some assistance should be sought as follows.

**1) Inserting the fixed lower mast section into the base casing in the ground:**

Two ropes are tied to the top of the mast's lower section, using the holes drilled for the pivot bolt. The section can then be placed with the lower end over the base casing and the top supported on a pair of stepladders, or similar. The section can be raised using the ropes, at the same time the lower end is guided into the casing using a section of angle iron.

**2) Installing the upper rotatable mast section:**

Place the clevis bracket at the top of the mast and fit the bolt. Then insert the mast into the oversize piping used as the tilt-over thrust bearing. These tasks can be eased by using a gin-pole with a pulley and rope. The gin-pole can be constructed from steel angle-iron and clamped to the mast with additional angle iron pieces or steel straps.

## Other Fold-over Masts

## Wooden Masts

In the early days of amateur radio, wood was a very popular material for constructing masts and even beam antennas. This material has become less popular because of the cost of quality seasoned timber and the lack of sensible fold-over designs that can carry the payload of a medium size beam antenna to a height of 10 to 15m (30 to 45ft). The selection of timber and weather treatment requires specialist knowledge, which is beyond the scope of this chapter.

## Commercial Masts

There is a range of commercial fold-over masts available (see RadCom, amateur radio suppliers or online suppliers for sources). Most of these masts have a lattice structure, with sections of the fold-over lattice mast telescoping into each other. This design enables a fairly large mast to be erected in a relatively small garden.

# TREES AS ANTENNA SUPPORTS

If there is a suitable tree in the garden, then this can be a very good support for a wire antenna that does not require planning permission. As antenna supports, trees are unstable in windy conditions, except in the case of very large trees where the antenna support is well down from the top branches. Therefore, good practice is to construct tree supported wire antennas more sturdily than is necessary with stable supports. To this end, the preferred method is to use a halyard and pulley shown in **Fig 13.22**.

The use of a halyard with a mast is shown in **Fig 13.22**(a). Here the halyard end can be lashed to a bracket. When a tree is used as the support, a weight could used to take up the movement of the tree, this concept is shown in **Fig 13.22**(b). The endless loop allows greater control when raising and lowering the antenna.

An example based on the technique described in **Fig 13.22**(a) was used by G0JMI and is shown in **Fig 13.23**. A short mast was secured to the top of a fir tree (whose top had been lopped off),

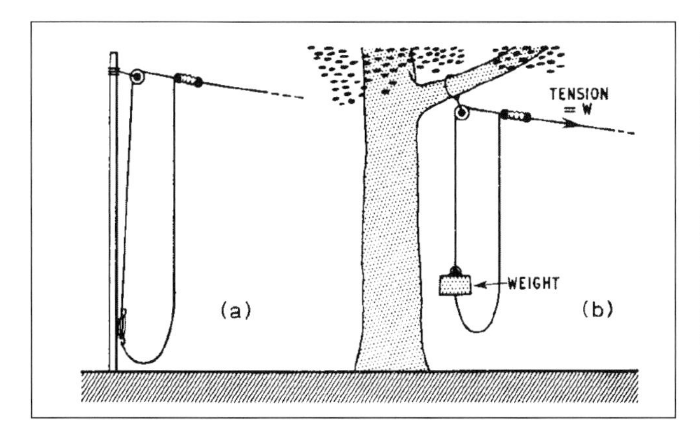


Fig 13.22: Halyard connections to (a) a pole and (b) a tree. The weight is equal to the required antenna wire tension

Fig 13.24: A compact 2 element multi-band beam

to this was fixed a pulley wheel to enable the antenna to be raised and lowered using a length of nylon cord.

## Fixing the Halyard Pulley

If the point where the pulley is to be attached can be reached using a ladder then fixing it to a branch, pole or building can be relatively straightforward. If it is not possible to reach the pulley fixing point then a line has to be thrown or propelled over the anchor point. A missile or weight should be used that will not cause damage or injury if things go wrong. The best object for this purpose is a squash ball or a plastic practice golf ball filled with wood filler. The best sort of pilot line is the cord used for electric strimmers because this line is strong and resists kinking. The line can be stretched out straight on the ground or 'zig-zagged' across the ground. If snarling of the line is a problem (eg from garden plants and bushes), then stationary reel can be made up by driving eight nails, arranged in a circle, through a 20mm (3/4inch) thick board. After winding the line around the circle formed by the nails, the line should reel off readily at lift-off. The board should be tilted at approximately right angles to the path of the line's travel. If it is necessary to retrieve the line to start again, the line should be drawn back slowly to avoid the weight from swinging and wrapping it around a branch making its retrieval difficult. The pilot line can then be used to pull a heavier line over the tree. This line is used to haul a pulley up into the tree after the antenna halyard has been threaded through the pulley. The line that holds the pulley must be capable of withstanding considerable chafing against the tree. A metal ring, of around 70 to 100mm (3 to 4inches) in diameter, could be used instead of a pulley. The wire antenna's halyard is passed through the ring, then once the ring is in place this is used to pull the end of the antenna up. This technique has more friction than a pulley but it will tend not to jam. This method of running a pilot line to enable a heavier line to be pulled up can also be used to place an antenna support over a roof.

Fig 13.23: Example of a tree used to support a wire antenna used by G0JMI

**SAFETY**

Safety should be a primary consideration when erecting antenna masts.

NEVER erect an antenna and mast that could possibly come in contact with electric power lines.

NEVER rush this sort of work.

ALWAYS stop to consider the implications of the next move, particularly when dealing with heavy sections of steel tubing.

DO NOT use an antenna support structure that requires the joint efforts of, for example, all members of the local radio club to raise and lower it.

Although help in the construction stages is always welcome.

# WIND LOADING

Most antennas have the potential to be brought down by very strong winds, therefore it is important to consider the effect of windloading on the antenna installation. There are some informative articles on the subject [6] [7] and an overview is given as follows to give an idea of the forces involved.

G3SEK [8] suggests that at approximate wind speed of 100mph, then every square foot of exposed area suffers a sideways pressure of around 25 pounds. Using metric units, at 45-50m/s the wind force is about 150kgf/m$^2$ (kilograms force per square metre). Therefore, to determine the wind-loading on an antenna, the area of the antenna needs to be known. The best method to calculate the antenna's area is to simplify the parts of the elements to a 'flat slab' area. For example, a pole 3m long and 50mm diameter can be represented by a flat slab 0.15m$^2$.

## Assessing the exposed areas

To calculate an antenna's exposed area to the wind, a compact HF beam antenna has been used as an example. This antenna was mounted on the roof of the house and is shown in **Fig 13.24**.

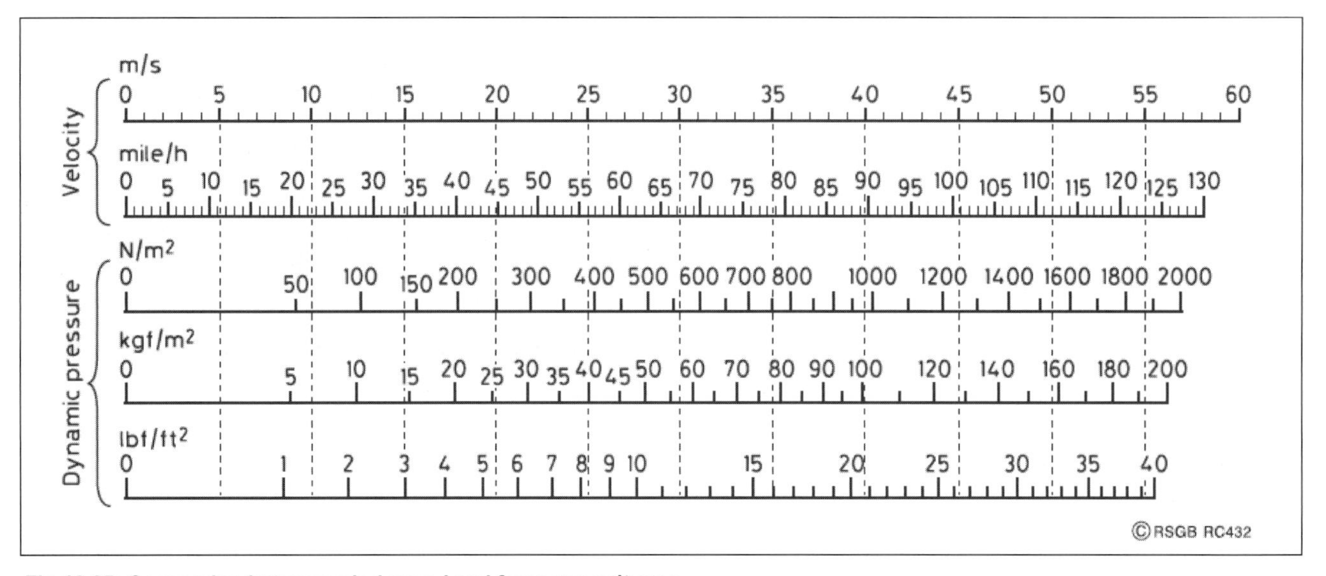


**Fig 13.25: Conversion between wind speed and force per unit area**

The individual components comprising the antenna, along with its mount, have areas exposed to the wind calculated as follows:

- This beam has 2 elements and these are 4.49m long and 25mm in diameter. The flat slab area of the elements is 4.49 x 0.025 x 2 = 0.2245m$^2$.
- The boom is 2.1m long by 35mm in diameter giving 2.1 x 0.035 = 0.0735m$^2$.
- The loading coils and spokes at the ends of the elements are more complicated. Therefore, they have been modelled as four cylindrical objects 0.3m long by 0.06m in diameter, giving an area of 0.3 x 0.06 x 4 = 0.072m$^2$.
- The mast fixed to the chimney is around 1m long and 0.05m in diameter, giving 0.05m$^2$.
- The rotator's area has been taken as 0.15m x 0.13m = 0.0195m$^2$.

With the antenna directly sideways on to the wind, the exposed area is given by: 0.2245m$^2$ + 0.072m$^2$ + 0.05m$^2$ + 0.0195m$^2$ = 0.366m$^2$. This **Fig 13.13**.has been rounded up, giving 0.4m$^2$. The boom supporting the two elements has not been included in the calculation because it would be end on to the wind if the beam were facing into wind, therefore its contribution would be negligible.

If the beam were now to be rotated by 90$^o$ into the wind, then the boom will now contribute to the windloading. The loading coils and spokes at each element's end now have to be modelled differently as eight 'slabs' of 0.125m long by 0.06m wide, or 8 x 0.125 x 0.6 = 0.06m$^2$. The antenna's area facing the wind end-on is given by: 0.0735m$^2$ + 0.06m$^2$ + 0.05m$^2$ + 0.0195m$^2$ = 0.203m$^2$, or 0.2m$^2$ when rounded down (ie around half the area of when the antenna is directly sideways on to the wind).

## Calculating the Wind Pressure

Using **Fig 13.25** it is possible to make a reasonable assessment of the force acting on the antenna structure by the wind. This antenna has a total area of 0.4m$^2$, the sideways force acting on this antenna with a 45m/s (100mph) wind, according to **Fig 13.25**, is 128kgf/m$^2$ x 0.4m$^2$ ~ 51kg. By turning the antenna so that the elements are sideways to the wind the area facing the wind is reduced to 0.2m$^2$ and the sideways force is reduced to ~ 25.5kg.

There are leverage forces that need to be taken into consideration with unsupported masts that extend above the supporting structure, such as a wall bracket, chimney bracket or guys on a mast. That is why an antenna should be fixed as close to the support or rotator as possible. If you have several antennas on the same mast that is turned by one rotator, then a rotator cage is recommended.

The example calculations above are simplifications of the situation in reality. The wind can come in gusts and there can be a lot of turbulence over the roof. However, the method of estimating wind forces by G3SEK does give a reasonable indication of the forces that could be encountered and enables the antenna structure to be engineered accordingly.

# LIGHTNING PROTECTION

The characteristics of thunderstorms in the UK were described by Alan Martindale, G3MYA, in two articles some years ago [9] [10]. However, the information provided by Alan Martindale is still relevant and a summary of the details and statistics follows:

By scientific observation it has been established that, on a nationwide basis, there are ten days in the year when there are thunderstorms in the vicinity of any one place. These thunderstorms do not have to be directly overhead and still are a risk when within clear audible range. The reasoning to follow is based on an average and will be subject to:

- Differences in the location involved
- The thunderstorm count from year to year.

However, these figures have been calculated over a long period of time and from a large variety of locations. Allowing for the fact that not all thunderstorms are overhead and that most storms consist of more than one storm centre, it has been calculated that on average ten storm centres pass directly overhead in any one year. However, seasonal variations, regularly occurring storm paths and other stray phenomena could double this figure. Therefore, the risk of lightning strokes has been calculated on the basis of twenty storm centres passing overhead per year.

In the UK and similar temperate regions, each thunderstorm centre produces about 20 or 30 lightning strokes during its average life of 30 to 60 minutes. Of these strokes, less than half (about 40%) are cloud to ground strokes and of these, about 95% ground strokes are of the negative type. Each storm centre covers a ground area of about 4km². If the storm centre were to remain stationary during its active life, it would produce 2.5 ground strokes per square kilometre (ie 25 strokes x 0.4 divided by 4km²). However, on average a storm centre travels across the ground at a speed of 50km/h, therefore its ground strokes are distributed over an area of 100km² assuming a duration of 30 minutes. This gives an average of 0.1 ground strokes per storm centre per square kilometre (ie 25 strokes x 0.4 divided by 100km²). Multiply this figure by twenty storm centres per year and this gives 2 ground strokes per square kilometre per year.

| OD | Wall | Weight |
|---|---|---|
| mm | g/m | |
| 6 | 1 | 42 |
| 8 | 1 | 60 |
| 10 | 1 | 76 |
| 12 | 1 | 93 |
| 13 | 1 | 103 |
| 14 | 1 | 110 |
| 16 | 1 | 127 |
| 19 | 1.5 | 227 |
| 20 | 2 | - |
| 20 | 1.5 | 235 |
| 22 | 2 | 339 |
| 22 | 1.5 | 261 |
| 25 | 2.5 | 477 |
| 25 | 2 | 398 |
| 25 | 1.5 | 298 |
| 28 | 1.5 | 336 |
| 30 | 3 | 687 |
| 30 | 2 | 484 |
| 32 | 1.5 | 387 |
| 35 | 2 | 564 |
| 36 | 1.5 | 438 |
| 40 | 5 | 1495 |
| 40 | 2 | 644 |
| 40 | 1.5 | 489 |
| 44 | 1.5 | 541 |
| 45 | 2 | - |
| 48 | 1.5 | 603 |
| 50 | 5 | 1923 |
| 50 | 2 | 820 |

**Table 1: Metric-size alloy tubing available in continental Europe. Material F22 (AlMgSo 0.5%). Tensile strength 22kg/mm2. Standard lengths are 6m**

Taking an average domestic property of about 20 x 50m or an area of 1,000m² as an example, one thousand such properties would fit into 1km². This means that the chances of a direct strike on a single property works out at around once every 500 years.

At this stage, it is shrewd to put everything into perspective. Radio amateurs tend to install antennas as high up as is practical, with good electrical connections and, therefore, are not too dissimilar from an earth potential. Consequently, there is an increased risk of a lightning strike on the property arising from the desire for the best radio communications possible. Therefore, if such electrically attractive devices are installed, then it is obvious that something has to be done to protect them and the property from lightning strikes.

One of the best lightning protection items may already exist at the QTH and this is the metal antenna mast. Ideally it should be in the centre of the property and be 5m (16ft) higher than any other part of the installation. To increase the mast's effectiveness as a lightning conductor, it should have a short length of copper rod attached to the top. This rod should not be less than 25mm² in section and have a sharp point at one end. The rod should be securely fitted to the mast's top 'point up' and there should be a solid electrical connection between them that is fully protected against corrosion. All joints on the mast should also have good electrical connections as well. G3MYA also recommends that the base of the mast should be connected to an earth rod driven into the ground of 0.5in galvanised pipe, hardened copper rod or 'T' section earth rod. The mast arrangement described tends to provide a cone of protection for all equipment within a radius of about 20m (60ft). In the event of a direct lightning strike on the antenna mast, the EMP generated in all electrical conductors inside and outside the house would be very high. On the approach of a severe electrical storm, the best protection from lightning is to disconnect all antennas from equipment and disconnect the equipment from the power lines and microphones.

Antenna feed lines should be 'earthed' to safely bleed off any static that could build up. Rotator cables and other control cables from the antenna location should also be disconnected and earthed where possible. In areas where the AC mains supply comes via overhead lines, the probability of lightning surges entering homes via the line is much higher. As stated earlier, most electrical storms pass by without causing physical damage. However, even fairly distant lightning can produce electrical and magnetic fields that can be coupled into power lines, telephone lines and antenna systems which can damage solid-state devices in the front-ends of receiving and computing communications equipment. An antenna selector switch, which allows all the antennas to be earthed when not in use, is a wise precaution. However, computer equipment and particularly telephone lines connected to routers can be susceptible, particularly if the telephone line uses an overhead distribution system. Disconnect all equipment connections from rigs as described above when going on vacation or not using the equipment for long periods of time.

When an electrical spark arcs across a distance of 25mm in dry air, the voltage potential across this distance is around 33kV. During one thunderstorm, a spark was seen to arc across a distance of 10cm from the end of a PL259 plug terminating the antenna's coaxial cable to a metal hinge on the window frame. The author is relieved that he was not holding this plug at the time.

## REFERENCES

[1] 'Eurotek', Erwin David, G4LQI, RadCom, Nov 1999

[2] Protection Against Atmospheric Corrosion, Karel Barton, published John Wiley, 1976

[3] 'In Practice', Ian White, G3SEK, RadCom, Mar 1995

[4] 'In Practice', Ian White, G3SEK, RadCom, Nov 1998

[5] 'The Paul Buyan Whip', Alfred W.Hubbard, KOOHM, QST, Mar 1963

[6] 'Wind loading' by DJ Reynolds, G3ZPF. RadCom, Apr and May 1988 (reprinted in The HF Antenna Collection, RSGB)

[7] 'Ropes and Rigging for Amateurs - A Professional Approach' by JM Gale, G3JMG, RadCom, March 1970 (reprinted in HF Antenna Collection and in the RSGB Microwave Handbook, Volume 1)

[8] 'In Practice' column, Ian White, G3SEK, RadCom, Jan,1995

[9] 'Lightning', Alan Martindale, G3MYA, RadCom, January 1984

[10] 'Lightning', Alan Martindale, G3MYA, HF Antenna Collection, Erwin David, G4LQI, RSGB



***Mike Parkin BSc(Hons), G0JMI,*** *obtained his first amateur radio licence in 1977 gaining his current licence in 1988. His working background has included telecommunications, radio engineering, mobile-communications, electronics, lecturing and consultancy. Mike has constructed much of his radio equipment covering the bands from 472kHz through to 47GHz. Mike's current business interests are the design and production of antennas covering the range from HF to UHF. Mike is the author of the Antennas Column in RadCom and has had various articles published in other journals. He is a Member of the Institute of Engineering & Technology, a Member of the City & Guilds Institute and a Chartered Engineer.*

# Transmission Lines

# 14

*Mike Parkin, G0JMI*

For an antenna to function efficiently it should be installed as high and clear of buildings, telephone lines and power lines as is practically possible. On the other hand, the transmitter that generates the RF power for driving the antenna is usually located in the shack, some distance from the antenna feedpoint. The connecting link between the two is the RF transmission line or feeder. Its sole purpose is to carry RF power from the transmitter to the antenna or received signals from the antenna to the receiver as efficiently as possible.

Any conductor of appreciable length compared with the wavelength will radiate power if it is carrying RF current; in other words it becomes an antenna. The transmission line must be designed so that RF power being carried to the antenna does not radiate.

Radiation loss from transmission lines can be prevented by using two conductors so arranged and operated that the electromagnetic field from one is balanced everywhere by an equal and opposite field from the other. In such a case the resultant field is zero; ie there is no radiation. This is illustrated in **Fig 14.1**.

## TRANSMISSION LINE BASICS

### Characteristic Impedance

A transmission line with its two conductors in close proximity can be thought of as a series of small inductors and capacitors distributed along its whole length. Each inductance limits the rate at which each immediately following capacitor can be charged when a pulse of electrical power is fed to one end of a transmission line. The effect of the LC chain is to establish a definite relationship between current and the voltage of the pulse. Thus the line has an apparent impedance called its characteristic impedance or surge impedance, whose conventional symbol is Zo. Transmission line characteristic impedance is unaffected by the line length. A more detailed description of impedance and transmission lines is described later.

### Velocity Factor

With open wire air-spaced lines the velocity of an electromagnetic wave is very close to that of light. In the presence of dielectrics other than air used in the construction of the transmission line (see below) the velocity is reduced because electromagnetic waves travel more slowly in dielectrics than they do in a vacuum. Because of this the wavelength as measured along the line will depend on the velocity factor that applies in the case of the particular type of line in use. The wavelength in a practical line is always shorter than the wavelength in free space.

### Mismatch and SWR

The feedpoint impedance of an antenna may not be exactly the same as the characteristic impedance of its associated feeder. The antenna is then said to be mismatched to the feeder.

When a wave travelling along a transmission line from the transmitter to the antenna (incident wave) encounters impedance that is not the same as Zo (discontinuity) then some of the wave energy is reflected (reflected wave) back towards the transmitter. The ratio of the reflected to incident wave amplitudes is called the reflection coefficient, designated by the Greek letter ρ (Rho).

$$|\rho| = | Z_L - Zo / Z_L + Zo |$$

Where $Z_L$ is the load impedance and Zo is the characteristic impedance of the transmission line. It follows that the magnitude of ρ lies between 0 and 1, being 0 for a perfectly matched line.

The reflectometer is an instrument for measuring ρ and comprises two power meters, one reading incident power and the other reflected power. Power detector directivity is possible because the incident wave voltage and current are in phase and in the reflected wave, 180° out of phase. The construction of a reflectometer meter is described in the test equipment chapter.

Whenever two sinusoidal waves of the same frequency propagate in opposite directions along the same transmission line, as in any system exhibiting reflections, a static interference pattern (standing wave) is formed along the line, as illustrated in **Fig 14.2**.

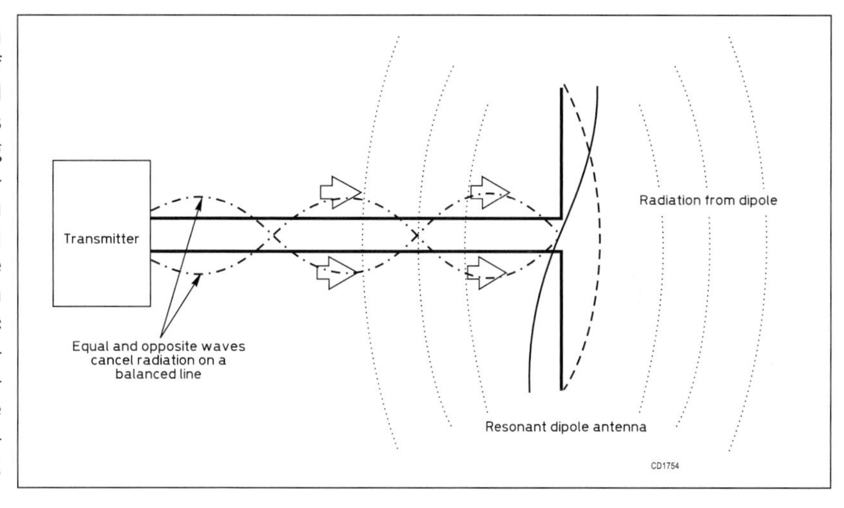


**Fig 14.1: RF energy on a transmission line connected to an antenna. No radiation occurs on the line provided the RF energy on each of the lines is equal and opposite. Once the RF energy reaches the antenna there is no opposition to radiation**

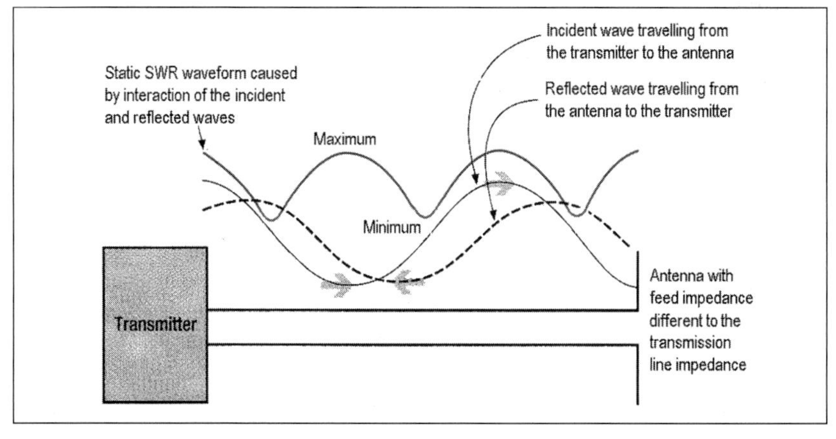


**Fig 14.2: Creation of a standing wave on a transmission line**

For the purposes of quantifying reflection magnitude, however, we are interested in the amplitude of the voltage or current nodes and nulls. Standing Wave Ratio (SWR) is defined as the ratio of the voltage or current maximum to the voltage or current minimum along a transmission line, as follows:

$$SWR = V_{max} / V_{min} = I_{max} / I_{min}$$

SWR can be measured using either a current or voltage sensor. The maximum must always be greater than the minimum, thus SWR is always greater than or equal to one. If no reflections exist, no standing wave pattern exists along the line, and the voltage or current values measured at all points along the transmission line are equal. In this case impedance match is perfect, the numerator and denominator of the equation are equal, and SWR equals unity.

As can be seen from Fig 14.2 the direct measurement of SWR must be made at two positions, one quarterwave apart. However, by using the reflectometer the SWR can be measured indirectly as:

$$SWR = 1 + |\rho| / 1 - |\rho|$$

The reflectometer, calibrated in SWR, has become the standard amateur radio tool for measuring transmission line mismatch.

It is often thought that a high SWR causes the transmission line to radiate. This is not true provided the power on each line is equal and opposite as shown in Fig 14.1.

## IMPEDANCE TRANSFORMATION

Impedance can be defined by the ratio of current and voltage. This will be familiar to you when looking at the current and voltage distribution of the standing wave on a dipole antenna as shown in Fig 14.1. The voltage is high and the current zero at the end of the dipole (high impedance) while at the centre of the dipole the voltage is low and the current high (low impedance). The centre is obviously the best place to feed the antenna when using low impedance transmission line.

If the transmission line is terminated with a short circuit, the impedance at that point will be very low as shown in **Fig 14.3**. It can be seen that the voltage is zero and the current is high at that point. The standing wave pattern shows that this very low impedance is repeated at every half-wave point down the line. On the other hand the impedance will be high a quarter of a wavelength down the line. This characteristic of transmission lines is often used as an impedance transformer and is used

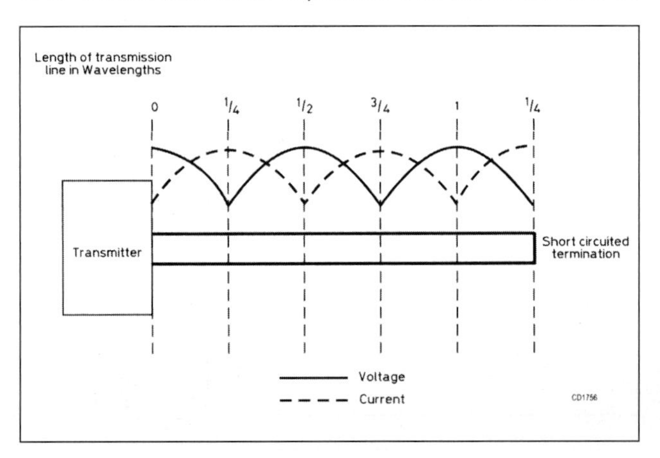


**Fig 14.3: Transmission line terminated with a short circuit, showing the patterns of voltage and current SWR. Note that the low impedance caused by the short circuit is reflected at half wavelengths from the short**

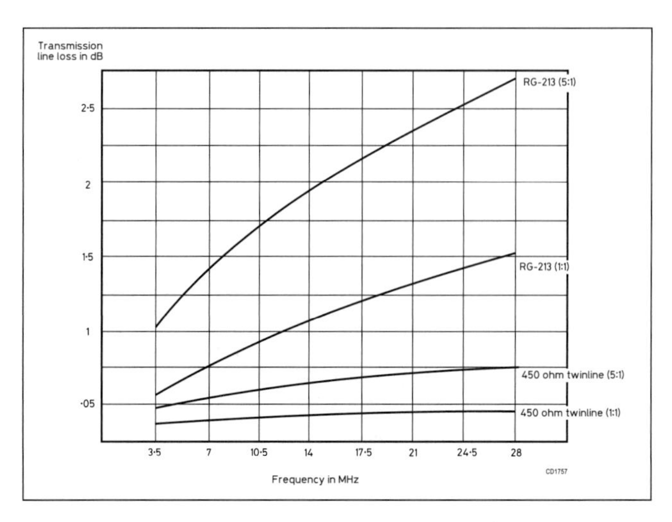


**Fig 14.4: Graph showing losses on 10m (30ft) of 450Ω twinline and RG-213 coaxial cable at an SWR of 1:1 and 5:1**

with the G5RV antenna described in the chapter on practical HF antennas.

The impedance transform effect is described later under Smith Chart.

## LOSSES IN TRANSMISSION LINE

Practical transmission line has losses due to the resistance of the conductor and the dielectric between the conductors. As in the case of a two-wire line, power lost in a properly terminated coaxial line is the sum of the effective resistance loss along the length of the cable and the dielectric loss between the two conductors. Of the two losses, the resistance loss is usually the greater; since it is largely due to the skin effect and the loss (all other conditions remaining the same) will increase directly as the square root of the frequency.

### Measurement of Coaxial Cable Loss

The classic method of measuring coaxial cable loss is to terminate the cable with a dummy load that is equal to the Zo of the line. Then use a power meter, first at the transmitter end and the load end ensuring that the transmitter power is maintained at a constant level during the test. Then calculate the loss from the difference in power readings using the formula:

$$dB\ loss = 10 \log_{10} (P1/P2)$$

where P1 is the power at the transmitter end and P2 is the power at the dummy load.

### Losses Due to SWR

As described above, a transmission line has losses due to the resistance of the conductor and the dielectric between the conductors. Losses at higher frequencies can also result from a poor quality outer conductor. **Fig 14.4** shows approximate losses for 450Ω twinline and RG-213 coaxial cable. Additional losses occur due to antenna/transmission line mismatch (SWR), also shown in Fig 14.4. These losses are for a transmission line over 30m (100ft) long. SWR losses on the HF bands are not as great as is often thought, although at VHF and UHF it is a different matter. As you can see from Fig 14.4, even an SWR of 5:1 on a 30m length of RG-213 coax at 28MHz, the attenuation is only just over 1dB over the perfectly terminated loss.

A reading of SWR due to a mismatch at the transmitter end of the transmission line will be lower than if the measurement were taken at the load (antenna) end. The reason is that the losses on the line attenuate the reflected wave. This means you

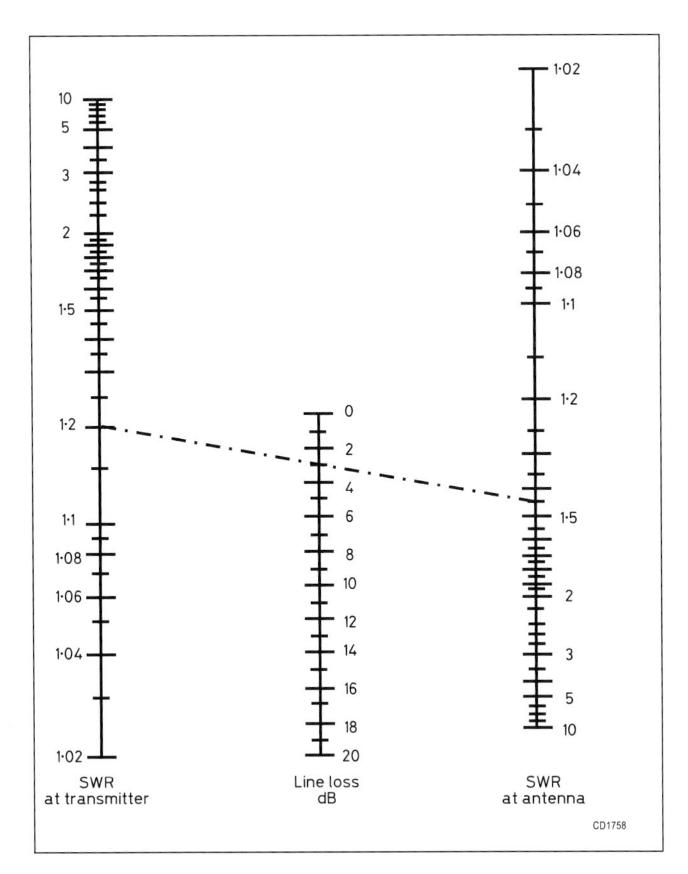


**Fig 14.5: Nomograph for calculating transmission line loss using an SWR meter**

can use an SWR meter to measure transmission line loss, using the power meter method (use a load that creates a mismatch, say 100Ω) described above. Measure the SWR at the transmitter and at then at the antenna. Use the graph in **Fig 14.5** to determine the cable loss.

## Low-loss Coax - Is it Worth it?

The attenuation factors of various correctly terminated coax cables is shown in **Fig 14.6**. These attenuation figures are for 30m (100ft) lengths and indicate that for frequencies below 30MHz there is not much to be gained by using expensive low-loss coax feeders. The method of construction of these cables to reduce loss is described and illustrated below.

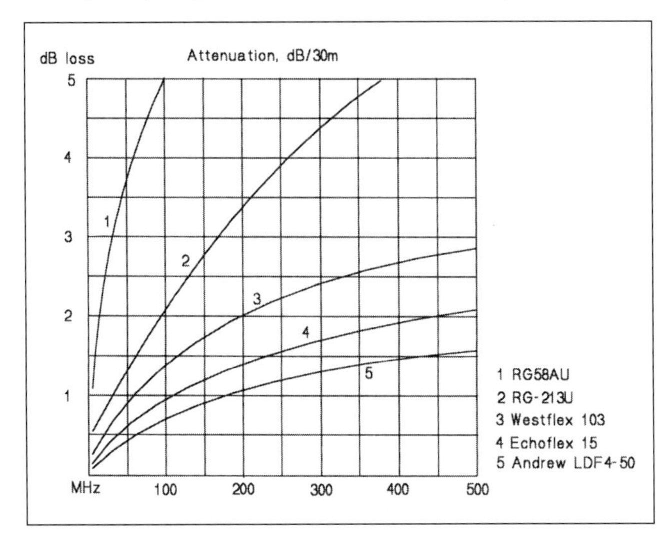


**Fig 14.6: Attenuation characteristics of various 30m (100ft) lengths of correctly terminated coax cables. With the exception of the thin RG58 coax the attenuation differences of the various cables below 30MHz are not significant**

At VHF, and particularly UHF frequencies, it is a different matter. Good quality coax can really enhance a station's performance. On a typical UHF installation at least a 3dB increase should be possible by replacing RG-213 with, say, Ecoflex. If this doesn't sound much remember that generally the size of a VHF/UHF antenna array has to be doubled to get 3dB gain.

# TRANSMISSION LINE CONSTRUCTION

Two types of transmission line have been used to construct antenna systems described in the antenna chapters. These are twin-line feeder and coaxial cable.

## Twin-line Feeder

Twin-line feeders can be constructed from two copper wires supported at a fixed distance apart using insulated spacers as shown in **Fig 14.7**. This type of construction is often known as 'open-wire feeder'. Spacers may be made from insulating material, such as plexiglas, polyethylene or plastic. The spacers shown in Fig 14.7 are specifically made for the job. The characteristic impedance of such a line can be calculated with the formula:

$$Zo = (276/ \varepsilon_r).\mathrm{Log}_{10}\,(2S/D)$$

Where $\varepsilon_r$ is the relative permittivity of the insulation.
Where S is spacing between the wire centres and d is the wire diameter.

The construction uses 1.5mm diameter copper wire, and spacers hold the wire around 75mm apart. Using the formula this gives a Zo of 550Ω. If 1mm diameter wire had been used, Zo would have been close to 600Ω.

The 300Ω twin line (the light coloured line shown in Fig 14.7) is constructed by moulding the conductors along the edges of a ribbon of polyethylene insulation and for this reason is sometimes known as ribbon line. This type of feeder is convenient to use but moisture and dirt tend to change the characteristic of the line.

A further variation of commercial twin-line feeder is 'window-line', which has windows cut in the polythene insulation at regular intervals. This reduces the weight of the line, reduces the loss due to the dielectric and breaks up the surface area where dirt and moisture can accumulate.

## Coaxial Cable

Coaxial cable transmission line is used in most amateur installations. The two conductors of the transmission line are arranged coaxially, with the inner conductor supported within the tubular outer by means of a semisolid low-loss dielectric.

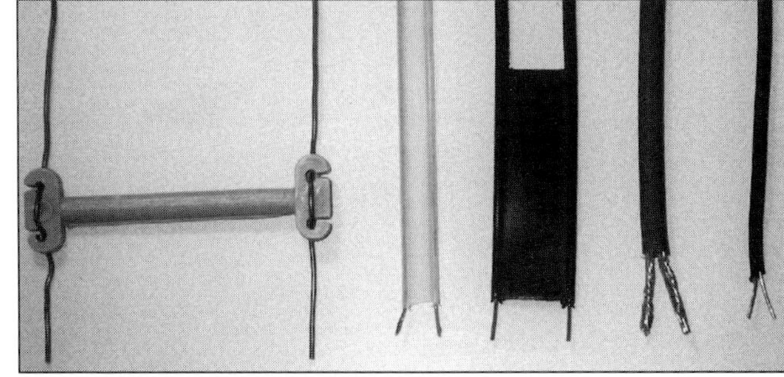

**Fig 14.7: Open wire line constructed using 1mm diameter copper wire and spacers. 300Ω twin line with polyethylene insulation, 450Ω 'window' twin line, 75Ω heavy duty twin line, 75Ω light weight twin line**

The characteristic impedance for concentric circular conductors is given by:

$$Zo = (138/\sqrt{\varepsilon_r}) \log10 (D/d)$$

Where D is the inside diameter of the outer conductor; d is the diameter of the inner conductor, and $\varepsilon$ is the dielectric constant of the insulator.

Coaxial cable has advantages that make it very practical for efficient operation in the HF and VHF bands. It is a shielded line and has a minimum of radiation loss. Since the line has little radiation loss, nearby metallic objects have minimum effect on the line because the outer conductor serves as a shield for the inner conductor.

Electromagnetic waves tend to propagate along the surface of conductors, rather than inside, due to the phenomenon of skin effect. Coaxial cable performance depends upon the conductivity and size of the outer surface of the inside conductor and the inner surface of the outer conductor.

The centre conductor of a coaxial cable may consist of either a single wire of the desired outer diameter, or from a twisted bundle of smaller strands. Stranded centre conductors improve cable flexibility while solid centre conductors provide the greatest uniformity of outer diameter dimension, which contribute to stable electrical characteristics.

The outer conductor of coax cable ideally should be made from a solid conductive pipe but this construction makes the cable difficult to bend. The flexibility and bend radius of such cables can be improved by corrugating the outer conductor; examples are shown in **Fig 14.8**.

Nearly all of the popular flexible coaxial cables employ braided outer conductors. These are not as effective electrically as solid outer conductors because gaps in the woven outer conductor permit some signal leakage or radiation from the cable, increasing the attenuation at higher frequencies. This effect can be minimised by adding a layer of copper foil under the braid.

The dielectric material that separates the outer conductor of a coaxial cable from its centre conductor determines the intensity of the electrostatic field between conductors and maintains the physical position of the inner conductor within the outer conductor. Common dielectric materials for coaxial cable include polyethylene, polystyrene and PTFE.

The least lossy dielectric material is a pure vacuum, which is totally impractical for use as a cable dielectric. However, the electromagnetic properties of air or gaseous nitrogen are very similar to a vacuum and can be used by mixing low-cost polyethylene with low-loss nitrogen. This is accomplished by bubbling nitrogen gas through molten polyethylene dielectric material before the polyethylene solidifies. This material is variously known as cellular polyethylene dielectric, foam dielectric, or poly-foam. It has half the dielectric losses of solid polyethylene at a modest increase in cost.

The characteristic impedance of most coax cable used in amateur radio installations is usually 50-ohms. Other impedance cable is used for impedance transformers and baluns (described later). The impedance of coaxial cable is often printed on the protective vinyl sheath.

In order to preserve the characteristics of the flexible, coaxial line, special coaxial fittings are available. These, and methods of fixing them, are described later.

Fig 14.8: Five examples of coaxial cable. From left to right, RG58A/U, 50-ohm RG213/U, Westflex 103 and Andrew LDF4-50 with outer sheath removed

## THE SMITH CHART

The Smith Chart was invented by Phillip H Smith and described as a Transmission-Line Calculator [1]. While transmission line calculations can be done using a computer with appropriate software such as TLW [2] the Smith chart is described here because it shows very clearly the action of a transmission line as an impedance transformer and the relationship between impedance and SWR .

### The Cartesian Impedance Chart

Impedance comprises resistance and reactance and is always expressed in two parts, R+jX. An impedance having a resistance of 75Ω and an inductive reactance 50Ω is conventionally written as:

75 +j50

For our consideration of impedance j can simply be regarded as a convention for reactance. The '+j' indicates inductive reactance and a '-j' indicates capacitive reactance. When the antenna is at its resonant frequency the +j and -j parts are equal and opposite so only the resistive part remains.

Impedance can be represented using a chart with Cartesian coordinates as shown in **Fig 14.9**.

This method of plotting and recording the impedance characteristics of antennas is rather like a Mercator Projection map, with the latitude and longitude of R and jX respectively plotted to define an impedance 'location'. Resonance, where the inductive and capacitive reactances in a tuned circuit or antenna element are equal and opposite, exists only on the zero reactance vertical line.

The impedance chart can be used to plot a series of measurements at various frequencies, which produces an impedance signature of the antenna or antenna system. These measurements are done with an impedance bridge and a professional

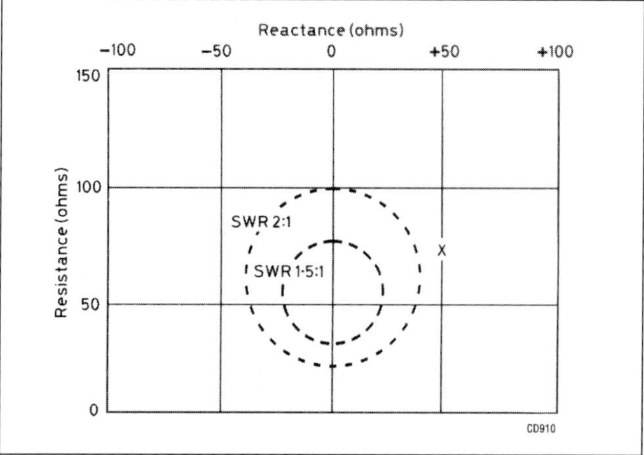


Fig 14.9: Impedance Chart, with X showing an impedance of 75R +50jX. The circles represent SWRs of 1.5:1 and 2:1 respectively for 50-ohm coaxial cable

Fig 14.10: General Radio 1606 RF impedance bridge. The indicated reactance value is valid for 1MHz and must be divided by frequency to get the true reactance. Inductive or capacitive reactance is established with the use of the switch located between the two dials

Fig 14.11: Basic simplified restricted range Smith chart

impedance bridge is shown in **Fig 14.10**. As you can see there are two calibrated controls, one for R and the other for j. Information from the calibrated dials on the instrument can be used to establish the impedance position on the chart when a measurement is made. An impedance noise bridge is described in the test equipment chapter, and other methods of measuring impedance are described in [3].

## The Smith Chart

The Smith Chart shown in **Fig 14.11** is an impedance map similar to the ones shown in Fig 14.9. It can be considered as just a different projection, just as maps have different projections, such as the Mercator Projection or the Great Circle projection. The most obvious difference with the Smith chart is that all the co-ordinate lines are sections of a circle instead of being straight. The Smith chart, by convention, has the resistance scale decreasing towards the top. With this projection the SWR circles are concentric, centred on the 50Ω point, which is known as the prime centre.

One of the advantages of the Smith impedance map projection is that it can be used for calculating impedance transforms over a length of coaxial feeder. Because the reflected impedance varies along the feeder it follows that you need to know the electrical length of your coaxial feeder to the antenna. The measured impedance, using an impedance bridge as previously described, is then modified by the impedance transform effect of the length of the feeder.

The impedance transformation Smith Chart is illustrated in **Fig 14.12**. An additional scale is added around the circumference, calibrated in electrical wavelength. Halfway round the chart equals 0.25 or quarter wavelength, while a full rotation equals 0.5 or half wavelength.

Two lengths of 50Ω coaxial feeder are shown superimposed around the circumference of a Smith chart; one length a quarter wave long and the other 3/8 wavelength). Both lengths are connected to a load having an impedance of 25 +j0. The quarter wave length of line (0.25) gives a measured impedance of 100

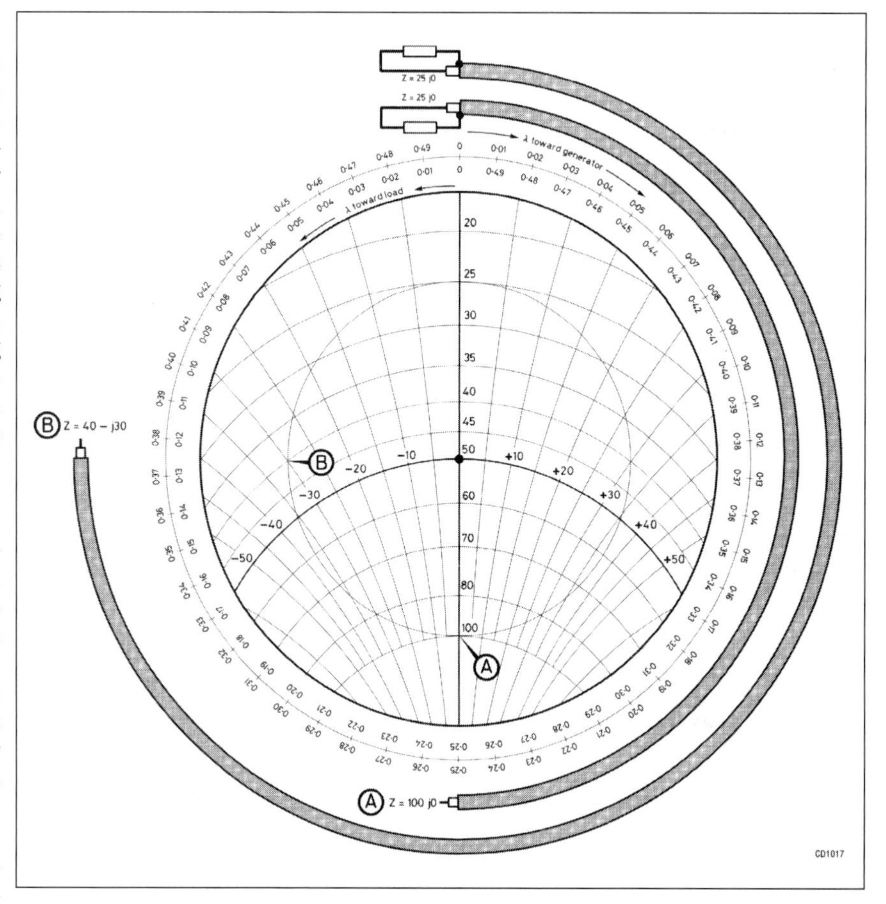


Fig 14.12: Smith chart, with transmission line electrical length scale, superimposed on two lengths of coaxial cable

+j0 at the other end while the 3/8 section (0.375) gives an impedance of 40+j30. It can also be seen from Fig 14.12 that a half wave length of coaxial would transform the impedance back to 25+j0.

## A Practical Smith Chart Calculator

You can make a Smith Chart calculator using the charts shown at the end of this chapter. The first one has a prime centre of 50Ω and a restricted impedance range. This makes it easier to use but the impedance excursions are limited. The other is the standard 50Ω chart which covers impedances from (theoretically) zero to infinity.

For this exercise we will make an impedance calculator using the restricted range chart as shown in **Fig 14.13**.

Make a photocopy of the chart (at the end of this chapter), enlarging it if necessary to bring it to a usable size. The chart is then glued to a circular sheet of stiff cardboard or thin aluminium. A small hole is drilled in the chart and backing material at the 50 +j0 point.

From a piece of very thin perspex or transparent plastic or celluloid cut a circle the same size as the chart to make an overlay. A hole is then drilled exactly at the overlay centre. Identifying the centre point should be no problem if a pair of compasses are used to mark the overlay before cutting.

Make a cursor by drawing a line along the radius of the overlay, using a fine tipped marker pen. Cover the line with a strip of transparent sticky tape to prevent the line rubbing out. Trim off the excess tape.

Fix the transparent overlay to the chart with a nut and bolt, with the tape covered line against the chart. Adjust the nut and bolt so that the overlay can be easily rotated, as shown in Fig 14.13.

The uses to which this calculator can be put are numerous and just three examples are described.

### Measuring coaxial cable electrical length

It is often important to know the electrical length of transmission line, either for making antenna feedpoint impedance measurements or constructing phasing lines for stacked beams or phased verticals. You can find the electrical length of coaxial cable by physically measuring its length and multiplying it by the cable velocity factor or by using a dip meter.

A more accurate method is to measure the electrical length directly using an RF impedance measuring instrument and the Smith Chart. It also assumes that the transmission line losses are low; in practice this means that the procedure will only work with relatively short lengths of fairly good quality coaxial cable..

1. Terminate the load (antenna) end of the cable with a 22Ω resistor.
2. Measure the impedance at the other end of the feeder.
3. Move the cursor so that it intersects the measured impedance point. The cursor will now point to the electrical wavelength of the feeder marked on the outer scale 'wavelengths towards generator'.

The cable may be several half wavelengths and part of a half wavelength long. The Smith chart will only register the part of a half wavelength, which is all we are interested in regarding the impedance transform effect.

### Calculating antenna impedance from measured impedance

This is a method of calculating antenna impedance from a measured impedance value, using coaxial cable whose electrical length has already been determined.

1. Connect the cable to the antenna.
2. Measure the impedance at the other end of the coaxial cable.
3. Move the cursor over the measured impedance point and mark the point on the overlay with a wax pencil.
5. Follow the cursor radially outwards to the scale marked wavelengths towards load. Write this number down.
6. Add the length of cable in wavelengths to this number.
7. If the number is larger than 0.5, subtract 0.5.
8. Rotate the overlay until the cursor points to this number on the wavelengths towards load scale.
9. The antenna impedance will be found on the cursor directly under the wax pencil mark.

**Example:**

The measured impedance is 35+j20 ohms and the cursor points to 0.407 on the wavelengths towards load scale.

The cable electrical length was measured as 0.13 wavelengths.

Then 0.407 + 0.13 = 0.537 wavelengths. Off scale - too big!

Subtract 0.5 wavelengths = 0.037 wavelengths.

Rotate the overlay until the cursor points to 0.037 on the wavelengths towards load scale.

The antenna impedance is shown as 28 -j8 ohms under the cursor at the same radius as the measured impedance.

### Calculation of SWR

Calculation of SWR is very simple using the Smith chart. The result is useful for correlating impedance measurements with SWR measurements. To measure SWR:

1. Move the cursor over the measured impedance point.
2. Mark the point on the overlay with a wax pencil.
3. Move the cursor to the 0 point on the outside scales.
4. The SWR can be read off as 50 divided by the mark on the cursor. The impedance measured above gives a reading of 27 +j0. 50 divided by 27 equals 1.85; the SWR in this case is 1.85:1.

You can, of course, calibrate the cursor in SWR. Just place the cursor in the vertical zero position and place marks on the cursor at the 33.3, 25 and 20 resistance points to give SWR marks at 1.5:1, 2:1 and 2.5:1 respectively.

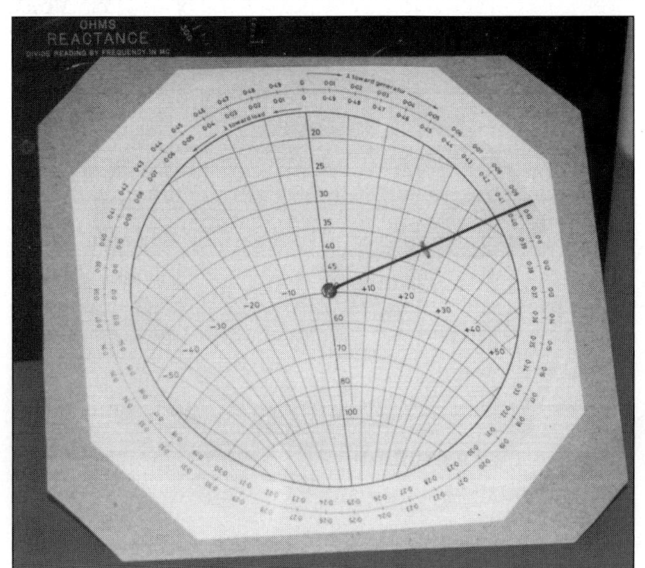


Fig 14.13: General construction of the Smith chart calculator

## ADMITTANCE

The method of calculating the total value of several resistors in series is add their individual values. And the simplest way of calculating resistors in parallel is to add their reciprocals; the answer is also a reciprocal and has to be converted back to R to be in a form which we are more familiar:

$$1/R1 + 1/R2 + 1/R3 = 1/R$$

The reciprocal of R is Conductance (symbol G) and you could work in Conductances if you were dealing with calculations involving lots of parallel circuits.

The reciprocal (or the dual) of Impedance is Admittance (symbol Y).

The reciprocal of reactance is Susceptance (symbol B). The unit of Conductance, Susceptance and Admittance is the Siemen.

It is important to know just what it is that your RF bridge is measuring. If the bridge is an Admittance bridge then the result, like the calculation of parallel resistors described above, will need to be converted into the more familiar ohms impedance.

### The Normalised Smith Chart

Most Smith charts are normalised so that they can be used at any impedance and not restricted to 50Ω, as are the ones so far described. This is achieved by assigning 1 to the prime centre; other values, for example, are 0.5 for 25Ω and 2 for 100Ω in a 50Ω system. Normalisation also extends the use of the chart to convert impedance to admittance (see sidebar) and vice versa. A chart for constructing a normalised Smith Chart calculator is also shown at the end of this chapter. The construction of the chart is the same as described above. To use the procedure described below, the cursor must be extended from the centre, ie the cursor line is extended from a radius to a diameter.

**To convert admittance to impedance:**

1. Convert the admittance to normalised admittance by multiplying each component of the admittance value by the prime centre, usually 50.
2. Move the cursor over the measured admittance point.
3. Mark the point on the overlay with a wax pencil.
4. Move the cursor 180 degrees so that the unmarked section of the cursor lies over the measured admittance point.
5. The mark on the cursor from 3 gives the normalised impedance reading.
6. Convert to actual impedance ohms by multiplying by 50.

## FITTING COAXIAL CONNECTORS

Fitting coaxial connectors to cable is something we all have to do at some time or other.

If you have had trouble in the past fitting connectors, you should find the methods described here by Roger Blackwell, G4PMK [4] helpful. Although specific styles of connector and cable are mentioned, the methods are applicable to many others.

### Cables and Connectors

The main secret of success is using the right cable with the right connector. If you're buying connectors, it is important to be able to recognise good and bad types, and know what cables the good ones are for. Using the wrong connector and cable combination is sure to lead to problems. Any information you can get, such as that from old catalogues, is likely to prove useful, especially if you can get the cable cutting dimensions and equivalents lists.

Cables commonly used in amateur radio are the American 'RG' (RadioGeneral MIL specification) types. RG213 is 10.5mm in diameter and is the most common cable used with type N and PL259 connectors. RG58 (5mm OD) is one usually used with BNC connectors. If there is any doubt about the quality of the cable, have a look at the braid. It should cover the inner completely.

There is another useful type of coax cable and that is M-RG8 (often known as Mini-8), which is a compromise between RG213 and RG58. This cable has an outside diameter of 6.5mm.

The three most popular connector types are the UHF, BNC and N ranges. If you can, buy connectors from a reputable manufacturer. There are some good surplus bargains about, so a trawl through the boxes at the local rally may prove worthwhile.

The PL259 UHF connector is not very good beyond 200MHz, because the 50 ohms impedance through the plug-socket junction is not maintained. The suitability of N and BNC connectors for use at UHF and beyond is due to their maintaining the system impedance (50 ohm) through the connector. PL259 plugs should have PTFE insulation. The plating should be good quality and there should be two or more solder holes in the body for soldering to the braid. There should be two small tangs on the outer mating edge of the plug, which locate in the serrated ring of the socket and stop the body rotating. If you are going to use small-diameter cable with these plugs, get the correct reducer. It is advisable to buy the reducers at the same time as buying the plugs because some manufacturers use different reducer threads.

With BNC, TNC (like the BNC, but threaded) N and C (like N, but bayonet) types, life can be more complicated. All these connectors are available in 50 and 75-ohm versions. Be sure you get the right one! All of these connectors have evolved over the years, and consequently you will meet a number of different types. The variations are mostly to do with the cable clamping and centre pin securing method.

If you are buying new connectors, then for normal use go for the pressure-sleeve type, which are much easier to fit.

All original clamp types use a free centre pin that is held in place by its solder joint onto the inner conductor. Captive contact types have a two-part centre insulator between which fits the shoulder on the centre pin. Improved MIL clamp types may have either free or captive contacts. Pressure sleeve types have a captive centre pin. As an aid to identification, **Fig 14.14** shows these types. Pressure clamp captive pin types are easy to spot; they have a ferrule or 'top hat' that assists in terminating the braid, a two-piece insulator and a centre pin with a shoulder. Unimproved clamp types have a washer, a plain gasket, a cone-ended braid clamp and a single insulator, often fixing inside the body. Improved types have a washer, a thin ring gasket with a V-groove and usually a conical braid clamp with more of a shoulder. There are variations, so if you can get the catalogue description it helps!

### Tools for the Job

To tackle this successfully, you really need a few special tools. While they may not be absolutely essential, they certainly help. Most of them you probably have anyway, so it's just a matter of sorting through the toolbox. First and foremost is a good soldering iron. If you never intend to use a PL259, a small instrument type iron is sufficient. If you use PL259s, something with a lot more heat output is required. Ideally a thermostatically-controlled iron is best; as with most tools a little extra spent repays itself handsomely in the future.

A *sharp* knife is another must. A Stanley-type is essential for larger cables, provided that the blade is sharp. For smaller cables, you can use a craft knife or a very sharp penknife. Use sharp blades, cut away from you, and keep the object you're cutting on the bench, *not in your hand*. Although sharp, the steel blades are

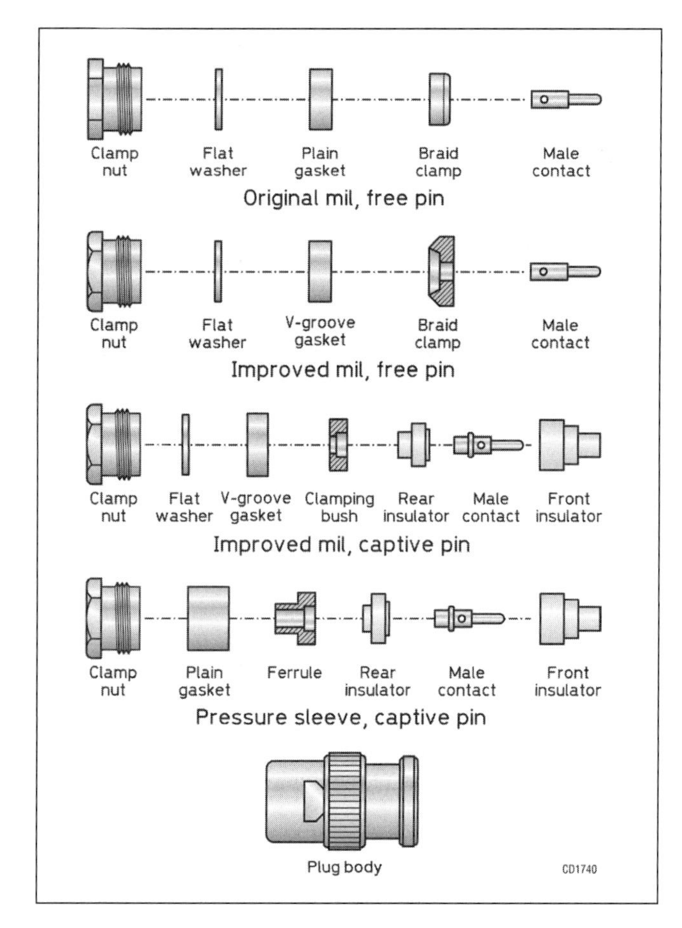


Fig 14.14: Types of BNC/N cable clamps

brittle and will shatter if you apply excessive force or bend them, with bits of sharp blade shooting all over the place. Dispose of used blades in a box or plastic jar. Model shops have a good range of craft knives, which will also do an excellent job.

A pair of small sharp scissors is needed for cutting braids, and a blunt darning needle (mounted in a handle made from a piece of wood dowelling) is useful for unweaving the braid. A scriber is also useful for this job. You will find a small vice a great help as well. For BNC, TNC and N type connectors, some spanners are essential to tighten the gland nuts. The BNC/TNC spanners should be thin 7/16in AF. Those for type N need to be 11/16 x 5/8 AF. A junior hacksaw is needed to cut larger cables. Finally, if you intend to put heatshrink sleeves over the ends of plugs for outdoor use, some form of heat gun helps, although the shaft of a soldering iron may work. (A hot-air paint stripper can be used for this purpose - with care). See the chapter on 'The Great Outdoors' for more on weatherproofing.

## Preparing Cables

Fitting a plug requires you to remove various bits of outer sheath, braid and inner dielectric. The important knack to acquire is that of removing one at a time, without damaging what lies underneath. To remove the outer sheath, use a sharp knife or scalpel. Place the knife across the cable and rotate the cable while applying gentle pressure. The object of doing this is to score right round the cable sheath. Now score a line from the ring you just made up to the cable end. If you have cut it just enough, it should be possible to peel away the outer sheath leaving braid intact underneath. If this is not something you've tried before, practise on a piece of cable first. For some connectors, it is important that this edge of the sheath is a smooth edge at right angles to the cable, so it really is worth getting right.

Braid removal usually just requires a bit of combing out and a pair of scissors. Removal of the inner dielectric is most difficult with large-diameter cables. Again, it is important that the end is a clean, smooth cut at right angles to the cable. This is best achieved by removing the bulk of the dielectric first, if necessary in several stages. Finally the dielectric is trimmed to length. There is a limit to how much dielectric you can remove at one go; 1-2cm is about as much as can be attempted with the larger sizes without damaging the lay of the inner. For the larger cables, it is worthwhile to pare down the bulk of the unwanted material before trying to pull the remainder off the inner. If you can, fit one plug on short cables before you cut the cable to length (or off the reel if you are so lucky). This will help to prevent the inner sliding about when you are stripping the inner dielectric.

## Fitting PL259 Plugs

### Without reducer, RG213 type cable (also URM-67)

First, make a clean end. For this large cable, the only satisfactory way is to use a junior hacksaw. Chopping with cutters or a knife just spoils the whole thing. Having got a clean end, refer to **Fig 14.15** for the stripping dimensions. First, remove the sheath braid and dielectric, revealing the length of inner conductor required. Do this by cutting right through the sheath and braid, scoring the dielectric, then removing the dielectric afterwards. Next carefully remove the sheath back to the dimension indicated, *without disturbing the braid*. Examine the braid; it should be shiny and smooth. If you have disturbed it, or it looks tarnished, start again a little further down.

With a hot iron, tin the braid carefully. The idea is to do it with as little solder as possible; a trace of a non-corrosive flux such as Fluxite helps. Lightly tin the inner conductor also at this stage.

Now slide the coupling piece onto the cable (threaded end towards the free end). Examine the plug body. If it isn't silver-plated, and you think it might not solder easily, apply a file around and through the solder holes. Now screw the body onto the cable, hard. When you've finished, the sheath should have gone into the threaded end of the connector, the inner should be poking out through the hollow pin, and the end of the exposed dielectric should be hard up against the inside shoulder of the plug. Look at the braid through the solder holes. It should not have broken up into a mass of strands; that's why it was tinned. If it has, then it is best to start again.

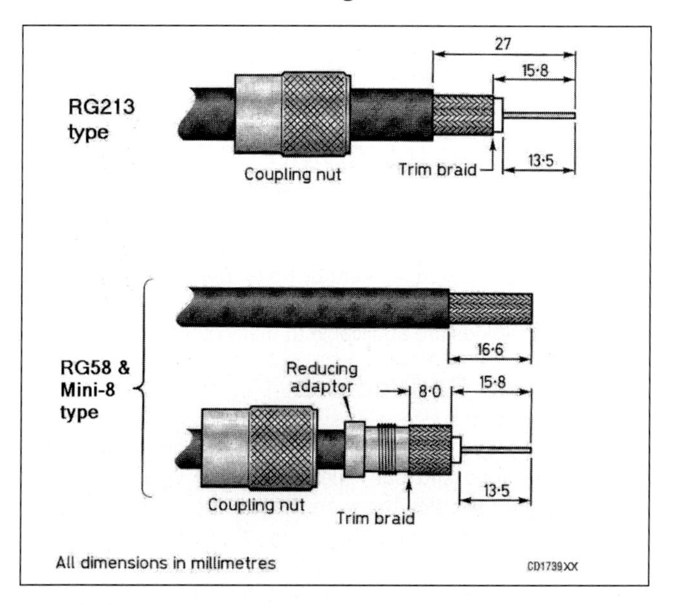


Fig 14.15: PL259 plug assembly

If all is well, lightly clamp the cable in the vice, and then apply the iron to the solder holes. Heat it up and then apply solder. It should flow into the holes; if it stays there as a sullen blob, the body isn't hot enough. Now leave it undisturbed to cool before soldering the inner by heating the pin and feeding solder down the inner. Finally, when it is all cool, cut any excess protruding inner conductor and file flush with the pin, then screw down the coupling ring. Merely as a confidence check, of course, test for continuity on both inner and outer from one end of the cable to the other, and check that the inner isn't shorted to the braid.

An alternative method of fitting PL-259 type connectors is used by military contractors. The outer insulation sheath is removed and the braid is folded back. The plug is then screwed onto the braid. This method needs no soldering of the braid which often results in shorts, and the pulltests on correctly fitted connectors show that at least 25kg is needed to dislodge the braid. The joint is finished by adding a short piece of heat-shrink sleeving on the tail of the cable to seal it from moisture.

### With reducer, RG58 and Mini 8 type cable

First, slide the outer coupler and the reducer onto the cable. Next, referring to Fig 14.15, remove the outer sheath without nicking the braid. Now, using a blunt needle, gently unweave the braid a bit at a time until it is all straight and sticking out like a ruff around the cable. Remove the inner dielectric, without nicking the inner conductor; so as to leave the specified amount of dielectric. Tin the inner conductor. Bring up the reducer until the end of the reducer is flush with the end of the outer sheath. Fold the braid back so it lies evenly over the shank of the reducer, then cut off the excess braid with scissors so that it is not in danger of getting trapped in the threads. Smooth it down once more, then offer up the plug body and, while holding the reducer and cable still, screw on the plug body until it is fully home. The only really good way of doing this is with two pairs of pliers. Now hold the assembly in the vice and ready the soldering iron. There has been a spirited discussion from time to time about the advisability of soldering the braid through the holes. Professional engineers use soldered connections or compression types.

## Fitting BNC and Type N plugs

These are 'constant impedance' connectors; that is, when correctly made up, the system impedance of 50Ω is maintained right through the connector. It is vital that the cable fits the connector correctly, therefore check that each part fits the cable properly after you prepare it. Refer to **Fig 14.16** for BNC dimensions, and **Fig 14.17** for N types.

### Original or unmodified clamp types

Slide the nut, washer and gasket onto the cable in that order. With the sharp knife, score through the outer sheath by holding the knife and rotating the cable, without nicking the braid. Run the knife along the cable from the score to the end, and then peel off the outer sheath.

Using a blunt needle, for example, start to unweave the braid enough to enable the correct length of dielectric to be removed. Now slip the braid clamp on, pushing it firmly down to the end of the outer sheath. Finish unweaving the braid, comb it smooth then trim it with scissors so that it just comes back to the end of the conical section of the clamp. Be sure that the braid wires aren't twisted.

Now fit the inner pin and make sure that the open end of the pin will fit up against the dielectric. Take the pin off and lightly tin the exposed inner conductor. Re-fit the pin and solder it in place by placing the soldering iron bit (tinned but with the solder wiped off) on the side of the pin opposite the solder hole. Feed a small quantity of solder (22SWG or so works best) into the hole. Allow the connector to cool and then examine it. If you've been careful enough, the dielectric should not have melted. Usually it does, and swells up, so with the sharp knife trim it back to size. This is essential, as otherwise the plug will not assemble properly. Remove any excess solder from around the pin with a fine file. Now push the gasket and washer up against the clamp nut, check the braid dressing on the clamp, and then push the assembly into the plug body. Gently firm home the gasket with a small screwdriver or rod and then start the clamp nut by hand. Tighten the clamp nut by a spanner, using a second spanner to hold the plug body still; *it must not rotate*. Finally, check the completed job with the shack ohmmeter.

### Modified or improved clamp types

In general, this is similar to the technique for unmodified clamp types described above. There are some important differences, however. The gasket has a V-shaped groove in it, which must face the cable clamp. The clamp has a corresponding V-shaped profile on one side; the other side may be conical or straight sided, depending on the manufacturer. If the clamp end has straight sides, the braid is fanned out and cut to the edge of the

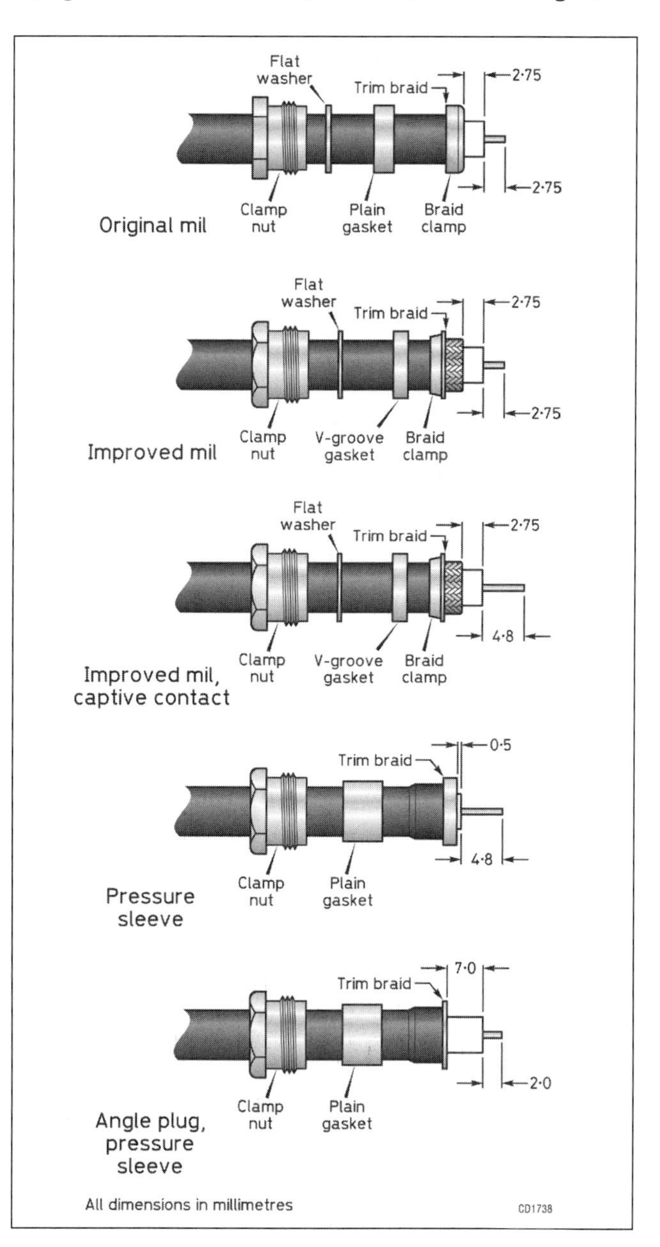


Fig 14.16: BNC dimensions, plugs and line sockets

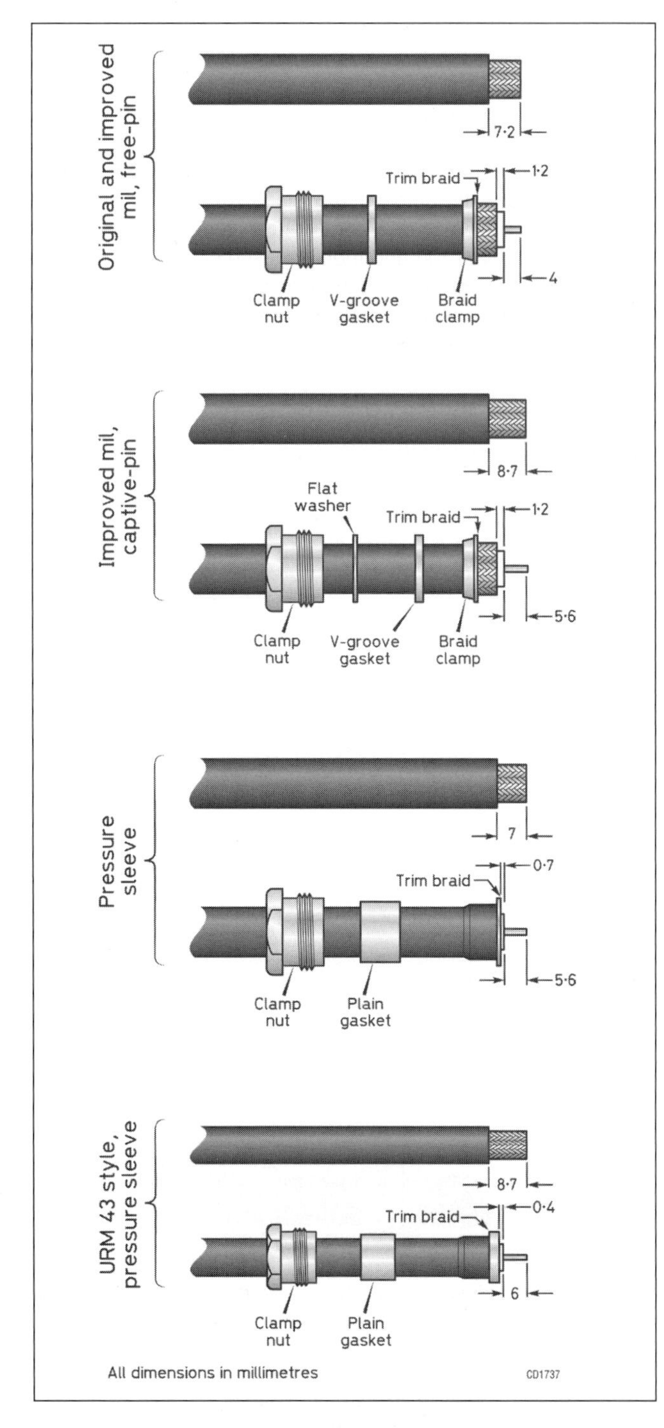


**Fig 14.17: N-type dimensions, plugs, angle plugs and line sockets**

**Fig 14.18: Partially assembled N-connector used with Ecoflex 15 coax. Note that the centre pin is a tight push fit over the coax centre conductor**

clamp only, not pushed down the sides. Some types have a small PTFE insulator, which is fitted before the pin is put on (common on plugs for the small RG174 cable). You now appreciate why having the assembly instructions for your particular flavour of plug is a good idea! Still, by using these instructions as a guide, it shouldn't be too difficult to get it right, even if it does not fit the first time. One important point - if the plug has been assembled correctly and tightened up properly, the clamp will have (intentionally) cut the gasket, which is then rather difficult to re-use. This thin gasket will not stand a second attempt. The thicker gasket types will often allow careful re-use.

### Captive contact types

These have a small shoulder on the pin, and a rear insulator, which fits between the pin and the cable. Most types use a thick gasket and a ferrule, although some use a V-grooved braid clamp and thin gasket. The ferrule type is described first because these are the most commonly available, and the easiest to fit. First, slip the nut and gasket onto the cable then strip off the correct amount of outer sheath by rotating the cable, producing a neat scored circle. Score back to the end of the cable and peel off the unwanted sheath. Comb out the braid, and with it fanned out evenly around the cable, slide the ferrule (small end first) on to the dielectric-covered inner conductor. Push it home so that the narrow portion of the ferrule slides under the outer sheath, and the end of the outer sheath rests against the ferrule shoulder.

Trim the braid with scissors to the edge of the ferrule. Slide up the gasket so that it rests gently against the ferrule shoulder, which will prevent the braid from being disturbed. Using the sharp knife, trim the dielectric back to the indicated dimension, without nicking the inner conductor.

Fit the rear insulator, which will have a recess on one side to accommodate the protruding dielectric. Incidentally, if you don't have the size for your particular plug, trim the dielectric until it fits; but don't overdo it! Now trim the exposed inner conductor to length and check by fitting the pin, whose shoulder should rest on the rear insulator unless the inner has been cut too long. Tin the inner lightly, then fit the pin and solder it by applying the iron tip (cleaned of excess solder) to the side of the pin opposite from the solder hole and feed a small amount of solder into the hole.

Allow to cool, and then remove the excess solder with a fine file. Now fit the front insulator (usually separate from the body) and push the whole assembly into the body. Push down the gasket gently into the plug body with a small rod or screwdriver. Start the nut by hand, and then tighten fully with one spanner, using the other to prevent the body from rotating. Check with the ohmmeter, then start on the other end - remember to put the nut and gasket on first!

### Variations

Angle plugs generally follow a similar pattern to the straight types, except that connection to the inner is via a slotted pin, accessed via a removable cap screw. Tighten the connector nut before soldering the inner. Line sockets are fitted in the same way as plugs. Interestingly, many connectors used on high-grade low loss cables appear to be solderless. This applies to the N-type male and female connectors used with Ecoflex15 (a high grade coax used at VHF/UHF), see **Fig 14.18**. This connector proved to be very easy to fit.

## SPLICING COAXIAL CABLE

The radio engineer's method of joining two lengths of coax together is to use coaxial connectors. However, in the description of coax cable splicing by G3SEK [5] that follows, you will see that a splice can be made entirely without connectors. A splice

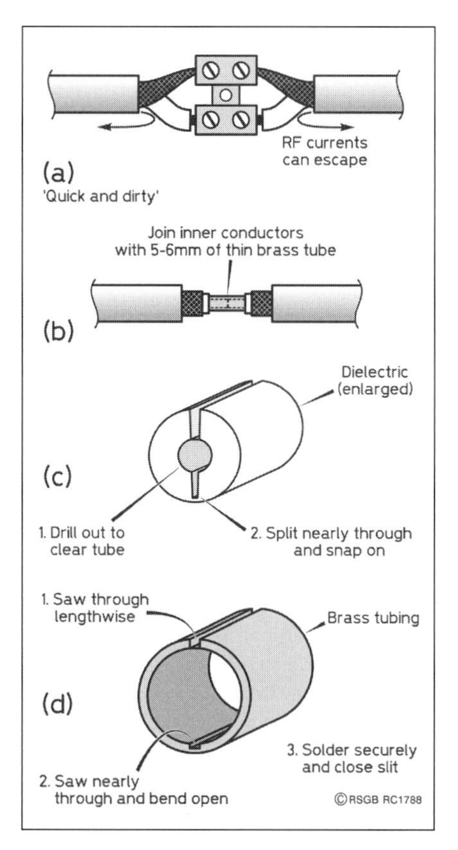


**Fig 14.19: Methods of splicing coaxial cable**

in coaxial cable needs to be as close as possible to an uninterrupted run of cable. In practice this requires four things:

- Constant impedance through the splice.
- As short an electrical length as possible, if it is not possible to make the impedance quite constant.
- Continuous shield coverage.
- Good mechanical properties: strong and waterproof.

At low frequencies coax can be spliced with a two-pole connector block as shown in **Fig 14.19(a)**. Tape over the joint and it's done. Even though this creates a non-constant impedance, the electrical length of the splice is so short that it's most unlikely to have any significant effect.

The main drawback is that the break in the shield cover provides an opportunity for RF currents to flow out from the inside of the shield and onto the outer surface (the skin effect makes RF currents flow only on surfaces). This may undo all your good efforts to keep RF currents off the feedline, using baluns or feedline chokes. For a truly coaxial splice you need to join and insulate the inner conductor, and then replace the outer shield. Avoid making a big blob of twisted inner conductors and solder if you can, because that will create an impedance bump - a short section of line with a different impedance from the coax itself.

The neatest and electrically the best way to join the inner conductors is to use a 5-6mm (¼in) sleeve of thin brass tubing, see **Fig 14.19(b)**. This is available from good hobby shops in sizes from 1,6mm (1/16in) outside diameter up to 12.7mm (½in), in steps of 0.8mm (1/32in); these sizes telescope together, by the way.

To replace the dielectric, take a piece of the original insulation, drill out the centre to fit over the sleeve, and split it lengthways so that it snaps over the top, see **Fig 14.19(c)**.

To complete the shield on braided coax, one good way is to push the braid away from each end while you join the inner conductor, and then pull it back over the splice. Solder the braid quickly and carefully to avoid melting the dielectric underneath. For mechanical strength you can tape a rigid 'splint' alongside the joint as you waterproof it. Alternatively the splice can be made using a very short length of air-insulated line of the same characteristic impedance. The inner conductor is joined using tubing as already described in **Fig 14.19(b)**. The outer is made from a short length of brass or copper tubing.

The outer tube is 'hinged' to fit over the joint as shown in **Fig 14.19(d)** shows how to then solder the whole thing up solidly. This method makes a very strong splice with excellent RF properties.

For 50Ω air-spaced coax, the ratio of the inner to outer conductor diameters is 0.43, so all you need to do is to choose the right diameters of tubing for the inner and outer conductors. Remember that the relevant dimensions are the outside diameter of the inner conductor, and the inside diameter of the outer conductor. It so happens that air-spaced line needs a larger inner diameter than solid-dielectric, semi-air spaced or foamed line, which conveniently accommodates the wall thickness of the inner sleeve. For UR67, RG213 and RG214, the best available choices are 8mm (5/16in) and 4mm (5/32in) outside diameters. These coaxial splices will be at least as good as a splice using coaxial connectors.

## A QUESTION OF BALANCE

At the beginning of this chapter, Fig 14.1 shows a dipole antenna fed with twin line feeder carrying transmitter power to a dipole antenna. The current flow in each conductor of the feeder is equal and opposite so no radiation from the feeder takes place; neither does the feeder pick up electromagnetic signals on receive. This balanced mode of transmission is often referred to as differential mode.

In the earlier days of amateur radio twin wire feeder was the only practical feeder available and the design of amateur feed methods was influenced by commercial radio practice.

Commercial HF radio stations often use large antennas such as rhombics. These antennas take up a considerable amount of space so transmission lines have to be very long. Furthermore there can be several of them in close proximity. It can be appreciated that such an arrangement requires that the feed lines have to be well balanced (equal RF current in each conductor) to prevent radiation loss and cross-talk (mutual interference between sets of lines). To achieve this balance an ATU is used that enables the currents in each transmission line conductor to be adjusted so that they are equal.

From this we often get idealistic images in textbooks showing a simple dipole; fed in the centre, with electric field lines neatly connecting the opposite halves, and lines of magnetic flux looping around the wires. **Fig 14.20** is a typical version of this pretty picture, showing only the electric field lines for clarity. Everything is symmetrical, with the system 'balanced' with respect to ground.

The reality of a typical installation is very different [6]. As **Fig 14.21** shows, the electric field lines connect not only with the

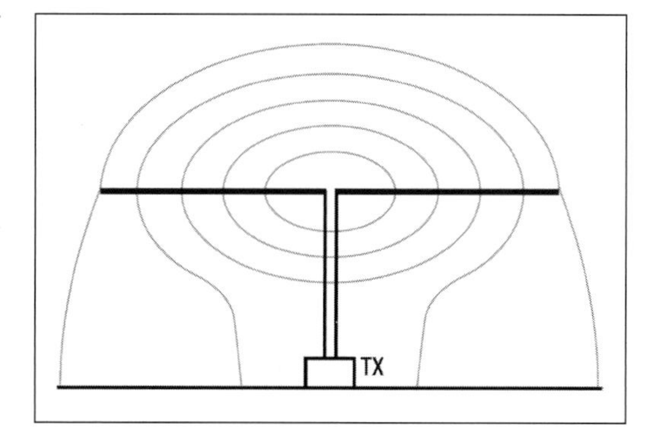


**Fig 14.20: A highly idealised picture of electric fields around a symmetrical antenna**

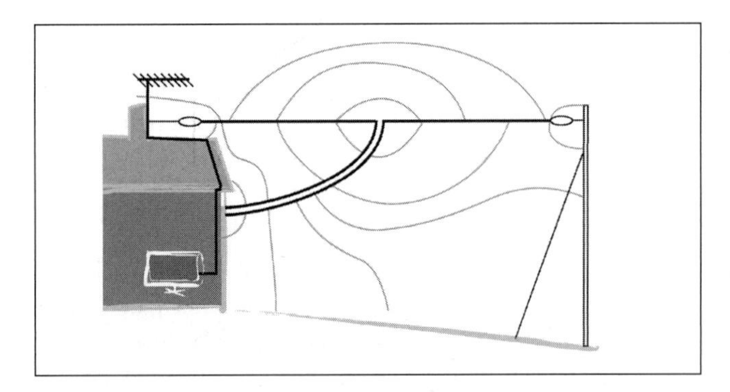

Fig 14.21: Typical reality, the distorting effects of nearby asymmetrical surroundings

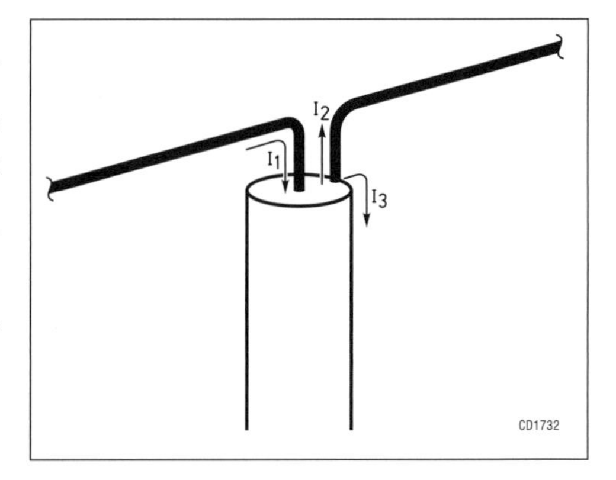


**Fig 14.22: Where currents on either side of the feedpoint are unequal, the difference I3 = I1 - I2 will flow down the outside of the cable**

opposite half of the dipole, but also with the feedline, the ground, and any other objects nearby. The currents induced into each conductor of the feedline from the antenna are similar in amplitude, but not opposite as shown in Fig 14.1. These are referred to as common-mode currents.

Although the electromagnetic coupling between the opposite halves of a horizontal dipole makes the antenna 'want' to be balanced, the coupling has to compete with the distorting effects of the asymmetrical surroundings. As a result, practical antennas can be very susceptible to the way they are installed, and are hardly ever well balanced.You might think that using coaxial cable would overcome these common mode effects.

The currents on the centre core (I1) and the inside of the shield (I2) are equal and opposite, ie 180° out of phase. The two conductors are closely coupled along their entire length, so the equal and antiphase current relationship (differential mode) is strongly enforced. Also, what goes on inside the cable is totally independent of the situation outside. This is due to the skin effect, which causes HF currents to flow only close to the surfaces of conductors and the inner and outer surfaces of the coaxial shield to behave as two entirely independent conductors.

Coax cable, unlike twin line feeder, is unaffected by nearby metal objects. It can be taped to a tower or even buried; yet the voltages and currents inside the cable remain exactly the same. About the only things you can do wrong with coax cable are to let water inside, or bend it so sharply that it kinks. That's why coax is popular - it is so easy to use.

However coax cable construction is not symmetrical, which means that it is inherently unbalanced. Broadly speaking the unbalance with coax line is caused by the fact that the outside of the braid is not coupled to the antenna in the same way as the inner conductor and inner surface of the outer braid

This results in a situation where there is a difference between the currents flowing in the antenna at either side of the feedpoint. This difference current is shown in **Fig 14.22** as I3, and is equal to (I1-I2). The current I3 has to flow somewhere. It cannot flow down the inside of the cable because I1 and I2 must be equal, so instead it flows down the outside of the outer sheath.

As a result, the feedline becomes part of the radiating antenna. This causes distortion of the radiation pattern, RF currents on metal masts and Yagi booms, and possible problems with 'RF in the shack' when running high power.

Common mode current I3 is further exacerbated if the feed to the antenna is asymmetrical as shown in Fig 14.21 and even more so if the feeder length is an electrical multiple of $\lambda/2$.

## BALUNS

The word 'balun' is short for 'balanced to unbalanced'. A balun is a device, which is used to connect a balanced load to an unbalanced coaxial line. There are two types of balun. The first is one that converts an inherently unbalanced source to a balanced load often known as a transformer or voltage balun; the second is a choke or current balun, which places a large series impedance on the outside of the feedline to choke off I3 currents that result from imbalance.

### Transformer or Voltage Balun

A balanced antenna feedpoint will present an equal impedance from each side to ground. The accusation levelled at choke baluns is that they treat the symptom (the surface current I3) without attempting to correct the imbalance that causes it. A transformer balun, on the other hand, does create equal and opposite RF voltages at its output terminals, relative to the grounded side of its input.

Voltage baluns may or may not involve a deliberate impedance transformation. You are probably familiar with the basic 1:1 and 4:1 baluns. **Fig 14.23(a)** shows a wire-wound 1:1

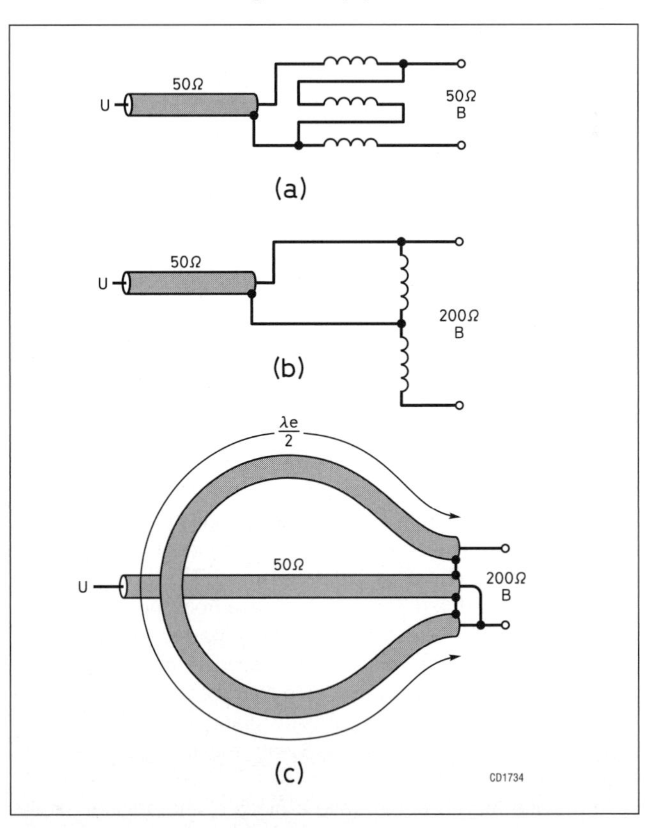


**Fig 14.23: Transformer/voltage - type baluns. Windings shown separately are bifilar or trifilar wound, usually on a ferrite rod or core. (a) wire-wound 1:1 (trifilar). (b) Wire-wound 4:1 (bifilar). (c) Coaxial 4:1, the half-wavelength must allow for the velocity factor of the cable**

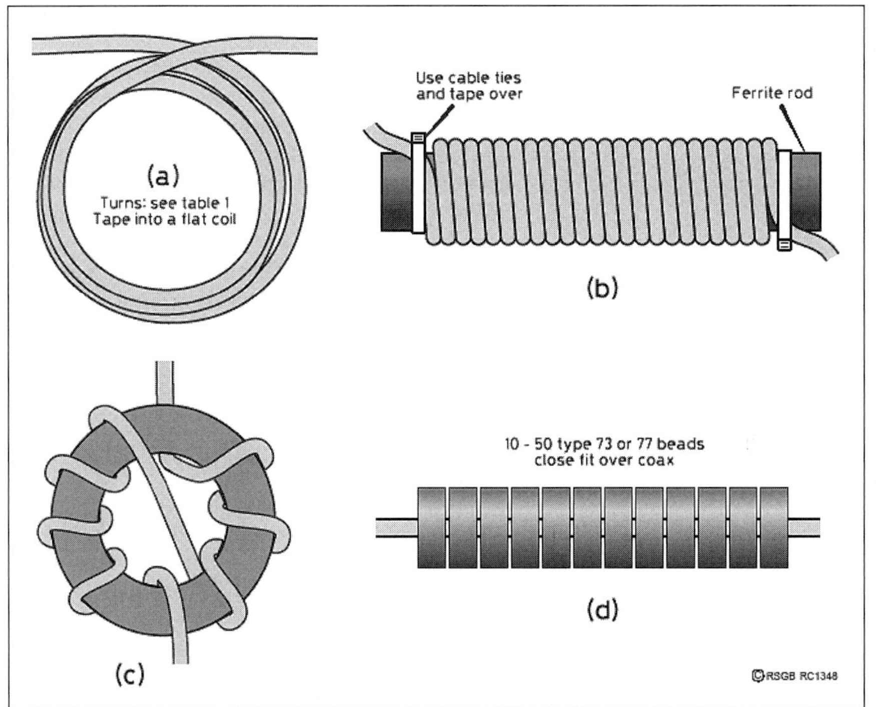


**Fig 14.24: Four varieties of choke-type balun. (a) Coil of coaxial cable, typically six to ten turns - see Table 14.1 - (only three turns are shown here). (b) Coaxial cable on ferrite rod. (c) Coaxial cable on toroid ring. (d) Ferrite sleeve. These chokes are designed to present a high impedance to unwanted currents on the outside of the coaxial cable**

| | RG213/UR67 | | RG58/UR76 | |
|---|---|---|---|---|
| **MHz** | **feet (m)** | **turns** | **feet (m)** | **turns** |
| **3.5** | 22 (6.7) | 8 | 20 (6.1 | 6-8 |
| **7** | 22 (6.7) | 10 | 15 (4.6) | 6 |
| **10** | 12 (3.7) | 10 | 10 (3.5) | 7 |
| **14** | 10 (3.5) | 4 | 8 (2.4) | 8 |
| **21** | 8 (2.4) | 6-8 | 6 (1.8) | 8 |
| **28** | 6 (1.8) | 6-8 | 4 (1.22) | 6-8 |

**Table 14.1: Lengths and number of turns of coax required to make a single-band coiled-coax choke**

balun, and both wire-wound and coaxial cable versions of the 4:1 type.

These all have the common property of forcing balance at their output terminals by means of closely coupled windings within the transformer itself. The 4:1 balun in **Fig 14.23(b)** is the easiest to understand. Typically, it transforms 50Ω unbalanced to 4-x-50 = 200 ohms balanced. This is achieved simply by a phase inversion.

An applied voltage v at one side of the feedpoint is converted by the transformer action into a voltage -v at the other side. These two voltages 180° out of phase represent the balance we are seeking to achieve. The 4:1 impedance transformation arises as follows. If the original voltage on the feedline was v, the voltage difference between opposite sides of the feedpoint is now 2v. Since impedance is proportional to voltage squared, and $(2v)^2 = 4v^2$, the impedance is stepped up by a factor of 4.

In the wire-wound 4:1 balun shown in Fig 14.23(b), the 180° phase inversion is achieved by the connection of the windings, while the coaxial equivalent in **Fig 14.23(c)** does it by introducing an electrical half-wavelength of cable between opposite sides of the feedpoint. Strong coupling between the windings and inside the coax forces the whole system into balance. This is often the balun of choice for VHF/UHF beams, see below.

A transformer balun is generally not the best type to be connected between coax feeder and a HF antenna. Any unbalance caused by external factors or the design of the antenna mean that any current difference between the two sides of the antenna has no where else to flow but down the outside of the coax. That does not mean that the balun is no better than a direct connection; when you insert the balun, it forces the currents in the antenna to readjust and become more symmetrical.

Although the transformer balun may do a good job, it may still not achieve total equality between the currents on opposite sides of the feed point. That may leave a residual difference current I3 to flow down the outside of the coax, as noted above. Unfortunately, having made its effort to minimise that difference current, the transformer balun does nothing to prevent it from flowing onto the coax. And there is still the possibility of additional currents being induced further down the surface of the feedline. This implies that a transformer-type balun may require additional RF chokes at the balun itself and possibly further back down the line.

The transformer balun is useful where it is necessary to connect an unbalanced ATU to balanced two wire transmission line. It is also essential in the comudipole arrangement, see the chapter on Practical HF Antennas.

## Choke or Current Baluns

As stated earlier a choke or current balun is a device that places large series impedance on the outside of the feedline to choke off I3 common currents. At its very simplest, a choke balun can be just a few turns of cable in a loop of diameter 300 to 600mm (6in to 12in), see **Fig 14.24(a)**.

We tend to think of these coils as inductors, but their high-frequency performance is actually dominated by the distributed capacitance between the turns. For example, take about 2.2m of thin coax like RG8X or RG58 and wind it into a five-turn bundle of about 125mm average diameter. This has an inductance of about 6µH, but the capacitance between the turns is equivalent to about 9pF in parallel with the 6µH. So instead of an inductor, what we actually have is a high-Q parallel resonant circuit with the measured impedance characteristics shown in **Fig 14.25**.

This parallel resonant circuit does not make a dependable RF choke. The impedance is only high around the resonant frequency, and much lower elsewhere. The resonant frequency is also quite sensitive to small changes affecting the capacitance between the turns, even how tightly the turns are taped together. The disadvantage of these chokes is that their performance is very dependent on the situation in which they're being used. This is because the impedance of the choke consists almost entirely of either inductive or capacitive reactance, at all frequencies except the very narrow region close to resonance as shown in Fig 14.25. Ideally these types of choke are only suitable for monoband antennas.

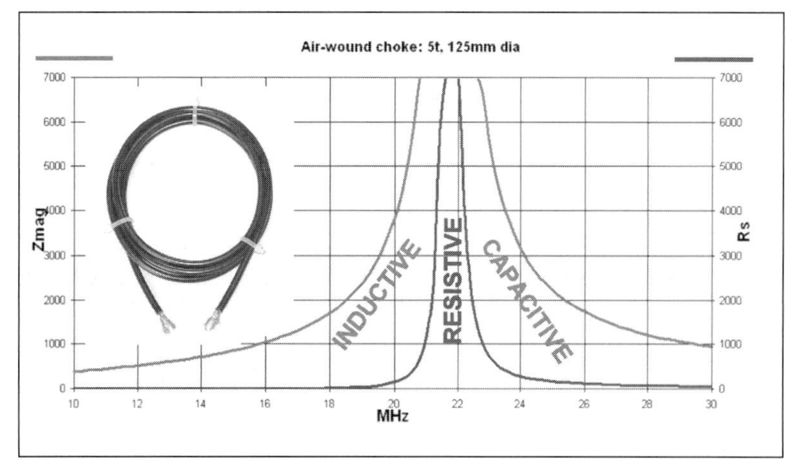


**Fig 14.25: Performance of an air-wound choke: notice the very sharp resonance at 21MHz**

## Ferrite loaded chokes

To overcome the problem of uncertain reactive impedance, the impedance of a dependable RF choke needs to be both large and predominantly resistive. The advantages of resistive impedance are that it cannot be cancelled out and it also tends to broaden the useful bandwidth of the choke. Any practical choke will also have some useful reactance, but resistive impedance is the only solid foundation for dependable performance. Earlier literature suggested that a resistive impedance of 500Ω for a choke balun would be adequate. More recently designs are aiming for an R value of several thousand ohms, rather than this lower value.

A high resistive impedance is achieved by engineering a certain amount of loss into the choke, and this is achieved using ferrite. Unlike many other radio engineering situations, resistive loss in an RF choke is a desirable entity because it appears as a very high value of R in the series impedance, $Z_{CHOKE} = (R \pm jX)$.

The simplest type of a ferrite choke can be seen in **Fig 14.24(b)** with 20 or more turns of RG58 wound on a thick ferrite rod. A more traditional method of building a choke is to wind RG58 on a ferrite toroid as shown in **Fig 14.24(c)**.

Another alternative, popularised by W2DU, is to feed the cable through tubes or beads to form a sleeve as shown in **Fig 14.24(d)**. Many commercial baluns use this method.

Strings of ferrite beads are not that cost-effective. Ferrite beads can usually take only one 'turn' of cable (one pass through the centre hole = one turn) and each individual bead generates quite a low impedance, so a high impedance will need a lot of beads in series. Ten or 20 beads will give enough impedance to handle minor EMC; for dependable performance, 40 or 50 beads are necessary. A commercial (Unadilla) balun (described as a W2DU HF 1:1 inline isolator) using this technique has around 45 beads and gives a resistive impedance at resonance of well over 1000 ohms at resonance, see **Fig 14.26**.

High common mode currents on coax feeders can be a problem when running high power. If the resistive impedance of a choke balun is too low it can allow a residual level of common-mode current to flow, causing overheating. The resistive (heat) loss in the choke equals $I_{CM}^2R$, where $I_{CM}$ is the residual level of common-mode current that remains after the choke has been inserted. If the choke has successfully suppressed the common mode current then the residual value of $I_{CM}$ will be very low and it is unlikely that there will be significant heating in the ferrite.

Ferrite chokes with a resistive impedance less than 1000 ohms are at a greater risk of under performing and overheating when using high power. Many of these chokes were designed to meet that inadequate target of 500 ohms, and some commercial examples have also suffered further cost-cutting, eg by using smaller quantities of ferrite and failing to use the correct materials. If a ferrite loaded choke begins to overheat, the ferrite may reach the Curie temperature at which its magnetic permeability collapses, allowing $I_{CM}$ to increase and causing further overheating.

To make a good ferrite choke the right grade of ferrite must be used; one that actually has some loss at the operating frequency. Additionally the choke must have the right amount of coupling between the ferrite material and the magnetic field around the cable.

GM3SEK [7] investigated choke construction methods, inspired by designs in the 2010 *ARRL Handbook*, which includes some new choke designs. These designs use a small number of relatively low-cost ferrite cores threaded onto a coil of cable as shown in **Fig 14.27**.

These cores have an oval central hole, 26 x 13mm, which will take several turns of thin transmitting coax like RG8X, or similar-sized cable of any other type. And although they are made by Fair-Rite in the USA, these particular cores don't have to be specially imported; they are readily and inexpensively available as stock items from Farnell UK [8].

This design concept has opened the way to a range of cost-effective ferrite chokes that can tackle the large majority of balun and other EMC problems across the HF spectrum. The three chokes shown in Fig 14.27 are only examples of what can be done; each choke delivers a high

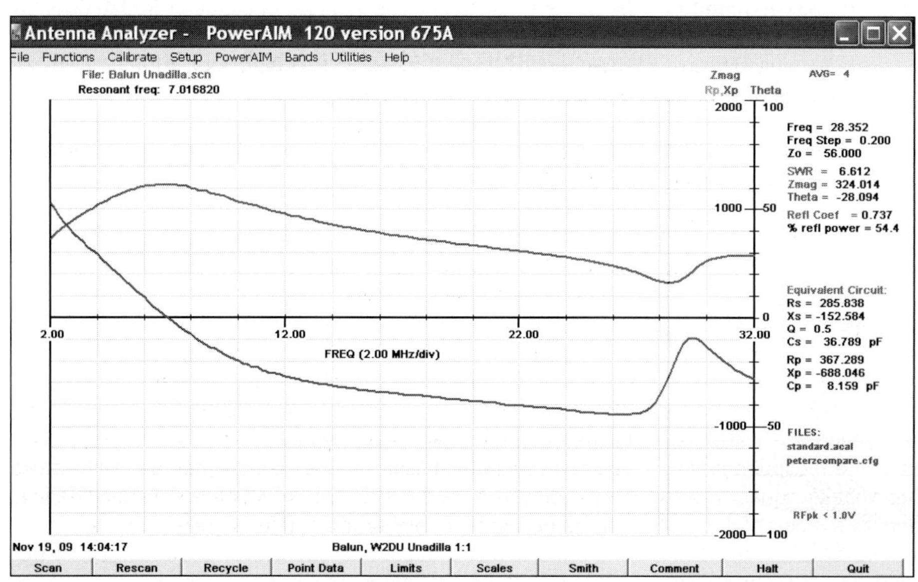


**Fig 14.26: Impedance measurements (using the AIM4170) of a commercial (Unadilla) balun (described as a W2DU HF 1:1 inline isolator). This balun uses around 45 beads**

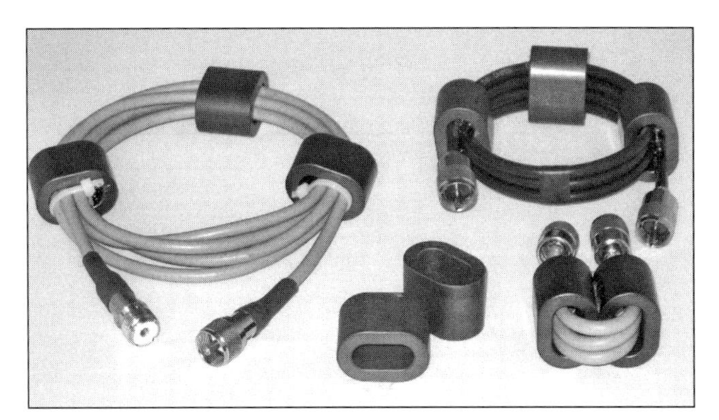

**Fig 14.27: Clockwise from the left: Low-bands ferrite choke, mid-bands choke, high-bands choke, the ferrite cores**

| | Turns | Mean diameter | Cores |
|---|---|---|---|
| **Low bands** | 5 | 125mm | 3 |
| **Mid bands** | 4 | 85mm | 3 |
| **High bands** | 3 | Close wound | 2, glued side by-side |

**Table 14.2: Dimensions of the three HF ferrite chokes using Fair-Rite 2643167851 or Farnell 1463420 ferrites**

resistive impedance over at least a 2:1 frequency range using only two or three of the oval Fair-Rite cores. The key dimensions for the three HF-band chokes are given in **Table 14.2**.

**Low Bands:** When two or three ferrite cores are threaded onto the flat five-turn coil as described earlier (Fig 14.27 left) the narrowband 21MHz choke from Fig 14.25 is transformed into a broadband choke covering 1.8 - 3.8MHz.

**Mid Bands:** To cover 5/7/10MHz, reduce the coil diameter and the number of turns but still use three cores (Fig 14.27 top right).

**High Bands:** For the 14 - 30MHz coverage, two of the same cores are superglued together side-by-side as shown in Fig 14.27 lower right. Three turns will make quite a respectable choke for a 20 - 10m beam. The impedance isn't quite as high as the lower-frequency chokes at their very best, but it is substantially resistive across the whole 14 - 30MHz range. In terms of 'value for ferrite' this two-core choke will at least equal a straight string of 40 to 50 ferrite beads.

Higher impedance or a wider bandwidth can be achieved by cascading any of these chokes in series along the cable.

While the GM3SEK recommended ferrite baluns produce high quality baluns of known characteristics there is nothing to be lost by trying balun construction with what you might already have. For example a choke constructed using eight turns of RG58 on a ferrite ring, as shown in Fig 14.24(c), has an impedance greater than 2000 ohms from 6 to 35MHz.

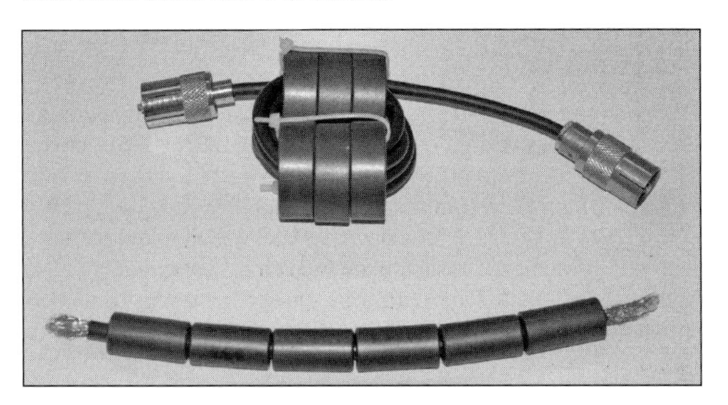

**Fig 14.28: Top, low band balun constructed from six 35mm OD ferrite rings and five turns of RG58: below, experimental W2DU balun using ferrite tubes**

A low band choke using six smaller ferrite rings, shown in **Fig 14.28**, has an impedance of greater than 2000Ω over a frequency range of 1.5 to 8MHz and greater than 500Ω up to 31MHz.

The level of common mode current on a coax feeder, before and after the insertion of a choke, can be checked using the clip-on RF current meter described in the chapter on Measurements.

## VHF and UHF Baluns

At VHF and UHF there are generally fewer difficulties with imbalance created by the antenna's surroundings. The problem is usually the difficulty of making a symmetrical junction at the feedpoint, because the lengths of connecting wires and the necessary gap at the centre of a dipole become significant fractions of the wavelength.

For all its popularity the gamma match is not a balun. It does nothing to create balance between the two sides of a dipole. On the contrary, it relies entirely on electromagnetic coupling between the opposite sides of the dipole to correct the imbalance of the gamma match itself. When used with an all-metal Yagi, the direct connection of the coax shield to the centre of the dipole invites the resulting imbalance currents to travel along the boom as well as the outer surface of the feedline.

The coaxial half-wave balun is definitely the 'best buy' for all VHF/UHF bands up to at least 432MHz, see Fig 14.24(c). It strongly enforces balance, yet it does not introduce an impedance mismatch unless the cable or its length is markedly different from a true electrical half-wavelength. The problem with using this balun is that the feedpoint impedance of the antenna must be transformed up to 200Ω. Fortunately, this is often very simple. For example, the highly successful family of DL6WU long Yagis [9] have a feedpoint impedance which is close to 50Ω at the centre of the dipole driven element; this impedance can be raised to the necessary 200Ω simply by converting the driven element into a folded dipole. Other alternatives for creating a symmetrical 200Ω feedpoint impedance include the T match and the delta match.

The impedance of the coaxial cable used in a half-wave balun is not important, though characteristic impedance of one-half the load impedance (ie in most cases 100Ω) has been shown to give optimum broadband balance. Low-loss 100Ω coax is difficult to obtain, though, and it is perfectly adequate to use good-quality 50Ω cable carefully cut to length with an allowance for the velocity factor.

# MATCHING THE ANTENNA TO THE TRANSMISSION LINE

Wire HF antennas are often used with an Antenna Tuning Unit, described in detail in the chapter on Practical HF Antennas. Another method is to use a matching arrangement at the antenna, particularly with beam antennas. Some of the more popular matching arrangements are described below.

## The Direct Connection

The halfwave dipole has a theoretical centre feedpoint impedance of 73Ω at resonance. In practice this value is less, particularly at HF, because of the presence of ground. Generally the centre of a dipole can be connected directly to 50Ω coax cable as shown in the Practical HF Antennas chapter, and will almost always provide a good match. A current balun, described earlier, may be necessary at higher transmitter powers.

## The Folded Dipole

A halfwave antenna that is used as the driven element in a parasitic array such as a Yagi will normally have a feedpoint

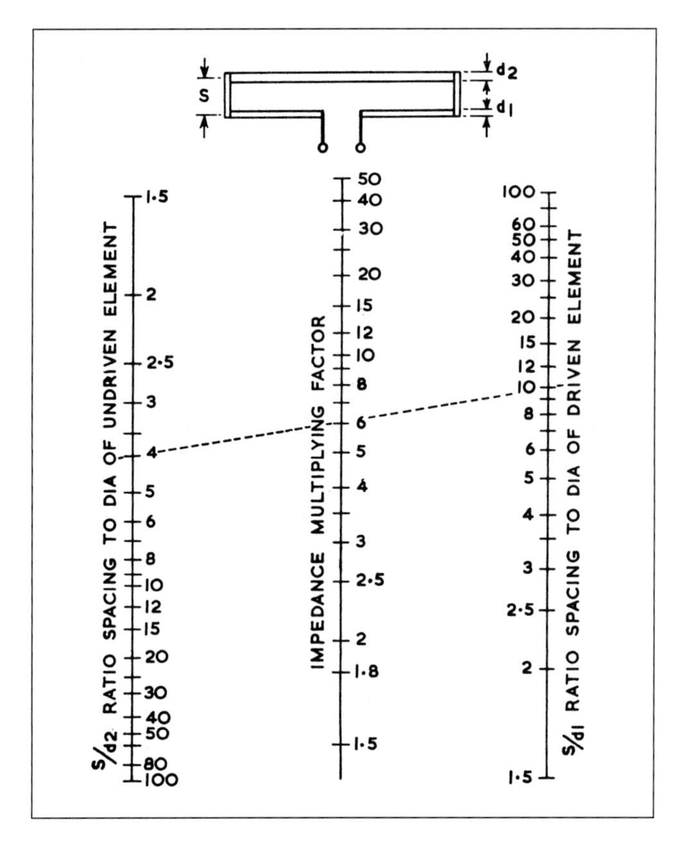


**Fig 14.29: A nomogram for folded dipole impedance ratio calculations. A ruler laid across the scales will give pairs of spacing / diameter ratio for any required multiplier. In the example shown the driven element diameter is one-tenth of the spacing and the other element diameter is one-quarter of the spacing, resulting in a setup of 6:1. This shows an unlimited number of solutions for a given ratio**

impedance much lower that 50Ω. This is due to the coupling between the driven element and the parasitic elements. In this case some impedance transformation is required.

A transformer can be used to step the antenna impedance up to the correct value but this can have the effect of reducing the bandwidth. It has been found that by folding the antenna a 4:1 impedance step-up can often be accomplished with an increase in impedance bandwidth.

Other ratios of transformation than four can be obtained by using different conductor diameters for the elements of the radiator. When this is done, the spacing between the conductors is important and can be varied to alter the transformation ratio. The relative size and spacing can be determined with the aid of the nomogram in **Fig 14.29**.

## The Gamma Match

The Gamma match is an unbalanced feed system suitable for matching coax transmission line to the driven element of a beam. Because it is well suited to plumber's delight construction, where all the metal parts are electrically and mechanically connected to the boom, it has become quite popular for amateur arrays.

A short length of conductor (often known as the gamma rod) is used to connect the centre of the coax to the correct impedance point on the antenna element. The reactance of the matching section can be cancelled either by shortening the antenna element appropriately or by using the resonant antenna element length and installing a series capacitor C, as shown in **Fig 14.30**.

Because of the many variable factors - driven-element length, gamma rod length, rod diameter, spacing between rod and driven element, and value of series capacitors - a number of combinations will provide the desired match. The task of finding a proper combination can be sometimes be tedious because the settings are interrelated.

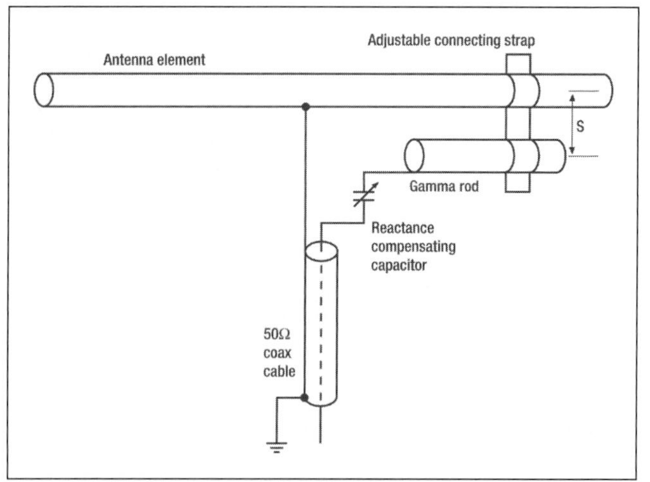


**Fig 14.30: Diagram of a Gamma match. Matching is achieved by altering the position of the gamma rod adjustable connecting strap point on the antenna element. The series capacitor C also has to be adjusted to cancel the inductance of the gamma rod. See text for approximate dimensions**

For matching a multi-element array made of aluminum tubing to 50Ω line, the length of the gamma rod should be 0.04 to 0.05 wavelengths long and its diameter 1/3 to 1/2 that of the driven element. The centre-to-centre gamma rod / driven element (S in Fig 14.30) is approximately 0.007 wavelengths. The capacitance value should be approximately 7pF per metre of wavelength at the operating frequency. This translates to about 140pF for 20 metre operation.

The exact gamma dimensions and value for the capacitor will depend on the radiation resistance of the driven element, and whether or not it is resonant. The starting-point dimensions quoted are for an array having a feed-point impedance of about 25Ω, with the driven element shortened approximately 3% from resonance.

### Adjustment

After installation of the antenna, the proper constants for the gamma generally must be determined experimentally. The use of the variable series capacitor, as shown in Fig 14.30, is recommended for ease of adjustment.

With a trial position of the tap or taps on the antenna, measure the SWR on the transmission line and adjust C1 for minimum SWR. If it is not close to 1:1, try another tap position and repeat.

### Construction

The gamma rod is made from thin aluminium tube whose diameter recommended in most publications is 1/3 to 1/6th of the antenna element diameter. However, it is worth trying what is to hand. The Simplified Gamma match by G3LDO, uses hard drawn copper wire as the gamma rod whose connection to the antenna element is achieved using a hose clamp. Note, though, the potential for electrolytic corrosion caused by using dissimilar metals. The traditional method of making a gamma match is to use an air-spaced variable capacitor and enclose it in a weatherproof metal box. Corrosion to the capacitor can still occur because of condensation.

The gamma match shown in **Fig 14.31** uses a fixed capacitor whose value is determined by experiment with a variable capac-

**Fig 14.31: Simplified Gamma match by G3LDO uses hard drawn copper wire as the gamma rod. Connection of the gamma rod to the antenna element is achieved using a hose clamp. A Philips capacitor is used as a reactance correction capacitor, which can be replaced with a fixed mica capacitor of the correct value (see text) once the adjustments are complete.**

itor. The value of the variable capacitor is then measured and a fixed capacitor (or several series/parallel combinations) substituted. This arrangement will handle 100W without breakdown and only requires a smear of grease to achieve weatherproofing.

## The Omega Match

The Omega match is a slightly modified form of the gamma match. In addition to the series capacitor, a shunt capacitor is used to aid in cancelling a portion of the inductive reactance introduced by the gamma section. This is shown in **Fig 14.32.** C1 is the usual series capacitor.

The addition of C2 makes it possible to use a shorter gamma rod, and makes it easier to obtain the desired match when the driven element is resonant. During adjustment, C2 will serve primarily to determine the resistive component of the load as seen by the coax line, and C1 serves to cancel any reactance.

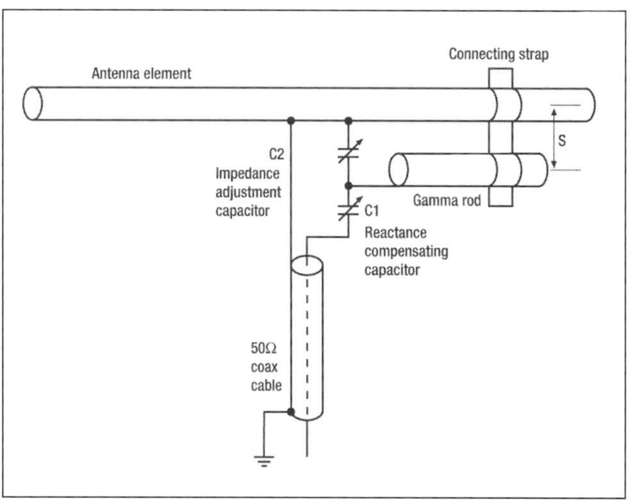


**Fig 14.32: Diagram of an Omega match. Matching is achieved by adjusting the parallel capacitor C2 and the series capacitor C1**

Fixed capacitors can be used to replace the variable ones once the matching procedure is complete. In general the dimensions are the same as for the Gamma match but the gamma rod can be shortened up to 50%.

The maximum value of C2 is approximately 1.4pF per metre of the operating frequency.

## REFERENCES

[1] 'Smith Radio Transmission-Line Calculator'. Phillip H Smith, *Electronics*, Jan 1939
[2] *The ARRL Antenna Book*, 20th edition, ARRL
[3] *The Antenna Experimenter's Guide*, 2nd Edition, Peter Dodd, G3LDO
[4] 'Fitting Coaxial Connections', Roger Blackwell, G4PMK, *RadCom* May 1988
[5] 'In Practice', Ian White, G3SEK, *Radcom*, Jun 1998
[6] 'Balanced to Unbalanced transformers', Ian White, G3SEK, *RadCom*, Dec 1989
[7] 'In Practice', Ian White, GM3SEK, *Radcom* May 2010.
[8] Farnell UK: http://uk.farnell.com/
[9] 'High performance long Yagis', Ian White, G3SEK, *RadCom*, Apr 1987

**The following three pages contain Smith charts that may be copied and enlarged to make a Smith chart calculator as described earlier in this chapter.**

***Mike Parkin BSc(Hons), G0JMI,*** *obtained his first amateur radio licence in 1977 gaining his current licence in 1988. His working background has included telecommunications, radio engineering, mobile-communications, electronics, lecturing and consultancy. Mike has constructed much of his radio equipment covering the bands from 472kHz through to 47GHz. Mike's current business interests are the design and production of antennas covering the range from HF to UHF. Mike is the author of the Antennas Column in RadCom and has had various articles published in other journals. He is a Member of the Institute of Engineering & Technology, a Member of the City & Guilds Institute and a Chartered Engineer.*

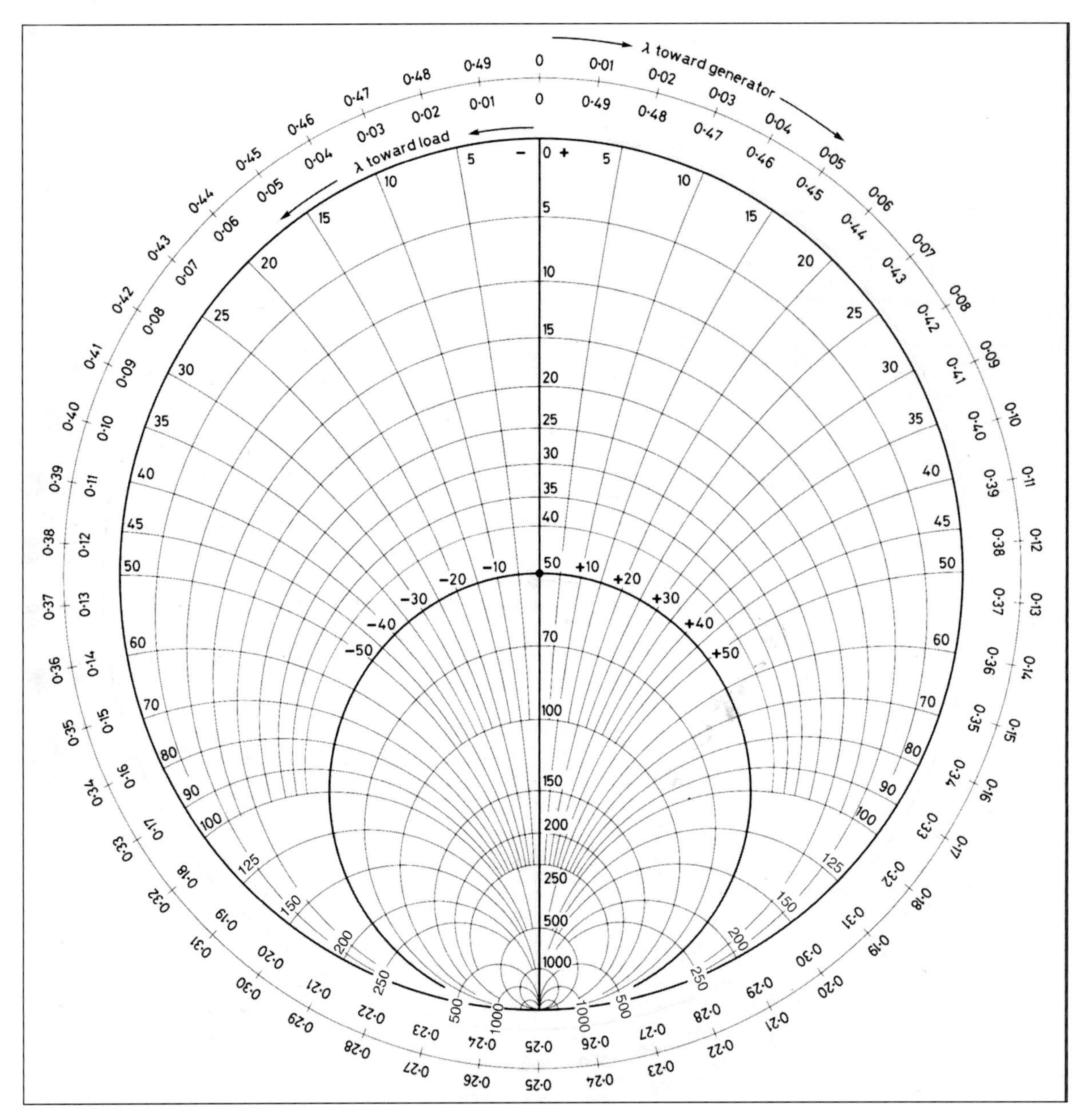


**Fig.14.31: Smith chart for constructing a 50-ohm impedance calculator**

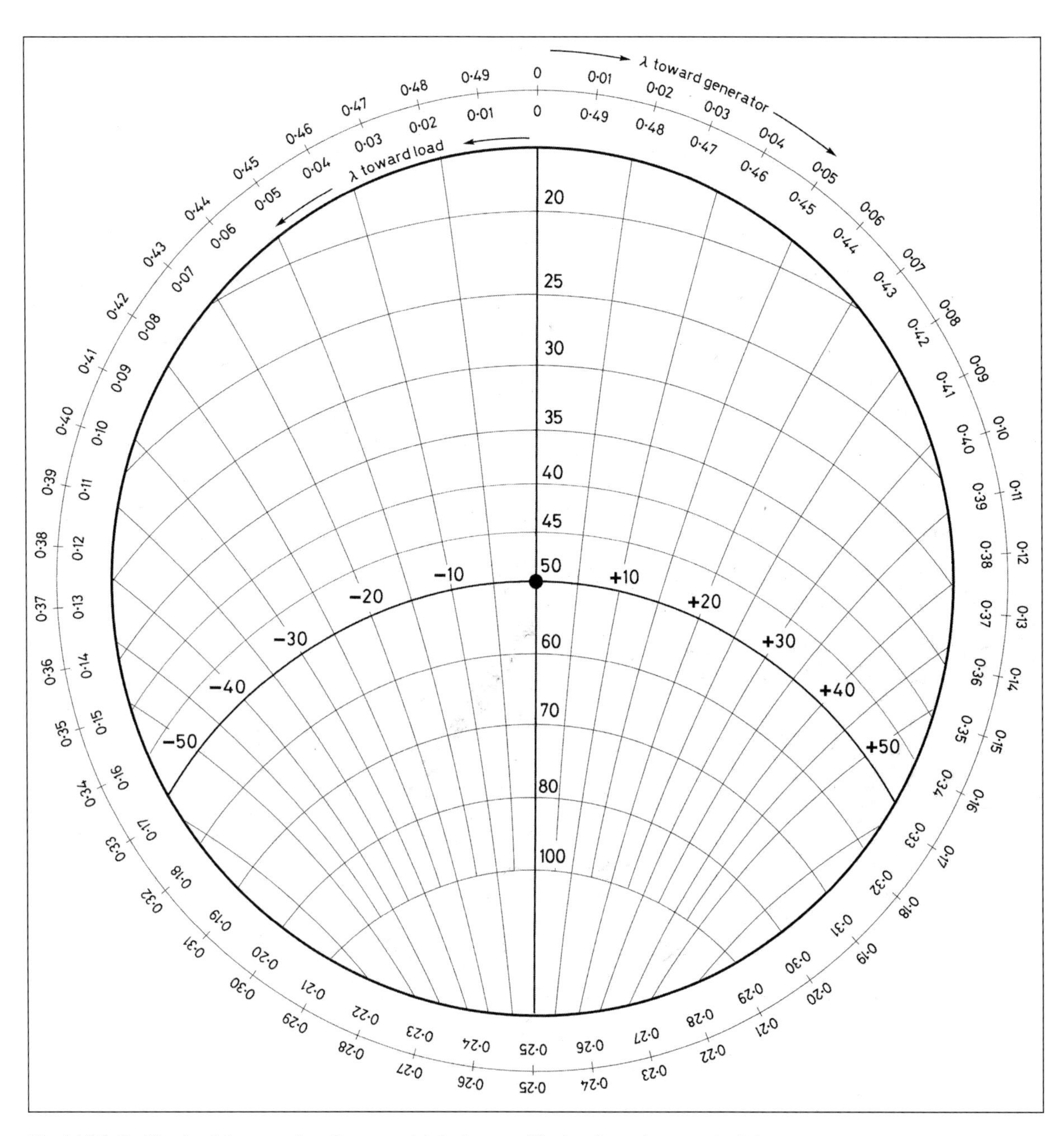


Fig.14.32: Smith chart for constructing a restricted range 50-ohm impedance calculator

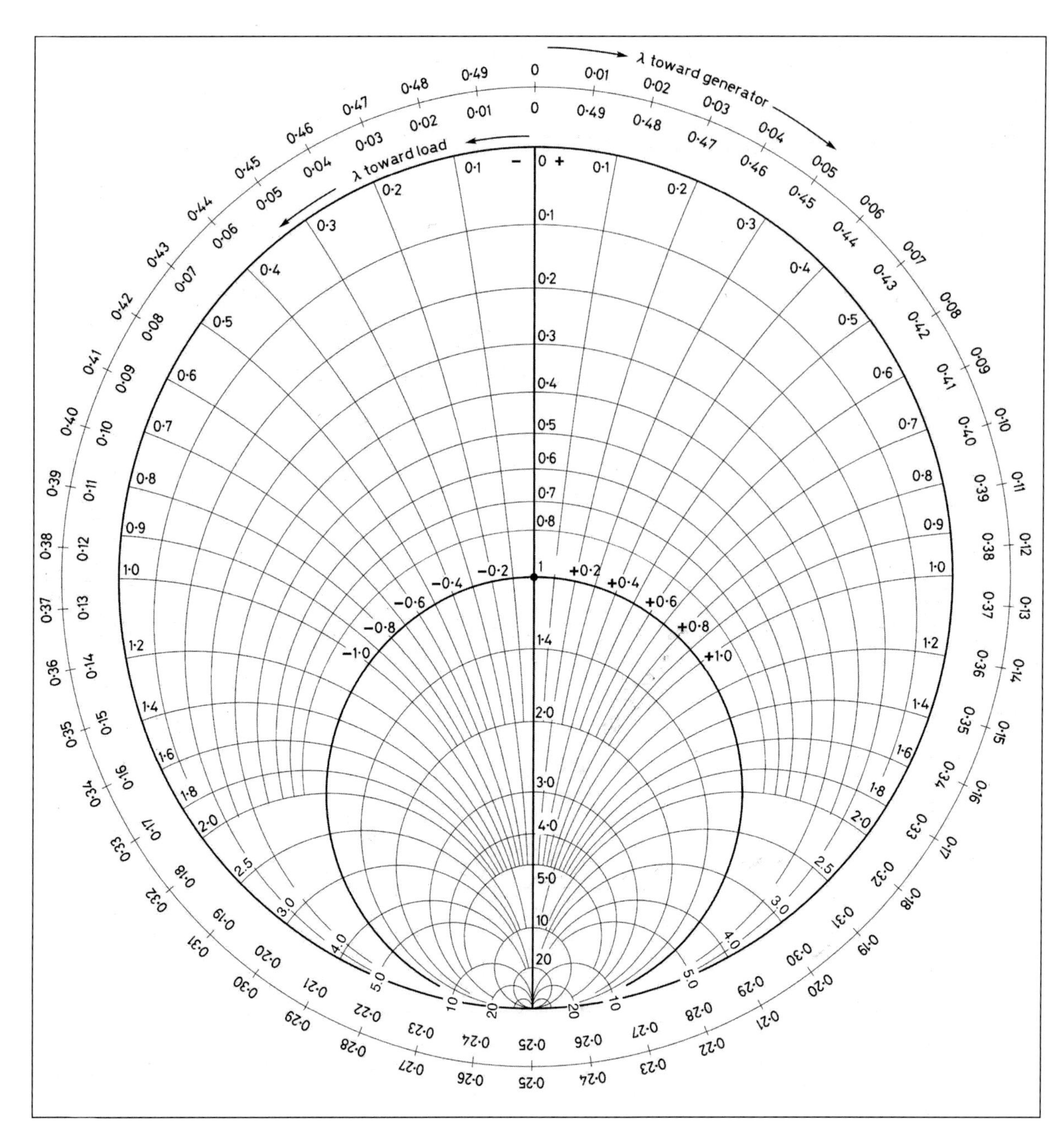


**Fig.14.33: Smith chart for constructing a normalised impedance calculator**

# Practical HF Antennas

# 15

*Mike Parkin, G0JMI*

## INTRODUCTION

A High Frequency (HF) radio station comprises of three fundamental components, these being:

- The HF transceiver and its power supply
- The interconnecting cable between the transceiver and the HF antenna that is referred to as the transmission line or feeder
- The HF antenna.

The HF antenna can be considered as the essential link between free space and the transmitter or receiver. Therefore, the HF antenna's role is two-fold:

- To ensure that as much of the accessible RF power as possible supplied by the transmitter is radiated as an RF signal
- To ensure that as much of the received RF signal as possible is made accessible as RF power for the receiver to function successfully.

Consequently, the HF antenna is one of the primary factors that determines the characteristics and effectiveness of the complete system. Therefore, the design of the antenna within its working environment is vital to the overall success of a radio system.

This chapter addresses the practical construction of HF antennas and Antenna Tuning Units (ATU). The scale and range of HF antennas is extensive and so this chapter is confined to the description and construction of the more commonly used antennas.

## END-FED WIRE ANTENNAS

### The End-fed Antenna

The most straightforward of all HF antennas is a length of wire suspended as high as possible between two anchor points, with one end of the wire connected directly to the radio equipment. An example of an end-fed antenna is shown in **Fig 15.1**. This may seem a rather a minimal antenna, however an end-fed wire can achieve surprisingly good results and can be fairly inconspicuous compared to other types of antenna. When the antenna tends to be longer than a wavelength (λ), it is often referred to as a long wire as well as an end-fed antenna.

With the antenna's wire span secured at each end at height, the equipment is connected to the wire span's nearer end using a short length of wire run to the Antenna Tuning Unit's (ATU) unbalanced termination. The ATU then provides the mechanism to match the antenna to the transceiver at the operating frequency. **Fig 15.1** shows an example of this arrangement and this form of inverted L end-fed antenna is often referred to as a Marconi antenna.

However, one of the undesirable aspects of the end-fed wire is that it can present a wide range of impedances at its feed point and many ATUs have difficulty accommodating such a range. The choice of end-fed wire's length may alleviate some of the matching problems and this is discussed in detail later within this chapter. In addition, with the wire being directly connected to the ATU, a significant amount of the transmitted RF power can be radiated in the shack. This could result in various undesirable effects including RF burns and interference to other equipment for example.

### Practical HF End-fed Antennas

The length of the wire span determines the lowest band that is practical to use. An eighth wavelength (λ/8) span of wire is about the minimum practical length usable and tends to present a low impedance at the wire's end. The same length of wire will also work as quarter wavelength (λ/4) antenna and similarly tends to present a low impedance at its end. However, when the wire is used as a half wavelength (λ/2) antenna the impedance presented is very high and can result in problems when trying to match the antenna to the transceiver using a conventional ATU. Similarly, when the same wire is used as a wavelength (λ) long antenna, then the impedance presented becomes high. When the wire becomes close to two wavelengths long (2λ), the impedance presented tends to be low at its end.

Guidance for selecting an optimum length for an end-fed wire antenna was described by Alan Chester, G3CCB, based on matching an end-fed wire using a conventional ATU that uses variable inductance/capacitance to tune out the mismatch [1]. **Fig 15.2** illustrates the wire lengths covering the nine HF bands starting with 160m. The heavy lines running horizontally indicate areas where the impedance presented at the wire's end might exceed the matching capabilities of many ATUs. To assess if a particular length of wire is suitable, a vertical line is dropped down the diagram. Where the vertical line encounters a heavy line, then this wire

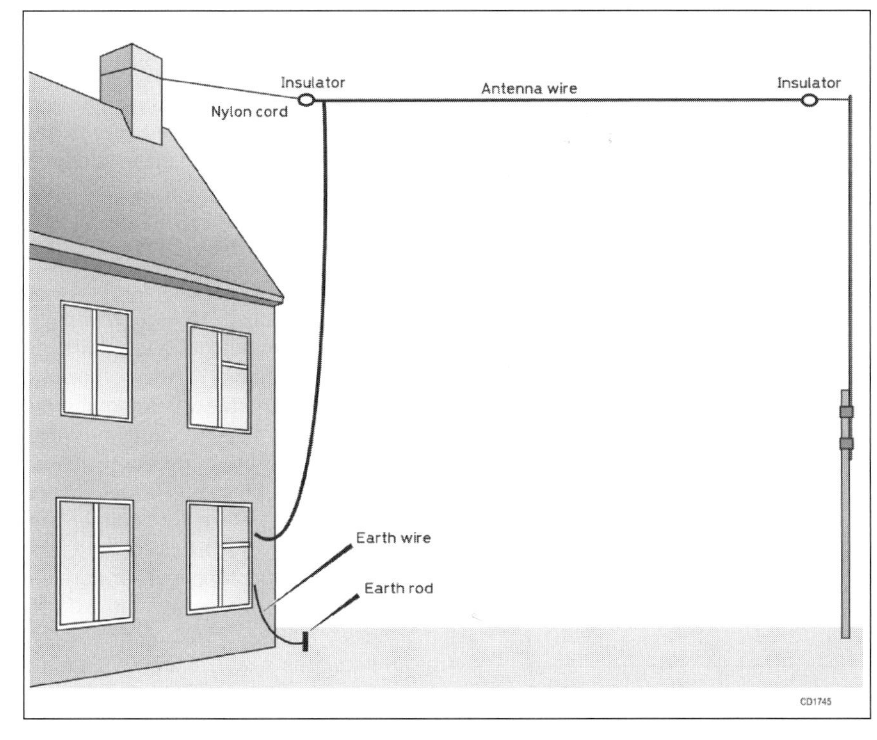


**Fig 15.1: The end-fed antenna, the most straightforward of all multi-band antennas**

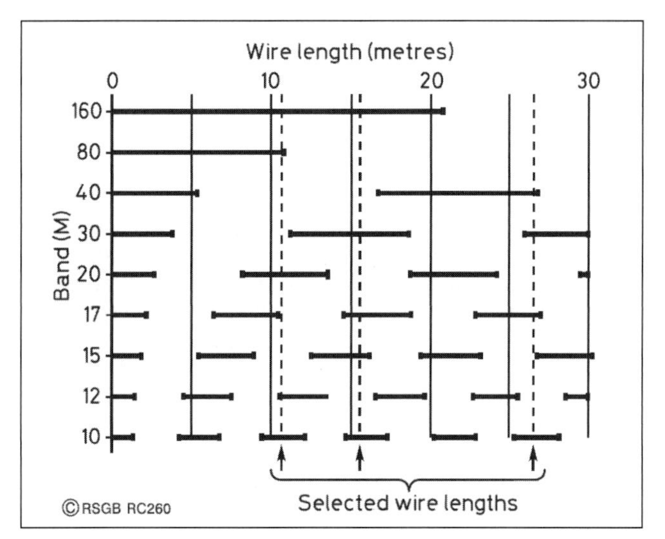


**Fig 15.2: Aerial wire lengths, showing 'no-go' lengths for various bands**

length could be expected to be difficult to match using a conventional ATU. Within **Fig 15.2**, three broken vertical lines have been included to indicate practical lengths that cover more than one band [2]. Using the broken vertical lines in **Fig 15.2** as examples:

- The vertical broken line for 10.5m passes through the 160m 'no-go' line. Then it just passes through the end of the 80m 'no-go' line. The line then goes through the middle of the 40m 'safe sector', passing through the 30m gap. At 20m the broken vertical line is blocked, however there are then openings at 17m, 15m and 12m.
- The next possibility is with a wire length of 15.5m. The line starts off by passing through the 160m 'no-go' line. However, there are then openings at 80m, 40m and 20m. If some tolerance is permitted, then at 17m and 15m are feasible. The line then passes through the clearance at 12m.
- Using the broken line for a 26.5m wire length, this allows potentially eight bands to be covered from 160m excluding 10m where difficulties in matching could be encountered using a conventional ATU.

### Remote End-fed Antenna

Situating the antenna's feed point remotely from the house/ building avoids and minimises the undesirable effects associated from RF radiation in the shack. Remotely locating the end-fed antenna's feed point also has the benefit that Electromagnetic Compatibility (EMC) electrical noise problems can be minimised on both transmit and receive. In addition, the unpredictable effect on the antenna caused by possible conduit, wiring and water pipe resonances associated with the house can be reduced. The concept of this arrangement is shown as **Fig 15.3**.

A disadvantage of this arrangement is that the ATU is located at a distance from the transceiver. This can be inconvenient when it comes to making adjustments to the ATU, for example when tuning up between bands. An ATU that can be remotely controlled provides a convenient method to overcome this problem of matching the antenna at a distance from the 'shack'. An example of a remotely controlled ATU used to tune an end-fed antenna is described later in this chapter (see Matching and Tuning).

### Transformer End-fed Antenna Matching

A technique to resolve situations where the impedance presented by an end-fed wire exceeds the ATU's capabilities is to transform the impedance to bring it into the range of most ATUs. A method to do this is to use an unbalanced radio frequency auto-transformer connected between the antenna and the ATU. This arrangement has the advantage that a length of coaxial cable can be run between the ATU and the transformer allowing the antenna's feed point to be remotely located outside, away from the shack. Assuming the antenna is situated at a suitable distance, then this arrangement has the advantage of minimising RF power from being radiated within the shack, so avoiding any undesirable effects that this may cause. However, it may be necessary to earth the shield of the coaxial cable to minimise any common mode currents. The concept of this arrangement is shown in **Fig 15.4**.

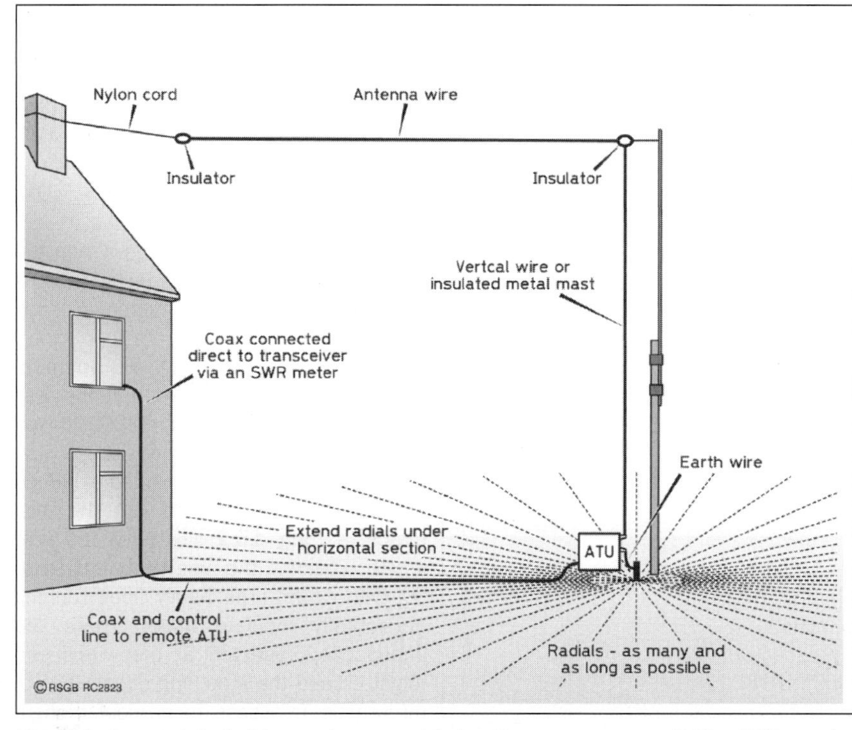


**Fig 15.3: A remotely fed long wire or end-fed antenna arrangement. The ATU can be either preset or automatic and may require a control cable in addition to the coaxial cable feeder**

## The Un-Un transformer

Transformers using iron powder toroid cores can be constructed to cover frequency ranges from about 2 to 40MHz. To enable the transformer to work efficiently, it is important that the wires forming the windings are laid side-by-side to maximise the magnetic field coupling between them. There will always be some leakage inductance associated with a transformer and this increases in proportion to the transformer's self-inductance. Therefore, a transformer that works well at 3.5MHz may not work as well at 28.5MHz and this often shows as a worsening SWR as the frequency is increased. The operation of toroid core transformers is discussed in further detail later within this chapter.

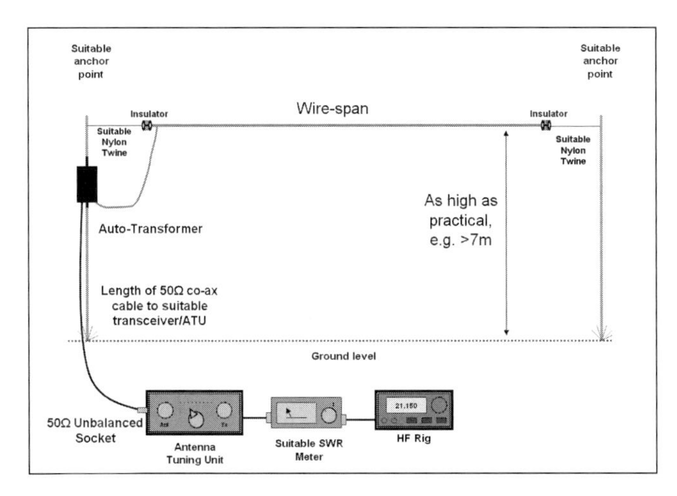


**Fig 15.4: Concept of an end-fed antenna arrangement using an auto-transformer to match the antenna to the feeder**

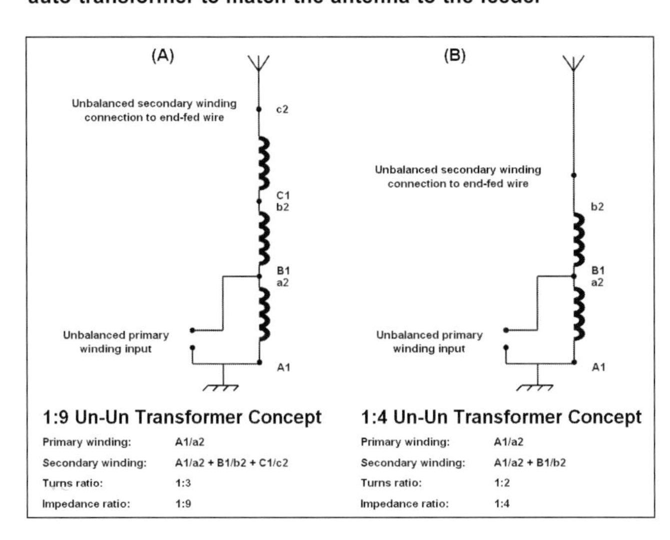


**Fig 15.5: Concept of 1:9 (A) and 1:4 (B) Unbalanced-Unbalanced (Un-Un) transformers**

The concept of a 1:9 and a 1:4 impedance transforming ratio unbalanced-to-unbalanced auto-transformer (or Un-Un) is shown in **Fig 15.5**. If the end-fed wire is thought of as being connected to the auto-transformer's secondary winding, then the ATU's unbalanced 'antenna' socket is connected to the auto-transformer's primary winding. The antenna's impedance presented at the auto-transformer's secondary winding is transformed down to present a lower impedance at the primary winding, enabling it to be brought within the capabilities of most ATUs.

Peter Miles, VK6YSF, and John Parfrey, M0UKD, have published several toroid iron powder core Un-Un 1:9 and 1:4 impedance ratio auto-transformer designs for use with end-fed wires, with details made available online [3] [4]. These designs used both the T130-2 and larger diameter T200-2 Micrometals Iron Powder Toroids obtainable from some radio rally traders, radio component retailers and online suppliers. These cores are typical of those used by radio amateurs for transformers used to match antennas to feeder cables.

A 1:9 impedance ratio auto-transformer can be made by winding 9 trifilar 3-wire turns on to a T130-2 core. A similar 1:9 auto-transformer consisting of 18 trifilar wound 3-wire turns can be made up using a larger T200-2 core. A lower 1:4 impedance ratio transformer can be made using T130-2 and T200-2 cores in a similar manner. The 1:4 T130-2 transformer comprised 9 bifilar wound 2-wire turns while the 1:4 T200-2 transformer

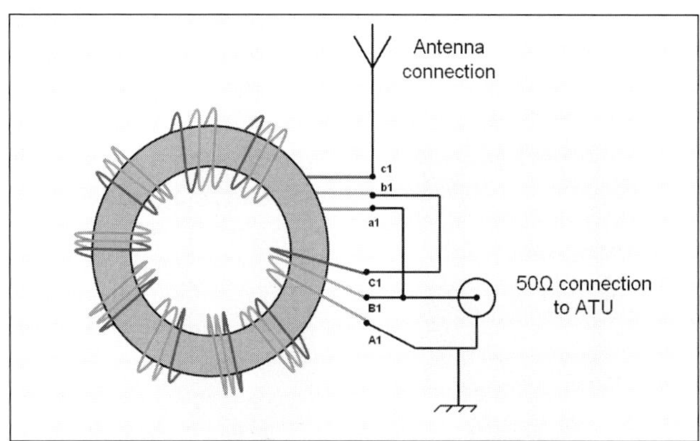


**Fig 15.6: Un-Un 1:9 transformer connection details**

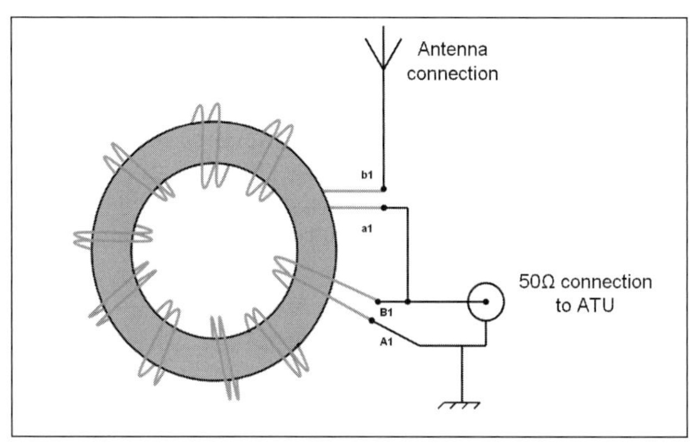


**Fig 15.7: Un-Un 1:4 transformer connection details**

comprised 18 bifilar wound 2-wire turns. **Figures 15.6** and **15.7** summarise the connection arrangements for these transformers for both types of toroid core used.

The performance of both 1:4 and 1:9 transformers using both types of core was assessed by G0JMI and the results are shown in **Tables 15.1** and **15.2** [5].

The assessments were made with approximately 10m of RG58 50Ω coaxial cable connected to the unbalanced primary winding to simulate the situation where the transformer is used remotely from the transmitter. To gain an indication of the performance of both auto-transformer types referred to 50Ω, the SWR was measured with each secondary winding terminated in a suitable dummy load (ie 450Ω for the 1:9

| | Transformer wound as 1:9 impedance ratio | |
|---|---|---|
| | Toroidal Core Type | |
| | T200-2 | T130-2 |
| MHz | SWR | SWR |
| 1.90 | 1.5:1 | 4:1 |
| 3.65 | 1.3:1 | 3:1 |
| 5.28 | 1.2:1 | 2.5:1 |
| 7.10 | 1:1 | 1.2:1 |
| 10.12 | 1:1 | 1.05:1 |
| 14.15 | 1.05:1 | 1.15:1 |
| 18.12 | 1.05:1 | 1.05:1 |
| 21.20 | 1.4:1 | 1.2:1 |
| 24.95 | 1.8:1 | 1.2:1 |
| 28.50 | 1.5:1 | 1:1 |
| 50.20 | 4:1 | 1.1:1 |

**Table 15.1: Comparison of the measured SWR for T200-2 and T130-2 toroidal cores tested for 1:9 transformer turns configuration**

| | Transformer wound as 1:4 impedance ratio | |
|---|---|---|
| | Toroidal Core Type | |
| | T200-2 | T130-2 |
| MHz | SWR | SWR |
| 1.90 | 1.5:1 | 4:1 |
| 3.65 | 1.3:1 | 3:1 |
| 5.28 | 1.2:1 | 2.5:1 |
| 7.10 | 1:1 | 1.2:1 |
| 10.12 | 1:1 | 1.05:1 |
| 14.15 | 1.05:1 | 1.2:1 |
| 18.12 | 1.1:1 | 1.05:1 |
| 21.20 | 1.4:1 | 1.2:1 |
| 24.95 | 2.2:1 | 1.2:1 |
| 28.50 | 1.8:1 | 1:1 |
| 50.20 | 5:1 | 1.1:1 |

**Table 2: Comparison of the measured SWR for T200-2 and T130-2 toroidal cores tested for 1:4 transformer turns configuration**

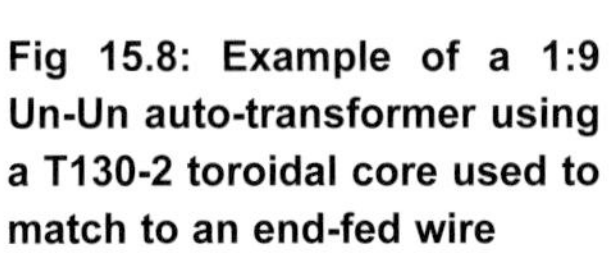

**Fig 15.8: Example of a 1:9 Un-Un auto-transformer using a T130-2 toroidal core used to match to an end-fed wire**

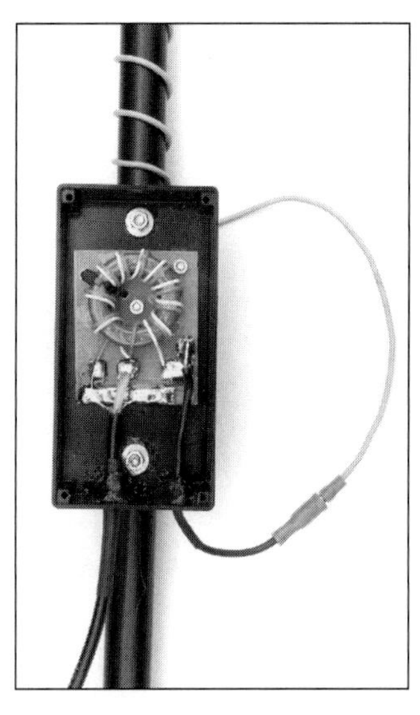

**Fig 15.9: Example of a 1:4 Un-Un auto-transformer using a T130-2 toroidal core used to match to a vertical helical 20m antenna**

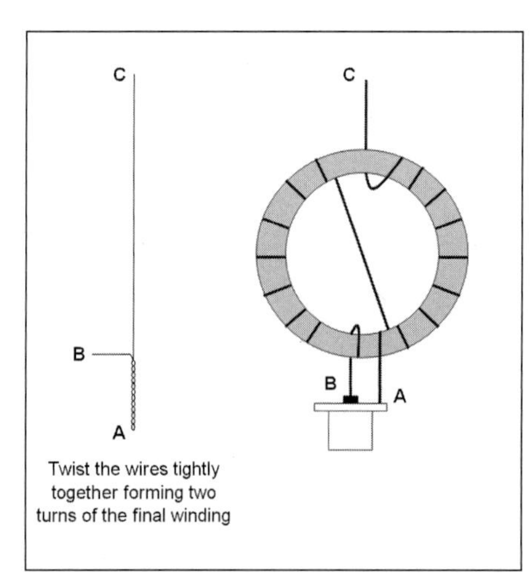


**Fig 15.10: Un-Un construction details. A capacitor of about 150pf is connected between A and B to improve the HF performance of the auto-transformer. The core comprised of two FT140-43 toroids that were stacked to improve the RF power handling capability**

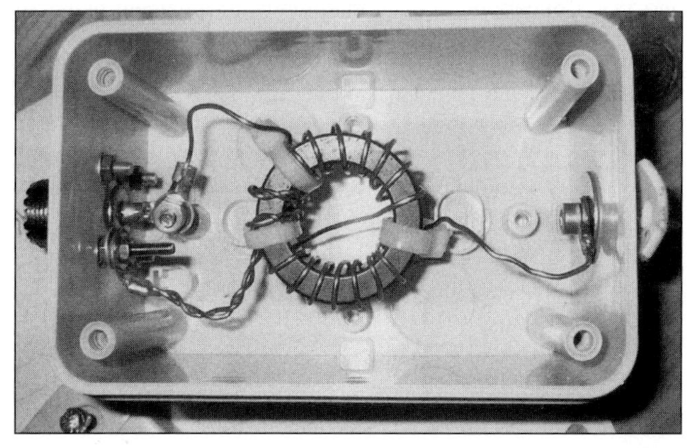

**Fig 15.11: The UN-UN transformer mounted in a suitable ABS box. The capacitors connected across the primary winding were added later**

and 200Ω for the 1:4 transformer). The two types of transformers assessed used plastic insulated 0.7mm diameter copper single core wire for the windings.

The results indicated that when assessed:

- The more inductive T200-2 1:9 and 1:4 transformer configurations had a tendency to be more effective between 1.9 and 21.2MHz compared to the T130-2 transformers.
- However, the lesser inductive T130-2 1:9 and 1:4 transformer configurations tended to be more effective from 7.0 to 50MHz compared to the T200-2 transformers.

## Examples of Un-Un Auto-transformers used to Match Antennas

**Fig 15.8** illustrates an example of a 1:9 impedance ratio auto-transformer using a Micrometals Iron Powder Toroid T130-2 core that was used to match an end-fed wire covering the 40m to 10m bands. This auto-transformer comprised 9 trifilar wound 3-wire turns with a small gap left between each turn as shown. This antenna had a wire span of about 10m and this was mounted at a height of around 7m. The concept of this antenna was previously illustrated as **Fig 15.4**.

**Fig 15.9** illustrates an example of a 1:4 impedance ratio auto-transformer similarly using a Micrometals Iron Powder Toroid T130-2 core that was used to match a helical vertical antenna tuned to the 20m band. This transformer comprised 9 bifilar wound 2-wire turns with a small gap left between each turn as shown. Helical vertical antennas are described in more detail later in this chapter.

Within **Fig 15.8** and **Fig 15.9** it can be seen where the coaxial cable was connected to the primary winding and the connection of the antenna to the secondary winding of the Un-Un auto-transformer. These transformer arrangements allowed RF powers of up to 100w to be handled using an ATU to provide a good match to the transceiver.

## PA1ZP Multi-band End-fed Antenna Covering 80m to 10m

Another example of an Un-Un auto-transformer used to match an end-fed wire antenna covering the 80m to 10m bands was by Jos van Helm, PA1ZP [6]. The following is a summary of the antenna's construction by Jos van Helm:

First the Un-Un auto-transformer was wound using two stacked FT140-43 toroid cores. Two stacked cores were used because this improved the RF power handling capabilities of the auto-transformer. The winding was made up from 6+6+2 turns, totalling 14 turns as illustrated in **Fig 15.10**. 1mm diameter single core enamelled copper wire was used for the windings. The initial 2 turns of the auto-transformer's winding were made up by tightly twisting together a sufficiently long length of the wire for the section between A and B as shown in **Fig 15.10**. The winding was secured in place on the toroid core using cable-ties. The Un-Un auto-transformer was housed in a suitable ABS box, with a SO259 socket mounted at one end and a connection pillar terminal at the other. Solder ring-tabs were attached to the SO259 socket and to the pillar terminal to enable the auto-transformer's primary and secondary windings to be connected. **Fig 15.11** shows the auto-transformer housed in the ABS box

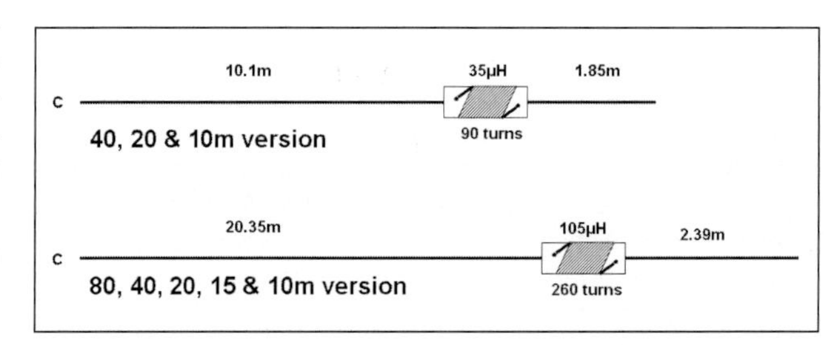


**Fig 15.12: Details of the construction of the two versions of the antenna. Point C is connected to point C of Fig 15.10**

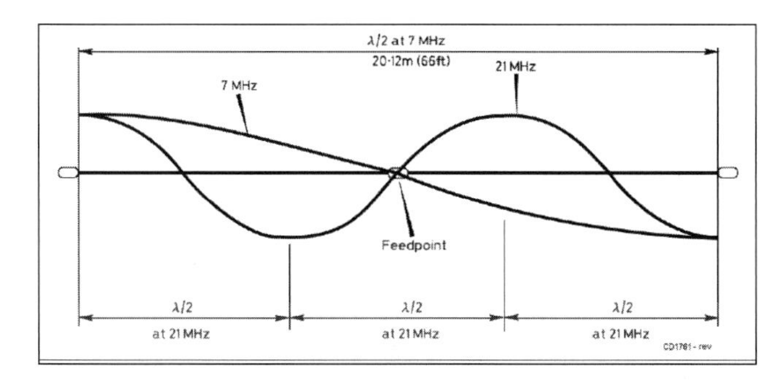


**Fig 15.13: Concept of the voltage distribution on a 7MHz half wavelength dipole. The feeder cable is connected to the antenna at a point where the feed impedance is low (where the voltage is low). The 7MHz dipole will also have a low impedance at the centre on the third harmonic, at 21MHz**

before the lid was attached and screwed down. A 68pf and an 80pf capacitor were connected in parallel across points A and B (across the primary winding) to enable the auto-transformer to be more effective on the higher HF bands [7]. These capacitors were rated at above 6kV to accommodate the RF voltages that could be expected to be handled by the antenna.

Two versions of the antenna were made up covering 80m to 10m (5 band version) and 40m to 10m (3 band version). Both antennas used an approximate λ/4 (quarter wavelength) long wire at the lowest band of operation. This wire was terminated in a loading coil, with a short length of wire extending beyond the loading coil that was used to tune the antenna. The wire used for the antenna was 18AWG insulated copper stranded wire (approximately 18 to 19SWG). The 5 band version of the antenna (80m to 10m) used a 105μH loading coil (260 turns) and the 3 band version (40m to 10m) used a 35μH loading coil (90 turns) as shown in **Fig 15.12**. The former used for each loading coil was a length of 30mm diameter plastic water pipe.

To tune up the 5 band version of the antenna, the antenna was first installed at a height of about 8m. The length of the longer wire section of the antenna was altered until a reasonable match was found on 40m, 20m, 15m and 10m by monitoring the SWR. Then, the length of the antenna's shorter wire section was altered for a resonance point on the 80m band by monitoring the SWR.

A similar tuning process was followed for the 3 band version of the antenna starting with the longer section. The length of the wire was altered until a reasonable match was found on 20m, 15m and 10m. Then the shorter wire section was adjusted to give a reasonable match on 40m.

The wire lengths found for both versions are illustrated in **Fig 15.12** for reference. However, these wire lengths may vary where the ground conditions, the wire used and the height of the antenna are different. This is because these factors can influence the velocity factor for the antenna affecting its length.

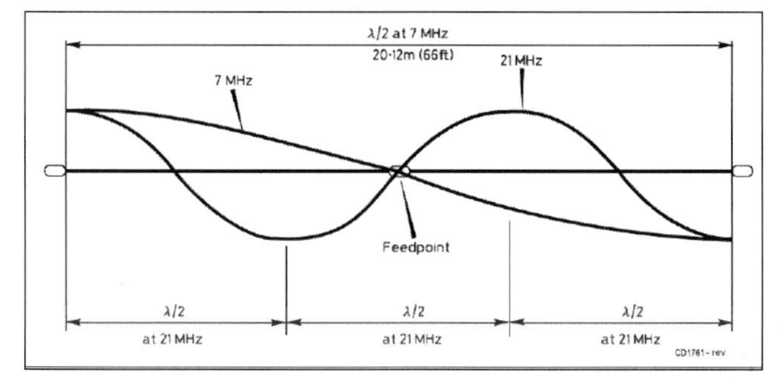


**Fig 15.14: Antenna (a) shows a quick fix installation for a dipole. With minimal effort the dipole height (b) can be raised substantially**

Both versions of the antenna were able to handle CW RF powers of up to 100w. When tuned, the following SWR measurements for the 5 band version of the antenna were:

- 80m: SWR was 1:1 at 3.590MHz, with SWR rising to 1.5:1 at +/- 50kHz either side of this frequency. However, the antenna could be tuned to another frequency within the 80m as desired.
- 40m: SWR was 1:1 at 7.090MHz. The SWR was under 1.5:1 across the 40m band.
- 20m: SWR was 1:1 at 14.150MHz. The SWR was under 1.5:1 across the 20m band.
- 15m: The lowest SWR was 1.21:1 at 21.250MHz, with the SWR rising to 1.7:1 at 21.450MHz and 3:1 at 21.000MHz. Therefore, it is suggested that the antenna should be used with an ATU on this band.
- 10m: The lowest SWR was 1.1:1 at around 28.0MHz, with the SWR rising to 1.5:1 at 28.750MHz. Shortening the longer section of the antenna's wire should allow a wider bandwidth of operation on 10m. However, this will worsen the SWR on the 15m band and change the minimum SWR point on the 40m and 20m bands.

PA1ZP did not give details of any insulators used to support the antennas. However, an installation as previously illustrated in **Fig 15.4** could provide a suitable solution using dog-bone insulators.

### Earth Arrangements and End-Fed Antennas

For an end-fed antenna to operate efficiently a good RF earth is usually required. The resistance of the earth connection is in series with the radiation resistance of the antenna and it is important that the ground resistance is as low as possible to maximise the efficiency of the end-fed antenna. A poor RF earth can result in an undesirably high RF potential developing on metal surfaces, including the cases of radio equipment and any metal shelving used for the equipment. In addition, the microphone, morse key or headset leads are also 'hot' with RF so that RF feedback and Broadcast Interference (BCI) problems could occur. Moreover, the circuitry of modern communications equipment could be electrically damaged under these circumstances. Antenna earthing techniques are described later within this chapter.

## CENTRE FED WIRE ANTENNAS

### The Centre-fed Dipole

The half wavelength (λ/2) dipole antenna is an uncomplicated and reliable antenna whose operation does not depend on an ATU or the connection of an earth to work effectively. The current and voltage distributions in one dipole leg are matched by those in the opposite leg as shown in **Fig 15.13**. This means that the half wavelength dipole is a balanced single band antenna that can be directly fed using a balanced feeder. The dipole has a low feed point impedance enabling it to be readily connected to a 50Ω coaxial feeder cable run to the radio equipment. However, if an unbalanced coaxial feeder cable is directly connected to the feed point, then good practice is to use a balanced-to-unbalanced (balun) transition to avoid the effects that arise from common mode currents (as described in the Transmission Lines chapter and later within this chapter). In most cases, the dipole has an efficiency that exceeds 95% making it an effective antenna that is practical to construct, as described later.

The half wavelength dipole also presents a low feed point

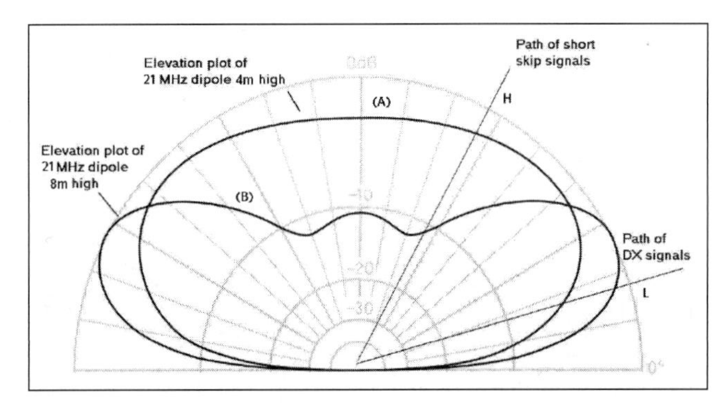


**Fig 15.15: Elevation radiation pattern for a dipole at different heights above ground, see Fig 15.14. (A) is for a dipole around 4m high. (B) Is from the antenna at 8m. Increasing the dipole's height from (A) to (B) indicates a theoretical DX path gain of 3dB and a reduction in short skip signals of up to 8dB**

impedance on its third harmonic. This characteristic means the antenna can be used as a third-harmonic antenna. For example, a 7MHz dipole will also work on the 15m band because it is close to resonance on 21MHz. However, an ATU may be required to tune the antenna in most cases to obtain a good match.

A practical dipole antenna for the higher HF bands is shown in **Fig 15.14**. This antenna could be installed by suspending it from the eaves of the house to a suitable pole situated at an appropriate distance from the house (eg an existing clothes line post). This arrangement gives the antenna an average height of around 4m (15ft) above the ground as illustrated in **Fig 15.14**(A). As can be seen by the elevation radiation pattern illustrated in **Fig 15.15**(A), the antenna is most effective for signals having a high angle of radiation tending to make it suitable for short skip contacts. However, raising this antenna to around a height of 8m (23ft), as shown in **Fig 15.14**(B), gives the antenna a lower angle of radiation as shown in **Fig 15.15**(B). This lower radiation angle tends to make the antenna more suitable for working DX contacts. Generally, these comments regarding the antenna's height and radiation characteristics apply to any horizontal antenna and not just to the HF band dipole summarised above.

There a number of methods to connect the coaxial cable to the antenna and two examples are shown in **Fig 15.16**. Both these arrangements take the strain of the coaxial cable off the connections. Also, having the ends of the cable facing downwards assists in preventing water from entering the coaxial cable. However, it is important to protect the junction against the entry of water by using either self-amalgamating tape or a non-corrosive sealant. As outlined previously, connecting an unbalanced coaxial cable to a balanced antenna can result in undesirable common mode currents flowing on the coaxial cable screen's outer surface. Therefore, some form of balun is required to minimise this effect.

The dipole can be supported using 2mm or 3mm diameter nylon rope with 'dog-bone' insulators at the ends of the dipole's legs. When measuring up the wire for each dipole leg, an extra length of about 150mm (6 inches) should be allowed at each end to give an allowance for connecting the dipole leg to the centre insulator and the end insulator. If each wire end is threaded through the insulator to form a loop, then this allows the length of each dipole leg to be adjusted for minimum SWR by folding the wire back on itself. However, it is important that the dipole is kept symmetrical during this procedure. The wire ends can be temporally held in position using tape and once the optimum length has been determined, cable ties can be used to secure each wire end in place. This method of connecting the dipole's legs to the insulators is shown in **Fig 15.16**(a). It is recommended not to use egg insulators and wire as the dipole's leg end supports because this can increase the end capacitance causing the antenna's performance to be unpredictable.

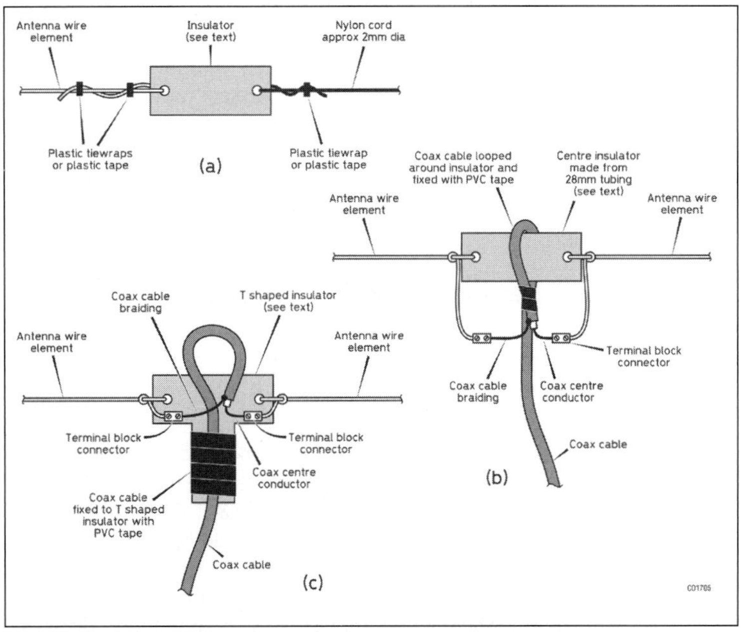

**Fig 15.16: (a) A convenient arrangement for constructing a dipole so that the element lengths can be adjusted to make the element longer than shown in Table 3 to allow for tuning. The excess is taped back along the element. (b) Method of connecting coaxial cable to the centre of the dipole using a short length of tubing or a dog-bone insulator. (c) Method of connecting coaxial cable to the centre of a dipole using a specially constructed T insulator. A sealant should be used to prevent water ingress at the exposed coaxial cable's end**

The dipole is described as a half wavelength ($\lambda/2$) antenna. However, in practice the dipole's length is slightly shorter than a half wavelength because of the influence of 'end-effect' and the wave's velocity factor as it travels along the wire (as described in the Antenna Principles chapter). How to calculate the length of a $\lambda/2$ dipole wire antenna was described in the Antenna Principles chapter (Equations 2a and 2b) using:

L (m) = 143 / f (MHz)

or

L (ft) = 468 / f (MHz)

Where f is the frequency in MHz.

This equation gives a close enough approximation on the higher frequency bands but may be a bit short for the lower bands. For example, in metres, the formula gives a dipole length of 40.63m for 3.52MHz while using MMANA-GAL a length of 40.96m is predicted for the same frequency (with the antenna modelled at a height of 10m using 2mm diameter copper wire. See the Antenna Principles chapter for further details of the MMANA-GAL application).

It should be noted that this is the total dipole length and the wire has to be cut in half at the centre, giving two legs of $\lambda/4$ each, where the feeder cable is connected. The gap between the two legs also forms part of the whole dipole length. Using the formula above for a dipole length at 7.10MHz gives 20.14m and compar-

| Frequency | Free Space | | Derived from 143/f MHz | | Predicted Lengths | |
|---|---|---|---|---|---|---|
| | λ | λ/2 | λ/2 | λ/4 | λ/4 2mm diameter copper wire | λ/4 25mm diameter copper tube |
| MHz | Metres | Metres | Metres | Metres | Metres | Metres |
| 1.83 | 163.93 | 81.97 | 78.14 | 39.07 | 40.00 | - |
| 1.90 | 157.89 | 78.95 | 75.26 | 37.63 | 38.50 | - |
| 3.52 | 85.23 | 42.62 | 40.63 | 20.32 | 20.48 | - |
| 3.65 | 82.19 | 41.10 | 39.18 | 19.59 | 19.74 | - |
| 7.10 | 42.25 | 21.13 | 20.14 | 10.07 | 10.08 | - |
| 10.12 | 29.64 | 14.82 | 14.13 | 7.07 | 7.19 | 7.09 |
| 14.05 | 21.35 | 10.68 | 10.18 | 5.09 | 5.24 | 5.20 |
| 14.20 | 21.13 | 10.57 | 10.07 | 5.04 | 5.19 | 5.14 |
| 18.12 | 16.56 | 8.28 | 7.89 | 3.95 | 4.01 | 3.94 |
| 21.05 | 14.25 | 7.13 | 6.79 | 3.36 | 3.42 | 3.35 |
| 21.20 | 14.15 | 7.08 | 6.74 | 3.40 | 3.40 | 3.32 |
| 24.95 | 12.02 | 6.01 | 5.73 | 2.87 | 2.90 | 2.85 |
| 28.05 | 10.70 | 5.35 | 5.10 | 2.55 | 2.60 | 2.56 |
| 28.50 | 10.53 | 5.27 | 5.02 | 2.51 | 2.56 | 2.52 |
| 29.50 | 10.17 | 5.09 | 4.85 | 2.43 | 2.48 | 2.44 |

**Table 3: Wavelengths and half wavelengths in free space together with the predicted resonant quarter wavelength length for each leg of a dipole relative to frequency for the HF amateur bands**

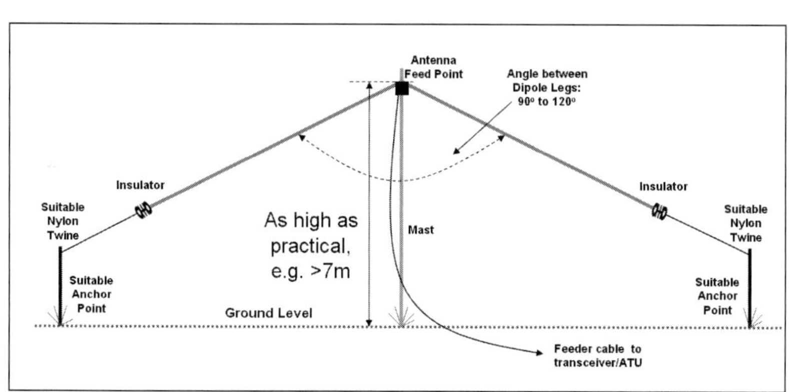


**Fig 15.17: Concept of the inverted V dipole antenna**

ing this length with a true half wavelength (λ/2) on 7.10MHz gives 21.125m.

Dipole dimensions for each amateur band are shown in the **Table 3** for comparison, where the wire lengths have been given as:

- The wavelength (λ) and the half wavelength (λ/2) in free space
- The half wavelength (λ/2) and quarter wavelength (λ/4) calculated using 143/f MHz
- The predicted quarter wavelengths (λ/4) at 10m agl for element dimensions of 2mm or 25mm as appropriate using MMANA-GAL.

The calculated and predicted quarter wavelength (λ/4) leg lengths have been included in Table 3 as a guide when making dipoles for the amateur bands for reference.

When making up a dipole using a larger diameter conductor for the antenna element, the conductor's length has to be reduced by an amount known as the K factor (based on the length to diameter ratio). For example, using MMANA-GAL, the calculated length for a dipole for 21.20MHz is 6.8m using a wire diameter of 2mm. If the conductor diameter is increased from 2mm to 25mm, the total length should be reduced to 6.64m to take account of this effect.

In practice, tubular elements are best constructed using different diameter telescopic sections. This makes it easy to adjust the length when adjusting the antenna to obtain the minimum SWR. Dipoles for 80m and 160m are long and should be made of hard drawn copper wire to reduce stretching and sagging due to the weight of the antenna and the coaxial cable. The feed point impedance of a dipole at resonance can vary either side of the nominal 73Ω, depending on the antenna's height above ground, the proximity of buildings and any electromagnetic obstacles, together with any bends in the wire. As a result of this, for a practical antenna it is difficult to obtain an SWR of 1:1 when the antenna is fed with 50Ω coaxial cable.

The dipole is a balanced symmetrical antenna and should ideally be fed with a balanced two-wire feeder. However, because almost all transmitters use a 50Ω coaxial line antenna socket, coaxial cable is almost universally used to feed the dipole antenna. Connecting unbalanced coaxial cable to a balanced antenna can affect the performance of the antenna because the unbalance can cause common mode currents to flow on the coaxial cable screen's surface. This effect can be minimised by ensuring that the coaxial cable is not a multiple of an electrical quarter wavelength and by bringing the coaxial cable away from the antenna element at as close to 90° as possible. Common mode currents flowing on the outer of the coaxial cable are undesirable because they can cause the line to radiate (leading to EMC problems). However, this effect should not be confused with the SWR on the transmission line because even a high SWR does not cause the line to radiate. A balun should also be used to reduce common mode currents and several choke baluns have been described in the Transmission Lines chapter. Descriptions of other forms of balun are included later within this chapter.

The dipole antenna normally requires two supports and this may be a problem at some locations. The solution may be to mount the antenna so that it is vertical or sloping. Where the space to install the antenna is restricted, the dipole's elements can be arranged in an inverted "V" configuration. An inverted V dipole has its two wire legs sloping down towards the ground from the central feed point, typically creating a 120 o or 90° angle between the legs of the dipole. This arrangement has the advantage that only a single support is required to hold up the central feed point, with this usually mounted at the top of a mast

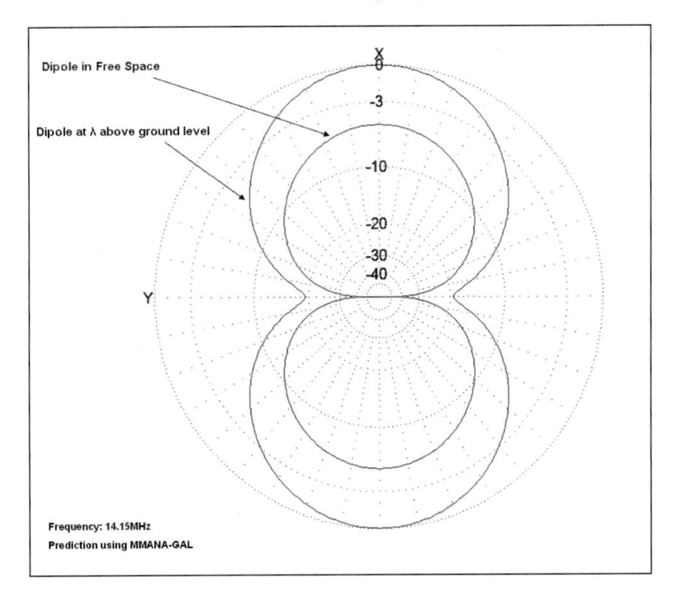


**Fig 15.18: Comparison polar diagrams (in dBd) for a dipole antenna in free space (inner trace) and the same antenna mounted at a wavelength (λ) above the ground (outer trace) modelled using MMANA-GAL**

or pole. This configuration of dipole can considerably reduce the ground footprint of the antenna without significantly impacting on its performance. When viewed from the side, the antenna looks like the letter "V" turned upside down and consequently from where it gets its name. **Fig 15.17** illustrates the concept of an inverted V dipole for reference. Where a metal mast is used to support the antenna, the feed point needs to be isolated from the mast to avoid it impairing the performance of the antenna through induction. One method to isolate the feed point is to attach a length of plastic tube to the top of the mast with about 15 to 30cm (6 to 12 inches) of the tube protruding above it. The feed point can then be fastened to the top of this tube giving at least 100mm of clearance. When the dipole is fed with a balanced feeder cable, a metal mast can also affect the feeder cable through induction and so impair the performance of the antenna. There should be a be gap left of at least 25mm (1inch) to isolate the balanced feeder cable from the mast to avoid this effect.

Another method to fit a dipole into the space available is to equally bend over the ends of a dipole's legs for a short length up to about 15%. This should not significantly affect the performance of the antenna provided the configuration of the dipole is kept symmetrical.

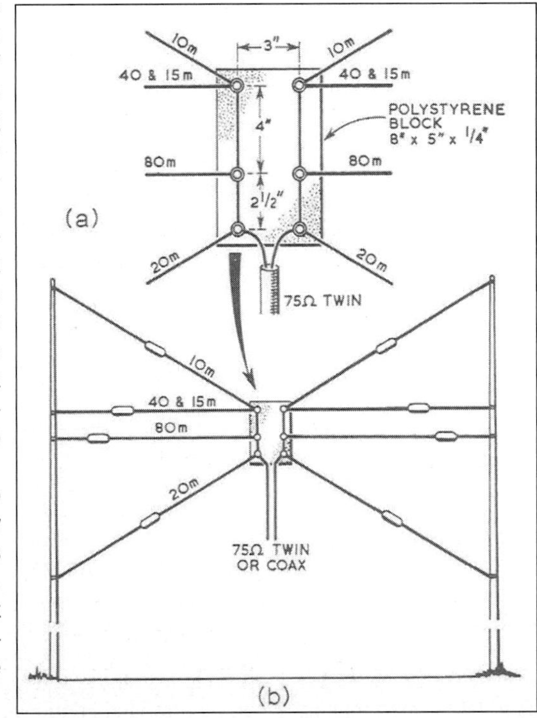


**Fig 15.19: F7FE Multi-Band Dipole Array**

### Dipole Polar Diagrams

The diagram usually shown for a dipole is the familiar azimuth polar 'Fig 15.of eight' diagram where there are clear nulls at the ends of the dipole's legs (as discussed in the Antenna Principles chapter). It can come as a surprise when stations are successfully worked apparently off the ends of the dipole, especially when the information in many antenna theory books implies that this should not be possible. The reason why this occurs can be explained using **Fig 15.18** as follows. The inner pattern shown in **Fig 15.18** is for a theoretical dipole in free space and shows that the nulls at the ends of the dipole are close to -40dB down from the maximum 0dBd axis. When the dipole is installed at about a wavelength (λ) above the ground level, then the gain increases by about 5dB relative to the dipole in free space as shown in **Fig 15.18**. This increase in gain is due to ground effect and depends on the quality and conductivity of the ground below the antenna. With the antenna at this height, the nulls are now close to -20dB rather than -40dB relative to the 0dBd maximum. Therefore, the signals are effectively increased by 20dB at the ends of the antenna and this explains why it is possible to often work stations off the ends of a dipole. Alternatively, if the antenna in this example had been turned directly towards these stations, then their signal strengths would have increased by 20dB as indicated in **Fig 15.18**.

However, for a practical antenna, any radiation from the feeder or re-radiation from nearby electromagnetic obstructions will further fill in the nulls at the antenna's ends. This makes it very difficult to predict how the antenna may perform and there can be a significant ambiguity between the same type of antenna when installed at different locations.

## MULTI-BAND ANTENNAS

Operating a station on several bands often involves installing a number of antennas, each with its own feeder cable. This can result in several coaxial feeder cables being run in parallel and their combined weight may become a problem, eg when raising/lowering the mast. Solutions to operating on multiple-bands where one feeder could be used include:

- Dipoles in their fundamental and harmonic modes
- Parallel dipoles
- Trap dipoles
- Multi-band doublet using tuned lines
- Multi-band doublet with an ATU
- The off-centre fed dipole.

Most multi-band systems can be improved by using an Antenna Tuning Unit (ATU). Therefore, it is probably a good decision to either purchase or to build an ATU. Most ATU designs are generally based on using straightforward RF circuitry and this technology tends not to date, unlike some modern equipment or computers. The construction of several ATU designs has been described later within this chapter.

### Parallel Dipole Arrays

One multi-band technique often used is to connect several dipole antennas together at their centres forming an array. Dipole arrays provide a convenient technique to avoid the need to use separate feeders for each antenna or an arrangement to switch the feeder cable between antennas.

An example of this technique was devised by F7FE using four

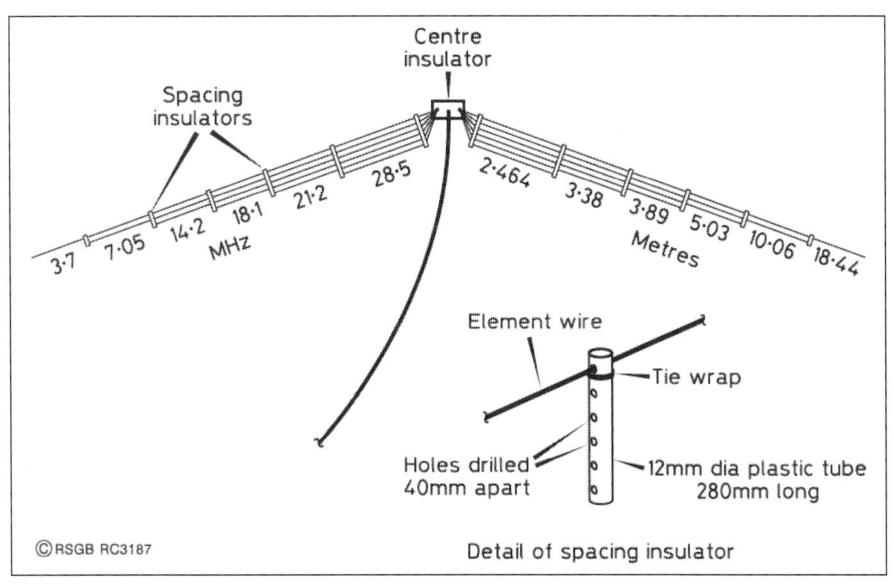


Fig 15.20: Concept of a parallel dipole array using suspended dipoles

Fig 15.21: The GM4UTP multi-band parallel dipole array showing the inverted V configuration. The antenna also supports a single quad element for 10m

dipoles covering the 80m, 40m, 20m and 10m bands as illustrated in **Fig 15.19** [8]. This arrangement also allowed operation on 15m using the 40m dipole as a third harmonic antenna. F7FE used two masts to support the array of dipoles, that fanned out either side of where the dipole centres shared a common termination and where the feeder cable was connected. Each dipole's length was determined using the mid-band frequency f (MHz) using the equation described previously:

Dipole Length (m) = 143 / f (MHz)

Or

Dipole Length (ft) = 468 / f (MHz)

The principle of operation of a dipole array relies on each dipole presenting a relatively high reactive impedance at frequencies other than its own resonant frequency. This results in small RF currents flowing in all the dipoles except the resonant dipole and this becomes the primary radiator.

The dipole array described above, along with the parallel dipole antenna to follow, is of the balanced variety and should be used with a balun if coaxial cable is used as the feeder. The balun used could be of the choke variety (as described in the Transmission Lines chapter) or of a transformer type. The description and construction of transformer type baluns are described later within this chapter.

A parallel dipole array is shown in **Fig 15.20** covering 80m, 40m, 20m, 17m, 15m and 10m where the dipoles are successively suspended off each other using insulated spacers made from electrical conduit. The arrangement excluded the 12m band, however this can be added with an additional dipole of 5.68m in length, suspended between the 10m and 15m dipoles and connected in the centre in the same manner as the other dipoles.

The antenna is best configured as an inverted 'V' with the weight of the centre insulator supported by a central mast or pole. Low centre band SWRs are possible by the careful tuning of each dipole. This can be achieved by arranging the ends of the elements so that they are clear of their support insulators by about 200mm. The dipole lengths can be reduced or increased by folding back the ends and securing them with plastic tape. The dipoles should be tuned starting with the highest frequency dipole and working downwards from there. The resonance of these dipoles can be interactive and when one dipole is adjusted, then this can affect the resonance of the others. The process can be iterative and the dipoles may have to re-resonated several times to achieve a good match on all bands. When using coaxial cable to feed the antenna, a 1:1 balun should be used and this can be situated on the mast below the central insulator.

An example of the parallel dipole constructed by GM4UTP is shown as **Fig 15.21**. The antenna was configured as an inverted 'V' with the weight of the centre insulator and the 1:1 balun mounted on a 10m high aluminium scaffold pole. The lowest frequency dipole supports the higher frequency dipoles using

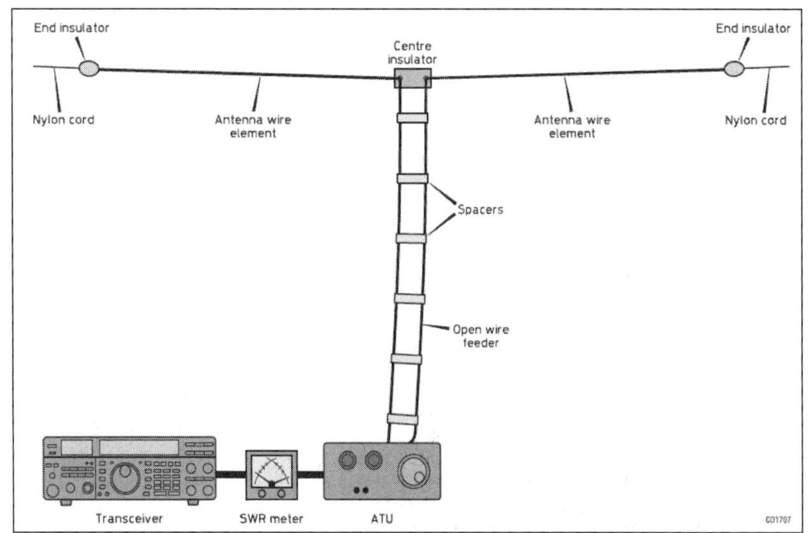


Fig 15.22: The tuned open-wire dipole using a tuned transmission line. When space is restricted, the antenna could be cut for 3/8 of a wavelength on 7MHz and it will tune all bands from 7 to 28MHz. The dipole length is not critical because the tuner provides the impedance match throughout the entire antenna system

spacing insulators made from 11mm diameter plastic electrical conduit.

The principle of operation of the parallel dipole array is the same as the dipole array previously described. This antenna also relies on each dipole presenting a relatively high reactive impedance at frequencies other than its own resonant frequency. This results in small RF currents flowing in all the dipoles except the resonant dipole and this becomes the primary radiator.

Dipole arrays and parallel dipoles provide a convenient technique for multi-band operation and avoid the need to use separate feeders for each antenna or an arrangement to switch the feeder cable between antennas. Where the space allows, many stations install a dipole array in the loft and very good results can be achieved using this arrangement.

### The Tuned Doublet or Random-Length Open-wire Tuned Dipole

This open-wire tuned dipole antenna is an effective and efficient antenna for multi-band use whose construction is comparatively straightforward. The antenna is usually fed with a tuned open-wire twin-line feeder, however ladder-line cable (slotted twin-line feeder) could be used. An ATU is used to tune the twin-line feeder to accommodate the wide variations of feed impedance encountered on the different bands. The concept of the antenna is shown in **Fig 15.22**.

This antenna should be at least a quarter wavelength long at the lowest frequency of operation, where it radiates with an effectiveness of approximately 95% relative to a $\lambda/2$ dipole. However, the high impedances associated with the shortness of this antenna results in SWR values of around 300:1 when using a 450Ω feeder cable. While the antenna is quite efficient, the impedances at the end of the tuned feeder will be outside of the matching capabilities of the many commercial ATUs that use a toroid based balun to provide the balanced match to the tuned feeders.

Using a doublet with a length of about 3/8 wavelength on the lowest frequency can overcome this high impedance problem. This length is a compromise between a quarter and half wavelength and will work on the 80m band, for example, if a $\lambda/2$ dipole is difficult to install at a given location. A 3/8 wavelength dipole has an effectiveness of greater than 98% relative to a $\lambda/2$ dipole and the SWR values can be far easier to match. For example, the SWR encountered is around 25:1 on 600Ω feeder, 24:1 on 450Ω feeder and 25:1 on 300Ω feeder lines. A 3/8 wavelength dipole at 3.5MHz is approximately 30m long and this means that a length from 27m to 30m makes an good radiator on the HF amateur bands from 80m to 10m.

If there is not enough room to install a 27m length of straight wire for operation on the 80m band, a 3 to 5m portion of each wire-end may be dropped vertically from each end support. There will not be a significant change in radiation pattern on the 80m and 40m bands although there will be a minor change in polarisation in the radiation at higher frequencies. However, the effect on propagation will be negligible. The twin-line feeder could be affected by the close proximity of metal objects and should be run with these taken into consideration (eg metal window frames or guttering).

An example of a lightweight multi-band doublet intended for portable use is shown in **Fig 15.23**. The main advantage of this antenna is that the doublet's span length is not critical, an advantage when intending to use the antenna at an unfamiliar location. In practice, usually a 20m span length, fed at its centre works, well for all bands HF bands from 40m to 10m.

The antenna can be supported at the centre using telescopic fibreglass poles. These poles are made for fishermen, are called roach poles and weigh around 500g. They are usually 1.2m long and when telescoped can extend to 6m, although sizes vary. The antenna's performance is improved by installing it as high as possible. The doublet's span can be made from 2mm diameter plastic covered insulated wire, fed in the centre using 300Ω ladder-line cable. The antenna's centre insulator can be made from a terminal block using screw connectors. This is fixed to the top of the support pole with plastic tape as shown in **Fig 15.23**. If the antenna is intended to be installed on a more permanent basis, then the centre insulator used needs to be a more weather resistant variety. Using 300Ω ladder-line cable is convenient because it can be run through the gaps in doors or windows when operating from a temporary premises. Dog-bone end insulators should be used at the antenna span's ends.

When using this type of an antenna, a suitable ATU is required that has the provision for feeding a balanced antenna using an integral balun. An example is the MFJ-901B because it has been found to work well with this arrangement. This ATU has the advantage that it weighs about 600g and is small (135 x 150 x

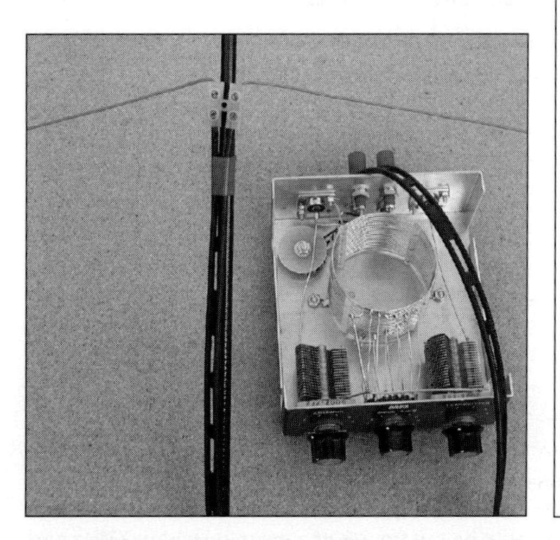

Fig 15.23: The multi-band doublet can provide a flexible and practical solution for portable operation

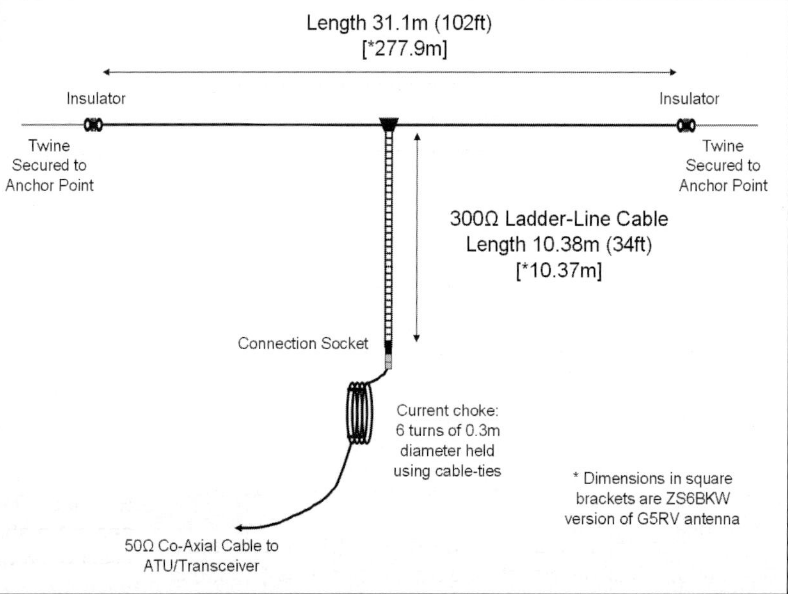


Fig 15.24: Construction of the full-size G5RV antenna. The dimensions shown in square brackets are for the ZS6BKW version - see text

60mm). As can be seen from **Fig 15.23**, this ATU uses a large low-loss air-spaced tapped inductor.

## The G5RV Antenna

The G5RV antenna [9] remains a popular HF bands antenna widely used by radio amateurs around the globe. This antenna was devised by Louis Varney, G5RV, around 1946 when he was looking for a solution to getting his station back on the air following the restoration of amateur transmitting licenses in the UK. One of Louis' interests was operating on 20m and his design was based on this requirement. However, the antenna was found to be effective on the other HF bands giving a multi-band antenna capable of use from 80m to 10m.

The G5RV antenna's geometry can be altered by converting it into an inverted 'V' or by bending the ends down to enable the antenna to be fitted into a smaller space without modification to the length. Generally, up to 15% of the length of each leg can be bent over without significantly changing the radiation pattern of the antenna on the lower and mid HF bands. However, any changes to the antenna's legs should be symmetrical to maintain the balance of the antenna.

The G5RV antenna was devised to work with older valve based radio equipment using a Pi-network as a means of matching to the antenna. The impedances presented by the antenna can result in fairly high SWRs being encountered depending upon the HF band that is in use. Therefore, an ATU is necessary to allow the antenna to be matched to solid-state transceivers that otherwise would not be able to tolerate the range of SWR met. Several examples of ATU are described at the end of this chapter.

### The Full and Half-Size G5RV Antennas

A G5RV antenna is often thought of as a wire-span of 31.1m (102ft) long that is centrally fed using about 10.38m (34ft) of air-spaced ladder-line cable. This configuration is often referred to as the full-size G5RV and covers the bands from 80m to 10m. The concept of the full-size G5RV antenna is shown in **Fig 15.24**. Another version of the antenna is the half-size G5RV, where the wire-span's length is 15.55m (51ft) and is centrally fed using about 5.17m (17ft) of air-spaced ladder-line cable and covers the bands from 40m to 10m.

The G5RV design is based on a wire-span that is three electrical half wavelengths ($3\lambda/2$) long at a desired frequency, centrally fed by a balanced line that is half an electrical wavelength ($\lambda/2$) long at the same frequency [10]. The length of the wire-span (allowing for the velocity factor) is given by:

Length of wire-span (m) = 150.(n - 0.05) / f (MHz) (1)

For the full-size G5RV, where f = 14.15MHz and n = 3 half wavelengths, then Equation (1) gives:

Length of wire-span (m) = 150.(3 - 0.05) / 14.15MHz = 31.27m (102.6ft)

In the interests of keeping things straightforward and since the whole system is brought into resonance using an Aerial Tuning Unit (ATU), the antenna's wire-span is cut to 31.1m (102ft).

The length of the air-spaced ladder-line feeder at 14.15MHz (using a 0.98 velocity factor) is given by:

Half Wavelength Feeder (m) = (0.98).(150 / 14.15MHz) = 10.38m (34ft) (2)

The wire-span's centre impedance at 14.15MHz is about 100Ω and the half wavelength ladder-line acts as a 1:1 impedance transformer presenting this impedance at its end. Therefore, if a 50Ω feeder cable is connected to the ladder-line there will be a mismatch, however this will be low at about 2:1 and be within the tuning capabilities of the ATU enabling a good match to be seen by the transceiver.

Fig 15.25: Mast mounted antenna-centre

Connecting an unbalanced coaxial cable to the end of the balanced ladder-line will require an arrangement to minimise common mode currents that flow on the outer of the co-axial cable's screen conductor. There are several balanced-unbalanced techniques available to overcome common mode current problems using a choke balun as described in the Transmission Lines chapter. However, one uncomplicated method is to make a current choke from about ten 300mm diameter loops of the coaxial cable (held together with cable-ties) and located this close to the coaxial cable's connection with the balanced ladder-line as possible. The description and construction of transformer type baluns are described later within this chapter.

The calculations for the half-size G5RV are similar to the larger version outlined above. However, using a frequency of 28.5MHz gives a wire-span of 15.55m (51ft) in length with an air-spaced ladder-line of 5.16m (17ft) long.

For both the full-size and half-size G5RV antennas, the ladder-line feeder may have a shorter length if other types of balanced line feeders are used because their velocity factors will be different compared to an air-spaced ladder-line. For example, often the length of the balanced line can reduce to typically 9.3m for the full-size G5RV and 4.65m for the half-sized G5RV antenna when other types of balanced line are used (eg commercially available polythene insulated 300Ω ladder-line).

The G5RV design is really only resonant at its design frequency (eg 14.15MHz for the full-size G5RV and 28.5MHz for the half-size G5RV). Often, the antenna is considered as acting more like a doublet-type antenna when not used on its resonant band.

### The Quarter-Size G5RV Antenna

Both the full-size and half-size G5RV antennas summarised above are still comparatively large antennas and may not fit within the space available. However, it is possible to use the G5RV's calculations to make a smaller antenna that is based on 6m rather than 20m. Taking 50.15MHz (6m band) as the design frequency, then using formulas (1) and (2) gives:

Length of wire-span (m) =150.(3 - 0.05) / 50.15MHz = 8.82m (28.94ft)

Half Wavelength Feeder (m) = (0.98).(150 / 50.15MHz) = 2.93m (9.81ft)

This antenna is sometimes called a quarter-sized G5RV and can be constructed using 5amp rated insulated stranded copper wire and 300Ω ladder-line. This antenna's wire-span was set up about 4m above the ground with a mast supporting the centre. Fishing-line was attached to the wire-span ends and fastened to anchor points to keep the antenna taught. The central insulator used for the antenna was made from a sheet of PVC plastic [11] and is illustrated in **Fig 15.25**.

When running the ladder-line away from the central insulator, if a metal mast is used to support the central insulator, then a gap of at least 25mm (1inch) should exist between the ladder-line and the mast. This is to minimise interactions between the metal mast and the ladder-line. The ladder-line could be run downwards at slight angle from the central insulator allowing a gap for this purpose. A good practice is to use a short length of plastic conduit tube secured vertically to the top of a metal mast. The central insulator can then be attached to the top of the conduit tube providing at least 100mm of isolation between the metal mast and the antenna as shown in **Fig 15.25**.

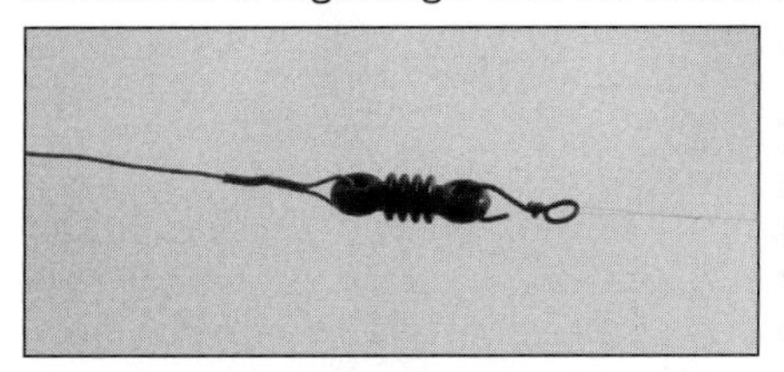

Fig 15.26: Wire-span end termination

When tested using 50.155MHz, the wire-span was found to be too long and the ends were equally folded back on themselves until the lowest SWR was observed. Then dog-bone insulators were added, the ends trimmed to length and terminated as illustrated in **Fig 15.26**. To further improve the match, the length of the 300Ω ladder-line was altered until the best SWR was achieved at 1.2:1. This final length of the wire-span was 8.41m and ladder-line was 2.65m in length, **Fig 15.27** illustrates the antenna.

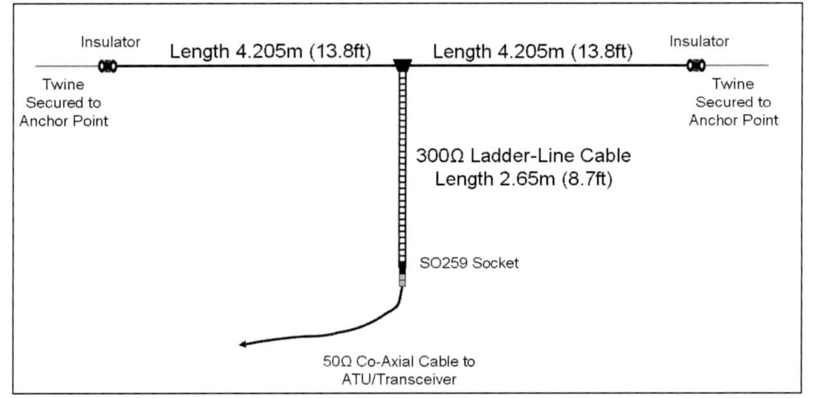


Fig 15.27: λ/4 G5RV Antenna Concept Covering 20m to 6m. A balun should be used when connecting coaxial cable to the ladder-line

Fig 15.28: Completed ladder-line SO259 termination

To terminate the end of the ladder-line, a short length of 20mm diameter electrical conduit tube was passed over the end. Then the ladder-line's conductors were soldered to centre and outer of an SO259 single-hole socket. The conduit tube was slid back over the soldered joints, vertically held firmly against the base of the SO259 socket using insulation tape and sufficient epoxy resin was poured into the conduit tube's interior to form a waterproof joint once set. **Fig 15.28** illustrates this arrangement. When the connecting the ladder-line socket to the radio equipment feeder cable's plug, the joint should be weatherproofed using self-amalgamating tape for example.

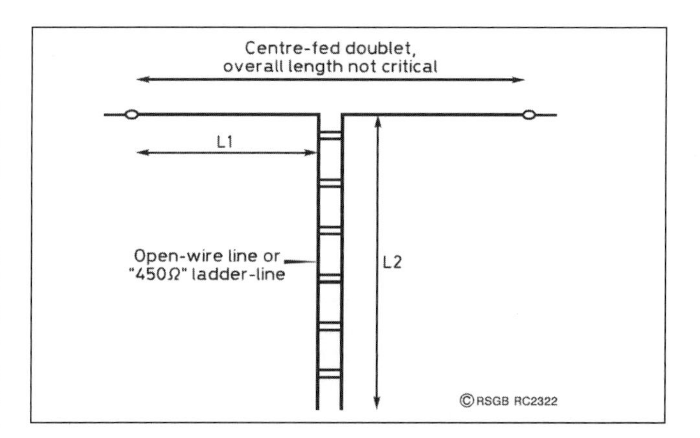


Fig 15.29: The HF Multi-band doublet fed with ladder line can have a wide range of feed impedances. To estimate the impedance, measure half the length of the doublet shown as L1, plus the electrical length of the feed-line shown as L2, (allow for velocity factor of L2)

Similarly to the previous antennas described, when a coaxial cable is connected to the ladder-line a balun should be used. The balun should be located as close to the co-axial cable's connection with the ladder-line as possible.

This antenna has a span-length of 8.41m (28ft) allowing it to be installed in a relatively small space compared with its larger versions. Use of the antenna on 40m and 80m is not really practical, however it does allow operation from 20m through to 6m.

### The ZS6BKW Modified Full-size G5RV Antenna

ZS6BKW (G0GSF) developed a computer program to determine the most advantageous length and impedance of the ladder-line matching section and the top length of a G5RV-type antenna. He arranged that his antenna should match as closely as possible into standard 50Ω coaxial cable and so be more useful to the user of modern solid-state equipment. The G5RV antenna total top length of 31.1m was reduced to 27.9m, and the ladder-line section was increased to 10.37m (ignoring the velocity factor). This ladder-line section must a characteristic impedance of 400Ω and can be made up from two 18SWG wires spaced at 250mm (10inch) apart. The ZS6BKW version of the antenna gives improved impedance matching over the original G5RV, but still cannot be used without an ATU with solid-state PA transmitters.

## The W6RCA Multi-band Doublet

Cecil Moore, W6RCA, devised a doublet antenna covering all the HF bands from 3.6 to 29.7MHz that did not need an ATU. This antenna is based on a conventional centrally fed wire span with a balanced ladder-line feeder run down to the radio equipment. The physical length of the ladder-line feeder is changed to suit the band in use so that the current maximum always coincides with the bottom of the ladder-line [12]. The ladder-line's input impedance at this point will be low and non-reactive giving a low SWR. This allows the direct connection of the transceiver to the ladder-line feeder, using a length of coaxial cable and a 1:1 choke balun to minimise any imbalance between the twin-line feeder and equipment. The concept of the antenna is shown

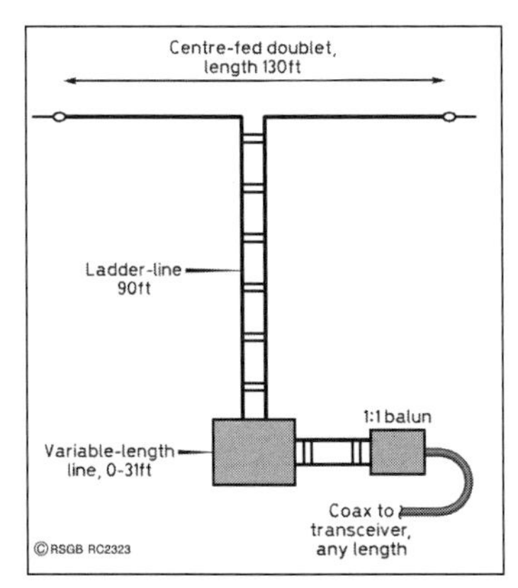


in **Fig 15.29** where the length of each doublet leg is L1 and the ladder-line feeder is L2 (allowing for the ladder-line's velocity factor (v)). A reasonable match can be obtained when the total length of L1 plus L2 is an odd multiple of an electrical quarter wavelength on each band, where the length can be estimated from:

**Fig 15.30(a): W6RCA's ladder-line switch-system allows this 39.62m (130ft) doublet to cover all eight HF bands with no ATU. Optimum dimensions will depend on several factors, however the line length can be changed to compensate**

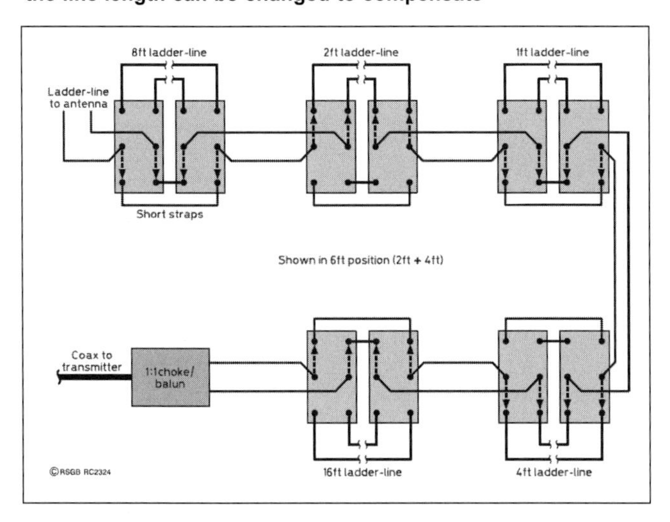


**Fig 15.30(b): Concept of W6RCA's ladder-line switch-system. This uses five pairs of surplus DPDT (changeover) relays. A suitable 1:1 choke balun is described in the Transmission Lines chapter**

$L1 + (L2 \times v) = n.\lambda/4$

Where v is the velocity factor and n is 3, 5, 7,9 and so on.......

W6RCA used a doublet of 39.62m (130ft) in length that was fed at its centre with 27.5m (90ft) of 450Ω ladder-line as shown in **Fig 15.30**(a). This length of the doublet was chosen because it is approximately a half wavelength at 3.5MHz and a full wavelength at 7MHz. The 27.5m (90ft) 450Ω ladder-line brings the current maximum to the bottom at 7.2MHz. The advantage of this particular combination is that all the other HF bands can be matched using a relatively small range of additional 450Ω ladder-line lengths. The longest additional ladder-line length required is 9.5m (31ft) for 3.6MHz to extend the line to an electrical half wavelength. All the other HF bands require ladder-line extensions somewhere between zero and 9.5m (31ft 2inches). To alter the overall ladder-line length, W6RCA built the variable ladder-line switch-system shown in **Fig 15.30**(b) consisting of 300mm (1ft), 600mm (2ft), 1.22m (4ft), 2.44m (8ft) and 4.88m (16ft) loops of 450Ω ladder-line. These can be individually switched in or out using DPDT (double-pole double-throw or double change over) relays, giving any line length from zero to 9.5m in 300mm steps. W6RCA found he could cover all amateur bands from 3.6 to 29.7MHz with a SWR of better than 2:1 using this arrangement.

However, the dimensions of the doublet, the main ladder-line and the extensions will depend on a number of factors. These include antenna's height, local earth properties, the use of other doublet configurations such as an inverted 'V' and the exact type of ladder-line used. The attributes of 450Ω ladder-line can vary in characteristic impedance, velocity factor and the quality of the insulation between different brands. Therefore, when constructing an antenna based on the W6RCA design, there will be a need to vary the antenna's dimensions to minimise the SWR for each HF band. A practical approach to tuning the antenna is to measure the SWR using a suitable antenna analyser and alter the ladder-line lengths to minimise the SWR.

To start with, increase the length of the permanent ladder-line by about 1m (3ft) from the recommended length and trim this for a minimum SWR at around 7.05MHz. This length may also be found to work adequately for the 15m and 12m bands as well. Secondly, determine the maximum additional ladder-line length required to tune down to 3.5MHz with an acceptable SWR. This extra length should be about 9.5m and the optimum lengths for all the other bands will be shorter than this.

For a shorter doublet covering 7.0 to 29.7MHz, a 20.12m (66ft) doublet and an 18.3m (60ft) ladder-line makes a reasonable starting-point, again with a zero to 9.5m variable ladder-line section. Note also that this shorter antenna can still be used as a shortened dipole on the 80m band, however an ATU will be required to resonate the arrangement and this may introduce more loss.

When using this design of antenna, it is important to use a 1:1 choke balun. This is because any low-impedance path to ground from either side of the ladder-line is likely to result in very strong unbalanced radiation from the ladder-line itself. This is a consequence of the 'odd quarter-wavelength' principle used in selecting the ladder-line length. A suitable choke balun was described in the Transmission Lines chapter.

## Baluns for use with HF Antennas

The concept of how connecting an unbalanced coaxial cable to a balanced antenna can cause common mode currents to flow along the outside of the cable's shield was described in the Transmission Lines chapter. Common mode currents are undesirable because they can affect the antenna's radiation performance, be a source of interference or a cause of problems from RF in the 'shack'. There are three main types of 'balanced-to-unbalanced' transitions (baluns) that can be used to minimise the effects of common mode currents:

### Tuned Baluns

This type of balun includes the Sleeve Choke Balun and Pawsey Balun. Both of these baluns rely on the properties of a λ/4 transmission line to minimise common mode currents. These baluns are often used for VHF and UHF applications, however their physical length tends to make them impractical for use at HF. These types of balun have been described in the Transmission Lines chapter.

### Choke Baluns

This type of balun relies on introducing a high inductive impedance to reduce common mode currents to make their effects negligible. Choke Baluns can be made from loops of the coaxial cable or by threading suitable ferrite cores over the coaxial cable used to feed the antenna. Both these balun techniques need to be situated as close as possible to where the coaxial cable is connected to the twin-line feeder or to the antenna's feed point. Choke baluns are often used for HF applications and are able to provide a wider operational bandwidth compared to tuned Sleeve or Pawsey Baluns. Several examples of Choke Baluns were described in the Transmsission Lines Chapter.

## Transformer-type Baluns

This type of balun uses a transformer to provide a match between an unbalanced coaxial cable and a balance load (eg twin-line cable or an antenna). Their use is often encountered with HF applications and many types of ATU use this technique to provide a balanced connection for a balanced twin-line cable.

Transformer-type (or voltage-type) baluns using iron powder toroid cores can be constructed to cover frequencies up to about 40MHz. These transformers have their origins in the work performed by C.L. Ruthruff [13] and function in the same manner as their lower-frequency counterparts. However, the low magnetic permeability of the toroid core allows these transformers to operate at much higher frequencies. To enable these transformers to work efficiently, it is important to maintain a high coefficient of coupling between the windings. To achieve this, the wires forming the windings are laid side-by-side to allow the magnetic field surrounding one wire to encompass the second wire so maximising the coupling. For efficient transformer action, the inductance presented must be large enough to ensure that connecting the transformer has no effect on the input other than that due to the load when connected to the secondary winding. However, there will always be some leakage inductance and this is usually considered as in series with the load. The leakage inductance increases in proportion to the transformer's self-inductance and these are functions of the coupling coefficient between the windings. Essentially, a transformer that works well at 3.5MHz may not work as well at 28.5MHz and this often shows up as a worsening SWR as the frequency is increased.

To achieve a high coupling between the windings, bifilar or trifilar winding techniques are used where the turns of the windings are effectively wound tightly together with the wires usually in parallel. The concept of two transformer baluns for impedance ratios of 1:1 and 1:4 are shown as **Figures 15.31(a)** and **15.31(b)**. The trifilar (1:1 ratio) arrangement tends to increase the leakage inductance and this can limit the upper frequency range limit when compared to the bifilar arrangement (1:4 ratio).

Peter Miles, VK6YSF, has published several toroid iron powder core transformer designs based on the concept shown in **Fig 15.31**. These have included 1:1 and 1:4 impedance transformers with their details made available online [14]. **Fig 15.32** illustrates one of these designs for a 1:1 impedance ratio transformer balun using a Micrometals Iron Powder Toroid T200-2 core. This balun comprised 17 trifalar wound 3-wire turns with a small gap left between each turn as shown. Another design is shown in **Fig 15.33** for a 1:4 impedance ratio transformer balun using a Micrometals Iron Powder Toroid T130-2. This balun comprised 17 bifilar wound 2-wire turns with a small gap left between each turn.

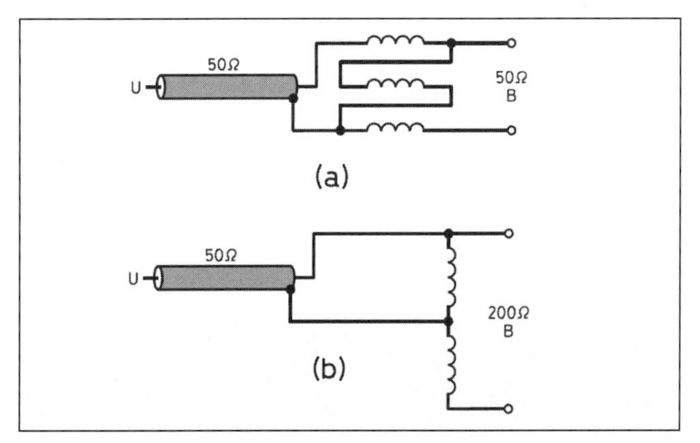


**Fig 15.31: Concept of transformer-type baluns. Windings (shown separately) are bifilar or trifilar usually wound on the core. (a) wire-wound 1:1 (trifilar). (b) Wire-wound 4:1 (bifilar). "U" for unbalanced, "B" for balanced**

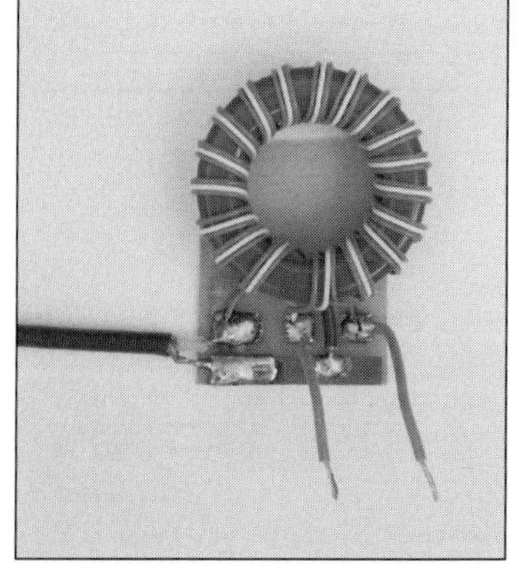

**Fig 15.32: 1:1 impedance ratio transformer balun wound using 17 trifilar turns on a T200-2 core**

**Fig 15.33: 1:4 impedance ratio transformer balun wound using 17 bifilar turns on a T130-2 core**

The performance was assessed by G0JMI for both the 1:1 and 1:4 types of balun design [15]. This assessment was made with approximately 10m of RG58 coaxial cable connected to the unbalanced primary winding to simulate the situation where the balun is to be used remotely from the transmitter, at the antenna's feed point for example. To gain an indication of the performance for both transformer balun types referred to 50Ω, the SWR was measured with each secondary winding terminated in a suitable dummy load (ie 50Ω for the 1:1 and 200Ω for the 1:4 balun). Tests were made at several HF frequencies with the results shown in **Table 4** for both types of balun.

It was found that the SWR for the 1:1 balun could be improved at lower HF frequencies by using a 200pf capacitor connected in parallel across the transformer's primary winding where the coaxial cable was terminated [16]. However, this was found not to be necessary for the 1:4 balun in this case.

The results indicated that the trifilar 1:1 balun (T200-2 core) had a tendency to be more effective between 3.5 and 14MHz provided it was used in association with a parallel 200pf capacitor (with a rating over 500 volts) across the primary winding. The bifilar 1:4 balun (T130-2) tended to be more effective from 3.5 to 28.5MHz.

| | 1:1 Balun (T200-2 core) | | | 1:4 Balun (T130-2 core) |
|---|---|---|---|---|
| Frequency (MHz) | No capacitor connected | Capacitor connected across primary | Capacitor | SWR with secondary load of 200Ω |
| | SWR | SWR | | SWR |
| 3.65 | 1.2:1 | 1:1 | 200pf | 2.5:1 |
| 5.2775 | 1.2:1 | 1:1 | 200pf | 1.4:1 |
| 7.1 | 1.2:1 | 1.3:1 | 200pf | 2.7:1 |
| 10.12 | 1.7:1 | 1.4:1 | 200pf | 4:1 |
| 14.25 | 2.1:1 | 1.5:1 | 200pf | 1.1:1 |
| 18.14 | 2.5:1 | 2:1 | 50pf | 2:1 |
| 21.25 | Over 4:1 | - | - | 2.5:1 |
| 24.95 | Over 4:1 | - | - | 1.7:1 |
| 28.5 | Over 4:1 | - | - | 1.7:1 |

**Table 4: SWR comparison for a 1:1 balun using a T200-2 and a 1:4 balun using a T130-2 core at HF**

HF transformer baluns tend to be physically smaller than their corresponding HF choke type baluns. Therefore, their size and lighter weight makes their use at height feasible close to the antenna's feed point. However, a core should be chosen that is able to withstand the RF transmit power to be run and there are a range of toroid cores available from several suppliers or online sources. Most single T200-T core baluns are capable of handling RF powers up to about 100w from 1.81MHz upwards. However, single T130-2 core baluns tend not to be able to handle RF powers of up to 100w below 10MHz and the RF power capability tends to lower to around 70w.

To improve the power handling capabilities for a transformer balun, it is possible to stack two cores on top of each other forming a larger core. When using enamel insulated single core copper wire for the balun's windings, the toroid core should be wrapped with a layer of PTFE tape to insulate it. This is to provide electrical isolation if the enamel insulation breaks allowing the wire to make contact with the core. The windings can be held in place on the toroid core using cable-ties.

When used outside, a transformer balun should be housed in a suitable weatherproofed box with the cables passed through sealed holes to the balun. The balun should be fixed inside the box using glue or cable-ties to stop it moving around in turbulent weather conditions.

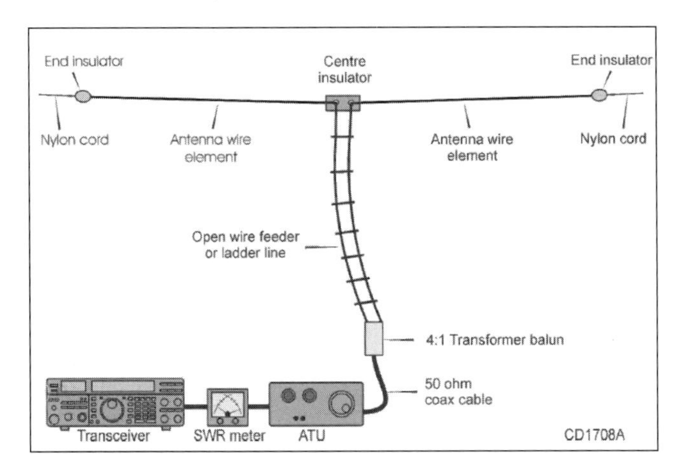


**Fig 15.34: The Comudipole feed arrangement for a multi-band doublet antenna**

## The Comudipole (Coaxial Cable Fed Multi-band Dipole)

In many locations there can be problems when trying to bring an open wire feeder into the shack. One solution to this could be the multi-band Comudipole antenna that was first described by Dick Rollema, PA0SE [17], as shown in **Fig 15.34**. This antenna arrangement was used to feed an inverted 'V' dipole with each leg 19m long (ie total length of 38m) mounted on the roof of a five-storey apartment building from a second floor shack. The antenna is not that much different from the tuned open wire dipole arrangement shown earlier. The twin-line feeder was brought down to a point where it was still clear of any metal objects to avoid the affects that these could cause from induction (eg metal guttering, metal down-pipes, metal windows and similar structures). At this point the twin-line feeder was connected to a 4:1 balun. Then a length of RG-213 coax was run to the shack where an L-network ATU was used to match the antenna to the transceiver. In practice the balun can be placed anywhere along the transmission line section from the antenna to the ATU. However, the feeder system should comprise as much of the twin-line feeder, or ladder-line, as possible because the losses with a high SWR on this type of line are much lower than with coaxial cable. If the space available restricts the length of the dipole, then a shorter dipole could be used. However, this shorter dipole should not be less than 15m (45ft) and may need to be used with a 1:1 balun. To provide the matching to the transceiver, a transmatch type ATU and balun could be used. An example of a suitable tuner is the G3TSO ATU and this is described later within this chapter.

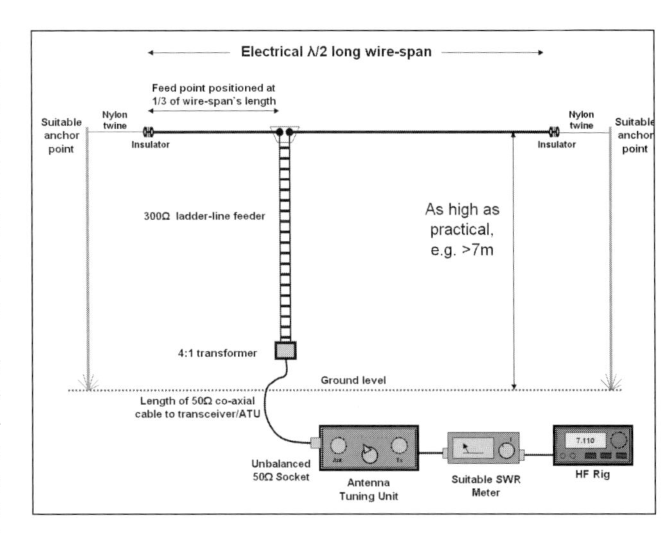


**Fig 15.35: Concept of the Off Centre Fed Dipole**

### The Off Centre Fed Dipole (OFCD)

The impedance at the centre of an electrical wavelength dipole is close to 50Ω when installed at a reasonable height above the ground. However, the impedance at about a third of the length from one end of the antenna is about 280Ω. Therefore, at this point, this makes it feasible to feed the antenna using 300Ω ladder-line. When the frequency (f) is increased to the second or fourth harmonic (ie 2 x f or 4 x f), the impedance at this point a third of the way along the antenna is also close to 280Ω allowing a reasonable match to 300Ω ladder-line at these frequencies.

The antenna can be fed with 300Ω ladder-line at either of the triple-crossing points and it will work on all three frequencies as a multi-band antenna. For example, if the wire span's length is an electrical half wavelength at 3.5MHz then the antenna works on both on 7.0MHz and 14.0MHz, provided the feed point is located at one third of the wire span's length from one end where it can be fed using 300Ω ladder-line. This arrangement is called an Off Centre Fed Dipole (OCFD) and the concept of this antenna is shown in **Fig 15.35**. If the feeder is not connected at a third of the wire span's length, the antenna presents an impedance of between 150 and 600Ω.

The OCFD antenna has a low SWR at its resonant frequency f, 2f and 4f enabling it to be used with a transceiver having an internal auto-ATU where higher SWRs may be difficult to be handled. However, this arrangement of antenna is likely to be unsuitable for the WARC bands without using an ATU with a matching capability able to handle much wider impedance ranges. This antenna arrangement will have some imbalance on the 300Ω ladder-line. Therefore, a current choke or balun is required to prevent common mode currents from flowing along the coaxial section of the feeder.

It is worth experimenting with this antenna to optimise its performance. This can be achieved through altering its overall length to move the position of the feed point by making one end longer than the other end. These variables can be made easier to adjust

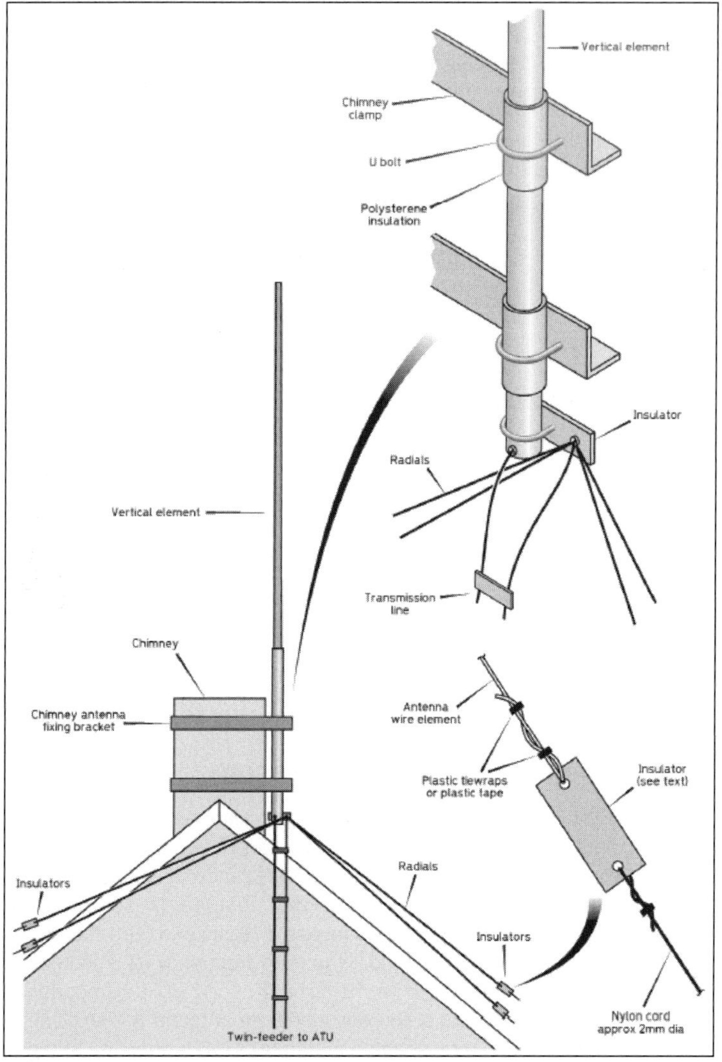

Fig 15.36: Concept of the Ground Plane antenna

by making the top section longer than required and folding the excess length at the end insulators back along the elements. The excess lengths can be temporarily held in place using tape while tuning the antenna. Once the optimum lengths have been found, the antenna's ends can be trimmed as appropriate before making a more permanent weatherproofed termination.

As a guide, for an OCFD antenna fed using 300Ω ladder-line feeder covering:

- 80m, 40m, 20m and 10m, the antenna is 42m long and fed 14m from one end
- 40m, 20m and 10m, the antenna is 21m long and fed 7m from one end.

## THE GROUND PLANE ANTENNA

The Ground Plane antenna is a vertically polarised aerial consisting of a vertical element with a counterpoise arrangement that is usually made from at least three radial wires as shown in **Fig 15.36**. The length of the vertical element and radial wires is an electrical quarter wavelength (λ/4) at the operating frequency. The vertical element is usually made of aluminium tube to make it self supporting and so does not require an end insulator. This has the effect that the velocity factor is higher compared to other antenna types at about 0.98. Therefore the electrical lengths of both the vertical element and radials are close to:

Vertical element and radial lengths (m) = (λ/4) x (0.98)

However, when constructing a Ground Plane antenna it is good practice to cut the vertical element and radials slightly longer and then trim their lengths until the SWR is minimised.

The radials are often made to slope downwards from the feed point although the angle is not critical. If the radials are at $90^{O}$ to the vertical element the feed impedance is around 30Ω. When the radials are sloped downwards at an angle of about 45o, the feed point impedance is around 45 to 55Ω making it feasible to match the antenna to 50Ω coaxial cable. When using coaxial cable, the central conductor is connected to the underneath of the vertical element and the outer shield is connected to where the radial wires fan-out from. In this configuration, the Ground Plane is a single-band antenna. Using twin-line cable to feed the antenna allows the antenna to be used for multi-band operation, although an ATU is required. However, the physical length of the vertical element will govern how effective the antenna is when this becomes less than a quarter wavelength.

**Fig 15.37** shows the predicted vertical radiation pattern for a 15m band Ground Plane antenna centred on 21.2MHz. This antenna was modelled using MMANA-GAL and comprised a vertical element (20mm diameter tube) and four radials (2mm diameter wire) each of 3.52m in length. The radials were symmetrically fanned-out around the vertical element and set at $90^{O}$ to it. Two patterns are shown in **Fig 15.37** with the antenna at 3.5m (about λ/4) and 14m (about λ) above the ground (agl). The pattern at 3.5m agl has a low angle of radiation of about 16o, while at 14m agl this angle becomes lower at about $10^{O}$ to the horizontal. This indicates, for 'dx' working, at 14m agl there should be an advantage over installing the antenna at 3.5m. At 14m agl, the radiation pattern also becomes divided into three lobes allowing the antenna to be also more effective for shorter-skip contacts.

## The Helical Vertical Antenna

A compact vertical antenna can be constructed by helically winding an electrical half wavelength (λ/2) of insulated single

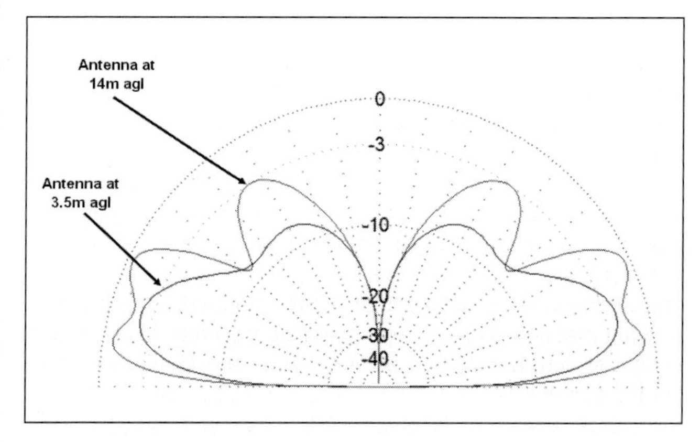


Fig 15.37: MMANA-GAL predicted vertical radiation pattern for a 15m band Ground Plane antenna at 3.5m and 14m agl

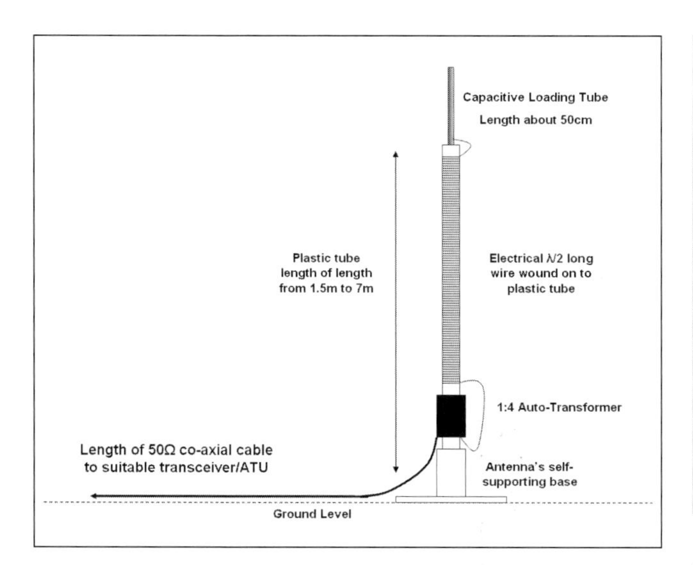


**Fig 15.38: Concept of the Vertical Helical antenna**

core copper wire onto a suitable length of plastic pipe of at least 20mm in diameter. A short length of metal tube is connected to the wire-end at the top of the antenna and acts as a capacitive load. The lower wire-end forms the antenna's feed point and the antenna is worked against ground as effectively a quarter wavelength system. An advantage of this vertical antenna system is that it does not require the use of radial wires, however the feeder cable should be run over the ground to the shack to maximise the antenna's performance. The antenna's voltage and current distribution is more linear than when lumped-inductance is used and is a possible reason why it can be an effective antenna. This vertical antenna was described in the ARRL 1976 Handbook where a variable inductive tuning arrangement was used to match it to the feeder cable [18].

This antenna is particularly useful for limited-space applications for the lower HF bands from 1.8 to 14MHz for example. The antenna can be used for the higher HF bands, although its use tends to be desirable only when an antenna of shorter than a natural quarter wavelength is required. The concept of this antenna is shown in **Fig 15.38** where a Un-Un auto-transformer was used to match the antenna to the coaxial feeder cable.

A 20m band version of the helical vertical antenna was made using a 2m length of 20mm diameter plastic electrical conduit as the antenna's former. This type of electrical conduit is available from many electricians' suppliers and DIY outlets, however a robust variety of conduit should be chosen that does not bend easily. Onto this was wound 8m of single core insulated wire of 1.5mm diameter (giving an electrical λ/4 in this particular case). The winding started about 40cm along the electrical conduit with approximately 100 helically wound turns extending out over the remaining 1.6m length to form the antenna's inductive section. Five equally spaced cable ties were used to hold the windings in place onto the conduit. A 57cm length of aluminium tube of 12mm diameter was secured to the top of the conduit using self-tapping screws. 52cm of tube was left protruding above the top of the conduit. The inductor's upper wire-end was attached to the tube just above where it came out of the conduit by soldering it to a ring-tab held in place with a self-tapping screw.

A 1:4 Un-Un auto-transformer was used to match the antenna to the 50Ω coaxial feeder. The Un-Un comprised 9 bifilar wound 2-wire turns on a T130-2 toroid core. The Un-Un was housed in small ABS box and two 20mm saddle-camps were fixed to the box to allow it to be secured to the lower end of the conduit. The coaxial cable feeder and antenna wire were passed through holes in the ABS and soldered to the Un-Un as described previously in this chapter (see 'Examples of Un-Un Auto-transformers used to Match Antennas' previously described for end-fed antennas). The ABS box and the cable access holes were sealed with external grade sealant to weatherproof it. Previously in this chapter the circuit details for a 1:4 Un-Un were shown in **Fig 15.7** and the Un-Un housing arrangements used for the 20m helical vertical were shown in **Fig 15.8**.

**Fig 15.39: 20m Helical Vertical antenna at use at G3FYX's QTH**

A base was made using 30cm length of electrical conduit of 25mm diameter vertically held on to a wooden board using angle-braces and jubilee clips. Four horizontal supports were attached to the bottom of the base to allow the base to stand upright on the ground when the antenna was slotted into the 25mm conduit.

To tune the antenna to 14.15MHz, 10m of coaxial cable was connected to the antenna's Un-Un auto-transformer and the antenna was set up on the ground clear of local objects. For testing, an antenna analyser was connected to the end of the coaxial cable to monitor the SWR. Initially, the antenna's inductive section was formed by helically winding 10m of wire on to the 20mm conduit. The inductance was reduced by trimming the wire and the SWR monitored until a good match was obtained. The gaps between the helical winding were also altered as part of the tuning process. The antenna was resonated at 14.15MHz where it gave an SWR of 1.3:1. Under a load of 100w CW the antenna maintained an SWR of 1.3:1 from 14.0 to 14.3MHz, although an ATU should be used. A comparison with a G5RV and λ/2 end-fed wire was made and signals received using the 20m helical vertical were about 1 S-Point lower compared to the other antennas. Using a MMANA-GAL antenna model, the antenna has a predicted low angle radiation pattern of about 25o to the horizontal and a gain of about -0.35dBi at ground level.

The 20m helical vertical antenna has been used by Roy Emery, G3FYX, when set up at ground level at his QTH. This antenna has enabled many contacts to be made throughout Europe and across the Atlantic using 100w SSB/CW on the 20m band. **Fig 15.39** illustrates the antenna set up at G3FYX's garden.

A 40m version of this antenna has also been constructed using 19m of insulated single core copper wire to form the inductor that comprised about 260 helically wound turns. The construction arrangements were the same as for the 20m helical vertical antenna and it used the same 1:4 Un-Un auto-transformer design. The antenna was tuned on 7.1MHz using the same technique as used for the 20m antenna. Received signals were about 1 to 2 S-Points lower compared to the G5RV and λ/2 antennas. However, this was to be expected for a very short length antenna on 40m. Stations have been worked all over the UK and Europe on the 40m band using this antenna with an ATU using 70w SSB/CW.

## THE CONTROLLED FEEDER RADIATION (CFR) DIPOLE

When an unbalanced coaxial cable is connected to a balanced load, the imbalance between them causes common mode cur-

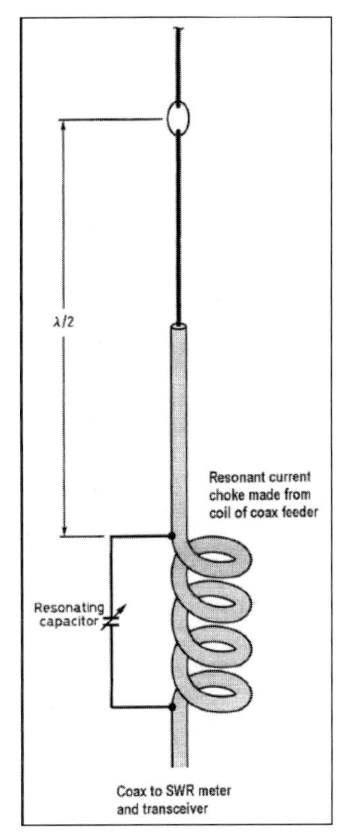


Fig 15.40: Controlled Feeder Radiation (CFR) dipole using a tuned parallel tuned choke made from the coaxial feedline

rents to flow along the coaxial cable screen's outer surface. An explanation of how common mode currents can occur due to an imbalance is described within the Transmission Lines chapter.

Common mode currents are usually undesirable because they can cause RF interference on transmit and an increase in noise on receive. However, their occurrence can be made use of because the coaxial cable's outer shield radiates and receives RF signals in a similar way as the wire connected to the cable's centre conductor. In other words, the coaxial cable becomes part of the radiation and reception characteristic of the antenna rather than its designated use as the transmission line.

If a high impedance choke is placed a quarter wavelength (λ/4) down from the feed point, the coaxial cable's outer screen will function as the counterpoise to the vertical element connected to the cable's centre conductor. This is the basis of the Controlled Feeder Radiation (CFR) antenna. The concept of the CFR dipole is shown in **Fig 15.40**.

This antenna's main advantage is mechanical simplicity and because it is a resonant antenna, there is no requirement for a separate ATU. The antenna can be stored using a cord reel and unwound when required for use. This makes the CFR dipole a practical and compact antenna for portable operation.

However, the CFR dipole is a single band antenna and the impedance of the choke is an important factor affecting the antenna's performance. Unfortunately, ferromagnetic ferrite baluns have been found to have an insufficiently high impedance

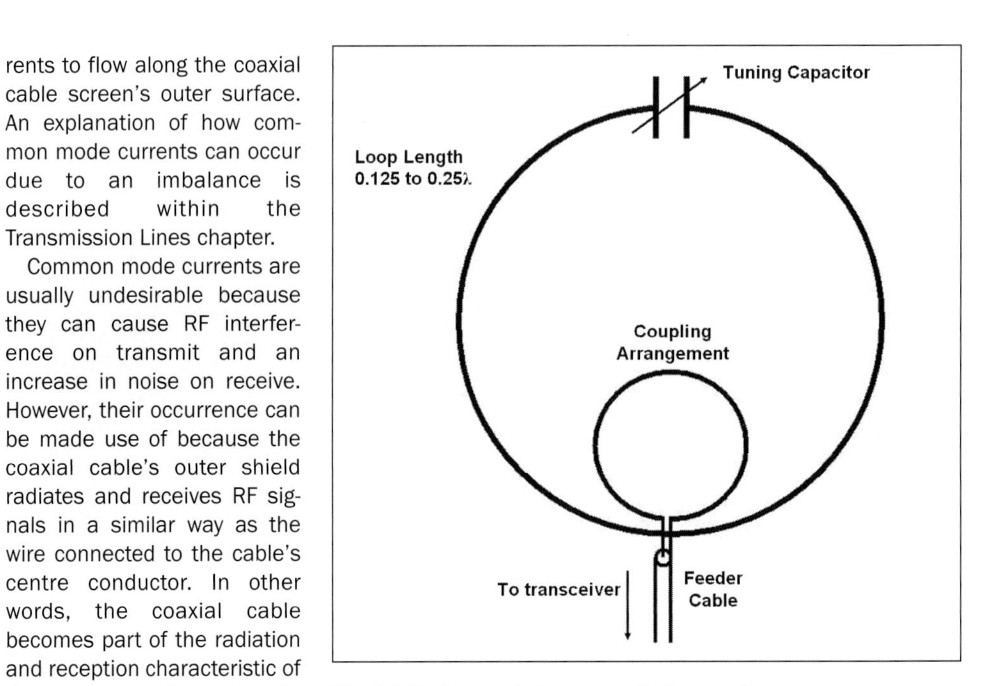


Fig 15.41: Concept of a magnetic loop antenna

for this application (this form of choke is described in the Transmission Lines chapter). An improved choke can be constructed by winding the coaxial cable feeder into a coil and using a capacitor to ensure the correct resonant point, see **Fig 15.40**.

A 14MHz version of this antenna can be constructed by taping 5.03m (16.5ft) of 16SWG multi-stranded insulated copper wire to a fibreglass support. The bottom end of this wire is connected to the centre conductor of the coaxial cable feeder (eg RG super eight or RG58). A suitable coaxial current choke for 14MHz can be constructed from nine turns of the coaxial cable wound on a length of 11cm plastic pipe and tuned with a 15pF air-spaced capacitor. This choke exhibits a high impedance of over 12kΩ at resonance and is placed 5.2m (17ft) from the feed point.

## HF LOOP ANTENNAS

If the space at the location is restricted or a wire HF antenna is undesirably conspicuous, then this can make it difficult to install a suitable aerial. Possibly, a loop antenna may be a potential solution.

### The Magnetic Loop

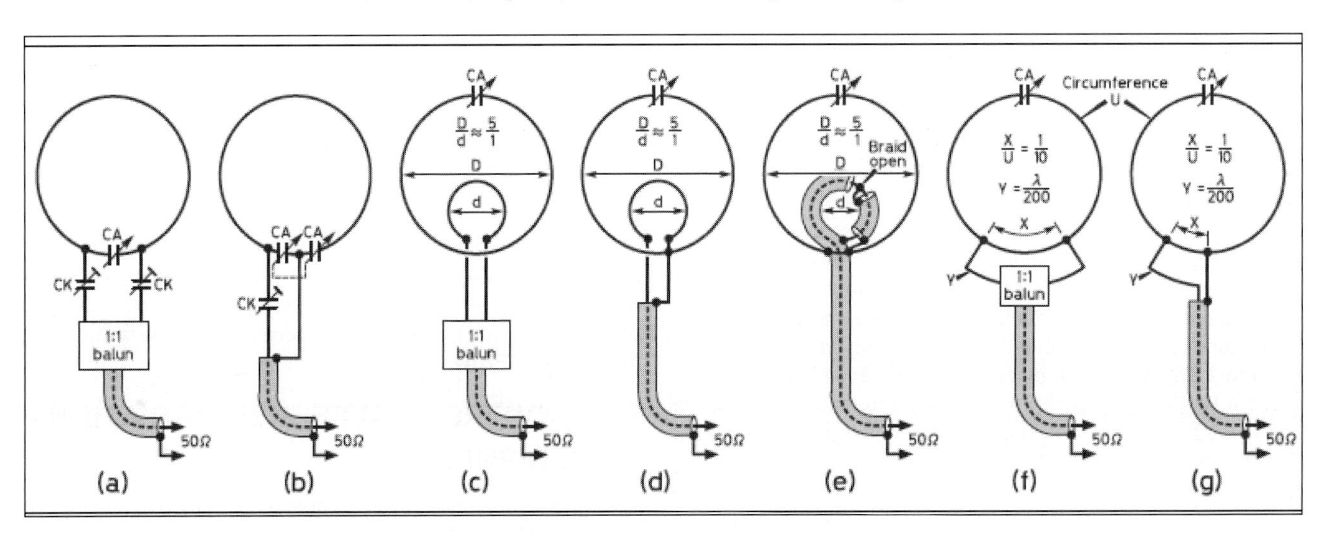


Fig 15.42: Techniques to match a magnetic loop antenna to a coaxial cable

A magnetic loop antenna comprises of a conductive loop, a series tuning capacitor and an arrangement to couple the antenna to the transceiver. The loop's circumference should be between 0.125λ to 0.25λ and this is brought to resonance using the tuning capacitor giving an operating range ratio of typically 1:2 (eg 3.5 to 7MHz, 7 to 14MHz or 14 to 28MHz). The length of the loop is important because if it is longer than 0.25λ, then the antenna tends to behave as an electric field rather than as a magnetic field antenna. Loop lengths under 0.25λ can make it difficult to accurately tune the antenna because the bandwidth becomes small.

The radiation resistance is a function of the area enclosed by the loop and is a maximum for a circular loop. To achieve a good radiation efficiency it is important to minimise the ratio between RF ohmic losses and the radiation resistance. The RF ohmic losses comprise the resistance of the loop and that of the tuning capacitor, although the latter tends to be much lower. Skin-effect causes the RF current to flow on the outer surface of the loop. Therefore, to minimise the loop's RF ohmic losses, a large diameter copper conductor (eg 20mm diameter or larger) should be used to attain a highly-conductive surface area. If it is possible to use a silver-plated large diameter conductor for the loop, then this can further improve the surface-area's conductivity and reduce the loop's ohmic losses. However, a well-constructed loop just 0.15m high can have a radiation efficiency of close to that of a ground plane antenna and does not require a ground plane to work.

The magnetic loop antenna is a tuned 'LC' circuit and it has a high Q-factor. Therefore, the RF voltage across the tuning capacitor is high even when using low transmit powers and it is directly proportional to the power. A good quality transmitting-type capacitor should be used that is able to cope with the high RF voltages involved. Good practice is to use a split-stator type capacitor of about 120pf per section. Each section is connected in series to eliminate rotor contact losses associated with conventional capacitors. Many magnetic loop antennas use a vacuum capacitor to tune the antenna because this type of capacitor is intended to be capable of handling high RF voltages across it. A magnetic loop's high-Q characteristic also results in a narrow effective bandwidth and requires accurate retuning for quite small changes in frequency. This can be overcome by the use of a remote control system designed to fine tune the capacitor to bring the antenna to resonance. The concept of a magnetic loop is shown in **Fig 15.41** for reference.

Several arrangements can be used to couple the transceiver to the magnetic loop, as shown in **Fig 15.42**. However, the usual technique is to use a Faraday loop constructed from a length of coaxial cable (eg RG58) as shown in **Fig 15.42**e. The diameter of the coupling-loop is about one-eighth of the main loop.

For reception a 'magnetic' antenna is much less susceptible to the electric component of nearby interference sources. The reduction of man-made noise is particularly important on the lower-frequency bands and is further enhanced by the directional properties of a loop. The high-Q characteristics of a magnetic loop provide an excellent filter in front of a receiver, reducing overload and cross-modulation from adjacent strong signals that can impair the received signal. On transmit, these high-Q properties dramatically reduce harmonic radiation and may reduce some forms of TVI and BCI (television and broadcast interference).

The magnetic loop utilises the near-field magnetic component of the electromagnetic wave and this results in much less RF absorption in nearby objects. This gives the magnetic loop the advantage that it can be used successfully indoors or on a balcony.

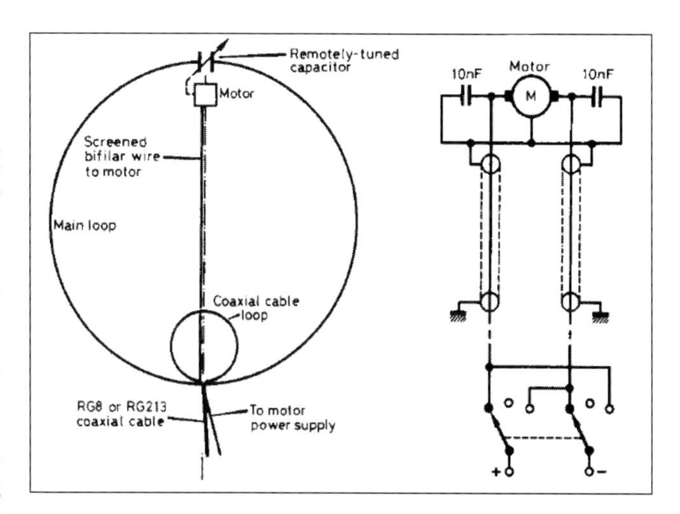


**Fig 15.43: Concept of the I1ARZ loop antenna including the tuning motor connections**

**SAFETY**

When using a magnetic loop antenna in transmit mode, it is good practice for the operator to be situated at a suitable distance from the antenna due to the strong radiated field strengths involved. Therefore, usually several metres of coaxial cable are necessary to connect the antenna to the transceiver.

## A Magnetic Loop for 14 to 29MHz

This magnetic loop design is by Roberto Craighero, I1ARZ [19] and he began experimenting with small diameter transmitting loops in 1985. I1ARZ is convinced that the loop is a thoroughly practical antenna that should not be considered as a compromise as an antenna. The main characteristics of I1ARZ's antenna include:

- 1m diameter circular loop, made from 22mm diameter copper pipe giving a circumference of 3.14m
- Split-stator 'butterfly' type capacitor of about 16 to 120pF per section
- A small loop made from coaxial cable used as the inductively coupled feed arrangement
- Maximum power capability of about 100w
- Remote control of tuning of capacitor using electric DC motor and reduction gearing.

The antenna's electrical characteristics were calculated using the formulas given by W5QJR [20] and are illustrated in **Table 5**. The overall design of the antenna is shown in **Fig 15.43**.

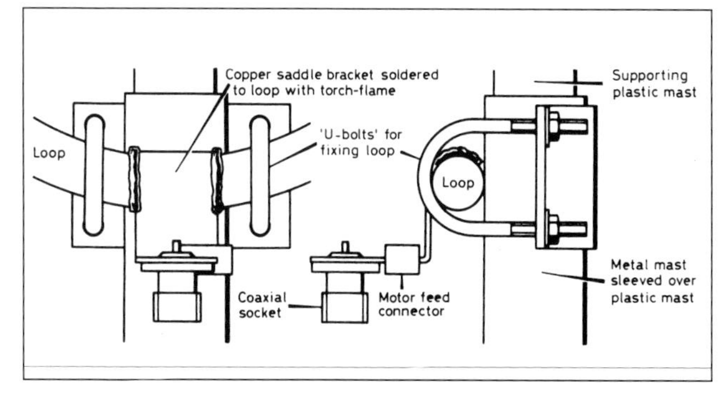


**Fig 15.44: Details of the bottom of the loop - front view**

## The loop

22mm outside diameter copper pipe was used to form the antenna's circular loop using a pipe-bending tool. Alternatively, the pipe may need to be taken to a professional to be bent. With the loop held vertically, each pipe end was longitudinally sawn making a 5cm cut along the vertical diameter of the pipe. Then for both ends, the upper half was cut away and the remaining lower half was flattened to form a strip. Later, these strips are to be passed through the insulated board that supports the tuning capacitor and suitably bent to allow them to be soldered to each stator section of the capacitor.

At the bottom of the loop, opposite where the tuning capacitor was mounted, a small copper bracket was soldered to the loop as shown in **Fig 15.44**. This bracket was used to attach the coaxial connector and the twin-lead connector for powering the tuning motor. The bracket should be soldered to the loop to ensure good electrical contact.

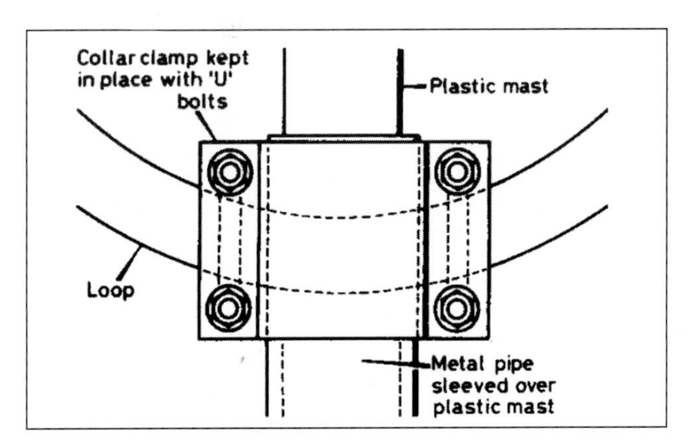


**Fig 15.45: Fixing the lower point of the loop to the support mast**

## External loop supporting mast

A stout 1.5m length of PVC pipe of about 40-50mm in diameter was used for the loop's support. About 200mm at the top of the PVC pipe was used for fixing the plastic board carrying the tuning capacitor. The remaining length, at the bottom, was used to attach the loop to the short mast. It is important not to use a metallic pipe as the loop's support because this will affect the RF performance of the antenna.

The lower section of the loop was fixed to the mast, using two U-bolts at the bottom, as shown in **Fig 15.45**. The ends at the top of the loop were held in place by two cast aluminium TV collar-clamps. The clamps were secured using stainless steel nuts and bolts to the back of the plastic board supporting the capacitor as shown in **Fig 15.46**. The bolts should sufficiently long to act as adjustable spacers to enable the loop to be mounted upright. The plastic supporting mast is fixed to the back of the board using two semi-circular clamps secured using suitable stainless steel nuts and bolts passed through the plastic board. The two copper strips of the loop must be bent at 90° and inserted through suitable holes in the board to reach the stators of the capacitor on the front side of the board as shown in **Figures 15.46** and **15.47**. Later these holes should be waterproofed using silicone compound.

## Tuning capacitor's board and cover

The size of the plastic board used depends upon the dimensions of the variable capacitor and motor. Suitable board materials for high-power operation are Teflon or Plexiglas (Perspex) of 10mm thickness. When calculating the size of the board, allow space for fixing the clamps of the loop and for a waterproof box to house the tuning capacitor, gearing and motor to protect them. The weatherproof box's base had an opening cut that was just wide enough to take the capacitor, gearing and motor. A layer of soft rubber was inserted between the surface of the supporting board and the waterproof box to act as a seal. The box was then fixed in place on the board using small stainless nuts and bolts that were tightened to compress the rubber layer making a watertight seal. The plastic box's lid was then pushed on and kept in place with a tight nylon lashing. Silicone compound was applied all round to seal the lid making the arrangement watertight. To prevent gradual UV deterioration of the cover from sunlight, the box and lid were painted with white external grade paint to protect them.

## The tuning capacitor

It is important to use a good quality transmitting-type variable capacitor that is capable of handling the high RF voltages encountered. With an RF power of 100w, the RF voltage can be between 4 - 5kV and it is recommended that a split-stator (or 'butterfly') capacitor of about 120pF per section is used. The advantage of this arrangement is the two sections are connected in series, thereby eliminating rotor contact losses that occur in conventional capacitors. Assuming that the loop is intended for use with a transmitter RF power of up to 100w, the spacing between the vanes should be at least 1.5-2mm. A vacuum capacitor would also be a good choice, although the high loop

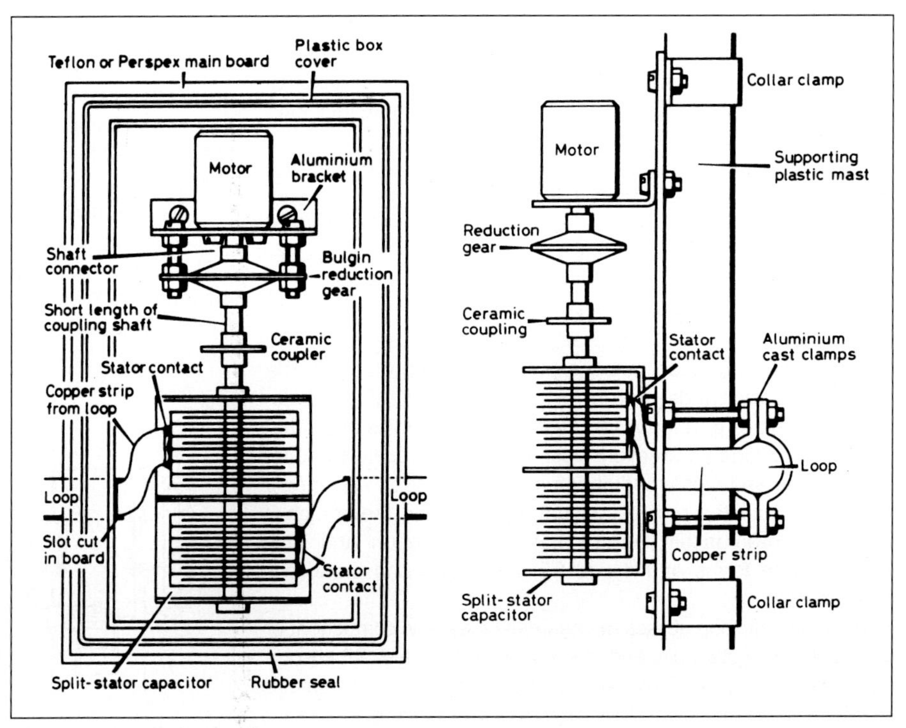


**Fig 15.46: The tuning capacitor's board's front and side views**

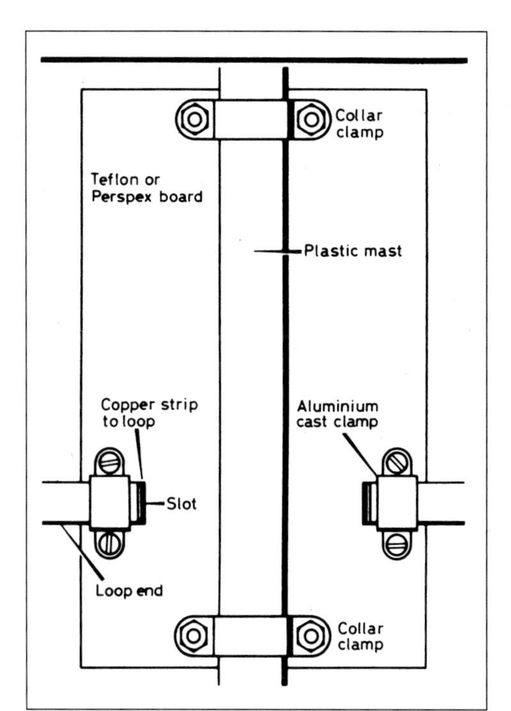


**Fig 15.47: Rear view of the tuning capacitor's board showing loop and support mast mountings**

currents tend to heat and distort the capacitor's thin metal. Consequently, this effect can cause detuning of the loop [21]. Experimenting with a vacuum capacitor tuned low-frequency loop showed that with an SSB signal (of 60w PEP), with its low power factor, there was no need to retune the antenna. However, when used with a CW signal, with its greater power factor, this usually resulted in the antenna needing to be retuned.

## The tuning capacitor's motor

To finely tune the antenna, a DC motor with a reduction gear capable of rotating the capacitor's shaft at a rate of about one turn, or less, per minute was used. The variable speed motor used was able to provide a slow rotation for accurate tuning while having a faster rate for changing bands. The motor I1ARZ used could operate between 3 and 12V, running slowly at the lowest voltage and more quickly at 12 volts. If it is not possible to find a suitable motor incorporating a suitable reduction gear, then it is possible to use a receiver-type slow-motion tuning drive with a ratio of about 25:1.

## Construction of the tuning system

When estimating the dimensions of the insulated supporting board, the following should be considered:

- The space required for the watertight cover
- The aluminium bracket for mounting the motor and the external reduction gearing
- The various couplings between the capacitor spindle and the motor.

The first step is to mark the centre line of the board, the capacitor is then bolted to the board ensuring that the shaft is aligned with the centre line marked on the board. The split-stator capacitor was also placed with the respective stator contacts symmetrically aligned in the vertical plane. This was to allow the loop's copper strips coming from the rear of the board to have the same lengths to each side of the capacitor (one strip was bent upwards, the other downwards). With a 'butterfly' or conventional capacitor, the copper strips must be bent horizontally as both contacts are the same height. Once the capacitor has been bolted to the board, the distance of the board from the centre of the capacitor shaft was accurately measured. This dimension was transferred to the centre line of the vertical side of the L-shaped aluminium bracket used for supporting the motor and an external reduction gear. A pilot hole was drilled just large enough to take the motor shaft. This assembly was carried out carefully because it was vital to the accurate alignment of the system. Once it had been verified that motor and capacitor shafts were in alignment, the motor was permanently fixed to the bracket, the pilot hole enlarged and holes drilled for the motor's fixing bolts. However, the bracket was not attached to the board at this stage. The next step was to adapt the motor shaft to take a shaft extension taking care not to introduce any eccentricity. If the motor used does not require external reduction gearing, the motor shaft can be inserted into a ceramic coupler (circular shape with ceramic ring and flexible central bush) ensuring everything is aligned. The lower flange of the aluminium bracket was then fixed to the board using nuts and bolts.

A Bulgin drive external reduction gear was used between the motor and the capacitor. This had two 4mm holes drilled in the bracket either side of the motor at the same distance as the mechanical connections of the gear and at the same level as the centre of the motor shaft. If the size of the motor used is wider, it may be necessary to add two short strips of brass or aluminium to the Bulgin gear to extend the fixing points of the drive. Two long 4mm diameter brass bolts were used to hold the gear in position and secured with nuts. The Bulgin gear was fixed to the motor shaft, with the other side of the gear connected to the ceramic coupler by means of a short shaft. A check of the tuning system was made by temporarily connecting the power supply to ensure that everything was working smoothly. The loop's copper strips were then soldered to the stator capacitor vanes. This required the use of a large wattage soldering iron, taking care that the best possible electrical contact was achieved.

## Motor feed-line

The feed-line was made from screen Hi-Fi twin-cable. The braid was connected using a tag attached to the motor's aluminium bracket. The motor was bypassed for RF using two 10nF ceramic capacitors connected to the braid. The feed-line cable was run along the supporting plastic mast and secured using cable-ties. At the base of the loop, a connector was soldered onto the small copper bracket and feed-line's braid soldered to the bracket.

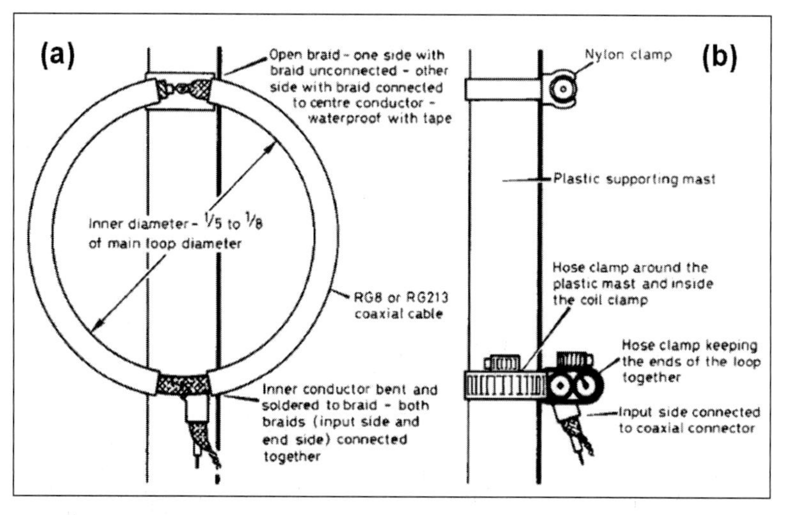


**Fig 15.48: Detail of (a) construction of the coaxial coupling-loop feed and (b) the plastic support mast**

Normal electrical twin cable was then run from this point to the operating position. There have been suggestions that feedline could be inserted inside the loop pipe, however this could reduce the efficiency of the antenna. A small box containing the DC power supply and the switch for reversing polarity of the supply was made up to allow operation from the shack.

## Coupling-loop and matching procedure

I1ARZ found that the most satisfactory coupling method was using a small single-turn Faraday coupling-loop made from coaxial cable (RG8 or RG213). This loop had a diameter one-eighth of the main loop. In practice, the optimum diameter of the coupling-loop may vary slightly and it may be worthwhile experimenting with slightly different sized loops until lowest SWR over a wide frequency is found. I1ARZ found the coupling-loop's optimum diameter was 18cm rather than the theoretical 12.5cm.

The coupling-loop should have the braid open at the top-centre, at this point one side is connected to the inner conductor of the coaxial cable. At the base of the coupling-loop, the inner conductor and braid are connected together and jointed to the braid on the input side of the coil as shown in **Fig 15.48**(a). The ends and braid of the coupling-loop are held together using a stainless steel hose clamp. This arrangement, in turn, is fixed to the mast at 90° to another hose clamp on the plastic mast as shown in **Fig 15.48**(b). This provides a method to adjust the coupling-loop by sliding it up or down the mast to find the best SWR position. The upper opening was protected with tape and, to avoid movement of the coil, it was fixed to the mast by means of cable-ties.

Final matching of the antenna was carried out after determining the final position of the installation. An SWR bridge was connected at the base of the loop close to the input coax connector. Alignment tests were made on 18MHz and then on 21MHz by applying minimum RF power. After the loop's resonance was found, the coupling-loop was moved up and down for the lowest SWR. The coupling coil must be maintained in the same plane as the loop. After finding the lowest SWR, the hose clamps and nylon clamps were tightened to keep the coupling-loop in position. The coaxial cable and tuning motor power line were kept vertical for at least 1m from their connectors at the base of the loop to avoid undesirable coupling with the loop itself and any subsequent difficulty in achieving a good match. The minimum SWR should be better than 1.5:1 on all bands.

## Installing and using the loop

The loop can be conveniently installed on a terrace, or concrete floor or roof.

An option to use as a stable base for the antenna is a large sun umbrella pedestal of the type that can be weighted down by filling it with water or sand. Light nylon guy lines can be used to minimise the risk of the loop falling over in high winds.

A transmitting loop operates effectively at heights of 1 to 1.5m above ground, and little will tend to be gained by raising it any higher than say 2 or 3m at most. I1ARZ tested the loop using a telescopic mast up to 9m above ground and found little difference in its performance. Therefore, I1ARZ uses the antenna at about 3m high. With a garden, the loop could be fixed directly to a short metallic mast driven into the ground. A small TV rotator could be used but this is not essential. The maximum radiation is in the plane of the loop and the minimum off the sides of the loop. Close large metallic objects can reduce the efficiency of the antenna in the direction of the plane of the loop (eg fences, piping, posts). The radiation is vertically polarised at all vertical angles making the loop suitable for DX, medium and short range contacts. There is nothing particularly complicated about operating with a small loop antenna other than the need to tune it to resonate it at the operating frequency. Initially, tune the loop for maximum signals and noise in 'receive' mode, this will bring the loop close to the tuning point for transmission. Then using the SWR meter, tune carefully with the aid of the polarity-reversing switch to the precise point where minimum reflected power is achieved. The 'receive-mode' procedure should always be used when changing bands.

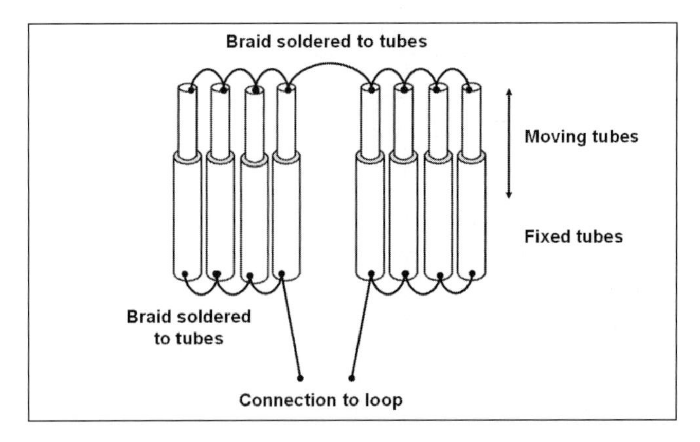


**Fig 15.49: Concept of G8CQX's tuning capacitor constructed using copper pipes and sleeving**

## G8CQX Magnetic Loop

John Hawes, G8CQX, has constructed a magnetic loop able to handle transmit powers up to 100 watts [22]. G8CQX, a keen HF bands operator, needed an inconspicuous antenna for use at his alternative QTH. Therefore, his thoughts turned to building a magnetic loop able to handle higher transmit powers.

Magnetic loops require a tuning a capacitor that is capable of handling the high RF voltages involved, even with comparatively low transmit RF powers.

G8CQX decided to construct a capacitor able to withstand higher RF voltages using parts obtained from a local hardware store. He was daunted by the prospect of fabricating numerous capacitor plates, so he had the idea of using a 'telescopic' approach for the tuning capacitor rather than using parallel plates. He made up a tuning capacitor using lengths of 15mm and 22mm diameter copper pipe. The 15mm diameter pipe was first insulated using three layers of white plastic sleeving (this is sold as central heating pipe covering). This arrangement gave a snug sliding fit inside the 22mm diameter pipe forming a capacitor. When measured, this 'telescopic' capacitor arrangement gave about 15pF to 70pF when the inner pipe was moved in and out.

A split-stator capacitor was constructed using eight 'telescopic' capacitors. Four 'telescopic' capacitors with their stationary sections commonly connected formed each section of the split-stator and the eight sliding inner pipes were commonly connected. This arrangement is shown in **Fig 15.49**. The loop was connected to the capacitor's stationary sections with the arrangement tuned by sliding the inner pipes up and down. This allowed the capaci-

**Fig 15.50: Feeder coupling arrangement to the loop**

Fig 15.51: The G8CQX magnetic loop

tance to vary from under 15pf to more than 150pf (as measured). Essentially, the capacitor required two 2m lengths of copper pipe to construct it.

To hold the capacitor in position, a frame was made using plastic water pipe. This was drilled to hold the static 22mm pipes in place that were then hot glued into position. The capacitor's inner sliding 15mm pipes were held in position using a similar arrangement. Dowels were pushed into the pipe ends and then these fixed into a suitably drilled plastic pipe. These pipes were then connected together by soldering braid to form a common low-resistance connection to accommodate higher RF currents (a 100 watt soldering-iron was necessary for this).

Using more plastic water pipe, some useful pipe fittings and a thick polystyrene sheet, the frame for the loop antenna was completed by gluing it together. A length of surplus URM67 coaxial cable, with its centre and braid conductors shorted at each end, formed the loop and this was connected to the capacitor's fixed stator-sections. The loop's length was trimmed to resonate the antenna from 40m to 17m.

The feed coupling was made using two ferrite rings of about 25mm diameter taped together and slipped over the loop. A reasonable match was found when 4 turns were wound around the ferrite cores. The antenna matched almost perfectly at the lowest frequency and it was still a reasonable match at the highest frequency with an SWR of about 2:1. The feed coupling is illustrated as **Fig 15.50**.

Finally, a motor was glued to the frame that drove a M3 threaded shaft. This engaged with a M3 nut attached to the capacitor's moving stator. A couple of limit-of-travel micro-switches were added in series with the motor to stop the moving stator from moving up and down too far. Three wires were brought back from this arrangement to enable the moving stator to be remotely moved up and down to tune the antenna using a double-pole double-throw toggle switch from the operating position. **Fig 15.51** illustrates the finished antenna.

G8CQX found that the magnetic loop antenna could comfortably handle a transmit 100w signal. As expected the tuning was sharp, however the reduction-drive using the M3 shaft drive made it possible to carefully peak the antenna remotely, with the transition from maximum capacity to minimum taking about 40 seconds. There were no problems from overrunning with the end limit switches suitably stopping the motor. As a comparison, G8CQX tested the magnetic loop against his 120ft long sky wire (about 40ft up) and found the loop about 10dB down. The loop has some directivity and it did exhibit a null (as expected), so it would need to be lined up as desired when in use. As a general receive test, one Saturday when the conditions on 20m were not good, G8CQX used the loop with his SDR# receiver. Most of the stations were around S5, so they could have been workable using CW. The strongest station received was more than +60dB up compared to the system noise floor.

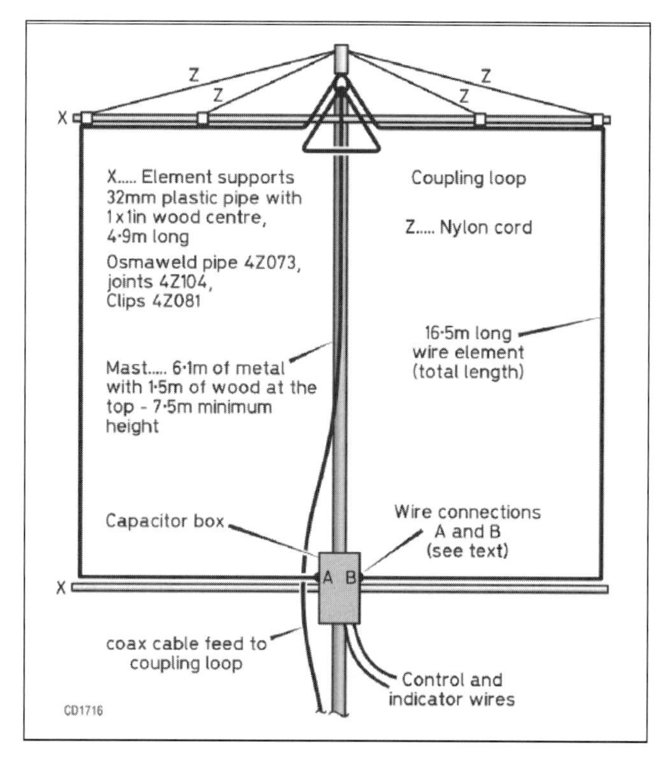


Fig 15.52: Overall view of the lower HF bands magnetic loop

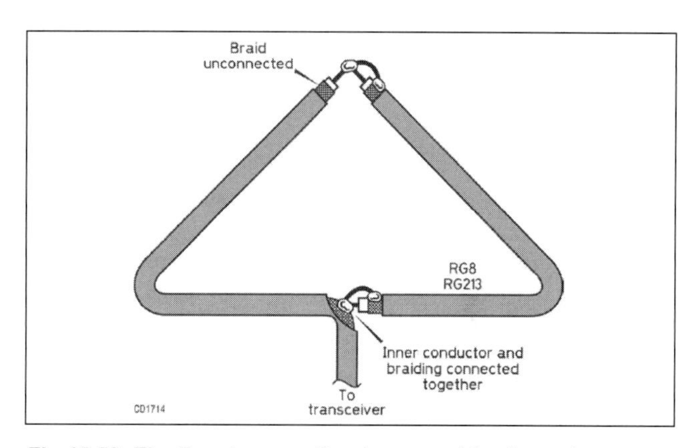


Fig 15.53: The Faraday coupling-loop used for the antenna

This magnetic loop antenna has provided G8CQX with a means to operate at his alternative premises using an inconspicuous antenna that can be readily taken to pieces enabling it to be transported fairly easily.

It is worth considering the RF voltage generated across the ends of the loop, especially when operating with a transmit power level of 100w. When the loop is at resonance, the circuit's Q-factor means that the RF voltage across the capacitor can be of the order of several kilo-volts. An advantage from using a split-stator, compared to a conventional capacitor, is each section shares the RF voltage reducing the likelihood of voltage breakdown between the plates. Such high RF voltages need to be taken seriously from a safety aspect.

## Wire Loop Antenna for the Lower HF Bands

C R Reynolds, GW3JPT, has constructed a different design of a practical loop antenna for the 160m, 80m and 40m bands using a large square loop as shown in **Fig 15.52** [23]. If this antenna was made from copper tube it would be very heavy, therefore GW3JPT used 19.5m (64ft) of plastic covered wire for the loop that was tuned using a 250-300pF tuning capacitor. A

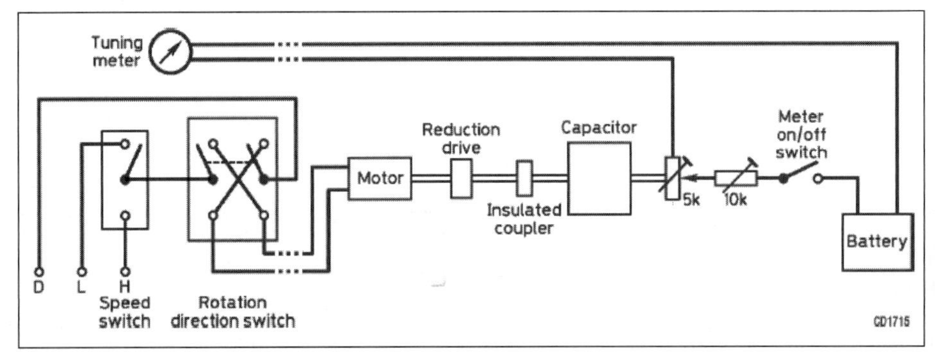


**Fig 15.54: Control and indicator system for the magnetic loop antenna**

Faraday -loop was used as the feed coupling and is shown in **Fig 15.53**. This was close coupled for about 0.77m (30in) each side of the centre of the triangle section of the element as shown in **Fig 15.52**. The wire loop will also work on 40m using a switch (or relay) to disconnect the capacitor at points A and B as shown in **Fig 15.52**. The loop is then tuned by the stray capacitance of the switch with the antenna element's length adjusted for correct matching using an SWR meter because the stray capacity cannot be varied.

The antenna and mast were fitted to a good ground post, did not need any support from guy wires and could be raised or lowered easily. For portable use, the antenna could be raised in a few minutes using three or four guy wires.

### Capacitor drive motor

A barbecue spit motor was used as the drive to tune the capacitor. This motor was chosen because it was geared down, however it did require extra reduction using a 6:1 or 10:1 epicyclical drive for more precise tuning. This motor rotated slowly when energised by a 1.5V battery and with 3V applied, it would run much faster. By switching from 1.5 to 3 volts a slow or fast tuning speed could be selected. A motor control unit was built and is summarised in **Fig 15.54**. The positive lead of the 3V battery is connected to H and the positive lead of the 1.5V battery is connected to L. The negative leads of both batteries are connected to D. The direction of rotation is achieved using a two-pole three-way switch. When the switch is set to the centre position the motor is disconnected from the battery ('OFF' position). The battery polarity to the motor is selected by the two other positions of the switch and should be labelled 'DOWN' or 'UP'. The drive mechanism must be electrically isolated from the high RF voltages present at the capacitor and an insulated coupler was made from plastic petrol-pipe to do this. This pipe size should be chosen so that it is a push fit onto the drive mechanism and capacitor shafts. The pipe was secured to the shafts by wrapping single strand copper wire around the ends of the pipe and tightening with a pair of pliers.

The tuning capacitor was made by GW3JPT and had the spindle extending out from both sides of the capacitor. One spindle was used to couple the capacitor to the drive mechanism and the other was used to turn a position indicator. The indicator circuitry was also electrically isolated from the capacitor as described above. The control unit was housed in a plastic box with the fast/slow and rotation direction switches fixed to the front, together with the capacitor position meter.

### Capacitor unit housing

A major problem when constructing any electrical circuits associated with antennas is protecting them from the weather. GW3JPT made a box to house the loop's tuning arrangements from exterior grade plywood. The bottom and sides of the box were fixed together using 25mm (1inch) square strips of timber. Glue and screws were then used to make the box's joints waterproof. The top of the box was made to be removable to allow access to the equipment inside. The box was painted with several coats of paint to protect it from the weather.

### Construction of capacitors

Capacitors that are able to handle the high RF voltages involved when tuning loop antennas are not easy to obtain. Therefore, GW3JPT made up a capacitor as shown in **Fig 15.55**. GW3JPT used aluminium and double-sided circuit board for the vanes, with nuts and washers used for the spacers. Insulating material was used for the end plates. The centre spindle and spacing rods were constructed from 6mm-threaded plated steel rod.

**Fig 15.55: GW3JPT's home constructed capacitor**

The end-plates were made first and were 76mm x 76mm (3 x 3in) as shown in **Fig 15.56**. These were taped together, back-to-back, for marking and drilling. The same was done with the vanes. Masking tape was used so the surface was not scratched around drill holes that were drilled to clear 6mm with the centre hole acting as a bearing. The number of vanes required dictated the length of the 6mm spindle. For double-sided board, washer/nut/washer spacers were used so that copper sides did not need to be electrically bonded first. The resulting spacing was about 6mm (0.25inch). The plate shape illustrated in **Fig 15.56**(a) was used because this was straightforward to cut out. The fixed vane was a simple rectangle as shown in **Fig 15.56**(b). This plate could be modified to reduce the minimum capacity as shown as the dotted line in **Fig 15.56**(c). For the size shown, six pairs of vanes with 6mm (0.25inch) spacing work out to about 150pf. Units using both printed circuit board and aluminium vanes have been used for about two years and both were still in good working condition at the time of writing.

### Operation

Loop tuning needs to be adjusted precisely for minimum SWR and should coincide with maximum RF power out. The tuning is critical because a few kilohertz off tune caused the SWR to rise dramatically. The best way of finding the correct position of the tuning capacitor was to listen for maximum noise, or signals, on receive whilst tuning the loop, then fine-tune using an SWR meter. The performance of this antenna on 80m was at least as good as a G5RV. It tuned all of 160m and gave good results when compared with local signals on the club nets.

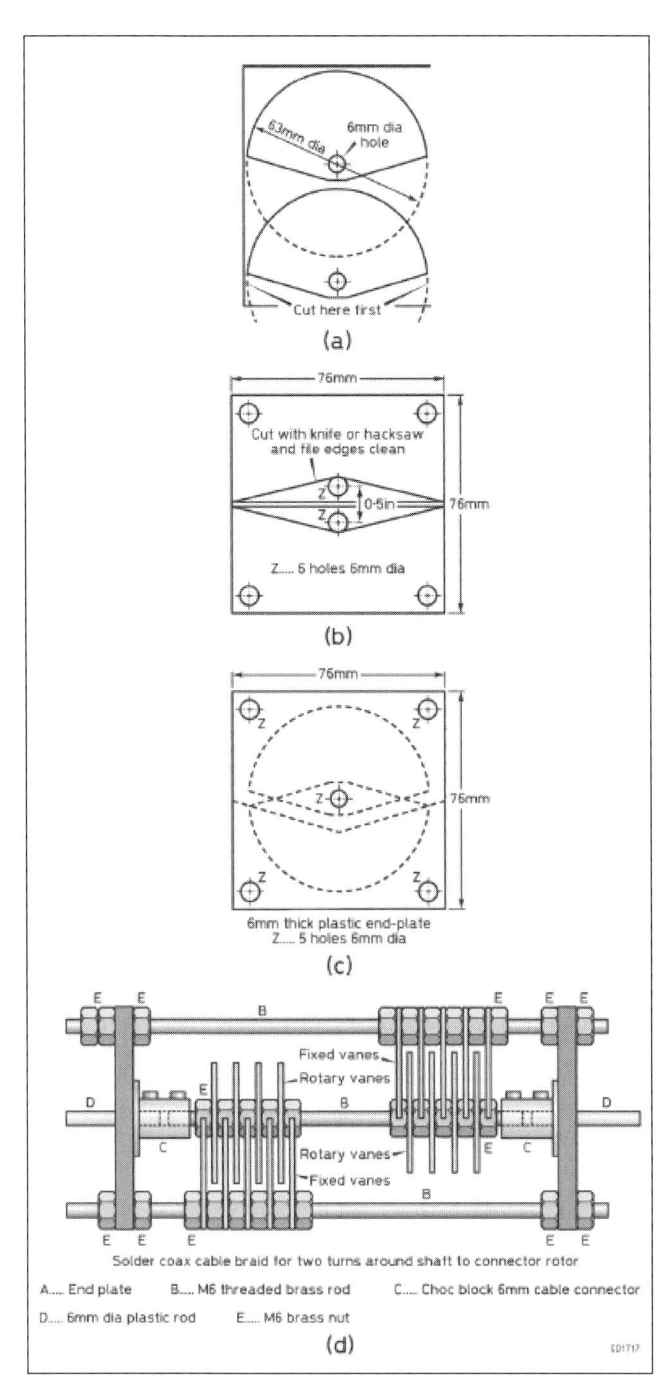


**Fig 15.56: Details of home constructed capacitor: (a) moving vanes, (b) fixed vanes, (c) fixed and moving vanes geometry showing minimum capacitance, (d) capacitor assembly**

## A Magnetic Loop for QRP Operation

John Corkett, M0XXF, has constructed a magnetic loop for portable use based on the G4ILO 'Wonder Loop' design [24] [25]. M0XXF was interested in an antenna that could be readily constructed using standard tools, required a minimum of space to accommodate it and could be taken out portable as and when required. After looking at several designs, M0XXF decided to build a magnetic loop mainly for use on 20m and this resulted in the antenna having a primary loop diameter of about 80cm and a smaller inner coupling-loop diameter of about 16cm.

The primary loop was made using a modified 2.5m length of RG213 co-axial cable with the outer braid forming the antenna.

**Fig 15.57: M0XXF Magnetic Loop based on the G4ILO design set-up for operation**

A short section of the inner conductor/insulation was removed at each end and the ends insulated. Then, the outer braid was drawn over the ends and a ring terminal soldered to each end. Using a suitable plastic box to hold the antenna's tuning capacitor, pillar-terminals were added to the outside to enable the primary loop to be connected. Internally, the pillar-terminals' ends were connected to the capacitor's terminals using heavier gauge insulated wire (able to cope with about 20 amps), with these terminations soldered.

The inner coupling-loop carries a lower current compared to the primary loop and was made from 5amp rated copper single core insulated wire as shown in **Fig 15.42**(d). This loop was connected to a length of RG174 50Ω coaxial cable that was run through the plastic box and terminated on a BNC socket mounted on the side. This arrangement allowed the antenna to be connected to the transceiver using a length of coaxial cable.

A central support was added made from white plastic conduit to support the primary loop and inner loop that were hung from a plastic hook as shown in **Fig 15.57**. For ease of portability, the antenna was attached to a camera tripod. A length of RG58 coaxial cable was used to connect the antenna with the transceiver. This arrangement ensured that the operator and the transceiver were at a suitable distance from the antenna when in transmit-mode. A heavier duty air-spaced tuning capacitor was used for the antenna and this handled CW RF signals up to 20 watts on 20m and 17m during testing. The capacitor had a range from about 20 to 350pF when measured. With the antenna tuned, the RF power was kept to 10w enabling numerous contacts to be made on 20m and 17m.

This magnetic loop antenna is a good example of how home-brewed equipment can achieve good results while being built using standard tools and a good soldering iron. The physical size of this HF antenna means that the space required to set it up is not really an issue.

## Comments on Magnetic Loop Antennas

The compact magnetic loop antenna is a potential solution where it is not possible to install an HF wire antenna. The magnetic loop by I1ARZ, in the worst case, is about 4 or 5dB down on a dipole at a height of a half wavelength. This is equates to less than one S-point. However, under conditions of normal HF fading difficulties might be experienced and it becomes difficult to make meaningful comparisons. GW3JPT's compact wire loop is about 60% efficient on160m. Making antenna assessments on this band is even more difficult because of the difficulty of finding a 160m dipole at a height of half a wavelength to compare it with. There has been some comment that the Faraday coupling-loop connections found in most descriptions of loops are incorrect (including the I1ARZ and GW3PJT designs above). The coaxial inner and braid at the top or apex of the loop is joined as shown previously in **Fig 15.42**(e), apparently making it a Faraday half loop. The inner to braid connection should be

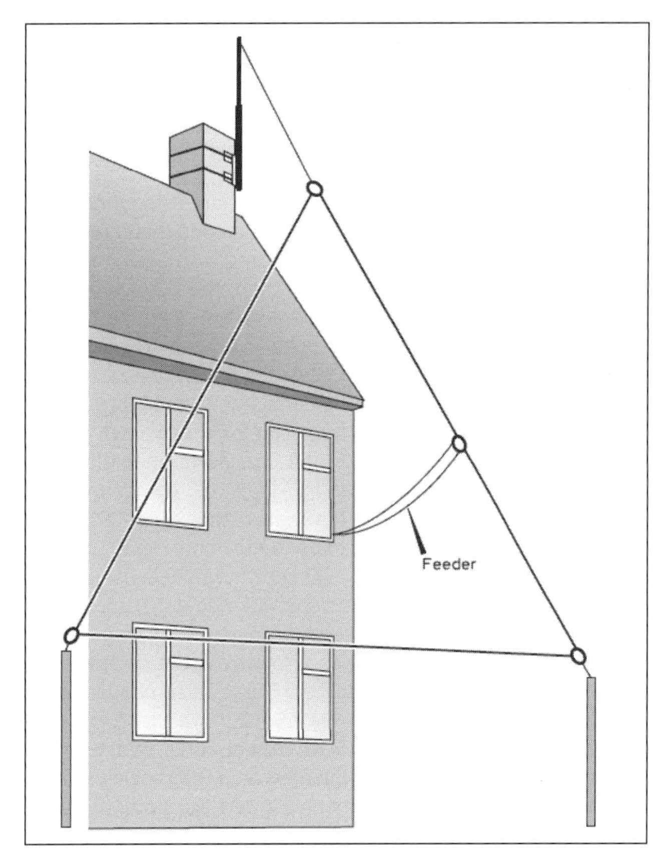


**Fig 15.58: Loop antenna one wavelength circumference on 7MHz**

removed, but the gap in the braid halfway round the loop should remain. The matching methods in **Fig 15.42**(a) and (b) have been used with loops of 10m or more diameter for 136kHz, for both transmit and receive.

### Multi-band Delta Loop Antenna

If there is the space available and the QTH can accommodate a larger loop, then it may be worthwhile trying one. If the loop is larger than 0.25λ it will lose its predominant 'magnetic' characteristic and become an 'electric' field antenna of the quad or delta type. In a similar manner to descriptions of magnetic loop antennas already described, the antenna's efficiency improves with an increase in size because the resistive losses still remain low. A full wave loop on 7MHz can be fed with coaxial cable and will also operate on the 14 and 21MHz bands without an ATU, provided that a transformer and balun are connected between the coaxial cable and the antenna. The shape of the loop is not too important either.

If a loop antenna in the form of an equilateral triangle is used, then only one high support is required as shown in **Fig 15.58**. This support could be a mast fixed to the chimney and, depending upon the situation, may not require planning approval. Part of this antenna is close to the ground and there is a possible danger of receiving an RF burn if the antenna was touched when the transmitter is on. Therefore, insulated wire for the lower half of the antenna is recommended, although the top half of the antenna could be constructed with bare copper wire.

A loop antenna of this type is not a high-Q device. Therefore very high voltages do not occur as would be found at the tips of a dipole for example.

The top half of the antenna can be constructed from bare copper wire, or insulated wire could be used for the entire the loop. However, using lightweight wire for the upper half of the loop, supported by a lightweight mast, has a lower visual impact. Using lightweight thin wire does not affect the antenna performance because the radiation resistance of the loop is fairly high.

The first experiments were carried out for this antenna design with the coaxial cable connected directly to the loop and gave an SWR of over 3:1 because the feed point impedance was greater than 100Ω. A 4:1 balun was fitted, enabling the antenna to be fed directly with 50Ω coaxial cable with little mismatch. The best results occurred when the antenna was fed about one third up from the bottom on the most vertical of the triangle sides as illustrated in **Fig 15.58**. This antenna gave good results even when the lowest leg of the triangle was only 0.6m above the ground. **Fig 15.58** shows the corner insulators fixed to the ground using tent pole type fixtures. However, the lower wire could be run along a wooden fence using suitable insulators to provide isolation.

The apex support used for the experimental loop antenna was a 2.5m length of scaffolding pole fixed to the chimney with a double TV lashing kit. The top of the chimney was about 9m above the ground and the pole gave the antenna enough height to have a reasonable clearance above the roof. The loop proved to be a good DX transmitting antenna on 7MHz. However, this antenna did tend to pick up electrical noise from the house and this had an effect on the reception of some weaker signals. To overcome this problem, this antenna could be used in conjunction with a smaller receive loop to help reduce electrical noise.

## ROTARY BEAM ANTENNAS

The rotary beam has become a popular antenna used for the upper HF amateur bands and many stations using three-element Yagi beam antennas are often encountered. The beam antenna offers power gain, compactness, a reduction in interference from undesired directions and the ability to change the beam-direction with relative ease using some form of rotator. All these aspects have many advantages compared to a fixed type antenna.

All the types of beam described in this chapter are parasitic beam antennas. The optimum element dimensions and their spacing can only be obtained over a narrow frequency range. Therefore, the parasitic beam works only over a relatively restricted bandwidth. However, in most cases, the bandwidth of a parasitic beam antenna is compatible with the width of the HF amateur band in use. For its physical and compact size, the performance of a parasitic beam antenna exceeds that of any comparable antenna and outweighs any disadvantages associated with it.

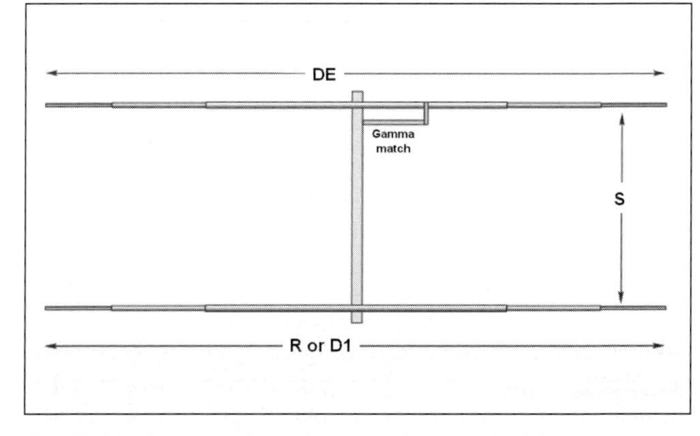


**Fig 15.59: Construction of a two-element Yagi beam antenna with dimension references given for the 20, 18, 15, 12 and 10m bands (see table in text)**

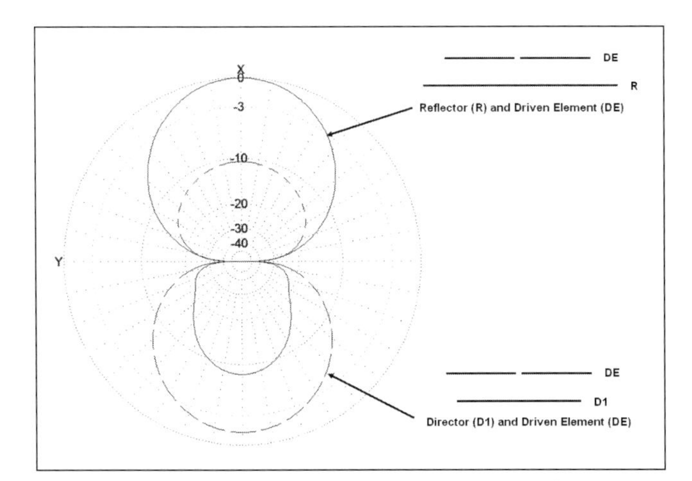


**Fig 15.60: MMANA-GAL predicted horizontal radiation patterns of a two-element beam comprising a driven element (DE) with the parasitic element as a reflector (R) and then as a director (D1)**

An explanation of the operation of the beam antenna that uses parasitic elements is provided within the Antenna Principles chapter for reference.

## Two-element beam antenna

The concept of a two-element beam antenna is shown in **Fig 15.59**. The parasitic element (R or D1) is energised by radiation from the driven element (DE). The parasitic element then re-radiates most of the RF energy. The phase relationship between the driven element's radiated signal and the signal re-radiated by the parasitic element establishes a concentrated signal that is 'beamed' away from the antenna. When the parasitic element is shorter, it operates as a director (D1) causing power gain in the direction towards D1. Conversely, when the parasitic element is longer than the driven element, it operates as a reflector (R) and causes power gain in the direction away from R [26].

When the parasitic element is to be used as a director, the spacing between it and the driven element is around 0.1 wavelength. The spacing when using the parasitic element as a reflector is approximately 0.13 wavelength. The effect of these options can be seen in the MMANA-GAL computer simulation in **Fig 15.60** for a two-element beam. There is only about a 1dB difference in the performance of a two-element beam when the parasitic element is used either as a director (D1) or reflector (R), although sometimes there can be a marginal improvement in the front-to-back ratio (F/B) when the parasitic element is a director. An advantage of a two-element beam where the parasitic element is a director, is that this arrangement is slightly smaller and lighter of the two configurations.

Guidance for the dimensions of a two-element beam is shown in **Table 6** (R & DE) and **Table 7** (DE & D1) for the 20m to 10m bands referred to **Fig 15.59**. These dimensions have been calculated using MMANA-GAL for a two-element beam giving a free space gain of about 4dBd, a F/B ratio of about 9dB and a feed point impedance of around 30Ω. These calculations assume an average tube diameter of 20mm on 28.5MHz, 24mm on 24.95 - 21.20MHz and 30mm on 18.12 - 14.15MHz. In practice the diameter of the tube used is not critical and its diameter should be sufficient to make the antenna mechanically stable. The predicted polar diagrams for both configurations of the two-element beam are shown in **Fig 15.60** where the 180$^{0}$ change in the direction of the radiated main lobe is illustrated.

The elements should be made of sections that telescope into each other as shown in **Fig 15.59**. Aluminium scaffolding pole 50mm (2inch) in diameter is useful material for the antenna's boom for any of the HF bands. The construction of the elements and methods of fixing the boom to the mast are described in the Antenna Construction Techniques chapter. As with all parasitic beams, the dimensions of the parasitic elements determine the antenna's performance. The length of the driven element determines the feed point impedance and this is usually less than 50Ω for a two-element beam. Therefore to feed the antenna with 50Ω coaxial cable requires a matching arrangement (eg the gamma match as described in the Antenna Principles chapter).

| | Frequency (MHz) | | | | |
|---|---|---|---|---|---|
| **Element (metres)** | **14.15** | **18.12** | **21.20** | **24.95** | **28.5** |
| **Reflector (R)** | 10.56 | 8.25 | 7.06 | 5.98 | 5.23 |
| **Driven Element (DE)** | 9.78 | 7.64 | 6.54 | 5.54 | 4.86 |
| **Spacing (S)** | 2.76 | 2.15 | 1.84 | 1.56 | 1.37 |
| **Element Diameter (mm)** | 30 | 30 | 24 | 24 | 20 |

**Table 6: Guidance for HF 2-element beams comprising a Reflector (R) and Driven Element (DE)**

| | Frequency (MHz) | | | | |
|---|---|---|---|---|---|
| **Element (metres)** | **14.15** | **18.12** | **21.20** | **24.95** | **28.5** |
| **Director (D1)** | 9.66 | 3.72 | 6.34 | 5.42 | 4.72 |
| **Driven Element (DE)** | 10.46 | 8.08 | 6.90 | 5.88 | 5.14 |
| **Spacing (S)** | 2.12 | 1.66 | 1.42 | 1.20 | 1.05 |
| **Element Diameter (mm)** | 30 | 30 | 24 | 24 | 20 |

**Table 7: Guidance for a HF 2-element beams comprising a Driven Element (DE) and Director (D1)**

| | Frequency (MHz) | | | | |
|---|---|---|---|---|---|
| **Element (metres)** | **14.15** | **18.12** | **21.20** | **24.95** | **28.5** |
| **Director (D1)** | 9.26 | 7.28 | 6.20 | 5.22 | 2.62 |
| **Driven Element (DE)** | 10.30 | 8.00 | 6.80 | 5.88 | 5.04 |
| **Reflector (R)** | 10.66 | 8.36 | 7.14 | 6.18 | 5.36 |
| **Spacing (S1)** | 3.24 | 2.53 | 2.16 | 1.80 | 1.61 |
| **Element Diameter (mm)** | 30 | 30 | 24 | 24 | 20 |

**Table 8: Guidance for HF 3-element beams comprising a Reflector (R), a Driven Element (DE) and a Director (D1)**

## The Three-element Yagi beam antenna

A three-element parasitic beam antenna consists of a reflector (R) behind the driven element (DE) and a director (D1) in front of it as shown in **Fig 15.61**. This arrangement of antenna is named after its developers Mr. Uda and Mr. Yagi and is know as a Yagi beam antenna. Yagi beam antennas are designed and constructed to maximise the performance of the antenna in terms of its gain and F/B ratio while keeping any side-lobes at a minimum [27]. However, this is a compromise and beam antennas tend to be built to give the highest practical gain for the best F/B

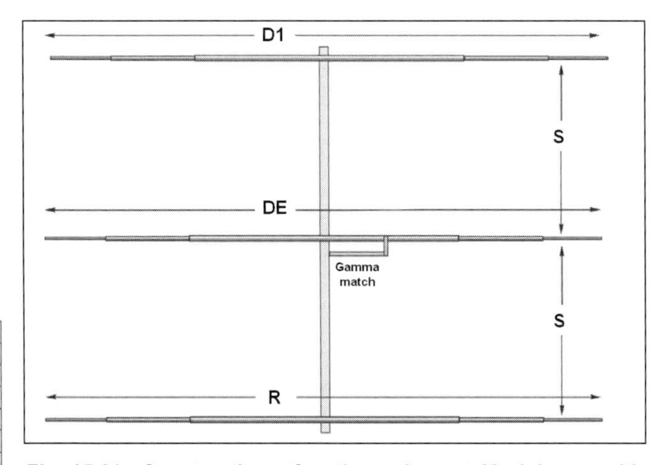


**Fig 15.61: Construction of a three-element Yagi beam with dimension references for the 20, 18, 15, 12 and 10m bands. See table in text for the dimensions**

ratio while keeping any side-lobes as minimal as possible. For a three-element beam, the spacing between the reflector and the driven element is about 0.15 to 0.2 wavelength, while the spacing between the driven element and the director is about 0.15 wavelength.

Guidance for the dimensions of a three-element beam is shown in **Table 8** for the 20m to 10m bands referred to **Fig 15.61**. These dimensions have been calculated using MMANA-GAL for an antenna giving a free space gain of about 5.3dBd and a F/B ratio of about 16.5dB. These calculations assume an average tube diameter of 20mm on 28.5MHz, 24mm on 24.95 - 21.20MHz and 30mm on 18.12 - 14.15MHz. In practice the diameter of the tube used is not critical and its diameter should be sufficient to make the antenna mechanically stable. The feed point impedance of this three-element Yagi beam design is around 40Ω. Therefore, to feed the antenna with 50Ω coaxial cable a matching arrangement is necessary (eg the gamma match as described in the Antenna Principles chapter).

## The Cubical Quad

The Cubical Quad beam is a parasitic array whose elements consist of closed loops having a circumference based on one-wavelength at the design frequency. The concept of a quad beam's construction for the HF bands is shown in **Fig 15.62**. Guidance for the dimensions for a two-element quad is shown in **Table 9** for the 20m to 10m bands referred to **Fig 15.62**.

The parasitic element is normally tuned as a reflector,

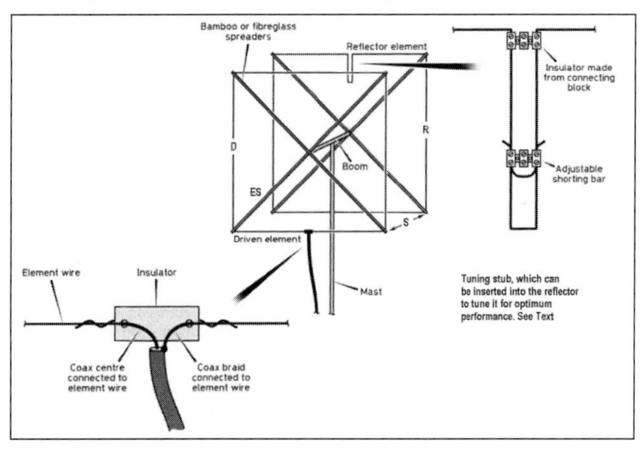


**Fig 15.62: Construction of a two-element wire quad. See table in text for the dimensions**

although this element could be tuned as a director. However, using the parasitic element as a director tends to make the gain and F/B ratio more inferior, consequently making the optimum settings even more difficult to obtain. The reflector (R) can be

| Frequency MHz | 14.1 | 18.1 | 21.2 | 24.9 | 28.5 |
|---|---|---|---|---|---|
| (S) Element Spacing (m) | 2.98 | 2.34 | 1.99 | 1.70 | 1.49 |
| (R) Reflector length (m)* | 5.56 | 4.38 | 3.73 | 3.17 | 2.77 |
| (ES) Element support length (m) | 3.93 | 3.1 | 2.64 | 2.24 | 1.96 |
| (D) Driven Element (m)* | 5.33 | 4.18 | 3.57 | 3.04 | 2.65 |
| *Note: These dimensions are for one side of the quad. The total length of the element is four times this figure. | | | | | |

**Table 9: Dimensions for a two-element quad beam for the upper HF bands. Refer to Figure 62 for D, S, R and ES. These dimensions have been calculated using the EZNEC application for a non-critical design to give a free-space gain around 6dBd and a front-to-back ratio greater than 20dB**

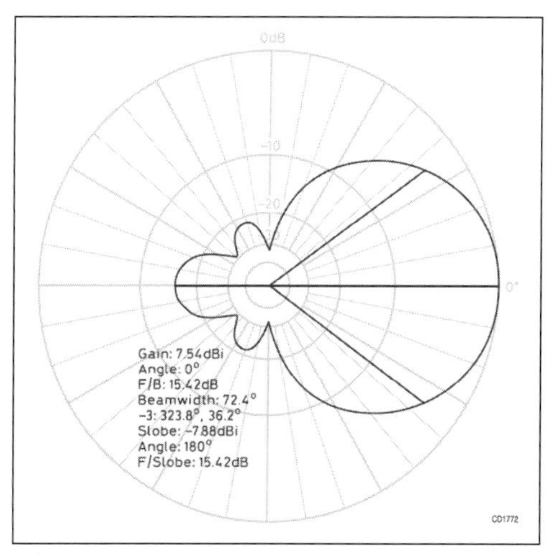


**Fig 15.63: Computer analysis of a two-element wire element quad with 0.14 wavelength element spacing**

constructed using the same dimensions as the driven element (D). However, the reflector needs to resonate at a slightly lower frequency and this can be done using a tuneable stub. This stub can also be adjusted to enable the best front-to-back (F/B) ratio of the beam to be obtained. The dimensions given in **Table 9** are for a quad using an element spacing of 0.14 wavelength. The computed free-space performance is shown in **Fig 15.63** where a gain of around 5.5dBd (7.5dBi) and a F/B ratio of about 15dB are predicted.

The lengths of the element supports (ES) are also given within **Table 9**. The lengths of these supports should be made slightly longer because dimension ES is the point where the element is connected to the support.

The feed impedance of the quad using the dimensions shown in **Table 9** is around 65Ω, therefore the driven element (D) can normally be connected directly to a 50Ω feed line with a low mismatch. The 0.14 wavelength spacing (S), given in **Table 9**, was chosen because it is the most prevalent in antenna literature.

The quad can be made into a multi-band antenna by interlacing quad loops for the different bands onto a common support structure. In this case the element support lengths (ES) should be the length for the lowest frequency band. The disadvantage of this arrangement is that the wavelength spacing (S) between the driven element (D) and the reflector (R) is different on each band. This problem can be overcome by using an element support structure with a modified geometry. This structure, that is devised to hold the element supports in place at the correct angles, is often referred to as a 'spider' and an example is shown in **Fig 15.64**. A multi-band quad using this type of geometry is often referred to as a 'boom-less' quad. Dimension (ES) will have

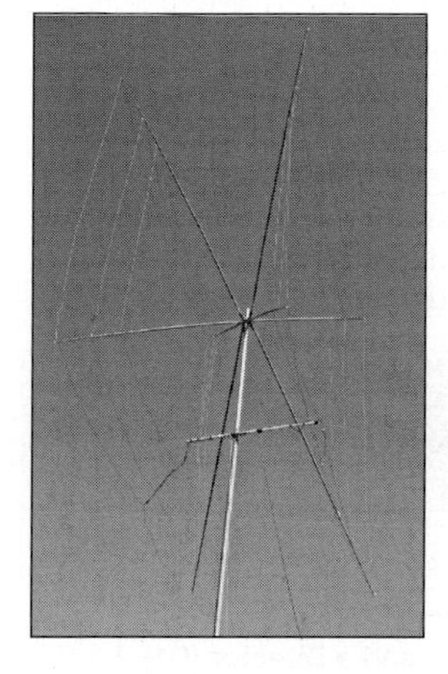

**Fig 15.64: A boom-less multi-band quad for the 14, 18 and 21MHz bands using the aluminium lightweight Labgear boom-less spider. An extra long stub is used on the 14MHz reflector because the element supports were too short to accommodate the size of the element required**

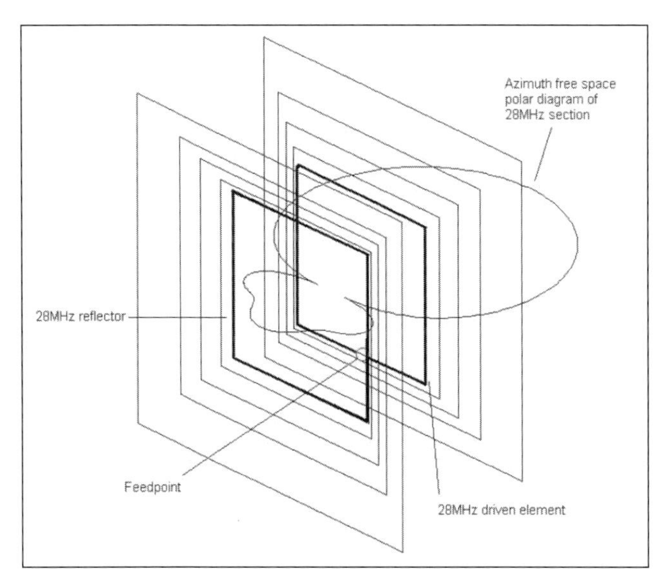


**Fig 15.65: Computer model of a five-band quad covering the 14 to 28MHz bands. Only the 28MHz band (highlighted with thick lines) is shown energised and the resultant horizontal free space polar diagram**

to be increased by around 5% with this boom-less arrangement. Methods of fixing cane or fibreglass element supports to booms are given in the Antenna Construction Techniques chapter.

Some designs of multi-band quad beams have the driven elements all fed in parallel. L.B. Cebik, W4RNL, has investigated multi-band quad beam antennas fed using this method and has found that this can potentially affect the antenna's performance [28]. When there is a harmonic relationship between any of the elements, then these elements will radiate RF signals that collectively can distort the radiation pattern of the beam. For example, taking a three-band quad for 14, 21 and 28MHz that is energised on 28MHz. The 14MHz driven element also presents good matching impedance to the feeder line because it is a two-wavelengths loop at 28MHz. The effect of this is a degradation to the desirable quad directivity pattern on 28MHz.

W4RNL constructed a computer model of a five-band quad using the boom-less configuration as shown in **Fig 15.65**. Each band was energised in turn and this demonstrated that the performance on each individual band could be as good as a single band quad. Therefore, the interpretation of this is to use a separate feeder for each band for a multi-band quad. However, to feed a five bands quad using this method would require five separate feeder cables to be run and this tends to become impractical. Therefore, to feed a multi-band quad, usually a single feeder cable arrangement is used that involves a remotely operated changeover switch at the feed point to change between the bands.

The quad beam shown in **Fig 15.64** for the 14, 18 and 21MHz bands was originally constructed with the driven elements connected and fed in parallel, however this proved unsuccessful. The driven elements were then fed using a length of 300Ω ladder-line feeder with coaxial cable used as the feeder to the 21MHz element as shown in **Fig 15.66**. This temporary arrangement was replaced with 450Ω ladder-line and soldered connections used instead of connector blocks. An Antenna Analyser AIM 4170 was used to assess the impedance characteristics of the quad antenna when fed using this ladder-line feeder arrangement. It was found that there were clear points of low SWR on the 14, 18 and 21MHz bands indicating that this ladder-line based technique provided a good match. This feeding technique has weathered winter storms without any problems. However, using this technique to feed all the upper HF bands has not yet been fully investigated.

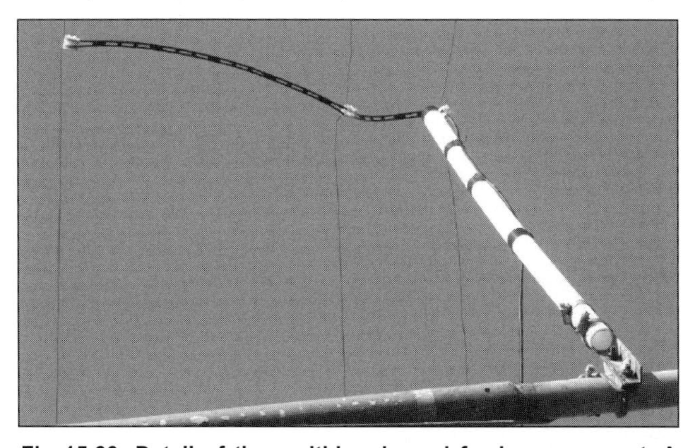

**Fig 15.66: Detail of the multi-band quad feed arrangement. A plastic tube is fixed to the mast to provide support for coaxial cable feeder. Connector blocks were originally used for connections to the elements, however these have been replaced as described in the text**

## COMPACT BEAMS WITH BENT ELEMENTS

Although many factors can affect an HF beam's installation at a given location, a key consideration is the physical space available. This is because this influences the actual size of the antenna and so the HF bands that are feasible. However, for a two-element beam it is possible to reduce the width of the antenna by bending the elements reducing the antenna's size by about a half. This arrangement was first suggested by John Reinartz, W1QP, in October 1937. Burton Simson, W8CPC, constructed such an antenna for the 20m band using 6mm copper tubing and adjustable brass rods to tune the antenna [29]. The elements were supported on a wooden frame and this allowed the element ends to be folded towards each other.

### The VK2ABQ two-element beam

A wire version of the W1QP/W8CPC two-element antenna was described, in 1973, by Fred Caton,VK2ABQ [30]. The VK2ABQ beam is essentially a 2 element beam that has been folded back on itself forming a square with sides of about λ/4 long as shown

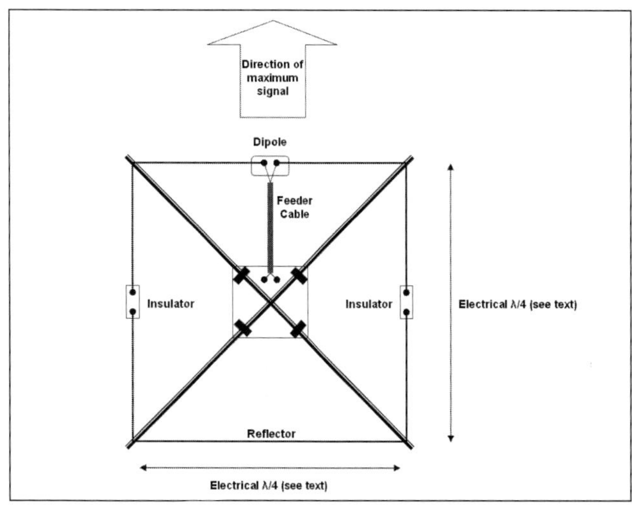


**Fig 15.67: Concept of the VK2ABQ 2-element beam antenna**

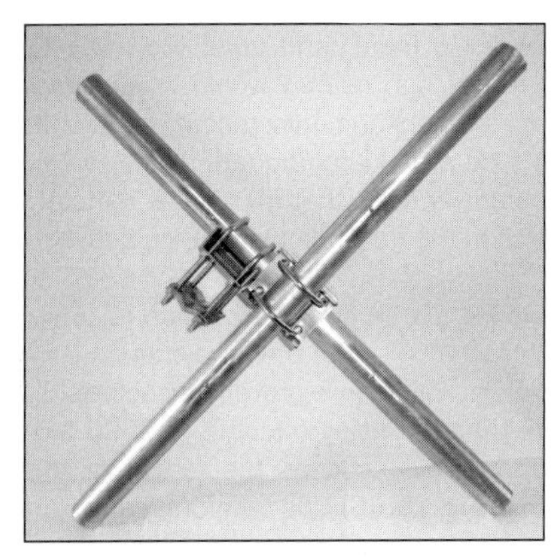

**Fig 15.68: Example of an "X" frame central cross-centre made by clamping two sections of aluminium tubing together**

in **Fig 15.67**. In this configuration, the beam consists of a symmetrical dipole and reflector whose ends support each other using insulators, with the antenna mounted on an "X" shaped frame. The insulators are constructed to give about a 6mm gap between the element tips, however this gap is 'not critical'. This arrangement results in an antenna that has about half the turning radius and area of an equivalent 2 element beam for the same band [31]. The length of each wire-side forming the loop is given by:

Wire-side (m) = 75.64 / f

Where f is the frequency in MHz.

A value of f towards the lower end of the band should be chosen. This results in a longer wire-length allowing the wire to be trimmed for tuning, with this done at about 1.5m above the ground. The impedance at the dipole's feed point can vary with height. This is due to the effect of the ground and the impedance varies from about 25 to 100Ω depending upon the height of the antenna above the ground.

Although a VK2ABQ beam is about half the size of a conventional 2 element beam, the antenna can be of a significant size depending upon the band required (eg about 5.3m square on 20m). Therefore, the central "X" frame used to support the wire loop needs to be sturdy. An "X" frame can be made using 40mm diameter PVC pipes and 16SWG aluminium tubing of 38mm diameter. The central cross-former was made from two 750mm long lengths of the aluminium tubing clamped together as shown in **Fig 15.68**. A second mast clamp was used to attach the antenna to its mast when completed.

Four lengths of 40mm diameter PVC pipes were used as the "X" frame's legs because they were a good sliding-fit over the central cross-former. Hose clips were used to secure each leg in place. Depending upon the band in use, the PVC pipes may need to be extended by joining two pipes together. This can be done by gluing a 400mm section of the PVC pipe that was slit longitudinally and slid inside between the pipes. Each leg's length was made adjustable to allow the wire loop to be tensioned using a telescopic arrangement. This consisted of a 1m length of 25mm PVC pipe slid inside each larger pipe through a modified 40mm end cap. The end of the 25mm diameter tube was enlarged using insulation-tape to enable the tube to be a sliding-fit inside the 40mm PVC pipe. A hose-clip was used to secure the 25mm pipe in place. **Figures 15.69a** and **15.69b** illustrate the adjustable telescopic arrangement. Push-on 25mm plastic couplings were modified with holes drilled to hold the wire loop in place on the end of the "X" frame. The wire used to make the beam's loop was external-grade insulated 17amp rated stranded copper wire, with dog-bone insulators used between the dipole and reflector ends.

The dipole/feeder connection was housed in an ABS box with holes drilled to take the dipole's wire ends and the coaxial feeder cable. To match the coaxial cable to the antenna, a current choke balun comprising ten turns of the coaxial feeder cable was used (as described in the Transmission Lines chapter).

10m and 20m versions of the beams were modelled using the MANNA-Gal antenna analysis application to predict the antennas' performance. This gave a free-space gain of about 3dBd and an F/B ratio of around 8dB.

Fine tuning adjustments of the antenna at the desired operating frequency gave an SWR of better than 1.2:1 using a 50Ω coaxial cable as the antenna's feeder. When the antenna's tuning was completed, the power level was increased to 100w for operational use.

A Multi-band version of this antenna can be constructed without any known difficulty by nesting one antenna within the other and using a common feed for the driven element. This concept of this technique is shown in **Fig 15.70**.

## The Mini-Beam

The Mini-beam [32] is a derivative of the VQ2ABQ antenna where the feed impedance is made less variable while still maintaining the performance of the antenna. This is done by modifying the antenna through making the reflector about 9% longer than the dipole's length. The length of each wire-side forming the loop is given by:

Wire-side (m) = 75 / f

Where f is the frequency in MHz (ie the loop's length is a wavelength).

When tuning the antenna the overall dimensions of the square loop are still maintained, however the reflector's length is increased as the dipole's length is shortened until close to a 50Ω match is found at the dipole's feed-point. This can be done by bending the wires back upon themselves through the insulators until a good match is found. Once the optimum dimensions have been found, the wires can be trimmed, the ends soldered and weatherproofed. This beam design was modelled using the MANNA-Gal antenna analysis application to predict the antennas' dimensions and performance. This gave a free-space gain of about 3dBd and an F/B ratio of around 8dB.

The dimensions for 20m, 17m, 15m, 12m and 10m versions of the antenna are shown in **Table 10**. An "X" shaped frame can be constructed for

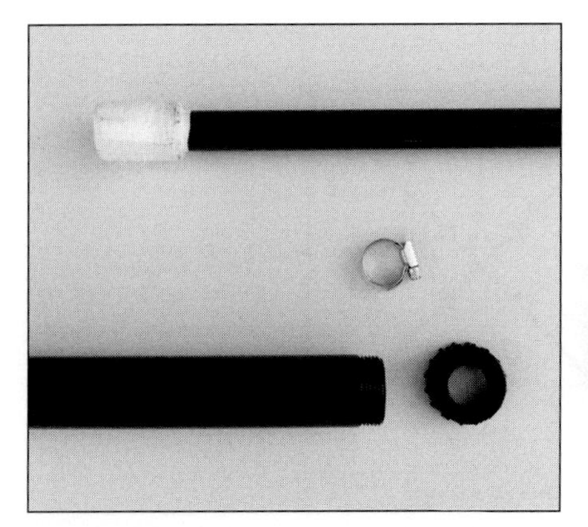

**Fig 15.69a: Concept of "X" frame's telescopic arrangement**

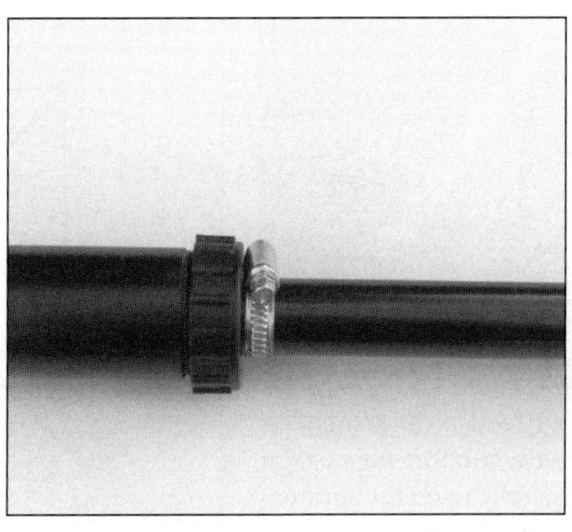

**Fig 15.69b: Concept of "X" frame's assembled telescopic arrangement**

| Frequency MHz | 14.15 | 18.12 | 21.20 | 24.95 | 28.50 |
|---|---|---|---|---|---|
| Length of each side | 5.30 | 4.14 | 3.54 | 3.01 | 2.63 |
| (R) Reflector length (m)* | 10.96 | 8.56 | 7.32 | 6.22 | 5.64 |
| (DE) Driven Element (m)* | 10.24 | 8.00 | 6.84 | 5.82 | 5.03 |

**Table 10: Mini-beam loop and element dimensions for the upper HF bands**

this antenna as previously described above for the VK2KBQ beam. Fine adjustments of a Mini-beam can result in an SWR of better than 1.1:1 being obtained using 50Ω coaxial cable as the antenna's feeder. When the antenna's tuning is completed, the power level can be increased to 100 watts for operational use. **Fig 15.71** illustrates an example of a 10m 2-element Mini-beam mounted on a mast.

## The G6XN Antenna and W4RNL 'Moxon Rectangle'

Les Moxon, G6XN, changed the structure of the VK2ABQ beam from a square to a rectangle by reducing the centre section spacing of the elements from 0.25λ to 0.17λ [33]. This modification improved the antenna's free space gain to about 4.5dBd and F/B ratio by over 20dB. L.B. Cebik, W4RNL found that reducing the element spacing to 0.14λ further improved the gain and F/B ratio [28] and this antenna he called the 'Moxon Rectangle', an antenna that has now become well known. Both antenna variations have a feed point impedance around 55Ω enabling a reasonable match to a 50Ω feeder cable. The concept of the geometries of these antenna designs is illustrated in **Fig 15.72**.

Peter Dodd, G3LDO, made an assessment of the MOXON Rectangle [34] and the equations below are derived from this work. The equations provide a guide to the wire lengths for the elements, where f is the frequency in MHz:

Wire conductor diameter: 1 to 2mm

Driven element (DE) length (m) = 142 / f

Reflector (R) length (m) = 149.6 / f

Then using the equations above to determine DE and R, the lengths A, B, C, D and E shown in **Fig 15.73** can be calculated, where:

Length A (m) = 108 / f

Length E (m) = 41.32 / f

Length B (m) = (DE - A) / 2

Length D (m) = (R - A) / 2

Length C (m) = E – (B+D)

The wire lengths should be cut slightly longer than calculated. This will allow the antenna to be tuned by bending the wires back

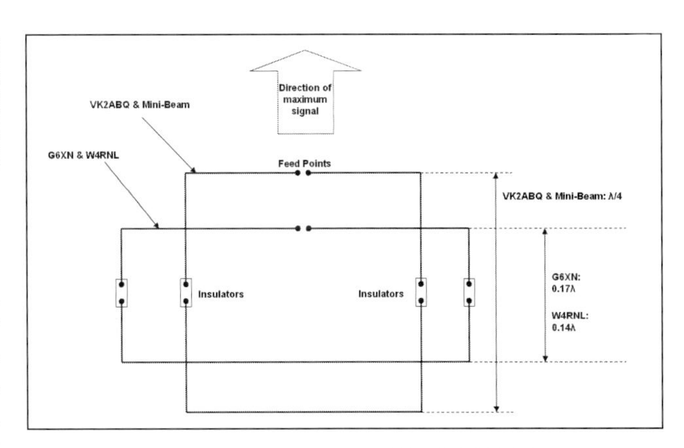


**Fig 15.72: Comparison between the VK2ABQ/Mini-beam and the G6XN/W4RNL antenna structures**

upon themselves through the insulators until a good match is found. Once the optimum dimensions have been found, the wires can be trimmed, the ends soldered and weatherproofed. An "X" shaped frame could be constructed for this antenna as previously described above, however this will need to be rectangular rather than square in shape.

## G3LDO suggested that the:

- MOXON Rectangle can be difficult to tune and can take time to do successfully.
- Dimension of C is critical and can significantly influence the performance of the antenna. This can result in different element lengths for antennas operating on the same band.
- The antenna's high performance can lead to difficulties when making a multi-band version due to the interaction between the elements. With care a multi-band MOXON Rectangle antenna can be constructed whose rectangular layout is similar to that shown in **Fig 15.70**.

## The G3LDO Double D Antenna

It is possible to make the profile of a two-element beam even smaller by folding the bent elements' ends back towards the mast in the vertical plane. This effectively allows the antenna to be accommodated in a smaller area compared to that needed for the VK2ABQ, G6XN or Moxon Rectangle configurations. The G3LDO antenna has a pyramidal configuration and its construction is shown in **Fig 15.74** [35]. To support the wire elements, the antenna used a frame made from canes. The wire elements were then held in position on the canes using plastic-tape. The elevated ends of the elements were held in place using nylon guy-lines whose ends were attached to the central mast. This antenna arrangement provides a strong and lightweight antenna.

Using the EZNEC antenna modelling application to assess the performance of this antenna, this predicted a free-space gain of 3.1dBd and an F/B ratio of better than 15dB. This predicted performance is not as good as the 'Moxon Rectangle', however the antenna is relatively straightforward to construct.

Referring to **Fig 15.74**, the approximate wire length can be calculated from:

Length A+B (m) = 79.00 / f

Length C (m) = 55.89 / f

Length D (m) = 16.41 / f

Length E (m) = 31.41 / f

The total element length (m) = 141.783 / f

Where f is the frequency in MHz.

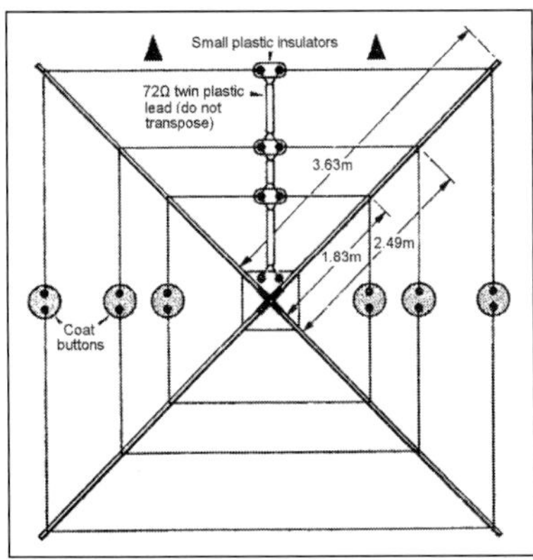


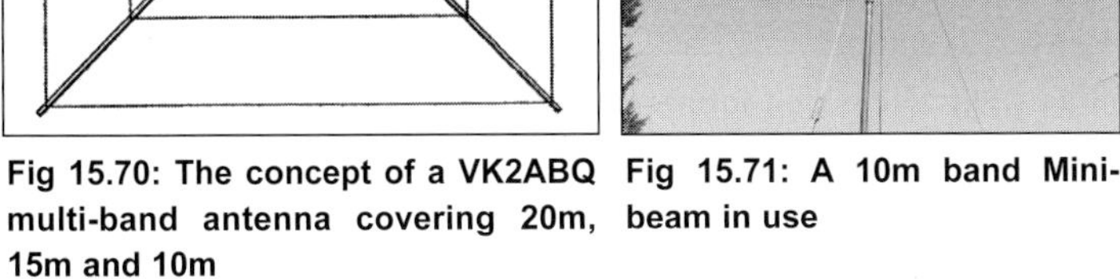

**Fig 15.70: The concept of a VK2ABQ multi-band antenna covering 20m, 15m and 10m**

**Fig 15.71: A 10m band Mini-beam in use**

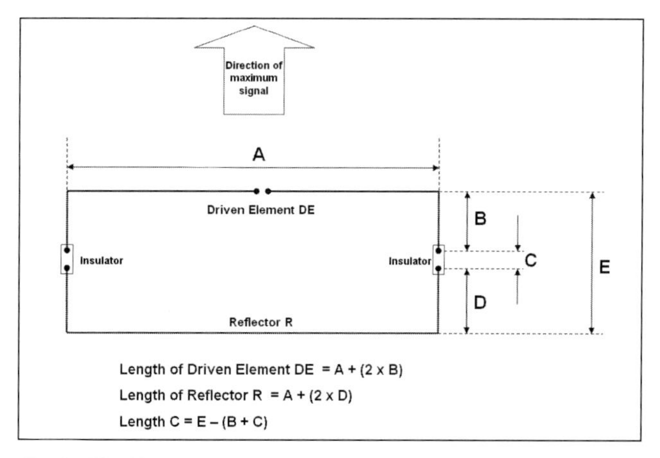


Fig 15.73: 'Moxon Rectangle' antenna key dimensions

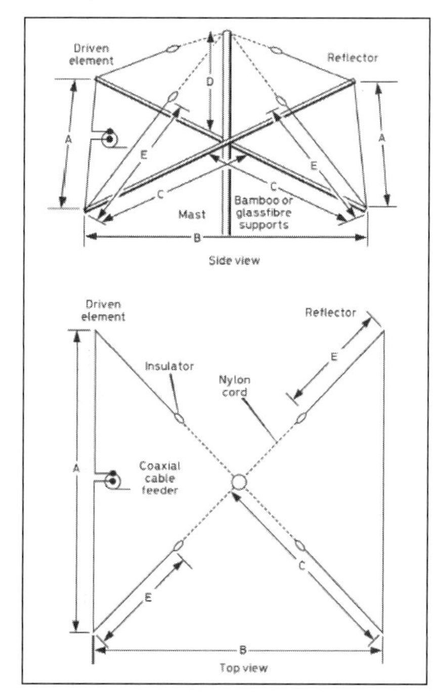


Fig 15.74: The wire Double D Antenna with approximate dimensions

To optimise the antenna, the element lengths were altered by bending the wires back upon themselves through the insulators until a good match was found. The reflector was adjusted to maximise the F/B ratio, while the driven element was adjusted for minimum SWR using 50Ω coaxial cable as the feeder. Cable-ties were used to hold the wire-ends in place and tightened once the most effective lengths had been found.

Often it is possible to increase the gain and improve the SWR by reducing dimension B by altering the angles of the fixing supports relative to the mast.

The Double-D antenna is also amenable to multi-banding by adding other antennas for different bands mounted on the same support. The same method of feeding can be used as with the quad and the VK2ABQ antenna.

## MOBILE ANTENNAS

The antenna is crucial to the success of mobile operation. The shape of the vehicle, space limitations and the vehicle's slipstream make the vertical whip antenna the most popular choice of mobile antenna. The easiest way to feed a mobile antenna is to make it a quarter wavelength (λ/4) long at the frequency in use. The resonant quarter wavelength is a function of frequency and is about 1.49m long on 50.2MHz and 2.63m on 28.5MHz (not allowing for the velocity factor). This length becomes progressively shorter on the higher VHF bands. Quarter wave antennas on the 28MHz band and higher bands are practical. However, the antenna's length on the lower HF bands becomes a different matter. Even on 21.2MHz a quarter wavelength antenna is about 3.5m long and on 14.15MHz is 5.3m (not allowing for the velocity factor). Therefore, a mobile antenna for the HF bands has to be shorter than a quarter wavelength for practical reasons. When the antenna's length is kept constant, as the frequency of operation is lowered, the feed point exhibits a decreasing resistance in series with an increasing capacitive reactance. In order to feed power into the antenna it must be brought to resonance making the feed point resistive. This is achieved by inductively loading the antenna by adding a loading coil. A loading coil for a mobile antenna must be rugged, be able to withstand the weather and the mechanical strain of a fast moving vehicle's slip stream. The following examples represent suitable solutions.

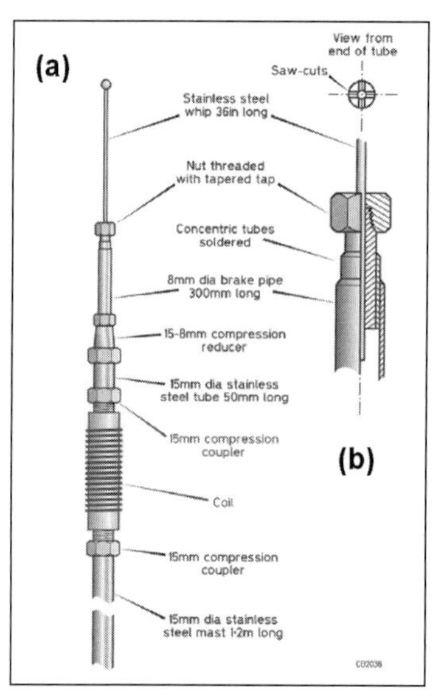


Fig 15.75: (a) Section of the G3MPO mobile antenna. (b) Detail of stainless steel whip / 8mm diameter brake pipe clamp

### The G3MPO Coil and Antenna

John Robinson, G3MPO, designed a single antenna structure that used a different loading coil for each HF band [36]. G3MPO has used this design during several thousands of miles of motoring and it has proved secure. Many of the components were sourced from a local plumbing supplier who stocked ready-made brass, stainless steel and plastic fittings.

The bottom section of the antenna was made from a length of 15mm stainless steel central heating tubing. Each coil former was made from white polypropylene waste pipe, 15mm plumber's brass compression couplers were used as the coil's terminations and provided the method of fixing the completed coil into the antenna. The construction of the antenna and coils is shown in **Fig's 75** and **76**.

The thread of a brass 15mm compression coupler was tightly screwed into the end of a length of 20mm diameter tubing making a strong joint (the ends of the tube can be pre-heated in hot water if necessary). A second coupler was screwed into the other end in a similar manner. This arrangement gives a coil former with ready-made 15mm connections at each end. This enables the coil former to be fitted and clamped directly onto the 15mm diameter lower mast section. Varying lengths of former were used for the higher frequency coils. Where a greater diameter is needed for the lower frequencies, the same 20mm former was used as a spine. Then a larger diameter polypropylene tube was slid over this tube and was attached by packing the space between the tubes with small pieces of car-

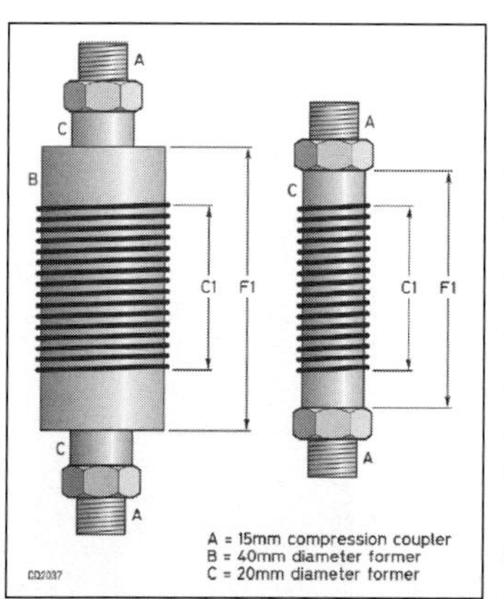


Fig 15.76: Coil former construction and dimensions of G3MPO antenna. F1 is the coil former length and C1 the coil winding length, see Table 11

| COIL DATA | | | | | | | | |
|---|---|---|---|---|---|---|---|---|
| F | D | F1 | Wire | N | C1 | L | Rr | Rf |
| MHz | mm | mm | SWG | | mm | μH | Ω | Ω |
| **29.0** | 20 | 77 | 18 | 9 | 28 | 0.9 | 35 | 48 |
| **24.9** | 20 | 90 | 18 | 15 | 44 | 1.7 | 29 | 48 |
| **21.2** | 20 | 115 | 18 | 23 | 64 | 3.0 | 22 | 47 |
| **18.1** | 20 | 140 | 18 | 34 | 92 | 4.5 | 17 | 43 |
| **14.25** | 20 | 146 | 20 | 45 | 96 | 8.4 | 11 | 34 |
| **10.13** | 40 | 115 | 20 | 31 | 72 | 19 | 6 | 26 |
| **7.05** | 40 | 165 | 20 | 58 | 117 | 41 | 3 | 20 |
| **3.65** | 40 | 280 | 22 | 130 | 157 | 153 | 0.8 | 21 |
| **1.9** | 40 | 305 | 28 | 294 | 263 | 558 | 0.2 | 37 |

**Table 11: Read in conjunction with Figure 76. F = frequency (MHz), D = coil former diameter (mm), F1 = Length of coil former tube (mm), N = number of turns, C1 = Length of coil winding (mm), L = coil**

repair glass mat soaked in resin at each end. The coil assembly was then waterproofed with a silicone rubber sealant.

The whip above the coil former comprised of a short length of small diameter copper tube, fixed to the top of the coil using its 15mm coil coupler. A further length of 8mm tubing was connected using a 15mm-to-8mm compression reducer and a length of stainless steel whip was slid inside this tubing. Suitable tubing is commonly used in refrigeration units and micro-bore central heating systems and was sourced from metal stockists. Vehicle brake-pipe components could also be used as shown in **Fig 15.75**.

A locking system was used to hold the whip in position once its length was set. This was made by cutting a thread along the last 12mm of the tube's end using a thread-cutting die and making cross cuts down its length with a mini-hacksaw. A locking-nut with a tapered thread was made from a short length of brass rod. This locking-nut was suitably drilled and taper-tapped to enable it to close the tube's end down onto the whip. This arrangement locked the whip in place in the required position, as shown in **Fig 15.75**. The whip structure was completed by connecting the telescopic section on top of the coil using a 15mm-to-8mm (micro-bore) brass reducer and a 51mm length of 15mm brass tubing, as shown in **Fig 15.75**(b).

The wire used for the coil was attached to the end couplers by passing each wire end through a small hole drilled at both ends of the polypropylene pipe just beyond where the end of the coil is to lie. The wire was passed into the inside of the tube and drawn through the coupler. Before fitting each coupler to the polypropylene tube, a hairpin loop of wire was soldered onto the inside of each coupler. The wire was then soldered onto this hairpin loop at the appropriate time (as described below).

Two or three lengths of double-sided tape were stuck to the coil former. A sufficient length of enamelled copper wire was cut for the particular coil to be made up. **Table 11** summarises the details for each coil and the approximate wire length required to be cut is given by:

Length ~ 7 x Number of turns x Diameter of the former

This allowed enough wire to be cut to wind the coil with some to spare.

The wire spacing was achieved by winding two lengths of wire onto the former side by side and subsequently removing one of them. The double-sided tape fixed to the former holds the remaining winding in position. The coil was wound beyond the holes through which the wire ends are to be taken and, after the spacing wire was removed, the winding was coated with polyurethane varnish. When dry, the coil was then wound back at each end to give the required number of turns for the HF band in use leaving sufficient wire to make the connections (about 10 to 20mm). The wire ends were then fed through into the former, out through the end couplings, cut to length and soldered to the coupling hairpins. The two small holes in the former were sealed with varnish and the winding was bound with a double layer of self-amalgamating tape to protect it from being damaged. The soldered connections were pushed down into the coupling, to keep them out of the way, and the coil was given two further coats of polyurethane varnish.

**Fig 15.77: Detail of the tuning section of the 'screwdriver' antenna, showing the coil and fingers that short the turns as it emerges. This photo, courtesy of Waters & Stanton plc, is of the WBB-3 derivative**

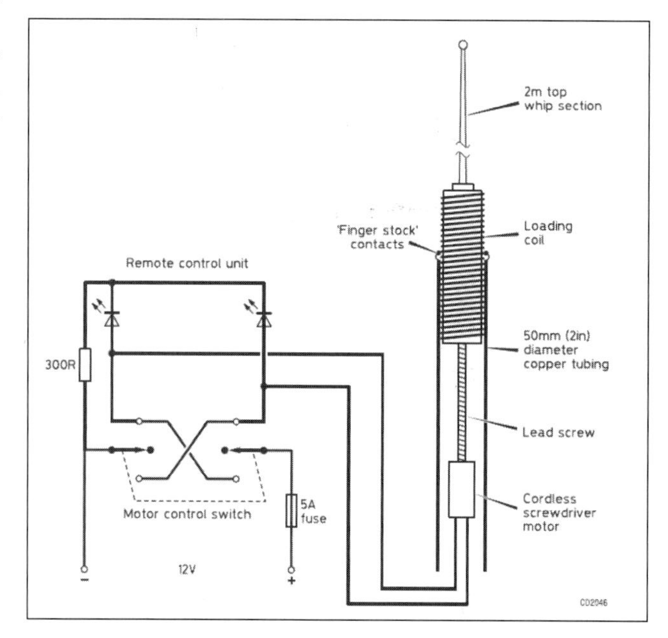


**Fig 15.78: W6AAQ's DK3 mobile antenna (not to scale). The control box is located by the driver and power obtained from either the rig supply or the cigar lighter socket. The original had relay switched capacitors, selected from the drivers control box, to match the antenna on the lower frequencies**

The W6AAQ Continuous Coverage HF Mobile Antenna

In this design the antenna resonance is adjusted from the drivers/operator's position using a cordless screwdriver electric motor. This motor rotates a brass lead-screw via a nut fixed to the coil to cause the coil to move up or down inside in a 1m long 50mm diameter aluminium, brass or copper tube as shown in **Fig 15.77**. As the motor is rotated the coil is raised or lowered so that more or less of the coil is contained within the lower tube section. Finger stock connectors are used to short the coil to the tube to obtain the appropriate resonance. A circuit of the antenna and the control box is shown in **Fig 15.78**. The antenna is tuned to resonance by first by listening for an increase in receiver noise, then applying transmit power and fine tuning for the lowest SWR.

## Matching a Mobile Antenna to the Feeder

All the mobile antennas described so far are fed with 50Ω coaxial cable. The usual arrangement is for the coaxial cable's centre to be is connected to the antenna and the braid to the vehicle body. However, the radiation resistance of the antenna will generally be lower than 50Ω and depends on the operating

| Frequency (MHz) | 1.8 | 3.6 | 7.05 | 10 | 14.2 | 18 | 21.3 | 25 | 28.5 |
|---|---|---|---|---|---|---|---|---|---|
| Radiation Resistance (Ω) | 0.2 | 0.8 | 3 | 6 | 12 | 17 | 21 | 28 | 36 |

**Table 12: Radiation resistance of a typical mobile antenna**

| Frequency (MHz) | 1.90 | 3.65 | 7.05 | 10.13 | 14.25 | 18.10 | 21.25 | 24.95 | 28.50 |
|---|---|---|---|---|---|---|---|---|---|
| Capacitance (pF) | 1000 | 1000 | 544 | 300 | 150 | 74 | 37 | 27 | 18 |

**Table 13: Suggested values for capacitive shunt feeding of a mobile antenna**

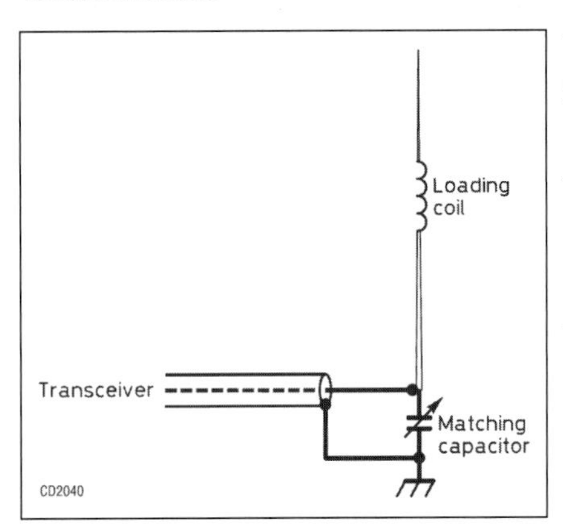


**Fig 15.79: Capacitor matching. In practice, the variation in capacity is achieved by switching-in the appropriate combination of fixed capacitor values**

frequency for a given antenna size. Typical radiation resistance figures for a 2.4m (8ft) antenna are shown in **Table 12**.

In practice, the feed impedance will include the RF resistance of the loading coil, the radiation resistance of the antenna and the total resistance loss. The total resistance loss is usually much greater than the radiation resistance at the lower operating frequencies. For example, the radiation resistance of an 80m antenna is around 1Ω, the loading coil RF resistance may be around 10Ω and the ground loss will be between 4 to 12Ω and depends on the size of the vehicle. This gives a total feed impedance in the region of 20Ω. When the antenna is fed with a 50Ω coaxial cable, this will give an SWR of 2.5:1 at resonance. As the transceiver is tuned off the antenna's resonance, the SWR will rapidly worsen and soon fall outside the range that is acceptable to a solid-state transceiver's 50Ω PA stage (unless the transceiver has a built-in ATU). At the other end of the HF spectrum, the radiation resistance is much higher and the coil losses become lower allowing the transceiver to be connected directly to the antenna using a length of 50Ω coaxial cable. There are several ways of matching the nominal 50Ω transceiver output to the impedance encountered at the base of a resonant mobile antenna. Of these the most common are:

**1. Capacitive shunt feeding**

This is the addition of a shunt capacitor directly across the antenna feed point as shown in **Fig 15.79**. Suggested capacitor values calculated by G3MPO are shown in **Table 13**. Exact values can be determined experimentally and will need to be switched for multi-band operation. The way that this works can be seen by referring to **Fig 15.80**. The curve A represents the feed impedance of a Pro Am antenna in the frequency range 3.55 to 3.65MHz, measured using a suitable antenna impedance analyser. At the lower frequency the impedance was about (10 - j50)Ω, while at the higher frequency this was (70 + j70)Ω. Nowhere on the curve is the SWR better than 2:1. By slightly increasing the inductance of the loading coil and compensating with a capacitor across the feed point, the curve can be shifted to B to achieve an improved match.

**2. Inductive shunt feeding**

This is achieved by using a small tapped matching coil inductor at the base of the antenna. With the loading coil adjusted to take into account the effect of the matching coil, the antenna

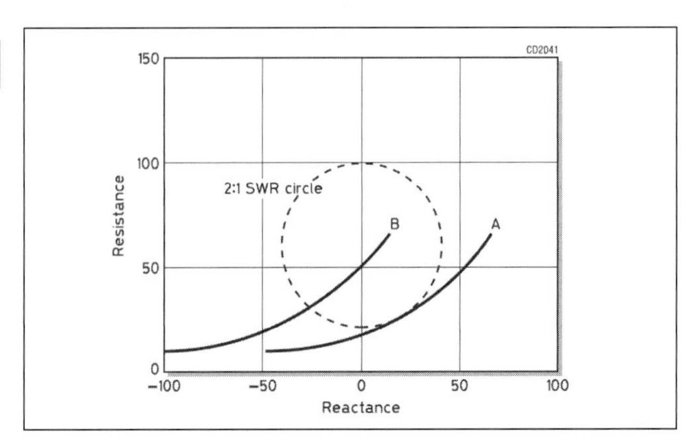


**Fig 15.80: Curve A is the feed impedance of an 80m mobile antenna in the frequency range 3.55 to 3.65MHz. On no part of the curve is the standing wave ratio better than 2:1. An improved match is achieved by increasing the inductance of the loading coil slightly, and compensating with a capacitor across the feed point, thereby moving the curve to B**

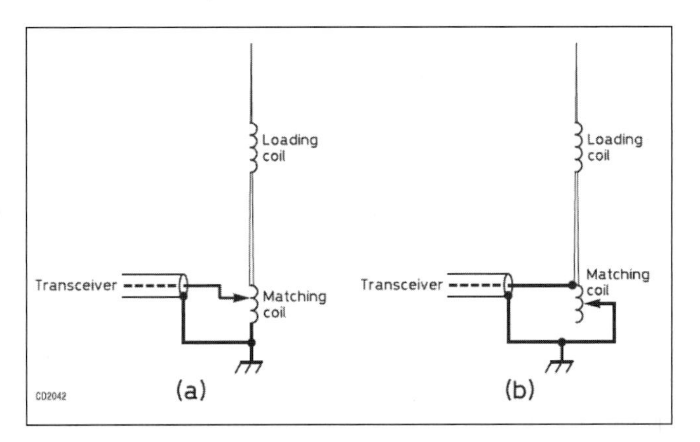


**Fig 15.81: Two methods of using a tapped inductance for matching at the base of the antenna**

base impedance is raised in proportion to the size of the base inductance. There are two antenna feeding techniques that could be used to do this:

- Selecting an inductor coil that results in a value greater than 50Ω at the lowest point of the antenna impedance/frequency curve. The transceiver connection is then tapped down the matching-coil to obtain the best match as shown in **Fig 15.81**(a).
- Using a variable inductance across the feed point as shown in **Fig 15.81**(b). The inductance value or tapping point must be changed when the frequency band is changed.

**3. Transformer matching**

This matching arrangement uses an RF transformer wound on a toroid core as shown in **Fig 15.82**. A commercial or home-brew matching transformer could be used. Mike Grierson, G3TSO, has designed an adjustable matching arrangement that used a Amidon T157-2 toroid core. This core was wound with 20 bifilar 2-wire turns using 18SWG (1.2mm) wire. Both windings were connected in series, in phase, with the second winding tapped every other turn as shown in **Fig 15.83**. With the loading coil adjusted to take into account the inductance of the transformer windings, antenna impedances from 50Ω down to 12Ω could be matched.

The three matching arrangements described were located part way between the transceiver and the antenna. If the matching arrangement is made remotely switch-able from the driver's seat, then this represents a greater degree of operator convenience compared to a manually adjusted matching arrangement.

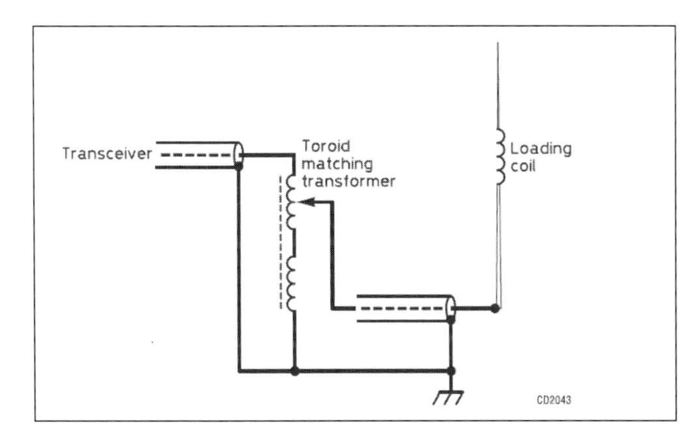


Fig 15.82: Matching arrangement for a mobile antenna using a variable ratio RF transformer

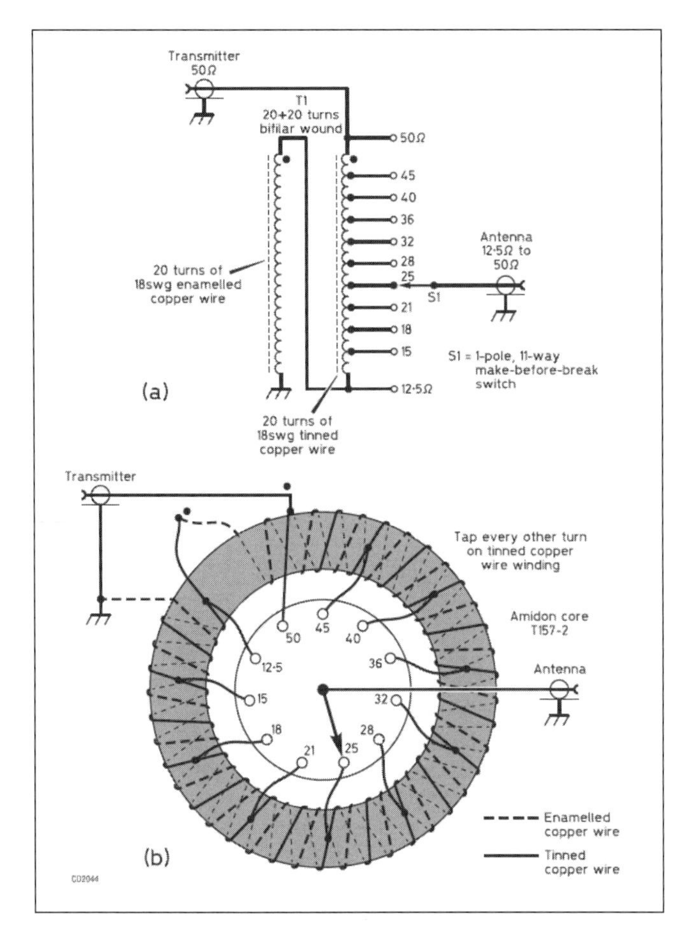


Fig 15.83: Details of the G3TSO RF transformer. (a) The circuit and impedance matching range. (b) Construction details

As described earlier, the feed impedance comprises the radiation resistance, earth resistance and coil RF resistance all in series. This allows the transceiver to be connected directly to the antenna without any matching at 14MHz and above. On the lower frequency bands the coaxial feeder cable is comparatively electrically very short for a mobile installation. This has the effect that losses caused by a higher SWR are minimal.

## Further Mobile Antenna Information

Details of mobile antenna construction, mounting and matching can be found in The Amateur Radio Mobile Handbook 2nd edition, RSGB [37].

# FIXED LONG WIRE HF BEAM ANTENNAS

Very occasionally stations are worked using large HF beam antennas that require a significant area of ground for their installation. Unlike smaller HF beam antennas, these large HF beams tend not to be able to be rotated and are set-up to beam a signal in a particular direction. For example, several years ago, a station in Zimbabwe using a V-beam HF antenna aimed in a northerly direction could often be worked on the upper HF bands from Europe. This large 'V' wire antenna was strung along on two lines of telephone poles that happened to be in the right orientation to beam a signal north. Sadly, termites ate the wooden telephone poles resulting in the demise of this antenna.

If access to a significant area of ground is available, then a fixed large HF beam antenna could be a possibility.

## V-beams

The V-beam consists of two wires made in the form of a V and fed at the apex with twin-wire feeder, as shown in **Fig 15.84**. A long-wire antenna, two wavelengths long and fed at the end, has four lobes of maximum radiation at an angle of 36 degrees to the wire. If two such antennas are erected horizontally to form a V that has an included angle of 72 degrees, then some lobes will tend to be in alignment and are additive in the direction of the bisector of the apex. The other lobes will tend to be out of alignment and cancel. The result is a pronounced bi-directional beam. The directivity and gain of V-beams depends on the length of the legs and the angle at the apex of the V. This is likely to be the limiting factor in most amateur installations and is the first point to be considered when designing a V-beam. The correct angle and the gain to be expected in the most favourable direction are given in **Table 14**.

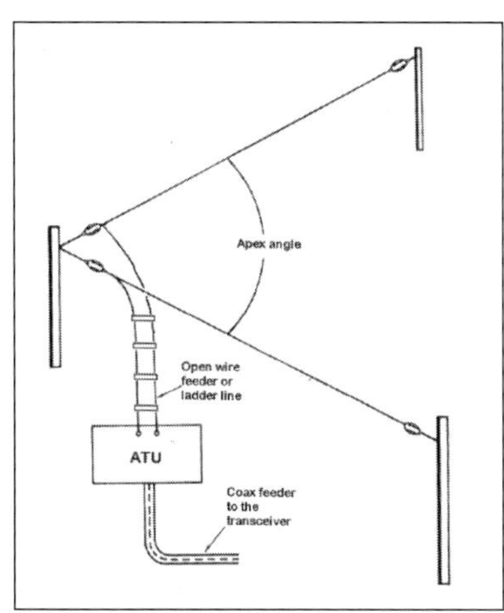


Fig 15.84: A practical resonant V-beam for the HF bands 14 to 29MHz. Ideally a balanced line ATU should be used to ensure an equal current level in each element

A practical V-beam capable of producing bi-directional high gain on the higher frequency bands (14 to 29MHz) is shown in **Fig 15.84**. V-beams are often constructed so that the apex is placed as high as possible with the ends close to the ground. This arrangement means that only one mast is required for the installation and, if space is available, several V-beams pointing in different directions could use a common support. The V-beam described is a resonant antenna. The input impedance of this antenna may rise to 2000Ω in a short V, but will be between 800 and 1000Ω in a longer antenna. Therefore, 400 to 600Ω feed lines can be used, matched to the transceiver with a suitable balanced ATU.

| Leg length (Wavelength) | Gain (dBd) | Apex Angle (Degrees) |
|---|---|---|
| 1 | 3 | 108 |
| 2 | 4.5 | 70 |
| 3 | 5.5 | 57 |
| 4 | 6.5 | 47 |
| 5 | 7.5 | 43 |
| 6 | 8.5 | 37 |
| 7 | 9.3 | 34 |
| 8 | 10 | 32 |

Table 14: V-beam gain and apex angles for given lengths of element

The V-beam can be

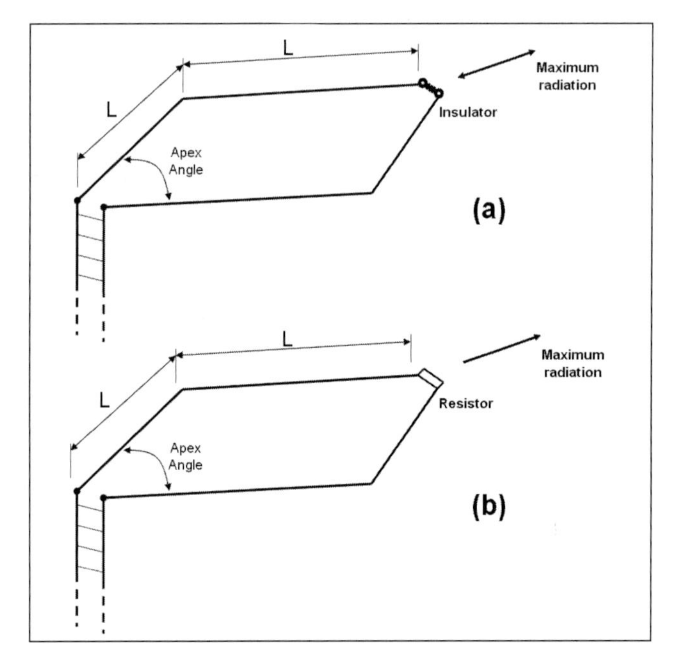


**Fig 15.85: The Rhombic antenna. Drawing (a) shows the resonant un-terminated version, and (b) the terminated version**

made unidirectional if it is terminated with resistors inserted approximately a quarter wavelength from the ends of the elements, so that final quarter-wavelengths act as artificial earths. A suitable value of resistor would be 500Ω for each leg. Termination resistors have the effect of absorbing some of the lobes produced and the elements become travelling wave devices.

### The Rhombic

The rhombic antenna is a V-beam with a second V added as illustrated in **Fig 15.85**. In a similar manner as the V-beam, the rhombic antenna uses the same lobe addition principle. However, there is an additional factor to be considered because the lobes from the front and rear halves must also add in phase at the required elevation angle. This introduces an extra degree of control in the design so that considerable variation of pattern can be obtained by choosing various apex angles and heights above ground. The rhombic gives an increased gain but takes up a lot of room and requires at least one extra support compared to the V-beam. As with the V-beam, the resonant rhombic has a bidirectional pattern as shown in **Fig 15.85**(a). The terminated version of the Rhombic, as shown in **Fig 15.85**(b), has a unidirectional pattern and is a travelling wave antenna. The terminating resistance absorbs noise and interference coming from the rear direction as well as transmitter power, that otherwise would be radiated backwards. This means signal-to-noise ratio is improved by up to 3dB without affecting signals transmitted in the wanted direction.

The use of tuned feeders enables the Rhombic, like the V-beam, to be used on several amateur bands. The non-resonant Rhombic differs from the resonant type in being terminated at the far end by a non-inductive resistor comparable in value with the characteristic impedance, the optimum value being influenced by energy loss through radiation as the wave travels outwards. An average termination will have a value of approximately 800Ω. It is essential that the terminating resistor be as near a pure resistance as possible, ie without inductance or capacitance ruling out the use of wire-wound resistors. The power rating of the terminating resistor should not be less than one-third of the mean power input to the antenna. For medium powers, suitable loads can be assembled from series or parallel combinations of resistors (eg combinations of 5 watt carbon resistors).

The terminating resistor may be mounted at the extreme ends of the rhombic at the top of the supporting mast. Alternatively the resistor may be located near ground level and connected to the extreme ends of the Rhombic via twin-wire feeder. The impedance at the feed point of a terminated Rhombic is

700-800Ω and a suitable feeder to match this can be made up of 16SWG wire spaced 300mm apart. The design of Rhombic antennas can be based on **Table 14**, considering them to be two V-beams joined at the free ends.

The design of V and Rhombic antennas is quite flexible and both types will work over a 2:1 frequency range or even more, provided the legs are at least two wavelengths long at the lowest frequency. For such wide-band use the angle is chosen to suit the element length at the mid-range frequency.

Generally the beamwidth and wave angle increase at the lower frequency and decrease at the upper frequency, even though the apex angle is not quite optimum over the whole range. In general, leg lengths exceeding 10 wavelengths are impractical because the beam is then too narrow. Advantages of the Rhombic over a V-beam are that it gives about 1-2dB increase in gain for the same total wire length and its directional pattern is less dependent on frequency. The Rhombic can be easier to terminate compared the V-beam. The Rhombic antenna's main disadvantage is that it requires four masts to support it.

## ANTENNA EARTHING TECHNIQUES

For some antenna designs to operate efficiently a good RF earth is often required. The end-fed wire antenna is a good example of where an RF earth can potentially improve the performance of the antenna.

### Using Real Earth

In practice a good RF earth connection is difficult to find and is only practicable from a ground floor room. The problem with the 'earth stake' is that the ground has resistance and the cable connecting the earth stake to the radio has reactance. Many ways have been tried to reduce the ground resistance and, in general, the more copper that can be buried in the ground the better. An old copper water tank, connected to the radio with a short length of thick copper wire can make a good earth. An RF earth can also be made from about 60m$^2$ of galvanised chicken wire. This is laid on the lawn early in the year and pegged down with large staples made from hard-drawn copper wire. The grass will grow up through the chicken wire netting and the wire will tend to disappear into the ground over a period of about two months. In the early stages, the lawn has to be mown so that the lawnmower does not cut too close and chew up the carefully laid wire netting (and in so doing, potentially damaging the lawnmower). Low band DXers tend to use multiple radials where many wires are buried that radiate out from the earth connection. The rule seems to be the more wires the better. These types of direct connection to earth arrangements can also provide an electrical safety earth to the radio equipment in the shack.

### Artificial Earths

When the shack is located above the first floor, engineering a low-impedance earth connection to ground level becomes difficult using the method described above. This is because the distance that the shack is located above ground level becomes a significant fraction of a wavelength on the higher HF bands and above. At frequencies where the earth cable's length is close to

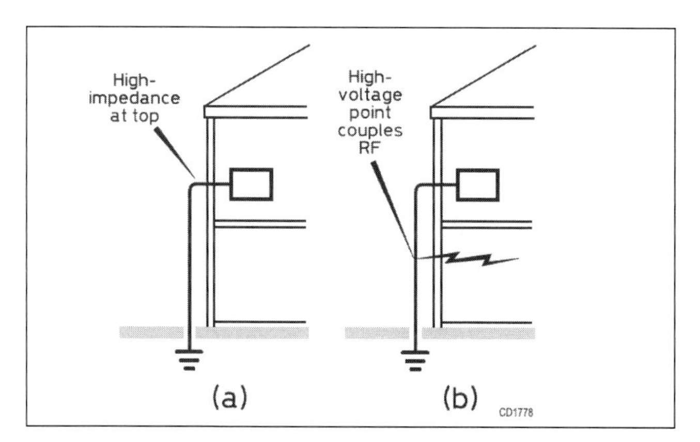


**Fig 15.86: Concept of why RF ground cable runs from upper floor are effective. (a) Ground cable run with quarter-wavelength resonance (or odd multiple) is ineffective because very little current flows into it. (b) Ground cable run with half-wavelength resonance (or multiples) will have high-voltage points which could couple RF currents into the domestic wiring**

one or three quarters of a wavelength, the earth connection will present a high impedance at RF frequencies. Effectively, the earth connection acts like an RF insulator and this is opposite to that required as shown in **Fig 15.86**(a). This situation can happen on one or more of the nine amateur HF bands.

Alternatively, a situation that can potentially occur on the bands typically above 10MHz is when the length of the earth cable allows it to resonate as a half-wavelength. The earth cable may act as a good RF earth. However, it also has a high-voltage point halfway along it and this may allow RF currents to be coupled into the domestic electrical wiring as shown in **Fig 15.86**(b). This situation can occur if the earth cable is run generally parallel to the electrical wiring within the wall of the building. This means that an earth wire run from the radio equipment located in an upstairs shack to an earth stake will provide a safety earth, however its usefulness as an RF earth is uncertain and not assured. In general, the earth cable run from an upstairs shack should be located as far from house wiring as possible.

Another technique of obtaining a good RF earth is to connect a quarter-wave (λ/4) radial for each band to the transceiver and ATU earth connector, then running the free ends outside, away from the transceiver. The impedance at the end of the radial is high and the current will be low, conversely where the radial is connected to the radio equipment its impedance will be low and the RF potential close to zero. The problem is then where to locate each radial for each band to be used. This arrangement will require some experimenting to find the best position, however the radials can be bent or even folded, although the length may have to be altered to maintain resonance. The radials are best located outside the building and run in the horizontal plane to minimise any coupling with the domestic electrical wiring. If the radial(s) are used indoors (eg round the skirting board), then wire with thick insulation should be used. The end(s) of the radial(s) should be insulated using several layers of insulating tape to protect against any high RF voltages that can exist when the transmitter is on. The best way to check the resonance of a radial is to connect it to the radio earth, make a loop in the radial and use a dip meter to check its resonance. If such an instrument is unavailable then use an RF current meter and adjust the radial length for maximum current.

Alternatively, one single length radial can be tuned on any band to place a zero RF potential at the transceiver by inserting a LC series tuning circuit between the transmitter and the radial.

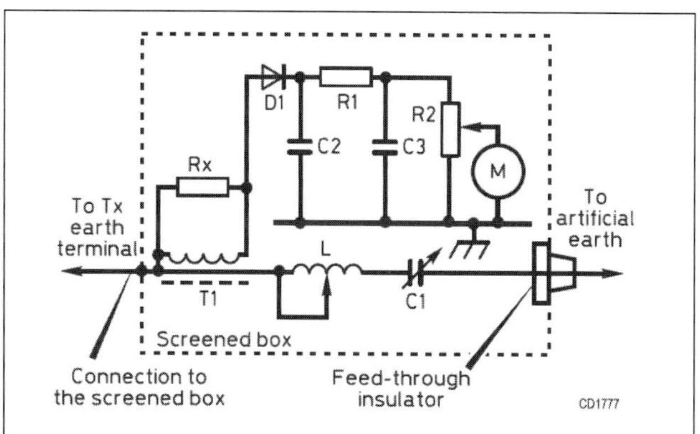


**Fig 15.87: Concept of SM6AQR's design for an earth lead tuner. T1 = Amidon T-50-43 ferrite toroid core. The transformer's primary is simply the earth lead through the toroid centre. The secondary = 20t small gauge enamelled wire. L = 28μH roller-coaster or multi-tapped coil with 10-position switch as described in the text. C1 = 200pF or more air-spaced variable capacitor with >1mm spacing that is insulated from panel and case. C2 , C3 = l0nF ceramic. D1 = AA119, R1 = 1k, R2 = 10k pot, Rx see text. M = 100μA or less**

These units are commercially available and have a tuneable LC tuning circuit. These units often have a through-current RF indicator to help tuning the radial or earth lead to resonance (maximum current). A unit could also be made and an example designed by SM6AQR [38] is shown in **Fig 15.87**. This design used a 28μH 'roller-coaster' inductor and a 200-300pF air spaced tuning capacitor with a plate spacing of at least 1mm. The capacitor and its shaft must be insulated from the metal cabinet used to house the circuitry. Alternatively, a multi-tapped fixed coil comprising as many taps as possible could be used as the inductor. The tuning indicator consists of a current transformer, rectifier, smoothing filter, sensitivity potentiometer and DC microammeter. The 'primary' of the current transformer is the artificial earth lead and this passes through the centre of the T1 ferrite toroid core. Onto this core the secondary was wound using 20 turns of thin enamelled wire. The resistor Rx across the T1 secondary should be non-inductive and be between 22 and 100Ω. Resistor Rx is selected to enable a convenient meter deflection to be set with the sensitivity control R2 on each required frequency and for the RF power used.

Caution: A separate electrical safety earth should always be used, in addition to the RF earth described above.

## MATCHING AND TUNING

An Antenna Tuning Unit (ATU) is a device used to interface the impedance transition between the transceiver and the transmission line whose load is the antenna. The purpose of the ATU is to maximise the RF power transfer between the transceiver and the transmission line/antenna to ensure that:

- As much of the accessible RF power as possible that is supplied by transmitter is radiated by the antenna as an RF signal
- As much of the received RF signal as possible is made accessible as RF power for the receiver to function successfully.

Many of the antennas described within this chapter will present an impedance that is not a good match with the transmission line. Similarly, the transmission line will present an impedance that is not a good match with the transceiver. The impedance presented by the transmission line will be influenced by the antenna's impedance and will often comprise a resistive and a reactive component (R +/- jX). Therefore, the ATU is used:

- As a means to tune out and so neutralise the reactive component of the impedance connected at its input
- To transform the resistive component of the impedance presented at its input.

The effect of these actions is the ATU presents an acceptable output impedance to the transceiver that, in most cases, is ideally 50Ω resistive.

An ATU has to accommodate at least four antenna arrangements to interface the line/antenna with the transceiver:

- Direct connection of a balanced or unbalanced antenna
- End-fed wire antenna worked against earth
- Antenna fed with coaxial cable
- Antenna fed with twin-line feeder or ladder-line.

As the function of the ATU is to match the impedance presented by the line/antenna system to usually 50Ω, the term AMU (Antenna Matching Unit) might be a more accurate description, however "ATU" is much more commonly used.

## Commercially available automated ATUs

Many stations now use an automated ATU, or auto-tuner, to match the transmission line/antenna to the transceiver. These 'tuners' provide a very convenient and compact approach to match an HF band antenna to the transceiver. Increasingly, many base-station transceivers tend to include an integral auto-tuner making the process to match the line/antenna to the transceiver even more straightforward. However, many auto-tuners are not able to accommodate higher SWRs that may be encountered with some types of antenna. This may necessitate the capabilities of a heavier-duty ATU and these tend to be manually tuned.

The basis of an auto-tuner is a matching network comprising a series of fixed inductors and capacitors. Using relays, these components are selected and configured in various combinations to obtain a good match between the transceiver and the line/antenna. The switching of the relays is controlled by a microprocessor based management system running an embedded tuning algorithm. The management system receives SWR and impedance samples from an RF head and uses these to select an appropriate combination of capacitance and inductance to obtain a good match. This selection is often controlled on an iterative basis and the auto-tuner can often be heard switching in and out various combinations of capacitors and inductors to obtain a low SWR. The construction of an automatic tuner is beyond the scope of this chapter, however Peter Rhodes, G3XJP, has designed the PicATUne auto-tuner [40] whose details can be found online [41].

An automatic ATU, that can be remotely controlled, provides a convenient method to match an antenna at a distance from the 'shack'. This concept was previously shown in **Fig 15.3**. As described earlier, this technique avoids the disadvantages associated with bringing the end of an antenna into the shack.

An example of an auto-tuner is the SG-235. This auto-tuner is understood to be capable of handling 500w. A simplified version of this auto-tuner is shown in **Fig 15.88**. Essentially, the SG-235 is a Pi-network, although if one of the banks of capacitors were to be switched out then it would be an L-network. On the left hand side of **Fig 15.88** are the capacitors associated with the low impedance 50Ω input from the transceiver. This left-hand bank of capacitors, with a total capacity of over 6000pF, can be switched in or out using relay contacts with a resolution of 100pF. On the antenna side of the ATU, the right-hand bank of capacitors has a total of nearly 400pF that can be switched in with a resolution of 12.5pF. To obtain a safe working voltage, the capacitance used for tuning is made up of groups of series/parallel capacitors.

The inductor section is made up of eight inductors giving a total inductance of 15.75μH with a switching resolution of 0.125μH. This inductor is connected between the capacitor banks and forms a 'Pi-network' with them. Antennas that are shorter that a quarter wavelength tend to present a low capacitive reactive impedance. These are tuned using the 4μH inductors (top right) that can be switched in or out to load the short antenna as required.

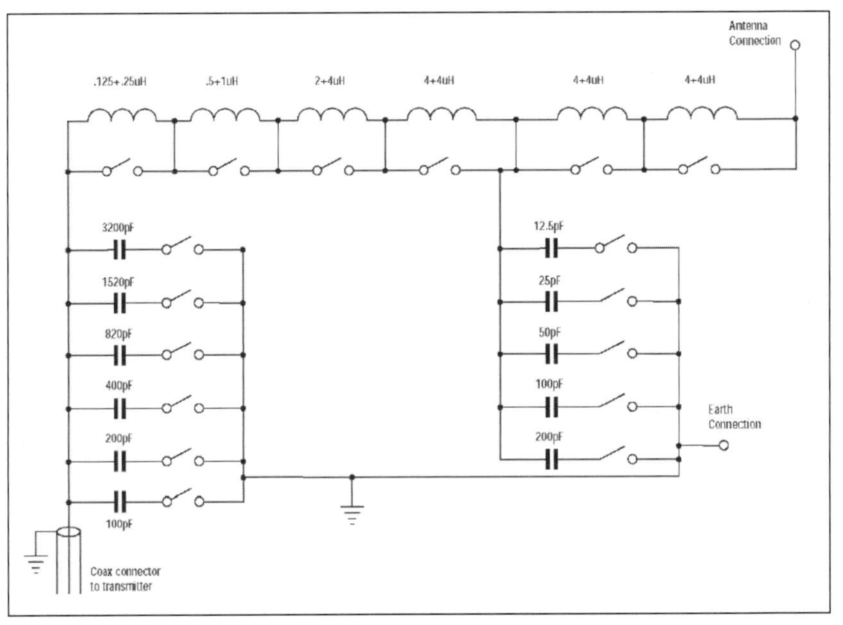


**Fig 15.88: Simplified diagram of the SG-235 auto-ATU. The Pi section inductor (the section between the two capacitor banks) is actually made up of eight inductors. The inductors (top right) are switched in for short antennas. Switching relays are controlled by a SWR/microprocessor circuit (not shown)**

## The Pi and T-network Antenna Tuner

The Pi-network ATU comprising an inductor connected between two capacitors was often used with older transceivers and has been included because many stations still use this form of ATU. The Pi-network has been the basis of for several ATU designs, as previously illustrated with the SG-235 auto-tuner in **Fig 15.88**. This circuit is theoretically capable of matching any transmitter to any antenna impedance (resistive or reactive). However, in practice the matching range is dependent on the component values used. For the widest step-up and step-down transformations, high-voltage variable capacitors are needed having low minimum and very large maximum capacitance values. Nowadays, obtaining such capacitors is not as straightforward as it used to be. The Pi-network possesses the advantage that it not only transforms impedance but also forms a low-pass filter, providing additional attenuation of undesirable harmonics and higher frequency spurii. The Pi-network was frequently encountered with valve radio transmitters and the concept of this circuit is shown in **Fig 15.89** [42]. C1 and C2 are the tuning capacitors

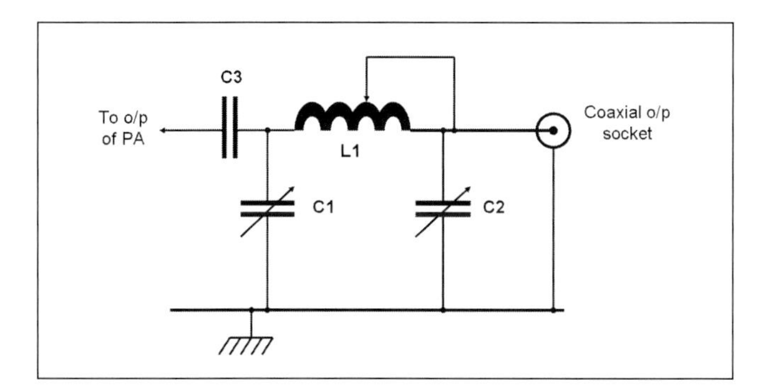


**Fig 15.89: Concept of the Pi-network used with older valve transceivers**

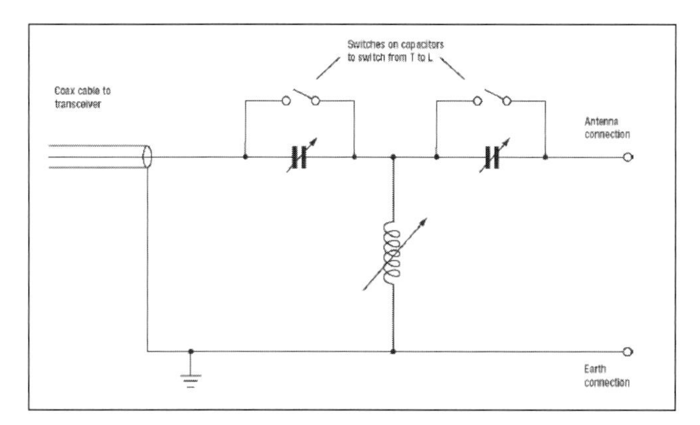


**Fig 15.90: Concept of the series capacitor Unbalanced-to-Unbalanced T-network. This arrangement forms the basis of many modern ATUs. The shorting switches across the capacitors allow the unit to be switched to an L network to reduce losses**

while L1 was often a 'roller-coaster' type variable inductor. C3 is a blocking capacitor connecting the Pi-network to the output of the PA stage of the transmitter.

Modern solid-state transceivers include integral low-pass filtering tailored to the individual bands, with the result that there is less of a need for the ATU to attenuate undesirable harmonics as was previously required. This has resulted in the greater use of the T-network that can provide an acceptably wide range of impedance transformations without a requirement for large-value variable capacitors. The T-network is a variety of the transmatch, however it does not use the split-stator type of capacitor used by other transmatch variants. The T-network's input and output ports are both unbalanced making its use suitable with coaxial feeder cables and unbalanced antennas. When the use of older valve transceivers was the norm, the T-network was regarded as a disadvantage because it forms a high-pass filter rather than a low-pass filter. However, now that low-pass filtering tends to be provided by the transceiver then this is less of a problem. While the T-network has become popular, it does suffer from losses at some transformation ratios on the higher frequencies. It is not uncommon for T-network to 'sink' around 20% of the RF power supplied to it depending upon the impedances it is expected to handle. This lost RF power is primarily 'sunk' by the inductor and care must be taken to ensure the T-network is not damaged when handling higher RF powers. These losses can be minimised by a simple modification of the T-network to an L-network using switches connected across the ends of the capacitors. The concept of a T-network that can be configured as an L-network is shown in **Fig 15.90** [43] [44].

## The G3TSO Transmatch

The following is an overview of the general purpose Transmatch ATU designed and constructed by Mike Grierson, G3TSO [45], who decided to base his ATU design on the T-network as described previously. The ATU's circuit details are illustrated as **Fig 15.91** and the design includes:

- The capability to match lines/antennas in the HF bands from 1.8 to 28MHz
- The ability to match over a wide impedance range
- The selection of different antennas
- The facility to ground all inputs when the station is not in use
- The inclusion of an SWR meter
- A balun allowing the unit to be capable of handling balanced lines

## Component selection

The values of capacitors required are not too critical and almost any high quality wide-spaced variable capacitor could be used. Ideally the variable capacitors used should:

- Be between 200pF and 400pF
- Be capable of working to 2,000V DC
- Be of the type that use ceramic end plates
- Have a plate spacing of at least 1.5mm between the stator and rotor plates.

These requirements are necessary to cope with the high voltages that can be developed across the capacitors' plates when matching high-impedance long-wire antennas. Surplus variable capacitors that are in good working condition and meet the above criteria should be suitable for use.

The inductors can be either fixed, with a number of taps selected by a rotary switch, or variable such as the 'roller-coaster' type that allows the maximum flexibility in matching.

All switches used are of the 'Yaxley' type and use ceramic

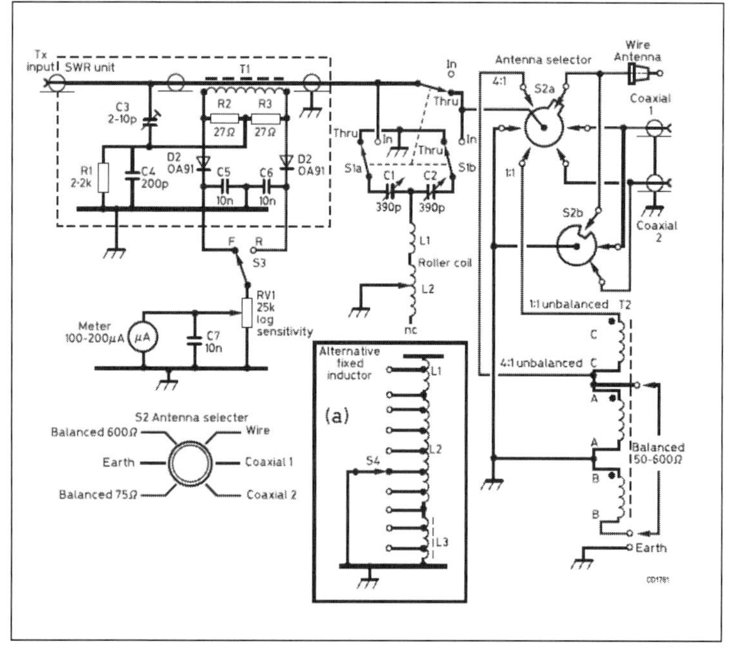


**Fig 15.91: Circuit details of the G3TSO ATU. This ATU was developed from the T-network variety of the Transmatch**

wafers. Paxolin wafers could be used, though they are not as good as the ceramic type. The antenna selector switch used a double-spaced switch unit giving six stops per revolution rather than the usual twelve. The switch wafers were modified by removing alternate contacts. This modification reduced the likelihood of arcing between them.

Many of the components required for the ATU could be sourced from radio rallies, from online suppliers or from radio component surplus suppliers, although the latter are becoming fewer-and-fewer.

### Balanced feeders

The T-network's input and output are both unbalanced. Therefore, to accommodate balanced feeders/antennas a balun transformer was included in the ATU's design. A balun that could be switched from a 1:1 ratio to a 4:1 ratio was used to enable impedances up to 600Ω to be matched. However, if a low impedance feeder from either a G5RV or W3DZZ type of antenna was to be matched, this could be done using the balun's 1:1 ratio option to reduce the losses. This balun's turns ratio was switched from the antenna selector switch. This provided a range of balanced inputs from about 45 to 600Ω without introducing too many losses into the system.

### Balun construction

The balun transformer was wound on a Micrometals T200-2 powdered-iron core, colour coded red (from Amidon). For sustained high-power operation, eg 400w, two cores could be taped together using plumbers' PTFE tape. This tape also provides an added layer of insulation between the core and the windings. The balun consisted of 14 turns of 16SWG enamelled-copper wire, wound 3-wire trifilar fashion onto the toroidal core. Care was taken that neither the core nor enamel covering was scratched during construction. 14 turns required approximately 100cm of 16SWG (1.6mm) wire. Three equal and slightly longer lengths of 16SWG cut wire were passed through the core until they reached about halfway to become the centre of the winding. It was found easier to wind from the centre to either end rather than from one end to the other. The T200-2 size core accommodated 14 trifilar wound turns without any overlapping of the start and finish of the winding. Close spacing occurs at the inside of the core and a regular spacing interval was maintained on the outside. A small gap should be left where the two ends of the winding come close together. When the balun was connected, the opposite ends of the same windings were identified using a continuity meter and marked for reference.

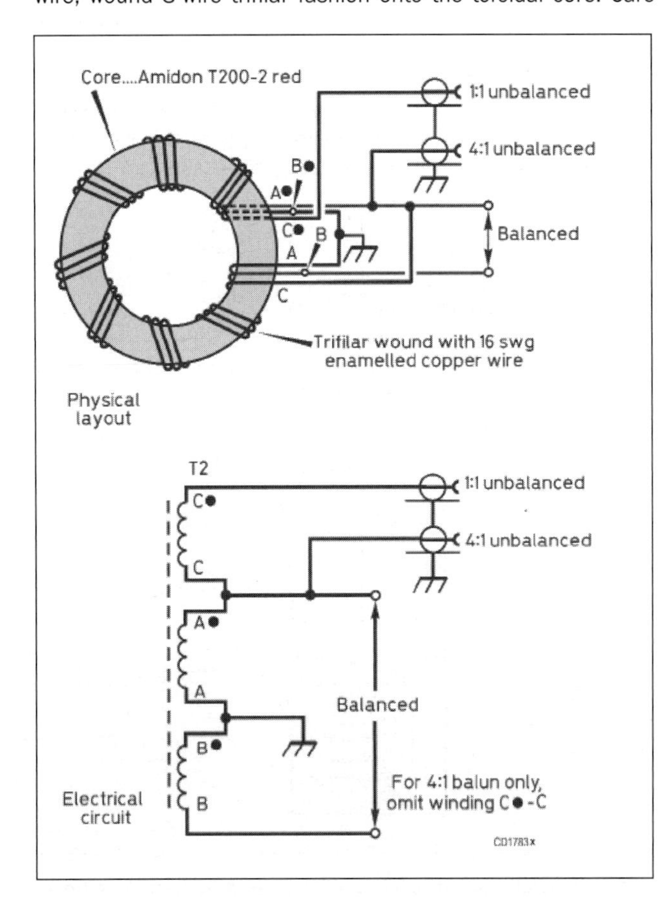


**Fig 15.92: 1:1 and 4:1 balun transformer arrangements**

Referring to the circuit diagram in **Fig 15.91**, a dot is used to signify the same end for each separate winding. It is essential that the various windings are correctly connected if the balun is to work properly. Details of how the balun transformer was wound and connected are shown in **Fig 15.92**. To hold the balun in place, it was directly supported by soldering the balun to the balanced input terminals. A sheet of 8mm Perspex was used to insulate the balun from the aluminium case. The construction of the 4:1 balun configuration is simpler and requires only two 2-wire bifilar windings.

### SWR measurement

To provide ease of matching, the ATU design included an SWR bridge whose circuit is shown in **Fig 15.91**. A conventional circuit design was used and this was a current-sampling bridge because this tends be less affected by frequency. The current transformer T1 used a ferrite ring of about 12mm diameter. Ferrite having a μ value of at least 125 should be used. The Amidon FT50-43 ferrite core is suited to this application. However, alternative ferrite rings are the FT23-43 (6mm diameter) and the FT37-43 (9.5mm) that have a μ of around 850. A short length of coaxial cable was passed through the ferrite core to form the primary after the 18 turn secondary had first been wound on. The braid of the cable was earthed at one end to form an electrostatic screen. However, both ends of the braid must not be earthed because it will form a shorted turn. D1 and D2 should be a matched pair of germanium diodes. These were selected from a number of similar-type diodes through comparing their forward and reverse resistances using a multi-range meter. **Fig 15.93** shows a suggested PCB layout for the SWR bridge.

The completed SWR bridge was tested away from the antenna tuner by placing it in line between a suitable transmitter and a 50Ω dummy load. The trimmer capacitor was adjusted to produce a zero-reflected reading with the forward reading at full scale. The bridge was then connected the reverse way around to check the diode balance by comparing the meter deflections in the opposite direction. The forward and reverse switch selection will be reversed if the signal direction through the bridge is reversed. It is advisable to check the bridge balances on a number of different bands, this is because C3 may be more sensitive at the higher frequency end of the operating range.

### Construction of the antenna tuner

The complete tuner layout is illustrated for guidance in **Figures 15.94** and **15.95**, with the components list provided as **Table 15**. A suitable case to house the ATU could be purchased, or prefabricated using 16 or 18SWG aluminium sheet bent into two interlocking 'U' shapes. 12mm aluminium angle provides suitable stiffening for a 'homebrewed' case as well as a method

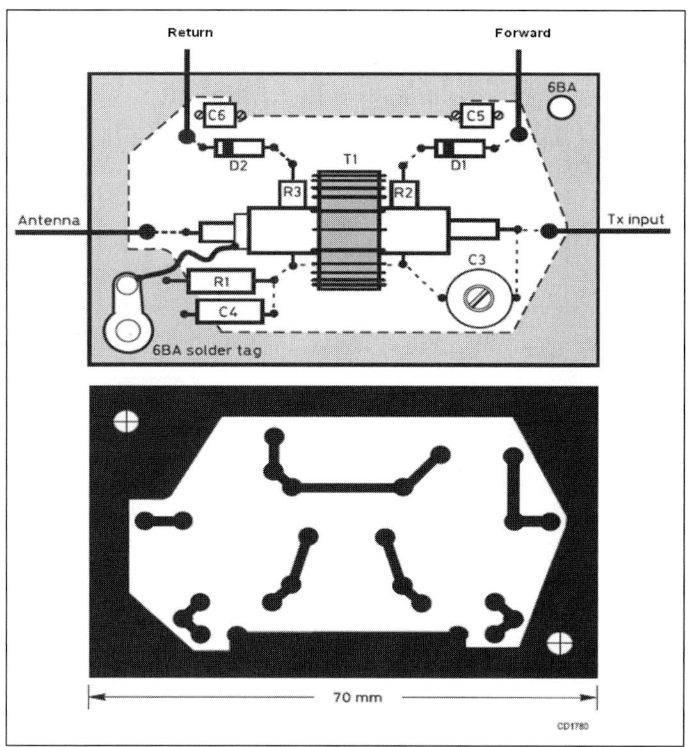


Fig 15.93: PCB and layout for the SWR bridge

of joining the sections together. The actual components used may well differ in size compared to those used for the original and this will determine the actual dimensions of the case required. The ATU's layout is not overcritical, however it is advised to minimise the lead lengths to avoid any unnecessary stray capacitance that could, for example, make the ATU's operation on 28MHz difficult.

The roller-coaster inductor's connections were arranged so that the minimum inductance was located at the end closest to the connections, ideally at the rear of the unit. A small heavy-duty coil (L1) was included for ease of 28MHz operation. As an alternative to the roller-coaster inductor, a switched inductor arrangement could be used. The switch used should be a ceramic type with substantial contacts. An example of a switched inductor was described by Hector Cole, G3OHK, and this is shown in **Fig 15.96**. A third toroidal inductor was included to permit operation on 1.8MHz (see **Fig 15.91**(a)). It is recommended that the bottom end of this inductor should be shorted to ground to prevent the build-up of high voltages which could arc over.

The capacitors C1 and C2 were 'electrically' above ground and must be mounted on insulators (a problem greatly reduced if the capacitors are constructed using ceramic end-plates). Ceramic pillars or Perspex could be used for mounting capacitors with metal end-plates. Additionally the shafts of the capacitors must be insulated and the use of insulated spindle couplers is recommended. To ease the rather sharp tuning characteristics that may be encountered on 21 and 28MHz, slow-motion drives were tried. However, they made tuning on the lower frequencies an arduous and lengthy process, therefore their use is not advised. A turns counter on the roller coaster makes for much simpler operation. However, something as simple as a slot in the cabinet with a Perspex window would allow the position of the 'jockey' wheel to be monitored.

Antenna switching could introduce excessive lead lengths as well as stray capacitance. For this reason the antenna selector switch was located on an extension shaft at the rear of the unit adjacent to the antenna inputs and the balun transformer. The wiring of the antenna switch was done to achieve the minimum lead lengths rather than to provide the front-panel selections in any logical order.

A separate IN/THROUGH switch enabled the tuner to be bypassed and the antennas routed directly to the transmitter. This switch was located on the rear panel adjacent to the input socket to minimise the lead length. This arrangement should not present too much of an operational problem because it intended for only for occasional use. It is necessary to ground the tuning components in the THROUGH position to minimise capacitance effects. Wiring of the tuner should be commenced after mounting all of the components using a heavier wire gauge, eg I6SWG tinned wire, coaxial cable braid or copper strip.

The SWR bridge was located directly adjacent to the transmitter input socket and all meter leads were kept away from the tuning components. It was also found necessary to screen the SWR bridge.

| ATU | |
|---|---|
| C1, C2 | 390pF 2,000VDC wkg, ceramic end-plates, eg Eddystone or Jacksons |
| L1 | 3t 10SWG, 25mm (1in) ID, 25mm (1in) long |
| L2 | Roller-coaster 36 turns, 38mm (1.5in) dia, 16SWG |
| T2 | Amidon T200-2 (red); 14 turns trifilar 16SWG enamel |
| S1 | Three-pole two-way ceramic Yaxley |
| S2 | One-pole six-way double-spaced ceramic Yaxley, one-pole six-way shorting water (one pole open) |
| **Alternative ATU circuit** | |
| L1 | 2.5t 14SWG 25mm (1in) ID tapped at 1 .5t |
| L2 | 14t 16SWG 1.25in ID tapped at 1, 2, 6, 9 and 14t |
| L3 | Amidon T1 57-2; 31t I8SWG enamel tapped at 6 and 27t |
| **SWR bridge** | |
| R1 | 2.2kΩ |
| C3 | 2-l0pF trimmer |
| C4 | 200pF mica |
| C5, C6, C7 | 10nF disc ceramic |
| R2, R3 | 27Ω |
| RV1 | 25kΩ log |
| D1, D2 | Matched OA91 etc (germanium diodes) |
| T1 | 18t 22SWG 13mm (0.5in) OD ferrite ring (Amidon FT50-43, Fairite 26-43006301). Primary: 38mm (1.5in) coaxial cable, braid earthed one end only to form electrostatic shield. |
| Meter | 100-200μA |
| S3 | SPCO miniature toggle |

Table 15: Components list for the G3TSO Transmatch

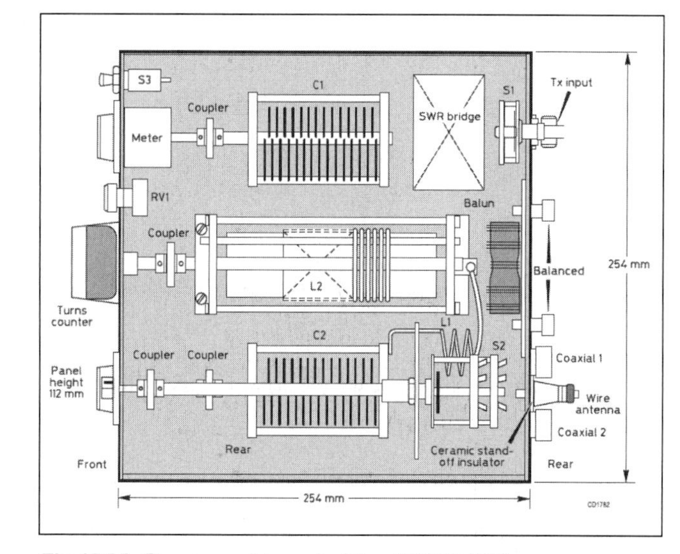


Fig 15.94: Component layout of the G3TSO ATU

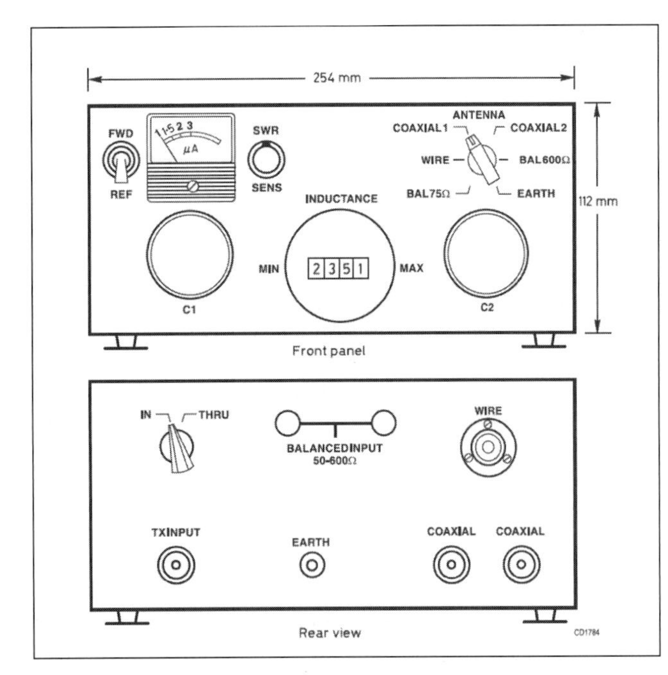


**Fig 15.95: The ATU's front and rear panels**

The antenna selector switch had two ceramic wafers and was arranged so that every other contact was removed to give double spacing. The second wafer was used for shorting and provided a ground for all unbalanced antennas that are not in use. This was largely to avoid capacitive coupling to other antennas. The balanced input is grounded to DC through the balun.

Balun switching was achieved by either taking the input from one side of the balanced input, giving a 4:1 ratio, or by selecting the third winding, giving a 1:1 ratio. An earth position enabled the transceiver input to be grounded to prevent static discharge into the receiver.

## Operation of the antenna tuning unit

If the SWR bridge is included in the design, it should be checked and balanced independently of the ATU, using a dummy load. Ideally it should be compared and calibrated against an SWR measuring device of known accuracy.

To use the ATU, start by selecting the required antenna and ensure that the THROUGH/IN switch is switched to the IN position. Set both C1 and C2 to their halfway positions, adjust the inductance for maximum signal on receive, and adjust C1 and C2 one at a time for maximum received signal. Using a low CW transmitter power, further adjust C1, C2 and the inductance to eliminate any reflected reading on the SWR meter. All tuning controls are interdependent and settings may need to be adjusted several times before minimum SWR is achieved. Often, more than one setting may give a matched condition and the settings required for the highest value of C1 should be used.

Once the transmitter is matched on low power, increase the operating power for any final adjustments. Never attempt to tune the ATU initially on full power

or with a valve power amplifier that has not been tuned up. Generally, the higher the frequency the lower the value of inductance required, but exceptionally high impedances may require more inductance than expected. Capacitance values may vary considerably, and it is not uncommon on the higher frequencies for one capacitor to be very sharp and require a minimum value while the other is flat and unresponsive. Using the components recommended, it is possible to match a wide range of impedances from 1.8 to 28MHz, but operation on 1.8MHz may

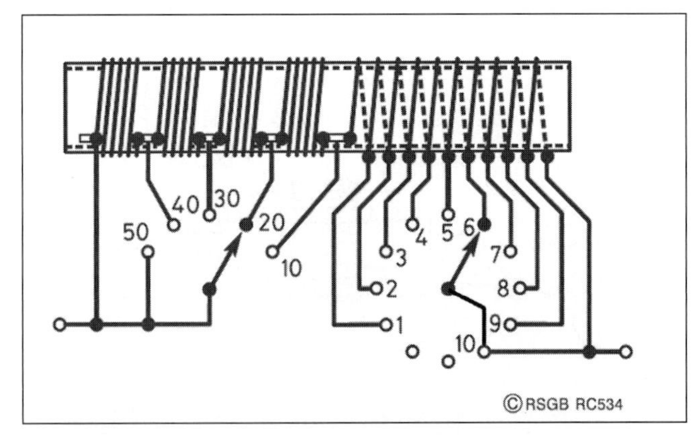


**Figure 15.96: A variable inductance ATU coil described by Hector Cole, G3OHK. This arrangement uses two switches and 14 taps to permit selection of from one to 50 turns of a 50-turn coil. This arrangement can be reset to any number of turns previously found suitable without a turns-counter as required for roller coaster coils**

become impossible if lower values of capacitance are used. However, fixed silver mica capacitors may be switched across C1 and C2 to compensate for this. Higher values of capacitance will almost certainly prevent operation on 28MHz and maybe 14MHz.

## Conclusion

The antenna tuner described is not a new revolutionary design, however it provided great flexibility as a solution to matching. The tuner's performance was good and was not inhibited by a lack of balanced input or restricted to a very narrow range of low impedances. The power handling capability of the tuner depends on the impedances encountered and the plate-spacings of the capacitors. As a rule, very high impedances should be avoided because arcing could occur in the switches and the efficiency of the unit may well suffer. Adjustment of the antenna or feeder length may remove any exceptionally high impedances that could be encountered.

G3TSO used this tuning unit with a 60m (180ft) doublet fed with an unknown length of 300Ω slotted-line feeder. The tuning unit could be tuned to give a 1:1 SWR on all amateur bands from 1.8 to 28MHz. Using Eddystone capacitors of the type recommended, the tuning unit should be capable of handling 100w into a fairly wide range of impedances up to several thousand ohms, and the full 400w into impedances up to 600Ω.

Two versions of the tuner have been built using the same basic circuit. One was for base station operation using a roller-coaster inductor and a smaller version using a range of switched inductors for portable use. The portable version has a slightly different layout and was a result of trying several other designs. This version also combined the IN/THROUGH facility on the inductor switch that necessitated several wafers. The balun used in this version was also the simpler 4:1 type and was connected with a 'flying' lead.

## Balanced ATUs

Many of the antennas that have been described previously are balanced and so require a balanced feeder to avoid problems arising from common mode currents. Therefore, when an ATU is necessary to match a balanced feeder/antenna to the transceiver, the antenna connection terminal to the ATU also needs to be balanced. When the ATU does not have a balanced antenna

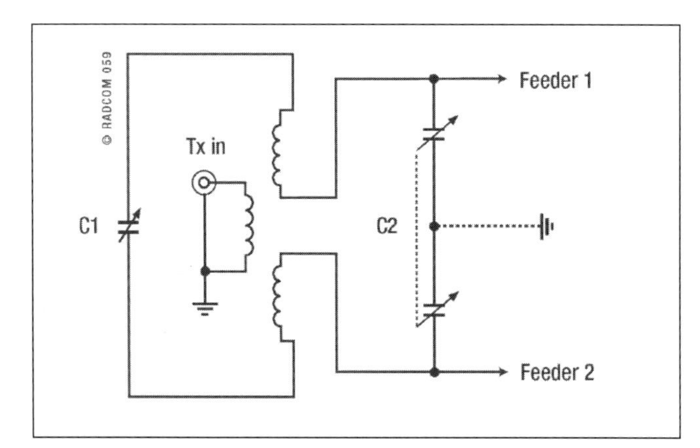


Figure 15.97: Concept of a balanced matching unit

connection, eg a T-network ATU, then a straightforward approach suggested by L.B. Cebik, W4RNL, is to use a 4:1 balun between the feeder/antenna and the ATU [46]. However, this system does have its limitations when the balanced impedance presented to a 4:1 balun is a low impedance. This is because the impedance presented to the ATU may then fall outside of its capabilities, although a 1:1 balun could be used as described earlier for the G3TSO transmatch ATU.

Another approach is to link-couple or inductively couple the feeder/antenna to the ATU's main inductor. However, this approach usually does not allow all the HF bands to be covered because all the best matching ratios can be difficult to provide. Without the provision for coil tapping and series connections, the most efficient operating mode may be inaccessible, even when the impedance presented by the feeder/antenna is 50Ω.

An ATU that overcomes some of these problems, based on established balanced matching techniques, was designed and built by Brian Horsfall, G3GKG [47]. The basis of the G3GKG design is essentially a link-coupled balanced Pi-coupler that did not require match selection through taps on the coils. This balanced input/output ATU design used one variable capacitor, C1, to 'tune' the network and a second one, C2, to 'tune out' any reactance at the feed point and match the overall impedance. The concept of the ATU circuitry is shown in **Fig 15.97**. The two controls may interact, however it is possible to obtain a perfect match by rotating them alternately. Tune up was achieved by rotating each control until a decrease was observed in the reflected power (or SWR) reading with the initial tune up using low power or with the aid of an antenna analyser. With practice it was possible to obtain zero reflected power, coincident with maximum forward power.

In general, the lower the impedance of the load the more capacitance will be required in C2 and, as the two capacitors are effectively in series as regards resonating the inductor, the lower will be the capacitance of C1. However, for either capacitor, the lower the capacitance when loaded, the higher will be the voltage across it at any given frequency and power. Furthermore, the lower the reactance in the load, the broader will be the tuning. It is only with very high impedances and highly reactive loads that the tuning becomes relatively sharp and critical.

## Practical considerations

There are three possible circuit arrangements of the two variable capacitors:

- C1 could be a twin-gang type with the frame earthed, to provide an 'electrical' centre about which the feeders are balanced.
- C2 could be a twin-gang type with the frame earthed, to provide an 'electrical' centre about which the feeders are balanced.
- C1 and C2 could be single-gang types that are completely isolated from earth with the whole of the secondary circuit, including the antenna system itself, 'electrically' floating. For this arrangement it is important that the control shafts are insulated.

In many amateur installations, the antenna will tend towards being unbalanced. Therefore, the latter floating method could be used where the whole feeder/antenna system finds its own 'balance'. For any of the cases above, high-value resistors should be connected to earth from each side of the feeder to prevent static voltage build-up.

C2 will need to have fairly wide-spaced plates, although its capacitance does not have to be too high provided higher feed point impedances are avoided. There is technique that may improve the match where a few extra metres of feeder could be inserted in series between ATU and the feeder/antenna on the troublesome band(s). This cable could be 300 or 450Ω ladder-line, rolled up in the shack. This technique was described earlier in this chapter as a method of matching used by W6RCA. However, if it is intended to use the antenna on all the HF bands, it may be difficult to avoid at least one band, or more, where the impedance presented is high resulting in a high voltage across the feeder/antenna. If the feed/antenna presents a very low impedance to the ATU at the lower frequencies, then the required value of C1 can be very high. This could necessitate either using a multi-gang variable capacitor for C1, with the sections in parallel, or additional fixed capacitors that could be switched in. A vintage receiver twin-gang air-spaced variable capacitor could be used for C1 where the two sections are connected in series to double the voltage rating.

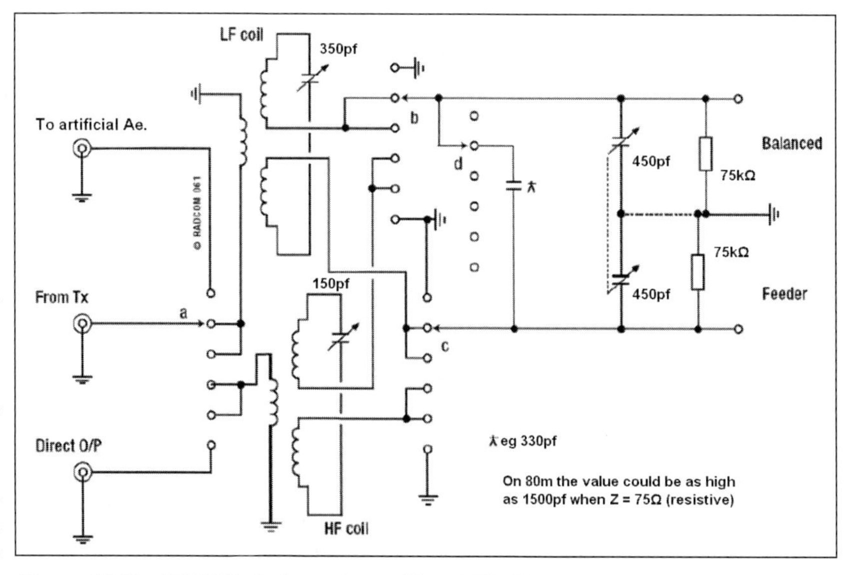


**Figure 15.98: G3GKG's balanced matching ATU. The two coil assemblies are: LF (used for 80, 60 & 40m bands) - (9 + 9) turns of 16SWG tinned copper wire on 2.25inch diameter former, with 3 turn link of PTFE-coated wire. HF (used for 20 to 10m bands) - (4 + 4) turns of B&W 1.75inch stock with 1 turn link between windings. A large, 4-gang, 6-way ceramic switch assembly a, b, c, d is used to select the bands**

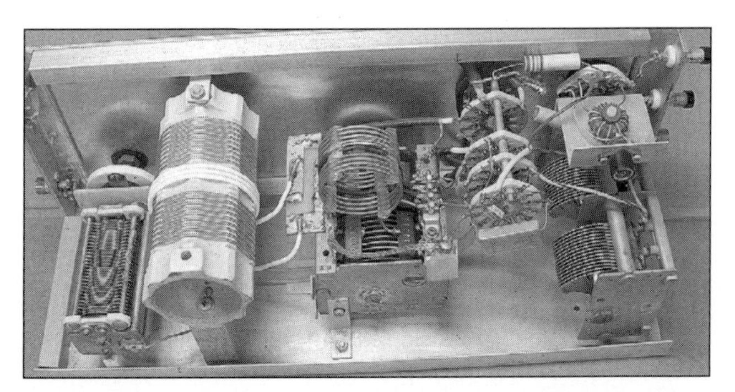

**Figure 15.99: Internal view of the G3GKG all-band balanced ATU. The LF coil for the 80, 60 & 40m bands is shown to the left and the coil for the higher frequencies to the right. The toroid/ switch assembly (top right) is to cater for an alternative feed directly to the receiver**

## The G3GKG All Band ATU

The full schematic diagram of the G3GKG all-band ATU is shown in **Fig 15.98.** Using this arrangement, it was not possible to make an ATU to cover all the HF bands using one coil and so two separate coils were used. Using two separate coils avoided switching the inductor 'hot' ends by having separate C1 tuning capacitors for each inductor (C1 as referred to **Fig 15.97**).

The internal construction and layout can be seen in **Fig 15.99.** The LF coil for the 80, 60 and 40m bands is shown to the left and the coil for the higher frequencies to the right. On one or more bands, the total capacity of C2 required to match the particular antenna in use was made up by switching in an extra fixed capacitor (C2 as referred to **Fig 15.97**). The value of this capacitor was determined to suit the feed impedance by temporarily substituting a variable capacitor and then testing using low power. C2 needs to be capable of providing coverage for each of the individual bands.

The low frequency coil was wound on a surplus Eddystone ribbed ceramic former, whose construction determined the turn spacing to about one turn's width. The two main windings were spaced as far apart as the former allowed giving a gap of about 25mm between the coils. Into this gap was wound the link-coil that was close-wound using thicker PTFE insulated wire. The coupling proved closer than it needed to be and was kept fairly loose to minimise the capacitive coupling effect (that otherwise could produce an undesirable in-phase current in the feeder). For virtually complete elimination of the capacitive effect, an earthed Faraday shield around the link-coil could have been used. The two 75k resistors provide a bleed to earth to prevent static building up on the antenna and feeder.

### Two-band version

If the requirement is for an ATU to cover only the 40 and 80m bands, then a simpler design could be used requiring one coil. This 2-band version of the G3GKV ATU is shown in **Fig 15.100.** The switching required was provided by using a ceramic 2-pole switch to add the padding capacitors across both variable capacitors on 80m for the reasons previously described above.

**Figure 15.100: The interior view of the two-band version of the G3GKV ATU**

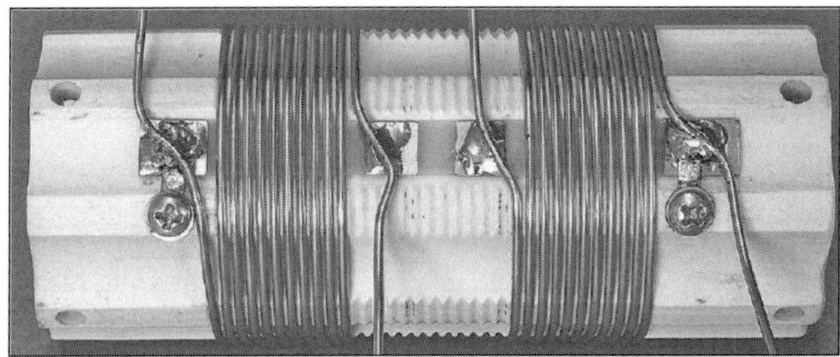

**Figure 15.101: Method of fixing the windings to a fluted ceramic coil former**

In this particular ATU a slightly smaller ceramic former was used (compared with **Fig 15.99**) that was salvaged from one of the tuner units from a BC-375 transmitter. The optimised windings for this inductor were 11 + 11 turns, with a 3-turn link-coil.

### ATU components

G3GKG devised a method of providing anchor points for the inner ends of the windings. These consisted of strips of fibreglass PCB, each cut to fit closely into one of the flutes in the former, with all the copper removed except for small pads at the ends that are used as anchor points. The PCB anchor points were fixed in place by using solder tags secured by screws to the two existing threaded holes in the former as shown in **Fig 15.101.**

The ends of the main windings were soldered to the copper PCB anchor points, leaving sufficient extra wire for the connections to the capacitors. A strip of thin polythene sheet (cut from the lid of a scrap plastic box) was wrapped around the inter-winding space and held in place using polythene adhesive tape before the link-coil was wound over it using heavy duty PTFE insulated silver plated stranded wire. The ends of this winding were twisted together before a short length of heat-shrink tubing was slipped over them, as close to the former as possible. Then a thin coat of acrylic varnish was applied to hold the heat-shrink tubing in place.

Obtaining appropriate fixed value capacitors suitable for an ATU can be difficult. There are many types of capacitor available that are claimed to have high kilovolt ratings. However, only some are suitable for high power RF applications. The best capacitors to look out for are the large 'mushroom-shaped' type of capacitor specifically designed for RF applications and rated at about 10kV. These capacitors were made by Plessey and TCC among others. They typically come in values between 10 and 1300pF and, unfortunately, tend to be only found at rallies nowadays. Some examples are shown in **Fig 15.102** (left and centre). These capacitors are ideal for RF powers up to the legal limit, although 1000pF or similar value capacitors of this type can be very large. Another glazed-ceramic encased type of

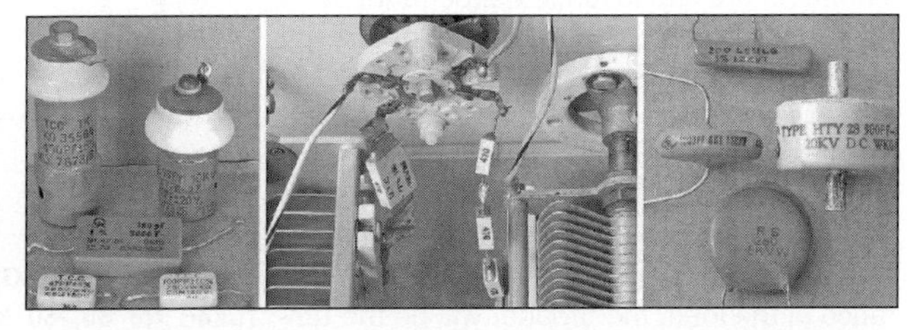

**Figure 15.102: The capacitors left and centre were found to be suitable for the ATU. However the other capacitors to the right are to be avoided**

capacitor, used by G3GKG in both ATU designs, came in two sizes that were rated at 1kV and 2kV RF. They were very stable and produced no perceivable heat when working at full power. However, these capacitors were only available in values of around 500pF and so parallel or series combinations of these capacitors were used as required.

Some capacitors, as shown in **Fig 15.102** (right) are designed for used in pulse applications and will tend to break down if subjected to the typically high RF voltages encountered in an ATU. The ceramic disc and 'doorknob' types also shown in **Fig 15.102** (right) may or may not withstand the voltage strain required. These capacitors are intended for coupling and decoupling situations and can be prone to large capacitance changes with temperature. Surprisingly, the moulded mica capacitors from the American TU units rated at up to 5kV may not be suitable. G3GKG has found that the internal construction of some of these capacitors to be poor (this is tends to be problem with 'vintage' components).

## The Z-match

Another well established link-coil coupled ATU design is the Z-match. Originally, this ATU was designed as a tank circuit of a valve PA [48]. The anode was connected to the top (or 'hot') end of the multi-band tuned circuit. It was fed directly from the PA valve, with its internal (source) impedance of several thousand ohms. When the circuit was adopted as an ATU [49], the tank circuit was fed directly from a source that required a 50Ω load via a 350pF variable coupling capacitor connected to the top (or 'hot') end of a multi-band parallel-tuned LC circuit.

Despite the difference between the required 50Ω load for the transmitter and the relatively high impedance of the tank circuit, the Z-match became very popular due to its straightforward design. Z-match ATUs were produced commercially and they are described here because they are still readily available and tend not to be expensive. An example of such a unit is shown in **Fig 15.103**.

The design of the Z-match was improved by Louis Varney, G5RV, and to follow is a summary of the article that he published [50]:

Referring to **Fig 15.104**(a), on the 3.5, 7 and 10MHz bands the inductor L1 is connected in parallel with the two sections of C1 (C1 is a double-ganged capacitor comprising C1a and C1b that are parallel connected). The much smaller inductor L2 can be thought of as a connecting lead between the top of C1a and the top of C1b. This assumption is valid at the low frequencies of 3.5 and 7MHz because the inductance of L2 is much less than that of L1. For these bands, the combination of L1, C1a plus C1b can be considered as a simple tuned circuit with one end earthed. Provided the capacitance range of C1a plus C1b is sufficient, the circuit will also tune to 10MHz. Although, it may be necessary to reduce the inductance of L1 by one to two turns to achieve resonance on 10MHz. However, it is important to avoid the occurrence of harmonic resonance between the two circuits comprising the multi-band tuned circuit. Therefore, the values of the inductors L1 and L2 must be selected to prevent this effect.

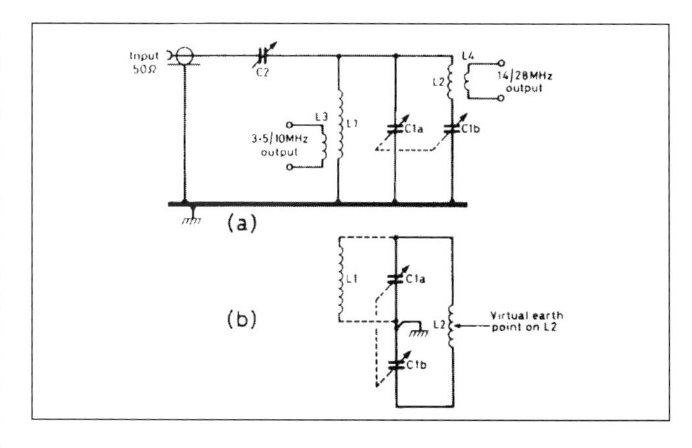


Figure 15.104: (a) The basic Z-match circuit. (b) The 14-28MHz tuned circuit shown in a more conventional form

On the 14, 18, 21, 24 and 28MHz bands the active tuned circuit consists of the two variable capacitor sections C1a and C1b working as a split-stator capacitor (with the moving vanes earthed) with L2 connected between the two sets of stator vanes. The inductance of L1 is much greater than L2, therefore L1 may be considered as an HF choke coil connected in parallel with C1a. L1 has no appreciable effect on the performance of the split-stator tuned circuit comprising L2, C1a and C1b. This can be demonstrated by first tuning this circuit to any band from 14 to 28MHz and noting the dial reading of C1a, C1b. Then the top of L1 is disconnected and the circuit retuned for resonance. It will be noticed that the effect of L1 is negligible. **Fig 15.104**(b) shows the effective 14 to 28MHz tuned circuit in a more conventional manner.

The relatively high impedance LC circuits of L1, C1a, C1b (tuning for 80m to 30m bands) and L2, C1a, C1b (tuning for the 20m to 10m bands) must be detuned slightly off resonance at the frequency in use. The effect of this is to present an impedance with an inductive component (ie R + jX). This impedance, in conjunction with the coupling capacitor C2, functions as a series resonant input circuit that, when correctly tuned, presents a 50Ω non-reactive load to the transmitter's output.

Figure 15.103: An early KW E-ZEE MATCH showing the general construction of a Z-match ATU

## Modifying the Z-match

Supplying RF power from the transmitter's output, requiring a 50Ω load, to the top of a high impedance parallel-tuned LC circuit tends not to be an efficient approach for matching. Therefore, G5RV redesigned the circuit to improve the match efficiency by introducing tapping points on coils L1 and L2. The circuit of the modified Z-match is shown in **Fig 15.105**.

### The final modified Z-match

The final design is shown in **Fig 15.106** and a list of components is in **Table 16**. The circuit incorporated switching for the appropriate coil coupling taps and selects the appropriate output-coupling coil (L3 or L4) to the feeder. For maximum output coupling efficiency on the 7MHz and 10MHz bands, a tap on L3 was selected by S1b. Provision was made for coaxial cable antennas to be

fed either directly or through the Z-match. The transmitter

| Item | Details |
|---|---|
| C1a-1b | Split-stator variable capacitor 20-500pF per section |
| C2 | 500pF single-section variable capacitor (shaft insulated) |
| L1 | 10t 4cm ID C/W 14SWG enamel copper wire. Tap T1 4t from earth end |
| L2 | 5t 4cm ID turns spaced wire dia 14SWG enamel copper wire. T2 1.5t from centre of coil (virtual earth point) |
| L3 | 8t 5cm ID C/W enamel copper wire over L1. T3 at 5t from earth end |
| L4 | 3t 5cm ID C/W over L2. 14SWG enamel copper wire |
| S1 | Ceramic wafer switch. All sections single-pole, five positions |
| S2 | Ceramic wafer switch. Single-pole, three positions |

**Notes:**

**(1)** C1a and C1b. A suitable 250 + 250pF (split-stator or twin-ganged) variable capacitor can be used since the capacitance required to tune L1 to 3.5MHz is approximately 420pF, and for 7.1MHz approximately 90pF. If C1a, C1b (paralleled) have a combined minimum capacitance of not more than 20pF, it should be possible also to tune L1 to 10MHz. Otherwise it may be necessary to reduce L1 to nine turns, leaving T1 at four turns from the 'earthy' end of L1. A lower minimum capacitance of C1a, C1b as a split-stator capacitor would also be an advantage for the 28-29.7MHz band.

**(2)** Taps on L1 and L2 are soldered to inside of coil turn. Tap on L3 soldered to outside of coil turn

**Table 16: Components list for the G5RV Z-match**

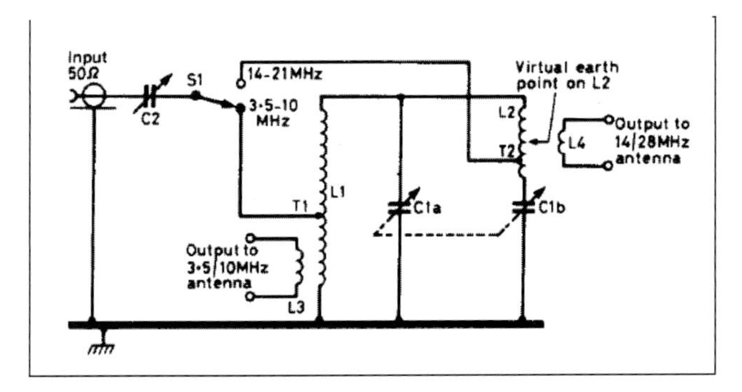


**Figure 15.105: The basic Z-match circuit showing the tapped-down feed arrangement**

output can be connected directly to a suitable 50Ω dummy load if necessary.

The layout is not critical, however it is advisable to mount the coils L1 and L2 with their axes at right angles to prevent undesirable mutual coupling between them. All earth leads should be kept as short as possible and the metal front panel should be earthed as a matter of course. The coupling capacitor, C2, should be mounted on an insulating sub-panel and its shaft fitted with an insulated shaft coupler to isolate it from the front panel, preventing hand-capacitance effects.

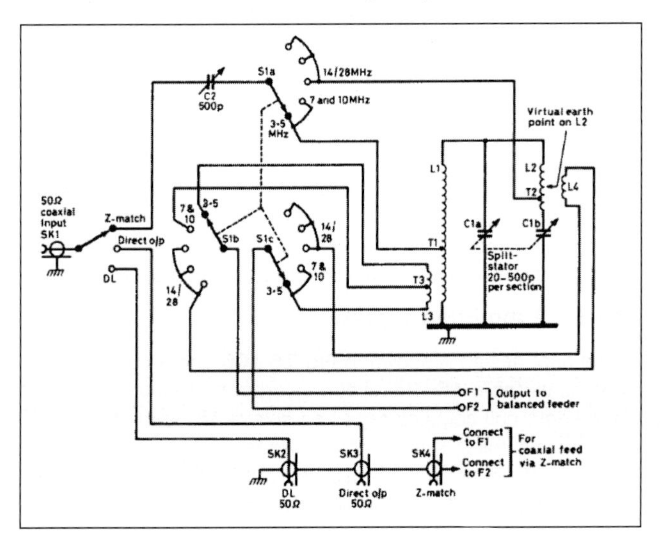


**Figure 15.106: G5RV's final modified Z-match circuit**

The receiving-type variable capacitors used in the experimental model Z-match had adequate plate spacing for CW and SSB (peak) output powers of up to 100w. For higher powers it would be necessary to use a transmitter-type split-stator capacitor (or two ganged single-section capacitors) for C1a, C1b. However, C2 required only receiver-type vane spacing even for high power operation. Tests with additional feed point taps on both L1 and L2 in the modified Z-match circuit showed no practical advantage. However, the tap on the output coupling coil L3 was found to be essential on 7MHz and 10MHz. The very tight coupling between L1/L3 and L2/L4, tended to reduce the operating Q value of the LC circuits. This rendered them more 'tolerant' of the complex reactive loads presented at the input end of the feeder(s) connected to the antenna(s) used.

G5RV noted that the efficiency of a conventional link coupled antenna tuning unit was better than that of either form of Z-match designs. In addition, by virtue of its design, the Z-match cannot satisfy all the required circuit conditions for all the HF bands. However, in its original form the Z-match does provide the convenience of covering the 3.5 to 28MHz bands without the necessity for switched or plug-in coils. Nevertheless, the inclusion of the simple switching shown in **Fig 15.106** is an advantage.

## Remote Controlling of an ATU

An ATU that can be remotely controlled provides a convenient method to match an antenna at a distance from the 'shack'. As previously described in this chapter, this technique avoids the disadvantages associated with bringing the end of an antenna into the shack. The concept of a remotely controlled ATU was shown in **Fig 15.3**.

## G3UCE Remote-Controlled ATU

L B Uphill, G3UCE, devised a remote-controlled ATU for use with an end-fed wire antenna that has proved satisfactory on all the HF bands. The concept of this remote controlled ATU is shown in **Fig 15.107**.

The ATU was located in a rear porch and was connected to the shack using 10m of coaxial cable. Control of the unit was provided through an 8-way multi-core cable connected to a control box in the shack. A good RF earth or a counterpoise close to the ATU was also necessary. The ATU could have been located outside, however this would have required the ATU to be weatherproof and locating the unit in the porch avoided this.

Once set up, the ATU provided instant selection of pre-selected settings. However, the ATU did take several hours to set up and should be sited in an accessible position. The ATU was tested with several different lengths of antenna from 18m (60ft) to 61m (200ft). Some antenna wire lengths proved difficult to tune when operation on all six bands was tried. The best antenna lengths for all six bands were 30m (100ft) and 40m (132ft). For the inductor, around 40 turns were required if 160m is the lowest band to be used. When 80m is the lowest band to be used, then 20-25 turns were found sufficient.

A junk box coil could be used for the inductor provided the turns are of a reasonably heavy copper wire. The wire spacing should allow the use of an instrument-type crocodile clip with narrow jaws to be clipped to any turn during setting-up, without shorting an adjacent turn. The inductor used was a 40 turn coil wound on a 190mm length of 45mm diameter plastic pipe using 14 to 16SWG tinned copper wire. One wire-end was fastened to a nut and bolt located at the pipe's end, then the inductor was

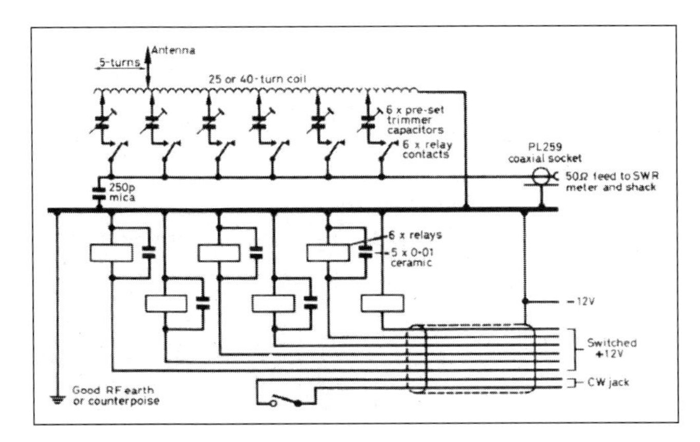


**Figure 15.107: Concept of the G3UCE remote-controlled ATU**

tightly wound using a similar thickness of string as spacing until 40 turns were made. The wire-end was then fastened to the pipe's other end to a second nut and bolt. The string spacer was then carefully removed and 3 or 4 strands of adhesive glue, such as epoxy resin, were applied across the turns to hold them in place.

The capacitors used were 100pF air spaced types for the higher frequency bands and 500pF 500V working mica presets for the lower bands. A capacitor may not be required for 160m, where a direct connection is made from the coaxial feeder cable to the coil. The relays were 12 volt types with contacts able to carry about 5A AC. A small control box with a 2-pole, 6-way Yaxley switch controlled the switching of the relays from the shack. One pole switched the relays and the other pole switched LED indicator lamps indicating the band selected. An 8-core miniature, screened cable was used to connect the ATU to the control box located in the shack.

With six bands to select, this left two wires spare and these were used to switch the transmitter on and off via the CW socket during adjustments from the ATU end. The setting up procedure was as follows:

- Start with the lowest frequency, then tune up the transmitter on a dummy load to the centre of the band. Initially a low transmit power was used for testing
- An SWR meter was inserted at each end of the feeder ready for testing (ie one at the shack end and one at the ATU end)
- Connect the feeder cable, energise the appropriate relay, and pass a small amount of RF power to the ATU
- Monitoring the SWR, find a tapping point on the coil by working from the aerial tap until the SWR reduces
- Then adjust the appropriate capacitor until a combination is found which gives the lowest SWR (ensure that the transmitter is switched off whilst manually adjusting the tapping point, to prevent the possibility of an RF burn to your fingers)
- Now check the shack SWR meter and if both SWR meters have similar readings, increase the transmit power then recheck the SWR. If the SWR remains low, the tap can be soldered permanently in place on the coil
- Now carry on to the next lowest frequency, remembering to switch in the appropriate relay.

On the highest frequencies, (15, 12 and 10m) the tap should not need to be more than four or five turns from the antenna. When all the bands have been satisfactorily set up, no further alterations must be made to the antenna length or the earth system or all adjustments will need to be repeated.

## REFERENCES

[1] 'Taming the End-Fed Antenna', Alan Chester, G3CCB, Radcom September 1994.

[2] RSGB HF Antennas for Everyone, edited by Giles Read, M1MFG: Chapter 1, page 2.

[3] VK6YSF Projects, Peter Miles, web link *http://vk6ysf.com/unun_9-1.htm*

[4] MOUKD Amateur Radio Projects, web link https://m0ukd.com/homebrew/baluns-and-ununs/91-magnetic-longwire-balun-unun/

[5], 'Antennas' Pages 34 to 36, Mike Parkin, GOJMI, RadCom January 2017.

[6]: 'A 3 or 5 Band End-Fed Antenna', Jos van Helm, PA1ZP. RSGB RadCom, February 2016 Pages 54 to 59.

[7] Radio Communication Handbook 5th edition. Section 12 HF Aerials Page12.42. Published by RSGB 1988.

[8] Radio Communication Handbook 5th edition. Published by RSGB 1988. Section 12 HF Aerials, Page 12.63, Fig 15.12.97.

[9] Radio Communication Centenary Issue. Published by RSGB July 2013. Pages 32 to 33.

[10] HF Antenna Collection, Edited by Erwin David G4LQI. Chapter 1 Single-Element Horizontally Polarised Antennas Pages 9 to 13, Louis Varney G5RV. Published by RGSB 1991.

[11] 'Antennas' Pages 30 to 31, Mike Parkin GOJMI, RadCom February 2016.

[12] 'In Practice', Radcom, Aug 1999.

[13] Some Broadband Transformers, C.L. Ruthfoff MIRE, published by the Institute of Radio Engineers 1959.

[14] VK6YSF Projects, Peter Miles, web link *http://vk6ysf.com/projects.htm, July 2012.*

[15] 'Antennas' Pages 60 to 32, Mike Parkin GOJMI, RadCom October 2016

[16] Radio Communication Handbook 5th edition. Section 12 HF Aerials Page12.42. Published by RSGB 1988.

[17] Electron (December 1992) and reported in 'Technical Topics', RadCom, May 1993.

[18] The ARRL 1976 Handbook 53rd edition. Chapter 21 HF Antennas Pages 606 to 607.

[19] 'Electrically Tunable Loop', Roberto Craighero, I1ARZ, Radcom February 1989

[20] 'Small high-efficiency antennas - alias the loop', 100-page booklet published by Ted Hart, W5QJR

[21] 'Eurotek', DK5CZ's comments on the use of vacuum cap, Radcom October 1991, p39

[22] 'Antennas' Pages 34 to 36, Mike Parkin GOJMI, RadCom November 2016

[23] 'Experimental Magnetic Loop Antenna' C R Reynolds, GW3JPT, Radcom February 1994

[24] The G4ILO Wonder Loop, *http://www.g4ilo.com/wonderloop.html, Julian Moss G4ILO*

[25] 'Antennas' Page 59, Mike Parkin GOJMI, RadCom March 2016

[26] 'Antennas' (2-Element Beam), Pages 56 to 58, Mike Parkin GOJMI, RadCom May 2016

[27] 'Antennas' (3-Element Yagi Beam), Pages 40 to 52, Mike Parkin GOJMI, RadCom June 2016

[28] L.B. Cebik, W4RNL, http//www.cebik.com/

[29] 'Concentrated Directional Antennas for Transmission and Reception', John Reinartz, W1QP and Burton Simson W8CPC, QST Oct 1937

[30] 'VK2ABQ Antenna', Fred Caton VK2ABQ, Electronics Australia, Oct 1973

[31] HF Antenna Collection, edited by Erwin David G4LQI. Published by RSGB 1991. Section 2 Horizontally Polarised Beams, Page 23.

[32] 'Collapsible 6m 2-element Beam Antenna', Pages 44 to 46, Mike Parkin, G0JMI, Radcom September 2012

[33] HF Antennas for all Locations, 2nd edition, Les Moxon, G6XN, RSGB

[34] 'Antennas', Radcom, Mar 2002

[35] 'Wire Beam Antennas and the Evolution of the Double-D', Peter Dodd, G3LDO, QST Oct 1984, also RadCom June/July 1980 (RSGB)

[36] 'An All-Band Antenna for Mobile or Home', John Robinson, G3MPO, Radcom, Dec 1992 and Jan 1993.

[37] The Amateur Radio Mobile Handbook, 2nd edition, Peter Dodd, G3LDO, RSGB

[38] 'Eurotek', Radcom, September 1993.

[39] EMC 07 Protective Multiple Earthing, RSGB. Weblink: *http://rsgb.org/main/files/2012/06/emc-leaflet-07.pdf*

[40] 'PicATUne - the Intelligent ATU', Peter Rhodes, G3XJP, RadCom, September 2001 to January 2002

[41] G3XJP PicATUne weblink: *http://sp-hm.pl/attachment.php?aid=413*

[42] Radio Communication Handbook 5th edition. Published by RSGB 1988. Section 17 Interference, Page 17.8.

[43] 'Modifying the MFJ989 Versatuner', David Knight, G3YNH, weblink: t*http://www.g3ynh.info/atu/mfj989c.html*

[44] 'Save your tuner for two pence', Tony Preedy, G3LNP, Radcom, May 2000

[45] 'A General-Purpose Antenna Tuning Unit', M J Grierson, G3TSO, RadCom, January 1987

[46] 'Link Coupled Antenna Tuners: A Tutorial', L B Cebik, W4RNL, *http://www.cebik.com/*

[47] 'The doublet de-mystified', Brian Horsfall, G3GKG, RadCom, January 2004

[48] 'To Turrets - Just Tune', King, W1CJL, QST, Mar 1948

[49] 'The Z-match Antenna Coupler', King, W1CJL, QST, May 1955

[50] 'An improved Z-match ATU', Louis Varney CEng MIEE AIL, G5RV, Radcom Oct 1998. HF Antenna Collection, edited by Erwin David G4LQI. Published by RSGB 1991. Section 6, Pages 116 to 119

***Mike Parkin BSc(Hons), G0JMI,*** *obtained his first amateur radio licence in 1977 gaining his current licence in 1988. His working background has included telecommunications, radio engineering, mobile-communications, electronics, lecturing and consultancy. Mike has constructed much of his radio equipment covering the bands from 472kHz through to 47GHz. Mike's current business interests are the design and production of antennas covering the range from HF to UHF. Mike is the author of the Antennas Column in RadCom and has had various articles published in other journals. He is a Member of the Institute of Engineering & Technology, a Member of the City & Guilds Institute and a Chartered Engineer.*

# Practical VHF/UHF Antennas

16

*Peter Swallow, G8EZE*

VHF and UHF antennas differ from their HF counterparts in that the diameter of their elements are relatively thick in relationship to their length and the operating wavelength, and transmission line feeding and matching arrangements are used in place of lumped elements and ATUs.

## THE (VHF) DIPOLE ANTENNA

At VHF and UHF, most antenna systems are derived from the dipole or its complement, the slot antenna. Many antennas are based on half-wave dipoles fabricated from wire or tubing. The feed point is usually placed at the centre of the dipole, for although this is not absolutely necessary, it can help prevent asymmetry in the presence of other conducting structures.

The input impedance is a function of both the dipole length and diameter. A radiator measuring exactly one half wavelength from end to end will be resonant (ie will present a purely resistive impedance) at a frequency somewhat lower than would be expected from its dimensions. Curves of 'end correction' such as **Fig 16.1** show by how much a dipole should be shortened from the expected half wavelength to be resonant at the desired frequency.

The change of reactance close to half-wavelength resonance as a function of the dipole diameter is shown in **Fig 16.2**.

In their simplest form, dipole antennas for 2m and 70cm can be constructed from 2mm diameter enamelled copper wire and fed directly by a coaxial cable as shown in **Fig 16.3**. The total element length (tip to tip) should be 992mm for 145MHz operation and 326mm to cover the band 432 to 438MHz. The impedance will be around 70 ohms for most installations, so that a 50-ohm coaxial cable would present a VSWR of around 1.4:1 at the transceiver end.

A more robust construction can be achieved using tubing for the elements and moulded dipole centre boxes, available from a number of amateur radio antenna manufacturers and at radio rallies. The dipole length should be shortened in accordance with Fig 16.1 to compensate for the larger element diameters.

Note that this simple feed may result in currents on the outside of the cable, and consequently a potential to cause interference to other electronic equipment when the antenna is used for transmitting. This can be reduced or eliminated by using a balun at the feed point.

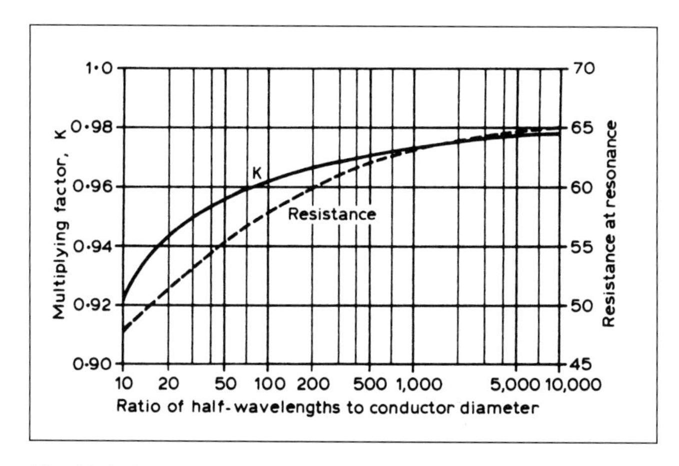


**Fig 16.1: Length correction factor for half-wave dipole as a function of diameter**

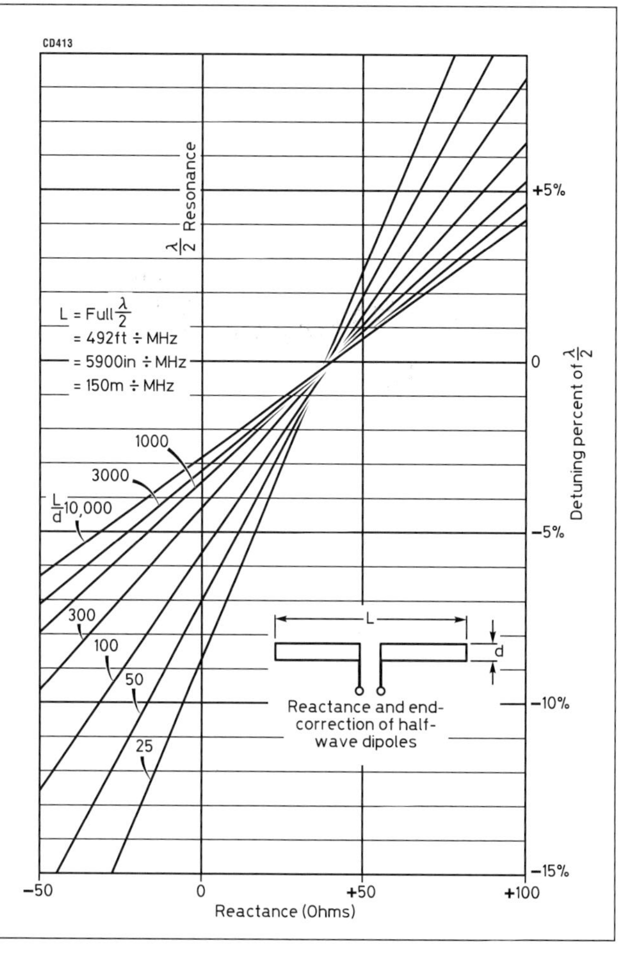


**Fig 16.2: Tuning and reactance chart for half-wave dipoles as a function of diameter**

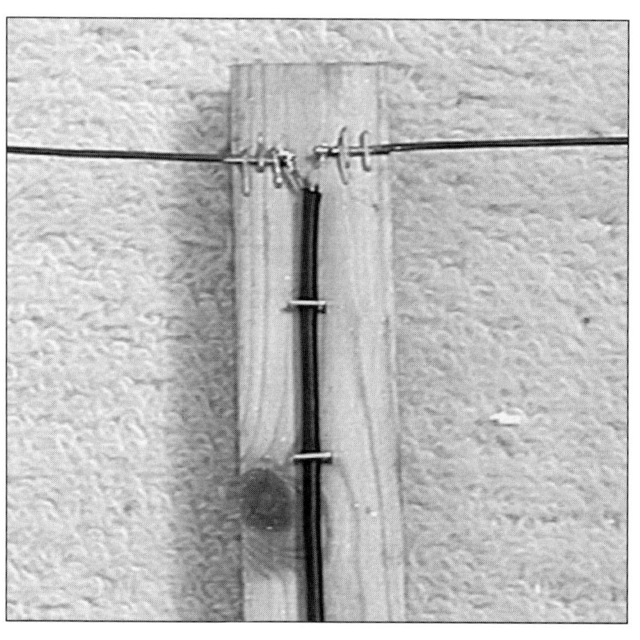

**Fig 16.3: Simple dipole construction for 2m and 70cm**

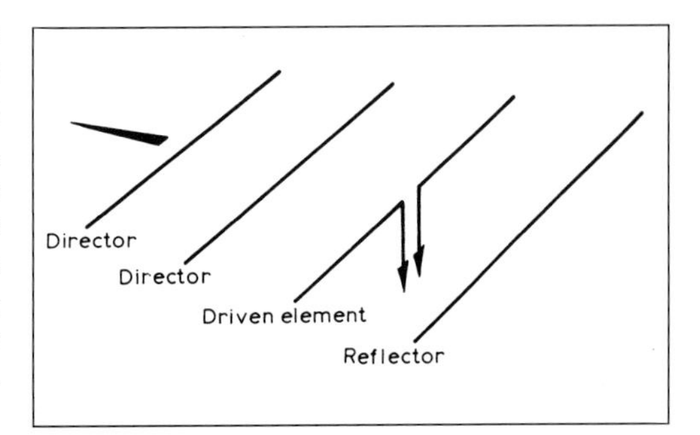


**Fig 16.4: Simple Yagi antenna structure, using two directors and one reflector in conjunction with a driven element**

# THE YAGI AND ITS DERIVATIVES

## The Yagi Antenna

The Yagi antenna was originally investigated by Uda and subsequently brought to Western attention by Yagi in 1928 in a form similar to that shown in **Fig 16.4**. It consists of a driven element combined with an in-line parasitic array. There have since been many variations of the basic concept, including its combination with log-periodic and backward-wave techniques.

To cover all variations of the Yagi antenna is beyond the scope of this handbook. A great number of books and many articles have been published on the subject, and a wide range of theoretical and practical pages can be found on the Internet with a simple search.

Many independent investigations of multi-element Yagi antennas have shown that the gain of a Yagi is directly proportional to the array length. There is a certain amount of latitude in the position of the elements along the array. However, the optimum resonance of each element will vary with the spacing chosen. With Greenblum's dimensions [1], in **Table 16.1**, the gain will not vary more than 1dB from the nominal value. The most critical elements are the reflector and first director as they decide the spacing for all other directors and most noticeably affect the matching. Solutions may be refined for the materials and construction methods available using one of the many software tools now freely available from the Internet, and discussed elsewhere in this handbook. These tools can be used to assess the sensitivity of a given design to alternative diameter elements and dimensions.

The optimum director lengths are normally greater the closer the particular director is to the driven element. (The increase of capacitance between elements is balanced by an increase of inductance, ie length through mutual coupling.) However, the length does not decrease uniformly with increasing distance from the driven element. **Fig 16.5** shows experimentally derived element lengths for various material diameters. Elements are mounted through a cylindrical metal boom that is two or three diameters larger than the elements.

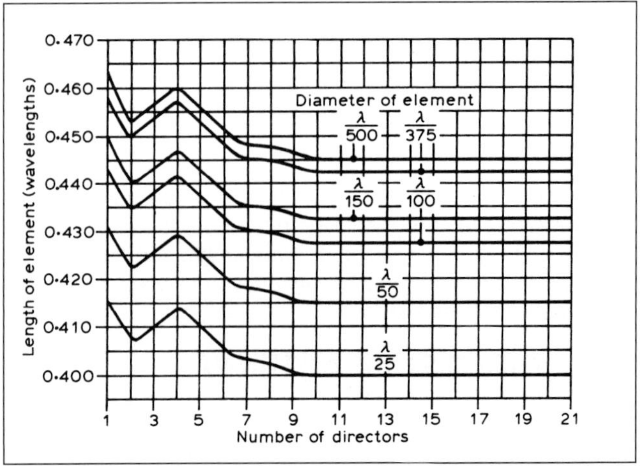


**Fig 16.5: Length of director versus position in the array for various element diameters (*ARRL Antenna Book*)**

Some variation in element lengths will occur using different materials or sizes for the support booms. This will be increasingly critical as frequency increases. The water absorbency of insulating materials will also affect the element lengths, particularly when in use, although plastics other than nylon are usually satisfactory. **Fig 16.6** shows the expected gain for various numbers of elements if the array length complies with **Fig 16.7**.

The results obtained by G8CKN using the 'centre spacing' of Greenblum's optimum dimensions shown in Table 16.1 produced identical gains to those shown in Fig 16.6. Almost identical radiation patterns (**Fig 16.8**) were obtained for both the E and H

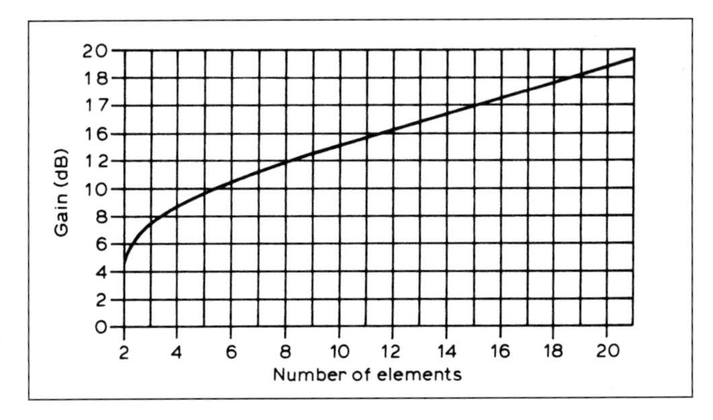


**Fig 16.6: Gain in dBd versus the number of elements of the Yagi array (*ARRL Antenna Book*)**

| Number of elements | R-DE | DE-D1 | D1-D2 | D2-D3 | D3-D4 | D4-D5 | D5-D6 |
|---|---|---|---|---|---|---|---|
| 2 | 0.15 -0.20 | | | | | | |
| 2 | | 0.07 -0.11 | | | | | |
| 3 | 0.16 -0.23 | 0.16 -0.19 | | | | | |
| 4 | 0.18 -0.22 | 0.13 -0.17 | 0.14 -0.18 | | | | |
| 5 | 0.18 -0.22 | 0.14 -0.17 | 0.14 -0.20 | 0.17 -0.23 | | | |
| 6 | 0.16 -0.20 | 0.14 -0.17 | 0.16 -0.25 | 0.22 -0.30 | 0.25 -0.32 | | |
| 8 | 0.16 -0.20 | 0.14 -0.16 | 0.18 -0.25 | 0.25 -0.35 | 0.27 -0.32 | 0.27 -0.33 | 0.30 -0.40 |
| 8 to N | 0.16 -0.20 | 0.14 -0.16 | 0.18 -0.25 | 0.25 -0.35 | 0.27 -0.32 | 0.27 -0.32 | 0.35 -0.42 |

*DE = driven element, R = reflector and D = director. N = any number. Director spacing beyond D6 should be 0.35-0.42*

**Table 16.1: Greenblum's optimisation for multielement Yagis**

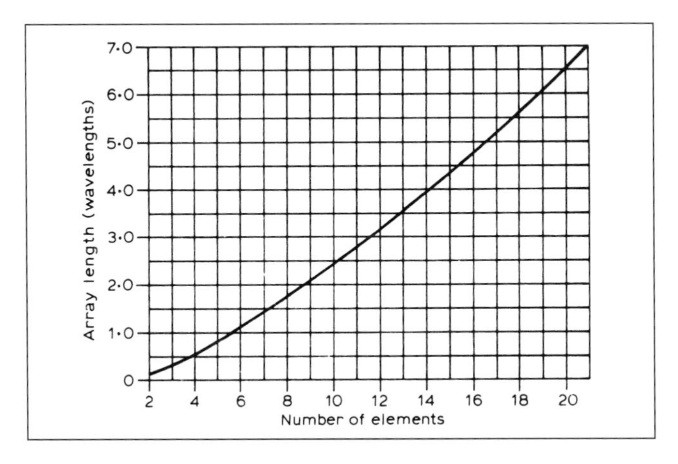


Fig 16.7: Optimum length of a Yagi antenna as a function of the number of elements (*ARRL Antenna Book*)

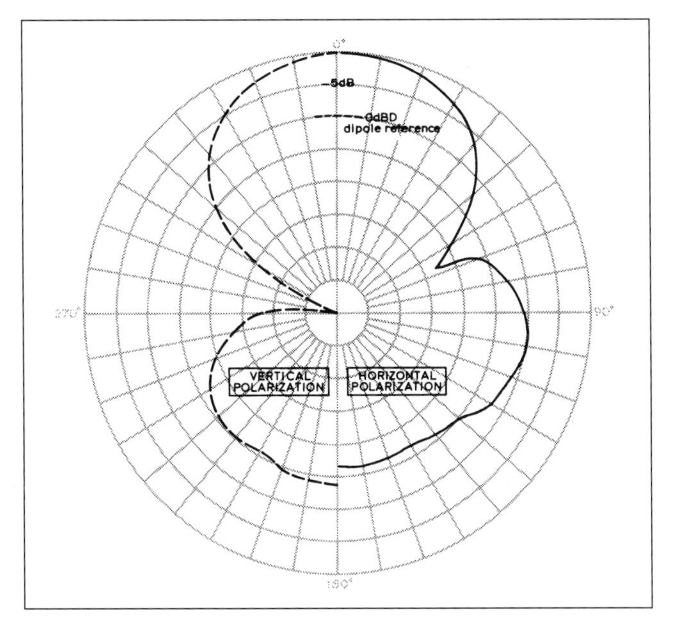


Fig 16.8: Radiation pattern for a four-element Yagi using Greenblum's dimensions

planes (V or H polarisation). Sidelobes were at a minimum and a fair front-to-back ratio was obtained.

Considerable work has been carried out by Chen and Cheng on the optimising of Yagis by varying both the spacing and resonant lengths of the elements [2].

**Table 16.2** and **Table 16.3** show some of their results obtained in 1974, by optimising both spacing and resonant lengths of elements in a six element array.

Table 16.3 shows comparative gain of a six element array with conventional shortening of the elements or varying the element lengths alone. The gain figure produced using conventional shortening formulas was 8.77dB relative to a λ/2 dipole (dBd). Optimising the element lengths produced a forward gain of 10dBd. Returning to the original element lengths and optimising the element spacing produced a forward gain of 10.68dBd. This is identical to the gain shown for a six-element Yagi in Fig 16.6. Using a combination of spacing and element length adjustment obtained a further 0.57dBd, giving 11.25dBd as the final forward gain as shown in Table 16.3.

A publication of the US Department of Commerce and National Bureau of Standards [3], provides very detailed experimental information on Yagi dimensions. Results were obtained from measurements to optimise designs at 400MHz using a model antenna range.

The information, presented largely in graphical form, shows very clearly the effect of different antenna parameters on realisable gain. For example, it shows the extra gain that can be achieved by optimising the lengths of the different directors, rather than making them all of uniform length. It also shows just what extra gain can be achieved by stacking two elements, or from a 'two-over-two' array.

The paper presents:

(a) The effect of reflector spacing on the gain of a dipole.
(b) Effect of different equal-length directors, their spacing and number on realisable gain.
(c) Effect of different diameters and lengths of directors on realisable gain.
(d) Effect of the size of a supporting boom on the optimum length of parasitic elements.
(e) Effect of spacing and stacking of antennas on gain.
(f) The difference in measured radiation patterns for various Yagi configurations.

| | $h_1/\lambda$ | $h_2/\lambda$ | $h_3/\lambda$ | $h_4/\lambda$ | $h_5/\lambda$ | $h_6/\lambda$ | Gain (dBd) |
|---|---|---|---|---|---|---|---|
| **Initial array** | 0.255 | 0.245 | 0.215 | 0.215 | 0.215 | 0.215 | 8.78 |
| **Length-perturbed array** | 0.236 | 0.228 | 0.219 | 0.222 | 0.216 | 0.202 | 10.00 |

$b_i1 = 0.250\lambda$, $b_i2 = 0.310\lambda$ ($i$ = 3, 4, 5, 6), $a = 0.003369\lambda$

Table 16.2: Optimisation of six-element Yagi-Uda array (perturbation of element lengths)

| | h1/λ | h2/λ | h3/λ | h4/λ | h5/λ | h6/λ | b21/λ | b22/λ | b43/λ | b34/λ | b35/λ | Gain (dBd) |
|---|---|---|---|---|---|---|---|---|---|---|---|---|
| **Initial array** | 0.255 | 0.245 | 0.215 | 0.215 | 0.215 | 0.215 | 0.250 | 0.310 | 0.310 | 0.310 | 0.310 | 8.78 |
| **Array after spacing perturbation** | 0.255 | 0.245 | 0.215 | 0.215 | 0.215 | 0.215 | 0.250 | 0.289 | 0.406 | 0.323 | 0.422 | 10.68 |
| **Optimum array after spacing and length perturbations** | 0.238 | 0.226 | 0.218 | 0.215 | 0.217 | 0.215 | 0.250 | 0.289 | 0.406 | 0.323 | 0.422 | 11.26 |

Table 16.3: Optimisation for six-element Yagi-Uda array (perturbation of element spacings and element lengths)

| Length of Yagi (λ) | 0.4 | 0.8 | 1.20 | 2.2 | 3.2 | 4.2 |
|---|---|---|---|---|---|---|
| **Length of reflector (λ)** | 0.482 | 0.482 | 0.482 | 0.482 | 0.482 | 0.475 |
| **Length of directors (λ):** | | | | | | |
| 1st | 0.424 | 0.428 | 0.428 | 0.432 | 0.428 | 0.424 |
| 2nd | - | 0.424 | 0.420 | 0.415 | 0.420 | 0.424 |
| 3rd | - | 0.428 | 0.420 | 0.407 | 0.407 | 0.420 |
| 4th | - | - | 0.428 | 0.398 | 0.398 | 0.407 |
| 5th | - | - | - | 0.390 | 0.394 | 0.403 |
| 6th | - | - | - | 0.390 | 0.390 | 0.398 |
| 7th | - | - | - | 0.390 | 0.386 | 0.394 |
| 8th | - | - | - | 0.390 | 0.386 | 0.390 |
| 9th | - | - | - | 0.398 | 0.386 | 0.390 |
| 10th | - | - | - | 0.407 | 0.386 | 0.390 |
| 11th | - | - | - | - | 0.386 | 0.390 |
| 12th | - | - | - | - | 0.386 | 0.390 |
| 13th | - | - | - | - | 0.386 | 0.390 |
| 14th | - | - | - | - | 0.386 | - |
| 15th | - | - | - | - | 0.386 | - |
| **Director spacing (λ)** | 0.20 | 0.20 | 0.25 | 0.20 | 0.20 | 0.308 |
| **Gain (dBd)** | 7.1 | 9.2 | 10.2 | 12.25 | 13.4 | 14.2 |

*Element diameter 0.0085λ. Reflector spaced 0.2λ behind driven element. Measurements are for 400MHz by P P Viezbicke.*

**Table 16.4: Optimised lengths of parasitic elements for Yagi antennas of six different boom lengths**

The highest gain reported for a single boom structure is 14.2dBd for a 15-element array (4.2λ long and reflector spaced at 0.2λ, with 13 graduated directors). See **Table 16.4**.

It has been found that array length is of greater importance than the number of elements, within the limit of a maximum element spacing of just over 0.4λ.

Reflector spacing and, to a lesser degree, the first director position affects the matching of the Yagi. Optimum tuning of the elements, and therefore gain and pattern shape, varies with different element spacing.

Near-optimum patterns and gain can be obtained using Greenblum's dimensions for up to six elements. Good results for a Yagi in excess of six elements can still be obtained where ground reflections need to be minimised.

Chen and Cheng employed what is commonly called the long Yagi technique. Yagis with more than six elements start to show an improvement in gain with fewer elements for a given boom length when this technique is used.

As greater computing power has become available, it has been possible to investigate the optimisation of Yagi antenna gain more extensively, taking into account the effects of mounting the elements on both dielectric and metallic booms, and the effects of tapering the elements at lower frequencies.

Dr J Lawson, W2PV, carried out an extensive series of calculations and parametric analyses, collated in reference [4], which although specifically addressing HF Yagi design, explain many of the disappointing results achieved by constructors at VHF and above. In particular, the extreme sensitivity of some designs to minor variations of element length or position are revealed in a series of graphs which enable the interested constructor to select designs that will be readily realisable.

| | Length 70.3MHz | 145MHz | 433MHz |
|---|---|---|---|
| **Driven elements** | | | |
| Dipole (for use with gamma match) | 79 (2000) | 38 (960) | 12 3/4 (320) |
| Diameter range for length given | 1/2 - 3/4<br>1/8 - 1/4<br>(12.7 - 19.0) | (6.35 - 9.5) | 1/4 - 3/8<br>(3.17 - 6.35) |
| *Folded dipole 70-ohm feed* | | | |
| *l* length centre-centre | 77 1/2 (1970) | 38 1/2 (980) | 12 1/2 (318) |
| *d* spacing centre-centre | 2 1/2 (64) | 7/8 (22) | 1/2 (13) |
| Diameter of element | 1/2 (12.7) | 1/4 (6.35) | 1/8 (3.17) |
| *a* centre/centre | 32 (810) | 15 (390) | 5 1/8 (132) |
| *b* centre/centre | 96 (2440) | 46 (1180) | 152 (395) |
| Delta feed sections (length for 70Ω feed) | 22½ (570) | 12 (300) | 42 (110) |
| Diameter of slot and delta feed material | 1/4 (6.35) | 3/8 (9.5) | 3/8 (9.5) |
| **Parasitic elements** | | | |
| Element | | | |
| Reflector | 85 1/2 (2170) | 40 (1010) | 13 1/4 (337) |
| Director D1 | 74 (1880) | 35 1/2 (902) | 11 1/4 (286) |
| Director D2 | 73 (1854) | 35 1/4 (895) | 11 1/8 (282) |
| Director D3 | 72 (1830) | 35 (890) | 11 (279) |
| Succeeding directors | 1in less (25) | 1/2in less (13) | 1/8in less |
| Final director | 2in less (50) | 1in less (25) | 3/4in less |
| One wavelength (for reference) | 168 3/4 (4286) | 81 1/2 (2069) | 27 1/4 (693) |
| Diameter range for length given | 1/2 - 3/4<br>(12.7 - 19.0) | 1/4 - 3/8<br>(6.35 - 9.5) | 1/8-¾<br>(3.17 - 6.35) |
| **Spacing between elements** | | | |
| Reflector to radiator | 22 1/2 (572) | 17 1/2 (445) | 5 1/2 (140) |
| Radiator to D1 | 29 (737) | 17 1/2 (445) | 5 1/2 (140) |
| D1 to D2 | 29 (737) | 17 1/2 (445) | 7 (178) |
| D2 to D3, etc | 29 (737) | 17 1/2 (445) | 7 (178) |

*Dimensions are in inches with millimetre equivalents in brackets.*

**Table 16.5: Typical dimensions of Yagi antenna components. Dimensions are in inches with metric equivalents in brackets**

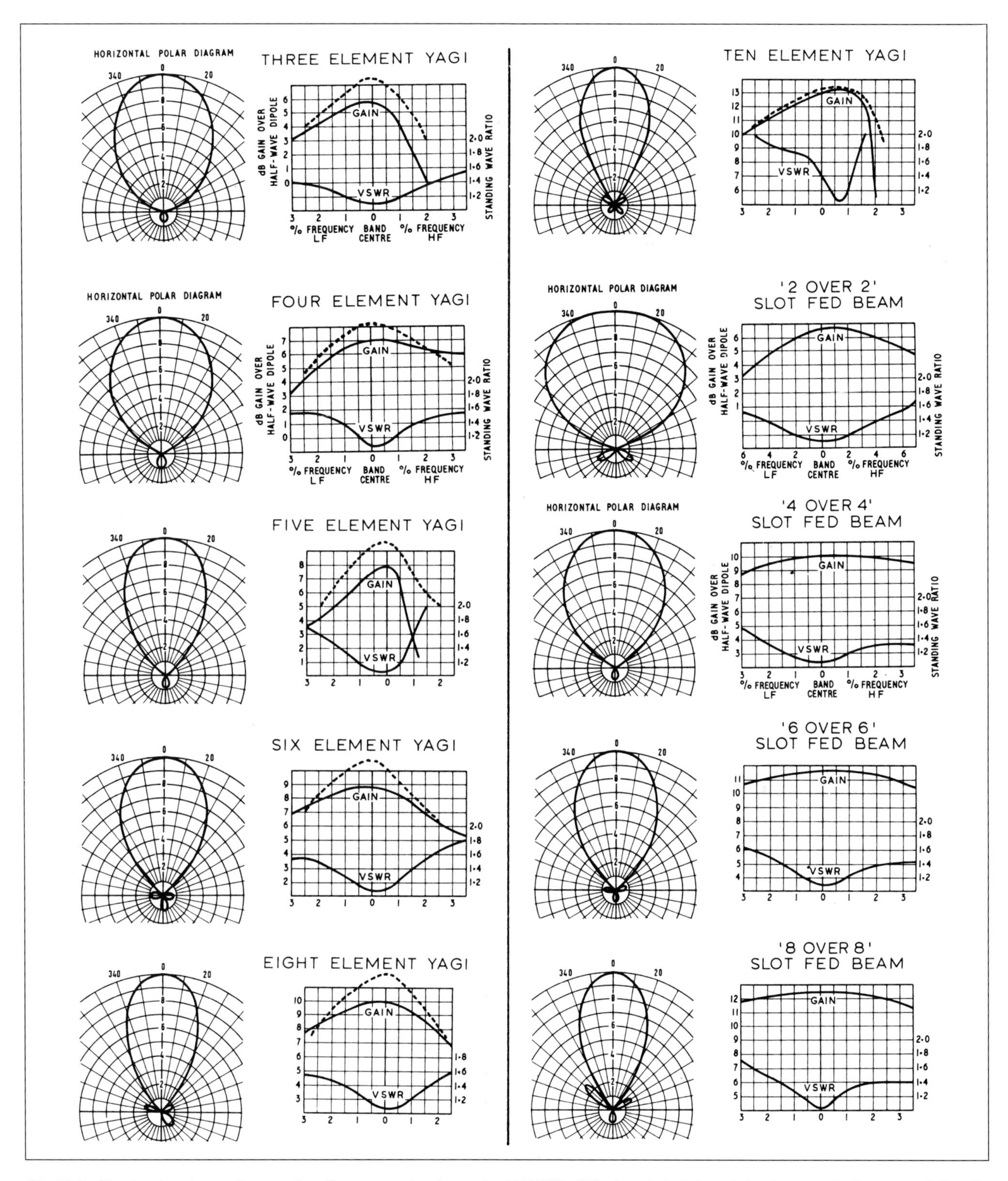


**Fig 16.9: Charts showing voltage polar diagram and gain against VSWR of Yagi and skeleton-slot antennas. In the case of the six Yagi antennas the solid line is for conventional dimensions and the dotted lines for optimised results discussed in the text**

The keen constructor with a personal computer may now also take advantage of modelling tools specifically designed for optimisation of Yagi antennas and arrays, eg [5], although some care is needed in their use if meaningful results are to be assured. The Internet is a good source for Yagi antenna design and optimisation programmes, many of which can be obtained free of charge, or for a nominal sum.

From the foregoing, it can be seen that several techniques can be used to optimise the gain of Yagi antennas. In some circumstances, minimisation of sidelobes is more important than maximum gain, and a different set of element spacings and lengths would be required to achieve this. Optimisation with so many independent variables is difficult, even with powerful computing methods, as there may be many solutions that yield comparable results.

Techniques of 'genetic optimisation' have been developed and widely adopted, which can result in surprising, but viable designs [6], [7]. The technique requires the use of proven computer-

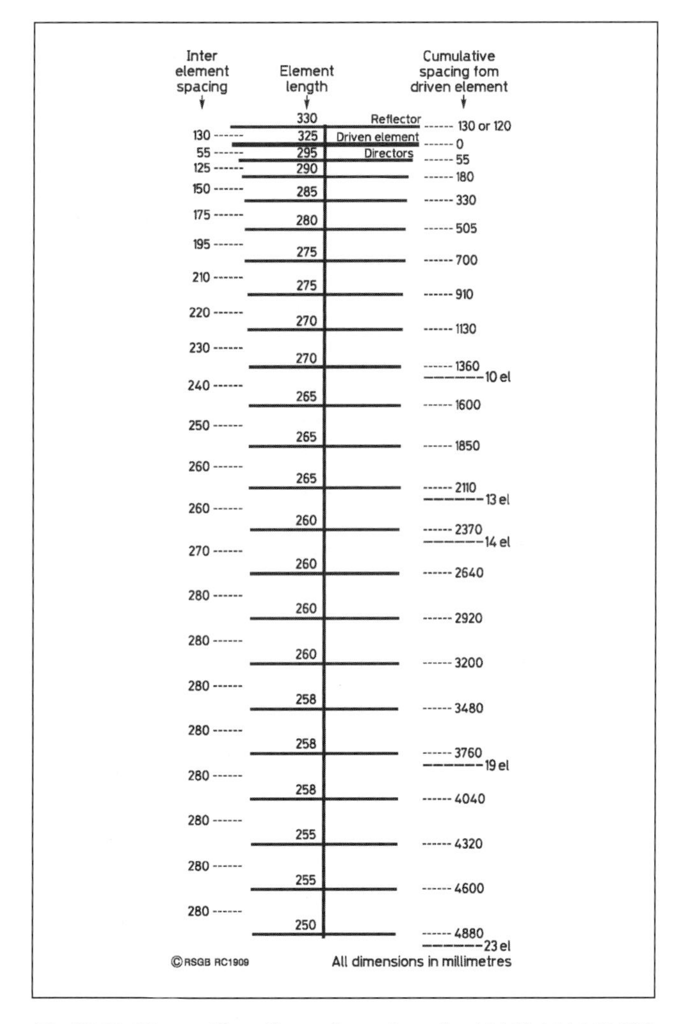


**Fig 16.10: Element lengths and spacings for 10 / 13 / 14 / 19 / 23 element 435MHz Yagi**

| Number of elements | 10 | 13 | 14 | 19 | 23 |
|---|---|---|---|---|---|
| Gain (dBd) | 11.7 | 13 | 13.3 | 15 | 16 |
| Horizontal beamwidth | 37° | 30.5° | 30° | 26.5° | 24° |
| Vertical beamwidth | 41° | 33° | 32° | 28° | 24.5° |

**Table 16.6: Performance of 10 / 13 / 14 / 19 / 23 element 435MHz Yagis**

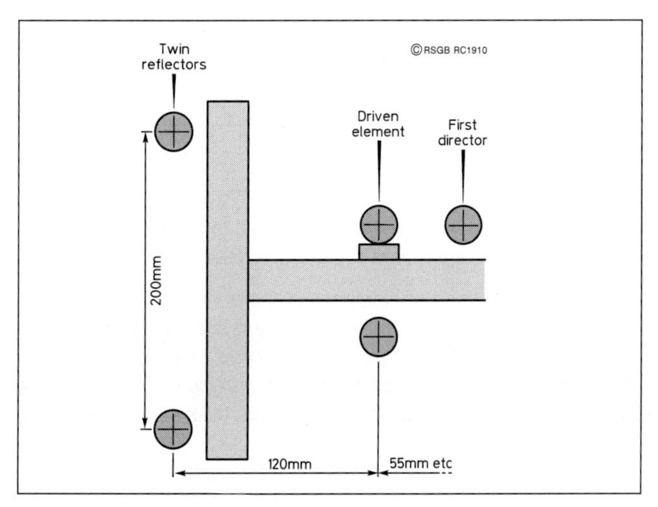


**Fig 16.11: Twin reflector details**

based analysis tools such as NEC, MININEC or their derivatives. The required parameters (gain, sidelobe levels, input impedance) are described and weighted according to their importance to the designer, together with the permitted variables.

A figure of merit is defined, which incorporates the weighting of the desired parameters. An initial structure is input, which is then analysed, its performance recorded, and an incremental change made to one of the variables. The process is repeated whilst the figure of merit continues to improve. However, unlike conventional optimisation methods, where local optimisation may obscure a better result that may also be available, a random process selects the variable(s) to be changed until a reasonably large seed population has been generated. Selection, crossover and mutation processes are then used to filter out poor designs and retain better ones, with each successive generation possibly containing better designs than the preceding one, if the selection algorithms have been well constructed. This technique is readily available to amateurs with home computers [8], [9].

Dimensions for Yagi antennas for 70, 145 and 433MHz are shown in **Table 16.5**. The table also includes dimensions for feeding two stacked Yagi antennas with a *skeleton slot feed*, described later in this chapter.

Typical radiation patterns, gains and VSWR characteristics for a range of different Yagi antennas are shown in **Fig 16.9**. The figure also contains information on skeleton slot Yagis, discussed later in this chapter.

## Long Yagi Antennas

The NBS optimisation described above has been extended by American amateurs [10]. Tapering of the spacing was studied by W2NLY and W6QKI who found [11] that, if the spacing was increased up to a point and thereafter remained constant at 0.3-0.4λ, another optimisation occurred. Both these are *single optimisation* designs.

Günter Hoch, DL6WU, looked at both techniques and decided that they could be applied together. The director spacing was increased gradually until it reached 0.4λ and the length was tapered by a constant *fraction* from one element to the next. The result is a highly successful *doubly optimised* antenna [12], [13].

Great care is required in constructing these antennas if the predicted gain is to be realised. This means following the dimensions and fixing methods *exactly* as laid out in the designer's instructions. Details for building a number of long Yagi antennas for VHF and UHF can be found through links at GM3SEK's website [14].

### F5JIO long Yagi for 435MHz

This antenna can be built with 10, 13, 14, 19 or 23 elements according to the space available (**Fig 16.10**). Its performance is shown in **Table 16.6**.

An extra 0.2dB gain and some reduction of backlobes can be obtained by fitting twin reflectors (**Fig 16.11**), but note that the spacing between the driven element and the reflector is reduced from 130mm to 120mm.

The 23 element version requires a boom length in excess of 5010mm, and must be solidly constructed and supported. The boom is made from 20 x 20mm square aluminium tubing, and all elements from 8mm diameter (round) tubing. All elements except the driven element must be *insulated* from the boom and mounted so that their centres are 8mm above its upper surface. Dimensions for the driven element and cable balun construction to provide a feed impedance of 50 ohms are shown in **Fig 16.12**.

Balun cable lengths and calculations are shown in **Table 16.7**. The cable should be a 75-ohm miniature PTFE insulated type, such as URM111 or equivalent. A small weatherproof

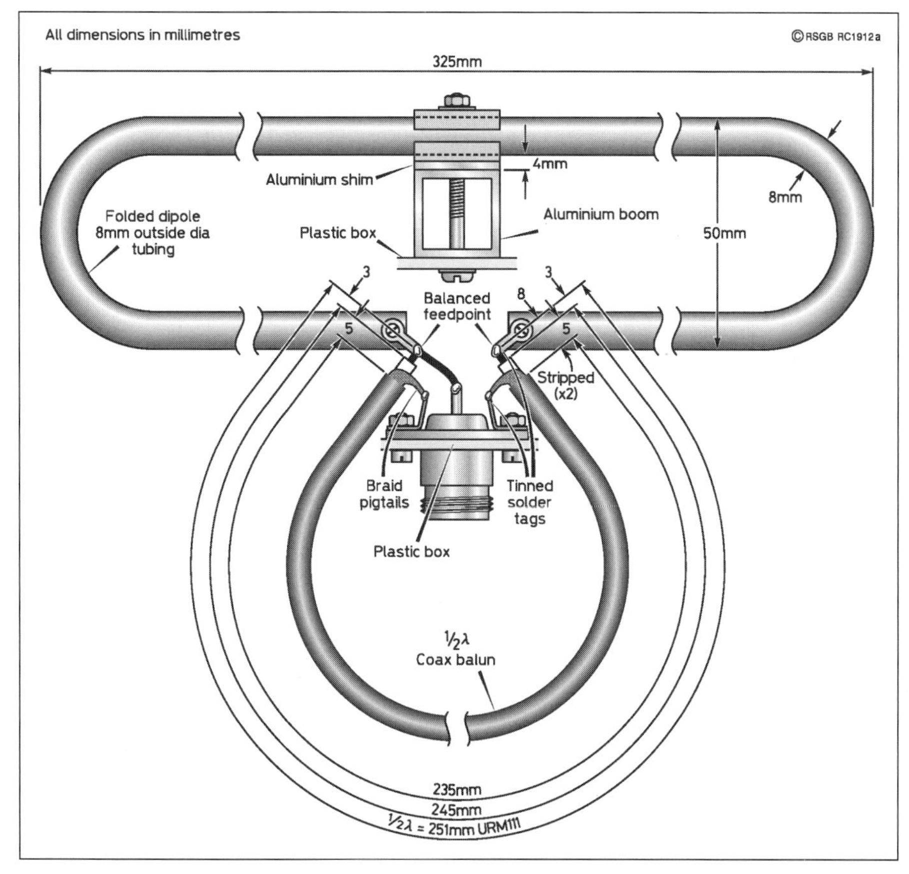


Fig 16.12: 435MHz long Yagi driven element and balun

λ/2@435MHz = 300,000/435 x 2 = 345mm (in air)
In URM111: 16mm of stripped end (@v=0.9) =18mm (electrical)

Cutting length = 345mm-18mm (@v=0.72 PTFE insulation) = 235mm (unstripped)

*Note: Use v=0.66 for Polyethylene insulation*

**Table 16.7: Balun cable length calculation**

box should be fitted over the ends of the element, inside which the balun cable may also be coiled. The driven element may be made from 9.5 x 1.6mm flat aluminium bar which is easier to bend and drill.

## Quad Antenna

The Quad antenna can be thought of as a Yagi antenna comprising pairs of vertically stacked, horizontal dipoles with their ends bent towards each other and joined, **Fig 16.13**. The antenna produces horizontally polarised signals, and in spite of its relatively small physical size a forward gain of 5.5 to 6dB can be obtained with good front-to-back ratio. Additional quad or single element directors can be added to the basic two element array in the same manner as the Yagi.

Typical dimensions for lightweight wire 51, 71 and 145MHz Quad antennas are given in **Table 16.8**, and a photograph of the 145MHz version is shown in **Fig 16.14**. This variant has equal size loops and uses a stub to tune the reflector. The boom is made from 15mm copper tubing with a T-piece in the centre for fixing to the mast or rotator. The element supports are made from 10 or 12mm square wooden dowelling fixed to square pieces of plywood using nuts and bolts. The plywood centres are fixed to the boom using L-brackets and hose clamps. A 50-ohm coaxial cable can be connected directly

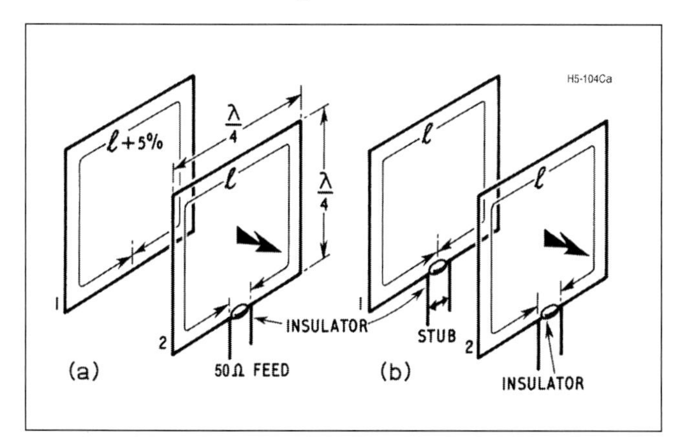


Fig 16.13: Quad antenna structure and electrical dimensions

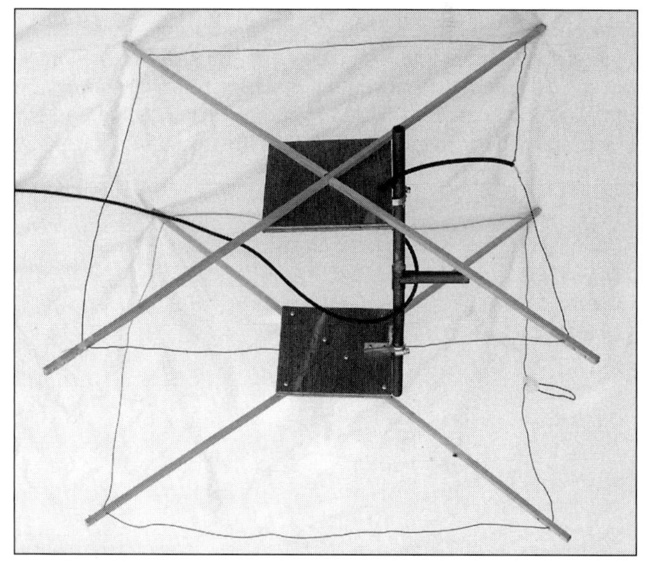

Fig 16.14: Wire quad antenna for 145MHz

| Band | Element spacing, mm | Reflector sides, mm | Driven element sides, mm |
|---|---|---|---|
| 51MHz | 840 | 1560 | 1500 |
| 71.5MHz | 600 | 1210 | 1080 |
| 145MHz | 294 | 548 | 524 |

**Table 16.8: Design dimensions for 51, 70 and 144MHz quad antennas**

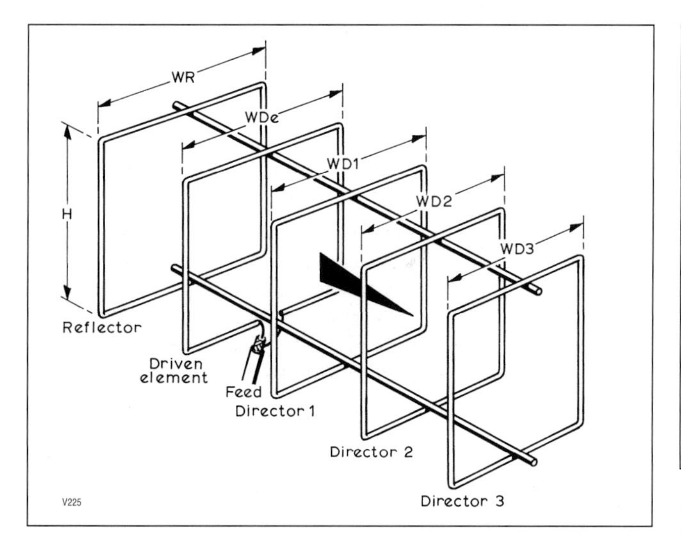


Fig 16.15: General arrangement for a multi-element quad antenna

to the driven element. A 1:1 balun will minimise currents on the outer of the cable, preventing distortion of the radiation pattern and potential EMC problems.

Quads may be stacked or built into a four square assembly in the same way as the basic Yagi (see below).

## Multi-element quad (Quagi)

The multiple element quad antenna or "Quagi" can offer a better performance with reduced sidelobes compared with the average simple Yagi, whilst retaining a simple robust form of construction (**Fig 16.15**). Dimensions for a four element, 145MHz antenna are given in **Table 16.9**. Generally the maximum number of elements used is five. Where more gain is needed, a pair may be stacked vertically or horizontally, although for maximum mechanical strength the vertical arrangement is to be preferred.

The whole structure may be made up of aluminium tube (or solid rod for the elements). The only insulator necessary is at the feed point of the driven element. In construction, it is best to make each element from one piece of material. A 3/8in aluminium rod will bend to form corners much more readily than tube that would also need a 'filler'. The corner radius should be kept small, and allowance must be made for the resultant 'shortening'

| | | | | |
|---|---|---|---|---|
| Height H | 21 (533) | 21 | 21 | 21 |
| Width reflector WR | 24½ (622) | 24½ | 24½ | 24½ |
| Driven WDe | 20½ (520) | 20½ | 20½ | 20½ |
| Director 1 WD1 | - | 18 (457) | 18 | 18 |
| Director 2 WD2 | - | - | 16 (406) | 16 |
| Director 3 WD3 | - | - | - | 14 (356) |
| Spacing | | | | |
| Reflector to Driven | 7 (178) | 19 (483) | 20 (508) | 20 |
| Driven to Director 1 | - | 12 (305) | 14½ (368) | 14½ |
| Director 1 to Director 2 | - | - | 14½ | 14½ |
| Director 2 to Director 3 | - | - | - | 14½ |
| Approx gain (dBd) | 5 | 7 | 10.5 | 12.5 |

*Element diameters all 3/8in (9.35mm). Feed impedance in all cases is 75 . Dimensions are in inches with millimetre equivalents in brackets.*

Table 16.9: Dimensions for a multi-element quad antenna for 144MHz

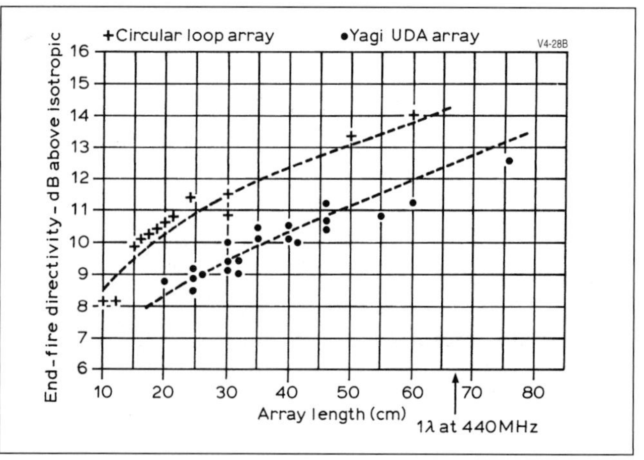


Fig 16.16: Comparative directivity of the Yagi and Quad as a function of overall array length. Although measured with circular loops, performance with square loops is comparable (*ARRL Antenna Book*). Note the gains are in dBi, not dBd

of element length, ie side of the quad element. For mechanical simplicity (and appearance) it is a good idea to arrange for all the element heights to be the same, and vary the width.

Fixing the elements to the boom and the boom to the mast is conveniently done with standard TV antenna fittings. Although suitable blocks or clamping arrangements can be made by the constructor, they often tend to be unnecessarily heavy. Purchased TV fittings can be more cost-effective than obtaining raw materials and there is also much less effort involved in construction. There are also several antenna manufacturing companies catering for the radio amateur who sell tubing, mast clamps and small components for securing elements to booms. They can often be found at rallies and amateur radio events, or advertise in the pages of *RadCom*.

If preferred, the reflector may be made the same size as the driven element, and tuned with a suitable stub. If vertical polarisation is required, instead of horizontal, then the feeder can be attached to the centre of one of the vertical sides of the driven element. (The same 'side' must always be used for correct phase relationship within stacked arrays.)

The relative performance of multi-element quad and Yagi antennas is shown in **Fig 16.16**, demonstrating that the shorter quad structures can provide gains comparable with longer Yagi antennas. This may be of benefit if turning space is limited (eg inside a loft). However, there is no such thing as a free lunch, and in general, the weight and wind loading of the multi-element quad antenna will be slightly greater than its Yagi counterpart.

## The loop Yagi

At frequencies above 433MHz, the construction of multiple quad antennas can be considerably simplified by bending the elements into circular loops. High gains can be achieved by using large numbers of elements, and the relatively simple construction allows gains up to around 20dBi to be realised with manageable boom lengths [15].

A practical horizontally polarised four element loop Yagi antenna for 435MHz is shown in **Fig 16.17**. 2mm diameter enamelled copper wire elements are fixed to a tubular metal boom using hose clamps. A three terminal, plastic mains power connector block is used to connect the coaxial cable and provide the method for fastening the driven element to the boom. the enamel insulation is removed from the ends of the driven element to a distance of 20mm at one end and 50mm at the other. The 50mm end is folded into a loop and passed back into the

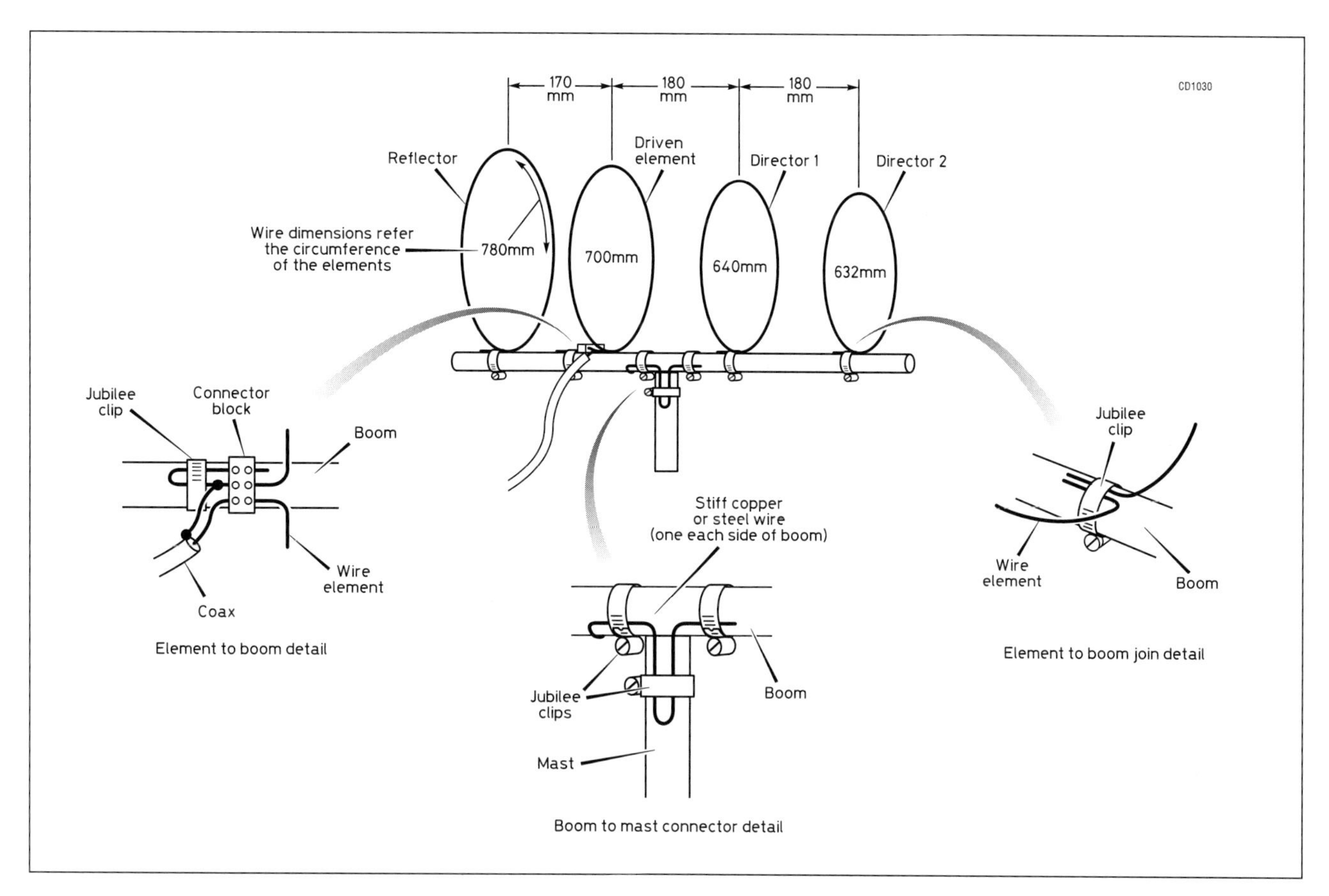


**Fig 16.17: A four element quad loop Yagi antenna for the 435MHz band**

connector block. The parasitic elements should be made 40mm longer than the dimensions shown, and the enamel removed from the last 20mm at each end. These ends should be bent at right angles and the remaining wire formed into a loop. The bent ends should be soldered together to simplify assembly. The boom and mast can be connected together using thick wire loops as shown, or the boom could be made from copper water pipe with a T-piece to connect to the support mast if preferred. A gain of around 9dBi should be achieved.

## GW3YDX Super Moxon Antenna

This antenna is essentially a pair of Moxon Rectangle antennas, arranged in tandem, with the second antenna parasitically coupled to the first (see 'Compact beams with bent elements' in the Practical HF Antennas chapter). It may also be thought of as a four element Yagi antenna with the ends of each pair of elements bent towards each other so that they can be mechanically joined by insulators (**Fig 16.18**). The result is a very compact and structurally sound directional antenna with a gain of around 6dBi and a front to back ratio of 26.5dB (**Fig 16.19**). A VSWR of better than 1.5:1 in 50 ohms was achieved between 50.0 and 50.3MHz [16]. The horizontal dimensions of the 2m version are just 770 x 640mm (**16.20**).

Dimensions for 6, 4 and 2m variants are shown in **Table 16.10**, all dimensions are in mm measured from the tubing centres. Dimensions K, L and M are measured along the boom from the position of the longest element J=0 in Fig 16.18. The main elements are made from 1/2" aluminium alloy tubing with 3/8" tubing or rod used for the corners. A 1" square alloy boom is sufficient for the 6m version. Insulators for joining the element ends and the driven element centres can be made from 3/8" fibreglass rod. The element end tubes can be slit and fitted with small stainless steel hose clamps to allow fine tuning. All four element

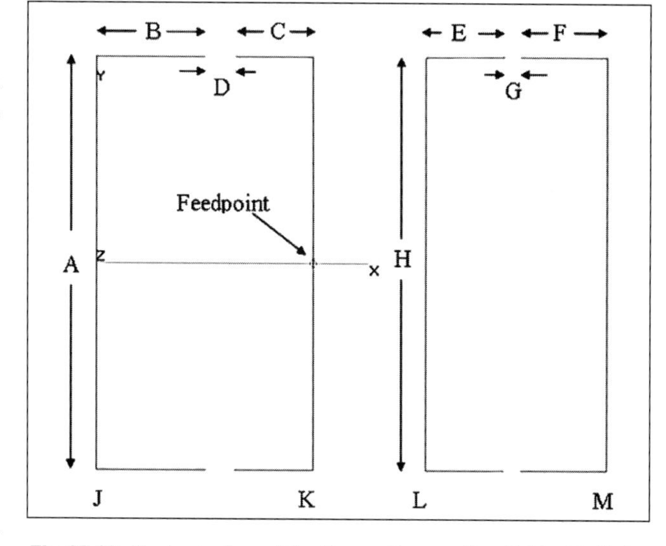


**Fig 16.18: Basic design of the Super Moxon. See Table 16.10 for 6m, 4m and 2m versions**

lengths should be altered by the same amount for each round of adjustments. A 1:1 50 ohm balun (see the Transmission Lines chapter) is required to feed the driven element.

The 4m version has been tested with input powers up to 1kW. The ends of the elements for the 2m version are separated by only 12mm, and has only been tested up to 50W input power. Appropriate precautions should be taken if high power operation is to be considered.

Note that the design is registered, and commercial manufacture requires the permission of the design owner [17]. However, radio amateurs may freely construct the antenna for their personal radio stations.

**Table 16.10: Tubing lengths for 6m, 4m and 2m versions of Fig 16.18. All dimensions are in mm, measured to tubing centres**

| | A | B | C | D | E | F | G | H | J | K | L | M |
|---|---|---|---|---|---|---|---|---|---|---|---|---|
| **6m** | 2160 | 395 | 280 | 105 | 290 | 310 | 60 | 2140 | 0 | 780 | 1201 | 1861 |
| **4m** | 1572 | 275 | 175 | 110 | 195 | 202 | 43 | 1572 | 0 | 560 | 860 | 1310 |
| **2m** | 730 | 135 | 86 | 55 | 82 | 90 | 12 | 730 | 0 | 276 | 434 | 615 |

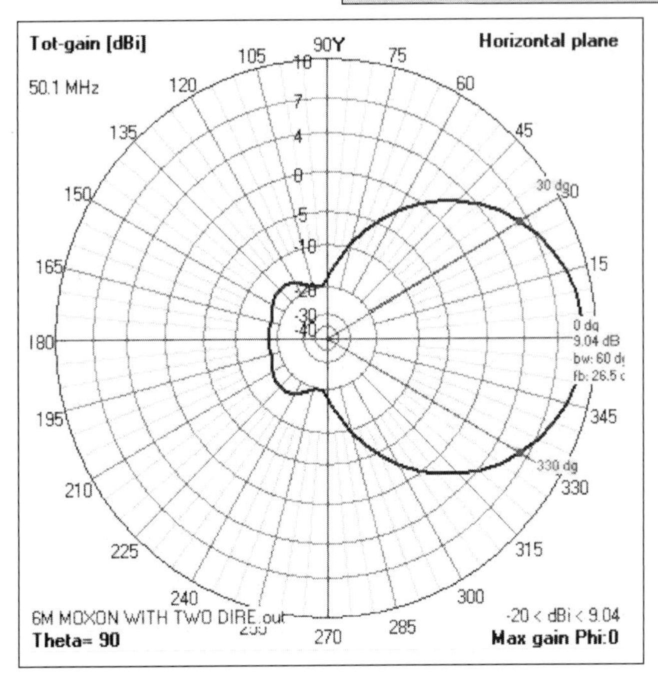


Fig 16.19: Azimuth radiation pattern of 6m Super Moxon antenna calculated using *4NEC2*

Fig 16.20: The 2m version of the Super Moxon is just 30" x 25"

# ANTENNA ARRAYS

## Array Principles

The gain achievable with any antenna structure is ultimately limited by the fundamentals of its operation. However, higher gains can be achieved by using several antenna elements in an *array*. The array can comprise antennas *stacked* vertically above each other, or arranged side by side in *bays*, or a combination of both. These are *broadside* arrays, where most of the radiated power is projected at right angles to the plane in which the array elements lie.

An array can also be formed where the main beam is projected along the array of elements; these are *endfire* arrays, of which the HB9CV and Yagi antennas are examples.

An array of elements has a narrower beamwidth, and hence a higher gain than the individual antennas. The maximum achievable gain could be N times greater than one element fed with the same power ($10\log_{10}$ N decibels) if there are N elements in the array. However, more complex feed arrangements can reduce the VSWR bandwidth and introduce losses, reducing the array gain. Arrays need care in construction and attention to detail, especially at UHF and above, but the results reward the effort expended.

Antenna array theory can be found in almost any book devoted to antennas. However, a good treatment with many radiation pattern examples can be found in Refs [18] and [19].

## Disadvantages of Multi-element Arrays

High gain cannot be achieved by simply stacking many elements close together. If we consider a dipole collecting power from an incident field for delivery to a load (receiver), it can be thought of as having a collecting area or effective aperture that is somewhat larger than the dipole itself. The higher the directivity of the antenna, the larger the effective aperture, as given by the relationship:

$$A_{eff} = \frac{\lambda^2}{4\pi} D$$

where D is the directivity of the antenna
$\lambda$ is the working wavelength

If the effective apertures of adjacent antennas overlap, the incoming RF energy is shared between them, and the maximum possible directivity (or gain) of the elements cannot be attained.

The generalised optimum stacking distance is a function of the half power beamwidth of the elements in the array, and is given by:

$$S_{opt} = \frac{\lambda}{\left[2 \sin\left(\frac{\phi}{2}\right)\right]}$$

where $\phi$ is the half power beamwidth and $S_{opt}$ is in wavelengths. Note that this is usually different for the E and H planes, so that the spacing of the elements is also usually different in each plane.

Also, when antennas are placed close together, *mutual coupling* between elements occurs. This leads to changes in the current distribution on the elements, changing both the radiation pattern and the feed point impedance of each element. The changes to the feed impedance often result in unequal powers being fed to the elements of the array, with consequential loss of gain.

Optimum stacking rules are based on the assumption of minimum mutual influence, which can be difficult to predict for complex antennas such as Yagis. However, antennas with low sidelobe levels are less susceptible than those with high sidelobes, as might be expected intuitively.

The coupling and effective aperture overlap problems cannot simply be solved by arbitrarily increasing the separation of the elements.

As the element spacing increases beyond one half wavelength, *grating sidelobes* appear, which can reduce the forward gain. The grating lobes are due solely to the array dimensions, and can be seen by plotting the array factor for the chosen configuration.

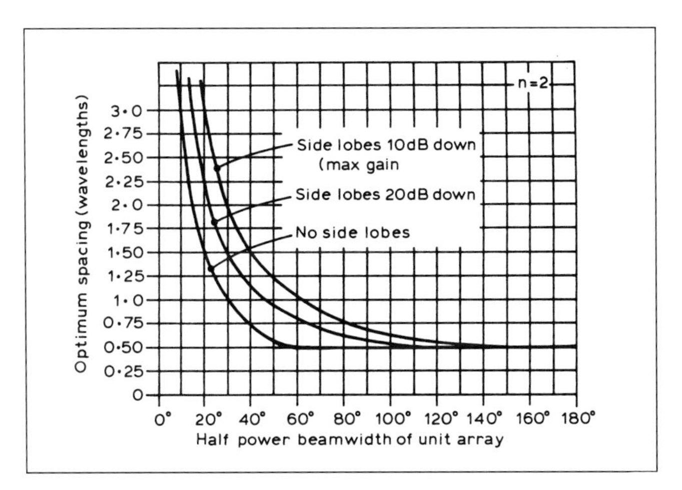


**Fig 16.21: Optimum stacking spacing for two-unit arrays. The spacing for no sidelobes, especially for small beamwidths, may result in no gain improvement over a single array element (*ARRL Antenna Handbook*)**

## Arrays of Identical Antennas

A parasitic array such as the Yagi can be stacked either vertically or horizontally to obtain additional directivity and gain. This is often called *collinear and broadside* stacking.

In stacking it is assumed that the antennas are identical in pattern and gain and will be matched to each other with the correct phase relationship, that is, 'fed in phase'. It is also assumed that for broadside stacking the corresponding elements are parallel and in planes perpendicular to the axis of the individual arrays.

With vertical stacking it is assumed the corresponding elements are collinear and all elements of the individual arrays are in the same plane.

The combination of the radiation patterns can add but can also cancel. The phase relationships, particularly from the side of the Yagi, are very complex.

Because of this complexity the spacing to obtain maximum forward gain does not coincide with the best sidelobe structure. Usually maximum gain is less important than reducing signals to the sides or behind the array.

If this is the case, 'optimum spacing' is one that gives as much forward gain as possible as long as the sidelobe structure does not exceed a specific amplitude compared with the main lobe. There will be different 'optimum' spacings according to the acceptable sidelobe levels.

**Fig 16.21** gives typical optimum spacing for two arrays under three conditions:

(a) optimum forward gain with sidelobe down 10dB,
(b) sidelobe 20dB down and
(c) virtually no sidelobe.

The no-sidelobe case can correspond to no additional forward gain over a single antenna. **Fig 16.22** shows the optimum stacking spacing for four-unit arrays.

The maximum forward gain of two stacked arrays is theoretically +3dB, and +6dB for four stacked arrays. More complex arrays could produce higher gain but losses in the matching and phasing links between the individual arrays can outweigh this improvement. When stacking two arrays, the extra achievable gain is reduced at close spacing due to high mutual coupling effects. With two seven-element arrays a maximum gain of about 2.5dB can be achieved with 1.6$\lambda$ spacing; with two 15-element arrays it is also possible to achieve the extra 2.5dB but the spacing needs to be 2$\lambda$.

The use of four arrays, in correctly phased two-over-two systems, can increase the realisable gain by about 5.2dB. Using seven-element Yagis produced a total gain of 14.2dB. With 15-element optimised Yagis a total gain of 19.6dB was obtained. (This was the highest gain measured during the experiments by Viezbicke [3].) The effects of stacking in combination with the physical and electrical phase relationship can be used to reduce directional interference.

Further information on antenna stacking can be found on GM3SEK's web site [14].

An improvement in front-to-back ratio can be accomplished in vertical stacking by placing the top Yagi a quarter-wavelength in front of the lower Yagi as shown in **Fig 16.23**. The top antenna is fed 90° later than the bottom antenna by placing additional cable in the upper antenna feed run. The velocity factor of the cable must be taken into account.

## A Coaxial Cable Harness for Feeding Four Antennas

Four identical antennas such as Yagis can be mounted at the corners of a rectangle as a stacked and bayed array as shown in **Fig 16.24**, with separations determined by their beamwidths as described above. Whilst feed harnesses can be purchased with the antennas, they can also be constructed using standard coaxial cables and connectors, as shown in **Fig 16.25**. Each antenna and all cables must have an impedance of 50 ohms. The two feeders L1 and L3 connected in parallel result in 25Ω at Point A.

This is transformed to 100 ohms by the cable between A and B, which must be an odd number of quarter wavelengths long. The two 100-ohm impedances connected in parallel at point B result

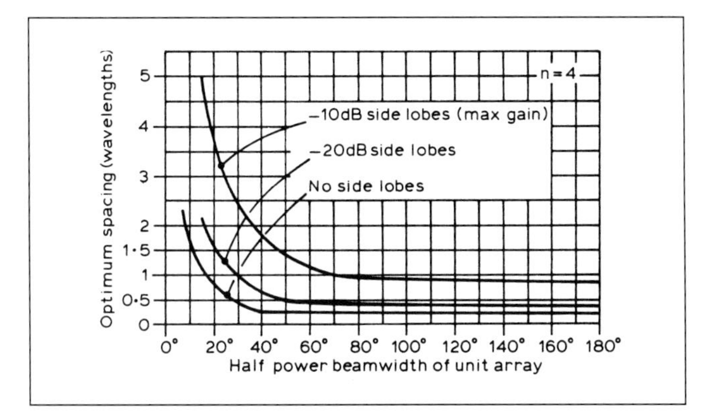


**Fig 16.22: Optimum stacking spacing for four-unit arrays (ARRL *Antenna Handbook*)**

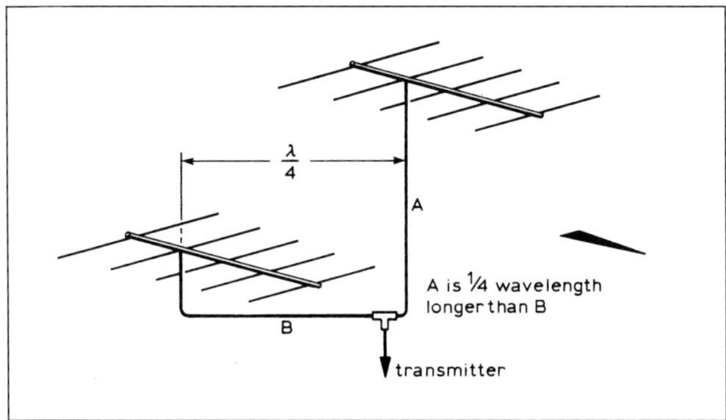


**Fig 16.23: Improving front to back ratio of stacked Yagi antennas with offset vertical mounting**

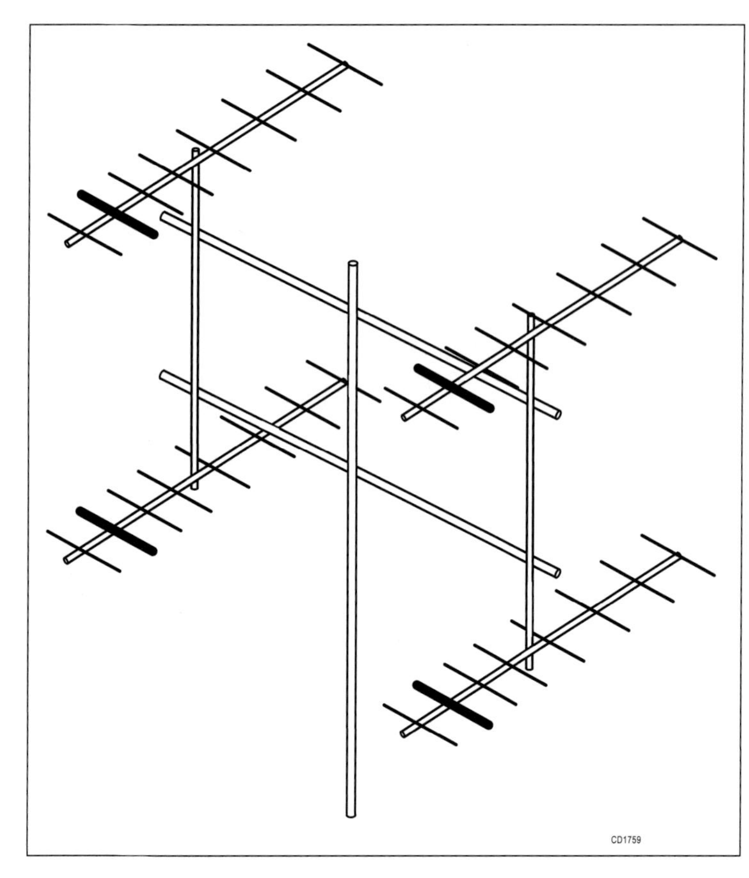


Fig 16.24: Four Yagi antennas stacked and bayed

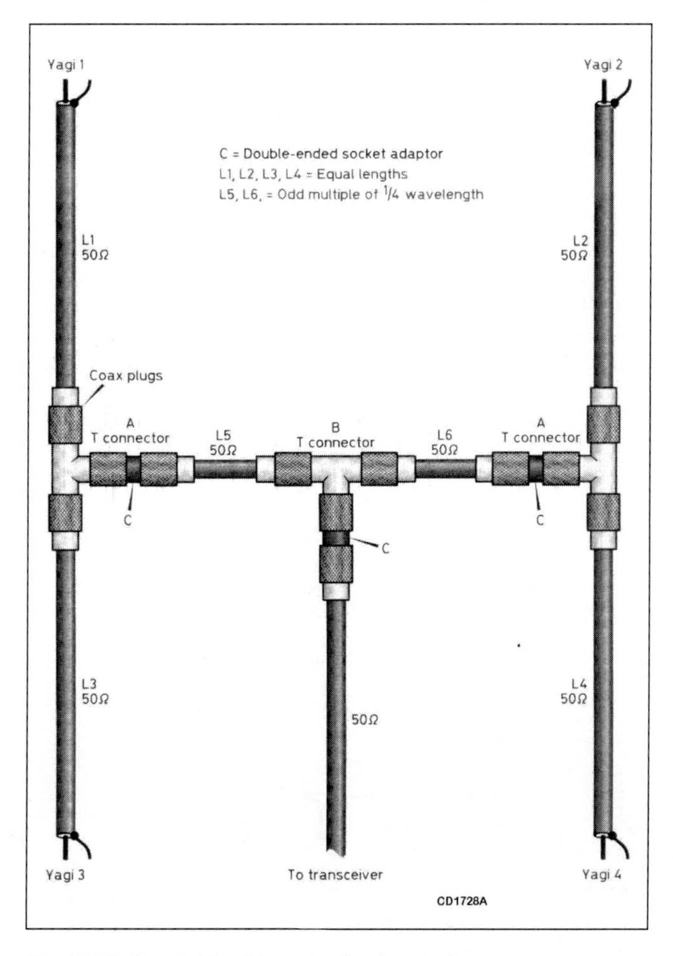


Fig 16.25: Coaxial feed harness for four antennas

in a 50-ohm impedance presented to the transceiver feeder. Feeders L1-L4 may be any convenient length, provided that they are all the same electrical length, preferably phase matched using measuring equipment..

## Skeleton Slot Feed for Two Stacked Yagis

A serious disadvantage of the Yagi array is that variation of the element lengths and spacing causes interrelated changes in the feed impedance. To obtain the maximum possible forward gain experimentally is extremely difficult. For each change of element length it is necessary to readjust the matching either by moving the reflector or by resetting a matching device.

However, a method has been devised for overcoming these practical disadvantages. It involves the use of a radiating element in the form of a skeleton slot. This is far less susceptible to the changes in impedance caused by changes in the length of the parasitic elements. A true slot would be a slot cut in an infinite sheet of metal. Such a slot, when approximately $\lambda/2$ long, would behave in a similar way to a dipole radiator. In contrast with a dipole, however, the electric field produced by a vertical slot is horizontally polarised.

The skeleton slot was developed during experiments to find how much the 'infinite' sheet of metal could be reduced without the slot antenna losing its radiating property. The limit was found to occur when there remained approximately $\lambda/2$ of metal beyond the slot edges. However, further experiments showed that a thin rod bent to form a 'skeleton slot' (approximately $5\lambda/8$ by $5\lambda/24$) exhibited similar properties to those of a true slot.

The way a skeleton slot works is shown in **Fig 16.26**. Consider two $\lambda/2$ dipoles spaced vertically by 5 /8. Since the greater part of the radiation from each dipole takes place at the current maximum (ie the centre) the ends of the dipoles may be bent without serious effect.

These 'ends' are joined together with a high-impedance feeder, so that 'end feeding' can be applied to the bent dipoles. To radiate in phase, the power should be fed midway between the two dipoles.

The high impedance at this point may be transformed down to that of the feeder cable with a tapered matching section/

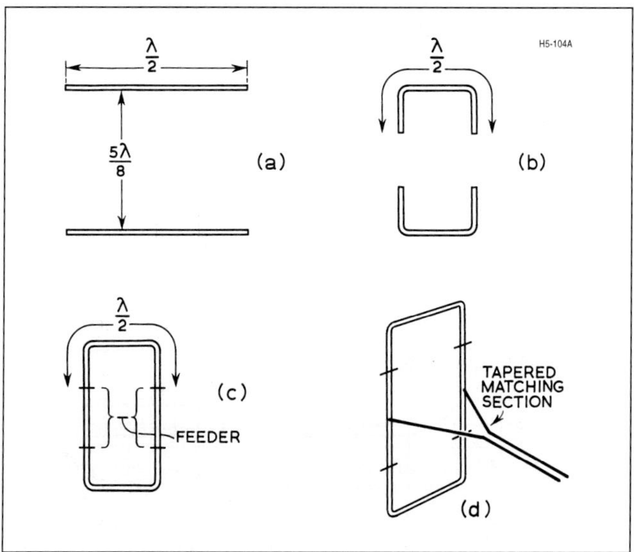


Fig 16.26: Development of the skeleton-slot radiator

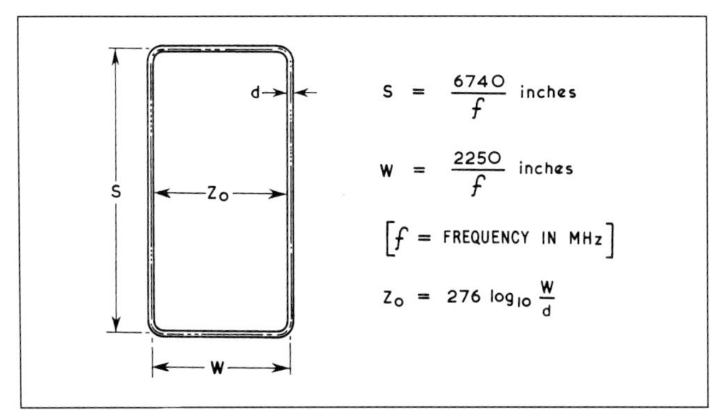


**Fig 16.27: Dimensional relationships of a skeleton-slot radiator. Both S and W may be varied experimentally, and will not change the radiation characteristics of the slot greatly. See text**

transmission line (ie a delta match). Practical dimensions of a skeleton-slot radiator are given in **Fig 16.27**.

These dimensions are not critical, and may be varied over a modest range without affecting the radiating characteristics of the slot. However, the feed impedance is very sensitive to dimensional changes, and must be properly matched by altering the length and shape of the delta section after completing all other adjustments.

It is important to note that two sets of parasitic elements are required with a skeleton-slot radiator and not one set as with a true slot. One further property of the skeleton slot is that its bandwidth is greater than a pair of stacked dipoles.Radiation patterns and VSWR data for some typical slot-fed Yagi antennas are shown in Fig 16.9 earlier in this chapter. Details of a practical 'six-over-six' skeleton slot Yagi antenna are shown in **Fig 16.28** with essential dimensions listed in **Table 16.11.**

## Skeleton Slot Yagi Arrays

Skeleton-slot Yagi arrays may be stacked to increase the gain but the same considerations of optimum stacking distance as previously discussed apply. The centre-to-centre spacing of a pair of skeleton-slot Yagi arrays should typically vary between

| Element | Length, in (mm) | Element spacing, in (mm) |
|---|---|---|
| A | 34 (864) | A - B, 14.25 (362) |
| B | 34 (864) | B - C, 14.25 (362) |
| C | 34 (864) | C - D, 14.25 (362) |
| D | 34 (864) | D - E, 34.25 (870) |
| E | 40 (1016) | D - Slot, 14.5 (368) |
| Boom | 70 (1720) | |

**Table 16.11: Dimensions for six-over-six slot fed 145MHz Yagi**

**Fig 16.29: A high gain 432MHz antenna consisting of four eight-over-eight slot-fed Yagi antennas arranged in a square formation**

1λ and 3λ depending on the number of elements in each Yagi array. A typical 4 x 4 array of 8-over-8 slot-fed Yagis for 432MHz is shown in **Fig 16.29**.

## Quadruple Quad for 144MHz

This collapsible antenna was designed for portable use [19], but is equally useable as a fixed antenna for use indoors or in a loft, and can achieve gains of between 10 and 11dBi (8 - 9dBd) on 2 metres. It is effectively a stacked quad using mutual coupling instead of a phasing harness to excite the outer elements. Constructional details are shown in **Fig 16.30**. Each section has a circumference of around 1.04 wavelengths, which is not as would be expected for conventional quads. The dimensions are the result of experiments to obtain the best front to back ratio and least sensitivity to adjacent objects, which can be important for portable or loft operation, ensuring that the antenna will

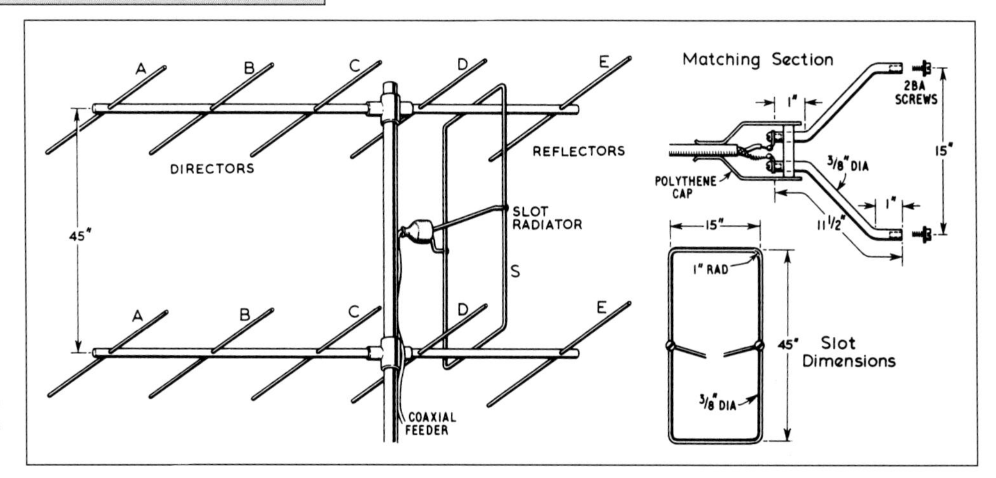


**Fig 16.28: Six-over-six skeleton slot fed Yagi antenna for 145MHz**

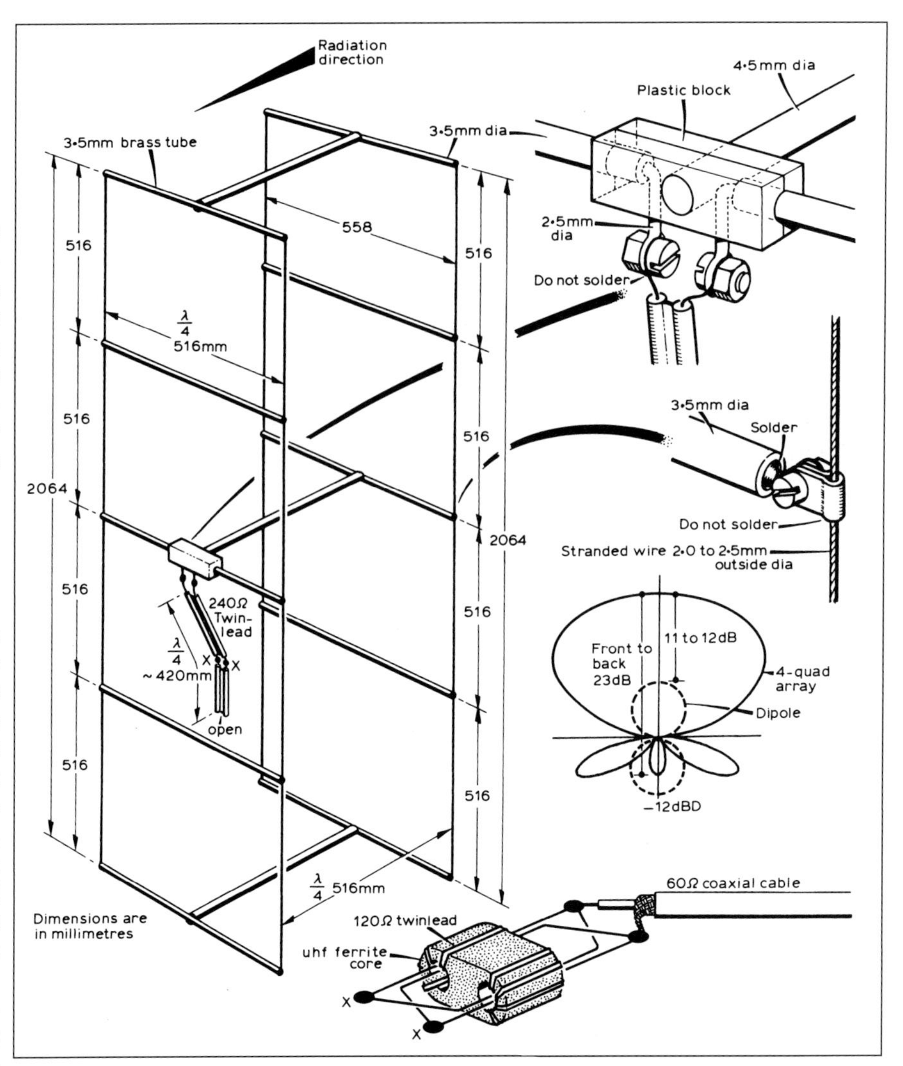


**Fig 16.30: Quadruple Quad. The match point xx should be found experimentally and will be approximately 200mm from the open end (*VHF Communications*)**

work without extensive adjustment. Note that the antenna was designed for low-power (1 watt) operation; the ferrite bead must not be allowed to saturate magnetically, or harmonic generation may occur. The bead may also become hot and shatter. For higher power operation, ferrite rings could be considered for the balun transformer, or a sleeve balun constructed as appropriate.

## Yagi Antenna Mounting Arrangements

The performance of Yagi antennas can be greatly degraded if they are not correctly installed and the feeder routed to minimise unwanted interaction with the antenna. Many commercial designs have fixed clamping positions which have been optimised to minimise coupling into the support mast. However, there are a number of precautions that can be taken when installing any Yagi antennas, whether operating in the same band as an array, or operating in several different bands.

Antenna performance may be completely destroyed if the mast is installed parallel, and through the Yagi antenna as in **Fig 16.31a**. The mast should be mounted at right angles to the antenna elements to minimise coupling between the elements and the mast, **Fig 16.31c**. If the antenna is to provide vertical polarisation, it should be offset from the main support mast with a stub mast if possible, **Fig 16.31b**. Mechanical balance can be restored (and the Yagi-mast separation increased) by using a symmetrical stub mast and a second antenna for the same band (bayed array), or for another frequency.

The mounting clamp should be placed midway between elements, and well away from the driven element. This is usually achieved by clamping near the mechanical balance point of the antenna. However, it is more important to keep the mast and clamp away from the adja-

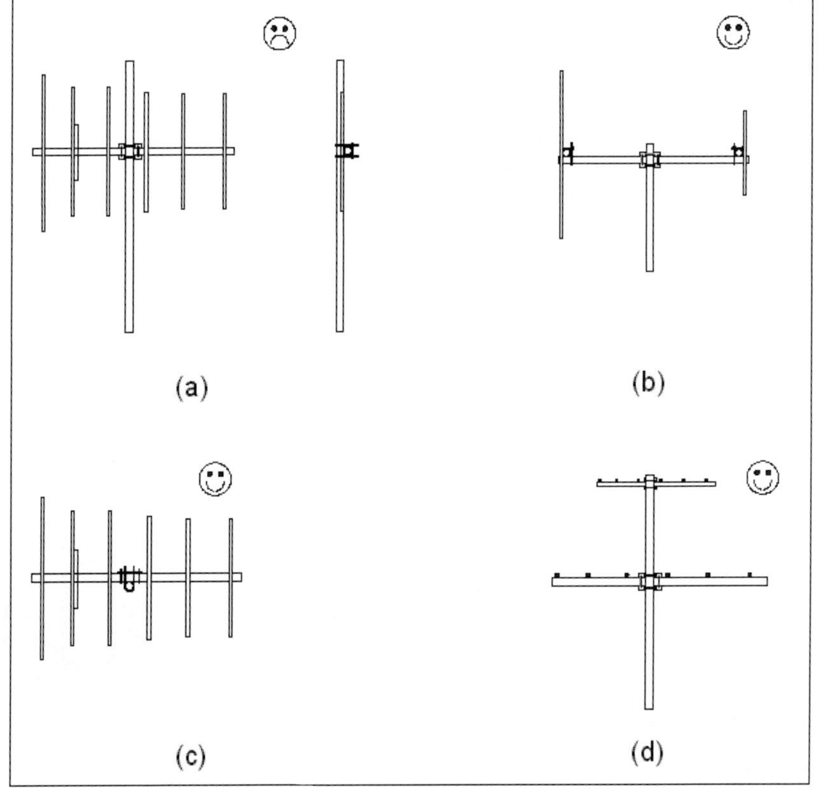


**Fig 16.31: Yagi mounting methods (a) the wrong way, antenna couples strongly with the mast, destroying performance (b) best arrangement for vertically polarised antennas, offset from the mast at least λ/4 with a horizontal tube (c) mast should be fastened mid way between elements, but not next to the driven element (d) put highest frequency antenna at top of a multi-band stack**

cent elements than to mechanically balance the antenna. In theory, the mast could be clamped to the antenna behind the reflector element(s). This is rarely done with antennas operating at wavelengths greater than 23cm because of mechanical constraints.

At 23cm and above, a 50mm (2in) pipe mast running through the antenna will seriously degrade its performance, even if the elements are at right angles to the pipe. Performance will not be so badly affected if the (horizontally polarised) antenna is right at the top of the mast, with the minimum amount of pipe required for clamping projecting through the elements. A smaller diameter pipe, eg 25mm (1in) for the mount will also reduce these effects, and is generally mechanically adequate to support higher frequency antennas.

The feed cable should be arranged to lie in a plane at right angles to the Yagi elements, or be taped below and along the boom until it can be run down the support mast. In the case of circularly polarised antennas, for example, crossed Yagi antennas fed in phase quadrature, the elements should be arranged at 45 degrees to the support mast when the antenna is viewed along its boom. There will be some degradation of circularity, but it can be minimised if the support mast is not an odd multiple of quarter wavelengths long. The feed cable should be taped to the boom and dressed on to the support mast with the minimum bend radius for which the cable is designed.

## Stacking Yagi Antennas for Different Bands

The optimum spacing between identical antennas to create higher gain arrays is discussed earlier in the chapter. However, in many cases, it may be desired to put several antennas for different bands on a common rotating mast. If the antennas are all pointing in the same direction on the mast, they should be separated sufficiently to ensure that their effective apertures do not overlap (see formula earlier in this chapter) to avoid interaction and mutual degradation of their radiation patterns. To a first approximation, the antenna gain may be used in place of the directivity in the formula. Yagi antennas may be stacked more closely together if alternate antennas point in directions at 90 degrees to each other. The separation may then be reduced so that the effective aperture of the lowest band antenna of any pair

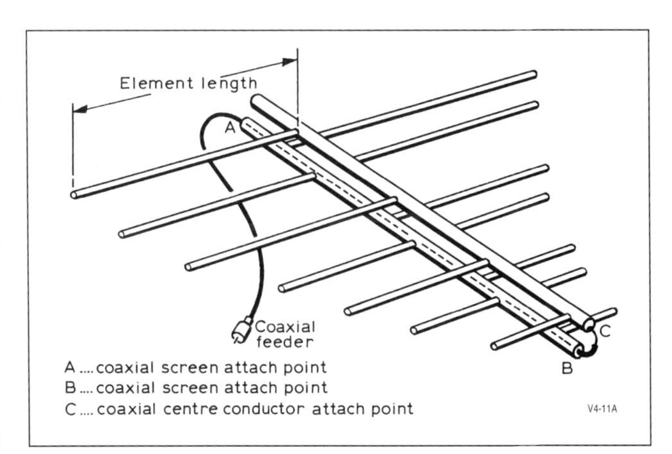


**Fig 16.32: Typical log-periodic antenna. Note that the bottom transmission line is fed from the coaxial outer while the top line is fed from the centre conductor (*Ham Radio*)**

is not physically encroached by the higher frequency antenna. Closer spacings may be possible without excessive interaction, but need to be investigated on a case by case basis using antenna modelling software or careful experiment.

Further information on antenna stacking can be found on GM3SEK's web site [14].

## THE LOG PERIODIC ANTENNA

The log-periodic antenna **Fig 16.32** was originally designed and proven at the University of Illinois in the USA in 1955 [21].Its properties are an almost infinite bandwidth, governed only by the number of elements used and mechanical limitations, together with the directive qualities of a Yagi antenna [22].

**Table 16.12** and **Table 16.13** show typical dimensions for element spacing and length for log-periodic arrays. These are derived from a computer-aided design produced by W3DUQ [22]. Other frequency bands can be produced by scaling all dimensions. The tabulated parameters have a 5% overshoot of the working frequency range at the low end and a 45% overshoot at the high-frequency end. This is done to maintain logarithmic response over the complete frequency range specified as the

| | 21-55MHz array | | | | | | 50-150MHz array | | | | | | 140-450MHz array | | | | | |
|---|---|---|---|---|---|---|---|---|---|---|---|---|---|---|---|---|---|---|
| Ele- | Length | | Diameter | | Spacing | | Length | | Diameter | | Spacing | | Length | | Diameter | | Spacing | |
| ment | (ft) | (mm) | (in) | (mm) | (ft) | (mm) | (ft) | (mm) | (in) | (mm) | (ft) | (mm) | (ft) | (mm) | (in) | (mm) | (ft) | (mm) |
| 1 | 12.240 | 3731 | 1.50 | 38.1 | 3.444 | 1050 | 5.256 | 1602 | 1.00 | 2.54 | 2.066 | 630 | 1.755 | 535 | 0.25 | 6.7 | 0.738 | 225 |
| 2 | 11.190 | 3411 | 1.25 | 31.8 | 3.099 | 945 | 4.739 | 1444 | 1.00 | 2.54 | 1.860 | 567 | 1.570 | 479 | 0.25 | 6.7 | 0.664 | 202 |
| 3 | 10.083 | 3073 | 1.25 | 31.8 | 2.789 | 850 | 4. 274 | 1303 | 1.00 | 2.54 | 1.674 | 510 | 1.304 | 397 | 0.25 | 6.7 | 0.598 | 182 |
| 4 | 9.087 | 2770 | 1.25 | 31.8 | 2.510 | 765 | 3.856 | 1175 | 0.75 | 19.1 | 1.506 | 459 | 1.255 | 383 | 0.25 | 6.7 | 0.538 | 164 |
| 5 | 8.190 | 2496 | 1.25 | 31.8 | 2.259 | 689 | 3.479 | 1060 | 0.75 | 19.1 | 1.356 | 413 | 1.120 | 341 | 0.25 | 6.7 | 0.484 | 148 |
| 6 | 7.383 | 2250 | 1.00 | 25.4 | 2.033 | 620 | 3.140 | 957 | 0.75 | 19.1 | 1.220 | 372 | 0.999 | 304 | 0.25 | 6.7 | 0.436 | 133 |
| 7 | 6.657 | 2029 | 1.00 | 25.4 | 1.830 | 558 | 2.835 | 864 | 0.75 | 19.1 | 1.098 | 335 | 0.890 | 271 | 0.25 | 6.7 | 0.392 | 119 |
| 8 | 6.003 | 1830 | 0.75 | 19.1 | 1.647 | 500 | 2.561 | 781 | 0.50 | 12.7 | 0.988 | 301 | 0.792 | 241 | 0.25 | 6.7 | 0.353 | 108 |
| 9 | 5.414 | 1650 | 0.75 | 19.1 | 1.482 | 452 | 2.313 | 705 | 0.50 | 12.7 | 0.889 | 271 | 0.704 | 215 | 0.25 | 6.7 | 0.318 | 97 |
| 10 | 4.885 | 1489 | 0.75 | 19.1 | 1.334 | 407 | 2.091 | 637 | 0.50 | 12.7 | 0.800 | 244 | 0.624 | 190 | 0.25 | 6.7 | 0.286 | 87 |
| 11 | 4.409 | 1344 | 0.75 | 19.1 | 1.200 | 366 | 1.891 | 576 | 0.50 | 12.7 | 0.720 | 219 | 0.553 | 169 | 0.25 | 6.7 | 0.257 | 78 |
| 12 | 3.980 | 1213 | 0.50 | 12.7 | 1.080 | 329 | 1.711 | 522 | 0.375 | 9.5 | 0.648 | 198 | 0.489 | 149 | 0.25 | 6.7 | 0.231 | 70 |
| 13 | 3.593 | 1095 | 0.50 | 12.7 | 0.000 | | 1.549 | 472 | 0.375 | 9.5 | 0.584 | 178 | 0.431 | 131 | 0.25 | 6.7 | 0.208 | 63 |
| 14 | | | | | | | 1.403 | 428 | 0.375 | 9.5 | 0.525 | | 0.378 | 115 | 0.25 | 6.7 | 0.187 | 57 |
| 15 | | | | | | | 1.272 | 388 | 0.375 | 9.5 | 0.000 | | 0.332 | 101 | 0.25 | 6.7 | 0.169 | 52 |
| 16 | | | | | | | | | | | | | 0.290 | 88 | 0.25 | 6.7 | 0.000 | |
| Boom | 25.0 | 7620 | 2.0 | 50.8 | 0.5 | 12.7 | 16.17 | 5090 | 1.5 | 38.1 | 0.5 | 152 | 5.98 | 1823 | 1.5 | 38.1 | 0.5 | 152 |

**Table 16.12: Spacing and dimensions for log-periodic VHF antennas**

| Element | Length | | Diameter | | Spacing | |
|---|---|---|---|---|---|---|
| | (ft) | (mm) | (ft) | (mm) | ( f t ) | (mm) |
| 1 | 0.585 | 178 | 0.083 | 2.1 | 0.246 | 75 |
| 2 | 0.523 | 159 | 0.083 | 2.1 | 0.221 | 67 |
| 3 | 0.435 | 133 | 0.083 | 2.1 | 0.199 | 61 |
| 4 | 0.418 | 127 | 0.083 | 2.1 | 0.179 | 55 |
| 5 | 0.373 | 114 | 0.083 | 2.1 | 0.161 | 49 |
| 6 | 0.333 | 101 | 0.083 | 2.1 | 0.145 | 44 |
| 7 | 0.297 | 91 | 0.083 | 2.1 | 0.131 | 40 |
| 8 | 0.264 | 80 | 0.083 | 2.1 | 0.118 | 36 |
| 9 | 0.235 | 72 | 0.083 | 2.1 | 0.106 | 32 |
| 10 | 0.208 | 63 | 0.083 | 2.1 | 0.095 | 29 |
| 11 | 0.184 | 56 | 0.083 | 2.1 | 0.086 | 26 |
| 12 | 0.163 | 50 | 0.083 | 2.1 | 0.077 | 23 |
| 13 | 0.144 | 44 | 0.083 | 2.1 | 0.069 | 21 |
| 14 | 0.126 | 38 | 0.083 | 2.1 | 0.062 | 19 |
| 15 | 0.111 | 34 | 0.083 | 2.1 | 0.056 | 17 |
| 16 | 0.097 | 30 | 0.083 | 2.1 | 0.000 | 0 |
| Boom | 1.99 | 607 | 0.5 | 12.7 | | |

**Table 16.13: Spacing and dimensions for log-periodic UHF antenna (420 - 1350MHz)**

log-periodic cell is active over approximately four elements at any one specific frequency. The logarithmic element taper ($\alpha$) is 28° for all three antennas. They have a forward gain of 6.55dBd, with a front-to-back ratio of typically 15dB and a VSWR better than 1.8:1 over the specified frequency range.

Construction is straightforward. The element lengths for the highest-frequency antenna allow for the elements to be inserted completely through the boom, ie flush with the far wall. The two lower-frequency antennas have element lengths calculated to butt flush against the element side of the boom, and a length correction factor must be added to each element if through-boom mounting is used.

The supporting booms are also the transmission line between the elements for a log-periodic antenna. They must be supported with a dielectric spacer from the mast of at least twice the boom-to-boom spacing. Feed-line connection and the arrangement to produce an 'infinite balun' is shown in **Fig 16.33**. Any change in the boom diameters will require a change in the boom-to-boom spacing to maintain the transmission line impedance. The formula to achieve this is:

$$Z_0 = 273 \log_{10} D/d$$

where $D$ is the distance between boom centres and $d$ the diameter of the booms. Mounting arrangements are shown in

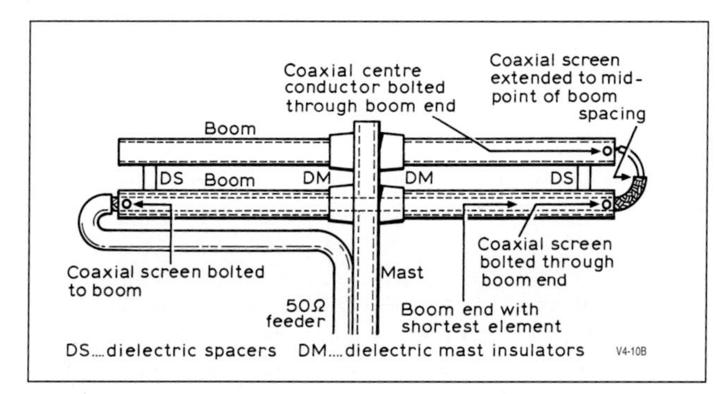


**Fig 16.33: Log-periodic antenna mast mounting and feeder arrangements (*Ham Radio*)**

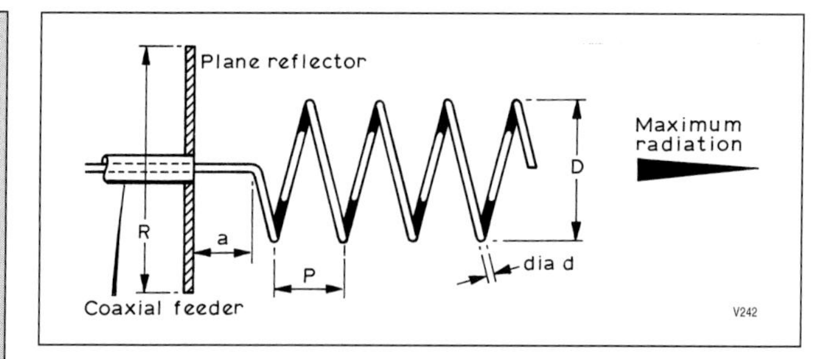


**Fig 16.34: The axial mode helix antenna. The plane reflector may take the form of a dartboard type wire grid or mesh. The dimensions given in Table 16.13 are based on a pitch angle of 12 degrees. The helix tube or wire diameter is not critical, but it must be supported by low loss insulators**

Fig 16.33. The antenna can be oriented for either horizontal or vertical polarisation if a non-conductive mast section is used. The horizontal half-power beamwidth will be typically 60° with a vertical half-power beamwidth of typically 100°.

## THE AXIAL MODE HELIX

The axial mode helix antenna provides a simple means of obtaining high gain and a wide-band frequency characteristic. When the circumference of the helix is of the order of one wavelength, axial radiation occurs, ie the maximum field strength is found to lie along the axis of the helix. This radiation is circularly polarised, the sense of the polarisation depending on whether the helix has a right or left-hand thread. The polarisation can be determined by standing behind the antenna. If a clockwise motion would be required to travel along the helix to its far end, the helix will generate and receive Right Hand Circularly Polarised (RHCP) waves.

A helix may be used to receive plane or circularly polarised waves. When signals are received from a transmitting helix care must be taken to ensure that the receiving helix has a 'thread' with the same hand of rotation as the radiator, or significant signal will be lost due to *polarisation mismatch*.

The properties of the helical antenna are determined by the diameter of the helix $D$ and the pitch $P$ (see **Fig 16.34**). It is also dependent on radiation taking place all along the helical conductor. The gain of the antenna depends on the number of turns in the helix.

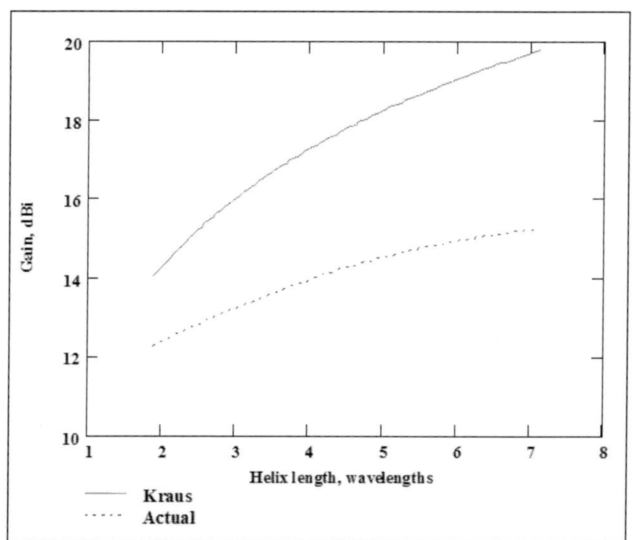


**Fig 16.35: Kraus's theoretical gain and realisable gain for a helix antenna of different lengths. The antenna has a circumference of 1.06$\lambda$ and $\alpha$ = 13°**

The diameter of the reflector $R$ should be at least $\lambda/2$, with the diameter of the helix $D$ about $\lambda/3$ and the pitch $P$ about $0.24\lambda$. A detailed description of the way in which the antenna radiates, and the relationships between pitch and diameter for different antenna characteristics are described by its inventor, J D Kraus in [23].

A helix of this design will have a termination / feed impedance of about 140 ohms. A 50-ohm impedance can be obtained by shaping the last quarter turn from the feedpoint to lie close to the reflector by reducing the pitch of the helix over the last turn.

Gain of the antenna is proportional to the number of turns in the helix, and may be enhanced slightly by tapering the open end towards the centre.

At higher frequencies an additional 1dB can be obtained by replacing the flat reflector with a cup that encloses the first turn. However, the theoretical gains published by Kraus and others are optimistic. (**Fig 16.35**) Maximum realisable gains are given by following formula for helix lengths between 2 and 7 wavelengths.

$$G_{max} = 10.25 + 1.22L - 0.0726L^2 \text{ dBi}$$

where $L$ is the length of the antenna in wavelengths.

A typical antenna with a seven turn helix has a gain of approximately 12dBi over a 2:1 frequency range. To fully utilise this gain it is necessary to use a circularly polarised antenna (eg a helix of the same sense) for both transmission and reception. If a plane-polarised antenna, such as a dipole, is used there will be an effective loss of 3dB due to polarisation mismatch. General dimensions for helix antennas are shown in **Table 16.14**.

## A Practical Helix Antenna for 144MHz

The greatest problem to be overcome with this type of antenna for 144MHz operation, with a helix diameter of 24½in, is the provision of a suitable support structure.

**Fig 16.36** shows a general arrangement in which three supports per turn (120° spacing) are used. Details of suitable drilling of the centre boom are given in **Fig 16.37**.

The helix may be made of copper, brass, or aluminium tube or rod, or coaxial cable. This latter alternative is an attractive material to use, being flexible with the braid 'conductor' weatherproofed. If coaxial cable is used the inner conductor should be

| Band | Dimensions D | R | P | a | d |
|---|---|---|---|---|---|
| General | 0.32 | 0.8 | 0.22 | 0.12 | |
| 144MHz | 25.5 (648) | 64 (1626) | 17.75 (450) | 8.75 (222) | 0.5 (12.7) |
| 433MHz | 8.75 (222) | 22 (559) | 6 (152) | 3 (76) | 0.375-0.5 (4.8 - 12.7) |
| 1296MHz | 3 (76) | 7 (178) | 2 (50) | 1.125 (28) | 0.125-0.25 (3.2-6.4) |
| Turns | 6 | 8 | 10 | 12 | 20 |
| Gain | 12dBi | 14dBi | 15dBi | 16dBi | 17dBi |
| Beamwidth | 47° | 41° | 36° | 31° | 24° |

Dimensions in inches, millimetres are given in brackets. The gain and beamwidth of the helical antenna are dependent upon the total number of turns as shown above.

Bandwidth $= 0.75 \text{ to } 1.3\lambda$

Feed impedance $= 140 \times \dfrac{\text{circumference}}{\lambda}$ ohms

(Note: λ and circumference must be in the same units.)

Beamwidth (degrees) $= \dfrac{12{,}300}{\text{No of turns}}$

Table 16.14: General dimensions for 144, 433 and 1296MHz helix antennas

Fig 16.36: General arrangements of support structure for a five-turn helical antenna for 144MHz. The antenna is right hand circularly polarised

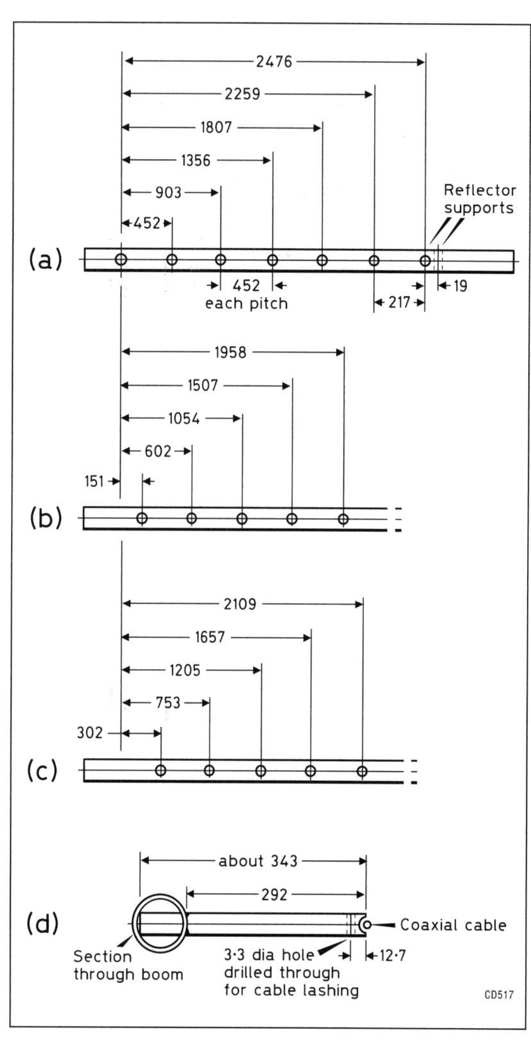


Fig 16.37: First side drilling dimensions for the 144MHz band helix antenna (a) reflector holes are drilled at right angles; (b) and (c) are drilled at intervals of 120 and 240 degrees respectively from (a); (d) cutting and filing dimensions for the element stand-offs

connected to the outer at each end, and the jacket well sealed to prevent moisture ingress and corrosion.

The reflector is located at a distance *a* behind the start of the first turn, and is supported by crossed supports from the central boom. Material for the reflector can be any kind of metal mesh such as chicken netting. Radial spokes alone are not sufficient and will reduce the gain by 1-2dB, unless connected together with wires in dartboard fashion.

It is not essential that the central boom should be constructed of non conductive material. Metal booms may be used provided that they are centrally placed along the axis of the helix. This can lead to a simple construction using square aluminium tubing, as sold for self-construct shelving in some DiY centres. Corners and end fixtures can also be used to fasten the boom rigidly to the reflector without having to resort to machining or fabricating brackets.

A square section also simplifies the mounting of the insulators, which can be made from Delrin or other plastic rod, and secured through the boom by a long screw or bolt with a single hole at the far end through which to thread the helix. The number of insulators will depend upon the rigidity of the helix material. At 433MHz, supports every 1.25 turns are adequate for a helix made from copper tubing.

Although probably too heavy for 144MHz designs, copper tubing for small-bore central heating is suitable for 433MHz helices. It is readily obtainable in DiY centres in malleable coils that can be easily shaped over a suitable former. Draw a line with a wax pencil or paint along adjacent turns whilst they are still on the former, which if sized correctly, will allow the turns to be drawn out to the correct positions whilst the marks remain in a straight line. This helps considerably when ensuring the turns diameter and pitch are maintained along the length of the helix. The ends of the tube should be hammered flat and soldered up to prevent the ingress of water.

The last fractional turn of the helix closest to the reflector should be brought very close to the reflector as it approaches the connector, to bring the impedance of the helix to 50 ohms. Helix antennas for higher frequencies are easier to construct than Yagi antennas of comparable gain and require little adjustment. Detailed instructions for building 435MHz and 1296MHz helix antennas for satellite communications have been published in [25] and [26].

## HAND-HELD AND PORTABLE ANTENNAS

### Normal Mode Helix

The normal mode helix antenna comprises a length of spring wire wound such that the diameter of the spring is less than 0.1λ, and typically of order 0.01λ. Such antennas become resonant when their axial length is around 0.1λ, and can be designed to offer manageable impedances at the base. The resonance occurs over a relatively narrow band, and is heavily influenced by any jacket or sleeve fitted over the helix, and by the nature of the groundplane (if any) against which it is fed. The current distribution along the length of the helix is similar to that of a whip antenna, but compressed into the much shorter length of the helix.

For hand-held radios, the length of the helix is dimensioned such that the current distribution is similar to that expected on a 5/8λ whip, ie the current maximum occurs about one third of the overall height above the feed point. This helps to maximise the radiation efficiency of the antenna, whilst also minimising effects of the variability of the ground plane (hand held radio) and body proximity on both the input impedance and radiating efficiency.

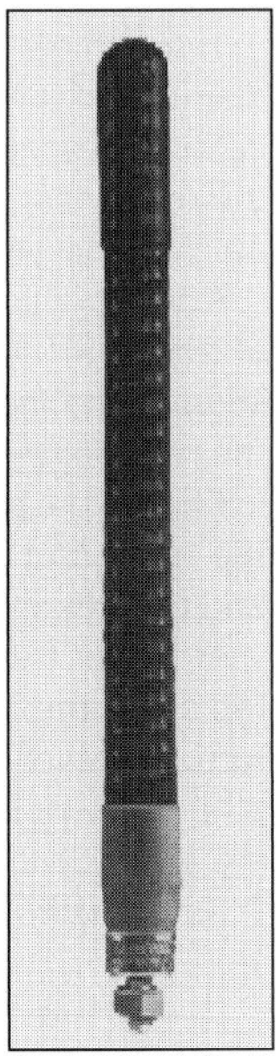

**Fig 16.38: A typical commercial helical antenna with screw mounting facility**

A 3/4λ whip over a moderate ground plane has a resistive match very close to 50 ohms. If this whip is coiled into a helical spring it will match to approximately 50 ohms and resonate at a lower frequency, partly due to capacitance between the coil turns.

If the spring is trimmed to the original frequency the result will be an antenna of about 0.1λ overall height. The actual length of wire is between 1/2λ and 5/8λ at the working frequency.

Electrically it is still a resonant 3/4λ antenna. Near-base capacitance also modifies the matching under certain frequency and ground plane conditions.

If the turns are very close together, the helical antenna will resonate at a frequency approaching the axial length, because of strong coupling between the turns. There is an optimum 'spacing' between turns for best performance. A 145MHz helical antenna typically has a spacing between turns equal to twice the diameter of the wire used.

The helical whip is very reactive off-resonance. It is very important that it is resonated for the specific conditions that prevail in its working environment.

Fortunately, it is often only necessary to change the number of turns to resonate the spring over such diverse conditions, ie a large ground plane or no ground plane at all. The resistive part of the impedance can vary between 30 and 150 ohms at the extremities.

Under typical 'hand-held' conditions (**Fig 16.38**), although to a small extent depending on the frequency of operation, the spring can offer something close to a 50-ohm impedance match. **Fig 16.39** shows the number of turns required for a typical 9mm diameter helix for 3/4λ resonance.

As the helical is reduced in length two effects occur. First, the radiation resistance is lower than the equivalent linear whip so the choice of a good conducting material is important to reduce resistive losses.

A plain steel spring compared with a brass or copper-plated helix can waste 3dB of power as heat. Secondly, the physical aperture of the helical whip is around one third that of a λ/4 whip, which would imply a loss of 4.77dB.

Results obtained from copper-plated, Neoprene-sheathed helical antennas, correctly matched to a hand-held transmitter at 145MHz, provided signals at worst 3dB and at best +1dB compared with a λ/4 whip. A λ/4 whip with minimal ground plane would offer signals about -6dB compared to a λ/2 dipole.

A helical antenna, resonant and matched, on a λ/2 square ground plane can give results 2-3dB below a λ/2 dipole. An alternative arrangement using a bifilar-wound helix gives identical results (within 0.2dB) to a λ/2 dipole.

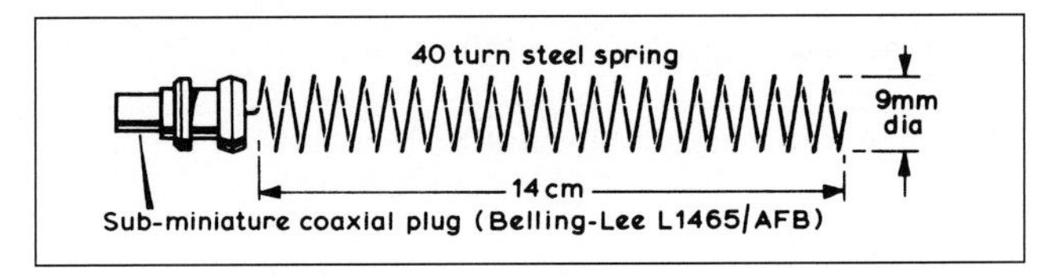


**Fig 16.39: Details of a home-made helical whip for 145MHz. A BNC plug could also be used**

## A Vertical Dipole for Portable Operation

A practical dipole for portable operation on in either the 2m or 6m band [27] is shown in **Fig 16.40**. The upper and lower sections together form a centre-fed half wavelength dipole. The feed cable is wound into a resonant choke to present a high impedance to the lower end of the dipole to reduce currents on the outside of the feed cable. Constructed from RG58C/U cable or similar, the antenna can be rolled up into a small space for travel, then unrolled and suspended from a suitable support for operation.

3870mm of RG8C/U cable is required for 145MHz operation. 470mm of the outer sheath and braid is stripped off, leaving the insulator and inner core to form the upper radiator. Measure out the length of the lower radiator from where the insulator is exposed to mark the starting point of the choke. Wind 4.6 turns on 32mm diameter PVC pipe to form the choke.

Feeding the cable through holes in the centre of end caps on the pipe allows the antenna to hang tidily. A ring terminal or solder tag soldered to the tip may be used for hoisting the antenna on nylon line or similar.

Tuning should be done outdoors, with the antenna positioned well away from objects that could affect the resonant frequency. Trim short pieces from the tip of the antenna to obtain a VSWR better than 1.3:1 (in 50 ohms) across 144 - 146MHz. If a longer feeder is required, the length below the choke should be a multiple of one half wavelength (680mm to compensate for the velocity factor of the dielectric) to minimise de-tuning.

A 6m variant can be constructed using 7280mm of RG58C/U cable. 11.8 turns of cable should be wound on a 50mm diameter PVC tube to form the choke. Any additional feeder should be a multiple of 1980mm.

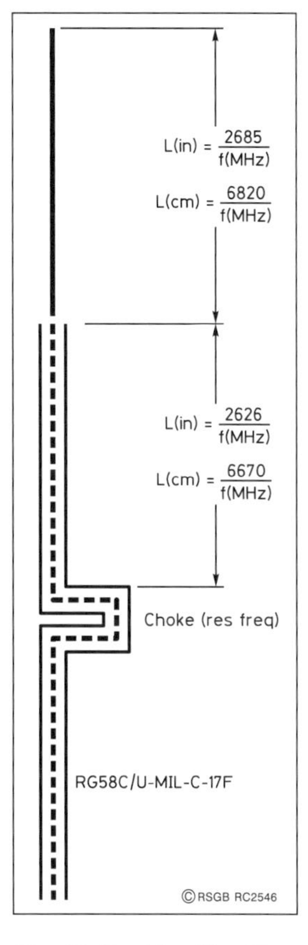


**Fig 16.40: The "feedline vertical" antenna**

## Collapsible 6m two-element beam

This compact 50MHz antenna (**Fig 16.41**) designed by G0JMI will fold down into a space of approximately 25 x 25cm by 1metre long. The radiating elements comprise a driven dipole and a reflector made from 0.7mm diameter single strand PVC insulated copper wire with their ends folded towards each other and separated by a 5mm spacer, **Fig 16 42**.

**Fig 16.41: Collapsible 6m 2-element beam antenna**

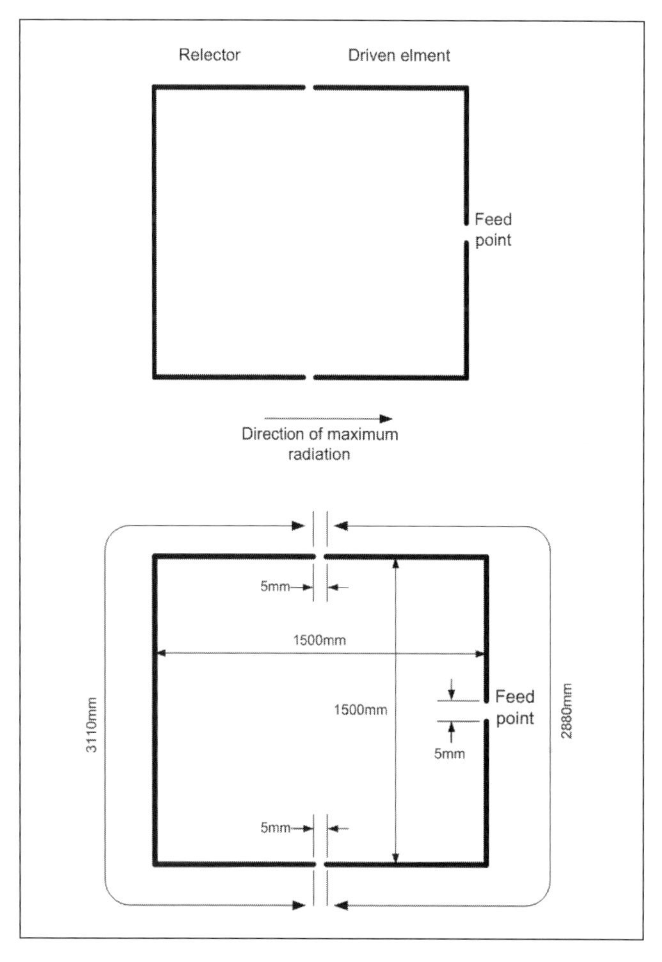


**Fig 16 42: Folded 2 element Yagi concept and dimensions**

The elements are spread by four lengths of plastic electrical conduit running to the corners, and fastened to a central plywood support by hinges that allow them to fall into the correct position under gravity and to be secured with a locking pin or bolt as required, see **Fig 16 43**. The central support is 150mm square, with a 50mm diameter central hole to take the top of a support mast. The mast is fastened in place by a hose clip and four suitable right angle brackets as shown in the photograph. The hinges are made from metal or wooden battens arranged so that the pivot is a little beyond each corner of the central support.

The wire is threaded through plastic conduit jointing sections that slip over the ends of the conduit, and the lengths adjusted to that the diagonals of the antenna are 2.12m to the wire elements. Paint marks on the wire are used to help position it correctly in the jointing sections on deployment. The 5mm gaps between the element ends and the feed point insulator are each made from 25mm x 5mm pieces of single sided glass fibre PCB with a 5mm gap in the copper at the centre.

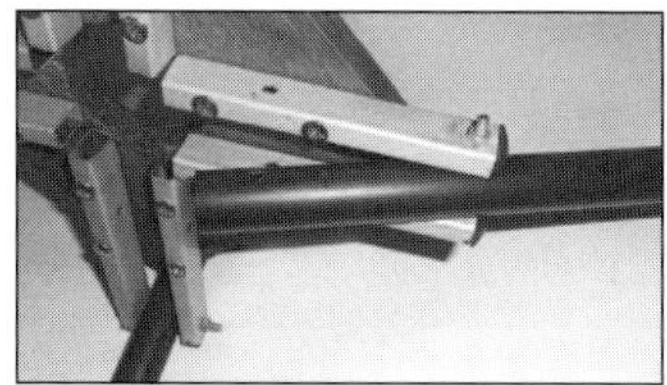

**Fig 16.43: Hinge assembly and pole mounting arrangements**

The ends of the dipoles are soldered in place and the two side insulators are covered and sealed using heat shrink tubing. Note that these insulators are at a point of maximum voltage, so a film of water across the gap will de-tune the antenna at least!

Two brass screw connectors taken from 'chocolate block' connector strip and soldered to the feed point insulator provide a removable connection for a quarter wavelength balun made from RG58 coaxial cable, see **Fig 16. 44**. The length of the balun was 1.4m, but may vary slightly with different thicknesses of cable jacket. The two legs of the balun are held together with insulation tape at about 12 cm intervals along its length. A plastic box with a snap-on lid protects the feed point from the weather.

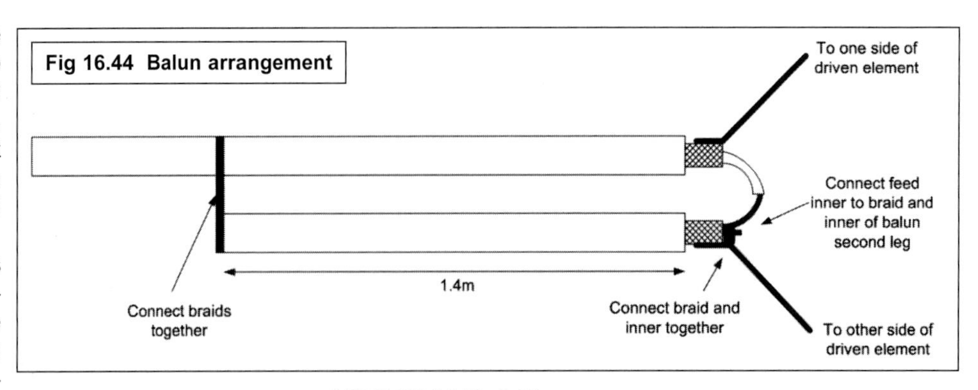


The antenna was found to present a VSWR better than 1.2:1 across the band when raised to 4m above ground level. The radiation pattern was measured and compared with that of a commercial two element Yagi, as shown in **Fig 16.45**. It is recommended that the transmitter power into this antenna does not exceed 50W. The antenna, folded down, is shown in **Fig 16.46**. Full construction details for the antenna can be found in [28].

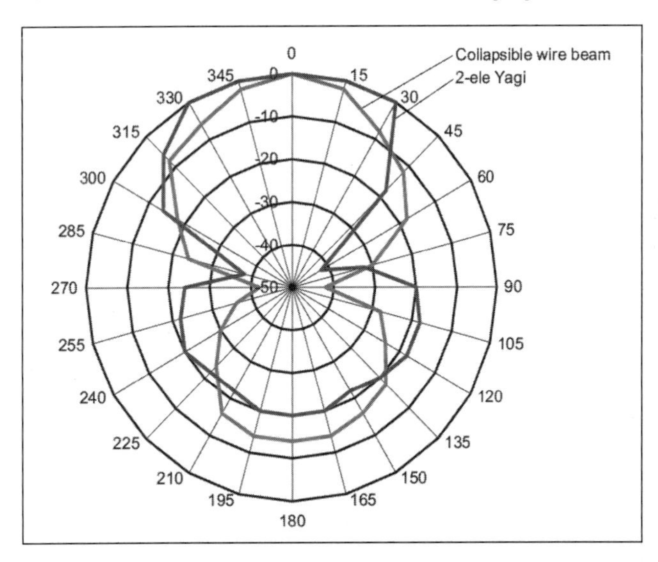


**Fig 16.45: Polar patterns of the 6m wire beam antenna compared with a commercial 2 element Yagi. Scales are degrees and 0 to -50dB**

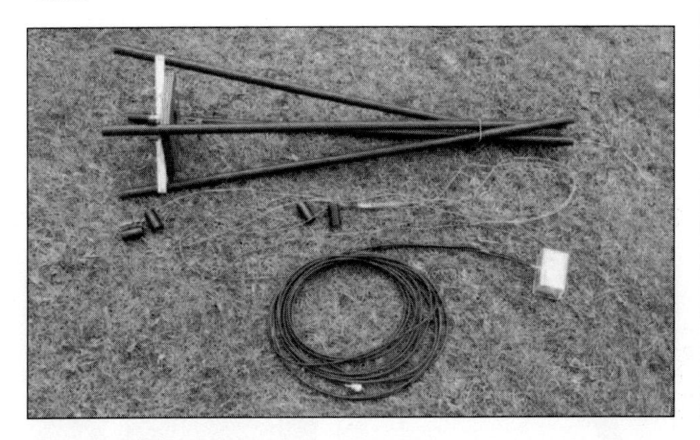

**Fig 16.46: 6m beam antenna folded down for transport and storage**

## HB9CV Mini Beam

The HB9CV mini-beam, because of its compact and straightforward construction, is suitable for both base station and portable use, and can be particularly useful in confined spaces such as lofts. Similar antennas are the lazy-H and ZL Special often used on the HF bands. The HB9CV version has one or two mechanical advantages that make it particularly suitable for VHF portable use. **Fig 16.47** [29] and **Fig 16.48** show two methods of construction for the HB9CV antenna. Note that a series capacitor of 3-15pF is required to adjust the gamma match/phasing combination to a VSWR of about 1.3:1 referred to 50Ω. The element spacing, and in particular the transmission line spacing (5mm in this case), is critical for optimum impedance matching and phasing, and therefore gain and front-to-back ratio.

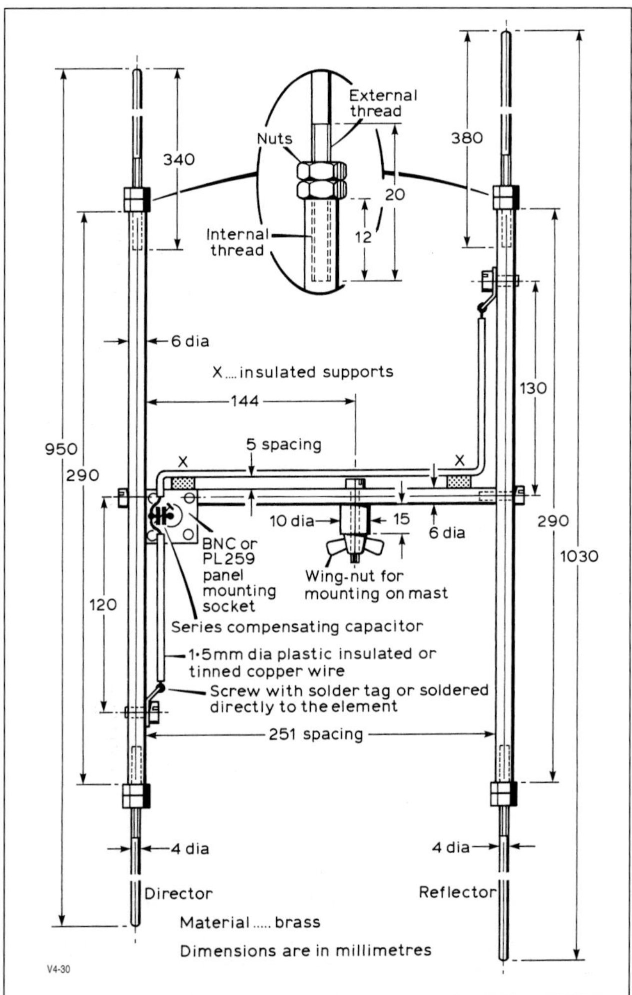


**Fig 16.47: A collapsible HB9CV antenna for the 144MHz band (*VHF Communications*)**

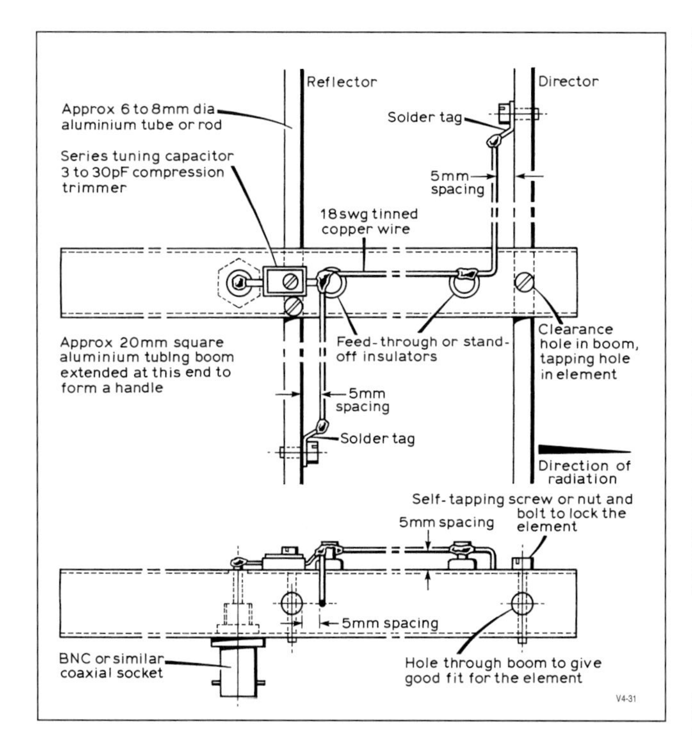


Fig 16.48: Alternative boom and feed arrangement for the 144MHz HB9CV antenna

The principle of operation is as follows. If two dipoles are close spaced (typically 0.1 - 0.2λ) and fed with equal currents with a phase difference corresponding to the separation of the dipoles, 'end-fire' radiation will occur along the line between the dipoles in one direction, and almost no radiation will occur in the reverse direction as explained earlier in this chapter in the section on arrays.

The different element lengths found on most HB9CV antennas improve the VSWR bandwidth, not the directivity as might at first be thought by comparison with a two element Yagi antenna.

The end at which the beam is fed defines the direction of radiation. A theoretical gain in excess of 6dBd should be possible. Depending on construction techniques, gains of 4 to 5dBd with front-to-back ratios of 10 to 20dB tend to be obtained in practice.

The radiation patterns shown in **Fig 16.49** and **Fig 16.50** are for the antenna of Fig 16.47. This antenna has a typical gain of 5dBd. Note the difference obtained when mounted at 10m

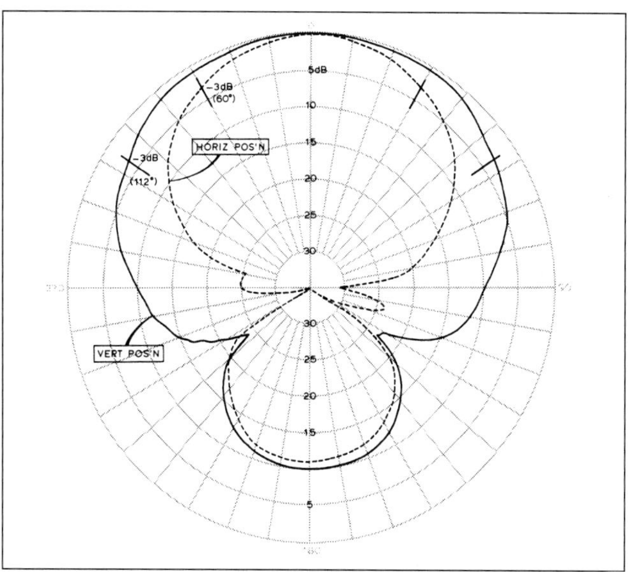


Fig 16.49: HB9CV antenna radiation patterns. Antenna 10m above ground

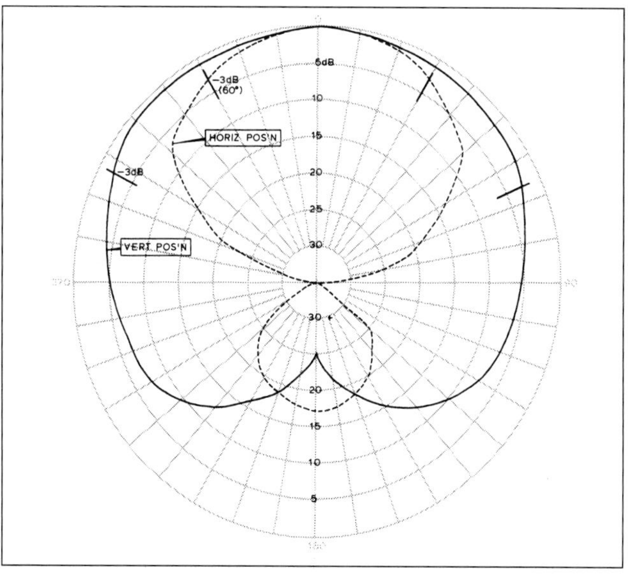


Fig 16.50: HB9CV antenna radiation patterns. Antenna hand-held, 1-2 metres above ground

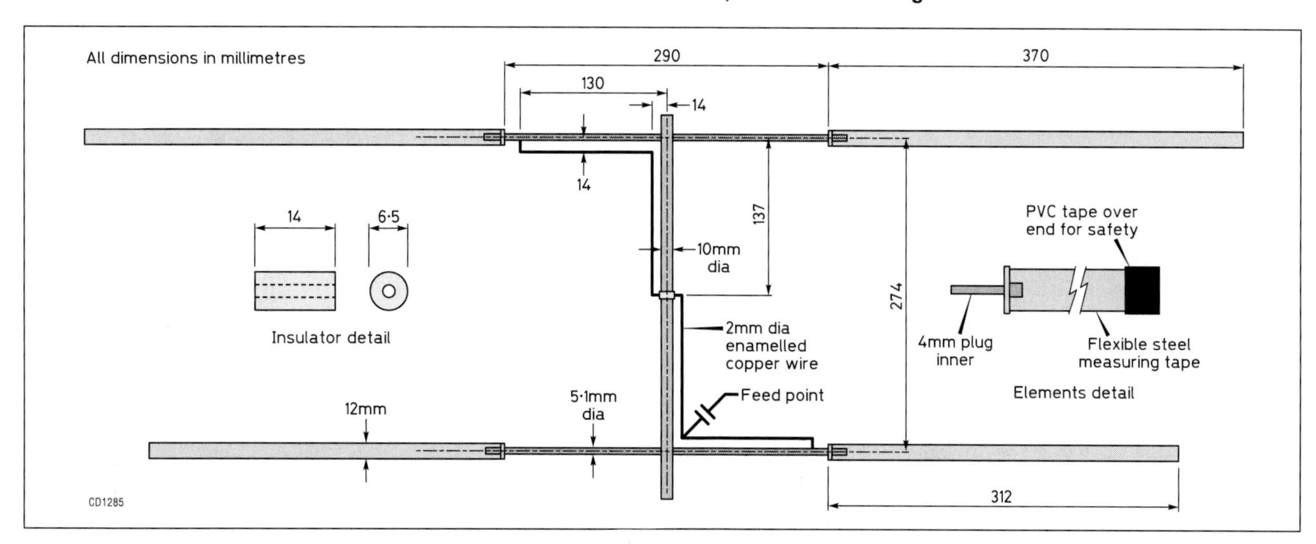


Fig 16.51: Construction details of the lightweight HB9CV antenna for 144MHz

(30ft) above the ground compared with hand-held measurements 1-2m above the ground. The latter height is typical for the antenna being used for direction finding.

## Lightweight HB9CV for 144MHz

The compact size of the HB9CV makes it eminently suitable for direction finding contests, EMC or interference probing, and portable work. The need for a very lightweight directional antenna for EMC investigations led to the design shown in **Fig 16.51**.

The boom and stub elements are made from thin walled brass tubing, soft soldered or brazed together. The removable elements are made from an old 12mm wide spring steel measuring tape soldered on to 4mm 'banana' plugs, although replacement tapes without housings can be purchased from good tool shops. The sharp ends must be protected by at least one layer of PVC tape or similar material. The feedline insulator where it passes through the boom can be made from Delrin or a scrap of solid polythene insulator from coaxial cable. The series matching capacitor in the example shown is 13pF, but should be adjusted for minimum VSWR, and the end of the coaxial cable and exposed connection to the capacitor should be sealed with silicone rubber compound if outdoor use is envisaged. The antenna can be supported on a simple wooden mounting using small 'Terry' spring clips to grip the boom.

## Foldable Four-element Yagi for 2m

This folding 4-element Yagi (**Fig 16.52**) is based on a lightweight FM broadcast receiving antenna [30]. The original antenna had a plastic insulator supporting a folded driven element, and three tubular parasitic elements made from rolled aluminium with an un-welded seam along the length of each tube.

The parasitic elements were supplied in two halves each fastened by a bolt and wing nut to a piece of shaped aluminium sheet that holds the elements at right angles to the boom when the wing nut is tightened, but allows them to lie folded along the boom when slackened off.

The key dimensions of the antenna modified for 2m operation are shown in **Fig 16.53**. The boom was re-drilled to take the original bolts and fittings.

The driven element was cut in half and re-joined with round aluminium bar that was a sliding fit inside the element tube. The curved ends were cut short to operate on 2m and made good with more of the aluminium rod bent into a "U" shape, **Fig 16.54**. The ends should be left loose until the antenna has been adjusted for optimum VSWR.

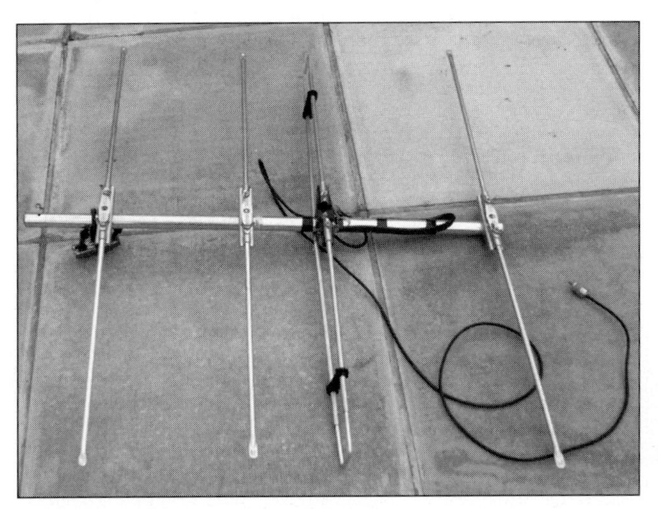

**Fig 16.52: General view of the foldable 2m antenna fully open and ready to use**

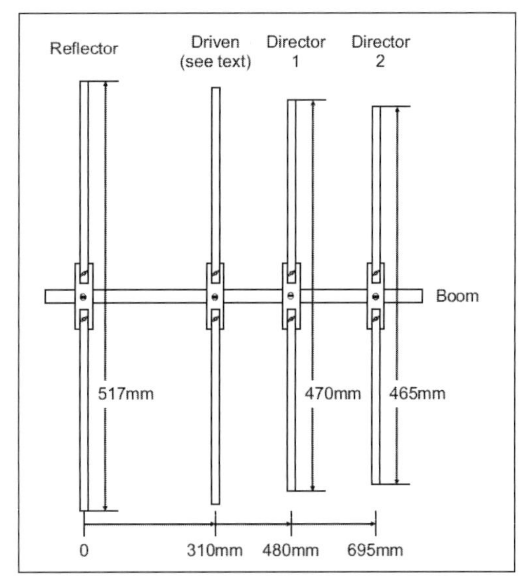


**Fig 16.53: Dimensions of antenna modified for two metre operation**

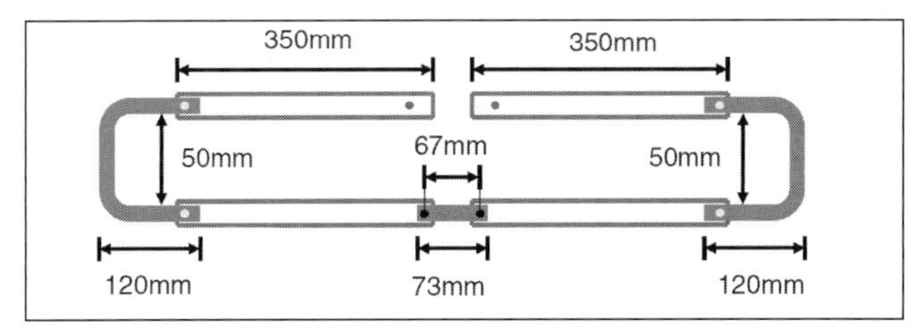


**Fig 16.54: Re-construction of the driven element. Materials shown shaded are aluminium bar. The feed point is at the top and the hinge at the bottom**

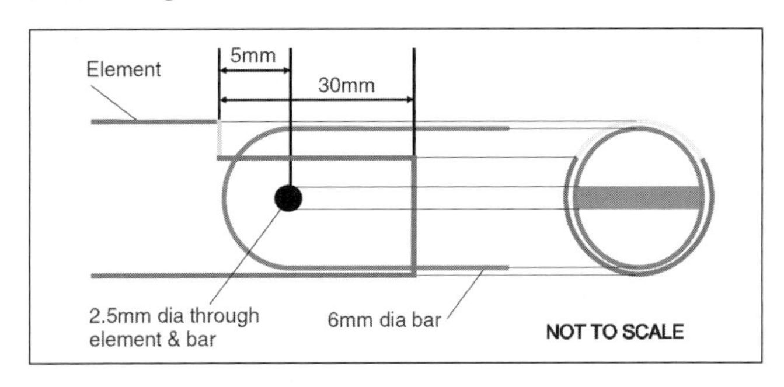


**Fig 16.55: Dimensions and drilling details for the driven element hinges**

**Fig 16.56: The driven element hinges partially folded. Note the insulating tape wrapped round the central bar to prevent a short circuit to the feed assembly fastener**

**(below) Fig 16.57: The antenna completely folded**

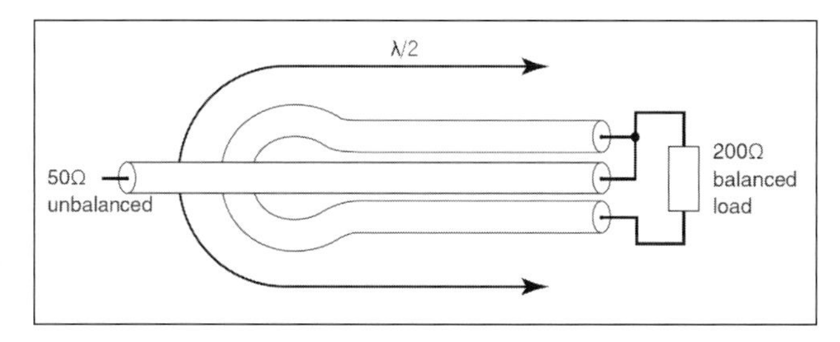


Fig 16.58: Balun details

The centre hinge is made by carefully cutting away part of the tube and rounding the end of the aluminium bar so that it can rotate around a 2mm bolt run through both the element and bar, **Figs 16.55 and 16.56**. The parasitic elements are shortened as necessary (Fig 16.c) and the ends flattened and de-burred for safety. The antenna can be folded to a small size to enable easy transportation (see **Fig 16.57**).

A 4:1 balun (**Fig 16.58**) made from coaxial cable was constructed. The author used 75 ohm cable for the phasing line, although 50 ohm cable would also be suitable if the antenna is tuned accordingly. The length of the phasing line must be one half wavelength long taking the velocity factor of the cable into account. Terminal tags are used to make connection with the driven element whilst allowing rotation of the wing nut bolts. The balun cables should be arranged symmetrically under the boom and taped to it, and the ends sealed with silicone rubber to prevent the ingress of moisture.

To tune the antenna, it should be raised above the ground and the U sections of the driven element adjusted symmetrically for lowest VSWR. When the optimum positions have been found, the elements and U sections should be drilled through and permanently bolted together.

## MOBILE ANTENNAS

The choice of an antenna for mobile VHF and UHF use is dependent on several factors. As the frequency increases the aperture of the antenna decreases, and propagation losses increase. This means that higher antenna gains are required for UHF than VHF to overcome the losses of both aperture and path.

Considerable reduction of beamwidth in the vertical plane is needed to achieve gain whilst retaining an omnidirectional pattern in the horizontal plane. A compromise has to be made to obtain maximum gain in the best direction that gives minimum disruption of signals when mobile.

For example an omnidirectional antenna of 6dBd gain will have a typical half-power beamwidth in the vertical plane of under 30 degrees. The narrow disc shaped beam that is produced can result in considerable variation in transmitted and received signal strength as the vehicle or antenna tilts. This is particularly the case where signals are reflected from nearby objects.

The choice of polarisation is not only dependent on compatibility with stations being received and the optimum for the propagation path. The aesthetics, mechanical complexity, safety and the mounting position of the antenna on the vehicle must be considered.

High-gain, relatively large, antennas suffer gain reductions with probable loss of omnidirectionality if the antenna is not roof mounted. The difference in mounting an antenna on the wing or boot of a car compared with mounting it on the top dead centre of the car roof can be at least 3dB. Variation of the radiation pattern can occur due to close-in reflections and surface-wave effect across the vehicle, as well as restriction of the 'line of sight'.

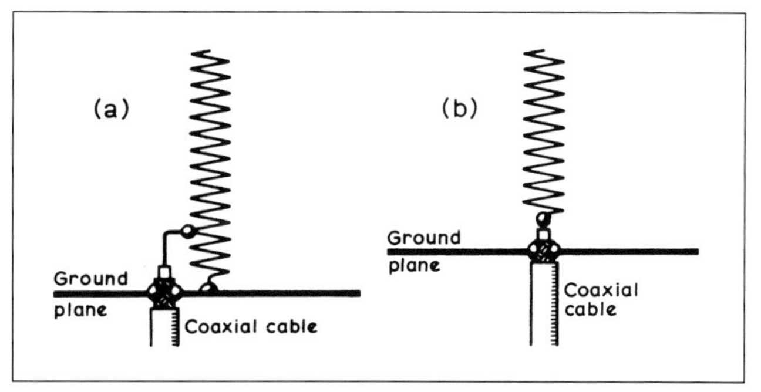


Fig 16.59: Two ways of feeding a helical antenna on a groundplane: (a) shunt feed, (b) series feed

### Normal Mode Helix on a Groundplane

The normal-mode helical (spring) antenna, when vehicle mounted, offers a gain approximately 2 to 3dB less than a dipole, but the overall height is reduced to the order of 0.1λ. An acceptable match to 50 ohms can often be obtained by simply adjusting its resonant length. Alternatively, a small inductance or capacitor across the base or an input tapping on an 'earthed' helical, as shown in **Fig 16.59**, will provide the required matching. The design and limitations of the normal mode helix were discussed earlier in this chapter under the heading of hand-held antennas.

### Quarter-wave Whip

This is the simplest and most basic mobile antenna. The image of the vertical λ/4 section is 'reflected' in the ground plane, producing an antenna that is substantially the same as a dipole, provided that the ground plane is infinitely large and made of a perfectly conducting material (**Fig 16.60**). In this case, all of the radiation associated with the lower half of the dipole is radiated by the top half, resulting in a 3dB improvement in signal strength in a given direction for the same power input to the antenna.

In practice the size of the ground plane and its resistive losses modify the pattern. The full 3dB is never realised. Measurement of a 5GHz monopole on an aluminium ground plane of 40 wavelengths diameter showed a gain of 2.63dBd. **Fig 16.61** and **Fig 16.62** show optimum patterns of a λ/4 whip measured on a ground plane of λ/2 sides and 1λ sides. Although the pattern is raised from the horizontal, on a medium sized ground plane the loss of horizontal gain is relatively small (20° and 1dB at 0° in Fig 16.48, but 40° and 6dB at 0° in Fig 16.49). However, as the groundplane size increases, the main lobe continues to rise until the situation of **Fig 16.63** pertains.

When a vertical radiator is mounted over a ground plane as described, the input impedance is typically halved. For the λ/4 whip or monopole, the input impedance is typically 36Ω - jX, that is to say approximately half the resistance of the dipole but with

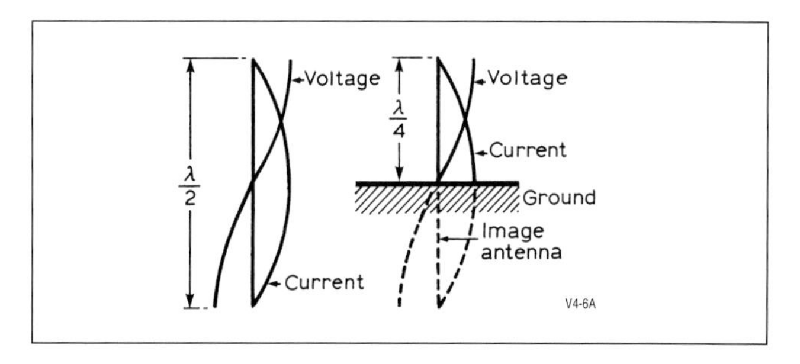


Fig 16.60: The λ/2 antenna and its grounded λ/4 counterpart. The missing λ/4 can be considered to be supplied by the image in ground of good conductivity

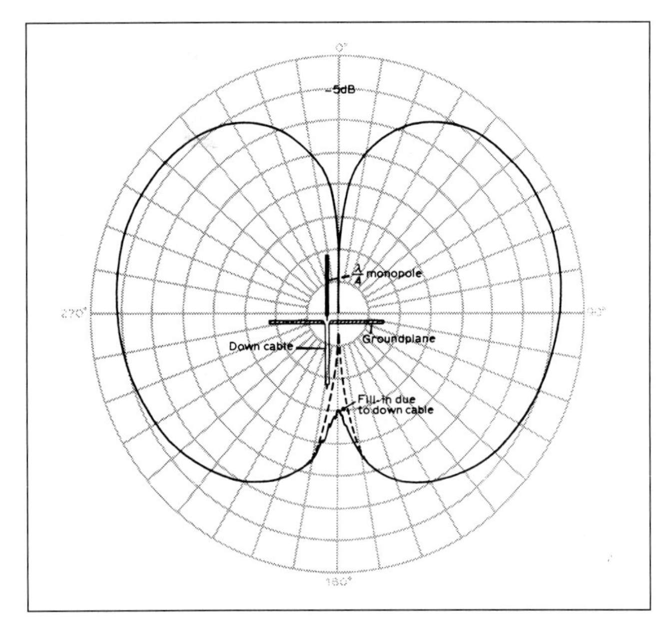


**Fig 16.61: Decibel radiation pattern of a λ/4 monopole over a λ/2 square groundplane at 145MHz**

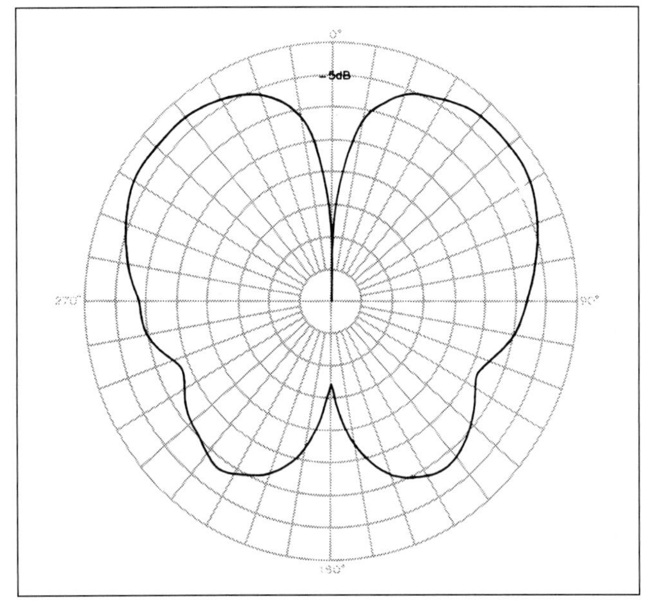


**Fig 16.62: Decibel radiation pattern of a λ/4 monopole over a 1λ square groundplane at 145MHz**

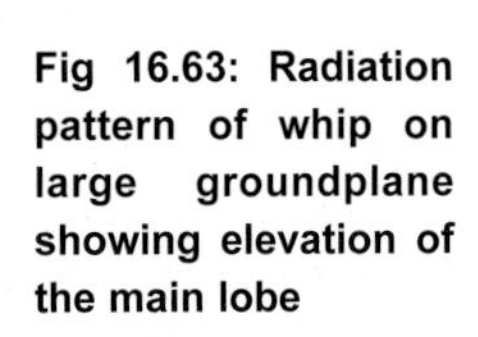

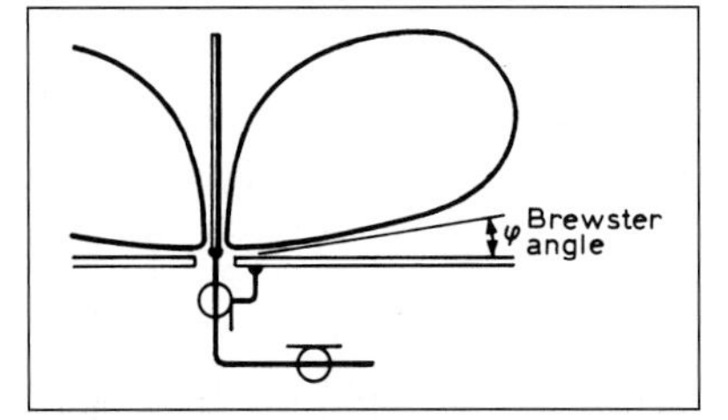


**Fig 16.63: Radiation pattern of whip on large groundplane showing elevation of the main lobe**

an additional reactive component. With 50 ohm cable impedance this would produce a standing wave ratio at the antenna base of about 1.5:1.

The simplest way to overcome this mismatch is to increase the length of the whip to produce an inductive reactance to cancel the capacitive reactance normally obtained. In practice an increase in length also raises the resistive value of the whip and a close match can usually be obtained to a 50-ohm cable.

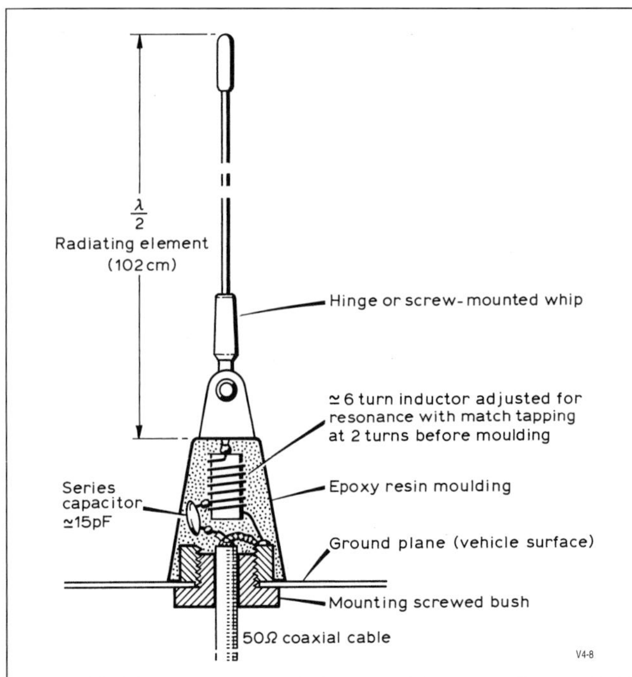


**Fig 16.64: A home-built mobile antenna and mount**

At VHF (145MHz) the λ/4 whip's simplicity and limited height (about 49cm/19in) is often an accepted compromise. At 70MHz the physical dimensions (about 102cm/40in) are such that size is the usual limit, making a 1/4λ whip preferable to a 'gain' antenna. The effective aperture of the antenna at this frequency is compatible with path loss conditions, and the ground-plane size, when roof-mounted on a vehicle, is such that the radiation angle is fairly low. However, the shape of the radiation pattern can result in a gain reduction of 3dB to each side of the vehicle.

## Half-wave and Five-eighths-wave Antennas

Ground-plane techniques described for the 1/4λ whip can be used for vertical gain antennas. If the 1/2λ dipole is extended in length, maximum forward gain (before the pattern divides into several lobes) is obtained when the dipole is about 1.2λ. This corresponds to a maximum length of 5/8λ for a ground-plane antenna.

A natural extension to the 1/4λ whip is the 1/2λ whip. However, such a radiator fed against a ground plane has a high input impedance. On the other hand, a 3/4λ radiator fed against a ground plane has a resistive input close to 50 ohms. Unfortunately, the resultant radiation pattern in the elevation plane is less than optimum.

If the 1/2λ whip could be made to look like a 3/4λ radiator then it would be possible to obtain a 50-ohm resistive input. A series coil at the ground-plane end of a 1/2λ radiator can be used to resonate at 3/4λ, but the input is still of fairly high impedance and reactive. If, however, the coil is shorted to the ground plane, tapping up the coil will provide the required match/input point. The addition of a capacitor in series with the input will compensate for the remaining reactive component. **Fig 16.64** shows details of such an antenna.

As the aperture of the antenna has been doubled compared with the 1/4λ whip, the gain over the whip approaches 3dB. Achievement of this figure requires minimum losses in the radiating element, ie it must be copper-plated or made from a good conducting material.

The maximum radiator size of 5/8λ for a single-lobe pattern can also make use of the impedance characteristics of the

3/4λ radiator. Construction is simpler than for a 1/2λ antenna. If the radiating element is made 5/8λ long, and a series coil is placed at the groundplane end, an input impedance very close to 50 ohms can be obtained. With correct materials a gain close to 4dBd can be achieved from the further increase in effective aperture. The radiation pattern is raised more than that of a 1/2λ antenna, so the improved gain of the 5/8λ may not always be realised. However, the simplicity of construction is an advantage.

**Fig 16.65** gives details of the series 5/8λ whip. One other advantage of this antenna is that over a wide range of mounting and ground-plane conditions it will self-compensate for impedance and resonance changes. It is preferable for both the 1/2λ and 5/8λ antennas to be 'hinged', particularly if roof-mounted, to enable folding or 'knock down' by obstructions, eg trees and garages.

Various gain figures have been reported for the 5/8λ whip antenna. Unfortunately not all antennas use optimum materials. Resistive steel wires or rods produce heating loss, and the use of a glass fibre-covered wire changes the resonant length by as much as 20%. The radiator therefore has to be cut shorter than 5/8λ, with an accompanying loss of aperture.

The construction of the series coil is important. Movement of the coil turns will change the antenna's resonance, giving apparent flutter. Some transceivers with VSWR-activated transmitter close-down will be affected by change of resonance of the antenna. This can make the power output of the transmitter continually turn down or be switched off, producing what appears as extremely severe 'flutter' on the transmission.

Several of the '5/8λ ground-plane antennas' discussed in various articles are in fact not truly antennas of this nature.

One of these devices worth considering for its own merits is that shown in **Fig 16.66**. It consists of a 5/8λ vertical element with a reactive sleeve of 0.2λ at the ground-plane end. The gain obtained from this antenna is typically 1.8dBd. As can be seen, the actual radiating element A-A is shorter than a 1/2λ antenna.

Another antenna family, with similar properties but different in construction, includes the 'J' and Slim Jim. These are described later in this chapter.

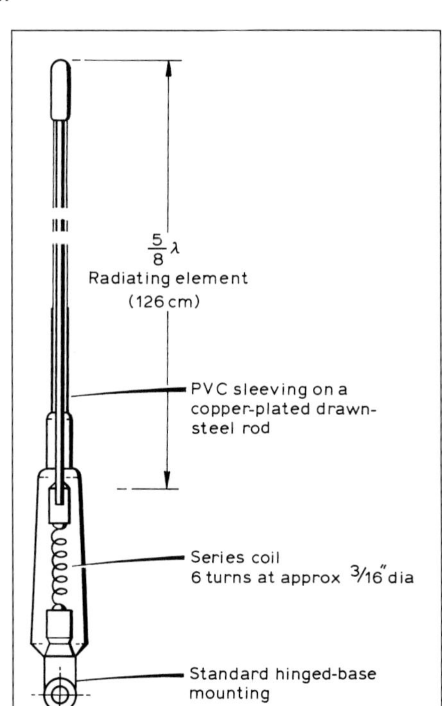


**Fig 16.65: Typical commercial 5/8λ mobile antenna and mount**

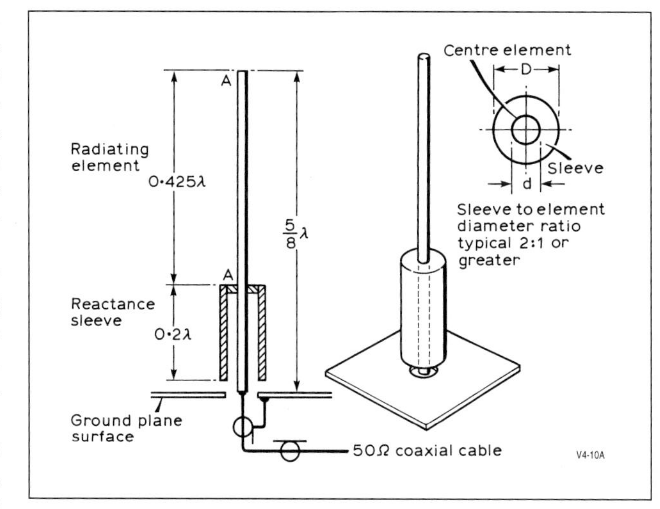


**Fig 16.66: The reactance-sleeve 5/8λ monopole (*Ham Radio*)**

## Seven-eighths-wave Whip

This mobile antenna is derived from the Franklin collinear shown later in this chapter. It consists of two 1/2λ elements coupled by a series 'phasing' capacitor. One effect of the capacitor is to resonate the combined elements at a lower frequency than that of a single 1/2λ element. However, reducing the length of the top element tunes the arrangement back to the original frequency.

The base impedance above a perfect ground plane is 300-400 ohms with some capacitive reactance. A series loading coil in combination with an L-matching section gives a good match to 50-ohm coaxial feeder. The match is maintained with quite modest groundplane size (1/4λ radials or 1/2λ diameter metal surface). This makes the 7/8λ whip suitable for vehicle mounting or for use as a base-station antenna.

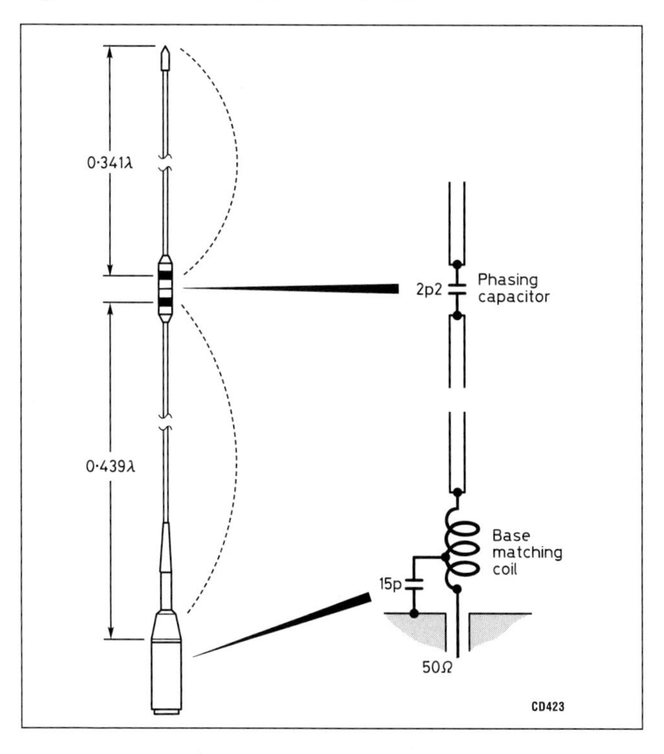


**Fig 16.67: The 7/8λ whip antenna. This is effectively two shortened half-wave elements in series with a series phasing capacitor between them. The assembly shown is that of a commercial form of the antenna. The dotted lines show the approximate current distribution**

The final length of the two radiator elements is somewhat dependent on their diameters and the design of the series capacitor and matching unit. **Fig 16.67** shows the general appearance and dimensions of a commercial version of the 7λ/8 whip, together with typical circuit components and current distribution in each element.

The theoretical gain of this antenna is 4.95dBi (2.8dBd) over a perfect ground plane. The professionally measured gain, with the whip on a 1m ground plane, was slightly over 4.7dBi for the full 144MHz band. The radiation pattern in the E (vertical) plane was predominantly a single lobe (torus/doughnut) peaking at 4° above the horizon and with a 3dB beamwidth of 38.5°.

## OMNIDIRECTIONAL BASE STATION ANTENNAS

### 144MHz Horizontally Polarised Omni-V

This antenna consists of a pair of λ/2 dipoles. The centres of the dipoles are physically displaced to produce quadrature radiation with the ends of each dipole supported on a λ/4 shorted stub. A pair of Q-bars are tapped down the stubs to a point where the impedance is 600Ω as shown in **Fig 16.68.** When the two units are fed in parallel, they produce an impedance of 300Ω at the centre. A 4:1 balance-to-unbalance coaxial transformer is fitted to the centre point of the Q-bars to enable a 75Ω coaxial feeder cable to be used. A 50Ω feed can be arranged by repositioning the Q bars on the antenna stubs to provide a tap at 400Ω on the stubs.

This can be achieved by monitoring the VSWR on the coaxial feeder whilst adjusting the Q bar position by small but equal amounts on both stubs. The balun should, of course, be constructed from 50Ω coaxial cable (transforming from 200Ω in the balanced section) for a match to 50Ω. The general arrangement is illustrated in **Fig 16.69** showing how the antenna may be arranged to give either an omnidirectional or bi-directional radiation pattern, and typical radiation patterns for either case are shown in **Fig 16.70**. **Fig 16.71** shows the gain and VSWR of these antennas as a function of the centre frequency.

### Quarter-wave Groundplane Antenna

This is one of the simplest omnidirectional antennas to construct and usually yields good results. However, some unexpected effects may occur when the antenna is mounted on a conductive mast, or if radio frequency current is allowed to flow on the outside of the feeder.

In its simplest form, the groundplane antenna comprises a quarter-wavelength extension to the inner of a coaxial cable,

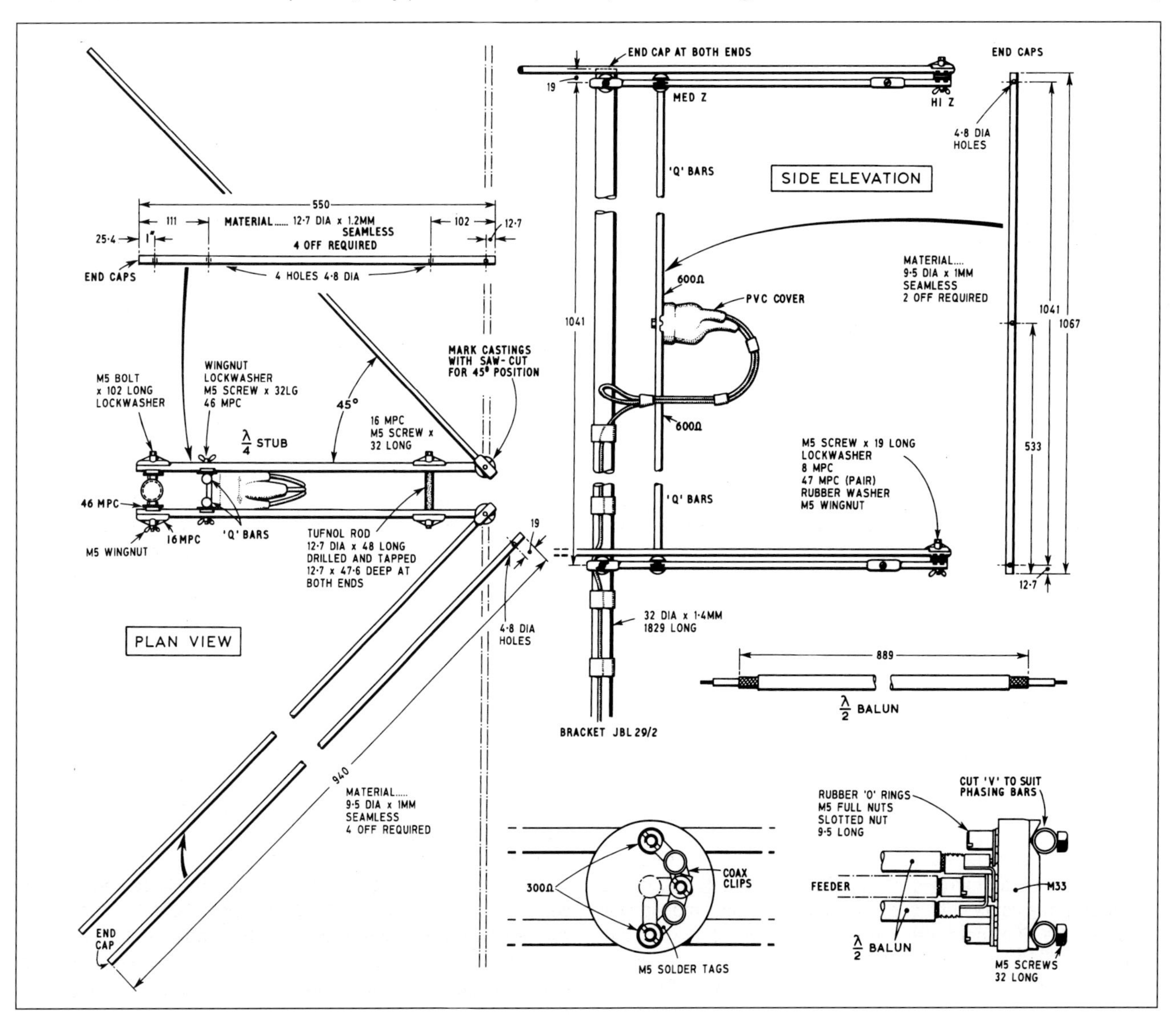


**Fig 16.68: The mechanical construction of the Omni-V (dimensions are in millimetres)**

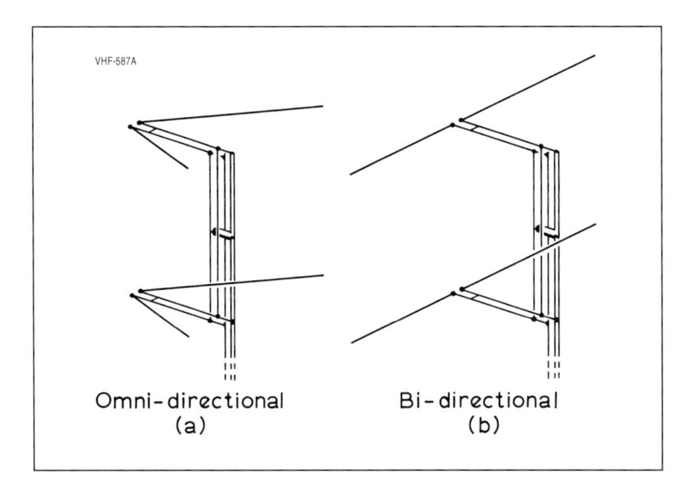


**Fig 16.69: Formation of the Omni-V antenna**

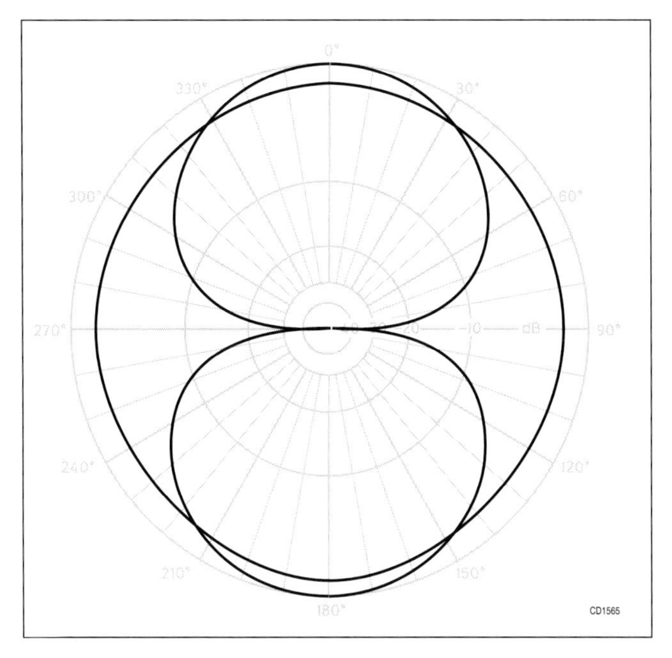


**Fig 16.70: The horizontal radiation patterns for an Omni-V antenna and its bi-directional counterpart**

with several wires extending radially away from the end of the outer of the coaxial cable, **Fig 16.72(a)**. The input resistance will be quite low, in the order of 20 ohms, although this may be transformed to a higher impedance by using a folded monopole radiator as shown in **Fig 16.72(b)**.

Equal diameter elements provide a 4:1 step-up ratio to around 80 ohms, and a smaller diameter grounded leg can reduce the input impedance to 50 ohms.

The feedpoint impedance can be modified by bending the groundplane rods downwards from the horizontal, **Fig 16.72(c)**. If the radiating element and the groundplane rods are all $\lambda/4$ in length, the input resistance is approximately:

$$R = 18(1+\sin\theta)^2 \quad \text{ohms}$$

where $\theta$ is the groundplane rod angle below the horizontal, in degrees. A 50 ohm resistance is achieved when $\theta$ is 42 degrees.

The ends of the groundplane rods are sometimes joined together with a conductive ring to provide additional mechanical stability. The ring increases the electrical size of the groundplane, and the length of the radials can be reduced by about 5%.

The few rods forming the groundplane usually do not prevent current flowing on any conductive supporting mast, or on the outside of a coaxial feeder. The mast or feeder can become a

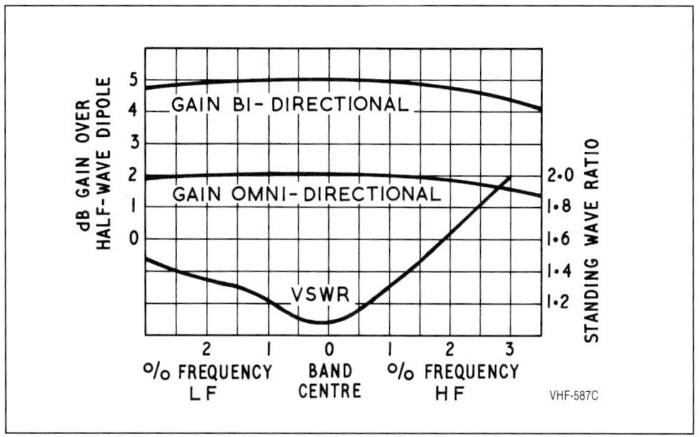


**Fig 16.71: Chart showing gain and VSWR as a function of frequency for the Omni-V and the bi-directional antenna**

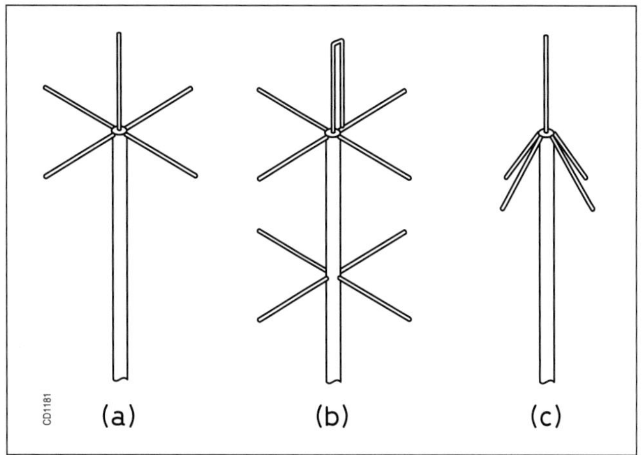


**Fig 16.72: Quarter wave groundplane antennas**

long radiating element which may enhance or destroy the radiation pattern of the antenna, dependent upon the magnitude and phase of the mast currents relative to that on the antenna. An example of this is shown in **Fig 16.73(a)**, where the monopole and groundplane is mounted on a $5\lambda$ mast (about 10 metres). The corresponding radiation patterns without mast or cable influences are shown in **Fig 16.73(b)**. The effects of ground reflections have been ignored in both cases.

Some antenna designs make use of these currents to enhance the gain of the monopole, and sometimes have a second set of groundplane rods further down the mast, tuned to present a high impedance to reduce currents flowing below that point. The mast currents can be reduced a little by using more radials in the groundplane or extending their length to around $0.3\lambda$.

An open circuited choke sleeve can be more effective than radial wires for mast current control. This technique is used in the skirted antenna described later in this chapter.

## Dual-band 145/435MHz Whip

This base station antenna, devised by Bert Veuskens, PA0HMV acts as an end fed antenna on 145MHz with a gain of 0dBd, and two stacked $5\lambda/8$ radiators with a gain of about 5dBd at 435MHz [31]. The radiator is made from 2mm copper wire, formed as shown in **Fig 16.74**. The coil at the base and a series capacitor provides the matching network at 145MHz. The folded stub section at the centre ensures the upper and lower $5/8\lambda$ sections radiate in phase at 435MHz. A groundplane comprising four short radials resonant at 435MHz is fixed to the input connector, see **Fig 16.75**. No groundplane is required for operation at 145MHz, although the coaxial feed can be coiled close to the

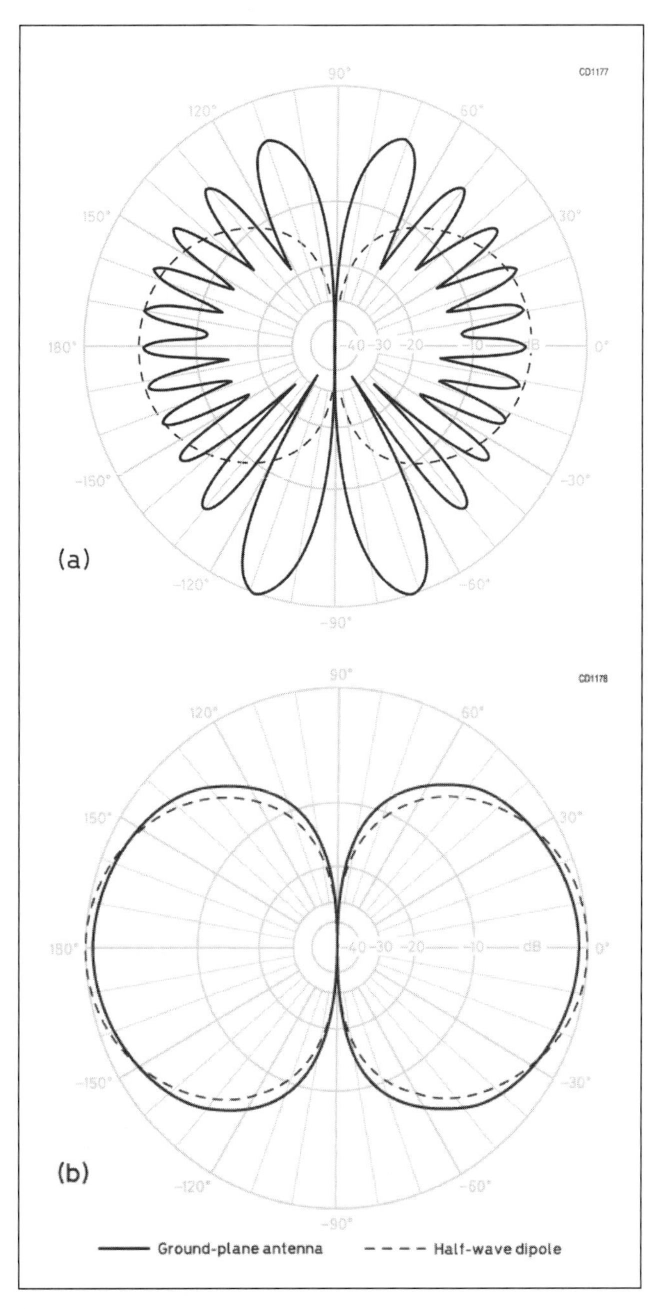


Fig 16.73: Radiation patterns of quarter-wave groundplane antenna, (a) with and (b) without a metallic mast

(right) Fig 16.74: Shape and dimensions of the radiator wire

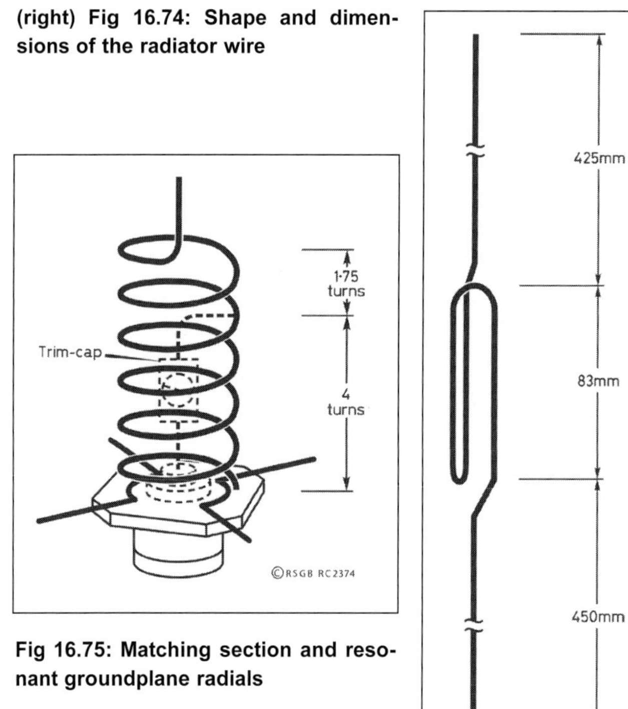


Fig 16.75: Matching section and resonant groundplane radials

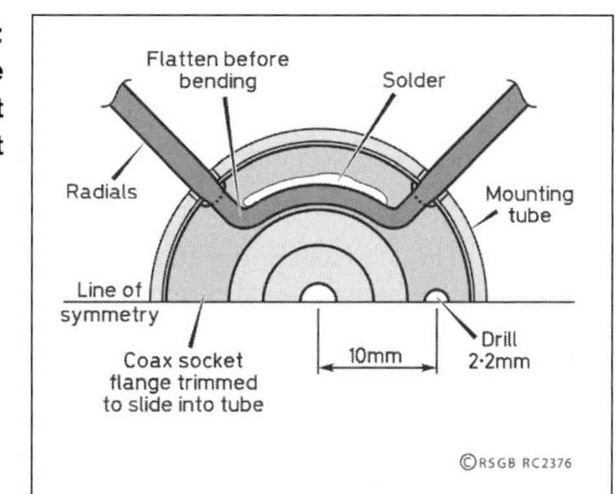


Fig 16.76: Details of the coaxial socket and attachment of radials

| Table 16.15: Materials required for the 145/435MHz omni antenna |
|---|
| Copper or brass mounting tube, 28mm OD x 220mm |
| PVC tube, 32mm OD, 28mm ID, x 1.2m |
| Jubilee clip, 32mm, preferably stainless steel |
| N-type 50Ω socket, PTFE dielectric, with square flange |
| Brass rod or tubing, 3mm OD x 720mm for radials |
| Bare or enamelled copper wire, 2mm (14SWG) 1.6m |
| Copper/brass tube or inner from connector block to butt-joint radiating elements (discard screws) |
| Tubular trimmer capacitor, 10pF or RG58 cable per Fig 16.74. Ensure that the capacitor can handle the transmitter power safely. |
| Silicone rubber sealing compound or bathroom sealant |

connector to choke off any current on the outside of the cable. The whole assembly is fitted inside a section of PVC tube for support and protection from the weather.

## Assembly

A list of materials is given in **Table 16.15**. The corners of the coaxial socket are filed off to make it a snug fit in the 28mm copper pipe, and a hole is required for the earthy end of the matching coil as shown in **Fig 16.76**. The radials should be cut slightly too long to allow for trimming after assembly. A jig to hold the radials and connector in place during soldering is essential, and can be made from a piece of chipboard. After soldering, each radial should be trimmed to 173mm measured from the centre of the socket.

The rotor terminal of the trimmer capacitor, or the centre conductor of the RG58 coaxial cable capacitor (**Fig 16.77**) should be soldered to the centre pin of the connector next. Prepare the matching coil by straightening and stretching 60cm of the

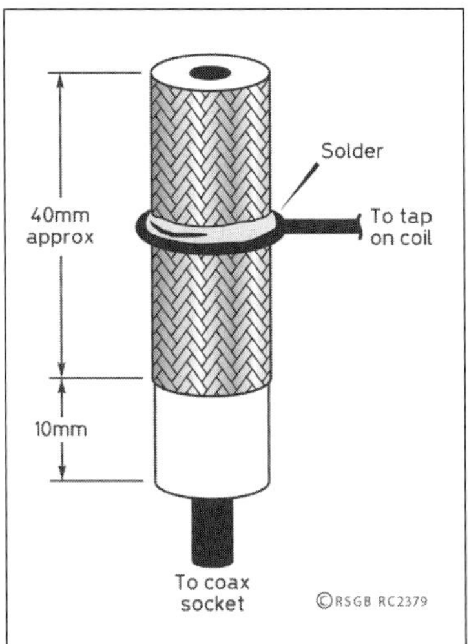


**(left) Fig 16.77: RG58 coaxial cable capacitor. Note that the length may need to be adjusted for best match**

**(below) Fig 16.78: Phasing stub bending detail**

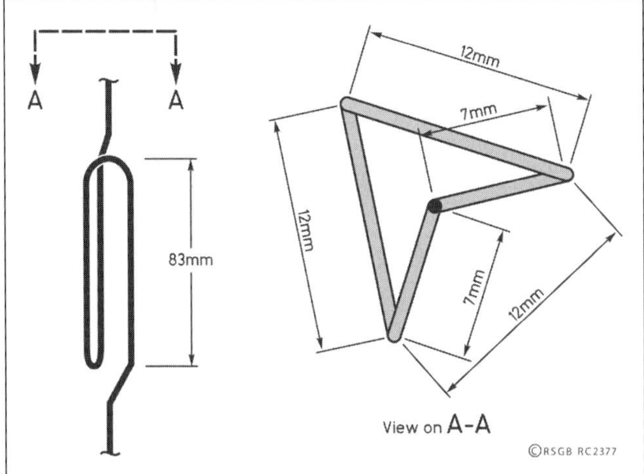


**(right) Fig 16.79: Mounting tube and PVC cover details showing slots to clear radials**

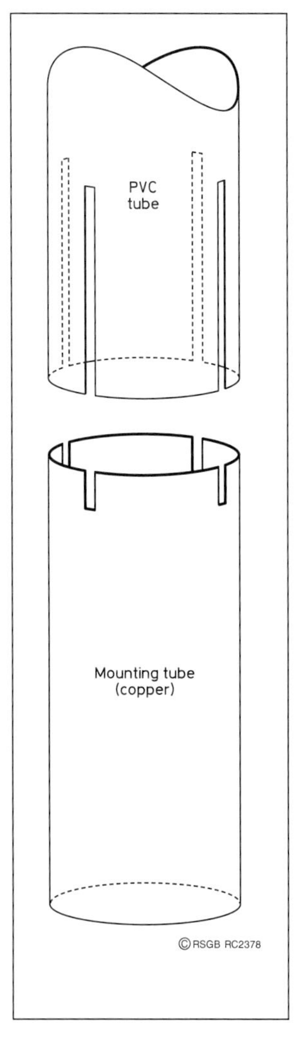

**Fig 16.80: The finished antenna installed on a mast**

antenna wire. If enamelled wire is used, scrape off the enamel at one end and at the capacitor tapping point before winding up and shaping on a 19mm former of tube, rod or dowel as shown in Fig 16.62. Solder the end of the coil into the hole drilled in the flange of the coaxial socket and connect the free end of the matching capacitor to the coil, four turns above the soldered end. Approximately 200mm of the lower radiator should be projecting upwards, coaxially with the coil.

With the remaining wire, shape the phasing section as in **Fig 16.78** using a 9.5mm drill bit as a former. Trim the lower wire end so that it makes up the 450mm length shown in Fig 16.61 with the wire on the coil when the ends are butted together. Prepare a polystyrene foam disc to centre the wire within the PVC support tube and slide onto the radiator wire below the phasing stub.

Butt-splice the two wires together by soldering them into a short piece of copper tubing or the inner part of a small cable connector with the screws removed.

Cut the top wire to 460mm. Fit a second foam centering disk on to the top wire. Four slots, 90 degrees apart, 7mm deep and 4mm wide are cut at one end of the copper pipe to clear the groundplane radials. This pipe will be used to clamp the finished antenna to the top of its mast with U-bolts and saddles.

### Tuning

Pass a short 50-ohm cable through the copper pipe and connect it to the N socket. Push the pipe over the socket until the groundplane radials are resting in the bottom of the four slots in the tube.

Set up the antenna without its cover tube well clear of objects that could de-tune it, but low enough for the top to be accessible. Trim the top element for minimum VSWR in the 70cm band at a frequency 3MHz higher than required, eg 438MHz for 435MHz operation. The PVC cover tube will reduce the frequency by this amount when it is installed.

Fix the two centering discs in place with a drop of epoxy glue, at 170mm below the end of the top wire and halfway between the top of the coil and the bottom of the phasing stub for the lower wire.

Slide the PVC cover tube (**Fig 16.79**) over the antenna until the radials are fully seated in the slots, check that the minimum VSWR is below 1.5:1 and at the desired frequency in the 70cm band. Raise the PVC tube just enough to allow access to the trimmer/cable capacitor and adjust for minimum VSWR at 145MHz. Push the PVC tube down again and secure with the Jubilee clip below the radials. Cap and seal the top and weatherproof the slots with sealant to keep rain out, whilst leaving a hole to prevent condensation from being trapped. The finished antenna is shown in **Fig 16.80**.

## The Skirted Dipole Antenna

The skirted dipole antenna (**Fig 16.81**) does not require groundplane radials, and can be mounted in a cylindrical radome for better appearance and lower wind induced noise. The skirt forms the lower part of a half wave dipole, and being one quarter wavelength long, presents a high impedance at its lower end, reducing unwanted currents on the mast. The current is further reduced by a second choke, with its open, high impedance end

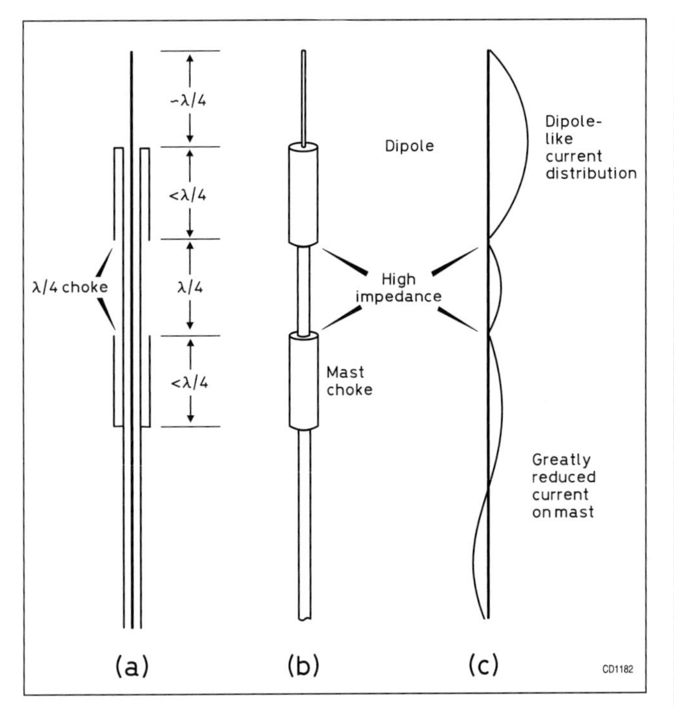


**Figure 15.81: Skirted dipole antenna with mast choke**

placed one quarter wavelength below the dipole skirt for best effect. The radiation pattern of this antenna closely resembles that of a half-wave dipole in free space.

## Gain Sleeve Dipole

The *gain sleeve dipole* (**Fig 16.82**) is derived from the 1.8dBd shunt-fed 5/8λ antenna described in the section on mobile antennas.

The radiating element B-B is a 1 element fed part way along its length with a 0.2λ series short circuited stub to provide the transformation to 50 ohms and phasing. The impedance of asymmetrical antennas was investigated by R W P King [32], and has a number of applications in the design of groundplane antennas with elevated feedpoints. Having approximately twice the aperture of the λ/2 dipole, a gain of typically 2.5-3dBd is achieved. Mechanical construction is open to interpretation but 'beer can' or plastic water pipe formats offer two solutions. Note that the mounting point should be at A-A and not on the 0.25λ sleeve.

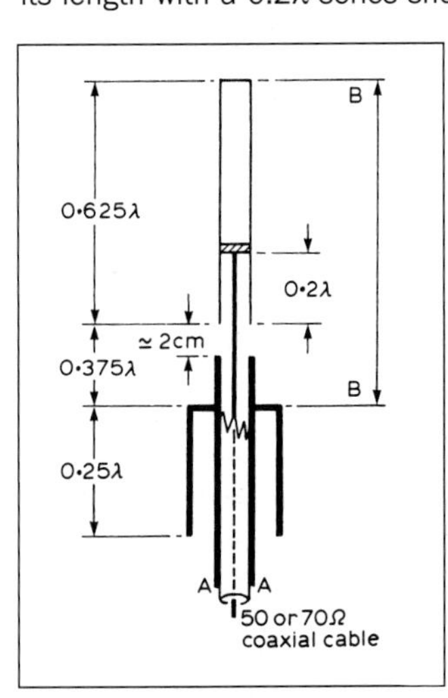


**Fig 16.82: Gain sleeve dipole**

## The Discone

The discone is often used where a single omnidirectional antenna covering several VHF/UHF bands is required. A single antenna is capable of covering the 70, 144 and 432MHz bands or 144, 432 and 1296MHz. However, as the antenna can operate over roughly a 10:1 frequency range, it will more readily radiate harmonics present in the transmitter output. It is therefore important to use a suitable filter to provide adequate attenuation. The radiation angle tends to rise after the first

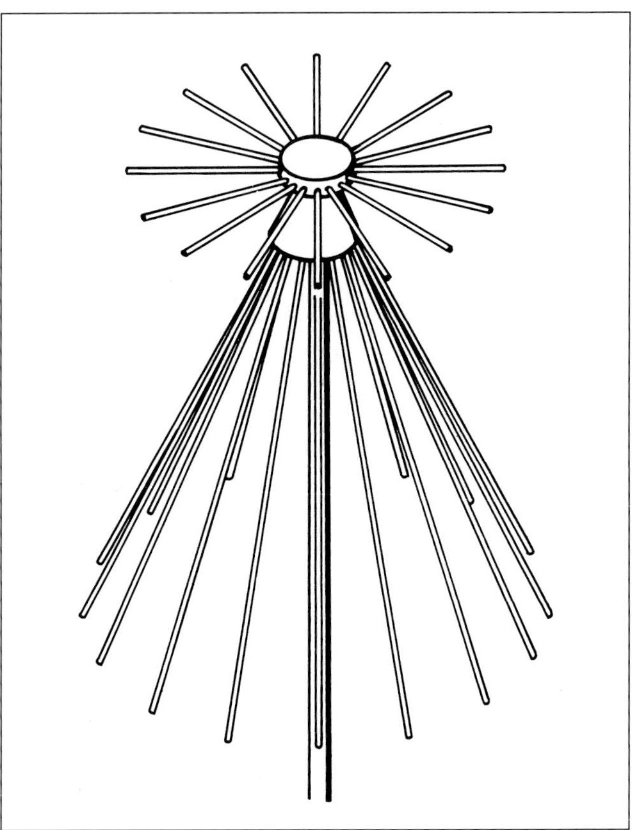

**Fig 16.83: General arrangement of skeleton form of discone**

octave of frequency and this is the normal acceptable working range. If correctly constructed, a VSWR of less than 2:1 can be obtained over the octave range. One characteristic of the basic discone is a very sharp deterioration of the VSWR at the lowest frequency of operation.

The discone consists of a disc mounted above a cone, and ideally should be constructed from sheet material. However, with only a little loss of performance the components may be made of rods or tubes as illustrated in **Fig 16.83**. At least eight or preferably 16 rods are required for the 'disk' and 'cone' for reasonable results. Open mesh may be used as an alternative to sheet metal or rods.

The important dimensions are the end diameter of the cone and the spacing of this from the centre of the disc. These are instrumental in obtaining the best termination impedance, ie 50 ohms [33].

**Fig 16.84** shows the key dimensions, which must satisfy the following requirements:

*A* The length of the cone elements are λ/4 at the *lowest* operating frequency (2952/*f* MHz inches).

*B* The overall disc diameter should be 70% of λ/4.

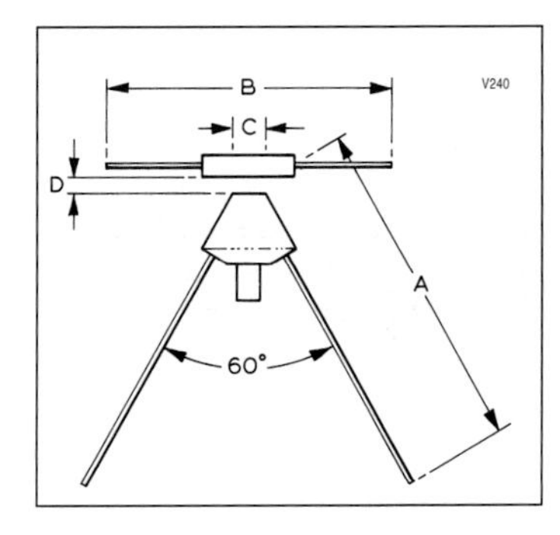


**Fig 16.84: Key dimensions of discone antenna**

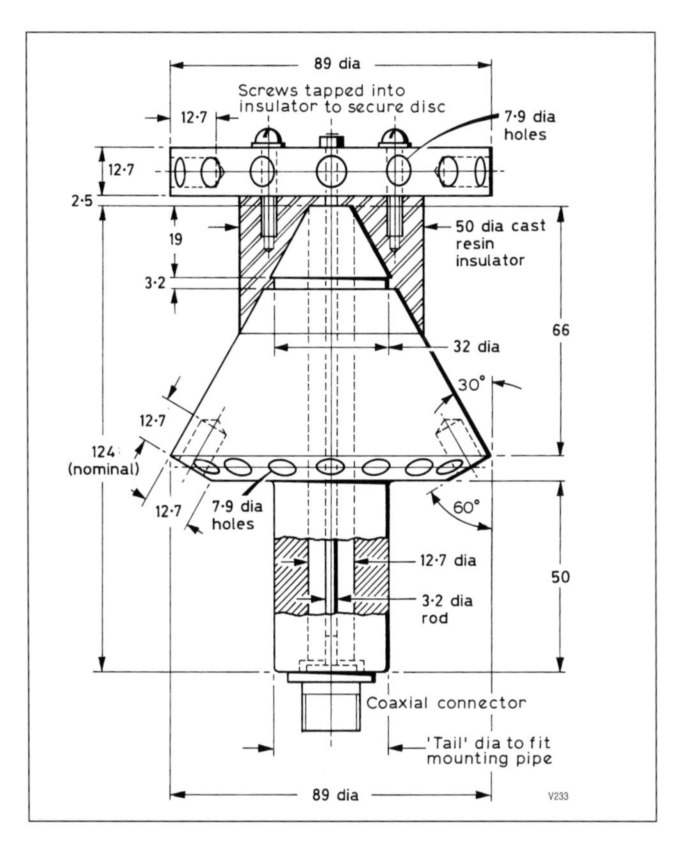


Fig 16.85: Details of discone hub assembly

*C* The diameter of the top of the cone is mainly decided by the diameter of the coaxial feeder cable. For most purposes 0.5in will be suitable.

*D* Spacing between top disc and the cone top is 20% of *C* or 0.1in for 50 ohms.

The detail given in **Fig 16.85** of the 'hub' construction will be suitable for any design using a 50-ohm feeder cable and may be taken as an example. A suitable insulator can be made with a potting resin or turned from nylon, PTFE or other stable low loss material.

The low frequency coverage can be extended by fitting a whip antenna to the centre of the disk. The antenna then operates like a quarter wave whip on a skeleton groundplane (the cone providing the groundplane). The VSWR can be optimised at a particular frequency below discone cut-off by adjusting the length of the whip, or several whips can be fitted, resonating at different frequencies, in the style of multi-band dipoles. This can be useful if the antenna is used for other purposes in addition to amateur band coverage. **Fig 16.86** shows the calculated elevation radiation patterns for a conventional discone at 145, 435 and 1296MHz, together with those of a discone of the same dimensions fitted with a 850mm whip to provide 70MHz coverage.

## Collinear Dipole Arrays

Communication with mobile stations is best achieved with a vertically polarised omnidirectional antenna, as there is no need to point the antenna in the direction of the mobile. However, a fixed station is not as constrained by mechanical considerations as a mobile, and can thus be fitted with longer, higher gain antennas.

This can be achieved by stacking dipoles vertically above one another in a collinear array, and feeding them with cables of equal lengths, as shown for the GB2ER repeater antenna later in this chapter. Another method of achieving gain with simpler feed arrangements is discussed below.

The current on a length of wire several wavelengths long will be distributed as shown in **Fig 16.87(a)**. The wire shown is 2λ long. Radiation at right angles to the wire will be poor, as the successive half wavelength current maxima are in opposite phases, and if the currents were equal, there would be perfect cancellation of the radiation from the oppositely phased pairs of current maxima. However, if all the current maxima were in phase, the radiated fields would add, and a high gain could be achieved. **Fig 16.87(b)**.

There are several ways of achieving this phase reversal. The simplest is to insert an anti-resonant network or a non-radiating half-wavelength of transmission line as a phasing section between the half-wave radiating elements, **Fig 16.87(c)**. The half wavelength transmission line can be realised as a quarter wavelength of ribbon cable, which can be wound around the insulator between the radiating elements (see the section on mobile antennas).

A more subtle approach uses radiating elements that are a little longer or shorter than one half wavelength. This helps the feeding arrangements, as end feeding a half wave dipole is difficult because of its very high impedance. The self reactance of the longer or shorter dipole is then used in the design of the phasing network between the elements to achieve the desired overall phase shift. The non-radiating transmission line can then often be replaced by a capacitor or an inductor in series with the residual element reactance, **Figs 15.88(a) and (b)**.

Again, a transmission line stub can be used to synthesise the required reactance, which may be more convenient or cheaper than a lumped component if significant RF power handling is required, **Fig 16.88(c)**. Sometimes a parallel tuned circuit is realised as an inductor resonated by the self-capacitance of the insulator separating the radiating elements, and upon which it is wound.

A technique devised by Franklin that has been attractive to VHF antenna manufacturers folds parts of the radiating element to provide the phasing section as shown in **Fig 16.89(a)**. Provided that the folded sections are significantly shorter than the radiating elements, the gain is not significantly degraded, although the whole structure is sensitive to capacitive loading by any housing and insulators required.

The radiation pattern is frequency sensitive, and the main lobe will squint upwards or downwards as the frequency changes from the nominal. Whilst these folded element designs look attractive for home construction, adjustments to optimise both the radiation pattern and input impedance are very difficult without proper measuring facilities. Poor gain and broken radiation patterns result if the sections are not properly excited and phased.

All these designs are end-fed, which have practical disadvantages for longer, multi-element arrays. If identical sections are used, the end elements carry less current than those close to the feed, reducing the overall efficiency of the antenna. Whilst different length radiators and phasing elements can be used to equalise the current distribution, the design and adjustment is lengthy, and definitely requires good radiation measurement facilities. If the array can be centre fed, any residual phasing errors tend to cancel out, and for a given length, the performance tends to be better because of a more uniform current distribution.

**Fig 16.89(b)** shows one means of achieving centre feeding with a Franklin array. Note the use of the quarter wave choke section at the base of the array, which is essential to prevent

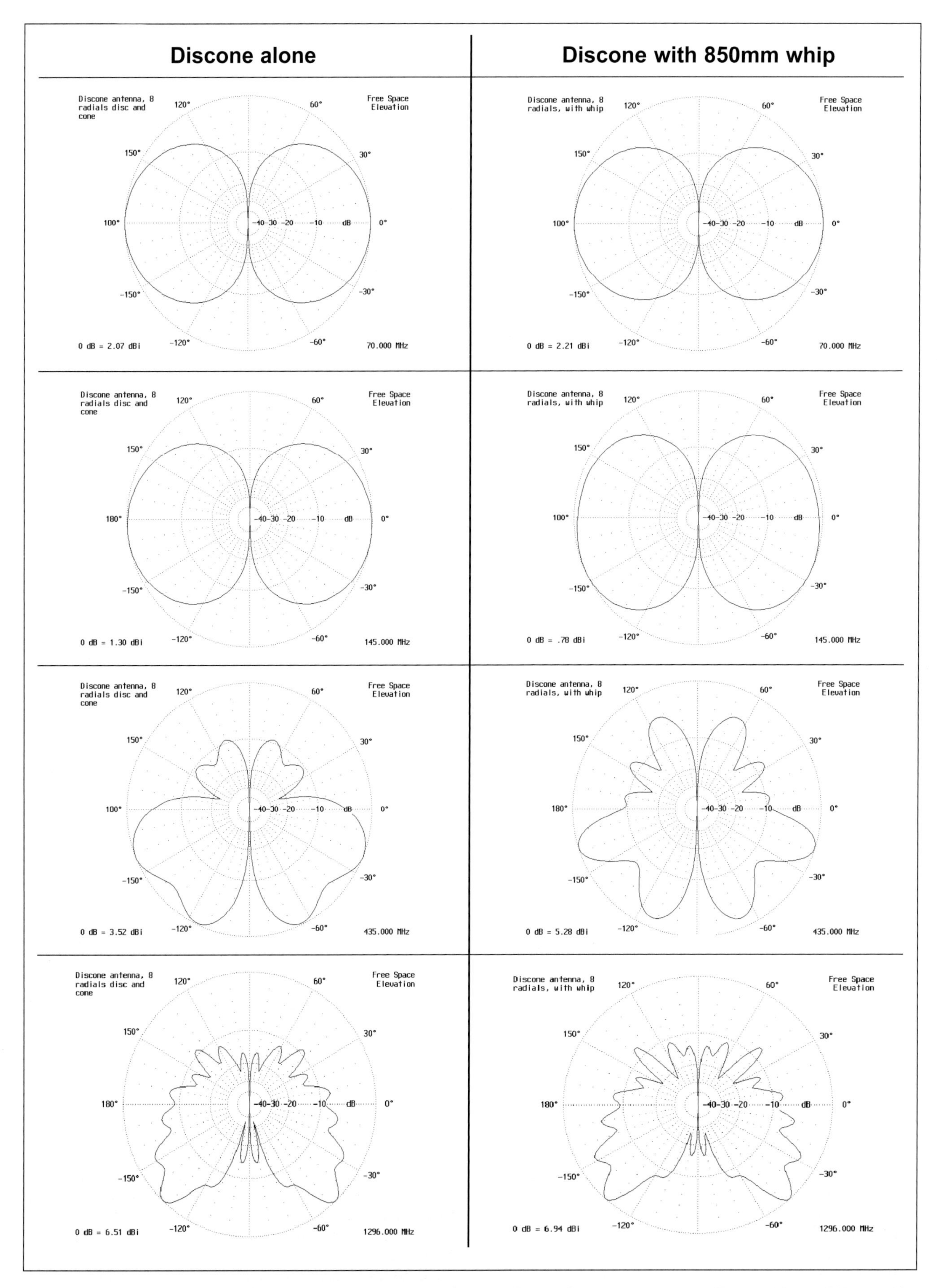


**Fig 16.86: Discone elevation radiation patterns, with and without whip for low-frequency operation. Top radials are 305mm, skirt radials 915mm and whip 850mm in length. The whip length was not optimised for 70MHz**

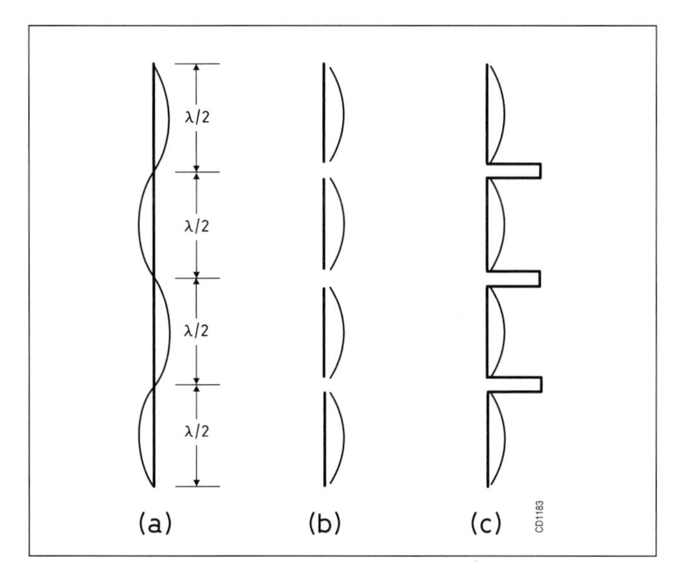


Fig 16.87: Current distribution on a wire, and derivation of the collinear antenna

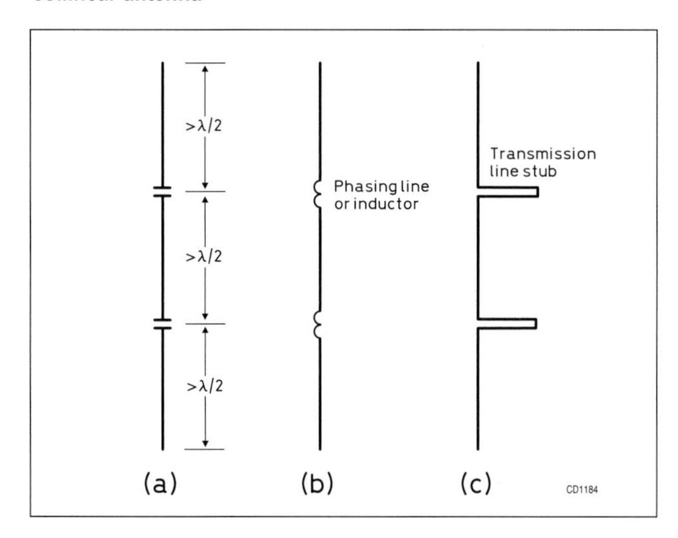


Fig 16.88: Realisation of collinear antennas

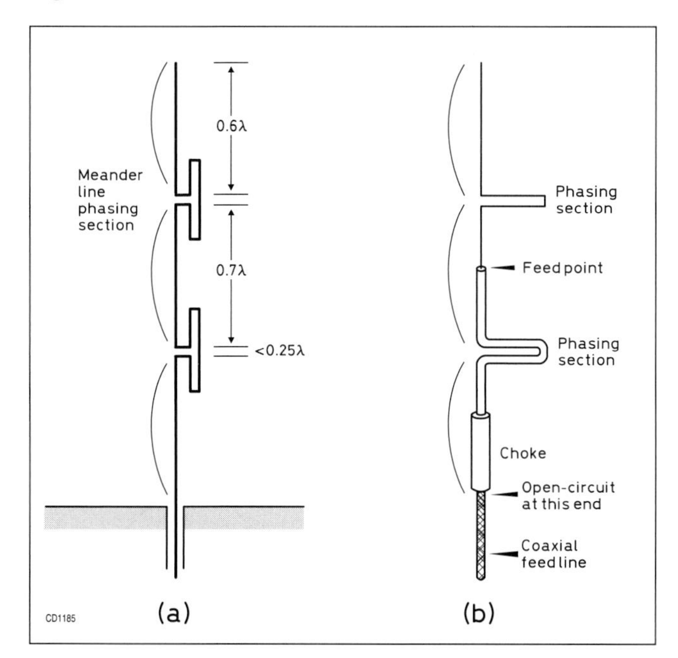


Fig 16.89: Franklin collinear antennas, end-fed and centre fed

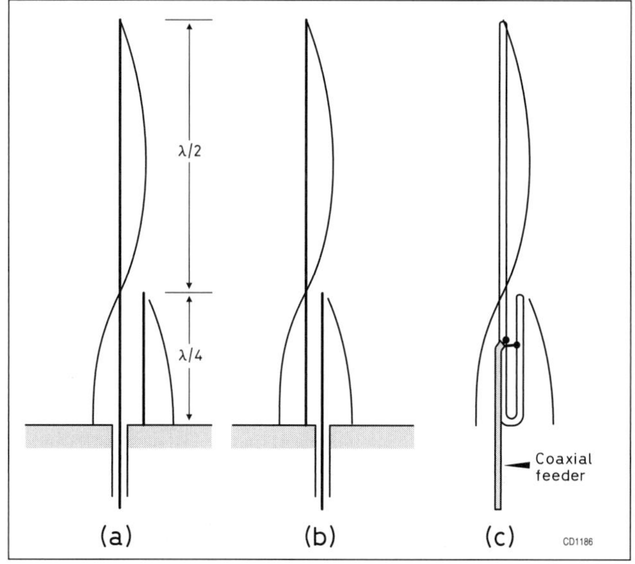


Fig 16.90: The J antenna

current flowing down the outer of the coaxial cable and destroying the performance of the antenna. The practical gain limit of the singly-fed collinear antenna is around 10dBi.

Practical collinear antennas in radio amateur use tend to use variations on Fig 16.88. The radiating elements may comprise combinations of lengths up to 5/8λ, with or without ground planes. The presence of a good groundplane increases the gain as the image or reflection effectively doubles the length of the array (see also the section on mobile antennas). However, good results can be achieved with collinear antennas directly mounted on pipe masts, especially if care is taken to minimise unwanted currents flowing on the mast.

## The J Antenna

Collinear antennas, by virtue of their operation as end-fed structures, have high feed point impedances. A good feed arrangement, valid for both ground-plane and mast mounted antennas, uses a quarter-wave short-circuited transmission line, as described in the chapter on antenna fundamentals.

**Fig 16.90** shows such an arrangement to end feed a half wave dipole mounted over a groundplane. The matching section should not radiate, and the overall effect is that of a half wave radiator raised λ/4 above the groundplane. Either leg of the quarter wave section can be fed, leading to the structure in **Fig 16.90(b),** which is identical to **Fig 16.90(a)** in terms of current distribution, and hence radiation performance. The evolution can be taken a stage further by removing the groundplane and feeding either leg of the quarter wave section as in **Fig 16.90(c)**; this is the 'J' or 'J-pole' antenna, which may use different diameters of tubing for the radiator and stub.

The 'Slim Jim' antenna provides an elegant solution for a simple, mechanically robust antenna made from a single piece of tubing as shown in **Fig 16.91**.

This antenna, described in [34] comprises a folded, open circuit 1/2λ radiator above a 1/4λ transformer section, and is a derivative of the 'J' antenna. The folded stub characteristics of the radiator provide some control over the reactive element of the input impedance.

The two ends of the tube can be joined by an insulator, eg a piece of stiff plastic tubing, to provide weather proofing and enhanced mechanical rigidity. Either balanced or unbalanced feeds can be used, tapped on to the 1/4λ transformer section at the point that provides the best match to the feeder. Coaxial feeders should be strapped or bonded to the quarter wave section to reduce unwanted currents on the outer of the cable.

The antenna has a maximum gain of around 2.8dBi (0.6dBd) in free space, although the main lobe is tilted up about 10 degrees. The main lobe can be brought to the horizontal by reducing the length of the upper section to about 0.4 wavelengths. This reduces the peak gain to around 2.5dBi (0.3dBd), and can make the feed impedance capacitive.

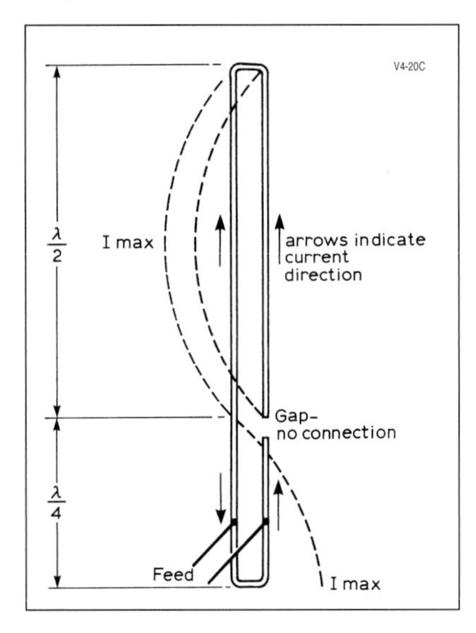


**Fig 16.91: The basic Slim Jim, showing direction of current flow and phase reversal (*Practical Wireless*)**

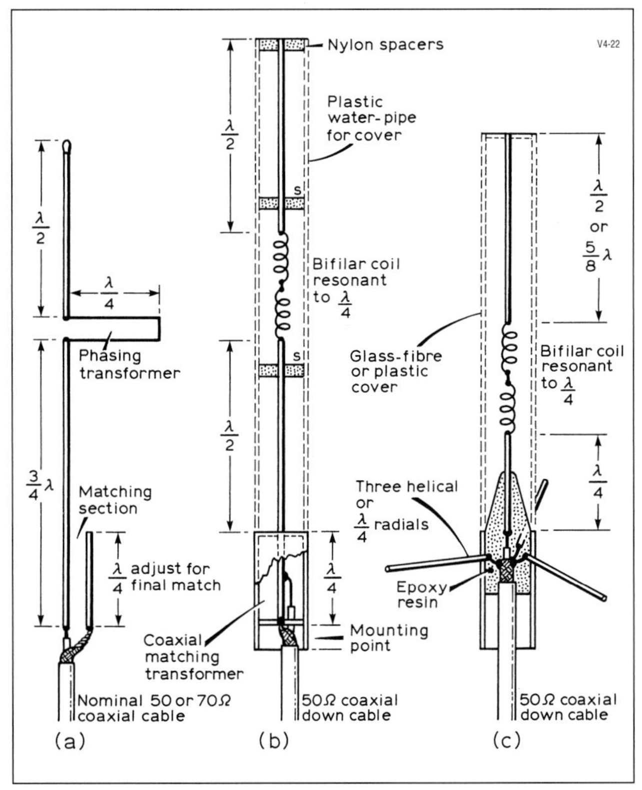


**Fig 16.92: A collinear form of 'J' antenna: (a) the addition of λ/4 sections as suggested by Franklin; (b) Use of a coaxial short-circuit λ/4 transformer to give an unbalanced input. The tapping point of the matching transformer is approximately 0.15λ from the 'earthy' end; (c) A variant of (b) with radials, With both (b) and (c) the λ/4 phasing transformer has been 'wound up' as a bifilar coil (each coil being wound in the opposite hand). While the inductive component is cancelled, the mutual capacitance of the windings makes them physically shorter than a quarter wavelength**

Phasing sections and additional elements can be combined to produce a collinear form for the 'J' as shown in **Fig 16.92(a)**. This antenna and that of **Fig 16.92(c)** have been used successfully to produce low-angle radiation for the GB3SN 144MHz repeater.

A variation of the techniques described, using coils as with the original Marconi concept, is shown in **Fig 16.93(a)** for 432MHz and **Fig 16.93(b)** for 144MHz. The expected gain is between 6 and 7dBd. Materials required for Fig 16.80(a) are as follows:

- One 2.5cm diameter 10cm long glassfibre tube.
- One 4.0mm diameter 1.2m long glassfibre rod.
- Four 2.0mm diameter 20cm long glassfibre rods.
- Length of braiding from 'junk' large coaxial or multicore cable.
- Length of 1.2mm wire for matching coils.
- Approximately 5cm square of singled sided PCB.

First, adjust the bottom 5/8λ element to give minimum VSWR by moving the tapping point on the bottom coil (approximately four turns). A fine adjustment can be made by altering the length of the first 5/8λ element. Next fit the centre matching coil and the top element. To obtain the best results, both elements should be the same length and approximately 5/8λ. Further improvement in VSWR is obtained by adjusting the centre matching coil (the coil is spread over 1/4λ).

The matching coil provides the phase change necessary to feed the top element and so adjustment is quite critical. If the matching coil has to be 'squeezed up' to obtain a good VSWR, the coil has too many turns.

The opposite is true if the coil has to be greater than 1/2λ for a good VSWR.

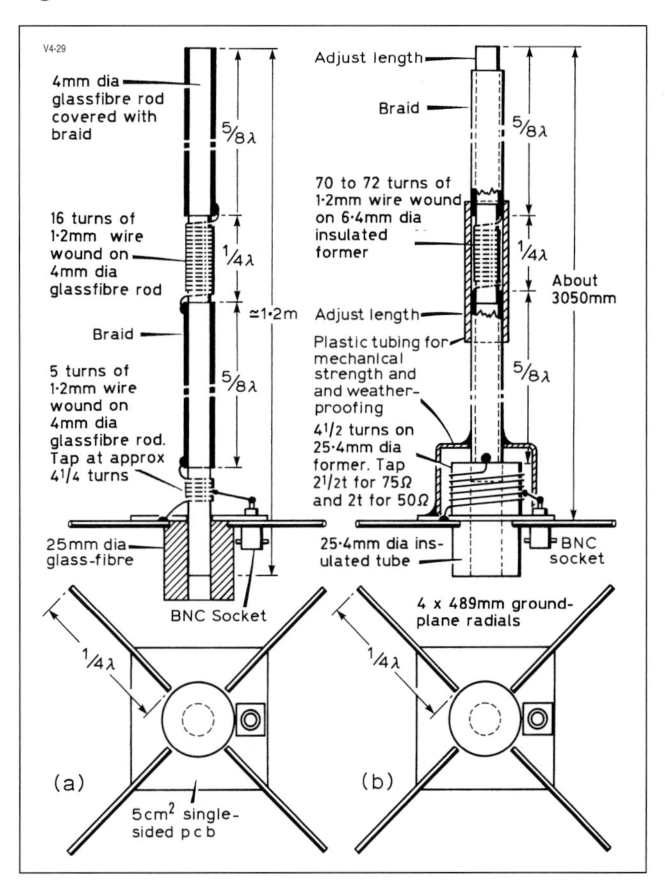


**Fig 16.93: (a) A 432MHz collinear antenna (b) A 144MHz collinear antenna (*UK Southern FM Journal*)**

To prevent the collinear going 'off tune' once set up, the elements are secured to the centre glassfibre rod and the matching coil taped with self amalgamating tape. Provided care is taken in setting up, a VSWR close to 1.1:1 can be obtained.

Materials required for Fig 16.80(b) are as follows:

- Two 12.7mm diameter by 1206 ±12mm, 5/8λ elements (adjustable).
- Four 495mm rods for the ground plane.
- One 6.4mm diameter by 762mm insulated rod.
- One 25mm diameter insulated tube (a cotton reel can be used).
- 1.6mm wire for matching and phasing coils.

The diagram shows extra insulated tubing over the matching and phasing coils to give more mechanical strength and weatherproofing. Setting up is carried out as follows. First, adjust the length of the bottom 5/8 element to give minimum VSWR.

Secondly, fit the phasing coil and the top element. The top element must be the same length as the set-up, bottom element. Next obtain the best VSWR by 'adjusting' the turns of the phasing coil. The coil provides the phase change necessary to 'feed' the top element. It consists of a length of 1.6mm wire, (about 1λ), coiled up to give 70-72 turns on a 6.4mm diameter former. The λ/4 spacing between the two elements is more critical than the number of turns. 68 turns gave a satisfactory VSWR with the prototype.

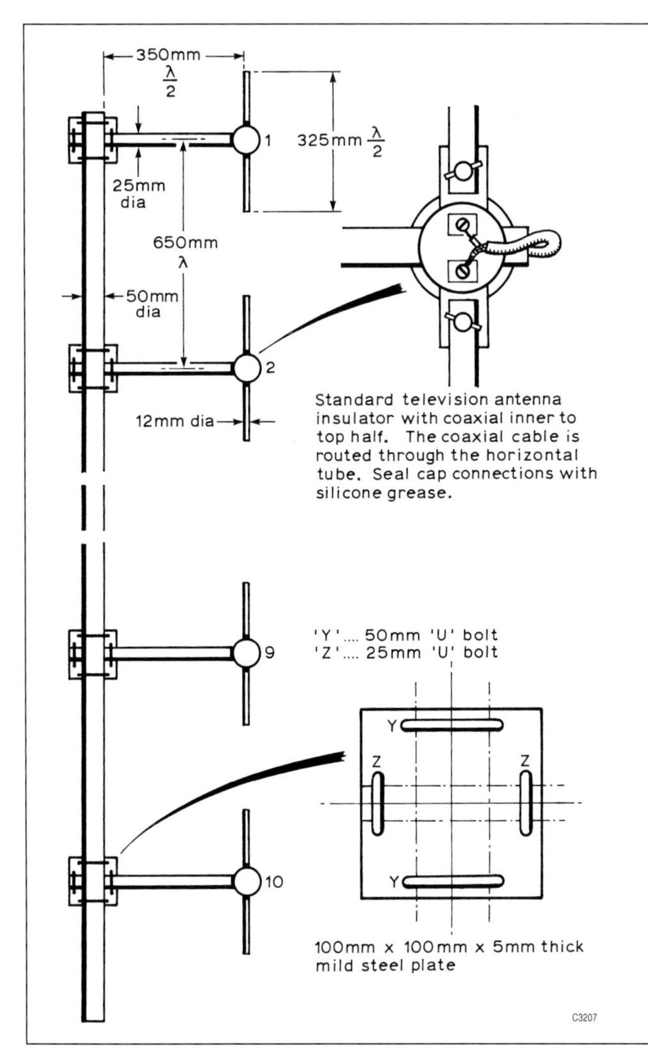


Fig 16.94: Mechanical details of GB3ER collinear

Some difficulty may occur in setting up the phasing coil. If more than seven turns have to be removed, go back to the first adjustment stage to ensure the bottom 5/8λ element is correctly matched. If the bottom element is not correctly set up the collinear will not tune up.

Careful adjustment should produce a VSWR of 1.1:1 at the chosen operating frequency.

A technique, widely used for commercial systems, combines λ/2 dipoles fed in phase from a single source, or alternatively with an appropriate variation of feeder cable length between dipoles to provide phasing.

The disadvantage with this form of antenna array is that some interaction occurs between cables and radiating elements. However, the disadvantage is balanced by the ability to modify the radiation pattern shape by simple adjustment of dipole spacing or phasing cable length.

The example given in **Figs 16.94, 16.95 and 16.96** is probably the simplest to set up and was devised for the GB3ER

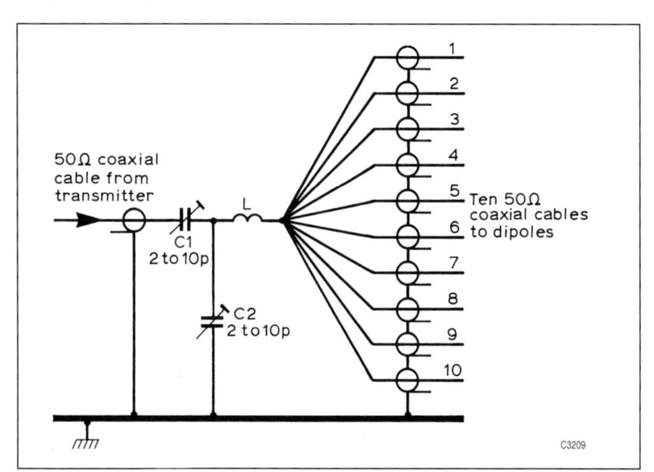


Fig 16.95: Matching Unit of GB3ER collinear

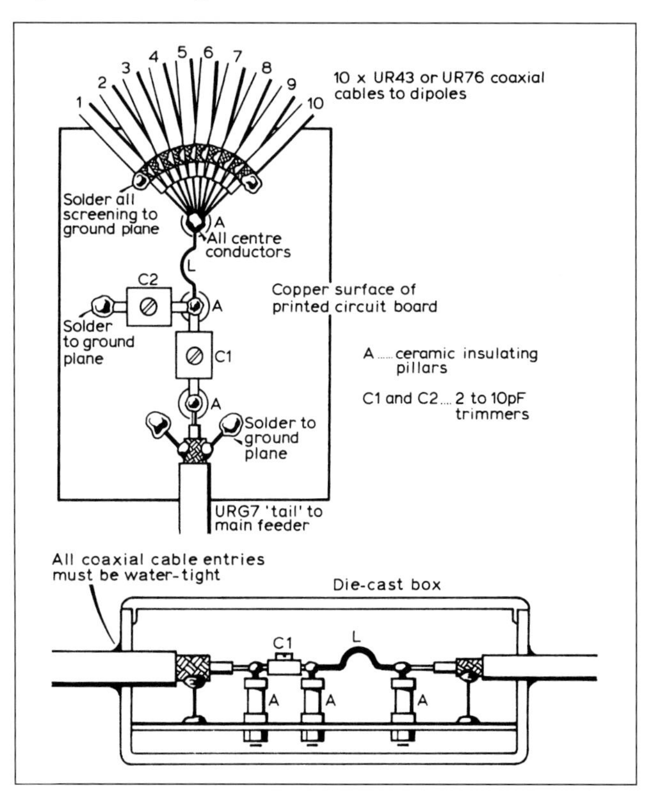


Fig 16.96: Matching unit layout for GB3ER collinear

Fig 16.97: (left) The completed antenna with G0EVV for scale. (right) A completed dipole mounted on the pole with part of the power splitter on the right

Fig 16.98: One completed dipole of the 70cm repeater antenna

432MHz band repeater. If the cables are made to be an odd number of quarter wavelengths long, an equal current feed to each dipole is assured.

## 70cm Repeater Antenna

A four element 70cm D-Star repeater antenna made from copper water pipe and plumber's fittings with excellent VSWR at 432 and 438MHz is shown in **Figs 16.97 and 16.98**. Each element comprises a folded dipole fed by coaxial cable run through the tubing and a tee piece at the electrically neutral point on the tube opposite the feed point.

The short piece of tubing attached to the tee piece is used to clamp the dipole to its support mast. A four-way power divider made from standard coaxial cable, N type connectors and tee pieces feeds the dipoles in phase and transforms their combined impedance to 50 ohms. This antenna requires care in construction and adjustment with access to impedance measuring equipment to make it perform well. A list of materials and details for construction are given in [35].

# ANTENNAS FOR SATELLITE COMMUNICATION

For the radio amateur, satellite ground station antennas fall into two groups. The first group comprises *steerable antennas*, which enable the passage of the satellite to be tracked across the sky. The second group consists of *fixed antennas*, which have essentially hemispherical radiation patterns to receive the satellite signals equally from any direction. These antennas do not need to be steered to receive signals during the satellite's passage.

The tracking antennas are usually of high gain, while the fixed antennas are usually of low gain, due to their hemispherical coverage. Fortunately, signal losses between ground and the satellite in line-of-sight are relatively low. With no obstructions, low-gain antennas of the fixed variety are often acceptable for reception of amateur or weather satellites, helped by the higher radiated powers available from current low-earth orbiting amateur satellites.

As many satellites rotate or change their orientation with respect to the ground station, both groups of antennas are designed to provide circular polarisation. Amateur convention calls for the use of Right-Hand circular polarisation for earth/space communications. However, the downlink may have either Left or Right handed polarisation according to the satellite and frequency in use. Of the higher-gain tracking antennas, the crossed Yagi and the helix antenna are the main ones used. The crossed Yagi is probably the easiest to construct and most readily available commercially. Construction details for these antennas were described earlier, and polarisation switching schemes are discussed at the end of this chapter.

## Crossed Dipole or Turnstile Antenna

**Fig 16.99** shows a simple arrangement of crossed dipoles above a ground plane for 145MHz. This type of antenna can be scaled for use at 29, 145 or 432MHz. Mechanical problems may make the reflectors inadvisable in a 29MHz version. The height above ground can be about 2m for 145MHz and 3m for 29MHz.

Typical dimensions are:

| | | | |
|---|---|---|---|
| 29MHz | driven elements ($\lambda/2$) | 188in | 4775mm |
| 145MHz | driven elements ($\lambda/2$) | 38in | 965mm |
| | reflectors | 40.5in | 1030mm |
| | spacing ($0.3\lambda$) | 24.5in | 622mm |

The phasing line comprises $\lambda/4$ of 72-ohm coaxial cable. The matching section for a 72-ohm feed is $\lambda/4$ of 50-ohm cable. When calculating the length of the $\lambda/4$ sections, the velocity factor of the cable must be taken into account. Typically this is 0.8 for cellular and semi-airspaced cables, and 0.66 for solid dielectric cables, but verification of the figure for the particular cable used should be obtained. As an example, a matching section of RG59/U would be 13in (330mm) in length. Omit the transformer section for 50-ohm operation. This will result in an input VSWR of less than 1.4:1. For a centre-fed crossed dipole, it is advisable to have a 1:1 balun to ensure a consistent pattern through 360° of azimuth. Dependent on the spacing between the dipoles and ground plane, the radiation pattern can be directed predominantly to the side, for satellites low on the horizon, or upwards for overhead passes.

By drooping the dipole elements at 45°, with a spacing of approximately $0.4\lambda$ between the dipole mounting boss and the

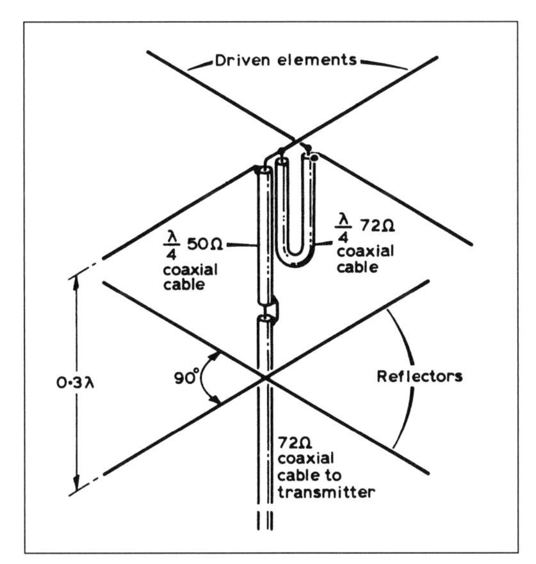


Fig 16.99: A crossed dipole antenna for 145MHz

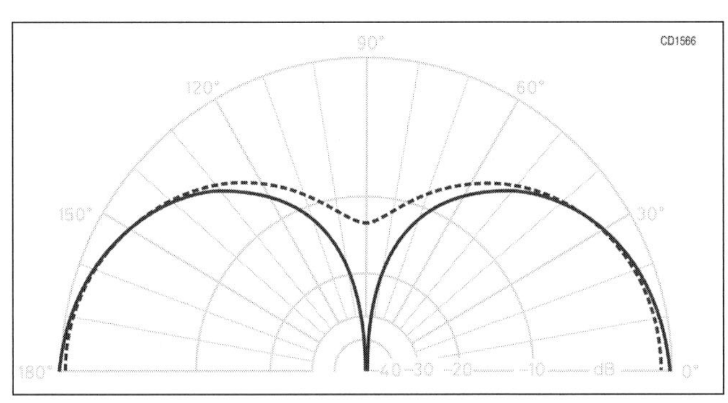


Fig 16.100: Elevation radiation pattern for a whip antenna sloping 30 degrees on a large groundplane. The pattern for a vertical whip is also shown for comparison

ground plane/reflectors, better coverage towards the horizon can be obtained. As ground reflections affect horizontal and vertical polarisation differently, low-to-horizon flight paths will not produce circular polarisation. This is due to 'ground scatter' of the satellite signal when it is low on the horizon, and ground reflections locally at the ground-based antenna. Circular polarisation is normally produced by feeding one dipole 90° out of phase to the second dipole, and can be achieved by having an extra λ/4 of cable on one side of a combining harness.

An alternate approach to this method of phasing is to use the phasing properties of a capacitive or inductive reactance.

Suppose, for example, that the length and diameter of the dipoles are set to give a terminal impedance of 70Ω - j70Ω (capacitive). If a second, crossed dipole is set to be 70Ω + j70Ω (inductive) the combined terminal impedance of the arrangement becomes 35Ω ± j0, ie 35 ohms resistive.

As the two dipoles are connected in parallel, the current in each dipole is equal in magnitude. However, due to the opposite phase differences of 45° produced by the capacitive and inductive reactances, the radiated fields are in phase quadrature (a 90° phase difference) which results in circularly polarised radiation.

## Hairpin or Sloping Antennas

As technology has advanced, so has the radiated power from low earth orbiting amateur radio satellites increased, to the extent that communications can be established using handheld radios if the satellite is well above the horizon. Accordingly, it can be worth constructing very simple antennas to get started with satellites.

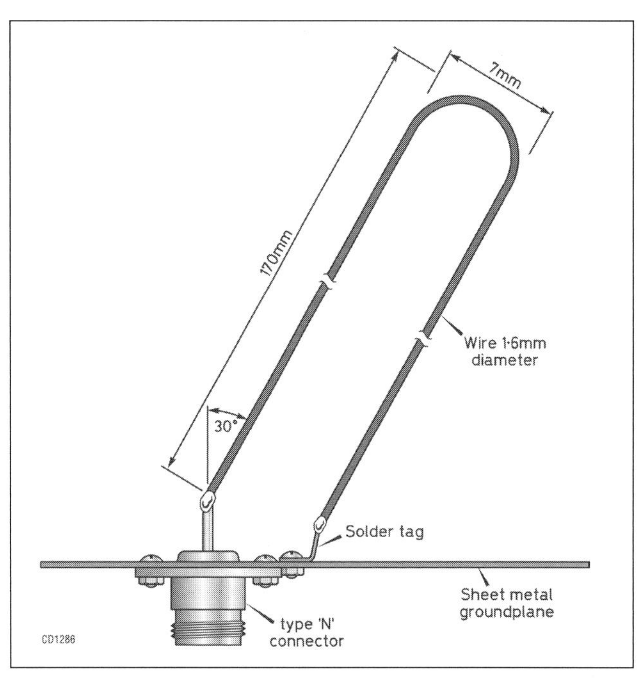


**Fig 16.101: Construction of 435MHz hairpin antenna and groundplane**

A vertical whip on a ground plane produces a null along the axis of the whip. However, if the whip is bent over by about 30°, the overhead null is filled without degrading the azimuth pattern too severely (**Fig 16.100**). For 145MHz operation, the whip should be around 480mm in length, and adjusted for best match according to the size of the groundplane.

At 435MHz, a *hairpin* construction from 1.6mm diameter wire (or even a wire coat hanger) is somewhat more rugged, and can offer wider bandwidths because of the transforming action of the folded dipole. The connector/feed point joint should be potted or sealed with silicon rubber or flexible epoxy resin if the antenna is to be used outdoors, to prevent moisture entering the cable (**Fig 16.101**). The antenna can be tuned, without seriously degrading the radiation pattern, by adjusting the spacing between the wires or by changing the slope of the hairpin a few degrees, or a combination of both.

## Eggbeater Antennas

This omnidirectional antenna produces excellent circularly polarised signals over a wide range of elevation angles if carefully constructed. It comprises a pair of crossed circular loops, slightly over one wavelength in circumference, and fed in phase quadrature. The only disadvantage is the high terminal impedance of each loop (approximately 140 ohms, which requires a transformer section to match to 50 ohms [36].

Performance has been further optimised for satellite operation by K5OE by using square loops and adding tuned reflectors as a groundplane [37]. This antenna can be built with an excellent match to 50 ohms without recourse to special impedance cables.

The general arrangement is shown in **Fig 16.102**. The loops are constructed of rod, tubing, or 2mm enamelled copper wire for 70cm, bolted to a 22mm plastic water pipe coupler. Critical dimensions are shown in **Table 16.16**. The tops of the loops are supported by a section of (plastic) water pipe that fits into the coupler, and capped to prevent water ingress. One loop should be set 5 - 10mm higher than the other so that the top of the loops can cross without touching. Instead of the 93-ohm RG62 cable described by K5OE, the phasing section can be made from two

**Fig 16.102: General construction arrangements for K5OE Eggbeater II antenna**

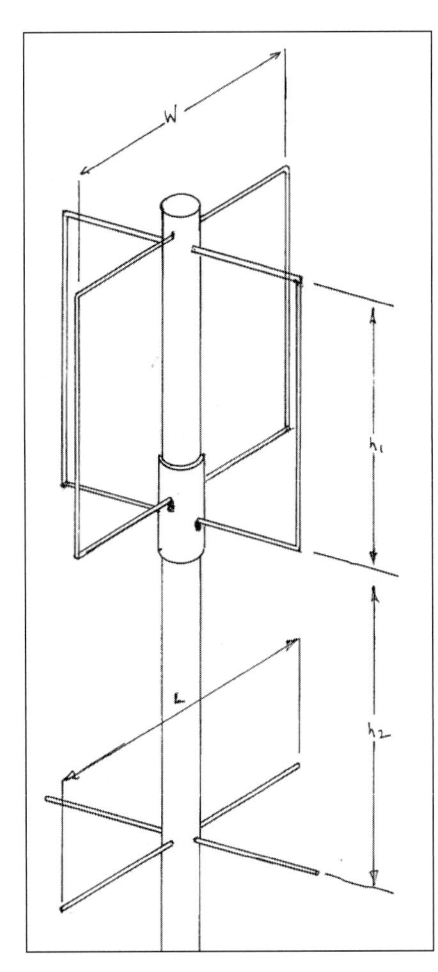


| Design frequency | w | h1 | L | h2 | Loop wire dia | Phasing cable length |
|---|---|---|---|---|---|---|
| 137MHz | 540 | 665 | 1065 | 1080 | 5 | 390 |
| 145.5MHz | 510 | 625 | 1000 | 1010 | 5 | 370 |
| 435MHz | 170 | 210 | 335 | 330 | 2 | 124 |

**Table 16.16: K5OE Eggbeater antenna critical dimensions in mm. Phasing cable lengths are for solid PTFE dielectric, and should be measured between where the inner dielectric clears the braided coaxial outer**

pieces of 50-ohm cable, each one quarter-wavelength long, with the outers connected together at each end. This makes a 100-ohm balanced wire transmission line, which can be coiled or folded up if required.

All four braids can be connected together if the connections are conveniently close together (e.g. the 70cm variant). The antenna is Left Hand Circularly Polarised (LHCP) if the connections are as shown in **Fig 16.103** when the antenna is viewed from below. To change the polarisation to RHCP, transpose the two phasing cable connections at *one* end only.

For many purposes, a 50-ohm coaxial feeder can be connected directly as shown. Five or six turns of feeder closely wound can provide a balun and choke against current on the outside of the cable if required.

A ferrite bead fitted on the feed cable close to the connection can be effective, and there is space for a low noise amplifier inside the water pipe between the coupler and the tuned reflectors if the antenna is to be used only for reception.

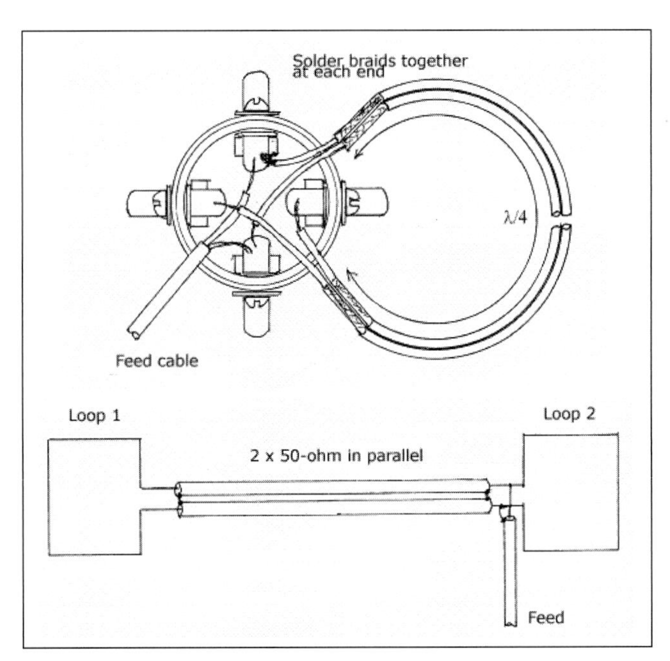


**Fig 16.103: Feed point construction detail and phasing line connections for Eggbeater antenna**

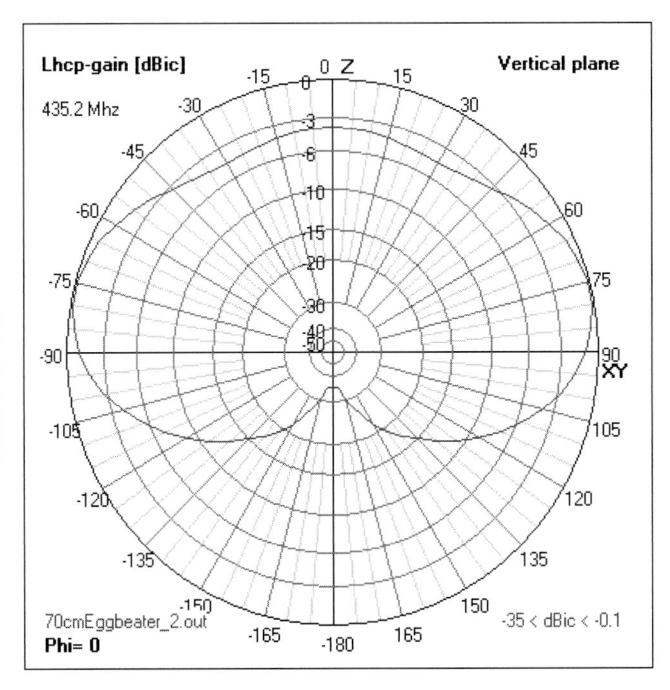


**Fig 16.104: Elevation radiation pattern for 70cm Eggbeater antenna. The azimuth pattern is essentially omnidirectional. Note the gain figures are in dBi**

Calculated radiation patterns for the 435MHz variant are shown at **Fig 16.104**.. **Fig 16.105** shows the antenna construction.

## Quadrifilar Helix or Volute Antenna

The quadrifilar helix (QFH) or volute is a four-element helical antenna which can be used to give either directional gain or hemispherical circular polarised coverage as originally described by Kilgus [38] [39]. The general form of a QFH is shown in **Fig 16.106**.

Radiation patterns produced for several combinations of turns and resonant lengths are shown in **Fig 16.107(a) to (d)**, and generally produce better circularly polarised signals at low elevation angles than a turnstile with drooping arms.

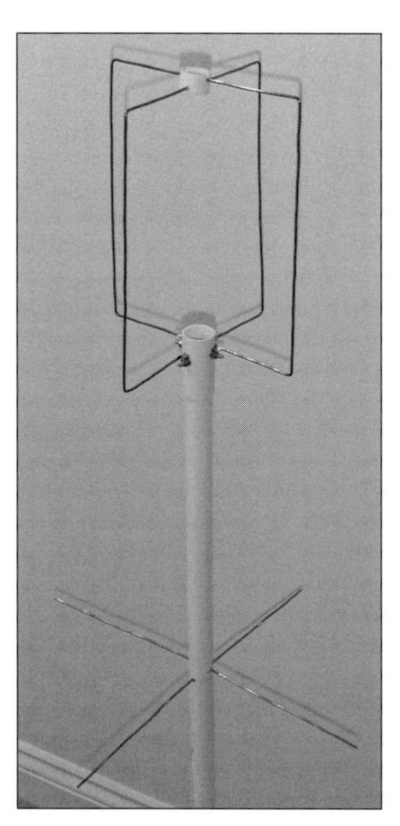

**Fig 16.105: K5OE Eggbeater II without weatherproofing**

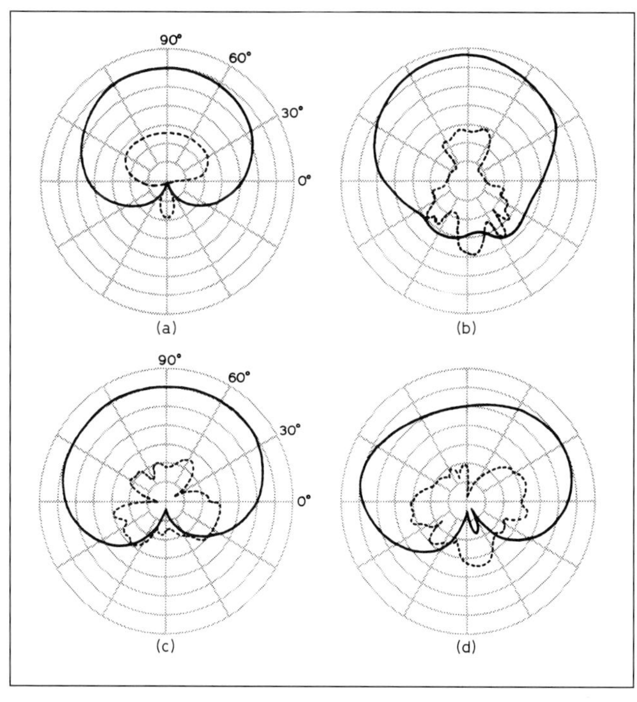


**Fig 16.107: Volute radiation patterns, co- and cross- circular polarisation (a) 3/4 turn λ/4 volute. (b) 3/4 turn λ/2 volute. (c) 3/4 turn 3/4λ volute. 3/4 turn 1λ volute *(Microwave Journal)***

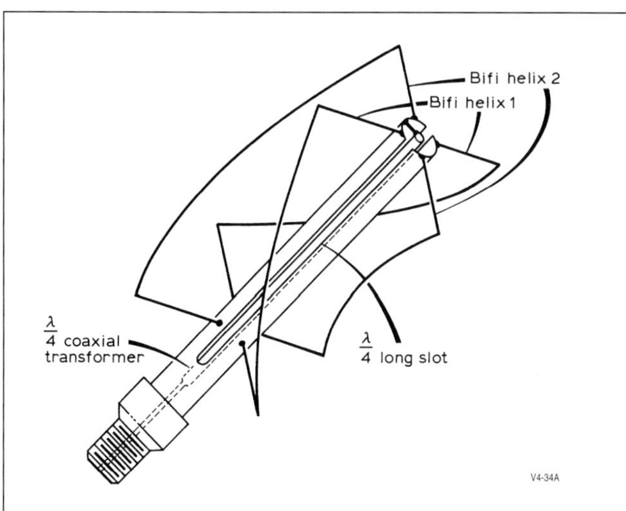


**Fig 16.106: A quarter turn volute with split sheath or slot balun (Microwave Journal)**

The QFH can make use of a phasing line or the reactance method (as previously discussed for the turnstile antenna) to produce circular polarisation. The number of 'turns' or part turns of the radiating elements combined with their length can be used to produce various radiation patterns. Elements that are multiples of λ/4 have open-circuit ends, while the elements that are multiples of λ/2 can be short-circuited to the mounting structure.

Good radiation patterns for communication with low earth orbiting satellites (LEOs) can be produced by QFHs with diameters less than 0.1λ and one quarter turn per loop. This makes antennas for 145MHz possible without large support structures or incurring high wind load penalties [40] [41].

Constructional details must be followed exactly if impedance measuring test equipment is not available to adjust the element lengths.

The radiation pattern of a quarter turn QFH antenna is shown in **Fig 16.108**.

'Long' QFH with multiple turns can produce even better low angle coverage, but are generally only suitable for 435MHz or above because of their length.

## CONTROL OF ANTENNA POLARISATION

Vertical polarisation is the most popular for FM and mobile operation in the UK. This means that a fixed station with antennas optimised for horizontal polarisation can only operate effectively if two antennas or a means of changing polarisation is available.

Space communication, where control of polarisation at the spacecraft end of the link can be difficult, has stimulated the use of circular polarisation. Its fundamental advantage is that, since reflections change the direction of polarisation, there can be far less fading and 'flutter' from reflections.

The use of circular polarisation at only one end of the link, with horizontal or vertical linear polarisation at the other end of the link, will result in a loss of 3dB. However, changes of polarisation caused by propagation often result in better reception of linearly polarised signals with a circularly polarised antenna. Conversely, if circular polarisation is used at both ends of the link, but the two stations use oppositely polarised antennas, the loss of signal due to polarisation mismatch can be 10dB or more.

It has been usual to standardise on Right-Hand Circular polarisation (RHCP) in the northern hemisphere for fixed communications. Space communications may use either Right or Left Hand Circular polarisation according to the frequency used and the satellite of interest.

The helix antenna produces Right or Left-Handed circular polarisation dependent upon whether the antenna element is wound clockwise or anti clockwise. Horizontal or vertical linear polarisation is also possible from helix antennas by using two helices and suitable phasing arrangements.

A compromise arrangement for receiving circular-polarisation signals is to use slant polarisation. To obtain this, a single Yagi

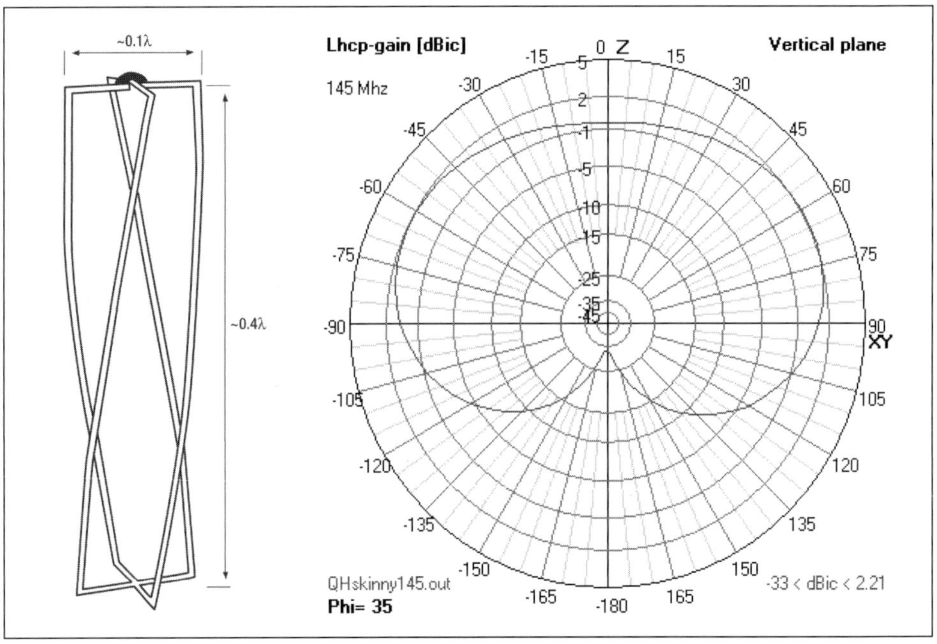

**Fig 16.108: Quarter turn slim quadrifilar helix antenna and elevation radiation pattern. Each loop is approximately 1λ in circumference. Support pole not shown**

is set at an angle of 45°. This enables horizontal and vertical signals to be received almost equally. At first sight one would expect a loss of 3dB for H and V polarised signals compared with an appropriately aligned Yagi. However, long-term practical measurements have shown when averaged that this arrangement gives a 6dB improvement with typical mixed polarised signals. In addition this arrangement it is only a little affected by the mounting mast, unlike a vertical polarised Yagi.

The simplest way of being able to select polarisation is to mount a horizontal Yagi and a vertical Yagi on the same boom, giving the well-known *crossed Yagi* antenna configuration (**Fig 16.109**). Separate feeds to each section of the Yagi brought down to the operating position enable the user to switch to either horizontal or vertical.

It is perhaps not generally realised that it is quite simple to alter the phasing of the two Yagis in the shack and obtain six polarisation options. These are two slant positions (45° and 135°), two circular positions (clockwise and anti-clockwise) and the original horizontal and vertical polarisation.

The presence of the mast in the same plane as the vertical elements on a Yagi considerably detracts from performance, but a simple solution is to mount the antennas at 45° relative to the vertical mast. With appropriate phasing, vertical and horizontal polarised radiation patterns can be obtained that are unchanged by the presence of the mast.

If a crossed Yagi is mounted at 45°, with individual feeders to the operating position, the polarisation available and the phasing required is as follows:

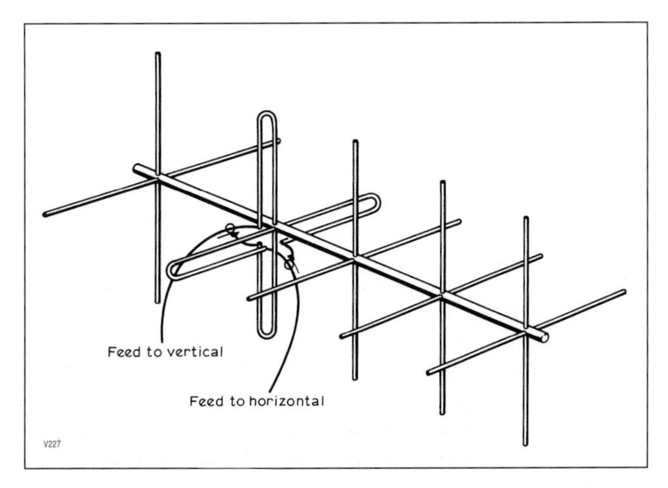


**Fig 16.109: General arrangement of a crossed Yagi antenna**

(a) Slant position 45° and 135°. Antennas fed individually.
(b) Circular positions clockwise and anti-clockwise. Both antennas fed 90°+ or 90° - phase relationship respectively.
(c) Horizontal and vertical polarisation. Both antennas fed with 0° or 180° phase relationship respectively.

Phasing is simply the alteration of the length of the feeders of each crossed Yagi to change the polarisation. Where a 90° phase shift is required, λ/4 of cable is inserted and, where a 180° phase shift is required, λ/2 cable is inserted. The polarisation switch must switch in the appropriate λ/4 'impedance transformer' and correct phasing by connecting the appropriate length(s) of cable.

## Mast-head Multiple Polarisation Switch for 145MHz

There are several disadvantages in locating the polarisation switch in the shack. Two feeders are required from the antenna to the equipment, and RF losses may be significant if the antennas are positioned on a mast. Whilst it is technically feasible to improve signal to noise performance by installing twin preamplifiers at the

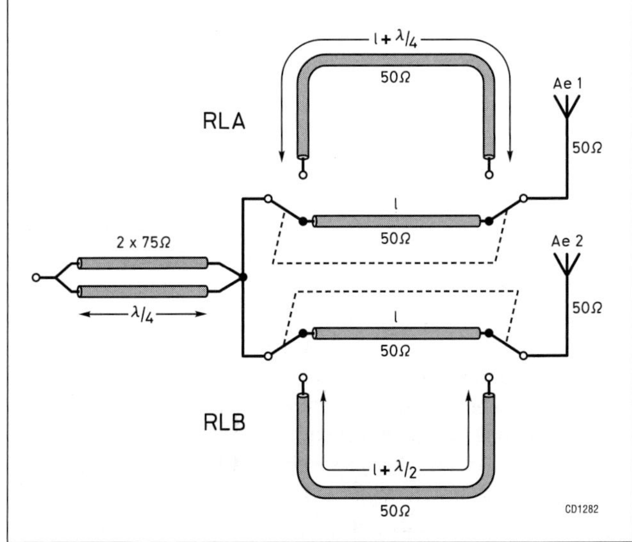


**Fig 16.110: Masthead antenna polarisation switch**

| Characteristic | 145MHz | 435MHz |
|---|---|---|
| Insertion loss | <0.2dB | <0.3dB |
| Mismatch | <1.03:1 | <1.2:1 |
| Isolation between normally open and moving contact | 42dB | 30dB |
| Isolation between moving contacts | 39dB | 28dB |

**Table 16.17: Half crystal-can relay measured characteristics**

antennas, the problems of equalising gain and ensuring relative phase stability are formidable. An alternative solution is to place the polarisation switch close to the antennas, which minimises the feeder losses, and to utilise a single preamplifier (with transmitter bypass arrangements if required) immediately after the switch to improve the system signal to noise ratio before despatching the signals to the receiver via a single feeder.

The polarisation switching circuit (**Fig 16.110**) offers a fair compromise between cost, complexity and performance. A pair of half crystal-can double pole changeover relays (**Table 16.17)** insert extra lengths of line into the feeds of a 50-ohm crossed Yagi antenna mounted at 45 degrees to its support mast, to provide four polarisations (**Fig 16.111**).

The antenna cable lengths must be cut to compensate for any phase differences between the antennas due to physical offset in the direction of propagation. Under ideal conditions with perfectly matched antennas, the four main polarisations will be achieved. If the antennas are not well matched to 50 ohms, the power division will not be equal, and mixed polarisation will result for each relay setting. If the VSWR is less than 1.2:1, the depolarisation will not be serious.

The relays are mounted on 1.6mm double sided copper clad glass fibre PCB with 50-ohm microstrip tracks for the RF circuits (**Fig 16.112**). RF interconnection and phasing lines are made from 3mm PTFE cable (URM 111 and T3264) which provides a good trade-off between losses and size for coiling up within a diecast weatherproof box. Polythene dielectric cables could possibly be used if soldered very quickly, but this is not recommended. The cable ends must be kept as short as possible as shown in Fig 16.99. The polarisation unit will handle 100 watts of CW or FM, provided that the antenna system VSWR at the relay is better than about 1.5:1. Miniature PTFE cables in small quantities can be obtained from [42].

If a transient suppression diode is to be fitted across the relay coil, it should be bypassed with a 1nF capacitor, or rectified RF pickup during transmission may cause the relay to change state. Experience has shown that it is better to dispense with the diode at the relay altogether if high transmitter powers are involved.

## Circular Polarisation Feed / Reversing Switch

The following simple arrangement generates remotely selectable Right or Left hand circular polarisation from a crossed Yagi antenna, or two separate, identical Yagi antennas with their elements set at right angles to each other. It will work well provided that both antennas are closely matched to 50 ohms, and it uses readily obtainable cables, adaptors and connectors throughout (**Fig 16.113**). It should be mounted close to the antenna(s).

The electrical length of the two antenna cables must be an *odd* number of quarter wavelengths. λ/4 should be sufficiently long to make the connections for 2m operation, and 3λ/4 for 70cm. Higher multiples of λ/4 will reduce the VSWR and polarisation bandwidths of the antenna. If longer antenna cables are necessary, equal lengths of 50-ohm cable should be used to connect the antennas to the λ/4 75-ohm sections at the relay.

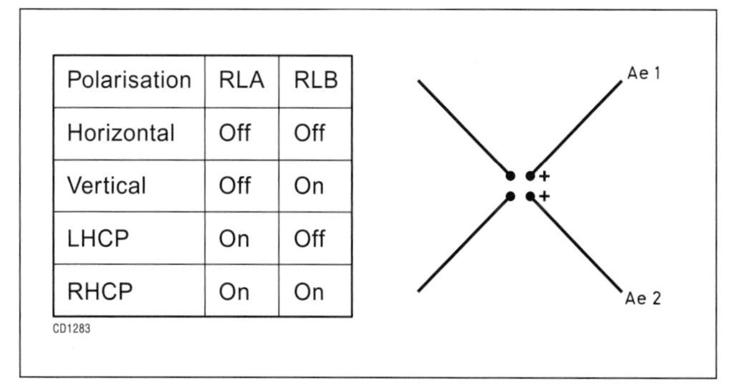

| Polarisation | RLA | RLB |
|---|---|---|
| Horizontal | Off | Off |
| Vertical | Off | On |
| LHCP | On | Off |
| RHCP | On | On |



**Fig 16.111: Rear view of crossed antenna with polarisation table. The radiated wave propagates into the paper**

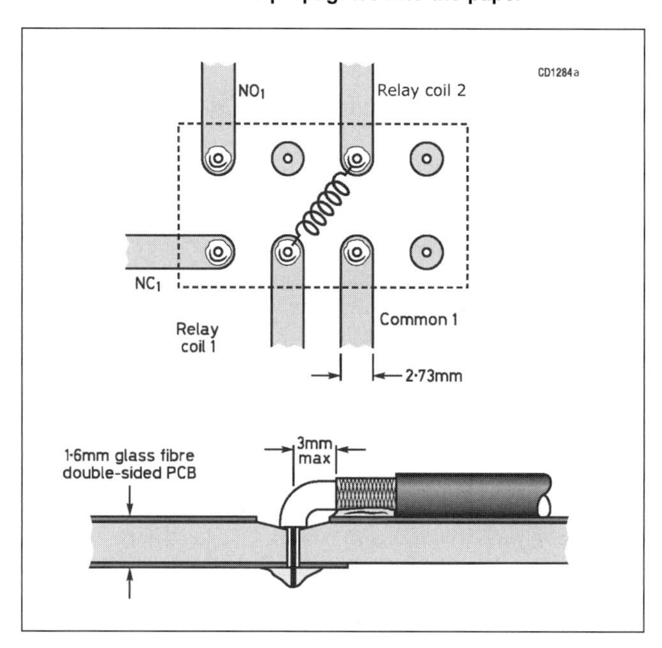


**Fig 16.112: Relay and phasing/matching cable mounting details**

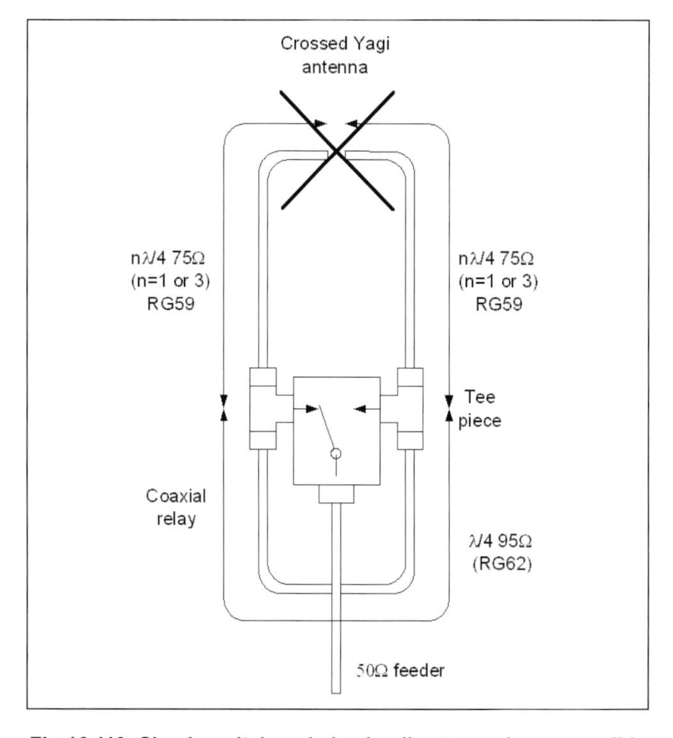


**Fig 16.113: Simple switch and phasing line to produce reversible circular polarisation**

The 50-ohm impedance of the antenna is transformed by the quarter wave sections to 100 ohms at the T-piece. A quarter wavelength of RG62 (93-ohm) cable provides the phasing needed to produce circular polarisation, and the common relay contact, in either position, 'sees' two 100-ohm loads in parallel, ie 50 ohms, for presentation to the downlead.

Suitably weatherproofed PL239-series connectors will work for 2m operation, but type N connectors should be used for 70cm. The electrical length of the connectors and tee-pieces must be taken into account when cutting the quarter wavelength cables.

For 1.3GHz and above, use separate circularly polarised antennas and a change-over relay if selectable polarisations are required.

## REFERENCES


[1] 'Notes on the Development of Yagi arrays', C Greenblum, *QST* and other ARRL publications, edited by Ed Tilton, W1HDQ

[2] 'Yagi-Uda Arrays', Chen and Cheng, *Proc IEEE*, 1975

[3] 'Yagi Antenna Design', P Viezbicke, *National Bureau of Standards Technical Note 688*, 1976

[4] *Yagi Antenna Design*, Dr J L Lawson, W2PV, ARRL, 1986

[5] '4NEC2' NEC2 based antenna modeller and optimiser http://www.qsl.net/4nec2/

[6] 'How good are genetically designed Yagis?', R A Formato, WW1RF, *VHF Communications* Vol. 2/98 pp87-93

[7] 'A genetically designed Yagi'" R A Formato, WW1RF, *VHF Communications* Vol, 2/97 pp 116-123

[8] YGO2, YGO3 Yagi Genetic Optimiser software may be downloaded from http://nec-archives.pa3kj.com/

[9] *Genetic Algorithms in Electromagnetics*, Randy L Haupt and Douglas H Werner, Wiley, 2007

[10] 'How to Design Yagis', W1JR, *Ham Radio*, Aug 1977, pp22-30

[11] 'Long Yagis', W2NLY and W6KQI, *QST* Jan 1956

[12] 'Extremely long Yagi Antennas', G Hoch, DL6WU, *VHF Communications*, Mar 82, pp131-138; Apr 1977, pp204-21

[13] 'DL6WU Yagi Designs', *VHF/UHF DX Book,* ed I White, G3SEK, DIR Publishing / RSGB 1993

[14] www.ifwtech.co.uk/g3sek/diy-yagi. Various long Yagi designs with construction details and other information

[15] 'The JVL LoopQuad', *Microwave Handbook,* Vol 1, M W Dixon, G3PFR, (ed) p4.7ff, RSGB

[16] 'The GW3YDX Super Moxon' Ron Stone, GW3YDX, RadCom July 2010 pp 70 -71

[17] GW3YDX@yahoo.com

[18] *Antennas*, J D Kraus, 2nd edn, McGraw-Hill, 1988, Chap 4

[19] *Antenna Theory, Analysis and Design*, Constantine A Balanis, Harper and Row, 1928 pp204-243

[20] 'A Quadruple Quad Antenna - an Efficient Portable Antenna for 2 Metres', M Ragaller, DL6DW, *VHF Communications* 2/1971, pp82-84

[21] 'Analysis and Design of the Log-Periodic Dipole Antenna' *Antenna Laboratory Report No.52*, University of Illinois, USA

[22] 'Log-Periodic Dipole Arrays', D E Isbell, *IRE Transactions on Antennas and Propagation* Vol AP-8, May 1960, pp260-267

[23] W3DUQ, *Ham Radio*, Aug 1970

[24] *Antennas*, J D Kraus, 2nd edn, McGraw-Hill, 1988 pp265-319

[25] 'Helical Antennas for 435MHz', J Miller, G3RUH, *Electronics and Wireless World* Jun 1985, pp43-46

[26] 'A Helical antenna for the 23cm Band', Hans-J Griem, DJ1SL, *VHF Communications* 3/83 pp184-189

[27] 'Feedline Verticals for 2m & 6m', Rolf Brevig LA1IC, *RadCom* Mar 2000 p36

[28] 'Collapsible 6m 2-ele beam antenna', Mike Parkin, G0JMI, *RadCom*, September 2012, pp44-46

[29] 'The HB9CV Antenna for VHF and UHF', H J Franke, DK1PN, *VHF Communications* Feb 1969

[30] 'Folding 2m Yagi', Eric Lake, 2W0WXM, *RadCom*, January 2013 pp42-44

[31] 'Eurotek', E David, G4LQI, *RadCom,* Sep 1999 pp31-32

[32] 'Asymmetrically Driven Antennas and the Sleeve Dipole', R W P King, *Proc IRE,* Oct 1955 pp1154-1164

[33] 'Designing Discone Antennas', J J Nail, *Electronics* Vol 26, Aug 1953, pp167-169

[34] 'Slim Jim Antenna', F C Judd, G2BCX, *Practical Wireless*, Apr 1978, pp899-901

[35] 'Designing and building a 70cm repeater aerial' David Stansfield G0EVV, RadCom October 2010 pp54-56

[36] 'Eurotek' *RadCom* May 1996 p59 & 61, Circular loop Eggbeater antenna

[37] http://wb5rmg.somenet.net/k5oe/Eggbeater_2.html. Eggbeater II omni LEO antenna

[38] 'Resonant Quadrifilar Helix Design' C C Kilgus, *Microwave Journal*, Vol 13-12, Dec 1970, pp49-54

[39] 'Shaped-Conical Radiation Pattern Performance of the Backfire Quadrifilar Helix', C C Kilgus, *IEEE Trans Antennas and Propagation* May 1975, pp392-397

[40] http://metsat.gogan.org/ant_qha.htm Construction details for a very slim quadrifilar helix antenna for 2m or 137.5MHz. Note that the elements are twisted just a quarter of a turn The .zip file associated with the link can be found at https://web.archive.org/web/20060422092048/http://www.pilotltd.net/qha.zip

[41] http://www.jcoppens.com/ant/qfh/index.en.php Description of 137MHz QFH construction with calculator for other frequencies turn numbers

[42] W H Westlake Electronics: www.whwestlake.co.uk. Polythene and PTFE cables in many sizes and impedances, including miniature types


*__Peter Swallow__, G8EZE, was directly involved with the design and development of fixed and mobile antennas for 14 years, and continues to experiment with antennas for the VHF and microwave bands to support his interest in amateur satellite communications.*

# Practical Microwave Antennas

17

*Peter Swallow, G8EZE*

The advantage that microwave antennas have over those used for other amateur bands is that they are relatively small. We can let our imaginations loose and have antennas with gains that are not achievable at lower frequencies and mount them at heights, in wavelengths, that users of the lower frequency bands can only dream of.

This chapter has been organised by type of antenna rather than by band usage, the types being:

- **Patch antennas:** These are easily constructed because they can be etched onto printed circuit board. They are small and have found many applications in commercial electronics such as mobile phones, cordless phones and local area networks. If you need a small antenna with low directivity, perhaps for feeding a parabolic dish, the patch may suit your needs.
- **Slot antennas:** These are omni-directional and can be used for repeaters or mobile use.
- **Helical antennas:** The big advantage of the axial-mode helical antenna is that it is circularly polarised. This has made it widely used for satellite operation.
- **Yagi antennas:** The Yagi is probably the best known antenna. With the reduced element size on the microwave bands some impressive gains can be achieved.
- **Horn antennas:** As the frequency increases, the horn antenna comes into its own at 10GHz and above.
- **Dish antennas:** Ideal for high gain antennas but more difficult to build. With the availability of offset dishes for the satellite TV market they are a good choice, but calculating where they are pointing is more difficult. W1GHZ has all the answers for using an offset dish. Amazingly, Michael Kohla, DL1YMK uses a dish for portable moonbounce DXpeditions, his design shows what can be achieved if you are really dedicated to amateur radio.

## PATCH ANTENNA

### Patch Antenna Theory

The patch or microstrip antenna comprises a very thin metallic plate mounted parallel and very close (typically <0.01 wavelengths) to a conducting ground plane (**Fig 17.1**). They can be designed to produce linear or circularly polarised signals, and a single element will have a typical gain of around 4-9dBi and 3dB beam widths of 35-80 degrees, dependent on the shape and size of the patch and the size of the ground plane.

The plate is a little shorter than a half wavelength, and may be fed at one edge, or from below through the ground plane, which

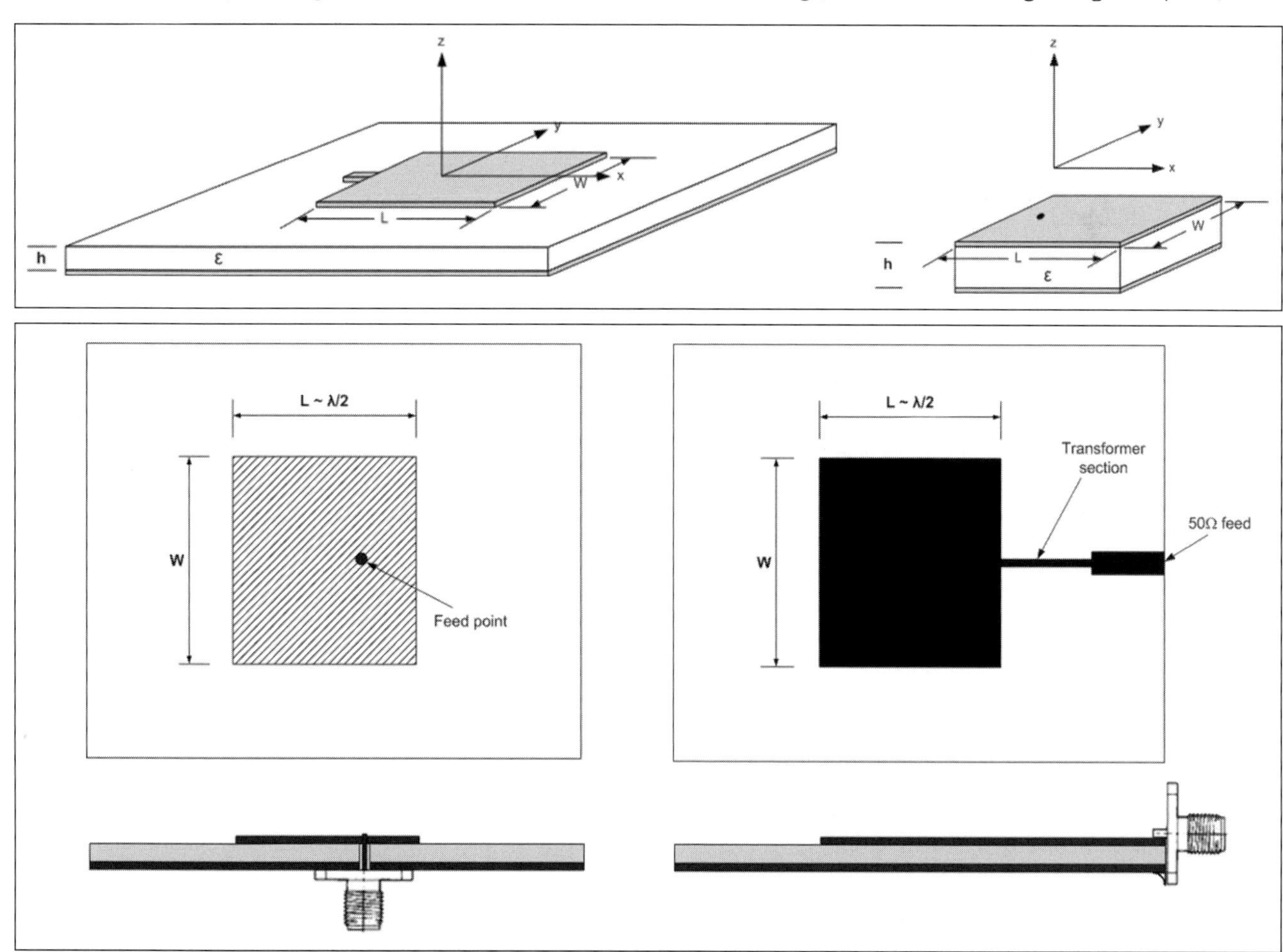


**(Top) Fig 17.1: Two types of simple patch antenna with essential dimensions**
**(Bottom) Fig 17.2: Two ways of feeding a microstrip patch antenna**

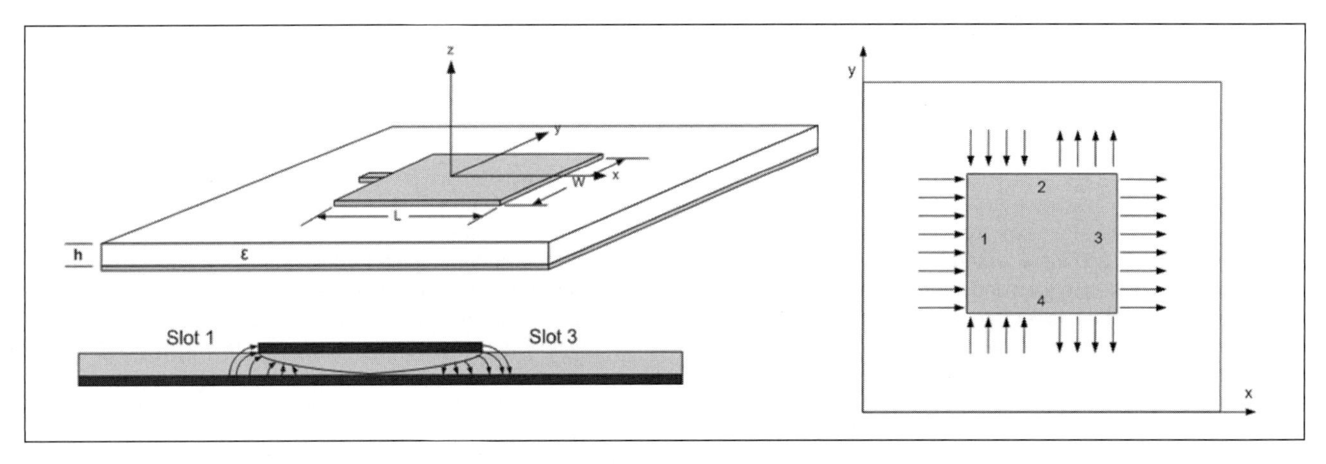


**Fig 17.3: Electric field distributions in and around a linearly polarised patch antenna**

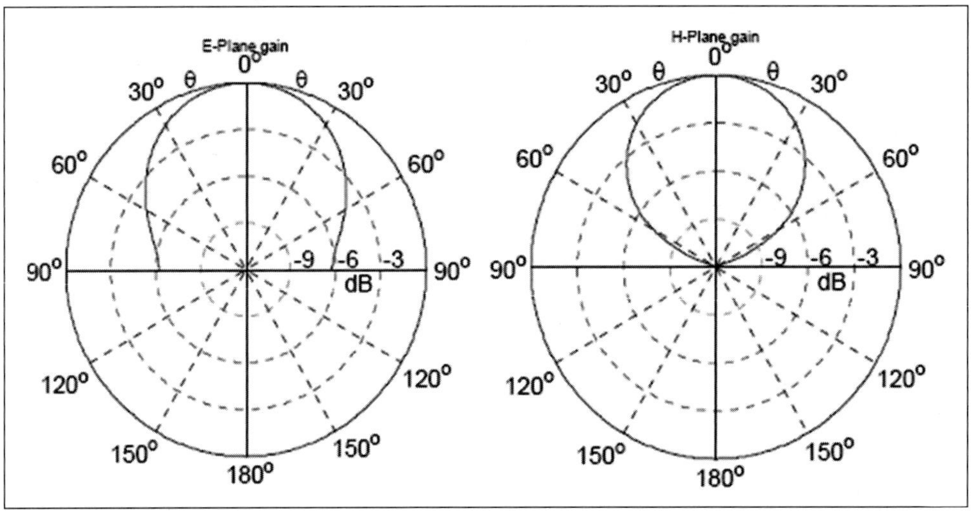

makes it suitable for manufacture in quantity using printed circuit techniques. **Fig 17.2** shows the feeding options.

Unfortunately, the simple structures shown in Fig 17.1 have narrow bandwidths because they behave as cavity resonators. They can support multiple modes, and much development has focussed on using parasitic elements and novel feed networks to extend the radiation bandwidth whilst keeping the structure simple and repeatable for manufacture.

The high Q nature of the cavity makes the antenna tuning sensitive to variations in the dielectric constant (ε) of the substrate material between the patch and the ground plane. This means that GRP PCB materials such as FR-4, though usable, may not give repeatable results at higher frequencies. A range of electrically uniform PCB-like materials with dielectric constants between about 1.5 and 15 are available from specialist manufacturers, and a wide range of datasheets and supporting material on microwave circuit board design is available from [1].

In general, the higher the dielectric constant, the more sharply tuned the patch will be and the more critical the dimensional accuracy. However, simple antennas can be made without PCB etching facilities using a steel rule and scalpel, and by peeling the copper away using a hot soldering iron to loosen the adhesive.

**Fig 17.4: Linear polarised patch radiation pattern on an infinite ground plane**

Fortunately, the bandwidths required for amateur applications are relatively narrow, and the design of simple patch antennas can be relatively straightforward, although the iterative numerical work can be tedious. It is better, where possible, to use software tools to obtain a basic design, and fine tune one or more prototypes to achieve the desired centre frequency and impedance match.

## Linearly Polarised Patch

The structure shown in **Fig 17.3** behaves as two slots, each of width w and height h, separated by a very low impedance parallel plate transmission line of length L. The length L is approximately $\lambda_g/2$ where $\lambda_g$ is the wavelength inside the transmission

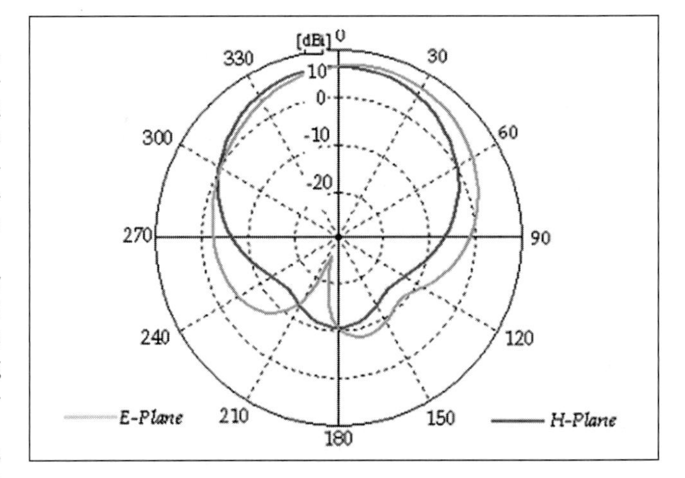


**Figure 17.5: Typical radiation patterns of a patch antenna on a limited ground plane**

line. The slots form a two-element phased array, where the fields from Slots 1 and 3 add in phase to produce maximum radiation normal to the patch, and the fields from Slots 2 and 4 effectively cancel each other. The overall electric field lies in the x-z plane.

If the patch is placed over an infinitely large groundplane, all radiation must be directed in the +z direction (**Fig 17.4**). With a finite ground plane there will always be radiation to the sides and rear of the antenna due to scattering and diffraction around the ground plane edges (**Fig 17.5**). In the extreme case, the ground plane barely extends beyond the patch dimensions (Fig 17.1), allowing radiation over very wide angles as required for mobile devices such as GPS receivers. Calculation of the scattering and diffraction effects on the radiation patterns is complex, and is not generally available in 'free' software.

## Circularly Polarised (CP) Patch

A square patch can be fed at two adjacent edges to enable it to radiate in both the x-z and the y-z planes simultaneously. With some adjustment to the feed lengths, the resultant fields can be arranged to have relative phases 90 degrees apart to produce a circularly polarised field in the z direction. The same result can be achieved by making the patch asymmetrical or feeding it from below the groundplane from a point that generates the correct cavity modes between the patch and the groundplane. According to simple theory, this would lie on the diagonal between opposite corners; in practice, the locus of feed points producing circular polarisation follows an S-shaped curve between the corners, and is dependent on the dielectric constant of the supporting material and the ground plane dimensions. Design is not straightforward and requires microwave modelling tools such as *Ansys HFSS* (no longer available in a freeware version) or *Sonnet* [2, 3] whose 'Lite' version does not include far-field radiation pattern capability. Circular or slightly elliptical patches can also produce CP, but are difficult to produce without photographic PCB etching facilities.

The radiation patterns of well-polarised CP antennas are the same in the x-z and y-z planes, and their comparatively broad beam widths makes a single patch a good candidate as a feed antenna for dishes.

## Software Design Tools for Patch Antennas

With the demise of *Windows XP*, many freely available software tools will not run on later operating systems, although *Windows 7* XP emulation does offer a solution. Whilst the interface is not as slick as later *Windows*-based software, there are a number of DOS-based programs that are both powerful and freely available, and can be run on *Vista*, *Windows 7* and *Windows 8* using *DOSBox* emulator software [4]. One such is *PUFF*, Computer Aided Design for Microwave Integrated Circuits, developed at Caltech and still available [5]. It has also been ported to *Linux* as free software complete with the original manuals under GPL licence.

For those wanting to run software natively under *Windows 7 or 8*, there are cut-down versions of professional microwave design tools such as *Sonet Lite* and *FEKO Lite* [6] that are available as freeware. These tools generally require some effort to learn to use effectively, although free tutorials are obtainable from the manufacturers' websites.

## Antenna for 5.8GHz

This design was produced by Gunthard Kraus. DG8GB, and is offered as an example because of the detailed design notes in the literature [7, 8, 9] for anyone wanting to repeat the design process for other frequencies or PCB materials. The techniques can be used to design patch antennas for any of the lower frequency amateur microwave bands, ie 23cm to 3cm. The majority of the design work used PUFF CAD software [10, 11], although the same techniques can be used with other microwave CAD software tools.

The patch antenna was developed for the 5.8GHz ISM (Industrial, Scientific and Medical) band for monitoring video feeds The 5.8GHz ISM band contains 16 channels at 9MHz intervals. between 5732MHz and 5867MHz. The antenna was thus required to have a self-resonant frequency of 5800MHz and a bandwidth of approximately 140MHz (2.4%). A 50-ohm feed was needed using semi-rigid cable and an SMA connector. The circuit board was made from Rogers R04003 [1], which is a PTFE and woven glass fibre material that is stable both electrically and mechanically, is easy to machine and has the following characteristics:

- $\varepsilon_r = 3.38$
- Printed circuit board thickness = 0/031 inches (0.813mm)
- Dielectric loss factor = 0.001
- Copper coating = 35μm

Good quality FR-4 glass-fibre PCB could probably have been used for this antenna, but tuning and impedance might not be repeatable from one antenna to another. The dimensions would also be very different because of the different dielectric constant of FR-4, about 4.2 at 5GHz. Variations in board thickness and copper glue thicknesses, especially with thin boards, become critical at higher frequencies with sharply resonant (high Q) antennas such as these where repeatability is required. For radio amateurs prepared to spend some time trimming an antenna to resonance, this might not be so important.

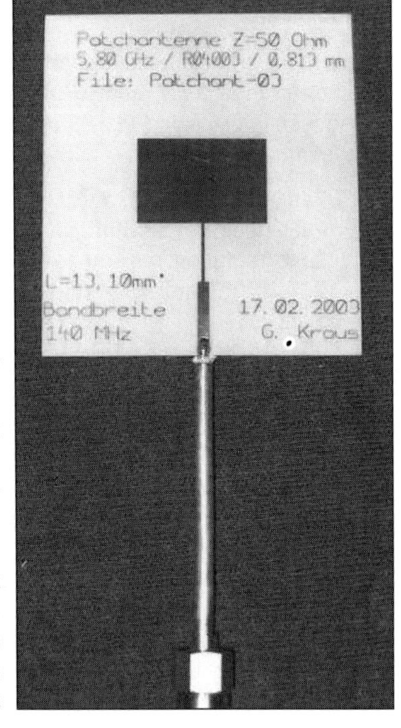


**Fig 17.6: The finished patch antenna for 5.8GHz**

A narrow, high impedance λ/4 line transforms the antenna feed point resistance to the required 50 ohms. The circuit board size was 50mm x 50mm and a short 50-ohm microstrip is also needed from the transformer line to the cable connection at the edge of the circuit hoard. This can be seen clearly in the photograph of the finished device in **Fig 17.6**.

## Design Procedure

The design procedure comprises:

- Obtain a basic patch design
- Model the design to verify the resonant frequency and input impedance of the patch structure on the desired circuit board material
- Design a matching circuit to provide a 50-ohm input impedance at the connector
- Build and measure the performance of a prototype
- Use the measurements to adjust / correct the model, re-design with the new parameters and build a second prototype
- Repeat until the measured results converge on the desired performance

An initial design was produced using the *patch16* program [12], with a centre frequency of exactly 5800MHz, while the bandwidth was deliberately increased to 2.9%. The dimensional inputs and outputs of *patch16* are in inches.

```
These are the design parameters:

 Length (L) =   .5271  inches
 Width (W) =  .8  inches
 Height (H) =   .032  inches
 Dielectric Constant (D) =  3.38
 Loss Tangent (T) =   .001
 Feedpoint Distance (F) =  0  inches

 Do you wish to edit any value? (Y/N):
```

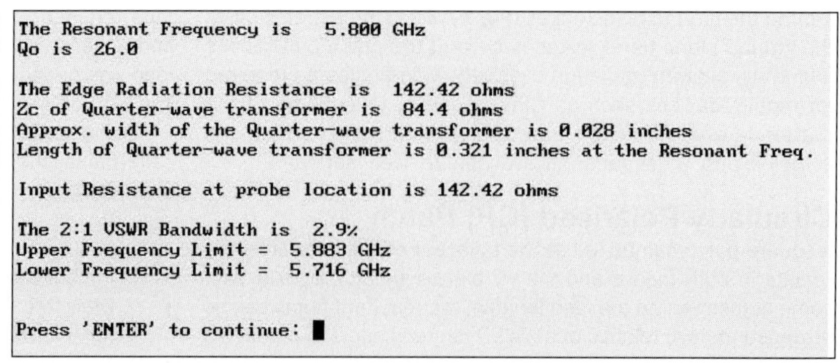


**(above) Fig 17.7: The input parameters for *Patch16***

**(right) Fig 17.8: Simulation results for the patch antenna**

The input and calculated characteristics of the antenna can be seen in **Figs 17.7 and 17.8**. The values required for the subsequent work using *PUFF* [5] must first be converted from inches into millimetres:

- Patch width = 20.32mm
- Patch length = 13.39mm
- Edge radiation resistance = 142.4Ω, equivalent to two 284.8Ω radiators at either patch edge connected in parallel

The patch was modelled in *PUFF* as a wide, lossy transmission line terminated at each end by the individual radiator resistances, 284.8Ω. The modelling frequency was set to exactly 5.8GHz and the impedance of the transmission line adjusted until the known width of 20.32mm was achieved, **Fig 17.9** (*PUFF* will not take transmission line width inputs directly in this mode, although other tools do).

The modelled electrical length of the transmission line was then varied until the structure was as near as possible to resonance at 5.8GHz ($s_{11}$ = 0°), using the smallest swept frequency range and the highest amplitude resolution possible. The results are shown in Fig 17.9. For a patch width of 20.32mm, the microstrip line will have a characteristic impedance of 7.34 ohms, and a mechanical length 14.36m to have an electrical length of λ/2 at 5.8GHz.

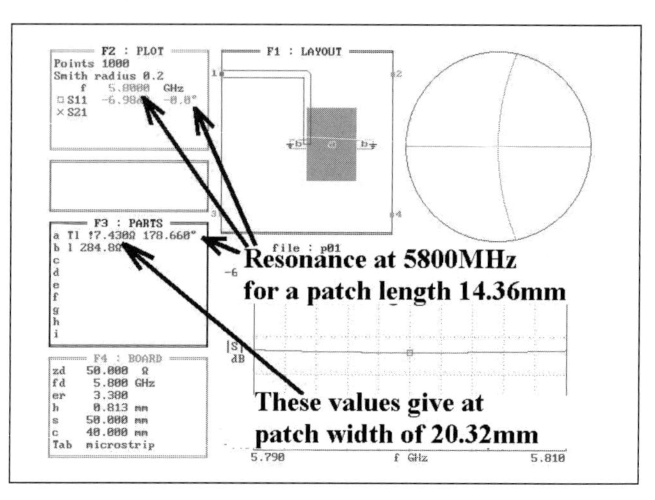


**Fig 17.9: Simulation of the patch antenna using *PUFF***

Microstrip lines operate in mixed dielectric media, ie air above and dielectric below the strip. This results in fringing fields which effectively extend the ends of the line by a factor dependent on the dielectric constant and thickness of the substrate, and the width of the line. Formulas to calculate the extension have been developed for a number of cases [13], but it is often easier to use correction curves such as those provided in the *PUFF* manual, see **Fig 17.10**. Fringing also accounts for a further difference between the calculated λ/2 length and its resonant length of about 1%. Software with more comprehensive features than *PUFF* can carry out these extension calculations and corrections automatically.

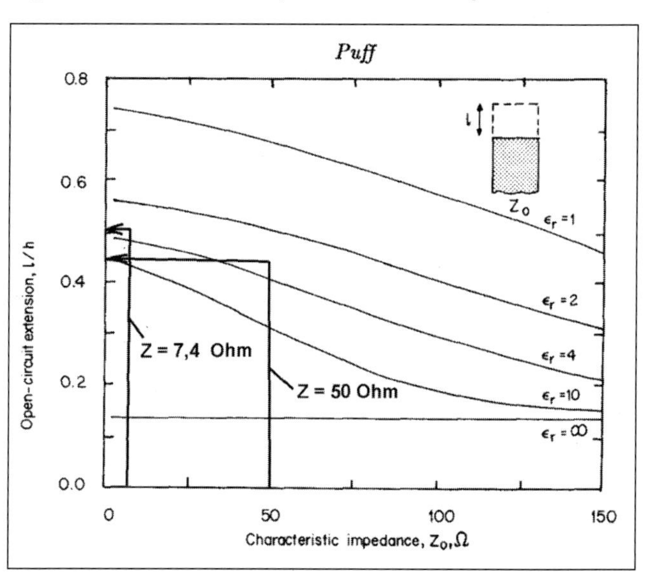


**Fig 17.10: Determining the open end extension using the graph published in the *PUFF* manual**

Overall, the resonant patch length is reduced by the fringing field extension at each end (2 x 0.41mm) and a further 1%, to (14.36 - 2 x 0.41) x 0.99 = 13.40mm as originally calculated by the *patch16* software. So, the real patch will be 13.40mm long, but the simulated patch used by *PUFF* will be 14.36mm long to obtain the correct impedance in the model to allow the matching section to be designed. The input resistance of the patch must be transformed to 50 ohms, which is done with a λ/4 transformer as shown in **Fig 17.11**. The transformer characteristics are:

- Line length = 8.18mm
- Line width = 0.765mm ~0.77mm

A short length of 50-ohm transmission line is needed to carry the signal from the transformer to the edge of the PCB. The transformer section is the highest impedance (narrowest) section of line in the circuit, and its nominal length (8.18mm) must be shortened to take into account the electrical extension effects described above.

Using the curves in Fig 17.10:

- The patch impedance is 7.4Ω, $\varepsilon_r$ =3.38, l/h = 0.51 and the board thickness is 0.813mm, so the patch extension l is 0.51 x 0.813 = 0.415mm.
- The 50Ω line l/h = 0.45, so the extension l is 0.45 x 0.813 = 0.366mm

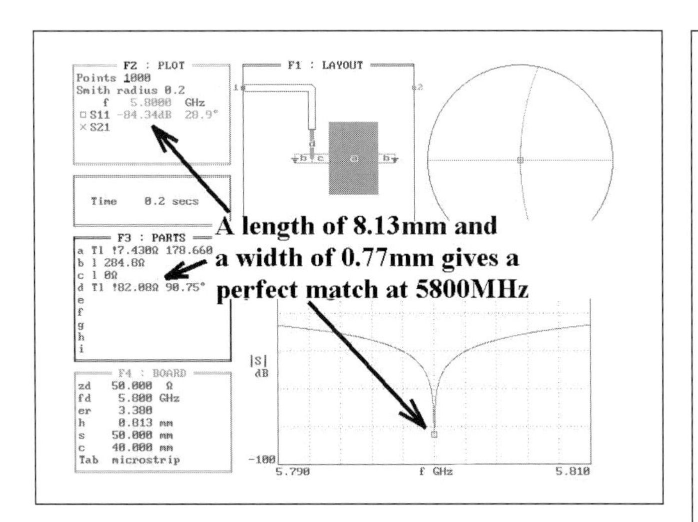


Fig 17.11: Matching the patch to 50 ohms with a quarter-wave transmission line

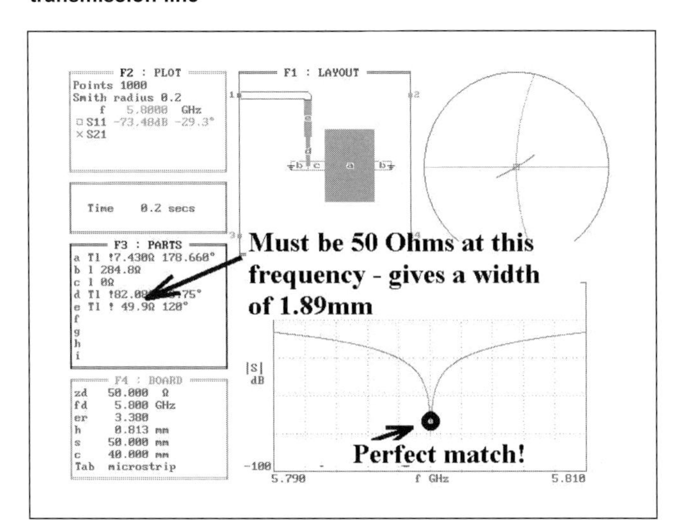


Fig 17.12: Bringing the signal to the edge of the board with a 50-ohm microstrip line

The extension factor of a step impedance discontinuity between two different strip widths is given by [9], [13] as:

$$\Delta L = \left(1 - \frac{w2}{w1}\right) \cdot l$$

The patch radiator end of the transformer must be extended by:

$$\Delta L = \left(1 - \frac{w2}{w1}\right) \cdot l = \left(1 - \frac{0.77}{20.32}\right) \cdot 0.42 = 0.41mm$$

The 50 ohm end of the transformer must be extended by:

$$\Delta L = \left(1 - \frac{w2}{w1}\right) \cdot l = \left(1 - \frac{0.77}{1.89}\right) \cdot 0.37 = 0.22mm$$

So, the corrected length of the transformer section is 8.18 + 0.41 + 0.22 = 8.81mm, and the width remains 0.77mm.

The microstrip feed line between the transformer section and the edge of the board must have a characteristic impedance of 50 ohms, and the overall structure (patch radiator, transformer and feed line) should present an input impedance of 50 ohms at the centre frequency at this point. The width of the feed line is 1.89mm, **Fig 17.12**.

Having calculated all the necessary dimensions, the antenna can be drawn using any PCB CAD tool, such as EAGLE [14], which is freeware and can generate Gerber files if you want to have the board produced professionally.

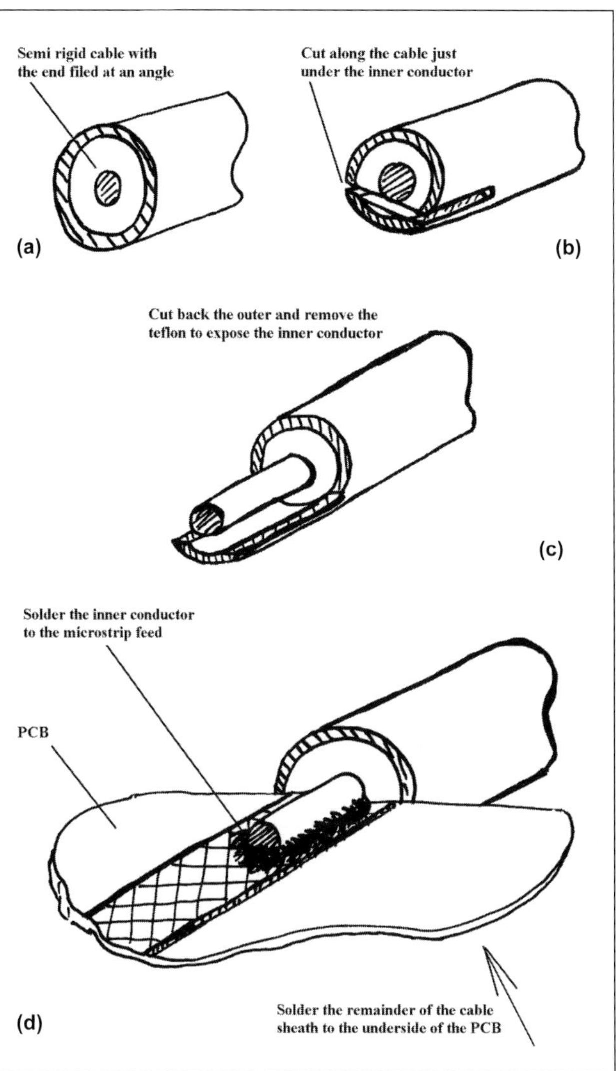


Fig 17.13: Details of how to fit the semi-rigid feeder: (8a) Semi -rigid cable with the end filed at an angle; (8b) Cut along the cable just under the inner conductor; (8c) Cut back the outer and remove the teflon to expose the inner conductor; (8d) Solder the inner conductor to the microstrip feed and the cable sheath to the underside of the PCB

The best way to draw this type of structure is to set out the circuit board and ground plane, then centre the patch and draw the remaining elements around it.

## Transition to Coaxial Cable

This technique can be used to produce a low VSWR transition between coaxial cable and stripline for a range of PCB thicknesses, by choosing the diameter of the semi-rigid cable and suitable coaxial connector (**Fig 17.13**).

- First, saw off the cable and carefully file the end flat at an angle. Do not forget to trim it
- Using a fine saw (eg a jeweller's or craft saw), make a cut parallel to the cable and a few millimetres long, precisely following the inner conductor
- Carefully cut perpendicular to the cable to expose the inner conductor completely and remove the internal Teflon insulation

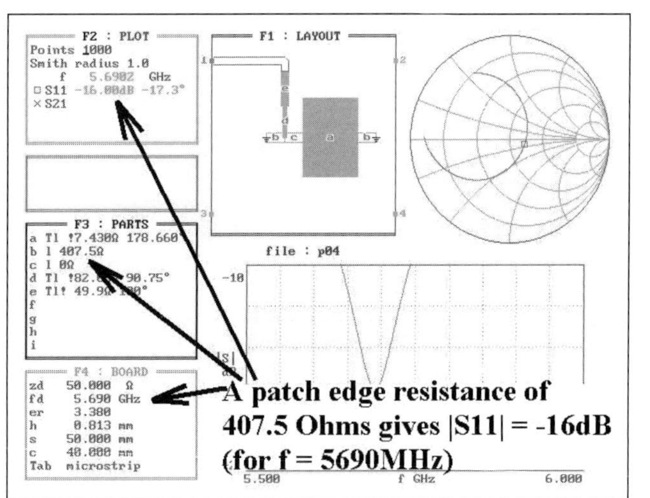


Fig 17.14: *PUFF* simulation showing the patch edge resistance of 407.5Ω and $|S_{11}|$ of -16dB

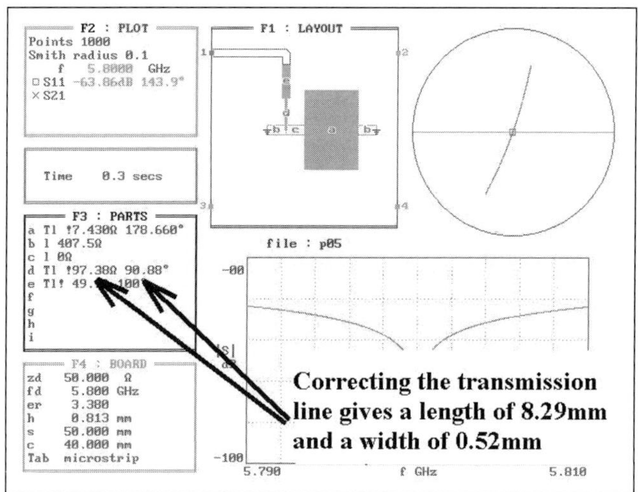


Fig 17.15: Correcting the matching transformer

- Push the circuit board into this cut, solder the inner conductor to the 50Ω microstrip feed line. Carefully solder the remainder of the cable sheathing to the underside.

## Prototype Antenna Performance

Measurements gave the resonant frequency as 5690MHz, with a return loss, $|S_{11}|$, of -16dB.

A closer examination of the input impedance showed that the input resistance was higher than 50 ohms and consequently the radiation resistance of the patch must have been higher than calculated initially. Entering the measured values into *PUFF* showed that the radiation resistance at each edge of the patch should have been 407.5 ohms instead of 284.8 ohms (**Fig 17.14**).

Changing the resonant frequency to the required 5800MHz gives an adjusted patch length of 14.36mm in *PUFF*, or 13.1mm on the PCB. The width is not altered. Using the same methods as before, a better match to 50 ohms is obtained by changing the transformer section to 0.52mm wide and 8.29mm length (**Fig 17.15**). A full description can be found in [7]. Tests on a second prototype using the revised dimensions resulted in a resonant frequency of 5802MHz, very close to the 5800MHz target.

The need to iterate more than once through the design and prototyping stages is typical of patch antenna fabrication, even with very good design tools. In the articles [7, 8, 9], DG8GB goes on to compare these calculations with equivalents using *Mstrip-40* (no longer available) and *Sonnet Lite* [2], which is available for download free of charge.

## SLOT ANTENNA

The slot antenna that has become known in the amateur world as the *Alford slot* actually derives from work by Alan D Blumlein of London, and is detailed in his patent number 515684 dated 7 March 1938. The work by Andrew Alford was carried out during the mid 1940s and 50s, and not applied to microwave bands but to VHF/UHF broadcasting transmitters in the USA.

Development by M Walters, G3JVL, was carried out during 1978 when designs for the GB3IOW 1.3GHz beacon were being investigated. The initial experiment was carried out at 10GHz as the testing was found to be much easier, especially when conducted in a relatively confined space. A rolled copper foil cylinder produced results close to those suggested in the original work. Initially, it was thought that the skeleton version would be best used at the lower frequencies only. However, several models for use on 144MHz have been constructed and they performed very well, but the design was, at the time, not regarded by G3JVL as being of interest.

Further developments have resulted in working models being constructed for the 50MHz, 432MHz, 900MHz and 1.3GHz bands. Early models operated below the designed frequency. A working theory to compensate for this was developed with assistance from G3YGF.

The 2.3GHz version developed by G3JVL fulfils the need for an omni-directional horizontally polarised antenna. This makes it particularly useful for beacons, fixed station monitoring purposes and mobile operation. Mechanical details of this antenna are shown in **Fig 17.16**. The prototype was made from 22mm outside diameter copper water pipe. Material is removed to produce a slotted tube with an outside diameter of 18.5mm and a slot width of 2.6mm. The tube is best formed around a suitable diameter mandrel to maintain circularity along its length. Small tabs are soldered at the top and bottom of the tube to define the slot length of 229mm. A plate is soldered across the bottom of the tube to strengthen the structure.

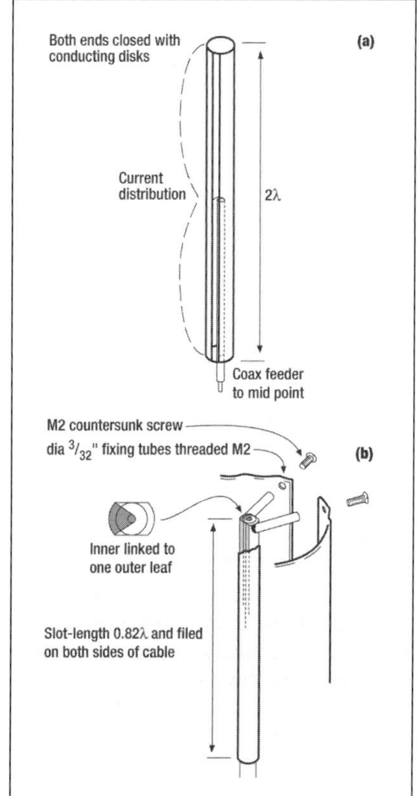


**Fig 17.16: Construction of the 2.3GHz Alford slot antenna using a dual slotted cylinder. The feed point impedance is 200Ω. (a) Dimensions for 2,320MHz are: slot length 280mm, slot width 3mm, tube diameter 19mm by 18SWG. (b) Construction of a suitable balun. The balun slots are 1mm wide and 26mm long**

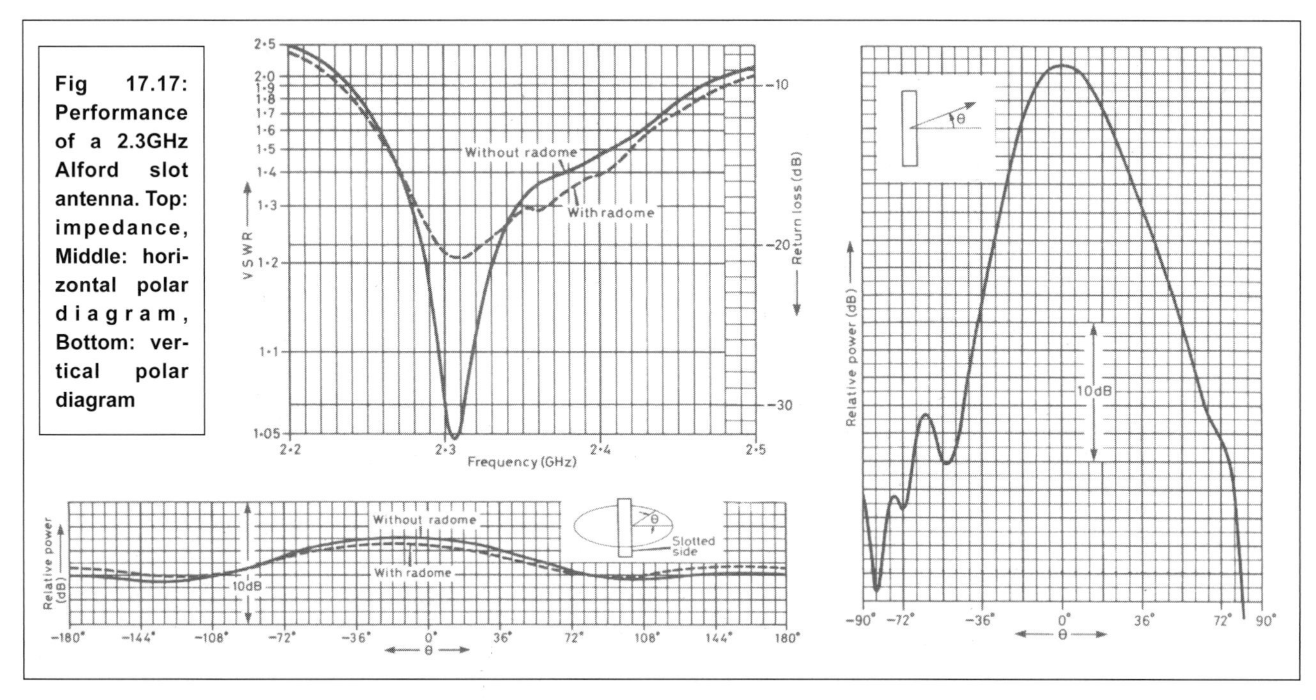


**Fig 17.17: Performance of a 2.3GHz Alford slot antenna. Top: impedance, Middle: horizontal polar diagram, Bottom: vertical polar diagram**

The RF is fed via a length of 0.141in (3.6mm) UT-141 semi-rigid coaxial cable up the centre of the tube to the centre of the slot via a 4:1 balun constructed at the end of the cable. The detailed construction of the balun is shown in Fig 17.16. The two diametrically opposite slots are cut carefully using a small hacksaw with a new blade.

The inner and outer of the cable are shorted using the shortest possible connection and the balun is attached to the slot using two thin copper foil tabs. If suitable test gear is available, the match of the antenna can be optimised by carefully adjusting the width of the slot by squeezing the tube in a vice, or by prising the slot apart with a small screwdriver. Typical antenna characteristics are shown in **Fig 17.17**. The gain of the antenna has been measured to 6.4dBi.

## HELICAL ANTENNA

Paolo Pitacco, IW3QBN, designed this array of 4 x 16 turn helix antennas for 2402MHz [15]. Such a project is an interesting challenge for design, manufacture and testing. The designer had no space for a parabolic reflector and required easy mounting and dismounting in case of strong wind. The idea was to design a system capable of operating as a satellite and terrestrial antenna with discrete gain, good capture area and rapid installation.

As a starting point a simple 16-turn helix antenna [16] was used to hear signals from the DOVE satellite (2401MHz), and to work with homemade ATV systems. The endpoint is an array of four of these helices, shown in **Fig 17.18**.

Radio amateurs today have many 'ready to use' computer programs available for helical antenna design, from simple DOS to complex *Windows* or *Linux* applications. It is possible to choose from any of the following techniques:

- Reproduce the design by G3RUH [17]
- Make a new design using KA1GT's program [18]
- Use Peter Ward's *Excel* worksheet, 'RF2' [19]
- Follow design formulas from antenna engineering literature [20, 21]

This excludes all commercial and costly programs. A mix of these ways was followed to check the results from outputs of several programs, and validate and refine using formulas [20]. The same input data (central frequency, number of turns and mechanical dimension) were used in each case.

The design was centred on 2402MHz, to obtain sufficient bandwidth to cover AO-40 and ATV frequencies. Using this centre frequency, a support diameter of 38mm was chosen, together with 31mm spacing between turns. Copper wire of 3mm diameter was used because it is self-supporting without the need for any type of plastic tube which would cause up to 15% detuning of the helix due to dielectric action.

Matching a single helix is not difficult. It is possible to achieve a broadband match without use of a network analyser or other measuring instruments, but multiple helix arrays are more difficult. Multiple antennas require a coupling system and matching to the transmission line, and the first is related to the second.

The aperture area of each antenna must be evaluated, and a method found to couple the array to a single cable.

**Fig 17.18: The completed 4x16 turn helical antenna for 2402MHz**

There are four methods:

1) Each antenna matched to 50Ω with a 4:1 coupler to the feeder
2) Unmatched (Z~140Ω) with a coaxial quarter-wavelength transformer and a 4:1 coupler to the feeder
3) Unmatched (Z~140Ω) with a quarter-wavelength open wire transformer to the feeder
4) Unmatched (Z~140Ω) with a coaxial quarter-wavelength transformer to the feeder

Solution 1 is easy and is used by G6LVB. Good connectors and cable, and great precision is needed to make four equal helix antennas and cables.

Solution 2 is difficult. Mechanical tools and a lot of connectors and cable are needed to make a good transformer.

By using Solution 3, the helix antennas and feeder can be matched in a single pass because a quarter-wavelength transformer with an output impedance equal to 4 x 50Ω is easy, but a wireline is a potential antenna.

Solution 4 has the advantages of 3, but by using coaxial line you avoid any problem of radiation and the matching line has a more stable impedance. It is a mechanically complex solution, but requires only one connector.

For this project, Solution 4 was chosen, and this drove subsequent mechanical and electrical issues. The helices must be placed at the correct spacing, and a quarter wavelength transformer must be constructed. For 2402MHz, λ is 12.48cm and λ/4 is 3.12cm, but calculations indicate an aperture area of 1.4λ or 17.5cm. A mechanical solution had to be found, using an odd multiple of a quarter wavelength, exactly 3λ/4 (93mm) to maintain transformation properties and distance between helices.

Impedance matching with coaxial transformer is straightforward when mechanical machining is not a problem, using the equation:

$$Z_m = \sqrt{Z_a \, x \left(Z_1 \text{ x } N_a\right)} \qquad (1)$$

Where $Z_a$ is the antenna impedance, $N_a$ is the number of antennas, $Z_l$ is the impedance of main feeder and $Z_m$ the required impedance for a match.

Using the equation below, it is possible to determine tube and conductor diameter to build the matching section.

$$Z_{coax} = \frac{138}{\sqrt{\varepsilon_r}} \cdot \log\left(\frac{D_g}{D_p}\right) \qquad (2)$$

Where $D_g$ stands for inner diameter of tube, $D_p$ is diameter of wire, $\varepsilon_r$ is 1.001 for air).

Because the computed helix impedance is 153Ω, a characteristic impedance of 173Ω is needed for this line. The quarter wavelength transformer was made from a commercially available aluminium tube with an external diameter of 12mm (10mm inside), and a 0.6mm diameter silvered copper wire. This represented a good compromise (about 170Ω).

The wire is held in position with a couple of thin (3mm) centre drilled nylon plugs, this means the transformer's dielectric will be near that of air (see **Fig 17.19**).

All four transformers are locked in position (on the upper side of the reflector) by means of two U-shaped brackets also used as a ground connection. One side is soldered on to an N connector (see **Fig 17.20**). The reflector plate is a made of a 3mm thick square of aluminium sheet with 30cm sides.

Each helix is made with 3mm diameter copper wire, wound on a 35mm diameter support, then gently relaxed and loaded using 11 spacers (every 1.5 turns), as shown in **Fig 17.21**. The

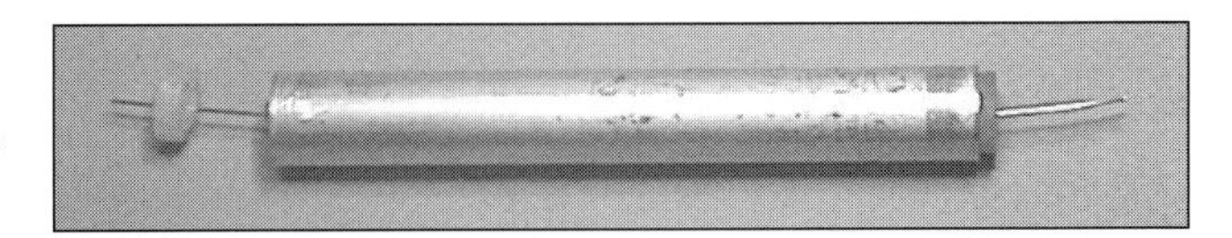

**Fig 17.19: Quarter wavelength transformer for 2402MHz 4 x 16 turn helical antenna**

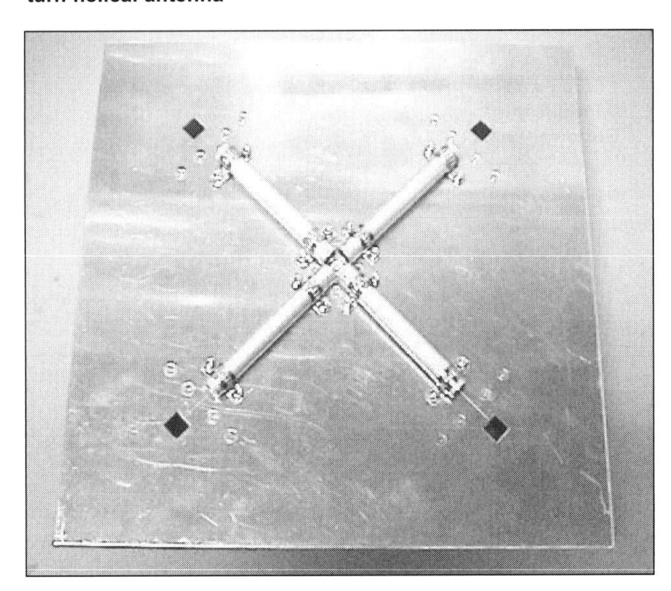

**Fig 17.20: Mounting of the four quarter wavelength transformers for 2402MHz 4 x 16 turn helical antenna**

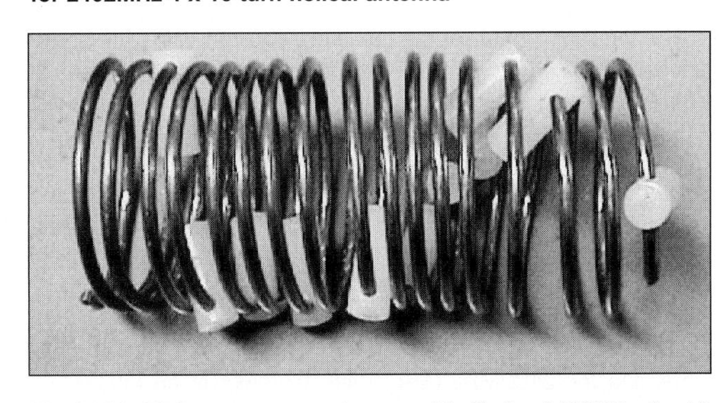

**Fig 17.21: Fitting spacers onto wound helix for 2402MHz 4 x 16 turn helical antenna**

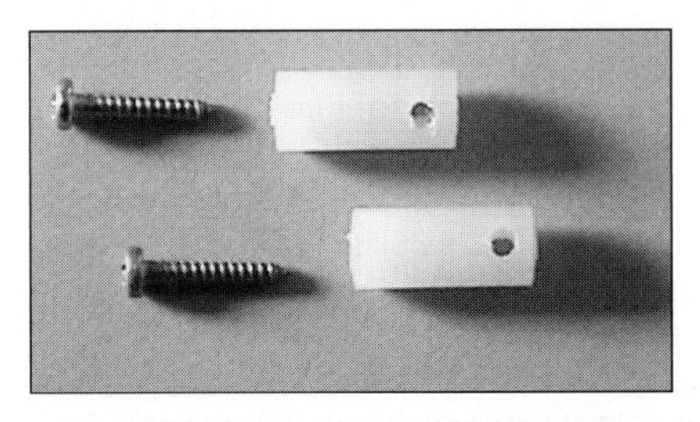

**Fig 17.22: Spacers for the helix of the 2402MHz 4 x 16 turn helical antenna**

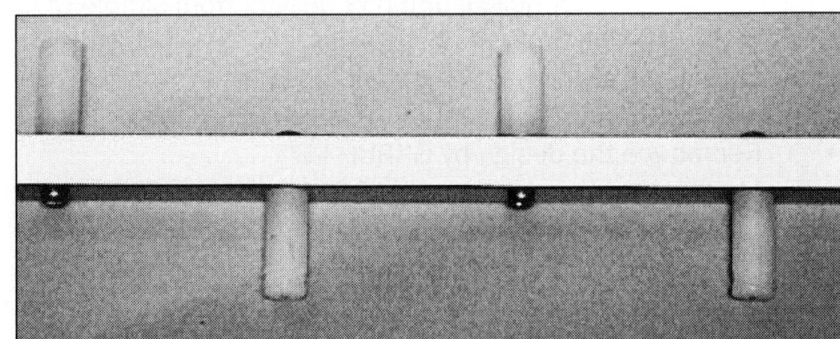

**Fig 17.23: Aluminium support tube for 2402MHz 4 x 16 turn helical antenna**

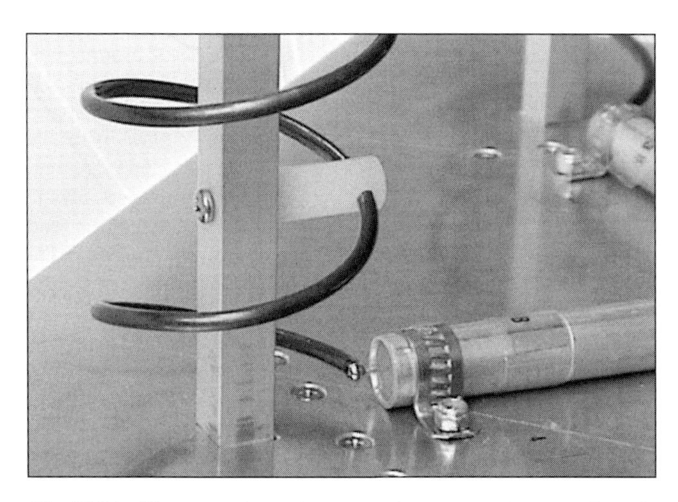

Fig 17.24: Close up view of connection to quarter wavelength transformer for 2402MHz 4 x 16 turn helical antenna

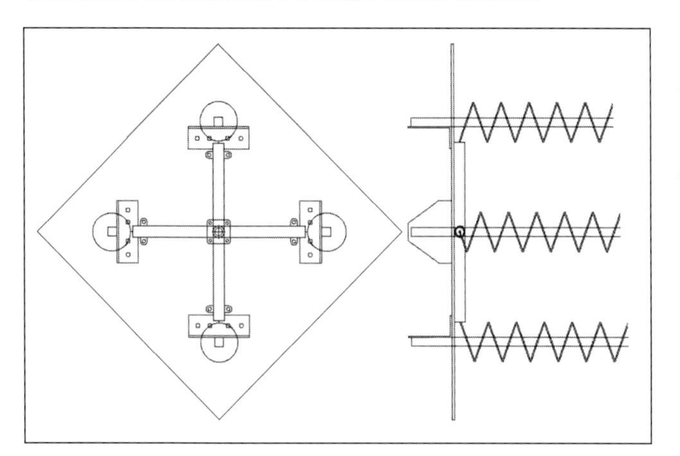

Fig 17.25: Complete mechanical drawing for 2402MHz 4 x 16 turn helical antenna

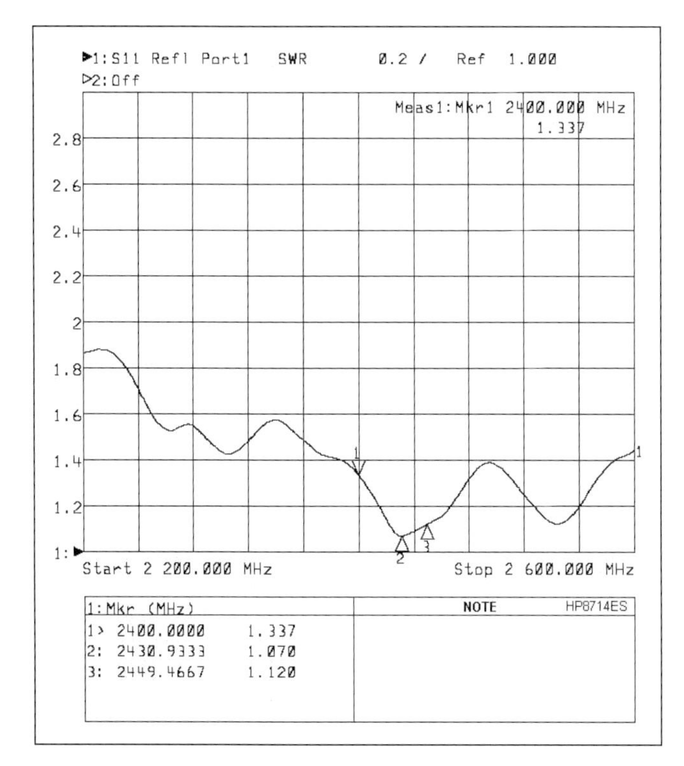


Fig 17.26: Network analyser measurement results for 2402MHz 4 x 16 turn helical antenna

spacers are small cylinders of nylon, 20mm long and 8mm diameter, each pre-drilled with a 3.2mm hole 15.6mm from the bottom (**Fig 17.22**). The first spacer is located at the end of the helix and the last, one turn before the feed point. The spacers are subsequently locked on a square aluminium tube (10mm side, 1mm depth) pre-drilled with 3mm holes at distances of 47mm (one each 1.5 turns) as shown in **Fig 17.23**.

This tube, which is 57cm long, has no effect on performance, and is used to hold the helix in position by means of L-shaped aluminium brackets measuring 25 x 50mm.

Another U-shaped bracket is used to hold converters, preamplifiers or other devices, and as a mounting point (eg a tripod).

All helices are wound in the same direction (counter clockwise as seen from rear) in order to maintain the phase of the feed point. **Fig 17.24** shows the electrical and mechanical connection between the helix and the coaxial transformer. Attention should be given to soldering the inner transformer conductor; pre-solder the 3mm copper wire beforehand. **Fig 17.25** is a complete mechanical drawing of this array.

The first measurement was made with the aid of an Agilent 8714ES Network Analyser, and the results were close to what was expected as shown in **Fig 17.26**. A low SWR was obtained on 2430MHz (near the centre of the S band), but the antenna gives good results over the entire band. Other lengths of transformer (92 and 94mm respectively) have been tried, but 93mm represents the best compromise for amateur use. The calculated gain is 18dB (including losses), and tests on AO-40 signals demonstrate this figure. A comparison was made with a 80cm dish with three-turn helix, and this gave 6-8dB higher gain.

## 2.4GHz Circularly Polarised Feed Antenna

Mode-S operation on the late AO-40 satellite stimulated interest in using small TV dish antennas to receive right hand circularly polarised downlink signals on 2.4GHz, as a higher gain could be obtained than with a single helical antenna. Whilst the patch antenna has been used as a feed by various experimenters with 600mm and larger offset-fed dishes, a short helical antenna feed is also suitable. Although circularity and overall performance is not as good as a well-designed patch, it is easy to construct and needs little adjustment. Note that the polarisation of the feed is reversed on reflection by the dish, and so must be wound Right Handed for a Left Handed CP antenna and vice-versa (see **Fig 17.27**).

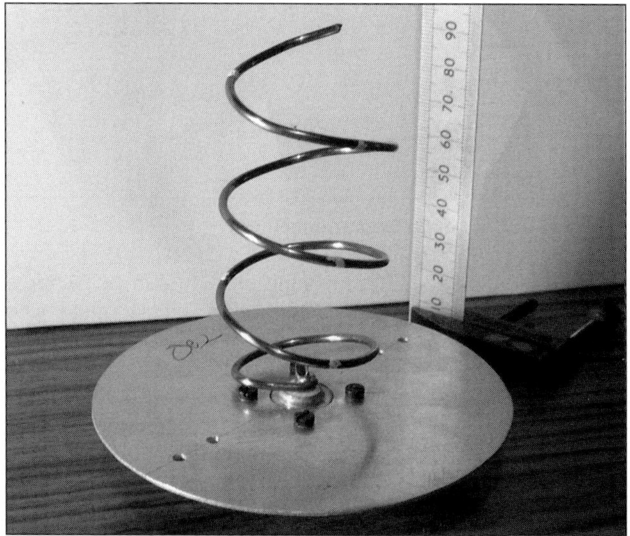

Fig 17.27: Left Hand Circular Polarisation (LHCP) helical feed antenna for 2.4GHz. This results in RHCP on reflection by the dish

For a 600mm dish, 3.25 turns of 2mm diameter enamelled copper wire close wound on a 39mm former and stretched to 80mm in length producing a helix with an internal diameter of about 38mm. The wire is stiff enough to be self-supporting, or 1.5 inch (38mm) plastic water pipe can be inserted down the middle. Do not use a thick plastic pipe cap, as it will affect performance!

A 100mm diameter ground plane with a Type N connector at the centre will support the helix. Form the connector end of the helix into the centre of the helix, remove the enamel and wrap the end round the connector pin so that the wire is parallel and close to the ground plane, to improve the impedance match. Solder in place and make minor adjustments to its position to optimise the match. The helix should work fairly well without adjustment if you do not have appropriate test equipment to optimise the match.

## Dual band feed for Es'hail-2/QO-100

QO-100 is the first amateur radio transponder to be placed in geostationary orbit on the Es'hail-2 satellite at 25.9°East. It uses a right hand circularly polarised uplink on 2.4GHz and a transponder-dependent linear vertical or horizontal polarised downlink at 10.49GHz. Whilst separate helix uplink and dish downlink antennas can be used, co-located feed antennas on a single dish are also popular.

G0MJW, PA3FYM and M0EYT have designed and constructed a number of dual band feed variants based on a left hand circularly polarised patch antenna for the 2.4GHz uplink through which the waveguide of a modified satellite TV LNB passes to receive the 10.49GHz downlink [1], [2]. The assembly is located at the focal point of a suitably sized parabolic dish, which can be either prime focus or offset fed. The reflector reverses the sense of the circular polarisation of the 2.4GHz antenna.

A phase locked loop (PLL) TV LNB can be made sufficiently stable for narrow band operation on QO-100 [3]. These generally have a feed horn of about 55mm diameter, which is too large to place in front of the patch antenna without destroying its performance. The feed horn of the LNB is cut off, and its (circular) waveguide extended with 22mm diameter copper water pipe, to which a copper or brass ground plane for the patch

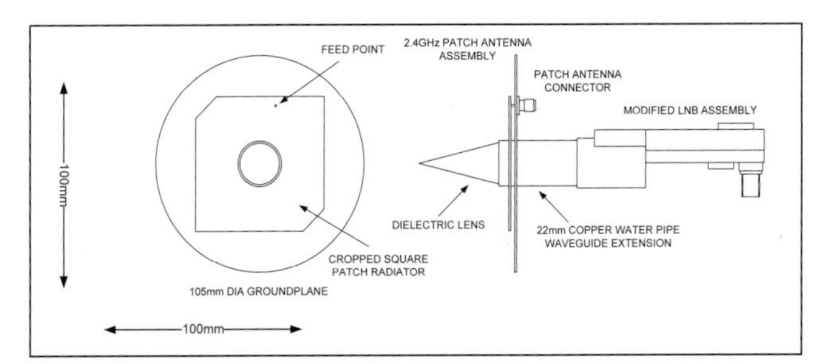


**Image 17.2: Dual band feed arrangement**

antenna is soldered. The radiating element can be either a square plate with cropped corners, or a circular plate with notches cut in it. This is soldered to the copper water pipe exactly 3mm above the groundplane and fed by an appropriate connector soldered to the back of the groundplane, **Image 17.1**. The patch dimensions and its feed point location are critical as it has to resonate in orthogonal planes at two separate frequencies either side of the uplink frequency to provide the required 90° phase difference needed for circular polarisation.

The copper pipe waveguide is attached directly to the LNB. Techniques employed include standard plumbing fittings, see [**Image 17.2**]. Although its open end will illuminate dishes with

**Image 17.2: Dual band feed with square patch for 2.4GHz.**

an F/D of around 0.6, the match is poor. A dielectric lens salvaged from a low cost TV "rocket" LNB (which has a shaped plastic cap in place of the feed horn) or turned from Nylon-6 inserted into the open end of the pipe will improve performance. Alternatively, a short horn made from a 22 - 42mm copper pipe adaptor can be added to the end of the waveguide if the notched circular patch is used [35], [36].

## WIDE BAND DIPOLE ARRAY FOR 23CM

Noel Hunkeler, F5JIO, designed an array of six half-wave dipoles over a reflector plane for a 23cm packet radio link [21]. It provides 10dBd gain and an SWR not exceeding 1.1:1 throughout the band.

Curtains of phased dipoles have been used by amateurs since before WWII, examples are shown in the 1937 Frank C Jones *Antenna Handbook*. F5JIO consulted *Rothammel*, the German antenna 'bible', which gives the following guidelines for sizing the reflector:

> *For best F/B ratio, it should extend at least a half wave beyond the perimeter of the curtain on all sides. If made of wire mesh instead of solid sheet to reduce windage, the wire pitch should be $\lambda/20$ or less. A reflector plane spaced $5\lambda/8$ behind the radiator adds maximum gain, up to 7dB, but a spacing of 0.1 - 0.3$\lambda$ gives better F/B ratio. If spaced at least 0.2$\lambda$ behind the curtain, the reflector plane does not affect the feed point impedance of the array.*

With the dimensions given in **Fig 17.28** and the parts shown in **Table 17.1**, the feed point impedance of each dipole pair is approximately 600 ohms balanced. Three pairs in parallel give 200 ohms balanced. The 4:1 re-entrant line balun transforms this to 50 ohms unbalanced. Note that each dipole is supported at its voltage node; hence the standoffs need not be high quality insulators.

For 23cm this antenna design is small, so a solid aluminium reflector is practical; this plate supports all other components. Bend the phasing rods slightly so they do not touch at the crossovers. For weather protection, a plastic food container serves as a radome. Its RF absorption will be negligible (especially if it is marked "microwavable") and it is much cheaper than Teflon. Although precision is required, construction of this antenna is not difficult.

## YAGI ANTENNAS

### Yagi Antenna for 13cm

Leo Lorentzen, OZ3TZ, designed these 10 and 22 element Yagi antennas for 13cm [22]. The design comes from a popular antenna designer, DL6WU, who has been much copied. While very popular, his writings are also courtesy of Prof H Yagi from Japan, dating right back to 1928 (see the other antenna chapters for more on Yagis).

The Danish version of this Yagi (**Fig 17.29**) is made of standard dimension brass materials that should be readily available. Accurate tools are required for construction at these frequencies, and care needs to be taken over setting out dimensions and cutting. Mark out the dipole positions on the antenna boom from a single datum point and check the marking an extra time before positioning the twenty centre punch marks and then drilling twenty 2mm diameter holes for the directors, which need to be soldered into the boom. Solder the folded dipole on top of the boom.

**Fig 17.30** shows the dimensions of the elements and **Fig 17.31** shows the detail of the dipole.

Drill two 10mm diameter holes in the 80 x 80mm, 1mm thick brass reflector plate, soldering it firmly to the antenna boom.

Behind the reflector, drill two 4mm diameter holes, spaced 75mm apart. Two 4mm diameter nuts are soldered on top of these holes so that the antenna can be secured to a mast fixture or something else. Mount a BNC or TNC socket in the reflector plate to accept the 44/65mm RG178s from the dipole/balun, see **Fig 17.32**.

The really important thing is to solder the screen directly onto the antenna socket and allow the screen to go right up to the inside pin on the socket, where the inner conductor in the cable will be soldered.

If the antenna is to be sited outdoors, Araldite can be used to make 50Ω cabling etc watertight.

The finished antenna fares well in practical use. To measure the gain, an absolutely stable test signal was received on a half-wave dipole connected to a 13cm converter. With the dipole antenna, an approximately S1 signal was received.

A 10-element yagi was then substituted for the dipole and an approximately S3 signal was received. Then the 22-element antenna was connected, and an impressive S7 signal was received.

According to the original designer of the antenna, a boom length of seven wavelengths and approximately twenty elements makes for a gain of something over 15dBd.

## Quad Loop Yagi for 9cm

This antenna was originally designed by Mike Walters, G3JVL. The design is shown in **Fig 17.33** with the critical dimensions shown in **Table 17.2**.

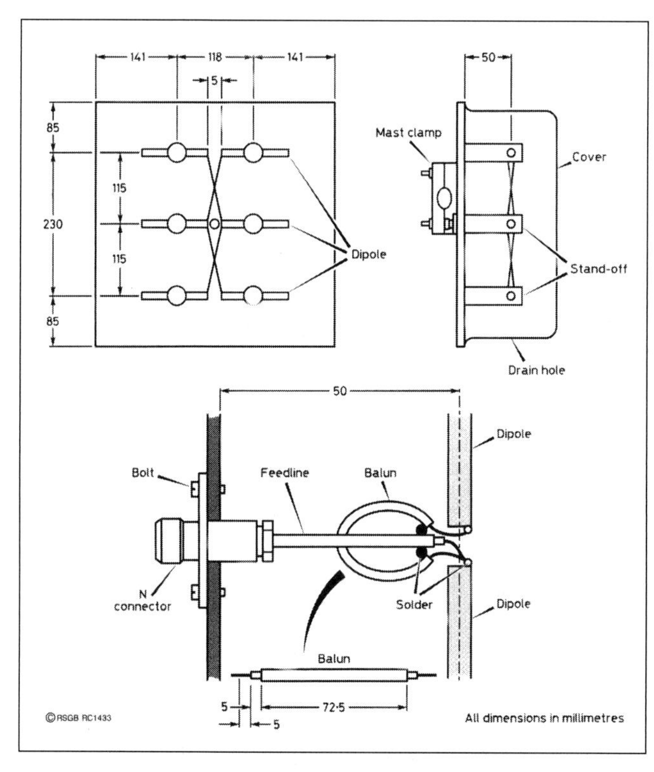


**Fig 17.26: Details of wide-band dipole array for 23cm**

| | |
|---|---|
| **Reflector** | 400 x 400 (340 min) 2.5 thick aluminium sheet (qty 1) |
| **Stand-off** | Teflon (or PVC) 60L x 20D (qty 6) |
| **Dipole** | Brass, silvered, 108L x 6D (qty 6) |
| **Rod, phasing** | Wire, silvered, 2D (qty 4) |
| **N connector** | (qty 1) |
| **Feedline** | Semi rigid coax, 50Ω, approximately 4D (qty 1) |
| **Balun** | as above, 92.5L (qty 1) |
| **Bolt** | M3 x 8, SS (qty 4) |
| **Cover** | Plastic food container (qty 1) |
| **Mast clamp** | From TV antenna (qty 1) |

**Table 17.1: Component list for wideband 23cm dipole array. All dimensions are in mm**

The design is a scaled version of the 1,296MHz design, adapted to the narrowband segment at 3,456MHz. There are 61 directors giving a boom length of 2m. The construction of the antenna is quite straightforward providing care is taken in the marking out process. In marking the boom for instance, measurements of the position of the elements should be made from a single point rather than marking out spacing between individual elements.

All elements are made from 1.6mm diameter welding rod cut to the lengths shown in the table, then formed into a loop as shown in **Fig 17.34a**. The driven element is brazed to a M6 x 25 countersunk screw drilled 3.6mm to accept the semi rigid coaxial cable. All other elements are brazed onto the heads of M4 x 25 countersunk screws. All elements, screws and joints should be protected with a coat of polyurethane varnish after assembly. If inadequate attention is paid to weatherproofing the antenna, its performance will gradually deteriorate as a result of corrosion.

Provided the antenna is carefully constructed, its feed impedance will be close to 50Ω. If a suitably rated power meter or impedance bridge is available, the match may be optimised by carefully bending the reflector loop toward or away from the driven element.

The antenna can be mounted using a suitable antenna clamp. It is essential that the antenna be mounted on a vertical support, as horizontal metalwork in its vicinity can cause severe degradation in its performance.

# HORN ANTENNAS

## 3cm Horn

Large pyramidal horns can be an attractive form of antenna for 10GHz and above. They are fundamentally broadband devices showing virtually perfect match over a wide range of frequencies, certainly over an amateur band. They are simple to design, toler-

**Fig 17.29: Picture of the 22 element Yagi antenna for the 13cm band**

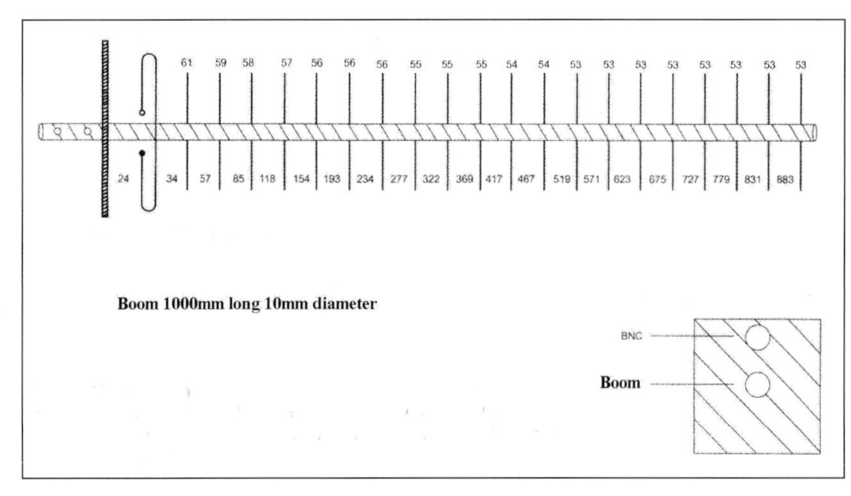


**Fig 17.30: Element dimensions for the 22 element 13cm Yagi**

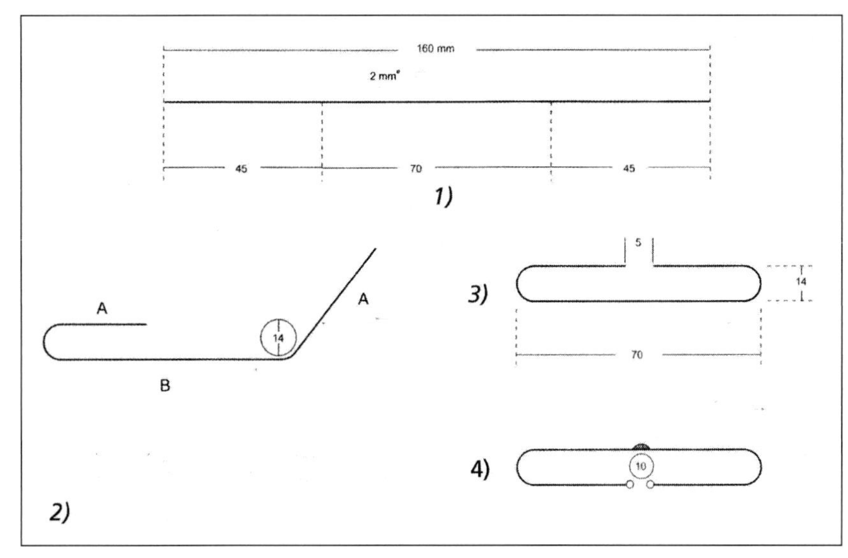


**Fig 17.31: Details of the dipole for the 22 element 13cm Yagi**

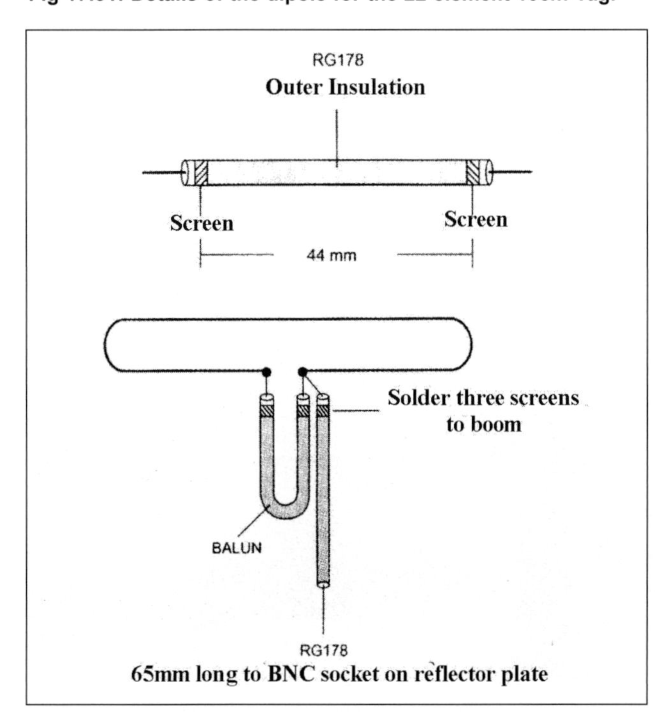


**Fig 17.32: Details of the balun for the 22 element 13cm Yagi**

ant of dimensional inaccuracies during construction and need no adjustment. Their gain can be predicted within a dB or so (by simple measurement of the size of the aperture and length) which makes them useful for both the initial checking of the performance of systems and as references against which other antennas can be judged. Their main disadvantage is that they are bulky compared with other antennas having the same gain.

Large (long) horns, such as that illustrated in **Fig 17.35**, result in an emerging wave which is nearly planar and the gain of the horn is close to the theoretical value of $2\pi AB/\lambda^2$, where A and B are the dimensions of the aperture. For horns which are shorter than optimum for a given aperture, the field near the edge lags in relation to the field along the centre line of the horn and causes a loss in gain.

For very short horns, this leads to the production of large minor lobes in the radiation pattern. Such short horns can, however, be used quite effectively as feeds for a dish. The dimensions for an optimum horn for 10GHz can be calculated from the information given in **Fig 17.36** and, for a 20dB horn, are typically:

- A = 5.19in (132mm)
- B = 4.25in (108mm)
- L = 7.67in (195mm)

Inevitably, there is a trade off between gain and physical size of the horn. At 10GHz this is in the region of 20dB or perhaps slightly higher. Beyond this point it is better to use a small dish. For instance a 27dB horn at 10GHz would have an aperture of 11.8in (300mm) by 8.3in (210mm) and a length of 40.1in (1,019mm) compared to a focal plane dish that would be 12in (305mm) in diameter and have a length of 3in (76mm) for the same gain.

Horns are usually fabricated from solid sheet metal such as brass, copper or tinplate. There is no reason why they should not be made from perforated or expanded metal mesh, provided that the size of the holes is kept below about $\lambda/10$. Construction is simplified if the thickness of the sheet metal is close to the wall thickness of WG16, ie 0.05in or approximately 1.3mm. This simplifies construction of the transition from the waveguide into the horn.

The geometry of the horn is not quite as simple as appears at first sight since it involves a taper from an aspect ratio of about 1:0.8 at the aperture to approximately 2:1 at the waveguide transition.

For a superficially rectangular object, a horn contains few right angles, as shown in **Fig 17.37** an approximately quarter scale template for a nominal 20dB horn at 10.4GHz. If the constructor opts to use the one piece cut and fold method suggested by this figure, it is strongly recommended that a full sized template be drafted on stiff card that can be lightly scored to facilitate bending to final form. This will give the opportunity to correct errors in measurement before transfer onto sheet metal and to prove to the constructor that, on folding, a pyramidal horn is formed!

The sheet is best sawn (or guillotined) rather than cut with tin snips, so that the metal remains flat and undistorted. If the

Fig 17.33: The 9cm quad loop Yagi. Dimensions are shown in Table 17.2

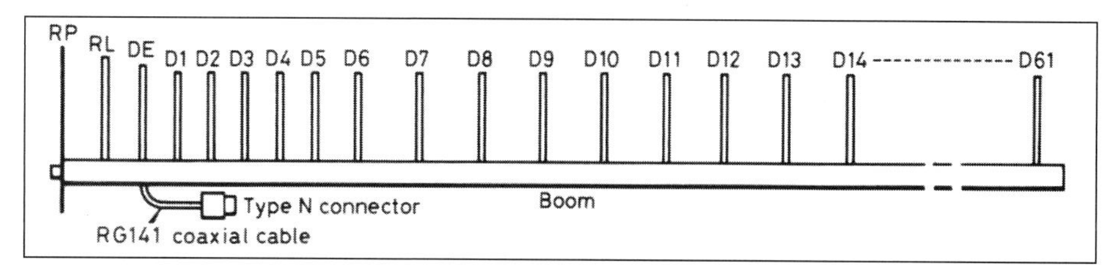


**Boom:**

| | |
|---|---|
| Boom diameter | 12.5mm od |
| Boom length | 2.0m |
| Boom material | aluminium alloy |

**Elements:**

| | |
|---|---|
| Driven element | 1.6mm diameter welding rod |
| All other elements | 1.6mm diameter welding rod |

**Reflector size:**

| | |
|---|---|
| Reflector plate | 52.4 x 42.9mm |

**Element lengths:**

| Length 1 (millimetres) (see Fig 7.34a) | | | |
|---|---|---|---|
| | | Directors 13-20 | 81.2 |
| | | Directors 21-30 | 78.2 |
| Reflector loop | 99.2 | Directors 31-40 | 76.1 |
| Driven element | 92.7 | Directors 41-50 | 75.1 |
| Directors 1-12 | 83.9 | Directors 51-60 | 74.6 |

**Cumulative element spacings (mm):**

| | | | |
|---|---|---|---|
| RP | 0.0 | RL | 29.5 |
| DE | 38.6 | D1 | 49.2 |
| D2 | 57.2 | D3 | 74.1 |
| D4 | 91.1 | D5 | 103.0 |
| D6 | 125.0 | D7 | 158.9 |
| D8 | 192.8 | D9 | 226.7 |
| D10 | 260.6 | D11 | 294.5 |
| D12 | 328.4 | D13 | 362.3 |
| D14 | 396.2 | D15 | 430.1 |
| D16 | 464.1 | D17 | 498.0 |
| D18 | 531.9 | D19 | 565.8 |
| D20 | 599.7 | D21 | 633.6 |
| D22 | 667.5 | D23 | 701.4 |
| D24 | 735.3 | D25 | 769.2 |
| D26 | 803.1 | D27 | 837.0 |
| D28 | 871.0 | D29 | 904.9 |
| D30 | 938.8 | D31 | 972.7 |
| D32 | 1006.6 | D33 | 1040.5 |
| D34 | 1074.4 | D35 | 1108.3 |
| D36 | 1141.2 | D37 | 1176.1 |
| D38 | 1210.0 | D39 | 1244.0 |
| D40 | 1277.9 | D41 | 1311.8 |
| D42 | 1345.7 | D43 | 1379.6 |
| D44 | 1413.5 | D45 | 1447.4 |
| D46 | 1481.3 | D47 | 1515.2 |
| D48 | 1549.1 | D49 | 1583.1 |
| D50 | 1617.0 | D51 | 1650.9 |
| D52 | 1684.8 | D53 | 1718.7 |
| D54 | 1752.6 | D55 | 1786.5 |
| D56 | 1820.4 | D57 | 1854.3 |
| D58 | 1888.2 | D59 | 1922.1 |
| D60 | 1956.1 | D61 | 1990.0 |

Table 17.2: Dimensions of 3.4GHz loop quad

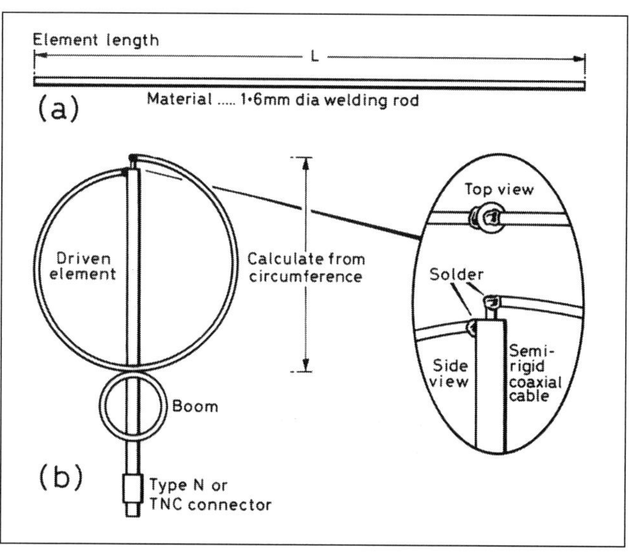


Fig 17.34: (a) Detail of element construction. (b) Assembly of the driven element for the 9cm quad loop Yagi

constructor has difficulty in folding sheet metal, the horn can be made in two or more pieces, although this will introduce more soldered seams which may need jigging during assembly and also strengthening by means of externally soldered angle pieces running along the length of each seam. Alternative methods of construction are suggested in **Fig 17.38**.

It is worth paying attention to the transition point that should present a smooth, stepless profile. The junction should also be mechanically strong, since this is the point where the mechanical stresses are greatest. For all but the smallest horns, some form of strengthening is necessary.

Another method of construction would be to omit the WG16 section and to mount the horn directly into a modified WG16 flange. In this case the thickness of the horn material should be a close match with that of WG16 wall thickness and the flange modified by filing a taper of suitable profile into the flange.

Whichever method of fabrication and assembly is used, good metallic contact at the corners is essential. Soldered joints are very satisfactory provided that the amount of solder in the horn is minimised. If sections of the horn are bolted or riveted

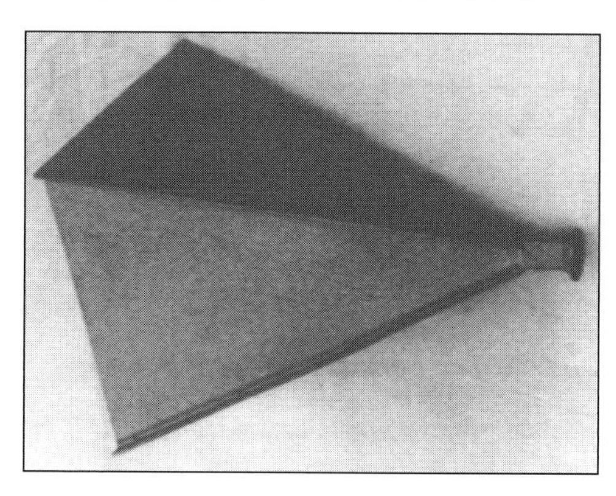

Fig 17.35: A large 10GHz horn

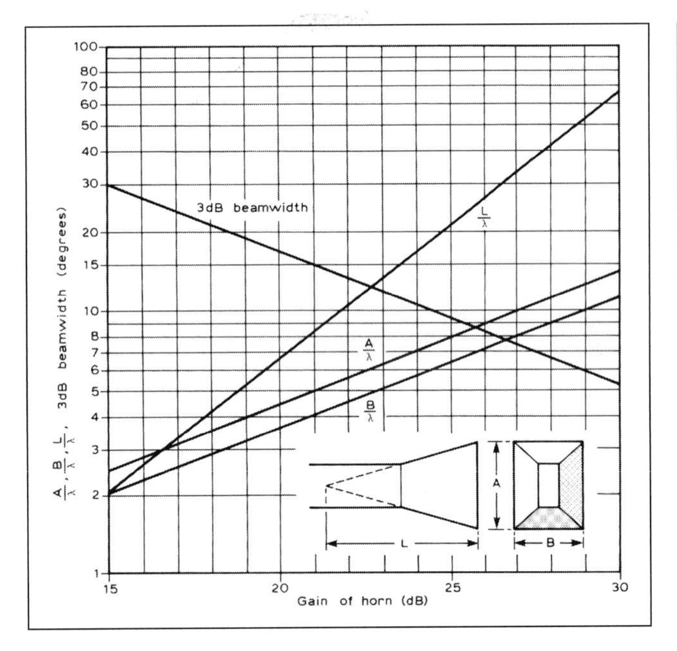


Fig 17.36: Horn antenna design chart

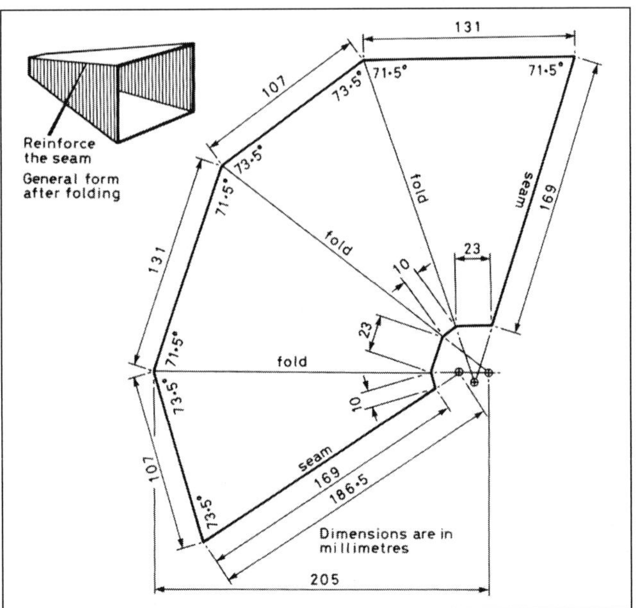


Fig 17.37: Dimensioned template for single piece construction of a 20dB horn

together, it is essential that many, close spaced bolts or rivets are used to ensure good contact. Spacing between adjacent fixing points should be less than a wavelength, ie less than 30mm at 10GHz.

## Rotatable linear polarisation dish feed for 10GHz

G3WDG has devised a rotatable linear polarisation dish feed for EME signal reception, where the signals between stations can be optimised if one of them can adjust its polarisation. The feed, based on an idea by Chapparal Communications Inc. [1] comprises a circular waveguide and feed horn soldered to the broad wall of a section of WG16. A 1.25mm diameter wire hook is placed in the circular guide with its handle passing through the broad wall of the WG16 as an E-field probe, **Image 17.3**. The handle is supported by a PTFE rod that passes through both broad walls of the rectangular waveguide, which allows it to be rotated about its axis and sets the probe depth accurately. RF current induced in the hook is maximum when the plane of the hook is aligned with the E-field in the circular guide. VSWR measuring facilities are needed to adjust the shape of the hook for best performance. G3WDG added a 180° model servo motor to control the angle of the probe remotely, and describes the construction of the feed and control circuit in detail in [38].

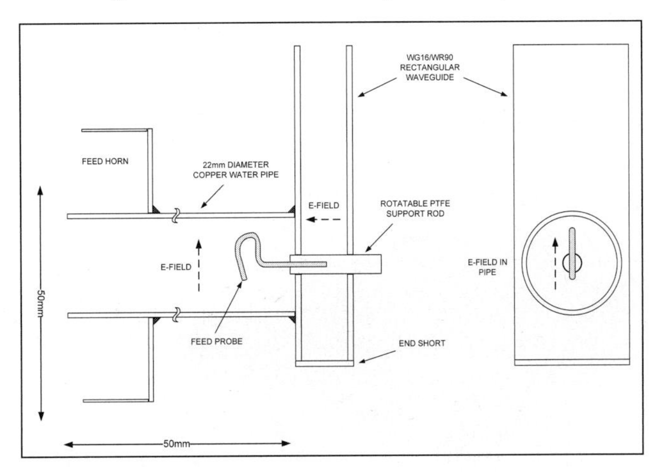


Image 17.3: Elements of rotatable polarisation dish feed

## 24GHz Horn Antenna

A 24GHz horn is a very easy antenna to construct, as its dimensions are not critical and it will provide a good match without any tuning. The dimensions, as per Fig 17.36, for optimum gain horns of various gains at 24GHz are shown in **Table 17.3**. Above 25dB the horn becomes very long and unwieldy and it becomes more practical to use a dish.

Suitable materials are PCB material, brass or copper sheet, or tin plate. If PCB material is used, the double-sided type will enable the joins to be soldered both inside and out for extra strength. An ideal source of tin plate is an empty oil can.

The transition from the guide to the horn should be smooth.

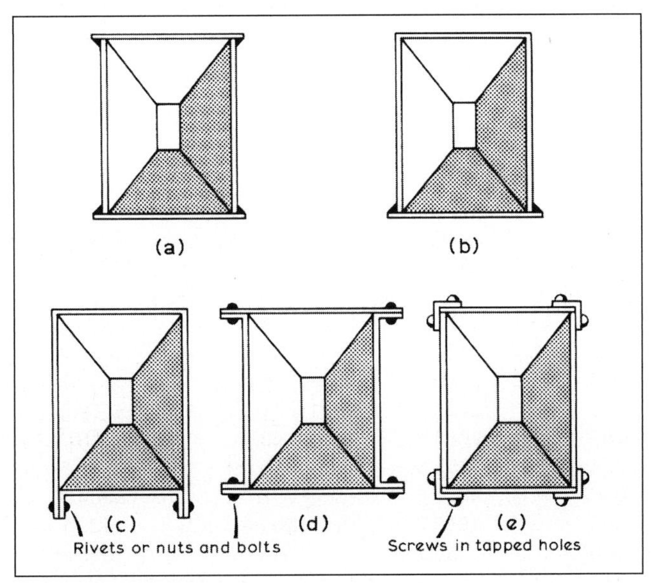


Fig 17.38: Alternative construction methods for constructing horn antennas

| Gain (dB) | Beamwidth (degrees) | Length (mm) | A (mm) | B (mm) |
|---|---|---|---|---|
| 15 | 30 | 26 | 31 | 25 |
| 20 | 17 | 81 | 55 | 45 |
| 25 | 9 | 270 | 98 | 79 |
| 30 | 5 | 819 | 174 | 141 |

Table 17.3: Gain of 24GHz horn antennas

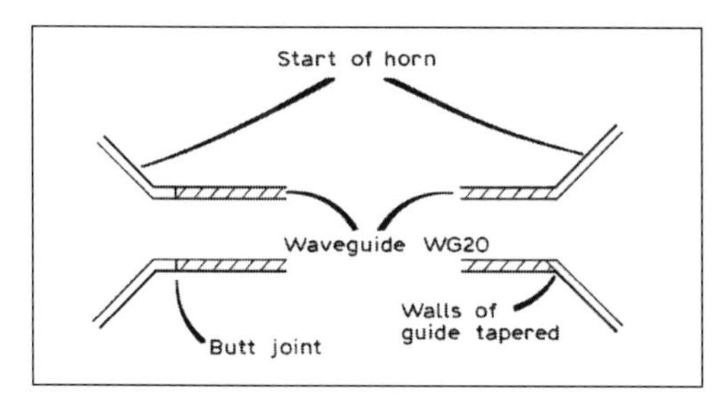


**Fig 17.39: Joining a 24GHz horn directly to a waveguide**

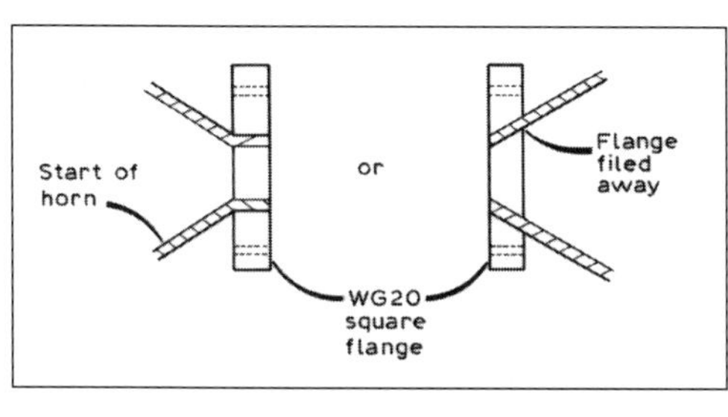


**Fig 17.40: Joining a 24GHz horn directly to a flange**

Use either a butt joint, or file the waveguide walls to a sharp edge, see **Fig 17.39**. The material should be cut to size and then soldered together and onto the end of a piece of waveguide.

Alternatively the horn may be coupled directly to the inside of a flange. In this case the material should be the same thickness as the waveguide would be and it is bent as it enters the flange, or the flange may be filed to be part of the taper, see **Fig 17.40**.

## DISH ANTENNAS

### Offset Feed Parabolic Dish

The spread of consumer satellite TV means that good offset feed dishes can be obtained easily and for reasonable prices. These can be found on *eBay* (or even from a local TV antenna installer) and are known as TVRO (TV Receive Only) or DSS (Digital Satellite System) antennas with sizes from 45cm (18in). They can offer excellent performance up to at least 10GHz [23].

An offset feed dish antenna has a reflector that is a section of a normal parabolic reflector, as shown in **Fig 17.41**. If the section does not include the centre of the dish, then none of the radiated beam is blocked by the feed and support structure. For small dishes, feed blockage in an axial feed dish causes a significant loss in efficiency. Thus, we might expect an offset feed dish to have higher efficiency than a conventional dish of the same aperture.

In addition to higher efficiency, an offset feed dish has another advantage for satellite reception. The dish in **Fig 17.42**, aimed upward toward a satellite, has its feed pointing toward the sky. A conventional dish would have the feed above it, pointing toward the ground, as shown in **Fig 17.43**. Any spillover from the feed pattern of the conventional dish would receive noise from the warm earth, while spillover from the offset dish would receive less noise from the cool sky. Since a modern low noise receiver has a noise temperature much lower than the earth, the conventional dish will be noisier. This affects the figure of merit G/T. The offset dish offers higher gain, G, since the efficiency is higher, plus reduced noise temperature, T, so both terms in the G/T ratio are improved.

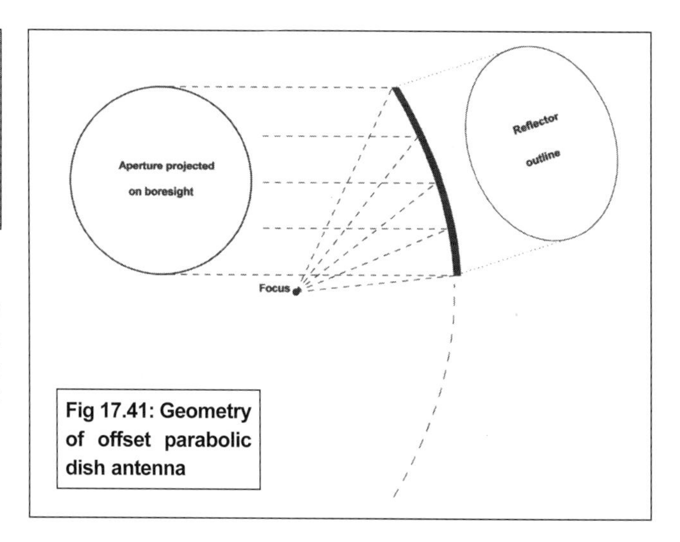


**Fig 17.41: Geometry of offset parabolic dish antenna**

The higher gain means more signals may be received from a source, and the lower noise temperature means that less noise accompanies it, so a higher G/T offers a higher signal-to-noise ratio

The real incentive in the USA to use an offset feed dish was provided by Zack Lau, KH6CP [24] (now W1VT) who pointed out that the 18in RCA DSS dishes were available by mail order for a few dollars. The RCA reflector is oval shaped, but the dish aperture should appear circular when viewed on its boresight (direction of maximum radiation) as shown in Fig 17.41. Thus the dish must be tilted forward for terrestrial operation. The angle, feed point location, and the rest of the dish geometry can be calculated as follows using the technique of Paul Wade, W1GHZ, or by reference to [25]:

If a dish has a strut and fittings to support the feed, the new feed should point along a line at right angles to, and centred on, the feed supporting clamp (usually circular inside). The phase centre for the feed will lie on this line, and be close to, or even inside the original feed support clamp. If the feed support is missing, some measurements and calculations are necessary to locate the focus of the dish and to position the feed.

We need to determine the tilt angle of the reflector so that the beam is horizontal (for terrestrial operation), then do some curve fitting calculations for the dish surface, calculate the focal length, and finally determine the focal point in relation to the offset reflector.

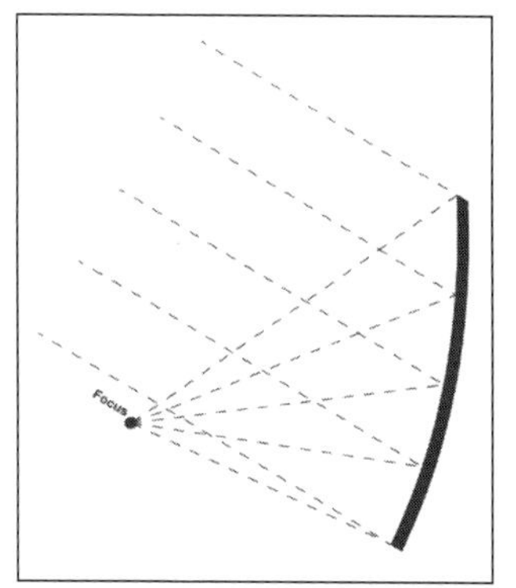


**Fig 17.42: Offset parabolic dish antenna aimed at a satellite**

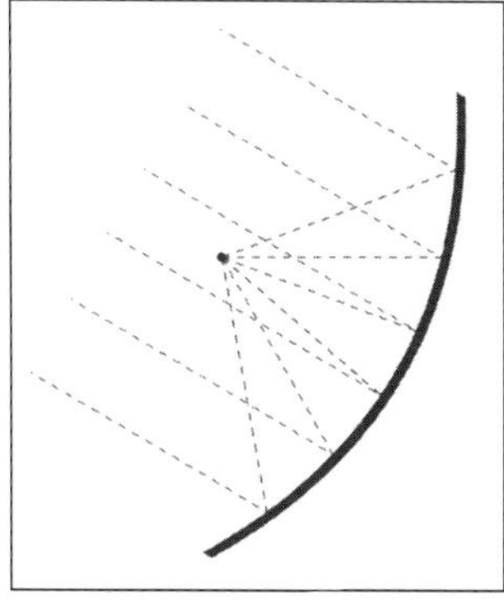

**Fig 17.43: Parabolic dish antenna aimed at a satellite**

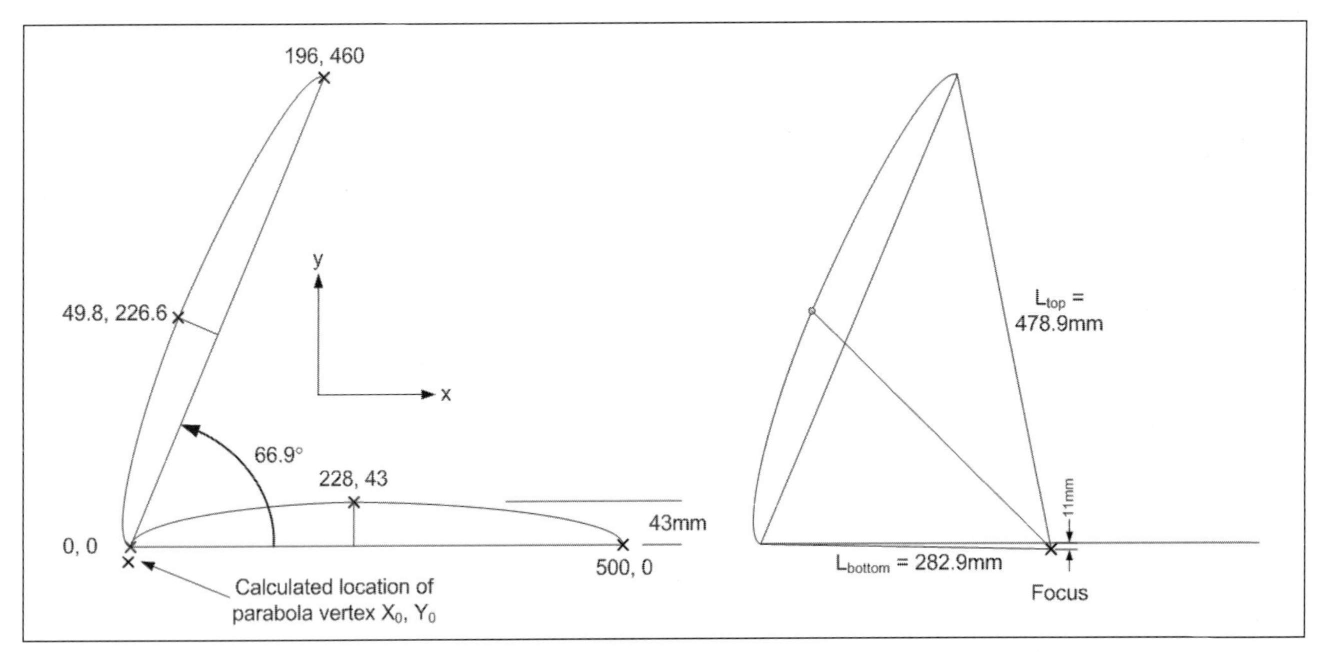


**Fig 17.44: The geometry of an RCA dish as described by W1GHZ**

A common type of offset parabolic reflector has an oval shape, with a long axis from top to bottom and a shorter axis from side to side. With the feed at the bottom, tilt the top of the reflector forward, until it appears circular from a distance, and it will be in the correct orientation to operate with the beam on the horizon. The tilt can be determined much more accurately with a simple calculation:

*Tilt angle (from horizontal) = arcsin (short axis / long axis)*

(Note: *arcsin* is called $\sin^{-1}$ on some scientific calculators)

For the RCA 18in dish, the short axis is 460mm and the long axis is 500mm. Therefore, the tilt angle = arcsin (460/500) = 66.9 degrees above horizontal. At 10GHz, one millimetre is sufficiently accurate for most dish dimensions, so using millimetres for calculations eliminates a lot of tedious decimals.

If the offset reflector is not oval, the essential dimensions can be found by placing it on the ground with the reflecting surface upward and filling it with water. The surface of the water in the dish should be an oval just touching the top and bottom rims, the length of the other axis of the oval provides the smaller dimension of the dish.

The third dimension needed is location and depth of the deepest point in the dish. This is probably not at the centre, but somewhere along the long axis. Using a straight-edge across the rim for an oval dish, or the water depth for other shapes, locate the deepest point and measure its depth and distance from the bottom edge on the long axis.

For the RCA dish, the deepest point is 43mm deep at 228mm from the bottom edge on the long axis. The three essential dish dimensions in x-y coordinates are shown in **Fig 17.44** and listed in **Table 17.4**.

When the dish is rotated around the bottom edge to the tilt angle (66.9°) above horizontal, the new x-y coordinates are calculated as:

$$x' = x*\cos(66.9) - y*\sin(66.9)$$

$$y' = x*\sin(66.9) + y*\cos(66.9)$$

If we assume that the bottom edge is *not* at the axial centre

| | Dish face down | | Dish rotated around bottom edge by 66.9° | |
|---|---|---|---|---|
| **Position** | x | y | x' | y' |
| **Bottom edge** | 0 | 0 | 0 | 0 |
| **Max depth** | 228 | 43 | 49.8 | 226.6 |
| **Top edge** | 500 | 0 | 196 | 460 |

**Table 17.4: RCA dish coordinates before and after rotation**

of a full parabola of rotation (the equivalent conventional dish of which the offset dish is a section), but is offset from the centre by an amount $X_0$, $Y_0$, then all three points must fit the equation:

$$4*f*(X-X_0) = (Y-Y_0)^2$$

The unknowns are $X_0$ and $Y_0$, and f, the focal length. The three x', y' points produce three equations and three unknowns which can be solved by matrix inversion or by substitution:

$$4*f*(0-X_0) = (0-Y_0)^2$$

$$4*f*(49.8-X_0) = (226.6-Y_0)^2$$

$$4*f*(196-X_0) = (460-Y_0)^2$$

f = 282.8mm = 11.13in

$X_0$ = 0.1mm, behind bottom edge

$Y_0$ = 11mm below bottom edge, so the feed doesn't block the aperture at all.

The position of the focus can be found by calculating its distance from the top and bottom of the dish using the rotated dish coordinates:

$$L = \sqrt{(x'-(f-X_0))^2 + (y'-Y_0)^2}$$

$$L_{top} = \sqrt{(196-282.7)^2 + (460+11)^2} = 478.9mm$$

$$L_{top} = \sqrt{(0-282.7)^2 + (0+11)^2} = 282.9mm$$

A knot tied in a piece of string or lacing cord run from the top

Fig 17.45: Using a knot tied in a piece of string to accurately place the feed horn at the focus of an offset dish

and the bottom of the dish accurately locates the focus, which is where the phase centre of the feed should be positioned (**Fig 17.45**).

Version 2 of the HDL_ANT program by W1GHZ will do the calculations and produce these figures. It is available at [26]. W1GHZ conjectured that the bottom edge of the dish should be at the centre of the full parabola, so that $X_0 = 0$ and $Y_0 = 0$. The calculations above can be repeated with slightly different tilt angles until $X_0 = 0$ and $Y_0 = 0$; for the RCA dish, the new tilt angle is 68.3°, a small change from the original estimate of 66. 9°. The focal length is then calculated to be 291mm. The later Version 3 of the HDL_ANT program [27] will do these modified calculations automatically. HDL_ANT. Note that this modified geometry is not necessarily usable with all antennas. The dimensions of a 650mm ChannelMaster dish which had a feed strut and LNB mounting clamp were measured using the techniques described above. The calculated vertex was about 12mm below the lower edge of the dish. Hand calculation and Version 2 of HDL_ANT placed the focus exactly where the LNB feed phase centre would have been. Version 3 of HDL_ANT resulted in an offset of the focus by 8.6 and 4.6mm in the x and y planes respectively. As W1GHZ comments, some empirical adjustment of the feed location may be necessary.

The calculations show the focal length of the dish to be 280mm. If it were a full parabola rather than just an offset section, the diameter would be 929mm, for an f/D = 0.31. However, a feed horn need only illuminate the smaller angle of the offset section, a subtended angle of about 78°.

This subtended angle is the same as a conventional dish with an f/D of 0.69, so a feed horn designed for a 0.69 f/D conventional dish should be suitable.

Rectangular feed horns of two different lengths were designed using G3RPE's graph [28] and *HDL_ANT* were used to design suitable rectangular horns, then two of different lengths were made from flashing copper.

*HDL_ANT* includes an approximation to G3RPE's curves so that the program can be used to design feed horns for both offset and conventional dishes as well as generate templates for constructing them.

The actual reflector geometry has an f/D of 0.31, making focal distance quite critical (it becomes more critical as f/D decreases). At f/D =0.31, the phase centre of the feed horn should be positioned within a quarter wavelength of the focus, ie ±8.5mm for 3cm operation.

To locate the focus accurately, the distance to both the top and bottom of the rim was calculated and a knot tied in a piece of string taped to the rim so that the knot was at the focus when the string is pulled taut, as demonstrated in Fig 17.45. Then a sliding plywood holder for the feed horn was made and taped in place then adjusted so that the knot in the string was at the phase centre of the horn, as shown in **Fig 17.46**. Materials aren't critical when they are not in the antenna beam!

Fig 17.46: Using a plywood holder to position the feed for an offset dish

Where should the feed horn be aimed? On a conventional circularly symmetric dish it is obviously at the centre (vertex) of the parabola. However, an offset feed is much closer to one edge of the dish, so that edge will be illuminated with much more energy than the opposite edge.

The article in reference [29] contains a detailed analysis of the various aiming strategies, concluding that small variations have little effect, so aiming at the centre of the reflector is close enough.

W1RIL, WB1FKF, N1BAQ, and N1BWT set up an antenna range and made some of the measurements to see how well the RCA offset dish performed. With a simple rectangular feed horn they measured 63% efficiency at 10GHz, significantly higher than ever measured on an 18in primary focus dish. Varying the focal distance showed that the calculations were correct and that this dimension was critical.

Methods for setting up an antenna test range and carrying out measurements are described in Chapter 9 of [23]. Note that the source and test antennas must be separated by a significant distance in terms of wavelength if meaningful feed optimisation is to be achieved.

The effects of feed beam width, phase centre and orientation on radiation pattern and gain for circularly symmetric and offset-fed parabolic antennas can also be investigated with the reflector antenna modelling tool ICARA [30] which calculates radiation patterns that can be compared directly as the reflector and/or feed geometry are varied.

## Parabolic Preloaded Stressed Dish for Moonbounce

Above 1GHz, the most favourable antenna form has proved to be the parabolic dish, as it combines high gain derived from its unrivalled directivity with low noise pick-up from the ground by well-suppressed sidelobes when properly designed. This is extremely important for EME applications, where every tenth of a dB signal/noise ratio makes the difference in completing a contact. Moreover, the dish can be used for different amateur bands by just changing the feed.

The following description of a lightweight, fully collapsible parabolic dish antenna (**Fig 17.47**) with a diameter of 4.1m is useful for those amateurs interested in moonbounce, who cannot set up a larger dish as a permanent installation due to confinements of their location.

This so-called 'preloaded stressed dish' was developed by Michael Kohla, DL1YMK, and was successfully used in his portable multiband moonbounce DXpeditions to Ireland, Madeira and Iceland in 2005-2007 on 70cm to 13cm. The idea of a quick to install, and even quicker to dismantle, dish is based on an earlier article by K2RIW in *QST* magazine [31]. The stressed dish does not use parabolically shaped, preformed struts, but generates the (close to) parabolic curvature by bending originally straight spokes with guywires fixed to a central point in front of the focal plane.

Although the parabolic shape will not be perfect with such a stressed dish, f/D ratios up to 0.45 are feasible, which allows circular polarized square feed horns to be used, as described by OK1DFC [32]. This dish was used on 13cm with just such a CP feedhorn made by OK1DFC with surprisingly good performance during the TF/DL1YMK DXpedition.

Except for screws and bolts, which are all stainless steel for durability, all other parts are made from aluminium to save weight. This description need not necessarily be followed in every detail; it should just give an incentive to the enthusiastic moonbouncer.

The central part of the dish is the hub that consists of a solid aluminium disc, 350mm diameter and 25mm thick. The centre of the disc has a 32mm hole with a bush welded to the centre. This bush is 70mm long with an outer diameter of 50mm and a 32mm bore. The bush has a dual purpose:

- the central pole used for the feed support is passed through it and fixed with two screws,
- the bush is the axis for a second aluminium disc with a central hole of 50.5mm and a diameter of 350mm (just the same as the first disc), but only 10mm thick. Details can be seen in **Fig 17.48**.

This second disc is an adapter plate for the two axis rotator system, but, above all, gives essential handling assistance for setting up the dish and fixing the upper mesh segments. It enables the front plate and spokes to be rotated when the four screws that fix it to the adapter plate are not fitted. One person can perform all of the assembly operations (although a helping hand always is welcome).

**Fig 17.47: Parabolic preloaded stressed dish for moonbounce used by Michael Kohla, DL1YMK, in Iceland**

**Fig 17.48: Bush arrangement at the centre of the dish**

It is not advisable to reduce the thickness of the hub or adapter plate because considerable mechanical stress has to be absorbed by the hub assembly when bending the spokes to the desired f/D ratio.

The thicker hub disc has 18 radial holes on the outer rim, all have a diameter of 12.5mm and a depth of 120mm. These holes will accept the 18 spokes of the dish that consist of 18 aluminium tubes 2000mm x 12mm x 1mm. The angle between the radial holes has to be 360° / 18 = 20°. These holes are preferably drilled on a CNC machine to guarantee symmetry.

A special conical anchor piece was made on a lathe; see **Fig 17.49**, to fix the 18 front guy wires for stressing the spokes to the central feed pole. Again, it has an inner diameter of 30.5mm for the protruding feed support pole, a standard 30mm aluminium tube with 3mm wall thickness for improved rigidity. The face of the cone has 18 screw-threaded eyes for easily fixing the guy wires with little steel hooks. A second identical anchor piece provides the fixing of the rear guy wires to the feed pole.

The rear guy wires cannot be omitted under any circumstances because - as experienced by DL1YM on his first portable EI Dxpedition [33] - a dish crash can be induced by high winds from the back of the dish. The guys will not only stabilise the structure against the wind pressure from the rear, but will also

**Fig 17.49: Conical anchor piece for preloaded stressed dish**

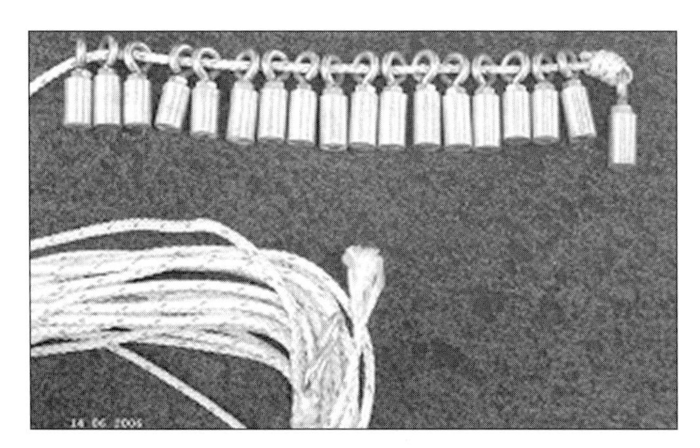

Fig 17.50: Front guy wires for preloaded stressed dish

retain the dish's shape at any elevation from the load of the feed pole with the feed horn and preamp attached.

The front guy wires are all equal and about 1930 millimetres long, the ones on the rear are about 2530mm. The exact length is not critical, as long as a dish depth of 450mm can be adjusted giving an f/D of 0.43 at a focal length of 1760mm. Remember: f = 16 / (depth x diameter). However, the guys have to be made of a very rigid polymer fibre like Polyester (PET, ie Dacron™) because they must not stretch under stress, nor lengthen in humid conditions (Polyamide absorbs up to 10% w/w of water, therefore is completely unsuitable!).

The front guys are fixed to the end of the spokes by means of aluminium dowels (10mm dia x 15mm, see **Fig 17.50**), again with thread eyes at their centre for easy fixing of steel hooks attached to the wires. The dowels can be easily inserted into (and removed from) the open ends of the 12mm spoke tubes. These dowels are strung onto another Dacron line, running through all thread eyes as the circumference of the dish in the stressed state of the struts. This line is 12.90m long giving a dish diameter of 4.1m and the preloading of the struts because they cannot relax into their straight shape, even if the guywires are removed [34].

The feed pole is a standard aluminium tube 1950mm x 30mm x 3mm, extended at the feed end by a glassfibre reinforced polyester tube of 1000mm x 40mm x 4mm. The glass fibre portion must not be shorter than this because a metal structure close to the mouth of the waveguide feeder would severely interfere with the electromagnetic field illuminating the dish, resulting in loss of proper circular polarisation (CP).

The mesh segments are all identical, made from aluminium wire mesh 5mm x 5mm x 0.7mm (galvanized chicken wire is also suitable, but at the expense of additional weight). They are pre-cut from a roll in a trapezoidal shape (1800mm long, 180mm small end, 760mm large end) ensuring an overlap of 40mm when the segments are mounted. The mesh panels are tied to the struts by flower binding wire, this is effective, reusable and cheap.

With this dish, any waveguide type of feeder may be used for 23cm and higher frequency bands; however, the easiest way of mounting a proper CP feed to the support pole is by attaching the original OK1DFC square WG feeder by a set of aluminium clamps, as can be seen in **Fig 17.51**. When stressing the dish to 0.45 f/D or deeper, no flare or choke ring is required on the feed for optimum illumination.

This stressed dish was also used on 70cm with convincing receive capabilities because a dish picks up less ground noise than a comparable yagi array of approximately 21dBd gain. The 70cm feed used was a modified version of the one-wavelength loop feed described by CT1DMK [35] The length of the feed line

Fig 17.51: Square waveguide feeder built by Zdenek Samek, OK1DFC, fitted to the preloaded stressed dish

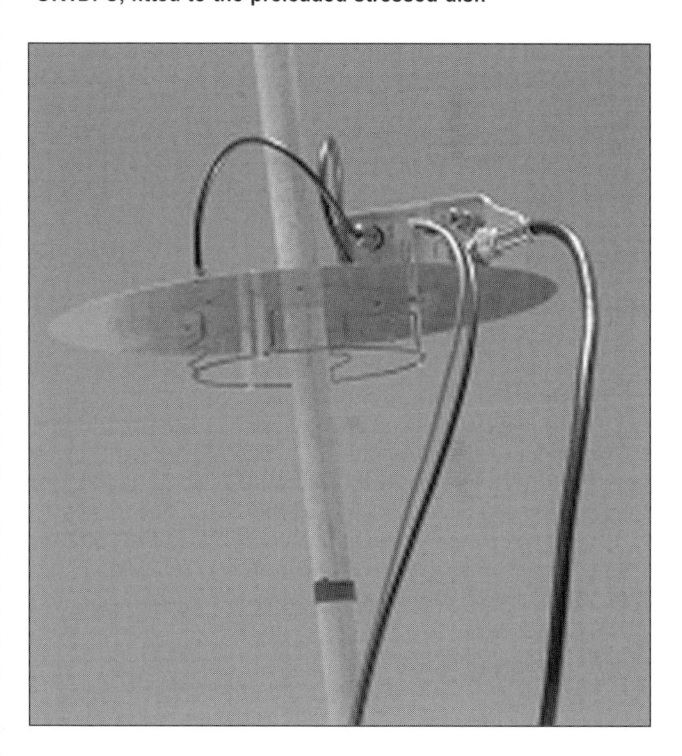

Fig 17.52: 70cm feed from Luis Cupido, CT1DMK, fitted to the preloaded stressed dish

was changed from the original quarter-wavelength to three-quarter-wavelength because the relays used for H/V polarization switching were shorted to ground on NO (the common CX-520D relay), the reflector has a five-eighths-wavelength diameter, which is a trade-off between minimised dish blockage and the desired 3dB beam-width of the feed, see **Fig 17.52**.

## REFERENCES


[1] Rogers Corporation: www.rogercorp.com

[2] Sonet Lite: www.sonnetsoftware.com/products/lite/

[3] 'An interesting program. Sonnet Lite 9.51', G Kraus, DG8GB, *VHF Communications Magazine* 3/2004, pp 156 -178

[4] www.dosbox.com or http://sourceforge.net/projects/dosbox/

[5] www.its.caltech.edu/~mmic/puffindex/puffE/puffE.htm
[6] FEKO Lite: www.feko.info/download/feko-lite
[7] 'Practical project: A patch antenna for 5.8GHz', Gunthard Kraus, DG8GB, *VHF Communications Magazine* 1/2004, pp 20 - 29
[8] 'Modern patch antenna design', Part 1, G Kraus, DG8GB, *VHF Communications Magazine* 1/2001, pp 49 - 63
[9] Modern patch antenna design', Part 2, G Kraus, DG8GB, *VHF Communications Magazine* 2/2001, pp. 66 - 86
[10] 'PUFF - a CAD program for microwave stripline circuits', Robert E Lentz, DL3WR, *VHF Communications Magazine* 2/1991, pp 66 - 69
[11] 'PUFF 2.1 - improved and expanded version', Dipl Ing A Gerstlauer, DG5SEB, *VHF Communications Magazine* 2/1998, pp 97 - 101
[12] Search for PATCH16.zip by WB0DGF on the Internet. This is a DOS program. One source is: www.nic.funet.fi /pub/ham/antenna
[13] *Microstrip Lines and SlotLines*, K.C.Gupta, Ramesh Garg, I.J.Bahl, Artech House
[14] Easily Applicable Graphic Layout Editor (EAGLE), CadSoft Computer, www.cadsoftusa.com
[15] 'An array of 4x16 turn helix antennas for 2402MHz', P Pitacco, IW3QBN, *VHF Communications Magazine* 1/2004, pp 2 - 6
[16] 'Antenne a elica, esperienze utili per tutti', Paolo Pitacco, IW3QBN, *AMSAT-I News* Vol8, N3, p13
[17] 'Un'elica di 16 spire per la banda S', G3RUH (translated I0QIT), *AMSAT-I News* Vol5, N1, p19
[18] 'HELIX.BAS Basic program for Helix design', KA1GT, *UHF/Microwave Experimenters Manual* (ARRL), pp 9-49
[19] "RF2.xls" spreadsheet by Peter Ward AX2VCI, The Numerical Electrical Codes Archive, http://nec-archives.pa3kj.com/
[20] *Antennas*, J D Kraus, R J Marhefka, McGraw Hill - 3th edition.
[21] Contribution by Noel Hunkeler, F5JIO, to 'Technical topics', *RadCom* July 1997, p 72
[22] From an article by Leo Lorentzen, OZ3TZ, *OZ* 1/2001, pp 18 -20 and *OZ* 11/2000, pp. 633 - 635
[23] *Paul Wade Antenna Book*, chapter 5; 'Offset-Fed Parabolic Dish Antennas', www.w1ghz.org/antbook/chap5.pdf
[24] Z Lau, KH6CP/1, 'RF', *QEX*, March 1995, p24
[25] 'Calculating the Focal Point of an Offset Antenna', Ing. Jiri Otypka, CSc, *VHF Communications Magazine*, 1/1995, pp 25 - 30, Note that there are printing errors in some of the formulas
[26] http://www.arrl.org/qexfiles. Click on 1995 tab, then download HDLANT21.zip
[27] *HDL_ANT* ver 3b4: www.w1ghz.org/software/hdl_3b4_all.zip
[28] D Evans, G3RPE, 'Pyramidal horn feeds for paraboidal dishes', *RadCom*, March 1975
[29] V Jamnejad-Dailami and Y Rahmat-Samii, 'Some Important Geometric Features of Conic-Section-Generated Offset Reflector Antennas', *IEEE Transactions on Antennas and Propagation*, AP-28, Nov 1980, pp 952 - 957
[30] Induced Current Analysis of Reflector Antennas ICARA V1.2 Student Edition, software and operating manual. Although advertised for Windows XP, it is compatible with Windows 7, 8 & 10 in both 32 and 64 bit versions: www.com.uvigo.es/ant
[31] 'A Twelve-Foot Stressed Parabolic Dish', R T Knadle, K2RIW, *QST*, Aug. 1972
[32] Z Samek, OK1DFC, paper in *Proc 10th EME-Conference in Prague, 2002*; www.ok1dfc.com/eme/emeweb.htm
[33] Dr M Kohla, DL1YMK, paper in *Proc 12th EME-Conference in Würzburg, 2006*; www.ok1dfc.com/peditions/EI-DL1YMK/index.html
[34] L Cupido, CT1DMK, '70cm Deep Dish Loop Feed', *Dubus* 3/2002
[35] M.Willis G0MJW, Remco den Besten PA3FYM, Paul Marsh M0EYT "Simple dual band dish feed for Es'hail-2/QO-100" Oscar News No.225 March 2019 pp 34-38 (AMSAT-UK).
[36] https://uhf-satcom.com/blog/patch_antenna
[37] e.g. Luiz EA3DOM and Dieter DF9MP "Low cost TVsat LNB Disciplined to a 10MHz reference" Oscar News No.212 December 2015 pp18 – 20 (AMSAT UK). A web search will produce a number of variants dependent on LNB type.
[38] http://www.google.co.uk/patents/US4414516
[39] C.Suckling G3WDG "A rotatable linear polarisation dish feed for 10GHz" Dubus Vol.46 Q3 pp8-18

***Peter Swallow**, G8EZE, was directly involved with the design and development of fixed and mobile antennas for 14 years, and continues to experiment with antennas for the VHF and low microwave bands to support his interest in amateur satellite communications. He is currently a systems engineering consultant with a company specialising in communications technologies.*

# Outdoor Portable Operating

18

*Jamie Davies, MM0JMI and Richard Marshall, G4ERP*

One aspect of amateur radio that has seen huge growth in recent years is that of outdoor operating. By this we mean the use of a portable or mobile station without recourse to mains power and where the station is carried in its entirety to the intended location by the operator. This style of operating includes Summits on the Air (SOTA) [1] activations, Worked all Britain [2] operations from remote areas and even long-distance backpacking with amateur radio.

A number of factors have led to this rise in popularity. Amongst them are the availability of small multiband radios, the success of the SOTA programme and perhaps the difficulty many face in operating a radio station effectively in a modern urban environment. Done well, it is an excellent way to combine two hobbies. A love of walking and the countryside enhances the operation of the radio station and vice versa. Whether this is a local walk to your nearest hill, a night spent in the mountains while climbing bigger SOTA summits or backpacking along one of our National Trails with a radio for company, the enjoyment can be immense. Done badly, though, it can be the ruination of both the walk and the operating session; the difference between doing well and badly come down to attitude, design and planning.

Outdoor operating poses many interesting challenges to the technically-minded. It presents numerous opportunities for the construction of equipment that would perhaps be bought from commercial sources given more normal operating constraints.

Inevitably, when weight is the most serious consideration, compromises in performance may have to be made. It is the balance between performance, weight and ease of set-up, and the effect of that balance on the engineering of an appropriate station, that are the main topics of this chapter.

## ALTITUDE AND ATTITUDE

Before committing to an outdoor setup, ask yourself what you most enjoy about radio. If it is the ease with which you crank up the linear and 5-element beam in your home station to spend hours breaking all pileups and getting 5,9 reports from the other side of the planet, then SOTA may not be for you. Whatever you manage to lug up the hill, with an effort that takes all the fun out of the walk, will be a disappointment by the standards of that massive home rig. If, on the other hand, you enjoy being surprised by how far a few watts and a bit of wire can get you on a lucky day, then you can take your radio to the hills with very little extra investment of energy and be delighted to gain a few contacts, even if some are from distant towns you can actually see on the landscape below. The temptation to carry just a bit more to do just a bit more will always be there but beware: in the graph between weight of equipment and overall fun, there is a definite peak, beyond which the fun falls away quite quickly. Be modest in expectations, and enjoy the fact that low power transceivers and simple antennae can be small and light. Even a 2m or 70cm handheld FM unit that will fit in the pocket can gain contacts 100 miles away from a high summit, and QRP CW or digital modes can reach thousands of miles on a good day, especially if the receiving station is also somewhere far from QRM.

## RELIABILITY

There are two reasons to focus on reliability: one is that outdoor environments and the shaking inherent in being carrier puts extra stress on components, and the other is that it is much more annoying to have something fail when you have carried it all the way up a mountain than when you are at home with tools and a kettle. There are two main strategies; to attempt to engineer very high reliability, or to carry enough equipment to mitigate against failures.

### Redundancy

A simple approach to reliability is to take duplicate equipment, but this approach can greatly increase the load that has to be carried. A reasonable compromise might be either to carry duplicate of only particularly vulnerable items, or to take a small FM handheld and halfwave vertical as a backup for a larger main station. This can be useful anyway, when conditions on a mountain top make erecting an antenna impossible.

### Identifying vulnerabilities

In attempting to minimise the chances of a failure halting operations, it is sensible to review each piece of equipment and estimate the likelihood and consequence of a failure would be to the station. Then it can either be re-engineered to reduce the possibility of a failure or backup provided that would at least allow operating to continue. The most common causes of failures are power (usually simply forgetting to recharge a battery), cables and connectors, be they microphone, control or coaxial. These are potentially fragile items anyway, but using them 'in the field' dramatically increases the chance of a failure. Use a few types of cable as possible, so that one or two spares can stand in for a range of cables in the station: keep this standardization of cables and connectors in mind if home-brewing or modifying equipment, for example by choosing BNC throughout instead of a mixture of BNC, PL259 and N-type. Standardizing on connector types does, however, carry the risk of making wrong connections: you pay for the ease of substitution of a bad cable with the need to think carefully whenever you assemble your station.

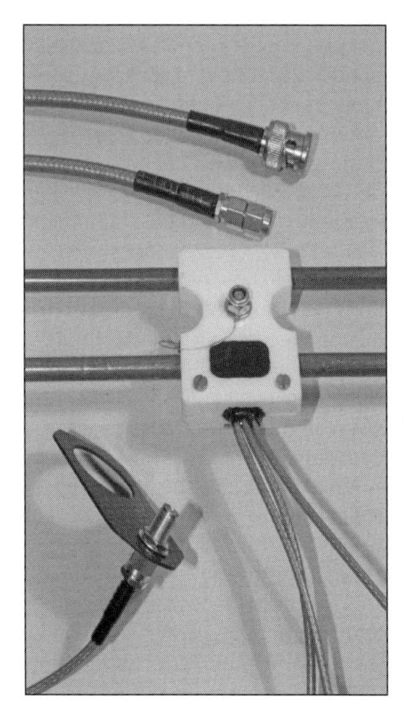

**Fig 18.1: Strain relief applied to antenna components. From the top: 1) Stiff Raychem Viton heatshrink sleeving has been applied to the crimped connections on the coax. This is glued in place using special RT125 two-part epoxy adhesive. 2) The coax balun joints and cable exit from the folded dipole have been encapsulated using RT125. 3) A carbon fibre strain relief plate has been added to prevent the weight of the coax downlead from damaging the more fragile RG316 of the dipole feed**

**Fig 18.2: Strain relief applied to control cables. 1) Top centre shows how the cable into the centre of this paddle key is trapped and held by the Tufnol block. Each connection has a small length of Kynar heatshrink sleeving covering the solder joints. 2) Top left and right are examples of heatshrink boots being used to protect the cable exits from connectors. 3) A headphone splitter made from three moulded cables. The splice contains sleeved, Kynar-covered joints with repair loops and there is a heatshrink covering for overall protection of the split. 4) Bottom left is an example of 'repair loops' - a small loop allowing the wire to be stretched slightly without damage. Note the crimped pins. 5) Deutsch Autosport connector and Raychem boot prior to shrinking. The connector is the intended housing for the crimped pins to their left**

Antenna supports can bend or snap - a small roll of duct tape can be useful for make-shift repairs, especially if used in connection with a splint improvised from twigs at the site.

## Making equipment more durable.

The topic of reliability is high on the agenda in military, aerospace and in motor racing circles and it is not surprising that similar techniques are used in all three. These techniques are mostly based on applying strain relief and waterproofing to both cables and equipment. Damage caused by tension on cables, or corrosion caused by water ingress, are two very common causes of failure. Equipment designed for use in the home shack or car is unlikely to have been designed to survive the climatic extremes of outdoor operating. **Fig 18.1** shows several instances where strain relief has been applied to antenna components. **Fig 18.2** shows how control and other cables can be engineered to last. Generally, crimped connectors are more reliable than the soldered equivalent if done correctly and are usually lighter. Solder joints generally fail at the sharp transition between the solder joint and wire. Kynar sleeving placed over a joint can dramatically reduce the chance of a failure. Heatshrink sleeving is an exceptionally useful product for protecting and strain-relieving electrical interconnections. It should, however, be used with care. In its intended environment it is used with matching high-temperature wires and cables. It should be used carefully with PVC cables where the heat that is required to shrink the sleeving can also cause the PVC to melt and the wires to fuse together. If a sound is heard when the cable is flexed after assembly, this is a good indication that it has been damaged.

## ANTENNAS

This is one aspect of outdoor operating where the keen constructor can make a significant difference to the efficiency of the station. Antennas intended for long-term use at home are likely to be far too bulky and heavy for the outdoor operator to carry. Efficient antennas for all the amateur bands can be constructed that will be far lighter than anything available from the usual commercial sources. This section explores a selection of designs that have been proven to work. Brief descriptions and constructional details are given for each type along with approximate weights and performance figures.

## VHF/UHF

Many new entrants to SOTA begin with VHF, because the investment is modest and many will already have a handheld FM unit, adequate for hills within a few tens of miles of centres of population. A diverse range of antenna designs can be used on the VHF bands. They all fall into one of two categories - beams and omnidirectional verticals.

Verticals are simple to erect and are often rucksack-mounted for ease of use and lightness. The handheld / 'rubber duck' antenna combination is the ultimate expression of this approach and is perfectly adequate in populous areas. Beams include both Yagis and quads (see the chapter on practical VHF antennas) although the latter are a heavier and more bulky solution.

The choice of a suitable antenna, of course, is entirely down to the philosophy of the operator and, in the SOTA context, the difficulty of qualifying a summit. This section offers a range of antennas from both groups which should satisfy most situations and tastes in operating styles. Having a range of antennas available allows the weight-conscious operator to choose the most appropriate solution for a particular expedition. Whilst the handheld / 'rubber duck' approach may well allow a summit qualification in some parts of the country, it is unlikely to succeed in more remote areas where a high-gain beam and 50 watts would be more suitable.

## Verticals

Those venturing to more remote areas need to consider vertical antennas somewhat more seriously. Whilst the standard 'rubber duck' type of antenna supplied with all handheld radios is useful, the efficiency is poor in comparison with a resonant antenna such as a quarter-wave ground plane or collinear. For those who do not wish to carry a beam or specifically want omnidirectional radiation from their antenna, some of the options are described in this section.

Although it is not in the same league as a beam, a rucksack-mounted vertical can at least be used in almost any weather conditions. It can also be used whilst on the move to keep in touch with other expeditions; I have made numerous contacts with stations on other summits by this means. It is necessary, however, for the user to remember that she is carrying it when approaching trees! A useful secondary effect of using a halfwave vertical is that a reduction in IMD products may be noticed thanks to the frequency selectivity created by using a resonant antenna and its matching network.

## A rucksack-mounted vertical for 2m

This variation on the halfwave theme employs a standard telescopic antenna salvaged from a broadcast radio as the radiating element. A length of around 3/8 wavelength or 700mm is ideal, resulting in a practical Q in the tuned circuit and easier matching than with a full halfwave. A 50 ohm match to the transceiver can be achieved by adding a tap part-way up the coil as shown in **Fig 18.3**.

Telescopic antennas often have a hinged joint and this can be employed to good effect to achieve a polarity change. In my example, a modification was required to allow the whip to be locked in the vertical position, the normal friction lock proving inadequate. This was achieved by adding a small aluminium collar that slides down over the joint locking it solid. This collar can be seen in **Fig 18.4** and **Fig 18.5**. A small modification in a lathe was also required to the whip section to produce a parallel end to the lower section.

The matching network is housed in a small aluminium enclosure whose lower portion fits over one of the mast sections which is in turn mounted on the rucksack. It is necessary to prevent the whip from rotating with respect to the housing as this would cause the connection to the matching network to fail. This is achieved with a small insulating block which is a tight fit inside the housing. The whip is attached by a screw which locates in a tapped hole in the block. Under the head of the screw (sized to match the hole in the whip) is a solder tag which is the top connection for the parallel tuned circuit. The whip is insulated from the housing by a PTFE bush which is an interference fit in the housing. A second solder tag and screw at the base of the housing connects the opposite end of the matching network to ground.

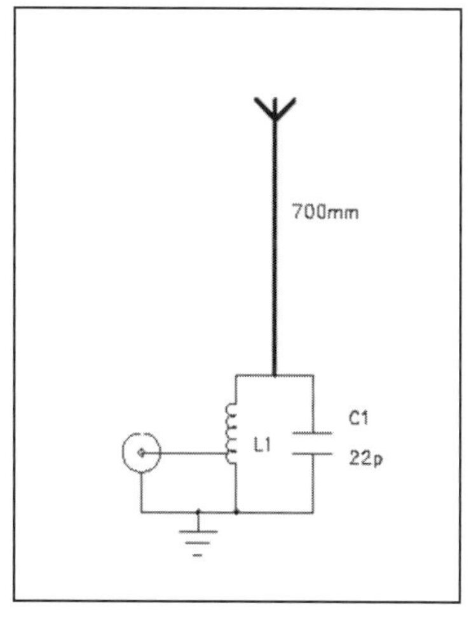


**Fig 18.3: Circuit representation of the rucksack antenna**

The capacitor is a 22pf tubular ceramic of surplus origins. The inductor comprises four turns of 1mm enamelled copper wire wound on a 5.0mm former and in my version is tapped one and a half turns from the bottom. Exact values will depend on the construction and the length of the whip. A fixed capacitor is used in preference to a variable type as these can have a tendency to creep with repeated temperature cycling or with vibration.

The cavity is covered by a small 0.7mm thick aluminium plate. The assembly is waterproofed using a small amount of non-corrosive sealant. A detailed diagram of the housing is shown in **Fig 18.6**. The completed antenna weighs 72g and the RG316 coax a further 21 grams. Measurements on a test range indicate a gain approaching 6db over the type of antenna typically fitted to handheld units. **Fig 18.7** shows the SWR plot.

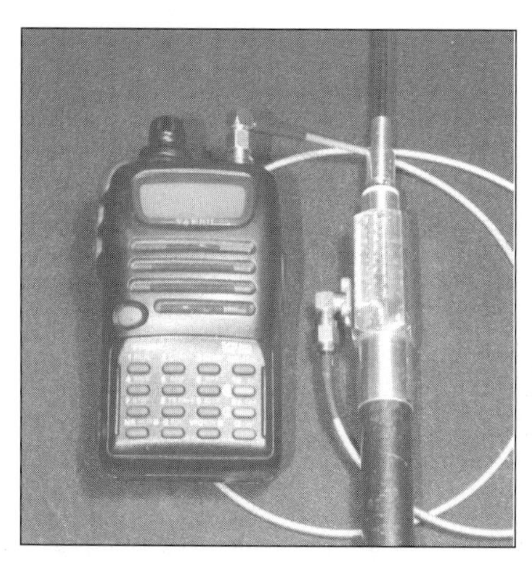

**Fig 18.4: The mount and matching network**

Assuming a similar length of whip is used, tuning of the assembled antenna should consist of adjusting the spacing of the turns on the coil for resonance. In my version, the plate made no difference to the tuning and so this procedure can be carried out with the lid removed. A non-conductive tool should be used to adjust the spacing. A useful adjusting tool can be made from a short length of FR4 PCB material with the copper foil removed.

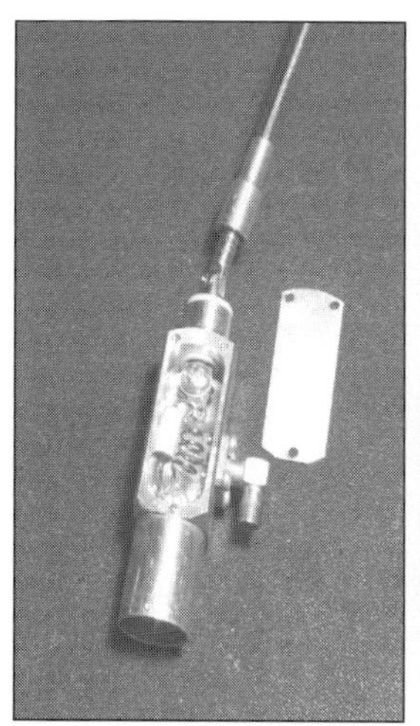

**Fig 18.5: Close-up of the mount and matching network**

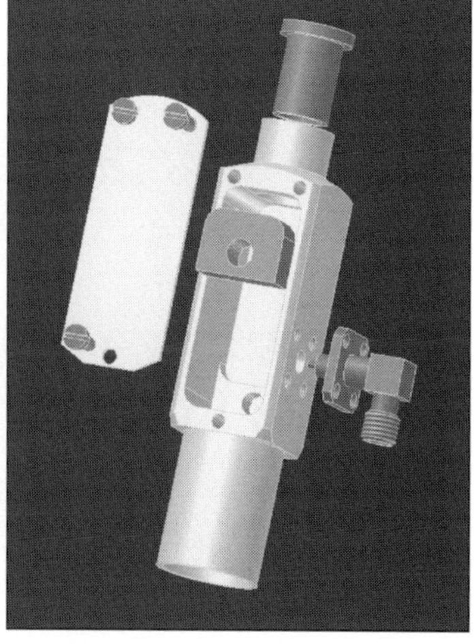

**Fig 18.6: Exploded view of the mechanical assembly**

## Collinears and J-poles

Other possibilities in this category are a collapsible collinear antenna constructed from coaxial cable, and the J-pole. Of the two, the J-pole is much more compact and can easily be mounted on a small mast. Suitable designs that can be adapted for use by the outdoor operator may be found in the chapter on practical VHF/UHF antennas. A ready-made version is available from MFJ.

## A 4m ground plane antenna

The length of a resonant 4m antenna renders it impractical for the outdoor operator to carry without it being broken down into shorter lengths which then require assembly on site. A traditional rigid antenna is also likely to prove heavy. Wishing to participate in the activity on 4m FM, I put together the collapsible

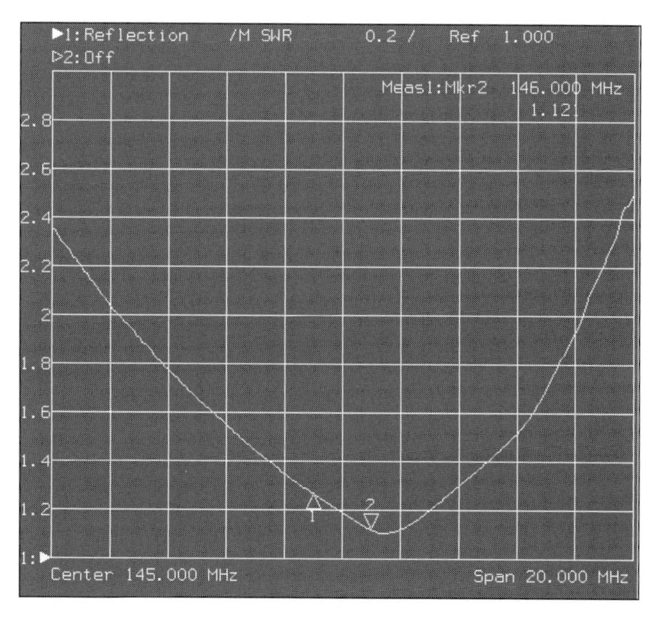


**Fig 18.7: SWR plot of the finished 2m vertical**

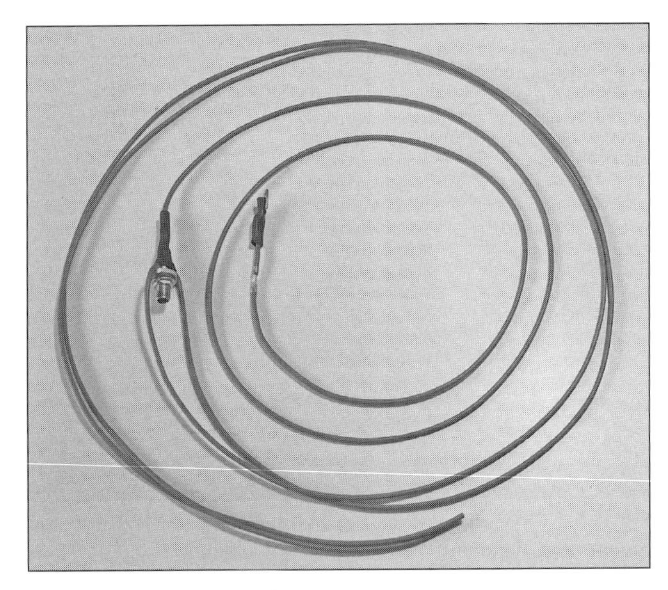

Fig 18.8: The 4m ground plane antenna

antenna shown in **Fig 18.8**. It is a quarter-wave ground plane with the elements comprising lengths of 16/0.2 hookup wire soldered to an SMA connector. The antenna is suspended from the top of a fibreglass mast using a hook fabricated from a short length of heavy gauge tinned copper wire. The hook is electrically isolated from the antenna thanks to the insulation on the PVC-coated wire and it is held in place by a short length of heat-shrink tubing. By experimentation, the best lengths for the elements were found to be 940mm and 920mm for the radiating element and ground plane elements respectively. When deployed, the ground plane elements are allowed to hang down from the SMA connector. The complete antenna weighs 31g.

## Yagis

Without doubt, the Yagi is the most appropriate antenna for the VHF and UHF bands when system performance is important. Not only will a beam mounted on a mast have more gain than a vertical, its directional properties may prove beneficial on sites where there are co-located commercial transmitters and the receiver is in danger of being overloaded (a band-pass filter on the antenna line also helps with this problem).

Most commercial Yagi designs intended for home use are generally bulky and heavy - they have to survive the ravages of our climate for long periods. For the outdoor operator, weight and ergonomics are the most important factors. The designs presented in this section have proved their worth in countless outings and despite their lightweight construction, perform at least as well as the equivalent commercial design.

## A 6-element Yagi for 2m

This no-compromise design has survived several hundred outings without a breakage and gives outstanding performance for an antenna weighing 0.5kg. The design criteria were:

- Lightweight
- Driven element can be used alone
- Quick assembly / disassembly (no nuts)
- Rugged - to survive the occasional crash-landing.
- High but stable forward gain.
- Readily-available materials.
- No part longer than 1m.

Having had considerable success with Yagis based on DL6WU's double taper principle [3], this was used as the starting point for the electrical design. It was optimised using EZNEC within the physical constraints defined above. Conventional designs use a reflector of greater than 1m in length. However, suitable tubing is not easily available and so a reduction in front-to-back ratio was accepted in order to maintain this goal. The azimuth plot of the finished design is shown in **Fig 18.9**.

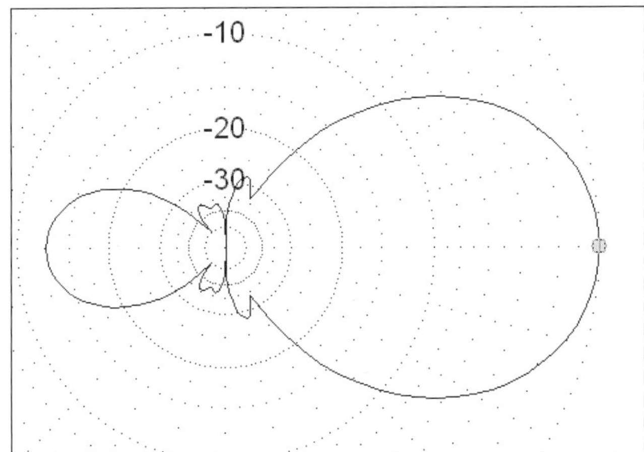


Fig 18.9: *EZNEC* simulated polar plot of the 6 element 2m Yagi. Predicted gain is 9.1dBd, F/B 12.6dB and -3dB beamwidth 50°

The elements are all insulated from the boom. This removes any possibility of a variation in performance caused by an intermittent contact in this important area. The passive elements are held to the boom using Terry clips. Nylon bushes on each element locate and insulate them from the boom. This method of attachment provides for very quick assembly and allows them to come free in the event of an accident - the elements simply popping off the boom.

The driven element (DE) is a folded dipole. An increase in weight has been accepted here in order to achieve a near-50-ohm match. This allows it to be used alone when conditions dictate - perhaps due to high winds on a summit. The mounting for the DE is replicated on the mast clamp to permit a direct fit without using the boom.

Dimensions for the antenna are shown in **Table 18.1**. The

Fig 18.10: Assembly drawing of the passive elements

| Element | Position (mm) | Length (mm) |
|---|---|---|
| R | 0 | 1000 |
| DE | 450 | 930 (see text) |
| D1 | 700 | 920 |
| D2 | 1050 | 900 |
| D3 | 1450 | 890 |
| D4 | 1980 | 880 |

**Table 18.1: Dimensions for the 6 element 2m Yagi**

folded dipole has an overall length of 930mm and the two sides have a centre-to-centre spacing of 30mm.

Aluminium for the antenna elements and boom can be obtained from larger DiY stores in 1m lengths. The boom comprises two lengths of 10mm square tube and the elements are all constructed from 6mm tubing. Each passive element has a pair of nylon bushes bonded in place at its centre. Two different lengths (4mm and 6mm) are needed to exactly match the 10mm boom. As an alternative, they can be machined as one piece from nylon. Initially, they are held in place with a drop of superglue. Once this has set, a bead of Araldite should be placed around the junction between the nylon flange and the tube. Application of a small amount of heat using a heatgun will flow the glue and help it form a better joint.

**Fig 18.10** shows assembly details of the passive element mount. The passive elements clip onto the boom using 8mm Terry clips. These fit into recesses in the boom and are held in place using 2.5mm (3/32") rivets. The recessed mounting protects the clips and the semicircular details in the sides of the boom locate the element accurately. These semicircular cutouts are machined using an 8mm ball-nosed cutter. Below the boom can be seen a strengthening plate. The original design did not use these and the soft aluminium of the boom eventually suffered after a fall. The plate was added to passive elements one and two and this has since proved reliable despite many mishaps. Both the bushes and clips are available from Farnell - one of the UK's major mail-order electronics companies [4].

The driven element is a folded dipole fed via a 4:1 coaxial balun for a near-50-ohm match. It is insulated from the boom by the plastic housing and secured to it by a titanium stud and 'R' clip which fits through a hole in the end of the stud. The distance quoted in the table is from the reflector to the mounting hole at the centre of the DE.

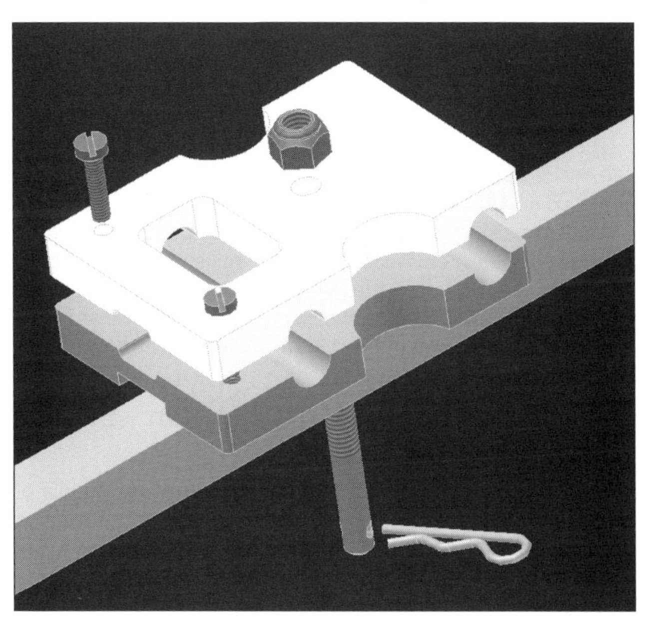

**Fig 18.12: Assembly drawing for the dipole centre**

Construction of the balun housing is shown in **Fig 18.11** and an assembly drawing in**Fig 18.12**.

In this particular case, a solid rod is used and the balun attached using M3 stainless steel screws and solder tags into threaded flats on the ends of the DE. The insulator comprises two pieces of 36mm x 51mm 6mm thick acrylic plate. The elements are located 30mm apart in semicircular grooves machined in each half using a 6mm ball-nosed cutter. The two halves are held together by two M2.5mm screws (seen at the bottom of the photo) and the titanium locating stud. (The temporary fixing in the photo shows an M4 standard screw.) A 10mm wide longitudinal groove on the bottom stops the driven element from rotating with respect to the boom. Once the performance has been verified, the cavity is filled with Araldite to achieve both waterproofing and mechanical strength.

The 'R' clip that locates the DE on the boom is held captive to the DE using a short length of fishing line. In case of loss due to breakage of the line, I take a spare. To date, this has not been needed.

The 4:1 balun comprises a halfwave of RG-316 connected as shown in **Fig 18.11** and **Fig 18.13**. It does not matter whether the coaxial connection is located in the centre of the balun or to

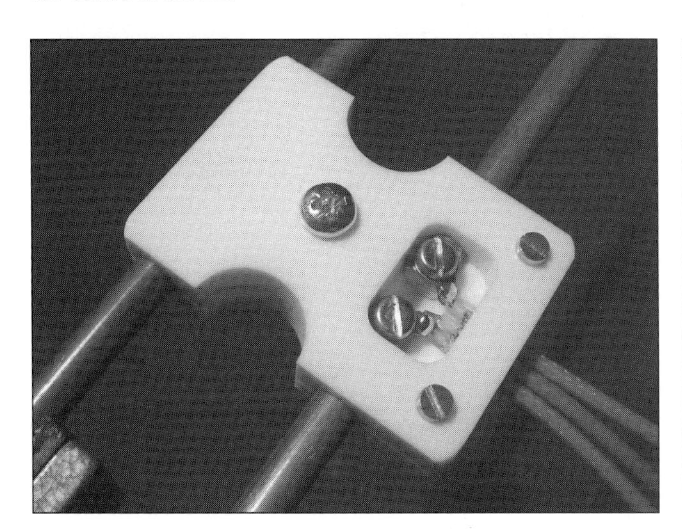

**Fig 18.11: Fully-insulated dipole centre**

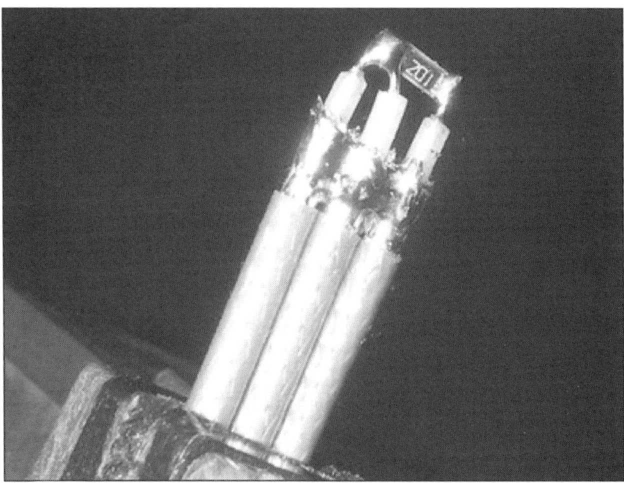

**Fig 18.13: Adjusting the balun for lowest VSWR**

one side. Allowing for the velocity factor, the length of the coax forming the balun is approximately 700mm. A diagram of the balun arrangement can be found in the chapter on practical VHF/UHF antennas.

The only tuning that is required is to cut the coaxial balun for the desired frequency. This can be done prior to connecting it to the dipole. Attach a 200-ohm surface mount or other non-inductive resistor as shown in **Fig 18.13**. Using an antenna analyser or other return loss indicator, adjust the length of the balun for best match at the desired centre frequency (usually 145MHz). Once this is done, the balun and feed can be attached to the dipole as shown in **Fig 18.11** and the assembly potted.

## A 4-element compact Yagi for 2m - the 'R' Clip Special

There are occasions when carrying a one metre-long bundle of poles and antenna elements is inconvenient. Having struggled to maintain control over both antenna and rucksack when using public transport, I developed this solution. The entire antenna and mast fit in a small stuff sack measuring 550 x 150mm. This can be carried in one of the side pockets on a rucksack - one that would normally carry trekking poles. The antenna and 3m mast in its bag weigh 672g. A separate bag containing the guys, pegs and coax weighs a further 160g.

The antenna has a forward gain of 7.3dBd which compares well with its larger cousins whilst offering a substantial saving in weight and bulk. **Table 18.2** shows the dimensions.

Conventional high-performance Yagi designs place great emphasis on the electrical integrity of the elements at their centres where the currents are highest. An inescapable consequence of a design that dismantles into parts shorter than one metre is the scope for poor contacts in this important area. The outdoor operator can take advantage of the transient nature of his operations and ignore the major source of this problem - corrosion. In this design, the act of assembling the antenna causes the electrically mating surfaces to be wiped clean of any surface build-up. In this case, the elements are all held in place using 'R' clips and whilst fiddly to assemble, this design has proved its worth on long treks where minimising weight and bulk are the primary considerations.

This four-element Yagi employs a folded dipole as the driven element and so, like the 6-element version in the previous section, the DE can be used in isolation should conditions dictate.

Fig 18.15: The 'R'-Clip Special driven element and mast clamp

| Element | Position (mm) | Length (mm) |
|---|---|---|
| R | 0 | 1015 |
| DE | 450 | 980 |
| D1 | 700 | 961 |
| D2 | 1050 | 913 |

Table 18.2: Dimensions of the 'R'-Clip Special

In this design, the passive elements are electrically connected to the boom. The design was optimised using EZNEC to give the highest stable gain using boom and element lengths that would fit inside the available sack. After simulation, the element lengths were corrected for the 10mm boom material. The EZNEC polar pattern is shown in **Fig 18.14**.

The elements are 4mm aluminium rod and the boom 10mm square-section tubing as can be seen in **Fig 18.15**. Both types of material are readily available in the larger DIY shops although better quality rod that is easier to machine would be an improvement. Referring to **Fig 18.16** the element rods are held captive in their aluminium mounts by 'R' clips and are located by radial grooves. The mounts are riveted to the boom using M2.5 rivets. M3 countersunk screws could equally well be used.

The driven element follows similar principles to the design of the 6-element but in common with the passive elements uses

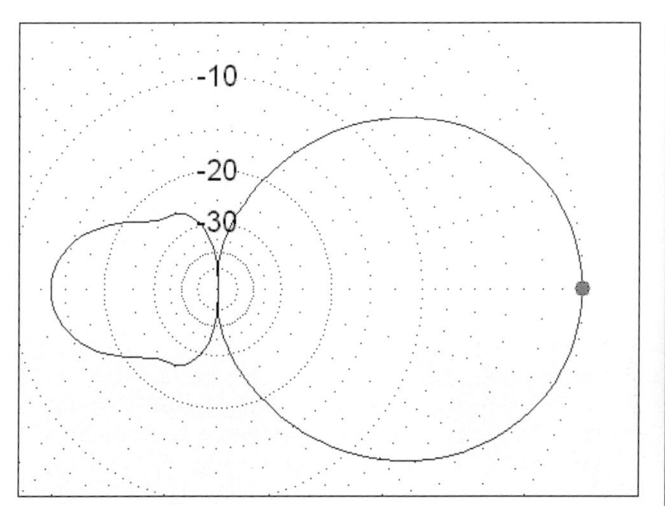


Fig 18.14: Simulation of the polar pattern for the 'R'-Clip Special. Predicted gain is 7.3dBd, F/B 13.5dB and -3dB beamwidth 57°

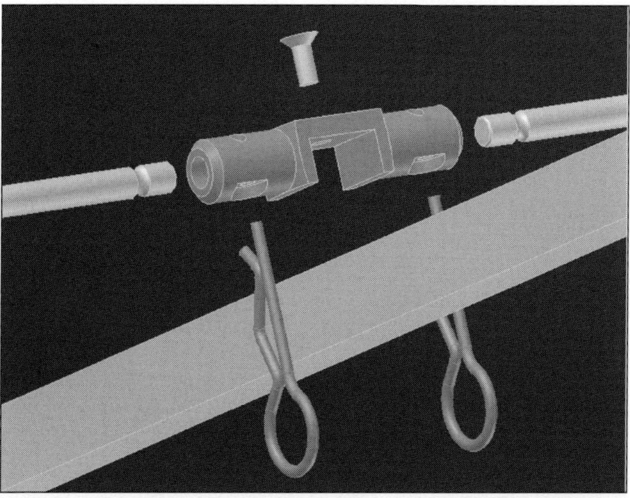

Fig 18.16: Assembly drawing showing the passive element mounts

| Element | Position (mm) | Length (mm) |
|---|---|---|
| R | 0 | 341 |
| DE | 133 | 324 |
| D1 | 193 | 312 |
| D2 | 317 | 304 |
| D3 | 460 | 296 |
| D4 | 625 | 293 |
| D5 | 832 | 292 |
| D6 | 1034 | 289 |
| D7 | 1246 | 288 |
| D8 | 1488 | 285 |
| D9 | 1736 | 283 |
| D10 | 1981 | 282 |

**Table 18.3: Dimensions of the 12 element 70cm Yagi**

the 'R' clip method of attachment. Two small machined bushes provide the transition between the rods and the coax of the balun and feeder. The fully insulated assembly is potted in Araldite whose properties are adequate for short-term exposure to moisture.

As supplied, the 'R' clips have sharp ends which can dig into the aluminium. It is worth spending time with a needle file chamfering them to smooth their passage through the mounts and past the grooves in the elements.

Tuning of the assembly is identical to that of the 6-element design above and the reader is referred to this section.

## A 12 element Yagi for 70cm

Whilst considerably smaller than its 2m relatives, a 70cm Yagi is still too bulky to be carried with the elements attached to the boom and for this reason a design using similar principles to those described for the 6-element 2m design is presented here. It is based on the DJ9BV series of 70cm Yagis [5] which in turn can trace their roots back to the work done by DL6WU, originally published in 1977 [3].

It has a total of 12 elements with dimensions as shown in **Table 18.3**. All are insulated from the boom which breaks down into two sections of approximately 1m length for ease of transportation. As **Fig 18.17** shows, the gain is approximately 13dBd with a F/B ratio of around 20dB (noticeably better than the 2m version, thanks to the lack of restriction in the length of the reflector). It weighs 385g without the mast clamp which is common to the 2m versions.

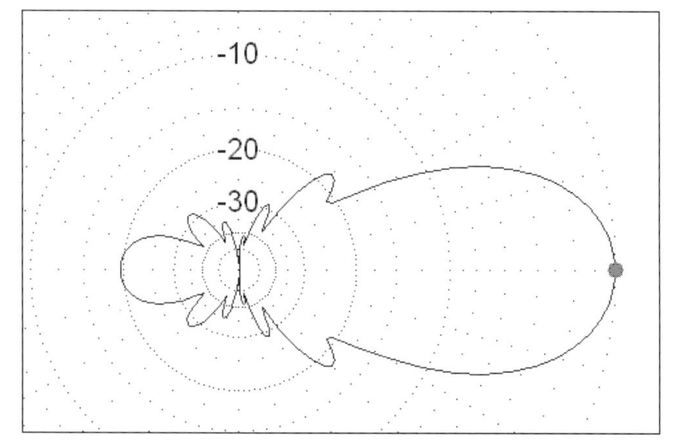


**Fig 18.17: Simulated polar pattern of the 12 element 70cm Yagi. Predicted gain is 13dBd, F/B 20.6dB and -3dB beamwidth 33°**

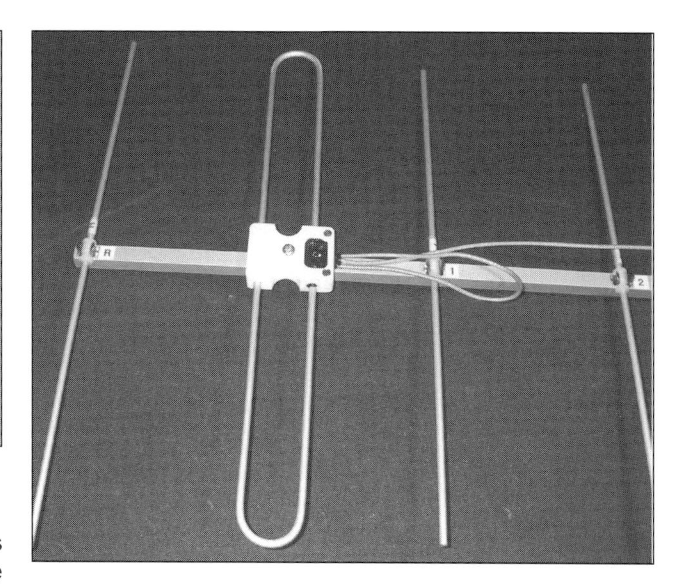

**Fig 18.18: Showing the construction of the first four elements of the 70cm Yagi**

Unlike the 2m version, the elements in this design are made of 4mm solid rod, however, the boom still uses 10mm square tube. The nylon bushes used to insulate the elements are all 5mm wide and are used in pairs to straddle the boom. In this case, the semicircular locating details in the boom are machined using a 6mm cutter. Notice that both elements and boom are labelled to avoid time being wasted during assembly of the antenna.

The driven element is 324mm overall in length and the two sides are spaced by 30mm from centre-to-centre. It can be made from a single rod (**Fig 18.18**).

The balun and feed are attached to the DE via small solder tags and M2.5 screws which fit into tapped holes. The distance quoted from the reflector to the DE is to the centre line of the side to which the balun is attached.

The coaxial cable used for the balun and for the feeder exit is RG-316. The total length is 250mm. Allowing 5mm at each end for termination, this gives an active length of 240mm.

## HF

Two types of HF antenna dominate lightweight portable operation: the inverted-V and the inverted-L (there's more about these in the chapter on practical HF antennas). The latter normally requires the use of some form of antenna matching unit and probably a counterpoise as well. It has the advantage that it can be fed directly with only a short length of coax and the mast can be used as part of the antenna. It does, however, require a matching unit. The inverted-V needs no counterpoise or matching unit and so potentially promises the lighter solution. It is also less susceptible to changes in tuning caused by variations in ground conductivity.

Space can be a problem in some situations and this may make the inverted-L a more attractive proposition, however, the radiation from an inverted-V is predominantly high-angle and this is ideal for most outdoor operations where the majority of listeners are in Europe (by choice: limited power makes DX operation difficult). The high-angle radiation particularly favours 60m with its NVIS properties and this largely explains the success that SOTA has enjoyed on this band.

On 80 and 160m, the space required to erect an HF antenna is often a problem but, by using loading coils, the overall length may be reduced at the expense of a reduction in bandwidth. For

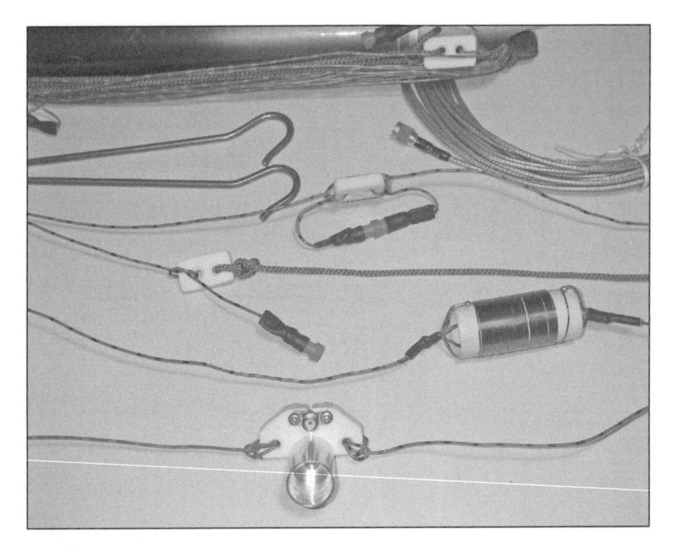

Fig 18.19: Components of the compact multiband inverted-V linked dipole

many types of outdoor operation such as SOTA and WAB, specific frequencies are used and so this lack of bandwidth is unlikely to cause a problem since the antenna can be pre-tuned for the desired part of the band.

## A shortened multiband inverted-V

Apart from ease of deployment, a shortened antenna is most useful in situations where space is limited - which is often the case on mountain tops. The design describe in **Fig 18.19** needs approximately 30% less real estate than a full-sized antenna, thanks to the loading coils. The reduction in bandwidth that these cause is small on 40m but more noticeable on 60m. If necessary, tuning can be achieved by adding short plug-in extensions to the ends.

The centre of the inverted-V is supported on a glass fibre pole 6m above ground. The height of the centre is not particularly critical to the tuning, but the height of the ends and hence the capacitance of the ends to ground has a large effect. The antenna should be tuned for the anticipated height of the ends above ground allowing for droop. The dipole centre can be seen at the bottom of **Fig 18.19**. The cup fits over the top of the mast making for quick attachment.

The loading coils are wound on 50mm lengths of 22mm diameter hot water pipe. This material appears to have excellent RF properties even up to VHF, thanks to a mix that includes PTFE. The coils comprise 55 turns of 0.56mm (24AWG) enamelled copper wire. Note the stirrup-shaped 18SWG tinned copper wire pigtails used for attachment of the antenna wire. These pass through 2mm holes drilled in the sides of the tube. Glued heatshrink sleeving is used to ruggedise the potentially fragile junctions between the coils and the dipole wires.

Band selection is achieved by means of the links that can be seen in **Fig 18.19**. They are connected to achieve 60m operation and left open for 40m. Each link comprises a 2mm plug and socket pair - again, ruggedised using glued heatshrink. When open, the ends of the active part of the dipole carry high voltages and the small insulators shown in the photograph are essential. These blocks are made from 22 x 12mm pieces of 6mm thick acrylic sheet although Tufnol or Perspex could equally well be used. The holes in these insulators are countersunk and deburred on both sides to reduce the possibility of a failure in the wires caused by sharp edges. There is a 2mm socket at the end of the 60m legs to allow an 80m extension to be added.

|  |  |
|---|---|
| Dipole centre to loading coil: | 4.4m |
| Loading coil to 40/60m link: | 1.6m |
| 60m leg: | 1.4m |
| Guy: | 2.5m |

Table 18.4: Table 18.1: Critical dimension of the multiband inverted-V

The wire is standard 7/0.2 plastic-covered equipment wire. Different colours have been used on either side of the dipole to aid deployment should a tangle occur. To a large extent this annoyance can be minimised by careful winding of the dipole onto a suitable 'former' when not in use and the author's example can be seen at the top of the photo. It is a 300mm length of a glass fibre tube with the ends slotted. A piece of hardboard is equally suitable - if heavier.

The antenna is fed via 7m of RG316. Whilst loss is not significant on HF, RG316 is used in preference to RG174 due to its higher voltage specification. This becomes important at higher powers but RG316 is also mechanically superior and so is to be preferred.

The dipole centre comprises a 40 x 20mm piece of insulating material which is attached using a single M4 screw to an aluminium cup. In operation, this is held onto the mast by gravity. Note the method of securing the wires to the dipole centre in a fashion that helps prevent failure at this point due to flexing.

The guy ropes shown in the photo are 2.5mm woven polypropylene cord. As an alternative, Liros manufacture a range of 2mm cords in high visibility colours that stand out well when buried amongst vegetation. The completed antenna weighs 282g including the coax, storage former and titanium guy pegs.

Tuning a trapped dipole from scratch can prove frustrating and simulation programs such as EZNEC do not accurately model the end capacitance, making the task doubly difficult. With this antenna, the no-traps design means that setting up the lengths is simply a matter of tuning the length for the highest band then adding in the links and cutting the next lowest band legs to length. A network analyser - even a simple one will make this an easy task. Should such a device not be available, the dimensions given in **Table 18.4** should prove very close and certainly in-band, allowing the SWR indication in a transceiver to be used to complete the operation. To allow for variations on-site, the kit could include pairs of small lengths of wire each fitted with 2mm plugs. These need be no more than 200mm long to have a significant effect on the resonant frequency.

Magnetic loop antennae offer some advantages for outdoor use. They can operate close to the ground, which is useful in windy locations, and designs based on a loop of co-ax are easy to transport. On the other hand, efficiency is not high, and bandwidth very narrow, making their use at a fixed frequency (for example, a QRP centre or a centre for FT8) the most suitable application.

## MICROWAVES

Nowhere is there more scope for optimisation of antenna systems than on the microwave bands. Normal portable operation on the bands above 13cm generally involves the use of a dish. Here weight is an advantage as it counters the effects of the wind on an antenna whose -3dB beamwidth is usually in single figures. Unfortunately there is no substitute for size in the quest for gain and when a lightweight system is to be constructed, compromises will have to be made.

This section explores different approaches to what is a relatively new area of outdoor operating (the author and G3CWI having made the first 3 and 6cm SOTA QSOs as recently as 2008).

## Yagis

On 23cm and above, Yagis become small enough to be transported with the elements in place. Unfortunately, the dimensional tolerances become difficult to maintain and so 13cm is probably the practical limit for their use. Above 13cm most microwave operators have switched to dishes and so reference designs for Yagis are not available.

## Dishes

A parabolic reflector (or dish) is the standard antenna for use on the microwave bands. On 3cm, the 60cm satellite TV offset-feed dishes are widely used - as are surplus 2ft dishes from Andrew Corporation and others. Both of these are impractical for portable operations due to their bulk and weight. Probably the practical limit is the small portable metallised plastic dish sold as part of caravan TV systems. These are approximately 30cm in diameter, yielding a gain of approximately 26 dBd - some 6dB less than a 2ft dish. Beyond this size, carrying the antenna on a rucksack would prove difficult. Experience has shown that even a dish this size can prove difficult to align when mounted on a lightweight mast in typical summit conditions. Perforated dishes are the traditional answer to this problem but none of a suitable size currently exists. There is plenty of information available on the construction and use of dishes and I would recommend reading the International Microwave Handbook [6] as a starting point.

## Horns

At present, the horn antenna offers the most effective solution on 3 and 6cm and whilst low in gain compared with the 30cm dish, is robust and lighter. By packing system components inside the mouth of the horn for transportation, the overall volume can be kept to a minimum.

## A 6cm horn and coax transition

Whilst the use of horns and Gunn oscillators, was common in the early days of 3cm they have fallen out of use except for dish feeds. These designs are not optimised for gain but for the specific F/D ratio needed for the dish. The design presented here is intended for use on 6cm where it has a gain of approximately 15dBd. Tapering it further down to WG-16 waveguide dimensions would allow it to be used on 3cm, where it would yield a gain of approximately 20dBd.

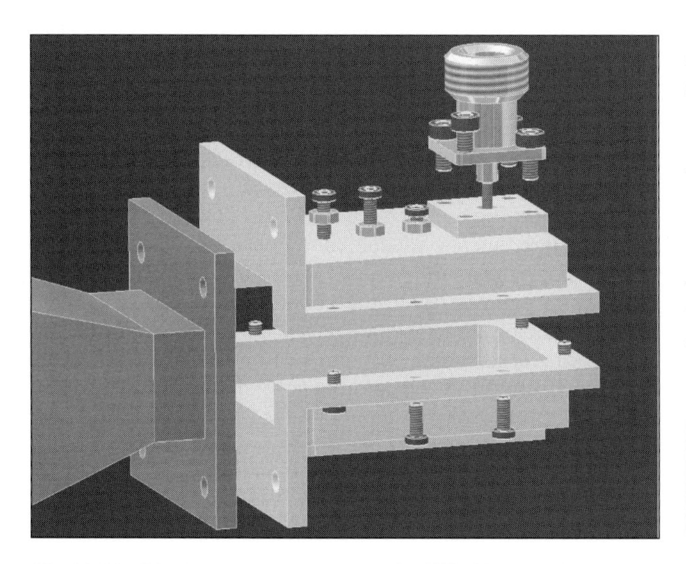

**Fig 18.20: Mechanical assembly of the WG-14 transition**

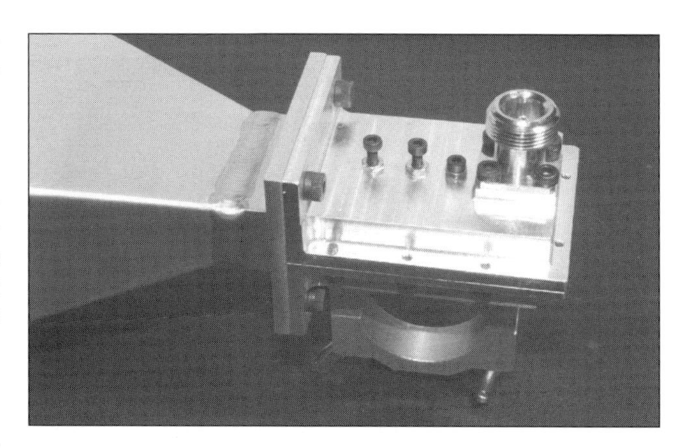

**Fig 18.21: The finished 6cm antenna and mount**

The purpose of a horn is twofold. Firstly it must give gain, and secondly it must smooth the transfer of the RF from waveguide to free space. Thus it acts as an impedance transformer. Whilst it is possible to use a shorter horn, the angle affects the efficiency of this transition. Too great an angle incurs a loss of gain. Too small an angle does little other than increase the weight and volume. This design is the best compromise between gain and weight.

The coax transition provides a second translation - this time from the 50-ohm impedance of the transverter to that of the waveguide. Despite a free choice of waveguide dimensions in this design, those for WG-14 were chosen to allow standard test equipment to be used. RF is launched into the waveguide by a probe attached to the centre pin of the 'N'-type connector. Matching is achieved by the three M2.5 screws seen inline with the connector (**Fig 18.20**).

The antenna and transition are mounted onto the aluminium mast coupler and the transverter mounts directly to the transition. The complete antenna with its mast clamp weigh 352g.

The horn was fabricated from a single piece of 1mm aluminium sheet with the joint welded. The walls were then welded to a flange to complete the antenna. The slot in the flange is 34.8 x 15.8mm - the dimensions of WG-14. The mouth of the horn measures 150 x 115mm and the length of the taper is 160mm. The finished unit is shown in **Fig 18.21**.

As an alternative, the horn could be constructed from 1.6mm copper-clad FR-4 PCB material or sheet brass with the joints soldered. A brass plate could be used for the flange. Suitable brass can be obtained from model shops.

The transition comprises two pieces of aluminium bolted together with eight M2.5 screws around the perimeter. Whilst not ideal from the RF point of view, this method of construction has been verified in back-to back experiments. The probe comprises a length of 2.0mm brass rod that projects 12.4mm into the cavity. Critical dimensions inside the cavity are given in **Table 18.5**:, see the materials section below for suitable specifications of metals. A 3cm version of the transition could be manufactured on similar lines. For dimensions, see [6].

| | |
|---|---|
| Probe to end wall: ........... | 7.0mm |
| Probe to first tuning screw:.... | 4.0mm |
| Tuning screw spacing:........ | 9.5mm |

**Table 18.5: Critical dimensions inside the cavity of the horn**

| Cable | Weight (g) of 4m | Loss (dB) 144MHz | Loss (dB) 432MHz | Loss (dB) 1296MHz | Max power (W) 432MHz |
|---|---|---|---|---|---|
| RG-174 | 34 | 1.68 | 3.43 | Forget it! | 26 |
| RG-316 | 46 | 1.68 | 2.96 | Ditto | 210 |
| RG-58C | 100 | 1.10 | 2.22 | High | 90 |
| RG-400 | 256 | 0.72 | 1.43 | 2.82 | 1050 |
| RG-214 | 500 | 0.44 | 0.92 | 1.92 | 330 |
| FSJ1-50A | 280 | 0.29 | 0.52 | 0.94 | 610 |

**Table 18.6: Coax cable characteristics for typical lengths**

## Patch antennas

A potential source of cheap ready-made lightweight antennas for 6cm is the Wi-Fi market. For example Solwyse advertise a 200mm square patch antenna whose specification shows a gain of 17dBd and a weight of 500g. It is believed that these antennas will handle powers up to 5W. More on patch antennas can be found in the chapter on microwave antennas.

## Feed Systems

Whilst feeder runs are relatively short in most outdoor applications, the correct choice of coaxial cable is still important. An inappropriate choice could either add unnecessary weight to the system or waste valuable RF power. There is little point in carrying a higher power station only to lose a large percentage of that extra power in the feeder.

## Cable

**Table 18.6** gives the characteristics of some commonly-used cables. For such short lengths, the loss can be ignored on the HF bands but it becomes significant on VHF and above. As an example, putting 50 watts of FM on 2m through RG-316 causes a noticeable increase in temperature of the coax - which is not surprising since 30% of the power is wasted as heat. The loss figures in the table relate to a 4m length - which is typical of outdoor systems. The weights quoted are just for the coax. Whilst it would appear that RG-174 is a serious candidate, it is not a cable that I would consider; the voltage rating is too low for high power operation and it is mechanically inferior to RG-316.

RG-58 is a sensible choice. However, if UHF operation is contemplated the lower-loss RG-400 will give better performance but at over twice the weight. At 23cm FSJ1-50A Heliax is a good choice although it is mechanically stiff and more difficult to use.

## Connectors

The choice of connectors may well be dictated by whatever transceiver is to be used. Unfortunately, most commercial transceiver manufacturers persist in fitting SO-239 sockets for the HF bands. Many do at least fit an 'N'-type for VHF and above. Whilst there are now good quality clamp-style PL-259 plugs available for RG-58 and RG-400 cables, I still prefer to use a BNC adaptor and a crimped BNC plug on the cable. This gives a mechanically-superior solution. Clamp-type connectors have a habit of coming loose over time, eventually causing a failure of the braid connection. Correctly assembled, a crimped connector is far superior - and lighter.

The FT817 offers the choice of two RF connectors and although the front panel-mounted BNC is less convenient operationally than the SO-239 rear socket, it is to be preferred from both the weight and reliability aspects.

Where a free choice is available, I always use crimped SMA connectors and reinforce the screen crimp with a length of heat-shrink sleeving. If the socket is mounted on a plate as shown in (**Fig 18.1**) I find it easy to mate the free plug even with cold or gloved hands. Microwave connectors are discussed above. Despite it often being described as a microwave band, coaxial cable is normally used on 23cm to connect the antenna to a ground-mounted transverter. If FSJ1-50A is to be used then BNC or 'N'-type connectors are really the only viable options.

In general, the narrow beam of microwave communication precludes the technique of randomly sweeping about hoping that there is a contact out there somewhere. Using the internet to advertise your intentions to perform an activation, and have people listening out for you, works much better.

## Antenna Supports

Whatever the antenna, and not withstanding the fact that the station may be on top of a mountain, some form of antenna support is generally needed. There are exceptions to the rule and the rucksack-mounted vertical is one - provided the operator is prepared to stand whilst operating. Where HF antennas or any form of directional VHF/UHF antenna is to be used, a mast of some description will be needed. The traditional amateur mast involves aluminium tubing and whilst a possibility for outdoor operating, it is a heavy solution. Two very different solutions are offered in this section. One follows the minimalist approach compromising height and functionality in exchange for lightness. The other is a traditional portable mast engineered for the lightest solution commensurate with ease of use and performance.

## The 'walker's approach'

There is an adage that says that: "nothing weighs less than nothing". Many walkers use a pair of walking poles to aid stability and ease the load on hard-worked knees etc. The average walking pole is approximately 1.35 metres in length when fully extended and so two poles end-to-end with a suitable joiner can be used as a compromise mast. For 60m in particular, this may be an adequate height for the centre of an inverted 'V' dipole. When erected over dry, rocky ground such as is found on many summits, the difference in performance between 2.7m and the average 6m mast is small. For the minimalist approach, all that is required other than the poles is a small coupler to connect them end-to end and a set of guys. The wrist loop on the top pole can be used as an attachment point for the centre of the dipole and for the guys. A more ergonomic solution is to fit an inverted cup over the top pole's grip to which are attached the antenna and guys.

Many summits afford such an excellent take-off that this solution can also be used on VHF. It does, however, assume that the terrain at the summit is known in advance.

The coupler: The bottom section of most poles is tapered. The length of this tapered section varies between models so it is worth choosing ones with a longer one if this technique is to be used. In its simplest form, the coupler can comprise a short length of tubing, the resultant flexibility in the centre of the 'mast' being accepted as part of the compromise. A better solution is to machine the coupler from solid, matching the taper on the end of the poles to tapers inside the coupler. This will achieve a straight mast but will not necessarily prevent the top one from rotating on its own when it is used for supporting a Yagi in windy conditions.

Most walking poles have a basket just above the tip and there is often a cutout in it's perimeter. By arranging pegs to locate in

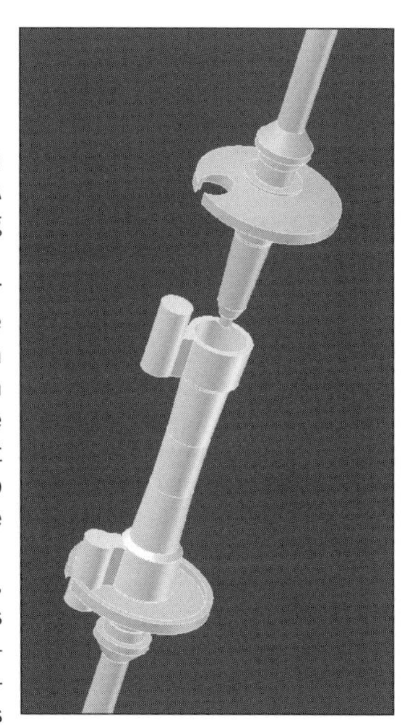
**Fig 18.22: A coupler for two walking poles**

these holes, an anti-rotation mechanism can be made. A suitable design weighing 39g is shown in **Fig 18.22**.

Attaching a Yagi or other directional antenna to the top of the 'mast' is more of a challenge and the solution very much depends on the type of pole. A small bracket clamped just below the grip of the upper poles is the most secure answer.

Whilst they are very strong, the use of walking poles in this manner may in exceptional circumstances cause them damage, rendering them useless for the return journey. The risks should be considered before adopting this approach.

## Plastic water pipe

A simple, light sectional mast can be made by cutting 22mm plastic water pipe into sections about 50cm long, and gluing pipe connectors to one end of each length. These can be slotted into one another at the summit, and are strong enough to lift a light antenna. They are not suitable for heavy yagis though, because they tend to buckle under wind strain.

Modified 'roach poles'

For a no-compromise solution to the problem of supporting antennas, there is no better starting-point than the 7 and 9m glass fibre fishing rods known as 'roach poles'. These are readily available from a number of suppliers and they provide an excellent source of lightweight material. They can be used in their unmodified form, but it is worth bearing in mind that the sections lock together by friction alone. That works well in their intended application as fishing rods where the pole is close to horizontal. When used vertically they have a tendency to collapse. The greater the head load and the larger the antenna the greater the chance.

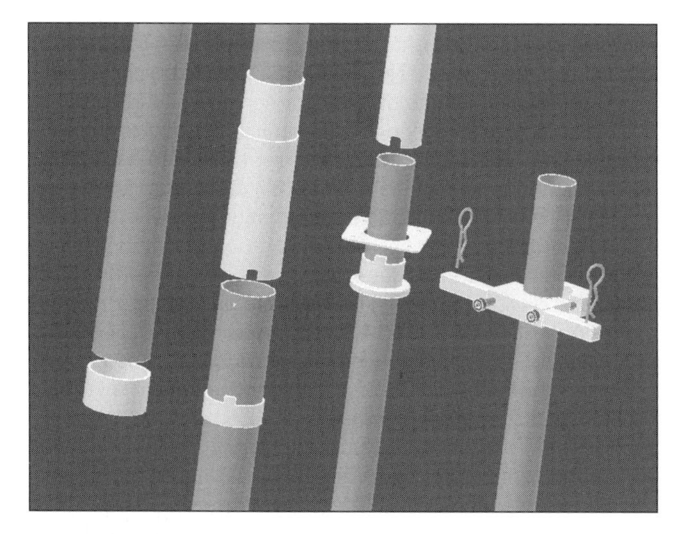
**Fig 18.23: Exploded view of the mast components**

For transportation, the rods retract into the handle section, making a package that is around 1.2m long. In my opinion, this is a bit long and when attached to a rucksack makes it prone to getting caught in overhanging branches or on rocks when descending steep mountainsides.

This design discards the handle (which is heavy anyway) and uses six middle sections from two rods, all reduced to 1m in length - or the length of an element of a 2m Yagi.

For VHF usage, the antenna is mounted at the top of the fourth section, locating the antenna 4m above ground. For HF, the remaining two sections are added, bringing the height up to nearly 6m. The first four sections lock using small spigots to allow the beam to be turned. The remaining two are left free to swivel. The mast is guyed at the top of the third section, where a square carbon fibre plate sits on top of a collar on the coupler.

The glass fibre pole sections fit together using aluminium tubes and locking rings (**Fig 18.23**). The tubes prevent the glass tubing from splitting and protect the bottom ends from damage on rocks. When made out of good quality aluminium, the wall thicknesses can be reduced to 1mm without fear of damage in use. A collar at the base of the mast protects the bottom pole where it touches the ground. The guying plate sits on top of a collar formed on the locking ring on the third pole. Space is left above the boom clamp on the fourth pole to allow the fifth and sixth sections to be added for HF operation.

The aluminium rings and tubes are bonded to the glass fibre poles using Araldite. Prior to bonding, the epoxy paint is removed from the poles to expose the bare glass fibre tubing and all parts are degreased using Methylated Spirit or Methanol.

The mast and boom clamp weigh 460g when used in the VHF configuration and 600g when extended for HF. The 8mm square spigots either side of the boom clamp form a universal fixing that can take any VHF or UHF antenna that uses the 10mm square tubing as a boom. The halves of the antenna boom are slid onto the spigots and secured using 'R' clips. This makes for very rapid assembly and allows a polarisation swap in just a few seconds (the boom has holes in both planes).

## Guys and pegs

It is true that a mast will stay erect with only three guys but, in my experience, the addition of a fourth greatly increases both the stability and ease of erection of the mast. The angles between the guys are much less critical and uneven terrain causes much less of a problem. The weight of the fourth guy is negligible and the fourth peg comes 'for free' since a spare would always have to be carried with a three-guy system to cater for accidental loss. A target weight for a complete four-guy system is 80g.

In this context, guy 'rope' is hardly an accurate description - and would lead to confusion when enquiring in boating shops. The term cord is more appropriate for the 2 to 3mm guys which are perfectly adequate for use with outdoor masts and antennas. The most readily available and cheapest rope is 2.5mm diameter polypropylene cord. This is usually available in 50m reels. Whilst the breaking strain is not quoted, it has proved adequate with the loads presented by the 6-element 2m antenna in high winds. It does, however, have a tendency to stretch in use so some re-tightening of guys may be necessary.

The best is the 2mm diameter Dyneema-based cord produced by Liros. This has a specified breaking strain of 240daN and very low stretch of less than 3%. It has proved highly abrasion-resistant in use. It is available in a fluorescent yellow or blue. The added visibility of the yellow is an advantage, highlighting the location of such thin guys. To some extent, the nightmare of the tangle can be overcome by the addition of short lengths of low-temperature heatshrink sleeving over the knots - as shown in **Fig 18.8**. These help prevent the loose ends catching.

In my experience there is only one type of guy peg to consider and that is the 1/8" (3.2mm) diameter Titanium pegs that are often available from the high street retailer Blacks in sets of six. They do not bend, retain their points and have an incredible capacity to find their way into cracks between rocks on the stoniest summit without damage. As supplied, the loops are rather open which could lead to a guy slipping off the peg. It is possible to close them up slightly in a strong vice but this must be done with the utmost care to avoid injury should one escape during the process. A set of four pegs weighs 28g. Aluminium and steel pegs are more readily available but neither has the same strength and they are nearly twice the weight. Whichever pegs are chosen, it is a good idea to add short lengths of yellow heatshrink sleeving to aid recovery if one is accidentally dropped. They can be very difficult to spot even in the shortest of vegetation and usually end up some considerable distance from where dropped.

A guying plate is important to allow the mast to rotate smoothly when required and not to damage the guys. **Fig 18.24** shows a 42mm square, 1mm thick carbon fibre plate that has been carefully deburred to prevent chafing of the guys. Even after several hundred activations, the guys are unmarked.

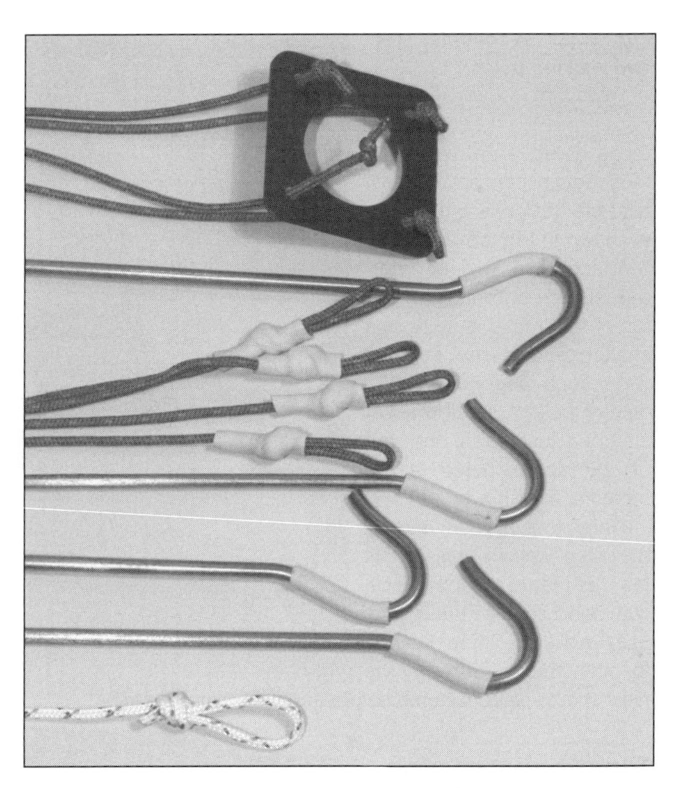

**Fig 18.24: Lightweight guy set including the guying plate, peg loops and guy pegs. An example of the fluorescent yellow chord is shown at the bottom**

## RADIO EQUIPMENT

There are many factors to consider when making a choice of transceiver for outdoor operating. These include:

- Weight
- Size
- Modes and bands of interest
- Output power
- Ruggedness
- Power requirements

There is no correct answer to the often-discussed question of which transceiver is best. Yaesu's FT-50, FT-847 and FT-817 plus Icom's IC-706 are all possibilities. So is a Tuna Tin Two knocked up from a few spare transitors and coils. On a remote summit the extra power of the FT857 may be useful - but not if the expedition is also a camping trip when weight and volume become the overriding consideration. High power means larger batteries and also larger AMUs. A specific SOTA expedition to an outlying mountain will usually warrant taking a spare transceiver as a backup. This might be impossible when backpacking. Anyway, in this case radio might not necessarily be the primary objective!

**Table 18.7** provides a comparison between some of the transceivers currently in use by SOTA operators. It focuses on parameters of interest to the outdoor enthusiast. Receive current was measured on SSB (or FM for the handheld) at a comfortable audio level into a headset. This was done with the radio powered by an external 13.8V PSU. Note that the weights quoted for the FT-857 and IC-706 include CW filters at approximately 20g.

The IC-706 consumes 1.55A on receive. This can be reduced to a slightly more manageable 1.36A if the display contrast is set to low. At this current, operating time will be noticeably less than with the FT857.

| Transceiver | Wt (kg) | Int batts | RX current (A) | Bands | Input voltage (V) |
|---|---|---|---|---|---|
| IC-706 Mk1 | 2.63 | No | 1.36 | HF/2m | 11.7 - 15.8 |
| FT-817 | 1.53 | Yes | 0.304 (min) | HF -70cm | 8.0 - 16.0 |
| FT-857 | 2.36 | No | 0.522 (min) | HF-70cm | 11.7 - 15.8 |
| FT-50 | 0.36 | Yes | 0.082 | 2m/70cm FM | 4.0 - 16.0 |

**Table 18.7: Comparison of popular transceivers for portable use**

Both the FT-817 and FT-857 allow the operator to choose the colour of the display backlight. The factory setup for the FT-817 results in a receive current of 333mA on external batteries. If the backlight is switched to 'off', this can be reduced to 304mA. Similarly, the FT-857 takes 594mA. The default setting is actually one of the lower current colours. Some will increase this to 630mA. By changing the display mode to 'Auto1', the consumption can be reduced to 522mA until a button is pressed when it temporarily increases. That represents a useful saving.

One of the most popular HF bands for SOTA use in the UK is 60m (currently available to all Full UK licensees). All three of the above HF radios can be modified to cover this allocation.

The FT-817, FT-857 and FT-50 can be configured to display the supply voltage on the front panel. I find this invaluable. Given a knowledge of the discharge curve of the battery pack being used, the state of charge of the pack can be established. This estimation is accurate enough to determine when to change the battery pack or call a halt to operating. It can also help guard against either damaging the rig with the low voltage or the battery pack by a deep discharge. Information on battery performance is given in a later section.

The transceivers mentioned above are all general purpose multi-band devices, however, none covers one of the emerging bands in the SOTA world - 4m. The Wouxun KG-699E/4m, is a 5W handheld which has opened up easy access to the FM portion of the band. The use of an external antenna is recommended and a suitable lightweight design can be seen in **Fig 18.23**.

As an alternative to this handheld, I have pressed a Tait T199 crystal-controlled mobile transceiver into service. This aluminium-cased unit is roughly the size of an FT-857 and weighs 1.6kg. It will happily generate 15W on transmit. From a well-located

| | Baofeng UV-3R | Yaesu FT-50R | Yaesu FT-817 |
|---|---|---|---|
| **Sensitivity (dBm)** | | | |
| **2m** | -129.5 | -126.5 | -127.5 |
| **70cm** | -130 | -124 | -125 |
| | | | |
| **IP3IP (dBm)** | | | |
| **2m close** | -21.5 | -33.5 | -24.5 |
| **2m far** | -21.5 | -27.5 | -3 |

**Table 18.8: V/UHF handheld radio RX performance. Sensitivity was measured for 10dB SNR with 1kHz modulation. IP measurements were with signals at 145.525 & 145.550MHz (close), 150.5 & 155.5MHz (far)**

summit, it is advisable to ensure that at least two working channels are available in addition to the FM calling channel of 70.45MHz.

## V/UHF Handhelds for FM Use

Anyone who has tried to operate on 2m near a commercial mast is likely to have suffered the consequences of a receiver that is unable to cope with overloading. The majority of the problems are caused by IMD products - usually a combination of one or more in-band signals and pager signals from the mast. For handhelds in particular this can render the band unusable for seconds or even minutes at a time.

**Table 18.8** lists the sensitivities and susceptibilities of three popular portable radios. For the single sample of each type tested, the UV-3R came out on top for both sensitivity and also for overloading when both interfering signals were in-band. If, however, one of the interfering signals were placed above 2m in the pager band, the FT-817 would cope best due to its superior internal filtering. Field experiments back up these results with the FT-817 and FT-857 coping very well even when used with high gain antennas.

See the chapter on receivers for more on measuring resistance to out-of-band signals.

## Signal Quality

The potential effect that a 'sagging' generator can have on a contest station is well-known and although outdoor operations tend to last for much less than the average day's contesting, it is still our responsibility to ensure that the signal being radiated is clean.

This is particularly so on VHF and above where the station is likely to radiate a considerable distance and where, in many cases, it is co-located with commercial radio stations.

If access to a spectrum analyser is not possible, simple on-air tests with a local station should be able to confirm the quality of the signal. These tests should be carried out using the complete station including the intended battery pack in varying states of charge to establish what happens when the pack becomes depleted. Ideally, apply a two-tone audio signal to the microphone input of the transceiver and monitor the RF signal on a spectrum analyser. Reduce the power supply voltage and repeat the measurements.

I found that there was no noticeable change in transmit IMD performance with either the FT-857 or the IC-706 when the voltage was taken down to 10V.

# ANCILLARY EQUIPMENT

Despite the maxim that "less is more", there are several ancillary items that the outdoor operator is likely to need.

## Lightweight Morse Key

On HF and on the microwave bands, CW is a very popular mode - and indeed is often essential for microwave operating. Most modern transceivers have an inbuilt keyer, but a standard paddle key intended for the home station is generally fragile in construction and difficult to use if not placed on a stable surface. They are usually made intentionally heavy to impart further stability when sending high-speed Morse. A neat solution to the fragility of the paddles is to make them retractable for transit. The stability needed to operate a conventional paddle key can be achieved by putting the transceiver's metallic case to good use and embedding magnets in the base of the paddle.

Unless you are intending to send Morse at over 20WPM, the solution in **Fig 18.25** (*see more on morse keys in chapter 19*) may appeal. By reverting to the up-down motion of a straight key much more stability will be achieved, and by using the rig's built-in keyer Morse can be sent at a reasonable speed, even with a gloved hand. I am no CW expert but I find this type of key much more usable in adverse weather conditions. It is also light, weighing in at 100g including 0.5m of cable.

The base comprises a 50 x 78mm piece of 5mm aluminium

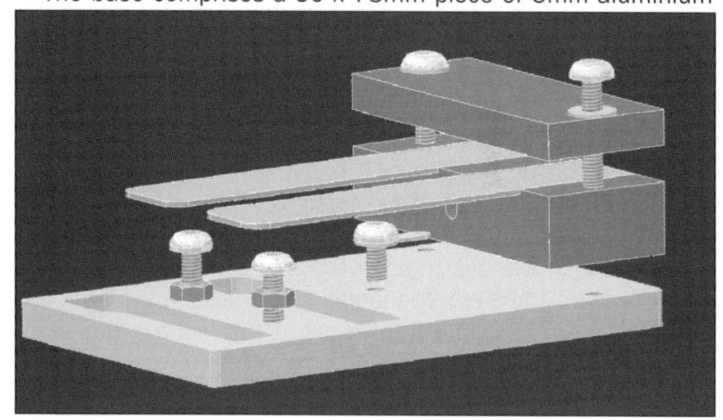

**Fig 18.25: Mechanical assembly of the paddle key**

plate which is lightened by machining pockets a shown. A layer of tape on the underside helps prevent it from moving laterally during use. The paddles are 14 x 66mm plates of 0.8mm brass and they are held in blocks made of Tufnol or any other insulating material. Grooves are milled in the lower block to clamp and locate the paddles and to prevent them rotating.

Five 2.5mm holes in the base are tapped at M3 to take the two paddle stops, the screws to hold the insulator blocks and an M3 solder tag for the earth connection. Further tags are soldered to the underside of the paddles and used to make electrical connections. The three soldered connections to the cable are reinforced with short lengths of Kynar sleeving. The cable is located in a hole through the bottom insulator block and is glued in place, strain-relieving the cable connections. Once adjusted, the screws are locked in place using M3 nuts.

As an aside, it is worth remembering how to switch the transceiver between its keyer and straight key modes when operating on the microwave bands. The key is normally used to send a series of long dashes during the process of antenna alignment.

## Tablet or smartphone

If you wish to use digital modes, the usual method is to connect a smartphone or tablet running digital mode software to the transceiver. For modern transceivers, this can be done via specific interfaces; for older, connection can be made via generic interfaces to microphone and headphone sockets. The experience of using these is similar to a base station, especially on a tablet-sized screen. Cases that carry keyboards can be particularly useful for PSK31 and other 'chatty' digital modes, but are less necessary for the stereotypes exchanges of FT8. If you are relying on your smartphone as a safety device, take care not to run its battery down too far using it for amateur radio.

## Logbook

Unlikely though it may sound, considerable effort has been expended by many outdoor operators over the years in pursuit of the lightest and most ergonomic logging method. Only when an attempt has been made to operate a rig and rotate a mast by hand, whilst simultaneously trying to stop a logbook flying away in the wind does the difficulty become apparent.

Short of taking a table or other substantial flat surface the outdoor operator will inevitably be balancing the logbook on an unstable surface. **Fig 18.26** shows three very different solutions.

At the top are two versions of a conventional paper logbook. In both cases, the log sheets are the same and comprise a number of laser-printed 100gsm sheets cut and bound using salvaged wire binding. The slightly shiny finish on this paper repels the occasional drop of rain long enough for it to be removed. Under most conditions a ballpoint pen can be used but when temperatures are excessively low or humidity is exceptionally high, a 2B pencil is more useful and water-resistant. Take both so that each acts as a spare for the other. To the left is the minimalist approach using a thin aluminium sheet as the backing. Measuring 95 x 125mm, it is too narrow to rest conveniently on most surfaces but as it weighs 55g, it is ideal for the long distance backpacker. The QSO start time can be taken from any number of devices that have a clock built-in - eg mobile phone, camera, GPS etc. To the right is a heavier and larger, but more ergonomic, solution. It weighs 105g without the stopwatch. The extra size of the backing allows it to rest more securely in your lap. As an alternative to standard Laser printer paper, some outdoor operators use waterproof paper and the accompanying writing implements. 'Toughprint' [7] from Memory Map and 'Rite in the Rain' [8] are two of these.

Below the paper logbooks are two completely different solutions. To the left is a small MP3 recorder weighing a total of 64g. With this, it is possible to log the audio during the activation, making sure to speak the time at the start of each recording. Given that these devices are stereo, microphone audio could be recorded on one side and rig audio on the other rather than trying to capture both sides of the contact using its internal microphone.

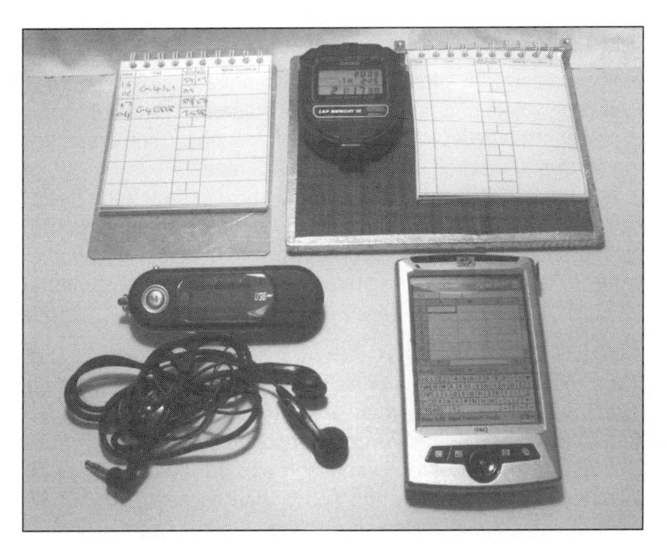

**Fig 18.26: A selection of logbook solutions**

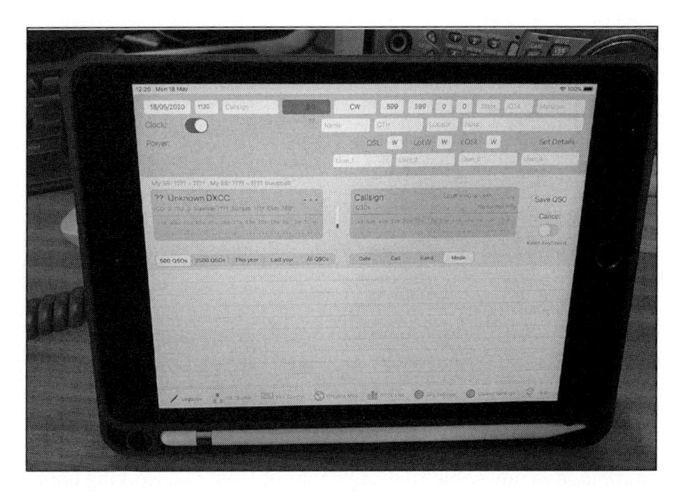

**Fig 18.27: iPad version of the RUMlogNG2Go Logging App by DL2RUM**

At least one experienced outdoor operator uses a Windows CE-based PDA for logging. Attempts at simulating data entry in a real environment have also not been very successful, paper logging proving much quicker. As a PDA often has a voice recorder built in, this may be a useful backup when time is short. Some devices store data into battery-backed RAM and so a flat battery could result in data loss. It is worth checking this before deciding on a purchase. Also nowadays the iPhone and iPad along with Android smartphones have apps such as *RUMlogNG2Go* which is suitable for logging and recording contacts **Fig 18.27**.

## Headphones

If it's windy you should use headphones to help you hear over the noise. If it isn't, you should wear headphones out of courtesy to other walkers.

In the former case, headphones will make a big difference to the overall S/N ratio. For the minimalist, earpieces of the type that come with an MP3 player are adequate. Slightly more ergonomic are a set of lightweight stereo 'phones. These are generally fragile and the plastic has a tendency to fracture - especially in very cold weather. The TM-201 headset from was originaly from Maplin has swivelling earpieces which allow it to stow flat, reducing the chances of damage in transit.

## Microphones and Headsets

In the context of ultra-lightweight backpacking, to carry a fist microphone weighing 170g makes little sense - and that is the weight of the MH-31 unit that comes with the FT-817. Inside this

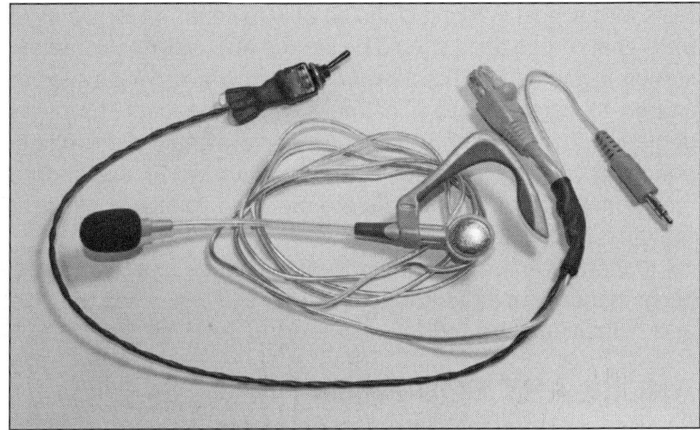

**Fig 18.28: Lightweight headset for the FT-817**

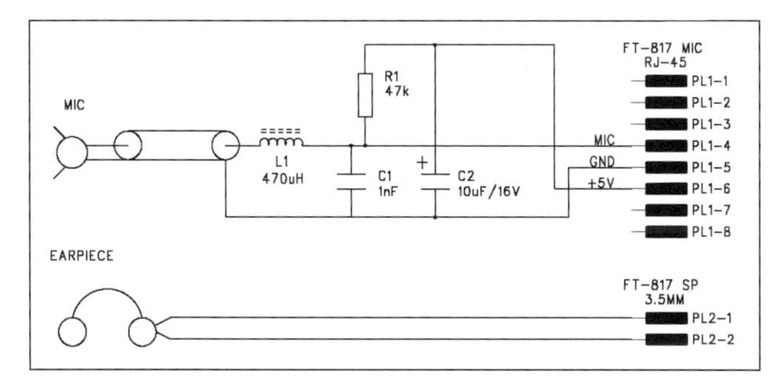


**Fig 18.29: Circuit diagram of L/W FT-817 headset**

bruiser of a mic is a steel weight, which when removed, reduces it's weight to 147g. This is still over 10% of the weight of the radio.

The headset shown in **Fig 18.28** was developed specifically for backpacking, but it has proved generally useful on VHF SOTA activations in adverse conditions when a beam antenna, a logbook, a microphone and an umbrella would have been too much for one pair of hands. The finished headset weighs 36g and should prove reliable if built along the principles described at the start of this chapter.

It is based on the BEP-55NC In-ear Headset Microphone which used to be available in the UK from Maplin [9], Maplin is no longer trading but there are other electronic supplers. This has a noise-cancelling insert which can be improved further by adding a foam sock as shown in **Fig 18.28**. The PTT comprises a standard SPST toggle switch. I would normally prefer a momentary switch but in this application there is little danger of it being accidentally left on transmit.

This microphone is not directly compatible with the FT-817 and so several discrete components are needed both to set the level and to prevent the RF feedback that was prevalent on HF prior to their inclusion. These components are all hidden in the heatshrink transition between the RJ-45 connector and the headset / PTT cables.

The circuit of the headset is shown in **Fig 18.29**. L1 and C1 form a LPF that prevents the RF feedback that was noticed when operating on HF whilst seated directly under the antenna. Resistor R1 is used to apply DC bias to the microphone insert and its value can be used to control the microphone gain. Values between 22k and 68kΩ were found to give a large adjustment range. Capacitor C2 provides decoupling of the +5V supply, preventing noise on this line from being fed into the microphone.

The BEP-55ND headset conductors comprise cores constructed of self-fluxing multi-strand enamelled copper wire and there is no conventional insulation between them so care must be taken not to damage this by applying too much heat. Tinning the wires quickly with a hot soldering iron was found to give good results. For reliability, the moulded 3.5mm plug that was part of the original headset was re-used and for the same reason, an RJ-45 cable with stranded wires was used for the connection to the FT-817's microphone input.

## Filters

The VHF-orientated outdoor operator is likely to find himself co-sited with all manner of radio installations. Many of these services have the potential to overload the front end of the average amateur radio transceiver. Particularly vulnerable are those with a broadband receiver such as can be found in most handhelds. In extreme cases the radio is rendered useless for minutes at a time. A model with selectivity in the front end of the receiver is much less likely to be susceptible to this overloading.

In the UK, a major cause of receiver overload is the pager traffic located just above 2m. Being so close in frequency, it is not an easy task to design a bandpass filter sharp enough to reject out-of-band signals whilst having reasonably low passband loss and in a package that is manageable for the outdoor operator.

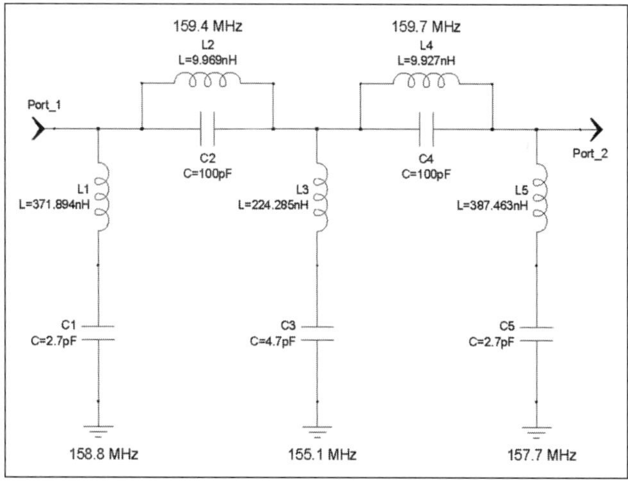


**Fig 18.30: Circuit diagram of the PMR Notch Filter. All inductors are wound with 0.71mm enamelled copper wire. L2 and L4 are 1t, 5mm av dia, 3.5mm long; L1 is 12t, 6mm av dia, 14mm long; L3 is 9t, 6mm av dia, 10mm long; L5 is 13t, 6mm av dia, 13mm long**

For near-guaranteed immunity, a filter such as the DCI 2m BPF is needed. At 300 x 150 x 75mm and weighing 1.37kg, this is a little large and heavy for the outdoor operator to carry.

Since these services are vertically polarised, the use of a directional, horizontally polarised antenna will often solve the problem. For the determined FM operator who needs to use vertical polarisation there is the MFJ-713. This is a compact 2m filter for use with low power rigs that may provide adequate protection. Simulation of this topology with realistic values indicates that it should have a useful 30dB rejection of the pager frequencies but at the expense of at least 3dB passband loss which would be unacceptable on transmit. This unit overcomes the problem by using RF-switching to bypass the filter on transmit but to achieve the switching requires an external power source.

Switching adds complexity and weight. An alternative to the conventional BPF is to introduce a notch at the frequency of the interfering signals. A suitable design for such a bandstop filter is shown in **Fig 18.30**. This topology produces a lower loss at 2m than other designs - the prototype being better than 2dB across the whole of 2m. The attenuation in the pager band is only 25dB but as each dB reduction in the interfering signal attenuates the

**Fig 18.31: PMR Notch filter - the finished unit**

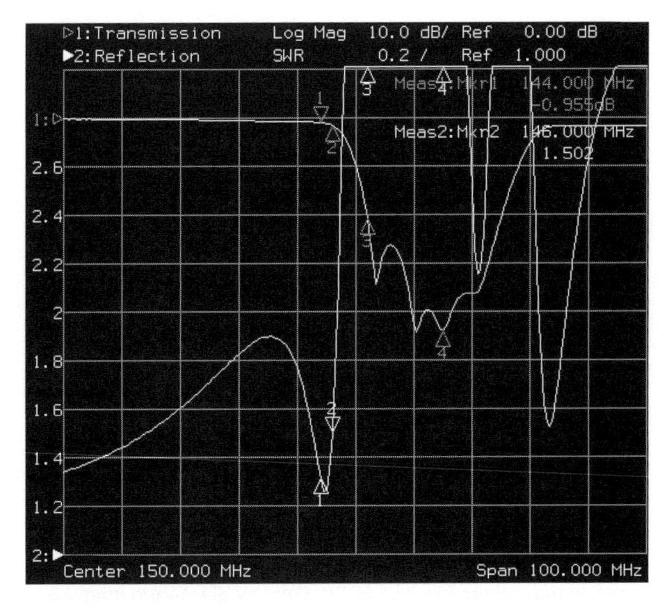


**Fig 18.32: Fig 18. 34: Frequency response and matching of the PMR Notch filter**

IMD product by 3dB it produces a noticeable improvement in receiver performance.

The final unit is shown in **Fig 18.31** and it's frequency response can be seen in **Fig 18.32** The unit measures 43 x 29 x 15mm and weighs 30g. For those without access to a network analyser adequate performance can be achieved by adjusting the coils for minimum loss on 2m and a good VSWR at the transceiver port. For those with a signal generator and detector, the notch frequencies are shown on the circuit diagram.

## HOMEBREW HF EQUIPMENT

Listen for a while on 7.032MHz and it will become apparent what success can be obtained by using low power CW into an inverted-V antenna. Whilst the FT-817 is very popular, a number of outdoor operators are using home-built QRP transceivers.

One possibility here is the design sold in kit form by K1SWL under the name Small Wonder Labs [10]. The SW40 is a 2.5W output CW transceiver covering up to 40kHz of the 40m band. When boxed, it would be approximately half the volume and a fraction of the weight of an FT-817. With a receive current of 20mA, it would last twice as long on a set of batteries. The limited selectivity may cause it to struggle on 40m at weekends but during quieter periods, it may be an ideal backpacker's radio.

## MICROWAVE EQUIPMENT

Traditional designs and equipment for the bands above 13cm have been relatively bulky and heavy. Bearing in mind the usual need for a talkback station as well, this amounts to an unacceptable load for the outdoor operator.

The separate talkback station permits full duplex cross-band operation during setup and dramatically reduces the time spent in this phase of the contact. When battery life and the effects of adverse weather on the hilltop activator are taken into account, this has to be the best method.

My combined 2m and microwave station fits into a 30 litre rucksack and the overall weight is approximately 13kg including all safety equipment, food and water for an expedition. Battery life is adequate for over one hour's operating. This station comprises an FT-857 on 2m running 50W to a 6-element Yagi and an FT-817 as the prime mover for a 3 or 6cm transverter. The FT-817's output is reduced to 0.5W to drive the transverter and so it runs on its own internal battery for the duration of the activation. Two NiMH battery packs are taken. One powers the transverter and the other the FT-857. At 'half-time' the packs are swapped to equalise usage. A headphone splitter cable feeds audio from the transceivers to separate earpieces. Arranging the transceivers so that the LH earpiece corresponds with the LH rig helps avoid potential confusion. Also, it is probably worth labelling the microphones if they are identical!

### Compact transverters for 3 and 6cm

At present, SOTA operators on the microwave bands are using transverters based on the Kuhne Electronics [11] modules by DB6NT (see the chapter on microwave transmitters and receivers). Whilst it is possible to buy some of the units as kits at a lower price, they are normally supplied aligned and ready to use. A glance at their catalogue will show a vast array of equipment for all of the microwave bands.

From a decent site, the basic transverter with around 200mW output and a 20dBd horn will allow many contacts to be made however, adding a 2W PA will vastly improve the range and put the station equipment on a par with many home stations. In considering an enclosure for the basic transverter, it is worth allowing room for a PA module to be added at a later date.

The transverters and modules come with full instructions and circuit diagrams so this description is limited to a few overall comments. **Fig 18.33** shows the inside of the 3cm unit. The two large modules are the transverter and 2W PA. Transmit/receive switching is performed by the microwave coaxial relay below the PA (the black rectangle). All microwave connections use RG-402 coax, the antenna connection being the 'N'-type right of centre.

The PCB that can be seen bottom right provides power sup-

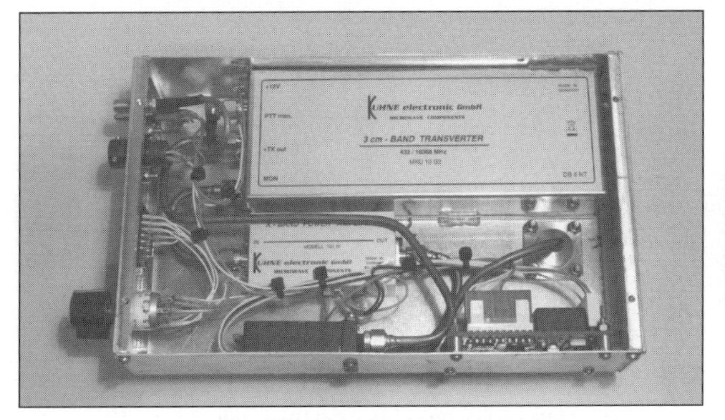


**Fig 18.33: Completed 3cm 2W DB6NT transverter**

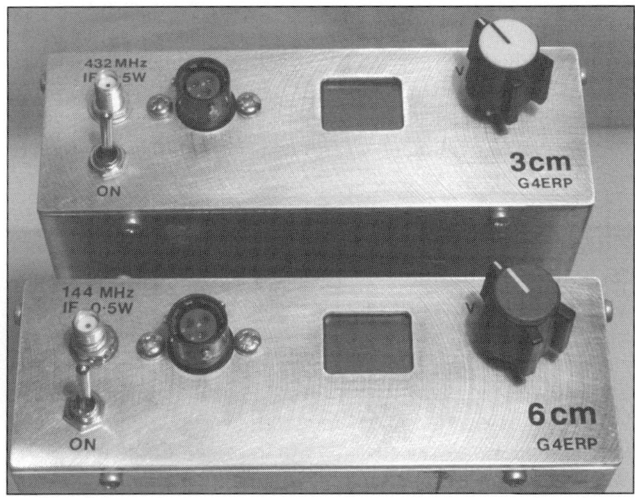


**Fig 18.34: Front panels of the 3 and 6cm transverters**

Fig 18.35: Transverter and horn mounted on the mast, aligned with the 2m talkback antenna

plies for the 28V relay and provides sequenced switching between transmit and receive. This is necessary to ensure that there is no damage caused to the sensitive receive semiconductors or to the PA transistor during switchover. If just the basic transverter is used, this is not necessary. A suitable comparator-based design can be found in [6]. The board is my own design but Kuhne electronic sell a similar sequencer.

The transverter cases shown in **Fig 18.34** comprise rigid front and rear panels machined from solid aluminium and a case bent from 0.8mm sheet. Front panel metering shows battery voltage, RF output and current drawn. Each complete assembly weighs 705g.

The mast-mounted transverter, microwave horn and VHF antenna are shown in **Fig 18.35**.

## POWER SOURCES

For outdoor operating, power sources almost inevitably mean batteries - perhaps augmented by on-site recharging using

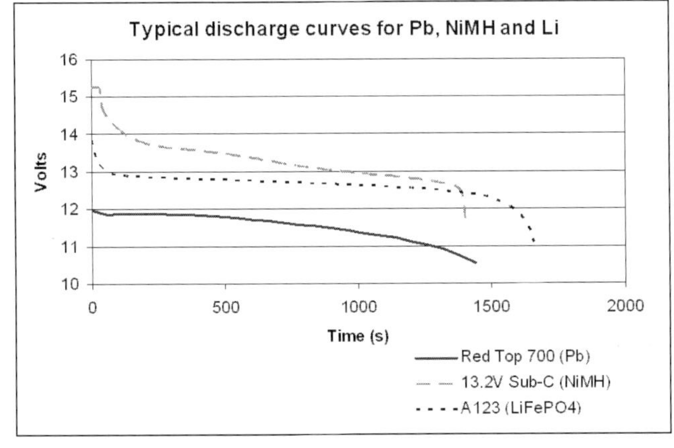


Fig 18.36: Typical battery discharge curves

solar or wind power. Recent developments in chemistries now provide us with a bewildering array of batteries to choose from. For more information, see the chapters on principles and power supplies.

The capacity of a battery is quoted as current multiplied by time in hours. The units are either mAh for small batteries or Ah for larger ones. Manufacturers often quote the capacity assuming that discharge takes place over 10 hours. Batteries that provide sufficient voltage to power transceivers such as the FT-817 will comprise several cells in a series and possibly parallel combination. The configuration of such a pack is often described in advertisements using shorthand terms such as 3S, 4S or maybe 4S2P. In the above examples, 3S and 4S describe battery packs made of 3 and 4 cells in series. 4S2P describes a pack of two 4S stacks in parallel to increase the capacity.

### Battery Chemistries

Battery technology is split between those types that can be recharged and those that are used once and then thrown away. Both environmentally and financially, rechargeable types are to be preferred but there are good reasons why the non-rechargeable types occasionally have a place in the outdoor operator's rucksack. For example, I use Lithium AA cells in my GPS unit when backpacking due to the problem of recharging sufficient batteries using solar power while away for long periods.

There is a vast range of rechargeable batteries but they all fall into one of three basic types which are explored here.

### Lead acid

In their wet form, as used in cars, they are not very attractive for outdoor operating, but when the electrolyte is contained in a gel to form the Sealed Lead Acid (SLA) battery they are more interesting. Unlike the wet variety, these batteries can be used in almost any orientation and cannot leak unless damaged. (If you wish to use one inverted, check the data sheet to ensure that this is allowed.) There are subtle differences in their construction based on the intended application which can range from starting cars to standby lighting. The manufacturer's data sheets should mention this.

In all cases, the individual cell voltage is a nominal 2V. Commonly available capacities range from 2Ah to 40Ah and whilst single cells are available, the most common format is the 12V pack. Generally speaking, they will happily source high currents for short periods but the internal resistance is such that the terminal voltage droops under load. As can be seen in **Fig 18.36**, the discharge curve of an SLA battery has a greater slope than the other types and so some of the capacity may be wasted if the voltage drops below the minimum allowable voltage for the transceiver. It is also the heaviest of the three types.

Lead acid batteries do not take well to being stored. Those that are used regularly and reasonably hard, tend to last the longest. A brand new battery that had been stored and top-up charged to the manufacturer's specification proved completely useless when attempts were made to bring it into service. In general an SLA battery has a life of about three years from manufacture. After this period, the available capacity may be seriously reduced. This means that for the occasional activator, SLA is probably a poor choice.

### Nickel

The first rechargeable battery packs that appeared in handheld transceivers used Nickel Cadmium cells (NiCd). In recent years, Nickel Metal Hydride (NiMH) cells have all but replaced NiCd,

Fig 18.37: Fig 8.39: Normal (left) and failed LiPo packs

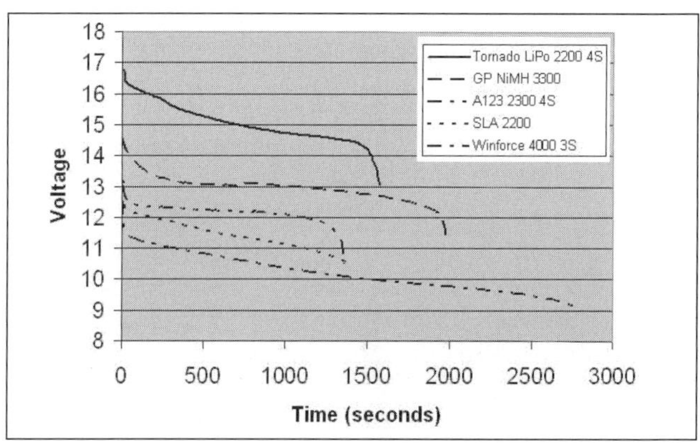


Fig 18.38: Typical battery discharge profiles

and recent legislation relating to disposal of used cells will probably soon spell the end of NiCd technology. We will concentrate on NiMH cells here. They are very similar in their characteristics to NiCd - if a little more difficult to charge correctly.

NiMH cells are available in an array of sizes and capacities. An individual cell has a nominal 1.2V terminal voltage and common capacities range from 200mAh to 4Ah. They are readily available in sizes from button cells up to 'D' cells. Considerable development has gone into the sub-C cell for high current radio control applications and these characteristics make them equally good for powering transceivers. Although available in lower capacities than SLA packs, the internal resistance is lower and so they can deliver higher relative currents without the voltage drooping. The energy density is also higher and so Nickel has been the choice of Radio Control enthusiasts for many years.

After an initial dip, the discharge characteristic is relatively flat. It then drops rapidly when the battery is depleted so most of the energy is available at a sensible terminal voltage. When the endpoint is reached, it is necessary to cease operating immediately to prevent damaging the battery by a deep discharge.

Unlike SLAs, NiMH batteries do not mind being stored. They have a low self-discharge rate and so can be stored for several days after charging without significant loss of capacity. They should be stored fully charged and if unused for a longer period, given a top-up charge before use. For longer periods of inactivity, it will be necessary to check them regularly - say every two months - and recharge them. They can be destroyed by a deep discharge and should never be taken below 0.8V per cell under load. For longer periods of inactivity, it will be necessary to check and recharge them regularly. This has recently become less onerous with a range of very low self-discharge cells from Energizer. These are available in both AA and AAA sizes. [12]

## Lithium

Lithium cells have been the biggest growth area in battery technology in recent years. It is the lightest technology currently available and cells with capacities of several Ah are now available. This performance comes at a cost in terms of their user-friendliness. Their SOA (safe operating area) is much smaller than for other types and they are capable of failing spectacularly if abused. If, however, their characteristics and limitations are understood and respected, they offer by far the best performance for the outdoor operator and they are perfectly safe. The different chemistries within the Lithium family of cells have slightly different voltages but all are in the range 3.3 to 3.8V. Those at the upper end of the range pose potential problems when used with popular mobile transceivers.

The discharge curve for Lithium is similar in characteristic to that of Nickel but it is even flatter which means that more of the cell's energy can be used. The end-of-use knee is very abrupt and so considerable care must be used to ensure that operation stops before this point is reached. Again, a deep discharge below this point will almost certainly destroy the cell.

All forms of Lithium cells are excellent at retaining their charge when stored. Unlike Nickel packs which require checking every couple of months, they can be safely left for many months without attention. This is only true when talking about the basic cells. Any battery pack with active electronics inside it should be checked as often as possible since this electronics draws current, eventually leading to a complete self-discharge.

It is not a good idea to store Lithium batteries fully-charged as subjecting them to cold in this condition can cause damage to the cells. The capacity of a Lithium cell is proportional to temperature and so cooling a cell that has been fully-charged at room temperature will result in an effective over-charge. Transportation of cells over a certain capacity by aircraft is problematical and strictly controlled by law. If they are to be stored or transported, then they should be discharged to a 50% SOC (state of charge). Most modern intelligent chargers have a function to discharge a pack to a user-defined SOC.

Three types of Lithium batteries are of interest:

**Lithium Polymer (LiPo)** is the lightest, most available but least stable of the variants. Compared with the equivalent NiMH battery, it is approximately two-thirds of the weight whilst having a similar volume. It requires careful charging and a specialist charger should always be used. Most of the packs that are available are intended for Radio Control applications where they are used in electrically powered vehicles. Many of the alarmist comments that have been made about these cells relate to the after-effects of crashes. They can be discharged at very high rates but can only be charged relatively slowly. They have acquired a reputation for failing spectacularly and although the author has never experienced anything quite so extreme he has recently had two packs fail. In both cases they failed in storage, the batteries self-discharging rapidly. The heat generated was sufficient to expand the packs (**Fig 18.37**).

**Lithium Ion (LiIon)** cells are most commonly found in the laptop PCs, cameras and any number of portable consumer appliances that can be found around the home. Most consumer packs have built-in electronics to safeguard the cells. One other major application for LiIon cells is set to outstrip all others in volume - HEVs (Hybrid Electric Vehicles). Here, large banks of cells in series / parallel totalling many hundreds of volts are used to store energy under braking using an onboard motor / generator. That energy is then used to help power the car under acceleration. LiIon cells have been chosen for this application because of the high charge and discharge rates they can withstand. In Formula 1 motor racing, KERS (Kinetic Energy Recovery Systems) packs are typically 3Ah capacity. The cells are charged and discharged at around 400A for short periods. SOTA operators are now starting to use LiIon cells as the voltage is more attractive than that of LiPo.

**Lithium Iron Phosphate (LiFePO4)** is not as light as LiPo or LiIon but it is much more user-friendly. The cell voltage is also slightly lower than the others and so a pack comprising four cells is ideal for use with most transceivers. Until recently, they were not readily available [13] and some users resorted to salvaging cells from the battery packs supplied with cordless drills. These particular cells are universally referred to as A123 but this is actually the name of the company.

There are two part numbers that you may see - 18650 and 26650. The latter is the more common and is the useful one for powering transceivers. Capacity is 2.3Ah which is marginal for most outdoor operations but I carry two or more packs which is in keeping with my policy on failure management (see below for details of charging).

## Lithium Polymer packs and mobile transceivers

While LiPO packs are undoubtedly attractive for powering mobile transceivers such as the FT-857, the discharge curves in **Fig 18.38** show that a 3S pack gives too low a voltage and a 4S pack exceeds the maximum voltage specification of the radio when fully charged. The Yaesu specification is quoted as 13.8V ±15% (or 11.7 to 15.9V) and the 100% SOC voltage for a 4S LiPo pack is approximately 16.8V.

A number of outdoor operators have overcome this problem by using the forward voltage drop of silicon diodes to reduce the input voltage of the transceiver. The preferred configuration is to use two bridge rectifiers connected as shown in **Fig 18.39**. Each bridge rectifier contributes a voltage drop of approximately 0.75V at low current depending on the specification of the device.

The second rectifier is simply linked or switched out of circuit once the pack voltage has fallen to a safe level. Although it may seem desirable to be able to switch out both bridge rectifiers, having one permanently in circuit guards against an operational error.

This technique does have a cost in terms of weight and an estimate including a suitable heatsink is 50g. This puts the complete power pack on a par with a LiFePO4 battery. A simpler solution is to limit charging to 80% SOC where the voltage is within limits and to accept the wasted capacity.

## Non-rechargeable Cells

Non-rechargeable cells still have a place in the outdoor operator's rucksack but their applications are very specific and are restricted to AA and AAA cells.

I carry spare batteries of this type for use in my GPS navigation device and my headtorch. There are two reasons for this: Firstly, non-rechargeable cells have a very long shelf life and hence only need checking periodically. Secondly, if the main rechargeable cells fail due to some external influence such as very low temperature, having cells of a different chemistry as spares guards against a double failure.

There are two types of non-rechargeable cell chemistries that are of interest namely Alkaline and Lithium. The best known of the first type is sold under the Duracell name and Lithium cells are manufactured by Energizer under the Ultimate Lithium brand (L91 and L92).

Lithium cells have better overall performance particularly at higher discharge currents where they have over twice the capacity. They are also considerably lighter. The FT-817 comes with an AA cell carrier and these are the obvious choice. **Table 18.9** gives their weights.

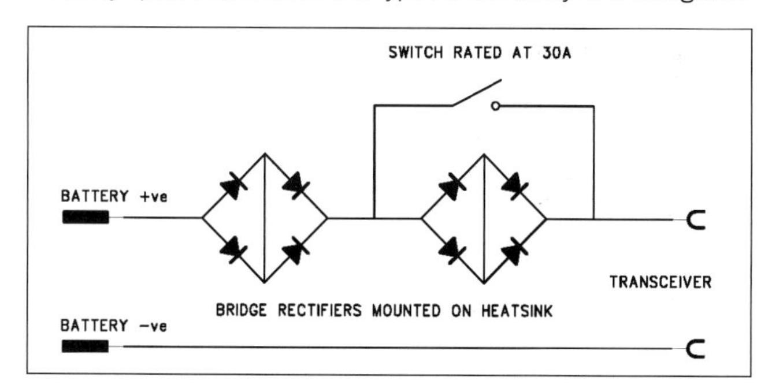


**Fig 18.39: Voltage reducing circuit for LiPo 4S packs**

| Type | Alkaline Weight per cell (g) | Lithium Weight per cell (g) |
|---|---|---|
| AAA | 12 | 7 |
| AA | 25 | 14.5 |

**Table 18.9: Non-rechargeable cell weights**

| | Lead Acid | NiMH | Lithium |
|---|---|---|---|
| **Cell voltage** | 2.0V | 1.2V | 3.5V approx |
| **Capacity** | 2Ah - 40Ah | 0.2Ah - 4Ah | 0.3Ah - 20Ah |
| **Weight** | Heavy | Medium | Lightest |
| **Discharge curve** | Sloped | Flat | Flat |
| **Charging** | Easy | Medium | Complex |
| **Safety (relative)** | Good | Medium | Care |

**Table 18.10: Battery performance compared**

## Battery Selection

Batteries are probably the heaviest single item that the outdoor operator has to carry. There is no obvious answer to the question of which battery is correct for a particular application but a correct choice will have a significant impact on the overall weight of the station.

This section explores some of the factors to consider and **Table 18.10** summarises these. The first and most important consideration is the voltage range of the equipment that will be powered from the battery.

Check the specification carefully to establish both the upper and lower limits. The input voltages specified in the handbooks of some commonly-used transceivers were shown earlier in (Table 18.7) as a guide. Too high an input voltage could damage a transceiver. Too low a voltage could be responsible for poor

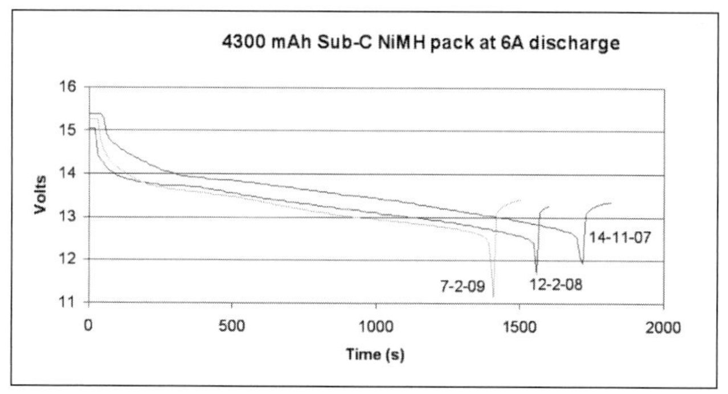


**Fig 18.40: Typical NiMH pack discharge test results showing the capacity loss with use**

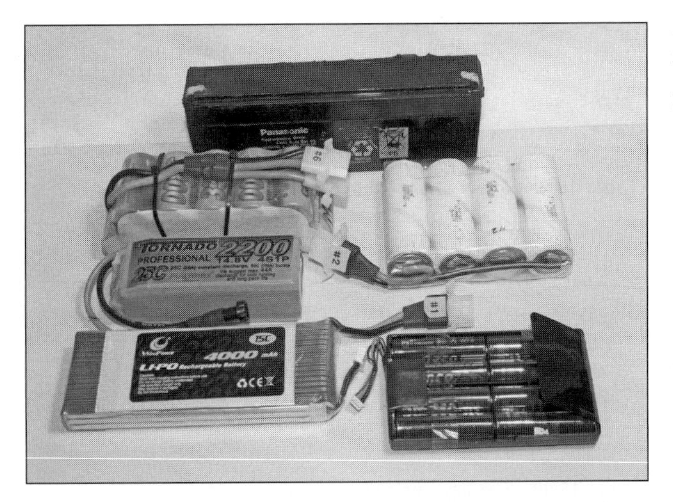

**Fig 18.41: Selection of batteries for outdoor operations. Clockwise from the bottom left: 4.0 Ah 3S LiPo pack (265g); 2.2 Ah 4S LiPo pack (242g); 3.3Ah 13.2V NiMH Sub-C pack (796g); 2.2Ah 12V SLA battery (824g); 2.3Ah 4S LiFePO4 pack (314g); 2.6Ah 9.6V NiMH AA cell pack (258g). It is no coincidence that of the six batteries shown, half are based on Lithium. For robustness and ease of use, NiMH is still to be preferred, but for the ultimate in performance, LiPo is the clear favourite. The two packs at the front are intended for use with the FT817 - indeed the one to the right uses the FT817's AA cell carrier. The voltage of the 3S LiPo is ideal for the FT-817 and this pack will power the rig for a multi-summit expedition without needing to be recharged. The remaining packs are for use with mobile transceivers such as the FT-857 which can be powered directly from all but the 4S LiPo pack. Here, the 100% SOC voltage exceeds the maximum voltage specified by most equipment manufacturers and so the diode dropper described earlier in this section should be used**

transmitted signal quality, and in some cases failure of the PA. The discharge curves in **Fig 18.40** will help with this decision and the higher initial voltage of a newly-charged pack needs to be taken into account.

When it comes to deciding which cells to use for a pack, NiMH is the most versatile technology. With a cell voltage of 1.2V it is easy to assemble a pack whose voltage is within the range of the transceiver.

Formerly, my choice for powering a mobile transceiver such as the FT-857 is a pack comprising 11 NiMH Sub-C cells. The fully-charged voltage is approximately 15.2V and for the entire usable part of the charge, the voltage is well over 12V. One such pack is shown in **Fig 18.41**.

Experience shows that a 3.3Ah pack weighing 720g will last over 45 minutes on SSB at 50 watts. More recently, many SOTA activators (including myself) have moved to Lithium battery packs and all three variants are now in regular use on the hills. This change was hastened by the availability of new A123 cells and sources of reliable, high capacity LiIon and LiPo packs from the Far East.

I now use different packs for one-day outings and for multi-day backpacking trips. Experience has shown that a single 4S LiFePO4 will last over 30 minutes on SSB at 50W, so for a day's multi-summit outing I take three packs. For backpacking trips I take a 3S 4Ah LiPo pack.

Charging is an important but often overlooked part of the battery selection process. In order to maximise the energy available and the longevity of a battery, it is necessary to ensure that charging is carried out correctly.

For optimum battery life and performance, a specialist charger is a must. At present, most LiPo cells come with a warning about safety when charging and with a suggestion that they should not be left unattended. One particular manufacturer suggests that they should be charged outdoors on a concrete block. Whilst this may be excessively cautious, I do not feel confident enough to leave a used LiPo pack recharging in my car whilst away on an activation whereas it is my normal practice with NiMH packs.

## Care, Use and Testing

Having made a significant investment in suitable batteries, it obviously makes sense to look after them and this section shows how to maintain them in the best possible condition.

## Charging

Just as it is possible to damage a battery during use, it is possible to do so when re-charging. Although sometimes difficult to find amongst their advertising information, the manufacturers data sheet will usually provide the correct parameters. Data sheets for many common batteries can be found on the Farnell website [4]. What follows here are generic comments relating to each type of chemistry.

It is now easy to find specialist chargers for the different chemistries - indeed many chargers will cater for all three types if not all sizes of battery and these units are strongly recommended. A selection of chargers is shown in **Fig 18.42**.

## Lead-acid chargers

The method of charging all lead acid batteries is the same but SLAs require additional care to prevent overcharging which can cause venting of the cells. An intelligent charger for a 12V battery will have several stages to the charge. These are:

- A bulk charge where the current is limited to a suitable value for the capacity of the battery. This is usually 0.25C so the 4A charger in **Fig 18.34** would be ideal for a 16Ah battery.
- When the voltage rises to 14.5V, the current is allowed to decrease and the voltage held at 14.5V.
- When the current drops to a low value the battery is fully charged.
- The charger then lowers its voltage to 13.8V and continues to trickle charge the battery. It can be left indefinitely in this state.

In the absence of a specialist charger, the above procedure can be executed using a current-limited bench PSU. Not all PSUs like being used in current-limit for extended periods so this technique should be used with care.

## NiMH chargers

The cheap plug-top chargers that come with so many portable devices are trickle-chargers that will take many hours to charge a pack. Because of the voltage that they give and method they use, there should be no chance of overcharging. The Radio Control fraternity have been fast-charging NiCd and NiMH cells for many years and there are a number of 12V-powered units available that will speed up the charging process. In this category are the Kokam and Overlander chargers shown in **Fig 18.43**. Most are multi-chemistry chargers that can be programmed for the desired current and then left to complete the charge unattended.

Whilst handy in an emergency, fast charging a battery is likely to shorten its life. Most cell manufacturers define a 'fast charge' as one that is greater than 0.5C - or 1.5A in the case of a 3Ah pack - and so this is my preferred setting for my Sub-C cell

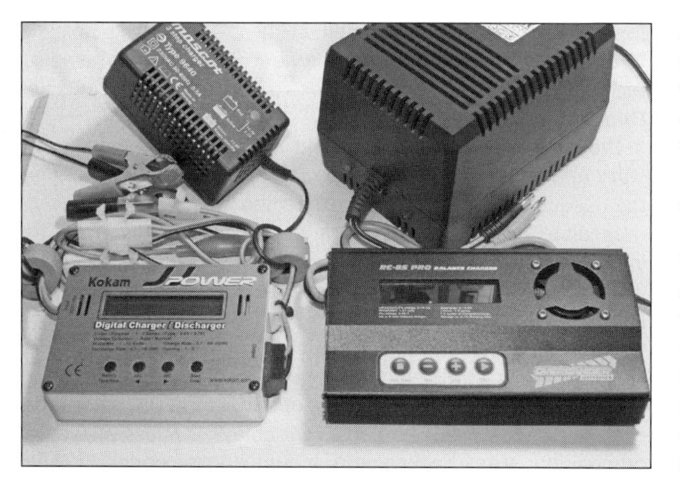


**Fig 18.42: A selection of intelligent battery chargers. The two chargers at the rear are mains-powered units specifically for SLA batteries. The large unit to the right is a Yuasa YCP4A12 three-stage charger delivering 4A. The Mascot plugtop unit to the left is for smaller batteries and charges at a maximum of 0.5A. On these units, LEDs indicate the status while charging. The two 12V-powered chargers at the front will cope with a range of chemistries. Both will cope with Lithium, NiMH and NiCd batteries, however, the unit to the right will also charge SLA batteries and all three types of Lithium batteries. Note the ferrite chokes that have been added to the leads on the Kokam unit to reduce spurious emissions. Most chargers of this type have a small alphanumeric display which is essential for their setting up and operation**

packs. A glance at the data sheet for the Ansmann AA and AAA cells indicates that a standard charge is 0.1C and 0.2C is referred to as a quick charge. I use the quick charge rates for my AA and AAA cells.

Trickle charging is usually carried out at 1.5V per cell, current limited to 0.1C, a full charge taking around 12 hours. Fast charging requires a specialist charger to detect when the cell reaches full charge. This point is indicated by a small drop in terminal voltage and an increase in temperature. The voltage drop is detected by the charger periodically removing the charge current and measuring the terminal voltage. Some chargers also measure the temperature rise although this is not a common feature.

## Lithium chargers

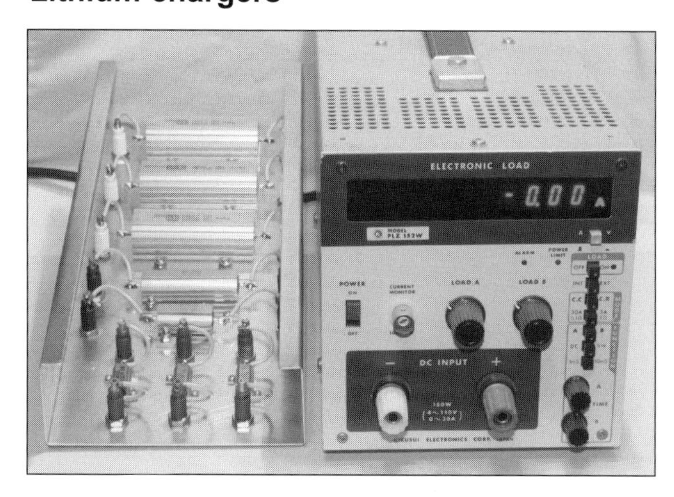


**Fig 18.43: Battery health monitoring equipment**

All multi-cell packs can become unbalanced to the point that some of their potential capacity is unavailable. This happens when one cell becomes fully charged before the others and charging has to stop at that point. With Lithium cells, it is critical that this is detected. Unless each cell's voltage is measured, it is possible to overcharge some of the cells in a pack, leading to damage and possible thermal runaway.

For this reason, it is advisable to use a charger with this cell balancing facility. The small LiPo batteries in Fig 18.38 have two connectors. The smaller is a multi-way connector which allows the charger to access the individual cell voltages. By this means it is possible to re-balance a pack. The pack is charged then individual cells are discharged until they all have the same terminal voltage. The battery can then be re-charged to its full capacity. At the very least, a charger that monitors the individual cell voltages should be used to prevent overcharging.

## Load testing

As has already been noted, batteries do not last forever. A regular check of their capacity should be made to determine when they should be replaced. There are two methods of doing this. One involves a complete charge-discharge cycle. The other involves the application of a load whilst measuring the cell voltage. Both are described in this section.

In both cases, similar equipment is needed and examples are shown in **Fig 18.43**. On the left is a homebrew load comprising a range of power resistors with values between 0.47 ohms and 100 ohms. These are mounted on an aluminium plate with 4mm sockets. This method also requires external current and voltage meters, and fans will be required to keep it cool if a discharge test is to be performed. On the right is an Electronic Load of the type that sometimes appears on the surplus market. It can be used in a variety of modes but the constant current setting is of interest in this application. The instrument varies its impedance to maintain a constant current regardless of applied voltage - something that cannot be done with a fixed resistor. Having a fixed current makes capacity checking much simpler since no additional equipment is needed. A fully hands-off test can be performed if the voltage is recorded using a data logger such as a PicoScope [14] to detect the end voltage.

## Capacity test

The battery is fully charged and allowed to settle for a few hours. A representative load is then applied and the voltage logged against time. The battery state must be monitored carefully and the procedure stopped immediately the battery reaches its end voltage. The capacity at that load current is calculated from the area under the curve of the resulting time / current graph.

$$Capacity = \sum (I * \delta t) \text{ aH}$$

If a measurement is taken every 30 seconds, the capacity in Ampere Hours is the sum of all the individual current and time measurements.

A glance at the data sheet for any cell will show that the measured capacity is related to the load. For this reason, the test should be carried out at a representative current. As experience shows that a 4.3Ah pack gives approximately 45 minutes of operating, I use 6A as my test current.

The disadvantage of this method is that in doing the test, one charge cycle is used up. However, if done on an annual basis and interspersed with the step method described in the next section, this overhead is probably acceptable. A series of tests

**Fig 18.44: Solar power being used to float charge a LiPo battery**

showing how the capacity of a battery pack degrades with time and usage can be seen in Fig 18.42. (The pack was first used in July 2006 but this data is not available.)

## Step test

The health of a battery can also be determined by applying a known load to a battery for a short period and measuring the quasi-steady voltages before loading and when the load is applied. This gives a measure of the internal resistance of the battery and hence its state of health. Unfortunately, few manufacturers supply data for their cells and so the only way to use this technique is to measure a battery when it is new and use this data as a benchmark for future tests.

With careful measurement, good correlation with discharge test results can be obtained. It should be noted that this technique is not suitable for use with packs that contain thermal or other protection circuitry as the characteristics of these components tend to dominate in the calculation. In order to get accurate results, it is essential to use exactly the same procedure each time. In particular, the timing of the three phases is critical and for this, a data logger such as a PicoScope will allow the voltage for the entire test to be recorded. The data for the calculation can then be extracted after the test has been completed.

If neither of the above methods can be followed, it is possible to get an approximate idea of the health of a pack by using the calculation built into chargers such as the Kokam unit shown in Fig 18.39. This displays the charge that has been applied to the pack. However, it should be noted that this value is optimistic as it includes the time taken to detect the end point and this is beyond the point where the battery is fully-charged.

## Solar Power

For the serious backpacker wanting to recharge batteries during a multi-day outing, solar power would seem to offer the ideal solution. It is a technology that is attracting plenty of development and interest at present, so although the performance is currently a little disappointing, this may change over the course of the next few years.

A solar panel converts light into electricity. In simple terms, the number of solar cells dictates the maximum terminal voltage but the light level has a big influence on the actual voltage and current that is available. As a result, solar power is highly unlikely to be a viable energy source on its own but it can be used to recharge the primary power source when it has become discharged.

The alternative to solar-powered charging, of course, is to carry additional battery packs and not worry about recharging the used ones. For a weekend expedition, this may be the sensible approach, but for a longer outing, this may be impractical. Let us compare the merits of the two options.

The major deciding factor is the amount of available sunlight. Practically, if you can't see a distinct shadow it's not worth getting it out of the rucksack. On many American Long Distance Trails, sufficient sunlight can be guaranteed to make solar-powered charging a realistic option. In the UK, where the climate can best be described as variable, it is a less attractive proposition. Many backpackers mount a solar panel on their rucksack whilst walking to maximise the available time for charging. In order for this to be practical, the size of the solar panel needs to be carefully considered as does the method and flexibility of the mounting so that output can be optimised. On a linear walk where the backpacker's relationship to the sun is reasonably constant, it is possible to benefit from solar charging from this technique.

A search on the web will indicate a variety of devices. Some can immediately be discounted for portable use unless the application is purely to recharge a couple of AA cells for a GPS unit, a head torch or an MP3 player. If the intended use is the recharging of batteries for powering radio equipment in the field, then a realistic recharge time has to be achieved. Panels with an output of at least 5W should be considered.

Currently there are three units available for the hiking market that are worthy of consideration: the Flexcell [15] Sunpack 7W and the Reware Powerpocket 6.5W and 12W Folding Solar Panels. The former weighs 700g but appears sturdy and capable of being successfully mounted on a rucksack. It rolls up into a cylinder of approximately 35 x 6cm. The latter units weigh 200g and 360g respectively but the 12W version is too large for rucksack mounting and so would need to be used statically during an activation to recharge an already spent pack or during the evening after the day's operating is finished. I have successfully used the PowerPocket 6.5W on multi-day trips (**Fig 18.44**) and the results presented here were obtained with this unit.

## Solar panel performance

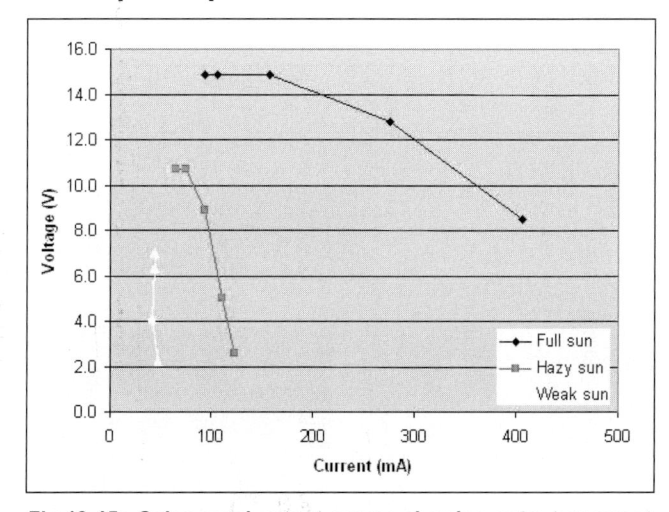


**Fig 18.45: Solar panel output curves showing output power at different light levels**

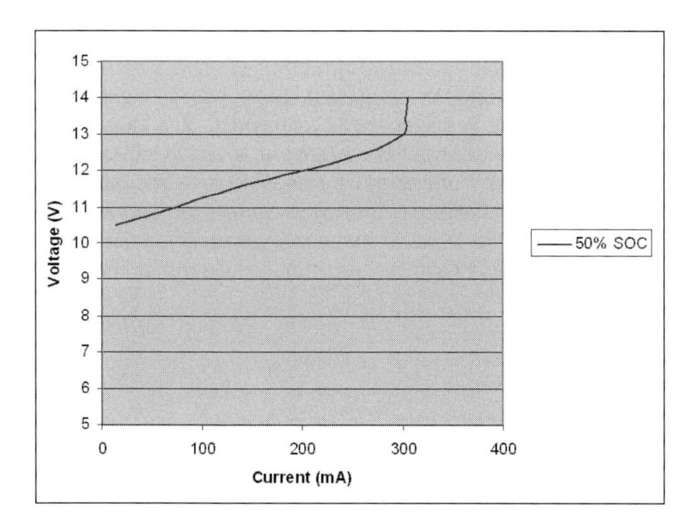


**Fig 18.46: FT817 charge curve**

So, what can you expect in the UK? The Reware 6.5W PocketPower solar panel data sheet specifies a maximum output of 433mA maximum at 12V. Can this be achieved?

The output of a panel is strongly dependant on light level and whilst hazy sun will provide enough power to charge a couple of AA cells, if the object is to replenish a discharged battery pack for the FT-817 then sunlight bright enough to give a strong shadow will be needed. The graph of **Fig 18.45** indicates that the maximum output that could be obtained in full sun in the south of England was 406mA but at this current draw, the voltage had dropped to 8.5V (an output of 3.45W). The best performance curve will only be achieved if the entire panel is in full sun and nothing must be allowed to mask the active area. The panel must also be kept pointing at the sun. If the error is kept to less than 30 degrees, almost full output will be obtained.

Charging batteries from solar power

In the UK, it is safe to say that there is one application that is ideally suited to solar charging. The FT-817 is the ideal radio for backpacking trips where summit activations are spaced over the course of several days. Maintaining charged batteries under these conditions using solar power is a practical proposition given reasonable weather.

Charging the FT-817. Tests on an FT-817 indicate that the internal charger limits the current into the battery at 300mA but that a useful charge is still achieved at lower voltages. These tests were performed with the battery at a 50% SOC. (See **Fig 18.45**) Comparing the graphs in **Fig 18.47** with that of**Fig 18.46**, it will be seen that in full sun, a charge current of over 250mA will be achieved.

Charging LiPo packs. My preference when using the FT-817 is to power it from an external higher voltage battery whilst the internal NiMH pack is kept as a spare. The current-limited 400mA output of the panel is ideal for charging a 4Ah LiPo 3S pack. All that is then required is a low voltage dropout linear regulator to limit the charging voltage to a safe level. While it would be possible to 'push the limits' and charge at the LiPo maximum allowable voltage of 12.6V, there is little to be gained and so a fixed voltage regulator is used. The specification for ST's LF120ABV indicates a maximum voltage drop of 350mV at 400mA load but experience shows that in full sun, a charge current of around 250mA will be achieved.

I used the circuit shown in **Fig 18.47** for this purpose. It allows the battery to be float charged during an activation as well as on the move. If there is insufficient sun, the FT-817 is then powered from the battery alone. The system is shown in **Fig 18.48**. The LDO regulator is potted into the boot of the connector to the right of the solar panel using RT-125 (see the Heatshrink section), construction being 'dead-bug' style with strain-relief on all the wires. The FT-817 power connector can be seen at the bottom of the photo.

By way of a demonstration of its effectiveness, I used this system on a six-day walk which included five SOTA activations. Thanks to topping up the battery using solar charging I returned home with the battery at a 75% SOC.

When there is insufficient sunlight to charge the LiPo battery due to the reduced terminal voltage, ancillary items such as phones and cameras can still be charged. For this, I have built similar adaptors and these are shown in **Fig 18.49**. The adaptor at the top allows a compact camera LiIon battery to be charged (it is held in place using the rubber band). The potting on the top of the nylon housing conceals a TO-220 voltage regulator. The contacts are spring-loaded terminals salvaged from a handheld radio microphone sideclip connector. The second adaptor is for charging a mobile phone and uses a regulator salvaged from a commercial car charger.

## MATERIALS

In the context of engineering the ultimate lightweight station, the choice of materials is critical. Thanks mostly to research in the field of aerospace, there is now a huge selection of advanced materials available. Knowing their properties is the first stage to the ultimate in station engineering.

Taking aluminium as an example, the range of strengths available from different grades is large and achieving the same durability in the weakest grade may require a part that is twice as heavy as using the strongest.

VHF antennas, of course, are primarily aluminium and so considerable weight savings can be made here by the correct choice of grade.

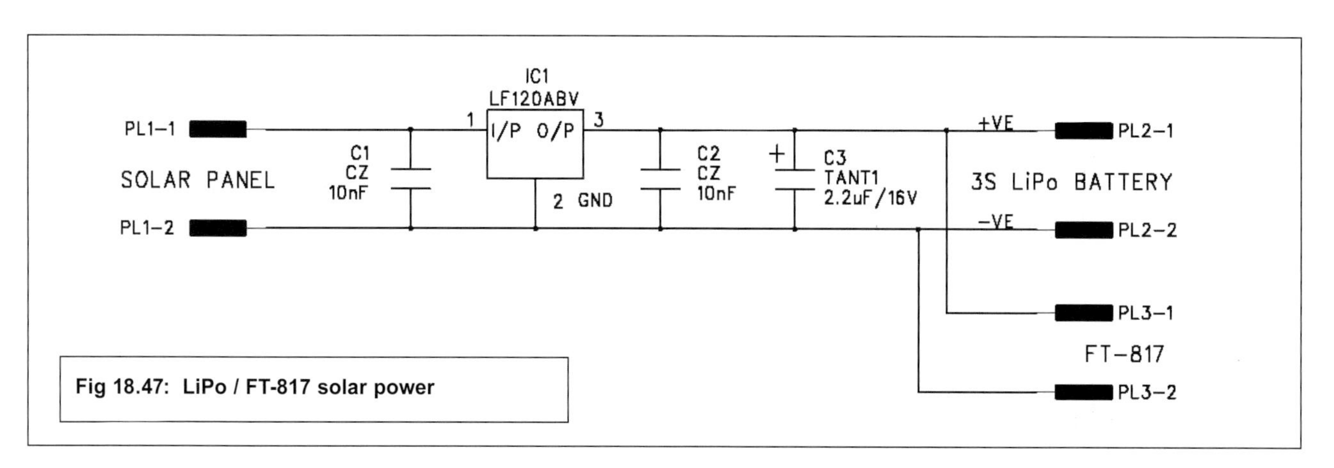


**Fig 18.47: LiPo / FT-817 solar power**

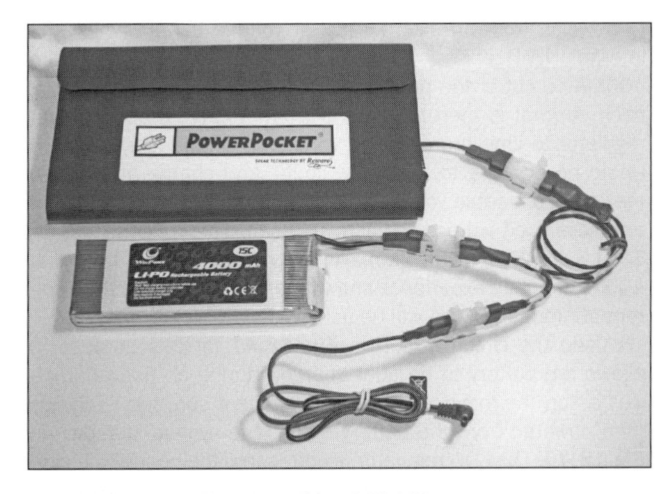


Fig 18.48: Solar float charging of 3S LiPo

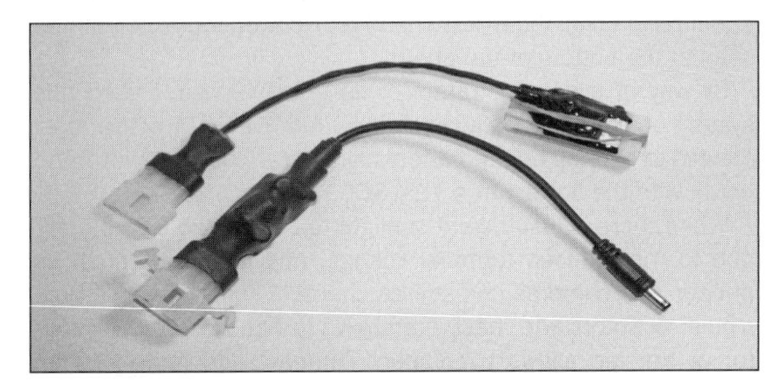

Fig 18.49: Fig 10.51: Ancillary charge adaptors

## Composites

Composites is the generic term for a large range of materials that are produced by combining widely differing substances. Examples include carbon fibre, glass fibre, chipboard, plywood and MDF. In each of these cases, resin is used to lock fibres in a defined orientation to each other, imparting a strength to the final product that is greater than that of the constituents.

When working with composite materials and carbon in particular, it is advisable to take precautions to avoid inhalation of dust. Any splinters should be removed immediately as the resins can cause the onset of an infection.

## Carbon fibre

Carbon fibre materials have limited use in outdoor stations. However, in certain applications they are an appropriate choice. In the raw state, it is supplied as woven sheets impregnated with adhesive. When laid into a mould and baked in a vacuum, it can produce a 3D part that is incredibly strong and much lighter than the fabricated equivalent.

Another application where carbon fibre is an appropriate material is as a microwave antenna such as a dish where it is covered in aluminium foil. Carbon fibre could be used to manufacture an ultra-light and dimensionally stable antenna. It can be bought in sheet form, ready-cured. As flat sheet, it is still stiffer and slightly lighter than the aluminium equivalent and so I have used it for logbook backings and coaxial cable supports.

Carbon fibre tubing is also available - at a cost - and weight-for-weight it is stiffer than a glass fibre equivalent. Be aware that many plastics have been made to look like carbon by clever printing. Carbon fibre is very hard on cutting tools. Top quality files and drills last well, but lesser quality tools are likely to have a very short life.

## Glass fibre

Glass fibre materials come in many forms but one is particularly useful for the outdoor enthusiast looking for a lightweight antenna support. A seven metre long roach pole is a useful source of cheap, lightweight material. It is not as strong as a carbon equivalent but it is much less expensive and is readily available. These poles are mostly covered in an epoxy-based paint which should be removed prior to any bonding operation. This can be done by using a sharp, flat file followed by fine abrasive paper.

# METALS

## Steels

Being generally heavy and prone to corrosion, steels would not normally find a place in the construction of lightweight equipment and if strength is required, titanium is generally a better option.

If the superior strength of steel is required then my favourite is EN24. This is reasonably easy to machine and is more commonly available than the stronger grades that have higher EN numbers. Mild steel has no practical use for the outdoor station.

## Aluminium

Most of the aluminium offered in DiY shops and at engineering shows is of indeterminate grade and generally soft. As such, it is a poor choice for most applications but it is ideal for antenna element construction since a small bend can more easily be straightened in the field.

For machined parts, there is nothing better than HE15. This is a high-strength alloy that can be reduced to a wall thickness of 1mm or less without worry about its longevity. My 1mm thick mast couplers have withstood numerous encounters with rocks over the years without sustaining any damage. There are several other specifications that describe the same material, so offcuts stamped with L168 or 2014A are equally valuable. As an even tougher alternative, L160 or 7075 can be used. None of these can be welded. If machined parts are to be welded, such as might be the case for the flange and walls of a microwave horn antenna, then HE30 is the preferred grade. Equivalents are L111 or 6082. It is possible to weld many types of aluminium sheet that can be found on the surplus market, but for the best results, the NS4 specification sheet is preferred and gauges down to 1mm can be worked reasonably easily.

## Titanium

Titanium is an exceptional material that has the strength of steel, does not rust yet is approximately the density of aluminium. As with other metals, there are many grades.

For machined parts, Ti6-4 is the most common material and it can be machined reasonably easily. For sheet metal work, it is essential to remember that there are two distinct grades. BSTA2 can be bent whereas BSTA10 (Ti64) definitely cannot.

## Brass

Despite its habit of tarnishing, its high density and its lack of strength, brass is still an exceptionally useful material for small components where weight is less of a problem. It can be machined easily, takes threads well and does not corrode. It is ideal for small bushes and studs. It is also easy to soft solder.

### Ropes

There are three types of rope. Those that stretch and are supposed to, those that stretch and are not supposed to and ropes that do not stretch. For guying portable antennas, Liros Dyneema-based Magic Gold is the author's favourite. It has a guaranteed stretch of less than 3% and has good abrasion resistance. The 2mm cord has a breaking strain of 240daN which has proved adequate, surviving some severe mountain-top conditions. It is available in a highly-visible luminous Neon Yellow which helps the thin cords stand out against most backgrounds. It can be obtained from most yachting shops. Tendon supply similarly well-specified cords for the climbing market. These start at 4mm diameter and are available from outdoor shops.

Shock cords are elastic ropes and they are very useful for the outdoor operator or backpacker. The author uses 3mm diameter shock cord in conjunction with button-operated toggles to hold together bundles of poles and antenna elements and to secure loads to rucksacks etc. The cord can be obtained from many sources, including caravanning and outdoor shops. Good outdoor shops also hold stocks of spare toggles in different colours.

## SEALANTS AND ADHESIVES

When used appropriately, both adhesives and sealants have an important part to play in the construction of the outdoor station. A visit to any DiY shop will reveal shelves full of tubes containing different compounds intended for different purposes.

### Tubed sealants

In general, there are two families of sealants - those that use acetic acid as a solvent and those that do not. Those which are often referred to as Silicone Sealants and which have the unmistakable smell of pear drops should be used near any metallic part with caution. They will often cause severe corrosion. Steels are particularly badly affected - even plated ones. Their main advantage is a faster cure time but, however tempting, that feature should be ignored.

Sealants are useful in a number of applications, including waterproofing box lids - such as the tuning network on the 2m vertical described earlier. As they do not form an adhesive bond, the parts can be relatively easily dismantled when needed.

When used to seal a box, a thin uniform layer should be applied - just enough to fill the largest gap. Gentle, slow tightening of the screws should then cause the two surfaces to extrude any excess which can then be cleaned away.

Applying a blob of sealant between a wire or a wire bundle and a solid surface such as an enclosure wall will help prevent a vibration-induced fatigue failure of a solder joint by damping any motion.

In the commercial world, Dow Corning's RTV744 has been proven to be compatible with all types of electronic equipment and metals. A more readily-available alternative is Unibond's Waterproof All-Purpose Sealant which also has reasonable adhesive properties. It causes a slight reaction with brass, producing a Verdigris-looking deposit after an extended period.

### Thread locking

Outdoor equipment is subjected to repeated thermal cycling and vibration in a way that home equipment never is. In the long-term, this can cause threaded components to come loose. Thread lock compound applied to the parts during assembly will prevent this problem. The most commonly used compounds are made by Loctite [16] who produce a range of products for different applications. For general low-temperature use, their 242 (Nutlock) compound is ideal. It does not require heat to permit disassembly. For applications that require a small amount of gap-filling, 641 can be used to good effect.

### Conformal coating

By far the best way to protect PCBs is to apply a conformal coat such as Concoat's HumiSeal 1B31 (available from Farnell etc). This can be carefully applied using a small artist's brush, covering small components as well as the board. Apart from preventing corrosion, it will wick under parts, damping any movement due to vibration. It is a good idea to avoid applying the coating to any open-frame connectors as these same wicking properties can cause problems.

### Adhesives

Adhesives do not generally like being subjected to moisture for long periods but for the length of the average outdoor operation, this is not a problem and so they can be used for bonding many types of materials together. Aluminium mast couplers can be bonded to glass fibre poles. Nylon bushes can be bonded to aluminium rods.

The most readily-available adhesive is Araldite which is a two-part epoxy. Provided the surfaces are cleaned and de-greased first, the resultant bond will last well. Adhesives work best where the parts overlap, giving a large bond area. Butt joints usually fail.

Hot melt adhesives of the type that are readily available in DiY shops are widely used in the manufacture of battery packs where they are used to hold the cylindrical cells together. They have many other uses in the outdoor station but should be used sparingly to avoid adding unnecessary weight to the assembly.

### Heatshrink

The reader who has made it thus far, will have probably have realised that I am a big fan of heatshrink sleeving. No longer a specialist material, anyone looking in a distributor's catalogue will note the large variety of types that are now available.

Unless heatshrink sleeving is to be used with the companion high temperature wires and cables it is wise to use the low temperature types. These are usually cheaper and have higher shrink ratios anyway. They are more versatile whilst being less likely to cause damage to PVC insulation. The Pro-Power range of 3:1 tubing available from Farnell shrinks at about 100°C compared with Raychem's DR-25 at 175°C.

If Raychem [17] 44 or 55 Specification wire - or indeed any high temperature wire - can be acquired, then the ultimate in lightweight flexible control and audio cables can be made by twisting the wires together by hand then covering in heatshrink. In this case DR-25 heatshrink can be used as this provides the toughest jacket.

To use this properly with the matching boots, it is necessary to glue the parts together using a two-part adhesive such as ResinTech's RT-125. These are the techniques that are used to manufacture the ultra-light yet robust looms that are used on racing cars.

An alternative to using plain heatshrink boots and two-part adhesives is to use pieces of adhesive-lined heatshrink to join sleeving to connectors and to cover inline splices. This is not as durable, but is quick to implement and does not require such specialist techniques.

For the protection of the soldered joints of in-line splices, Kynar semi-rigid sleeving is recommended. The complete splice is then further covered with adhesive-lined tubing to strengthen the transition. Heatshrink products should be used in a well-ventilated area as small amounts of fumes are released during the shrinking process.

## SAFETY AND WALKING EQUIPMENT

SOTA and other outdoor operating spans a huge range of environments, from a 5-minute amble from a car park to multi-day serious mountain traverses in remote country. The amount of effort that has to be spent on planning, and the amount of money that may have to be spent on equipment, follows a similar wide range (though anyone contemplating one of the more adventurous types of expedition will presumably be appropriately experienced and equipped already). In any environment, the outdoor operator will be at greater risk than the average hill walker, for the following reasons;

- A higher risk of hypothermia due to the length of time spent stationary.
- A higher danger from thunderstorms (when carrying poles and erecting antennas).
- Instability caused by carrying a larger, heavier and perhaps less well-balanced load.
- The distraction caused by an intense concentration on the radio (missing deteriorating weather).
- A greater than normal determination to reach a summit whatever the conditions.

To this list should be added the hazards facing the more 'normal' hill walker such as trips, navigation and the vagaries of the British weather.

This section attempts to indicate 'best practice' and to show what equipment is available to help reduce the risks to a minimum. Following this to the letter will not guarantee your safety. Experience built up over increasingly challenging expeditions, backed by formal training or tutoring by a mentor is probably the best way to learn safely.

### Clothing

*"There is no such thing as bad weather, only the wrong clothes"* says the old proverb. The importance of using the correct clothing cannot be stressed enough. The marketplace is full of outdoor garments of different types and a discussion of the various options could fill a book in its own right. If you intend heading off into the great outdoors for the first time you are strongly recommended to read the specialist walking magazines available from high street newsagents. They publish excellent reviews of equipment and offer good advice. What follows here are no more than a few hints.

- Go to a specialist outdoor shop and ask for their advice. Most such shops are staffed by knowledgeable enthusiasts.
- Don't be tempted to wear Denims - ever. At best, they restrict leg movement and waste energy. Worse, they get wet and stay wet, leading to heat loss.
- Always take spare layers, even when the weather shows no sign of it being necessary. In particular, take a spare hat and gloves. Loss of a glove could result in frostbite.
- Good supportive boots are essential - despite what you might read. We are hikers - not fell runners!
- Buy the best you can afford. The best will keep you warmer or will be lighter - or both.
- Look after your investment. Apart from the cost of replacement, it could be important for your continued wellbeing.

Here are some of my favourites:

**Extremities Sticky Windy Gloves**. They will allow you to write, operate a rig and even assemble SMAs. Not as warm as some, though. Take them as your spare pair.

**Montane Atomic Pants and Jacket**. They weigh very little, pack up into tiny stuff sacks and whilst not as breathable as some waterproofs still work very well.

**TrekMates Merino T-shirt**. A great base layer - especially for long treks where the wool's anti-odour properties are appreciated.

**Scarpa Ranger GTX boots**. A good hill walking boot. Having leather uppers, they require more care and maintenance than others. They will take a Grivel G10 Crampon for short periods despite their rating.

**Lowa Renegade GTX Lo trail shoes**. Excellent for backpacking on defined paths.

**Balaclava**. It will stop your face from freezing and so will help prevent a Q4 report on SSB!

### Navigation

In the UK we are extremely lucky to have the most incredibly detailed maps at our disposal. The Ordnance Survey's 1:25,000 series maps are a treasure trove of information for the trained eye to use. When used in conjunction with a compass, a GPS receiver and even satellite imaging, it helps the outdoor operator to understand and anticipate the terrain. There have been one or two cases where SOTA activators have activated the wrong summit due to errors in navigation so quite apart from making sure you get back safely, accurate navigation is very important.

A selection of navigation tools is shown in **Fig 18.50**. Their correct use requires practice and details of training courses can be found in the walking magazines.

### Maps and compass

GPS receivers usually provide a compass facility - examples can be seen in **Fig 18.50**. Some work by using your GPS-calculated direction and some have built-in Hall-effect sensors. They all use electronics.

It is a wise precaution to take a traditional compass based on a magnetic needle and to know how to use it in conjunction with a map. Silva manufactures a good range of plastic-bodied compasses of varying sizes from the tiny to the classic Silva Ranger 3.

Even when your GPS receiver is still working, there can be occasions when a map and compass are preferable. A compass can be operated in arctic conditions when fingers pushing buttons would freeze in seconds.

### GPS

Many personal computing devices incorporate a GPS receiver these days, however none is waterproof and most do not have the features that are available in devices specifically aimed at the walker.

This is not the place to discuss the workings of GPS, so this section concentrates on the basics, on those features that matter most and how to get the best from a receiver.

A GPS receiver calculates its location by clever mathematics based on the time taken for time-critical signals to travel from satellites in known locations in space. It could be viewed as DF in reverse. It is unaffected by cloud cover - or even rain, the system working at 1.6 GHz. It is important to have a clear view of as many satellites as possible to get the most accurate calculation of position.

Tree cover used to be a big problem, however, the latest high sensitivity receiver chip sets are very much better than the older generation ones, which lost lock very easily. Whilst users have been known to place their GPS receiver in their hats, a more

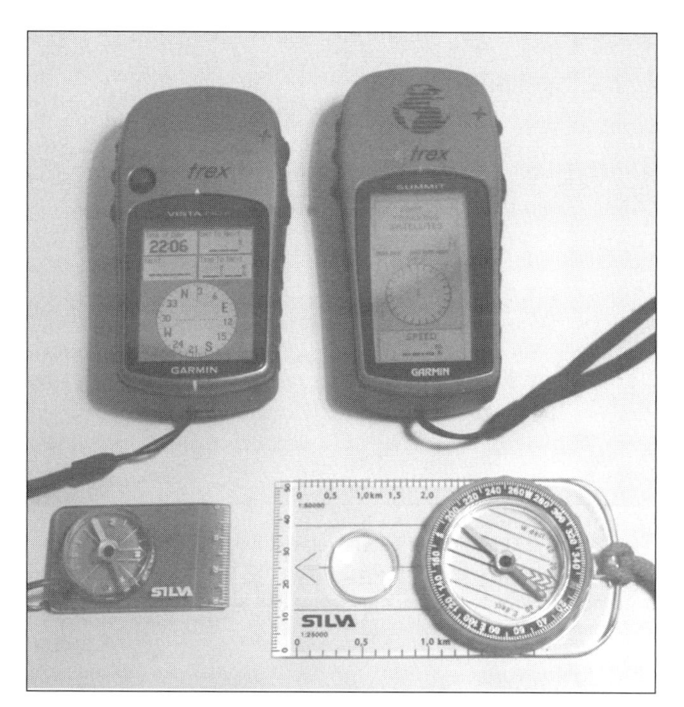


**Fig 18.50: A selection of navigation tools: The newer Etrex Vista HCx on the left has a colour screen and higher sensitivity receiver than the older Summit on the right. The small Silva compass below it may be adequate as a backup. The Silva Ranger 3 on the right is preferable**

normal and more readily accessible location is on a rucksack strap where the top part of the unit can see the sky. Garmin sell a carrying case which has a useful clip. If the receiver is supplied with a lanyard, it is a good idea to attach it to the rucksack to prevent its accidental loss.

Most GPS receivers run on two AA cells and the life is often no more than 20 hours. The receiver then automatically switches off when its preset end voltage is reached. Often a large percentage of the cell's capacity is wasted. Although they have a lower initial voltage, rechargeable NiMH cells will give longer use than Manganese Dioxide non-rechargeable cells due the poor slope of the latter's discharge curve. The latest receivers allow the user to select the type of battery that has been fitted and this should help to maximise the life of a set of cells.

Typically, the user will decide on some key points on a map that define the intended journey (track junctions, turns, landmarks, etc). These waypoints are then linked to form a route which is downloaded to the receiver. Whilst it is possible to enter all this information directly into the device, this is time-consuming to say the least. Most come with a PC interface and software which allow this to be done more efficiently. With the route downloaded, the receiver is then told to follow it from one waypoint to the next. In so doing, it records a log of the actual path taken (a so-called breadcrumb trail). This can be used later in the walk to allow the user to retrace his steps or it can be downloaded to the PC for analysis. It is fascinating to overlay a tracklog on an OS map and see the exact route.

**Fig 18.51: iPhone and iWatch both have built in compass**

In choosing a receiver there are several important features to consider.

- **Sensitivity**. The latest generation receivers such as Garmin's Vista HCx are infinitely better in tree cover than older models such as the Etrex Summit.
- **Routes**. Some units are able to store a maximum of only 20 routes. For a long multi-summit expedition this may be a problem.
- **Reversible routes**. Both the Summit and the Vista allow you to follow a route in either direction, however, the Vista is less intuitive to set up and is less well-documented.
- **Maps**. Later units like the HCx can store and display real maps if a micro-SD card is installed. Cross-referencing the GPS position to a location on a paper map is made much simpler.
- **Tracklog memory**. Older devices record into a limited-size internal flash memory. Depending on the logging mode chosen, this is generally adequate for a weekend expedition but would not cope with a Long Distance Path walked over a period of a week. Modern units like the Vista have the ability to log onto a micro-SD card. The same weekend expedition that would fill two-thirds of the internal flash memory would use only 100kB of a 1GB card.

Also shown in **Fig 18.51** is an iPhone and iWatch the later vesions both have a built in compass, some Android devices also have compass features, the photo shows both with different readings this is because they were photographed inside, but seem to be fairly accurate outside compared to my standard compass and both had the same readings.

## Rucksacks and Waterproofing

Another difference between the outdoor operator and the aver-

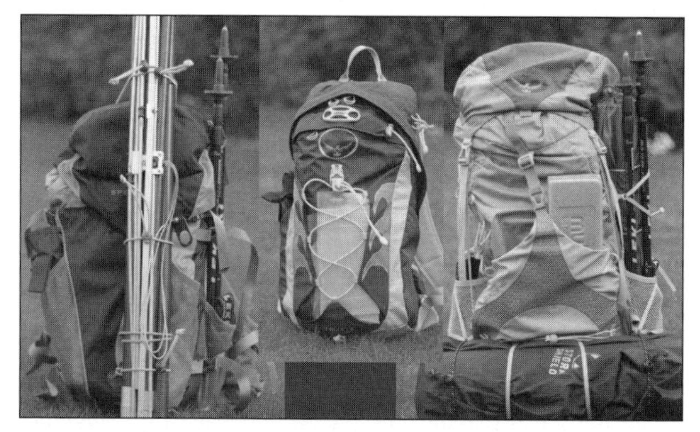

**Fig 18.52: Rucksacks suitable for different radio-orientated activities: (left) Deuter Speed Lite 30 litres, 830g. Hardwearing and just large enough to cope with an overnight stop if you are a minimalist. Shown complete with mast and antenna for 2m. (middle) Osprey Talon II 10 litres, 630g. Excellent for a minimalist activation or mountain biking. Will carry an FT-817 and essential walking equipment. (right) Osprey Exos 46 litre, 1030g. Slightly less robust. Shown ready for a week's expedition with camping equipment, FT-817 and spare batteries. The 'R'-Clip Special and mast are in the bag on the right with the walking poles.**

age hill walker is in the type of equipment he needs to carry and the relative fragility of this equipment. These factors have a bearing on the choice of a suitable rucksack.

The primary specifications you will see for a rucksack are its capacity and its weight. The term Daysack is applied to something generally less than 25 litres and often without a hip belt. A hip belt adds stability and transfers weight to the hips, making a large or heavy sack more comfortable to carry. For a single activation or even a day's expedition, a Daysack is usually big enough, although the wearer's shoulders may suffer when carrying a heavy load and so one with a hip belt is to be recommended. A 30 litre rucksack could be adequate for an overnight stop but this is specialist ultra-light territory and probably not for beginners. A 45 litre rucksack is a better bet. Again, for the ultra-lightweight enthusiast a 45 litre rucksack could be adequate for longer expeditions but 55 litres would be less spartan.

Whatever the application, the quality of the material is important. Here, it is very much a case of "if it feels strong enough, it is". For this reason it is best to visit a good outdoor shop and see for yourself - especially if straying into ultra-lightweight territory. The nature of the loads to be carried could result in damage to some of the fabrics used in ultra-lightweight rucksacks. That said, recent advances in materials and design techniques mean that even a strong 45 litre rucksack can now weigh less than 1kg.

When choosing a rucksack for outdoor operating, the following questions are worth asking yourself.

- What type of expeditions do I want to cover?
- Has it got attachment points for antennas?
- Can I mount a GPS receiver in a convenient place?
- Has it got lots of separate compartments? This helps with segregating radios from walking necessities.
- Is the back system going to protect me from awkward-shaped objects? Will it be difficult to pack?
- Is it comfortable when fully-loaded? Good shops will load up a rucksack so you can make a realistic assessment.

Generally-speaking rucksacks are not waterproof. It is normal practice to stow any critical items inside specialist waterproof bags which can be bought from outdoor shops. They are useful for larger items like sleeping bags or maybe a transceiver but the outdoor operator will need to carry many small articles - such as coax adaptors, batteries etc and these are best organised into several small, separate groups. A cheap way of doing this is to use Baco Zip'n'Seal freezer bags. They are readily available, cheap, will last many activations before needing replacement whilst coming in a range of sizes. For the outdoor operator, the added advantage of keeping radio items in waterproof bags is that the bags help protect the rucksack from damage due to sharp corners. When the bag is worn it can be cheaply replaced.

**Fig 18.52**: shows three rucksacks that have proved suitable for different types of expeditions. All three have hip belts. An alternative source of suitable rucksacks Is the camera market. Since the last edition of the Manual was published several of the camera bag manufacturers have produced sacks that combine space for camera equipment with capacity for hill walking items. The padded pockets are adjustable which allows radio equipment to be accommodated securely. They generally come supplied with a rain cover. manufacturers to consider include Lowepro, Manfrotto and Think Tank / MindShift Gear.

## Weight

The most valuable piece of equipment the outdoor operator can acquire is a good set of digital scales. Using these, the activator can keep a check on the weight of his load. I maintain a spreadsheet listing the weights of every piece of equipment I intend carrying and from this can calculate the overall weight of the rucksack. The weight that can be carried depends on your own physique, level of fitness and the nature of the expedition. Some activators will happily carry 15kg all day. I am of slight build and have found my limit to be around 12kg for a moderately strenuous expedition or 13kg for backpacking.

**Fig 18.53: A selection of safety-related items for the hill walker: Laser Competition tent, Storm Shelter 100, two head torches and a carabiner with a whistle, mini-compass and single LED torch**

## Safety Equipment

In addition to the clothing and navigation equipment mentioned in the preceding sections, it is wise to carry several other safety-related items in case of emergencies.

Hypothermia is a real possibility at altitude even in summer. Any form of shelter will make a big difference to your chances of survival should you be unfortunate enough to suffer a forced stay on a summit. Survival bags come in different sizes but Vango's one-person Storm Shelter 100 weighs 380g and its deployment promotes an instant increase in body temperature thanks to the shelter it provides from a chilling wind. The alternative is to carry a small tent that can also be used for more extended operating periods as well as for emergencies. It should be said, however, that an exposed summit is not the best environment for a tent. Merely putting in the guys may be impossible on rocky terrain. The best of the lightweight tents is Terra Nova's Laser Competition at 990g. The material is relatively fragile and care is needed to avoid damage - especially on rocky ground.

To summon help should it be needed, a whistle should always be carried - and possibly a mobile phone as well. Six blasts on the whistle is the universal distress call and this should always be tried first before 'phoning for help. Mobile phones often have no coverage and should never be relied upon. It is also best to leave them switched off unless they are needed as battery drain is drastically increased when areas of poor coverage are entered.

Walking after dark in an urban environment is relatively easy. There is usually so much light pollution. This is not the case in the mountains and navigating on a cloudy night can be completely disorientating. For emergency use, it is a good idea to carry a small head torch such as the Petzl Tikka Plus. It weighs 80g and has four LEDs which run on AAA cells.

Whilst it could never be used for fell running, it is fine in an emergency (and ideal for in-tent use at night). If a night-time walk is contemplated, a more powerful and versatile torch will be required. For mountain marathons or night time activations, then something similar to Black Diamond's Icon is more appropriate. It runs on three AA cells and has both a single 3W LED and four additional LEDs for a choice of high intensity spot or broad beam. As battery changing on any equipment in the dark is difficult, I also carry a small single LED torch to help in such situations. Naturally, spare cells should always be carried. **Fig 18.53** shows a selection of safety-related items.

## COURTESY TO OTHERS

On all but the remotest hills, you are likely to meet other people who have come to the same summit, to enjoy the views and the peace and tranquility of the mountain top. Finding the place festooned with trip-hazard guy ropes and a man yelling 'CQ SOTA' into a microphone over and over again is unlikely to be welcomed. The SOTA rules have a generous definition of 'summit', allowing the radio operator to be stationed far enough away from the actual peak not to be a nuisance to other walkers. Please make use of this rule, so that you are seen as at worst a harmless eccentric, rather than as a selfish nuisance.

## SUPPLIERS

| | |
|---|---|
| Aluminium - antennas: | Major DiY shops (B&Q etc) |
| Carbon sheet: | Demon Tweeks |
| 'R' Clips: | Trident Racing Supplies [18] |
| NiMH packs: | Overlander Technologies |
| SLA packs; | DMS Technologies |
| Cyclon SLA cells: | Farnell [4] |
| Ropes: | Cotswold Outdoors [19] |
| Glass Roach poles: | SOTA Beams [20] |
| Coax cables | Times [21] |

## REFERENCES


[1] SOTA Websites: *www.sota.org.uk* and *www.sotawatch.org*
[2] Worked All Britain: *http://wab.intermip.net*
[3] 'More gain with Yagi antennas', Gunter Hoch DL6WU, VHF Communications, 4/1977
[4] Farnell: *http://uk.farnell.com/*
[5] 'Opt70 Yagis', Rainer Bertelsmeier, DJ9BV, Dubus Technik V.
[6] International Microwave Handbook, Andy Barter, G8ATD, RSGB
[7] Toughprint waterproof paper. *www.toughprint.com/*
[8] Rite in the Rain waterproof paper. *www.riteintherain.com*
[9] Ebay (Maplin has alas disappeared now~0
[10] Small Wonder Labs: *www.smallwonderlabs.com/*
[11] Kuhne electronic gmbh: *www.kuhne-electronic.de/en*
[12] Energizer batteries: https://energizer.eu/uk/blog/product/energizer-recharge-power-plus-aa/
[13] Lithium Ion Phosphate packs: *http://www.buddipole.com/portablepower.html*
(14] PicoScope: *www.picotech.com/picoscope-oscilloscope-software.html*
[15] Flexcell Solar Panels: *www.flexcell.com*
[16] Loctite: *www.loctite.co.uk*
[17] Raychem: *http://raychem.te.com/*
[18] Trident Racing Supplies: *www.tridentracing.co.uk*
[19] Cotswold Outdoors: *www.cotswoldoutdoor.com*
[20] SOTA Beams: *www.sotabeams.co.uk*



***Richard Marshall, G4ERP**, has had a passion for radio from an early age. After many years of successful contesting on VHF and above, his interest turned to the Summits on the Air programme and he is one of a select band to have achieved the coveted Mountain Goat award. After 17 years in defence electronics he has spent time as Head of Electronics or Chief Electronics Engineer for the Renault F1 Team.*

***Jamie Davies, MM0JMI**, is a radio amateur keen on home construction, QRP, and enjoying simple rigs. Living close to the Southern Uplands, he enjoys that chance to get on the air from hilltops and is author of the RSGB book SOTA Explained.*

*By profession, he is a scientist-engineer working in synthetic biology, and is a professor at the University of Edinburgh.*

# Morse Code

# 19

*Roger Cooke, G3LDI*

In the UK and an increasing number of countries, it is no longer necessary to pass a Morse test in order to gain access to the bands below 30MHz. This might be thought to lead to an overall decline in Morse activity, but there has been a welcome influx of new operators trying their hand at Morse, at their own speed, and without the daunting hurdle of a formal test to overcome.

The Morse broadcast service of the RSGB, using the call GB2CW, has been rejuvenated and there are already a number of volunteers broadcasting on a regular basis, on VHF, UHF and also HF. Some Clubs are even offering their own Morse Proficiency Certificates.

The IARU band plans [1] reserve the bottom portion of most bands for Morse, with the faster ('QRQ') signals generally occupying the lowest 30kHz of each HF band, and slower stations higher up. By agreement within IARU, the 10MHz band is not to be used for SSB or other wide bandwidth modes, and is almost exclusively given over to Morse code.

Radio amateurs tend to use the terms 'Morse' and 'CW' interchangeably. This may seem strange, since CW stands for Continuous Wave, whereas Morse code is anything but continuous. However, almost all amateur Morse is sent by on-off keying of a continuous carrier so CW is widely used as a synonym for Morse.

Operators come to use Morse for a variety of reasons, and compulsion is no longer one of them. **Table 19.1** lists ten arguments in favour of the use of Morse.

## HISTORY

Morse code isn't a code, and it wasn't invented by Morse. Strictly speaking, a scheme in which each letter is represented by a symbol (combination of dots and dashes in this case) is referred to as a cypher, not a code, and the basic idea of using dots and dashes was invented by Samuel Morse's assistant, Alfred Vail.

### The Early Days

Experiments were being carried out with early telegraph systems since the discovery of the relation between electricity and magnetism, following the experiments of Oersted in 1819 and continued by Ampere in 1820. Ampere suggested using a wire and magnetized needle for each letter and by 1825 Schilling had constructed a single needle system based on the galvanometer invented by Schweigger in 1820, using a paper disk attached to the needle such that either a white or black face showed depending on the direction of the current. Schilling's code showed a considerable advance over earlier schemes.

This was followed by work on the magnetization of soft iron by Arago and Faraday in 1830 and the work of Henry in 1830 on electromagnetism. In 1834 Gauss and Weber constructed a crude electromagnetic telegraph by stringing a wire over the housetops in Gottingen to connect The Astronomical Observatory, Physical Cabinet and the Magnetic Observatory. The slow oscillations of magnetic bars to the left or right, according to the current, were viewed through a telescope. Obviously dreadfully slow, an improved version was introduced by Steinheil in 1837, incorporating many of the features of the later methods of Morse, Vail, Cooke and Wheatstone.

**Table 19.1: Why operating using Morse is still popular, despite learning the code no longer compulsory in the UK**

### Advantages of using Morse code

**1. Simple equipment.** It is possible to build a simple CW transmitter using fewer than a dozen parts, and a direct conversion receiver can give acceptable performance on CW. The art of 'homebrewing' is alive and well, and building a CW transmitter and/or receiver can get an amateur onto the air on CW very easily and cheaply.

**2. International.** It is easy to get around the language barrier by the use of abbreviations such as *QTH* and *73*, so that two amateurs can have an elementary contact without knowing each other's language, and without accent or phonetic problems that may arise on phone.

**3. Silent operation.** Wearing headphones and using a straight key or bug, it is possible to operate CW silently, at night time without disturbing others sleeping in the house. Similarly, holiday operation on CW from a hotel can be done in 'stealth' mode.

**4. Morse gets through.** Cross-mode contacts aren't very common, perhaps because they're not valid for most awards, but when struggling to copy a weak phone station on a crowded band it's surprising how often a switch to CW will enable the contact to be completed.

**5. Spectrum efficiency.** The minimum bandwidth needed to copy an SSB signal is about 1.8kHz, and to sound natural the requirement is more like 2.5kHz. On the other hand IF filters for CW operation are typically only 500Hz wide, and if necessary CW can be copied through filters of 100Hz or less. Put simply, at least five times as many CW contacts can fit in the bandwidth required for SSB.

**6. Less breakthrough.** Those who operate both modes know that breakthrough problems are worse on SSB than CW. It is always better to try to resolve a TVI or RFI problem but if this is not possible it may be that switching to CW enables operation to continue when it is impossible on phone. And if power has to be reduced, CW comes into its own.

**7. More competitive.** The difference between 'big gun' and 'little pistol' seems to be accentuated on phone. The low power or antenna-limited operator may struggle for contacts on phone, whereas CW is a great leveller.

**8. Morse is a skill.** It's wrong to say that phone operation doesn't require skill, but the basic skill required is that of speech, which just about everyone has. On the other hand, a new skill has to be learned in order to be able to communicate using CW, and the sense of achievement can be considerable.

**9. Automation.** There are computer programs which make it possible to automate the transmission and reception of Morse code. This makes it possible under some conditions to engage in CW contacts without knowing the code, but this would be to miss the whole point: it would be much better to use a more efficient automated mode such as PSK31. In any case, a good human operator can easily out-perform most computer techniques when copying a CW signal on a channel with even a moderate amount of interference or fading.

**10. Morse is easy.** The general public thinks that Morse code is a language, and at special event stations Morse operation always proves fascinating to visitors. All that is needed to acquire Morse skills is to learn the symbols for 26 letters, ten numbers and a few special characters. This is a very great deal easier than learning a foreign language complete with all its grammar and vocabulary.

Out of the sixty or more different early methods proposed for telegraphy, there emerged two most widely known ones; the needle telegraph of Cooke and Wheatstone, introduced in England and the electromagnetic telegraph of Morse and Vail, originating in the USA.

## Dots and Dashes

Starting in 1837 and building on the work of Joseph Henry, the artist Samuel Finley Breese Morse worked to develop a practical electric telegraph.

Morse's idea was that common words in a message would be represented by numbers, and these numbers would be transmitted as a series of dots. At the receiving end, the dots would be marked by the equipment on paper tape.

The operator would then have to decode the dots to work out which numbers were being sent, and look up the number combinations in a dictionary to find the word. Less common words would be spelled out, with numbers again representing letters. At about the same time in England, Cooke and Wheatstone were also working on the development of an electric telegraph.

While Morse was working on his system in New York City, his partner Alfred Vail thirty miles away in Morristown, New Jersey made numerous improvements to Morse's equipment.

It was Vail's idea that instead of using a series of dots for numbers, the code could be made up of combinations of dots, dashes and spaces to represent letters directly. In this way a message could be spelled out without the need to look up code numbers in a book. Samuel Morse was an incurable self-publicist and failed to give Vail credit in his lifetime as the true inventor of the code that bears Morse's name. When developing his code, Vail visited a local printer in Morristown to find out which letters occurred most frequently in English text, and he assigned the shortest combinations of dots and dashes to these letters. At the time of the first public demonstrations of Morse code, the code was made up of four elements in addition to the standard space: dot, dash, long dash and long space. The code underwent further changes but remained in use as American Morse code until well into the 20th century.

The electric telegraph and Morse code started to be used in Europe in the mid to late 1840s. A German, Frederick Gerke, addressed the problem of distinguishing between standard and long dashes, and standard and long spaces, by reducing the code to the elements of dot and dash that we are familiar with today.

More changes were made and in 1865 the 'continental' or 'international' Morse code came into being, in very much the form that exists today. Small changes have been made to the code since then, the most recent being the adoption in 2004 of a new Morse code symbol for the '@' used in Internet addresses.

Both Vail and Morse intended that Morse characters should be printed onto paper tape and read as dots and dashes from there, but it soon became apparent that operators were learning to decode from the clicks of the machinery marking the tape. This led to the development of the sounder and now whenever Morse is written down it is always written down directly as letters, never as individual dots and dashes.

A modern exception to this is QRSS, or extremely slow Morse decoded using a FFT program (as used, for instance, on 136kHz and for very low power experiments) which is displayed as dots and dashes on a computer screen and decoded by eye.

## THE CODE

Although Morse is composed of dots and dashes it is better to think of the elements as 'dits' and 'dahs'. These are in the correct proportions so that saying 'di-dah-di-dit' out loud is the correct

| Alphabet | |
|---|---|
| A | di-dah |
| B | dah-di-di-dit |
| C | dah-di-dah-dit |
| D | dah-di-dit |
| E | dit |
| F | di-di-dah-dit |
| G | dah-dah-dit |
| H | di-di-di-dit |
| I | di-dit |
| J | di-dah-dah-dah |
| K | dah-di-dah |
| L | di-dah-di-dit |
| M | dah-dah |
| N | dah-dit |
| O | dah-dah-dah |
| P | di-dah-dah-dit |
| Q | dah-dah-di-dah |
| R | di-dah-dit |
| S | di-di-dit |
| T | dah |
| U | di-di-dah |
| V | di-di-di-dah |
| W | di-dah-dah |
| X | dah-di-di-dah |
| Y | dah-di-dah-dah |
| Z | dah-dah-di-dit |

| Accented letters | |
|---|---|
| à, á, â | di-dah-dah-di-dah |
| ä | di-dah-di-dah |
| ç | dah-di-dah-di-dit |
| ch | dah-dah-dah-dah |
| è, é | di-di-dah-di-dit |
| ê | dah-di-di-dah-dit |
| ñ | dah-dah-di-dah-dah |
| ö, ó, ô | dah-dah-dah-dit |
| ü, û | di-di-dah-dah |

| Abbreviated numerals | | | |
|---|---|---|---|
| 1 | di-dah | 6 | dah-di-di-di-dit |
| 2 | di-di-dah | 7 | dah-di-di-dit |
| 3 | di-di-di-dah | 8 | dah-di-dit |
| 4 | di-di-di-di-dah | 9 | dah-dit |
| 5 | di-di-di-di-dit | 0 | daaah (long dash) |

| Full length numerals | |
|---|---|
| 1 | di-dah-dah-dah-dah |
| 2 | di-di-dah-dah-dah |
| 3 | di-di-di-dah-dah |
| 4 | di-di-di-di-dah |
| 5 | di-di-di-di-di |
| 6 | dah-di-di-di-dit |
| 7 | dah-dah-di-di-dit |
| 8 | dah-dah-dah-di-dit |
| 9 | dah-dah-dah-dah-dit |
| 0 | dah-dah-dah-dah-dah |

| Punctuation and other codes | |
|---|---|
| Full stop (.) | di-dah-di-dah-di-dah |
| Comma (,) | dah-dah-di-di-dah-dah |
| Colon (:) | dah-dah-dah-di-di-dit |
| Question mark (?) | di-di-dah-dah-di-dit |
| Apostrophe (') | di-dah-dah-dah-dah-dit |
| Hyphen or dash (-) | dah-di-di-di-di-dah |
| Fraction or slash (/) | dah-di-di-dah-dit |
| Brackets - open [(] | dah-di-dah-dah-dit |
| - close [)] | dah-di-dah-dah-di-dah |
| Double hyphen (=) | dah-di-di-di-dah |
| Quotation marks (") | di-dah-di-di-dah-dit |
| Error | di-di-di-di-di-di-di-dit |
| Message starts (CT) | dah-di-dah-di-dah |
| Message ends (AR) | di-dah-di-dah-dit |
| End of work (VA) | di-di-di-dah-di-dah |
| Wait (AS) | di-dah-di-di-dit |
| Understood (SN) | di-di-di-dah-dit |
| The @ in e-mails | di-dah-dah-di-dah-dit |

**Spacing and length of signals**

1. A dash is equal to three dots.
2. The space between the signals which form the same letter is equal to one dot.
3. The space between two letters is equal to three dots.
4. The space between two words is equal to seven dots.

**Table 19.2: The Morse code and its sound equivalents**

representation of the letter 'L'. Spacing and rhythm are essential to good Morse code. The inter-element space is the same duration as a dit length, while the length of a dah is three times the dit length. The space between letters of a word is three dit lengths and the space between words is seven dits.

As well as the letters and numbers, **Table 19.2** lists commonly used punctuation and procedural symbols ('prosigns') in common use on the amateur bands.

## SPEED

For many years it was necessary in order to gain an amateur licence, to pass a Morse test at 12 words per minute (WPM). This was, in the earlier days of amateur radio, followed by an enforced year on the air using Morse only, and proof had to be given of a number of contacts using Morse in order to be allowed to use 'Phone', as speech modes were called in those days. This is one reason why so many of the older licensees are excellent CW operators. In practice 12WPM is the minimum speed that is generally heard on the amateur bands. A word is defined as 50 dit lengths, and this is referred to as the Paris standard since the word 'PARIS', including the seven dit lengths at the end of the word, is exactly the right length. By coincidence the word 'MORSE' is also exactly the right length.

A simple method of measuring speed is repeatedly to send PARIS, including the correct inter-word gap, and count how many are sent in a minute. At a speed of 12WPM the word can be sent just once in five seconds. Another method of measurement which is simple to do but overstates the speed by 4 percent is to count the number of dashes in a five second period.

## LEARNING MORSE CODE

Prior to trying to copy any Morse, it really is essential to commit the code to memory. The essential characters are obviously the alphabet and numbers, together with a few punctuation marks. One of the easiest ways of achieving this is to have a number of small pieces of paper, with the character on one side and the Morse equivalent in dit-dah mode, on the other. Picking these at random and then converting will make the learning curve that much easier, plus it can be done anywhere and at any time.

A later chapter in this book shows the extent to which the personal computer has become an integral part of the radio amateur's shack. There is a method of learning Morse code which goes back to the 1930s but has come into its own in the age of the computer. That is the Koch method.

Ludwig Koch was a German psychologist who carried out a systematic study of skilled Morse operators, and then conducted a series of trials to find the most efficient way of teaching Morse to new students. Existing methods at that time were based on learning the letters either visually, or audibly at a slow speed and then increasing speed by constant practice. This approach to learning Morse is fundamentally flawed. Learning all the letters at slow speed requires a conscious mental translation from the combination of dits and dahs to the corresponding letter. After all letters have been memorised, the student hears a combination of dits and dahs and searches in a mental look-up table for the right one.

With a great deal of practice, speed increases but everyone who learns by this method experiences a 'plateau' at 8-10WPM where no more progress seems possible. Many aspiring CW operators give up in frustration at this point.

What is happening is that the process of responding to individual dits and dahs and looking them up in a mental table is hitting a natural upper speed limit. The student who can move beyond that limit has made the unconscious transition from the old 'lookup' method to a stage where the whole Morse character is recognised by reflex. Koch's insight was to see that it was essential that each character retain its acoustic wholeness and never be broken down by the student into dits and dahs. The essence of the Koch method is that the letters are learned one at a time, but with the correct rhythm and at full speed.

There is no process of decoding the dits and dahs, but from the beginning a reflex is built, so that when 'dah-di-di-dah' is heard, it is automatically recognised as X without any conscious decoding process. It's the same way that a touch typist knows where the letters are on a keyboard without having to think about it.

Another key feature of the Koch method is that it provides positive reinforcement. Once a letter has been learned, it has been learned at full speed and it is possible to measure one's progress towards mastery of the full alphabet.

Koch's technique was difficult to implement for classes of students in the 1930s but has come into its own in the day of the personal computer, when each student can learn at his or her own pace. A program written by Ray Goff, G4FON, implements the Koch method and can be downloaded from [2]. Much of the credit for rediscovery and promotion of the Koch method must go to N1IRZ, and the G4FON program follows his suggested implementation.

The student starts by learning just two letters, at a speed of 15WPM. The program sends random combinations of these letters for a fixed period, typically three to five minutes, and the student must copy them onto paper. At the end of the run the copy is compared with what the computer sent, as displayed on the screen. If 90 percent or greater correct copy is received, a third character is added to the set and another run started. It is up to the student how many runs to do in a day. When Ludwig Koch carried out his investigations in the thirties he was seeking to train commercial operators but found that too many sessions in a day gave diminishing returns and the amateur student is more likely to have to fit in training sessions around other activities.

As the training programme progresses and more and more Morse characters are added, the student gains a real sense of achievement, and proof that it is possible to master the copying of Morse characters at full speed. G4FON's program introduces new characters in the order suggested by N1IRZ which mixes easy and difficult letters, numbers and punctuation in an apparently random order.

Once all 43 characters have been learned the program lets the student practice with text files (**Fig 19.1**), or with simulated contacts. At this stage, most people would also have started to listen to real contacts on the amateur bands, though the quality of 'real' Morse code can be rather variable.

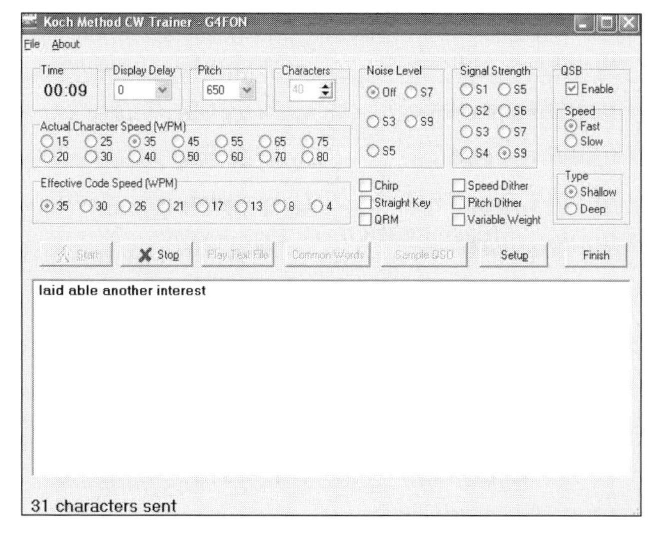


**Fig 19.1: Part of the screen of the G4FON Koch Morse trainer program**

VOLUNTEERS ARE ALWAYS NEEDED by the GB2CW service, so if you think you could help with this very worthwhile scheme, please give it some thought. It just takes one hour per week of your time and you will be promulgating the use of CW to a new generation of operators who probably don't even realise what they are missing.

Transmissions take place on both the 80 metre and 2 metre bands, so you can choose which band you wish to use. The uptake on HF seems to be slow and it looks like two metres might be a better option. If you have a lot of newly licensed members of your club, 2m FM would probably be a good choice, as it would be localised to your particular area. If you would like to take part, all you would need is a 2m transceiver and a computer. Contact the G2CW Coordinator via the RSGB [3, 4].

Another way of learning is the Farnsworth Method. With this method, you learn each character at 15 words per minute with large spacing in between characters. This has been proven to be the best method for long-range development. Once the characters are learned, copying speed is easily increased by decreasing the spacing between each character.

No matter if you learn quickly, or slowly, the key to learning is practice. With enough practice, just about anyone can learn Morse code. Sometimes, skipping a day or two of practice is helpful, and can get you back on track.

The ubiquitous computer can have its dangers, as well. The Koch method is the quickest means of learning the code but some students in their impatience have resorted to use of Morse decoding software as an aid to getting on the air on CW more quickly. This approach leads to reliance on the computer and the temptation must be resisted. However, there are small and portable Morse practice devices that can be carried in the pocket. They can be used anytime, anywhere.

## METHODS OF LEARNING

### The GB2CW Service

GB2CW has been around for a lot of years, started in the days when the Morse test was mandatory. Now that the Morse test has been abolished, the decision to learn Morse is down to the individual. The GB2CW scheme has been rejuvenated and we now have around 24 volunteers all over the UK tutoring both the newcomer to the code and some of the older licensees who wish to rekindle their Morse on the air. This service is run by volunteer CW operators transmitting on a regular schedule, using VHF, UHF, and HF including 60m. They are usually interactive, and will accommodate all speeds, according to the needs of the listener. The broadcast schedule is published by the RSGB [3, 4].

Radio Society of Great Britain
Advancing amateur radio since 1913

RADIO SOCIETY OF GREAT BRITAIN
MORSE PROFICIENCY PROGRAMME

*This is to certify that*

*passed a Morse Code Proficiency Test on*

*sending and receiving at a speed of ___wpm*

Regional Manager

**Fig 19.2: The RSGB's Morse certificate**

### The CWOps Club

This is more suited to the well-organised and regimented among us! It is on-line so does not require a radio, but it does require more commitment.

The CwOps club (*www.cwops.org*) runs 8 week, internet-based Morse training courses three times a year. Organised by the club's CW Academy these commence each January, April and September and cater for complete beginners, improvers and competent CW operators wishing to improve further. For each course a maximum of 5 students is allocated to an advisor who is a club member and will be responsible for the conduct of the course; students can be geographically spread but will normally reside in the same or neighbouring time zones. The courses are free but students must commit to about 30 minutes practice each day as well as be able to take part in twice weekly face to face sessions over Skype during which progress can be gauged and issues discussed. No formal assessments are involved and self motivation is the key to progress. Further information can be obtained from the website: *www.cwops.org*

Many newcomers to amateur radio do not realise the potential of CW, and look upon it as a mode that requires a lot of study and practice. In fact, Morse can give a lot of pleasure: a pride in the skill that it takes and the ease with which DX can be worked. If every radio club in the UK automatically had a GB2CW teaching program, there would soon be a new generation of good CW operators. All clubs are urged to instigate such a scheme and maintain it, encouraging all new licensees to take part. See the sidebar for how to volunteer as a GB2CW tutor.

Volunteer Morse Test Assessors are also needed. Full details of how to become an Assessor can be found in the Morse section of the RSGB website [5].

The RSGB also offer a Certificate of Proficiency which is in the form of an award which you can gain by taking a Morse examination **Fig 19.2**.This has recently been redesigned. Tests start at 5WPM and go as high as you want to go. The intention of the Certificate of Proficiency is to provide a form of recognition for those who wish to have something to record their achievement in learning Morse code. It is intended to provide incentive for further study and practice to attain a standard that the student can be proud to display. More information is available from [5]. Phillip Brooks is the RSGB Morse Test Coordinator and is quite dedicated to this task.

### Morse Readers

It can be tempting to be lazy and read Morse without all that practice. This is possible, and there are Morse readers available. However, it should be emphasised that absolutely nothing can substitute for the human brain. There is no Morse reader on this planet that can do the same job as the brain, and that would be upheld by any good CW operator that you might speak to about this. However, in the interest of unbiased writing, take a look at the advertising 'blurb' for a Morse reader [6].

*Relax and place this tiny pocket size portable Morse Code /CW Reader near your receiver's speaker. Then watch Morse code signals turn into text messages as they scroll across an easy-to-read LCD display. No cables to hook-up, no computer, no interface, no other equipment needed! Use it as a backup in case you miscopy a few characters.*

This can help with your practice if you can find a suitable CW station on a reasonably clear frequency. However, the problem with these devises is that it can also detract from making the operator read the CW, always relying on the reader. It is also a fact that trying

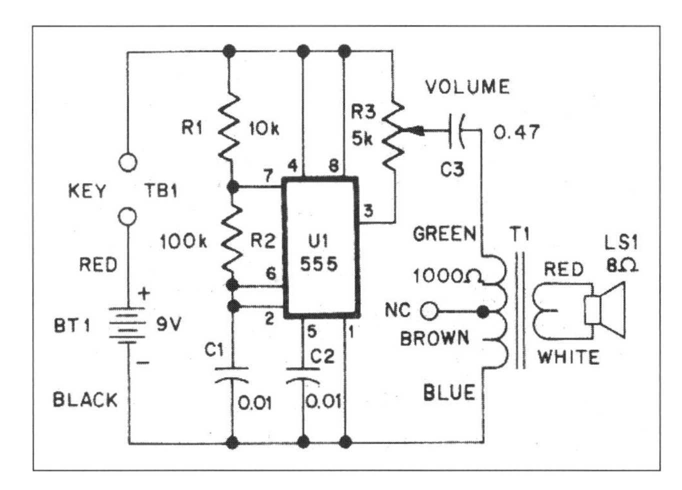


Fig 19.3: Simple Morse practice oscillator

to get a machine to copy a weak, fading signal surrounded by interference is just about impossible, whereas the brain can pick out the 'needle in the haystack'. This skill is irreplaceable and it is what makes CW so attractive to the DX operator.

## Learning to Send

Although the PC makes it possible for a student to learn Morse code without assistance, when it comes to sending by far the best approach is to get an experienced operator to listen to you and correct any bad habits at the earliest opportunity. This cannot be emphasised enough. Bad sending will lead to zero contacts because nobody will wish to converse with somebody sending badly, whereas properly sent Morse, from a very nicely keyed transmitter is a pleasure to listen to. The student who has learned using the Koch method may have fewer bad habits than most.

If no-one is available to listen in person then one method that can work is to make tape recordings and play them back to try and discover weaknesses. Only when the student's sending sounds perfect should it be tried on the air.

## Morse Practice Oscillator

In order to practice sending off the air, a pleasantly keyed oscillator is needed. You can make one fairly easily on a piece of Vero board, using a 555 timer chip. A suitable circuit containing easily sourced components is shown in **Fig 19.3**. It could form the basis of a club project and should not take too long to construct.

## Modern way of encouraging learning, the BOOTCAMP!

Norfolk ARC started this method a few years ago now and it has proved to be very popular, such that it has been copied by the Essex CW Club, to great success. It is the CW Bootcamp. The originally scheme was started three years ago with a seven-day

Fig 19.4: A modern straight key [photo: RA Kent (Engineers)]

Fig 19.5: A mechanical bug key

Bootcamp. This was successful, but it really did prove to be difficult to maintain, both for students and tutors alike! Now they have two Bootcamps a year, spring and autumn and just for a day. The Essex Club attended our Bootcamp and decided to emulate that formula for their Club. Whilst it is a serious day of Morse, a lot of fun can be had at the same time, with an exchange of ideas, display of Morse keys and discussions of the various paddles and other keys that are available.

This is a scheme that other clubs can adopt, no matter how small the club is. Concentrated practice for a day really does pay dividends and can be infectious! Breaks are taken for coffee and lunch and in Norfolk we are lucky to be supported by the wives with a very nice selection of cakes, biscuits and so on. This has been such a success that we are now asked when the next one will be. One added advantage of the Bootcamp is that we can address the sending technique of each student, and correct any errors they may be making. This is necessary, because a bad technique will lead to bad Morse. Encouraging each student to have an actual QSO on the air can also be very rewarding, plus it lessens the nerves that some suffer from when actually making a "real" QSO!

## MORSE KEYS

It is important to start by getting a good quality straight key ('pump handle') such as the one in **Fig 19.4**, and position it correctly on the table. Adjust the height of the seat so that when not sending, the arm is resting horizontally, and it requires only a small lift of the forearm to hold the knob of the Morse key. This should be held with the forefinger on top of the knob, the thumb to the left and slightly underneath the knob, and the second finger either on top or to the right of the knob (assuming a right-handed operator).

Dots are made by a small wrist movement whereas dashes require a more pronounced downward movement of the wrist. Think of the keying contacts being closed by the wrist joint. The hand is just the lever which happens to pull the key contacts together. Do not tap the key or push down on the key with the hand. The technique is best shown and then practised at length.

The forearm and the upper arm should make an angle of approximately 90º. When sending, movement of the key should come from a combination of the elbow and the wrist. Often the key is mounted at the edge of the table but if it is further back, the forearm should be slightly above the table, not resting on it.

It is possible to make your own key. Dick Biddulph, M0CGN designed a simple one that should be possible for those that enjoy construction [7].

## Bugs and Elbugs

With practice it is possible to send at speeds over 20WPM on a straight key, but the operator who aspires to high-speed (QRQ) working will sooner or later wish to learn how to use a bug.

Fig 19.6: A paddle for use with an electronic keyer

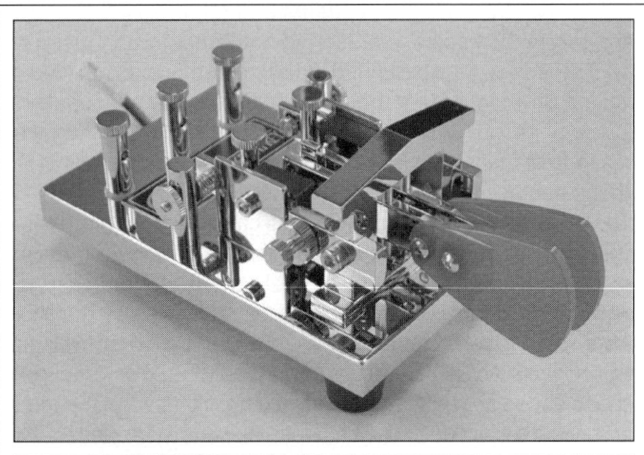

Fig 19.7: Two GHD keys by Toshihiko Ujiie

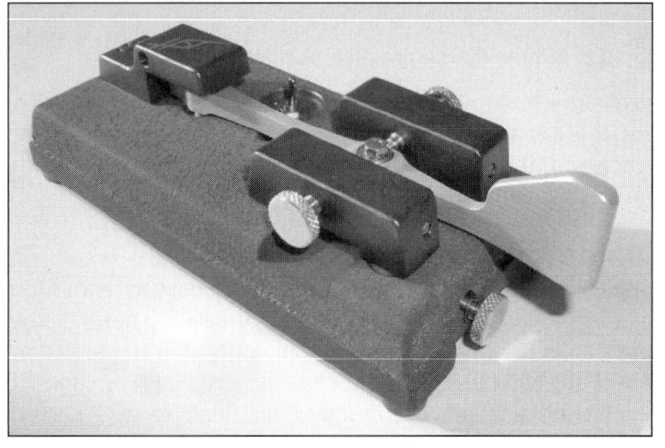

19.8: Two of the keys manufactured by Begali

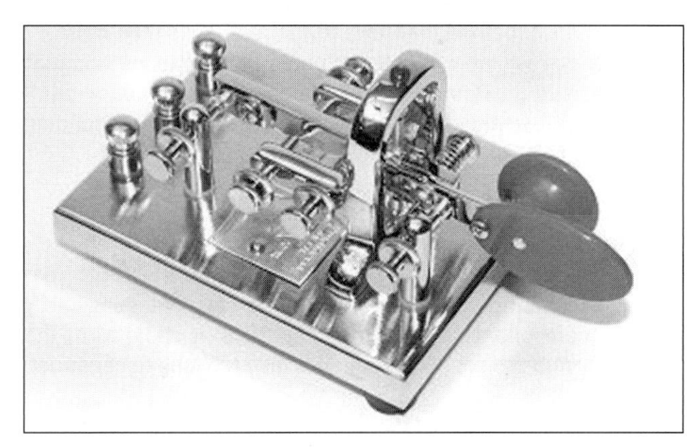

Fig 19.9: A presentation key made by Vibroplex

The semi-automatic key (**Fig 19.5**), generally known as the bug after the trade mark of the original maker, is a mechanical key in which the arm moves from side to side instead of up and down. It has two pairs of contacts; dashes are made singly by moving the knob to the left, thereby closing the front contacts. A train of dots is produced by similarly moving the paddle to the right against a stop. This causes the rear portion of the horizontal arm to vibrate and close the rear pair of contacts. A properly adjusted bug key will produce at least 25 dots.

There are not many in use today, although they are the last resort of being able to identify the sender by just listening to his keying, or his 'fist'. It was often common practice to add emphasis to a CQ for example by sending extra long dashes. With an electronic keyer, this type of personality on the key is not possible. Semi-automatic paddles, such as the old Eddystone Bug, are now collectors items and have increased in value over the last few years.

Bugs were in widespread use until the advent of the electronic keyer, or 'elbug'. The first elbugs were cumbersome affairs using combinations of relays and valves. Transistor types followed, giving way to IC-based keyers and now PIC and other microprocessor keyers.

Most modern transceivers incorporate an electronic keyer, and many station logging and computer logging programs provide a CW keyer as standard. Many programs support both paddle input and keyboard input so the operator can choose whether to use the program either as a keyer taking input from a paddle, or as a keyboard sender.

## Paddles

The key or 'paddle' (**Fig 19.6**) used with an electronic keyer is a derivative of the mechanical bug key, with movement from side

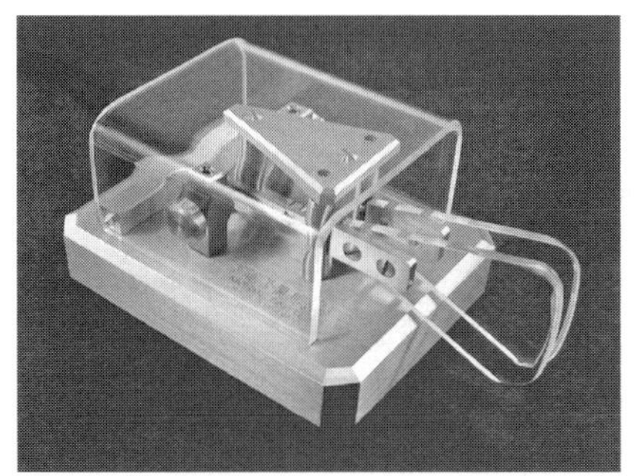

Fig 19.10: A paddle by Gerhard Schurr

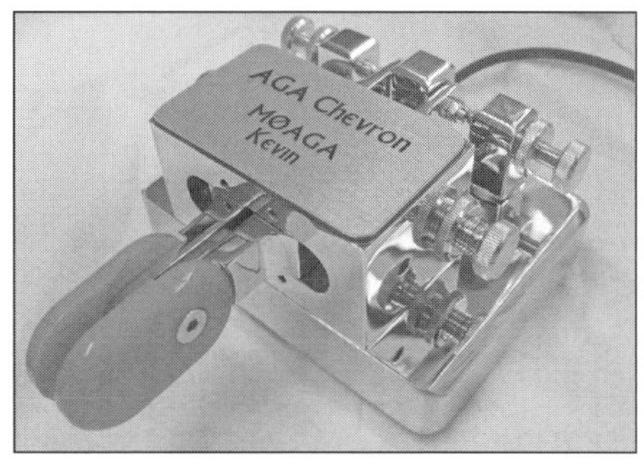


Fig 19.12a: A UK-made paddle from Kevin Gunstone, M0AGA

to side instead of up and down. For the right-handed operator, moving the paddle to the right with the thumb produces a train of dots. Moving the paddle to the left with the side of the index finger produces a train of dashes.

An iambic or 'squeeze' keyer requires a twin paddle key, and as well as movement of each paddle separately; they can be squeezed together, hence the name. According to which contact is closed first, a train of either di-dah-di-dah or dah-di-dah-dit is produced.

Fig 19.11: This innovative touch paddle, designed by Peter Raven, G4KLM, has no moving parts

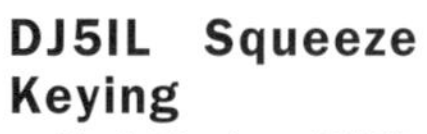

## DJ5IL Squeeze Keying

Karl Fischer DJ5IL has written a very informative article on Squeeze Keying. Anybody thinking of buying a paddle would do well to do a lot of research with various paddles, both single and double lever, to ascertain which feels most comfortable. It would also be a good idea to read this lengthy article first and learn all about squeeze keying.

*http://cq-cq.eu/DJ5IL_rt007.pdf*

It is possible to make a paddle if you have good engineering skills, indeed some amateurs have produced some very nice home made paddles. Access to an engineering workshop is necessary in order to produce a well-engineered and attractive paddle.

As with the straight key, the newly acquired bug or electronic key should not be tried on the air until the operator has gained confidence through practice. Some electronic keys provide auto-spacing in an attempt to enforce the three or seven dit gaps between characters and words, but it is still just as easy to send bad Morse on an elbug as on a straight key. Again, lots of practice is the answer and is very advisable.

Commercial

Fig 19.12b: 9a5n paddle

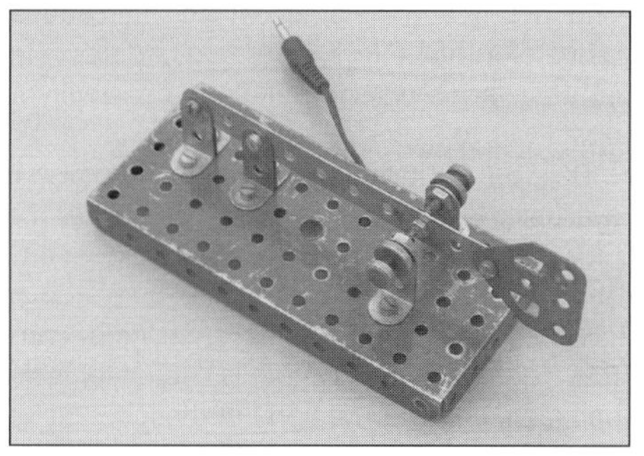

Fig 19.12c: Meccano flex 8

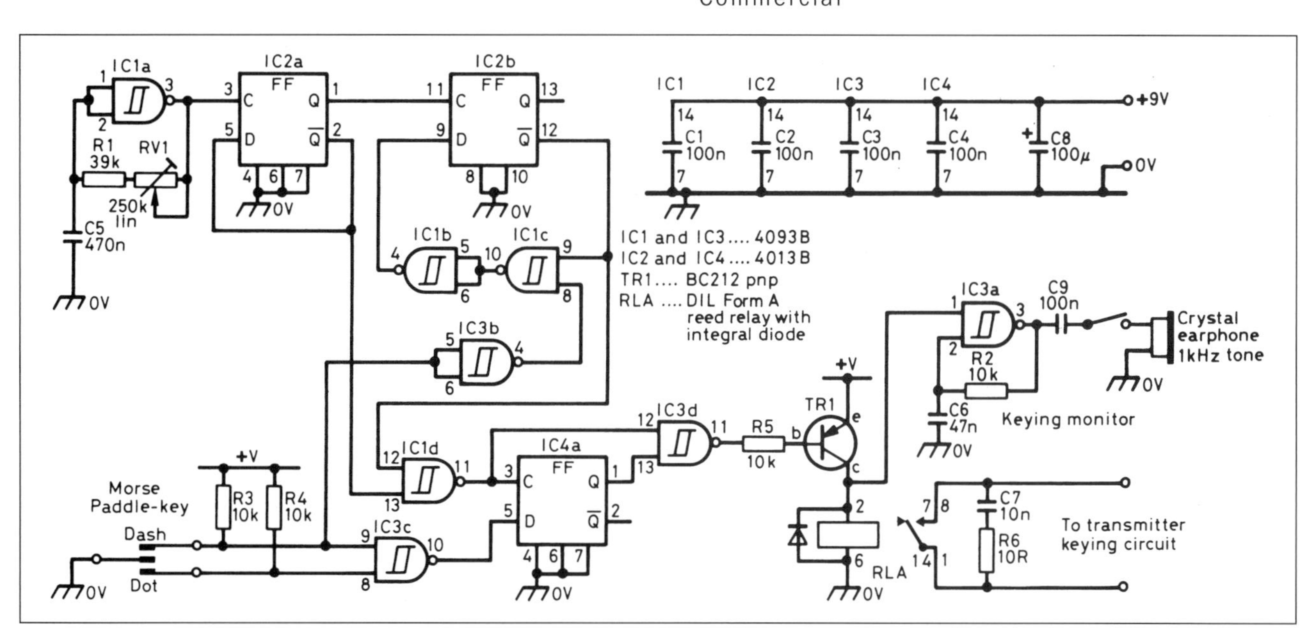


Fig 19.13: The G3BIK simple keyer uses four ICs for precision Morse

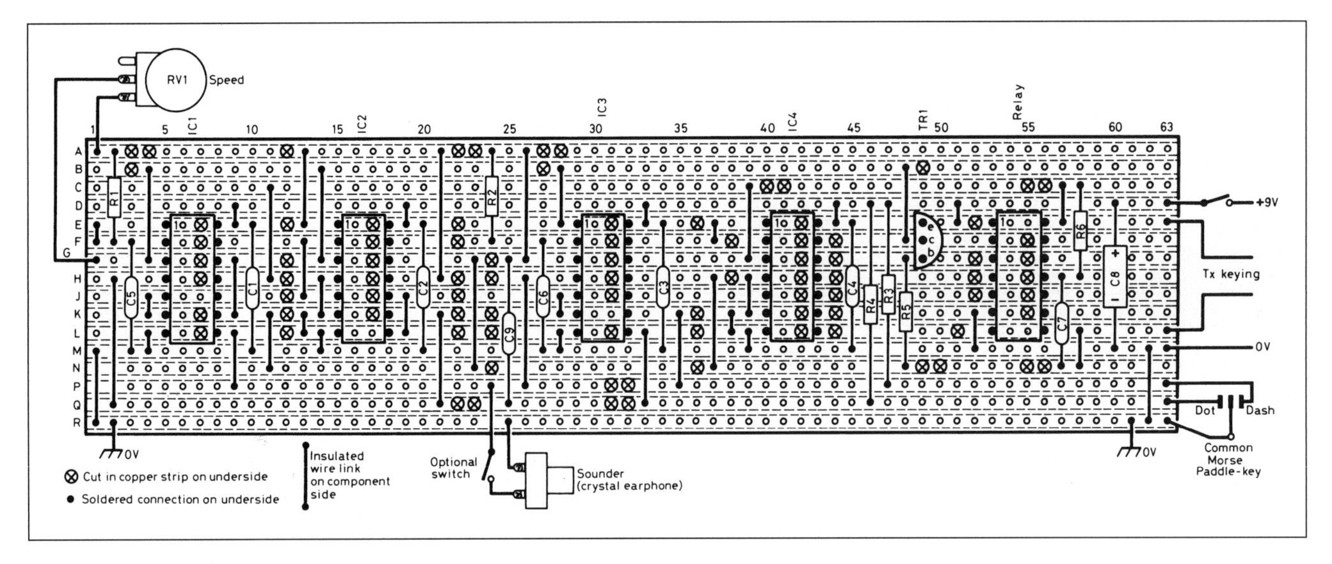


**Fig 19.14: Construction of the G3BIK keyer is easy using strip board. The component side is shown**

paddles are the only route for most people and the quality of a paddle is, of course, proportional to its price. However, if you are keen on CW and intend to use the mode a lot, it is much like a pianist owning a piano. If he has an old, out-of-tune upright with some notes missing, there is no incentive to play. The better quality paddle leads to a much better quality Morse, as it will always be a pleasure to use it. Here are some examples.

GHD keys, made by Toshihiko Ujiie (JA7GHD) [8] in Sendai City, are beautiful, practical and innovative. There is a large range of straight keys, bugs and paddles (**Fig 19.7**) An interesting innovation on one paddle is the ability to convert it from a dual lever to a single lever model.

Begali is an Italian manufacturer who also has several paddles available. **Fig 19.8** shows a popular one, the Sculpture, and a recent addition, the High Speed single lever model. These are available from [9].

Vibroplex is an old established American manufacturer and they, too, have lots of models available, including a single lever one [10]. I have owned one of these for 51 years now! The presentation model is shown in **Fig 19.9**.

Schurr keys are not so well known in the UK, but again are a quality product. Made by Gerhard Schurr, DH2SAA, [11], his motto is "If you stop trying to be better, soon you will stop being good." This is a good motto to adopt when learning and practising Morse! His Profi paddle can be seen in **Fig 19.10**

Peter Raven, G4KLM, manufactures a different type of paddle, with no moving parts (**Fig 19.11**). It is a touch paddle and relies on the capacitive effect to operate. It does require care when sending in that the user must not touch the paddle unintentionally. Peter has also made a memory keyer with a built-in touch paddle. He has sold several keyers and paddles and those that have them like them very much. The touch principle does take some practice and you either like it or you don't. It's always best to try several different types before you decide.

Shown in **Fig 19.12a** it is a UK 'Rolls Royce' of paddles. It is not cheap, but the quality is superb. It is hand made by Kevin Gunstone, MOAGA, and his engineer Alan, G4HCD, of Chevron Morse Keys [12]. It is shown in **Fig 19.12a**. Unfortunately they do not make a single lever paddle.

The latest innovative paddle to hit the market is the solid state paddle from Neno 9A5N. This is a new concept. These paddles have no mechanical parts such as bearings, contacts, springs, magnets, or adjustment screws. Instead, they use solid state silicon sensors to detect the minute deflection of the levers that the finger pieces are attached to. A microprocessor detects the output from the sensors and creates the dot and dash outputs. Single lever and twin lever paddles are available. See **Fig 19.12b**.

## Keys by RA1AOM

Valery Pavlov RA1AOM, from St. Petersburg is a Morse operator and a skilled engineer who has an appreciation of the Morse Key as a work of art, as well as a tool with which to send Morse. Val was a professional operator in the soviet army, so he does appreciate a well made key himself.

He is similar to Kevin Gunstone MOAGA with his Chevron paddle However, Val has taken this to another level and produced not only superb keys, but they take on the look of a work of art. He uses mineral stones, such as Jasper, serpentinite, obsidian, granite for the bases and highly polished stainless steel.

Manufacturing methods have changed a lot over the years, and most manufacturers use neodymium magnets instead of springs. A couple of his keys are shown in the pictures below.

## Cootie Keys, or Side-Swipers

I first built a side-swiper from a hacksaw blade, mounted on a wooden base, with contacts arranged with brass brackets and bolts as contacts. There is a certain skill in using a Side-Swiper or Cootie Key as they are called in the US. Careful control of the paddle is necessary in order to maintain a good Morse character formation, otherwise it can sound very joined up! Martin G4ZXN is a keen exponent of the Side-Swiper and has a web site devoted to their construction and use. He has made several from Meccano and one is shown in the picture **Fig 19.12c**. If you are thinking of trying one for yourself, it would pay to practice off-air for a while to master the technique, because a badly used one will produce bad Morse! There is a lot of interest in Cooties as can be seen from the web site and there is also a Side-Swiper Net.

*http://www.sideswipernet.org/keys/g4zxn-keys.php*

## Simple Electronic Keyers

### BIK simple electronic key

This keyer designed by Ed Chicken, G3BIK, is of basic design in that it uses only four integrated circuits and does not include the iambic facility. According to which side the paddle is moved, a train of dots or dashes is produced. The speed is adjustable in the range 5-35WPM. A small sounder is included as a side tone (keying monitor).

The circuit diagram is shown in **Fig 19.13**, and the stripboard layout is in **Fig 19.14** . Details of the paddle key are contained in the full article [13].

### K1EL Winkey

WinKey [14] is a single chip Morse keyer using a PIC IC. Powered from the PC, the keyer is designed to attach to a PC's external port and provide accurate transmitter keying for numerous Windows$^{TM}$-based logging programs and other ham radio software.

Due to timing latency inherent in the multi-threaded Windows operating system, it is difficult to generate accurately timed Morse. The host PC communicates to WinKey over a simple interface. Letters to send, along with operational commands, are sent from the host to the keyer over the serial link.

WinKey buffers ASCII characters sent by a Windows$^{TM}$-based software application. It then translates them to Morse, directly keying a transmitter or transceiver. In addition, WinKey has paddle inputs so that an operator can break-in and send using paddles at any time. A speed potentiometer interface is provided so that an operator can instantly dial any speed desired. The user can tailor WinKey's keying characteristics precisely to a particular transmitter.

A PCB board with a component kit is available from [14]. Don't be fooled by the appearance (**Fig 19.15**), good things come in small packages!

## CW TRANSMISSION AND RECEPTION

A crystal-controlled QRP transmitter for CW can be made with fewer than a dozen components, and give many contacts under favourable conditions. There are many designs for simple CW-only homebrew transmitters or transceivers and several kits are available that offer CW-only transmission.

Almost all commercial transceivers are designed with SSB in mind, and sometimes give the impression that CW has been added as an afterthought. Even top of the range Yaesu transceiver, the FT1000MP, is known for its key clicks and several modifications have been developed - by individuals and not the manufacturer - to address this shortcoming in the design of an otherwise excellent piece of equipment.

**Fig 19.15: Winkey single chip Morse keyer**

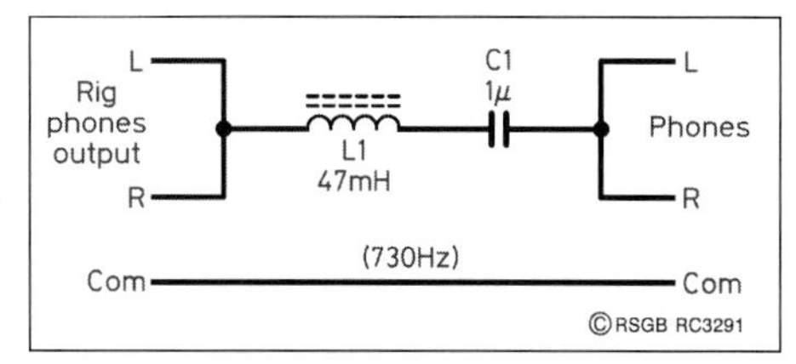


**Fig 19.16: Circuit diagram of the two-component CW filter**

In many transceivers a dedicated IF filter for CW is an optional extra. Some rigs incorporate passband tuning which can be used to narrow the receive bandwidth without the expense of buying additional internal filters. There are occasions under crowded band conditions when it is extremely difficult to separate the many CW signals that can pass together through an SSB filter, which may have a bandwidth of up to 2.5kHz. IF filters most commonly available to fit in a transceiver for CW reception have a bandwidth of 500Hz though 250Hz filters can sometimes be obtained.

In cases where IF filtering is not available, CW filtering can be done at audio frequencies, and combinations of IF and AF filtering can be extremely effective. The problem with audio filtering on its own is that any filtering done after the AGC detector will result in 'pumping' whereby strong signals which are not heard by the operator because of the audio filter nevertheless are let through by the IF filter and cause desensitisation of the receiver. More recent designs using IF DSP (digital signal processing) can help to address this problem by implementing the filter before detection, though the dynamic range of the DSP system may then become an issue.

Different operators have different styles, and some of the best contest and DXpedition operators like to use relatively wide filters and do most of the separation of CW signals in their head: it is sometimes said that the best CW filter is between the ears. This practice is necessary in contests because often the calling station will be way off frequency. Nevertheless there are times when a narrow IF filter is essential for pileup and weak signal CW work.

A very simple add-on audio filter which provides some selectivity as well as helping to clean up the hiss, clicks, hum and thumps which spoil the audio of some rigs, has been described by Fraser Robertson, G4BJM [15]. In its simplest form it consists of a simple series-resonant circuit and with the inductor and capacitor values shown in **Fig 19.16** it resonates at around 730Hz. The components can simply be wired inline as a headphone extension lead, and covered with self amalgamating tape or heat-shrink tubing to provide some mechanical stability. Alternatively, it can housed in a small box as in **Fig 19.17**.

One of the more sophisticated filters around is the DSP-599zx by Timewave, **Fig 19.18**. This is a very versatile digital signal processor and for CW use it is extremely useful for enhancing a weak signal, especially under adverse conditions, such as high noise. Using this on top band where my noise level is usually around S-8 to S-9, coupled with the filters on the transceiver can mean the difference between a contact or no contact. It takes some time to set it up properly and also to become accustomed to using it, but once done a filter of this type is indispensable.

Most transceivers incorporate semi break-in or 'VOX' keying on CW. This means that the user does not have to operate a separate transmit/receive switch, but transmission starts automatically almost at the instant that the key is pressed. The equipment returns to receive a short period after the key is opened, and this period can usually be varied. The intention is to ensure the rig stays in transmit between Morse characters, but returns again to receive without the operator having to throw a switch.

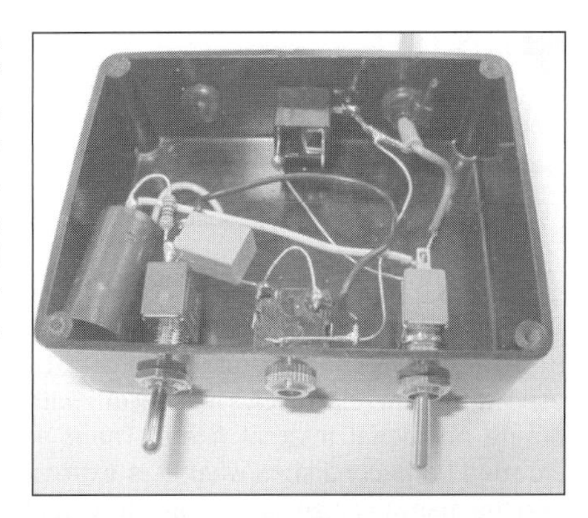

Fig 19.17: The deluxe version of G4BJM's two-component filter includes switches to connect a loudspeaker, and to bring the filter in and out of circuit

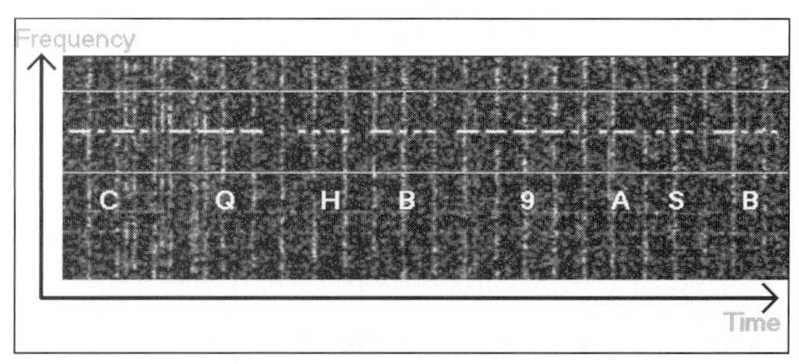


Fig 19.20: A QRSS signal on a curtain display. The translated letters have been added to the picture; they are not present on the original display. The vertical lines are static crashes [pic: ON7YD]

Fig 19.18: The Timewave DSP-599zx is an add-on DSP filter which can improve CW reception

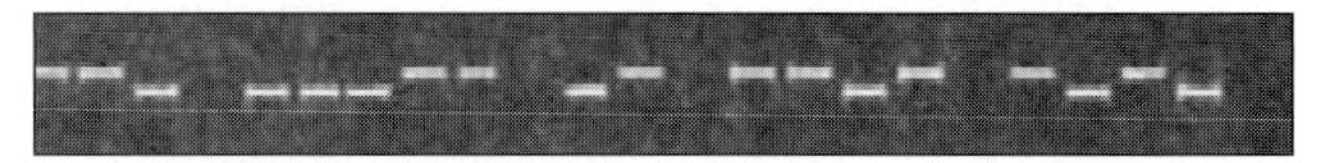

Fig 19.21: DFCW uses different frequencies instead of dots and dashes. This says "G3AQC" [pic: ON7YD]

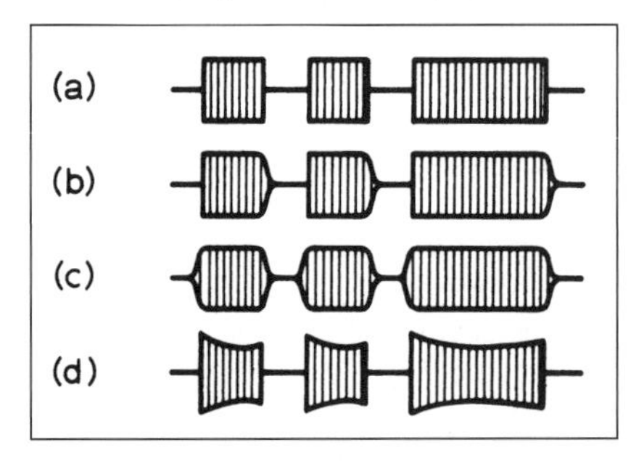


Fig 19.19: Keying envelope characteristics. (a) Click at make and break; (b) click at make, click at break suppressed; (c) ideal envelope with no key clicks; (d) affect on keying envelope of poor PSU regulation

Full break-in ('QSK') is also frequently implemented, with a varying degree of success. Full break-in means that the operator can listen between the individual dits and dahs of a Morse character, right up to 40WPM and beyond. With careful transceiver design this goal can be achieved, but design compromises and shortfalls generally mean that a number of manufacturers' implementations are far from perfect. Ten-Tec and Elecraft have, however, really understood the design principles involved. Although it is achievable, QSK is not commonly used and can be very distracting so is a personal choice.

When examining the keying waveform of many amateur transceivers, the correct 1:1:3 ratio between spaces, dits and dahs is not always maintained or if it is, this is at the expense of reduced receive time when using full break-in. A common feature of poorly designed rigs when using semi break-in is a short first dit, because of time taken in the transceiver to switch between transmit and receive. If this is used to drive an amplifier which itself has a slow T/R relay, the overall effect in extreme cases can be to lose the first dit of a transmission altogether.

If an oscilloscope is available it can be used to display the outline of the pattern made by the signal, this is called the keying envelope. The keyed RF signal is fed to the Y-plates or vertical amplifier of a slow-scan oscilloscope, with the timebase set in synchronism with the keying speed. The square shape in **Fig 19.19(a)** is a very 'hard' signal and will radiate key clicks over a wide range of frequencies. The hard leading edge in **Fig 19.19(b)** will also cause clicks.

The transceiver design should ensure that the rise and fall times are lengthened such that interference is no longer objectionable, but without impairing intelligibility at high speed. The goal should be to achieve rise and fall times of about 10% of the dit length (**Fig 19.19(c)**).

The characteristics of the power supply may contribute to the envelope shape, as the voltage from a power unit with poor regulation will drop quickly each time the key is closed, and rise when it is released. This can lead to the shape in **Fig 19.19(d)**.

## QRSS AND DFCW

Not all Morse used on the amateur bands is intended for reception by ear. When a signal is transmitted over fairly stable paths, such as on 136kHz, it can be received well below the threshold of audibility by using a computer's sound card and software [16] that displays the signal as dots and dashes on the screen.

QRSS is extremely slow speed CW, with dot lengths typically from three to 120 seconds. The name is derived from the Q-code QRS (reduce your speed). To take advantage of the very narrow bandwidth of the transmitted signal an appropriate filter at the receiver end is needed. Making a 'software filter' using Fast Fourier Transform (FFT) [17]] has some advantages over the old fashioned hardware filter. One of the main advantages, when using it for reception of slow CW signals, is that FFT does not provide a single filter but a series of filters with which it is possible to monitor a complete spectrum at once. This means that it is not necessary to tune exactly into the signal, which can be very delicate at sub-Hertz bandwidths. Also it is possible to monitor more than one QRSS signal at the same time.

At first glance it looks as if it is complicated to do this, even if FFT presents you this nice multi-channel filter it might be difficult to monitor all these channels. Further the long duration of the dots and dashes is unfavourable for aural monitoring.

A solution to the above problems is to show the outcome of the FFT on screen rather than making it audible. The result is a graphic where one axis represents time, the other axis represents frequency and the colour represents the signal strength. If the vertical axis represents time it is called a waterfall display, while a curtain display is where the horizontal axis represents time (**Fig 19.20**).

A variant of QRSS is Dual Frequency CW (DFCW) where the dots are transmitted on one frequency, and the dashes on a slightly higher frequency (**Fig 19.21**). This saves time as the dashes can be the same length as the dots. In fact, a time reduction of better than 50% can be achieved, which can be very important at these extremely slow speeds, especially when trying to have a contact in the relatively short duration ionospheric propagation window.

## QRQ OR HIGH SPEED CW

Some amateurs spend a lot of time using QRQ, or high speed CW. The main use for this mode is communications using the ionised trails of meteorites. These hit the Earth's atmosphere and radio waves can be reflected off the ionised tail. This phenomenon only lasts a finite period of time, but high speed Morse can be used to communicate with others involved in the same activity.

High speed CW can also be used for conventional communications. The human brain has limits, however, and for general conversation on the air, speeds of more than 40WPM become very difficult. 'Rag-chewing' at around 30WPM is commonplace however.

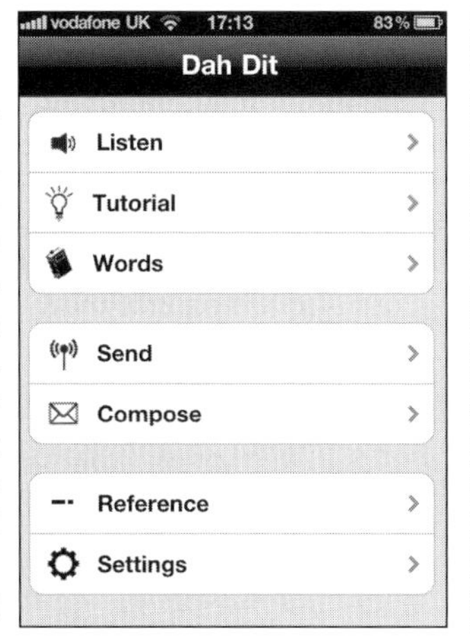


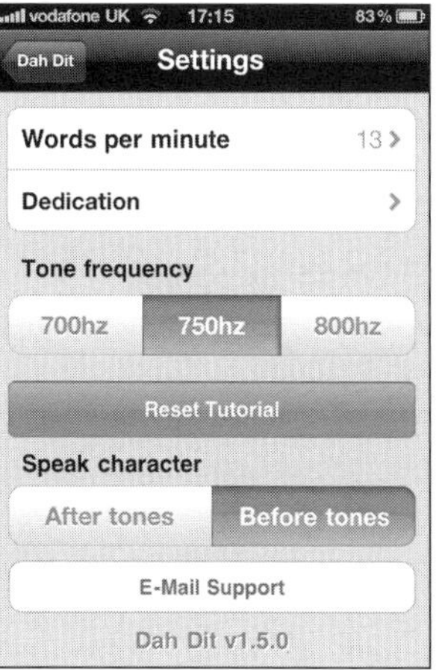

**Fig 19.22: Three screens from Stephen Breen's Morse code app, *Dah Dit*, for the Apple IPhone**

Then there are those whose interest is to test their ability to the limit. There are annual speed contests in which proponents can show off their expertise. The world record is in *The Guinness Book of Records*. There are several levels of achievement, measured according to the number of errors made (see sidebar).

## USING COMPUTERS . . .

The ubiquitous computer is now an essential part of an amateur radio station. It not only helps with functions such as all the modern digital modes, logging, DXCC record-keeping, satellite tracking, and so on, it is also used for contesting, whatever mode is used.

Some stations have several computers, networked together, all performing different tasks but available on a Local Area Network (LAN ).

## . . . for Learning

Learning Morse has become a whole lot easier with the PC. There is a plethora of Morse programs aimed at the beginner through to the advanced CW operator who just wishes to hone his skills with a bit of practice. Before the computer era, it usually meant a weekly trip to night school or a local amateur, and then listening to whatever could be found on the receiver.

The two learning methods discussed earlier are catered for by several programs. A computer is ideal for learning by using the Koch method and suitable software can be found at [19]. If you wish to try the Farnsworth method, take a look at programs by ZL1AN or AA9PW. The ZL1AN software [20] is useful for plain text, has a very nice keying tone, and you can enter your own mixed text into a file and use that. AA9PW [19] allows you to set up a huge number of options, including call signs, and punctuation.

### Mobile phone app

One of the many applications (or apps) for the Apple Iphone is very useful for someone trying to learn Morse. This Morse code tutor for the IPhone/IPod touch is called *Dah Dit - Morse Code Tutor*. It is written by Stephen Breen and is available in the online app store at modest cost.

Most people have their mobile phone with them all of the time, so wherever you are and at whatever time, you can use this app on your phone. It will give you the spoken character then the Morse code or the Morse code and then the spoken character. It has a progressive tutorial feature that gradually introduces new characters into the letters/numbers/punctuation you are learning until you have learned them all. Once you have achieved that, the program has a 1000 common word list to send for you to copy and type back into it. It will even give statistics on which characters you are struggling with! The screens you see on your phone are illustrated in **Fig 19.22**.

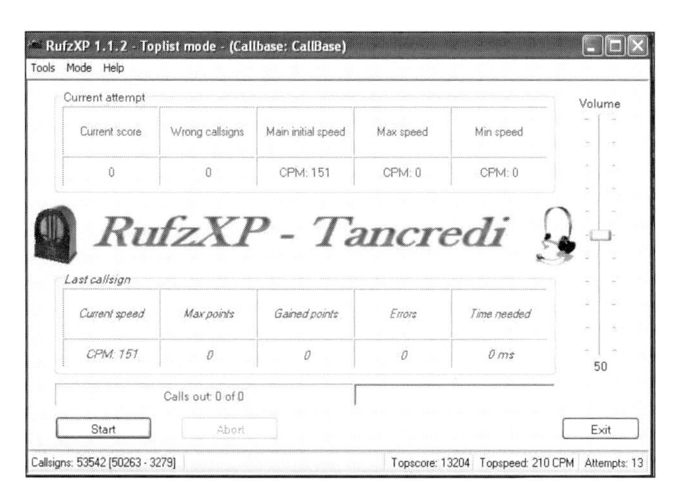


**Fig 19.23: Morse practice program *RufzXP***

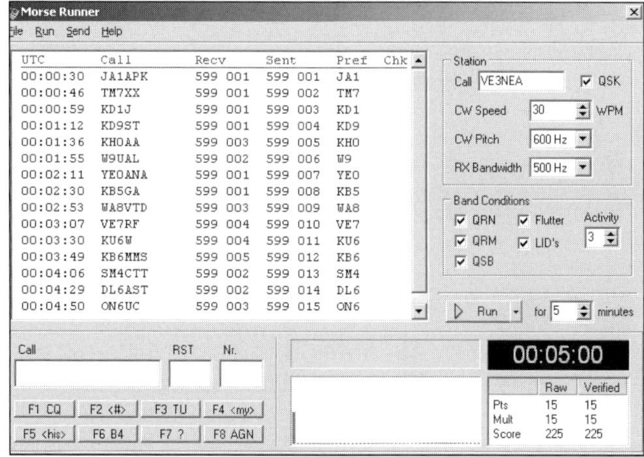


**Fig 19.24: A *MorseRunner* screen**

## RufzXP

*RufzXP* is another Morse program designed for practice at reception. Written by DL4MM and IV3XYM, it is by Tancredi and is free to download (from [21]) and use. It sends a chosen number of true amateur radio calls, randomly selected and the user has to type the call on the keyboard. In effect, it is good for two purposes, receiving the call and also improving your typing skill on the keyboard. If the call has been typed in correctly, the speed then increases. If the call is typed in with errors, the speed decreases.

The last call heard can be repeated by hitting F6, for which you receive a 50% penalty. The number of points for each call is primarily related to the speed that the call is sent, the number of errors and the length of the call. The time you take to type the call also has an influence.

At the end of the test you are given a score and your call is inserted into a rate table. There are four modes of operation, toplist, trainer, contest and HST, the high speed trainer.

Just like *Morserunner* (see below) it can be addictive. It is quite interesting to see the score improving each time you use the program. A photograph of the *RufzXP* screen can be seen in **Fig 19.23**.

### High Speed Morse World Records

Guinness World Records Ltd has recognized the high-speed telegraphy achievement of Andrei Bindasov, EU7KI.

"On 6 May 2003 Andrei Bindasov (Belarus) transmitted 216 marks of mixed text per minute during the 5th International Amateur Radio Union World Championship in High Speed Telegraphy in Belarus," the Guinness database listing states.

Witnessing the accomplishment in Minsk were HST International Referee Oscar Verbanck, ON5ME, Region 1 Executive Committee member Panayot Danev, LZ1US, and IARU Region 1 HST Coordinator Oliver Tabakovski, Z32TO.

Bindasov also sent 271 letters per minute and 230 figures per minute during the phases of the 2003 HST competition [18]

## Learn CW Online

This is another complete package whereby the student can sign in and use the facilities without downloading any program at all. The training is similar to other programs, with letter and number groups, words, call signs, plain text, words in other languages, practice files and so on. You can select speed, tone, Farnsworth and all the usual parameters to what suits you.

There is also a high scores list and you can compete with friends doing a similar thing, a Forum where discussions take place, and a neat text to CW converter, so you can input your own text and play it back in Morse. This is repeatable too, so you can fill in the gaps the second time.

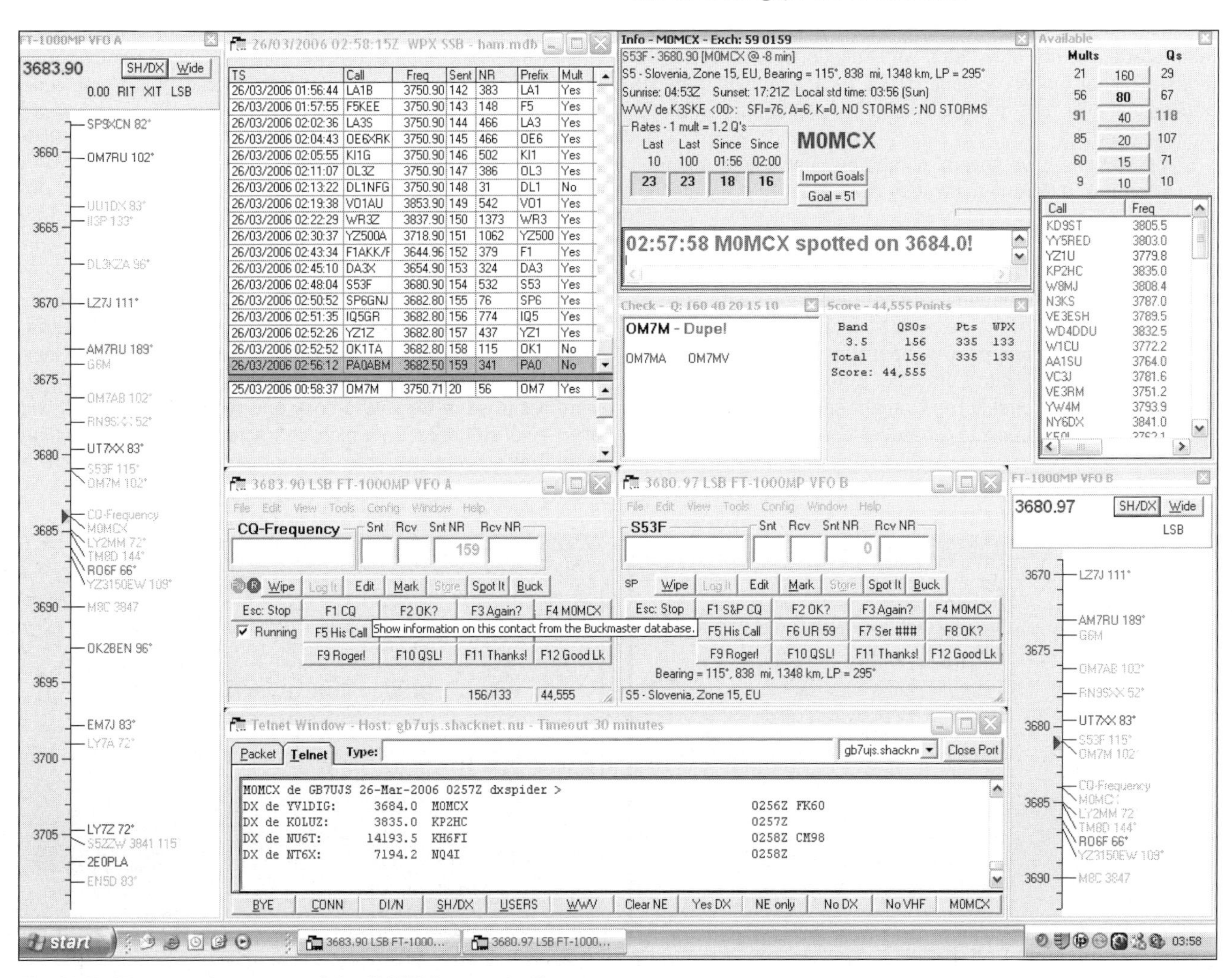


Fig 19.25: The complex screen of the *N1MM* free contesting program

## . . . for Contesting

### Getting started

Once you become more proficient with Morse code, you might wish to enter a few contests. The RSGB Cumulative Club contests occupy just 90 minutes of an evening. They use three modes, CW, RTTY and SSB. In the CW section, there is a QRS corral, where beginners can try their luck. Being thrown into the deep end of the 30WPM+ brigade is what dissuades a lot of new licensees, but the QRS corral can encourage more activity because they find it is not impossible.

Practice being the name of the game, it would be a good idea to obtain some before taking part in a contest. Try downloading a program called *MorseRunner* [22]. It is a very addictive program however, enabling the user to take part in a pseudo contest on the computer. It provides a window on the screen (**Fig 19.24**) that is very similar to the *N1MM* contest program. Using *MorseRunner* improves your operating skills as well as your speed.

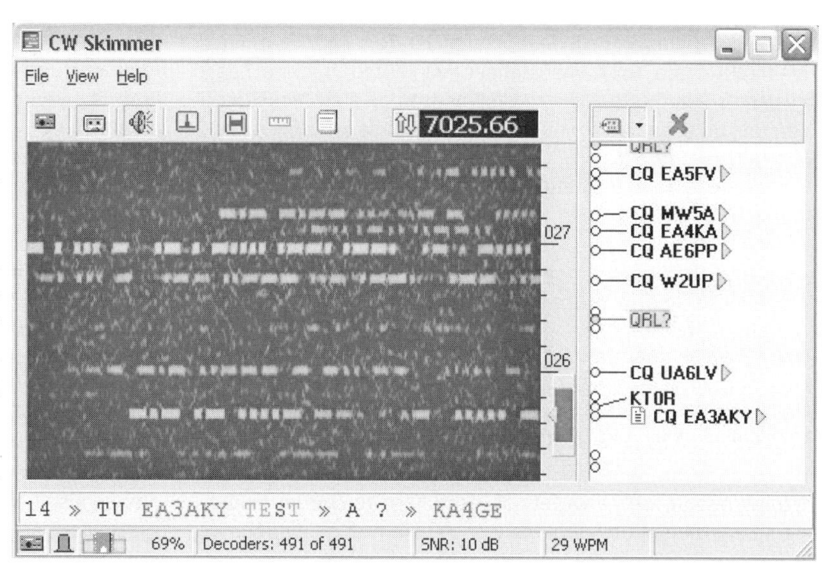


Fig 19.27: CW Skimmer from Afreet Software Inc can decode a pile-up or, with the right receiver, an entire band of Morse stations

### Advanced

There are several contest programs available, for instance *N1MM* [23], *SD* [24], and *Writelog* [25]. *N1MM* is gaining popularity and is used by a large number of contesters. It is free but takes some setting up. There may be somebody who can help you to do that. The on-screen display can get complicated (**Fig 19.25**) but that is variable and up to the individual as to how many windows are used. Other programs are listed at [26].

*Wintest* is a very popular contest program. Although it is not free, it is used by quite a number of serious contesters, and DX-peditions. It can be networked quite easily and using it this way with several operators, either locally or networked on the Internet, it provides functions that some other contest loggers don't have. **Fig 19.26** shows the program being used in a major contest.

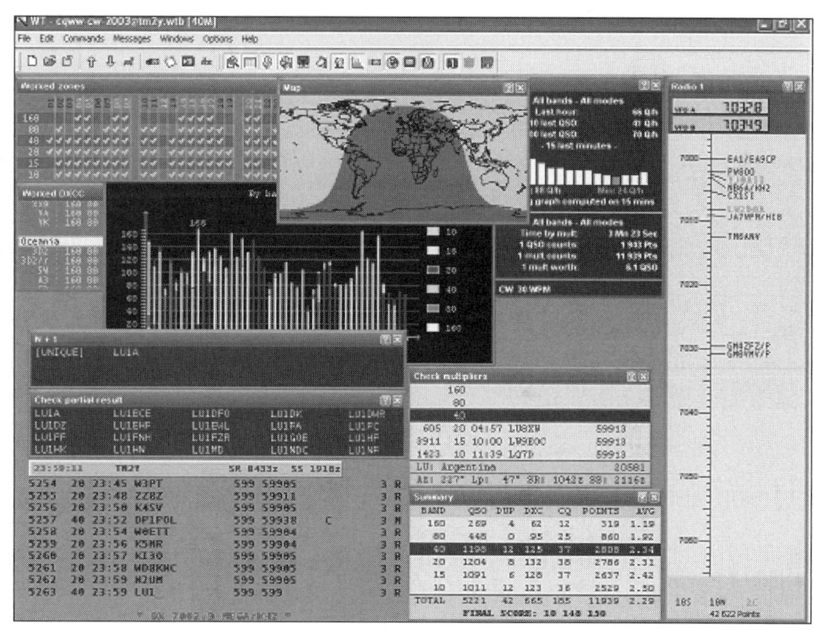

Fig 19.26: The *Wintest* logging program

One program that has caused quite a stir among the avid contesters is called *CW Skimmer*. This is a new development by Afreet Software Inc. This is software created by Alex Shovkoplyas, VE3NEA, who also produced *Morse Runner*, *DX Atlas* and so on. *CW Skimmer* is a Morse reader with a difference. It uses a very sensitive CW decoding algorithm based on the methods of Bayesian statistics and is capable of decoding all stations within the receiver passband. Not only that, but as you see from **Fig 19.27**, it produces a waterfall type display with all the calls displayed as well. Love it or hate it, it is yet another addition to the stable of software available to make life easier for the operator [27]

Although the contest programs do send your CW for you, it is still necessary to be able to use the paddle, so it is a good idea to have one wired in parallel to enable you to interject comments.

Contest programs are fine for logging, 'duping' and so on, but nothing replaces the brain, so use the practice programs on a regular basis and you will maintain your own level of speed and will gradually increase it too.

## Welcome to the reverse beacon network!

The Reverse Beacon Network is a revolutionary new idea. Instead of beacons actively transmitting signals, the RBN is a network of stations listening to the bands and reporting what stations they hear, when and how well. How does this have anything to do with Morse? More than you might imagine!

Well, to begin with, you can see band openings in near-real time on an animated map. You can call a quick CQ, and see which reverse beacons hear you, and how strong you are.

But the real breakthrough is in the database of past "spots". You can instantly find out what stations, from a given country or zone, have been heard, at what times and on what frequencies. You can see when you've been spotted, who spotted you, and how loud you were.

There is more! Now, for the first time, you can compare your signal with those of your friends and competitors, in near real time or historically. If you wonder how your signal stacked up during last weekend's contests, the Signal Comparison Tool will give you real, quantitative data. Tell it what stations you want to compare, based on signals heard by a given reverse beacon on a certain band at a certain time, and there you'll have it. Of course, whether you like what you see is up to you.

It really is an extremely useful tool and takes no effort on your part in order to use it, other than a simple TEST or CQ message. The Reverse Beacon Network is also run by volunteers and there are a huge number taking part these days. It can be a useful propagation tool, an antenna checking tool, and you can see if your improvements have taken effect.

*http://www.reversebeacon.net/*

## MORSE AS AN ASSISTIVE TECHNOLOGY

Morse code has been employed as an assistive technology, helping people with a variety of disabilities to communicate. Morse can be sent by persons with severe motion disabilities, as long as they have some minimal motor control. Morse code can be translated by computer and used in a speaking communication aid. In some cases this means alternately blowing into and sucking on a plastic tube ('sip-and-puff' interface). An important advantage of Morse code over row column scanning is that, once learned, it does not require looking at a display.

Also, it appears faster than scanning. People with severe motion disabilities in addition to sensory disabilities (eg people who are also deaf or blind) can receive Morse through a skin buzzer. In one case reported in the magazine QST, an old shipboard radio operator who had a stroke and lost the ability to speak or write could communicate with his physician (a radio amateur) by blinking his eyes in Morse. Another similar case was a lady who was paralysed completely with the exception of the ability to move one finger. Her son noticed that she was desperately trying to move her finger in a rhythm and, knowing she was a qualified Morse operator, made up an oscillator and she was able to communicate using Morse.

It has also been speculated that learning and using Morse is a great exercise for the brain and can help stem the onset of Alzheimers. Perhaps that is why there are so many active G3 stations! It all helps!

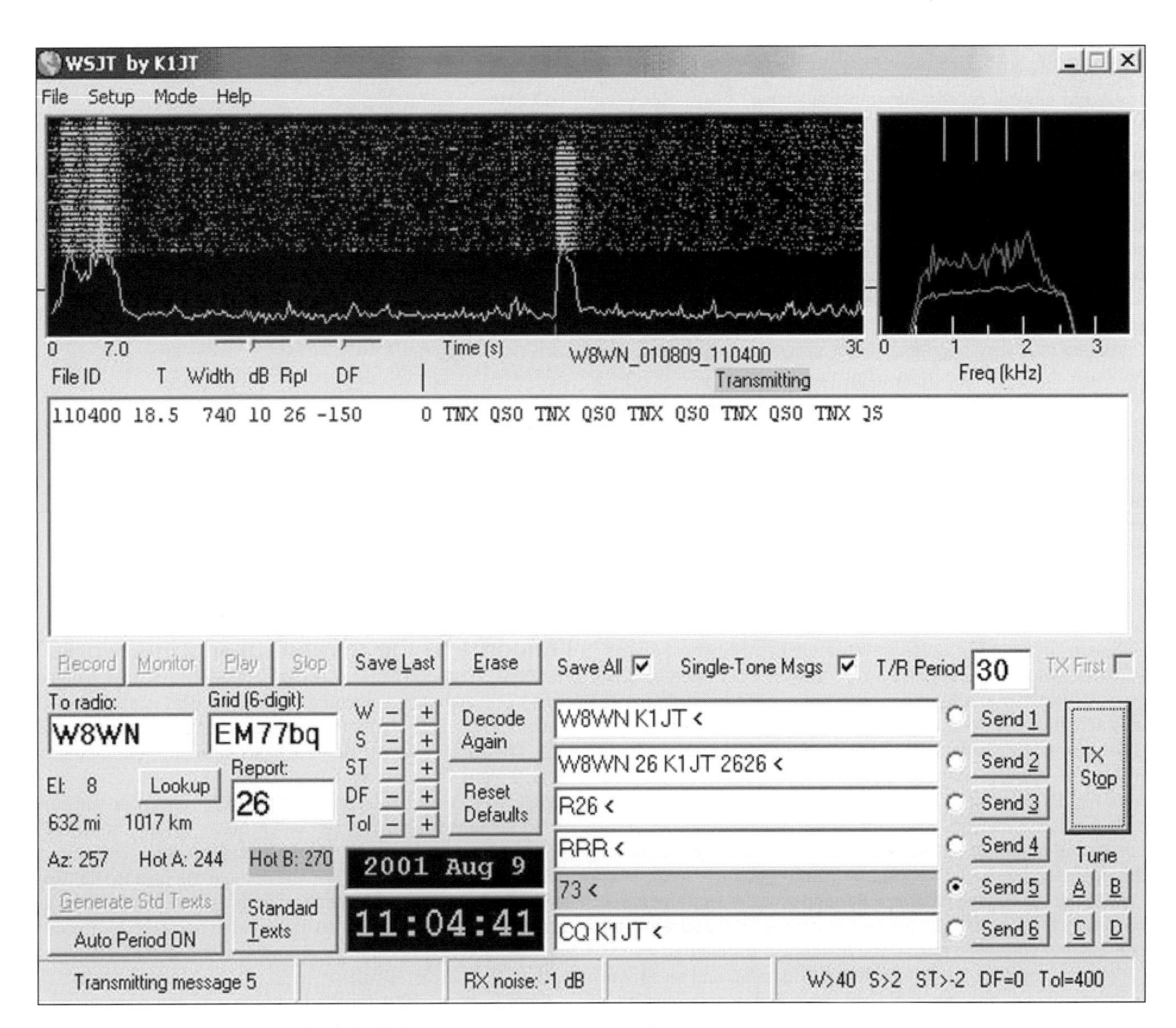


**Fig 19.28: The *WSJT* high speed CW program**

## GOING FASTER

A very useful program is one that is regularly used for MS (Meteor Scatter) communications. It is called *WSJT* and can be downloaded for your own use from [28].

It can decode fraction-of-a-second signals reflected from ionized meteor trails, as well as steady signals more than 10 dB weaker than those required for conventional CW. One of its operating modes is optimized for amateur EME (Earth-Moon-Earth) communications (see the chapter on space communications).

*WSJT* is open source software and is licensed under the GNU General Public License. To use the program you will need a computer equipped with a sound card and running the Microsoft Windows, Linux, or FreeBSD operating systems.

Pings from meteor trails with 'underdense' ionization are nearly always available in usable numbers. Even 100W, single-Yagi stations at suitable distances can usually hear several pings from each other in a 10 to 20 minute period. At typical high speed CW speeds, around 8000 letters per minute, a ping lasting 0.1 second contains about 13 characters-just about enough for your callsign, the other station's call, and perhaps a signal report.

With coordinated timing, good frequency calibration, and some diligence, operators who take the time to learn the technique can easily complete contacts this way. It's a fascinating way to work a bunch of new grid locators, or DXCC entities. It can also work wonders for fattening your multiplier total in a VHF contest. You do not need an EME-class station, and best of all, you don't need to wait for a meteor shower or for one of those all-too-elusive band openings.

The screen-capture in **Fig 19.28** shows the main *WSJT* screen during a meteor scatter contact between K1JT and W8WN. Thunderstorms were present to the west of K1JT at the time, explaining the two static crashes near the beginning of the displayed 30-second recording as well as the noisier-than usual baseline of receiver background noise (the green line). The signal about 18 seconds into the record is a ping from an underdense meteor trail, and the message it conveyed is displayed in the central text box.

## SUMMARY

In summary, Morse should not be seen as daunting. Millions of people have mastered the use of Morse code over the years and many amateurs continue to use it on a daily basis both for routine and state-of-the-art communication methods.

## REFERENCES


[1] IARU bandplans listed by region: www.iaru.org

[2] Computer program for learning the Morse code. www.g4fon.net/CW%20Trainer.htm

[3] *RSGB Yearbook*, published annually by, and available from, the Radio Society of Great Britain

[4] RSGB web site: www.rsgb.org

[5] Morse Certificate. www.rsgb.org/morse/

[6] The MFJ-461 Pocket Morse Reader: www.mfjenterprises.com//Product.php?productid=MFJ-461

[7] 'A Simple Morse Key', Dick Biddulph, M0CGN, *RadCom*, April 2001

[8] GHD Morse keys: www.mtechnologies.com/ghd/

[9] Begali Paddles and keys: www.i2rtf.com/html/hst.html

[10] Vibroplex keys: http://vibroplex.com/vibrokeyer.html

[11] Schurr paddles: www.mtechnologies.com/schurr/

[12] Chevron paddles: http://chevronmorsekeys.co.uk/

[13] 'The BIK Simple Electronic Key', Ed Chicken, G3BIK, *Radio Communication,* Aug 1993

[14] Winkey details: www.k1el.com/

[15] 'A Two-Component CW Filter' Fraser Robertson, G4BJM, *RadCom*, Aug 2002

[16] Viewers for QRSS and DFCW can be downloaded from www.weaksignals.com

[17] QRSS and DFCW information at ON7YD's web site www.qsl.net/on7yd/136narro.htm

[18] High speed CW information: www.iaru-r1.org/index.php?option=com_content&view=category&id=46

[19] AA9PW Morse practice program: www.aa9pw.com/radio/morse.html

[20] ZL1AN Morse program: www.qsl.net/zl1an/Software/teach4software.html

[21] *RufzXP* Morse Practice program: www.rufzxp.net/

[22] MorseRunner by VE3NEA: www.dxatlas.com/

[23] N1MM logging program: http://n1mm.hamdocs.com/

[24] *SD* by EI5DI: www.ei5di.com/

[25] *Writelog*: www.writelog.com/

[26] Contest Programs: www.ac6v.com/logging.htm#CON

[27] Skimmer CW program: www.dxatlas.com/CwSkimmer/

[28] *WSJT* Meteor Scatter program: www.vhfdx.de/wsjt/


***Roger Cooke*** *spent 32 years as a TV/Video technician. He has written numerous articles for Practical Wireless, Radcom, QST and 73 magazine, on various subjects. Author of Morse Code for Radio Amateurs and RTTY and PSK31 for Radio Amateurs (both RSGB). Edited the 'Data' column in RadCom a few years ago. He now writes a regular bi-monthly column in Practical Wireless called "The Morse Mode".*

*Taught Morse for the amateur radio test for decades, both at home and at night school. Still teaching Morse, at home with a class known as The Bootcamp, twice per year to prepare them for working DX, contests and enjoying the mode.. He is also the RSGB's GB2CW co-ordinator and has arranged five on-air classes using GB2CW on two metres with four tutors each week. Now 81 years young, still enjoying Morse after 65 years using it on the air.*

# Digital Communications

20

*Andy Talbot, G4JNT*

First, a little history. Amateurs have been transmitting data, usually typed or printed text, to each other for many decades. The first datacomms used mechanical teleprinters - a sort of early printer with a mechanical keyboard that generated a code for each letter made up of five bits. These bits were then transmitted serially using one of two tones to represent a 'one' or a 'zero'. Decoding was by purely mechanical means with a constant speed motor and cams, and much time was spent by the operators maintaining and oiling their machines. The use of teleprinters led to the term RTTY, for Radio TeleTYpe. RTTY is still in widespread use on the amateur bands today, although most now use computers for generating and receiving the signals

Once computers had become established in many shacks, but before the advent of the Internet, a widespread network of interconnected stations and nodes allowed operators to send messages to each other and to exchange files. The Packet Radio network allowed one station to 'connect' to another station by specifying the callsign, and the network was able to correctly route traffic through multiple nodes around the country and the world. In its heyday, the Packet network was a reasonably reliable messaging system, although in busy periods it could take tens of minutes or even hours for some messages to get through.

Once the Internet came along to do the same job a lot quicker and more robustly, the packet network mostly died, although a few nodes and stations can still be found. Packet radio still has a place on a more local and personalised level for emergency communications where it allows short uncorrupted messages to be routed over locally set up, ad hoc, radio networks when other infrastructure has failed

With the advent of the personal computer (PC), the whole amateur data communications field has grown to encompass a lot more than exchanging hand typed messages. What has made this possible is the rapid growth in what are usually referred to as 'soundcard modes'. The computer's audio ports are interfaced to the radio so that the PC generates audio that is subsequently upconverted to RF, usually using the radio in SSB mode. On receive, audio is fed into the PC and after digitation of the analogue waveform, the power of Digital Signal Processing is used to do all of the signal processing. transmit/receive switching is also usually placed under computer control with another interface line, allowing datamode operation to take place without having to make regular changes on the radio.

Details of computer interfacing, including a number of pitfalls for the unwary, are given in the Computers in the Shack chapter.

Before going on to look at some of the modern digital communications techniques in detail, we will need to take a look at some of the theory behind the transmission of data over a radio link.

## DIGITAL COMMUNICATIONS PRINCIPLES

### Modulation

Data consists of binary information, usually referred to as 'ones' and 'zeros'. To send these over a radio link we need to modulate the RF carrier in a way such that the receiver can differentiate between a 1 and a 0 being sent. The easiest and most obvious way is to switch a carrier on for a '1' and off for a '0' and is known as Amplitude Shift Keying or ASK. The modulation rate, or the

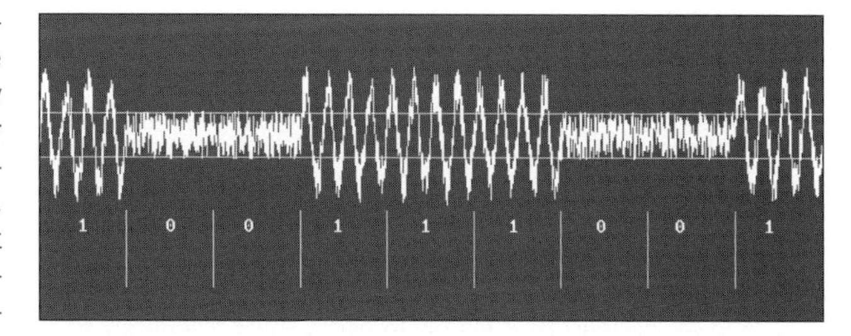


**Fig 20.1: Amplitude Shift Keyed (ASK) carrier with added noise**

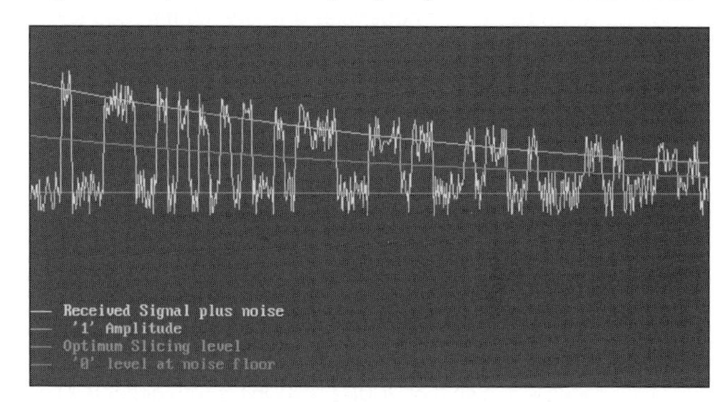


**Fig 20.2: Fading amplitude shift keyed (ASK) signal**

rate at which the bits change, is usually referred to as the Symbol or Baud Rate. For simple modulations like ASK and FSK, Baud rate is equal to the Bit rate

The problem with ASK is knowing where to make the decision for 0 or 1. Where the signal is varying in amplitude (fading) this 'slicing' or decision level, which is usually placed half way between zero and the maximum received amplitude, will have to move correspondingly. Automatic gain control (AGC) can be used to track the amplitude, but the AGC time constant will have to match the fading on the RF channel, and in addition we need to know the number of ones and zeros being transmitted as this will affect the average signal strength. A '0' will consist only of noise, which may itself have large random spikes on it that could mistakenly be measured as a '1'. **Fig 20.1** shows a typical ASK signal corrupted with noise. The optimum slicing level for the best 0 / 1 decision in the presence of fading can be seen in **Fig 20.2**.

The problem with selecting the optimum slicing level for on-off keying can be mitigated by forcing the modulation to have a fixed duty cycle of 50% so the slicing level is always exactly half the amplitude of the peak signal. One way of doing this is to multiply, or exclusive-OR, the data with its own clock. Each '0' bit then corresponds to, say, a high/low transition and each '1' bit to a low/high transition as shown in **Fig 20.3**. This is commonly called Manchester Coding, and has been adopted for the Opera beacon signal format described later in this chapter.

Frequency Shift Keying, or FSK, overcomes the amplitude threshold problem. The simplest route is to use two tones and switch them alternately - one tone for a '1', and the other tone for a '0'. Demodulation is then performed by looking at each tone separately and comparing the relative amplitudes of the signals, including added noise, to decide which element was sent. This particular technique is know as Frequency Exchange Keying, or wide shift Frequency Shift keying (W-FSK) and relies on the two tones being widely spaced in frequency. This is only

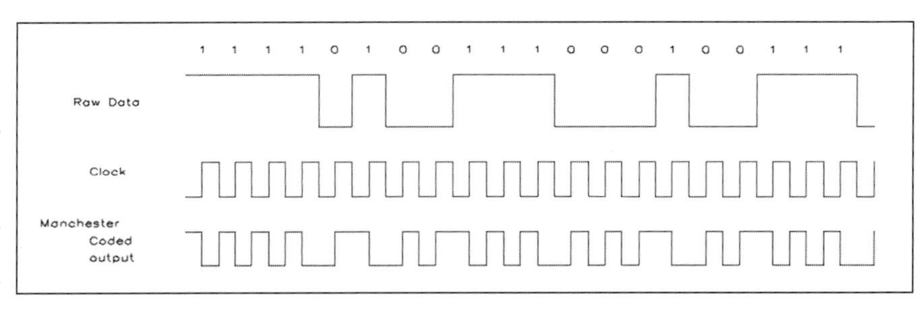


**Fig 20.3: The principle of Manchester encoding**

the case if the data rate is significantly less than the difference between the two tones because the frequency spread from each tone being switched causes the sidebands from each to merge as data rate rises. RTTY with data at 45 or 50 bits per second (bit/s) and 170Hz shift is a typical example of this mode.

Wide shift FSK has a poor bandwidth efficiency, using significantly more width than is needed to send the data. This is not very socially acceptable in the congested amateur bands! The way to lower bandwidth is simply to use a lower frequency shift for a given data rate. This has the effect of merging the sidebands from the two tone frequencies into one continuous spectrum. It means that using two tuned circuits to separate out which tone was transmitted no longer works. By converting the varying audio frequency to a voltage it can be fed to a comparator to decide whether a One or Zero was sent. The Phase Locked Loop is the most popular means of doing this in hardware with other techniques used in DSP based demodulators. A PLL generally needs a higher signal to noise ratio to stay locked and to track the changing frequency reliably than a two tone demodulator would have been able to cope with.

Narrow shift FSK was employed with good results in the early days of telephone modems where originally 300bit/s was available using two tones with a shift of 200Hz. 1200bit/s with either 800 or 1000Hz shift followed shortly and this worked reliably enough over the 25 - 40dB S/N capability of the analogue telephone network in its day. The 1200bit/s standard, with its switching sidebands, takes up much of the available 3kHz audio bandwidth, and data rate could not be further raised by using FSK techniques. Simple low cost modem chips appeared to cater for these modes and the packet radio community adopted these telephone standards in the 1980s due to their simplicity of implementation.

Unfortunately, the greater S/N requirement for Narrow Shift FSK means that, for radio communication, high error rates frequently occur. If an FM link is used for transmission of the two audio tones, the inherently good quality of this mode when signals are strong allows Narrow Audio-FSK to work successfully for short range local contacts. The vast majority of the Packet network relies on this medium of 1000Hz shift audio carrying 1200bit/s data in a single FM channel, but it is hardly an efficient modulation scheme. On HF, attempts to use 200Hz shift for 300bit/s data were fraught with failure and lost data; HF packet forwarding was for some time limited to the 50bit/s or less of AMTOR or RTTY. A better way of modulating an RF carrier is needed if we are to improve on this.

FSK does, however, have its uses where frequency scattering is prevalent. This is often the case with narrowband amateur microwave links, and when using very narrow bandwidths at HF.

MSK, or Minimum Shift Keying is a special case of FSK where the shift is exactly half of the symbol rate. A Gaussian filter is often employed to shape the frequency transition and reduce the bandwidth of the signal, leading to the term GMSK. GMSK is used on the GSM Mobile phone network, and in the G3RUH 9600 Baud packet radio modem and some amateur satellite links. GMSK requires coherent decoding to work properly and a quite complex decoding algorithm if the best use is to be made of the mode.

## Coherent Schemes

A solution to improving the bandwidth efficiency and S/N performance is Binary Phase Shift Keying, BPSK, often just referred to as PSK. In its simplest form the RF carrier is reversed in phase by 180 degrees for each '1' bit of data and zero degrees for a '0' bit. This is accomplished very simply by multiplying the RF carrier by 1 or -1 (for a '0') in a balanced modulator. It is obvious that at the receiver there will be an ambiguity as to which phase shift is a '1' and which corresponds to a '0'. One way around this is to transmit a preamble of a known pattern of '1's and '0's. If the preamble is received inverted then we know the whole demodulation process is upside down and needs to be swapped.

The other way is by differential coding. Instead of transmitting the absolute phase, a '1' is sent by changing the phase from one bit interval to the next, for a '0' the phase is not changed. Therefore a string of zeros will appear as an unchanging plain carrier, and a string of '1's as a repeated pattern of 0 / 180 degree phase changes.

The merits of coherent schemes can best be appreciated by looking at their analogue counterparts, comparing SSB voice with plain AM. SSB generally gives about 10dB advantage in noise over AM but at the expense of a more complicated receiver demodulator having to reinsert the carrier. PSK shows a similar advantage over FSK or ASK.

The spectrum of PSK depends on the symbol rate, and if the phase is hard switched then keying sidebands can extend to some considerable distance either side, as shown in **Fig 20.4**. To keep the bandwidth within reasonable limits, the amplitude of a PSK waveform is ramped at the transition point so the actual phase change occurs at zero amplitude as illustrated in **Fig 20.4**.

The rate of ramp is a trade off between bandwidth and S/N degradation due to loss of net energy in the entire symbol interval. PSK31 is an example at one extreme of this trade-off, where

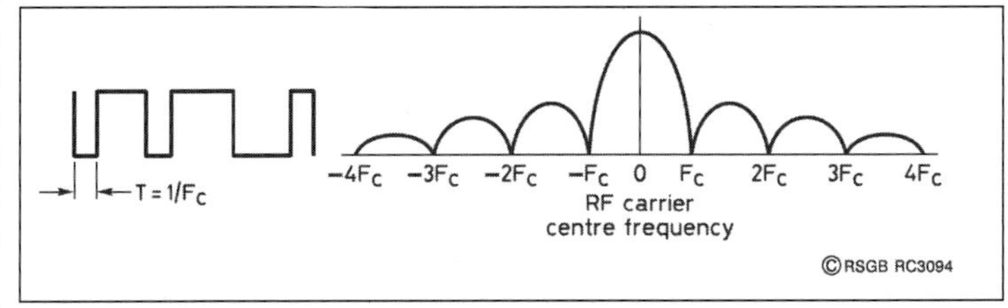


**Fig 20.4: Keying sidebands of hard-switched PSK**

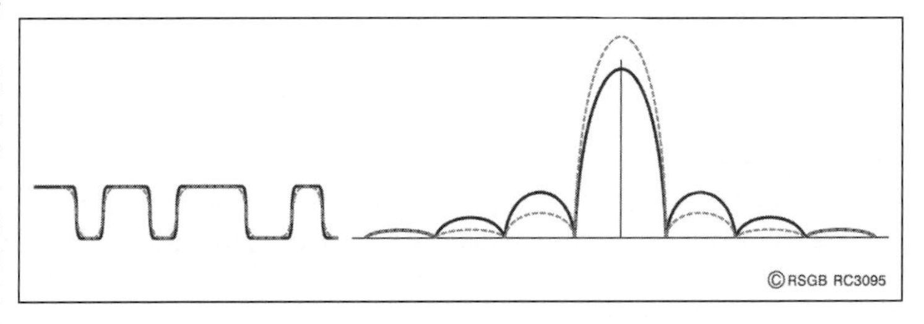


**Fig 20.5: Reducing PSK bandwidth by reducing the amplitude at the phase transition point**

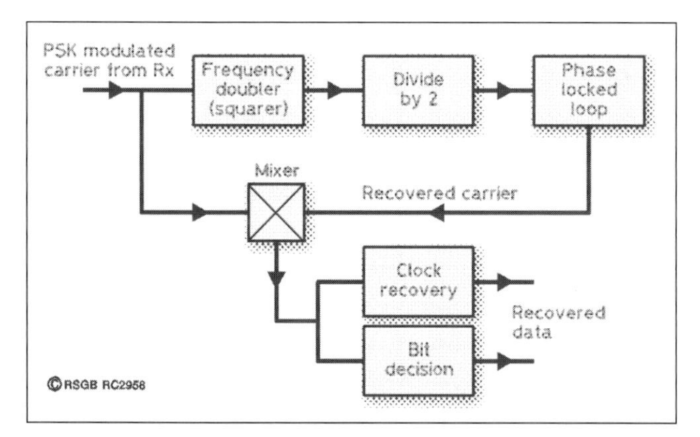


Fig 20.6: The first of two decoding schemes for PSK modulation. Suitable for non-differentially-coded signals. It uses a frequency-doubler to resolve the 0° / 180° ambiguity before regenerating the carrier via a phase-locked loop

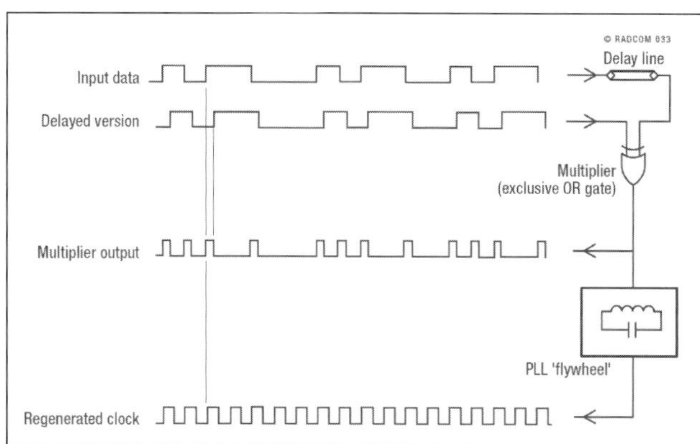


Fig 20.8: Clock regeneration from a random data stream

the entire symbol is shaped to a half sine-wave at the signalling rate. This leads to a very narrow bandwidth equal to the data rate, but sacrifices a couple of dB in S/N over a hard switched version with high levels of 'keyclicks'.

## Decoding PSK

PSK is more complicated to decode than is ASK or FSK with particular areas of complexity being those of carrier and bit clock regeneration, as well as controlling the spectrum of the signal. To demodulate a PSK signal, it is necessary first to generate a local oscillator - equivalent to the BFO in an SSB receiver - in order to mix with the incoming signal in order to recover the phase shift information. Unlike SSB, however, it has to be phase locked to the incoming waveform otherwise the recovered phase would slowly drift at the difference frequency, making resolution of the 0/180 degree phase shift impossible to achieve.

One way of doing this is to use a special phase locked loop (PLL) which can lock up to the signal when it is in either of its two phase states by squaring (frequency doubling) the incoming waveform which has the result of generating a constant phase continuous carrier at twice the input frequency. This can then be filtered out, divided by two and used to lock a PLL to recover the transmitted phase information. See **Fig 20.6** Another way is to use a variant of the third method SSB generator, called a Costas Loop, where a PLL works on the In-phase and Quadrature components of the signal and maintains lock, with the added bonus of demodulation as an inherent part of the carrier recovery. Costas loops are rather complex to build in hardware, but can be more easily programmed using DSP.

For differentially coded PSK demodulation of the binary data is slightly easier without having to resort to a PLL. Provided the incoming frequency is known with sufficient accuracy, by comparing the instantaneous signal with that received exactly one bit interval earlier using a mixer, the differentially modulated code can extracted (**Fig 20.7**) Provided the frequency error is low enough such that the phase drift during a bit interval is significantly less than the 180 degree data change, errors are minimised. To detect a 180 degree phase shift we need an error of less than 90 degrees. To achieve this phase shift during one bit interval the frequency needs to be set to within 90°/ 360° of the signalling rate. So, for a 300 bit/s DPSK waveform, the frequency has to be known to within 75Hz. For noisy signals, higher accuracy is needed and a figure of a tenth the bit rate is usually taken as a rule of thumb. The delay and comparison needed for DPSK has to be made at the bit transition point, where the phase may or may not have changed, so the data clock needs to be recovered for optimal decoding. In most cases a clock is also needed in order to be able to feed the recovered data into subsequent processor circuitry.

Clock regeneration is probably the most important part of a PSK demodulator, as the accurate timing needed for optimum demodulation of noisy signals is derived directly from this. The clock not only has to be locked to the frequency of the data bits, it has to be synchronous with the transitions so maximum use can be made of the full symbol period for making the vital 1/0 decision. Any clock jitter here degrades demodulation as the precise changeover point cannot be found.

One method is to use a similar technique as that for carrier recovery, a non linear operation such as a pulse generator triggered from transitions or zero crossings, followed by a narrow bandwidth Phase Locked Loop to extract just the clock signal from the jittery pulse output. **Fig 20.8** shows the clock extraction. Other techniques that can be used are to examine the amplitude of filtered or band-limited PSK. The waveform will have a maximum at the middle of the symbol period and knowledge of the time at which the maximum occurs allows a loop to be locked. A simpler technique is just to look for any phase transition, assume this is at the correct point and synchronise a locally generated clock from there. Periodical comparisons can be made to keep the clock locked. Other more complex schemes are possible, including those that combine clock recovery with carrier regeneration, such as the Costas loop.

One variation on BPSK is Quadrature PSK, with four phases separated by 90 degrees as shown in **Fig 20.9** This allows two bits to be sent at a time so Baud rate is now one half of the bit rate. QPSK gives better bandwidth efficiency than BPSK, but needs a 3dB higher S/N for reliable copy. This is a story that recurs repeatedly in the field of digital communications.

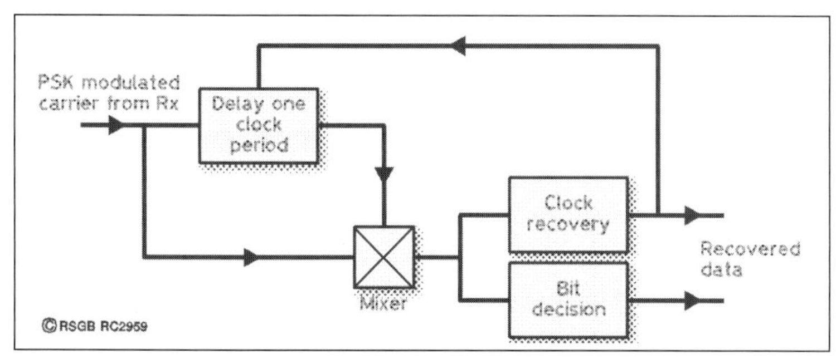


Fig 20.7: This scheme is used for differentially-coded modulation where the data is coded as a change from one bit to the next. Note the need for recovery of the bit clock in order to make the phase comparison accurately

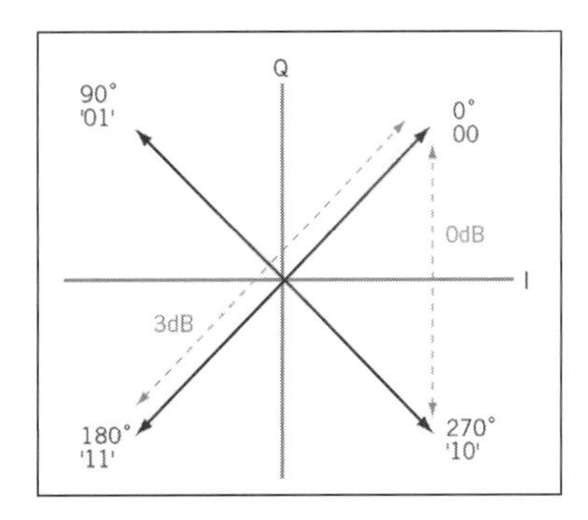


Fig 20.9: The four symbol states of a QPSK waveform showing the amplitude differences between states

## Combined Schemes

Many modern digital modulations combine or expand on these basic types. For example, Multiple Frequency Shift Keying, or MFSK, uses one of several tones for each interval. If, for example, eight tones were used then each tone could represent three bits at a time, binary '000' through to '111'. MFSK offers a lower bandwidth than FSK for a given data rate, and in DSP implementations is almost as straightforward to decode as binary FSK.

In fact, a 32 tone MFSK signal, where each tone represents one of the 32 possible RTTY codes, was developed over half a century ago by the UK Foreign Office - where it was called Piccolo because it sounded a bit like the musical instrument.

Quadrature Amplitude Modulation is another combined scheme. A carrier is altered both in amplitude and phase to cram several bits into one symbol. 256 QAM, for example, has 256 states with each symbol carrying 8 bits of information. As each state is only separated from its adjacent one (in phase and amplitude space) by a small amount compared with the overall signal amplitude, a high Signal to Noise ratio is required to prevent a small amount of noise from corrupting one symbol into another. **Fig 20.10** shows a simulation of a 16-QAM signal with 18dB S/N

And here lies the big trade-off for digital communications. Simple schemes like FSK and BPSK work on low S/N links but occupy large bandwidth. Making each modulation symbol carry more bits reduces the bandwidth requirement but increases the received S/N needed, and needs greater transmit power.

## Parallel Tone Modulation

All the modulation types considered so far have been formed by modulating a single tone or a carrier, but there is now a whole family of digital modulation types that are built up from multiple simultaneous tones. One problem with single tone modulations is that when multipath is present - which is often the case on radio links - a receiver may pick up two or more copies of the transmitted signal with different time delays. If the amplitudes of each component are not too far apart, the resulting demodulated bits will overlap each other leading to errors. The overlapping bits cause intersymbol interference and especially on the HF bands multipath can be very bad, with delays of up to several milliseconds being typical. The overlap severely limits usable modulation rates to 200 Baud, or even lower. This limitation was the reason why, at HF, RTTY was so popular for so long; its 45 Baud signalling was robust enough to overcome multipath if signals were strong enough. At VHF/UHF, multipath delays of tens to hundreds of microseconds are typical.

The need for higher data rates over multipath-prone channels forced designers to come up with new schemes and rethink modulation completely. If a number of tones are transmitted simultaneously within a channel, these can each be modulated at a symbol rate, that is low compared with multipath delays. If each tone is modulated independently (with BPSK or QPSK for example), the data carried on each can be merged after demodulation, leading to a net data rate comparable with what might have been achieved in the same bandwidth with a single tone scheme. But now, as the symbol rate is much lower, multipath is not an issue. As an example, consider a 32 tone scheme with a spacing of 75Hz leading to a signal bandwidth of 2400Hz. This bandwidth could, ideally, support a data rate of up to 4800bit/s using BPSK, although for any typical implementation 2400bit/s would be more realistic. But, any multipath delays of more than a few tens of microseconds would damage the demodulation process. If the 32 tones were modulated at a data rate sufficient for the sidebands not to spread into the adjacent tones, which might be around 50bits/s each, then the net data rate becomes 32 * 50 = 1600bit/s which is not quite as good as a single tone scheme but much more multipath-resistant. Such parallel tone modulation types have been used for military communications for many years now.

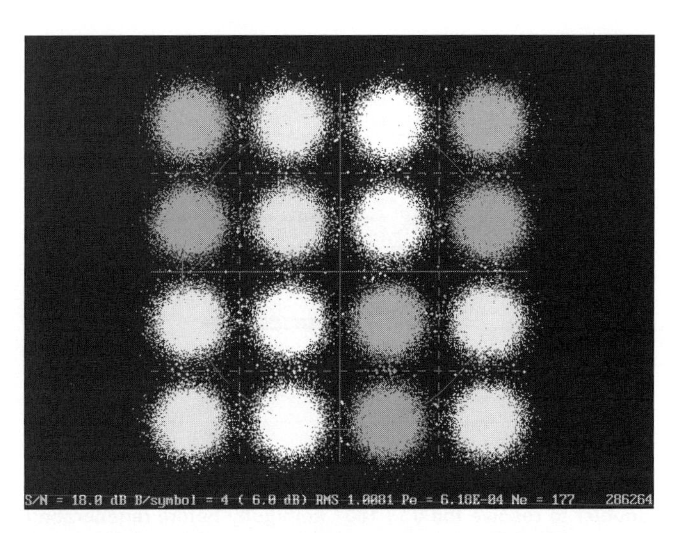


Fig 20.10: Simulation of a 16-QAM signal with 18dB S/N

DSP techniques allow this concept to be pushed further. By synchronising the data switching with tone spacing, it is possible to ensure that although the sidebands from each modulated carrier overlap they do so in a controlled way that prevents interference. The 32 tone example will now be modulated at 75 symbols/s, leading to a net data rate of 2400bit/s - exactly that possible with a single tone. Synchronising the tone spacing with symbol rate to avoid mutual interference leads to the name Orthogonal Frequency Division Multiplex OFDM - which is now one of the most widespread data modulations, being used for terrestrial and cable digital television transmission, DAB radio,

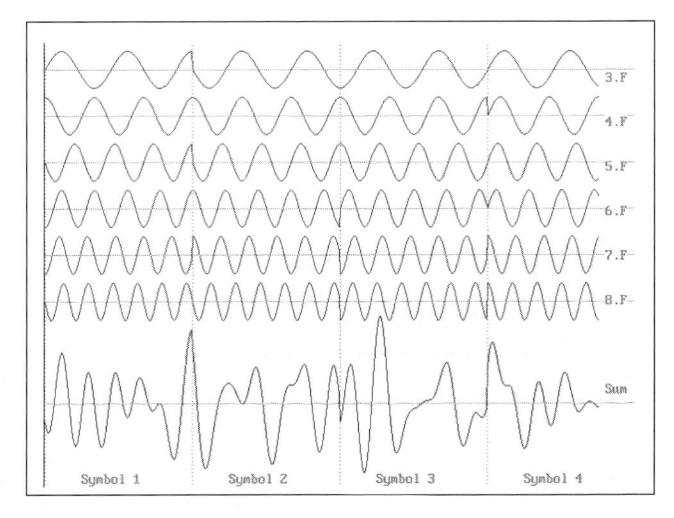


Fig 20.11: The problem of summing QPSK modulated carriers

Digital Shortwave Broadcasting (DRM) some military communications, 3G mobile phones and Wimax Internet. It is just beginning to make inroads to amateur radio.

A downside to OFDM is that a highly linear transmitter is required, so the resulting power efficiency is not very good as PAs have to be run considerably backed-off. Where many frequencies are sent in parallel, at certain times their amplitudes can add up coherently (voltages add) and a brief power output many times the mean level can be seen. **Fig 20.11** illustrates this, where six QPSK modulated carriers are shown with their resultant voltage summation. The problem can be lessened by staggering the start phases in a pseudo-random manner to reduce linearity requirement.

## TRAFFIC OR TYPE OF COMMUNICATION

### Keyboard to Keyboard Modes

This is the simplest type of amateur data communications and one of the most 'friendly' in operation. RTTY and PSK31 are popular examples of Keyboard to Keyboard modes where everything that is typed on the sender's keyboard is transmitted directly over the air and printed or shown character by character at the receiving station. Transmitted codes will often include control characters such as backspace, and it can be quite intriguing to watch a distant typist correct a spelling mistake by backspacing and retyping.

Keyboard to Keyboard modes often do not include any error correction, so are prone to interference or bit errors which result in corrupted characters and garbled text. If interference is slight, such as just the occasional burst, the meaning of garbled text can often be inferred from the surrounding characters or words. Attempts have been made in some schemes to add in Forward Error Correction (FEC), but this nearly always leads to delays in timing, making quick break-in (or Tx/Rx changeover) time consuming, so FEC is rarely used.

### File Transfer

In contrast to keyboard-keyboard modes, transferring data files requires that absolutely no errors are to be introduced, so a protocol has to used to check the validity of the received data. There are two ways of doing this, and both are often used together. Forward Error Correction adds redundancy to the transmitted data, so that if a few bits are corrupted on transmission, there is enough additional data to be able to regenerate the missing data.

Very strong FEC is needed for one way error free traffic, and in practice data integrity can never be guaranteed so handshake operation is usually employed. Here the transmitting station appends a checksum to short bursts of data. The receiver checks each burst with the checksum, and transmits an acknowledgement back to say either “send again” or “OK. Next please”. Protocols have to be developed to allow for lost bursts, and lost or corrupted acknowledgements, and are referred to as Automatic ReQuest (ARQ) protocols.

AMTOR (AMateur Teleprinter Over Radio), known as SITOR in the commercial world, adds simple error checking to teletype transmissions with an ARQ protocol for greatly reducing errors on short messages.

The latest generation of File Transfer modes may incorporate Internet Protocol (IP) type messages for compatibility with networking or Internet connection.

### Visual Schemes - or Fuzzy Modes

These are a sort of half-way-house between data transmission and slow scan television. Signals are sent in a form that can be displayed on a PC screen and read by visual inspection. Frequency or amplitude (usually on/off) modulation can be used, and there is at least one case of a phase shift visual modes available in software

SlowCW or QRS(S) is the simplest of these visual schemes. A carrier is switched on or off using normal Morse code symbols, but slowed down to such an extent that a single dot interval can take anything from one second to minutes. Digital Signal Processing, uses the FFT routine to filter the signal to a narrow bandwidth commensurate with the dot interval and display the result on a waterfall plot. This type of plot shows frequency along one axis (usually the vertical one) with time is along the other. Signal intensity is shown by the colour or brightness of the display. The advantage of a waterfall plot is that a band of frequencies can be observed simultaneously, and several QRSS transmissions can be decoded at the same time

The first instance of QRSS being used on the amateur bands was in 1997 when G4JNT used it to transmit on 73kHz to G3PLX at a distance of 393km. The dot interval used for that transmission was 40 seconds, and the complete message took 3 hours to send. **Fig 20.12** shows a plot of the received message at G3PLX.

DFCW, or Dual Frequency CW is an alternative to QRSS. Instead of 'dashes' being coded as a transmission three times longer than a 'dot', they are sent for the same duration as a dot, but on a slightly different frequency. Also, since each symbol is now of an equal duration, intersymbol gaps are redundant and can be removed. Sometimes, to aid readability, they are kept but made shorter. DFCW typically shortens messages by about 20 - 40 percent.

DFCWi (the I is for 'idle') carries the process a stage further and introduces a third tone for the intersymbol and, more importantly, the inter letter and word gaps. In noisy conditions a long series of symbols can get broken up, and give the appearance of multiple words. The third idle tone shows an operator if there really is a gap there, or if the signal has faded out, in which case the message contents can be inferred more easily than if DFCW were used. DFCWi transmissions often have a 100 percent duty cycle, although small intersymbol off periods can be inserted to aid readability. **Fig 20.13** Shows the DFCWi received from the 10GHz beacon GB3SCX over a rain-scattered path.

Software for generating DFCW can be obtained from [1] and for DFCWi [2] More on using QRSS and DFCW can be found in the Morse chapter.

Hellschreiber is a visual scheme where individual letters are sent to be read off a screen. Traditional Hellschreiber (often abbreviated to just Hell) uses rapid on-off keying with a raster scanned display such as that shown in **Fig 20.14.** Later versions adopted frequency shifting so that the result can be read on a waterfall display. There are two types of FSK Hell. Sequential Multitone (SMT) Hell transmits the vertical components making up the pixels of each letter sequentially so only one tone is transmitted at a time. This allows a high efficiency non-linear transmitter to be employed, often the case on the LF bands, but means the letters will be shown with a slant to them. Suitable choice of

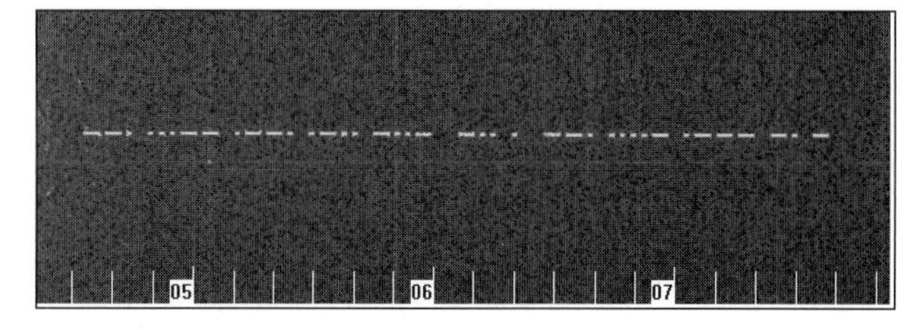


**Fig 20.12: The first ever QRSS reception was on the old 73kHz band. The time axis shows that the transmission took three hours to complete**

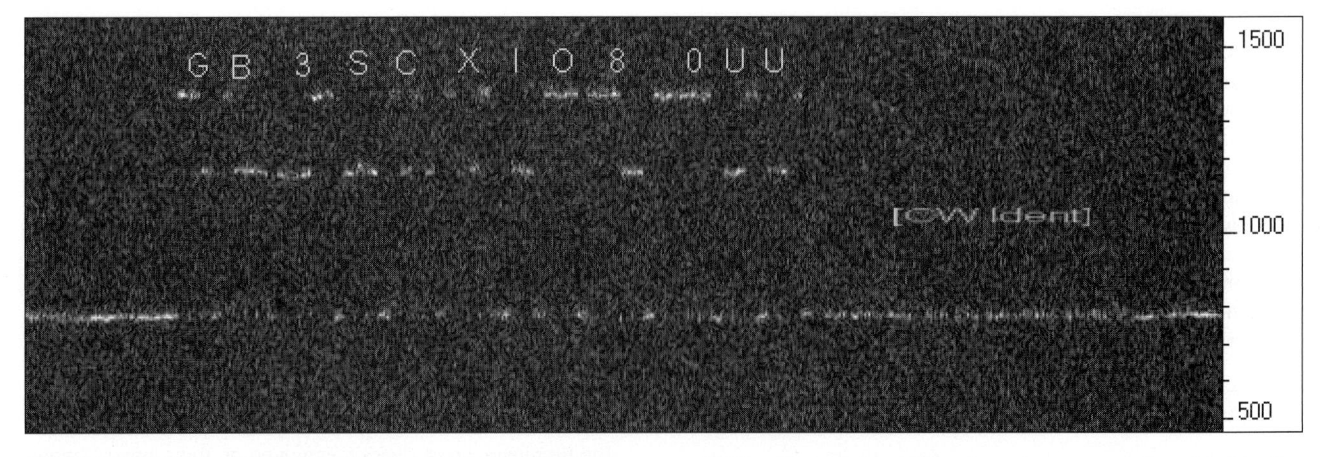


**Fig 20.13: Weak signal DFCWi reception of the GB3SCX beacon on 10GHz (the callsign letters have been manually added to the picture for clarity). Dashes are on the top line, dots on the middle and the idle state on the bottom**

font can make the result quite acceptable and in practice one of the best fonts for SMT-Hell is a simple 5x7 dot-matrix such as used on the earliest computer printers and screens.

Parallel tone MT Hell generates all the tones corresponding to each vertical element making up each letter simultaneously, so a linear transmitter is required. This gives a more pleasing and possibly easier to read display, but is generally constrained to HF where linear SSB transmitters are the norm. Software for generating and receiving all the Hell modes can be found at [3].

Modern Slow Scan TV is nowadays usually transmitted digitally, and is often included within the integrated data communications software packages. More details on SSTV can be seen in the chapter on visual modes.

## MODERN DATA MODES

With the ready availability of powerful and fast DSP using PCs and soundcards, whole new families of advanced data modes have been designed. Some of these are just extensions of straightforward ASK / FSK or PSK with error correction, while others have been designed with specific target applications.

One aspect of any new datamode that has to be considered is that of matching it to the channel characteristics. This has to involve considering the bandwidth of the signal, the level and duration of multipath, any frequency scattering or Doppler effects that may be encountered, and if transmissions need to be in short bursts, or continuous.

Linear transmitters as used at HF and for weak signal working at VHF/UHF allow waveforms with an amplitude component such as parallel tone signalling. This usually allows optimum bandwidth and S/N to be considered. Where non linear transmitters are in use, modulation is restricted to single tone constant amplitude (although on-off is usually possible). This might be the case at LF and also on the microwave bands where the transmitted signal is generated by multiplying up from a lower frequency source. A few examples are shown in **Table 20.1**.

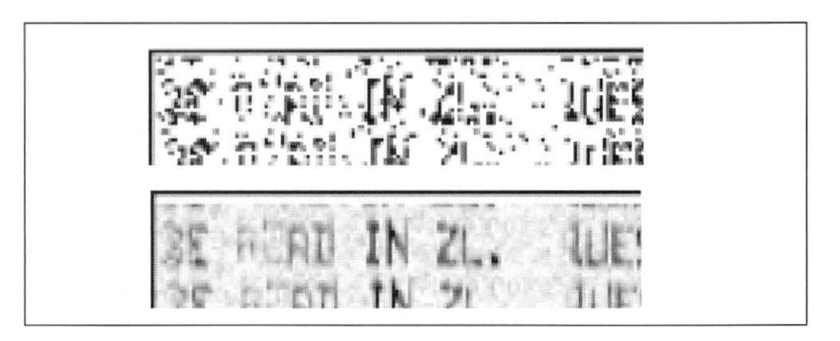

**Fig 20.14: Hellschreiber in black and white (top) and greyscale (bottom)**

## ERROR CORRECTION & ITS IMPLICATIONS

Forward Error Correction can be strong or weak, or roughly equivalent to soft or hard. Strong FEC can expand the data to such an extent that six times as many bits are transmitted as actually make up the message; weak FEC may just double the number of bits sent. Error correction can be a complex mathe-

| | |
|---|---|
| **LF 136kHz - 2MHz** | Multipath quite low - if any usually due to a single ionospheric hop. Low bandwidths, due to narrow allocations, typically a few Hertz or tens of Hertz maximum. Often non linear transmitters are used for high power / efficiency.<br>*ASK, FSK, (including MFSK) and PSK are usual* |
| **HF Skywave 3 - 50MHz** | Bad multipath - up to several milliseconds. Deep fading of several seconds. Bandwidths typically a few hundred Hertz to a 3kHz SSB channel width. ionospheric Doppler shifts of a few Hertz.<br>*FSK, PSK, parallel tone are usual, with extended interleaving to cope with fades* |
| **V/UHF 144MHz - 2GHz** | Severe multipath from multiple reflections, typically tens to hundreds of microseconds. Doppler shifts if communicating with moving vehicles. Bandwidths either SSB 3kHz for weak signals / DX work, or based on an FM channel of 12.5 / 25kHz for local working. Meteor Scatter requires short bursts of fast data. Moonbounce introduces scattering, and works with very weak signals over extended durations.<br>*FSK/MFSK , PSK, MSK are all used* |
| **Microwaves 3GHz and up** | Massive multipath from moving objects leads to scattering. Very high levels of spectral spreading due to troposcatter and rain scatter Bandwidth not a major issue, but for weak signals, a 3kHz SSB channel is preferred.<br>*FSK / MFSK with a shift matching the spread is the only practical modulation for badly scattered paths* |

**Table 20.1: Matching data modes to the radio frequency medium**

matical process and many of the better FEC schemes can rarely be described without recourse to a thorough understanding of the maths behind the process. FEC always introduces a delay between transmission and decoding, and on the strongest schemes this delay can take up the complete duration of the message if data has been interleaved over its whole length.

Soft error correction makes an attempt to correct as many bits as possible with limited redundancy, presenting the best it can do to turn what would otherwise be garbled text into something readable. Although some errors may still be present, hopefully these will have been reduced enough to make it possible to infer or guess the message contents. Hard error detection involves such a high degree of processing that the received message is either perfect or nothing. Occasionally a badly garbled message, or impulsive noise, can be decoded and a message presented as valid, but the contents are usually so ridiculous that it is quite clear a false decode has occurred.

Examples of both are seen in amateur service. Soft FEC can be seen in QPSK31 (see below) where the majority of errors due to impulsive type interference can be coped-with. Hard FEC with deep source coding is encountered in some of the WSJT modes for very weak signalling like EME.

Source coding is the technique of reducing the unwanted redundancy in transmitted data to lower the number of bits required for a given message. The varicode alphabet used in PSK31 and a few other amateur modes is an example, where fewer bits are used for the common letters in normal text, such as 'e' and 't', with corresponding more for 'z', 'q' and 'j'. More complex source coding can compress amateur callsigns to a few tens of bits making use of their known structure of letters/numbers only, as well as locators and other numeric information.

A downside to source coding it that it removes the possibility to make use of contextual correction or guesswork on the part of the receiving operator. So source coding tends to be used with high levels of usually strong error correction. This all leads to a big trade off; is it better to heavily source code then expand with strong FEC, or leave in the natural redundancy use soft or no FEC and let the operator make the best of the result? Examples of all these can be found in modern amateur data communications.

Soft Decision Decoding is often included as part of the FEC process. Traditionally a '1' or '0' is decided at the receiver by determining if the signal lies on one side or the other of a threshold, such as amplitude, frequency etc. But, in a fully DSP implemented modem, more information is available to the decoder than a single threshold, and the actual level of the signal and noise is known. A soft decision decoder can make use of this information, in conjunction with a particular type of signal encoding adding redundancy and interleaving known as convolutional coding, to make a best-guess at what was most likely to have been transmitted. Viterbi soft-decoding of convolutionally encoded signals can give one of the strongest FEC schemes for continuous data there is, and when combined with additional block error correction on the source data (for example Reed Solomon error correction) can lead to a completely all-or nothing solution.

Combined strong error correction schemes are seen commercially for Digital TV broadcasting. Amateur radio applications can be found in the WSJT WSPR and JT65 modes.

## READY-TO-GO SOUNDCARD BASED DATAMODES

Several software packages exist that allow nearly all of the datamodes currently in use to be implemented with a common soundcard interface. Many of them are free public domain software and can be downloaded from the Internet and installed. Others offer a limited free trial period then ask for payment. Some offer additional facilities once payment has been sent. Some of the most popular packages are shown in **Table 20.2**. There are many others that are specific to particular modes, they are described in the relevant section.

## RTTY

The oldest and most well established keyboard to keyboard mode it uses FSK with a shift of, usually, 170Hz in the amateur service. A five bit code with letters/numbers shift, allows only a restricted alphabet of upper case letters, numbers and punctuation. Most operation at HF is at a speed of 45.45 Baud (symbol period 22ms), although at VHF 50 Baud is often encountered. At UHF and above, higher shifts may be needed to cope with frequency scattered signals, and at 10GHz 850Hz shift 50 baud RTTY was used for a while on the GB3SCX beacon. The use of a letters/numbers shift makes the system quite error prone as if the shift character is lost; whole sentences can be garbled. A number of software packages can be used for RTTY, including *MMTTY*, *MultiPSK* and *FlDigi*. MMTTY allows operator intervention to repair corruption to the letter/number shift and apply this to whole sections of garbled received text.

## AMTOR

This is a derivation of RTTY that introduces error correction by ARQ. Signalling speed is increased to 100 Baud, and characters are combined into groups of three at a time for transmission.

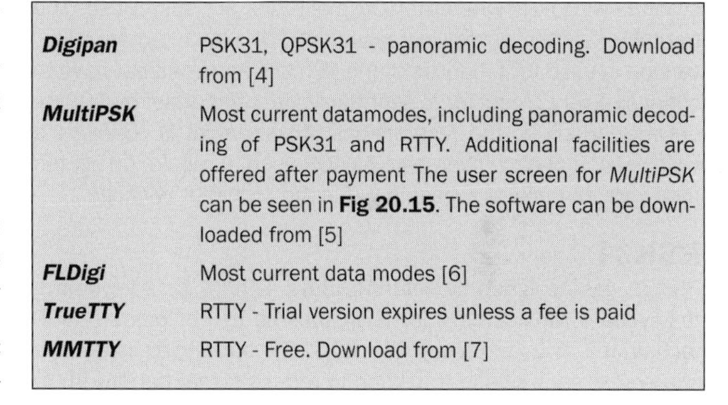

| | |
|---|---|
| ***Digipan*** | PSK31, QPSK31 - panoramic decoding. Download from [4] |
| ***MultiPSK*** | Most current datamodes, including panoramic decoding of PSK31 and RTTY. Additional facilities are offered after payment The user screen for *MultiPSK* can be seen in **Fig 20.15**. The software can be downloaded from [5] |
| ***FLDigi*** | Most current data modes [6] |
| ***TrueTTY*** | RTTY - Trial version expires unless a fee is paid |
| ***MMTTY*** | RTTY - Free. Download from [7] |

**Table 20.2: Examples of multimode data software**

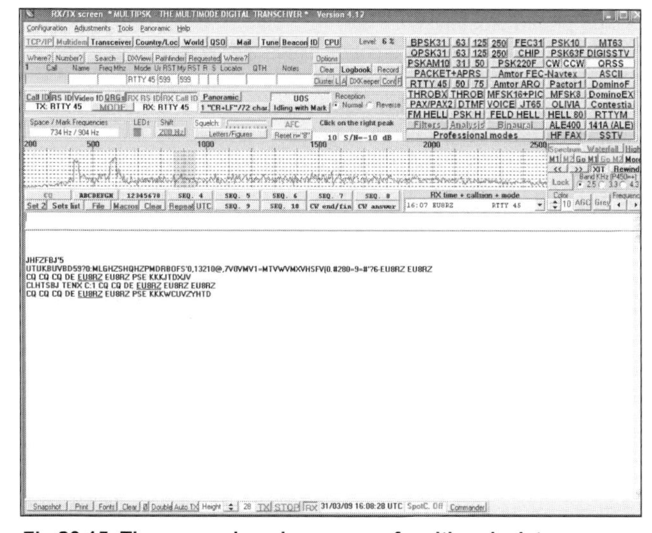

**Fig 20.15: The comprehensive screen of multimode data program *MultiPSK***

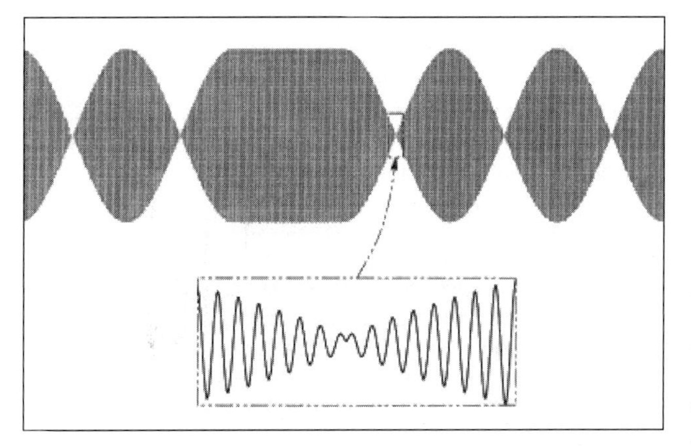

Fig 20.16: A typical BPSK symbol showing the amplitude reduction at crossover

The five-bit code is expanded to seven bits to introduce error detection (but not correction) by only using those seven-bit codes that have exactly three ones (there are 35 of these out of the possible total of 128). On receive, if the number of bits in any character is not exactly three, an error is assumed and the receiving station requests a retransmission. The error correction / ARQ is not perfect, but does result in a mostly error free connection. AMTOR is a connected mode where one station has to connect to another to set up a link.

It is a handshaking mode and needs continuous, rapid receive/ transmit switching. Its regular chirp-chirp can often be heard around 14100kHz where AMTOR mailboxes are still in use. The frequency shift is usually increased to 200Hz. An alternative FEC version is used for broadcasts and CQ calls and does not have to connect. Every character is sent duplicated and offset by 200ms between them, so the receiver has two attempts at receiving a valid version of each character. AMTOR used to require dedicated hardware, but now is available within the *MultiPSK* package.

## PSK31

PSK31 was designed by Peter Martinez, G3PLX ,as a keyboard-to-keyboard replacement for RTTY making use of modern DSP techniques. It uses Binary PSK at 31.25 baud with amplitude shaping at each phase transition to reduce signal bandwidth to the absolute minimum possible and results in a bandwidth of just 32Hz. A typical PSK31 symbol can be seen in **Fig 20.16** showing the amplitude detail at the phase crossover.

The full set of 256 ASCII characters is available, but they have been re-coded into a variable length alphabet called Varicode. Here, the most popular characters (assuming lower case English language plain text) such as '[space]' 'e' and 't' have been allocated codes with a lower bit count than the less common letters like 'z' and 'x'.

The narrow bandwidth and inherently better signalling efficiency of coherent PSK over non-coherent RTTY means that PSK31 will work at far weaker signal strengths than RTTY can. In practice, it offers around 10 - 13dB improvement over 170Hz shift 45 baud RTTY. Also, something like five PSK31 stations can typically occupy the spectrum taken up by just one RTTY signal

An experimental version of PSK31 introduced FEC with convolutional coding on a QPSK waveform, produced by G3PLX at the same time. The FEC was aimed at improving the performance in the presence of interference bursts. The change to QPSK from BPSK introduced a 3dB S/N penalty so some of the advantages to the 'mild FEC' were lost. The narrowband signal can get corrupted when more than a few Hertz of signal scattering is present - such as on trans-polar paths.

Subsequently, other writers modified PSK31 to work at both lower and higher data rates. A quarter speed version, PSK08 was tried on 73 and 137kHz, and PSK62 and PSK125 have been designed for faster hand typed operation when conditions are good, or to cope with scattered signal paths.

Nearly all datamode software packages include PSK31 and QPSK31; many packages also include its later variants. Most popular is *Digipan* which provides a spectral display of up to 3kHz of spectrum and decodes all the PSK31 signal in this band up to a maximum of around 32 different stations. A simple point and click process allows the operator to reply to any one of the multiple decodes, and selects the transmit tone frequency appropriately. **Fig 20.17** shows the *Digipan* screen in Panoramic Mode.

*MultiPSK* also offers this panoramic feature; Most other datamode packages just allow one signal at a time to be included

## OLIVIA

This is a teletype protocol that transmits a stream of ASCII (7-bit) characters, sent in blocks of five at a time. Each block takes two seconds to transmit, giving an effective data rate of 2.5 character/second or equivalent to about 25bits/s. It is a Multi-FSK mode with the default mode being 32 tones within a 1000Hz audio bandwidth. The tones are spaced by 1000Hz / 32 = 31.25Hz and their amplitude is shaped to minimise the amount of energy sent outside the nominal bandwidth. The baud rate is 31.25 MFSK tones/second. To accommodate for different conditions and for the purpose of experimentation, the bandwidth and the baud rate can be changed.

Strong Forward Error Correction is built in by the use of Walsh functions so decoded text is of the all-or-nothing nature within the structure of the two-second five-character blocks.

## THROB

THROB is a DSP sound card mode that uses Fast Fourier Transform technology to decode a five tone signal. The THROB program has been described as an attempt to push DSP into the area where other methods fail because of sensitivity or

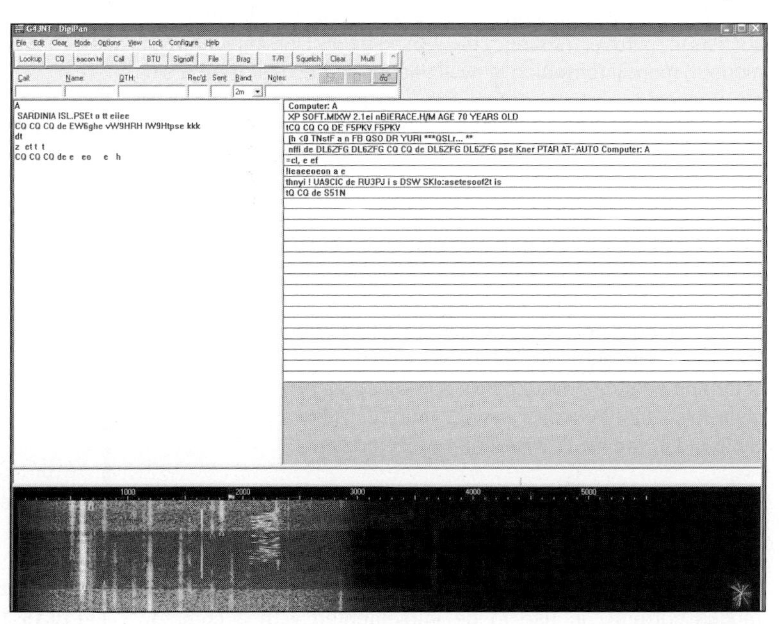

Fig 20.17: The *Digipan* screen

propagation difficulties and at the same time work at a reasonable speed. The text speed is slower than other modes and runs from a dedicated software package or as one of the modes within *MultiPSK* and *FlDigi* The name comes from the sound the signal makes - a slow throbbing tone

## MFSK16 / MFSK8

MFSK16 is an advancement on the THROB mode and encodes 16 tones with Constant Phase Frequency Shift Keying to minimise bandwidth. Continuous FEC sends all data twice with an interleaving technique to reduce errors from impulse noise and static crashes. A new improved Varicode is used to increase the efficiency of sending extended ASCII characters, making it possible to transfer short data files between stations under fair to good conditions. The relatively wide bandwidth of 316Hz for this mode allows faster baud rates with a typing speed equivalent to about 42WPM, and greater immunity to multipath interference. A second version called MFSK8 is available with a lower baud rate (8) but greater reliability for DXing when trans-polar interference is a major problem. Both versions are available in a package written by IZ8BLY, and also as options in *MultiPSK* and *FlDigi*.

## MT63

This is a robust signalling scheme to keyboard text over paths that experience fading and interference from other signals. It is accomplished by a complex scheme to encode text in a matrix of 64 tones over time and frequency. This overkill method provides a 'cushion' of error correction at the receiving end while still providing a 100WPM rate. The wide bandwidth of 1kHz for the standard version makes this mode less desirable on crowded bands such as 14MHz. MT63 is available in *MultiPSK* and *FlDigi*.

Other datamodes can be found within some of the multi-purpose packages. In many cases these are variants of those described, operating at different speeds / bandwidths or with different error correction. Some are adaptations for specific purposes, like digital Slow Scan TV or image transfer. Such applications are specific to the software package employed, for instance *MultiPSK* .

## ROS

A relatively new mode ROS is named after its designer José Ros. it is described as a "Digital Spread Spectrum Mode, but with a narrow total occupied bandwidth of about 2.2kHz". The first contact using ROS took place on 18 Feb 2010 from Vitoria in Spain to the University of Twente in the Netherlands, covering a distance of 1265Km on 7.065 MHz.

ROS can be downloaded free of charge from [8]. Some limited documentation is supplied as a PDF file. Not much technical information is provided about the internals of the mode but ROS is a 16-tone MFSK waveform arranged so the frequency is stepped pseudo-randomly over the occupied 2200Hz bandwidth to reduce the effects of interference.

At first launch two rates were supported, 16 Baud and 1 Baud to cope with different band conditions and signal strengths. The decoding software automatically synchronises to any symbol rate. Subsequently lower rate and bandwidth options were added, especially to cater for LF signalling. The latest version of the software now offers the option of I/Q stereo output mode for directly driving direct quadrature upconverters.

The relatively wide bandwidth and low data rate means that ROS offers weak signal advantages over many current established HF datamodes. In some cases signals that are so weak they don't show up on the waterfall display can be correctly decoded. The ROS User screen is shown in **Fig 20.18**.

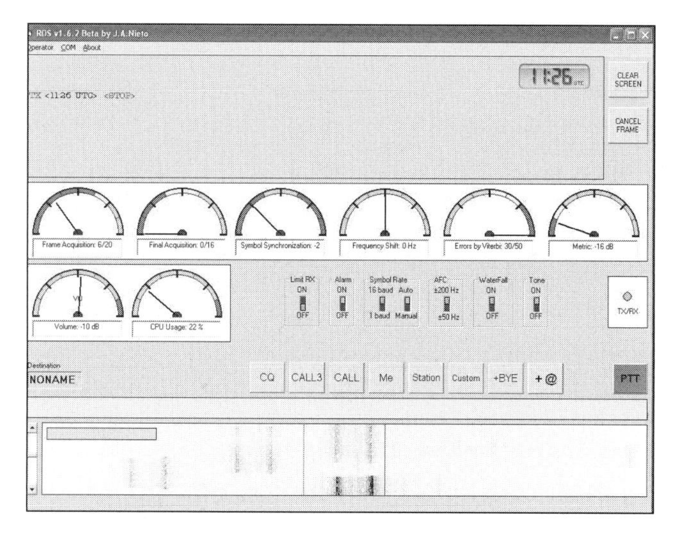


Fig 20.18: The ROS user screen

## WIDEBAND DATA MODES

ROS is a typical example of modern HF digital communications practice in that it spreads the signalling energy well outside the necessary bandwidth needed for communication. Necessary bandwidth is equal to or higher than the symbol rate; imagine a continuous 101010 . . .. pattern being sent, the alternate high and low parts of the waveform, after filtering, each constitute a half cycle of a sine wave.

As a rule of thumb necessary bandwidth is often taken as being equal to the symbol rate - for the standard ROS mode, it is 16 baud or 16Hz. For multi-symbol modes like MFSK or QPSK, several bits are sent for each symbol so the data rate is proportionately higher.

In environments like HF where multipath, scattering, Doppler shift and interference are prevalent the transmitted signal is usually expanded over a significantly wider bandwidth than necessary to minimise the damage done by the interference mechanisms. These can often be observed as the classic HF fading, often noticeable as a deep audio null drifting through an SSB or AM voice channel over a period of a few seconds. By spreading out the data out over the bandwidth occupied by a typical voice signal, 2 - 3kHz, the effects of the moving null are reduced.

There are many different ways of spreading the energy. Single tone modes like Stanag 4285 / MIL-STD -110A waveforms use an underlying 2400 baud symbol rate modulated with 8 PSK (3 bits per symbol) then use massive redundancy, error correction and interleaving to encode a lower data rate robustly. Other waveforms adopt a lower symbol rate, but hop a single tone over the band in different ways.

ROS uses a pseudo random spreading code with an underlying simple 16 FSK modulation. Contrast with K1JT's JT65 code in *WSJT*, where 65 tone slots are directly used. Reed Solomon error correction is directly applied to the tones to be able to cope with several being lost in QRM. Modern schemes such as OFDM transmit thousands of tones simultaneously, each one modulated at a low rate with the total being the combination of all individual data streams summed. All these techniques appear on the amateur bands, at some times.

## Social Issues

Commercial and military users usually allocate their frequencies in 3kHz blocks corresponding to a single voice channel. With a guaranteed clear channel, all the advantages of the relatively

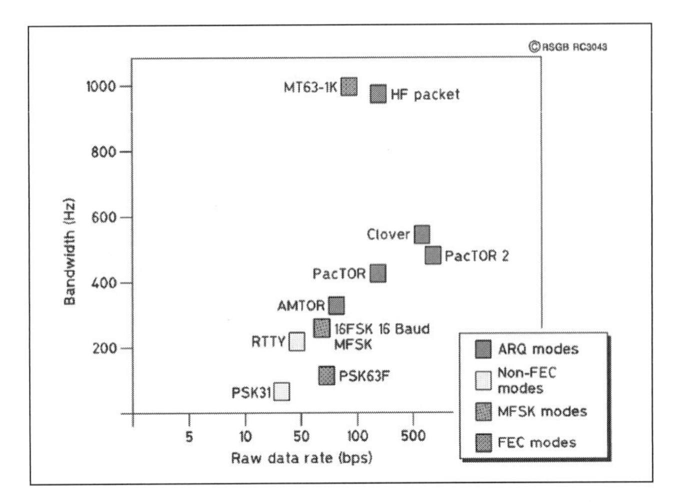


**Fig 20.19: Comparison of the performance and speed of various data modes [BARTG]**

wide band can be used for robust low data rate communications, often with the overall data throughput being adjusted automatically in response to band conditions. Stanag 4285 automatically adjusts over the range 4800 down to 75 Bits/second as propagation alters. On a crowded amateur band, with many different and incompatible modes each vying for space, a plethora of different wideband energy spreading modes doesn't work. If users try to stay in the segments allocated for data modes then narrowband modes like PSK31, RTTY and MultiPSK have to fight for their patch with the wideband modes obliterating several of the narrow ones.

Ideally, experimenters wanting to use the full 3kHz bandwidth for their data should really use the SSB part of the band but problems can occur here as well - not least SSB operators objecting to data transmission in 'their' segments and the all-too-frequent heavy occupancy of this part of the band during contests.

So, while wideband datamodes have their place in the experimental scheme of things in amateur radio, and offer very clear advantages in communications reliability over their narrowband cousins, their place on the narrow and mostly de-regulated amateur bands really needs to be considered carefully. And should a new wideband mode be suddenly released for free download, knowing many operators will immediately want to try it out without fully appreciating the implications for interference.

## MODES COMPARED

A summary of performance and speed of many of the amateur data modes was printed in the BARTG magazine, and is reproduced in **Fig 20.19**.

## LOW BANDWIDTH SLOW DATAMODES FOR LF/MF

Operation on the LF and MF bands (137 and 500kHz) has generated a few datamodes customised to the requirement for narrow bandwidth extremely weak operation. LF propagation is usually stable with little ionospheric scattering. These are specialised, and usually need their own dedicated software for operation rather than being included as options within the multi-mode software suites.

### Weak-signal Operation on LF (WOLF)

WOLF was written by Stewart Nelson, KK7KA, with a Windows graphical user interface by Wolfgang Büscher, DL4YHF. Wolf stands for Weak-signal Operation on Low Frequency, and can operate over a wide range of signal levels.

The original WOLF implementation by KK7KA sends ten symbols per second using BPSK at 10bits/s. Five of these symbols carry data bits, the other five bits contain a pseudo-random sequence for synchronisation. In the WOLF GUI, this original WOLF speed is called 'normal' speed. A few slower and faster variants were subsequently added. The slower variants may help with extremely weak signals and the WOLF GUI [9] offers:

- Normal speed: 10 symbols per second . One frame = 96 seconds; copy may be possible after 24 seconds
- Half speed: 5 symbols per second. One frame = 192 seconds = 3' 12"
- Quarter speed: 2.5 symbols per second. One frame = 384 seconds = 6' 24"
- Double speed: 20 symbols per second ("Fast"). One frame = 48 seconds
- Fourfold speed: 40 symbols per second ("Turbo"). One frame = 24 seconds

The example shown in **Fig 20.20** is the first trans-Atlantic signal received on the now-defunct 73kHz band. It illustrates how a weak signal is built up after many iterations:

Using 10 Bit/second BPSK with no amplitude shaping of the waveform, WOLF is a rather 'antisocial' modulation to use on narrow LF bands and its use has fallen off, being replaced with more modern soundcard modes.

### Jason

This is a keyboard-to-keyboard communication program, tailored for very low S/N ratios using only 4Hz of band. The coding scheme is based on the ideas about Incremental Frequency Keying, initially proposed by Steve Olney, VK2ZTO, where the information is coded in the absolute value of the difference between two frequencies sent sequentially. This has the advantage of not needing a precise initial tuning and a tuning error of a few Hertz is perfectly acceptable. Another characteristic is that as the frequencies are sent one at a time there is no need for a linear amplifier.

**Signal received by John Andrews, W1TAG, on 19 March 2001 at Worcester, Massachusetts**

```
C:\wolf>wolf -f 799.892 -r 8001.95 -t 0.02 -q bmu2.wav -s 1200
WOLF version 0.51
t:  24 f:-0.020 a: 1.0 dp: 99.4 ci:11 cj:391 94R.7A8A4???UWC ?
t:  48 f:-0.029 a:-1.5 dp:103.7 ci: 5 cj: 12 6A.S69BGZ//LK8B ?
t:  96 f: 0.019 a:-1.4 dp: 98.7 ci: 4 cj:207 JFWWUL???N*E .Y ?
t: 192 f: 0.029 pm: 118 jm:119                LQ5HQ*2G569R2MW ?
t: 288 f:-0.029 pm: 150 jm: 52                /S SYS7FZSV XXZ -
t: 384 f:-0.029 pm: 175 jm: 52                ??????AWI2N 20Y ?
t: 480 f:-0.029 pm: 195 jm: 52                /KR/F 2U4X8ZSMT -
t: 576 f:-0.029 pm: 198 jm: 52                VMEVXXDPY2J4RTJ ?
t: 672 f:-0.029 pm: 205 jm: 52                .L5DKCA9GEDS.AX -
t: 768 f:-0.029 pm: 226 jm: 52                3J0FRV/XG6S7D7X ?
t: 864 f:-0.029 pm: 238 jm: 52                RJSWH TQU4IJ6NS ?
t: 960 f:-0.029 pm: 256 jm: 52                PE41I9K3DDS9ZHM -
t:1056 f:-0.010 pm: 321 jm:211                H96XY075JF4B YU ?
t:1152 f:-0.010 pm: 357 jm:211                Q*CW RQAU447BNT ?
t:1248 f:-0.010 pm: 447 jm:211                6D 6FLYJTCVY0.N ?
t:1344 f:-0.010 pm: 472 jm:211                Q*C*GBR88N0/5*T ?
t:1440 f:-0.010 pm: 482 jm:211                CQ M0BMU M0BMU  -
t:1536 f:-0.010 pm: 499 jm:211                CQ M0BMU M0BMU  -
```

**Fig 20.20: Decoded text from the first WOLF transatlantic transmission, showing how the text is extracted by coherently integrating 18 seconds blocks of received data**

The frequency deltas can assume one of 16 different values. After sending one tone, the next is shifted by the appropriate amount, up or down depending on the setting of the USB/LSB switch. With 16 deltas there are 17 tones, and any overflow causes a wraparound.

With 16 possible deltas, each baud (change in frequency) encodes four bits (a nibble) at a time which are not enough for a reasonable alphabet, so each character takes two nibbles to send. But now a problem arises: how to get character synchronisation? In other words, which is the high-order and which is the low-order nibble?

The high order bit of each nibble is used to encode this info meaning the high-order nibble is of the form '1xxx'b, while '0xxx'b is the low-order one with xxx standing for the actual information transmitted. So now there are 6 bits available to encode our alphabet, enough for a shortened 64 character alphabet of ASCII code goes from 0x20 (the blank) to 0x5F allowing the transmission of all upper case letters, the ten digits, and practically all the punctuation symbols normally used.

A dedicated software package, Jason [10] is needed for transmission and reception.

## Weak Signal QSO Mode (WSQ)

A new development by ZL2AFP, and ZL1BPU, is WSQ, or Weak Signal QSO mode. Designed specifically for hand typed two way contacts on LF, as opposed to mainly beacon type operation as offered by WSPR and Opera, WSQ uses differential Multi Frequency Shift keying, where information is encoded by the change of frequency from one tone to the next in a similar way to the older Jason mode (see above). This use of incremental keying makes the mode more tolerant of tuning errors and slow drift. Tone spacing is 2000/1024 or approximately 1.95Hz and 32 tones are used, giving signal frequency span of about 65Hz

Variable length coding is used whereby a character is encoded by transmitting either one or two sequential tones. A single tone is used for lower case letters and the commonly used symbols of space, carriage return, and full-stop. For the less common letters and symbols, two successive tones are sent. **Table 20.3** lists the character / tone symbols. For each symbol sent, the value for VAR is added to the value of the previous tone number, and then incremented by one. If the result is greater than 32, then 32 is subtracted for a final tone number in the range 0 to 31. The increment by one means that when no data is being transmitted the tone is slowly changing. Expressed mathematically, $F_N = (F_N\text{-}1 + V + 1)$ MOD 32 where $F_N$ and $F_N\text{-}1$ are the current and previous tone frequencies, and V is the increment value taken from the table.

There is no error correction as this would greatly slow down real time operations. The use of differential, or incremental frequency encoding coupled with a symbol rate of 2.048 seconds reduces the S/N needed for reliable copy. The WSQ software with full documentation can be downloaded from [11].

| Char | ASCII | Var | Char | ASCII | Var |
|---|---|---|---|---|---|
| SPACE | 32 | 0 | T | 84 | 20, 29 |
| ! | 33 | 11, 30 | U | 85 | 21, 29 |
| " | 34 | 12, 30 | V | 86 | 22, 29 |
| # | 35 | 13, 30 | W | 87 | 23, 29 |
| $ | 36 | 14, 30 | X | 88 | 24, 29 |
| % | 37 | 15, 30 | Y | 89 | 25, 29 |
| & | 38 | 16, 30 | Z | 90 | 26, 29 |
| ' | 39 | 17, 30 | [ | 91 | 1, 31 |
| ( | 40 | 18, 30 | \ | 92 | 2, 31 |
| ) | 41 | 19, 30 | ] | 93 | 3, 31 |
| * | 42 | 20, 30 | ^ | 94 | 4, 31 |
| + | 43 | 21, 30 | _ | 95 | 5, 31 |
| , | 44 | 27, 29 | ` | 96 | 9, 31 |
| - | 45 | 22, 30 | a | 97 | 1 |
| . | 46 | 27 | b | 98 | 2 |
| / | 47 | 23, 30 | c | 99 | 3 |
| 0 | 48 | 10, 30 | d | 100 | 4 |
| 1 | 49 | 1, 30 | e | 101 | 5 |
| 2 | 50 | 2, 30 | f | 102 | 6 |
| 3 | 51 | 3, 30 | g | 103 | 7 |
| 4 | 52 | 4, 30 | h | 104 | 8 |
| 5 | 53 | 5, 30 | i | 105 | 9 |
| 6 | 54 | 6, 30 | j | 106 | 10 |
| 7 | 55 | 7, 30 | k | 107 | 11 |
| 8 | 56 | 8, 30 | l | 108 | 12 |
| 9 | 57 | 9, 30 | m | 109 | 13 |
| : | 58 | 24, 30 | n | 110 | 14 |
| ; | 59 | 25, 30 | o | 111 | 15 |
| < | 60 | 26, 30 | p | 112 | 16 |
| = | 61 | 0, 31 | q | 113 | 17 |
| > | 62 | 27, 30 | r | 114 | 18 |
| ? | 63 | 28, 29 | s | 115 | 19 |
| @ | 64 | 0, 29 | t | 116 | 20 |
| A | 65 | 1, 29 | u | 117 | 21 |
| B | 66 | 2, 29 | v | 118 | 22 |
| C | 67 | 3, 29 | w | 119 | 23 |
| D | 68 | 4, 29 | x | 120 | 24 |
| E | 69 | 5, 29 | y | 121 | 25 |
| F | 70 | 6, 29 | z | 122 | 26 |
| G | 71 | 7, 29 | { | 123 | 6, 31 |
| H | 72 | 8, 29 | | 124 | 7, 31 |
| I | 73 | 9, 29 | } | 125 | 8, 31 |
| J | 74 | 10, 29 | ~ | 126 | 0, 30 |
| K | 75 | 11, 29 | DEL | 127 | 28, 31 |
| L | 76 | 12, 29 | CRLF | 13/10 | 28 |
| M | 77 | 13, 29 | IDLE | 0 | 28, 30 |
| N | 78 | 14, 29 | ± | 241 | 10, 31 |
| O | 79 | 15, 29 | ÷ | 246 | 11, 31 |
| P | 80 | 16, 29 | ° | 248 | 12, 31 |
| Q | 81 | 17, 29 | × | 158 | 13, 31 |
| R | 82 | 18, 29 | £ | 156 | 14, 31 |
| S | 83 | 19, 29 | BS | 8 | 27, 31 |

**Table 20.3: WSQ character to tone mapping**

## CMSK

MSK (Minimum Shift Keying) is very similar to PSK, but instead of changing the phase to signal the data bits, the frequency is advanced or retarded a very small amount (exactly half the symbol rate), sufficient to exactly achieve a 180 degree phase shift in one bit period. Because the resulting phase change is produced smoothly, without any sudden changes in phase, the signal does not require any amplitude modulation for spectrum management. For PSK modes, such modulation must be employed to drop the output to zero at the phase change, in order to reduce the keying sidebands. The MSK spectrum is very similar to PSK, but the phase relationship between the carrier and the data is different. MSK is little used on HF, but has been widely used on LF, notably at 100 baud and 200 baud for Differential GPS beacons, and at 50 baud and 100 baud for VLF submarine communications.

The main advantage of MSK over PSK is that because there is no amplitude information on the signal, the transmitting amplifier need not be linear. The transmitter duty cycle is always 100% so the average signal strength is greater. In other respects the mode is similar to PSK, and in fact the same receiver demodulator can be used, although a different means of recovering symbol sync is required.

The LF and MF Amateur bands are characterized by relatively stable carrier phase on received signals, accompanied by low

Doppler shift. These bands have very strong lightning interference, but mostly local, not the background of random impulse noise typical of lower HF. There can also be considerable man-made interference. While there is multi-path reception, especially on 160m, the path changes are slow. The slow fades can be very deep, and signals can be quite weak, so in order to have a conversation at typing speed on these bands, we need a mode that is very sensitive, narrow band, has excellent impulse noise tolerance, but need not have strong phase or Doppler tolerance.

It is no accident that MSK is used by military and commercial services on VLF and LF, or by the MF DGPS beacons around 300kHz, although all of these are broadcast, rather than chat (QSO) systems. A well designed MSK chat mode can however provide all the advantages just described. In the ZL2AFP CMSK chat mode software, several important design features have been added which further improve the already excellent robustness and sensitivity of MSK. The CMSK software with full documentation can be downloaded from [12].

The mode can be described as Correlated, Convolved, Chat-mode MSK, but for short just CMSK. It uses a very sensitive cross-correlator in the receiver to exactly identify a pseudo-random sequence which marks the interleaver start position and the FEC dibit order; it uses a convolutional FEC system with interleaver; and uses direct MSK modulation (not differentially coded as in PSK31). The ITU Definition for the mode (which is typically transmitted as an audio subcarrier by an SSB exciter) is J2B, the same definition as PSK31.

## WSJT MODES

Joe Taylor, K1JT, has introduced a whole suite of customised digital communications modes for use across the entire amateur radio spectrum. The name derives from 'Weak Signals by k1JT' They are not keyboard-to-keyboard modes and rely on a strict protocol being set up with regard to timing and information exchanged. Most of them include:

- Source coding and compression
- Massive Forward Error Correction
- A strong synchronisation scheme
- High levels of data redundancy
- A modulation suited to the path in use such as narrowband LF, badly scattered EME reflections or short duration meteor pings
- A decoder that searches in time and frequency to allow for tuning / Doppler offsets and incorrectly set clocks
- A timing protocol that relies on UTC being known to within a few seconds, and a QSO exchange based on time slots.

Source coding first involves reducing the components of a QSO to its basic form including some or all of callsign, locator, report, transmit power and handshake signals such as OK and fail. By making use of the redundancy and limited character sets in callsigns and locators a total QSO 'over' can be compressed into a few tens or a hundred bits.

"The reduction of the source data consid¬erably shortens the message which now allows it to follow different concepts for subsequent processing. One is to transmit it with little correction at a fast data rate over good quality channels. This is the solution seen in the ISCAT communications waveform aimed primarily at aircraft and meteor reflections. These are usually of good quality and high signal strength, but of length typically of a few hundreds of milliseconds.".

Alternatively, the short source data can be expanded in a controlled way by interleaving it repeatedly, adding error correction and synchronisation bits to make a very robust weak signal mode capable of surviving bad QRM. This, in different forms, can be seen in JT65, and WSPR. As all the WSJT modes are designed for weak and scattered signal paths, they all make use of the various types of frequency modulation rather than coherent modulation like PSK. But, all are deliberately designed so that timing and frequency shift / spacing are synchronous and related directly to the soundcard sampling rates.

All the WSJT modes are available as options within one package [13] although WSPR has its own dedicated software tool for beaconing.

A complete rewrite of the WSJT suite which became WSJT-X updated the user interface and introduced several new modes using a new form of very powerful error correction known as LDPC, or Low Density Parity Check Coding. This is a more modern technique than that used in the earlier WSJT modes but in now only possible as computer power has increased to levels that make the decoding possible. The resulting error correction is so powerful that LDPC codes can get closer to the Shannon limit than any other type of correction can offer. The penalty is the need for vastly increased processor usage during the decoding period.

## FSK441

This is a wide fast shift modulation designed for exploiting the short duration strong returns from meteor reflections. FSK441 uses four-tone frequency shift keying at 441 baud, the frequencies of the audio tones are 882, 1323, 1764, and 2205Hz. Each encoded character uses three tone intervals and takes 3/441 seconds or around 2.3ms for transmission. FSK441 accommodates an alphabet of 43 characters, the same ones used in the PUA43 system developed by Robert Larkin, W7PUA. No error correction is used, and text can be anything typed in.

The four possible 'single-tone' character codes, namely 000, 111, 222, and 333, are reserved for special use as shorthand messages. When sent repeatedly, these reserved characters generate pure single-frequency carriers. Their pings are easily recognised by the human ear and also by appropriate software. The present definition of the shorthand messages is respectively "R26", "R27", "RRR", and "73" for the four tones - messages that are frequently encountered in amateur meteor scatter communications.

Timing is based around slots of fixed length transmit/receive periods which have to be agreed beforehand by both parties, and are then specified in the setup screen. **Fig 20.21** shows the FSK441 user screen during operation.

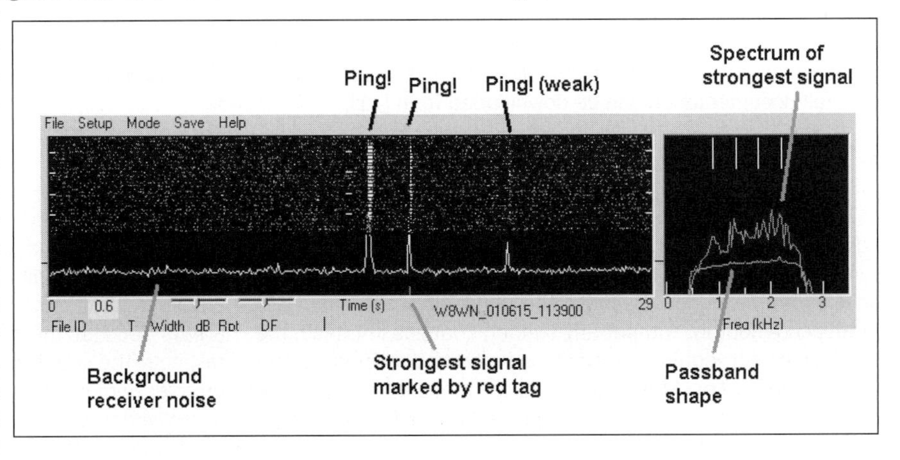


**Fig 20.21: FSK441 in action**

The tone frequencies and data rates are derived from the standardised soundcard sampling rate of 11025Hz. 441 Baud = 11025 / 25 and the tone frequencies are respectively two, three, four and five times this.

FSK441 is no longer a part of the latest WSJT-X suite, having been replaced by MSK144. However it remains popular amongst MS operators using software packages written by other authors. A web search on 'FSK441' should find several hits.

## MSK144

This is the only WSJT-X mode not to adopt multiple frequency shift keying modulation. Instead, MSK144 uses Minimum Shift Keying which is a sort of cross between FSK and PSK. The frequency shift is exactly half the symbol rate and 144 symbols are transmitted in a 72ms duration frame at 2000 baud with a frequency shift of 1000Hz. Frames are continuously repeated for the duration of a transmit period which may be anything up to 30 seconds.

The error correction uses the new LDPC technique which should make MSK144 a better performer than FSK441

## JT65

This is a weak signal mode operating in alternate one minute time slots, starting on the minute boundary. Each transmission takes about 52 seconds. Massive source coding is used, and a message consists only of Callsign, four digit locator (for example IO90) and a few acknowledgement and reporting codes customised for EME operation. The compressed source data is expanded six times and all are interleaved, modified by adding error correction bits and spread out to form 64 separate symbols in a one of 63 code.

The signalling makes use of sequential MFSK in 65 individual tone slots. 63 of the tones are used in the 1 of 63 coding, one is left empty and the lowest tone frequency is used for synchronisation. There are a total of 128 time slots within a 47.5 second window during which one of the tones is always present. Timing derives from 11025Hz /4096, or approximately 0.32 seconds each. Half of the slots are used by the one-of-63 MFSK code and the other half, which are interspersed in a pseudo-random manner over the message duration, are allocated to the single lowest frequency tone used for synchronisation. This strong synchronisation code allows the decoding software to search over time to find the vector, even when the EME delay and computer clocks may introduce errors adding up to several symbol durations.

Three variants of JT65 are in use, depending on the frequency band and expected spreading. The 65 tones can have three different spacings and signal bandwidths:

**JT65A** is designed for HF and 50MHz terrestrial operation with the narrowest tone spacing of 11025 / 3096 = 2.69Hz, giving a total signal width of 175Hz.

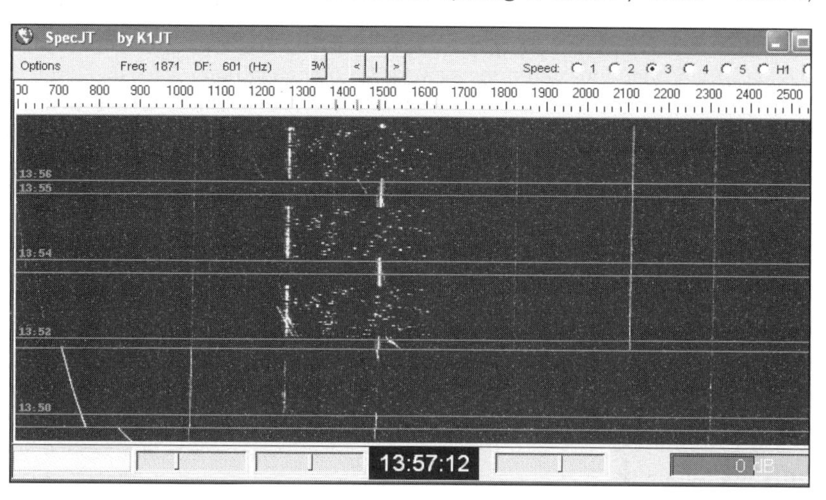


Fig 20.22: Spectrum of JT65B transmission

**JT65B** is targeted at 50 - 144MHz terrestrial and EME, with a tone spacing twice that of the A variant for a signal bandwidth of 350Hz

**JT65C** doubles tone spacing, again, for use at UHF with 700Hz bandwidth.

**Fig 20.22** shows the spectrum of a JT65B received off-air from the GB3VHF 144MHz beacon, with the lowest synchronisation tone forming a clearly visible pattern.

Operation runs in alternate one-minute slots of transmit and receive, the use of even and odd minutes has to be agreed beforehand and is always specified as part of the EME setup scheduling.

The decoder stores the entire received data and searches in time and frequency for the sync vector. Once found, it knows the tuning and timing error and can then do a best estimate of each symbol using a sophisticated soft-decision process developed by K1JT. With a symbol time of 0.3s, and tone spacing of 2.7Hz the noise bandwidth of the decoder is of the order of 3Hz. This directly gives a 10dB S/N advantage over the (typical) 30 - 50Hz bandwidth needed for aural copy of CW. When normalised to the standard 2.5kHz bandwidth, reliable JT65 decoding can be made at S/N ratios of 26dB, equivalent to +3dB S/N in the 3Hz actual signal bandwidth. The reason that reliable decoding of an FSK signal can be made in a S/N so low is due solely to the massive encoding redundancy and error correction coupled with soft decision decoding.

The final message cannot be directly equated to a total effective-number of bits due to the use of a 63 symbol set. But, had a 1 of 64 symbol scheme had been used instead of 63, this could be said to represent 6 bits as $2^6 = 64$. In that case, the total effective number of bits could be said to be $64^{64} = 4096$. With the synchronisation vector adding another 64. This is a considerable expansion over the few tens of bits making up the compressed source data, and explains the extreme robustness of this protocol. **Fig 20.23** shows the JT65 screen during operation.

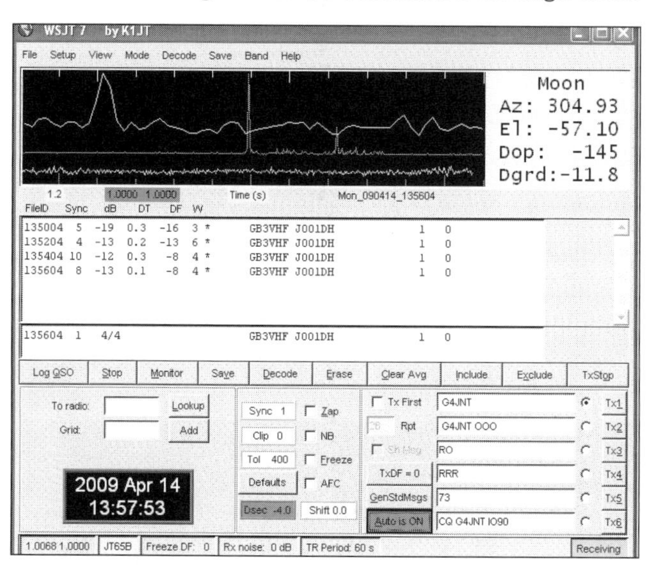


Fig 20.23: The JT65 screen in operation

## QRA64

QRA64 is used in a similar manner to JT65, but also adopts LDPC coding. IT offers a wider range of tone spacings than JT65 so is more adaptable for frequency bands from LF to microwaves. Unlike JT65, it doesn't have a synchronising tone which makes it more difficult to spot a weak signal on a waterfall display

## FT8

FT8 was introduced after HF users had adopted JT65 and discovered its amazing ability to work weak signals where even CW couldn't cope and signals were barely audible, if at all. The one-minute cycles were considered too slow for HF operating practice, so FT8 was introduced with a 15 second Tx/Rx cycle, meaning a QSO could be exchanged using the same rigid message protocol in a period of just over one minute. Changes were subsequently made to the encoding to allow longer callsigns for contest and DXpedition operations, and autofill of the message fields was added to assist and speed up operator interaction.

FT8 is not quite as sensitive as JT65 or QRA64 but is still probably 10dB - 15dB better than CW. As with the other latest modes in the WSJT-X suite, LDPC error correction coding is used. The massive take up of this mode often means that a band may appear to be deserted of SSB and CW activity, but there is almost certainly going to be some FT8 activity in its small narrow sub-segment

## WSPR

WSPR or Weak Signal Propagation Reporter is the latest of the WSJT modes. Unlike the others it has been designed primarily for beacon type operation (although the latest WSJT software does offer a 'QSO mode). The source data is compressed to callsign, 4 digit locator and transmit power specified to one of twenty levels ranging from -30 to +30dBm. The total again occupying a few tens of bits. This is expanded to a one-of-four tone MFSK signal, with 162 symbols spread over a 110.6 second interval. Effective total number of bits is therefore 324. Unlike JT65, the synchronisation code does not form a separate tone, but is interleaved with the data, again in a pseudo random manner.

The 110s transmission fits into a two minute time slot (there is just time for a CW ident at the end for when licence conditions require this). The WSPR software is designed for simultaneous transmission and reception of other signals, so to manage this transmit slots are allocated in a pseudo random manner as a fixed percentage of the total. The percentage can be defined by the user as 33%, 25%, 20% etc of the total. By randomising transmit periods in this way, the likelihood of repeated clashes between two stations so that they would never hear each other is minimised.

The tone spacing is 1.46Hz, derived from the later 12000Hz soundcard sampling rate / 8192. The symbol duration is the reciprocal of this, or 682.7ms. This very low spacing and bandwidth, just 6Hz wide in total, means that WSPR primarily finds use at LF and HF. Every amateur band up to 144MHz now has a designated WSPR spot (200Hz wide) and on 500kHz, 7 and 10MHz in particular stations can always be found participating in 'WSPRing' as it is known.

As WSPR was designed for signal reporting, some facilities were built into the software to give additional benefits. The WSPR software monitors a 200Hz wide chunk of spectrum corresponding to tone frequencies from 1400 to 1600Hz and can decode every WSPR signal within it, searching over time as well to allow for errors in PC clock setting. The operating screen showing signals received on the 7MHz band can be seen in **Fig 20.23**. The S/N is automatically measured in the decoding process and reported along with the message contents.

Now as signal power, location and S/N are known there is enough information to determine path details. The software allows successful decodes to be automatically uploaded to a central WSPR database [14] where every 'hit' is logged and a complete propagation map between all participating stations can be generated.

## Opera

Opera is a complete calibrated engineering beacon / data transmission and reception system, capable of assessing propagation and determining the suitability of a path to support traffic by providing an average Signal / Noise assessment and the percentage fade below a 3dB margin.

It uses on-off keying (OOK) of a single carrier, employing Manchester Coding (see Fig 20.3) to force a duty cycle of exactly 50%. Opera is completely free running, no time locking being required, and is designed to deliver full performance using normal HF amateur radio equipment and sound cards. The system utilises web linking to distribute spot data to other users and is linked to the *PSK-Reporter*, A local copy of decodes and a web-linked copy of other users' spots is displayed in real time on the *Windows* software.

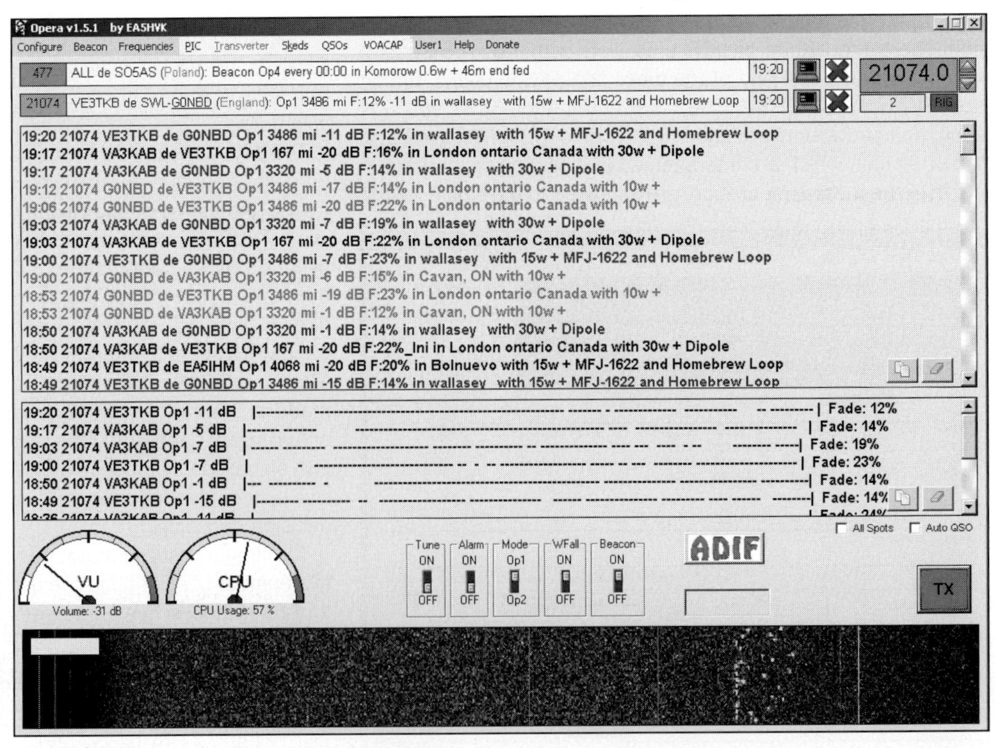


**Fig 20.25: The Opera user screen showing (from the top) 'spots', received signals and fading, the control panel and a waterfall display**

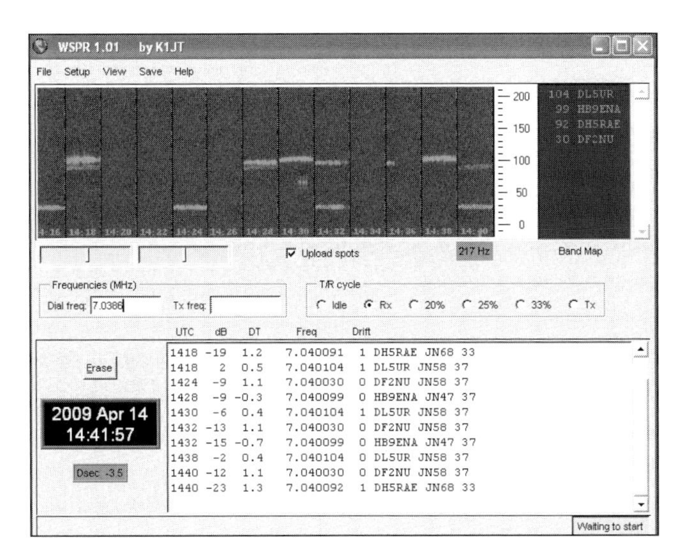


Fig 20.24: The WSPR screen

The message is encoded into a unique 239 bit binary data word with strong Forward Error Correction (FEC) to allow successful decode of the message with only 50% of the signal being intercepted. The 50% loss may be within a block, or as bits lost in QSB / QRM that are randomly distributed along the time line.

Two versions of the encoded message are available:

- Callsign only 'Beacon Mode'
- 15-characters plain text or 'QSO Mode' giving 20dB S/N in a 120 second transmit period (this is not active at the time of writing, but the software exists)

The system is compatible with both CW and SSB transmitters in that both COM-port ON/OFF key and audio tone transmitter drive is offered by the software. Trials have also proven compatibility with FSK systems, using recovered tones either inside or straddling the decoder pass band.

Using SSB transceive enables the full capabilities of the system to be exploited; the audio drive via the sound card provides agile Tx tone frequency placement, There is no gain to be made from the deployment of high stability equipment or narrow filtering; to the contrary, narrow filtering reduces the achievable S/N level.

### Running Opera

In the period between transmit demand and commencement of transmission, the received audio spectrum is scanned. The transmit carrier is placed in the lowest S/N position, allowing a guard band of a few Hertz to prevent collisions with other users, QRM lines and areas of increased noise. This maximises link efficiency.

Reception / decoding is provided by a *Windows* based software package which contains both transmit and receive components of the system. An option is also provided to show the 239 bit binary code, for PIC and other types of systems to offer transmit-only, fixed beacon use.

Opera comes with a wide choice of symbol lengths, effectively these equate to the noise bandwidth used in the decoder. The mode name roughly corresponds to the beacon transmission period so that, for instance, OP.5 has a 30 second transmission, and OP4H takes four hours.

The minimum decode signal/noise level is set by transmit cycle time/symbol length, ranging from -20 dB S/N (all S/N values are normalised to a 2.5kHz bandwidth), for the fastest OP0.5 (0.5 minute or 30 second transmission), to -50 dB S/N for Op4H with its 4 hours transmission period and roughly one minute long symbols. The expected minimum S/N values achievable are listed for each mode in **Table 20.4**.

| Mode | S/N |
|---|---|
| OP05 | -20dB |
| OP 1 | -23dB |
| OP 2 | -26dB |
| OP 4 | -30dB |
| OP 8 | -34dB |
| OP16 | -37dB |
| OP32 | -41dB |
| Op64 | -44dB |
| OP4H 8.9kHz sound card SDR | -49dB |
| 16kHz SDR optical test version | N/A |
| QSO Mode (all bands) | -20dB |

**Table 20.4: A comparison of signal to noise ratios for the various Opera modes**

Modes normally used at VHF and HF vary from OP0.5 to OP4, with their 30s to 4 minute transmission periods. OP4/OP8 are generally in use on 477kHz and OP8/OP32 has become the standard for the 136kHz band. OP16/OP64 has been added to support sub-100kHz experimental stations in the USA.

OP4H provides the user with an SDR by using the sound card as a direct 8kHz transmitter and receiver, as does the 'Experimental Light' version which provides a 16kHz carrier via the sound card, and allows direct reception of a keyed 16kHz modulated carrier.

Measurement of percentage fade is displayed via the provision of a visual representation of the signal level along the time line and presented as a % fade value. This measurement is a real time indication of the path's ability to support a data transmission.

**Fig 20.25** shows the Opera user screen with the fade indication on the lower panel and the web-delivered other user / band activity recorded on the upper panel.

The text bars display the transmit web beacons and decodes from other users; checking the 'All Spots' tick-box, at the lower right-hand side allows local recording of the band in use or 'all bands'. Opera can be downloaded from [15].

## SPECIALIST MODES

### EbNaut BPSK at Low Frequencies

Coherent signalling experiments on 137kHz (and lower) have been taking place with some good success software written by a group of amateurs determined to push the weak signalling capabilities of the bands to the limit. The name EbNaut derives from the term Eb/No meaning energy per bit, used as a normalised measure of signal to noise ratio for digital communications systems. The EbNaut software [18] is not plug and play; it requires that you have an intimate knowledge of your signal structure, the format and timing of the message and, for most users, how to drive Spectrum Lab [19] as well as the innards and meaning of the terms used in the driver and receiver software. All the communication link parameters MUST be agreed in advance - except its actual contents, of course! The mode requires a very accurate and stable frequency source based on GPS or Rubidium locking and timing on the PC accurate to 100ms. The latter can usually be achieved with a network time setting programme.

| Callsign | Locator | Frequency MHz | Modulation types |
|---|---|---|---|
| GB3RAL | IO91IN | 40.0500<br>50.0500<br>60.0500<br>70.0500 | CW, JT65B, (four VHF freqs, phase locked) |
| G4JNT/P | IO80UU59 | 70.031 | PSK31 Telemetry |
| GB3VHF | JO01EH | 144.4300 | CW, JT65B, phase reversals |
| GB3NGI | IO85VB | 144.4820 | CW, JT65B |
| GB3WGI | IO64BL | 144.4870 | CW, JT65B |
| GB3CSB | IO75XX57 | 1296.9850<br>2320.9850<br>3400.9850 | FSK-CW, JT4G |
| GB3SCS | IO80UU59 | 2320.9050 | CW, JT4G |
| GB3SCX | IO80UU59 | 10368.9050 | CW, JT4G |
| GB3SCK | IO80UU59 | 24048.905 | CW, JT4G |
| GB3SCF | IO80UU59 | 3400.9050 | CW, JT65C |
| GB3SEE | IO91VG | 10368.8500<br>24048.9600 | FSK-CW, JT4G<br>FSK-CW, JT4G |
| GB3CAM | IO92WI | 24048.8700 | FSK-CW, JT4G |

**Table 20.5:**

## How It Works

Modulation is Binary Phase Shift Keying, BPSK, but unlike previous amateur use of PSK, the receiver does not attempt, itself, to recover carrier or symbol timing. It assumes the transmitting station is "good enough". Typical symbols rates used at 137kHz have ranged from 0.25 seconds per symbol to 2s, using messages that last up to 20 minutes, so the carrier must not drift by no more than would result in about 45 degrees of phase shift over this Tx period. 45 degrees in 20 minutes at 137kHz requires a frequency stability of about one part per billion which is well within the capability of GPS or rubidium standards.

The message is convolutional coded to add a very high level of error correction. EbNaut uses eight or more feedback registers for rate 1/8 or longer coding. It is this extreme level of added redundancy that allows the system to get within a dB of the Shannon limit for some paths. In addition, an outer checksum is added as a further check of correct decoding as it is quite normal, and statistically inevitable, that completely random noise will occasionally get through the decoding process and generate garbage messages.

## Transmitting

The EBNaut encoder uses the PC clock for symbol timing, so this has to be maintained accurately using a time server or GPS. Specify the sub-mode (the code rate and level of checksum strength) the symbol period and the start time. The symbols are generated and sent to a transmitter by the simple expedient of toggling the RTS and DTR lines on a COM port. It is up to the user to use these to generate a 180° phase shift; something as simple as a changeover relay and centre tapped transformer could suffice. DDS driver PIC code can be modified to reprogramme the chip's phase shifter in response to an external input. Diode ring mixers are another possibility. A soundcard and upconversion is not an option for this mode however; the accuracy and stability of soundcards are just not good enough.

## Receiving

This is more complicated. The EbNAutRx software comes in two packages, compiled for 32 bit or 64 bit Windows (also available in Linux) and such is the computational overhead that if you have a modern machine, use the 64 bit version to speed the decoding

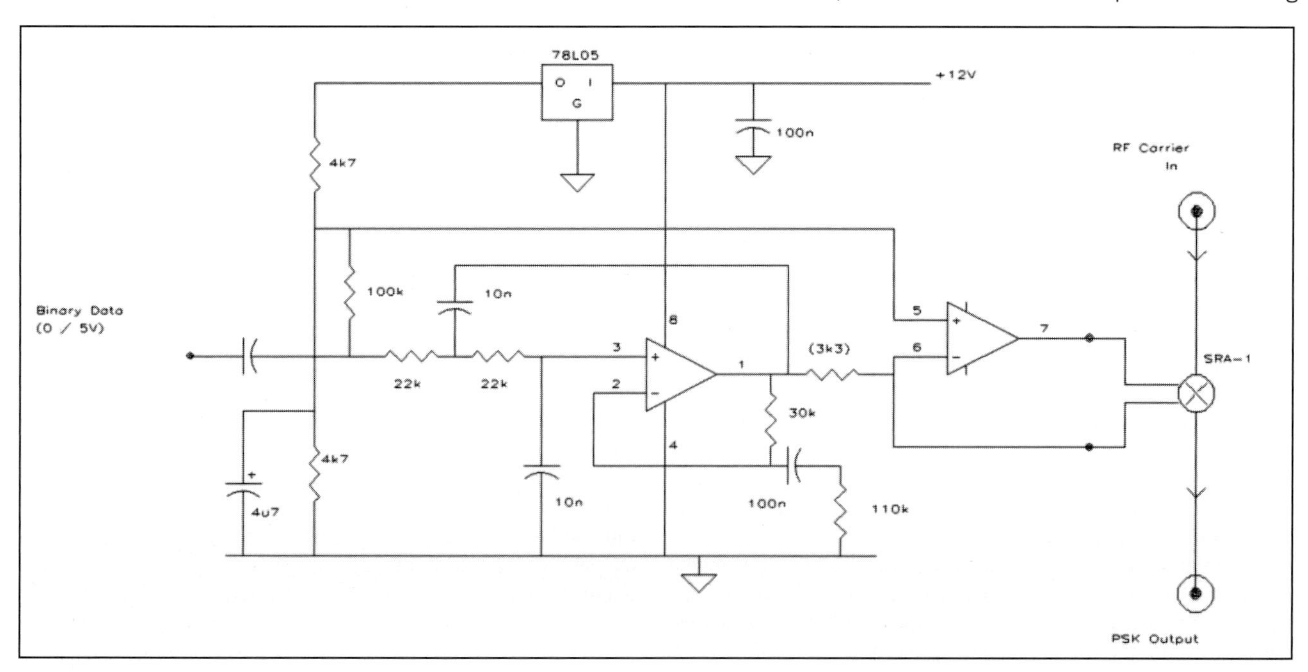


**Fig 20.26: Direct PSK generation at 400 Baud**

process. It only works off-line, with .WAV files that have been recorded and saved. Extreme frequency stability means normal sound card recording is not stable enough and another solution is needed. So the first thing is to generate a properly synchronised (and ideally time stamped) .WAV file of the received audio. Spectrum Lab can do this using GPS derived 'clicks' added to the audio to take-out soundcard inaccuracy and drift. Custom .WAV files with addition of a chunk labelled 'inf1' allow arbitrary sample rates, time stamp and other information to be included.

To run the EbNautRx software . first browse to find the recorded file. It is essential that the exact start time of the recording is known so that any offset, deliberate or otherwise, can be entered into the start offset window. The message length, symbol period and sub-mode are entered and the decode started. Since the absolute phase cannot possibly be known in advance, the decoder has to search over a range of phases to determine this – you can see it choosing settings as the software runs. It is doing an enormously complex computational task, so don't be surprised to see the CPU capacity at 100% and CPU cooling fans come on! It is running thousands of Viterbi decoding sessions, and matching checksums in order to get a valid one. If you are fortunate, a message will soon pop up after just a few percent of the overall progress. Decoding even a short message can easily take several minutes, so just leave it to get on with the job. Even after a message appears, it will go on searching for better solutions. All the results are stored in an accompanying .TXT file generated by the decoder **Fig 20.27**.

If there is uncertainty in start times – like the Tx station is a bit 'out' for example - multiple instances of EbNautRx can be started in parallel, all with identical parameters entered apart from various offsets in start time. It may take hours to run, but the chances are, if there is a message to be decoded it will find it!

A few datamodes exist that derive from the days before PCs and soundcards were ubiquitous. They are based around dedicated hardware and use proprietary, or at least customised, protocols. Although little used by amateurs now, they are often is use by commercial organisations.

**Pactor** is packet radio over Tor. The AMTOR signalling protocol forms the base modulation and a packet radio error correction/ detection layer is overlaid.

**Clover** is a proprietary four tone MFSK system designed by the Hal corporation who supply the dedicated DSP modems. Hal modems can also be used for RTTY and Pactor.

## DIRECTLY GENERATING DATAMODE WAVEFORMS

Instead of using a soundcard plus upconverter for transmission, some datamodes lend themselves to direct generation at RF using suitable hardware.

Any simple FSK mode can be generated from a voltage tuneable oscillator, although it will be necessary to set the shift correctly. This simple scheme is directly applicable to DFCW / DFCWi where the keying lines can drive a varicap across the crystal determining the output frequency.

Binary PSK can be generated at low level with a double balanced mixer, by feeding the data as a bi-directional current into the IF port as shown in **Fig 20.26**.

The diagram shows an optimised filter for a 400 Baud signal as used for the telemetry on some amateur satellites. A mixer with an IF port response to DC is needed; one where both IF port connections are floating from ground is ideal as no negative power rail is then required for generating a bi-directional current. For use up to UHF frequencies the SRA-1 or SBL-1 type are

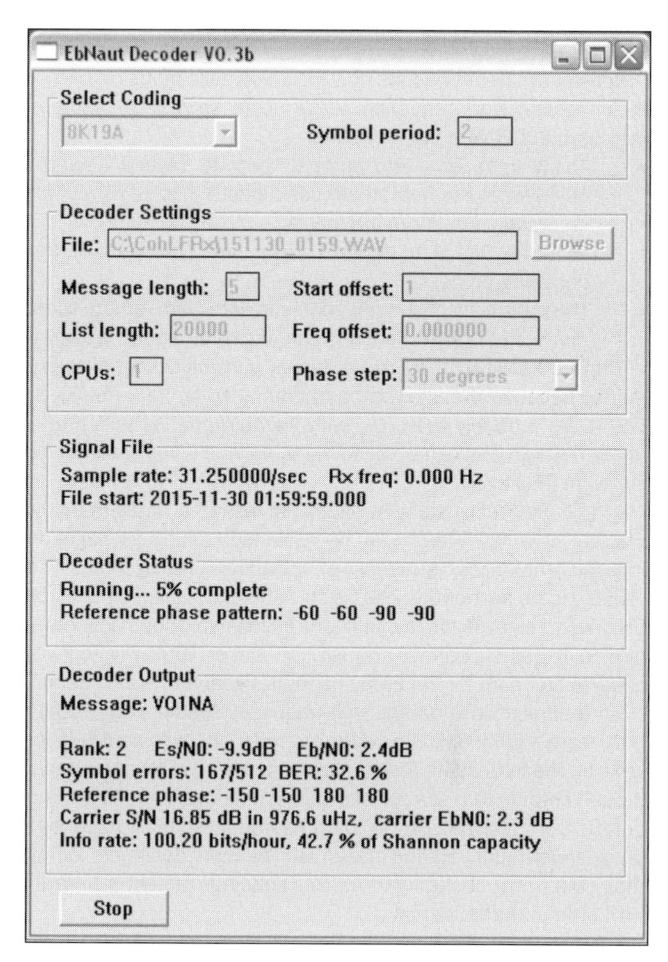


**Fig 20.27: The EbNaut Receiver screen showing a successful decode of VO1NA running about 20 Watts on 137kHz**

perfectly adequate.

For more complex schemes a Direct Digital Synthesiser can be programmed from a microcontroller such as a PIC, in real time. Devices like the AD9852 [16] offer frequency, amplitude and phase programmability so for the lower frequencies where no further frequency multiplication is needed, this technique can be used to generate some quite advanced modulations. For details of direct PSK31 generation using a PIC and AD9852 DDS see [17].

Beacon GB3VHF uses the same DDS chip followed by a X2 multiplier to generate JT65B on 144MHz, the VHF beacon cluster at GB3RAL directly generates JT65 on 40/50/60/70MHz coherently phase locked to a master GPS locked reference.

## DATAMODES ON BEACONS

An experimental beacon GB3SSS used first on 1.9MHz then on 3.6MHz generated a multi purpose beacon transmission containing CW ident, Stepped power and PSK31 all under precise GPS timed control.

For microwave beacons where frequency multiplication is in use, DDS devices followed by multipliers are not the preferred route due to excessive spurious outputs. Several beacons of the GB3SC# cluster use a DDS as part of a phase locked loop with a crystal source forming the variable element. By reprogramming the DDS in real time, slow frequency shift type datamodes can be generated, provided the PLL is fast enough to follow the keying.

For microwave beacons where more complex modulations are needed, an upconverter approach can be used with the DDS generating at VHF which is then upconverted to the final microwave frequency. G6GXK adopted this route for the GB3XGH 10GHz beacon.

Several amateur beacons now include a data mode as part of their transmission sequence giving users several advantages over normal CW or aural reception:

- Ability to receive and correctly decode several decibels below what the human ear can detect
- Automatic detection and logging
- Accurate Signal to Noise measurement as part of the decoding process
- The possibility of sending real time data from remote sites.

A summary of the UK beacons carrying data modes is shown in **Table 20.5**. At HF there are not many possibilities for different modes because the allowed bandwidth is so limited. PSK31 or WSPR are probably the only really suitable options that give a low bandwidth and can be decoded by anyone using several free software packages.

At VHF, frequency stability issues are less of a problem and a stability of a few Hertz can be assumed, taking propagation induced anomalies and receiver instability into account. The JT65B mode, part of the WSJT suite with its 65 tones spaced by 5Hz, was selected for the first 2m beacon to adopt new data-modes. It was chosen for consistency since JT65B is already a de-facto standard for 2m EME and weak signal terrestrial comms.

On the microwave bands, with frequency scatter and receiver drift more significant, the wider spaced JT4G was selected for several beacons as its four tones spaced by 315Hz can punch-though chronic rain scatter and tuning drift over the 48 second duration of its signal. So far, data has been added to beacons on a pretty-much *ad-hoc* basis with beacon designers doing their own thing, choosing tones and spacings almost arbitrarily with little standardisation.

For example, JT4G on the Bell Hill beacons, GB3SCS and GB3SCX, use the lowest of the four tones as the nominated reference (10368.905000MHz in the case of GB3SCX) meaning listeners have to set an SSB receiver to 10368.9042MHz to correctly decode it. A total of four tones are used.

Taking a different approach, GM6BIG, on the GB3CSB beacon carefully defines the centre of the tones as being the reference point where carrier and the mark tone for the FSK CW lies. CW space, and the four JT4 tones mean that beacon transmits six different tones.

## PACKET RADIO

The use of packet radio has declined in recent years, but it is still used for specialist purposes such as APRS (Automatic Packet Reporting System) and DXCluster.

## REFERENCES

[1] DFCW software. www.qsl.net/on7yd/136khz.htm
[2] DFCWi software. www.g4jnt.com/DFCW.htm
[3] Hellschreiber modes: www.k3pgp.org/software.htm
[4] Digipan. www.digipan.net/
[5] MultiPSK. http://f6cte.free.fr/index_anglais.htm
[6] FlDigi. www.w1hkj.com/
[7] MMTty http://hamsoft.ca/
[8] ROS software: http://rosmodem.wordpress.com/
[9] DL4YHF's WOLF GUI. http://www.qsl.net/dl4yhf/wolf/wolf_gui_manual.html
[10] I2PHD Jason. www.weaksignals.com
[11] www.qsl.net/zl1bpu/SOFT/WSQ.htm
[12] www.qsl.net/z/zl1bpu/CMSK/cmsk.htm
[13] K1JT WSJT and WSPR. http://physics.princeton.edu/pulsar/K1JT/
[14] WSPR Site and Report database. http://wsprnet.org/drupal/
[15] For the download page for the Opera mode go to: http://rosmodem.wordpress.com/
[16] Analog Devices AD9852 DDS. www.analog.com/en/rfif-components/direct-digital-synthesis-dds/ad9852/products/product.html
[17] Direct PSK31 generation. www.g4jnt.com/beacons.htm
[18] EbNaut Coherent signalling http://abelian.org/ebnaut/
[19] Spectrum Lab audio processing tool http://www.qsl.net/dl4yhf/spectra1.html

***Andy Talbot*** *BSc, G4JNT, is a retired professional electronics engineer working in the field of radio communications and signal processing. Andy's main interests being at the two opposite ends of the spectrum - microwaves and the LF bands. He has always been a home constructor, designing and building equipment from scratch right from the early days. Working on the new 73 and 137kHz bands started off his interest in DSP, and in conjunction with G3PLX pioneered the first use of ultra- narrow-band techniques and low data-rate signalling on the amateur bands. Later, in conjunction with G4GUO, he took part in the first amateur use of digital voice at HF. He has written many articles for various journals, and writes two regular columns for the RSGB's* RadCom *magazine.*

# Computers in the Shack

# 21

*Andy Talbot, G4JNT*

Computers have been part of a well-equipped radio shack for some years, mostly for data communications and logging/contesting. More recently, cheap online connections have brought Internet resources into many amateur stations.

This chapter aims to show that there are many other uses for the shack computer, making it an essential tool for the constructor.

## INSIDE A COMPUTER

Before discussing what a computer can do for you as a radio amateur, it is useful to take a brief look at how it works.

For the purposes of this book, it will be assumed that the computer is a 'PC' operating under Windows. Other computers such as the Apple Mac could be pressed into service for some applications with the appropriate software. Other operating systems such as LINUX can also be used.

The essential components of a computer are a processor to do the work, memory to store information, inputs and outputs for communication between the computer and the operator (eg keyboard and screen), software to give it instructions and a power supply.

### Central Processing Unit

The CPU is the engine of a computer and is measured by the number of cycles of work it can carry out in one second. Thus a computer carries out 2000 million instructions per second. Modern CPU chips get very hot and have heat sinks or even their own cooling fans.

### Memory

Just like the human brain, a computer cannot work without being able to store information. There are several types of memory, divided into their function, the storage medium and the amount of time that information can be stored.

#### Bootstrap

This is a tiny, permanent, memory on a chip. It is known as Read Only Memory (ROM) as it cannot be over-written with new data. Its function is to give the CPU the very basic information it needs to start up and function as a computer.

#### RAM

In contrast to ROM, Random Access Memory (RAM) is designed to be continuously re-used. It is the temporary storage used to hold all of the data required during processing. RAM is located on chips so it can be 'written to' and 'read' very rapidly. It is commonly described in 'Megs', though this is Megabytes, not Megahertz.

#### Hard disk

Most of a computer's storage is done on a magnetic disk. Although reading and writing is nowhere near as fast as RAM, it has the advantage that it keeps its information indefinitely, even when the computer is not powered up.

#### Removable memory

This refers to the disks that can be taken out of a computer for future reading by the same, or another, computer at a later date. Older machines have 3.5in so called 'diskettes' which can store about 1 megabyte of data, whereas all modern computers use CDs capable of storing up to 800MB or DVDs which can store several gigabytes - that's thousands of megabytes. Additionally it is now possible to plug in an external memory device (flash stick) capable of storing many tens of gigabytes or more in a small space and with rapid read and write. External hard disks can be added, and these can store hundreds of gigabytes.

### Input / Output

Abbreviated to I/O, these devices are what is needed for human beings to interact with the computer. They include the keyboard, screen, mouse and sound card.

### Operating System

Usually stored on the hard disk, this is the permanently installed software, that makes the CPU into a usable computer. The most commonly used operating system is Microsoft Windows, although there are amateur radio programs that run under DOS (not used as often now) LINUX or MAC OS. The operating system defines how the various parts of the computer work together and how it connects to real people.

### Power Supply Unit

Like any piece of electronic equipment, the computer has a PSU, to run from the mains. Additionally lap-top computers have hefty batteries capable of running the unit for an hour or two.

## SOFTWARE

Although the operating system is an essential piece of software, it cannot do anything other than make a computer. To perform any useful task, such as word processing or sending e-mail, additional software known as programs must be installed.

A new computer will usually come with some programs, usually an Internet browser and some office functions, but there is an almost infinite number of additional programs that can be added to perform specialist functions. Although some programs, especially those for commercial applications such as producing this book, are very expensive, many are quite cheap or even free. Fortunately, many amateur radio programs are in the latter category.

### Operating Aids

Computers are used to enhance the shacks of many keen DXers. Facilities available to the operator include:

- Logging
- Contest aids
- Rig control
- Maps
- Data communications
- DXCluster
- Propagation information
- Maps and locators

Most of these are outside the scope of this *Handbook*, but detailed descriptions of all of the above can be found in [1].

### Drawing

If you are not good at drawing circuits, good quality illustrations, such as many of those in this book, may be drawn using 'com-

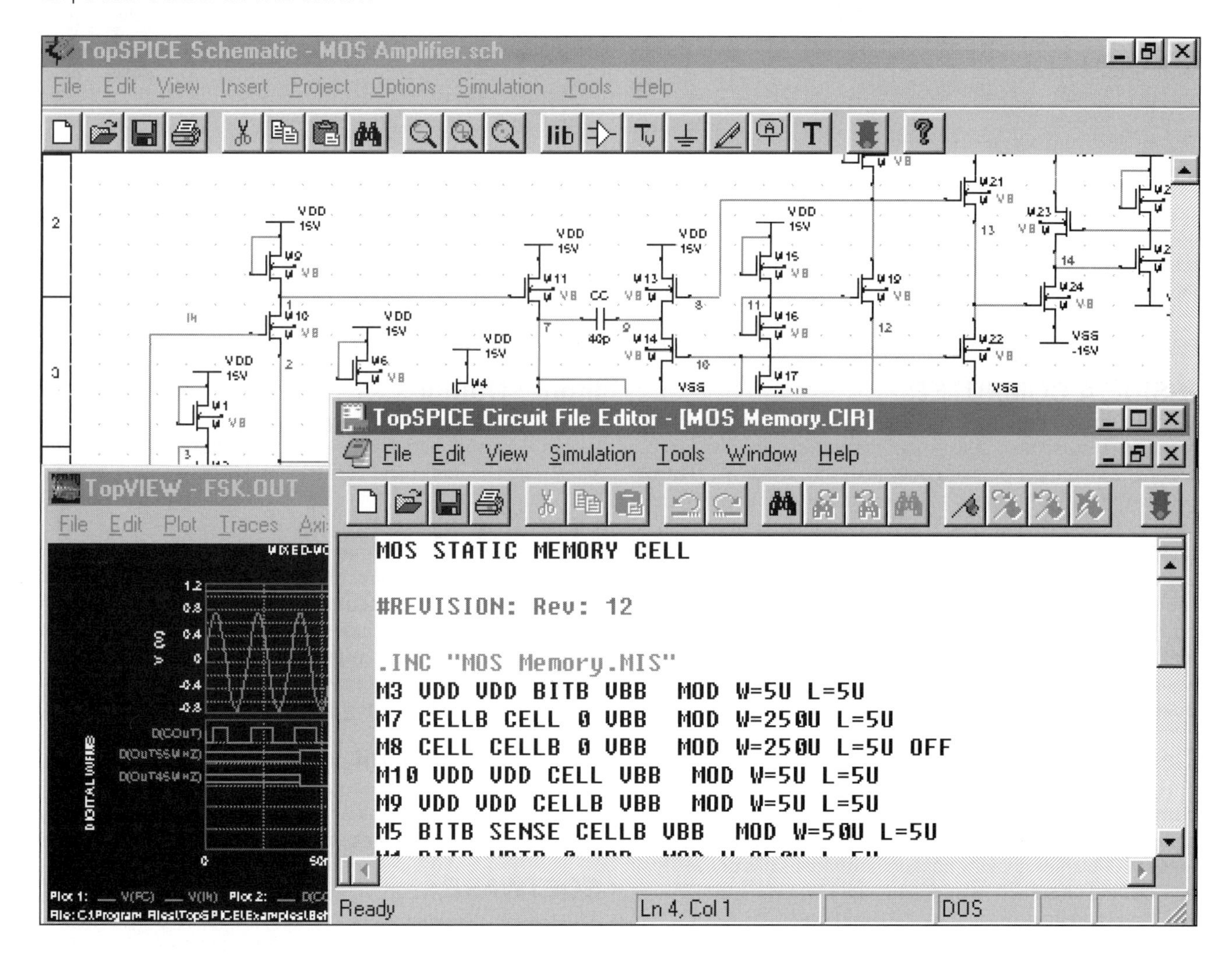


**Fig 21.1: A typical circuit simulator / analysis screen. [Source http://penzar.com/topspice/topspice.htm]**

puter aided design' (CAD) software. There are many generic drawing packages, from professional quality software such as Adobe Illustrator and CorelDraw to inexpensive or even free programs available for download on the Internet.

These save the work of producing neat straight lines, boxes and circles but components must be individually drawn. An alternative is to use a CAD program specially tailored for electronic circuit design. These have the facilities of a generic program, but also have a library of component symbols. Most of the simulator programs described below include schematic drawing facilities.

Basic drawing facilities quite suitable for producing block diagrams and simple circuit diagrams are also included within a number of office type programs such as word processors and spreadsheets.

## Circuit Drawing and PCB Layout Packages

A number of packages aimed at electronics design allow circuit diagrams to be drawn, with most components types stored in libraries. In many cases, schematic design is coupled with Printed Circuit Board layout, with the ability to autoroute a PCB directly from the original circuit diagram.

PCB layouts can be exported to files in standard formats that can be sent to PCB manufacturing companies or machinery. Other facilities like parts lists, design rule checking are often included in these software packages, as is the facility to export net lists to circuit simulators.

One package that is widely used by amateurs is EAGLE Light, by CadSoft Solutions [2]. This free program provides a limited subset of that provided by the full, professional, version. Restrictions include: The maximum board area is limited to 100 x 80mm; Only two signal layers can be used (top and bottom); The schematic editor can only create one sheet; Use is limited to non-profit applications or evaluation purposes.

## Circuit Simulators

These allow circuits to be drawn and then analysed, and most are based on the industry standard SPICE. Several are available to try out as free demo versions, usually with restrictions. A useful list can be found at [3].

Some simulators incorporate printed circuit board (PCB) design, whilst others can export data to a dedicated design program. The result can be printed and used in producing the PCB itself.

It is possible to simulate both analogue and digital circuits, or even a mixture of the two. Several of the projects in this book have been initially designed by this type of program, most notably *PUFF*.

The user starts by drawing the circuit diagram, made from graphical elements provided with the software. The result can be analysed by the program's 'virtual' test equipment, to check how well (or whether) it works.

The information displayed can include amplitude vs frequency response, phase vs frequency, group delay vs frequency, gain or

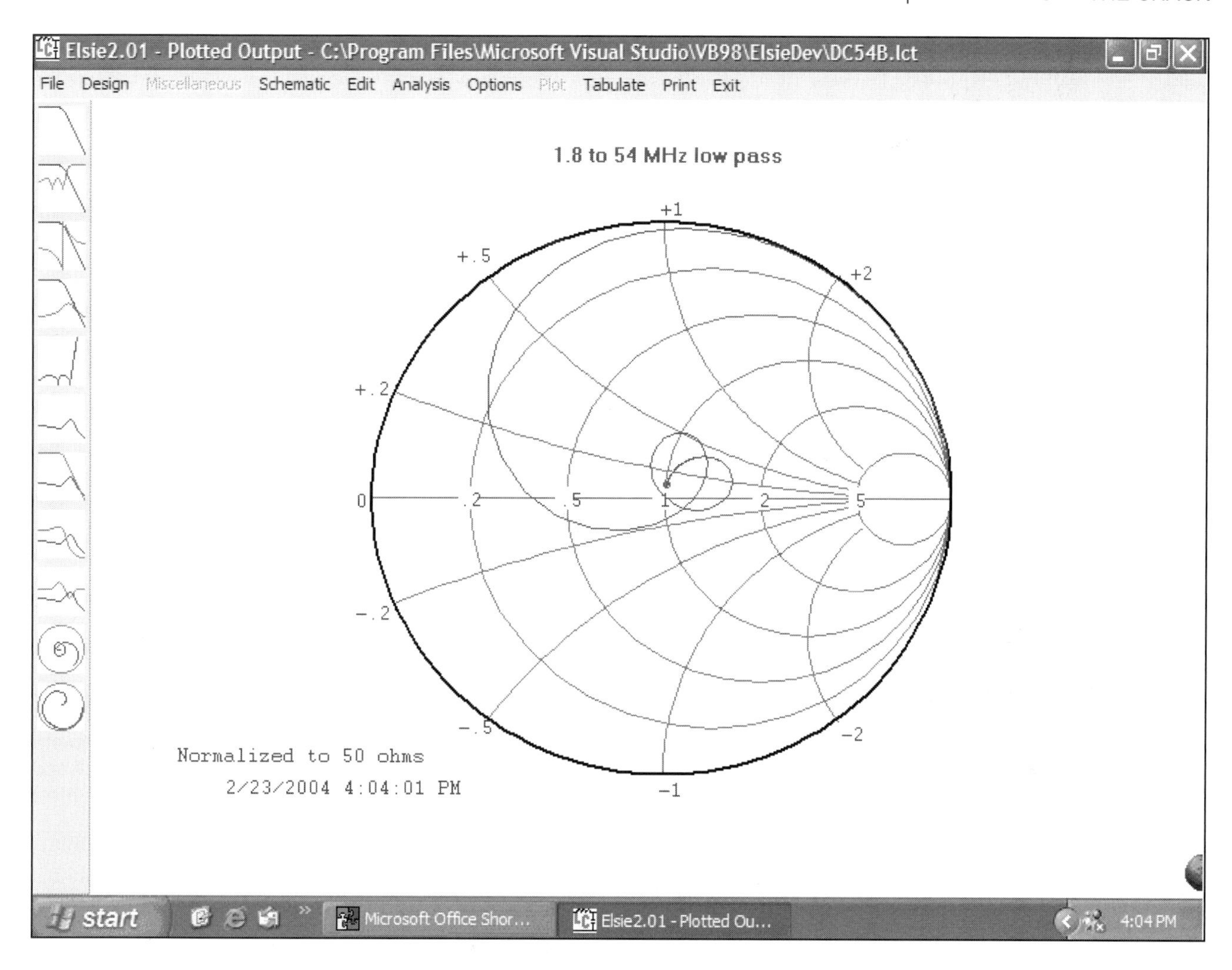


**Fig 21.2: A Smith Chart produced by *ELSIE*, the filter design program from Tonne Software**

loss, and the impedance at input and output. A typical display is shown in **Fig 21.1**.

Changes can be made and the results tested, all before any real construction takes place. As with all software simulations, it is no substitute for knowledge and errors can be reduced by having a basic understanding of how real electronic circuits work.

Simulators can usually predict performance quite accurately (mostly for analogue circuits), although they do require substantial effort to learn to use effectively.

A circuit simulator won't do the designing for you, but it does help a lot in finding out what is wrong with a design, and predicting what will happen before you actually build the circuit. Note that the cheaper programs will not operate with time-varying inputs, eg inter-modulation analysis.

## Spreadsheets

Many aspects of circuit design can be calculated, so a spreadsheet program is useful for making the most complex and interactive calculations. Examples are the optimisation of inductors and filters, Yagi dimensions and linear amplifiers. See also the chapter on Software Defined Radio for a spreadsheet-based tutorial.

The 'number crunching' abilities of modern spreadsheet software can remove the need to write small custom programs for solving many design problems. In days gone by, special software written in a high level language was often needed to solve long equations or to optimise solutions. Now, spreadsheets such as Excel have optimisation and data analysis tools available for uses as well as all the basic mathematical functions. Spreadsheets are the ideal solution to quickly plotting and displaying results.

## Word Processors

As well as their obvious use for producing written text and documents, modern word processing software permits seamless integration of graphics and pictures such as circuit diagrams exported from circuit design software and digital photographs, allowing a one-stop solution for producing complete, professional looking, documents.

In the past, constructors would use rub-down lettering or adhesive characters to produce professional-looking legends on a front-panel. Nowadays a word-processor can generate text in many fonts, and with a little practice it can also be used to produce arrows, meter scales, etc.

## Other Software

A useful tool is AADE Filter Design which can be downloaded free from [4]. It will calculate gains, impedances, group delay, phase, return loss and more. Filter types handled include Butterworth, Chebyshev, elliptic (Cauer), Bessel, Legendre and linear phase low-pass, high-pass, band-pass, and band-reject filters, as well as coupled resonator band-pass and crystal Ladder band-pass filters using identical crystals. A tutorial is included for those wishing to know more about filters.

A large range of programs can also be downloaded from the website originally produced by Reg Edwards, G4FGQ [5]. These include calculators for the design of antennas, transmission lines, inductors, filters and amplifiers. Also included are propagation path-loss calculators.

Another useful source is Tonne Software [6] which has programs for filter design, meter scales, customised maps, antennas and matching.

Most are totally free, or available free in a cut-down version. The filter design program, *ELSIE* (**Fig 21.2**), was used in the 'PIC-A-STAR' project which features on the CD accompanying this book.

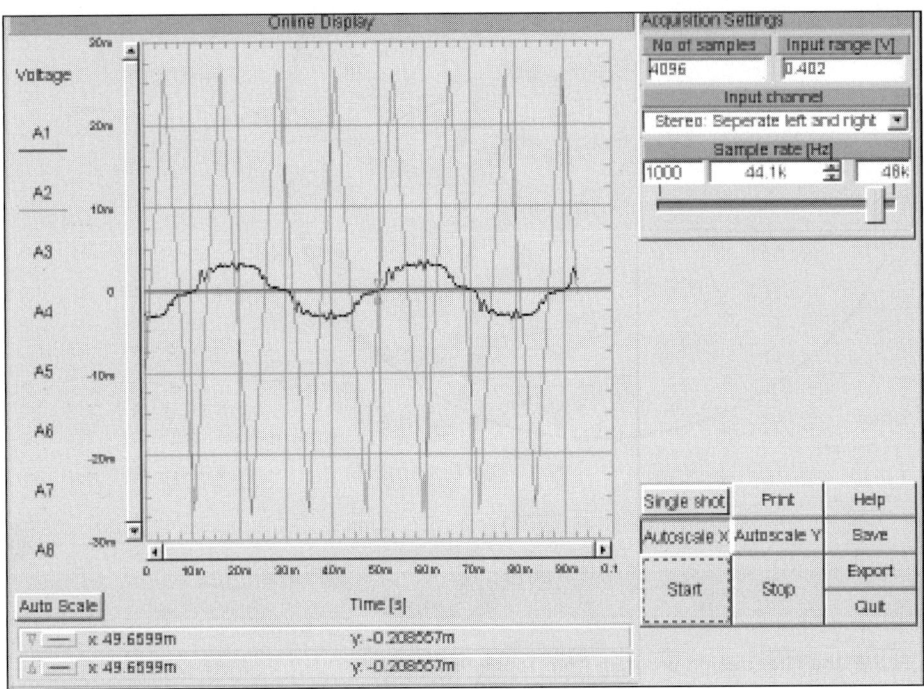


**Fig 21.3: A sound-card-based signal analyser with data logging facilities [From http://www.hacker-technology.com]**

## Test and Measurement

A computer can be used to make all sorts of measurements with the appropriate software, and either a sound card or additional hardware (eg an analogue to digital converter).

Audio analysis software such as Spectrum Lab [7] uses the computer's sound card and can be used for such things as measuring distortion, intermodulation, noise, oscillator drift and modulation. It has a built-in tone generator.

Also using the sound card are programs that can measure and display audio signals (in fact up to several tens of kHz), including those at the output of a radio, and may also be used as a data logger for monitoring beacons. One of these is shown in **Fig 21.3**.

## Antenna Simulation

It is no longer essential to dismantle your antenna in order to experiment with ideas for a new one. Thanks to software originally developed for the US military, and made available in a usable form by amateurs, it is possible to make a 'virtual' antenna on the computer. You can then measure its performance, alter it and measure it again until something practical is found - or the project is abandoned as a failure. Work on a practical antenna need only take place when the 'virtual' one has been optimised.

As with all computer programs, it is possible to get the wrong answer, but this is usually a case of 'garbage in, garbage out'. A good knowledge of antenna theory and practice will help you avoid the major pitfalls. Information on how to obtain antenna simulation software can be found at [8].

Much more about the use of antenna simulation programs can be found in the chapter on Antenna Basics.

## The Internet

For very little money, you can have facilities at your fingertips that only a few years ago would have been the envy of the best equipped library in the world, together with communications facilities that have almost totally eclipsed the postal service.

Although the Internet is often used to describe the vast array of web sites, it is actually the network on which sits the web, e-mail and many other facilities. The most common are:

### World Wide Web

The 'web' is a repository for many millions of documents, including text, pictures, sound and video. Amateur radio is well represented with information on just about every aspect of the hobby.

There are two main ways to find the information you want. You could start with a large site that

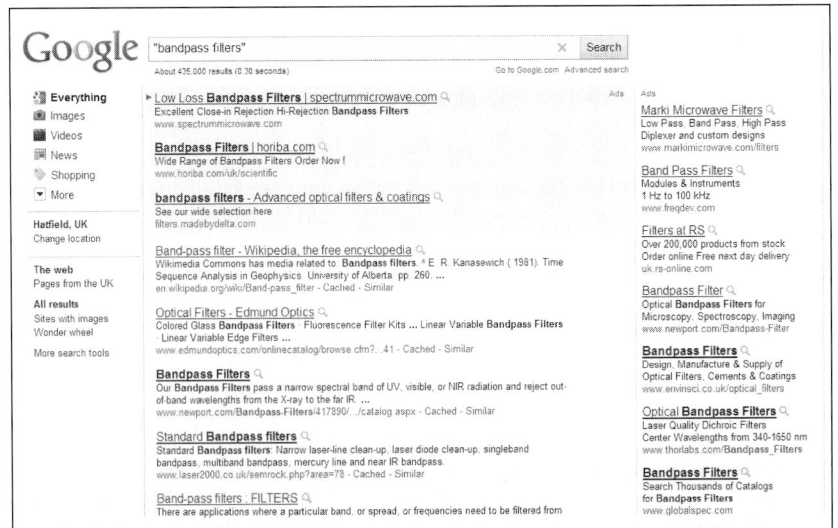


**Fig 21.4: An Internet search for the phrase "bandpass filters" produces nearly half a million results, including advertisements for commercial filters, technical papers, descriptions and advice by radio amateurs, and lists of where to buy components**

covers most aspects of amateur radio, such as those run by the Radio Society of Great Britain [9] or the American Radio Relay League [10], and follow the links to sites carrying additional information.

Alternatively, use a 'search engine' such as Google [11] or Yahoo [12], type in one or more keywords (eg circuit simulation) to get a large list of sites that may be useful (see **Fig 21.4**).

The sort of information that can be found on the web includes:

- News and events diaries
- Calculators for component values
- Equipment modifications
- Component catalogues
- Advice from experienced amateurs
- DXpedition information and logs
- Circuits and construction projects
- Propagation and solar data

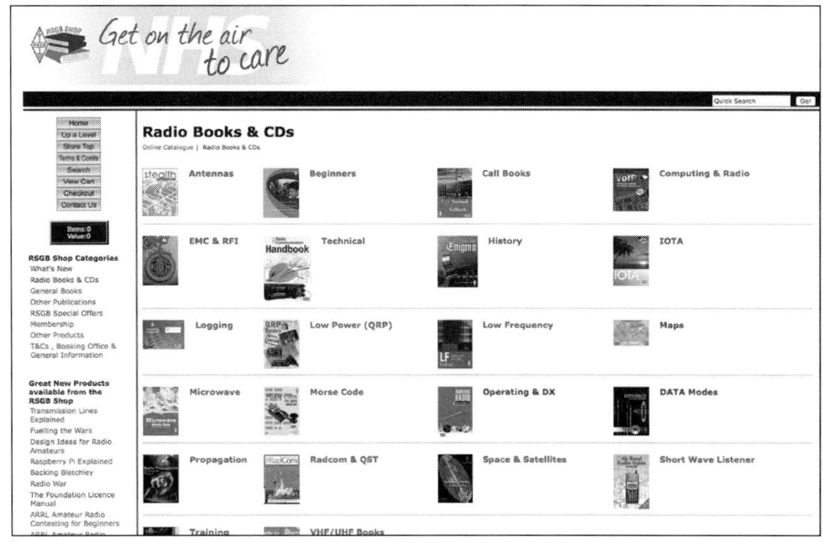


**Fig 21.5: Part of the RSGB's e-commerce site where books, CD-ROMs, maps etc can be bought by credit/debit card. It includes a search facility and detailed descriptions for those who like to browse before they buy**

## E-commerce

Most shops have a presence on the world wide web, and some are available only via the 'web'. This includes major component suppliers as well as much smaller specialist outlets. All display their wares, and most encourage electronic sales by credit card. This has two advantages, you can browse without leaving your home and you can buy from overseas shops - note, though that VAT, import duty and other charges may be payable on entry to the UK.

The on-line catalogues of Farnell [13], Maplin Electronics [14}, and RS Components [15] and many others are not only places where components can be purchased, but also a really good source of information such as data sheets.

Auction sites such as eBay [16] are where second-hand (and new) bargains may be found. Simply enter a search term (eg "ATU") and choose from an array of items. As with all 'blind' purchases, such as a newspaper advertisement, the buyer should take precautions to prevent fraud. If possible visit the seller and check before you buy.

One of the first examples of e-commerce was book selling, and the Internet makes it possible to search for what you want from millions of books, including many specialist publications that you will never see in a high street shop. Amateur radio books can be bought direct from the ARRL [10] and the RSGB [9] (**Fig 21.5**).

All sorts of radio and electronics books - even technical papers -can be searched for (by keyword, author or title) on big sites such as Amazon [17] which often has second-hand books listed alongside the new ones. Again, auction sites are a good source of second-hand and antique books.

### E-mail

Although your main contact with other amateurs may well be on the air, or at the local radio club, e-mail can still be useful. It can be used to maintain a dialogue with like-minded amateurs all over the world, and allows the exchange not only of text but also pictures (including circuit diagrams), sound files and programs.

### Groups

Similar to e-mail are newsgroups and reflectors. These allow groups of amateurs who have something in common, for instance an interest in VHF contesting or the city they live in, to share news and information with all of the other members. if you are new to a particular aspect of amateur radio, this is often the place to get advice from those with much more experience than you.

Some groups, such as those hosted by Yahoo or Google require you to enter a password (available free) before gaining access. A good example of a special interest group is https://groups.yahoo.com/neo/groups/picastar-users/ which is used by those interested in the PIC-A-STAR project to compare notes and get help.

## CDs and DVDs

Publications such as the ARRL's *QST*, RSGB;s *RadCom* (including archives going back to pre-WWII days) and books such as this one are also available in full on CD and DVD. In addition to saving shelf space, digital books and magazines are usually fully-searchable (ie any word or phrase can be searched for - much more useful than a conventional Index). Visually impaired people can considerably magnify each page on screen if required, or even use a program that reads the publication out loud.

## Developing External Hardware

The PC can be used to develop software that will eventually be copied to a chip in order to drive stand-alone equipment. This is often called embedded software, or firmware. An example of this is development of software for running on microcontrollers such as the family of PIC devices, described later on in this chapter.

Other external programmable hardware includes gate arrays and Field Programmable Gate Arrays (FPGA) which are the modern replacement for the old TTL and CMOS chips.

A wide range of these devices from those equivalent to a few logic gates, to chips with millions of programmable cells, made by Xilinx and Altera, is available. Both of these companies provide free development software, which after a bit of practice will allow both simple and complex logic (and even some analogue) functions to be built using just the low cost chip and a few external components. There is even a choice of ways FPGA circuitry can be designed - from use of a pseudo high level language such as VHDL or by thinking in terms of logic gates and drawing out the functions schematically as if they were going to be made from separate logic gates.

Programming and development using FPGAs is outside the scope of this Handbook, but full details of two companies that

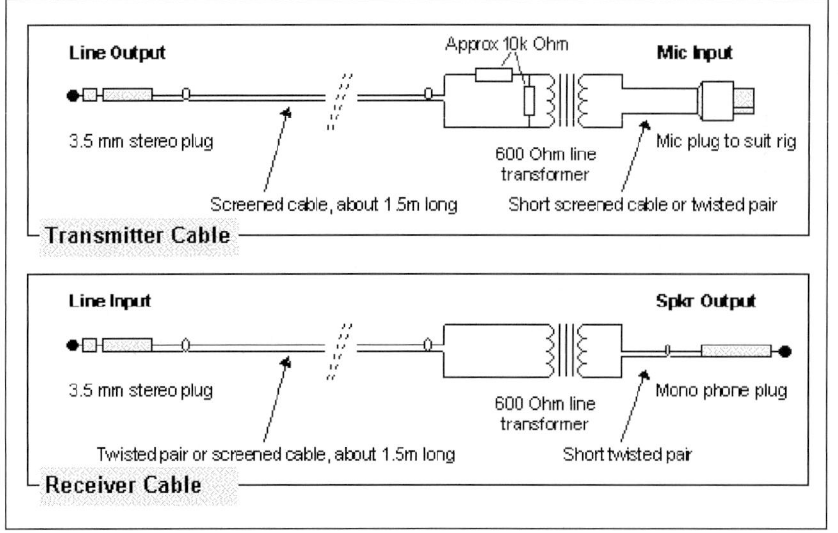


**Fig 21.6: The simple cables used to connect PC to radio**

provide devices and support hardware and software can be found at [18, 19].

The chapter on Software Defined radio also deals with the use of Programmable Logic.

# CONNECTING TO THE REAL WORLD

## Sound and Audio Interfacing

To operate digital modes, a modern computer with sound card is required (computer specification depends on software, but an old 1GHz laptop will provide a lot of pleasure). Some simple cables, easy to make, are also required, or a commercial interface such as the RIGBLASTER™ can be purchased. A conventional HF SSB transceiver is used, connected via the receiver audio output and microphone, or ideally line, audio input. It is best if this is a modern solid-state unit, with good filters and low drift, but many operators use older rigs for RTTY, MT63, Hellschreiber and SSTV with no particular problems. These are the modes least affected by drift and poor frequency netting.

Some of the newer modes require very high stability, and very low frequency offset is necessary between transmit and receive. Most synthesised transceivers will suffice. A transceiver that drifts less than 5Hz per over will operate the newer modes very successfully. Unfortunately offset cannot be accurately corrected by using the Receiver Incremental Tuning (RIT).

Older VFO-controlled transceivers with poorer stability can still be used with wider bandwidth modes such as RTTY and MT63. DominoEX, while a relatively narrow-band mode, is very effective with older rigs, as it was specifically designed to be immune to drift and frequency offset.

The connections between computer and transceiver are quite straightforward. Most amateurs should be able to build the required cables. **See Fig 21.6.**

The resistors in the transmit cable are used to attenuate the sound card signal so that it does not overload the transceiver. If the microphone socket is used, a lower value of resistor may be necessary across the transformer to further attenuate the audio. If an accessory socket is used, the values shown may suffice.

While the receiver cable is shown with a connector to be directly plugged into the external speaker socket on the receiver, with many rigs this will disconnect the speaker, which isn't helpful. It is best in this case to use an adaptor allowing both the PC cable and an external speaker to be connected.

There is a very good reason for not using the computer speakers instead of an external speaker on the transceiver. Computer speakers receive their audio from the sound card output, and by connecting the LINE IN signal from the radio to the LINE OUT or SPEAKER OUT and the speakers, the receiver output signal will also be sent to the microphone input of the transceiver. This causes feedback problems, especially if VOX is used.

### The Importance of isolation

The transformers shown in Fig 21.6 provide complete DC isolation between the computer and the radio transceiver. The most compelling reason to do this is to prevent serious damage to the radio and computer.

Most power supplies are grounded for safety reasons. If the power supply cable to the transmitter becomes loose, the full 20A transmitter current can pass through the microphone circuit, down the cable and through the computer sound card to ground via the PC power cable. Even if the transmitter power cable is considered reliable, significant current could still flow through the sound card cable, causing instability, hum and RF feedback. The simple expedient of isolating the connections also reduces the risk of RF in the computer, and computer noises in the radio.

### External soundcards

Although the computer's internal soundcard will fulfil the majority of requirements for operating most digital or data modes (see the chapter on Digital Communications) there are circumstances where an additional external audio interface would be useful.

For instance, a second soundcard module can be useful if a

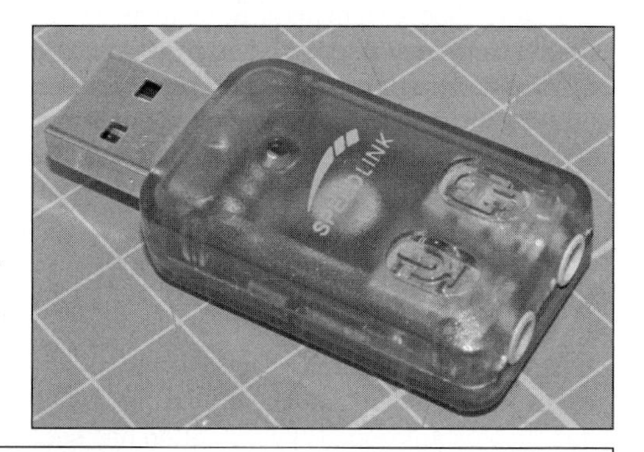

**Fig 21. 7: (top) USB headphone dongle; (bottom) high-end external USB soundcard**

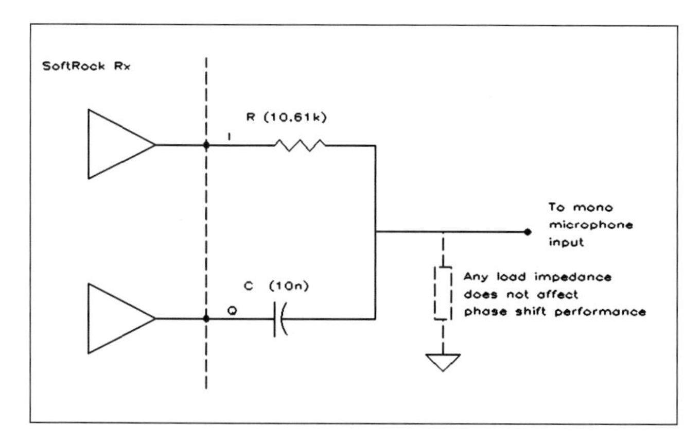


**Fig 21.8: Combining I and Q channels to achieve sideband rejection over a narrow frequency range**

Software Defined Radio normally requires the computer's own soundcard for its audio, but another input / output is still required for running data modes (see the chapter on SDR).

While there are ways of transferring audio in software such as the Virtual Audio Cable software package [20] it may be easier to use a second soundcard and cross-connect audio cables between the two. Using a second, or even multiple soundcards can also make it easier to connect multiple receivers, modems and test equipment without having to continuously unplug and rearrange connectors. Modern operating systems will accept multiple soundcards without even blinking.

A second soundcard can be installed into one of the computers extension slots, but these days the USB interface offers a more versatile route.

There is a wide range of external USB sound adapters on the market; these range from low cost headphone / microphone adapters such as the top unit shown in **Fig 21.7** up to the high quality top end external line interface such as that shown below it.

The high-end units offer proper line inputs and outputs and can usually be connected in exactly the same way as described above. Being designed for high end audio and music processing, some include audio isolation and proper decoupling.

Low cost headphone adapters, often referred to as 'headphone dongles' usually have to be treated differently as they only have a single mono microphone input and a stereo output jack for directly feeding headphones. To avoid the need for large value electrolytic capacitors inside the module, the headphone output is usually delivered as a directly coupled signal from the audio amplifier with all connections, including the common - the barrel of the 3.5mm jack socket - sitting above ground at typically 2.5V. The headphone signal lines swing either side of this to give a bi-directional drive across the headphone. The presence of this mean DC level gives us a problem if we want to use the audio output for wrapping round to another audio input, or to go off to a modem. If the common point is connected directly to ground, this DC will be shorted, probably resulting in damage to the headphone adapter unit.

A simple solution is just to insert DC isolation in the form of capacitors in series with both ground and audio paths. However, transformer isolation as shown in Fig 21.6 is the preferred solution as it offers additional noise immunity by separating audio and computer ground connections.

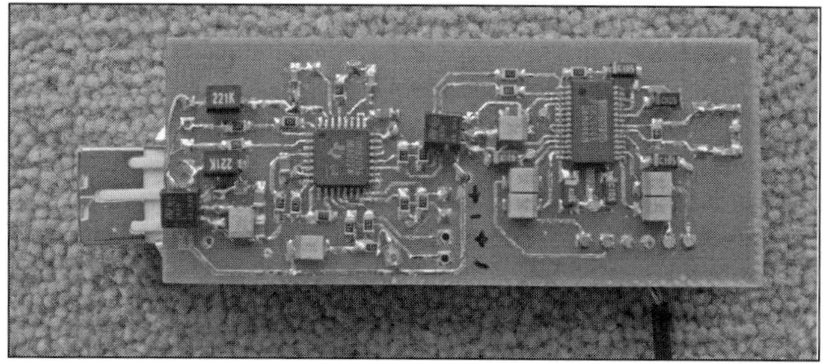

**Fig 21.9: USB Codec and Hub for a stand alone soundcard module suitable for building into homebrew projects. An FTDI232 chip on the reverse side of the PCB gives a serial COM port**

Also, beware that some very low cost units (such as 'eBay specials') provide an unfiltered Pulse-Width Modulated digital drive direct to the headphones. These will require extra filtering to remove the PWM switching components at a few tens of kHz and are best avoided, but the PWM signal may be useful as a drive to an EER type transmitter (see the LF chapter).

The mono microphone input is ground-referenced so no harm will come with a direct connection, although transformer coupling will help in reducing RF and digital induced noise into the audio path. Having a mono input prevents a full stereo I/Q input from SDR front ends such as the Softrock from being able to make use of such adapters.

If only a narrow bandwidth is wanted, for example of no more than a couple of hundred Hertz wide to use with a narrowband mode like WSPR, a simple phase shift circuit can be used to combine I/Q channels into one for sideband rejection. The circuit of **Fig 21.8** will give sideband rejection over a narrow range around 1500Hz when driven by the I/Q audio outputs from an SoftRock or similar SDR.

Another use of low cost headphone adapters is to build them into the actual equipment, such as an SDR or modem, for which audio interfacing is required so that a single USB interface cable is all that is needed to connect to a PC.

The five volt signal present on the USB can be tapped off to power external hardware by opening up the adaptor casing and connecting to the two outer pins of the USB socket. A continuity check will show which of these is connected to ground and which must be the 5V supply.

A low cost pair of computer speakers - especially battery operated ones, will usually function with a 5V supply, and a headphone 'dongle' built into a pair of scrap speakers can make a low cost additional audio output for an SDR, while keeping the computer's own soundcard free for other purposes.

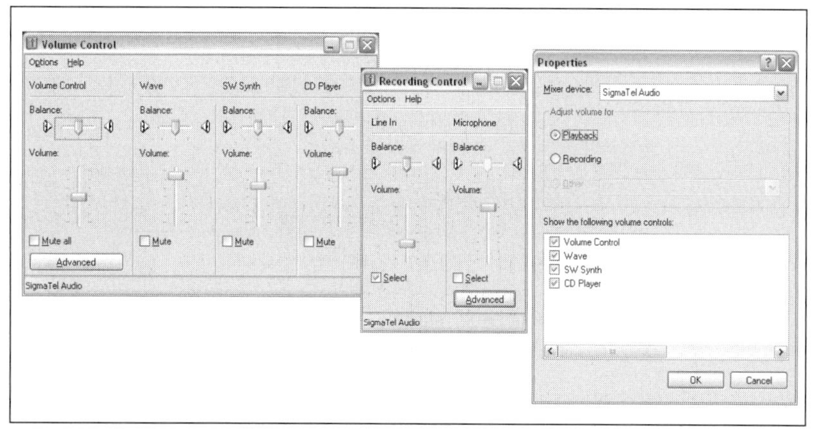


**Fig 21.10: Some of the controls available in the Windows 'volume control' software**

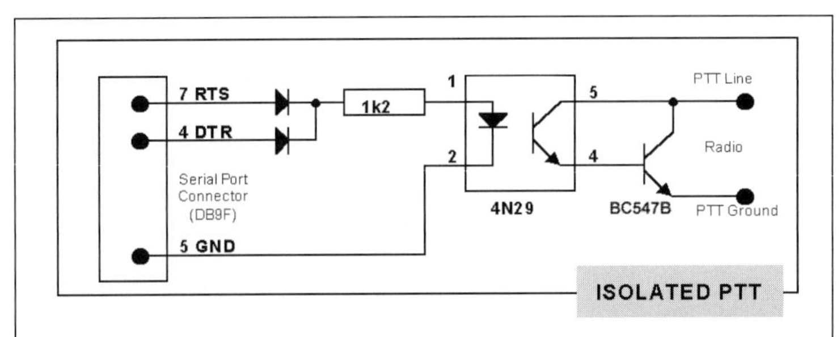


Fig 21.11: An opto-isolated PTT circuit

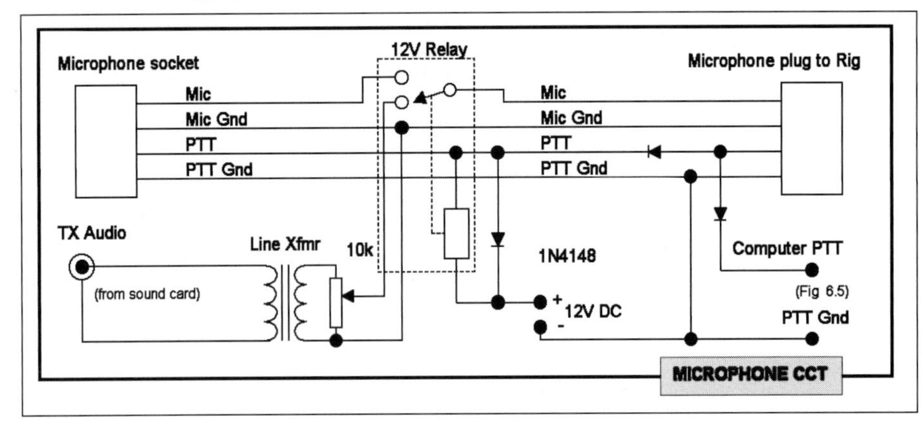


Fig 21.12: A simple way to connect microphone and computer

## Single chip USB soundcard

Another solution is to build a dedicated USB soundcard into any new receiver or transmitter project. The PCM2900 Stereo Audio Codec with USB interface is a single chip 28 Pin SSOP package audio input / output facility. With the addition of little more than a 12MHz crystal, a few coupling/decoupling capacitors and a 3.3V regulator, a high specification dual line input / line output becomes available as an additional soundcard. As the USB port supplies 5V at up to 100mA a USB powered plug-and-play project is feasible.

The PCM2900 is available from several major component suppliers such as Farnell [13]. The photograph **(Fig 21.9)** shows one of these chips mounted on a PCB along with a TUSB2036 three port USB hub chip. Not shown, on the underside of the board is an FTDI FT232R chip providing a USB COM port.

## Soundcard controls

Recording and playback levels, as well as input / output selection are made using the computer's own operating system, and invariably accessed on Windows machines by clicking on the loudspeaker icon in the lower right taskbar. This usually brings up a 'Sound Mixer' or 'Volume Control' window similar to that in **Fig 21.10**. The menu options vary between manufacturers, but it will always be possible to select between all the Input and output connections, to set recording and playback (volume) levels and to select which soundcard is the default. Hidden settings, or those not on immediate display, can usually be found in 'Advanced' or the Record menu options.

## VOX and PTT Control

Most operators find VOX operation of digital modes quite appropriate and reliable, although the delay may need to be set longer than for Morse or SSB. If for some reason direct control of the rig is necessary, the transmit control must also use an isolated circuit. An opto-coupler does this nicely, driving the Press-to-Talk (PTT) directly without requiring a relay or any further power supplies.

The digital mode software usually controls the transceiver via a serial port, by driving RTS or DTR (often both) positive on transmit, with an appropriate delay before sending tones out from the sound card. The design in **Fig 21.11** is an appropriate PTT circuit for a transceiver with positive voltage on the PTT line and a current when PTT is closed of up to 100mA or less.

Many transceivers include an 'accessory socket', offering line-level audio inputs and outputs for transmit and receive. Using these instead of the microphone socket and speaker socket can be really convenient, but can lead to a range of unexpected problems.

Sometimes PTT is not available from the accessory socket, and sometimes the VOX does not operate from this socket. The signal levels can also be quite different to the speaker and microphone connections. Even more troublesome, some transceivers leave the microphone operating while sending data through the accessory socket, so coughs, mutterings and keyboard clatter go out over the air!

A simple home-made adaptor (see **Fig 21.12**) provides a way to operate voice and data modes interchangeably without disconnecting anything. Using this design, the data transmit cable is connected by default, but when the microphone PTT switch is

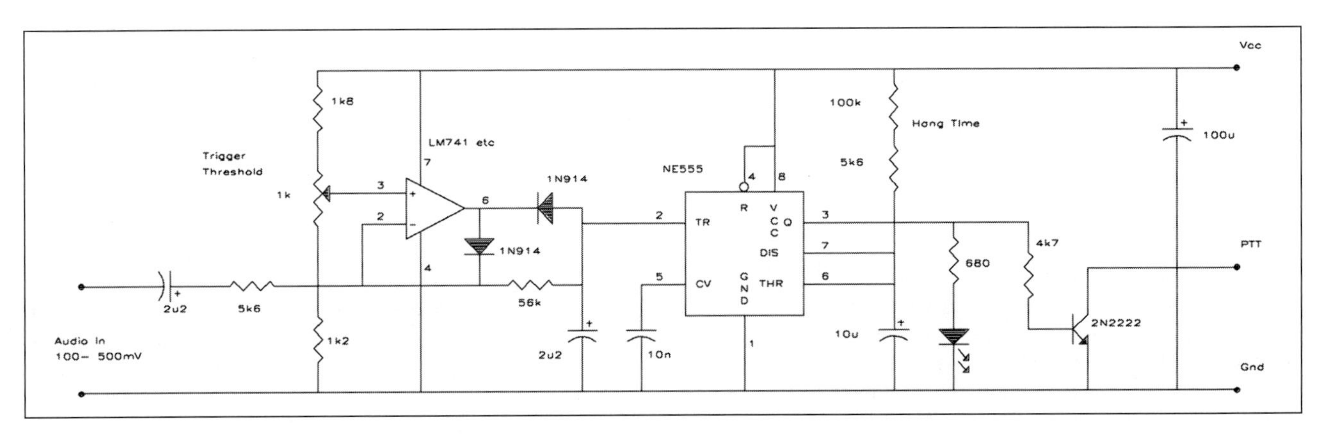


Fig 21.13: An audio-operated VOX switch is useful for when direct PTT control is not possible

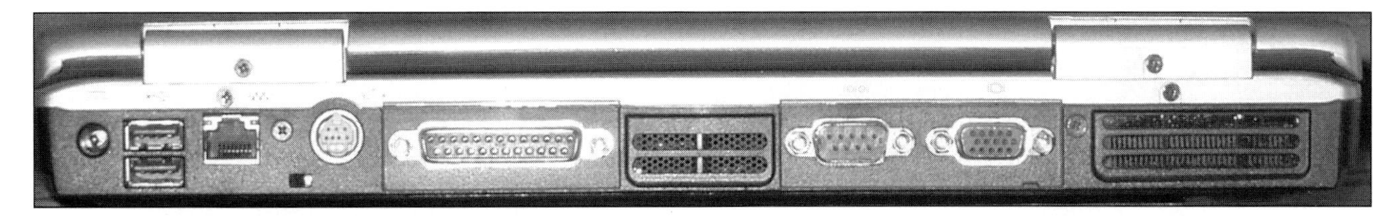

**Fig 21.14: Input and output connectors on a laptop computer. From left: Two USB ports, network cable, video out, parallel port (25-way), serial port (9-way) and external monitor**

depressed, the relay switches over the audio input and normal microphone use occurs. Operation is simple, and feels natural. The isolated PTT circuit of Fig 21.11 can be built into the same box, and the whole assembly replaces the transmit cable in Fig 21.6. There are several similar designs offered as kits [21].

There are several commercial interface designs available for users not disposed to building a home-made or a kitset interface, but not all provide full isolation. There are also USB interfaces suitable for laptop computers and others with no serial port or sound card.

An area that causes confusion among beginners is the business of setting up and adjusting the sound card. The adjustments are all performed in software, mostly using an application provided with the operating system, and once set for one mode or program, the settings should be correct for all the rest. There are two main software adjustments, for transmit and for receive, and it is not very obvious where to find these, especially the receiver adjustments. The better applications provide direct access to the adjustments. In addition to the gain settings, you need to select the correct inputs and outputs, and disable those not being used. The procedure and these adjustments are described in detail in the RSGB publication *Digital Modes for All Occasions* [22], a reference work recommended for both novice and experienced operators.

| Pin No. 25-Pin | 9-Pin | Function | Signal Name | Direction of signal flow |
|---|---|---|---|---|
| 1 | - | Protective Ground | | |
| 2 | 3 | Transmit Data | TXD | DTE > DCE |
| 3 | 2 | Receive Data | RXD | DTE < DCE |
| 4 | 7 | Request To Send | RTS | DTE > DCE |
| 5 | 8 | Clear To Send | CTS | DTE < DCE |
| 6 | 6 | Data Set Ready | DSR | DTE < DCE |
| 7 | 5 | Signal Ground | GND | |
| 8 | 1 | Data Carrier Detect | | |
| 20 | 4 | Data Terminal Ready | DTR | DTE > DCE |
| 22 | 9 | Ring Indicator | | |

**Table 21.1:RS232 Port connections and signals**

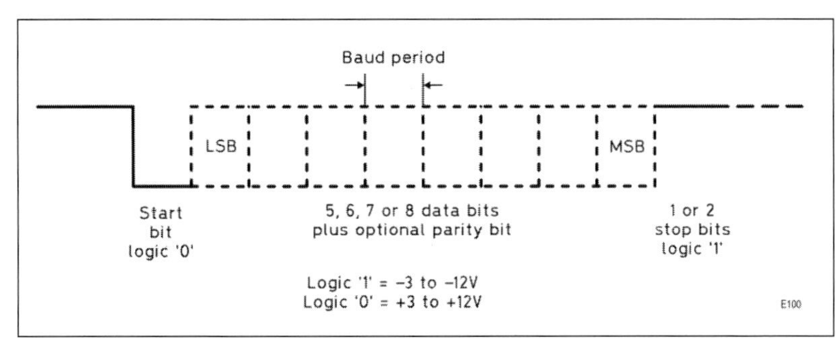


**Fig 21.15: RS-232 Serial bit format**

Sometimes it is not desirable, or feasible, to use direct PTT control. Many modern PCs do not have a serial port, and those that do sometimes pulse the control signals on it which can cause unexpected transmit switching. The audio VOX circuit shown in **Fig 21.13** can be used to overcome this problem. The audio from the soundcard, which must be activated only during transmit, is rectified in a precision rectifier and fed to a comparator. When its level exceeds a threshold it triggers a timer to holdover short breaks in the audio. The output from the timer interfaces to the PTT control. Adjust delay and threshold settings to best suit your preferences and audio drive level.

## I/O Sockets

On modern computers the I/O sockets offer mainly 4 USB ports and on PCs a serial port. On some computers such as the apple laptop, only offer a couple of type C ports which can be converted with a lead to a standard USB port. Previously computers used to offer one or two serial ports for connections to a modem or plotter for example, and a parallel port for a printer, most printers nowadays use a USB cable.

Serial means that the eight bits in a byte of data are sent one at a time down a single wire, whereas a Parallel port sends the bits simultaneously but needs eight wires for each byte, plus other lines for control. Both of these were slow, and could not provide an easy 'plug-and-play' solution for the growing range of computer accessories becoming available. To cope with this, the more modern USB interface was developed to overcome the limitations, and USB has now taken over just about all aspects of computer interconnectivity **Fig 21.14** shows the range of input and output ports available on a typical older laptop computer.

### The serial ports

These are often called 'COM' or RS-232 ports, and were traditionally used for external modems (including packet radio TNCs), connecting to other computers, plotters and some early printers and other media COM ports are rarely fitted to modern PCs, but they are widely available as external USB adapters. Units offering multiple COM ports on a single USB connection are available. Ref [23] has more details of USB interfaced COM ports.

**Table 21.1** gives the pin connections for 9-pin and 25-pin serial connectors. Data is transmitted serially using stop and start bits to synchronise each end of the link, as shown in **Fig 21.15**.

The speed of transmission can be set over a wide range of values to suit the equipment and data transfer requirements, but 9600 bits per second (otherwise known as 9600 baud) is a useful mid-range value for many activities. Higher speeds up to 115200 baud (and in some cases depending on hardware, even higher rates), can be set. The number of bits sent can also be adjusted over the range 5, 6, 7 or 8, although these days 8 bits is the only real practically useful setting. Finally, simple error checking is provided for by the

**Fig 21.16: FDTI USB module**

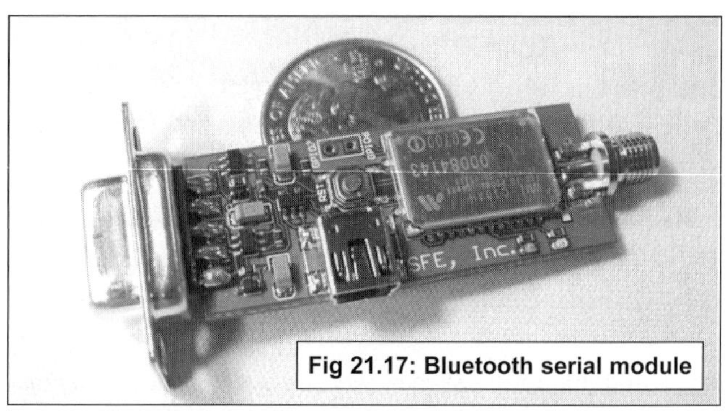

**Fig 21.17: Bluetooth serial module**

use of an optional parity bit.

This huge range of speeds and data formats, with its limited speed made the serial port a nightmare to set up in some cases, and contributed to its fall from favour for modern usage. However, there is one very big advantage, that makes the serial port concept live on. It is absolutely universal, everyone understands it and can implement the standard whatever computer language, operating system or hardware is in use!

Every computer made has always, in some manner or other, incorporated a serial port. This makes it the most widely supported format, ever, for interfacing between external hardware and in spite of its complexities in setting up, continues to live on. In many cases the inelegance of the multitude of serial formats is hidden from users, who just see a connection, but the serial RS232 format is there, hidden beneath a layer of software that has done the configuring automatically. A prime example of this is seen in GPS modules that communicate with their host over a Bluetooth radio interface. With suitable Bluetooth hardware the serial data can be seen and decoded to a form for connection to a true serial port. Serial interfaces, or UARTS, are also built into just about every microcontroller there is.

The other advantage of the serial port, is that, unlike the parallel port (see below) it is still fully supported in all programming languages for the latest operating systems. So writing your own software for communicating with the outside world does not require special drivers specific to the operating system.

Serial port support can be integrated seamlessly, usually from a pull down menu or by a line or two of high level programming code.

## Parallel port

Parallel connectors, or printer ports, were commonly used for connecting printers before USB ports became universal. They are characterised by having parallel 0/5V TTL level outputs (eight data and four handshaking) and five status inputs on a 23-pin connector. The use of logic levels made the parallel port a favourite for interfacing to external logic hardware, and a number of the older chip device programmers, such as for PICs and CPLDs, used the printer port for their interface.

Since the advent of Windows 95 and later operating systems for PCs, direct access to the printer port became more difficult, requiring special software drivers, so its use as an optional I/O port has gradually diminished. Most modern machines no longer even offer a parallel port, with printers now going via USB or Network connections.

## The Universal Serial Bus (USB)

Nowadays, the USB port reigns supreme on modern computers. It has solved many of the compatibility problems experienced with the older I/O systems, and offers a plug-and-play solution, having the ability to interrogate peripherals as they are connected and prompt for (named) drivers if new equipment is introduced. Furthermore, the USB port carries its own 5 volt power, allowing most low power accessories to be self-powered and removing the need for separate power supplies.

But the USB bus is not easy for amateurs to use directly - the USB specification runs to several hundred pages. However, the situation has improved recently as a range of chips have become available that allow users to interface to PCs via USB, and to make the connection completely seamless and transparent.

FTDI Chip [23] supply two integrated circuits, the FT232 and FT245 devices with very useful properties. The FT232 (as its type number might suggest) allows a virtual serial port to be established. The chip connects directly to a USB cable on one side of the interface and can extract the USB port's own power to use for accessories. The user side of the chip supplies the serial lines that would be seen on a COM port, receive and transmit data and two handshaking lines.

Drivers supplied with the device (in practice these can be downloaded from the manufacturer's web site) configure the interface as a new COM port which can be opened from any PC software capable of communicating with a serial port - which can be your own software written in any language of choice.

The FT245 chip adds another level of interfacing capability. As well as implementing the same Virtual Com Port mode, this device adds a high speed parallel interface with FIFO (first in, first out) suitable for connecting to custom logic and microprocessors; sending and receiving data at megabit per second rates. Full details, datasheets and programming details can be found in [23].

Complete USB-RS232 leads and modules are available that plug into a USB port on a PC and provide a 9-pin serial interface with correct signalling voltages, all powered from the USB port itself. These can provide a quick and easy solution for adding a serial port to a laptop or modern desktop. One such module can be seen in (**Fig 21.17**).

USB Parallel port interfaces are also produced, but unlike the COM port there is no universal way of addressing these. In practice they are only useful for adding an older style printer without USB capability. However, it may be possible to make some exter-

**Fig 21.18: XPort Ethernet interface module**

nal devices or hardware 'look' like a printer, in which case such an interface could prove useful.

### Bluetooth

Another option is to use the short range radio interfacing offered by Bluetooth. Working at 2.4GHz with a few milliwatts of power, this bi-directional radio link gives wireless communications over ranges of a few tens of metres. The protocols permit networking, printers, serial communications and audio to be carried between suitable peripherals. A Bluetooth serial module allows pseudo COM ports to be added in the same way as USB ones.

A Bluetooth serial interface is shown in **Fig 21.17**. Modern laptops often come with Bluetooth capability already built in, and interfacing to such a module is merely a matter of 'pairing' the devices, then opening the COM port that is then allocated. For older computers and desktops without Bluetooth capability, a Bluetooth 'dongle' can be inserted into a spare USB port, or a Bluetooth card can be installed. Again, the serial protocol offers wide versatility when it comes to programming and using your link with the outside world.

### Network or Ethernet

Normally associated with inter-computer linking, the Ethernet port is increasingly being used as a high speed interface in its own right. With a data capability up to 100 MBits/s (with 1GB/s systems now appearing), it exceeds the capacity possible with USB and is the interface of choice for several Software Defined Radios.

A stand alone module such as the XPort from Lantronix shown in **Fig 21.18** permits a pseudo COM port (again!) to be set up, operating at speed up to a few MB/s. The module is set to a unique IP address on the network, and data is communicated to and from it as if it were any other network node, such as another computer or a modem.

In fact, the module contains a web page in its own right, and can be accessed using browser software such as Internet Explorer. Being a networked accessory, by suitably setting up IP addresses in your router or modem, it is possible to directly access this module from anywhere in the world via the Internet. This opens up the possibility for very long distance remote control.

**Fig 21.19: The Raspberry Pi single board computer model-B rev 1**

## Controlling your Radio

Many modern base-station transceivers incorporate sockets so that they can be controlled by a computer using special software. Station control programs, such as those used by DXpeditions and contesters, have facilities to control at least transmit and receive frequency.

This can be extended to being able to control your station remotely over a modem (with the appropriate approval, if necessary), as described in [24].

Most rigs use a serial RS232 protocol, so the connection methods mentioned above can all be used. There are considerable differences between the control protocols from different manufacturers, but all are well documented in their respective manuals, and most transceiver control software can cope with all types, which can be changed in the setup menu.

| **Raspberry-PI Specifications for the Model A and Model B** | |
|---|---|
| **SoC:** | Broadcom BCM2835 (CPU, GPU, DSP, SDRAM, and single USB port) |
| **CPU:** | 700MHz ARM1176JZF-S core (ARM11 family, ARMv6 instruction set) |
| **Graphics Processor:** | Broadcom VideoCore IV @ 250 MHz; OpenGL ES 2.0 (24 GFLOPS); MPEG-2 and VC-1 (with licence), 1080p30 h.264/MPEG-4 AVC high-profile decoder and encoder |
| **Memory (SDRAM):** | Model A 256MB (shared with GPU); Model B 512MB (shared with GPU) as of 15 October 2012 |
| **USB 2.0 ports:** | Model A 1 (direct from BCM2835 chip);. Model B 2 (via the built in integrated three-port USB hub) |
| **Video input::** | A CSI input connector allows for the connection of a custom designed camera module |
| **Video outputs:** | Composite RCA (PAL and NTSC), HDMI (rev 1.3 & 1.4), raw LCD Panels via DSI |
| **14 HDMI resolutions:** | From 640×350 to 1920×1200 plus various PAL and NTSC standards. |
| **Audio outputs:** | 3.5 mm jack, HDMI, and, as of revision 2 boards, $I^2S$ audio (also potentially for audio input) |
| **Onboard storage:** | SD / MMC / SDIO card slot (3.3V card power support only) |
| **Onboard network:** | Model A None; Model B 10/100Mbps Ethernet (8P8C) USB adapter on the third port of the USB hub |
| **Low-level peripherals:** | 8 × GPIO, UART, $I^2C$ bus, SPI bus with two chip selects, $I^2S$ audio +3.3 V, +5 V, ground |
| **Power ratings:** | Model A 300 mA (1.5 W); Model B 700 mA (3.5 W) |
| **Power source:** | 5 volt via MicroUSB or GPIO header |
| **Size:** | 85.60 mm × 53.98 mm |
| **Weight:** | 45g |
| **Operating systems:** | Arch Linux ARM,[2] Debian GNU/Linux, Gentoo, Fedora, FreeBSD, NetBSD, Plan 9, Raspbian OS, RISC OS, Slackware Linux |

**Table 21.2: Specifications for the Models A and B of the tiny Raspberry Pi single board computer**

| Mnemonic. | Operands | Description | Cycles | Status Bits Affected | Notes |
|---|---|---|---|---|---|
| **Byte Oriented File Register Commands** | | | | | |
| ADDWF | *f* , d | Add W to *f*, place result in d (*f* or W) | 1 | C,DC,Z | 1,2 |
| ANDWF | *f* , d | AND W with *f*, place result in d (*f* or W) | 1 | Z | 1,2 |
| CLRF | *f* | Clear *f* | 1 | Z | 2 |
| CLRW | | Clear W | 1 | Z | |
| COMF | *f* , d | Complement *f* place result in d | 1 | Z | 1,2 |
| DECF | *f* , d | Decrement *f* | 1 | Z | 1,2 |
| DECFSZ | *f* , d | Decrement *f*, skip next instruction if zero | 1 or 2 | | 1,2,3 |
| INCF | *f* , d | Increment *f* | 1 | Z | 1,2 |
| INCFSZ | *f* , d | Increment *f*, skip next instruction if zero | 1 or 2 | | 1,2,3 |
| IORWF | *f* , d | Inclusive OR W with f | 1 | Z | 1,2 |
| MOVF | *f* , d | Move *f* to destination | 1 | Z | 1,2 |
| MOVWF | *f* ,d | Move W to *f* | 1 | | |
| NOP | | No Operation | 1 | | |
| RLF | *f* , d | Rotate Left *f* through carry | 1 | C | 1,2 |
| RRF | *f* , d | Rotate Right *f* through carry | 1 | C | 1,2 |
| SUBWF | *f* , d | Subtract W from *f* | 1 | C,DC,Z | 1,2 |
| SWAPF | *f* , d | Swap nibbles in *f* | 1 | | 1,2 |
| XORWF | *f* , d | Exclusive OR W with *f* | 1 | Z | 1,2 |
| **Bit Oriented File Register Commands** | | | | | |
| BCF | *f* , b | Clear bit B of register *f* | 1 | | 1,2 |
| BSF | *f* , b | Bit Set *f* | 1 | | 1,2 |
| BTFSC | *f* , b | Bit Test *f*, skip next instruction if clear | 1 or 2 | | 3 |
| BTFSS | *f* , b | Bit test *f*, skip next instruction if set | 1 or 2 | | 3 |
| **Literal and Control Operations** | | | | | |
| ADDLW | k | Add literal to W | 1 | C,DC,Z | |
| ANDLW | k | AND literal with W | 1 | Z | |
| CALL | k | Call Subroutine | 2 | | |
| CLRWDT | | Clear Watchdog timer | 1 | | |
| GOTO | k | Go to address | 2 | | |
| IORLW | k | Inclusive OR literal with W | 1 | Z | |
| MOVLW | k | Move literal to W | 1 | | |
| RETFIE | | Return from Interrupt | 2 | | |
| RETLW | k | Return with literal in W | 2 | | |
| RETURN | | Return from Subroutine | 2 | | |
| SLEEP | | GO into standby / sleep mode | 1 | | |
| SUBLW | k | Subtract W from Literal | 1 | C,DC,Z | |
| XORLW | k | Exclusive OR literal with W | 1 | Z | |

*NOTES:*
1) *When an I/O register is modified as a function of itself (eg MOVF PORTB, f) the value used will be that value present on the pins themselves rather than that stored in the output registers*
2) *Clears prescalar when executed on the TMR0 register*
3) *If the Programme Counter (PC) is modified or a conditional test is true, the instruction requires two cycles. The second cycle is executed as a NOP*

**Table 21.3: PIC 16Fxxx instruction set**

## Communications

Amateurs have been using home computers for making contacts ever since they became available - RTTY programs were available for the 'BBC-B' computer, for example. The chapter on Data Communications gives much more information.

Towards the end of the 20th century, ways were developed to link amateur radio and the Internet, so that radio amateurs can communicate even though one of them has no radio, or to interlink repeaters via the Internet. Three systems are in use at the time of writing: Echolink [25], eQSO [26] and IRLP [27].

## RASPBERRY PI

The Raspberry Pi is a single-board computer about the size of a credit-card (see **Fig 21.19**), developed in the UK by the Raspberry Pi Foundation with the intention of promoting the teaching of basic computer science in schools. It is manufactured in two board configurations. The specification is shown in **Table 21.2**.

The Pi has a Broadcom BCM2835 system on a chip (SoC), which includes an ARM1176JZF-S 700MHz processor, VideoCore IV GPU, and was originally shipped with 256 megabytes of RAM, later upgraded to 512 MB. It does not include a built-in hard disk or solid-state drive, but uses an SD card (the memory used in the majority of digital cameras) for booting and persistent storage.

The Raspberry-Pi Foundation provides Debian and Arch Linux ARM distributions for download. Tools are available for Python as the main programming language, with support for BBC BASIC (via the RISC OS image or the Brandy Basic clone for Linux), C, Java and Perl.

A number of Raspberry Pi specific peripheral devices and cases are available from third-party suppliers.These include the Raspberry Pi Foundation sanctioned Gertboard, which is designed for educational purposes, and expands the Raspberry Pi's GPIO pins to allow interface with and control of LEDs,

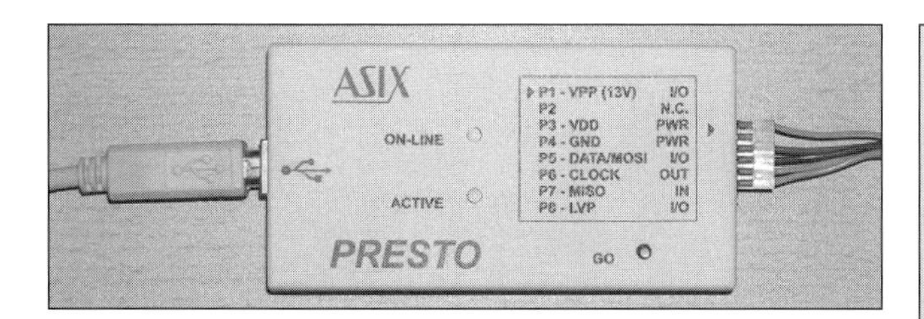


**Fig 21.20: The Presto PIC programmer**

switches, analogue signals, sensors and other devices. It also includes an optional Arduino compatible controller (see later in this chapter) to interface with the Pi.

There is no conventional analogue audio input or output. Most users use a USB "headphone dongle" or other USB audio device. Details of such devices are given elsewhere in this chapter.

More about the Raspberry-Pi can be found on the foundation website [28]. Many amateur radio applications have been written, and one place to discuss them is the Raspberry_Pi_4-Ham_RADIO group on Yahoo [29].

## MICROCONTROLLERS

Small microcontroller chips such as the PIC appear in several projects throughout this Handbook. Here a short overview of these is described, together with how you can program these for your own personalised tasks. Microcontrollers are single chips that can probably be considered the simplest to run stored programmes and execute actions based on the results. A microcontroller chip is usually a small processor that includes its own programme and data memory, a range of peripherals such as a serial interface and timers, and drivers for input and output pins; they are often designed to be used in completely stand-alone applications.

Programming a PIC for any task is similar to writing in any other programming language, although much more knowledge of the target processor is needed. Many programmers use a high level language such as C or Basic for microcontrollers, and many users prefer this route for the ease of programming it offers. The best compilers can 'hide' the complexities of the chip, but these versions can be quite expensive - low cost, and even free or shareware C language compilers do exist, but using these often gives little advantages over programming in assembler.

There is no doubt that for optimum processing speed and keeping resultant code size small, as well as understanding all the device's idiosyncrasies, the native assembler is more efficient than a high level language such as C. For some tasks however, such as string handling, a high level language can prove more suitable. **Table 21.3** shows the assembler commands available for the mid range PIC family

To develop PIC software (more correctly called firmware as it is embedded into a final application) a Personal Computer of some sort is invariably used to develop the code via various software tools supplied, usually free of charge, from the PIC manufacturers.

A programmer driven from the PC is then used to load the developed code into the chip. Programmers of different sorts and levels of sophistication can either be purchased from most electronic component suppliers, or simpler ones capable of programming many, but not necessarily all, of the PIC processor types can be quite easily built. A modern USB controlled programmer is shown in **Fig 21.20**. Most programmers run off either the parallel port or the serial port, and now most of the commercially manufactured units employ a USB interface.

There are four stages to producing a fully functional PIC design:

**Microchip PICkit 3**
A small versatile USB programmer produced by Microchip. It will handle all the modern devices although the obsolete 16F84 and a few other older devices cannot be programmed with it. As well as PIC programming, it can also be used with serial EEPROMS and offers in-circuit debugging and several other useful utilities. See the microchip website [30] for more details. The PICkit is stocked by most of the major component suppliers, and often comes bundled with a PIC development board - sometimes as part of a special offer.

**The WISP648 Programmer**
A design for home construction that requires a programmed PIC as part of the design. A chicken and egg situation arises here if this is to be the first programmer, but ready programmed PICs are available from the designer in [32].

**The Presto Programmer** [33]
A modern USB powered device. It can handle all microcontrollers and CPLDS as well as most serial EEPROM and Flash memory chips.

**Table 21.4: PIC programmers**

1. **Design the circuit and hardware**, making use of the PIC data sheets to determine which pins are to be used for each function - many of the special peripheral functions requires their interfaces to be on specific pins.
2. **Write the assembler code** - the list of instructions to perform the function required. This in the form of an assembler, or source, file in text format and usually generates a file with a name like xxx.ASM where xxx is the file name. There is usually a section that defines memory locations and register names and sets up peripheral functions, followed by the operating code itself. A range of 'include' files are provided by Microchip for each processor type to make naming programming easier by naming all the registers and individual control bits. On a PC, a text editor such as *Notepad* or the DOS 'EDIT' command can be used to write the source code.
3. **Assemble the source code into machine code**. The assembler generates a file xxx.HEX which contains the machine code in a form useable by the programmer. It also generates a full listing, xxx.LST which contains a repeat of the source code with all associated memory maps, variables and labels shown. Any errors in the source code such as illegal calls, mistyped labels or variable names or illegal values appear in an error file, xxx.ERR; this will be empty for a 'perfect' assembly. Warnings are also produced when certain operations that may cause potential problems in some circumstances are made. An excellent assembler is that produced by Microchip, called 'MPASM', which can be downloaded from [30] .
   MPASM is called from the command line by invoking it along with the filename of the .ASM file eg 'MPASM TESTPROG' After assembly is complete, it reports back on the number of errors encountered in the process along with any messages or warnings about the code. Details of these are stored in the. ERR file.
4. **Program the chip**. Most modern devices are programmed serially by making use of three of the pins plus ground. The master reset pin is raised to +13V while the chip is powered normally from its 5V rail, and the data is clocked in serially using two pins, one for clock and the other for data. Some

programmers have a Zero Insertion Force socket for programming raw chips; others just provide the four connections needed for in-circuit programming. Microchip supply a range of programmers and debugging tools and details of a several programmers suitable for home constructors are included in **Table 21.3**. This list is by no means exclusive and many more programmer designs can be found by searching the web.

An intermediate stage between steps 3 and 4 is available - that of simulation.

Host computer software takes the .ASM or .HEX files and simulates the functioning of the PIC hardware with the user code which can be single stepped if required, allowing all intermediate values, the state of all registers and I/O lines to be examined at any time for debugging. Alternatively, break points can be set and registers checked at this point. It is only a simulation, and the functioning of peripherals such as A/D converters has to be assumed. Timing and clock cycle count is shown directly, a function which can be very tedious if undertaken manually on the source code.

As an alternative, stages 1 to 3 can be replaced by a high level programming language. Special versions of Basic and C can be used to write code for the PICs (as well as most other types of microcontroller). These produce the.HEX file directly by the compiler and users rarely see the assembler code produced.

Programming in a high level language can be considerably easier for many people as it removes most of the need for a detailed understanding of the device's architecture. However, the amount of object code produced by most high level languages, particularly low cost or free ones, is generally larger than when writing the same functions in assembler. Therefore a high specification, more expensive, chip with more memory has to be employed.

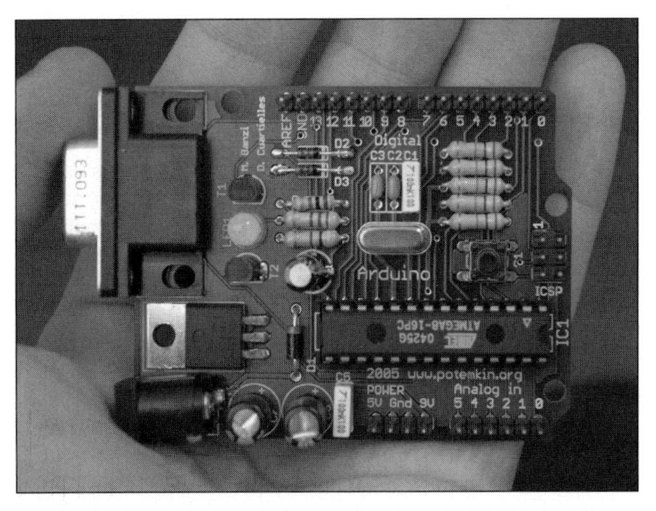


Fig 21.21: An official Arduino board with descriptions of the I/O locations

- **Microcontroller:** Atmel ATmega168
- **Operating voltage:** 5V
- **Maximum supply voltage:** 6 to 20V
- **Digital I/O pins:** 14 (of which 6 provide PWM output)
- **DC current per I/O pin:** 40 mA
- **DC current for 3.3V pin:** 50 mA
- **Flash memory:** 16kb of which 2kb is used by Boot loader
- **SRAM:** 2kb
- **EEPROM:** 1KB
- **Clock speed:** 16MHz
- **Power source:** Via the USB connection or with an external power supply
- **Communications & Networking**

Table 21.5: Features of the A000066 Uno board

Fig 21.22: The Pinguino [http://hackinglab.org/]

Also, writing time-critical code, such as that to be used in some simple DSP functions is impossible as the high level languages do not take account of processing time and clock cycles.

## DSPic

The latest addition to the Microchip stable is the new DSPic range which has a more complex instruction set aimed at digital signal processing as well as microcontroller functions. While being aimed mostly at the automotive and power supply / conditioning market, the range of instructions makes them suitable for many audio and SDR related tasks. [34] has more details.

## Other Microcontrollers

The Microchip PIC is not the only microcontroller available for experimenters, although it is probably the most well known. It is not the fastest and for significantly higher operation speed, Atmel's AVR microcontrollers come in a range of sizes and capabilities, and are the microcontroller of choice for many. More details can be found in [35].

## Arduino

The Arduino (**Fig 21.21**) is a single-board microcontroller, intended to make control and programming more accessible to everyone. The hardware consists of an open-source hardware board designed around an 8-bit Atmel AVR microcontroller, or a 32-bit Atmel ARM.

Pre-programmed into the on-board microcontroller chip is a boot loader that allows uploading programs into the microcontroller memory without needing a chip (device) programmer. Arduino boards can be purchased pre-assembled or as kits. Hardware design information is available for those who would like to assemble an Arduino by hand.

The A000066 Arduino Uno is a microcontroller board based on the ATmega328. It has 14 digital IO pins, six analogue inputs, a 16 MHz crystal oscillator, a USB connection, power jack, programming header and a reset button. The board contains everything needed to support the microcontroller, simply connect it to a computer with a USB cable or supply DC power to get started.

The main features of the Arduino are shown in **Table 21.5**, and more information can be found at [36].

## Pinguino

The Pinguino is similar to the Arduino electronics prototyping platform but makes use of the PIC controller family instead of the Atmel devices. It supports different 8- and 32-bit Microchip microcontrollers (8-bit: PIC18Fx550, PIC18Fx5K50, PIC18Fx6J50 and PIC18Fx7J53 family. 32-bit: PIC32MX Mips

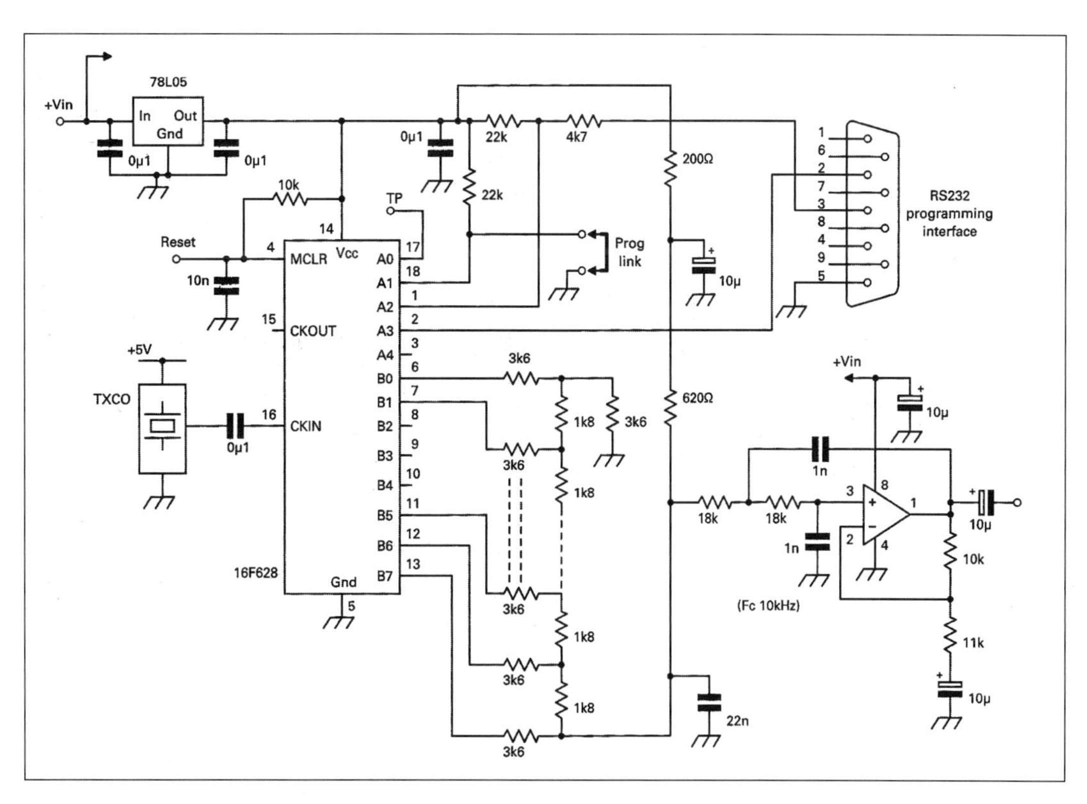


**Fig 21.23: Complete circuit diagram of the low frequency Direct Digital Synthesizer using a PIC, discrete R-2R ladder D/A converter and op-amp filter**

family) all with built-in USB module. It comes with a USB bootloader. This small program running inside the microcontroller is responsible for transferring your application from your PC to the microcontroller memory and handing over the control to this program afterwards.

No programmer in normally needed, unless the board is being built from scratch. In normal use, once the PIC has been initially programmed, the microcontroller then can be reprogrammed over the USB port. Pinguino boards can be used in various different ways depending on your skills: using the Pinguino IDE and the Pinguino Language (based on C and almost compatible with Arduino language); with the Pinguino 32-bit MIPS-elf GCC toolchain (C/C++) or 8-bit SDCC/GPutils toolchain (C only); with Microchip MPLAB X IDE toolchain.

Pinguino is an Integrated Development Environment (IDE) which gives everyone the ability to write, compile and upload programs on a Pinguino board.

The programming language is an Arduino-like or Arduino-influenced rather than Arduino-compatible language. Users can use the same keywords but as a different processor type is employed, cannot include Arduino's libraries in the code, although adapted libraries are available. Pinguino's Design Environment, libraries and compilers are all available for GNU/Linux, Windows and Mac OS X. All the free open source software is released under the terms of the GNU GPLv2 (General Public Licence version 2).

Pinguino Boards are Open Hardware. More details can be found at [37]. One use of this board is illustrated in Chapter 8.

## Project: PIC Based Audio Source

The circuit of **Fig 21.23** is for a PIC based audio source that implements a low rate Direct Digital Synthesizer (DDS) running on a standard 'workhorse' PIC 16F628.

This device contains a counter-timer that can be configured to generate an interrupt when it overflows. An interrupt, as its name suggests, is a routine that pauses whatever the processor is currently doing and sends it away to do something else; when complete, the processor returns to its original task - which for the PIC can just be sleeping. Interrupts are ideal for timing and Digital Signal Processing tasks as their timing can be accurately controlled.

On the Microchip PIC range [30], the counter timer can be run from the processor's internal clock signal which in turn is derived from the crystal or externally supplied oscillator. On the 16F family, the specification states this can be up to 20MHz, which is divided by four internally for a maximum processor clock of 5MHz. For LF and audio generation we need to be able to generate at least the range of frequencies that a PC soundcard can manage, which these days means a sampling rate of at least 48kHz although a sampling rate closer to 100kHz would be desirable. To be able to generate audio tones of virtually any arbitrary frequency, a 32 bit DDS architecture is needed.

A DDS is no more than a counter that is incremented by a fixed value for each clock input, rolls-over when it reaches maximum and starts counting again. The counter value at each instant is used as the input to a sine lookup table. The resulting value looked-up from the table is sent to a Digital to Analogue (D/A) converter whose output, after filtering to remove clock and alias products is the wanted audio waveform. The value the counter is incremented each time is related to the wanted output frequency as follows

A counter of length 32 bits rolls over every 232 counts, or each 4294967296 clocks. This defines the underlying resolution or step size for the DDS.

For a 100kHz input clock it is therefore 100kHz / 232 = 0.00002328Hz (23.3µHz) The value by which the counter has to be incremented each time is then given by N = Fout / Fclock * 232. For an output of 8.98kHz with a 100kHz clock N is 8980/100000 * 232 = 385688063 or expressed in hexadecimal 0x16FD21FF.

The PIC's internal clock can run at up to 5MHz which is the rate at which instructions can be executed. If we want to generate a waveform by sampling at 100kHz this means an absolute maximum of just 50 instructions can take place during each sampling instant. Several instructions are used-up just processing the interrupt and forcing the timer-counter to divide by 50 without actually contributing to waveform generation.

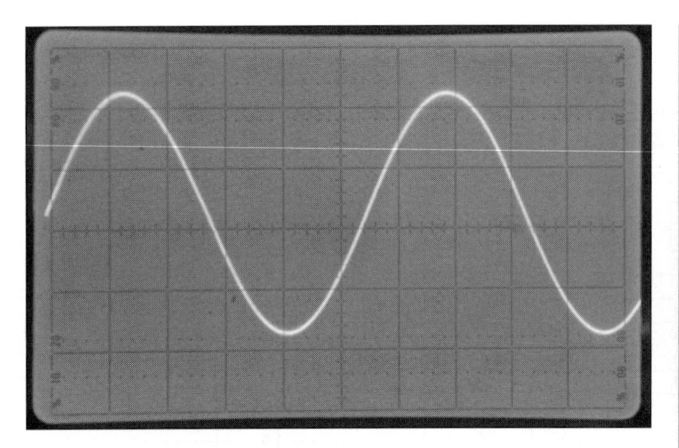

Fig 21.24: Oscilloggram of the PIC DDS sinewave output

Fortunately is it just possible using a PIC's internal 8 bit architecture to implement a 32 bit addition, table lookup and output to a D/A converter in less than 40 clock cycles.

## The algorithm

The registers in the PIC are eight bits wide so four are concatenated into a 32-bit accumulator - we'll call these D3 through to D0. Four more registers, F3/2/1/0, hold the value N determining the frequency, and for a single wanted tone these remain unaltered. Every interrupt cycle (at 100kHz for a 20MHz oscillator) the value in the F registers is added to the accumulated D3 - D0, which just overflows back to zero when its maximum count is reached.

The most significant byte stored in D3 alone is then used as the address for the sine look-up table. The resulting 8-bit value is placed on the PICs 8-bit parallel output Port-B.

The 32 bit addition, table lookup and output can be managed in about 35 clock cycles within the interrupt service routine, leaving a small amount of spare capacity. The complete listing can be found at [38],

## D/A converter

A ladder made up of resistor pairs in the ratio of 1:2 driven directly from PORTB. CMOS outputs, especially those on PIC devices switch between 0V and Vcc (usually 5 volts) with about 10 ohms of internal resistance so the 8-bit PORTB output from the PIC can directly drive the sixteen 1.8k and 3.6k resistors making up the chain at even lower cost than a custom D/A chip.

## Amplitude resolution and spurii

The restricted table-lookup capabilities of this baseline PIC family mean that tables with more than 256 entries and more than an 8-bit output word are complicated to arrange, and would not fit into the limited amount of clock cycles allowed, but for simple audio and VLF purposes the 8-bit DDS table is adequate. The rule of thumb for all DSP that says spurious and quantising noise levels will lie at roughly 6.N dB below full scale, where N is the number of bits. suggests an 8-bit lookup table ought to give about -48dBc spurious.

Fig 21.26: The finished PIC based Low Frequency DDS Source

The DDS concept works best where output frequency is kept lower than a quarter of the clock. It cannot, in any case ever go above half the clock rate due to the Nyquist sampling criteria. For a simple 8-bit table like this it is probably better to stay below one-sixth of the clock.

## Output filter

The D/A output now has to be filtered to remove alias products and spurii above 10kHz. A single stage, third order opamp filter can do the job admirably. The R-2R ladder maintains a constant output impedance of 1.8kΩ whatever the output voltage.

Biasing the output opamp filter and output levels ideally needs the output from the ladder to have its DC level raised and its 0-5V amplitude range attenuated. This is done by adding a 620Ω pull up resistor to 5V, through another 200 ohms decoupled to prevent noise of the 5V rail appearing directly in the audio path.

The result is an output impedance of 461 ohms and a peak - peak waveform amplitude of 1.28V superimposed on a mean DC level of 3.44V. The first stage of filtering comes from the 22nF capacitor from between the output of the resistor ladder to ground for a first stage cutoff at 15.6kHz. The opamp filter, whose values are taken from standard tables has an underlying gain of 2.1 times, resulting in an output of 2.7V peak-peak, or very close to 1V RMS.

The values shown give a cut-off at 10kHz, but a higher frequency would be possible by proportionally reducing the value of the two 1nF capacitors and/or the 18k resistors if some degradation in spurii at the upper frequency band is accepted.

## Results

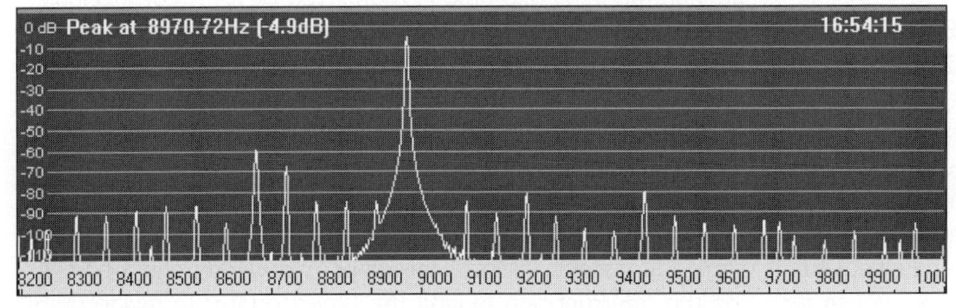


Fig 21.25: Output spectrum of the simple PIC-based audio frequency DDS

**Fig 21.24** shows the generated and filtered waveform at 8.9kHz with its close in spectrum in **Fig 21.25**. Spurii are typically -50 to -70dB individually, but as with any DDS, their actual frequency and spread is unpredictable. All we can really give is an upper limit to their magnitude, and here they are comfortably below the -48dBc predicted from the 6.N dB rule-of-thumb. The relative power of all the spurii when combined over the full audio band probably do amount to about the expected value.

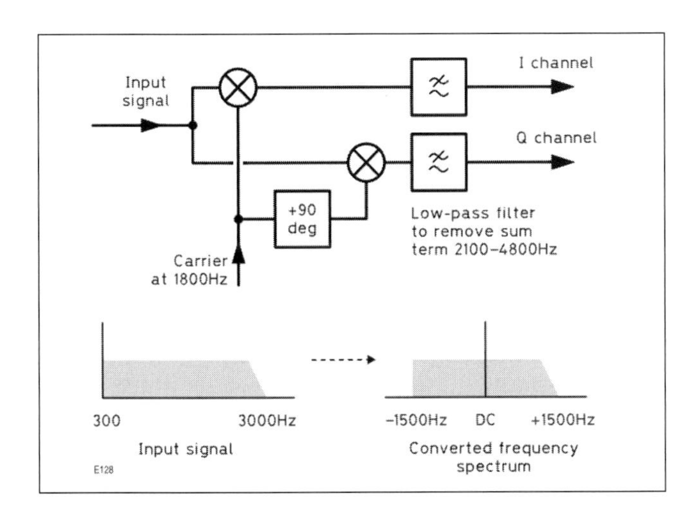


Fig 21.27: Frequency spectrum folding in I/Q mixing

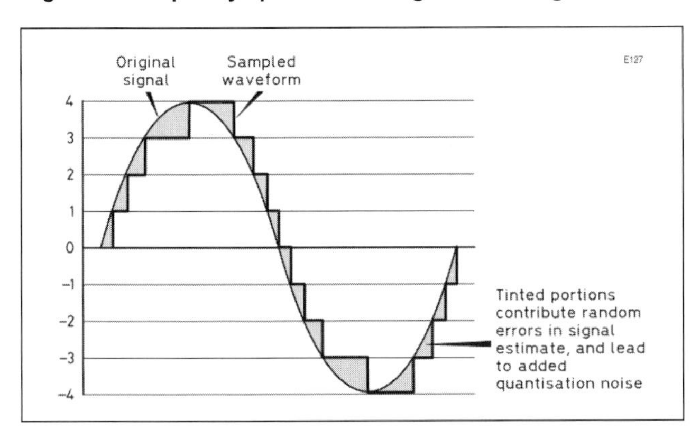


Fig 21.28: Generation of quantisation noise

### Software

The PIC code for the DDS to generate a single frequency can be found at [38]. The archive also contains a more complex version of the software with two stored frequencies that can be selected externally for frequency shift keying, These can be updated under RS232 serial control with auto baud rate determination allowing for arbitrary choice of PIC clock. Full details are supplied in the accompanying documentation, including the PCB layout as shown in **Fig 21.26.**

## DIGITAL SIGNAL PROCESSING

DSP is the latest aspect of computing that makes all the soundcard based data modes and Software Defined Radio possible (see also the chapter on SDR).

Essentially, DSP requires an analogue signal to be converted to a succession of samples whose value represents the instantaneous value at the instant of sampling. The rate at which the waveform is sampled is called the sampling rate.

## Sampling and the Nyquist Limit

An audio signal has to be sampled at a sufficient rate, the Nyquist limit, to minimise alias products and distortions, as described in the chapter on Principles. The other aspect to be considered is the number of discrete levels at the output of the A/D converter.

Too few, and the errors introduced by the quantising process contribute an unacceptable level of quantising noise; too many, and the A/D converter becomes complex and expensive.

The number of levels is related to the width of the output data in bits. An 8-bit A/D has 256 quantising levels, a 16-bit A/D (as in the majority of soundcards) offer $2^{16}$ levels = 65536.

**Fig 21.27** shows the quantising process graphically. As a rule of thumb, the maximum dynamic range (or Signal to quantising Noise) that an A/D converter offers is given approximately by:

S/N (dB)= 6 x N where N is the number of bits

So a 16-bit soundcard gives approximately 96dB of dynamic range, a simple 8-bit converter only a maximum of 48 dB dynamic range.

## Signal Processing

Once the waveform has been turned into a series of numbers, the real part of DSP can start. Several functions available in the analogue domain have direct equivalents in DSP. Frequency mixing is performed by multiplication of each waveform sample with a sample of a numerically generated Local Oscillator waveform (another series of numbers).

In practice, nearly all DSP is performed using two parallel channels consisting of 0 and 90 degree phase shifted signals, as this allows cancellation of unwanted mixer sidebands similar to the phasing method of SSB generation. In analogue processing terms, this can be thought of as separating the two image frequencies output from a mixer. This quadrature sampling of two parallel streams is usually referred to as In-phase and Quadrature, or I/Q processing. **Fig 21.28** shows this graphically.

I/Q data is referred to as a Complex signal, and consists of both negative and positive frequency components. Complex signals and negative frequencies are an aspect of DSP that many amateurs fail to visualise, but so long as it is thought of as only a mathematical short cut, and that negative frequencies are every bit as real as positive ones in the final result, you can't go far wrong.

A local oscillator in DSP consists of Sine and Cosine samples of a Numerically Controlled Oscillator, generated from a sine look-up table stored in memory..

Filtering is another task that is common with the analogue world, and consists of multiplying a block of successive samples with a set of filter coefficients, then summing all the results into a single, filtered, sample. The block of input samples is then shifted one sample to include the latest one and rejecting the earliest, and the whole multiplication / summation performed again to give the next filtered sample. When the DSP process is complete, the samples can be passed to a Digital to Analogue converter to generate real audio, for example.

The processes just described are all that is needed for a Software Defined radio for SSB or CW reception. A simple soundcard-only SDR doing just these processes is that designed for listening to VLF transmissions such as SAQ. SAQRx is a soundcard receiver written by SM6KLM.

It uses the soundcard sampling at 44kHz to receive radio signal up to 20kHz, which are then filtered to 3000, 1000, or 300Hz bandwidth and downconverted to audio for replaying. The LO can be tuned with the mouse or keyboard, and the IF bandwidth set the same way. SAQRx can be obtained from [39]

## The FFT and Visual Displays

But DSP can do a lot more than replace analogue processing. The Fast Fourier Transform is a widely used example of a technique only available in DSP. The FFT takes a block of samples of any waveform, and performs a mathematical process on them to give a set of samples representing the spectrum of that waveform in the Frequency Domain.

The resulting data can be used to show the spectrum of the

signal (up to half the sampling rate), or for other more complex tasks such as data demodulation.

For the process to work, the block has to be a binary power of two, so block lengths of 256, 512 all the way up to 262144 (and sometimes higher) are used. The resulting spectrum is generated to a resolution of the sampling rate divided by the block length. So, for example, an 8kHz sampled waveform passed through an 8192 point FFT will output a series of values consisting of the spectrum of the waveform in units of 8000/8192 = 0.977Hz multiples. These successive frequency samples are usually referred to as bins (as in waste bin). Bin zero contains the DC component of the waveform, bin 1 the amplitude of the frequency components at 0.977Hz, all the way up to bin 4095 that contains the amplitude of the frequencies between 3998 and 3999Hz approximately.

Above this Nyquist limit, the values are repeated, mirrored. If complex, or I/Q values are put into the FFT process, this mirrored spectral output is modified to show the complex spectrum above and below the zero term.

There are many spectral analysis programs around now, and several are described in the Data Communications chapter.

Other aspects of DSP are too numerous to mention, and beyond the scope of this Handbook. More details of DSP and some practical examples can be found in [40].

## REFERENCES


[1] *The Amateur Radio Operating Manual*, 7th ed, Don Field, G3XTT, and Steve Telenius-Lowe, 9M6DXX, RSGB
[2] www.cadsoftusa.com/downloads/libraries
[3] www.terrypin.dial.pipex.com/ECADList.html
[4] www.aade.com/filter.htm
[5] www.zerobeat.net/G4FGQ/index.html
[6] www.tonnesoftware.com/
[7] www.qsl.net/dl4yhf/spectra1.html
[8] www.cebik.com
[9] RSGB web site, www.rsgb.org
[10] ARRL web site, www.arrl.org
[11] Google Internet search engine, www.google.co.uk, or www.google.com
[12] Yahoo Internet search engine, www.yahoo.co.uk, or www.yahoo.com
[13] http://uk.farnell.com
[14] www.maplin.co.uk/
[15] http://rswww.com/
[16] www.ebay.co.uk/
[17] www.amazon.co.uk, or www.amazon.com
[18] www.xilinx.com/
[19] www.altera.com/
[20] VAC (Virtual Audio Cable). www.screenvirtuoso.com/vac.html
[21] The BARTG publish designs from time to time. See www.bartg.org.uk/
[22] Appendix D, *Digital Modes for All Occasions*, Murray Greenman ZL1BPU, RSGB
[23] www.ftdichip.com/
[24] 'There's a remote possibility . . .', David Gould, G3UEG, *RadCom*, Aug/Sep 2005
[25] www.echolink.org/
[26] http://en.wikipedia.org/wiki/EQSO
[27] www.irlp.net/
[28] www.raspberrypi.org/
[29] https://groups.yahoo.com/neo/groups/Raspberry_Pi_4-Ham_RADIO/info
[30] www.microchip.com
[31] www.melabs.com
[32] www.voti.nl/wisp648/
[33] www.asix.net
[34] Microchip PIC Microcontrollers and DSPic. www.microchip.com
[35] www.atmel.com/
[36] www.arduino.cc/
[37] www.pinguino.cc/
[38} www.g4jnt.com/PIC_DDS.zip
[39] http://sites.google.com/site/sm6lkm/saqrx
[40] *"Command" - Computer, Microcontrollers and DSP for the Radio Amateur*. Andy Talbot, G4JNT, RSGB 2003


***Andy Talbot*** *BSc, G4JNT, is a professional electronics engineer currently working in the field of radio communications and signal processing. He has been a radio amateur for 29 years, with his main interests being at the two opposite ends of the spectrum - microwaves and the LF bands. He has always been a home constructor, designing and building equipment from scratch right from the early days. Working on the new 73 and 137kHz bands started off his interest in DSP, and in conjunction with G3PLX pioneered the first use of ultra-narrowband techniques and low data-rate signalling on the amateur bands. Later, in conjunction with G4GUO, he took part in the first amateur use of digital voice at HF. He has written many articles for various journals, and writes the 'Data' column for the RSGB's* RadCom *magazine.*

# Electromagnetic Compatibility 22

*Robin Page-Jones, G3JWI*

Electromagnetic compatibility (always abbreviated to 'EMC') is the ability of various pieces of electrical and electronic equipment to operate without mutual interference. So far as amateur radio is concerned, the object is to achieve good EMC performance: that is, not to suffer from received interference or to cause interference to others.

In practice, the amateur must endeavour to minimise interference caused by his (or her) station and, where appropriate, increasing the immunity of susceptible local domestic radio and electronic equipment. Complementary to this is interference to amateur reception. In recent years this has become increasingly important.

National administrations have enacted legislation defining minimum EMC standards which products on sale to the public must meet.

These standards lay down maximum permitted emitted interference, and also the minimum immunity which equipment must have to unwanted signals. In the UK the standards are issued by the British Standards Institute, and are harmonised to the common standards of the European Community. In general, EMC standards are framed round a normal domestic or industrial radio environment, and fall short of what would be ideal from the radio amateur's point of view. Amateurs tend to generate higher field strengths and attempt to receive smaller signals than other radio users in a typical residential area.

So, for the radio amateur, interference breaks down into two major categories

- Interference caused by operation of the transmitter.
- Interference to reception, usually simply called radio frequency interference or RFI.

In recent years, the number of cases of serious interference to neighbours' equipment has declined markedly. This is due to a number of factors one of which is the coming of the EMC regulations which have made EMC a major factor in the design of all types of domestic electrical and electronic equipment. An even more important factor, however, is the appreciation by amateurs themselves of the need to design and operate their stations to minimise interference to neighbours. At the same time, the "digital age" has meant that interference to amateur reception (RFI) has become more of a problem.

Interference to and from an amateur station involve the same basic principles but dealing with them requires different procedures. For this reason this chapter has been divided into two parts.

## INTERFERENCE FROM TRANSMITTING

Interference from the transmitter falls into two categories.

- Interference caused by the legitimate amateur signal, on the normal operating frequency, breaking through into some piece of susceptible equipment. This is usually called breakthrough to emphasise that it is not really a transmitter fault, but rather a defect in the equipment which is being interfered with.
- Interference due to unwanted emissions from the amateur station. The general name for this is spurious emissions.

Nowadays, breakthrough is much more likely to be a cause of interference than spurious emissions, but before jumping to any conclusions carry out a few simple checks.

First consider what sort of equipment is affected. If it does not use radio in any way, then the cause must be breakthrough. If the susceptible equipment makes use of radio in some way (usually broadcast radio or TV but it could be some form of communication or control device) then the cause could be either breakthrough or a spurious emission from the transmitter.

The next step is to find out if the interference only occurs when the equipment being interfered with is tuned to specific frequencies. If so, check whether similar equipment in your own house (or in neighbours' houses) is affected when tuned to the same frequencies. In effect, if your transmitter is radiating a spurious signal, then anything tuned to that frequency is likely to be affected.

### Breakthrough

All electronic equipment is to some extent vulnerable to strong radio frequency fields and is a potential EMC threat. Any amateur who lives in close proximity to neighbours will have to give some thought to the avoidance of breakthrough. This can be tackled in two ways:

- By installing and operating the station so as to reduce the amount of radio frequency energy reaching neighbouring domestic equipment. The term good radio housekeeping has been coined to cover all aspects of this activity.
- By increasing the immunity of the affected installation to the amateur signal.

#### Good radio housekeeping

The essence of good radio housekeeping is keeping your RF under reasonable control, putting as much as possible where it is wanted (in the direction of the distant station) and as little as possible into the local environment.

Installations designed to achieve this will also minimise the pick-up of locally generated noise, so that much of the advice in this section applies to reception as well as to transmission. Where a station is to be installed in a typical suburban environment, in close proximity to neighbours, it is essential to plan with EMC in mind.

#### The antenna

It is always good practice to erect any antenna as far from houses as possible, and as high as practical, but for HF operation the relatively long wavelengths in use give rise to special problems. In locations where breakthrough is likely to be a problem, HF antennas should be:

- Horizontally polarised. House wiring and other leads tend to look like antennas working against ground, and hence are more susceptible to vertically polarised signals.
- Balanced, to minimise out-of-balance currents, which can be injected into the house wiring, particularly in situations where a good earth is not practical. These currents will also give rise to unwanted radiation.
- Compact, so that the whole of the radiating part of the antenna can be kept as far from house wiring as possible. Try to avoid antennas where one end is close to the house while the other end is relatively far away. This encourages direct coupling between the near end and the house wiring, inducing RF currents which will be greater than would otherwise be the case.

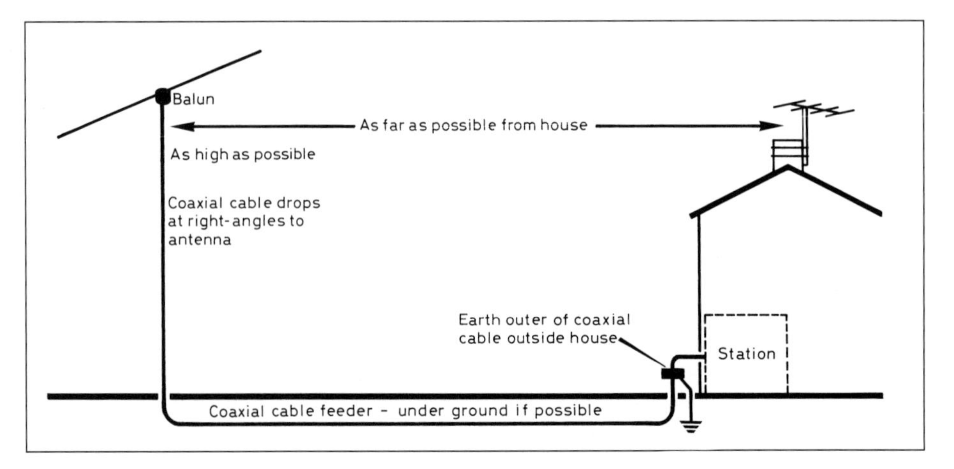


**Fig 22.1: Antenna and feeder system with EMC in mind**

The sort of thing to aim for is a dipole, or small beam, located as far as possible from the house (and neighbouring houses), and fed with coaxial cable via a balun (**Fig 22.1**). This is not too difficult to achieve above 10MHz, but at lower frequencies a compromise is usually necessary. A balanced feeder can be used, but coaxial feeder is convenient and allows the use of ferrite chokes to reduce any unwanted currents which may find their way on to the braid.

## Earths

From the EMC point of view, the purpose of an earth is to provide a low-impedance path for RF currents which would otherwise find their way into household wiring, and hence into susceptible electronic equipment in the vicinity. In effect, the RF earth is in parallel with the mains earth path as in **Fig 22.2**. Good EMC practice dictates that any earth currents should be reduced to a minimum by making sure that antennas are balanced as well as possible. An inductively coupled ATU (*see the chapter on Practical HF Antennas*) can be used to improve the isolation between the antenna/RF earth system and the mains earth. The impedance of the mains earth path can be increased by winding the mains lead supplying the transceiver and its ancillaries onto a stack of ferrite cores as described below for breakthrough reduction.

Antennas which use the earth as part of the radiating system - ie antennas tuned against earth - should be avoided since these inevitably involve large RF currents flowing in the earth system. If this type of antenna must be used, arrange for it to be fed through coaxial cable so that the earth, or better some form of counterpoise, can be arranged well away from the house. A remotely tuned ATU can be used to make a simple multi-band antenna (*see the chapter on Practical HF Antennas*).

The minimum requirement for an RF earth is several copper pipes 1.5m long or more, driven into the ground at least 1m apart and connected together by thick cable. The connection to the station should be as short as possible using thick cable or alternatively flat copper strip or braid.

Where the equipment is installed in an upstairs room, the provision of a satisfactory RF earth is a difficult problem, and sometimes it may be found that connecting an RF earth makes interference problems worse. In such cases it is probably best to avoid the need for an RF earth by using a well-balanced antenna system - but don't forget lightning protection. (*see the chapter on Antenna Basics*).

For more on choosing antennas and earths, see the antenna chapters in this book.

## Operating from difficult locations

If there is no choice but to have antennas very close to the house or even in the loft, then it will almost certainly be necessary to restrict the transmitted power. Some modes are more likely to give breakthrough problems than others, and it is worth looking at some of the more frequently used modes from this point of view.

### *WARNING: Protective Multiple Earthing (PME)*

Many houses, in the UK are wired on what is known as the PME system. In this system the earth conductor of the consumer's installation is bonded to the neutral close to where the supply enters the premises, and there is no separate earth conductor going back to the sub-station. Under certain rare supply fault conditions a shock or fire risk could occur where external conductors such as antennas or earths are connected. For this reason the supply regulations require additional bonding and similar precautions in PME systems.

Many houses in UK were wired on the old TN-S system, where a separate earth goes back to the sub-station. In such systems there were no problems with connecting an external radio antenna or earth. It has recently become evident that changes to maintenance and installation practice mean that the inherent safety of the old TNS systems can not always be guaranteed. The situation is being reviewed. Until further information is available all installations should be treated as if they were PME.

**Read RSGB EMC lealet EMC 07 Earthing and the Radio Amateur before connecting any earth or antenna system to equipment inside the house.**

**If in doubt consult a qualified electrician**

Leaflet *EMC 07* is available on request from RSGB, or from the RSGB EMC Committee web site [1].

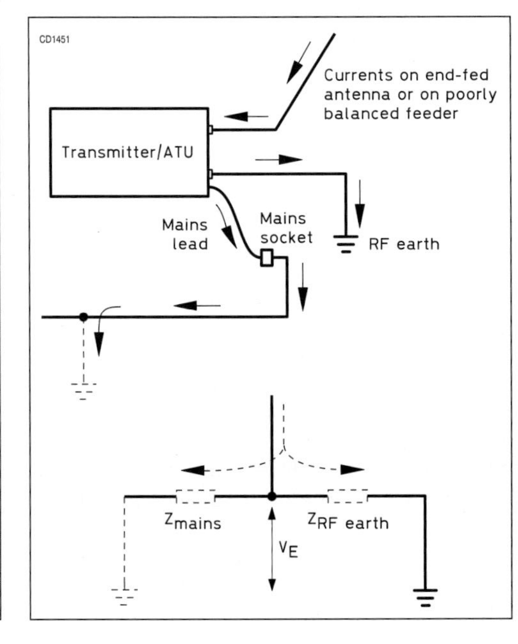


**Fig 22.2: Earth current divides between RF earth and mains. The current down each path will depend on the impedances. The transmitter earth terminal will be at $V_E$ relative to 'true' earth potential**

**SSB** is a popular mode but also the most likely to give breakthrough problems, particularly where audio breakthrough is concerned.

**FM** is a very EMC friendly mode, mainly because in most cases the susceptible equipment sees only a constant carrier turned on and off every so often.

**CW** has two big advantages. First, providing the keying waveform is well shaped, the rectified carrier is not such a problem to audio equipment as SSB. The slow rise and fall gives relatively soft clicks which cause less annoyance than SSB. The second advantage is that it is possible to use lower power for a given contact than with SSB.

**Data modes** used by amateurs seem to be EMC friendly. Some data modes are very efficient so that low power can be used if necessary.

## Passive intermodulation products (PIPs)

This phenomenon is a fairly unusual aspect of good radio housekeeping but one which has turned up from time to time since the early days of radio. Traditionally it is known as the rusty bolt effect. It occasionally causes harmonic interference to be generated by amateur transmissions, but far more often it simply degrades receiver performance without being identified as a problem.

Mixing and harmonic generating circuits use non-linear elements such as diodes to distort the current waveform and hence to generate the required frequency components. A similar effect will be produced whenever the naturally produced semiconductor layer in a corroded metal joint forms an unwanted diode. These unwanted diodes are usually most troublesome in the antenna system itself, particularly in corroded connectors. In the case of a single transmitter the effect simply causes excessive harmonic radiation, but where two or more transmitters are operating in close proximity the result can be spectacular intermodulation product generation.

On receive, the result is much-reduced receiver intermodulation performance, and in severe cases there will be a noticeably high background noise level which consists of a mishmash of unwanted signals. The best way to avoid troubles of this sort is to keep the antenna system in good repair and to examine all connections every few months. PIPs can be generated in corroded metal gutters and similar structures not directly associated with the antenna system. The solution is to clean up or remove the corroded metalwork or, where this is not possible, to short-circuit the corroded junctions with a conductive path. It is obviously good practice to keep antennas away from doubtful metalwork.

## Breakthrough at the receiving end

Breakthrough is simply unwanted reception, and the basic mechanism by which signals are picked up is the same as for any other reception. Breakthrough can occur to such a wide range of equipment that for the sake of general discussion it is simpler to assume some non-specific device, in other words a black box.

For signals to get into our black box, they must be picked up on a wire which is a significant fraction of a wavelength long, which means that on HF the external leads are the most common mode of entry and direct pick-up by the equipment itself is unlikely unless the transmitter field strength is very high. At frequencies above about 50MHz pick-up by wiring inside the black box becomes more likely but only if the box is made of non-conducting material.

**Fig 22.3** shows how unwanted RF signals might get into the black box - in this case, some sort of alarm circuit is assumed with a sensor connected by twin cable, and an amplifier and power supply inside the box.

The sensor lead acts as a crude earthed antenna so that electro-

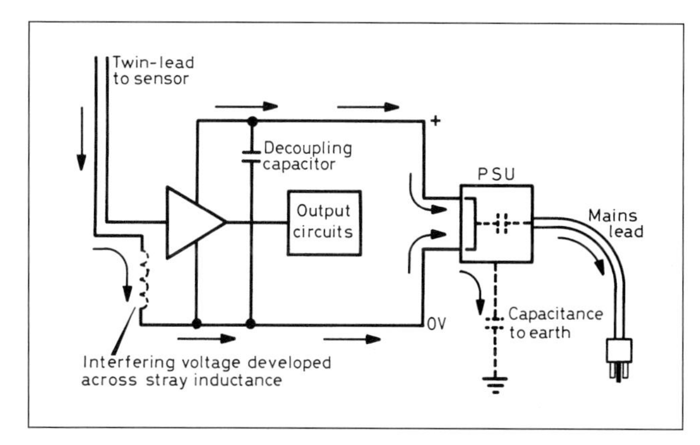


**Fig 22.3: Path of RF signal in a typical sensor / alarm device**

magnetic energy from the transmitter causes currents to flow in the antenna formed by the sensor lead, through any stray impedance between the input connection and the 0V rail, and through the power supply to earth. These currents are called common-mode currents because they flow on both conductors in the cable in phase - in effect they act as if the pair of wires were one conductor. The wanted signals are differential, flowing in one direction on one wire, and the reverse direction in the other. Common-mode and differential currents are illustrated in **Fig 22.4**.

The key to avoiding breakthrough is to prevent the unwanted signals picked up on the external lead from getting into the circuits inside the box. There are two ways of doing this:

- Bypass the unwanted RF by providing a low-impedance path across the vulnerable input circuit.
- Use a ferrite choke to increase the impedance of the unwanted antenna where it enters the black box, effectively reducing the currents getting into the sensitive internal circuitry.

## Bypassing the unwanted RF

The principle of bypassing is to arrange for a potential divider to be formed in which the majority of the unwanted signal is dropped across a series impedance, as shown in **Fig 22.5**. In some instances the series element may be a ferrite bead or, where circuit conditions permit, a low-value resistor, but in many cases no series element is used and the stray series impedance of the lead provides the series element. A ceramic capacitor with a value in the region of 1 to 10nF would be typical. It is important to keep the leads as short as possible, and to connect the 'earthy' end of the capacitor to the correct 0V point - usually the point to which the amplifier 0V is connected, and to which the inputs are referred. ('0V' in this context has much the same meaning as

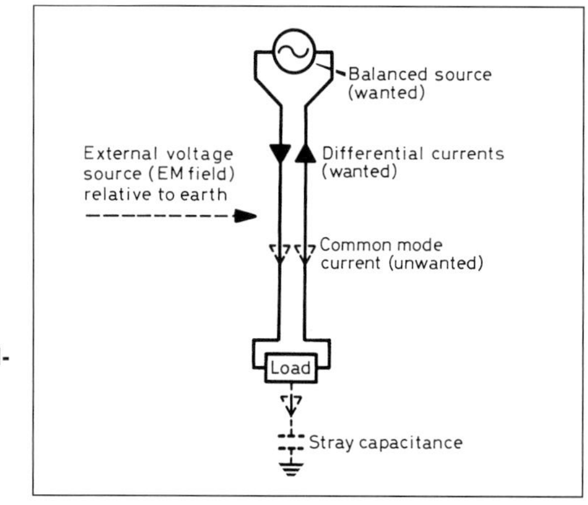


**Fig 22.4: Differential-mode and common-mode currents**

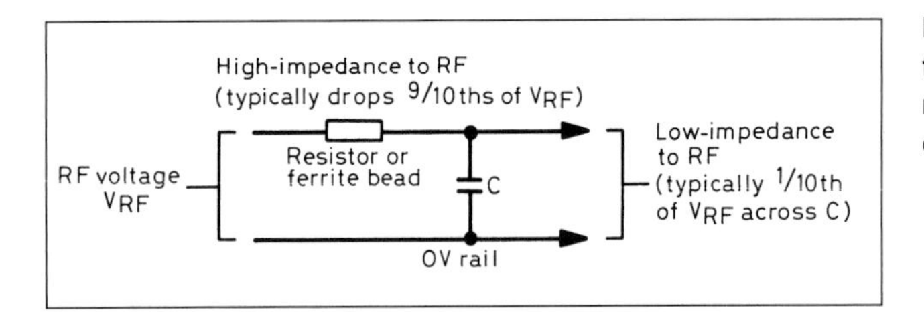


Fig 22.5: The principle of by-passing or decoupling

'ground', but avoids any confusion with other meanings of 'ground' used in radio discussions such as in 'capacitance to ground' etc.)

Generally, bypassing is not very practical in domestic situations and is included as food for thought for anyone building their own equipment. It is inadvisable for the amateur to attempt to modify commercial equipment unless he (or she) has expert knowledge - in particular, never attempt to modify equipment belonging to someone else.

## Ferrite chokes

Ferrite chokes are also used to form a potential divider to the unwanted RF currents, though in this case the series impedance is increased outside the black box, and the stray capacitance or resistance inside the box forms the shunt element (**Fig 22.6**). The great benefit of this technique is that it does not involve any modification inside the box. All that is required is to wind the susceptible lead on to a suitable ferrite core. Ferrite cores which are designed for EMC purposes have a reasonably high permeability, which is combined with a relatively high loss at the frequencies of interest. This enables a high impedance to be achieved without resonance effects becoming dominant. Further information on core types will be found in [1].

A very important feature of this type of ferrite choke is that it acts only on the common-mode interfering currents - differential-mode wanted signals will not be affected. The go and return currents of the differential signal are equal and opposite at any instant, and so their magnetic fields cancel out (except in the space between the wires) and there is no external field to interact with the ferrite.

The most popular core for EMC use is the toroid or ring, because the ring shape means that the magnetic field is confined inside the core, giving a relatively high inductance for a given material. Where a simple toroid cannot be fitted because the connectors cannot be removed a split core can be used. Cores with quite large diameters are advertised on the web. Make sure that the core is suitable for EMC purposes in the required frequency range. The core must be tightly closed with no gap otherwise inductance will be greatly reduced.

The inductance of a ferrite ring choke is proportional to the length of ferrite through which each turn passes (ie the thickness

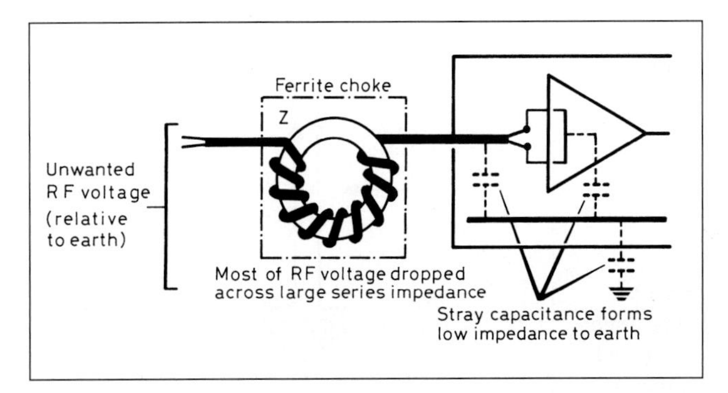


Fig 22.6: The series choke. The impedance of Z forms a potential divider with stray capacitance

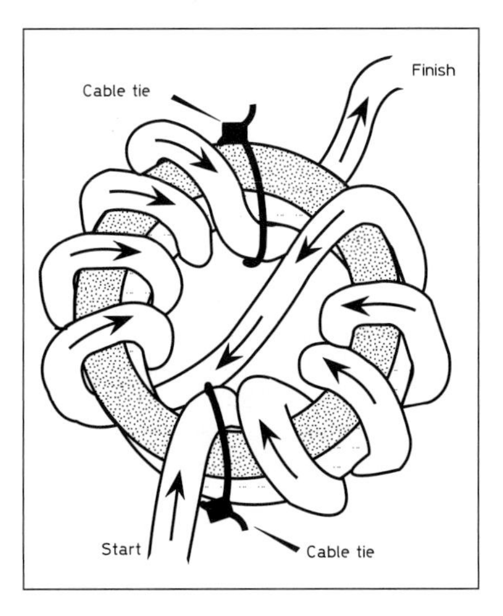


Fig 22.7: Winding a ferrite choke for minimum shunt capacitance

or 'depth' of the ring) and also to the square of the number of turns. Traditionally cores about 6.5mm thick have been used, two of these being stacked to give an effective thickness of 13mm or so. To make a choke for HF frequencies, 12 to 14 turns on two such rings should be satisfactory. For VHF, fewer turns should be used - about seven turns on a single ring will be effective at 144MHz. At the time of writing the ferrite cores available from the RSGB are about 12.7mm thick so that one of these rings is equivalent to two of the thinner rings. It is most important to use the correct type of ferrite ring.

There is a very wide range of ferrite materials, and many rings are unsuitable for EMC purposes. The simplest way to ensure that the correct rings are obtained is to purchase them from a reliable supplier, advertising rings specifically for EMC purposes. Ferrite rods, from old AM radios, can be used but will require more turns (20 or more) to make an effective choke, and are generally only effective above 20MHz.

The shunt capacitance across a ferrite ring choke can be minimised by winding the core as in **Fig 22.7**.

## Breakthrough to TV and video equipment

The advent of digital TV and the use of digital video recorders, in place of analogue TV and video cassette recorders, has changed the way that interference becomes evident. The main difference is that interference to digital TV results in 'freezing' and audio break-up or even loss of picture and sound. The traditional indicators such as picture patterning and audible effects which were discernible as specific to amateur radio interference do not occur with digital TV. Similar effects can be due to any cause which makes the effective signal to noise ratio to fall below the level required for adequate reception. In common with most digital situations, the change from a perfect picture to complete loss of picture takes place over the range of a few decibels.

This makes investigation of alleged interference more difficult, but fortunately, the underlying causes and remedies for interference to TV and video have not changed. The requirement is to prevent the amateur signal from getting in to the various units comprising the domestic installation.

Many households receive TV programmes via the Internet - Internet Protocol Television (IPTV). So far breakthrough to IPTV has not been a problem but this could change as more households take up the faster internet access systems. (See Breakthrough to internet access systems - below.)

The complexity of modern domestic TV and video installations makes it impractical to list problems and fixes for specific installations, but the principles of preventing ingress of unwanted RF by good installation practice and by the use of common-mode

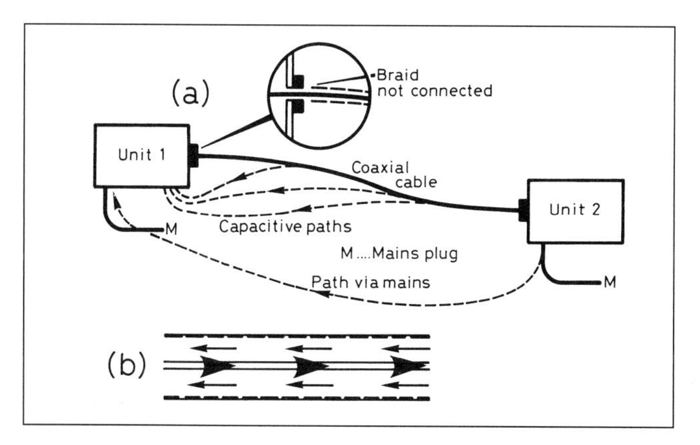


**Fig 22.8: (a) Broken braid connection causes the current which would normally flow on the inner surface of the braid to return by random paths. (b) In a correctly connected coaxial cable the currents balance, giving no external field**

chokes has not changed. In a terrestrial TV installation, the most vulnerable point is the antenna input, and in the majority of cases eliminating unwanted signals at this point will clear up the problem. The small size of UHF TV antennas makes it unlikely that large interfering signals will be picked up on the antenna itself except possibly above 100MHz. Corrosion of the connector at the end of the cable coming down from the antenna may indicate that water has got into the cable; the commonest cause is chafing or splitting of the sheath, though it could be a failure of the seal at the antenna itself. In a complicated installation comprising several units, make sure that all connectors are correctly mated, and that coaxial leads are correctly 'made-off' with the braid properly connected. In many cases, equipment will work with the braid of an interconnecting cable disconnected. In this case the return path for the signals in the cable is via random earth paths through other parts of the system. This reduces the immunity of the installation to breakthrough, and allows the leakage of radio frequency interference (**Fig 22.8**)

There have not been many reported cases of interference to satellite TV and it is not practical to say more than to check for obvious problems such as loose or damaged connections. The best place to look for the latest information is the *RadCom* 'EMC' column or the RSGB EMC web site [1].

## TV Antenna Amplifiers

A well-designed masthead amplifier which is correctly installed, close to the TV antenna, is unlikely to give problems at HF because the length of cable, before the amplifier, is short compared to the wavelength of the amateur signal, but at VHF the conditions could be very different.

Problems arise if the unwanted signal reaching the input to the amplifier is large enough to drive it into non-linearity. Masthead amplifiers get their DC power through the coaxial feeder, so that any filters or braid-breakers must pass (and not short-circuit) the required DC supply. Unless specific information is available, it is best to use only ferrite-ring chokes in the down leads from such amplifiers.

Indoor amplifiers are much more likely to cause problems, especially where HF is concerned, but fortunately they can be easily removed from the circuit for test purposes. Many amplifiers have several outputs, allowing more than one set to be operated from the same TV antenna. If the amplifier has to be retained in the installation it should be considered as part of the TV set (or sets), and the standard techniques, using chokes and filters, on the antenna side of the amplifier will still apply.

TV amplifiers vary in input bandwidth. Some will accept all signals from about 40MHz and below up to the top of the UHF TV band. Others are designed for UHF only. In countries such as the UK where all terrestrial TV is on the UHF bands it is obviously better to choose a UHF only amplifier.

## Chokes and filters for TV installations

The simplest way of choking-off signals travelling down the braid is to use a ferrite ring choke; used in this way it is often called a ferrite-ring braid-breaker. The low-loss cable commonly used for UHF TV is not suitable for winding on to the core. It is best to make the choke separately, using a length of smaller-diameter coaxial cable, with suitable TV connectors at each end. **Fig 22.9** shows a choke for HF breakthrough. The coaxial cable used for the choke should, of course, be 75 ohms , but there is no need to use low-loss cable since the length involved is quite short.

Braid-breakers and filters are available from a number of commercial manufacturers, and these are usually effective provided that the correct type is used. Some devices are simply braid-breakers, with no filtering of signals travelling down the inside of the coaxial cable, while others include high-pass or band-stop filters optimised for specific bands, so it is important to study the specification before making a decision. In some instances it may be necessary to use two devices (for instance, a braid-breaker in series with a high-pass filter) to achieve the desired rejection. Further information on braid-breakers and filters can be found in [1]. The RSGB stocks suitable ferrite rings and filters. Information can be found on the Society's web site [2].

Try a choke or filter on the antenna lead first, then try chokes on other vulnerable leads. Do not remove a device if it does not have much effect. In many cases of breakthrough, the unwanted signals enter the installation by more than one route, and the observed interference is the result of the signals adding or subtracting.

If a ferrite choke is only partially successful, another should be tried in series to increase the rejection, but it is usually not worthwhile to go beyond two. When the sources have been identified, chokes which are not needed can be removed.

Chokes for mains leads are made by winding the cable onto ferrite rings in the same way as for a braid-breaker. It goes without saying that in this case it is essential to avoid damaging the cable in any way. If the cable is unusually thick it may be necessary to use fewer turns and more rings. Due to the closed magnetic circuit in a toroidal core, it is not necessary to pull the cable tight onto the ferrite material - all that is required is for the winding to be secure and the ends not too close together.

Leads to external loudspeakers may need to be choked as described below for audio equipment.

## Radio and audio equipment

In general the approach to radio and audio problems is the same

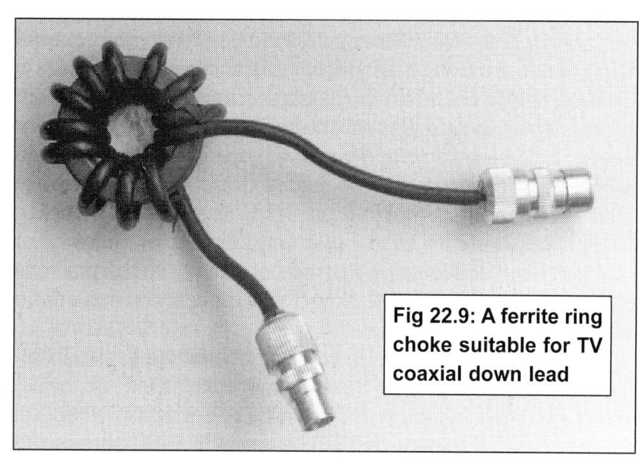

**Fig 22.9: A ferrite ring choke suitable for TV coaxial down lead**

as that for TV, though there are differences of emphasis. Interference caused by the braid of the antenna feeder acting as an unwanted antenna can still occur, but has become less common as many radios now have built-in antennas. As with digital TV, digital radios will not exhibit the traditional interference symptoms. The leads to the remote speakers of a domestic audio system frequently pick up interference and this can usually be cured by winding the appropriate number of turns on to a ferrite ring (or pair of rings) to form a choke at the amplifier end of the lead. Where this is not possible, a split core or a ferrite rod can be used,

Image interference can occur when a strong signal on the 'wrong side' of the local oscillator beats with it to give the IF. It is fairly common on the 1.8MHz band, where amateur signals give image responses on medium-wave receivers. For instance, a receiver with a 455kHz IF, tuned to 990kHz (303m) would have a local oscillator of 1445kHz. This would beat with a strong amateur signal on 1.9MHz, which would be tuned in on the medium-wave band like any other signal. This is a case of breakthrough caused by poor image rejection of the susceptible receiver.

Amateur signals can also beat with the harmonics of the local oscillator or other oscillators in the susceptible equipment, causing spurious responses. These give rise to interference which is tuneable at the receiver, but is nevertheless a case of breakthrough.

Where breakthrough to a portable radio is experienced, there is not very much that can be done from the outside of the set since there are no external leads to choke. However, it is often possible to move the set to a position where the interference is negligible. Frequently portables only give trouble when operated from the mains, and in this case a ferrite choke in the mains lead is likely to be effective.

### Security systems

Breakthrough to security systems has become less common in recent years even though increasing numbers of such systems are being installed. Probably this is due to better EMC awareness by manufacturers and better installation practice by installers. Experience has shown that the standard breakthrough measures such as ferrite chokes and by-pass capacitors, are rarely effective.

The most usual cause of breakthrough to security systems is poor immunity in the passive infra red sensors (PIRs). The solution is to replace the offending units with ones which are more immune to RF signals. Occasionally the control panel has been found to have insufficient immunity, but this is unusual. More information will be found in the RSGB's leaflet EMC 03 *Dealing with security alarm EMC problems* [3]. The best place to find up to date information on security system problems is the 'EMC' column of *RadCom*.

Most security systems are professionally installed, and it is up to the installer to ensure that it is sufficiently immune to operate in the environment in which it is installed. A well installed system with suitably immune PIRs is unlikely to be affected by an amateur station practising reasonable radio housekeeping.

If a complaint of breakthrough to a security system is received, it is advisable to be open with the complainant and to carry out checks to see exactly what the problem is. As in all breakthrough problems check your station and your radio housekeeping, before going any further.

Once you are sure of the problem, advise the complainant that the installer should be approached, and give them a copy of the RSGB's Leaflet EMC 02 *Radio Transmitters and Home Security Systems* [4].

All new installations should be using CE marked units, including the PIRs.

### Breakthrough to telephones

The main problem with telephones is breakthrough, caused by rectification and amplification of RF currents, and the solution is to avoid the RF getting into the equipment in the first place. There are two courses of action:

- First check your radio housekeeping.
- Prevent common-mode interfering currents getting into vulnerable circuits by fitting ferrite chokes as close to the susceptible units as possible.

Most households have extension telephones connected by plug and socket to the 'line jack' provided by the telephone company and these may involve quite long lengths of interconnecting cable. At the start of an investigation, unplug any extension units, leaving only the instrument near the line jack connected.

Telephones with additional features powered from a mains power unit are more likely to have breakthrough problems because of the earth path through stray capacitance in the power unit, so use a telephone powered from the line for initial tests. Clear any breakthrough on this, and then reconnect the extension leads and telephones one by one, dealing with breakthrough problems as they arise. In some cases it may be necessary to re-route vulnerable extension leads, but in most cases liberal use of ferrite chokes will prove effective.

Some telephones may have unusually poor immunity. If this is suspected check by substituting a known good unit. In such cases replacing the telephone may be the most practical option.

Breakthrough may be caused by signals being picked up on the lines before they come into the house. Commercial filters intended for the UK telephone installations are difficult to obtain and are not particularly effective on HF. However some types of 'microfilters' sold for use with self-install ADSL have characteristics which may make them useful in cases of HF breakthrough. Information on selecting and using these filters can be found in [5]. Where it is suspected that the problem is due to a fault on the lines themselves, contact the authority responsible for the lines, but before doing this it is important to ensure that your radio housekeeping is in order and that you are operating reasonably with regard to all the circumstances.

### Breakthrough to internet access systems

There are three types of access system:

- Fibre and cable systems
- Radio systems using frequencies above 1GHz.
- Systems which use the telephone lines.

The first two of these do not appear to give rise to many problems. It remains to be seen whether there will be problems should the use of these systems become more widespread.

The dial-up modem which was once the most popular means of internet access has been replaced by DSL systems. DSL stands for Digital Subscriber Line and there is a whole family of DSL systems.

Until recently the most common one in the UK was ADSL but this is being superseded by VDSL There are millions of ADSL installations in the UK but there have been relatively few cases of breakthrough reported. Further information on breakthrough to ADSL can be found at [6].

Internet access based on VDSL2 is now available In much of the UK. This uses frequencies in the HF band up to about 17MHz. To date breakthrough to VDSL2 has not been a problem but this could change as VDSL becomes more widespread, particularly if "self- install" become popular. Emissions from VDSL2 can cause interference to amateur reception. See Interference to Reception, below. The best place to look for the latest information is in RadCom and on the RSGB EMC Committee web site [1].

## Interference Caused by Unwanted Emissions from the Transmitter

As mentioned above, where interference is caused by spurious emissions from a transmitter, the interference can only be to equipment which uses radio in some way, and will be evident on similar equipment in the vicinity that is tuned to the same frequency. This is in contrast to breakthrough which is dependent on the susceptibility of the 'victim' equipment rather than any defect in the transmitter.

Spurious emissions fall into two major categories:

- Predictable spurious emissions which are generated as part of the process of carrier generation, and hence are (at least in principle) predictable.
- Oscillations caused by faults in the design or construction of a transmitter. These give rise to unexpected emissions which can occur on almost any frequency. This is known as instability.

### Predictable spurious emissions

The commonest predictable spurious emissions are harmonics of the carrier. These harmonics are simply a measure of the distortion of the sine wave which constitutes the carrier; an absolutely pure sine wave would have no harmonic content. In practice the object is to generate a carrier which is as undistorted as possible, and where necessary to filter it so that harmonics are kept to a negligible level.

The other major class of spurious emissions is the products caused by multiplication and mixing processes in the transmitter. In almost all transmitters the carrier is generated by mixing the outputs of oscillators (fixed and variable) to produce the final frequency. In most cases there are several mixer stages and at each stage unwanted products will be generated as well as the wanted signals.

The secret of good design is to make sure that unwanted products are at frequencies widely separated from the wanted signals so that they can be readily filtered out. As with harmonics, the object is to ensure that everything except the wanted output is attenuated to a negligible level. How small 'negligible' is depends on the frequency bands in use and the power of the transmitter. As a general rule on HF, harmonics should be at least 50dB, and other spurii at least 60dB, below the carrier. On VHF the figures should be 60dB and 80dB respectively.

There is another class of predictable spurious emissions, and this is splatter - the generation of unwanted intermodulation products by the modulation process in an SSB transmitter. The important intermodulation products are the odd-order products because they appear close to the carrier and hence cannot be filtered out. (*See also the Building Blocks chapters.*)

There will always be some intermodulation products generated in any SSB transmitter, and the object is to design and operate the transmitter in such a way that they are kept within reasonable bounds. 35 or 40dB down on the carrier is a reasonable target to aim for (see **Fig 22.10**).

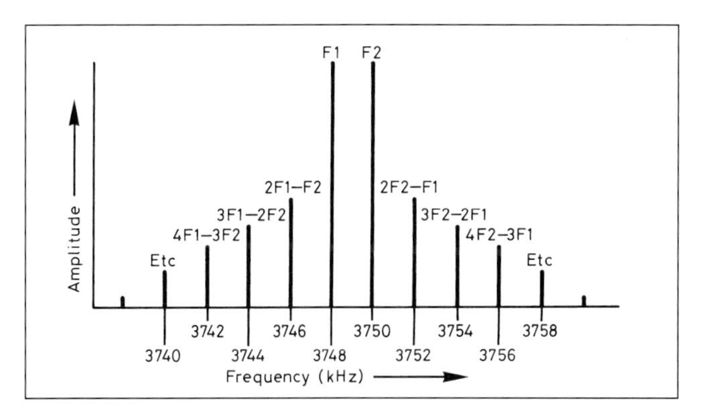


**Fig 22.10: Odd-order 'intermods' from a two-tone SSB signal on the 3.5MHz band**

## Identifying and Rectifying Spurious Emissions

Interference caused by harmonics of a transmitter can usually be identified by consideration of the frequencies on which interference is evident and the frequency on which the transmitter is operating. For instance, if the problem is interference to a VHF broadcast signal on 100.5MHz when the transmitter is operating on 50.25MHz, then it is very likely that second-harmonic radiation is the culprit. Where excessive harmonic radiation is suspected, first check the transmitter and its adjustment and, if the problem persists, fit a low-pass filter as in **Fig 22.11**. Note that the filter comes after the VSWR meter (whose diodes can generate harmonics) and, where an ATU is used, it should come after the filter.

With non-harmonic spurious emissions the situation is more difficult. The best procedure is to monitor the output from the transmitter to see if any excessive spurious signals are present using a spectrum analyser. Once the spurious signal has been identified, the transmitter should be investigated; the frequency of the spurious signal will often indicate the stage which is likely to be at fault. Where the problem is not too severe, using an ATU on HF or a suitable band-pass filter on VHF may be all that is required.

The spectrum analyser should be connected up as in **Fig 22.12**. It is good practice to start with more attenuation than is expected to be required and to remove it to establish the desired level. Remember that the attenuator connected to the transmitter must be capable of handling the full transmitter power.

If no spectrum analyser or measuring receiver is available, a general coverage receiver can be pressed into service. Set up the receiver with a short antenna some distance from the transmitter operating into a dummy load. In such an arrangement the magnitude of any spurious signals cannot be determined but the technique will nonetheless be adequate to show up serious spurious emissions.

### Spurious emissions due to instability

This is caused by unintentional feedback in some stage of a transmitter causing oscillation. Instability also falls into two categories:

- Direct instability where feedback in one of the amplifying stages of the transmitter causes oscillation on or near the normal operating frequency.
- Parasitic oscillation, where incidental inductance and capacitance cause oscillation at a frequency far removed from that at which the circuit would normally operate.

Both these effects are serious but direct instability is perhaps the most dangerous because it can result in very large spurious signals being generated at a frequency where the antenna can be expected to be a good radiator. In very bad cases the spurious signal can be comparable with the normal output power, causing widespread

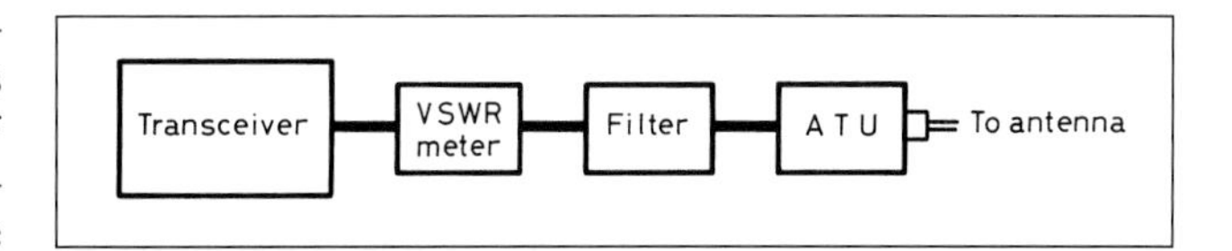


**Fig 22.11: Position of the filter. The ATU ensures that the filter sees a 50-ohm load**

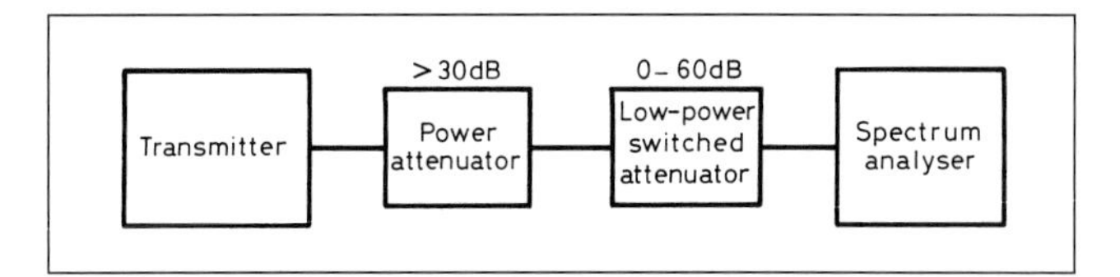


**Fig 22.12: Connecting a spectrum analyser to a transmitter**

interference. Fortunately both types of instability are rare in modern, well-designed, transceivers but care should be exercised when repairs have been carried out. This is particularly true where power transistors have been replaced with 'near-equivalents'.

The way to avoid instability is by employing good design and construction, with particular reference to layout, screening and decoupling. Where appropriate, parasitic stoppers should be included. (*See the chapters on transmitters*).

The two main indicators of instability are bad signal reports and erratic operation of the transmitter. If instability is suspected and suitable test gear is not available, the best procedure is to arrange for a report from someone who can receive you at good strength, and who knows what they are looking for. Ask them to look as far either side of your signal as possible, looking for unstable signals coming and going in sympathy with your main signal.

It is important to remember that a transmitter may be stable when operating into a good 50 ohm load but unstable when connected to an antenna.

Erratic operation usually means that the RF output and power supply current varies unpredictably as the drive is varied or as the ATU is tuned but, before jumping to the conclusion that the transmitter is faulty, make sure that the problem is not caused by poor station lay-out. It is not uncommon for RF from the transmitter to find its way into the power supply, or into ancillaries.

This can cause very erratic operation. In a modern commercial transceiver, which is normally well behaved, this is more likely to be the cause of erratic operation than instability in the transceiver itself. Check the antenna/earth arrangements, and review your radio housekeeping.

### Frequency halving

This is a spurious emission which doesn't come into either of the above categories. It occurs in bipolar transistor power amplifiers, when the input and output conditions are such that the parameters of the device are modified between one cycle and the next due to the internal parasitic capacitances acting like varactor diodes.

The commonest cause is too little or too much drive to the amplifier stage. The result is a significant output at half the input frequency. The practical effect is that an output at half frequency appears at certain tuning conditions. Divisions other than two are possible but less common.

Harmonics are also present in the output and often the most obvious symptom of this effect, is a 'harmonic' at one and a half times the nominal frequency. When tuning a VHF power amplifier into a dummy load, it is advisable to make sure that it is tuned well away from any frequency halving condition otherwise a change of load, such as connecting an antenna, may cause it to change to the halving condition.

## INTERFERENCE TO RECEPTION

## Background Interference Level

There are a many potential sources of radio frequency interference (RFI) and space does not permit discussing them all in this chapter.

The RSGB leaflet EMC 04 *Interference to Amateur Radio Reception* [7] gives a comprehensive list along with tips on tracking down and identifying sources, This can be found on the RSGB EMC Committee website [1].

### The radio noise environment

On HF there will always be noise picked up by the antenna, however well it is sited - this is called the ambient radio noise. This noise may be either man-made or natural, and unless the receiving antenna is extremely inefficient, will be much greater than the thermal noise generated in the front-end of the receiver. The 'bottom line' of this noise is the natural noise from atmospheric discharges which are always taking place somewhere in the world and are propagated by the ionosphere. Added to this is the man-made noise from electrical and electronic devices of various sorts.

There may also be local atmospheric static but this is usually not a serious problem in the UK. As frequencies rise into the VHF region, the natural noise decreases, until above about 100MHz the thermal noise in the receiver front-end becomes significant.

There is a great deal of misunderstanding about the ambient noise level on the HF bands. A casual measurement using the 9kHz bandwidth, usually used for HF EMC measurements, could give the impression that the noise level is so high that relatively high levels of broadband emissions would not cause a significant problem.

Arguments such as this have been used in attempts to justify relaxations of the EMC standards. However this is a complete misunderstanding of the HF electromagnetic environment.

### Man-made noise

Sources of man-made noise in a typical residential location tend to fall into two broad categories. Firstly there are the traditional sources such as motors and other sparking devices. Sometimes these can cause quite high levels of interference but in most cases they are in use for only a limited time.

The other category is interference from digital electronic equipment, particularly the ubiquitous switch-mode power supply (SMPU). All such devices are 'unintentional emitters' and are generally only a problem if they are not compliant with the EU EMC regulations.

In recent years a new class of digital RF noise sources has become important. This is where RF energy is deliberately generated for the transmission of data down cables, originally Intended for other uses such as telephone or electricity services, but where radio communication is not intended. These are also classed as unintentional emitters. With such systems it is possible to manage the frequency spectrum of the RFI to avoid certain frequency bands. The most prominent example of this is the Powerline Adaptors used in In-House PLT to link computers and to distribute video signals. This is sometimes called 'managed interference'.

### The ambient noise floor

For our purposes the noise floor is defined as the noise which exists when specific local sources of noise (sometimes called 'incidental noise') have been eliminated. If the ambient noise is observed in an amateur band on a reasonably sited horizontal dipole antenna such as that illustrated in Fig 22.1, a 'floor' will be seen consisting of the natural noise, mentioned above, plus an aggregate of distant man-made sources. This is usually fairly constant in any particular location. While this is still a useful concept, the practical application has been confused by the growing number of 24/7 RFI sources in residential locations. The term noise floor is sometimes used to refer to the ambient noise in general, so care is needed in comparing figures from different sources.

Further information can be found in Leaflet EMC 16 "Background Noise on the HF Amateur Bands", available on the RSGB EMCC web site.

### Antennas and noise

In general the good radio housekeeping rules for transmitting antennas apply to receiving antennas too, and for basically the same reason. Just as house wiring tends to pick up vertically polarised transmissions it also tends to emit vertically polarised local noise. It has been suggested that a significant proportion of this may be surface wave. Where HF operation is concerned, there is a difference between transmission and reception antenna requirements. Since the ambient noise sets a limit to small-signal reception it may be better to mount a relatively poor antenna in a low noise location than to have an efficient antenna mounted where the man-made noise is high.

In normal amateur HF operating, signals arrive by ionospheric propagation and will be a mixture of vertical and horizontal polarisation. With this in mind it is worth looking at antennas from the point of view of optimising the ratio between signal and ambient noise.

- **The dipole.** The horizontal half wave dipole has always been a popular antenna. Apart from its simplicity and low cost it is has the major advantage of receiving primarily horizontally polarised signals (except off the ends). It is insensitive to vertically polarised noise,
- **The vertical antenna tuned to resonance on one or more amateur bands.** These have become popular of late because they are convenient and unobtrusive. Verticals have a reputation of picking up more noise than a horizontal dipole and this seems to be deserved. They are sensitive to vertically polarised noise and in particular they will respond to surface wave noise.
- **The tuned loop antenna.** Tuned loops used as the main antenna for transmission and reception are not very common and there is no generally accepted view on their noise performance. Receiving tuned loops are often used where a small but sensitive antenna is needed. It is possible to make a small tuned loop feeding directly into a high impedance amplifier providing gain and also transforming the high impedance to 50 ohms. The author has used diamond shaped loops of about 30cm side and from one to four turns for EMC investigations. The sensitivity is remarkably good and it is possible to see the noise floor even in a quiet location. Caution is needed in interpreting noise levels measured on loops. The relationship between E and H noise fields will be different in different electromagnetic environments.

## Suppressing Interference at Source

### Impulsive interference

Where mechanical contacts are concerned, there are a number of well-tried remedies based on the principle of absorbing the energy which would otherwise be released when the contact is broken. The energy is initially stored in the magnetic field, due to the normal operating current flowing in any inductance which may be present in the circuit, and in many cases this will be considerable. When the contact is broken, the magnetic field collapses and a large voltage appears for a short period as the contacts open, causing a spark. Radio frequency currents are exchanged between the inductance and capacitance in the vicinity of the contact, using the ionised air of the spark as a bridge.

The traditional way of absorbing the energy is to connect a low value resistor and a capacitor across the contacts as in **Fig 22.13**. This effectively quenches the spark, by dissipating the unwanted energy in the resistor. The capacitor should be between 0.01 and 0.1µF and the voltage rating must be several times the voltage being switched. Special capacitors rated for use on AC mains are available, and these must always be used where mains voltages are involved. Units containing a resistor and capacitor in one encapsulated unit can be purchased from component suppliers.

Small, low-voltage DC motors can be suppressed by using a shunt capacitor of between 0.05 and 1µF, and a series choke with adequate current rating and a value of between 10 and 30µH, as in **Fig 22.14**. The chokes are more effective at higher frequencies, and may not be required if only low frequencies are involved. Mains motors are best dealt with by purchasing a suitable mains filter. This should be installed as close to the machine as practical.

### Interference from digital equipment

It is relatively easy to reduce the leakage of interference from digital equipment at the design stage - it is really a matter of good engineering practice. Good decoupling and the provision of a substantial ground plane for the common 0V rail is a good start. It is important to prevent external leads having energy coupled into them from shared return paths, and so acting as antennas (**Fig 22.15**).

Ideally, interference-generating circuits should be completely screened. The screen should be connected to the common 0V point through a path which has the lowest possible impedance. Leads should be decoupled where they pass out through the screen.

In situations where interference reduction is a major factor, for instance where digital control circuits are actually part of the receiver, special attention should be paid to screening and feedthrough capacitors should be used on all leads except those carrying RF.

It is important to choose the correct value - too large a capacitance will distort fast digital signals. So far as possible, the screen should be continuous and, where there are any joints, there must be good electrical contact along the mating surfaces.

### Interference from computer installations

Modern computer installations are put together from units complying with the requirements of the relevant EMC standards. If the various units are connected together using, good quality screened leads, then the EMC performance should at least be reasonable. At HF frequencies most RFI problems are caused by common-mode currents flowing on the interconnecting leads, as illustrated in **Fig 22.16**. Where a screened lead is used to connect two units together, the screen has two functions.

As might be expected, it does reduce radiation by forming an electrostatic screen around the conductors, but more importantly it provides a low-impedance path back to the computer for the common-mode currents which would otherwise leak back by devious routes, as illustrated in **Fig 22.17**.

As with all electronic equipment the 0V arrangements on the computer are a compromise, and this sometimes leads to emissions along the lines of Fig 22.15, even if the lead is screened. In such cases a ferrite choke on the lead can be effective in reducing RFI.

### Mains filters

Many mains extension leads for use with computers are advertised as containing surge or spike suppressors - generally these will not contain RFI filter components. To a large extent these have taken the place of traditional mains interference filters using capacitors and inductors. Mains RFI filters can still be purchased though they are harder to find than they were.

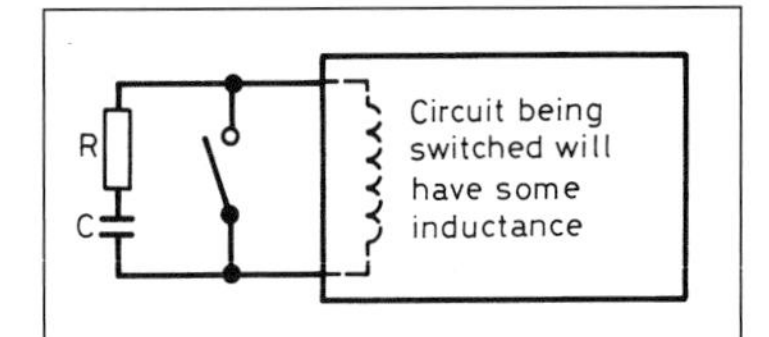


**Fig 22.13: A resistor and capacitor used to absorb the energy released when contact is broken**

When considering a mains filter, make sure that it contains components to suppress both differential and common-mode currents. Some mains filters are sold with a simplified circuit diagram, and if this indicates that it contains an arrangement similar to **Fig 22.18**, then the filter will probably give a reasonable degree of attenuation to both modes. If a mains filter is being considered as an interference reduction measure, it is worth trying a ferrite choke made by winding the mains lead on a ferrite ring, before buying a mains filter. A ferrite choke of this sort has the advantage of reducing common-mode currents on all three conductors (line, neutral and earth). It may be more effective, particularly at higher frequencies and will certainly be much cheaper. See Ferrite Chokes earlier in this chapter.

### Specific RFI concerns

As well as the RFI sources listed in Leaflet EMC04 [7] it is worth mentioning a few specific sources which are causing concern.

Solar panels: These are becoming widespread in the UK and while many do not radiate noise in the amateur bands there are enough which do to make this a serious issue. Noise from solar panels can be identified by the fact that it varies with sunlight and drops away at night. It seems likely that the noise emanates from the control and transforming electronics.

**Wind farms**: There have not been many complaints of RFI to date but there is concern that the wider adoption of switch mode conversion techniques may give rise to more interference complaints.

**High efficiency lights**: This is an ongoing situation. Most lamps do not cause any problems but there has been a number of serious cases. These mainly come from devices which do not comply with the appropriate EMC regulations. Even if a noisy lamp is simply scrapped and replaced with a compliant unit it is still important that information be sent to the RSGB through the EMC forum, accessible via the EMCC web site [1].

**Plasma TVs**: Plasma TVs usually do not cause significant RFI when new, but as they age emissions can become big enough to cause serious interference on the HF amateur bands. Most of the emissions are directly from the screen of the TV set, so that remedies such as ferrite rings do not have much effect.

Fortunately plasma displays are being superseded by other techniques in domestic TV installations.

## NON-RADIO SPECTRUM USERS

In the previous sections we have been dealing with interference accidentally generated by various types of electrical activity such as commutation in electromagnetic devices or high-speed switching in digital products. In recent years an entirely different type of RFI has become a major issue. This has arisen from the use of the existing cables such as electricity supply cables and telephone wires to carry high speed data. At radio frequencies

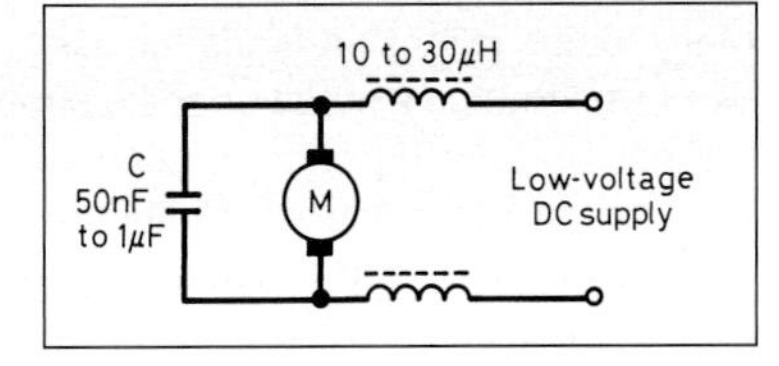


**Fig 22.14: Suppressing a small low-voltage DC motor**

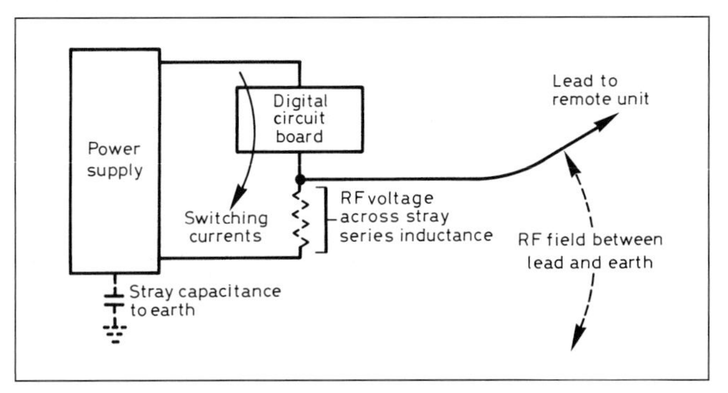


**Fig 22.15: External lead, nominally at 0V, acting as an unwanted antenna**

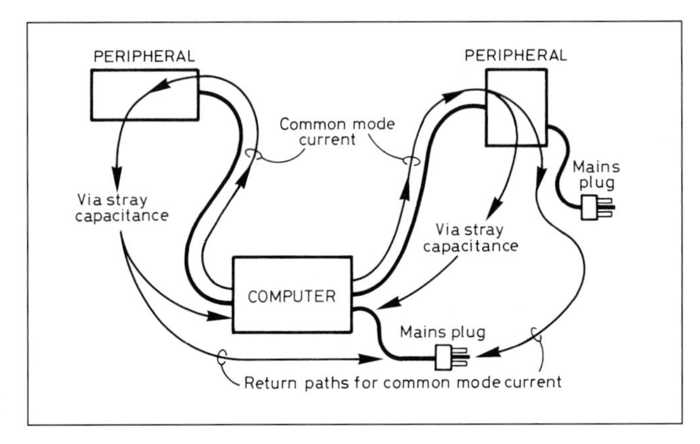


**Fig 22.16: Common-mode current paths**

such systems are imperfect transmission lines so RFI leakage is inevitable. The Recommendations and Standards relating to such systems require that the spectrum of the signals injected into the cable (called the "mask") should be designed so that the power can be significantly reduced on vulnerable frequencies, such as the amateur bands. This is known as "notching". For some technologies the notches are fixed and obligatory while in others the requirement is to build into the system the ability to notch should the need arise. The RSGB's interest in these techniques goes back to the late nineteen nineties. A Technical Working Group was set up by the Radio Communication Agency (RA) "The compatibility of VDSL and PLT with radio services in the range 1.6 to 30MHz". The final report was issued in 2002. At the time of writing this document is available on the archive

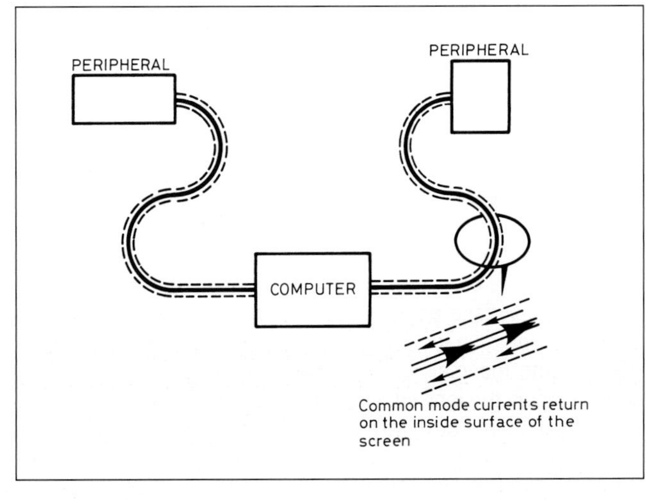


**Fig 22.17: Screening confines common mode currents**

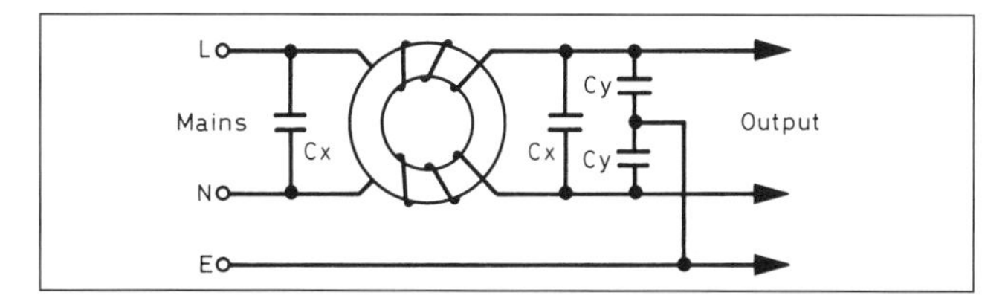


**Fig 22.18: Mains filter with balanced inductor (the action of the coil is similar to a very large number of turns of two-core lead on a ferrite ring)**

pages of the OFCOM web site and is useful background reading for anyone interested in the history of the subject.

## Systems using telephone lines - DSL

DSL (Digital Subscriber Line) is a general term for a family of techniques whereby data is passed over the telephone lines to a customer's premises. The term 'xDSL' is sometimes used when referring generally to all DSL systems. The RSGB has been following the progress of DSL systems since they were first introduced in the in the mid 1990s. Very few cases of interference to or from DSL systems were reported until 2013 when significant emissions from an upgraded version of VDSL2 were noted. Interference to and from DSL is known in the industry as ingress and egress.

**ADSL (Asymmetric Digital Subscriber Line):** Until about 2010 most households in the UK had ADSL broadband internet access. This uses frequencies up to about 1.1MHz, though an upgraded version ADSL2+ goes up to 2.2MHz. In a typical ADSL system a modem in the local exchange feeds the signal out to a modem at the customer's premises. The throughput depends on the distance from the exchange, the maximum distance being in the region of 3 miles. Despite the many millions of connected households there have been very few cases of either ingress or egress to ADSL services. It has always been assumed that this is due to the relatively good balance on the lines at lower frequencies and the fact that high levels of broadcast signals and of ambient noise at medium and long wave frequencies effectively masked any emissions.

**VDSL (Very high speed Digital Subscriber Line):** In a VDSL system the internet signal comes to the street cabinet by fibre and is distributed over the telephone lines to the customer's premises. Since the path over the twisted pair is relatively short - up to about a kilometre - much higher data rates are possible. Download speeds in the region of 40Mb or more are achievable. This arrangement is also known as FTTC (fibre-to-the-cabinet). In many locations in the UK the telephone lines come underground to the distribution pole and are then fed out to the customer's premises on a "dropwire". The dropwire may be quite short but in some cases it can be over 50m. Almost all cases of reported Interference are from overhead installations.

The system in the UK is VDSL2 using frequencies up to about 17MHz. There are three upstream bands (going from customer to the cabinet) and three downstream bands (going from the cabinet to the customer). The upstream bands are 25kHz to 138kHz, 3.75 to to 5.2MHz and 8.5 to 12MHz and the downstream bands are 138kHz to 3.75, 5.2 to 8.5MHz and 12 to 17.6MHz. How the bands are used in any particular installation depends on the distance from the street cabinet and the type of service. It had been expected that interference from VDSL would be most likely to be from the upstream bands, where the signals, are generated at the customer's premises, This seems to be the case though interference from the downstream bands also occurs.

Since about 2013, the appreciation of the radio interference potential of VDSL has changed. Around that time the system in the UK was upgraded and higher frequency bands brought into use. Since then there have been a steady trickle of reports of interference. In 2015 self-install began to be used. It seems likely that this will be a major factor in VDSL interference.

## Identifying and Mitigating VDSL Interference

VDSL interference is very like white noise and the effect is simply that the receiver seems to be insensitive. The band will be abnormally quiet because small signals and the fluctuations of the normal ambient noise are being "drowned", but the S meter reading will be high. VDSL interference can be confirmed by listening on the band edges; 3.75, 5.2, 8.5, 12.0 and 17.6 MHz These frequencies are nominal and the actual step change of noise will be slightly off these frequencies. The steps are easy to see on a Software Defined Radio (SDR). **Fig 22.19** is an SDR plot of emissions in the 8.5 to 12MHz upstream band. Broadband antennas sensitive enough to enable the ambient noise floor to be clearly seen are difficult to achieve on HF. In this case a simple 10.1MHz inverted Vee dipole was used giving rise to the curved shape of the plot. The plot was taken at night when the 12MHz broadcast band was "closed" but the 9MHz band was still open. For diagnostic purposes it is easier to measure the steps separately using a narrow frequency span.

VDSL emissions tend to be constant 24/7 but transient interference has been reported where the VDSL is "retraining".

Work is in hand to find out the cause of emissions from VDSL and to evaluate common-mode chokes and similar measures. In many cases the interference is not related to a VDSL installation in the amateur's house. In such cases the interference is coming either from a source in the neighbourhood or possibly from an accumulation of small emissions from many local dropwires. At the time of writing "notching" is not available in the UK. More information can be found in Leaflet EMC 15 "VDSL Interference to HF Radio" [9] and in [10],

## FutureTechnologies

A related technology called G.fast is already being evaluated.. This brings the fibre nearly to the customer's premises - the local distribution pole for instance - and uses the twisted pair for the last few tens of metres. This would use frequencies up into the VHF region.

The situation on VDSL is changing rapidly and up to date information can be found In RadCom or on the RSGB EMC Committee web site [1].

## Systems using electricity supply cables

PLT (PLC) (Powerline Telecommunication or Communication) is the use of electricity wiring to carry data signals. Low frequency signalling on electricity supplies has quite a long history, but this involves very low frequencies, and is completely different from the internet access systems which use frequencies in the HF

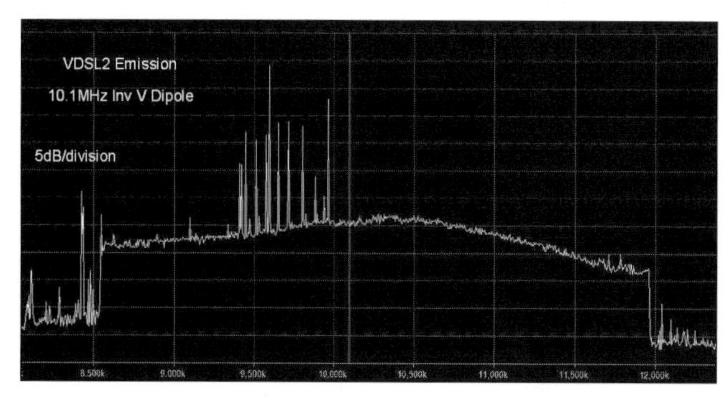


**Fig 22.19: Emissions from VDSL2. The 8.5 to 12MHz upstream band**

band. PLT is not really a single identifiable technology but a name for systems which use the electricity cabling to carry high speed data signals. There are two categories of PLT, Access PLT which provides internet access via the electricity supply cables and In-House PLT which uses the electricity wiring to distribute data between equipment in the same building.

**Access PLT**: The proposed systems vary in detail, but in the UK they involve injecting the data signals at the electricity sub-station. The signal travels up the cable and into the houses supplied by the sub-station. Over the years there have been a number of trials in the UK but none has proved satisfactory. Apart from EMC issues, Access PLT is limited by the fact that the data stream is effectively shared by a number of subscribers and it is difficult to see how such systems could deliver the download speeds now being offered by other technologies.

**In-house PLT:** This links computers and other devices using the electricity wiring in the house as the transmission medium. Devices of this sort have been about for some time, but at first they were mainly used for linking computers using plug-in modems known as Powerline Adapters (PLAs). These early PLAs only generated interference when files were being passed, so interference was not a serious problem.

IIn-house PLT came to prominence with the large scale use of PLAs for video distribution in domestic IPTV installations. Older PLAs operate in the HF band between about 4MHz and 28MHz but some newer devices operate up to 87MHz. PLAs avoid transmission in the international amateur bands and this results in 'notches' where the transmitted power is about 30dB down on the rest of the spectrum. This gives fairly good protection to the amateur bands but the rest of the spectrum including the UK 5MHz amateur band is unprotected. A new standard, EN50561-1, includes notches for the amateur bands and other vulnerable services but permits high levels of emissions elsewhere in the HF bands. It also gives some degree of protection to the HF broadcast bands. More information will be found in RadCom and on the EMCC web site [1].

In-house PLT is relatively easy to identify by listening inside and outside the amateur bands. The notching should be clearly heard, with interference getting much worse as the receiver is tuned away from the amateur band [11].

A technology known as "MIMO" is starting to be used for in-house PLT. MIMO stands for Multiple In Multiple Out and involves injecting the data, not only into the live and neutral, but also onto the earth wire. It has been suggested that this could raises the potential for interference but so far as is known there have been no confirmed reports of increased interference in the amateur bands. The RSGB, along with other National Societies, are watching the situation closely.

## Reducing the Effects of Interference at the Receiver

Where it is not possible to suppress the interference at source, the only solution is to attempt to mitigate its effects at the receiver. There are two traditional techniques available: the first is some form of noise limiter or blanker and the second is by cancellation.

The noise limiter has been around since the early days of radio, and the modern equivalent, the noise blanker, does a similar job in a rather more complex way, and much more effectively.

Cancellation is a very powerful tool for dealing with received interference and deserves to be much more widely known. The principle is to receive both the wanted signal and the interference on two antennas and to adjust conditions so that the interference is cancelled, leaving the wanted signal largely unaffected. This is not as difficult as it sounds, and is well worth considering in cases of severe interference. More information on noise blankers and cancellation (null steering) can be found in the chapter on *HF receivers* and in [1].

Digital signal processing (DSP) techniques are now being used in receivers to enhance the extraction of wanted signals from interference. These techniques can be very powerful, but determining what can be achieved in any particular situation is a matter of practical testing. If you are attempting to combat specific interference, study all the information, and if possible arrange a test on the interference in question.

## REPORTING INTERFERENCE

In the UK complaints of interference can be made to OFCOM. Before making a complaint read Leaflet EMC 09.[12]. It is important to make sure that the source is not in your own home and if possible to give some indication of where the RFI might be coming from. Complaints must be reasonable bearing in mind that, in a typical residential location, some man-made noise is to be expected.

## FURTHER ASSISTANCE

Where an EMC problem fails to yield to the procedures that have been discussed in this chapter, it is time to call for some help. In most cases this means contacting your National Society.

The RSGB's EMC Committee has a Helpdesk for EMC advice and also a forum where interested parties can post EMC related queries and solutions This can be accessed through the RSGB EMC Committee web site [1].

## REFERENCES


[1] RSGB EMC Committee (EMCC) web site. Access from the RSGB web site

[2] RSGB web site www.rsgb.org

[3] Dealing with Security Alarm EMC Problems Leaflet EMC 03. Available on the EMCC web site.

[4] Radio Transmitters and Home Security Systems Leaflet EMC 02. Available on the EMCC web site.

[5] EMC Column RadCom Feb and April 2005, Oct 2007

[6] EMC Column RadCom June 2009

[7] Interference to Amateur Radio Reception. Leaflet EMC 04. Available on the EMCC web site.

[8] Background Noise on the HF Amateur Bands. Leaflet EMC 16. Available on the EMCC web site.

[9] VDSL Interference. Leaflet EMC 15. Available on the EMCC web site.

[10] EMC Column RadCom June 2016.

[11] Interference from In-House PLT. Leaflet EMC 14. Available on the EMCC web site.

[12] Handling Inbound Interference. Leaflet EMC 09. Available on the EMCC web site.


*With thanks to **Robin Page-Jones**, G3JWI, and the EMC team for updating this chapter.*

# Power Supplies

# 23

*Stuart Swain, G0FYX*

Amateur radio equipment normally derives its power from one of four sources:

- The public AC mains which is nominally 230 volts at 50Hz in the EU, though 240 volts still exists in the UK.
- Batteries, which are either primary (non-rechargeable) or secondary (rechargeable)
- Engine driven DC generators or alternators, whether separate or incorporated in a vehicle.
- Renewable sources such as wind, possibly water turbine or pedal driven generators, solar cells or rarely, thermocouples. As these are intermittent in nature, a secondary battery with a regulator must be used in conjunction.

For fixed stations the AC mains is readily available, is relatively cheap and is almost always used. It can be converted by transformers, rectifiers, smoothing circuits or switch mode (high frequency) circuits to a wide range of direct voltages and currents necessary for amateur equipment use.

Batteries, both primary and secondary (accumulators) have always provided a convenient though relatively expensive source of power, especially for low powered or portable rigs, or test equipment.

Discrete engine driven generators can give an output of DC or AC, but the most popular give an output of 240/230V AC at a nominal 50Hz, matching the domestic mains. Car alternators at present provide charging for a 12V accumulator, but there is no standard for electric cars.

Renewable sources provide AC or DC according to type, and are discussed in detail later.

## SAFETY

The operation of all power supplies (except perhaps low-voltage, low current primary batteries) can be dangerous if proper precautions are not taken.

The domestic mains can be lethal. High voltage power supplies for valve equipment can also be lethal and great care must be taken with them. There is a case on record of 12 volts proving fatal, for it is not voltage which kills you but current, and the law of the late German doctor Georg Simon Ohm applies. Having wet hands is asking for trouble; if you must handle high voltages, do so with one hand in your pocket.

### Commercial Equipment Hazards

Some manufacturers use the main on/off switch in one leg of the mains and put the fuse in the other (**Fig 23.1**). It is common to see this potentially lethal arrangement in the UK. The fuse must come before the switch and not the other way around. Often when looking at a schematic the error can be missed unless you trace out the wiring. In some cases, the schematic shows it wired the correct way but the actual transceiver wiring is different!

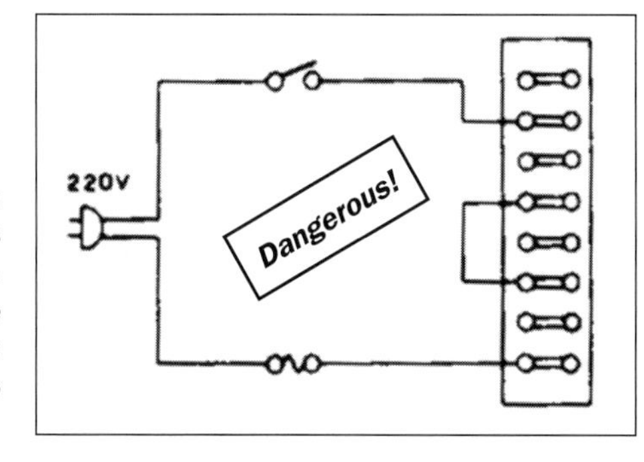


**Fig 23.1: Unsafe mains input wiring. The mains fuse should be in series with the switch**

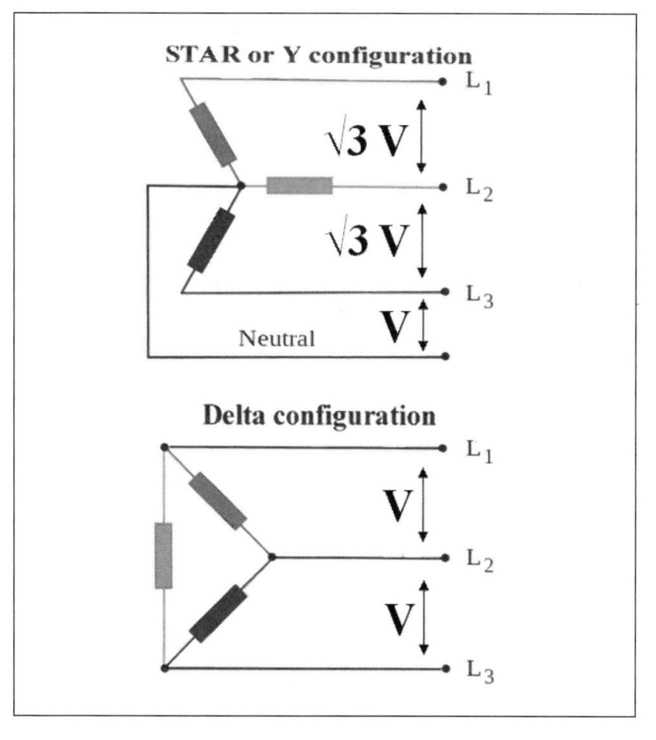


**Fig 23.2: Differences between the UK and US three-phase distribution systems. $L_1$, $L_2$ and $L_3$ are the physical wires and V the voltage between them**

The other common problem found is that the early Japanese transceivers and linears only had a two-core mains lead. As well as this they fit capacitors between the Line and ground and Neutral and ground. Without a three-core earthed mains cable the metalwork floats at half the mains supply voltage and if you touch the cabinet and another piece of equipment which is earthed you get a mighty big slap. This arrangement is illegal in the UK, and many other countries, as all exposed metalwork should be connected to the public mains earth conductor.

### Three-phase Differences

Although not directly of relevance to amateur radio, some workshop tools may use a three phase mains supply. It is important to appreciate that the UK arrangement differs from that used in the US and many other countries.

**Fig 23.2** shows the STAR type distribution system used in the UK, and the DELTA type used in the USA and elsewhere. Note the significant voltage differences between the two systems.

### Fuel Hazards

Where petrol or liquefied gas is used for your generator's fuel, there is a fire hazard, particularly when re-fuelling.

If you persist in re-fuelling while the engine is running, you may have a fire.

## SUPPLIES FROM THE PUBLIC MAINS

It is assumed that the supply will be a nominal 230 volts at 50Hz in accordance with EU regulations.

A rectifier (or rectifiers) will be needed to convert AC to DC. Rectifiers are nearly always silicon PN junction or Schottky diodes but Silicon Carbide diodes will be met occasionally (see the chapter on semi-conductors for more on rectifier diodes). These are all very efficient in that they have a very low forward voltage drop and a very high reverse resistance, within their rating. Copper oxide, selenium and germanium diodes are obsolete for this purpose. The diode has a conventional symbol (**Fig 23.3**) in which the arrow points in the direction of current flow, not electron flow - so the arrow head is the anode, and the plate the cathode. By convention it is the cathode, which is marked +, banded or coloured red because it is positive when rectifying AC.

**Fig 23.4** shows three types of rectifier circuits which are widely used in amateur Power Supply Units (PSUs), together with waveforms of the current supplied by the rectifier(s). In all cases this can be resolved into a DC and an AC component. The latter is called ripple, and requires removal. It can be seen that the half-wave circuit of **Fig 23.4(a)** has a worse ripple than either of the full wave circuits, and has the lowest frequency component (50Hz). Also notice that DC flows through the transformer, which may cause saturation and over-heating if the transformer is not designed for this purpose. Consequently it is not much used.

The full-wave circuit of **Fig 23.4(b)** needs a centre-tapped transformer while the bridge circuit of **Fig 23.4(c)** does not. The bridge incurs two diode (voltage) drops but uses the transformer winding more efficiently. Microwave oven transformers are not designed to use bridge rectifiers, and the insulation (if any) of the low potential end is inadequate, but the circuit of Fig 23.4(a) may be used if the high leakage reactance of the transformer can be tolerated. The oven power supply cannot be used without modification, as it is neither smooth enough nor of the right polarity to feed a valve amplifier.

In most cases the rectifier diodes feed a large value capacitor (the reservoir capacitor), which stores energy during the part of the cycle when the diodes are conducting and releases it when the diodes are not conducting, so maintaining a relatively constant output voltage. The difference in voltage between the diodes conducting and not conducting causes an AC ripple voltage to be superimposed on the average DC output voltage (see below re smoothing circuits). In a few cases the diodes feed an inductor (choke) which is then followed by a capacitor. The choke is heavy and expensive, but does lower the peak current in both diodes and capacitor.

A rectifier diode for mains frequency has three important parameters:

- Maximum mean forward current
- Maximum peak forward current
- The peak inverse voltage, which is encountered by the diode when it is not conducting. This is made up of the instantaneous voltage of the transformer, ie when it is at its negative peak, added to the voltage appearing on the smoothing capacitor. Placing diodes in series is a means of increasing peak inverse voltage, but shunt (voltage sharing) resistors need to be added across each of the diodes to ensure equal voltage across each diode.

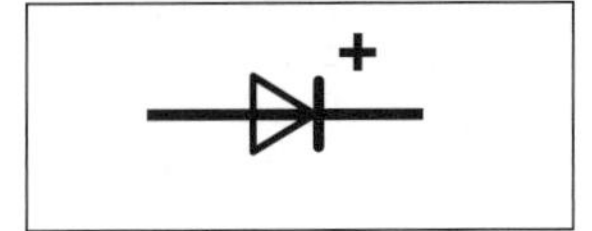


Fig 23.3: Rectifier symbol

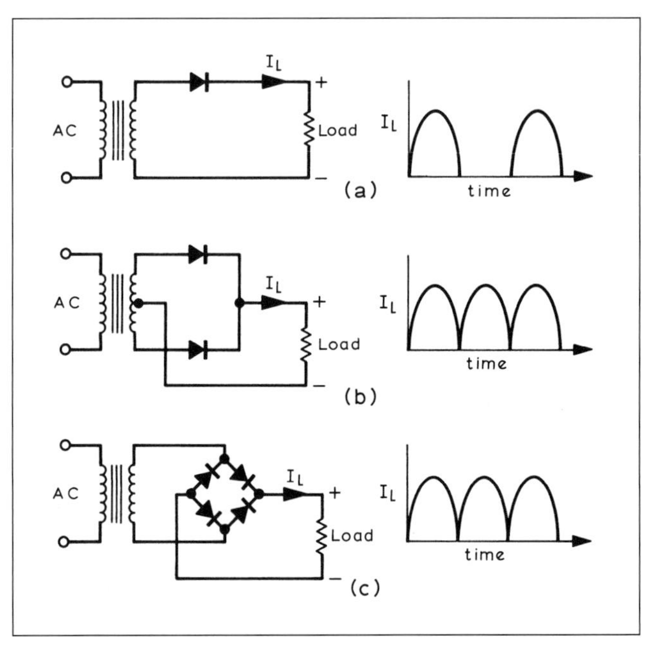


**Fig 23.4: Rectifier circuits showing the output current waveforms with resistive loads. (a) half-wave, (b) full-wave or bi-phase half-wave, (c) bridge**

**Table 23.1** shows the voltages and currents associated with various configurations. **Table 23.2** gives the parameters for some of the more common of the many diodes now available.

Other categories are fast and ultra-fast recovery diodes which are used along with Schottky (low forward voltage and fast recovery time) diodes in switch mode power supplies and soft recovery diodes, which because they switch relatively slowly, cause less RF interference. Avalanche diodes break down on over-voltage throughout the silicon chip and not a localised spot.

There are many packages of four diodes as a bridge, ready to mount on a heat-sink. These are relatively cheap, but watch the manufacturer's peak inverse voltage rating; does it apply to the individual diodes or to the bridge? Table 23.2 also lists some common bridge packages.

### Voltage Multipliers

When a DC voltage greater than the peak of the available AC is needed, a voltage multiplier circuit can be used. These can give a large voltage multiplication, but with poor regulation (decrease of output voltage with increase of load current). The operation may be visualised roughly by thinking of the diodes as 'ratchets'. The mechanical ratchet passes motion in one direction only, and in the multiplier, each stage 'jacks up' the voltage on the following stage. **Fig 23.5** shows some of the possible circuits.

### Smoothing Circuits

These are low pass filters which follow the rectifier diodes, and the behaviour of the circuit depends on the input element of the filter.

This is usually a capacitor in mains frequency circuits and a choke in switch mode PSUs. There may be further components where a greater ripple reduction is needed (see later in this chapter).

#### Capacitor input

An example is the bridge rectifier of **Fig 23.6** in which $R_s$ is the effective resistance of the transformer (resistance of the secondary, plus the square of the turns ratio times the primary resistance).

| Circuit | Average DC output voltage | PIV across diode | Average diode current | Diode peak current | Secondary RMS current |
|---|---|---|---|---|---|
| 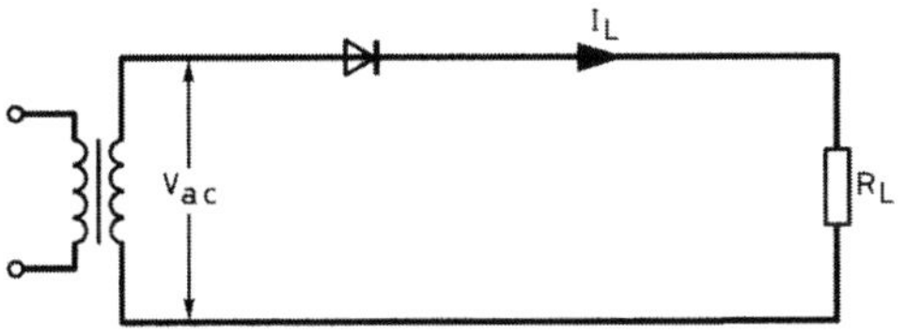 $V_{ac}$, $I_L$, $R_L$ | $0.45V_{ac}$ | $1.4V_{ac}$ | $I_L$ | $3.14I_L$ | $1.57I_L$ |
| 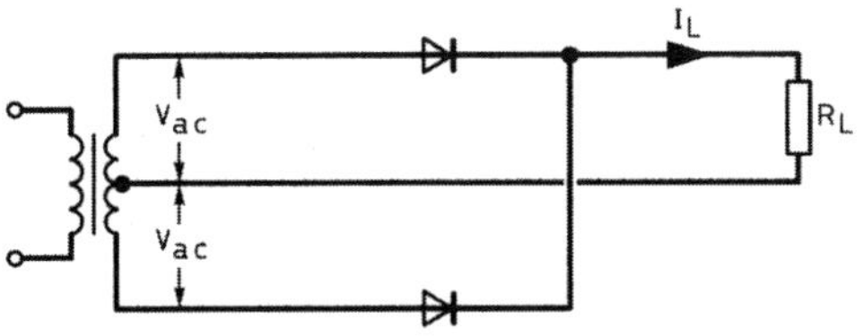 $V_{ac}$, $V_{ac}$, $I_L$, $R_L$ | $0.9V_{ac}$ | $2.8V_{ac}$ | $0.5I_L$ | $1.57I_L$ | $0.785I_L$ |
| 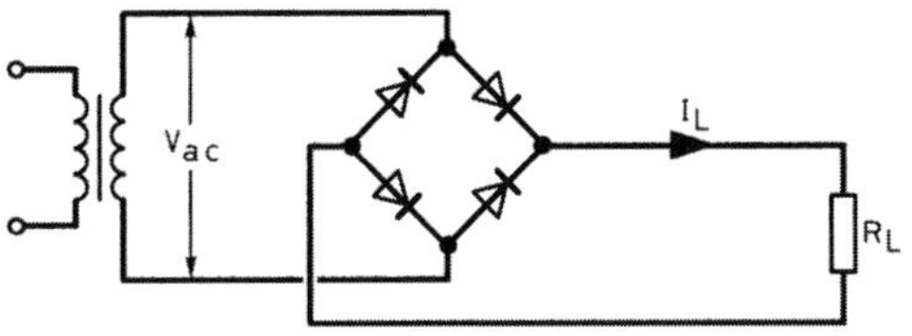 $V_{ac}$, $I_L$, $R_L$ | $0.9V_{ac}$ | $1.4V_{ac}$ | $0.5I_L$ | $1.57I_L$ | $1.11I_L$ |
| 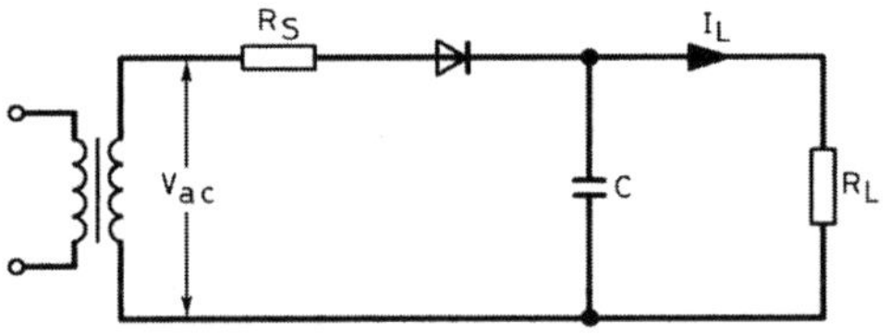 $R_S$, $V_{ac}$, C, $I_L$, $R_L$ | $1.4V_{ac}$ (no load) | $2.8\ V_{ac}$ maximum | $I_L$ | See Fig 24.6 | = Diode RMS current See Fig 24.7 |
| 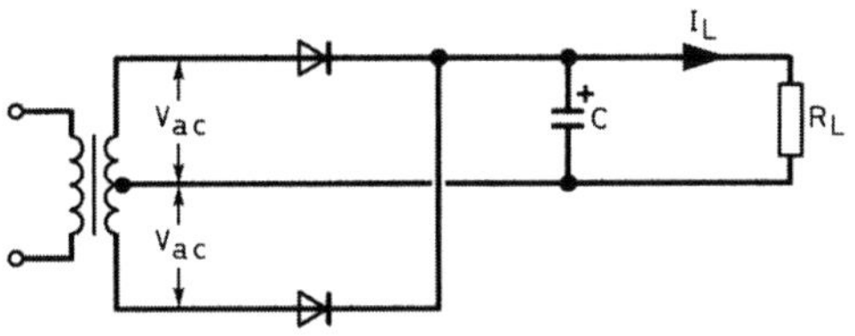 $V_{ac}$, $V_{ac}$, +C, $I_L$, $R_L$ | $1.41V_{ac}$ | $2.82V_{ac}$ | $0.5I_L$ | See Fig 24.6 | $I_L$ |
| 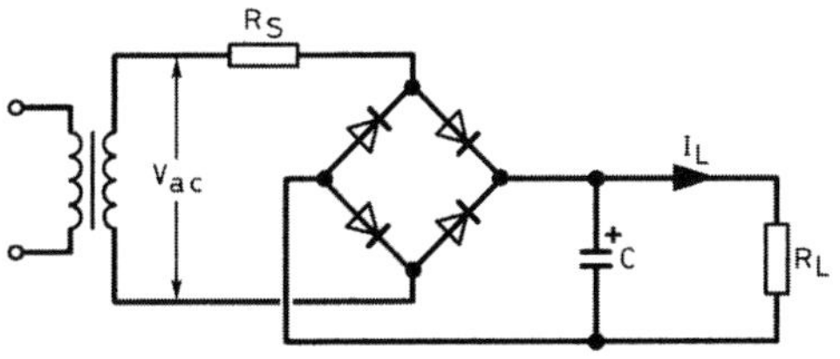 $R_S$, $V_{ac}$, +C, $I_L$, $R_L$ | $1.4V_{ac}$ (no load) See Fig 15.6 | $1.4V_{ac}$ maximum | $0.5I_L$ | See Fig 24.6 | = Diode RMS current x 1.4 See Fig 24.7 |
| 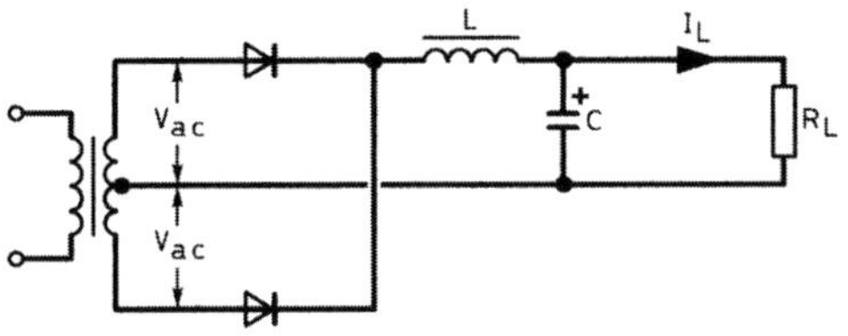 $V_{ac}$, $V_{ac}$, L, +C, $I_L$, $R_L$ | $0.9V_{ac}$ | $1.41V_{ac}$ | $0.5I_L$ | $2I_L$ when $L = L_C$ | $0.65I_L$ |
| 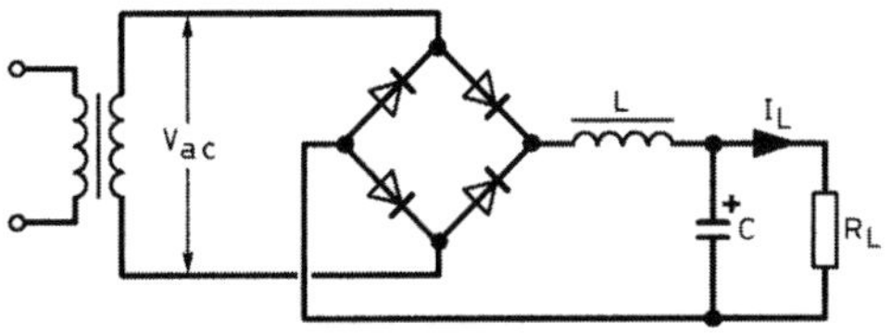 $V_{ac}$, L, +C, $I_L$, $R_L$ | $0.9V_{ac}$ | $1.4V_{ac}$ | $0.5I_L$ | $2I_L$ when $L = L_C$ | $1.22I_L$ when $L = L_C$ |

CD1665

**Table 23.1: Operating conditions of single-phase rectifier circuits**

| Type | VRRM | Iav | IFRM | IFSM |
|---|---|---|---|---|
| **Diodes** | | | | |
| 1N4001* | 50 | 1.0 | 10 | 20 |
| 1N4007* | 1000 | 1.0 | 10 | 20 |
| 1N5401† | 100 | 3.0 | 20 | 60 |
| 1N5408† | 1000 | 3.0 | 20 | 60 |
| BY98-300 | 300 | 10 | 50 | 100 |
| BY98-1200 | 1200 | 10 | 50 | 100 |
| BY96-300 | 300 | 30 | 100 | 200 |
| BY96-1200 | 1200 | 30 | 100 | 200 |
| **Bridge rectifiers** | | | | |
| 1KAB10E | 100 | 1.2 | 25 | 50 |
| 1KAB100E | 1000 | 1.2 | 25 | 50 |
| MB151 | 100 | 15 | 150 | 300 |
| MB156 | 600 | 15 | 150 | 300 |
| GBPC3502 | 200 | 35 | 200 | 400 |
| GBPC3506 | 600 | 35 | 200 | 400 |

*Note: The diodes marked * and † are wire ended - the rest are mounted on screwed studs. $V_{RRM}$ is the maximum reverse voltage or peak inverse voltage, Iav is the average output current in amps, $I_{FRM}$ is the maximum repetitive peak current in amps, $I_{FSM}$ is the maximum non-repetitive peak forward current with a maximum duration of 5ms.*

**Table 23.2: Electrical characteristics of some common diodes and bridge rectifiers**

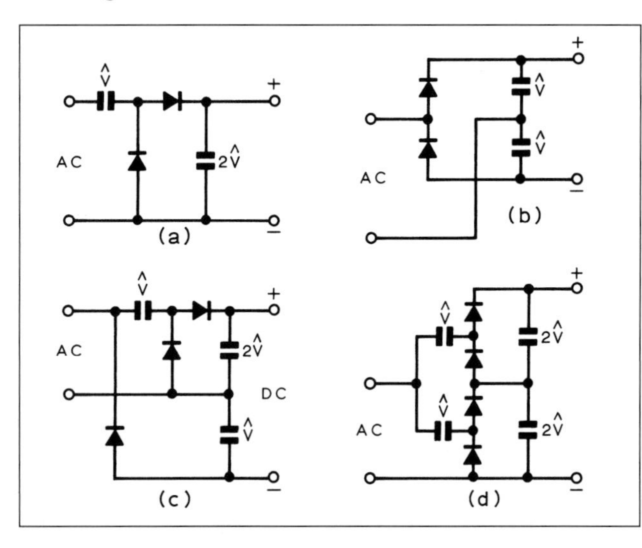


**Fig 23.5: Voltage multiplier circuits. (a) Half-wave voltage doubler; (b) full-wave voltage doubler; (c) voltage tripler; (d) voltage quadrupler. V^ = peak value of the AC input voltage. The working voltages of the capacitors should not be less than the values shown. The diodes have to withstand 2V, but to avoid problems with surges it is advisable to double these voltages**

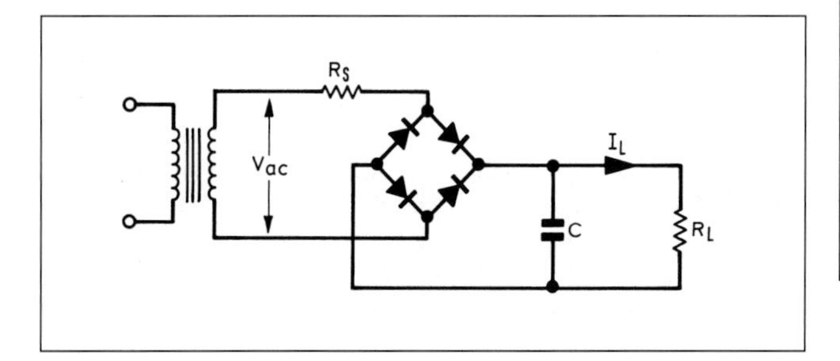


**Fig 23.6: Bridge rectifier with capacitor input filter**

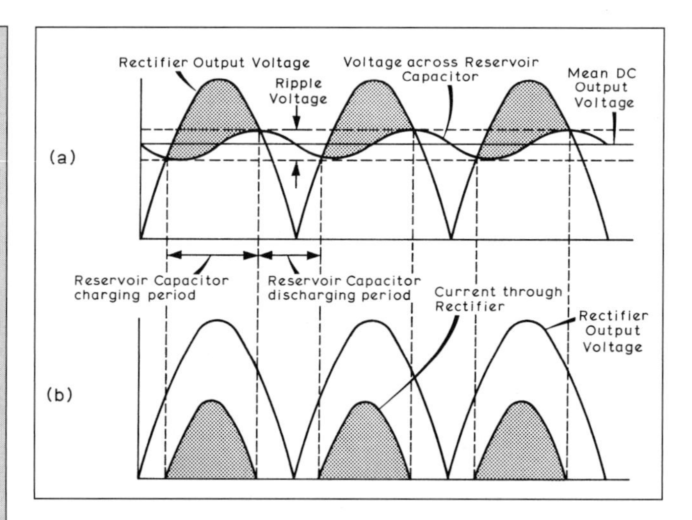


**Fig 23.7: The output voltage and current waveforms from a full-wave rectifier with capacitor input filter. The shaded portions in (a) represent periods during which the rectifier input voltage exceeds the voltage across the reservoir capacitor, causing charging current to flow into it from the rectifier**

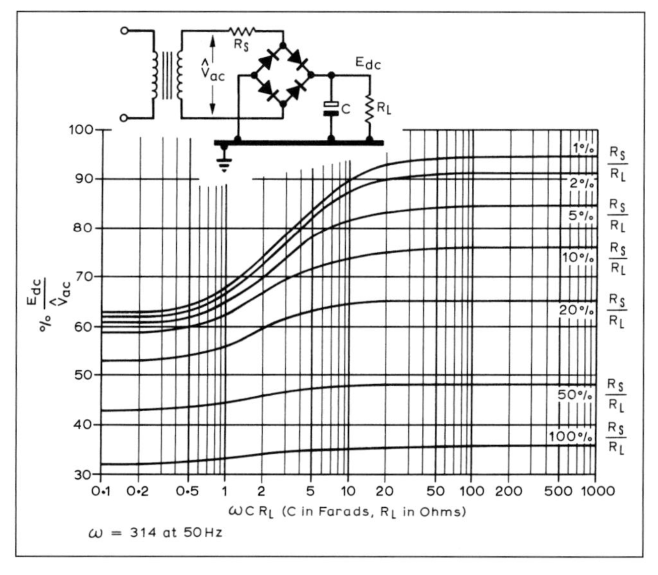


**Fig 23.8: Output DC voltage as a percentage of peak AC input voltage for a bridge rectifier with capacitor input filter**

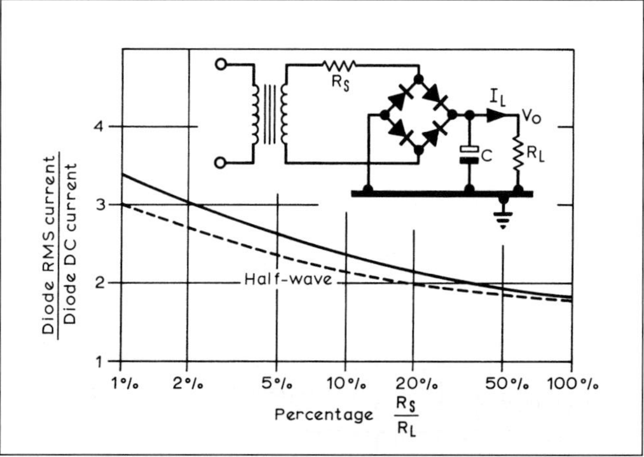


**Fig 23.9: Relationship between diode RMS current and percentage $R_S/R_L$ for values of $\omega CR_L$ greater than 10. ($\omega$ = 314 for 50Hz mains). The dotted line applies to half-wave rectifiers**

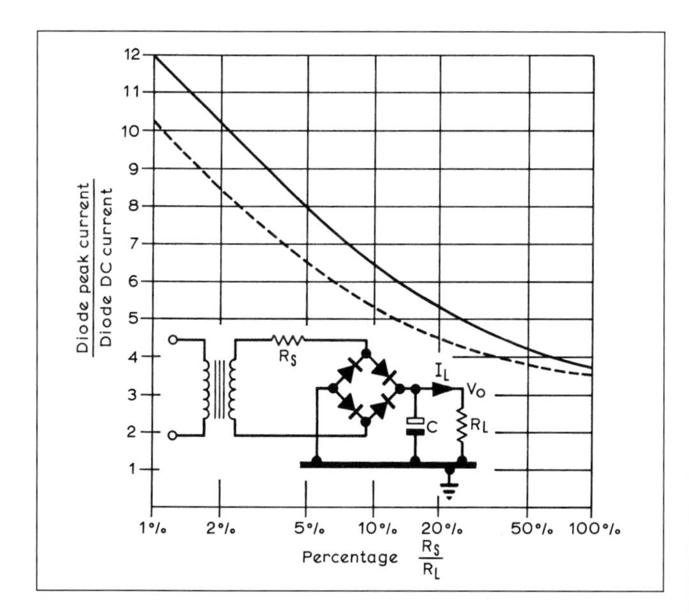


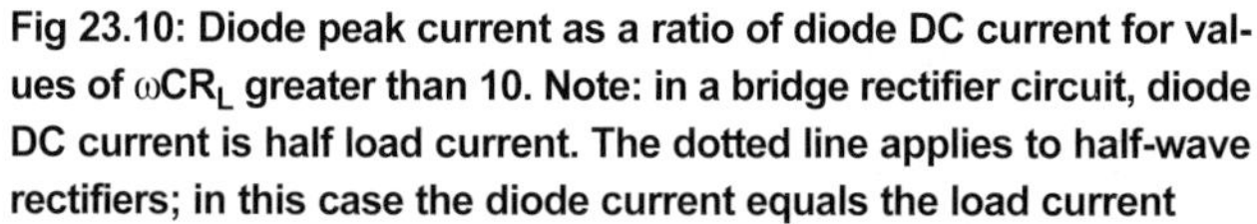
Fig 23.10: Diode peak current as a ratio of diode DC current for values of $\omega CR_L$ greater than 10. Note: in a bridge rectifier circuit, diode DC current is half load current. The dotted line applies to half-wave rectifiers; in this case the diode current equals the load current

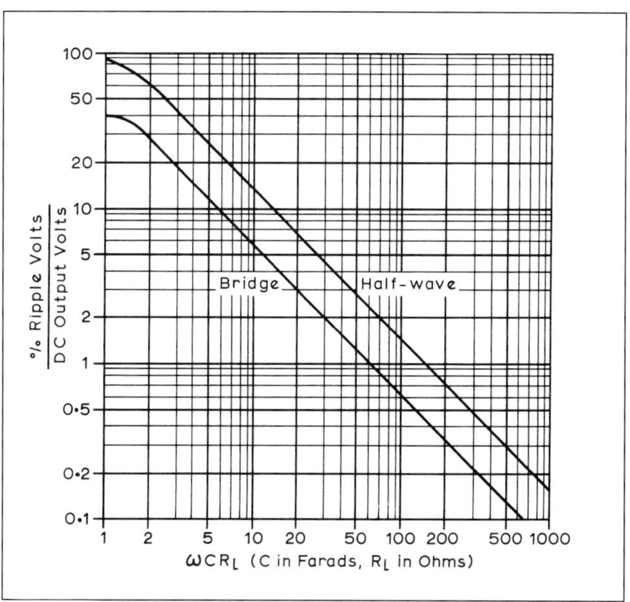


Fig 23.11: Percentage ripple voltage (RMS) against values of $\omega CR_L$ ($\omega = 2\pi f$ where f is the mains supply frequency)

**Fig 23.7** shows voltage and current waveforms of the capacitor as it charges up during part of the cycle and discharges during the rest. The ratio of output voltage to peak input voltage depends on the value of the capacitor, the load and effective resistance in series with the rectifier. **Fig 23.8** shows the relationship graphically being $2\pi$ times the input frequency.

The charging and discharging of the reservoir capacitor constitutes the ripple current and all electrolytic capacitors have a maximum ripple current rating (see the chapter on passive components). Exceeding the maximum will overheat the capacitor and shorten its life. Ripple current is difficult to calculate, but can be measured by using a true RMS ammeter, or estimated at three times the DC rating of the PSU.

As the effective resistance of the transformer becomes a smaller and smaller fraction of the load resistance, $R_L$, the peak rectifier current increases (see **Figs 23.9 and 23.10**). With increasing C (Fig 23.6) the peak rectifier current increases and the capacitor ripple current increases, although the capacitor (output) ripple voltage decreases (**Fig 23.11**). The efficiency also decreases, meaning more transformer and diode heating. A simple idea to avoid destroying the diodes by excessive current, is to limit the maximum peak current by adding a series resistor to augment the $R_s$ of Fig 23.6. The minimum value of $R_s$ is given by:

$$R_s = V/I_{FRM} \qquad (1)$$

where V is the output voltage of the transformer and $I_{FRM}$ the maximum peak current for the diode.

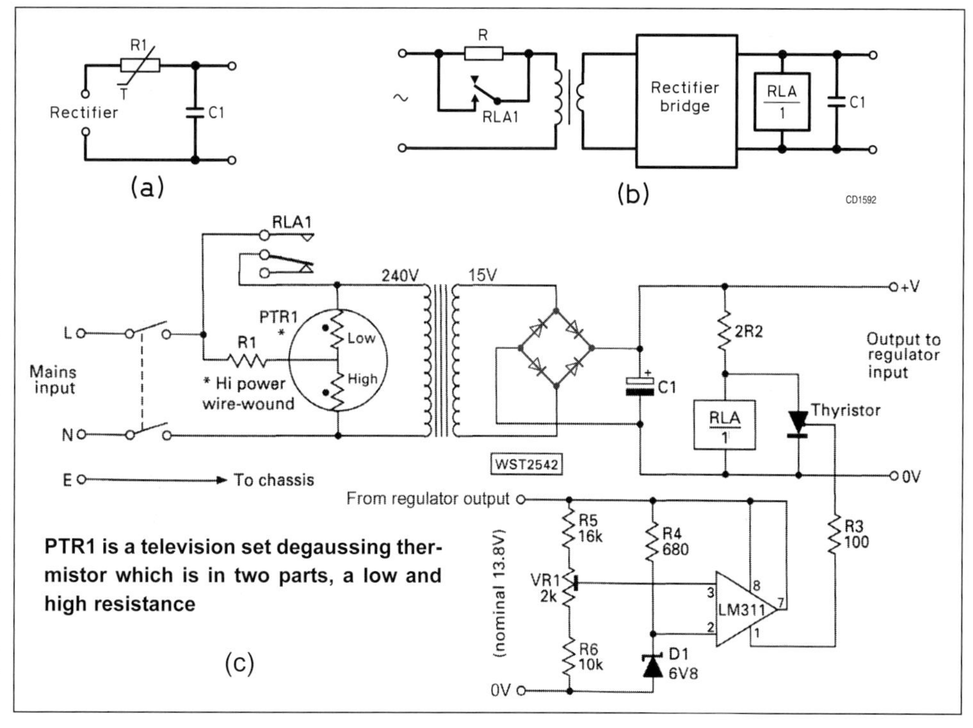


Fig 23.12: Soft starting circuits. C1 is the smoothing capacitor. (a) Using R1, an NTC thermistor. (b) Using a relay to short out a series resistor. (c) A more sophisticated circuit. for a 12V supply. When power is applied, C1 charges up slowly because R1 and the low thermistor limit the inrush current. When C1 is sufficiently charged to operate RLA, the relay closes and RLA1 puts full mains voltage across the transformer primary. The high resistance part of the thermistor remains hot and keeps the 'low' part high. The circuit also provides over-voltage protection by 'crow-barring' C1 through the 2R2 resistor. An enhancement is to place a neon lamp in series with a 150kΩ resistor across the relay contact; this flashes briefly at switch-on and stays on if overvoltage occurs or an attempt is made to switch on with a load connected. [The use of the circuit in (c) is by permission of *Practical Wireless*]

Then calculate the effective resistance of the transformer as already explained. If it is more than the value calculated above, no augmentation is necessary. If it is less, add a series resistor to make up the difference, bearing in mind the necessary power rating.

### Soft starting

When first switched on, the capacitor's inrush current may be excessive and there are means of limiting it.

A simple way is to connect an NTC thermistor (see the chapter on passive components) of correct current rating in series with the primary of the transformer, as in **Fig 23.12(a)**. Another way is to have a resistor in series with the primary, which is shorted by a relay whose coil is in parallel with the reservoir capacitor (**Fig 23.12(b)**).

This relay must be chosen so that it closes at about 75% of the normal output voltage. By putting the limiter in the primary, the doubling of the magnetising current and possible magnetic saturation when the transformer is switched on at a zero crossing of the mains, is avoided.

A refinement of this circuit is shown in **Fig 23.12(c)**, where over-voltage protection is also provided.

### Inductor (choke) input

Here the situation is different (see **Fig 23.13**) - if the inductor's value is above a certain limit (see below), current flows during the whole time (**Fig 23.14**), much reducing the peak value The critical value, ($L_c$) for the inductance in a full wave circuit is:

$$L_c = \frac{R_S + R_L}{6\pi f} \qquad (2)$$

Where $L_c$ is in Henrys, f is the supply frequency in Hertz, and resistances are in ohms. If $R_S$ is much less than $R_L$ and the frequency of the supply is 50Hz, for a full wave rectifier, this reduces to:

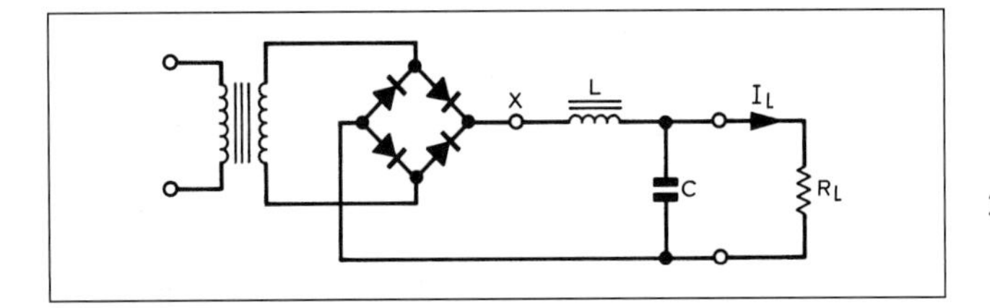


**Fig 23.13: Bridge rectifier with choke input filter**

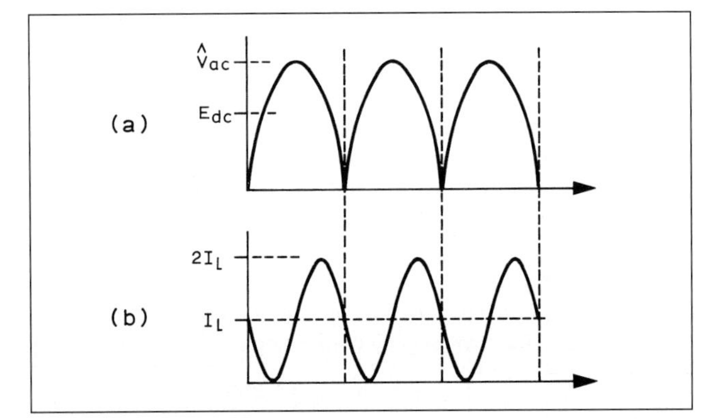


**Fig 23.14: Waveforms at rectifier output (point X in Fig 23.13) in a choke input circuit. (a) Voltage waveform. (b) Current waveform ($L = L_C$)**

$$L_c = \frac{R_L}{940} \qquad (3)$$

It will be seen that the inductance required increases as the load current decreases, (the load resistance $R_L$ increases), so it may be necessary to provide a minimum current by means of a bleeder resistor if the inductor input is to remain effective (the output voltage will rise if it is not).

The value of this resistor in ohms is 940 times the maximum value of the inductor in henrys. The inductance of an iron cored 'swinging choke' filter depends on the current through it.

## Choice of Components

The ideas given here are for power supplies which will always work but which may not be the most economical in components. The reason for this is that generally components are cheap, but time is important and troubleshooting can be difficult. Components should always be chosen on a 'worst case' basis. That is assuming that:

- The mains voltage can fluctuate from its nominal value by ±10%. This does not allow for drastic load shedding during a very hard winter!
- Electrolytic capacitors generally have a tolerance of +50% to -20%, so that a capacitor marked 100µF can have any value between 150µF and 80µF
- Rectifier diodes should have a peak inverse voltage rating of at least three times the output voltage of the transformer (if one is used), when the mains voltage is 10% higher than nominal. This does not allow for spikes on the mains, and either an even higher rating should be adopted or some means of spike reduction installed.
- Choose a diode or diodes with an average current rating of at least twice the required value.

### Capacitor input

1. The secondary resistance of the transformer is assumed to be 0.1Ω and the primary resistance 5Ω. The turns ratio (see below) is 0.0729 to 1 so the effective resistance of the primary transferred to the secondary will be 0.027Ω and the total effective resistance of the transformer is 0.127Ω.
2. The bridge rectifier is a type SKB25/02, which is rated at an average current of 10A, a peak current of 359A and peak inverse voltage of 200V. The RMS transformer output voltage is 17.7V (see below). The minimum value of $R_s$ is 17.7/350 = 0.047Ω, which is well below the effective resistance of the transformer, so no added resistance will be necessary.
3. $R_s/R_L$ = 0.047 /2.76 = 0.017 or 1.7%. The average diode current is 2.5A (two diodes share the 5A).
4. Referring to Fig 23.8, the peak diode current is 12 times this ie 30A and the RMS current from Fig 23.9 is 3.2 times the mean ie 8A.
5. Assume $\omega CR_L$ = 100 (this is an arbitrary choice based on the need for low ripple voltage - see later), so $E_{DC}/E_{peak}$ = 0.95 (from Fig 23.8 which also shows that there is not much to be gained by increasing $\omega CR_L$ further). Therefore the secondary voltage = 13.8/0.95 = 14.5. However this does not include the voltage drop in each diode of about 1.5V (a total of 3V) so the secondary voltage required is 17.5V.
6. $R_L$ is 2.76 so C is 100/(2π50 x 2.76) which equals 115,000µF, the mains frequency being 50Hz. Electrolytics have a tolerance of -20 to +50% so this would be scaled

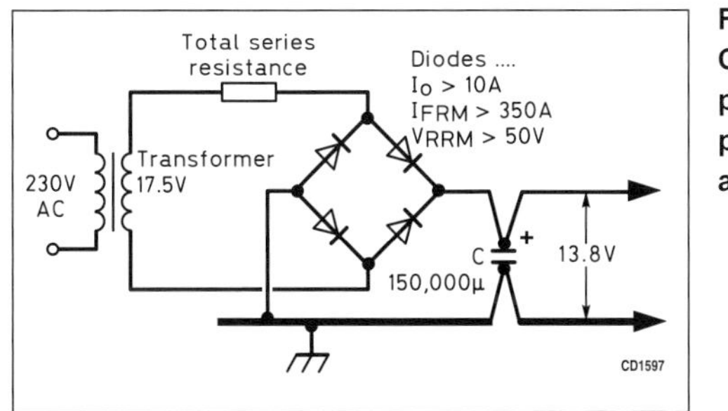


**Fig 23.15: Circuit of a power supply for 13.8V at 5A**

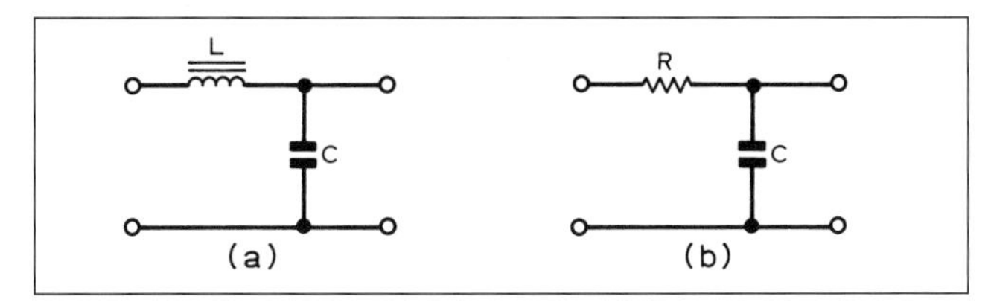


**Fig 23.16: Additional smoothing sections to follow the circuits of Fig 23.6 or 23.13**

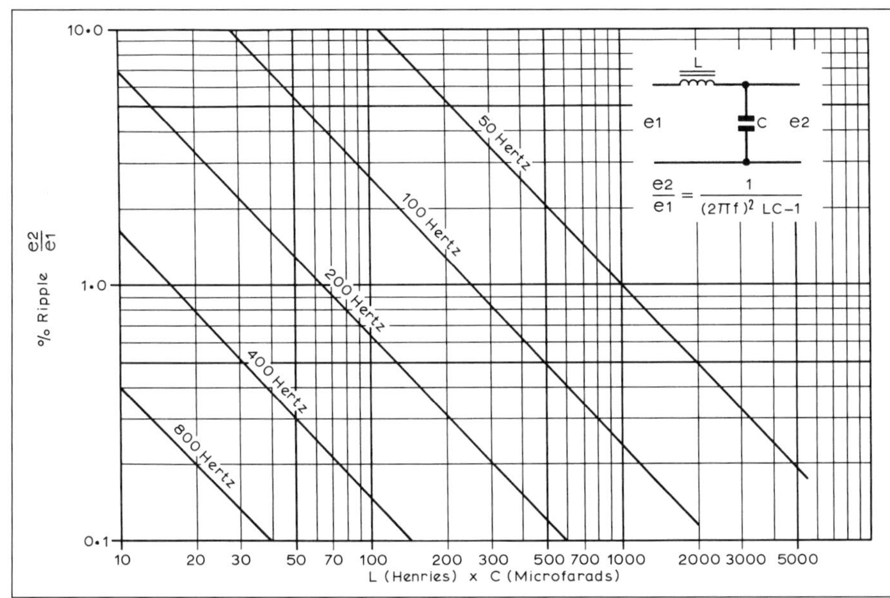


**Fig 23.17: Relationship between large ripple and product of LC**

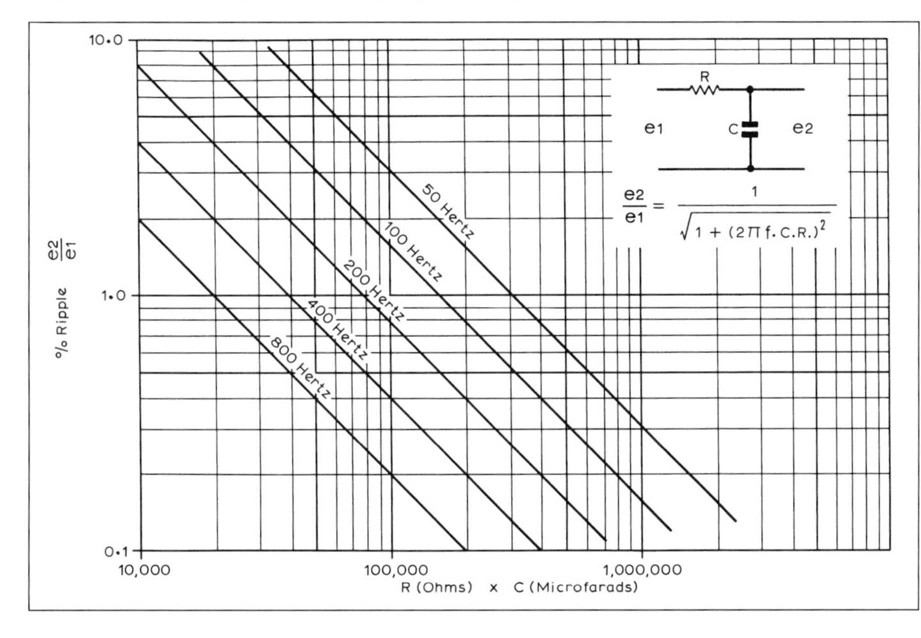


**Fig 23.18: Relationship between percentage ripple and product of RC**

up to 150,000µF, an available value if somewhat expensive. Because of this, a higher voltage is often rectified, a smaller capacitor used, and the ripple removed by a voltage regulator circuit - see below.

7 Fig 23.11 gives the ripple as about 0.6%.

The input and output leads to the capacitor should have as little in common as possible to avoid the introduction of ripple. Take them independently to the capacitor terminals as in **Fig 23.15**.

The design of all other power supplies working from the mains, no matter what voltage or current, follows the same rules except the following:

- When an input inductor (choke) is used (see below).
- Where a 12 or 24V secondary battery (usually a vehicle battery) is 'floated' across a DC supply and takes the place of the reservoir capacitor.

## Inductor (Choke) input

Here the inductor is connected directly to the rectifier diodes (Fig 23.13) and is followed by a smoothing capacitor. Note that it is equally possible to use a centre-tapped transformer and just two diodes to make a full wave rectifier.

As mentioned above in equations (2) and (3), the inductor must have a certain minimum value for a given load. This must be calculated for the smallest current expected, which may only be that of a bleeder, if fitted.

The voltage and current values and waveforms for a circuit using at least this value of inductance are shown in Fig 23.14. It is clear that the peak rectifier current is only double the mean current. Under these conditions the transformer RMS current is 1.22 times the load current and the average current per diode is half the load current.

The output voltage is 0.9 of the transformer RMS output minus voltage drops in winding resistance of both transformer and choke less the diode(s) drop. The diode drop can be neglected for power supplies above say 100V and estimated from manufacturer's data where significant. (Power diodes do not just drop 0.6V!)

The value of the filter capacitor (the first if more than one) is arranged to give the wanted ripple voltage ER from the equation:

$$ER = \frac{\text{output voltage}}{0.8LC} \qquad (4)$$

where L is in henrys, C in microfarads and the supply frequency 50Hz (full wave rectification).

Further smoothing can be added to either capacitor or inductor input by means of LC or RC circuits as in **Figs 23.16, 23.17 and 23.18**.

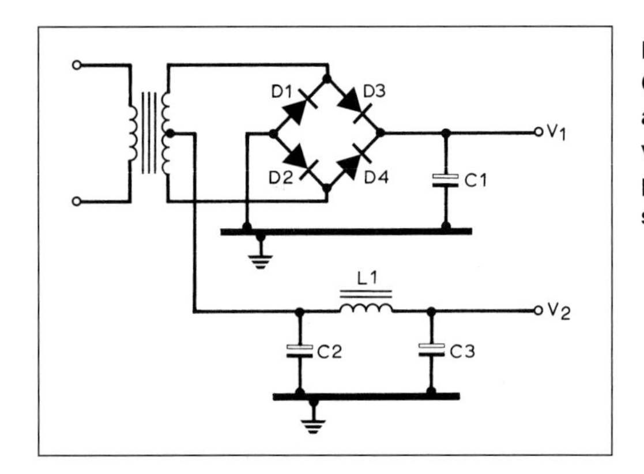


Fig 23.19: Circuit of a dual-voltage power supply

The RC circuit is unsuitable for high currents because it drops the voltage.

Care must be taken that L and C do not resonate at the ripple frequency or any harmonics thereof. For a full wave circuit on 50Hz, the lowest ripple component is at 100Hz and the LC product is 2.53 for resonance. Normally LC is very much higher than this (see Figs 23.17 and 23.18). Note that the choke should be designed so that it does not become magnetically saturated at full DC output. An air gap in the core will help in this respect. The 'swinging choke' is so designed that it does approach saturation at full output and has a higher inductance at low currents where it is needed to satisfy equation (3).

## Dual Power Supply

The circuit of **Fig 23.19** shows a dual voltage supply from a centre-tapped transformer. On analysis, although it uses a bridge rectifier, the two halves of the bridge feed the two supplies separately. The earth point may be chosen to give V and V/2 or +V and -V as required. In the example shown, output V2 is half of V1.

## Voltage Regulators or Stabilisers

These are circuits that give a virtually constant voltage output regardless of load or input, within certain limits. They are necessary for supplying variable frequency oscillators (VFOs), DC amplifiers and some logic circuits. At higher voltages they are necessary for supplying screen and/or grid bias for valve amplifiers.

### Shunt regulators - zener diodes

These are diodes with a sharp breakdown voltage, (see the chapter on semi-conductors). If fed from an un-regulated source through a resistor (**Fig 23.20**) it forms a simple regulator for more or less constant loads, with the advantage of being able to source or sink current. There are reference diodes whose breakdown voltage is nearly independent of temperature, usually available in the 8-10 volt region, at a specified current. Below 5 to 6 volts the zener diode has a negative temperature coefficient and above, a positive one. **Table 23.3** gives some figures for typical zener diodes made by Philips.

The series resistor value ($R_s$) is calculated so that the diode provides regulation when the input voltage ($V_s$) is at its minimum when the load current ($I_L$) is a maximum. It is important to check that the diode's maximum dissipation is not exceeded when these conditions are reversed, ie when $V_s$ is at its maximum and $I_L$ a minimum. The expression for the series resistor is:

$$R_s = \frac{V_{s(min)} - V_{zener}}{I_{L(max)} + I_{zener(min)}}$$

The resistor must be rated for

$$\frac{(V_{s(max)}) - V_{zener})^2}{R_s} \text{ watts.}$$

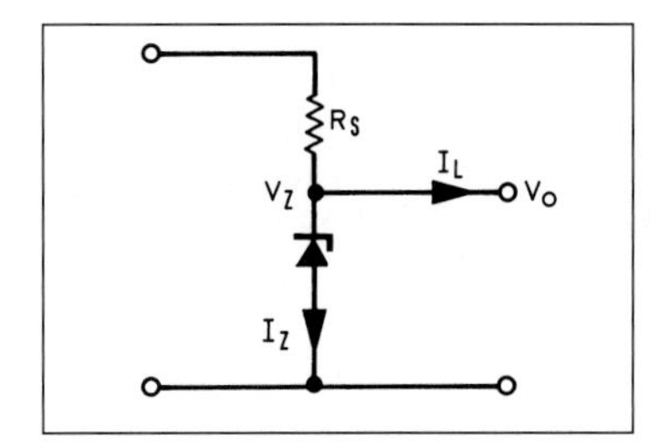


Fig 23.20: A simple zener diode voltage regulator

| Type | Zener voltage (V) | Normal operating current (A) | Zener slope resistance (Ω) | Maximum dissipation (W) | Temp coeff (mV/°C) |
|---|---|---|---|---|---|
| BZX79C2V4 | 2.4 | 0.005 | 100 | 0.4 | 3.5 |
| BZX79C6V2 | 6.2 | 0.005 | 10 | 0.4 | +2.0 |
| BZX79C75 | 75 | 0.002 | 255 | 0.4 | +80 |
| BZT03C7V5 | 7.5 | 0.1 | 2 | 3.25 | +2.2 |
| BZT03C270 | 270 | 0.002 | 1000 | 3.25 | +300 |
| BZY91C10 | 10 | 5.0 | NA | 75* | NA |
| BZY91C75 | 75 | 0.5 | NA | 75* | NA |

** On a heatsink.*
*'NA' means 'not available'.*
*While these are all Philips devices, all semiconductor manufacturers make zener diodes in one or more sizes. The data on Philips devices were used for this table because they were to hand and not because they are recommended above other makes. This table represents the extremes of size, power dissipation and voltage.*

**Table 23.3: Electrical / thermal characteristics of some zener diodes**

### Shunt regulators - integrated circuits

There are many voltage reference devices made for shunt regulator purposes, ranging from 1.22 up to about 36V. Some have three terminals to allow fine adjustment. They are used in the same way as the zener diodes previously described

### Shunt regulators - high voltage

Beam tetrodes used for linear power amplifiers need a screen supply, and the 4CX250 series in particular can sink or source screen current. A series string of zener diodes could be used to cope with this, but would use expensive high wattage diodes. The circuit of **Fig 23.21** transfers the problem to a cheaper power transistor, but does not remove the need for a series string of diodes. These can now be of low wattage and so cheaper. The higher the $h_{FE}$ of the transistor or Darlington pair, the cheaper will be the diodes. Formerly gas filled voltage regulators were used, and are still available. As long as you allow for the difference between the striking and running voltages, the procedure is the same as that for zener diodes.

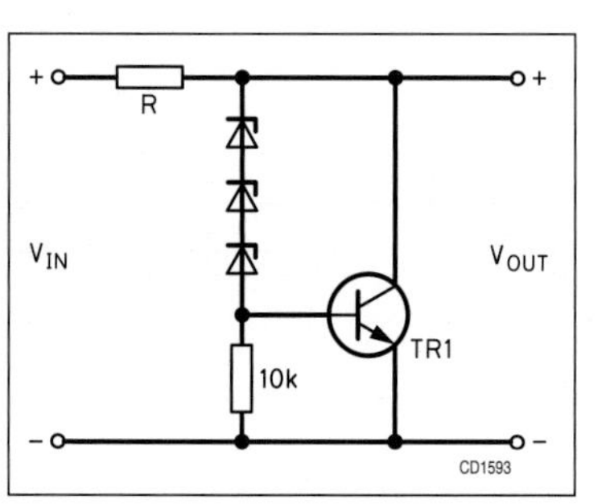


Fig 23.21: Stabiliser for screen of beam tetrode

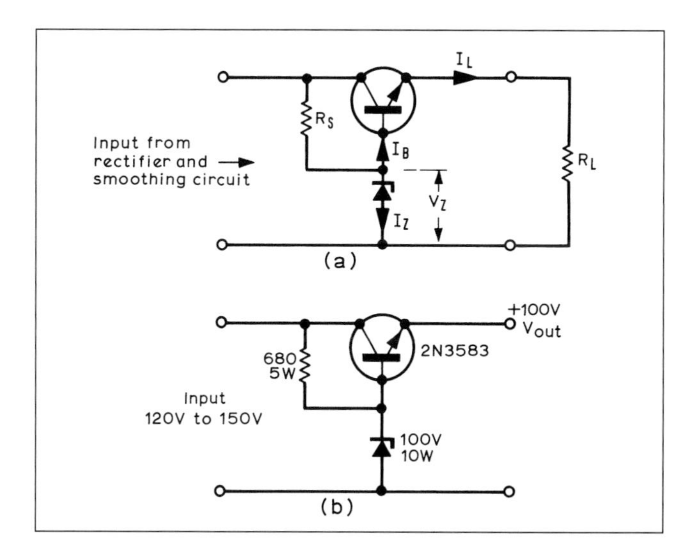


Fig 23.22: A series transistor regulator

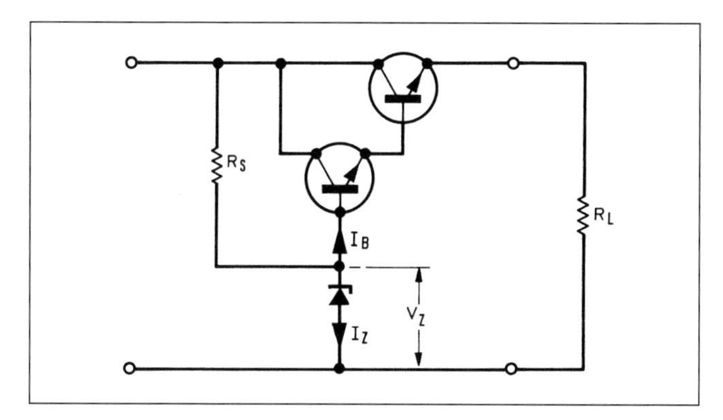


Fig 23.23: A series regulator using two compounded transistors as the series element

## Series regulators

In these an active device is placed in series with the supply, and negative feedback applied in such a way that the output voltage remains constant in spite of varying load and input voltage. The output voltage, or a definite fraction of it, is compared with a reference voltage and the difference amplified to control the series pass element in such a way as to minimise the difference. The greater the gain of the amplifier, the better will be the final result, provided the circuit is stable. A single transistor may be good enough in less demanding situations. The pass element can be either a bipolar or field effect transistor. Protection from failure of the pass transistor is advised to avoid damage from over-voltage.

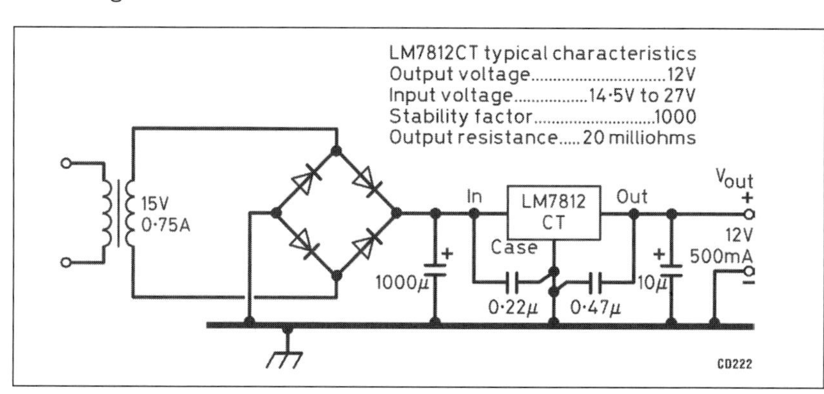


Fig 23.24: A 12V 500mA power unit using a regulator type LM7812CT

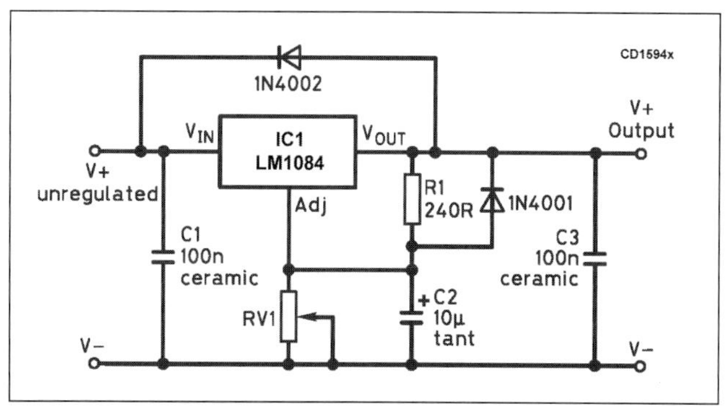


Fig 23.25: Adjustable voltage regulator using an IC. The value of RV1 is ($V_{out}$ - 1.25)R1/1.25 ohms

**Figs 23.22 and 24.23** show the simplest type of regulator using only two semiconductors. As previously said, they are only suitable in less demanding situations. There is not much excuse for building voltage regulators out of discrete components, as integrated circuits are cheap and often better.

## IC voltage regulators

There are many IC voltage regulators, of which only linear types will be considered here. They all have the elements so far described incorporated into a single chip, and often include various over-load and over-temperature protection. There are four main types:

1. Positive fixed voltage eg 5, 12, 15 and 24V
2. Positive adjustable voltage, adjustable by external resistors
3. Negative fixed voltage
4. Negative adjustable voltage, adjustable by external resistors

All need capacitors connected close to their input and output pins to prevent oscillation, see **Fig 23.24**, and it is advisable to fit diodes to prevent capacitors in the load from discharging back through the IC.

This is particularly important for LM317 and LM1084 type adjustable types (**Fig 23.25**) - take the maker's advice! There is no excuse for attempting to add components to a fixed regulator to get an increased voltage; the adjustable ones cost little more.

**Table 23.4** gives some typical examples. The circuits for adjustable types do vary from manufacturer to manufacturer (Table 23.4 again).

Most IC voltage regulators have internal current limiting and some have 'fold back' current limiting in which the voltage falls sharply if you attempt to exceed the current limit, see **Fig 23.26**.

The input voltage must exceed the output voltage by a stated amount; this is called the drop-out voltage.

An external pass transistor, connected as shown in **Fig 23.27** for a positive regulator, may be used to increase the output current. The resistor is chosen so that the transistor is turned on (by 0.6V on its base) just before the integrated circuit's current maximum is reached.

This applies particularly to the 723 type of IC where its ability can be extended almost indefinitely (note that the correct compensation capacitor must be connected according to the data sheet) [1]

| Type | Voltage | Current | Polarity | Vin(min) | Vin(max) |
|---|---|---|---|---|---|
| **Fixed voltage** | | | | | |
| MC78L05APC | 5.0 | 0.1 | + | 6.9 | 30 |
| MC79L05APC | 5.0 | 0.1 | | 6.9 | 30 |
| LM78 12CT | 12.0 | 1.0 | + | 14.6 | 35 |
| LM79 12CT | 12.0 | 1.0 | | 14.5 | 35 |
| LT1086CT12 | 12.0 | 1.5 | + | 13.5 | 25 |
| MC78T 15CT | 15.0 | 3.0 | + | 17.8 | 40 |
| 78P 05SC | 5.0 | 10.0 | + | 8.25 | 40 |
| LM2931A | 5.0 | 0.4 | + | 5.65 | 25* |
| **Adjustable voltage** | | | | | |
| LM317LZ | 1.2-37 | 0.1 | + | NA | 40 |
| TL783C | 1.25-120 | 0.7 | + | NA | 125 |
| LM317T | 1.2-37 | 1.5 | + | Vo + 3 | Vo + 40 |
| LT1086CT | 1.2-29 | 1.5 | + | Vo + 1.5 | Vo + 30 |
| 79HGSC | 2.1-24 | 5.0 | | NA | 35 |
| MAX 667 | 5-12 | 0.25 | + | NA | 18* |

** Low 'drop-out' type, ie it has a low voltage drop across the series transistor. More of these are now available.*

*Notes. There are many other different voltage and current rated stabilisers and they are made by many different manufacturers. All the high-powered devices must be fixed to a suitable heatsink. All 'fixed voltage' devices can have their voltage adjusted upwards by adding a resistor, a diode or a zener diode in series with their 'common' lead. The value of resistor varies with the device and the manufacturer, and the latter's literature should be consulted.*

**Table 23.4: IC voltage regulators**

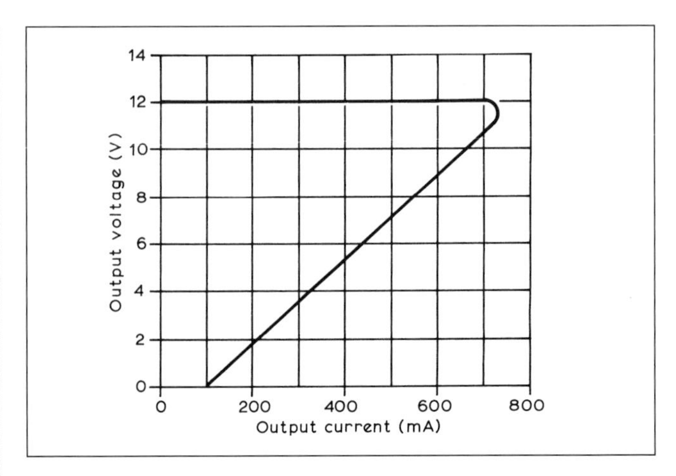


**Fig 23.26: Voltage current characteristic showing 'fold-back'**

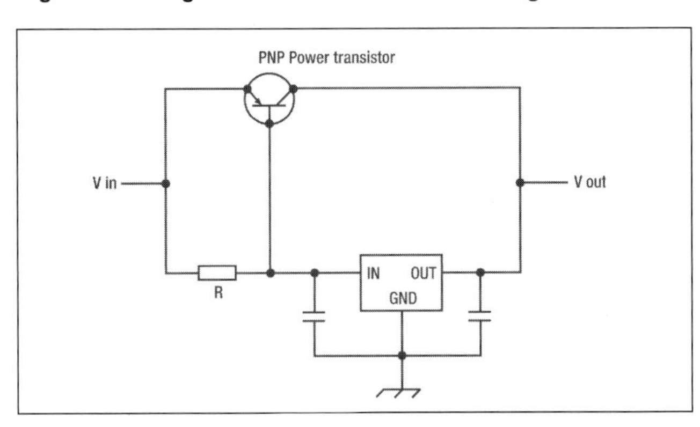


**Fig 23.27: Connecting an external pass transistor to increase the output current**

### Field effect transistors (FETs)

Instead of using bipolar transistors for the pass element, a MOSFET (see the chapter on semiconductors) can be used with advantage. They are now cheap, and have the advantages that they are not subject to thermal run-away, have a reasonably constant gain with drain current, and a very low on-resistance. A disadvantage is that as they are only made in enhancement versions, the gate requires a voltage somewhere between 2 and 8V (according to type) above the source. For efficiency this should be derived from a low power auxiliary rectifier and transformer or winding.

As MOSFETs have an integral reverse diode, protection from damage due to charge stored in capacitors after the regulator is not required.

**Fig 23.28: The TV de-gaussing thermistor used in Fig 23.12(c)**

## Over-voltage Protection

If the pass transistor fails to a short circuit condition, the whole input voltage (perhaps 18-20V for a 13.8V supply) will reach the output. Many rigs do not like this very much, and to avoid expense, some form of over-voltage protection ought to be included. The most common is called a crow-bar, so called because it short circuits the supply and either blows a fuse or operates a relay to open circuit the rectifier (see Fig 23.12(c)). A type of thermistor which can be used here comes from the de-gaussing circuit of a TV set, and is shown in **Fig 23.28**. Varistors and diodes can be specially made for this purpose.

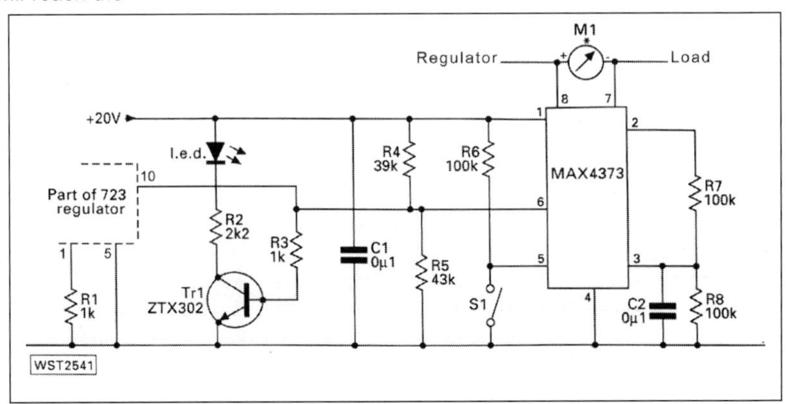


**Fig 23.29: Current limiting and protection [*Practical Wireless*]**

## Over-current Protection

Most ICs have internal current limiting, but where augmentation of the ICs current is used, often with the 723, the internal limit may not work or may

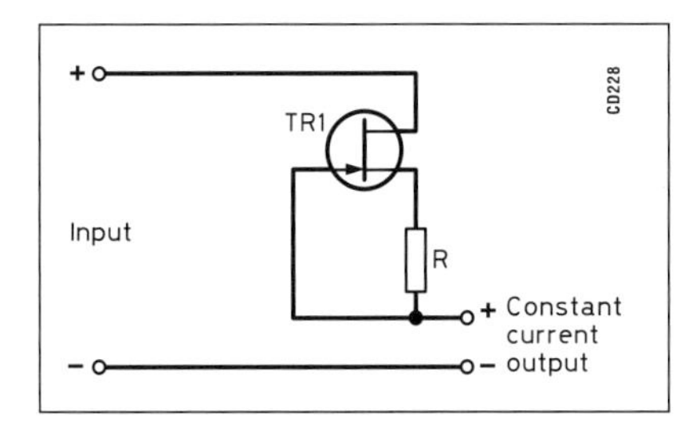


**Fig 23.30: Principle of the constant-current circuit. TR1 small JFET; R to set current - depends on $G_m$ of FET**

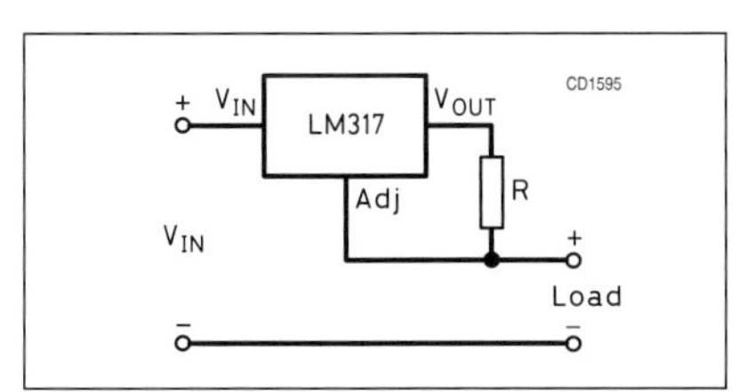


**Fig 23.31: LM317 used as a constant-current regulator $I_{LOAD}$ = 1.25/R**

need too many millivolts in a current sensing resistor. The Maxim MAX 4373 is a very useful IC that needs very few millivolts from the current sensing resistor, and provides a latched over- current flag. **Fig 23.29** shows it being used to sense the voltage across an ammeter, and shut down a 723 IC by crowbarring an internal amplifier transistor. If you use this circuit, you must put a 1k resistor between pin 1 (Current Sense) of the 723 IC and ground, to prevent destruction of the transistor by excess current. You restore the latch by grounding pin 5 of the MAX 4373. Full details of the Max 4373 series are available from Maxim.

Quick-acting Positive-Temperature Coefficient (PTC) thermistors are now made with very short time constants for current limiting in series connection.

## Constant Current Circuits

From time to time, a constant current source is needed, for example to charge a secondary battery (see below). A depletion mode FET with a source bias resistor will do the trick (see **Fig 23.30**), but the $I_{DSS}$ of a junction transistor is only loosely specified. Corresponding to drop out voltage, the knee voltage of the FET has to be overcome before constant current is achieved. To save trouble, FETs with built in source resistors can be bought.

A similar arrangement with a bipolar transistor, whose base is held at a fixed potential, can provide a constant current up to several amps if required.

The LM317 adjustable regulator can be used in the circuit of **Fig 23.31**. The LM317 passes enough current to maintain 1.25V between the output and Adj pins, so the current in amps will be 1.25/R where R is in Ohms. As the circuit only has two connections, it can be used for either positive or negative supply, having due regard to polarity.

For a really high current supply, put a suitable choke in series with the primary of the transformer of a normal full wave circuit. Achieving the correct choke is easier done by trial and error!

## SWITCH MODE POWER SUPPLIES

All the previously mentioned DC supplies transformed the mains, rectified it and perhaps regulated it. At 50Hz, the transformer is bulky and heavy. If a higher frequency could be used, the transformer and smoothing arrangements could be made much smaller.

The switch mode device does just this, and was used in the 1930s as a means of effectively transforming DC. With the advent of transistors, much higher frequencies, up to the MHz region could be used, also making efficient regulation possible, if needed.

The following two sections are extracts from the *Power Supply Handbook* [2].

## Introduction to Switch-Mode Supplies

Today almost every item of domestic and other electronics seems to contain some sort of switch-mode supply. Cellular telephone handsets utilise switch-mode supplies to generate the positive and negative supply required by the RF power amplifier stages and positive voltages for the processor ICs, as well as chargers for the battery. Personal computers have for many years used off-line switchers, even ham radio equipment has had them included in some way or another.

Many people who don't know any better think that putting a switch-mode supply inside a sensitive radio is crazy as they generate horrific interference because of the rapidly changing switching currents. However, with care in the design and construction, adequate screening and filtering they can be made surprisingly quiet. Like it or not the 'switcher' is here to stay and when one understands the subtleties of these supplies most of the bad feelings go away. For compact, lightweight supplies switchers are hard to beat, and they often cost far less than a conventional linear supply of the same power rating.

The easiest way to design switch mode converters is to get data sheets of suitable ICs and use the circuits offered. Anyway, make sure that input and output are adequately filtered.

### How much better is switch-mode than linear?

An excellent example of the differences between the two types of supplies was presented in the SGS Thomson Micro-Electronics book *Power Switching Regulators - Designer's Booklet 1st Edition* (1993). The company name has since changed to ST-Microelectronics but the data given is timeless:

Switching regulators are more efficient than linear types so the transformer and heatsink can be smaller and cheaper. But how much can you gain?

We can estimate the savings by comparing equivalent linear and switching regulators. For example suppose we want a 4A / 5V supply.

For a good linear regulator the minimum dropout voltage will be at least 5V at 4A. The dropout voltage is given by:

$$Vi_{(min)} = V_o + V_{drop} + 1/2\ V_{ripple}$$

Using a 60Hz supply with a 10,000µF smoothing capacitor and assuming a minimum mains voltage of 80% of nominal we need a minimum of 13.25V DC at the input of the regulator. It is prudent to raise this a bit so we will choose 14V.

Power dissipated in series element is:

$P_d = (V_{in} - V_o) \times I_o$ = 36W    Note: This is a 20W supply!

The heatsink will need to have a thermal resistance of 0.8°C/W or better.

The transformer needs to supply a power of 14V x 4A = 56W. It must therefore be dimensioned for about 62VA to take care of the assumed transformer efficiency factor of 90%.

In contrast, a switching regulator IC using the same nominal input voltage of 14V will dissipate a maximum of 7W. This power is divided more or less equally between the IC and the recirculation diode. It follows that the transformer needs to be about 30VA and the heatsink needs to have a thermal resistance of about 11°C/W.

As can be seen from **Table 23.5**, the approximate cost savings are 50% on the transformer and around 80% on the heatsink.

| | Linear | Switching |
|---|---|---|
| **Transformer** | 62VA | 30VA |
| **Heatsink** | 0.8°C/W | 11°C/W |

**Table 23.5: Comparison between linear and switch mode supplies**

| Output Current | Output Power | Dissipation in output transistor | Efficiency |
|---|---|---|---|
| 1A | 13.8W | 16.2W | 46% |
| 2A | 27.6W | 38W | 42% |
| 5A | 69W | 56W | 55% |
| 10A | 138W | 74W | 65% |
| 15A | 207W | 63W | 76% |

**Table 23.6: Dissipation in pass transistors in a 13.8V linear power supply**

As well as the obvious cost savings one other point should be noted, and that is the lower operating temperatures within the equipment allow a smaller and cheaper box to be used without as many cooling slots.

If you now consider a typical linear regulator for a 12A power supply, we can draw some inference of how much more current we could get if the supply was converted to a switching type.

In a 13.8 volt nominal 12 amp supply the typical pass transistor dissipation is shown in **Table 23.6**. At the maximum output current of 15 amps the total power supplied by the transformer / rectifier / smoothing capacitor bank to the regulator is 207 + 63 Watts. This is a total of 270W.

Assuming an efficiency of 85% we should be able to achieve 16.6A by using the same transformer, rectifier and smoothing capacitor, and changing the regulator from a linear type to a step-down switchmode type.

## 12V to 24V Boost Converter

For those applications where a simple step-up DC converter is required, the Boost Converter is a good choice when the output current required isn't too high. A Boost Converter can deliver an efficiency of around 90%.

The output is not inherently short circuit proof because of the topology and hence adequate fusing should be fitted to the input. The circuit shown is dimensioned to provide a maximum of 1A of output current.

A basic Boost Converter is shown in **Fig 23.32** using the UC3843B and a low cost mosfet. The UC3843B is an 8-pin DIL IC containing everything except the power switch. The mosfet chosen is the International Rectifier IRFZ-24N that has a suitably low on-resistance and can safely handle up to 17A of RMS current and 68A of peak current.

At 24W output it requires little heatsinking. The circuit will operate from approximately 9V to 16V and provide an efficiency of about 90% at 1A output current.

**Fig 23.33: Boost converter inductor wound on T106-26 core**

The UC3843B features an under-voltage lockout threshold of 8.4V so if the input supply voltage falls to less than this level the circuit will be shut down and restart when the supply voltage rises above the threshold. In automotive applications this provides protection should the battery voltage fall below 8.4V whilst the engine is being cranked. Another device in the series is the UC3842B and this has an under-voltage lockout threshold of about 16V so it is unsuitable for 12V automotive applications.

The Boost Converter operates in a flyback mode by alternatively storing energy in the form of magnetic flux in an inductor when the mosfet is turned on and then releasing it to the output when the mosfet is turned off. Because the switching frequency is high, the value of inductor required is quite small and occupies very little volume.

The inductor is wound on a powdered iron core made by Micrometals and able to supply up to 1A of output current. The core material is type 26 and its diameter is 1.06in (27mm) - part number T106-26, colour coded white/yellow. The storage inductor is wound with 22SWG enamelled copper wire with 33 turns (**Fig 23.33**). The boost converter operates at about 100kHz, which is set by R1 and C2 on the oscillator input (pin 4)

Feedback resistors R3 and R6 set the output voltage. The internal reference voltage is 2.5V. The converter features a soft start circuit to reduce the inrush current on switch on. Components D1, C1, D2 and R2 form the soft start circuit. D1 and D2 are 1N4148 or 1N914 types. C1 should be a low leakage tantalum capacitor. The inductor peak current, and hence the maximum power output, is set by the low value resistor Rs in the source of TR1. The voltage developed across Rs is a replica of the triangular waveform drain current, and is filtered by R5 and C3 then fed to the current limit input of IC1. Rs needs to be a high wattage resistor and a 2W wire-wound would be the best choice for this component. The gate drive to TR1 is via R4 to suppress ringing caused by the Miller effect in the mosfet.

The output rectifier diode D3 should be a Schottky for the best efficiency. A type rated at 3A average current is the optimum

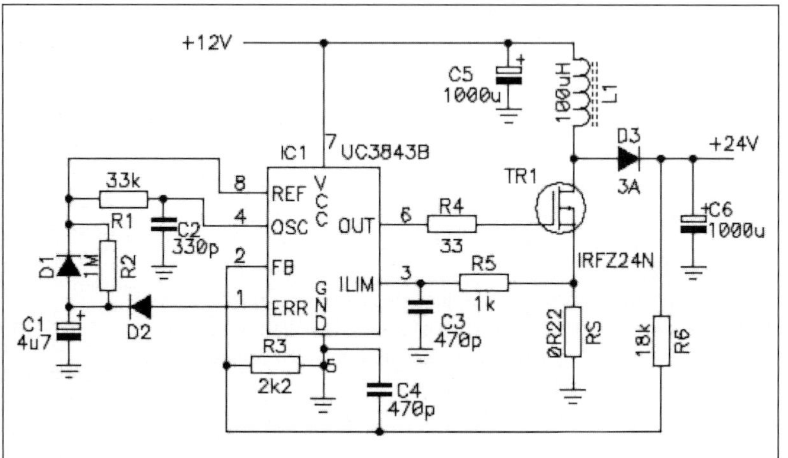


**Fig 23.32: (right) A 12 volt to 24 volt boost converter circuit, together with (above) a photo of an amateur built version [photograph by G3AAT]**

choice with a PIV rating of at least 30V. A suitable device would be the International Rectifier 31DQ03. Ultra-fast diodes will also work with a slightly lower efficiency. A suitable device would be a BYV28-50 or MUR310. Normal silicon rectifier diodes such as 1N4000 or 1N5400 series are far too slow to work effectively at 100kHz and should not be used.

The output smoothing capacitor C6 needs to be a type designed for switchmode supply applications and having a low equivalent series resistance (ESR) to reduce the output ripple voltage and able to sustain high peak ripple current. An ESR of less than 0.25Ω is needed if the output ripple voltage is to be less than 100mV p-p. The peak charging current into C6 is approximately the same as the average input current and so a diode and smoothing capacitor capable of handling high peak current is required. A suitable type would be a Nichicon PJ series rated at 35V and having a value of at least 1000µF. The input capacitor C5 can also be a Nichicon PJ type or a normal electrolytic because the ripple current it experiences is not as severe as C6.

All the resistors except Rs can be quarter watt or half watt types. TR1 will require a small heatsink with an insulating washer and mounting bush to prevent the drain shorting to ground. Because of the high circulating currents the grounding of components around IC1 and the mosfet is critical. Although the frequency is only 100kHz the construction should use short leads and low inductance grounding similar to VHF construction techniques.

One or two points about boost converters need to be appreciated. The first is that the power to be converted has to come from the input supply. For a converter with an input voltage range of 9 to 16V the input current will be highest when the input voltage is 9V. With an efficiency of 90% the input power required is 1.1 x 24 = 26.4W and with a 9V supply this is an average current of 3A. When the input voltage is 16V the average input current is about 1.7A. If the input supply cannot deliver the required current the converter will not operate correctly.

Many linear regulated power supplies, although capable of supplying the average current, are not able to supply the high input current pulses that a boost converter draws. Although the average input current is only 3A the peak current can be several times this figure for a short period when the mosfet turns on. If the linear supply runs into current limit the converter will hiccup and behave erratically. When testing a boost converter using a linear regulated bench supply as the input source, it is often necessary to connect a large electrolytic capacitor (10,000µF or more) across the bench supply output to help supply the pulsing peak current.

If the output ripple voltage needs to be less than 100mV, then rather than increasing the value of C6 it is better to insert an additional filter LC network in series with the output (**Fig 23.34**). This can be a small inductor also wound on 26 material (T50-26), and having a value of about 10 to 20% of the main switching inductor. This with an electrolytic capacitor of about 100µF will reduce the ripple voltage to about 10mV.

For applications requiring more output current the value of Rs can be reduced. If Rs is set to 0.1Ω the converter can supply up to 2A. The rectifier diode will need to be a 5A device and the value of C6 will need to be larger; a value of 2 x 1000µF PJ types in parallel will suffice.

In **Fig 23.35** the Boost Converter is running at approximately 40% duty cycle and supplying about 35W output. The mosfet drain voltage swings between approximately 0.25V above ground up to 25V and the steep transient seen is when the drain current is switched off. This peaks at about +40V. This transient is clamped by the reservoir capacitor to less than 1V peak. The second stage L-C output filter reduces this to approximately 5mV peak.

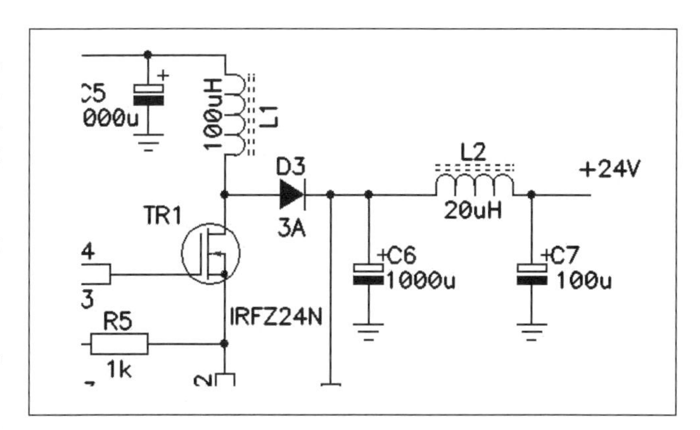


**Fig 23.34: Additional output filter**

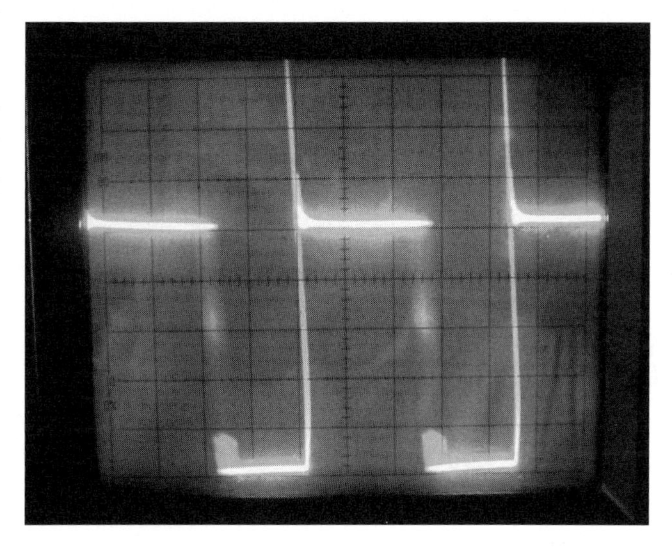

**Fig 23.35: Drain waveform of boost converter**

# BATTERIES

The basic types of batteries have been described in the chapter on Principles. There is more on the use of batteries in the chapter on operating out of doors.

There are two types, primary or one shot, and secondary or rechargeable. Strictly speaking, a 'battery' is an assembly of two or more 'cells', but a single cell is commonly also called a battery.

## Primary Batteries

At present there are two main varieties, that are suitable for amateur use, based on zinc or lithium. The battery derives its energy from the metal used as the negative electrode. The positive electrode has an effect as well and has to be able to dispose of the hydrogen, which would otherwise be liberated there. A depolariser surrounds the electrode if it is unable to do this. For watches and similar purposes zinc-mercury oxide and zinc-silver oxide cells are available at some cost.

### Zinc-carbon

These form the oldest and cheapest primary cells and are called dry cells as the electrolyte, although not dry, is immobilised so that it cannot spill.

Three different electrolytes are used, an aqueous solution of ammonium chloride (sal ammoniac) in the cheapest, zinc chloride in 'high power' cells, and sodium hydroxide (caustic soda) in manganese-alkaline cells. **Table 23.6** gives some parameters of typical types.

Manganese-alkaline cells are made in the same sizes and **Table 23.7** gives details.

| Type | Voltage (V) | Weight (g) | Maximum size (mm) | Current range (mA) |
|---|---|---|---|---|
| AAA | 1.5 | 9 | 45 10.5 | 0-25 |
| AA | 1.5 | 18 | 50.5 14.5 | 0-40 |
| C | 1.5 | 48 | 50 26.2 | 20-60 |
| D | 1.5 | 110 | 61.5 34.2 | 25-100 |
| PP3 | 9 | 38 | 48.5 17.5 26.5 | 1-10 |
| PP9 | 9 | 410 | 81 52 66 | 5-50 |
| C (HP) | 1.5 | as 'C' above | | 0-1000 |
| D (HP) | 1.5 | as 'D' above | | 0-2000 |

*Note. Where there are two dimensions, the first is the length and the second the diameter. The current range is that which gives a reasonable life. Manufacturers do not often give capacities in ampere-hours. The shelf life of either type of cell is about a year although it can be improved by keeping it cold. 'HP' is the high-powered type.*

**Table 23.6: Characteristics of typical zinc-carbon cells and batteries**

| Type | Voltage | Weight (g) | Capacity (Ah) |
|---|---|---|---|
| AAA | 1.5 | 11 | 1.2 |
| AA | 1.5 | 22 | 2.7 |
| C | 1.5 | 67 | 7.8 |
| D | 1.5 | 141 | 18.4 |
| PP3 | 9 | 45 | 0.55 |

*Note. The dimensions are as above and the capacity is in ampere-hours. They have a shelf life of several years.*

**Table 23.7: Characteristics of typical manganese-alkaline batteries**

### Zinc-air

These are similar to zinc-carbon, but use the oxygen of the air as the depolariser. You buy them sealed and they only 'come to life' when the seal is removed. Potassium hydroxide (caustic potash) is used as the electrolyte, and slowly absorbs carbon dioxide from the air, ending the life of the cell. They must be used in a well-ventilated housing. They have a higher energy density than zinc-carbon cells ie they pack more energy into a given weight or space.

### Lithium

The negative electrode is the highly reactive metal lithium, so the electrolyte contains no water. The positive electrode is either iron disulphide (1.5V) or manganese dioxide (3V on load). The electrolyte is either an organic liquid or thionyl chloride (2.9V).

These cells have a long shelf life making them good for battery back up, and have a low internal resistance. They also present a fire hazard if broken or an attempt to charge is made.

## Secondary Batteries

Lead-acid, Nickel-Cadmium (Ni-Cad), Nickel-Metal Hydride (Ni-MH) and Lithium are the only types in amateur use which will be described.

### Lead-acid

This the earliest cell, with lead plates and dilute sulphuric acid electrolyte. Vehicle batteries nearly always use these, as the alternatives (Ni-Cad or Ni-MH, see below) are too expensive. They are heavy, but have very low resistance. One feature is that if left discharged for some time, the plates sulphate irreversibly and the battery is almost always ruined. The usual charging is at a constant voltage of 13.8V, with some form of current limiting to prevent too large a current flowing initially. Do not try to charge a 12V battery from a 13.8V PSU. If the mains fails or is disconnected, the battery will discharge into the PSU, possibly damaging it. Any charger should therefore contain a diode to prevent reverse current flow. Over-charging a sealed battery will result in explosive gases being generated and may burst the vent.

| Size | Voltage (V) | Capacity (Ah) | Weight (g) |
|---|---|---|---|
| AAA | 1.2 | 0.18 | 10 |
| AA | 1.2 | 0.5 | 22 |
| C | 1.2 | 2.2 | 70 |
| D | 1.2 | 4.0 | 135 |
| PP3 | 8.4 | 0.11 | 46 |
| PP9 | 8.4 | 1.2 | 377 |

*Note. The dimensions are the same as those for the zinc-carbon cells/batteries above.*

**Table 23.8: Characteristics of typical nicad (NiCd) cells / batteries**

### Nickel-Cadmium

Ni-Cads (1.2V) have an electrolyte of potassium hydroxide in water, which is attacked by carbon dioxide in the air. Today the cells are sealed to prevent this, and as such are much used by amateurs. **Table 23.8** gives data on a selection. Cadmium compounds are toxic and Ni-Cads should be disposed of with care; many local authorities make provision for this and should be consulted.

Charging can be done at a constant current of C/10 amps where C is the capacity in ampere-hours. This will be complete in about 14 hours (allowing for inefficiency) and moderate over-charging at this rate does not result in harm. This long charging time is a nuisance, so fast charge circuits have been developed. When fully charged, the Ni-Cad cell voltage actually decreases with time. Maxim make a charger IC for Ni-Cads, the MAX 713. This detects the decrease and stops fast charge, **Fig 23.36**.

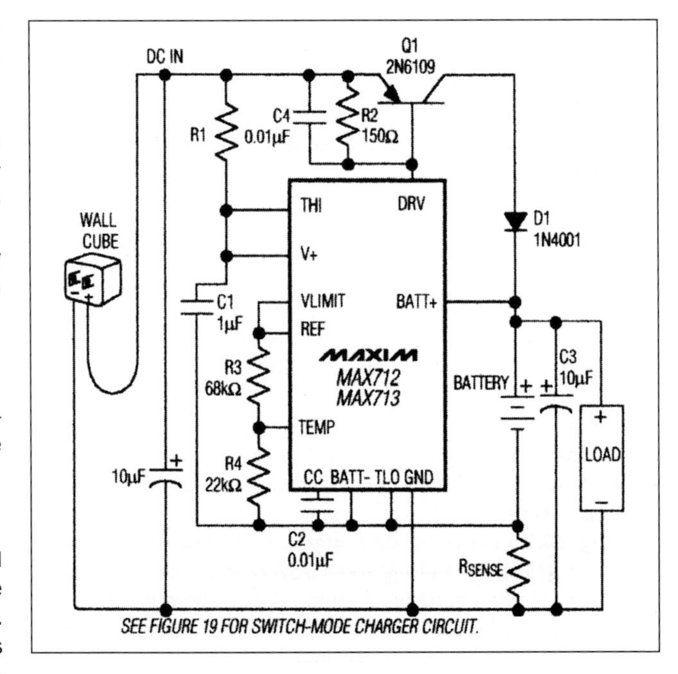


**Fig 23.36: The Maxim MAX 712/713 Ni-Cad and Ni-MH fast charger ICs**

| | Metal-hydride | Nicad |
|---|---|---|
| Capacity (Ah) | 3.5 | 2.0 |
| Voltage (V) | 1.2 | 1.2 |
| 'Memory'* | None | Severe |
| Toxics | None | Cadmium |
| Discharge rate (A) | <12-15 | <15 |
| Overcharge capability | Cont. at C/5 | Cont. at C/5 |

** 'Memory' is an alleged effect [4] which shows up if a cell is only partially discharged before recharging. It is said to reduce the capacity of the cell. This has been disputed [5].*

**Table 23.9: Comparison of metal hydride and nicad 'C' size cells [3]**

There is a memory effect with Ni-Cads if re-charged before being completely discharged; the cell 'remembers' that it was not fully discharged and will not be able to discharge fully after subsequent charge. Some authorities dispute this, believing that it only occurs after repeated discharge to less than complete. If your cell suffers from this, short out the cell only when discharged as much as it will. Beware of reversing the current through the cells of a battery; this damages them. Connecting a resistor across each cell of a battery is recommended, but not usually possible.

Opinion also differs with the procedure for storing cells that are not required. Unlike the lead-acid cell, Ni-Cads may be left discharged, and some have found this better than leaving them fully charged. Maker's advice is not readily available.

### Nickel-metal hydride cells

They have advantages over nicads in that they have a higher energy density and they do not contain cadmium. A simple comparison between one of these and the same size in nicad is given in **Table 23.9**. Ni-MH cells are very similar to Ni-Cads, the voltage is the same, but capacity for the same size is higher. There is no memory effect, but the self-discharge rate of earlier cells is higher. Ni-MH cells do not exhibit the decrease of voltage with time at full charge (like Ni-Cads do); the voltage remains constant. The MAX 712 (Fig 24.36) detects this and stops fast charge. Cells such as IMMEDION made by the MAHA company (kindly supplied for review by Nevada Radio), and MaxE made by the Ansmann company, do not have this problem.

### Lithium cells

Development of lithium cells is proceeding fast. There are now lithium-ion, lithium-polymer and lithium-iron phosphate. All are relatively costly, but their energy density is about twice that of NiMH cells. As a direct result, life prolonging techniques have become important.

To get maximum storage life, the cells should be charged at

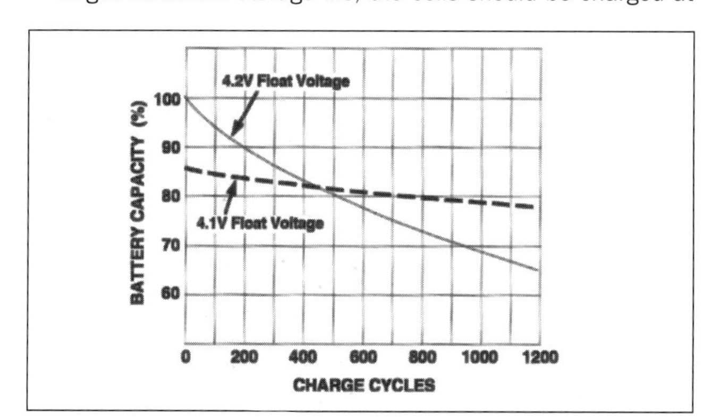


**Fig 23.37: Cycle life and capacity vs 4.1V and 4.2V float voltages**

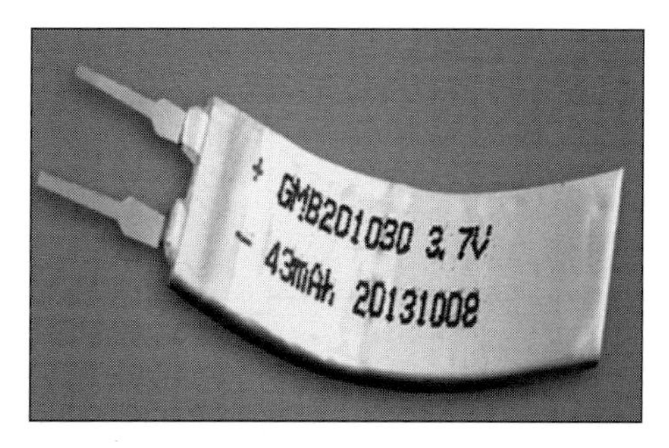


**Fig 23.38: Thin film lithium polymer (LiPo) battery [photo: GMB]**

3.6 V, to 40% and kept refrigerated at 4.4 degrees Celsius. If you use a lithium battery in your laptop, you could not ask for a worse environment. It will run hot and be kept nearly fully charged.

There is an unavoidable loss of about 5% capacity in the first 24 hours after full charge, and about 3% per month thereafter, assuming that the temperature is at or below 20 degrees C. High rates of charge or discharge also cause loss of life, which is very dependent on charging voltage, as shown in **Fig 23.37**. The best way to charge is at a voltage limited constant current, and there are many ICs on the market which do just this. Consult IC manufacturer's data sheets for more detail.

Because of their generally small size and flexibility, rechargeable thin film lithium polymer (LiPo) batteries (**Fig 23.38**) could be used where space is limited. More info can be found at [6].

### Hybrid cells

Various manufacturers now produce hybrid cells which combine the advantages of alkaline primary cells with those of nickel metal hydride. They are designed as cheap replacements for alkaline primary cells, being rechargeable up to 1000 times. The 9v 'PP3' size battery should be very useful for test instruments (and smoke detectors).

### Battery safety

As mentioned above, the electrolyte of lead-acid cells is sulphuric acid, and should be treated with the greatest respect, and not allowed to touch the skin. If it does, it should be immediately washed off with running water. In particular eyes should be protected. If it gets on clothes, if left, it will slowly make a hole. The explosive nature of the gas evolved by unsealed cells has already been noted. Sparks from the terminals on connection or disconnection can ignite the gas, and the entire cell could explode.

The electrolytes of the other types are undesirable also, and should be washed off if they contact the skin. Lithium is an extremely reactive metal, and if a cell containing it bursts, a fire may start. Do not attempt to destroy such a cell by burning.

A secondary battery, or low voltage high current power supply, will raise a ring or metal watch strap to red heat quickly if it forms a short circuit.

## REVERSE BATTERY PROTECTION

Applying power with the wrong polarity can damage equipment. Here are four simple ways to prevent this:

1. Put a power diode in series with the load. This has the disadvantage of wasting a volt or so across the diode, but a Schottky diode would be better (0.31V at 8A for the 95SQ015). **See Fig 23.39(a)**.
2. Put a power diode in parallel with the load and a fuse in series. If the power is incorrectly applied, the diode conducts and the fuse will blow, see **Fig 23.39(b)**.
3. Use a relay to switch the power with a diode in series with the relay, which will then only operate if the power is correctly applied, **Fig 23.39(c)**.

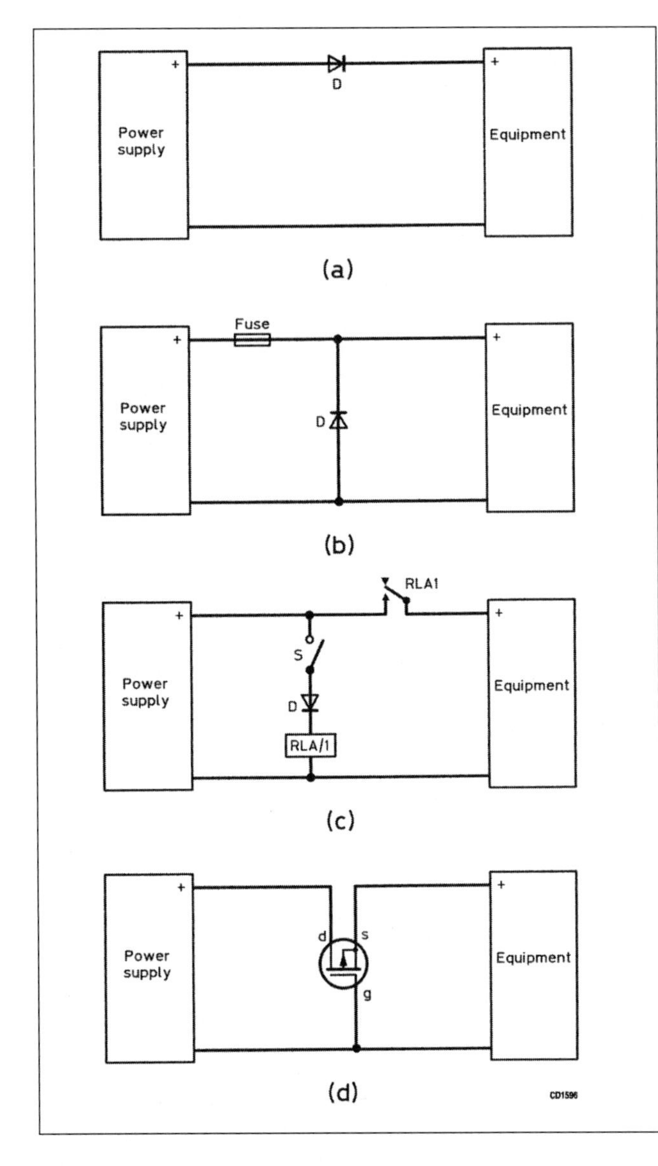


Fig 23.39: Four methods of reverse-polarity protection. D1 is a silicon diode; in (a) it must carry the whole current. In (c), S is the on-off switch, A/1 is the relay operating coil and A1 is the normally open relay contact. (d) Using a p-channel MOSFET

4 Use a power MOSFET in the circuit reported by G4CLF. The MOSFET is only turned on by the correct power supply polarity. The 'on' resistance of available MOSFETs is so low that it may be neglected. See **Fig 23.39(d)**.

## RENEWABLE ENERGY SOURCES

### Wind, Water and Pedal Generators

These may not find much application at home, but could be useful in portable operation. As all of them supply power intermittently, a rechargeable battery will be needed. This implies a regulator, and a simple shunt regulator is all that is necessary, plus a means of preventing reverse current. Commercial practice uses buck-boost switched mode regulators to provide an output voltage even when the solar panel is producing a lower voltage.

A wind generator is shown in **Fig 23.40**, and one type of regulator is shown in **Fig 23.41**.

**Fig 23.40: A typical wind generator *[Marlec]***

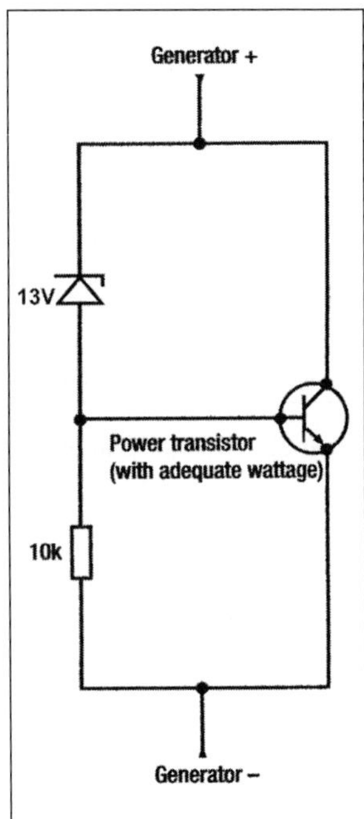


**(above right) Fig 23.41: A shunt regulator suitable for intermittent generators**

### Solar Cells

Solar cells come in various sizes. You cannot estimate the available power by exposing it to bright sunlight, then measuring the no-load voltage and the short circuit current. Thévenin's theorem does not apply, but many makers will provide you with performance curves.

Use of a buck/boost converter (see above and [2]) will extend the effective range of useable solar flux. However, you should beware of possible electromagnetic interference from the converter.

## REFERENCES


[1] LM723 data sheet. Available on the Internet at http://pdf.datasheetcatalog.com/datasheet/nationalsemiconductor/DS008563.PDF

[2] *Power Supply Handbook*, John Fielding, ZS5JF, RSGB, 2006

[3] *Encyclopedia of Chemical Technology*, Vol 3, Kirk-Othmer, 4th edn, John Wiley, New York, 1992, p1070

[4] *Art of Electronics*, Horowitz and Hall, 2nd edn, Cambridge University Press, 1989, p927

[5] 'Technical Topics', *Radio Communication* June 1989, p34

[6] http://www.gmbattery.com/


## ACKNOWLEDGEMENTS


Grateful thanks are extended to the following who provided information and/or illustrations during the revision of this chapter:

Amberley Working Museum; Analog Devices; Ansmann AG; Chloride Power Protection; Cliff Keys, G8IML, Editor in Chief of *Power Systems Design Europe* magazine; Dallas Maxim; Future Electronics; GM batteries; Marlec Engineering; National Semiconductor; Nevada Radio; *Practical Wireless;* Vishay.

# Measurement and Test Equipment

# 24

*Philip Lawson, G4FCL*

TO THE RADIO AMATEUR and electronics enthusiast, test equipment is a 'must have' for activities such as checking and demonstrating compliance with the licence conditions, servicing equipment and for testing self-build projects. In recent years, basic test equipment of quite remarkable accuracy and functionality has become available on the market, which can be bought cheaply and acquired quickly via the internet. This has largely dispensed with the need to build such equipment, so that time and money can be spent on buying or building more exotic equipment or accessories. This trend is reflected here, where there is more emphasis on the operation and correct use of basic test equipment, with self-build being confined to more specialist equipment and accessories.

→ *For a deeper delve into test equipment, including many bits of test equipment and test accessories that you can make, see Test Equipment for the Radio Amateur, 5th Edition, available from the RSGB bookshop.*

Remember! A user who understands the operation and limitations of his or her test equipment, has the ability not only to make measurements, but the ability to make the right measurements, and to correctly interpret their meaning. Amateur Radio is a leisure activity, and there is great joy to be had from the skilful use of such equipment.

## THE MULTIMETER

This is a basic measuring instrument that every electronics enthusiast should have. Available as a digital or an analogue type, both sorts have their pros and cons, and so ideally you would have one of each (see **Fig 24.1**).

**Fig 24.1 An analogue and a digital multimeter side-by-side**

You can now buy a really good, brand new, multimeter of either variety for as little as £10-£20, which is fantastic value for money, although as with most things, the more you pay, the more features, better precision, and higher quality you get. For example, a basic digital multimeter will measure voltage, resistance and current. A better one could measure inductance, capacitance, temperature, and frequency as well, and might even incorporate an AF signal generator (**Fig 24.2**). Let's look more closely at what can be measured, how, and things to look out for......

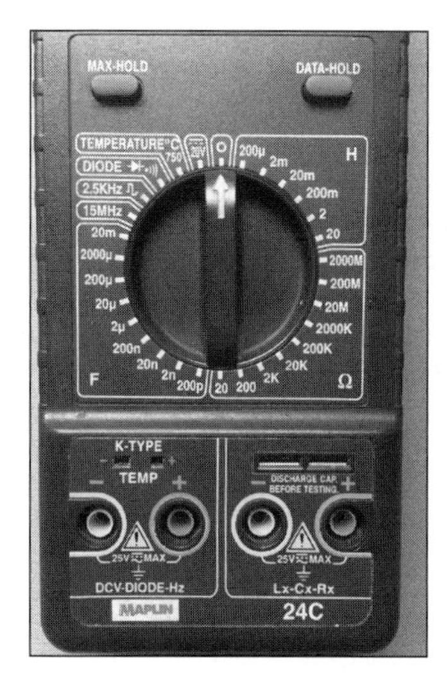


**Fig 24.2 Close-up of functions available on a typical DMM**

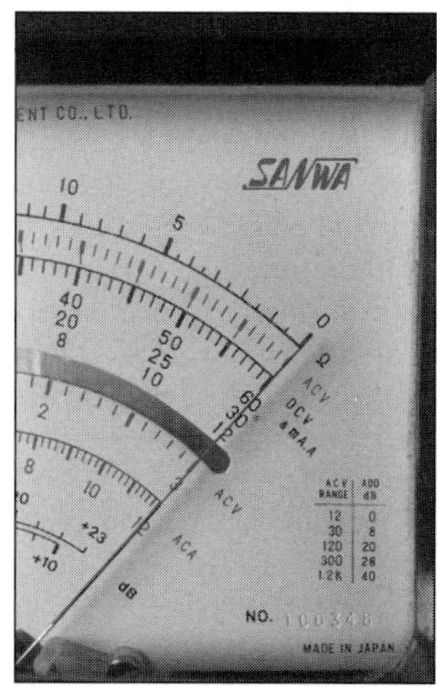


**Fig 24.3 Avoiding parallax error with a mirror behind the scale an analogue multimeter**

**Voltage**: A modern digital multimeter is capable of measuring a steady AC or DC voltage extremely accurately because its 'loading' effect on the circuit is normally negligible, and because it presents the result clearly on a display with large numbers. The analogue type in contrast will usually 'load' the circuit more, and should have a sensitivity of at least 20kΩ/V to reduce this effect. As the needle position can be hard to interpret, better analogue multimeters have a mirror behind the needle so that they can be read 'head on' (ie with no parallax) to minimise the scale interpretation error (**Fig 24.3**). That said, the analogue multimeter is often superior where the voltage is slow moving, as fluctuations can be followed far more easily. For the repair of valve equipment, I would prefer the latter for voltage measurements for the reasons given, and additionally because if there is AC (mains hum) on the DC (HT) supply, a digital multimeter can get very confused and give a variable and misleading readout. Beware also, that the working range of many digital multimeters may extend to only to a few tens, rather than hundreds of volts, because they are intended to test modern transistorised equipment, and so are not suitable for the task. Either way, when you pick up those two probes, know the highest voltage likely to be encountered and set the multimeter to a suitable range *before* making a measurement. If you don't, you may trigger the 'quick-blow' fuse, damage the instrument, and (if analogue) permanently bend the needle or burn out the movement. Also be aware that the digital multimeter contains a clock oscillator to process the display, and the pulses may appear as interference transmitted through the test probes.

**Current**: Similar arguments apply-ideally the working range should extend to at least a few amps.

**Resistance**: A digital multimeter really comes into its own here and can give a quick and accurate result. However, when measuring large resistances (say >1MΩ) ensure that the multimeter's internal resistance (shown in its specification) is at least 10x the resistance that you're trying to measure (and preferably a lot more) to minimize shunt errors. Cheaper multimeters may have an internal resistance of only 10MΩ, the best ones in the order of 10GΩ.

**CARE!** Never attempt to measure resistance with the power on, and ensure that one end of the resistor is disconnected from the circuit to avoid a false reading caused by other components shunted across it. To avoid errors when measuring low resistances, ensure that your probes are clean and well fitted at the other end. Bringing the two probes together, an analogue multimeter should go roughly full scale: use the 'zero' knob to manually set the pointer to precisely '0 ohms' on the dial. Make good use of the parallax mirror if provided. A digital multimeter should auto-zero, or show a small resistance (<1Ω) that ought to be deducted from any low resistance measurement.

**Inductance and Capacitance**: Connect the isolated component to the probes or slots provided. Remember that any test leads inherently have a small self-inductance along them and a small self-capacitance across them. This 'parasitic' value should be noted and subtracted from the item being tested if it significant in comparison.

**Temperature**: Sometimes a thermocouple on a twisted-pair cable is provided which plugs into the multimeter. Note the thermocouple temperature range, and secure with tape, and possibly heatsink compound, to the item to be measured. If you use a different thermocouple, make sure that it has the right type letter (eg 'J'), otherwise readings will be incorrect.

**Frequency measurement**: An in-built frequency counter may allow limited in-circuit frequency measurement via the test probes. Do not exceed the maximum specified voltage between the probes when connecting to a powered set in this mode. Normally intended for probing transistorised equipment, you may need an external capacitor (DC block) with a high DC working voltage in order to protect the multimeter.

**Signal-Generator**: This may usefully provide a basic AF waveform (perhaps sine, square, triangular, or sawtooth) to the probes. Again, do not exceed the maximum specified voltage between the probes when injecting a signal into a powered-up set; use an external DC block if necessary.

Handling a nice multimeter is one of the joys of electrical construction and maintenance. If you understand how it works and use it correctly, your joy will be complete, and you will have mastered a basic competence essential to any electrical work.

## The Multitester

One can now buy a microprocessor-based hand-held tester for £10-20 (or a kit to make one for half that) which can accurately measure the properties of many active and passive components. Not just inductance, capacitance, and resistance (and typically over a wider range than a multimeter), but capacitor ESR and diode and transistor characteristics (polarity, leakage, barrier potential, gain etc) as well. It may also check the operation of some common ICs - make sure that it covers the ones you're interested in. They are so clever that they will even work out which leg is which, so you don't have to! They are quick and easy to use, and give valuable information on a wide variety of component - very handy if you have a lot of unidentified components of uncertain history to test! They represent extremely good value for money, and are another 'must' for the shack (**Fig 24.4**).

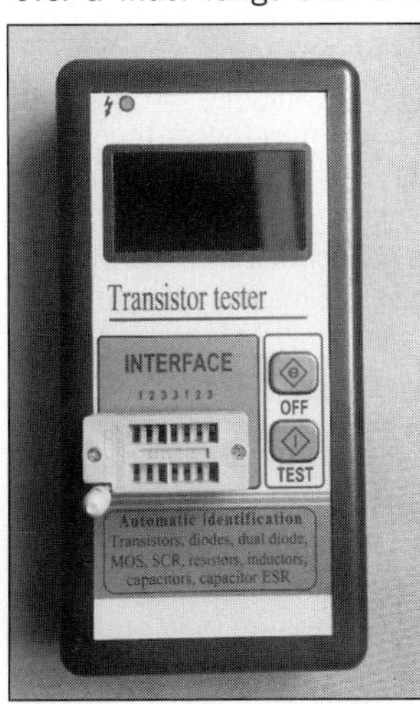


**Fig 24.4 A typical multitester**

## Leads & Probes

These connect the circuit under test to the test equipment. They need to be robust (to withstand constant use), should be quick and easy to attach, have minimal impact on the circuit under test, and present the signal to the test equipment with as little degradation as possible. Ensure they are kept in good condition and are adequately rated, especially if working with high voltages or currents.

**DC/LF leads**: Typically attached to multimeters and the like, these are characterised by having a push-fit plug at the test instrument end, and either a probe point, or a press & release grip to the equipment under test at the other. These are connected by a length of flexible, multi-stranded, insulated copper wire. With use, the torsion can gradually break the strands, causing an open circuit. Depending on the design, it is often possible to remove the plug, probe, or grip cover to resolder the joint, which should be done carefully and neatly so that joint resistance is minimised, the bond is strong, and the solder does not foul the cover when you try to reattach it (**Fig 24.5**). Always try to keep the probes clean, as any dirt or grease will act to increase resistance causing measurements to be inaccurate and erratic. Some typical leads and probes are shown in **Fig 24.6**.

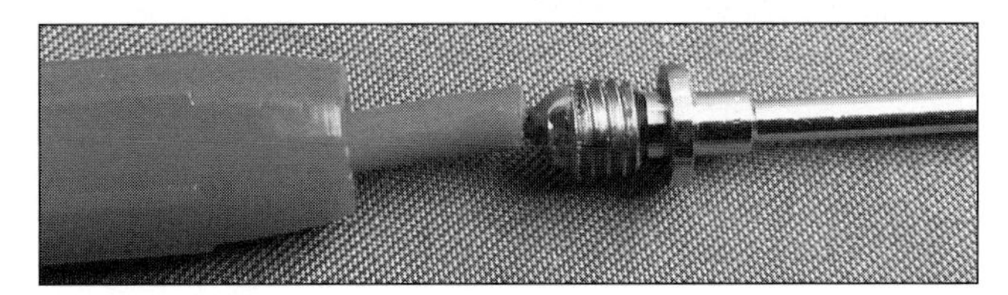

**Fig 24.5 When re-soldering the probe to its cable – make a neat, strong, and compact joint like this**

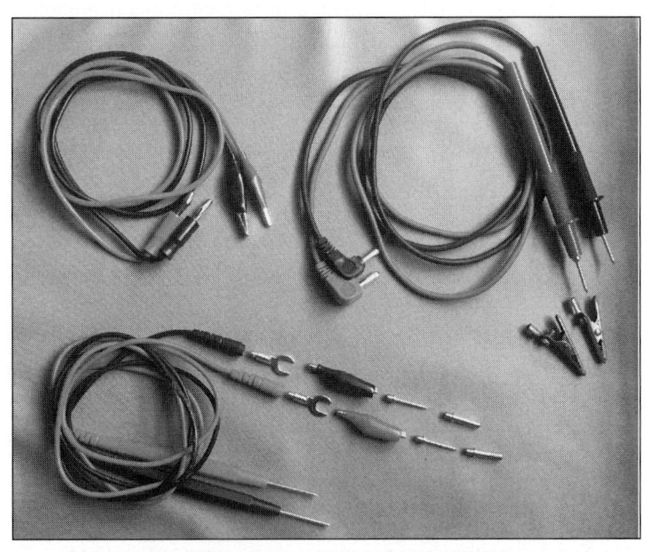

**Fig 24.6 A selection of typical DC/LF test leads and probes**

**RF Leads & Probes**: At low frequencies, a coaxial lead terminated in two crocodile clips may suffice for the injection or monitoring of a signal (**Fig 24.7**). One of the crocodile clips is connected to the centre conductor of the coax, whilst the other is on a short length of flexible wire connected to the braid. The braid connection is the earth, whilst the actual signal being examined is connected via the centre conductor. This forms a very basic (one might say crude) 'probe', which suffers from the fact that the crocodile clips are large and clumsy and often uninsulated. Beware of damage to the circuit by the clips shorting out, and to you, if there is a large voltage present!

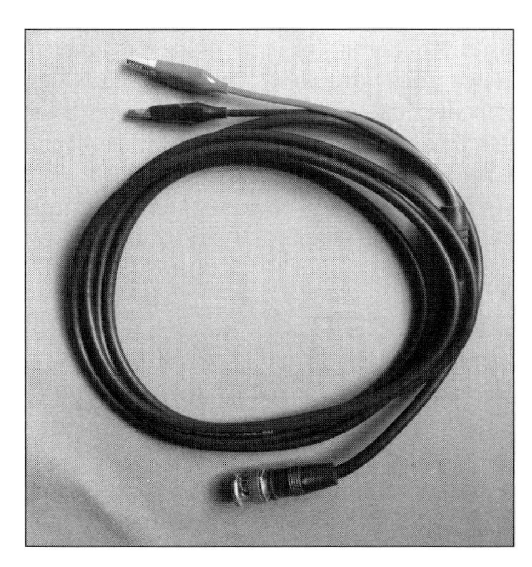

Fig 24.7 A low-frequency RF test lead / basic probe

If connected to an oscilloscope with a 1MΩ input impedance, the input impedance presented by the circuit is typically in the order of 1MΩ, shunted by about 60pF or more due to the cable and the capacitive nature of the flying leads. Handy for quickly and simply injecting or monitoring a signal in low frequency work, this type of probe is unsuitable for accurate RF measurements because of its poor frequency response.

A much better arrangement is a lead with a calibrated probe (**Fig 24.8**). The probe is usually plastic covered (ie insulated), has several attachment options available to connect to the circuit, and incorporates a small earth lead with miniature covered crocodile clip. A typical 'times one' (x1) probe as it is called, when connected to an oscilloscope with a 1MΩ input impedance, will present the circuit with an impedance of about 1MΩ shunted by about 40pF.

A variant on this type is one which has a divide-by-ten arrangement in it, known as a 'times ten' (x10) probe, which attenuates the input voltage by a factor of 10 and so allows a larger value to be measured. Importantly, it also raises the impedance across the circuit by ten times (to 10MΩ in this example) shunted by less capacitance (about 15pF), both of which helpfully decrease the loading on the circuit being measured. A more expensive probe will have a slide-switch which allows selection of either x1 or x10 and may also provide a 'reference' position which grounds the probe, and is useful for calibration purposes when making a measurement (**Fig 24.9**). The probe will often be useful to 60 or 100MHz and should come with other information, such as its input capacitance and voltage derating with frequency.

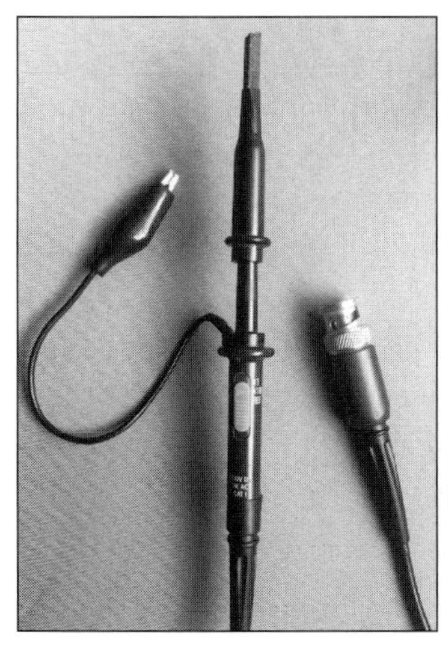

Fig 24.8 A 'proper' RF probe

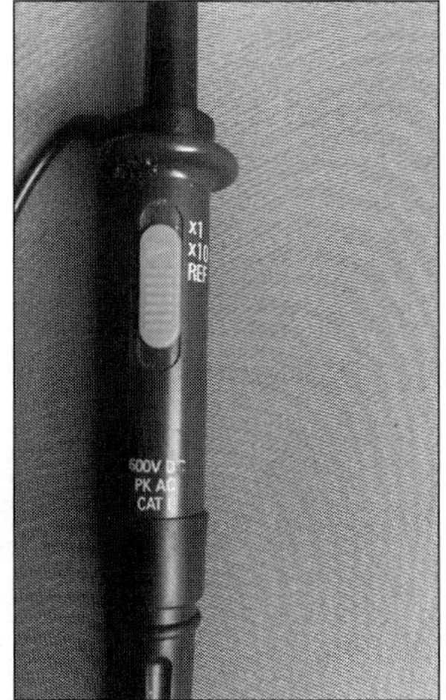


Fig 24.9 Close-up of the RF probe's switch positions and its low frequency voltage rating

An RF probe will often incorporate a 'compensation' trimmer which can be adjusted for best frequency response (**Fig 24.10**). A plastic adjuster may be provided so that the capacitance setting is not influenced by a metal screwdriver.

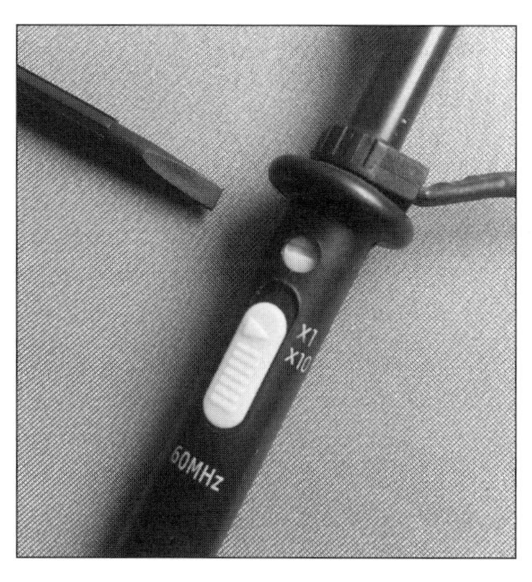


Fig 24.10 Close-up of the compensation screw and its non-metallic adjuster

Again, probe leads need to be keep be kept clean, and care taken not to strain the cable, especially at the two ends where it enters the housing, as damage and poor measurements will result.

As to interconnecting equipment, for HF work, these leads are normally coaxial (for low-loss RF transmission) and will usually have BNC plug at each end to connect, for example, a signal generator or an oscilloscope to the equipment under test. At higher frequencies, this may change to perhaps N-type, or SMA.

**Logic Probes**: These help to investigate the operation of digital ICs and associated circuitry. They will normally detect logic hi-, lo-, and indeterminate state, and display the result by means of a green, red, and amber LED respectively, and/or with audible support. They may have a switch to swap from continuous running to 'single-shot' mode, and another to change from TTL to CMOS. Contact with the circuit under test is via a pointy metal probe, and power may be drawn through a couple of wires which come out of the body of the probe and terminate in crocodile clips so that they can be connected to the digital equipment's ground and supply rail. They may also have facility for internal power. Such probes are inexpensive to purchase (costing around £10 upwards), and simple to make, perhaps being housed in the shell of a pen.

**RF Diode Probes**: Measurement of RF voltage can be problematic as multimeters are not designed to measure at radio frequencies, and many oscilloscopes struggle beyond perhaps 20 or 30MHz. An alternative is to make a simple RF detector, comprising a diode, capacitor and resistor housed in a slim metal cylinder with a metal probe to the circuit and a coaxial output to a moving coil meter or a digital voltmeter. If more sensitivity is required, the RF can be amplified before detection by an IC.

## THE OSCILLOSCOPE

THE OSCILLOSCOPE (**Fig 24.11**) is an extremely useful piece of diagnostic equipment to have in the workshop, not least because it has a pictorial rather than a numerical display, which allows signs of distortion, noise, DC offsets, and modulation on a signal to be easily identified. The instrument permits the amplitude of a signal, and it variation, to be easily measured, and despite working in the time domain, it is usually possible to at least estimate the frequencies involved.

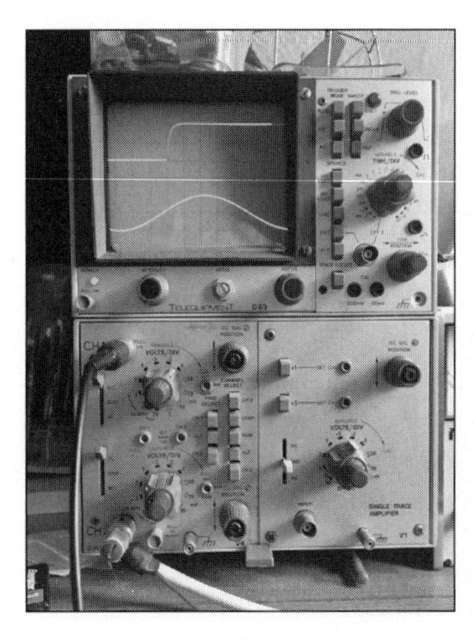

Fig 24.11 A cheap second-hand dual-trace CRT oscilloscope – old, but perfectly good for many test measurements

A basic cathode ray oscilloscope (CRO) with a bandwidth to a few tens of MHz can be bought second-hand for just a few tens of pounds, and can be ideal for AF, IF, or low-frequency RF work. Alternatively, you could purchase a modern LCD version for a few hundred pounds, or a small box which will convert your PC into an oscilloscope for a price somewhere in between the two. Digital oscilloscope kits are available for just £10-30 online, although you should check that the quality and functionality is adequate for your needs before you buy.

The modern equivalent to the old CRO is more expensive, and is likely to have all sorts of 'bells & whistles' - the question is: do you really need them? There is a lot of satisfaction to be had from skilfully using a cheap CRO; however, unless you are prepared to maintain it (for example deal with an electrically noisy y-shift control) a modern version might better.

## Operation

**Fig 24.12** shows a block diagram for an oscilloscope reduced to its four fundamental elements:

1) A power supply: to suit the needs of its amplifiers, timebase, display, and other internal electronics.
2) A Y-amplifier: The signal of interest (whose amplitude could range from mV to tens of volts) is applied to the Y-channel of the oscilloscope and amplified to produce a vertical movement on the screen. This Y-deflection therefore represents the signal's amplitude at a point in time, so to facilitate measurement on the vertical axis, the amplifier's Y-gain is made variable by a control calibrated in V/division or mV/division.
3) A Timebase: An oscillator which usually drives an X-amplifier to cause the waveform to traverse the display in the horizontal plane. At the end of the transit, the spot is made to return (flyback) to the starting position ready for the next display without being seen (blanking). The horizontal X-axis deflection represents time, and so the speed of the transit is made variable by a control calibrated in s/division, ms/division and μs/division.
4) A Display: There are three types, which, at the end of the day, tend to determine the shape and size of the overall equipment:

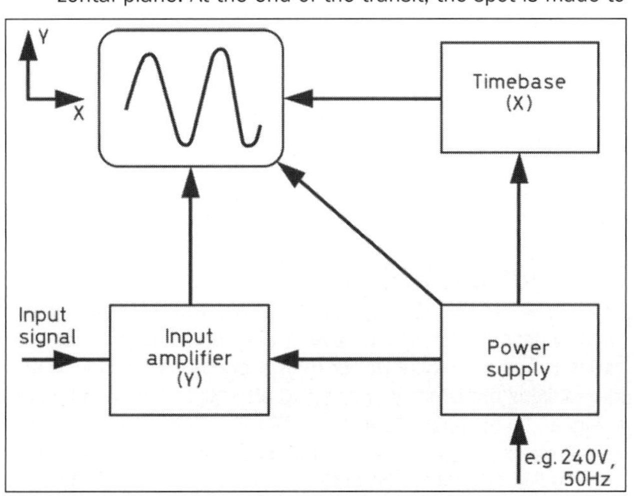


Fig 24.12 Block diagram of an oscilloscope

## Cathode Ray Tube (CRT)

With a single-trace CRO, an electron gun fires a beam of electrons at the phosphor coating on the inside of the glass screen, causing it to illuminate. The focus, brightness, and astigmatism controls set electro-potentials in the CRT and thus enable the user to form a small, bright, round spot that can be deflected vertically by the signal applied to the Y-amplifier, and horizontally by the signal from the timebase and X-amplifier.

In a dual-trace CRO, control of this beam is split in turn between the two Y-channels, giving two traces from one beam. Less common is the dual-beam CRO which utilises two independent electron guns and therefore permits comparison of traces driven from two separate timebases.

CROs are generally analogue, with a few specialist units (such as sampling 'scopes) using digital techniques. Unless the display is continually refreshed, or there is persistence (storage) the trace will just fade away.

Horizontal and vertical measuring divisions are usually marked on a clear plastic sheet (a graticule) placed in front of the display.

#### Liquid Crystal Display (LCD)

These types generally use an ADC (analogue-to-digital converter) to convert incoming signals into a digital value which can then be stored in RAM and used to plot the signal on a display under microprocessor control. Each trace may be a different colour so that the signal in each channel can be easily identified. Most of these can easily store waveforms as well as communicate directly with a computer for further storage, printing, and analysis. They may also have digital voltmeters and frequency counters incorporated into them that will show digital values on the display as well.

A modern tablet-like touch-screen oscilloscope (**Fig 24.13**) may boast a bandwidth of 100MHz, and the ability to display at least four bright and colour-coded channels. Moreover, it may automatically calculate and display the parameters of common waveforms and have the capacity to transmit results wirelessly, or via Ethernet LAN, to a remote location.

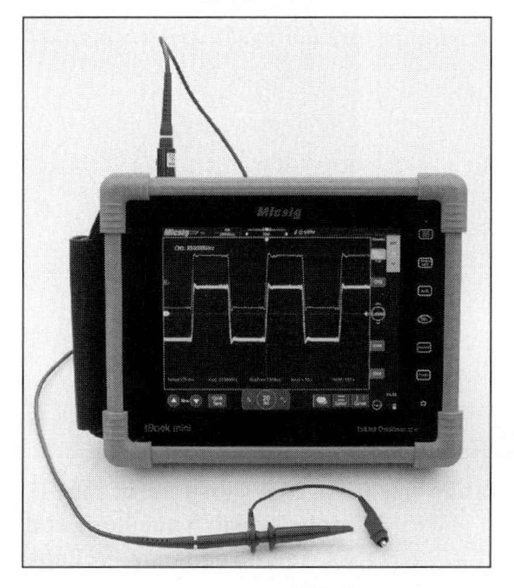


Fig 24.13 A modern touch-screen oscilloscope

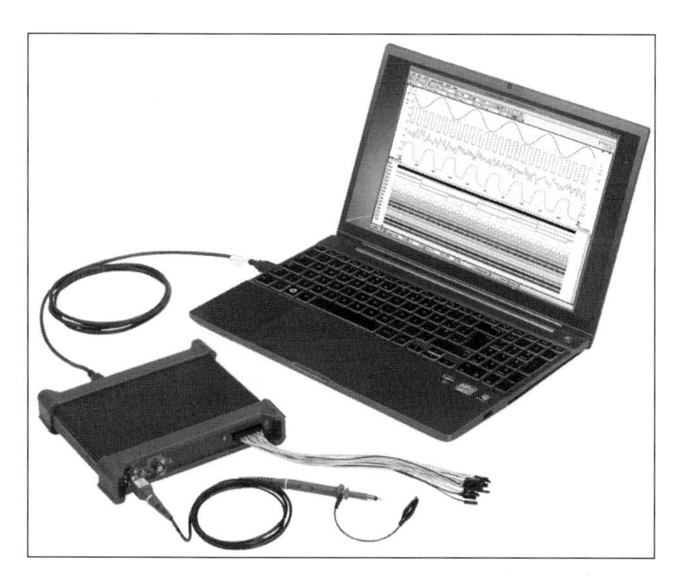

Fig 24.14 A modern PC oscilloscope

**Personal Computer (PC) Screen – Hardware Interface**

These types also use a high-performance ADC to convert the incoming signals into a digital value, but rely on the power and functionality of the ubiquitous personal computer to do the processing and display the output. They consist of a box which interfaces to the PC through a USB or similar transport cable, and usually connects with the circuit under test using compensated probes (**Fig 24.14**).

## Personal Computer (PC) Screen – Software Interface

Here the signal to be examined is simply applied via the USB, jack, or other input to the user's PC soundcard. A computer program is then run which processes the signal to form the oscilloscope display. Note however, that the bandwidth of this 'soft' oscilloscope is limited to that of the PC sound card. Various programs are available which can perform this function, and may provide other facilities such as audio spectrum analysis. For a long list of options covering Windows, Linux, Mac, OS, and Android platforms see: www.dxzone.com/catalog/Software/Oscilloscope

## Controls

The following are typical:

- Y-Gain (volts/div): changes the gain of the vertical amplifier so that a suitable size signal can be displayed on the screen.
- Y-position: allows position of the trace to be varied vertically in order to place it at a convenient point on the screen.
- AC/DC/GND: selects the type of coupling to the input signal. 'DC' allows the AC signal plus any DC to be displayed; 'AC' removes the DC component from display; 'GND' allows input to be grounded internally in the oscilloscope in order to zero-position the trace.
- CH1/CH2/ALT/CHOP/ADD: selects how the channels will be displayed. Channel 1 or 2 separately; both channels on an alternate basis; both channels, but chopping between them during a single scan; Channel 1 and 2 added together - there may also be a channel-invert function, which allows subtraction instead.
- Timebase (Time/Div): changes the speed at which the trace traverses the screen and thus affects the number of cycles of an input signal being displayed on the screen.
- X-position: allows horizontal shift of the trace in order to put it in a convenient position for taking measurements.
- X-magnification: normally allows a x10 magnification if required, equivalent to a faster time on the timebase.
- Trigger Level: determines the trigger level of the input signal. This can be put to AUTO or set to a given level. The input signal triggers the timebase so that a stationary waveform appears on the screen.
- Source: selects the source for the trigger circuit: Internal, Channel 1, Channel 2, Line (mains) or External.
- Intensity: controls the brightness of the image displayed on the screen.
- Cal: an output for X- and Y- amplifier calibration purposes, usually at a frequency of 1kHz with a fixed amplitude, often 0.1V peak.
- CRT displays will usually have a Focus, and possibly Astigmatism and Rotation controls as well to adjust the sharpness, shape, and orientation of the image respectively.
- LCD and PC oscilloscopes usually have the added capability of automatic measurement functions, which can make the determination of key parameters such period, frequency, peak-to-peak voltage, and waveform rise-and-fall times so much easier. These measurements are displayed on the screen in real time, perhaps with the option to average or hold the value. As well as enabling quick display of the key characteristics of different types of waveform, there are usually markers that can be placed on the waveform, so that parts of it can be investigated and measurements taken.

## Compensated Probes

Although one may use a common (uncompensated) probe to connect the oscilloscope to the circuit under test, its bandwidth is significantly limited by its internal resistance. A dedicated oscilloscope probe in contrast, incorporates a parallel capacitance which can be adjusted by a screwdriver to achieve a flat response in the bandwidth of the oscilloscope (**Fig 24.15**).

This is normally achieved by connecting the compensated probe to the oscilloscope's 'Cal' output and adjusting the probe's compensating capacitor until an undistorted squarewave is obtained (**Fig 24.16**). **Fig 24.17** shows possible responses when the probe is connected to the oscilloscope's calibrator jack.

Fig 24.15 Adjusting a compensated probe

Fig 24.16 A correctly compensated probe

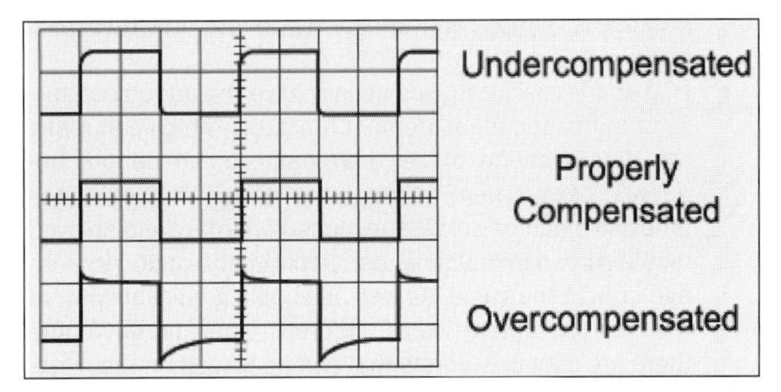


Fig 24.17 Under-, over- and properly compensated probes (reproduced with kind permission of the ARRL)

The coaxial lead on a compensated probe should not be cut, as it may not be possible to compensate for the new length, and reflections caused by impedance mismatch may also arise.

The shortest ground lead possible should be used from the probe to the circuit ground, as long ground leads are inductors at high frequencies, causing 'ringing' and other undesirable effects.

## Voltage Measurements

The easiest voltage measurement to take is the peak-to-peak value. For example in **Fig 24.18** if the vertical displacement (Y) is six divisions from peak-to-peak and the Y-gain is 0.5V/div, then the peak-to-peak voltage is 6x0.5=3V. The peak value is half of this (=1.5V), and the RMS value is $1/\sqrt{2}$ (or 0.7071) times this value (=1.06V).

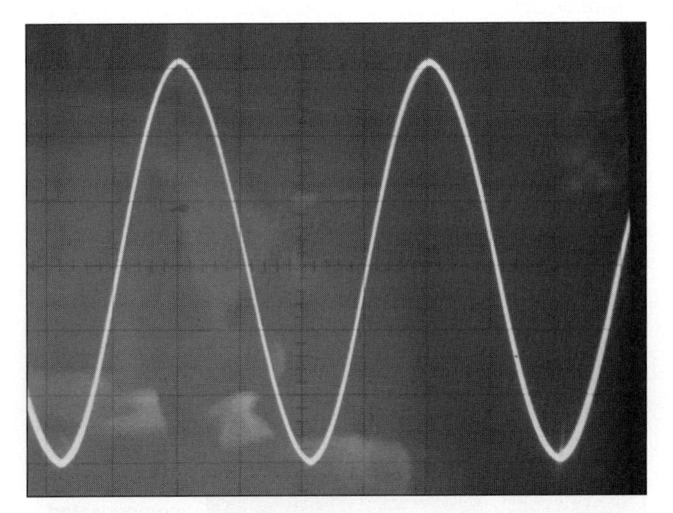

Fig 24.18 Measuring a sine wave

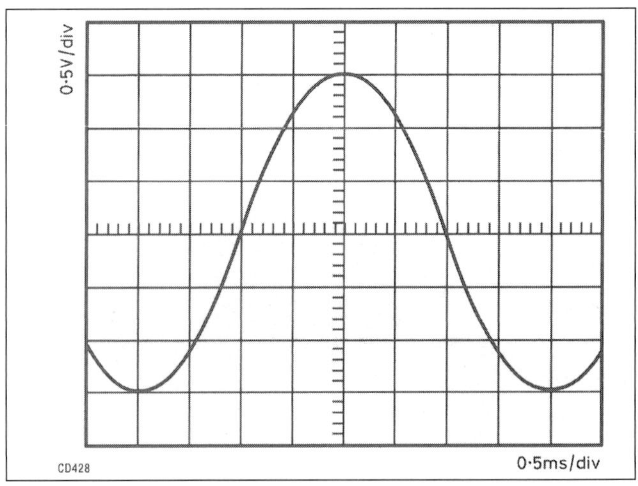


Fig 24.19 Measurement of time and frequency

Accuracy of voltage measurement is unlikely to be better than about 5-10% when taking readings from a CRO trace. However with LCD and PC equipment, the on-screen displayed value will be much more precise.

## Time and Frequency Measurements

For a sinewave, it is best to measure from like-point to like-point, such as between two adjacent negative peaks. For example, in **Fig 24.19** this distance is eight units. If the X or timebase setting is 0.5ms/div, this represents a period of 8x0.5=4ms. The frequency is the reciprocal of this, ie 1/4ms or 250Hz.

The same principle applies to a rectangular waveform, whose edges are clearer, and whose amplitude, period, and frequency are much easier to estimate. For a pulse waveform, another parameter that can be measured is the mark-to-space ratio, ie the ratio of time when it is high to when it is low. In the case shown (**Fig 24.20**), it is a measure of the durations on the screen and is 4:2 (which is the same as 2:1). The duty cycle can also be ascertained, which is the ratio of it being high to the duration of the whole pulse cycle, in this instance 4:6, ie 2:3 or, expressed as a percentage, 66.7%.

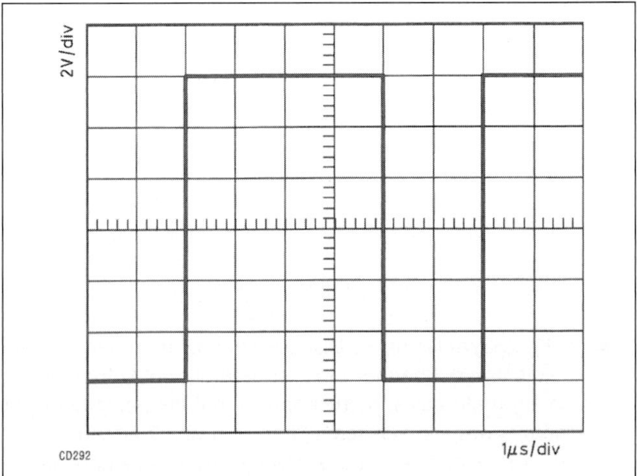


Fig 24.20 Measurement of mark-space ratio and duty cycle

## Differential Phase Measurement

There are a couple of methods that can be used here. The first (and probably most common) allows two signals of interest (which we shall call 'A' and 'B') to be displayed simultaneously together, with the associated phase difference. This method can cope with any shape of waveform, the upper frequency limit being dependent on the oscilloscope and any small phase shifts introduced by the two input-channel amplifiers (Y1 and Y2).

Signals A and B are applied to Y1 and Y2 respectively and the gains adjusted so that the waveforms are approximately equal - **Fig 24.21** shows a typical display. For most accurate results, the input selection is set to CHOP. The phase shift can then be calculated from the amount of offset between the traces.

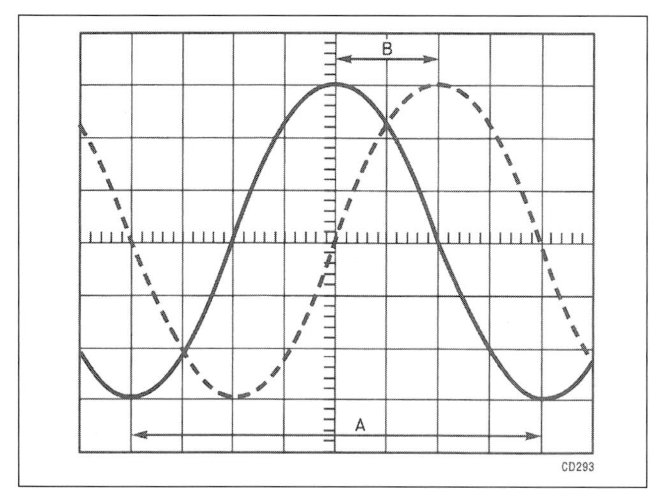


**Fig 24.21 Phase measurement – first method**

Remembering that one cycle is equivalent to 360°, the phase shift (in degrees) = 360*B/A.

The second method is most suitable for the audio frequency range or just above. It assumes a sine wave input, and only requires a single-channel (rather than dual-channel) oscilloscope. This method relies on one signal being applied to the X-input - see **Fig 24.22**.

To start with, both channels (X and Y) are simultaneously connected to the input signal and adjusted so that a suitable sized display is obtained and there is a straight line at approximately 45°. Next, the Y-channel is connected instead to the output signal of the circuit under test. Various displays may be obtained. **Fig 24.23** shows how these can be interpreted and the phase-angle obtained.

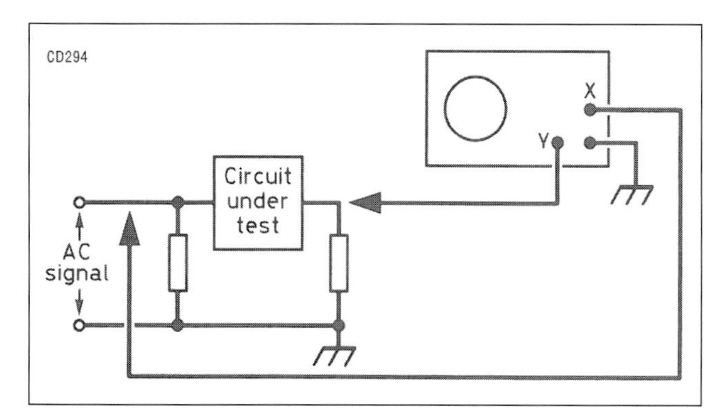


**Fig 24.22 Phase measurement - second method**

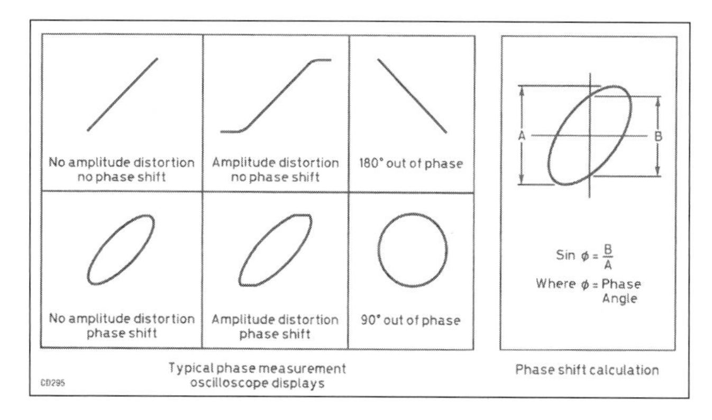


**Fig 24.23 Displays and calculation of phase shift**

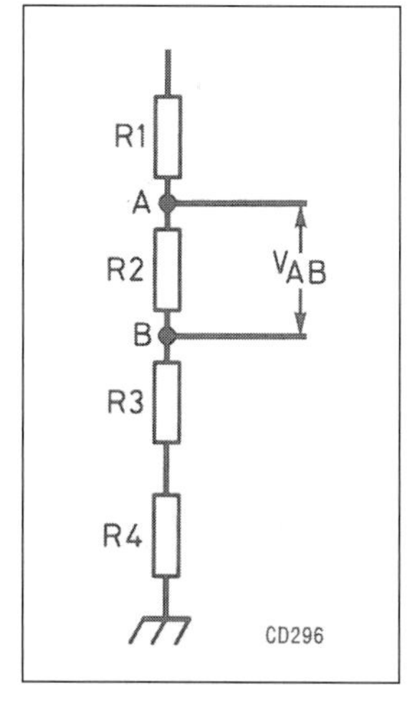


**Fig 24.24 Differential voltage measurements**

## Differential Voltage Measurements

Occasionally one may have to make a voltage measurement between two points in a circuit, neither of which is grounded; for example the voltage VAB in **Fig 24.24**. This can be accomplished in two ways. The first is to measure A with respect to ground and then B with respect to ground. The difference between these two values is equal to VAB. This is satisfactory for DC and if any AC waveforms at A and B are identical in shape. It is also the only method available for a single trace oscilloscope.

The second method is applicable for both dual-trace and dual-beam oscilloscopes. There will be controls on the oscilloscope that will allow display of Y1+Y2 and Y1-Y2. In the latter case, one must use the invert control on channel Y2 and then the ADD facility; this will then give Y1- Y2. Make sure that the gain settings on both Y1 and Y2 are the same, the Y1-Y2 display is set up, and for the dual-beam oscilloscope only: that both timebases are set identically. Put one probe (Y1) on A and the other (Y2) on B; the display will then show VAB. The magnitude and time readings are then as described earlier.

## Equipment Limitations

Three things to watch out for:

Frequency response: This is limited by the Y-amplifier (plus CRT) bandwidth and by sampling rates. Particular care must be taken with rectangular waveforms, as these consist of a series of harmonics in fixed ratios, and the higher harmonics will exceed the oscilloscope's bandwidth. The effect of this is that the displayed waveforms will become progressively more rounded as their frequency rises, eventually tending to a sinewave.

Loading: Where the capacitance presented by the oscilloscope probe (perhaps 20-40pF) is of the same order as the circuit under test, the circuit will be disturbed and may be especially noticeable with tuned circuits, as their frequency $f = 1/(2\pi\sqrt{LC})$. A divide-by-10 probe will help to reduce loading effects on the circuit, but reduce (rather than eliminate) is the operative word.

Safe Input Voltage: The oscilloscope's maximum input voltage is typically 400V DC plus the peak AC signal that can be displayed. Exceeding this will damage internal components of the oscilloscope. Although a divide-by-10 probe can be used to extend the voltage range, these have a voltage limit (typically 600V DC), but this may need to be derated as frequency rises - check the specification. For most semiconductor applications the voltage limit does not cause a problem, however for high-voltage valve circuits one must pay due regard to these limitations for one's safety.

## THE SPECTRUM ANALYSER

Unlike the oscilloscope, which displays the signal in the time domain, the spectrum analyser presents it in the frequency domain, which is extremely helpful for assessment of factors such as modulation, intermodulation, noise, and harmonics. By addition of a tracking generator, the spectrum analyser can also display the response of filters and networks.

### Commercial Products

Historically a very expensive item, they are rapidly becoming more affordable as second-hand CRT-based models increasingly come on to the market, and as LCD software-based models (either standalone or, using an external PC or oscilloscope for processing and display) take over. **Fig 24.25** shows a typical spectrum analyser.

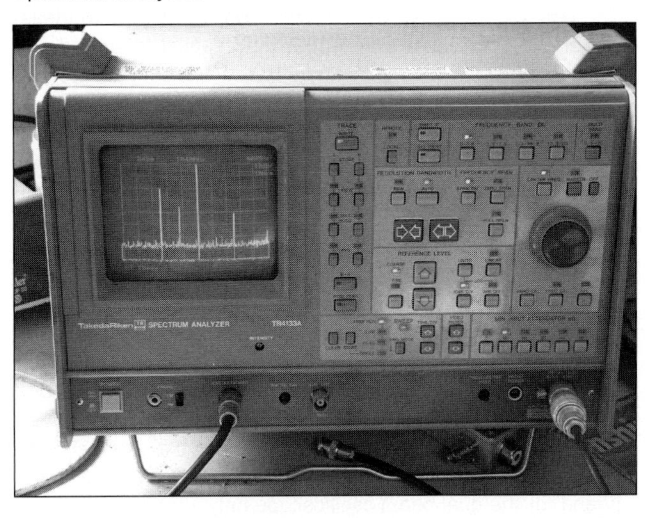


**Fig 24.25 A typical spectrum analyser**

→ To build your own, see for example:
'SA3000 100kHz to 3GHz Spectrum Analyser', Rubens Fernandes, VK5FE, RadCom, August 2018.
'The simple man's Spectrum Analyser' *http://www.hanssummers.com/spectrumanalyser.html*

### Operation

There are various types of spectrum analyser.

**Self-Contained Swept-tuned**

This is the most common and sweeps a desired frequency range sequentially in time to display its results on a screen. Good for displaying periodic and random signals, it fails to capture transient responses.

**Self-Contained Real-time**

Real-time analysers sample the whole frequency range simultaneously and thus preserve the time dependency between signals as well as the capabilities of a swept-tuned type.

**Hardware Interface - PC/Osclloscope Screen**

A potentially cheaper and more flexible alternative to a self-contained spectrum analyser is to use an add-on unit (a hardware interface) which uses an oscilloscope or PC as the display medium. Such units have a lower frequency limit (eg 400kHz), and an upper limit being dependent on type, eg 100, 250, 500MHz. They come in a range of profiles, from a large probe, to a stand-alone box. The bandwidth of the oscilloscope or PC is not usually a limitation as the high-frequency processing is taken care of within the add-on unit.

**Software Interface - Personal Computer (PC) Screen**

As with the 'soft' oscilloscope, the signal to be examined is simply applied via the USB, jack, or other input to the user's PC soundcard. A computer program is then run which processes the signal to form the spectrum analyser display. Again, the frequency limit of the spectrum analyser display is that of the PC sound card, creating an analyser for the audio spectrum only. However, that limitation could be overcome by adding a down-converter to baseband, which would enable a 20kHz or so slice of spectrum about a carrier or harmonic to be examined, for example when performing a two-tone test. A block diagram for such a system is shown in **Fig 24.26**.

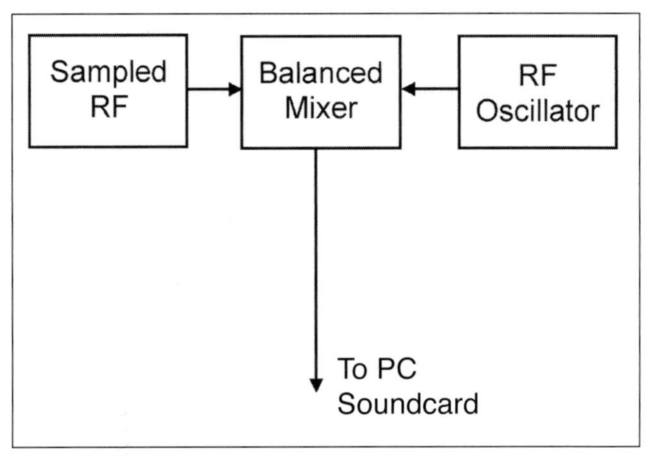


**Fig 24.26 Typical arrangement for PC soundcard spectrum**

→ For software based spectrum analysers covering most operating platforms, see a long list at: www.dxzone.com/catalog/Software/Spectrum_analyzers

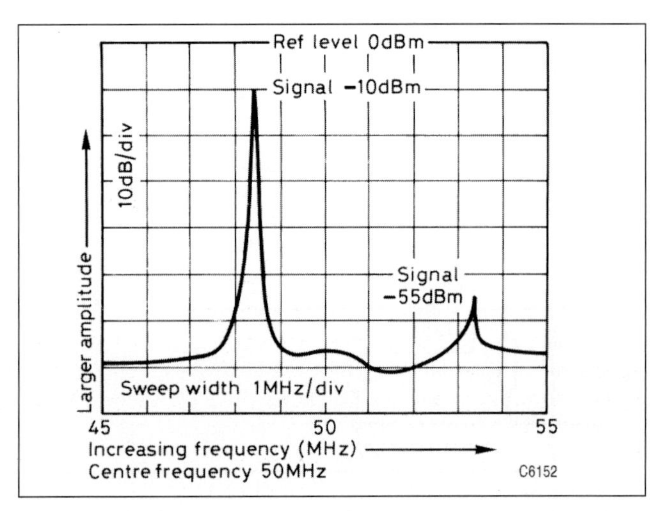


**Fig 24.27 Typical screen display of a spectrum analyser**

## General Use

The horizontal axis is linearly calibrated in frequency with the lower frequency to the left (which is never zero because of the way the analyser works!) and the higher frequency to the right-hand side of the display. The operator can either choose these upper and lower frequency limits, or alternatively scan over a range centred about a specific frequency.

The vertical axis is amplitude and can be either linear (V/div) or logarithmic (dB/div). For most applications, a logarithmic scale is chosen (typically 10dB/div) because it enables signals over a much wider range to be seen on the spectrum analyser. As the scale is normally calibrated in dBm (decibels relative to 1 milliwatt) absolute power levels as well as differences in levels can be seen and measured. **Fig 24.27** shows a typical display at 10dB/div vertically and 1MHz/div horizontally.

A spectrum analyser will typically be rated for a maximum input power of +20dBm (0.1W), beyond which damage to the instrument will occur. In a similar manner, there will be a maximum permissible DC input voltage.

→ For a good tutorial on spectrum analysers see: *http://www.radio-electronics.com/info/t_and_m/spectrum_analyser/rf-analyzer-basics-tutorial.php*

→ See also a 'Guide to Spectrum and Signal Analysis': https://www.anritsu.com/

## Measurement of Harmonics

**Fig 24.28** shows a typical display and depicts a 30MHz signal with harmonics at 60, 90, 120 and 150MHz. The third harmonic (for example) is shown 48dB down on the fundamental but the display does show outputs at the 2nd, 4th and 5th harmonics as well. If the coupler and spectrum analyser had been adjusted so that 0dB represents 100W, then the third harmonic power content is -48dB relative to 100W or 1.6mW. Measurement of the harmonic output of a transmitter is the obvious application here.

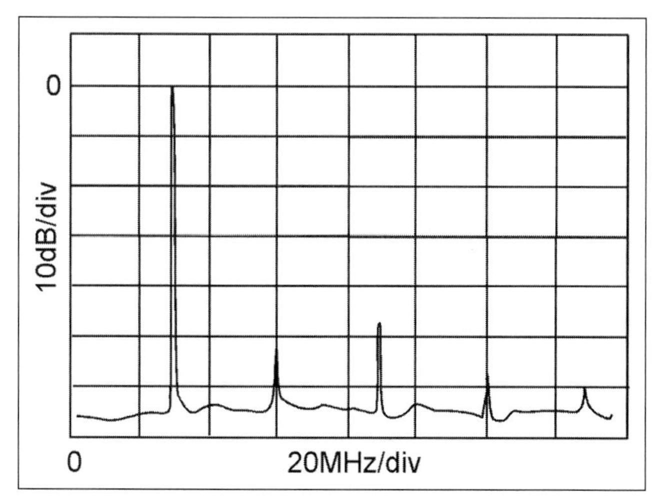


**Fig 24.28 Spectrum analyser display showing harmonics**

## Measurement of Amplitude modulation

Transmitter modulation depth can be determined by spectral analysis as well as from the oscilloscope display; however the spectrum analyser will generally permit observation at much higher frequencies. **Fig 24.29** shows a typical spectrum analyser plot for an amplitude modulated carrier modulated at 1kHz.

The double sideband structure is immediately obvious. Note that, unusually, the vertical axis is in linear mode at V/div. The

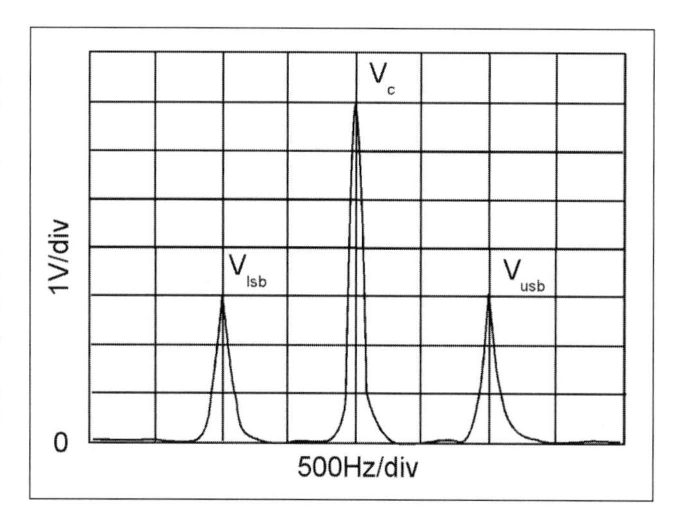


**Fig 24.29 Modulation depth measurement using a spectrum analyser**

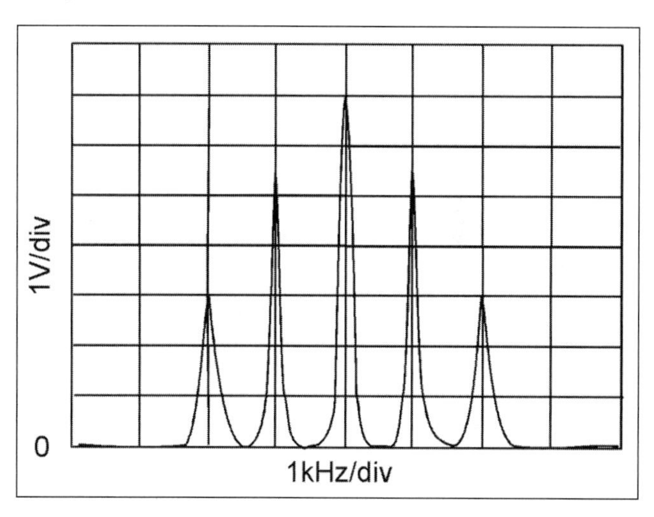


**Fig 24.30 Additional sidebands caused by over-modulation**

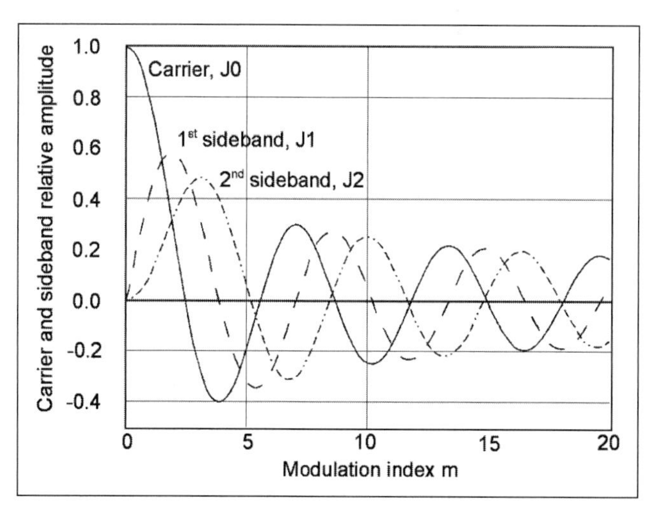


**Fig 24.31 Bessel function curves showing relationship of carrier and sideband amplitudes**

percentage modulation depth 'D' is given by:

D = 100(Vlsb+Vusb)/Vc

For the case shown, the modulation depth is 86%. For low levels of modulation it is better to use a logarithmic display, but the calculation becomes more complex.

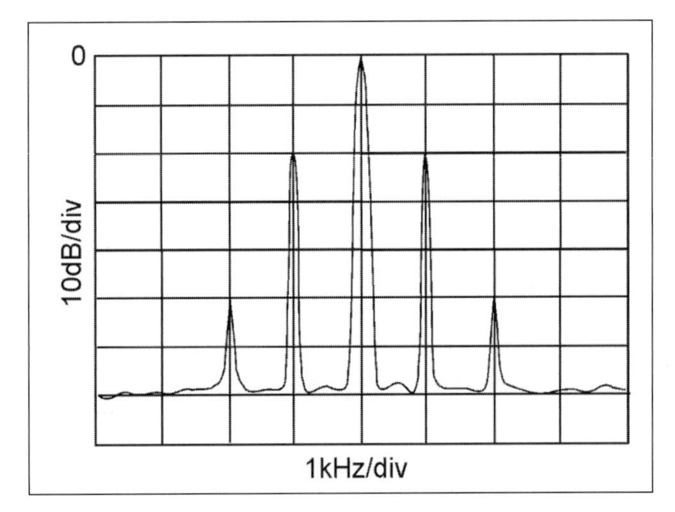


Fig 24.32 Typical narrow band FM spectrum analyser plot

When the value of each sideband reaches half the carrier level, 100% modulation depth has been reached. Proceeding beyond this, distortion begins to occur with the number of additional sidebands rising as the distortion level worsens. **Fig 24.30** shows a typical display - obviously this is a situation to be avoided.

## Measurement of Frequency modulation

### Theory

The Bessel function curves of **Fig 24.31** show the complex relationship between carrier and sideband amplitudes of an FM wave as a function of modulation index 'm'. Observe that the carrier component J0, and the various sidebands JN, all go to zero amplitude (cross the x-axis) at specific values of m.

Looking at the signal in the frequency domain, the associated sidebands will be spaced about the carrier at the modulation frequency. However, for measurement and calculation purposes, 'significant' sidebands are only those which have a voltage of at least 1% (-40dB) of that of the unmodulated carrier – hence the higher order sidebands can be ignored for practical purposes. **Fig 24.32** shows a typical spectrum, and corresponds to a modulation index of 1.2. Clearly only the first two sidebands are 'significant' and of interest here.

### Measurement of Deviation

From the foregoing, the carrier and first order sidebands independently go to zero at certain values of m (**Fig 24.33**). The following table shows the specific values of m for which this occurs:

| Carrier (J0) | 1st sideband (J1) |
|---|---|
| 2.40 | 3.83 |
| 5.52 | 7.02 |
| 8.65 | 10.17 |

If a tone of modulation frequency (fm) is applied to the transmitter such that the carrier or sidebands are nulled, the value of m is now known. Furthermore, the peak frequency deviation of the carrier (Δf) can be now be calculated since the modulation index (m), is the ratio of carrier frequency deviation to the frequency of the modulating signal, that is:

Δf = m x fm

This process is called 'the Bessel Zero Method', and can also be used for checking the calibration of deviation meters.

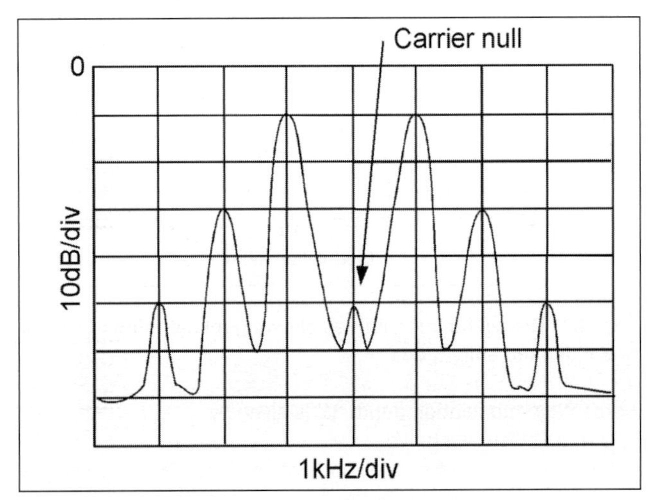


Fig 24.33 Spectrum analyser plot for first carrier null

### Measurement of Bandwidth

In narrowband FM voice communication, (speech quality) bandwidth is limited and the number of higher order sidebands is restricted, so all sidebands with less than 10% of the carrier amplitude (-20dB) can be ignored. This means that the bandwidth (B) can be calculated approximately by using Carson's Rule, the formula for which is:

B = 2(Δf + fm)

An example showing both deviation and bandwidth calculations is given in **Table 24.1**.

**Example**

A signal generator or transmitter is adjusted to obtain the first carrier null of the FM signal and a plot similar to the **Fig 24.33** will occur (for practical reasons the carrier may not disappear completely). This first null appears at a modulation index of 2.4 and the sidebands are at 1kHz spacing (fm).

The peak deviation Δf is therefore 2.4 x 1kHz = 2.4kHz. Using the formula to calculate B, this results in an approximate bandwidth of (2 x 2.4kHz) + (2 x 1kHz) = 6.8kHz.

A similar analysis can be undertaken by finding other nulls for the carrier or sidebands.

The situation is slightly more complicated with wideband FM signals, but there are good tutorials on the Internet.

Table 24.1 Example of FM deviation and bandwidth calculation

# ELECTROMAGNETIC INTERFERENCE & COMPATIBILITY (EMI/EMC) MEASUREMENTS

EMI: A spectrum analyser with an appropriate transducer attached, such as an antenna or a pickup coil, is capable of measuring either conducted or radiated EMI. **Fig 24.34** illustrates a simple setup used for measuring field strengths.

The antenna is used to convert a radiated field into a voltage for the analyser to measure. The field strength will be the analyser reading modified by the antenna correction factor, as the antenna may not have a flat frequency response.

EMC: This is an important factor for high-frequency circuits in close proximity to each other. For example, in a multistage circuit, parasitic oscillation from one stage can couple to a nearby stage and cause unpredictable behaviour. A common technique used to search for spurious radiation is with an inductive-loop probe. The loop probe is simply a few turns of wire that attaches to the spectrum analyser with a flexible coaxial cable (**Fig 24.34** above). Various parts of the circuit can then be 'probed' to identify the location, as well as the frequency and relative amplitude

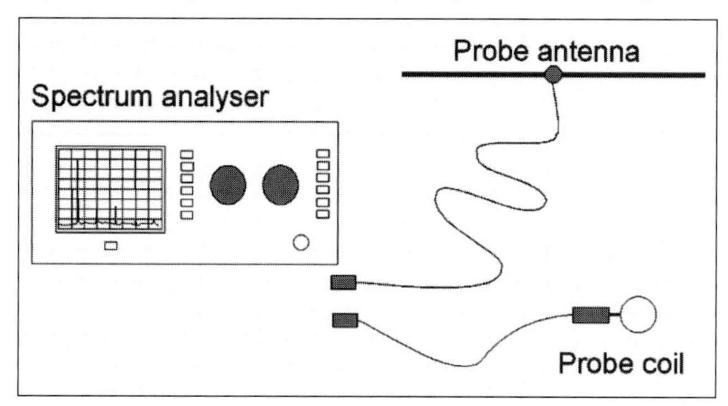


**Fig 24.34 Field strength measurement test set-up**

of an unwanted signal. Once the unwanted signal has been identified, shielding and/or design techniques can be used to reduce or eliminate the cause of the interference.

On a wider point, many Radio Amateurs suffer from RFI caused by modern electronic devices. A particular problem can be interference from wind farms, solar PVs, and VDSL. A useful discussion of how the amateur bands are affected by these emissions, and what might be done to eliminate, or at least mitigate their effects, is contained in the following article:

→ Radio Frequency interference', John Rogers, MOJAV, *RadCom*, March 2017

## Signal Tracing & Spectrum Analyser Probes

A really useful feature of the spectrum analyser is the ability to identify multiple signals (ie different frequencies) and to follow the one of interest around the circuit. A simple E- or H-Field probe can be made for this purpose. Search the internet, or see Test Equipment for the Radio Amateur, 5th Edition (available from the RSGB bookshop) for examples you can make.

## Antenna System Analysers

These can measure various features of both antennas and transmission lines, either separately, or when connected together as an antenna and feed system. Typical parameters that can be measured are:

- Impedance (resistance and reactance)
- VSWR (or return loss)
- Resonant frequency
- Bandwidth
- Velocity factor
- Cable loss
- Cable length
- Distance to fault

Measurements are usually undertaken to see how well the antenna or feed-point impedance compares to a system value, often 50Ω. Improving the 'match', improves forward power transfer to the antenna, and reduces reflected power.

Antenna analysis is a specific application of the Scalar Network Analyser (SNA) or vector network analyser (VNA). The former just measures the magnitude of the reflection coefficient while the latter also measures its phase. So while a scalar analyser can measure impedance, the vector network analyser can go a step further and display the values of the resistive and reactive elements that make up that 'complex' impedance.

Such equipment can also be used to check input and output impedances of filters, amplifiers, matching networks, resonance of traps, attenuation, stub performance etc. Simple analysers may just give a readout, whereas those with a PC interface can generally provide Smith-chart plots of complex impedance, phase, etc.

→ Antenna Analysers, SNAs, and especially VNAs can be expensive; so check-out internet review sites to see how those that have already bought one have fared with their purchase. A really good comparison site is eHam.net at:

https://www.eham.net/reviews/products/31

→See also W8WWV - 'Analyzing Three Antenna Analyzers':

*http://www.seed-solutions.com/gregordy/Amateur%20 Radio/Experimentation/*

EvalAnalyzers.htm#Autek%20RF-1

→ You can also build your own from a kit; examine the following websites:VK5JST (Apparently, over 17,000 have built; note the excellent comments on this build by M5POO):

*http://users.on.net/~endsodds/aamk7.htm*

*http://www.m5poo.co.uk/vk5jst-aerial-analyser/N2PK*

*http://n2pk.com/*

DG8SAQ and DG5MK (some new products here):

«https://sdr-kits.net/DG8SAQ-VNWA-models and FA-VA5 vector antenna analyzer

A simpler and cheaper way to measure impedance is with an impedance bridge. There are two important types:

### The Noise Bridge

Often used to set up antennas, it is especially useful as it can determine both the 'real' (resistive) part and the 'imaginary' (reactive) part of an impedance. The impedance to be measured is simply connected to the 'unknown' socket, the noise generator switched on, and the receiver tuned to the frequency at which the test is to be made. Two controls (one which varies a resistance and the other a capacitance) are then adjusted to obtain a minimum noise reading on the receiver S-meter. This null occurs when the resistive and reactive parts of the impedance being measured coincide with those inside the bridge, and the values of R and X can be read-off the instrument. Note that the reactance X could be positive (inductive) or negative (capacitive) so the control will be marked accordingly (**Fig 24.35**). The value of L or C at the frequency in use can now be calculated since:

$L=X/\omega$ and $C=1/\omega X$ where $\omega=2\pi f$

**Fig 24.35 A commercial noise bridge for the Amateur market**

### The Return Loss Bridge

A return loss bridge (RLB) is usually a wideband resistive bridge network which can be used to measure the impedance of coaxial cables, antennas, tuning stubs, filters, duplexers, cavities etc. It compares an unknown impedance with a known impedance (typically 50Ω) and a voltage is produced which is related to the impedance mismatch. The directivity of the bridge is a measure of the quality of the bridge and must be much better than the return loss that you expect to measure. The directivity and insertion loss of the bridge will vary with frequency.

RLBs are simple to build and have a large bandwidth, typically 1 to 500MHz - somewhat wider than a typical VSWR meter. **Fig 24.36** shows a typical arrangement for using an RLB.

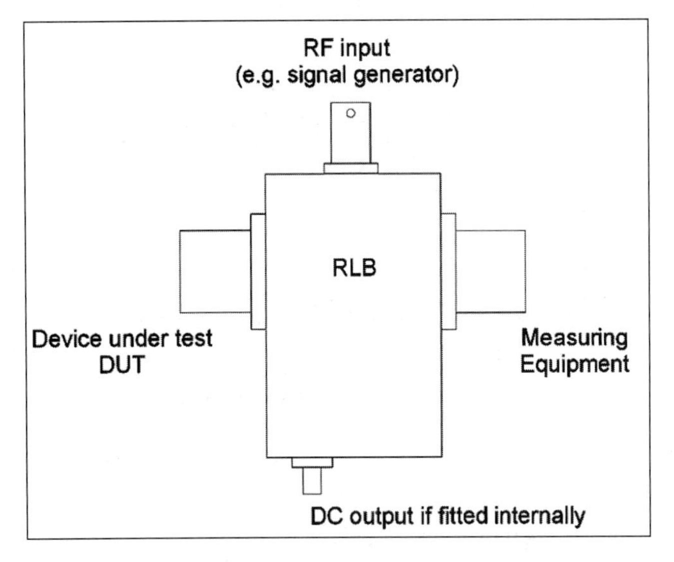


**Fig 24.36 Typical connections when using a return loss bridge**

→ For a few short practical articles on RLBs see:
*http://www.qsl.net/n9zia/rlb/index.html*
*http://www.vk2zay.net/article/179*
www.wetterlin.org/sam/Reflection/ManualReturnLoss.pdf
'Design Notes', Andy Talbot, G4JNT, RadCom, January, 2018

| | | | |
|---|---|---|---|
| R1, R2 | 4k7 | C1, C2 | 330n |
| R3 | 10k | C3, C4 | 33n |
| R4 | 560R | C5, C6 | 3n3 |
| RV1 | 1k trimmer | C7, C8 | 330p |
| RV2 | 47k dual gang log pot. | C9, C10 | 100μ, 25V |
| RV3 | 10k lin. pot. | C11 | 47μ, bipolar |
| B1 | 28V, 40mA bulb | IC1 | LM358 |

**Table 24.2: Low frequency oscillator components list**

## SIGNAL SOURCES

Sources of controlled frequency and amplitude are necessary for setting up both transmitters and receivers. Ideally, for receiver adjustment, it is desirable to have an RF source covering from a few hundred kilohertz up to the highest frequency used at the station. The amplitude should be known from a fraction of a microvolt up to tens of millivolts. In a good signal generator both frequency and amplitude are accurately known but such instruments are costly and certainly difficult to make and calibrate in an amateur workshop.

Many good signal generators appear on the surplus market, although with the cheaper ones the frequency calibration is sometimes not too accurate; however, this is not too important as the amateur almost always has a means of checking frequency. In selecting an instrument, the quality of the internal attenuator, and the effectiveness of the internal screening, is all important, as at very low levels, a poorly screened oscillator will emit sufficient signal to bypass the attenuator and prevent low-microvolt output levels being attained.

Simple oscillators can be constructed for tuning over a limited range, such as VFOs and VXOs. There are many designs on the Internet as well as from manufacturer's data sheets. A number of kits and modules for oscillators based on direct digital synthesis (DDS) are available, via the Internet, but a good attenuator will be needed.

An audio frequency generator is useful for testing audio amplifiers and for checking the performance of transmitters. The design given here provides a sinewave output and a frequency range well in excess of the audio range.

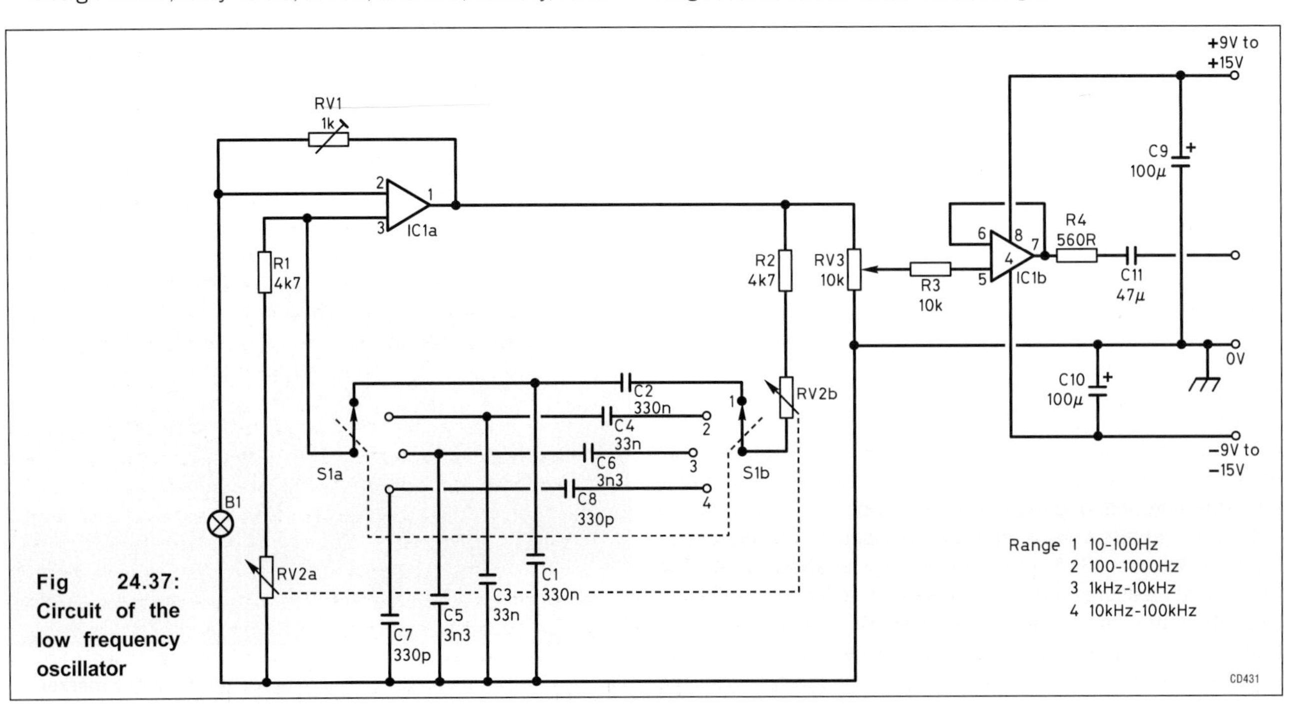


**Fig 24.37: Circuit of the low frequency oscillator**

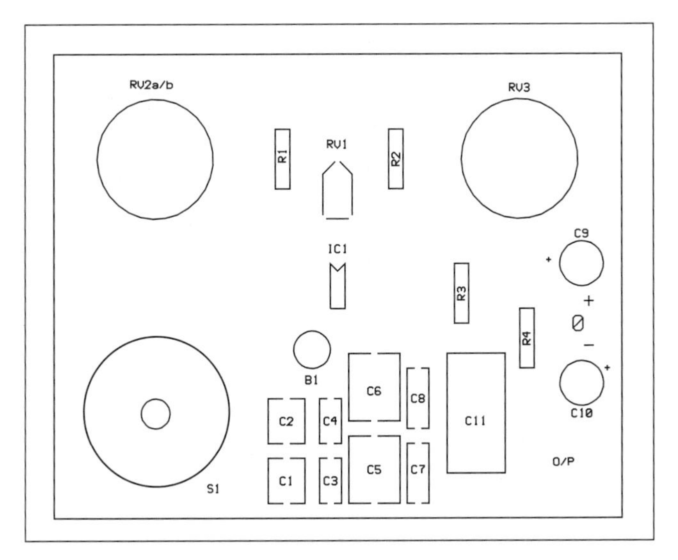


**Fig 24.39: Component layout of the low frequency oscillator (not to scale)**

## A Low-Frequency Sinewave Oscillator for 10Hz-100kHz

The circuit diagram for this oscillator is shown in **Fig 24.37**. It is based on a Wien bridge oscillator formed around IC1a and buffered by IC1b. The main frequency-determining components are R1/R2 and RV2 with capacitors C1 to C8. In the configuration shown, stable oscillation can occur only if the loop gain remains at unity at the oscillation frequency. The circuit achieves this control by using the positive temperature coefficient of a small lamp to regulate the gain as the oscillator varies its output. Potentiometer RV3 forms the output level control, with R4 giving a defined output resistance of approximately 600Ω and C11 providing DC isolation. Capacitors C9 and C10 provide power supply line decoupling. The approximate ranges provided are:

1: 10Hz-100Hz
2: 100Hz-1kHz
3: 1kHz-10kHz
4: 10kHz-100kHz

The exact range is dependent on the tolerance of the particular components used and ambient temperature variations. The circuit requires a symmetrical plus and minus supply between 9 and 15V. A components list is given in **Table 24.2**. The layout of the circuit is not critical, but a PCB pattern (**Fig 24.38**) is given in Appendix B and a component layout in **Fig 24.39**. If some ranges, or the output level control, are not required then the layout can be tailored accordingly. The feedback resistor RV1 should be adjusted so that the output on all ranges is just below the clipping level. No frequency calibration is required but it would be wise to check with a frequency counter that the ranges are as suggested. An oscilloscope is required for setting up the adjustment of RV1.

## A Crystal-Based Frequency Marker

The purpose of this unit is to produce a 'comb' of output frequencies which are all based on a crystal. The unit described here gives outputs at harmonics of 1MHz, 100kHz, 25kHz, 12.5kHz and 10kHz with an additional output of a sine wave at 1kHz which may be useful as an accurate modulation signal. The sine wave output has an output resistance of approximately 600Ω and maximum amplitude of approximately 2.5V peak to peak.

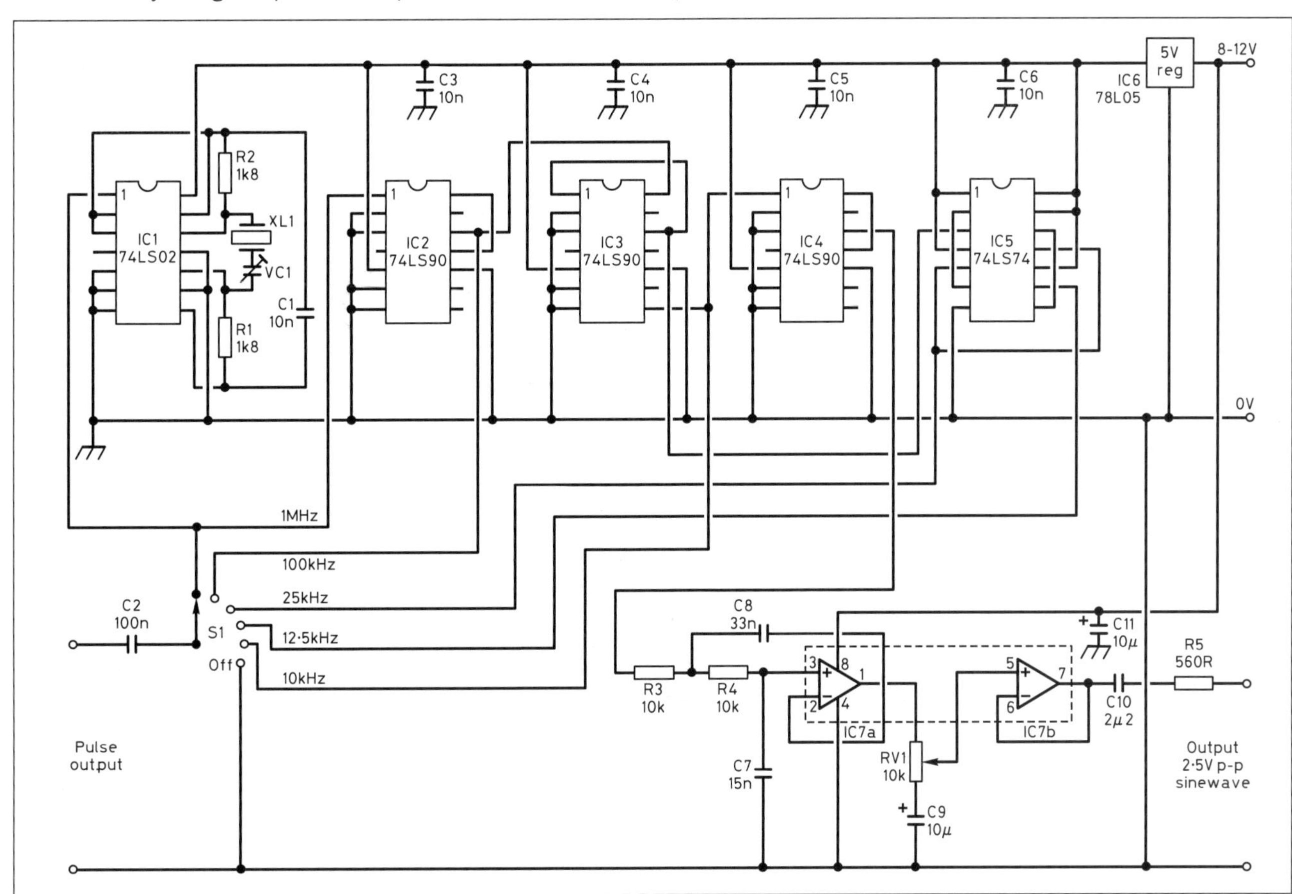


**Fig 24.40: Circuit of the crystal-based frequency marker**

| | | | |
|---|---|---|---|
| R1, R2 | 1k8 | C10 | 2µ2 |
| R3, R4 | 10k | IC1 | 74LS02 |
| R5 | 560R | IC2, IC3, IC4 | 74LS90 |
| VR1 | 10k lin pot. | IC5 | 74LS74 |
| C1 | 10n, ceramic | IC6 | 78L05 |
| C2 | 100n, ceramic | IC7 | LM358 |
| C3, C4 | 10n, ceramic | SW1 | 2p, 6w rotary switch |
| C5, C6 | 10n, ceramic | XL1 | 1MHz crystal, HC6U |
| C7 | 15n | VC1 | 30p trimmer |
| C8 | 33n | IC Sockets | 14p, 5 off; 8p, 1 off |
| C9, C11 | 10µ, 25V tant bead | | |

Resistors are 0.25W/0.5W, 5% unless specified otherwise.

**Table 24.3: Crystal-based frequency marker components list**

### Circuit Description

The circuit diagram is shown in **Fig 24.40**. The signal is derived from a 1MHz crystal-controlled oscillator formed by XL1 and IC1 plus various components. Capacitor VC1 allows a slight variation of the crystal frequency for calibration as described later. This 1MHz signal is divided by 10 by IC2 to give a 100kHz signal. This signal is then passed to IC3 which has a 50kHz output and also a 10kHz output. The 50kHz output is divided by dual flip-flop IC5 to give a 25kHz and 12.5kHz output. The 10kHz signal from IC3 is divided by 10 by IC4 to give a 1kHz square-wave output. The 1kHz square wave is then filtered by an active low-pass filter formed by IC7a. The variable-amplitude sine-wave output is then buffered by IC7b. R5 forms the output resistance of the buffer.

### Construction

A components list is given in **Table 24.3**. The layout for this circuit is not critical but the completed circuit should be housed in a metal box to prevent unwanted radiations. The output should be via a coaxial socket to a small antenna when in use. It requires a supply of 8 to 12V DC at about 50mA. If the voltage regulator IC6 is omitted the circuit can be fed straight from a 5V supply but ensure there is a supply to the 1kHz filter IC7. A PCB pattern (**Fig 24.41**) is in Appendix B and component placement in **Fig 24.42**.

### Calibration

The frequency of the 1MHz crystal oscillator can be adjusted by a small amount by VC1. The output from the oscillator or a harmonic should be checked against an accurate frequency reference.

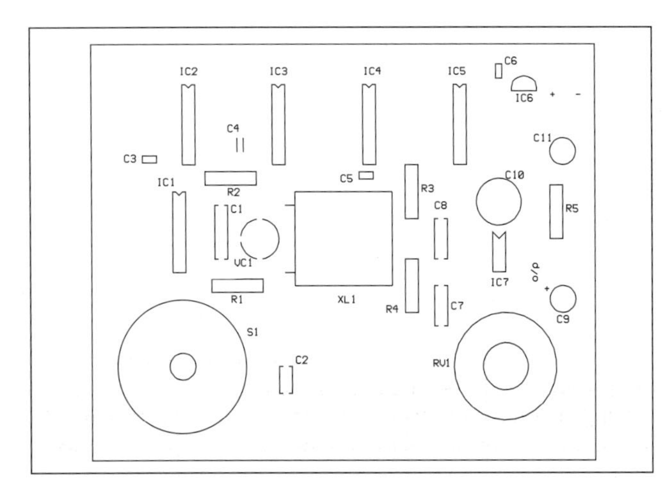


**Fig 24.42: Component placement details for the crystal-based frequency marker (not to scale)**

## THE FREQUENCY COUNTER

A frequency counter is an invaluable instrument for alignment, test, and calibration purposes. Among other things, it allows a

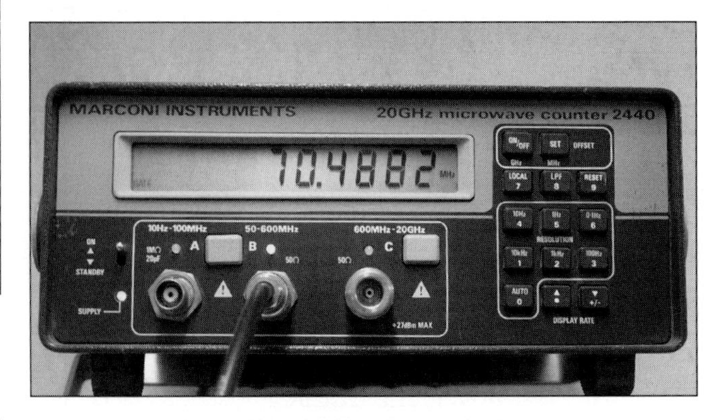


**Fig 24.43 A Marconi 2440, 20GHz frequency counter**

transmitter frequency to be checked and verified as in-band; and a receiver to be aligned so that the dial readout is correct.

The cost of frequency counters varies considerably. For example, £150 or so will buy a 3GHz hand-held unit with a short whip aerial which by which to pick up or 'sniff' for a local RF field; it may also decode and display any CTSS tone modulating the signal. Although workbench units can be much more expensive it is possible to buy something for less (including kits), but operating frequency range, resolution, and accuracy are likely to be compromised. **Fig 24.43** shows a commercial frequency counter capable of counting to 20GHz.

### Operation

**Fig 24.44** shows a block diagram of a basic frequency counter. In essence, the counter counts the number of input cycles for a given unit of time and displays the result in Hz (or its multiples: kHz, MHz etc). To achieve this, a wave-shaping circuit takes the input signal, amplifies it, and converts it into a rectangular waveform of sufficient magnitude to operate the counting circuits. Meanwhile a clock produces a series of pulses which determine the counting period of the frequency counter. These pulses are typically 10ms, 100ms, or 1s long, and derived from a crystal-controlled clock-oscillator. They are thus of high accuracy, and are applied to a gate, which can be considered as an on/off switch operated by the clock. When the clock-pulse opens the gate, a train of pulses from the wave-shaping circuit is sent to the counting circuit which counts the number of pulses for the duration that the gate is open. This count is then frozen, decoded, and used to drive a digital readout, perhaps in the form of an LED or LCD display.

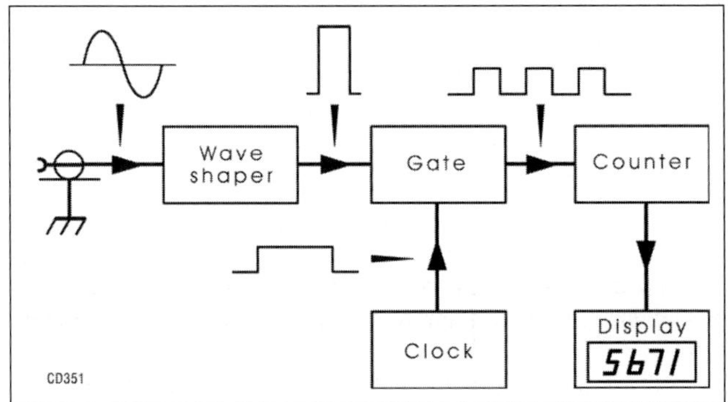


**Fig 24.44 Simplified block diagram of a digital frequency counter**

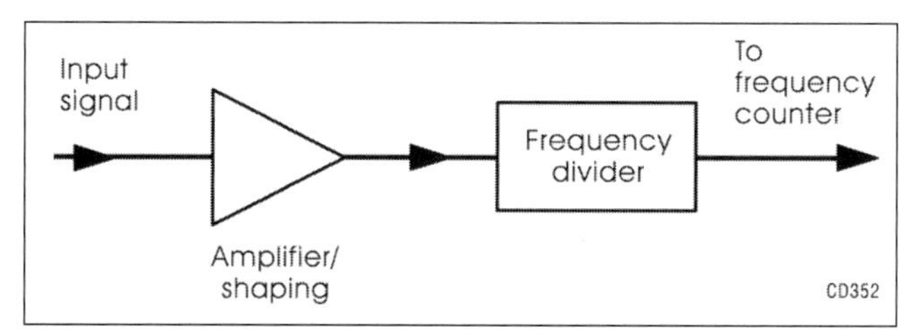


Fig 24.45 Pre-scaling

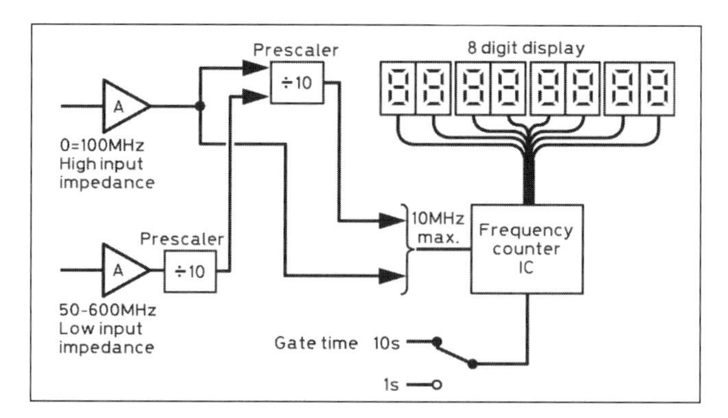


Fig 24.46 Block diagram of a frequency counter with pre-scaling

More expensive frequency counters will allow measurement of the period, usually in microseconds or milliseconds. This is useful for frequencies less than about 100Hz and will generally give a more accurate result, though you will need to calculate the frequency (f), which is the reciprocal of the period (T). That is: (f=1/T).

### Accuracy and Resolution

The accuracy of a frequency counter clearly depends on the accuracy of the clock signal. It can be increased if the clock oscillator is housed in a thermostatically controlled crystal oven and compared with a standard frequency source for calibration. The resolution of a frequency counter, which is the smallest digit that it displays, is typically 1Hz. However this will change according to frequency range and the length of the clock signal.

## Pre-scalers

To increase the range of a frequency counter it is common to put ahead of the input some components that perform pre-scaling. **Fig 24.45** shows that this consists of an amplifier, a wave-shaping circuit, and a high-speed frequency divider (the pre-scaler).

Pre-scaling will divide the input frequency by a known amount (eg 2, 10, 64, or 100) with the resulting signal being applied to the basic frequency counter (**Fig 24.46**). It should be borne in mind however, that pre-scaling reduces the resolution, so that whereas a frequency counter without pre-scaling may measure to 1 Hz, if a pre-scale of 10 is introduced, then the same counter will only read to 10Hz.

### Making measurements

There is usually a high (1MΩ) impedance, and a low (50Ω) impedance input for the lower and higher frequency ranges respectively. The input can be taken from a test probe, direct pick-up off-air using an antenna, or from an RF sampler or coupler.

- Know the maximum input ratings of your frequency counter, and take care not to exceed them
- Keep RF input levels low (only a few tens of mV of RF is needed).

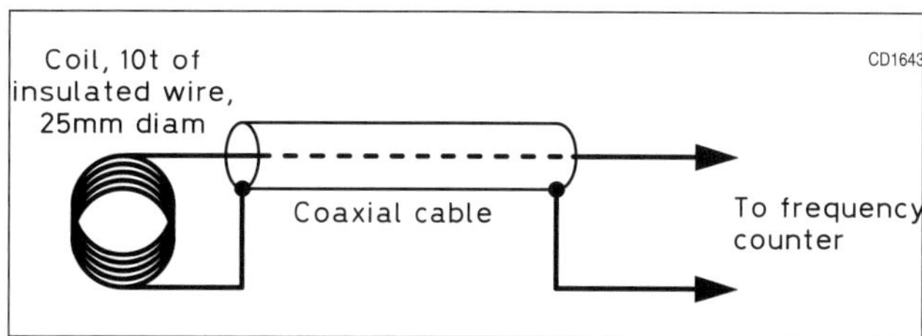


Fig 24.47 Pick-up loop

- Consider whether any AC or DC offset voltage is safe. Use AC coupling, especially in higher-voltage valve circuits. Add an external DC block if necessary.

When making measurements of transmitter frequency, the carrier should be unmodulated to avoid confusing the counter; hence:

- With AM and FM transmitters, do not speak into the microphone or provide any other form of modulation.
- With SSB sets, the CW mode should be used instead, with key down. Take care not to exceed the thermal dissipation limit of the PA.
- With digital transmissions, no superimposed data should be transmitted, just a carrier.

### Probes

Be aware that while a high-impedance probe generally helps to minimise loading of the counter on the circuit under test, if connected directly to the tuned circuit in an oscillator, it may well change the frequency of operation of the circuit by 'pulling' of the oscillator, and may stop it working altogether. This is because the probe can present a shunt capacitance in the order of 10pF at 144MHz, which is a reactance of only 110Ω. The probe may also affect the DC operating conditions. Therefore:

- Do not attach the probe to frequency-determining elements.
- Measure the oscillator frequency after the buffer amplifier.
- Put a 1kΩ resistor in series with the probe.

Another alternative is to use a pick-up loop, as shown in **Fig 24.47**. Ensure only loose coupling, as this minimises influence on the circuit and damage to the counter by input power overload.

## Frequency Standards

It is fine having test equipment but how accurate is it? Has it been calibrated recently? It is unlikely that an atomic frequency standard is owned or the cost of periodic calibration can be justified. So how can items such as frequency counters be calibrated?

In the past the best frequency standard available to most amateurs was a crystal oscillator carefully adjusted to zero-beat with a station of known frequency, such as one providing a standard frequency service. These standard frequencies are maintained to an accuracy of typically one part in $10^{11}$. However, if the sky-wave is used there could be a large error in reception due to Doppler shift and there will be fading of the signal. These problems can be avoided by using a low-frequency transmission such as those from MSF or WWVB. Timing information is also impressed on these signals in GMT / UTC. In addition, in the UK, the BBC maintains the accuracy of the Droitwich 198kHz (formerly 200kHz) carrier to high accuracy - on a long-term basis being 2 parts in $10^{11}$.In the past, accuracies of the order of 0.1ppm were good enough to ensure keeping transmissions within band edges but for EME

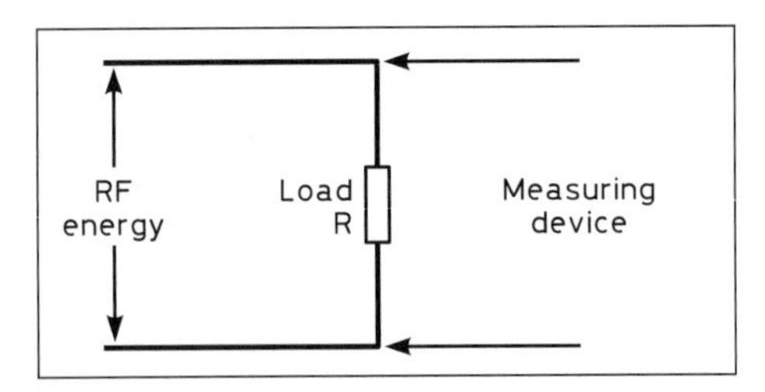


Fig 24.48: Power measurement arrangement

and weak signal VHF work this is not really good enough. Now there is available a high accuracy clock available from space - the GPS satellites. All that is required is equipment and a method to decode the signal and produce a usable output. They all use various GPS modules which extract and output the 1pps signal and in some cases a 10kHz signal. These can often be found on eBay.

## POWER OUTPUT MEASUREMENTS

The UK Amateur Licence requires that you should be able to measure transmitter output power in order to comply with the licence conditions. The oscilloscope can be used up to about 30MHz to monitor modulated waveforms and measure output power, but above this it becomes an expensive item and may provide unwanted loading effects on the equipment being monitored. The familiar VSWR meter monitors forward and reflected signals and the scale can be made to represent power in a 50Ω line. It is possible to use an RF voltmeter across a given load to measure power. The higher you go in frequency, the more difficult it becomes to measure the modulation and power with relatively cheap equipment. Yet it is a condition of the licence that these parameters can be monitored. It is more difficult to measure PEP than average carrier power.

### Constant amplitude Signals

In a carrier-wave situation (CW, FM or unmodulated AM), the output is of constant amplitude and so it is relatively easy to measure the steady output power. To measure these signals using the circuit as shown in **Fig 24.48**, the power output is given by:

$$P_{out} = V^2 / R$$

where V must be the RMS value of the voltage.

This voltage measurement can be carried out using an oscilloscope or RF voltage probe. A VSWR meter can also provide this value.

### Amplitude modulated Signals

These pose more of a problem; two cases are dealt with below.

### Amplitude modulation (A3E)

With no modulation, the problem reverts to the measurement of power of a constant carrier as described above. If the carrier is amplitude modulated (A3E) then the overall output power increases. The power is divided between the sidebands and the carrier component. With 100% modulation the average output power increases to 1.5 times the unmodulated condition - the power contained in each of the two sidebands being one-quarter that in the carrier. It is suggested that for this form of modulation the carrier power is measured (ie no modulation) and multiplied by 1.5 to give the maximum output power.

If an exact value for the output power is required it is necessary to determine the modulation index. This can be carried out using an oscilloscope of adequate frequency response. Set the oscilloscope as shown in **Fig 24.49** and

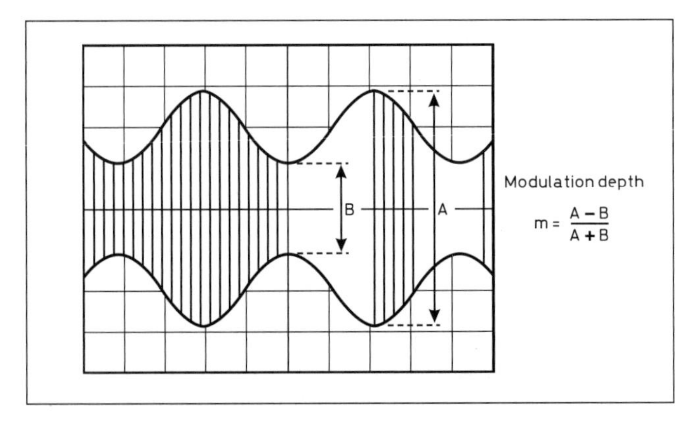


Fig 24.49: Modulation depth measurement

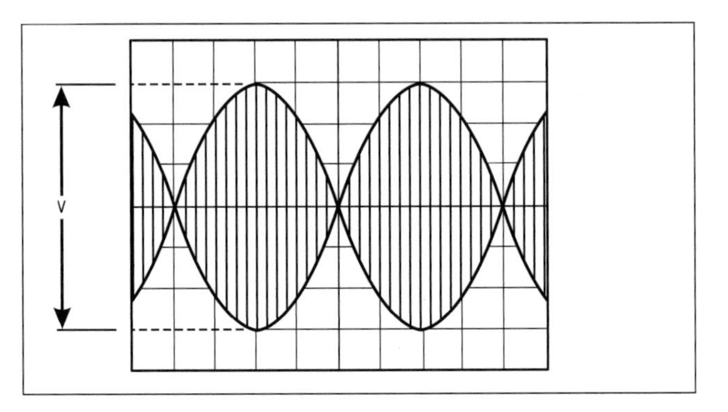


Fig 24.50: Two-tone test display

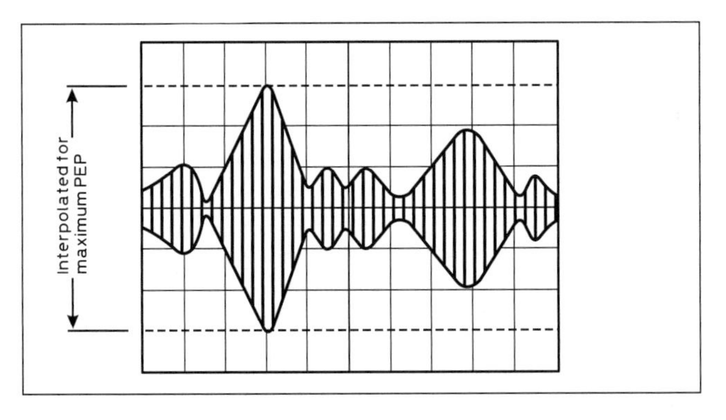


Fig 24.51: Speech waveform and interpolated maximum PEP level

calculate the modulation depth m. The output power is then given by:

$$P_{out} = \frac{V^2}{R}\left(1 + \frac{1}{2}m^2\right)$$

where V is the RMS value of the unmodulated carrier and R the load.

For test purposes a constant amplitude audio signal can be fed in at the microphone socket of the transmitter using the low-frequency oscillator described earlier.

### Single sideband (J3E)

With single sideband, no power is output until modulation is applied. The output envelope is non-sinusoidal in appearance. The normal method for measuring output power is by observation of the modulation envelope and determination of the peak envelope power - this is the parameter defined by the UK licensing authority. This can be accomplished using an oscilloscope of suitable frequency response as described below, or using a SWR meter that will respond to peak envelope power.

**Fig 24.50** shows the display when a two-tone test signal is fed in via the microphone socket. If the peak-to-peak voltage V at the crest of the envelope is measured across a load of value R, then the PEP is given by:

$$P_{out} = \frac{V_{pk-pk}^{2}}{8R}$$

The equivalent peak-to-peak voltage reading can then be interpolated for the maximum allowable PEP and the position noted on the display. **Fig 24.51** shows a typical display for a speech waveform and the interpolated maximum PEP level.

## POWER AND VSWR METERS

### Power Meters

There are two main types of power meter:

#### Through-line

Based on RF samplers or couplers they sample the power on the transmission (feed) line and indicate a power level. They are usually directional, so that both forward and reverse power can be measured.

#### Absorption

These incorporate a sensor which absorbs the power, and generates a signal proportional to this power to drive a meter or similar. Sensors involving heating use detectors such as thermistors, thermocouples and the RF (hot-wire) ammeter. They are good at measuring average power in systems because the heating effect produces an averaging, and so they measure power irrespective of waveform. Good for signals such as CW, AM, FM and pulse waveforms they are unable to measure instantaneous power in signals such as SSB.

For instantaneous measurements, a diode-based detector is normally used, and this is the sensor arrangement behind virtually all VSWR and power meters that the Radio Amateur will buy.

#### Using Other Instruments

Oscilloscopes and spectrum analysers can also be used to measure power; the oscilloscope has a relatively low frequency limit compared to the spectrum analyser, however the latter cannot cope with high power and relies on attenuators to reduce the input power to a suitable level.

**Fig 24.52** shows the ubiquitous Bird power meter with its plug-in elements that give different power and frequency ranges.

**Fig 24.52 The Bird power meter**

Ready-made power meters and kits for home-construction are available on the internet.

### VSWR Meters

These combine a directional coupler and detector to enable power to be measured in both directions. From this, the reflected power and the VSWR can be deduced and displayed. It is common to have two analogue meters, or one analogue meter with two needles, so that forward power and VSWR can be displayed simultaneously (**Fig 24.53**). This gives the operator a valuable indication of the power being delivered into the aerial feeder, the quality of the power transfer or 'matching' and a visual indication of any fault condition, such as a short, or an intermittent caused by weather affecting the integrity of the aerial-feeder arrangement.

**Fig 24.53 A typical commercial power & VSWR meter**

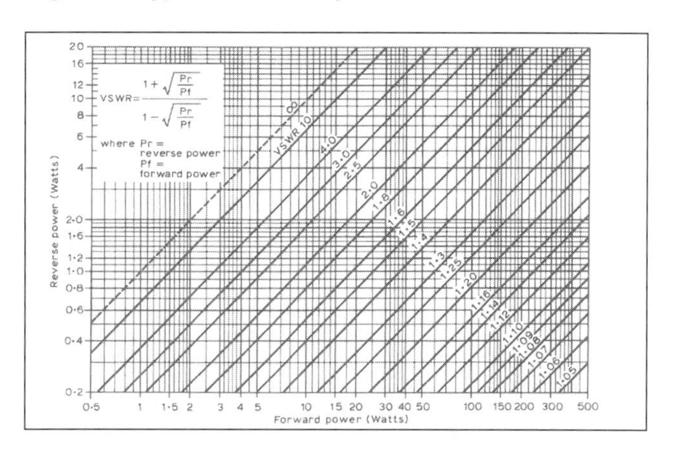

**Fig 24.54 Relationship between VSWR and forward/reflected power**

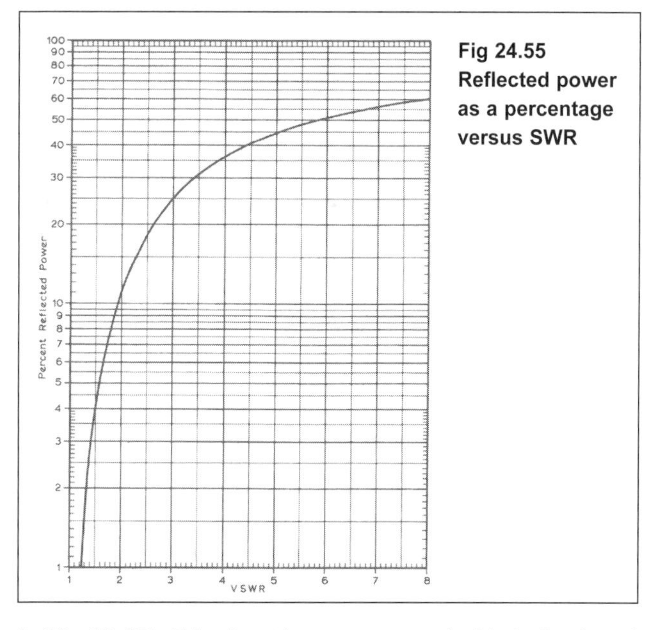

**Fig 24.55 Reflected power as a percentage versus SWR**

**Fig 24.54** shows the relationship between forward and reflected power and VSWR, which are the parameters of practical interest, while **Fig 24.55** demonstrates how reflected power increases dramatically with increasing VSWR. A 1:1 VSWR is ideal, but from a practical viewpoint, it is worth trying to get a SWR better than 2:1 (better than 11% reflected). The guidelines of **Table 24.4** are suggested practical conditions and the actions that should be taken.

# FIELD STRENGTH AND RF CURRENT METERS

## Field Strength Meters

These can be used to produce polar plots (radiation patterns) of antennas. That said, Field Strength Meters are often not used by amateurs to make field strength measurements, but rather, just to indicate the presence of RF. Commonly, the transmitter system is tweaked for maximum indication on the meter to maximise the radiated power. Costing just a few pounds, these units usually rely on some form of telescopic antenna to pick-up the near-field and a simple tuned or untuned detector arrangement (**Fig 24.56**). The absorption wavemeter, or a dip oscillator in absorption mode, can be used for a similar purpose.

→ In connection with this, there is an excellent Windows program 'PolarPlot' (free for non-commercial use) by Bob Freeth, G4HFQ, which involves a receiver, an antenna on a rotator, and a PC with a soundcard, which will measure and display the radiation pattern of a beam antenna (**Fig 24.57**). It relies on an antenna receiving a signal, and the audio output (with AGC disabled) plotted with rotation. The accompanying help files assist configuring the setup for taking measurements. Download PolarPlot at: *http://www.g4hfq.co.uk/download/PolarPlotSetup.exe*

→ For a short tutorial on the relationship between field strength and received power, see:
*http://www.giangrandi.ch/electronics/anttool/rx-field.shtml*

## RF Current Meters

A basic clip-on version of this handy device takes about 10 minutes to tack together (**Fig 24.58**). When you're convinced how useful it is, you can then go on to build a more permanent version.

The clip-on RF current meter has a long history, early versions involved breaking a ferrite ring into two equal pieces - which takes some doing! The constructional breakthrough was G0SNO's idea to use a large split ferrite bead intended for HF interference suppression. This clamps around the conductor under test, to form the one-turn primary of a wideband current transformer. The secondary winding is about 10 turns, and is connected to a load resistor, R1-R2, and the diode detector.

The load resistor, R1-R2, is important because it creates a low series impedance when the current transformer is effectively inserted into the conductor under test. For the values shown in **Fig 24.58**. (10-turn secondary, 2 x 100Ω) this is 50/102 = 0.5Ω. Some circuits omit this resistor, but that creates a high insertion impedance - exactly the opposite of what is needed. Also, more secondary turns create a lower insertion impedance, but at the expense of HF bandwidth.

The other components in **Fig 24.58** are discussed in G0SNO's article which is reproduced on the 'In Practice' website (see below). Component types and values are critical only if you want to make a fully calibrated meter with switchable current ranges. However, for a first try, and for most general RFI investigations, the meter is almost as useful without any need for calibration. Make R4 about 4.7-10kΩ, and omit R3 and S1. If the meter is either too sensitive or not sensitive enough, either change R4 or change the HF power level.

Just about any split ferrite core intended for RFI suppression will do the job, but there are a few practical points. Choose a large core, typically with a 13mm diameter hole. This allows you to clip the core onto large coax, mains and other multicore cables while still leaving enough space for the secondary winding (which should be made using very thin enamelled or other insulated wire). It is important that the core closes with no air gap - and that can be a problem.

A major disadvantage of the basic split ferrite core in its plastic housing is that the housing is not meant to be repeatedly opened and closed, so the hinge will soon break. By all means try out this gadget in the basic form but you will soon be thinking about something more permanent. The classic way to do this is using a clothes-peg **Fig 24.59** but there are now better alternatives.

For example, while the first pho-

| VSWR | % Reflected Power | Comment |
|---|---|---|
| 0.0 - 2.5 | 0-18 | solid-state rig VSWR-protection starts to operate; try looking for an improvement at higher SWRs |
| 2.5 - 5.0 | 18-45 | valve equipment will probably survive - look to improve the SWR to nearer 2:1 |
| 5 upwards | 45-100 | check the feed/antenna system - there is a problem! |

**Table 24.4: Guidelines for various SWRs**

**Fig 24.56 A typical field strength meter for Amateur use**

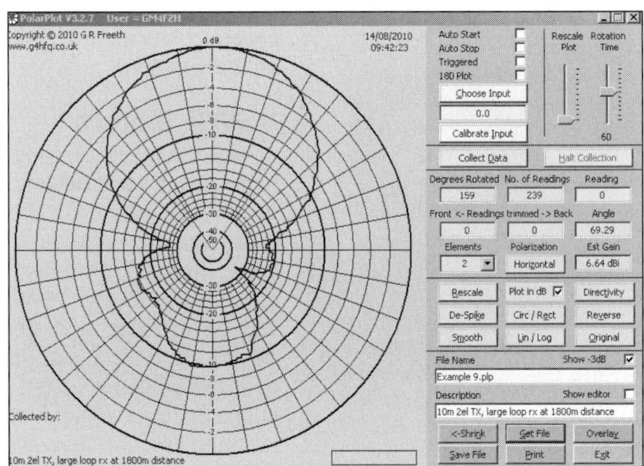


**Fig 24.57 Typical display using PolarPlot**

tograph (**Fig 24.59**) shows a rather heavy-duty version using two strong clothes-pegs, fibreglass sheet and epoxy glue, the second photograph shows GI0XAC's neat and simple version using a giant plastic paper-clip, with a small plastic-cased meter stuck on the side. The only requirement of the clip is that it must be non-metallic, and can hold the two halves of the core accurately together while the whole weight of the meter is dangling from the cable.

Another option worth investigating would be using pliers-style plastic work-clamps that are sold in a range of sizes by hobby shops. Whatever you use, it is vital that you glue the two halves of the core to the clip in such a way that they always close tightly together with no air gap. Hint: glue one half of the core to one side of the clip first, and let that side set; don't try to glue the second half until the first is good and solid.

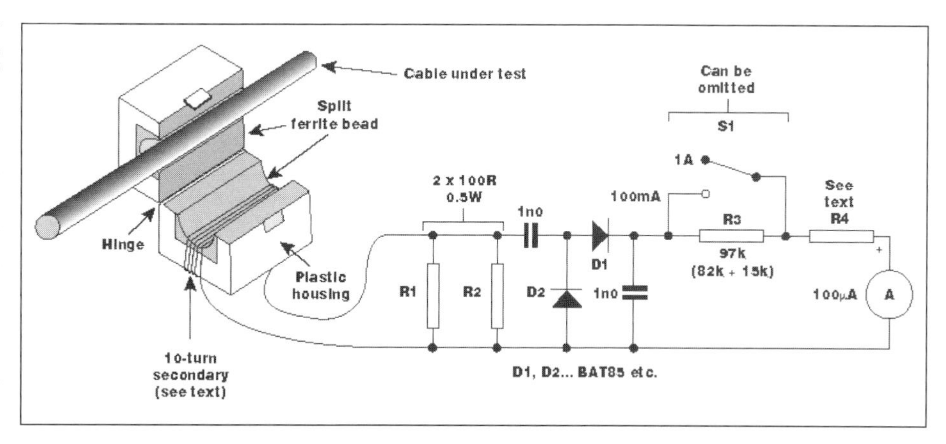


Fig 24.58: G0SNO's clipon RF current meter

A clip-on RF current meter could hardly be simpler to build, and it's an ideal project for beginners and clubs.

→ See, 'Clamp-on RF Current Meter - The most useful tool for RF interference troubleshooting!' on GM3SEK's 'In Practice' website. Design based on G0SNO's original article in RadCom, April 1993, page 74. *http://www.ifwtech.co.uk/g3sek/clamp-on/clamp-on.htm*

→ See also: 'Clip-on RF Current Probe', David Lauder, G0SNO, *RadCom*, October 2016.

## ABSORPTION AND HETERODYNE WAVEMETERS

An absorption wavemeter is usually a hand-held unit which absorbs RF energy from a source close by and uses this to drive some form of indicator such as a moving coil meter (**Fig 24.60**). Generally consisting of little more than a diode and a tuned circuit, they are quite simple in design, can be easily made, and require no DC supply. A commercial Grid-Dip Oscillator (GDO), or its transistorised equivalent, can usually be used as a wavemeter simply by using the instrument in absorption mode. An earpiece socket may be provided, so that AM audio can be demodulated and checked for quality. Two important uses of an absorption wavemeter are:

- Checking that the correct harmonic is selected for driving the next stage in circuits such as multipliers;
- Checking for the presence of harmonics in the output of a transmitter stage.

An absorption wavemeter provides a course, but unambiguous identification of harmonic frequency, and a rough idea of relative amplitude.

### Operation

A good absorption wavemeter should have a continuous frequency range of at least 2:1. For example, a wavemeter might be tuneable between 60 and 150MHz, which would allow idetification of the fundamental output of a 4m transmitter on 70MHz, and assessment of the level of its 2nd harmonic radiating at 140MHz. The hand-held unit is placed within a few feet or a few inches of the source depending on the power output, and the tuning-knob adjusted for maximum indication. Tuning can be quite sharp, and care must be taken not to miss the frequency of interest or overdrive the meter when it's located.

The absorption wavemeter shown in **Fig 24.60** is an easily built unit covering 65-230MHz.

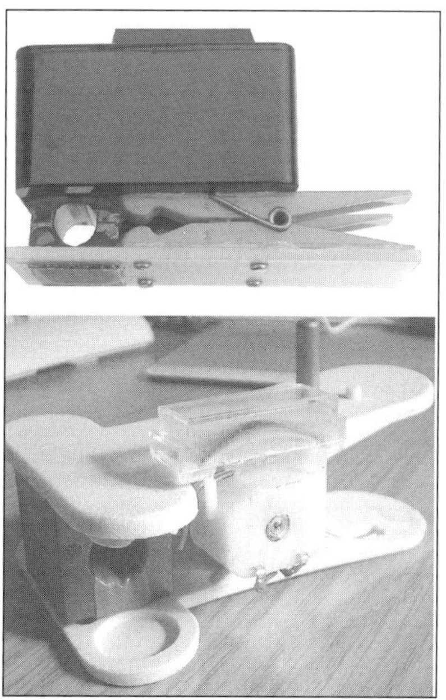

Fig 24.59: Two versions of the clip-on RF current meter

Construction is straightforward and all the components, apart from the meter, are mounted on a Perspex plate of thickness 3 or 4mm and measuring 190 x 75mm. Details of the tuned circuit are shown in **Fig 24.61**(a) and should be closely followed. The layout of the other components is not critical provided they are kept away from the inductor. RFC1 and 2 are made by winding 80T of 40SWG ECW on a 10k, 0.5W resistor; D1 is an OA91, BAT85 or equivalent; C1 is a 4-50pF variable and C2 a 470pF ceramic capacitor. For accurate calibration, a signal generator should be used but, provided the inductance loop is carefully constructed and the knob and scale are non-metallic, the dial markings can be determined from **Fig 24.61**(b).

In operation the unit should be loosely coupled to the circuit under test and the capacitor tuned until the meter indicates resonance (a maximum). For low-power oscillators etc a more sensitive meter should be used (eg 50μA

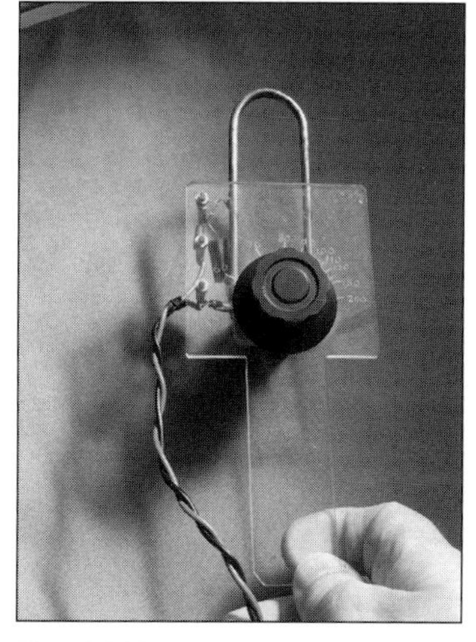

Fig 24.60 A home-made VHF absorption wavemeter

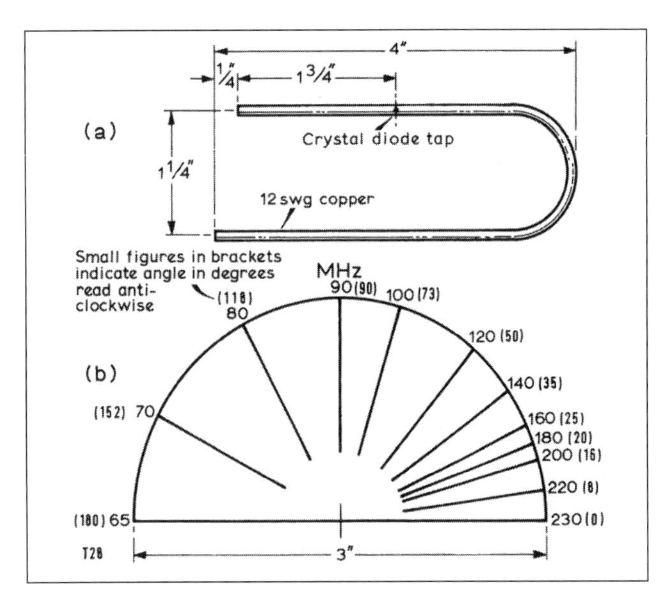


Fig 24.61 Constructional details of simple absorption wavemeter

or 100μA). The wavemeter can also be used as a field strength indicator when making adjustments to VHF antennas. A single-turn coil should be loosely coupled to the wavemeter loop and connected via a low-impedance feeder to a dipole directed towards the antenna under test.

A heterodyne wavemeter in contrast, is basically an absorption wavemeter, but with a very accurate signal generator attached. The absorption wavemeter picks up or 'sniffs' the transmitted signal, and the signal generator is adjusted until it is on the same frequency or 'zero-beat'. The equipment often has an internal crystal calibrator and look-up tables attached so that the received frequency can be better interpreted from the dial. Heterodyne wavemeters, rapidly fell out of use with the widespread availability of frequency counters, and with the advent of synthesized transceivers having digital readouts. However, at UHF and microwave, old cavity-tuned instruments can still be put to good use. **Fig 24.62** shows such a unit.

## TRANSMITTER / RECEIVER MEASUREMENT

The previous sections have dealt with the test equipment that can be used for making various measurements; this section looks specifically at the type of tests and measurements that can be performed on equipment. Many of these measurements require quality test equipment which only a few will either own or have access to. Some of this equipment is obtainable at radio rallies, surplus stores or Internet auction sites.

→ A good source of reference, and with further details, is Peter Hart's book, specifically the first chapter: 25 Years of Hart Reviews, Peter Hart, G3SJX, RSGB, 2005.

### Receiver Measurements

In very simple terms, there are broadly two sets of parameters which define the effectiveness of any receiver - how well it receives wanted signals and how well it rejects unwanted signals. The types of receiver measurements can be expanded into various types as suggested below.

- Sensitivity and SINAD
- Noise figures
- AGC response
- Intermodulation and blocking
- Receiver reciprocal mixing

#### Sensitivity

**Fig 24.63** shows the test arrangement for making measurements of sensitivity, spurious response rejection and selectivity. A suitable resistive load is connected to the external loudspeaker socket of the radio and the audio output monitored using a true RMS voltmeter. It is important that this indicator shows the true RMS level of the noise output as well as mixed noise and sine wave output. A dB scale is by far the most convenient. The audio output level should be low enough to avoid audio distortion at all times. The audio monitor is just a simple amplifier and speaker so that the receiver audio can be heard.

As an alternative to the voltmeter, an audio spectrum analyser or FFT analyser can be used with software to compute signal-to-noise ratios directly. This can have a number of advantages with improved accuracy and ease of separating signals and noise.

Sensitivity measurements on SSB or CW are made with the RF generator set to give a 1kHz audio beat note. With the RF generator switched off the audio output level on noise alone is set to a convenient value. The generator is then turned on and the RF level set to give a 10dB increase in audio output. This level gives the sensitivity for 10dB (signal + noise) to noise ratio (s + n ): n.

The receiver noise floor is 9.5dB lower than this figure. If the

Fig 24.62 A UHF heterodyne wavemeter

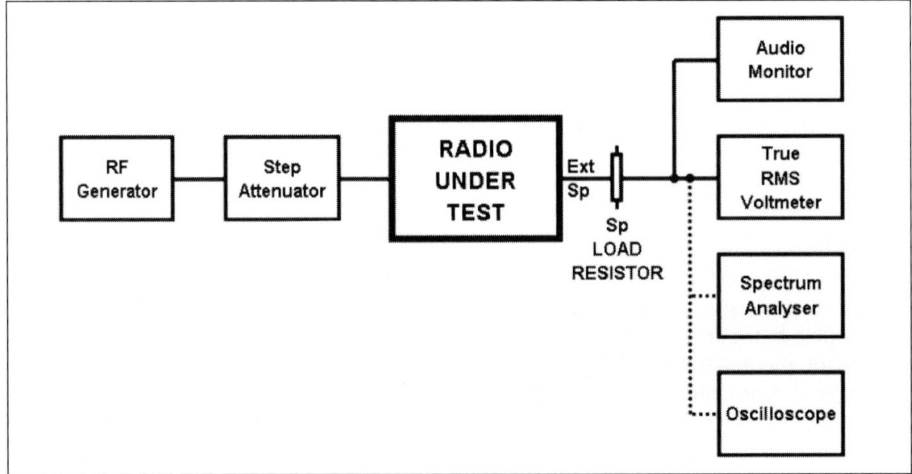


Fig 24.63: Sensitivity related receiver measurements

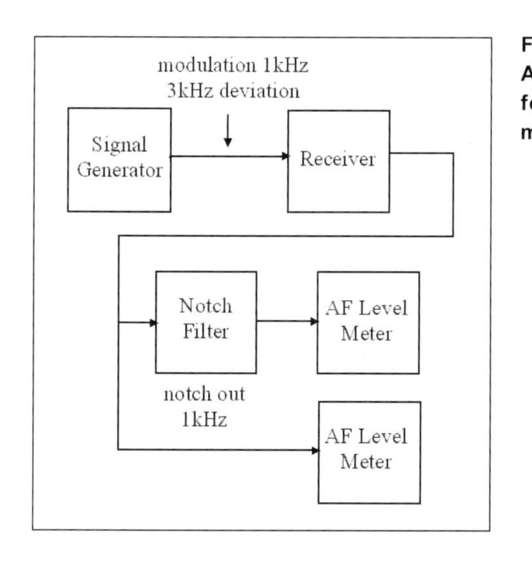


**Fig 24.64: Arrangements for SINAD measurement**

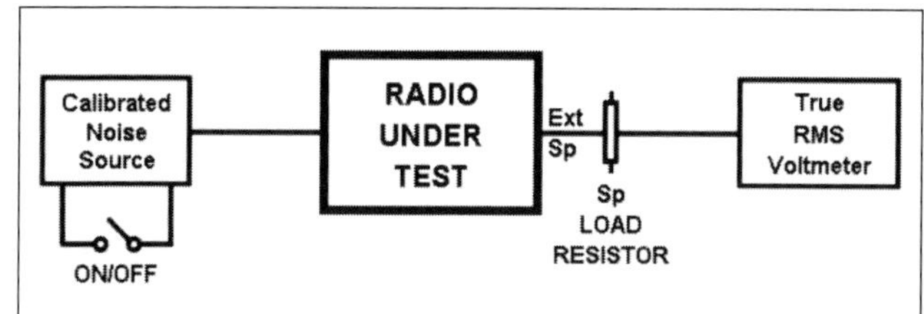


**Fig 24.65: Noise figure measurements**

generator had been set to give a 3dB increase in audio output, the level would be equal to the noise floor of the receiver. Measurements of (s + n) : n for AM and FM signals are made, not by turning the RF generator on and off, but by switching the 1kHz modulation on and off: 30% modulation depth is usually used for AM and 60% of the maximum deviation is used FM (eg for a 25kHz channel spacing, which uses a maximum of 5kHz peak deviation, the modulation is set to 3kHz).

## Sinad

Another measurement that can be made to assess and specify the sensitivity performance of a radio receiver is SINAD (signal plus noise plus distortion to noise plus distortion). It is a common performance measurement in many applications including many two-way FM radio systems especially at VHF and above.

SINAD is a measurement that can be used with any radio communication equipment and expresses the degradation of the signal by unwanted or extraneous signals including noise and distortion. Defined as the ratio of the total signal power level (Signal + Noise + Distortion) to the unwanted signal power (Noise + Distortion) it is normally expressed in decibels (dB) and is typically used for measuring or specifying the sensitivity of a radio receiver: The higher the figure, the better the quality of the audio signal.

$$\text{SINAD} = 10\ \text{Log}_{10} \left\{\frac{S+N+D}{N+D}\right\} \text{dB}$$

where: S+N+D = combined Signal + Noise + Distortion power level; N+D = combined Noise + Distortion power level. NOTE: SINAD is a power ratio and not a voltage ratio in this calculation.

**Fig 24.64** shows a typical arrangement for measuring SINAD. An RF signal modulated with a 1kHz audio tone is fed to the radio receiver. Two audio measurements are then made, one with the 1kHz modulation and one with it notched out. Once the figures for the signal plus noise plus distortion and the noise plus distortion are obtained it is then possible to calculate the value of SINAD for the radio receiver.

The notch filter characteristics for SINAD measurements are of importance and ETSI (European Telecommunications Standards Institute) defines such a notch filter in ETR 027. Briefly this specifies that for the standard modulating frequency of 1kHz, the 1kHz should be attenuated by at least 40dB, at 2kHz the attenuation should not exceed 0.6dB and the filter characteristic shall be flat within 0.6dB over the ranges 20Hz to 500Hz and 2kHz to 4kHz. Without modulation, the filter shall not cause more than 1dB attenuation of the total noise power of the audio frequency output of the receiver under test.

Whilst measurements of SINAD can be made using individual items of test equipment, there are a number of commercial SINAD meters. These meters connect directly to the radio receivers for the measurements. There is also a piece of software called SINAD which uses the PC soundcard.

Sensitivity performance of a radio receiver can be assessed by determining the RF antenna input level to achieve a SINAD figure of 12dB (a typical figure that is used commercially). This equates to a distortion factor of 25% with a modulating tone of 1kHz. (Note: the typical ETSI - European Telecommunications Standards Institute - specification states that a deviation level of

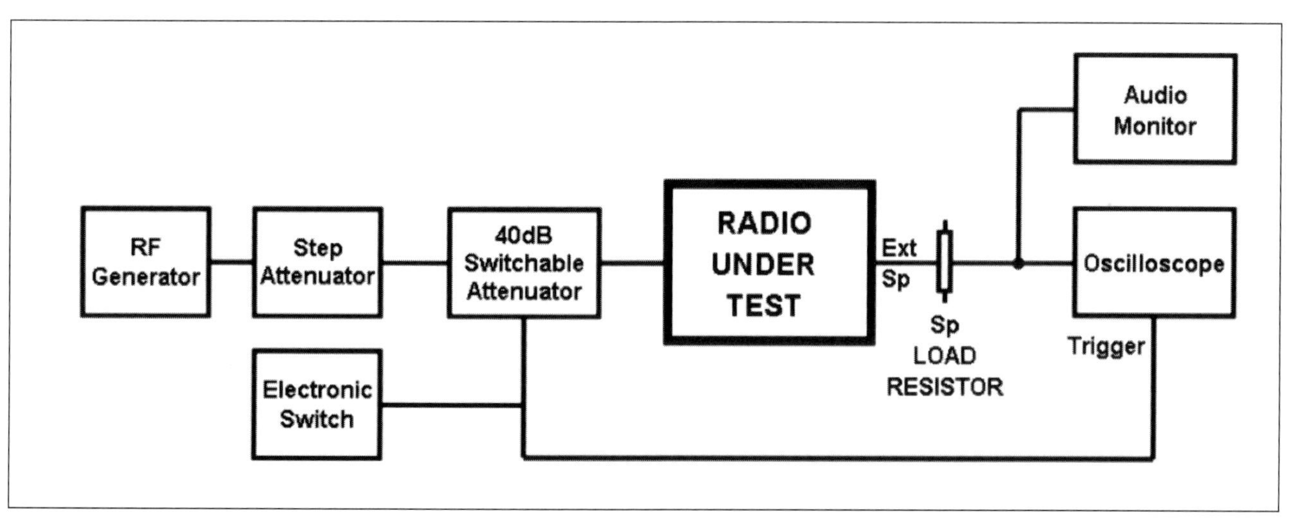


**Fig 24.66: AGC measurements**

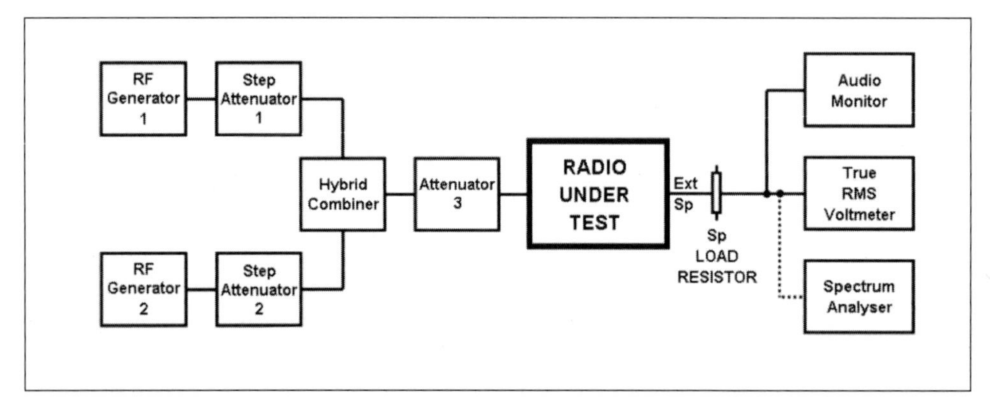


Fig 24.67: Intermodulation and blocking receiver measurements

12.5% of the channel spacing should be used, with AM the modulation depth needs to be specified).

A typical specification for a narrow band FM receiver might state that a receiver has a sensitivity of 0.25µV for a 12dB SINAD. The lower the RF input voltage needed to achieve the given level of SINAD the better the receiver performance. A 12dB SINAD figure is considered the maximum acceptable level of noise that will not swamp intelligible speech.

### Noise Figure

As an alternative to measuring (s + n) : n or SINAD, the noise figure of a receiver may be measured using a suitable calibrated noise source. This technique gives more accurate results with low-noise VHF or UHF receivers and has the advantage that noise figure is independent of receiver bandwidth. Noise figure, noise floor, and bandwidth may be related for an SSB receiver as follows:

noise floor (dBm) = noise figure (dB) + 10 $\log_{10}$ (bandwidth) - 174

where the bandwidth is in Hertz

**Fig 24.65** shows the test configuration for making noise figure measurements. A detailed description is beyond the scope of this book but, essentially, two readings are taken, one of the receiver output with the noise source off and a second with the calibrated noise source switched on. It is important that the impedance presented to the receiver antenna connector does not change between on and off states. If the noise source is variable, it is set to give a 3dB increase in noise level and the receiver noise figure is simply read from the calibrated scale on the instrument. With fixed noise sources, the noise figure of the receiver is calculated from the on and off readings.

→ See 'The Measurement of Noise', Dave Roberts, G8KBB, *RadCom* January 2007

### AGC response

**Fig 24.66** shows the test configuration for making AGC measurements. The drive source from the generator includes a home-constructed 40dB attenuator which can be switched quickly and cleanly in or out of circuit by an electronic trigger signal. This signal is also used to trigger an oscilloscope scan to view the audio output from the receiver. The attack and decay characteristics for a 40dB change in level over a range of signal inputs can be observed together with the threshold level at which the AGC starts to operate

### Intermodulation and blocking

The test arrangement for making measurements on intermodulation and blocking is shown in **Fig 24.67**. This requires the use of two signal generators.

To make intermodulation measurements, it is most important to ensure that intermodulation products are not produced within the generators by one generator coupling into the other. A hybrid coupler and not just a resistive combiner is necessary as this will give some 30dB extra isolation. This extra isolation will only be achieved if the coupler is properly terminated at the output port and this requires that attenuator 3 is not reduced too low in value. Typically use a 10dB attenuator.

Even with a suitable coupler, it is quite difficult to keep test equipment intermodulation within limits and get accurate results when receiver dynamic ranges exceed 100dB. Some generators are better than others and a careful balance between the generator output is often necessary to achieve best results.

To measure the third-order intermodulation, the levels of the two generators are increased equally by using the step attenuators until the amplitude of the intermodulation product generated in the receiver gives a measured s + n : n ratio of 10dB.

**NOTE**: if the AGC is operative at this level, a spectrum analyser or notch filter needs to be used as for sensitivity measurements under this condition.

The difference between the amplitudes of either input signal as measured at the antenna input and the on-tune level or sensitivity figure for similar s + n : n is termed the intermodulation ratio. It is more convenient to quote intermodulation performance in terms of Intercept Point (IP),

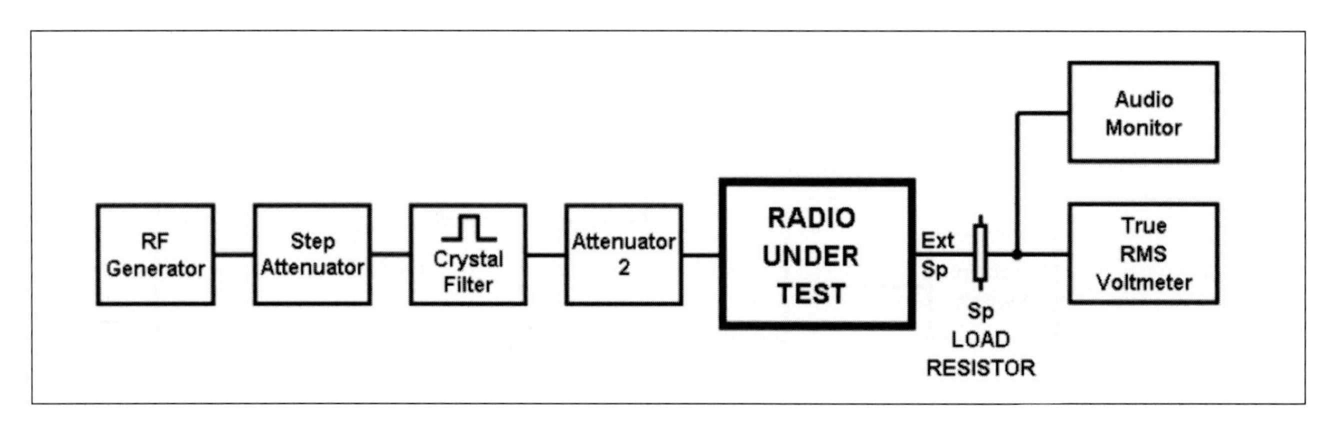


Fig 24.68: Reciprocal mixing measurements

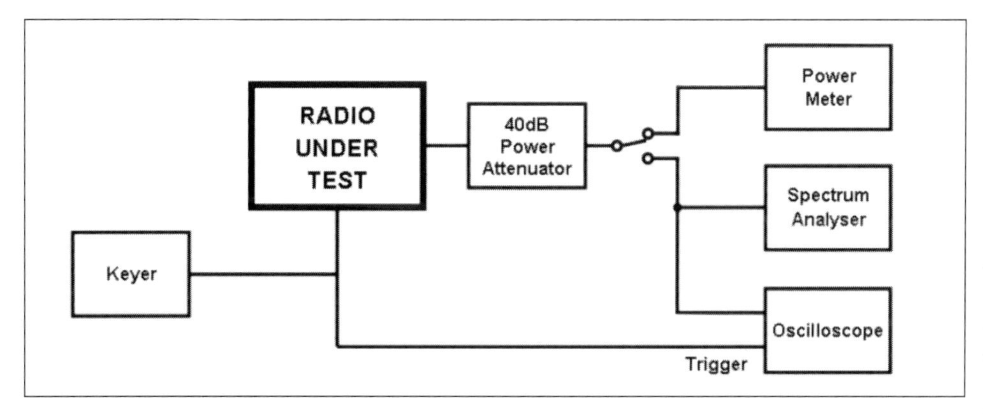


Fig 24.69: CW transmitter measurements

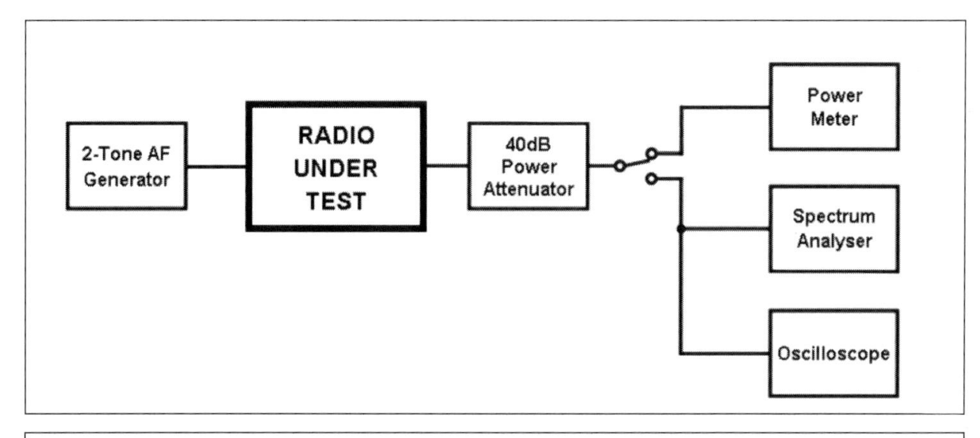


Fig 24.70: Two-tone SSB transmitter measurement

as this is independent of measurement signal-to-noise ratio and bandwidth.

$$\text{3rd-order IP (dBm)} = \frac{(3S - I)}{2}$$

where: S is the amplitude in dBm of each input signal and I is the amplitude in dBm of intermodulation product generated when related to the receiver input.

Having measured the third order Intercept (IP3) and the noise floor (NF) of the receiver from the sensitivity measurement, the two-tone Spurious Free Dynamic Range or Intermodulation-Limited Dynamic Range is calculated from the expression:

Dynamic Range (dB) = 0.667(IP3 - NF)
where IP3 and NF are expressed in dBm

For every 1dB increase in generator levels, the third-order intermodulation products in the receiver increase by 3dB. If this does not happen, intermodulation may be occurring simultaneously in more than one stage in the receiver, front-end AGC may be operative, or intermodulation products are being generated within the measuring setup.

The above expressions for IP3 and dynamic range assume that the 3:1 ratio holds and cannot be accurately applied if this is not the case.

IP3 measurements are made over a range of frequency spacings, with and without the front-end preamplifiers. Second-order measurements are also made to assess front-end filter effectiveness. In-band linearity is assessed by setting the generators 200Hz apart, centred in the receiver passband, and observing the audio output on a spectrum analyser. The resulting intermodulation products observed are a good indication of the linearity of the total signal path right through to the loudspeaker. It is quite noticeable that a really clean-sounding receiver generally shows intermodulation products some 40 to 50dB down on either of the two tones and with a less-clean-sounding receiver these levels are only 20 to 30dB down. The results usually hold over a wide range of signal levels and are often improved by manually reducing the RF gain control.

Sometimes slow AGC gives better results than fast. Front-end blocking is caused by gain compression in the receiver front-end stages ahead of the main IF filters. Generator 1 is set on-tune at a defined S-meter level (typically S9). Generator 2 is offset from the on-tune frequency and the level increased until the S-meter reading drops by 1dB. The level of generator 2 related to the receiver input is taken as the blocking level. Measurements at different offset frequencies are usually made. If reciprocal mixing is poor then it may not be possible to measure blocking by this method, but then it is probably irrelevant as well. AGC is not normally applied to the front-end of modern receivers but, if it is, the measured blocking level will be dependent on the on-tune signal level. Measuring at lower on-tune levels may not be possible due to reciprocal mixing.

### Receiver reciprocal mixing

**Fig 24.68** shows the test arrangement for making reciprocal mixing measurements. It is most important that the sideband noise spectrum of the RF generator is considerably lower than that of the receiver local oscillator. In order to achieve this, measurements are made at a single frequency by inserting a narrow bandwidth crystal filter together with suitable matching components and attenuators between the generator and the receiver. For example, a 21.4MHz filter could be used and then all measurements relate to the 21MHz band.

The generator should be tuned to the passband of the filter and as close to the roll-off as possible on the measurement side. The total loss of the filter, matching components and attenuators are noted for calibration. The receiver is tuned away from the generator frequency, noting the generator level required to give a 3dB increase in noise output from the receiver. Hence the noise due to reciprocal mixing is then equal to the noise floor of the receiver.

The phase-noise-limited dynamic range is the difference between the generator level as seen at the receiver input and the receiver noise floor. Measurements are made in SSB bandwidths for compatibility with the other measurements, although CW bandwidths may really be better for close-in measurements. The phase noise of the receiver synthesiser in dBc/Hz can be calculated at the specified offset with the equation:

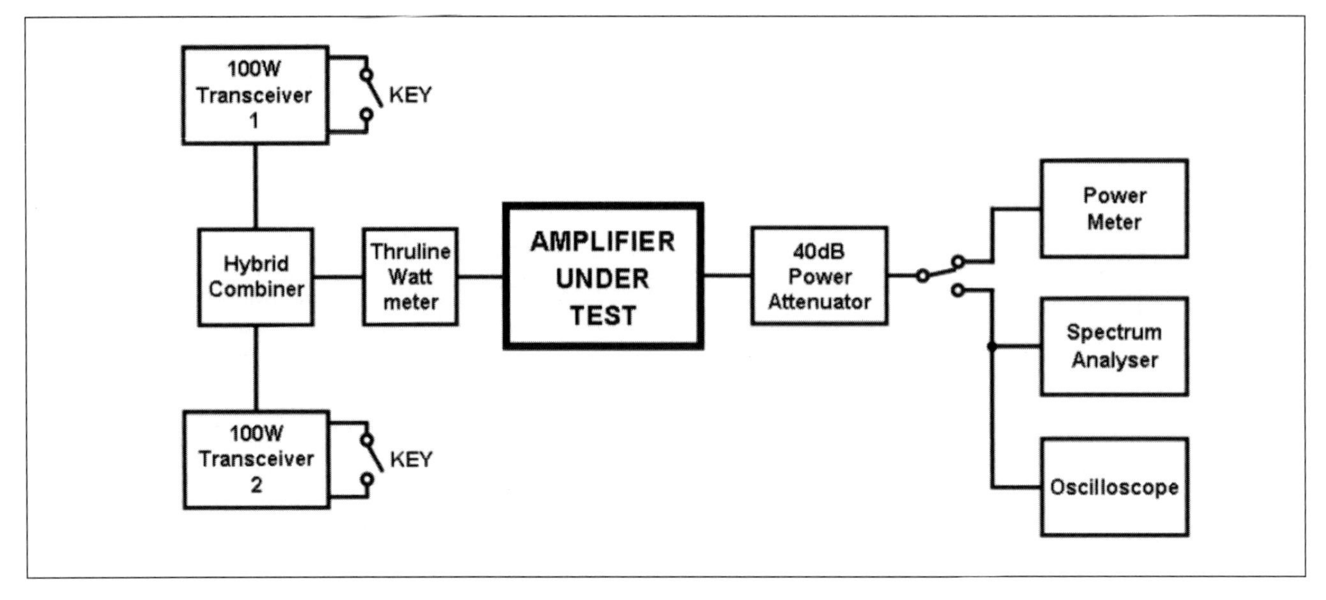


Fig 24.71: Two-tone linear amplifier transmitter measurements

Oscillator noise (dBc/Hz) = - (PNDR + 10 $\log_{10}$ B)

where: PNDR is the phase noise limited dynamic range in dB; and B is the receiver noise bandwidth in Hz (typically 2500 for SSB).

## Transmitter Measurements

With transmissions, one is interested in spectral purity so as not to cause undue interference to other users of the spectrum and have what many call a 'clean' signal. This involves both the initial transmitter and any following amplifiers used.

The test arrangement for carrying out CW transmitter measurements is shown in **Fig 24.69**. The measurement of power, harmonic and spurious outputs is largely self-explanatory. It is important to keep input attenuation levels on the spectrum analyser as high as possible to avoid the possibility of inaccurate harmonic measurements due to harmonics being generated within the analyser itself.

A critical test is to insert an additional 10dB attenuation and make sure that the relative levels do not change, just a change in the noise floor. CW keying is checked at 40 words per minute, which gives a 31ms dot length on the oscilloscope. Rise time, fall time, delays or distortions and any first-character differences are all noted, checking at both full break-in and semi-break-in.

The test arrangement for carrying out SSB transmitter measurements is shown in **Fig 24.70**. First, the transmitter is driven to full rated power output using a single audio tone. The amplitude of the waveform displayed on the oscilloscope is noted. Then the transmitter is driven by two equal-level audio tones (700Hz and 1700Hz) to the same peak amplitude level on the oscilloscope. The PEP level is then the same as the power out-

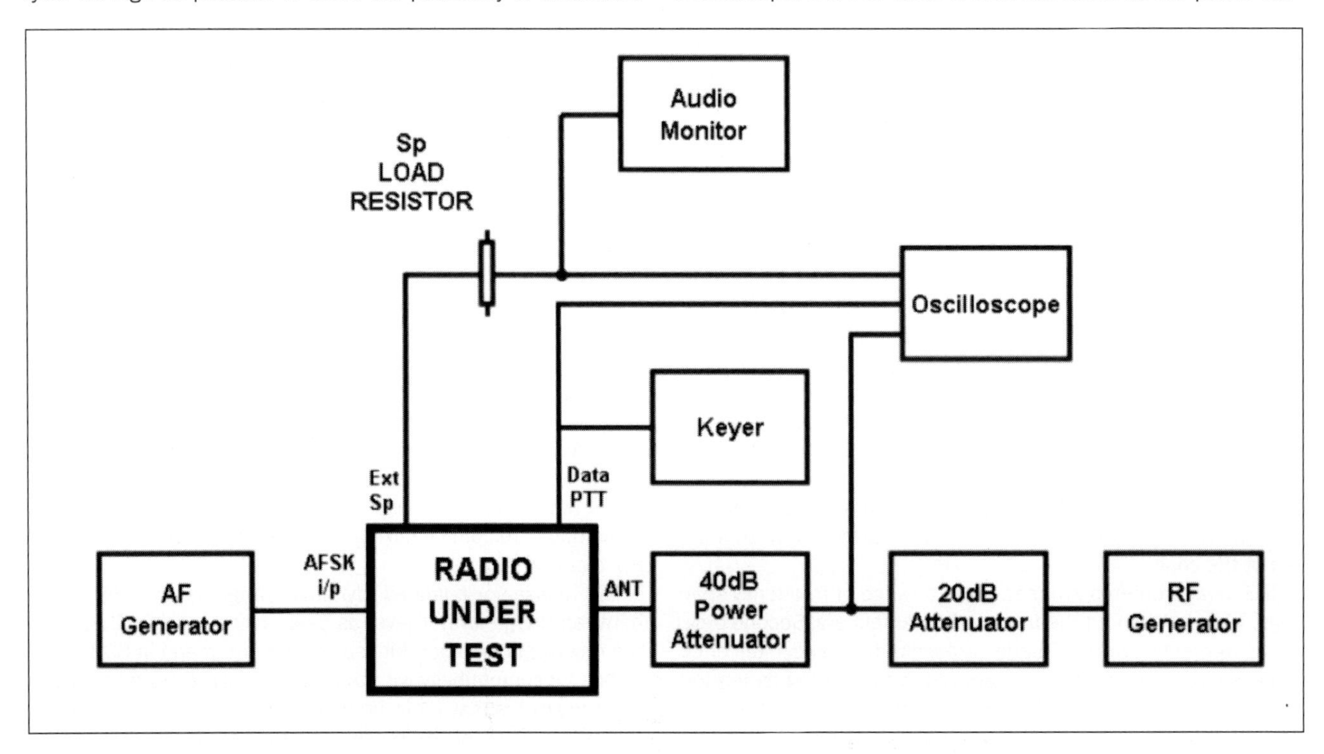


Fig 24.72: T/R switching speed measurement

put on a single tone. If an accurate and reliable PEP meter is available, the oscilloscope transfer method is not needed and the power level can be read directly from the meter scale. The levels of the 3rd and 5th order intermodulation products are measured using a spectrum analyser. It is common to quote levels with respect to PEP as this is universally adopted for all amateur radio products, reviews and specifications. With the transmitter driven by a single audio tone the levels of carrier and unwanted sideband are measured using the spectrum analyser and an estimate made of audio distortion.

### Linear Amplifiers

**Fig 24.71** shows the specific test arrangement used to measure the two-tone distortion performance of SSB linear amplifiers. Two 100W HF transceivers are coupled together using a power hybrid coupler. This is a homebrew unit using ferrite-cored transformers, such designs are described in many books on RF design. This couples two 50Ω sources to a 25Ω load with half the power dissipated in a 100Ω resistor. This resistor needs to be air-blown hard by a suitable fan. The 25Ω load comprises the linear amplifier under test shunted by an additional 50Ω load.

Hence only one quarter of the original power is available to drive the linear, but this is sufficient, giving 40 - 50W mean power or 80 - 100W PEP. The residual level of intermodulation products is around -50dB to -60dB, so that the distortion level of the drive source is significantly better than that of the amplifier to be measured. This test setup must be used with great care. The transceivers are used on CW and both must be switched to transmit before either key is pressed. Also both keys must be released before either transceiver is switched back to receive. This is most important otherwise leakage through the hybrid coupler from one transmitter could damage the receiver in the other transceiver. The measurement of power output and distortion follows the same practice as for SSB transmitter measurements.

### Transmit-Receive Switching Speed

A reasonably fast and a clean switchover between receive and transmit and back to receive is needed on some data modes. The test arrangement is shown in **Fig 24.72**. An audio generator at 1kHz drives the transmitter in AFSK mode via the AFSK or microphone input.

The receive signal generator is set to give a reasonable on-tune signal level and coupled into the antenna connector via 60dB of attenuation. This ensures that the amount of transmit signal entering the signal generator is insufficient to cause any damage. The data PTT line on the radio is keyed from a pulse generator and the resulting RF output, receiver audio and keying waveform observed on a multichannel oscilloscope.

The times for the receiver to be muted, the transmitter to reach full power, the transmitter to be muted and the receiver to regain full sensitivity are all read from the oscilloscope display. Any anomalous behaviour can also be observed. Receiver and transmitter mute times are usually very fast. Transmit and receive enable times less than 20ms are quite fast and fully acceptable for all modes.

***Philip Lawson, G4FCL*** *Licensed in the early 1970s, initially as G8ING and later as G4FCL. After university, Philip took up a professional career in electronic engineering and is now a Chartered Engineer with over 30 years' experience, leading RF, Microwave and Optical product development in telecoms, aerospace, and defence. He has published over 40 internal and external reports and his achievements include several world firsts.*

| Function | Brief Description | Filename |
|---|---|---|
| **DTMF, CTCSS, tones** | Produced by ComTekk. This is a trial version but nonetheless appears very useful, some functions are partly disabled. It allows DTMF, CTCSS, paging and other tones to be generated for testing radio equipment. The software allows auto-calibration and has an audio oscilloscope and spectrum analyser. | *ComTekkSetup106_EU.exe* |
| **DTMF, CCIR, ZVEI and CTCSS decoder** | This is a multi-tone decoder with recording log. Typical tonesets are DTMF, CCIR, ZVEI etc) plus CTCSS. It could be useful to those testing and repairing equipment. It is an older program and not so easy to find on the web but use filename as search word. | *Wintn22.zip* |
| **Bench top calculators** | 1. A very useful calculator with many engineering and radio functions and conversions. Free.<br>2. Gcalc (Linux) at https://wiki.gnome.org/action/show/Apps/Calculator. Free | *Benchtop-calc.zip* |
| **Intermodulation** | 1. A free download for calculating intermodulation products up to the fifth.order<br>2. Free program courtesy of TCS Consultants at www.tcstx.com/software/intermod/intermodulation.htm | *imodbeta.msi* |
| **Meter dial maker** | 1. This is a basic version and will allow meter scales to be designed and printed. Useful when wanting to make your own scales to fit an analogue meter.<br>2. *Galva* from www.radioamateur.org/download/index.html | *MeterInstall221.exe*<br>*Galva_185-3L.zip* |
| **Oscilloscope** | 1. There are various programs that perform this function along with other facilities such as audio spectrum analysis. See for instance www.dxzone.com/catalog/Software/Oscilloscope<br>2. *Xoscope* at http://xoscope.sourceforge.net. Linux. Free | various |
| **Audio signal generator, including two tones for SSB testing** | There are several programs that perform this function, see for instance www.dxzone.com/catalog/Software/Signal_Generator | various |
| **Polar plot** | This program will measure and display the radiation pattern of a beam antenna. It relies on an antenna receiving a signal and the audio output (assuming AGC not operative) plotted with rotation. | *PolarPlotSetup.exe* |
| **Real time audio spectrum analyser** | Although the title suggests it is an audio spectrum analyser, it also functions as a complex signal generator and oscilloscope. | *TrueRTA_se.exe* |
| **RF toolbox (1)** | A DOS based program with various calculation options including noise and gain calculations, R, L and C calculations (including attenuators), resonance and VSWR. | *rftbox.zip* |
| **RF toolbox (2)** | This program allows various calculations to be made as well as antenna design. Trial version by Black Cat Systems. | *RFToolBoxSetup.exe* |
| **SINAD measurement** | A program from ComTekk using the latest DSP technology to measure some key receiver performance parameters by analysing the spectral content of received audio signal. A trial version. | *CTSinadSetup120.exe* |
| **Spectrum analysers** | Various programs perform this function in the audio range, the programs normally offer additional functions See for instance www.dxzone.com/catalog/Software/Spectrum_analyzers | various |
| **VSWR calculator** | This is a calculator and contains calculators for Directivity Error, Mismatch Error and Ratio to dB. | *vna.exe* |
| **RF system analysis** | A program for modelling the noise, gain, intermodulation/spurious<br>performance, and compression point of a system of cascaded components where each component is characterized by known parameters (as may have been determined by measurements). The Lite version of the program is freeware. | *Scw5Lite.exe* |

**Table 24.5: Typical software test equipment**

# Construction and Workshop Practice 25

*Original chapter by Eamon Skelton, EI9GQ*

There are many reasons why radio enthusiasts might choose to build at least some of their own equipment. Even in a world where commercially made equipment is readily available, there are some advantages to rolling your own. The most obvious advantage is the potential to save a lot of money by building your own rig instead of buying off-the-shelf equipment.

It is possible to build useful radio equipment using components that were recovered from scrap equipment. Surplus components can often be bought at very reasonable prices from specialist suppliers or from Internet auction sites. A well stocked junk-box is a very valuable asset to the radio constructor.

Cost is not the only consideration for the home constructor, there are other advantages to building your own. The home constructor enjoys a great deal of flexibility when he/she builds a new item of equipment. In most cases, the designer, builder and the end-user will be one and the same person. This usually means that the operator will end up using equipment which is built to his specifications rather than generic equipment which may be less suitable for the task.

The home constructor can also decide where to concentrate the available resources. For example: he might decide to use a £10 transistor in the front-end of a homebrew receiver, even if it only gives a small advantage over a £1 transistor. This small advantage may well be worth £9 extra to someone who wants to have the best possible receive performance. This kind of flexibility is not usually available to the commercial manufacturer.

There are many situations where commercially made equipment is not readily available and home made equipment is the only option. Radio equipment for LF (low frequency) and for the higher microwave bands is usually home made.

Homebrew is not always the cheapest option, some specialised components can be difficult to find and expensive to buy. If you value your time, homebrew can be very expensive indeed. It is not unusual for complex projects to take dozens or even hundreds of hours to complete.

So why would anybody want to waste time and money by building amateur radio equipment when second-hand commercially made equipment is quite cheap and readily available? The most compelling reason for building your own equipment is the knowledge that you will gain from building it and the great sense of satisfaction that goes with operating a homebrew station.

Homebrew radio is not the exclusive preserve of a small number of technical gurus. Home construction is accessible to all radio enthusiasts, young and old, the newcomer or the experienced constructor, the able-bodied and the disabled. Anybody with a soldering iron and a handful of components can build a working circuit. Basic components like transistors, diodes, resistors and capacitors are very cheap and easy to find.

The cost of failure is very low. If a circuit doesn't work first time, you can easily modify it or even completely re-design it and try again. You will probably learn more from the projects that fail to work at the first attempt than you will learn from the projects that work perfectly the first time you apply power to them.

Not every homebrew project involves electronic components and solder smoke. A lot of radio equipment and test instruments are software based. If you use a software defined radio (see the chapter dedicated to this subject), you could completely redesign and rebuild it without ever modifying the hardware or even taking the lid off the box!

Your project might be as simple as measuring and cutting a wire dipole or as complex as building a multi-band transceiver. It is a good idea to start with a few simple projects before you progress to more complex projects. It would be possible for the new constructor to build a very complicated circuit as a first project, but it can be very discouraging for the newcomer when a circuit just refuses to work. The experience gained from building and debugging a few simple projects will be very valuable when you start to build more elaborate circuits.

## THE WORKSHOP

Only a small minority will be fortunate enough to have a separate radio shack for operating and a dedicated workshop for construction projects. Most of us have to make the best of limited resources. The radio shack usually serves as the workshop and increasingly as a computer workstation. The shack may have to be shared with other family members or be used as a centre for other hobbies and activities.

My first radio shack was a small table in the corner of the bedroom. I still have painful memories of standing on upturned IC pins in my bare feet! Even now, despite the fact that I have a dedicated outdoor radio shack, some of my homebrew activities take place on the kitchen table.

A garden shed or a bench at the back of the garage can be pressed into service as a radio shack. An attic, basement or spare bedroom can make a good radio shack. Caravans, campers or even boats (subject to local planning regulations) can be used as a radio shack. If space is very limited, a small cupboard, a wardrobe or a roll-top desk can be used as the radio shack.

## TOOLS AND INSTRUMENTATION

Given unlimited resources, the home constructor can build just about anything that could be built in a factory or a professional machine shop. A few amateurs will have the facilities and the necessary skills for both electronic construction and metalworking. Most of us have to work in less ideal circumstances. A typical

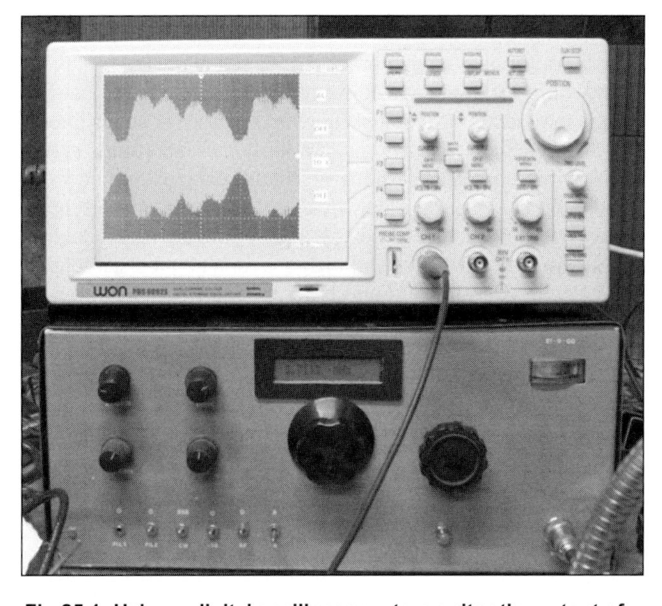


**Fig 25.1: Using a digital oscilloscope to monitor the output of a home made SSB transmitter**

radio shack will contain a basic electronics tool-kit and a few very basic metalworking tools.

The simplest practical tool-kit will consist of a few basic hand tools like a soldering iron, screwdrivers, pliers, side-cutters etc.

A simple multi-meter is also an essential part of the home constructors tool-kit. I have owned many different multi-meters over the years and have found that the ability of a meter to survive a fall onto a hard floor is inversely proportional to the price of the instrument. Even if you have an expensive multi-meter, it is a good idea to keep a cheap one in the shack for use in risky environments.

The choice between analogue and digital meters is largely a matter of personal preference. Digital meters will offer greater accuracy and resolution; analogue meters tend to be better at tracking slowly changing values of voltage, current or resistance. Very cheap digital meters tend to be almost as accurate as more expensive models, so there is little to be gained by paying extra for a high quality instrument. My favourite digital meter cost less than £4 from a local supermarket. When buying an analogue multi-meter, it is worth spending a little more on a meter with a sensitivity of at least 20kΩ per volt. There is much more on meters in the chapter on Measurements and Test Gear.

Most general purpose soldering irons can be used for electronics. A temperature controlled soldering station is desirable, but not strictly necessary for most construction work. A relatively high powered iron is useful for heavier work and for desoldering. A lighter iron in the 15-18 watt range is more suitable for delicate electronics work. A 25 watt iron is a good compromise between the two.

More advanced test equipment like oscilloscopes, signal generators, and spectrum analysers are very useful, but they are not absolutely essential (see the Measurements and Test Equipment chapter).

The basic tool-kit described above is enough to get you started. Once you have gained some construction experience, you can buy or even build any other instruments that are required. Surplus and second-hand test equipment from industrial, education and military sources can sometimes be found at radio rallies and Internet auction sites like eBay. Some specialist suppliers offer surplus and used test equipment. Buying from a specialist will be a lot more expensive than taking pot luck at a rally or auction, but the extra expense can often be justified, especially if the equipment comes with a guarantee. A specialist supplier will usually be able to offer a repair and calibration service for the test equipment that they sell.

The more experienced radio constructor's shack will usually have a few electronic test instruments. An oscilloscope is one of the most useful instruments. Professional laboratory grade oscilloscopes are quite expensive, though surplus and second-hand scopes are quite widely available. Digital storage scopes were once considered to be the state-of-the-art in test equipment. The ready availability of high speed, high precision digital to analogue converter ICs and other recent advances in digital electronics have greatly reduced the cost of digital scopes. **Fig 25.1** show a digital oscilloscope used to monitor the output of a home made SSB transmitter.

Digital oscilloscopes which use a standard PC (personal computer) to process and display sampled signals are now available ready built and in kit form.

Another instrument which is of great value to the amateur constructor is the digital frequency counter. The cost of new frequency counters has been falling steadily for many years. The availability of modern micro-controllers like the PIC micro have made the job of building a home made frequency counter [1] a lot easier (**Fig 25.2**).

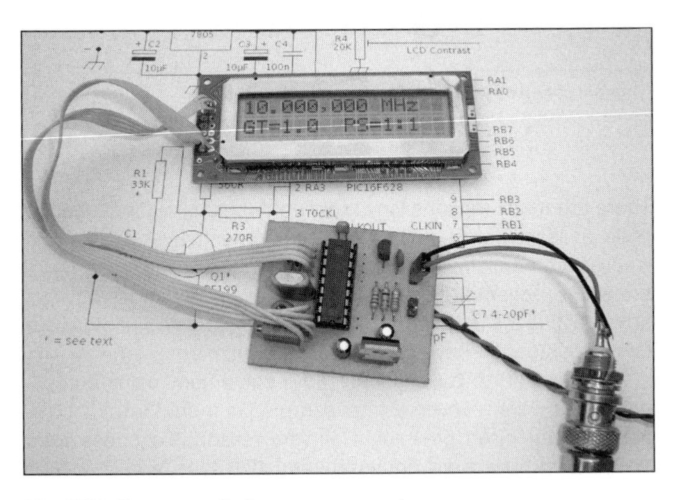


**Fig 25.2: Home-made frequency counter**

A good signal generator is a useful tool for the home constructor. Again, lab grade signal generators are very expensive, but a simple RF signal generator is a fairly easy construction project. Sophisticated test instruments like spectrum analysers and network analysers are extremely expensive and rarely found in the amateur's shack. This situation is changing gradually as faster, better and cheaper digital instruments are developed.

A standard PC fitted with a sound card and some clever software makes a very useful audio signal generator, oscilloscope and spectrum analyser. Unfortunately, PC sound cards have an upper frequency limit of about 22kHz. A relatively simple RF to AF converter can be used to mix an RF signal down to audio frequencies so that is can be sampled by the PC. Most, if not all, PC sound cards use input coupling capacitors. This means that a sound card based oscilloscope will not be able to display DC signals or very low frequencies below a few Hz. Oscilloscope and spectrum analysis software is available for most types of home computer and operating systems. Much of this software is open-source and/or available free of charge.

## METALWORKING

Metalworking or 'chassis bashing' is a very important part of amateur radio construction. Commercially made metal and plastic enclosures are available at reasonable prices from most component suppliers. You may find that such enclosures are not exactly suited to your needs and you will need to modify them. For some projects, a home made enclosure will be the only option.

Surplus or scrap equipment is another good source of high quality metal enclosures, but you may find there is just as much work involved in modifying an ex-equipment enclosure to suit your project as would be required to build a new one from scratch. 'Murphy's Law' says that all of the holes and openings on the front and rear panels of a second-hand enclosure will be the wrong size, wrong shape and in all the wrong places.

The successful home-brewer will need to master the art of mechanical engineering, metalworking and fabrication. Most amateur constructors will not have access to machine shop tools like lathes and milling machines. However, a basic tool-kit is all that is needed for most construction projects.

You will need a selection of basic hand tools. The tools required for metalworking are a little different to those in our electronics tool-kit mentioned earlier in this chapter. A minimal tool-kit would include a good hacksaw, a drill, an engineer's rule (steel), a bench vice and a selection of basic hand tools like screwdrivers, files, spanners etc.

A more complete metalworking kit might include a few more specialised items like sheet metal shears, a metal nibbling tool, a drill stand and an assortment of reaming tools, hole cutters and chassis punches.

An old fashioned hand drill can be useful for delicate jobs where an electric drill would be too fast and difficult to control. A small 12V DC drill is useful for drilling small holes in PCBs. Dremel and similar types of rotary multi-purpose tools can be very useful for drilling, grinding and polishing. A bending brake is useful for making nice sharp and straight bends in sheet metal. You can easily make your own bending brake from a couple of lengths of angle steel and a few bits of metal hardware.

Obviously there will be a limit to how much you can spend on your hobby, but you should try to buy reasonably good quality tools. You can cut corners (if you will pardon the pun) in some areas. In most respects, a cheap hammer is just as good as a more expensive one. On the other hand, you certainly shouldn't compromise on the quality of screwdrivers, hacksaws, drill bits or side-cutters.

## SAFETY

Safety should always be the first and most important consideration for the home constructor. Some of the dangers facing the home constructor are quite obvious. Every radio amateur will know that you shouldn't work on live mains-powered equipment because of the risk of electric shock. The dangers of working on low voltage equipment might not be so obvious. Never wear jewellery when working on electrical equipment. Even a few volts from a high current source will melt a metal ring or bracelet.

Always wear eye protection when working with power tools. You should always wear a dust mask when cutting or grinding hazardous materials like fibreglass PCB laminate.

Only a fool would climb a tower without a proper safety harness, but much lower platforms such as chairs or tables can be dangerous if they collapse while you are standing on them.

The risk of injury to yourself and others can be greatly reduced if you apply a little bit of common sense. Make sure that live equipment is properly insulated and housed in a safe enclosure. Even if you feel that you know what you are doing and feel comfortable about working on live equipment, there is always the danger that someone else will be electrocuted by it. A visitor to the shack, a young child or a stray animal may not be aware of the danger.

You should also take great care when using and storing chemicals. Trays and tanks containing etching or cleaning chemicals can be very hazardous to children and small animals. Don't leave exposed etching or developing trays unattended. Don't place dangerous chemicals near the edge of the workbench where they can be reached by children or accidentally knocked over. Always wear rubber gloves and eye protection when working with chemicals.

Take particular care to keep PCB developer and other caustic chemicals away from your eyes. Remember that paint, thinner, paint remover and other solvents are usually highly flammable and could pose a serious fire hazard. It is worth noting that many of the things that we keep in our radio shacks would not be allowed in an industrial or commercial environment because of health and safety regulations and conditions imposed by insurers.

Fumes from soldering and chemicals can pose a health hazard. As hand soldering is usually done very close to the operator's face, it is very important to ensure that your workbench is well ventilated. Batteries should not be charged in a confined space because of the risk of fire or explosion due to the build up of hydrogen gas. Hot soldering irons should be placed in a proper stand to reduce the risk of burns and damage to equipment.

Amateur radio is just a hobby. There is no need to take unnecessary risks to save time or to reduce cost. If you need professional help or expert advice, don't be afraid to ask for it. Experienced radio builders are usually willing to help out if you have a query. If you are a licensed amateur, you can just ask on your local homebrew net frequency. Your local radio club may be able to offer help and advice. There are several homebrew mailing lists, web forums and Usenet groups. If you are concerned about the safety of a particular tool or product, ask the manufacturer for advice or a safety data-sheet.

More safety advice can be found at the end of this chapter.

## MATERIALS

The first requirement is an elementary knowledge of the materials useful and normally available to constructors.

### Metals

#### Aluminium and aluminium alloys

*Typical uses: cabinets, boxes, panels, masts, beam antennas, heatsinks.*

These are good electrical conductors and are the lightest in weight of the normal metals available. They are non-magnetic, of medium to high cost, and available in sheet, rod, tube and other forms. The annealed quality usually bends and machines easily but soldering requires special fluxes and solders, and often the soldered joint is not good electrically. With the unknown quality of aluminium that most amateurs encounter, soldering is best left alone or experimented with before committing the project to this method of fastening.

Long-term corrosion is also a problem with soldered joints in aluminium unless specialist methods are applied to prevent this. Adhesive bonding is very good but such joints are not usually reliable electrically. Aluminium is non-corroding in normal use, but direct contact with brass or copper should be avoided as these react and encourage corrosion with the resultant troubles of poor or non-existent electrical contact (see the section on corrosion.) Nickel-plated or stainless-steel screws, nuts and washers, rather than brass or steel, should be used to reduce the chances of corrosion around this type of fastening.

A metal that looks similar to aluminium, and which may come into the hands of the unwitting, is magnesium alloy. Filings and chippings from this metal are highly flammable. They can burst into a glaring flame with the heat generated by filing and drilling, and trying to put out such a fire with water only makes matters worse. (Factories where this alloy is machined use a chemical fire extinguisher, one type of which goes under the title of DX powder!) It is difficult to tell the difference by just looking, but size-for-size magnesium alloy sheet weighs less than aluminium and usually on bending there is a 'granulating-cum-scrunching' feel.

For reliable bending of sheet aluminium up to 1.5mm thick, it should be possible to bend a sample strip back on itself and hammer the fold flat without breakage. Annealing aluminium is possible, but under specialised conditions. Before any heating is attempted it is essential to degrease the part thoroughly.

A workshop 'dodge' which sometimes works, and is certainly worth a try if bending or forming proves difficult, is to heat the aluminium slowly (the slower the better) to such a temperature that a small pine stick, when rubbed on the heated surface, just chars, but does not ignite (equivalent to approximately 300°C) Care needs to be taken to avoid overheating, for this will destroy the properties of the aluminium. This method of localised annealing has proved effective when bending tube or rod for the folded dipole elements of beams. In this case the area to be annealed is small and the 'dodge' can be applied easily. Another method, suitable for larger areas, is to rub soap on the surface and warm the aluminium until the soap turns black. However, it is essential that it is the aluminium which is warmed by the blow torch or hot air gun and not the soap. This is ensured by applying the heat to the non-soaped face. Any bending or forming should be carried out as soon as the annealed piece has cooled, because most grades of aluminium alloys start to re-harden and may actually improve in strength after such heating. Do not attempt to speed up the cooling process by quenching in water. Allow the part to air cool. Note that during the heating of the component there must be no visible change in the surface appearance of the aluminium, otherwise the aluminium will lose its properties and disintegrate. This fact can be verified, if you are so minded, by attempting to bring a small piece of aluminium sheet to red heat.

Heat treatment of aluminium and its alloys is a specialised process used to improve a particular characteristic of the metal, and is carried out in controlled conditions not usually available in the home workshop. Technically the above techniques are more of a part softening process rather than true annealing.

For interest, the usual method of commercially annealing most aluminium alloys is that the part is 'soaked' for about two hours at 420°C, air cooled, and then soaked for a further two to four hours at 225°C. All this is performed in a furnace or salt bath which has its temperature closely controlled. Most, if not all, of the aluminium used in extruded tube or sections has undergone this type of treatment immediately prior to extrusion, ie in the billet form.

### Brass

*Typical uses: Morse keys, terminal posts, weatherproof boxes, extension spindles, waveguides and other microwave components.*

This is an expensive but good non-magnetic electrical conductor, and is available in sheet, rod, tube and other forms. It can be soldered easily but adhesive bonding can prove difficult. For work involving bending or forming, the most suitable grade is ductile brass. For panels and non-formed parts, the half-hard and engraving brasses are adequate. It can be annealed like copper, though care is necessary as brass is nearing its melting point when heated to bright red. Brass is non-corroding in normal use but reacts with aluminium and zinc.

### Copper

*Typical uses: heatsinks, coils, antenna wires, PCBs, tuned lines, waveguides, earthing stakes and straps.*

This is a very good, though expensive, electrical conductor. It is non-magnetic and is available in sheet, rod, tube and other forms. Before work of a forming or bending nature is attempted, this metal should be annealed by heating as uniformly as possible to a bright red heat and air or water cooled. If considerable bending is required, this annealing should be repeated as soon as the metal begins to resist the bending action. In the annealed state, copper bends very easily. Soft or hard soldering present no problems but adhesive bonding can be troublesome. It is non-corroding in normal use but does react with aluminium and zinc, especially in an out-of-doors environment.

### Steels

*Typical uses: Masts and tabernacles, screw fixings, guy wire stakes and parts subject to high wear or heavy loads.*

These are electrical conductors, which are magnetic except for the expensive stainless types. They are cheap and available in numerous forms and qualities. The common grades are called mild steel or GCQ (good commercial quality) steel. The black quality steels are usually cheaper and used for such things as the stakes for mast guys, or similar 'rough' work. Silver steel is a special grade suitable for making tools, pivot pins and other items which require the parts to be tough or hardened and tempered. Most steel sheet forms commonly available will bend, solder and machine easily, and can be annealed by heating to bright red and allowing to cool slowly in air. Do not quench in water or in any way cool rapidly, for this may cause some steels to harden. Corrosion is a problem unless plated or well painted. For outdoor use the commercial process of galvanising is perhaps the best form of protection. The next best thing is a few coats of paint.

### Tin plate

*Typical uses: Boxes, screening cans and plates, light-duty brackets, retaining clips, spacers.*

This is a good electrical conductor. It is very thin steel coated with an even thinner layer of tin on each side, and is magnetic, cheap and available in sheet form up to 0.5mm thick. It can be easily soldered, bonded, bent and machined, and is non-corroding in normal use. Cut edges should be re-tinned with solder if the full benefits of the non-corroding properties are required. It can be annealed but this will destroy the tin coating. This is a 'friendly' metal to use and is normally readily available in the form of biscuit tins and similar containers, hence the old timers' expression 'an OXO tin special'.

### General comments on metals

All of the above metals work-harden and will break if repeatedly flexed at the same point. Annealing removes the effects of work-hardening, providing it is carried out before the part is over-stressed. There are professional standards which classify the above metals and each is given a specific identifier code. Fortunately, amateur constructors do not normally need to enter this maze of professional standards, and metal suppliers usually understand that to most of us steel is steel and brass is brass!

## Plastics

Plastics are electrical and thermal insulators, and are not suitable as RF screens unless the plastic is specially metal coated or impregnated. The insulation can normally be considered as excellent for most amateur purposes.

The following is a brief description of the more commonly available and useful plastics, and is a very small selection of the many plastics in use today.

### Laminates

*Typical uses: PCBs, matrix boards, coil formers, insulating spacers.*

Various base fabrics such as paper, cotton, glass, asbestos etc, are bonded together by selected resins, and usually compacted and cured under pressure. The combination of the resin and the base fabrics produce laminates which may be used for many applications. Most are available in sheet, rod and tube form. The cured laminate cannot be formed easily. Normal machining is possible, particularly if attention is paid to the lay of the base material. Drilled or tapped holes should be arranged so that they go through at right-angles to, and not in the same plane as, the laminations.

Where components made from this material are exposed to the elements and expected to insulate, the glass-based laminates should be preferred. This also applies where the dielectric properties are important (VHF converters, RF amplifiers etc). The normal heat generated by valves and similar electronic components will not harm these laminates but the glass or asbestos fabric-based laminates should be used for higher temperatures (100-140°C). Costs range from expensive for the paper and cotton bases to very expensive for the nylon and glass bases. Glassfibre repair kits for cars are a useful laminate for weatherproofing antenna loading coils and making special covers or insulators. The filler putty supplied with these kits may contain metal.

### ABS (acrylonitrile butadiene styrene)

*Typical uses: antenna insulators, coil formers, handles, equipment enclosures.*

This is expensive and is available in natural white coloured sheet. It machines easily and can be formed by heating, similar to Perspex®. Bonding requires proprietary adhesives. A tough plastic and a good insulator.

### Acetal copolymer

*Typical uses: bearing bushes for rotating beams, Morse key paddles or electrical parts such as insulators or feeder spacers.*

This is a medium to low cost plastic and is available in white or black rod and sheet. It machines very easily without specialist tools and is a useful plastic to have available in the workshop. It cannot be formed easily. Bonding requires proprietary adhesives and such joints are usually the weakest part of any assembly.

### Acrylics (Perspex®)

*Typical uses: decorative and protective panels, dials, Morse key paddles, insulated fabricated boxes and covers.*

A medium-cost plastic which is available in clear or coloured sheet, rod or tube. It is non-flexible and can shatter or crack under shock or excessive loads, although it is often used for see-through machine guards. The clear sheet is ideal for covering and protecting the front panels of equipment. Perspex® may be formed by heating but not with a flame as this plastic is combustible and gives off unpleasant fumes if burnt. If placed in a pan of water and simmered (or in an oven) at around 95 °C, the plastic softens and can be formed or bent very easily. Forming should be stopped and the work re-heated the moment hardening or resistance to bending is felt, otherwise breakage will occur.

Bonding requires proprietary adhesives. A properly made bond is structurally sound and can be transparent. Normal drilling, sawing and filing are straightforward, providing the work is adequately supported. Most sheet forms are supplied covered with protective paper and it should be worked with this left in place. It is not very heat resistant and should not be placed in direct contact with any heat source such as lamps, heatsinks and valves. Where transparency is not required a better plastic to use would be Acetal or, for particularly tough applications, ABS.

### Nylon 66

This is a cheap to medium-cost plastic, and is available in sheet, rod or tube. It is usually supplied in its natural creamy white colour. It can be machined but it does tend to spring away from any cutting edge, making tapping or threading difficult. It is not easily formed. Proprietary adhesives are available which claim to bond nylon successfully. There are other types of nylon but most of these are expensive and intended for special applications, such as bearings, gears etc. For most amateur purposes the Acetal copolymer, mentioned previously, is generally an easier material to use. Note that nylon is susceptible to moisture absorption and ultra-violet rays.

### Polyethylene (ultra high molecular weight)

A medium- to high-cost plastic, normally available in sheet and rod. It is usually supplied in its natural white colour and it can be machined easily but not formed. Proprietary adhesives are available for bonding. This is an ideal plastic for outdoor components such as insulators, feeder separators etc as it is virtually rot-proof.

### Polycarbonate

This is an expensive material, normally available in transparent sheet form. It can usually be bought at builders' merchants where it is sold as vandal-proof glazing. A very tough plastic, virtually unbreakable (bullet proof!) and, though it can be machined, it will wear out normal tools very quickly. It is ideal for making an insulated base for a vertical mast/antenna and in other areas where impact, high loads and temperature changes would rule out other less-durable plastics. Polycarbonate is particularly sensitive to ammonia and clear sheet will turn white, become brittle and fail if this substance is used for cleaning it. A common trade name for polycarbonate is Lexan.

### Polypropylene

This is an expensive plastic, normally available in opaque-coloured rod form. It can be machined and formed but not bonded. It can be sensitive to prolonged frictional contact with metals, particularly copper, and disintegration can occur in these circumstances. Because of its strength and resistance to atmospheric attack, it is usually used by amateurs in its rope form for halyards, mast guys etc.

The twisted strands are normally melt-welded together to prevent fraying, using a soldering iron. A naked flame should not be used as a substitute for the soldering iron as this plastic burns and melts, and burning droplets can go anywhere, even on the hands!

### Polystyrene

This is a relative cheap plastic and is available in a variety of types, shapes and colours from black to transparent. It can be formed, machined, painted and bonded very easily - a model maker's delight! Usually used by radio amateurs for coil formers, insulated extender spindles and in other areas requiring insulation. It is a particularly tough flexible material, although some transparent types can be brittle. Some forms are also heat sensitive. A paper/card laminate of this plastic is available and this is very useful for making mock-ups of cabinets, boxes etc. Model-making suppliers usually stock extruded polystyrene sections, some shapes of which can be utilised in making bezels and other cabinet embellishments.

### PTFE

This is an expensive material, noted for its excellent dielectric performance and low frictional properties. It is available in sheet, tube and rod, and is normally supplied in its natural off-white colour. PTFE is easy to machine with normal wood or metal turning tools. However it is extremely difficult to grip the PTFE in a normal lathe chuck or a vice as it has a consistency like putty and deforms easily under pressure. Round PTFE rod is normally turned by gripping in a collet chuck. Bondable PTFE in sheet form is supplied with one side burned by liquid sodium that allows normal contact adhesives to be used. This sheet has a white side and a dark brown side. PTFE is usually used for low-friction bearings, insulators (up to UHF), capacitor dielectric and the nozzles of de-soldering guns. The fumes from overheated PTFE are very toxic.

### PVC

This cheap material is available in many forms including rod, tube and sheet. It is usually grey or black in colour, and can be easily machined, formed and bonded. Proprietary adhesives should be used (although hot-air welding with a filler rod is also possible) but skill is required to produce structurally sound joints. Certain of the building types of PVC encountered seem to have some conducting capabilities which can lead to problems if used in electrical or RF applications.

A suggested test for this is to try to cook a small sample in a microwave oven (alongside a cup half full with water), and if no metal is present the PVC should stay cool to just warm. However, note that microwave ovens operate on the agitation of water molecules so this test is not reliable for some drainage pipes made from PVC which contain carbon black or other filler materials that have poor RF properties even at lowish frequencies. 40m traps wound on PVC can exhibit very low Q factors!

PVC insulating tapes are strong, cheap and normally self-adhesive, and are supplied in a variety of colours and widths intended for wrapped insulation. Some of the poorer-quality tapes do not weather very well and suffer adhesion failure with the first frost.

## Adhesives

Many modern adhesives are hazardous and it is essential to follow meticulously the manufacturer's instructions when mixing, using and curing them. Most are insulators and unsuitable for electrical joints. Five general rules should be applied for bonding:

- Degrease the parts thoroughly; even finger marks impair results.
- Roughen the joint faces unless a transparent joint is required.
- Do not place bonded joints under a peeling type of load.
- Ensure that the work is dry and warm.
- Wear protective glasses and gloves and have the necessary first aid chemicals readily to hand.

### Epoxides

A group of medium to expensively priced, cold- or heat-setting resins (usually self-generated heat) that can be used for bonding, surface coating, laminating or encapsulation. Air and gas bubbles are the biggest problem with encapsulation (this work is carried out professionally under vacuum). The problem can be minimised by warming the work and the resin to around 40°C and providing a generous shrinkage allowance with a large pouring area which can be cut off from the cured encapsulation. Careful thought should be given to the necessity of encapsulation, for once completed, the encapsulated module cannot be altered or repaired. Encapsulation is usually used when circuits are subjected to harsh vibrational and environmental conditions. (See also 'silicone sealants'). The correct choice of encapsulating epoxides is critical. Incorrect types when cured shrink and can compress components so much that they change shape and often fail, eg electrolytic capacitors.

These are usually two-part adhesives and require careful mixing just prior to use. A structural joint should not be over-clamped during bonding, and a bonding gap of typically around 0.05mm is required for the joint to be made properly. In other words, don't squeeze out all of the glue! Surface coatings can be applied by dipping, spraying or brushing. Flexible resins are usually used for this type of work and are ideal for protecting beams, traps etc.

### Cyanoacrylates ('Superglue')

These expensive adhesives are available in various grades, each intended for bonding a particular set of materials. The low- to medium-viscosity grades are suitable for most amateur work. They are scientific marvels of bonding and as such require correct and proper application to ensure success. Releasing or debonding agents are available and it is a wise precaution to keep some of this handy in case of accidental bonding of fingers etc. The household type of superglue is unsuitable for outdoor use as the bond fails when moisture is present. If you accidentally glue your fingers together holding them under a cold water tap and gently easing them apart causes the bond to peel away. These adhesives should be used and cured in well-ventilated conditions. It is advisable to wear protective glasses when using this adhesive.

### Toughened acrylics

These are expensive, fast-curing adhesives intended for structural joints. Various types are available and are usually supplied in two parts - the 'glue' and the primer/activator. The glue is applied to one side and the primer to the other. They are suitable for use on most of the materials already mentioned, but some may not be used with certain plastics as they dissolve the material and eventually the joint fails. They are usually easier to use than the cyanoacrylates.

## Other Materials

### Silicone rubber compounds

These are medium-priced materials, available as paste in squeeze tubes and as a liquid in tins. When cured, they normally set to give a white or translucent silicone rubber finish. They are ideal for encapsulation and the sealing or weatherproofing of antenna connecting boxes and similar out-of-door items. The electrical insulating characteristics are excellent and can be used to prevent parts from vibrating in equipment used for mobile or portable work. One type of this compound emits acetic acid during curing and this may damage some insulators and component connections.

Though not normally sold as such, one type of this compound has been used successfully as a resilient adhesive for structural and pressure-sealing joints on metal, plastic and glass.

### Self-amalgamating tapes

These are a form of insulating tape which, when stretched and overlap-wrapped around cables, coaxial plugs etc, will amalgamate or flow together as one. They are reasonably priced and available in widths up to 50mm either with or without a self-adhesive face. Excellent weather-resistant properties. The self-adhesive form of this tape is ideal for waterproofing antenna traps, joints and connectors.

### Nickel silver

*Typical uses: Boxes, screens, nuts, bolts and washers etc where the non-corroding properties are required.*

An expensive non-magnetic electrical conductor which is corrosion resistant. Available in sheet and rod. It is often used by railway modeller's for it is a very 'friendly' metal to use, soldering and bending easily. It has some resistivity.

### Wood

*Typical uses: Aerial masts, booms, spreaders, storage boxes, and other parts which do not require RFI protection.*

A common material, not usually used today for radio work as it has no ability to conduct electricity or to screen against RFI. Its insulating properties are marginal unless dried and treated to prevent ingress of water/moisture. The radios of the 1920s and 1930s were cabinetmaker's delights but this was before the 'enlightened' days of electromagnetic compatibility. Other joinery-type materials such as chip board, MDF, plywood etc are equally unsuitable for RF applications but may be utilised to make mock-ups of proposed designs, especially for the panels and cabinets.

## Corrosion

There are two main processes of metal corrosion. The first relies on environmental conditions such as rain or condensation which results in an acidic electrolytic liquid being formed on the surface of the corroding metal. As the metal corrodes, the acidity of the liquid increases until the electrolytic process of corrosion

| | | | |
|---|---|---|---|
| 1. | Magnesium alloys | 8. | Lead |
| 2. | Zinc | 9. | Tin |
| 3. | Aluminium alloys | 10. | Nickel |
| 4. | Cadmium | 11. | Brass |
| 5. | Mild steel | 12. | Copper |
| 6. | Stainless steel 18/8 | 13. | Silver |
| 7. | Lead-tin solders | 14. | Gold |

*Note: the numbers are item numbers only and are not values!*

**Table 25.1: Galvanic series**

becomes almost self-sustaining. The second occurs due to the electrolytic action occurring between dissimilar metals in contact, and is referred to as galvanic action. Both processes change the metal into a different form, which in the case of steel or iron we know as 'rust', and often refer to the process as oxidisation.

**Table 25.1** shows the galvanic relationship between metals. The numbers are item numbers only to show the position of each metal in the galvanic series. The actual values depend on several factors such as temperature, radiation and acidity etc. This list enables metals to be selected which will have the minimum galvanic corrosion effect on one another. The greater the list separation, the greater will be the possibility of corrosion.

For example, brass and copper are adjacent and would therefore not cause problems if in direct contact with each other. Brass and aluminium alloy are widely separated and corrosion occurs if these metals are in direct contact with each other. This state can be reduced by tinning the brass with lead-tin solder, which falls about halfway between the other two metals (galvanic interleaving). The higher item-numbered metal will normally promote the corrosion of the lower item-numbered one.

Dissimilar-metal galvanic action corrosion is avoided by ensuring that the galvanic series separation is minimal. If widely separated metals do need to contact each other, then a suitable interleaving material should be used to reduce the galvanic separation level.

Corrosion by moisture or rain is more difficult to combat effectively. The commonly accepted anti-corrosion treatment consists of protection by paint and, providing the paint coat remains intact or is renewed regularly, this is a very effective treatment. Steel and iron can also be coated by protective metals such as zinc (galvanising) or nickel (plating), both methods being normally outside the range of the home workshop. Aluminium is slightly different, for this metal forms its own protective oxide barrier which, providing it is not disturbed, will prevent further corrosion. It is this oxide barrier which makes soldering aluminium difficult. Unfortunately, this natural protective oxide layer can be disturbed by stress or galvanic action, and the corrosion process bites deeper into the metal. Anodising is a process on aluminium which forms a controlled layer of oxide on the surface and presents a toughened surface finish which can be coloured by dyeing. Brass and copper can be considered as corrosion resistant for most amateur purposes. However, their surfaces do oxidise and this can impair good electrical contact. Certain platings, such as silver, nickel or chromium, can reduce this. The platings themselves can also corrode but usually at a much slower rate than the parent metal.

If not adequately protected, corroding metal will gradually lose strength and the device from which it is made will fall apart. This is usually seen as collapsed masts, broken antennas and similar expensive disasters.

Corrosion also affects RF and electrical connectors, particularly feeder-to-antenna connections, and causes a gradual decline in the overall performance of the system. Signals become weaker and calls to DX stations which used to be answered are ignored. Most observed lowering of performance in this area is usually (or eventually), discovered to be due to corrosion. This can be practically eliminated by first ensuring that no dissimilar metals are in contact to cause galvanic corrosion, and second that water is excluded from all connections. Copper coaxial-cable inners should never be connected directly to the aluminium elements of a beam but should be tinned first as previously stated. Coaxial plugs and sockets should be fitted with heat-shrink sleeving and wrapped with amalgamating tape to prevent the ingress of water. Connector boxes can be filled with silicone-rubber compound for a like purpose. Stainless steel antenna fittings are the least affected by corrosion but even these would benefit from a coating of protective lacquer, particularly on screwed fittings. The position of stainless steel in the above list is interesting, for sometimes the metal in contact with the stainless steel will corrode due to galvanic action, particularly in wet, smoggy or salt-laden conditions, and some stainless steel fittings are supplied with plastic interleaving spacers and washers to prevent this.

Condensation is also a problem with outdoor enclosed or boxed-in items such as rotators, remote tuners, antenna traps, coaxial cable and the like - where there is condensation, there will be corrosion. The simplest, but not necessarily the easiest, solution is to allow the enclosed area to 'breathe' by introducing suitable weather-resistant holes as drainage vents, while ensuring that these are at the lowest point of the enclosed area and cannot be inlets for insects, rain or the run-off water from the outside of the enclosure. Complete sealing usually makes matters worse, for a fully water and air-tight enclosure still produces internal moisture by condensation. Full hermetic sealing is difficult to apply for it normally requires the ability to pressurise the enclosure with an inert gas, as with some military or maritime equipment. If the items within the enclosure can be protected as if they were exposed to the elements, much of the corrosive effect of condensation is reduced.

In the case of rotators, attempts at filling the voids with grease does not help, for the grease forms small pockets which can hold water. It is better to lacquer or wax oil the moving parts and the inside of the housing, and to use grease for its intended purpose of lubrication. Any electrical items within the rotator should be sprayed with a commercial, non-insulating, waterproofing liquid.

The threads, screws, nuts, bolts etc should be given a light coating of anti-seize compound on assembly and, after assembly, sprayed over with lacquer or wax oil, or coated with a water-repellent grease such as lanolin. This makes for easier maintenance and reduces the possibilities of corrosion.

Corrosion is not limited to outdoor items. Corrosion of connector pins of microphones, plug-in PCBs, computers etc is not uncommon. Careful selection of mating materials to avoid galvanic action, combined with appropriate painting, plating etc and regular maintenance, will reduce the effects of corrosion.

## CONSTRUCTION TIPS

Before you start a construction project, you should have a definite plan of action. Changing your mind part way through the project can be expensive in terms of wasted time and materials. When building an enclosure for a project, you should measure the exact dimensions of any printed circuit boards or other types of circuits used and the height of the tallest components mounted on the board to make sure that the size and shape of the enclosure will be suitable for the job. Make allowance for any inter-stage screens or compartments within the main enclosure. You should also consider the possibility that you may want to introduce modifications or add new features at a later time.

## ENCLOSURES

At one time, most amateur radio construction projects were enclosed in a home made aluminium box. Aluminium sheet was relatively cheap to buy and it made economic sense to build your own enclosures. In recent years, the price of aluminium sheet in small quantities has risen to the extent that it is almost as cheap to buy a ready-made box as it is to build one. Commercially made boxes are available in a wide range of shapes and sizes. Most of the smaller boxes are made from sheet aluminium. A few of the more expensive types are made from die-cast aluminium.

Die-cast boxes are stronger and more rigid than boxes made from aluminium sheet. The greater rigidity of die-cast boxes makes them very suitable for use as VFO enclosures or for any other project that requires good mechanical stability. Larger boxes are often made from a combination of aluminium and steel. The bottom half of the box which includes the front and back panels is made from aluminium and the top cover is made from vinyl coated or painted steel. The bottom half of the case is easily drilled or machined to make any holes or other openings that are required on the front and rear panels. The superior strength of the steel top cover makes this type of enclosure very strong and durable.

If your project calls for a size or shape of enclosure that is not commercially available or, like me, you just prefer to make your own, a home made box can easily be made from sheet metal. If you have a bending brake or a simple set of bending brackets, it is a very easy job to bend the aluminium sheet into the required shape. Even if you don't have any special metalworking tools, it is possible to make a respectable looking enclosure using only a hacksaw, a straight edge, a sharp carpet knife and a few very basic hand tools (pencil, measuring-tape etc).

The box shown in **Fig 25.3** was made as an enclosure for a VHF transverter. The top cover is a simple U shape made from a rectangular piece of aluminium sheet. The bottom half of the box is also made from the same 1mm aluminium sheet.

Before you start cutting and bending the aluminium, you should make a drawing of the box. The exact dimensions should be marked on the drawing. For more complex projects, I use a computer with drawing or CAD (computer aided design) software to create plans and templates for metal boxes. Suitable software is readily available for most personal computers. I use "Xfig" vector drawing software on a Linux PC. For more about using computer aided design, see the chapter on Computers in the Shack.

For very simple projects, a pencil and ruler is usually faster and easier. When you are drawing the plans, you should remember to account for the thickness of the sheet. The inside measurements

Fig 25.3: Home made aluminium box

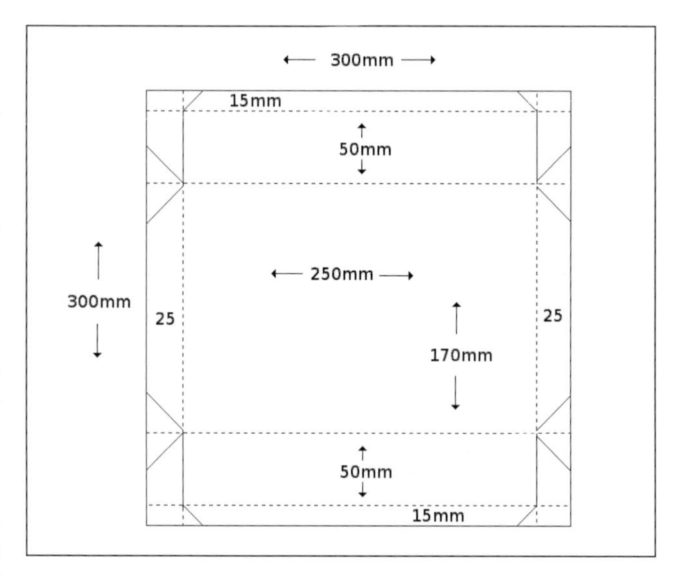


**Fig 25.4: It is advisable to make a detailed drawing before making a box**

of a box made from 1mm sheet will be at least 2mm less than the outside measurement. The bottom half of the simple project box was made from a 300mm x 300mm square of 1mm thick aluminium sheet. A detailed drawing is shown in **Fig 25.4**.

Since this box was made without the aid of any special tools, it was necessary to find a way of making perfectly straight 90° bends without using a bending brake. This was achieved by cutting a deep score along each line that was to be cut or bent. A very sharp carpet knife (Stanley or similar) was used to score the metal. Be very careful when you are cutting the score. Use a good quality metal-bodied knife and not a cheap plastic one which might break under pressure. A steel engineers rule can be use as a guide for the knife blade.

The 300mm square was cut from a larger sheet of aluminium. As this was literally a 'kitchen table' project, a sheet metal cutting guillotine was not available. It is very difficult to cut a perfectly straight edge with a hacksaw and even more difficult with an electric jigsaw.

The 300mm square was carefully marked out at one corner of the sheet with a felt-tip pen. After double checking my measurements, the carpet knife was used to cut a deep score along lines that were to be cut. The sheet was then carefully bent over a straight edge. Our kitchen table was not suitable for this job because it has a bevelled edge. I used a surplus sheet of veneered panel that was left over from a DiY project as a suitably square edge and to protect the kitchen table or counter-top from damage. A thick sheet of steel or the edge of a metal workbench can also be used. You will find that it is quite easy to bend the sheet along the line of the knife score. With a little practice, you will get a perfectly straight bend or cut first time, every time.

When you want to cut rather than just bend the sheet, you should bend it to 90° and then partially straighten it again. Move the bent section back and forward a little more each time until it is so weak that it breaks off cleanly. The type of soft aluminium sheet that is sold by hardware and DIY stores is quite flexible and it probably won't break until it has been bent to almost 180° several times. Be patient at this stage. Don't apply too much force or you will end up with a slight curve at the edge of the sheet.

Once the sheet has been cut out, you should mark the bends and cuts with a felt-tip pen. A fine laundry marker or a CD/DVD marker pen is ideal for this job. The bends should then be scored with the carpet knife. Since you only want to bend the metal rather that cut it, you should only cut the score deep

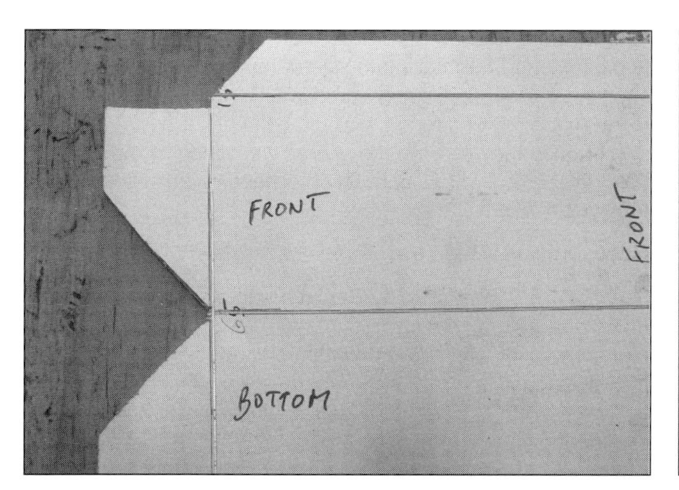


Fig 25.5: The cut sheet ready for bending

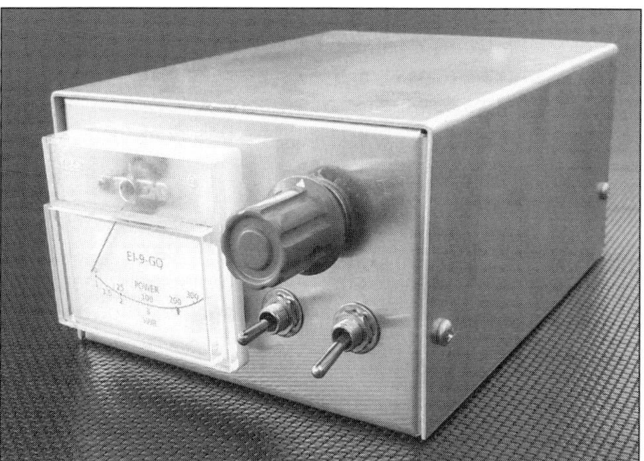

Fig 25.6: SWR meter enclosure using the score and bend method

enough so that the metal bends cleanly and easily. If the score it too deep, it will weaken the structure of the box or in the worst case, it might break off altogether! At this stage, you should cut off any unwanted parts at the edge of the sheet. You will usually need to cut out a V shape where there will be a 90° bend. **Fig 25.5** shows the sheet when it is ready for bending.

The bottom half of the box should now be bent into its final shape. This job is not as easy as the first cutting/bending operation because we are now dealing with a three-dimensional structure. The task becomes progressively more difficult as each part of the box is bent into shape. A small bench vice and a few pieces of timber will make this task a lot easier.

The 15mm lip at the very edge of the sheet is particularly tricky. Since this lip will not be visible when the lid is on the box, a small amount of damage in this area can be tolerated. A vice grip or large pliers can be used to tidy up the edges, provided that you take care not to damage the front panel. You can use a fine file or a medium grade of sandpaper to remove any sharp edges. This is not critical for the bottom half of the case because all of the edges will be inside.

The top half of the case is very easy to make. A simple U shaped cover can be made from a rectangular sheet of aluminium or other metal. Only two bends are required.

The most critical part of the project is the front edge of the top half of the box. This is the part of the box that will be most visible to the user. Take great care when you are cutting this edge, or preferably cut this panel from the edge of the original sheet of aluminium so that the guillotine-cut edge is towards the front of the enclosure. The aluminium top cover can be replaced by a steel cover. Steel is much harder to cut than sheet aluminium, but it is more durable and easier to paint. Fancy perforated steel is easier to bend than plain steel sheet. The perforated sheet will allow some air-flow through the enclosure which will help to keep high power circuits cool. This type of construction is often used for power supplies and RF power amplifiers.

The score and bend method of construction has served me well over the years, I have used this method to build cases for many projects of all different sizes. **Fig 25.6** shows a SWR bridge which was built using score and bend.

There are many other ways of making metal cases and each individual builder will have their own favourite method. One approach is to make every panel of the box from separate pieces of sheet metal. The panels are screwed together at the corners. Angle aluminium or angle steel can be used as corner supports. The panels can be drilled and secured to the angle pieces with nuts and bolts. As an alternative to the strip of angled metal, a solid length of square metal bar can be drilled and tapped and used as the corner pieces. Square plastic, fibreglass or even wood can be used for the corners, but metal is to be preferred for screened enclosures.

## Painting the Enclosure

Most of my enclosures are painted with aerosol spray paint. Aluminium is notoriously difficult to paint. It helps if the metal is absolutely clean before you apply the primer coat. Special etching primers are available which are used for painting motor caravans or aircraft. You can get away with using ordinary car primer, provided you apply it as several light coats. After it has dried thoroughly, the primer should be lightly rubbed down with very fine wet sandpaper. Be careful, the fresh primer will rub off the aluminium panels very easily.

You may need to apply several coats to fill in any scratches in the metal. Larger scratches or pinholes in the surface of the primer can be covered with a filler/stopper compound before sanding. Once you have a clean and smooth coat of primer, you can paint the enclosure with an aerosol enamel. Aerosol paints use a lot of paint thinner and they tend to run very easily. Do a few tests on a scrap sheet before you tackle the real project. Remember that aerosol paints use thinners and propellants that are highly flammable. Keep well away from sparks and flames. Wear a mask and work in a ventilated area.

Letraset or similar rub-on lettering can be used to label the controls on your equipment. Several coats of clear lacquer should be applied over the finished panel to protect the letters from damage. The first coat of lacquer should be applied very lightly so that it will dry quickly. If you use a heavy coat, the thinners used in the aerosol will tend to melt the plastic lettering. **Fig 25.7** shows the front panel of a home made SSB/CW transceiver. For more information, see the finishing and painting sections near the end of this chapter.

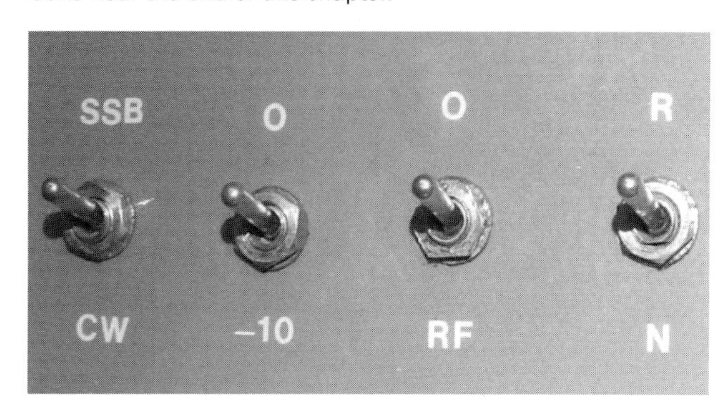


Fig 25.7: Smart lettering on a front panel

## Two Useful Home Made Tools

One of the problems most of us have in metalwork and the like is cutting or filing things square and straight. Also, marking out to ensure that parts will fit together correctly is sometimes not as easy as it appears. The few extras described here should simplify things.

### Bending bars

See **Fig 25.8**. These are normally home-made. The things to check when obtaining the steel angle are its straightness and squareness. If these are not 'true' the corrective actions needed will require skills which may be beyond those so far acquired. Old bed-frame angles, provided they have not rusted too badly, usually make very good bending bars. One bed frame can be used to make several bending bars of different lengths. The choice of length and distance between the clamping bolts of the bars is related to the maximum width of metal or other sheet purchased and the maximum size of panel worked. Aluminium sheets, for example, usually come in 1.8 x 1.2m size and trying to cut this is rather difficult. It would be better, if buying in these sort of quantities, to have the sheets guillotined by the supplier to widths which will fit between the bending bar clamp bolts. (Suggest 482mm.) The length of the sheet will not matter.

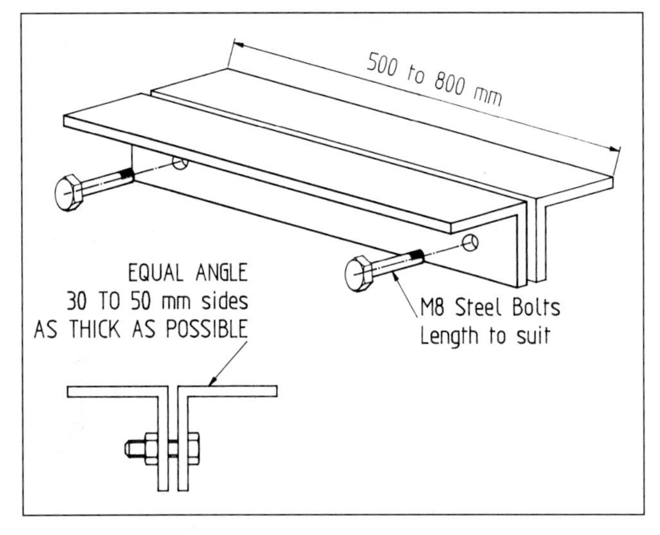


**Fig 25.8 Bending bars**

### Hole embossing tool

See **Fig 25.9 and 25.10**. This very simple home-made tool forms holes along the edges of metal panels to accept self-tapping screws of about 2.5mm diameter (depending on metal thickness), and with an engagement length of about 12mm. It facilitates the rapid manufacture of cabinets, boxes and screens, eliminating much of the accurate metal bending normally associated with this type of work.

The working parts and the two side plates of the tool are made from 1in wide by 0.125in thick-ground gauge plate, which is a tough but workable steel capable of being hardened and tempered. It is not necessary to harden the finished tool, unless it is intended to form mild steel but, if facilities are available to do this, it will improve the durability. The base and top plates are made from 1in wide by 0.375in thick mild steel.

The vee points must be on the centre-line of the tool. The vee notches are not so critical for they play no part in the shape of the hole produced, but they should allow sufficient clearance for the formed metal to flow into. The point and notch sizes shown are suitable for most metal thicknesses up to 1.6mm (16SWG). The vee points can be made as a set, with the four plates clamped together. The fully shaped points should be polished with an oilstone to obtain square, burr-free, sharp edges along each face of the point. The edges of each point are the working parts of the tool and should be made with care. It is essential that the flat faces on both sides of the notch and point ends of each set of plates are square, parallel and level with each other.

Tapping-size holes should be marked out and drilled on each top plate only, and these are then used as a template to drill the remaining holes on each plate. The holes in each bottom plate should be tapped 2BA and the holes in the remaining plates opened up to 0.1875in diameter - a good fit for the high tensile steel 2BA screws. ISO M5 may be used in place of the 2BA screws and tapped holes. The outside faces of the holes in the top plates should be countersunk deep enough to ensure that the screw heads are slightly under-flush. It is good practice to lightly countersink both sides of every hole to remove burrs and facilitate assembly.

The back edges should also be square and level to ensure that the vice pressure is applied evenly during forming. The 0.125in thick side plates and the 0.375in thick top and bottom plates form the bearing for the sliding part of the tool, which must slide easily with the minimum of play in any direction. Failure to get this right will ruin the action of

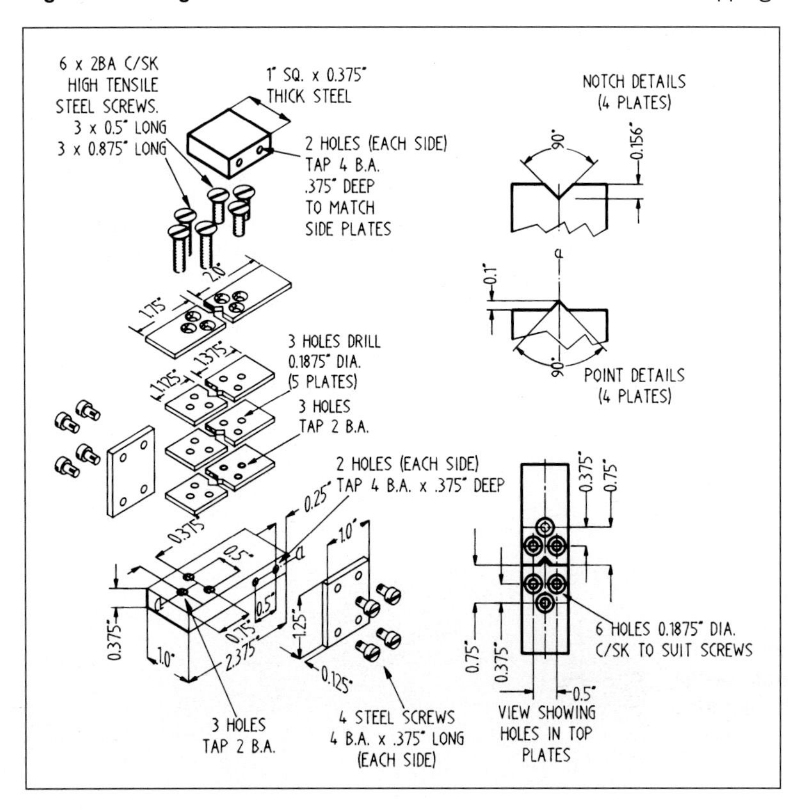


**25.9: Hole embossing tool**

**Fig 25.10: Holes made by the embossing tool**

the tool, for each point must just slide over the opposing one or two points with the minimum of clearance.

To operate, the tool is placed between the jaws of a 4in vice and supported by the extended top plates. The metal to be holed is rested on the base of the tool and the marked out hole position is aligned with the tip of the visible point. The vice is then closed with the minimum of force until each face of the working ends of the tool just contacts the metal sheet. It is pointless trying to go beyond this, for squeezing the tool and the work by excessive pressure from the vice will only ruin the tool and the work. The vice is then opened and the metal sheet gently prised away from the points. This method of releasing the formed sheet could be improved upon, but the tool would become more complex to make and, as a spot of oil on the working faces eases the problem, this extra complexity is not worthwhile.

The tool performs very well on aluminium, tin plate and annealed brass or copper in thicknesses up to 0.0625in (16SWG). It is not recommended for use on sheet steel unless the points have been hardened and tempered.

## Hole Making

Drilling is a straightforward operation providing a few key guidelines are observed. The drill speed should be adjusted to match the size of drill and the material being drilled. As a general rule, the smaller the drill, the higher the speed. A correctly sharpened drill should not require heavy pressure to cut. If it does, either the material is too hard or the drill needs to run slower. There is a tendency for normally sharpened drills not to cut some brass properly. This is due to the incorrect rake angle of the cutting edges of the drill. The problem can be solved by sharpening the helix edge on the face of the flute to give a rake angle of about 15 degrees, ie leaning backwards away from the normal cutting edge (**Fig 25.11**). Drilling some plastics can also present problems, such as chipping of the edges and breakaway of the material as the drill breaks through. Some improvement is possible by sharpening the drill to produce an included point angle of around 80 degrees. The standard included point angle is 118 degrees.

When drilling steel, it is advisable to use a coolant such as soluble cutting oil to keep the drill and the work cool. It also saves having to re-sharpen the drill so often. Paraffin is a good coolant for aluminium and copper. Brass and most plastics do not normally need a coolant, providing the drilling speed is correct, but treat as steel if necessary. These coolants may be applied either by an oilcan, or a brush. (Old liquid soap squeeze bottles make good coolant dispensers and any left-in detergent is not detrimental.) Soapy water is also a good coolant and certainly better than no coolant at all - it is best for most plastics.

Holes should always be centre-popped before drilling and, if a BS1 centre drill is available, this should be used next to provide an accurate location for the drill.

The holes drilled in thin sheet are often anything but round because the drill does not have enough depth of metal to round the hole properly before it breaks through. There are at least two ways round this. The easiest is to drill the hole undersize and bring to size using the taper reamer, which must be allowed to cut without forcing, otherwise another fancy-shaped hole will be produced (another reason for scribing-in the holes beforehand.) Another way is to alter the cutting angle of the drill to an included angle of about 140 degrees and thin the chisel edge of the drill to a point. It is worthwhile keeping a set of drills sharpened in this manner especially for thin sheet drilling. These sheet metal drills normally require a much slower speed and the holes should also be pre-drilled using the centre drill.

Whatever is being drilled should be well supported and clamped to prevent rotation and lifting. Failure in this direction can lead at best to a broken drill and at worst to serious personal injury, for drills have a habit of picking-up just as they are about to break through the hole. (Plastic and copper are particularly susceptible to this.) A panel whirling round on the end of a 10mm diameter drill is a frightening sight!

All drilling will produce burrs around the hole edges and it is good engineering practice to remove these using the hand-held countersink bit referred to previously. A file will only scratch the rest of the surrounding surface and bend the burrs into the hole. The use of a large drill for hole de-burring is not recommended, particularly on the softer metals, unless the 'touch' for this method has been acquired.

It is usual to step drill holes larger than 10mm diameter, that is a hole of smaller diameter (3mm) is drilled first, then another slightly larger (+1mm), and so on until the finished size is reached. Step drilling is unsuitable for sheet material. There are special, though expensive, stepped drills available for drilling holes up to 40mm diameter in sheet material.

Making large round holes can be tackled in at least two ways. The first requires a washer or tank cutter (**Fig 25.12**) and an extra slow speed drill (a joiners' brace is effective). The biggest snag with this method is trying to obtain an even cut around the full circle. By clamping the work to a block of wood and drilling through into the wood for the centre pilot of the tank cutter, a guide is provided which improves things a little. The main thing with this method is not to be in a hurry.

The second method can be applied equally well to non-round holes. Contiguous holes of about 5mm diameter are drilled on the waste side of the hole or cut-out and 1 to 2mm away from the finished size markings. The waste is removed, using tin snips. The hole is then carefully filed to size, using the bending bars for support, and as a guide for any straight portions.

To de-burr large holes, a small half-round needle file can be used in the 'draw' fashion. The file is held at both ends and drawn round the edges to be de-burred in a manner similar to

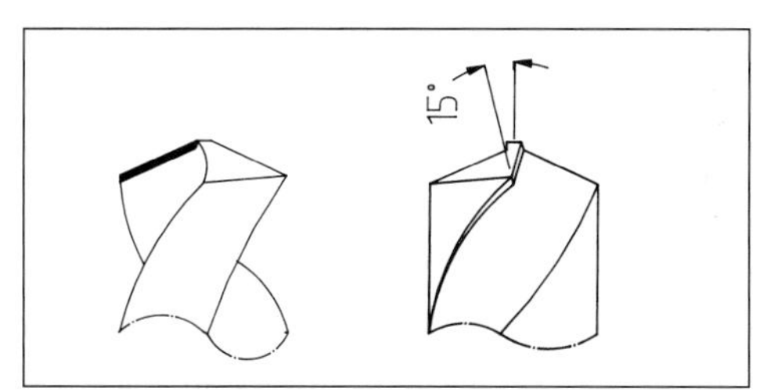


**Fig 25.11: Modified drill rake angle to improve the drilling of brass**

**Fig 25.12: Large round hole making tools. Clockwise: deluxe home-made tank cutter, commercial tank cutter, screw-up punch, hole saw**

using a spoke shave. The file should be held at an angle to produce a small 45° chamfer around each edge of the hole.

Machine countersinking of holes requires a very slow speed drill (60RPM). They can be produced using the handle-mounted countersink bit referred to previously. Even the multi-toothed rose-countersink bit will chatter and leave a very unsightly surface if too high a speed is used. Countersunk screws will not sit properly on such holes. There are countersink file burrs available which do a similar job and can be used in the DIY drill, providing the work is held securely with the burr square to the work. They do have a tendency to skid over the surface, leaving a trail of deep scratches, if not located properly.

Drilling accurately positioned holes can be a problem and templates can be used with success. Keeping the small holes for multi-pin DIL ICs in-line and at the correct spacing is difficult, but the job is made easy by using a piece of the correct pitch matrix board as a drilling template. One of the holes (pin 1 is suggested) is drilled first in the required position. The template is located from this by passing a drill through the hole and the matrix board.

The matrix board is then aligned with the rest of the marked-out hole positions and clamped using the toolmakers' clamps. The remaining holes are drilled through using the matrix board as a template. It helps to mark each row of holes on the matrix board template. If a considerable amount of such accurate drilling is to be done, it is worth making a set of metal templates for each size of IC using this method. Sometimes the component itself can be used as a template to reduce the risk of error, eg slide switches, coaxial sockets, dial fixing holes etc.

## Filing

Accurate filing is a skill which can only be acquired by practice. General rules are:

- Always use a file handle. This eliminates the risk of running a file tang into the wrist and enables the file to be guided properly.
- Use a sharp file. It is normal practice to use new files for brass and, as their sharpness wears off, use them for filing steel or aluminium. New files should be kept separately or otherwise identified.
- Do not force the file to cut. Only a light relaxed pressure is required, which also aids the accuracy of filing.
- Keep the file clean by brushing with a file card or by rubbing a piece of soft brass or copper along the teeth grooves.
- Support the work properly. Trying to file with the work held in one hand and the file in the other is a guarantee of failure.

The height of the vice should be such that when the filer's arm is bent to place the fisted hand under the chin, the point of the elbow should just rest on the top of the vice jaws. An old but effective form of ergonomics!

A strip of emery cloth wrapped along the file can be used to obtain a better surface finish. This method was frowned upon in apprentice schools but it works and saves time. The draw filing technique referred to previously imparts the final touches to the filed edges and removes the filing burrs. The bending bars can be used as a filing guide (**Fig 25.13**) and a set of bars can be made especially for this purpose, checking them regularly to see that the guiding surfaces are not bowed by too much one-spot filing.

## Bending and Forming

The first essential is to ensure that the material can be bent. This sounds obvious but it is better to check first than to find out after all but the bending work has been done. Annealing can be applied as mentioned previously under the section on materials.

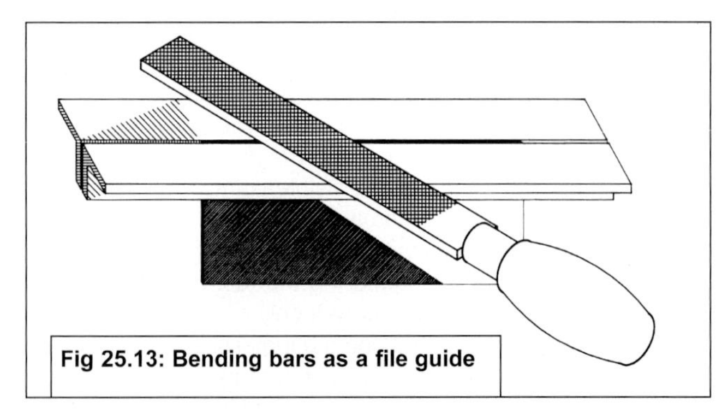

**Fig 25.13: Bending bars as a file guide**

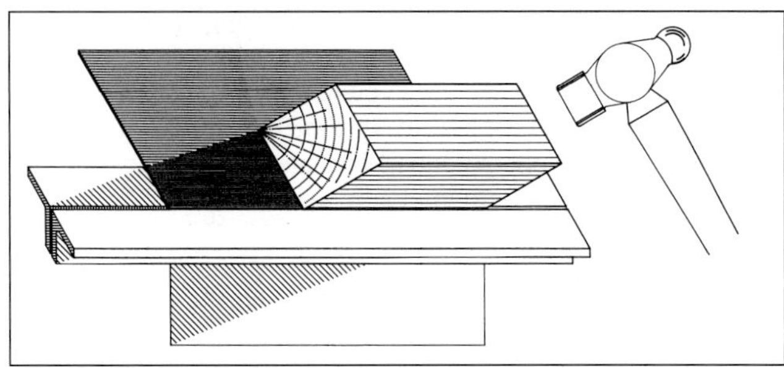

**Fig 25.14: Method of bending in the bending bars. Note that the cutter block is held into the point of the bend**

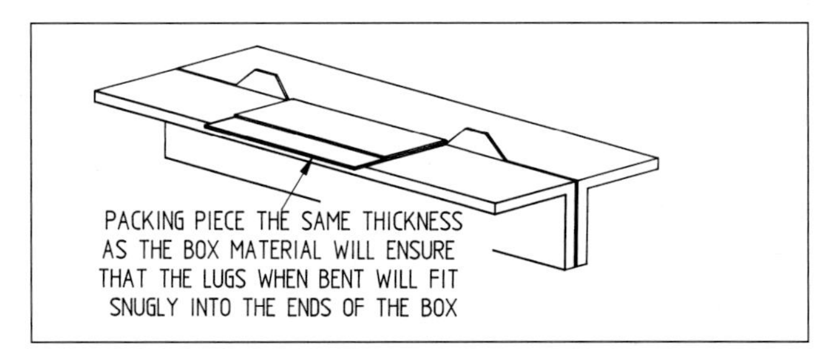


**Fig 25.15: Making allowance for bending the end lugs of the box**

A metal hammer should never be used directly on the metal when bending or forming. Either use a soft-faced hammer or a block of wood or plastic as a buffer for the hammer blows (**Fig 25.14**). This prevents all the humps and hollows which would otherwise occur. Do not use any metal as a buffer block as there is a danger of the metal chipping and flying into the face or eyes. The block should be kept into and near the point of bending. It seems easier and quicker to try to use the sheet as a lever and hammer as far away from the point of bending as possible. This will only produce a bend which curves up and back again. Where three sides are to be formed by bending, the point of intersection of the marked-out bend lines should be drilled before bending, with a hole diameter three times the thickness of the metal. This prevents corner bulge. **Fig 25.15** shows how to ensure that the end lugs of the box fit snugly inside the box. **Fig 25.16** shows one method of bending the ends of a similar box.

Tube bending for beams, tuned lines etc is not difficult and flattening or kinking can be reduced by observing the following:

- Ensure that the tube is suitable for bending. Anneal as necessary and re-anneal as soon as resistance to bending is felt.
- Unless skilled or equipped with specialised tooling, do not attempt to bend to a radius of less than three times the outside diameter of the tube, eg 12mm OD tube should have a 36mm minimum bend radius.
- Always bend round a former shaped to the required radius (**Fig 25.17**).

- Pack the tube tightly with fine sand (clean birdcage or builders' sand is ideal). Wet the sand and cork both ends. This will minimise the risk of kinking during bending, and the sand can be washed out afterwards. The tube can be re-annealed if necessary with the sand left in place but the corks should be removed to let the hot gases escape.

A low-melting-point, lead-like material is available which can be used in place of sand, and is known as bending metal. It is poured molten into the blocked, ended tube, allowed to set and, when the bending has been completed, is melted out to be used again. This process should be carried out with the tube at an angle to allow the hot air to escape. The molten bending metal should be poured very slowly for it has a tendency to 'blow-back'. Normal lead is not a substitute for bending metal and should not be used, particularly with aluminium tubes, for the melting temperature required for lead may destroy the aluminium's properties.

Modelling wax (also known as American wax) can be used in place of sand or the bending metal. The wax is heated, not boiled, until molten and then poured into the tube in the same manner as the bending metal. The wax should be allowed to set hard before bending. Unfortunately it is not possible to re-anneal with this wax left in place.

Another tube-bending tool is the bending spring. This is used by plumbers and pipe fitters. Each tube size should have a matching bending spring, for it is all too easy to use the wrongly matched spring and have it permanently trapped inside the bent tube!

## Cutting

Cutting long strips of metal with tin snips or shears is an expert's job. The cut edges usually produced by non-experts are anything but straight and they require flattening to remove the cutting curl. Tin snips are best used where a one-snip cut will remove the required amount of metal, such as 45-degree corners, or the trimming to length of narrow strips. Snips should not be used on plastics, laminates or copper-clad board (PCBs), for the cutting action can cause de-lamination or shattering of the edges.

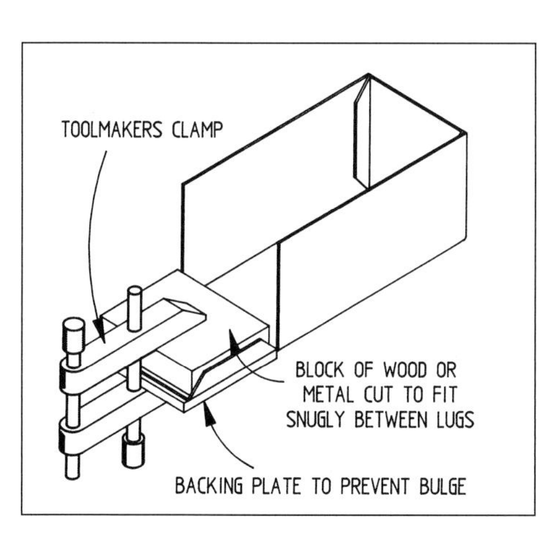


**Fig 25.16: Bending the ends of the box using a toolmaker's clamp to secure the forming block**

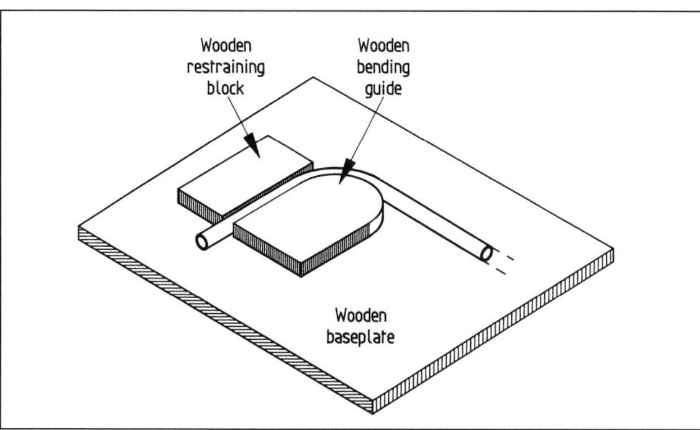


**Fig 25.17: Tube or rod bending set-up**

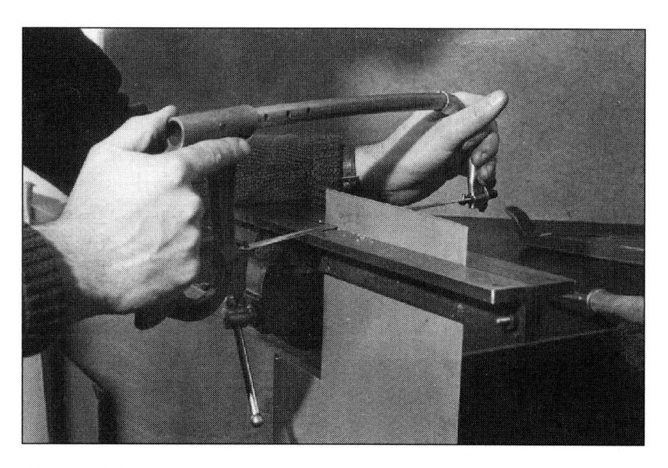

**Fig 25.18: Bending bars as a sawing guide**

A guillotine is by far the quickest way of cutting most sheet material but it is not a normal home-workshop tool. For the home workshop the cutting of nearly all materials is best done by sawing, using a hacksaw and the bending bars as a guide and support (**Fig 25.18**). The hacksaw should be fitted with a 24 to 32 teeth per inch blade for most sheet work, and coarser blades for cutting blocks of material. The blades known as ding-dong style, ie with their cutting teeth arranged along a wavy edge, are very good for most work. When sawing copper-clad laminates such as PCBs, it is advisable to score through the copper beforehand using a sharp chisel. This will bevel the edge of the copper and prevent delamination due to sawing. The coolants recommended for drilling can be used if required. A fine-toothed roofers' saw is very useful to cut sheets which are too big to pass inside the hacksaw frame. This is similar to a joiners' saw but shorter with a stiffer fine-toothed blade. A pad saw or fine-toothed machine hacksaw blade which is handle mounted is equally effective in these circumstances. Some constructors use a fretsaw with either a metal-cutting blade or an Abrafiler to overcome the problem, particularly when making large holes.

Hand-operated nibblers are available for cutting most thin sheet materials, including laminates, although they are sometimes difficult to guide properly. The power-operated nibblers are not normally a home-workshop tool. The DiY jig-saw fitted with a fine-toothed, metal-cutting blade is very good but it is essential to support the work very well and even then it is a noisy process. Similarly the DiY router-cutter is extremely useful, especially for large holes, but it can be a dangerous tool to use.

Cutting and sawing are areas where gadgets are forever appearing, each claiming to save time and produce a better job! The hacksaw has not yet been replaced as a good general-purpose cutting tool.

# SOLDERING

Whether they are made in the radio amateur's workshop or commercially by a large electronics company, almost all electronic circuits will use soldered electrical connections. The type of soldering that is usually used for assembling electronic circuits is known as soft soldering.

To make a solder joint between two metal conductors, a metal alloy (the solder) which has a relatively low melting point is applied between the two conductors and heat is applied to the joint. The heat causes the solder to melt and flow around the joint. When the joint is allowed to cool, the solder makes a good electrical and mechanical joint between the two conductors.

For a typical hand assembled circuit, the heat is provided by a hand-held soldering iron. The main difference between soldering and welding is the much lower temperatures used. A solder joint is formed at a temperature which is well below the melting point of the metals to be joined. The term soft soldering usually refers to joints that are formed at a temperature of less than 450°C. Higher temperature soldering methods like hard or silver soldering are rarely used for electronic assembly work because the much higher temperatures involved would damage the PCB and electronic components. Until recently, most solders used for electronic work were made from a tin/lead alloy, the most common type used is a mix of about 60% tin and 40% lead. The introduction of the new RoHS (reduction of hazardous substances) directives in Europe have lead to a gradual phasing out of lead solder over the last few years. From 2006, electronic circuits are required to be lead-free. These new regulations have led to the development of lead-free solders with a high percentage of tin and a small amount of copper and/or silver. Lead-free solders generally have a higher melting temperature than tin/lead solder. The 99.3% tin, 0.7% copper alloy that is often used to replace lead solders has a melting point of 227°C and a recommended soldering iron tip temperature of 350°C.

The use of a soldering flux prevents oxidation of the metals to be joined and helps to 'wet' the solder so that it flows freely and makes a better joint. Tins of solder flux are available from most component suppliers, but most solders that are designed for electronics work have a built-in flux core so that no additional flux is required. Traditional solder fluxes are made from rosin which is produced from a naturally occurring resin found in pine trees and some other plants. Some types of flux are very acidic and are not suitable for electronics work. Most types of flux will leave some residue around the soldered joint. This flux is usually removed from mass produced circuit boards by washing them in a solvent.

New regulations that are intended to reduce the amount of VOCs (volatile organic compounds) released into the environment have also led to changes in the formulation of soldering fluxes. Water soluble fluxes can be removed without the use of environmentally harmful solvents.

## Rules for Soldering

- All surfaces must be scrupulously clean. This applies equally well to the bit of the soldering iron. These bits are often iron plated to prevent rotting, and this plating should not be removed but wiped clean using a damp cloth or sponge. The bit should be kept 'tinned' with the solder and flux in use. Note that some fluxes will rapidly corrode the iron plated soldering bits. In the case of PCBs, the copper cladding should be cleaned using a fine abrasive eraser block. Component wires and pins etc should be similarly cleaned for, although many brand-new parts come with the wires dipped in a solder-through or flux coating, it is still worthwhile just making certain. There must be no dirty or greasy patches on any of the surfaces to be soldered and corrosion spots must be removed. Even finger marks can spoil the soldering, so "clean it and when you think it's clean, clean it again!"
- Ensure that there is just sufficient heat to make the joint. Too much heat is harmful to electronic work and, in the case of structural work, it causes distortion. Not enough and the solder will not bond correctly.
- Heat and apply solder to the greater conductor of heat first, which in the case of PCBs is usually the copper pad and in the case of switches etc is the solder tag, and then finally the wire, (the lesser conductor of heat) in one continuous process.

## Soldering Technique

After cleaning the parts, the following technique should be used to make a sound mechanical and electrical joint on PCBs.

Apply the clean, tinned hot iron to the pad adjacent to the component wire to be soldered, and tin the copper surface by applying just sufficient multi-cored solder to the point where the tip of the iron is in contact with the copper. When the solder and flux starts to wet the copper surface, slide the iron into contact with the wire and apply a little more solder. As soon as both the wire and the copper surface are blended together in a small pool of molten solder, draw the iron up the wire away from the work.

This whole process should take seconds, not minutes. The less time any component is under this form of heat stress the better. In the case of particularly heat-sensitive components, it is advisable to heatsink the component's leads during soldering. A good soldering heatsink can be made from a crocodile clip, on to each jaw of which is soldered a small strip of copper. A small set of metal forceps will also serve as a clamp-on heatsink. The clamp-on action is required so that both hands are left free to do the soldering. Commercial soldering heatsink clips are available.

This method of soldering applies equally well to the joining of wires to components such as switches, potentiometers etc. In this instance, good practice demands that the wires should be wrapped and secured mechanically before soldering, but bear in mind that the joint may need to be unsoldered later.

Good soft-soldered joints should look neat and smooth, with no draw-off points, and be continuous around or along the joint. Large blobs of solder do not ensure that a joint is well-made. The cored solder should always be applied to the work and the iron at the same time, not carried on the iron to the work.

The results of incorrect soldering procedures are 'dry joints' which are either open-circuit or high-resistance points and will prevent the circuit from functioning correctly. Often the effects of such soldering will appear after several hours of operation, and this really leads to frustration especially to the novice constructor. Commercial equipment has exactly the same problems, if this is any consolation! In most instances it is difficult to locate a dry joint visually. Sometimes a perfectly sound looking joint turns out to be 'dry' immediately at the contact point, which is hidden under the solder. As a rough guide most 'good' joints appear shiny and devoid of any 'craters' or occlusions, whereas some 'dry' joints appear dull and pitted. There is no easy way to identify 'dry' joints but there is an easy way of avoiding them - follow the rules and guidelines set out above.

To join structural parts together by soft soldering, each mating face should be pre-tinned, ie given a thin coat of solder using the appropriate flux to ensure proper 'wetting'. The tinned faces must be clean and shiny with no burnt or blackened flux spots.

The two tinned faces are then painted with flux, brought together, heated to just melt the solder, and clamped together until the solder has cooled. There is no need to apply any more solder. If the job has been done properly, a neat, clean solder fillet will be seen along the mating edges. The flux residue must be washed away immediately after soldering to prevent long-term corrosion. It is not good practice to have long soldered joints on PCB laminates. The expansion caused by the necessary heat will de-laminate the copper. In these circumstances it is better to tack solder along the required structural joint, placing 2mm long solder-tacks at about 10mm intervals. Any such joints on copper laminates will be as mechanically strong as the laminate-to-copper bond.

It is better practice to secure mechanical parts such as angle brackets, shaft housings etc by screws or clips which pass right through the laminate, and to use solder only to make a sound electrical connection.

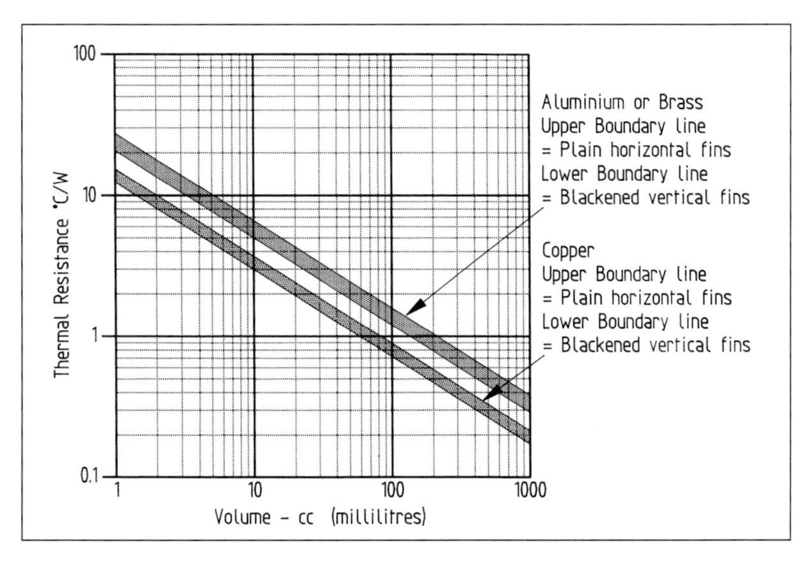


**Fig 25.19: Heatsink thermal resistance / size chart**

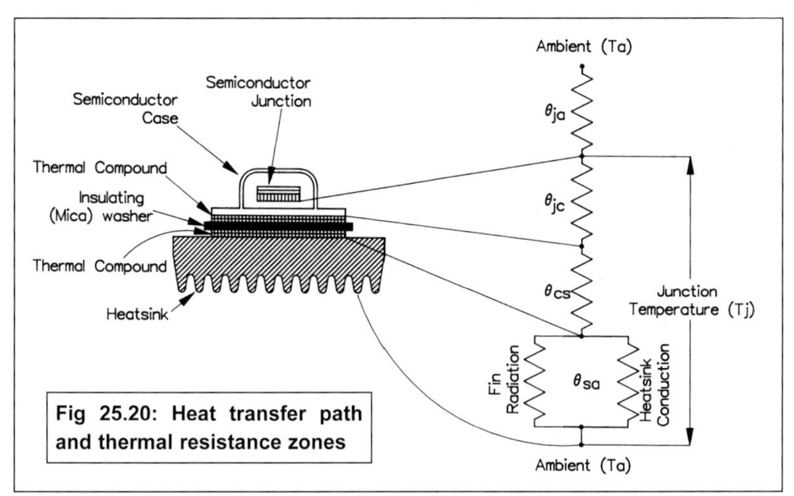


**Fig 25.20: Heat transfer path and thermal resistance zones**

## HEATSINKS

Heat levels for semiconductors are usually defined as 'not to be exceeded' junction temperatures. The power dissipation at these junction temperatures is specified in watts, and can range from milliwatts to several hundred watts. Whenever the device is operating, heat is being generated. Unless some means of heat dissipation is used, the recommended operational limits will be exceeded very rapidly and the device will fail. A semiconductor junction takes microseconds to reach its operating temperature and only a few more microseconds to destroy itself if heat is not dissipated. Electronic protection circuits can help safeguard the device but will not eliminate the requirement for heat dissipation. Semiconductor devices generate heat very rapidly and it is essential to ensure proper cooling. The heat must be dissipated at the same rate as it is being generated in order to ensure that the device remains within its specified temperature operating limits. The usual system consists of a heat-conducting material which takes the heat from the device and transfers it to the atmosphere. The materials used for heatsinks should have high thermal conductivity to conduct the heat away from the heat source, and high emissivity to radiate this heat to the atmosphere. Emissivity depends on the surface finish and texture of the radiating elements, rather than the type of material, but thermal conductivity depends on the type of material. Most commercial heatsinks are made from aluminium, for it is lighter in weight, cheaper and can be extruded or cast into the complex shapes required for compact, efficient heatsinks. Aluminium can also be anodised and dyed black to increase its emissivity. Heatsinks should be made and positioned such that the fins radiate to atmosphere. Fan cooling enables heatsinks to be smaller for the same thermal resistance, and as a guide, a substantial airflow over the fins can reduce the existing thermal resistance by up to 40%, depending on the number of fins receiving air.

Painting a natural metallic finished heatsink with a very thin coating of matt black paint will also produce a reduction in overall thermal resistance by increasing the emissivity. The actual amount of reduction depends on the mattness of the black paint and the thickness of the coating but a 10 to 25% improvement could be expected.

**Fig 25.19** shows the thermal resistance possible at various volumes for different metals, finishes and fin positioning. The bands indicate the effects of a matt black finish and the vertical or horizontal positioning of the fins. The top edge of each band represents the thermal resistance of natural-finished horizontal fins. The bottom edge represents matt-black-finished vertical fins. The band for aluminium or brass is an approximation, for the thermal conductivity of either metal varies widely depending on the alloying metals, but it can be considered reasonably accurate for most amateur purposes.

The heat transfer path is considered as flowing from the semiconductor (heat source) to ambient air, with each junction or transitional point treated as a thermal resistance (**Fig 25.20**). The thermal resistances of fins and heatsink to ambient are considered in parallel to give the total heatsink value. Similarly the thermal resistance of the device junction to air is considered to be in parallel with the total thermal resistance.

*Heatsink formulas:*

$$P_D = \frac{T_j - T_a}{\theta_{jc} + \theta_{cs} + \theta_{sa}} \qquad (1)$$

and

Total thermal resistance x Power dissipation = Temperature rise in °C above ambient of the transistor junction (2)

from which:

$$\theta_{sa} = \frac{T_j - T_a}{P_D} - (\theta_{jc} + \theta_{cs}) \qquad (3)$$

also:

$$T_j - T_a = P_D \times \theta_{ja} \qquad (4)$$

where:

$T_j$ = Maximum allowable junction temperature (°C)
$T_a$ = Ambient temperature (°C)
$\theta_{ja}$ = Thermal resistance, junction to air (°C/W)
$\theta_{jc}$ = Thermal resistance, junction to case (°C/W)
$\theta_{cs}$ = Thermal resistance, case to heatsink, plus any insulating washer and heat-conducting compound (°C/W). (Can be assumed to be between 0.05 and 0.2°C/W).
$\theta_{sa}$ = Thermal resistance, heatsink to ambient air (°C/W).
$P_D$ = Power dissipation (W)

Readily available semiconductor data do not always specify all of the above details but approximations may be used to design a heatsink suitable for most amateur applications. The usual data includes $P_{tot}$ in watts ($P_D$) at a case temperature of usually 25°C. This can be used to derive the junction temperature ($T_j$) for substitution in formula (1). Because $\theta_{ja}$ = 25°C/$P_{tot}$ and $T_j$ = 25°C/$\theta_{ja}$ then $T_j = P_{tot}$°C.

Example: A 2N3055 is to be used as the pass transistor in a 13.8V DC regulated PSU. The available data shows $P_{tot}$ as 115W at 25°C, thus $T_j$ = 115°C and $\theta_{jc}$ = 0.2°C/W (25°C/$P_{tot}$). It is estimated that, for most applications, $\theta_{cs}$ = 0.1°C/W, assuming thermally conducting compound and a mica insulating washer are used. (If no insulating washer is used, $\theta_{cs}$ = 0.05°C/W but the heatsink may be 'live'.) Substituting in formula (3):

$$\theta_{sa} = \frac{115 - 25}{115} - (0.2 + 0.1)$$

$$= 0.3°C/W$$

which is the heatsink thermal resistance required to ensure correct thermal operation.

The above example assumes that the 2N3055 is working at its limit, and this should seldom be the case. In this example it would be usual to supply the pass transistor with 18V DC to allow adequate regulation, and the transistor would be run at 5A, or half the rated amperage. The difference between the supply voltage and the output voltage is 18 - 13.8 = 4.2V DC, hence the power to be dissipated will be 4.2 x 5 = 21W. Substituting this in formula (3) gives:

$$\theta_{sa} = \frac{115 - 25}{21} - (0.2 + 0.1)$$

$$= 4.0°C/W$$

This is the thermal resistance required for this power supply. If necessary a further allowance could be made for the inefficiency of the semiconductor and the calculated thermal resistance decreased to allow for this. In this example this factor has been ignored.

From Fig 25.19 it can be seen that about 6cm$^3$ of blackened or 9cm$^3$ of natural finished copper (vertical fins) will be a suitable minimum size. **Fig 25.21** shows a typical home-made heatsink based on the above calculations. Ideally the two bent plates should be soldered together to minimise any thermal resistance at this junction. Thermal conducting compound should be used if this joint is not soldered. The copper should be annealed before bending.

Using a heatsink with a lower-than-calculated thermal resistance will not affect the operation of the cooled semiconductor.

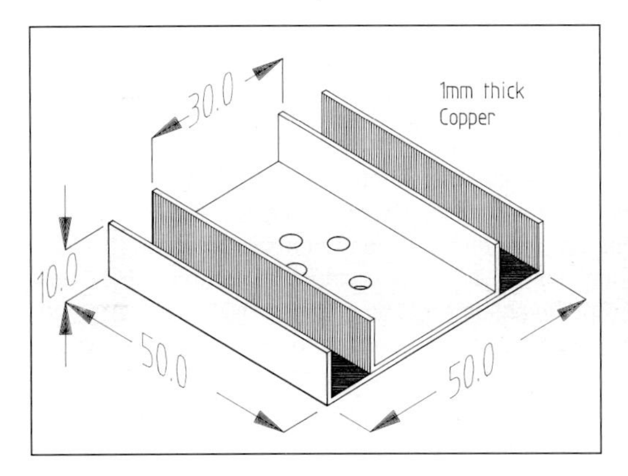


**Fig 25.21: Heatsink based on the calculations of the example in the text**

**Fig 25.22: Power mosfet on a substantial heatsink**

It would be unwise to use a heatsink with a higher-than-calculated thermal resistance. In other words, a larger heatsink is better than one too small (**Fig 25.22**)!

In use the heatsink will become warm to the touch but it should not become untouchable. If it does, then the semiconductor providing the heat is probably being overworked. It would be worthwhile checking the calculations and ensuring that each mating joint is made properly. Some semiconductor devices are designed to run at high temperatures and a hot heatsink would be expected. This heat represents wasted energy, and the necessity of large cooling systems, though seeming to be the norm, indicate inefficiency.

Professional designers use very sophisticated computer programs to design their heatsinks. Even then they rely on trials and measurements to tune-up their results! The above formulas and chart are accurate for most amateur purposes, eliminating the need for complex equations or computers.

Heat can also be dissipated from the connecting pins, together with the case, of semiconductor devices, and this can be useful if trying to keep the finished unit compact. A heatsink can be attached to the pin side of the board by a layer of thermally conducting, electrically insulating, elastomer and mechanically fastened to the board with screws. This method can provide about a 5 to 10% reduction in the required overall size of the main heatsink. Large soldering pads for the pin connections also act as heatsinks, and these can be used in conjunction with a suitable heatsink for such devices as audio amplifiers, power regulators, rectifiers etc.

The duty cycle of the device also affects the size of heatsink required. A 50% duty cycle can allow about a 20% reduction in the size of heatsink, depending on the heatsink design. During the OFF cycle, the heatsink must be able to dissipate all of the heat generated during the ON cycle, and it is usual in these circumstance to provide forced-air (or water) cooling to ensure that this is achieved. Many solid-state transceivers specify ON and OFF times for continuous full-power carrier (FSK, AM etc) operation. The small handheld transceivers start to get very warm if they are transmitting for longer than they are receiving. In the case of handheld transceivers, the batteries usually limit the transmit time but, if such transceivers were operated from a mains PSU, it would be essential to ensure that the heatsink arrangements were adequate.

## CIRCUIT CONSTRUCTION

Every electronic circuit consists of a number of point-to-point electrical connections. Modern mass produced electronic circuits are usually built on some form of printed circuit board. Older production methods use a variety of point to point connection methods, usually involving tagstrip or stand-off terminals and insulated wire.

Many amateur circuits need to be functional rather than reproducible. If the project is to be housed in a screened enclosure, it won't really matter whether it is visually ugly or attractive. For portable operation, weight or size can be the most important factor. There are many different construction methods used by home constructors. Before you start building your project, you should consider the various options available and try to choose a method that best suits your needs.

The following list suggests some of the factors worth considering before starting on any circuit design and making:

1 Permanent or experimental?
2 How will the circuit/s be housed and mounted?
3 What power supplies are required (battery, mains, internal, external etc)?
4 What types of inputs/outputs are required?
5 What controls are required?
6 What safeguards are required (eg accidental switch-on, wrong polarity, over-voltage etc)?
7 Methods of construction. (Can you make it?)

In the rush to make the circuit, it is all too easy to discover afterwards that some of the above factors were important! Many solid-state devices are susceptible to damage by static discharge. Handling and soldering such components requires care. It is safer to assume that all semiconductor devices are prone to static discharge damage and treat accordingly. Assembly and repair should be carried out using the normal anti-static precaution of connecting the soldering iron, PCB and operator to a common point. This equalises the static level and is further improved if the common point is properly grounded.

Commercial equipment is available, consisting of an anti-static mat, wrist strap and connecting leads for the other tools. Semiconductors which are very sensitive to static are usually supplied mounted in a metal strip or wire clip (or a carbon-conductive foam in the case of ICs) which shorts together every pin of the device. This anti-static protection should be left in place until the device is plugged or soldered in position, and then carefully removed. A suitable size 'Bulldog' clip connected to the ground mat by a length of flexible wire makes an excellent anti-static connector for ICs.

## Experimental, Temporary or Prototyping Circuits

One of the most important features in an experimental or prototype circuit is the ability to make quick and easy changes to the circuit. The effect of the changes is then observed and further changes are easily made if necessary. If you take a wrong turn in the development of your circuit and the modifications are unsuccessful, the circuit can easily be restored to it's original configuration.

The word 'breadboard' is often used to describe an experimental circuit. At one time, it was common practice to use a wooden board and conductive nails or pins to keep the components in place. The components and interconnecting wires could be soldered to the nails or simply wrapped around them. Since the days of the original wooden breadboard, both amateur and professional constructors have devised many ingenious methods of circuit prototyping. We will examine some of the most popular methods in detail.

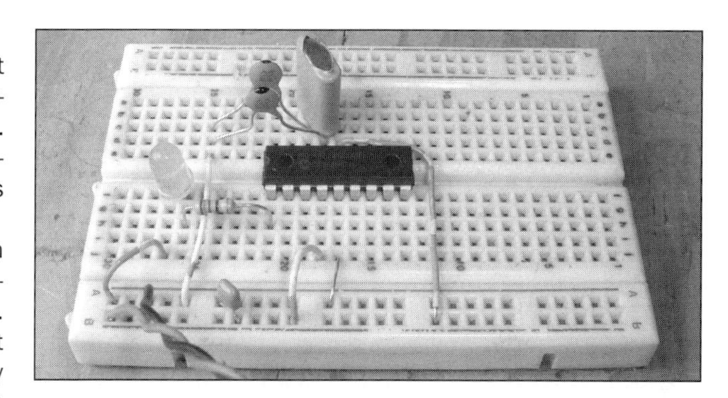

**Fig 25.23: Modern breadboarding system**

The modern equivalent of the old fashioned breadboard is still known by the same name as it's predecessor. Breadboard prototyping systems are available from most component suppliers. The most common type of breadboard is shown in **Fig 25.23**.

This type of prototyping system uses a 0.1in matrix of holes for mounting the components. This is ideal for mounting DIP ICs and other components with a 0.1in lead spacing. Each row of holes is numbered and connected internally to the other holes in the same row. There is usually a gap at the centre of the breadboard to facilitate the mounting of ICs. Such breadboards are available in a variety of sizes and some of the more expensive types come with a built in power supply.

One of the simplest and most popular construction methods uses a sheet of copper-clad PCB laminate or other conductive material as a base or ground plane for the other components. Any components that require a connection to ground can be soldered directly to the metal ground plane. Other components can be soldered together using copper wire or mounted on insulating stand-off terminals. If PTFE or ceramic stand-off terminals are not available, high value resistors can be used instead. A 10MΩ resistor is not likely to have any significant effect on the operation of a typical circuit.

This method is widely known as 'dead bug construction', probably because of the resemblance of an inverted IC to an insect lying on it's back. This is one of the most flexible of all prototyping systems. Apart from a soldering iron, no special tools are required for this construction method. Changes or additions to the circuit are trivially easy and this technique probably offers the fastest and simplest practical method of getting your circuit from a paper schematic to an assembled circuit. **Fig 25.24** shows a dead bug circuit of an audio preamp and power amplifier.

There are several types of point-to-point construction which are based on a collection of isolated pads or 'islands'. One such method uses a grid of square or rectangular pads that are cut or

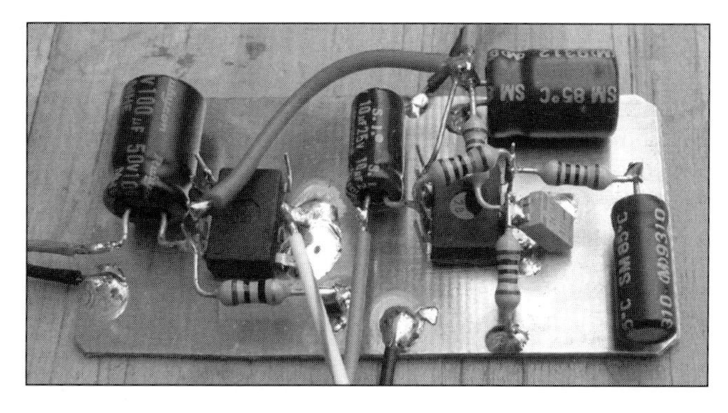

**Fig 25.24: 'Dead bug' construction**

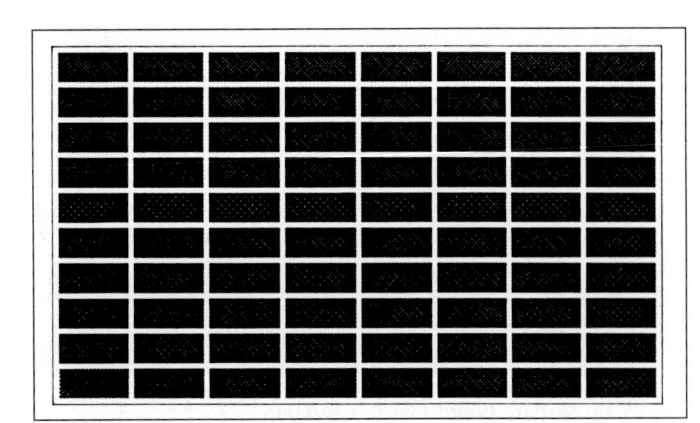

Fig 25.25: Grid pattern PCB for experimental circuits

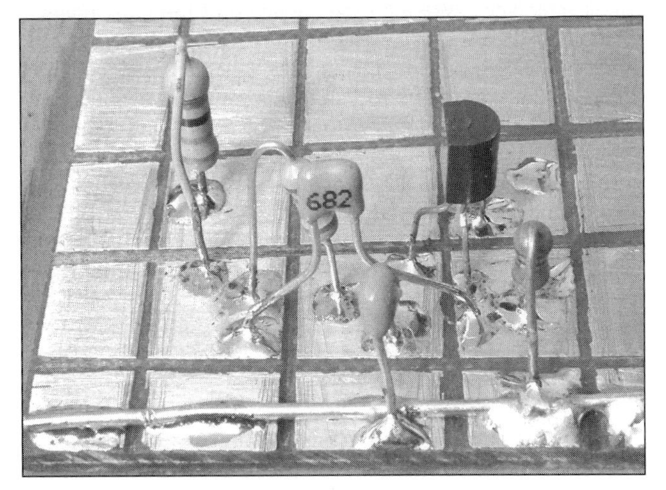


Fig 25.26: 'Pad' or 'island' construction

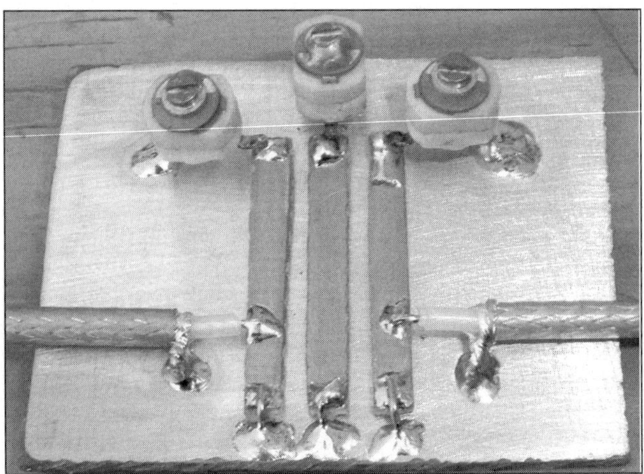

Fig 25.27: Microwave filter built using glue-on copper laminate

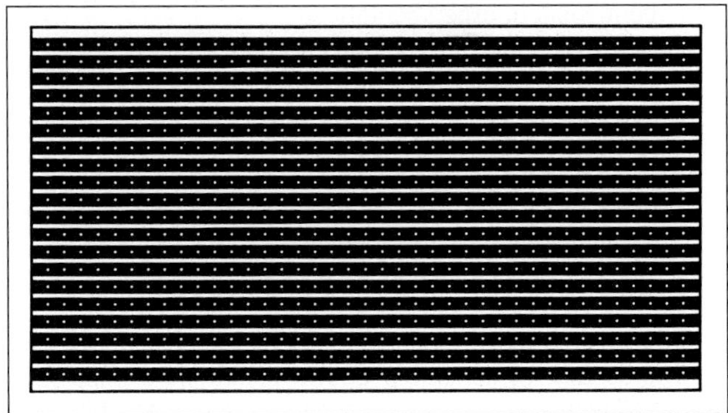

Fig 25.28: Strip board PCB suitable for experimental or permanent circuits

etched on the copper side of copper-clad PCB laminate (**Fig 25.25**). The component leads are soldered directly to the copper pads. Connections to adjacent pads can be made with a blob of solder. Connections to more distant parts of the circuit can be made using insulated wire. **Fig 25.26** shows an active low pass filter built on a square grid which was cut into the face of a copper-clad board using a hacksaw blade. The components can be surface mounted on the copper side of the board or if the board is drilled, the components can be mounted on the opposite side of the board. This is probably the easiest way of making a double-sided circuit board.

A few variations of the pad/island method have evolved over the years. Some builders use small square or circular pads made from PCB laminate which are glued to a copper ground plane. One example of this type of construction is known as 'Manhattan Construction', presumably because it resembles an aerial view of city blocks. Each pad is glued to the copper plane using super glue. Each component can be mounted on a collection of small generic pads or alternatively, special dedicated pads can be made for each type of component. Special pads for IC, transistors, etc can be cut from PCB laminate. At least one company sells ready made pads for this type of construction. As with dead bug style construction, ground connections are soldered directly to the copper plane.

Dead bug and Manhattan style construction are excellent ways of making UHF and microwave circuits. Resonant sections of transmission line can be formed by glueing rectangular strips of single-sided PCB to a ground plane. Microstripline circuits made by this method are much more forgiving of errors in construction or design than conventional etched PCBs. If a line is found to be slightly off resonance, it can be lifted off the board and re-trimmed or replaced by an new one. A similar problem with an etched PCB would usually require the replacement of the entire board. Glue-on microstripline circuits are very easily interfaced with dead bug and other similar methods of construction. **Fig 25.27** shows a UHF band-pass filter made from three narrow strips of copper laminate.

Strip board offers many of the advantages of conventional PCB construction. A typical strip board is made from SRBP (synthetic resin bonded paper) or other composite material. The board has a 0.1in x 0.1in matrix of holes and parallel copper tracks with a 0.1in spacing on one side of the board (**Fig 25.28**). The components are mounted on the top side of the board and the component leads are soldered to the tracks on the bottom of the board. A special cutting tool is used to cut any required gaps in the tracks. A sharp knife or drill bit makes a good substitute for this tool. Strip board is also available as 0.1 matrix board without any copper tracks. This type of matrix board is used as a framework for building circuits. The component leads and connecting wires are soldered together at the back of the board. Both types of strip board are often used in combination with wire wrap pins and tools. Wire wrap is a method of assembling circuits which uses special pins, wire and a wire wrap tool (rather than solder) to make electrical connections. **Fig 25.29** shows a circuit built on stripboard.

One method we have not considered so far is virtual construction. Instead of actually building a prototype, it is possible to create a mathematical model of the circuit and use electronic simulation software to test the circuit. Suitable software is available for most computer platforms and operating systems.

Simulation can be a very valuable tool for the circuit designer. It can sometimes allow you to test circuits that would otherwise

Fig 25.29: Stripboard construction

be difficult to design and build. Testing on a simulator can give the designer the opportunity to experiment with component values when fine-tuning a design. Changes can easily be made to a circuit in a matter of seconds. All but the most complex simulations will take no more than a few seconds to run on a modern computer.

Electronic simulation software can give very accurate results, but it is important that the user has a good understanding of the software and it's limitations. A simulator is only as good as the mathematical algorithms used by the programmer that created it. Even if the simulator is very accurate, it is only going to produce accurate results if it is given the right data. Even a seemingly small error in creating the model can have a dramatic effect on the simulation results. A capital 'M' instead of a lowercase 'm' or vice versa could turn millivolts into megavolts or megohms into milliohms. Computer software often lacks any kind of sanity check so it might allow to to put 100 megavolts across a 10 ohm resistor without asking if this is really what you intended to do.

**Fig 25.30** shows a simulation of a VHF power divider and **Fig 25.31** shows the predicted frequency response of the circuit. The simulation software used is QUCS (quite universal circuit simulator) [2] which is free and open source software for Windows, Linux and other Unix based systems like BSD and Mac OSX.

## Permanent Circuits

Although they are described as experimental or prototyping circuits, all of the circuit construction methods mentioned so far are quite suitable for use in permanent circuits. Some of my dead-bug circuits have given more than 20 years of reliable service and will hopefully continue to work for many more years.

## Printed Circuit Boards

Almost all modern electronic circuits are built on some form of printed circuit board. The most common type type of PCB consists of one or more layers of copper foil which is bonded to an insulating substrate. The substrate is usually made from a composite material like fibreglass/epoxy or SRBP. Some specialised PCBs use a ceramic substrate for use at high temperatures or as part of a hybrid module.

Most of the PCBs made by home constructors will be either single-sided with a single conducting layer on one side of the board, or double-sided with a copper layer on both sides of the board. Multi-layer boards with more than two layers are commonplace in commercially made equipment, but they are rarely seen in home made equipment.

Fibreglass PCB is probably the best choice for home constructors. Fibreglass is very durable and it has very predictable electrical characteristics. The great strength and durability of fibreglass PCB makes it quite difficult to drill and cut. Always wear a mask and work in a well ventilated area when cutting, drilling or

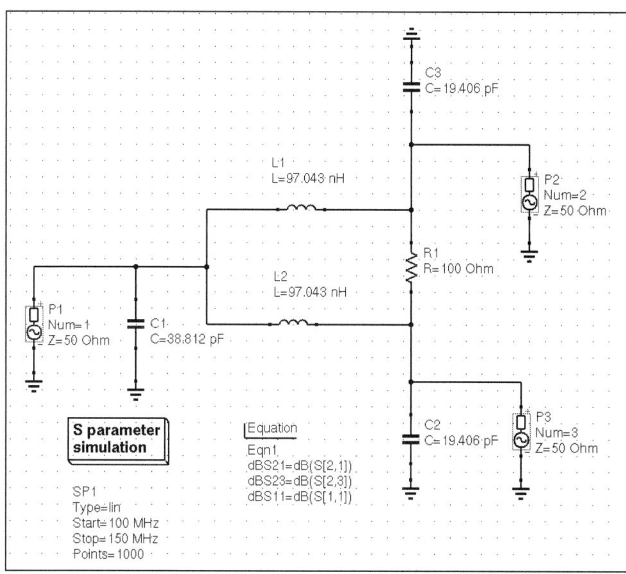


Fig 25.30: Computer simulation of a VHF power divider

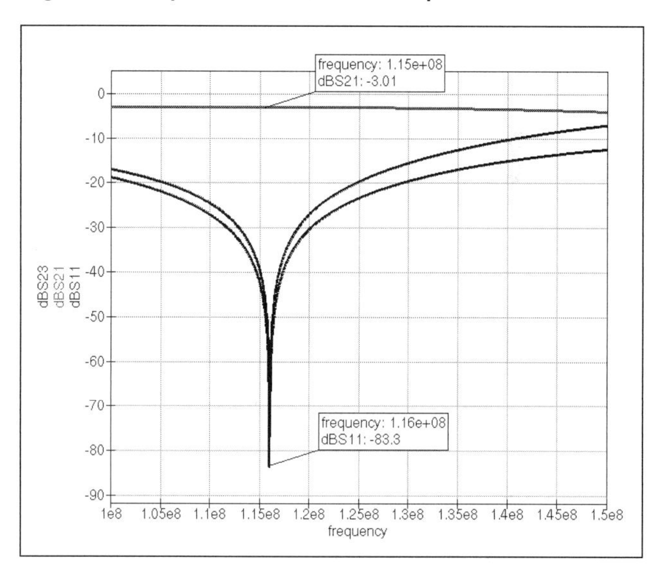


Fig 25.31: Computer-predicted frequency response of simulated circuit from Fig 25.30

grinding fibreglass. SRBP board is cheap and readily available. It is easier to drill than fibreglass, but it can also be difficult to cut with a saw because it cracks very easily.

Before we can make a PCB, we must first design a suitable track layout for the circuit. The rules for designing a PCB layout are much the same as for any other type of circuit construction. The designer must ensure that adequate grounding, decoupling and screening are used at critical points in the circuit. The input and output of high gain amplifiers must be well isolated from one another. Ground connections should be short and direct. One approach that is recommended for radio frequency or high-speed digital PCBs it to use double-sided PCB and use most of, or even all of one side of the board as a ground plane.

For a simple one-off PCB, the track pattern can simply be drawn directly on the copper PCB with an etch-resist pen. Some types of permanent waterproof pens can be used as etch-resist pens. I have had good results with fine laundry markers and CD/DVD marking pens. The board is then immersed in an etchant which will dissolve the exposed copper, leaving only the tracks that are covered by the etch-resist ink. Etch-resist pens and suitable etchant are available from many component suppliers. After the etch resist is removed from the board, it can be drilled and the components mounted on the board and soldered into place.

## General PCB guidelines

1. Keep mains and high-voltage circuits well away from other circuits, preferably by using separate boards. In the interests of safety, any tracks etc carrying high voltages should be covered by an insulating material to prevent accidental contact.
2. Avoid high current density points. Rapid changes of line width should be avoided, particularly in circuits carrying several amperes. The recommended current density I in amps is:

   $$I = 3(\sqrt[3]{w^2})$$

   where w is the width of the copper track in millimetres. Circuit board tracks make excellent fuses if incorrectly designed!
3. The capacitance C in picofarads between each side of a double-sided board is:

   $$C = 0.0885 \times K \times (A/h).$$

   where A is the area (sq cm) of the smallest side; h is the thickness (cm) of the laminate excluding copper, and K is the dielectric constant (relative permittivity) of the laminate material, which for glassfibre is about 4.5.

## Creating the PCB artwork

There are many different ways of creating the track pattern for PCBs. Instead of drawing the pattern directly onto the copper board as described above, the pattern can be drawn by hand on white paper and then photocopied onto a clear acetate overhead projection film which can be used to produce a PCB by photographic methods. Some of my first PCB layouts were created using rub on dry transfers for the component pads and rub-on tape for the tracks.

Such techniques are rapidly becoming obsolete due to the ready availability of cheap personal computers and PCB CAD (computer aided design) software.

The design of simple PCB layouts is not very demanding in terms of computer resources. A simple PCB CAD program will run on a very humble PC. One of the computers that I use for PCB design has an 850MHz CPU and 384MB of RAM. This computer was rescued while it was on its way to the recycling centre. Such a computer would be considered by many as little more than a doorstop and could be acquired at little or no cost. The Linux operating system and PCB [3] CAD software are both free and open source, so the total expenditure on hardware and software was somewhere in the region of zero.

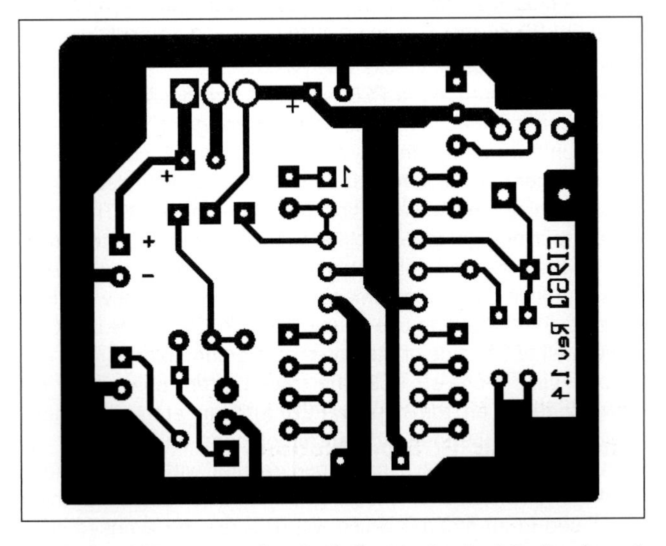

**Fig 25.32: PCB drawn using the industry standard Gerber format for CAD**

If you don't have a PC in the shack, you really should consider getting one. Even a very basic model can be a very useful design tool. See the chapter on computers in the shack for many other amateur radio applications for a computer.

The size and complexity of PCB design software varies from the very simplest CAD program which can be used to draw single or double sided track layouts up to very large and complex EDA (electronic design automation) packages costing thousands of pounds. If you don't require advanced features like schematic capture, auto routing and circuit simulation, an ordinary vector drawing program can be used to draw the track layout. However, it is probably worth spending a little time and effort to learn how to use proper PCB layout software.

You should beware of "Free!" CAD software because it often comes with strings attached. Some commercial PCB houses provide free CAD software for download. This free software often uses a proprietary file format that can only be used by the same PCB manufacturer that gave you the 'free' software. Unless you want to be tied to a single PCB supplier, you should make sure that your PCB CAD software can produce output files in the industry standard Gerber format. **Fig 25.32** shows a PCB layout drawn using PCB CAD software.

Once the PCB artwork has been created, we must find a way of getting the track pattern from the computer screen onto the copper-clad board. Commercial PCB makers use a special high resolution plotter to print the layout onto a clear film which can be used as a master artwork for the PCB production process. Alternatively, the artwork is printed on paper and then photographed with a special camera. It is possible to make PCBs at home using similar methods.

Before we consider these methods, we will look at an even easier way of making prototype PCBs [4] The plastic toner used in ordinary laser printers and photocopiers is quite waterproof and makes a very good etch-resist. Special PCB transfer films are available from specialist PCB suppliers and some electronic component suppliers. The film is available in A4 size sheets. The process for making the PCB is quite simple. First, a mirror image of the PCB track pattern is printed onto the transfer film using a laser printer or photocopier. After the copper PCB has been carefully cleaned, the transfer sheet is placed toner side down in contact with the copper PCB and the toner is ironed onto the copper using a standard clothes iron. The transfer sheet is then carefully peeled away from the copper board, leaving the toner behind. At this stage, any flaws or gaps in the track pattern can be carefully touched up with an etch resist pen. After the board has cooled down, it can be etched in the same manner as a hand drawn or photographically produced board.

Many home constructors have found that the type of glossy photo paper that is normally used in inkjet printers makes a good substitute for the special toner transfer film. I have had particularly good results with Epson branded Glossy Photo Paper. Before we progress to the printing stage, a few words of warning are in order.

Putting anything other than the manufacturers recommended paper or film through your laser printer or photocopier could damage it. Using the printer in this manner will probably invalidate the manufacturers warranty. The printer/copier fuser unit is a high temperature roller which is used to melt the toner into the surface of photocopy paper. Some types of plastic OHP film or plastic coated paper could make a horrible mess of it. If in doubt, you can check the heat resistance of your paper by running a hot clothes iron over it. If you have any reservations about putting inkjet paper through your laser printer, then you should buy the 'proper' transfer film instead. You have been warned! I have had good results with TEC200 film in the past. The most popular type of film currently on the market is called Press-N-Peel.

The track pattern should be printed onto the transfer film using an ordinary black laser printer or photocopier. Since most PCBs will be much smaller than an A4 sheet, you should be able to make several PCBs for each sheet of film. If you use transfer film, you should pay close attention to the manufacturers instructions. If you are using glossy photo paper, the track pattern should be printed on the glossy side of the paper using a laser printer and not an inkjet printer.

Remember that the process of ironing the toner onto the surface of the board will reverse the image. The image printed on the film must be a mirror image of the track pattern that will appear on the finished PCB. use the print mirror image function of your PCB CAD software. If you are using ordinary drawing software, you will find that most if not all drawing software will have a 'flip horizontal' function that will produce the desired effect. Before you commit your track pattern to an expensive sheet of transfer film, you should first print a copy on ordinary A4 paper. Hold the paper up to the light and make sure that the track pattern is correct. Check the size of the image carefully. If the pattern contains pads for an IC, you can use a spare IC or an IC socket to check that the pin spacings are correct. Once you are satisfied that the track pattern is ok, the image can be printed on the film. When the film has been printed, you should avoid handling it. Fingerprints on the film will lead to poor adhesion of the toner to the PCB.

The edges of the film or glossy paper should be trimmed with a scissors so that the remaining film is just slightly bigger than the track pattern. The PCB should be cleaned with very fine steel wool, rinsed thoroughly and dried with a lint-free cloth. The transfer film should then be placed toner side down on the copper PCB. The film can be secured to the PCB using a single length of paper masking tape along one edge only. Vinyl tape is not suitable for this job. The PCB and transfer film should be placed on a heat resistant platform before ironing. I use the back of a telephone directory for this purpose. The instructions below are for glossy photo paper. For special toner transfer films, you should follow that manufacturer's instructions instead.

The toner is then ironed onto the surface of the board using a hot clothes iron. The amount of time and pressure required will depend on the temperature of the iron and the type of toner used in the printer or copier. I use a hot iron at close to maximum setting. The iron is placed on the back of the photo paper

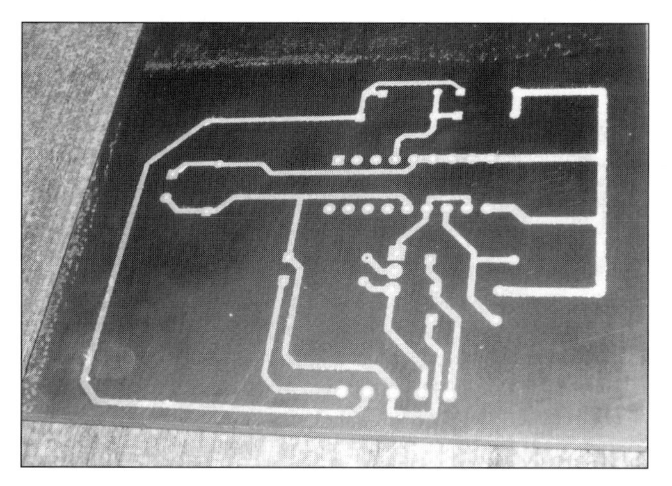

Fig 25.33: Home made PCB after ironing

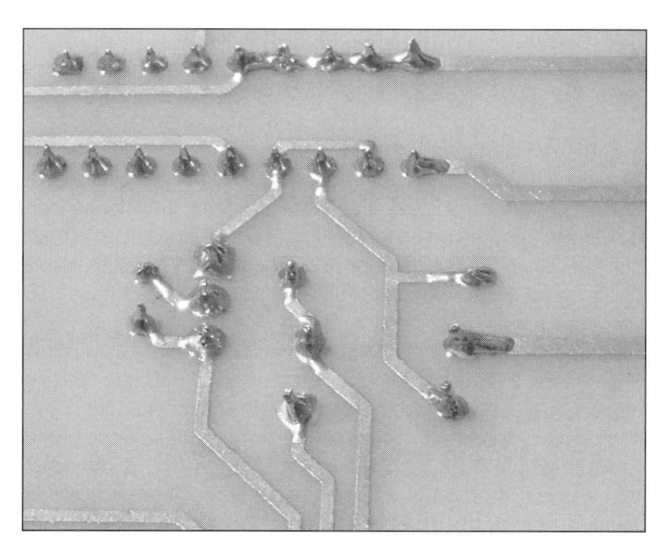

Fig 25.34: Home produced PCB

and pressed firmly down for about 10 seconds while the toner sticks to the PCB surface. The iron is then moved slowly back and forth for a further 15 to 20 seconds. The amount of pressure required will depend to some extent on the complexity and size of the PCB pattern. Heavy pressure on the iron will tend to flatten the toner and tends to make lines slightly thicker and pads slightly larger. Do not attempt to peel the photo paper off the PCB at this stage. The board should be placed in warm soapy water which will soften the photo paper. You should then unplug the iron and leave the room immediately. All homebrew radio enthusiasts have a natural curiosity and a tendency to poke at things, even when we know we shouldn't.

Leave the board in the warm water for at least twenty minutes before you try to remove the photo paper. By this time, the paper will be very wet and should be easy to remove. Any remaining paper residue can be removed by brushing it off with your finger tip while holding the board under running water. Very stubborn residue can be removed with a coarse sponge. Don't be too afraid of damaging the track pattern, the toner is usually quite durable. Any small gaps in the toner can be repaired with an etch-resist pen. **Fig 25.33** shows the board after ironing. The board is now ready to be etched.

The toner transfer method is a good way of making one-off PCBs. The resulting boards are not quite as good as photographically produced boards from commercial PCB makers, but as you will see from **Fig 25.34** they can come quite close and are certainly good enough for most amateur projects. Most PCB CAD software can be used to generate one or more silk-screen layers. The silk-screen layer can be used to show the outline of components, component values and other text like PCB revision information.

If you want to give your home made PCBs the professional look, you can use the toner transfer method to print the silk-screen layer on the top of the PCB. Simply iron a mirror image of the silk-screen layer onto the top of the PCB. The PCB should be drilled first so that the component holes can be used as guide for positioning the transfer film. **Fig 25.35** shows the silk-screen layer of a home made PCB. The term silk-screen is usually used to describe this layer, even when a printing technology other than silk-screen printing is used.

## Photographic PCB production

The best way to make very high quality PCBs is to use the same photographic methods that are used by commercial PCB makers. One of the biggest advantages of the photographic method is the ability to make many identical PCBs from a single master artwork.

Most photo sensitive PCBs are coated with a positive photo-resist. This means that we will need to use a positive artwork in which the copper areas are opaque and the copper-free areas are transparent. Note that some PCB production systems require a negative master artwork. The PCB artwork is created using a CAD system as used for the toner transfer method above. The artwork is then printed onto a clear film. A laser printer or photocopier can be used to print the artwork onto a clear acetate overhead projection film. You must use the type of overhead projection film that is designed for use in photocopiers and laser printers. I have had very good results with Staedtler brand OHP laser film.

Photo sensitive PCB is coated with a special UV light sensitive photo-resist. To transfer the track pattern to the PCB, the positive artwork is placed in close contact with the photo-resist and the board is exposed to a source of UV light. After exposure the board is developed in a bath of developing solution. The photo-resist that was exposed to UV at the exposure stage of the process is washed away by the developer, leaving only the photo-resist that was protected by the opaque parts of the master artwork.

### Exposure

The first, and probably the most critical step, in the process is the exposure of the photo-sensitive PCB laminate to a UV source. A high quality master artwork is the key to success. The printed image from most laser printers and copiers is not quite opaque enough to make a good master artwork. You should adjust the printer settings for maximum toner density for best results. I have found that it is often necessary to stack two identical copies of the printed artwork to obtain the required opacity. The two sheets must be very carefully aligned. Most PCB CAD software will produce alignment marks which will help you to achieve precise alignment. When you cut out the two copies of the master artwork from the A4 sheet, one copy should be cut slightly larger than the other. The larger copy should be placed on a flat surface and the smaller copy placed on top of it. When the two sheets have been perfectly aligned, A strip of Sellotape can be placed along one edge of the upper sheet to secure it in position. Hold the artwork up to the light and check it for any flaws or alignment errors. Note that the toner should be printed on the side of the OHP film which will be closest to the PCB. If the toner is printed on the wrong side of the film, some UV light will be able to creep around the edges. This can lead to thinning and fuzzy edges of the PCB tracks.

Some home constructors have reported that they can make good master artwork using an inkjet printer and the type of OHP film that is specially designed for use in inkjets. My earlier attempts to make PCB artwork using a standard low cost inkjet printer were far from successful. Even though the printed artwork looked quite good, the black ink from my old inkjet printer seemed to be almost completely transparent to UV light. My more recent efforts have been much more successful.

I now use a photo quality HP D5160 inkjet and Stephens brand inkjet film from a local newsagent. This produces a very high quality positive artwork which is at least as good as the best results I have seen from the laser printer. With a B&W print resolution of 1200dpi and even higher resolution in photo mode, this type of printer can produce artwork which is good enough for the very fine tracks and pads used for surface-mount components. You should check that the type of film you are using is compatible with your inkjet printer. However, inkjets are much less prone to damage than laser printers because they don't use a high temperature pressure roller (toner fuser). Even in the worst case scenario where the print head is destroyed, HP and several other brands of inkjet have the print head incorporated into the ink cartridge. If you do manage to damage the print head, a new cartridge will probably cure the problem. As with the artwork produced by the laser printer, I often use two stacked sheets of film to achieve the best possible results.

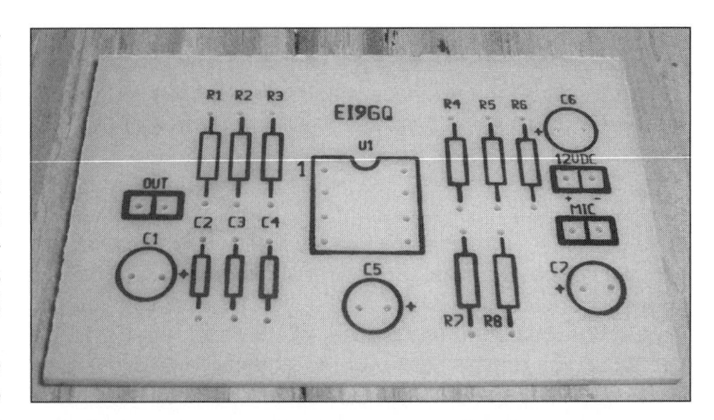


**Fig 25.35: Silk screen layer on home made PCB**

The PCB can be exposed using a commercial or home made UV exposure unit. A typical unit will consist of a lightproof box containing UV fluorescent tubes, a sheet of glass for mounting the PCB and a timer. UV light can be very harmful. A home made exposure unit would have to be built in a totally enclosed box and the lid should be fitted with a safety switch to ensure that the tubes can never be illuminated while the box is open. A simple timer mechanism should also be included. Exposure to high intensity UV can lead to eye damage or even total blindness. The type of UV tubes used in EPROM erasers and the germicidal tubes used in sterilising units produce short wavelength UV which is considered particularly dangerous.

Ultra violet fluorescent tubes of the kind used for disco lighting (black light), banknote scanners, suntan lamps or insect zappers can be used for PCB exposure.

Ordinary fluorescent tubes can also be used, but exposure times will be much longer. A single Philips 30 Watt TLD 30W/33 tube gave good results with an exposure time of one hour at a distance of 10cm. Many constructors use photographic equipment to expose PCBs. A photographic contact frame can be used to hold the PCB and artwork and a photoflood bulb used as the UV light source. Optimum exposure times will vary depending on the power of the bulb and the distance of the PCB.

One very obvious source of UV light is the sun. The PCB exposure frame is simply exposed to direct sunlight for a few minutes. This method can sometimes produce excellent PCBs, but the unpredictability of the weather makes it difficult to get consistent results. I have found that the optimum exposure time to direct sunlight is somewhere between two and five minutes. My tests were performed under a clear blue sky in the early afternoon.

After many years of experimenting with different types of UV lamps, I have found that the best UV light source can be purchased for no more than a few pounds from any DIY or electrical supplier. Quartz-halogen lamps of the type used for garden lights and security lighting produce enough UV light to expose a PCB in a few minutes. I use a 500W halogen lamp and a very simple exposure frame made from a small sheet of glass. The glass from a small picture frame is ideal for this purpose. The OHP film master artwork is secured to the glass with a strip of Sellotape and the PCB is held in place by a small clamp made from a strip of aluminium. The 500W halogen lamp produces a lot of heat so the artwork should not be placed too close to it. A 28-30cm spacing between the lamp and the exposure frame ensures that the PCB doesn't get too hot and gives an exposure time of about 10 minutes. The lamp should not be left unattended because of the risk of fire or heat damage. **Fig 25.36** shows the halogen lamp and exposure frame.

**Fig 25.36: Simple exposure unit using 500W security lamp**

A strip of the photosensitised board should be used to assess the exposure time required. **Fig 25.37** shows a suggested exposure testing strip, which should be made of the same materials as those used for the master transparency. Each section of the test strip is progressively uncovered until the complete strip has been exposed. For example, if the timer is set to, say, 20s then each step would be 20s greater than the previous one. This is normal photographic practice. The fully exposed strip is now developed. The image of the test strip should be visible, at least in sections, and at this stage the board is washed thoroughly in running water, dried and then etched. It is possible to assess the correct exposure without etching but etching is more reliable and much easier to see.

The developing is usually carried out by using sodium hydroxide dissolved in water in the ratio of between 7 and 16g per litre, depending on the resist type. Cooled boiled water is preferred. During mixing, the sodium hydroxide tends to produce a boiling effect and great care should be taken to prevent contact with the eyes etc. The usual safety precautions should be observed: glasses, gloves, no children or pets etc. The mixed solution has a limited life and a fresh solution should be made for each batch of developing.

## Developing

The two most commonly used PCB developers are sodium hydroxide and sodium metasilicate. Sodium hydroxide is widely available from most hardware stores as caustic soda or drain cleaner. Both types of developer can give excellent results. I have found that caustic soda is a bit more difficult to use. To get good results, the strength and temperature of the caustic solution must be carefully controlled. Sodium metasilicate based developer is much less critical and will deliver consistent results over a wider temperature range.

PCB developer is a caustic solution which should be kept away from the eyes and skin. You should wear eye protection and rubber gloves when working with the developer. One teaspoon (5gm) of sodium hydroxide crystals (caustic soda) dissolved in 500ml of water make an excellent PCB developing solution. Wait until the crystals have fully dissolved before you use the developer. Sodium metasilicate developer is available in liquid or crystal form from most electronic component suppliers. You should follow the suppliers instructions for mixing and safe handling.

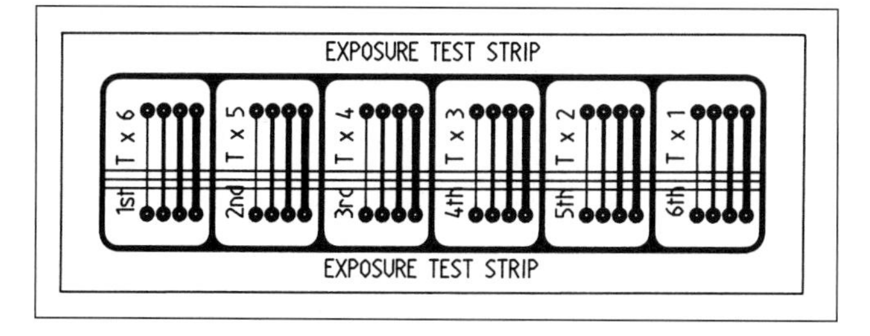


**Fig 25.37: Exposure test strip for photosensitised PCB**

I use a plastic margarine or ice-cream tub as a PCB developing tray. The PCB should be placed resist side up in the developer. Make sure that the PCB is not exposed to direct sunlight or strong fluorescent lights before it is developed. The developing tray should be rocked gently from side to side to while the etch resist is washed away. The developing time will vary depending on the UV exposure time, the strength of the developing solution and the thickness of the photo-resist layer.

After about one minute in the developer, you will see the outline of the PCB tracks appearing on the board. After two minutes, you should start to see bare copper. At three to four minutes, the board should be almost completely developed. The board should be left in the developer for a further minute to ensure that all of the exposed etch resist is washed away. The board should now be removed from the developer and thoroughly rinsed with running water. If your developing time is much longer than 4-5 minutes, you probably need to increase the UV exposure time or slightly increase the strength of the developing solution.

The remaining photo resist should be an exact replica of the master artwork. The board should be handled very carefully at this stage. The resist must not be scratched before the board is etched. **Fig 25.38** shows a PCB in the developing solution.

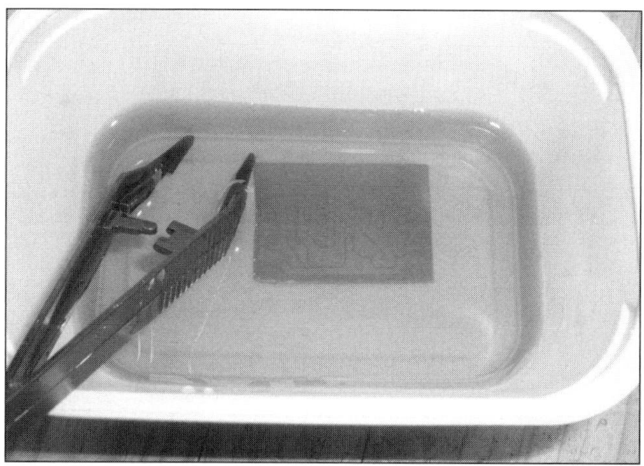

**Fig 25.38: PCB in the developing solution**

## Etching

The two most common types of PCB etchant are ferric chloride and sodium persulphate.

The optimum temperature for PCB etching solutions is higher than room temperature. Commercial PCB etching tanks are fitted with heating elements and mechanical agitation systems to speed up the etching process. Sodium persulphate doesn't work very well at room temperature, so ferric chloride is probably the best etchant for home PCB construction unless you have a heated etching tank. Ferric chloride is corrosive and quite toxic in concentrated form. However, it is probably no more dangerous than some household chemicals like bleach or drain cleaner. Wear rubber gloves and eye protection. Wear old clothes, ferric chloride stains won't wash out easily. Etching time depends on the temperature, strength and condition of the etchant, the thickness of the copper foil and the area of copper to be removed. Boards that are upside down (with the copper facing downwards) will etch faster than boards that are lying copper

Fig 25.39: PCB in the etchant

side up. Double sided boards can be etched standing on edge if the etchant is deep enough. Typical etching times range from five minutes for a heated/agitated tank up to about an hour for an unheated etching tank on a cold day. A few tests with off-cuts of scrap PCB will help you to determine the optimum etching time, temperature and etchant strength.

**Fig 25.39** shows a PCB in ferric chloride etchant. Note that the PCB is floating face-down on top of the etchant. Also note the length of insulated wire that allows for handling of the board without coming into contact with the etchant. The board should be lifted and wiggled from side to side several times during the etching process. This board took about 35 minutes to etch at a room temperature of 22°C.

I would recommend that you use a deep plastic tub rather than the shallow trays that are supplied with most etching kits. Metal tubs should be avoided for obvious reasons! Once the board is etched it should be rinsed with clean water. If you find that there are still a few traces of unetched copper, you can put the board back in the etchant for a few minutes before you rinse it again. After etching, the resist can be removed from the board with paint/varnish remover or steel wool.

## Double-sided boards

There are a few different ways of making double sided boards using photographic methods. It is possible to process both sides of the board at the same time by using a double sided exposure frame and a pair of UV light sources. This requires extremely accurate alignment of the artwork on both sides.

I find it is a lot easier to do one side at a time. The first side of the board is exposed, developed, etched and drilled as it would be for a single sided board. Use a sharp new drill to reduce the risk of damaging the photo-resist and protective coating on the opposite side. When the first side is finished, it can be covered with strips of PCV tape to protect it while you are etching the second side. Now you can remove the plastic protective film from the second side of the board. The artwork for the second side should be aligned carefully with the holes in the board. This ensures that both sides of the board are properly aligned. The second side of the board can now be exposed, developed and etched. Be very careful to protect the first side from the etchant while you are etching the second side. After etching, the protective tape and photo-resist can be removed, hopefully to reveal a perfect double sided PCB.

Double sided boards can also be made by etching a pair of single sided boards and sticking them together with epoxy resin. If you use 0.8mm laminate, the finished product won't be much thicker than standard 1.6mm board.

SMT (surface mount technology) boards are made using exactly the same methods as boards for leaded components. PCBs with leaded components are often referred to as PTH (plated through hole) technology. This is a bit of a misnomer in the case of home made boards because most amateurs will not have the equipment to make boards with electroplated holes. I have used the photographic methods described here to make boards using surface-mount ICs and 0603 type chip capacitors and resistors. The main differences between boards for leaded components and boards for SMT are the methods used for mounting and soldering the components. It is not too difficult to solder most SMT components provided that you have a good magnifying glass and a small tip for your soldering iron. Some of the newer SMT packages like the BGA (ball grid array) package have all of the connections underneath the chip so that no pins are visible. Such technologies will provide a new and difficult challenge for the home constructor.

## Drilling the PCB

Hand cranked, egg-beater style drills are just about usable for hand drawn PCBs. To drill the tiny holes required by CAD produced PCB layouts, a 12V miniature PCB drill or a mains powered Dremel or similar type of rotary tool will do a much better job.

After many years of breaking drill bits, I have at last discovered a reliable way of drilling holes in PCBs. I use a 12V PCB drill with a reduced supply voltage of around 6 volts. Using a lower supply voltage reduces the kick caused by the torque reaction when the drill starts turning. The lower voltage also reduces the speed of the drill which results in longer life from HSS drill bits. I use a push-to-make foot switch to operate the drill. This leaves both hands free to hold the drill and PCB. You should wear safety goggles when using the drill.

To avoid drilling holes in the table, drill the board on on old telephone book or paperback. The exact procedure I use is as follows: Place the tip of the drill in the centre of the pad to be drilled. I use a 100W light so that I can see what I'm doing. Taking care to keep the drill absolutely vertical, press the foot switch. Don't hold the drill too firmly. The hole in the centre of the pad will act as a guide for the drill. Even if you are a little off centre, the drill will tend to centre itself with the hole in the copper pad. When the hole has been drilled, withdraw the drill vertically while it is still turning and release the foot switch. I use a drill size of 0.7mm to 1.0mm for most PCB work.

Commercial PCB makers use tungsten carbide drills for drilling fibreglass PCB. I have found that these drills are easily broken if they are used in a hand drill. My own preference is for HSS (high speed steel) drills which are relatively cheap and readily available. Most constructors recommend using a drill stand for PCB work. I prefer to use a hand-held drill and the above technique. I have drilled thousands of holes since the last time I

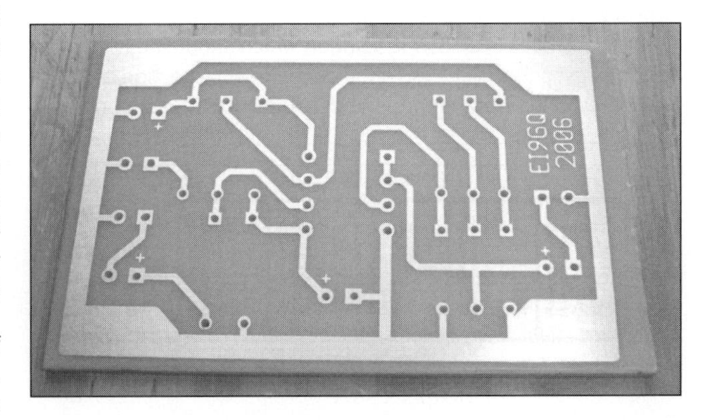


Fig 25.40: Completed home made PCB

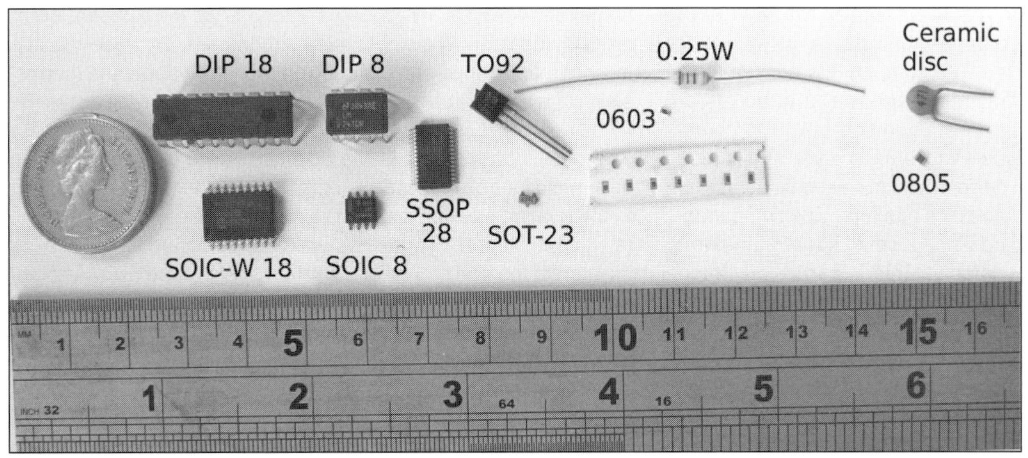


Fig 24.41: A selection of some of the most widely used SM package types showing their size

broke a drill-bit. Once the PCB has been etched and drilled, the components can be mounted on the board and the leads soldered to the component pads. Commercial PCBs usually have a protective layer of solder resist applied to the board. This is not strictly necessary for home produced boards, but if you want to preserve the shiny new appearance of your PCB, it can be coated with a clear lacquer after soldering. Special lacquers are available which will allow re-soldering of the board for future modifications or repairs. **Fig 25.40** shows the track side of a home made PCB after etching and drilling.

## SMT

Boards for surface-mount components are made in much the same way as boards for through-hole components, but there are a few obvious differences. SMT boards are usually much smaller than their through-hole counterparts. As well as occupying a smaller area of PCB, SM components have a much lower profile than other component types. This makes them particularly well suited to double sided or multi-layer PCBs where components are mounted on both sides of the board. SMT layouts tend to use narrower track widths and pad sizes. SM devices usually use much smaller lead spacings than through-hole components. For example: SOIC (small outline integrated circuit) ICs have a pin-to-pin spacing of 0.05in (1.27mm) or 20 pins to the inch. This is exactly half the pin spacing of a standard DIP package. Smaller SM ICs use 40 or even 50 pins to the inch. **Fig 25.41** shows a selection of some of the most widely used SM package types.

Running tracks between IC pins is common practice for DIP ICs. This is not so easy with SOIC packages and just about impossible for smaller packages. This restriction greatly reduces the routing options available to the PCB designer. Regardless of whether the track routes are chosen by human brain power or auto-routing PCB CAD software, at some point you will find you are completely surrounded by tracks and there is no clear path to the next pad you want to reach. In this situation, the only way to avoid an ugly wire link is to use a double sided or multi-layer PCB.

Inter-layer connections are made by small plated holes called vias. Because vias are not soldered to the copper tracks, they are almost flush with the surface of the PCB. In many circuits, the vias can actually be hidden underneath surface mounted components. This is not easy to achieve with home made PCBs. Most home made PCBs use either a single sided or double sided layout. Vias are typically made by drilling through the PCB and soldering short lengths of wire between the top and bottom layer. Multi-layer (more than two copper layers) boards with electro-plated vias are not easily made in the amateur workshop.

The two main differences between SMT and through-hole PCBs are the way that the components are mounted on the board and the way the components are soldered to the copper pads. Surface mount components are as the name suggests, mounted on one surface of the PCB. Unlike single layer through-hole PCBs which have a 'component side' and a 'track side', single layer SMT PCBs have the components and the copper pads and tracks on the same side of the board.

The most common type of board used for home construction uses a double-sided board. Most of, or quite often all of the components are mounted on one side only. Hand soldered vias are used to make connections to the tracks on the other side of the board. Such boards are reasonably easy to make and offer a reasonable range of routing options to the layout designer. By convention, the component side is considered to be the 'top' copper layer during the design process. Commercially produced PCBs will usually have a solder mask layer and silk screen layer above the top copper later.

## SMT Soldering

The basic principles of soldering for SMT boards is just the same as for any other type of construction. The solder joints are made by applying flux and solder to the metals to be joined and heating the joint to a temperature above the melting point of the solder. The methods employed for most SMT boards are often quite different to the methods used for through-hole boards. Solder and flux can be applied to the PCB pads in a paste form.

There are several different ways of applying the solder paste. One method used by commercial board makers is to apply the solder paste to the PCB through a screen. This is similar to silk screen printing except that a solder and flux paste is used instead of ink. The other method is to apply the paste to each PCB pad individually via a small nozzle. For one-off prototyping, reworking and repairs, the solder paste can be applied by hand using a tool similar to a syringe which has a metal nozzle at one end and a plunger at the other end. It is also possible to use a conventional soldering iron to solder SM devices.

**Fig 25.42** shows a six pole SSB filter which uses HC49/SMT crystals and 0805 chip capacitors. This board was soldered by hand using a pointed tip in the soldering iron.

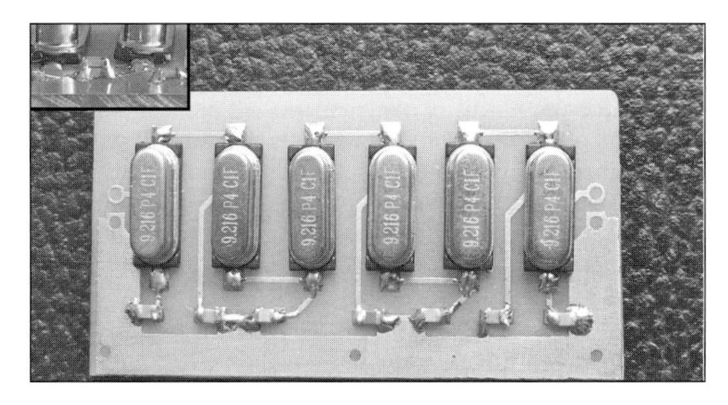

Fig 25.42: Hand-soldered SMT components in a filter

Once the solder/flux paste has been applied to the PCB, the components can be mounted in position using a computer-controlled pick-and-place machine. The components are fed to the pick-and-place machine in much the same way that bullets are fed to a machine gun using an ammunition belt. SM components are supplied on tape reels which have perforations along one edge of the tape. These perforations are used as a guide for the feed system of the pick-and-place machine. The components may also be glued to the board using a small dot of adhesive to keep them in place during soldering. This is particularly important if the components are to be wave soldered. For prototyping, reworking and repair, a simple vacuum operated hand tool can be used to place the components on the PCB.

There are several different methods of soldering the components to the PCB. Wave soldering is sometimes used for SMT boards or for boards that use a mix of SM and through-hole components. Wave soldering equipment works by running a wave of molten solder along the surface of the PCB. This method is ideal for through-hole PCBs because there are no components on the track side of the board. Without any components to get in the way, it is possible to solder the entire board in a single operation using a shallow wave of solder. Wave soldering is not particularly well suited to SMT soldering. The most common soldering method for SMT boards is known as reflow soldering. There are several methods of applying heat to the board. Infra-red, hot gas, and laser heating are all in common use. Each component of the board will have it's own specification for maximum soldering temperature and duration. Most modern components are able to survive the high temperatures encountered in lead-free reflow soldering. In a typical system, the entire board is pre-heated to a temperature of about 150°C at a controlled rate of about 2°C per second. After this 'preheat zone', the board is held at this elevated temperature for the second stage of the process. During this 'heat soak zone' the solder flux is activated. The next stage is the 'reflow zone' where the temperature is increased to above the melting point of the solder. This is typically 200-250°C.

The solder must remain liquid for long enough to flow properly around each joint. Surface tension will make the solder flow onto the solder pads and 'wick' along component leads. Temperatures much above 200°C are close to the maximum temperature limits for most semiconductors. Many components like plastic capacitors, switches, plastic connectors and even the PCB laminate can only survive such high temperatures for a short period of time. The final stage is the 'cooling zone' where the PCB is cooled to room temperature.

So how do we go about bringing all this space-age technology into the amateur's shack? How much expensive equipment will we need? You might be surprised to discover that SMT reflow soldering techniques are well within the capabilities of the amateur constructor. In some respects, reflow soldering is easier, cheaper and more consistent than conventional soldering of through-hole PCBs using a soldering iron.

A humble pair of tweezers is a good alternative to the expensive pick-and-place machine. Metal tweezers are suitable for most PCB work. Plastic or bamboo tweezers are also available for the more delicate jobs. The tweezers might be a bit slower than its industrial counterpart, but it will get the job done eventually. Component placement is not quite as critical as you might imagine. Surface tension during soldering tends to pull the components into line with the PCB pads, even when they have been slightly misplaced before soldering.

Specialised soldering equipment like vapour-phase ovens are staggeringly expensive. However basic infra-red ovens are not too expensive, although they are probably still beyond the reach of all but the most enthusiastic home constructor. The most popular home reflow oven is a domestic toaster oven of the type that might normally be used to make toasted sandwiches. Because the normal operating temperature of such an oven is far above the thermal soak zone temperature and the maximum setting is well above the reflow temperature of the solder, this approach requires a great deal of skill on the part of the operator. Some constructors have built microcontroller based systems to control the 230V AC current to the heating element so that the oven temperature can be accurately controlled. The control system can be programmed to provide the correct thermal profile for the reflow soldering process. A home-brew oven of this type can give excellent results.

Another method of reflow soldering uses a stream of hot gas to heat the solder. Simple systems use hot air, some more expensive systems use a stream of hot nitrogen. The simplest form of hot gas soldering equipment is the hot air pencils that are used for SMT rework and repairs. For this type of work, the entire board is often placed on a hot plate or over a hot air 'bath' so that the entire board is at an elevated temperature. The hot air pencil is then used to reflow solder individual components on the top side of the board. Hot air rework stations are available at reasonably low prices. Basic units cost less than £100. Ordinary hot air guns of the type sold at DIY stores are not ideal for SMT repair and reworking because the air temperature it too high and the large nozzle makes it impossible to heat a small area of the board without reflowing the entire board. However, this type of hot air gun can be used to reflow solder a complete new PCB. Most guns have a high and low setting. The low setting is typically about 300°C which is still a bit too hot. The board temperature can be controlled by keeping the gun at a suitable distance from the board. There are no absolute rules to determine the correct distance. Different guns will have different temperatures, flow rates, nozzle velocity etc. Some practice with a suitable thermometer and a few scrap SMT boards will help you to find the ideal settings.

It is just about possible to draw a track layout for some of the larger SM components by hand. For more complicated layouts using SM IC's, computer CAD software is the only practical way of producing the track layout. All modern PCB CAD software has libraries of SM component footprints. In fact, most CAD software has better support for SM components than through-hole components because SMT has been the most widely used construction method over the last couple of decades. **Fig 25.43** shows a simple PCB layout for a microphone amplifier and active LPF for use in a home made SSB transmitter.

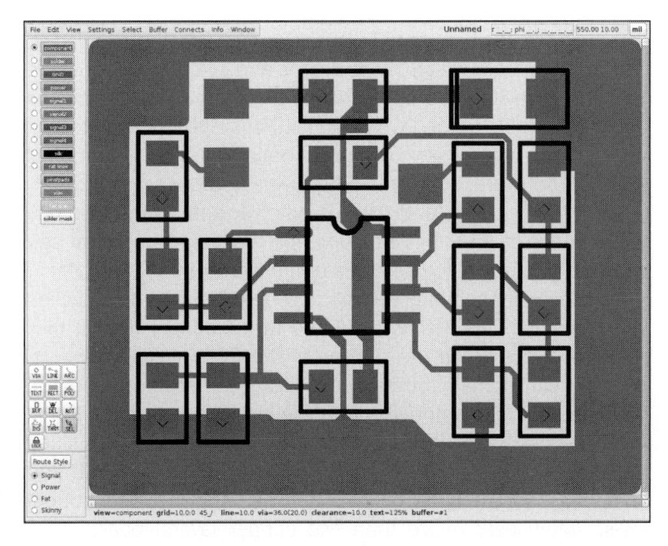

**Fig 25.43: A simple PCB layout for parts of a home made SSB transmitter produced using a CAD program**

**Fig 25.44: The printed circuit after etching**

I will build two of these amplifier modules to demonstrate some of the reflow soldering techniques described above.

The layout was drawn using PCB [3] running on a Linux PC. The artwork was printed onto a sheet of Stephens transparency film using my HP D5160 inkjet printer. Two identical copies of the artwork were printed on a single A4 sheet. The sheet was then cut in half so that the two track patterns could be stacked to make a more opaque master artwork. This was used as the master artwork for the photographic PCB process which is described in detail earlier in this chapter. The artwork and pre-coated photo-sensitive PCB were clamped in my exposure frame and exposed to the halogen lamp for exactly ten minutes. After exposure, the PCB was developed in sodium metasilicate photo-resist developer, rinsed with clean water and then etched in a ferric chloride bath. **Fig 25.44** shows the PCB immediately after etching, before the photo-resist was removed.

After etching, the two separate PCB patterns were carefully cut out using a hacksaw. Next the photo-resist was removed using fine steel wool. The signal track width used in this board is only 10 mils (0.254mm) so don't scrub too hard with the steel wool or you might lose some of the tracks. The use of powerful solvents like paint stripper to remove the resist is best avoided because the solvent may undermine the thin copper tracks and destroy the adhesive bond between the copper foil and the board substrate. An alternative way of removing the photo-resist is to re-expose the board without using the artwork as a mask and then place the board back in the developer tray for a few minutes.

There are several ways of applying the solder/flux paste to the PCB pads. Unless special equipment is available, the best way is to use a small pointed tool to apply a small amount of paste to each component pad. Alternatively, you can use one of the special syringes that are sold for this purpose. Solder paste supplied in syringes is usually water based and needs to be stored in a refrigerator. This is rather inconvenient for the occasional constructor who keeps solder in storage for long periods of time. This type of paste is relatively expensive and it needs to be shipped on-ice from the supplier using overnight delivery. In this part of the world, 'overnight' usually means several days. The other thing that this paste has in common with a choc-ice, is that it has an expiry date printed on the container. Solder paste is also sold in small plastic tubs. This type of paste doesn't usually have an expiry date or any special storage requirements. However, this paste is too thick to use with a syringe type of dispenser. Both lead free and tin/lead pastes are readily available.

I use Lodestar type L309 paste. This paste can be applied to the PCB using a small pointed tool like a toothpick or cocktail stick. **Fig 25.45** shows how the solder paste was applied to one of the microphone amplifier PCBs using a cocktail stick. The amount of paste required for each pad is very small. A 50g tub of paste will last for a very long time.

After the paste was applied, I used small metal tweezers to place the components on the board and another wooden cocktail stick to push the components down into the paste and poke them into correct alignment with the pads. A good illuminated magnifier is very useful for this kind of work (**Fig 25.46**).

For those who only have experience of through-hole PCB soldering, the next stage is really the fun part. The soldering process only takes a couple of minutes. The whole board is soldered in a single operation and you won't have to worry about dirty oxidised iron tips, dry joints, solder bridges etc.

The board was mounted in a metal 'helping hands' type of holder and placed at the edge of the bench so that I would have free access to the top and bottom of the board. I am a great advocate of 'kitchen table technology', but this is a job for a rough wooden or metal bench.

The hot air from the heat gun is likely to damage a more delicate surface. I used a standard DIY hot air gun on the low heat setting. The high setting of 500°C - 480L/min would just toast the board and blow the components away! The low setting gives an air temperature of 300°C and a flow rate of 280L/min. This is still too hot, so don't allow the nozzle to get too close to the PCB.

My exact procedure is as follows:

1 Heat the underside of the board with the gun at a distance of 0.3 to 0.4m until the board reaches a temperature of 150°C. I used a heat resistant thermocouple and digital multimeter as a thermometer. Even at this distance, it takes less than one minute to reach the desired temperature. This is a bit faster than the recommended temperature ramp rate, but it works well in practice.

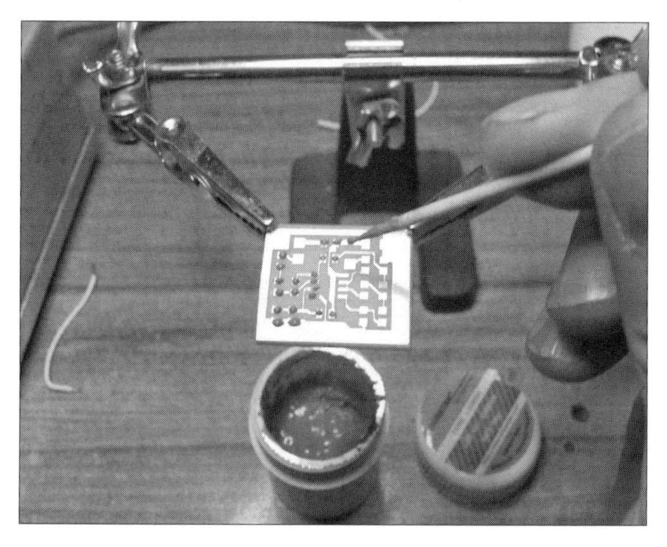

**Fig 25.45: Applying the solder paste to one of the PCBs using a cocktail stick**

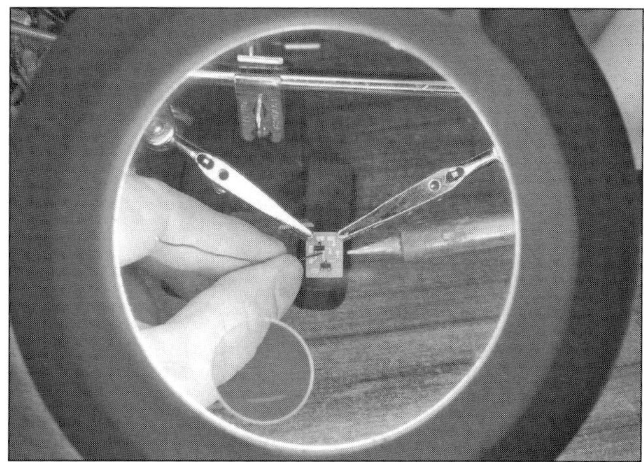

**Fig 25.46: The job is made much easier by using a large illuminated magnifier and a jig to hold the work**

**Fig 25.47: (above) Close-up view of one of the boards**

**Fig 25.48: Finished SMT board compared with a DIP version**

2 Quickly move the gun to the top side of the board and continue to heat the board until the temperature reaches about 200°C. This is the critical phase of the operation. Too much air will tend to dislocate the components when the solder flows. Too much heat will damage the components or the laminate. Move the gun closer to the board (about 10-20cm). You should find that the solder paste begins to flow after a few seconds. Keep the heat at this level until all of the solder paste has melted.
3 A few seconds after full reflow, withdraw the heat gun and allow the board to cool.
4 Check the board carefully under a strong light using a magnifier. I have found that this technique rarely produces solder bridges. If you do find any bridges, they can be removed using a hot iron and desoldering braid. Any dry or 'pasty' joints can be reflowed using a fine pointed soldering tip.

Both of the amplifier boards turned out quite well. No solder bridges were found. Both ICs were perfectly centred on the pads with good solder connections at the pins. One of the boards had a couple of 'pasty' looking joints which indicates that I didn't heat the entire board to a uniform temperature or the reflow zone temperature was a little too low. It only took a few seconds to rework these joints using a soldering iron. **Fig 25.47** shows a close-up view of one of the boards. **Fig 25.48** shows a finished board compared to a through-hole version of the same circuit using a DIP package.

The cosmetic appearance of the boards is reasonably good. A close examination of the board shows that there is a small amount of visible flux residue around the joints. As this type of paste uses an non-corrosive 'no-clean' flux, this is of no great consequence and can be left on the board.

## Hand soldering SMT ICs.

It is also possible to solder SM ICs using an ordinary soldering iron. There are several techniques in common use. You can solder the pins one-by-one using a fine tip and very fine cored solder. This is reasonably easy to do for SOIC packages, but it is extremely difficult for smaller devices with very close pin spacing. One approach that is popular with radio amateurs is to tack a few corner pins to keep the IC in place, then solder all of the remaining pins without worrying too much about solder bridges. The basic idea is to lash on excessive amounts of solder and flux during the initial soldering stage. The excess solder and any solder bridges are then removed using desoldering braid. This technique can produce excellent results, even with high pin density packages.

A similar technique called 'drag soldering' can be used to solder high pin density ICs. The board is cleaned and then fluxed using SMT flux from a syringe or a flux pen. I use a CircuitWorks CW8100 no-clean flux dispensing pen (Maplin N63AA). As with the previous method, a few corner pins of the IC are tacked with solder to keep the IC in place. Make sure that all of the IC pins are sitting properly on the PCB pads before you start the final soldering. Using a relatively large and flat iron tip and copious amounts of flux, solder the IC pins, several at a time while gradually dragging the tip along the IC pins using cored solder. This is not quite as difficult as it sounds because surface tension will tend to draw the solder to the pads and IC pins. The hot iron will easily flow the solder from pin-to-pin so that excess solder tends to disappear and solder bridges are avoided.

Once soldering is complete, do a visual check and remove any bridges using desoldering braid. The drag soldering stage of the operation is easier if the board is mounted vertically during soldering. Start soldering at the top of each row of pins so that gravity will assist the flow of solder as you move the iron across the pins. This is known as 'vertical drag soldering'.

For quad packages with pins on four sides, solder the two opposing vertical sides first and then rotate the board 90 degrees before soldering the remaining two sides. YouTube viewers can get a close up view of the process by searching for "SMT soldering" or "drag soldering". Be warned! Watching the experts at drag soldering is a lot like watching the TV Chef cutting a tomato into ten thin slices in a couple of seconds. Trying to repeat this performance with a blunt knife in the EI9GQ kitchen tends to result in a messy pile of tomato purée. However, with a little bit of practice, drag soldering techniques are not too difficult to master.

**Fig 25.49** shows a drag soldered IC on a commercially made PCB. The IC was removed from the board using a hot air pen. The PCB pads were cleaned up using a hot iron and desoldering braid. The IC was then re-soldered to the PCB using only a flux pen, soldering iron and cored solder. The following statement might seem counter-intuitive but a fine pointed soldering iron tip is not the correct tool for drag soldering. I normally use a

**Fig 25.49: A drag soldered IC**

temperature-controlled iron with a pointed tip for most soldering work. My preferred tool for drag soldering SM ICs is an ordinary Antex 25W iron with a standard flat tip. During drag soldering, several IC pins are soldered at once. The flat iron tip lies across the top of the pins and should never intrude into the gap between pins. Use of a pointed tip will slow down the soldering process and is likely to lead to solder bridges. The key to success is to use plenty of flux. Only a small amount of solder is required. If excessive solder is used, it will tend to gather at the end of each row of pins so that it is very easy to remove with desoldering braid or some ordinary stranded copper wire. The flux residue can be removed using PCB cleaner or most alcohol based cleaners.

### Rework and repairs

The ability to remove and replace SM components is of great importance to the the amateur constructor. A few simple SMT reworking tools allows the amateur to repair or modify existing home made or commercial equipment. The usual goal is to remove and replace a single SM device without damaging or disturbing the surrounding components. There are several methods in common use. Small chip capacitors and resistors are relatively easy to remove and replace. There are special iron tips for desoldering such devices. Even with a standard type of tip, anyone with a keen eye and a quick hand can flow the solder at both ends of the device and slide it off it's PCB pads. Removing SM ICs is a little more difficult. There are special desoldering tools which will fit over the pins of most types of IC package, but you will need a different head for each package type you are likely to encounter. Hot air reflow is the most popular method of SMT reworking and repair. The general purpose heat gun as used for our earlier experiment is a rather blunt instrument for this kind of work. A small temperature controlled heat pencil is a much more suitable tool. A heat pencil has a small nozzle which can focus the hot air on one individual component rather than heating the entire board. It is a good idea to pre-heat the entire board from below using an air-bath or hot plate heater. This reduces the time that it takes for the air pencil to reflow the top of the board. It also reduces the potential for stress and damage that could result from heating one small area of a cold board. Localised expansion and contraction can lead to cracking of the very hard ceramics used for chip devices. In extreme cases, it could even result in a warped PCB.

**Fig 25.50: The bottom of a BGA style IC from a modern mobile phone. It has 244 connections, including the array of 64 pins in the centre used for conducting heat**

**Fig 25.51: Suggested ergonomic strategy for some controls**

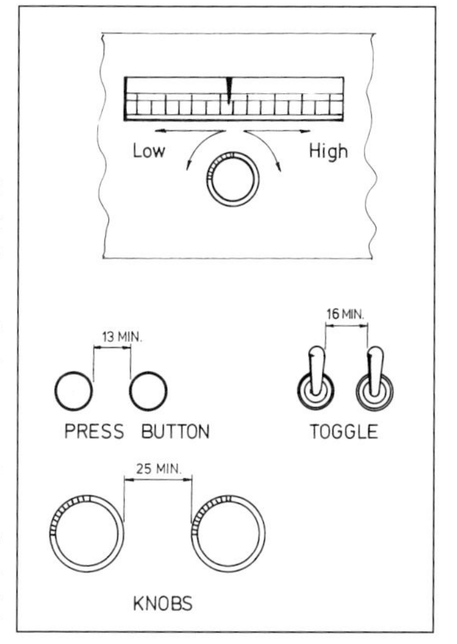


Some of the newer IC package types will present an interesting challenge for the home constructor. Many of these devices have no connections at the edge of the package. Electrical connections are made via an array of pads on the underside of the package. These pads are reflow soldered to their corresponding pads on the PCB during PCB assembly. One such package type is the BGA or Ball Grid Array. BGA devices have an array of solder balls on the bottom of the chip carrier package. The solder balls flow onto a matching array of PCB pads during reflow soldering. This type of chip can achieve a much higher pin density than conventional IC packages. New BGA devices are supplied pre-balled and ready for soldering.

Reworking or repairing BGA circuits without using new replacement devices is tricky because of the need to re-ball the chip before it can replaced on the board. **Fig 25.50** shows the bottom of a BGA style IC which was removed from a modern mobile phone. This is a 1cm square package with 244 connections hidden under the chip carrier. The array of 64 pins in the centre is used for conducting heat away from the chip core, leaving a mere 180 pins for normal electrical connections. A quad style chip with pins on four sides would need to have more than 150 pins to the inch (6pins/mm) to achieve the same pin density.

## DESIGN THOUGHTS

Home construction allows individual ideas to be designed into the project being made. With commercial equipment 'you pays your money' but you don't always get exactly what you would like. Some controls you find unnecessary and others which you want are not there or are hidden inside the case. These factors are often overlooked until you make something for yourself and then aspects of commercial designs take on a different meaning. The quality or lack of it becomes apparent and it is very difficult to acquire this appraisal ability without having made, or having tried to make, something for yourself.

Over the years there has become an awareness of the value of ergonomics (see **Fig 25.51**). Most of us now know that controls should be positioned according to their purpose and that this purpose should be self-evident from either the type of control or the area in which it is placed. For example, most transceiver tuning knobs appear adjacent to the tuning display and, when rotated clockwise, the frequency increases. Usually, it is evident by looking at most transceivers which is the tuning knob.

Unfortunately the same cannot be said about press buttons and other controls. The use of press buttons needs great care, especially where accidental operation could cause damage. If correctly designed, it should be impossible to accidentally press two or more buttons at the same time. Also, if such an action is done accidentally or deliberately, it should not destroy the device, or send it into an unusual operating or locked-up state. Ideally, any button-type control should be accompanied by an indicator which shows what mode the button is in. Power supply switches are particularly important in this respect, especially with portable equipment, where it is all too easy to switch it on and not be aware of this fact.

Rotary switches are less susceptible to accidental switching but they must be positioned and have suitable style knobs to facilitate switching. Most rotary switches occupy more space than push buttons and it is this which usually restricts their use on the compact rigs of today. Rotary switches are ideal for such controls as mode and band selection. One look at the switch position shows immediately what mode or band is selected. It is impossible to attempt to select two modes or two bands and there is no need for any other form of indicator: a power-saving factor also.

Toggle and slide switches are an equally effective form of self-indicating switches but toggle switches are prone to accidental operation. Potentiometric rotary controls, such as AF volume, Morse speed and RF gain etc should all be self-indicating. Also, clockwise rotation should increase the function. Rotary concentric controls should give each control a related and easily identified function. Knob styles can also affect the quality of presentation of the finished work. Unfortunately, knobs are expensive but ex-equipment knobs can often be used as a cheaper alternative. The size of knob should relate to the accuracy required from the function it controls. For example, a tuning knob would not feel right if it was less than say 40mm diameter, but the same diameter for a volume control would be unsuitable and usually unnecessary. Slider controls are very good, especially when simultaneous operation is required on such as faders and mixers, but this facility is not usually required on radio equipment. These controls also take up space if they are to be accessed easily. Some commercial amateur radio equipment uses the smaller type of this control for the less-utilised or preset controls.

Many more factors can be found, and it is discovering these which adds to the pleasure of home construction

## REFERENCES


[1] 'Homebrew', *RadCom*, October 2006

[2] http://qucs.sourceforge.net/

[3] http://pcb.gpleda.org/

[4] 'Homebrew', *RadCom*, July and August 2006


### Safety recommendations for the amateur radio workshop

1. All equipment should be controlled by one master switch, the position of which should be well known to others in the house or club.

2. All equipment should be properly connected to a good and permanent earth (but see the box about PME in the chapter on Practical HF Antennas, and Note A).

3. Wiring should be adequately insulated, especially where voltages greater than 500V are used. Terminals should be suitably protected.

4. Transformers operating at more than 100V RMS should be fitted with an earthed screen between the primary and secondary windings or have them in separate slots in the bobbin.

5. Capacitors of more than 0.01μF capacitance operating in power packs etc (other than for RF bypass or coupling) should have a bleeder resistor connected directly across their terminals. The value of the bleeder resistor should be low enough to ensure rapid discharge. A value of 1/C megohms (where C is in microfarads) is recommended. The use of earthed probe leads for discharging capacitors in case the bleeder resistor is defective is also recommended. (Note B). Low-leakage capacitors, such as paper and oil-filled types, should be stored with their terminals short-circuited to prevent static charging.

6. Indicator lamps should be installed showing that the equipment is live. These should be clearly visible at the operating and test position. Faulty indicator lamps should be replaced immediately. Gas-filled (neon) lamps and LEDs are more reliable than filament types.

7. Double-pole switches should be used for breaking mains circuits on equipment. Fuses of correct rating should be connected to the equipment side of each switch in the live lead only. (Note C.) Always switch off before changing a fuse.

8. In metal-enclosed equipment install primary circuit breakers, such as micro-switches, which operate when the door or lid is opened. Check their operation frequently.

9. Test prods and test lamps should be of the insulated pattern.

10. A rubber mat should be used when the equipment is installed on a floor that is likely to become damp.

11. Switch off before making any adjustments. If adjustments must be made while the equipment is live, use one hand only and keep the other in your pocket. Never attempt two-handed work without switching off first. Use good-quality insulated tools for adjustments.

12. Do not wear headphones while making internal adjustments on live equipment.

13. Ensure that the metal cases of microphones, Morse keys etc are properly connected to the chassis.

14. Do not use meters with metal zero-adjusting screws in high-voltage circuits. Beware of live shafts projecting through panels, particularly when metal grub screws are used in control knobs.

15. Certain chemicals occur in electronic devices which are harmful. Notable amongst these are the polychlorinated biphenyls (PCBs) which have been used in the past to fill transformers and high-voltage capacitors and beryllium oxide (BeO) which is used as an insulator inside the case of some high-power semiconductors. In the case of PCBs, the names to look out for on capacitors are: ARACLOR, PYROCHLOR, PYRANOL, ASBESTOL, NO-FLAMOL, SAF-T-KUL and others [3]. If one of these is present in a device, it must be disposed of carefully. The local Health and Safety Authority will advise. In the case of beryllium oxide, the simple rule is DON'T OPEN ANY DEVICE THAT MAY CONTAIN IT.

*Note A. - Owing to the common use of plastic water main and sections of plastic pipe in effecting repairs, it is no longer safe to assume that a mains water pipe is effectively connected to earth. Steps must be taken, therefore, to ensure that the earth connection is of sufficiently low resistance to provide safety in the event of a fault. Checks should be made whenever repairs are made to the mains water system in the building.*

*Note B. - A 'wandering earth lead' or an 'insulated earthed probe lead' is an insulated lead permanently connected via a high-power 1kΩ resistor or a 15W 250V lamp at one end to the chassis of the equipment; at the other end a suitable length of bare wire with an insulated handle is provided for touch contacting the high-potential terminals to be discharged.*

*Note C. - Where necessary, surge-proof fuses can be used.*

# QRP - Low Power Operating

# 26

*Steve Hartley, G0FUW*

Operating on the amateur bands with low power, or QRP, continues to be one of the more popular aspects of amateur radio; the G-QRP Club has had a steady membership of around 4000 for the last decade.

This chapter is intended to provide an overview of QRP and to signpost the reader to further sources of information if this proves to be of interest. The author has been licensed for over thirty years and most of his operating has been QRP, but he has been known to 'turn up the wick' on occasion; being keen on QRP is not like some ethical commitment, it is just good fun and can be extremely rewarding.

The origins of QRP clearly rest in the Q-code, where an operator can send 'QRP?' in Morse to ask the other station if they are able to reduce power. No specific power limits are inferred, just to reduce power to prevent overloading, maybe. The notion of using the minimum power necessary to make a contact remains good practice but is often not observed and using the maximum power available is taken as the norm.

As higher powers became easier to generate and licence conditions allowed hundreds of watts on most bands, many operators decided to voluntarily reduce power to see what could be achieved with minimal equipment and 'flea power'. QRP became a way of operating.

Over the years there have been various attempts to define what is recognised as QRP operation. The current internationally agreed levels are 5W RF output for CW, 10W pep for SSB and data. It is worth noting that in the UK, every Foundation Licence holder is by default, a voice or data QRP operator.

As with the licence limits, these power levels are maximum limits and some use much less power electing to operate with no more than one watt. Transmitting with milliwatt levels is generally referred to as QRPp operating, like extreme QRP.

## WHAT CAN BE ACHIEVED?

There is a common jibe that goes along the lines of 'Life's too short for QRP' which infers that making contacts with low power takes too long and more power is needed to make operating worthwhile. Well, there is no denying that many contacts are easier to make with high power, in the same way that it is easier to catch fish in a trawler's net, than with a fly on a thin line. However, with many QRP stations achieving awards and contest results not too far behind the big guns, there can be no doubt that you can make contacts with low power.

Probably the most prolific QRP operator was George Burt, GM3OXX, now a silent key. George rattled up over 300 countries with just one watt or less, and wire antennas. The author has cherished QSL cards from New Zealand, Australia and the USA, all made with a Yaesu FT817 and a dipole. On 50MHz he has worked many countries with 500mW. Many also have fun with QRP in VHF contests and are amazed at what can be worked from a good high location.

Propagation conditions clearly play an important part in what can be worked, especially on HF and 50MHz, but skilful planning of operating times, and good operating techniques can keep the contacts flowing.

For those who doubt the claims, it is worth noting that ten watts of SSB is less than 18dB down on a 400-watt transmitter. That is the 'plus plus' part of the very very strong signal, or just under three 'S' points down for a '59' report. That is, of course,

**Fig 26.1: George, G3RJV, speaking at the Yeovil QRP Convention**

**WHAT IS MEANT BY QRP?**

of no consolation when the received signal is only 3 'S' points above the noise, but the physics of RF power cannot be denied; QRP signals CAN make contacts.

## WHY DO IT?

If your Licence allows you to operate with high power, why would you use less? There are a number of reasons, some perfectly practical, others down to personal choice.

First of all, if you are going to operate from a remote site, like a mountain top, where you need to carry all your gear with you, having a low power radio that can work from a small battery, maybe with a solar panel to keep it topped up, suddenly looks attractive. With improved battery technology, a small sealed lead acid battery or a Li-Po power pack allows the operator to be active for many hours.

Similarly, if you are travelling by public transport or via airlines, an effective QRP station can be contained in a single carry-on flight case or rucksack making it easy to take part in events like Islands on the Air, Summits on the Air, or to activate a rare square or island location.

Costs can be much less with QRP. It is true that some high quality commercial QRP radios command a high price but there

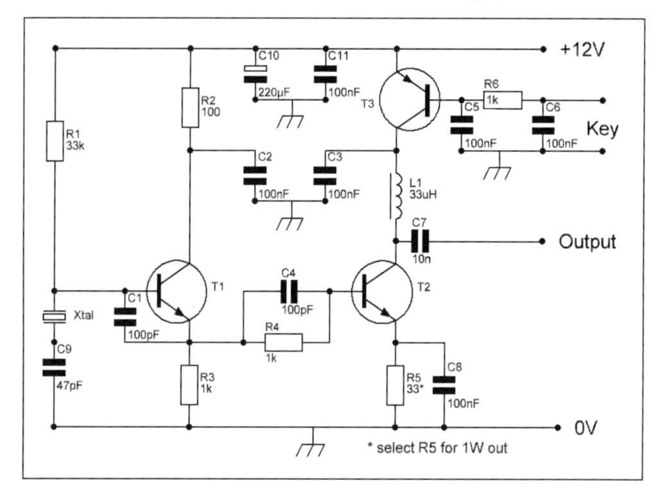


**Fig 26.2: Circuit of the now classic OXO transmitter, designed by George GM3OXX**

is no need for heavy duty power supplies, SWR meters or antenna systems. A home brew approach can have a useful QRP station on the air for very little outlay so low power operation is attractive for those with limited funds.

As previously mentioned, it is recognised good practice to use only enough power to make the contact and with many stations exchanging '59 plus plus' reports, there is scope for many others to back off the RF output control. That minimises interference to adjacent channels and allows better use of the spectrum. Reducing power is also one of the best ways to reduce the potential for your amateur radio signals to cause breakthrough or other EMC issues, especially in the urban environment; QRP signals are friendly to the RF environment in more ways than one.

The final, and probably the most common reason for amateurs opting to operate with low power is the challenge it brings. We can turn around the aforementioned common jibe and consider whether life is too long for QRO (high power operating). It is not uncommon for new Full Licence holders to work a hundred countries over a weekend, or certainly within a year of being licensed and be left wondering 'what's next?'. Some will seek new challenges in VHF/UHF/microwave operating, some may try bouncing signals off the moon, or maybe aim for a hundred countries via satellites, but many simply choose to reduce power and seek to work the world with QRP.

## WHAT ARE THE DOWN SIDES?

There aren't many negatives to operating with low power but it would be wrong to exclude some mention if the downsides here.

As mentioned earlier, there is no question that operating QRP will deny the operator the chance to bag some DX stations and that can be frustrating. Part of the challenge is to figure out how to overcome that, maybe by waiting for a quieter time, changing to another band, or improving the antenna system. There are many ways to work it out through 'self-training' and experimentation, without resorting to the 'sledgehammer' approach. That said, taking up QRP is not like making a pledge to some religious order and if you really need that DX in the log, there is no shame in increasing your power by just enough to make the contact.

If you make it known that you have taken up QRP operating, you will, inevitably, hear the 'life's too short' jibe. Some take this to heart and get angry but the best retort is to point to your log and explain your rationale for 'going QRP'. Amateur radio is a diverse hobby and as the saying goes, 'each to their own'; there is room for many special interests.

## DO PEOPLE STILL BUILD KITS?

Kits are not always used, many QRPers gather components and build from scratch. SPRAT, the journal of the G-QRP Club, is full of circuits and construction advice to enable readers to source parts, make PCBs and maybe modify the designs for their own use. This can be a very economical way to go but when you are just starting out kits provide all the parts you need along with full building instructions.

Whilst the availability of electronic components and amateur radio kits is no longer a feature of the average high street, there are still a number of kit suppliers in the UK and many of them include QRP equipment. Being able to order from overseas via the internet has also opened up a whole new range of sources for buying kits and parts.

Walford Electronics has an ever-changing range of kits for receivers, transmitters, transceivers and accessories, all very QRP-friendly. Their kits come without enclosures with the focus on sound electronic design and repeatability. The range usually includes direct conversion receivers, Morse and voice transmitters accessories and test equipment. Many of the receiver and transmitter kits can be combined to form integrated transceivers.

**Fig 26.3: a homebrew valve station by PA0WDW**

Kanga UK have recently shifted emphasis towards surface mount components but still supply some 'traditional' kits with through-hole components. Some come with enclosures. They have kits for a direct conversion receiver with a digital dial, an SDR receiver kit, an SSB two-tone tester kit (**Fig 26.4**), K1EL keyers, and many more. They have also been supplying the Mountain Topper ultra-portable transceivers designed by Steve Weber, KD1JV, that will literally slip into a pocket but still produce a respectable level of QRP power.

If antennas, baluns and antenna tuners are your thing, SOTA Beams provide a range of kits to cover your needs. They are aimed more at portable use on the fells and mountains but by association, that tends to be QRP and there is nothing stopping their kits being used at the home station.

Spectrum Communications provide a range of kits but they are perhaps best known for their VHF transverters. They also sell kits for 'building blocks' which can be used in a mix and match kind of way, or used in conjunction with projects sourced from other suppliers.

The G-QRP Club has a limited range of kits for sale to members. These include neat enclosures and can be built to form a matching line up of receiver, transmitter and ATU.

Looking overseas, HF Signals are based in India but have developed a global reputation for solid designs that are 'hackable': modifications are encouraged. Their BITX series of QRP

**Fig 26.4: Kanga UK Two Tone Test kit under construction**

SSB transceiver kits have allowed many to experience the joy of homebrew QRP without breaking the bank. At the time of writing the Micro BITX, multiband SSB transceiver was available at version 6 but the single band 40m version had been retired. They also had a very competitively priced antenna analyser kit available.

QRP Labs are based in Turkey but ship from the Far East. They list an impressive range of kits at competitive prices. Some are complete transceivers, like the QCX single band Morse transceiver, which has sold by the thousands. Other kits are more of the 'building block' type and a modular approach can be used to built transceivers or QRP WSPR beacons for testing propagation or comparing antennas.

This is not a comprehensive list, and suppliers do come and go, but is intended to show that kit building is not a relic of the past and modern techniques, like SDR, Direct Digital Synthesisers

**Fig 26.5: Yaesu FT818 is a very portable QRP 'shack in a box**

and surface mount construction are very must part of QRP in the 21st century. Contact details are provided below.

## IS IT ALL ABOUT HOMEBREW?

QRP is often taken to be synonymous with home brew and it is certainly easier to build a QRP transceiver, or transverter, than a full legal power linear amplifier. There are many classic designs that can be built with a fairly minimal tool kit and little more than a multi-meter for test equipment. In recent years the K series of kit transceivers from Elecraft, the BITX series by Ashar Farhan and the now legendary QCX transceiver kit by Hans Summers' QRP Labs have taken home brew to a new level that perform as well as many commercial rigs.

However, if wielding the soldering iron is not your thing than there are a growing number of commercial transceivers on the market. The Yaesu FT817 (**Fig 26.5**), and its successor the FT818, set the bar quite high in terms of having a 'QRP shack in a box' covering most bands from 160m to 70cm at five watts on just about all modes. The Elecraft KX3 has proved very popular with fewer bands than the FT817 but a much higher specification receiver, more QRP power on SSB and the option of a built-in ATU. At the time of writing the Icom IC-705 is eagerly awaited with full SDR operation in a radio shaped box.

Another recent shift in the market has been the availability of lower priced radios from China. Initially, these were limited to VHF/UHF FM handhelds and a few mobile FM transceivers but we are now seeing HF Morse and SSB radios giving a good account of themselves at bargain prices. For example, the Xiegu G90 is an all band all mode HF SDR transceiver with a built-in ATU for around half the price of the nearest equivalent from the other, more well-established, manufacturers.

For VHF or UHF and certainly for the GHz bands, many operators will use transverters and these are often designed for QRP power levels with optional external power amplifiers to achieve higher output powers.

There are also some bargains to be had on the second hand market. Some 'vintage' QRP radios can still give a good account of themselves on the air. The Kenwood TS120 remains popular for HF with all band coverage and ten watts output on SSB and Morse. For VHF the Yaesu FT290 and FT690 models provide three watts of FM, SSB and Morse on 2m and 6m respectively. Ten Tec produced a very well-respected range of QRP radios for many years and they often pop up at rallies or on-line auction sites. Clearly, there are questions about reliability with older gear but these old workhorses should not be dismissed out of hand.

The easier, but maybe less obvious, QRP option is to simply turn down the RF Power control on your QRO transceiver. In the days of valves this was not so easy to achieve but with modern solid state power amplifiers it is normally very easy to set the output power to QRP levels. Not all transceivers will go down to five watts but many do. So, if you are 'QRP curious' it is possible to see what is possible without changing equipment or melting solder.

## IS IT ALL ABOUT MORSE CODE?

A common misconception is the QRP operation is all carried out with the Morse key but voice contacts have been part of the QRP game for many years. Older readers may recall that all UK Licence holders were limited to ten watts on top band at one time.

SSB is very popular for QRP operation and with the advent of more digital modes being made available using small laptops via accessory cables the magic of weak signal modes is ideal for the QRPer. There are many Foundation Licence holders that will testify to the joy of QRP voice and data communications, even if they did not realise their ten watts of SSB made them QRP operators.

There are many contests where QRP SSB is welcomed. Some are specific to QRP operation and those that take part are often amazed at what can be achieved when there are stations keen to get others in the log. For example, the RSGB Backpackers series of contests encourages portable VHF operation with low power and simple antennas and Practical Wireless has been running a 2m QRP contest every June for over thirty years.

Morse code has its place in the QRP world but it is not the only mode of operation!

## WHAT OPERATING SKILLS ARE NEEDED?

Regardless of mode, there are some operating skills that will be useful to practice if you are to up your success rate.

As with all amateur radio operation, good listening skills are essential. Listening to establish if the frequency is clear, or listening to establish who is using the frequency are simple

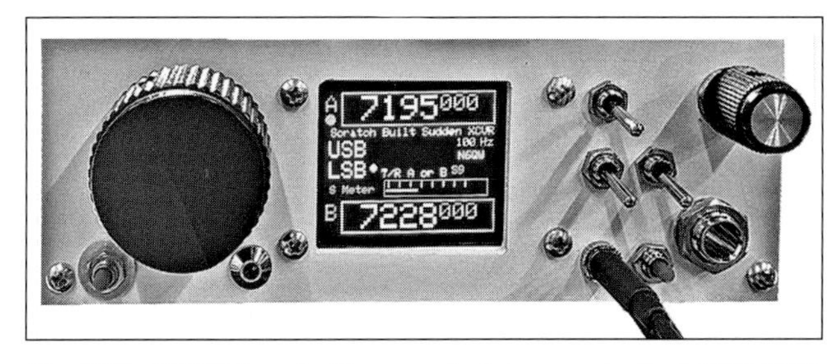


**Fig 26.6: N6QW has built many Arduino controlled QRP SSB rigs**

examples that help to establish if a call is appropriate; if the station you are hearing is only working stations on the other side of the globe, there is little point trying to attract their attention.

Being conscious of propagation conditions is another skill that applies to all aspects of HF and some VHF amateur operation. The RSGB provides propagation forecasts and looking at websites like the WSPR reports or DX spots will give a clue as to which band is best to use at any time.

SPRAT
THE JOURNAL OF THE G QRP CLUB
DEVOTED TO LOW POWER COMMUNICATION

| ISSUE Nr. 150 | © G-QRP-CLUB | SPRING 2012 |
| --- | --- | --- |



In This Issue:
Valve Transceiver for 40m ~ Receiver RF Amplifier
Michigan Mighty Mite ~ Over the Horizon on 481 THz
Calibrating Simple Noise Sources ~ Know Your Phones & Ears
Power Beeper ~ Simple Bipolar Transistor Tester
G4FQN 80m DC Receiver ~ Micro-80 after 20 years ~ Valve Day
Membership News ~ New Sales Items
Antennas, Anecdotes, Awards ~ Communications and Contests
Member's News ~ Club Sales

THIS IS SPRAT NUMBER 150!

**Fig 26.7: The quarterly journal SPRAT is now available in full colour**

Sitting in the middle of the band calling 'CQ' is unlikely to return a large number of responses, even if the band is open; those tuning through the band are usually looking for strong signals to hook up with. However, there are QRP centres of activity in most amateur bandplans and calling 'CQ' on, or near, those frequencies can be worthwhile as other QRP operators will be listening there.

A more productive technique is to 'tail end' other contacts. In other words, find a couple of stations in conversation and wait for them to complete their contact. Once they have signed off with each other call one of them. That often will bring a very positive response as the other station does not have to go hunting for another contact or even call 'CQ'. Occasionally the station you call will be closing down but the other may still be listening and away you go.

Operating in contests is a good way to build your countries/squares worked without too much hunting. If you are going to join in a contest, even if you have no intention of submitting an entry, it is well worth finding out what exchange information is required. Some will require just a report and a serial number but some will be looking for your RF power, your locator square, post code, year of birth, etc. and you will not endear yourself to the serious contest operator if you don't know it in advance. Also remember that the serious contest operator will not be interested in your name, location or your equipment, unless it is part of the required information for the contest, so best to just make the contact and move on.

## WHAT AWARDS ARE AVAILABLE?

If contesting is not your thing, and it is not to everyone's taste, you can look to gain operating awards. These are usually based on contacts accumulated over an extended period of time, sometimes over a lifetime. The most popular is the ARRL's DXCC Award for working one hundred countries and doing that with QRP is a real achievement.

For the Foundation Licence holder, the RSGB has a Foundation Award based on a much smaller number of countries worked across a number of bands, which is an excellent way to get the new licence holder active with a purpose.

The G-QRP Club has a number of awards for its members. The basic one being awarded for just 25 countries with endorsements for each additional 25 countries. There is one for working 20 Club members again with endorsements for each additional 20 members worked and another award for working 10 countries with two-way QRP. The QRP Master Award is gained by working 60 members, 75 countries and 20 countries with two-way QRP. There is also a trophy for the best log submitted for operating on World QRP Day, the 14th of June each year.

**Fig 26.8: G-QRP members building kits at the Club Convention**

The QRP Amateur Radio Club International has an award for its members for making a contact that equates to one thousand miles per watt, for example a 3000 mile contact with three watts would qualify, as would a 500 mile contact with 500 milliwatts.

As you can see, there is no shortage of challenges for the QRP operator to take on!

## WHAT IS THE G-QRP CLUB?

In October 1974 the Reverend George Dobbs, G3RJV, wrote to Short Wave Magazine seeking amateurs who were interested low power operation with a view to forming a Club to share ideas and technical information. The Club was formed soon after and it was not long before a members' journal (SPRAT) was being produced and distributed from George's vicarage. Forty-odd years later, the Club is still going and SPRAT has become a very well-respected publication with a number of RSGB books flowing from it.

In that time thousands of people have been members and, whilst some members move on, the numbers remain steady at about four thousand members each year. Sadly, George became a silent key in 2019 but the Club continues as a not-for-profit organisation, devoted to low power communication. Membership is currently just £6 a year and in addition to the quarterly journal SPRAT members have access to Club Sales, which include electronic components, kits and books, all at very reasonable prices. The Club also hosts an annual Convention with kit building workshops and QRP-related talks. As noted above, the Club has an active awards and trophy scheme.

## IS IT JUST A UK THING?

There are QRP Clubs in many countries, not just the UK. The QRP Amateur Radio Club International is maybe one of the biggest. Based in the USA it really does have an international membership and produces a quarterly journal, aptly named QRP Quarterly. The Club also has awards, runs a number of QRP contests and hosts the world-famous Four Days in May Convention, which runs alongside the massive Dayton Hamvention, now held in Xenia, Ohio.

In Europe there are QRP Clubs in many countries, Italy, Spain, the Benelux countries, Russia and the Czech Republic to name just a few.

Fig 26.9: The Peregrino miniature QRP SSB transceiver from the EA-QRP Club

Down under, Peter Parker, VK3YE, and Charlie Morris, ZL2CTM, regularly post YouTube videos about QRP. Peter is a regular contributor to SPRAT in the UK.

Again, there many more national QRP organisations and the ones mentioned above are just to illustrate how far the interest in low power operation has spread. It is also worth noting that these national Clubs and societies do not exist in isolation. Here is just one an example of how the world-wide QRP community has collaborated. In 2014 the Spanish EA-QRP Club produced the Peregrino SSB transceiver and made a kit available. In 2019 the construction information was translated into English and published in SPRAT by Nigel, G0EBQ, which was then picked up by Pete Juliano, N6QW, over in California, who modified the design are those mods were shared in SPRAT.

## WHERE CAN I FIND OUT MORE?

### Clubs

G-QRP Club: *http://www.gqrp.com/*
QRP Amateur Radio Club International: *https://www.qrparci.org/*

**Books** (all available from RSGB Bookshop)
QRP Basics, ISBN: 9781 9050 8684 9
International QRP Collection, ISBN: 9781 9050 8655 9
Low Power Spratbook, ISBN: 9781 9050 8686 3
QRP Scrapbook, ISBN: 9781 9101 9379 2
Low Power Communication, ISBN: 978 0 87259 582 8
More QRP Power, ISBN: 978 0 87259 965 9

### International QRPers

QRP School: *http://qrpschool.com/*
Peter Parker, VK3YE: *http://vk3ye.com/*
Charlie Morris, ZL2CTM: *http://zl2ctm.blogspot.com/*
Pete Juliano, N6QW: *http://www.n6qw.com*

### Kits

Kanga UK: *https://www.kanga-products.co.uk/*
Walford Electronics: *https://walfords.net/*
Spectrum Communications: *http://spectrumcomms.co.uk/*
QRP Labs: *https://www.qrp-labs.com/*
HF Signals: *https://www.hfsignals.com/*
QRP Guys: *https://qrpguys.com/*

# QRP 80M CW TRANSCEIVER

This transceiver was developed as a project that would hopefully appeal to the relatively inexperienced constructor, although it is probably not ideal for the complete beginner. The ability to handle a soldering iron and identify components is necessary, as is the ability to read and work from a circuit diagram, but a lot of detail is included to make construction as straightforward as possible.

## DESIGN PHILOSOPHY

The aim was to produce a basic transceiver with minimum component count that required little in the way of tools or test gear to get working, but which would give creditable performance.

A tall order, perhaps. The first step was to decide exactly what was required. The obvious choice was a single-band QRP rig. On the receiver side, the author opted for direct conversion, as a superhet design would be far more complex. For the same reason, he decided on CW rather than SSB.

Now, if you are going to operate QRP CW for the first time, the ideal place to do it is on 80m. On this band, a watt or two will give you QSOs all over the UK and well into Europe, There are many circuits for this kind of transceiver around. A lot of them seem to be severely limited by over-simplification. First consider that VFO (variable frequency oscillator) control is highly desirable, but that RIT (receiver incremental tuning) is absolutely essential. If you are using crystal-control for transmit and receive, you have to rely on the other station being slightly off your frequency in order to produce the necessary beat note, or be able to shift the frequency of your oscillator slightly. With RIT and a VFO, it is a simple matter to 'net' onto another signal, and then adjust your RIT for a comfortable note. Another highly desirable feature is a sidetone oscillator. This enables you to hear what you are sending, and does not add greatly to the overall complexity.

The method of construction is at least as important as the electronics in a practical project.

I opted to use double-sided copper-clad board to make a chassis / panel arrangement, with a rear apron and a screened enclosure for the VFO. Cutting and drilling are kept to a minmum,

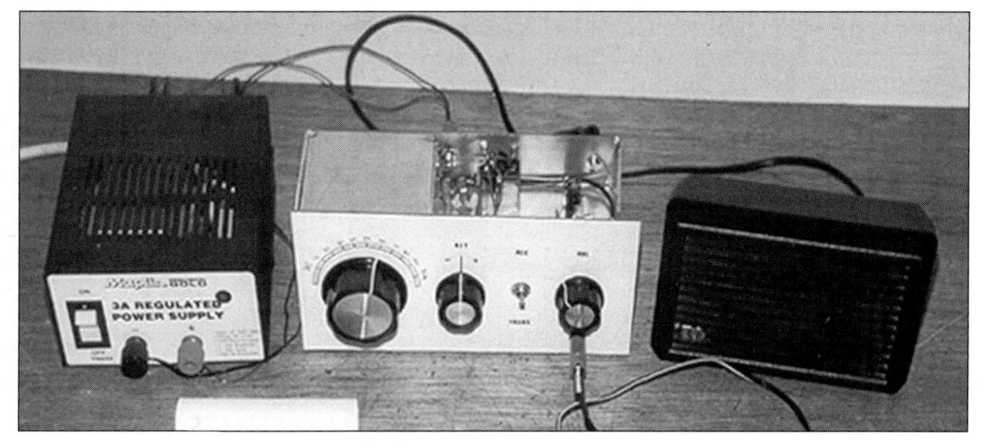


Fig 26.10: The completed transceiver (centre)

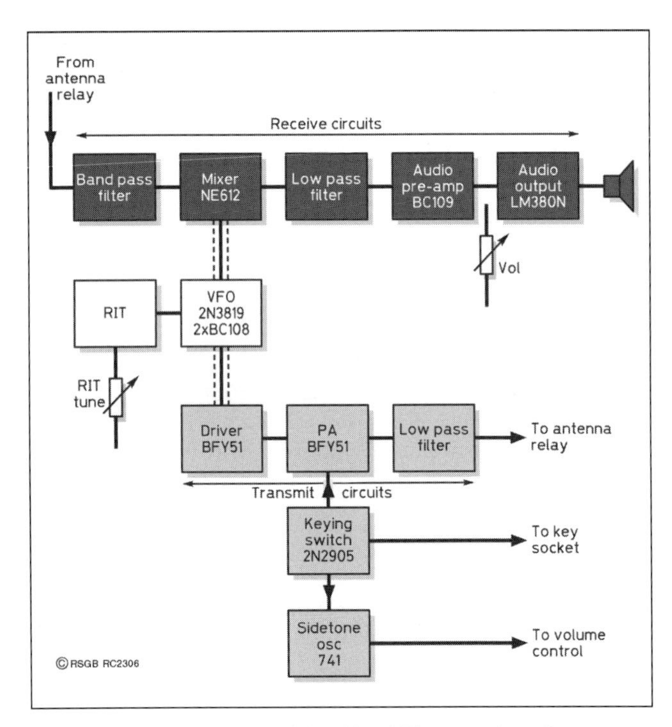


Fig 26.11: Block diagram of the 80m CW transceiver. Power supplies have been omitted for clarity

and the need to manufacture printed circuit boards avoided by mounting most components (with the exception of those in the VFO, which are mounted on a small piece of matrix board) in a 'dead bug' fashion. More on that later.

## DESCRIPTION

The block diagram of the transceiver is shown in **Fig 26.11**. The VFO is common to transmit and receive.

On receive, the incoming signal is passed from the aerial socket by the changeover relay to the band-pass filter (which greatly attenuates out-of-band signals) and then to the mixer where it is mixed with the VFO signal to produce an audio beat note. Most operators favour a note of 800 - 1000Hz, so the VFO will be offset by this amount. Because there will be many in-band but unwanted signals at the mixer input, there will be many more at the output after the mixing process. Although most of these will be beyond the range of hearing, some may be strong enough to overload the audio stages.

The low-pass filter will all but remove those at radio frequencies, and also attenuate the higher audio frequencies. The audio preamp now amplifies the wanted signal to a sufficient level to drive the output stage. On transmit, the VFO signal is amplified by the driver and then the PA to give an output in excess of 1W. Harmonics are attenuated by the low-pass filter and the signal is passed to the aerial socket by a change-over relay. Keying is by means of a 'keying switch' (electronic, not mechanical) which keys the 13V supply to the PA. This keyed 13V also supplies the sidetone oscillator, the output of which passes to the audio output stage via the sidetone level control.

The RIT in this design allows the operator to tune approximately ±1.5kHz of the frequency on receive. The position of the RIT control does not, of course, affect the transmit frequency.

## THE VFO

There is nothing remarkable about the VFO (**Fig 26.12**). Similar circuits are to be found in many pieces of equipment. The oscillator transistor is a 2N3819 field-effect transistor (FET). This operates at a low power level so the generation of excess heat, the enemy of VFO stability, is minimised. The supply is stabilised at +5V by IC1. The buffer stage consists of two BC108s, chosen because they are cheap, common, and most amateurs must have a few lying around. The coil L1 is wound on a T50-2 toroidal core. Wound with the number of turns specified, you should have a VFO that works on the correct frequency, which might not be the case if any old former and core were used (I don't know of any current source of new coil formers). In practice, frequency stability is quite adequate.

Mention must be made of the tuning capacitor, VC1. The tuning range is determined by the value of this component. 12pF allows coverage of the whole of the CW segment with a little to spare at each end, while 10pF will just cover the required range. The author prefered to have a little overlap, and used 12pF. With the amount of bandspread that this provides, the tuning rate is

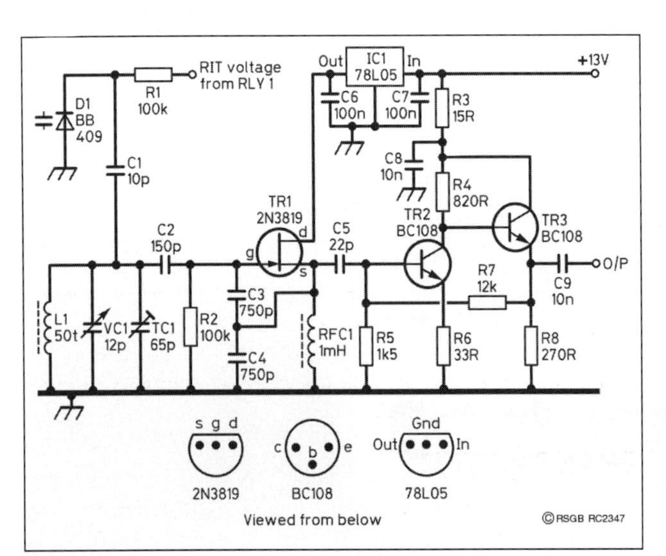


Fig 26.12: The VFO consists of a buffered Colpitts oscillator.

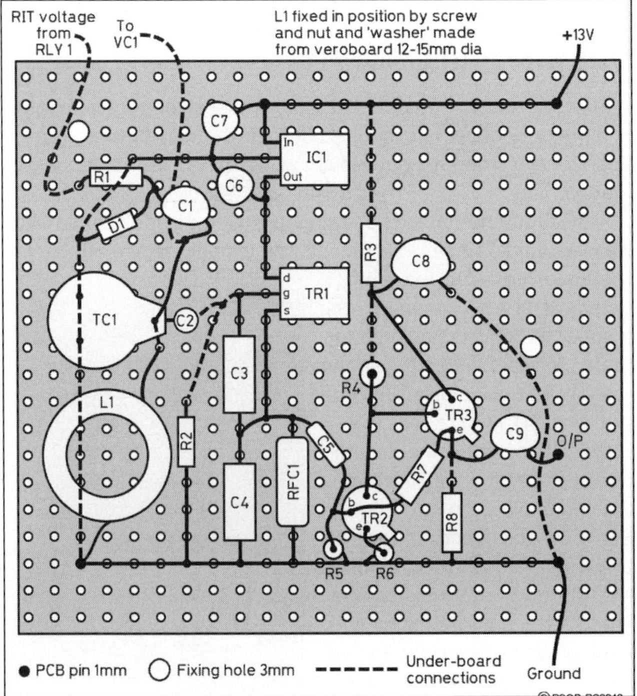


Fig 26.13: The VFO is built on matrix board.

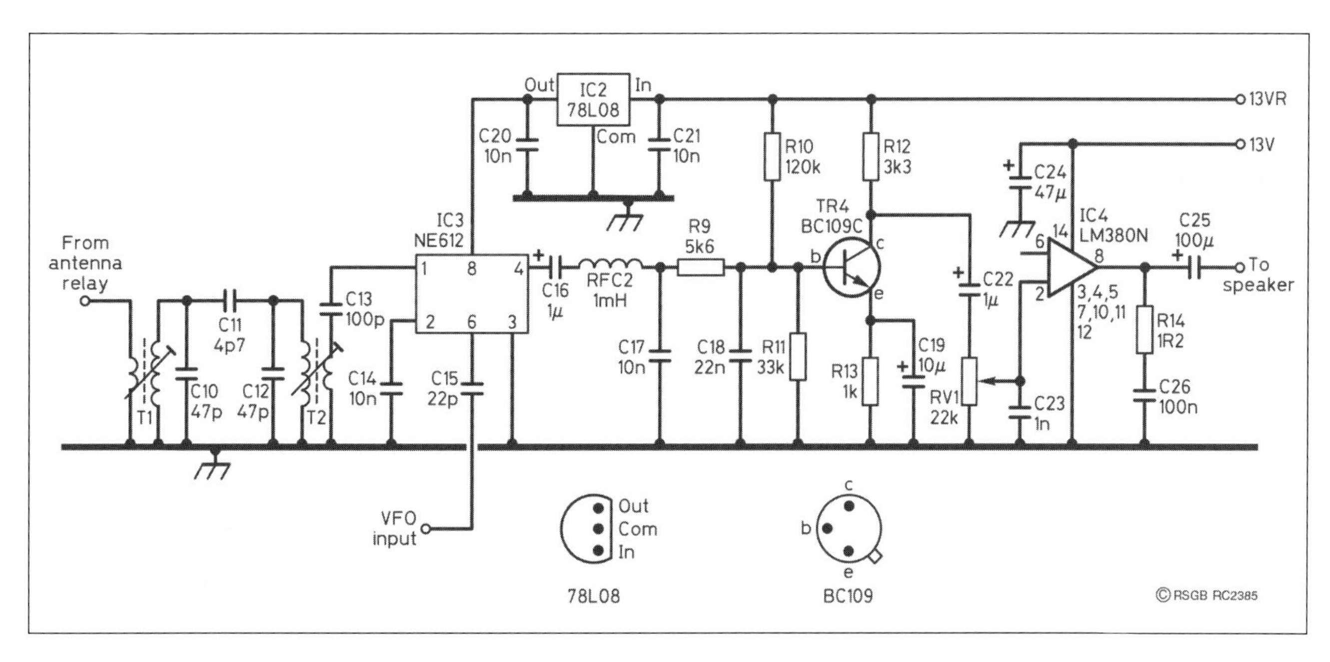


**Fig 26.14: Circuit of the receiver. The VFO input is via a short length of miniature coax from C9**

comfortable without a slow-motion drive. This, of course, simplifies mechanical construction. The value of a variable capacitor can be reduced by removing some of the plates. Actually, it is the number of gaps that determines the value, so if you divide this number into the value, you will know how many picofarads each gap contributes. The excess plates should be carefully removed with long-nosed pliers, taking care not to damage the remaining ones or over-stress the bearings or shaft. So, if you have a component with a value that is a little too high, you can still use it, although the author wouldn't recommend trying to reduce the value by more than 50% as the results become less predictable.

It was decided not to build the VFO 'dead bug' fashion for two reasons: (a) the rigidity (very important in a VFO) wouldn't be as good as it could be; and (b) thermal stability. With the frequency-determining components in contact with the chassis, they would be more prone to sudden changes in temperature than if they

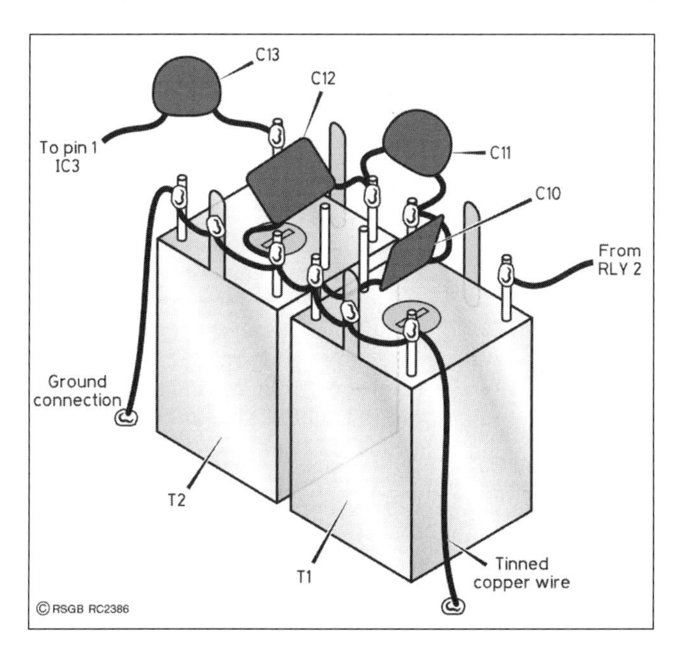


**Fig 26.15: T1 and T2 are mounted upside down on the earth plane and wired as shown.**

were mounted independently. As printed circuits had been ruled out, the VFO was built on a 21/4in × 21/2in piece of 0.1in matrix board. This, in turn, was mounted using two 12mm M2.5 screws, with extra nuts as spacers to keep the board clear of the chassis. See **Fig 26.13** for the layout of the VFO board.

It turned out to be essential that the VFO be fully enclosed, not for the obvious reason of RF screening but because the slightest draught caused unacceptable drift. Holes are drilled in the screens to allow connections to pass through. Obviously this should be done before assembly. Initially the lid can be left off. Decoupling capacitor C40 should be mounted close to where the lead carrying the RIT voltage enters the enclosure.

## THE RECEIVER

Referring to **Fig 26.14** and the detail in **Fig 26.15**, the band-pass filter comprises T1, T2, C10, C11 and C12. The transformers are from Toko, which make circuits such as this easily reproducible.

Next comes the mixer. The NE612 is a useful device, containing an oscillator as well as a mixer. The author decided not to make use of the oscillator in this case. The NE612 operates from an 8V supply provided by IC2, a 78L08.

As mentioned previously, the output contains many signals, audio and RF. The RF signals are unwanted and are removed by RFC2 and C17. The higher audio frequencies are progressively attenuated by R9 and C18, which form a low-pass filter. This gives 3dB attenuation at 1.3kHz and 6dB per octave thereafter. TR4 is the audio preamplifier. A BC109 was chosen rather than a BC108 because it gives more gain, which is required here.

The audio output stage is an LM380N. As shown in **Fig 26.16**, it is mounted upside down in the 'dead bug' style.

Unlike the previous stages, it is supplied with 13V on transmit as well as on receive, as it is needed to amplify the sidetone signal to loudspeaker level.

## TRANSMITTER

Referring to **Fig 26.17**, when the key contacts are closed, RLY2 switches the aerial to the PA stage and 13V from the receiver to the transmit circuits. The RIT control on the front

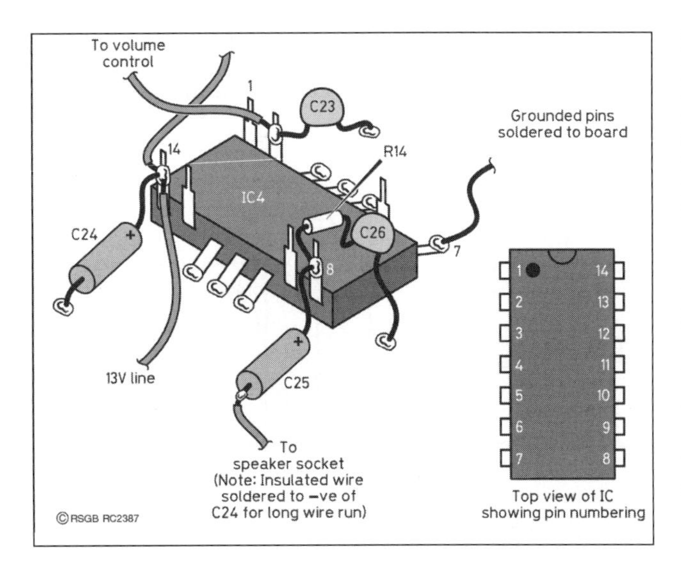


Fig 26.16: IC4, the audio output chip, is mounted 'dead bug' style on the ground plane and wired as shown.

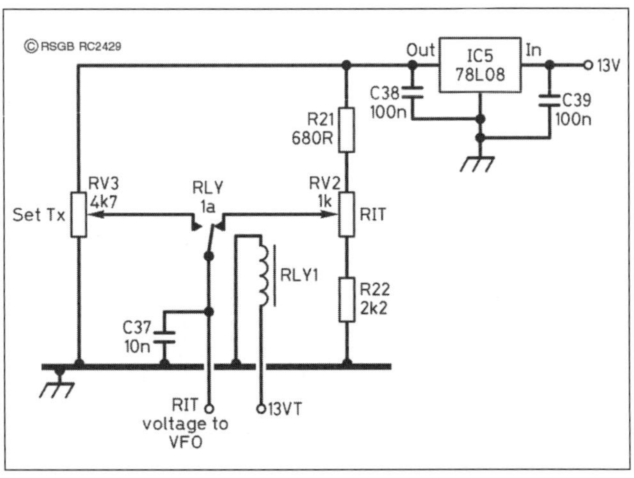


Fig 26.18: How the RIT voltage to the VFO is switched between transmit and receive.

panel becomes no longer operative, and the VFO offset voltage now comes from RV3.

The VFO signal is fed to the input of the driver stage, a BFY51. The collector is coupled to the PA by L2, a toroidal transformer. It is more usual to tune this with a trimmer but (on 80m at least) the tuning is very flat, so the number of turns is optimised for a standard-value capacitor, in this case 150pF (C29). There is adequate drive over the whole of the tuning range.

Low-frequency parasitic oscillations, normally below 100kHz, sometimes occur in the driver stages of QRP transmitters, and can go unnoticed by the operator. The author remember coming across a rough CW signal on about 3630kHz. and identified the station concerned, tuned down the band and also found him on 3560kHz. The two signals were 70kHz apart and, as expected, found another rough CW signal at 3490kHz. The station concerned was only running 3W and was nearly 200 miles away.

This problem is normally caused by poor decoupling and is easily cured. The output from TR5 is taken via a link winding on L2 to the base of TR6, the PA. This is a BFY51 and produces in excess of 1W output. The DC feed choke, RFC3, is not critical.

Some variation in wire thickness and number of turns shouldn't cause any problems, so just try what you have available. Take care, though, not to damage the enamel insulation when passing the wire through the bead, as this is easy to do.

The output is coupled by C32 to the low-pass filter, comprising L3, L4, C33, C34 and C35. It then passes via the changeover relay RLY2 to the aerial socket.

TR7 is the keying switch. Its use is preferable to directly-keying an RF stage such as the PA or driver, which can be unpredictable because RF has a tendency to find its way into keying lines. Also, there is the added advantage that the key or keyer only has to cope with the base current of TR7, a few milliamps.

The disadvantage of this kind of circuit is that it appears to be very good at injecting RF into supply lines and causing a shift in VFO frequency on 'key down' – something which is all too common in simple QRP equipment. The solution, again, is adequate decoupling, which is taken care of by C30 and C31.

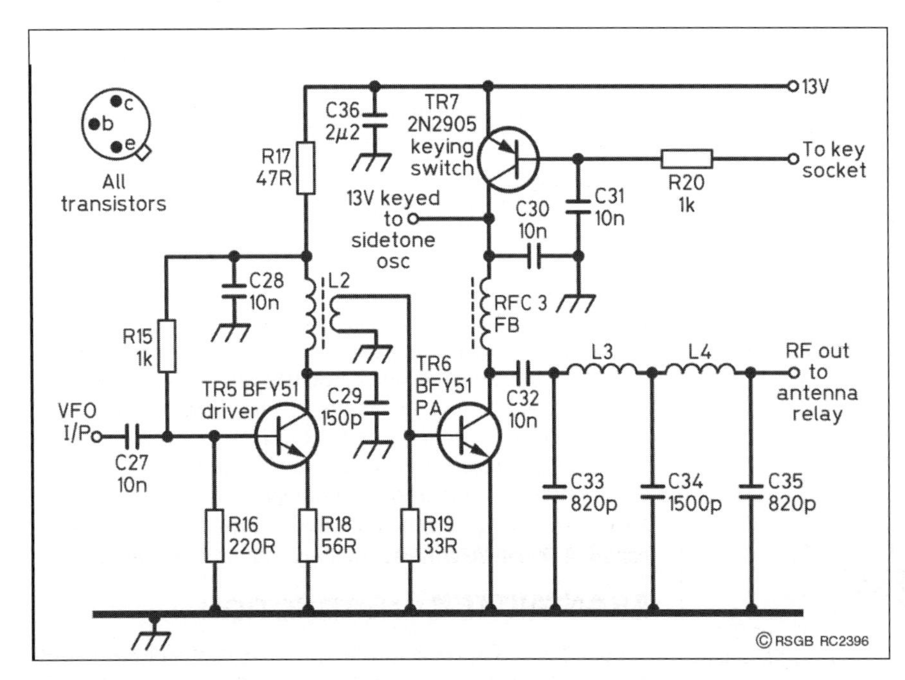


Fig 26.17: The transmitter produces over 1W output and, thanks to adequate decoupling, is quite stable.

## RIT

The requirement here is that the VFO frequency can be varied over a small range without moving the main tuning control. As you can see from **Fig 26.18**, RV2, the RIT control, is active on receive only. On switching to transmit, the RIT control has no effect and the frequency reverts to that set by the main tuning.

The actual shift in frequency is accomplished by a BB409 varicap (variable capacitance) diode coupled to the VFO tuned circuit. When this is reverse-biased (ie positive to the cathode) it exhibits a capacitance which is dependent on the voltage so, by varying the voltage, it will tune the VFO over a small range. A fixed, stable voltage is required on transmit. This is provided by RV3. On receive, a variable voltage is provided by RV2, the front-panel-mounted RIT control. The supply to RV2 and RV3 is stabilised by IC5. Relay RLY1 selects the voltage from RV2 or RV3, depending on whether the transceiver is in the receive or transmit mode.

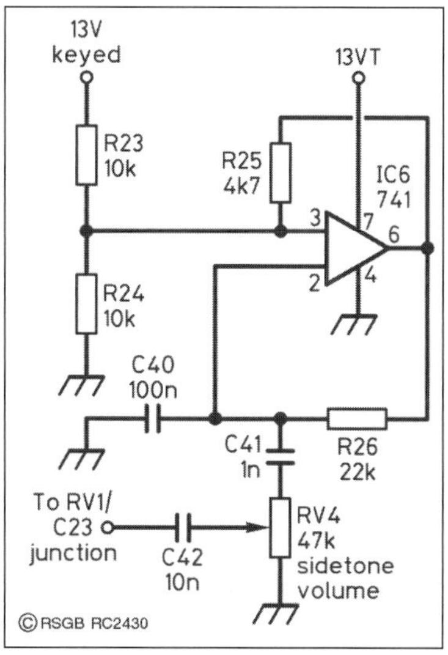

## SIDETONE

The sidetone oscillator (**Fig 26.19**) uses a 741 IC. It draws its supply from the 13V line, but is only activated on 'key down' by 13V from the keying switch, TR7. The output is taken to the volume control, RV1, via the sidetone level control, RV4, which is used to reduce the level of signal reaching the volume control. This is for operator comfort.

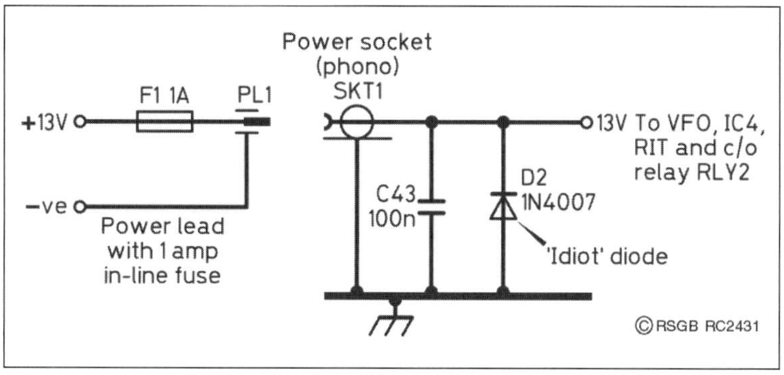

**Fig 26.19. The sidetone oscillator and its associated level control.**

**Fig 26.20: The connection to the supply contains protection against power of the wrong polarity being applied accidentally.**

## POWER SUPPLY

Power is supplied to the transceiver via a phono socket on the rear apron (see **Fig 26.20**). This is decoupled by C43. There is a diode, normally reverse-biased, connected directly across the supply input. This is for protection so, if the supply is accidentally connected the wrong way round, the diode will be forward-biased and will conduct, blowing the 1A fuse, which should be installed in a holder in the power lead.

## AERIAL SWITCHING

This is accomplished using a double-pole changeover relay RLY2 (see **Fig 26.21**).

## MECHANICAL CONSTRUCTION

As stated previously, this project was designed for easy construction. Mechanical work has been minimised; nonetheless some is required. The copper-clad board should be double-sided glass fibre but SRBP could be used. This would be easier to cut, but the end result is not as rigid. If you have, or if you know someone who has, access to a workshop guillotine, this part of the project could be very easy.

If not, you will have to cut the board by hand. Possibly the best way to do this is clamp it in a vice between two pieces of angle iron which are then used as a cutting guide.

Using a hacksaw, cut the pieces very slightly oversize, then file down to the required dimensions. Refer to **Fig 26.22** and **Fig 26.23** for the dimensions.

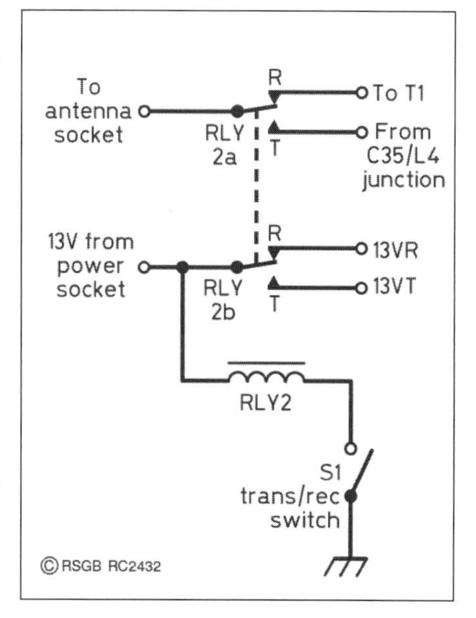


**Fig 26.21: Transmit / receive switching.**

When you have cut the base, front panel, rear apron, VFO enclosure and the two triangular supports, the holes can be drilled. First, mark their locations accurately and use a centre punch.

Start by drilling all holes 3mm in diameter. The larger holes can then be drilled to 6mm. At 10mm, more care is needed, as the drill can easily bite into the PCB. If you can, clamp the board firmly and drill through as slowly as possible.

The aerial socket can be any type you choose. If you want to fit an SO-239, a 16mm hole will be needed. This can be made by marking out, drilling to 10mm and enlarging with a file. Alternatively, drill a series of small holes (say, 2mm) inside the circumference of a 16mm circle, cutting out the centre and finishing off with a file.

Finally, drill the two 3mm holes in the base for the screws which support the VFO board. In this instance, there is no need to measure - just place the matrix board that you will be using in position, and mark through the holes.

## ASSEMBLY

With everything cut out and drilled, assembly can begin. Start by mounting the front panel on the base. Initially, use just one blob of solder in the centre. Then fix the side pieces of the VFO enclosure, again soldering lightly. After this, solder the rear apron in place. Finally, solder the two triangular supports in place. Do not mount the rear of the VFO enclosure at this stage. Inspect your work and if happy apply more solder to the joints and also solder in more places.

The end result will look more pleasing if the front panel is faced with card or thick paper, bearing labelling for the controls, along with a tuning scale for the VFO, and zero and ± marks for the RIT control. This can be done at any stage and fixed with adhesive.

## ELECTRICAL CONSTRUCTION

This is best done in a logical order, rather than haphazardly. Refer to **Fig 26.24**, and start by mounting the large components: controls, sockets, switches, relays (the latter can be fixed in place with a spot of Super Glue®). Draw pencil marks on the

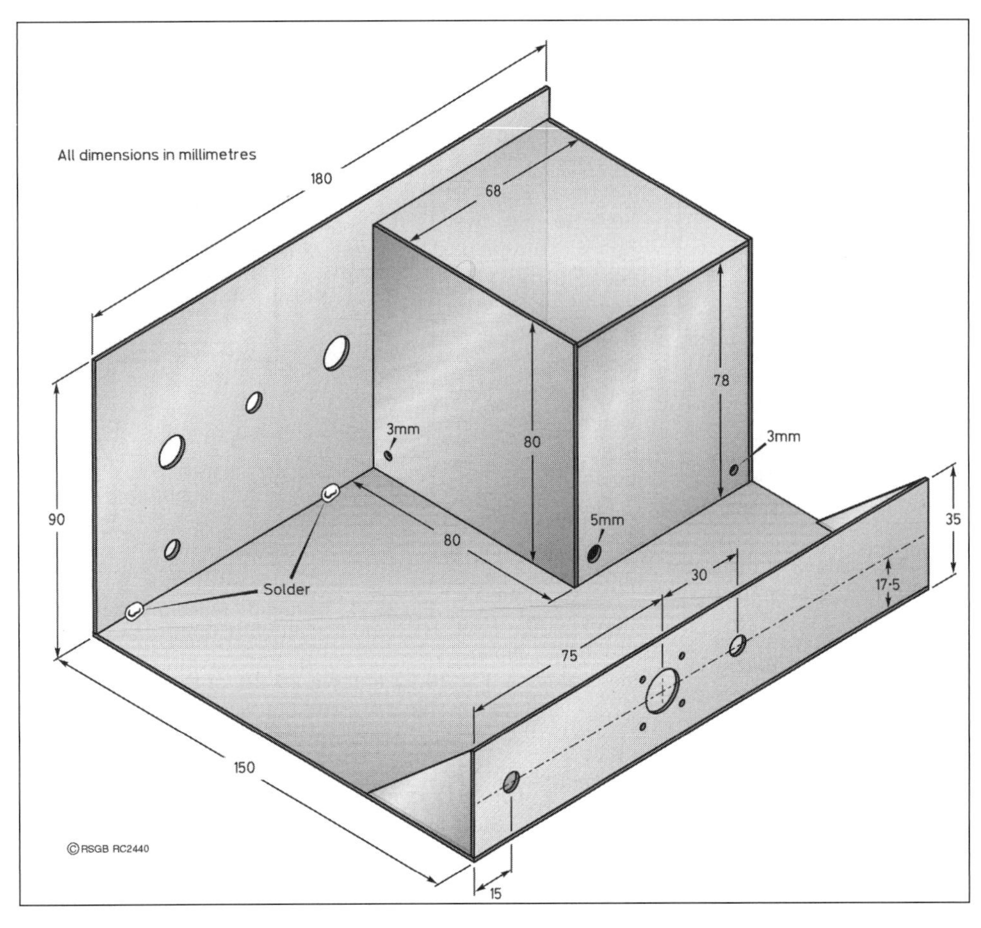


Fig 26.22: The chassis, seen from the back.

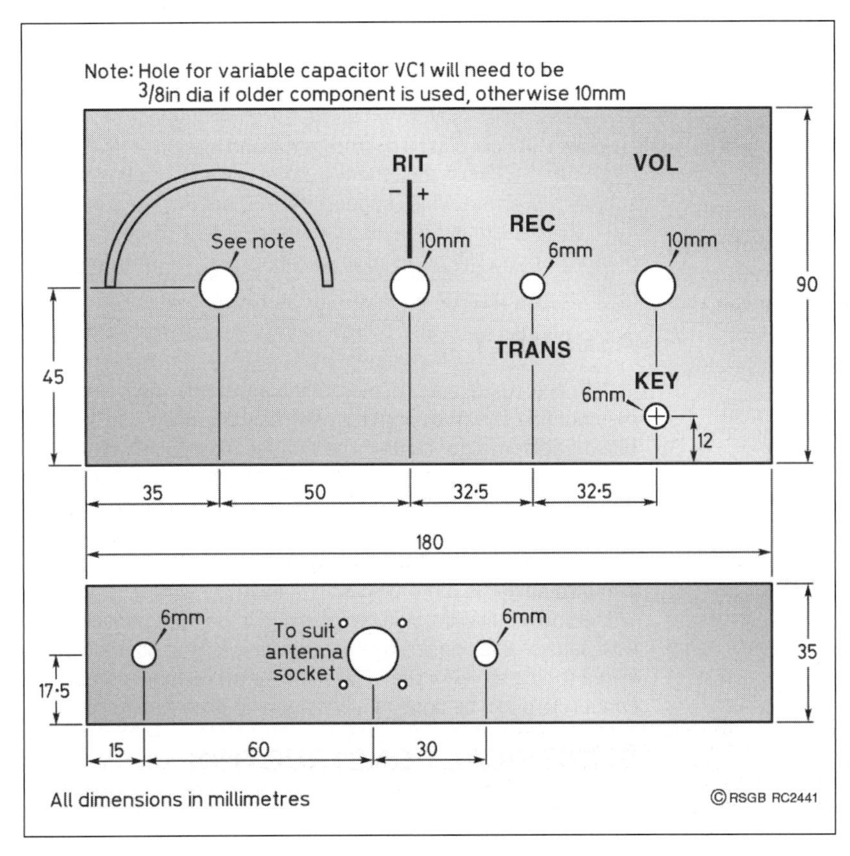


Fig 26.23. Front panel and rear apron layouts.

base, along the line of the receiver (30mm from the side edge) and transmitter (30mm from the back edge). These are used as guides for the smaller components.

The first major job is the VFO. Fit all components and make all connections, taking great care to get the pin connections of the semiconductors correct. Check and double-check everything. Mount the VFO board, ensuring that the connections on the underside are clear of the base. Connect the variable capacitor using stiff wire (1.25 or 0.9mm). The lead to the RIT pin can be grounded for now. Connect a 13V supply and check that there is 5V at the output of IC1. All being well, the tuning range can now be roughly set. With VC1 at half-mesh, connect a short length of wire to the output (C9). Using a test receiver tuned to 3550kHz, adjust TC1 until a signal is heard. This is all that is needed at the moment. Don't be concerned if the VFO is not very stable at this stage.

Now for the receiver. Referring to **Fig 26.24**, start by assembling the band-pass filter (**Fig 26.25**) and fix it in place. Continue with IC3, IC2 and the smaller components. IC3 is mounted in the same manner as IC4 ('dead bug', ie legs upwards), but only pin 3 is grounded. Where grounding is necessary, leads are soldered directly to the base. Carry on until the receiver is complete. The volume control is connected with miniature screened audio cable, and the output from IC4 is connected to the loudspeaker socket. The receiver can now be tested using a temporary connection from the VFO to IC3, with 13V to IC4, TR4 and IC2, and a length of wire as an aerial to the input of the band-pass filter. Sensitivity will be low, but it should be possible to hear something by tuning the VFO up and down its range.

Continue with the auxiliary circuits - the RIT, sidetone, changeover relay RLY2 [1] and the power connector. Permanent power connections can be made to all stages, a length of miniature coax connected to C15, and the wiring tied down neatly as shown. Now carry on and complete the transmitter and make a thorough check of everything.

## TESTING

If at any stage a problem appears, turn off the power and investigate it. Connect a loud-

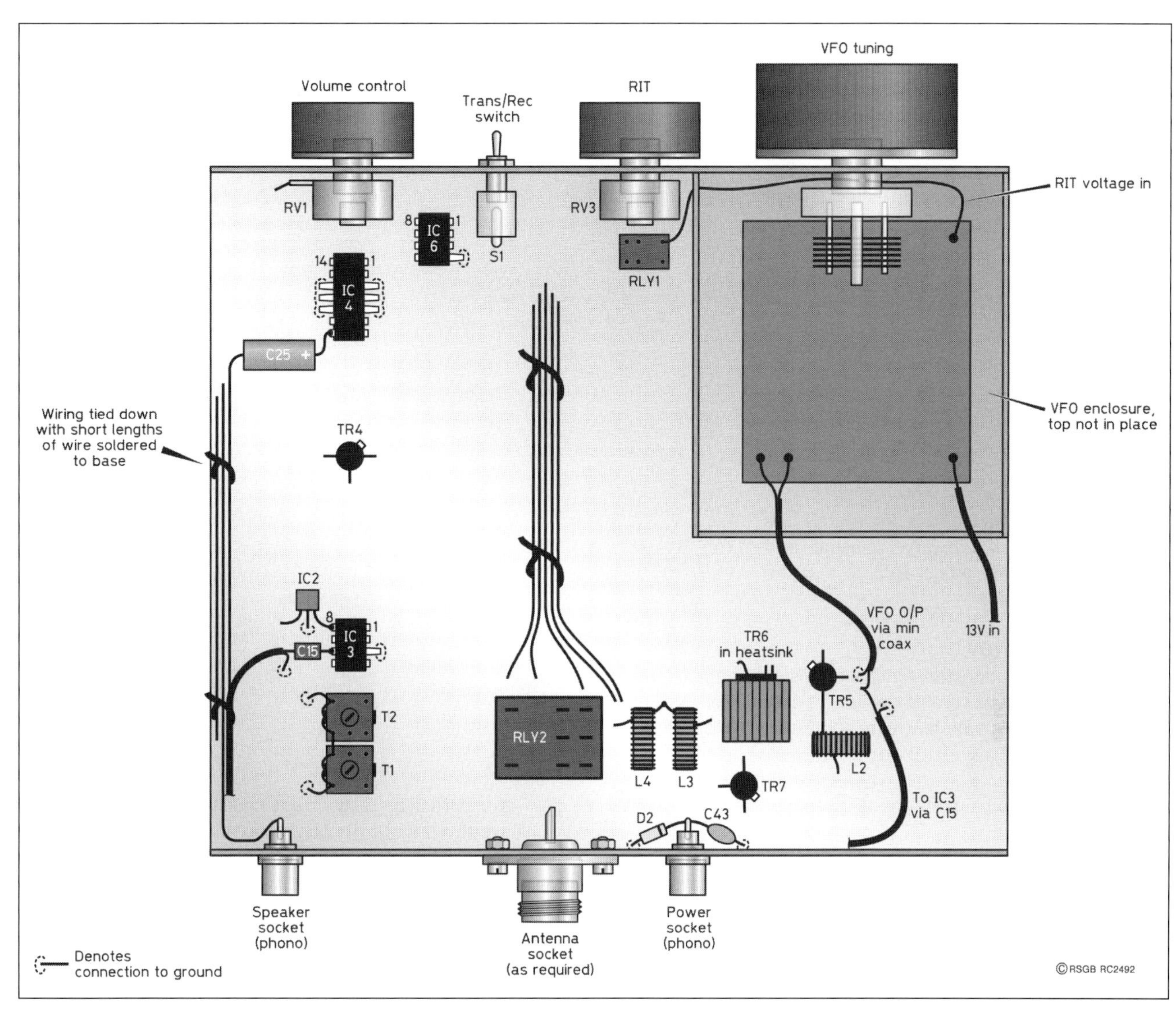


**Fig 26.24: Layout of the major components. Smaller components are soldered between the larger ones.**

speaker and an aerial, and turn on. Advancing the volume control should result in noise from the loudspeaker. Set the RIT control to zero. Tune the VFO and search for a steady signal at the centre of the tuning range. When one is found, using a suitable tool, adjust the cores of T1 and T2 for maximum volume. Now check the function of the RIT control.

Disconnect the aerial and connect a Morse key and a dummy load [2]. Set RV3 to the centre of its travel. Switch to transmit and press the key. The sidetone should be heard in the speaker. Holding the key down, tune the test receiver until the signal is found, somewhere near 3550kHz. If you have a power meter, it can be connected between the transceiver and dummy load. It should show an output of at least 1W.

**Table 1. RIT, sidetone, switching and power input circuits components list**

| | |
|---|---|
| **Resistors** | |
| R21 | 680R |
| R22 | 2k2 |
| R23, 24 | 10k |
| R25 | 4k7 |
| R26 | 22k |
| RV2 | 1k |
| RV3 | 4k7 |
| RV4 | 47k |
| **Capacitors** | |
| C37, 42 | 10n ceramic disc |
| C38–40, 43 | 100n ceramic disc |
| C41 | 1n ceramic disc |
| **Semiconductors** | |
| IC5 | 78L08 |
| IC6 | 741 |
| D2 | 1N4007 |
| **Miscellaneous items** | |
| RLY1 | 12V single-pole change over relay |
| RLY2 | 12V double-pole change over relay |
| F1 | 1A fuse with in-line holder |
| S1 | SPST switch |
| PL1 | Phono plug |
| SKT1 | Phono socket |

**Table 2. Transmitter components list**

| | |
|---|---|
| **Resistors** | |
| R15, 20 | 1k |
| R16 | 220R |
| R17 | 47R |
| R18 | 56R |
| R19 | 33R |
| All resistors 0.6W metal film | |
| **Capacitors** | |
| C27, 28, 30–32 | 10n |
| C29 | 150p |
| C33, 35 | 820p |
| C34 | 1500p |
| C36 | 2μ2 |
| **Inductors** | |
| L2 | 42 turns of 0.375mm enamelled wire on a T50-2 core, plus a 5-turn link of PVC-covered wire over the main winding |
| L3, L4 | 22 turns of 0.56mm wire on T50-2 core |
| RFC3 | 20 turns of 0.19mm wire on ferrite bead |
| **Semiconductors** | |
| TR5, 6 | BFY51 |

## SETTING UP

There are only three adjustments to be made – the VFO, RIT and bandpass filter – and they have already been roughly set.

## VFO

Close the vanes of VC1 fully. The tuning knob should be set to exactly nine o'clock. Now set the test receiver to 3490kHz (if VC1 is a 12pF component) or 3500kHz if it is 10pF.

Adjust TC1 until the signal is heard. Set the tuning control to the three o'clock position and check the frequency of the VFO by finding the signal with the test receiver. If using 10pF for VC1, the frequency should be about 3600kHz; if using 12pF it should be about 3610kHz. The VFO enclosure can now be completed, but don't use too much solder on the lid - you might want to remove it at some future time. The VFO must be calibrated, but not yet. Since you have just heated everything up with a soldering iron, now is not a good time - allow several hours to elapse first.

## RIT

Set the VFO to the centre of its range, the RIT control to zero (centre), and tune the VFO to give a beat note on the test receiver. Switch to transmit and adjust RV3 for exactly the same note. Switch between transmit and receive and carefully adjust RV3 so that the note doesn't change.

**Table 4. VFO components list**

| | |
|---|---|
| **Resistors** | |
| R1, 2 | 100k |
| R3 | 15R |
| R4 | 820R |
| R5 | 1k5 |
| R6 | 33R |
| R7 | 12k |
| R8 | 270R |
| All resistors 0.6W metal film | |
| **Capacitors** | |
| C1 | 10p ceramic plate |
| C2 | 150p close tolerance polystyrene |
| C3, 4 | 750p close tolerance polystyrene |
| C5 | 22p polystyrene |
| C6, 7 | 100n ceramic disc |
| C8, 9 | 10n ceramic disc |
| VC1 | 10 or 12p (see text) |
| TC1 | 65p (Maplin WL72P) |
| **Inductors** | |
| L1 | 50 turns of 32SWG on T50-2 toroidal core |
| RFC1 | 1mH |
| **Semiconductors** | |
| TR1 | 2N3819 |
| TR2, 3 | BC108 |
| IC1 | 78L05 |
| D1 | BB409 |

**Table 3. Receiver components list**

| | |
|---|---|
| **Resistors** | |
| R9 | 5k6 |
| R10 | 120k |
| R11 | 33k |
| R12 | 3k3 |
| R13 | 1k |
| R14 | 1R2 |
| RV1 | 22k log pot |
| All resistors 0.6W metal film unless specified otherwise | |
| **Capacitors** | |
| C10, 12 | 47p ceramic plate |
| C11 | 4p7 ceramic plate |
| C13 | 100p ceramic plate |
| C14, 17 | 10n ceramic disc |
| C15 | 22p polystyrene |
| C16, 22 | 1μ, 16V electrolytic |
| C18 | 22n ceramic disc |
| C19 | 10μ, 16V electrolytic |
| C20, 21 | 10n ceramic disc |
| C23 | 1n ceramic disc |
| C24 | 47μ, 16V electrolytic |
| C25 | 100μ, 16V electrolytic |
| C26 | 100n ceramic disc |
| **Inductors** | |
| RFC2 | 1mH |
| T1, 2 | Toko KANK3333R |
| **Semiconductors** | |
| TR4 | BC109 |
| IC2 | 78L08 |
| IC3 | NE612 |
| IC4 | LM380N |

## Bandpass filter

With the aerial connected, find a steady signal in the centre of the tuning range and carefully adjust T1 and T2 for maximum signal strength. This time take care and make sure it is right.

## Calibration

It just remains to calibrate the VFO scale and the transceiver will be ready for use.

Before using it in earnest, though, it would be a good idea to get a local amateur to listen to your signal to make sure that all is well.

## AND FINALLY...

The transceiver is delightfully easy to use. To net onto a station, start with the RIT at its centre position and adjust the main tuning for zero beat. The RIT can now be set to give the desired note.

With a reasonable aerial (eg a G5RV) you shouldn't have any shortage of QSOs. You are unlikely to achieve WAC or DXCC (please prove me wrong) - on the other hand, you will almost certainly have no problems with TVI or BCI. Have fun!

## NOTES

[1] As a particular relay is not specified for RLY2, you will have to work out the connections for yourself. If no data are available, this can be done visually and confirmed with a test meter on a resistance range.

[2] This can be three 150Ω 0.5W resistors in parallel.

[3] The TOKO coils specified for T1 and T2 in the receiver circuit are no longer manufactured. Direct 45uH replacements are currently sold in the UK by Spectrum Communications of Dorset; http://spectrumcomms.co.uk/ and are available to G-QRP Club members via their Club Sales; http://www.gqrp.com/sales.htm

## REFERENCES

[1] 80m Project abridged from the RSGB's: *'50 Projects for Radio Amateurs'* by Mike Browne, G3DIH.

*__Steve Hartley, G0FUW,__ has been a member of the G-QRP Club since before he was licensed as a UK radio amateur. Attracted by the simple homebrew approach to radio communications, and the sagely writings of G3RJV, he passed the City & Guilds Radio Amateurs' exam in 1983. Steve was first on the air as G1KVY before passing the Morse test to become G0FUW. He has served the RSGB for many years in various capacities including four and a half years as a Board Director, one of those as Chairman. Appointed as Chairman of the G-QRP Club in 2018 Steve continues to build his own radio equipment and support newcomers through running Buildathons and amateur radio training.*

# *Appendix A:* General Data

## Capacitance

The capacitance of a parallel-plate capacitor is:

$$C = \frac{0.224\ KA}{d} \text{ picofarads}$$

where K is the dielectric constant (air = 1.0), A is the area of dielectric (sq in), and d is the thickness of dielectric (in).

If A is expressed in centimetres squared and d in centimetres, then:

$$C = \frac{0.0885\ KA}{d} \text{ picofarads}$$

For multi-plate capacitors, multiply by the number of dielectric thicknesses.

The capacitance of a coaxial cylinder is:

$$C = \frac{0.242}{\log_{10}(D/d)} \text{ picofarads per centimetre length}$$

where D is the inside diameter of the outer and d is the outside diameter of the inner.

## Capacitors in series or parallel

The effective capacitance of a number of capacitors in series is:

$$C = \frac{1}{\frac{1}{C_1} + \frac{1}{C_2} + \frac{1}{C_3} + \text{etc}}$$

The effective capacitance of a number of capacitors in parallel is:

$$C = C_1 + C_2 + C_3 + \text{etc}$$

## Characteristic impedance

The characteristic impedance $Z_0$ of a feeder or transmission line depends on its cross-sectional dimensions.

(i) Open-wire line:

$$Z_0 = 276 \log_{10} \frac{2D}{d} \text{ ohms}$$

where D is the centre-to-centre spacing of wires (mm) and d is the wire diameter (mm).

(ii) Coaxial line:

$$Z_0 = \frac{138}{\sqrt{K}} \log_{10} \frac{d_o}{d_i} \text{ ohms}$$

where K is the dielectric constant of insulation between the conductors (eg 2.3 for polythene, 1.0 for air), $d_o$ is the inside diameter of the outer conductor and $d_i$ is the diameter of the inner conductor.

## Decibel

The decibel is the unit commonly used for expressing the relationship between two power levels (or between two voltages or two currents). A decibel (dB) is one-tenth of a bel (B). The number of decibels N representing the ratio of two power levels $P_1$ and $P_2$ is 10 times the common logarithm of the power ratio, thus:

$$\text{The ratio } N = 10 \log_{10} \frac{P_2}{P_1} \text{ decibels}$$

If it is required to express voltage (or current) ratios in this way, they must relate to identical impedance values, ie the two different voltages must appear across equal impedances (or the two different currents must flow through equal impedances). Under such conditions the power ratio is proportional to the square of the voltage (or the current) ratio, and hence:

$$N = 20 \log_{10} \frac{V_2}{V_1} \text{ decibels}$$

$$N = 20 \log_{10} \frac{I_2}{I_1} \text{ decibels}$$

## Dynamic resistance

In a parallel-tuned circuit at resonance the dynamic resistance is:

$$R_D = \frac{L}{Cr} = Q\omega L = \frac{Q}{\omega C} \text{ ohms}$$

where L is the inductance (henrys), C is the capacitance (farads), r is the effective series resistance (ohms), Q is the Q-value of the coil and $\omega = 2\pi \times$ frequency (hertz).

## Frequency – wavelength – velocity

The velocity of propagation of a wave is:

$$v = f\lambda \text{ metres per second}$$

where f is the frequency (hertz) and $\lambda$ is the wavelength (metres).

For electromagnetic waves in free space the velocity of propagation v is approximately $3 \times 10^8$m/s and, if f is expressed in kilohertz and $\lambda$ in metres:

$$f = \frac{300{,}000}{\lambda} \text{ kilohertz}$$

$$\lambda = \frac{300{,}000}{f} \text{ metres}$$

$$\text{Free space } \frac{\lambda}{2} = \frac{492}{\text{MHz}} \text{ feet}$$

$$\text{Free space } \frac{\lambda}{4} = \frac{246}{\text{MHz}} \text{ feet}$$

Note that the true value of v is $2.99776 \times 10^8$m/s.

### Impedance

The impedance of a circuit comprising inductance, capacitance and resistance in series is:

$$Z = \sqrt{R^2 + \left(\omega L - \frac{1}{\omega C}\right)^2}$$

where R is the resistance (ohms), L is the inductance (henrys), C is the capacitance (farads) and $\omega = 2\pi \times$ frequency (hertz).

### Inductors in series or parallel

The total effective value of a number of inductors connected in series (assuming that there is no mutual coupling) is given by:

$$L = L_1 + L_2 + L_3 + \text{etc}$$

If they are connected in parallel, the total effective value is:

$$L = \frac{1}{\frac{1}{L_1} + \frac{1}{L_2} + \frac{1}{L_3} + \text{etc}}$$

When there is mutual coupling M, the total effective value of two inductors connected in series is:

$$L = L_1 + L_2 + 2M \text{ (windings aiding)}$$

$$\text{or } L = L_1 + L_2 - 2M \text{ (windings opposing)}$$

### Ohm's Law

For a unidirectional current of constant magnitude flowing in a metallic conductor:

$$I = \frac{E}{R} \qquad E = I\,R \qquad R = \frac{E}{I}$$

where I is the current (amperes), E is the voltage (volts) and R is the resistance (ohms).

### Power

In a DC circuit, the power developed is given by:

$$W = E\,I = \frac{E^2}{R} = I^2R \text{ watts}$$

where E is the voltage (volts), I is the current (amperes) and R is the resistance (ohms).

### *Q*

The Q-value of an inductance is given by:

$$Q = \frac{\omega L}{R}$$

where L is the inductance (henrys), R is the effective resistance (ohms) and $\omega = 2\pi \times$ frequency (hertz).

### Reactance

The reactance of an inductance and a capacitance respectively is given by:

$$X_L = \omega L \text{ ohms}$$

$$X_C = \frac{1}{\omega C} \text{ ohms}$$

where L is the inductance in henrys, C is the capacitance in farads and $\omega = 2\pi \times$ frequency (hertz).

The total reactance of an inductance and a capacitance in series is $X_L - X_C$.

### Resistors in series or parallel

The effective value of several resistors connected in series is:

$$R = R_1 + R_2 + R_3 + \text{etc}$$

When several resistors are connected in parallel the effective total resistance is:

$$R = \frac{1}{\frac{1}{R_1} + \frac{1}{R_2} + \frac{1}{R_3} + \text{etc}}$$

### Resonance

The resonant frequency of a tuned circuit is given by:

$$f = \frac{1}{2\pi\sqrt{LC}} \text{ hertz}$$

where L is the inductance (henrys) and C is the capacitance (farads).

If L is in microhenrys (μH) and C is picofarads (pF), this formula becomes:

$$f = \frac{10^3}{2\pi\sqrt{LC}} \text{ megahertz}$$

The basic formula can be rearranged thus:

$$L = \frac{1}{4\pi^2 f^2 C} \text{ henrys}$$

$$C = \frac{1}{4\pi^2 f^2 L} \text{ farads}$$

Since $2\pi f$ is commonly represented by $\omega$, these expressions can be written as:

$$L = \frac{1}{\omega^2 C} \text{ henrys}$$

$$C = \frac{1}{\omega^2 L} \text{ farads}$$

See Figs 23.1 and 23.2.

### Time constant

For a combination of inductance and resistance in series the time constant (ie the time required for the current to reach $1/\varepsilon$ or 63% of its final value) is given by:

$$t = \frac{L}{R} \text{ seconds}$$

where L is the inductance (henrys) and R is the resistance (ohms).

For a combination of capacitance and resistance in series, the time constant (ie the time required for the voltage across the capacitance to reach $1/\varepsilon$ or 63% of its final value) is given by:

$$t = C\,R \text{ seconds}$$

where C is the capacitance (farads) and R is the resistance (ohms).

### Transformer ratios

The ratio of a transformer refers to the ratio of the number of turns in one winding to the number of turns in the other winding. To avoid confusion it is always desirable to state in which sense the ratio is being expressed, eg the 'primary-to-secondary' ratio $n_p/n_s$. The turns ratio is related to the impedance ratio thus:

$$\frac{n_p}{n_s} = \sqrt{\frac{Z_p}{Z_s}}$$

where $n_p$ is the number of primary turns, $n_s$ is the number of secondary turns, $Z_p$ is the impedance of the primary circuit (ohms) and $Z_s$ is the impedance of the secondary circuit (ohms).

## COIL WINDING

Most inductors for tuning in the HF bands are single-layer coils and they are designed as follows. Multilayer coils will not be dealt with here.

The inductance of a single-layer coil is given by:

$$L\,(\mu H) = \frac{D^2 \times T^2}{457.2 \times D + 1016 \times L}$$

where D is the diameter of the coil (millimetres), T is the number of turns and L is the length (millimetres). Alternatively:

$$L\,(\mu H) = \frac{R^2 \times T^2}{9 \times R + 10 \times L}$$

where R is the radius of the coil (inches), T is the number of turns and L is the length (inches).

Note that when a ferrite or iron dust core is used, the inductance will be increased by up to twice the value without the core. The choice of which to use depends on frequency. Generally, ferrite cores are used at the lower HF bands and iron dust cores at the higher. At VHF, the iron dust cores are usually coloured purple. Cores need to be moveable for tuning but fixed thereafter and this can be done with a variety of fixatives. A strip of flexible polyurethane foam will do.

### Designing inductors with ferrite pot cores

This is a simple matter of taking the factor given by the makers and multiplying it by the square of the number of turns.

Example

A RM6-S pot core in 3H1 grade ferrite has a 'factor' of 1900 nanohenrys for one turn. Therefore 100 turns will give an inductance of:

$$100^2 \times 1900\text{nH} = 10000 \times 1900\text{nH} = 19\text{mH}$$

There are a large number of different grades of ferrite; for example, the same pot as above is also available in grade 3E4 with a 'factor' of 3300. Manufacturers' literature should be consulted to find these 'factors'.

**Table A.1**

| Diameter (mm) | Approx SWG | Turns/cm | Turns/in |
|---|---|---|---|
| **1.5** | **16–17** | **6.6** | **16.8** |
| **1.25** | **18** | **7.9** | **20.7** |
| **1.0** | **19** | **9.9** | **25** |
| **0.8** | **21** | **12.3** | **31** |
| **0.71** | **22** | **13.9** | **35** |
| **0.56** | **24** | **17.5** | **45** |
| **0.50** | **25** | **19.6** | **50** |
| **0.40** | **27** | **24.4** | **62** |
| **0.315** | **30** | **30.8** | **78** |
| **0.25** | **33** | **38.5** | **97** |
| **0.224** | **34–35** | **42.7** | **108** |
| **0.20** | **35–36** | **47.6** | **121** |

**Note: SWG is Imperial standard wire gauge. The diameters listed are those which appear to be most popular; ie they are listed in distributor's catalogues. The 'turns/cm' and 'turns/in' are for enamelled wire.**

**Table A.3: Wire table**

| Diameter (mm) | Approx SWG | Max current (A) | Fusing current (A) | Resistance at 20 deg C ohms/km |
|---|---|---|---|---|
| **2.5** | **12** | **7.6** | **325** | **3.5** |
| **2.0** | **14** | **4.9** | **225** | **5.4** |
| **1.5** | **16–17** | **2.7** | **147** | **9.7** |
| **1.0** | **19** | **1.2** | **81** | **22** |
| **0.71** | **22** | **0.61** | **46** | **43** |
| **0.5** | **26** | **0.30** | **28** | **87** |
| **0.25** | **32** | **0.076** | **10** | **351** |
| **0.20** | **36** | **0.049** | **7.1** | **541** |

**'Max current' is the carrying capacity at 1.55A/mm². This is a very conservative figure and can usually be doubled. The 'fusing current' is approximate since it depends also on thermal conditions, ie if the wire is thermally insulated, it will fuse at a lower current.**

**Table A.2: Coaxial cables**

| Type | Nominal impedance (ohms) | Outside diameter (mm) | Velocity factor | Capacitance (pF/m) | Attenuation per 30m (100ft) of cable 10MHz (dB) | 100MHz (dB) | 1000MHz (dB) |
|---|---|---|---|---|---|---|---|
| RG58-U/UR43 | 52 | 5.0 | 0.66 | 100 | 1.0 | 3.3 | 10.6 |
| RG213/UR67 | 50 | 10.3 | 0.66 | 100 | 0.68 | 2.26 | 8.0 |
| Westflex 103 | 50 | 10.3 | 0.85 | 78 | 0.27 | 0.85 | 2.7 |
| EchoFlex 15 | 50 | 14.6 | 0.86 | 77 | 0.26 | 0.28 | 2.9 |
| LDF4-50 | 50 | 16 | 0.88 | 88 | 0.21 | 0.68 | 2.5 |
| UR70 | 75 | 5.8 | 0.66 | 67 | 0.5 | 1.5 | 5.2 |
| RG59BU | 75 | 6.15 | 0.66 | 68 | 0.5 | 1.5 | 4.6 |
| RG62AU | 93 | 6.15 | 0.84 | 44 | 0.3 | 0.9 | 2.9 |

This short list of coaxial cables represents what is available in distributors' advertisements.
The data in this table has been obtained from these catalogues and from the Internet.
Note that there are various standards of measuring coax loss, such as dB/100m, dB/10m and dB/100ft and at various frequencies.
The data above uses dB/30m (approximately 100ft) representing the most common length from the shack to the antenna.

### Table A.4: Basic SI units

| Quantity | Name of unit | Unit symbol |
|---|---|---|
| **Electric current** | **ampere** | **A** |
| **Length** | **metre** | **m** |
| **Luminous intensity** | **candela** | **cd** |
| **Mass** | **kilogramme** | **kg** |
| **Thermodynamic temperature** | **kelvin** | **K** |
| **Time** | **second** | **s** |

### Table A.5: Derived SI units

| Physical quantity | SI unit | Unit symbol |
|---|---|---|
| **Electric capacitance** | **farad** | **F = A s/V** |
| **Electric charge** | **coulomb** | **C = A s** |
| **Electric potential** | **volt** | **V = W/A** |
| **Electric resistance** | **ohm** | **Ω = V/A** |
| **Force** | **newton** | **N = kg m/s²** |
| **Frequency** | **hertz*** | **Hz = s⁻¹** |
| **Illumination** | **lux** | **lx = lm/m²** |
| **Inductance** | **henry** | **H = V s/A** |
| **Luminous flux** | **lumen** | **lm = cd sr** |
| **Magnetic flux** | **weber** | **Wb = V s** |
| **Magnetic flux density** | **tesla†** | **T = Wb/m²** |
| **Power** | **watt** | **W = J/s** |
| **Work, energy, quantity of heat** | **joule** | **J = N m** |

*** Hertz is equivalent to cycle per second.**
**† Tesla is equivalent to weber per square metre.**

### Table A.6: Comparison of Celsius and Fahrenheit thermometer scales

| Celsius | Fahrenheit | Celsius | Fahrenheit |
|---|---|---|---|
| **–50** | **–58** | **+80** | **+176** |
| **–45** | **–49** | **+85** | **+185** |
| **–40** | **–40** | **+90** | **+194** |
| **–35** | **–31** | **+95** | **+203** |
| **–30** | **–22** | **+100** | **+212** |
| **–25** | **–13** | **+105** | **+221** |
| **–20** | **–4** | **+110** | **+230** |
| **–15** | **+5** | **+115** | **+239** |
| **–10** | **+14** | **+120** | **+248** |
| **–5** | **+23** | **+125** | **+257** |
| **0** | **+32** | **+130** | **+266** |
| **+5** | **+41** | **+135** | **+275** |
| **+10** | **+50** | **+140** | **+284** |
| **+15** | **+59** | **+145** | **+293** |
| **+20** | **+68** | **+150** | **+302** |
| **+25** | **+77** | **+155** | **+311** |
| **+30** | **+86** | **+160** | **+320** |
| **+35** | **+95** | **+165** | **+329** |
| **+40** | **+104** | **+170** | **+338** |
| **+45** | **+113** | **+175** | **+347** |
| **+50** | **+122** | **+180** | **+356** |
| **+55** | **+131** | **+185** | **+365** |
| **+60** | **+140** | **+190** | **+374** |
| **+65** | **+149** | **+195** | **+383** |
| **+70** | **+158** | **+200** | **+392** |
| **+75** | **+167** | | |

### Table A.7: Colour coding for glass fuses

| Colour | Rating (A) | Colour | Rating (A) |
|---|---|---|---|
| **Green/yellow** | **10** | **Green** | **0.75** |
| **Red/turquoise** | **15** | **Blue** | **1.0** |
| **Eau-de-Nil** | **25** | **Light blue** | **1.5** |
| **Salmon pink** | **50** | **Purple** | **2.0** |
| **Black** | **60** | **Yellow and purple** | **2.5** |
| **Grey** | **100** | **White** | **3.0** |
| **Red** | **150** | **Black and white** | **5.0** |
| **Brown** | **250** | **Orange** | **10.0** |
| **Yellow** | **500** | | |

**Note that this coding does not apply to the ceramic-bodied fuse commonly found in 13A plugs etc.**

### Table A.8: Conversion factors

| To convert | into | multiply by | Conversely |
|---|---|---|---|
| **Amp hours** | **Coulombs** | **3600** | **$2.778 \times 10^{-4}$** |
| **Atmospheres** | **Lb/sq in** | **14.70** | **0.068** |
| **Centigrade** | **Kelvin °C + 273 = K** | **K - 273 = °C** | |
| **Cubic inches** | **Cubic feet** | **$5.787 \times 10^{-4}$** | **1728** |
| **Cubic inches** | **Cubic metres** | **$1.639 \times 10^{-5}$** | **$6.102 \times 10^{4}$** |
| **Degrees (angular)** | **Radians** | **$1.745 \times 10^{-2}$** | **57.3** |
| **Dynes** | **Pounds** | **$2.248 \times 10^{-6}$** | **$4.448 \times 10^{5}$** |
| **Ergs** | **Foot pounds** | **$7.376 \times 10^{-8}$** | **$1.356 \times 10^{7}$** |
| **Feet** | **Centimetres** | **30.48** | **$3.281 \times 10^{-2}$** |
| **Foot pounds** | **Kilowatt hours** | **$3.766 \times 10^{-7}$** | **$2.655 \times 10^{6}$** |
| **Gausses** | **Lines per sq in** | **6.452** | **0.155** |
| **Grams** | **Dynes** | **980.7** | **$1.02 \times 10^{-3}$** |
| **Grams per cm** | **Pounds per in** | **$5.6 \times 10^{-3}$** | **178.6** |
| **Horse power** | **Kilowatts** | **0.746** | **1.341** |
| **Inches** | **Centimentres** | **2.54** | **0.3937** |
| **Kilograms** | **Pounds (lb)** | **2.205** | **0.454** |
| **Kilometres** | **Feet** | **3281** | **$3.048 \times 10^{-4}$** |
| **Kilometres** | **Nautical miles** | **0.540** | **1.853** |
| **Kilometres** | **Statute miles** | **0.621** | **1.609** |
| **Kilowatt hours** | **Joules** | **$3.6 \times 10^{6}$** | **$2.778 \times 10^{-7}$** |
| **Kilowatt hours** | **HP hours** | **1.341** | **0.7457** |
| **Knots** | **Miles per hour** | **1.1508** | **0.869** |
| **Lamberts** | **Candles per sq cm** | **0.3183** | **3.142** |
| **Lamberts** | **Candles per sq in** | **2.054** | **0.4869** |
| **Lumens per sq ft** | **Foot candles** | **1** | **1** |
| **Lux** | **Foot candles** | **0.0929** | **10.764** |
| **Metres** | **Feet** | **3.28** | **0.3048** |
| **Metres** | **Yards** | **1.094** | **0.9144** |
| **Miles per hour** | **Feet per second** | **1.467** | **0.68182** |
| **Nepers** | **Decibels** | **8.686** | **0.1151** |
| **Tons** | **Pounds** | **2240** | **$4.464 \times 10^{-4}$** |
| **Watts** | **Ergs per second** | **$10^{7}$** | **$10^{-7}$** |

### Table A.9: Greek alphabet

| Capital letters | Small letters | Greek name | English equivalent |
|---|---|---|---|
| Α | α | **Alpha** | **a** |
| Β | β | **Beta** | **b** |
| Γ | γ | **Gamma** | **g** |
| Δ | δ | **Delta** | **d** |
| Ε | ε | **Epsilon** | **e** |
| Ζ | ζ | **Zeta** | **z** |
| Η | η | **Eta** | **é** |
| Θ | θ | **Theta** | **th** |
| Ι | ι | **Iota** | **i** |
| Κ | κ | **Kappa** | **k** |
| Λ | λ | **Lambda** | **l** |
| Μ | μ | **Mu** | **m** |
| Ν | ν | **Nu** | **n** |
| Ξ | ξ | **Xi** | **x** |
| Ο | ο | **Omicron** | **o** |
| Π | π | **Pi** | **p** |
| Ρ | ρ | **Rho** | **r** |
| Σ | σ | **Sigma** | **s** |
| Τ | τ | **Tau** | **t** |
| Υ | υ | **Upsilon** | **u** |
| Φ | φ | **Phi** | **ph** |
| Χ | χ | **Chi** | **ch** |
| Ψ | ψ | **Psi** | **ps** |
| Ω | ω | **Omega** | **ō** |

### Table A.10: Useful twist drill sizes

| **Screw size** | **2** | **3** | **4** | **5** | **6** |
|---|---|---|---|---|---|
| **Clearance drill** | **2.10** | **3.10** | **4.10** | **5.10** | **6.10** |
| **Tapping drill** | **1.55** | **2.65** | **3.50** | **4.50** | **5.20** |

**Where all sizes are in millimetres.**

| Twist drill no | ¼in 0.250 in | 1 0.228 in | 9 0.196 in | 17 0.173 in | 24 0.152 in | 32 0.116 in | 43 0.089 in | 50 0.070 in |
|---|---|---|---|---|---|---|---|---|
| **Clearance for woodscrew no** | **14** | **12** | **10** | **8** | **6** | **4** | **2** | **0** |
| **Clearance for BA** | **0** | **1** | **2** | **3** | **4** | **6** | **8** | **10** |
| **Tapping size for BA** | **—** | **—** | **0** | **1** | **2** | **4** | **6** | **8** |

**Table A.11: Component colour codes**

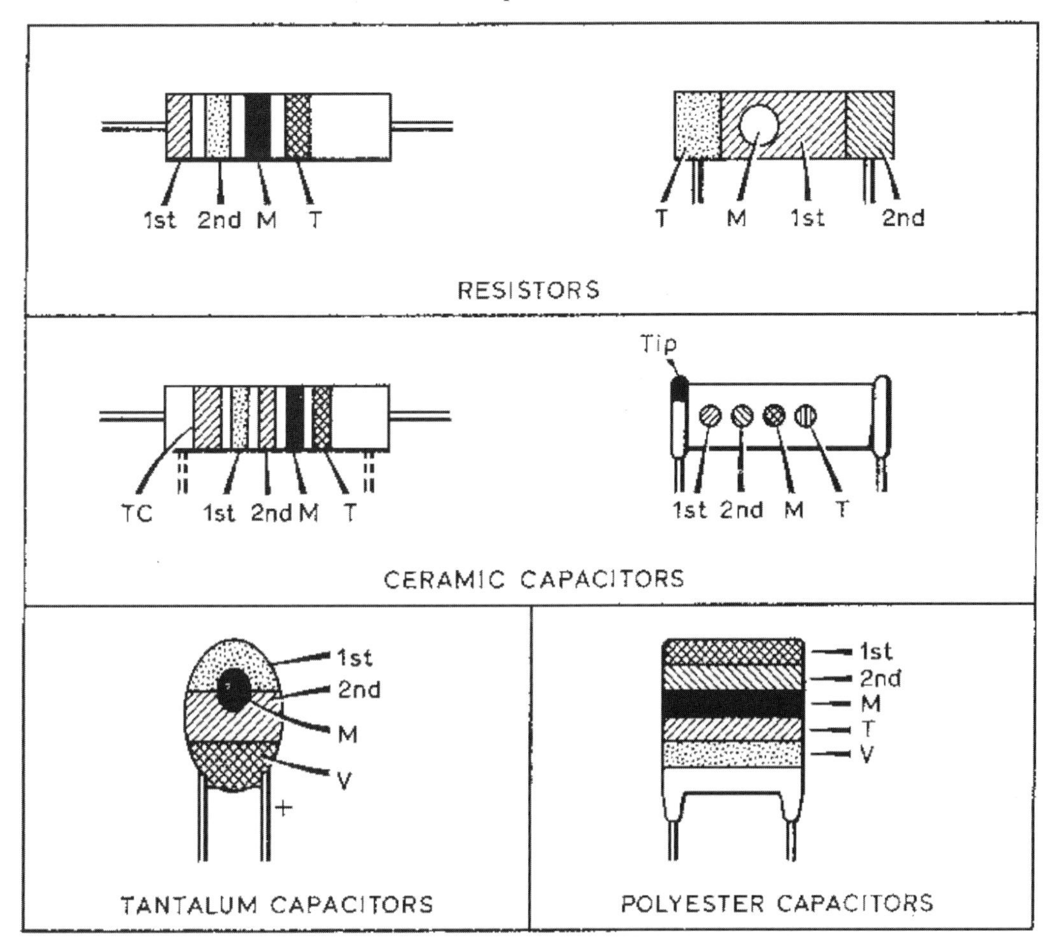


| Colour | Significant figure | Thermal multiplier | Tolerance (T) (percent) | Temp coeff (TC) (parts/10⁶/°C) | Volrtage (V) (tantalum cap) | Voltage (V) (polyester cap) |
|---|---|---|---|---|---|---|
| **Black** | **0** | **1** | **±20** | **0** | **10** | — |
| **Brown** | **1** | **10** | **±1** | **–30** | — | **100** |
| **Red** | **2** | **100** | **±2** | **–80** | — | **250** |
| **Orange** | **3** | **1000** | **±3** | **–150** | — | — |
| **Yellow** | **4** | **10,000** | **+100, –0** | **–220** | **6.3** | **400** |
| **Green** | **5** | **100,000** | **±5** | **–330** | **16** | — |
| **Blue** | **6** | **1,000,000** | **±6** | **–470** | **20** | — |
| **Violet** | **7** | **10,000,000** | — | **–750** | — | — |
| **Grey** | **8** | **100,000,000** | — | **+30** | **25** | — |
| **White** | **9** | **1,000,000,000** | **±10** | **+100 to –750** | **3** | — |
| **Gold** | — | **0.1** | **±5** | — | — | — |
| **Silver** | — | **0.01** | **±10** | — | — | — |
| **Pink** | — | — | — | — | **35** | — |
| **No colour** | — | — | **±20** | — | — | — |

**Units used are ohms for resistors, picofarads for ceramic and polyester capacitors, and microfarads for tantalum capacitors.**

| Letter symbol | Tolerance of capacitor | Letter symbol | Tolerance of capacitor |
|---|---|---|---|
| A | +/- 0.05% | K | +/- 10% |
| B | +/- 0.10% | M | +/- 20% |
| C | +/- 0.25% | N | +/- 0.05% |
| D | +/- 0.5% | Q | +30%, -10% |
| E | +/- 0.5% | P | +100% ,-0% |
| F | +/- 1% | S | +50%, -20% |
| G | +/- 2% | T | +50%, -10% |
| H | +/- 3% | Z | +80%, -20% |
| J | +/- 5% | | |

**Capacitor tolerance codes**

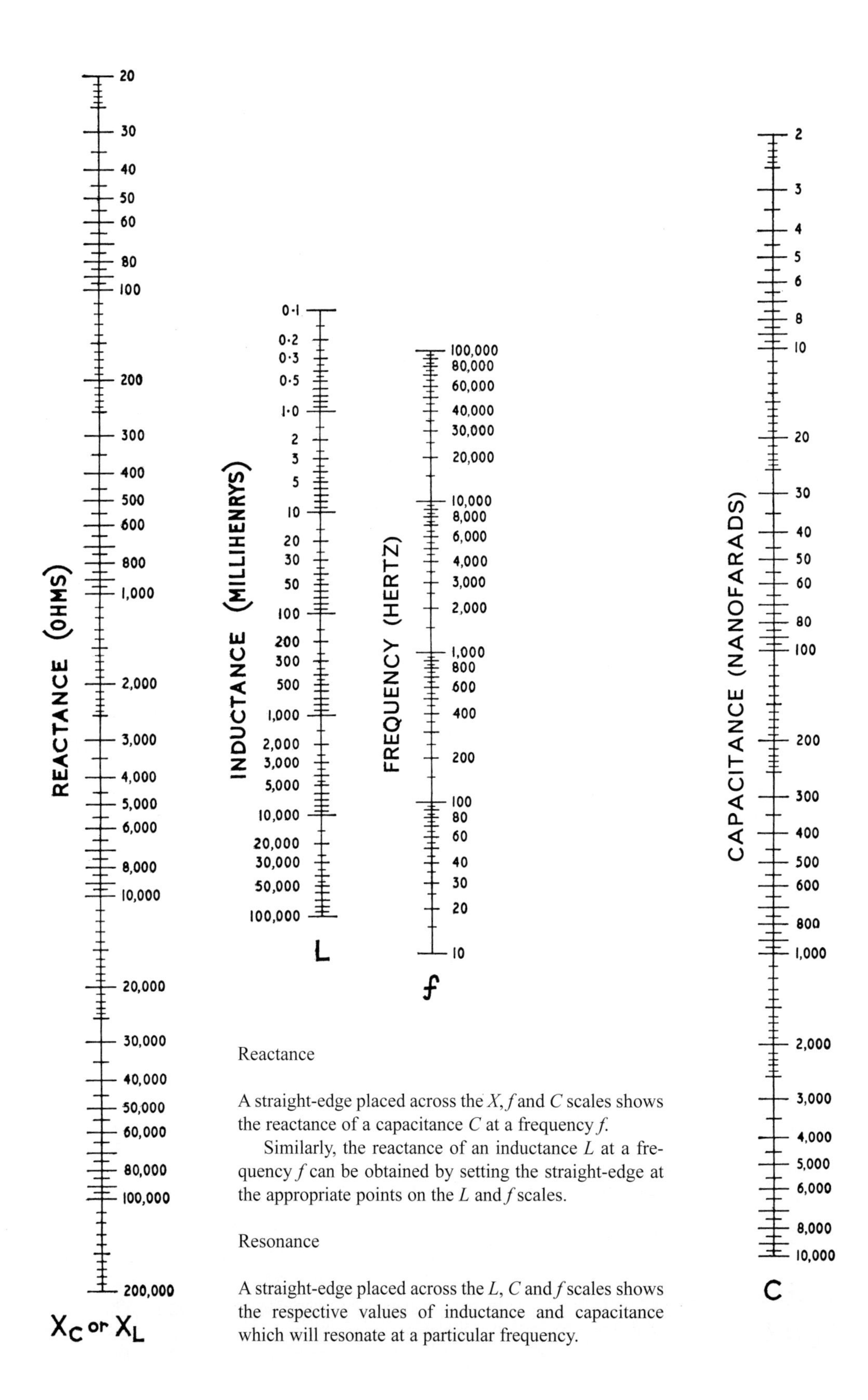


Reactance

A straight-edge placed across the *X*, *f* and *C* scales shows the reactance of a capacitance *C* at a frequency *f*.

Similarly, the reactance of an inductance *L* at a frequency *f* can be obtained by setting the straight-edge at the appropriate points on the *L* and *f* scales.

Resonance

A straight-edge placed across the *L*, *C* and *f* scales shows the respective values of inductance and capacitance which will resonate at a particular frequency.

**Fig A.1: Reactance and resonance chart - audio frequency**

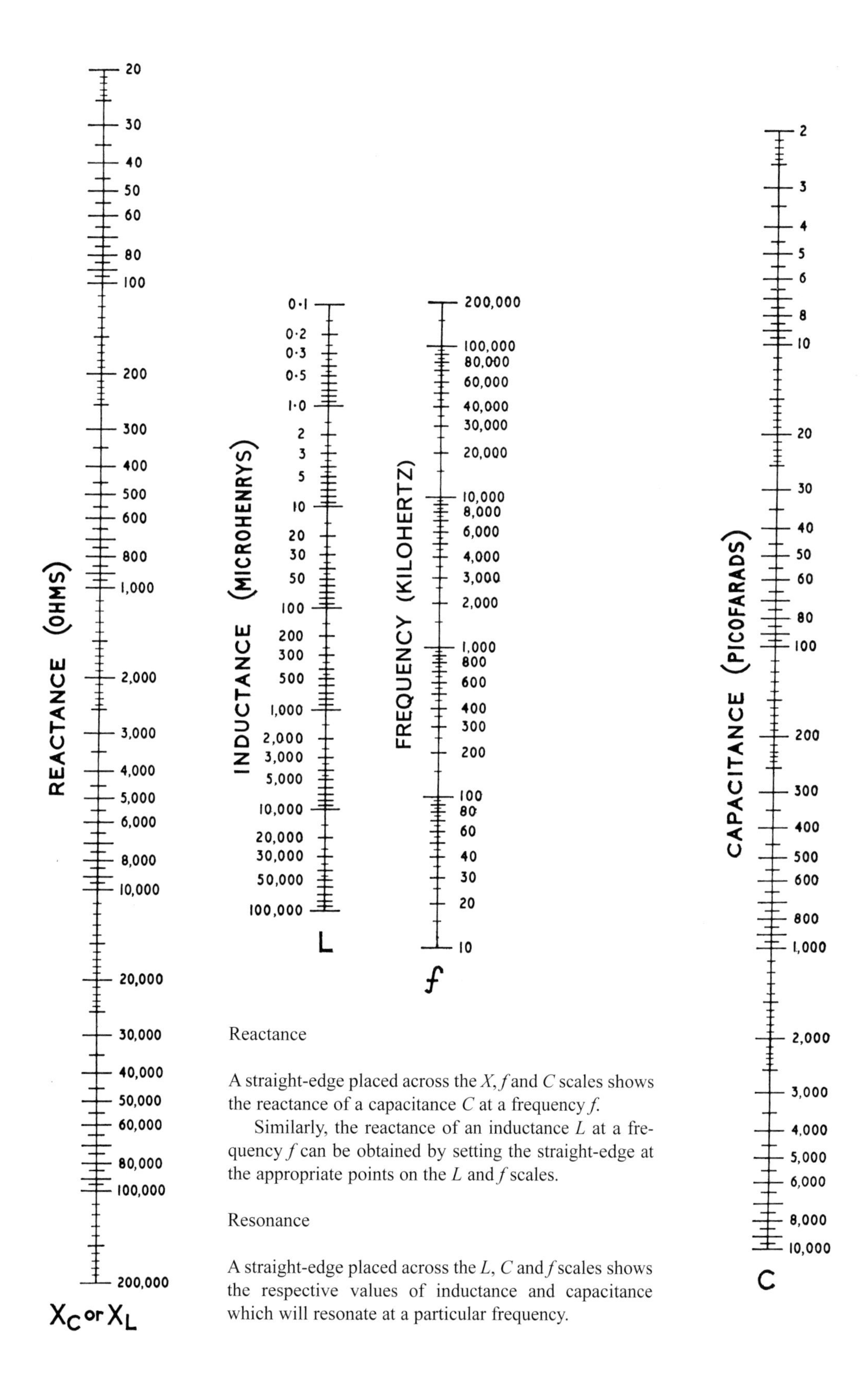


Reactance

A straight-edge placed across the *X*, *f* and *C* scales shows the reactance of a capacitance *C* at a frequency *f*.

Similarly, the reactance of an inductance *L* at a frequency *f* can be obtained by setting the straight-edge at the appropriate points on the *L* and *f* scales.

Resonance

A straight-edge placed across the *L*, *C* and *f* scales shows the respective values of inductance and capacitance which will resonate at a particular frequency.

**Fig A.2: Reactance and resonance chart - radio frequency**

## FILTER DESIGN CALCULATIONS

### Coupling between two resonant circuits tuned to the same frequency

The coupling coefficient is the ratio of the mutual inductance between windings to the inductance of one winding. This is true where the primary and secondary are identical; for simplicity, this is taken to be the case.

When the peak of the response is flat and on the point of splitting, the coupling is at its critical value, which is given by:

$$k_c = \frac{1}{Q} \quad (Q_p = Q_s)$$

Hence, the higher the Q, the lower the coupling required. In an IF transformer, the coupling is set at the critical value; however, for use in wide-band couplers it is convenient to have it slightly higher. The design formulae given below are based on a coupling/critical coupling ratio of 1.86, corresponding to a peak-to-trough ratio of 1.2:1, or a response flat within 2dB over the band.

The most convenient way of introducing variable coupling between two tuned circuits is with a small trimmer between the 'hot' ends of the coils (see Fig A.3). This is equivalent, except where phase relationships are concerned, to a mutual inductance of the value:

$$M = \frac{C_1}{C_1 + C} L$$

Hence the coupling coefficient is:

$$k = \frac{C_1}{C_1 + C}$$

The purpose of the damping resistors R in Fig A.3 is to obtain correct circuit Q; they should not be omitted unless the source or load provides the proper termination.

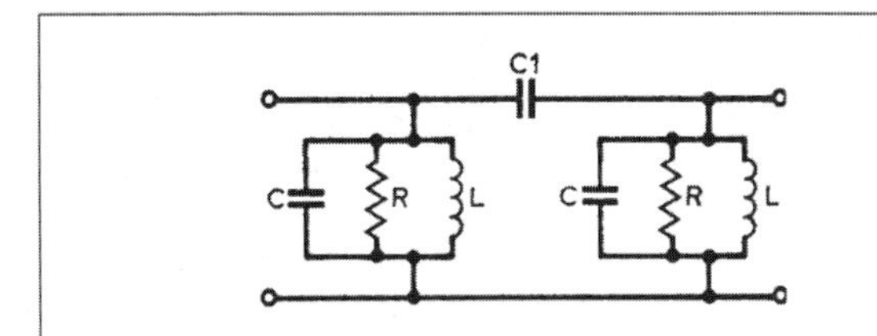


**Fig A.3: Basic coupler circuit**

Given set values of damping resistance, pass-band and centre frequency, all values may be calculated from the following formulae:

$$k = 0.84 \frac{\text{Bandwidth (kHz)}}{\text{Centre frequency (kHz)}}$$

$$Q = \frac{1.86}{k} \qquad L = \frac{R}{2\pi f Q} \qquad C = \frac{1}{L}\left(\frac{1}{2\pi f}\right)^2$$

where C is in microfarads, L in microhenrys and f is the centre frequency in megahertz. R is in ohms.

Note that C includes all strays: if the calculated value of C is less than the estmated strays on any band, a lower value of R should be used. The bandswitch can increase the strays to 20pF or more.

Coupling capacitance $C_1$ is given by:

$$C_1 = \frac{k}{1-k} C$$

### Elliptic filters

Using modern design procedure, a 'normalised' filter having the desired performance is chosen from a series of precalculated designs. The following presentation, originally due to W3NQN, uses normalisation to a cut-off frequency of 1Hz and termination resistance of 1Ω, and all that is required to ascertain the constants of a practical filter is to specify the actual cut-off frequency and termination resistance required and to scale the normalised filter data to those parameters.

The following abbreviations are used in these curves:

A = attenuation (dB),
$A_p$ = maximum attenuation in pass-band,
$f_4$ = first attenuation peak,
$f_2$ = second attenuation peak with two-section filter or third attenuation peak with three-section filter,
$f_6$ = second attenuation peak with three-section filter,
$f_{co}$ = frequency where the attenuation first exceeds that in the pass-band,
$A_s$ = minimum attenuation in stop-band,
$f_s$ = frequency where minimum stop-band attenuation is first reached.

The attenuation peaks $f_4$, $f_6$ or $f_2$ are associated with the resonant circuits L4/C4, L6,C6 and L2/C2 on the respective diagrams.

#### Applications

Because of their low value of reflection coefficient (P) and VSWR, Tables 1-1, 1-2 and 1-3 of Table A.12 and Tables 2-1 to 2-6 inclusive of Table A.13 are best suited for RF applications where power must be transmitted through the filter. The two-section filter has a relatively gradual attenuation slope and the stop-band attenuation level ($A_s$) is not achieved until a frquency $f_s$ is reached which is two to three times the cut-off frequency. If a more abrupt attenuation slope is desired, then one of the three-section filters (Tables 2-1 to 2-6 in Table A.13) should be used. In these cases the stop-band attenuation level may be reached at a frequency only 1.25 to 2 times $f_{co}$.

Tables 1-4 to 1-6 of Table A.12 are intended for AF applications where transmission of appreciable power is not required, and consequently the filter response may have a much higher value of VSWR and pass-band ripple without adversely affecting the filter performance. If the higher pass-band ripple is acceptable, a more abrupt attenuation slope is possible. This can be seen by comparing the different values of $f_s$ at 50dB in Tables 1-4, 1-5 and 1-6 which have pass-band ripple peaks of 0.28, 0.50 and 1.0dB respectively. The values of $A_s$ for the audio filters were selected to be between 35 and 55dB, as this range of stop-band attenuation was believed to be optimum for most audio filtering requirements.

It should be noted that all C and L tabular data must be multiplied by a factor of $10^{-3}$.

With one exception, all the C and L tabulated data of each table have a consistent but unequal increase or decrease in value, a characteristic of most computer-derived filter tables. An exception will be noted in Table 1-5, $A_s$ = 50, column C1. The original author points out that this is not an error but arose from a minor change necessitated in the original computer program to eliminate unrealisable component values.

#### How to use the filter tables

After the desired cut-off frequency has been chosen, the frequencies of $f_s$ and the attenuation peaks may be calculated by multiplying their corresponding tabular values by the required cut-off frequency ($f_{co}$). The component values of the desired

**Table A.12: Two-section elliptic-function filters normalised for a cut-off frequency of 1Hz and terminations of one ohm**

| REFLECTION COEFFICIENT, VSWR & Ap | As dB | fs Hz | f4 Hz | f2 Hz | C1 Farad | C3 Farad | C5 Farad | C2 Farad | L2 Henry | C4 Farad | L4 Henry |
|---|---|---|---|---|---|---|---|---|---|---|---|
| Table 1–1 | 70 | 3·24 | 3·39 | 5·42 | 110·4 | 235 | 103·5 | 4·34 | 199·0 | 11·72 | 187·5 |
| p = 4% | 65 | 2·92 | 3·07 | 4·88 | 109·6 | 233 | 101·0 | 5·39 | 197·9 | 14·67 | 183·7 |
| VSWR = 1·08 | 60 | 2·56 | 2·68 | 4·24 | 108·2 | 229 | 96·9 | 7·20 | 195·8 | 19·88 | 177·3 |
| Ap = 0·0069 dB | 55 | 2·37 | 2·48 | 3·90 | 107·2 | 227 | 93·8 | 8·57 | 194·3 | 23·9 | 172·7 |
| | 50 | 2·13 | 2·23 | 3·48 | 105·5 | 223 | 88·6 | 10·88 | 192·0 | 31·0 | 164·7 |
| Table 1–2 | 70 | 3·07 | 3·22 | 5·13 | 118·3 | 243 | 110·8 | 4·73 | 203 | 12·78 | 191·0 |
| p = 5% | 65 | 2·79 | 2·92 | 4·64 | 117·4 | 241 | 108·3 | 5·82 | 202 | 15·82 | 187·2 |
| VSWR = 1·11 | 60 | 2·46 | 2·57 | 4·06 | 116·0 | 237 | 104·0 | 7·67 | 200 | 21·2 | 180·7 |
| Ap = 0·011 dB | 55 | 2·28 | 2·39 | 3·75 | 115·0 | 234 | 100·8 | 9·07 | 198·5 | 25·3 | 175·9 |
| | 50 | 2·06 | 2·16 | 3·36 | 113·2 | 230 | 95·6 | 11·43 | 196·0 | 32·4 | 168·1 |
| Table 1–3 | 70 | 2·79 | 2·92 | 4·64 | 138·4 | 262 | 129·6 | 5·59 | 210 | 15·09 | 196·4 |
| p = 8% | 65 | 2·56 | 2·68 | 4·24 | 137·4 | 259 | 126·9 | 6·75 | 208 | 18·32 | 192·4 |
| VSWR = 1·17 | 60 | 2·28 | 2·39 | 3·75 | 135·9 | 255 | 122·4 | 8·72 | 206 | 23·9 | 185·7 |
| Ap = 0·028 dB | 55 | 2·06 | 2·16 | 3·36 | 134·2 | 251 | 117·4 | 10·98 | 204 | 30·6 | 178·4 |
| | 50 | 1·887 | 1·970 | 3·05 | 132·2 | 245 | 111·8 | 13·55 | 201 | 38·4 | 170·3 |
| Table 1–4 | 55 | 1·701 | 1·773 | 2·71 | 217 | 317 | 190·8 | 18·03 | 191·5 | 49·7 | 162·3 |
| p = 25% | 50 | 1·556 | 1·617 | 2·44 | 213 | 306 | 181·3 | 22·8 | 187·3 | 63·8 | 151·9 |
| VSWR = 1·67 | 45 | 1·440 | 1·493 | 2·22 | 209 | 295 | 170·6 | 28·3 | 182·7 | 80·9 | 140·5 |
| Ap = 0·28 dB | 40 | 1·325 | 1·369 | 1·988 | 203 | 279 | 155·8 | 36·4 | 176·0 | 108·0 | 125·1 |
| | 35 | 1·236 | 1·273 | 1·802 | 195·9 | 262 | 139·2 | 46·4 | 168·2 | 144·3 | 108·3 |
| Table 1–5 | 55 | 1·618 | 1·690 | 2·56 | 248 | 348 | 214 | 21·3 | 181·4 | 58·7 | 151·0 |
| p = 33% | 50 | 1·481 | 1·540 | 2·30 | 249 | 336 | 210 | 27·4 | 174·9 | 76·7 | 139·3 |
| VSWR = 2·00 | 45 | 1·369 | 1·416 | 2·08 | 244 | 318 | 197·5 | 34·7 | 169·2 | 99·8 | 126·5 |
| Ap = 0·50 dB | 40 | 1·270 | 1·308 | 1·878 | 238 | 299 | 177·3 | 44·4 | 161·7 | 133·7 | 110·8 |
| | 35 | 1·186 | 1·222 | 1·700 | 229 | 280 | 163·3 | 57·0 | 153·9 | 177·6 | 95·5 |
| Table 1–6 | 55 | 1·528 | 1·591 | 2·39 | 314 | 401 | 276 | 28·3 | 156·9 | 77·5 | 129·1 |
| p = 45% | 50 | 1·407 | 1·459 | 2·16 | 308 | 381 | 260 | 35·5 | 153·3 | 99·6 | 119·4 |
| VSWR = 2·67 | 45 | 1·245 | 1·313 | 1·898 | 306 | 365 | 247 | 46·6 | 150·7 | 135·0 | 108·9 |
| Ap = 1·00 dB | 40 | 1·217 | 1·250 | 1·755 | 296 | 341 | 227 | 59·2 | 138·9 | 176·2 | 92·0 |
| | 35 | 1·145 | 1·174 | 1·597 | 284 | 315 | 203 | 75·4 | 131·6 | 237 | 77·7 |
| | As dB | fs Hz | f4 Hz | f2 Hz | L1 Henry | L3 Henry | L5 Henry | L2 Henry | C2 Farad | L4 Henry | C4 Farad |

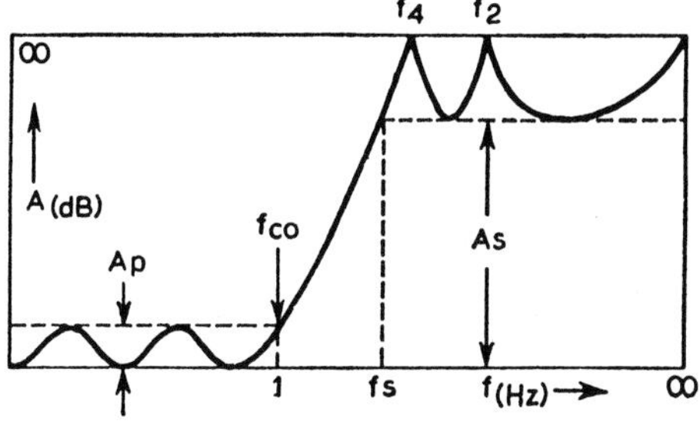


Figure 1 ✱✱

✱ All tabulated data of C and L must be multiplied by $10^{-3}$; for example, in Table 1–1, the normalized value of C1 is $110{\cdot}4 \times 10^{-3}$, for $A_s = 70$dB

✱✱ In the above tabulation, the top column headings pertain to Figure 1 while the bottom column headings pertain to Figure 2

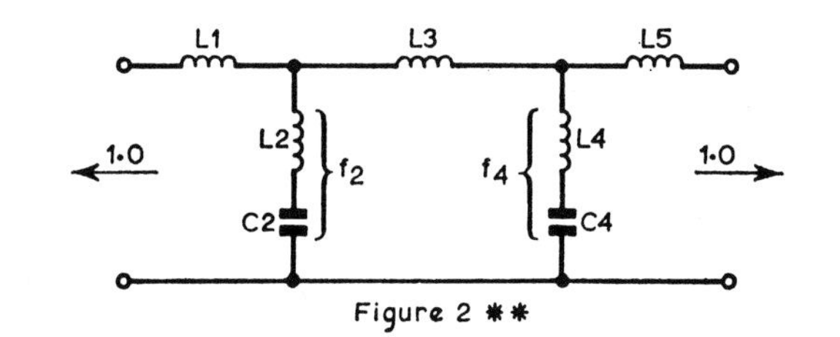


Figure 2 ✱✱

filter are then found by multiplying C and L values in the tables by $1/Rf_{co}$ and $R/f_{co}$ respectively.

*Example 1*

A low-pass audio filter to attenuate speech frequencies above 3kHz with a minimum attenuation of 40dB for all frequencies above 3.8kHz, and to be terminated in resistive loads of 1.63kΩ. (This odd value has been chosen merely for convenience in demonstrating the design procedure.)

The circuit of Fig 1 in the Tables is chosen because this has the minimum number of inductors, which are both more expensive and have higher losses than do capacitors. The parameters are:

$$A_s = 40\text{dB} \qquad f_{co} = 3\text{kHz} \qquad R = 1.63\text{k}\Omega$$

From Table 1-5 of Table A.12, $A_s = 40$dB, calculate $f'_s$, $f'_4$ and $f'_2$. (Numbers with the prime (′) are the frequency and component values of the final design: numbers without the prime are from the filter catalogue.)

(1) $f'_s = f_s(f_{co}) = 1.270 \times 3 = 3.81\text{kHz}$.

$f'_4 = f_4(f_{co}) = 1.308 \times 3 = 3.92\text{kHz}$

$f'_2 = f_2(f_{co}) = 1.878 \times 3 = 5.63\text{kHz}$

**Table A.13: Three-section elliptic-function filters normalised for a cut-off frequency of 1Hz and terminations of one ohm**

| REFLECTION COEFFICIENT, VSWR & Ap | As dB | fs Hz | f4 Hz | f6 Hz | f2 Hz | C1 Farad | C3 Farad | C5 Farad | C7 Farad | C2 Farad | L2 Henry | C4 Farad | L4 Henry | C6 Farad | L6 Henry |
|---|---|---|---|---|---|---|---|---|---|---|---|---|---|---|---|
| Table 2–1 | 70 | 2·00 | 2·04 | 2·49 | 4·35 | 79·6 | 209 | 201 | 63·1 | 7·42 | 180·2 | 30·9 | 196·4 | 26·3 | 155·2 |
| p = 1% | 64 | 1·836 | 1·876 | 2·27 | 3·95 | 78·3 | 204 | 194·8 | 58·2 | 9·10 | 178·4 | 38·4 | 187·6 | 33·0 | 148·3 |
| VSWR = 1·02 | 60 | 1·743 | 1·780 | 2·15 | 3·72 | 77·3 | 200 | 190·3 | 54·5 | 10·35 | 177·1 | 44·1 | 181·4 | 38·2 | 143·5 |
| Ap = 0·43 x $10^{-3}$dB | 55 | 1·624 | 1·657 | 1·990 | 3·41 | 75·8 | 194·2 | 183·5 | 48·5 | 12·42 | 175·2 | 53·8 | 171·4 | 47·2 | 135·6 |
| | 50 | 1·524 | 1·554 | 1·854 | 3·15 | 74·1 | 187·8 | 176·3 | 41·8 | 14·75 | 172·8 | 65·3 | 160·7 | 58·0 | 127·1 |
| Table 2–2 | 70 | 1·836 | 1·876 | 2·27 | 3·95 | 93·8 | 222 | 212 | 75·7 | 8·34 | 194·8 | 35·8 | 201 | 29·4 | 167·0 |
| p = 2% | 64 | 1·701 | 1·737 | 2·09 | 3·61 | 92·5 | 216 | 205 | 70·7 | 10·08 | 193·1 | 43·8 | 191·6 | 36·2 | 160·0 |
| VSWR = 1·04 | 60 | 1·624 | 1·657 | 1·990 | 3·41 | 91·5 | 212 | 200 | 67·1 | 11·35 | 191·6 | 49·8 | 185·1 | 41·3 | 154·8 |
| Ap = 1·7 x $10^{-3}$dB | 55 | 1·524 | 1·554 | 1·854 | 3·15 | 89·9 | 206 | 192·7 | 61·1 | 13·47 | 189·4 | 60·0 | 174·8 | 50·2 | 146·7 |
| | 50 | 1·414 | 1·440 | 1·702 | 2·86 | 87·5 | 196·9 | 182·1 | 52·2 | 16·70 | 186·1 | 76·4 | 160·0 | 64·8 | 135·0 |
| Table 2–3 | 70 | 1·743 | 1·780 | 2·15 | 3·72 | 104·2 | 230 | 219 | 84·7 | 9·06 | 203 | 39·7 | 201 | 31·8 | 172·5 |
| p = 3% | 65 | 1·624 | 1·657 | 1·990 | 3·41 | 102·8 | 224 | 211 | 79·7 | 10·84 | 201 | 48·1 | 191·8 | 38·7 | 165·4 |
| VSWR = 1·06 | 60 | 1·524 | 1·554 | 1·854 | 3·15 | 101·2 | 217 | 203 | 74·1 | 12·86 | 198·3 | 57·8 | 181·6 | 46·8 | 157·5 |
| Ap = 3·9 x $10^{-3}$dB | 55 | 1·440 | 1·466 | 1·737 | 2·92 | 99·5 | 211 | 194·8 | 67·9 | 15·12 | 195·9 | 69·0 | 170·8 | 56·3 | 149·1 |
| | 50 | 1·367 | 1·391 | 1·636 | 2·73 | 97·6 | 203 | 186·2 | 61·2 | 17·65 | 193·1 | 82·2 | 159·2 | 67·5 | 140·1 |
| Table 2–4 | 70 | 1·701 | 1·737 | 2·09 | 3·61 | 113·0 | 236 | 224 | 93·0 | 9·37 | 208 | 41·6 | 202 | 32·7 | 177·0 |
| p = 4% | 65 | 1·589 | 1·621 | 1·942 | 3·32 | 111·6 | 230 | 217 | 88·0 | 11·18 | 205 | 50·2 | 192·3 | 39·6 | 170·0 |
| VSWR = 1·08 | 60 | 1·494 | 1·523 | 1·813 | 3·07 | 110·0 | 224 | 208 | 82·4 | 13·20 | 203 | 60·0 | 181·9 | 47·6 | 161·9 |
| Ap = 6·9 x $10^{-3}$dB | 55 | 1·414 | 1·440 | 1·702 | 2·86 | 108·3 | 217 | 199·6 | 76·3 | 15·47 | 201 | 71·4 | 171·1 | 57·0 | 153·4 |
| | 50 | 1·325 | 1·347 | 1·576 | 2·61 | 105·6 | 206 | 187·5 | 67·3 | 18·94 | 196·9 | 89·7 | 155·6 | 72·2 | 141·3 |
| Table 2–5 | 70 | 1·662 | 1·696 | 2·04 | 3·51 | 120·6 | 242 | 229 | 99·9 | 9·77 | 211 | 43·9 | 201 | 33·9 | 179·4 |
| p = 5% | 65 | 1·556 | 1·586 | 1·897 | 3·23 | 119·2 | 235 | 221 | 94·9 | 11·61 | 209 | 52·7 | 191·1 | 40·9 | 172·0 |
| VSWR = 1·11 | 60 | 1·466 | 1·494 | 1·774 | 3·00 | 117·6 | 228 | 212 | 89·3 | 13·67 | 206 | 62·8 | 180·8 | 49·0 | 164·1 |
| Ap = 11 x $10^{-3}$dB | 55 | 1·367 | 1·391 | 1·636 | 2·73 | 115·2 | 219 | 199·7 | 81·0 | 16·81 | 203 | 78·8 | 166·2 | 61·9 | 152·7 |
| | 51·5 | 1·325 | 1·347 | 1·576 | 2·61 | 113·8 | 213 | 193·4 | 76·5 | 18·57 | 201 | 88·2 | 158·3 | 69·5 | 146·6 |
| | 50 | 1·305 | 1·327 | 1·548 | 2·55 | 113·1 | 211 | 190·2 | 74·1 | 19·51 | 199·7 | 93·2 | 154·4 | 73·7 | 143·5 |
| Table 2–6 | 70 | 1·556 | 1·586 | 1·897 | 3·23 | 139·7 | 252 | 237 | 116·2 | 11·30 | 214 | 52·0 | 193·4 | 39·1 | 180·0 |
| p = 8% | 65 | 1·466 | 1·494 | 1·774 | 3·00 | 138·1 | 245 | 228 | 110·9 | 13·30 | 212 | 61·9 | 183·5 | 46·6 | 172·5 |
| VSWR = 1·17 | 60 | 1·390 | 1·415 | 1·668 | 2·79 | 136·3 | 238 | 218 | 105·0 | 15·54 | 210 | 73·2 | 173·0 | 55·3 | 164·4 |
| Ap = 28 x $10^{-3}$dB | 55 | 1·325 | 1·347 | 1·576 | 2·61 | 134·4 | 230 | 208 | 98·6 | 18·05 | 207 | 86·3 | 161·9 | 65·4 | 155·8 |
| | 50 | 1·252 | 1·271 | 1·471 | 2·39 | 131·4 | 218 | 193·9 | 89·2 | 21·9 | 202 | 107·3 | 146·1 | 81·6 | 143·4 |
| | As dB | fs Hz | f4 Hz | f6 Hz | f2 Hz | L1 Henry | L3 Henry | L5 Henry | L7 Henry | L2 Henry | C2 Farad | L4 Henry | C4 Farad | L6 Henry | C6 Farad |

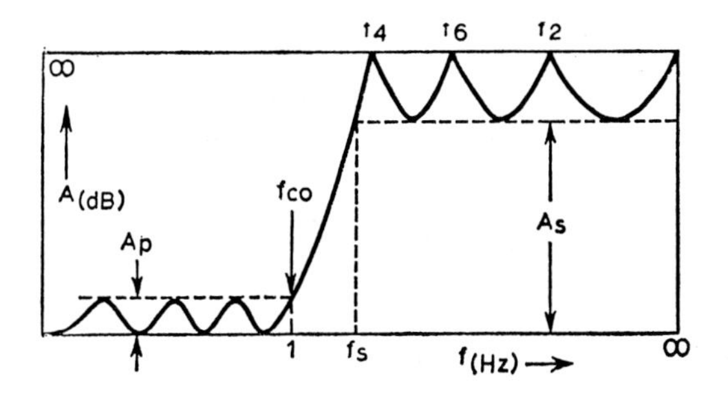


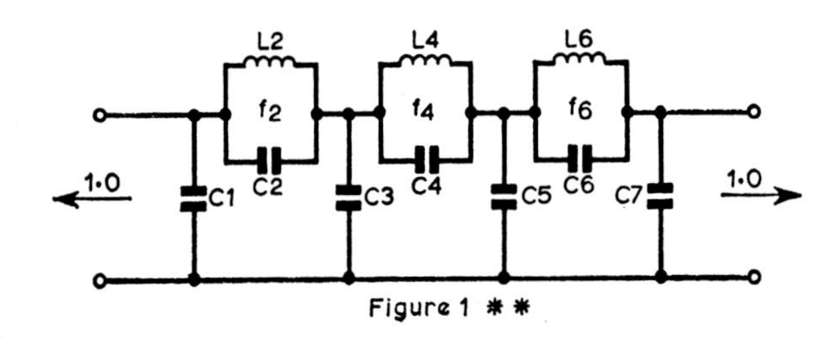


* All tabulated data of C and L must be multiplied by $10^{-3}$; for example, in Table 2–1, the normalized value of C1 is 79·6 x $10^{-3}$, for As = 70dB

** In the above tabulation, the top column headings pertain to Figure 1 while the bottom column headings pertain to Figure 2

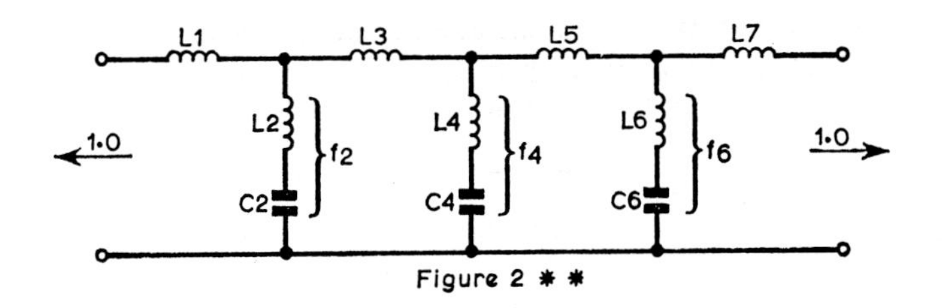


(2) Calculate factors $1/Rf_{co}$ and $R/f_{co}$ to determine the capacitor and inductor values.

$$1/Rf_{co} = 1/(1.63 \times 10^3)(3 \times 10^3)$$
$$= 1/(4.89 \times 10^6)$$
$$= 0.2045 \times 10^{-6}$$

$$R/f_{co} = (1.63 \times 10^3)/(3 \times 10^3) = 0.543$$

(3) Calculate the component values of the desired filter by multiplying all the catalogue tabular values of C by $1/Rf_{co}$ and L by $R/f_{co}$ as shown below:

$$C'1 = C1(1/Rf_{co}) = (238 \times 10^{-3})(0.2045)\,10^{-6} = 0.0487\mu F$$
$$C'3 = C3(1/Rf_{co}) = (299 \times 10^{-3})(0.2045)\,10^{-6} = 0.0612\mu F$$
$$C'5 = C5(1/Rf_{co}) = (177.3 \times 10^{-3})(0.2045)\,10^{-6} = 0.0363\mu F$$
$$C'2 = C2(1/Rf_{co}) = (44.4 \times 10^{-3})(0.2045)\,10^{-6} = 0.00908\mu F$$
$$C'4 = C4(1/Rf_{co}) = (133.7 \times 10^{-3})(0.2045)\,10^{-6} = 0.00273\mu F$$
$$L'2 = L2(R/f_{co}) = (161.7 \times 10^{-3})(0.543) = 87.8mH$$
$$L'4 = L4(R/f_{co}) = (110.8 \times 10^{-3})(0.543) = 60.1mH$$

These calculations, which may conveniently be performed with a pocket calculator, complete the design of the filter.

It should be noted that all the elliptic-function data is based

**Table A.14: Butterworth filters**

| K | $C_1$ $L_1$ | $C_2$ $L_2$ | $C_3$ $L_3$ | $C_4$ $L_4$ | $C_5$ $L_5$ | $C_6$ $L_6$ | $C_7$ $L_7$ | $C_8$ $L_8$ | $C_9$ $L_9$ | $C_{10}$ $L_{10}$ |
|---|---|---|---|---|---|---|---|---|---|---|
| **1** | **2.000** | — | — | — | — | — | — | — | — | — |
| **2** | **1.4142** | **1.4142** | — | — | — | — | — | — | — | |
| **3** | **1.000** | **2.000** | **1.000** | — | — | — | — | — | — | |
| **4** | **0.7654** | **1.8478** | **1.8478** | **0.7654** | — | — | — | — | — | — |
| **5** | **0.6180** | **1.6180** | **2.000** | **1.6180** | **0.6180** | — | — | — | — | — |
| **6** | **0.5176** | **1.4142** | **1.9319** | **1.9319** | **1.4142** | **0.5176** | — | — | — | — |
| **7** | **0.4450** | **1.2470** | **1.8019** | **2.000** | **1.8019** | **1.2470** | **0.4450** | — | — | — |
| **8** | **0.3902** | **1.1111** | **1.6629** | **1.9616** | **1.9616** | **1.6629** | **1.1111** | **0.3902** | — | — |
| **9** | **0.3473** | **1.000** | **1.5321** | **1.8794** | **2.000** | **1.8794** | **1.5321** | **1.000** | **0.3473** | — |
| **10** | **0.3129** | **0.9080** | **1.4142** | **1.7820** | **1.9754** | **1.9754** | **1.7820** | **1.4142** | **0.9080** | **0.3129** |

on the use of lossless components and purely resistive terminations. Therefore components of the highest possible Q should be used and precautions taken to ensure that the filter is properly terminated.

It will be noticed that some rather curious values of both capacitance and inductance may emerge from the calculations but these may be rationalised to the extent that the tolerance on the values of components need not be closer than some ±3%.

*Example 2*

A three-section low-pass filter to suppress harmonics at the output of a transmitter covering the HF bands up to a frequency of 30MHz with a matching impedance of 50Ω and a minimum attenuation in the stop-band of 50dB.

The parameters are, from Table 2-2 (circuit Fig 2) of Table A.13:

$A_s$ = 50dB  $f_{co}$ = 30MHz  R = 50Ω

From Table 2-2 (bottom line) of Table A.13, calculate $f'_s$, $f'_4$, $f'_6$ and $f'_2$.

(1) $f'_s = f_s(f_{co}) = 1.414 \times 30 = 42.4$MHz.
$f'_4 = f_4(f_{co}) = 1.440 \times 30 = 43.2$MHz
$f'_6 = f_6(f_{co}) = 1.702 \times 30 = 51$MHz
$f'_2 = f_2(f_{co}) = 2.860 \times 30 = 85.8$MHz

(2) Calculate factors $1/Rf_{co}$ and $R/f_{co}$ to determine the capacitor and inductor values respectively.

$1/Rf_{co} = 1/50(30 \times 10^{-6}) = 66 \times 10^{-11}$
$R/f_{co} = 50/(30 \times 10^{6}) = 1.67 \times 10^{-6}$

(3) Calculate component values of the desired filter by multiplying all tabular values of C by $1/Rf_{co}$ and L by $R/f_{co}$, remembering to multiply all values in the tables by $10^{-3}$.

$C'2 = C2(66 \times 10^{-11})$
$= (186.1 \times 10^{-3})(66 \times 10^{-11})$
$= 12{,}286.6 \times 10^{-14}$F
$= 12{,}282.6 \times 10^{-2}$pF
$= 122.8$pF

$L'1 = L1(1.67 \times 10^{-6})$
$= (87.5 \times 10^{-3})(1.67 \times 10^{-6})$
$= 146.1 \times 10^{-9}$H
$= 0.15\mu$H

$C'4 = (160 \times 10^{-3})(66 \times 10^{-11})$
$= 105.6$pF
$C'6 = 89.1$pF

$L'2 = 0.03\mu$H  $L'3 = 0.33\mu$H
$L'4 = 0.13\mu$H  $L'5 = 0.30\mu$H
$L'6 = 0.11\mu$H  $L'7 = 0.09\mu$H

As a check, it will be found that the combination C4, L4 tunes to 43.2MHz and that the other two series-tuned circuits tune to the other two points of maximum attenuation previously specified.

In order to convert the values in the filter just designed to match an impedance of 75Ω it is only necessary to multiply all values of capacitance by 2/3 and all values of inductance by 3/2. Thus C6 and L6 in a 75Ω filter become approximately 59.4pF and 0.17μH respectively.

## Butterworth filters

Frequency response curve:

$$A = 10 \log_{10}\left[1 + \left(\frac{f}{f_c}\right)^{2K}\right]$$

where A is the attenuation, f is the frequency for an insertion loss of 3.01dB, and K is the number of circuit elements.

### Low- and high-pass filters

Table A.14 is for normalised element values of K from 1 to 10 (number of sections) reduced to 1Ω source and load resistance (zero reactance) and a 3.01dB cut-off frequency of 1 radian/s (0.1592Hz). In both low-pass and high-pass filters:

$$L = \frac{R}{2\pi f_c} = L\ (1\Omega/\text{radian}) \qquad C = \frac{1}{2\pi f_c R} = C\ (1\Omega/\text{radian})$$

where R is the load resistance in ohms and $f_c$ is the desired 3.01dB frequency (Hz).

An example of a Butterworth low-pass filter is given in Fig A.4 (see Table A.14 for element values). In these examples of five-element filters (a) has a shunt element next to the load and (b) has a series element next to the load. Either filter will have the same response. In the examples of five-element filters given in Fig A.5, (a) has a series element next to the load and (b) has a shunt element next to the load. Either filter will have the same response.

### Butterworth band-pass filters

Centre frequency  $f_0 = \sqrt{f_1 f_2}$

Bandwidth  $BW = f_2 - f_1$

If the bandwidth specified is not the 3.01dB bandwidth ($BW_c$), the latter can be determined from:

$$BW_c = \frac{BW}{(10^{0.1A} - 1)/2K}$$

where A is the required attenuation at cut-off frequencies.

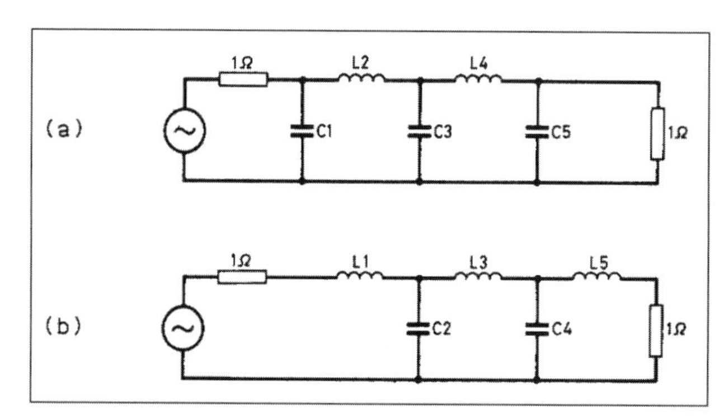


**Fig A.4: Butterworth low-pass filter**

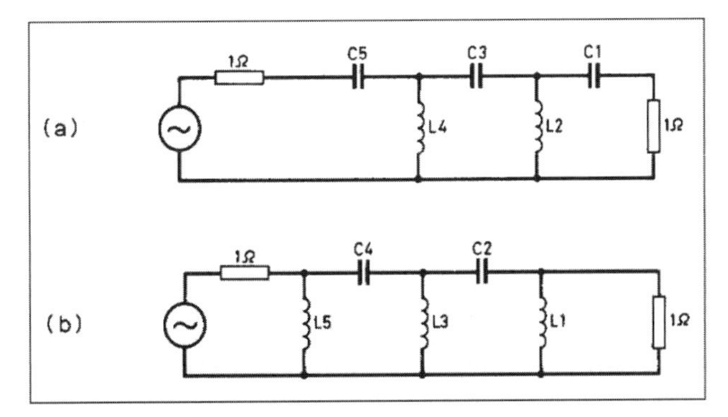


Fig A.5: Butterworth high-pass filter

Lower cut-off frequency:

$$f_{cl} = \frac{-BW_c + \sqrt{(BW_c)^2 + 4f_0^2}}{2}$$

Upper cut-off frequency: $f_{cu} = f_{cl} + BW_c$

An alternative, more-convenient method, is to choose a 3.01dB bandwidth (as wide as possible) around the desired centre frequency and compute the attenuation at other frequencies of interest by using the transformation:

$$\frac{f}{f_c} = \left[\left(\frac{f}{f_0} - \frac{f_0}{f}\right)\frac{f_0}{BW_c}\right]$$

## Chebyshev filters

Tables A.15 to A.18 provide the essential information for both high-pass and low-pass filters of T and π form. Figures are given for pass-band ripples of 1, 0.1, 0.01, and 0.001dB which respectively correspond to VSWR of 2.66, 1.36, 1.10 and 1.03.

The filters in this case are normalised to a frequency of 1MHz and an input and output impedance of 50Ω. This means that for any particular desired frequency the component values simply have to be divided by the required frequency in megahertz.

The 1MHz is the cut-off frequency; attenuation increases rapidly above the frequency for a low-pass filter and correspondly below for a high-pass type.

The filter data is also dependent on the impedance which as given is for 50Ω. For other impedances the component values need to be modified by the following:

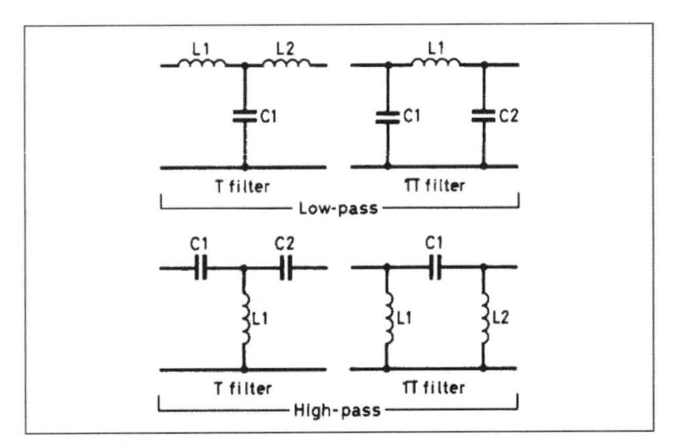


Fig A.6: Single-section three-pole filter elements

$\frac{Z_n}{50}$ for inductors $\qquad \frac{50}{Z_n}$ for capacitors

where $Z_n$ is the required impedance.

There is an advantage in using toroidal-form inductors due to their self-screening (confined-field) properties. Mica or silver mica capacitors are superior to other types for filter applications.

Practical filters for the amateur HF bands are given in Table A.19.

## Constant-*k* and *m*-derived filters

The filter sections shown in Fig A.7 can be used alone or, if greater attenuation and sharper cut-off is required, several sections can be connected in series. In the low-pass and high-pass filters, $f_c$ represents the cut-off frequency, the highest (for the low-pass) or the lowest (for the high-pass) frequency transmitted without attenuation. In the band-pass filter designs, $f_1$ is the low-frequency cut-off and $f_2$ the high-frequency cut-off. The units for L, C, R and f are henrys, farads, ohms and hertz respectively.

All the types shown are for use in an unbalanced line (one side grounded), and thus they are suitable for use in coaxial line or any other unbalanced circuit. To transform them for balanced lines (eg 300Ω transmission line or push-pull audio circuits), the series reactances should be equally divided between the two legs. Thus the balanced constant-k π-section low-pass filter

Table A.15: Chebyshev low-pass filter ('T' configuration)

| | Ripple (dB) | $L_1$ | $L_2$ | $L_3$ | $L_4$ | $L_5$ | $C_1$ | $C_2$ | $C_3$ | $C_4$ |
|---|---|---|---|---|---|---|---|---|---|---|
| **Single section (3-pole)** | **1** | **16.10** | **16.10** | — | — | — | **3164.3** | — | — | — |
| | **0.1** | **8.209** | **8.209** | — | — | — | **3652.3** | — | — | — |
| | **0.01** | **5.007** | **5.007** | — | — | — | **3088.5** | — | — | — |
| | **0.001** | **3.253** | **3.253** | — | — | — | **2312.6** | — | — | — |
| **Two-section (5-pole)** | **1** | **16.99** | **23.88** | **16.99** | — | — | **3473.1** | **3473.1** | — | — |
| | **0.1** | **9.126** | **15.72** | **9.126** | — | — | **4364.7** | **4364.7** | — | — |
| | **0.01** | **6.019** | **12.55** | **6.019** | — | — | **4153.7** | **4153.7** | — | — |
| | **0.001** | **4.318** | **10.43** | **4.318** | — | — | **3571.1** | **3571.1** | — | — |
| **Three-section (7-pole)** | **1** | **17.24** | **24.62** | **24.62** | **17.24** | — | **3538.0** | **3735.4** | **3538.0** | — |
| | **0.1** | **9.40** | **16.68** | **16.68** | **9.40** | — | **4528.9** | **5008.3** | **4528.9** | — |
| | **0.01** | **6.342** | **13.91** | **13.91** | **6.342** | — | **4432.2** | **5198.4** | **4432.2** | — |
| | **0.001** | **4.69** | **12.19** | **12.19** | **4.69** | — | **3951.5** | **4924.1** | **3981.5** | — |
| **Four-section (9-pole)** | **1** | **17.35** | **24.84** | **25.26** | **24.84** | **17.35** | **3562.5** | **3786.9** | **3786.9** | **3562.5** |
| | **0.1** | **9.515** | **16.99** | **17.55** | **16.99** | **9.515** | **4591.9** | **5146.2** | **5146.2** | **4591.9** |
| | **0.01** | **6.481** | **14.36** | **15.17** | **14.36** | **6.481** | **4542.5** | **5451.2** | **5451.2** | **4542.5** |
| | **0.001** | **4.854** | **12.81** | **13.88** | **12.81** | **4.854** | **4108.2** | **5299.0** | **5299.0** | **4108.2** |

**Inductance in microhenrys, capacitance in picofarads. Component values normalised to 1MHz and 50Ω.**

Table A.16: Chebyshev low-pass filter (Pi configuration)

| | Ripple (dB) | $C_1$ | $C_2$ | $C_3$ | $C_4$ | $C_5$ | $L_1$ | $L_2$ | $L_3$ | $L_4$ |
|---|---|---|---|---|---|---|---|---|---|---|
| **Single section (3-pole)** | 1 | 6441.3 | 6441.3 | — | — | — | 7.911 | — | — | — |
| | 0.1 | 3283.6 | 3283.6 | — | — | — | 9.131 | — | — | — |
| | 0.01 | 2007.7 | 2007.7 | — | — | — | 7.721 | — | — | — |
| | 0.001 | 1301.2 | 1301.2 | — | — | — | 5.781 | — | — | — |
| **Two-section (5-pole)** | 1 | 6795.5 | 9552.2 | 6795.5 | — | — | 8.683 | 8.683 | — | — |
| | 0.1 | 3650.4 | 6286.6 | 3650.4 | — | — | 10.91 | 10.91 | — | — |
| | 0.01 | 2407.5 | 5020.7 | 2407.5 | — | — | 10.38 | 10.38 | — | — |
| | 0.001 | 1727.3 | 4170.5 | 1727.3 | — | — | 8.928 | 8.928 | — | — |
| **Three-section (7-pole)** | 1 | 3538 | 5052 | 5052 | 3538 | — | 17.24 | 18.20 | 17.24 | — |
| | 0.1 | 3759.8 | 6673.9 | 6673.9 | 3759.8 | — | 11.32 | 12.52 | 11.32 | — |
| | 0.01 | 2536.8 | 5564.5 | 5564.5 | 2536.8 | — | 11.08 | 13.00 | 11.08 | — |
| | 0.001 | 1875.7 | 4875.9 | 4875.9 | 1875.7 | — | 9.879 | 12.31 | 9.879 | — |
| **Four-section (9-pole)** | 1 | 6938.3 | 9935.8 | 10,105 | 9935.8 | 6938.3 | 8.906 | 9.467 | 9.467 | 8.906 |
| | 0.1 | 3805.9 | 6794.5 | 7019.9 | 6794.5 | 3805.9 | 11.48 | 12.87 | 12.87 | 11.48 |
| | 0.01 | 2592.5 | 5743.5 | 6066.3 | 5743.5 | 2592.5 | 11.36 | 13.63 | 13.63 | 11.36 |
| | 0.001 | 1941.7 | 5124.6 | 5553.2 | 5124.6 | 1941.7 | 10.27 | 13.25 | 13.25 | 10.27 |

**Inductance in microhenrys, capacitance in picofarads. Component values normalised to 1MHz and 50Ω.**

Table A.17: Chebyshev high-pass filter ('T' configuration)

| | Ripple (dB) | $C_1$ | $C_2$ | $C_3$ | $C_4$ | $C_5$ | $L_1$ | $L_2$ | $L_3$ | $L_4$ |
|---|---|---|---|---|---|---|---|---|---|---|
| **Single section (3-pole)** | 1 | 1573 | 1573 | — | — | — | 8.005 | — | — | — |
| | 0.1 | 3085.7 | 3085.7 | — | — | — | 6.935 | — | — | — |
| | 0.01 | 5059.1 | 5059.1 | — | — | — | 8.201 | — | — | — |
| | 0.001 | 7786.9 | 7786.9 | — | — | — | 10.95 | — | — | — |
| **Two-section (5-pole)** | 1 | 1491 | 1060.7 | 1491 | — | — | 7.293 | 7.293 | — | — |
| | 0.1 | 2775.6 | 1611.7 | 2775.6 | — | — | 5.803 | 5.803 | — | — |
| | 0.01 | 4208.6 | 2018.6 | 4208.6 | — | — | 6.098 | 6.098 | — | — |
| | 0.001 | 5865.7 | 2429.5 | 5865.7 | — | — | 7.093 | 7.093 | — | — |
| **Three-section (7-pole)** | 1 | 1469.2 | 1028.9 | 1028.9 | 1469.2 | — | 7.160 | 6.781 | 7.160 | — |
| | 0.1 | 2694.9 | 1518.2 | 1518.2 | 2694.9 | — | 5.593 | 5.058 | 5.593 | — |
| | 0.01 | 3994.1 | 1820.9 | 1820.9 | 3994.1 | — | 5.715 | 4.873 | 5.715 | — |
| | 0.001 | 5401.7 | 2078 | 2078 | 5401.7 | — | 6.410 | 5.144 | 6.410 | — |
| **Four-section (9-pole)** | 1 | 1460.3 | 1019.8 | 1002.7 | 1019.8 | 1460.3 | 7.110 | 6.689 | 6.689 | 7.110 |
| | 0.1 | 2662.2 | 1491.2 | 1443.3 | 1491.2 | 2662.2 | 5.516 | 4.922 | 4.922 | 5.516 |
| | 0.01 | 3908.2 | 1764.1 | 1670.2 | 1764.1 | 3908.2 | 5.578 | 4.647 | 4.647 | 5.578 |
| | 0.001 | 5216.3 | 1977.1 | 1824.6 | 1977.1 | 5216.3 | 6.657 | 4.780 | 4.780 | 6.657 |

**Inductance in microhenrys, capacitance in picofarads. Component values normalised to 1MHz and 50Ω.**

Table A.18: Chebyshev high-pass filter (Pi configuration)

| | Ripple (dB) | $L_1$ | $L_2$ | $L_3$ | $L_4$ | $L_5$ | $C_1$ | $C_2$ | $C_3$ | $C_4$ |
|---|---|---|---|---|---|---|---|---|---|---|
| **Single section (3-pole)** | 1 | 3.932 | 3.932 | — | — | — | 3201.7 | — | — | — |
| | 0.1 | 7.714 | 7.714 | — | — | — | 2774.2 | — | — | — |
| | 0.01 | 12.65 | 12.65 | — | — | — | 3280.5 | — | — | — |
| | 0.001 | 19.47 | 19.47 | — | — | — | 4381.4 | — | — | — |
| **Two-section (5-pole)** | 1 | 3.727 | 2.652 | 3.727 | — | — | 2917.3 | 2917.3 | — | — |
| | 0.1 | 6.939 | 4.029 | 6.939 | — | — | 2321.4 | 2321.4 | — | — |
| | 0.01 | 10.52 | 5.045 | 10.52 | — | — | 2439.3 | 2439.3 | — | — |
| | 0.001 | 14.66 | 6.074 | 14.66 | — | — | 2837.3 | 2837.3 | — | — |
| **Three-section (7-pole)** | 1 | 7.159 | 5.014 | 5.014 | 7.159 | — | 1469.2 | 1391.6 | 1469.2 | — |
| | 0.1 | 8.737 | 3.795 | 3.795 | 8.737 | — | 2237.2 | 2023.1 | 2237.2 | — |
| | 0.01 | 9.985 | 4.552 | 4.552 | 9.985 | — | 2286.0 | 1949.1 | 2286.0 | — |
| | 0.001 | 13.50 | 5.195 | 5.195 | 13.50 | — | 2584.1 | 2057.7 | 2584.1 | — |
| **Four-section (9-pole)** | 1 | 3.651 | 2.549 | 2.507 | 2.549 | 3.651 | 2844.1 | 2675.6 | 2675.6 | 2844.1 |
| | 0.1 | 6.656 | 3.728 | 3.608 | 3.728 | 6.656 | 2206.5 | 1968.9 | 1968.9 | 2206.5 |
| | 0.01 | 9.772 | 4.410 | 4.176 | 4.410 | 9.772 | 2230.5 | 1858.7 | 1858.7 | 2230.5 |
| | 0.001 | 13.05 | 4.943 | 4.561 | 4.943 | 13.05 | 2466.3 | 1911.8 | 1911.8 | 2466.3 |

**Inductance in microhenrys, capacitance in picofarads. Component values normalised to 1MHz and 50Ω.**

would use two inductances of a value equal to $L_k/2$, while the balanced constant-k π-section high-pass filter would use two capacitors of a value equal to $2C_k$.

If several low- (or high-) pass sections are to be used, it is advisable to use m-derived end sections on either side of a constant-k section, although an m-derived centre section can be used.

**Table A.19: Practical Chebyshev low-pass filters (3 section, 7 pole)**

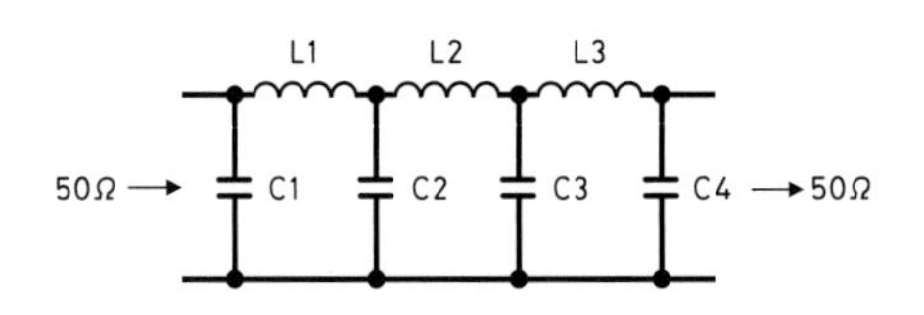


| Amateur band (MHz) | 28 | 21 | 14 | 7 | 3.5 | 1.8 | |
|---|---|---|---|---|---|---|---|
| **$F_c$** | **30.9** | **21.69** | **15.16** | **7.98** | **4.11** | **2.05** | **MHz** |
| **VSWR** | **1.10** | **1.06** | **1.09** | **1.08** | **1.07** | **1.09** | |
| **C1, C4** | **82** | **100** | **160** | **300** | **560** | **1200** | **pF** |
| **C2, C3** | **180** | **240** | **360** | **680** | **1300** | **2700** | **pF** |
| **L1, L3** | **0.36** | **0.49** | **0.72** | **1.37** | **2.62** | **5.42** | **µH** |
| **L2** | **0.42** | **0.59** | **0.85** | **1.62** | **3.13** | **6.41** | **µH** |

**Ripple is 0.01dB.**

The factor m relates the ratio of the cut-off frequency and $f_\infty$, a frequency of high attenuation. Where only one m-derived section is used, a value of 0.6 is generally used for m, although a deviation of 10 or 15% from this value is not too serious in amateur work. For a value of m = 0.6, f will be $1.25f_c$ for the low-pass filter and $0.8f_c$ for the high-pass filter. Other values can be found from:

$$m = \sqrt{1 - \left(\frac{f_c}{f_\infty}\right)^2}$$

for the low-pass filter and:

$$m = \sqrt{1 - \left(\frac{f_\infty}{f_c}\right)^2}$$

for the high-pass filter.

The filters shown should be terminated in a resistance R, and there should be little or no reactive component in the termination.

## Microstrip circuit elements

In the calculation of microstrip circuit elements it is necessary to establish the dielectric constant for the material. This can be done by measuring the capacitance of a typical sample.

$$\text{Dielectric constant } e = \frac{113 \times C \times h}{a}$$

where C is in picofarads, h is the thickness in millimetres and a is the area in square millimetres. This should be done with a sample about 25mm square to minimise the effects of the edges. Having found the dielectric constant, it is now necessary to calculate the characteristic impedance ($Z_0$) of the microstrip. There are many approximations for this but the following is simple and accurate enough (±5%) for amateur use since microstrip is fairly forgiving of small errors:

$$Z_0 = \frac{131}{\sqrt{(e + 0.47)}} \times \log_{10}\left(\frac{13.5h}{w}\right)$$

where e is the dielectric constant, h is the dielectric thickness (see Fig A.8), and w is the conductor width. Note that h and w must be in the same units, eg both in centimetres. The formula assumes that the conductor is thin relative to the dielectric.

An accurate plot of $Z_0$ against w/h for dielectric constants between 2 (approximately that of PTFE) through 4 (approx that of epoxy-glassfibre) to 6 is given in Fig A.9. The next operation is to determine the velocity factor, the ratio of the velocity of electromagnetic waves in the dielectric to that in free space. Here, too, the equations are complex but the factor only changes slowly as w/h changes. Fig A.10 gives figures for the above range of dielectric constants.

A microstrip resonator has the form of a strip on one side of a double-sided PCB with the other side as a ground plane. It is usually a quarter-wavelength long. The length is calculated from the free space length multiplied by the velocity factor as estimated above. The line width depends on the required $Z_0$ and a starting point for experiment would be 50–100Ω. If it is necessary to tune the resonator accurately, it should be made shorter

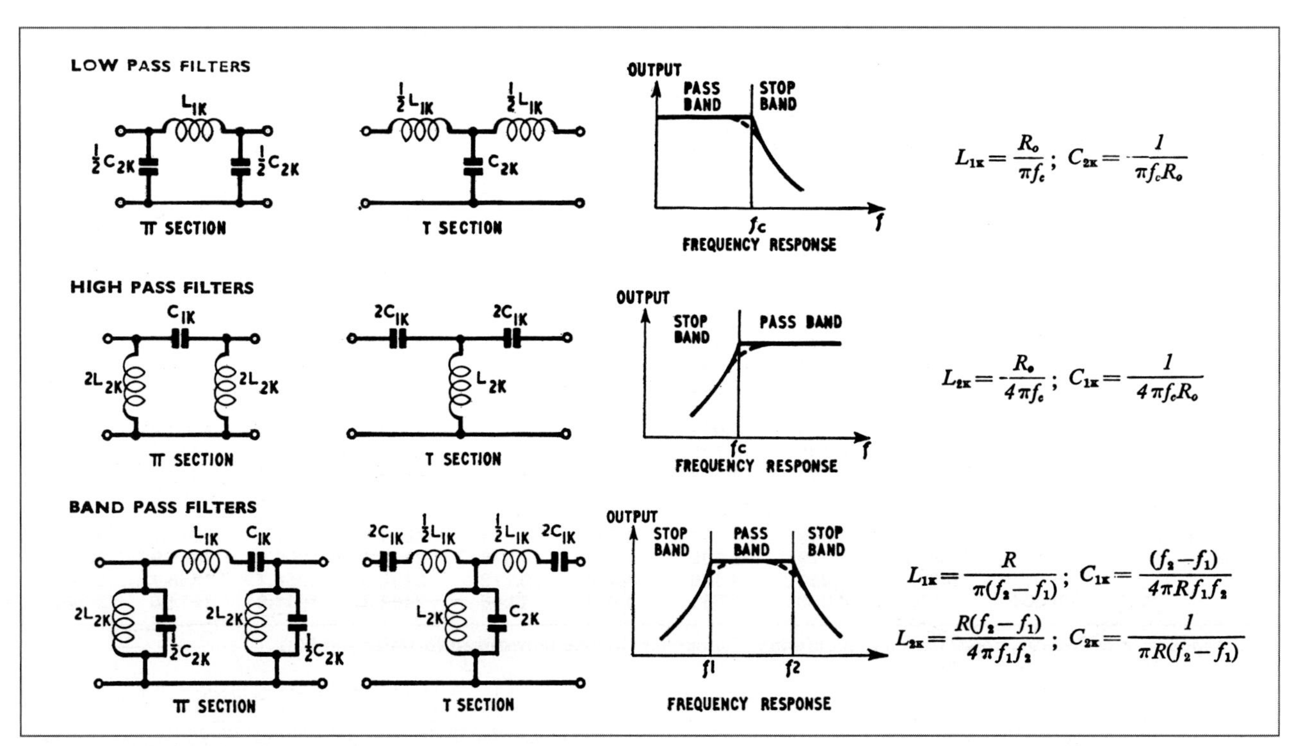


Fig A.7(a): Constant-k filters

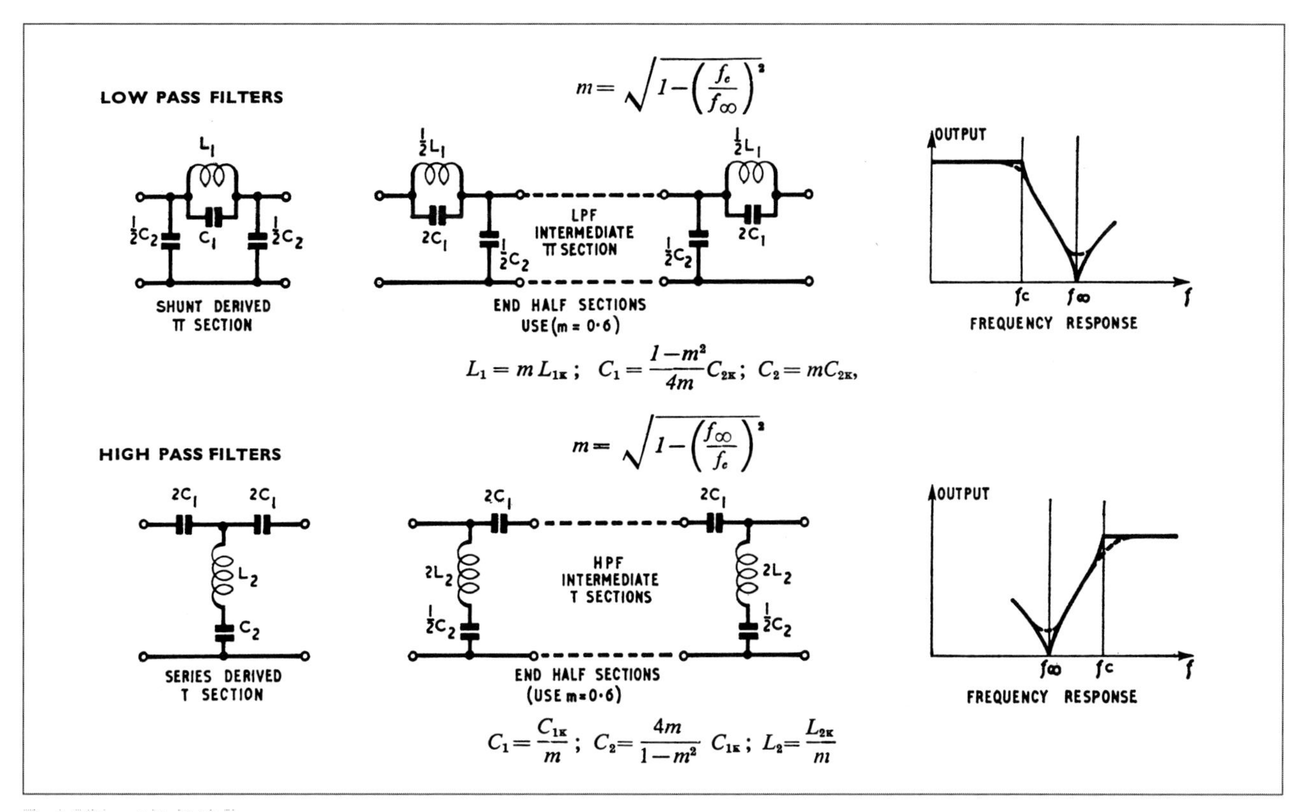


Fig A.7(b): m-derived filters

than calculated above and a trimmer capacitor connected between the 'hot' end and the ground plane. The new length can be calculated from:

$$l = 0.0028\lambda \times \tan^{-1}\left(\frac{\lambda}{0.188CZ_0}\right)$$

where l is the length in centimetres, λ is the wavelength in centimetres, C is the capacitance (say at half maximum) in picofarads, $Z_0$ is the characteristic impedance in ohms and $\tan^{-1}(*)$ is the angle in degrees of which * is the tangent. * represents the figures in the bracket.

Coupling into and out of the line may be directly via tapping(s) or by additional line(s) placed close to the tuned line. A spacing of one line width and a length of 10–20% of the tuned line would be a starting point for experiment.

## Materials

The most-used material for amateur purposes is glassfibre reinforced epoxy double-sided PCB. It has a dielectric constant of 4.0–4.5, depending on the resin used. Accurate lines may be made by scoring through the copper carefully with a scalpel or modelling knife and lifting the unwanted copper foil after heating it with a soldering iron to weaken the bond to the plastic.

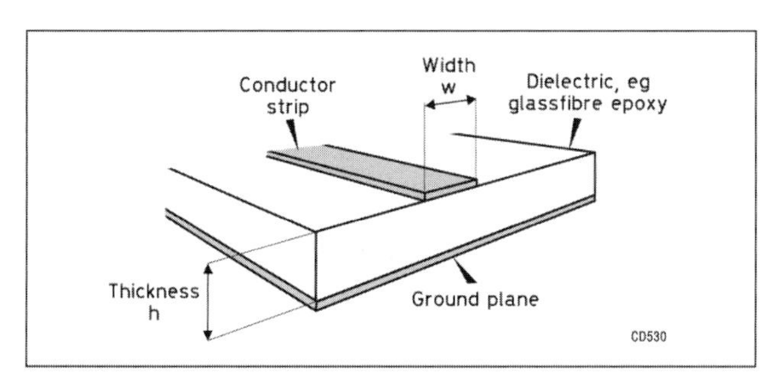


Fig A.8: Dimensions involved in calculating the characteristic impedance of microstrip

For microwave use, glassfibre-reinforced PTFE with a dielectric constant close to 2.5 is the preferred material. Further information is given in the Microwave Handbook [4].

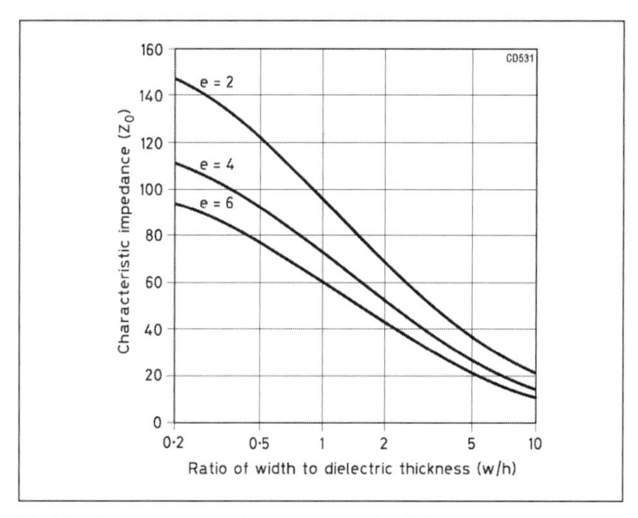


Fig A.9: Characteristic impedance versus w/h for dielectric constants between 2 and 6

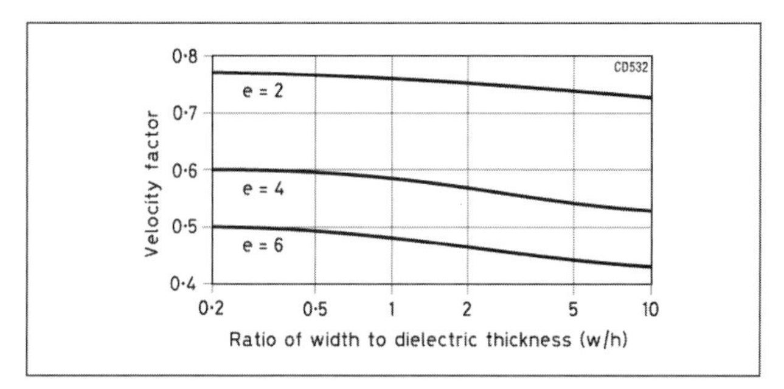


Fig A.10: Velocity factor versus w/h for dielectric constants between 2 and 6

## Op-amp-based active filters

Design information (taken, by permission, from reference [1]|) will be given for four common filter configurations. All are based on inexpensive op-amps such as the 741 and 301A (or their duals or quads in one package) which are adequate when frequencies are in the voice range, insertion gain is between unity and two (0–6dB), signal input and output voltages are in the range between a few millivolts and a few volts, and signal (input, feedback and output) currents between a microamp and a milliamp. This covers the bulk of common amateur applications. No DC supplies to the op-amps are shown.

### 2nd order 'Sallen and Key' Butterworth low-pass filter

Referring to Fig A.11, the cut-off (–3dB) frequency $f_c$ is:

$$f_c = \frac{1}{2\pi\sqrt{(R_1 R_2 C_1 C_2)}}$$

Choosing 'equal components', meaning $R_1 = R_2 = R$ and $C_1 = C_2 = C$, then:

$$f_c = \frac{1}{2\pi RC}$$

For a second-order Butterworth response, the pass-band gain must be 4dB or ×1.586. This is achieved by making $(R_A + R_B)/R_A = 1.586$. This is implemented with sufficient accuracy with 5% standard-value resistors of $R_A = 47k\Omega$ and $R_B = 27k\Omega$. This means that a 1V input generates an output of 1.586V at a frequency in the pass-band and 0.707 × 1.586 = 1.12V at the –3dB cut-off frequency; the roll-off above $f_c$ is 12dB/octave or 20dB/decade.

Example. Design a two-pole 'equal component' Butterworth low-pass filter (Fig A.12) with $f_c$ = 2700Hz.

Choosing for C a convenient value of 1nF and solving:

$$R = \frac{1}{2\pi f_c C} = 59k\Omega$$

This can be made up from 56kΩ and 2.7kΩ in series.

Should R come out below 10kΩ, choose a larger C; if R would be larger than 100kΩ, select a smaller C; then recalculate R.

### The multiple-feedback bandpass filter

Providing two feedback paths to a single op-amp, a band-pass filter can be made with Q up to 10. To get reasonably steep roll-off at low Q, from two to four identical sections (Fig A.14) are cascaded. The centre frequency is given by:

$$f_0 = \frac{1}{2\pi C}\sqrt{\frac{1}{R_3} \cdot \frac{R_1 + R_2}{R_1 R_2}}$$

for which the three resistors can be calculated from:

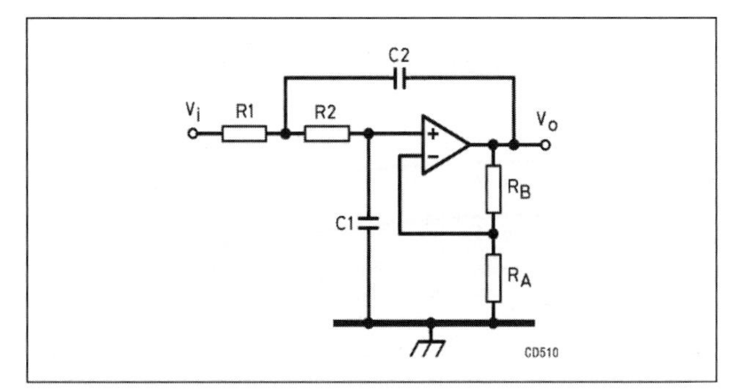


Fig A.11: "Sallen & Key" Butterworth low-pass filter

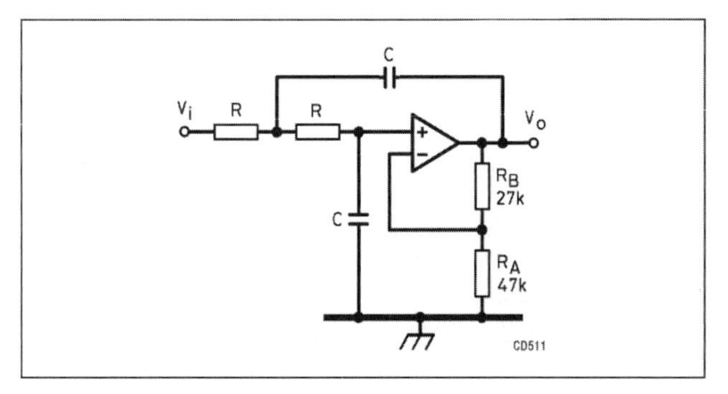


Fig A.12: Equal component low-pass filter

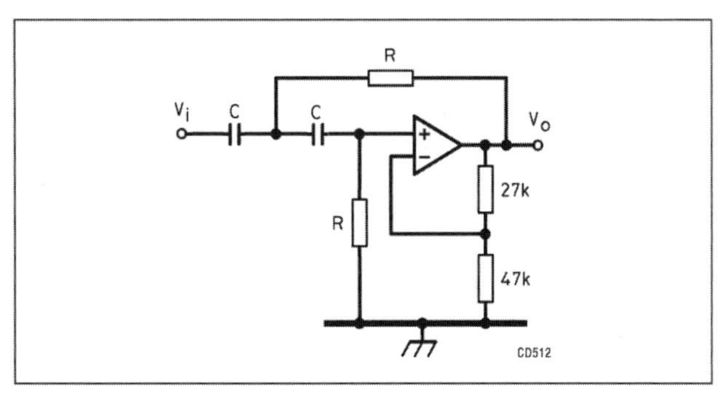


Fig A.13: Equal component high-pass filter

$$R_1 = \frac{Q}{2\pi f_0 G_0 C}$$

$$R_2 = \frac{Q}{2\pi f_0 C(2Q^2 - G_0)}$$

$$R_3 = \frac{Q}{\pi f_0 C}$$

The equations for $R_1$ and $R_3$ combine into:

$$G_0 = R_3/2R_1$$

Also, the denominator in the formula for $R_2$ yields:

$$Q > \sqrt{(G_0/2)}$$

Example. Design a band-pass filter with centre frequency 800Hz, –6dB bandwidth of 200Hz and centre-frequency gain of 2.

A two-section filter, with each section having a 200Hz –3dB bandwidth, is indicated.

$$Q = 800/200 = 4$$

$$G_0 = \sqrt{2} = 1.4$$

Select a convenient C, say 10nF.

$$R_1 = \frac{4}{6.28 \times 800 \times 1.4 \times 10^{-8}} = 56.9k\Omega, \text{ (use } 56k\Omega)$$

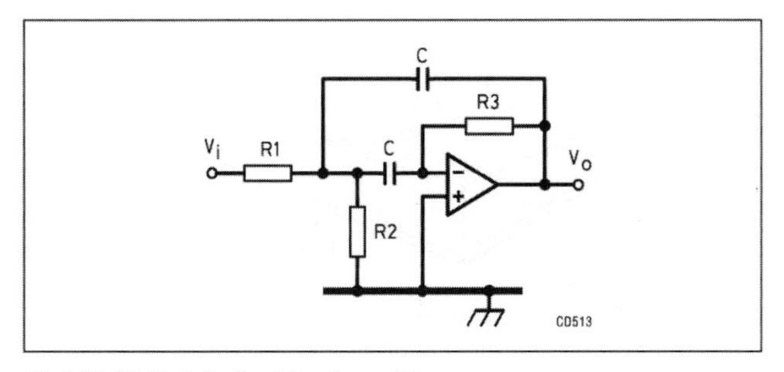


Fig A.14: Mulitple feedback band-pass filter

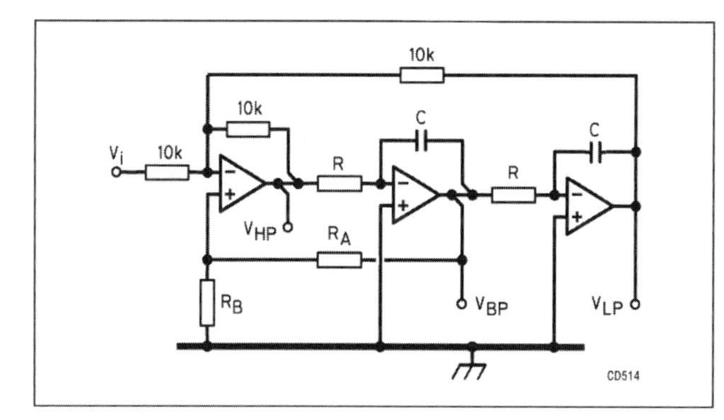


Fig A.15: State variable filter

$$R_3 = 2 \times 56.9 \times 1.4$$
$$= 159\text{k}\Omega \text{ (use } 100\text{k}\Omega + 56\text{k}\Omega)$$
$$R_2 = \frac{4}{6.28 \times 800 \times 10^{-8} \times (2 \times 4^2 - 1.4)}$$
$$= 2.60\text{k}\Omega \text{ (use } 5.1\text{k}\Omega \text{ in parallel with } 5.1\text{k}\Omega)$$

Note that the centre frequency can be shifted up or down at constant bandwidth and centre frequency gain by changing $R_2$ only, using ganged variable resistors for cascaded sections:

$$R_2' = R_2 \left( \frac{f_0}{f_0'} \right)^2$$

### The state-variable or 'universal' filter

Three op-amps, connected as shown in Fig A.15, can simultaneously provide second-order high-pass, low-pass and band-pass responses. The filter is composed of a difference amplifier and two integrators. The common cut-off/centre frequency is given by:

$$f_{cL} = f_{cH} = f_0 = \frac{1}{2\pi RC}$$

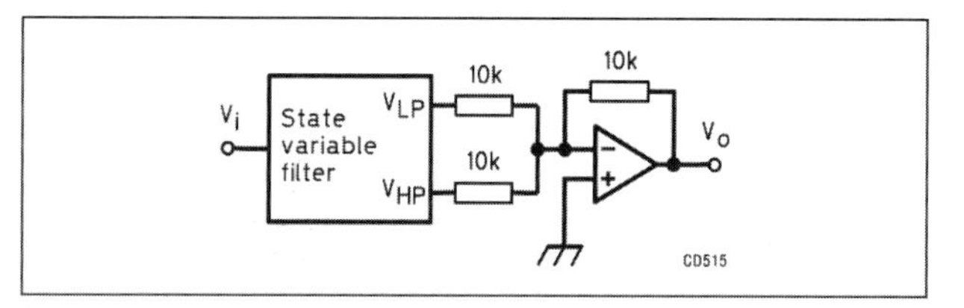


Fig A.16: Using the state-variable filter to obtain a notch response

The filter's Q depends only on $R_A$ and $R_B$:

$$R_A = (3Q - 1)R_B$$

There is no way to simultaneously optimise the performance of high-/low-pass and band-pass performance. For a Butterworth response, Q must be 0.7 and even for a second-order 3dB-ripple Chebyshev response the Q is no more than 1.3, obviously too low for good band-pass response. No DC voltage should be applied to the input of this filter and there should be no significant DC load on its outputs.

By adding the low-pass and high-pass outputs from a variable-state filter in a summing amplifier, a notch response is obtained. See Fig A.16.

For an application of the variable-state filter refer to the active filters section in Chapter 5.

## REFERENCES


[1] *The Design of Operational Amplifier Circuits, with Experiments*, Howard M Berlin, W3HB, E&L Instruments Inc, Derby, Conn, USA, 1977.

# *Appendix B:* Printed Circuit Board Artwork

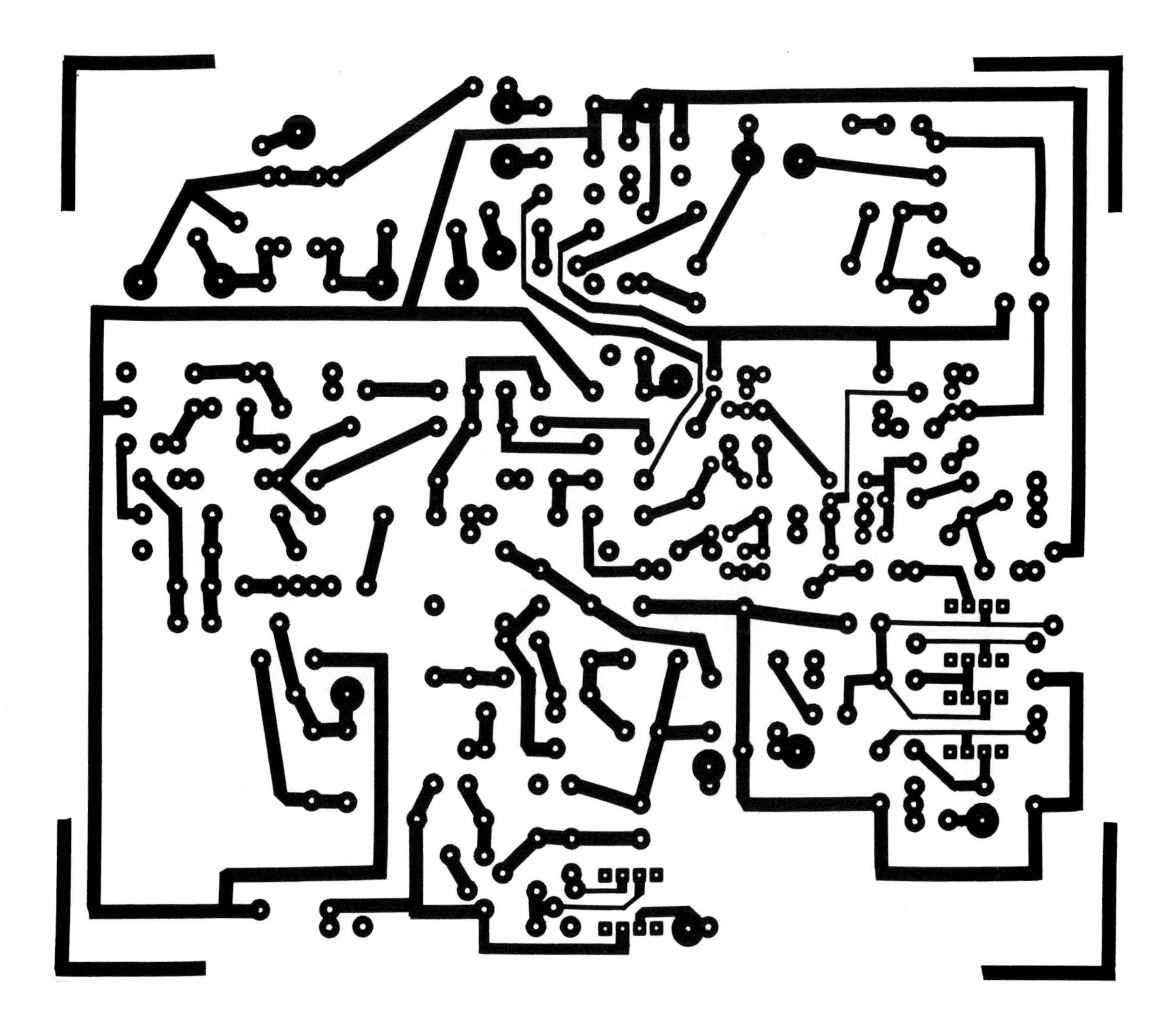

Fig 7.36: 1.8MHz QRP transceiver PCB layout

**IMPORTANT NOTICE**

**Whilst every effort has been made to ensure that the artwork in this appendix is displayed accurately, it is the responsibility of the reader to check that he/she is using the correct drawing, that it is accurate and the correct size (some may need re-scaling on a photocopier) and whether the artwork is 'mirror image'. It is important to read the associated text in the relevant chapter.**

**The RSGB cannot be held responsible for any errors or consequential losses incurred by the use of this artwork.**

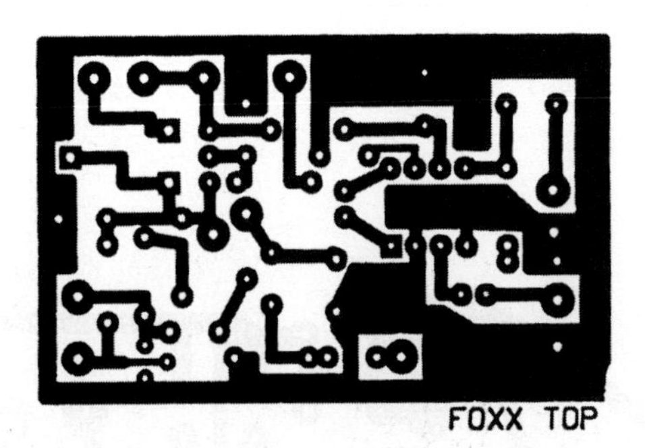


Fig 7.43: FOXX2 PCB layout

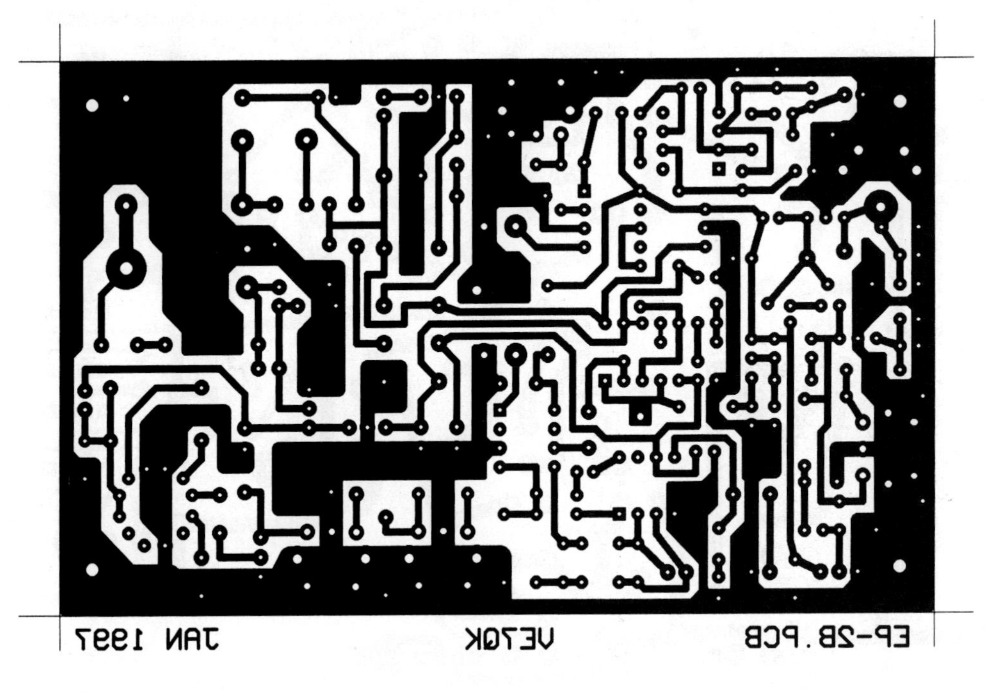


Fig 7.49: Epiphyte-2 PCB layout. This image is reversed

**Before using this artwork, please read the IMPORTANT NOTICE on page B.1 of this Appendix**

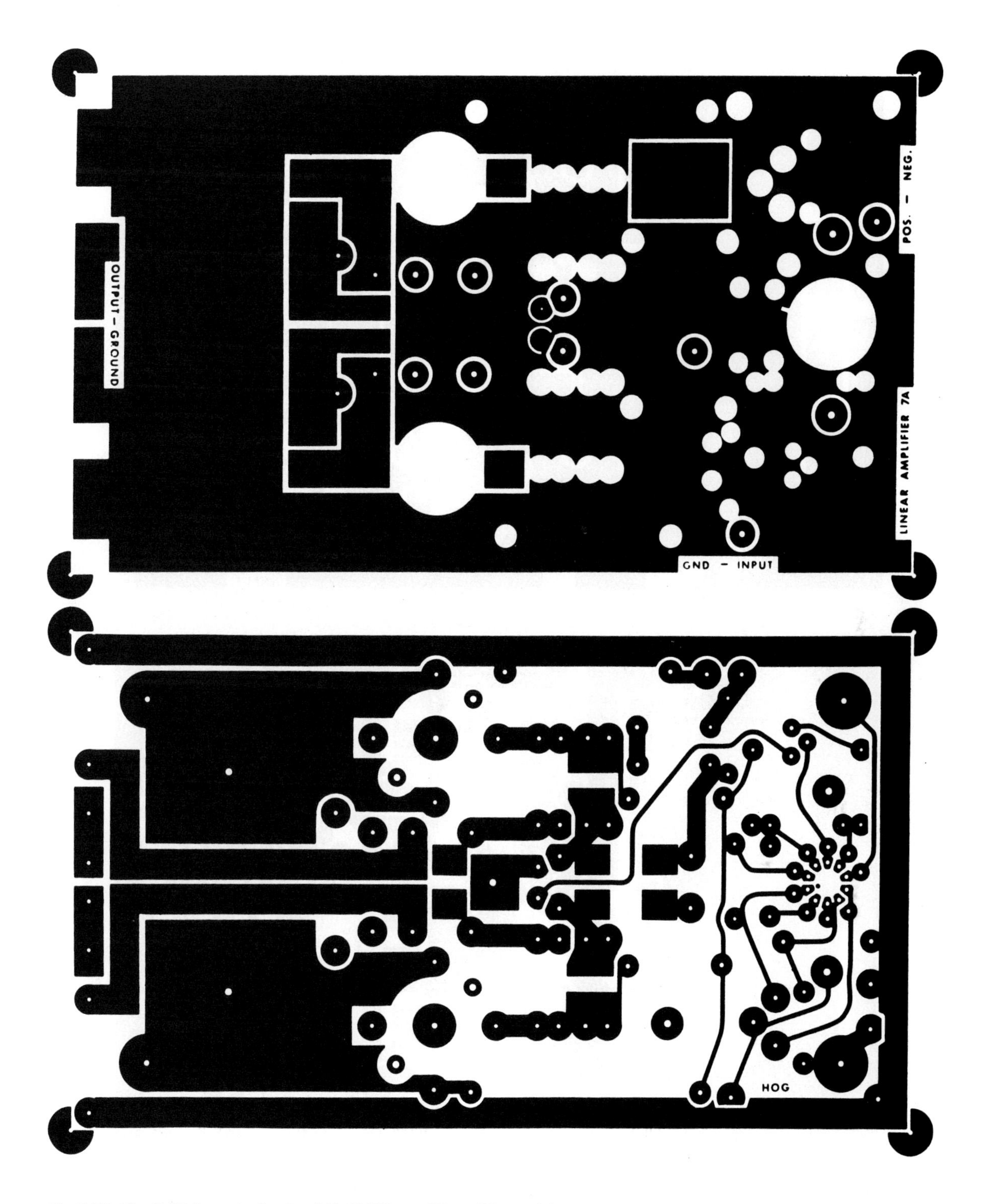


**Fig 7.56: The PCB layouts for the 140-300W amplifiers (Motorola)**

**Before using this artwork, please read the IMPORTANT NOTICE on page B.1 of this Appendix**

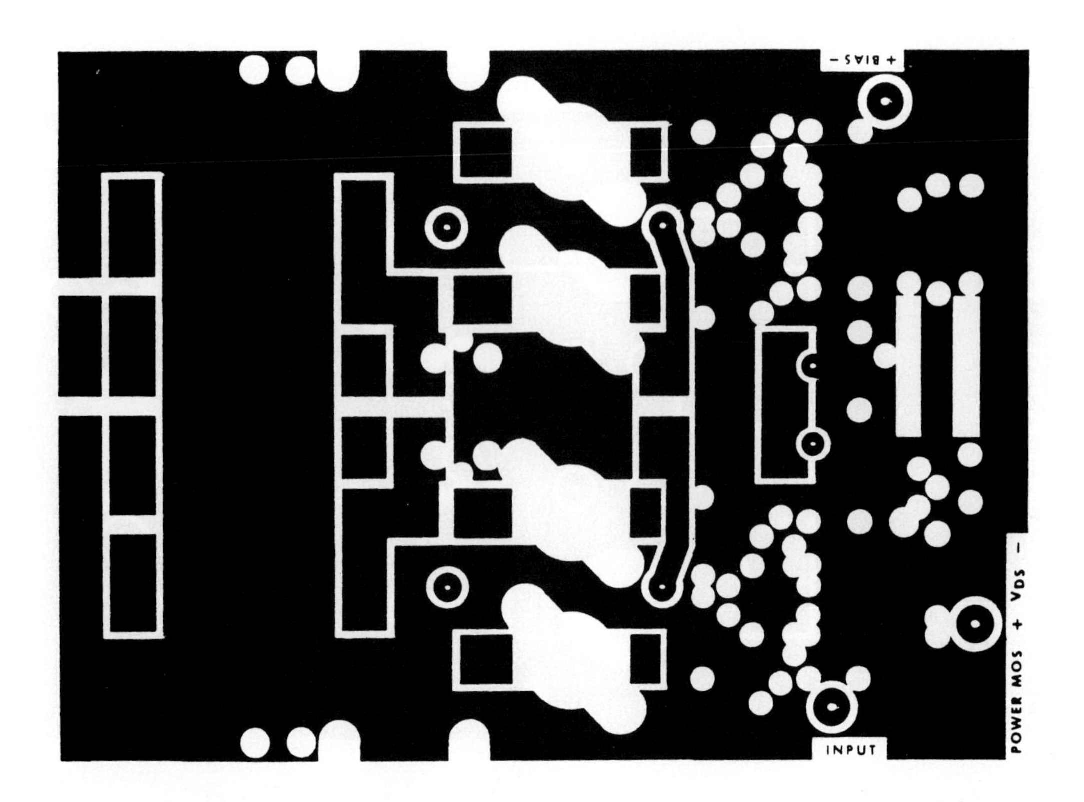


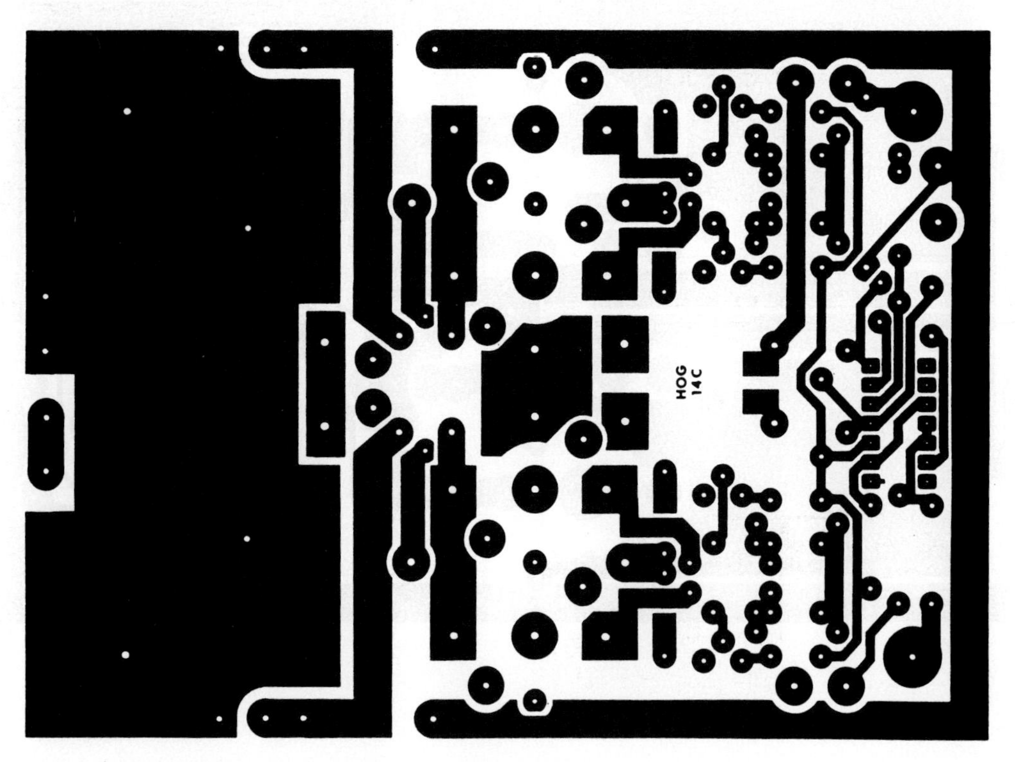


Fig 7.63: PCB layouts for the 600W amplifier (Motorola)

**Before using this artwork, please read the IMPORTANT NOTICE on page B.1 of this Appendix**

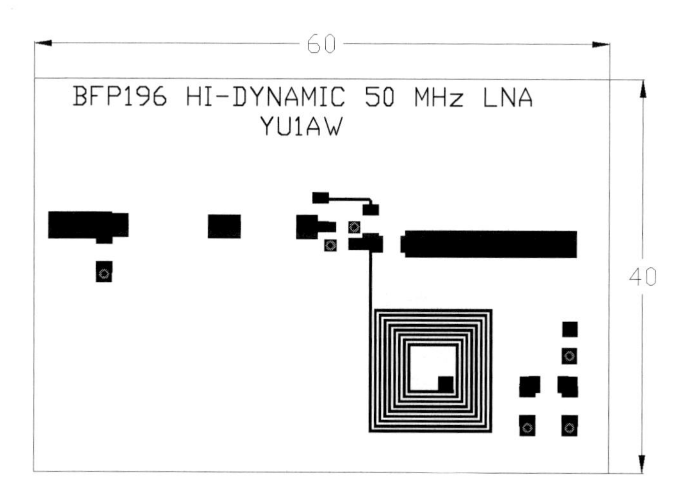


Fig 9.41: PCB layout for the 6m low noise preamplifier. The illustration should be re-scaled as shown

60
BFP196 HI-DYNAMIC 70 MHz LNA
YU1AW
40

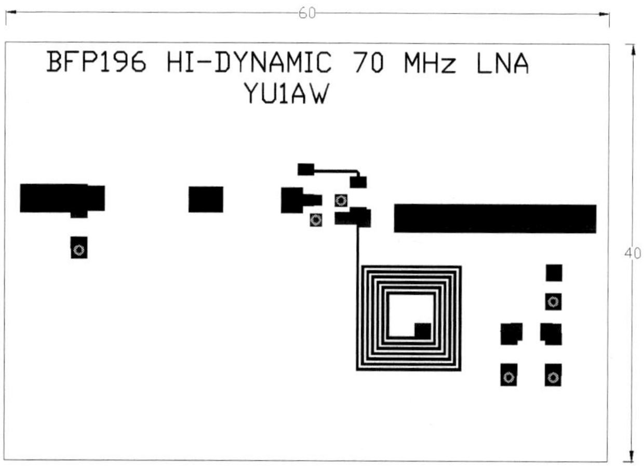

Fig 9.42: PCB layout for the 4m low noise preamplifier. The illustration should be re-scaled as shown

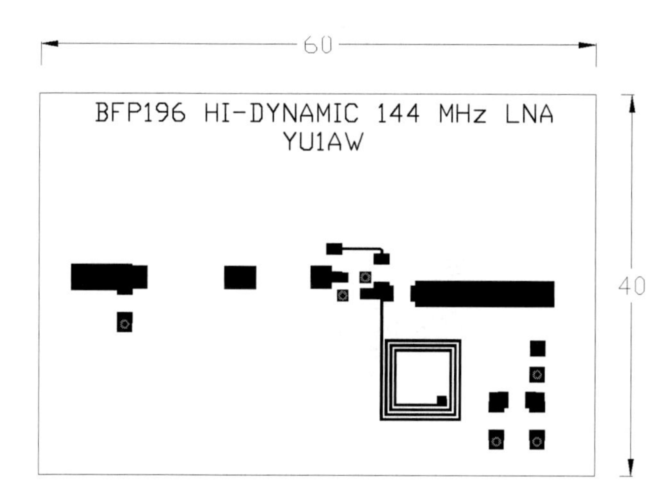


Fig 9.43: PCB layout for the 2m low noise preamplifier. The illustration should be re-scaled as shown

BFP196 HI-DYNAMIC 50 MHz LNA
YU1AW
Ant
L
Ctrimer
Rx
n=10zov.
D=8mm
l=10mm
d=0.7mm CuI
10-30 pF
(24pF)

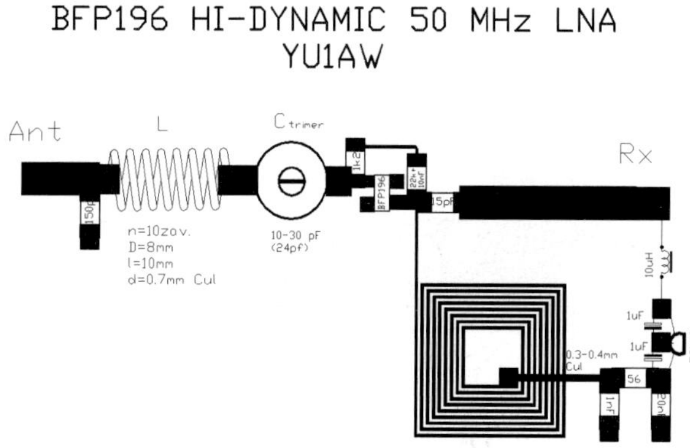

Fig 9.51: Component layout for the 6m low noise preamplifier

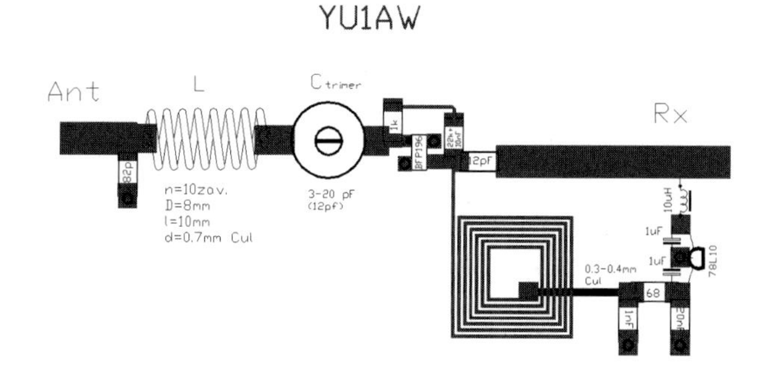


Fig 9.52: Component layout for the 4m low noise preamplifier

Ant
L
Ctrimer
Rx
n=8zov.
D=8mm
l=10mm
d=1mm CuAg
2-10pF
(4.3pF)

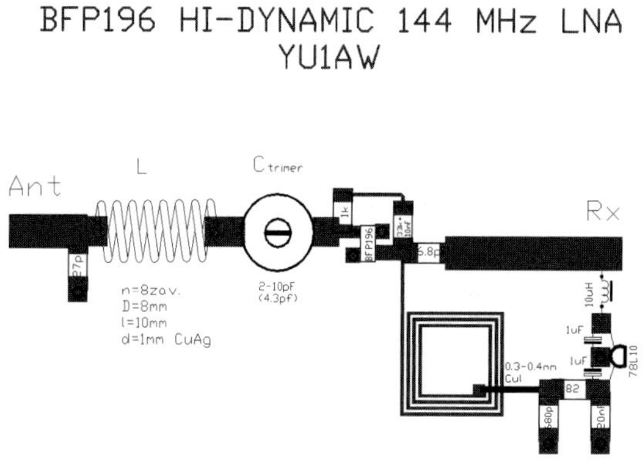

Fig 9.53: Component layout for the 2m low noise preamplifier

**Before using this artwork, please read the IMPORTANT NOTICE on page B.1 of this Appendix**

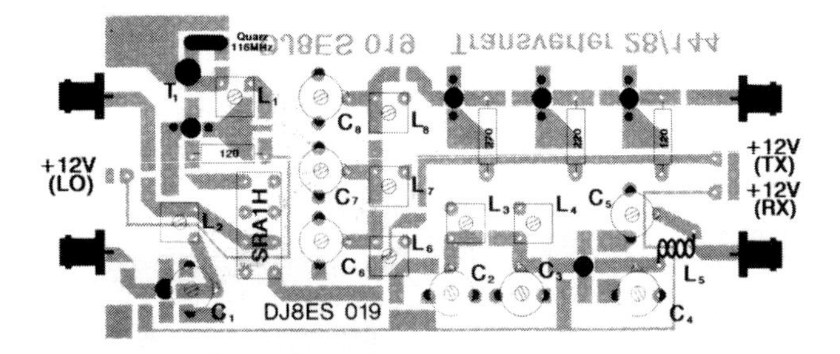


Fig 9.116: Component layout for the 2m transverter showing the component side of the PCB. The board size is 54 x 108mm

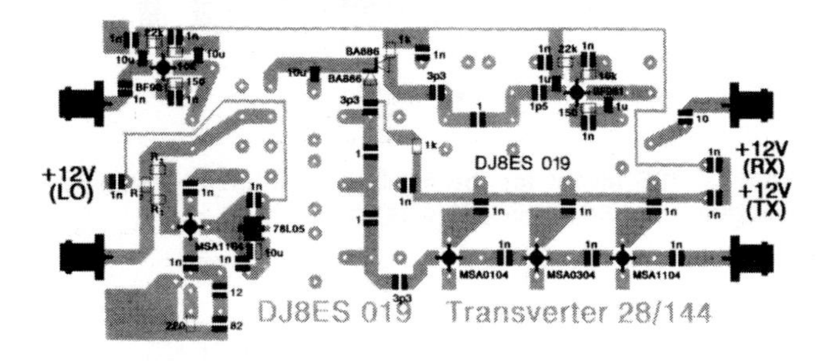


Fig 9.117: Component layout for the 2m transverter showing the track side of the PCB and the positions of the SMD components

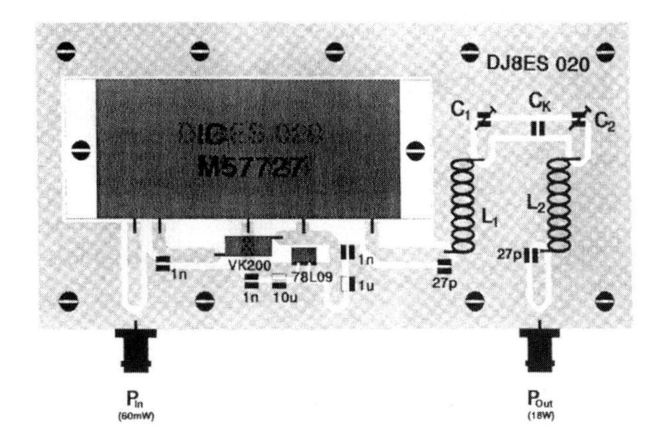


Fig 9.123: Component layout of the 2m power amplifier to be used with the 2m transverter. The board size is 54 x 108mm

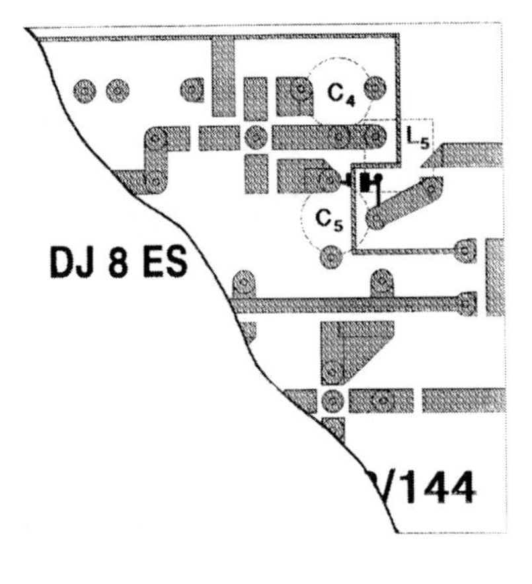


Fig 9.125: Details of modifications to the 2m transverter PCB for use as 6m transverter

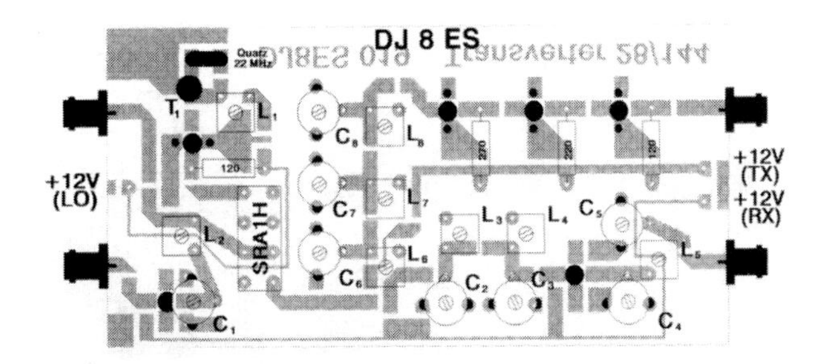


Fig 9.126: Component layout for the 6m transverter showing the component side of the PCB

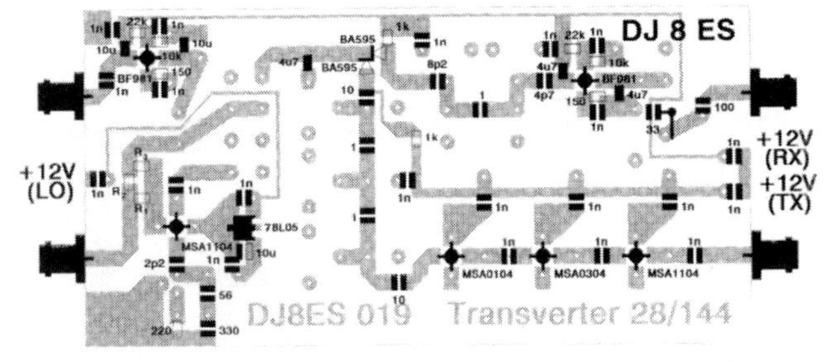


Fig 9.127: Component layout for the 6m transverter showing the track side of the PCB and the positions of the SMD components

**Before using this artwork, please read the IMPORTANT NOTICE on page B.1 of this Appendix**

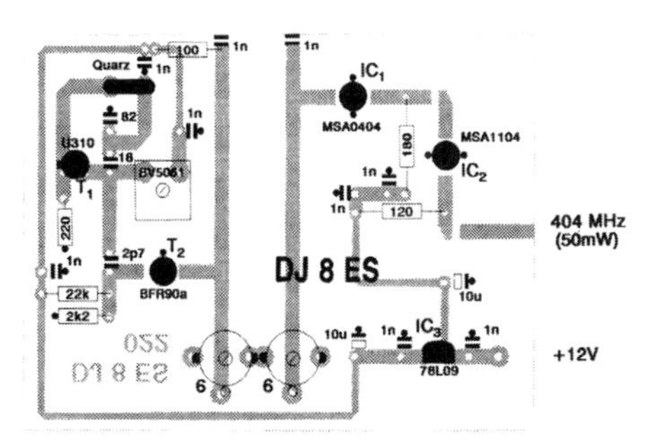


Fig 9.133: Component layout of the local oscillator used on the 70cm transverter showing the component side of the PCB.

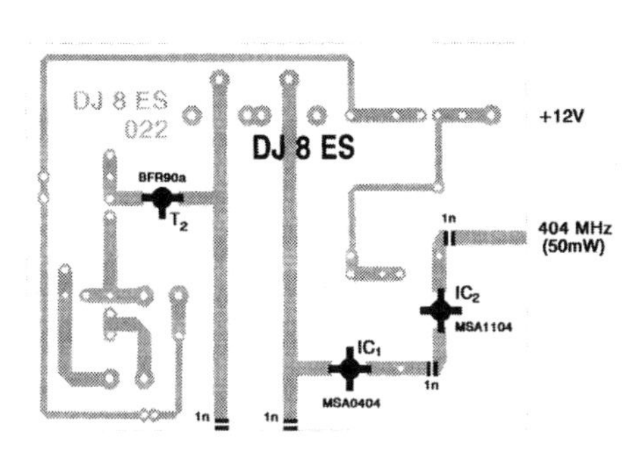


Fig 9.134: Component layout of the LO on the 70cm transverter showing the track side of the PCB and the positions of the SMD components

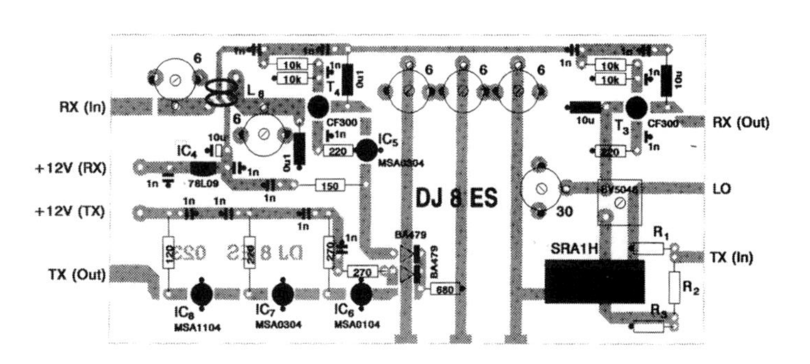


Fig 9.135: Component layout for the 70cm transverter showing the component side of the PCB

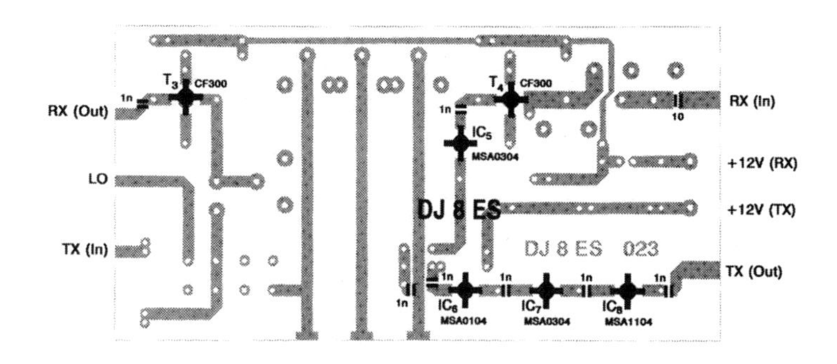


Fig 9.136: Component layout for the 70cm transverter showing the track side of the PCB and the positions of the SMD components

**Before using this artwork, please read the IMPORTANT NOTICE on page B.1 of this Appendix**

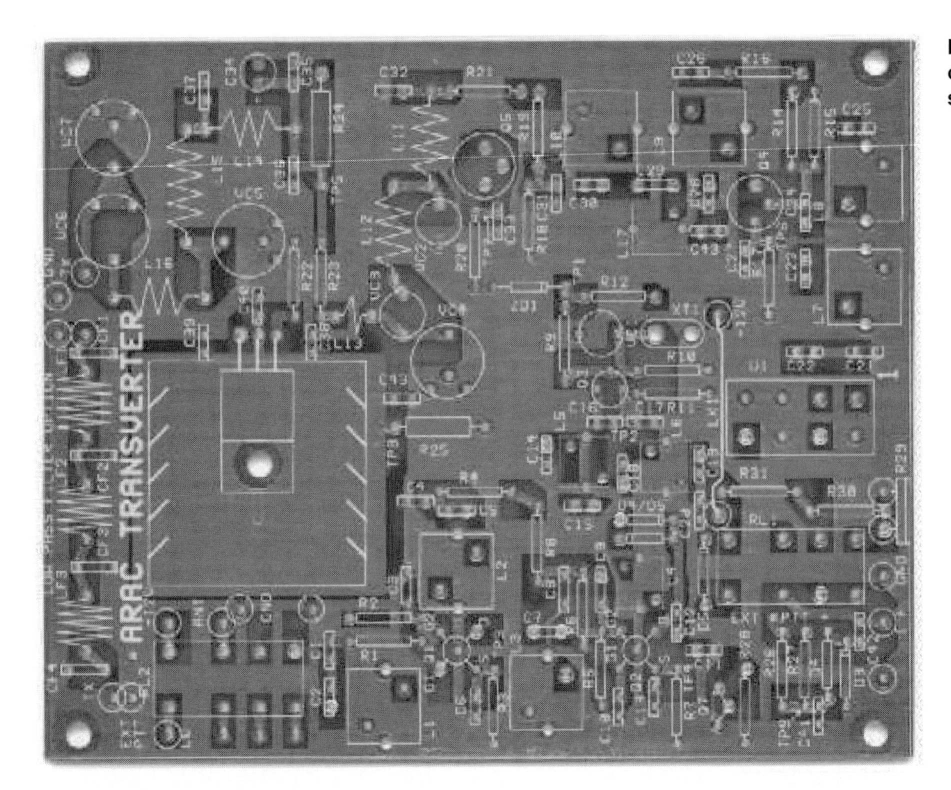

**Fig 9.141: Fig 9.143: Component side of PCB for the 4m transverter. Not to size**

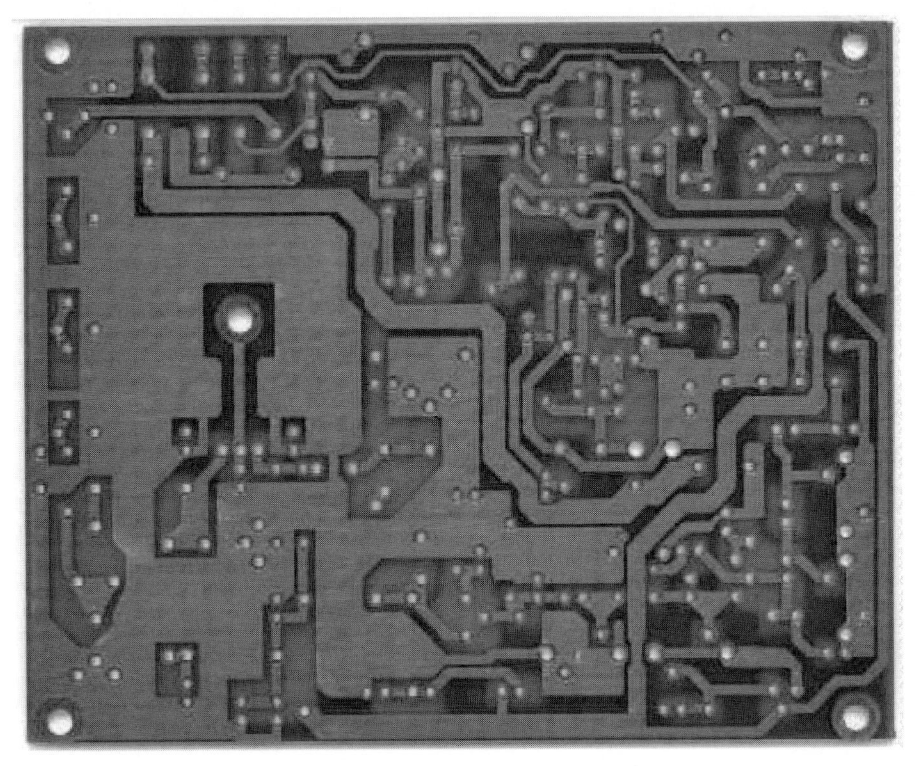

**Fig 9.142: Track side of PCB for the 4m transverter. Not to size**

**Before using this artwork, please read the IMPORTANT NOTICE on page B.1 of this Appendix**

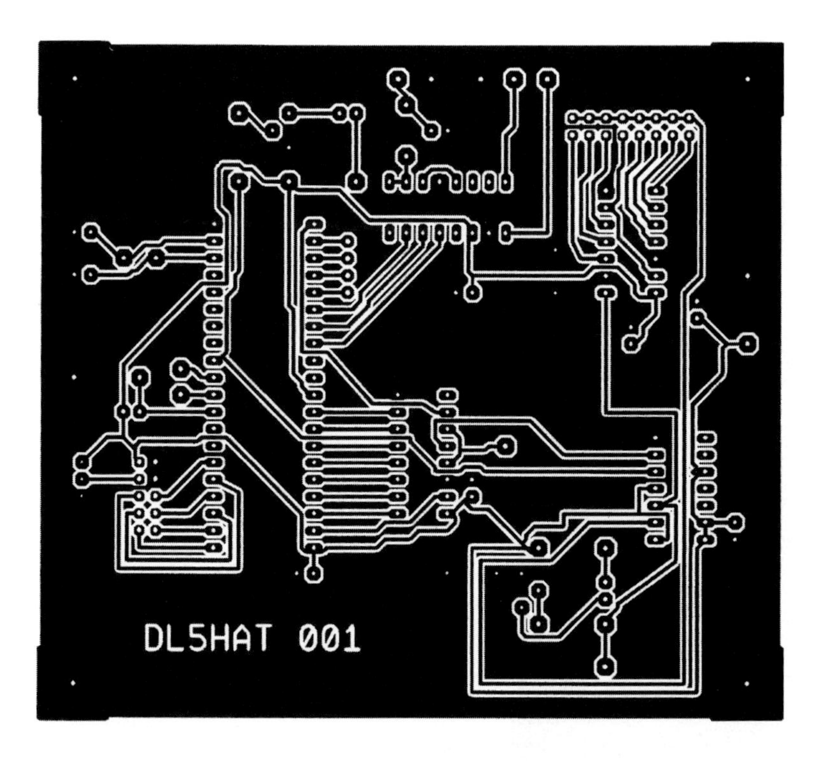


Fig 11.41: PCB layout (DL5HAT-001) for GPS control stage of the high precision frequency standard for 10MHz. The finished PCB should be 100mm x 100m

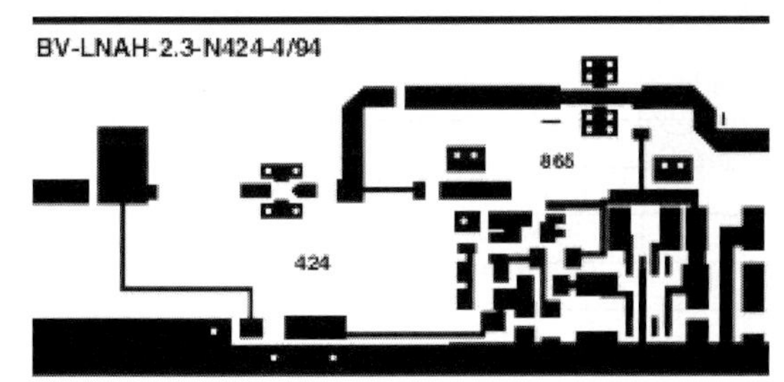


Fig 11.50: PCB layout for 13cm PHEMT. PCB dimensions are 34 x 72mm

**Before using this artwork, please read the IMPORTANT NOTICE on page B.1 of this Appendix**

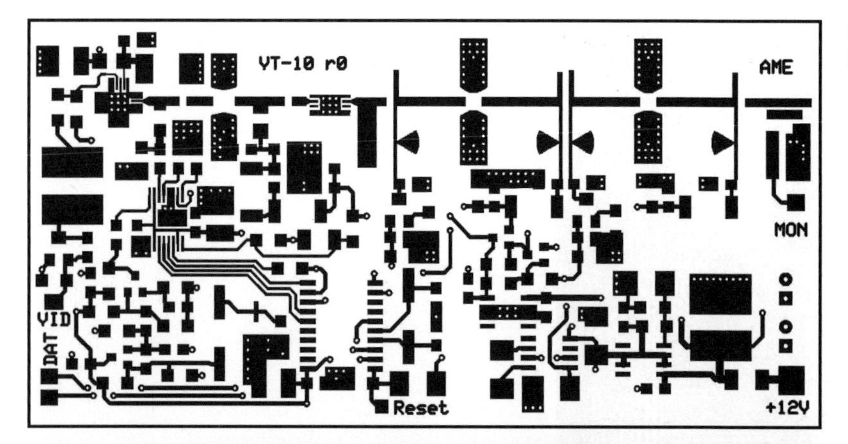


**Fig 11.103: Top side PCB layout for the ATV transmitter**

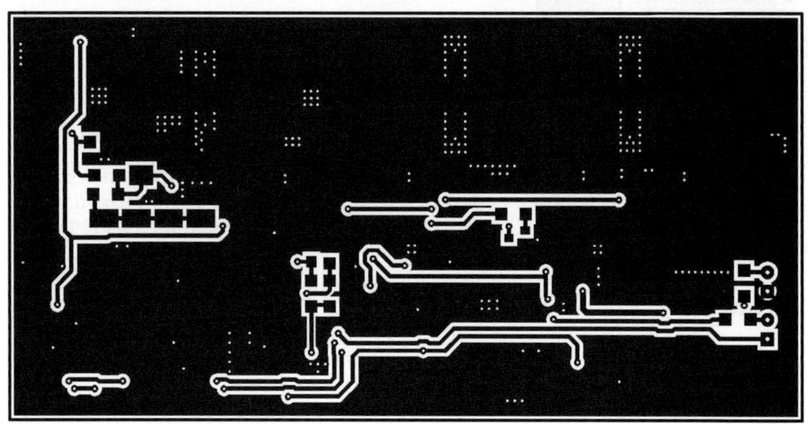

**Fig 11.104: Bottom side PCB layout for the ATV transmitter**

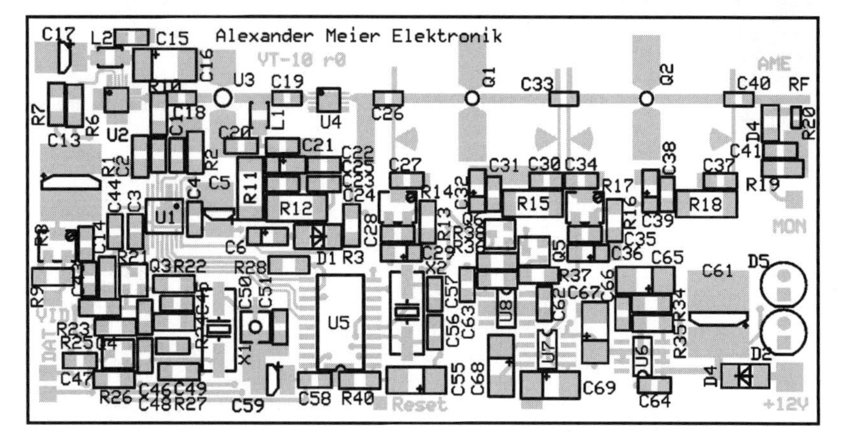


**Fig 11.105: Top side component layout for the ATV transmitter**

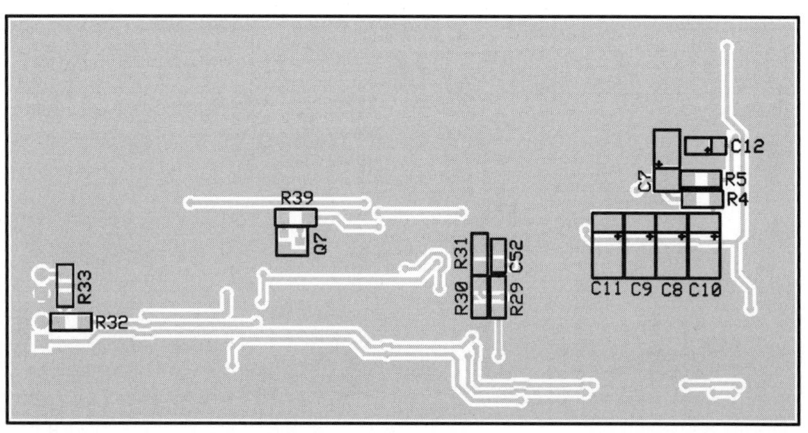


**Fig 11.106: Bottom side component layout for the ATV transmitter**

**Before using this artwork, please read the IMPORTANT NOTICE on page B.1 of this Appendix**

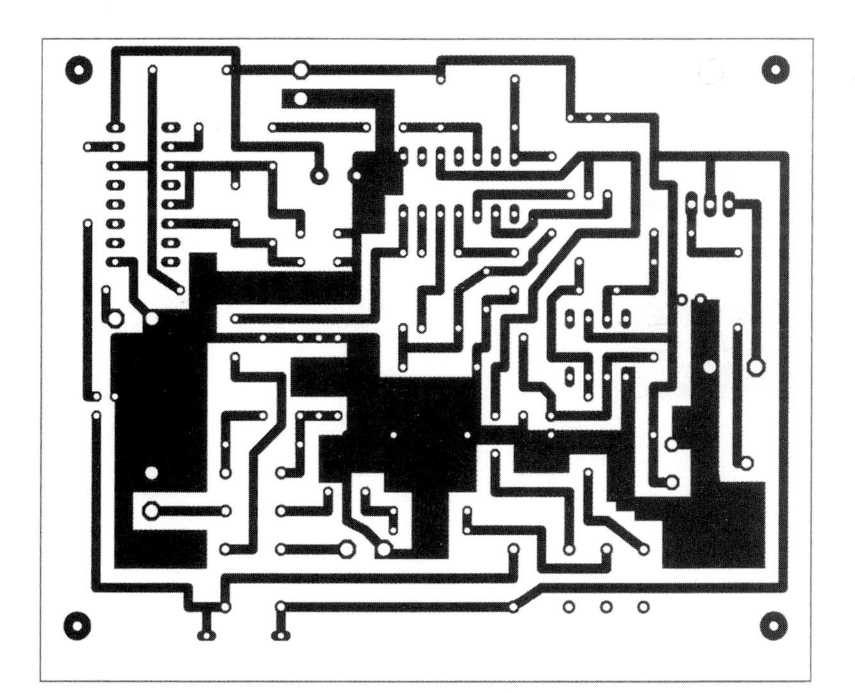

**Fig 11.154: Transverter PCB foil pattern, full size**

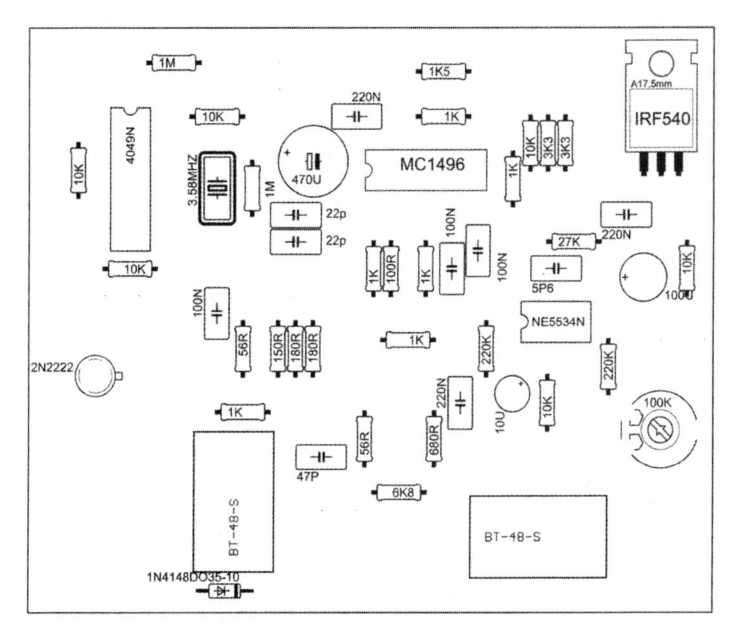


**Fig 11.155: Transverter component placement**

**Before using this artwork, please read the IMPORTANT NOTICE on page B.1 of this Appendix**

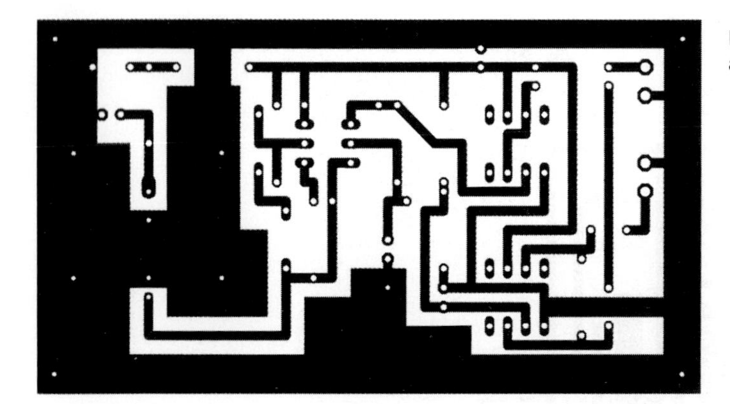

**Fig 11.162: Receive head PCB foil pattern, full size (two op-amp version)**

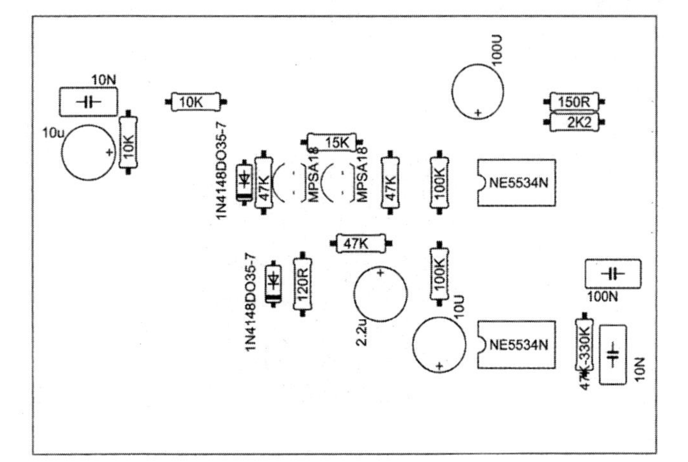


**Fig 11.163: Receiver head component placement (two op-amp version)**

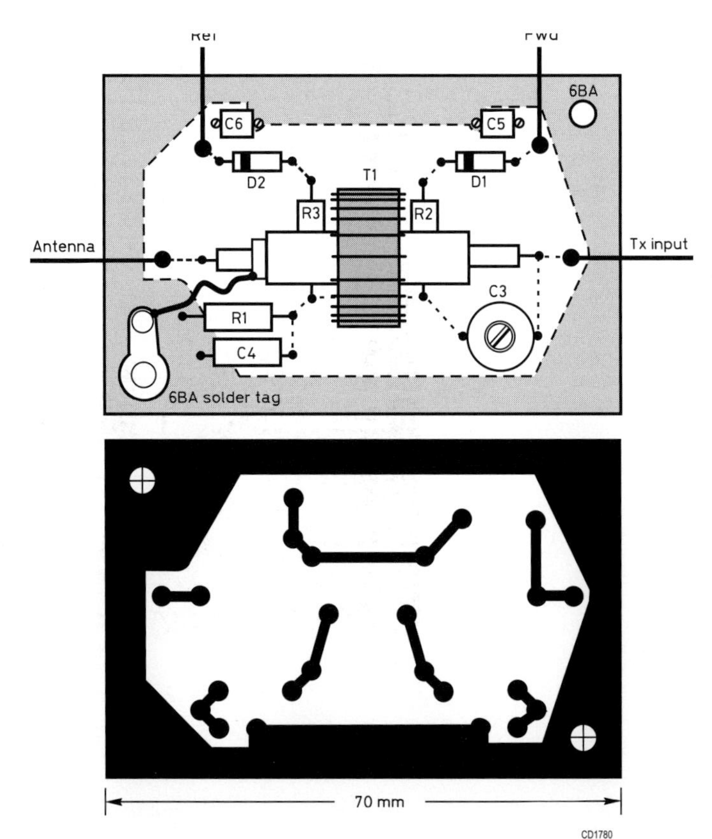


**Fig 15.37: PCB and layout for the SWR bridge**

**Before using this artwork, please read the IMPORTANT NOTICE on page B.1 of this Appendix**

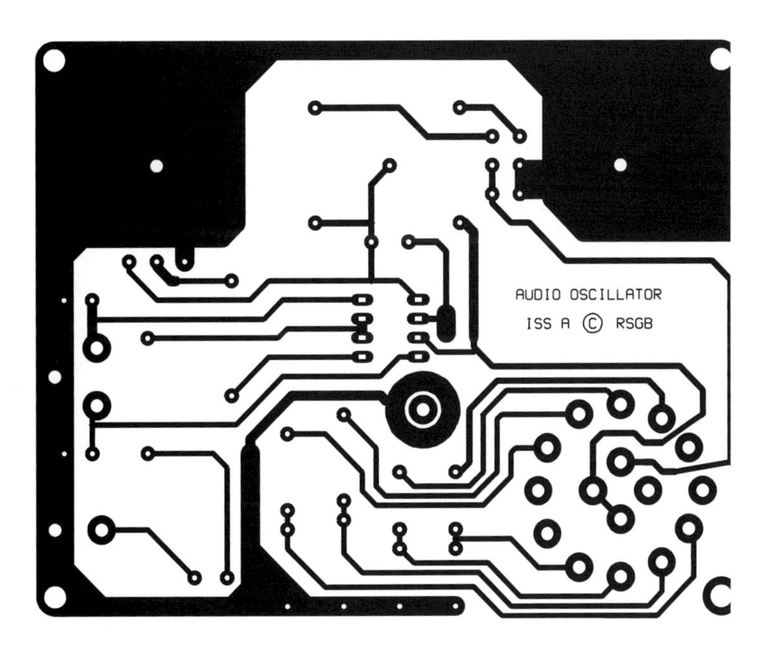


Fig 24.38: Low frequency oscillator PCB layout

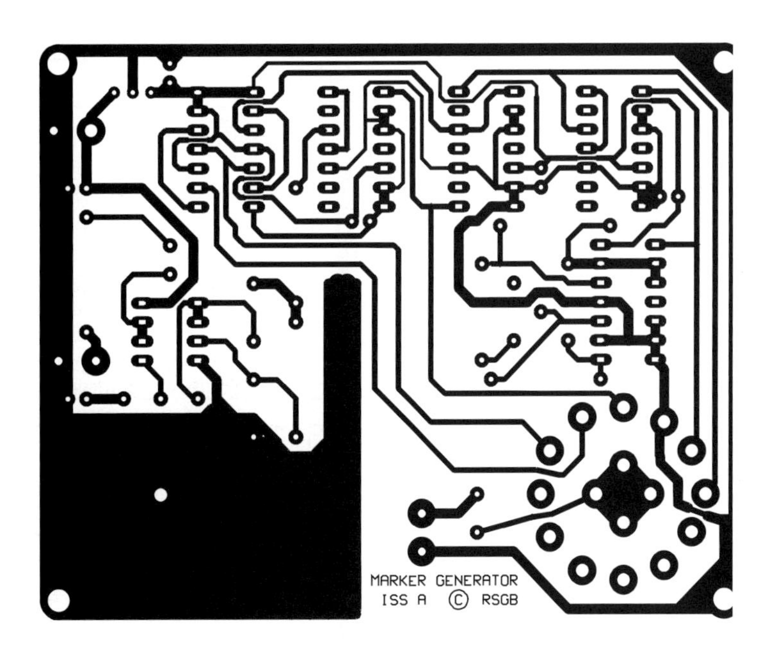


Fig 24.41: Frequency marker PCB layout

**Before using this artwork, please read the IMPORTANT NOTICE on page B.1 of this Appendix**

# Index

## A

## E



## F



## G



## H